CONCISE DICTIONARY OF

BIOMEDICINE AND MOLECULAR BIOLOGY

SECOND EDITION

CONCISE DICTIONARY OF BIOMEDICINE AND MOLECULAR BIOLOGY

SECOND EDITION

PEI-SHOW JUO, PH.D.
PROFESSOR OF BIOLOGY EMERITUS
STATE UNIVERSITY OF NEW YORK
COLLEGE AT POTSDAM
POTSDAM, NEW YORK

CRC PRESS

Boca Raton London New York Washington, D.C.

Library of Congress Cataloging-in-Publication Data

Juo, Pei-Show.
Concise dictionary of biomedicine and molecular biology / Pei-Show Juo.--2nd ed.
p. cm.
ISBN 0-8493-0940-9 (alk. paper)
1. Medicinal sciences--Dictionaries. 2. Molecular biology--Dictionaries. I. Title.

R121 .J86 2001
610′.3—dc21 2001043892

Direct all inquiries to CRC Press LLC, 2000 N.W. Corporate Blvd., Boca Raton, Florida 33431.

International Standard Book Number 0-8493-0940-9
Library of Congress Card Number 2001043892
Printed in the United States of America 1 2 3 4 5 6 7 8 9 0
Printed on acid-free paper

DEDICATED TO MY WIFE

Phyllis Tsou Juo who worked tirelessly to word process all entries as well as draw and proofread the chemical structures. Without her generous help, the completion of this dictionary would have been impossible.

PREFACE

The rapid advance and accumulation of knowledge in modern life sciences has created the need for a dictionary that integrates terminology and abbreviations from diversified disciplines so the reader can grasp quickly the meaning of the terms without lengthy searching in many sources. This dictionary fills the need for such a handy reference volume; it provides simple, clear, up-to-date definitions of terms commonly used in cell biology, bacteriology, virology, immunology, biochemistry, genetics, biomedicine, and related fields.

This dictionary also provides chemical structures and molecular weights of commonly used chemicals, drugs, antibiotics, naturally occurring compounds, products of DNA recombinant technology, and substances of environmental concern and explains enzymatic reactions and specific activities of restriction endonucleases. Brand names and generic names of common drugs or antibiotics are cross-referenced with their chemical structures so the reader can easily interrelate chemical structures with specific drugs or antibiotics. Scientific names of unicellular organisms and their unique characteristics are presented in the dictionary to give the reader a brief overview of the bacteria or protozoa frequently encountered in the literature.

This dictionary consists of over 30,000 entries, including approximately 4,000 chemical structures and their functions, 1,200 equations of enzymatic reactions, 600 restriction endonucleases and their modes of action, a large number of commonly used drugs and antibiotics and their mechanisms of action and medical applications. Entries were drawn from various scientific dictionaries, reference handbooks, research journals, and a large number of textbooks in diversified disciplines of the life sciences.

The terms and chemical structures included in this dictionary have been selected with a "quick access" philosophy in mind, for an audience that includes students of the life sciences, professionals in the allied health fields who often encounter unfamiliar scientific terms and chemical structures, or anyone who is simply interested in knowing specific chemical structures.

In preparing my manuscript, I consulted a number of standard dictionary sources including: Bowker, R. R., *The Dictionary of Cell Biology*, 2nd ed., Lackie, J. M. and Dow, J. A. T., Eds., Academic Press, New York, 1995. Budavari, S., Ed., *The Merck Index*, 11th ed., Merck & Co. Inc., Rahway, NJ, 1989. Coombs, J., *Dictionary of Biotechnology*, 2nd ed., Stockton Press, New York, 1992. Dox, I. G., Melloni, B. J., and Eisner, G. M., *Melloni's Illustrated Medical Dictionary*, 3rd ed., Parthenon Publishing Group, Pearl River, NY, 1993. Fasman, G. D., *Practical Handbook of Biochemistry and Molecular Biology*, CRC Press, Boca Raton, FL, 1990. Glanze, W. D., *The Mosby Medical Encyclopedia*, rev. ed., Penguin Group, New York, 1992. King, R. C. and Stansfield, W. D., *A Dictionary of Genetics*, 4th ed., Oxford University Press, New York, 1990. Singleton, P. and Sainsbury, D., *Dictionary of Microbiology and Molecular Biology*, 2nd ed., John Wiley & Sons, New York, 1993. Stanley, L. et al., *Nursing 93 Drug Handbook*, Springhouse Corporation, Spring House, PA, 1993. Stenesh, J., *Dictionary of Biochemistry and Molecular Biology*, 2nd ed., John Wiley & Sons, New York, 1989. Urdang, L. and Swallow, H. H., Ed., *Mosby's Medical & Nursing Dictionary*, C. V. Mosby, St. Louis, MO, 1983. Webb, E. C., *Enzyme Nomenclature 1992, Recommendations of the Nomenclature Committee of the International Union of Biochemistry and Molecular Biology on the Nomenclature and Classification of Enzymes,* Academic Press, New York, 1992. Mitchell-Hatton, S. L., *The Davis Book of Medical Abbreviations*, F. A. Davis Company, Philadelphia, PA. DeSousa L. R. et al., *Common Medical Abbreviations*, Delmar, Albany, NY, 1995. Fathman, L., Ed., *Medical Drug Reference*, Mosby, St. Louis, MO, 2001. *Mosby's Medical Dictionary,* 5th ed., Mosby, St. Louis, MO. Karch, A. M., *Lippincott's Nursing Drug Guide,* Lippincott, Philadelphia, PA, 2000. *The Bantam Medical Dictionary*, Market House Books, New York, 2000. Pease, Jr., R. W., *Merriam-Webster's Medical Desk Dictionary*, Merriam-Webster, Springfield, MA, 1996. White, J. S. and White, D. C., *Source Book of Enzymes,* CRC Press, Boca Raton, FL, 1997. Smith, A. D., *Oxford Dictionary of Biochemistry and Molecular Biology*, 2000.

I also consulted the following journals: *Arch. Biochem. Biophys., Biochem. J., Biochim. Biophys. Acta, J. Biological Chem., Biochem. Biophys. Res. Comm.*

Pei-Show Juo

CONTENTS

A

α See alpha.

A Abbreviations for 1. absorbance, 2. adenine or adenosine, 3. alanine, 4. ampere. Symbols for 1. Helmholtz free energy and 2. mass number (in chemistry).

Å An angstrom unit. Å = 10^{-1} nm (nanometer) = 10^{-4} μ (micron) = 10^{-7} mm (millimeter) = 10^{-8} cm (centimeter) = 10^{-10} m (meter).

A^- Common symbol for anions.

A- A prefix meaning without.

A68 A protein found in the brains of patients with Alzheimer's disease.

A_{260} ($A_{260\ nm}$) Absorbance at the wavelength of 260 nm.

$A^{1\%}_{260}$ ($A^{1\%}_{260\ nm}$) Absorbance of a 1% solution at the wavelength of 260 nm.

A_{280} ($A_{280\ nm}$) Absorbance at the wavelength of 280 nm.

$A^{1\%}_{280}$ ($A^{1\%}_{280\ nm}$) Absorbance of a 1% solution at the wavelength of 280 nm.

$A^{1\%}_{1\ cm}$ Absorbance of a 1% solution of a given substan ce measured at a specific wavelength in a cuvette with a light path of 1 cm.

A1 Abbreviation for 1. Apolipoprotein A1. 2. Angiotensin I.

AII Abbreviation for angiotensin II.

AIII Abbreviation for angiotensin III.

$[A]_{0.5}$ Symbol for the enzyme kinetic value of the concentration of a substrate at which the velocity of the reaction is half of the maximum velocity.

A23187 An ionophore that can transport divalent ions, particularly Ca^{++}, across membrane lipid bilayers into the cell or cell organelles.

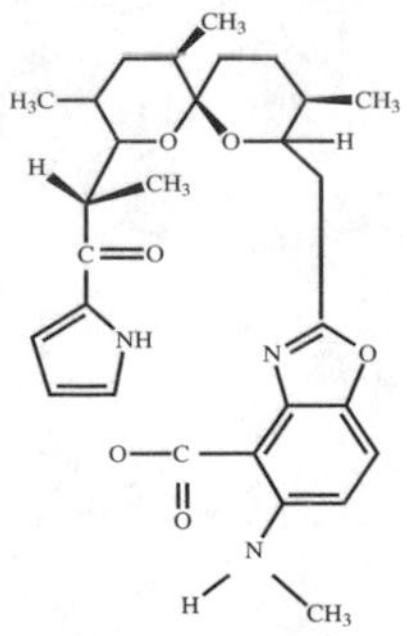

2,5A An oligoadenylate in which adenine nucleotides are linked together through 2′ and 5′ positions of ribose. It acts as an endonuclease activator in interferon-treated cells.

A Antigen Referring to blood group antigen A in the ABO blood group system.

A Band A transverse myosin-containing dark band in the sarcomere of a striated muscle fibril as seen under the electron microscope.

A9 Cells Established heteroploid mouse fibroblasts that are deficient in HGPRT.

A Chain Referring to 1. The shorter polypeptide chain of insulin. 2. The heavy chain of immunoglobulin.

A RNA Referring to double helical RNA with conformation resembling A-DNA (also known as RNA 11).

A Site Referring to aminoacyl-tRNA binding site on ribosome (also known as acceptor site).

A Type Inclusion Body A type of inclusion body formed in cells infected with certain pox viruses.

A Type Particle An intracellular, noninfectious, retrovirus-like particle.

AA Abbreviation for 1. Arachidonic acid. 2. Australia antigen. 3. Acetic acid. 4. Amino acid.

AAA 1. Abbreviation for alpha-aminobutyric acid. 2. A genetic code (codon) for the amino acid lysine.

AAA Pathway Abbreviation for aminoadipic acid pathway.

AAAD Abbreviation for aromatic amino acid decarboxylase.

AAAE Abbreviation for amino acid-activating enzyme.

AAAF Abbreviation for albumin auto-agglutinating factor.

aa-AMP Abbreviation for amino acid-AMP complex or aminoacyl-AMP complex.

αAAN Abbreviation for alpha amino acid nitrogen.

AAC A genetic code (codon) for the amino acid asparagine.

AacI (BamHI) A restriction endonuclease isolated from *Acetobacter aceti* sub. *liquefaciens* and *Bacillus amyloliquefaciens* with the following specificity:

```
   ↓
5′..........GGATCC..........3′
3′..........CCTAGG..........5′
                ↑
```

AACE Abbreviation for antigen-antibody crossed immunoelectrophoresis.

αADA Abbreviation for alpha amino adipic acid.

AaeI (BamHI) A restriction endonuclease isolated from *Acetobacter aceti* sub. *liquefaciens* having the same specificity as BamHI.

AAF Abbreviation for acetic-alcohol-formalin mixture

AAG A genetic code (codon) for the amino acid lysine.

αAIBA Abbreviation for alpha amino isobutyric acid.

α-Amylase An endo-amylase that catalyzes the hydrolysis of starch to dextrins.

AAN Abbreviation for alpha amino nitrogen.

α_2-Antiplasmin Plasma protein that regulates fibrinolysis.

α_1-Antitrypsin Protein in the blood plasma that inhibits serine proteases.

AAO Abbreviation for amino acid oxidase.

aa-O-AMP Abbreviation for aminoacyl-O-adenosine monophosphate.

AAP Abbreviation for 1. Alanine aminopeptidase. 2. Arginine aminopeptidase. 3. Aspartate aminopeptidase. 4. Alpha-2 antiplasmin.

AAR Abbreviation for antigen-antibody reaction.

AAS Abbreviation for 1. Atomic absorption spectrophotometry. 2. Atomic absorption spectrometer.

aa-S-CoA Abbreviation for aminoacyl-S-CoA.

AASH Abbreviation for adrenal androgen-stimulating hormone.

AAT Abbreviation for α_1-antitrypsin.

AatI A restriction endonuclease isolated from *Acetobacter aceti* with the following specificity:

```
              ↓
5′..........AGGCCT..........3′
3′..........TCCGGA..........5′
              ↑
```

AatII A restriction endonuclease isolated from *Acetobacter aceti* with the following specificity:

```
                ↓
5′..........GACGTC..........3′
3′..........CTGCAG..........5′
            ↑
```

AAT Medium Abbreviation for adenine-aminopterin-thymine medium.

AA-tRNA Abbreviation for aminoacyl-tRNA.

AAU A genetic code (codon) for the amino acid asparagine.

AAV Abbreviation for adeno-associated virus.

Ab Abbreviation for antibody.

Ab- A prefix meaning from, off, or away from.

AB Toxin Referring to toxin that has two major components, an active A component and a binding component B responsible for binding to the target cell.

ABA Abbreviation for abscisic acid.

Abacavir Sulfate (mol wt 671) A reverse transcriptase inhibitor that inhibits the HIV viral replication.

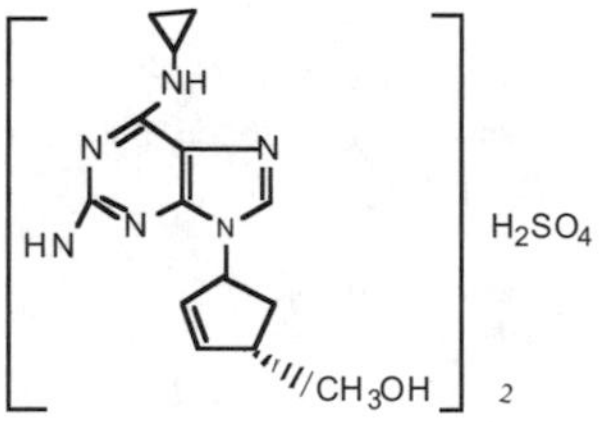

Abacterial Free from bacteria.

Abamectin (Avermectin B_1) An anthelmintic agent or insecticide. It consists of two components: avermectin B_{1a} and B_{1b}.

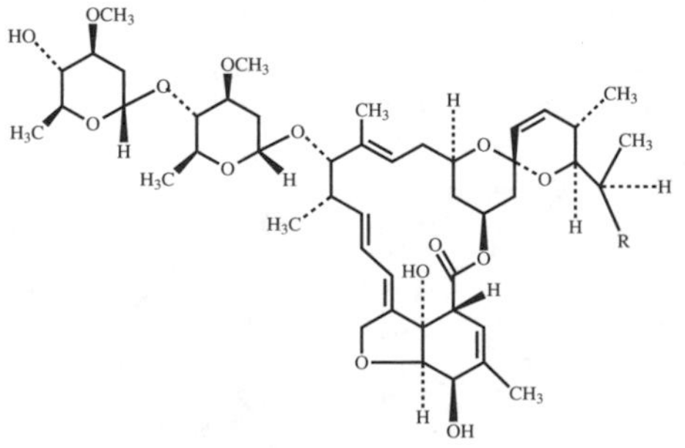

Component B_{1a}, $R = C_2H_5$
Component B_{1b}, $R = CH_3$

Abarticular Pertaining to a set or structure remote from the joint or a condition not affecting the joint.

Abasia The inability to walk properly owing to the paralytic condition of the leg muscle.

Abaxial Pertaining to a position directed away from the axis.

Abbe Condenser A device placed beneath the microscope stage to obtain illumination.

Abbe Refractometer A device used for the direct measurement of the light-retarding property of a solution.

Abbokinase A trade name for urokinase that catalyzes the conversion of plasminogen to plasmin.

Abbot Pump A small, portable pump used for delivery of a precise quantity of medication in solution through an intravenous infusion device.

Abbot's Staining Method A method that stains the bodies of bacteria red and spores blue.

ABC Abbreviation for 1. Absolute basophil count. 2. Acid-buffered citrate. 3. Antigen-binding capacity. 4. Antigen-binding cells. 5. ATP-binding cassette.

ABC Excinulease Abbreviation for the enzyme complex produced by *uvrA, uvrB, uvrC* genes of *E coli* that mediates incision and excision steps of DNA repair.

ABC Immunoperoxidase Method An immunological method that uses preformed avidin-biotin-peroxidase for detection of antigen-antibody reaction.

ABC Transporters A family of transport proteins that are involved in the transport of amino acids, sugars, inorganic ions, polysaccharides, peptides and proteins.

ABCD Abbreviation for a combination drug containing adriamycin, bleomycin, CCNU and dacarbazine.

Abciximab An anti-platelet agent that interferes with platelet membrane function and inhibits platelet aggregation and prolongs bleeding time.

ABCM Abbreviation for a combination drug containing adriamycin, bleomycin, cytoxan and mitomycin C.

ABD Abbreviation for a combination drug containing adriamycin, bleomycin and DTIC.

Abdominal Actinomycosis Abdominal diseases caused by infection of *Actinomyces*, e.g., *A. israelii*. (See also Actinomycosis.)

Abdominalgia Abdominal pain.

Abdominocentesis Surgical puncture of the abdominal wall for diagnostic purposes.

Abdominoplasty Surgical removal of excess fat from the abdomen.

Abdominoscopy Inspection or examination of the abdominal cavity or organs by an endoscope.

Abduct Movement of the body structure or appendage in a direction away from the midline or median plane.

Abductin An insoluble, rubber-like protein from the internal triangular hinge ligament of scallops.

Abductor A muscle that pulls a structure away from the axis of the body.

Abe Abbreviation for abequose.

ABE Abbreviation for a mixture containing acetone, butanol and ethanol.

ABE Process A fermentation process for the production of acetone, butanol, and ethanol by *Clostridium acetobutylicum* from carbohydrate, e.g., molasses.

Abecarnil (mol wt 404) An anxiolytic agent.

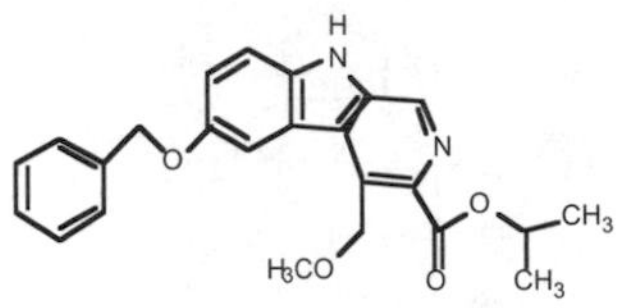

Abelcet A trade name for amphotericin B complex, an antifungal agent.

Abelson Murine Leukemia Virus A replication-defective *v-onc*$^{+}$ murine leukemia virus isolated from a prednisolone-treated BALB/c mouse inoculated with Moloney murine leukemia virus.

Abembryonic Located away from the embryo.

Abenol A trade name for acetaminophen, an antipyretic and analgesic agent.

Abequose (3,6-Dideoxy D-Galactose, mol wt 147) An unusual sugar found in the lipopolysaccharide of a bacterial cell wall.

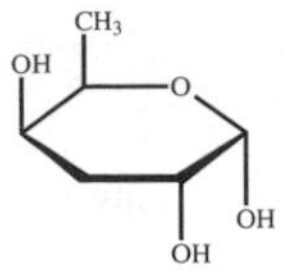

Abernethy's Sarcoma A malignant neoplasm of fat cells usually occurring on the trunk.

Aberrant Deviation from the normal.

Abetalipoproteinemia An inherited disorder characterized by the absence of plasma low density lipoprotein (betalipoprotein) and the presence of acanthocytes in the blood (acanthocytosis).

ABH Antigens Referring to the blood group antigen A, B and H. The H antigen is the precursor of blood group antigen A and B. Individuals having neither A nor B antigen express the H antigen.

Abient Having a tendency to move away from the stimuli.

Abietic Acid (mol wt 302) The principal constituent of colophony rosin which is capable of stimulating growth of lactic and butyric bacteria.

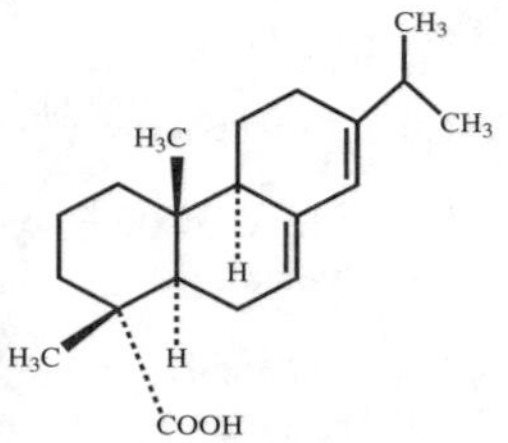

Abikoviromycin (mol wt 161) An antiviral antibiotic produced by *Streptomyces abikoensis* and *Streptomyces rubescens*.

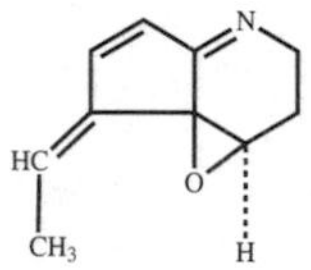

Abiogenesis The theory of spontaneous generation of a living organism from nonliving matter.

Abiogenic Pertaining to abiogenesis.

Abiosis The absence of life.

Abiotic Pertaining to substances that are of nonbiological origin or an environment characterized by the absence of biological organisms.

Abiotrophy The loss of function of certain cells and tissues, possibly due to a latent inherited trait.

Abirritant An agent that relieves irritation.

Abirritation A reduced responsiveness to irritating stimuli.

***abl* Gene** An oncogene in mouse pre-B cell leukemia that encodes protein kinase (tyrosine).

Ablactation The weaning of a child from the breast.

Ablastins Substances or agents that inhibit and prevent the reproduction or cell division of microorganisms.

Ablation Surgical removal of an organ or part of the organ, e.g., amputation.

Ablepharia Congenital absence of the eyelid (partial or total).

Ablepsia Blindness.

Abluent Agent or substance with purifying property.

ABLV Abbreviation for Abelson leukemia virus.

ABM Paper The aminobenzyloxy methylcellulose paper capable of covalently binding single-stranded DNA.

ABMA Abbreviation for anti-basement membrane antibody.

ABMT Abbreviation for autologous bone marrow transplantation.

AbMuLV Abbreviation for Abelson murine leukemia virus.

Abnerval Current Pertaining to the electric current passing from a nerve terminal into a muscle.

Abneural Away from the central nervous system.

Abnormal Hemoglobin A hemoglobin that differs from normal hemoglobin in function, amino acid sequence, and electrophoretic mobility.

ABO Blood Group System A human blood group system in which there are two antigens on the red blood cell surface denoted A and B. The four major blood types, A, B, AB, and O, are named based upon the presence or absence of these antigens. Type A possesses antigen A, type B possesses antigen B, type AB possesses both antigens A and B, and type O possesses neither antigen A nor B. The plasma of type A blood contains anti-B (antibody to antigen B); type B blood contains anti-A (antibody to antigen A); the type O blood contains both anti-A and anti-B; the blood type AB contains neither anti-A nor anti-B.

Aboaggregin B A protein that binds to glycoprotein 1b of the platelet membrane.

Aboral In a direction away from the mouth.

Aborticide Agent or substance that causes abortion.

Abortifacient Agent or substance that causes abortion.

Abortive Complex Any enzyme-substrate complex in which the substrate is bound to the enzyme in a manner that renders the catalysis inactive.

Abortive Infection A viral infection that does not lead to the formation of infectious progeny virions.

Abortive Transduction Bacterial transduction in which the DNA from the donor cell fails to integrate into the chromosome of the recipient bacterium.

Abortus An aborted fetus.

ABOS Abbreviation for a combination drug containing adriamycin, bleomycin, oncovin and streptomycin.

ABP Abbreviation for 1. A combination drug containing adriamycin, bleomycin and prednisone. 2. Androgen-binding protein. 3. Arterial blood pressure. 4. Actin-binding proteins.

ABP-50 Abbreviation for actin-binding protein-50, a 50 kDa protein from *Dictyostelium* that cross-links actin filaments into tight bundles.

ABP-67 Abbreviation for actin-binding protein-67 encoded by SAC6 gene, mutations in which lead to disruption of the actin cytoskeleton.

ABP-120 Abbreviation for actin-binding protein-120 from *Dictyostelium* capable of cross-linking filaments.

ABP-280 Abbreviation for actin-binding protein-280 from *Dictyostelium* with an actin-binding domain similar to that in ABP-120.

ABPC Abbreviation for avidin-biotin-peroxidase complex.

Abortus Fever A form of brucellosis caused by *Brucella abortus*.

AbrI A restriction endonuclease isolated from *Azospirillum brasilense* with the following specificity:

↓
5′..........CTCGAG..........3′
3′..........GAGCTC..........5′
↑

Abrin A protein phytotoxin (toxic lectin or toxalbumin) obtained from seeds of jequirity, *Abrus precatorius* (Leguminosae). It inhibits protein synthesis and possesses antitumor activity.

Abrine (mol wt 218) A compound obtained from seed of *Abrus precatorius*.

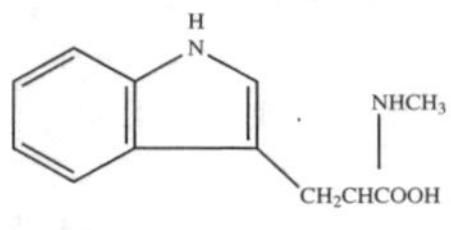

Abrism A morbid condition resulting from the ingestion of the seeds of *Abrus precatorius* that contain phytotoxin abrin.

Abrosia A condition caused by fasting or abstaining from food.

Abruptio Placentae Premature separation of the placenta from the wall of the uterus.

Abrus A genus of papilionaceous plants (e.g., *Abrus precatorius*) producing toxic lectin or phytotoxin.

Abs Abbreviation for absorption or absolute.

Abscess A localized accumulation of pus or a cavity containing pus and surrounded by the inflamed tissue.

Abscisic Acid (mol wt 264) A plant hormone produced by plants that promotes dormancy and abscission.

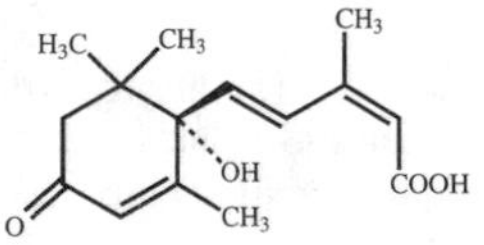

Abscisin A group of plant hormones that accelerate abscission of plant parts, induce and maintain dormancy in seeds.

Abscissa The horizontal axis (x-axis) in a plane rectangular coordinate system.

Abscission Separation of leaves, flowers, and other plant organs from a plant due to the formation of an abscission zone or layer.

Abscission Layer (Abscission Zone) The cells at the base of a leaf, flower, or fruit that form an abscission corky layer leading to the separation of the plant part from the plant.

Absinthe 1. Wormwood or its essence. 2. A green liqueur prepared by steeping herbs of anise and wormwood in alcohol.

Absinthin (mol wt 497) The chief bitter part of wormwood (*Artemisia absinthium*), which was used to flavor alcohol beverages.

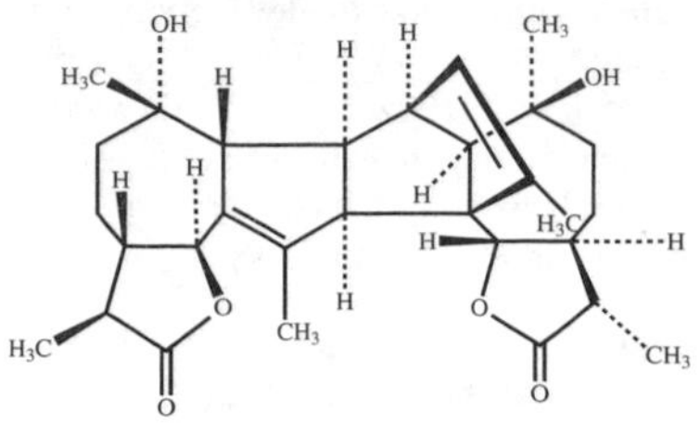

Absinthism An addiction to absinthe.

Absinthium The common wormwood. *Artemisia absinthium*, a bitter plant used as a stomachic tonic, anthelmintic, and flavoring in alcohol beverage.

Absolute Alcohol Dehydrated alcohol or anhydrous ethyl alcohol.

$$CH_3\text{-}CH_2OH$$

Absolute Catabolic Rate The mass of protein catabolized per day, which is determined by multiplying the fractional turnover rate by the volume of the plasma pool.

Absolute Configuration The actual spatial arrangement of the atoms around the asymmetric carbon atoms in a molecule.

Absolute Counting The radiation count that includes all disintegration that occurs in the sample and is expressed as disintegrations per minute (dpm).

Absolute Deviation The numerical difference between an experimental value and the true (or the best) value of the quantity being measured.

Absolute Plating Efficiency The percentage of cells that give rise to colonies when inoculated into nutrient medium.

Absolute Reaction Rate The rate or velocity of a chemical reaction that is proportional to the concentration of energy-activated reaction complex.

Absolute Refractory Period The time period in which sodium channels of a nerve cell are inactivated and the cell is incapable of responding to any stimulus regardless of its strength.

Absolute Scale A temperature scale based on absolute zero.

Absolute Specificity The extreme selectivity of an enzyme that catalyzes only the reaction with a single specific substrate (in a monomolecular reaction) or a single specific pair of substrates (in a bimolecular reaction).

Absolute Temperature The temperature measured on the absolute scale and expressed in degrees above absolute zero (approximately –273° C).

Absolute Temperature Scale (Kelvin Temperature Scale) A temperature scale on which the zero point is absolute zero (–273° C) and the degrees denoted T or K match those of the centigrade scale. Zero degrees on the centigrade scale (0° C) equals 273° on the Kelvin scale.

Absolute Zero The zero point on the absolute temperature scale that is –273°C. It is the temperature at which all atomic motion stops.

Absorbance (A) The measurement of the amount of light absorbed by a solution. It is mathematically defined as

$$A = \log \frac{I_o}{I} \quad \text{or} \quad A = \log \frac{1}{T}$$

where I_o is intensity of incident light, I is intensity of transmitted light and T is percent of light transmitted (% transmittance or %T).

Absorbance Index See absorption coefficient.

Absorbance Unit The amount of light-absorbing material contained in 1 ml of a solution that has an absorbance of 1.0 when measured with a 1-cm cuvette.

Absorbancy Variant spelling of absorbance.

Absorbate A substance that is absorbed by another substance.

Absorbed Antiserum An antiserum from which antibodies have been removed by the addition of antigens or other antibody-absorbing substances.

Absorbed Dose The energy imparted by ionizing radiation per unit mass of irradiated material. The unit of absorbed dose is the rad (radiation absorbed dose), which equals 100 ergs per gram.

Absorbefacient Agents or substances that promote absorption.

Absorber A material used to absorb radioactive radiation.

Absorptiometer Instruments used to measure 1. the amount of gas absorbed by a liquid, 2. the thickness of a liquid layer between parallel glass plates, or 3. the color intensity or color difference.

Absorption 1. The uptake of one substance by another substance. 2. The passage of materials across a biological membrane. 3. The transfer of the energy of incident radiation to the matter through which it passes.

Absorption Band A portion of the electromagnetic spectrum in which a molecule absorbs radiant energy.

Absorption Coefficient The proportionality constant in Beer-Lambert's Law

$$A = \varepsilon cl \quad \text{or} \quad \varepsilon = \frac{A}{cl}$$

where ε is the absorption coefficient and A is the absorbance. l is the length of light path (usually 1 cm) and c is the concentration of a substance. The absorbance at a given wavelength of a 1 M solution in a 1-cm cuvette (light path = 1 cm) is termed molar absorptivity, molar absorption coefficient, or molar extinction coefficient.

Absorption Line Synonym of absorption band.

Absorption Optical System An optical system used for measuring molecular boundary movement in a solution during the centrifugation or electrophoresis.

Absorption Ratio The ratio of the concentration of a substance in solution to its absorptivity.

Absorption Spectrometry The process of measuring absorption spectrum of a substance with a spectrometer.

Absorption Spectrum The extent to which a substance absorbs light of different wavelengths. It can be obtained by plotting the absorbance of a substance at different wavelengths.

Absorptive Lipemia Transitory accumulation of excessive lipid in the blood following the ingestion of lipid.

Absorptivity See absorption coefficient.

Abstergent Having the cleaning or purgative property.

Abstraction The removal of either an atom or an electron from a compound.

Abstriction The formation of spores in fungi by successive cutting of sections of the sporophore and the development of end wall or septum at the constriction point.

Abterminal Moving from the end toward the center.

Abu Abbreviation for aminobutyric acid.

A_2bu Abbreviation for 2,4-diaminobutanoic acid.

ABV Abbreviation for a combination drug containing adriamycin, bleomycin and velban.

Abz Abbreviation for aminobenzoic acid or aminobenzyl.

Abzymes Nonenzyme substances that are capable of catalytic activity, e.g., catalytic antibody or catalytic RNA.

AC Abbreviation for 1. A combination drug containing adriamycin and cyclophosphamide. 2. Adenylate cyclase or adenylyl cyclase. 3. Alternating current. 4. Anti-cholinergic. 5. Anticoagulant. 6. Anti-complementary. 7. Artificial chromosome.

Ac 1. Abbreviation for an Acetyl group or CH_3CO^- radical. 2. Symbol for the chemical element Actinium (atomic weight 227, valence 2).

AC Calorimetry A technique for the measurement of heat capacity of the sample on both cooling and heating and monitoring of its isothermal time-dependence.

ACA A genetic code (codon) for the amino acid threonine.

7-ACA (7-aminocephalosporanic acid, mol wt 272) Hydrolytic product of the antibiotic cephalosporin C, a potent inhibitor of bacterial β-lactamase.

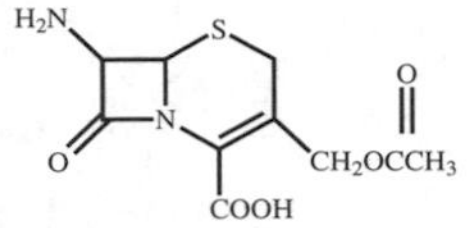

AcaI (AsuII) A restriction endonuclease isolated from *Anabaena catenula* with the following specificity:

↓
5′..........TTCGAA..........3′
3′..........AAGCTT..........5′
↑

AcaII (BamI) A restriction endonuclease from *Anabaena catenula* having the same specificity as BamHI.

AcaIII (MstI) A restriction endonuclease from *Anabaena catenula* with the following specificity:

↓
5′..........TGCGCA..........3′
3′..........ACGCGT..........5′
↑

AcaIV (HaeIII) A restriction endonuclease from *Anabaena catenula* with the following specificity:

↓
5′..........GGCC..........3′
3′..........CCGG..........5′
↑

Acacia 1. A genus of leguminous shrubs or trees consisting of several economically and medically important species, e.g., *A. senegal* yielding acacia (gum arabic) and *A. catechu* producing catechu. 2. Gum arabic, the dry gummy exudate from the stems and branches of *A. senegal* and other African species of *Acacia*.

Acacic Acid (mol wt 489) A compound isolated from pods of *Acacia concinna*.

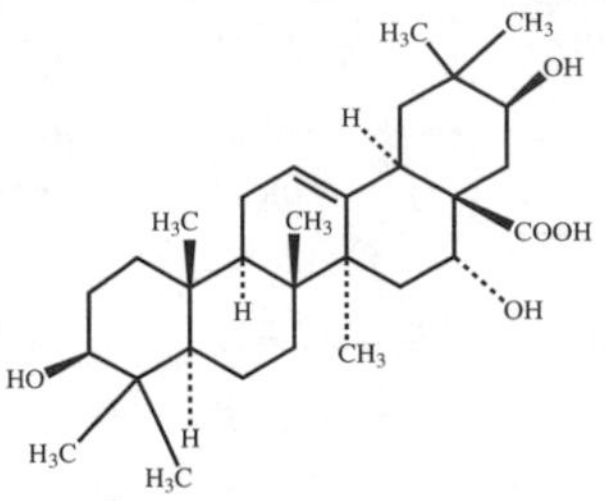

Acacin Gum arabic.

Acacine Variant spelling of acacin.

Acadesine (mol wt 258) A nucleoside analog and a cardioprotective agent.

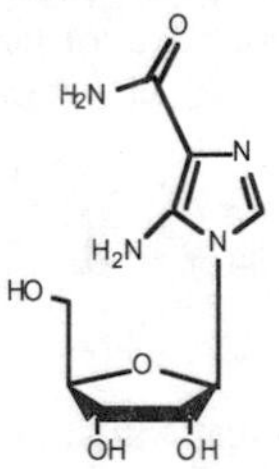

Acalcerosis A condition in which the body is deficient in calcium.

Acalcicosis A disorder caused by a deficiency of calcium in the diet.

Acamprosate Calcium (mol wt 400) A substance used for treatment of alcoholism.

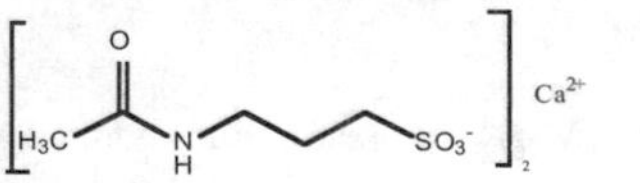

Acampsia A condition in which a joint becomes rigid and inflexible.

Acantha A spine or a spinous projection.

Acanthamebiasis Infection caused by *Acanthamoeba castellani.*

Acanthamoeba A genus of free-living amoebae (order Amoebida).

Acantho- A prefix meaning thorny or spiky.

Acanthocyte An abnormal red blood cell having several protoplasmic projections that give it a thorny appearance.

Acanthocytosis The presence of acanthocytes in the circulating blood system, most commonly associated with abetalipoproteinemia.

Acanthoid See acantha.

Acantholysis Destruction of the epidermis.

Acanthoma Carcinoma of the epidermis or tumor of the skin.

Acanthosis The thickening and warty growth of the prickle-cell layer of the skin, e.g., eczema and psoriasis.

Acapnia A marked decrease of CO_2 in the blood.

Acapsular Without a capsule.

Acarbose (mol wt 646) A pseudotetrasaccharide isolated from *Actinoplanes*. It inhibits α-glucosidase and reduces sugar absorption in the gastrointestinal tract.

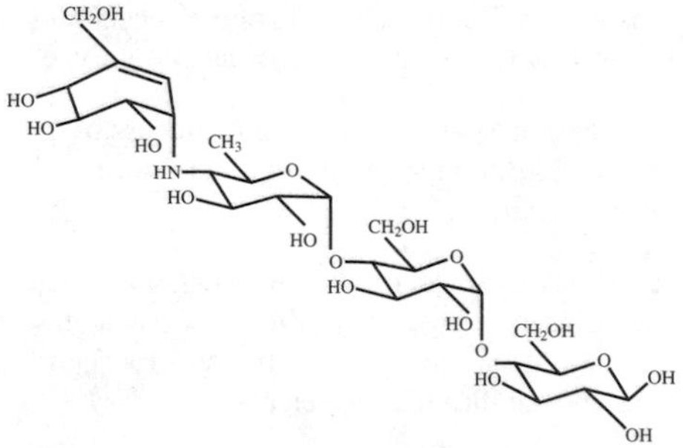

Acardia Congenital absence of the heart.

Acariasis Any disease caused by an acarid, e.g., scrub typhus.

Acaricide Any chemical agent that kills mites and ticks.

Acarid Any member of the order Acarida that includes a great number of parasitic and free-living mites.

Acaro- A prefix meaning pertaining to mites.

Acarology The science that deals with mites and ticks.

Acarus 1. A genus of small mites causing itch, mange, and other skin diseases. 2. A mite.

Acaryote (Akaryote) Cell without a nucleus.

ACAT Abbreviation for acyl-CoA cholesterol transferase, the enzyme that catalyzes the formation of cholesteryl esters from cholesterol.

Acatalasia A metabolic disorder characterized by the congenital absence of the enzyme catalase.

ACB Abbreviation for antibody-coated bacteria.

ACBP Abbreviation for acyl-CoA binding protein.

ACC A genetic code (codon) for the amino acid threonine.

AccI A restriction endonuclease from *Acinetobacter calcoaceticus* with the following specificity:

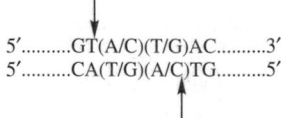

5′..........GT↓(A/C)(T/G)AC..........3′
5′..........CA(T/G)(A/C)↑TG..........5′

AccII (FnuDII) A restriction endonuclease from *Acinetobacter calcoaceticus* with the following specificity:

5′..........CGCG..........3′
3′..........GCGC..........5′

AccIII (BspMII) A restriction endonuclease from *Acinetobacter calcoaceticus* with the following specificity:

5′..........TCCGGA..........3′
3′..........AGGCCT..........5′

Acc16I A restriction endonuclease from *Acinetobacter calcoaceticus* 16 with the following specificity:

5'........TGCGCA........3'

Acc65I A restriction endonuclease from *Acinetobacter aceti* 655 with the following specificity:

5'........GGTACC........3'
3'........CCATGG........5'

Acc113I A restriction endonuclease from *Acinetobacter calcoaceticus* with the following specificity:

5'........AGTACT........3'
3'........TCATGA........5'

AccB1I A restriction endonuclease from *Acinetobacter calcoaceticus* B1 with the following specificity:

5'........GGYRCC........3'
3'........CCRYGG........5'

R = A or G Y = C or T

AccB2I A restriction endonuclease from *Acinetobacter calcoaceticus* B2 with the following specificity:

5'........RGCGCY........3'
3'........YCGCGR........5'

R = A or G Y = C or T

AccB7I A restriction endonuclease from *Acinetobacter calcoac* B7 with the following specificity:

5'........CCANNNNNTGG........3'
3'........GGTNNNNNACC........5'

AccBSI A restriction endonuclease from *Acinetobacter calcoaceticus* BS with the following specificity:

5'........CCGCTC(-3/-3)........3'

AccEBI (BamHI) A restriction endonuclease from *Acinetobacter calcoaceticus* having the same specificity as BamHI.

Accelerating Voltage Voltage responsible for accelerating electrons prior to their emission from the electron gun, e.g., in an electron microscope.

Acceleration An increase in speed or velocity of an object or a reaction.

Acceleration Gravity The acceleration of a freely falling object caused by the force of gravity that is expressed in term of cm/sec^2 or ft/sec^2.

Accelerator 1. A substance that speeds up any chemical reaction. 2. An instrument that speeds up small particles for bombarding the nuclei of atoms. 3. A muscle or nerve that speeds up the performance of a motion.

Accelerator Globulin Synonym of blood coagulation factor V.

Accelerin Synonym of blood coagulation factor Va.

Accentuator Substances or physical agents that increase the intensity of the microorganism staining reaction, e.g., phenol or heat.

Acceptor Arm See acceptor stem.

Acceptor Control The dependence of the respiratory rate of mitochondria on the ADP concentration (also known as respiratory control).

Acceptor Control Ratio The rate of respiration, in terms of oxygen uptake per unit time, in the presence of ADP, divided by the rate in the absence of ADP.

Acceptor End Referring to the CCA-terminal of the tRNA.

Acceptor Junction See acceptor-splicing site.

Acceptor RNA Outdated term for transfer RNA (tRNA).

Acceptor Site See A-site.

Acceptor-Splicing Site The segment at the 3′ end of an intron in a genome (also known as acceptor junction).

Acceptor Stem The arm or stem on the 3′ end of the tRNA where the amino acid is covalently linked.

Accessory Cells Cells of predominantly monocyte and macrophage lineage, e.g., macrophage, dendritic cells, and Langerhans cells that cooperate with T and B lymphocytes for the expression of humoral and/or cell-mediated immunity.

Accessory Chromosome An unpaired chromosome.

Accessory DNA The surplus DNA present in certain cells or during certain stages of cell development owing to the gene amplification.

Accessory Factors The blood clotting factors that serve to enhance the rate of proteolytic activation of other blood clotting factors.

Accessory Pigments Photosynthetic pigments such as carotenoids, phycobilins, and chlorophyll b that harvest and transfer light energy to the photosynthetic reaction centers mediated by the primary pigment chlorophyll a.

Accessory Protein The protein whose action accelerates the activity of other proteins.

AcCh Abbreviation for acetylcholine.

AcChR Abbreviation for acetylcholine receptor.

Acclimation (Acclimatization) The physiological adjustment of an organism to a new environment.

AcCoA Abbreviation for acetyl-CoA.

Accolate A trade name for zafirkulast a leukotriene receptor antagonist and an antiasthmatic agent.

Accommodation 1. Adaptation or adjustment to surrounding environments in order to maintain a state of homeostasis. 2. Automatic adjustment of the eye to focus on objects at different distances.

Accrementition 1. Reproduction by budding. 2. Growth by gradual external addition.

Accretion See accrementition.

Accumulation Coefficient The rate of increase in concentration of molecules adsorbed on a surface, compared to the concentration of the same type of molecule in the phase in contact with the surface.

Accumulation Theory A theory of aging that states aging is due to the accumulation of toxic substances.

Accumulation Time The time necessary for the accumulation of a quantum of radiant energy before it can be released.

Accumulator Organism An organism capable of absorbing and retaining large amounts of specific substances.

Accupril A trade name for quinapril, an antihypertensive agent that prevents the production angiotensin II.

Accutane A trade name for isotretinoin, an antiacne agent.

ACD Solution Acid-citrate-dextrose solution. (See also Alsever's solution).

ACE Abbreviation for 1. Mixture of alcohol-chloroform-ether. 2. Angiotensin-converting enzyme. 3. Amplification control element (a DNA sequence that functions as the origin for amplification). 4. A combination drug containing adriamycin, cyclophosphamide and etoposide.

-acea A suffix in animal taxonomic nomenclature denoting a family.

-aceae A suffix in plant taxonomic nomenclature denoting a family.

Acebutolol (mol wt 336) A β_2-adrenergic blocking agent with antihypersensitive, antianginal and antiarrhythmic activity.

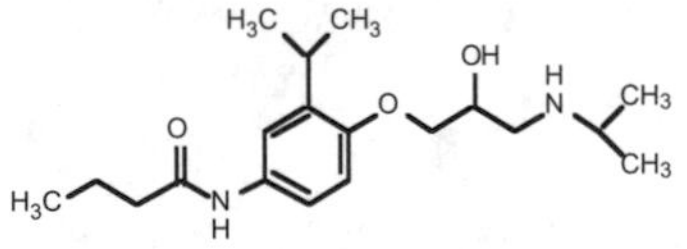

Acecainide (mol wt 277) A cardiac depressant (antiarrhythmic).

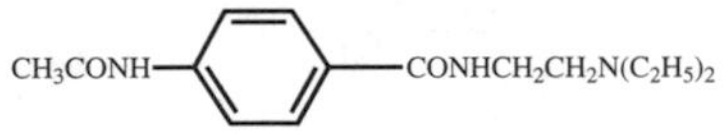

Acecarbromal (mol wt 279) A sedative and hypnotic agent.

$(C_2H_5)_2$-CBr-CO-NH-CO-NH-$COCH_3$

Aceclofenac (mol wt 354) An anti-inflammatory agent.

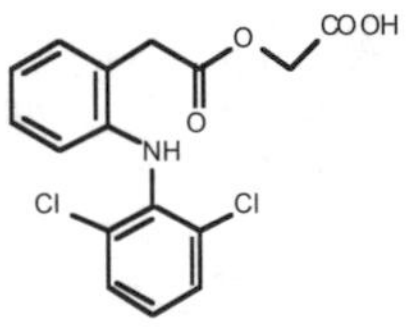

Acedapsone (mol wt 332) An antimalarial and antibacterial agent.

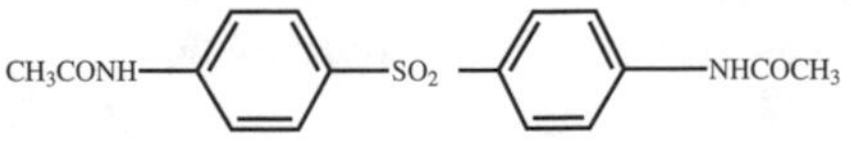

Acediasulfone (mol wt 306) An anti-bacterial agent.

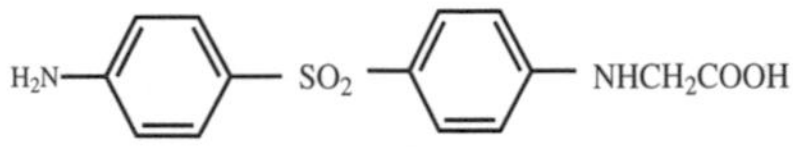

Acefylline (mol wt 238) A diuretic, cardiotonic agent, and a bronchodilator.

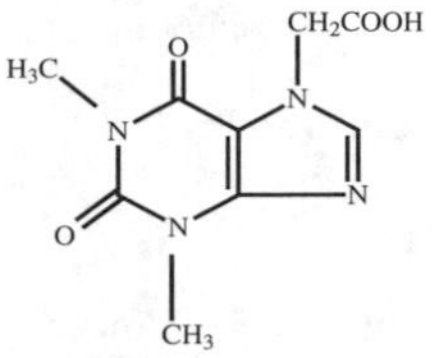

Aceglatone (mol wt 258) An antineoplastic agent.

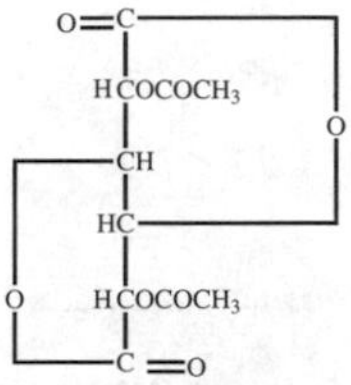

Aceglutamide (N-acetyl-L-glutamine, mol wt 188) An amino sugar derivative; its aluminum complex can be used as an antiulcerative agent.

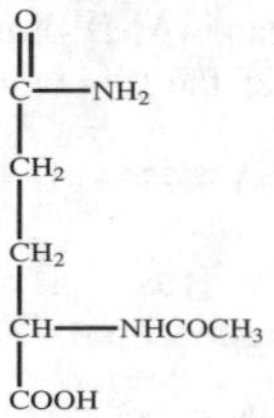

ACEI Abbreviation for angiotensin converting enzyme inhibitor which prevents the formation of angiotensin-II.

Acel-Imune A trade name for diphtheria and tetanus toxoids and acellular pertussis vaccine.

Acellular Containing no cell, e.g., viruses, viroids, prions, or plasmodium.

Acemetacin (mol wt 426) An anti-inflammatory agent.

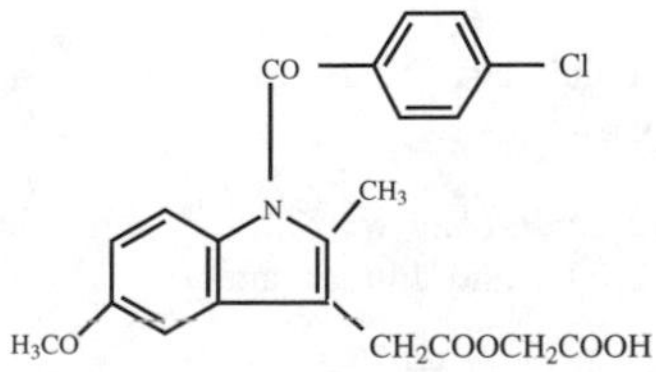

Acenaphthene (mol wt 154) A compound that occurs in coal tar possessing insecticide and fungicide activity.

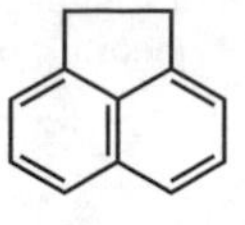

Acenocoumarol (mol wt 353) A synthetic anticoagulant and vitamin K antagonist.

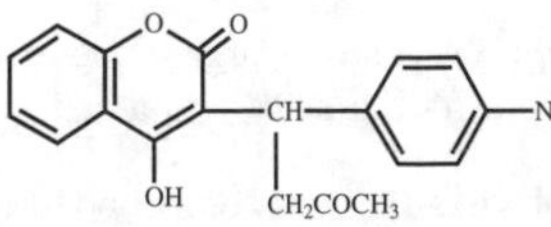

Acentric 1. Not located in the center. 2. A chromosome or chromosome fragment without a centromere.

Aceon A trade name for perindopril, an antihypertensive agent.

Acephalobrachia A congenital defect in which a fetus lacks both arms and a head.

Acephaly (Acephalia, Acephalism) A congenital abnormality in which the head of a fetus is absent or not properly developed.

Acephate (mol wt 183) A systemic insecticide and a cholinesterase inhibitor.

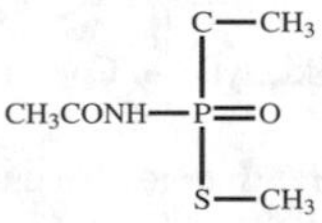

Acephen A trade name for acetaminophen, an analgesic and antipyretic agent that inhibits the synthesis of prostaglandins (pain mediators).

Acepromazine (mol wt 326) A tranquilizer used in veterinary medicine to immobilize large animals.

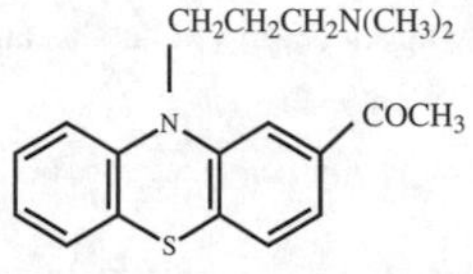

Acerola The ripe fruit of West Indian cherry fruit (*Malpighia punicifolia*), the richest natural source of ascorbic acid (vitamin C:1690 mg/100 g of pitted fruit).

Acervuline Occurring in cluster form.

ACES (mol wt 182) N-(2-Acetamido)-2-aminoethanesulfonic acid; used for the preparation of buffers in the pH range of 6.0 to 7.5.

$NH_2COCH_2NHCH_2CH_2HSO_3$

Acesulfame (mol wt 163) A nonnutritive artificial sweetener.

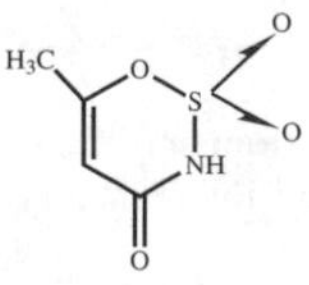

Acet- A combining form meaning vinegar.

Aceta A trade name for acetaminophen, an analgesic and antipyretic agent.

Acetabularia A genus of large single-celled green algae having a foot, a stalk, and a cap.

Acetal (mol wt 118) 1. A compound formed from acetaldehyde and alcohol in the presence of anhydrous calcium chloride.

$$CH_3CH{-}(OC_2H_5)_2$$

2. A compound formed between aldehyde and two alcoholic OH groups.

Acetaldehyde (mol wt 44) The product of the oxidation of ethanol.

$$CH_3CHO$$

Acetaldehyde Dehydrogenase The enzyme that catalyzes the following reaction:

$$\text{Acetyl-CoA} + \text{NADH} \rightleftharpoons \text{Acetylaldehyde} + \text{CoA} + \text{NAD}^+$$

Acetaldehyde Syndrome Accumulation of acetaldehyde in the blood.

Acetamide (Acetic Acid Amide, mol wt 59) A solvent for many organic and inorganic compounds.

$$CH_3CONH_2$$

Acetamidocaproic Acid (mol wt 173) An anti-inflammatory agent (zinc salt acts as antiulcerative agent)

$$CH_3CONH(CH_2)_5COOH$$

Acetamidoeugenol (mol wt 277) An anesthetic agent.

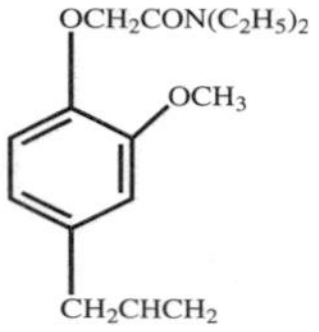

Acetaminophen (mol wt 151) An analgesic and antipyretic agent. It inhibits the synthesis of prostaglandins, which act as mediators for pain and fever.

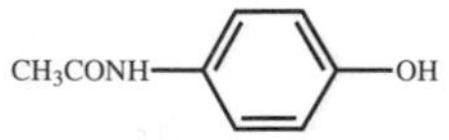

Acetaminosalol (mol wt 271) An anti-pyretic, analgesic, and anti-inflammatory agent.

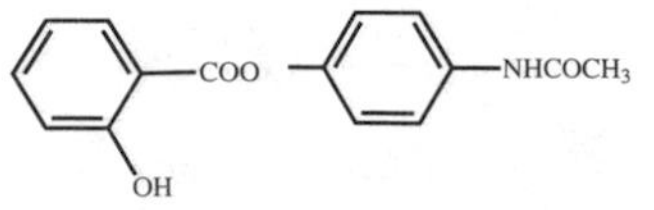

Acetanilide (mol wt 135) An analgesic and antipyretic agent.

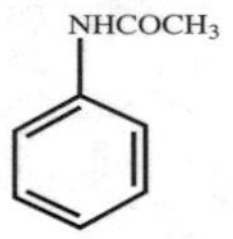

Acetarsone (mol wt 275) An antiprotozoal *(Trichomonas)* and antibacterial agent.

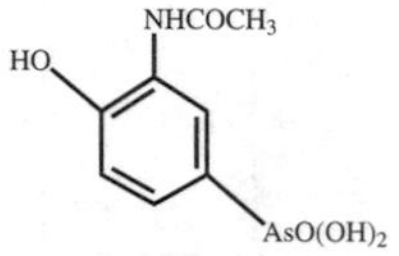

Acetate A salt of acetic acid.

Acetate-CoA Ligase (ADP-Forming) The enzyme that catalyzes the following reaction:

$$\text{ATP} + \text{acetate} + \text{CoA} \rightleftharpoons \text{ADP} + \text{Pi} + \text{acetyl-CoA}$$

Acetate-CoA Ligase (AMP-Forming) The enzyme that catalyzes the following reaction:

$$\text{ATP} + \text{acetate} + \text{CoA} \rightleftharpoons \text{AMP} + \text{PPi} + \text{acetyl-CoA}$$

Acetate CoA-Transferase The enzyme that catalyzes the following reaction:

$$\text{Acyl-CoA} + \text{acetate} \rightleftharpoons \text{A fatty acid anion} + \text{acetyl-CoA}$$

Acetate Kinase The enzyme that catalyzes the following reaction:

$$\text{ATP} + \text{acetate} \rightleftharpoons \text{ADP} + \text{acetyl-phosphate}$$

Acetate Kinase (Pyrophosphate) The enzyme that catalyzes the following reaction:

$$\text{PPi} + \text{acetate} \rightleftharpoons \text{Pi} + \text{acetyl-phosphate}$$

Acetate Thiokinase (Acetyl-CoA Synthetase) See acetate-CoA ligase.

Acetazolamide (mol wt 222) A carbonic anhydrase inhibitor and diuretic agent.

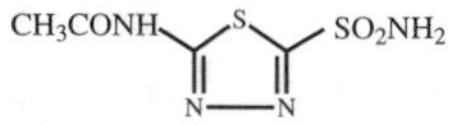

Acetest A method used to test for the presence of abnormal quantities of acetone in the urine of patients with diabetes mellitus or other metabolic disorders.

Acetiamine (mol wt 366) A fat-soluble derivative of vitamin B_1. It acts as enzyme cofactor in a number of biochemical reactions.

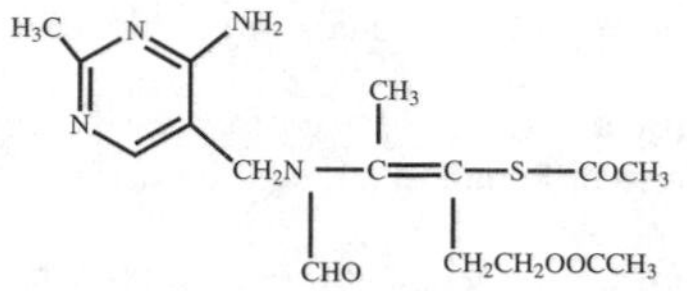

Acetic Acid Bacteria Any bacteria capable of acetification (e.g., *Acetobacter* spp and *Gluconobacter* spp).

Acetic Acid Glacial (mol wt 60) A clear, colorless, pungent liquid found in vinegar that is miscible with water, alcohol, glycerin, and ether.

CH_3COOH

Acetic Anhydride (mol wt 102) A compound derived from two molecules of acetic acid by removal of one molecule of water.

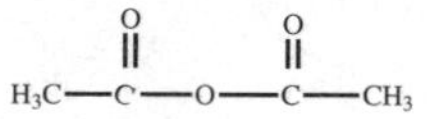

Acetic Fermentation The production of acetic acid or vinegar from a weak alcoholic solution by microorganisms.

Aceticlastic Capable of catabolizing acetate.

Acetification The aerobic conversion of ethanol to acetic acid by bacteria (e.g., *Acetobacter* spp).

Acetin A mixture of acetic acid and glycerin.

Acetivibrio A genus of bacteria (family Bacteroidaceae).

Acetoacetate Salt of acetoacetic acid.

Acetoacetate Carboxylase See acetoacetate decarboxylase.

Acetoacetate-CoA Ligase The enzyme that catalyzes the following reaction:

$$\text{ATP} + \text{acetoacetate} + \text{CoA} \rightleftharpoons \text{AMP} + \text{PPi} + \text{acetoacetate-CoA}$$

Acetoacetate Decarboxylase The enzyme that catalyzes the conversion of acetoacetate to acetone and carbon dioxide.

Acetoacetic Acid (mol wt 102) A ketoacid produced from bacterial fermentation or metabolism of fatty acid; one of the ketone bodies.

CH_3COCH_2COOH

Acetoacetyl-ACP A complex of acetacetate and acyl-carrier protein, an intermediate in the synthesis of fatty acid (see also acetoacetyl-S-ACP).

$CH_3COCH_2CO-S-ACP$

Acetoacetyl-CoA The condensation product from acetyl-CoA.

$CH_3\text{-}CO\text{-}CH_2\text{-}CO\text{-}S\text{-}Co\text{-}A$

Acetoacetyl-CoA Hydrolase The enzyme that catalyzes the following reaction:

$$\text{Acetoacetyl-CoA} + H_2O \rightleftharpoons \text{CoA} + \text{acetoacetate}$$

Acetoacetyl-CoA Reductase The enzyme that catalyzes the following reaction:

$$\text{3-Hydroxyacyl-CoA} + \text{NADP}^+ \rightleftharpoons \text{3-keto-acyl-CoA} + \text{NADPH}$$

Acetoacetyl-CoA Synthetase See acetoacetate-CoA ligase.

Acetoacetyl-CoA Thiolase See acetyl-CoA acetyltransferase.

Acetoacetyl-S-ACP A variant writing of acetoacetyl-ACP (see also acetoacetyl-ACP).

Acetobacter A genus of Gram negative bacteria of the family Acetobacteraceae.

Acetobacterium A genus of Gram-negative, obligate anaerobic bacteria occurring in marine and freshwater sediments.

Aceto-Carmine A stain used in the preparation of chromosome squashes consisting of 5% carmine in 45% acetic acid.

Acetogen Any bacterium capable of producing acetate as the main product from CO_2 and H_2 and/or from certain sugars (e.g., *Acetobacterium woodii*; *Clostridium aceticum*).

Acetogenesis Acetate formation by microorganisms.

Acetohexamide (mol wt 324) A sulfonylurea, an oral antidiabetic agent that stimulates insulin release from pancreatic beta cells and reduces glucose output by the liver.

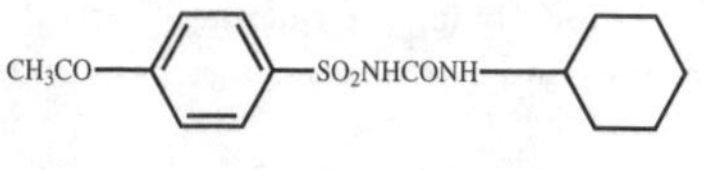

Acetohydroxamic Acid (mol wt 75) An antiurolithic and antibacterial agent that prevents formation of renal stones by inhibiting bacterial urease activity (a urease inhibitor).

$CH_3CONHOH$

Acetoin (mol wt 88) 2-Keto-3-hydroxybutane, a product of microbial fermentation.

$CH_3CH(OH)COCH_3$

Acetoin Dehydrogenase The enzyme that catalyzes the following reaction:

$$\text{Acetoin} + NAD^+ \rightleftharpoons \text{Diacetyl} + NADH$$

Acetoin Racemase The enzyme that catalyzes the following reaction:

$$\text{(S)-Acetoin} \rightleftharpoons \text{(R)-Acetoin}$$

Acetokinase See acetate kinase.

Acetolactate Decarboxylase The enzyme that catalyzes the following reaction:

$$\text{(S)-2-Hydroxy-2-methyl-3-keto-butanoate} \rightleftharpoons \text{(R)-2-acetoin} + CO_2$$

Acetolactic Acid (mol wt 134) A product formed from pyruvic acid during the biosynthesis of the amino acid valine.

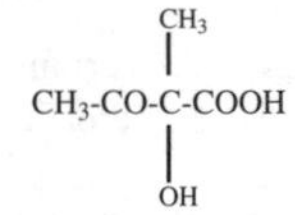

Acetomeroctol (mol wt 465) A topical anti-infective agent.

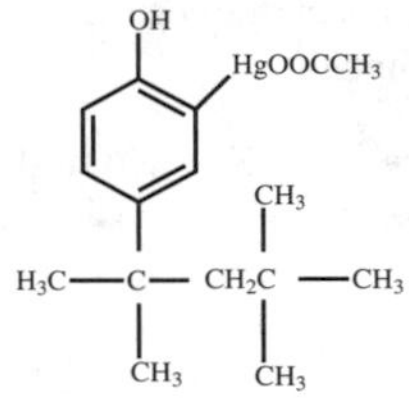

Acetone (mol wt 58) A ketone that can be formed either from condensation of acetyl CoA or from bacterial fermentation. It is found in considerable quantities in the blood and urine of the diabetic patient.

$CH_3—CO—CH_3$

Acetone Body See ketone body.

Acetone Butanol Bacteria Bacteria that are capable of acetone-butanol fermentation (production of acetone and butanol), e.g., *Clostridium acetobutylicum.*

Acetone Ethanol Bacteria Bacteria that are capable of production of acetone and ethanol, e.g., *Bacillus macerans*

Acetone Powder A powder preparation that is obtained by the removal of acetone from tissue-acetone homogenate through vacuum filtration. The powder contains proteins that can be isolated and purified subsequently.

Acetonemia The presence of excessive amounts of acetone or ketone bodies in the blood.

Acetonuria The presence of excessive amounts of acetone or ketone bodies in the urine.

Aceto-Orcein A reagent used in preparation of chromosome squashes consisting of 1% orcein in 45% acetic acid.

Acetophenazine (mol wt 412) A phenothiazine tranquilizer and antipsychotic agent.

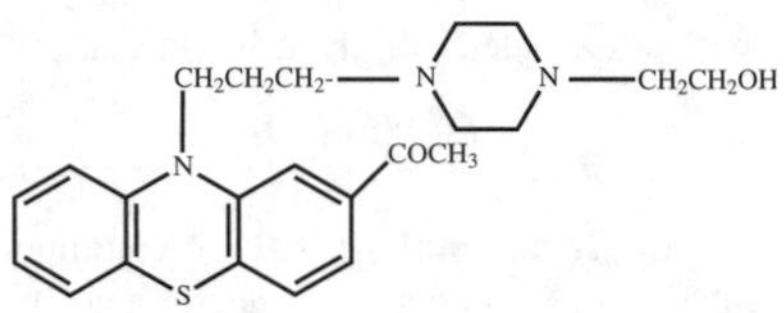

Acetophenone (mol wt 120) A hypnotic agent.

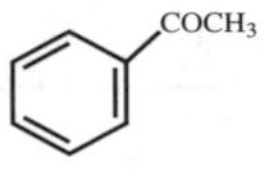

Acetosulfone Sodium (mol wt 391) An antibacterial *(Leprostatic)* agent.

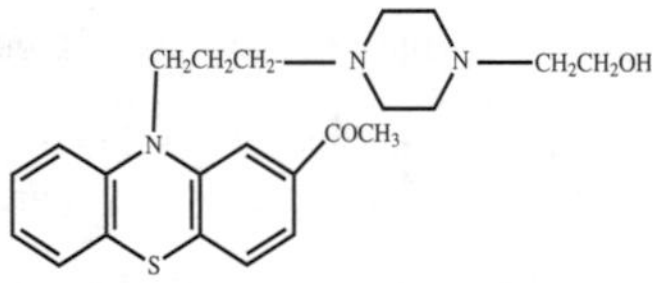

Acetosyringone (mol wt 196) Compound found in the wounded but metabolically active plant cells that can activate the virulent genes on the Ti plasmid of *Agrobacterium tumefaciens.*

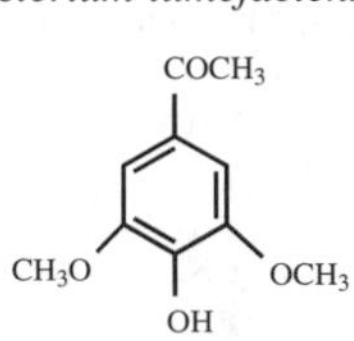

Acetoxan A high molecular weight polysaccharide produced by *Acetobacter xylinum.* It consists of glucose, mannose, ribose, and rhmanose.

Acetoxolone (mol wt 513) An antiulcerative agent.

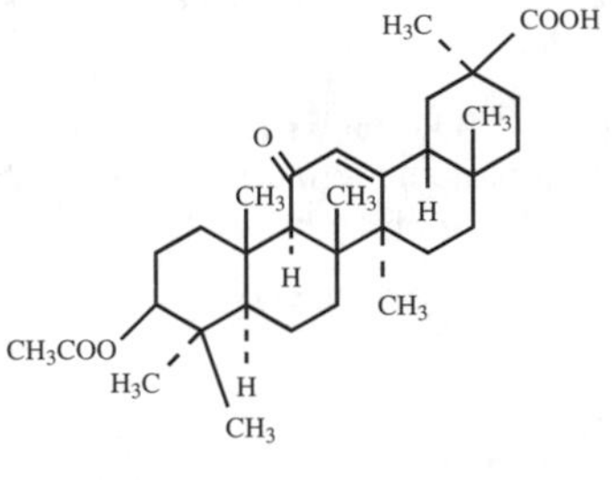

Acetoxypregnenolone (mol wt 375) An anti-inflammatory and antiarthritic agent.

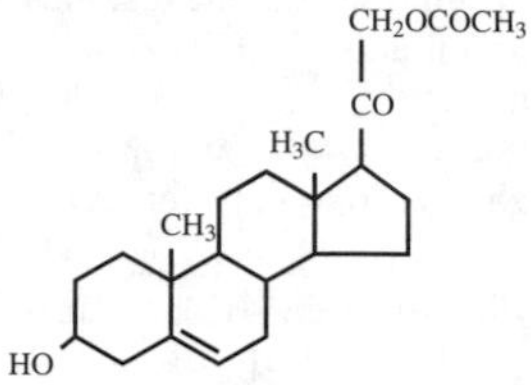

Acetozone (mol wt 180) A germicide used in the bleaching of flour and food oils. It can cause severe skin burns.

$C_6H_5CO\text{-}O\text{-}O\text{-}CO\text{-}CH_3$

Acetrizoate Sodium (mol wt 579) A reagent used as a radiopaque medium.

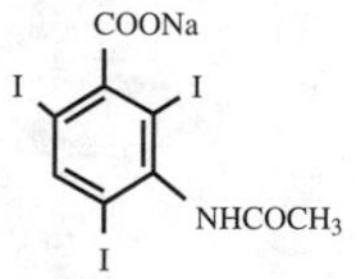

Acetyl Referring to CH_3CO^-.

Acetylation A reaction in which an acetyl radical CH_3CO^- is introduced into an organic compound.

Acetylcarnitine (mol wt 203) A carnitine acetyl ester and a nootropic agent.

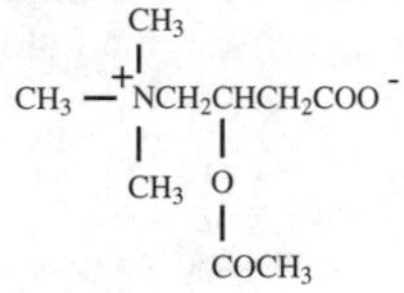

Acetylcholine (mol wt 146) A neurotransmitter responsible for transmission of nerve impulses.

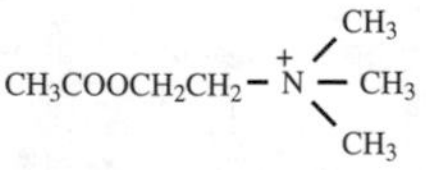

Acetylcholine Chloride (mol wt 182) A cholinergic and mitotic agent.

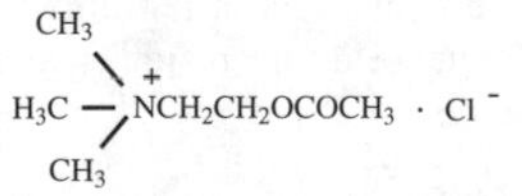

Acetylcholine Esterase The enzyme that hydrolyzes acetylcholine to choline and acetic acid.

Acetylcholine Hydrolase See acetylcholinesterase.

Acetylcholine Transporter Protein An integral membrane protein of the synaptic vesicles of cholinergic neurons. It transports newly synthesized acetylcholine molecules into the synaptic vesicles.

Acetyl-CoA (Acetyl Coenzyme A, mol wt 809) A condensation product of coenzyme A and acetic acid. It is the entry compound for the Krebs cycle.

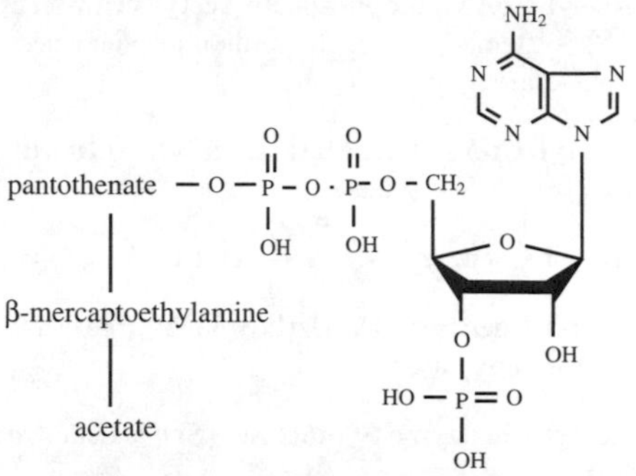

Acetyl-CoA Acetyltransferase The enzyme that catalyzes the formation of acetoacetyl CoA.

$$\text{Acetyl CoA} + \text{Acetyl CoA} \rightleftharpoons \text{Acetoacetyl CoA} + \text{CoA-SH}$$

Acetyl-CoA Carboxylase The enzyme that catalyzes the following reaction:

$$\text{ATP} + \text{acetyl-CoA} + CO_2 + H_2O \rightleftharpoons \text{ADP} + \text{Pi} + \text{malonyl-CoA}$$

Acetyl-CoA Carboxylase Kinase The enzyme that catalyzes the following reaction:

$$\text{ATP} + \text{acetyl-CoA carboxylase} \rightleftharpoons \text{ADP} + \text{acetyl-CoA carboxylase phosphate}$$

Acetyl-CoA Carboxylase Phosphatase The enzyme that catalyzes the following reaction:

$$\text{Acetyl-CoA carboxylase phosphate} + H_2O \rightleftharpoons \text{Acetyl-CoA carboxylase} + \text{orthophosphate}$$

Acetyl-CoA Carnitine O-Acetyltransferase The systematic name for carnitine O-acetyl transferase.

Acetyl-CoA Chloramphenicol O-Acetyl-Transferase The systematic name for chloramphenicol O-acetyltransferase.

Acetyl-CoA Choline Acetyltransferase The systematic name for choline acetyltransferase.

Acetyl-CoA Hydrolase The enzyme that catalyzes the following reaction:

$$\text{Acetyl-CoA} + H_2O \rightleftharpoons \text{CoA} + \text{acetate}$$

Acetyl-CoA Kanamycin 6'-N-Acetyl Transferase The systematic name for kanamycin 6'-N-acetyltransferase

Acetyl-CoA Orthophosphate Acetyl Transferase The systematic name for orthophosphate acetyl transferase.

Acetyl-CoA Synthetase (Acetyl Activating Enzyme) See acetate-CoA ligase.

Acetyl-Coenzyme A See acetyl-CoA.

Acetyl-Coenzyme A Carboxylase See acetyl-CoA carboxylase.

Acetyl-Coenzyme Synthetase See acetate-CoA ligase.

Acetylcysteine (mol wt 163) A derivative of cysteine, a mucolytic agent, and an antidote for acetaminophen poisoning. It increases production of respiratory tract fluids to help liquefy and to reduce the viscosity of tenacious secretions.

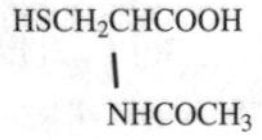

Acetyldigitoxin (mol wt 807) A cardiotonic agent obtained from the enzymatic hydrolysis of lanatoside A. It consists of aglycone digitoxigenin, digitoxose, and acetylated digitoxose.

Acetylene (mol wt 26) A substrate used for assaying nitrogenase activity.

HC≡CH

Acetylene Reduction Assay An assay for nitrogen fixation based upon the conversion of acetylene to ethylene by nitrogenase in nitrogen fixation.

Acetylesterase The enzyme that hydrolyzes an acetic ester to an alcohol and an acetate.

N-Acetyl-D-Galactosamine (mol wt 221) An amino sugar present in various polysaccharides.

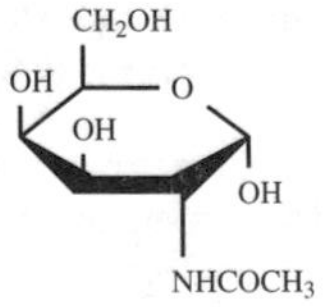

α-N-Acetylgalactosaminidase The enzyme that catalyzes the hydrolysis of the terminal nonreducing N-acetyl-D-galactosmine residues in N-acetyl-α-D-galactosaminides.

β-N-Acetylgalactosaminidase The enzyme that catalyzes the hydrolysis of the terminal nonreducing N-acetyl-D-galactosamine residues in N-acetyl-β-D-galactosaminides.

N-Acetyl-D-Glucosamine Kinase The enzyme that catalyzes the following reaction:

ATP + N-acetyl-D-glucosamine

⇌

ADP + N-acetyl-D-glucosamine 6-phosphate

N-Acetylglucosamine 6-Phosphate (mol wt 301) The phosphorylated form of N-acetylglucosamine.

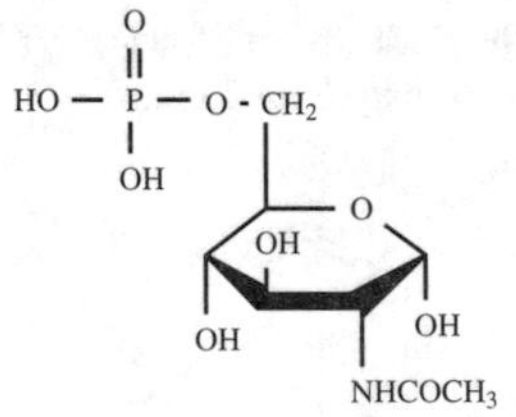

N-Acetylglucosamine 6-Phosphate Deacetylase The enzyme that catalyzes the following reaction:

N-Acetyl-D-glucosamine 6-phosphate + H_2O

⇌

D-Glucosamine 6-phosphate + acetate

N-Acetylglucosamine 6-Phosphate 2-Epimerase The enzyme that catalyzes the following reaction:

N-Acetylglucosamine 6-phosphate

⇌

N-Acetylmannosamine 6-phosphaten

N-Acetylglucosamine Phosphomutase The enzyme that catalyzes the following reaction:

N-Acetyl-D-glucosamine 1-phosphate

⇌

N-Acetyl-D-glucosamine 6-phosphate

α-N-Acetylglucosaminidase The enzyme that catalyzes the hydrolysis of terminal non-reducing N-acetyl-D-glucosamine residues in N-acetyl-D-glucosaminides

β-N-Acetylglucosaminylglycopeptide β-1,4-Galactosyltransferase The enzyme that catalyzes the following reaction:

UDP-galactose
+
N-acetyl–β-D-glucosaminylglycopeptide

⇌

UDP + β-D-galactosyl
1,4-N-acetyl-β-D-glucosaminylglycopeptide

6-Acetylglucose Deacetylase The enzyme that catalyzes the following reaction:

6-Acetylglucose + H_2O ⇌ D-Glucose + acetate

N-Acetylglutamate (mol wt 189) An acetylated form of glutamic acid. It is a cofactor for carbamoylphosphate synthetase.

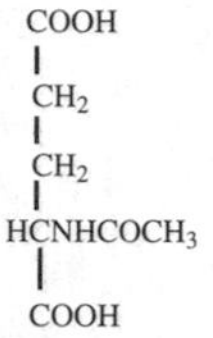

N-Acetylglutamyl-Phosphate Reductase The enzyme that catalyzes the following reaction:

N-Acetyl-L-glutamate 5-semialdehyde + Pi + $NADP^+$
⇅
N-Acetyl-5-glutamylphosphate + NADPH

N-Acetylglutamate Synthetase The enzyme that catalyzes the synthesis of N-acetylglutamate from glutamate and acetyl-CoA.

N-Acetyl-γ-Glutamylphosphate Dehydrogenase The enzyme that catalyzes the following reaction:

N-Acetylglutamate 5-phosphate + NADPH
⇅
$NADP^+$ + Pi + N-acetylglutamate 5-semialdehyde

N-Acetylglutamate Kinase The enzyme that catalyzes the following reaction:

N-acetyl-L-glutamate +ATP
⇅
N-Acetyl-L-glutamate 5-phosphate + ADP

β-N-Acetyl-D-Hexosaminidase The enzyme that catalyzes the hydrolysis of terminal nonreducing N-acetyl-D-hexosamine residues in the N-acetyl-β-hexosaminides.

β-N-Acetyl-D-Hexosaminide N-Acetyl Hexosaminohydrolase The systematic name for β-N-acetyl-D-hexosaminidase.

N-Acetyl-Hydroxyproline (mol wt 173) A derivative of hydroxyproline and an antirheumatic agent.

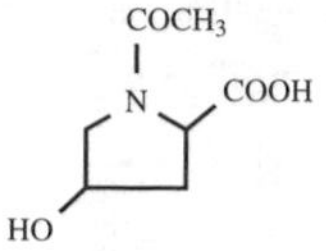

N-Acetyl 5-Hydroxytryptamine (mol wt 218) A metabolite of serotonin and an inhibitor of hydroxyindole-O-methyltransferase.

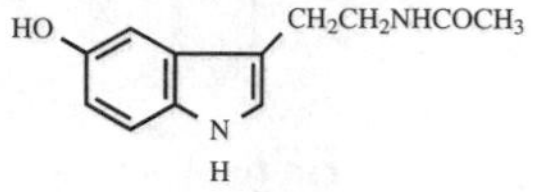

N-Acetylimidazole (mol wt 110) An acetylating agent specific for tyrosyl residues in a protein.

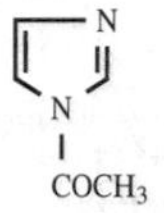

Acetyl-Kinase (Acetate Kinase, Acetokinase) The enzyme that catalyzes the transfer of a phosphate group from ATP to an acetate.

N-Acetyllactosamine Synthetase The enzyme that catalyzes the following reaction:

UDP-galactose + N-acetylglucosamine
⇅
UDP + N-acetyllactosamine

Acetylleucine Monoethanolamine (mol wt 234) An antivertigo agent.

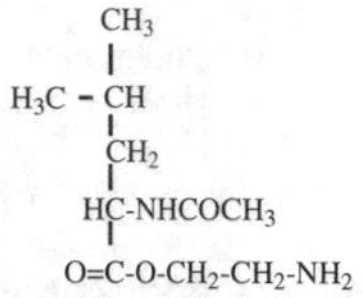

Acetyllipoamide (mol wt 235) An enzyme cofactor.

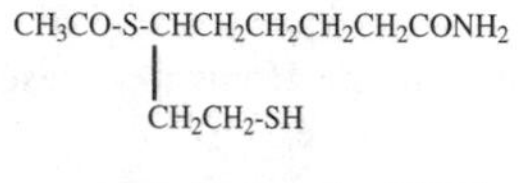

N-Acetylmannosamine (mol wt 221) An amino sugar found in the mucopolysaccharide.

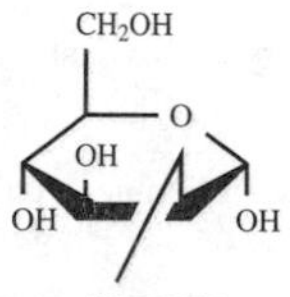

N-Acetylmannosamine 6-Phosphate A compound required for synthesis of N-acetyl neuraminic acid.

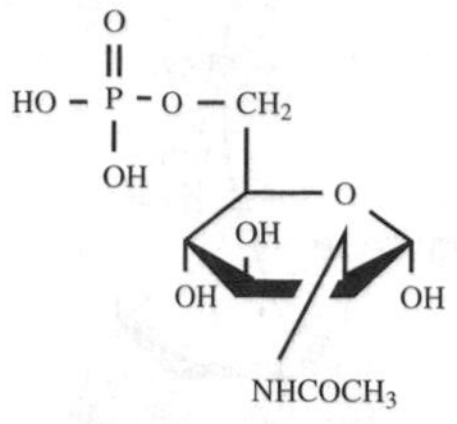

N-Acetylmethionine (mol wt 191) A derivative of methionine and a lipotropic agent.

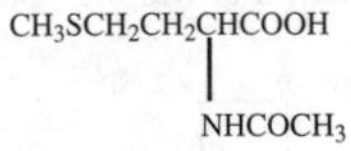

N-Acetylmuramic Acid (mol wt 293) A compound derived from acetic acid, glucosamine, and lactic acid. It is a major building block of bacterial cell walls.

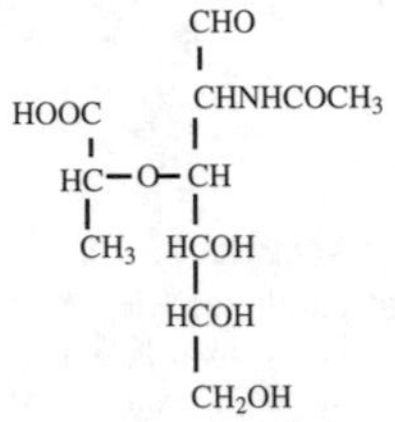

N-Acetylmuramidase Synonym for lysozyme.

N-Acetylmuramoyl-L-Alanine Amidase The enzyme that catalyzes the hydrolysis of the link between N-acetylmuramoyl residues and L-amino acids, e.g., L-alanine.

N-Acetylneuraminate Lyase The enzyme that catalyzes the following reaction:

N-Acetylneuraminate

$\rightleftharpoons$

N-Acetyl-D-mannosamine + pyruvate

N-Acetylneuraminate 9-Phosphatase The enzyme that hydrolyzes N-acetylneuraminate 9-phosphate to N-acetylneuraminate and the inorganic phosphate.

N-Acetylneuraminate Pyruvate-Lyase See N-acetylneuraminate lyase.

N-Acetylneuraminate Synthetase The enzyme that catalyzes the following reaction:

N-Acetylneuraminate + Pi

$\rightleftharpoons$

N-Acetyl-D-mannosamine + phosphoenol pyruvate + H_2O

N-Acetylneuraminic Acid (mol wt 309) A compound derived from acetic acid, mannosamine, and pyruvic acid and a major building block of animal cell coats.

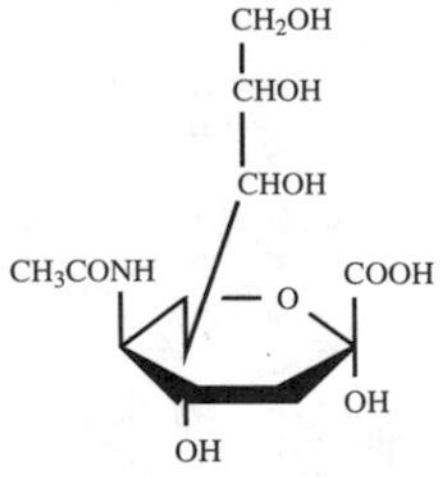

N-Acetylneuraminic Acid Aldolase Synonym of N-acetylneuraminate lyase.

Acetyl-Number The number of milligrams of potassium hydroxide required to neutralize the acetic acid in 1 gram of acetylated fat.

N-Acetylornithine (mol wt 174) A derivative of ornithine and an intermediate in the biosynthesis of ornithine.

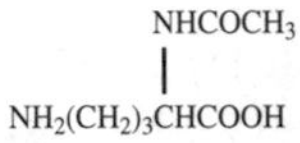

Acetylornithine Cycle A major pathway in bacteria and plants for the synthesis of ornithine from glutamic acid and N-acetylornithine.

N-Acetylornithine Deacetylase The enzyme that hydrolyzes α-N-acetyl-L-ornithine into acetate and L-ornithine.

N-Acetylornithine + H_2O $\rightleftharpoons$ Acetate + ornithine

N-Acetylornithine Transferase The enzyme that catalyzes the following reaction:

N-Acetyl-L-ornithine + 2 α-ketoglutarate

$\rightleftharpoons$

N-Acetyl-L-glutamate semialdehyde + glutamate

Acetyl-Pheneturide (mol wt 248) An anticonvulsant agent.

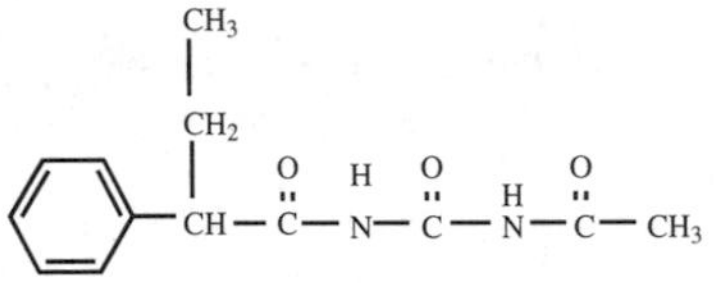

Acetylphenylhydrazine (mol wt 150) An antipyretic agent.

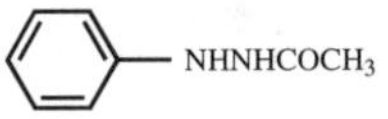

Acetylputrescine Deacetylase The enzyme that catalyzes the following reaction:

N-Acetylputrescine + H_2O

$\rightleftharpoons$

Acetate + putrescine

Acetylpyruvate Hydrolase The enzyme that catalyzes the following reaction:

Acetylpyruvate + H_2O $\rightleftharpoons$ Acetate + pyruvate

Acetyl-S-ACP A complex of acetate and acyl carrier protein and a substrate for the biosynthesis of fatty acid (see also acetyl-ACP).

$CH_3CO\text{-}S\text{-}ACP$

Acetylsalicylate (Aspirin, mol wt 180) An analgesic, antipyretic, and anti-inflammatory agent.

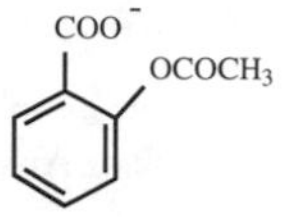

Acetylsalicylsalicylic Acid (mol wt 300) An analgesic agent.

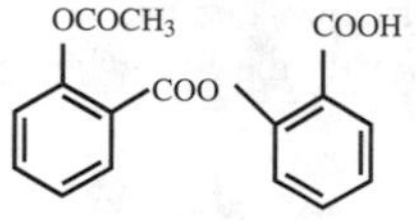

Acetyl-S-CoA Acetyl coenzyme A.

$$H_3C-\overset{O}{\overset{\|}{C}}-S-CoA$$

N-Acetyl-Serine The acetylated form of the amino acid serine.

CH_2OH
|
$CHNHCOCH_3$
|
$COOH$

Acetylserotonin N-Methyltransferase The enzyme that catalyzes the following reaction:

S-adenosyl-L-methionine + N-acetylserotonin

⇅

S-adenosyl-L-homocysteine + N-acetyl-5-methoxytryptamine

Acetylspermidine Deacetylase The enzyme that catalyzes the following reaction:

Acetylspermidine + H_2O ⇌ acetate + spermidine

Acetyl-Sulfamethoxypyrazine (mol wt 322) An antibacterial agent.

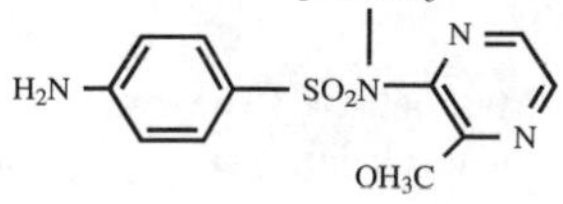

Acetylsulfisoxazole (mol wt 309) An antibacterial agent.

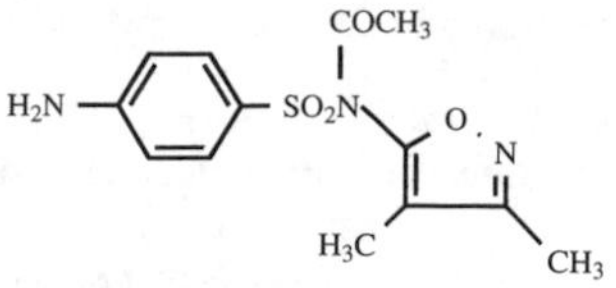

Acetyltransacylase The enzyme that catalyzes the following reaction:

Acetyl-CoA + ACP-SH

⇅

Acetyl-S-ACP + CoA-SH

AcFuCy Abbreviation for a combination drug containing actinomycin-D, fluorouracil and cytoxan.

ACG A genetic code (codon) for the amino acid threonine.

AcG Abbreviation for accelerator globulin.

Ac-Gly Abbreviation for acetyl glycine.

ACH Abbreviation for adrenocortical hormone.

ACh Abbreviation for acetylcholine.

Δ_2Ach Symbol for the (all-Z)-eicosa-8,11-dienoyl group.

Δ_3Ach Symbol for the (all-Z)-eicosa-5,8,11-trienoyl group.

Δ_4Ach Symbol for the (all-Z)-eicosa-5,8,11,14-tetraenoyl group.

Achalasia A disorder characterized by the inability of a muscle to relax, particularly in the gastrointestinal tract.

A-Channel Type of potassium-selective ion channel that is activated by depolarization after a preceding hyperpolarization.

Achard-Thiers Syndrome A hormonal imbalance disorder in postmenopausal women with diabetes, characterized by growth of body hair in a masculine distribution.

Achatin-1 An endogenous neuro-excitatory tetrapeptide (Gly-Phe-Ala-Asp) isolated from ganglia of African snail.

AChE Abbreviation for acetylcholinesterase.

Acheiria The congenital absence of one or both hands.

Aches-N-Pain A trade name for ibuprofen, a nonsteroidal anti-inflammatory agent.

Achlorhydria A disorder characterized by the absence of hydrochloric acid in the gastric juice.

Achlorophyllous See achlorotic.

Achloropsia The inability to perceive the color green.

Achlorotic Lacking chlorophyll.

Acholeplasma A genus of facultatively anaerobic, urease-negative bacteria (family Acholeplasmataceae).

Acholeplasmaviruses Bacteriophages that infect *Acholeplasma* species, (e.g., MV-L3 phage of the family plasmaviridae).

Acholia The absence of biliary secretion (lack of bile).

Acholous Lacking bile.

Acholuria The absence of bile pigment in the urine.

AChR Abbreviation for acetylcholine receptor.

AChRAb Abbreviation for acetylcholine receptor antibody.

Achrodextrin Small molecular weight dextrin that does not give color reaction with iodine.

Achromacyte Decolorized red blood cell.

Achromasia 1. Lack of pigment in the skin. 2. Lack of a staining reaction in a cell.

Achromatic Free from color.

Achromatin Tissue that cannot be readily stained.

Achromatism Total color blindness.

Achromatocyte Variant spelling of achromocyte.

Achromatophil A cell that has little or no affinity for cytoplasmic stain.

Achromatopsia Complete color blindness.

Achromatopsy See achromatopsia.

Achromatosis A disorder characterized by the absence of normal pigment in the skin.

Achromaturia Excretion of colorless urine as a consequence of diuresis or chronic renal failure.

Achromia A congenital condition characterized by the deficiency of natural pigment.

Achromic Devoid of color.

Achromic Point A point in the hydrolysis of starch at which the addition of iodine fails to produce a blue color.

Achromocyte A sickle-shaped, hypochromic erythrocyte (also called achromatocyte).

Achromoderma The absence of pigment in the skin.

Achromogenic An organism incapable of producing pigment.

Achromotrichia The absence of pigment in the hair.

Achromycin A trade name for tetracycline, an antibiotic that inhibits bacterial protein synthesis.

Achromycin V A trade name for tetracycline hydrochloride, an antibiotic that inhibits bacterial protein synthesis.

Achylia Deficiency of hydrochloric acid and pepsinogen in the stomach. Also called achylosis.

ACI Abbreviation for adenylate cyclase inhibitor.

AciI A restriction endonuclease from *Arthrobacter citreus* with the following specificity:

5'........C↓CGC........3'
3'........GGC↑G........5'

Acid A substance that is capable of forming hydrogen ions (H^+) when dissolved in water.

-acid A combining form meaning an acid or pertaining to an acid.

Acid Alcohol A reagent used for decolorization in the acid-fast staining of microorganisms, e.g., *Mycobacterium tuberculosis*. It consists of 3 ml of concentrated hydrochloric acid in 100 ml of alcohol.

Acid Anhydride Referring to a compound that contains two acyl groups bound to an oxygen atom.

Acid Base Balance The maintenance of a constant, optimum internal pH environment in the various fluid compartments of the body.

Acid Base Catalysis The catalysis in solution in which the catalysts are 1. free protons and/or free hydroxy ions or 2. various acidic and/or basic species that serve as proton donors and/or proton acceptors.

Acid Base Indicator A substance in a weak acid or weak base that has a different color in acid or base solution.

Acid Base Metabolism The metabolic processes that maintain the balance of acids and bases in body fluids.

Acid Base Titration A titration in which either acid or base is added to a solution and the change in pH is followed by means of a pH measurement.

Acid Citrate-Dextrose Solution A solution of sodium chloride, sodium citrate, and dextrose that is used as an anticoagulant in the collection and storage of blood (see also Alsever's solution).

Acid CoA Ligase The enzyme that catalyzes the following reaction:

$$GTP + \text{an acid} + CoA \rightleftharpoons GDP + Pi + \text{acyl-CoA}$$

Acid Curd The coagulant formed from coagulation of milk proteins by an acid.

Acid Dyes The anionic dyes that contain an acidic organic component that stain positively charged structures.

Acid Fast The property of bacteria with lipid-rich cell walls, e.g., mycobacteria that resist decolorization by acid-alcohol after staining with basic dyes.

Acid Fast Stain A staining method used to demonstrate the acid-fast property of certain bacteria, e.g., *Mycobacterium tuberculosis*.

Acid Fuchsin (mol wt 586) A pH indicator and biological stain.

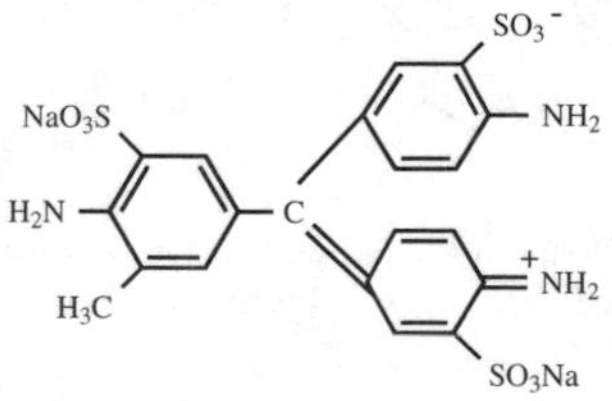

Acid Glycoprotein A plasma protein whose concentration increases following trauma, acute inflammation, and malignancy.

Acid Hematin A hematin formed by treatment of hemoglobin with acid below pH 3.

Acid Hydrolase Any hydrolase that is active in mildly acidic conditions (pH 5-6).

Acid Metabolism The metabolic process of certain photosynthetic organisms in which carbon dioxide is absorbed at night, stored in the form of acid and released for use in daytime when the acid is broken into carbon dioxide. It permits the photosynthesis with stoma closed.

Acid Mucopolysaccharide A group of heteropolysaccharides that contain N-acetylated hexosamine in its characteristic repeating disaccharide unit.

Acid Number The number of milligrams of potassium hydroxide required to neutralize the free fatty acids in 1 gram of fat.

Acid Perfusion Test An experimental animal-model test to determine the sensitivity of the esophagus to acid in which 0.1 N hydrochloric acid and normal saline are dripped alternately into the esophagus via a nasal-esophageal tube. A positive response is pain with acid but not with saline.

Acid pH A pH value below 7.

Acid Phosphatase A nonspecific phosphomonoesterase with an optimum pH below 7. It catalyzes the hydrolysis of an orthophosphoric monoester to an alcohol and a H_3PO_4.

Acid Phosphomonoesterase Synonym of acid phosphatase.

Acid Poisoning A toxic condition caused by the ingestion of a toxic acid such as hydrochloric, nitric, phosphoric, or sulfuric acid.

Acid Precipitation 1. The rain, fog, or snow with high acidity caused by the pollutants (e.g., SO_2, or nitrogen oxides) from industry, motor vehicle exhausts, and other sources. 2. Precipitation of DNA, RNA, or proteins by an acid.

Acid Protease or Acid Proteinase Proteolytic enzymes that are active only under acid conditions.

Acid Violet 7B (mol wt 706) A dye.

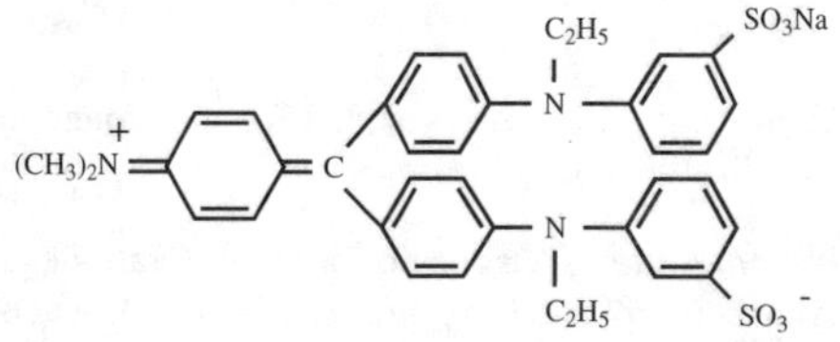

Acidemia A condition characterized by an abnormal increase in the hydrogen-ion concentration in the blood or loss of bicarbonate as a result of increase of hydrogen ion concentration.

Acidic Pertaining to 1. an acid or 2. a pH value less than 7.

Acidic Amino Acid An amino acid with one amino group and two carboxyl groups, e.g., aspartic acid.

Acidic Dye An anionic dye capable of binding to and staining positively charged macromolecules or structures.

Acidic Proteins Proteins that are rich in acidic amino acids.

Acidify To lower the pH or to make acid.

Acidimetry 1. The chemical analysis of solutions by means of titrations. 2. Determination of the amount of an acid by titration against a standard alkaline solution.

Acidity The acid content of a fluid.

Acidocyte See eosinophil.

Acidogen Agent or drug capable of stimulating the release of HCl in the stomach.

Acidolysis Hydrolysis by means of an acid (acid hydrolysis).

Acidophil 1. Cells or microorganisms that have affinity for acidic dyes. 2. An organism that grows only under acidic conditions.

Acidophilic Having an affinity for acidic dye or staining readily with acidic dyes.

Acidophilic Adenoma A tumor of the pituitary gland, characterized by cells that can be stained red with an acid dye.

Acidophilism The overactivity of the acidophilic cells in the pituitary gland

Acidophilus Milk A medicinal beverage produced by fermentation of heat-treated skimmed milk with *Lactobacillus acidophilus*. It contains a high concentration of lactic acid.

Acidosis A condition in which there is an abnormal increase in hydrogen ion concentration in the blood owing to the imbalance of bicarbonate/carbonic acid ratios.

Acidosome A non-lysosomal vesicle found in the ciliate protozoan *Paramecium*.

Acidothermus A genus of aerobic, Gram-negative, thermophilic (growing at 37-70° C), acidophilic (growing at pH 3.5-7.0), cellulolytic, nonmotile bacteria isolated from acidic hot springs.

Aciduria A condition characterized by the presence of an excessive amount of acid in the urine.

Aciduric 1. Tolerant to an acidic condition. 2. Capable of growth under acid condition, e.g., *Lactobacillus spp.*

Acifluorfen (mol wt 362) An herbicide.

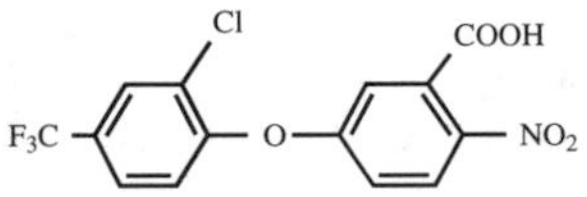

Acifran (mol wt 218) An antihyperlipoproteinemic agent

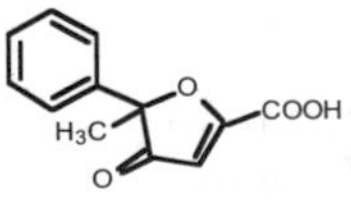

Acinetobacter A genus of oxidase-negative, catalase-positive Gram-variable aerobic bacteria of the family Neisseriaceae, which occur in soil, water, and in clinical specimens as an opportunist pathogen.

Acinus (Plural: Acini) 1. Any small saclike structure in the body (also called alveolus). 2. A subdivision of the lung consisting of the tissue distal to a terminal bronchial. 3. A small lobe of a compound gland or a saclike cavity at the termination of a passage. 4. One of the minute grape-shaped secretory portions of an acinous gland.

Acipimox (mol wt 154) An antihyperlipoproteinemic agent.

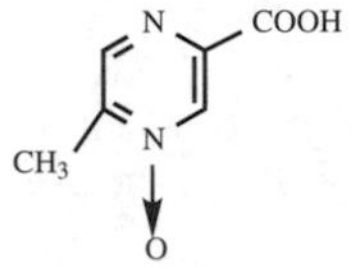

Acirculatory Without a circulatory system.

Acitretin (mol wt 326) An antipsoriatic agent.

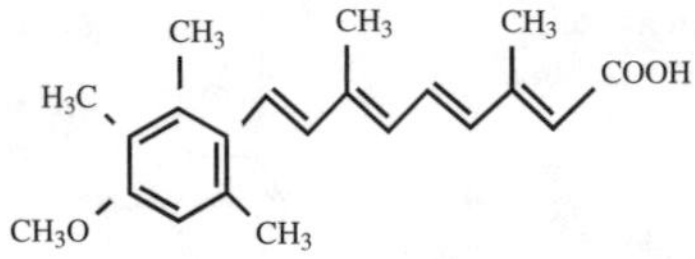

AclI A restriction endonuclease from *Acinetobacter calcoaceticus* M4 with the following specificity:

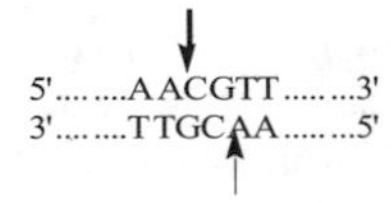

Aclacinomycins An antitumor antibiotic complex of the anthracycline group, produced by *Streptomyces galilaeus*. A number of different aclacinomycins have been identified.

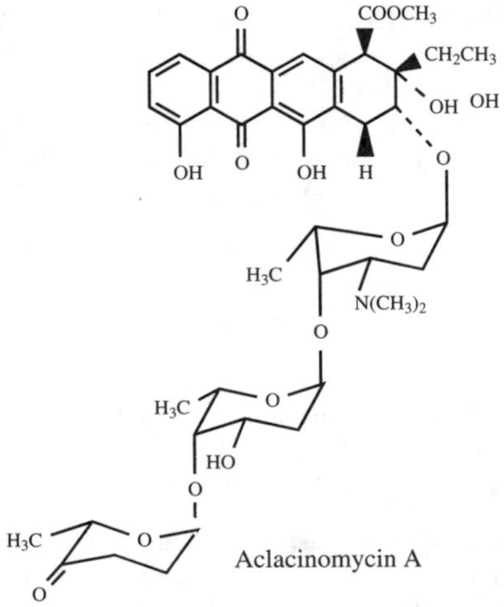

Aclacinomycin A

Aclatonium Napadisilate (mol wt 723) A cholinergic and spasmolytic agent.

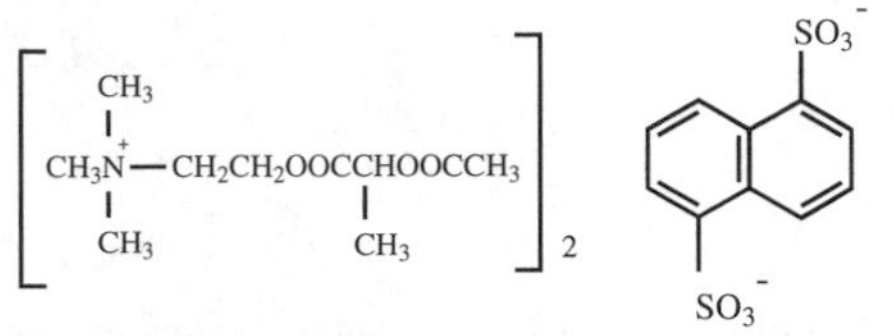

AclNI A restriction endonuclease from *Acinetobacter calcoaceticus* N20 with the following specificity:

5'.........ACTAGT.......3'
3'.........TGATCA.......5'

Aclovate A trade name for alclometasone, a topical corticosteroid.

AclWI A restriction endonuclease from *Acinetobacter calcoaceticus* W2131 with the following specificity:

5'.........GGATC(4/5).......3'

ACM Abbreviation for 1. A combination drug containing adriamycin, cyclophosphamide and methotrexate. 2. A mixture containing albumin, calcium and magnesium.

Acne A skin disorder characterized by papulopustular skin eruption caused by the inflammation of the subaceous glands. It may occur as pimples on the face, neck, shoulder, and upper back.

AcNeu Abbreviation for N-acetylneuraminic acid.

AcNPV Abbreviation for *Autographa californica* nuclear polyhydrosis virus.

Ac_2O Abbreviation for acetic anhydride.

AcOEt Abbreviation for ethylacetate.

AcOH Abbreviation for acetic acid.

Acology Medical science that deals with cures and remedies.

Acon A trade name for vitamin A.

Aconine (mol wt 500) An antipyretic agent obtained by hydrolysis of aconitine.

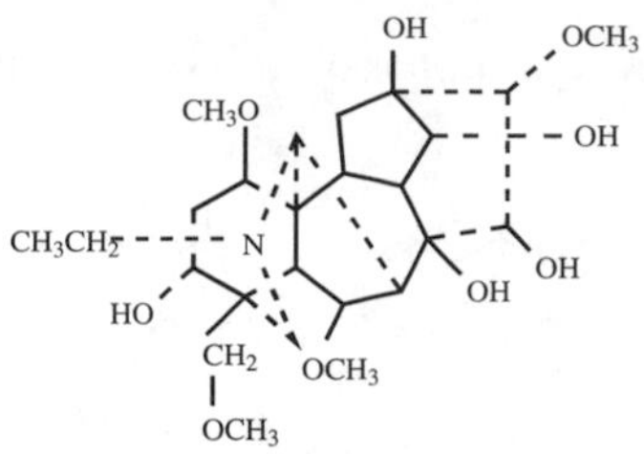

Aconitase (Aconitate Hydrolase) The enzyme that catalyzes the following reaction:

Citrate ⇌ *cis*-Aconitate ⇌ Isocitrate

Aconitate Decarboxylase The enzyme that catalyzes the following reaction:

cis-Aconitate ⇌ Itaconate + CO_2

Aconitate Hydrase The enzyme that catalyzes the following reaction:

Citrate ⇌ cis-Aconitate + H_2O

Aconitate Δ-Isomerase The enzyme that catalyzes the following reaction:

trans-aconitate ⇌ *cis*-aconitate

Aconite The dried tuberous root of *Aconitum napellus* (Ranunculaceae). It consists of aconitine, aconine, napelline, picraconitine, aconitic acid, itaconic acid, succinic acid, malonic acid, fat, and levulose. It has been used as an antihypertensive and antipyretic agent.

***cis*-Aconitic Acid (mol wt 174)** A tricarboxylic acid formed from citric acid in the Krebs cycle.

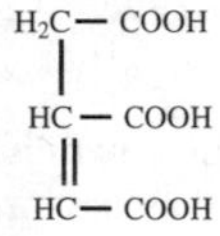

Aconitine (mol wt 646) A constituent of aconite and an antipyretic agent.

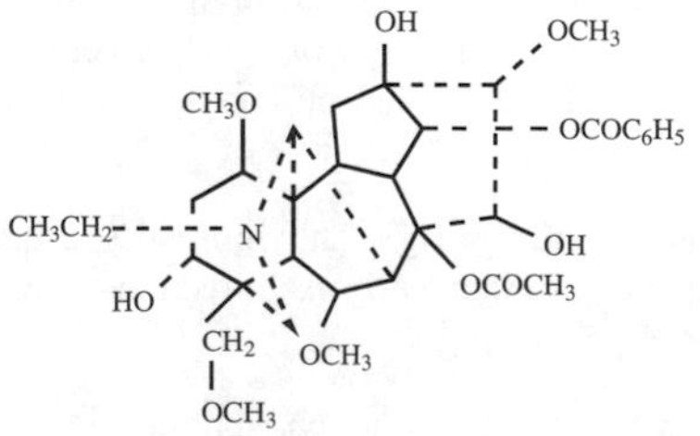

Aconitine Amorphous Mixture of amorphous alkaloids from *Aconitum napellus* (Ranunculaceae) consisting of aconitine, mesaconitine, hypaconitine, neopelline, ephedrine, sparteine, neoline, and napelline.

Aconuresis Lack of control of urination.

ACOP Abbreviation for a combination drug containing adriamycin, cyclophosphamide, oncovin and prednisone.

Acorn-Tipped Catheter A flexible catheter with an acorn-shaped tip used in various diagnostic procedures.

Acou- A combining form that means hearing.

Acoumeter An instrument used for determining the acuteness of hearing.

Acousmatagnosis Failure to recognize sound.

Acoustic Pertaining to sound and the sense of hearing.

Acoustic Gene Transfer Transforming cells by using ultrasound.

Acoustic Microscope A microscope in which the object being viewed is scanned with sound waves and the image is reconstructed with light waves on a video screen.

Acoustics The science that deals with sound and hearing.

ACOX Abbreviation for acyl-CoA oxidase.

ACP Abbreviation for 1. Acyl-carrier protein, an essential component for synthesis of fatty acid. 2. Alanine carboxy-peptidase. 3. Arginine carboxy-peptidase. 4. Aspartate carboxy-peptidase.

ACPP Abbreviation for adreno-cortico-polypeptide.

ACP-SH Variant writing of ACP (acyl carrier protein).

Acquired Hemolytic Anemia An autoimmune disease characterized by the formation of antibodies to one's own red blood cells.

Acquired Immune Deficiency Syndrome (AIDS) An infectious disease caused by HIV and characterized by diminished immune responsiveness due to the deficiency of T helper cells or the impaired function of helper T cells following infection by HIV.

Acquired Immunity The immunity acquired through exposure to antigens, infections, or vaccination.

AcrI (AvaI) A restriction endonuclease from *Anabaenopsis circularis* with the following specificity:

↓
5′.........CPyCGPuG.........3′
3′.........GPuGCPyC.........5′
↑

AcrII (BstEII) A restriction endonuclease from *Anabaenopsis circularis* with the following specificity:

↓
5′.........GGTNACC.........3′
3′.........CCANTGG.........5′
↑

Acranil (mol wt 461) An antiprotozoal and antiviral agent with radioprotective and interferon-inducing activity.

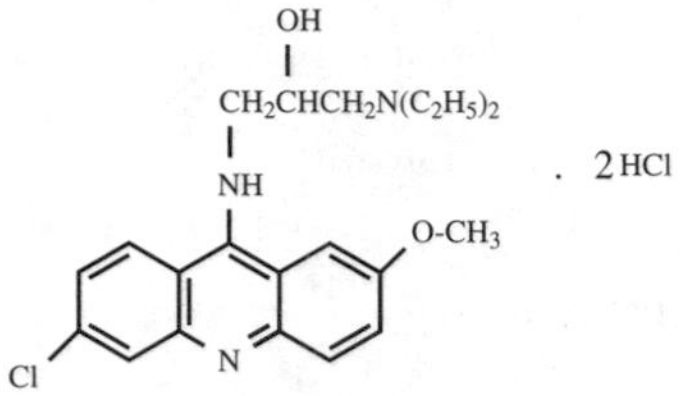

Acrasin A substance produced by slime mold to serve as a chemotactic factor for cell aggregation and the formation of a fruit body.

Acridine (mol wt 179) A benzopyridine compound used in the synthesis of dyes and drugs.

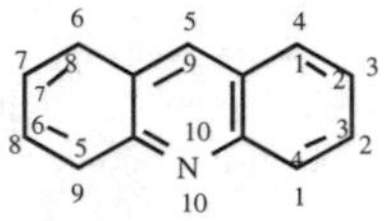

Acridine Orange (mol wt 265) An acridine dye used to stain nucleic acids (ssDNA or ssRNA fluoresces orange-red; dsDNA or dsRNA fluoresces green). It is also mutagenic.

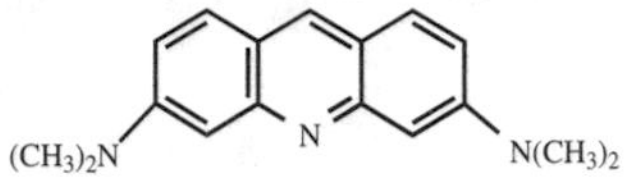

Acriflavine An acridine dye that functions as an antiseptic agent. It also inhibits mitochondriogenesis.

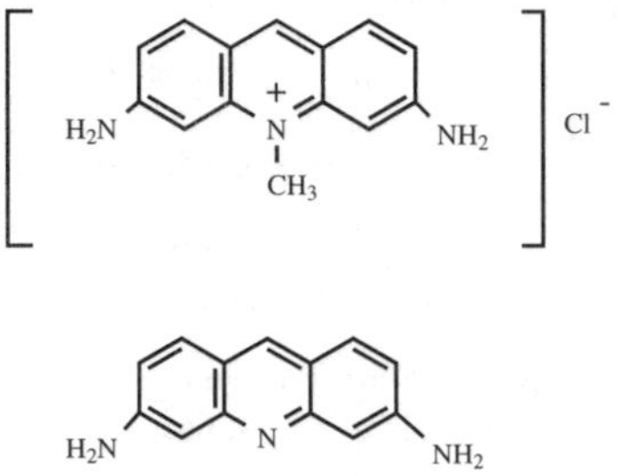

Acrisorcin (mol wt 388) A synthetic antifungal agent.

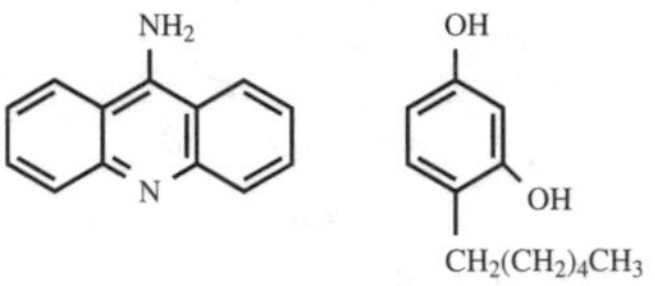

Acrivastine (mol wt 348) A histamine H-1 receptor antagonist with antihistaminic activity.

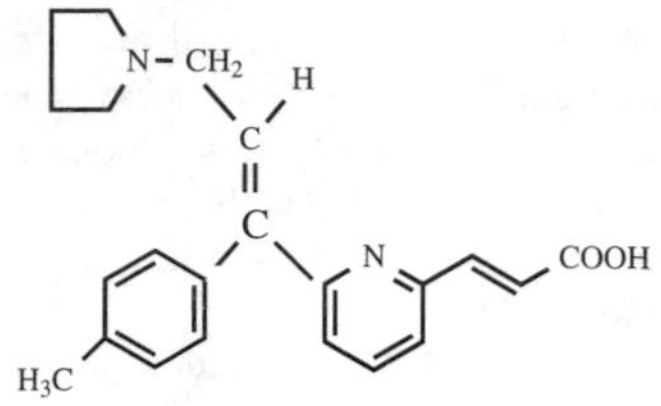

Acro- A prefix meaning tip or outermost part.

Acroarthritis Arthritis in the arms or legs.

Acroataxia The lack of muscular coordination of the fingers and toes.

Acrocentric A chromosome whose centromere is located close to one end of the chromosome.

Acrocyanosis A disorder characterized by the cyanotic discoloration, coldness, and sweating of the extremities (especially the hands) caused by arterial spasm.

Acrohyperhidrosis Excessive sweating of the hands and feet.

Acrokeratosis The warty growth on the hands and feet.

Acrolein (mol wt 56) An aldehyde used as an aquatic herbicide.

$$CH_2 = CH - CHO$$

Acrolein Test A qualitative test for glycerol, based upon the dehydration and oxidation of glycerol to acrolein by heating with potassium bisulfate.

Acromegaly A disorder characterized by the overgrowth of skeletal structures due to the excessive production of growth hormone after the normal growth period has ended.

Acronematic Referring to an eukaryotic *Flagellum* that is smooth and tapers to a fine point.

Acropathology The science that deals with diseases which affect the extremities.

Acropeptide A protein fraction obtained by heating protein above 140° C in a nonaqueous solvent.

Acroposthitis Inflammation of the prepuce.

Acropurpura Purpura of extremities.

Acrosclerosis Scleroderma of the extremities or thickening of the skin and subcutaneous tissue of the hands and feet due to the swelling and thickening of fibrous tissues.

Acrosin A protease in spermatozoa that preferentially cleaves peptide bonds involving amino groups of arginine or lysine.

Acrosomal Process The narrow channel through the surface of the egg coat formed by the polymerization of a pool of actin located behind the acrosomal vesicle of the sperm cell.

Acrosomal Reaction Exocytotic release of enzymes from the acrosomal vesicle when the sperm makes contact with the egg.

Acrosomal Vesicle Vesicle in the sperm head containing enzymes that catalyze the breakdown of the egg surface coat.

Acrosome The membrane-bound structure at the anterior end of a sperm cell. It contains digestive enzymes that enable the sperm to penetrate the protective layers around the oocyte.

Acrosomin A complex of lipoglycoprotein present in the acrosome of spermatozoa.

Acrospore A spore developed at the tip of the fungal hypha.

Acrotism Absence of a pulse.

Acrylamide (mol wt 71) A substance used for the preparation of polyacrylamide gel for electrophoresis. It is a potential cancer causing agent.

$$CH_2 = CH - \overset{\displaystyle O}{\overset{\|}{C}} - NH_2$$

ACS Abbreviation for acetyl-CoA synthetase.

AcsI A restriction endonuclease from *Arthrobacter citreus* with following specificity:

```
          ↓
5'........(A/G)AATT(T/C).......3'
3'........(T/C)TTAA(A/G).......5'
                ↑
```

ACT Abbreviation for activated coagulation time.

ACT A trade name for sodium fluoride used for bone remineralization.

Acta-Char A trade name for activated charcoal used to bind drugs and chemicals within the GI tract.

Actaplanin A glycopeptide antibiotic produced by *Actinomyces missouriensis*, different actaplanins contain different sugars.

Actarit (mol wt 193) An anti-arthritic agent.

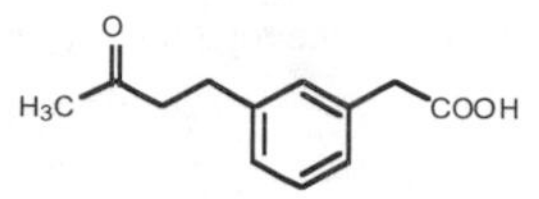

Act-C Abbreviation for actinomycin C.

Act-D Abbreviation for actinomycin D.

ACTH (Adrenocorticotropic Hormone, Adrenocorticotropin, or Corticotropin) A peptide hormone from the anterior pituitary gland that stimulates the production of glucocorticoids.

ActHIB A trade name for *Haemophilus influenzae* b vaccine conjugated with tetanus toxoid.

Acthrel A trade name for corticorelin, a synthetic hormone and an analog of human CRH that stimulates the release of ACTH from anterior pituitary.

Acticort A trade name for hydrocortisone, an immunosuppressor and anti-inflammatory agent.

Actidil A trade name for triprolidine hydrochloride, an antihistaminic agent.

Actidione (mol wt 281) An inhibitor that binds to 80S ribosomes, thus inhibiting protein synthesis.

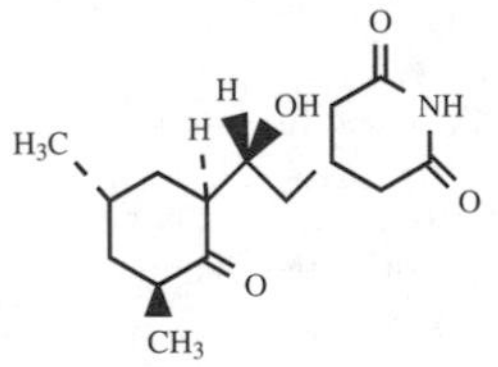

Actidose-Aqua A trade name for activated charcoal used as an antidote.

Actifed A trade name for a combination drug containing an adrenergic bronchodilator and vasoconstrictor (pseudoephedrine hydrochloride) and an antihistamine (triprolidine hydrochloride).

Actifed Allergy A trade name for a combination drug containing pseudoephedrine and diphenhydramine used as a decongestant.

Actifed Sinus Daytime A trade name for a combination drug containing pseudoephedrine and acetaminophen used as a decongestant.

Actifed Sinus Nighttime A trade name for a combination drug containing pseudoephedrine, diphenhydramine and acetaminophen used as a decongestant.

Actigall A trade name for ursodiol that suppresses hepatic synthesis of cholesterol and the intestinal absorption of cholesterol.

Actilyse A trade name for alteplase, a plasminogen activator.

Actimmune A trade name for interferon gamma lb with phagocyte-activating property.

Actin A major protein found in most types of eukaryotic cells. The monomeric form of actin (G-actin) can be polymerized to form noncontractile microfilaments (F-actin). The interaction between F-actin and myosin causes microfilaments to slide to one another, thereby bringing about movement and contraction. α-Actin is found in differentiated muscle cells; β-actin or γ-actin is present in all nonmuscle cells. Actins are involved in a variety of cellular events: e.g., chromosome movement, cytokinesis, phagocytosis, exocytosis, and cytoplasm streaming.

Actin-Binding Domain A structure feature found in the actin-binding protein.

Actin-Binding Proteins Proteins that bind to actin microfilaments and regulate the length or assembly of microfilaments.

Actin-Filament The two-stranded helical polymer of protein actin.

Actinic Pertaining to light rays or radiation energy.

Actinic Dermatitis A skin inflammation or rash resulting from exposure to sunlight, X-ray, or atomic particle radiation.

Actinic Keratosis A localized thickening of the outer layers of the skin caused by excessive exposure to the sun.

Actinidain A protease isolated from the kiwi fruit or Chinese gooseberry with activity similar to papain

***Actinidia* Anionic Protease** Synonym of acitnidain.

Actinidin Synonym of actinidain.

Actinidine A monoterpenoid alkaloid occurring in *Actinidia polygama.*

Actinin An actin-binding protein found in a Z line or Z disk of striated muscle. It plays an important role in anchoring the thin filaments to Z lines. Two forms of actinin, denoted α and β, have been identified.

Actinium (Ac) A radioactive metallic element, with atomic weight 227, valence 2. It occurs in some ores of uranium.

Actino- A prefix signifying a ray or rays.

Actinobacillosis Infection or disease caused by a species of *Actinobacillus*.

Actinobacillus A genus of Gram-negative bacteria of the Pasteurellaceae.

Actinobolin (mol wt 300) An antibiotic produced by *Streptomyces grieoviridus var atrofaciens* that inhibits protein synthesis.

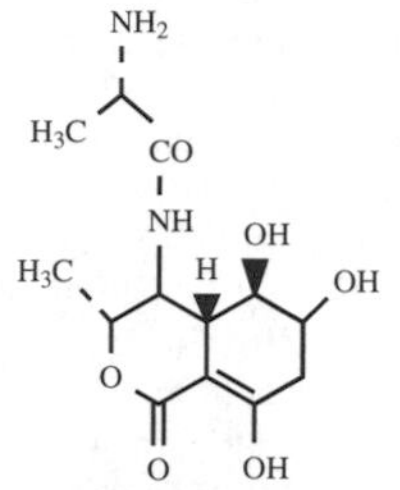

Actinochemistry The science that deals with the effects of visible radiation.

Actinogelin A protein factor that effects the Ca^+-sensitive gelation of actin filaments.

Actinogen A substance that gives off radiation.

Actinoidin See vancomycin.

Actinomadura A genus of bacteria (order Actinomycetales).

Actinometer A device for the measurement of the absorbed light by means of a photochemical reaction of known quantum yield.

Actinomyces A genus of asporogenous bacteria (order Actinomycetales).

Actinomycete Any member of the order Actinomycetales.

Actinomycetin Cell-free culture filtrate of Actinomycetes (e.g., *Streptomyces albus*); contains bacteriolytic substances that are protein in nature and capable of dissolving dead Gram-positive bacteria.

Actinomycin A peptide antibiotic produced by *Streptomyces*. It consists of two identical cyclic peptides joined to a phenoxazone ring system. It exists in many different forms, e.g., Actinomycin C, D, F.

Actinomycin D (mol wt 1255) An antibiotic, produced by *Streptomyces chrysomallus*, that inhibits the transcription of DNA to RNA by binding to DNA and thereby preventing it from being an effective template for synthesis of RNA.

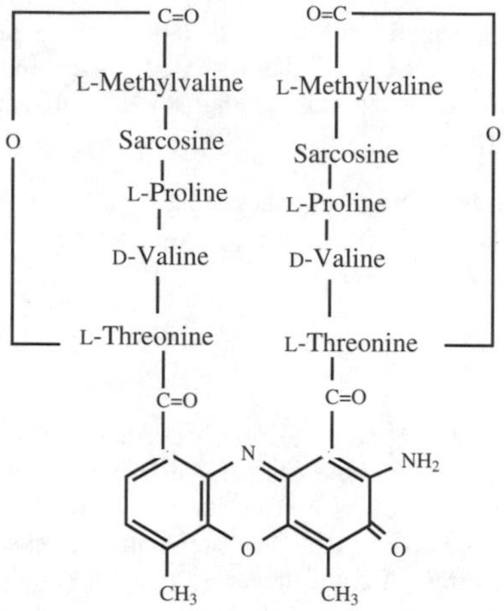

phenoxazone ring

Actinomycoma A swelling due to infection by Actinomycete.

Actinomycosis Infection or disease caused by a species of *Actinomyces*.

Actinomyosin The contractile element in the muscle consisting actin and myosin.

Actinon An isotope of radon having a half-life of about four seconds.

Actinoneuritis Inflammation of nerves due to the excessive exposure to X-rays or other radioactive radiation.

Actinophage Bacteriophages of Actinomycetales, e.g., øEC and VP5.

Actinoplanes A genus of aerobic, asporogenous bacteria (order Actinomycetales) that occurs in soil, plant litter, and aquatic habitats.

Actinopolyspora A genus of bacteria (order Actinomycetales).

Actinorhodine (mol wt 635) An antibiotic produced from *Actinomyces sp.*

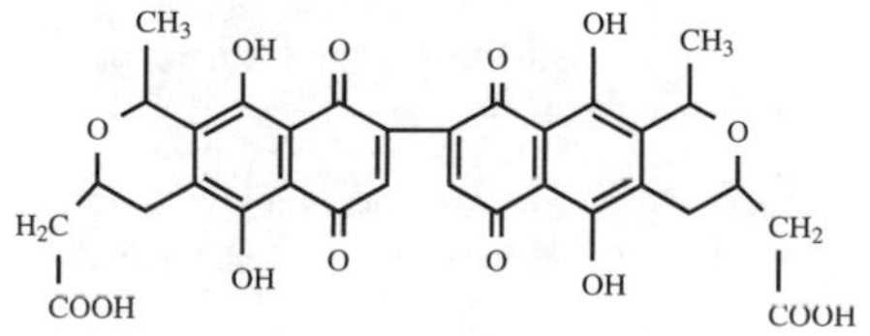

Actinorrhiza A bacterium-plant root association in which nitrogen-fixing root nodules are formed in certain nonleguminous angiosperms infected by the strains *Frankia.*

Actinosynnema A genus of bacteria (order Actinomycetales) that occurs on vegetable matter in aquatic habitats.

Actinotherapy Treatment of illness by sunlight, UV light, or X-ray.

Action Current The electric current produced in the cell membrane of a nerve by the electrical activity in the tissue. This current serves to depolarize adjacent membrane areas and thereby initiates a repetition of the action potential along the nerve fiber.

Action Potential A localized change of electrical potential across the membrane of a nerve or muscle fiber that serves as the means of transmission of a nerve impulse. In the absence of an impulse, the inside is electrically negative and the outside is positive (the resting potential). During the passage of an impulse at any point on the fiber, the inside becomes positive and the outside, negative.

Action Spectrum The extent of radiation of the different wavelengths on a chemical, biochemical, or physiological response. It is a plot of quantitative responses as a function of wavelength.

Actiphenol (mol wt 275) A metabolic product found in the culture filtrate of cycloheximide-producing *Streptomyces albulus.*

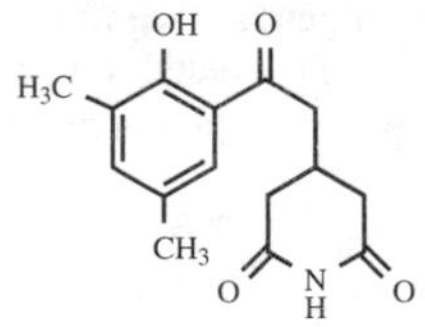

Actiprophen A trade name for ibuprophen, a nonsteroidal anti-inflammatory and analgesic agent.

Actiq A trade name for fentanyl, a narcotic agonist analgesic agent.

Activase A trade name for alteplase, a thrombolytic enzyme and tissue plasminogen activator.

Activated Alumina Thoroughly dried alumina.

Activated Amino Acid Referring to the amino acid-AMP complex (aa-AMP).

Activated Carbon Material prepared by distillation of plant material used for adsorption of gases and decolorization of solutions.

Activated Charcoal A form of carbon that readily adsorbs organic material.

Activated Christmas Factor Referring to coagulation factor IXa.

Activated CO_2 The CO_2 that is carried by biotin or biotin-enzyme complexes, e.g., carboxybiotin.

Activated Hageman Factor Referring to coagulation factor XIIa.

Activated Lymphocytes Lymphocytes that have been stimulated by specific antigen or mitogen.

Activated Macrophages Macrophages expressing elevated metabolic and phagocytic activity following stimulation by agents such as lymphokines.

Activated Protein C A protease that degrades blood coagulation factors Va and VIIIa.

Activation A process of 1. initiating a chemical or biochemical reaction, 2. initiating development of an egg or an organ, 3. converting an inactive components of complement to a functionally active form, 4. initiating differentiation and proliferation of immunoactive cells.

Activation Energy The energy required to elevate molecules from one energy level where they are nonreactive to a higher energy level at which they can react spontaneously. It is the difference in energy between that of the activated complex and that of the reactants.

Activator 1. Substance or ion that can serve as a cofactor for an enzymatic reaction. 2. Substance that is capable of turning on a chain reaction. 3. Substance that causes another substance to become active.

Active Acetaldehyde An acetaldehyde molecule attached to a thiamine pyrophosphate.

Active Acetate Referring to acetyl CoA.

Active Acetyl Referring to 1. acetyl-CoA or 2. acyllipoic acid.

Active Amino Acid Referring to an amino-acyl-AMP complex.

Active Ammonia Referring to carbamoyl phosphate, a substance that serves as initial reactant for the biosynthesis of pyrimidine nucleotides.

Active Carbohydrate Referring to UDP-sugar or GDP-sugar.

Active Carbon Dioxide See carboxybiotin.

Active Center See active site.

Active Enzyme Centrifugation A method for the determination of sedimentation and diffusion coefficients of the enzyme-substrate complex.

Active Fatty Acid Referring to the fatty acid and CoA complex (acyl-CoA complex).

Active Formaldehyde Referring to the N^5,N^{10}-methylenetetrahydrofolate.

Active Formimino A formimino group (NH=CH-) attached to THFA (tetrahydrofolic acid).

Active Formyl A formyl group (O=CH) attached to tetrahydrofolic acid.

Active Fructose Referring to fructose 1-6-diphosphate or fructose 1-6-bisphosphate.

Active Gas Gas that combines readily with other substances.

Active Glucose Referring to ADP-glucose, GDP-glucose, or UDP-glucose.

Active Glycolaldehyde A glycolaldehyde group (CH_2OH-CO-) attached to thiamine pyrophosphate.

Active Immunity Immunity acquired by an individual as a result of his/her own reactions to pathogenic microorganisms or their products or as a result of vaccination.

Active Immunization The induction of an active state of immunity by administration of a specific antigen.

Active Immunotherapy Treatment of disease by immunization of the patient with an immunostimulant to augment his/her immunological activity.

Active Iodine Iodine, e.g., iodinium ion (I^+), capable of reacting with tyrosine to form iodotyrosine complexes.

Active Mediated-Transport An active transport that requires transport protein and energy.

Active Methionine Referring to S-adenosyl methionine.

Active Methyl Referring to either 5-methyltetrahydrofolic acid or S-adenosyl methionine.

Active One Carbon Unit Any of the one carbon unit carried on the tetrahydrofolate involed in a variety of biosynthetic reactions.

Active Phosphate Referring to adenosine 5′-triphosphate or guanosine 5′-triphosphate.

Active Phospholipid A cytidine 5′-diphosphate derivative of either a phospholipid or a component of phospholipids.

Active Pyruvic Acid Referring to the complex of pyruvate and thiamine pyrophosphate.

Active Site Region of an enzyme molecule at which the substrate binds and the catalytic event occurs (also called catalytic site).

Active Site-Directed Irreversible Inhibitor An artificially designed inhibitor giving irreversible inhibition for a given enzyme. It consists of a functional group for binding onto the active site, a nonpolar fragment to interact with the nonpolar region outside the active site, and an active group capable of alkylating a functional group of the enzyme just outside the nonpolar region.

Active Sulfate Referring to either 3′-phosphoadenosine 5′-phosphosulfate or adenosine 5′-phosphosulfate.

Active Translocation See active transport.

Active Transport The transport of a substance across a biological membrane against its concentration gradient or electrochemical gradient that requires energy and specific transport proteins.

Activin A polypeptide hormone found in the ovarian follicular fluid that selectively stimulates secretion of FSH (follicle-stimulating hormone).

Activity Coefficient The ratio of the activity of a given substance to its molar concentration.

Activity Stain Any reagent that is capable of color development after reacting with a particular enzyme.

Actobindin A monomeric protein that is capable of binding two molecules of monomeric actin.

Actomyosin A complex of the muscle protein actin and myosin.

Actonel A trade name for risedronate sodium, a biophosphonate.

ACTP Abbreviation for adrenocorticotropic polypeptide.

Actron Caplets A trade name for ketoprofen, a non-narcotic analgesic agent.

ACU A genetic code (codon) for the amino acid threonine.

Acular A trade name ketorolac tromethamine ophthalmic, an anti-inflammatory agent.

Aculeacin-A A lipopeptide antibiotic that inhibits the formation of yeast cell wall glucan.

Aculeacin-A Deacylase The enzyme that catalyzes the hydrolysis of the amide bond in aculeacin-A and related neutral lipopeptide antibiotics, releasing long-chain fatty acid side chains.

Acumentin An actin-modulating protein isolated from rabbit alveolar macrophages.

Acupressure A therapeutic technique of applying digital pressure in a specified way at designated points on the body to relieve pain.

Acupuncture A method of producing analgesia or altering the function of a system of the body by inserting fine, wire-thin needles into specific sites on the body along a series of lines or channels, called meridians.

Acute 1. A disease that has a rapid onset and persists for a relatively short period of time. 2. An exceptionally severe or painful condition.

Acute Abdomen An abnormal condition characterized by the acute onset of severe pain within the abdominal cavity.

Acute Hemorrhagic Conjunctivitis (AHC) An infectious form of conjunctivitis, caused by an enterovirus, characterized by subconjunctival hemorrhages.

Acute Hypoxia A condition of rapid loss of available oxygen.

Acute Intermittent Porphyria An inherited liver disease, characterized by increase of concentrations of δ-aminolevulinate and porphobilinogen in liver and urine due to the deficiency of enzymes in the metabolism of δ-aminolevulinate.

Acute Lymphocytic Leukemia A malignant disease of the immune system characterized by a failure of lymphocyte maturation.

Acute Myelocytic Leukemia (AML) A malignant neoplasm of blood-forming tissues characterized by the uncontrolled proliferation of immature granular leukocytes.

Acute Phase Proteins Serum proteins that increase rapidly in the blood and remain prominent during early stages of infection and inflammation. They are nonimmunoglobulin factors (e.g., C-reactive protein) and are important in innate immunity.

Acute Radiation Exposure Exposure to an intense ionizing radiation within a short period.

Acute Serum A serum obtained shortly after the onset of a disease.

Acute Test A toxicity test on laboratory animals that requires only a single dose of chemical administration.

Acute Transfection The short term infection of cells with DNA.

Acutrim A trade name for phenylpropanolamine, an appetite suppressant and decongestant. It acts as an agonist of dopamine.

ACV (mol wt 363) A biosynthetic precursor of penicillins and cephalosporins produced by *penicillium chrysogenum*.

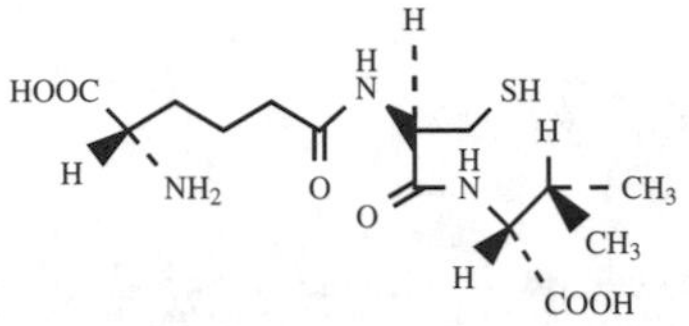

ACVD Abbreviation for acute cardiovascular disease.

Acyanoblepsia Inability to identify the color blue.

AcyI A restriction endonuclease isolated from *Anabaena cylindrica* with the following specificity:

↓
5′..........GPuCGPyC..........3′
3′..........CPyGCPuG..........5′
↑

Acycloguanosine Synonym of acyclovir.

Acyclovir (mol wt 225) An antiviral agent active against several herpes viruses. It interferes with viral DNA synthesis.

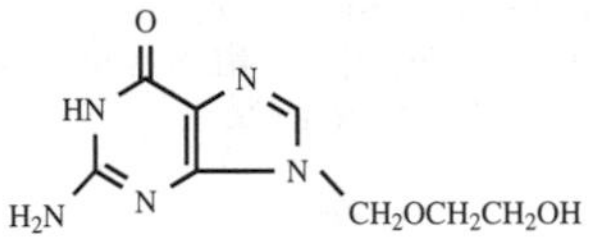

Acyl-Activating Enzyme Synonym of acetate-CoA ligase.

Acyl[Acyl-Carrier Protein] Desaturase The enzyme that catalyzes the following reaction:

$$\text{Stearoyl-[acyl-carrier protein]} + AH_2 \rightleftharpoons \text{Oleoyl-[acyl-carrier protein} + A + 2\ H_2O$$

Acyl[Acyl-Carrier-Protein] Synthetase The synonym of long chain fatty acid-[acyl-carrier-protein] ligase.

Acyl-Adenylate A fatty acid–AMP complex or an amino acid–AMP complex.

Acyl-Agmatine Amidase The enzyme that catalyzes the following reaction:

$$\text{Benzoylagmatine} + H_2O \rightleftharpoons \text{Benzoate} + \text{agmatine}$$

Acylamide Amidohydrolase The systematic name for amidase.

N-Acylamino L-Acid Amidohydrolase The systematic name for aminoacylase.

Acylamino Acid-Releasing Enzyme Synonym of acy-laminoacyl Peptidase.

Acylase Synonym of amidase.

Acylase I Synonym of aminoacylase.

Acyl-Aminoacyl Peptidase The enzyme that catalyzes the cleavage of an N-acetyl or N-formyl amino acid from the N-terminal of a peptide.

Acylation The introduction of an acyl radical (RCO^-) into an organic compound.

Acyl-Carnitine A complex of fatty acid and carnitine in the transfer of fatty acid from cytoplasm into mitochondria.

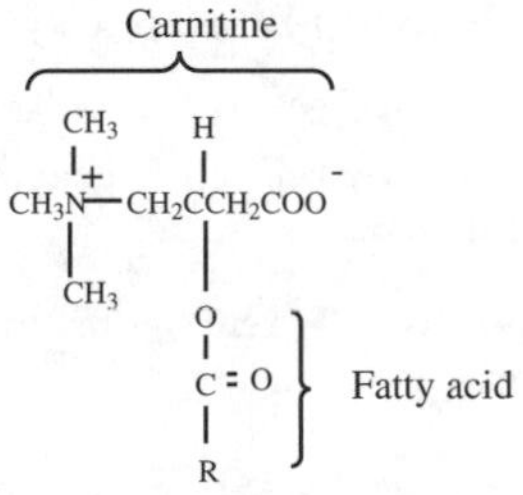

Acyl-Carnitine Hydrolase The enzyme that catalyzes the following reaction:

$$\text{Acetylcarnitine} + H_2O \rightleftharpoons \text{A fatty acid} + \text{L-carnitine}$$

Acyl-Carrier Protein A protein that constitutes part of the fatty acid synthetase complex and serves as a carrier of acyl groups during the fatty acid biosynthesis. Abbreviated as ACP or ACP-SH.

Acyl-Carrier Protein Acetyltransferase The enzyme that catalyzes the following reaction:

$$\text{Acetyl-CoA} + \text{ACP} \rightleftharpoons \text{CoA} + \text{acetyl-ACP}$$

Acyl-Carrier Protein Malonyl Transferase The enzyme that catalyzes the following reaction:

$$\text{Malonyl-CoA} + \text{ACP} \rightleftharpoons \text{malonyl-ACP} + \text{CoA}$$

Acyl-Choline Acylhydrolase The systematic name for cholinesterase.

Acyl-CoA Referring to fatty-acid-CoA complex.

Acyl-CoA Acetyl-CoA C-Acyltransferase The systematic name for acetyl-CoA C-transferase.

Acyl-CoA Dehydrogenase The enzyme that catalyzes the following reaction:

$$\text{Acyl-CoA} + \text{acceptor} \rightleftharpoons \text{2,3-dehydroacyl-CoA} + \text{reduced acceptor}$$

Acyl-CoA Dehydrogenase (NADP⁺) The enzyme that catalyzes the following reaction:

$$\text{Acyl-CoA} + NADP^+ \rightleftharpoons \text{2,3-dehydroacyl-CoA} + \text{NADPH}$$

Acyl-CoA Desaturase The enzyme that catalyzes the following reaction:

$$\text{Stearyl-CoA} + AH_2 + O_2 \rightleftharpoons \text{Oleyl-CoA} + A + 2\,H_2O$$

Acyl-CoA Hydrolase The enzyme that catalyzes the following reaction:

$$\text{Acyl-CoA} + H_2O \rightleftharpoons \text{CoA} + \text{a carboxylate}$$

Acyl-CoA Oxidase The enzyme that catalyzes the following reaction:

$$\text{Acyl-CoA} + O_2 \rightleftharpoons \text{trans 2,3-dehydroacyl-CoA} + H_2O_2$$

Acyl-CoA Oxygen 2-Oxidoreductase The systematic name for acyl-CoA oxidase.

Acyl-CoA Reductase The enzyme that catalyzes the following reaction:

$$\text{Long chain aldehyde} + \text{CoA} + NADP^+ \rightleftharpoons \text{Long chain acyl-CoA} + \text{NADPH}$$

Acyl-CoA Synthetase The enzyme that catalyzes the following reaction:

$$\text{Fatty acid} + \text{ATP} + \text{CoA} \rightleftharpoons \text{Acyl-CoA} + \text{AMP} + \text{PPi}$$

Acyl-Enzyme An enzyme that forms covalently linked acyl-enzyme intermediate with release of the product during the enzymatic hydrolysis of an ester or amide bond. The release of acyl group from the acyl enzyme intermediate is achieved by a second step called deacylation.

N-Acyl-Glucosamine 2-Epimerase The enzyme that catalyzes the following reaction:

N-Acetyl-D-glucosamine
⇅
N-Acetyl-D-mannosamine

N-Acyl-D-Glucosamine 6-Phosphate 2- Epimerase The enzyme that catalyzes the following reaction:

N-Acyl-D-glucosamine 6-phosphate
⇅
N-Acyl-D-mannosamine 6-phosphate

Acyl-Glycerol Glycerol with fatty acid attached to it. See also glyceride.

2-Acyl-Glycerol O-Acyltransferase The enzyme that catalyzes the following reaction:

Acyl-CoA + 2-acylglycerol
⇅
CoA + diacylglycerol

Acyl-Glycerol Kinase The enzyme that catalyzes the following reaction:

ATP + acylglycerol
⇅
ADP + acylglycerol 3-phosphate

Acyl-Glycerol Lipase The enzyme that catalyzes the hydrolysis of glycerol monoesters of long chain fatty acids.

1-Acyl-Glycerol 3-Phosphate O-Acyltransferase The enzyme that catalyzes the following reaction:

Acyl-CoA + 1-acyl-sn-glycerol 3-phosphate
⇅
CoA + 1,2-diacyl-sn-glycerol 3-phosphate

1-Acyl-Glycerophosphocholine O-Acyltransferase The enzyme that catalyzes the following reaction:

Acyl-CoA + 1-acyl-*sn*-glycero-3-phopshocholine
⇅
CoA + 1,2-diacyl-*sn*-glycero-3-phosphocholine

Acyl-Lysine Deacylase The enzyme that catalyzes the conversion of ε-N-acyl-L-lysine into a fatty acid and L-lysine.

Acyl-Malonyl-S-ACP Condensing Enzyme The enzyme that catalyzes the following reaction:

Acetyl-S-ACP + malonyl-S-ACP
⇅
Acetoacetyl-S-ACP + ACP + CO_2

N-Acyl-Mannosamine Kinase The enzyme that catalyzes the following reaction:

ATP + N-acyl-D-mannosamine
⇅
ADP + N-acyl-D-mannosamine 6 phosphate

Acyl-Migration The movement of acyl group from one functional group to another during the intermolecular rearrangement.

N-Acyl-Muramoyl-Alanine Peptidase The enzyme that catalyzes the cleavage of the N-acylmuramoyl-L-alanine bond.

N-Acyl-Neuraminate 9-Phosphatase The enzyme that catalyzes the following reaction:

N-Acyl-neuraminate 9-phosphate + H_2O
⇅
N-Acylneuraminate + Pi

Acylneuraminyl Hydrolase Synonym of neuraminidase.

N-Acylpeptide Hydrolase Synonym of acylaminoacyl peptidase.

Acyl-Phosphatase The enzyme that catalyzes the following reaction:

An acyl-phosphate + H_2O ⇌ Acetate + Pi

Acyl-Phosphate A mixed anhydride of phosphoric acid and a carboxylic acid.

Acyl-Phosphate-Hexose Phosopho-Transferase The enzyme that catalyzes the following reaction:

Acyl phosphate + D-hexose
⇅
An acid + D-hexose phosphate

5′-Acyl-Phosphoadenosine Hydrolase The enzyme that catalyzes the following reaction:

Acyl-CoA + $NADP^+$
⇅
2,3-dehydroacyl-CoA + NADPH

Acyl-Pyruvate Hydrolase The enzyme that catalyzes the following reaction:

3-Acyl-pyruvate + H_2O
⇅
A carboxylate + pyruvate

Acyl-S-CoA (Acyl coenzyme A) A complex formed from the combination of a fatty acid and coenzyme A.

N-Acyl-Sphingosine (Ceramide) A compound formed by enzymatic transfer of acyl group from acyl-CoA to sphingosine.

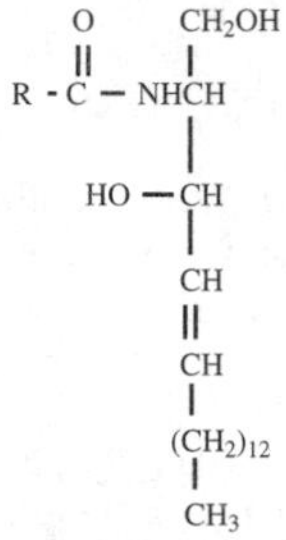

Acyl-Sphingosine Kinase See ceramide kinase.

Acyl-Transferase The enzyme that catalyzes the transfer of an acyl group from an acyl-CoA to another compound.

Acystia Congenital absence of a bladder.

Acystinervia Paralysis of the bladder.

Acytostelium A genus of cellular slime molds (class Dictyosteliomycetes).

AD Abbreviation for 1. Activation domain. 2. Alzheimer's disease. 3. Actinomycin D. 4. Analgesic dose. 5. Alcohol dehydrogenase.

ADA Abbreviation for enzyme adenosine deaminase.

***ada* Protein** A protein involved in DNA repair in cells of *E. coli* induced by exposure to a low concentration of certain alkylating agents. It can reverse the effects of a methylation agent.

Adacrya Insufficient lacrimal secretion or absence of tears.

Adactyla A congenital absence of one or more digits of the hand or foot.

Adagen A trade name for adenosine deaminase used for treatment of severe combined immunodeficiency diseases.

Adair Equation An equation used for the calculation of the average number of bound ligand molecules per molecule of total protein from binding data.

Adalate A trade name for nifedipine, a calcium channel blocker, coronary vasodilator, antianginal, and antihypertensive agent.

Adamalysin A protease.

Adamkiewicz Reaction The formation of a violet color upon treatment of a protein solution with acetic acid and sulfuric acid.

Adansonian Taxonomy A method of biological classification proposed by Michel Adanson in which relationships between organisms are defined by the number of common characteristics they possess.

Adapalene (mol wt 412) A modulator for the processes of cellular differentiation, keratinization and inflammation and used for the treatment of acne vulgaris.

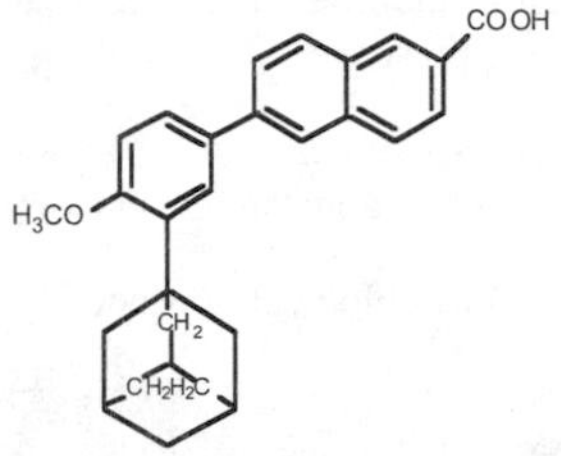

Adapin A trade name for doxepin hydrochloride, an antidepressant that blocks the uptake of norepinephrine and serotonin by the presynaptic neurons.

Adaptation 1. The gradual modification of an individual, species, or a population of organisms for survival in a particular environment. 2. Adjustment of the pupil of the eye to variations in light intensity.

Adapter A synthetic, single-stranded, noncomplementary oligonucleotide used in conjunction with a linker to add cohesive ends to a DNA molecule.

Adapter RNA Referring to tRNA.

Adaptin A major coat protein of the clathrin-coated vesicles.

Adaptive Enzyme Enzymes that are synthesized only in response to the presence of substrates or substrate-related compounds. Also called an inducible enzyme.

Adaptive Immunity An antigen-specific defense mechanism against infectious agents that are acquired as a result of immunization, infection, or natural exposure to antigen.

Adaptive Response A DNA repair system induced in cells of *E. coli* in response to exposure of low concentrations of certain alkylating agents.

Adaptor Variant spelling of adapter.

Adaptor tRNA Variant spelling of adapter tRNA.

Adavite A trade name for an oral multiple vitamin preparation containing both fat-soluble (e.g., A, D, E) and water-soluble (e.g., B, C) vitamins and folic acid.

ADB Abbreviation for a combination drug containing adriamycin, dacarbazine and bleomycin.

ADBC Abbreviation for a combination drug containing adriamycin, DTIC, bleomycin and CCNU.

ADC Abbreviation for 1. Alanine decarboxylase. 2. Albumin-dextrose-catalase. 3. Arginine decarboxylase. 4. Aspartate decarboxylase.

ADCC (Antibody-Dependent Cellular Cytotoxicity) The cytotoxic or cytolytic activity of an effector cell in the immune system that is dependent upon the presence of antibody.

Adderall A trade name for a combination drug containing dextroamphetamine and amphetamine.

Addiction Physiological dependence on a substance.

Addictology The science that deals with addiction.

Addis Count A method for counting red blood cells, white blood cells, epithelial cells, casts, and protein content in the sediment of a 12-hour urine sample.

Addison's Disease A life-threatening disease caused by the failure of adrenocortical function.

Additive 1. A substance added to another to improve the desirable property or to suppress the undesirable feature. 2. An agent added to food to improve color, flavor, texture, or quality.

Address Sequence The region of amino acid sequence in a polypeptide hormone in which the hormone is considered to bind to its specific receptor.

Addressins Ligands on the mucosal endothelial cells that serve as specific homing receptors on lymphocytes derived from Peyer's patches.

Adduct To move toward the midline (opposite of abduct).

Adductin A membrane skeleton protein that interacts with junctional complex and links the spectrin assemblies.

Adductor A muscle that draws a structure toward the median line of the body.

Ade Abbreviation for adenine.

Adelomorphous Without a clearly defined form.

Adelphoparasite Parasite that is closely related to its host.

Adenalgia Pain in a gland.

Adenase See adenine deaminase.

Adendritic Neurons without dendrites.

Adenectomy The surgical removal of a gland.

Adenia The enlargement of a gland or the enlargement of the lymph nodes without leukemic changes in the blood.

Adenic Pertaining to a gland.

Adenine (mol wt 135) A nitrogen-containing aromatic base that serves as a component of an informational monomeric unit in nucleic acids.

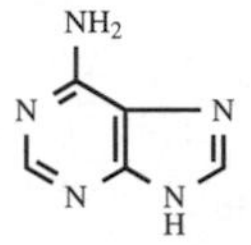

Adenine Aminohydrolase See adenine deaminase.

Adenine Arabinoside (mol wt 267) A synthetic purine nucleoside possessing antitumoral and antiviral activity.

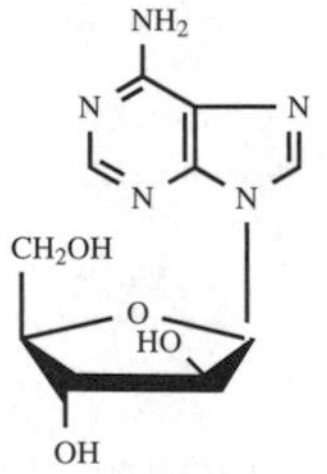

Adenine Deaminase The enzyme that catalyzes the following reaction:

$$\text{Adenine} + H_2O \rightleftharpoons \text{Hypoxanthine} + NH_3$$

Adenine Nucleotide A nucleotide containing a pentose sugar, an adenine base, and a phosphate.

Adenine Nucleotide Carrier See ATP-ADP translocase.

Adenine Phosphoribosyl Transferase The enzyme that catalyzes the following reaction:

$$\text{AMP} + \text{pyrophosphate} \rightleftharpoons \text{Adenine} + 5'\text{-phospho-ribosyl pyrophosphate}$$

Adenitis Inflammation of a gland or lymph node.

Adeno- A combining form meaning pertaining to a gland.

Adenoacanthoma A malignant neoplasm derived from glandular tissue, e.g., cancer of the uterus.

Adeno-Associated Virus A defective virus (family of parvoviridae) whose reproduction depends upon the presence of adenovirus.

Adenoblast Any embryonic cell from which a glandular cell is derived.

Adenocarcinoma A malignant epithelium tumor of glandular structure.

Adenocard A trade name for adenosine; used as an antiarrhythmic agent.

Adenocele A cysticlike tumor of a gland.

Adenochondroma A neoplasm of cells derived from glandular and cartilaginous tissues, e.g., tumor of the salivary glands.

Adenochondrosarcoma Any malignant tumor composed of glandular and cartilagelike tissue.

Adenocorticotropins A group of adenocorticotropic hormones (e.g., ATCH) from human, sheep, and pig that differ in their amino acid sequences.

Adenocyst Tumor composed of glandular and cystic elements.

Adenocystic Carcinoma A malignant tumor of epithelium tissue occurring frequently in the salivary glands, breast, and mucous glands of the upper and lower respiratory tract.

Adenocytes The secretory cells of a gland.

Adenodiastasis The presence of glandular tissue in locations other than their normal sites.

Adenofibroma A tumor of connective tissue found in the ovary and composed of glandular and fibrous elements.

Adenofibrosis The formation of fibrous tissue in a gland.

Adenogenesis The process of development of a gland.

Adenohypophysis The anterior lobe of the pituitary gland that secretes growth hormones.

Adenoid 1. Resembling a gland. 2. Tonsil (lymphoid gland).

Adenoid Hyperplasia A disorder characterized by partial respiratory obstruction caused by enlargement of adenoid glands.

Adenoidectomy Surgical removal of the adenoid.

Adenoiditis Inflammation of the adenoid.

Adenolipoma A neoplasm consisting of elements of glandular and fatty tissue.

Adenology The science that deals with glands.

Adenoma A benign tumor of glandular epithelium in which the cells form glands or glandlike structures.

-adenoma A combining form meaning a "tumor composed of glandular tissue or glandlike structure."

Adenomatoid Resembling adenoma.

Adenomatosis The presence of multiple adenomas.

Adenomegaly Enlargement of the lymph node.

Adenomyomatosis A disorder characterized by the formation of benign nodules resembling adenomyomas, found in the uterus or in parauterine tissue.

Adenomyosis 1. A benign neoplastic disorder characterized by tumors composed of glandular tissue and smooth muscle cells. 2. A malignant neoplastic disorder characterized by the invasive growth of uterine mucosa in the wall of the uterus or the oviducts.

Adenopathy Swelling or enlargement of glands.

Adenosarcoma A malignant glandular tumor composed of connective tissue and glandular elements.

Adeno-Satellite Viruses Referring to adeno-associated virus.

Adenoscan A trade name for adenosine used as an antiarrhythmic agent.

Adenosinase See adenosine nucleosidase.

Adenosine (mol wt 267) A nucleoside composed of adenine and D-ribose.

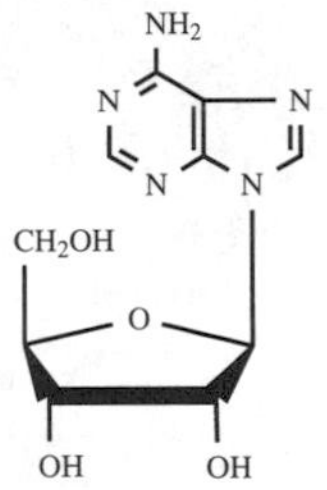

Adenosine Aminohydrolase See adenosine deaminase.

Adenosine 2′,3′-Cyclic Monophosphate (mol wt 329) A cyclic nucleotide similar to adenosine 3′-5′ cycle monophosphate.

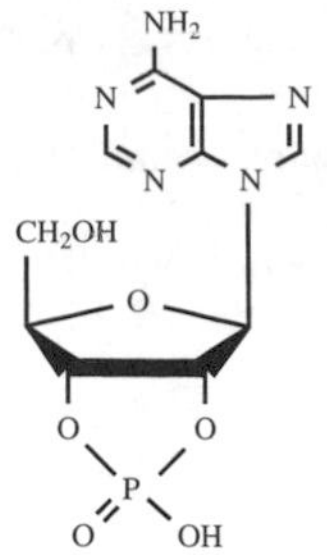

Adenosine 3′,5′-Cyclic Monophosphate (mol wt 329) The key intracellular regulator for a number of cellular processes in bacteria, plants and animals. Abbreviated as cAMP.

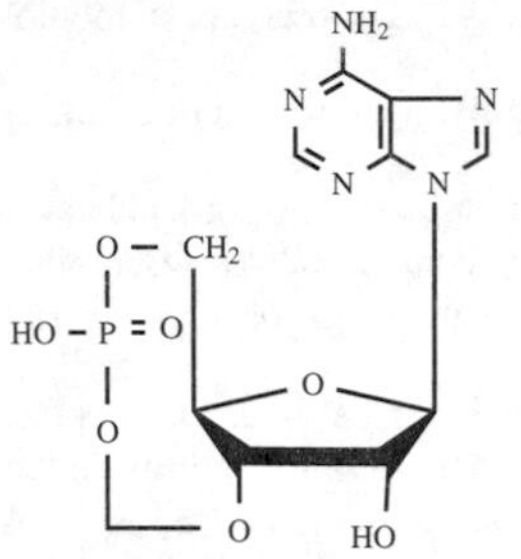

Adenosine Deaminase The enzyme that catalyzes the following reaction:

$$\text{Adenosine} + H_2O \rightleftharpoons \text{Inosine} + NH_2$$

Adenosine Deaminase Deficiency A disorder characterized by deficiency of the enzyme adenosine deaminase leading to the accumulation of adenosine within the cell and the impairment of function of helper T cells.

Adenosine Diphosphatase See apyrase.

Adenosine 5′-Diphosphate (mol wt 427) A diphosphate adenosine nucleotide with the phosphate groups linked to the carbon 5 position of the ribose.

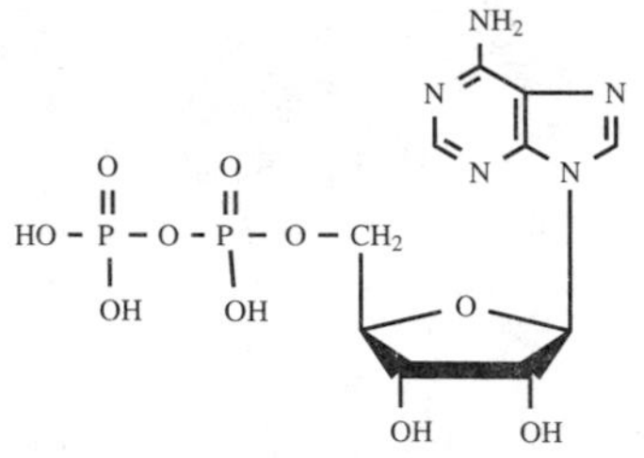

Adenosine Hydrolase The enzyme that catalyzes the conversion of adenosine into adenine and ribose.

Adenosine Kinase The enzyme that catalyzes the transfer of a phosphate group from ATP to adenosine.

$$\text{ATP} + \text{adenosine} \rightleftharpoons \text{ADP} + \text{AMP}$$

Adenosine Monophosphate (mol wt 347) The monophosphate form of the adenine nucleotide. The phosphate group may be linked to the 2′, 3′, or 5′ position of the ribose to form adenosine 2′-monophosphate, adenosine 3′- monophosphate, and adenosine 5′-monophosphate, respectively (also called adenylic acid or adenylate).

Adenosine 2′-Monophosphate (mol wt 347) An adenosine monophosphate nucleotide with a phosphate group linked to the carbon 2 position of the ribose.

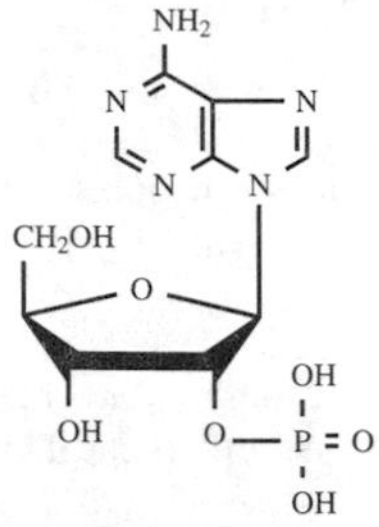

Adenosine 3′-Monophosphate (mol wt 347) An adenosine monophosphate nucleotide with a phosphate group linked to the carbon 3 of the ribose.

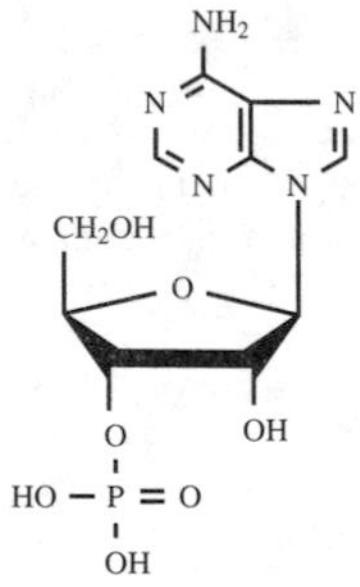

Adenosine 5′-Monophosphate (mol wt 347) An adenosine monophosphate nucleotide with the phosphate linked to the carbon 5 position of the ribose.

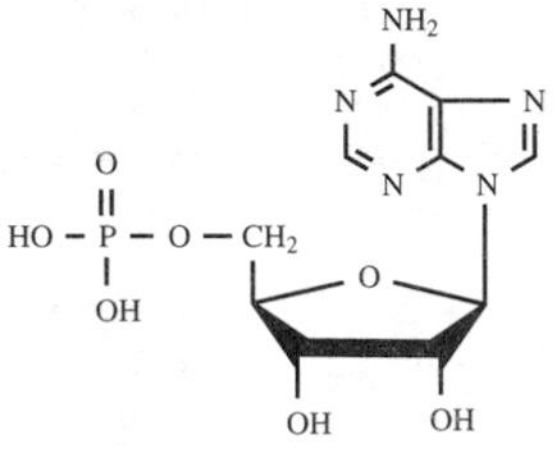

Adenosine Nucleosidase The enzyme that hydrolyzes adenosine to adenine and ribose.

$$\text{Adenosine} + H_2O \rightleftharpoons \text{Adenine} + \text{D-ribose}$$

Adenosine Phosphate The adenine nucleotide that may consist of one, two, or three phosphate groups (also called adenylic acid or adenylate).

Adenosine Phosphate Deaminase The enzyme that catalyzes the following reaction:

$$5'\text{-AMP} + H_2O \rightleftharpoons 5'\text{-IMP} + NH_3$$

Adenosine 5′-Phosphosulfate (mol wt 427) An active sulfate.

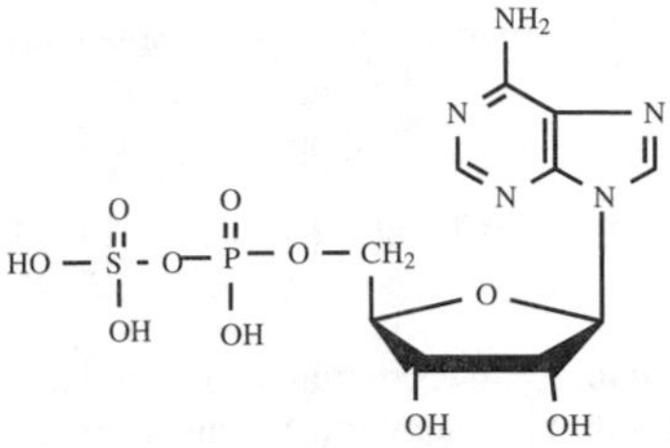

Adenosine Tetraphosphatase The enzyme that catalyzes the following reaction:

Adenosine tetraphosphate + H_2O ⇌ ATP + Pi

Adenosine Triphosphatase (ATPase) The enzyme that catalyzes the following reaction:

ATP + H_2O ⇌ ADP + Pi

Adenosine 5′-Triphosphate (ATP, mol wt 507) A triphosphate form of adenosine nucleotide with 3 phosphate groups linked to the carbon 5 position of the ribose. It is the major form of energy in cells.

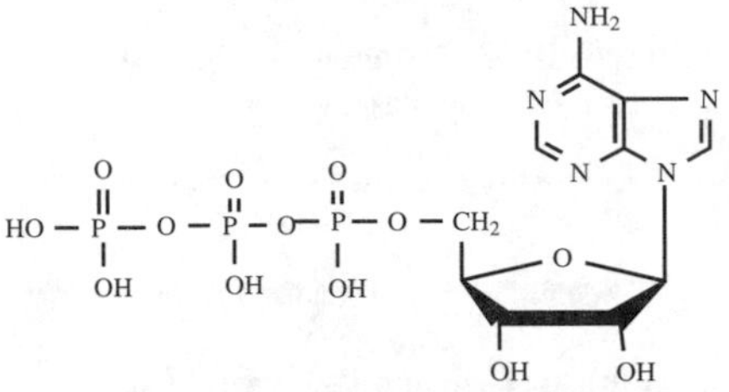

Adenosis 1. Disease of a gland, especially a lymphatic gland. 2. An abnormal development or enlargement of glandular tissue.

Adenosylhomocysteinase The enzyme that hydrolyzes S-adenosyl-L-homocysteine to adenosine and L-homocysteine.

S-Adenosyl-L-homocysteine + H_2O

⇅

Adenosine + L-homocysteine

S-Adenosylhomocysteine (mol wt 384) An intermediate in the biosynthesis of methionine.

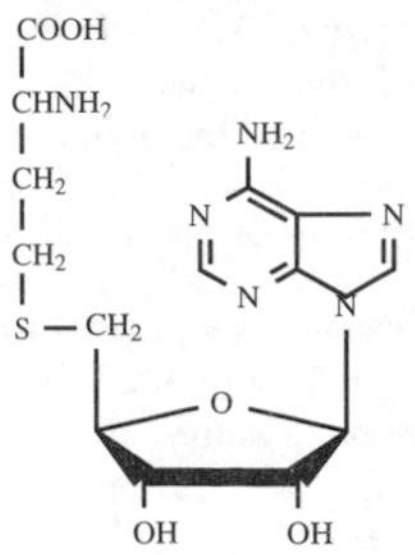

S-Adenosylhomocysteine Deaminase The enzyme that catalyzes the following reaction:

S-Adenosyl-L-homocysteine + H_2O

⇅

S-Inosyl-L-homocysteine + NH_3

S-Adenosylhomocysteine Hydrolase See S-adenosylhomocysteinase.

S-Adenosylhomocysteine Nucleosidase The enzyme that catalyzes the following reaction:

S-Adenosylhomocysteine + H_2O

⇅

Adenine + S-D-Ribosyl-L-homocysteine

S-Adenosylmethionine (mol wt 400) A methyl group donor involved in enzymatic transmethylation reactions in living organisms.

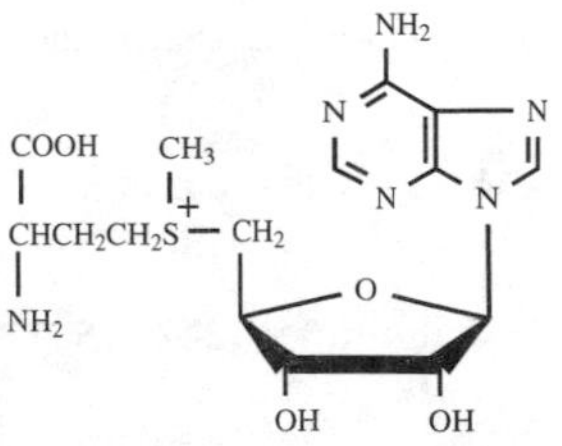

S-Adenosyl L-Methionine Catechol O-Methyl Transferase Synonym of catechol O-methyltransferase.

Adenosylmethionine Cyclotransferase The enzyme that catalyzes the following reaction:

S-Adenosylmethionine + H_2O

⇅

Methylthioadenosine + L-homoserine

S-Adenosyl-L-Methionine Histamine N-*tele*-Methyltransferase The systematic name for histamine N-methyltransferase.

S-Adenosyl-L-Methionine Phenyl-Ethanolamine N-Methyl Transferase The systematic name for phenylethanolamine N-methyltransferase.

S-Adenosyl-L-Methionine Protein L-Glutamate O-Methyl Transferase The systematic name for Protein glutamate O-methyltransferase.

Adenotonsillectomy The surgical removal of the palatine tonsils and adenoids.

Adenoviridae A family of nonenveloped icosahedral DNA-containing animal viruses.

Adenovirus Any one of the viruses in the family of Adenoviridase.

Adenovirus-Associated Virus A small, naked, and icosahedral virus that contains single-stranded DNA. Its replication depends upon the presence of adenovirus.

Adenyl Cyclase See adenylate cyclase.

Adenylate Referring to adenylic acid or adenosine monophosphate.

Adenylate Cylase The enzyme that catalyses the synthesis of cyclic AMP (cAMP) from ATP.

$$\text{ATP} \rightleftharpoons \text{cAMP} + \text{PP}$$

Adenylate Kinase The enzyme that catalyzes the addition of phosphate groups to adenine nucleotide.

Adenylate Pool The total content of the intracellular AMP, ADP, and ATP.

Adenylic Acid Referring to adenosine monophosphate.

3′-Adenylic Acid See Adenosine 3′-monophosphate.

5′-Adenylic Acid (mol wt 347) See Adenosine 5′-monophosphate.

Adenylic Acid Deaminase See AMP deaminase.

Adenylosuccinate (mol wt 463) An intermediate in the purine nucleotide biosynthetic pathway.

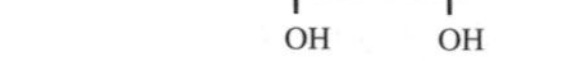

Adenylosuccinate AMP-Lyase See adenylosuccinate lyase.

Adenylosuccinate Lyase The enzyme that catalyzes the following reaction:

$$\text{Adenylosuccinate} \rightleftharpoons \text{Fumarate} + \text{AMP}$$

Adenylosuccinate Synthetase The enzyme that catalyzes the following reaction:

$$\text{GTP} + \text{IMP} + \text{L-aspartate} \rightleftharpoons \text{GDP} + \text{Pi} + \text{adenylosuccinate}$$

Adenylpyrophosphatase Synonym of adenosine triphosphatase.

Adenylyl Referring to the adenosine monophosphate group, the acyl group derived from adenylic acid.

Adenylyl Transferase Synonym of nucleotidyl transferases.

Adenylylate Introduction of adenylyl group into a compound through the action adenylyl transferase.

Adenylylcyclase Synonym of adenylate cyclase.

Adenylylsulfatase The enzyme that catalyzes the following reaction:

$$\text{Adenylylsulfate} + H_2O \rightleftharpoons \text{AMP} + \text{sulfate}$$

Adenylylsulfate Kinase The enzyme that catalyzes the following reaction:

$$\text{ATP} + \text{adenylylsulfate} \rightleftharpoons \text{ADP} + 3'\text{-phosphoadenylyl sulfate}$$

Adenylylsulfate Reductase The enzyme that catalyzes the following reaction:

$$\text{AMP} + \text{sulfite} + \text{acceptor} \rightleftharpoons \text{Adenylylsulfate} + \text{reduced acceptor}$$

Adermin See vitamin B_6 or pyridoxine.

ADF Abbreviation for actin depolymerizing factor.

ADH Abbreviation for 1. alcohol dehydrogenase and 2. antidiuretic hormone.

Adhalin A glycoprotein of skeletal muscle sarcolemma.

Adherent Cells Cells that can adhere to the surface of glass and plastics, e.g., macrophages.

Adhesin Compound used for adhesion to other cells or to inanimate surfaces.

Adhesion Molecule Molecules expressed on the surface of a cell that mediate the adhesion of the cell to other cells or to the extracellular matrix.

ADI Abbreviation for acceptable daily intake.

Adiabatic Process 1. A process conducted without either gain or loss of heat. 2. A system conducted in an isolated surrounding.

Adiaphoresis Absence of sweat or deficiency of perspiration.

ADIC Abbreviation for a combination drug containing adriamycin and dimethylimidazolecarboxamide.

Adinazolam (mol wt 352) An antidepressant.

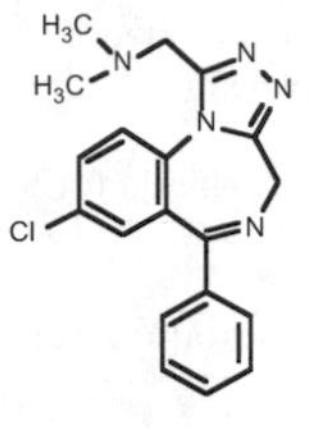

Adipex-P A trade name for phentermine hydrochloride, used as a cerebral stimulant.

Adiphenine Hydrochloride (mol wt 348) An anticholinergic agent. It also possesses antimuscarinic and antispasmodic activities.

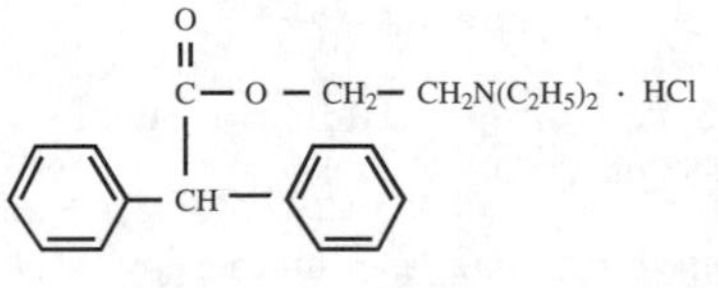

Adipo- A combining form "meaning pertaining to fat."

Adipocyte A fat cell; a cell of adipose tissue.

Adipofibroma A fibrous neoplasm of connective tissue in which there are fatty components.

Adipokinetic Fat-mobilizing or lipotropic.

Adiponecrosis Necrosis or death of fatty tissue in the body.

Adipose Fat or fat-storing tissue.

Adipose Cell A cell filled with a large quantity of lipid.

Adipose Tissue Tissue in which fat is extensively deposited.

Adiposis A disorder characterized by the excessive accumulation of fat in the body.

Adiposity Synonym for obesity.

Adipost A trade name for phendimetrazine tartrate, used as a cerebral stimulant.

Adipsin A serine protease homolog synthesized by mammalian adipocytes. It is a glycoprotein with molecular weight about 37–44 kD.

Adjunct A substance, treatment, or procedure used for increasing the efficiency or safety of the primary substance, treatment, or procedure.

Adjuvant Substance capable of enhancing or potentiating an immune response to an antigen when administered together with the antigen.

Adjuvant 65 A water-in-oil emulsion adjuvant made by emulsifying peanut oil with mannide monooleate and stabilized with aluminum monostearate.

Adjuvant Peptide Synonym of muramyl dipeptide used as adjuvant.

Adjuvanticity The ability of a substance to function as an adjuvant.

AdK Abbreviation for adenylate kinase.

Adler Test A test used for the identification of *Leishmania spp.*

Adlumidine (mol wt 367) An alkaloid present in the plants of *Adlumia fungosa*, *Corydalis thalictrifolia*, and *C. incisa*.

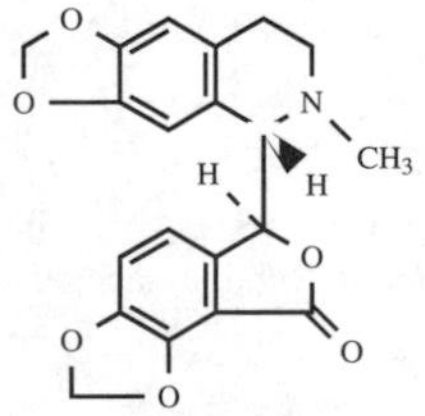

d-adlumidine

Adlumine (mol wt 383) An alkaloid present in some plants of the family Fumariaceae, e.g., *Adlumina fugosa*.

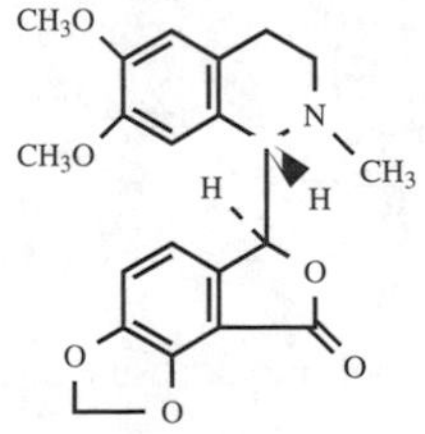

AdML Promoter Abbreviation for adenovirus major late promoter.

ADN Abbreviation for anti-deoxyribonuclease.

A-DNA (A Form DNA) Right-handed and double-stranded DNA containing about 11 residues per turn. The planes of the base pairs are tilted 20 degrees away from the perpendicular to the helix axis. It is formed through the dehydration of B-DNA.

Ado Abbreviation for adenosine.

AdoCbl Abbreviation for adenosylcobalamin

AdoHcy Abbreviation for adenosyl-L-homocysteine.

-adol A suffix referring to a drug that relieves pain.

Adolescence The period between puberty and maturity.

Adolescent Pertaining to or characteristic of adolescence.

AdoMet Abbreviation for S-adenosyl-L-methionine.

Adonitol (mol wt 152) A sugar alcohol derived from ribose (see also ribitol).

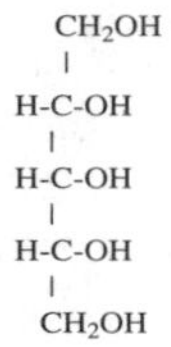

Adonitoxin (mol wt 551) A toxin isolated from *Adonis vernalis* of Ranunculaceae.

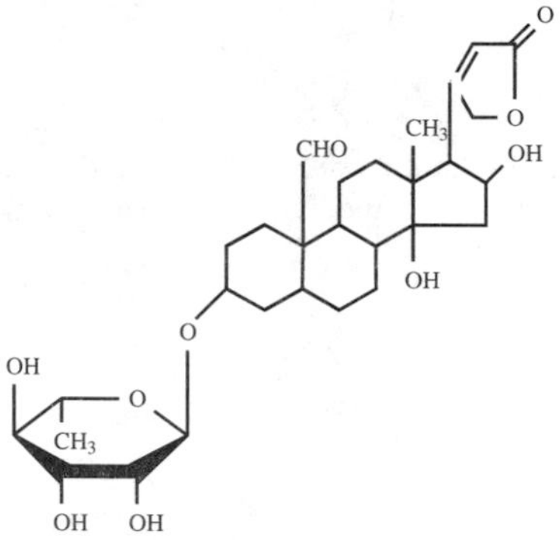

Ado2'P Symbol for adenosine 2'-phosphate.

Ado2'3'P Abbreviation for adenosine 2',3'-phosphate.

Ado3'P Abbreviation for adenosine 3'-phosphate.

Ado3'5'P Abbreviation for adenosine 3',5'-phosphate.

Ado5'P Abbreviation for adenosine 5'-monophosphate.

AdoP[CH$_2$]P Symbol for adenosine 5'-[α,β-methylene]diphosphate.

Ado5'PP Abbreviation for adenosine 5'-diphosphate.

AdoPP[CH$_2$]P Symbol for adenosine 5'-[β,γ-methylene]-triphosphate.

AdoPPGlc Abbreviation for adenosine diphosphate glucose.

AdoPP[NH]P Abbreviation for adenosine 5'-[β,γ-imido]triphosphate.

Ado5'PPP Abbreviation for adenosine 5'-triphosphate.

AdoPPPS Abbreviation for adenosine 5'-γ-thiotriphosphate.

AdoPPRib Abbreviation for adenosine diphosphoribose.

AdoPPS Abbreviation for adenosine 5'-β-thiodiphosphate.

Ado3'P5'PS Abbreviation for adenosine 3'-phosphate 5'-phosphosulfate.

Ado5'PS Abbreviation for adenosine 5'-thiophosphate.

Adoptive Cellular Immunotherapy The treatment by transfer of cultured immune cells that have antitumor activity into a tumor-bearing host.

Adoptive Immunity See adoptive transfer.

Adoptive Tolerance The immunological tolerance acquired by receiving lymphocytes from tolerant animals.

Adoptive Transfer Transfer of immunity from an immune individual to a nonimmune individual by the transfer of immunocompetent cells.

ADP Abbreviation for adenosine diphosphate or adenosine 5′-diphosphate.

ADP-Aminohydrolase See ADP deaminase.

ADPase Abbreviation for adenosine diphosphatase.

ADP-ATP Carrier Protein An integral membrane protein of the inner mitochondrial membrane responsible for transport of ADP and ATP across the membrane.

ADP-Deaminase The enzyme that catalyzes the following reaction:

$$ADP + H_2O \rightleftharpoons IDP + NH_3$$

ADP[α,β-CH2] Abbreviation for α,β-methylene adenosine 5'-diphosphate.

ADPG Abbreviation for ADP-glucose.

ADP-Glc Abbreviation for ADP-glucose.

ADP-Glucose An intermediate in the synthesis of glycogen or starch.

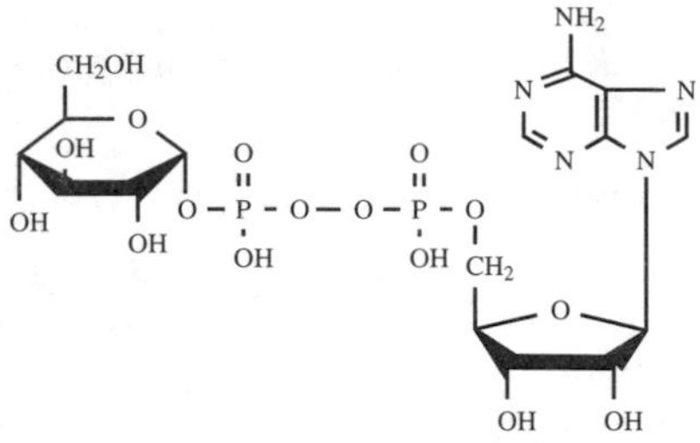

ADP-Glucose-Glycogen Glucosyl Transferase The enzyme that catalyzes the synthesis of glycogen.

$$\text{ADP-glucose} + (\text{glycogen})_n \rightleftharpoons (\text{Glycogen})_{n+1} + \text{ADP}$$

ADP-Glucose Pyrophosphorylase The enzyme that catalyzes the following reaction:

$$\text{ADP} + \text{glucose 1-phosphate} \rightleftharpoons \text{ADP-glucose} + \text{Pi}$$

ADP-Phosphoglycerate Phosphatase The enzyme that catalyzes the following reaction:

$$\text{3-(ADP)-2-phosphoglycerate} + H_2O \rightleftharpoons \text{3-(ADP)-glycerate} + \text{Pi}$$

ADPR Abbreviation for ADP-ribose or adenosine diphosphate ribose or adenosine diphosphate ribosyl.

ADP-Rib Abbreviation for ADP-ribose.

ADP-Ribose Phosphorylase The enzyme that catalyzes the following reaction:

$$\text{ADP} + \text{ribose 5-phosphate} \rightleftharpoons \text{ADP-ribose} + \text{Pi}$$

ADP-Ribose Pyrophosphatase The enzyme that catalyzes the following reaction:

$$\text{ADP-ribose} + H_2O \rightleftharpoons \text{AMP} + \text{D-ribose 5-phosphate}$$

ADP-Ribosyl Abbreviation for adenosine diphosphoribosyl group.

ADP-Ribosyl Cyclase Synonym of NAD^+ nucleosidase.

ADP-Ribosylation The process of transfer of an ADP-ribosyl group from NAD^+ to a protein.

ADP-Ribosylation Factor A protein that acts as activator for ADP-ribosyltransferase.

ADP-Ribosyltransferase The enzyme that catalyzes the transfer of ADP-ribosyl group from NAD to protein.

$$NAD^+ + \text{protein} \rightleftharpoons \text{ADP-ribose protein comples} + \text{nicotinamide}$$

ADPRT Abbreviation for ADP-ribosyl transferase.

ADPS Abbreviation for adenosine 5'-β-thiodiphosphate.

ADP-Sugar Pyrophosphatase The enzyme that catalyzes the following reaction:

$$\text{ADP-sugar} + H_2O \rightleftharpoons \text{AMP} + \text{sugar 1-phosphate}$$

ADP-Sulfurylase The enzyme that catalyzes the following reaction:

$$\text{ADP} + \text{sulfate} \rightleftharpoons \text{Pi} + \text{adenylylsulfate}$$

ADP-Thymidine Kinase The enzyme that catalyzes the following reaction:

$$\text{ADP} + \text{thymidine} \rightleftharpoons \text{AMP} + \text{thymidine 5}'\text{-phosphate}$$

Adr Abbreviation for adrenaline.

Adrafinil (mol wt 289) An α-adrenergic agonist used for treatment of depression.

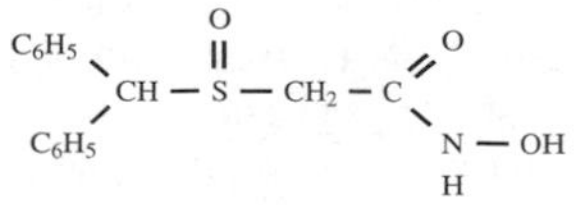

Adrenal Referring to the endocrine glands located near the kidneys.

Adrenal Androgen Any of the C19 steroid hormones produced in the cortex of the adrenal gland.

Adrenal Cortex Part of the adrenal gland that secretes adrenal cortical hormones.

Adrenal Cortical Carcinoma A malignant neoplasm of the adrenal cortex.

Adrenal Cortical Hormone Hormones secreted by the adrenal cortex.

Adrenal Cortical Steroid Steroids produced by the adrenal cortex.

Adrenal Gland Two endocrine glands located near the kidney, each consisting of two parts: a medulla that secretes epinephrine and norepinephrine and a cortex that secretes the adrenal cortical hormones.

Adrenal Medulla The inner part of the adrenal gland that secretes the hormones epinephrine and norepinephrine.

Adrenalectomy The surgical removal of one or both adrenal glands.

Adrenaline See epinephrine.

Adrenaline Oxidase Synonym of amine oxidase (flavin-containing).

Adrenalinemia The presence of abnormal amounts of epinephrine in the blood.

Adrenaline Tolerance Test A test for the diagnosis of glycogen storage disease by measuring the level of blood glucose as a function of time following the injection of adrenaline.

Adrenalize To stimulate or to excite.

Adrenalone (mol wt 181) An intermediate in the manufacturing of epinephrine.

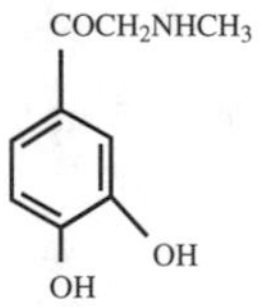

Adrenergen Substances with physiological action resembling that of adrenaline.

Adrenergic 1. Pertaining to motor nerve fibers that release either epinephrine or norepinephrine at the nerve ending when nerve impulses arrive there. 2. Pertaining to drugs or hormones that elicit the effects of epinephrine or norepinephrine.

Adrenergic Blocking Agent See antiadrenergic.

Adrenergic Drug Drugs that elicit the effects of epinephrine or norepinephrine.

Adrenergic Receptor A site on an effector cell that reacts to adrenergic stimulation. There are two types of adrenergic receptors: α-adrenergic and β-adrenergic receptor. The actions of α-adrenergic receptors inhibit adenylate cyclase and stimulate smooth muscle contraction in the blood vessels, while that of β-adrenergic receptors activate adenylate cyclase and stimulate glycogenolysis, gluconeogenesis in liver, and accelerate heart beat and bronchodilation.

Adrenergic Synapse A chemical synapse that uses norepinephrine or epinephrine as the neurotransmitter.

-adrenia A combining form meaning adrenal activity.

Adrenoceptor Synonym of adrenergic receptor.

Adrenochrome (mol wt 179) A pigment obtained by the oxidation of epinephrine that acts as a psychotomimetic agent.

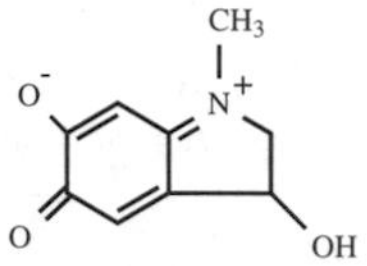

Adrenocortical Steroid The 21-carbon steroid hormones derived from the adrenal cortex (see corticosteroid).

Adrenocorticotrophic Variant spelling of adrenocorticotropic.

Adrenocorticotrophic Hormone Variant spelling of adrenocorticotropic hormone.

Adrenocorticotropic Pertaining to the stimulation of the adrenal cortex.

Adrenocorticotropic Hormone (ACTH) A polypeptide hormone secreted by the anterior lobe of the pituitary gland that stimulates the synthesis and secretion of adrenal cortical hormones by the adrenal cortex.

Adrenocorticotropin Synonym for adrenocorticotropic hormone.

Adrenodoxin An iron-containing protein that functions as an electron carrier in microsomal, nonphosphorylating electron transport systems.

Adrenodoxin Reductase Synonym of ferredoxin-$NADP^+$ reductase.

Adrenogenic Originating from the adrenal gland.

Adrenogenital Syndrome A metabolic disorder resulting from hyperactivity of the adrenal cortex.

Adrenoglomerulotrophin Variant spelling of adrenoglomerulotropin.

Adrenoglomerulotropin (mol wt 216) An aldosterone-stimulating hormone secreted by the adrenal cortex.

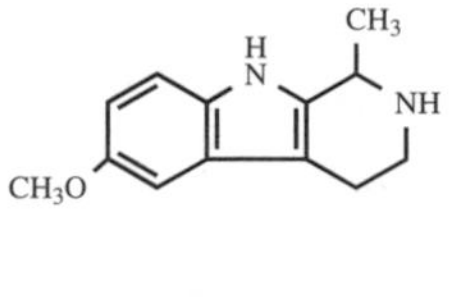

Adrenosterone (mol wt 300) An androgen (19 carbon steroid) produced by the adrenal gland.

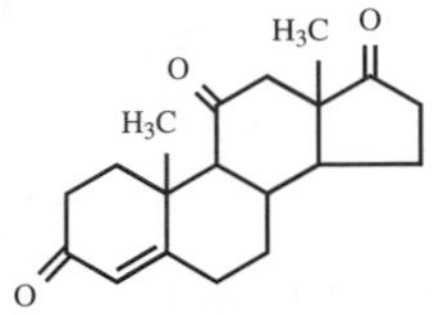

Adriamycin A trade name for doxorubicin, an antibiotic that interfers with RNA synthesis.

Adrucil A trade name for fluorouracil, which inhibits DNA synthesis.

ADS Abbreviation for antibody deficiency syndrome.

Adsorbate A substance that is adsorbed to the surface of another substance.

Adsorbent A substance used to adsorb another substance.

Adsorbocarpine A trade name for pilocarpine hydrochloride, used as a miotic agent.

Adsorbonac Ophthalmic Solution A trade name for NaCl (5%) solution, used for removal of excess fluid from coronea.

Adsorption 1. The attachment of ions or substances onto the surface of another substance. 2. The adhesion of molecules to the surface of a solid. 3. The attachment of phages or virions onto the cell surface. 4. Attachment of antigen onto the surface of the antibody molecule.

Adsorption Chromatography A form of chromatography in which molecules are separated on the basis of their adsorption and deadsorption properties on a solid adsorbent, e.g., column chromatography.

Adsorption Isotherm A plot of amount of solute adsorbed by an absorbent versus the concentration of free solute at constant temperature.

Adsorption Protein (A Protein) A specific protein on the phage particle responsible for adsorption of phage particles onto the bacterial cell surface.

ADT Abbreviation for agar-gel diffusion test.

AdV Abbreviation for adenovirus.

Advil A trade name for ibuprofen, used as an anti-inflammatory, analgesic and antipyretic agent.

Adx Abbreviation for adrenodoxin, a ferredoxin isolated from adrenal cortex mitochondria.

AE Abbreviation for anion exchanger.

AEA Abbreviation for a mixture containing alcohol, ether and acetone.

AEBS Abbreviation for anti-estrogen-binding site.

AEBSF Abbreviation for aminoethylbenzene sulphonyl fluoride.

AEC Abbreviation for 1. Alveolar epithelial cells. 2. Aminoethylcarbazole.

AE-Cellulose Aminoethylcellulose, an anion exchanger used for ion exchange chromatography.

AED Abbreviation for aminoethyldextran.

Ades Aegypti Mosquito that transmits yellow fever and dengue.

Aegyptianella A genus of Gram-negative bacteria of the family Anaplasmataceae.

AE-HPLC Abbreviation for anion-exchange high performance liquid chromatography.

Aequorin A calcium-dependent photo-protein from luminescent jellyfish *Aequorea forskaolea.*

Aero- (Aer-) A prefix meaning pertaining to air or to gas.

Aerobes Oxygen-dependent organisms or microorganisms.

Aerobic Capable of using molecular oxygen for oxidation, respiration, and growth.

Aerobic Bacteria Bacteria requiring oxygen for growth and respiration.

Aerobic Glycolysis The pathway in which glucose is converted to pyruvic acid and NADH. The NADH thus produced is transferred into the mitochondrial electron transport chain for oxidative phosphorylation.

Aerobic Respiration Metabolic reactions or pathways using oxygen as the terminal electron acceptor for breakdown of nutrient molecules (e.g., glucose) to yield energy and produce CO_2 as waste product.

Aerobid A trade name for flunisolide, a respiratory corticosteroid.

Aerobiosis 1. The state or condition in which oxygen is present. 2. Life in the presence of oxygen or air.

Aerococcus A genus of Gram-positive bacteria of the family Streptococcaceae.

Aerogel The highly porous gel that has maintained its original structure despite the loss of solvent.

Aerogenic Capable of producing gas.

Aerolate A trade name for theophylline, a bronchodilator and phosphodiesterase inhibitor.

Aerolone A trade name for isoproterenol, used as a bronchodilator.

Aerolysin A channel-forming protein secreted by the *Aeromonas hydrophila*.

Aeromonad Any species or strain of *Aeromonas*.

Aeromonas A genus of Gram-negative bacteria of the family Aeromonadaceae.

Aeromonolysin A proteolytic enzyme from *Aeromonas proteolytica*.

Aeroscope A device for gathering particles or microorganisms from the atmosphere for microscopic examination.

Aeroseb-Dex A trade name for dexametasone, a bronchodilator used for treatment of skin rash and inflammation.

Aerosol Minute particles of liquid and/or solid dispersed in a gas or air.

Aerosporin A trade name for polymyxin B sulfate, used as an anti-infective agent.

Aerotaxis The migration of cells or microorganisms toward oxygen.

Aesculin A derivative of 6,7-dihydroxycoumarin-6-glucose used for identification of certain aesculin-hydrolyzing bacteria (e.g., group D *Streptococci*). Hydrolysis of aesculin gives a brown coloration in the presence of ferric salt.

AET (mol wt 281) Abbreviation for 2-(2-aminoethyl)-2-thiopseudourea dihydrobromide, a radioprotective agent.

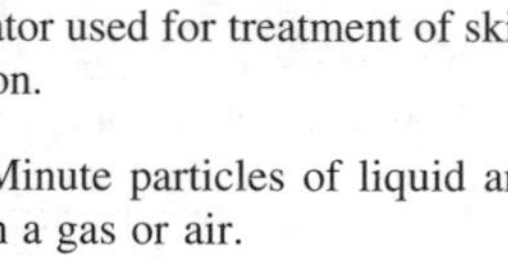

Aetiology The science that deals with causation.

AeuI (EcoRII) A restriction endonuclease from *Achromobacter eurydice* with the following specificity:

5′..........CC(A/T)GG..........3′
3′..........GG(T/A)CC..........5′

AEV Abbreviation for Avian erythroblastosis virus (see Avian Acute Leukaemia virus).

AEX Abbreviation for anion exchange.

AF Abbreviation for 1. Activation function. 2. Albumin free. 3. Amniotic fluid. 4. Angiogenesis factor.

AfaI A restriction endonuclease from *Acidiphilium facilis* 28H with following specificity:

5'.... .....GTAC...........3'
3'.........CATG...........5'

Afamin A vitamin-binding protein in mammalian serum.

AFAR Abbreviation for aflatoxin-B_1-aldehyde reductase.

AFB Abbreviation for 1. Acid-fast *Bacillus* or acid-fast bacteria. 2. Aflatoxin-B

AFB_1 Abbreviation for aflatoxin-B1.

AFC Abbreviation for antibody-forming cell.

AfeI A restriction endonuclease from *Alcaligenes faecalis* with following specificity:

5'.....AGCGCT.....3'
3'.....TCGCGA.....5'

Afebrile Without fever.

Afferent Proceeding toward a center.

Afferent Nerve A nerve which transmits impulses to the central nervous system.

Afferent Neurons Neurons that carry impulses to the central nervous system.

Affinin (mol wt 221) A lipid amide isolated from *Heliopsis longipes* of Compositae, with local anesthetic properties.

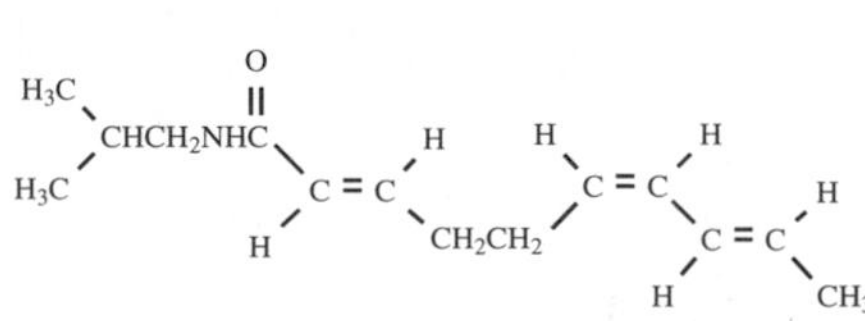

Affinity Chromatography A chromatographic technique in which a substance with a selective binding affinity is coupled to an insoluble matrix such as dextran, sepharose, or polyacrylamide. The coupled insoluble matrix is then used to bind or

trap its complementary substance. These trapped molecules can be selectively eluted with an appropriate eluting solution.

Affinity Constant The equilibrium constant for association (complex formation), it is the reciprocal of the dissociation constant for the complex.

Affinity Electrophoresis Electrophoresis performed on a support medium containing immobilized ligands that are capable of binding with components of the proteins to be electrophoresed. The binding of ligands to the proteins lead to a change in the electrophoretic mobility of the proteins.

Affinity Elution A chromatographic technique in which a mixture of compounds are adsorbed nonspecifically to a column, and the compound of interest is selectively eluted through its binding to a ligand in the eluting solvent.

Affinity Labeling A technique for labeling and identification of the active site of an enzyme or protein. It involves noncovalent binding of a bifunctional reagent to the active site. The bound bifunctional reagent is then linked covalently through its second chemically reactive group to an amino acid residue near or on the active site.

Affinity Matrix Any supporting material to which the bio-specific reagent is attached in the affinity chromatography.

Affinoelectrophoresis See affinity electrophoresis.

Affinophore A macromolecular polyelectrolyte bearing affinity ligands for specific protein and used in electrophoresis for alteration of the electrophoretic mobilities of the proteins to be electrophoresed.

Affinophoresis Synonym for affinoelectrophoresis.

AFG Abbreviation for aflatoxin-G.

Afibrinogenemia A genetic disorder characterized by the absence of plasma fibrinogen in the blood.

AFID Abbreviation for alkali flame ionization detector.

AFL Abbreviation for aflatoxin-L.

AflI (AvaII) A restriction endonuclease derived from *Anabaena flos-aquae* with the following specificity:

↓
5′..........GG(A/T)CC..........3′
3′..........CC(T/A)GG..........5′
↑

AflII A restriction endonuclease derived from *Anabaena flos-aquae* with the following specificity:

↓
5′.........CTTAG.........3′
3′.........GAATC.........5′
↑

AflIII A restriction endonuclease from *Anabaena flos-aquae* with the following specificity:

↓
5′.........CTTAG.........3′
3′.........GAATC.........5′
↑

Aflate A trade name for tolnaftate, an anti-fungal agent.

Aflatoxicosis A mycotoxicosis caused by ingestion of aflatoxins.

Aflatoxin B One of the aflatoxins produced by *Aspergillus flavus* and *A. parasiticus*, it is carcinogenic and capable of inhibition of salt-induced conversion of B-DNA to Z-DNA.

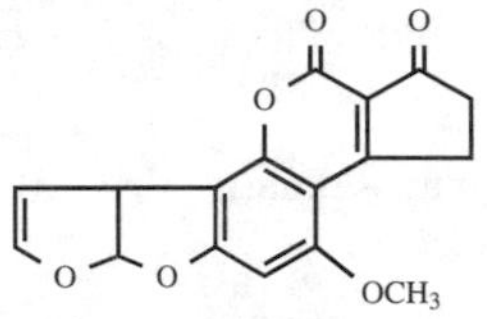

Aflatoxin B_1

Aflatoxin G A carcinogenic toxic metabolite from *Aspergillus flavus* and *A parasiticus.*

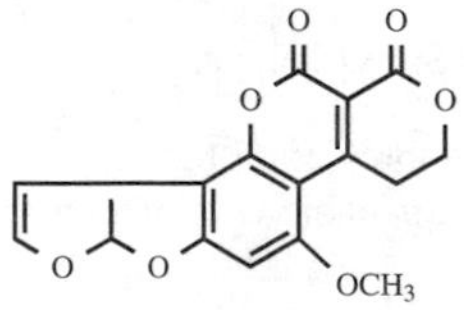

Aflatoxin G_1

Aflatoxin M An aflatoxin B derivative found in the milk of cows fed toxic meal.

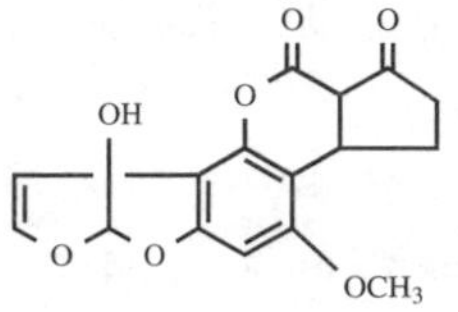

Aflatoxin M_1

Aflatoxins A group of mycotoxins produced by strains of *Aspergillus flavus* and *A. parasiticus*. There are many different types of aflatoxins, e.g., aflatoxin B_1, B_2, G_1, G_2, M_1, and M_2. Aflatoxins are capable of binding with purines leading to the inhibition of DNA replication and RNA transcription.

Afloqualone (mol wt 283) A skeletal muscle relaxant.

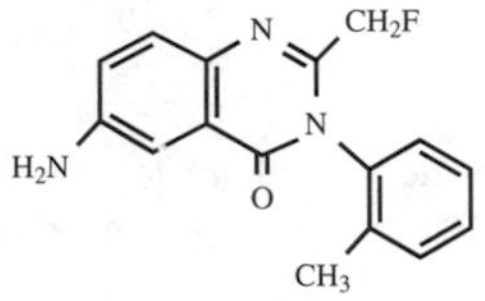

AFLP Abbreviation for amplified fragment length polymorphism.

AFM Abbreviation for 1. Aflatoxin-M. 2. Atomic-force microscope.

AFP Abbreviation for alpha-fetoprotein (α-fetoprotein).

αFP Abbreviation for alpha-fetoprotein.

AFQ Abbreviation aflatoxin-Q

African Sleeping Sickness A disease of the nervous system caused by the parasite *Trypanosoma gambiense* or *T. rhodesiense*. It is transmitted to humans by the bite of the testse fly (also known as African trypanosomiasis).

African Trypanosomiasis See African sleeping sickness.

Afrin A trade name for oxymetazoline hydrochloride, an adrenergic vasoconstrictor used as a decongestant.

AFT Abbreviation for aflatoxin.

Afterstain See counterstain.

Ag 1. The abbreviation for antigen. 2. The symbol for the element silver with atomic weight 108, valence 1.

A/G Referring to the albumin and globulin ratio.

AgAb Abbreviation for antigen-antibody complex.

Agalactia Absence or failure of the secretion of milk.

Agalactosis See agalactia.

Agalactosuria Absence of galactose from urine.

Agamete 1. Any of the unicellular organisms that reproduce asexually, e.g., bacteria. 2. Any asexual reproductive cell that forms a new cell without fusion with another cell.

Agammaglobulinemia An inherited immunological disorder characterized by a failure to produce γ-globulins or immunoglobulins.

Agamogenesis Asexual reproduction.

Agar (Agar-Agar) A complex of galactan (polysaccharide) extracted from red algae (Rhodophyceae) used as a solidifying agent for preparation of agar medium or agar gel. It consists of two main components: agarose (about 70%) and agaropectin (about 30%).

Agar Diffusion Method A method for determining the sensitivity of a microorganism to an antibiotic or antimicrobial drug based on the measurement of the size of the zone of growth inhibition when an antibiotic or a drug is placed in a hole, or a filter paper disk on an agar medium plate that has been seeded with the microorganism.

Agar Gel Diffusion Test See Ouchterlony gel diffusion test.

Agar Gel Electrophoresis Zonal electrophoresis in which agarose gel is used as the supporting medium.

Agarase The enzyme that hydrolyzes agar into the reducing sugar galactose.

β-Agarase I The enzyme that hydrolyzes agarose into neoagarooligosaccharide. It is a useful enzyme for recovering DNA fragments from agarose.

Agaricic Acid (mol wt 417) A compound extracted from the fungus *polyporus afficinalis*, used for treatment of diarrhea, reducing bronchial secretion, and night sweats.

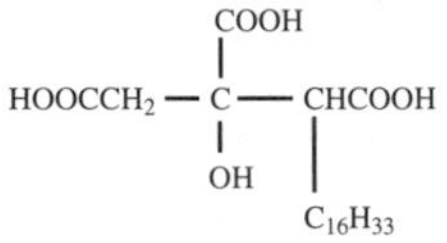

Agaritine (mol wt 267) A compound found in the commercial edible mushroom *Agaricus bisporus*.

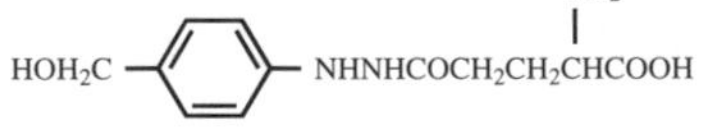

Agarobiose (mol wt 144) A disaccharide obtained from the degradation of agar.

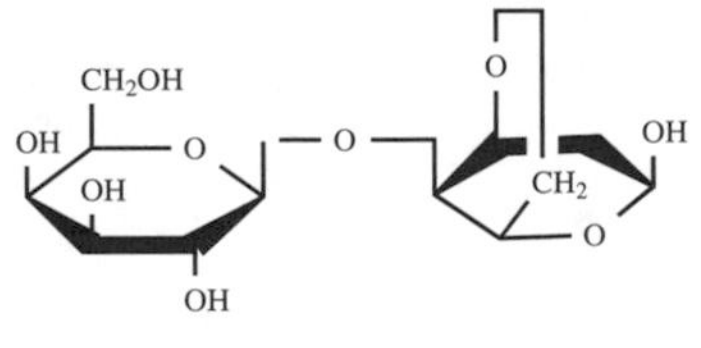

β-D-galactose anhydrogalactose

Agaropectin The sulfated and nongelling fraction of agar. It contains sulfated galactan and glycuronic acid.

Agarophyte Any agar-producing seaweed.

Agarose The neutral gelling component of agar, it contains a linear polymer of galactose and anhydrogalactose in an alternating α-(1,3) and β-(1,4) linkage.

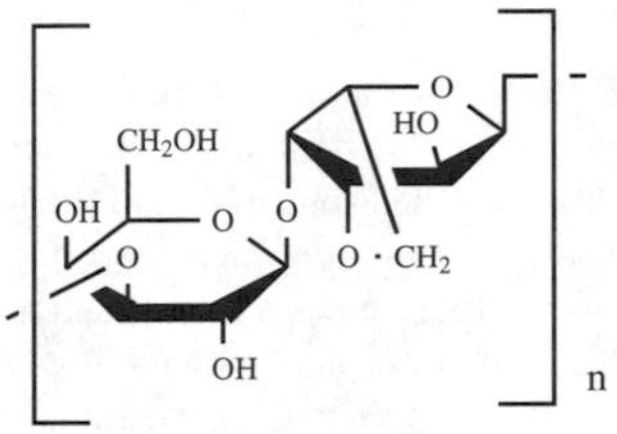

galactose anhydrogalactose

Agarose 3-Glycanohydrolase Synonym of agarase.

AGC A genetic code (codon) for the amino acid serine.

AGD Abbreviation for agar gel diffusion.

AGE Abbreviation for agar gel electrophoresis.

AgeI A restriction endonuclease from *Agrobacterium gelatinovorum* with following specificity:

5'........A↓CCGGT........3'
3'........TGGCC↑A........5'

Agent Anything capable of producing a physical, chemical, or biological effect.

Agent Orange A mixture of 2,4-dichlorophenoxyacetic acid and 2,4,5-trichlorophenoxyacetic acid, used as a defoliation agent.

AGG Abbreviation for 1. A genetic code (codon) for amino acid arginine. 2. Alpha-gamma globulinemia.

Agglomeration Group of cells that form a mass.

Agglutinating Antibody Antibody capable of mediating agglutinating reactions, e.g., antibody to bacteria.

Agglutination Inhibition Test A serologic test for the identification of certain agglutinating antigens (e.g., protein or virus) based on the inhibition of agglutination (formed by the agglutinating antigen) by the known antibody.

Agglutination Reaction An antigen-antibody reaction in which antigens of particulates combine with soluble antibody to form visible antigen-antibody complexes (e.g., clumping of bacterial cells by antibody).

Agglutination Test See agglutination reaction.

Agglutinin An antibody or substance capable of causing the clumping or agglutination of bacteria, cells, or inanimate particles.

Agglutinogen Any cell surface antigen that can induce the production of agglutinins and bind to them to produce an agglutination reaction.

Agglutinoid An agglutinin that has lost its ability to produce an agglutination reaction but is still capable of combining with its corresponding agglutinogen.

Aggrastat A trade name for tirofiban, an antiplatelet agent.

Aggrecan A major proteoglycan of cartilage.

Aggressin Referring to the product of a pathogenic microorganism that promotes the invasiveness of the organism in the host, e.g., hyaluronate lyase and streptokinase.

AGL Abbreviation for acute granulocytic leukemia.

Aglandular Lacking glands.

Aglaukopsia Inability to distinguish the color green.

Aglycone The noncarbohydrate moiety of a glycoside (e.g., methyl group in a methyl-glucoside).

Aglycosuria Absence of glucose in the urine.

Agmatinase The enzyme that catalyzes the following reaction:

$$\text{Agmatine} + H_2O \rightleftharpoons \text{Putrescine} + \text{urea}$$

Agmatine (mol wt 130) A compound formed by decarboxylation of arginine.

$$H_2N\underset{\underset{NH}{\|}}{C}NH(CH_2)_4NH_2$$

Agmatine Deiminase The enzyme that catalyzes the following reaction:

$$\text{Agmatine} + H_2O \rightleftharpoons \text{N-Carbamoylputrescine} + NH_3$$

Agmatine Kinase The enzyme that catalyzes the following reaction:

$$\text{ATP} + \text{agmatine} \rightleftharpoons \text{ADP} + \text{N-phosphoagmatine}$$

AGMK Abbreviation for African green monkey kidney.

Agnoprotein A basic peptide encoded by a gene in SV40 (Simian virus 40). It plays an important role in viral assembly.

Agonist A structural analog that is capable of stimulating a biological response like a natural ligand by occupying the cell receptor.

AGP Abbreviation for 1. Acid glycoprotein. 2. Arabinoglycan protein

αGP Abbreviation for alpha glycerophosphate.

AGPase Abbreviation for ADP-glucose pyrophosphorylase.

AGPT Abbreviation for agar gel precipitation test.

Agranular Reticulum Referring to the reticulum without attached ribosomes.

Agranulocyte Any white blood cell that has a nongranular cytoplasm (e.g., a lymphocyte).

Agranulocytosis An abnormal condition characterized by a severe reduction in the number of granulocytes (basophils, eosinophils, and neutrophils) in the blood.

Agravic Lack of gravity (zero G).

Agretope Part of the antigen that combines with the desetope of an MHC molecule on an antigen-presenting cell.

Agrin A component of the synaptic basal lamina that causes aggregation of acetylcholine receptors and acetyl cholinesterase on the surface of muscle fibers of the neuromuscular junction.

Agrobacterium A genus of Gram-negative bacteria of the Rhizobiaceae capable of causing tumor formation in plants.

Agrocinopines A class of sugar phosphodiester opium found in plant crown gall.

Agrocins Antibiotics that are produced by certain strains of *Agrobacterium* and active against other strains of the same genus.

Agroclavine (mol wt 238) A nonpeptide ergot alkaloid obtained from cultures of fungi parasitic to *Elymus mollis*.

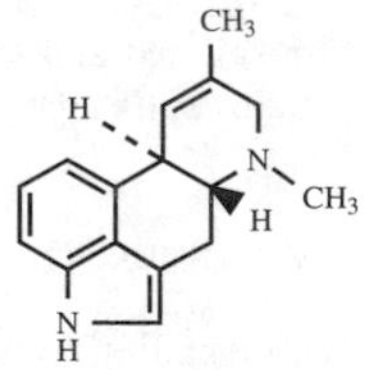

Agrocybin (mol wt 147) An antibiotic produced by basidiomycete *Agrocybe dura*.

$$CH_2OHC{\equiv}C\text{-}C{\equiv}C\text{-}C{\equiv}C\text{-}O\text{-}CO\text{-}NH_2$$

Agroinfection A method of introducing foreign DNA into a plant in which the foreign DNA is incorporated into a Ti-plasmid, and the plasmid is then introduced into *Agrobacterium tumefacians*. The transfer of foreign DNA into a plant is achieved by the infection of the plant with the *A. tumefacians* containing the constructed Ti-plasmid.

Agromyces A genus of microaerophilic to anaerobic, catalase-negative, asporogenous bacteria (order Actinomycetales).

Agronomy The science that deals with crop production.

Agropine (mol wt 415) A rare amino acid derivative produced by the Ti plasmid of certain plant crown-gall tumor.

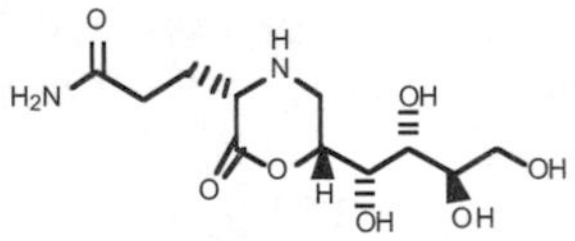

Agrostin A protein from the seeds of *Agrostemma githago*. It is a ribosome-inhibiting protein.

Agrylin A trade name for anagrelide, an antiplatelet agent.

Agrypnia Insomnia.

Agrypnotic An agent or drug that prevents sleep.

AGT Abbreviation for antiglobulin test.

AGTH Abbreviation for adrenoglomerulotropin hormone.

AGTT Abbreviation for abnormal glucose tolerance test.

AgTX Abbreviation for agitoxin.

AGU A genetic code (codon) for the amino acid serine.

AGV Abbreviation for gentian violet aniline.

AH Abbreviation for 1. Anti-hyaluronidase. 2. Aromatic hydrocarbon

AHA A trade name acetohydroxamic acid, a urinary tract agent.

AhaI (CauII) A restriction endonuclease from *Aphanothece halophytica* with the following specificity:

5′..........CC↓(CG)GG..........3′
3′......... GG(GC)↑CC..........5′

AhaII (AcyI) A restriction endonuclease isolated from *Aphanothece halophytica* with the following specificity:

5′...........GPu↓CGPyC...........3′
3′...........CPyGC↑PuG...........5′

AhaIII A restriction endonuclease isolated from *Aphanothece halophytica* with the following specificity:

5′.........AAA↓TTT.........3′
3′.........TTT↑AAA.........5′

AHBD Abbreviation for alpha-hydroxy-butyric dehydrogenase.

AHBDH Abbreviation for alpha-hydroxy-butyric dehydrogenase.

AHC Abbreviation for acute hemorrhagic conjunctivitis.

AhdI A restriction endonuclease from *Aeromonas hydrophilia* (NEB 724) with following specificity:

5'........GACNNN↓NNGTC........3'
3'........CTGNN↑NNNCAG........5'

AHF Abbreviation for antihemophilic factor.

AHG Abbreviation for 1. antihemophilic globulin and 2. antihuman globulin.

AHH Abbreviation for aryl-hydrocarbon hydroxylase.

Ahistan (mol wt 284) A synthetic antihistamine agent.

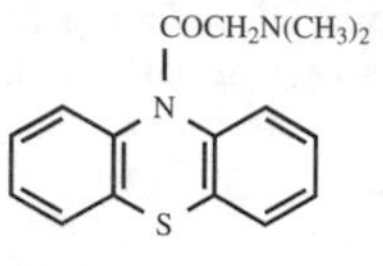

AHLS Abbreviation for anti-human lymphocyte serum or anti-human lymphocytic serum.

AHR Abbreviation for aryl hydrocarbon receptor.

3α-HSD Abbreviation for 3-α-hydroxysteroide dehydrogenase.

AHTG Abbreviation for anti-human thymocytic globulin or anti-human thymocyte globulin.

AHTP Abbreviation for anti-human thymocytic plasma.

AHx Abbreviation for 6-aminohexanoic acid.

AI Abbreviation for 1. Amylase inhibitor. 2. Anaphylatoxin inhibitor.

AIA Abbreviation for 1. Amylase inhibitor activity. 2. Anti-immunoglobulin antibodies. 3. Anti-insulin antibody

AIAb Abbreviation for anti-insulin antibody.

AIB Abbreviation for aminoisobutyrate.

AIBA Abbreviation for amino-isobutyric acid.

AIC Abbreviation for aminoimidazole carboxamide.

AICA Abbreviation for aminoimidazole carboxamide.

AICAR Abbreviation for aminoimidazole carboxamide ribonucleotide.

AICF Abbreviation for auto-immune complement fixation.

AICR Abbreviation for aminoimidazole carboxamide ribonucleotide.

AIDS Abbreviation for acquired immune deficiency syndrome. A transmissible, fatal human disease that affects the immune system and is caused by HIV (human immune deficient virus).

AIF Abbreviation for apoptosis-inducing factor.

AIHA Abbreviation for autoimmune hemolytic anemia.

AIHD Abbreviation for auto-immune hemolytic disease.

AILA Abbreviation for angio-immunoblastic lymph-adenopathy.

AinI (PstI) A restriction endonuclease from *Anabaena inqualis* having the same specificity as PstI.

AinII (BamHI) A restriction endonuclease from *Anabaena inqualis* having the same specificity as BamHI.

α_2-Interferon A trade name for interferon apha-2.

AIP Abbreviation for auto-immune precipitation.

AIR (mol wt 295) Abbreviation for 5′-amino-imidazol ribonucleotide. An intermediate in the biosynthesis of a purine nucleotide.

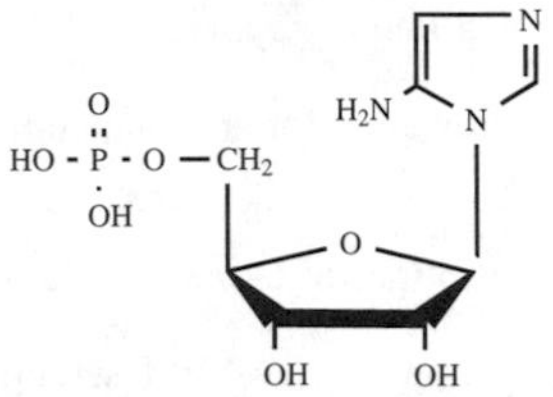

Air Embolism The presence of air in the cardiovascular system, resulting in the obstruction of the flow of blood through the vessel.

Air Fermentor A fermentor in which the circulation of culture is achieved by pumping sterile air in at the bottom of the culture vessel.

Air Peak The peak that is produced when a small amount of air is injected together with the sample into the gas chromatographic column.

Airet A trade name for albuterol, a bronchodilator.

AIS Abbreviation for anti-insulin serum.

AitAI (XhoII) A restriction endonuclease from *Aquaspirillum itersonii* having the same specificity as XhoII.

AITT Abbreviation for arginine insulin tolerance test.

Ajmaline (mol wt 326) An antihypersensitive and an antiarrhythmic agent from the roots of *Rauwolfia serpentina* (Apocynaceae).

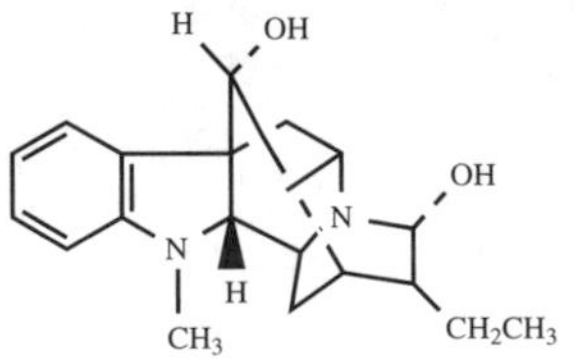

Ajoene (mol wt 234) An antithrombotic agent found in garlic (*Allium satiuum).*

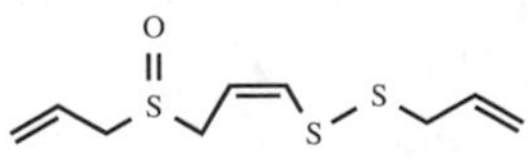

Ajowan oil Oil from the seeds of *Carum copticum* (Umbelliferae). It is toxic to earthworms, *Staphylococci* and *E. coli.*

Ajugarins Five related compounds isolated from the leaves of *Ajuga remota* (e.g., Ajugarin I, II, III, IV, and V).

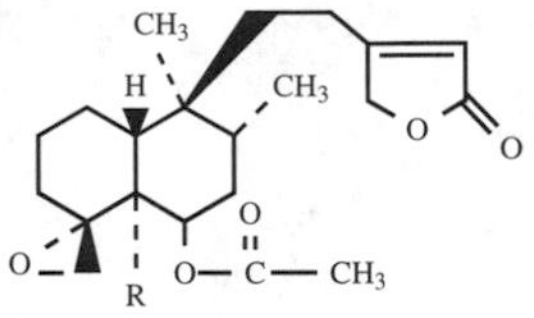

Ajugarin I: R = CH_2COOCH_3

Ajugose A polysaccharide from the root of *Ajuga nipponensis* and *Verbascum thapsiforme.* It contains galactose, glucose, fructose, and verbascose.

AK Abbreviation for 1. Arginine kinase. 2. Adenylate kinase

Akabori Reaction The formation of an alkamine by the reaction of an aldehyde with the amino group of an amino acid.

AKAP Abbreviation for A-kinase anchoring protein, it is a cAMP-dependent protein kinase anchor protein.

Akarpine A trade name for pilocarpine ophthalmic.

Akaryocyte A cell without a nucleus.

Akaryote See akaryocyte.

AK-Chlor A trade name for chloramphenical, an antibiotic that inhibits bacterial protein synthesis.

AK-Con A trade name for naphazoline hydrochloride, used as an ophthalmic vasoconstrictor.

AK-Dex A trade name for dexamethasone sodium phosphate, used as a corticosteroid.

AK-Dilate A trade name for phenylephrine hydrochloride, used as a mydriatic agent.

AKG Abbreviation for α-keto-glutarate.

AKH Abbreviation for adipokinetic hormone.

A-Kinase The enzyme that phosphorylates the target proteins in response to a rise in intracellular cyclic AMP concentration.

Akinete 1. A specialized bacterial cell capable of resistance to desiccation and cold. 2. A thick-walled, nonmotile resting cell produced by certain algae. 3. A nonmotile spore.

Akineton A trade name for biperiden hydrochloride or biperiden lactate, an anticholinergic drug used as an antiparkinsonian agent.

Aklomide (mol wt 200) An antiprotozoal and antibacterial agent.

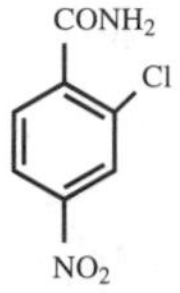

AK-Mycin A trade name for erythromycin, an antibiotic.

AK-Nefrin Ophthalmic A trade name for phenylephrine hydrochloride, used as a mydriatic agent.

Akne-Mycin A trade name for erythromycin, an antibiotic that inhibits bacterial protein synthesis.

AKP Abbreviation for alkaline phosphatase.

Ak-Pentolate A trade name for cyclopentolate hydrochloride, used to dilate the pupil.

AK-Pred A trade name for prednisolone, a hormone.

AKR Abbreviation for aldo-keto reductase.

AK-Spore A trade name for a combination ophthalmic drug containing neomycin, polymyxin-B and bacitracin.

Akt An oncogene of the transforming retrovirus AKT8.

AK-Tracin A trade name for bacitracin, an antibiotic.

Al Symbol for aluminum (atomic weight 27, valence 3).

AL Abbreviation for 1. active lipid. 2. argininosuccinate lyase.

AL 721 Abbreviation for an active lipid 721. A nutritional supplement, it is composed of neutral glyceride, phosphatidylcholine, and phosphatidylethanoamine at a ratio of 7:2:1.

Ala Abbreviation for alanine.

ALA Abbreviation for 5-aminolevulinic acid.

AlaAT Abbreviation for alanine aminotransferase.

A-Lac Abbreviation for alpha lactalbumin.

Alacepril (mol wt 407) An inhibitor for angiotensin-converting enzyme, used as an antihypertensive agent.

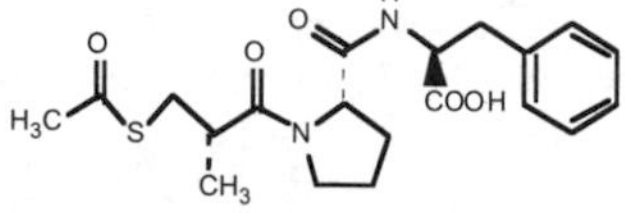

Alachlor (mol wt 270) An herbicide.

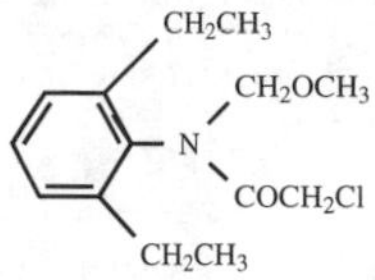

ALAD Abbreviation for 5-aminolevulinic acid dehydrogenase.

ALADH Abbreviation for 5-aminolevulinic acid dehydrogenase.

AlaDH Abbreviation for alanine dehydrogenase.

Alafosfalin (mol wt 196) A synthetic antibiotic which has a broad spectrum of activity.

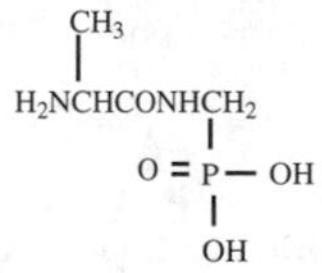

Alamethicin A peptide antibiotic produced by strains of *Trichoderma viride*. It acts as an ionophore.

Alanine (mol wt 89) A protein amino acid.

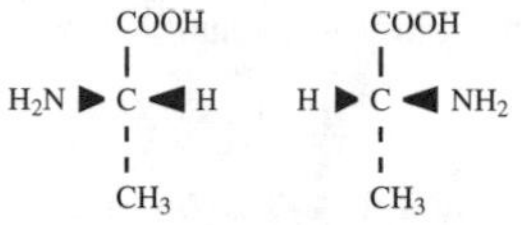

L-alanine D-alanine

β-Alanine (mol wt 89) A naturally occurring amino acid not found in proteins.

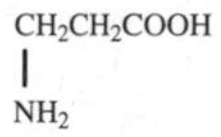

Alanine Aminopeptidase The enzyme that catalyzes the preferential cleavage of N-terminal alanine from a peptide.

Alanine Aminotransferase The enzyme that catalyzes the following reaction:

$$\alpha\text{-Keto-glutarate} + \text{alanine} \rightleftharpoons \text{pyruvate} + \text{glutamate}$$

Alanine Carboxypeptidase The enzyme that catalyzes the release of C-terminal alanine from a peptide.

Alanine Cycle A metabolic cycle in which alanine is formed in the muscle by the transamination of glucose-derived pyruvate.

Alanine Dehydrogenase The enzyme that catalyzes the following reaction:

$$\text{L-Alanine} + NAD^+ \rightleftharpoons \text{pyruvate} + NADH$$

Alanine α-Keto-Acid Transaminase The enzyme that catalyzes the following reaction:

L-Alanine + α-keto-acid

⇅

Pyruvate + a L-amino acid

L-Alanine NAD$^+$ Oxidoreductase The systematic name for alanine dehydrogenase.

L-Alanine 2-Oxoglutarate Aminotransferase The systematic name for alanine transaminase.

Alanine Racemase The enzyme that mediates the interconversion of D-form and L-form alanine.

L-alanine ⇌ D-alanine

Alanine Transaminase The enzyme that catalyzes the following reaction:

L-Alanine + 2-oxoglutarate

⇅

Pyruvate + L-glutamate

Alaninium The alanine cation.

$CH_3CH(NH_3^+)\text{-}COOH$

Alaninol The alcohol obtained by the reduction of carboxyl group of alanine.

$CH_3CH(NH_2)CH_2OH$

L-Alanosine (mol wt 149) An antibiotic produced by *Streptomyces alanosinicus*.

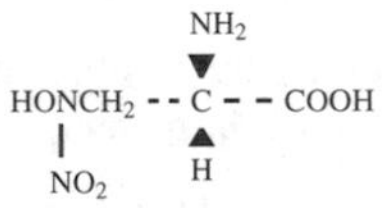

Alantolactone (mol wt 232) An anthelmintic (nematodes) agent obtained from the roots of *Inula helenium* (Compositae).

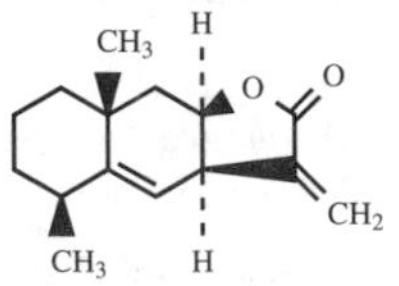

β-Alanyl-CoA Ammonia-Lyase The enzyme that catalyzes the conversion of β-alanyl CoA into acryl-CoA and NH_3.

Alanyl-Polyphosphoribitol Synthetase The enzyme that catalyzes the following reaction:

ATP + D-alanine + polyribitol-phosphate

⇅

AMP + pyrophosphate +
D-alanylpolyribitol polyphosphate

Alanyl-tRNA Ligase See alanyl-tRNA synthetase.

Alanyl-tRNA Synthetase The enzyme that catalyzes the following reaction:

ATP + L-alanine + tRNA^{ala}

⇅

AMP + pyrophosphate + L-analyl-tRNA^{ala}

Ala-Quin A trade name for iodochlorhydroxyquin used as an anti-infective agent.

Alarmone Any signal molecule that serves to reorient a cell's economy in response to stress.

ALAS Abbreviation for aminolevulinic acid synthetase.

Alastrim A viral disease similar to small pox, found in Brazil, West Indies, and Africa.

ALAT Abbreviation for alanine amino transferase.

Alazine A trade name for hydralazine hydrochloride, used as an antihypertensive agent.

Alazopeptin (mol wt 364) An antibiotic used as an antineoplastic agent and produced from *Streptomyces griseoplanus*.

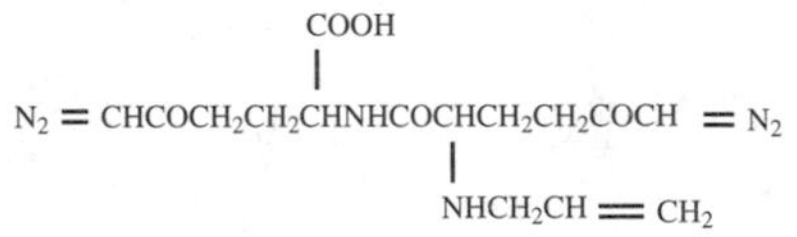

Alb- A prefix meaning white.

Albamycin Synonym for novobiocin.

Albendazol (mol wt 265) An anthelminic agent.

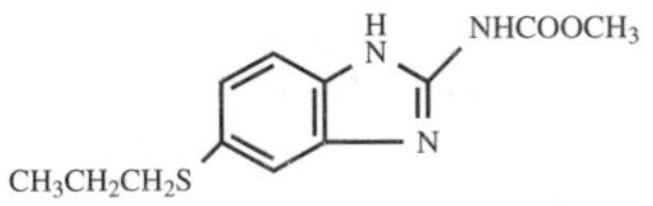

Albenza A trade name for albendazole, an anthelminic agent.

Albert Glyburide A trade name for glyburide, an anti-diabetic agent.

Albert Oxybutynin A trade name for oxybutynin, an urinary antispasmodic agent.

Albert's Stain A stain used to demonstrate the metachromatic granules in bacterial cells.

Albicidin An antibiotic produced by *Xanthomonas albilineans*, which inhibits DNA synthesis in *Escherichia coli*.

Albinism An inherited metabolic disorder in humans characterized by lack of skin pigmentation due to a deficiency of the enzyme tyrosinase.

Albino An individual that is deficient in skin pigmentation.

Albizziin (mol wt 147) A nonprotein amino acid occurring in some species of genus *Albizzia*. It is an antagonist of glutamine.

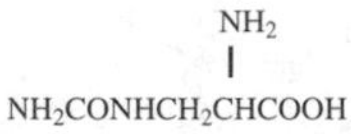

Albofungin (mol wt 521) An antifungal antibiotic produced by *Streptomyces albus* var *fungistaticus*.

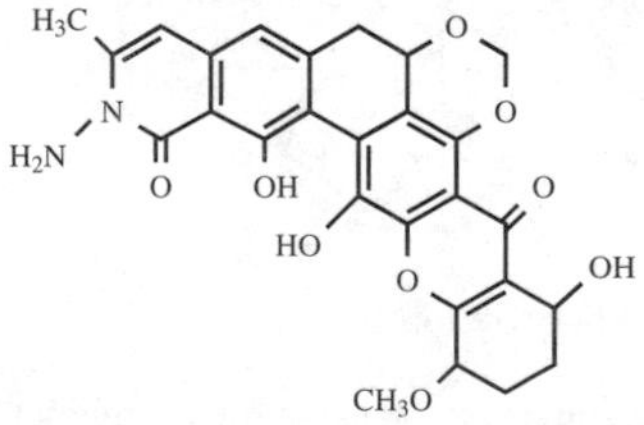

Albomycin An antibiotic produced by *Actinomyces subtropicus*. It is a cyclic peptide with a pyrimidine base (cytosine) and is effective against both Gram(+) and (–) bacteria.

ALBP Abbreviation adipocyte lipid-binding protein.

Albumen Egg white or dry egg white (pure protein is spelled as albumin).

Albumen Sac A two-layered ectodermal sac enclosing the albumen of chick egg.

Albumin A group of heat-coagulatable proteins that are soluble in water and in dilute salt.

Albumin A A blood serum constituent that occurs in cancer cells but does not circulate in the blood of cancer patients.

Albuminar A trade name for an albumin solution.

Albuminoids Synonym for scleroproteins.

Albuminorthostatic The presence of albumin in the urine due to long periods of standing.

Albumin Tannate A tannin-albumin complex containing about 50% tannin. It is an antidiarrheal and astringent agent.

Albuminuria The presence of excessive amounts of protein, mainly albumin, in the urine.

Albustix Test A rapid, semiquantitative test for protein in the urine by means of paper strips impregnated with buffer and protein indicator.

Albutein A trade name for an albumin solution.

Albuterol (mol wt 239) A bronchodilator and tocolytic agent that relaxes bronchial and smooth muscle of the uterine.

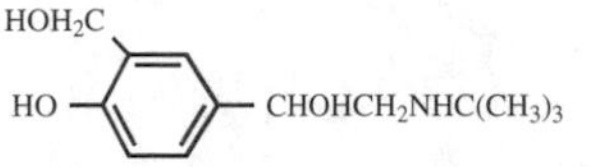

Albutoin (mol wt 212) An anticonvulsant.

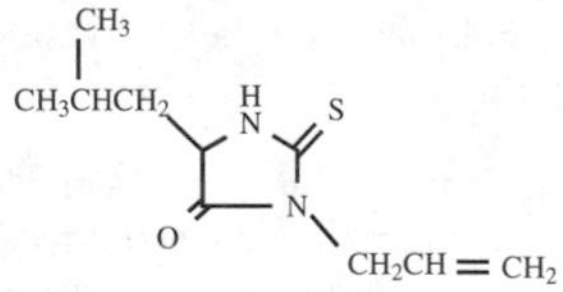

Alcalase A proteolytic enzyme.

Alcaligenes A genus of catalase-positive, oxidase-positive, Gram-negative bacteria that occur in soil, water, and in the alimentary tract of vertebrates.

Alcaptonuria A metabolic disease in which the homogentisic acid oxidase is absent leading to the secretion of homogentisic acid in the urine.

AlcDH Abbreviation for alcohol dehydrogenase.

Alcian Blue (mol wt is about 1300) A basic dye used for staining glycoproteins and polysaccharides.

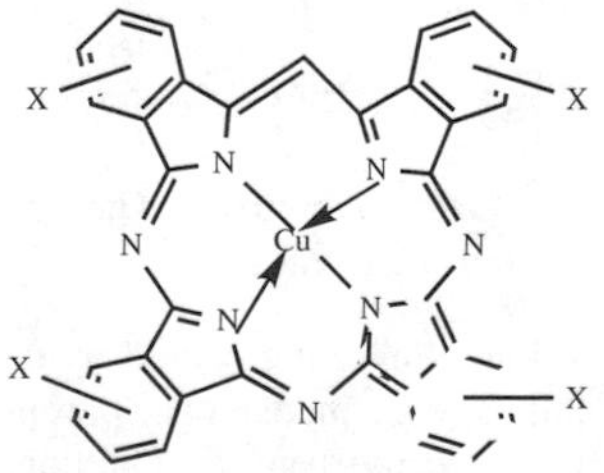

X = an onium group, e.g.,

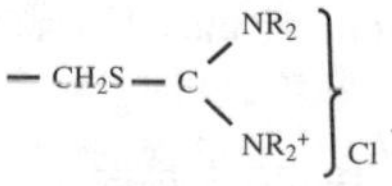

(R = alkyl or aryl)

Alclofenac (mol wt 227) An analgesic, antipyretic, and anti-inflammatory agent.

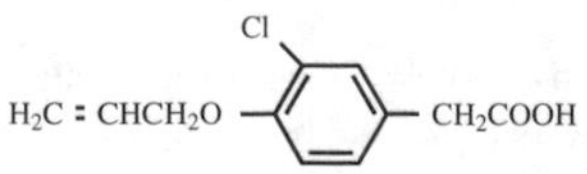

Alclometasone (mol wt 409) A corticosteroid and a topical anti-inflammatory agent that suppresses inflammation and normal immune responses.

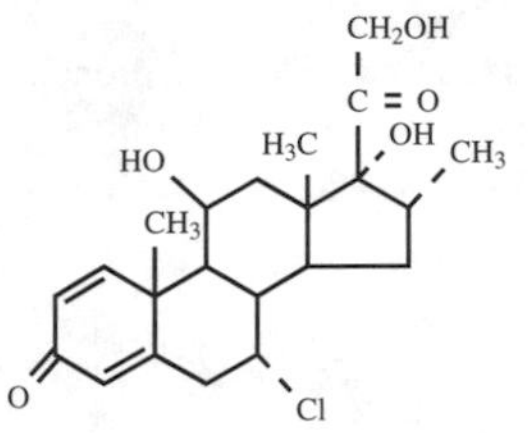

Alclovate A trade name for alclometasone dipropionate, used as a corticosteroid.

Alclox A trade name for the cloxacillin sodium, an antibiotic.

Alcohol 1. Organic compounds containing one or more hydroxyl groups attached directly to carbon atoms. Alcohols are designated as primary, secondary, tertiary depending on whether the hydroxyl group is linked to a carbon atom that is bound to one, two, or three other carbon atoms. 2. Ethyl alcohol; ethanol.

Alcohol Dehydrogenase A NAD-linked dehydrogenase that catalyzes the following reaction:

$$\text{NADH + acetaldehyde} \rightleftharpoons \text{Ethanol} + \text{NAD}^+$$

Alcohol NAD+ Oxidoreductase The systematic name for alcohol dehydrogenase.

Alcohol NADP+ Oxidoreductase The systematic name for alcohol dehydrogenase (NADP+).

Alcohol Oxidase The enzyme that catalyzes the following reaction:

$$\text{A primary alcohol} + O_2 \rightleftharpoons \text{An aldehyde} + H_2O_2$$

Alcohol Oxygen Oxidoreductase The systematic name of alcohol oxidase.

Alcohol Poisoning Poisoning caused by the ingestion of alcohols, e.g., methanol, isopropanol, and ethanol. It may cause nausea, vomiting, abdominal pain, blindness, or death.

Alcoholic Fermentation A microbial pathway for conversion of sugar to ethyl alcohol.

Alcoholic Hydroxyl Group A hydroxyl group linked to an aliphatic carbon chain.

Alcoholism A diseased condition caused by the excessive use of alcohol liquors.

Alcoholysis The cleavage of a covalent bond of an acid derivative by a reaction with an alcohol.

Alcomicin A trade name for gentamicin sulfate, an antibiotic.

Alconefrin A trade name for the phenylephrine hydrochloride, used to reduce blood flow and nasal congestion.

Alcuronium A skeletal muscle relaxant.

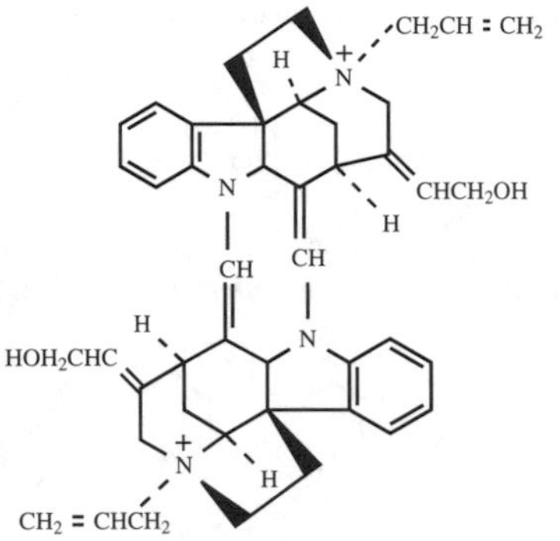

ALD Abbreviation for aldolase.

Aldactazide A trade name for a combination drug containing hydrochlorothiazide and spironolactone, used as a diuretic agent.

Aldactone A trade name for spironolactone, a diuretic agent.

Aldaric Acid A dicarboxylic sugar acid of an aldose in which both the aldehyde group and the primary alcohol group have been oxidized to carboxyl groups.

Aldase A trade name for the enzyme hyaluronidase.

Aldazine A trade name for thioridazine hydrochloride, used as an antipsychotic agent.

Aldecin A trade name for beclomethasone dipropionate, used as an anti-inflammatory agent.

Aldehyde An organic molecule that contains an aldehyde group.

Aldehyde Dehydrogenase (NAD+ Specific) The enzyme that catalyzes the conversion of an aldehyde to an acid using NAD+ as coenzyme.

$$\text{An aldehyde} + \text{NAD}^+ + H_2O \rightleftharpoons \text{An acid} + \text{NADH}$$

Aldehyde Dehydrogenase (NADP+ Specific) The enzyme that catalyzes the conversion of an aldehyde to an acid using NADP+ as coenzyme.

$$\text{An aldehyde} + \text{NADP}^+ + H_2O \rightleftharpoons \text{An acid} + \text{NADPH}$$

Aldehyde Group An organic compound containing the CHO- group joined directly onto another carbon atom.

Aldehyde NADP⁺ Oxidoreductase The systematic name for aldehyde dehydrogenase (NADP⁺).

Aldehyde Oxidase The enzyme that catalyzes the conversion of an aldehyde to an acid in the presence of oxygen.

$$\text{An aldehyde } H_2O + O_2 \rightleftharpoons \text{An acid} + O_2^-$$

Aldehyde Reductase The enzyme that catalyzes the following reaction:

$$\text{Alditol} + NADP^+ \rightleftharpoons \text{Addose} + NADPH$$

Alderase A trade name for tolrestat, an inhibitor for adlose reductase.

Aldesleukin A trade name for interleukin-2.

ALDH Abbreviation for 1. Alcohol dehydrogenase. 2. Aldehyde dehydrogenase.

Aldicarb (mol wt 190) An insecticide (acaricide, nematocide).

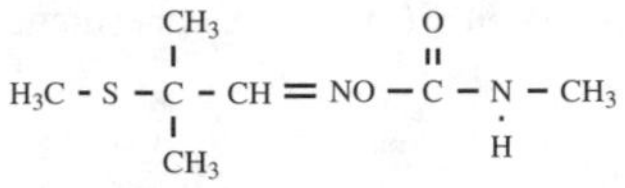

Aldimine An organic molecule of the general formula R-CH=NH.

Aldioxa (mol wt 218) An ulcerative agent.

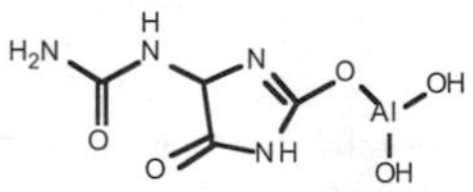

Alditol An acyclic polyhydroxyl alcohol (sugar alcohol).

Aldo- A prefix meaning aldose.

Aldoclor A trade name for a combination drug containing chlorothiazide and methyldopa, used as an antihypertensive agent.

Alditol NADP⁺ Oxidoreductase The systematic name for aldehyde reductase.

Aldoketomutase Synonym of lactoylglutathione lyase.

Aldoketose Any monosaccharide derivative containing both aldehydic carbonyl group and ketonic group.

Aldol (mol wt 88) An hypnotic and sedative agent.

$$CH_3 - CHOH - CH_2 - CHO$$

Aldol Condensation A condensation reaction of two ketones, two aldehydes, or an aldehyde and a ketone.

Aldolase The enzyme that catalyzes the following reaction:

Fructose 1,6-bisphosphate
⇅
Dihydroxyacetone 3-phosphate +
glyceraldehyde 3-phosphate

Aldomet M A trade name for methyldopa, an antihypertensive agent.

Aldonic Acid A sugar acid in which the aldehyde group of a monosaccharide is oxidized to a carboxyl group (COOH).

Aldopentose 5-carbon aldo-sugar.

Aldopyranose An aldose in a pyranose form.

Aldoril A trade name for a combination drug containing a diuretic (hydrochlorothiazide) and an antihypertensive (methyldopa).

Aldose A monosaccharide that has an aldehyde group.

Aldose 1-Dehydrogenase The enzyme that catalyzes the following reaction:

$$\text{D-Aldose} + NAD^+ \rightleftharpoons \text{D-Aldolactone} + NADH$$

Aldose 1-Epimerase See Aldose mutarotase.

Aldose Mutarotase The enzyme that catalyzes the interconversion of α-D-glucose to β-D-glucose.

Aldose 1-Phosphate Adenyltransferase The enzyme that catalyzes the following reaction:

ADP + aldose 1-phosphate
⇅
Pi + ADP-aldose

Aldose 6-Phosphate Reductase The enzyme that catalyzes the following reaction:

Sorbitol 6-phosphate + NADP⁺
⇅
D-Glucose 6-phosphate + NADPH

Aldose Reductase The enzyme that catalyzes the following reaction:

$$\text{Alditol} + NADP^+ \rightleftharpoons \text{Aldose} + NADPH$$

Aldosterone (mol wt 360) A highly active corticoid steroid hormone from the adrenal cortex that controls the metabolism of electrolyte and water.

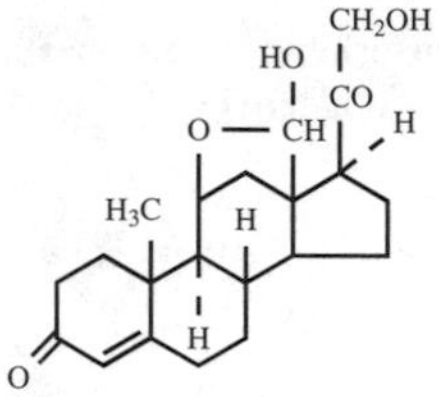

Aldosteronism A disorder characterized by the excessive production of aldosterone and the loss of blood potassium.

Aldosugar A sugar with an aldehyde group.

Aldotetrose Any 4-carbon aldosugar.

Aldotriose Any 3-carbon aldosugar.

Aldrin (mol wt 365) An insecticide.

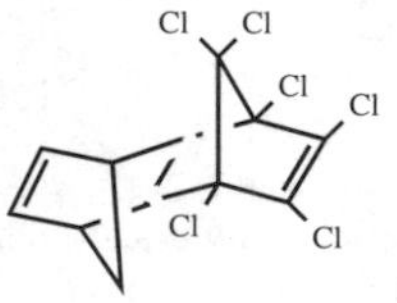

Alendronic Acid (mol wt 249) A suppressant of bone resorption.

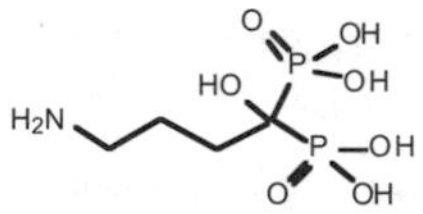

Aleukia An absence or reduced number of white blood cells in the blood.

Aleurone (Aleuron) Protein granules occurring in the outermost cell layer of the endosperm of wheat and other grains.

Aleuroplast Any colorless plastid.

Aleve A trade name for naproxen, a nonsteroidal anti-inflammatory and analgesic agent.

Alexan A trade name for cytarabine (cytosine arabinoside); inhibits DNA synthesis.

Alexidine (mol wt 509) An antimicrobial agent.

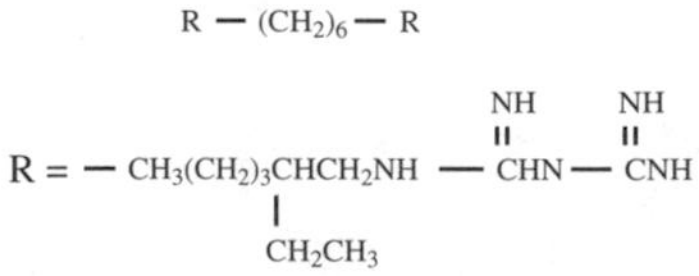

Alexin (Alexine) 1. A thermolabile and nonspecific factor that causes bacteriolysis in the presence of sensitizer (e.g., complement). 2. Synonym for complement.

Alexitol Sodium A complex of sodium polyhydroxyaluminum monocarbonate and hexitol used as an antacid.

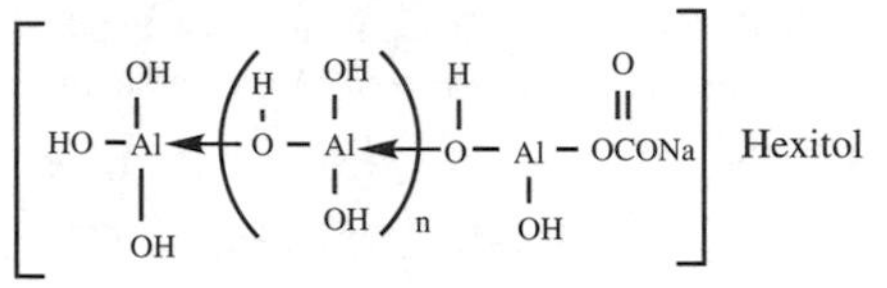

Alfadolone Acetate (mol wt 391) An anesthetic agent.

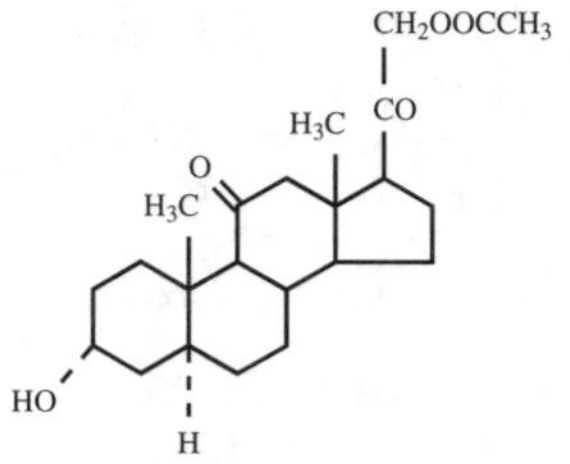

Alfalfa Mosaic Virus A ssRNA-containing plant virus that has a wide host range.

Alfaxalone (mol wt 332) An anesthetic agent.

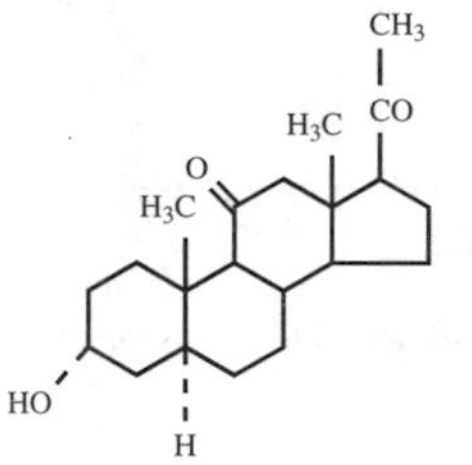

Alfenta A trade name for alfentanil hydrochloride, used as an opioid analgesic agent.

Alfentanil (mol wt 417) A narcotic and analgesic agent.

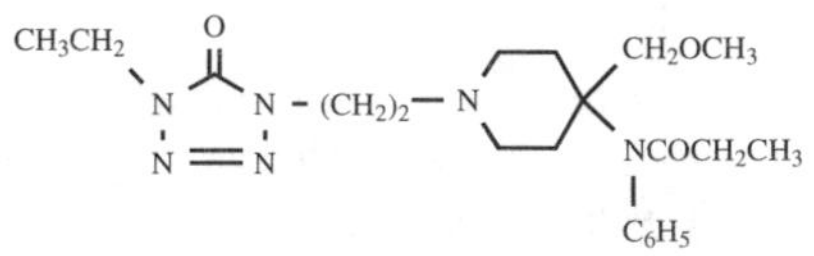

Alferon N A trade name for interferon alfa-n3.

Alfuzosin (mol wt 389) An antihypertensive agent.

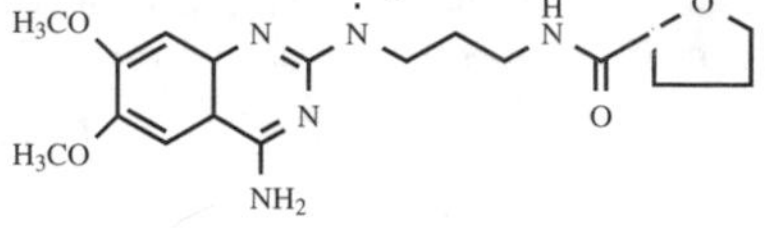

ALG Abbreviation for antilymphocyte globulin.

Alga A heterogeneous group of unicellular and multicellular eukaryotic photosynthetic organisms capable of photosynthesis.

Algae Plural of alga.

Algal Pertaining to algae.

Algal Bloom A sudden development of a heavy growth of algae.

Algesic Increased sensitivity to pain.

-algesic A combining form meaning pertaining to pain.

Algestone (mol wt 346) A topical anti-inflammatory agent.

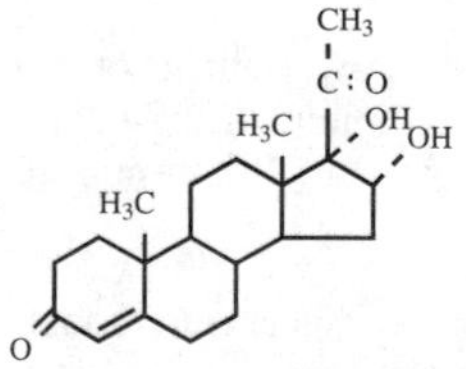

Algestone Acetophenide (mol wt 449) An agent used for treatment of acne.

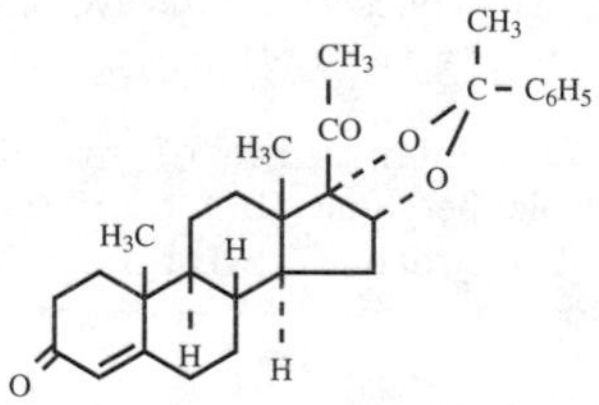

Algicides Chemical agents that kill algae, e.g., copper sulfate.

Algid Malaria A form of malaria caused by the protozoan *Plasmodium falciparum*; characterized by coldness of the skin, profound weakness, and severe diarrhea.

Algin Synonym of alginate, a gelling polysaccharide extracted from giant brown seaweeds (*Marcrocystis pyrifera*).

Alginate Salt form of alginic acid.

Alginate Lyase The enzyme that catalyzes the eliminative cleavage of polysaccharide containing β-D-mannuronic acid residues.

Alginate Synthetase The enzyme that catalyzes the following reaction:

$$\text{GDP-D-mannuronate} + (\text{alginate})_n \rightleftharpoons \text{GDP} + (\text{alginate})_{n+1}$$

Alginic Acid An algal polysaccharide consisting of β-1,4-linked D-mannuronic acid residues and α-1,4-linked L-guloronic acid residues.

Algivorous Feeding on algae.

Algology The science that deals with algae.

Ali122571 (BamHI) A restriction endonuclease from *Acetobacter liquefaciens* having the same specificity as BamHI.

Ali122581 (BamHI) A restriction endonuclease from *Acetobacter liquefaciens* having the same specificity as BamHI.

Ali28821 (PstI) A restriction endonuclease from *Acetobacter liquefaciens* having the same specificity as PstI.

AliI (BamHI) A restriction endonuclease from *Acetobacter liquefaciens* having the same specificity as BamHI.

AliAJI (PstI) A restriction endonuclease from *Acetobacter liquefaciens* having the same specificity as PstI.

Alibendol (mol wt 251) A choleretic and antispamotic agent.

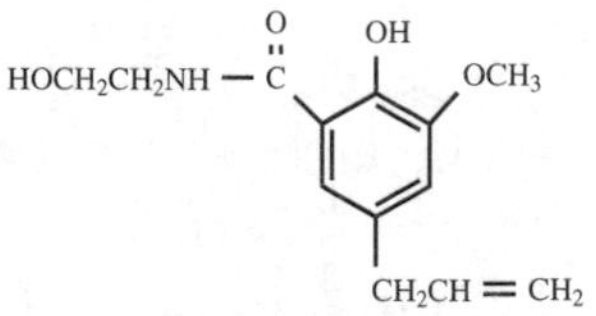

Alicyclic Denoting a compound derived from a saturated cyclic hydrocarbon.

Alien Substitution Replacement of one or more chromosomes of a species by those from another species.

Ali-Esterase See carboxylesterase.

Alimentary Pertaining to food.

Alimentary Glycosuria The increase of the level of glucose in the urine following a carbohydrate rich meal.

Alimentary Toxic Aleukia A severe mycotoxicosis caused by the ingestion of moldy grain contaminated with certain fungi, e.g., *Fusarium tricinctum*.

Alimentary Tract The food tract starting at the esophagus and ending at the rectum and anus.

Alimentation Nourishment.

Alimentology The science that deals with nutrition.

Aliphatic Pertaining to an organic compound that has an acyclic chain structure.

Aliphatic Acid An acyclic organic acid characterized by an open carbon chain.

Aliquot 1. A part of a whole that divides the whole evenly without a remainder. 2. Fraction of a whole.

Alitame (mol wt 331) A sugar substitute in NutraSweet.

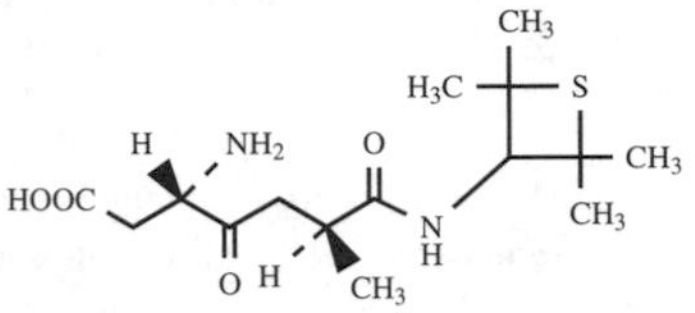

Alizapride (mol wt 315) An antiemetic agent.

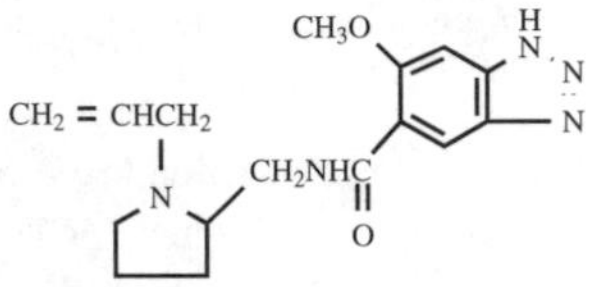

Alizarin (mol wt 240) A biological dye used in tests for aluminum, indium, mercury, and zirconium.

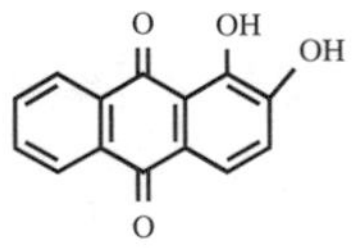

Alizarin Cyanine Green F (mol wt 623) A biological dye.

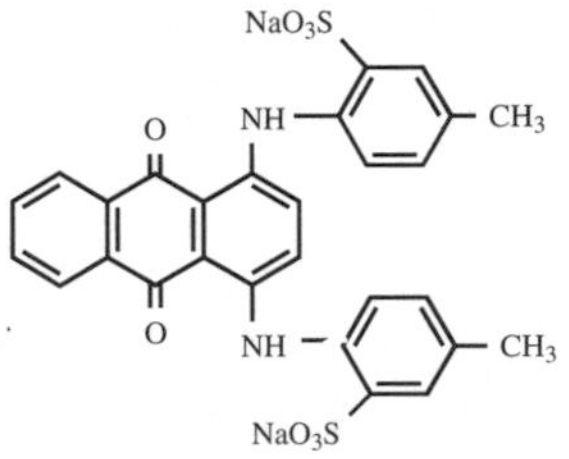

Alizarine Blue (mol wt 291) A pH indicator dye.

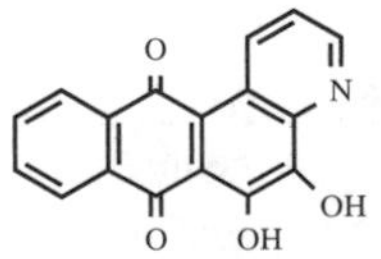

Alizarine Orange (mol wt 285) A pH indicator dye.

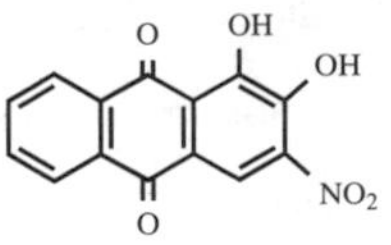

Alizarine Yellow R (mol wt 287) A pH indicator dye.

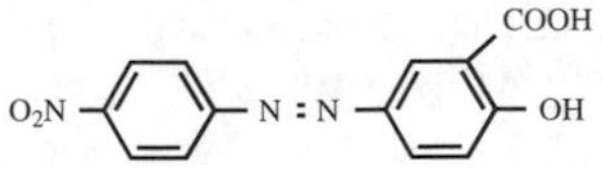

Alka Seltzer A trade name for a combination drug containing sodium bicarbonate, aspirin, and citric acid, used as an antacid.

Alkaban-AQ A trade name for vinblastine sulfate, used as an antineoplastic agent.

Alkalemia A disorder characterized by the high alkalinity in the blood.

Alkali A hydroxide of one of the alkali metals that can combine with an acid to form a salt that turns red litmus blue.

Alkali Metal Any element of group IA in the periodic table: lithium (Li), sodium (Na), potassium (K), rubidium (Rb), cesium (Cs), and francium (Fr).

Alkalimetry 1. The chemical analysis of a solution by means of titrations. 2. A determination of the amount of a base by titration against a standard acid solution or vice versa.

Alkaline The concentration of hydroxyl (OH^-) ions in a solution with a pH greater than 7.0.

Alkaline Earth An element of group IIA in the periodic table: beryllium (Be), magnesium (Mg), calcium (Ca), strontium (Sr), barium (Ba), and radium (Ra).

Alkaline Hematin A hematin formed from hemoglobin by treatment with alkali above pH 11.

Alkaline Lipase Synonym of triacylglycerol lipase.

Alkaline pH A pH value above 7.0.

Alkaline Phosphatase A nonspecific phosphomonoesterase with activity optima at alkaline pH.

$$\text{Orthophosphoric monoester} + H_2O \rightleftharpoons \text{Alcohol} + H_3PO_4$$

Alkaline Phosphomonoesterase See alkaline phosphatase.

Alkaline Protease Synonym of subtilisin.

Alkaloids Basic nitrogen-containing organic compounds produced by plants that are physiologically active in vertebrates. Many have a bitter taste and some are poisonous, e.g., nicotine, morphine, quinine, caffeine, and cocaine.

Alkalophile An organism that grows optimally under alkaline conditions.

Alkalosis A condition in which there is an excess amount of bicarbonate in the blood.

Alkanal-Reduced-FMN Oxygen Oxido-Reductase The systematic name for alkanal monooxygenase (FMN-linked).

Alkane A saturated aliphatic hydrocarbon of the general formula C_nH_{2n+2}.

Alkannin (mol wt 288) An astringent.

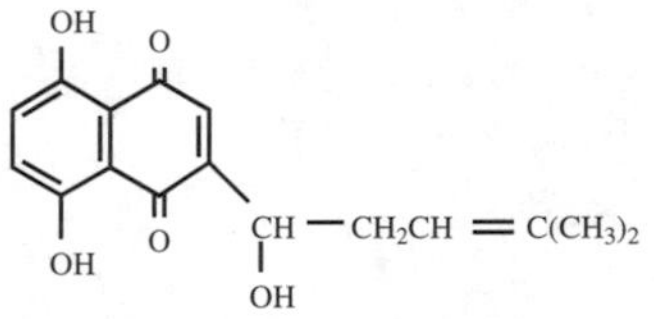

Alkaptonuria A hereditary disease characterized by the excretion of dark urine due to the lack of enzyme to degrade homogentisic acid.

Alkene An unsaturated aliphatic hydrocarbon containing one or more double bonds.

Alkenyl Group The radical derived from an alkene, or from a derivative of an alkene, by removal of a hydrogen atom.

2-Alkenyl-2-Acylglycerol Choline-Phospho-Transferase The enzyme that catalyzes the following reaction:

CDP-choline + 1-alkenyl-2-acylglycerol

⇅

CMP + plasmenylcholine

1-Alkenyl-Glycero-Phosphocholine O-Acyltransferase The enzyme that catalyzes the following reaction:

Acyl-CoA + 1-alkenylglycero-phosphocholine

⇅

CoA + 1-alkenyl-2-glycero-phosphocholine

Alkenyl-Glycerophosphocholine Hydrolase The enzyme that catalyzes the following reaction:

1-(1-Alkenyl)-*sn*-glycero-3-phosphocholine + H_2O

⇅

An aldehyde + *sn*-glycero-3-phosphocholine

1-Alkenyl-Glycerophosphoethanolamine O-Acyltransferase The enzyme that catalyzes the following reaction:

Acyl-CoA + 1-alkenylglycerophosphoethanolamine

⇅

CoA + 1-alkenyl-2-glycerophosphoethanolamine

Alkenyl-Glycerophosphoethanolamine Hydrolase The enzyme that catalyzes the following reaction:

1-(1-Alkenyl)-glycero-3-phosphoethanolamine + H_2O

⇅

An aldehyde + glycero-3-phosphoethanolamine

Alkeran A trade name for melphalan, an antineoplastic agent.

Alkofanone (mol wt 365) An antidiarrheal agent.

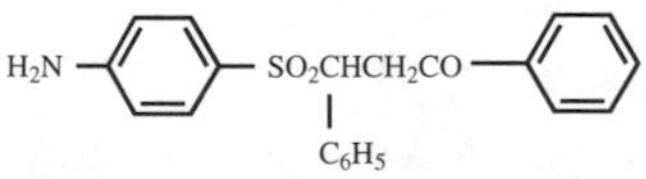

Alk-P Abbreviation for alkaline phosphatase.

Alk-Pase Abbreviation for alkaline phosphatase.

Alkyl Any group derived from alkane by removal of one hydrogen atom.

Alkylamidase The enzyme that catalyzes the following reaction:

N-Methylhexanamide + H_2O

⇅

Hexanoate + methylamine

Alkylated DNA Glycohydrolase The systematic name for DNA-3-methyladenine glycosidase (releasing methyladenine and methylguanine).

Alkylating Agents Agents capable of substitution of an alkyl group for an active hydrogen atom in an organic compound.

Alkylation The introduction of an alkyl group into an organic compound.

S-Alkylcysteine Lyase The enzyme that catalyzes the following reaction:

S-Alkyl-L-cysteine + H_2O

⇅

Alkyl thiol + NH_3 + pyruvate

Alkylglycerol Kinase The enzyme that catalyzes the following reaction:

ATP + alkylglycerol

⇅

ADP + alkylglycerol 3-phosphate

1-Alkylglycerophosphocholine O-Acetyl-Transferase The enzyme that catalyzes the following reaction:

Acetyl-CoA + 1-alkyl-*sn*-glycero-3-phosphocholine ⇅ CoA + 2-acetyl-1-alkyl-*sn*-glycero-3-phosphocholine

1-Alkylglycerophosphocholine O-Acyl-transferase The enzyme that catalyzes the following reaction:

Acyl-CoA + 1-alkyl-*sn*-glycero-3-phosphocholine ⇅ CoA + Phosphatidylcholine

Alkyl-Group The radical derived from an alkane, or from a derivative of an alkane, by the removal of a hydrogen atom.

Alkyl-Halide A derivative of a hydrocarbon in which one or more hydrogen atoms have been replaced by a halogen (F, Cl, Br, I).

Alkylhalidase The enzyme that catalyzes the following reaction:

Bromochloromethane + H_2O ⇅ Formaldehyde + bromide + chloride

Alkylmercury Lyase The enzyme that catalyzes the following reaction:

$RHg^+ + H^+ \rightleftharpoons RH + Hg^{2+}$

Alkyne An unsaturated aliphatic hydrocarbon that contains one or more triple bonds.

Alkynyl Group The radical derived from an alkyne, or from a derivative of an alkyne, by the removal of a hydrogen atom.

ALL Abbreviation for acute lymphocytic leukemia.

Allantoate Amidohydrolase See allantoinase.

Allantoate Deiminase The enzyme that catalyzes the following reaction:

Allantoate + H_2O ⇅ Ureidoglycine + NH_3 + CO_2

Allantoic Acid The carboxylic acid that is the end product of purine catabolism in some teleost fishes.

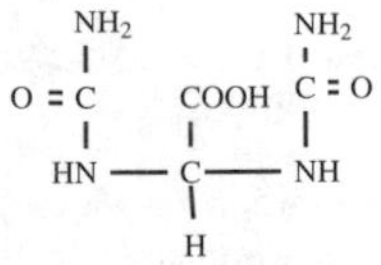

Allantoicase The enzyme that catalyzes the following reaction:

Allantoate + H_2O ⇌ Ureidoglycolate + Urea

Allantoin (mol wt 158) A heterocyclic product of purine catabolism, more prevalent in mammals than and in some reptiles. It has been used topically to stimulate healing of suppurating wounds.

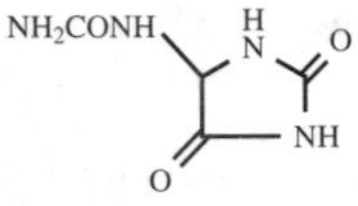

Allantoin Amidohydrolase The systematic name for allantoinase.

Allantoin Racemase The enzyme that catalyzes the following reaction:

(S)(+)-Allantoin ⇌ (R)(-)-Allantoin

Allantoinase The enzyme that catalyzes the following reaction:

Allantoin + H_2O ⇌ Allantoate

Allatum Hormones Hormones synthesized by the insect corpus allatum that influence the qualitative properties of each molt in the holometabolous insects.

Allegra A trade name for fexofenadine, an antihistaminic agent.

Allele One of two or more alternative forms of a given gene that control a particular characteristic occupying corresponding loci on the homologous chromosomes (e.g., brown-eye genes and blue-eye genes are alternative alleles for eye color).

Allele Specific Oligonucleotide An oligonucleotide that is constructed with a DNA sequence homologous to a specific allele.

Allelic Complementation The production of nearly normal phenotype in an organism carrying two different mutant alleles in the *trans* configuration.

Allelic Exclusion The phenotypic expression of a single allele in cells containing two different alleles for that genetic locus.

Allelo- A prefix meaning pertaining to another.

Allelopathic A substance produced by one organism that adversely affects another organism.

Allen Correction A method of correcting absorbance measurements due to interferring substances.

Allenolic Acid (mol wt 216) A compound used for the preparation of estrogenic compounds.

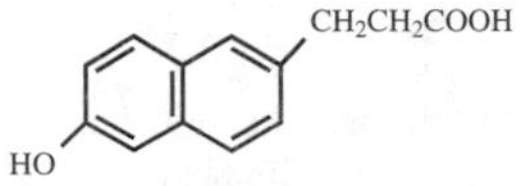

Aller-Chlor A trade name for chlorpheniramine maleate, used as an antihistaminic agent.

Allerdryl A trade name for diphenhydramine hydrochloride, used as antihistaminic agent.

Allergen An antigen capable of provoking production of IgE antibodies that bind to the mast cells. Subsequent exposure to allergen causes mast cell degranulation and release of histamine leading to allergic reactions.

Allergenic Pertaining to a substance capable of causing allergic reactions.

Allergesic A trade name for a combination drug containing phenylpropanolamine hydrochloride and chlorpheniramine maleate.

Allergex A trade name for chlopheniramine maleate, used an antihistaminic agent.

Allergic Asthma A form of asthma caused by the exposure of the bronchial mucosa to an inhaled airborne allergen. It is an IgE-mediated allergic reaction.

Allergic Contact Dermatitis An inflammation of the skin due to exposure to a chemical sensitizer.

Allergic Reaction The antigen-antibody mediated allergic reactions. There are four main types of allergic reactions, e.g., type I hypersensitivity (IgE mediated); type II (antibody-dependent cytotoxic reaction); type III (soluble antigen-antibody-complex mediated hypersensitivity); type IV (delayed type hypersensitivity).

Allergic Rhinitis An allergic disorder of humans caused by pollen, house dust, animal dander, or spores of fungi; characterized by wheezing, sneezing, coughing, copious flow of watery discharges, itching nose, mouth, excessive flow of tears, headache, and insomnia (also called hay fever and pollenosis).

Allergoids Chemically modified allergens that provoke the production of IgG antibody (not IgE), thereby reducing allergic symptoms.

Allergy An immunological hypersensitivity reaction resulting in tissue inflammation and organ dysfunction. Allergies are classified into four main types: Types I, II, III, and IV hypersensitivity. Types I, II, and III involve immunoglobulin antibodies and their interaction with different antigens. Type IV allergy is associated with T cells that react directly with antigen and cause local inflammation.

AllerMax A trade name for diphenhydramine hydrochloride, used as an antihistaminic agent.

Allicin (mol wt 162) An antibacterial agent from garlic (*Allium sativum*)

$$CH_2{=}CHCH_2\overset{\displaystyle O}{\underset{}{S}}{-}SCH_2CH{=}CH_2$$

Allidochlor (mol wt 174) An herbicide.

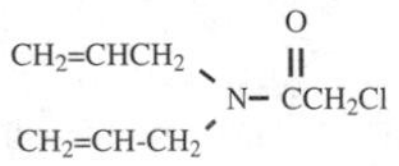

Alliin (mol wt 177) A compound isolated from garlic (*Allium sativum*); it possesses antibacterial activity similar to allicin.

$$CH_2{=}CHCH_2S({=}O){-}CH_2CH(NH_2){-}COOH$$

ALLN Abbreviation for N-acetyl-leucyl-leucyl-norleucinal.

Allo- 1. A combining form meaning dissimilar. 2. An isomeric form. 3. A dissimilar genome.

AlloAb Abbreviation for alloatibody.

AlloAg Abbreviation for allogeneic antigen.

Alloantibody 1. Antibody directed against alloantigen. 2. An antibody from one individual that reacts with antigen present in another individual of the same species.

Alloantigen Antigen from an individual that elicits the formation of antibody in another individual of the same species.

Allobarbital (mol wt 208) A hypnotic and sedative agent.

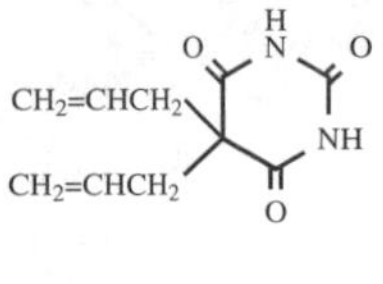

AlloBMT Abbreviation for allogeneic bone marrow transplantation.

Allochthonous An organism or substance foreign to a given environment.

Alloclamide (mol wt 311) An antitussive and antihistaminic agent.

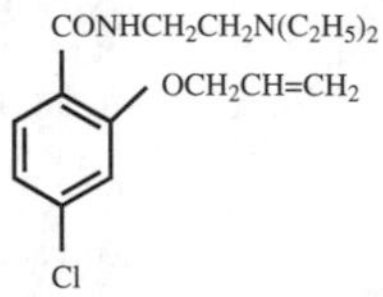

Allocupreide Sodium (mol wt 321) An antiarthritic agent.

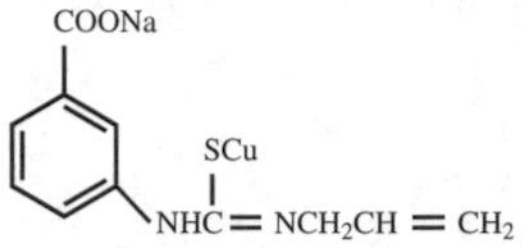

Allodiploid An individual, organism, or cell that has two genetically distinct sets of chromosomes derived from different ancestral species.

Allogamy See cross fertilization.

Allogeneic The relationship that exists between genetically dissimilar members of the same species.

Allogeneic Graft See allograft.

Allogeneic Inhibition *In vitro* damage to cells caused by contact with genetically dissimilar cells.

Allogenic Variant spelling of allogeneic.

Allograft A tissue or organ graft between two genetically nonidentical members of the same species (also known as homograft).

Allograft Rejection Rejection reaction against a transplant between genetically nonidentical individuals of the same species. The rejection is due to the cell-mediated immune response of the recipient against non-histocompatibility antigens of the donor.

Allolactose (mol wt 342) A disaccharde consisting of β-D-galactosyl-(1,6)-D-glucose. It is a minor product by the action of β-galactosidase on lactose and is a natural inducer of the *lac* operon in *E. coli.*

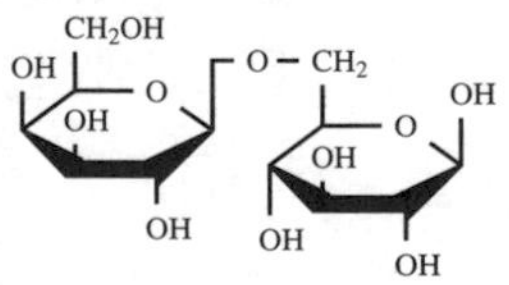

Allomerism The variation in the chemical composition of substances that have the same crystalline form.

Allometric Growth The variation in the relative rates of growth of the various parts of the body.

Allomonas A genus of bacteria (family Vibrionaceae).

Allomone Substance produced by one organism that influences the behavior of another organism of a different species.

Allomorphism The variation in the crystalline form of substances that have the same chemical composition.

Alloparasite A parasite that is not closely related taxonomically to its host.

Allophenate Hydrolase The enzyme that catalyzes the following reaction:

$$\text{Urea 1-carboxylate} + H_2O \rightleftharpoons 2CO_2 + 2NH_3$$

Allophycocyanin A red accessory pigment of algae that consists of a protein conjugated to a phycobilin.

Alloplex Interaction The reaction of a disordered protein molecule that undergoes refolding upon contact with another protein molecule.

Alloploid An organism that arises from the combination of two or more sets of chromosomes from different ancestral species.

Allopolyploid See alloploid.

Allopurinol (mol wt 136) A xanthine oxidase inhibitor and an antigout agent.

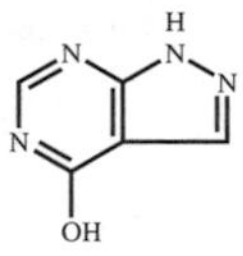

Alloremed A trade name for allopurinol, used as an antigout agent.

All-or-None Reaction A reaction that occurs either completely or not at all.

D-Allose (mol wt 180) A 6-carbon aldohexose.

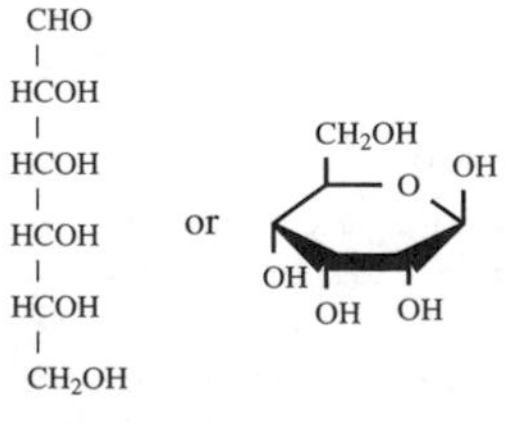

D-Allose

Allose Kinase The enzyme that catalyzes the following reaction:

$$\text{ATP} + \text{D-allose} \rightleftharpoons \text{ADP} + \text{D-allose 6-phosphate}$$

Allosteric Pertaining to the topologically distinct sites on a protein or an enzyme molecule.

Allosteric Activation The activation of an allosteric enzyme by binding of a positive effector onto the allosteric site of the enzyme.

Allosteric Activator Any substance or factor that activates or enhances the activity of an allosteric enzyme (also known as a positive effector).

Allosteric Effectors Substances (usually metabolites) that bind to the allosteric site of an allosteric enzyme leading to changes in shape and activity of the enzyme.

Allosteric Enzyme See allosteric protein.

Allosteric Inhibition The inhibition of an allosteric enzyme by binding of a negative effector onto the allosteric site of the enzyme.

Allosteric Inhibitor Substance that is capable of binding onto the allosteric site of an enzyme and acts as a negative effector leading to inhibition of the enzyme activity.

Allosteric Interaction The interaction of an allosteric protein or enzyme with an effector molecule leading to the changes in shape and activity of the protein or enzyme.

Allosteric Protein Proteins that possess topologically and functionally distinct sites. The three-dimensional structure and the biological properties can be altered by the binding of a specific effector molecule at sites other than the active site.

Allosteric Regulation The control of an enzymatic reaction by an effector-mediated change on an allosteric enzyme.

Allosteric Site A specific receptor site on an enzyme molecule remote from the active site. The binding of an effector molecule onto the allosteric site leading to changes in shape and activity of the allosteric protein or enzyme.

L-Allothreonine An isomer of the amino acid L-threonine.

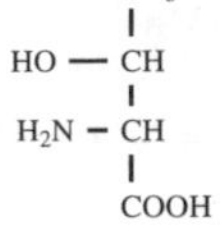

L-Allothreonine Acetaldehyde Lyase See L-allothreonine aldolase.

L-Allothreonine Aldolase The enzyme that catalyzes the conversion of L-allothreonine to glycine and acetaldehyde.

Allotopy The phenomenon of the formation of a membrane-enzyme complex resulting in alteration of the properties of both enzyme and membrane.

Allotrope Different forms of the same element that differ in physical and chemical properties.

Allotype One of a group of structurally and functionally similar proteins from the same species.

Alloxan (mol wt 142) An antimicrobial agent with antineoplastic activity.

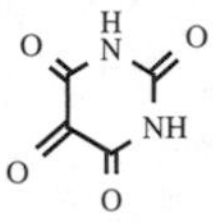

Alloy A mixture of two or more metals that are formed by mixing molten metals.

Allozymes A group of enzymes that are produced by different alleles of the same gene.

Allura Red AC (mol wt 496) A color additive for foods, drugs, and cosmetics.

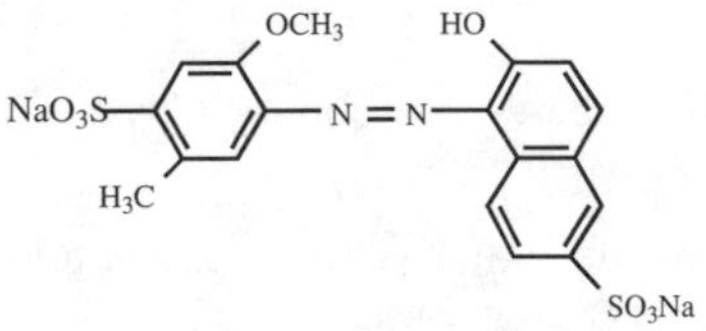

Allyl-Alcohol (mol wt 58) A colorless liquid with a pungent property.

$$CH_2 = CH - CH_2OH$$

Allylamine (mol wt 57) A highly active synthetic antifungal agent against dermatophytes.

$$CH_2 = CH - CH_2 - NH_2$$

Allylestrenol (mol wt 300) A progestogen.

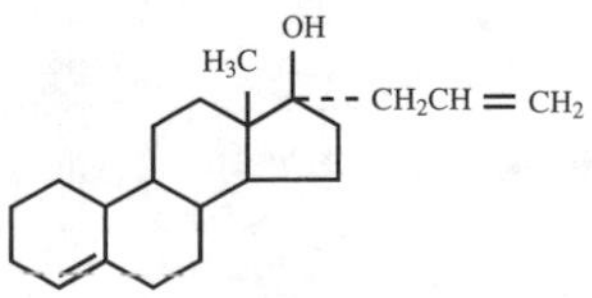

Allylisothiocyanate (mol wt 99) A counterirritant.

$$CH_2 = CH - CH_2 - N = C = S$$

Allylprodine (mol wt 287) A narcotic and analgesic agent.

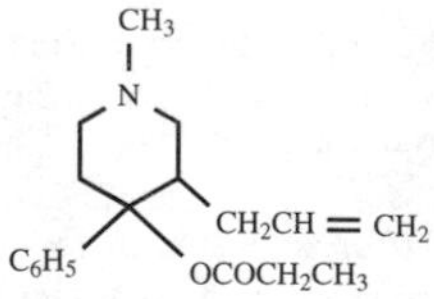

Allysine (mol wt 145) A non-protein amino acid that is formed from lysine.

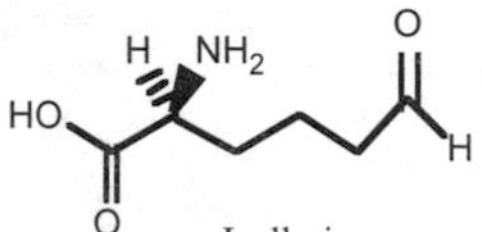

L-allysine

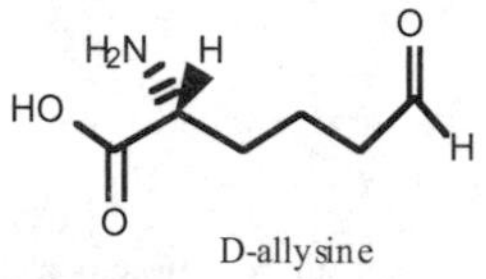

D-allysine

Almagate (mol wt 630) A compound used as an antacid.

$$Al_2Mg_6(OH)_{14}(CO_3)_2 \cdot 4\,H_2O$$

ALME Abbreviation for acetyl-lysine methyl ester.

Alminoprofen (mol wt 219) An anti-inflammatory and analgesic agent.

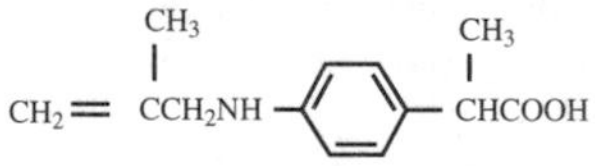

Almitrine (mol wt 478) A respiratory stimulant.

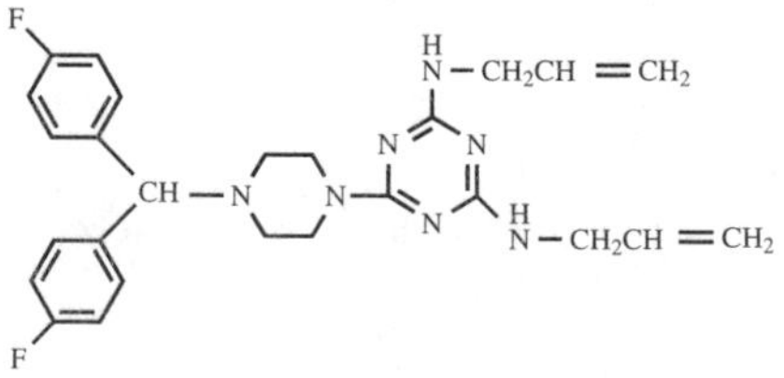

Aloe-Emodin (mol wt 270) A cathartic agent occurring in various species of *Aloe*.

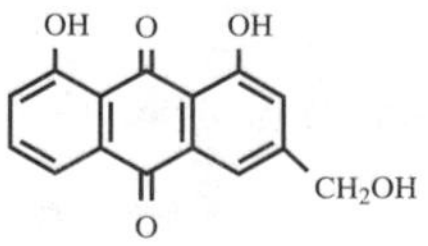

Aloin (mol wt 418) A purgative agent from various species of *Aloe*.

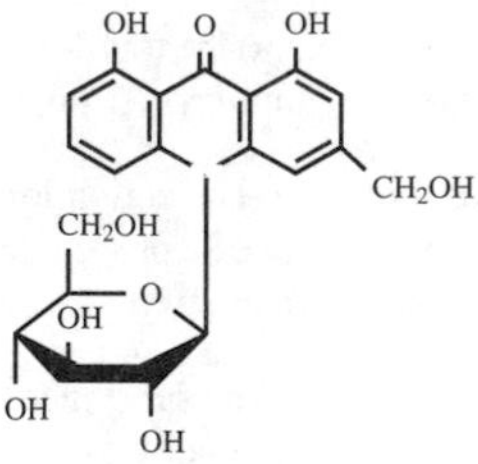

Alopecia Baldness.

Alophen Pill A trade name for phenophthalein, a laxative.

Alora A trade name for estradiol, a hormone.

Aloxidone (mol wt 155) An anticonvulsant.

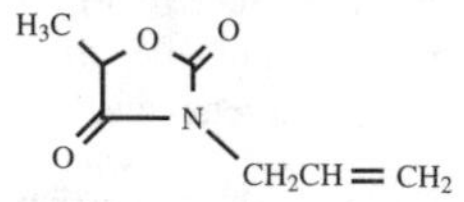

ALP Abbreviation for 1. Alkaline phosphatase. 2. Alpha lipoprotein.

αLP Abbreviation for alpha lipoprotein.

Alpha (α) Denotes 1. The first carbon atom next to the carbon atom that carries the principal functional group of the molecule. 2. A specific configuration of the substituents at the anomeric carbon in the ring structure of carbohydrates (e.g., α-D-glucose). 3. A type of protein structure (e.g., alpha helix). 4. A type of optical rotation.

Alpha-Amanitin (α-Amanitin) A toxic cyclic octapeptide produced by the poisonous mushroom (*Amanita phalloides*) that binds to the eukaryotic RNA-polymerase II and inhibits the synthesis of mRNA.

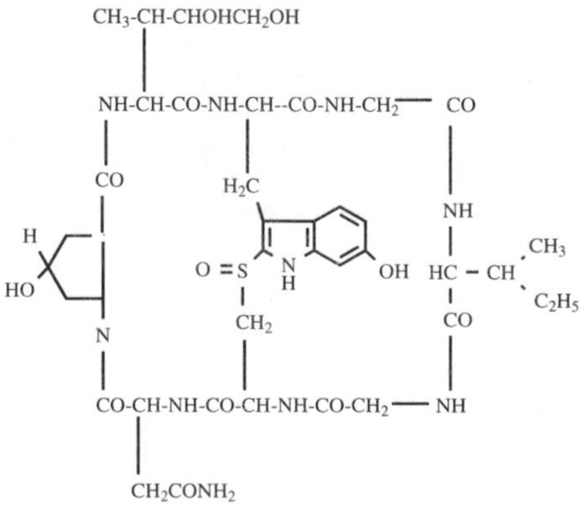

Alpha-Amino Acid (α-Amino Acid) An amino acid that has an amino group attached to the alpha (α) carbon.

Alpha-Amylase (α-Amylase) The enzyme that catalyzes the endohydrolysis of 1,4-α-D-glucosidic linkage in a polysaccharide containing three or more 1,4-α-linked D-glucose units.

Alpha-1-Antitrypsin (α_1-Antitrypsin) See alpha-1 protease inhibitor (α-$_1$ protease inhibitor).

Alpha-Blockers Agents or drugs that block the alpha adrenergic receptors at the nerve endings of the sympathetic nerve system by ephinephrine-like hormone.

Alpha-Bungarotoxin (α-Bungarotoxin) A protein neurotoxin produced by snakes of the genus *Bungarus* that blocks the binding of acetylcholine at the postsynaptic membrane.

Alpha-Chain (α-Chain) 1. One of the two types of polypeptide chains in hemoglobin. 2. One of the protein subunits in RNA-polymerase.

Alpha-Fetoprotein (α-Fetoprotein) A plasma protein found in the human fetus; its concentration decreases from 400 mg/100 ml to about 3 μg/100 ml in adult. Elevation of α-fetoprotein occurs in patients with certain malignancies.

Alpha-Globulin (α-Globulin) One of a group of proteins in the blood plasma with the electrophoretic mobility between α- and γ- globulins.

Alphagon A trade name for brimonidine tartrate, an anti-glaucoma agent.

Alpha-Helix (α-Helix) 1. Spiral-shaped secondary structure of a protein molecule. 2. Spiral arrangements of DNA or RNA.

Alpha-Hemolysis (α-Hemolysis) The development of a greenish zone around a bacterial colony (e.g., pneumococci) growing on the blood-agar medium due to the partial hemolysis of red blood cells.

Alpha-Interferon (α-Interferon) Interferon produced by leukocytes.

Alpha-Keratin (α-Keratin) The helical form of keratin in which the polypeptide chains are in the alpha helical configuration.

Alpha-Lactalbumin (α-Lactalbumin) 1. A component of lactose synthetase. 2. A heat-stable protein in the milk of mammals.

Alphalase A trade name for α-amylase.

Alpha-Lipoprotein (α-Lipoprotein) See high-density lipoprotein.

Alphanate A trade name for human antihemophilic factor produced by DNA recombinant technology and used for treatment of hemophilia A or factor VII deficiency individuals.

AlphaNine SD A trade name for a virus-filtered human coagulation factor IX, used for preventing and controlling bleeding with factor IX deficiency individuals,

Alpha-Oxidation (α-Oxidation) A minor pathway for the oxidation of fatty acids in germinating plant seeds.

Alpha-Particle (α-Particle) A subatomic particle consisting of two protons and two neutrons; frequently emitted by radioactive isotopes.

Alpha-Peptide (α-Peptide) A peptide that is cleaved from the N-terminus of the lacZ-encoded β-galactosidase of *E. coli* during autoclaving.

Alphaprodine (mol wt 261) A narcotic, analgesic agent.

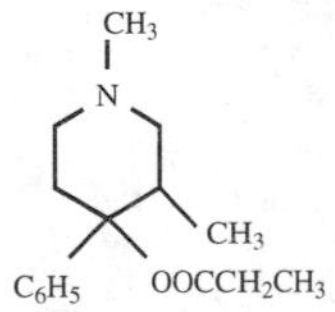

Alpha-1-Protease Inhibitor (α_1-Protease Inhibitor) A protease inhibitor that prevents the action of elastase on alveolar tissue in patients who have deficiency in α1-antitrypsin.

Alpha-Radiation (α-Radiation) Radiation consisting of alpha particles.

Alpha-Receptor (α-Receptor) One type of adrenergic receptor that is more sensitive to adrenaline than to isoproterenol.

Alpha-Tocopherol See vitamin E.

Alphatrex A trade name for betamethasone benzoate, a hormone used as an anti-inflammatory agent.

Alphavirus A genus of insect-transmitted (mosquito) viruses of the family Togaviridae capable of causing diseases in humans and animals (e.g., encephalitis).

Alpidem (mol wt 404) An anxiolytic agent.

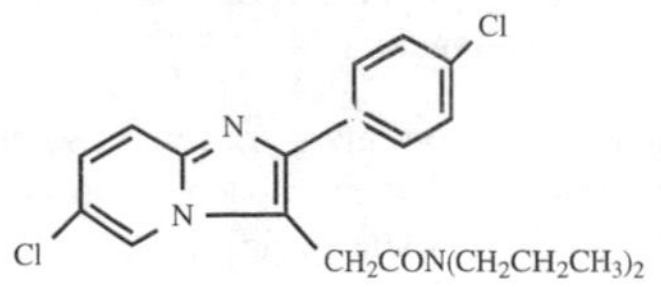

Alpiropride (mol wt 382) An antimigraine agent.

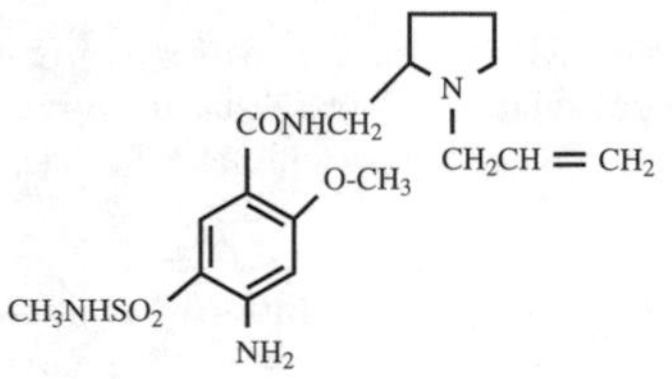

Alprazolam (mol wt 309) A sedative agent and tranquilizer. It is an inhibitory neurotransmitter.

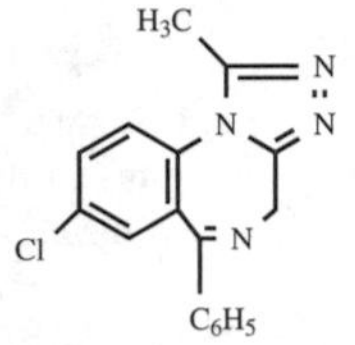

Alprenolol (mol wt 249) A β-adrenergic blocker used as an antihypertensive and antiarrhythmic agent.

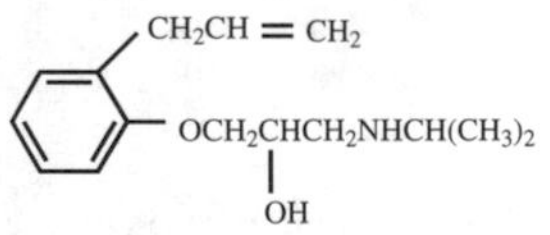

Alprin A trade name for trimethoprim, used as an anti-infective agent and inhibits the synthesis of folic acid in bacteria.

Alprostadil Referring to prostaglandin E_1, an agent capable of relaxing smooth muscle of the ductus arteriosus.

ALR Abbreviation for aldose reductase.

ALS Abbreviation for antilymphocyte serum.

Alsactide (mol wt 2120) A synthetic peptide analog of short-chain ACTH.

Alsever's Solution An anticoagulating solution containing D-glucose (20.5 g), sodium citrate dihydrate (8.0 g), and NaCl (4.2 g) per liter of water (pH 6).

ALT Abbreviation for 1. Alanine amino-transferase. 2. Autolymphocyte therapy.

Altace A trade name for ramipril, used as an antihypertensive agent.

Alteplase A tissue plasminogen activator that binds to fibrin and converts the plasminogen to plasmin to initiate local fibrinolysis.

Alterna-Gel A trade name for aluminum hydroxide, used as an antacid.

Alternaric Acid A complex compound, produced by *Alternaria solani*, which contains a diketotetrahydropyran group linked to a long-chain fatty acid; it inhibits germination of the spores of certain fungi and causes wilting and necrosis in the tissues of higher plants.

Alternate Generation A type of reproduction in which a sexual generation alternates with one or more asexual generations.

Alternate Host The secondary host for organisms that require two hosts to complete their life cycles.

Alternation Enzyme A T_4-encoded enzyme that is injected into the bacterial cell during phage infection. It modifies host RNA polymerase and renders the enzyme incapable of initiating transcription at the promoter site of the host DNA.

Alternative Complement Pathway A pathway for activation of complement. It involves properdin factor D, properdin factor B, and part of the classic pathway (also known as the properdin pathway).

Alternative Splicing A mechanism for generating multiple protein isoforms from a single gene consisting of nonconsecutive exons. The transcript resulting from the alternative splicing does not contain all the exons compared to that of constitutive splicing, which generates a transcript containing all exons.

Alteromonas A genus of aerobic, chemoorganotrophic, Gram-negative bacteria.

Altexide A trade name for a combination drug containing spironolactone and hydrochlorothiazide, used as a diuretic agent.

Althiazide (mol wt 384) A diuretic and antihypertensive agent.

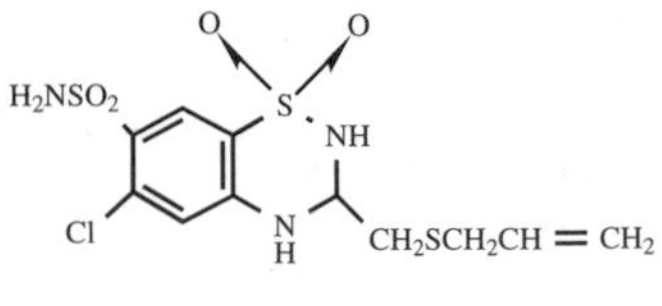

Altrenogest (mol wt 310) A synthetic oral progestogen.

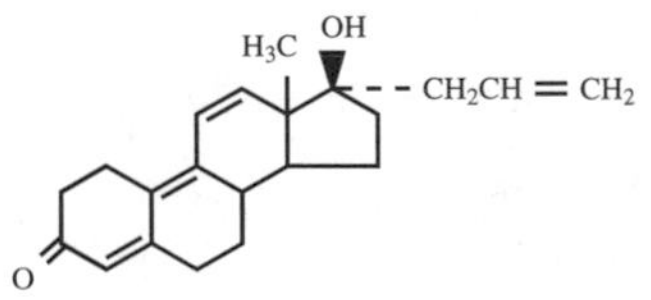

Altretamine (mol wt 210) An antitumor agent and a chemosterilant for house flies and other insects.

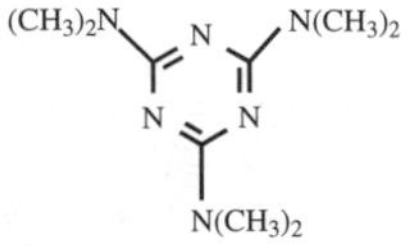

Altro-Heptulose Synonym of sedoheptulose.

D-Altrose (mol wt 180) A 6-carbon aldosugar.

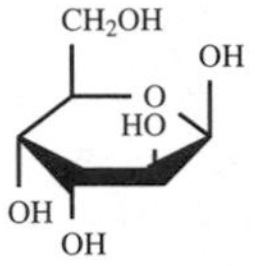

AluI A restriction endonuclease from *Arthobacter luteus* with the following specificity:

5′..........AGCT..........3′
3′..........TCGA..........5′

Alu-Cap A trade name for aluminum hydroxide, used as an antacid.

Aludrox Suspension A trade name for a combination drug containing aluminum hydroxide and magnesium hydroxide, used as an antacid.

Alum A double sulfate salt of aluminum and a monovalent metal used as an astringent and a styptic agent.

Alum Precipitation A technique in which a soluble immunogen is converted into a particulate form by mixing with a solution of alum, e.g., aluminum potassium sulfate.

Alumina Aluminum oxide used as an adsorbent in adsorption column chromatography.

Alumina Gel A gel prepared from ammonium sulfate and aluminum sulfate and used in the purification of proteins.

Aluminum (Al) A metallic element with atomic number 13, atomic weight 27, valence 3.

Aluminum Adjuvant An aluminum compound, such as aluminum hydroxide, aluminum phosphate, or alum, that is used as an adjuvant in immunization.

Aluminum Ammonium Sulfate (mol wt 237) An astringent and stypic agent.

$AlNH_4(SO_4)_2$

Aluminum Bis(acetylsalicylate) (mol wt 402) An analgesic and antipyretic agent.

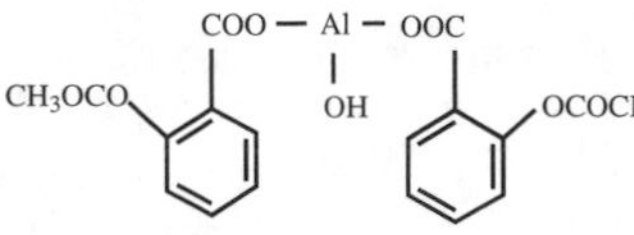

Aluminum Borohydride (mol wt 72) A reducing agent.

$Al(BH_4)_3$

Aluminum Hydroxide (mol wt 78) An antacid and vaccine adjuvant. It can be used to reduce acid load in the GI tract and to elevate gastric pH.

$Al(OH)_3$

Aluminum Magnesium Silicate (mol wt 262) An antacid.

$Al_2Mg(SiO_4)_2$

Aluminum Nicotinate (mol wt 393) A peripheral vasodilator and antilipemic agent.

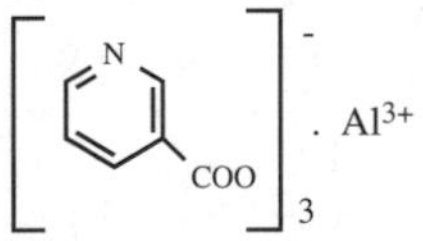

Aluminum Phosphate (mol wt 122) An antacid. It can be used to provide supplemental phosphate.

$AlPO_4$

Alupent A trade name for metaproterenol sulfate, a beta-adrenergic bronchodilator.

Alu-Sequences A family of closely related, dispersed sequences in human genomes. Each contains a common cleavage site for the restriction enzyme AluI.

Alurate A trade name for aprobarbital, a sedative and hypnotic agent.

Alu-Tab A trade name for aluminum hydroxide, used to reduce acid load in the GI tract.

ALV Abbreviation for avian leukosis virus.

Alverine (mol wt 281) An anticholinergic agent.

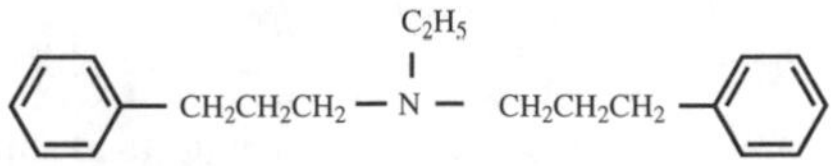

Alveolar Pertaining to alveolus.

Alveolar Cell Carcinoma A malignant pulmonary neoplasm that arises in a bronchiole and spreads along the alveolar surfaces.

Alveolar Macrophage Macrophages in the lung.

Alveoli Plural of alveolus.

Alveolitis Inflammation of the pulmonary alveoli, caused by inhalation of allergic substances.

Alveolus A small saclike structure, e.g., air sac of the lungs.

AlwI A restriction endonuclease from *Acinetobacter lwoffi* with the following specificity:

5′..........GGATC(N)$_4$..........3′
3′..........CCTAG(N)$_5$..........5′

(N)$_4$ = 4 identical nucleotides
(N)$_5$ = 5 identical nucleotides

Alw21I A restriction endonuclease from *Acinetobacter iwoffi* with following specificity:

5'........G(A/T)GC(A/T)C........3'
3'........C(T/A)CG(TA)G.........5'

AlwNI A restriction endonuclease from *Acinetobacter lwoffi* with the following specificity:

5′.........CAGNNNCTG.........3′
3′.........GTCNNNGAC.........5′

Alw26I A restriction endonuclease from *Acinetobacter iwoffi* RFL26 with following specificity:

5'........GTCTC(N)1........3'
3'........CAGAG(N)5........5'

Alw441 A restriction endonuclease from *Acinetobacter lwoffi* with the following specificity:

5′..........GTGCAC..........3′
3′..........CACGTG..........5′

Alymphocytosis The reduction of the total number of circulating lymphocytes in the blood.

Alysiella A genus of bacteria (*Cytophagales*), e.g., *A. filiformis* in the oral cavity of various vertebrates.

ALZ-50 Abbreviation for a monoclonal antibody to Alzheimer's brain tissue protein A-68.

Alzheimer's Disease A disease characterized by the formation of numerous plaques in the brain that consist of degenerated axons and neurites. The chief symptoms include confusion, disorientation, impaired judgement, and inability to carry out purposeful movement.

A_m A symbol for molar absorptivity.

AM Abbreviation for alveolar macrophage.

Am Symbol for 1. The chemical element americium. 2. An allotypic determinant or marker on the heavy chain of the human IgA molecule

A_2M Abbreviation for α_{-2}-macroglobulin.

Am Mutant A mutant that has an amber codon.

AMA Abbreviation for 1. Anti-mitochondrial antibody. 2. Anti-myosin antibody.

AmaI (NruI) A restriction endonuclease from *Actinmadura madurae* with the following specificity:

5′...........TCGCGA..........3′
3′...........AGCGCT..........5′

Amacodone A trade name for a combination drug containing acetaminophen and hydrocodone bitartrate, used as an opioid analgesic agent.

AMA-Fab Abbreviation for anti-myosin antibody with Fab fragment.

Amalgam An alloy of mercury and another metal or metals.

Amanita phalloides A genus of poisonous mushroom (*Agaricales*) that produces amantoxin (e.g., α-amanitin).

Amanitin Poisonous cyclic peptide derived from the mushroom (*Amanita phalloides*). It exists in α, β, and γ forms, the α amanitin inhibits eukaryotic RNA polymerase II.

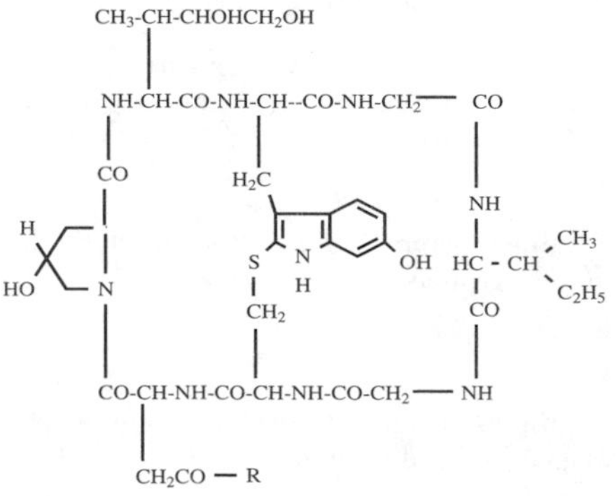

α-amanitin : R=NH_2

α-Amanitin See alpha-amanitin.

Amanozine (mol wt 187) A synthetic diuretic agent.

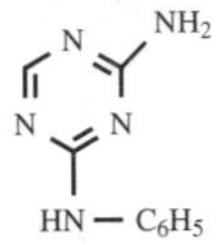

Amantadine (mol wt 151) A polycyclic, antiviral, and antiparkinsonian agent.

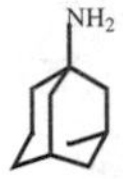

Amantoxin Toxic cyclic peptides produced by some species of *Amanitia,* e.g., *A. phalloides* and *A. verna.*

Amanullin A nontoxic substance with a similar chemical composition as amanitin, produced by *Amanita.*

Amaphen A trade name for a combination drug containing acetaminphen, caffeine, and butabital, used as an antipyretic and analgesic agent.

Amaryl A trade name for glimepiride, an antidiabetic agent.

Amastigotes Rounded protozoan cells lacking flagella.

Amatine A trade name for midodrine hydrochloride, an antihypotensive agent.

Amaurosis Blindness.

AMB Abbreviation for amphotericin B.

Ambazone (mol wt 237) An antibacterial agent.

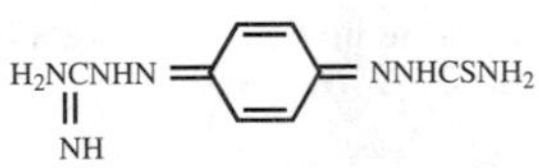

Ambenonium Chloride (mol wt 609) A cholinergic agent and cholinesterase inhibitor. It prevents the destruction of acetylcholine released from parasympathetic and somatic efferent nerves.

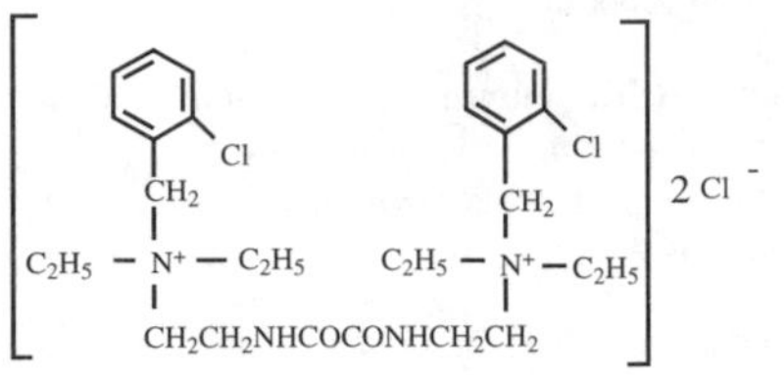

Ambenyl A trade name for a combination drug containing a narcotic analgesic-antitussive (codeine sulfate), two antihistamines (bromodiphenhydramine hydrochloride and dipenhydramine hydrochloride), expectorant (potassium guaiacolsulfonate), and ammonium chloride.

Amber Codon A termination codon (UAG) on the mRNA for termination of protein synthesis.

Amber Mutant A mutant that contains an amber codon in a vital gene.

Amber Mutation Mutation that produces an amber codon (UAG) and terminates the polypeptide chain prematurely.

Amber Suppression The suppression of an amber codon.

Amber Suppresser Any mutant encoding a tRNA whose anticodon is capable of inserting an amino acid at the amber codon termination site.

Ambien A trade name for zolpidem, a hypnotic agent that binds with gamma-aminobutyric receptor. Produces CNS depression.

Ambiguous Codon A codon that can lead to the incorporation of more than one amino acid into a polypeptide.

AmBisome A trade name for liposomal amphotericin-B complex produced by DNA recombinant technology and used for treatment of fungal infection.

Ambivalent Codon A codon that is expressed in some mutants but not expressed in other mutants.

Amblosis Abortion or miscarriage.

Amblyopia Reduced vision in the eye without any apparent structural abnormality.

Amboceptor Referring to 1. a bacteriolytic substance capable of interacting with complement or 2. hemolysin or antibody specific for surface antigens on erythrocytes.

Ambodryl A trade name for bromodiphenhydramine hydrochloride, an antihistaminic agent.

AMBP Abbreviation aminomethylene bisphosphate.

Ambroxol (mol wt 378) An expectorant.

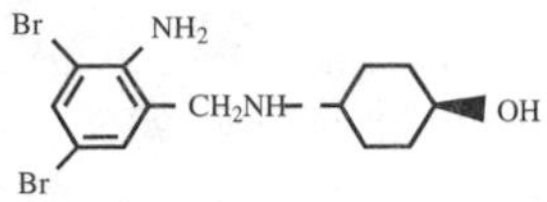

Ambruticin An antifungal antibiotic (a cyclopropylpolyene-pyran acid) isolated from a strain of

Polyangium cellulosum. It possesses activity against *Candida* spp and other pathogenic fungi.

Ambucaine (mol wt 308) A local anesthetic agent.

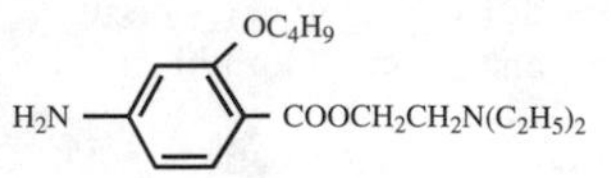

Ambucetamide (mol wt 292) An antispasmodic agent.

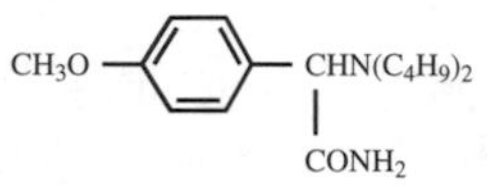

Ambuphylline (mol wt 269) A bronchodilator.

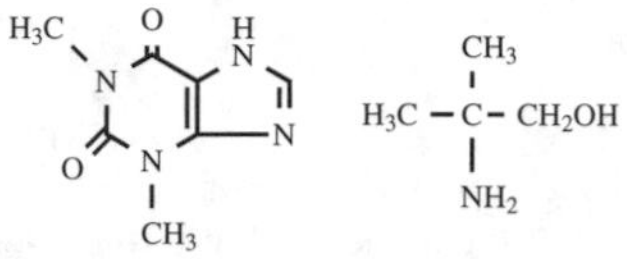

Ambuside (mol wt 394) A diuretic and antihypertensive agent.

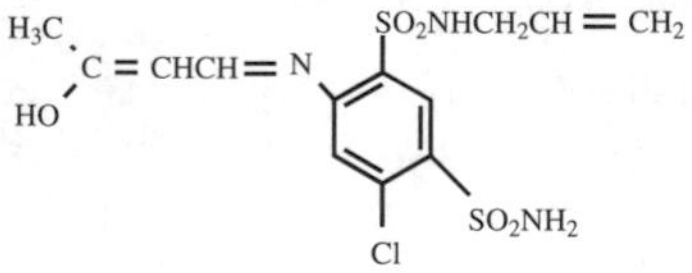

Ambutonium bromide (mol wt 391) An anticholinergic agent.

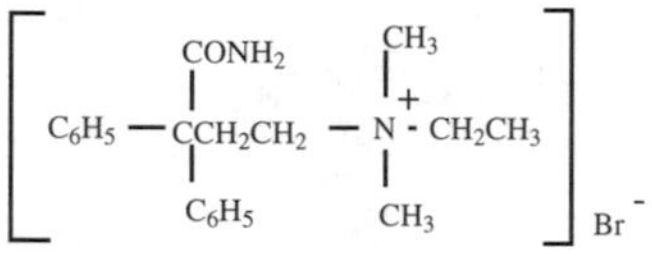

AMC Abbreviation for 7-amino-4-methyl coumarin.

Amcill A trade name for ampicillin, an antibiotic.

Amcinonide (mol wt 503) A topical glucocorticoid used as an anti-inflammatory agent.

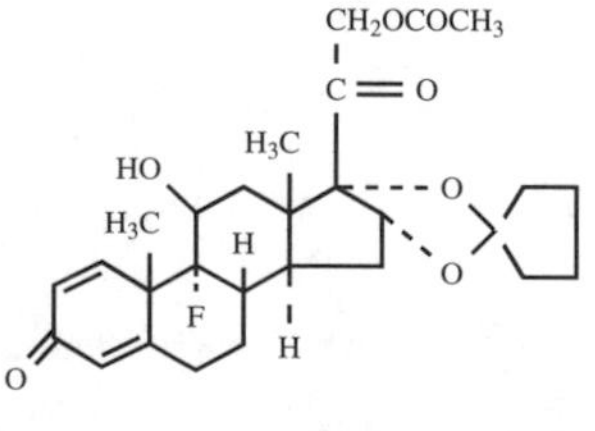

Amcort A trade name for triamcinolone diacetate, used as an anti-inflammatory agent.

AMD Abbreviation for alpha-methyl-dopa.

Amdinocillin (mol wt 325) A semisynthetic antibiotic related to penicillin.

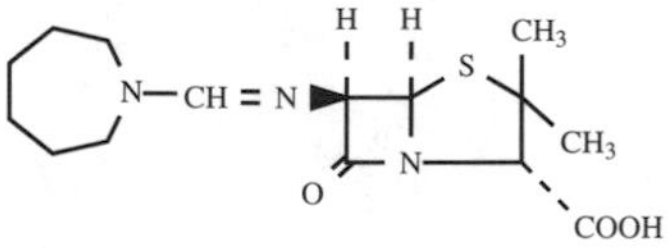

Amdinocillin Pivoxil (mol wt 440) A semisynthetic antibiotic related to penicillin.

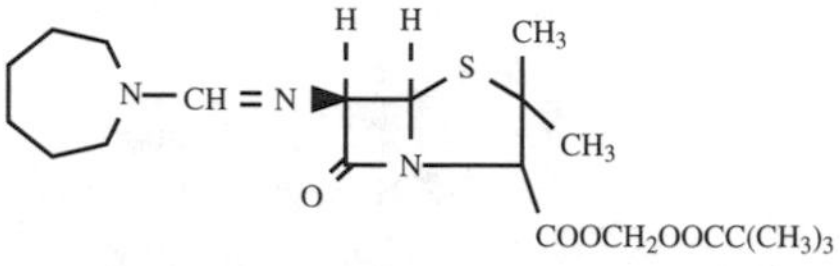

AME Abbreviation for acetoxylmethyl ester.

AmeI (ApaLI) A restriction endonuclease from *Aquaspirillum metamorphum* with the following specificity:

↓
5′..........GTGCAC..........3′
3′..........CACGTG..........5′
↑

AmeII (NaeI) A restriction endonuclease from *Aquaspirillum metamorphum* having the same specificity as NaeI.

Ameba 1. A unicellular organism with an indefinite changeable form. 2. Any protozoa of the genus *Amoeba*.

Amebae Plural of ameba.

Amebiasis Infection caused by a pathogenic amebae (e.g., infection of the intestine or liver by *Entamoeba histolytica*).

Amebic Abscess A collection of pus formed by disintegrated tissue in a cavity; caused by the protozoan *Entamoeba histolytica*.

Amebic Dysentery An inflammation of the intestine caused by infestation with *Entamoeba histolytica*.

Amebicide Any substance that kills amebae.

Amebocyte Any free cell capable of movement by pseudopodia.

Ameboid Movement The movement characteristic of amebae by projection of protoplasm (pseudopodia) toward which the rest of the cell's protoplasm flows.

Amelanotic Pertaining to tissue that lacks melanin.

Ameloblast An epithelial cell from which tooth enamel is formed.

Amelodentinal Pertaining to both the enamel and dentin of the teeth.

Amelogenesis The formation of the enamel of the teeth.

Amelogenesis Imperfecta A hereditary dental defect characterized by a brown coloration of the teeth, resulting from either severe hypocalcification or hypoplasia of the enamel.

Amelogenin A protein in dental enamel.

Amen A trade name for medroxyprogesterone acetate, used to suppress ovulation.

Amensalism A form of species interaction in which one organism is adversely affected, but the other is neither inhibited nor stimulated.

Americium A man-made radioactive element with atomic number 95, valences 3, 4, 5, 6.

Amersol A trade name for ibuprofen, used as an anti-inflammatory agent.

Ames Test A preliminary screening test for the identification of carcinogens that uses a mutant strain of *Salmonella* that cannot grow on a histidine-deficient medium unless the original mutation is reversed by a carcinogen.

Amethopterin (mol wt 454) A folic acid analog acting as an inhibitor for dihydrofolic acid reductase.

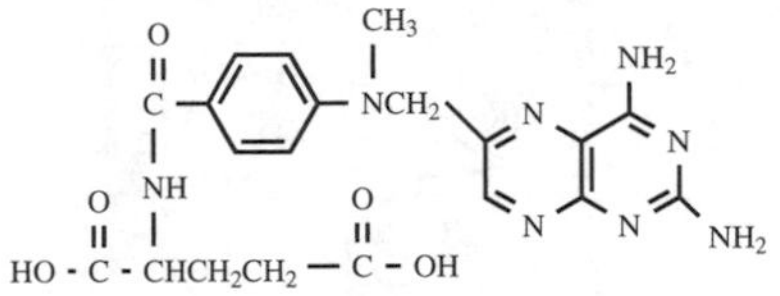

Ametropia A refractive disorder of the eye in which parallel rays of light do not focus on the retina.

Ametryn (mol wt 227) An herbicide.

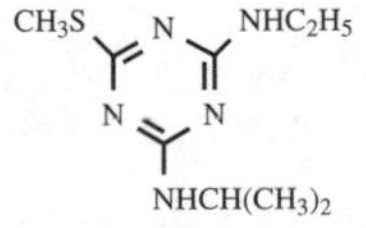

Amezinium Methyl Sulfate (mol wt 313) An antihypotensive agent.

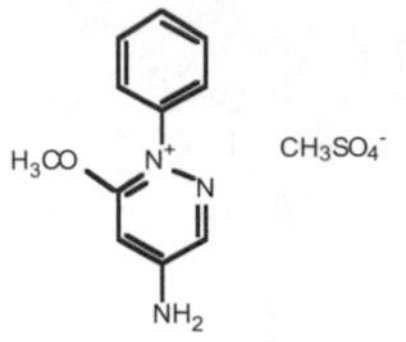

AMF Abbreviation for anti-muscle factor.

Amfenac (mol wt 255) An anti-inflammatory, antipyretic, and analgesic agent. It also inhibits platelet aggregation.

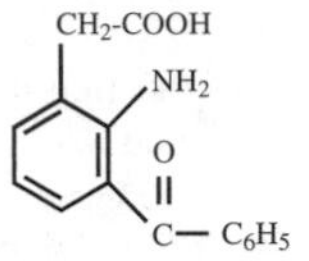

AMG Abbreviation for 1. Amyloglucosidase. 2. Anti-macrophage globulin.

α_2MG Abbreviation for α_2- macroglobulin.

Amicar A trade name for aminocaproic acid, a hemostatic agent used to inhibit plasminogen activating substances.

Amicetin (mol wt 619) An antibiotic substance produced by *Streptomyces vinaceus-drapp*.

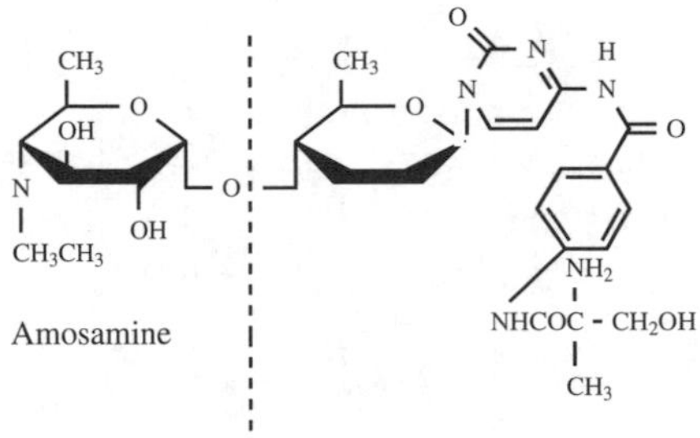

Amicibone (mol wt 357) An antitussive agent.

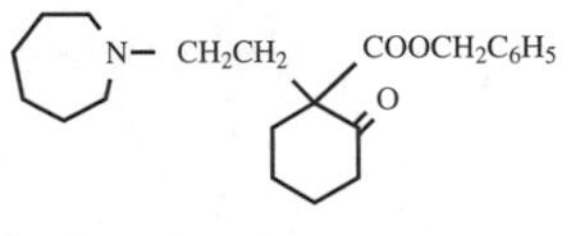

Amicil A trade name for ampicillin, an antibiotic.

Amicoumacin A (mol wt 423) A major component of a complex of antibiotics produced by *Bacillus pumilus*. It also exhibits anti-inflammatory activity.

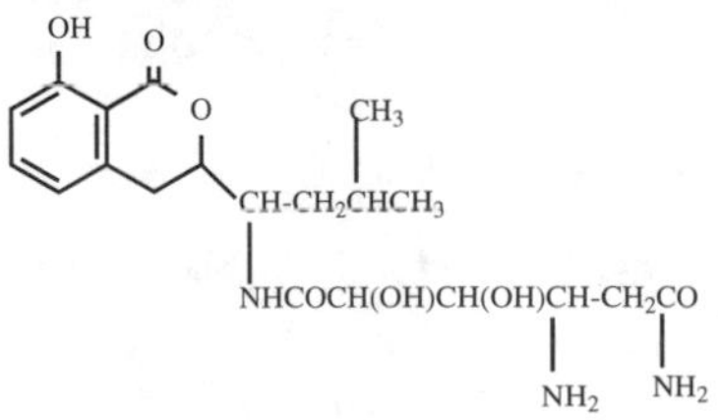

Amicyanin (mol wt 12,000) A blue protein present in certain methylamine-utilizing bacteria (e.g., *Pseudomonas*).

Amidase An enzyme that catalyzes the following reaction:

Monocarboxylic acid amide + H_2O ⇌ Monocarboxylate + NH_3

Amidation The introduction of an amide group into an organic compound.

Amide Any compound containing one or more acyl group attached to the nitrogen atom.

Amide Bond A covalent bond between an amino group of one molecule and a carboxyl group of another molecule (see also peptide bond).

Amide Group The $-CONH_2$ group derived from an acid by the replacement of the OH of the carboxyl group with an amino group.

Amidephrine (mol wt 244) A vasoconstrictor and nasal decongestant.

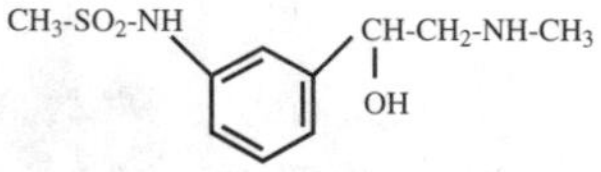

Amidinoaspartase The enzyme that catalyzes the following reaction:

N-Amidino-L-asparatate + H_2O ⇌ L-Aspartate + urea

Amidinomycin (mol wt 198) An antiviral antibiotic produced by *Streptomyces flavochromogenes*.

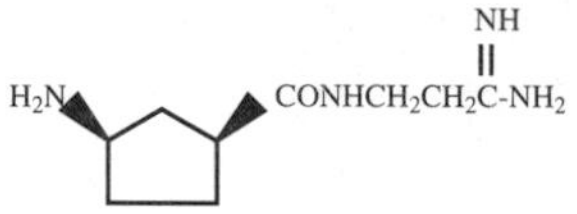

Amido Black (mol wt 617) A dye used for staining protein in polyacrylamide gel (also known as naphthol blue black).

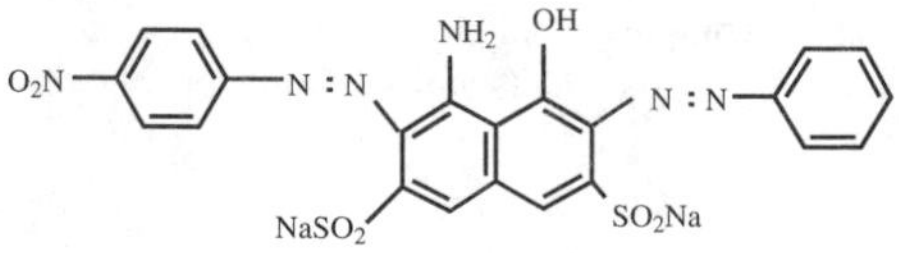

Amidobenzene See aniline.

Amidochlor (mol wt 297) A plant growth regulator.

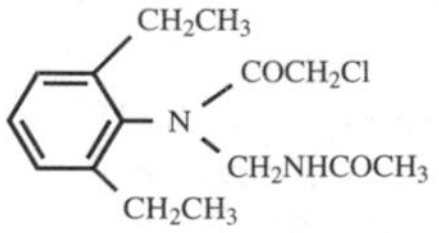

Amidolysis Any cleavage of an amide to the parent oxy acid and ammonia.

Amidomycin (mol wt 797) A cyclic antibiotic substance produced by *Streptomyces*. It consists of 4 moles each of D(-)-valine and D(-)-α-hydroxyisovaleric acid linked alternately by ester and amide bonds to form a 24-member ring.

Amidophosphoribosyl Transferase The enzyme that catalyzes the following reaction:

5-Phospho-β-D-ribosylamine + PPi + L-glutamate ⇌ L-glutamine + 5-phospho ribose 1-diphosphate + H_2O

Amidorphin An opioid peptide that is amidated at the C-terminus.

Amifloxacin (mol wt 334) An antibacterial agent.

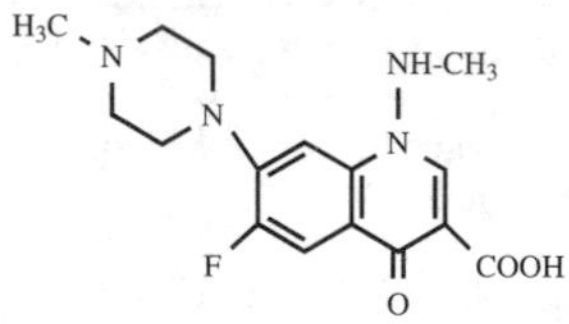

Amigesic A trade name for salsalate, a non-steroidal anti-inflammatory agent.

Amikacin (mol wt 586) A semisynthetic aminoglycoside antibiotic derived from kanamycin A. It binds to 30S ribosome, inhibiting bacterial protein synthesis.

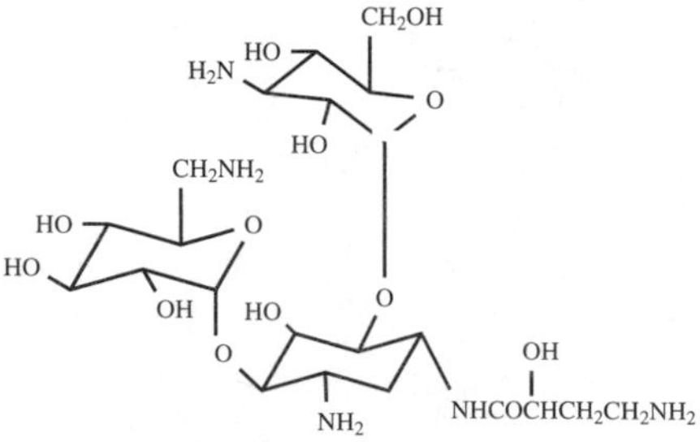

Amikin A trade name for amikacin sulfate, an antibiotic that inhibits bacterial protein synthesis.

Amiloride (mol wt 230) A potassium-sparing diuretic agent that inhibits sodium reabsorption and potassium excretion.

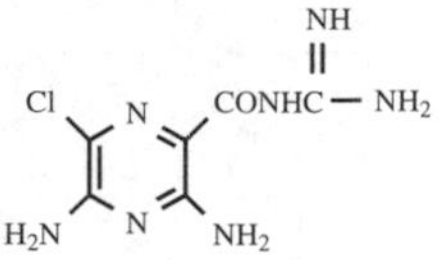

Aminacrine (mol wt 194) An anti-infective and antiseptic agent.

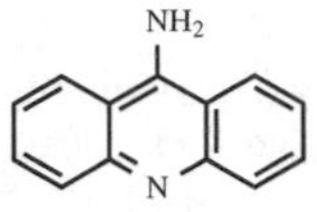

Amination The introduction of an amino group into an organic compound.

Amine A basic organic compound derived from ammonia by the replacement of one or more of its hydrogen atoms with hydrocarbon groups. The replacement of one, two, and three hydrogen atoms results in production of primary, secondary, and tertiary amines, respectively.

Amine 220 (mol wt 351) A fungicide, emulsifier, and soil stabilizer.

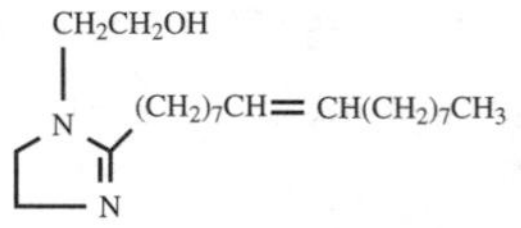

Amine Dehydrogenase The enzyme that catalyzes the following reaction:

$$RCH_2NH_2 + H_2O + \text{acceptor} \rightleftharpoons RCHO + \text{reduced acceptor} + NH_3$$

Amine N-Methyltransferase The enzyme that catalyzes the following reaction:

S-Adenosyl-L-methionine + an amine ⇌ S-Adenosyl-L-homocysteine + a methylated amine

Amine Oxidase (Copper-Containing) The enzyme that catalyzes the following reaction:

$$RCH_2NH_2 + H_2O + O_2 \rightleftharpoons RCHO + NH_3 + H_2O_2$$

Amine Oxidase (Flavin-Containing) The enzyme that catalyzes the following reaction:

$$RCH_2NH_2 + H_2O + O_2 \rightleftharpoons RCHO + NH_3 + H_2O_2$$

Amine Oxygen Oxidoreductase The systematic name for amidase.

Amine Sulfotransferase The enzyme that catalyzes the following reaction:

3-Phosphoadenylylsulfate + an amine ⇌ Adenosine 3′,5′-bisphosphate + a sulfamate

Amineptine (mol wt 337) A central nervous system stimulant.

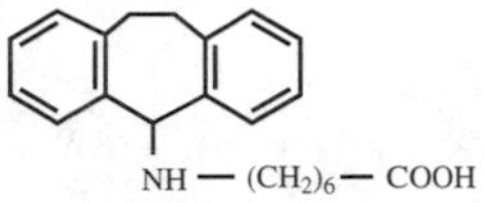

Amino Acid An organic acid that contains both a basic amino group and an acidic carboxyl group. The alpha amino acid, in which the amino group is attached to the alpha carbon, is the building block of peptides and proteins.

Amino Acid N-Acetyltransferase The enzyme that catalyzes the following reaction:

Acetyl-CoA + a D-amino acid ⇌ CoA + an N-acetyl-D-amino acid

Amino Acid Activating Enzymes A family of enzymes, at least one for each amino acid, that catalyzes the attachment of an amino acid to its specific tRNA molecule (also known as aminoacyl-tRNA synthetases).

Amino Acid Activation A process for linking an amino acid to its specific tRNA, catalyzed by the aminoacyl-tRNA synthetase as indicated below:

amino acid + ATP + tRNA ⇌ aminoacyl-tRNA + AMP + PPi

Amino Acid Analysis The qualitative and quantitative determination of amino acids in a peptide or protein.

Amino Acid Analyzer An instrument for the automated amino acid analysis in a peptide or protein.

Amino Acid Arm The base-paired stem of tRNA containing both 5' and 3' ends.

Amino Acid Composition The quantity and quality of amino acids in a peptide or protein.

D-Amino Acid Dehydrogenase The enzyme that catalyzes the following reaction:

$$\text{A D-amino acid} + H_2O + \text{acceptor} \rightleftharpoons \text{A keto acid} + NH_3 + \text{reduced acceptor}$$

L-Amino Acid Dehydrogenase The enzyme that catalyzes the following reaction:

$$\text{An L-amino acid} + H_2O + NAD^+ \rightleftharpoons \text{A keto acid} + NH_3 + NADH$$

α-Amino Acid Esterase The enzyme that catalyzes the following reaction:

$$\text{An } \alpha\text{-amino acid ester} + H_2O \rightleftharpoons \text{An } \alpha\text{-amino acid} + \text{an alcohol}$$

Amino Acid Nitrogen The nitrogen derived from amino acids.

Amino Acid Oxidase An enzyme that catalyzes the deamination of amino acids. There are two types of amino acid oxidase, namely D-amino acid oxidase and L-amino acid oxidase.

D-Amino Acid Oxidase The enzyme that catalyzes the following reaction:

$$\text{A D-amono acid} + H_2O + O_2 \rightleftharpoons \text{An } \alpha\text{-ketoacid} + HN_3 + H_2O_2$$

L-Amino Acid Oxidase The enzyme that catalyzes the following reaction:

$$\text{A L-amono acid} + H_2O + O_2 \rightleftharpoons \text{An } \alpha\text{-ketoacid} + HN_3 + H_2O_2$$

D-Amino Acid Oxidase (FAD-Specific) A flavin-dependent amino acid oxidase that catalyzes the oxidative deamination of D-amino acid:

$$\text{D-Amino acid} + H_2O + FAD \rightleftharpoons \alpha\text{-Ketoacid} + NH_3 + FADH_2$$

L-Amino Acid Oxidase (FAD-Specific) A flavin-dependent amino acid oxidase that catalyzes the oxidative deamination of L-amino acid:

$$\text{L-Amino acid} + H_2O + FAD \rightleftharpoons \alpha\text{-Ketoacid} + NH_3 + FADH_2$$

Amino Acid Oxygen Oxidoreductase Systematic name for amino acid oxidase.

Amino Acid Racemase The enzyme that catalyzes the following reaction:

$$\text{L-Amino acid} \rightleftharpoons \text{D-Amino acid}$$

Amino Acid Replacement The substitution of one amino acid for another at a given position in a polypeptide chain as a result of a mutation or biochemical manipulation.

Amino Acid Residue The amino acid present in a peptide or protein. It is the amino acid minus the atoms that are removed from it in the process of formation of a peptide bond.

Amino Acid Sequence Linear order of the amino acid residues in a peptide or protein.

Amino Acid Side Chain The atoms or group of atoms of an amino acid molecule other than the alpha amino group and the carboxyl group.

Amino Acid Site Synonym of aminoacyl site.

Amino Acid tRNA Ligase See aminoacyl tRNA synthetase.

Aminoacetic Acid Synonym for glycine.

Aminoaciduria The presence of excessive amounts of amino acids in the urine.

Aminoacyl- A combining form meaning an amino acid that is esterified through its carboxyl group with another molecule.

Aminoacyl Adenylate An amino acid that has been esterified through its carboxyl group to the phosphate group of AMP. It is an aminoacyl-AMP complex.

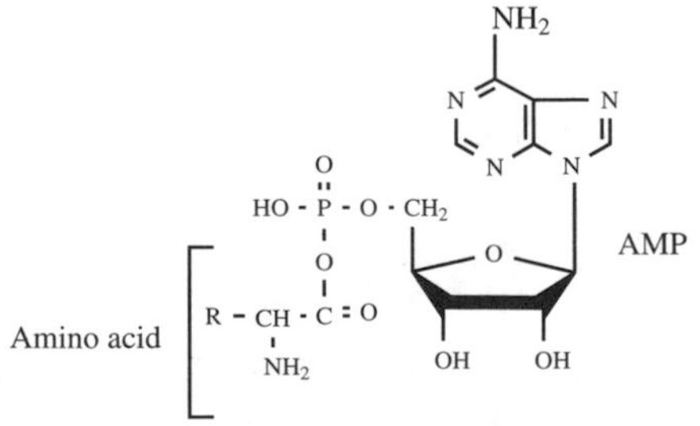

Aminoacyl Site Site on a ribosome where the incoming charged-tRNA binds (also known as A site).

Aminoacyl Synthetase See aminoacyl-tRNA synthetase.

Aminoacyl tRNA A tRNA molecule that has a covalently bound amino acid (also known as charged tRNA).

Aminoacyl tRNA Hydrolase The enzyme that catalyzes the following reaction:

N-Substituted aminoacyl-tRNA + H_2O

⇅

N-Substituted amino acid + tRNA

Aminoacyl tRNA Site See aminoacyl site.

Aminoacyl tRNA Synthetase The enzyme that catalyzes the linkage of an amino acid to its appropriate tRNA molecule by the formation of an ester bond, using energy provided by the hydrolysis of ATP.

amino acid + ATP + tRNA

⇅

aminoacyl-tRNA + AMP + PPi

Aminoacylase The enzyme that catalyzes the following reaction:

An N-acyl-L-amino acid + H_2O

⇅

A carboxylate + an L-amino acid

Aminoadipic Pathway A metabolic pathway for biosynthesis of lysine found in some algae

***p*-Aminobenzoic Acid (mol wt 137)** A factor of vitamin B that is essential for the growth of bacteria.

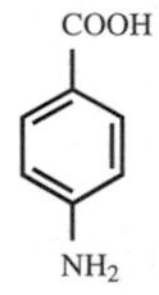

Aminobutyraldehyde Dehydrogenase The enzyme that catalyzes the following reaction:

4-Aminobutanal + NAD^+ + H_2O

⇅

4-Aminobutanoate + NHDH

γ-Aminobutyrate Bypass A reaction sequence for the conversion of α-ketoglutaric acid to succinic acid that differs from the normal sequence in the citric acid cycle. It occurs in brain tissue.

γ-Aminobutyrate Shunt A variation of tricarboxylic acid cycle in which a-keto-glutaric acid is converted to glutamate which is then decarboxylated to form g-aminobutyrate.

Aminobutyrate Transaminase The enzyme that catalyzes the following reaction:

4-Aminobutanoate + α-ketoglutarate

⇅

Succinate semialdehyde + L-glutamate

γ-Aminobutyric Acid (mol wt 103) A non-protein amino acid occurring in the central nervous system.

NH_2-CH_2-CH_2-CH_2-COOH

Aminochlorthenoxazin (mol wt 227) An antipyretic and analgesic agent.

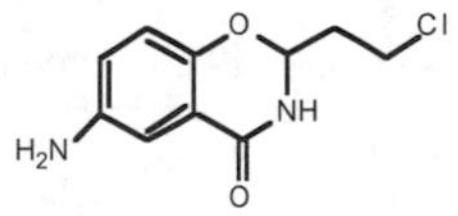

1-Amino-Cyclopropane-Carboxylic Acid (mol wt 101) A naturally occurring amino acid found in pears and apples.

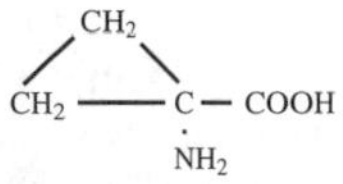

Aminoethyl The –CH_2CH_2-NH_2 group.

Aminoethylbenzenesulfonyl Fluoride (mol wt 202) An inhibitor of serine endopeptidase.

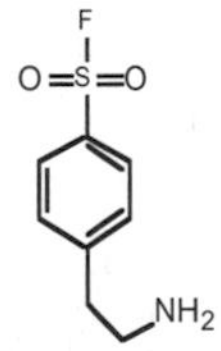

N-(2-Aminoethyl)-5-isoquinolinesulfonamide (mol wt 250) An inhibitor for casein kinase I and protein kinase A.

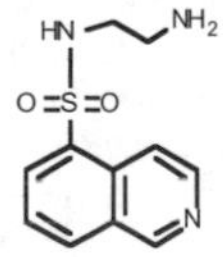

2-Amino-3-Formyl-Pentenoic Acid (mol wt 143) A naturally occurring amino acid found in mushrooms.

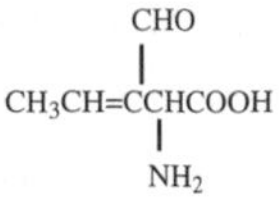

Amino-Group The -NH_2 group.

α-Aminoheptanoic Acid (mol wt 145) A naturally occurring amino acid found in fungi (e.g., *Claviceps purpurea*).

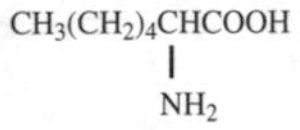

2-Amino-4,5-Hexadienoic Acid (mol wt 127) A naturally occurring amino acid found in mushrooms (e.g., *Amanita solitaria*).

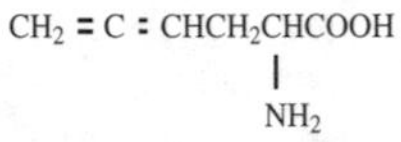

3-Amino-4-Hydroxybutyric Acid (mol wt 119) An anti-inflammatory and antifungal agent.

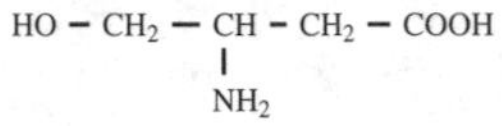

4-Amino-3-Hydroxybutyric Acid (mol wt 119) An anticonvulsant.

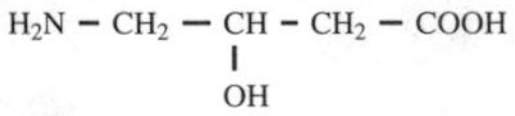

2-Amino-6-Hydroxy-4-Methyl-4-Hexenoic Acid (mol wt 159) The naturally occurring amino acid found in seeds of the plant *Aesculus Claifornica.*

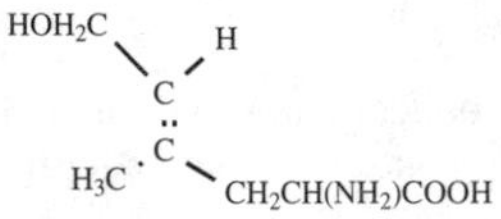

2-Amino-3-Hydroxy-Methyl-3-Pentenoic Acid (mol wt 145) A naturally occurring amino acid found in mushrooms.

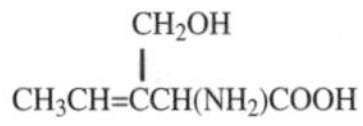

Aminoimidazole Ribonucleotide See AIR.

Amino-Imino Tautomerism A form of tautomerism in which the amino-form and imino-form of a compound are in equilibrium.

a-Amino-Isobutyric Acid (mol wt 103) An amino acid found in muscle protein of *Iris tingitana.*

$(CH_3)_2C(NH_2)COOH$

b-Amino-Isobutyric Acid (mol wt 103) A naturally occurring amino acid found in *Iris tingitana.*

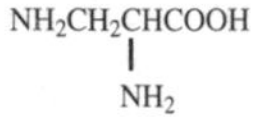

Aminolevulinate Dehydratase Synonym of porphobilinogen synthetase.

Aminolevulinate Hydro-lyase Synonym of porphobilinogen synthetase.

Aminolevulinic Acid (mol wt 130) An intermediate in biosynthesis of tetrapyrrole.

$NH_2\text{-}CH_2\text{—}CO\text{-}CH_2\text{-}CH_2\text{-}COOH$

Aminolysis Any hydrolytic deamination reaction in which an amino group is replaced by a hydroxyl group.

2-Amino-4-Methoxy-*trans*-3-Butenoic Acid (mol wt 131) A naturally occurring amino acid found in *Pseudononas aerugenosa.*

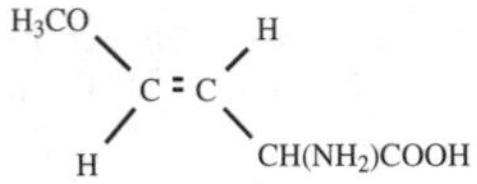

g-Amino-a-Methylene Butyric Acid (mol wt 115) A naturally occurring amino acid found in the groundnut plant (e.g., *Arachis hypogaea*).

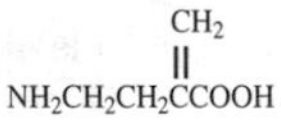

2-Amino-4-Methyl Hexanoic Acid (mol wt 145) A naturally occurring amino acid found in plant seeds of *Aesculus Californica.*

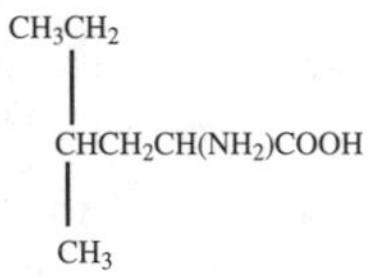

2-Amino-4-Methylthiazole (mol wt 115) An antihyperthyroid agent.

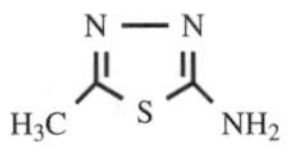

Aminometradine (mol wt 195) A diuretic agent.

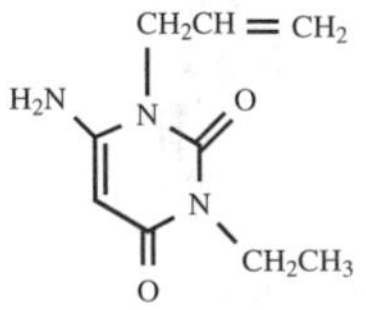

Amino-Oligopeptidase Synonym of membrane alanyl aminopeptidase.

Amino-Opti-C A trade name for vitamin P (bioflavonoid).

Amino-Opti-E A trade name for vitamin E

6-Aminopenicillanic Acid (mol wt 216) A derivative of natural penicillin and a precursor of many semisynthetic penicillins. It has little or no antibacterial activity by itself.

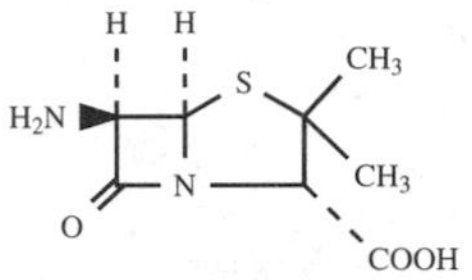

Aminopeptidase The enzyme that hydrolyzes the N-terminal amino acid residues from oligopeptides or polypeptides.

Aminophylline (mol wt 456) A bronchodilator and smooth muscle relaxant.

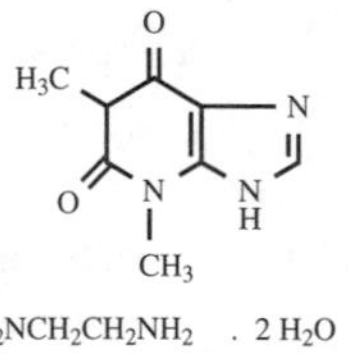

Aminopromazine (mol wt 328) An antispasmodic agent.

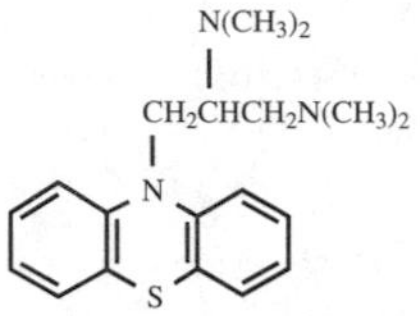

***p*-Aminopropiophenone (mol wt 149)** An antidote for cyanide.

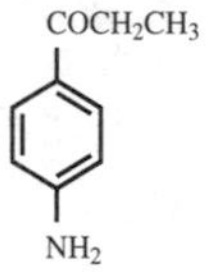

Aminopropylon (mol wt 302) An analgesic agent.

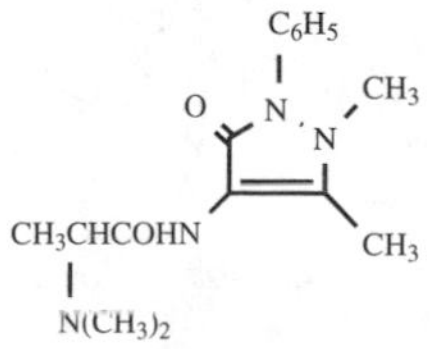

Aminopterin (mol wt 440) A folic acid analog that inhibits the enzyme dihydrofolate reductase and is used as a rodenticide.

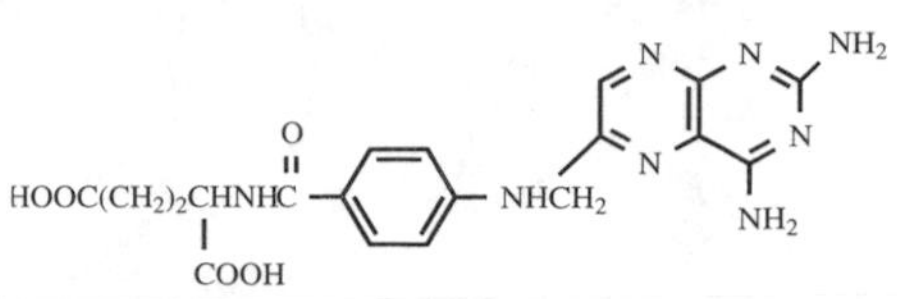

2-Aminopurine Synonym of adenine.

Aminopyrine (mol wt 231) An antipyretic and analgesic agent.

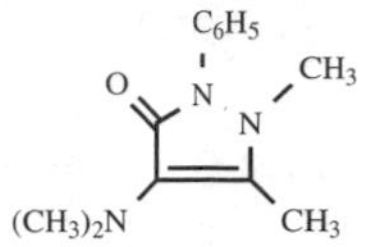

Aminoquinolines A group of antimalarial agents, e.g., 3-, 5-, 6-, and 8-aminoquinoline.

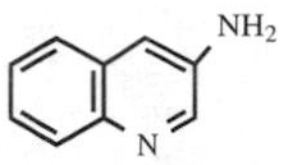

3-aminoquinoline

Aminoquinuride (mol wt 372) A substance used as an antiseptic agent.

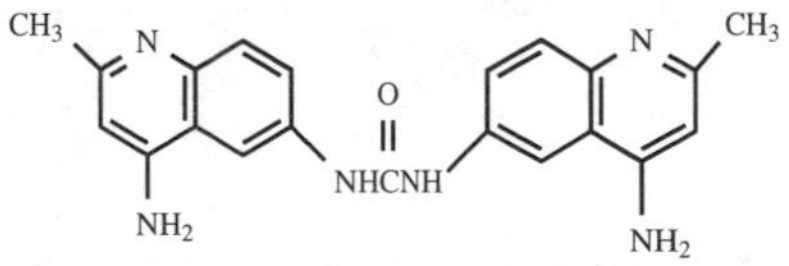

Aminorex (mol wt 162) An anorexic agent.

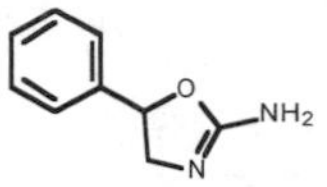

***p*-Aminosalicylic Acid** An antibacterial (tuberculostatic) agent.

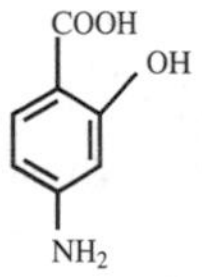

Aminosuccinic Acid Synonym for aspartic acid.

Aminotransferase The enzyme that catalyzes the transfer of an α-amino group from an α-amino acid to a keto acid.

Amiodarone (mol wt 645) An antiarrhythmic and antianginal agent that prolongs the refractory period and action potential duration and decreases repolarization.

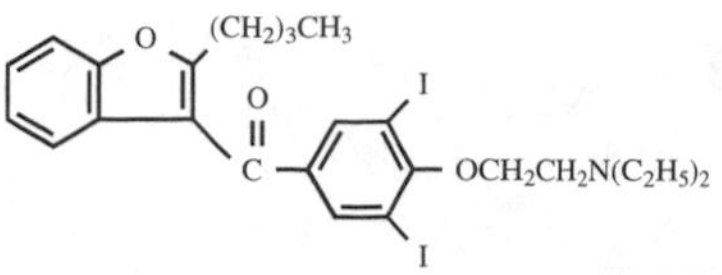

Amiphenazole (mol wt 191) A barbiturate and morphine antagonist.

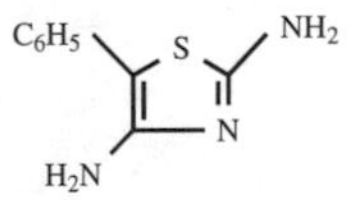

Amiprilose (mol wt 305) An immunomodulator.

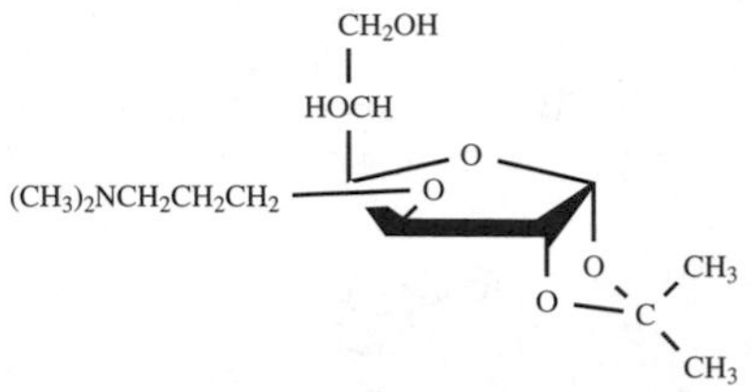

Amisometradine (mol wt 195) A diuretic agent.

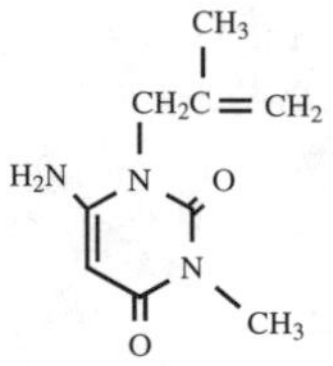

Amisulpride (mol wt 369) An antipsychotic agent.

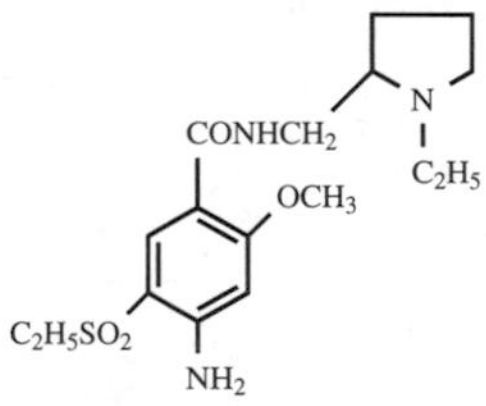

Ami-Tex LA A trade name for a combination drug containing phenylpropanolamine hydrochloride and guaifenesin used as nasal decongestant.

Amiton (mol wt 269) An insecticide (miticide).

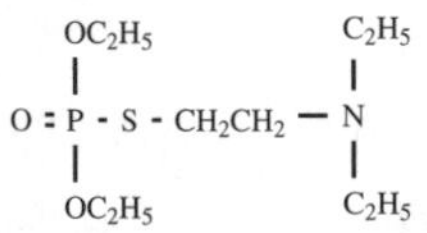

Amitosis A form of cell division in which the nucleus divides by constriction without participation of a mitotic apparatus.

Amitraz (mol wt 293) A scaricide and scabicide.

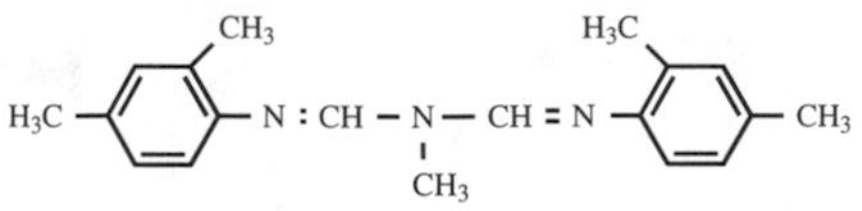

Amitril A trade name for amitriptyline hydrochloride, used as an antidepressant.

Amitriptyline (mol wt 277) An antidepressant that increases the amount of norepinephrine or serotonine in the CNS by blocking their retake by the presynaptic neurons.

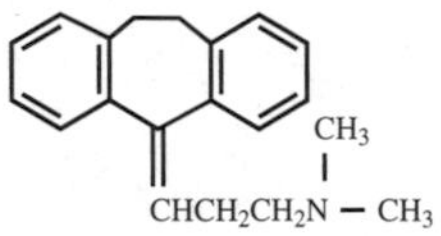

Amitriptylinoxide (mol wt 293) An antidepressant.

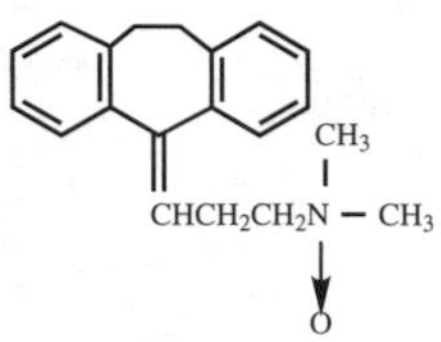

Amitrole (mol wt 84) An herbicide.

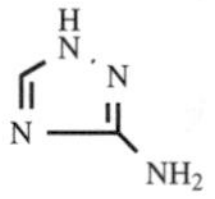

Amixetrine (mol wt 261) An anti-inflammatory and anticholinergic agent.

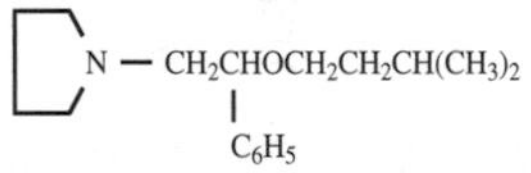

Amixis A type of reproductive cycle in which karyogamy and meiosis do not occur.

AML Abbreviation for acute myelocytic leukemia.

Amlexanox (mol wt 298) An antiallergic and antiasthmatic agent.

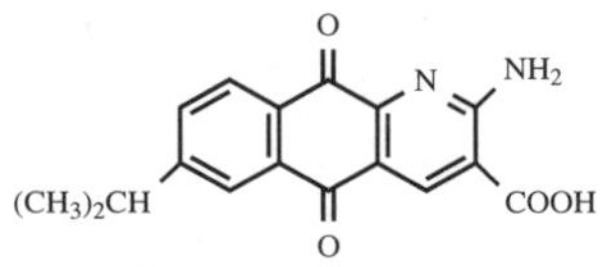

Amlodipine (mol wt 409) An antianginal and antihypertensive agent.

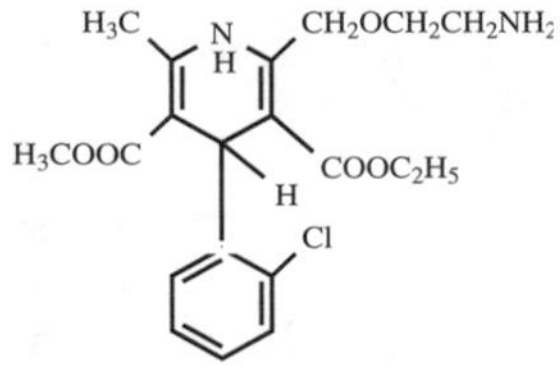

AMLR Abbreviation for autologous mixed lymphocyte reaction.

AMLS Abbreviation for anti-mouse lymphocyte serum.

A_2M-MA Abbreviation for methylamine-modified α_{-2} macroglobulin.

Ammoni- A combining form meaning pertaining to ammonium.

Ammonia A colorless gas, NH_3.

Ammonia Assimilation See ammonia fixation.

Ammonia Detoxification The formation of ammonium salt and nitrogen-excretion compounds from ammonia.

Ammonia Fixation The conversion of ammonia to glutamic acid, glutamine, or carbamoyl phosphate.

Ammonification Formation of ammonia through the decomposition of organic nitrogen compounds (e.g., amino acids and proteins) by microorganisms.

Ammonifying Bacteria Bacteria capable of ammonification.

Ammonium Acetate (mol wt 77) A diuretic agent.

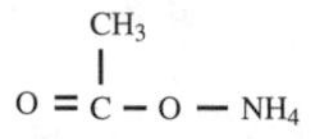

Ammonium Benzoate (mol wt 139) An anti-infective agent.

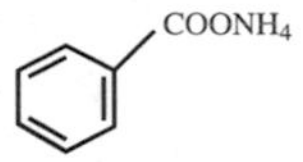

Ammonium Bicarbonate (mol wt 79) An expectorant.

NH_4HCO_3

Ammonium Bromide (mol wt 98) A sedative agent.

BrH_4N

Ammonium Kinase The enzyme that catalyzes the following reaction:

$$ATP + NH_3 \rightleftharpoons ADP + \text{phosphoramide}$$

Ammonium Mercuric Chloride (mol wt 379) An antifungal compound used in ointments for the treatment of chronic eczema.

$Cl_4H_8HgN_2$

Ammonium Persulfate (mol wt 228) A potent oxidizer used for promoting polymerization of acrylamide gels.

$(NH_4)_2S_2O_8$

Ammonium Salicylate (mol wt 155) An analgesic agent used topically to loosen psoriatic scales.

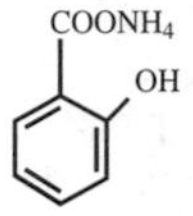

Ammonium Sulfate Fractionation The precipitation or fractionation of proteins by means of different concentrations of ammonium sulfate. Different proteins can be precipitated with different concentrations of ammonium sulfate.

Ammonium Valerate (mol wt 119) A sedative agent.

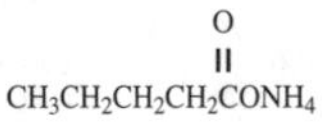

Ammonolysis The cleavage of a covalent bond of an acid derivative by reaction with ammonia so that one of the products combines with the hydrogen atom and the other combines with the amino group.

Ammonotelic Organism An organism that excretes ammonia as the principal form of nitrogenous waste from catabolism of nitrogenous compounds (e.g., amino acids and proteins).

Amnesia The loss of memory.

Amniocentesis A diagnostic procedure for removal of a sample of amniotic fluid or fetal materials from a pregnant woman with a hypodermic syringe needle inserted through the abdominal wall.

Amnion The fluid-filled sac surrounding the embryo.

Amnion Cell A cell that forms the fluid-filled sac in which the embryo of mammals and higher vertebrates develops.

Amniotic Fluid The fluid that fills the membranous sac enclosing the fetus.

Amniotic Sac A thin-walled membrane structure that contains the fetus and amniotic fluid.

AMO Abbreviation for alkene mono-oxygenase.

Amobarbital (mol wt 226) A barbiturate sedative-hypnotic agent and mitochondrial respiratory inhibitor.

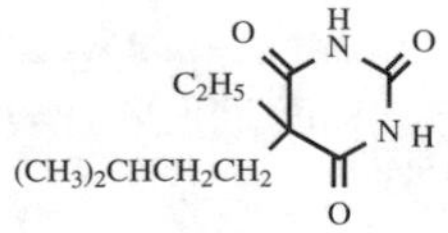

Amodiaquin (mol wt 356) An antimalarial agent.

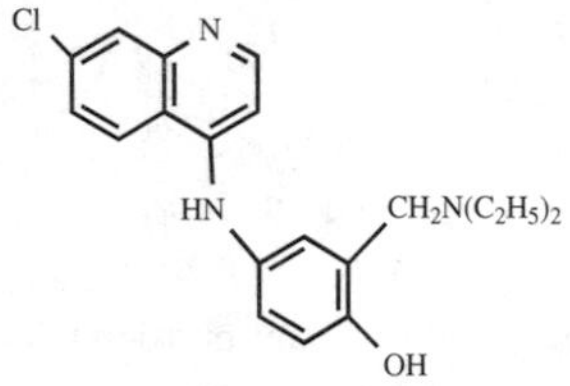

Amoeba A genus of free-living amebae (order Amoebida).

Amoebiasis Any disease or infection of humans or animals caused by a protozoon of Amoebida (e.g., dysentery and meningoencephalitis).

Amoebic dysentery A disease caused by an *amoeba*.

Amoebicide Any chemical agent that kills or inhibits amebae.

Amoeboid Characteristic of amoeba.

Amoeboid Movement A mode of locomotion that depends on pseudopodia and extensions of cytoplasm.

Amol Abbreviation for attomole (10^{-18} mole).

Amolanone (mol wt 309) A local anesthetic agent.

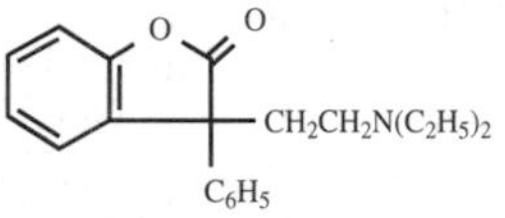

Amoproxan (mol wt 426) An antiarrhythmic agent.

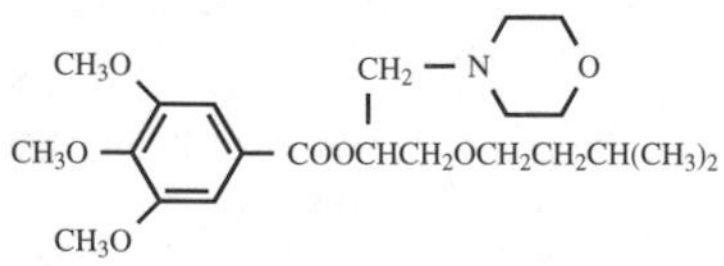

Amorolfine (mol wt 318) A topical antifungal agent.

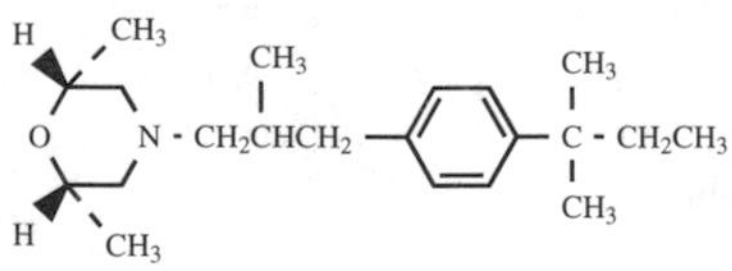

Amorph A gene or mutant allele that has little or no effect on the expression of a trait compared to that of the wild-type allele.

Amorphosporangium A genus of bacteria (order Actinomycetales, wall type II).

Amorphous A formless mass, devoid of regular structure.

Amoscanate (mol wt 271) An anthelmintic agent (*Schistosoma*).

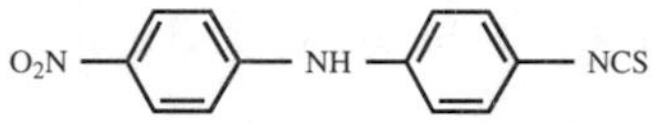

Amosulalol (molwt 380) An antihypertensive agent.

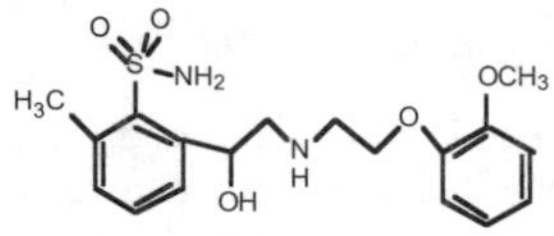

Amotriphene (mol wt 404) A coronary vasodilator.

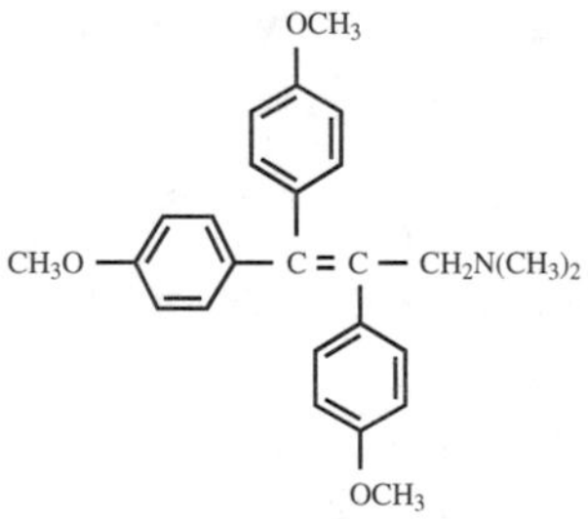

Amoxapine (mol wt 314) An antidepressant that increases the concentration of norepinephrine or serotonin in the CNS by blocking their uptake by the presynaptic neurons.

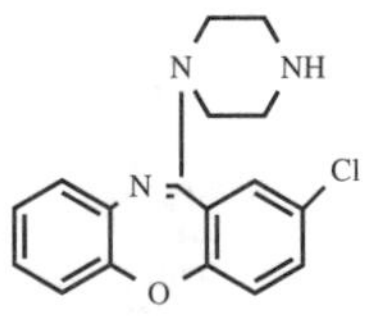

Amoxicillin A semisynthetic oral penicillin that inhibits bacterial cell synthesis.

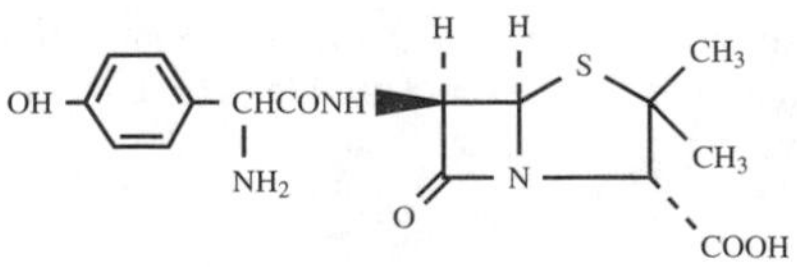

Amoxil A trade name for amoxicillin, an antibiotic.

Amp Abbreviation for ampere, an electric unit.

AMP Abbreviation for adenosine 5′-monophosphate or adenosine 3′-monophosphate.

2'AMP Abbreviation for adenosine 2'-monophosphate.

3'AMP Abbreviation for adenosine 3'-monophosphate.

5'AMP Abbreviation for adenosine 5'-monophosphate.

cAMP Abbreviation for adenosine 3′,5′-cyclic monophosphate.

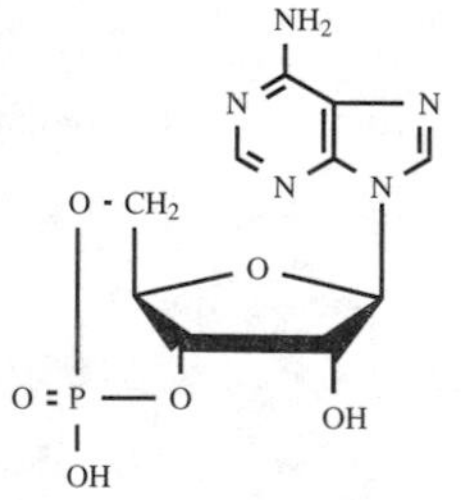

AMP Aminase See AMP deaminase.

AMP Deaminase The enzyme that catalyzes the following reaction:

$$\text{AMP} + H_2O \rightleftharpoons \text{IMP} + HN_3$$

AMP Nucleosidase The enzyme that catalyzes the following reaction:

$$\text{AMP} + H_2O \rightleftharpoons \text{Adenine} + \text{D-ribose 5-phosphate}$$

AMP Thymidine Kinase The enzyme that catalyzes the following reaction:

$$\text{AMP} + \text{thymidine} \rightleftharpoons \text{Adenine} + \text{thymidine 5-phosphate}$$

AMPase Abbreviation for adenosine monophosphatase.

AMPD Abbreviation for AMP deaminase.

cAMP-Dependent Protein Kinase Synonym of protein kinase.

Ampere A unit of electrical current. The amount of current passed through a resistance of one ohm by an electrical potential of one volt or the amount of current that deposits 0.001118 gm of silver per second when passed through a silver nitrate solution.

Amperometric Titration A titration in which either the titrant or the substance being titrated is electroactive.

AmpFLP Abbreviation for amplified fragment length polymorphism.

Amphecloral (mol wt 265) An anorexic agent.

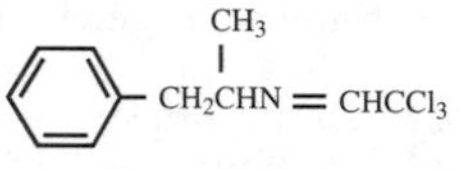

Amphenidone (mol wt 186) A sedative and hypotic agent.

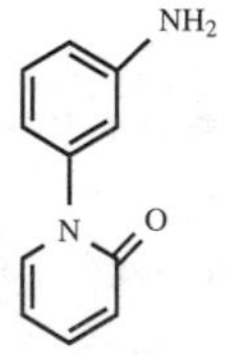

Amphetamine (Benzedrine, mol wt 135) A CNS stimulant and anorexic agent that promotes transmission of nerve impulses by releasing stored norepinephrine from the nerve terminals in the brain.

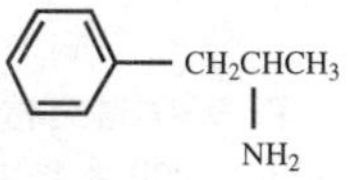

Amphetaminil (mol wt 250) A psychotropic agent.

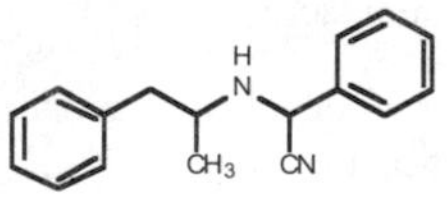

Amphibaric A pharmacologically active substance that can either lower or raise the blood pressure depending on the dose used.

Amphibolic Pathway A metabolic pathway that has both catabolic and anabolic functions.

Amphigony Referring to sexual reproduction.

Amphikaryon A nucleus that contains the diploid number of chromosomes.

Amphimixis The mixing of germinal substances during fertilization.

Amphion Any molecules that contain ionic groups of opposite charge.

Amphipathic Molecule A molecule having both hydrophilic and hydrophobic regions.

Amphiphilic See Amphipathic molecule.

Amphiphysin An acidic protein found in the synapses of avian and mammalian nervous system.

Amphiprotic Synonym of amphoteric.

Amphiregulin A bifunctional transmembrane glycoprotein. It inhibits growth of several human carcinoma cells in culture.

Amphitene Referring to a chromosome where one end is thick and the other end is thin.

Amphitrichous Having a single flagellum at each end of a cell (e.g., amphitrichous bacteria).

Amphitrophic An organism that can grow photosynthetically under the light or chemotrophically in the dark.

Amphocin A trade name for amphotericin B, an antifungal agent.

Amphoion Variant spelling of amphion.

Amphojel A trade name for aluminum salt used as antacid.

Ampholyte A molecule that has both acidic and basic groups and is capable of forming pH gradients. It is commonly used as a medium for electrofocusing.

Amphomycin A polypeptide antibiotic produced by *Streptomyces canus*, active against Gram-positive bacteria.

Amphotabs A trade name for aluminum hydroxide, used as an antacid.

Amphotec A trade name for amphotericin B, an antifungal antibiotic.

Amphoteric A substance capable of acting as either an acid or a base or a substance bearing either a net negative or a net positive charge.

Amphotericin B (mol wt 924) An antifungal agent from *Streptomyces nodosus* that alters cell membrane permeability.

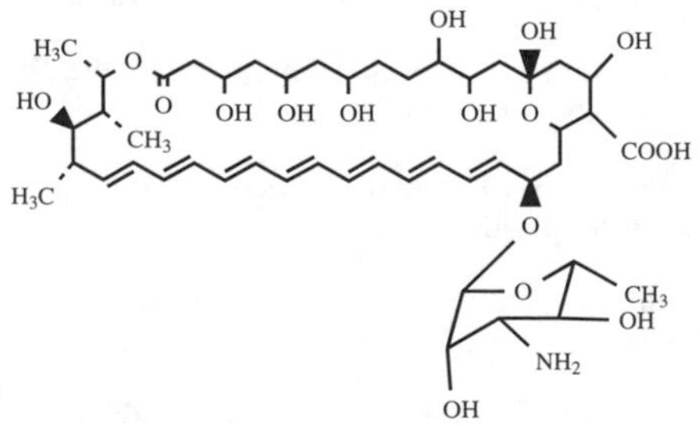

Amphotropic An organism capable of replicating both in its host of origin or in foreign host cells (e.g., a murine retrovirus that can replicate in both murine and nonmurine mammalian cells).

Ampicin A trade name for ampicillin, an antibiotic.

Ampicillin (mol wt 349) A semisynthetic derivative of penicillin. It inhibits bacterial cell wall synthesis or peptidoglycan cross linking.

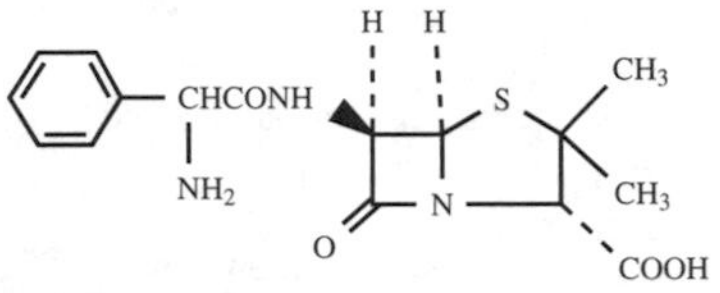

Ampilean A trade name for ampicillin, an antibiotic.

Ampiroxicam (mol wt 447) An anti-inflammatory agent.

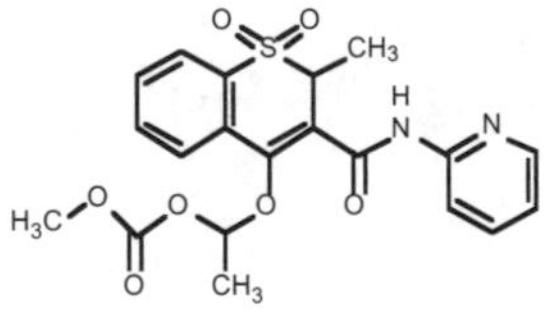

AMPK Abbreviation for AMP-activated protein kinase.

AMPKK Abbreviation for AMP-activated protein kinase kinase.

Amplicon A PCR (polymerase chain reaction) amplified DNA fragment.

Amplification 1. An increase in the number of copies of a particular gene resulting from the replication of the vector in which the gene has been cloned. 2. An increase in the number of copies of a plasmid by the exposure of its host cell to chloramphenicol, an antibiotic that inhibits chromosome replication.

Amplimer An oligonucleotide that serves as a primer in the PCR and defines the boundaries of an amplification product.

Amplitude of Accommodation The total accommodative power of the eye, determined by the difference between the refractive power for farthest vision and that for nearest vision.

Ampoule Variant spelling of ampule.

AMPr Gene in a vector that codes for β-lactamase and confers resistance of the vector-infected bacteria to ampicillin.

Amprace A trade name for enalapril maleate, used as an antihypertensive agent.

Amprolium (mol wt 279) A coccidiostatic agent.

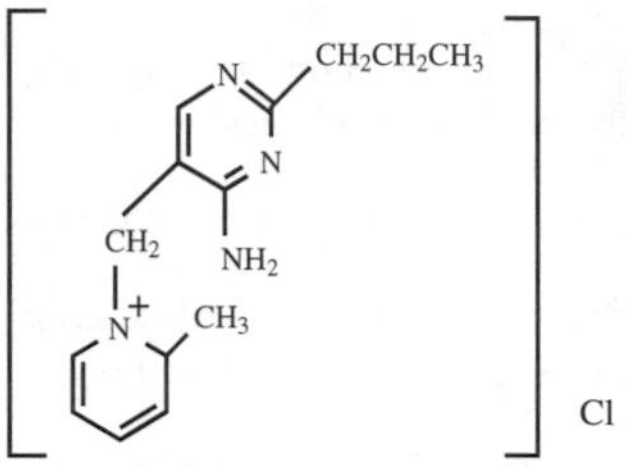

Amprotropine phosphate (mol wt 405) An anticholinergic agent.

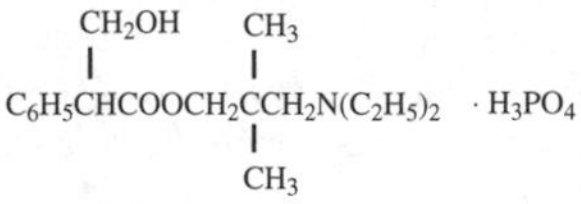

Ampule A small glass or plastic vial.

Ampullariella A genus of bacteria (order Actinomycetales occurring in soil and freshwater habitats.

Ampulliform Flask shaped.

Amrinone (mol wt 187) A cardiotonic agent.

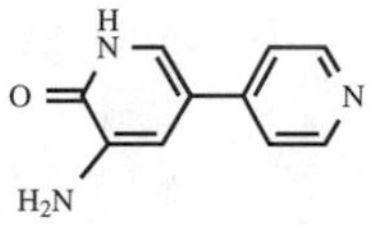

AMS Abbreviation for anti-macrophage serum.

Amsacrine (mol wt 393) An immune suppressant with antiviral and antineoplastic activity.

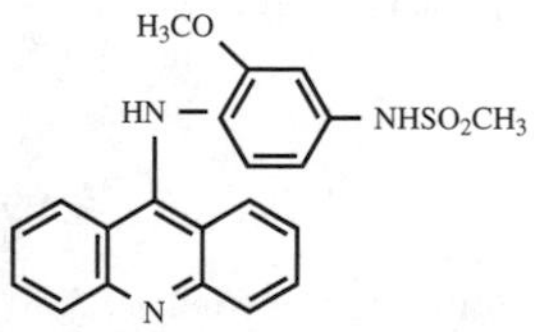

AMSH Abbreviation for alpha-melanocyte-stimulating hormone.

AMT Abbreviation for alpha methyltyrosine.

Amtolmetin Guacil (mol wt 420) An analgesic and anti-inflammatory agent.

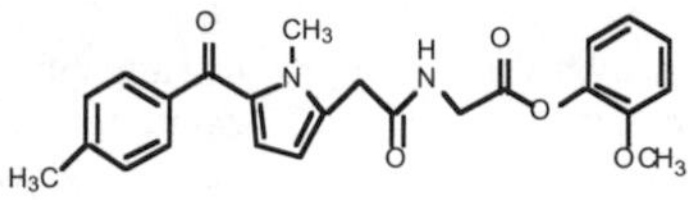

AMU Abbreviation for atomic mass unit.

A-MuLV Abbreviation for Abelson murine leukemia virus.

AMV Abbreviation for 1. avian myeloblastosis virus and 2. alfalfa mosaic virus.

AMyAb Abbreviation for antimyocardial antibodies.

Amyelination The failure to form myelin.

Amyelinic Neuroma A tumor that contains only nonmyelinated nerve fibers.

Amygdalin (mol wt 457) A β-glycoside, present in bitter almonds consisting of gentiobiose linked to mandelonitrile.

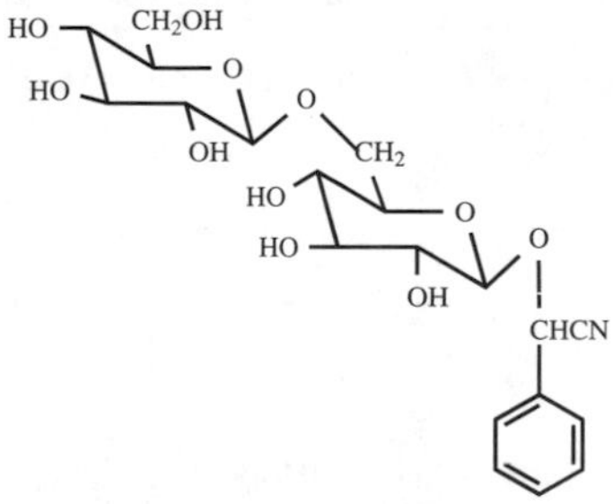

Amylase An enzyme that catalyzes the hydrolysis of starch and glycogen.

α-Amylase The enzyme that hydrolyzes mainly internal 1,4-α-D-glucosidic bonds adjacent to 1,6-α-D-glucosidic branch points yielding glucose, maltose, and branched small dextrin.

β-Amylase The enzyme that hydrolyzes α(1,4)-D-glucan from the nonreducing end yielding maltose units.

γ-Amylase The enzyme that catalyzes the hydrolysis of 1,4-glucosidic linkages from nonreducing ends of a glucan chain releasing β-D-glucose (also known as 1,4-α-glucosidase).

Amyl-Carbamate, Tertiary (mol wt 131) A hypnotic agent.

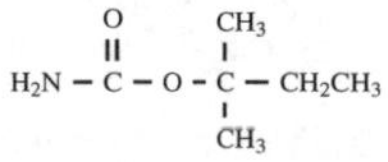

Amyl-Cresol (mol wt 178) An antiseptic, germicide, and mold preventive agent.

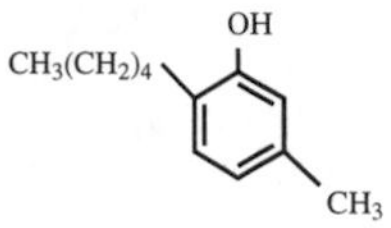

Amylene (mol wt 70) An asphyxiant.

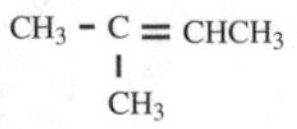

Amylin A 37-residue polypeptide found in the amyloid-rich pancreatic cells from patients suffered from type 2 diabetic,

Amyl-Nitrite (mol wt 117) A vasodilator and an antidote for cyanide.

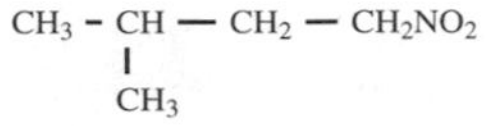

Amylo- A combining form meaning starch.

Amylobarbitone A trade name for amobarbital, an anticonvulsant, sedative and hypnotic agent.

Amylocaine Hydrochloride (mol wt 272) A local anesthetic agent.

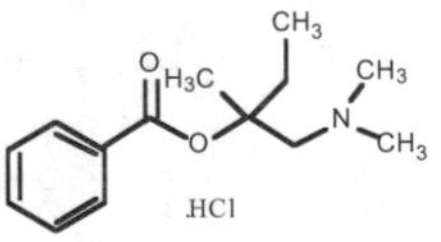

Amyloclastic See Amylolytic.

Amyloclastic Method A method for assaying amylase by determining the amount of unhydrolyzed starch that remains after incubation of the starch with the enzyme amylase.

Amylodextrin Soluble starch obtained from partial hydrolysis of starch.

Amyloglucosidase See γ-Amylase.

Amylo-1-6-Glucosidase The enzyme that catalyzes the endohydrolysis of 1,6-α-D-glucosidic linkages in amylopectin.

Amyloid A starchlike proteinaceous material believed to be a glycoprotein.

Amyloid Peptide β (β-Amyloid Peptide) A glycoprotein associated with Alzheimer's disease.

Amyloid-A Protein Referring to a serum protein synthesized by the stimulation of cytokine.

Amyloidosis A disorder characterized by the formation of amyloid deposits in tissues and organs.

Amylolysis The hydrolysis of starch.

Amylolytic Capable of hydrolyzing starch.

Amylometric Method A method for assaying amylase by determining the amount of starch that is hydrolyzed during the incubation of starch with the enzyme amylase.

Amylopectin A form of starch that consists of α(1,4) and α(1,6) glucosidic linkages.

Amylopectin 6-Glucanohydrolase Synonym of α-dextrin endo-1,6-α-glucosidase.

Amylophosphorylase See Phosphorylase.

Amyloplast Nonpigmented plastide specialized for synthesis and storage of starch or glycogen.

Amylo-Process A process of hydrolysis of starch to sugar by molds such as *Mucor rouxii* and *Rhizopus japonicus*.

Amylopsin Synonym for α-amylase.

Amylose A straight-chain form of starch consisting of α(1,4)-glucosidic bonds between successive glucose units.

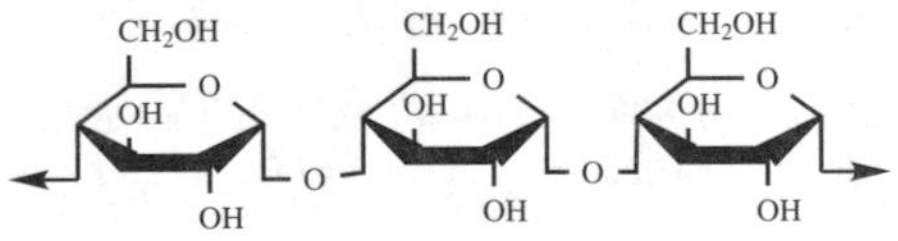

Amylose Synthetase The enzyme that catalyzes the synthesis of amylose from ADP-glucose or UDP-glucose.

Amylosucrase The enzyme that catalyzes the following reaction:

$$\text{Sucrose} + (1,4\text{-}\alpha\text{-D-glucosyl})_n \rightleftharpoons \text{D-fructose} + (1,4\text{-}\alpha\text{-D-glucosyl})_{n+1}$$

Amylovorin An extracellular polysaccharide produced by *Erwinia amylovora* that consists essentially of (1,3)- and (1,6)-β-linked galactan backbone with side chains containing residues of galactose, glucuronic acid, glucose, and pyruvate.

Amylozyme A trade name for a-amylase.

Amylpenicillin Sodium (mol wt 336) An antibiotic produced by *Aspergillus flavus*.

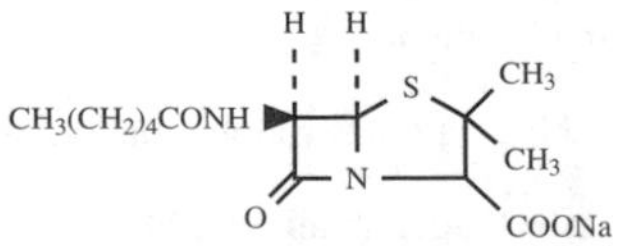

Amytal A trade name for amobarbital, a sedative.

ANA Abbreviation for antinuclear antibody, the antibodies that are directed against nuclear constituents. The presence of these antibodies is an indicator for certain autoimmune diseases, e.g., systemic lupus erythematosus.

ANAA Abbreviation for advanced nucleic acid analyzer.

Anabaena A genus of filamentous *Cyanobacteria.*

Anabasine (mol wt 162) An insecticide.

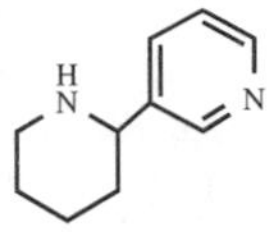

Anabolic Pertaining to anabolism.

Anabolic Pathway Series of reactions leading to the synthesis of one or more specific cellular components.

Anabolic Steroid Synthetic steroids used to promote muscle strength and body growth.

Anabolism A biosynthetic process by which simple substances are converted into more complex compounds.

Anacidity The lack of gastric hydrochloric acid.

Anacin A trade name for a combination drug containing aspirin and caffeine.

Anacin-3 A trade name for acetaminophen, used as an analgesic and antipyretic agent.

Anacobin A trade name for vitamin B_{12} (cyanocobalamin).

Anacystis A genus of cyanobacteria.

Anadrol-50 A trade name for oxymetholone, an anabolic hormone that promotes body tissue-building processes.

Anaerobe An organism capable of growing in the absence of molecular oxygen.

Anaerobic Pertaining to anaerobe.

Anaerobic Digestion The anaerobic breakdown of complex organic materials (e.g., animal and/or plant materials or sewage) to simple substances.

Anaerobic Fermentation Fermentation in the absence of molecular oxygen.

Anaerobic Glycolysis The pathway that converts glucose to lactic acid in the absence of molecular oxygen (also known as glycolysis).

Anaerobic Photosynthetic Bacteria Bacteria that carry out the photosynthetic reactions of photosystem I in the absence of molecular oxygen.

Anaerobic Respiration The energy-yielding metabolic process that uses substances (e.g., fumarate, nitrate, sulfur) other than oxygen as terminal electron acceptors.

Anaerobiospirillum A genus of Gram-negative bacteria (family Bacteroidaceae).

Anaerobiotic Life under anaerobic conditions.

Anaeroplasma A genus of obligately anaerobic, cell-wall-less, sterol-requiring bacteria.

Anaerovibrio A genus of Gram-negative bacteria (family Bacteroidaceae) that occurs in the rumen.

Anafranil A trade name for clomipramine hydrochloride, used as an antidepressant.

Ana-Gard A trade name for epinephrine hydrochloride, bronchodilator and antiglaucoma agent.

Anagrelide (mol wt 256) An antithrombotic agent and platelet aggregation inhibitor.

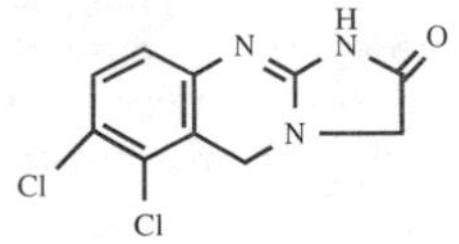

Analbuminemia A metabolic disorder characterized by an impaired synthesis of serum albumin.

Analeptics Drugs that stimulate central nervous system.

Analgesia The relief of pain without loss of consciousness.

Analgesic 1. Relieving pain. 2. A drug that relieves pain.

Analog 1. Chemical compounds that are similar in structure but nonidentical in composition. 2. Structures that are not homologous but have a similar function. 3. Structures that are similar in function and appearance but nonidentical in origin and development.

Analogous Enzyme Variants Enzyme variants that differ significantly in their molecular structures and catalytic properties.

Analogue Variant spelling of analog.

Analytical Biochemistry Biochemistry that deals with the qualitative and quantitative determination of substances in living systems.

Analytical Method Method that deals with the identification and characterization of specific substances (e.g., electrophoresis, analytical centrifugation, and HPLC).

Analytical Ultracentrifuge A high-speed centrifuge, equipped with optical systems, used for analytical analysis (e.g., determination of sedimentation coefficients and molecular weights).

Anamid A trade name for karamycin, an antibiotic.

Anamnesis See Anamestic reaction.

Anamnestic Reaction A heightened immunological response to a previously encountered antigen.

Anaphase A stage in mitosis in which the chromatids of each chromosome separate and move to opposite poles.

Anaphoresis The movement of charged particles or molecules toward the anode.

Anaphylactic Hypersensitivity An IgE-mediated type I hypersensitivity that involves the reaction of allergen with IgE-sensitized mast cells leading to mast cell degranulation; release of bioactive amines (e.g., histamine, serotonin); vasodilation; smooth muscle constriction; or acute asthma, bronchospasm, or death in severe cases. It is also known as immediate-type hypersensitivity.

Anaphylactic Response See Anaphylactic hypersensitivity.

Anaphylactic Shock See Anaphylactic hypersensitivity or type I hypersensitivity.

Anaphylactoid Reaction A clinical response similar to anaphylaxis but not IgE-mediated.

Anaphylatoxin Substances resulting from complement activation (e.g., C3a and C5a) that are capable of causing an anaphylactic reaction.

Anaphylatoxin Inactivator An α-globulin with a molecular weight of about 300,000 that destroys the biological activity of C3a and C5a.

Anaphylaxis See Anaphylactic hypersensitivity or Type I hypersensitivity.

Anaplasia 1. The loss of normal differentiation of cells. 2. The reversion of cells to a more primitive, embryonic state.

Anaplasma A genus of Gram-negative bacteria (family Anaplasmataceae).

Anaplasmosis Any disease or infection caused by tick-transmitted *Anaplasma*.

Anaplastic Pertaining to anaplasia.

Anaplasty Plastic surgery.

Anaplerotic A reaction that replenishes intermediates depleted by other metabolic pathways.

Anapolon A trade name for oxymetholone, an anabolic steroid.

Anaprox A trade name for naproxen sodium, used as an anti-inflammatory agent.

Anasarca A generalized, massive edema in the subcutaneous tissue.

Anaspaz A trade name for hyoscyamine sulfate, used as an anticholinergic agent.

Anastral Mitosis A type of mitosis that occurs in higher plants, characterized by the absence of an astral array of microtubules at the poles of a mitotic spindle.

Anastrozole (mol wt 293) An antineoplastic agent.

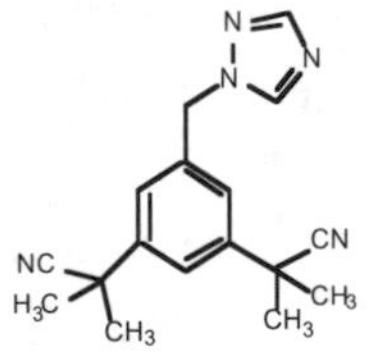

Anatensol A trade name for fluphenazine hydrochloride, used as an antipsychotic agent.

Anatomy The science that deals with structure and organization of plants or animals.

Anatus A combination drug containing phenylpropanolamine and dextromethorphan guaifenesin used as antitussive agent.

Anavar A trade name for oxandrolone, used as an anabolic steroid.

ANC Abbreviation for absolute neutrophil count.

ANCA Abbreviation for antineutrophil cytoplasmic antibodies.

Ancalomicrobium A genus of chemoorganotrophic, facultative anaerobic prosthecate bacteria found in aquatic habitats.

Ancasal A trade name for aspirin, used as an analgesic and antipyretic agent.

Ancef A trade name for cefazoline, an antibiotic that inhibits bacterial cell wall synthesis.

Anchorage-Dependent Cells Cells that will grow, survive, and maintain function only when attached to an inert surface such as glass or plastic.

Anchorage-Independent Cells Cells that are capable of growth and maintaining vital functions in suspension culture, e.g., transformed cells.

Ancitabine (mol wt 225) A cytostatic agent and intermediate in the synthesis of cytarabine. It is also an antineoplastic agent.

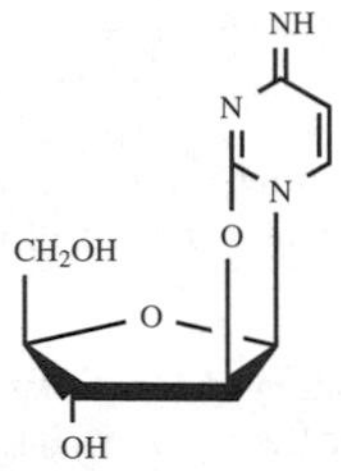

Ancobon A trade name for flucytosine, used as an antifungal agent.

Ancrod A protease from snake venom (e.g., Malaysian pit viper, rattlesnake) that cleaves peptide bonds involving the COOH group of arginine in fibrin. It is an anticoagulant.

Ancymidol (mol wt 256) A plant growth regulator.

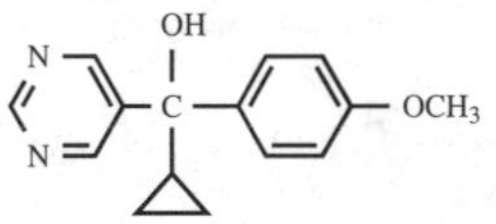

Andersen's Disease A glycogen-storage disease characterized by a genetic deficiency of amylo-(1,4:1,6)-transglucosidase.

Andro- A prefix meaning male.

Androderm A trade name for testosterone, a male hormone.

Androgamone A gamone secreted by the male gamete.

Androgen A general term for male hormones.

Androgenesis A fertilized egg that consists of only paternal chromosomes because of the disintegration of the maternal nucleus before syngamy.

Androgenic Producing masculine characteristics.

Androgenic Hormone Male hormones, such as testosterone.

Androgynous Having the characteristics of both sexes.

Android A trade name for methyltestosterone, a hormone that stimulates target tissue to develop normally in androgen-deficient males.

Androisoxazole (mol wt 329) An anabolic steroid.

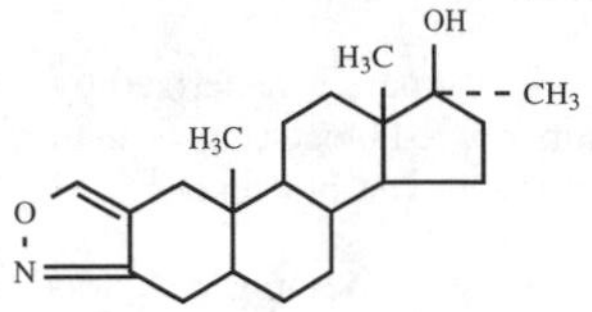

Androlone A trade name for nandrolone phenpropionate, an anabolic steroid.

Andronaq A trade name for testosterone, a male sex hormone.

Andronate A trade name for the hormone testosterone cypionate.

Androphage Male-specific bacteriophage that infects only host cells that contain a conjugative plasmid.

Andropository A trade name for testosterone, a male hormone.

Androstane The parent ring system of androgens.

Androstenediol (mol wt 290) An anabolic steroid.

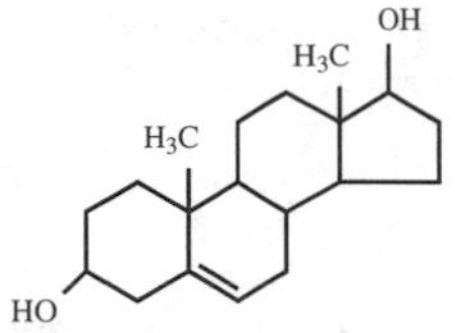

4-Androstene-3,17-Dione (mol wt 286) A male sex steroid hormone.

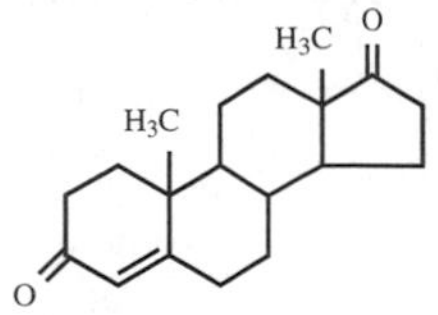

Androsterone (mol wt 290) A male sex hormone.

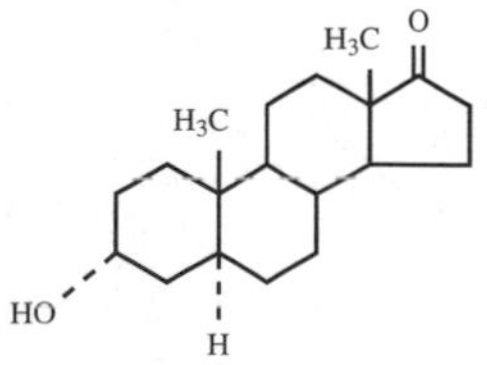

Andrumin A trade name for dimenhydrinate, used as an antiemetic agent.

Andryl A trade name for the hormone testosterone enanthate.

Anectine A trade name for succinylcholine chloride, a depolarizing agent.

Anemia A disorder characterized by a decrease in the number of red blood cells or in the hemoglobin content of the red blood cells.

-anemia A suffix meaning red blood cell deficiency or its remedy.

Anemic Of or pertaining to anemia.

Anemic Anoxia A disorder characterized by a deficiency of oxygen in body tissues.

Anemonin (mol wt 192) An antibiotic derived from *Anemone pulsatilla*.

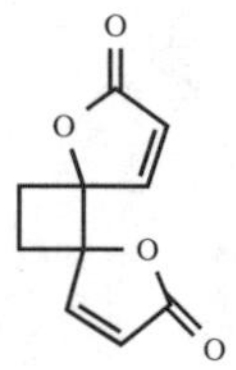

Anergan A trade namne for promethazine hydrochloride, used as an antihistaminic agent.

Anergy 1. Lack of an expected immune response. 2. Inability to react.

Anestacon A trade name for lidocaine hydrochloride, an antiarrhythmic agent.

Anesthesia The loss of sensation induced by an anesthetic substance or by hypnosis.

Anesthesiology The science that deals with anesthetics and their effects on organisms.

Anesthetic A substance that produces anesthesia.

Anesthetic Drug A drug that induces either a local or a total loss of sensation in the body.

Anethole (mol wt 148) Chief constituent of anise, star anise, and fennel oil. It has been used as a carminative agent.

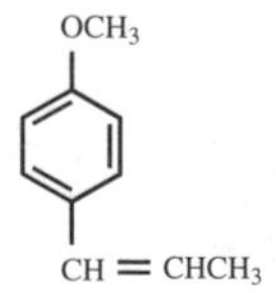

Anethole Trithione (mol wt 240) A cholinergic agent.

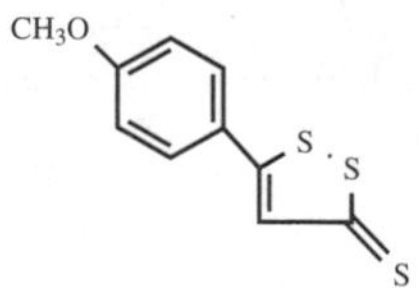

Aneucentric Referring to an aberration that generates a chromosome with more than one centromere.

Aneuploidy An aberration in which the chromosome number of an individual is not an exact multiple of the haploid set for the species.

Aneurysm 1. A blood-filled, saclike dilatation of the wall of an artery. 2. A saclike dilatation.

Aneusomatic Referring to an organism whose cells contain variable numbers of individual chromosomes.

ANF Abbreviation for 1. Atrial natriuretic factor. 2. Antinuclear factor. 3. Alpha naphthoflavone.

αNF Abbreviation for alpha naphthoflavone.

Ang-I Abbreviation for angiotensin-I.

Ang-II Abbreviation for angiotensin-II.

Angelic Acid (mol wt 100) Substance from sunbul root (*Angelia archangelica*).

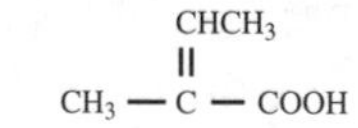

Angiitis An inflammation of blood vessels or lymphatic vessels.

Angina A disorder marked by severe strangling pain or choking feeling.

Angina Pectoris A cramping pain in the chest due to an oxygen deficiency in the heart.

Angio- A combining form meaning pertaining to a vessel (usually a blood vessel).

Angioblast Cells associated with the formation of blood cells and blood vessels.

Angioblastoma A tumor of blood vessels in the brain.

Angiocardiography X-ray examination of the heart and the vessels of the heart following injection of an opaque medium.

Angioedema An allergic reaction characterized by the localized swelling of the skin and subcutaneous tissues.

Angiofibroma An angioma containing fibrous tissue.

Angiogenesis The process of blood vessel formation.

Angiogenin A protein from human adenocarcinoma that induces angiogenesis.

Angiography The X-ray visualization of the internal anatomy of the heart and blood vessels following the intravascular administration of radiopaque contrast medium.

Angioid Resembling a blood vessel or lymphatic vessel.

Angioma Any benign tumor consisting primarily of blood vessels or lymph vessels.

-angioma A combining form meaning a tumor composed chiefly of blood and lymphatic vessels.

Angiomatosis A disorder characterized by the presence of numerous vascular tumors.

Angiomyoma A tumor composed of vascular and muscular tissue elements.

Angiomyosarcoma A tumor containing vascular, muscular, and connective tissue elements.

Angioneurotic Anuria A disorder characterized by an almost complete absence of urination caused by the destruction of tissue in the renal cortex.

Angioplasty A medical procedure to enlarge the narrowed or obstructed arteries.

Angiopoietic Causing the formation of blood vessels.

Angiospasma Spasmodic contraction of a blood vessel.

Angiostenosis Constriction or narrowing of a blood vessel.

Angiotensin A vasoconstrictor peptide derived from plasma globulin by the action of renin.

Angiotensin I The inactive decapeptide precursor of angiotensin II. It is produced from angiotensinogen by the action of the enzyme renin.

Angiotensin I Converting Enzyme The enzyme that converts angiotensin I to angiotensin II.

Angiotensin II The active octapeptide formed from angiotensin I by hydrolytic removal of two amino acids in a reaction catalyzed by angiotensin I converting enzyme.

Angiotensinase A Synonym of glutamyl aminopeptidase.

Angiotensinase C Synonym of lysosomal Pro-x carboxypeptidase.

Angiotensin Forming Enzyme Synonym of renin.

Angiotensinogen The α_2-globulin from which the decapeptide angiotensin I is cleaved in a reaction catalyzed by the enzyme renin.

Angiotensinogenase Synonym of renin.

Angiotonin See angiotensin.

Angle Head Rotor A fixed-angle centrifuge rotor used to hold centrifuge tubes during the centrifugation.

Angle Rotor See Angle head rotor.

Angstrom (Å) Unit of length: 1 Å = 0.1 nm = 10^{-4} micron = 10^{-7} mm = 10^{-8} cm.

Angular Methyl Group A methyl group linked to the perhydrocyclopentanophenanthrene ring system of steroids.

Angular Velocity The velocity of rotation expressed in terms of radians per unit of time (the central angle transversed per unit time during the course of rotation). One revolution equals 2π radians.

Anhaptoglobinemia The lack of sufficient amounts of haptoglobins in the blood.

Anhidrosis A disorder characterized by inadequate perspiration or the failure of the sweat gland to function.

Anhidrotic 1. Of or pertaining to anhidrosis. 2. An agent that reduces or suppresses sweating.

Anhydrase The enzyme that catalyzes the removal of water from a compound.

Anhydride A chemical compound formed by removal of one or more molecules of water from an acid or base.

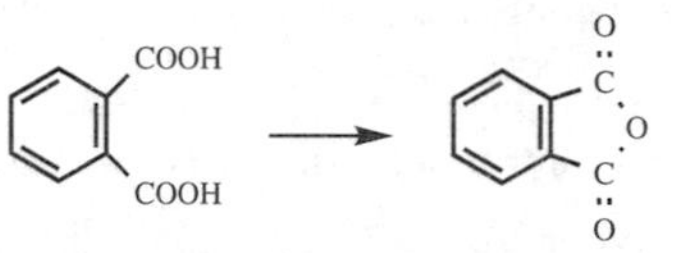

Anhydrobiosis Referring to an organism that can sustain the removal of cellular water and return to normal living upon addition of water.

Anhydrous Devoid of water.

Anicteric Absence of yellow discoloring of the skin and eyes (jaundice).

Anicteric Hepatitis A mild form of hepatitis in which there is no jaundice.

Anilazine (mol wt 276) A fungicide.

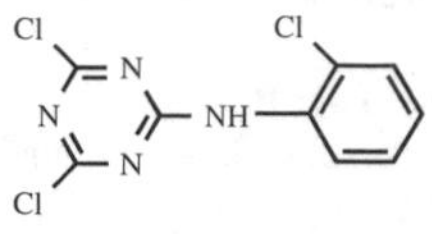

Anileridine (mol wt 352) A narcotic, analgesic agent.

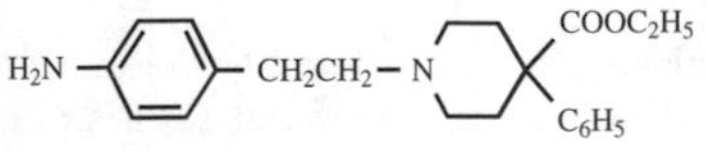

Aniline (mol wt 93) A poisonous, oily aminobenzene and a cancer suspectant.

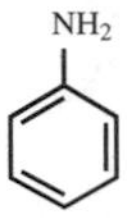

Aniline Dyes Any dye derived from aniline.

Aniline Mustard (mol wt 218) A useful substance in cancer research.

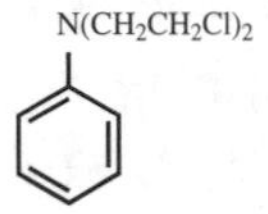

1-Anilino-8-Naphthalenesulfonate (mol wt 299) A hydrophobic fluorescent probe and visualization reagent for proteins.

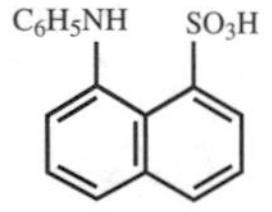

Animal Starch Referring to glycogen.

Animal viruses Viruses that infect and replicate in the cells of animals.

Anion A negatively charged ion or molecule.

Anion Exchanger A positively charged ion-exchange resin that binds and exchanges anions.

Anion Gap The difference between the concentrations of cations and anions in the serum.

Anion Transport Protein An integral membrane protein of the erythrocyte responsible for transport of anions across the membrane.

Anionic Detergent A surface-active agent that carries a negative charge.

Aniracetam (mol wt 219) A nootropic agent.

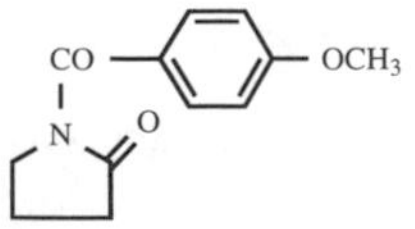

Anise The dried ripe fruit of the *Pimpinella anisum* plant (Umbelliferae). Extract of anise used in the preparation of carminatives and expectorants.

Aniseikonia An abnormal ocular condition in which each eye perceives the same image as being of a different size.

Anisindione (mol wt 252) An anticoagulant.

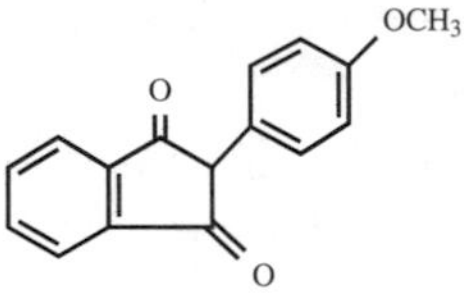

Anisochromasia The color variation of erythrocytes in which the central portion of the red blood cell is almost colorless due to the unequal distribution of hemoglobin.

Anisocytosis A blood disorder characterized by abnormal variation in size of the red blood cells.

Anisogamates Gametes differing in shape, size, and/or behavior.

Anisogamy The presence or conjugation of gametes that are of unequal size and structure.

Anisokaryosis Variation in the size of the nucleus of a cell of the same general type.

Anisometropia A disorder characterized by a difference in the refractive powers of the two eyes.

Anisomycin (mol wt 265) An antibiotic isolated from *Streptomyces griseolus* and *S. roseochromogenes* that inhibits protein synthesis through its inhibitory action on peptidyl transferase.

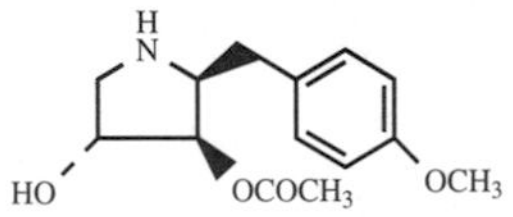

Anisopoikilocytosis A disorder of the blood characterized by the presence of the abnormal shape and large size of red blood cells.

Anisotropic Having different properties along different axes.

Anisotropic Inhibitor An agent that inhibits energy transduction in oxidative phosphorylation by binding to certain negatively charged sites on the outer surface of the mitochondrial inner membrane.

Anisotropine Methylbromide (mol wt 362) An anticholinergic agent.

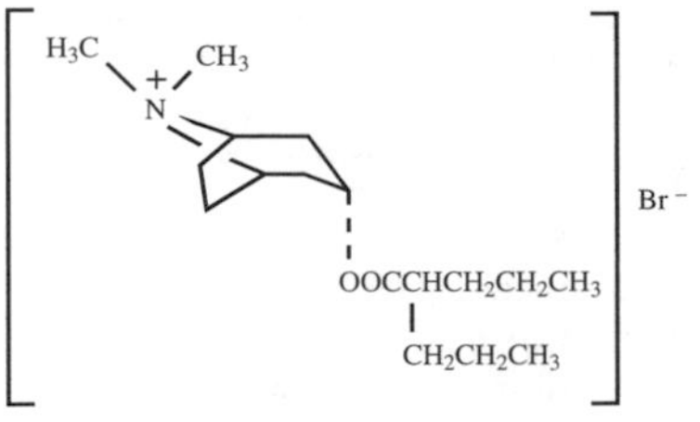

Anisotropy The variation in the physical properties of a substance as a function of the direction in which these properties are measured.

Anistreplase An anisoylated plasminogen streptokinase activator complex with thrombolytic enzyme activity. It consists of streptokinase and human plasminogen.

ANIT Abbreviation for alpha naphthylisothiocyanate.

αNIT Abbreviation for alpha naphthylisothiocyanate.

Ankyrin A peripheral membrane protein of the red blood cell that links the spectrin to the anion transport protein.

Anlage The undifferentiated cells from which a particular organ, tissue, or structure develops.

ANLL An abbreviation for acute nonlymphocytic leukemia.

Annealing The process in which two separate strands of nucleic acid interact to form a duplex molecule.

Annexin A Ca^{++}-dependent phospholipid-binding protein. There are different types of annexin, e.g., annex I, II, III, IV, V and VI.

Anode The positive terminal of an electric cell to which the negatively charged molecules or ions are migrating.

Anodic Of or pertaining to the anode.

Anolor DH A trade name for a combination drug containing hydrocodone bitartrate and acetaminophen used as an analgesic agent.

Anomer Referring to carbohydrate stereoisomers that differ from each other only in configuration at the anomeric carbon of the ring structure (e.g., α-D-glucose and β-D-glucose).

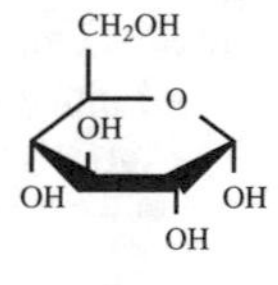

α-D-glucose

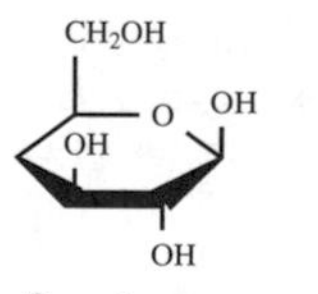

β-D-glucose

Anomeric Carbon The carbon atom of the carbonyl group in a carbohydrate with open chain projection.

Anonymous DNA Segment DNA segment of unknown function.

Anoquon A trade name for a combination drug containing butabarbital, acetaminophen and caffeine used as a non-narcotic analgesic agent.

Anorexia Lack of appetite.

Anorexiant Substances that induce loss of appetite.

Anosmia Lack of sense of smell.

Anovulation Failure of the ovaries to produce and release mature eggs.

Anoxemia A deficiency in blood oxygen content.

Anoxia A condition characterized by lack of oxygen in the tissue.

Anoxybiontic Incapable of using molecular oxygen for growth.

Anoxygenic Not producing oxygen.

Anoxygenic Photosynthesis Photosynthesis that does not split water and produce oxygen.

Anoxyphotobacteria Bacteria that can carry out only anoxygenic photosynthesis.

ANP Abbreviation for atrial natriuretic peptide.

AnPL Abbreviation for anionic phospholipid.

ANS Abbreviation for 1. 8-Anilino-1-naphthalene sulfate. 2. Autonomic nervous system.

Ansaid A trade name for flurbiprofen, used as an anti-inflammatory agent.

Ansamycins A group of structurally related antibiotics that consists of a macrocyclic ring (e.g., rifamycins, streptovaricins, and tolypomycins).

Anserinase The enzyme that hydrolyzes anserine, carnosine, homocarnosine, glycyl-leucine, and other dipeptides.

Anserine (mol wt 240) A dipeptide found in skeletal muscle of geese and other animals but not in man.

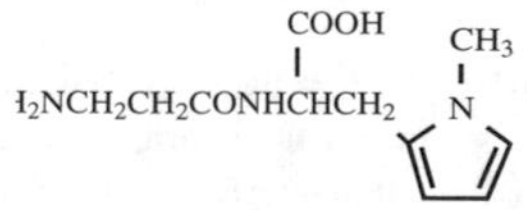

Anspor A trade name for cephradine, an antibiotic.

ANT Abbreviation for adenine nucleotide translocase.

Antabuse A trade name for disulfiram, which blocks the oxidation of alcohol at the acetaldehyde stage and is used as an alcohol deterrent.

Antacid An agent (e.g., drug or dietary substance) that buffers, neutralizes, or absorbs hydrochloric acid in the stomach.

Antadine A trade name for amantadine hydrochloride, used as anviral agent.

Antagonism The killing, injury, or inhibition of growth of one species of microorganism by another.

Antagonist A molecule, such as a drug, enzyme inhibitor, or hormone, that diminishes or prevents the action of another molecule.

Antamanide A cyclic peptide that acts as an antidote for phallatoxins and anatoxins.

Antazoline (mol wt 265) An antihistaminic agent.

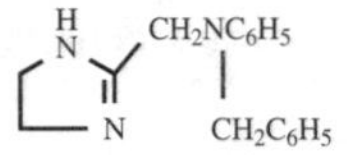

Ante-Iso Fatty Acid A fatty acid that is branched at the carbon atom preceding the penultimate carbon atom at the hydrocarbon end of the molecule.

Antenna Complex The light-harvesting pigment-protein complex in the photosynthetic organisms that funnel the absorbed incident radiation to the photochemical reaction centers.

Antepar A trade name for piperazine citrate, an anthelmintic agent.

Ante-Penultimate Carbon The third carbon atom from the end of a chain.

Anterior The front or ventral, the front of a structure.

Anterior Chamber-Associated Immune Deviation Selective suppression of delayed (Type IV) hypersensitivity reactions within the eye. Its occurrence can lead to adverse effects or infections on the visual system.

Anterior Pituitary Front part of the pituitary gland. It secretes a number of hormones that regulate other endocrines.

Anthel A trade name for pyrantel embonate, an anthelmintic agent.

Anthelmintic 1. A substance that destroys or prevents the development of parasitic worms. 2. An anthelmintic drug that interferes with the parasites' metabolism, blocks their neuromuscular action, or renders them susceptible to the destruction by the host's macrophages.

Antheraxanthin A type of xanthophyll found in green algae.

Antheridiol (mol wt 471) A plant sex hormone that controls and regulates plant fertility.

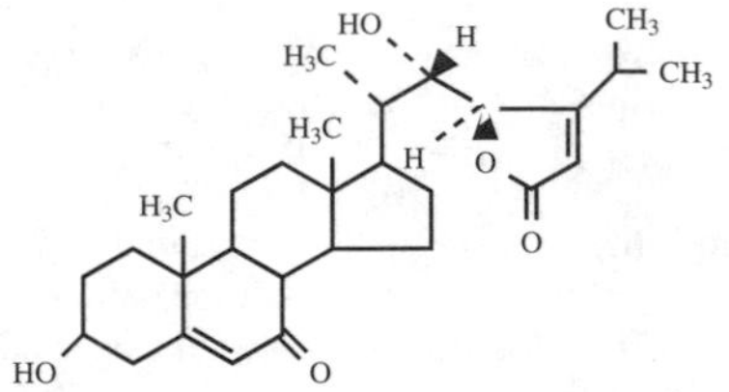

Anthiolimine (mol wt 605) An anthelmintic agent.

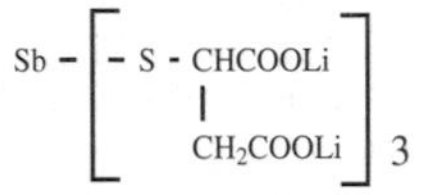

Anthocyanin Water-soluble plant flavonoid pigments responsible for red, pink, purple, and blue colors of higher plants.

Antho-K Amide A neuropeptide from the sea anemone *Anthopleura elegantissima.*

Anthracene (mol wt 178) A tricyclic aromatic compound used in scintillation cocktails.

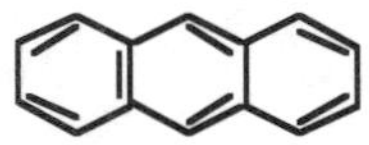

Anthracosis A chronic lung disorder caused by the deposit of coal dust in the lungs and characterized by the formation of black nodules on the bronchioles.

Anthracycline Antibiotics Antibiotics that have a tetrahydrotetracenequinone chromophore and function as intercalating agents that inhibit DNA and RNA synthesis.

Anthralin (mol wt 226) A topical antipsoriatic and antifungal agent.

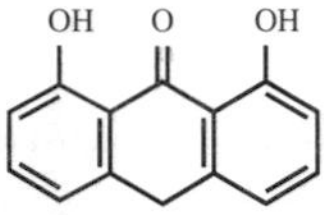

Anthramycin (mol wt 315) An antibiotic from *Streptomyces refuineus* that possesses antibacterial and antitumor activity. It binds covalently to nucleic acids and causes inhibition of DNA and RNA synthesis.

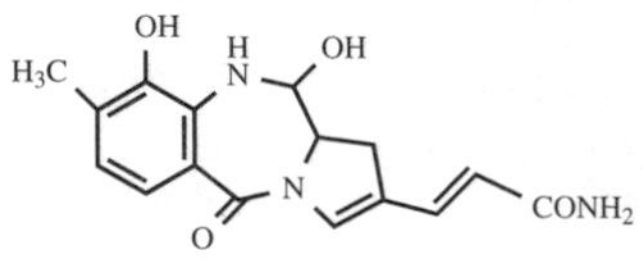

Anthranilate Adenyltransferase The enzyme that catalyzes the following reaction:

ATP + anthranilate
⇅
PPi + N-adenylanthranilate

Anthranilate N-Benzoyltransferase The enzyme that catalyzes the following reaction:

Benzoyl-CoA + anthranilate
⇅
CoA + N-benzoylanthranilate

Anthranilate CoA Ligase The enzyme that catalyzes the following reaction:

ATP + anthranilate + CoA
⇅
AMP + PPi + anthranilyl-CoA

Anthranilate 1,2-Dideoxygenase The enzyme that catalyzes the following reaction:

Anthranilate + NADPH + O_2 + H_2O
⇅
Catechol + CO_2 + $NADP^+$

Anthranilate 3-Monooxygenase The enzyme that catalyzes the following reaction:

Anthranilate + tetrahydrobiopterin + O_2
⇅
3-Hydroxyanthranilate + hydrobiopterin + H_2O

Anthranilate N-Methyltransferase The enzyme that catalyzes the following reaction:

S-Adenosyl-L-Methionine + anthranilate
⇅
S-Adenosyl-L-homocysteine + N-methylanthranilate

Anthranilate Phosphoribosyltransferase The enzyme that catalyzes the following reaction:

N-(5-Phospho-D-Ribosyl)-Anthranilate + PPi
⇅
Anthranilate + 5-phosphoribose 1-diphosphate

Anthranilate Synthetase The enzyme that catalyzes the following reaction:

Chorismate + L-glutamine
⇅
Anthranilate + pyruvate + L-glutamate

Anthranilic Acid (mole wt 137) An intermediate in tryptophan biosynthesis.

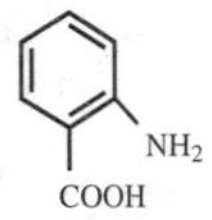

Anthrarobin (mol wt 226) A parasiticide.

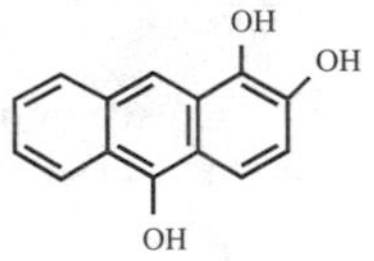

Anthrax An animal disease caused by *Bacillus anthracis*.

Anthrax Toxin A plasmid-encoded toxin, produced by *Bacillus anthracis*, consisting of oedema factor (factor I), protective antigen (factor II), and lethal factor (factor III). Each component alone has no toxic activity.

Anthrone (mol wt 194) A compound used for the colorimetric determination of carbohydrate.

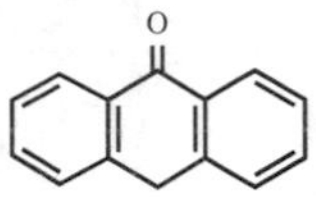

Anthrone Reaction A colorimetric reaction for carbohydrates, particularly hexoses, that is based on the production of a green color on the treatment of the sample with anthrone.

Anthropophilic Referring to a parasite or pathogen that preferentially infects humans.

Anti- A prefix meaning against, opposing, or counteracting.

Anti-AChR Abbreviation for anti-acetylcholine receptor.

Antiadrenergic 1. Substance that blocks the sympathetic or adrenergic activity. 2. Any substance that blocks the action of norepinephrine or the conduction at adrenergic nerve terminals.

Antiamebic Destructive or suppressive to amoebas.

Antianemic 1. Preventing or correcting anemia. 2. Any agent capable of preventing or correcting anemia, e.g., vitamen B_{12}.

Antianemic Factor Referring to vitamin B_{12}.

Antianginal 1. Capable of preventing or relieving angina pectoris by enlarging arteries of the heart and improving blood flow. 2. A drug that is capable of preventing or relieving angina pectoris.

Antiantibody An antibody directed against an antigenic determinant on an antibody molecule.

Antiantidote Substance that blocks the action of an antidote.

Antianxiety Capable of preventing or relieving anxiety.

Antiarachnolysin An antitoxin that blocks the action of arachnolysin (a hemolytic compound from spider venoms).

α-Antiarin (mol wt 567) A cardioactive glycoside isolated from the latex of the upas tree (*Antiaris toxicaria*). It causes vomiting, evacuation, protration, and death.

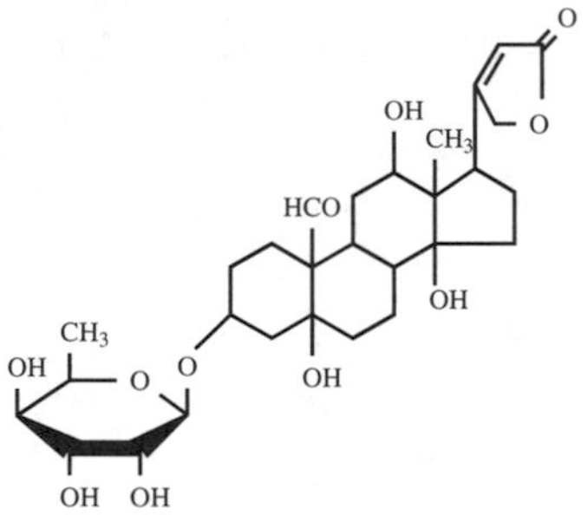

Antiarose (mol wt 164) An isomer of 6-deoxyglucose, a component in some cardiac glycosides.

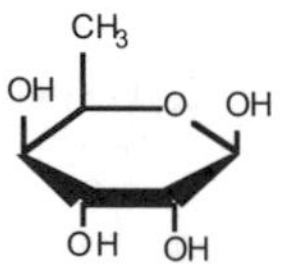

Antiarrhythmic 1. Capable of preventing, relieving or correcting heart rhythm problems. 2. A substance capable of suppressing, preventing, relieving or correcting heart rhythm problems.

Antiasthmatic Agent A drug that is used for treatment of asthma.

Antiauxin A substance that functions as a competitive inhibitor of auxin.

Antibacterial An action or substance that kills bacteria or inhibits their growth or replication.

Antibechic Substance having the property of relieving or curing a cough.

Antiberiberi Factor Any substance that has antiberiberi characteristics (e.g., thiamin).

Antibiotic A compound produced by a microorganism that kills or inhibits the growth of other microorganisms at low concentration. It includes natural, semisynthetic, or wholly synthetic antimicrobial substances.

Antibiotic Ear A trade name for a combination drug containing neomycin, polymyxin and hydrocortisone used for treatment of eye and ear infection.

Antibody A specific glycoprotein (immunoglobulin or Ig) produced by the immune system of the vertebrate in response to the exposure to an antigen and capable of reacting specifically with that antigen.

Antibody Binding Fraction Referring to the Fab fragment of an immuoglobulin molecule.

Antibody Combining Site The region on an antibody molecule that is capable of binding with a corresponding antigenic determinant.

Antibody Dependent Cell-Mediated Cytotoxicity A form of cell-mediated immunity in which immune effector cells kill the antibody-coated target cells.

Antibody Dependent Cytotoxic Hypersensitivity See antibody-dependent cell-mediated cytotoxicity (also known as type II hypersensitivity).

Antibody Diversity See Antibody heterogeneity.

Antibody Excess The high antibody-to-antigen ratio in an antigen-antibody reaction mixture.

Antibody Fixation Binding of antibody to a cell receptor.

Antibody Heterogeneity A population of antibodies to a given antigen that differ in size, structure, charge, and immunological properties.

Antibody-Mediated Immunity See Humoral immunity.

Antibody Repertoire The total collection of the antibodies available to an organism that are capable of recognizing virtually any foreign substance it will ever encounter.

Antibody Titer The highest dilution of an antiserum that will produce a detectable antigen-antibody reaction (e.g., precipitation, agglutination).

Antibody Valence The number of combining sites per antibody molecule. Most antibody molecules have a valence of two (two combining sites per antibody molecule).

Anticarcinogen Any substance possessing anticarcinogenic activity.

Anticephalagic Any substance capable of relieving a headache.

Anti-C-Gal-IgG Abbreviation for antibody against C-terminal sequence of α–galactosidase.

Anticholinergic 1. Counteracting the action of acetylcholine. 2. The blockage of acetylcholine receptors.

Anticholinergic Agent Any substance that is capable of competing with the acetylcholine for its receptor sites at the synaptic junctions.

Anticholinesterase An inhibitor of cholinesterase.

Anticoagulant A substance capable of preventing blood clotting.

Anticodon A trinucleotide sequence in the anticodon loop of a tRNA molecule that is complementary to the trinucleotide sequence (codon) on the mRNA molecule.

Anticodon Arm See Anticodon loop.

Anticodon Deaminase The enzyme that catalyzes the deamination of adenine that occurs at the first position (5′-end) of an anticodon.

Anticodon Loop The loop or arm on the tRNA that possesses anticodon.

Anticomplement Any substance or agent capable of inactivating a component of complement.

Anticomplement Fluorescent Antibody Technique A fluorescent antibody technique used for the detection of complement-Ag-Ab complexes by incubation with fluorescent-labeled antibody against complement.

Anticonvulsant An agent capable of preventing or controlling a convulsion.

Antidementia Drug A drug that slows the progression of Alzheimer's disease and related forms of mental deterioration.

Antidepressant A substance capable of preventing and relieving psychic depression.

Antidermatosis Vitamin Referring to pantothenic acid.

Antidiabetic Agent A drug used for treatment of diabetes mellitus.

Antidiarrheal Agent A drug used for treatment of diarrhea.

Antidiuresis A decrease in urine excretion.

Antidiuretic Any substance that decreases the excretion of urine.

Antidiuretic Hormone (ADH) A hormone (e.g., vasopressin) that decreases the production of urine by increasing the reabsorption of water by the renal tubules.

Anti-DNA Abbreviation for antibody against DNA.

Anti-DNase Abbreviation for antibody against DNase.

Anti-DNase B Abbreviation for antibody against DNase B.

Antidote An agent or substance capable of counteracting the action of a poison.

Anti-dsDNA Abbreviation for antibody against double-stranded DNA.

Antidysenteric Agent or substance that is effective against dysentery.

Antieczematic An agent used to treat eczema.

Antiedemic Agent capable of controlling edema.

Antiemetic Agent or substance that is capable of suppressing vomiting.

Anti-ENA Abbreviation for antibody against extractable nuclear antigen.

Antienzyme 1. A substance capable of inhibiting enzyme activity. 2. Antibody to an enzyme.

Antiepileptic Substance or drugs capable of preventing or suppressing epilepsy.

Antieurodonic Agent capable of preventing dental cavities.

Antifebrile Synonym for antipyretic.

Antifolate An antimetabolite of folic acid or an inhibitor of dihydrofolate reductase.

Anti-Freeze Protein A glycoprotein present in arctic and antarctic fish. It depresses the freezing point of water by inhibiting the formation of water crystals.

Antifungal Any substance capable of killing fungi or inhibiting their growth.

Antigalactic Reducing the secretion of milk.

Anti-GBM Abbreviation for antibody against glomerular basement membrane.

Antigen Any substance that is capable of eliciting the formation of antibodies in a vertebrate host or generating a specific population of lymphocytes reactive with that substance. Antigens are macromolecules (e.g., proteins and polysaccharides) that are foreign to the vertebrate host.

Antigen Antibody Complex The molecular complex formed from the reaction between an antigen and its complementary antibody molecule.

Antigen Antibody Reaction The interaction between an antigen and its complementary antibody.

Antigen Binding Cell A cell capable of binding specific antigen, e.g., T lymphocyte.

Antigen Binding Site That region of an antibody molecule, or T cell receptor, that binds an antigenic determinant.

Antigen Deletion The loss of a particular antigenic determinant due to a mutation or loss of a plasmid.

Antigen Dependent Differentiation The differentiation of a lymphocyte to effector cells that depends upon the presence of antigen or binding of antigen to the cell surface receptor (e.g., differentiation of B cell to plasma cell).

Antigen Determinant The small three-dimensional configuration on the surface of the antigen molecule that elicits the production of antibody.

Antigen Excess The high antigen-to-antibody ratio in an antigen-antibody mixture. Under this condition the antigen-antibody complexes formed are small and soluble.

Antigen Excess Zone A zone in the antigen-antibody reaction in which antigen is in excess.

Antigen Gain The formation of new antigenic determinant(s) due to a mutation or the acquisition of a plasmid.

Antigen Independent Differentiation Differentiation of T or B cells from lymphocyte precursors in the absence of antigen.

Antigen Presenting Cell A cell that is capable of presenting antigen to lymphocytes (e.g.,T cells and B cells) for initiation of an immune response.

Antigen Receptor Membrane-bound molecule on the surface of T lymphocyte, capable of binding of antigen.

Antigen Tolerance See Immunological tolerance.

Antigen Valence The number of antigenic determinants per antigen molecule.

Antigenic Competition The decrease in the immune response to one antigen due to the presence of a second antigen.

Antigenic Conversion 1. The change of bacterial cell surface antigen due to the presence of prophage. 2. The disappearance of antigenic determinants caused by the presence of antibody.

Antigenic Determinant See Antigen determinant or epitope.

Antigenic Drift Changes in antigenic specificity that occur over an extended period of time.

Antigenic Mimicry The acquisition or production of host antigens by a parasite, enabling it to escape the detection by the immune system of the host.

Antigenic Modulation 1. The alteration of antigenic sites on a cell surface due to the presence of bound antibody. 2. The suppression of cell surface antigen by the presence of homologus antibody.

Antigenic Variation The successive changes in cell surface antigens that give rise to an extensive range of antigenically distinct forms.

Antigenicity The ability of an antigen to elicit the formation of specific antibodies.

Anti-Glaucoma Agent A drug used to treat glaucoma, a disorder due to the buildup of excessive pressure in the eye.

Antiglobulin Antibody homologous to the antigenic determinants of serum γ-globulins.

Antiglobulin Test A test to detect immunoglobulin (Ig) by exploiting the ability of antiglobulin to agglutinate Ig-bound antigens or Ig-bound cells.

Antigonadotropic Agent capable of inhibiting the action of gonadotropin.

Antigonadotropin An agent that inhibits the action of gonadotropin.

Antigonorrheic An agent or substance used to treat gonorrhea.

Anti-Gout Agent A drug used for treatment of gout.

Anti-Gray-Hair Factor Referring to ρ-aminobenzoic acid.

Anti-HAA Abbreviation for antibody against hepatitis-associated antigen.

Anti-HAV Abbreviation for antibody against hepatitis A virus.

Anti-HBc Antibody to hepatitis B core protein.

Anti-HBV Abbreviation for antibody against hepatitis B virus.

Anti-HBs Antibody to hepatitis B surface antigen.

Anti-HBsAg Abbreviation for antibody against hepatitis B surface antigen.

Antihemagglutinin Any substance capable of inhibiting the action of hemagglutinin.

Antihemolysin Any substance capable of inhibiting the action of hemolysin.

Antihemolytic An agent capable of preventing hemolysis.

Antihemophilic Factor A protein factor participating in the cascade reaction that leads to blood coagulation (also known as factor VIII). It is encoded by a gene located at the end of the long arm of the X chromosome.

Antihemophilic Globulin See Antihemophilic factor.

Antihemorrhagic Agent capable of preventing bleeding.

Antihemorrhagic Vitamin Referring to vitamin K. Deficiency in vitamin K causes prolonged clotting time.

Antihidrotic Antiperspirant that is capable of preventing sweat excretion.

Antihistamine Any substance capable of reducing the physiological and pharmacological effects of histamine and of suppressing allergic symptoms.

Antihormone An antibody to a hormone.

Antihydrophic Agent capable of preventing the accumulation of fluid in the tissue.

Antihypercholesterolemic Agent capable of preventing elevation of the serum concentration of cholesterol.

Antihyperglycemic Agent capable of reducing blood glucose concentration.

Antihyperlipidemic Substance capable of preventing accumulation of lipid in the blood.

Antihypertensive Action or agent capable of reducing blood pressure.

Antihypotensive A drug or agent that elevates blood pressure in the individuals who have dangerously low blood pressure.

Anti-I Abbreviation for antibody against blood group antigen I.

Anti-idiotype Antibody to an idiotypic determinant (variable region domain) of an immunoglobulin molecule. It contains antibodies homologous to each of the individual determinants (idiotypes).

Anti-Immunoglobulin Antibody produced against foreign antibody.

Anti-infective Vitamin Referring to vitamin A.

Anti-inflammatory Action or agent capable of reducing inflammation.

Antiketogenic A substance capable of preventing acidosis or ketosis.

Antilipidemic Agent capable of reducing the amount of lipid in the serum.

Antilirium A trade name for physostigmine salicylate, used as an adjunct to anesthesia. It inhibits the destruction of acetylcholine released from parasympathetic and somatic efferent nerves.

Antilithic Agent A drug that promotes excretion of calcium and prevents the formation of calculi

Antiluretic Antisyphilitic.

Antilymphocyte Globulin The globulin fraction of an antilymphocyte serum.

Antilymphocyte Serum Serum which contains antibodies that are directed against lymphocytes.

Antimalarial Agent capable of destroying or suppressing the development of malaria.

Anti-Manic Drug A drug that relieves the mental and physical hyperactivity and incapacitating mania and mood elevation.

Antimetabolite A substance that possesses the structural similarity with a natural metabolite and competively inhibits the utilization of natural metabolites and exogenous substrates by an organism.

Antimicrobial Action or agent capable of killing or inhibiting the multiplication of microbes.

Antimicrobial Agents Chemical or biological agents that kill or inhibit the growth of microorganisms.

Antimicrobics Antimicrobial agents.

Antimigraine Drug A drug that relieves or prevents migraine headaches.

Antiminth A trade name for pyrantel pamoate, an anthelmintic agent that blocks neuromuscular action and paralyzes the worms.

Antimitochondrial Antibody Antibody against mitochondria.

Antimitotics Agents that inhibit mitosis.

Antimony Chemical element (Sb) with atomic weight 121, valence 3.

Antimony Poisoning Poisoning caused by ingestion or inhalation of antimony or antimony compounds; characterized by vomiting, sweating, diarrhea, and a metallic taste in the mouth.

Antimuscarinic Action or agent capable of inhibiting a mascarinic effect.

Antimutagen A substance that counteracts the action of a mutagen.

Antimycin A Antibiotics produced by *Streptomyces* species that inhibit the electron transport system between cytochrome-b and c_1.

Antimycin A_1 (mol wt 549) An antibiotic produced by *Streptomyces* species.

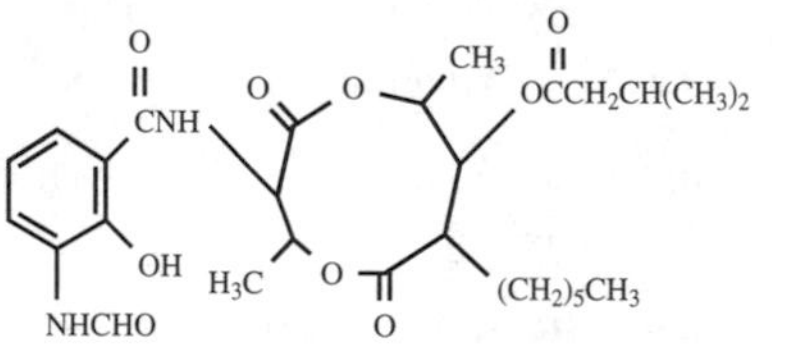

Antimycin A_3 (mol wt 521) An antifungal antibiotic produced by *Streptomyces blastmyceticus*.

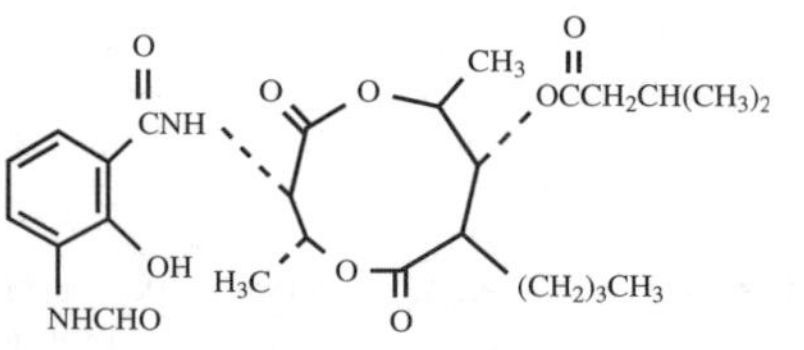

Antimycotic Antifungal.

Antinatriuresis The reduction of urinary excretion of sodium.

Antinaus 50 A trade name for promethazine hydrochloride, an antihistaminic agent.

Antinauseant Agent capable of preventing nausea.

Antineoplastic The action or agent capable of preventing proliferation of malignant cells.

Antineoplastic Hormone A hormonal substance produced by an endocrine gland or a synthetic analog of the naturally occcurring hormone used for cancer chemotherapy.

Antinephritic Action or agent capable of counteracting kidney inflammation.

Antineuritic Agent capable of preventing or relieving neuritis.

Antineuritic Factor Referring to vitamin B_1 or thiamine.

Antineuritic Vitamin Referring to thiamine or vitamin B_1.

Antinuclear Antibodies (ANA) Antibodies that are directed against nuclear constituents. The presence of such antibodies is an indicator of certain diseases, e.g., systemic lupus erythematosus and rheumatoid arthritis.

Antiobesity Agent A drug or agent that promotes weight loss.

Antioxidant 1. A substance that is capable of preventing oxidation. 2. A substance added to foods to inhibit oxidation of the food.

Antiparallel Two chains running in opposite directions, e.g., antiparallel strand arrangement in DNA.

Antiparallel Pleated Sheet The polypeptide chains in a pleated sheet running in opposite directions.

Antiparallel Spin The spin of two particles in opposite directions.

Antiparallel Strands Two polynucleotide strands running in opposite directions, one strand progressing from the 3′-terminal to the 5′-terminal, and the other progressing from the 5′-terminal to the 3′-terminal direction.

Antiparasitic Action or agent capable of inhibiting or killing parasites.

Anti-Parkinsonism Agent A drug used to relieve trembling and rigidity of Parkinson's disease and related disorders.

Antipediculotic Action or agent capable of preventing or inhibiting pediculosis.

Antipellagra Factor Referring to nicotinic acid.

Antiperistaltic Action or agent capable of inhibiting or diminishing peristalsis.

Antipernicious Anemia Factor Referring to vitamin B_{12}.

Antiperspirant Action or agent capable of inhibiting sweating.

Antiphthisic Action or agent capable of preventing tuberculosis.

Antiplasmin 1. Inhibitor for the enzyme plasmin. 2. Antibody to plasmin.

Antiplasmodial Destructive to plasmodia.

Antiplastic Action or agent capable of slowing cellular division or minimizing cicatrix formation.

Antiplatelet Agents capable of reducing the number of platelets in the blood.

Antipodagric Action or agent used to treat gout.

Antiport Coupled transport; a transport system that transports two solutes across a membrane in opposite directions.

Antiproliferative Agent A drug that suppresses the excess proliferation of cells.

Antipromoter Referring to 1. a substance that counters the action of a promoter factor in carcinogenesis or 2. a substance capable of preventing attachment of RNA polymerase to the promoter sequence in the DNA.

Antiprotozoal Agents Chemical agents capable of killing or inhibiting protozoas.

Antipruritic Action or agent capable of relieving itching.

Antipsoriotic Action or agent capable of counteracting psoriasis.

Antipsychotic Agent or procedure that counteracts or diminishes symptoms of psychosis.

Antipurine Any purine analog that acts as an antimetabolite in the metabolism of nucleic acid.

Antiputrefactive Acting against putrefaction.

Antipyogenic Agent capable of preventing the formation of pus.

Antipyretic Action or agent capable of reducing fever.

Antipyric Agent capable of preventing the formation of pus.

Antipyrimidine Any pyrimidine analog that acts as an antimetabolite in the metabolism of nucleic acid.

Antipyrine (mol wt 188) An antipyretic and analgesic agent.

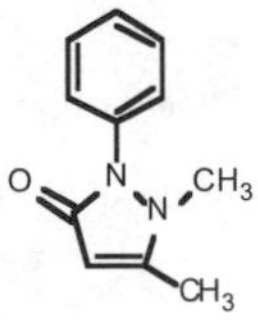

Antirabic Antirabies.

Antirachitic Action or agent capable of preventing rickets.

Antirachitic Vitamin Referring to vitamin D.

Antireflux A drug that alleviates gastroesophageal reflux (commonly known as heartburn)

Antirepressor Referring to the product of the *cro* gene in λ phage that prevents the synthesis of repressor leading to the replication of phage genome and lysis of the infected bacterial cells.

Antirheumatic Action or agent capable of counteracting the effects of rheumatic disease.

Anti-RNP Abbreviation for antibody against ribonucleoprotein.

Antiscorbutic Agent capable of counteracting scurvy, a disease due to the deficiency of vitamin C.

Antiscorbutic Factor Referring to vitamin C.

Antiscorbutic Vitamin Referring to ascorbic acid or vitamin C.

Antiself Referring to antibodies or lymphocytes that react with self-antigens and lead to the development of autoimmune disease.

Antisense RNA An ssRNA molecule that is complementary to a specific RNA transcript of a gene and capable of hybridizing with the specific RNA and blocking its function.

Antisense Strand Synonym for anticoding strand.

Antisepsis The destruction and prevention of growth of microorganisms that cause disease, decay, or putrefaction.

Antiseptic Action or agent capable of opposing sepsis, putrefaction, or decay by preventing or arresting the growth of microorganisms.

Antisera Plural of antiserum.

Antiserum Serum from an immunized individual that contains antibodies against a particular antigen.

Antiserum Anaphylaxis A type of hypersensitivity reaction caused by the injection of serum from a sensitized individual (also called passive anaphylaxis).

Antisialagogue Agent capable of suppressing the flow or formation of saliva.

Antisialic Suppressing or decreasing the secretion of saliva.

Antisigma Factor A protein produced by T_4-infected *E. coli* that prevents the recognition of the promoter by the sigma factor of host DNA-dependent RNA polymerase.

Anti-SM/RNP Abbreviation for antibody against smooth muscle ribonucleoprotein.

Antispas A trade name for dicyclomine hydrochloride, used as an anticholinergic agent.

Antispasmodic Agent or action capable of relieving spasms or convulsions.

Antispastic Agent capable of reducing or relieving spasticity.

Antistatic Agent capable of reducing charge accumulation.

Antisterility Factor Referring to vitamin E.

Antistreptolysin O Test A test used for detecting or quantifying serum antibodies to streptolysin O.

Antisympathetic Agents capable of blocking the activity of the sympathetic nervous system.

Anti-T_3 Abbreviation for triiodothyronine autoantibody.

Antitermination A transcriptional control mechanism in which the transcription termination signal on DNA fails to stop the action of the RNA polymerase leading to a continuous transcription beyond the inter-cistronic terminator sequence.

Antitermination Factor A protein that prevents the termination of RNA synthesis by DNA-dependent RNA polymerase.

Antithrombin A substance in blood capable of neutralizing thrombin and preventing the formation of a blood clot.

Antithyroid Agent A substance that inhibits thyroid function by preventing the synthesis, release, or utilization of thyroxine.

Antitoxin An antibody capable of neutralizing a toxin.

Antitrypanosomal Acting against *trypanosomes.*

α_1-Antitrypsin A group of antitrypsin glycoproteins migrating in the α_1 region in electrophoresis.

Antitrypsin Test A test based upon the capability of serum to inhibit the activity of trypsin. The antitrypsin power of serum is increased in patients with cancer, nephritis, and other diseases.

Antituberculous Destructive to *Mycobacterium tuberculosis*.

Antitumor Antibiotics An antibiotic that arrests or reverses the growth of a malignant tumor.

Antitumor Antimetabolite An antimetabolite that arrests or reverses the growth of a malignant tumor.

Anti-Tus A trade name for guaifenesin, used as an antitussive agent.

Antitussive Action or agent capable of preventing or relieving coughing.

Antityphoid Action or agent effective against typhoid.

Antiuratic Action or agent capable of preventing the deposition of urate in the tissue of the urinary tract.

Anti-Urolithic Agent A drug used for treatment of kidney stone.

Antivenin An antitoxin prepared from the serum of immunized horses and used to counteract animal venom.

Antivert A trade name for meclizine hydrochloride, an antiemetic.

Antiviral Destructive to viruses.

Antiviral Agents Compounds that inhibit the replication of viruses in cells, tissues, or organisms.

Antiviral Protein A protein, induced by interferon, that binds to ribosomes and inhibits the translation of viral RNA or messenger RNA derived from viral DNA.

Antivirin A virus inhibitory factor produced by various cells, e.g., Hela-S_3.

Antivitamin A substance capable of inactivating a vitamin.

Antixenic Responding to a foreign antigen or foreign substance.

Antixerophthalmic Agent effective in the treatment of xerophthalmia.

Antixerophthalmic Factor Referring to vitamin A.

Antrafenine (mol wt 589) An analgesic agent.

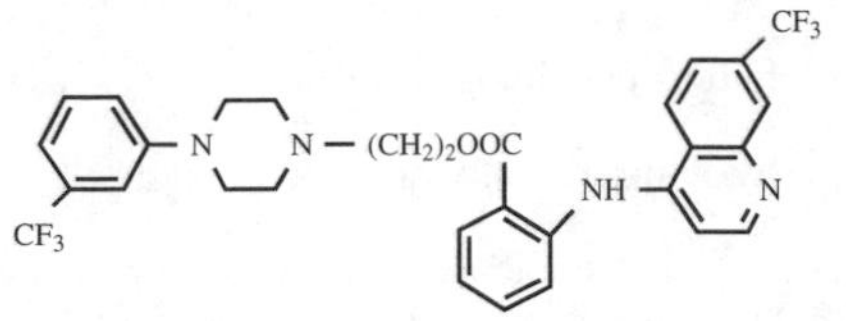

Antrenyl A trade name for xoyphenonium bromide, used as an anticholinergic agent.

αNTU Abbreviation for alpha naphthyl thiourea.

ANTU (mol wt 202) Abbreviation for α-naphthalenylthiourea.

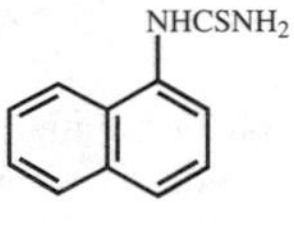

Antuitrin A trade name for gonadotropin, a hormone.

Anturane A trade name for a sulfinpyrazone, an antigout agent that blocks the tubular readsorption of uric acid.

Anucleate Lack of a nucleus.

Anucleolate Lacking a nucleolus.

Anucleolate Mutation A mutation that produces a cell lacking the nucleolus organizer.

Anuphen A trade name for acetaminophen, used as an analgesic and antipyuretic agent.

Anuresis Retention of urine in the bladder.

Anuria The absence of urine formation.

Anusol HC A trade name for hydrocortisone, a topical corticosteroid.

ANX Abbreviation annexin, a Ca^{++}-dependent phosphate-binding protein.

Anxanil A trade name for hydroxyzine hydrochloride, used as an antianxiety agent.

Anxiolytic Agent A drug that reduces anxiety.

Anzemet A trade name for dolasetron, an antiemetic agent,

AO Abbreviation for 1. Acridine orange. 2. Aldehyde oxidase. 3. Ascorbate oxidase.

AocI (SauI) A restriction endonuclease from *Anabaena* species with the following specificity:

5′..........CC↓TNAGG..........3′
3′..........GGANT↑CC..........5′

AocII (SduI) A restriction endonuclease from *Anabaena* species with the following specificity:

5′........G(A/G/T)GC(A/C/T)↓C.........3′
3′........C↑(T/C/A)CG(T/G/A)G.........5′

Aor51HI A restriction endonuclease from *Acidiphilium organovorum* 51H with following specificity:

5'........AGC↓GCT........3'
3'........TCG↑CGA........5'

AorI A restriction endonuclease isolated from *Acetobacter aceti sub orleanensis* with the following specificity.

5′.........CC↓(A/T)GG.........3′
3′.........GG(T/A)↑CC.........5′

AosI A restriction endonuclease isolated from *Anabaena oscillariodes* with the following specificity:

5′.........TGC↓GCA.........3′
3′.........ACG↑CGT.........5′

AosII A restriction endonuclease isolated from *Anabaena oscillariodes* with the following specificity:

5′.........GPu↓CGPyC.........3′
3′.........CPyGC↑PuG.........5′

AosIII (SacII) A restriction endonuclease from *Anabaena ascillarioides* with the following specificity:

5′.........CCGC↓GG.........3′
3′.........GG↑CGCC.........5′

AOX Abbreviation for 1. Alcohol oxidase. 2. Acyl-CoA oxidase. 3. Aldehyde oxidase.

AOX-1 Abbreviation for alcohol oxidase-1

AP Abbreviation for 1. aminopurine, 2. action potential, and 3. alkaline phosphatase.

AP-1 Abbreviation for activator protein-1.

AP2 Abbreviation for activator protein-2.

6-APA Abbreviation for 6-aminopenicillanic acid.

APA Abbreviation for 1. Amino-penicillanic acid. 2. Anthopleurin-A. 3. Anti-pernicious anemia.

ApaI A restriction endonuclease isolated from *Acetobacter pasteurianus sub pasteurianus* with the following specificity:

5′.........GGGCC↓C.........3′
3′.........C↑CCGGG.........5′

APAF Abbreviation for anti-pernicious anemia factor.

ApaLI A restriction endonuclease from *Acetobacter pasteurianus* with the following specifity:

5′..........G↓TGCAC..........3′
3′..........CACGT↑G..........5′

Apalcillin (mol wt 522) An antibiotic related to penicillin that is active against both Gram-positive and Gram-negative bacteria.

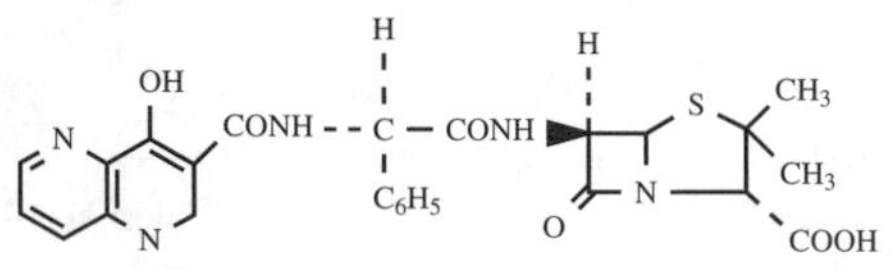

Apallesthesia The loss of the ability to detect vibration.

Apamide A trade name for acetaminophen.

Apamin A toxic neuropeptide from the venom of the honeybee.

Apancrea Without pancreas.

Aparalytic Without paralysis.

Aparathyrodism The absence or reduced function of the parathyroid gland.

Aparkane A trade name for trihexyphenidyl hydrochloride used as an antiparkinsonian agent.

Apathetic Pertaining to apathy.

Apathy Absence of feeling or emotion.

Apatite 1. A mineral containing the ions of Ca^{++} and PO_4^{-3} 2. Calcium phosphate present in the bone and teeth.

Apazone (mol wt 300) An anti-inflammatory and analgesic agent.

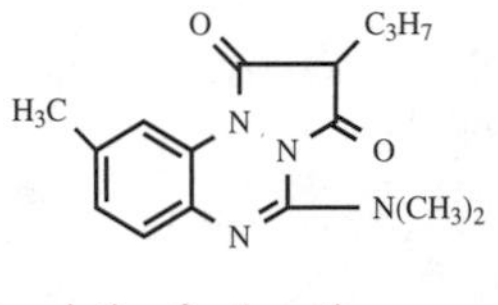

APC Abbreviation for 1. antigen presenting cells, 2. aphidicolin, 3. aspirin, and 4. phenacetin.

APE Abbreviation for acetone powder extract.

ApeI (MluI) A restriction endonuclease from *Achromobacter pestifer* with the following specificity:

5′..........A↓CGCGT..........3′
3′..........TGCGC↑A..........5′

ApeAI (NaeI) A restriction endonuclease from *Aquaspirillum peregrinum* having the same specificity as NaeI.

Apepsinia The absence of secretion of pepsin by the stomach.

A-Peptide The octadecapeptide cleaved from the a chain of a fibrinogen when it is converted into fibrin by thrombin.

Aperient A purgative or laxative.

Aperiodic Polymer A polymer consisting of nonidentical repeating units.

Aperitive 1. Having a stimulating effect on appetite. 2. Aperient (mild laxative).

Apex The top, the end, or the tip of a structure.

APF Abbreviation for anabolic-promoting factor.

APG Abbreviation for acid precipitable globulin.

APH Abbreviation for aminoglycoside phosphotransferase.

APHP Abbreviation for anti-*Pseudomonas* human plasma.

Aphanizomenon A genus of filamentous cyanobacteria.

Apheresis Process of removing blood or a blood element from the body.

Aphidicolin (mol wt 338) A novel tetracyclic diterpene antibiotic with antiviral and antimitotic property isolated from *Cephalosporium aphidicola*.

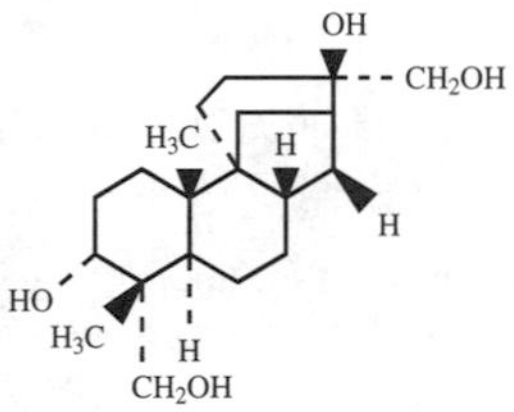

Apholate (mol wt 387) An insect chemosterilant and mutagen.

Aphosphagenic The absence of or insufficient production of phosphorus.

Aphotic Without light.

Aphotic Zone The part of the ocean beneath the photic zone where light does not penetrate enough for photosynthesis to take place.

Aphthous Fever See Foot and mouth disease.

Aphthovirus A genus of viruses (family Picornaviradae).

Apical Surface The surface of an epithelial cell that faces the lumen.

Apiculate Having a short, pointed projection at one end.

Apiculture The science that deals with honeybees.

Apicycline (mol wt 631) An antibacterial agent.

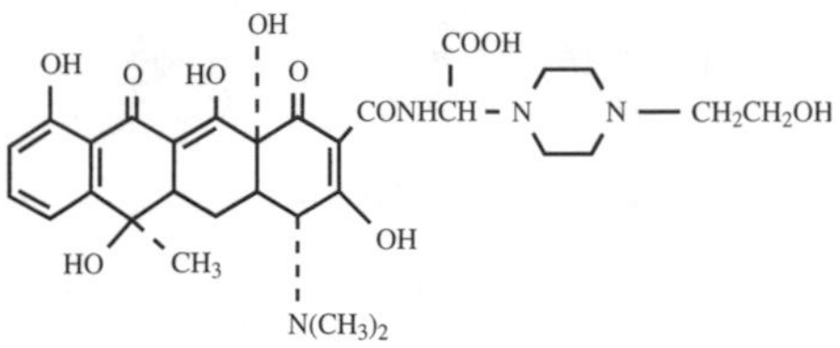

Apiose (mol wt 150) A sugar found in parsley.

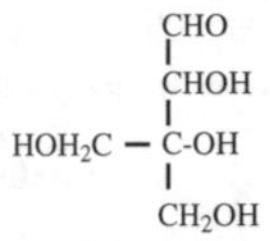

α_2PIPC Abbreviation for α_2-plasmin inhibitor-plasmin complex.

APL Abbreviation for acute pro-myelocytic leukemia.

Aplanogamete A nonmotile gamete.

Aplanospores Nonmotile sexual spores.

Aplasia Lack of an organ development.

Aplasmomycin (mol wt 799) An antibiotic produced by *Streptomyces griseus*.

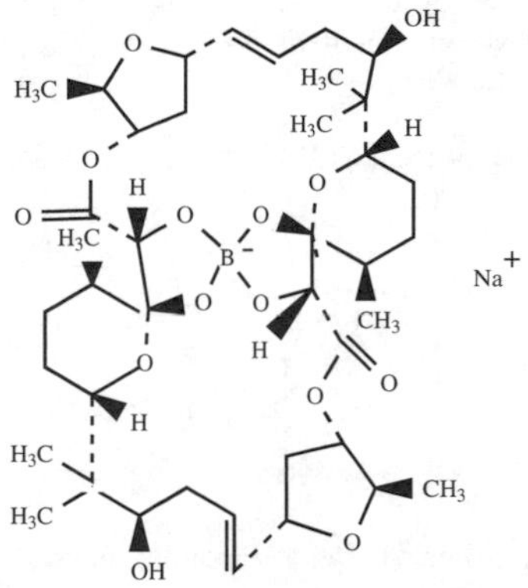

Aplastic Anemia A deficiency of all of the formed elements of the blood.

APLH Abbreviation for anterior pituitary-like hormone.

Apligraf A trade name for a skin collagen used for topical wound dressing.

Aplisol A trade name for tuberculin purified protein derivative, used for the diagnosis of tuberculosis.

APM Abbreviation for acid-precipitable material.

APMA Abbreviation for 4-aminophenylmercuric acetate.

APML Abbreviation for acute pro-myelocytic leukemia.

APM-Lactoferrin Abbreviation for aminopeptidase M-modified lactoferrin.

APMSF Abbreviation for *p*-amidinophenylmethylsulfonyl fluoride, an inhibitor for trypsin-like serine endopeptidases.

APO Abbreviation for a combination drug containing adriamycin, prednisone and oncovin.

Apo- A prefix denoting detached or separate.

ApoI A restriction endonuclease from *Acetobacter protophormiae* with the following specificity:

apoA Abbreviation for apolipoprotein A.

apoA-I Abbreviation for apolipoprotein A-I, the major protein of plasma high density lipoprotein.

ApoA-II Abbreviation for apolipoprotein A-II, an apolipoprotein associated with high density lipoprotein which it stabilizes.

ApoA-IV Abbreviation for apolipoprotein A-IV, A major component of high density lipoprotein and chylomicrons.

Apoatropine (mol wt 271) An antispasmodic agent.

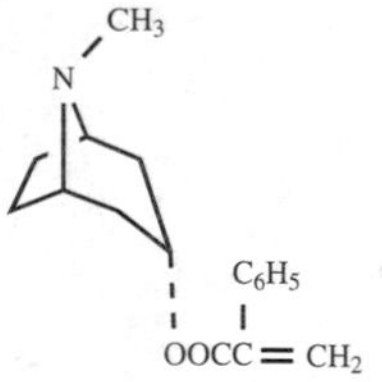

ApoB Abbreviation for apolipoprotein B.

ApoC-I Abbreviation for apolipoprotein CI, a component of chylomicrons and very low density lipoprotein.

ApoC-II Abbreviation for apolipoprotein CII, a component of very low density lipoprotein and an activator of acylglycerol lipase.

ApoC-III Abbreviation for apolipoprotein CIII, a major component of very low density lipoprotein, it binds sugar and sialic acid residues and inhibits lipoprotein and hepatic lipase.

Apochlorotic Colorless.

Apochromatic Lens An objective microscope lens in which chromatic aberration has been corrected for three colors and spherical aberration for two colors.

Apocodeine (mol wt 281) An emetic agent.

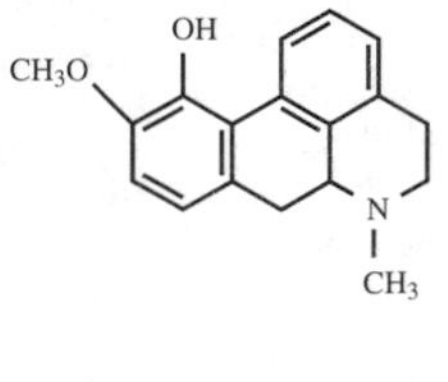

Apocrine A sweat gland.

ApoD Abbreviation for apolipoprotein D, an apolipoprotein that occurs in a macromolecular complex with lecithcin-cholesterol acyltransferase and in high density lipoprotein.

ApoE Abbreviation for apolipoprotein E, a component of very low density lipoprotein.

Apoenzyme The protein component of an enzyme; apoenzyme + coenzyme = holoenzyme.

Apoferritin The protein component of ferritin.

Apogamy Reproduction of an organism without involving the fusion of gametes.

ApoH Abbreviation for apolipoprotein-H, an lipoprotein that binds to anions such as heparin and phospholipids.

ApoJ Abbreviation for apolipoprotein-J.

Apolar Nonpolar.

Apolipoproteins The lipid-free portion of lipoprotein or the protein part of the lipoprotein.

Apomeiosis Nuclear division without meiosis.

Apomictic See Apomixis.

Apomixis (Apomictic) Reproduction without fertilization (e.g., asexual production of seed).

Apomorphine (mol wt 267) An emetic agent derived from morphine by removal of a molecule of water.

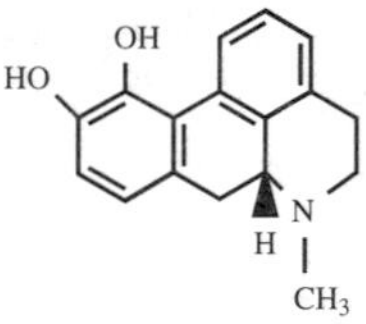

Aponia 1. Absence of pain. 2. Absence of physical exertion.

Aponic Analgesic or pertaining to the relief from fatigue.

Apophlegamatic 1. Secretion of an abnormal quantity of mucus. 2. An expectorant.

Apophysis Any outgrowth.

Apophysitis An inflammation or fragmentation of a bony apophysis.

Apoplastidic Lacking a plastid.

Apoplexy 1. A sudden impairment of cerebral function leading to coma. 2. A hemorrhage within an organ.

Apoprotein The protein part of a conjugated protein.

Apoptosis Programmed cell death.

Aporepressor A product of a regulator gene that regulates the operation of an operon system, but it does not function as a repressor in an operon system unless it binds with a copressor (e.g., end product of a pathway).

Aporinosis A disorder due to a nutritional deficiency.

Aposome A cytoplasmic inclusion that originates within the cell.

Aposthia The developmental absence of the prepuce.

Aposymbiotic An organism that has lost the symbionts it normally possesses.

APP Abbreviation for 1. Acute phase protein. 2. Alum-precipitated protein. 3. Acid-precipitated pyridine or acid precipitated proteins. 4. Amyloid precursor protein. 6. Alveolar proteinosis protein.

AP-P Abbreviation for aminopeptidase-P.

Apparent Specific Volume The change in volume per gram of solute when a known weight of solute is added to a known volume of solvent.

APPase Abbreviation for alkaline pyrophosphatase.

Appendage A filamentous structure, e.g., bacterial flagellum.

Appendicitis Inflammation of the appendix.

Appendix The accessory part attached to the main structure.

APPG Abbreviation for aqueous procaine penicillin G.

APP(NH)P Symbol for β,γ—imido-ATP, An ATP analog.

APPP Abbreviation for acute phase plasma protein.

AprI (NaeI) A restriction endonuclease from *Actinosynnema pretiosum* with the following specificity:

```
           ↓
5′..........GCCGGC..........3′
3′..........CGGCCG..........5′
              ↑
```

Apraclonidine (mol wt 245) An α_2-adrenergic agonist, used for treatment of post-surgical elevated intraocular pressure.

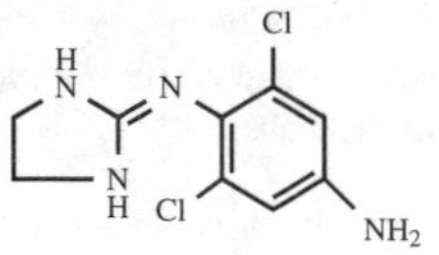

Apramycin (mol wt 540) An aminoglycoside antibiotic isolated from *Streptomyces tenebrarius*.

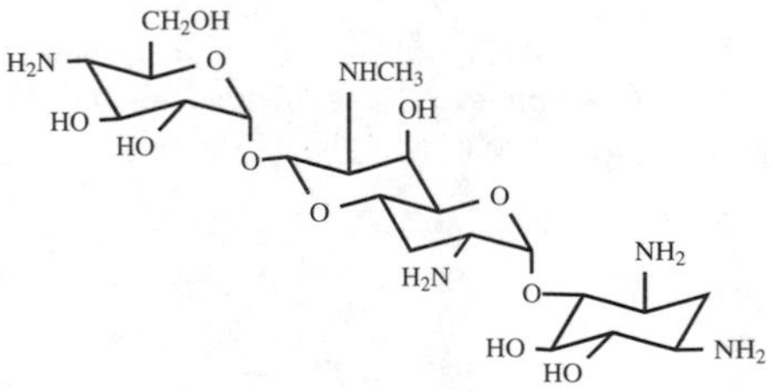

Apresazide A trade name for a combination drug containing hydralazine hydrochloride and hydrochlorothiazide hydrochloride, used as an antihypertensive agent.

Apresodex A trade name for a combination drug containing hydrochlorothiazide and hydralazine hydrochloride, used as an antihypertensive agent.

Apresoline A trade name for hydralazine hydrochloride, an antihypertensive.

Apresoline-Esidrix A combination drug containing hydralazine and hydrochlorothiazide, used as an antihypertensive agent.

Aprinox A trade name for bendroflumethiazide, a diuretic agent.

Aprobarbital (mol wt 210) A sedative and hypnotic agent.

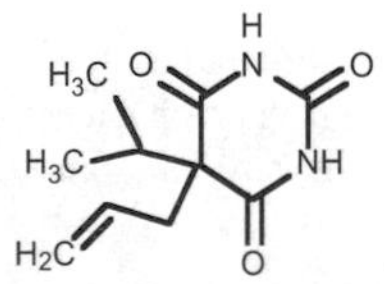

Apronalide (mol wt 184) A sedative and hypnotic agent.

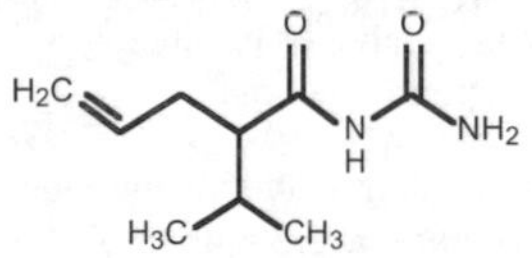

A-Protein 1. A protein of the capsid of TMV and M13 phage. 2. A protein subunit of the enzyme tryptophan synthetase.

Aprotic Solvent A type of solvent that acts neither as a proton acceptor nor as a proton donor with respect to the solute.

Aprotinin A basic polypeptide that inhibits many serine proteases, e.g., trypsin, chymotrypsin, and some bacterial proteases.

APRP Abbreviation for acute-phase reactant protein.

APRT Abbreviation for adenine phosphoribosyl transferase.

APS (mol wt 427) Abbreviation for adenosine 5′-phosphosulphate (also called adenosine 5′-sulphatophosphate or adenylylsulphate).

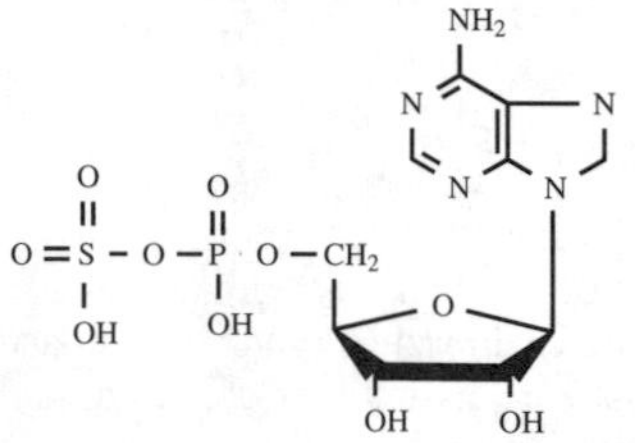

APSA Abbreviation for anisoylated plasminogen streptokinase activator.

APSAC 1. Abbreviation for anisoylated plasminogen streptokinase activator complex. 2. A trade name for anistreplase, a thrombolytic enzyme.

AP-Site Abbreviation for apurinic or apyrimidinic site. The site in DNA containing no purine or pyrimidine base.

APT Abbreviation for alum-precipitated toxoid.

APT Agar An agar-based medium containing tryptone, yeast extract, glucose, Tween 80, citrate, and various inorganic salts for growing organisms, such as *lactobacilli* and Brochothrix.

APT Paper Abbreviation for 2-aminophenylthioether blotting paper with a reactive diazo group that is capable of covalent binding of ssDNA, ssRNA, or proteins.

APT Test Abbreviation for aluminum precipitated toxoid test.

Aptitude A physiological state of lysogenic bacterium that is capable of producing infective bacteriophages upon induction.

ApuI (AsuI) A restriction endonuclease from *Alteromonas putrefaciens* with the following specificity:

5′..........G↓GNCC..........3′
3′..........CCNG↑G..........5′

APUD Abbreviation for amine precursor uptake and decarboxylation.

APUD-C Abbreviation for amine precursor uptake and decarboxylation cell.

Apurinic Acid A DNA molecule from which the purines have been removed by mild acid hydrolysis.

Apurinic Site The site on a DNA molecule containing no purine bases.

APW Abbreviation for alkaline peptone water (1% peptone and 1% NaCl in distilled water, pH 8.6-9.0) used for isolating *Vibrio cholerae*.

ApyI A restriction endonuclease isolated from *Arthobacter pyridinolis* with the following specificity:

5′..........CC↓(A/T)GG..........3′
3′..........GG(T/A)↑CC..........5′

Apyrase The enzyme that catalyzes the hydrolysis of ATP to AMP and two molecules of orthophosphate.

Apyrimidinic Acid A DNA molecule from which the pyrimidines have been removed by treatment with hydrazine.

Apyrimidinic Site The site on a DNA molecule containing no pyrimidine bases.

Aq Abbreviation for aqueous.

AquI A restriction endonuclease isolated from *Agmenellum quadruplicatum* with the following specificity:

5′..........C↓PyCGPuG.........3′
3′..........GPuGCPy↑C.........5′

Aqua Water.

Aquacare A trade name for urea, an osmotic diuretic agent, that elevates the osmolarity of the glomerular filtrate.

Aquachloral Supprettes A trade name for chloral hydrate, used as a sedative-hypnotic agent.

Aquacide Any substance capable of removing water molecules.

Aquacort A trade name for hydrocortisone, a hormone.

Aquamephyton A trade name for vitamin K.

Aquametry The quantitative determination of water.

Aquamox A trade name for quinethazone, used as a diuretic agent.

Aquaphyllin A trade name for theophylline, used as a bronchodilator.

Aquasol A A trade name for vitamin A.

Aquasol E A trade name for vitamin E.

Aquaspirillum A genus of Gram-negative, asporogenous bacteria.

Aquatag A trade name for benzthiazide, used as a diuretic agent.

Aquatenin A trade name for methyclothiazide, used as a diuretic agent.

Aquatensen A trade name for methyclothiazide, used as a diuretic agent.

Aquatic Relating to fresh or salt water.

Aquation The formation of aquoions.

Aquazide A trade name for trichlormethiazide, used as a diuretic agent.

Aquazym A trade name for α-amylase.

Aqueous Of or pertaining to water.

Aqueous Humor The lymphlike fluid circulating in the anterior and posterior chambers of the eye.

Aqueous Solution A solution in which water is the solvent.

Aquest A trade name for estrone, a hormone.

Aquocobalamin A vitamin B_{12} derivative (see also Cobalamin).

Aquo-Ion A complex ion that contains one or more water molecules.

Ar Symbol for the chemical element Argon.

AR Abbreviation for 1. Aldose reductase. 2. Allergic reaction. 3. Allergic rhinitis. 4. Analytic reagent. 5. Androgen receptor. 6. Auto-radiography.

Ara Abbreviation for arabinose.

ArA Abbreviation for arachidonic acid.

***ara* Operon** Abbreviation for arabinose operon, an operon that deals with the metabolism of arabinose.

Arabic Gum A water-soluble exudate from stems of *Acacia senegal* or related species.

Arabinan A polysaccharide consisting of arabinose.

Arabinan Endo-1,5-α-L-Arabinosidase The enzyme that catalyzes the endohydrolysis of 1,5-α-arabinofuranosidic linkages in 1,5-arabinans.

Arabinase Synonym of arabinan endo-1,5-α-L-arabinosidase.

Arabinitol Synonym for arabitol.

L-Arabinitol 2-Dehydrogenase The enzyme that catalyzes the following reaction:

$$\text{L-Arabinitol} + \text{NAD}^+ \rightleftharpoons \text{L-Ribulose} + \text{NADH}$$

D-Arabinitol 4-Dehydrogenase The enzyme that catalyzes the following reaction:

$$\text{D-Arabinitol} + \text{NAD}^+ \rightleftharpoons \text{D-Xylulose} + \text{NADH}$$

L-Arabinitol 4-Dehydrogenase The enzyme that catalyzes the following reaction:

$$\text{L-Arabinitol} + \text{NAD}^+ \rightleftharpoons \text{L-Xylulose} + \text{NADH}$$

α-N-Arabinofuranosidase The enzyme that catalyzes the hydrolysis of terminal nonreducing α-arabinofuranoside residues in α-arabinosides.

Arabinofuranosylcytosine (mol wt 243) A cytidine analog useful in cancer chemotherapy.

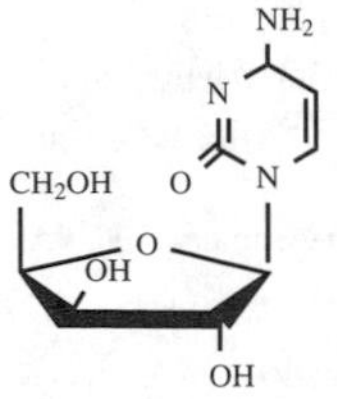

Arabinogalactan A water soluble polysaccharide extracted from the timber of western larch trees.

Arabinogalactan Endo-1,3-β-Galactosidase The enzyme that catalyzes the endohydrolysis of 1,3-D-galactosidic linkages in arabinogalactan.

Arabinogalactan Endo-1,4-β-Galactosidase The enzyme that catalyzes the endohydrolysis of 1,4-D-galactosidic linkages in arabinogalactan.

D-Arabinokinase The enzyme that catalyzes the following reaction:

$$\text{ATP} + \text{D-arabinose} \rightleftharpoons \text{ADP} + \text{D-arabinose 5-phosphate}$$

L-Arabinokinase The enzyme that catalyzes the following reaction:

$$\text{ATP} + \text{L-arabinose} \rightleftharpoons \text{ADP} + \text{L-arabinose 5-phosphate}$$

Arabinonate Dehydratase The enzyme that catalyzes the following reaction:

D-Arabinonate
⇅
2-Dehydro-3-deoxy-D-arabinonate + H_2O

D-Arabinonolactonase The enzyme that catalyzes the following reaction:

D-Arabinono-1,4-lactone + H_2O
⇅
D-Arabinonate

L-Arabinonolactonase The enzyme that catalyzes the following reaction:

L-Arabinono-1,4-lactone + H_2O
⇅
L-Arabinonate

Arabinose (mol wt 150) An aldopentose occurring as a component of hemicellulose and pectic polysaccharide.

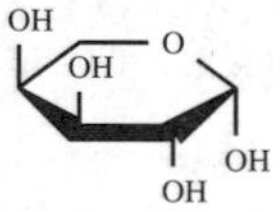

L-Arabinose

D-Arabinose 1-Dehydrogenase The enzyme that catalyzes the following reaction:

D-Arabinose + NAD^+
⇅
D-Arabinono-1,4-lactone + NADH

D-Arabinose 1-Dehydrogenase (NADP-specific) The enzyme that catalyzes the following reaction:

D-Arabinose + $NADP^+$
⇅
D-Arabinono-1,4-lactone + NADPH

L-Arabinose 1-Dehydrogenase The enzyme that catalyzes the following reaction:

L-Arabinose + NAD^+
⇅
L-Arabinono-1,4-lactone + NADH

D-Arabinose Isomerase The enzyme that catalyzes the following reaction:

D-Arabinose ⇌ D-Ribulose

L-Arabinose Isomerase The enzyme that catalyzes the following reaction:

L-Arabinose ⇌ L-Ribulose

Arabinose Operon An operan that deals with metabolism of arabinose.

Arabinose 5-Phosphate Isomerase The enzyme that catalyzes the following reaction:

Arabinose 5-phosphate
⇅
Ribulose 5-phosphate

β-L-Arabinosidase The enzyme that catalyzes the following reaction:

A β-L-Arabinoside + H_2O
⇅
An alcohol + L-arabinose

Arabinosuria The excretion of arabinose in the urine.

Arabinosylcytosine See Arabino-furanosylcytosine.

Arabinosyl Nucleosides A group of arabinonucleosides that inhibit activities of DNA polymerases and reverse transcriptase by acting as analogs of biological nucleosides or nucleotides.

Arabinoxylan A heteroglycan consisting of both xylose and arabinose.

Arabitol (mol wt 152) A sugar alcohol formed by the reduction of arabinose.

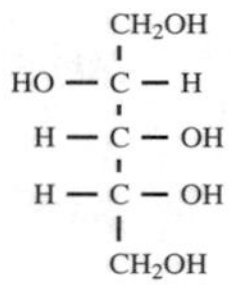

D-Arabitol

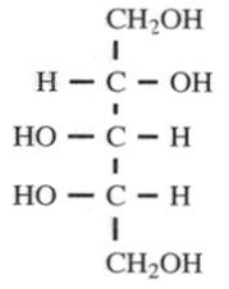

L-Arabitol

Araboflavin (mol wt 376) An antagonist of riboflavin.

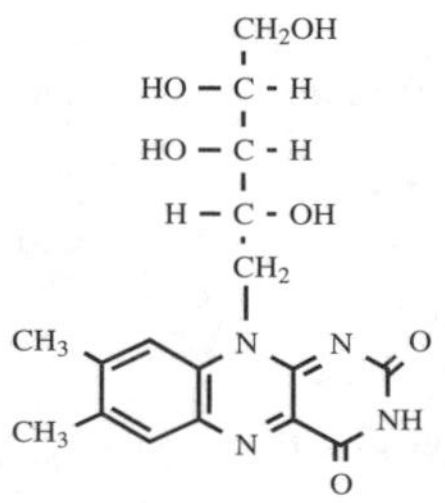

Ara-C Abbreviation for cytosine arabinoside.

Arachain A major protein from groundnut (*Arachis hypogaea*).

Arachidic Acid (mol wt 313) A saturated fatty acid found in peanut, vegetable, and fish oils (also called Eicosanoic acid).

$CH_3(CH_2)_{18}COOH$

Arachidonate-CoA Ligase The enzyme that catalyzes the following reaction:

ATP + arachidonate + CoA

⇌

AMP + PPi + Arachidonyl- CoA

Arachidonate 5-Lipooxygenase The enzyme that catalyzes the following reaction:

Aarachidonate + O_2

⇌

(6E,8Z,11Z,14Z)-(5S)-5-hydroperoxyicosa-6,8,11,14-tetraenoate

Arachidonate 12-Lipooxygenase The enzyme that catalyzes the following reaction:

Aarachidonate + O_2

⇌

(5Z,8Z,10E,14Z)-(12S)-12-hydroperoxyicosa-5,8,10,14-tetraenoate

Arachidonate 15-Lipooxygenase The enzyme that catalyzes the following reaction:

Aarachidonate + O_2

⇌

5Z,8Z,11Z,13E)-(15S)-15-hydroperoxyicosa-5,8,11,13-tetraenoate

Arachidonate Oxygen 5-Oxidoreductase The systematic name for arachidonate 5-lipooxygenase.

Arachidonate Oxygen 12-Oxidoreductase The systematic name for arachidonate 12-lipooxygenase.

Arachidonate Oxygen 15-Oxidoreductase The systematic name for arachidonate 15-lipooxygenase.

Arachidonic Acid (mol wt 304) A 20-carbon fatty acid with four double bonds. It serves as the precursor for the synthesis of prostaglandins, thromboxanes, and leukotrienes.

$CH_3(CH_2)_4(CH{=}CHCH_2)_4(CH_2)_2COOH$

Arachidonyl-CoA Synthetase Synonym of long chain fatty acid-CoA ligase.

Arachnia A genus of anaerobic catalase-negative, asporogenous bacteria (order Actinomycetales).

Arachnin A protein from the groundnut *Arachis hypogaea* that is soluble in water and alcohol and insoluble in ether.

Arachnoiditis Inflammation of the arachnoid membrane.

Arachnoid Membrane A thick, delicate membrane enclosing the brain and the spinal cord.

Arachnolysin An active hemolytic component of spider venom.

Aragonite A crystal form of calcium carbonate appearing in pearls.

Aralen A trade name for chloroquine hydrochloride, used as an antimalarial agent.

Araline A trade name for chloroquine phosphate, used as an antimalarial agent.

Aramine A trade name for metaraminol bitartrate, used as an adrenergic agent.

Aramite (mol wt 335) A miticide.

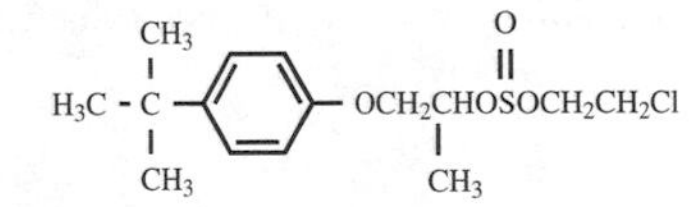

Aranidipine (mol wt 388) An antihypertensive agent.

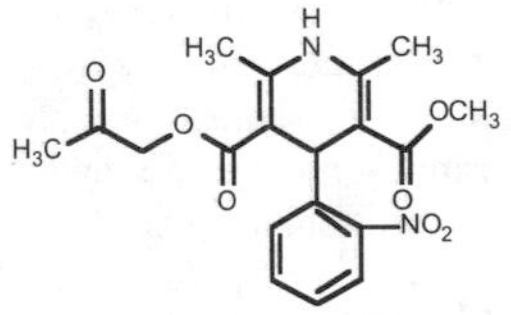

***ara*-Operon** An operon that deals with the metabolism of arabinose.

Ara-T Abbreviation for 1-β-D-arabinofuranosylthymine. An arabinosyl thymidine nucleoside isolated from marine sponges (*Cryptoethya crypta*). It has antiviral activity against herpes simplex viruses, varicella-zoster virus, and vaccinia virus.

ara-U Abbreviation for arabinosyl-uracil.

Arava A trade name for leflunomide, an enzyme inhibitor and an antiarthritis drug.

Arbaprostil (mol wt 367) An antiulcerative agent.

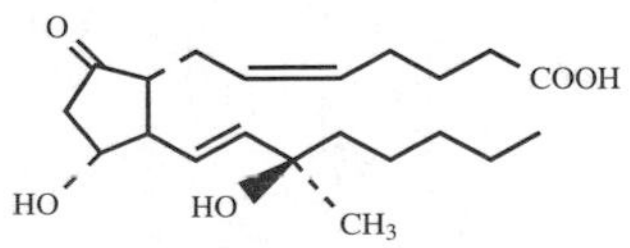

Arbekacin (mol wt 553) An antibacterial agent.

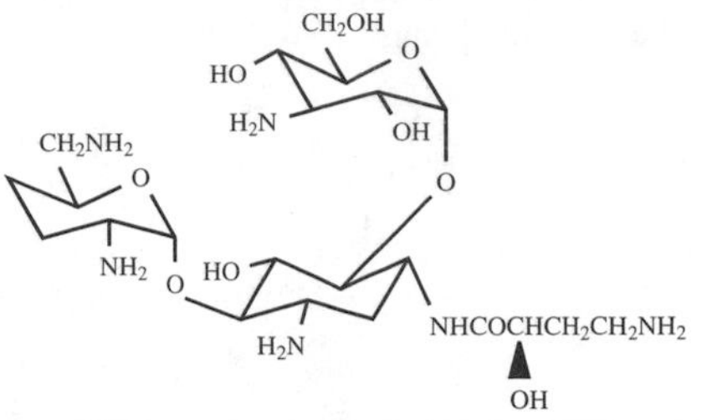

ARBOR Virus Abbreviation for arthropod-borne virus.

Arboviruses Arthropod-borne viruses. A nontaxonomic category of viruses that can replicate in both vertebrate hosts and arthropod vectors, e.g., viruses in the families Arenaviridae, Bunyaviridae, Reoviridae (Orbivirus), Togaviridae (Alphavirus), and Flaviviradae.

Arbutin (mol wt 272) A diuretic agent.

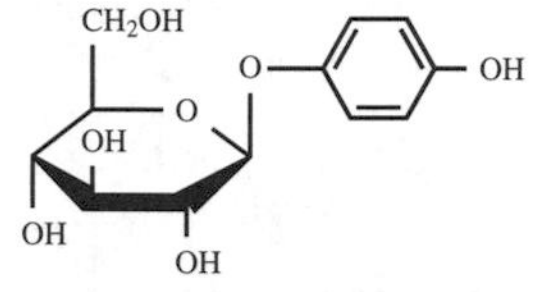

Arcanobacterium A genus of asporogenous bacteria (order Actinomycetales).

Arcella A genus of testate amoebae (order Arcellinida).

Archaebacteria A phylogenetic group of prokaryotes that differs from both eubacteria and eukaryotes in the sequence of its 16S rRNA and composition of it's membrane phospholipid. They represent a primary biological kingdom related to both eubacteria and eukaryotes.

Archibald Method A centrifugal method for determining molecular weights and assessing size homogeneity of macromolecules.

Architectural Gene A gene that determines the site of an enzyme within a cell.

ARD Abbreviation for acute respiratory diseases.

Ardeparin Sodium A low molecular heparin and an antithrombotic agent. It is a partially depolymerized porcine mucosal heparin.

Arecaidine (mol wt 141) A compound isolated from seeds of *Areca catechu* (betel nut palm).

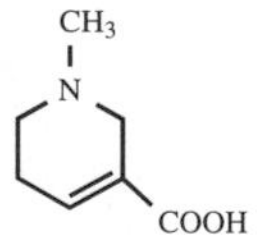

Arecoline (mol wt 155) An alkaloid from seeds of the betel nut palm *Areca catechu*. It has been used as an anthelmintic agent.

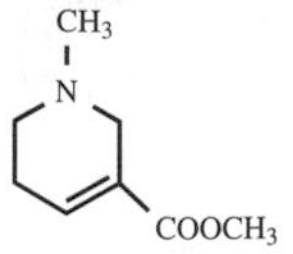

Aredia A trade name for pamidronate disodium, used as an antihypercalcemic agent.

Arenaviridae A family of enveloped animal viruses containing negative-stranded RNA and ribosomes (e.g., lymphocytic choriomenigitis).

Arene Any monocyclic or polycyclic aromatic hydrocarbon.

ARF Abbreviation for 1. Acute rheumatic fever. 2. ADP-ribosylation factor. 3. Audio-response frequency.

ARF-1 Abbreviation for ADP-ribosylation factor-1.

Arfonad A trade name for trimethaphan camsylate, used as an antihypertensive agent.

Arg Symbol for arginine.

ARG Abbreviation for auto-radiography.

***arg* Oncogene** An oncogene that encodes a tyrosine kinase.

Argatroban (mol wt 509) An antithrombotic agent.

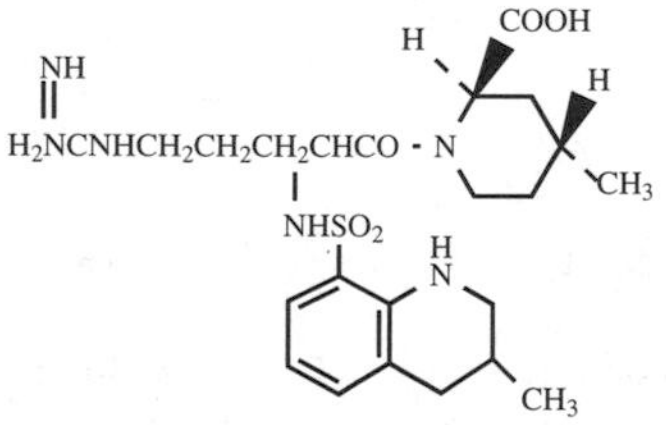

Argentaffin Cell A cell containing serotonin-secreting granules that are stained readily with silver solution.

Argentophilic 1. Staining well with silver stains. 2. Having affinity for silver, e.g., silver nitrate.

Argentum Latin name for silver.

Argesic A trade name for salsalate, an antipyretic, analgesic, anti-inflammatory and antirheumatic agent.

L-Arginase The enzyme that catalyzes the hydrolysis of arginine to ornithine and urea.

$$\text{L-Arginine} + H_2O \rightleftharpoons \text{L-Ornithine} + \text{urea}$$

Arginine (mol wt 174) A basic protein amino acid.

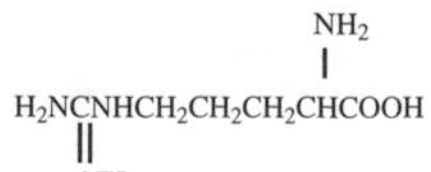

Arginine Amidinase See Arginase.

Arginine Amidinohydrolase The systematic name for arginase.

Arginine Aminopeptidase A protease that catalyzes the release of N-terminal arginine in a di- or tripeptide.

Arginine Carboxy-Lyase The systematic name for arginine decarboxylase.

Arginine Cycle Urea cycle.

Arginine Decarboxylase The enzyme that catalyzes the conversion of arginine to agmatine and CO_2.

$$\text{L-Arginine} \rightleftharpoons \text{Agmatine} + CO_2$$

Arginine Deiminase The enzyme that hydrolyzes arginine to citrulline and NH_3.

$$\text{L-Arginine} + H_2O \rightleftharpoons \text{L-Citrulline} + NH_3$$

Arginine Desiminase See arginine deiminase.

Arginine Kinase The enzyme that catalyzes the phosphorylation of arginine.

$$\text{ATP} + \text{arginine} \rightleftharpoons \text{ADP} + \text{N-phospho-L-arginine}$$

Arginine 2-Monooxygenase The enzyme that catalyzes the following reaction:

$$\text{L-Arginine} + O_2 \rightleftharpoons \text{4-Guanidinobutanamide} + CO_2 + H_2O$$

Arginine NADPH Oxygen Oxidoreductase (Nitric Oxide-Forming) The systematic name for nitric-Oxide synthetase.

Arginine Recemase The enzyme that catalyzes the following reaction:

$$\text{L-Arginine} \rightleftharpoons \text{D-Arginine}$$

Arginine tRNA Ligase The enzyme that catalyzes the following reaction:

$$\text{ATP} + \text{L-arginine} + \text{tRNA}^{arg} \rightleftharpoons \text{AMP} + \text{PPi} + \text{L-arginyl-tRNA}^{arg}$$

Arginine Vasopressin A vasopressin molecule in which the eighth amino acid residue has been replaced by an arginine residue.

Arginine Vasotocin A supraopticoneurohypophysial peptide of nonmammalian vertebrates.

Argininemia An autosomal recessive disorder characterized by an increased amount of arginine in the blood caused by a deficiency of arginase.

Argininium The monocation of arginine.

Argininosuccinase The enzyme that catalyzes the following reaction:

$$\text{Arginosuccinate} \rightleftharpoons \text{fumarate} + \text{arginine}$$

Argininosuccinate Lyase See arginosuccinase.

Argininosuccinate Synthetase The enzyme that catalyzes the following reaction:

$$\text{ATP} + \text{citruline} + \text{asparatate} \rightleftharpoons \text{AMP} + \text{PPi} + \text{arginosuccinate}$$

Argininosuccinic Acid (mol wt 290) An intermediate in the urea cycle.

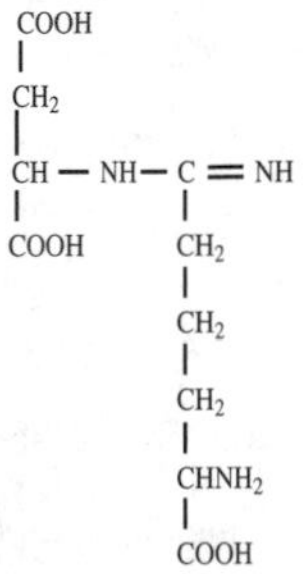

Argininosuccinic Acidemia An inherited metabolic defect in humans characterized by a deficiency of the enzyme argininosuccinase.

Argininosuccinuria An inherited metabolic disorder characterized by a large renal excretion of argininosuccinic acid.

Arginyltransferase The enzyme that catalyzes the following reaction:

$$\text{Arginyl-tRNA} + \text{protein} \rightleftharpoons \text{tRNA} + \text{L-arginyl-protein}$$

Arginyl-tRNA Synthetase See arginine tRNA ligase.

Argon (Ar) A colorless, odorless, chemically inactive gas with atomic weight 40.

Argon Detector An ionization detector employed in gas chromatography in which argon is used to ionize the organic compounds being separated. It is useful for trace analysis of steroids, fatty acids, and related compounds of relatively high molecular weights.

Argyremia The presence of silver in the blood.

Argyria A deep, ash-gray discoloration of skin and conjunctiva resulting from chronic exposure to silver or silver salt.

Argyrol Referring to silver-protein complexes.

ARIA Abbreviation for automated radio-immuno-assay.

Ariboflavinosis A condition caused by deficiency of vitamin B_2 in the diet and characterized by lesions at the corners of the mouth, on the lips, and around the nose and eyes.

Aricept A trade name for donepezil, an acetylcholinesterase inhibitor used for treatment of Alzheimer's disease.

Arimidex A trade name for anastrazole, an anti-estrogen and antineoplastic agent.

Aristocort A trade name for triamcinolone, used as an anti-inflammatory agent.

ARLD Abbreviation for alcohol-related liver disease.

Arlidin A trade name for nylidrin hydrochloride, used as a peripheral vasodilator.

ARM Abbreviation for ATP-regulated module.

Armed Macrophages Macrophages capable of antigen-specific cytotoxicity.

Armour A trade name for a thyroid preparation containing T_3, T_4 in their natural state and ratio.

Arogenic Acid (mol wt 227) A precursor in the biosynthesis of phenylalanine and tyrosine.

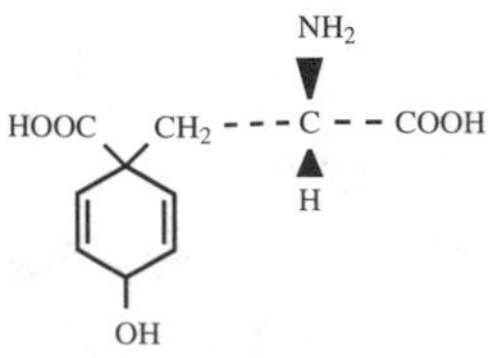

Aromatic 1. Of or pertaining to a carbocyclic organic compound that contains a benzene ring. 2. Having a smell or odor.

Aromatic Alcohol An alcohol that contains a phenyl hydrocarbon.

Aromatic Amino Acid An amino acid that contains a benzene ring, e.g., tyrosine and phenylalanine.

Aromatic L-Amino Acid Decarboxylase The enzyme that catalyzes the following reaction:

$$\text{L-Tryptophan} \rightleftharpoons \text{Tryptamine} + CO_2$$

Aromatic Amino Acid Glyoxylate Transaminase The enzyme that catalyzes the following reaction:

$$\text{An aromatic amino acid} + \text{glyoxylate} \rightleftharpoons \text{An aromatic keto acid} + \text{glycine}$$

Arotinolol (mol wt 371) An antihypertensiive, antianginal and antiarrhythmic agent.

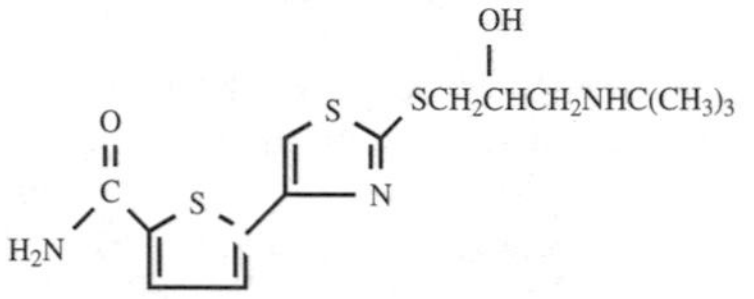

ARP Abbreviation for 1. Actin–related protein. 2. Assimilatory regulatory protein.

Arrestin A trade name for trimethobenzamide hydrochloride, an antiemetic agent.

Arrhenius Equation The equation relating the rate constant of a reaction to the absolute temperature.

$$\ln k = \ln A - E/RT \text{ where}$$

ln = natural logarithm,
R = gas constant,
A = Arrhenius factor,
E = activation energy of the reaction, and
T = Absolute temperature.

Arrhenius Plot A plot of the logarithm of the rate constant of a reaction versus the reciprocal of the absolute temperature; used for determining the activation energy of the reaction.

Arrhenoblastoma An ovarian neoplasm.

Arrhythmia Any deviation from the normal pattern of the heartbeat.

Arrhythmogenic Causing irregular heartbeats.

ARS Abbreviation for autonomously replicating sequence. A DNA sequence that is capable of conferring a nonreplicative DNA fragment to replicate in yeast cell when it is linked to the nonreplicative DNA fragment.

Arsacetin (mol wt 259) An antisyphilitic agent.

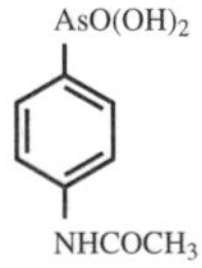

Arsenamide (mol wt 377) An anthelmintic agent.

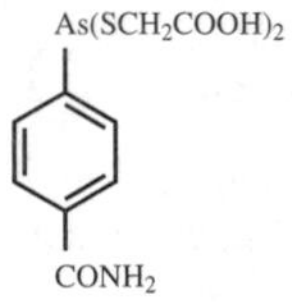

Arsenic (As) A chemical element with atomic weight 75, valence 3 or 5.

Arsenic Acid (mol wt 142) A poisonous compound.

AsH_3O_4

Arsenic Poisoning Poisoning caused by the ingestion or inhalation of arsenic or a substance containing arsenic element.

Arsenoacetic Acid (mol wt 184) An antistimulant for nervous diseases.

$H_2O_3AsCH_2COOH$

Arsenolysis The cleavage of a covalent bond of an acid derivative by reaction with arsenic acid (H_3AsO_4) so that one of the products combines with the H and the other product combines with the H_2AsSO_4 group of arsenic acid.

Arsphenamine (mol wt 439) An antisyphilitic agent.

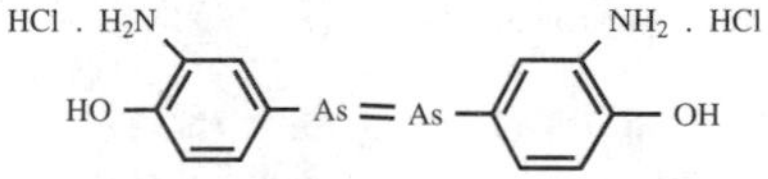

Arsthinol (mol wt 347) An antiamoebic agent.

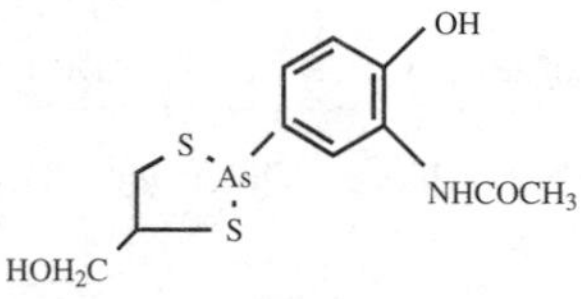

ART Abbreviation for automated reagin test, a standard test for syphilis.

Artane A trade name for trihexyphenidyl, used as an antiparkinsonian agent.

Arteether (mol wt 312) An antimalarial agent.

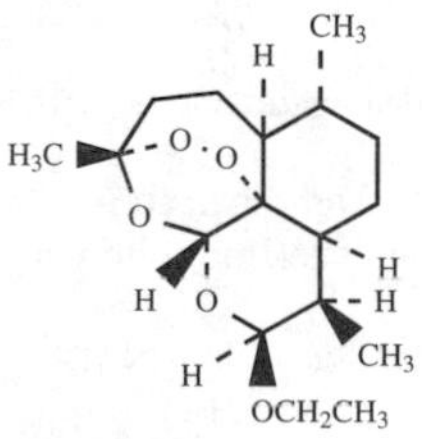

Artefact Any feature that does not occur in a specimen under natural conditions, but appears in that specimen after the experimentation. Artefacts are due to the disturbance introduced by the process of experimentation.

Artemether (mol wt 298) An antimalarial agent.

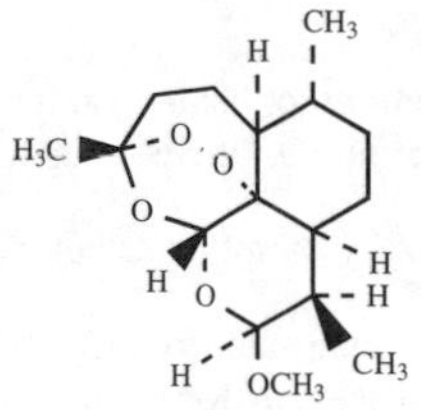

Artemisinin (mol wt 282) An antimalarial agent isolated from the Chinese medicinal herb *Artemisia annua* (Compositae).

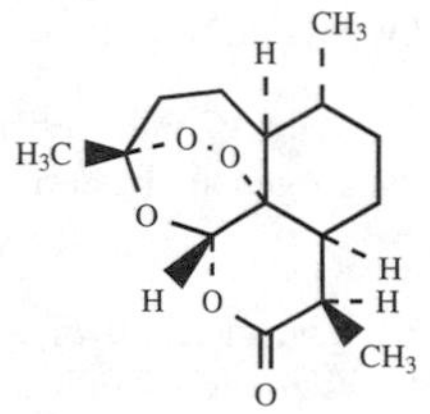

Arterial Of or pertaining to arteries.

Arterio- A combining form meaning pertaining to an artery.

Arterioflexin A trade name for clofibrate, used as an antilipemic agent.

Arteriography X-ray visualization of arteries after introduction of radiopaque contrast medium into the bloodstream or into a specific vessel.

Arterioles Branches of the arteries present within organs.

Arteriosclerosis A loss of elasticity or a hardening of an artery.

Arteritis Inflammation of an artery or arteries.

Arterivirus A genus of virus in the family Togaviridae, e.g., equinine arteritis virus.

Artery A blood vessel carrying blood away from the heart.

Artesunate (mol wt 384) An antimalarial agent and inhibitor of cytochrome oxidase.

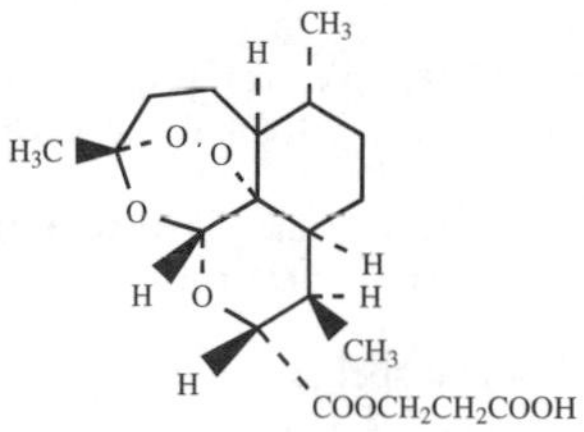

Arthr- A combining form meaning pertaining to a joint.

Arthra-G A trade name for salsalate, used as an analgesic and antipyretic agent.

Arthralgen A trade name for a combination drug containing acetaminophen and salicylamide, used as an analgesic and antipyretic agent.

Arthralgia Any pain that affects a joint.

Arthrexin A trade name for indomethacin, used as an anti-inflammatory, analgesic, and antipyretic agent.

Arthrinol A trade name for aspirin, used as an analgesic and antipyretic agent.

Arthritis Inflammatory condition of the joints, characterized by pain and swelling.

Arthrobacter A genus of obligately aerobic, catalase-positive, asporogenous bacteria (order Actinomycetales).

Arthrocentesis The puncture of a joint with a needle and the withdrawal of fluid for diagnostic purposes.

Arthropan A trade name for choline salicylate, used as an analgesic and antipyretic agent.

Arthropathy Any disease of the joints.

Arthropod Borne Virus Insect-transmitted viruses (see arbovirues).

Arthrospores Spores formed by the fragmentation of hyphae of certain fungi, algae, and cyanobacteria.

Arthrotec A trade name for a combination drug containing diclofenac and misoprostol used for treatment of rheumatoid arthritis.

Arthus Immune-Complex Reaction An inflammatory reaction characterized by edema, hemorrhage, and necrosis due to the formation of soluble immune complexes.

Arthus Phenomenon See Arthus immune-complex reaction.

Arthus Reaction See Arthus immune-complex reaction.

Articular Referring to a joint.

Articulin A cytoskeletal protein from the epiplasm of flagellate and ciliate organisms.

Articulose A trade name for prednisolone acetate, used as an anti-inflammatory agent.

Artifact Variant spelling of artefact.

Artificial Induction The induction of a prophage by a change in the conditions of the bacterial culture such that the immunity substance is either inactivated or not synthesized.

Artificial pH Gradient A linear pH gradient prepared manually or prepared with a gradient former.

Artificial Respiration The process of maintaining respiration by manual or mechanical means when normal breathing has stopped.

ARV Abbreviation for AIDS-associated retrovirus.

Aryl Any univalent organic radical derived from an arene by loss of one hydrogen atom.

Aryl-Acylamidase The enzyme that catalyzes the following reaction:

$$\text{An anilide} + H_2O \rightleftharpoons \text{A carboxylate} + \text{aniline}$$

Aryl-Acylamide Amidohydrolase The systematic name for aryl-arylamidase.

Aryl-Alcohol Dehydrogenase (NAD$^+$) The enzyme that catalyzes the following reaction:

$$\text{An aromatic alcohol} + NAD^+ \rightleftharpoons \text{An aromatic aldehyde} + NADH$$

Aryl-Alcohol Dehydrogenase (NADP$^+$) The enzyme that catalyzes the following reaction:

$$\text{An aromatic alcohol} + NADP^+ \rightleftharpoons \text{An aromatic aldehyde} + NADPH$$

Aryl-Alcohol Oxidase The enzyme that catalyzes the following reaction:

$$\text{An aromatic primary alcohol} + O_2 \rightleftharpoons \text{An aromatic aldehyde} + H_2O_2$$

Aryl-Aldehyde Dehydrogenase (NAD$^+$) The enzyme that catalyzes the following reaction:

$$\text{An aromatic aldehyde} + NAD^+ + H_2O \rightleftharpoons \text{An aromatic acid} + NADH$$

Aryl-Aldehyde Dehydrogenase (NADP$^+$) The enzyme that catalyzes the following reaction:

$$\text{An aromatic aldehyde} + NADP^+ + H_2O \rightleftharpoons \text{An aromatic acid} + NADPH$$

Aryl-Aldehyde Oxidase The enzyme that catalyzes the following reaction:

$$\text{An aromatic aldehyde} + O_2 \rightleftharpoons \text{An aromatic acid} + H_2O_2$$

Arylamine N-Acetyltransferase The enzyme that catalyzes the following reaction:

Acetyl-CoA + an arylamine

⇅

CoA + an N-acetylarylamine

Arylamine Glucosyltransferase The enzyme that catalyzes the following reaction:

UDP-glucose + lipopolysaccharide

⇅

UDP + α–D-glucosyl-lipopolysaccharide

Arylene Any bivalent organic group derived from an arene by loss of two hydrogen atoms.

Arylesterase The enzyme that catalyzes the following reaction:

Phenyl acetate + H_2O ⇌ Phenol + acetate

Aryl-Formanidase The enzyme that catalyzes the following reaction:

N-Formyl-L-kynurenine + H_2O

⇅

Formate + L-kynurenine

Aryl-Group Referring to an organic radical derived from an aromatic compound by loss of a hydrogen atom.

Arylsulfatase The enzyme that catalyzes the following reaction:

A phenol sulfate + H_2O

⇅

A phenol + sulfate

Arylsulfate Sulfohydrolase The systematic name for arylsulfatase.

Arylsulphatase Test A test used in the identification of *Mycobacterium* species possessing arylsulphatase.

As Symbol for arsenic with atomic weight 75, valence 3.

AS Abbreviation for 1. Ammonium sulfate. 2. Angeli's salt. 3. Antisense. 4. Anti-streptolysin. 5. Argininosuccinate synthetase. 6. Arthero-sclerosis. 7. Asparagine synthetase.

ASA Abbreviation for acetylsalicylic acid (aspirin).

5ASA Abbreviation for 5-aminosalicylic acid.

Asacol A trade name for mesalamine, a gastrointestinal anti-inflammatory agent.

ASAD Abbreviation for aspartate semialdehyde dehydrogenase.

ASAT Abbreviation for aspartate amino-transferase.

Asbestos An incombustible fibrous mineral of magnesium and calcium silicate; a suspected carcinogen.

Asbestosis A chronic lung disease caused by inhalation of asbestos fibers that results in the development of alveolar, interstitial, and pleural fibrosis.

Ascariasis An infection caused by a parasitic worm, *Ascaris lumbricoides*.

Ascaris A roundworm and intestinal parasite capable of inducing the production of IgE.

Ascarylose (mol wt 148) 3,6-Dieoxymannose, A component of glycolipids found in eggs of nematode.

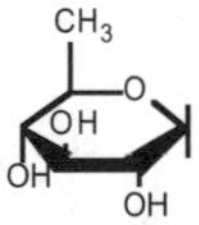

Ascending Boundary The electrophoretic boundary that moves upward in one of the arms of a Tiselius electrophoresis cell.

Ascending Chromatography A chromatographic technique in which the mobile phase moves upward along the support.

Aschaffenburg-Mullen Phosphatase Test A phosphatase test for milk.

Ascites The condition in which fluid accumulates in the peritoneal cavity.

Ascitic Of or pertaining to ascites.

Asclepain A protease from the latex of milkweed (*Asclepias syriaca*).

Ascoli Test A serological test for anthrax.

Ascomycetes A large group of true fungi with septate hyphae producing large numbers of asexual conidiospores and ascospores in asci.

Ascorbase See Ascorbate oxidase.

Ascorbate Cytochrome-b_5 Reductase The enzyme that catalyzes the following reaction:

2 L-Ascorbate + O_2

⇅

Dehydrioascorbate + $2H_2O$

Ascorbate 2,3-Dioxygenase The enzyme that catalyzes the following reaction:

$$\text{Ascorbate} + O_2 \rightleftharpoons \text{Oxalate} + \text{threonate}$$

L-Ascorbate Oxidase The enzyme that catalyzes the following reaction:

$$\text{Ascorbate} + O_2 \rightleftharpoons \text{Oxalate} + \text{threonate}$$

Ascorbate Oxygen Oxidoreductase The systematic name for L-ascorbate oxidase.

Ascorbate Peroxidase The enzyme that catalyzes the following reaction:

$$\text{L-Ascorbate} + H_2O_2 \rightleftharpoons \text{Dehydroascorbate} + 2H_2O$$

Ascorbemia The presence of ascorbic acid in the blood in amounts greater than normal.

Ascorbic Acid (mol wt 176) A vitamin, essential to man but not normally essential to microorganisms. It is a strong reducing agent.

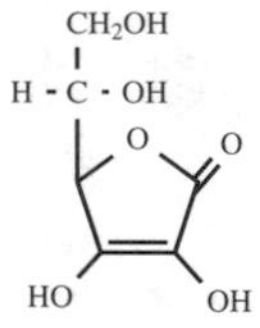

L-Ascorbic acid

Ascorbicap A trade name for ascorbic acid (vitamin C).

Ascorburia The presence of ascorbic acid in the urine in amounts greater than normal.

Ascospore A sexual spore of ascomycetes produced in a saclike structure known as an ascus.

Ascriptin A trade name for aspirin.

-ase A suffix denoting an enzyme.

AseI A restriction endonuclease isolated from *Aquaspirillum serpens* with the following specificity:

```
        ↓
5′.........ATTAAT.........3′
3′.........TAATTA.........5′
             ↑
```

AseII (CauII) A restriction endonuclease from *Aquaspirillum serpens* with the following specificity:

```
        ↓
5′..........CC(C/G)GG.........3′
3′..........GG(G/C)CC.........5′
                 ↑
```

Asemantic Molecule Any molecule that is not produced by organism.

Asendin A trade name for amoxapine, an antidepressant.

Asepsis 1. A condition of being aseptic. 2. A state in which potentially harmful microorganisms are absent.

Aseptic Free of microorganisms capable of causing infection or disease.

Aseptic Acid A mixture of boric acid, hydrogen peroxide, and salicylic acid in water, used as an antiseptic solution.

Aseptic Fever A fever not associated with infection.

Aseptic Technique Procedures used to prevent the introduction of fungi, bacteria, viruses, mycoplasm, or other microorganisms into the culture medium.

Asexual Any type of reproduction not involving the union of gametes, or meiosis.

Asexual Reproduction A type of reproduction without union of gametes.

ASG Abbreviation for anti-serum globulin.

ASGP Abbreviation for ascites sialoglycoprotein.

AsiAI A restriction endonuclease from *Arthrobacter* species A7359 with the following specificity:

```
          ↓
5'……..ACCGGT……..3'
3'……..TGGCCA……..5'
              ↑
```

Asialia Absence of saliva.

Asian flu A strain of influenza virus.

Asiderosis An abnormal decrease of iron in the body leading to the reduction of red cells in the blood.

A-Site The site on ribosomes for attachment of incoming charged tRNA in protein synthesis.

A-Site-P-Site Model The model of protein synthesis in which a ribosome possesses two binding sites: the aminoacyl site (A site) that binds the incoming aminoacyl-tRNA and the peptidyl site (P site) that binds the peptidyl-tRNA.

ASK Abbreviation for antistreptokinase.

ASL Abbreviation for 1. Anti-streptolysin. 2. Argininosuccinate lyase.

ASLT Abbreviation for antistreptolysin-O test.

ASM Abbreviation for 1. Acid-soluble metabolites. 2. Airway smooth muscle.

ASMA Abbreviation for anti-smooth muscle antibody.

Asmalix A trade name for theophylline, used as a bronchodilator.

Asmavent A trade name for albuterol, an adrenergic agonist used as a bronchodilator and antiasthmatic agent.

Asn Abbreviation for asparagine.

ASN Abbreviation for alkaline-soluble nitrogen.

AsnI A restriction endonuclease from *Arthrbacter* species NCM with the following specificity:

5'........ATTAAT........3'
3'........TAATTA........5'

ASO Abbreviation for 1. Antistreptolysin-O. 2. Allele-specific oligonucleotides.

ASO Test Abbreviation for antistreptolysin-O test.

ASOT Abbreviation for antistreptolysin-O test.

Asp Abbreviation for aspartic acid.

ASP Abbreviation for 1. Acid-soluble product or acid-soluble protein. 2. Acylation-stimulating protein.

AspI A restriction endonuclease from *Arthrobacter* species with the following specificity:

5'........GACNNNGTC........3'
3'........CCGNNNCAG........5'

Asp471 (XhoI) A restriction endonuclease from *Alcaligenes* species having the same specificity as XhoI.

Asp521 (HindIII) A restriction endonuclease from *Alcaligenes* species having the same specificity as HindIII.

Asp781 (StuI) A restriction endonuclease from *Alcaligenes* species having the same specificity as StuI.

Asp6971 (AvaII) A restriction endonuclease from *Achromobacter* species having the same specificity as AvaII.

Asp7001 (XmnI) A restriction endonuclease from *Achromobacter* species having the same specificity as XmnI.

Asp7031 (XhoI) A restriction endonuclease from *Achromobacter* species having the same specificity as XhoI.

Asp7071 (ClaI) A restriction endonuclease from *Achromobacter* species having the same specificity as ClaI.

Asp7081 (PstI) A restriction endonuclease from *Achromobacter* species having the same specificity as PstI.

Asp7181 (KpnI) A restriction endonuclease from *Acrhromobacter* species having the same specificity as KpnI.

Asp7421 (HaeIII) A restriction endonuclease from *Achromobacter* species having the same specificity as HaeIII.

Asp7481 (HpaII) A restriction endonuclease from *Achromobacter* species having the same specificity as HapII.

Asp7631 (ScaI) A restriction endonuclease from *Achromobacter* species having the same specificity as ScaI.

Asp700I A restriction endonuclease from *Arthrobacter* species 700 with the following specificity:

5'........GAANNNNTTC........3'
3'........CTTNNNNAAG........5'

Asp718I A restriction endonuclease from *Acinetobacter* species 718 with the following specificity:

5'........GGTACC........3'
3'........CCATGG........5'

AspAI (BstEII) A restriction endonuclease isolated from *Alcaligenes* species with the following specificity:

5′..........GGTNACC...........3′
3′..........CCTNTGG...........5′

L-Asparaginase The enzyme that catalyzes the hydrolysis of L-asparagine to L-aspartic acid and ammonia.

Asparagine (mol wt 132) A basic amino acid found in protein.

$NH_2COCH_2CH(NH_2)$-COOH

Asparagine Ketoacid Transaminase The enzyme that catalyzes the following reaction:

L-Asparagine + an α-ketoacid $\rightleftharpoons$ L-Aspartate + an amino acid

Asparagine Synthetase The enzyme that catalyzes synthesis of L-asparagine.

ATP + L-aspartate + NH_3 $\rightleftharpoons$ AMP + PPi + L-asparagine

Asparagine-tRNA Ligase See Asparaginyl-tRNA synthetase.

Asparaginyl-tRNA Synthetase The enzyme that catalyzes the following reaction:

ATP + L-asparagine + $tRNA^{asn}$ $\rightleftharpoons$ AMP + PPi + L-asparaginyl-$tRNA^{asn}$

Aspartame (mol wt 294) An artificial sweetener. It is dipeptide ester (160 times sweeter than sucrose). Also known as L-aspartyl-L-phenylalanine methyl ester.

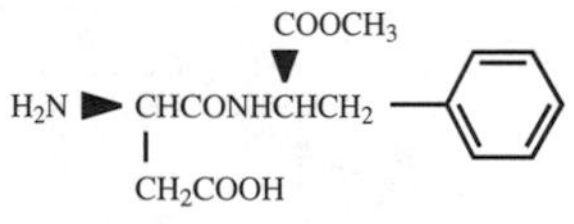

Aspartase See Aspartate ammonia-lyase.

Aspartate N-Acetyltransferase The enzyme that catalyzes the following reaction:

Acetyl-CoA + L-aspartate $\rightleftharpoons$ CoA + N-acetyl-L-aspartate

Aspartate Aminopeptidase The enzyme that catalyzes the release of N-terminal aspartate from a peptide.

Aspartate Aminotransferase The enzyme that catalyzes the following reaction:

L-aspartate + α-ketoglutarate $\rightleftharpoons$ oxaloacetate + L-glutamate

Aspartate Ammonia-Ligase The enzyme that catalyzes the following reaction:

ATP + L-aspartate + NH_3 $\rightleftharpoons$ Asparagine + PPi + AMP

Aspartate Ammonia-Ligase (ADP-Forming) The enzyme that catalyzes the following reaction:

ATP + L-aspartate + NH_3 $\rightleftharpoons$ Asparagine + Pi + ADP

Aspartate Ammonia-Lyase The enzyme that catalyzes the following reaction:

L-Aspartate $\rightleftharpoons$ Fumarate + NH_3

Aspartate Carbamoyltransferase The enzyme that catalyzes the following reaction:

Carbamoylphosphate + L-aspartate $\rightleftharpoons$ Pi + N-carbamoyl-L-aspartate

Aspartate 1-Decarboxylase The enzyme that catalyzes the following reaction:

L-Aspartate $\rightleftharpoons$ β-Alanine + CO_2

Aspartate 4-Decarboxylase The enzyme that catalyzes the following reaction:

L-Aspartate $\rightleftharpoons$ L-Alanine + CO_2

Aspartate β-Decarboxylase See Aspartate 4-decarboxylase.

Aspartate Kinase The enzyme that catalyzes the following reaction:

ATP + L-aspartate $\rightleftharpoons$ ADP + 4-phospho-L-aspartate

D-Aspartate Oxidase The enzyme that catalyzes the following reaction:

D-Aspartate + H_2O + O_2 $\rightleftharpoons$ Oxaloacetate + NH_3 + H_2O_2

L-Aspartate Oxidase The enzyme that catalyzes the following reaction:

L-Aspartate + H_2O + O_2 $\rightleftharpoons$ Oxaloacetate + NH_3 + H_2O_2

Aspartate 2-Oxoglutarate Aminotransferase The systematic name for aspartate transaminase.

Aspartate Racemase The enzyme that catalyzes the following reaction:

L-Aspartate ⇌ D-Aspartate

Aspartate Semialdehyde Dehydrogenase The enzyme that catalyzes the following reaction:

L-Aspartate β-semialdehyde + Pi + $NADP^+$

⇅

L-β-Aspartylphosphate + NADPH

Aspartate Transaminase An enzyme that catalyzes the transfer of an amino group from the aspartic acid molecule to another molecule.

Aspartate Transcarbamoylase The enzyme that catalyzes the following reaction:

Carbamoylphosphate + L-aspartate

⇅

Pi + N-carbamoyl-L-asparate

Aspartate tRNA Ligase See Aspartate-tRNA synthetase.

Aspartate tRNA Synthetase The enzyme that catalyzes the following reaction:

ATP + L-aspartate + $tRNA^{asp}$

⇅

AMP + PPi + L-aspartyl-$tRNA^{asp}$

Aspartic Acid (mol wt 133) An aliphatic, acidic, and polar alpha amino acid.

COOH
|
CH_2
|
$CHNH_2$
|
COOH

Aspartic Semialdehyde (mol wt 118) A derivative of aspartic acid in which one of the two carboxyl groups has been converted to an aldehyde group.

CHO
|
CH_2
|
$CHNH_2$
|
COOH

Aspartoacylase The enzyme that catalyzes the following reaction:

N-Acyl-L-aspartate + H_2O

⇅

A carboxylate + L-asparate

Aspartokinase Synonym of aspartate kinase.

Aspartyl-tRNA Ligase See Aspartate-tRNA synthetase.

Aspartyl-tRNA Synthetase See Aspartate-tRNA synthetase.

A-Spas A trade name for dicyclomine hydrochloride, an antidiarrheal and antispasmodic agent.

AspBI (AvaI) A restriction endonuclease from *Anabaena* species having the same specificity as AvaI.

AspEI A restriction endonuclease from *Auerobaccterium* species with the following specificity:

5'.........GACNNN↓NNGTC........3'
3'.........CTGNN↑NNNCAG........5'

Aspergillic Acid (mol wt 224) An antibiotic produced by *Aspergillus flavus*.

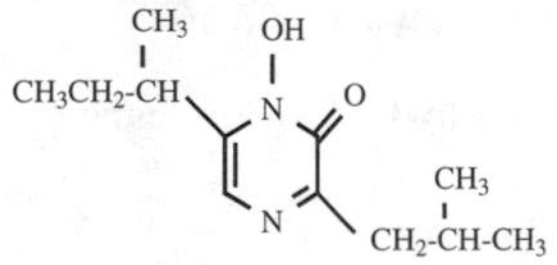

Aspergillin Any antibacterial agents produced by *Aspergillus*.

Aspergillopepsin I A protease from *Aspergillus* species that catalyzes the activation of typsinogen but does not clot milk.

Aspergillopepsin II A protease from *Aspergillus niger* that catalyzes the hydrolysis of peptide bonds between Asn-Gly; Gly-Ala; and Tyr-Thr.

Aspergillopeptidase A The enzyme that hydrolyzes peptide bonds involving carboxyl groups of arginine or leucine.

Aspergillosis Infection or disease caused by *Aspergillus* species.

Aspergillus A genus of ascomycetous fungi.

Aspergum A trade name for aspirin, used as an analgesic and antipyretic agent.

Asperlicin (mol wt 536) A naturally occurring nonpeptide cholecystokinin antagonist produced by *Aspergillus alliaceus*.

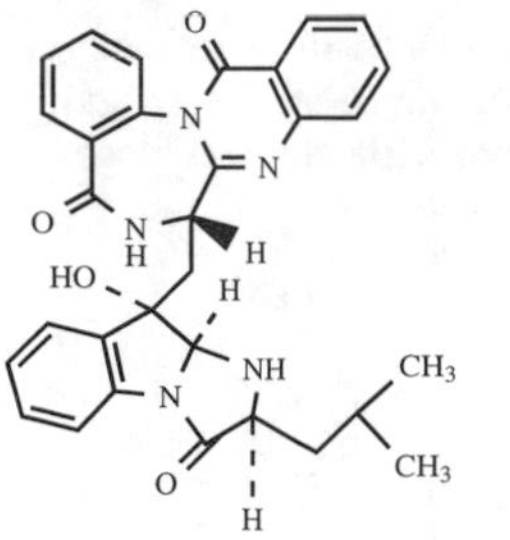

AspHI (HgiAI) A restriction endonuclease from *Achromobacter* species with the following specificity:

5′.........G(A/T)GC(A/T)↓C.........3′
3′.........C↑(T/A)CG(T/A)G.........5′

Asphygmia Temporary absence of pulse.

Asphyxia Loss of consciousness due to insufficient oxygen supply.

Asphyxiant Any agent capable of causing asphyxia.

Aspidin (mol wt 461) An anthelmintic agent.

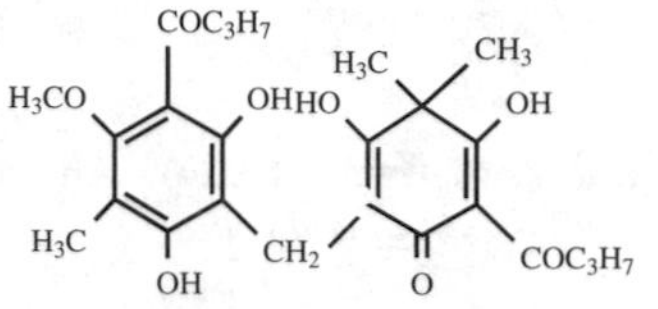

Aspidinol (mol wt 224) An anthelmintic agent.

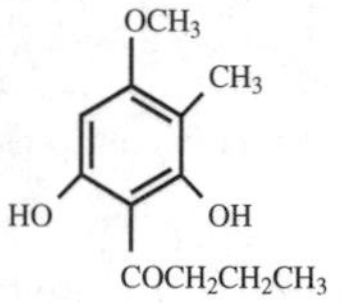

Aspidium Rhizome and stipes of *Dryopteris filix-mas* (Polypodiaceae) used as an anthelmintic agent.

Aspidosperma Dry bark of Aspidosperma *quebracho-blanco* (Apocynaceae) used as a respiratory stimulant.

Aspidospermine (mol wt 354) A diuretic and respiratory stimulant agent.

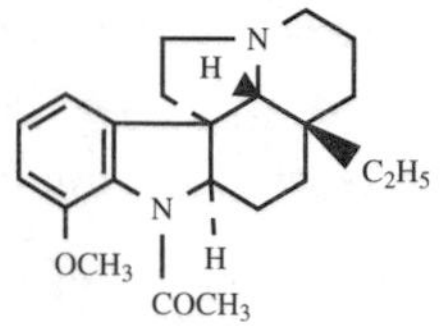

Aspiration To remove a liquid layer, such as a supernatant, from a sample using a pipet or equivalent attached to a vacuum source.

Aspirin (mol wt 180) An analgesic, antipyretic, and anti-inflammatory agent that inhibits the synthesis of prostaglandin and blocks the generation of pain impulses.

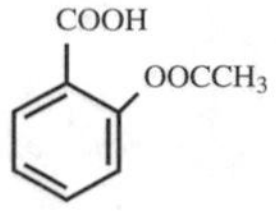

AspLEI A restriction endonuclease from *Arthrobacter* species LE3860 species with the following specificity:

5'........GCG↓C........3'
3'........C↑GCG.......5'

AspN Abbreviation for asparagine.

AspNH$_2$ Symbol for asparagine.

AspOMe Abbreviation for O-methyl aspartate.

Asporogenous Incapable of forming spores.

Aspoxicillin (mol wt 494) A semisynthetic penicillin; an antibiotic that inhibits bacterial cell synthesis.

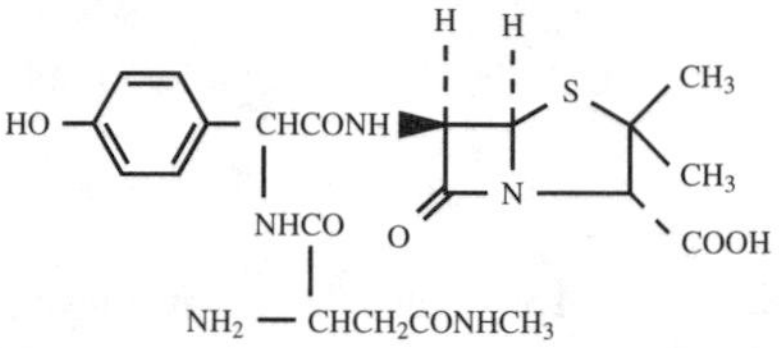

Aspro A trade name for aspirin, used as an analgesic and antipyretic agent.

Asproject A trade name for sodium thiosalicylate, used as an analgesic and antipyretic agent.

AspS9I A restriction endonuclease from *Arthrobacter* species S9 species with the following specificity:

5'........G↓GNCC........3'
3'........CCNG↑G........5'

ASS Abbreviation for argininosuccinate synthetase.

Assay To determine either the concentration or the activity of a substance.

Assimilation The process of conversion of nutrients by an organism into complex constituents of the organism.

Assimilatory Nitrate Reductase (NADH) The synonym of nitrate reductase.

Assimilatory Nitrate Reductase (NADPH) Synonym of nitrate reductase.

Assimilatory Nitrate Reduction The process of reduction in plants whereby nitrate is reduced to ammonium and then assimilated into cellular organic compounds.

Assimilatory Reduction The process of reduction in plants whereby substances, e.g., sulfate and nitrate, are reduced and then assimilated into cellular organic compounds.

Association Colloid A surface-active agent that tends to aggregate and to form micelles in solution.

Association Constant The equilibrium constant for the formation of a complex from simpler components, e.g., association of a proton and an anion to form an acid or the formation of complexes between enzyme and inhibitor or between antigen and antibody.

AST Abbreviation for angiotensin sensitivity test.

Astacin A protease from the crayfish.

Astatin A zinc-containing endopeptidase.

Astatine Radioactive halogen, one of the rarest elements in nature.

Astaxanthin (mol wt 597) A carotenoid pigment found mostly in animals.

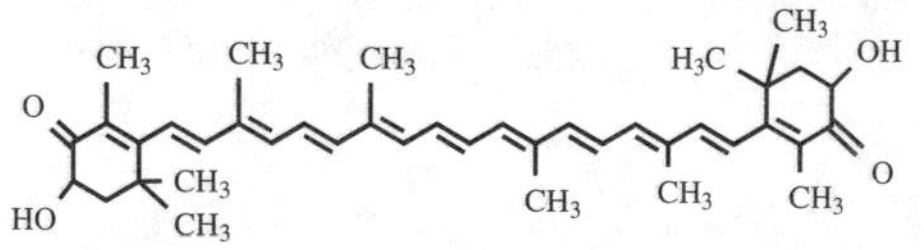

Astelin A trade name for azelastine, a histamine H_1 blocker.

Astemizole (mol wt 459) An anti-histaminic agent that blocks the effects of histamine at the H-1 receptor.

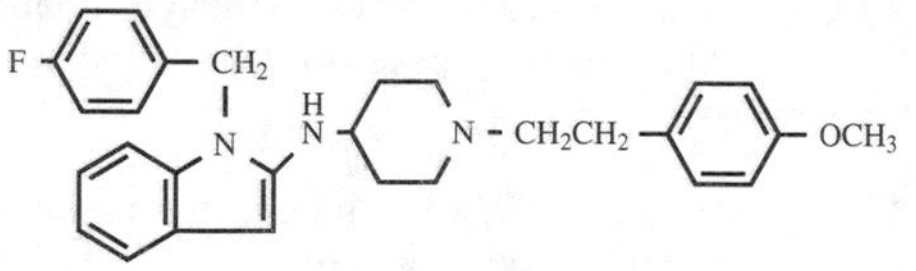

Asthma Type 1 hypersensitivity reaction that affects the respiratory tract; characterized by shortness of breath and wheezing.

-asthma A combining form meaning labored breathing.

AsthmaNefrin A trade name for epinephrine hydrochloride, a hormone.

Asticcacaulis A genus of chemoorganotrophic, strictly aerobic prosthecate bacteria.

Astral Mitosis Mitosis characterized by the presence of an astral array of microtubules at each pole of the mitotic spindle.

Astramorph A trade name for morphine sulfate, used as an opioid analgesic agent.

Astrin A trade name for aspirin, used as an analgesic and antipyretic agent.

Astringent 1. A medicinal substance that shrinks and hardens tissues. 2. Substance that causes contraction of tissue and arrests discharge.

Astroblast Primitive cell before it develops into an astrocyte.

Astroblastomata A malignant neoplasm of the brain and spinal cord.

Astrocyte The large neuroglial cell that has a star-shaped cell body with numerous processes radiating outward.

Astroviruses Spherical, ether-resistant viruses having a six-pointed star-shaped surface projection.

AstWI (AcyI) A restriction endonuclease isolated from *Anabaena* species with the following specificity:

↓
5′..........GPuCGPyC..........3′
3′..........CPyGCPuG..........5′
↑

AsuI A restriction endonuclease isolated from *Anabaena subcylindrica* with the following specificity:

↓
5′..........GGNCC..........3′
3′..........CCNGG..........5′
↑

AsuII A restriction endonuclease isolated from *Anabaena subcylindrica* with the following specificity:

↓
5′..........TTCGAA..........3′
3′..........AAGCTT..........5′
↑

AsuIII (Acyl) A restriction endonuclease isolated from *Anabaena subcylindrica* with the following specificity:

↓
5′..........GPuCGPyC..........3′
3′..........CPyGCPuG..........5′
↑

AsuHPI A restriction endonuclease from *Actinobacillus suis* HP with the following recognition sequence:

5'........GGTGA(8/7)........3'

It cleaves either within the recognition site or a short specific distance from it.

AsuNHI A restriction endonuclease from *Actinobacillus suis* NH with the following specificity:

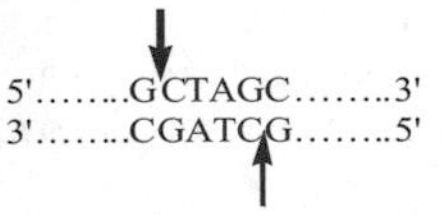

ASV Abbreviation for anti-snake venom.

ASW Abbreviation for artificial seawater.

As-x Abbreviation for amino acid aspartic acid or asparagine.

Asymmetric Lacking symmetry.

Asymmetric Carbon A carbon atom at which four different substituents are attached.

Asymmetric Synthesis The synthesis of only one of two optical isomers.

Asynaptic The failure of homologous chromosomes to pair during the first meiotic division.

Asynergy A condition characterized by faulty coordination among groups of organs or muscles that normally function harmoniously.

Asystole The absence of a heartbeat or muscular contraction of the heart.

At Symbol for the element astatine.

AT Abbreviation for 1. Adenine-thymine. 2. Amino-transferase. 3. Anaphylatoxin. 4. Angiotensin. 5. Anti-thrombin. 4. Antitrypsin. 5. Antitriptyline.

A/T Abbreviation for adenine and thymine ratio in DNA.

AT-III Abbreviation for antithrombin III.

ATA Abbreviation for 1. Anti-thyroglobin antibody. 2. Anti-toxoplasm antibodies. 3. Aurin tricarboxylic acid.

Atabrine A trade name for quinacrine hydrochloride, used as an anthelmintic agent.

Atacand A trade name for candesartan cilexitil, an angiotensin-II receptor antagonist used as an antihypertensive agent.

Atamet A trade name for a combination drug containing levodopa and carbidopa used as antiparkinsonism drug.

Atarax A trade name for hydroxyzine hydrochloride, an antianxiety agent.

Atasol A trade name for acetaminophen, used as an analgesic and antipyretic agent.

Atavism See Atavistic.

Atavistic A reversion to phenotype not present for several generations.

Ataxia Lack of muscular coordination.

ATC Abbreviation for 1. Acetylthiocholine. 2. Activated thymus cell. 3. Activated thymocyte.

ATCase Abbreviation for enzyme aspartate transcarbamoylase.

ATD Abbreviation for anti-thyroid drugs.

ATE Abbreviation for adipose tissue extract.

Atelectasis An abnormal condition characterized by the collapse of lung tissue; prevents the respiratory exchange of carbon dioxide and oxygen.

Atelo- A combining form denoting incomplete.

Atelocardia Incomplete cardiac development.

Atenolol (mol wt 266) An antihypertensitive, antianginal, and antiarrhythmic agent that blocks the response of beta-1 adrenergic receptors.

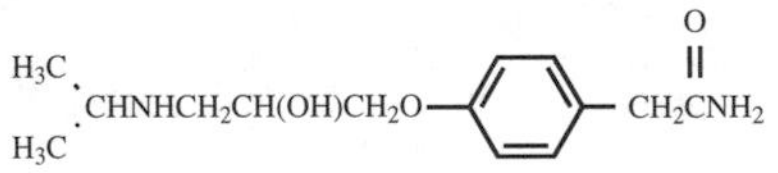

ATF Abbreviation for activating transcription factor.

ATF-2 Abbreviation for activating transcription factor 2.

ATG Abbreviation for 1. Anti-thymocyte globulins. 2. Anti-thrombocyte globin. 3. Anti-thyroglobin.

Atgam A trade name for lymphocyte immunoglobulin or antithymocyte globulin.

AT/GC Ratio The adenine-thymine to guanine-cytosine ratio in DNA.

αtGDP Abbreviation for GDP-bound transducin α subunit.

ATGG Abbreviation for anti-thymocyte gamma globulin.

αtGTP Abbreviation for GTP-bound transducin α-subunit.

αtGTP[S] Abbreviation for GTP[S]-bound transducin α-subunit,

Atherogenesis The development of atherosclerosis or the formation of atheromas.

Atheroma A lipid-containing deposit in the arteries undergoing hardening.

Atheromatosis The development of atheromas.

Atheromatous Of or pertaining to atheromas.

Atherosclerosis A chronic cardiovascular disorder in which plaques develop on the inner walls of the arteries and narrow the passageway.

Athiaminosis A condition resulting from lack of thiamine in the diet.

Athlete's Foot A disease caused by dermatophytic fungi affecting chronically wet feet with skin abrasions; tinea or ringworm of the feet.

Athrombin-K A trade name for warfarin, an anticoagulant.

Athymic Lacking a thymus.

Athymic Mouse Mouse without a thymus (nude mouse).

Ativan A trade name for lorazepam, an anti-anxiety agent.

ATL Abbreviation for adult T-cell leukemia.

ATLA Abbreviation for adult T-cell leukemia antigen.

ATLL Abbreviation for adult T cell leukemia-lymphoma

ATLV Abbreviation for adult T-cell leukemia virus.

Atm Symbol for atmospheric pressure; the amount of pressure exerted by the atmosphere at sea level (760 mm of mercury).

αTM Abbreviation for alpha-thrombomodulin.

Atom The smallest unit of a chemical element; composed of protons, neutrons, and electrons.

Atomic Absorption Spectrophotometry An analytical method for the spectrophotometric determination of elements. It is based on a measurement of the radiation absorbed by unexcited, nonionized, and ground-state atoms, which are produced when compounds are dissociated into atoms by means of a flame.

Atomic Mass The mass of a neutral atom expressed in terms of atomic mass units.

Atomic Mass Unit A unit of atomic mass: one-twelfth of the mass of a neutral ^{12}C atom or $1.6605655 \times 10^{-24}$ gram.

Atomic Number (Z) The number of protons in the nucleus of an atom, it also represents the number of orbital electrons surrounding the nucleus of the neutral atom.

Atomic Radius The distance between the nucleus and the outermost electron shell of an atom.

Atomic Volume The atomic weight of an element divided by its density.

Atomic Weight The sum of the protons and neutrons in an atomic nucleus as compared with C^{12}, which is assigned a value of 12.0000.

Atomic Weight Unit One-twelfth of the mass of the carbon isotope (1.67×10^{-24} gram).

Atomization The conversion of a liquid to a fine spray by breaking up a solution into fine drops.

Atomizer A device to produce a fine spray by breaking up a solution into fine drops.

Atonic Referring to 1. weakness, 2. lack of normal tone, or 3. lack of vigor.

Atopic Of or pertaining to a hereditary tendency to develop allergic reactions, e.g., asthma and atopic dermatitis (see also atopy).

Atopy An inherited tendency to develop immunological hypersensitivity states to common environmental allergens, e.g., hay fever, asthma, and urticaria. It is an IgE-mediated, type I hypersensitivity reaction.

Atorvastatin (mol wt 558) An antihyperlipoproteinemic agent used for lowering blood cholesterol.

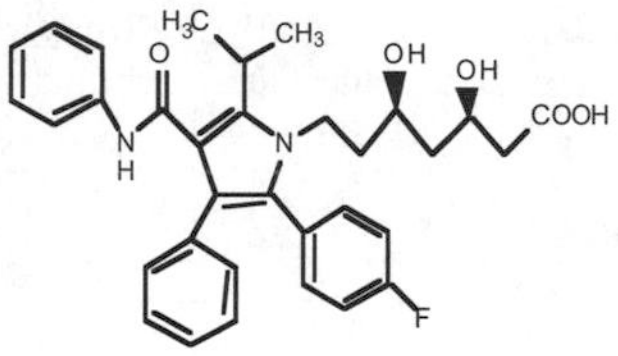

Atovaquone (mol wt 367) An antiprotozoal, antipneumocystis, and antimalarial agent.

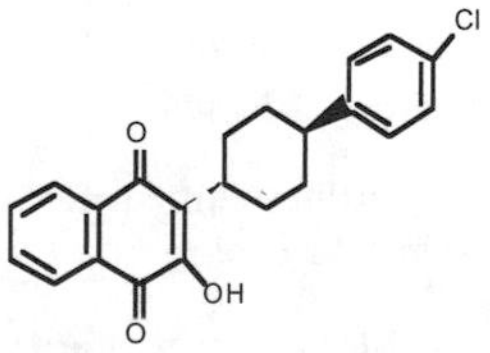

Atoxic Nontoxic.

ATP Abbreviation for adenosine 5′-triphosphate, the major form of energy in the cell.

ATP Acetate Phosphotransferase The systematic name for acetate kinase.

ATP ADP Carrier System A coupled system in mitochondria for outward transport of ATP and inward movement of ADP across the mitochondrial membrane.

ATP ADP Cycle The sum of the reactions for synthesis and hydrolysis of ATP.

ATP ADP Translocase A transport protein that mediates the transport of ATP and ADP across the inner mitochondrial membrane.

ATP AMP Phosphotransferase The systematic name for adenylate kinase.

ATP Arginine N-Phosphotransferase The systematic name for arginine kinase.

ATP Carbamate Phosphotransferase The systematic name for carbamate kinase.

ATP Choline Phosphotransferase The systematic name for choline kinase.

ATP Creatine N-Phosphotransferase The systematic name for creatine kinase.

ATP Deaminase The enzyme that catalyzes the following reaction:

$$ATP + H_2O \rightleftharpoons ITP + NH_3$$

ATP Diphosphatase The enzyme that catalyzes the following reaction:

$$ATP + H_2O \rightleftharpoons AMP + 2Pi$$

ATP 5'-Dephosphopolynucleotide 5'-phosphotransferase The systematic name for polynucleotide 5'-hydroxyl-kinase.

ATP 1,2-Diacylglycerol 3-Phosphotransferase The systematic name for diacylglycerol kinase.

ATP Diphosphohydrolase Synonym of ATP diphosphatase.

ATP Fructose 6-Phosphate 1-Phosphotransferase The systematic name for 6-phosphofructokinase.

ATP Galactose 1-Phosphotransferase The systematic name for galactokinase.

ATP Gluconate 6-Phosphotransferase The systematic name for gluconokinase.

ATP Glucose 6-Phosphotransferase The systematic name for glucokinase.

ATP Glycerol 3-Phosphotransferase The systematic name for glycerol kinase.

ATP dGMP Phosphotransferase The systematic name for guanylate kinase.

ATP Hexose 6-Phosphotransferase The systematic name for hexokinase.

ATP Kanamycin 3'-O-Phosphotransferase The systematic name for kanamycin kinase.

ATP Monophosphatase Synonym of adenosine triphosphatase.

ATP NAD$^+$ 2'-Phosphotransferase The systematic name for NAD$^+$ kinase.

ATP Nicotinamide Nucleotide Adenylyltransferase The systematic name for nicotinamide nucleotide adenylyltransferase.

ATP Nucleoside Diphosphate Phosphotransferase The systematic name for nucleoside diphosphate kinase.

ATP Nucleoside Phosphate Phosphotransferase The systematic name for nucleoside phosphate kinase.

ATP 3-Phospho-D-Glycerate 1-Phosphotransferase The systematic name for phosphoglycerate kinase.

ATP Phosphohydrolase The systematic name for adenosine triphosphatase.

ATP Phosphorylase-b Phosphotransferase The systematic name for phosphorylase kinase.

ATP Protein O-Phosphotransferase (Calmodulin-Dependent) The systematic name for Ca^{2+}/calmodulin-dependent protein kinase.

ATP Protein Phosphotransferase The systematic name for protein kinase.

ATP Pyrophosphate-Lyase (Cyclizing) The systematic name for adenylate cyclase.

ATP Pyruvate 2-O-Phosphotransferase The systematic name for pyruvate kinase.

ATP Regenerating System An enzymatic system linked to the mitochrondrial proton gradient for the synthesis of ATP.

ATP Ribose 5-Phosphate Pyrophosphotransferase The systematic name for ribose phosphate pyrophosphokinase.

ATP Ribose 5-Phosphotransferase The systematic name for ribokinase.

ATP Ribulose 5-Phosphate 1-Phosphotransferase The systematic name for phosphoribulokinase.

ATP Synthetase See ATPase.

ATP Synthetase Complex Referring to the F_oF_1-ATPase complex that is responsible for synthesis of ATP from ADP and inorganic phosphate.

ATPase Abbreviation for the enzyme adenosine 5′-triphosphatase (F_oF_1 complex). It catalyzes the following reaction:

$$ATP + H_2O \rightleftharpoons ADP + Pi$$

ATP(α,β-CH$_2$) Abbreviation for α,β-methylene-adenosine 5’-triphosphate.

ATP(β,γ-CH$_2$) Abbreviation for β,γ-methylene-adenosine 5’-triphosphate.

[ATP]$_e$ Abbreviation for extracellular ATP concentration.

[ATP]$_i$ Abbreviation for intracellular ATP concentration.

ATPS Abbreviation for adenosine 5’-γ-thiotriphosphate.

ATP[βS] Abbreviation for adenosine 5’-[β-thio]triphosphate.

Atractyloside (mol wt 803) A glucoside produced by the Mediterranean thistle (*Atractylis gummitera*). It inhibits the ADP/ATP exchange carrier system in the mitochondrial inner membrane.

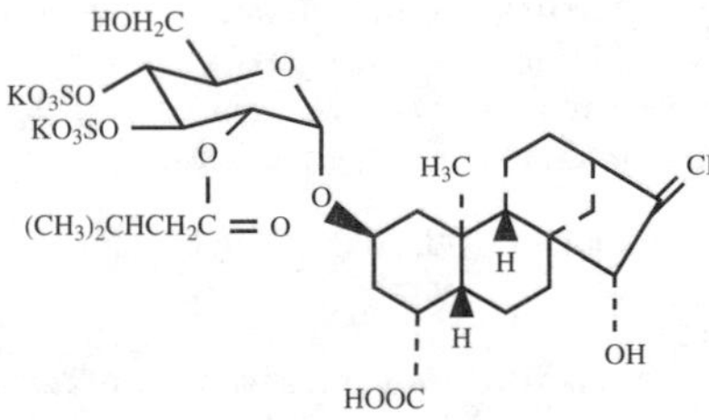

Atractyloside Barrier Referring to the blockage of ATP-ADP transport across the inner mitochondrial membrane produced by the action of atractyloside.

Atrazine (mol wt 216) An herbicide.

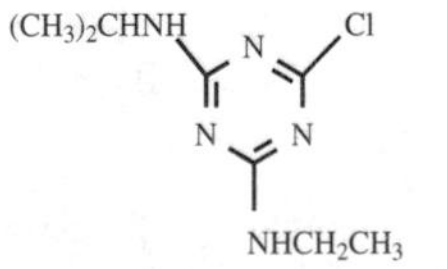

Atretol A trade name for carbamazepine, an anticonvulsant and analgesic agent.

ATR-FTIR Abbreviation for attenuated total reflection Fourier-transform infrared.

Atria The chambers of the heart that receive blood returning to the heart.

Atrial Natriuretic Peptide See natriuretic peptide.

Atrichous 1. Hairless. 2. Lacking flagella. 3. Lacking any filamentous appendages.

Atriopeptins A bioactive protein isolated from mammalian cardiac atria.

Atriposol A trade name for atropine, an anticholinergic, antispasmodic and an antimuscarinic agent.

Atrium 1. A chamber or cavity. 2. One of the two (right or left) chambers of the heart that receives the blood returning to the heart.

Atrohist A trade name for atropine sulfate, an eye muscle relexant and pupil enlarger.

Atrolactamide (mol wt 165) An anticonvulsant.

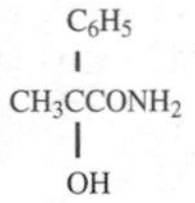

Atrolysin A protease from the venom of the western diamondback rattlesnake.

Atrolysin F A hemorrhagic endopeptidase from the venom of the western diamondback rattlesnake. It digests the γ-chain of fibrinogen.

Atromid-S A trade name for clofibrate, used as an antilipemic agent.

Atropar A trade name for atropine, an anticholinergic and antiarrhythmic agent.

Atrophy A wasting, progressive degeneration and loss of function of any part of the body.

Atropic Pertaining to atrophy.

Atropine (mol wt 289) An alkaloid derived from deadly nightshade (*Atropa*) and seeds of the thorn apple. It possesses antispasmodic activity and is an antidote for organophosphous insecticide poisoning.

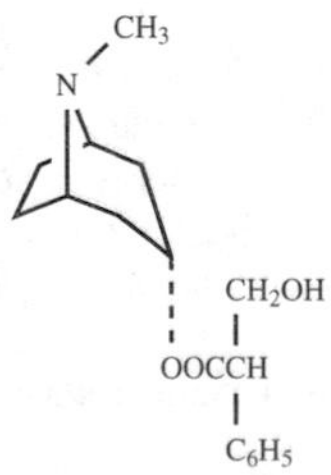

Atropine N-Oxidide (mol wt 305) An anticholinergic agent.

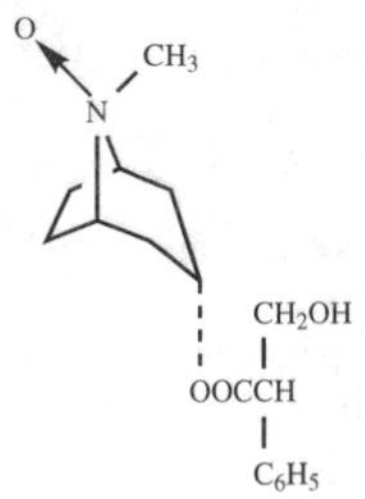

Atropinism Poisoning due to atropine.

Atropisol A trade name for atropine sulfate, an eye muscle relaxent and pupil enlarger.

Atrovent A trade name for ipratropium bromide, used as a bronchodilator.

Atroxase A protease from the venom of the western diamondback rattlesnake.

ATS Abbreviation for 1. Antithymocyte serum. 2. Anti-tetanic serum. 3. Arteriosclerosis.

AtsI A restriction endonuclease from *Aureobacterium testaceum* 4842 with the following specificity:

5'........GACNNNGTC........3'
3'........CTGNNNCAG........5'

ATT Abbreviation for arginine tolerance test.

AttB Site See Att site.

Attenuate 1. To decrease the virulence of a bacterium or a virus. 2. To decrease the attachment ability of enzymes or proteins.

Attenuated Vaccine A vaccine containing live pathogens whose virulence for a given host has been reduced or abolished by attenuation.

Attenuation Any procedure in which the virulence of a given (live) pathogen for a particular host is reduced or abolished without altering its immunogenicity.

Attenuator Region See Attenuator site.

Attenuator Site The site between the operator region and the first structural gene of an operon where most RNA polymerases stop transcription. The addition of specific antitermination factor will cause transcription to proceed.

Attenuvax A trade name for a live measles virus vaccine.

Atto- A prefix meaning 10^{-18} in the metric system of measurement.

Attomole A unit equals to 10^{-18} mole.

AttP-Site See Att site.

Att-Site A DNA sequence at which site-specific recombination occurs during integration of the genome of a temperate bacteriophage into the chromosome of its host; the site on the phage genome is designated attP, that on the host chromosome is designated attB.

AtuII (EcoRII) A restriction endonuclease from *Agrobacterium tumefaciens* having the same specificity as EcoRII.

AtuBI (EcoRII) A restriction endonuclease isolated from *Agrobacterium tumefaciens* with the following specificity:

5′.........CC(A/T)GG.........3′
3′.........GG(T/A)CC.........5′

AtuCI(Bc/1) A restriction endonuclease from *Agrobacterium tumefaciens* with the following specificity:

5′..........TGATCA..........3′
3′..........ACTAGT..........3′

ATX Abbreviation for *Anemonia sulcata* toxin.

A-Type Particle Intracellular, noninfectious, retroviruslike particles that possess reverse transcriptase activity and an RNA genome coding for the structural protein of the particles.

Au Symbol for gold, atomic weight 197, valence 1 and 3.

AUA A genetic codon for the amino acid isoleucine.

AuAg Abbreviation for Australian antigen.

AuBMT Abbreviation for autologous bone marrow transplantation.

AUC A genetic codon for the amino acid isoleucine.

Aucubin (mol wt 346) Substance isolated from leaves, roots, stalks, and seeds of *Aucaba japonica* (Cornaceae).

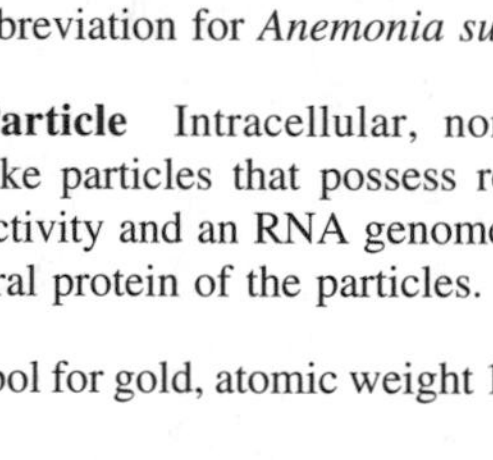

Audiogenic Sound producing; caused by sound.

Audiology The science that deals with hearing.

Audiometer A device for determining the acuteness of hearing or intensity of sounds.

AUG The initiation codon on mRNA.

Augmentin A trade name for a combination drug containing amoxicillin and clavulanic acid, used in the treatment of urinary tract infections.

AuHAA Abbreviation for Australian hepatitis-associated antigen.

AUL Abbreviation for acute undifferentiated leukemia.

AUO Abbreviation for amyloid of unknown origin.

Auracyanin A blue copper-containing bacterial outer membrane glycoprotein that donates electrons to cytochrome c_{554}.

Auralgan A combination drug containing antipyrine, benzocaine and dehydrated glycerin used to reduce inflammation, congestion and alleviates pain and discomfort in acute otitis media.

Auramine O (mol wt 304) A basic, yellow substituted diphenylmethane fluorochrome. It forms a highly fluorescent complex with horse-liver dehydrogenase.

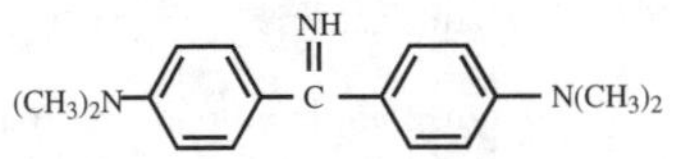

Auramine-Rhodamine Stain A mixture of auramine O and rhodamine B; a fluorescent acid-fast stain.

Auranofin (mol wt 678) An antirheumatic and anti-inflammatory agent.

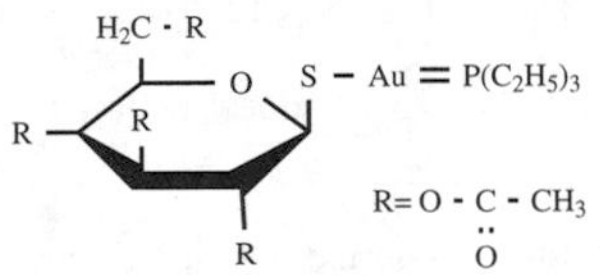

Aurantiogliocladin (mol wt 196) An antibiotic produced by a *Gliocladium* sp.

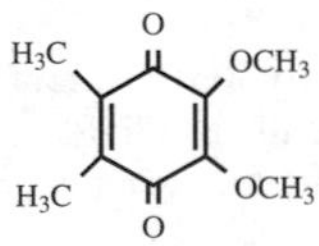

Aurate The anion of $Au(OH)_4^-$.

Aureo- A prefix meaning golden or yellow.

Aureofungin An antifungal antibiotic produced by *Streptoverticillium cinnamomeum var terricolum*. It is a golden-yellow powder that is insoluble in water but soluble in dilute alkali or ethanol.

Aureolysin A protease that catalyzes the cleavage of the B chain of insulin with specificity similar to that of thermolysin.

Aureomycin A trade name for the antibiotic chlorotetracycline that inhibits bacterial protein synthesis.

Aureothin (mol wt 397) An antibiotic isolated from cultures of *Streptomyces thioluteus*.

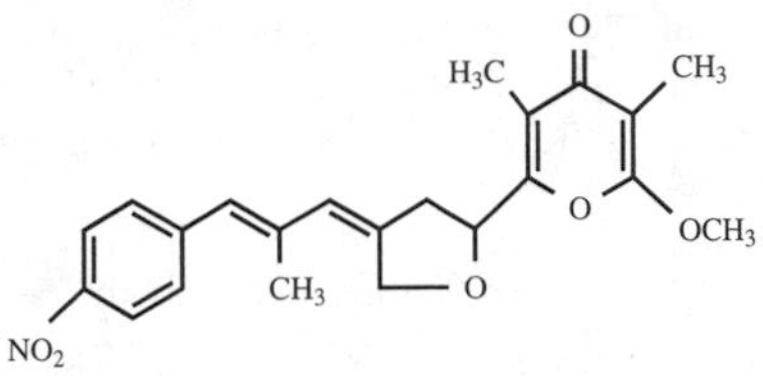

Aureothricin (mol wt 242) An antibiotic substance produced by *streptomyces* sp.

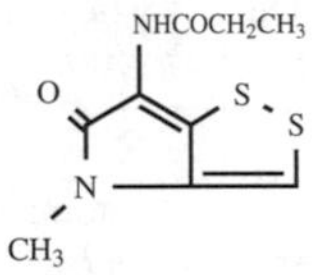

Auri- A prefix denoting 1. the ear or 2. gold.

Aurintricarboxylic Acid (mol wt 422) A dye that inhibits protein synthesis by blocking the attachment of messenger RNA to ribosomes. It is also a chelating agent with high affinity for beryllium, which has been used for treatment of beryllium poisoning.

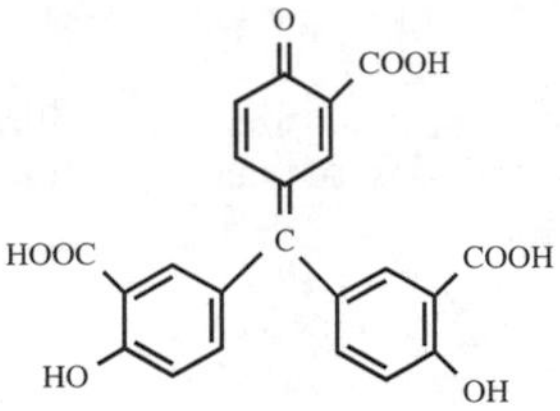

Auro- A prefix denoting gold.

Aurososme An artificially induced organelle occurring in animal cells cultured in the presence of gold.

Aurothioglucose (mol wt 392) An antirheumatic agent.

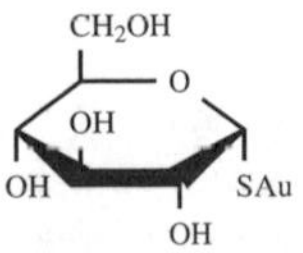

Aurothioglycanide (mol wt 363) An antirheumatic agent.

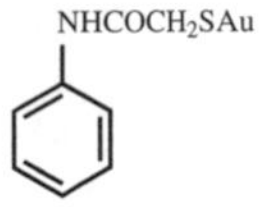

AuSH Abbreviation for Australian serum hepatitis.

Australia Antigen The serum hepatitis antigen (hepatitis B) that was originally discovered in an Australian aborigine.

Autacoid Substances, e.g., hormones, that are produced in one organ and act as a means to control the physiological processes in another part of the body.

AutoAb Abbreviation for autoantibody.

Autoactivation The activation of an endocrine gland by its own secretory product or products.

Auto-agglutination 1. The spontaneous agglutination of one's own cells or cellular particles by unimmunized plasma or serum. 2. The agglutination of bacteria in saline.

Autoagglutinin An agglutination produced by an autoantibody from one's own immune system.

Autoantibody An antibody produced by the immune system against one's own cellular constituents.

Autoantigen Self-antigen that becomes antigenic in one's own immune system.

Autocatalysis A process in which the product of a reaction acts as a catalyst to accelerate the reaction.

Autoclave Apparatus used for sterilization employing air-free saturated steam, pressure, and high temperature.

Autocoupling Hapten The formation of hapten-tissue-protein complexes when the hapten is injected into an animal and combines with tissue protein leading to the formation of antibody against the hapten.

Autocrine Effects of hormones on the cells that produces them.

Autocrine Cell A cell that possesses the receptors for hormones, growth factors, or other signaling substances that are produced by the same cell.

Autocrine Growth Pathway The cells that secrete their own growth-promoting factors (e.g., hormones and cytokines) for cell proliferation.

Autodiploidy The state or condition in which there are two genetically identical or nearly identical chromosome sets from the same ancestral species.

Autoecious Completing a life cycle on the same host.

Autoerythrocyte Sensitization An autoimmune hemolytic anemia characterized by the spontaneous appearance of painful, hemorrhagic spots on the body (e.g., arms and legs) due to the production of autoantibody against the red blood cell.

Autoerythrophagocytosis The ingestion of autologous erythrocytes by macrophages.

Autofluorescence The fluorescence of tissues due to molecules naturally present in the tissues.

Autogamy 1. Self-fertilization. 2. Fusion of haploid nuclei or gametes derived from a single cell.

Autogenesis 1. Spontaneous generation. 2. Self-reproduction originating from within the organism.

Autogenous Pertaining to autogenesis.

Autogenous Regulation The regulation of expression of a gene or an operon by its own product(s).

Autogenous Vaccine A virulent, immunogenic material prepared from a pathogenic microorganism for vaccination against the pathogenic organism.

Autogeny See Autogenesis.

Autograft A transplant from one site to another in the same individual or a tissue graft between genetically identical members of the same species.

Autohemagglutination Clumping of the red blood cells of an individual by the serum of the same individual.

Autohemolysin Autoantibody capable of destroying one's own red blood cells.

Autohemolysis Lysis or destruction of one's own red blood cells.

Autoimmune Immune reaction to one's cellular constituents.

Autoimmune Disease A disease produced by an autoimmune response in an individual.

Autoimmune Hemolytic Anemia Anemia produced by the lysis of an individual's own red blood cells due to the presence of autoantibodies against erythrocyte antigens.

Autoimmune Reaction Antiself immune reaction in which the immune system is incapable of distinguishing self from nonself.

Autoimmune Response Antiself immune response (see Autoimmune reaction).

Autoimmunity Antiself immune response in which the immune system is directed against one's own cellular constituents.

Autoinducible Enzyme An enzyme whose induction in a given organism is accomplished by the presence of a specific substance (autoinducer) that is produced by itself.

Autointoxicant A toxic substance generated within the body.

Autokinesis Spontaneous or voluntary activity or movement.

Autoleukoagglutinin Antibody capable of clumping one's own white blood cells.

Autologous Derived from same organism.

Autolysate The products of autolysis.

Autolysins Endogenous substances (e.g., enzymes) involved in the breakdown of certain structural components during particular phases of cellular growth and development.

Autolysis The lysis of a cell or tissue by one's own autolysins or enzymes.

Autolysosome A membrane-bound organelle resulting from fusion of a lysosome and an autophagosome.

Autolytic Of or pertaining to autolysis.

Automutagen A mutagenic compound that is produced by an organism and is mutagenic for the same organism.

Autonomic 1. Having the ability to function independently. 2. Pertaining to the autonomic nervous system.

Autonomic Drugs Drugs that mimic or modify the function of the autonomic nervous system.

Autonomic Nervous System The portion of the nervous system that regulates and governs the involuntary action of the body (e.g., cardiac and smooth muscle contraction and secretion from glandular structures).

Autoorganotroph Cell capable of utilizing CO_2 as a carbon source in the presence of an organic electron donor.

Autooxidation Spontaneous oxidation by atmospheric oxygen without the presence of a catalyst.

Autophagic Lysosome Secondary lysosome that contains hydrolytic enzymes involved in the digestion of intracellular materials.

AUT/PAGE Abbreviation for acid urea triton polyacrylamide gel electrophoresis.

Autophagic Vacuoles Vacuoles containing morphologically recognizable cytoplasmic components.

Autophagosome Phagosome or phagocytic vacuole formed by the process of autophagy.

Autophagy The intracellular digestion of endogenous material of the cell within a lysosome.

Autoplasty Repair of lesions with tissue from the same body.

Autoplex-T An anti-inhibitor coagulant complex, it consists of vitamin K-dependent clotting factor.

Autoploid Having two or more sets of chromosomes derived from the duplication of a haploid chromosome set.

Autopolyploid An organism having more than two genetically identical or near identical sets of chromosomes derived from a single set of chromosomes.

Autopsy The after-death dissection of a body to determine the cause of death or nature of a disease.

Autoradiography A technique in which X-ray film is used to localize a radioactive compound in a biological or biochemical preparation.

Autosomal Pertaining to those chromosomes that are not sex chromosomes.

Autosomal Inheritance A pattern of inheritance in which the transmission of traits depends on the presence or absence of certain genes on the autosomes.

Autosome Any chromosome that is not a sex chromosome.

Autotechnicon An automatic device used for transfer of tissue blocks through dehydrating and cleaning agents prior to embedding in a paraffin wax.

Autotoxin Toxic substance produced by the body.

Autotransfusion The transfusion of blood into an individual from whom it was obtained prior to the transfusion.

Autotransplantation Transfer of tissue from one part of the body to another.

Autotroph See autotrophic cells.

Autotrophic Pertaining to autotroph.

Autotrophic Cells Cells that use inorganic material as a source of nutrient and CO_2 as the sole source of carbon.

AUU A genetic code or codon for the amino acid isoleucine.

Auxanographic Technique A procedure used for identifying a substance or growth factor by employing an auxotroph, which requires a specific factor for growth.

Auxanology The science that deals with growth and development.

Auxesis Growth due to the increase in cell size rather than to the increase in number of cells.

Auxiliary Enzyme The enzyme that is added to a primary enzymatic assaying system in order to determine the enzymatic activity of the primary enzyme that is not readily assayed directly.

Auxiliary Pigment See Accessory pigment.

Auxin One of a group of plant hormones, such as indole acetic acid, that promotes cell division, chromosomal DNA synthesis, and longitudinal growth.

Auxo- A combining form meaning pertaining to growth, acceleration, or stimulation.

Auxocarcinogen The auxiliary group of atoms in the molecule of a carcinogen that influences the activity of the carcinogenophore.

Auxochrome A group of atoms that bind to a molecule containing a chromophore and intensifies the color of the chromophore.

Auxocyte Premeiotic germ cell (a cell destined to enter meiotic prophase.

Auxotroph Mutant microorganism that requires a growth factor not required by the original strain from which it was derived.

Auxotrophic Pertaining to auxotroph.

AV Abbreviation for a combination drug containing adriamycin, and vincristine.

AvaI A restriction endonuclease isolated from *Anabaena variabilis* with the following specificity:

```
         ↓
5′..........CPyCGPuG..........3′
3′..........GPuGCPyC..........5′
                  ↑
```

AvaII A restriction endonuclease isolated from *Anabaena variabilis* with the following specificity:

```
         ↓
5′.........GG(A/T)CC.........3′
3′.........CC(T/A)GG.........5′
                 ↑
```

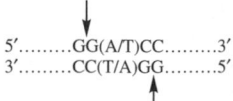

Avalide A combination drug containing angiotensin-II receptor antagonist irbesartan and hydrochlorothiazide used for treatment of hypertension.

Avandia A trade name for rosiglitazone maleate used as an oral antidiabetic agent.

Avapro A trade name for irbesartan, an antihypertensive agent.

AVC A trade name for sulfanilamide used for treatment of fungal infection.

Avenin The glutlin of oats.

Aventyl A trade name for nortriptyline hydrochloride, used as an antidepressant.

Average Intrinsic Association Constant The value of the association constant for the binding of a given antigen by the corresponding antibodies that is determined when one-half of all the binding sites are occupied by the antigen.

Average Molecular Weight The value of the molecular weight that is determined for a sample consisting of a mixture of molecules.

AvFd Abbreviation for ferredoxin from *Azotobacter vinelandii*.

AvFld Abbreviation for flavodoxin from *Azotobacter vinelandii*.

AvFld-1 Abbreviation for flavodoxin-1 from *Azotobacter vinelandii*.

AvFld-2 Abbreviation for flavodoxin-2 from *Azotobacter vinelandii*.

AVH Abbreviation for acute viral hepatitis.

AviI (AsuII) A restriction endonuclease from *Anabaena variabilis* having the same specificity as AsuII.

AviII (MstI) A restriction endonuclease from *Anabaena variabilis* with the following specificity:

```
             ↓
5′..........TGCGCA..........3′
3′..........ACGCGT..........5′
              ↑
```

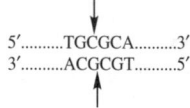

Avian Leukosis Virus A group of avian type C retroviruses that induce neoplastic disease after a long latent period.

Avian Sarcoma Viruses A group of v-onc$^+$ type C retroviruses that, after a short latent period, cause tumors in fowl and sometimes other animals.

Avianized Vaccine Any vaccine containing microorganisms whose virulence for a given host has been attenuated by adaptation in live chicks and/or serial passage through chick embryos.

Avicel An insoluble cellulose microcrystaline.

Avidin A protein (mol wt 68,000) present in the egg white, it consists of four subunits and is capable of binding with biotin.

Avidity 1. The tendency of an antibody to bind antigen. 2. Affinity. 3. The stability of an antigen-antibody complex.

Avirax A trade name for acyclovir, an antiviral agent that inhibits viral DNA replication.

Avirulent Lacking virulence or lack of a capability to cause diseases.

Avita A trade name for tretinoin, an antineoplastic agent.

Avitaminosis A disease due to lack of vitamins.

Avlosulfan A trade name for dapsone, a leprostatic agent.

Avogadro's Law A law stating that equal volumes of all gases at a given temperature and pressure contain the identical number of molecules.

Avogadro's Number 1. The number of molecules in a gram-molecular weight of a compound (6.023×10^{23} molecules per mole). 2. The number of atoms in a gram-atomic weight of an element (6.023×10^{23}).

Avoirdupois Weight The English system of weight in which there are 7,000 grains, 256 drams, or 16 ounces to one pound (One ounce = 28.35 grams; one pound = 453.59 grams).

Avonex A trade name for interferon beta-1 produced by DNA recombinant technology and used for treatment of relapsing forms of multiple sclerosis.

Avoparcin A glycopeptide antibiotic produced by *Streptomyces candidus*, which inhibits Gram-positive bacteria by interfering with peptidoglycan synthesis.

AVP Abbreviation for 1. [Arg^8]vasopressin or 8-arginine vasopressin. 2. Anti-viral protein. 3. A combination drug containing adriamycin, vincristine and procarbazine.

Avrl A restriction endonuclease from *Anabaena variabilis* with the following specificity:

↓
5′..........CPyCGPuG..........3′
3′..........GPuGCPyC..........5′
↑

AvrII A restriction endonuclease from *Anabaena variabilis* with the following specificity:

↓
5′..........CCTAGG..........3′
3′..........GGATCC..........5′
↑

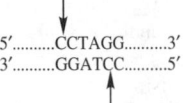

AWU Abbreviation for atomic weight unit.

Axanthopsia The inability to perceive the color yellow.

Axenic Free from other organisms.

Axenic Animal Germ-free animal.

Axenic Culture 1. A contamination-free culture. 2. A culture of one strain free from other strains.

Axerophthal Referring to vitamin A aldehyde.

Axerophthol Referring to vitamin A.

Axial Bond A bond in a ring structure that is at right angles to the plane of the ring.

Axial Substituent A substituent attached to an axial bond.

Axid A trade name for nizatidine, a histamine H2 blocker.

Axis of Polarity Gradient information that distinguishes one end of an organism from the other.

Axis of Rotational Symmetry An axis of symmetry such that rotation of a body will yield one or more structures that are identical to the structure before the rotation.

Axoaxonic Synapse A type of synapse in which the axon of one neuron comes in contact with the axon of another neuron.

Axocet A trade name for acetaminophen that contains butalbital.

Axodendritic Synapse A type of synapse in which the axon of one neuron comes in contact with the dendrites of another neuron.

Axodendrosomatic Synapse A type of synapse in which the axon of one neuron comes in contact with both the dendrites and the cell body of another neuron.

Axon One of the protoplasmic processes of a neuron, it carries the impulse away from the cell body.

Axonal Transport Movement of proteins and membrane-bounded vesicles along the axon of a nerve cell between the cell body and the synaptic knobs.

Axonemal Microtubules Microtubules present as a highly ordered bundle in the axoneme of a cilium or flagellum of the eukaryotic organism.

Axoneme The bundle of 9 + 2 complex of the microtubules in the cilia and flagella of the eukaryotice organism.

Axoplasm The cytoplasm of an axon.

Axosomatic Synapse A type of nerve synapse in which the axon of one neuron comes in contact with the cell body of another neuron.

Ayercillin A trade name for penicillin G procaine, an antibiotic that inhibits bacterial cell wall synthesis.

AYF Abbreviation for anti-yeast factor.

Aygestin A trade name for norethindrone acetate, used to suppress ovulation.

Aza- A prefix indicating the presence of nitrogen in a heterocyclic ring.

Azaadenine (mol wt 136) An andenine analog.

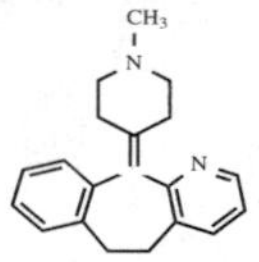

Azacitidine (mol wt 244) An antineoplastic agent.

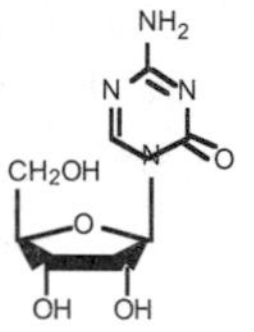

Azacosterol (mol wt 389) An avian chemosterilant used for control of nuisance birds.

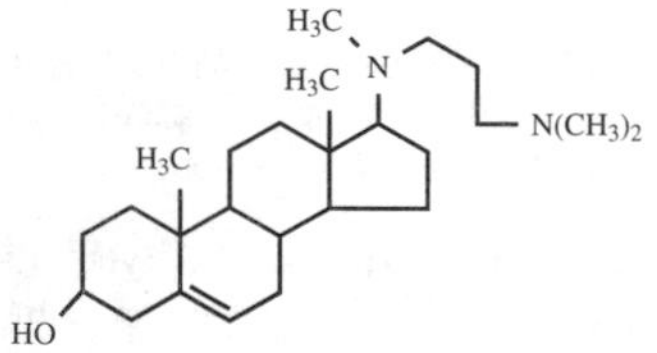

Azactam A trade name for aztreonam; inhibits synthesis of bacterial cell walls.

Azacyclonol (mol wt 267) A tranquilizer.

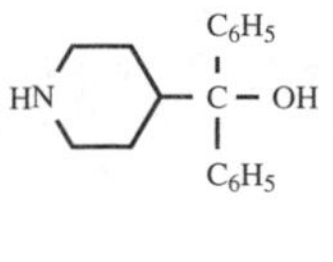

Azacytidine (mol wt 244) A pyrimidine antibiotic with antitumor activity. It disrupts the processes of transcription and translation.

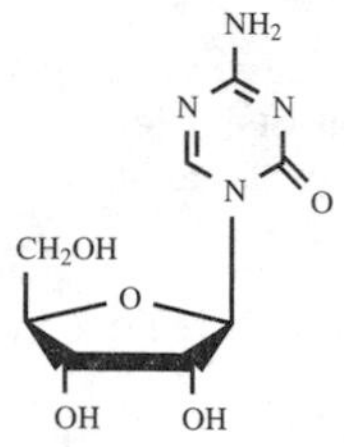

Azacytosine (mol wt 112) A cytosine analog.

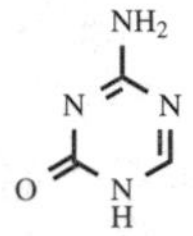

8-Azaguanine (mol wt 152) A purine analog that inhibits the growth of some cancer cells.

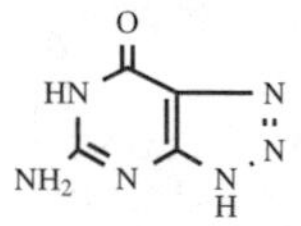

Azahypoxanthine (mol wt 137) An analog of xanthine.

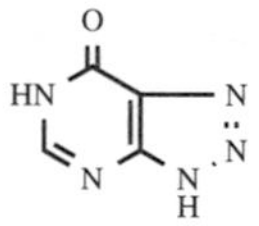

Azanidazole (mol wt 246) An antiprotozoal agent.

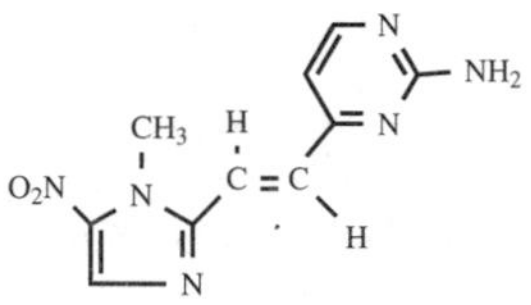

Azaperone (mol wt 327) A tranquilizer or sedative agent.

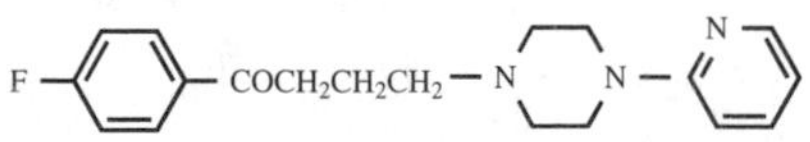

Azaserine (mol wt 173) An antibiotic and antitumor agent obtained from *Streptomyces*. It is an analog of glutamine that inhibits reactions in which glutamine acts as an amino group donor.

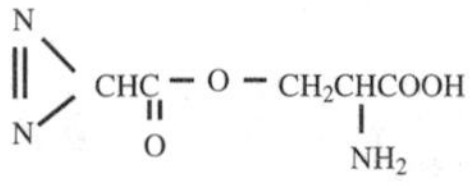

Azasetron (mol wt 350) An antiemetic agent.

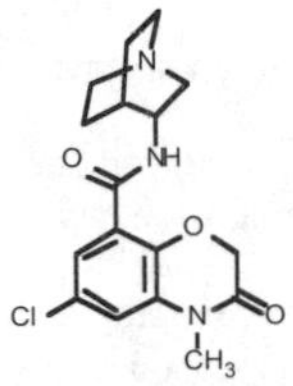

Azatadine (mol wt 290) An antihistaminic agent that competes with histamine for H-1 receptor sites on the effector cells.

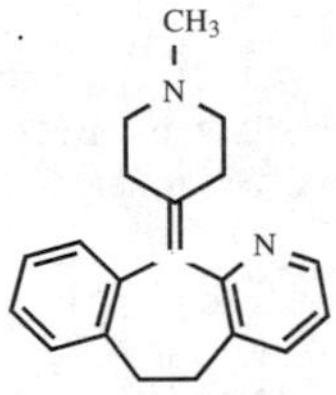

Azathioprine (mol wt 277) A purine antagonist commonly used for suppression of immune responses. It also inhibits purine synthesis.

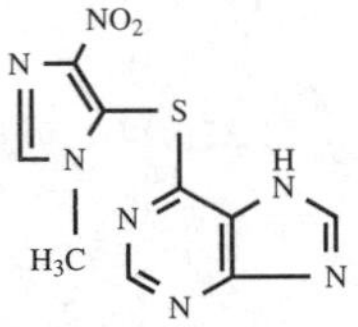

Azathiothymine (mol wt 143) A thymine analog.

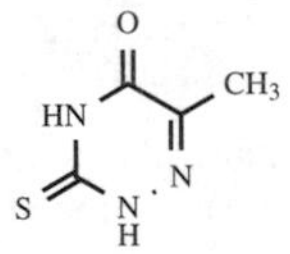

Azathiouridine (mol wt 261) An analog of uridine.

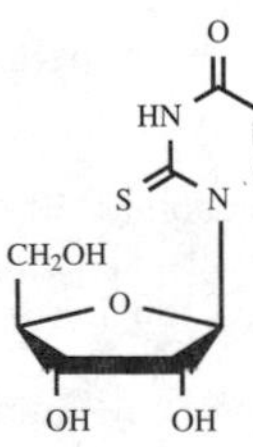

Azathymine (mol wt 127) A thymine analog, its deoxyriboside (azathymidine) inhibits the biosynthesis of DNA in neoplastic and normal cells *in vitro*.

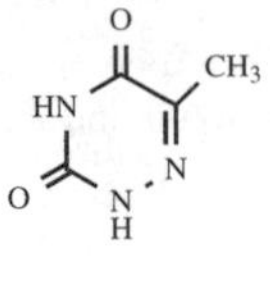

Azauracil (mol wt 113) An analog of uracil.

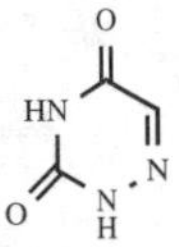

6-Azauridine (mol wt 245) A uridine analog that is useful in cancer chemotherapy.

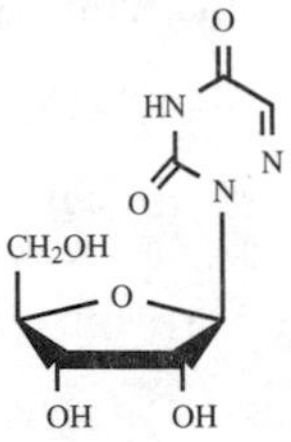

Azelaic acid (mol wt 188) An antiacne agent.

COOH
|
$(CH_2)_7$
|
COOH

Azelastine (mol wt 382) An antiasthmatic and antihistaminic agent.

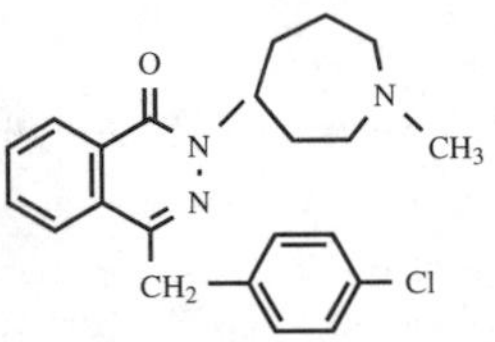

Azelex A trade name for azelaic acid, an antiacne agent.

Azeotrope A mixture of two or more liquids whose boiling point is below the boiling points of the component liquids.

AzG Abbreviation for azaguanine.

Azidamfenicol (mol wt 295) An antibacterial agent.

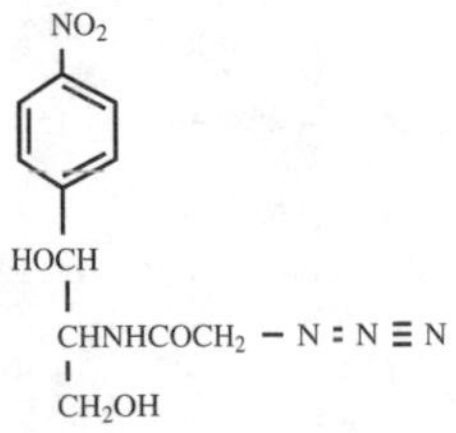

Azide (N_3^-) Any compound containing the characteristic formula $R(N_3)x$. R may be a metal, hydrogen atom, or halogen atom. Azide acts as a respiratory inhibitor and prevents the reduction of oxidized cytochrome oxidases.

Azide Group The $N = \overset{+}{N} = N^-$ group.

Azidocillin (mol wt 375) An antimicrobial agent related to penicillin.

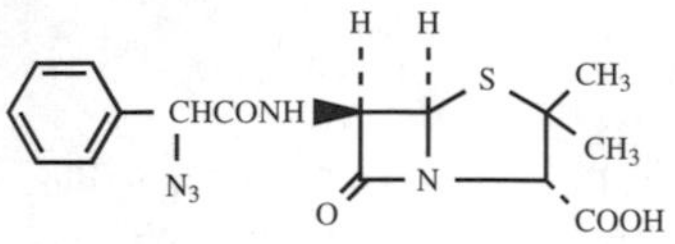

3'-Azido-3'-Deoxythymidine See AZT.

Azidothymidine See AZT.

Azithromycin (mol wt 749) An antibacterial agent that binds to 50S ribosomal subunits inhibiting bacterial protein synthesis.

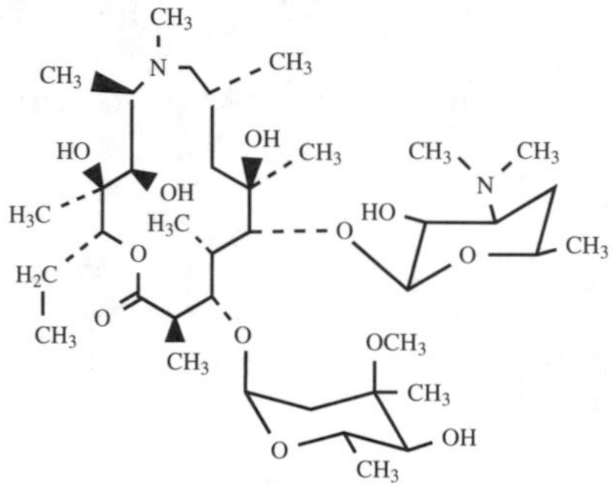

Azlin A trade name for azlocillin sodium, an antibiotic.

Azlocillin (mol wt 462) An antibacterial agent related to penicillin. It inhibits bacterial cell wall synthesis.

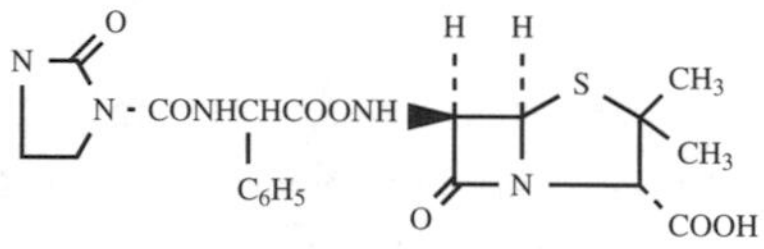

Azmacort A trade name for triamcinolone acetonide, a hormone with anti-inflammatory and immuno-suppressive activity.

Azo- A prefix meaning containing nitrogen.

Azo Compound A compound that contains an azo group (-N=N-).

Azo Dye A dye that contains the azo group.

Azo Gantanol A trade name for a combination drug containing sulfamethoxazole and phenazopyridine hydrochloride, used as an antibacterial agent.

Azo Gantrisin A trade name for a combination drug containing sulfisoxazole and phenazopyridine, used as an antibacterial agent.

Azo Group The grouping -N=N-.

Azo Protein A protein in which a tyrosine residue has been coupled to an aromatic diazo compound.

Azolid A trade name for phenylbutazone, used as an anti-inflammatory agent.

-azoline A suffix referring to substances capable of reducing allergic responses and narrowing blood vessels.

Azomonas A genus of Gram-negative or Gram-variable motile bacteria (family Azotobacteraceae) that occur in soil and water.

Azomycin (mol wt 113) An antibiotic produced by *Streptomyces* sp. It is effective against both Gram-positive and Gram-negative bacteria.

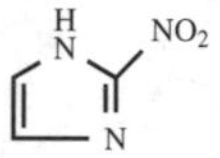

Azoospermia Lack of sperm in the semen.

Azopt A trade name for brinzolamide that decreases intraocular pressure in open-angle glaucoma

Azosemide (mol wt 371) A diuretic agent.

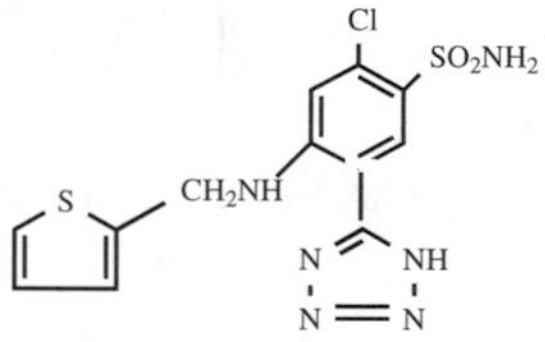

Azospirillum A genus of Gram-negative or Gram-variable, asporogenous, nitrogen-fixing bacteria.

AZO-Standard A trade name for phenazopyridine, a urinary analgesic agent.

Azote Nitrogen (so called because of its inability to sustain life).

Azotemia The retention in the blood of excessive amounts of nitrogenous compounds due to the failure of the kidneys to remove urea.

Azotobacter A genus of Gram-negative, cyst-forming bacteria (family Azotobacteraceae) that occurs in fertile soils of near-neutral pH.

Azotouria The presence of excess nitrogenous compounds in the urine.

AZT (mol wt 267) Abbreviation for 3′-azido-3′-deoxythymidine, a thymidine analog used for treatment of AIDS.

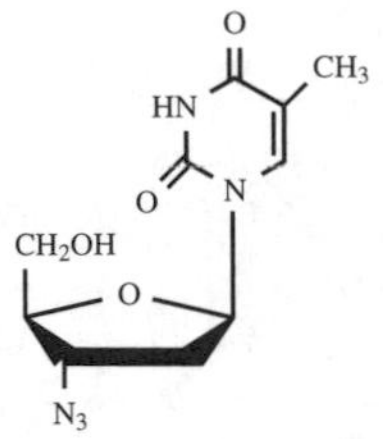

AZ-Test A pregnancy test using urine from a potentially pregnant woman.

Aztreonam (mol wt 435) A synthetic monocyclic β-lactam antibiotic that inhibits bacterial cell wall synthesis.

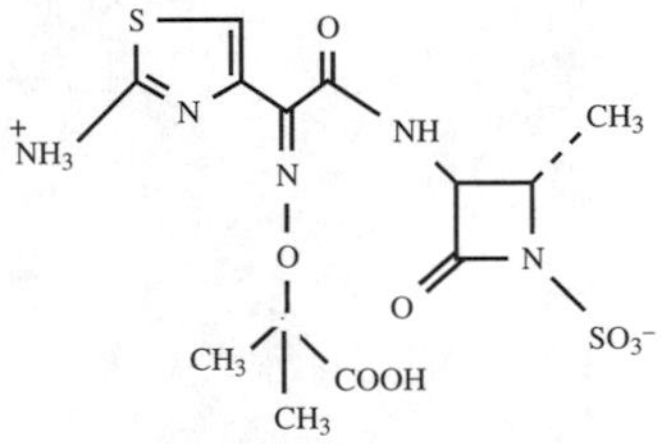

AzU Abbreviation for azauracil.

5AzU Abbreviation for 5-azauracil.

6AzU Abbreviation for 6-azauracil.

Azulene (mol wt 128) A deep blue bicyclic aromatic dye.

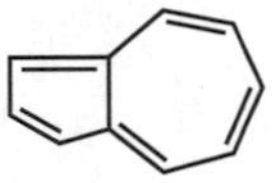

Azulfidine A trade name for sulfasalazine, an anti-infective and anti-inflammatory agent.

AzUR (AZUR) Abbreviation for 6-azauridine

5AzUR Abbreviation for 5-azauridine

6AzUR Abbreviation for 6-azauridine

Azure A (mol wt 292) A violet-blue basic dye.

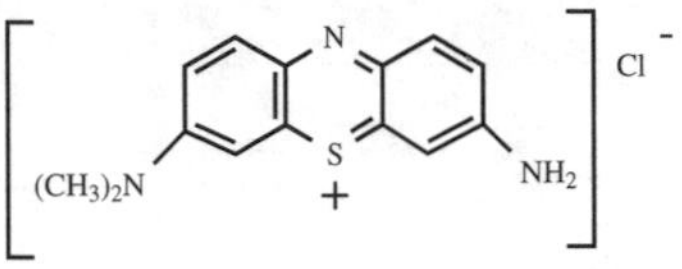

Azure B (mol wt 306) A violet-blue basic dye used in cytochemistry.

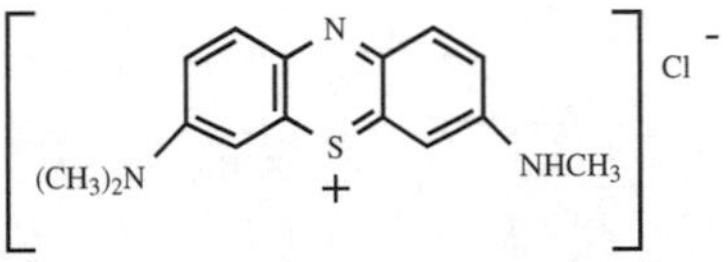

Azure C (mol wt 278) A violet-blue basic dye.

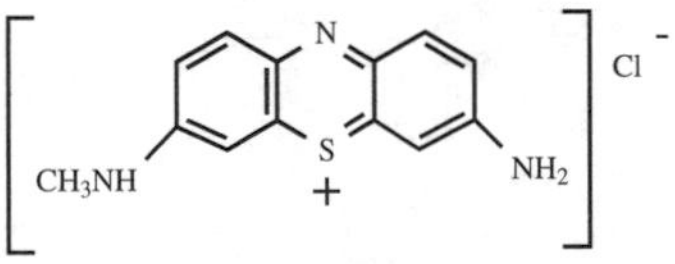

Azurin A blue copper-containing protein found in certain bacteria (e.g., *Alcaligenes denitrificans, Bordetella pertussis, Paracoccus denitrificans, Pseudomonas aeruginosa*).

Azurophilic Having an affinity for methylene azure blood stains.

Azygosperm See Azygospore.

Azygospore A spore that is produced directly from a gamete without conjugation.

Azygous Single, unpaired structure.

Azymic Not containing enzyme or not enzymatic.

B

B Abbreviation for 1. amino acid asparagine or aspartic acid; 2. *Bacillus*; 3. 5-bromouracil or 5-bromouridine; 4. chemical element boron.

β See beta.

B_1 Abbreviation for vitamin B_1 (thiamine).

B_2 Abbreviation for vitamin B_2 (riboflavin).

B_3 Abbreviation for vitamin B_3 (nicotinic acid).

B_6 Abbreviation for vitamin B_6 (pyridoxine).

B_7 Abbreviation for vitamin B_7 (carnitine).

B_8 Abbreviation for vitamin B_8 (adenosine phosphate).

B_9 Abbreviation for vitamin B_9 (folic acid).

B_{12} Abbreviation for vitamin B_{12} (cyanocobalamin).

B1, 2, 3, ... Referring to backcross generation, e.g., B1, the first backcross generation (mating of an individual with one of its parents or with an individual of the genotype identical to the parent); B2, the second backcross generation (mating of B1 individual with one of its parents or with an individual of the genotype identical to the parent).

B_{663} A lipid soluble substituted phenazine used as an antileprosy drug (see also clofazimine).

B Cell (B lymphocyte) Lymphocytes that are processed or derived from the Bursa of Fabaricius in avian species and from bone marrow in mammals.

B Cell Antigen Receptors The membrane-bound, surface immunoglobulin (sIg) displayed by B cells.

B Cell Differentiation Factors The lymphokines that are involved in the differentiation of B lymphocytes.

B Chromosome Supernumerary chromosomes that are not duplicates of any of the members of the basic complement of the normal or A chromosome. B chromosomes are usually devoid of structural genes and do not pair with A chromosomes during meiosis.

B DNA The classic form of Watson-Crick double helix DNA containing about 10 residues per turn; having a helix diameter of about 24 Å. The planes of the base pairs in the double helix are perpendicular to the helix axis.

B Granule Membrane-bound vesicles or granules found in the beta cell of pancreatic islets that contain stored insulin.

β Granules See B granule.

B Lymphocyte See B cell.

Ba The symbol for the chemical element barium, valence 2.

^{137}Ba Radioactive isotope of barium.

BA Abbreviation for 1. bacterial agglutinin; 2. benzyl-adenine; 3. bile acid; 4. blocking antibody; 5. blood agar; 6. boric acid; 7. bovine albumin.

BAA Abbreviation for benzoyl arginine amide.

Baa Helices See Baa helix.

Baa Helix A basic amphiphilic a helix that contains an array of basic amino acid residues on one side and hydropathic residues on the other. It occurs in the calmodulin binding region of many proteins.

Babes-Ernst Granules Metachromatic granules that occur in bacteria.

Babesia A genus of parasitic protozoa of erythrocytes in vertebrates that causes babesiosis.

Babesiasis Variant spelling of babesiosis.

Babesiosis Any infection or disease caused by a species of *Babesia*, e.g., red water fever (also known as babesiasis).

Babo's Law The law stating that the lowering of vapor pressure is proportional to the mole fraction of nonvolatile solute in a solution.

BABP Abbreviation for bile acid-binding protein.

BAC A trade name for a combination drug containing butalbital, aspirin, and caffeine.

Bac36I A restriction endonuclease from *Bacillus alcalophilus* 36 with the following specificity:

```
        ↓
5′..........GGNCC..........3′
3′..........CCNGG..........5′
               ↑
```

BacI (SacII) A restriction endonuclease isolated from *Bacillus acidocaldarius* with the following specificity:

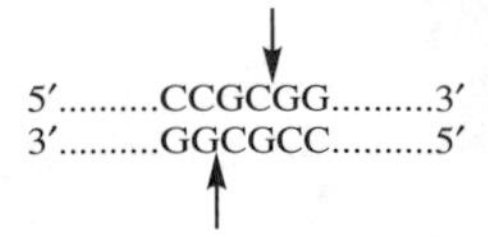

Bacampicillin (mol wt 466) A semisynthetic antibiotic related to penicillin that inhibits bacterial cell wall synthesis.

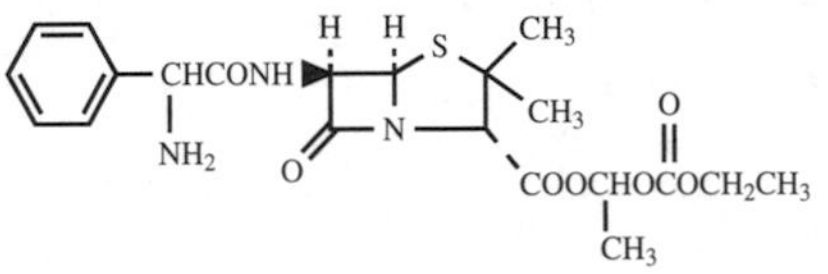

Bacarate A trade name for phendimetrazine tartrate that promotes transmission of nerve impulses.

Baciguent A trade name for bacitracin, an antibiotic.

Bacill- A prefix meaning rod-shaped bacterium.

Bacillary Dysentery A diarrheal disease caused by infection with *Bacillus.*

Bacille Calmette-Guerin (BCG) See Bacillus Calmette Guerin.

Bacillemia The presence of *Bacillus* in the blood.

Bacilli Plural of *Bacillus.*

Bacillicide Any substance capable of killing *Bacilli.*

Bacilliform The shape like a *Bacillus.*

Bacillin Synonym of bacilysin.

Bacillolysin A protease isolated from species of *Bacillus* that catalyzes the preferential hydrolysis of peptide bonds involving the amino group of leucine or phenylalanine.

Bacillosis Any infection or disease caused by the *Bacillus.*

Bacilluria The presence of *Bacilli* in the urine.

Bacillus A genus of Gram-positive, strictly aerobic or facultative anaerobic, catalase-positive, rod-shaped, and endospore-forming bacteria.

Bacillus anthracis A species of Gram-positive, facultative anaerobe; causal agent of anthrax.

Bacillus Calmette-Guerin (BCG) An attenuated form of *Mycobacterium bovis* used as a vaccine against tuberculosis.

Bacillus thuringiensis A Gram-positive spore-forming bacterium that causes the death of lepidopterous larvae. It is used as an insecticide for biological control.

Bacilysin (mol wt 270) An antibiotic produced by soil *Bacillus* that inhibits the bacterial cell wall synthesis.

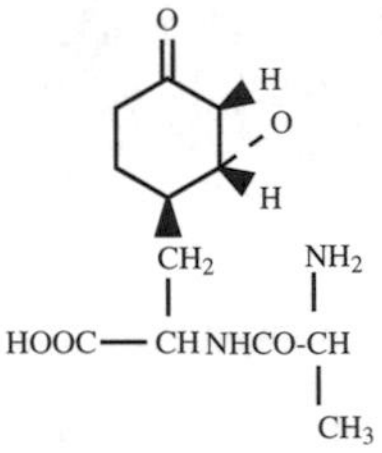

Bacimethrin (mol wt 155) An antibiotic produced by *Bacillus megatherium.*

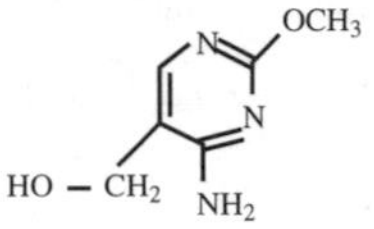

Bacitin A trade name for bacitracin, an antibiotic that inhibits bacterial cell wall synthesis.

Bacitracin A cyclic peptide antibiotic isolated from *Bacillus subtilis* that inhibits bacterial cell wall synthesis.

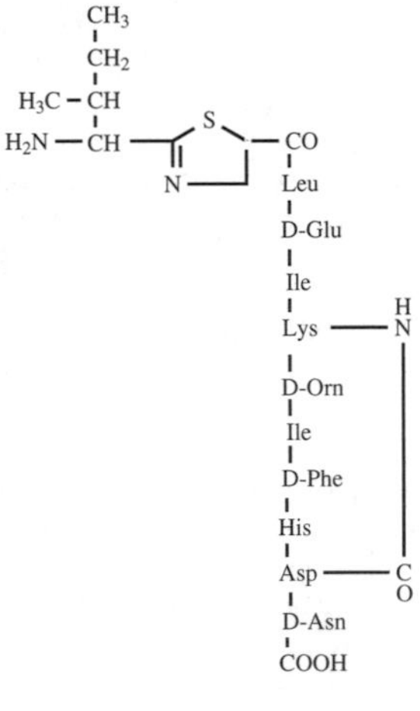

Backbone The linear chain structure of a polymer from which the side groups attach.

Backcross A cross between an offspring and one of its parents or an individual genetically identical to one of its parents.

Background Constitutive Synthesis The transcription of genes in a repressed operon due to the occasional dissociation of repressor from the operator, thus allowing RNA polymerase to bind to the promotor to synthesize gene product.

Background Radiation The amount of radioactivity registered by a radiation detector in the absence of a radioactive sample.

Back-Mutation A mutation that restores the original phenotypic character or restores the original nucleotide sequence.

Backward Flow The flow of the solvent of a solution of macromolecules in a direction opposite to the direction of the movement of the macromolecules.

Baclofen (mol wt 214) A muscle relaxant that reduces the transmission of the nerve impulse from the spinal cord to skeletal muscle.

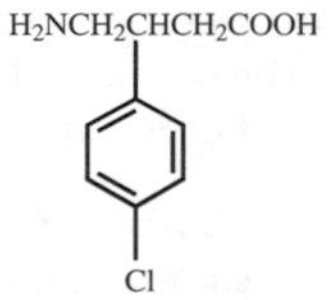

BACO Abbreviation for a drug combination containing bleomycin, adriamycin, CCNU, and oncovin.

BACON Abbreviation for a combination drug containing bleomycin, adriamycin, CCNU, oncovin, and nitrogen mustard.

BACONP Abbreviation for a combination drug containing bleomycin, adriamycin, CCNU, oncovin, nitrogen mustard, and prednisone.

BACOP Abbreviation for a combination drug containing bleomycin, adriamycin, cytoxan, oncovin, and prednisone.

BACT Abbreviation for a combination drug containing BCNU, ara-C, cytoxan, and 6-thioguanine.

Bacteraemia A variant spelling of bacteremia.

Bacteremia The presence of bacteria in the blood.

Bacteria (plural of bacterium) Unicellular, prokaryotic organisms. A diverse group of cellular organisms capable of causing diseases in plants and animals including humans.

Bacterial Chemotaxis The response of bacteria toward the chemical attractant.

Bacterial Chlorophyll A photosynthetic pigment of photosynthetic bacteria capable of extracting electrons from donors other than water.

Bacterial Dysentery Dysentery caused by bacterial infection, e.g., *Shigella* (shigellosis).

Bacterial Endocarditis A bacterial infection of the endocardium or the heart valves.

Bacterial Endotoxin Any bacterial toxin that is released only after cell lysis. It is part of the bacterial cell wall (see also endotoxin).

Bacterial Ferredoxin An iron-sulfur protein produced by bacteria; acts as an electron carrier.

Bacterial Flagellum A locomotory filament in bacteria.

Bacterial Lawn A continuous cover of bacteria on the surface of a solid growth medium.

Bacterial Vaginosis A disease (vaginitis) caused by bacteria and characterized by a malodorous vaginal discharge.

Bacterial Virus See bacteriophage.

Bactericholia The presence of bacteria in the bile.

Bactericidal Capable of killing bacteria.

Bactericide Any substance capable of killing bacteria.

Bactericidin 1. Bactericidal antibody. 2. A nonspecific bactericidal plasma factor. 3. An antibacterial protein produced by an invertebrate.

Bacterinia An unfavorable response to a bacterial vaccination.

Bacteriochlorophyll See bacterial chlorophyll.

Bacteriocidal A variant spelling of bactericidal.

Bacteriocide A variant spelling of bactericide.

Bacteriocin A protein produced by one species of bacterium that is toxic to another related species (also known as protein antibiotic).

Bacteriocin Factor A DNA plasmid or episome that carries the genetic information for the synthesis of bacteriocin.

Bacteriocinogen Referring to bacteriocin factor.

Bacteriocinogenic Capable of producing bacteriocin.

Bacteriocinogenic Factor Referring to bacteriocin factor.

Bacteriocin-Typing A method used to distinguish bacteria on the basis of the bacteriocins they produce or the bacteriocins to which they are susceptible.

Bacteriocuprein (CuZn-SOD) Referring to superoxide dismutase, a bacterial metalloenzyme that contains Zn and Cu.

Bacteriocyte A specialized cell that contains intracellular bacterial symbionts.

Bacteriological Filter A filter used to sterilize solutions by removing microorganisms through filtration.

Bacteriology The science that deals with bacteria.

Bacteriolysin 1. An antibody capable of lysis of bacterial cells in the presence of complement. 2. A substance capable of the lysis of a bacterial cell.

Bacteriolysis The lysis or disintegration of a bacterial cell.

Bacteriolytic Pertaining to bacteriolysis.

Bacterio-opsin The apoprotein of the bacteriorhodopsin.

Bacteriophaeophytin A phaeophytin derivative of bacterial chlorophyll, it acts as an electron carrier in photosynthetic reactions.

Bacteriophage Any virus capable of infecting bacteria. Bacteriophages are classified into ten major families:

Myoviridae (nonenveloped virions with a long contractile tail, containing dsDNA, e.g.,T_2, P_2).

Styloviridae (nonenveloped virions with a long noncontractile tail, containing dsDNA), e.g., λ, T_5).

Pedoviridae (nonenveloped virions with a short noncontractile tail, containing dsDNA, e.g., T_3, T_7).

Tectiviridae (nonenveloped virions containing dsDNA, e.g., PRD1).

Corticoviridae (nonenveloped virions containing dsDNA, e.g., PM2).

Plasmaviridae (enveloped virions containing dsDNA, e.g., MV-L2).

Inoviridae (filamentous virions containing ssDNA, e.g., MV-L1).

Microviridae (icosahydral virions containing ssDNA, e.g., ϕX174).

Leviviridae (icosahydral virions containing ssRNA, e.g., MS2, QB).

Cystoviridae (enveloped virions containing dsRNA, e.g., ϕ6).

Bacteriophage α 15 An icosahedral bacteriophage of the family Leviviridae, containing ssRNA. It infects enterobacteria.

Bacteriophage AP50 An icosahedral bacteriophage of the family Tectiviridae, containing dsDNA. It infects *E. coli, Salmonella typhimurium,* and *Pseudomonas aeruginosa.*

Bacteriophage β An icosahedral bacteriophage of the family Leviviridae, containing ssRNA. It infects enterobacteria.

Bacteriophage Bam 35 An icosahydral bacteriophage of the family Tectiviridae containing dsDNA.

Bacteriophage BPB1 A bacteriophage of the family Styloviridae with a long, noncontractile tail containing dsDNA. It infects *Bacillus* species.

Bacteriophage Conversion Acquisition or loss, by a bacterium, of one or more phenotypic characteristics as a result of infection by a bacteriophage.

Bacteriophage f1 A filamentous bacteriophage of the family Inoviridae containing single-stranded cccDNA. It infects enterobacteria.

Bacteriophage f2 An icosahedral bacteriophage of the family Leviviridae containing ssRNA. It infects enterobacteria.

Bacteriophage fd A filamentous bacteriophage of the family Inoviridae containing ssDNA. It infects enterobacteria.

Bacteriophage G4 An icosahedral bacteriophage of the family Microviridae containing single-stranded cccDNA.

Bacteriophage If1 A filamentous phage of the family Inoviridae containing single-stranded cccDNA. It infects enterobacteria.

Bacteriophage If2 A filamentous bacteriophage of the family Inoviridae containing single-stranded cccDNA. It infects enterobacteria.

Bacteriophage Ike A filamentous bacteriophage of the family Inoviridae, containing single-stranded cccDNA. It infects enterobacteria.

Bacteriophage λ A temperate bacteriophage of the family Styloviridae with long, noncontractile tail and isometric head containing dsDNA with cohesive sites on both ends. It infects *E. coli.* and has been used as a clonal vector.

Bacteriophage M13 A filamentous bacteriophage of the family Inoviridae containing single-stranded cccDNA. It infects enterobacteria and has been used as a clonal vector.

Bacteriophage MS2 An icosahedral bacteriophage of the family Leviviridae containing ssRNA. It infects *Pseudomonas and Caulobacter.*

Bacteriophage Mu A temperate bacteriophage of the family Myoviridae with a long contractile tail and isometric head containing linear dsDNA. It infects various enterobacteria.

Bacteriophage MV-L1 A filamentous bacteriophage of the family Inoviridae containing single-stranded cccDNA. It infects *Acholeplasma laidlawii.*

Bacteriophage MV-L2 A bacteriophage of the family Plamaviridae, an enveloped bacteriophage containing double-stranded cccDNA. It infects *Acholeplasma laidlawii.*

Bacteriophage MV-L3 A bacteriophage of the family Podoviridae with an isometric head and a short noncontractile tail containing linear dsDNA. It infects *Acholeplasma laidlawii.*

Bacteriophage N4 A bacteriophage of the family Podoviridae with short noncontractile tail and isometric head containing linear dsDNA. It infects *E. coli.*

Bacteriophage P1 A temperate bacteriophage of the family Myoviridae with a long contractile tail containing linear dsDNA. It infects *E. coli.*

Bacteriophage P2 A temperate bacteriophage of the family Myoviridae with a long contractile tail and isometric head containing linear dsDNA. It infects *E. coli.*

Bacteriophage P4 A satellite bacteriophage that infects *E. coli.* It requires all the late genes of a helper phage for completion of the replication cycle. It contains linear dsDNA with sticky ends.

Bacteriophage P22 A temperate bacteriophage of the family Podoviridae with a short noncontractile tail and isometric head containing circularly permuted dsDNA. It infects *Salmonella.*

Bacteriophage Packaging Insertion of a recombinant lambda DNA into *E. coli* for replication and encapsidation into plaque-forming bacteriophage particles.

Bacteriophage PBS1 A bacteriophage of the family Myoviridae with a long contractile tail and isometric head containing dsDNA with deoxyuridine nucleotide. It infects *Bacillus subtilis.*

Bacteriophage ϕ6 An enveloped bacteriophage of the family Cystoviridae containing segmented genome of three pieces of dsRNA. It infects *Pseudomonas.*

Bacteriophage ϕ29 A bacteriophage of the family Podoviridae with short noncontractile tail and an elongated head containing linear dsDNA. It infects *Bacillus subtilis.*

Bacteriophage ϕX174 A bacteriophage of the family Microviridae containing single-stranded cccDNA. It infects enterobacteria.

Bacteriophage PM2 An icosahedral bacteriophage of the family Corticoviridae containing doubled-stranded cccDNA. It infects *Pseudomonas.*

Bacteriophage PRD1 A bacteriophage of the family Tectiviridae containing dsDNA. It infects both Gram-positive and Gram-negative bacteria.

Bacteriophage QB An icosahedral bacteriophage of the family Leviviridae containing ssRNA. It infects *Pseudomonas* and *Caulobacter.*

Bacteriophage R17 An icosahedral bacteriophage of the family Leviviridae containing ssRNA. It infects enterobacteria.

Bacteriophage SPO1 A bacteriophage of the family Myoviridae with a long contractile tail and large isometric head containing double-stranded DNA with hydroxymethyl uracil nucleotides. It infects *Bacillus subtilis.*

Bacteriophage T1 A bacteriophage of the family Styloviridae containing dsDNA.

Bacteriophage T2 A bacteriophage of the family of Myoviridae with a long contractile tail and isometric head containing circularly permuted (terminal redundant) dsDNA. It infects *E. coli.*

Bacteriophage T3 A bacteriophage of the family Podoviridae.

Bacteriophage T4 A bacteriophage of the family Myoviridae with a long contractile tail and isometric head containing circularity permuted (terminal redundant) dsDNA. It infects *E. coli.*

Bacteriophage 5 A bacteriophage of the family Styloviridae containing dsDNA. It infects enterobacteria.

Bacteriophage T7 A virulent bacteriophage of the family Podoviridae with a short noncontractile tail and isometric head containing dsDNA. It infects enterobacteria.

Bacteriophage Typing A technique used for the typing of bacteria in which the strains of bacteria are distinguished on the basis of their susceptibility to a range of bacteriophages.

Bacteriophage Z A bacteriophage of the family Styloviridae.

Bacteriophage ZJ/2 A filamentous bacteriophage containing ssDNA.

Bacteriophagous Organisms capable of phagocytizing bacteria.

Bacteriopheophytin A variant spelling of bacteriophaeophytin.

Bacteriorhodopsin A hydrophobic protein pigment found in the membrane of *Halobacterium* and responsible for proton translocation.

Bacterioruberin A carotenoid pigment found in *Halobacterium.*

Bacteriostat Agent capable of bacteriostasis.

Bacteriostatic Capable of inhibiting the growth and reproduction of bacteria.

Bacteriostasis The inhibition of bacterial growth and reproduction without destruction.

Bacteristatic Synonym of bacteriostatic.

Bacterium A prokaryotic unicellular organism. A diverse group of prokaryotic cellular organisms capable of causing diseases in plants and animals including humans.

Bacteriuria The presence of bacteria in the urine.

Bacterivore Organism capable of ingesting bacteria as a source of nutrients.

Bacterization The process of coating seeds or tubers with bacteria prior to planting for the purpose of promoting plant growth.

Bacteroid A modified bacterial cell, e.g., *Rhizobium* cell in the root nodules of legume plant.

Bacteroidaceae A family of Gram-negative, anaerobic, chemoorganotrophic, asporogenous, rod-shaped bacteria.

Bacteroides A genus of Gram-negative bacteria of the family Bacteroidaceae.

Bacteroidosis A bacterial infection caused by *Bacteroides.*

Bacto Trade names for the products of Difco Laboratories, Inc.

Bactocill A trade name for oxacillin sodium, an antibiotic.

Bactopen A trade name for cloxacillin sodium, a penicillinase-resistant penicillin that inhibits bacterial cell wall synthesis.

Bactoprenol A lipid-soluble, membrane-bound polyprenol of bacterial origin.

$$H-\left[CH_2-\overset{CH_3}{\overset{|}{C}}=CH-CH_2\right]_{11}-OH$$

Bactrim A trade name for a combination drug containing sulfamethoxazole and trimethoprim, used as an antibacterial agent.

Bactroban A trade name for mupirocin, an anti-infective agent that inhibits bacterial protein and RNA synthesis.

BacTrx Abbreviation for thioredoxin from *Bacillus acidocaldarius.*

Baculoviridae A family of rod-shaped insect viruses containing double-stranded cccDNA. Baculoviruses have been engineered to synthesize foreign proteins.

Baculoviruses Any virus of the family Baculoviridae.

BAD Abbreviation for biological aerosol detector.

BaE Abbreviation for barium enema.

BAE Cell Abbreviation for bovine aortic endothelial cell.

BAEC Abbreviation for bovine aortic endothelial cell.

BAEE Abbreviation for benzoyl arginine ethyl ester or benzyl arginine ethyl ester.

Baeocyte A small reproductive cell formed by certain members of Cycanobacteria.

Bagasse A form of fibrous biomass formed as a by-product from the crushing of sugarcane. It contains approximately 50% cellullose, 25% pentosan, and 25% lignin.

Bagassosis A lung disease associated with inhalation of the dust of bagasse (the crushed, juiceless residue of sugarcane).

BAIB Abbreviation for beta amino isobutyric acid.

βAIB Abbreviation for beta amino isobutyric acid.

Baicalein (mol wt 270) An astringent isolated from roots of *Scutellaria baicalensis.*

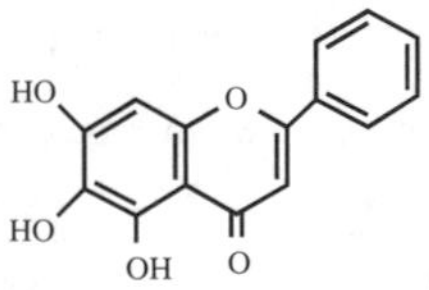

Baikiain (mol 127) A naturally occurring non-protein amino acid.

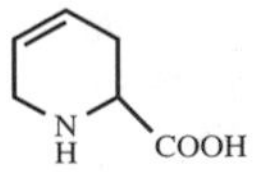

Bakankosin (mol wt 357) A nitrogenous glucoside from the seed of *Strychnos vacacoua.*

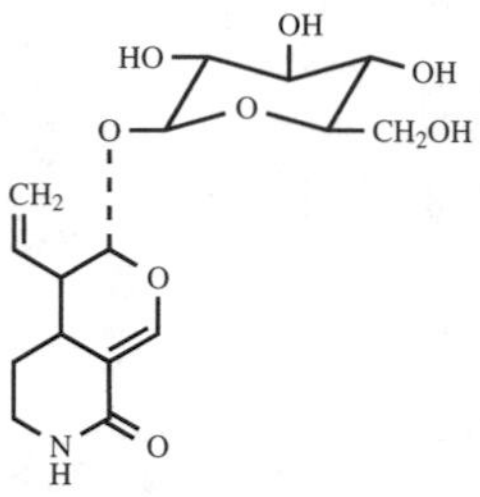

Baker's Yeast A strain of *Saccharomyces cerevisiae* used in the baking industry.

BAL (mol wt 124) Abbreviation for British Anti-Lewisite (2,3-dimercaptopropanol). It blocks the flow of electrons in mitochondrial electron transport chains.

$$HS\text{-}CH_2\text{-}CH(SH)\text{-}CH_2OH$$

BalI A restriction endonuclease from *Brevibacterium albidum* with the following specificity:

```
        ↓
5′.........TGGCCA.........3′
3′.........ACCGGT.........5′
        ↑
```

Balanced Growth The growth of cells in which all the cellular components increase by the same factor.

Balanced Polymorphism A type of polymorphism in which the frequencies of the coexisting forms do not change noticeably over generations.

Balanced Salt Solution (BSS) A salt solution used in tissue culture to provide satisfactory ionic strength, pH, and osmotic conditions for the maintenance and growth of cells.

Balanced Selection Selection that favors the heterozygotes that produce a balanced polymhorism.

β-Alanine (mol wt 89) A naturally occurring amino acid not found in proteins.

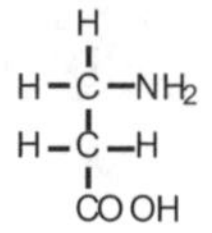

Balanitis Inflammation of the penis.

Balantidial Dysentery See balantidiasis.

Balantidiasis An infection caused by the protozoan *Balantidium coli.*

Balantidicidal Capable of destroying protozoans of the genus *Balantidium.*

Balantidium A genus of protozoa (order of Trichostomatida).

Balb/c Mouse An inbred mouse strain that is predisposed to form myelomas upon intraperitoneal injection of mineral oil.

Balbiani Chromosome A polytene chromosome discovered by E.G. Balbiani in *Chironomus* larvae.

Balbiani Ring An extremely large puff in the polytene chromosome of a salivary gland cell during a significant portion of larval development. It contains a large number of DNA loops upon which mRNAs are being transcribed.

Ball and Stick Model A molecular model in which the atoms and bonds are represented by spheres and sticks respectively. The bond angles are specified for each atom, and the bond length and atomic radii are fixed.

Ball Rebound Test A test for the determination of the elastic response of a polymeric material by the measurement of energy absorbed when a steel ball impacts the material from a fixed height and the height of rebound is measured. The more elastic the polymer, the greater the rebound height.

Ball Thrombus The coagulated mass of blood that contains platelets, fibrin, and cellular fragments. It can block a blood vessel or a heart valve.

Ballistocardiograph A device employed to measure the volume of the blood pumped by the heart and to calculate cardiac output.

Ballotini Small glass beads employed for the ballistic disintegration of cells.

Balm 1. An aromatic gum resin obtained from certain trees. 2. Fragrant aromatic oil or substance used for healing, soothing, and relieving pain.

Balminin DM A trade name for dextromethorphan, a non-narcotic antitussive agent.

Balneology The science that deals with the chemical composition of mineral waters and their healing characteristics.

Balsam Resinous saps from evergreens (see balm).

BALT Abbreviation for bronchus-associated lymphoid tissue.

Baltimore Classification A system of virus classification proposed by Baltimore based on the nature of the viral genomes and their expression. The viruses are classified into five main classes: 1. dsDNA viruses, 2. dsRNA viruses, 3. Positive stranded RNA viruses, 4. Negative stranded RNA viruses, and 5. ssRNA viruses that employ reverse transcriptase for the synthesis of DNA for cellular integration.

B_{am} Symbol for the amount of drug required to saturate a specific population of receptors on a membrane sample.

Bambermycin A flavophospholipid antibiotic complex from *Streptomyces bambergiensis,* consisting of moenomycin A, B_1, B_2, and C.

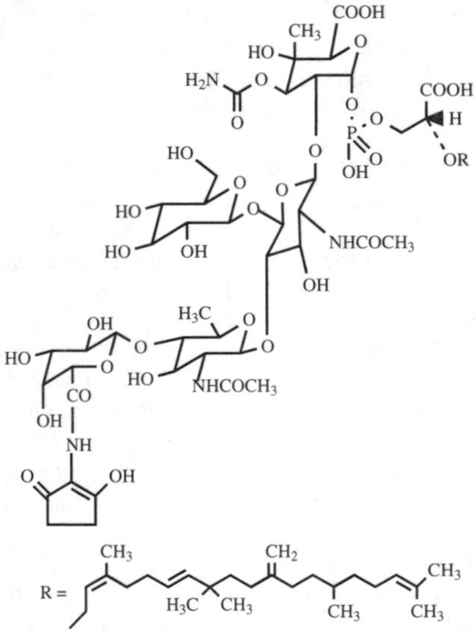

moenomycin A

Bambuterol (mol wt 367) A bronchodilator.

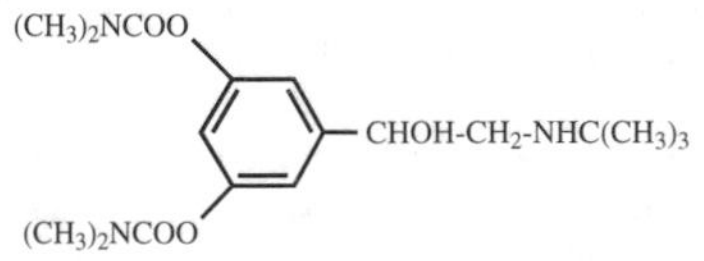

BAME Abbreviation for benzoyl arginine methyl ester.

Bamethan (mol wt 209) A vasodilator.

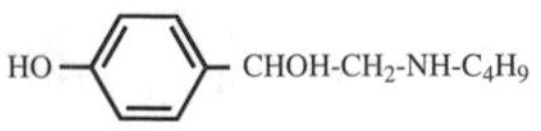

BamFI (BamHI) A restriction endonuclease from *Bacillus amyloliquefaciens* F with the same specificity as HamHI.

BamHI A restriction endonuclease from *Bacillus amyloliquefaciens* H with the following specificity:

5′.........GGATCC.........3′
3′.........CCTAGG.........5′

BamHI Methylase The enzyme that methylates the internal cytosine residue of the BamHI sequence.

Bamifylline (mol wt 385) A bronchodilator.

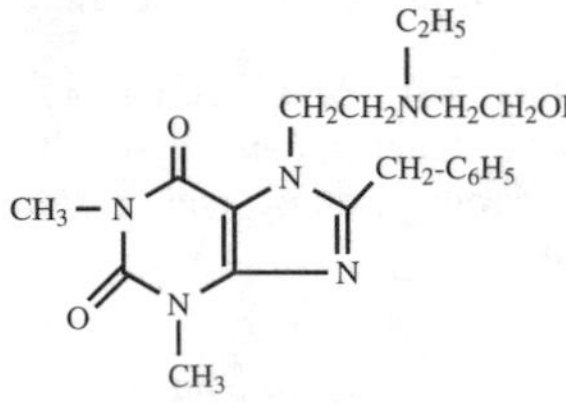

Bamipine (mol wt 280) An antihistaminic agent.

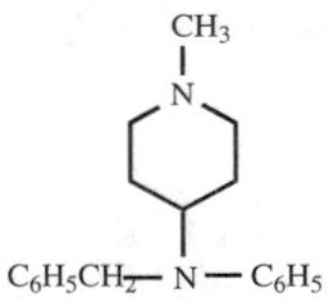

BamKI (HamH1) A restriction endonuclease from *Bacillus amyloliquefaciens* K with the same specificity as HamH1.

BamNI (BamH1) A restriction endonuclease from *Bacillus amyloliquefaciens* N with the same specificity as HamHI.

BamNxI (AvaII) A restriction endonuclease from *Bacillus amyloliquefaciens* N with the same specificity as Ava1I.

BAMON Abbreviation for a combination drug containing bleomycin, adriamycin, methotrexate, oncovin, and nitrogen mustard.

BanI A restriction endonuclease from *Bacillus aneurinolyticus* with the following specificity:

5′..........GGPyPuCC.........3′
3′..........CCpuPyGG.........5′

BanII A restriction endonuclease from *Bacillus aneurinolyticus* with the following specificity:

5′..........GPuGCPyC.........3′
3′..........CPyCGPuG.........5′

BanIII A restriction endonuclease from *Bacillus aneurinolyticus* with the following specificity:

5′..........ATCGAT.........3′
3′..........TAGCTA.........5′

Banana Oil 1. A solution of nitrocellulose in amyl acetate, so named because of its penetrating banana-like odor. 2. Synonym of amyl acetate.

Band 1. A zone of macromolecules obtained in density gradient centrifugation, electrophoresis, isoelectric focusing, or chromatography. 2. A bundle of protein fibers as seen in striated muscle. 3. An absorption band. 4. Patterns of contrast bands observed in chromosomes after staining.

Band III The membrane protein of human erythrocytes that acts as an ion-transport/exchange carrier.

Band 4.1 A protein found in the membrane skeleton of human erythrocytes.

Band 4.2 A protein associated with anion channels in the membrane of human erythrocytes.

Band Cells Immature neutrophils released from bone marrow.

Band Elimination Filter A filter that blocks the transmission of a given range of wavelengths.

Band Pass The range of wavelengths of a radiation that passes through a filter.

Band Width The width of an absorption band or a range of wavelengths.

Banding Techniques for staining chromosomes with fluorescent or chemical stains.

Banesin A trade name for acetaminophen, an analgesic and antipyretic agent.

Banflex A trade name for orphenadrine citrate, a skeletal muscle relaxant.

Bang Method A technique for determination of glucose in urine that employs alkaline copper thiocyanate.

BAO (mol wt 252) Abbreviation for bis-aminophenyl oxadiazole, a compound used for fluorescent staining of DNA.

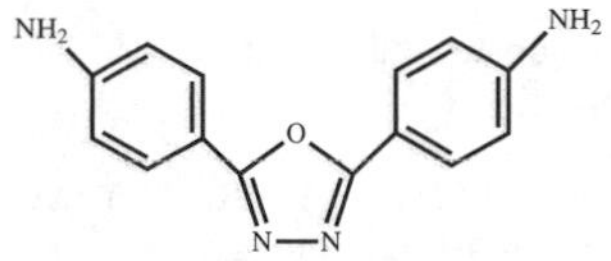

BAP Abbreviation for 6-benzylaminopurine and bacterial alkaline phosphatase.

BA/PL Ratio Abbreviation for bile-acid/phospholipid ratio.

BAPP Abbreviation for a combination drug containing bleomycin, adriamycin, platinol, and prednisone.

Baptisia An antiseptic alkaloid from the root of wild indigo *(Baptisia tinctoria)*.

Bar 1. A metric unit of pressure equal to 1,000,000 dynes per square centimeter or 750 mm mercury. 2. A sex-linked dominant mutation in *Drosophila melanogaster* that reduces the number of facets in the compound eye.

Barban (mol wt 258) A selective herbicide for wild oats.

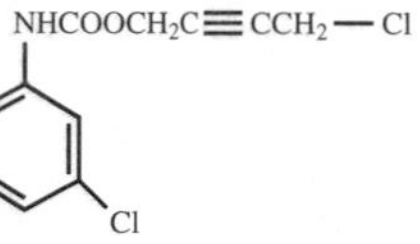

Barbased A trade name for butabarbital sodium, a sedative and hypnotic agent.

Barbilixir A trade name for phenobarbital, an anticonvulsant, sedative, hypnotic, and antiepileptic agent.

Barbita A trade name for phenobarbital, an anticonvulsant.

Barbital (mol wt 184) A sedative and hypnotic agent (also called veronal).

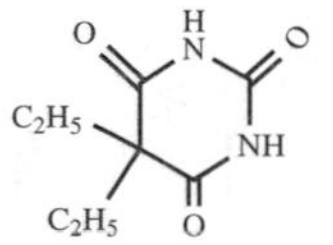

Barbitone Synonym of barbital.

Barbiturase The enzyme that catalyzes the following reaction:

$$\text{Barbiturate} + 2H_2O \rightleftharpoons \text{malonate} + \text{urea}$$

Barbiturates A group of derivatives from barbitaric acid that act as sedative or hypnotic agents. Unsubstituted barbituric acid has no hypnotic properties.

Barbituric Acid (mol wt 128) An acid used for the preparation of barbiturates.

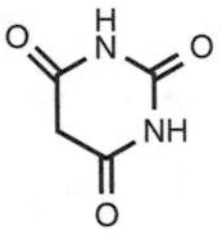

Barbiturism Addiction to a barbiturate or poisoning by derivatives of barbituric acid.

Barbloc A trade name for pindolol, an antihypertensive agent.

Barfoed's Test A colorimetric method to distinguish monosaccharides from disaccharides by means of cupric acetate in dilute acetic acid.

Bar-Graph A graph in which data are represented by bars at the ordinate or the abscissa.

Bariatrics The science that deals with obesity, its treatment, and disease associated with obesity.

Baridium A trade name for phenazopyridine hydrochloride, an analgesic and antipyretic agent.

Barium (Ba) A chemical element with atomic weight 137, valence 2.

Barium Enema The rectal infusion of barium sulfate (a radiopaque contrast medium) for X-ray examination of the intestinal tract.

Barium Ionophore I (mol wt 645) An ionophore for assaying Barium activity with solvent polymeric membrane electrodes.

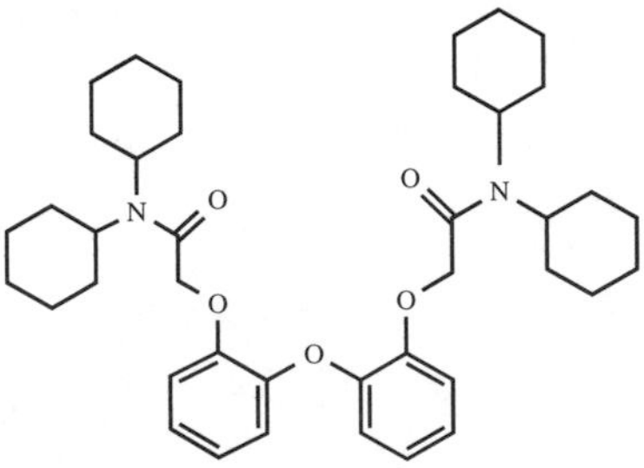

β-ARK Abbreviation for beta-adrenergic receptor kinase.

βARK1 Abbreviation for β-adrenergic receptor kinase-1.

βARK2 Abbreviation for β-adrenergic receptor kinase-2.

Barn A unit of area of the atomic nucleus equal to 10^{-24} cm^2.

Barnidipine (mol wt 492) An antihypertensive and an antianginal agent.

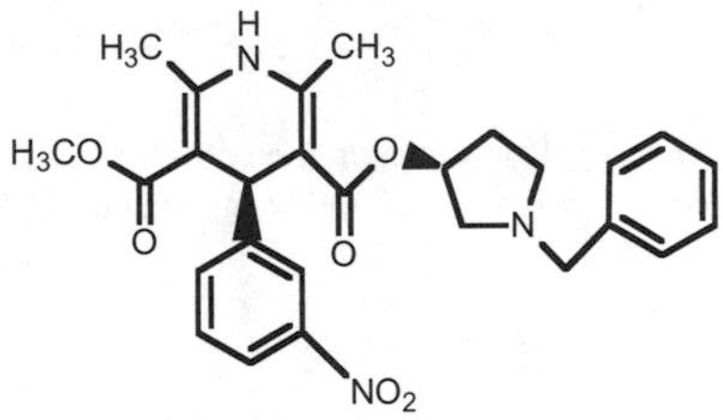

Baro- A prefix meaning pressure.

Baroceptor See baroreceptor.

Baroduric Barotolerant.

Barograph A device that monitors and records changes in barometric pressure.

Barometer An instrument for measuring atmospheric pressure.

Barophiles Organisms that grow optimally under the condition of high hydrostatic pressure.

Barophilic Shows the characteristics of a barophile.

Baroreceptor (Baroceptor) A receptor in the central nerve system that responds to changes in blood pressure.

Barosperse A trade name for the radiopaque medium barium sulfate.

Barostat A device for regulation or maintenance of pressure at a constant value.

Barotaxis A type of taxis in which pressure is the stimulus.

Barotolerant Organisms that can grow under the condition of high hydrostatic pressure.

Barotrauma A noninfective inflammatory disorder of the ear caused by marked changes in barometric pressure between environmental atmosphere and normal air within the cavity.

Barr Body An inactive, condensed X chromosome in the nuclei of somatic cells of female mammals.

Barrier Substance A substance applied to skin for protection against exposure to irritants or for preventing diffusion of moisture or gas.

Barthrin (mol wt 337) An insecticide.

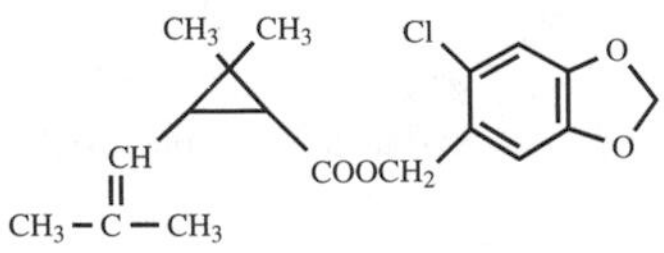

Bartonella A genus of Gram-negative bacteria of the family Bartonellaceae.

Bartonellaceae A family of bacteria of the order Rickettsiales.

Bartonellosis Any infection or disease caused by *Bartonella bacilliformis.*

Bartter's Syndrome A genetic disorder characterized by kidney enlargement, overactive adrenal gland, and an increased production of prostaglandin.

Baruria The excretion of urine with a high specific gravity.

Barwin A barley seed protein.

Basal Body A cell structure resembling a centriole that organizes and anchors the microtubule assembly of cilium or flagellum.

Basal Cell Carcinoma A malignant tumor of an epithelial cell.

Basal Cells The relatively undifferentiated cells in an epithelial sheet that give rise to more specialized cells, e.g., cells of the base layer of stratified epithelium.

Basal Enzyme The small quantity of an inducible enzyme produced in the absence of an inducer.

Basal Gel A trade name for basic aluminum carbonate gel, used as an antacid.

Basal Lamina A thin, noncellular layer of ground substance lying just under the epithelial surface.

Basal Membrane A sheet of tissue that forms the outer layer of the choroid and lies under the pigmented layer of the retina.

Basal Metabolic Rate (BMR) The amount of energy used in a unit of time by a resting organism to maintain its vital functions.

Basal Metabolism The metabolic activity required for maintaining the vital functions of an organism.

Base 1. A substance that turns litmus indicator blue; combines with acid to form salt; gives up hydroxyl ions (OH^-) or reduces hydrogen ion concentration. 2. The nitrogenous components of DNA or RNA. 3. A quantity that is being raised to an exponential power.

Base Analog A substance whose structure is similar to one of the bases of DNA or RNA.

Base Composition The relative concentration of purine and pyrimidine bases in DNA or RNA in terms of mole percent.

Basedow's Goiter An enlargement of the thyroid gland due to the hypersecretion of thyroid hormone following iodine therapy.

Basement Membrane The noncellular layer of extracellular matrix that underlies the epithelium.

Base Pair (bp) A pair of hydrogen-bound purine and pyrimidine bases in DNA or RNA, e.g., A:T pair, C:G pair, and A:U pair.

Base Pair Rules The rule states that adenine pairs with thymine or uracil and that guanine pairs with cytosine in double-stranded DNA or RNA.

Base Pair Substitution A type of mutation in which one purine is substituted by another purine or one pyrimidine by another pyrimidine (transition), or a purine is substituted by a pyrimidine or vice versa (transversion).

Base Piece The membrane-bound portion of ATPase (also known as the F_0 component of ATPase).

Base Ratio Referring to the adenine/thymine ratio, guanine/cytosine ratio, or purine/pyrimidine ratio.

Base Saponification Number The number of milligrams of KOH equivalent to the amount of acid required to neutralize the alkaline constituents present after saponifying one gram of sample.

Base Sequence The linear order of purine and pyrimidine nucleotides in a strand of DNA or RNA.

Base Stacking The arrangement of the base pairs in parallel planes in the interior of a double helical DNA structure.

Base Substitution The replacement of one base for another in DNA or RNA.

Basic Pertaining to a base or a solution having a pH greater than 7.

Basic Aluminum Carbonate Gel 1. A phosphorous-binding agent used to prevent recurrent renal phosphatic calculi. 2. An antacid gel.

Basic Amino Acid An amino acid with two or more amino groups and one carboxyl group, e.g., lysine or arginine.

Basic Blue 3 (mol wt 360) A basic dye.

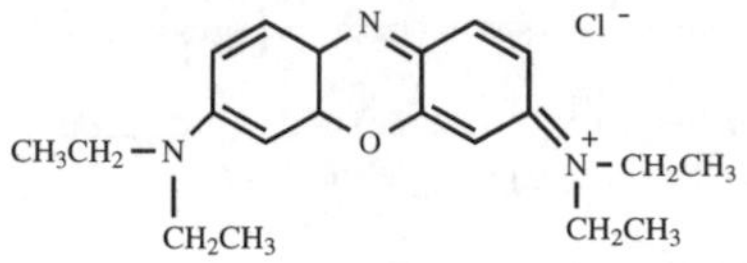

Basic Blue 47 (mol wt 371) A basic dye.

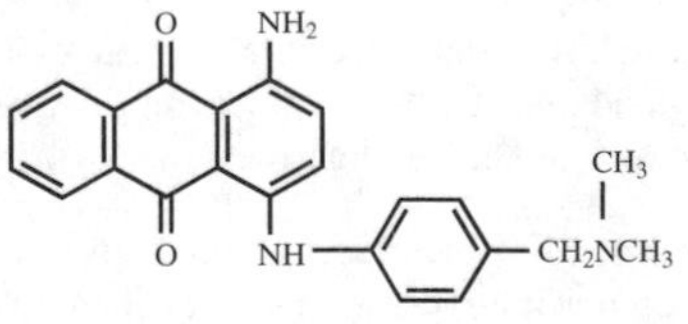

Basic Chromatic Readily stained with basic dye.

Basic Dye A cationic dye that binds to and stains negatively charged macromolecules or structures.

Basic Fuchsin (mol wt 338) A basic dye.

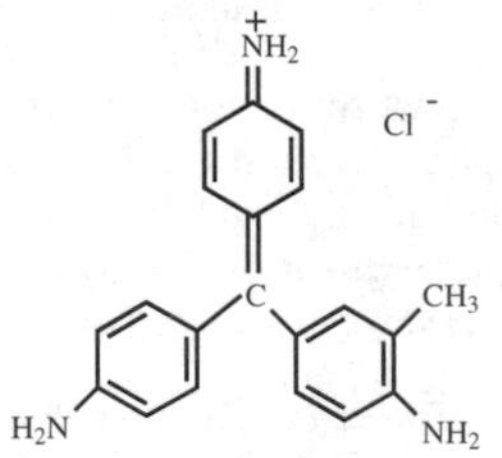

Basic Number Referring to the haploid chromosome number.

Basic Orange See acridine orange.

Basic Proteins Proteins that are rich in basic amino acids.

Basic Red 29 (mol wt 369) A basic dye.

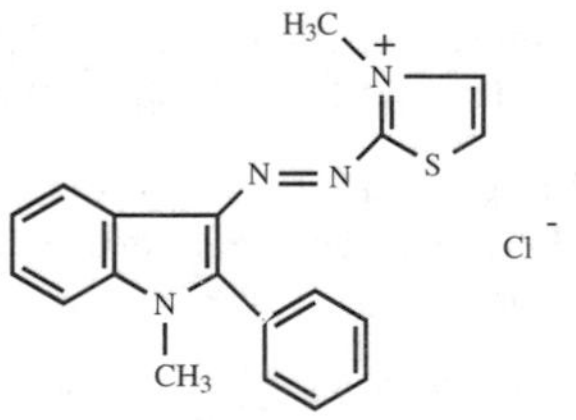

Basic Replicon The smallest part of a replicon that encodes all the functions necessary for replication.

Basic Stains See basic dye.

Basidia Plural of basidium.

Basidiocarps The fruit bodies of basidiomycetes that bear basidia.

Basidiolichen A lichen in which the mycobiont is a basidomycete.

Basidiomycete A group of fungi characterized by the formation of sexual basidiospores on a basidium.

Basidiospores The sexually produced spores formed on basidia (basidiomycete).

Basidium The basidiospore-bearing structure of basidiomycete.

Basiliximab A monoclonal antibody produced by recombinant DNA technology that acts as an interleukin-2 antagonist and used as an immunosuppressive agent.

Basillar membrane The cellular membrane of the mammalian inner ear.

Basiloma Tumor of basal cells.

Basocytosis The presence of an abnormally high number of basophils in the blood.

Basophil White blood cell that can be stained readily with basic dyes.

Basophilia The presence of an abnormally high number of basophilic leukocytes in the blood.

Basophilic Having an affinity for a basic dye or staining readily with a basic dye.

Basophilic Adenoma A tumor of the pituitary gland composed of cells that can be stained readily with basic dyes.

Basophilic Leukemia A malignant neoplasm of blood-forming tissues characterized by the presence of a large number of immature basophils.

Basosquamous Cell Carcinoma A malignant skin tumor composed of basal and squamous cells.

BAT Abbreviation for brown adipose tissue.

Batch Adsorption A technique for adsorption of a solute from a solution by stirring the solution with an adsorbent, e.g., DEAE-Sephadex. The adsorbed solute can be subsequently eluted from the adsorbent.

Batch Culture A type of cell culture in which cells are cultured in a volume of liquid medium.

Batch Elution The elution of the adsorbed solutes from an absorbent by stirring the absorbent with different eluent or different concentration of an eluent.

Bathochromic Group A group of atoms that is capable of causing a shift of the absorption spectrum of a compound when attached to the compound (e.g., shifting from a lower wavelength to a longer wavelength).

Bathochromic Shift A shift of the absorption spectrum of a compound (e.g., to a longer wavelength and lower frequency).

Batimastat (mol wt 478) An antineoplastic agent.

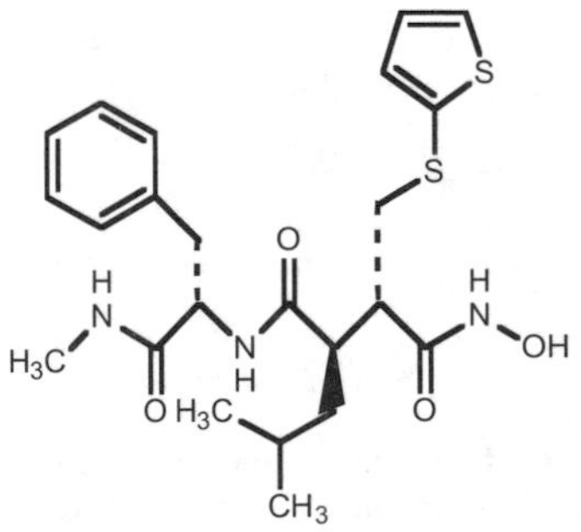

Batrachotoxin (mol wt 539) A toxic steroidal alkaloid extracted from the skin of neotropical poison-dart frogs (species of *Phyllobates*). It causes an irreversible increase in the permeability of the membrane to sodium ions.

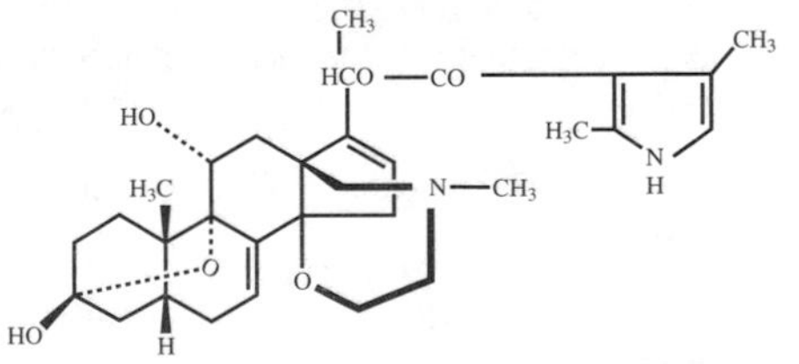

Batrachotoxinin A (Mol wt 417) A less toxic steroidal alkaloid (in comparison with bactrachotoxin).

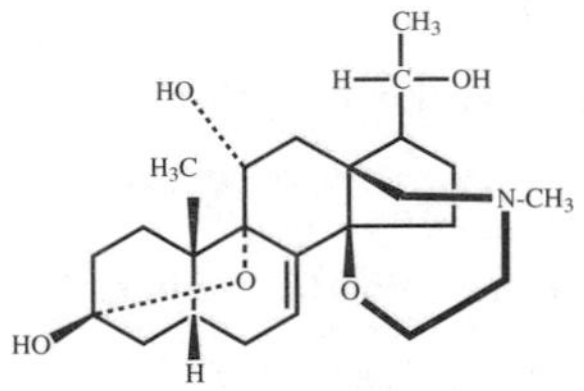

Battey Bacillus Strains of *Mycobacterium avium* or *M. intracellulare* that cause a chronic pulmonary disease resembling tuberculosis.

BavI (PvuII) A restriction endonuclease from *Bacillus alvei* with the following specificity:

↓
5′.........CAGCTG.........3′
3′......... GTCGAC.........5′
↑

BAVIP Abbreviation for a combination drug containing bleomycin, adriamycin, velban, imidazole, and prednisone.

Baycol A trade name for cerivastatin sodium, an HMG-CoA inhibitor used as an antihypertensive agent.

Bayer Aspirin A trade name for aspirin, an analgesic and antipyretic agent.

Bayer's Junction Sites of adhesion between the outer membrane and the cytoplasmic membrane of Gram-negative bacteria.

Bayhep B A trade name for Hepatitis B immunoglobulin (human).

Bayrab A trade name for rabies immunoglobulin (human).

Bayrho-D A trade name for Rh_o-D immunoglobulin (human).

Baytet A trade name for tetanus immunoglobulin (human).

Baytussin A trade name for guaifenesin, an antitussive agent.

Bb Activated factor B in the alternative pathway of complement activation.

BB Abbreviation for 1. binding buffer; 2. brush border.

BBB Abbreviation for blood-brain barrier.

BbeI (NarI) A restriction endonuclease from *Bifidobacterium breve* with the same specificity as NarI.

BbeAI A restriction endonuclease from *Bifidobacterium breve* S50 with the following specificity:

↓
5′..........GGCGCC..........3′
3′......... CCGCGG..........5′
↑

BbiI (PtsI) A restriction endonuclease from *Bifidobacterium bifidum* with the same specificity as PstI.

BbiII (AcyI) A restriction endonuclease from *Bifidobacterium bifidum* with the following specificity:

↓
5′..........GPuCGPyC..........3′
3′......... CPyGCPuG..........5′
↑

BbiIII (XhoI) A restriction endonuclease from *Bifidobacterium bifidum* with the same specificity as XhoI.

β-Blocker Referring to the inhibitor of adrenergic receptor.

BBM Abbreviation for brush-border membrane.

BBMV Abbreviation for brush border membrane vesicle.

BBP Abbreviation for bilin-binding protein.

BbrI (HindIII) A restriction endonuclease from *Bordetella bronchiseptica* having the same specificity as Hind III.

BbrPI A restriction endonuclease from *Bacillus brevis* with the following specificity:

5'........CACGTG........3'
3'........GTGCAC........5'

BbsI A restriction endonuclease from *Bacillus laterosporus* with the following specificity:

5'...........GAAGAC(2N).........3'
3'.......... CTTCTG(6N)..........5'

BbuI A restriction endonuclease from *Bacillus* species with the following specificity:

5'........GCATGC........3'
3'........CGTACG........5'

BbvI A restriction endonuclease from *Bacillus brevis* with the following specificity:

5′..........GCAGC(8N).........3′
3′......... CGTCG(12N)........5′

BbvII A restriction endonuclease from *Bacillus brevis* with the following specificity:

5′..........GAAGAC(2N)..........3′
3′...........CTTCTG(6N)..........5′

Bbv12I A restriction endonuclease from *Bacillus brevis* 12 with the following specificity:

5'........GWGCWC........3'
3'........CWCGWG........5'

W= A or T

Bbv16II A restriction endonuclease from *Bacillus brevis* 16 with the following recognition sequence. It cleaves either within the recognition site or at a short specific distance from it.

5'........GAAGAC(2/3)........3'

BC Abbreviation for 1. bacterial culture; 2 blood culture.

BCA Abbreviation for bicinchoninic acid.

BcaI (HhaI) A restriction endonuclease from *Branhamella catarrhalis* with the following specificity:

5′..........GCGC..........3′
3′..........CGCG..........5′

BCAA Abbreviation for branched-chain amino acid.

BCAF Abbreviation for basophil chemotaxis augmentation factor.

BCAP Abbreviation for balloon catheter angioplasty.

BCAVe Abbreviation for a combination drug containing bleomycin, CCNU, adriamycin, and velban.

BCB Abbreviation for brilliant cresyl blue.

BCC Abbreviation for basal cell carcinoma.

BCCP Abbreviation for biotin carboxyl carrier protein.

BCD Abbreviation for B-cell differentiation factor.

BCDF Abbreviation for B cell differentiation factor.

BCE Abbreviation for basal cell epithelioma.

Bce170 (PstI) A restriction endonuclease from *Bacillus cereus* with the same specificity as PstI.

Bce71I (HaeIII) A restriction endonuclease from *Bacillus cereus* with the same specificity as HaeIII.

BceFI (FnuDII) A restriction endonuclease from *Bacillus cereus* with the following specificity:

5′..........CGCG..........3′
3′..........GCGC..........5′

BceFI (FnuDII) Methylase A restriction methylase from *Bacillus cereus* that methylates or modifies external cytosine residue of the BceFI sequence.

CH_3
↓
5′..........CGCG..........3′
3′..........GCGC..........5′
↑
CH_3

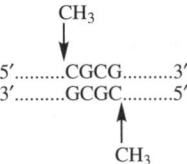

B-Cell Abbreviation for B lymphocyte (bone marrow derived lymphocytes).

B-Cell Antigen Receptor See B cell antigen receptor.

B-Cell Differential Factor See B cell differential factor.

BCF Abbreviation for basophil chemotactic factor.

BCG Abbreviation for *Bacillus* Calmette Guerin, an attenuated strain of *Mycobacterium bovis* that is used as a vaccine against tuberculosis.

βCG Abbreviation for beta subunit of chorionic gonadotropin.

BcgI A restriction endonuclease from *Bacillus coagulans* with the following specificity:

↓ ↓
5′..........(10N)CGA(6N)TGC(12N)..........3′
3′..........(12N)GCT(6N)ACG(10N)..........5′
↑ ↑

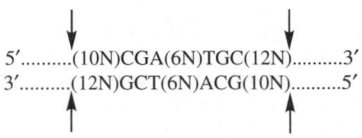

BCGF Abbreviation for B-cell growth factors.

BCGF-1 Abbreviation for B cell growth factor 1.

BCGF-2 Abbreviation for B cell growth factor 2.

B-Chain The large peptide chain of insulin.

BChl Abbreviation for bacteriochlorophyll.

BChl a Abbreviation for bacteriochlorophyll a.

BChl b Abbreviation for bacteriochlorophyll b.

BCHOP Abbreviation for a combination drug containing bleomycin, cytoxan, hydroxydaunomycin, oncovin, and prednisone.

B-Chromosome See B chromosome.

BCIP Abbreviation for 5-bromo-1-chloro-3-indolyl phosphate.

BCIP/NBT Abbreviation for 5-bromo-4-chloro-3-indolylphosphate/nitro-blue tetrazolium.

BCKAD Abbreviation for branched-chain α keto acid dehydrogenase.

BCKADH Abbreviation for branched-chain alpha-keto acid dehydrogenase.

BclI A restriction endonuclease from *Bacillus caldolyticus* with the following specificity:

↓
5′..........TGATCA..........3′
3′..........ACTAGT..........5′
↑

***bcl*2 Oncogene** An oncogene that encodes a plasma membrane protein.

***bcl*3 Oncogene** An oncogene associated with B-cell chronic lymphocytic leukemia.

BCLD Abbreviation for B cell lymphoproliferative disorder.

BCLL Abbreviation for B-cell lymphatic leukemia.

BCLT Abbreviation for blood clot lysis time.

BcmI (ClaI) A restriction endonuclease from *Bacillius* species with the following specificity:

↓
5′..........ATCGAT..........3′
3′..........TAGCTA..........5′
↑

BcnI (CauII) A restriction endonuclease from *Bacillus centrosporus* RFL1 with the following specificity:

↓
5′..........CC(C/G)GG..........3′
3′..........GG(G/C)CC..........5′
↑

BCNU Abbreviation for N,N′-bis(2-chloroethyl-N-nitrosourea, a reagent that crosslinks the strands of cellular DNA, interfering with RNA transcription (see also carmustine).

BcoI A restriction endonuclease from *Bacillus caogulans* with the following specificity:

↓
5'........CPyCGPuG........3'
3'........GPuGCPyC........5'
↑

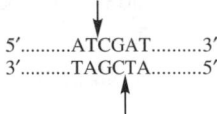

Bco33I (HaeIII) A restriction endonuclease from *Bacillus coagulans* with the same specificity as HaeIII.

B-Complex A large group of water soluble vitamins that include thiamine, riboflavin, nicotinic acid (niacin), pyridoxine, cyanocobalamin, biotin, choline, carnitine, folic acid, inositol, pantothenic acid, and para-aminobenzoic acid.

BCOP Abbreviation for a combination drug containing BCNU, cyclophosphamide, oncovin, and prednisone.

BCP Abbreviation for biotin carboxyl carrier protein.

BCR Abbreviation for a breakpoint cluster region on the human chromosome 22 involved in Philadelphia translocation.

BcrI (NlaIV) A restriction endonuclease from *Bacillus cereus* with the follow specificity:

```
           ↓
5′..........GGNNCC..........3′
3′..........CCNNGG..........5′
             ↑
```

BCVP Abbreviation for a combination drug containing BCNU, cyclophosphamide, vinblastine, and prednisone.

BCVPP Abbreviation for a combination drug containing BCNU, cyclophosphamide, vinblastine, prednisone, and procarbazine.

BD Abbreviation for binding domain.

BDB Abbreviation for bis-diazotized benzidine.

BD-cellulose Abbreviation for benzoylated diethylaminoethyl cellulose (an ion exchanger in ion exchange chromatography).

Bdellovibrio A genus of aerobic Gram-negative bacteria.

BDG Abbreviation for buffered deoxycholate glucose.

B-DNA See B DNA.

BDNF Abbreviation for brain-derived neurotrophic factor.

BDOPA Abbreviation for a combination drug containing bleomycin, DTIC, oncovin, prednisone, and adriamycin.

BdU Abbreviation for bromo-deoxy-uridine.

BDUR (BdUR) Abbreviation for bromodeoxyuridine.

Be Symbol for beryllium with atomic weight 9 and valence 2.

Beaker Cell Specialized cells that release mucus and form glands of the lining of the GI tract (also called goblet cell or mucus-secreting epithelium cell).

BEAM Abbreviation for a combination drug containing BCNU, etoposide, ara-C, and melphalan.

Beauvericin A cyclic peptide antibiotic from *Fusarium* species active against Gram-positive bacteria (see also enniatin).

Bebeerine (mol wt 595) An antimalarial agent from the plant *Nectandra rodioei*.

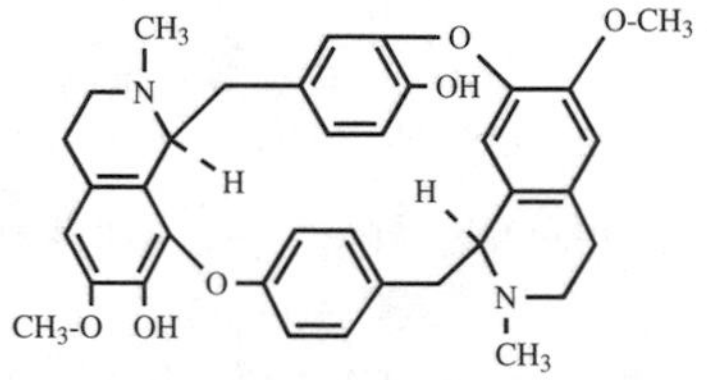

BEC Abbreviation for blood ethanol concentration.

Becanthone (mol wt 385) An anthelmintic agent.

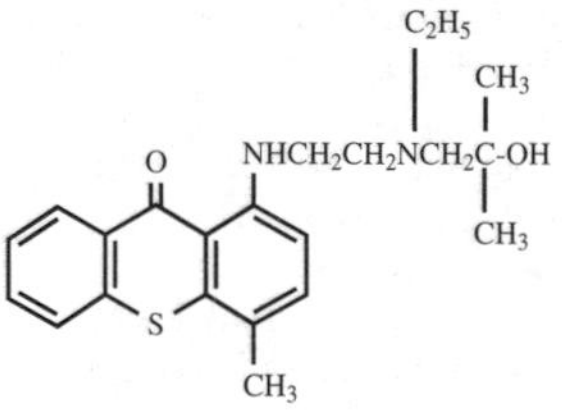

Bechic 1. Of or pertaining to coughing. 2. Any agent promoting coughing.

Becker's Muscular Dystrophy An inherited disorder of the muscle characterized by progressive muscular weakness.

Beckmann Thermometer A highly sensitive mercury thermometer with a large bulb. It is used to measure small changes in temperature.

Beckwith's Syndrome A genetic disorder of infants characterized by low blood sugar and overproduction of insulin.

Beclamide (mol wt 198) An anticonvulsant.

$Cl\text{-}CH_2CH_2\text{-}CO\text{-}NH\text{-}CH_2\text{-}C_6H_5$

Beclobrate (mol wt 347) An antihyperlipoproteinemic agent.

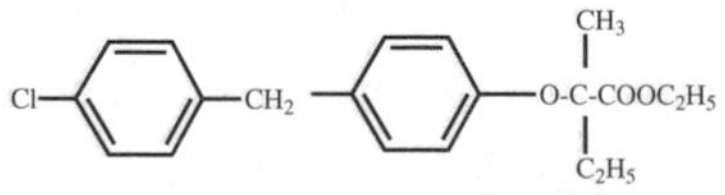

Beclodisk A trade name for beclomethasone dipropionate, a corticosteroid hormone.

Becloforte A trade name for beclomethasone dipropionate, a corticosteroid hormone.

Beclomethasone (mol wt 409) A glucocorticoid with antiallergic, antiasthmatic, and anti-inflammatory activity.

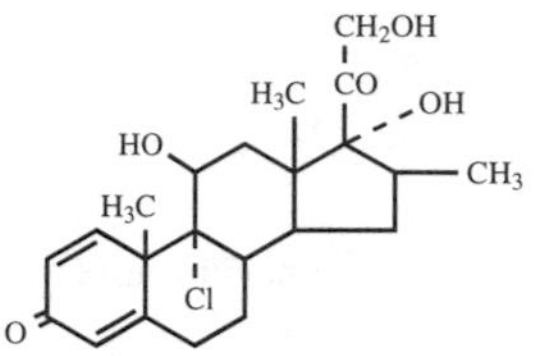

Beclotiamine (mol wt 319) An antithiamine compound derived from thiamine.

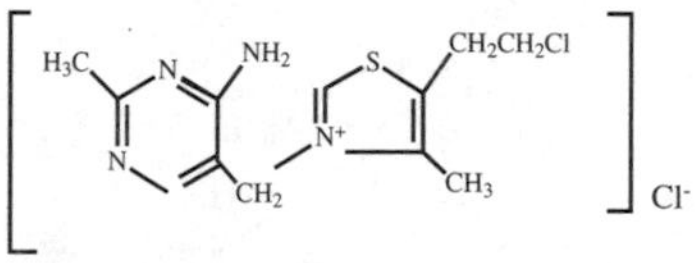

Beclovent A trade name for beclomethasone dipropionate, a corticosteroid that decreases inflammation.

Beconase A trade name for beclomethasone dipropionate, an anti-inflammatory agent.

Beconase AQ A trade name for beclomethasone dipropionate, a nasal inhalation aerosol.

Becquerel (Bq) A unit of radioactivity, 1 Bq = 27.03×10^{-12} Ci.

Bed Volume The volume occupied by the packing in a chromatographic column.

Bedbug A blood-sucking arthropod (*Cimex lectularius*). Its bite causes itching, pain, and redness.

Beef Tapeworm Infection An infection caused by the tapeworm *Taenia saginata*, transmitted to humans by eating contaminated beef.

Beelith A nutritional supplement containing magnesium and vitamin B_6.

Beepen-VK A trade name for penicillin V potassium.

Beer-Lambert Law The law states that the absorbance of a light-absorbing substance in solution is directly proportional to its concentration and the length of the light path.

Beer's law The law states that the absorbance of a light-absorbing substance in solution is directly proportional to its concentration.

Beesix A trade name for vitamin B_6.

Beet Sugar Referring to sucrose.

Beeturia A redish coloration of urine resulting from eating beets.

Befunolol (mol wt 291) An antiglaucoma agent.

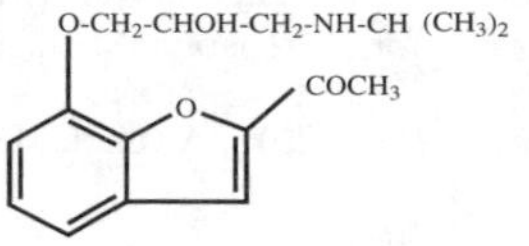

Beggiatoa A genus of gliding bacteria (*Cytophagales*).

Behenic acid (mol wt 341) A minor fatty acid in the fats of seeds, milk, and marine animal oils.

$CH_3(CH_2)_{20}COOH$

BEI Abbreviation for butanol extractable iodine.

Beijerinckia A genus of Gram-negative, aerobic, catalase-positive, chemoorganotrophic, asporogenous bacteria.

Bejel A nonvenereal form of syphilis prevalent among children, caused by the spirochete *Treponema pallidum.*

Beldin A trade name for diphenhydramine, used as an antihistaminic agent.

Belix A trade name for diphenhydramine, used as an antihistaminic agent.

Belladonna The dried leaves and fruits of *Atropa belladonna*; a poisonous plant containing hyoscyamine and atropine. It possesses anticholinergic activity.

Belling's Hypothesis The hypothesis states that crossover does not require breakage and reunion. According to the hypothesis, the genes replicate first, followed by the intergenic connections between the adjacent newly synthesized genes.

Bell-Shaped Curve A symmetrical bell-shaped curve in an X/Y plot.

Bemegride (mol wt 155) A central nervous system stimulant. It counteracts barbiturate poisoning.

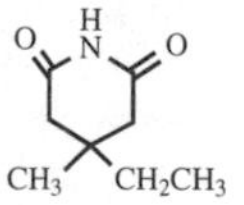

Bemote A trade name for dicyclomine hydrochloride, an anticholinergic agent.

Bena-D A trade name for diphenhydramine, an antihistaminic agent.

Benactyzine (mol wt 327) An antagonist of acetylcholine in the central and peripheral nervous system.

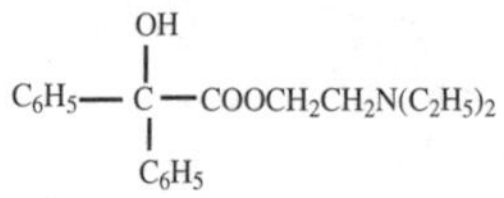

Benadryl A trade name for diphenhydramine hydrochloride, an antihistaminic drug. It competes with histamine for H_1 receptors on effector cells.

Benalaxyl (mol wt 325) A fungicide.

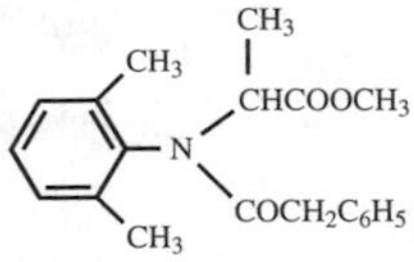

Benahist A trade name for diphenhydramine, an antihistaminic agent.

Benaphen A trade name for diphenhydramine, an antihistaminic agent.

Benapryzine (mol wt 341) An anti-anticholinergic agent.

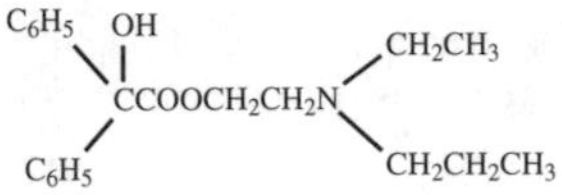

Benazepril (mol wt 425) An angiotensin-converting enzyme inhibitor used as an antihypertensive agent.

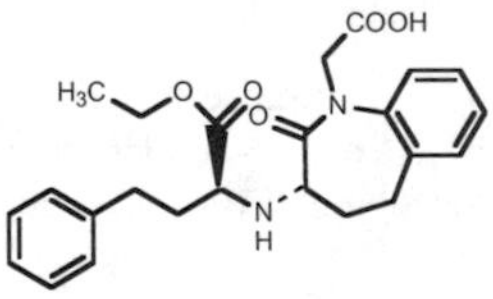

Bence-Jones Protein The immunoglobulin light chains present in the urine of patients with multiple myeloma.

Bencyclane (mol wt 289) A vasodilator.

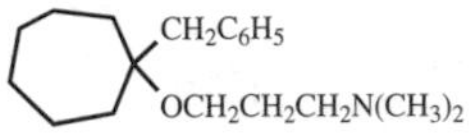

Bendazac (mol wt 282) An anti-inflammatory agent for treatment of dermatitis, e.g., eczema and other skin disorders.

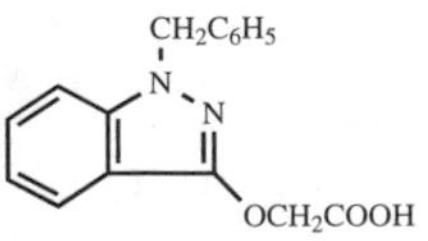

Bendazol (mol wt 208) A vasodialtor.

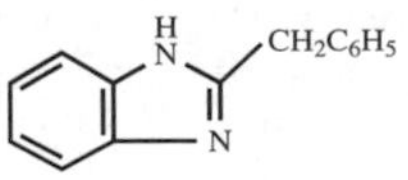

Bendiocarb (mol wt 223) A contact insecticide.

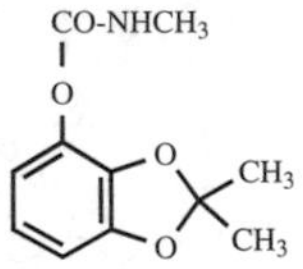

Bendrofluazide See bendroflumethiazide.

Bendroflumethiazide (mol wt 421) A diuretic agent that inhibits sodium reabsorption and increases urine secretion.

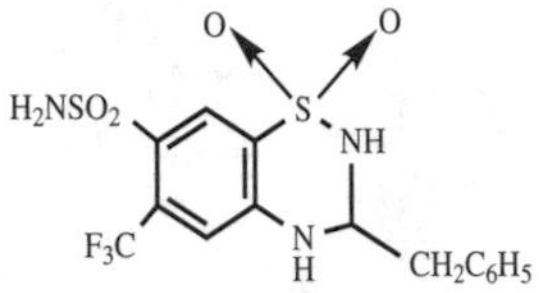

Benedict's Test A colorimetric test for reducing sugars based on the reduction of cupric ions by the reducing sugar in the alkaline solution.

Benedict's Reagent Reagent used for the determination of reducing sugars; it contains sodium citrate, sodium carbonate, and copper sulfate.

Benefix A trade name for coagulation factor IX produced by DNA recombinant technology and used for preventing and controlling bleeding in patients with factor IX deficiency.

Benemid A trade name for probenecid, an antigout agent that blocks renal tubular reabsorption of uric acid.

Benexate Hydrochloride (mol wt 446) An antiulcerative agent.

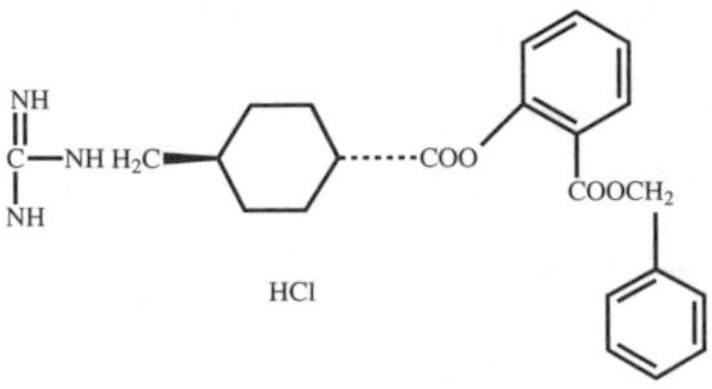

Benfluorex (mol wt 351) An antihyperlipoproteinemic agent.

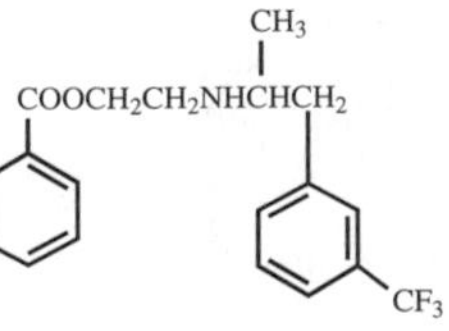

Benfotiamine (mol wt 466) A vitamin B_1 source.

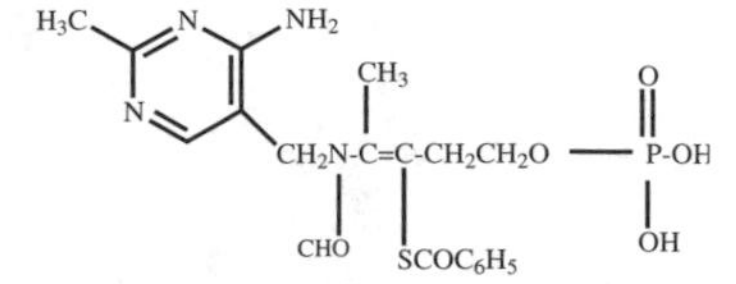

Benfuracarb (mol wt 411) An insecticide and inhibitor of cholinesterase.

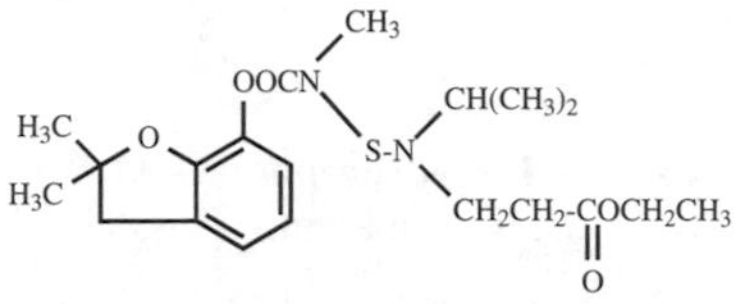

Benfurodil Hemisuccinate (mol wt 358) A cardiotonic agent and a vasodilator.

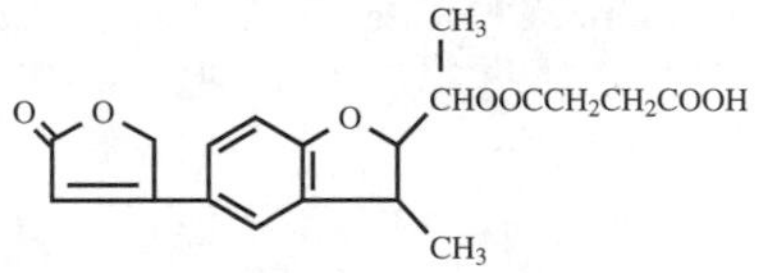

Ben-Gay A trade name for methyl salicylate that acts as a counterirritant to block pain.

Benidipine (mol wt 506) An antihypertensive agent.

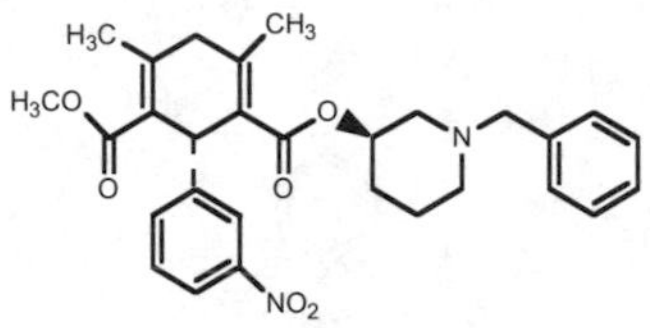

Benign 1. A nonmalignant tumor lacking the ability to invade surrounding normal tissue. 2. A mild self-limiting, nonrecurrent disease.

Benign Tumor A nonmetastasizing tumor.

Benisone A trade name for betamethasone benzoate, a corticosteroid.

Benmoxine (mol wt 240) An antidepressant.

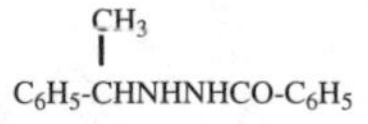

Benn A trade name for probenecid, an antigout agent that blocks the renal tubular reabsorption of uric acid.

Benoject A trade name for diphenhydramine hydrochloride, an antihistaminic agent.

Benomyl (mol wt 290) An anthelmintic and antifungal agent.

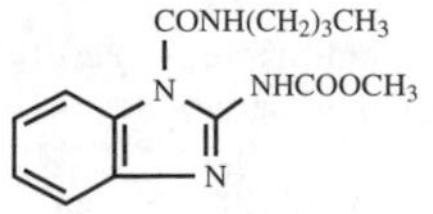

Benorylate (mol wt 313) An analgesic and anti-inflammatory agent.

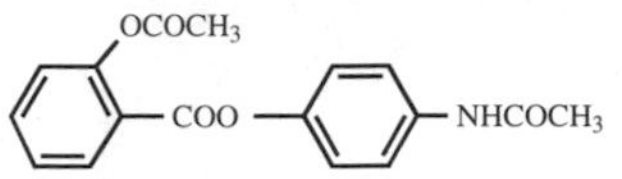

Benoxaprofen (mol wt 302) An anti-inflammatory and analgesic agent.

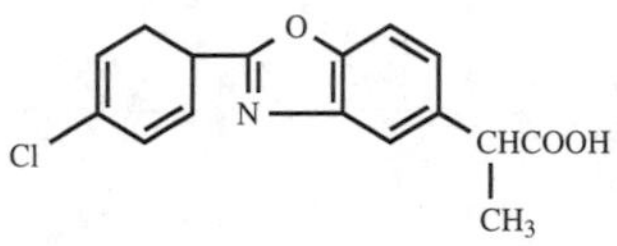

Benoxinate (mol wt 308) A topical anesthetic agent used to obtain surface anesthesia.

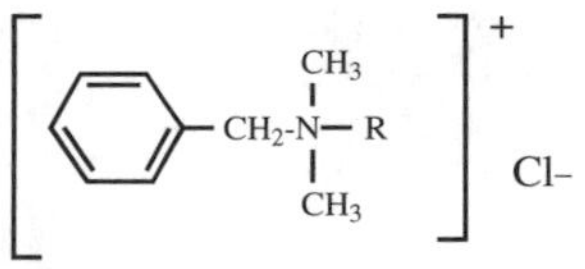

Benoxyl A trade name for benzoyl peroxide lotion used as an antimicrobial and comedolytic agent.

Benperidol (mol wt 381) An antipsychotic agent and a tranquilizer.

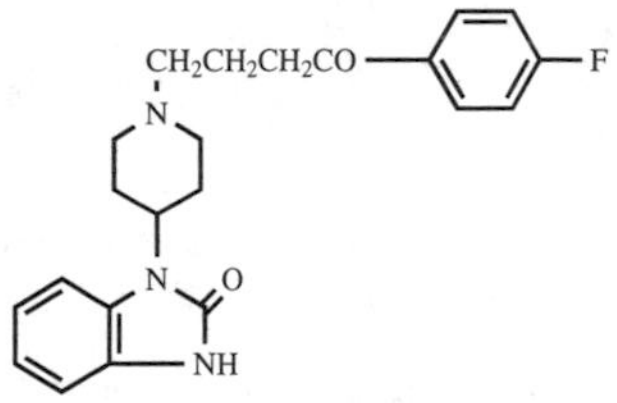

Benproperine (mol wt 309) An antitussive agent.

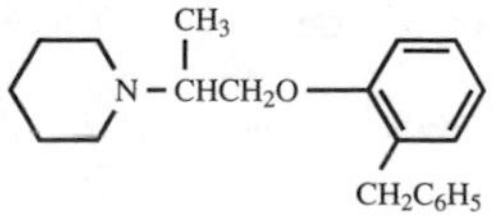

Benserazide (mol wt 257) An antiparkinsonian agent.

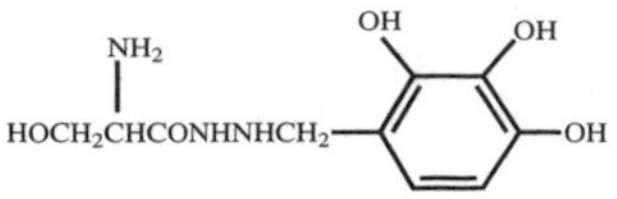

Bensylate A trade name for benztropine mesylate, an antiparkinsonian agent.

Bentazon (mol wt 240) An herbicide.

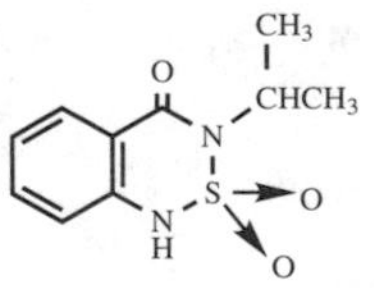

Bentiromide (mol wt 404) A synthetic chymotrypsin-labile peptide used in the diagnosis of exocrine pancreatic disease.

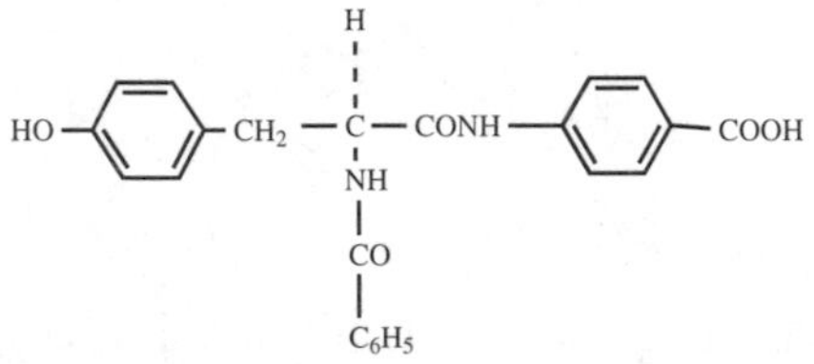

Bentonite A colloidal aluminum silicate containing chiefly montmorillonite; used as a suspending agent and as an inhibitor of nucleases.

$$Al_2O_3SiO_2H_2O$$

Bentyl A trade name for dicyclomine hydrochloride, used as an anticholinergic and antispasmodic agent.

Bentylol A trade name for dicyclomine hydrochloride, used as an anticholinergic and antispasmodic agent.

Benuryl A trade name for probenecid, used as an antigout agent that blocks the renal tubule reabsorption of uric acid.

Benylin A trade name for diphenhydramine hydrochloride, used as an antihistaminic agent.

Benza A trade name for benzalkonium chloride, used as a disinfectant.

Benzac A trade name for benzoyl peroxide gel, used as an antimicrobial and comedolytic agent.

Benzagel A trade name for benzoyl peroxide gel, used as an antimicrobial and comedolytic agent.

β-Benzalbutyramide (mol wt 175) An antilipoproteinemic agent that inhibits cholesterol biosynthesis.

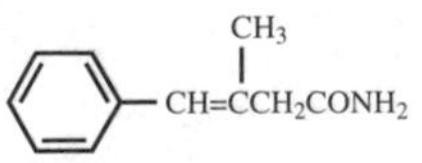

Benzaldehyde (mol wt 106) A compound in the oil of the kernels of bitter almond.

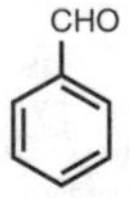

Benzaldehyde Dehydrogenase (NAD^+) The enzyme that catalyzes the following reaction:

$$\text{Bezaldehyde} + NAD^+ + H_2O \rightleftharpoons \text{Benzoate} + NADH$$

Benzaldehyde Dehydrogenase ($NADP^+$) The enzyme that catalyzes the following reaction:

$$\text{Benzaldehyde} + NADP^+ + H_2O \rightleftharpoons \text{benzoate} + NADPH$$

Benzalkonium Chloride A disinfectant and antibacterial agent used for treatment of wounds and burns.

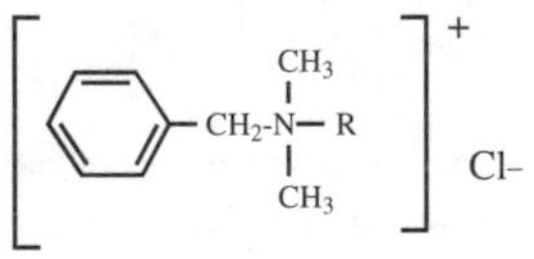

Benzamidase Synonym of aminoacylase.

Benzamidine (mol wt 121) An inhibitor of trypsin.

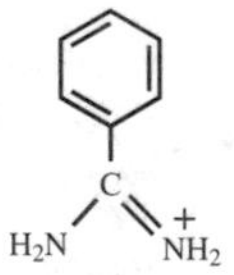

Benzamycin Gel A trade name for a combination drug containing erythromycin and benzoyl peroxide, used as an anti-infective agent.

Benzapyrene See benzopyrene.

Benzbromarone (mol wt 424) A uricosuric agent.

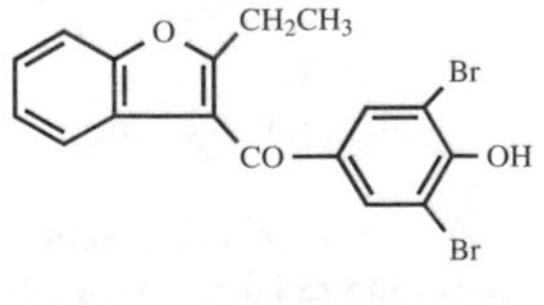

Benzene (mol wt 78) Compound consisting of a six-membered ring and three double bonds, it occurs as part of many biological molecules and is also a solvent and intermediate in the production of styrene, nylon.

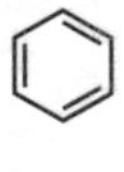

Benzene Poisoning A toxic condition caused by the ingestion of benzene or benzene-related products such as tolulene or xylene. It is characterized by nausea, headache, dizziness, and incoordination.

Benzenediol Oxygen Oxidoreductase The systematic name for laccase.

Benzenoid Having the property of benzene.

Benzestrol (mol wt 298) An estrogenic diphenol.

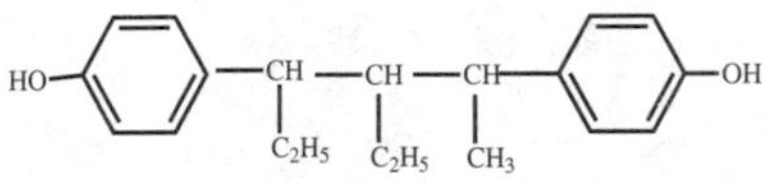

Benzethonium Chloride Mol wt 448) A topical anti-infective agent used for disinfecting the skin and for treatment of infections of the eye, nose, and throat.

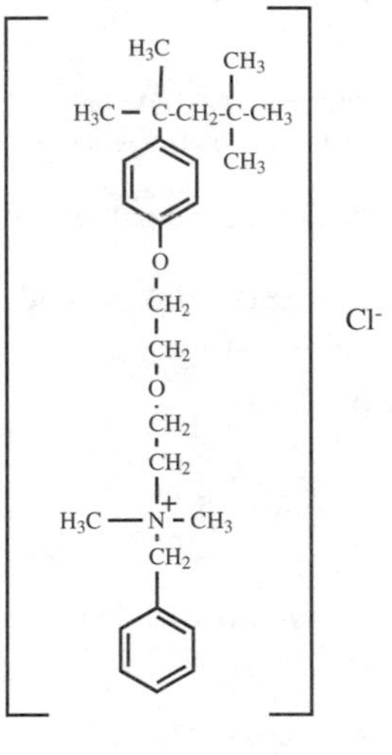

Benzide A trade name for bendroflumethiazide, used as a diuretic agent.

Benzidine (mol wt 184) A reagent used for detection of hydrogen peroxide in milk and blood and for detection of bacterial cytochrome.

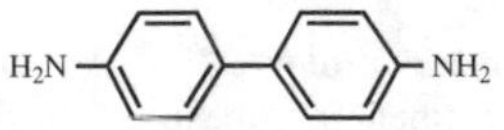

Benzidine Test A test used for detection of peroxidase and bacterial cytochrome based on the formation of benzidine blue upon treatment of the sample with benzidine and hydrogen peroxide.

Benzilonium Bromide (mol wt 434) An anticholinergic agent.

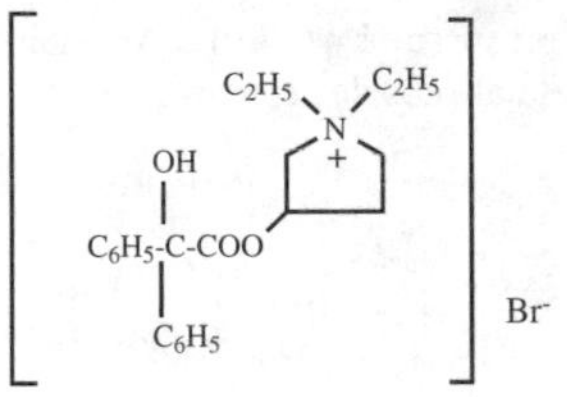

Benzimidazole (mol wt 118) An antifungal and antiviral agent.

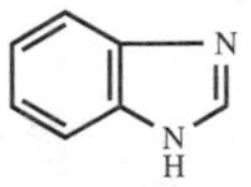

Benziodarone (mol wt 518) A coronary vasodilator.

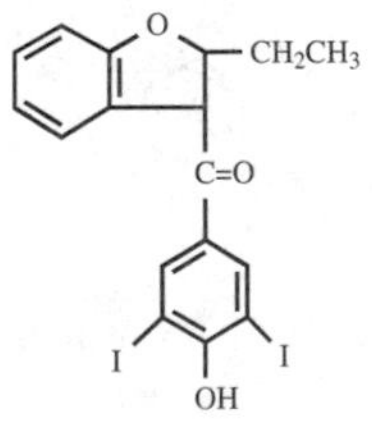

Benznidazole (mol wt 260) An antiprotozoal *(Trypanosoma)* agent.

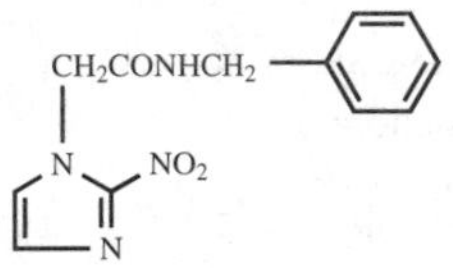

Benzoate A salt of benzoic acid acid.

Benzoate CoA Ligase The enzyme that catalyzes the following reaction:

ATP + benzoate + CoA

⇅

AMP + PPi + benzoyl-CoA

Benzoate 1,2-Dioxygenase The enzyme that catalyzes the following reaction:

Benzoate + NADH + O_2

⇅

Catechol + CO_2 + NAD^+

Benzoate Hydroxylase See benzoate 1,2-dioxygenase.

Benzocaine (mol wt 165) An anesthetic agent.

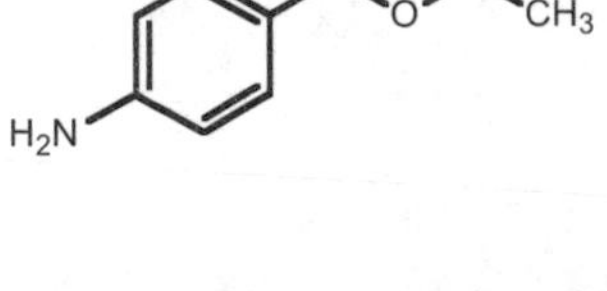

Benzoctamine (mol wt 249) An anxiolytic agent and a skeletal muscle relaxant.

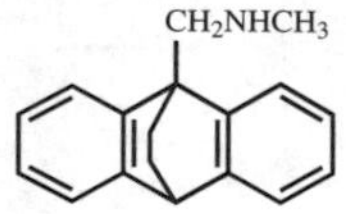

Benzodepa (mol wt 281) An antineoplastic agent.

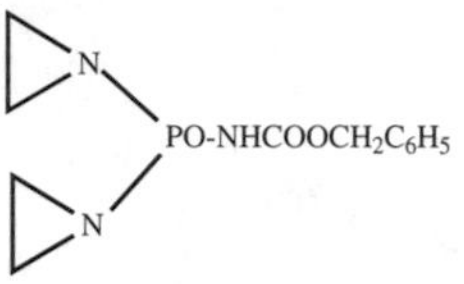

Benzoic Acid (mol wt 122) An aromatic acid, an antibacterial, and an antifungal agent.

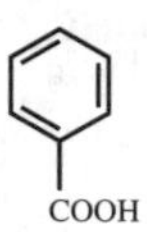

Benzoin (mol wt 212) A resin from the cut stem of species of Styracaceae with antiseptic activity.

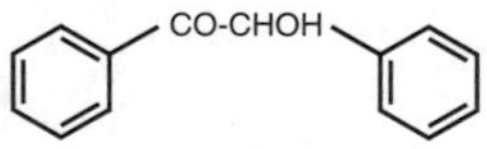

Benzoin Aldolase The enzyme that catalyzes the following reaction:

Benzoin ⇌ 2 Benzaldehyde

Benzolism Benzene poisoning.

Benzonatate (mol wt 604) An antitussive agent.

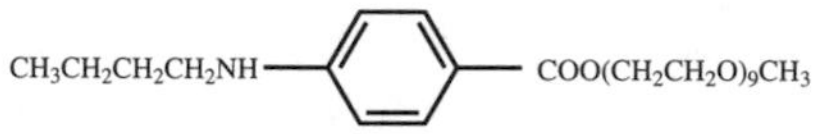

Benzopyrene (mol wt 252) A polycyclic, aromatic hydrocarbon and a potent mutagen and carcinogen.

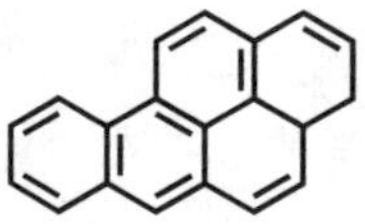

Benzoquinonium Chloride (mol wt 618) A synthetic neuromuscular blocking agent used as a skeletal muscle relaxant.

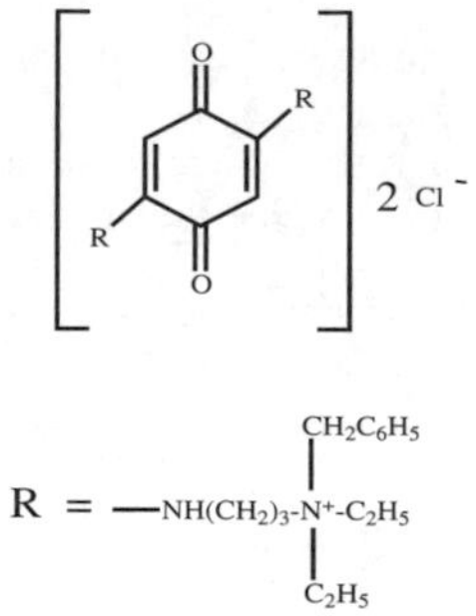

Benzoxiquine (mol wt 249) A disinfectant.

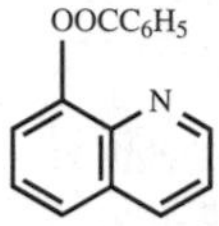

Benzoyl Referring to PhCO or C_6H_5-CO group.

Benzoylcholinesterase Synonym of cholinesterase.

Benzoylformate Decarboxylase The enzyme that catalyzes the following reaction:

Benzoylformate ⇌ Benzaldehyde + CO_2

Benzoyl-Isothiocyanate (mol wt 163) An antituberculostatic agent.

Benzoyl-Peroxide (mol wt 242) An antibacterial and keratolytic agent.

$(C_6H_5CO)_2O_2$

Benzoylpas (mol wt 257) An antibacterial agent.

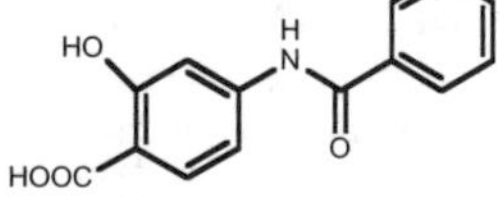

Benzphetamine (mol wt 239) An anorexic agent used for treatment of obesity. It also acts as a cerebral stimulant.

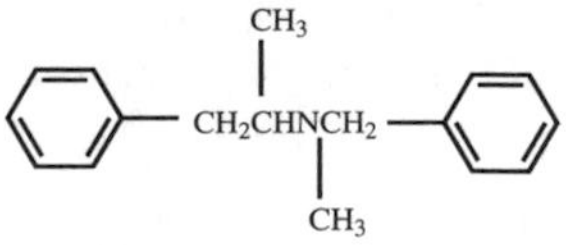

Benzpiperylon (mol wt 347) An anti-inflammatory agent.

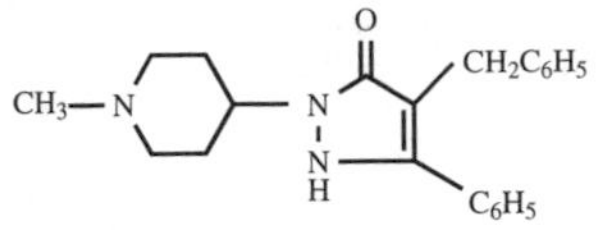

Benzpyrinium Bromide (mol wt 337) A cholinergic agent.

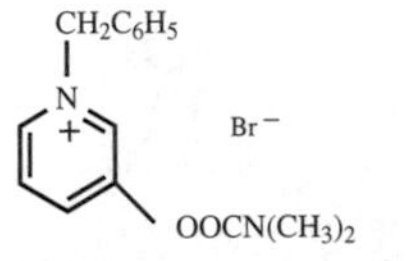

Benzquinamide (mol wt 404) An antiemetic agent used for treatment of postoperative nausea and vomiting.

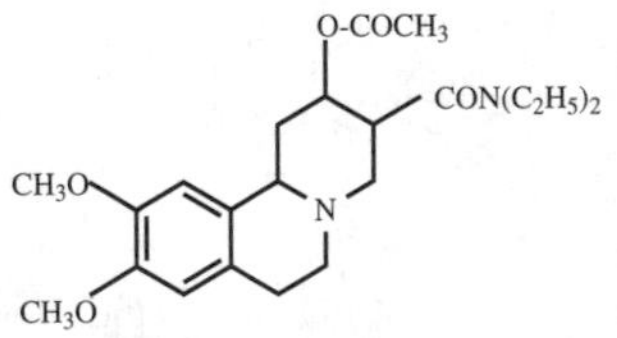

Benzthiazide (mol wt 432) A diuretic agent that increases urine secretion by inhibiting sodium reabsorption.

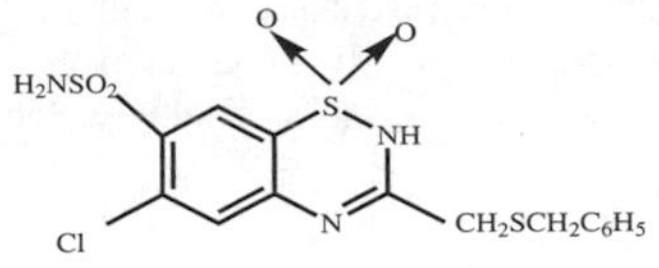

Benztropine Mesylate (mol wt 404) An anticholinergic and antihistaminic agent that blocks central cholinergic receptors.

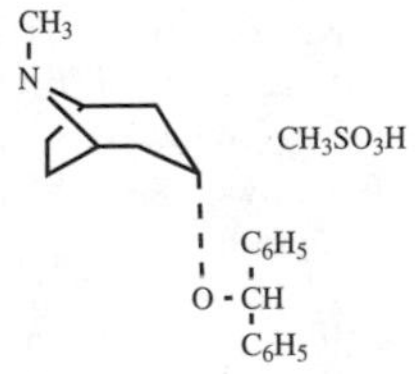

Benzydamine (mol wt 309) An analgesic, anti-inflammatory, and antipyretic agent.

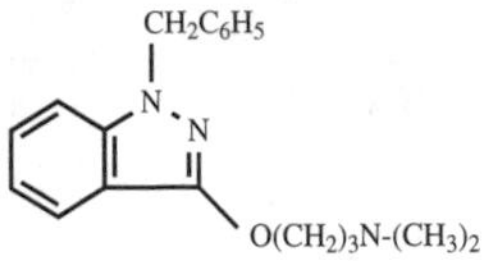

Benzyl Referring to $PhCH_2$ or C_6H_5-CH_2 group.

Benzyl-Alcohol (mol wt 108) An aromatic alcohol used as a topical anesthetic and as a bacteriostatic agent.

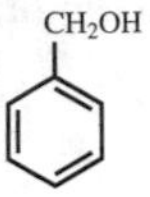

6-Benzylaminopurine (mol wt 225) A synthetic cytokinin.

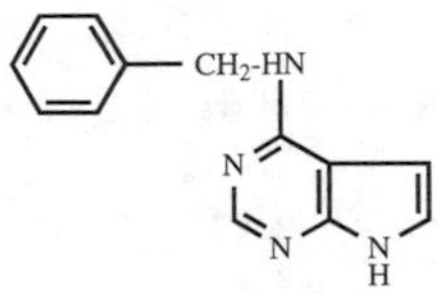

Benzyl-Benzoate (mol wt 212) An insecticide (scabicide).

$C_6H_5COOCH_2C_6H_5$

Benzyl-Benzoate Lotion A skin lotion for control of parasitic skin infestation (e.g., scabies, *Pediculus capitis, Phthirus pubis*).

Benzylhydrochlorothiazide (mol wt 388) An antihypertensive and a diuretic agent.

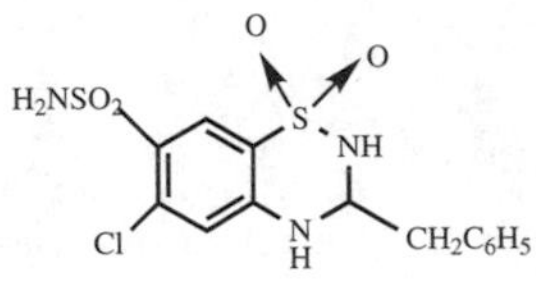

Benzylmorphine (mol wt 375) A narcotic analgesic agent.

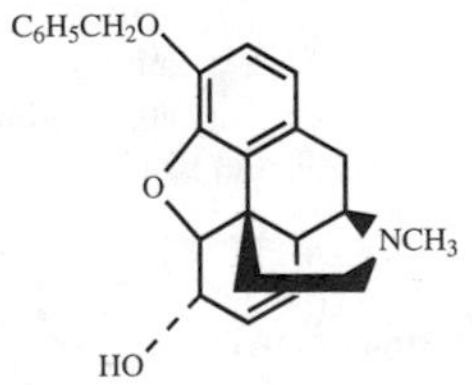

Benzyl-Oxycarbonyl Referring to the following group:

Ph-CH_2-O-CO-

Benzylpenicillinic Acid (mol wt 334) A free acid form of penicillin G that inhibits bacterial cell wall synthesis.

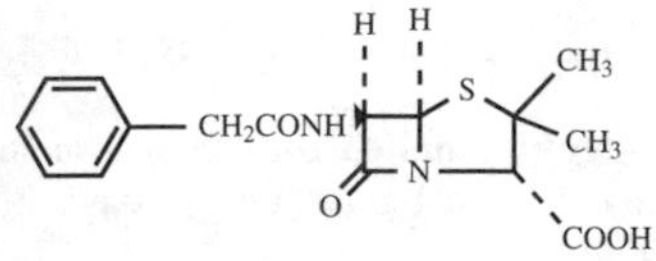

Benzylpenicillin Sodium (mol wt 356) The sodium form of penicillin G that inhibits bacterial cell wall synthesis.

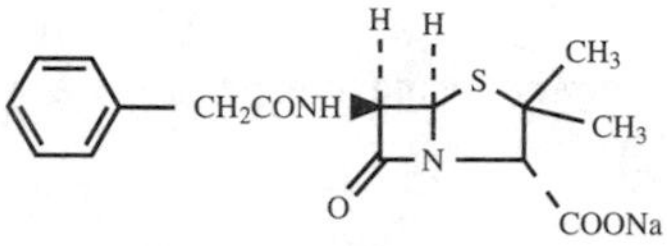

Benzylsulfamide (mol wt 262) An antibacterial agent.

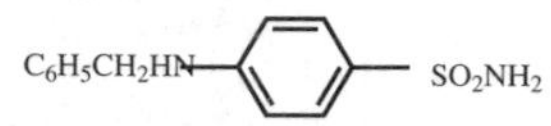

BEP Abbreviation for a combination drug containing bleomycin, etoposide, and platinol.

BepI (FnudII) A restriction endonuclease from *Brevibacterium epidermidis* with the following specificity:

```
          ↓
5′.........CGCG.........3′
3′.........GCGC.........5′
          ↑
```

Bepadin A trade name for bepridil hydrochloride, a calcium channel blocker used as an antianginal agent.

Bepen A trade name for betamethasone benzoate, a corticosteroid.

Bepridil (mol wt 367) An antianginal agent and a calcium channel blocking agent that inhibits calcium influx across the cardiac and smooth muscle cell.

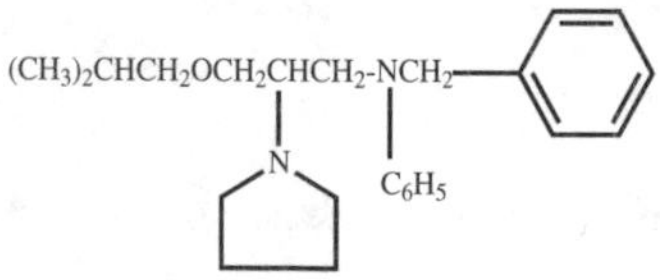

Beractant An extract from bovine lung used as a surfactant.

Berberine (mol wt 336) An alkaloid derived from *Dydrastis canadensis* and an antibacterial, antimalarial, and antipyretic agent.

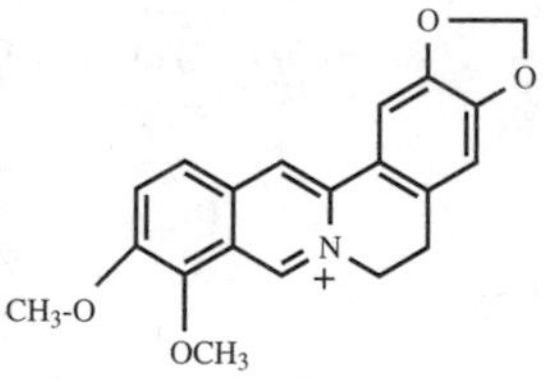

Bergaptene (mol wt 216) An antipsoriatic agent.

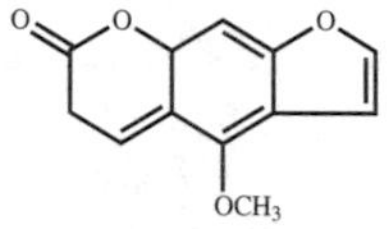

Bergenin (mol wt 328) An antitussive compound found in *Bergenia sibirica* (Saxifragaceae).

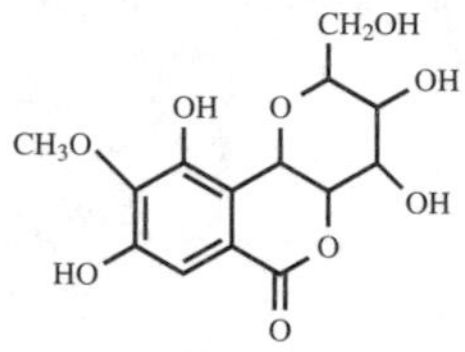

Berger's Disease A disorder of the kidney characterized by the periodic appearance of blood and protein in the urine and deposition of IgA in the central portions of the glomeruli. The disease may lead to kidney failure.

Beriberi A disease caused by a deficiency of thiamine (vitamin B_1). It is characterized by pain, paralysis, atrophy, and loss of body weight.

Berkefeld Filter A diatomaceous filter used to sterilize solutions.

Berkelium (Bk) A chemical element with atomic weight 247, valence 3 and 4.

Berk-Shap Technique A DNA mapping technique in which mRNA is hybridized with single-stranded DNA and the nonhybridized DNA is digested with specific endonuclease. The hybridized DNA is then dissociated from mRNA and analyzed by gel electrophoresis.

Berlock Dermatitis A skin disorder characterized by the hyperpigmentation and appearance of dark patches and sores on the skin. It is caused by a unique reaction to oil of bergamot, commonly used in perfumes.

Bermoprofen (mol wt 296) An anti-inflammatory, analgesic, and antipyretic agent.

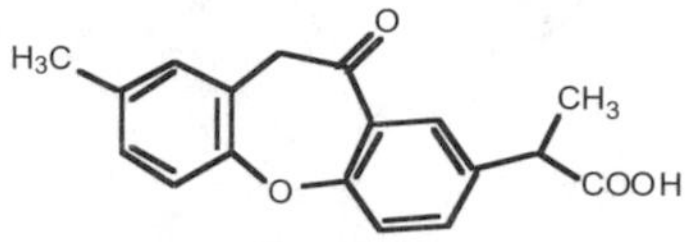

Bernard-Soulier Syndrome A blood-clotting disorder characterized by the inability of platelets to aggregate due to the deficiency of essential glycoprotein in the platelets.

Berninamycin A cyclic peptide antibiotic isolated from *Streptomyces bernesis.* It inhibits bacterial protein synthesis.

Bernoulli's Law A law stating that the velocity of a gas or a fluid flowing through a tube is inversely proportional to the pressure it exerts against the side of the tube. The greater the velocity, the lower the pressure.

Berthelot Reaction A colorimetric reaction for the determination of ammonia in the urine based on the production of blue indophenol upon treatment of urine with phenol and sodium hypochlorite.

Bertielliasis Infection by the tapeworm *Bertiella.*

Berylliosis Poisoning resulting from the inhalation of dusts or vapors that contain beryllium or beryllium compounds.

Beryllium (Be) A light-weight metallic element with atomic weight 9, valence 2.

BES (mol wt 213) Abbreviation for N,N-bis(2-hydroxyethyl)-2-aminoethanesulfonic acid, used for the preparation of buffers in the pH range of 6.2 to 8.2.

$$(HOCH_2CH_2)_2NCH_2CH_2SO_3H$$

Besipirdine (mol wt 251) A nootropic agent.

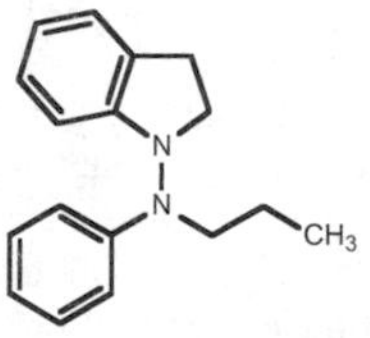

Besnoitia A genus of protozoa of the family Besnoitidae.

Bestrabucil (mol wt 721) An antineoplastic agent.

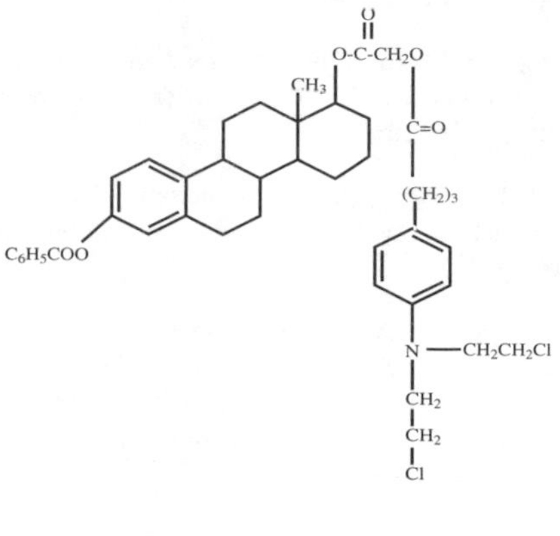

Beta (β) The second letter of the Greek alphabet employed to denote 1. the second carbon atom next to the carbon atom that carries the principal functional group of the molecule; 2. a specific configuration of the substituents at the anomeric carbon in the ring structure of carbohydrates, e.g., β-D-glucose; 3. type of protein configuration; 4. the symbol for buffer value; 5. a type of protein subunit.

Beta-2 A trade name for isoetharine hydrochloride, a bronchodilator.

Beta Adrenergic Blocking Agent Any agent that blocks the activity of the beta receptor (see also antiadrenergic).

Beta Adrenergic Receptor See beta receptor.

Beta Adrenergic Stimulating Agent See adrenergic.

Beta Alanine (β-Alanine, mol wt 89) A naturally occurring nonprotein amino acid.

$CHNH_2$
|
CH_2
|
COOH

Beta Alaninemia A genetic disorder characterized by seizures and drowsiness due to the deficiency in the enzyme (β-alanine-α-ketoglutarate aminotransferase) for the metabolism of the amino acid alanine.

Beta Barrels A barrel-shaped supersecondary structure formed by rolling up of the extended β sheet.

Beta Bend A tightly folded polypeptide chain in which the straight runs of the secondary structure are joined by stretches of polypeptide that change direction (reverse turn). The β bends are largely responsible for the formation of globular structure of the proteins.

Beta Blocker Referring to the inhibitors or antagonists that block adrenergic receptors.

Beta Carotene (mol wt 537) A precursor of vitamin A that yields two molecules of vitamin A per molecule of beta carotene upon cleavage.

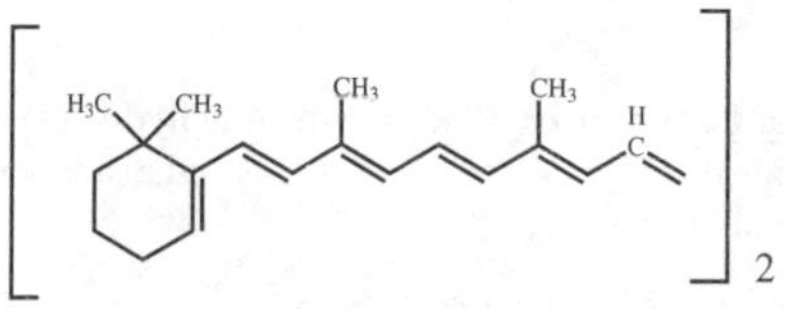

Beta Cells Insulin-producing cells situated in the islets of Langerhans of the pancreas.

Beta Chain A type of polypeptide chain occurring in protein, e.g., β chain of hemoglobin.

Beta Configuration A configuration of polypeptide chains in which the polypeptide chain is in a fully extended form.

Beta Decay The radioactive disintegration of an atomic nucleus that results in the emission of beta particles.

Beta Emitter A radioactive nuclide that emits beta particles.

Beta Fetoprotein A protein found in fetal liver and in some adults with liver disease.

Beta Galactosidase (β-galactosidase) The enzyme that catalyzes the cleavage of the β-galactosidic linkage in lactose.

Beta Galactoside Permease The enzyme that controls the rate of the entrance of β-galactoside (e.g., lactose) into the cell. It is one of the three enzymes encoded by the lac operon in *E. coli.*

Beta Globulin One of the plasma proteins associated with transportation of thrombin and prothrombin.

Beta Hemolysis The development of a clear zone around a bacterial colony growing on blood-agar medium.

Beta Hemolytic Streptococci The pyogenic *Streptococci* that hemolyze the red blood cells in blood agar (e.g., *Streptococcus* of group A, B,C, E, F, G, H, and K).

Beta Interferon Interferon produced by fibroblasts in mammalian connective tissue.

Beta Keratin A type of keratin in which the polypeptide chains are in the parallel pleated-sheet configuration.

Beta Lactam A four-member ring structure in which carbonyl group and a nitrogen are joined by an amide linkage. It occurs in the lactam antibiotics, e.g., penicillin.

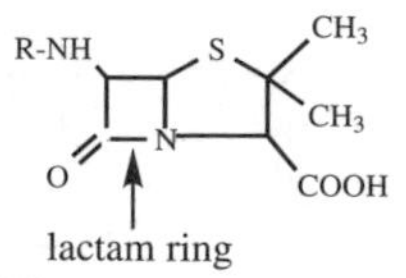

Beta Lactamase The enzyme that hydrolyzes the amide linkage in the lactam ring (also known as penicillinase).

Beta Lactam Inhibitor Substances that inhibit the inactivation of β-lactam antibiotics by β-lactamase or penicillinase.

Beta Lactoglobulin A type of protein in milk.

Beta Lipoprotein Referring to low-density lipoprotein.

Beta Lysin A highly reactive, heat-stable cationic protein that is bactericidal for Gram-positive bacteria.

Beta Meander A supersecondary structure of the protein consisting of an antiparallel β sheet formed by sequential segments of polypeptide chain that are connected by tight reverse turns.

Beta$_2$ Microglobulin (β$_2$-Microglobulin) An immunoglobulin-like molecule associated with class I histocompatibility antigen but not encoded by the major histocompatibility complex.

Beta Oxidation A pathway for oxidative degradation of fatty acid into acetyl-CoA by successive oxidations at the β-carbon atom of a fatty acid.

Beta Oxybutyria The presence of β-oxybutyric acid in the urine as in diabetes.

Beta Particle An electron or positron emitted from the nucleus of an atom during radioactive decay of the atom.

Beta Plateau The high potential region of the characteristic curve of a proportional radiation detector at which the count rate is almost independent of the applied voltage.

Beta Pleated Sheet (β Sheet) A type of protein configuration in which the polypeptide chains are partially extended and held together by the interchain hydrogen bonds.

Beta Ray A beam of beta particles.

Beta Receptor An adrenergic receptor on the membrane that responds to epinephrine. Activation of a beta receptor causes a variety of physiological reactions, e.g., vasodilatation, increase of heart beat, and muscle relaxation.

Beta Rhythm A low-voltage brain wave.

Beta Thalassemia An inherited disorder characterized by the reduction in rate of synthesis of the β chain of hemoglobin.

Beta Threshold The lowest potential at which beta particles can be detected with a proportional radiation detector.

Betabloc A trade name for metoprolol tartrate, a beta-1 adrenergic blocker used as an antihypertensive agent.

Betacellulin A potent mitogenic glycoprotein purified from conditioned medium of mouse pancreatic beta tumor cells, retinal pigment epithelial cells and vascular smooth muscle cells.

Betacin A bacteriocin-like, heat-resistant, proteinaceous substance produced by bacteriophage SPb-containing lysogens of *Bacillus subtilis.*

Betacort A trade name betamethasone valerate, corticosteroid hormone.

Betaderm A trade name betamethasone valerate, a corticosteroid hormone.

Betadine A trade name for povidone-iodine, a topical anti-infective agent.

Betagan A trade name for levobunolol, a nonselective beta blocker.

Betagen A trade name for povidone iodine, an anti-infective agent.

Betahistine (mol wt 136) A vasodilator.

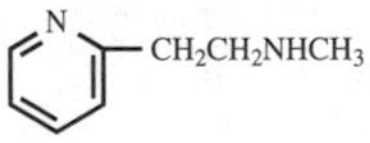

Betaine (mol wt 117) A methyl group donor that occurs in tissues of plants and animals.

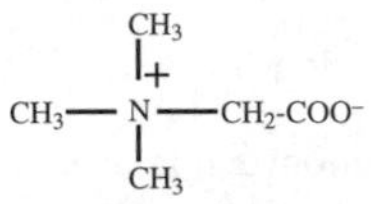

Betaine Aldehyde Dehydrogenase The enzyme that catalyzes the following reaction:

$$\text{Betaine aldehyde} + NAD^+ + H_2O \rightleftharpoons \text{Betaine} + NADH$$

Betaine Homocysteine Methyl-transferase The enzyme that catalyzes the following reaction:

$$\text{Betaine} + \text{L-homocysteine} \rightleftharpoons \text{dimethyl glycine} + \text{L-methionine}$$

Betalin 12 A trade name for vitamin B_{12}.

Betalin S A trade name for vitamin B_1 (thiamine hydrochloride).

Betameth A trade name for betamethasone sodium phosphate, a corticosteroid.

Betamethasone (mol wt 392) A glucocorticoid and a topical anti-inflammatory agent.

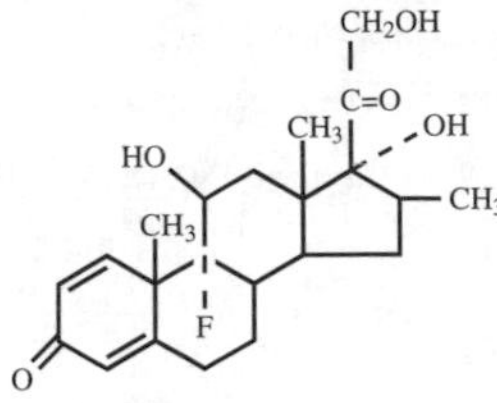

Betamin A trade name for vitamin B_1.

Betapen-VK A trade name for penicillin V potassium.

Betaseron A trade name for interferon beta 1B produced by DNA recombinant technology.

Betasine (mol 434) An iodine source.

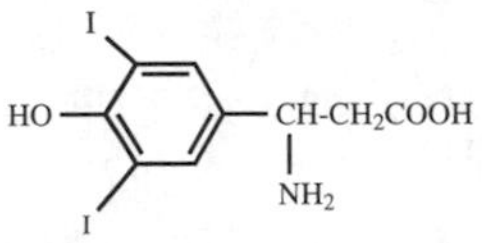

Beta-Tim A trade name for timolol maleate, a beta adrenergic blocker used as an antihypertensive agent.

Betatrex A trade name for betamethasone valerate, a corticosteroid.

Betatron An electron accelerator capable of producing electron beams of high energy as well as X-ray of extremely penetrating power.

Betaxin A trade name for thiamine hydrochloride (vitamin B_1).

Betaxolol (mol wt 307) A β_1-adrenergic blocker that is used as an antihypertensive and antiglaucoma agent.

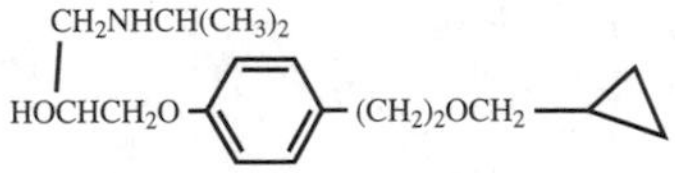

Betazole (mol wt 111) A stimulant for gastric secretion.

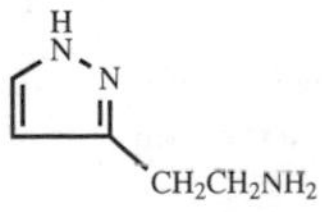

Betel Dried leaves of *Piper betle* (Piperaceae) that are used as counterirritant.

Bethanechol Chloride (mol wt 197) A cholinergic agent.

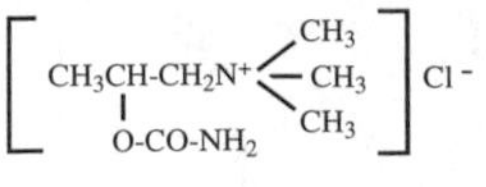

Bethanidine (mol wt 177) An adrenergic blocker used as an antihypertensive agent.

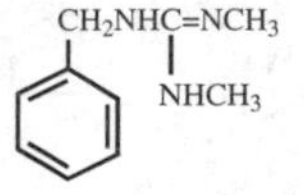

Betimol A trade name for timolol maleate, a beta adrenergic blocker used as an antihypertensive agent.

Betnesol A trade name for betamethasone sodium phosphate, a corticosteroid and hormonal agent.

Betnovate A trade name for betamethasone valerate, a corticosteroid and hormonal agent.

Betoptic S A trade name for betaxolol hydrochloride, a beta adrenergic blocking agent used as an antihypertensive.

Betoxycaine (mol wt 352) A local anesthetic agent.

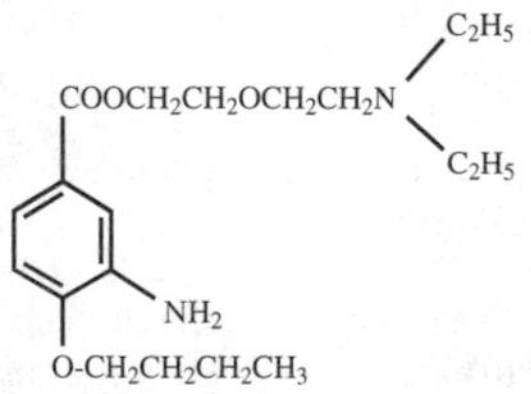

Betula A preparation from leaves and bark of *Betula alba* (white birch) containing birch tar oil, used as flavoring in pharmaceuticals.

BeV Abbreviation for 1,000,000,000 (10^9) electron volts.

Bevantolol (mol wt 345) A β1-adrenergic blocker used as antianginal, antihypertensive, and antiarrhythmic agent.

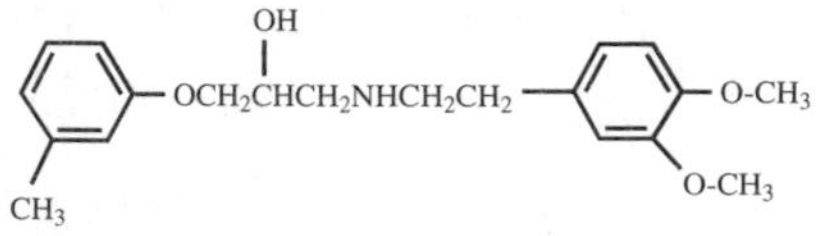

Bevitamel A trade name for a nutritional supplement containing melatonin and vitamin B_{12}.

Bevonium Methyl Sulfate (mol wt 466) An anticholinergic, antispasmodic, and bronchodilator.

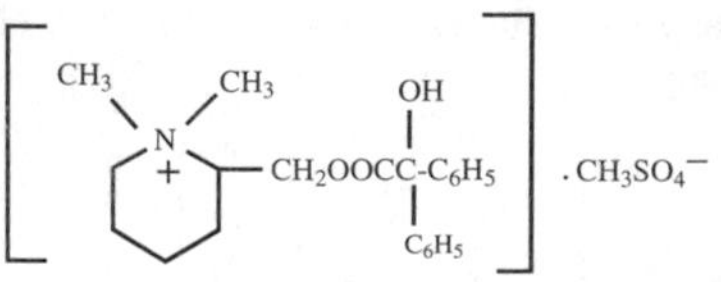

BEVS Abbreviation for baculovirus expression vector system.

Bex A trade name for aspirin.

Bezafibrate (mol wt 362) An antihyperlipoproteinemic agent.

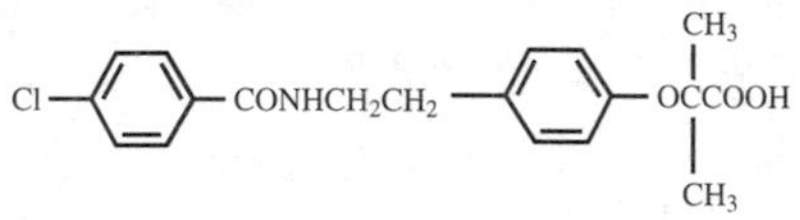

Bezitramide (mol wt 493) A narcotic analgesic agent.

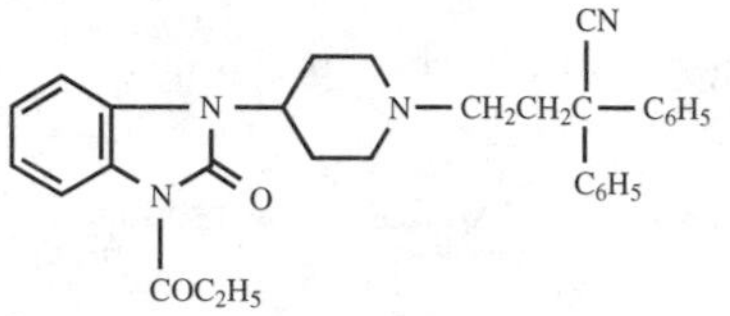

Bezoar An agglomeration of food or foreign material in the intestinal

BF 23 A bacteriophage (resembling T7) that infects *E. coli.*

B/F Ratio Abbreviation for bound-to-free ratio (in ligand binding assay). A plot of the B/F ratio vs. the concentration of free ligand is known as Scatchard plot.

BFA Abbreviation for brefeldin-A.

BfaI A restriction endonuclease from *Bacillus fragilis* NEB688 with the following specificity:

5'........C↓TAG........3;
3'........GAT↑C........5'

Bfas-PLA₂ Abbreviation for *Bungarus fasciatus* PLA_2.

BF_0F_1 ATPase A bacterial type of F_0F_1 proton ATPase.

BFGF Abbreviation for basic fibroblast growth factor.

BfmI A restriction endonuclease from *Bacillus firmus* S8336 with the following specificity:

5'........C↓TPuPyAG........3'
3'........GAPyPuT↑C........5'

BFP Abbreviation for biological false-positive.

BFPR Abbreviation for biological false-positive reaction.

BFR Abbreviation for bacterioferritin.

BfrI A restriction endonuclease from *Bacteroides fragilis* with the following specificity:

```
     ↓
5'........CTTAAG........3'
3'........GAATTC........5'
               ↑
```

BFT Abbreviation for bentonite flocculation test.

BFU Abbreviation for burst forming unit.

BG Abbreviation for blood glucose.

β-Gal Abbreviation for b-galactoside.

BGG Abbreviation for bovine gamma globulin (γ-globulin).

BGH Abbreviation for bovine growth hormone.

BglI A restriction endonuclease isolated from *Bacillus globigii* with the following specificity:

```
                         ↓
5′.........GCCNNNNNGGC.........3′
3′.........CGGNNNNNCCG.........5′
                     ↑
```

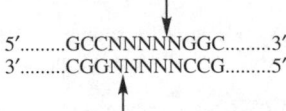

BglII A restriction endonuclease isolated from *Bacillus globigii* with the following specificity:

```
          ↓
5′.........AGATCT.........3′
3′.........TCTAGA.........5′
               ↑
```

BGlc Abbreviation for blood glucose.

BGlu Abbreviation for blood glucose.

BGP Abbreviation for 1. beta glycerophosphatase; 2. beta glycerophosphate.

βGP Abbreviation for 1. beta glycerophosphatase; 2. beta glycerophosphate.

β_2GP1 Abbreviation for beta-2 glycoprotein-1.

BGSA Abbreviation for blood granulocyte specific activity.

BH_4 Abbreviation for tetrahydrobiopterin.

BHA Abbreviation for butylated hydroxyanisole.

Bhang 1. A preparation from leaves and flowering tops of the marijuana plant (*Cannabis sativa*). 2. A toxicant product obtained from bhang.

BHBA Abbreviation for beta hydroxybutyric acid.

βHBA Abbreviation for beta hydroxybutyric acid.

BHC Abbreviation for benzene hexachloride.

BHCG Abbreviation for beta human chorionic gonadotropin.

βHCG Abbreviation for beta human chorionic gonadotropin.

BHCGH Abbreviation for beta human chorionic gonadotropin hormone.

βHCGH Abbreviation for beta human chorionic gonadotropin hormone.

BHD Abbreviation for a combination drug containing BCNU, hydroxyurea, and dacarbazine.

BHDV Abbreviation for a combination drug containing BCNU, hydroxyurea, dacarbazine, and vincristine.

BHI Abbreviation for biosynthetic human insulin.

BHIA Abbreviation for brain-heart infusion agar.

BHIB Abbreviation for brain-heart infusion broth.

BHK Abbreviation for baby hamster kidney.

BHK-21 An established cell line derived from baby hamster kidney. The cells are heteroploid and fibroblast like.

BHS Abbreviation for beta hemolytic *Streptococcus*.

βHS Abbreviation for beta hemolytic *Streptococcus*.

BHT Abbreviation for butylated hydroxytoluene.

Bi 1. Symbol for the chemical element bismuth with atomic weight 209, valence 3 and 5. 2. A prefix meaning two or twice.

BIAcore A trade name for an optical biosensor that measures the real-time interactions of macromolecules with ligands immobilized on a dextran-coated gold surface.

Bialaphos A linear tripeptide produced by *Streptomyces hygroscopicus*.

Bialamicol (mol wt 437) An antiamebic agent for treatment of chromic amebiasis.

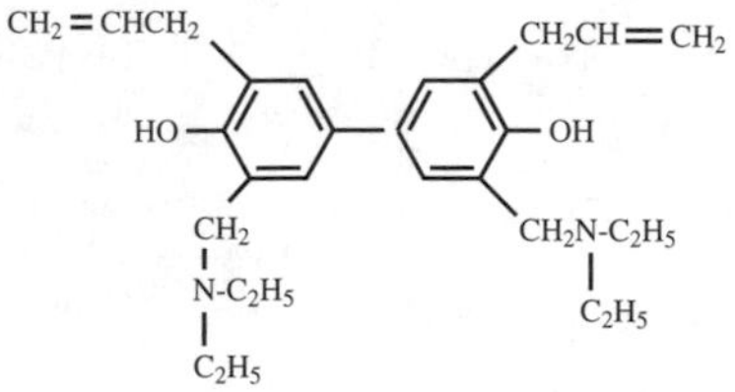

Bial's Reaction See orcinol reaction.

Biamine A trade name for vitamin B_1 (thiamin hydrochloride).

Biapenem (mol wt 430) An antibacterial agent.

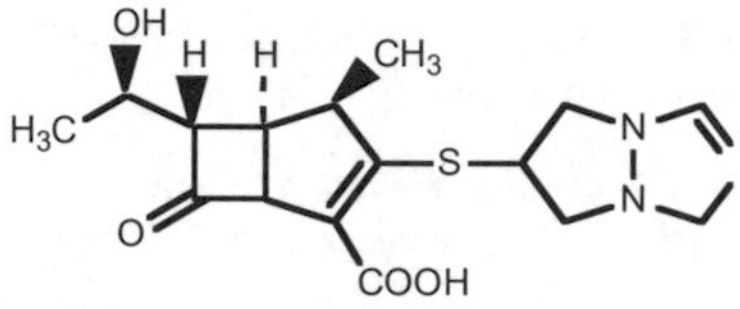

Biavex II A trade name for rubella and mumps virus vaccine (attenuated).

Biaxial Possessing two axes.

Biaxin Filmtabs A trade name for clarithromycin, an antibiotic derived from erythromycin that inhibits bacterial protein synthesis.

Bibasic Having two replaceable hydrogen atoms or a compound with two hydrogen atoms replaceable by a monovalent metal.

Bibenzonium Bromide (mol wt 364) An antitussive agent.

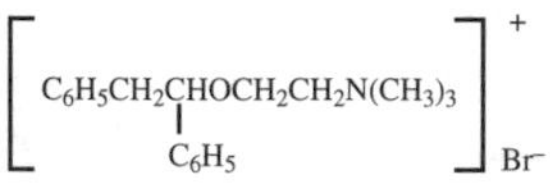

Bibrocathol (mol wt 650) A topical antiseptic agent.

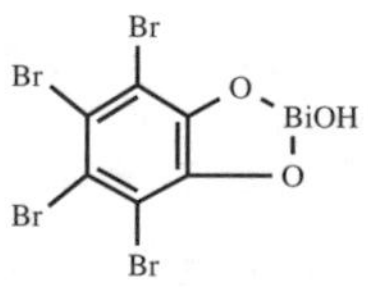

Bibrotoxin A vasoconstrictor peptide isolated from the venom of the burrowing wasp *Atractaspis bibroni*.

Bicalutamide (mol wt 430) A nonsteroidal agent that inhibits androgen uptake and cytosol binding of androgen in the target tissues.

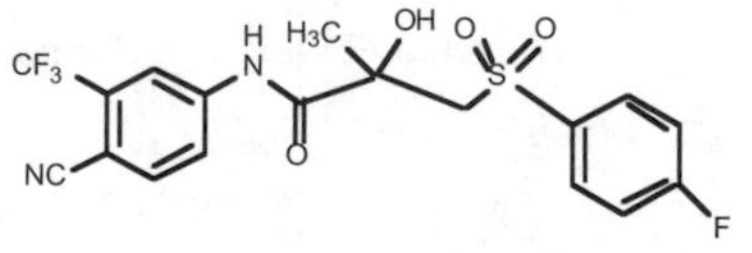

Bicameral Consisting of two chambers.

Bicarbonate The ion of HCO_3^- and its salt. It is the main form of carbon dioxide in solution at neutral pH.

Biceps A muscle possessing two heads.

Bi-Chinine A trade name for quinine bisulfate, an antimalarial agent

Biciliate Having two cilia.

Bicillin L-A A trade name for penicillin G benzathine, an antibiotic that inhibits bacterial cell wall synthesis.

Bicine (mol wt 163) A reagent used for the preparation of buffers in the pH range of 7.7 to 9.1.

$(HOCH_2CH_2)_2NCH_2COOH$

Bicistronic Messenger RNA An RNA that carries two translation initiation sites.

Bicitra A trade name for sodium citrate, an anticoagulant used for collection of blood.

BiCNU (BCNU) A trade name for carmustine that crosslinks strands of cellular DNA and interferes with RNA transcription.

Biconcave Referring to an optical lens having a depression on both sides.

Bicoid Protein A *Drosophila* morphogenetic protein that is required for the development of anterior structures in the embryo.

Biconvex Protruding from both sides or surfaces.

Bicozamycin (mol wt 302) An antibiotic isolated from *Streptomyces sapporensis.*

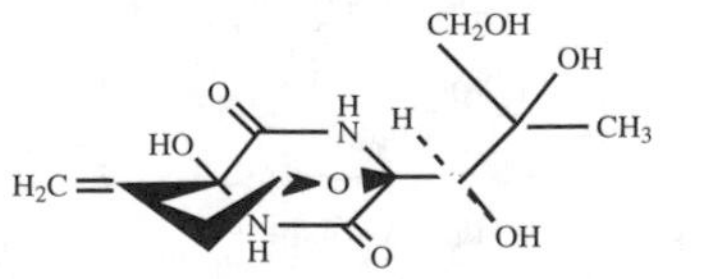

Bicuculine (mol wt 367) A GABA antagonist.

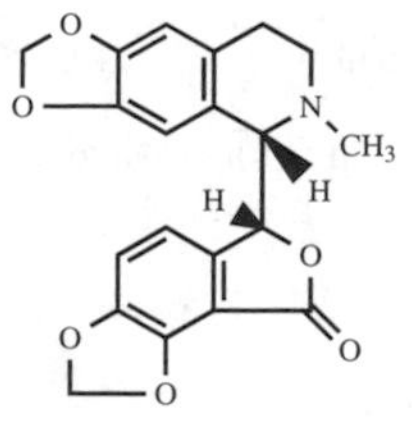

Bicyclic An organic compound in which only two ring structures occur.

Bicyclic Cascade A type of control mechanism for regulating enzyme-catalyzed reactions through a cyclic modification and demodification of the enzyme, e.g., the control of glycogen metabolism by the cyclic modification and demodification of both protein kinase and phosphorylase.

b.i.d. Abbreviation for *bis in die* meaning twice a day (in prescriptions).

Bidentate A ligand that is chelated to a metal ion by means of two donor atoms.

Bidirectional Genes A pair of open reading frames, one on the plus strand and the other on the minus strand of the same dsDNA.

Bidirectional Replication A theta mode of the double-stranded DNA replication in which two replication forks are moving in opposite directions.

Biduotertian Fever A form of malaria caused by infection with *Plasmodium* and characterized by overlapping attacks of chills and fever.

Biennial A plant that requires two years to complete its life cycle.

Bietamiverine (mol wt 318) An antispasmodic agent.

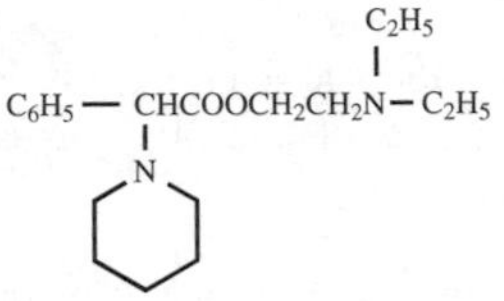

Bietanautine (mol wt 732) An antihistaminic, antiemetic, and antiparkinsonian agent.

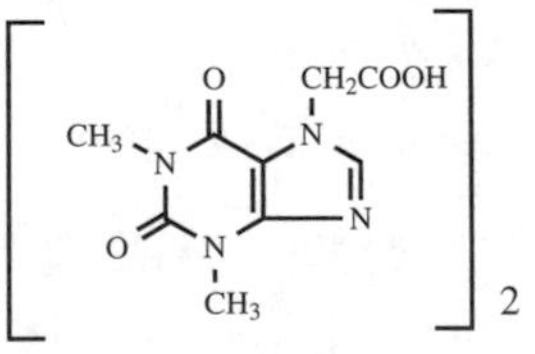

$(C_6H_5)_2CHOCH_2CH_2N(CH_3)_2$

Bietaserpine (mol wt 708) An antihypertensive agent.

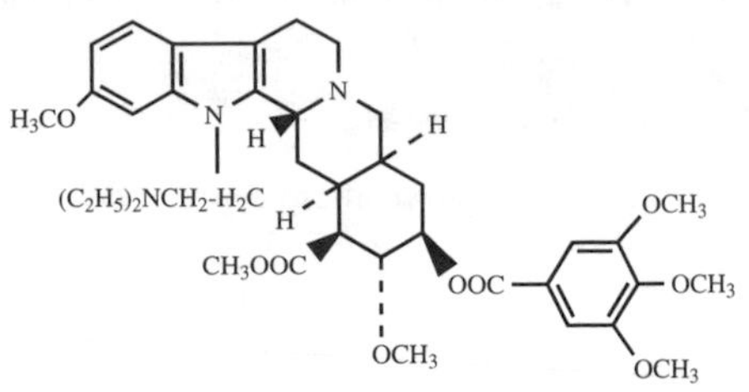

Bifactorial Heterothallism See heterothallism.

Bifenox (mol wt 342) An herbicide.

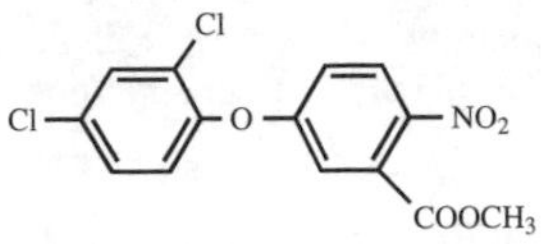

Bifenthrin (mol wt 435) An insecticide (acaricide).

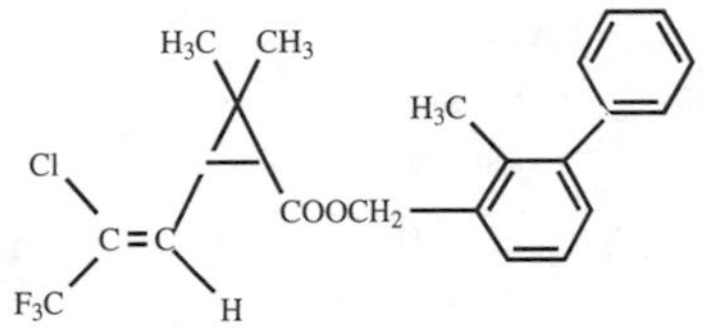

Bifermentolysin A group of thiol-activated cytolysins (oxygen-labile proteins) produced by certain bacteria (also called SH-activated cytolysin).

Bifid 1. Divided into two parts. 2. Bacterium that is forked at one or both ends.

Bifidobacterium A genus of Gram-positive, asporogenous, anaerobic bacteria.

Bifidus Factor A factor found in human milk that causes a predominant occurrence of *Lactobacillus bifidus* in the intestinal tract of breast-fed infants (also known as *Lactobacillus bifidus* factor).

Bifidus Pathway A pathway that ferments glucose to a mixture of lactic acid and acetic acid.

Biflagellate Having two flagella.

Bifluranol (mol wt 292) An antiandrogen substance.

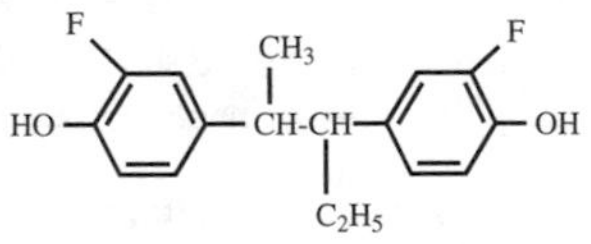

Bifocal 1. Pertaining to the characteristic of having two foci. 2. Having two areas of different focal lengths in a lens, e.g., one for near vision and one for far vision.

Bifonazole (mol wt 310) An antifungal agent.

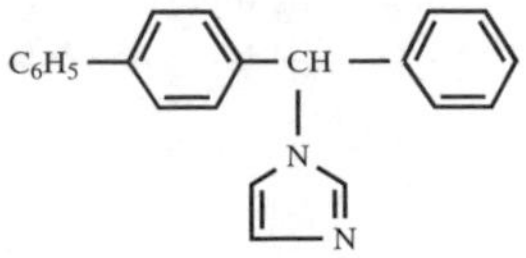

Bifunctional Having two functions.

Bifunctional Catalyst A catalyst that can provide both an acidic and a basic catalytic function.

Bifunctional Feedback A feedback mechanism that controls in two directions, e.g., the regulatory adjustment of pH when it becomes either above or below the normal value.

Bifunctional Reagent A compound that has two reactive groups that can interact either with two groups of one protein or with one group of each of the two different proteins.

Bifunctional Vector A cloning vector that is capable of replicating in two different organisms, e.g., in *E. coli* and yeast.

Big Bang Theory The concept that the universe was born in a gigantic explosion about 15 to 20 billion years ago.

Big T Antigen A viral encoded protein with molecular weight of 105,000 found in papovavirus-infected cells and involved in tumor formation (also known as large T antigen).

Biglycan A connective tissue glycoprotein of the extracellular matrix.

Bijou A small, screw-cap glass bottle.

Biken Test A precipitin test for detecting the production of heat-labile enterotoxin by *E. coli.*

Bikunin A plasma glycoprotein and a serine protease inhibitor.

Bilaminar Pertaining to, or comprising, two thin layers or plates.

Bilateral 1. Having two sides. 2. Occurring or appearing on two sides.

Bilateral Symmetry A body or a structure that can be halved in a single plane so that each half is approximately the mirror image of the other.

Bilateria Animals with bilateral symmetry.

Bilayer Referring to the lipid bilayer with a thickness of two molecules (see lipid bilayer).

Bilayer Lipid Membrane Referring to the lipid bilayer in the membrane (see membrane).

Bile Substance secreted by the liver that emulsifies fats and aids in the digestion and absorption of fats. It consists of cholesterol, bile salts, and lecithin.

Bile Acids Referring to 24-carbon steroid carboxylic acids such as cholic acid or deoxycholic acid that occur in the bile.

Bile Alcohol Referring to polyhydroxylated steroids.

Bile Duct The duct by which bile passes from the liver to the gallbladder.

Bile Pigment A degradation product of the heme portion of hemoglobin and other heme proteins, e.g., bilirubin and biliverdin.

Bile Salt The surface-active agents (salts of bile acids) in the bile that aid in the emulsification of fats during digestion.

Bile Salt Hydrolase Synonym of choloylglycine hydrolase.

Bile Solubility Test 1. Use of bile salt to differentiate bacteria or strains of bacterium based upon the solubility of bacteria in the bile salt (e.g., sodium deoxycholate). 2. A test used in differential diagnosis of pneumonoccal and streptococcal infections.

Bilharzioma A tumor of the urinary bladder.

Bili Abbreviation for bilirubin.

Bilianic Acid Any derivative of bile acid formed by the oxidative opening of one of the rings in the molecule with the formation of two carboxyl groups.

Biliary Referring to the bile, gallbladder, or ducts that transport bile.

Biliary Calculus A stone formed in the biliary tract that consists of bile pigments and calcium salts.

Biliary Cirrhosis An inflammatory condition in which the flow of bile through the liver is obstructed.

Biliary Duct A duct by which bile passes from the liver to the gallbladder.

Biliation Secretion of bile.

Bili-C Abbreviation for conjugated bilirubin.

Bilicyanin Blue pigment found in the gallstone that is formed by the oxidation of bilirubin or biliverdin.

Bilifecia The presence of bile in the feces.

Bilifuscin Brown pigment found in the human gallstone.

Biligenic Bile producing.

Biliprotein See phycobiliprotein.

Bilin A colored bile pigment, such as urobilin, that is formed by the oxidation of a colorless bilinogen.

Bilinogen A precursor of bile pigment, e.g., urobilinogen that forms a colored bilin (urobilin) pigment upon oxidation.

Biliprotein A chromoprotein in which the prosthetic group is a covalently linked pigment.

Bilirubin (mol wt 585) A bile pigment in the bile formed from breakdown of heme.

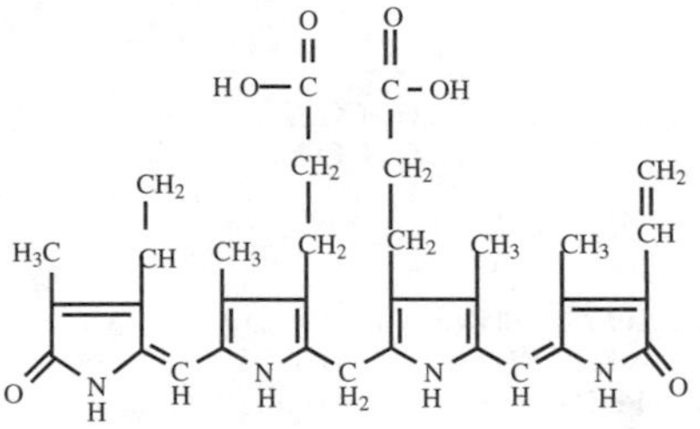

Bilirubin Diglucuronide A conjugated form of bilirubin that is formed in the liver by the esterification of two molecules of glucuronic acid to the two propionic acid residues of bilirubin.

Bilirubin Glucuronoside Glucuronosyltransferase The enzyme that catalyzes the following reaction:

2 Bilirubin-glucuronoside

⇅

Bilirubin + bilirubin-bisglucuronoside

Bilirubin Oxidase The enzyme that catalyzes the following reaction:

Bilirubin + O_2 ⇌ Biliverdin + H_2O

Bilirubin Oxygen Oxidoreductase The systematic name for bilirubin oxidase.

Bilirubinemia The presence of bilirubin in the blood.

Bilirubinuria The excretion of bilirubin in the urine that causes colorization of the urine (dark brown).

Bilitherapy The use of bile or bile salt preparations for treatment of digestive disorders.

Biliuria The presence of bile in the urine.

Biliverdin(e) (mol wt 583) The bile of amphibia and birds. It occurs in patients with carcinomatous obstruction of the bile duct.

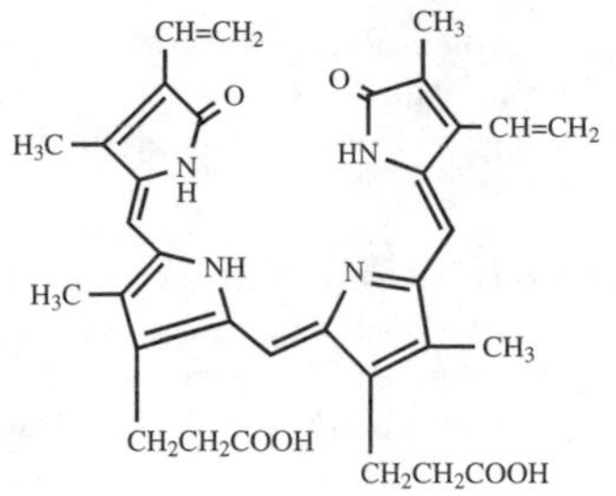

Biliverdin Reductase The enzyme that catalyzes the following reaction:

Bilirubin + $NADP^+$ ⇌ Biliverdin + NADPH

Bilobulate Having two lobules or two cells.

Biltricide A trade name for praziquantel, an anthelminic agent that alters the permeability of the cellular membrane.

Bimetal A type of thermometer in which the sensing element consists of two thin strips of metals having different expansion coefficients in response to changes in temperature.

Bimodal 1. A population in which the measurements of a given character are clustered around two values. 2. Having two distinct modes or peaks.

Bimolecular Lipid Membrane See membrane.

Bimolecular Reaction A chemical reaction that requires either two molecules of a single reactant or one molecule each of two different reactants to form products.

Binapacryl (mol wt 322) A fungicide and miticide.

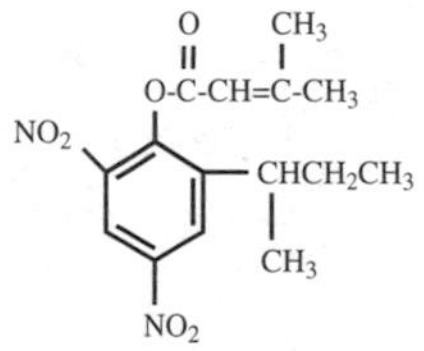

Binary Consisting of two parts.

Binary Acid An acid containing no oxygen.

Binary Alloy An alloy containing two major elements, exclusive of impurities.

Binary Fission A type of asexual reproduction in prokaryotes in which a cell divides into two identical cells by a nonmitotic process.

Binder Substance used to hold ingredients together or to increase the strength of the interaction force.

Bindin A protein found in the acrosome of the sea urchin spermatozoon that binds to the vitelline layer of the egg to confer specificity in the fertilization.

Binding Assay Any method for measuring protein-ligand interactions, e.g., measurement of binding of cAMP to protein kinase, aminoacyl-tRNA to ribosome, and DNA or RNA to nitrocellulose paper.

Binding Energy The energy for the noncovalent interactions between enzyme and substrate or between receptor and ligand.

Binding Factor A protein factor required for the binding of ligand to protein or carrier, e.g., binding of aminoacyl-tRNA to ribosomes.

Binding Proteins Proteins that are capable of binding onto other substances, e.g., sugars, ions, or nucleic acids.

Binding Site The region on a macromolecule that is responsible for specific binding of a ligand or a molecule.

Binedaline (mol wt 293) An antidepressant.

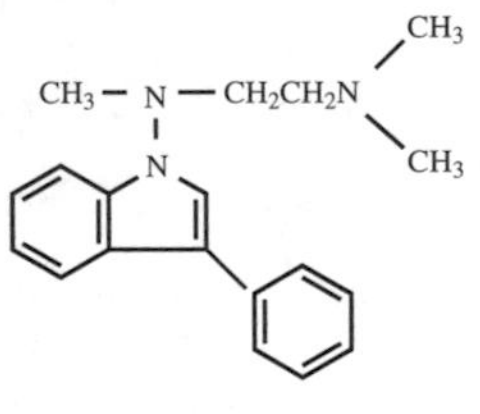

Binifibrate (mol wt 499) An antihyperlipoproteinemic agent.

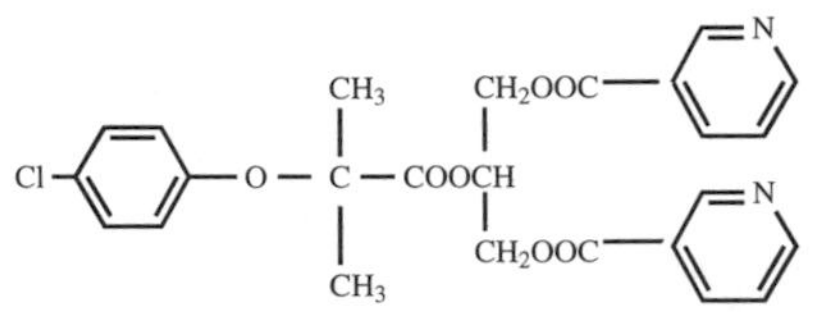

Binocular 1. Pertaining to both eyes. 2. A microscope or telescope that can accommodate viewing by both eyes.

Binocular Fixation The process of having both eyes directed at the same object at the same time.

Binomial Having two names, e.g., names of genus and species in the binomial nomenclature of organisms.

Binomial Nomenclature A system for naming organisms in which the organism consists of two names, the first designating the genus to which it belongs and the second, the name of the species.

BinSI (EcoRII) A restriction endonuclease isolated from *Bifidobacterium infantis* S76e with the following specificity:

↓
5′..........CC(A/T)GG...........3′
3′..........GG(T/A)CC...........5′
↑

BinSII (NarI) A restriction endonuclease isolated from *Bifidobacterium infantis* S76e with the following specificity.

↓
5′..........GGCGCC..........3′
3′..........CCGCGG..........5′
↑

Binuclear Having two nuclei.

Binucleolate Having two nucleoli.

Bio- A prefix meaning life or biological.

Bioactive Substance Substances that are capable of causing specific effects or reactions on target tissue or organisms.

Bioassay The laboratory measurement of either the activity of or the concentration of a substance using living cells or living organisms.

Bioautography A method used in conjunction with bioassay and chromatography for detection and identification of bioactive substances.

Bioavailability The quantity of an administered drug or other substance that becomes available for activity in the target tissue.

Bio-Bead S A polystyrene supporter used in gel filtration.

Bioblend A trade name for medically pure gases.

Biocatalyst Referring to the catalyst that is derived from the living system, e.g., an enzyme.

Biocef A trade name for cephalexin hydrochloride, a first generation cephalosporin antibiotic.

Biochanin-A (mol wt 284) An aromatic compound and substrate for biochanin-A reductase.

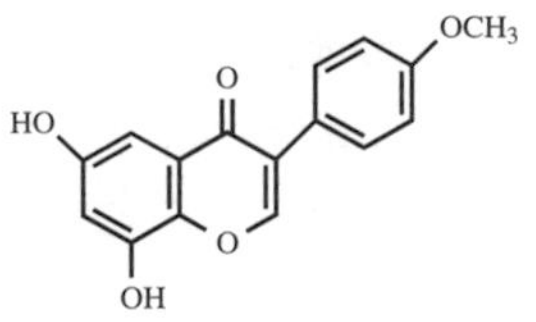

Biochanin-A Reductase The enzyme that catalyzes the following reaction:

Dihydrobiochanin-A + $NADP^+$

⇅

Biochanin-A + NADPH

Biochemical Energetics The free energy relationships of biochemical reactions.

Biochemical Genetics Genetics that deals with the chemical nature of the hereditary determinants and their mechanisms.

Biochemical Lesion A biochemical alteration that leads to a visible pathological condition.

Biochemical Marker A mutation that can be detected and identified by biochemical means.

Biochemical Mutant A mutant with a biochemical marker (see also auxotroph).

Biochemical Oxygen Demand (BOD) The amount of oxygen that is consumed by organisms for the oxidation of organic compounds. It is used to estimate the degree of water contamination.

Biochemistry The science that deals with the chemistry of living systems.

Biochip A device consisting of a biosensor with an integrated circuit.

Biochrome The naturally occurring colored compound in plants, animals, or microorganisms.

Biochronometry The science that deals with the temporal organizations and the timekeeping mechanisms of biological systems.

Biocide A substance capable of killing or arresting the growth of living organisms (e.g., bactericides, fungicides, and pesticides).

Bioclate A trade name for blood clotting factor VIII produced by DNA recombinant technology and used for treatment of hemophilia A.

Biocolloid A colloid or a mixture of colloids from plants or animals.

Biocomputer A computer in which the silicon in the microchips is replaced by a synthetic protein coated with a silver compound to act as a metallic semiconductor.

Biocontrol See biological control.

Bioconversion 1. Conversion of biomass to energy. 2. Conversion of substrate to product by an enzyme. 3. Conversion of substances to cell mass.

Biocytin A complex of biotin and protein in which biotin is covalently linked to the lysine residue of the protein. It also refers to biotin-lysine complexes.

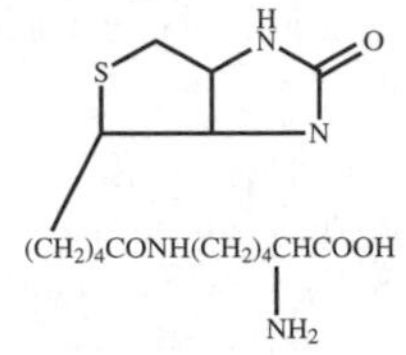

biotin-lysine complex

Biocytinase The enzyme that catalyzes the breakdown of biocytin to biotin and lysine or biotin and protein.

Biodegradable Capable of being decomposed by microorganisms.

Biodeterioration The deterioration (spoilage) of materials by microorganism or other biological activity.

Biodine A trade name for povidone-iodine that is active against bacteria, fungi, and viruses.

Biodynamic Pertaining to the dynamic relationship between an organism and its environment.

Biodynamics The science that deals with effects of dynamic processes on living organisms.

Bioelectric Current An electric current generated on the cell surface by the potential differences across the excitable cell membranes (e.g., nerve and muscle cells).

Bioelectrochemistry The application of principles and techniques of the electrochemistry to biological and medical problems.

Bioelectronics The application of electronic techniques to problems in biology and medicine. It combines biosensors or other biological activities with electronic circuits.

Bioemulsifier Biomolecules that function as emulsifying agents.

Bioenergetics The science that deals with energy changes in the living organisms.

Bioengineering Application of engineering principles to solve biomedical problems.

Bioequivalent Referring to a drug that has the same effect on the body as another drug.

Bioethics The interdisciplinary science that deals with the moral and social implications of the practice and developments in medicine and the life sciences.

Biofeedback A technique permitting an individual to gain control over internal physiological processes such as heartbeat, blood pressure, muscle tension, and brain wave activity.

Bioflavonoids 1. A group of colored phenolic compounds that contribute to the maintenance of normal blood vessel permeability and fragility. They also possesses antioxidant activity. 2. A large group of colored phenolic substances in plants, e.g., flavones and anthocyanins.

Biofuel Any solid, liquid, or gaseous fuel from biomass.

Biofuel Cell A device containing redox reactions, catalyzed by isolated enzymes or living cells for generating energy.

Biogas Referring to methane gas produced from animal manure by bacterial fermentation.

Biogel A trade name for a group of polyacrylamide and agarose gels used in gel filtration.

Biogenesis 1. The synthesis of a substance in a living organism. 2. The doctrine that living things arise only from preexisting living things.

Biogenic 1. Produced by the action of a living organism, e.g., fermentation. 2. Essential to life and the maintenance of health, e.g., food, water, and proper rest.

Biogenic Amine A group of amines that are produced by living organisms and are involved in many physiologically important functions, e.g., norepinephine, epinephrine, serotonin, and histamine.

Biogeochemistry The science that deals with the interaction of living organisms with the mineral environment of the earth's crust.

Biogeography The science that deals with the geographical distribution of organisms, e.g., plants, animals, fungi, and bacteria.

Bioginkgo An extract from the leaves of a *Ginko biloba* tree used as a dietary supplement to increase blood circulation.

Biokinetics The science that deals with growth changes and movements within the developing organisms.

Bioleaching Solubilization of metals by living organisms.

Biological Chemistry The science that deals with the chemistry of living things.

Biological Clock The periodicity of either a biological function or a biochemical reaction.

Biological Control Control of a pest population by introduction of predatory, parasitic, or disease-causing microorganisms.

Biological False Positive Reaction (BFPR) A positive serological reaction obtained as a result of the presence of disease(s) or condition(s) other than those being tested for.

Biological Half-Life The time required for the body to eliminate one-half of an administered dosage of any substance by regular physiological processes.

Biological Nitrogen Fixation The conversion of atmospheric nitrogen to ammonia by nitrogen-fixation organisms, e.g., *Rhizobium*.

Biological Oxidation Reduction The oxidation-reduction reactions occurring in a biological system, e.g., citric acid cycle and electron transport system.

Biological Oxygen Demand See biochemical oxygen demand.

Biological Stain A dye used for staining cellular structures of cells or tissues.

Biological Value The relative nutritional value of a protein that is based on the amino acid composition, digestibility, and availability of the digested products.

Biologicals Products produced from living organisms or their products.

Biology The science dealing with living systems.

Bioluminescence Emission of light by certain living organisms through a biochemical reaction, e.g., luciferase catalyzed ATP/luciferin system.

Biolysis Lysis by biological means.

Biomacromolecules High molecular weight bioactive molecules, e.g., DNA, RNA, and proteins.

Biomass Mass of biological material or mass of an organism or a group of organisms in a given habitat or a given area.

Biomechanics The application of mechanical principles to living organisms.

Biomedical Engineering Application of engineering principles and techniques to solve medical, industrial, and biological problems.

Biomembrane A biological membrane.

Biomere Structure formed from the arrangement of biomolecules.

Biomethanation The anaerobic fermentation for the production of biogas methane.

Biometry The science that deals with the application of statistics to biological systems.

Biomineral A mineral produced by biological organisms.

Biomineralization Mineral deposition due to biological activity, e.g., deposition of sulfur by photosynthetic bacteria.

Biomolecule Molecules occurring in a living system, e.g., proteins, nucleic acids, amino acid and sugars.

Biomonomer Monomer molecules occurring in living system, e.g., amino acids, nucleotides, and simple sugars, that can be enzymatically converted to polymers.

Biomox A trade name for amoxicillin, an antibiotic.

Bion Any living organism.

Bionic The application of biological principles to a man-made device or system, e.g., robots.

Bionics The science that deals with the application of principles and data of biological systems to solve engineering problems or to create a new engineering system.

Bionosis Any infection or disease caused by living organisms.

Biont An organism that has a specified mode of life (see symbiont).

Bioorganic The science that deals with the organic chemistry of biologically important substances.

Biophagous Feeding on living organisms.

Biophylaxis A nonspecific defense mechanism of the body, such as phagocytosis.

Biophysics The science that deals with the physics of living systems and their components.

Biopolymer Polymers that occur in living systems, e.g., proteins (polymer of amino acids), DNA (polymer of nucleotides), and polysaccharide (polymer of monosaccharides).

Bioprobe A type of biosensor in which the biocatalyst (e.g., enzyme) is immobilized with a detection system, e.g., enzyme electrode.

Biopsy The removal of tissue from an organ or part of the organ for microscopic examination to confirm or to establish a diagnosis or a cause.

Biopterin (mol wt 237) A growth factor for some insects and a coenzyme for hydroxylation of amino acids.

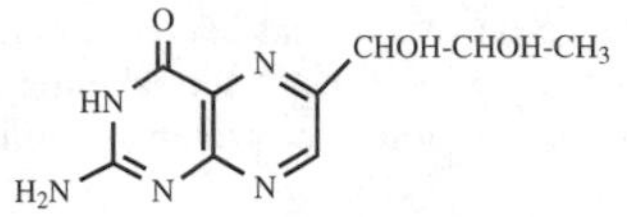

Bioreactor A fermenter or other apparatus used for carrying out bioconversion reactions.

Bioresmethrin (mol wt 338) An insecticide.

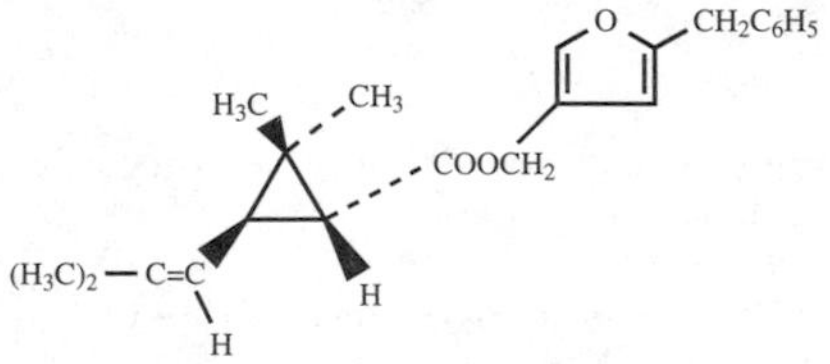

Biorhythm A cycle of change that occurs in an organism.

Bioscrubbing The employment of organisms immobilized on a supporting system for the removal of toxic material or pollutants.

Biosensor A device that employs biological material for detecting or measuring a chemical reaction or concentration of a particular substance.

Biosone A trade name for hydrocortisone acetate, a corticosteroid.

Biosphere The world of living organisms.

Biosurfactants Biological molecules that function as surfactants. They are capable of lowering surface tension.

Biosynthesis The process by which cells synthesize bioactive molecules from small and relatively simple components.

Biosynthetic Pertaining to biosynthesis.

Biosynthetic Pathway A pathway that leads to the synthesis of biomolecules (also known as anabolic pathway).

Biota A collective term for organisms living in a given region.

Biotechnology The scientific manipulation of living organisms or components of living organisms at the molecular level to produce useful products to improve human health, industrial efficiency, and food production, e.g., genetic engineering, DNA recombination, and monoclonal antibody.

Biotic Pertaining to life.

Biotin (mol wt 244) A vitamin of the vitamin B complex that acts as a coenzyme in carboxylation-decarboxylation reactions and as a growth factor for certain fungi and bacteria. It is also a compound commonly used for labeling macromolecules, e.g., biotinylated protein or biotinylated DNA.

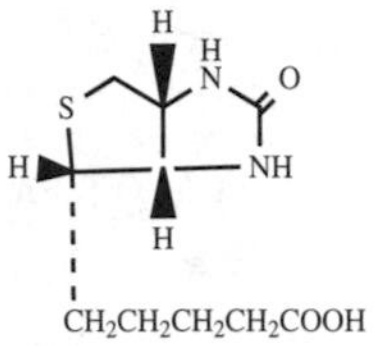

Biotin Carboxylase The enzyme that catalyzes the following reaction:

ATP + biotin-carboxyl-carrier protein + CO_2
⇅
ADP + Pi +
Carboxylbiotin-carboxyl-carrier Protein

Biotin Carboxyl Carrier Protein A 22 kD protein that is covalently linked to biotin. The carboxylation of the biotin by the biotin carboxylase occurs on this carrier protein.

Biotin-CoA Ligase The enzyme that catalyzes the following reaction:

ATP + Biotin + CoA
⇅
AMP + PPi + biotinyl-CoA

Biotinidase The enzyme that catalyzes the following reaction:

Biotin amide + H_2O ⇌ biotin + NH_3

Biotinyl-CoA Ligase See biotin-CoA ligase.

Biotinyl-CoA Synthetase See biotin-CoA ligase.

Biotinylated DNA A DNA probe labeled with biotin for the detection of complementary sequences. The biotinylated DNA can be visualized by complexing with streptavidin that is coupled to a color-generating agent.

Biotinylation of Nucleic Acids A nonradioactive method of labeling nucleic acid probes using nick translation to incorporate biotin-derivitized nucleotides.

Biotinyllysine Referring to biotin-lysine or biotin-protein complexes.

Biotope 1. The environment occupied by an organism or organisms. 2. The particular location of a parasite or pathogen. 3. The spatial distribution of a biomass in a cross-section of a river or lake.

Biotransformation The biochemical changes of a foreign substance that occur in the body (also see bioconversion).

Biotropin A trade name for somatropin, a hormone.

Biotroph An organism that derives nutrients from the living tissues of another organism.

Biotype 1. A physiologically distinct form within a species. 2. Any organism that can be distinguished by its metabolic and/or physiological properties.

BiP Abbreviation for binding protein.

Biparental Zygote A diploid zygote that contains equal genetic contributions from male and female parents.

Biparous Producing two individuals at one birth.

Bipartite Having two parts.

Bipartition Dividing into two products.

Biperiden (mol wt 311) An anticholinergic agent that blocks the cholinergic activity and restores the natural balance of neurotransmitters in the CNS.

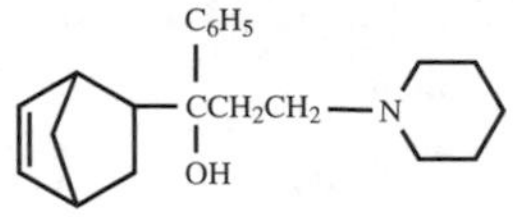

Biphenamine (mol wt 313) A topical anesthetic, antibacterial, and antifungal agent.

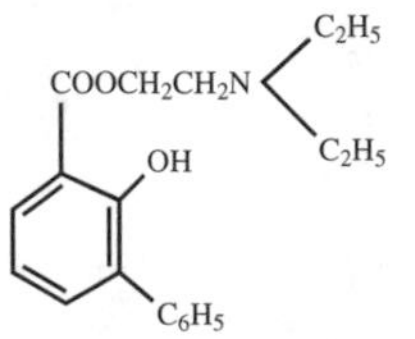

Bipiperidyl Mustard (mol wt 293) An obesifying agent that causes rapid lipid deposition.

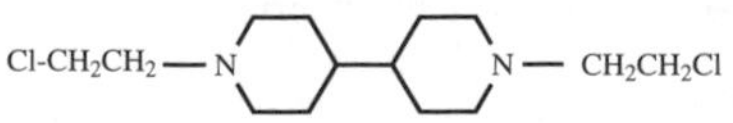

Bipolar Having two poles.

Bipolar Cell A group of retinal interneurons that receive input from photoreceptors and release it to ganglion cells.

Bipolar Filaments Filaments that have opposite polarity at the two ends, e.g., thick filament of the striated muscle.

Biradial Symmetry Object or structure with both radial and bilateral symmetry.

Birbeck Granule A rocket-shaped cytoplasmic granule found in Langerhans cells of the skin.

Birefringence The phenomenon in which a beam of light enters into a medium and splits into two polarized lights traveling in different directions.

bis in die (b.i.d.) Latin for twice a day.

Bisacodyl (mol wt 361) A cathartic agent that promotes fluid accumulation in the colon and small intestine; used to treat constipation.

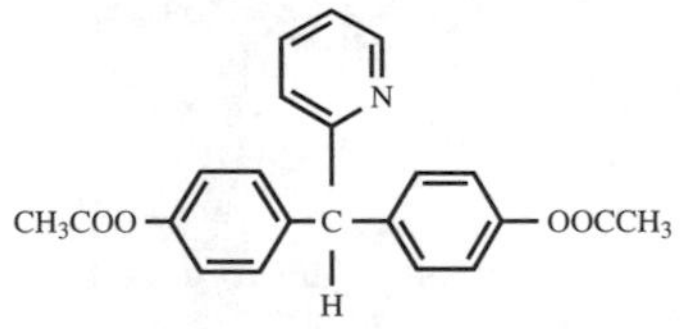

Bisacolax A trade name for bisacodyl, a laxative that promotes fluid accumulation in the colon and small intestine.

Bis-Acrylamide (mol wt 154) Abbreviation for N,N′-methylene-bisacrylamide. A reagent used as a crosslinking agent in the preparation of polyacrylamide gel.

$$(CH_2{=}CHCONH)_2CH_2$$

Bisalbuminemia A symptomless condition of humans in which double albumin bands are detected on electrophoreis.

Bisantrene (mol wt 398) An antineoplastic agent.

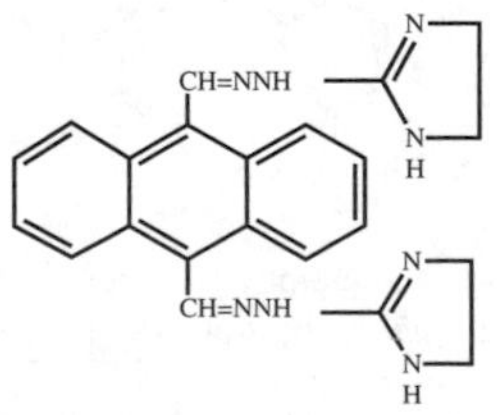

Bisbentiamine (mol wt 771) A vitamin B_1 source.

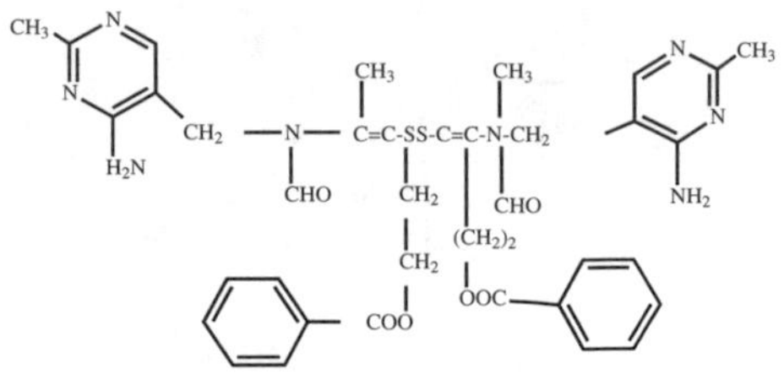

Bisdequalinium Chloride (mol wt 666) A disinfectant.

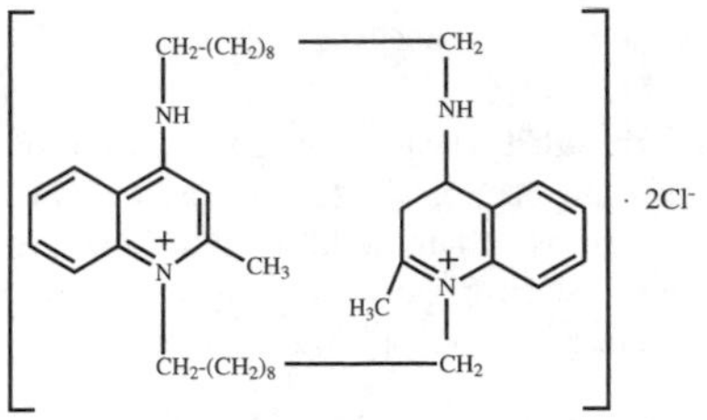

Bisdithiocarbamate An antifungal agent usually complexing with metal ions.

$$M\text{-}S\text{-}CSNHCH_2\text{-}CH_2\text{-}NHCSS\text{-}M.$$
(M = metal ions).

Bisexual 1. Producing both eggs and sperms. 2. Possessing physical or psychological characters of both sexes. 3. Engaging in both heterosexual and homosexual activity.

Bismark Brown R (mol wt 461) A biological stain.

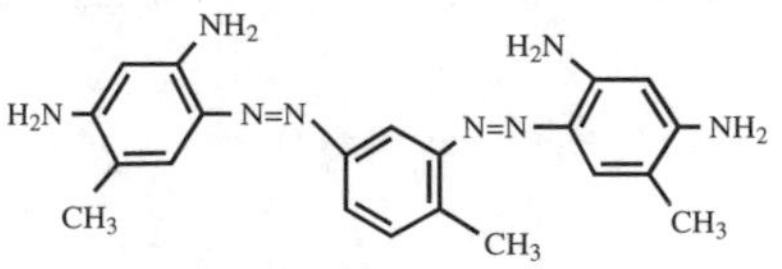

• 2HCl

Bismark Brown Y (mol wt 419) A brown, basic diazo dye used for vital staining, e.g., staining a vaginal smear.

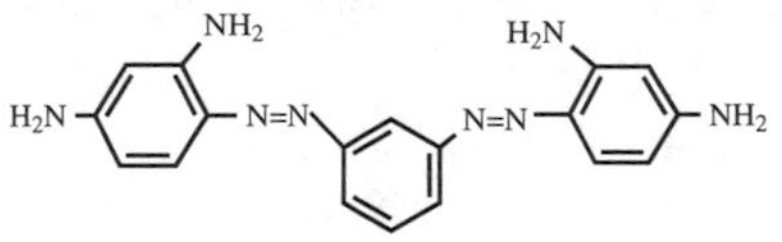

•2HCl

Bismatrol A trade name for bismuth subsalicylate, an antidiarrheal agent.

Bismuth (Bi) A chemical element with atomic weight 209, valence 3 and 5.

Bismuth Stomatitis A disorder characterized by the darkening of the inside cheek in the mouth due to a long period of using bismuth compound.

Bismuth Subgallate (mol wt 394) An antidiarrheal agent.

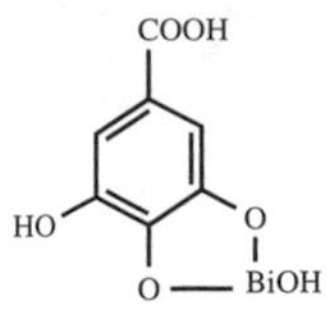

Bismuth Subsalicylate (mol wt 362) An antidiarrheal agent.

$HOC_6H_4COOBiO$

Bismuth Sulfite Agar An agar medium containing peptone, beef extract, glucose, disodium phosphate, ferrous sulfate, brilliant green, and bismuth sulphite; used for isolating strains of *Salmonella* (also called Wilson and Blair's agar).

Bisoprolol (mol wt 325) An antihypertensive agent that blocks the stimulation of beta-1 adrenergic receptors.

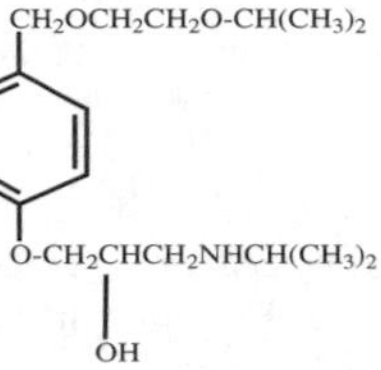

Bisorine A trade name for isoetharine hydrochloride, a bronchodilator that relaxes bronchial smooth muscle by acting on beta-adrengergic receptor.

Bisoxatin Acetate (mol wt 417) A cathartic agent.

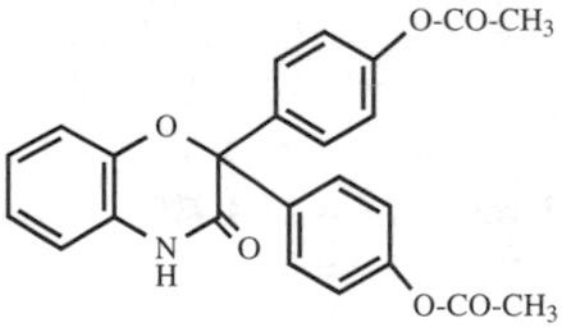

Bisphenols Antimicrobial agents that contain two phenolic residues.

Bisphosphate Two phosphate groups attached to different atoms in a molecule, e.g., fructose-1-6-bisphosphate.

1,2-Bisphosphoglyceric Acid (mol wt 226) A high energy 3-carbon acid and an intermediate in glycolysis.

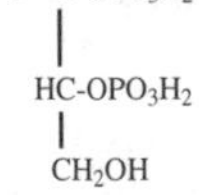

1,3-Bisphosphoglyceric Acid (mol wt 266) A high energy 3-carbon acid and an intermediate in glycolysis.

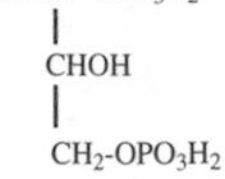

2,3-Bisphosphoglycerate Mutase The enzyme that catalyzes the following reaction:

2,3-Bisphosphoglycerate

⇅

1,3-Bisphosphoglycerate

2,3-Bisphosphoglycerate Phosphatase The enzyme that catalyzes the following reaction:

ADP + 2,3-bisphosphoglycerate + H_2O

⇅

3-Phosphoglycerate + ATP

Bistris (mol wt 209) A buffer substance.

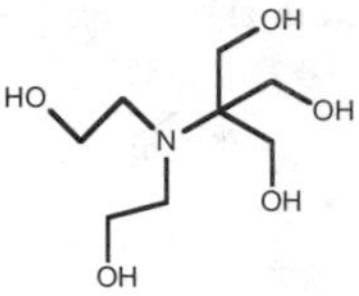

Bisubstrate Reaction An enzymatic reaction that involves two substrates.

Bitertanol (mol wt 337) An agriculture fungicide.

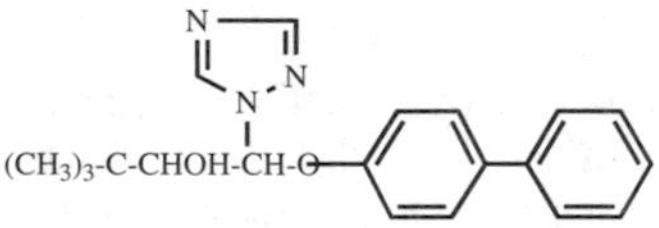

bITF Abbreviation for biotinylated ITF fusion protein.

Bithionol (mol wt 356) An anthelminic and antiseptic agent used for the treatment of infestation of the giant liver fluke.

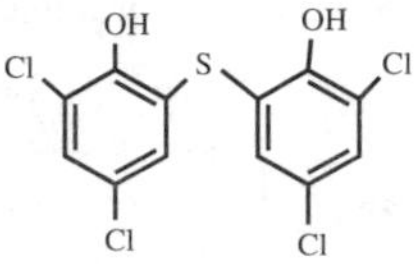

Bithorax Gene A homeotic gene in *Drosophila* that specifies the segmental identity of the body.

Bitolterol (mol wt 462) A bronchodilator that relaxes bronchial smooth muscle by acting on beta-2 adrenergic receptors.

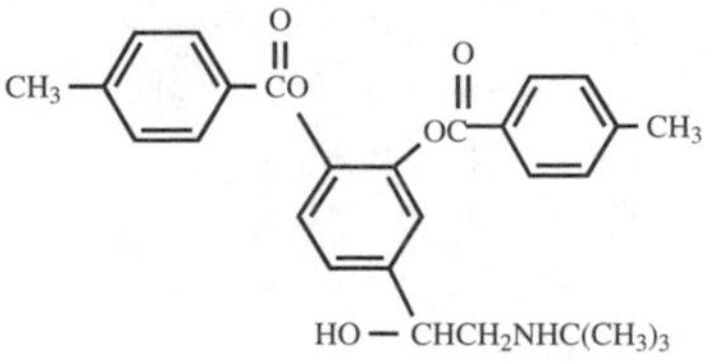

Bitoscanate (mol wt 192) An anthelminic agent (nematodes).

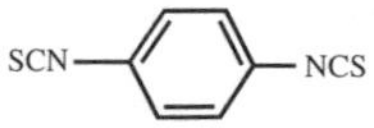

Bitot's Spots White or gray triangular deposits on the bulbar conjunctiva adjacent to the lateral margin of the cornea (a clinical sign of vitamin A deficiency).

Bitrate A trade name for a combination antianginal drug that contains pentaerythritol tetranitrate and phenobarbital sodium.

Bittner Mouse Milk Virus Referring to mouse mammary tumor virus.

Bittner Virus Synonym of mouse mammary tumor virus.

Biuret (mol wt 103) A compound formed by the condensation of two molecules of urea.

NH_2-CO-NH-CO-NH_2

Biuret reaction A colorimetric reaction for the qualitative and quantitative determination of proteins, based on the production of a purple color upon treatment of biuret, peptides, or proteins, with copper sulfate in an alkaline solution.

Biuret test See Biuret reaction.

Bivalent 1. A synapsed pair of homologous chromosomes seen during meiosis 2. A numerical expression of the capacity of a chemical element or compound to combine with other atoms or compounds, e.g., divalent Ca^{++}, monovalent Na^+, and divalent IgG.

Bixin (mol wt 394) A carotenoid carboxylic acid from the seeds of *Bixa brellana* (Bixaceae).

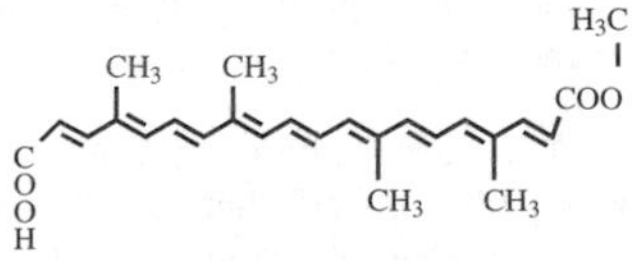

BJ Protein Abbreviation for Bence-Jones protein.

BJP Abbreviation for Bence Jones protein.

Bk A symbol for berkelium.

BK Cell Abbreviation for Sf9 cell expressing human B_2 bradykinin receptor.

BK Virus A virus of the family Papovaviridae.

BK-A Abbreviation for basophil-derived kallikrein-A.

BKHS Abbreviation for bovine kidney heparan sulfate.

BL Abbreviation for Burkitt's lymphoma.

BLA-36 Abbreviation for an antigen found in patients with Hodgkin's disease and B cell lymphomas.

Black Albumin Referring to bovine naphthol blue black.

Black Body Referring to a physical object that absorbs all the radiant energy falling upon it and emits it in the form of thermal radiation.

Black Cyanate A mixture containing 45% calcium cyanide; made from calcium cyanamide by heating it with sodium chloride and carbon.

Black Leaf 40 A trade name for an insecticide that consists of a 40% solution of nicotine sulfate.

Black Light Referring to the radiation of the wavelength of 300-420 nm.

Black Lung Disease A disorder due to the repeated inhalation of coal dust that may lead to the impaired lung function.

Black Membrane An artificially prepared molecular lipid bilayer; it appears black.

Black Mildews Plant parasitic fungi of the family Meliolaceae (order Dothideales) that form dark mycelia.

Black Piedra A human mycosis caused by infection with *Piedraia hortae* in which hard black nodules (masses of dark hyphae) adhere firmly to the infected hairs.

Black Powder A low explosive consisting of potassium nitrate, charcoal, and sulfur.

Black Stem Rust Disease of wheat caused by *Puccinia graminis*.

Black Tongue A fungal disease of the tongue characterized by the appearance of dark patches on the surface of the tongue.

Black Widow Spider A poisonous spider (*Latrodectius mactans*) whose bite causes sweating, stomach cramps, nausea, and intense pain.

Black Widow Spider Antivenom An agent used for treatment of black widow spider bite. It combines and neutralizes the venom.

Black Widow Spider Venom A potent neurotoxin that causes release of acetylcholine from presynaptic terminals of cholinergic chemical synapses.

Blackout A temporary loss of vision or consciousness.

Blackwater Fever A type of malaria characterized by the profound hemoglobinuria.

Bladder 1. A hollow sac serving as a receptacle for fluid or gas. 2. The urinary bladder.

Blanex A trade name for a combination drug containing chlorzoxazone and acetaminophen.

Blank A solution or a mixture of reagents that excludes the sample.

-blast A suffix meaning an embryonic state.

Blast Cell A precursor of a red blood cell in the earliest stage of development.

Blast Cell Transformation The differentiation of T lymphocytes to a larger, cytoplasmic-rich lymphoblast.

Blastema A group of cells that is capable of developing into an organized structure during regeneration or forming a new organism by asexual reproduction.

Blasticidin S (mol wt 422) A nucleoside antibiotic produced by *Streptomyces griseochromogenes.*

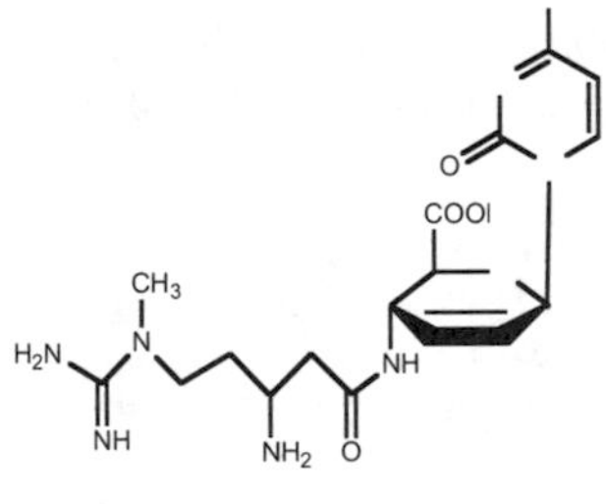

Blastin Any substance capable of stimulating the growth or proliferation of cells.

Blasto- A prefix meaning embryo.

Blastobacter A genus of rod-shaped to spherical, Gram-negative, budding bacteria occurring in ponds and lakes.

Blastocel The large central cavity of an embryo at the blastula stage.

Blastocoele Variant spelling of blastocel.

Blastocrithidia A genus of homoxenous parasitic protozoa (family Trypanosomatidae), which occurs in the gut of bugs and mosquitoes.

Blastocyst The embryo at the time of its implantation into the uterine wall; it gives rise to the tissues and organs of the developing mammalian organism.

Blastocystis A genus of protozoa (formerly known as fungi of Entomophthorales).

Blastocyte An undifferentiated very early embryonic cell.

Blastocytoma A tumor of undifferentiated or embryonic tissue.

Blastoderm The layer of cells forming the wall of the blastocyst in mammals and the blastula in lower animals during the early stages of embryonic development.

Blastodisc A disc-shaped layer of cells formed by the cleavage of a large yolky egg. Mitosis within the blastodisc produces the embryo.

Blastogenesis The production of blast cells caused by mitogen activation.

Blastokinin A protein secreted by the uterus in many mammals that stimulates and regulates the implantation process of the blastocyst in the uterine wall.

Blastolysis Destruction of a germ cell or blastoderm.

Blastoma A tumor consisting of undifferentiated or embryonic tissue.

Blastomatosis The development of tumors from embryonic tissue.

Blastomere A daughter cell resulting from cell division or cleavage of the fertilized egg.

Blastomyces A genus of fungus (*Hyphomycete*) that causes blastomycosis.

Blastomycosis An infectious skin disease caused by fungus *Blastomyces dermatitidis.*

Blastophthoria Destruction of germ cells.

Blastospore A spore produced by a budding process along the mycelium.

Blastula A hollow sphere structure enclosed within a single layer of cells that occurrs at an early stage of development in various animals.

Blastulation Formation of blastocyst or blastula.

Blattabacterium A genus of Gram-negative bacteria that occur as intracellular endosymbionts of the cockroach (*Blatta* species).

Bleaching 1. Loss of the chromophore or pigment from a biological organelle or loss of pigment(s) from an organism (e.g., loss of chlorophylls from a photosynthetic microorganism). 2. To whiten a textile or paper.

Bleaching Liquid A solution of either sodium or calcium hypochloride.

Bleb A blister or an accumulation of fluid under the skin, usually associated with lesions.

Bleed To lose blood from the blood vessels.

Bleeder An individual with a tendency to bleed or an individual who has blood clotting disease.

Bleeding The release of blood from the blood vessels.

Bleeding Diasthesis A tendency to have an abnormal blood clotting.

Blender An apparatus used for disintegration of tissues or cells.

Blending Vigorous agitation with high-shear forces.

Blenn- A prefix meaning mucus.

Blenna Mucus.

Blennadenitis Inflammation of mucous glands.

Blennorrhea Excessive discharge of mucus.

Blennostasis The cessation or reduction of mucous discharge.

Blennostatic Action or agent capable of causing blennostasis.

Blennothorax A large amount of mucus in the chest.

Blennouria The presence of mucus in the urine.

Blenoxane A trade name for bleomycin sulfate, an antineoplastic agent that inhibits DNA synthesis.

Bleomycin An antineoplastic antibiotic produced by *Streptomyces verticillus* that binds to DNA and causes it to degrade. It consists of peptide and disaccharide moietics.

Blephamide A combination drug containing sulfacetamide sodium, prednisolone acetate used as an anti-inflammatory and anti-infective agent.

Blepharal Pertaining to the eyelids.

Blepharitis An inflammatory condition of the eyelid characterized by swelling, redness, and crusts of dried mucus on the lids.

Blepharo- A prefix meaning eyelid or eyelash.

Blepharomelasma Pigmentation of the eyelids.

Blepharon Eyelid.

Blepharoplegia Paralysis of the eyelids.

Blepsopathy A disorder of eyesight.

Blewit The edible fruiting body of *Lepista saeva*.

bLF Abbreviation for bovine lactoferrin.

βLG Abbreviation for beta lactoglobulin.

Blight A wide range of unrelated plant diseases caused by various types of fungi, e.g., cane blight, chestnut blight, fireblight, and late blight.

Blighted Ovum A fertilized egg that fails to develop.

Blister A fluid-filled space either within or beneath the epidermis.

BLM Abbreviation for 1. bilayer lipid membrane, 2. bimolecular lipid membrane, and 3. black lipid membrane.

BlnI A restriction endonuclease from *Brevibacterium linens* with the following specificity:

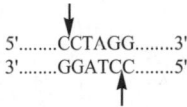

```
     ↓
5'........CCTAGG........3'
3'........GGATCC........5'
              ↑
```

Bloat A disorder of the abdomen resulting from the accumulation of intestinal gas or swallowing air.

Blocardren A trade name for timolol maleate, an antihypertensive agent.

Blochman Bodies Referring to the intracellular, prokaryotic symbiont in mycetomes in insects.

Block Electrophoresis A preparative electrophoretic technique employing a large supporting block, e.g., starch block or agar block, for handling of a large quantity of sample.

Block Copolymer A polymer consisting of a long stretch of two or more nonomeric units linked by chemical valences in one simple chain.

Block Polymer A high molecular weight polymer whose molecule is made up of alternating sections of onc chemical composition separated by sections of a different chemical nature or by a coupling group of low molecular weight.

Blocker Any agent that blocks a biological action.

Blocking Antibody 1. Antibody incapable of agglutinating homologous antigen (incomplete antibody). 2. Antibody capable of preventing anaphylaxis by combining with the allergen (e.g., IgG).

Blocking Factors Substances that are present in the serum of tumor-bearing animals and are capable of blocking the ability of immune lymphocytes to kill tumor cells.

Blocking Group See protecting group.

Blood The fluid circulating through the cardiovascular system; consists of plasma, red blood cells, white blood cells, and platelets.

Blood Agar An agar-based culture medium consisting of defibrinated blood and nutrient, used to cultivate certain microorganisms (e.g., *Staphylococcus epidermidis, Diplococcus pneumoniae,* and *Clostridium perfringens*) and to detect bacterial hemolysin.

Blood Bank A place or institution for storage of blood or plasma.

Blood Brain Barrier A specialized capillary arrangement in the brain that restricts the passage of most substances into the brain, thereby preventing dramatic fluctuations in the brain's environment.

Blood Buffers A buffer system responsible for maintaining the proper pH of the blood.

Blood Cerebrospinal Fluid Barrier See blood-brain barrier.

Blood Clot The insoluble complex of polymerized fibrin molecules and trapped cells.

Blood Clotting The biochemical reactions by which a blood clot is formed. It involves the conversion of fibrinogen to fibrin through a series of reactions mediated by the blood clotting factors.

Blood Coagulation See blood clotting.

Blood Count A computation of the number of blood cells (e.g., red and white blood cells) per cubic millimeter of blood.

Blood Culture A procedure for detecting the presence of viable bacteria in the blood.

Blood Doping The transfusion of blood to individuals (e.g., athletes) to increase the oxygen-carrying ability of the blood for the purpose of improving performance and endurance.

Blood Dyscrasis A pathologic condition in which any of the constituents of the blood are abnormal in either quality or quantity.

Blood Fluke A parasitic flatworm of the genus *Schistosoma.*

Blood Gas Gas dissolved in the blood, e.g., oxygen, carbon dioxide, and nitrogen.

Blood Gas Analysis A method for analysis of acid-base balance in the blood by determining the quantity of nitrogen, oxygen, and carbon dioxide in the blood. It is an important test for the evaluation of conditions of heart, kidney, or drug overdose.

Blood Gas Tension The partial pressure exerted by a gas in the blood.

Blood Glucose Concentration of glucose in the blood.

Blood Group The classification of blood based on the presence or absence of genetically determined antigens on the surface of the red blood cell (e.g., system of ABO or MN).

Blood Group A Red blood cells that possess antigen A on the cell surface.

Blood Group Antigen Referring to the antigens on the surface of red blood cells that are used for classification of blood types.

Blood Group AB Red blood cells that possess both antigen A and antigen B on the cell surface.

Blood Group B Red blood cells that possess antigen B on the cell surface.

Blood Group Chimerism The phenomenon in which dizygotic twins exchange hematopoietic stem cells while in utero and continue to form blood cells of both types after birth.

Blood Group O Red blood cells that possess neither antigen A nor antigen B on the cell surface.

Blood Group Substance See blood group antigens.

Blood Group System A blood classification system that classifies individuals into groups on the basis of the presence or absence of specific blood group substances.

Blood Island Hemopoietic center in the embryonic yolk sac where blood cells and vessels are forming (also known as a blood islet).

Blood Islet Synonym for blood island.

Blood Lactate The concentration of lactate in the blood as a result of metabolism without oxygen.

Blood Osmolarity The osmotic pressure of blood (normal value: 280 - 295 mOsm/L).

Blood Plasma Blood without cells.

Blood Platelets The minute, colorless protoplasmic disc of vertebrate blood. It plays an important role in blood coagulation.

Blood Poisoning Referring to the presence of pathogenic bacteria or fungi in the blood (also called septicemia).

Blood Pressure The force or pressure that the blood exerts on the wall of the vessels or heart.

Blood Serum Blood without cells and fibrin.

Blood Smear Smearing or spreading of a blood sample onto a microscope slide for diagnostic examination.

Blood Substitute Substance used to replace circulating blood or to extend blood volume, e.g., plasma, serum, and albumin.

Blood Sugars Sugars or sugar concentrations present in the blood, e.g., glucose, fructose, and galactose.

Blood Test Any test that determines the characteristics of the blood, e.g., blood sugar concentration or bacterial or viral infection.

Blood Transfusion The administration of whole blood or a component of blood to replace blood lost because of surgery or disease.

Blood Type The phenotype of erythrocyte antigen in a given individual, determined by an antigen-antibody reaction.

Blood Typing The identification of the blood types or blood group substances of an individual.

Blood Urea Nitrogen (BUN) Concentration of urea nitrogen in the blood. BUN is an important parameter for monitoring kidney function.

Blood Vessel The network of tubes and capillaries (e.g., artery or vein) through which blood circulates.

Bloodletting The removal of blood for therapeutic purpose.

Bloom Referring to the stage of abundant growth of planktonic algae.

Bloom Syndrome A type of dwarfism inherited as an autosomal recessive trait. Individuals with bloom syndrome have decreased immunity and show sun sensitivity.

Bloomed Lens A magnesium chloride coated lens used for the reduction of the amount of incident light refracted back from the surface of the lens.

Blot Transfer of electrophoretically separated components from a gel to chemically treated paper (e.g., nitrocellulose paper or membrane).

Blotting Cellulose Nitrate Membrane Cellulose nitrate membrane used for electroblotting of DNA and protein.

Blotting Hybridization See blotting technique.

Blotting Nylon 66 Membrane An amphoteric nylon cast on a polyester support and used for blotting DNA.

Blotting Technique A technique in which the chromatographically or electrophoretically separated protein, DNA, or RNA are transferred from the separating gel to the nitrocellulose membrane or other appropriate film for subsequent detection by enzyme-labeled or radioactively labeled probe.

Blowing Agent A substance incorporated in a mixture for the purpose of producing foam.

βLP Abbreviation for beta lipoprotein.

BlpI A restriction endonuclease from *Bacillus lentus* NEB819 with the following specificity:

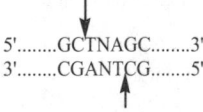

```
      ↓
5'........GC TNAGC........3'
3'........CGANT CG........5'
               ↑
```

BLS Abbreviation for B cell lymphoproliferative syndrome.

BluI (XhoI) A restriction endonuclease isolated from *Brevibacterium luteum* with the same specificity as XhoI.

BluII (HaeIII) A restriction endonuclease isolated from *Brevibacterium luteum* with the same specificity as HaeIII.

Bluboro A trade name for aluminum acetate, used as an astringent that produces cooling and vasoconstriction.

Blue Cotton Referring to phthalocyanine cellulose (cotton that bears linked trisulfophthalocyanine residues).

Blue Dextran A polymeric dye with approximate molecular weight of 2,000,000; used as a molecular marker in gel filtration.

Blue Fever Rocky Mountain spotted fever caused by *Rickettsia rickettsii* and characterized by the appearance of dark blue color on the skin following infection.

Blue Gel A trade name for pyrethrin, an anthelminic agent.

Blue Green Algae See blue-green bacteria.

Blue Green Bacteria Photosynthetic prokaryotes that contain chlorophyll and phycocyanin pigments.

Blue Mould Referring to the blue-spored species of *Penicillium* and *Peronospora* of fungi.

Blue Print Paper A paper that has been treated with ammonium ferric citrate and potassium ferricyanide solution. Upon exposure to light and wash, all black marks appear as white in a blue background.

Blue Proteins Copper-containing proteins with maximum absorption at 600 nm, e.g., plastocyanin, azurin, ascorbate, oxidase, and ceruloplasmin.

Blue Pus Pus formed from infections with *Pseudomonas aeruginosa* that has a bluish tinge due to the presence of pyocyanin.

Blunt End Terminus of double-stranded DNA that are completely base paired (also known as flushed end).

Blunt End DNA The linear double-stranded DNA with flush ends on both terminals.

Blunt-End Ligation The joints of DNA fragments that have blunt ends for specific DNA ligases, e.g., T_4 ligase.

Blush Localized density observed in X-ray examination of blood vessels due to increased vascularity in a tumor or due to leakage of blood.

Blutene A trade name for tolonium chloride.

BLV Abbreviation for bovine leukosis virus.

B-Lymphocyte (B Cell) Lymphocyte derived from bone marrow (in mammals) or from Bursa of Fabricius (in birds) that is responsible for production of antibody.

BM Abbreviation for 1. basal metabolism, 2. body mass, and 3. bowel movement.

B_2M Abbreviation for β_2-microglobulin.

β_2M Abbreviation for β_2-microglobulin.

B_{max} Abbreviation for 1. maximum number of binding sites; 2. maximum binding capacity.

BMCC Abbreviation for bacterial microcrystalline cellulose.

BMCP Abbreviation for bovine mast cell protease.

BME Medium Abbreviation for basal modified Eagle's medium.

βME Abbreviation for beta mercaptoethanol.

Bme18I A restriction endonuclease from *Bacillus magaterium* 18 with the following specificity:

5'........G(G/A/T)GC(C/T/A)↓C........3'
3'........C↑(C/T/A)CG(G/A/T)G........5'

B_2MG Abbreviation for beta-2 microglobulin.

BMM Abbreviation for bone marrow-derived macrophage.

BMMC Abbreviation for bone marrow-derived mast cell.

BMOPP Abbreviation for a combination drug containing bleomycin, methyldiamine, oncovin, procarbazine, and prednisone.

BMϕ Abbreviation for biogel-elicited macrophage.

BMP Abbreviation for bone morphogenetic protein.

BMR Abbreviation for basal metabolic rate.

BMT Abbreviation for bone marrow transplantation.

BMV Abbreviation for brome mosaic virus.

BmyI A restriction endonuclease from *Bacillus mycoides* with the following specificity:

5'........G(G/A/T)GC(C/T/A)↓C........3'
3'........C↑(C/T/A)CG(G/A/T)G........5'

BNC Abbreviation for brain Na^+ channel.

BND Abbreviation for benzoylated-naphthoylated DEAE.

BND-Cellulose Abbreviation for benzoylated and naphthoylated diethylaminoethylcellulose (an ion exchanger for ion exchange chromatography).

β-NF Abbreviation for β-naphthoflavone.

BNG Abbreviation for bromo-naphthyl β-galactoside.

BNGase Abbreviation for bromo-naphthyl β-galactosidase.

BNP Abbreviation for natriuretic peptide.

BoAChE Abbreviation for bovine acetylcholinesterase.

BOAP Abbreviation for a combination drug containing bleomycin, oncovin, adriamycin, and prednisone.

Boas' Test A test for hydrochloric acid content of the stomach (also known as resorcinol test).

Boat Conformation A conformation for cyclohexane and other six-membered ring structures that resembles an outline of a boat.

BOBA Abbreviation for beta-oxy-butyric acid.

βOBA Abbreviation for beta-oxy-butyric acid.

Boc Abbreviation for tertiary butyloxycarbonyl group.

$$(CH_3)_3\text{-C-O-CO-}$$

BOC-AA Abbreviation for tertiary butoxycarbonyl amino acid (also known as tBOC-AA).

BOC-Amino Acid (Butoxycarbonyl Amino Acid) An amino acid in which the amino group has been protected by attachment of a tertiary butoxycarbonyl group and used in solid peptide synthesis.

BOD Abbreviation for biochemical oxygen demand or biological oxygen demand. It is the amount of dissolved oxygen needed for microbial oxidation of biodegradable matter in an aquatic environment.

BOD Test A test that measures the amount of oxygen consumed per liter of sewage or per liter of its known dilution.

Bodo A genus of protozoa (suborder Bodonina) occurring in organically polluted water.

Body Microflora The characteristic microflora of various parts of the body, e.g., ear microflora, eye microflora, and gastrointestinal tract flora.

Bohr Effect The effect of pH on the affinity of oxygen for hemoglobin.

Boivin Antigens Synonym for O antigens.

Bolandiol (mol wt 276) An anabolic steroid that increases constructive metabolism and muscle strength.

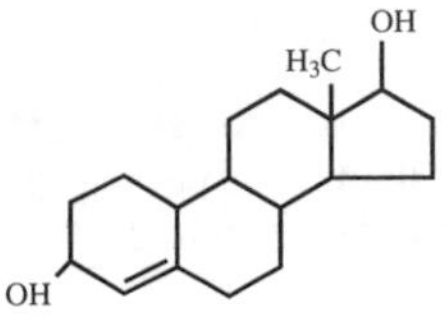

Bolasterone (mol wt 316) An anabolic steroid that increases constructive metabolism and muscle strength.

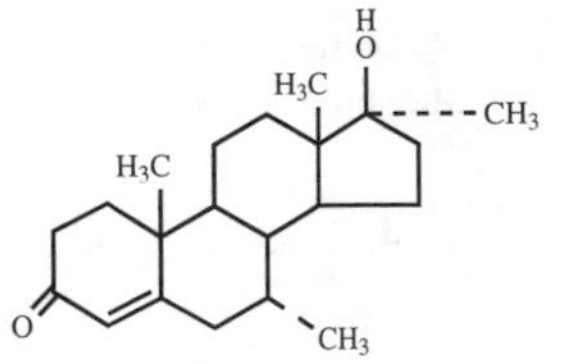

BOLD Abbreviation for a combination drug containing bleomycin, oncovin, lomustine, and dacarbazine.

Boldenone (mol wt 286) An anabolic steroid and an androgen that increases constructive metabolism and muscle strength.

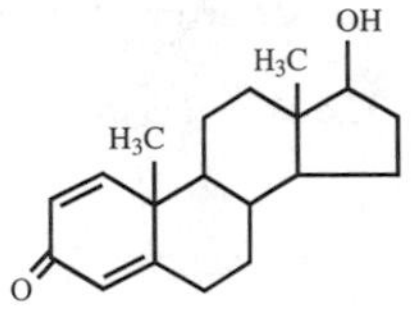

Bolivian Haemorrhagic Fever A viral hemorrhagic fever caused by Machupo virus (Arenaviridae).

Bollum's Enzyme See DNA nucleotidylexotransferase.

Bolometer A sensitive temperature transducer used for the measurement of minute quantities of radiant heat.

Boltzmann Constant The ratio of the molar ideal-gas constant to Avogadro's number.

Bomb Calorimeter An instrument used to obtain the caloric or thermal value of fuel or food.

Bombardment The impingement upon an atomic nucleus of accelerated particles such as neutrons or deutrons for the purpose of inducing fission or of creating unstable nuclei.

Bombay Blood Group A rare variant of the human ABO blood group that does not have A, B or O antigen. Individuals with Bombay blood type can not synthesize precursor H substance.

Bombesin A tetradecapeptide found in the skin of European amphibians of the family Discoglossidae. It is a potent stimulant for gastric and pancreatic secretion in mammals. It is also a mitogen.

5-Oxo-Pro-Gln-Arg-Leu-Gly-Asn-Gln-Trp

H_2N — Met-Leu-His-Gly-Val-Ala

Bombolitin Peptides isolated from venom of the bumblebee (*Megabombus pennsylvanicus*) that are rich in hydrophobic amino acids and capable of lysing erythrocytes and liposomes.

Bombyxin A brain secretory peptide of silkmoth that activates the prothoracic gland to produce ecdysone.

Bomyl (mol wt 282) An insecticide.

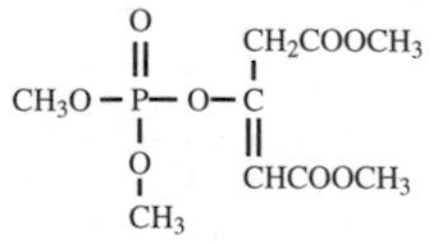

Bonamine A trade name for meclizine hydrochloride, an antiemetic agent.

Bond The linkage between two atoms, two groups of atoms, two ions, or two molecules.

Bond Angle The angle between two bonds in a molecule.

Bond Energy The energy required to break a chemical bond.

Bond Length The length of the bond between two atoms (the distance between the centers of the nuclei of the two bonded atoms).

Bond Radius One-half of the bond length.

Bone A hard, supporting connective tissue.

Bone Ash An ash composed principally of tribasic calcium phosphate with minor amounts of calcium fluoride, calcium carbonate, and magnesium phosphate.

Bone Black A black residue (impure animal charcoal) prepared from bone and used as a decolorizing agent or adsorbent.

Bone Marrow Specialized soft tissue filling the space in the cancellous part of most bone shafts; the primary lymphoid organ.

Bongkrekic Acid (mol wt 487) A toxic antibiotic produced by *Pseudomonas cocovenenans*; it inhibits the ADP/ATP exchange carrier system.

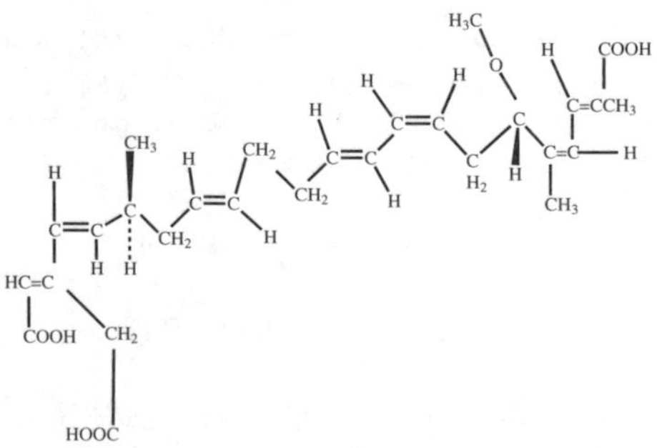

Bonine A trade name for meclizine hydrochloride, an antiemetic agent.

BONP Abbreviation for a combination drug containing bleomycin, oncovin, natulan, and prednisone.

BoNT Abbreviation for botulinum neurotoxin.

BoNT/E Abbreviation for botulinum neurotoxin-E.

Bontril PDM A trade name for phendimetrazine tartrate, a cerebral stimulant that promotes transmission of nerve impulses by release of stored norepinephrine from terminals in the brain.

Booster A second or subsequent dose of a vaccine administered for eliciting a higher rate of antibody production.

Booster Response See secondary immune response.

BOP Abbreviation for a combination drug containing BCNU, oncovin, and prednisone.

BOPAM Abbreviation for a combination drug containing bleomycin, oncovin, prednisone, adriamycin, and methotrexate.

Bopindolol (mol wt 380) An antihypertensive agent that promotes transmission of nerve impulses by release of stored norepinephrine from nerve terminals in the brain.

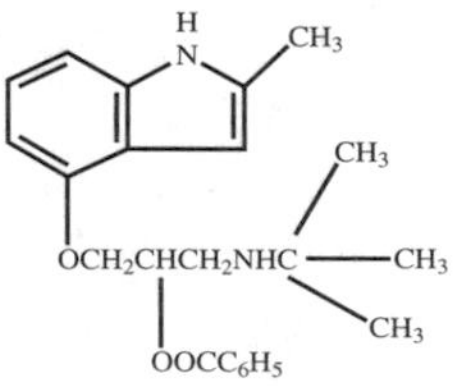

BOPP Abbreviation for a combination drug containing BCNU, oncovin, prednisone, and procarbazine.

Boracic Acid Synonym for boric acid.

Borate 1. Salt of boric acid. 2. Any polyatomic anion with boron as the central atom.

Borax A hydrated form of sodium borate.

$$Na_2B_4O_7 \cdot 10H_2O$$

Bordeaux Mixture An antifungal preparation first used for the treatment of downy mildew of vines. It is prepared by mixing copper sulfate and calcium hydroxide.

Bordeaux R (mol wt 502) A dye.

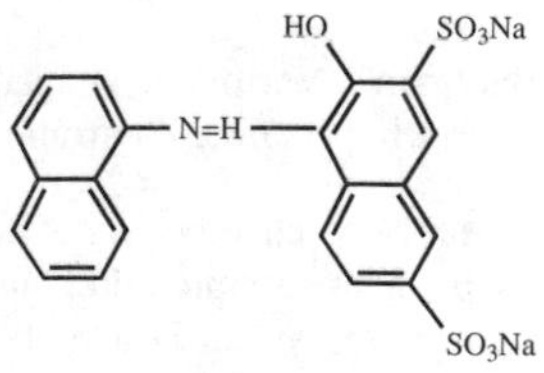

Bordet-Gengou Agar An agar-based medium containing glycerol, soluble starch, and fresh horse or sheep blood used for the isolation of *Bordetella* species.

Bordetella A genus of aerobic, chemoorganotrophic, Gram-negative bacteria.

Boric Acid (mol wt 62) Substance commonly used for the preparation of buffer (e.g., tris-borate buffer). It also possesses antimicrobial activity.

H_3BO_3

Boric Acid Gel A crosslinked polymer of dihydroborylanilino substituted methacrylic acid with 1,4-butanediol dimethacrylate used for chromatographic separation of mixtures whose components form complexes of varying stability with boric acid.

Borism Poisoning caused by inhalation or absorption of excessive doses of boric acid or borax.

Borneol (mol wt 154) A compound used in manufacturing esters and a substrate for borneol dehydrogenate.

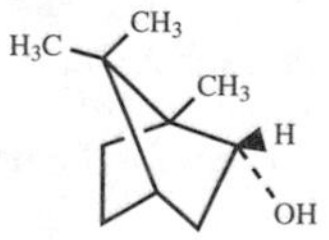

Borneol Dehydrogenase The enzyme that catalyzes the following reaction:

$$\text{Borneol} + \text{NAD}^+ \rightleftharpoons \text{Camphor} + \text{NADH}$$

Bornholm Disease A self-limiting disease in man caused by a group B Coxsackie virus.

Bornyl Chloride (mol wt 173) An antiseptic agent.

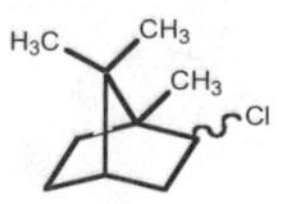

Boromycin (mol wt 880) An antibiotic from *Streptomyces antibioticus* that contains boron.

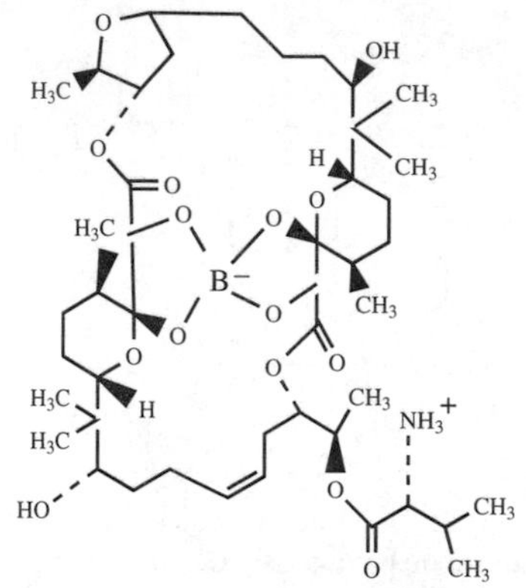

Boron (B) A chemical element with atomic weight 11, valence 3.

Borrelia A genus of Gram-negative bacteria (family Spirochaetaceae).

Borreliosis Any disease caused by a *Borrelia* species.

Bostrycoidin (mol wt 285) An antibiotic produced by *Fusarium bostrycoides*.

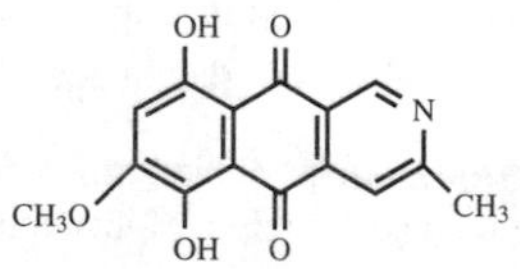

Botany The science that deals with plants.

Bothrolysin A protease that cleaves the peptide bonds between Gln-His, Ser-His, and Ala-Leu in the insulin B chain and the peptide bond between Pro-Phe in angiotension I.

Bothropasin A protease that cleaves the peptide bonds between His-Leu, Ala-Leu, Tyr-Leu, and Phe-Leu in the insulin B chain.

Botox A trade name for botulinum toxin type A, a purified neurotoxin complex.

Botryomycosis A chronic localized infection of the skin and subcutaneous tissues caused by *Pseudomonas aeruginosa* and *Staphylococcus aureus* (also known as bacterial pseudomycosis).

Bottleneck Effect A form of genetic drift resulting from reduction of a population by a natural disaster such that the surviving population is no longer genetically representative of the original population.

Bottromycins A group of antibiotics produced by *Streptomyces* species that inhibits bacterial protein synthesis.

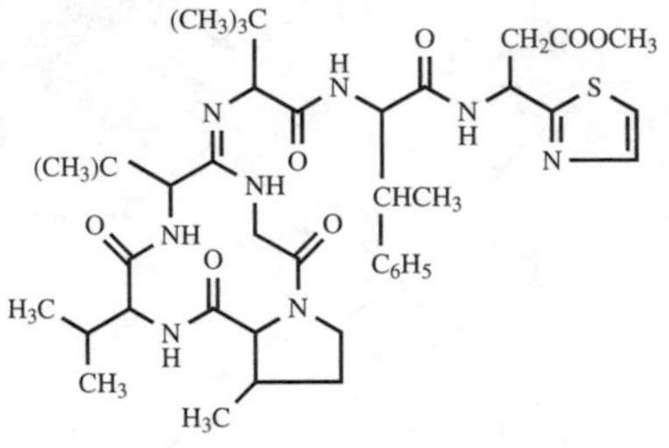

bottromycin A_2

Botulinolysin A thiol-activated cytolysin, e.g., butulinum toxin produced by *Clostridium botulinum.*

Botulinum C_2 Toxin A nonneurotoxin produced by certain strains of *Clostridium botulinum.*

Botulinum toxin A protein neurotoxin produced by strains of *Clostridium botulinum*; responsible for the symptoms of botulism.

Botulism Food poisoning due to the toxin from *Clostridium botulinum*. The disease is characterized by sight problem, muscle weakness, nausea, and vomiting and is often fatal.

Bouin's Fluid A fixative solution containing picric acid, formalin, and glacial acetic acid.

Bound Enzyme An insoluble form of enzyme prepared through an enzyme immobilization process.

Boundary Sedimentation A centrifugation technique used to determine the sedimentation coefficient of a macromolecule by measuring the boundary movement that forms during centrifugation.

Bound-to-Free Ratio The ratio of the amount of ligand bound to the receptors to the amount of ligand free in solution in a specified system. Plot of [B/F] vs. [F] yields a Scatchard plot.

Bouquet Fever See dengue.

Boutonneuse Fever A tick-transmitted infectious disease caused by *Rickettsia conorii*.

Bouvardin A cyclic hexapeptide isolated from plant (*Bouvarda ternifolia*) used as a drug against dysentery.

Bovine Pertaining to cattle, cow, ox, or other closely related species.

Bovine Tubercle Bacillus Referring to *Mycobacterium bovis*.

Bowie-Dick Test A method for testing the performance of an autoclave.

Bowman-Birk Inhibitor A trypsin inhibitor from soybean.

Box-like Bacteria Variously shaped, angular bacteria found in hypersaline environments.

Boyden Chamber A chamber that is divided into two sections by a membrane filter and used to study leukocyte migration and the activity of chemotactic substances.

Boyden Procedure A method for treating red blood cells with tanic acid to enable them to adsorb soluble antigen.

Boyle's Law The law states that the volume of a sample of gas varies inversely with the pressure at constant temperature.

BP Abbreviation for 1. base pair in double strand DNA or RNA, 2. boiling point, and 3. blood pressure.

BPA Abbreviation for biphenol A.

BPAF Abbreviation for bovine pulmonary arterial fibroblast.

BPB Abbreviation for bromphenol blue.

BpeI (Hind III) A restriction endonuclease isolated from *Bordetella pertussis* with the following specificity:

```
        ↓
5′...........A AGCTT...........3′
3′...........TTCGA A...........5′
                  ↑
```

BPG Abbreviation for 2,3-bisphosphoglycerate.

BPGM Abbreviation for bisphosphoglycerate mutase.

BPH Abbreviation for bacteriophaeophytin.

BPheo Abbreviation for bacteriopheophytin.

BPheo a Abbreviation for bacteriopheophytin a.

BPheo b Abbreviation for bacteriopheophytin b.

BpiI A restriction endonuclease from *Bacillus pumilus* SW 4-3 with the following specificity:

```
                          ↓
5'........GAAGAC(N)2........3'
3'........CTTCTG(N)6......5'
                    ↑
```

BPIG Abbreviation for bacterial polysaccharide immune globulin.

BPL Abbreviation for 1. benzylpenicilloyl-polylysine; 2. beta propio-lactone.

BPM Abbreviation for beats per minute.

BpmI A restriction endonuclease from *Bacillus pumlius* with the following specificity:

5'........CTGGAG(n)$_{16}$........3'
3'........GACCTC(N)$_{14}$........5'

BPN Abbreviation for a combination drug containing bacitracin, polymyxin-B, and neomycin.

BPS Abbreviation for benzyl penicillin sodium.

BPTI Abbreviation for bovine pancreatic trypsin inhibitor.

Bpu14I A restriction endonuclease from *Bacillus pumilus* 14 with the following specificity:

5'........TTCGAA........3'
3'........AAGCTT........5'

Bpu1102I A restriction endonuclease from *Bacillus pumilus* RFL 1102 with the following specificity:

5'........GCTNAGC........3'
3'........CGANTCG........5'

BpuAI A restriction endonuclease from *Bacillus pumilus* with the following specificity:

5'........GAAGAC(N)$_2$........3'
3'........CTTCTG(N)$_6$........5'

BPV Abbreviation for bovine papilloma virus.

BQ Abbreviation for biphenoquinone.

Br Symbol for bromine.

BR Abbreviation for 1. bacteriorhodopsin; 2. bilirubin.

b_5R Abbreviation for cytochrome-b_5 reductase.

Brachi- A prefix meaning arm.

Brachial Pertaining to the arm.

Brachialgia Pain in the arm.

Brachiomonas A genus of unicellular flagellated green algae that is closely related to *Chlamydomonas*.

Brachurin A protease from hepatopancreas of fiddler crab with broad specificity for peptides.

Brachyurin A collagenolytic proteinase.

Bracket fungi (Shelf Fungi) Fungi (Aphyllophorales) whose fruiting bodies resemble brackets or shelves on infected trees or rotting logs.

Bradford Method A method for protein determination based upon the shift of the absorption maximum of Coomassie blue upon binding to protein.

Bradshaw Test A test for the presence of Bence-Jones protein in urine by layering dilute, acidified urine over the concentrated hydrochloric acid.

Bradycardia An abnormal slowness of the heart beat (less than 60 contractions per minute).

Bradycor (mol wt 2805) A bradykinin antagonist.

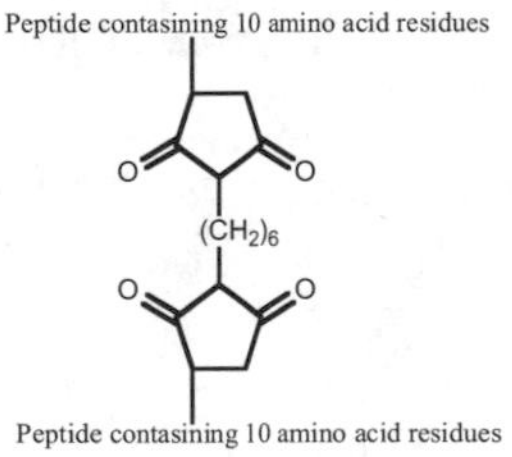

Bradycrotic Slow pulse.

Bradyecoia A mild deafness.

Bradykinesia Abnormal slowness of movement.

Bradykinin A bioactive peptide involved in regulation of body electrolyte balance, smooth muscle contraction, and vasodilation.

Arg-Pro-Pro-Gly-Phe-Ser-Pro-Phe-Arg

Bradyphagia Slowness in swallowing due to an organic impediment.

Bradypnea An abnormally slow rate of breathing.

Bradyrhizobium A genus of Gram-negative bacteria of the family Rhizobiaceae.

Bradytrophia Retarded growth and slow metabolic rate.

Bradytrophic Capable of causing decreased activity in living organisms.

Bradyuria The extremely slow passage of urine.

Bragg Curve A plot of specific ionization as a function of either distance or energy.

Bragg Scattering The scattering of radiation by a crystal.

Brain Extract (Bovine) A phospholipid-rich extract from bovine brain. It contains phosphatidylserine, phosphatidylcholine, phosphatidylethanolamine, sphingomyelin, cerebrosides, cholesterol, and fatty acid.

Brain Barrier System See blood-brain barrier.

Brain Fever Any inflammation of the brain or meninges (see also encephalitis).

Brain Heart Infusion Medium A medium containing infusions of calf-brain and beef-heart, proteose, D-glucose, NaCl, and Na_2HPO_4 for culturing bacteria and medically important fungi.

Brain Scan A diagnostic procedure employing radioisotope imaging techniques to identify intracranial masses, lesions, tumors, or areas where blood supply is blocked.

Brain Stem The portion of the brain that connects the forebrain and spinal cord.

Brain Tumor A neoplasm of the intracranial portion of the central nervous system.

Brallobarbital (mol wt 287) A sedative and hypnotic agent.

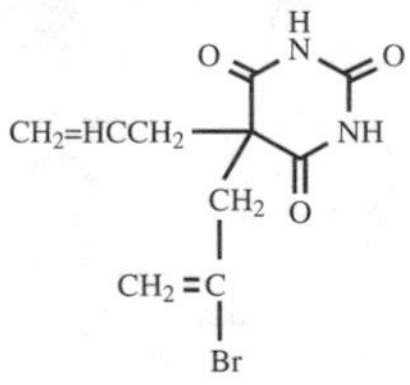

Bran A by-product from flour production. It consists mainly of the grain coat.

Branched Chain Amino Acids Amino acids with branched side chains, e.g., leucine and isoleucine.

Branched Chain Amino Acid Aminotransferase The enzyme that catalyzes the following reaction:

Leucine + α-ketoglutarate
⇅
4-Methyl-2-ketopentanate + L-glutamate

Branched Dextran Exo-1,2-α-Glucosidase The enzyme that catalyzes the hydrolysis of 1,2-α-D-glucosidic linkages at the branched points of dextran-producing free D-glucose.

Branched Chain Fatty Acid A fatty acid having a branched chain.

Branched Chain Fatty Acid Kinase The enzyme that catalyzes the following reaction:

ATP + 2-Methylpropanoate
⇅
ADP + 2-methylpropanoyl phosphate

Branched Chain Ketoaciduria A genetic disorder characterized by the presence of urinary ketoacids derived from branched-chain amino acids (isoleucine, valine and leucine).

Branched Metabolic Pathway A sequence of metabolic reactions that diverges and gives rise to two or more end products.

Branched Migration The movement of the branch point in branched DNA formed from two DNA molecules with identical sequences.

Branched Polymer The chain of a polymer consisting of side chain(s).

Branching Enzyme The enzyme that introduces branches into a linear polysaccharide.

Brassinolide (mol wt 481) A plant growth regulator.

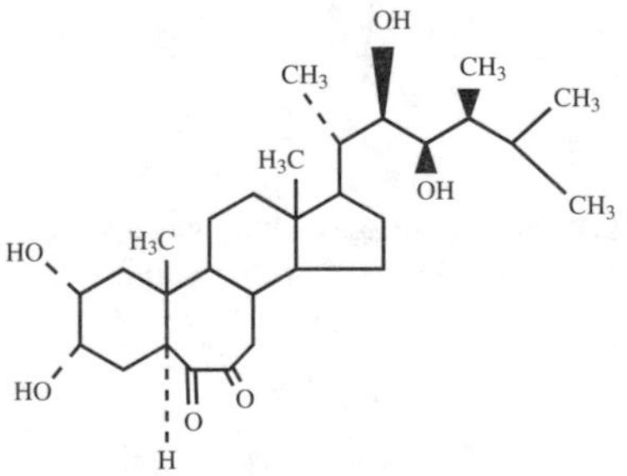

Braun Lipoprotein A rod-shaped murein lipoprotein with molecular weight of about 7500 and occurring in *E. coli.*

BRBC Abbreviation for bovine red blood cell.

BRCA1 A tumor-suppressive gene linked to familial breast and ovary cancers.

BRCA2 A tumor-suppressive gene linked to familial breast and ovary cancers.

Br-CBE Abbreviation for bromoconduritol-B epoxide.

BrdU Abbreviation for bromodeoxyuridine.

BrdUrd Abbreviation for bromodeoxyuridine.

Br-dUTP Abbreviation for bromodexoy uridine triphosphate.

Breakage and Reunion The mechanism for genetic recombination in which parts of two chromosomes are exchanged through crossing over, physical breakage, and reunion.

Breakbone Fever Synonym for dengue.

Breast Cancer A malignant neoplastic disease of breast tissue.

Breast Milk Jaundice A liver disease of the breast-fed infant characterized by jaundice and hyperbilirubinemia due to a metabolite in the mother's milk that inhibits the infant's ability to conjugate bilirubin to protein for excretion.

Breathalyzer A breath analyzer used to detect alcohol content in a breath sample.

Breda Virus An RNA virus causing diarrhea in young calves (Togaviridae).

Breed Count A microscopic counting of bacteria in a dried, stained film of the milk.

Breeder Reactor A nuclear reactor capable of converting nonfissionable uranium-238 into fissionable plutonium-239, which can be used as fuel.

Brefeldin A (mol wt 280) A fungal metabolite from *Penicillium brefeldianum* exhibiting a wide range of antibiotic activity.

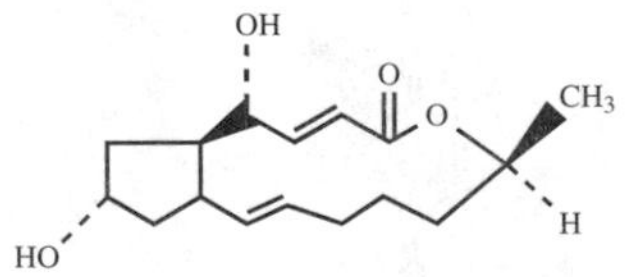

Brei Homogenized tissue used in biochemical analyses.

Breinl Strain A virulent strain of *Rickettsia prowazekii* that can be cultivated in human macrophage cultures.

Bremsstrahlung An electromagnetic radiation resulting from deceleration of high-energy beta particles in the electrostatic fields of atomic nuclei.

Brenner Tumor A benign ovarian neoplasm consisting of nests or cords of epithelial cells enclosed in fibrous connective tissue.

Breoneesin A trade name for guaifenesin, an antitussive agent.

Brepho- A prefix meaning embryo, fetus, or newborn infant.

Brephoplasty Tissue used in transplantation that originates from an embryo or from a newborn of the same species.

Brequinar (mol wt 375) An immuno-suppressant.

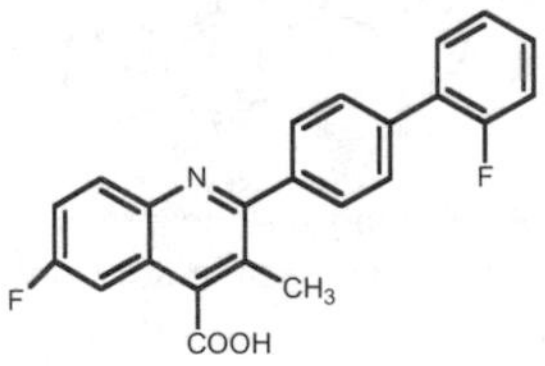

Brethaire A trade name for terbutaline sulfate, a broncodilator that relaxes bronchial smooth muscle by acting on beta-2 adrenergic receptors.

Brethine A trade name for terbutaline sulfate, a bronchodilator that relaxes bronchial smooth muscle by acting on beta-2 adrenergic receptors.

Brettanomyces A genus of yeast (class Hyphomycetes).

Bretylate A trade name for bretylium tosylate.

Bretylium Tosylate (mol wt 414) An antiarrhythmic and antiadrenergic agent.

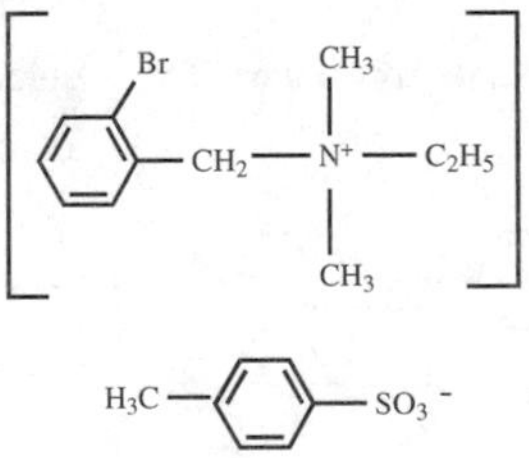

Bretylol A trade name for bretylium tosylate, an antiarrhythmic agent.

Brevetoxins Polycyclic toxins produced by the dinoflagellate *Ptychodiscus brevis* and capable of killing fish.

Brevibacterium A genus of obligately aerobic, catalase-positive, asporogenous bacteria (order Actinomycetales).

Brevibloc A trade name for esmolol hydrochloride, an antiarrhythmic agent.

Brevicollis An abnormal shortness of the neck.

Brevicon A trade name for a combination oral contraceptive drug containing ethinyl estradiol and norethindrone.

Brevinin A secreted amphibian peptide with antibacterial and hemolytic activities.

Brevital Sodium A trade name for methohexital sodium, used as a general anesthetic agent.

Brewer's Thioglycollate Medium A liquid medium (pH 7.0-7.2) used for the culture of anaerobes containing glucose, tryptone, agar, sodium thioglycollate, and redox indicator dye.

Brewer's Yeast One of various strains of yeast (*Saccharomyces cerevisiae*) used for brewing.

Brewing The preparation of beer or lager by the fermentation of an aqueous extract of malted barley containing essential oils and bitter resins of the dried female flowers (cones) of the hop (*Humulus lupulus*).

Bricanyl A trade name for terbutaline sulfate, a beta-adrenergic drug that relaxes bronchial smooth muscle by acting on beta-adrenergic receptors.

Bridging Atom 1. An atom that connects two groups in a molecule, such as the oxygen atom that connects two phosphoryl groups in ATP. 2. A metal ion in a metal bridge complex.

Brietal Sodium A trade name for methohexital sodium, an anesthetic agent.

Brij Polyoxyethylene ethers of high molecular weight aliphatic alcohols that are used as nonionic detergents and surfactants, e.g., Brij 30, 35,58, 78, and 99.

Brilliant Black BN (mol wt 868) A biological dye.

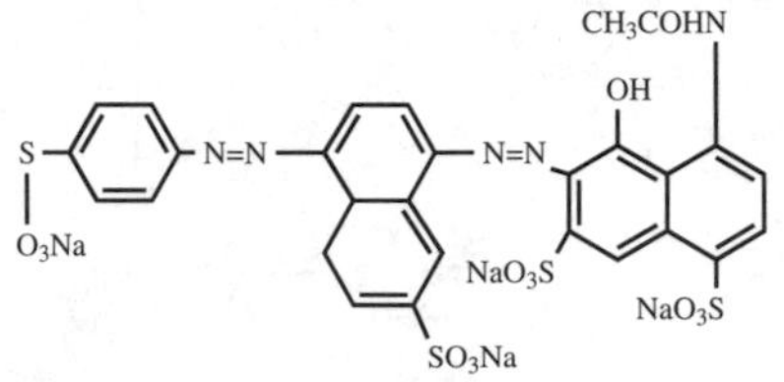

Brilliant Blue FCF (mol wt 793) A biological stain.

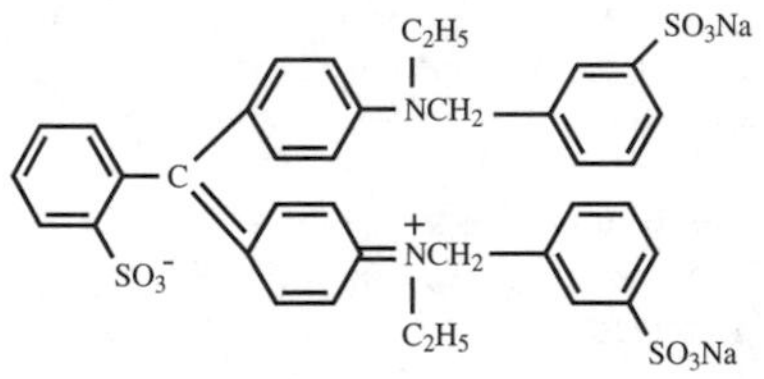

Brilliant Blue G (mol wt 854) A dye for staining proteins in gel electrophoresis.

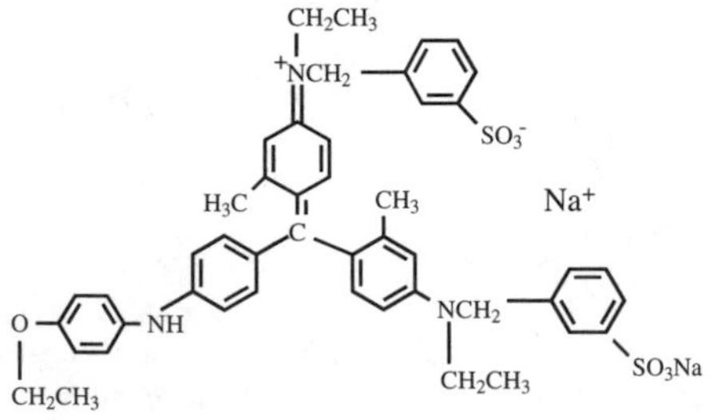

Brilliant Blue R (mol wt 826) A dye for staining proteins in gel electrophoresis.

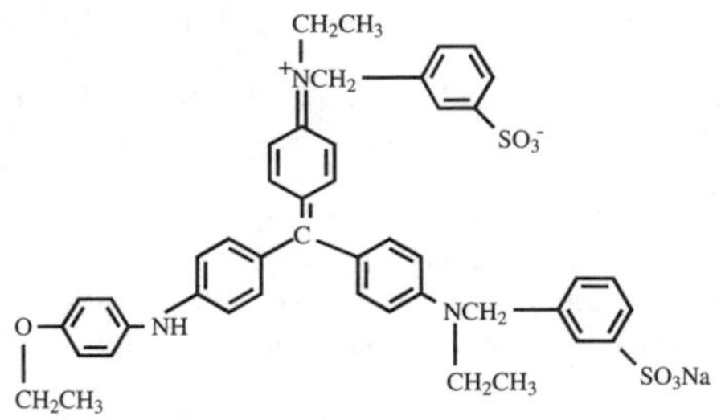

Brilliant Cresyl Blue ALD (mol wt 386) A biological dye.

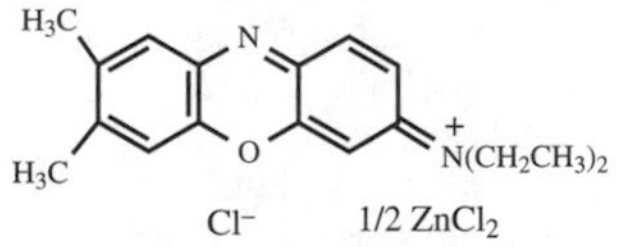

Brilliant Crocein MOO (mol wt 556) A biological dye.

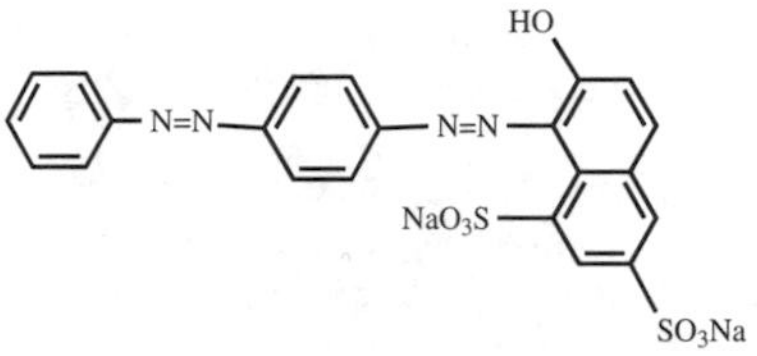

Brilliant Green (mol wt 483) A dye and an antiseptic agent.

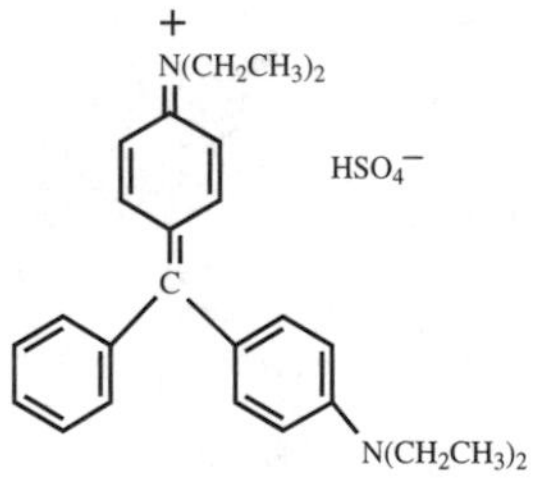

Brilliant Green Agar A selective medium used for selective staining of Salmonellae. It contains lactose, phenol red, and brilliant green (selective agent).

Brilliant Yellow (mol wt 625) A biological stain.

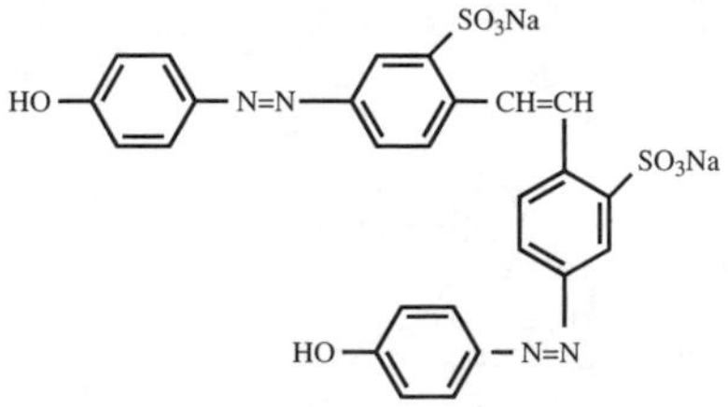

Brill-Zinsser Disease A mild form of typhus caused by *Rickettsiae typhi.*

Brimonidine (mol wt 292) An antiglaucoma agent.

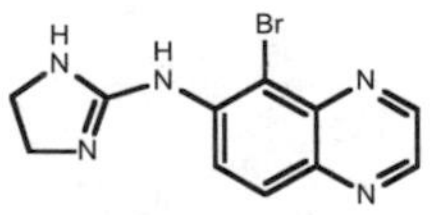

Brinase A fibrinolysin-like protease from *Aspergillus oryzae* capable of hydrolyzing fibrin and fibrinogen.

British Anti-Lewisite A sulfhydryl chelating agent (2,3-dimercaptopropanol) developed during World War II as a detoxicant for certain war poisons.

British Thermal Unit (BTU) The amount of heat required to raise the temperature of one pound of water by 1° F (from 63 to 64° F).

BRM Abbreviation for 1. biological response modifiers; 2. biuret reactive material.

Broad Beta Disease A type of hyperlipoproteinemia characterized by the accumulation of a lipoprotein containing high concentrations of cholesterol and triglycerides.

Broad Spectrum Antibiotic An antibiotic that has a wide range of antibacterial (Gram-positive and Gram-negative) activity, e.g., chloramphenicol or a tetracycline.

Broad Spectrum Pesticides Chemical pesticides capable of killing a wide range of pests.

Brochothrix A genus of Gram-positive, facultatively anaerobic bacteria.

Brodie's Abscess A form of osteomyelitis caused by Staphylococcal infection of bone.

Brodie's Solution A salt solution used in the Warburg manometer containing sodium chloride, Evan's blue, and sodium choleate.

Brodifacoum (mol wt 523) A rodenticide with anticoagulant activity.

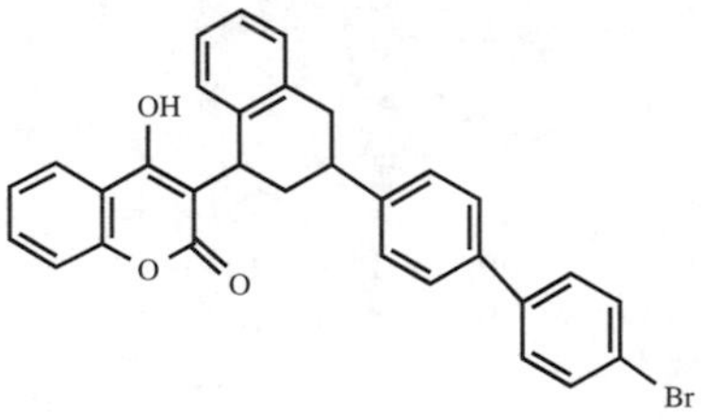

Brodimoprim (mol wt 339) A dihydrofolate reductase inhibitor and antibacterial agent.

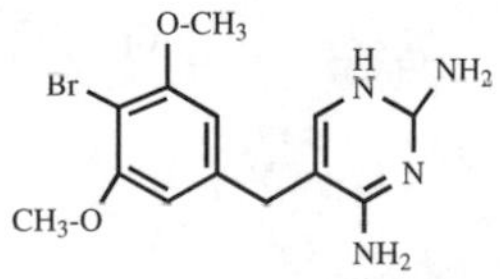

Bromacil (mol wt 261) An herbicide.

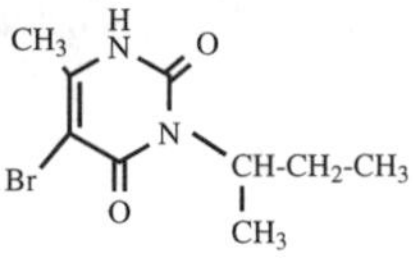

Bromadiolone (mol wt 527) A rodenticide with anticoagulant activity.

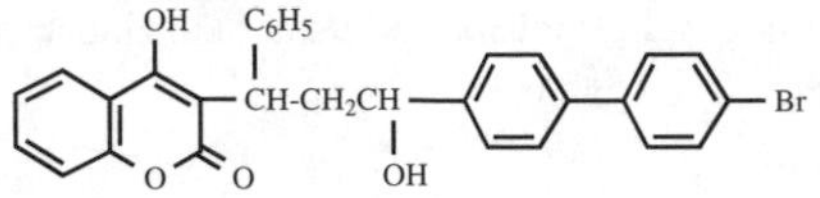

Bromates Salts of bromic acid.

Bromatology Science dealing with food (nutrition).

Bromazepam (mol wt 316) An anxiolytic agent.

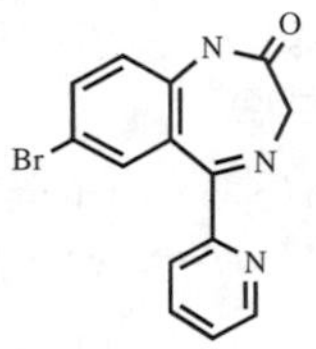

Bromcresol Green (mol wt 698) A dye.

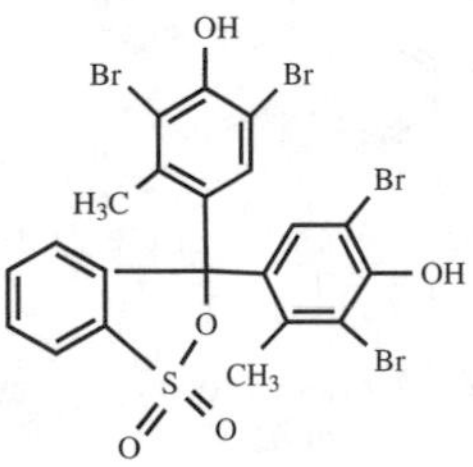

Bromcresol Purple (mol wt 540) A pH indicator (pH 5.2-6.8).

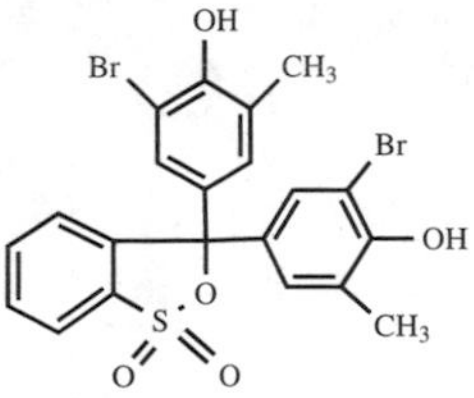

Bromelain Protein-digesting and milk-coagulating enzyme (protease) found in the pineapple fruit juice and stem tissue.

Bromelin Synonym for bromelain.

Bromethalin (mol wt 578) A rodenticide.

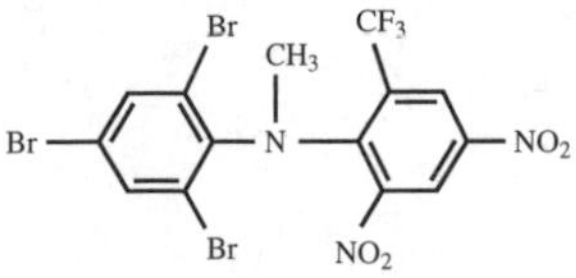

Bromfed-AT A trade name for a combination drug containing brompheniramine maleate, dextromethorphan hydrochloride, and pseudoephedrine hydrochloride.

Bromhidrosis An abnormal condition in which the apocrine sweat has an unpleasant odor.

Bromic Acid (mol wt 129) An oxidizing agent.

$Ba(BrO_3)_2$

Bromide A binary compound of bromine and another element or organic radical.

Bromidism Bromide poisoning characterized by headache, skin eruption, apathy, and muscular weakness.

Brominate The replacement of hydrogen with bromine in a compound.

Bromination The introduction of one or more bromine atoms into an organic molecule.

Bromindione (mol wt 301) An anticoagulant.

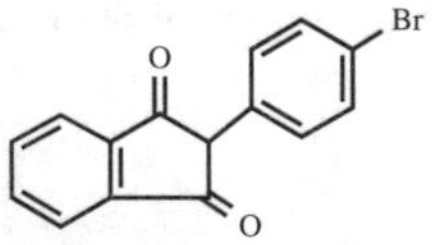

Bromine (Br) A chemical element with atomic weight 80, valence 1, 3, 5, and 7.

Bromine 82 A radioisotope of bromine emitting beta and gamma radiation.

Bromism Referring to bromide poisoning.

***p*-Bromoacetanilide (mol wt 214)** An analgesic and antipyretic agent.

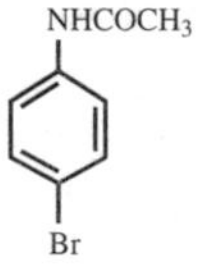

8-Bromoadenosine (mol wt 346) An analog of adenosine that inhibits binding of cAMP and adenosine to activated cAMP/adenosine binding protein.

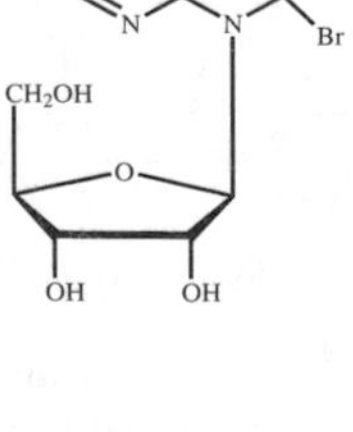

8-Bromoadenosine 3,-5′-Cyclophosphoric Acid (mol wt 408) A cyclic nucleotide analog for study of calcium-mediated pathways.

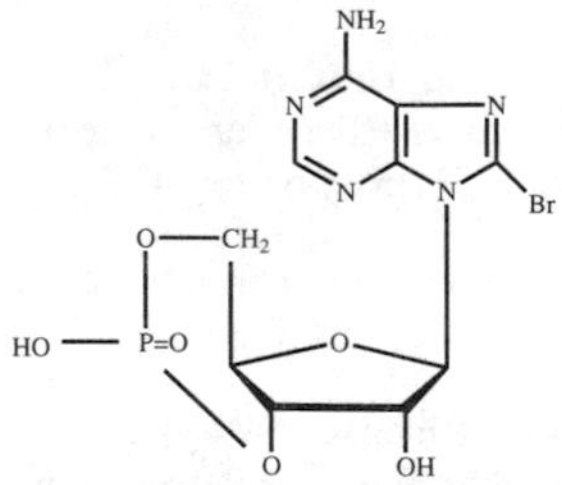

Bromocalcimycin A23187 (mol wt 603) A ionophore with a higher affinity for calcium than magnesium.

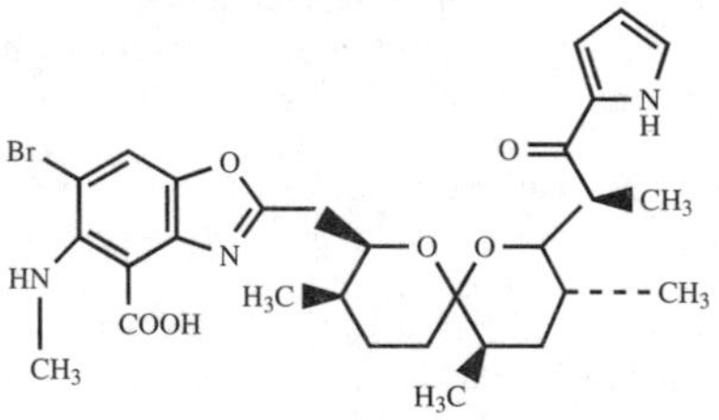

5-Bromo-4-Chloro-3-Indolyl β-D-Galactopyranoside (mol wt 409) A substrate for assaying β-galactosidase.

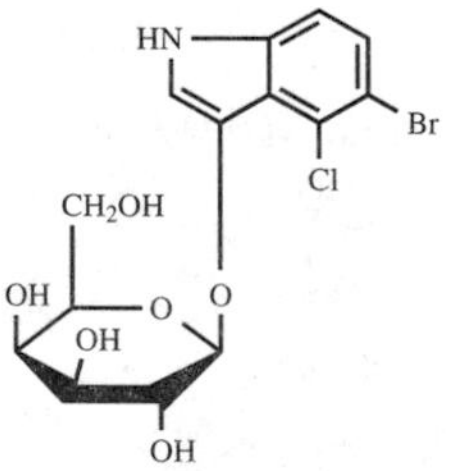

5-Bromo-4-Chloro-3-Indolyl-β-D-Glucopyranoside (mol wt 409) A substrate for assaying β-glucosidase.

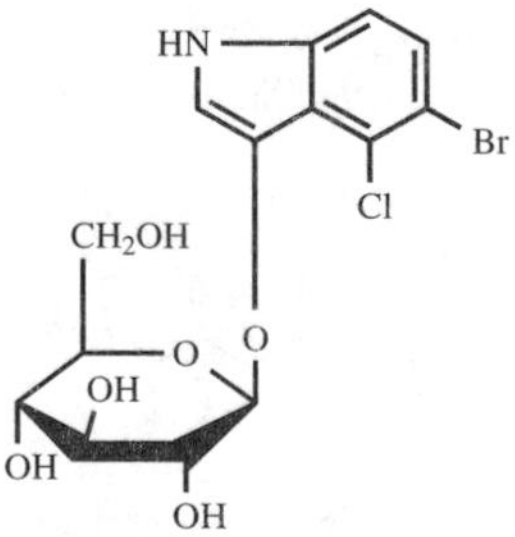

Bromocresol Green See bromcresol green.

Bromocresol Purple See bromcresol purple.

Bromocriptine (mol wt 655) An inhibitor of prolactin and an antiparkinsonian agent. It also inhibits prolactin secretion and acts as a dopamine-receptor agonist.

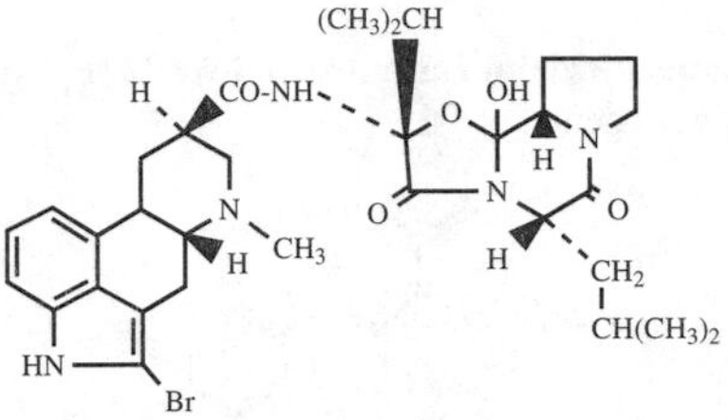

5-Bromocytosine (mol wt 190) An analog of cytosine.

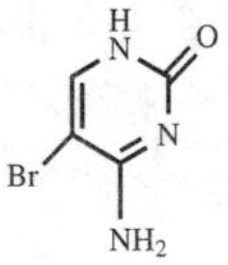

5-Bromo-2′-Deoxyuridine (mol wt 307) A thymidine analogue used as an antiviral drug.

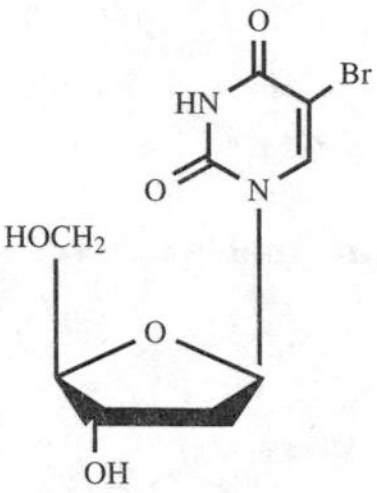

Bromoderma An acneiform, bullous, or nodular skin rash caused by allergic reactions to bromides.

Bromodiphenhydramine (mol wt 334) An antihistaminic agent.

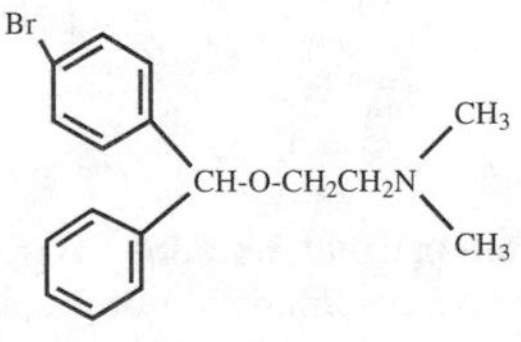

Bromofenofos (mol wt 581) A flukicide.

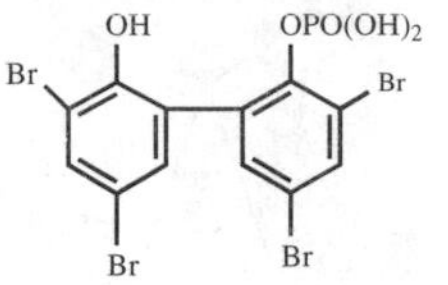

5-Bromoguanosine 3′,5′-Cyclophosphate Sodium salt (mol wt 446) A membrane-permeable cGMP derivative, it induces the HL-60 cells to differentiate.

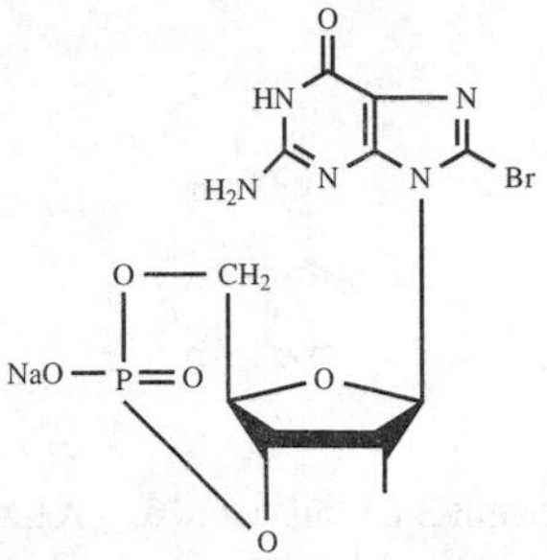

Bromoindoxyl Acetate (mol wt 254) A substrate for histochemical staining of esterases.

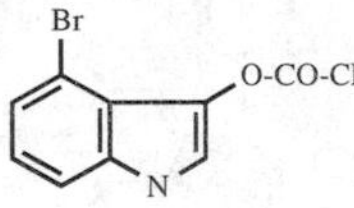

Bromoiodism Poisoning by bromine and iodine or their derivatives.

5-Bromo-3-Methyl-2(2-Nitrophenyl-mercapto)-3H-Indole (mol wt 363) A reagent for specific cleavage of proteins at the carboxylic side of the tryptophanyl residue.

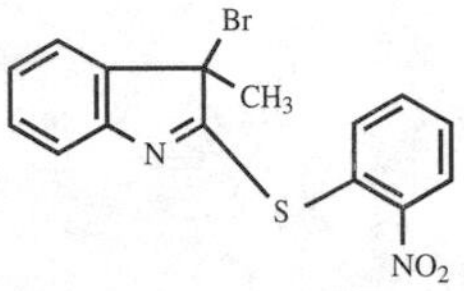

6-Bromo-2-Naphthyl-α-D-Galactopyranoside (mol wt 385) A high affinity inhibitor of lactose permease system of *E. coli.*

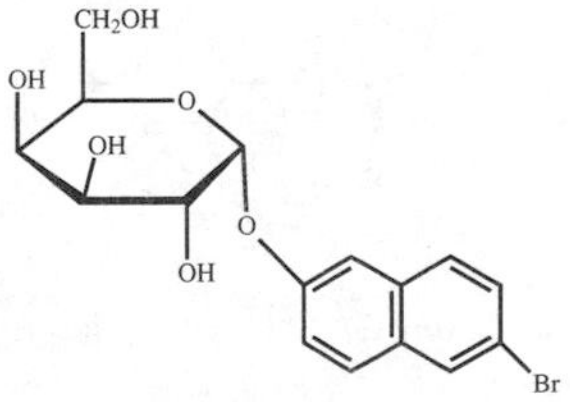

6-Bromo-2-Naphthyl-β-D-Galactopyranoside (mol wt 385) Substrate for assaying β-galactosidase activity.

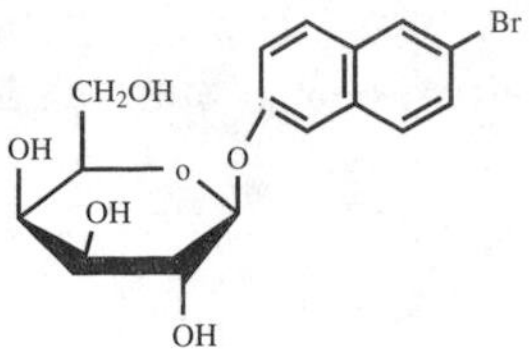

Bromoperoxidase Synonym of peroxidase.

Bromophenacyl Bromide (mol wt 278) A reagent used for modification of histidine residues in proteins.

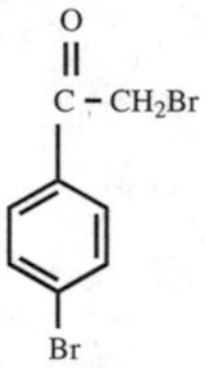

Bromophenol Blue (mol wt 670) A tracking dye in gel electrophoresis.

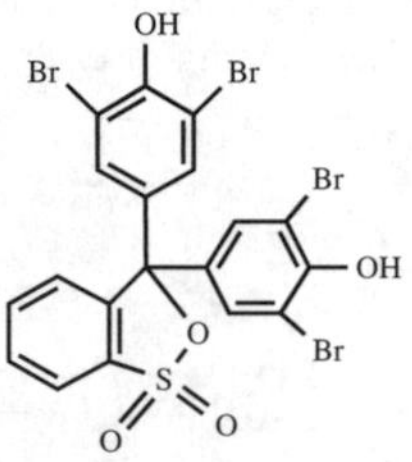

Bromophenol Red (mol wt 512) A dye and pH indicator.

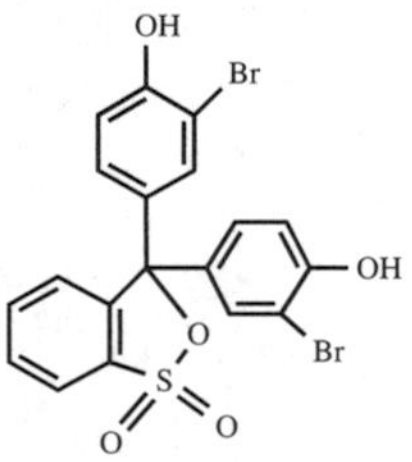

Bromophos (mol wt 366) An insecticide (acaricide).

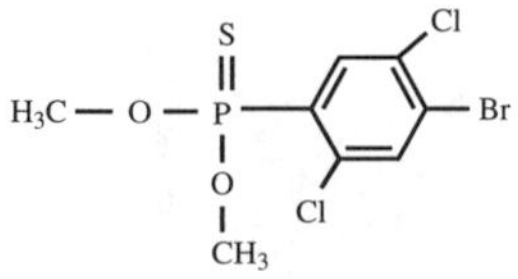

Bromopride (mol wt 344) An antiemetic agent.

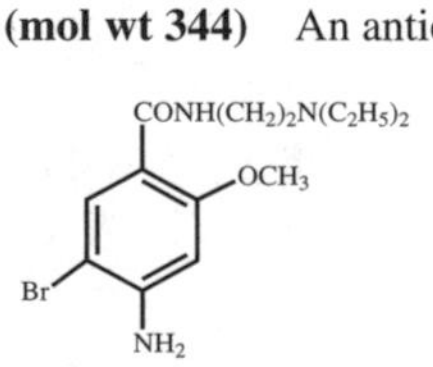

Bromopropylate (mol wt 428) An acaricide.

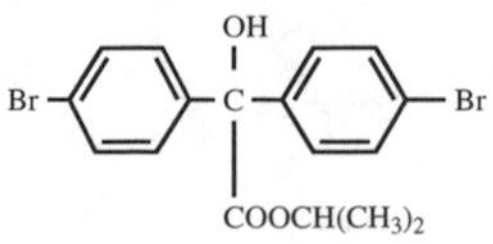

Bromopyruvic Acid (mol wt 167) A reagent used for labeling cysteine residues in a protein.

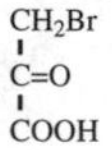

Bromosalicylchloranilide (mol wt 327) An antifungal agent.

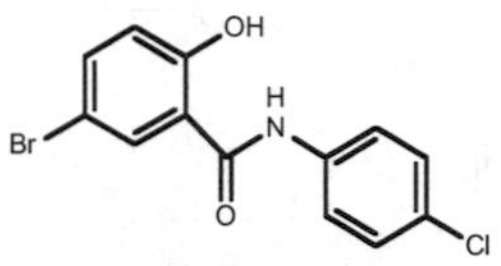

Bromosalicylhydroxamic Acid (mol wt 232) An antibacterial agent (tuberculostatic).

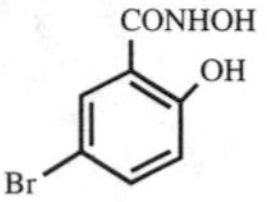

Bromosalicylic Acid Acetate (mol wt 259) An analgesic agent.

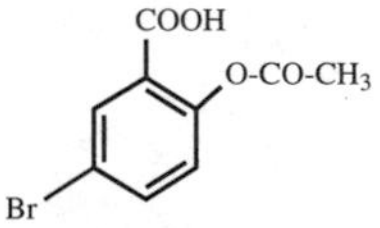

Bromosaligenin (mol wt 203) An anti-inflammatory agent.

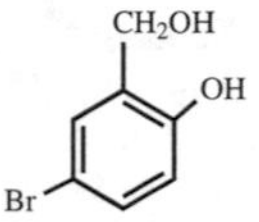

Bromosuccinimide (mol wt 178) A reagent used for specific oxidation of methionine and tryptophan.

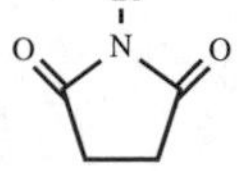

Bromosulfalein (mol wt 838) A reagent used for quantitative determination of proteins.

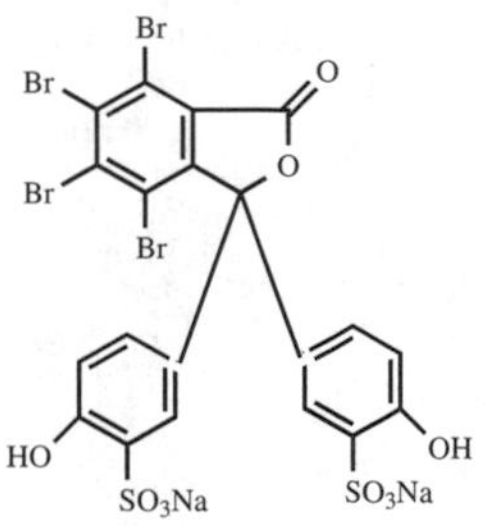

Bromothymol Blue (mol wt 624) A pH indicator dye.

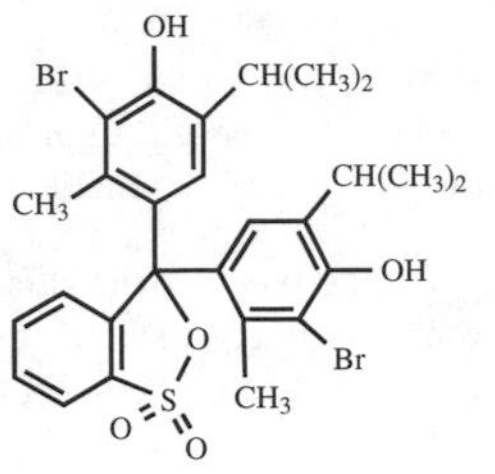

5-Bromouracil (mol wt 191) A mutagenic pyrimidine analog that is readily incorporated into DNA in place of thymine.

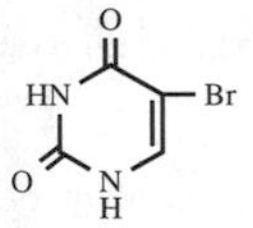

5-Bromouridine (mol wt 323) An analog of thymidine.

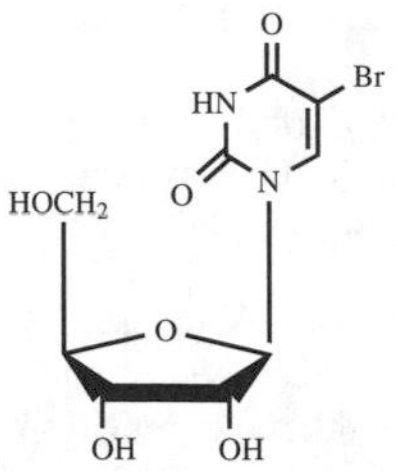

Bromoviruses An RNA plant virus with divided genome in the group bromoviruses.

Bromoxylenol Blue (mol wt 568) A pH indicator dye.

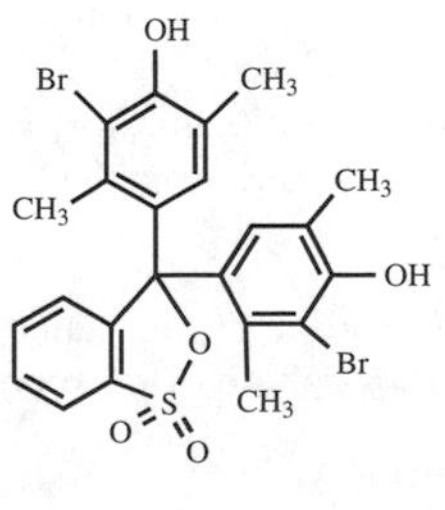

Bromoxynil (mol wt 277) An herbicide.

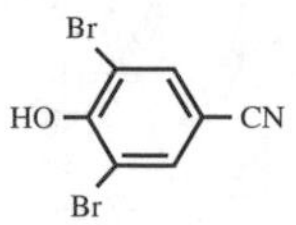

Bromoxynil Nitrilase The enzyme that catalyzes the following reaction:

3,5-Dibromo-4-hydroxybenzonitrile + H_2O

⇌

3,5-Dibromo-4-hydroxybenzoate + NH_3

Bromperidol (mol wt 420) An antipsychotic agent.

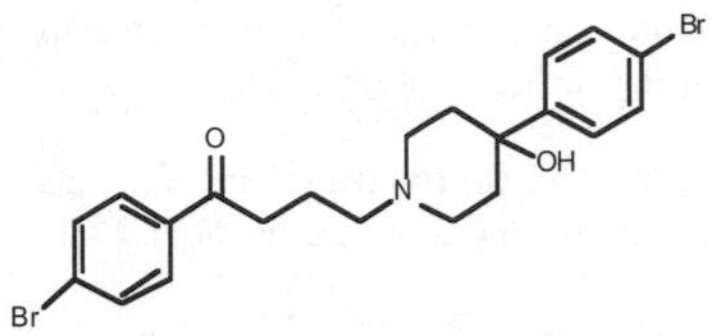

Brompheniramine (mol wt 319) An antihistaminic agent that competes with histamine for H_1 receptors.

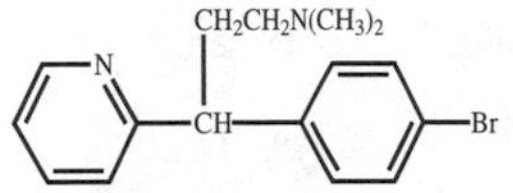

Bromphenol Blue See bromophenol blue.

Brompton's Cocktail An analgesic solution containing alcohol, morphine or heroin, cocaine, and phenothiazine for the control of pain in the terminally ill patient.

Bromsulphalein Test A test to measure the liver's ability to remove sulfobromophthalein from the blood.

Bromthymol Blue See bromothymol blue.

Bronalide A trade name for flunisolide, a corticosteroid.

Bronch- A prefix meaning the bronchus of the lung.

Bronchi Plural of bronchus.

Bronchial Pertaining to the bronchi and/or their branches.

Bronchial Asthma A form of allergy involving the lungs and bronchi. It is characterized by coughing, wheezing, choking, and shortness of breath.

Bronchial Capsule A trade name for a combination drug containing theophylline and guaifenesin, used as a bronchodilator.

Bronchiectasis A disorder of the bronchial tree characterized by irreversible dilatation and destruction of the bronchial walls.

Bronchiole A branch of the bronchus in a lung.

Branchiolectasis Abnormal dilatation of bronchioles.

Bronchiolitis Inflammation of the bronchioles.

Bronchitis Inflammation of bronchi.

Bronchobid Duracaps A trade name for a combination drug containing theophylline and ephedrine hydrochloride, used as a bronchodilator.

Bronchoconstrictor Any agent or substance capable of narrowing the lumen of a bronchus.

Bronchodilator Substance or agent capable of dilating the lumen of a bronchus.

Bronchogenic Carcinoma One of the malignant lung tumors originating in a bronchial tube.

Bronchography Radiographic examination of the bronchial tubes after injection of radiopaque material into the bronchus.

Broncholith A bronchial calculus (stone).

Broncholithiasis The presence of bronchial calculi.

Bronchomycosis A fungal infection of the bronchi.

Bronchopathy Disease of the bronchial tubes.

Bronchoplegia The paralysis of muscular fibers in the walls of the bronchi.

Bronchopneumonia Inflammation of the lungs and bronchioles characterized by chills, fever, high pulse and respiratory rates, cough with bloody sputum, severe chest pain, and abdominal distention.

Bronchopulmonary Of or pertaining to the bronchi and the lungs of the respiratory system.

Bronchoscope A lighted tubular device for examination of the interior of the trachea and bronchi.

Bronchoscopy The visual examination of the bronchi using a bronchoscope.

Bronchospasm The narrowing and obstruction of the respiratory airway due to spasmodic constriction of the bronchial tubes.

Bronchostenosis Narrowing of the lumen of a bronchial tube.

Bronchovesicular Pertaining to the tubes and air sacs in the lung.

Bronchus Any one of the several large air passages in the lungs through which pass inhaled air and exhaled waste gases.

Brondecon Tablets A trade name for a combination drug containing oxtriphylline and guaifenesin.

Broniten Mist A trade name for epinephrine bitartrate, an antitussive agent.

Bronkaid Mist A trade name for epinephrine, a bronchodilator.

Bronkephrine A trade name for ethyl norepinephrine hydrochloride, a bronchodilator.

Bronkodyl A trade name for theophylline, a bronchodilator that relaxes smooth muscle of the bronchial airways and pulmonary blood vessels.

Bronkometer A trade name for isoetharine mesylate, a beta-2 adrenergic agonist used as a bronchodilator and an antiasthmatic agent.

Bronkosol A trade name for isoetharine hydrochloride, a bronchodilator that relaxes bronchial smooth muscle by acting on beta-2 adrenergic receptors.

Bronopol (mol wt 200) An antibacterial and antifungal agent.

$$HOCH_2CBr(NO_2)CH_2OH$$

Bronsted Acid A molecule or an ion that acts as a proton donor.

Bronsted Base A molecule or an ion that acts as a proton acceptor.

Bronsted-Lowry Theory The theory states that acids act as proton donors and bases as proton acceptors.

Bronsted Plot A plot of the logarithm of the rate constant versus the negative logarithm of the dissociation constant.

Bronzed Diabetes A disease characterized by the presence of excessive amounts of glucose in the urine due to the accumulation of iron in the pancreas and liver.

Broparoestrol (mol wt 363) An estrogen.

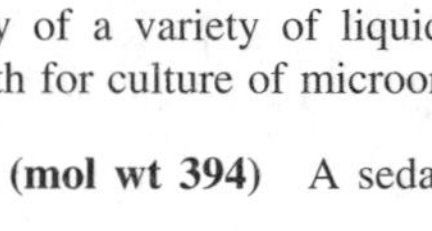

Brophylline A trade name for dyphylline that inhibits phosphodiesterase and increases cAMP concentration.

Broth Any of a variety of liquid media, e.g., nutrient broth for culture of microorganisms.

Brotizolam (mol wt 394) A sedative and hypnotic agent.

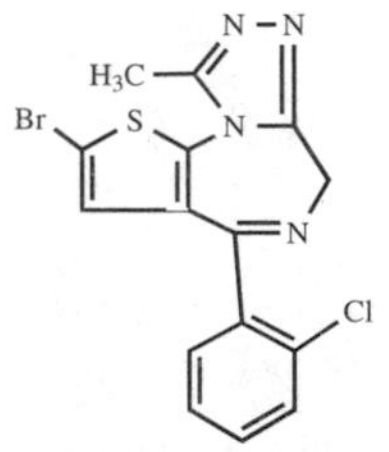

Brovincamine (mol wt 433) A peripheral vasodilator.

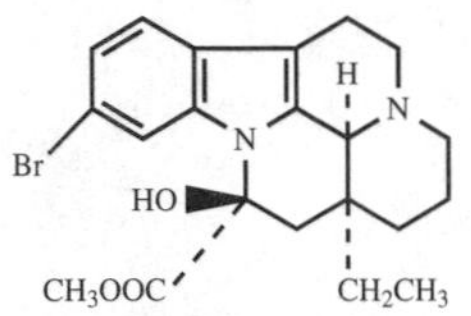

Brown Fat A heat-producing fat tissue stored in the body by hibernating animals (e.g., mammals and human).

Brown Hopps Modification Modification of Gram stains for staining bacteria in deparaffinized, hydrated tissue sections.

Brown Rust A disease of wheat caused by *Puccinia recondita.*

Brownian Motion The constant zigzag movement of small particles or organisms in a liquid medium due to molecular bombardment.

Brown's Tubes A set of glass tubes containing aqueous suspensions of barium sulphate in increasing concentrations, used for estimation of number of bacteria in a suspension by comparing optical density or turbidity of bacterial suspensions with concentrations of barium sulfate in the particular tube.

Broxyquinoline (mol wt 303) An antiseptic and disinfectant.

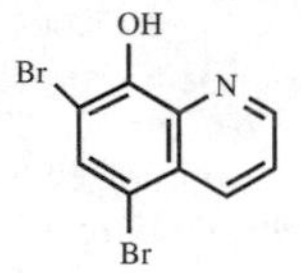

Brucella A genus of Gram-negative, aerobic, chemoorganotrophic, catalase-positive bacteria.

Brucellin An antigenic polysaccharide obtained from cultures of *Brucella* species.

Brucellin Test A skin test (analogous to the tuberculin test) used for detecting *Brucella*-induced hypersensitivity.

Brucellosis A disease caused by any of several species of *Brucella* and characterized by fever, chills, sweating, and weakness.

Brucine (mol wt 394) A CNS stimulant.

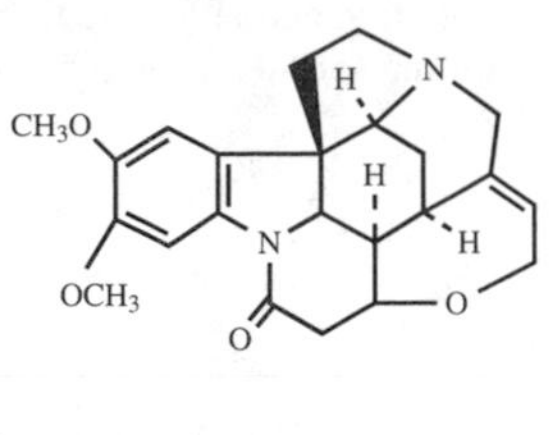

Brufen A trade name for ibuprofen, an anti-inflammatory agent that inhibits prostaglandin synthesis.

Bruise A blood clot in a tissue.

BrUrd Abbreviation for 5-bromouridine.

Brush Border Dense covering of microvilli on the apical surface of epithelial cells in the intestine and kidney. The microvilli play an important role in the nutrient absorption by increasing the surface area of the cell.

Bruton's Agammaglobulinemia A sex-linked, inherited disorder characterized by the absence of gamma globulin in the blood.

Bryonia Dry root of *Bryonia alba* used as a cathartic and purgative agent.

Bryrel A trade name for piperzine citrate, an anthelmintic agent that blocks neuromuscular action.

BS Abbreviation for 1. blood sugar; 2. blood agar.

BSA Abbreviation for 1. bismuth-soluble agar; 2. bis-trimethyl-silyl-acetamide; 3. bovine serum albumin.

Bsa29I A restriction endonuclease from *Bacillus* species 29 with the following specificity:

↓
5'........AT CGAT........3'
3'........TAGC TA........5'
↑

BsaAI A restriction endonuclease isolated from *Bacillus stearothermophilus* with the following specificity:

↓
5′..........PyAC GTPu ....3′
3′..........PuTG CAPy ....5′
↑

BsaBI A restriction endonuclease isolated from *Bacillus stearothermophilus* with the following specificity:

↓
5′..........GATNN NNATC ....3′
3′..........CTANN NNTAG..........5′
↑

BsaHI A restriction endonuclease from *Bacillus stearothermophilus* with the following specificity:

5'........GPuCGPyC........3'
3'........CPyGCPuG........5'

BsaJ1 A restriction endonuclease isolated from *Bacillus stearothermophilus* with the following specificity:

5′..........CCNNGG..........3′
3′..........GGNNCC..........5′

BsaMI A restriction endonuclease from *Bacillus stearothermophilus* with the following specificity:

5'........GAATGCN........3'
3'........CTTACGN........5'

BsaOI A restriction endonuclease from *Bacillus stearothermophilus* with the following specificity:

5'........CG(A/G)(T/C)CG........3'
3'........GC(T/C)(A/G)GC........5'

BsaPI (MboI) A restriction endonuclease isolated from *Bacillus stearothermophilus* with the same specificity as MboI.

BsaWI A restriction endonuclease from *Bacillus stearothermophilus* W with the following specificity:

5'........(A/T)CCGG(A/T)........3'
3'........(T/A)GGCC(T/A)........5'

BscI (ClaI) A restriction endonuclease isolated from *Bacillus species* with the following specificity:

5′..........ATCGAT..........3′
3′..........TAGCTA..........5′

Bsc4I A restriction endonuclease from *Bacillus schlegelii* 4 with the following specificity:

5'........CCNNNNNNNGG........3'
3'........GGNNNNNNNCC........5'

Bsc91I A restriction endonuclease from *Bacillus* species 91 with the following specificity:

5'........GAAGAC$(N)_2$........3'
3'........CTTCTG$(N)_6$........5'

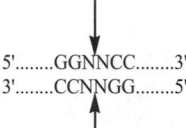

BscBI A restriction endonuclease from *Bacillus* species A11 with the following specificity:

5'........GGNNCC........3'
3'........CCNNGG........5'

BscFI A restriction endonuclease from *Bacillus* species with the following specificity:

5'........GATC........3'
3'........CTAG........5'

BSDL Abbreviation bile-salt-dependent lipase.

BseI (HaeIII) A restriction endonuclease isolated from *Bacillus stearothermophilus* strain 82 with the following specificity:

5′..........GGCC..........3′
3′..........CCGG..........5′

BseII (HpaI) A restriction endonuclease isolated from *Bacillus stearothermophilus* strain 82 with the following specificity:

5′..........GTTAAC..........3′
3′..........CAATTG..........5′

Bse8I A restriction endonuclease from *Bacillus* species 8 with following specificity:

5'........GATNNNNATC........3'
3'........CTANNNNTAG........5'

Bse21I A restriction endonuclease from *Bacillus* species 21 with the following specificity:

5'........CCTNAGG........3'
3'........GGANTCC........5'

Bse118I A restriction endonuclease from *Bacillus* species 118 with the following specificity:

5'........RCCGGY........3'
3'........YGGCCR........5'

R= A or G Y=C or T

BseAI A restriction endonuclease from *Bacillus stearothermophilus* with the following specificity:

5'........TCCGGA........3'
3'........AGGCCT........5'

BseCI A restriction endonuclease from *Bacillus stearothermophilus* species with the following specificity:

5'........ATCGAT........3'
3'........TAGCTA........5'

BseDI A restriction endonuclease from *Bacillus stearothermophilus* RFL 1434 with the following specificity:

5'........CCNNGG........3'
3'........GGNNCC........5'

BseNI A restriction endonuclease from *Bacillus species N* with the following specificity:

5'........ACTGGN........3'
3'........TGACCN........5'

BsePI A restriction endonuclease isolated from *Bacillus stearothermophilus* P6 with the following specificity:

5′...........GCGCGC...........3′
3′...........CGCGCG...........5′

BseRI A restriction endonuclease from *Bacillus* species R with the following specificity:

5'........GAGGAG$(N)_{10}$........3'
3'........CTCCTC$(N)_{8}$........5'

BSF Abbreviation for B cell-stimulating factor.

BSF-1 Abbreviation for B cell-stimulating factor 1

BSF-2 Abbreviation for B cell-stimulating factor 2.

BsgI A restriction endonuclease from *Bacillus sphaericus* NEB 581 with the following specificity:

5'........GTGCAG$(N)_{16}$........3'
3'........CTCGTC$(N)_{14}$........5'

Bsh1236I A restriction endonuclease from *Bacillus sphaericus* with the following specificity:

5'........CGCG........3'
3'........GCGC........5'

Bsh1285I A restriction endonuclease from *Bacillus sphaericus* RFL 1285 with the following specificity:

5'........CGPuPyCG........3'
3'........GCPyPuGC........5'

Bsh1365I A restriction endonuclease from *Bacillus sphaericus* RFL 1365 with the following specificity:

5'........GATNNNNATC........3'
3'........CTANNNNTAG........5'

BshAI (HaeIII) A restriction endonuclease from *Bacillus sphaericus* with the same specificity as HaeIII.

BshBI (HaeIII) A restriction endonuclease from *Bacillus sphaericus* with the same specificity as HaeIII.

BshCI (HaeIII) A restriction endonuclease from *Bacillus sphaericus* with the same specificity as HaeIII.

BshDI (HaeIII) A restriction endonuclease from *Bacillus sphaericus* with the same specificity as HaeIII.

BshEI (HaeIII) A restriction endonuclease from *Bacillus sphaericus* with the same specificity as HaeIII.

BshFI (HaeIII) A restriction endonuclease from *Bacillus sphaericus* with the same specificity as HaeIII.

BshNI A restriction endonuclease from *Bacillus sphaericus* Tk 4-5 with the following specificity:

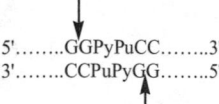

5'........GGPyPuCC........3'
3'........CCPuPyGG........5'

Bsi1854I A restriction endonuclease isolated from *Bacillus subtilis* with the following specificity:

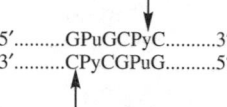

5'..........GPuGCPyC..........3'
3'..........CPyCGPuG..........5'

BsiBI A restriction endonuclease from *Bacillus species* with the following specificity:

5'........GATNNNNATC........3'
3'........CTANNNNTAG........5'

BsiC1 A restriction endonuclease isolated from *Bacillus* species with the following specificity:

5'............TTCGAA............3'
3'............AAGCTT............5'

BsiEI A restriction endonuclease from *Bacillus stearothermophilus* with the following specificity:

5'........CGPuPyCG........3'
3'........GCPyPuGC........5'

BsiHKAI A restriction endonuclease from *Bacillus stearothermophilus* with the following specificity:

5'........G(A/T)GC(A/T)C........3'
3'........C(T/A)CG(T/A)G.........5'

BsiLI A restriction endonuclease from *Bacillus* species with the following specificity:

5'........CC(A/T)GG........3'
3'........GG(T/A)CC........5'

BsiMI A restriction endonuclease from *Bacillus* species with the following specificity:

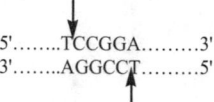

5'........TCCGGA.........3'
3'........AGGCCT.........5'

BsiQI A restriction endonuclease from *Bacillus* species with the following specificity:

5'........TGATCA........3'
3'........ACTAGT........5'

BsiSI A restriction endonuclease from *Bacillus stearothermophilus* with the following specificity:

5'........CCGG........3'
3'........GGCC........5'

BsiW1 A restriction endonuclease isolated from *Bacillus species* with the following specificity:

5'...........CGTACG...........3'
3'...........GCATGC...........5'

BsiXI A restriction endonuclease from *Bacillus* species with the following specificity:

5'........ATCGAT........3'
3'........TAGCTA........5'

BsiY1 A restriction endonuclease isolated from *Bacillus* species with the following specificity:

5'...CCNNNNNNNGG...3'
3'...GGNNNNNNNCC...5'

BsiZI A restriction endonuclease from *Bacillus* species with the following specificity:

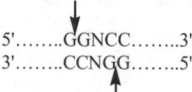

5'........GGNCC........3'
3'........CCNGG........5'

BSL Abbreviation for blood sugar level.

BslI A restriction endonuclease from *Bacillus* species with the following specificity:

5'........CCNNNNNNNGG........3'
3'........GGNNNNNNNCC........5'

BsmI A restriction endonuclease isolated from *Bacillus stearothermophilus* with following specificity:

5′.........GAATGCN........3′
3′.........CTTACGN.........5′

BsmA1 A restriction endonuclease isolated from *Bacillus stearothermophilus* with following specificity:

5′.........GTCTC(1N)........3′
3′........CAGAG(5N)........5′

BsmBI A restriction endonuclease from *Bacillus stearothermophilus* B61 with the following specificity:

5'........CGTCTC$(N)_1$........3'
3'........GCAGAG$(N)_5$.......5'

BsmFI A restriction endonuclease from *Bacillus stearothermophilus* F with the following specificity:

5'........GGGAC$(N)_{10}$........3'
3'........CCCTG$(N)_{14}$.......5'

BSO Abbreviation for buthionine sulphoximine.

BsoBI A restriction endonuclease from *Bacillus stearothermophilus* F with the following specificity:

5'........CYCGRG........3'
3'........GRGCYC........5'

R= A or G Y= C or T

BsoFI A restriction endonuclease from *Bacillus stearothermophilus* with the following specificity:

5'........GCNGC........3'
3'........CGNCG........5'

BSP Abbreviation for bromsulphalein.

Bsp2I (ClaI) A restriction endonuclease from *Bacillus* species RFL2 with the same specificity as ClaI.

Bsp4I (ClaI) A restriction endonuclease from *Bacillus* species RFL4 with the same specificity as ClaI.

Bsp5I (HpaII) A restriction endonuclease from *Bacillus* species RFL5 with the same specificity as HpaII.

Bsp6I (Fnu4HI) A restriction endonuclease from *Bacillus* species RFL6 with the following specificity:

5′..........GCNGC..........3′
3′..........CGNCG..........5′

Bsp12I (SacII) A restriction endonuclease from *Bacillus* species RFL12 with the same specificity as SacII.

Bsp13I A restriction endonuclease from *Bacillus* species 13 with the following specificity

5'........TCCGGA........3'
3'........AGGCCT........5'

Bsp16I (EcoRV) A restriction endonuclease from *Bacillus* species RFL16 with the same specificity as EcoRV.

Bsp17I (PstI) A restriction endonuclease from *Bacillus* species RFL17 with the same specificity as PstI.

Bsp18I (MboI) A restriction endonuclease from *Bacillus* species RFL18 with the same specificity as MboI.

Bsp19I A restriction endonuclease from *Bacillus* species 19 with the following specificity:

5'........CCATGG........3'
3'........GGTACC........5'

Bsp21I (Cfr10I) A restriction endonuclease from *Bacillus* species RFL21 with the following specificity:

5′..........PuCCGGPy..........3′
3′..........PyGGCCPu..........3′

Bsp30I (BamHI) A restriction endonuclease from *Bacillus* species RFL30 with the same specificity as BamHI.

Bsp63I (PstI) A restriction endonuclease isolated from *Bacillus sphaericus* 63 with the same specificity as PstI.

Bsp64I ((MboI) A restriction endonuclease isolated from *Bacillus sphaericus* 64 with the same specificity as MboI.

Bsp67I (MboI) A restriction endonuclease isolated from *Bacillus sphaericus* 67 with the same specificity as MboI.

Bsp68I A restriction endonuclease from *Bacillus megaterium* RFL 68 with the following specificity:

5'.........TCGCGA........3'
3'.........AGCGCT........5'

Bsp71I (HaeIII) A restriction endonuclease isolated from *Bacillus sphaericus* 71 with the same specificity as HaeIII.

Bsp74I (MboI) A restriction endonuclease isolated from *Bacillus sphaericus* 74 with the same specificity as MboI.

Bsp76I (MboI) A restriction endonuclease isolated from *Bacillus sphaericus* 76 with the same specificity as MboI.

Bsp78I (PstI) A restriction endonuclease isolated from *Bacillus sphaericus* 78 with the same specificity as PstI.

Bsp105I (MboI) A restriction endonuclease isolated from *Bacillus sphaericus* 105 with the same specificity as MboI

Bsp106I (ClaI) A restriction endonuclease isolated from *Bacillus sphaericus* 106 with the same specificity as ClaI.

Bsp119I A restriction endonuclease from *Bacillus* species RFL 119 with the following specificity:

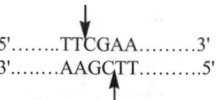

5'........TTCGAA.........3'
3'........AAGCTT..........5'

Bsp120I A restriction endonuclease from *Bacillus* species RFL 120 with the following specificity:

5'........GGGCCC........3'
3'.........CCCGGG.........5'

Bsp143I A restriction endonuclease from *Bacillus* species RFL 143 with the following specificity:

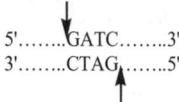

5'........GATC........3'
3'........CTAG........5'

Bsp1407I A restriction endonuclease from *Bacillus stearothermophilus* RFL 1407 with the following specificity:

5'........TGTACA........3'
3'........ACATGT........5'

Bsp211I (HaeIII) A restriction endonuclease isolated from *Bacillus sphaericus* 211 with the same specificity as HaeIII.

Bsp 1286I A restriction endonuclease isolated from *Bacillus sphaericus* IAM 1286 with the following specificity:

5′.......G(GAT)GC(CAT)C......3′
3′.......C(CTA)CG(GTA)G......5′

Bsp1720I A restriction endonuclease from *Bacillus* species 1720 with the following specificity:

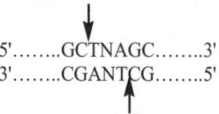

5'........GCTNAGC........3'
3'........CGANTCG........5'

BspAI (MboI) A restriction endonuclease isolated from *Bacillus sphaericus* with the same specificity as MboI.

BspA2I A restriction endonuclease from *Bacillus* species A2 with the following specificity:

5'........CCTAGG........3'
3'........GGATCC........5'

BspBI (PstI) A restriction endonuclease isolated from *Bacillus sphaericus* JL14 with the same specificity as PstI.

BspBII (AsuI) A restriction endonuclease isolated from *Bacillus sphaericus* JL14 with the following specificity :

5′..........GGNCC..........3′
3′..........CCNGG..........5′

BspCI A restriction endonuclease from *Bacillus* species C1 with the following specificity:

5'........CGATCG........3'
3'........GCTAGC........5'

BspDI A restriction endonuclease from *Bacillus* species with the following specificity:

5'........ATCGAT........3'
3'........TAGCTA........5'

BspEI A restriction endonuclease from *Bacillus* species with the following specificity:

5'........TCCGGA........3'
3'........AGGCCT........5'

BspHI A restriction endonuclease isolated from *Bacillus* species H with the following specificity:

5′........TCATGA........3′
3′........AGTACT........5′

BspLI A restriction endonuclease from *Bacillus* species RJ3-212 with the following specificity:

5'........GGNNCC........3'
3'........CCNNGG........5'

BspLU11I A restriction endonuclease from *Bacillus* species with the following specificity:

5'........ACATGT........3'
3'........TGTACA........5'

BspMI A restriction endonuclease isolated from *Bacillus species* M with the following specificity:

5′.......ACCTGC$(N)_4$.......3′
3′.......TGGACG$(N)_8$.......5′

BspMII A restriction endonuclease isolated from *Bacillus* species *M* with the following specificity:

5′..........TCCGGA..........3′
3′..........AGGCCT..........5′

BspRI (HaeIII) A restriction endonuclease isolated from *Bacillus sphaericus* R with the following specificity.

5′........GGCC........3′
3′........CCGG........5′

BspTI A restriction endonuclease from *Bacillus* species RFL 12651 with the following specificity:

5'........CTTAAG........3'
3'........GAATTC........5'

BspXI (ClaI) A restriction endonuclease isolated from *Bacillus sphaericus* X with the same specificity as ClaI.

BspXII (BclI) A restriction endonuclease isolated from *Bacillus sphaericus* X with the following specificity:

5′..........TGATCA..........3′
3′..........ACTAGT..........5′

BsrBI A restriction endonuclease from *Bacillus stearothermophilus* CPW 193 with the following specificity:

5'........GAGCGG........3'
3'........CTCGCC........5'

BsrBRI A restriction endonuclease from *Bacillus stearothermophilus* with the following specificity:

5'........GATNNNNATC........3'
3'........CTANNNNTAG........5'

BsrDI A restriction endonuclease from *Bacillus stearothermophilus* D70 with the following specificity:

5'........GCAATGNN........3'
3'........CGTTACNN........5'

BsrFI A restriction endonuclease from *Bacillus stearothermophilus* with the following specificity:

5'........PuCCGGPy........3'
3'........PyGGCCPu........5'

BsrGI A restriction endonuclease from *Bacillus stearothermophilus* with the following specificity:

5'........TGTACA........3'
3'........ACATGT........5'

BsrHI (BsePI) A restriction endonuclease isolated from *Bacillus stearothermophilus* H4 with the same specificity as BsePI.

BsrSI A restriction endonuclease from *Bacillus stearothermophilus* CPW 19 with the following specificity:

5'........ACTGG(1/-1)........3'

BSS Balanced salt solution used in cell cultures.

BssAI A restriction endonuclease from *Bacillus* species with the following specificity:

5'........RCCGGY........3'
3'........YGGCCR........5'

R= A or G Y= C or T

BssGI (BstXI) A restriction endonuclease isolated from *Bacillus stearothermophilus* G6 with the same specificity as BstXI.

BssGII (MboI) A restriction endonuclease isolated from *Bacillus stearothermophilus* G6 with the same specificity as MboI.

BssHI (XhoI) A restriction endonuclease isolated from *Bacillus stearothermophilus* H3 with the following specificity:

5′........CTCGAG........3′
3′........GAGCTC........5′

BssHII (BsePI) A restriction endonuclease isolated from *Bacillus stearothermophilus* PS with the following specificity:

5′........GCGCGC........3′
3′........CGCGCG........5′

BssKI A restriction endonuclease from *Bacillus stearothermophilus* with the following specificity:

5'........CCNGG........3'
3'........GGNCC........5'

BssSI A restriction endonuclease from *Bacillus stearothermophilus* S719 with the following specificity:

5'........CTCGTG........3'
3'........GAGCAC........5'

BssT1I A restriction endonuclease from *Bacillus stearothermophilus* T1 with the following specificity:

5'........CCWWGG........3'
3'........GGWWCC........5'

W= A or T

BST Abbreviation for bismuth sodium triglycolamate.

BstI (BamHI) A restriction endonuclease isolated from *Bacillus stearothermophilus* with the following specificity:

5′........GGATCC........3′
3′........CCTAGG........5′

Bst71I A restriction endonuclease from *Bacillus stearothermophilus* strain 71 with the following specificity:

5'........GGAGC$(N)_8$........3'
3'........CCTCG$(N)_{12}$........5'

Bst98I A restriction endonuclease from *Bacillus stearothermophilus* with the following specificity:

5'........CTTAAG........3'
3'........GAATTC........5'

Bst1107I A restriction endonuclease from *Bacillus stearothermophilus* RFL 1107 with the following specificity:

5'.......GTATAC........3'
3'........CATATG........5'

BstBI (AsuII) A restriction endonuclease isolated from *Bacillus stearothermophilus* B225 with the following specificity:

5′..........TTCGAA..........3′
3′..........AAGCTT..........5′

BstBAI A restriction endonuclease from *Bacillus stearothermophilus* BA with the following specificity:

5'........YACGTR........3'
3'........RTGCAY........5'

R=A or G Y= C or T

BstCI (HaeIII) A restriction endonuclease isolated from *Bacillus stearothermophilus* C1 with the same specificity as HaeIII.

BstDI (BstEII) A restriction endonuclease isolated from *Bacillus stearothermophilus* D428 with the same specificity as BstEII.

BstD102I A restriction endonuclease from *Bacillus stearothermophilus* D102 with the following specificity:

5'.......CCTGCTC(-3/-3).......3'

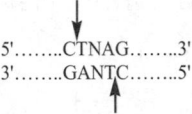

BstDEI A restriction endonuclease from *Bacillus stearothermophilus* DE with the following specificity:

5'........CTNAG........3'
3'........GANTC........5'

BstDSI A restriction endonuclease from *Bacillus stearothermophilus* DS with the following specificity:

5'........CCRYGG........3'
3'........GGYRCC........5'

R = A or G Y= C or T

BstEII A restriction endonuclease isolated from *Bacillus stearothermophilus* ET with the following specificity:

5′........GGTNACC........3′
3′........CCANTGG........5′

BstEIII (MboI) A restriction endonuclease isolated from *Bacillus stearothermophilus* ET with the same specificity as MboI.

BstFI (HindIII) A restriction endonuclease isolated from *Bacillus stearothermophilus* FH58 with the same specificity as HindIII.

BstF5I A restriction endonuclease from *Bacillus stearothermophilus* F5 with the following specificity:

5'........GGATG(2/0)........3'

BstGI (BclI) A restriction endonuclease isolated from *Bacillus stearothermophilus* with the following specificity:

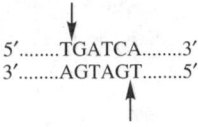

5′........TGATCA........3′
3′........AGTAGT........5′

BstHI (XhoI) A restriction endonuclease isolated from *Bacillus stearothermophilus* HI with the following specificity:

5′..........CTCGAG..........3′
3′..........GAGCTC..........5′

BstH2I A restriction endonuclease from *Bacillus stearothermophilus* H2 with the following specificity:

5'........RGCGCY........3'
3'........YCGCGR........5'

BstHPI A restriction endonuclease from *Bacillus stearothermophilus* HP with the following specificity:

5'.......GTTAAC........3
3'.......CAATTG........5'

BstJI (HaeIII) A restriction endonuclease isolated from *Bacillus stearothermophilus* J460 with the same specificity as HaeIII.

BstKI (BclI) A restriction endonuclease isolated from *Bacillus stearothermophilus* K554 with the same specificity as BclI.

BstLI (XhoI) A restriction endonuclease isolated from *Bacillus stearothermophilus* L95 with the same specificity as XhoI.

BstMI (ScaI) A restriction endonuclease isolated from *Bacillus stearothermophilus* M571 with the following specificity:

5′..........AGTACT..........3′
3′..........TCATGA..........5′

BstMCI A restriction endonuclease from *Bacillus stearothermophilus* MC with the following specificity:

5'........CGRYCG........3'
3'........GCYRGC........5'

R= A or G Y = C or T

BstNI (EcoRII) A restriction endonuclease isolated from *Bacillus stearothermophilus* with following specificity:

5′..........CC(A/T) GG..........3′
3′..........GG(T/A) CC..........5′

BstNSI A restriction endonuclease from *Bacillus stearothermophilus* 1161NS with the following specificity:

5'........RCATGY........3
3'........YGTACR........5'

R= A or G Y= C or T

BstOI A restriction endonuclease from *Bacillus stearothermophilus* with the following specificity:

5'........CC(A/T)GG........3'
3'........GG(T/A)CC........5'

BstPI (BstEII) A restriction endonuclease isolated from *Bacillus stearothermophilus* with the following specificity:

5′..........GGTNACC..........3′
3′..........CCANTGG..........5′

BstQI (BamHI) A restriction endonuclease isolated from *Bacillus stearothermophilus* Q407 with the same specificity as BamHI.

BstRI (EcoRV) A restriction endonuclease isolated from *Bacillus stearothermophilus* R463 with the same specificity as EcoRV.

BstSI (AvaI) A restriction endonuclease isolated from *Bacillus stearothermophilus* S183 with the following specificity:

5′..........CPyCGPuG..........3′
3′..........CPuGCPyC..........5′

BstSFI A restriction endonuclease from *Bacillus stearothermophilus* SF with the following specificity:

5'........CTRYAC........3'
3'........GAYRTC........5'

R=A or G Y=C or T

BstSNI A restriction endonuclease from *Bacillus stearothermophilus* SN with the following specificity:

5'.......TAC↓GTA........3'
3'........ATG↑CAT........5'

BstTI (BstXI) A restriction endonuclease isolated from *Bacillus stearothermophilus* T12 with the same specificity as BstXI.

BstUI A restriction endonuclease isolated from *Bacillus stearothermophilus* with the following specificity:

5′..........CG↓CG..........3′
3′..........GC↑GC..........5′

BstVI (XhoI) A restriction endonuclease isolated from *Bacillus stearothermophilus* V with the same specificity as XhoI.

BstWI (EcoNI) A restriction endonuclease isolated from *Bacillus stearothermophilus* W574 with the following specificity:

5′..........CCTNN↓NNNAGG..........3′
3′..........GGANNN↑NNTCC..........5′

BstXI A restriction endonuclease isolated from *Bacillus stearothermophilus* XI with the following specificity:

5′........CCANNNNN↓NTGG........3′
3′........GGTN↑NNNNNACC........5′

BstXII (MboI) A restriction endonuclease isolated from *Bacillus stearothermophilus* XI with the same specificity as MboI.

BstYI A restriction endonuclease isolated from *Bacillus stearothermophilus* Y406 with the following specificity:

5′...........Pu↓GATCPy..........3′
3′...........PyCTAG↑Pu..........5′

BstZI (XmaIII) A restriction endonuclease isolated from *Bacillus stearothermophilus* Z130 with the following specificity:

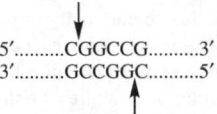

5′..........C↓GGCCG..........3′
3′..........GCCGG↑C..........5′

Bsu6I A restriction endonuclease from *Bacillus subtilis* 6V1 with following specificity:

5'........CTCTTC(1/4)↓........3'

Bsu15I A restriction endonuclease from *Bacillus subtilis* 15 with following specificity:

5'........AT↓CGAT........3'
3'........TAGC↑TA........5'

Bsu36I (SauI) A restriction endonuclease isolated from *Bacillus subtilis* with the following specificity:

5′..........CC↓TNAGG..........3′
3′..........GGANT↑CC..........5′

BsuBI (PstI) A restriction endonuclease isolated from *Bacillus subtilis* with the same specificity as PstI.

BsuEII (FnuDII) A restriction endonuclease isolated from *Bacillus subtilis* with the same specificity as FnuDII.

BsuFI (HpaII) A restriction endonuclease isolated from *Bacillus subtilis* with the following specificity:

5′..........C↓CGG..........3′
3′..........GGC↑C..........5′

BsuMI (XhoI) A restriction endonuclease isolated from *Bacillus subtilis* Marburg 168 with the same specificity as XhoI.

BsuRI (HaeIII) A restriction endonuclease isolated from *Bacillus subtilis* with the following specificity:

5′..........GG↓CC..........3′
3′..........CC↑GG..........5′

BTA Abbreviation for N-benzoyl tyrosine amide.

BTag Abbreviation for biotinylated tag control peptide.

BTB Agar Abbreviation for bromthymol blue agar. A medium that contains peptone, lactose, glucose, yeast extract, Maranil, thiosulphate, and bromthymol blue for culture of enterobacteria.

BTEE Abbreviation for N-benzoyl-L-tyrosine ethyl ester, a substrate for assaying chymotrypsin.

bTF Abbreviation for bovine serum transferrin.

BthI (XhoI) A restriction endonuclease isolated from *Bacteriaum thermophilum* RU326 with the following specificity:

5′..........C↓TCGAG..........3′
3′..........GAGCT↑C..........5′

BthII (BinI) A restriction endonuclease isolated from *Bacterium thermophilum* RU326 with following specificity:

5′..........GGATC(4N)↓..........3′
3′..........CCTAG(5N)↑..........5′

BTK Abbreviation for Bruton's tyrosine kinase.

BTRP Abbreviation for bovine TRP (transient receptor potential) protein.

BTS Abbreviation for Benedict's test for sugar.

BTSMC Abbreviation for bovine tracheal smooth muscle cells.

BTU Abbreviation for British Thermal Unit. The amount of heat required to raise the temperature of one pound of water one degree Fahrenheit.

***btu*B Protein** An outer membrane protein encoded by the btuB gene in *E. coli*. It acts as a receptor for bacteriophage BF23 and for various colicins, and it is also involved in the uptake of vitamin B_{12}.

BTX Abbreviation for benzene/toluene/xylene mixture.

BTX-B Abbreviation for brevetoxin-B.

B-type Inclusion Body A type of cellular inclusion body in the cells infected with pox virus.

Bu Abbreviation for butyryl.

BU Abbreviation for 5-bromouracil.

Bubble Column Fermenter A fermenter in which sterile air is bubbled into the bottom of a tall column for mixing and circulating of the culture.

Bubo An enlarged, inflamed lymph node.

Buboes Swollen lymph nodes.

Bubonic Plague A form of plague characterized by the painful swelling of lymph nodes.

Bucetin (mol wt 223) An analgesic and antipyretic agent.

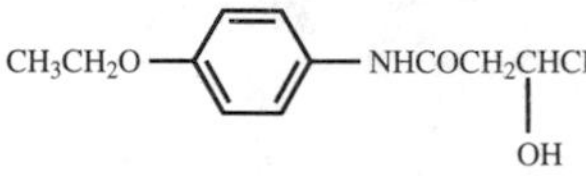

BuChE Abbreviation for butyrylcholine esterase.

Buchu Dried leaves of *Barosma betulina* used as a urinary antiseptic and diuretic agent.

Bucillamine (mol wt 223) An immunomodulator.

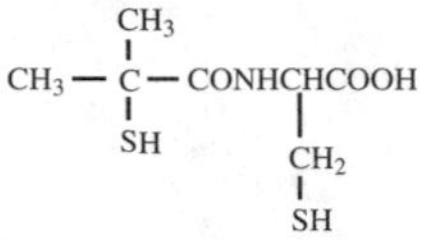

Bucladesine (mol wt 469) A cardiostimulant.

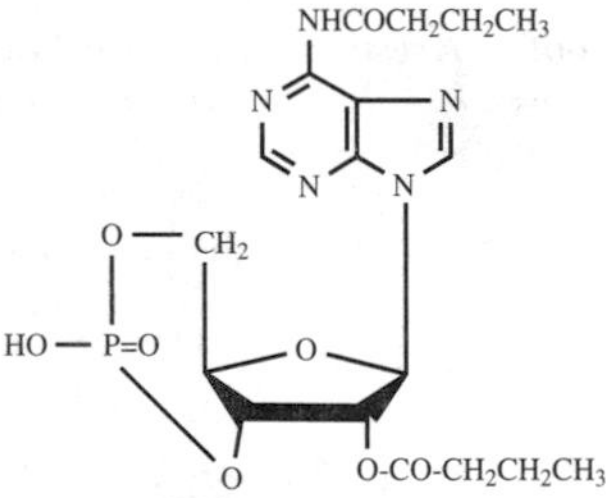

Bucladin-S A trade name for buclizine hydrochloride, an antiemetic agent.

Buclizine (mol wt 433) An antiemetic agent.

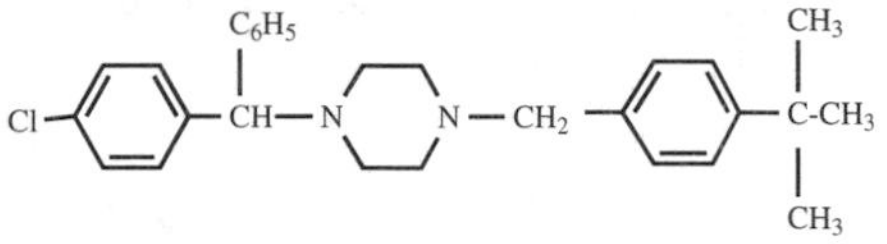

Buclosamide (mol wt 228) An antifungal agent.

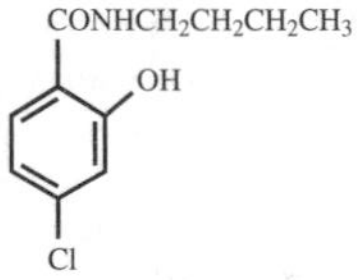

Bucloxic Acid (mol wt 295) An anti-inflammatory agent.

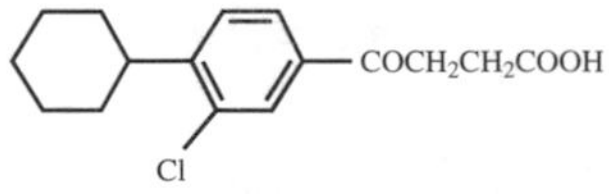

Bucolome (mol wt 266) An anti-inflammatory agent.

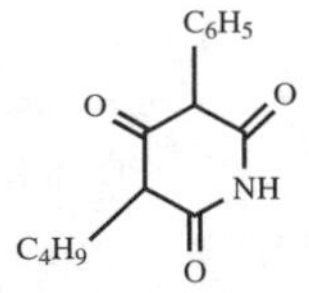

Bucumolol (mol wt 305) An antianginal, antiarrhythmic agent.

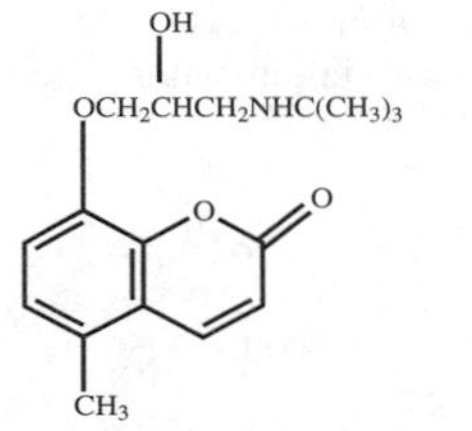

Budd-Chiari Syndrome A disorder of the hepatic circulation marked by venous obstruction and leading to liver enlargement.

Budding 1. A process of asexual reproduction in which a new organism is formed as an outgrowth from the parent organism. 2. A process of virus release without disruption of the cell.

Budesonide (mol wt 431) An anti-inflammatory agent.

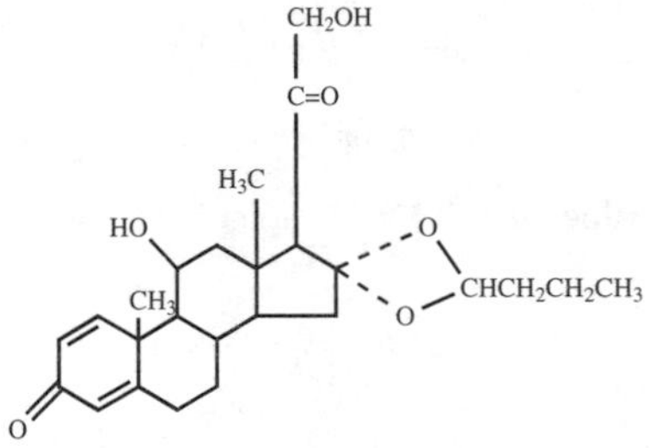

Budipine (mol wt 293) An antiparkinsonian agent.

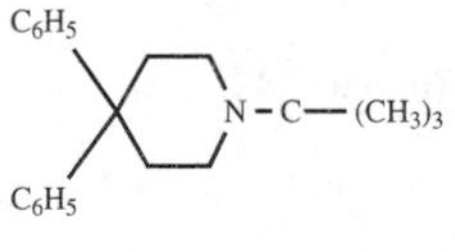

BUdr (BrdU) Abbreviation for 5-bromo-2-deoxyuridine.

Budralazine (mol wt 240) An antihypertensive agent.

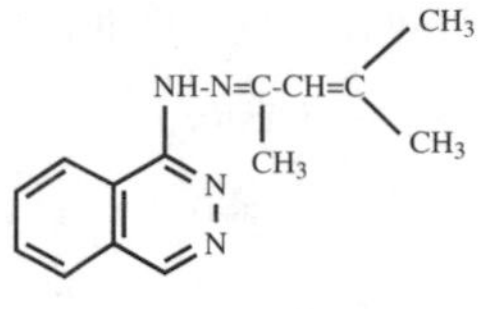

Bufalin (mol wt 387) A cardiotonic steroid from dried venom of the Chinese toad.

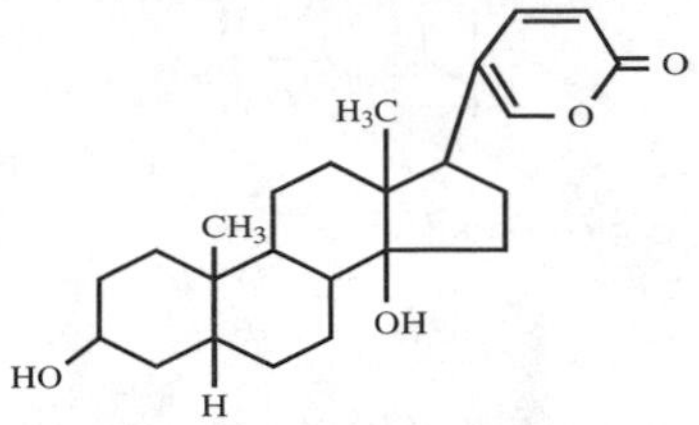

Bufeniode (mol wt 551) A vasodilator and an antihypertensive agent.

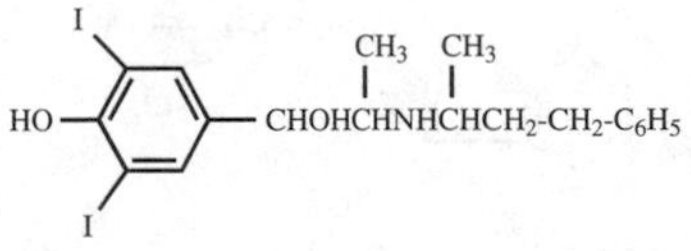

Bufexamac (mol wt 233) An anti-inflammatory and antipyretic agent.

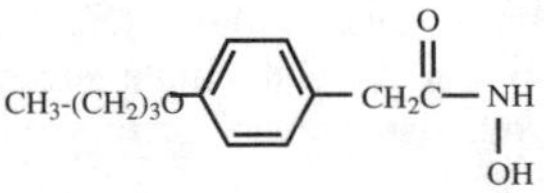

Buffer A solution containing a mixture of a weak acid and its conjugate weak base that is capable of resisting substantial changes in pH upon the addition of small amounts of acidic or basic substances.

Buffer Value The number of equivalents of either protons or hydroxyl ions that is required to change the pH of a buffer by one pH unit.

Bufferin A trade name for aspirin.

Buffex A trade name for aspirin.

Buffy Coat The thin layer of white cells formed at the surface of packed red cells when unclotted blood is centrifuged.

Buflomedil (mol wt 307) A peripheral vasodilator.

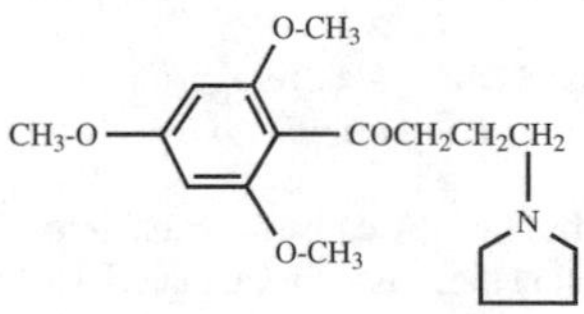

Buformin (mol wt 157) An antidiabetic agent.

$$CH_3CH_2CH_2CH_2NH\text{-}C(=NH)\text{-}NH\text{-}C(=NH)NH_2$$

Bufotoxin (mol wt 757) A toxin from the common European toad.

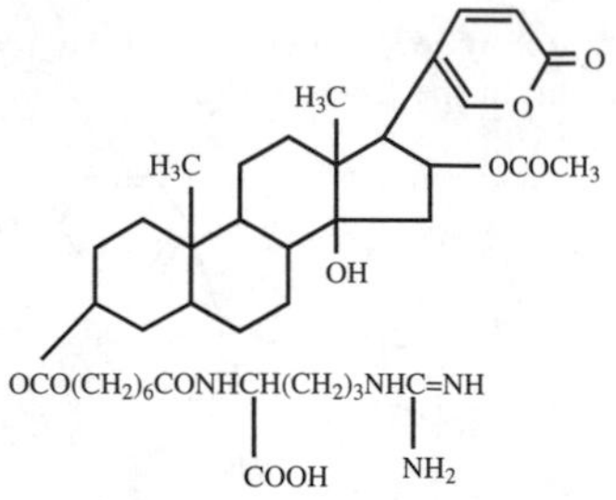

Bufuralol (mol wt 261) An antianginal and antihypertensive agent.

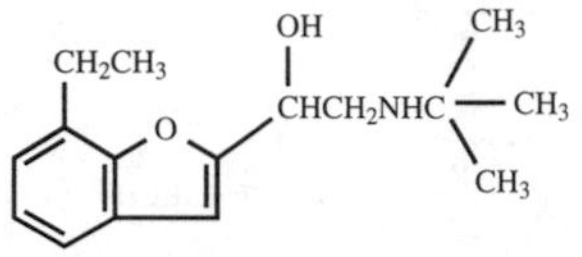

Building Block A molecule that serves as a structural unit in biopolymers, e.g., nucleotides, amino acids, or sugars.

Building Block Biomolecules Referring to nucleotides, amino acids, and simple sugars that are used for synthesis of biopolymers.

Builder Detergent A substance capable of increasing the effectiveness of a soap or a synthetic detergent by acting as a softener, or as a sequestering and buffering agent.

Bulbocapnine (nol wt 325) A peripheral dopamine-receptor blocking agent isolated from root of *Coyrdalis cava* (Papaveraceae).

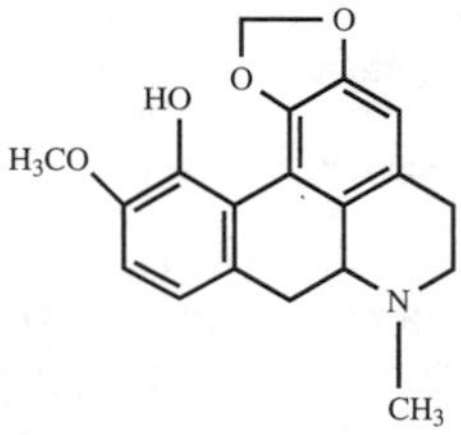

Bulgarian Buttermilk See bulgaricus milk.

Bulgaricus Milk A beverage made by fermenting milk with *Lactobacillus bulgaricus*.

Bulk Cathartic A cathartic capable of softening and increasing the mass of fecal material in the bowel.

Bulking Agent Chemically inert substances used to increase volume.

Bulla A thin-walled blister of the skin.

Bullate Bubble or blisterlike swellings.

Buller Phenomenon A form of dikaryotization in which a dikaryotic cell donates a nucleus to a monokaryotic (haploid) cell.

Bullgecin A low molecular weight glycopeptide that causes bulge formation in susceptible bacterial cells in the presence of β-lactam antibiotic.

Bullous Myringitis An inflammatory condition of the ear characterized by fluid-filled vesicles on the eardrum that cause sudden-onset and severe pain in the ear.

Bumadizon (mol wt 326) An analgesic, antipyretic, and anti-inflammatory agent.

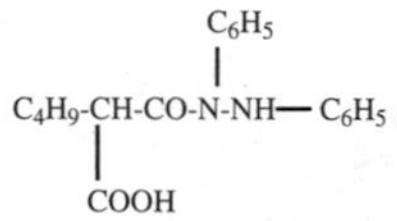

Bumetanide (mol wt 364) A diuretic agent that inhibits reabsorption of sodium and chloride.

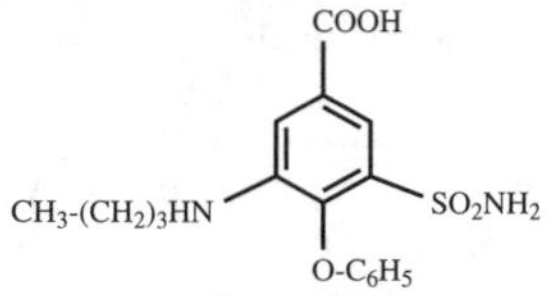

Bumex A trade name for bumetanide, a diuretic agent.

Buminate A trade name for human serum albumin, a blood-volume expander.

BUN Abbreviation for blood urea nitrogen.

Bunaftine (mol wt 326) An antiarrhythmic agent.

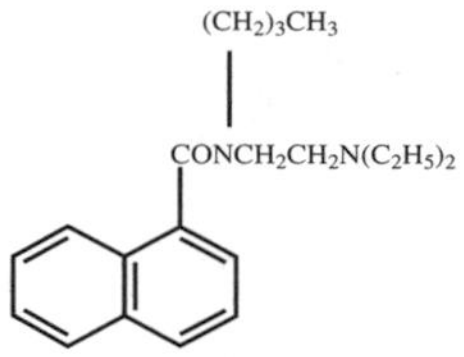

Bunamidine (mol wt 383) An anthelmintic agent.

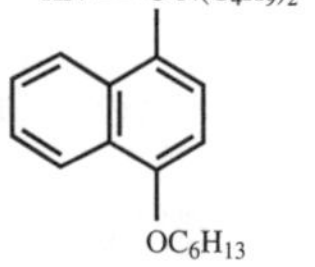

Bunamiodyl Sodium (mol wt 661) Compound used as radiopaque medium in diagnosis.

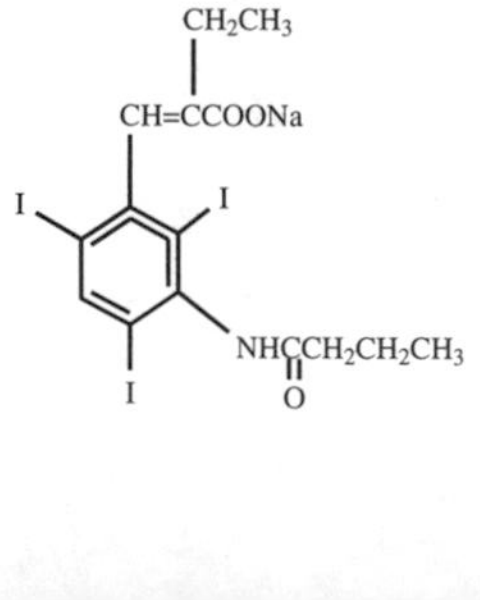

Bunazosin (mol wt 373) An antihypertensive agent.

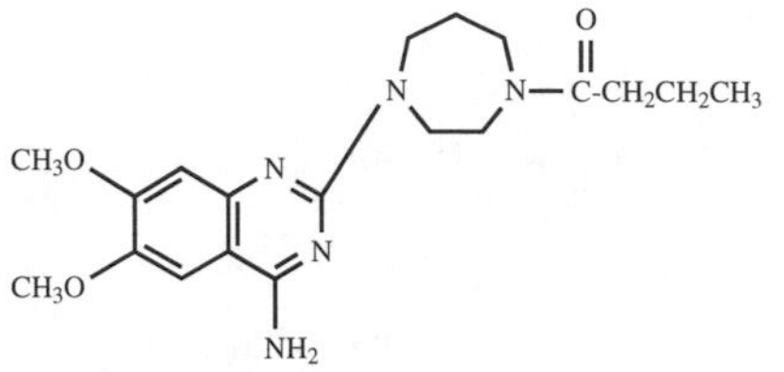

Bundle Protein Proteins that cause the association of actin molecules to form bundles.

Bundle Sheath A layer of cells surrounding the vascular bundle in a leaf of C_4 plants. The C_3 pathway is located in the bundle sheath cells.

Bungarotoxins Protein toxins of venom of the snake (*Bungarus multicinctus*). It binds with acetylcholinergic receptors.

Bunion An abnormal enlargement of the joint at the base of the great toe due to inflammation of bursa.

Bunitrolol (mol wt 248) An antihypertensive, antiarrhythmic, and antianginal agent.

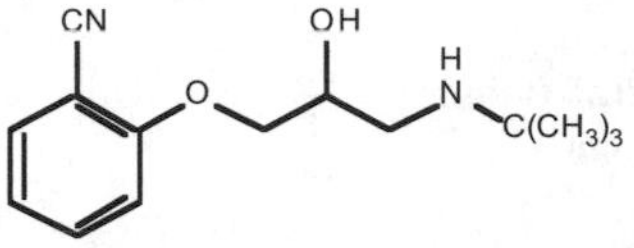

Bunyamwera Arbovirus One of a group of arthropod borne viruses that infect humans (carried by mosquitoes from rodent to human) and cause mild diseases, e.g., headache, low-grade fever, and rash.

Bunyaviridae A family of enveloped viruses with divided-RNA-genome (minus-stranded RNA), e.g., bunyamweravirus.

Bunyavirus A virus of the Bunyaviridae.

Buoyant Density The density of a macromolecule as determined by equilibrium density gradient centrifugation.

Buoyancy Factor The factor of $(1 - \bar{v}p)$ where v is the partial specific volume of the solute and p is the density of the solution.

Bupivacaine (mol wt 288) A local anesthetic agent.

Bupranolol (mol wt 272) An antihypertensive, antiglaucoma, antianginal, and antiarrhythmic agent.

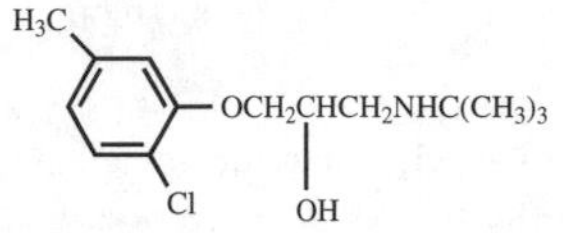

Buprenex A trade name for buprenorphine hydrochloride, an analgesic agent.

Buprenorphine (mol wt 468) A narcotic analgesic agent.

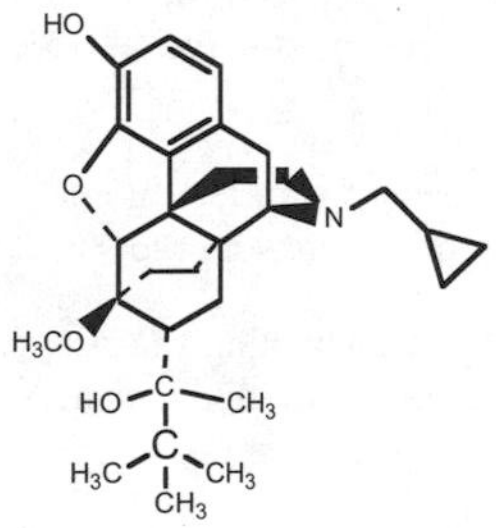

Bupropion (mol wt 240) An antidepressant.

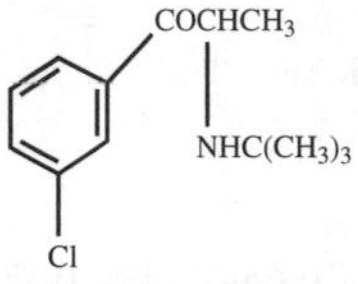

Buramate (mol wt 195) An anticonvulsant and antipsychotic agent.

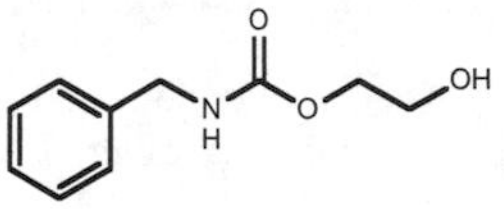

Burdon's Stain A stain used for staining bacterial intracellular lipid. A heat-fixed smear is first stained with sudan black B, washed, dried, and counterstained with safranin (lipid stains black; cytoplasm, red).

Buret A graduated tube with stopcock used for delivery of known volumes of liquid in titrations.

Burgundy Mixture An agricultural antifungal preparation consisting of a mixture of copper sulfate and sodium carbonate.

Burinex A trade name for bumetanide, a diuretic agent.

Burkitt's Lymphoma A malignant B cell lymphoma caused by Epstein-Barr virus; primarily affects children in Africa.

Burmate (mol wt 195) An anticonvulsant and antipsychotic agent.

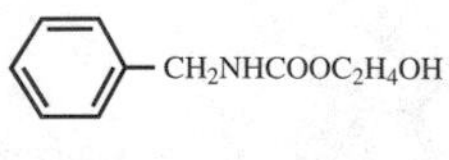

Burmex A trade name for bumetanide, diuretic agent that inhibits the reabsorption of sodium.

Burnet's Clonal Selection Theory See clonal selection theory.

Burow's Solution A preparation containing aluminum sulfate, acetic acid, calcium carbonate, and water; used as a topical astringent, antiseptic, and antipyretic for a wide variety of skin disorders.

Burr Cell One of a variety of cells that has spicules or tiny projections on the surface.

Bursa A fluid-containing sac between skin and muscle, between muscle and muscle, or around the joint that minimizes friction.

Bursal Equivalent The hypothetical organ or organs analogous to the bursa of Fabricius in nonavian species.

Bursa of Fabricius The hindgut organ located in the cloaca of birds that controls the ontogeny of B lymphocytes.

Bursectomy Removal of bursa by surgical procedure or by hormonal treatment.

Bursitis An inflammation of the bursa or connective tissue structure surrounding a joint.

Burst Size The average number of phage particles released per infected bacterial cell in a lytic infection.

Buruli Ulcer A chronic, progressive, granulomatous skin lesion caused by *Mycobacterium ulcerans.*

Buschke-Löwenstein Tumor A large cauliflower-like malignant but nonmetastasizing genital tumor of humans caused by Herpesvirus.

Buserelin (mol wt 1239) A synthetic nonapeptide, the gonad-stimulating agent used for the treatment of prostatic carcinoma.

Busodium A trade name for butabarbital solution, a sedative and hypnotic agent.

BuSpar A trade name for buspirone hydrochloride, an anti-anxiety agent.

Buspirone (mol wt 386) An anxiolytic agent.

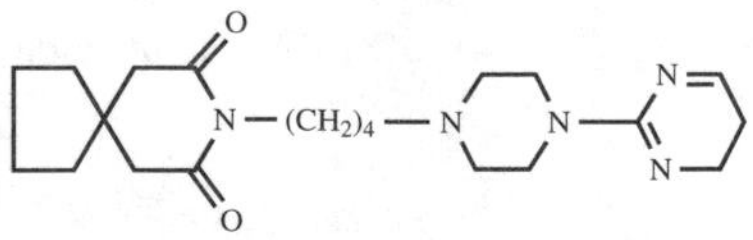

Busulfan (mol wt 246) An alkylating agent with antileukemic activity.

$CH_3SO_2O(CH_2)_4OSO_2CH_3$

Butabarbital sodium (mol wt 234) A sedative, hypnotic agent prescribed for the relief of anxiety, nervous tension, and insomnia.

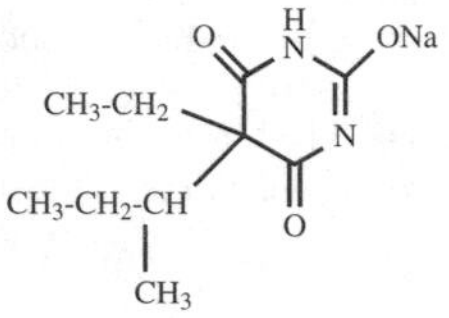

Butacaine (mol wt 306) A local anesthetic agent.

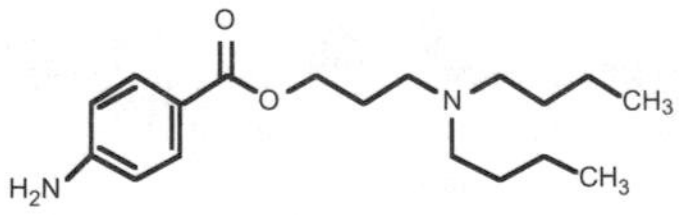

Butalbital (mol wt 224) A sedative and hypnotic agent.

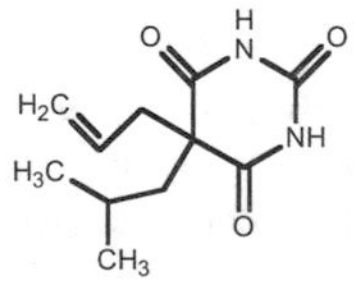

Butalan A trade name for butabarbital, a sedative and hypnotic agent.

Butallylonal (mol wt 303) A sedative and hypnotic agent.

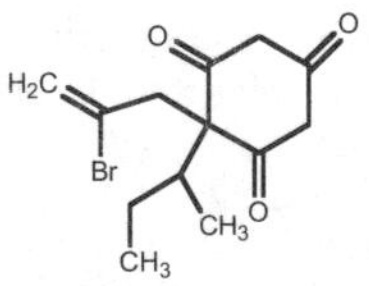

Butamben (mol wt 193) Anesthetic agent.

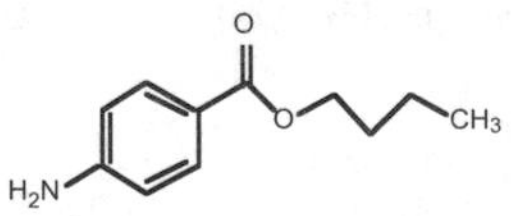

Butanal Dehydrogenase The enzyme that catalyzes the following reaction:

Butanal + CoA + $NADP^+$

⇅

Butanoyl-CoA + NADPH

Butanol (mol wt 74) An organic solvent (also known as butyl alcohol).

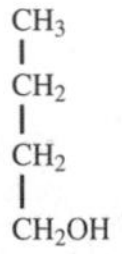

Butamirate (mol wt 307) An antitussive agent.

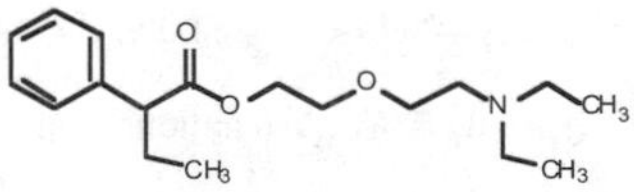

Butanilicaine (mol wt 255) An anesthetic agent.

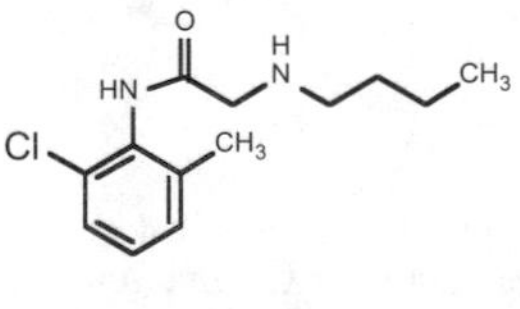

Butaverine (mol wt 289) An antispasmodic agent.

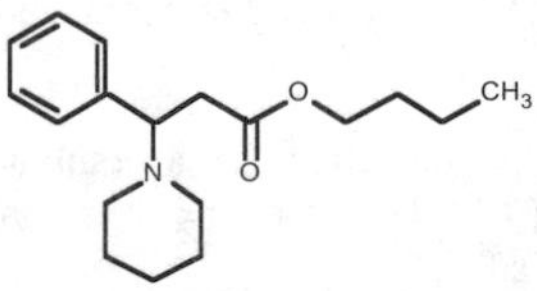

Butazolamide (mol wt 250) A diuretic agent.

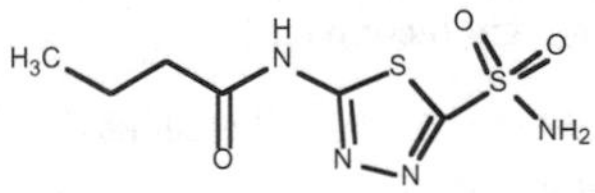

Butenafine (mol wt 317) An antifungal agent.

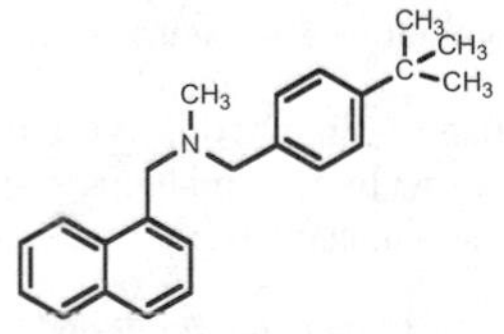

α-β-*trans*-Butenoyl-ACP An intermediate in the biosynthesis of fatty acid.

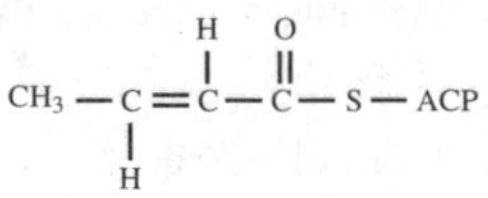

Butethal (mol wt 212) A sedative and hypnotic agent.

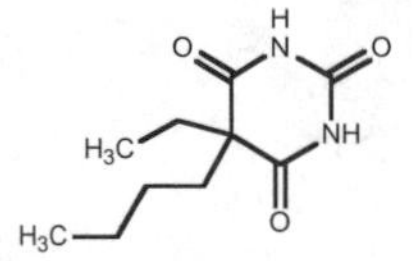

Butethamate (mol wt 263) An antitussive agent.

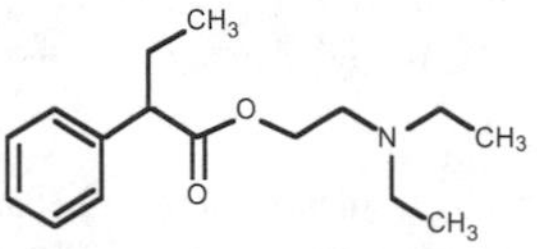

Buthiazide (mol wt 354) A diuretic and antihypertensive agent.

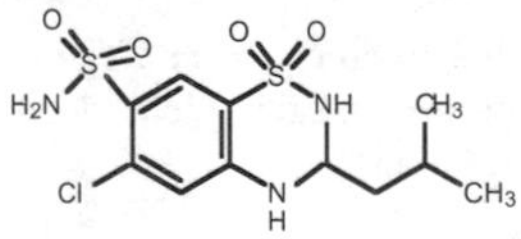

Butibufen (mol wt 220) An anti-inflammatory agent.

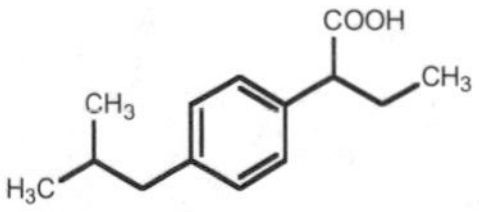

Butoconazole (mol wt 412) An antifungal agent.

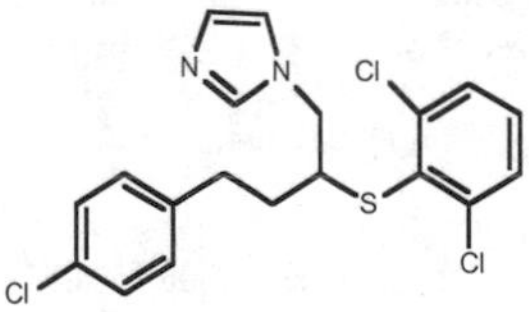

Butofilolol (mol wt 311) An antihypertensive agent.

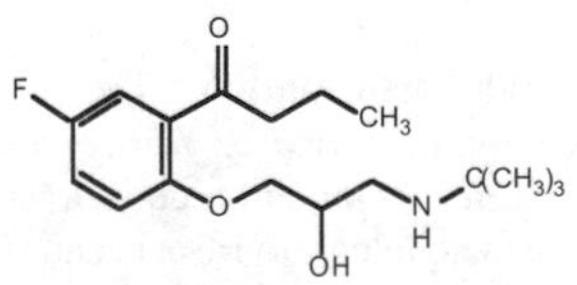

Butorphanol (mol wt 327) An analgesic and antitussive agent.

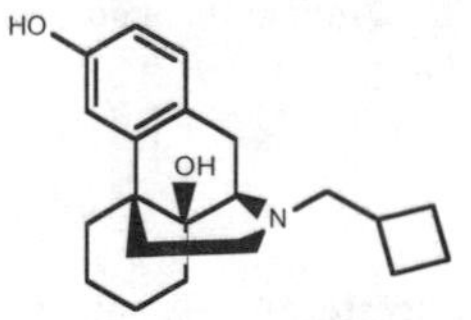

Butoxycaine (mol wt 293) An anesthetic agent.

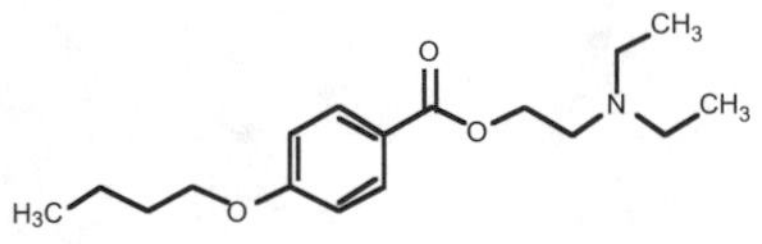

Butter Fat The oil portion of the milk of mammals.

Butterfly Mode of DNA Replication The replication of certain animal virus DNA in which the partially replicated DNA resembles the shape of a butterfly.

Buttermilk A fermented product obtained by fermentation of pasteurized skimmed milk with *Streptococcus lactis*, *S. cremoris*, and *Leuconostoc cremoris.*

Butyl Alcohol Referring to butanol.

Butyl Alcohol Fermentation The fermentation of glucose by bacteria that yields acetone, isopropanol, and n-butanol.

Butyloxycarbonyl Chloride (mol wt 136) A reagent used for protection of amino groups in the chemical synthesis of peptide.

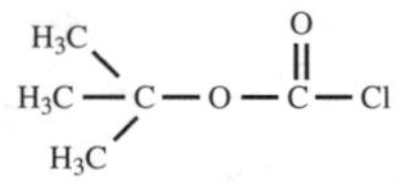

Butyrate-CoA Ligase See butyryl-CoA ligase.

Butyrate Kinase The enzyme that catalyzes the following reaction:

ATP + butanoate

⇅

ADP + butanoyl phosphate

Butyric Acid (mol wt 88) An organic acid.

$CH_3CH_2CH_2COOH$

Butyric Acid Fermentation The fermentation of glucose by *Butyrivibrio fibrisolvens, Fucobacterium nucleatum*, and certain species of *Clostridium* that yields butyric acid and acetic acid.

Butyrivibrio A genus of bacteria of the family Bacteroidaceae.

Butyryl-ACP An intermediate in fatty acid biosynthesis.

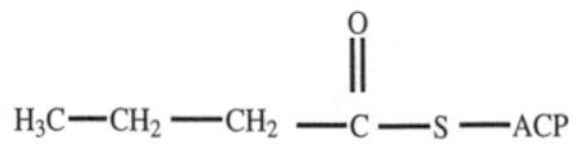

Butyrylcholineesterase A relative nonspecific plasma and liver acetylcholinesterase.

Butyryl-CoA Dehydrogenase The enzyme that catalyzes the following reaction:

Butanoyl-CoA + acceptor

⇅

2-Butenoyl-CoA + reduced acceptor

Butyryl-CoA Ligase The enzyme that catalyzes the following reaction:

ATP + butyrate + CoA

⇅

AMP + PPi + butyrate-CoA

Butyryl-CoA Synthetase See butyryl-CoA ligase.

Buzepide (mol wt 336) An anticholinergic agent.

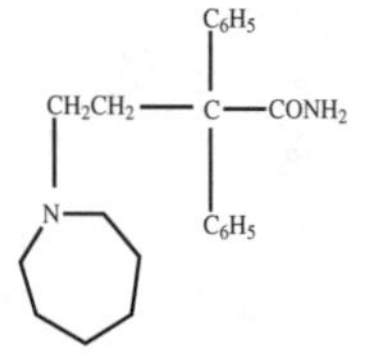

BV Abbreviation for 1. biliverdin; 2. baculovirus; 3. bacitracin-V; 4. bed volume.

BVAP Abbreviation for a combination drug containing BCNU, velban, adriamycin, and prednisolone.

BVCPP Abbreviation for a combination drug containing BCNU, velban, cytoxan, procarbazine and prednisone.

BVCPPB Abbreviation for a combination drug containing BCNU, velban, cytoxan, procarbazine, prednisone, and bleomycin.

βVLDL Abbreviation for beta-migrating very low density lipoprotein.

BVU Abbreviation for bromoisovaleryl urea.

BVX Abbreviation for bacitracin-X.

BXSB Mouse Abbreviation for a mouse strain genetically prone to develop lupus erythematosus-like disease spontaneously.

Byclomine A trade name for dicyclomine hydrochloride, an anticholinergic agent.

Bydramine A trade name for diphenhydramine hydrochloride, an antihistaminic, sedative, hypnotic, and cough suppressant.

Byssinosis A disorder associated with handling of cotton, flax, and hemp and characterized by airway obstruction.

Bx Abbreviation for biopsy.

Bz The benzoyl group.

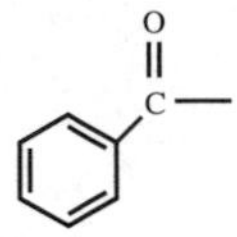

Bz_2-ATP Abbreviation for benzoyl-benzoyl-ATP.

Bzdz Abbreviation for benzoyldiazepine.

bZIP Abbreviation for basic leucine zipper.

Bzl The benzyl group.

BzOH Abbreviation for benzoic acid.

C

C Abbreviations for 1. carbon element, 2. complement, 3. cysteine, and 4. cytidine or cytosine.

° C Degrees centigrade; degrees Celsius.

^{14}C Radioactive isotope of carbon with an atomic weight of 14, half-life about 5700 years.

C Band A band (stained with giesma stain) appearing on the chromosome around the centromere.

C Genes The genes that encode the constant region of the light or heavy chain of an immunoglobulin molecule.

C Reactive Protein (CRP) A beta globulin found in the serum of patients with diverse inflammatory diseases. It reacts with the pneumococcal type-C polysaccharide.

C Region Referring to the constant region of the heavy chain or light chain of the immunoglobulin molecule.

C Value The amount of DNA in its haploid genome.

C1 The first component of the classical pathway of complement; it consists of three subcomponents C1q, C1r, and C1s.

C1 Inhibitor A complement regulatory protein that binds to C1r and C1s leading to inactivation of the classical complement pathway.

C2 The second component of complement.

C2 Photosynthesis Referring to the photorespiration that produces phosphoglycolic acid (a 2-carbon organic acid) from ribulose 1,5-bisphosphate.

C3 The third component of complement.

C3 Convertase A complex of C2b and C4b that acts as a protease converting C3 to C3a and C3b.

C3 Pathway See Calvin cycle.

C3 Photosynthesis Referring to C3 pathway.

C3 Plants Plants that employ only the C3 pathway for conversion of carbon dioxide to carbohydrate. The C4 pathway is absent from the C3 plant.

C4 The fourth component of complement.

C4 Pathway A pathway that catalyzed the conversion of carbon dioxide to the oxaloacetic acid employing the phosphoenolpyruvate system (also known as Hatch-Slack pathway).

C4 Plants Plants that employ both C3 and C4 pathways for conversion of carbon dioxide to carbohydrate. C4 plants also contain vascular bundle sheath cells around the vascular bundle which is absent from C3 plants.

C5 The fifth component of complement.

C5 Convertase A complex of subcomponents of complement (C4b-C2b-C3b) that acts as a protease converting C5 to C5a and C5b.

C6 The sixth component of complement.

C7 The seventh component of complement.

C8 The eighth component of complement.

C9 The ninth component of complement. Polymerization of C9 forms lytic holes in the cell membrane leading to cell lysis.

$C_{18:0}$ Abbreviation for octadecanoic acid, a saturated fatty acid with 18 carbons (also known as stearic acid or stearate).

CA Abbreviation for 1. capsid; 2. carbonic anhydrase; 3. cold acclimated; 4. cold agglutination; 5. common antigen; 6. a combination drug containing cytoxan and adriamycin.

Ca Symbol for calcium with atomic weight 40, valence 2.

CA-15-3 An antibody specific for an antigen frequently present in the serum of metastatic breast carcinoma patients.

CA-19-9 A tumor associated antigen found on the Lewis A blood group antigen.

CA-125 A cell surface glycoprotein detectable in higher concentration in the serum in patients with adenocarcinomas such as breast, gastrointestinal tract, and uterine cervix cancer.

C2a Abbreviation for a fragment of C2 component of complement.

C3a Abbreviation for a fragment of C3 component of complement.

C4a Abbreviation for a fragment of C4 component of complement.

C5a Abbreviation for a fragment of C5 component of complement.

Ca^{++} Calmodulin Complex The complex of calcium and calmodulin formed by the reversible binding of four calcium ions to calmodulin.

Ca^{2+} Calmodulin Dependent Protein Kinase The enzyme the catalyzes the following reaction in the presence of calcium and calmodulin:

$$\text{ATP} + \text{protein} \rightleftharpoons \text{ADP} + \text{phosphoprotein}$$

Ca^{2+} Mg^{2+}-ATPase The ATPase that requires both calcium and magnesium for activity.

CAA A codon or genetic code for the amino acid glutamine.

CAAT Abbreviation for computer-assisted axial tomography.

CAAT Box A nucleotide sequence about 75 base pairs upstream from the transcription initiation point in eukaryotic cells. It has the consensus sequence of GG(T/C)CAATCT.

Cabenegrin An orally active antidote against snake venom isolated from root of a South American plant called *Cabeca de Negra*. It consists of two active components (cabenegrin A1 and Cabenegrin A2).

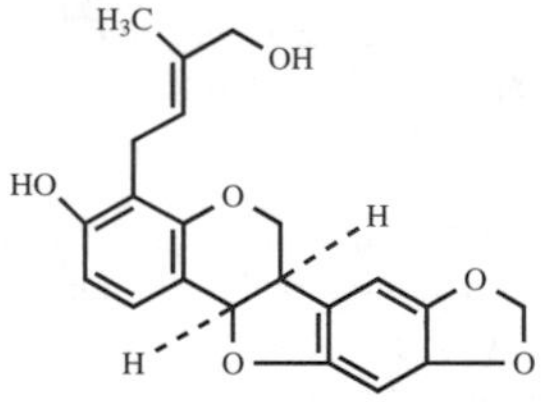

cabenegrin A-1

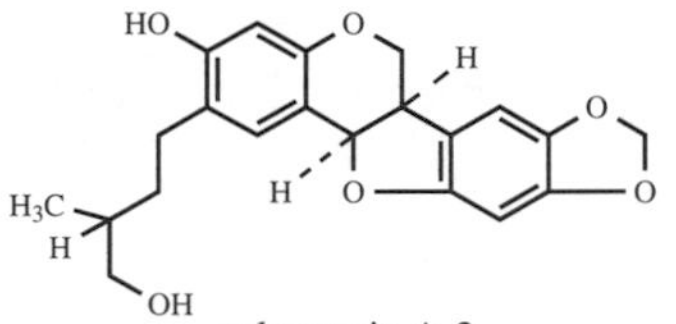

cabenegrin A-2

Cabergoline (mol wt 452) A prolactin inhibitor and dopamine receptor agonist.

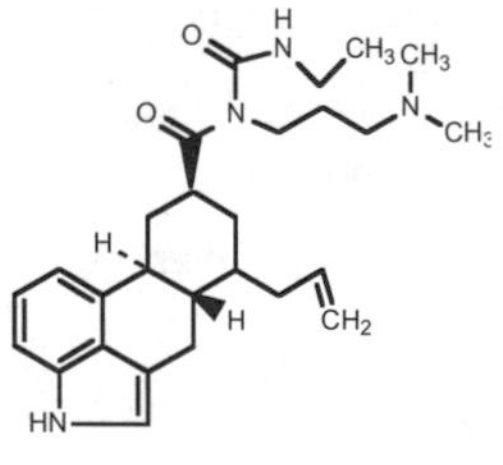

CABG Abbreviation for coronary artery bypass graft.

CABOP Abbreviation for a combination drug containing cyclophosphamide, adriamycin, bleomycin, oncovin, and prednisone.

CABP Abbreviation for 2-carboxyarabinitol 1,5-bisphosphate.

CaBP Abbreviation for calcium-binding protein.

CaBP3 A calcium-binding protein of endoplasmic reticulum lumen.

CABS Abbreviation for a combination drug containing CCNU, adriamycin, bleomycin, and streptozocin.

CAC A codon or genetic code for the amino acid histidine.

Cac8I A restriction endonuclease from *Clostridium acetobutyliticum* (NEB846) with the following specificity:

5'........GCN↓NGC........3'
3'........CGN↑NCG........5'

Ca^{2+}/Calmodulin Kinase II Synonym of Ca^{2+}/calmodulin-dependent protein kinase.

Cachet A flat capsule containing an unpleasant-tasting drug. The cachet is swallowed intact by the patient.

Cachexia Severe malnutrition, weaknesss, and muscle wasting due to chronic illness.

$CaCl_2$ Transformation See calcium phosphate transformation or calcium phosphate transinfection.

Cacodylic acid (mol wt 138) A substance used for treatment of chronic eczema and anemia. It has been also used as an herbicide.

$$(CH_3)_2As(O)OH$$

Cactinomycin An antibiotic complex isolated from *Streptomyces chrysomallus*. It is a mixture of actinomycin C1, C2 and C3.

Cactus grandiflorus A large flowered cereus. It contains a cardiotonic agent used as a circulatory stimulant.

CAD Abbreviation for a combination drug containing cytoxan, adriamycin, and dacarbazine.

CAD Protein A multifunctional protein found in many eukaryotes containing domains for carbamoyl phosphate synthetase, aspartate transcarbamylase, and dihydroorotase for pyrimidine biosynthesis.

Cadang-Cadang A viroid that infects coconut.

Cadaver Corpse, a dead body.

Cadaverine (mol wt 102) A biogenic polyamine and a homolog of putrescine produced by decarboxylation of lysine.

$NH_2\text{-}(CH_2)_5NH_2$

Cadexomer Iodine A hydrophilic, modified starch polymer produced by the reaction of dextrin with epichlorohydrin and iodine. It is used as a vulnerary agent.

Cadherin Referring to cell surface proteins that mediate calcium-dependent cell adhesion in animal tissues.

Cadmium (Cd) A chemical element with atomic weight 112, valence 2.

Cadmium Ionophore A selective ionophore for cadmium.

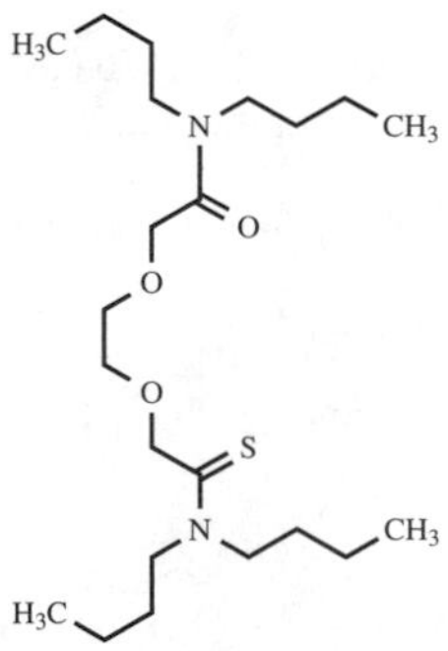

Cadmium Mycophosphatin A cadmium-binding phospho-glycoprotein rich in aspartic and glutamic acids and phosphoserine.

Cadmium Oxide (CdO, mol wt 128) An ascaricide.

Cadmium Salicylate (mol wt 387) An antiseptic agent.

$Cd(C_7H_5O_3)_2$

Cadmium Succinate (mol wt 228) A fungicide.

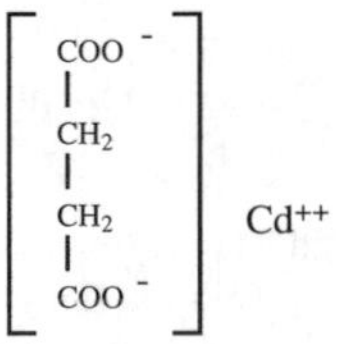

cADP Abbreviation for cyclic adenosinediphosphate.

cADPR Abbreviation for cyclic adenosine diphosphate ribose.

cADP-Ribose Abbreviation for cyclic adenosine diphosphate ribose.

Cadralazine (mol wt 283) An antihypertensive agent.

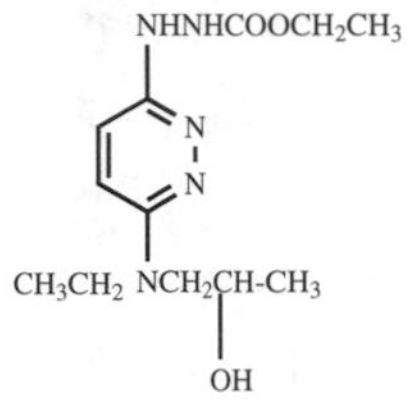

CAE Abbreviation for cellulose acetate electrophoresis.

Caecum A saclike extension or a blind diverticulum on the digestive tract.

Caedibacter A genus of Gram-negative bacteria that are endosymbionts in certain strains of *Paramecium aurelia.*

CAF Abbreviation for a combination drug containing cyclophosphamide, adriamycin, and fluorouracil.

Cafaminol (mol wt 267) A nasal decongestant.

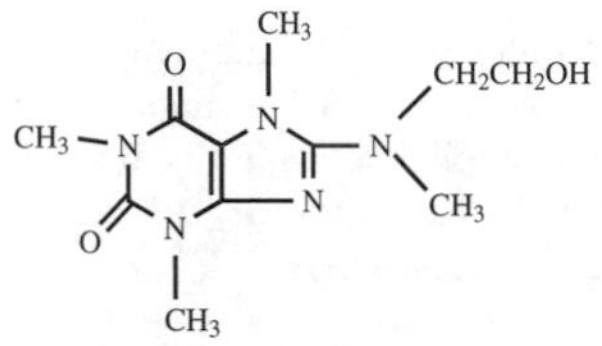

Cafergot A trade name for a combination drug containing ergotamine and caffeine, used as an adrenergic blocker.

Caffedrine A trade name for caffeine, a purine derivative that has been used as a cardiac and respiratory stimulant and diuretic agent. It also acts as an inhibitor for excission repair of DNA.

Caffeic Acid (mol wt 180) A lipooxygenase inhibitor.

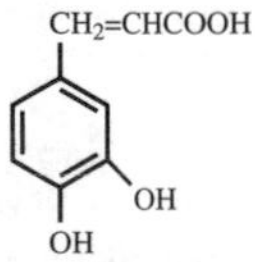

Caffeine (mol wt 194) A substance occurring in tea, coffee; a purine derivative that has been used as a cardiac and respiratory stimulant and diuretic agent. It also acts as an inhibitor for excission repair of DNA.

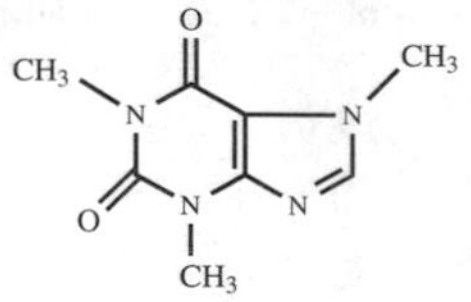

Caffeinism A chronic disorder due to excess consumption of beverages containing caffeine.

CAFVP Abbreviation for a combination drug containing cytoxan, adriamycin, fluorouracil, vincristine, and prednisone.

CAG A codon or genetic code for the amino acid glutamine.

Caged ATP A type of protected ATP analog, e.g., adenosine 5'-triphospho-1-(2-nitrophenyl) ethanol. It releases ATP upon photolysis by a short pulse of light of 360 nm wavelength.

Caged Compound An organic compound designed to change into an active form upon irradiation with specific wavelengths of light.

CalCaR Abbreviation for calcium-induced calcium release.

Cairns Mode of DNA Replication A mechanism of replication of double-stranded circular DNA in which the replication is initiated at a fixed point and proceeds bidirectionally or unidirectionally around the circular DNA.

cAK Abbreviation for cyclic AMP-dependent protein kinase.

CAK Abbreviation for CDK-activated kinase.

cal Abbreviation for small calorie(s).

Cal Abbreviation for large calorie(s) (1 Cal = 1000 cal).

Calamine A pink powder of zinc oxide and a skin protectant containing about 98% zinc oxide and 0.5% ferric oxide.

Calamus Dried rhizome of plant *Acorus calamus* (e.g., sweet flag, sweet cane) used as an anthelmintic agent.

Calan A trade name for verapamil hydrochloride, used as an antianginal agent.

Calberla's Solution A mixture of glycerol (5 ml), ethanol (10 ml), and water (15 ml) used as a microscopic mounting medium.

Calcein Blue (mol wt 321) A fluorescent indicator.

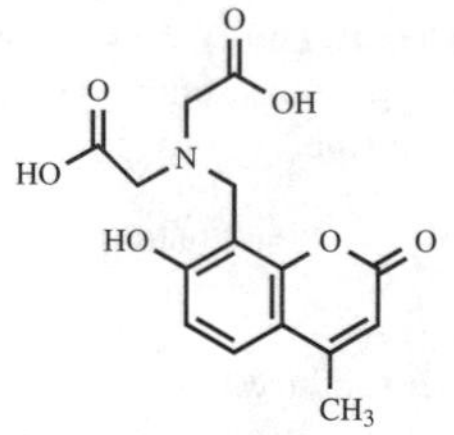

Calcemia Abnormally high level of calcium in the blood.

Calcet A trade name for a combination of calcium and vitamin D.

Calcibind A trade name for sodium cellulose phosphate, used to bind calcium in the GI tract and to decrease the amount of calcium absorbed.

Calcicolous Capable of growth in a calcium-rich environment.

Calcicosis A lung disorder due to the prolonged inhalation of dust of lime stone.

Calciday A trade name for calcium salt used as an antacid.

Calcifediol (mol wt 401) The circulating form of vitamin D_3 formed in the liver. It is a calcium regulator.

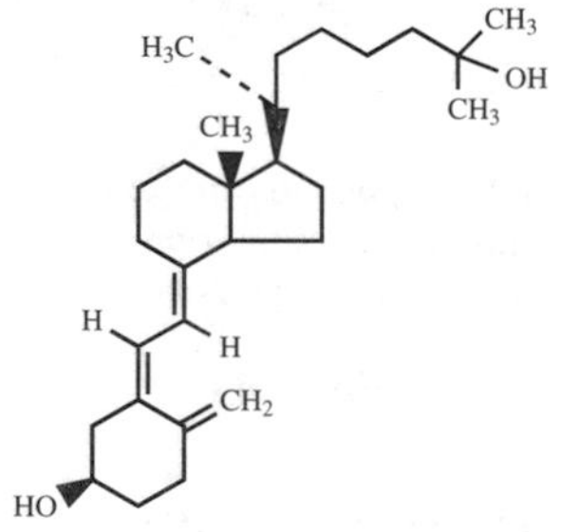

Calciferol Referring to vitamin D_2.

Calcification 1. Hardening by deposition of mineral salts in the bones and teeth. 2. Pathological hardening of an organic tissue by the deposition of calcium.

Calcijex A trade name for calcitriol, a calcium regulator.

Calcilac A trade name for calcium carbonate, used to reduce the acid load in the GI tract and to elevate gastric pH and pepsin activities.

Calcilean A trade name for heparin calcium, an anticoagulating agent that prevents the conversion of fibrinogen to fibrin.

Calcimar A trade name for calcitonin (from salmon), a calcium regulating peptide hormone.

Calcimax A trade name for calcium carbonate, used to reduce the acid load in the GI tract and to elevate gastric pH and pepsin activities.

Calcimedin A calcium-binding protein.

Calcimycin (mol wt 524) A polyether antibiotic isolated from *Streptomyces chartreusensis*. It is a divalent cation ionophore.

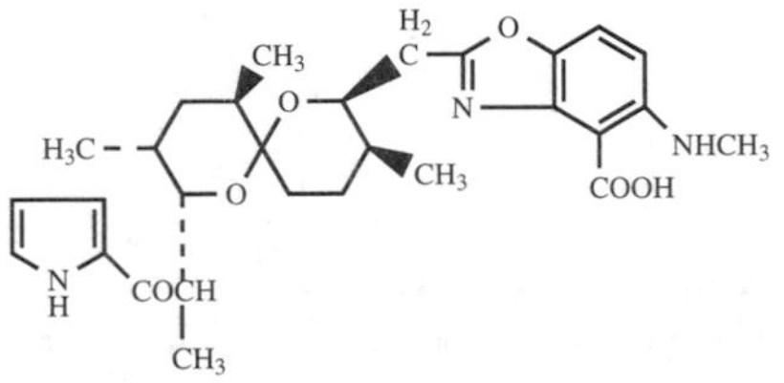

Calcineurin A calmodulin binding protein.

Calcinosis Deposition of calcium in the skin and muscle.

Calciosome A discrete cytoplasmic organelle in non-muscle cells, it contains a high content of calsequestrin-like protein.

Calciparine A trade name for heparin calcium, an anticoagulating agent that prevents the conversion of fibrinogen to fibrin.

Calcipenia Disorder characterized by the deficiency of calcium in the tissue.

Calciphilia A condition in which the tissue tends to absorb calcium and becomes calcified.

Calciphorin A calcium ionophore polypeptide isolated from the inner membrane of calf mitochondria.

Calcipotriene (mol wt 413) An antipsoriatic ointment used for the treatment of psoriasis.

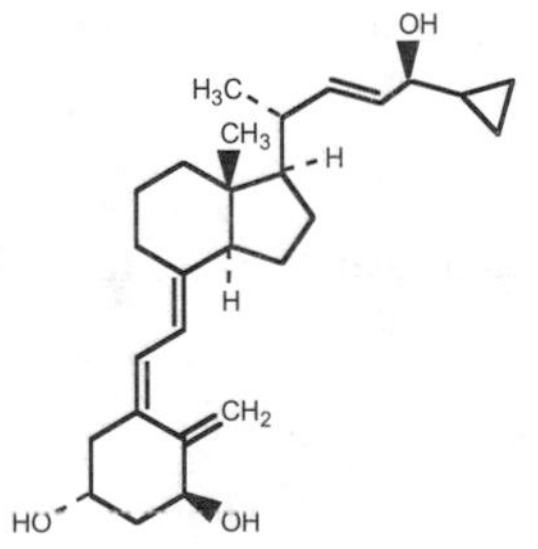

Calcisome An intracellular calcium reservior (a specific region of endoplasmic reticulum, e.g., sarcoplasmic reticulum in muscle cell) for the supply of cytosolic calcium.

Calcite 500 A trade name for calcium carbonate, an antacid.

Calcitonin A calcium-regulating peptide hormone secreted from the thyroid gland in mammals; it lowers the level of calcium in the blood, decreases osteoclastic activity, and reduces mineral release and collagen breakdown in bones.

Calcitonin Salmon Calcitonin from salmon used to decrease serum calcium concentration and promote renal excretion of calcium.

Calcitriol (mol wt 417) A biologically active form of vitamin D_3 and a calcium regulator.

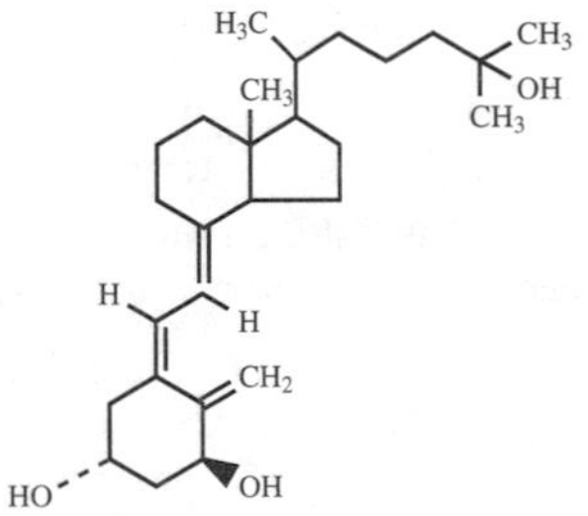

Calcium (Ca) A chemical element with atomic weight 40 and valence 2.

Calcium 45 (^{45}Ca) Radioactive calcium with a half-life of about 164 days.

Calcium Acetate (mol wt 158) A food stabilizer and corrosion inhibitor.

$Ca(CH_3COO)_2$

Calcium Acetylsalicylate (mol wt 398) An analgesic, antipyretic, and anti-inflammatory agent.

$Ca(OOCC_6H_4OCOCH_3)_2$

Calcium Antagonist A drug or substance that inhibits the influx of calcium into cardiac and smooth muscle cells.

Calcium Arsenate (mol wt 398) An insecticide.

$Ca_3(AsO_4)_2$

Calcium Bromide (mol wt 200) A sedative and an anticonvulsant.

$CaBr_2$

Calcium N-Carbamoylaspartate (mol wt 214) A psychostimulant.

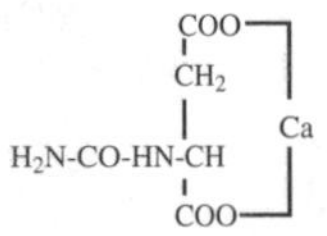

Calcium Carbonate (mol wt 100) A compound used to reduce the total acid load in the GI tract and to elevate gastric pH and pepsin activity.

$CaCO_3$

Calcium Channel   The membrane proteins that permit the controlled (gated) passage of calcium ions through the membranes.

Calcium Channel Blocker   Any agent or drug that prevents calcium from entering smooth muscle cells, causing smooth muscles to relax and reducing muscle spasms.

Calcium 2-Ethylbutanoate (mol wt 270)   A sedative agent.

$$\left[\begin{array}{c} CH_3CH_2CHCOO^- \\ | \\ C_2H_5 \end{array} \right]_2 Ca^{2+}$$

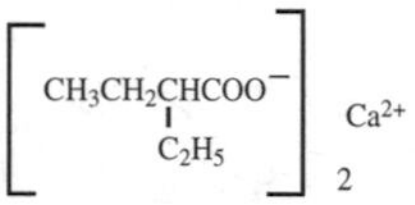

Calcium Ferrous Citrate (mol wt 514)   A hematinic agent.

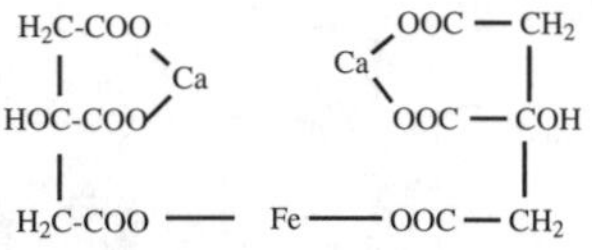

Calcium Formate (mol wt 130)   A food preservative.

$$Ca(HCOO)_2$$

Calcium Gluceptate   See calcium gluconate.

Calcium Gluconate (mol wt 430)   A calcium replenisher.

$$\left[\begin{array}{c} COO^- \\ H\,COH \\ HOCH \\ H\,COH \\ H\,COH \\ CH_2OH \end{array} \right]_2 Ca^{++}$$

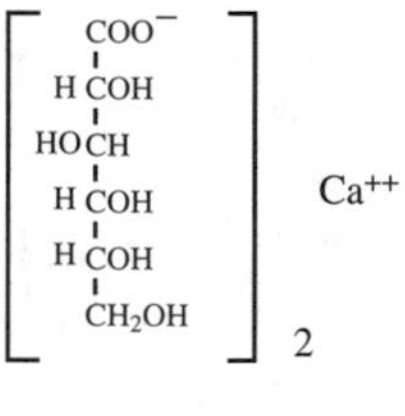

Calcium Iodate (mol wt 390)   An antiseptic agent and a nutritional source of iodine.

$$Ca(IO_3)_2$$

Calcium Ionophore I (mol wt 685)   A calcium ionophore used in membrane electrodes to determine calcium activity.

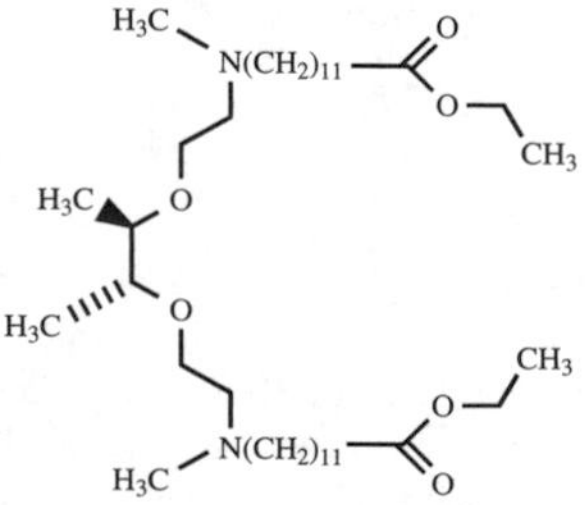

Calcium Ionophore II (mol wt 461)   An extremely efficient carrier of calcium ions.

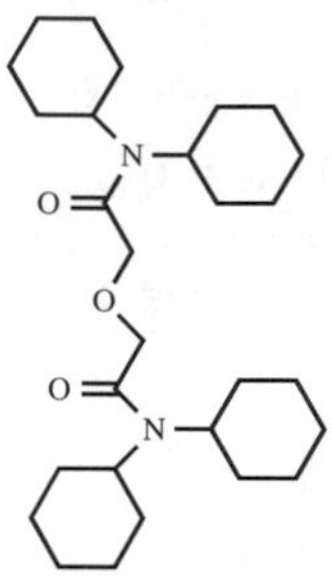

Calcium Ionophore III (mol wt 524)   A reagent used to prepare an optical chemical sensor for calcium and magnesium.

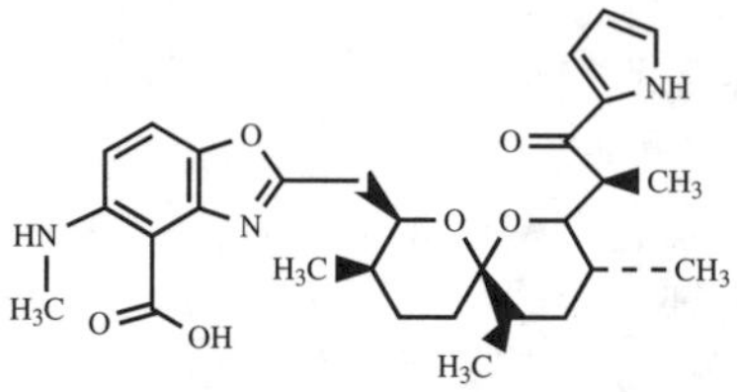

Calcium Ionophore IV (mol wt 801)   An extremely lipophilic calcium ionophore.

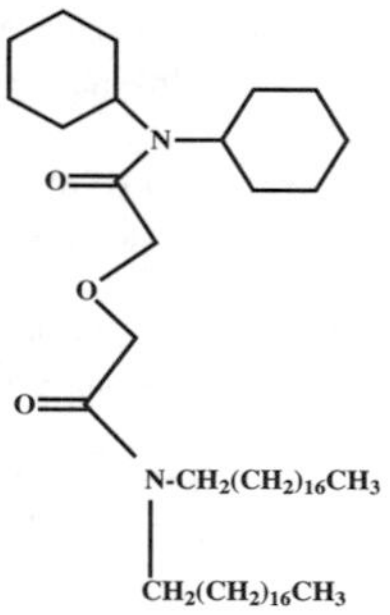

Calcium Lactate (mol wt 218)   A calcium replenisher.

$$Ca[CH_3CH(OH)COO]_2$$

Calcium Mesoxalate (mol wt 156)   An oral hypoglycemic agent.

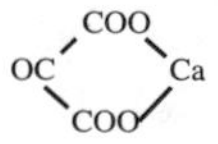

Calcium Pantothenate (mol wt 477)   A vitamin and enzyme cofactor.

$$[HOCH_2C(CH3)_2CHOHCONHCH_2CH_2COO]_2$$

Calcium Peroxide (mol wt 72)   An antiseptic agent.

$$CaO_2$$

Calcium Phosphate Gel A gel, prepared from calcium chloride and trisodium phosphate, used for purification of protein.

Calcium Phosphate Transfection Technique used to introduce DNA into mammalian cells with the aid of calcium phosphate.

Calcium Phosphate Transformation See calcium phosphate transfection.

Calcium Pump The calcium gradient across the membrane generated by the calcium-dependent ATPase at the expense of ATP.

Calcium Transporting ATPase A membrane ATPase that forms an essential component of the calcium pump.

Calciuria The presence of calcium in the urine.

Calculase An unclassified enzyme used for softening hard dental plaque.

Calculosis The presence of multiple calculi (stones) in the body.

Calculus Stony concretion resulting from the aggregation of mineral salts in various parts of the body.

CaldeCort A trade name for hydrocortisone.

Caldecrin A pancreatic protein that lowers serum calcium concentration.

Calderol A trade name for calciferol.

Caldesmon A calmodulin-binding protein.

Caldesmon Kinase The enzyme that catalyzes the following reaction:

$$\text{ATP + caldesmon} \rightleftharpoons \text{ADP + caldesmon phosphate}$$

Caldesmon Phosphatase The enzyme that catalyzes the following reaction:

$$\text{Caldesmon phosphate} + H_2O \rightleftharpoons \text{Caldesmon + phosphate}$$

Calelectrin A protein from the electric organ of *Torpedo marmorata*.

Calf Thymus Ribonuclease H A ribonuclease that catalyzes the cleavage of RNA to 5′-phosphomonoester.

Calfactant An extract of natural surfactant from calf lung, it contains phospholipids and proteins and is used to modify the surface tension.

Calglycine A trade name for calcium carbonate.

Caliciviridae A family of plus-stranded RNA viruses (formally classified as Picornaviridae).

Caliculus A cup-shaped structure.

Calicyclic A trade name for salicylic acid.

California Encephalitis An acute viral encephalitis caused by a mosquito-transmitted bunyavirus.

California Mastitis Test A test to estimate the number of white blood cells in the milk for the detection of mastitis (inflammation of mammary gland).

Californium (Cf) A radioactive element with atomic weight of 249 and half-life of about 45 minutes.

CALL Abbreviation for 1. common acute lymphoblastic leukemia; 2. common acute lymphocytic leukemia.

CALLA Abbreviation for 1. common acute lymphoblastic leukemia antigen; 2. common acute lymphocytic leukemia antigen.

Callose A linear 1,3—β-D-linked glucan.

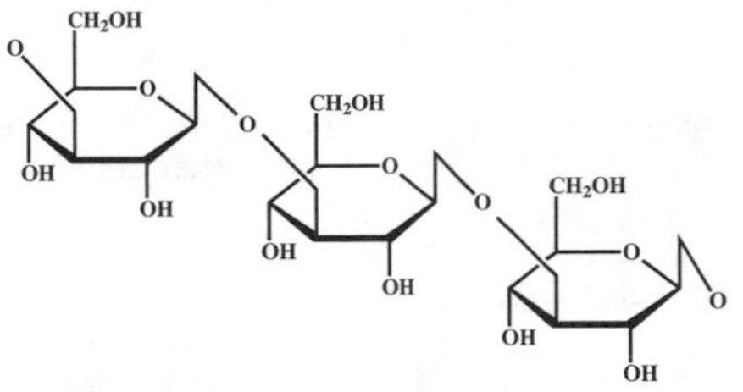

Callus Undifferentiated clone of plant cells.

Calm X A trade name for dimenhydrinate, used as an antiemetic agent.

Calmazine A trade name for trifluoperazine hydrochloride, used as an antipsychotic agent.

Calmodulin A small, heat-stable, acid-stable, calcium-binding protein involved in calcium-regulated biochemical processes in eukaryotic cells.

Calnexin A calcium-binding protein of the endoplasmic reticulum.

Calorie A unit of energy, the amount of heat required to raise the temperature of 1.0 g of water from 14.5° C to 15.5° C.

Calorific Heat generating.

Calorimeter A device to measure heat given off by an individual or by a chemical reaction.

Calothrix A genus of filamentous cyanobacteria.

Calotropin (mol wt 533) A poison isolated from the milk sap of *Calotropis procera* (Ascelpiadaceae).

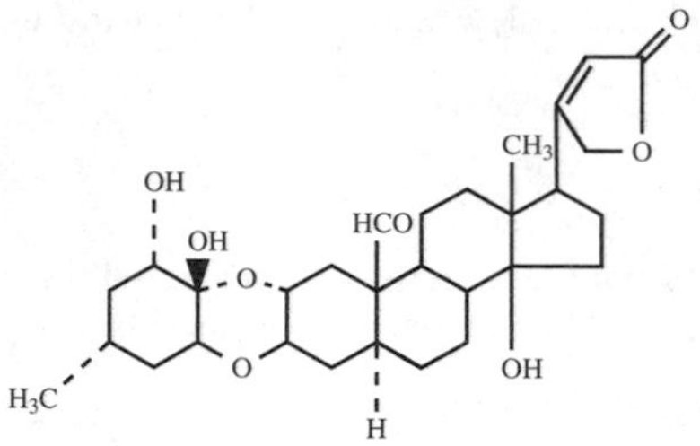

Calpain A calcium-dependent protease that catalyzes the hydrolysis of peptide bonds involving carboxyl group of tyrosine, methionine, or arginine.

Calpastatin A protein inhibitor for calpains (protease).

Calphotin A calcium-binding protein in the cytoplasm of photoreceptor cells of *Drosophila melanogaster.*

Cal-Plus A trade name for calcium carbonate; used to reduce the total acid load in the GI tract and to elevate gastric pH and pepsin activities.

Calponin A thin filament-associated protein implicated in the regulation and modification of smooth muscle contraction.

Calpromotin A cytoplasmic protein of the erythrocytes which activates calcium-dependent potassium transport.

Calregulin See calreticulin.

Calreticulin A calcium-binding protein of the endoplasmic reticulum lumen.

Calsan A trade name for calcium carbonate, an antacid.

Calsequestrin An acidic glycoprotein found in the sarcoplasmic reticulum; it binds calcium ion, serves as a calcium reservior, and releases calcium upon muscle contraction.

Calspectrin Synonym of fodrin, a protein from the bovine brain.

Calstorin A calcium-binding protein of the microsomal lumen of the rat brain.

Caltine A trade name for salmon calcitonin.

Caltrate A trade name for calcium salt used as an antacid.

Caltrin A small basic protein of the male seminal vesicle fluids which acts as a calcium transport inhibitor.

Calusterone (mol wt 316) An antineoplastic agent.

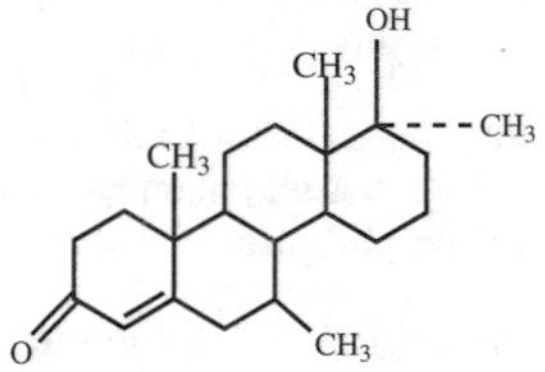

Calvin Cycle The pathway for the reduction of carbon dioxide to carbohydrate (e.g., phosphoglycerate, sugar) in photosynthesis employing the ribulose 1,5-bisphosphate carboxylase system (also called C_3 pathway or Calvin Benson cycle).

Calvinosome A structure in the prokaryotic photosynthetic cell containing enzymes of the Calvin cycle for reduction of carbon dioxide to carbohydrate.

Calymmatobacterium A genus of Gram-negative bacteria. It causes granuloma inguinale in humans.

CAM Abbreviation for 1. a combination drug containing cytoxan, adriamycin, and methotrexate; 2. cell adhesion molecule; 3. chorioallantoic membrane; 4. crassulacean acid metabolism; 4. constitutively active mutant.

CaM Abbreviation for calmodulin.

$[Ca^{2+}]_m$ Abbreviation for mitochondrial matrix Ca^{2+} concentration.

CaM Kinase II Synonym of Ca^{2+}/calmodulin-dependent protein kinase.

Camalox A trade name for a combination drug containing aluminum hydroxide, magnesium hydroxide, and calcium carbonate, used as an antacid.

CAMB Abbreviation for a combination drug containing cyclophosphamide, adriamycin, methotrexate, and bleomycin.

Cambendazole (mol wt 302) An anthelmintic agent.

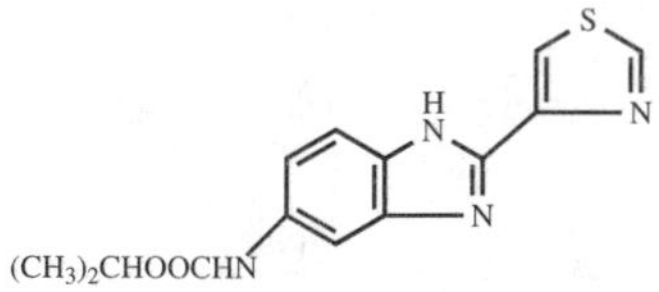

Cambium A layer of embryonic cells between the xylem and the phloem (vascular cambium) or between the cork and the phelloderm (cork cambium).

Camcolit A trade name for lithium carbonate, used to alter chemical transmitters in the CNS.

CAMEO Abbreviation for a combination drug containing cytoxan, adriamycin, methotrexate, etoposide, and oncovin.

CAMF Abbreviation for a combination drug containing cyclophosphamide, adriamycin, methotrexate, and fluorouracil.

CaMK1 Abbreviation for calmodulin-dependent protein kinase-1.

CAMKK Abbreviation for calmodulin-dependent protein kinase 1-kinase.

CAMLO Abbreviation for a combination drug containing cytoxan, adriamycin, methotrexate, leucovorin, and oncovin.

Camostat (mol wt 398) An orally active, nonpeptide protease inhibitor.

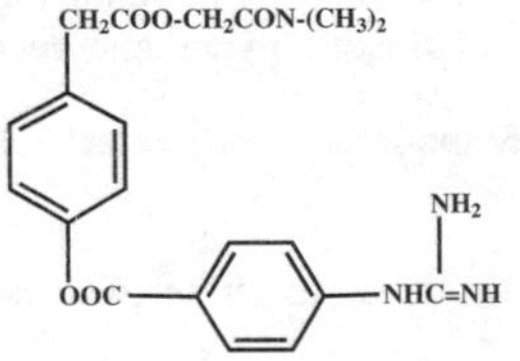

cAMP Abbreviation for cyclic adenosine monophosphate (cyclic AMP). It plays an important role in regulating various biochemical processes in cells.

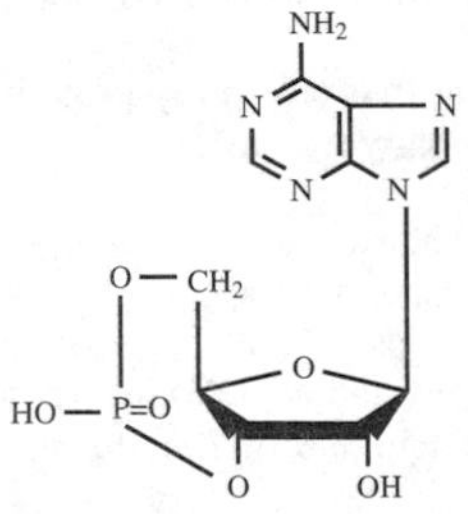

CAMP Abbreviation for a combination drug containing cytoxan, adriamycin, methotrexate, and procarbazine.

Campain A trade name for acetaminophen, used to block the generation of pain impulses.

Campbell Model A model which proposes that integration of plasmid or phage genome into a bacterial chromosome is accomplished by a single crossing over between two circular molecules.

Campesterol (mol wt 401) A typical plant steroid found in the rapeseed oil from *Brassica campestris*.

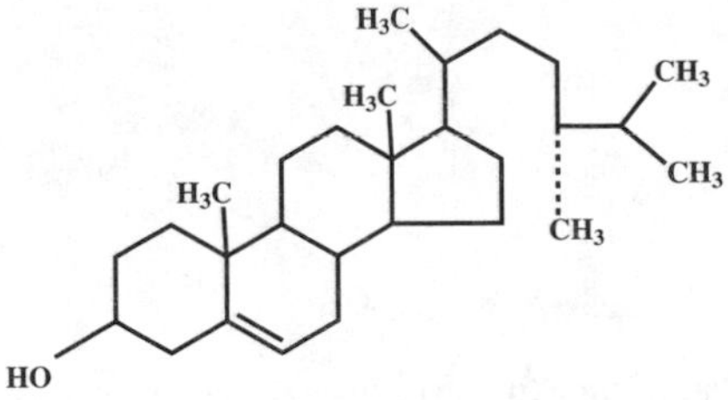

Camphor (mol wt 152) An anti-infective agent.

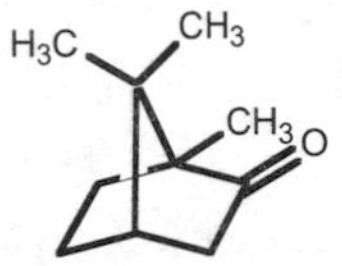

Camphothecin (mol wt 348) An antitumor alkaloid from the wood stem of the Chinese tree (*Camphotheca acuminata*).

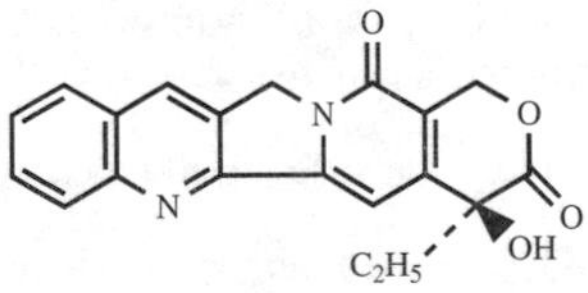

Camptosar A trade name for irinotecan hydrochloride, an antineoplastic agent.

Campylobacter A genus of Gram-negative asporogenous bacteria.

Campylobacteriosis Any infection or disease caused by *Campylobacter*.

CaMV Abbreviation for cauliflower mosaic virus.

Camylofine (mol wt 320) An anticholinergic agent.

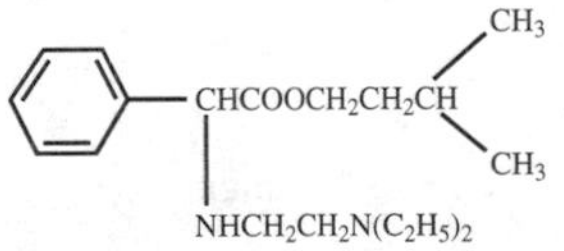

Canada Balsam A gum used in the preparation of a permanent mount of tissue slices or other specimens for microscopic examination.

Canaliculization The formation of small canals in the tissue.

Canaliculus A small canal.

Canaline (mol wt 139) A basic non-protein α-amino acid that inhibits pyridoxal-dependent enzymes.

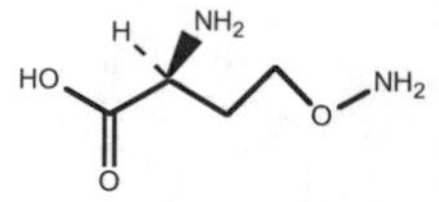

Canavanase See arginase.

Canavanine (mol wt 176) A naturally occurring nonprotein, basic amino acid.

$H_2NC(NH)NHOCH_2CH_2CH(NH_2)COOH$

Cancellous Spongy or honeycomb structure of some bone tissues.

Cancer Any disease in humans or animals in which the uncontrolled proliferation of cells leads to the formation of malignant tumors.

Cancer Gene See oncogene.

Cancer-Inducing Virus Viruses capable of inducing cancers (e.g., retroviruses).

Cancerocidal Capable of killing cancer cells.

Canceroid 1. A skin tumor of low malignancy. 2. Cancerlike.

Cancrum An ulcer that spreads rapidly.

Candesartan (mol wt 440) An angiotensin II receptor antagonist used as an antihypertensive agent.

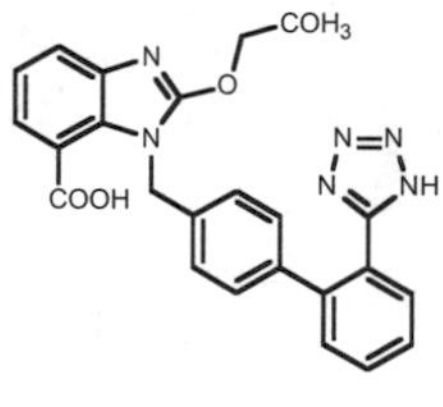

Candida A genus of fungi.

Candidapesin A protease that catalyzes the preferential cleavege of peptide bonds involving the carboxyl groups of hydrophobic amino acids.

Candidiasis Any fungal infection or disease caused by species of *Candida*.

Candistatin A trade name for nystatin, an antifungal agent.

Candling A procedure for the observation of the content of a developing chicken embryo.

Canesten A trade name for clotrimazole, used as a local anti-infective agent.

Canine Relating to dogs.

Canker A sore.

Canrenone (mol wt 340) An aldosterone antagonist and diuretic agent.

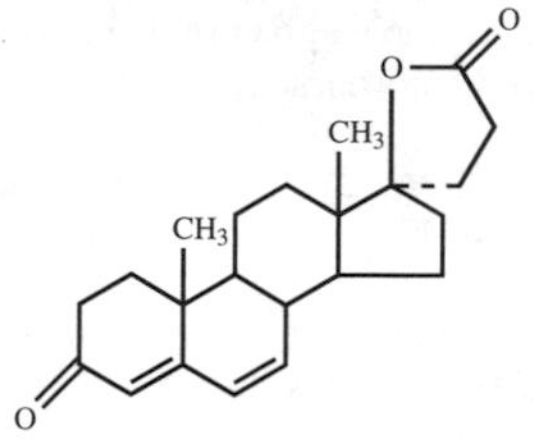

Cantil A trade name for mepenzolate bromide, used as an anticholinergic agent that competitively blocks acetylcholine leading to decrease of GI mobility and inhibition of gastric acid secretion.

CAO Abbreviation for a combination drug containing cytoxan, adriamycin, and oncovin.

$[Ca^{2+}]_o$ Symbol for extracellular calcium concentration.

CaOX Abbreviation for calcium oxalate.

Cap Referring to the 5′-cap of eukaryotic mRNA (see capped 5′-end).

CAP Abbreviation for 1. catabolite activating protein; 2. cellulose acetate phthalate; 3. cystine aminopeptidase; 4. a combination drug containing cyclophosphamide, adriamycin, and prednisone.

Cap Binding Protein Any protein that binds to the cap of eukaryotic mRNA.

Cap Site Abbreviation for 7-methylguanosine 5'-triphosphate site.

Capastat A trade name for capreomycin sulfate, an antibiotic and antimicrobial agent.

CAPBOP Abbreviation for a combination drug containing cytoxan, adriamycin, procarbazine, bleomycin, oncovin, and prednisone.

Capillary Blood vessel intermediate between the arteriole and venule.

Capillary Electrophoresis An electrophoresis performed in a long capillary tube with high resolution of component separation.

Capillary Viscometer An instrument for measuring the viscosity of a liquid, i.e., the time required for a given volume of liquid to flow through a capillary.

CAPK Abbreviation for ceramide-activated protein kinase.

Caplets Capsule-shaped tablets that are easier to swallow than round pills.

Capobenic Acid (mol wt 325) A cardiac depressant (antiarrhythmic).

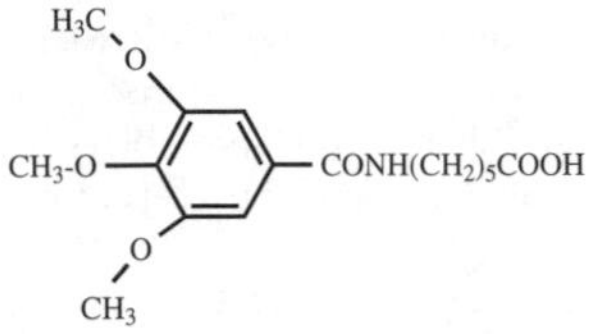

Capoten A trade name for captopril, used as an antihypertensive agent that inhibits the conversion of angiotensin I to angiotensin II.

Capozide A trade name for a combination drug containing hydrochlorothiazide and captopril, used as an antihypertensive agent.

CAPP Abbreviation for ceramide-activated protein phosphatase.

Capped 5′-End Methylated guanosine added posttranscriptionally to the 5′ end of a eukaryotic mRNA.

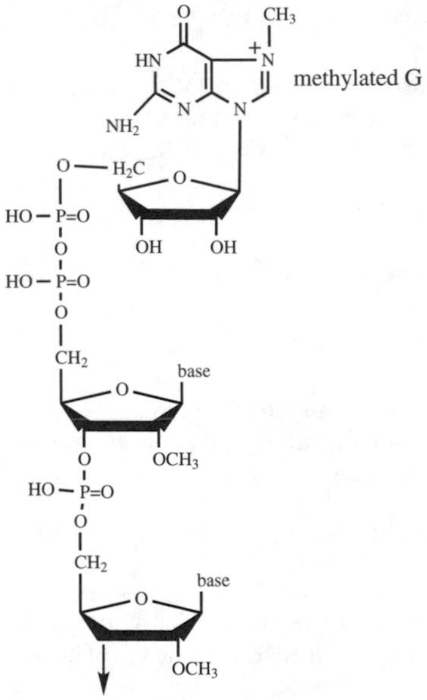

capped 5' end

Capreomycin A cyclic peptide antibiotic isolated from *Streptomyces capreolus*.

Capreomycin IA (mol wt 558) An antibacterial agent.

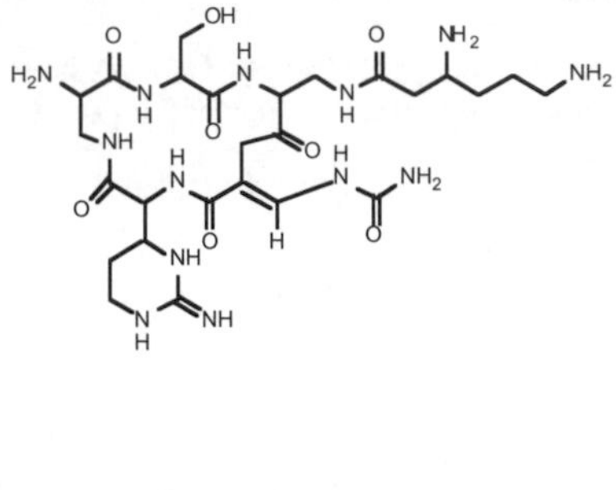

Capreomycin IB (mol wt 542) An antibacterial agent.

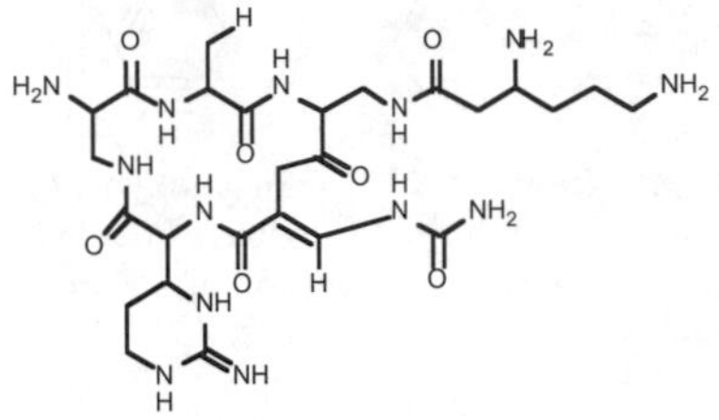

Caprin A trade name for heparin calcium, used as an anticoagulant that prevents the conversion of angiotension I to angiotensin II.

Caproic Acid (mol wt 116) A fatty acid.

$CH_3(CH_2)_4COOH$

CAPS (mol wt 221) 3-(Cyclohexylamino)-propanesulfonic acid, a biological buffer substance.

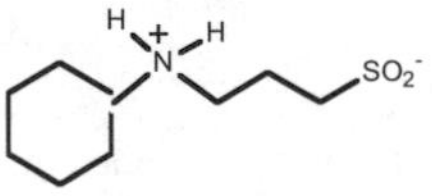

Capsaicin (mol wt 305) Pungent agent in fruit of various species of *Capsicum*. It possesses topical analgesic activity.

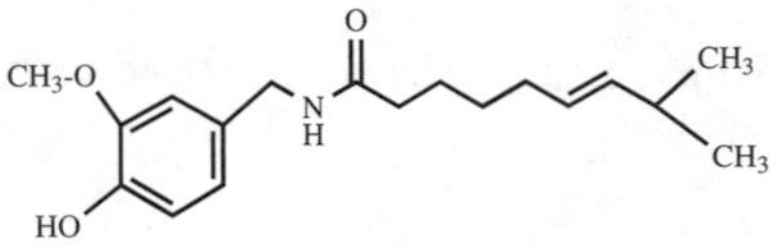

Capsid The protein coat that surrounds the nucleic acid of a virus.

Capsomers Structural units of the capsid of a virus.

Capsular Antigen Any antigen, usually polysaccharide in nature, that is located on the surface of the bacterial capsules.

Capsular Polysaccharide A polysaccharide component of a bacterial capsule.

Captafol (mol wt 349) An agricultural fungicide.

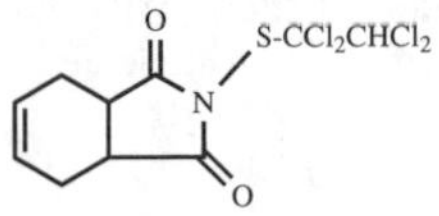

Captan (mol wt 301) An antifungal agent and a bacteriostat in soap.

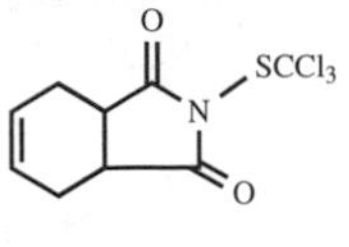

Captodiamine (mol wt 360) An anxiolytic agent.

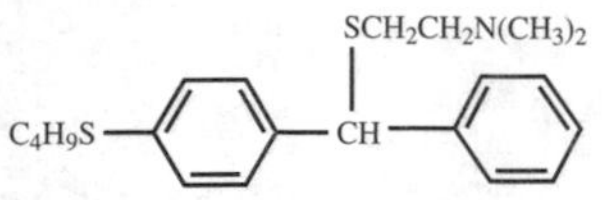

Captopril (mol wt 217) An antihypertensive agent and an inhibitor of angiotensin converting enzyme that prevents conversion of angiotensin I to angiotensin II.

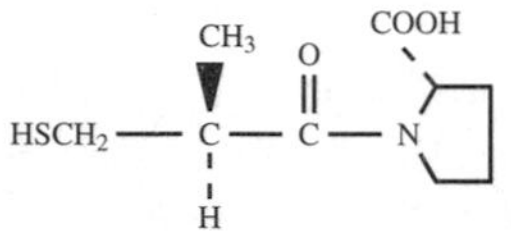

Capurate A trade name for allopurinol, used as an antigout agent; reduces uric acid production.

Capuride (mol wt 186) A hypnotic agent.

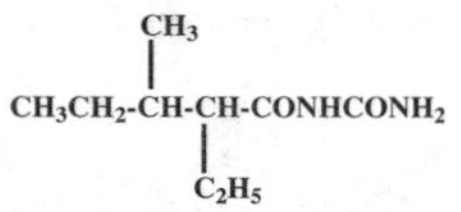

CaR Abbreviation for calreticulin.

C5aR Abbreviation for C5a (complement subcomponent 5a) receptor.

Carafate A trade name for sucralfate, used as an antiulcer agent.

Caramiphen Ethanedisulfonate (mol wt 769) An antitussive agent.

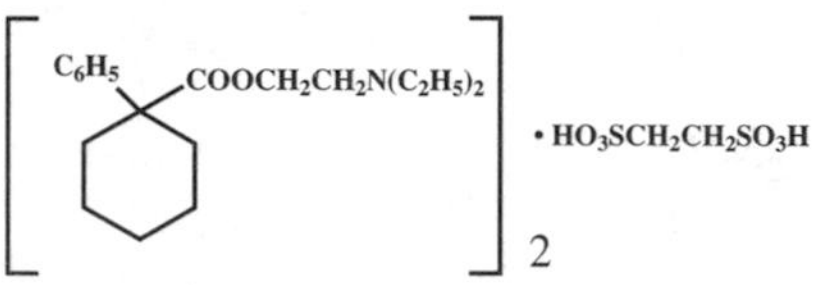

Caramiphen Hydrochloride (mol wt 325) An anticholinergic agent.

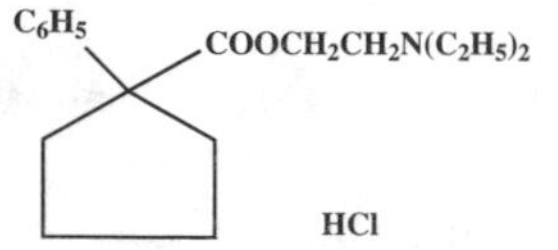

Carazolol (mol wt 298) An antihypertensive, antianginal, and antiarrhythmic agent.

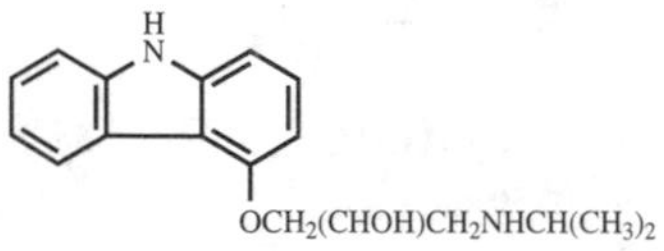

Carbacel A trade name for carbachol, used as a miotic agent.

Carbachol (mol wt 183) A substance that causes contractions of the sphincter muscle of the iris leading to miosis and produces ciliary spasm and vasodilation of conjunctival vessels of the outflow tract.

$$[NH_2COOCH_2CH_2N(CH_3)_3]^+Cl^-$$

Carbadox (mol wt 262) An antibacterial agent.

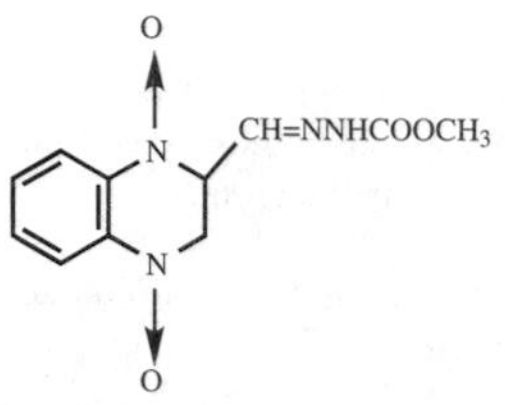

Carbamate Kinase The enzyme that catalyzes the following reaction:

$$\text{ATP} + NH_2 + CO_2 \rightleftharpoons \text{Carbamoyl phosphate}$$

Carbamazepine (mol wt 236) An analgesic and anticonvulsant that increases efflux or decreases influx of sodium ions across the cell membrane.

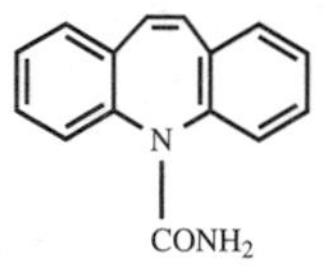

Carbamino Compound A compound formed by the reaction of carbon dioxide with a primary aliphatic amine.

Carbamino Group Referring to –NH-COO⁻ group.

Carbaminohemoglobin The carbamino compound that is formed by the reaction of hemoglobin with carbon dioxide.

Carbamoyl Referring to $-CO-NH_2$ group.

Carbamoylaspartotranskinase See aspartate carbamoyl transferase.

Carbamoyl-Methyl Group Referring to $-CH_2-CO-NH_2$.

Carbamoyl-Phosphate (mol wt 141) A substrate for synthesis of pyrimidine nucleotides.

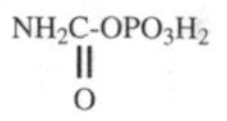

Carbamoyl-Phosphate L-Aspartate Carbamoyl Transferase The systematic name for aspartate carbamoyltransferase.

Carbamoyl-Phosphate L-Ornithine Carbamoyl Transferase The systematic name for ornithine carbamoyltransferase.

Carbamoylphosphate Synthetase (ammonia) The enzyme that catalyzes the following reaction:

$$2ATP + NH_3 + CO_2 + H_2O \rightleftharpoons 2ADP + \text{phosphate} + \text{carbamoyl phosphate}$$

Carbamoylphosphate Synthetase (glutamine hydrolyzing) The enzyme that catalyzes the following reaction:

$$2ATP + \text{L-glutamine} + CO_2 + H_2O \rightleftharpoons 2ADP + PPi + \text{carbamoyl phosphate}$$

Carbamoylputrescine Amidase The enzyme that catalyzes the following reaction:

$$\text{N-Carbamoylputrescine} + 2\ H_2O \rightleftharpoons \text{putrescine} + H_2O + CO_2$$

Carbamoylputrescine Amidohydrolase See carbamoylputrescine amidase.

Carbamoylsarcosine Amidase The enzyme that catalyzes the following reaction:

$$\text{N-Carbamoylsarcosine} + H_2O \rightleftharpoons \text{Sarcosine} + CO_2 + NH_3$$

Carbamoylsarcosine Amidohydrolase See carbamoylsarcosine amidase.

Carbamoylserine Ammonia-Lyase The enzyme that catalyzes the following reaction:

$$\text{Carbamoylserine} + H_2O \rightleftharpoons \text{Pyruvate} + 2NH_2 + CO_2$$

Carbamoylserine Deaminase See Carbamoylserine ammonia-lyase.

Carbamyl Aspartotranskinase Synonym of aspartate carbamoyltransferase.

Carbanion A negatively charged carbon.

Carbapen A trade name for carbenicillin sodium, an antibiotic inhibiting bacterial cell wall synthesis.

Carbapenem Antibiotics Any of the broad spectrum b-lactam antibiotics that inhibit peptidoglycan synthesis.

Carbarsone (mol wt 260) An antiamoebic agent.

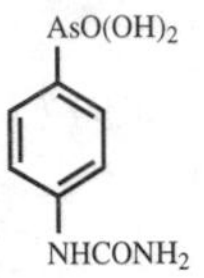

Carbaryl (mol wt 201) An insecticide.

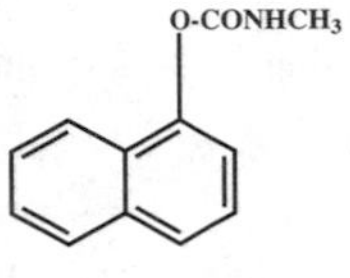

Carbatrol A trade name for carbamazepine, an antiepileptic agent.

Carbendazim (mol wt 191) A fungicide.

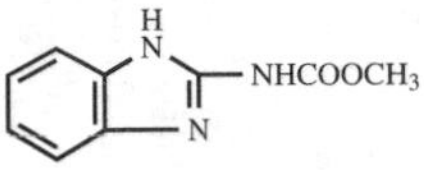

Carbenicillin (mol wt 378) A semisynthetic antibiotic related to penicillin, an antibiotic that inhibits bacterial cell wall synthesis.

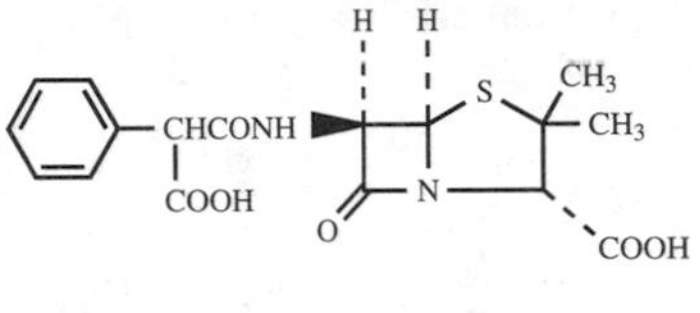

Carbenoxolone (mol wt 571) An antiulcerative agent.

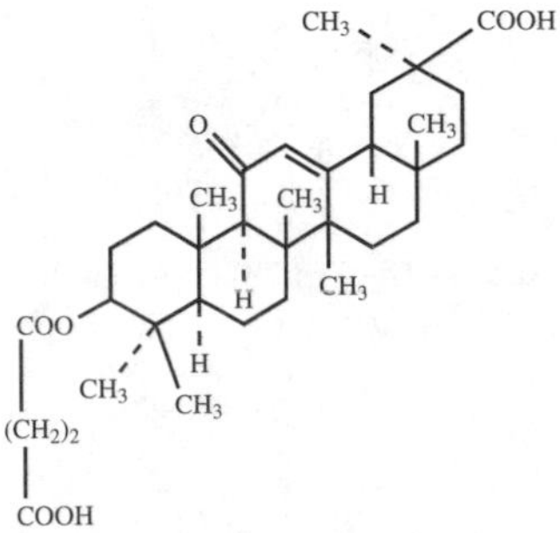

Carbetapentane (mol wt 333) An antitussive agent.

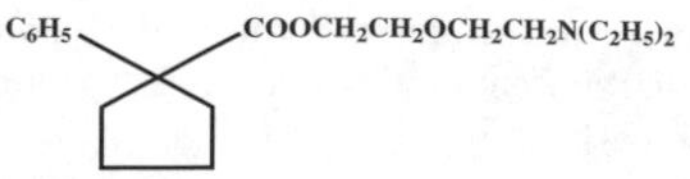

Carbetidine (mol wt 321) An analgesic agent.

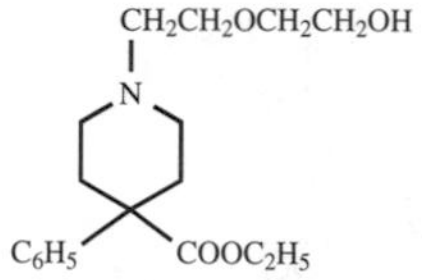

Carbetocin (mol wt 988) A synthetic analog of oxytocin possessing uterotonic and galactogogic activities.

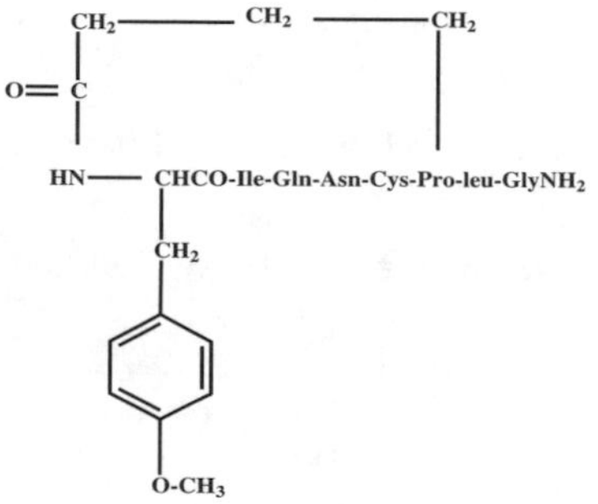

Carbex A trade name for selegiline hydrochloride, an irreversible inhibitor of monoamine oxidase and used for the treatment of Parkinson's disease.

Carbidopa (mol wt 244) An antiparkinsonian agent in combination with levodopa.

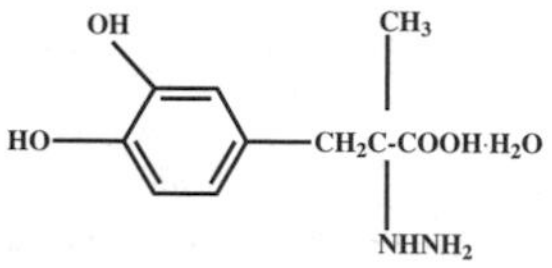

Carbimazole (mol wt 186) A thyroid inhibitor.

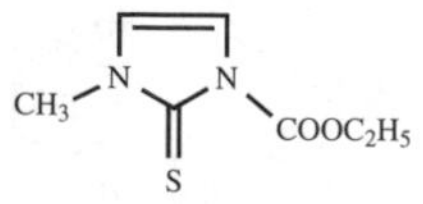

Carbinoxamine (mol wt 291) An antihistaminic agent.

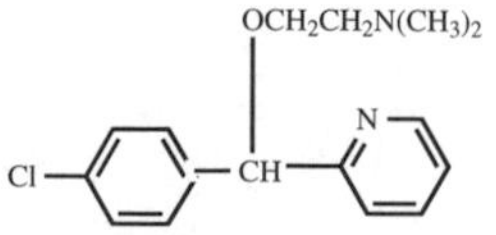

Carbiphene (mol wt 430) An analgesic agent.

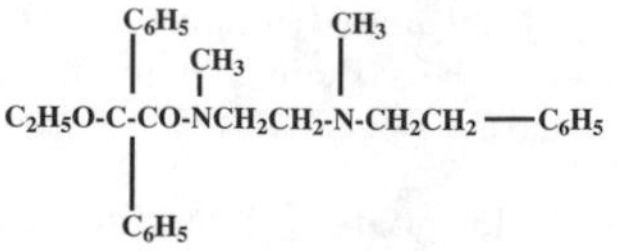

Carbobenzoxy Chloride (mol wt 171) An amino group blocker in peptide synthesis.

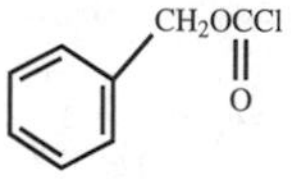

Carbocaine A trade name for mepivacaine hydrochloride, used as an anesthetic agent that interferes with sodium-potassium exchange across the nerve cell membrane.

Carbocation A positively charged carbon atom.

Carbocylic Any cyclic chemical structure that contains only carbon atoms.

Carbocyclic Pertaining to an organic compound that has a ring structure consisting of only carbon atoms.

Carbocysteine (mol wt 179) A mucolytic agent and expectorant.

$$HOOCCH_2SCH_2CH(NH_2)COOH$$

Carbodiimide Any organic compound with the general structure of

R-N=C=N-R

Carbohydrate An aldehyde or a ketone derivative of a polyhydroxy alcohol that is synthesized by living cells, e.g., sugar and starch. It is usually presented with the general formula $C_xH_{2x}O_x$.

Carboligase The enzyme that catalyzes the formation of acetoin from acetaldehyde.

Carbolith A trade name for lithium carbonate; it alters chemical transmitters in the CNS.

Carbometer An instrument that measures the carbon dioxide content of breath.

Carbomycin A 16-member ring macrolide antibiotic complex similar to leucomycin and erythromycin.

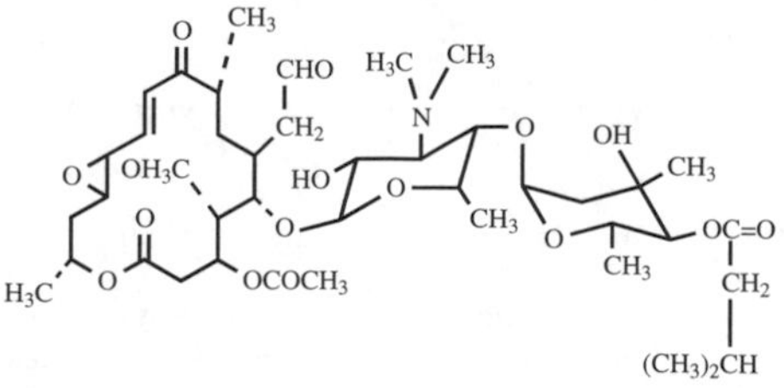

Carbon A chemical element with atomic weight 12 and valence 4.

Carbon-13 Nuclear Magnetic Resonance Spectroscopy A variant of nuclear magnetic resonance spectroscopy in which resonances in carbon-13 nuclei in natural carbon of organic molecules are examined instead of proton responses.

Carbon 14 Radioactive carbon with half-life of about 5700 years.

Carbon Assimilation See carbon dioxide fixation.

Carbon Clock The use of radiocarbon in carbon dating to establish the age of a biological sample.

Carbon Cycle The cyclic interconversion of carbon compounds, e.g., reactions where photosynthetic organisms reduce carbon dioxide to carbohydrates while the heterotrophic organisms oxidize the carbohydrates to carbon dioxide.

Carbon Dating See radiocarbon dating or carbon clock.

Carbon Dioxide Assimilation See Calvin cycle or C_3 pathway.

Carbon Dioxide Combining Power of Plasma The maximum amount of carbon dioxide that 100 mL of plasma can retain in the form of biocarbonate when the plasma is saturated with carbon dioxide.

Carbon Dioxide Effect See greenhouse effect.

Carbon Dioxide Fixation The photosynthetic conversion of carbon dioxide to carbohydrates via the Calvin cycle or C_3 pathway (see Calvin cycle or C_3 pathway).

Carbon Monoxide Dehydrogenase The enzyme that catalyzes the following reaction:

$$CO + H_2O + \text{racceptor} \rightleftharpoons CO_2 + \text{reduced acceptor}$$

Carbon Monoxide Hemoglobin See carboxyhemoglobin.

Carbon Monoxide Oxidase The enzyme that catalyzes the following reaction:

$$CO + H_2O + O_2 \rightleftharpoons CO_2 + H_2O_2$$

Carbon Reduction Cycle See Calvin cycle or carbon fixation.

Carbonate Dehydratase The enzyme that catalyzes the following reaction:

$$H_2CO_3 \rightleftharpoons CO_2 + H_2O$$

Carbonate Hydro-Lyase The systematic name for carbonate dehydratase.

Carbonate Ionophore I (mol wt 316) A lipophilic, neutral ionophore for carboxylate and hydroxide.

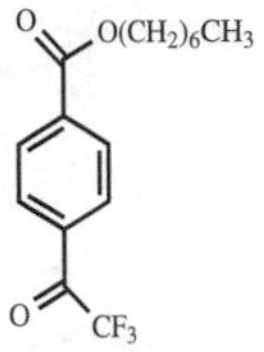

Carbonate Ionophore II (mol wt 407) A lipophilic, neutral ionophore for carboxylate and hydroxide.

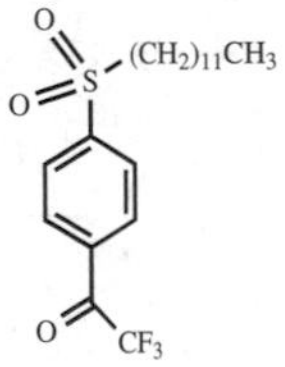

Carbonic Anhydrase A zinc-containing enzyme in erythrocytes that catalyzes the following reaction:

$$H_2CO_3 \rightleftharpoons CO_2 + H_2O$$

Carbonium Ion A positively charged carbon atom.

Carbonylcyanide 3-Chlorophenylhydrazone (mol wt 204) One of the potent uncouplers of oxidative phosphorylation.

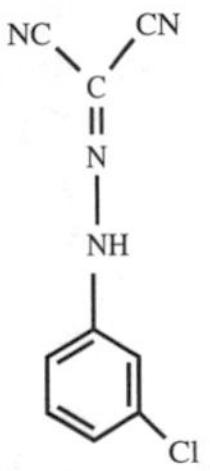

Carbonyl Group A pair of atoms consisting of a carbon atom linked to an oxygen atom by a double bond (C = O).

Carbonyl Reductase (NADP) The enzyme that catalyzes the following reaction:

$$\text{R-CHOH-R}' + NADP^+ \rightleftharpoons \text{R-CO-R}' + NADPH$$

Carboplatin (mol wt 371) An antitumor agent that produces crosslinking of cellular DNA and interferes with transcription.

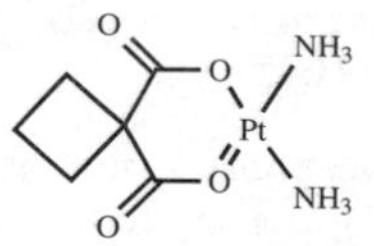

Carboprost (mol wt 369) Substance that produces strong, prompt contractions of uterine smooth muscle.

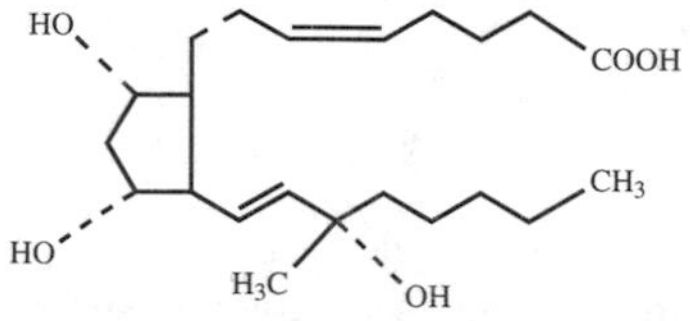

Carboquone (mol wt 321) An anticancer agent.

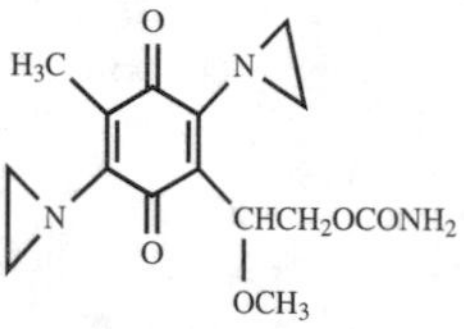

Carbowax A trade name for polyethylene glycol, a reagent used for cell fusion and protein purification.

Carboxamide Group Referring to a group of -CO-NH_2

Carboxin (mol wt 235) A systemic plant fungicide.

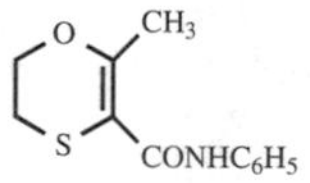

β-Carboxyaspartic Acid (mol wt 177) An amino acid found in ribosomal protein of *E. coli.*

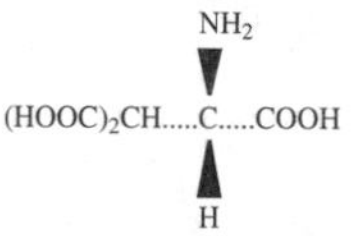

Carboxybiotin A biotin molecule with a molecule of carbon dioxide attached.

Carboxycathepsin A protease that releases dipeptide from the C terminus of a polypeptide. It also catalyzes the conversion of angiotensin I to angiotensin II.

Carboxydismutase See ribulose 1,5-*bis*-phosphate carboxylase.

γ-Carboxyglutamic Acid (mol wt 191) An amino acid found in blood coagulation proteins and in the proteins of calcified tissues.

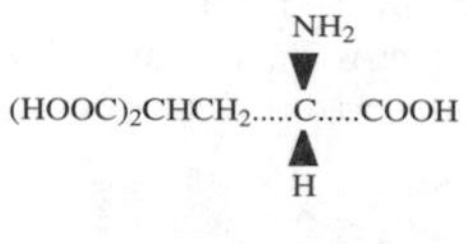

Carboxyhemoglobin A hemoglobin derivative formed by the union of hemoglobin with carbon monoxide.

Carboxyl-CoA Synthetase The enzyme that catalyzes the following reaction:

ATP + an α,ω-dicarboxylic acid + CoA

⇅

AMP + PPi + an ω-carboxyacyl-CoA

Carboxyl-Group The COOH of an organic acid.

Carboxylase Enzymes that catalyze the carboxylation or decarboxylation reactions.

Carboxylate Reductase The enzyme that catalyzes the following reaction:

An aldehyde + acceptor + H_2O

⇅

A carboxylate + reduced acceptor

Carboxylation The introduction of carbon dioxide into an organic compound.

Carboxylesterase The enzyme that catalyzes the following reaction:

A carboxylic ester + H_2O

⇅

An alcohol + carboxylate

Carboxylic Acid An organic compound containing a COOH group.

Carboxyl-Terminal The C terminus of a protein that has a free COOH group

Carboxymethyl Cellulose An ion exchanger used in ion exchange chromatography.

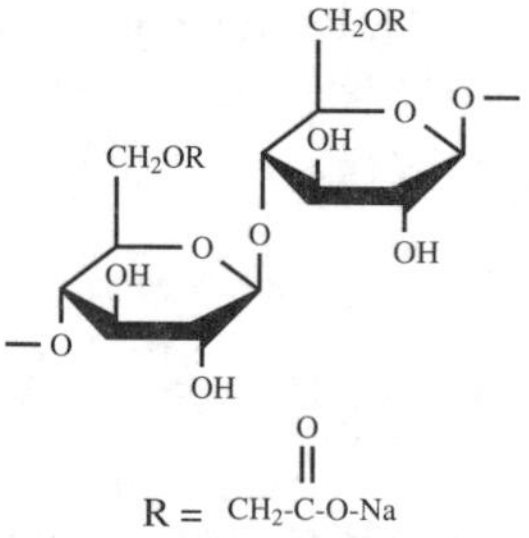

Carboxypeptidase An exopeptidase that catalyzes the release of a free amino acid from the C terminus of a polypeptide.

Carboxypeptidase A An exopeptidase that catalyzes the sequential hydrolysis of a peptide chain from the C terminus end but incapable of hydrolyzing peptide bonds involving aspartate, glutamate, arginine, lysine, and proline.

Carboxypeptidase B An exopeptidase that catalyzes the release of lysine or arginine from the C terminus of a polypeptide.

Carboxypeptidase H A carboxypeptidase from the storage granules of secretory cells with similar activity as carboxypeptidase M.

Carboxypeptidase M A membrane-bound carboxypeptidase that catalyzes the release of lysine or arginine from the C terminus of a polypeptide.

Carboxypeptidase P A membrane-bound carboxypeptidase that catalyzes the release of amino acids other than proline from the C terminus of a polypeptide at pH 4.

Carboxypolypeptidase Synonym of carboxypeptidase.

Carboxysome A polyhedral inclusion body observed in some blue-green algae containing the enzyme D-ribulose-1,5-bisphosphate carboxylyase.

Carbromal (mol wt 237) A sedative and hypnotic agent.

$$(C_2H_5)_2CBrCONHCONH_2$$

Carbubarb (mol wt 271) A sedative and hypnotic agent.

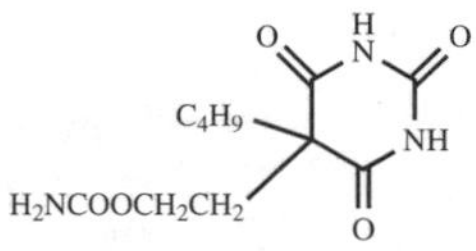

Carbutamide (mol wt 271) An antidiabetic agent.

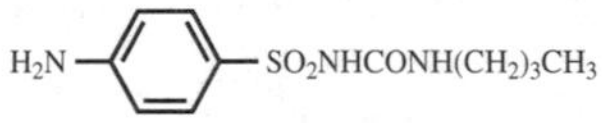

Carbuterol (mol wt 267) A bronchodilator with selectivity for airway smooth muscle receptors.

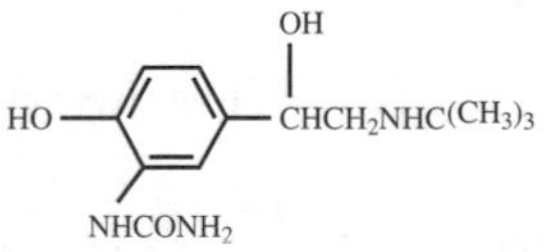

Carcino-Embryonic Antigen (CEA) An oncofetal glycoprotein antigen associated with certain tumors. It occurs in very small quantities in normal adults but its concentration increases significantly in patients with cancers of colon, breast, pancreas, and liver.

Carcinogen A cancer-inducing agent.

Carcinogenesis The development of cancer.

Carcinogenic Capable of inducing cancer.

Carcinogenicity The capacity to produce cancer.

Carcinogenophore The atoms or chemical groups in a chemical carcinogen responsible for the carcinogenic activity of the molecule.

Carcinoma A malignant tumor derived from epithelial cells, e.g., skin cancer or breast cancer.

Carcinomatosis Spread of carcinoma to multiple sites in the body.

Carcinostasis Inhibition of cancerous growth.

Cardene A trade name for nicardipine, used as an antianginal agent that inhibits calcium ion influx across cardiac and smooth muscle cells.

Cardia The esophageal opening of the stomach.

Cardiac Pertaining to the heart or esophageal opening.

Cardiac Glycoside A steroid glycoside such as oubain and digitalis capable of acting directly on the heart muscle and improving cardiac output. Cardiac glycosides are derived from plant tissue and contain steroid aglycone (either a C23 or C24 compund) complexed with sugar.

Cardiac Muscle The striated muscle of the heart consisting of individual heart muscle cells cross-linked by cell junction.

Cardiac Output The volume of blood pumped out of the heart in one minute.

Cardiac Puncture A method for withdrawing blood from an animal by inserting a syringe directly into the heart.

Cardialgia Pain in the heart.

Cardiatelia Incomplete development of the heart.

Cadioactive Having an influence on the heart or any agent that has an influence on the heart.

Cardiobacterium A genus of anaerobic, facultative aerobic, chemoorganotrophic, oxidase-positive, catalase-negative, Gram-negative bacteria.

Cardiography A technique for determination of the graphical movement of the heart using cardiographic instruments.

Cardiolipin The phospholipid (diphosphatidyl glycerol) used as an antigen in the Wasserman test for syphilis (see also diphosphatidyl glycerol).

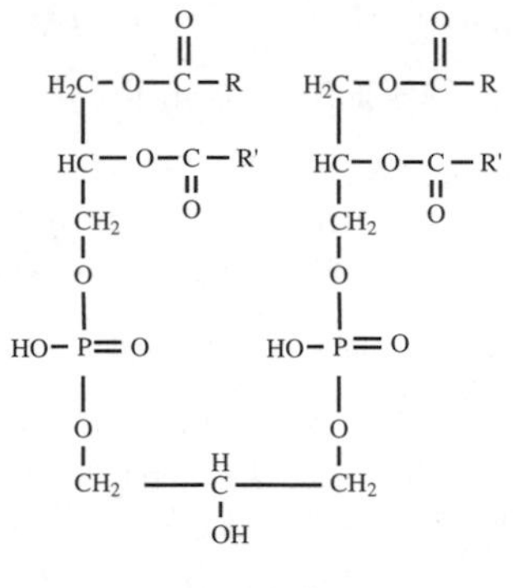

Cardiology The medical science that deals with the heart and diseases of the heart.

Cardiomegaly The enlargement of the heart.

Cardiomyoliposis The degeneration of fatty substances in the muscle of the heart.

Cardiomyoplasty A surgical technique to replace or reinforce the damaged cardiac muscle with skeletal muscle.

Cardiopathy Any disease of the heart.

Cardiopulmonary Pertaining to the heart and lung.

Cardiopulmonary Bypass A method by which the circulation is maintained while heart is deliberately stopped during heart surgery.

Cardiopulmonary Resuscitation An emergency procedure consisting of artificial respiration and manual external cardiac massage.

Cardioquin A trade name for quinidine polygalacturonate, used as an antiarrhythmic agent that prolongs the action potential.

Cardiorenal Pertaining to the heart and kidney.

Cardiorrhexis Rupture of the heart wall.

Cardioscope An instrument used for visual examination of the interior of the heart.

Cardiospasm The failure of the cardiac sphincter to relax during swallowing; results in esophageal obstruction.

Cardiotonic Substance with a favorable or tonic effect on the heart.

Cardiotoxic Substance with a toxic effect on the heart.

Cardiotoxin A toxic peptide from cobra venom capable of causing irreversible depolarization of cell membrane and contraction of skeletal muscle.

Cardiovascular Pertaining to the heart and blood vessels.

Carditis Inflammation of the heart.

Cardizem A trade name for diltiazem hydrochloride, used as an antianginal agent that inhibits calcium influx across the membrane of the cardiac and smooth muscle cells leading to the decrease of myocardial contractibility and oxygen demand.

Cardophyllin A trade name for aminophylline, used as a bronchodilator. It inhibits phosphodiesterase and prevents destruction of cAMP leading to the relaxation of smooth muscle of the bronchial airway and pulmonary blood vessels.

Carfecillin Sodium (mol wt 476) A semisynthetic antibiotic related to penicillin.

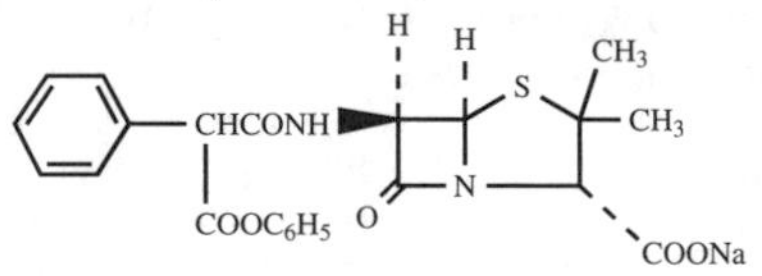

Carfimate (mol wt 175) A sedative and hypnotic agent.

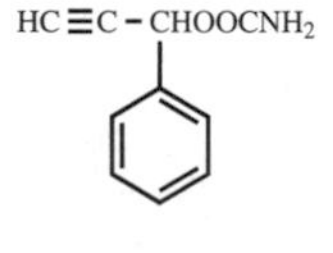

Cargutocin (mol wt 916) An oxytocic agent.

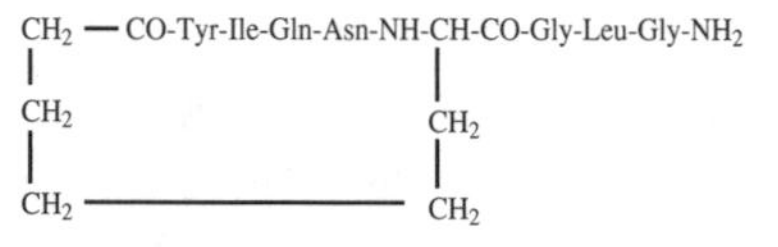

Caricain A protease with activity similar to that of papain.

Caries Muscular death or breakdown of bone.

Carindacillin (mol wt 495) A semisynthetic antibiotic related to penicillin that inhibits bacterial cell wall synthesis.

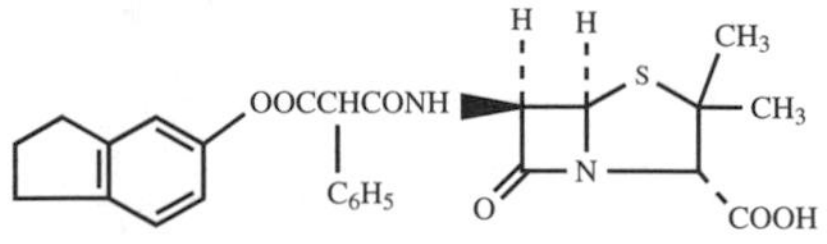

Cariogenic Any substance that promotes dental caries.

Cariostatic Any agent capable of inhibiting the progress of dental caries.

Carisoprodol (mol wt 260) A skeletal muscle relaxant capable of reducing transmission of impulses from the spinal cord to skeletal muscle.

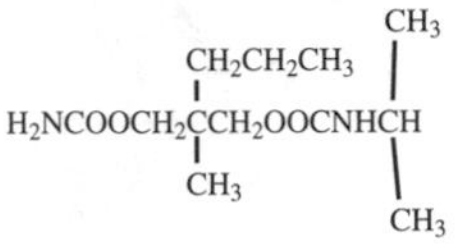

Carlsbad Salt A purgative solution consisting of sodium chloride, sodium bicarbonate, and anhydrosodium sulfate.

Carminitive Agent An agent that relieves flatulence used to treat gastric discomfort.

Carmofur (mol wt 257) An orally active derivative of fluorouracil possessing antineoplastic activity.

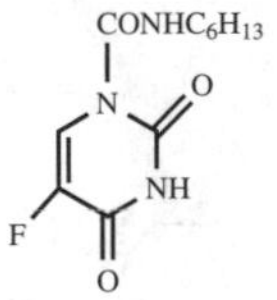

Carmol HC A trade name for hydrocortisone.

Carmustine (mol wt 214) An agent that cross-links strands of cellular DNA and disrupts the process of transcription leading to the cell death.

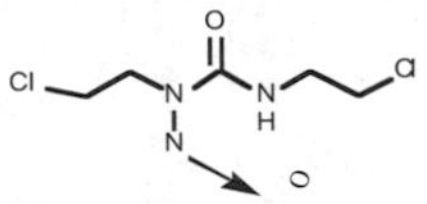

Carnidazole (mol wt 244) An antiprotozoal agent (trichomonacide).

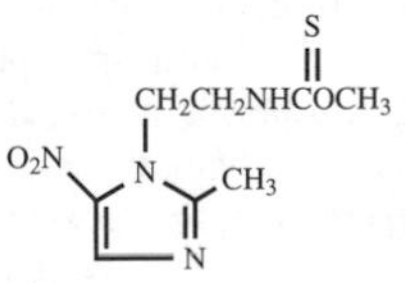

Carnitinamidase The enzyme that catalyzes the following reaction:

$$\text{L-Carnitinamide} + H_2O \rightleftharpoons \text{L-Carnitine} + NH_3$$

Carnitine An essential cofactor in fatty acid metabolism. It transports fatty acid from the cytoplasm to the mitochondria.

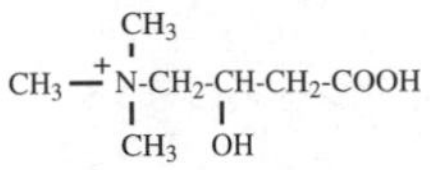

Carnitine Acetyltransferace The enzyme that catalyzes the following reaction:

$$\text{Acetyl-CoA} + \text{carnitine} \rightleftharpoons \text{CoA-SH} + \text{acetylcarnitine}$$

Carnitine Decarboxylase The enzyme that catalyzes the following reaction:

$$\text{Carnitine} \rightleftharpoons \text{2-Methylcholine} + CO_2$$

Carnitine Dehydrogenase The enzyme that catalyzes the following reaction:

$$\text{Carnitine} + NAD^+ \rightleftharpoons \text{3-Dehydrocarnitine} + NADH$$

Carnitine Palmitoyltransferace The enzyme that catalyzes the following reaction:

$$\text{Palmitoyl-CoA} + \text{carnitine} \rightleftharpoons \text{CoA-SH} + \text{palmitoylcarnitine}$$

Carnitor A trade name for L-carnitine, used to facilitate the transport of fatty acid into the mitochondria.

Carnivore An organism that feeds on other animals.

Carnosine (mol wt 226) A dipeptide of N-β-alanylhistidine found in the muscle of animals.

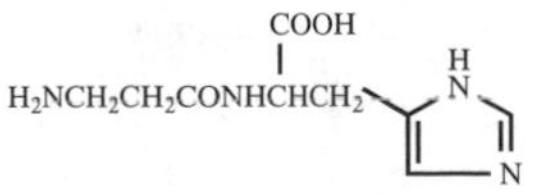

Carnosine Synthetase The enzyme that catalyzes the following reaction:

$$\text{ATP} + \text{L-histidine} + \beta\text{-alanine} \rightleftharpoons \text{AMP} + \text{PPi} + \text{carnosine}$$

α-Carotene (mol wt 537) An isomer of β-carotene found mainly in carrots.

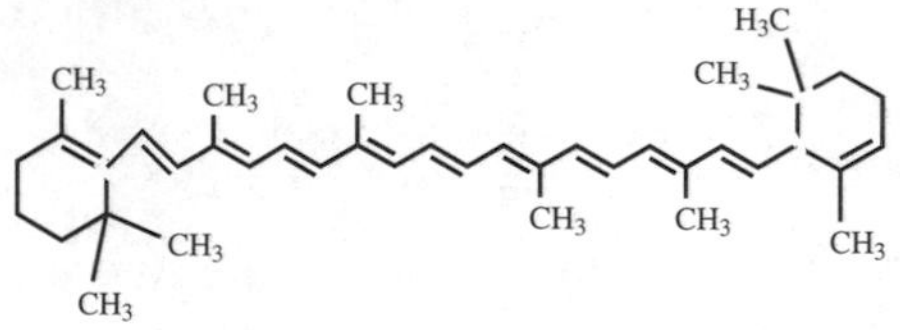

β-Carotene (mol wt 537) A widely distributed provitamin A found in various species of plants and animals, e.g., carrots.

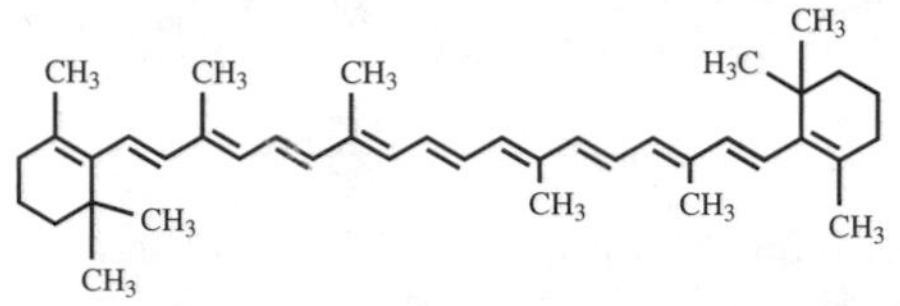

δ-Carotene (mol wt 537) A type of carotene that occurs in carrots and certain varieties of tomatoes.

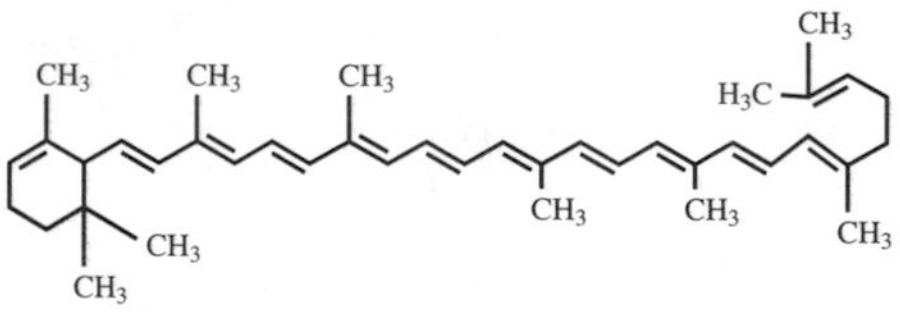

γ-Carotene A rare carotenoid found in *Penicillium sclerotiorum.*

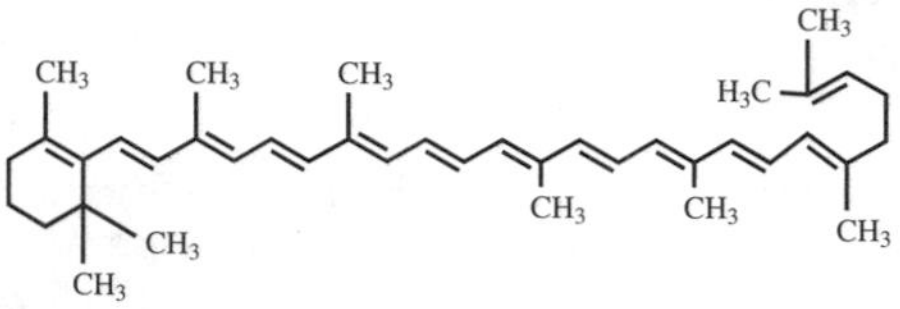

Carotenemia The increase of carotene in the blood causing the yellowish pigmentation of the skin (also known as carotinemia or xanthemia).

Carotenes A group of structurally related plant pigments consisting of a tetraterpene structure but lacking oxygen functionality such as hydroxyl groups.

Carotenoid A class of plant pigments based on a tertraterpene structure including carotenes and oxygen-containing derivatives of carotenes (e.g., xanthophylls). Carotenoids function as accessary pigments in photosynthesis and are widely distributed in microorganisms, plants, and animals.

Caroverine (mol wt 365) An antispasmodic agent.

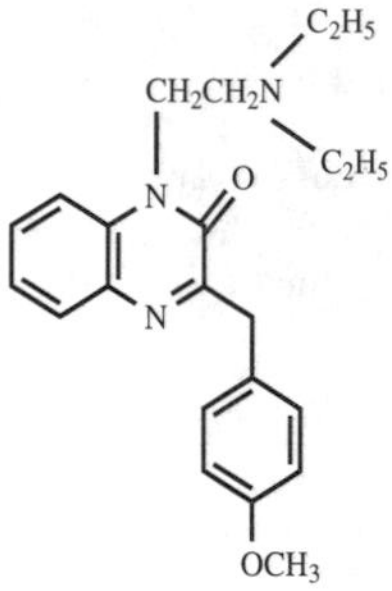

Caroxazone (mol wt 206) An antidepressant.

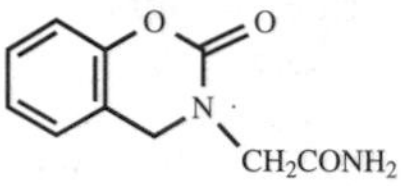

Carpetimycins An antibiotic related to thienamycin and olivanic acids and produced by *Streptomyces griseus.*

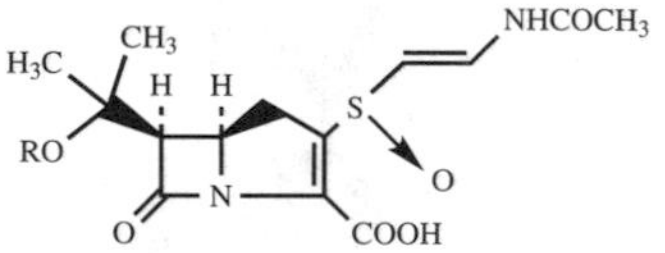

carpetimycin A: R = H
carpetimycin B: R = SO_3H

Carprofen (mol wt 274) An anti-inflammatory agent used for treatment of rheumatoid arthritis.

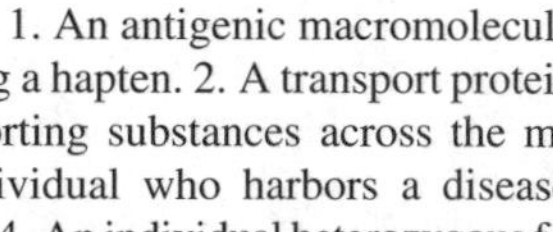

Carrageenan A mixture of sulfated galactans associated with cell walls of many rhodophycean algae, e.g., red seaweed (Rhodophyceae).

Carrier 1. An antigenic macromolecule capable of carrying a hapten. 2. A transport protein capable of transporting substances across the membrane. 3. An individual who harbors a disease-causing organism. 4. An individual heterozygous for a single recessive gene. 5. A stable isotope of an element used to mix with a radioisotope of an element to give a sufficient quantity for chemical experimentation.

Carrier Ampholyte The ampholyte that forms the pH gradient in isoelectric focusing.

Carrier Free Radioisotope Undiluted radioisotope.

Carrier Gas The inert gas that functions as the mobile phase in gas chromatography.

Carrier Protein A membrane protein that binds and transports a solute molecule across the membrane by undergoing conformational changes.

Carr-Price Reaction A colorimetric reaction for the determination of vitamin A.

Carsalam (mol wt 163) An analgesic agent.

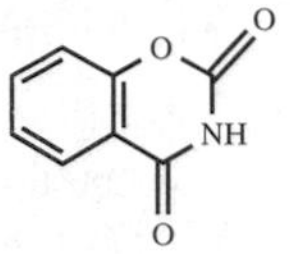

Carteolol (mol wt 292) An antihypertensive, antianginal, antiarrhythmic, and antiglaucoma agent that blocks beta-1 (myocardial) and beta-2 (pulmonary, vascular) receptor sites.

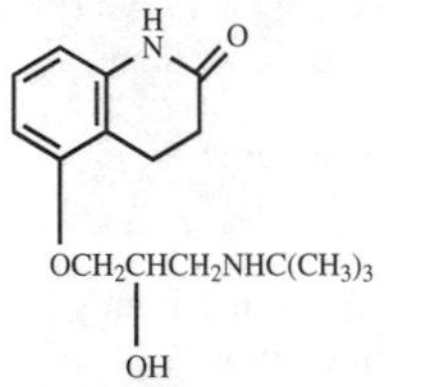

Carticaine (mol wt 284) A local anesthetic agent.

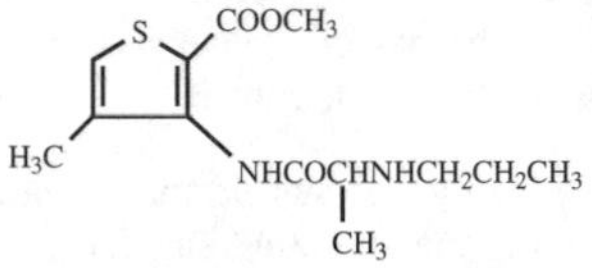

Cartrol A trade name for carteolol, a beta adrenergic blocking agent used as an antihypertensive drug.

Carubicin (mol wt 514) An antitumor antibiotic related to daunorubicin and doxorubicin.

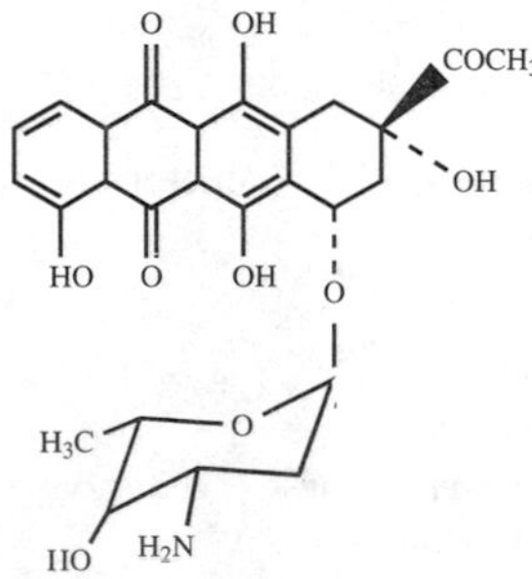

Carumonam (mol wt 466) A synthetic monocyclic β-lactam antibiotic that inhibits bacterial cell wall synthesis.

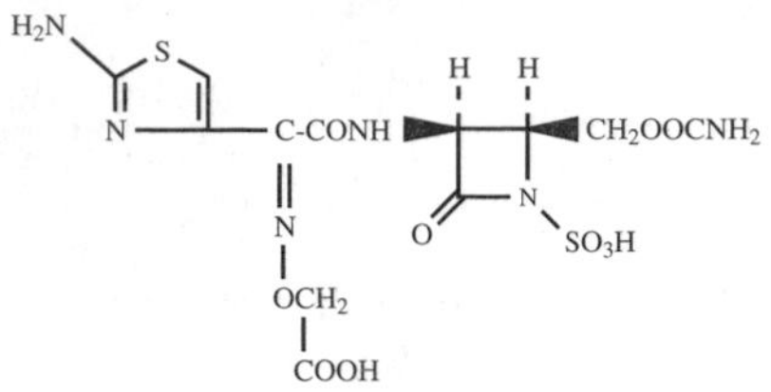

Carvacrol (mol wt 150) An insecticide (nematocide).

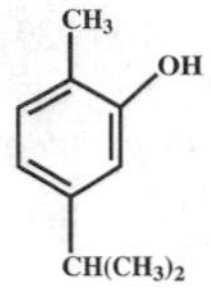

Carvedilol (mol wt 406) An antihypertensive and antianginal agent.

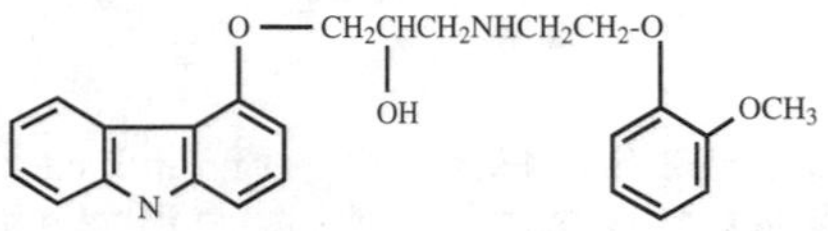

Caryo- A prefix meaning nucleus.

Cascade Reaction An interlinked series of enzymatic reactions in which the products of one reaction accelerate or catalyze a second reaction and so forth.

Cascara Sagrada Dried bark of *Rhamnus prushiana* used as a cathartic agent.

Casein A mixture of related phosphoproteins occurring in milk and cheese.

Casein Kinase Synonym of protein kinase.

Casodex A trade name for bicalutamide, an antiandrogen.

Cassaidine (mol wt 408) A cardiotonic agent from the bark of *Erythrophleum guineense* (Leguminosae).

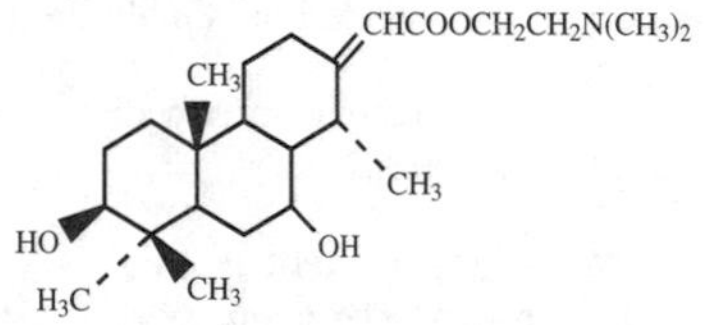

Cassaine Mol wt 406) A cardiotonic agent from the bark of *Erythrophleum guineense* (Leguminosae).

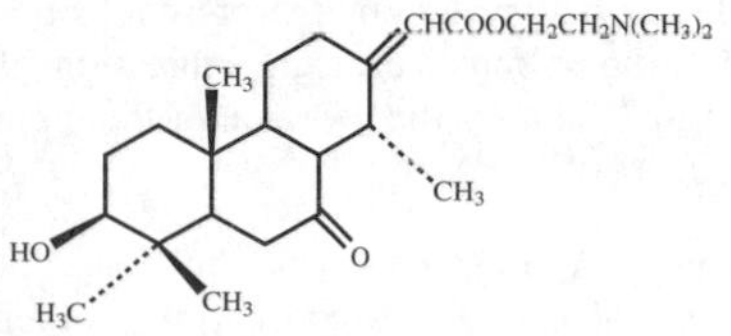

Cassette Loci containing functionally related nucleotide sequences that lie in tandem and can be substituted for one another, e.g., the mating type reversals observed in yeast (removing one cassette and replacing it by another containing a different nucleotide sequence).

Cassette Model A model to explain mating-type interconversion in yeast (see cassette).

Castanospermine (mol wt 189) A polyhydroxyl alkaloid isolated from the seeds of Australian leguminous tree (*Castanospermum australe*) and an inhibitor of glycosidase.

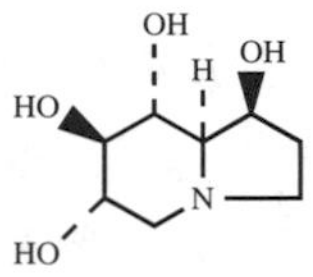

Castle's Intrinsic Factor A thermolabile mucoprotein capable of promoting vitamin B_{12} absorption. It occurs in normal gastric juice, but is deficient in patients with pernicious anemia.

Castor Oil A common organic solvent used as a water-repellent coating.

Castration Removal of testes or ovaries.

CAT Abbreviation for 1. chloramphenicol acetyltransferase; 2. computerized axial tomography or computer-assisted tomography; 3. a combination drug containing cytosine-arabinoside, adriamycin, and 6-thioguanine.

CAT Assay Abbreviation for chloramphenicol acetyl transferase assay.

CAT Scan Abbreviation for computerized axial tomography or computer-assisted tomography.

Catabolic Pertaining to catabolism.

Catabolism Metabolic breakdown or degradation of a complex molecule into simple products.

Catabolite Any metabolic intermediate generated in catabolic reactions.

Catabolite Activator Protein (CAP) A cyclic AMP-binding protein. The complex of CAP and cyclic AMP stimulates transcription by binding to certain promoter sites on DNA.

Catabolite Repression The repression of an inducible enzyme system by the presence of a preferred carbon source, e.g., the repression of *lac* operon in *E. coli* by the presence of the preferred glucose.

Cataflam A trade name for diclofenac potassium, an analgesic, antipyretic, and anti-inflammatory agent.

Catalase The enzyme that catalyzes the conversion of H_2O_2 to H_2O and O_2.

Catalysis A catalyst-mediated catabolic reaction.

Catalyst A substance that accelerates a chemical reaction without itself being consumed or changed by the reaction.

Catalytic Antibody An antibody that catalyzes a specific chemical reaction by lowering the free energy of activation.

Catalytic Center See active site or catalytic site.

Catalytic Site The region of an enzyme that binds substrate, forms enzyme substrate complex, and catalyzes the reaction.

Catalytic Subunit The subunit of the regulatory enzyme that binds substrate and possesses enzymatic activity.

Cataplasia The degenerative reversion of a cell or tissue to an embryonic state.

Catapres A trade name for clonidine, an antihypertensive agent.

Cataract Loss of transparency of the lens of the eye (opacity of the eye).

Catarrh The excessive secretion of thick phlegm or mucous by the mucous membrane of the nose, nasal sinuses, or nasopharynx.

Catatoxic Steroid A steroid that stimulates the activity of drug-metabolizing enzymes.

Catechin (mol wt 290) An astringent (diarrheal) agent.

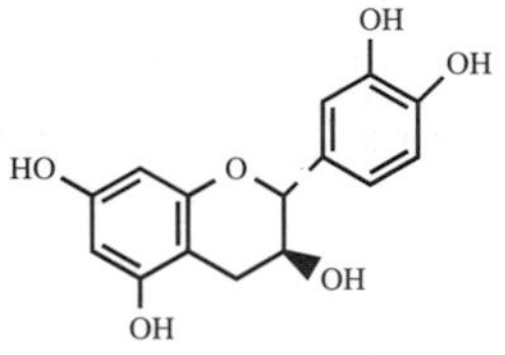

Catechol (mol wt 110) An aromatic alcohol.

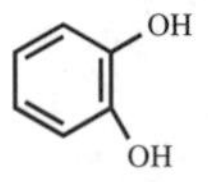

Catechol 1,2-Dioxygenase The enzyme that catalyzes the following reaction:

$$\text{Catechol} + O_2 \rightleftharpoons \text{cis-cis-Muconate}$$

Catechol 2,3-Dioxygenase The enzyme that catalyzes the following reaction:

$$\text{Catechol} + O_2 \rightleftharpoons \text{2-Hydroxymuconate semialdehyde}$$

Catechol Methyltransferase The enzyme that catalyzes the following reaction:

S-Adenosylmethionine + catechol

$\rightleftharpoons$

S-Adenosylhomocysteine + guaiacol

Catechol Oxidase The enzyme that catalyzes the following reaction:

$$2\ \text{O-diphenol} + O_2 \rightleftharpoons 2\ \text{O-quinone} + 2H_2O_2$$

Catecholamine A neurotransmitter derived from tyrosine, such as dopa, dopamine, epinephrine, or norepinephrine.

Catenane Structure consisting of two or more interlocking rings.

Catenated DNA Dimer A DNA molecule consisting of two interlocking circular DNAs.

Catharsis Purging or cleansing out of the bowels by giving a laxative.

Cathartic Any agent or drug that promotes movement and evacuation of intestinal content.

Cathepsin A proteolytic enzyme.

Cathepsin A A serine-type carboxypeptidase with broad specificity.

Cathepsin B An endopeptidase that preferentially cleaves peptide bonds between arginines.

Cathepsin C A dipeptidyl aminopeptidase that releases N-terminal depeptides from a polypeptide.

Cathepsin D A protease with activity similar to pepsin A.

Cathepsin G A protease that preferentially cleaves peptide bonds involving leucine, tyrosine, phenylalanine, methionine, tryptophan, glutamine, and asparagine.

Cathepsin H A protease with both aminopeptidase and carboxypeptidase activities.

Cathepsin L A protease that cleaves proteins with a preference for residues bearing a large hydrophobic side chain at the P2 position.

Cathepsis Protein hydrolysis by cathepsin.

Catheter A flexible slender tube used to introduce or remove fluid from a body cavity or body passageway.

Catheterization The introduction of a catheter into a body passage for medical or diagnostic purposes.

Cathode The negative electrode.

Catholyte The electrolyte that is in immediate contact with the cathode in the isoelectric focusing.

Cation A positively charged ion.

Cation Exchanger A negatively charged ion-exchange resin that is capable of binding and exchanging cationic molecules.

Cationic Detergent A positively charged surface-active agent.

Cationic Protein A positively charged antimicrobial protein present in the granules of phagocytic cells.

Ca^{++}-Transporting ATPase The enzyme that catalyzes the reaction in which hydrolysis of ATP is coupled with transport of calcium ions.

CAU A genetic code for the amino acid histidine.

CauI (AvaII) A restriction endonuclease from *Chloroflexus aurantiacus* with the following specificity:

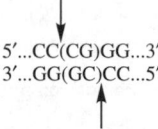

```
  ↓
5′...GG(A/T)CC...3′
3′...CC(T/A)GG...5′
           ↑
```

CauII A restriction endonuclease from *Chloroflexus aurantiacus* with the following specificity:

```
     ↓
5′...CC(CG)GG...3′
3′...GG(GC)CC...5′
          ↑
```

CauIII (PstI) A restriction endonuclease from *Chloroflexus aurantiacus* with the same specificity as PstI.

Caulobacter A genus of chemoorganotrophic, strictly anaerobic, prosthecate bacteria.

CAV Abbreviation for a combination drug containing CCNU, adriamycin, and vinblastine.

CAVe Abbreviation for a combination drug containing cyclophosphamide, adriamycin, and velban.

Caveola A closed plasmalemal vesicle.

Caveolin A protein that lines the cytoplasmic surface of the caveola.

Caverject A trade name for alprostadil, a prostaglandin used for the relaxation of vascular smooth muscle.

Cavity 1. Hollow space in the body. 2. Loss of tooth structure due to decay.

Cavity Slide A microscope slide with a circular depression on one side.

CAVP Abbreviation for a combination drug containing cytoxan, adriamycin, vincristine, and prednisone.

CB Abbreviation for cytochalasin B.

C_{2b} Abbreviation for a subcomponent of C2 complement.

C_{3b} Abbreviation for a subcomponent of C_3 complement.

C_{4b} Abbreviation for a subcomponent of C_4 complement

C_{5b} Abbreviation for a subcomponent of C_5 complement

CB Agar Abbreviation for chocolate blood agar.

CBA Mouse Abbreviation for a strain of inbred mouse.

CBA/N Mouse Abbreviation for a CBA mutant mouse.

CBB Abbreviation for Coomassie brilliant blue.

CBC Abbreviation for complete blood count.

CBD Abbreviation for 1. cellulose-binding domain; 2. cyclosporin A-binding domain.

CBE Abbreviation for a combination drug containing cytoxan, BCNU, and etoposide.

CBF Abbreviation for centromere-binding factor.

CBG Abbreviation for 1. corticosteroid-binding globulin; 2. cortisol-binding globulin.

CBH Abbreviation for cellobiohydrolase.

CbiI A restriction endonuclease from *Clostridium bifermentans* B4 with the following specificity:

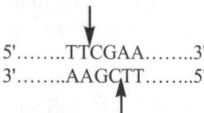

5'........TT↓CGAA........3'
3'........AAGC↑TT........5'

Cbl Abbreviation for cobalamin.

CBPPA Abbreviation for a combination drug containing cytoxan, bleomycin, procarbazine, prednisone, and adriamycin.

CBV Abbreviation for 1. a combination drug containing CCNU, bleomycin, and velban; 2. circulating blood volume.

CBVD Abbreviation for a combination drug containing CCNU, bleomycin, velban, and daxamethasone.

Cbz Abbreviation for carbobenzoxy.

CBZ Abbreviation for carbamazepine.

CBZ-Amino Acid An amino acid in which the amino group has been protected by attachment of a carbobenzoxy group.

cc Abbreviation for cubic centimeter (milliliter).

CC Abbreviation for chondrocalcin.

C3c Abbreviation for the subcomponent of C3 (third component of complement).

CCA A genetic code for the amino acid proline.

CCAT Abbreviation for 1. CCA-terminal in tRNA; 2. conglutinating complement adsorption test.

CCA-Terminal The 3′-end of the tRNA where the amino acid attaches during the process of protein synthesis.

CCAVV Abbreviation for a combination drug containing CCNU, cyclophosphamide, adriamycin, vincristine, and VP16.

CCB Abbreviation for calcium channel blockers.

CCC A genetic code for the amino acid proline.

CCCC Abbreviation for centrifugal counter-current chromatography.

cccDNA Abbreviation for covalently closed circular DNA.

CCCP Abbreviation for carbonyl cyanide 3-chlorophenylhydrazone.

C_2-Cer Abbreviation for N-acetyl sphingosine.

C_2-Ceramide Abbreviation for N-acetyl sphingosine.

C_6-Cer Abbreviation for N-hexanosyl-acetyl sphingosine.

C_6-Ceramide Abbreviation for N-hexanosyl-acetyl sphingosine.

CCD Abbreviation for charged couple device.

CCE Abbreviation for carboline-carboxylic acid ester.

CCF Abbreviation for 1. cardiolipin complement fixation; 2. cephalin-cholesterol flocculation.

CCF Agar Abbreviation for cefoxitin-cyclosterine fructose agar.

CCFE Abbreviation for a combination drug containing cyclophosphamide, cisplatin, fluorouracil, and estramustine.

CCG The genetic code for the amino acid proline.

CciNI A restriction endonuclease from *Curtobacterium citreum* N with the following specificity:

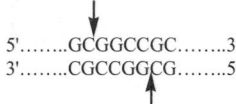

5'........GCGGCCGC........3'
3'........CGCCGGCG........5'

CCK Abbreviation for hormone cholecytokinin.

CCK-A Abbreviation for cholecystokinin-A.

CCK-OP Abbreviation for cholecystokinin octapeptide.

CCK-PZ Abbreviation for cholecystokinin pancreozymin.

CCL Abbreviation for carcinoma cell line.

CCM Abbreviation for a combination drug containing cyclophosphamide, CCNU, and methotrexate.

CCMA Abbreviation for a combination drug containing cyclophosphamide, CCNU, methotrexate, and adriamycin.

cCMP Abbreviation for cyclic cytidine monophosphate.

CCMT Abbreviation for catechol-O-methyl transferase.

CCNU Abbreviation for lomustine.

CCOB Abbreviation for a combination drug containing CCNU, cyclophosphamide, oncovin, and bleomycin.

CcP Abbreviation for cytochrome c peroxidase.

CcR Abbreviation for cytochrome c reductase.

CcRase Abbreviation for cytochrome c reductase.

CCSP Clara cell secretory protein.

CCU The genetic code for the amino acid proline.

CCV Abbreviation for a combination drug containing CCNU, cytoxan, and vincristinc.

CCVB Abbreviation for a combination drug containing CCNU, cyclophosphamide, vincristine, and bleomycin

CCVPP Abbreviation for a combination drug containing CCNU, cylophosphamide, vincristine, procarbazine, and prednisone.

^{115}Cd Abbreviation for radioactive cadmium.

CD Abbreviation for cluster of differentiation, a type of cell surface marker designated as CD1, 2, 3,......CD130 for identification of different groups of leukocytes. They are glycoproteins expressed on various cell types and act as cell markers, receptors, oncogenic markers, and cell to cell adhesion function-associated antigens. For example, CD4 (marker for helper T cells and delayed hypersensitivity T cells); CD8 (marker for suppressor T cells and cytotoxic T cells).

Cd Abbreviation for 1. chemical element cadmium with atomic weight 112, valence 2 and 2. candela, a unit of luminous intensity.

CD3 Cell surface marker on T cells associated with T cell receptors.

CD4 Cell surface marker of helper T cells and delayed hypersensitivity T cells.

CD8 Cell surface marker of suppresser and cytotoxic T cells

CD44 Glycoprotein that plays a general role in cell adhesion and lymphocyte homing.

CD59 A glycoprotein that is a potent inhibitor of complement membrane attack complex.

C_3d Abbreviation for a fragment of C3 complement.

CD Marker A series of antigenically distinct protein molecules occurring on the surface of leukocytes and other cell types and used in the characterization of leukocytes.

CDA Abbreviation for 1. chenodeoxycholic acid; 2. chloro-deoxy-adenosine; 3. completely denatured alcohol.

CdA A trade name for cladribine, an antineoplastic agent.

C^{14}-Dating A technique for estimating the age of carbon in biological remains using ^{12}C and ^{14}C (see also radiocarbon dating).

cdb3 Abbreviation for cytoplasmic domain of band 3.

CDC Abbreviation for 1. cell division cycle; 2. chenodexoycholate.

Cdc Genes Abbreviation for cell division cycle genes.

CDC Protein Abbreviation for cell division cycle protein.

CDDP A trade name for cisplatin, an alkylating agent used as an antineoplastic agent.

cDDP Abbreviation for *cis*-diamine dichloroplatin.

CDGS Abbreviation for carbohydrate deficient glycoprotein syndrome.

CDH Abbreviation for ceramide dihexoside.

CDI Abbreviation for collision-induced dissociation.

Cdi27I (EcoRII) A restriction endonuclease from *Citrobacter diversus* RFL27 with the same specificity as EcoRII.

C2-Dihydroceramide Abbreviation for N-acetyldihydroceramide.

CDK Abbreviation for 1. cell division protein kinase; 2. cyclin-dependent kinase.

CdmCl Abbreviation for guanidinium chloride.

cDNA Abbreviation for complementary DNA. The DNA synthesized by reverse transcriptase with an RNA template.

C-DNA Abbreviation for C form of DNA. It consists of a right-handed double helix with 9.3 nucleotide residues per turn.

cDNA Library A DNA library that contains DNA genes made from a population of mRNAs using reverse transcriptase. It is different from the normal gene library because it contains only transcribed DNA linked to a vector and cloned in a suitable host.

cDNA Probe Radioactively labeled or enzyme-linked cDNA (single stranded) used to determine the presence or absence of specific sequence in RNA or DNA.

CDNB Abbreviation for 1-chloro-2,4-dinitrobenzene.

CDP Abbreviation for cytidine diphosphate.

CDP-Abequose Epimerase The enzyme that catalyzes the following reaction:

CDP-3,6-dideoxy-D-glucose

⇅

CDP-3,6-dideoxy-D-mannose

CDP-Acylglycerol Arachidonyltransferase The enzyme that catalyzes the following reaction:

Arachdonyl-CoA + CDP-acylglycerol

⇅

CoA + CDP-diacylglycerol

CDPC Abbreviation for cytosine diphosphate choline.

CDP-Choline A cytidine diphosphate derivative of choline that serves as a donor of choline.

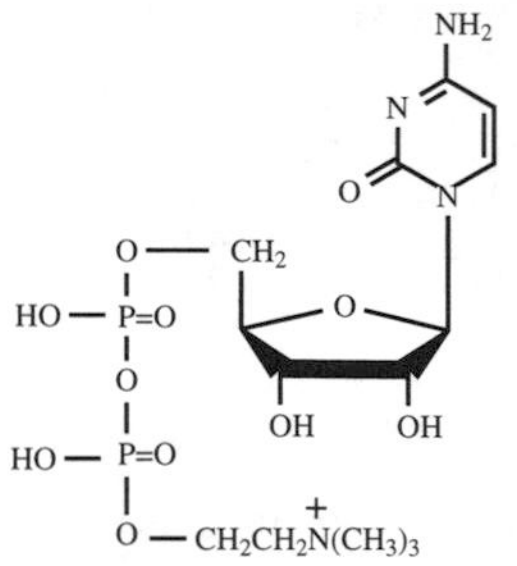

CDP-Diacylglycerol An intermediate in biosynthesis of some phosphoglycerides.

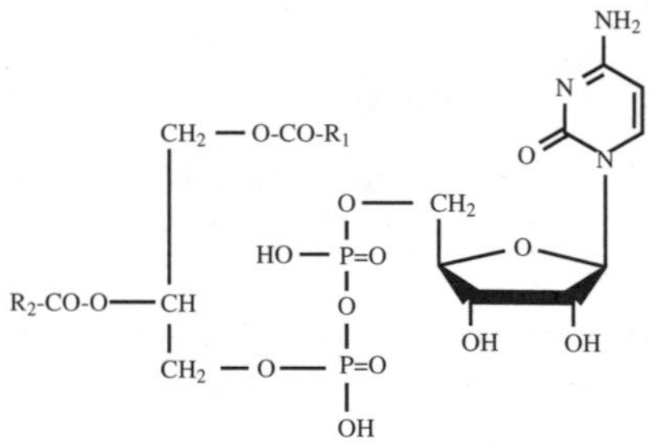

R = fatty acid

CDP-Diacylglycerol-Inositol 3-Phosphatidyl Transferase The enzyme that catalyzes the following reaction:

CDP-diacylglycerol + inositol

⇅

CMP + phosphatidylinositol

CDP-Etn Abbreviation for CDP-ethanolamine.

CDP-Glucose 4,6-Dehydratase The enzyme that catalyzes the following reaction:

CDP-glucose

⇅

CDP-4-dehydro-6-deoxy-D-glucose + H_2O

CDP-Glycerol Glycerophosphotransferase The enzyme that catalyzes the following reaction:

CDP-glycerol + $(\text{glycerophosphate})_n$

⇅

CMP + $(\text{glycerophosphate})_{n+1}$

CDP-Glycerol Pyrophosphatase The enzyme that catalyzes the following reaction:

CDP-glycerol + H_2O

⇅

CMP + glycerol 3-phosphate

CDPK Abbreviation for calmodulin-like domain protein kinase.

CDP-Ribitol Ribitolphosphotransferase The enzyme that catalyzes the following reaction:

$$\text{CDP-ribitol} + (\text{ribitol phosphate})_n \rightleftharpoons \text{CMP} + (\text{ribitol phosphate})_{n+1}$$

CDR Abbreviation for 1. calcium-dependent regulator protein; 2. complimentarily determining region.

CD2R Abbreviation for a marker molecule of 50 kDa, restricted to activated T cells and some NK cells.

CD-ROM Abbreviation for compact disc-read only memory.

Ce Symbol for the chemical element cerium with atomic weight 140, valence 2.

CE Abbreviation for 1. calexcitin; 2. chlorimuron ethyl; 3. capillary electrophoresis; 4. chick embryo; 5. cholesteryl ester; 6. cation exchanger; 7. cholera exotoxin; 8. cytopathic effect; 9. chloroform/ether mixture.

C_3e Abbreviation for a subcomponent of C3 (third component of a complement).

CEA Abbreviation for carcino-embryonic-antigen, the antigen present in certain cancers. It occurs in very small quantities in normal adults but its concentration increases significantly in patients with cancers of colon, breast, pancreas, and liver.

CEB Abbreviation for a combination drug containing carboplastin, etoposide, and bleomycin.

C/EBP Abbreviation for CCAAT/enhancer-binding protein.

Ceclor A trade name for cefaclor, an antibiotic that inhibits bacterial cell wall synthesis.

Cecon A trade name for vitamin C (ascorbic acid).

Cecropins A major class of antibacterial proteins in hemolymph of some insects (e.g., cecropia moth) as part of the immune response.

Cedax A trade name for ceftibuten hydrochloride, a third generation cephalosporin antibiotic.

Cedilanid A trade name for deslanoside, a drug that promotes movement of calcium from extracellular to intracellular cytoplasm and inhibits Na^+-K^+-activated ATPase.

Cedocard SR A trade name for isosorbide dinitrate, an antianginal agent.

CeeNU A trade name for lomustine, used as an antineoplastic agent that cross-links strands of cellular DNA and interferes with transcription.

Ceetamol A trade name for acetaminophen, an anti-inflammatory, antipyretic, and analgesic agent that inhibits prostaglandin synthesis.

CEF Abbreviation for chicken embryo fibroblast.

Cefaclor (mol wt 386) A semisynthetic cephalosporin antibiotic that inhibits bacterial cell wall synthesis.

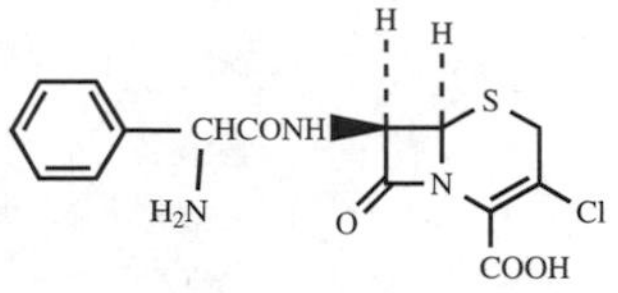

Cefadroxil (mol wt 381) A semisynthetic cephalosporin antibiotic that inhibits bacterial cell wall synthesis.

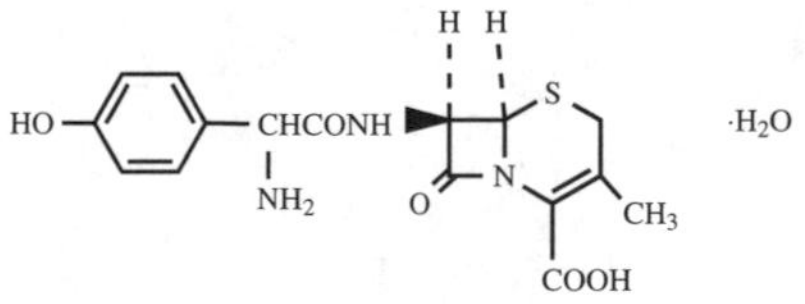

Cefadyl A trade name for cephapirin sodium, an antibiotic that inhibits bacterial cell wall synthesis.

Cefamandole (mol wt 463) A semisynthetic cephalosporin antibiotic that inhibits bacterial cell wall synthesis.

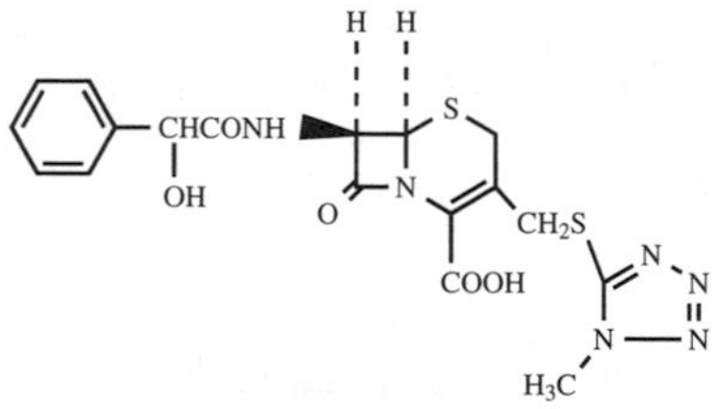

Cefatrizine (mol wt 463) A semisynthetic cephalosporin antibiotic that inhibits bacterial cell wall synthesis.

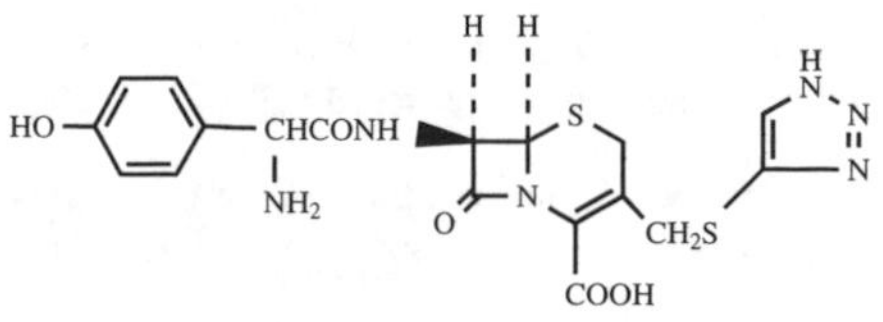

Cefazedone (mol wt 548) A semisynthetic cephalosporin antibiotic that inhibits bacterial cell wall synthesis.

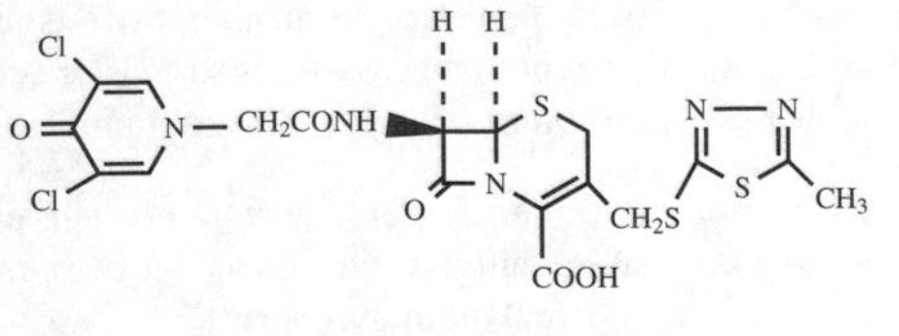

Cefazolin (mol wt 455) A semisynthetic antibiotic derived from 7-aminocephalosporanic acid that inhibits bacterial cell wall synthesis.

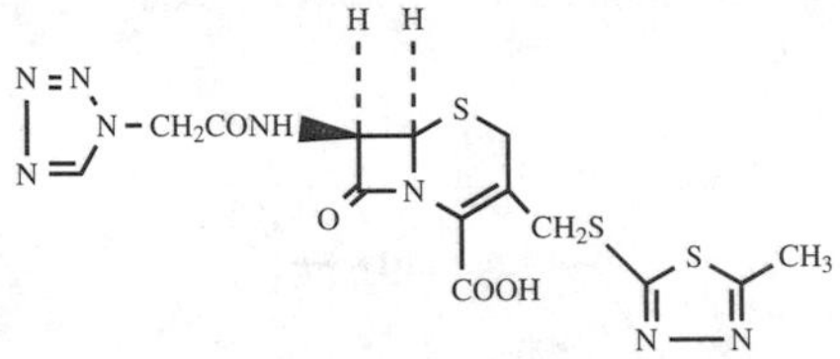

Cefbuperazone (mol wt 628) A broad spectrum antibiotic.

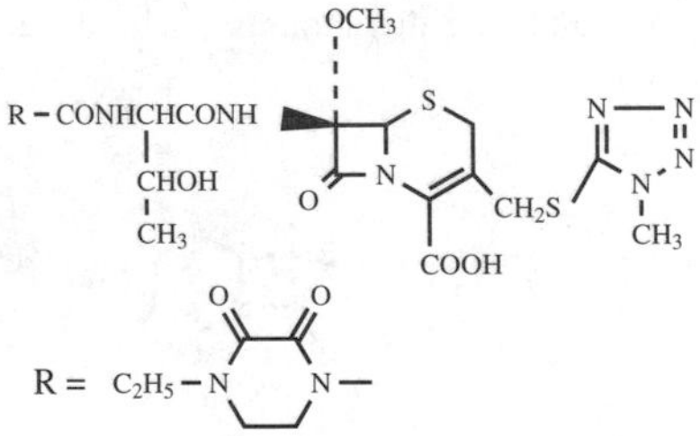

Cefepime (mol wt 481) An antibacterial agent.

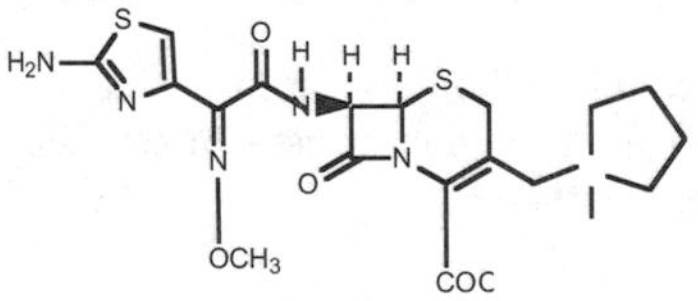

Cefixime (mol wt 453) A third-generation cephalosporin antibiotic that inhibits bacterial cell wall synthesis.

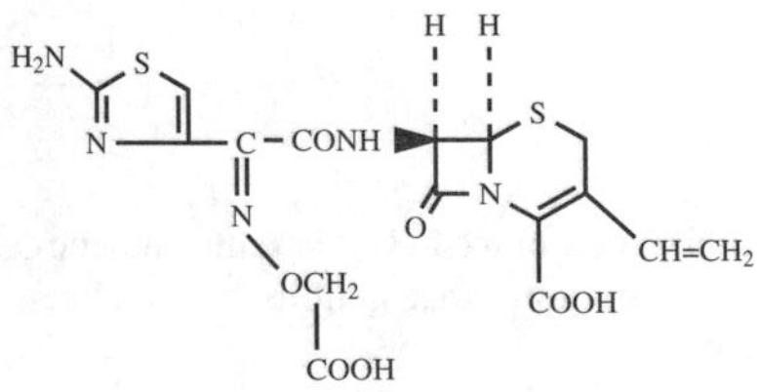

Cefizox A trade name for ceftizoxime sodium, an antibiotic that inhibits bacterial cell wall synthesis.

Cefmenoxime (mol wt 512) A third-generation cephalosporin antibiotic that inhibits bacterial cell wall synthesis.

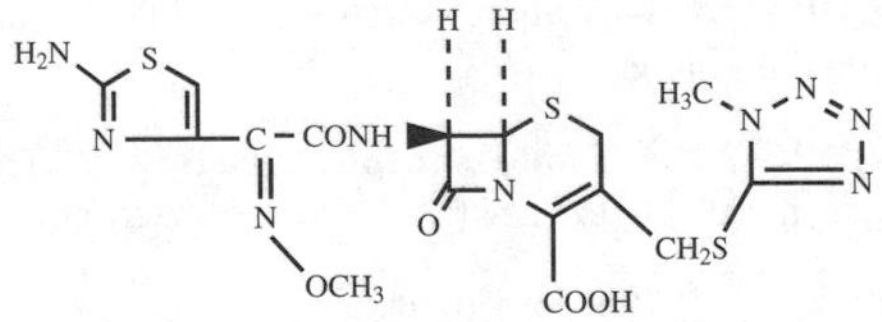

Cefmetazole (mol wt 472) A semisynthetic antibiotic derived from cephamycin, it inhibits bacterial cell wall synthesis.

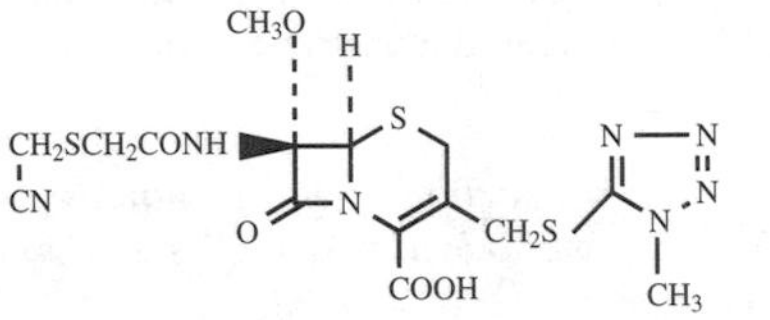

Cefminox (mol wt 520) A semisynthetic, broad spectrum cephamycin antibiotic that inhibits bacterial cell wall synthesis.

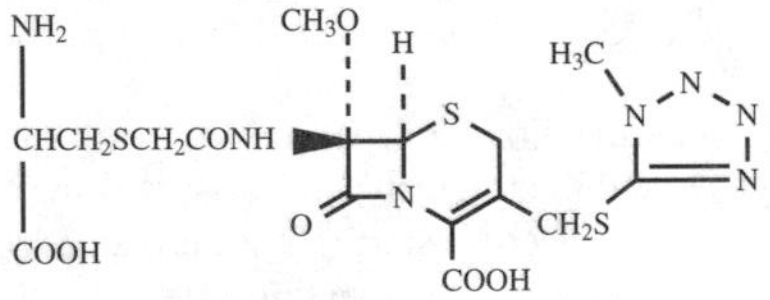

Cefobid A trade name for cefoperazone sodium, an antibiotic that inhibits bacterial cell wall synthesis.

Cefodizime (mol wt 585) A third-generation cephalosporin antibiotic derived from cefotaxime that inhibits bacterial cell wall synthesis.

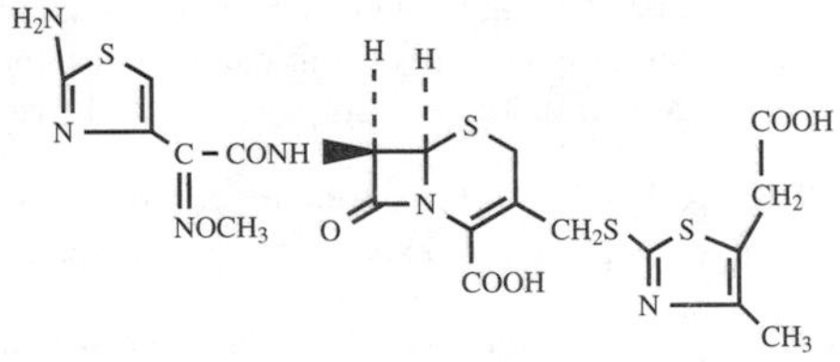

Cefonicid (mol wt 543) A semisynthetic cephalosporin antibiotic that inhibits bacterial cell wall synthesis.

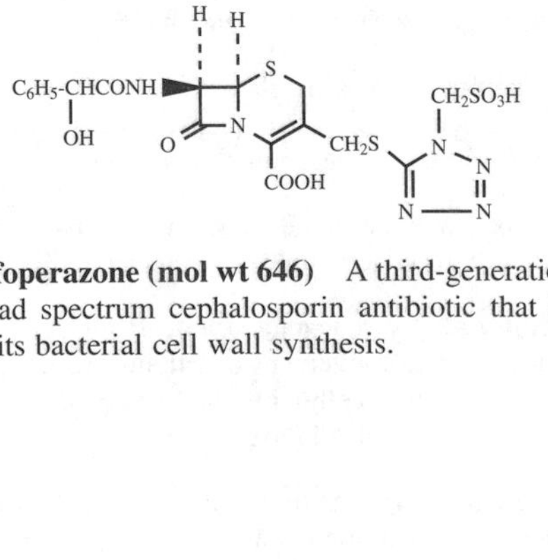

Cefoperazone (mol wt 646) A third-generation, broad spectrum cephalosporin antibiotic that inhibits bacterial cell wall synthesis.

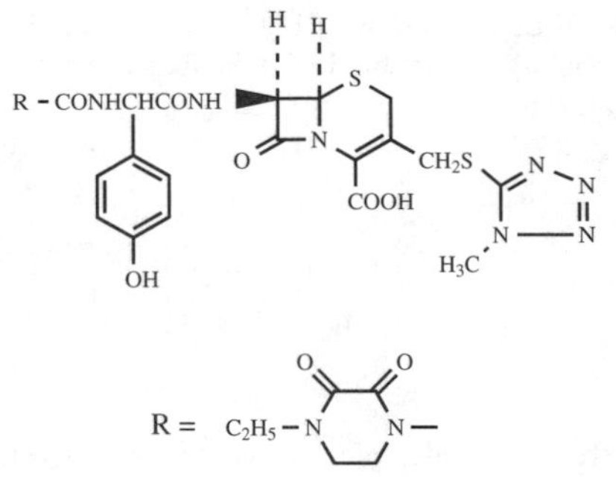

Ceforanide (mol wt 520) A semisynthetic cephalosporin that inhibits bacterial cell wall synthesis.

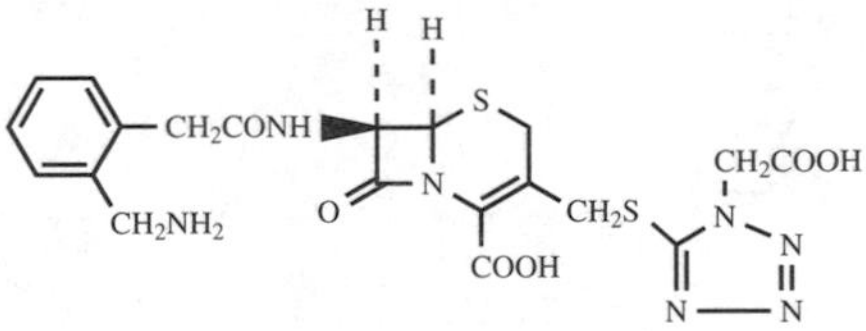

Cefotan A trade name for cefotetan, an antibiotic that inhibits bacterial cell wall synthesis.

Cefotaxime (mol wt 455) A broad spectrum third-generation cephalosporin antibiotic that inhibits bacterial cell wall synthesis.

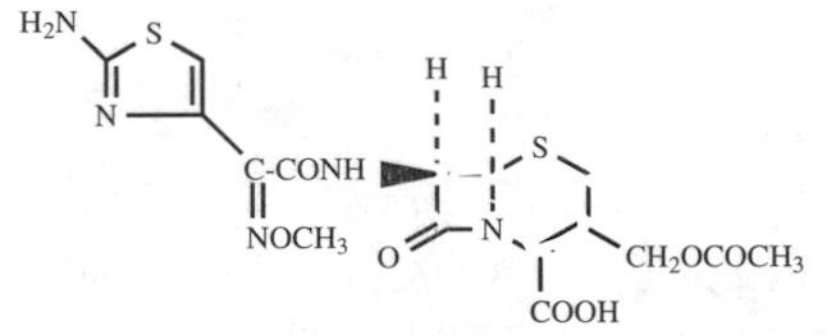

Cefotetan (mol wt 576) A semisynthetic antibiotic derived from cephamycin that inhibits bacterial cell wall synthesis.

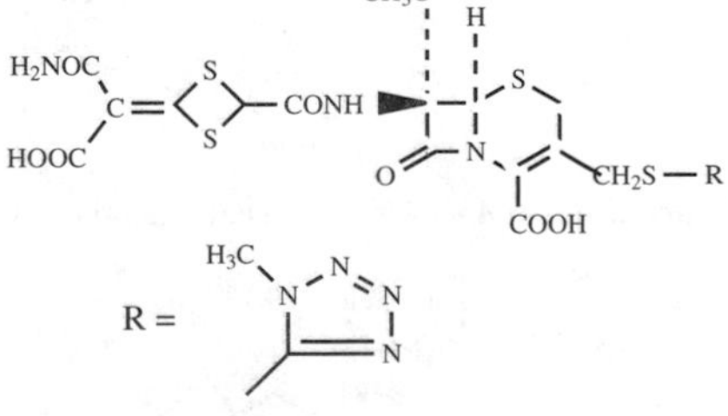

Cefotiam (mol wt 526) A semisynthetic cephalosporin antibiotic that inhibits bacterial cell wall synthesis.

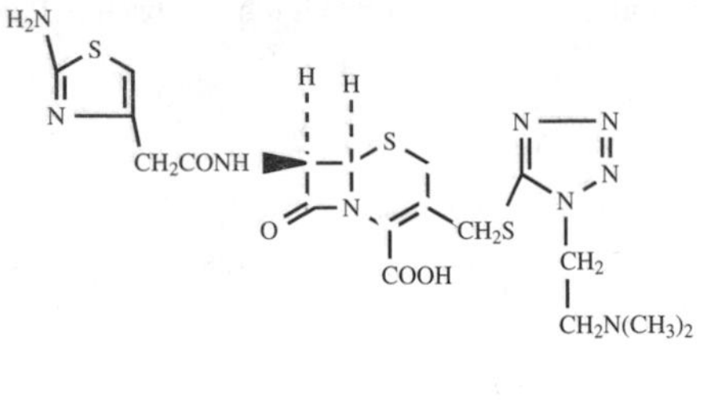

Cefoxitin (mol wt 427) A semisynthetic antibiotic derived from cephamycin that inhibits bacterial cell wall synthesis.

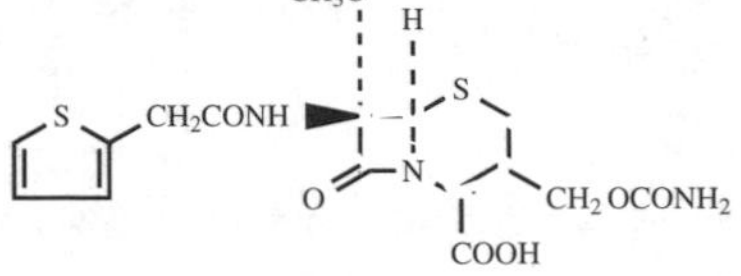

Cefpimizole (mol wt 671) A third-generation cephalosporin antibiotic that inhibits bacterial cell wall synthesis.

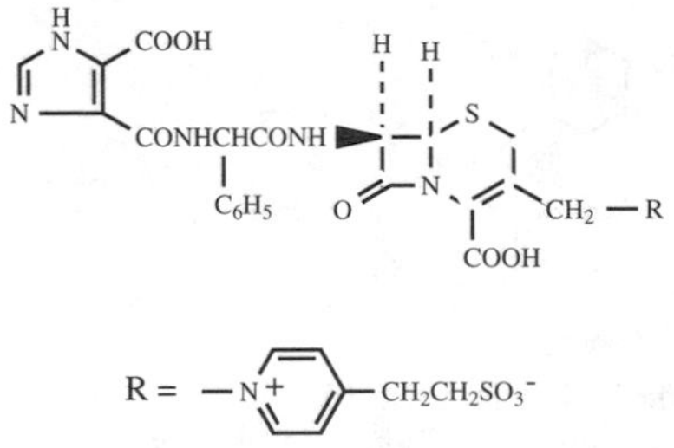

Cefpiramide (mol wt 613) A semisynthetic cephalosporin antibiotic that inhibits bacterial cell wall synthesis.

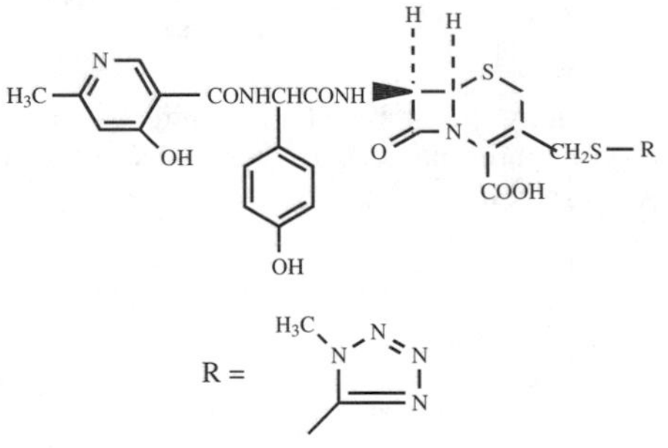

Cefpodoxime Proxetil (mol wt 558) A broad spectrum third-generation cephalosporin antibiotic that inhibits bacterial cell wall synthesis.

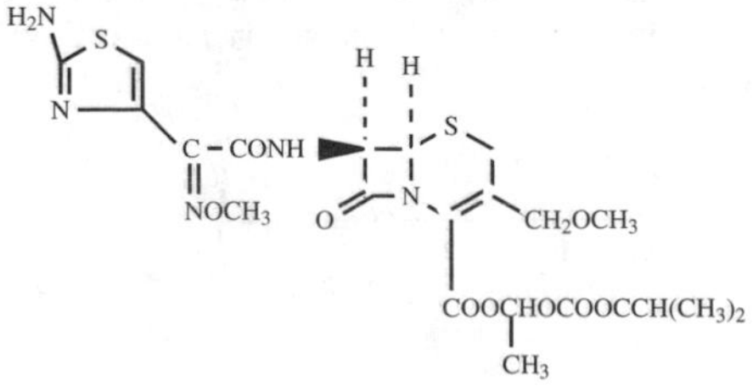

Cefprozil (mol wt 407) A cephalosporin-type antibiotic.

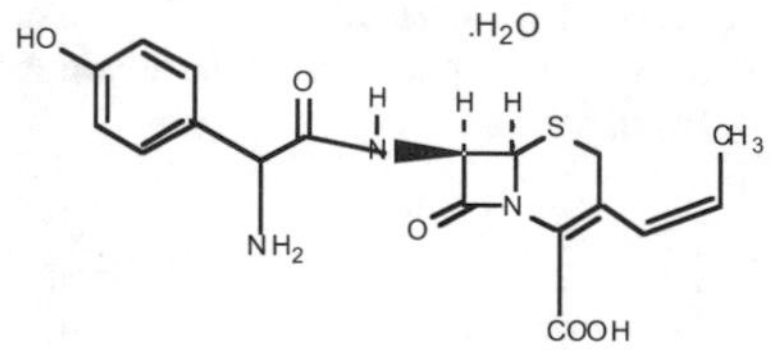

Cefracycline A trade name for tetracycline, an antibiotic.

Cefroxadine (mol wt 365) An orally active cephalosporin derivative that inhibits bacterial cell wall synthesis.

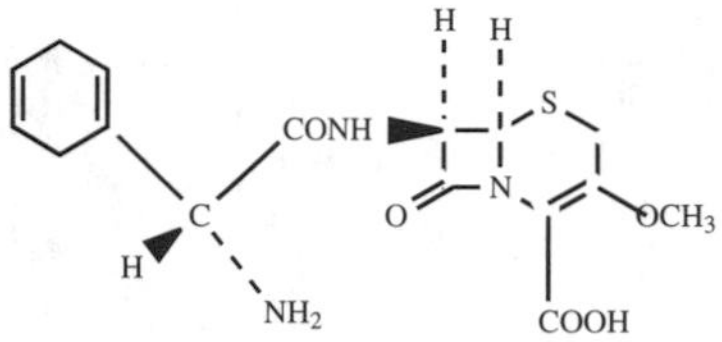

Cefsulodin (mol wt 533) A third-generation cephalosporin antibiotic that inhibits bacterial cell wall synthesis.

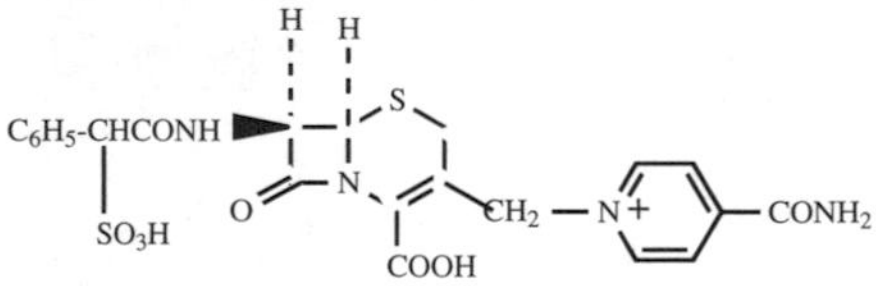

Ceftazidime (mol wt 547) A third-generation cephalosporin antibiotic that inhibits bacterial cell wall synthesis.

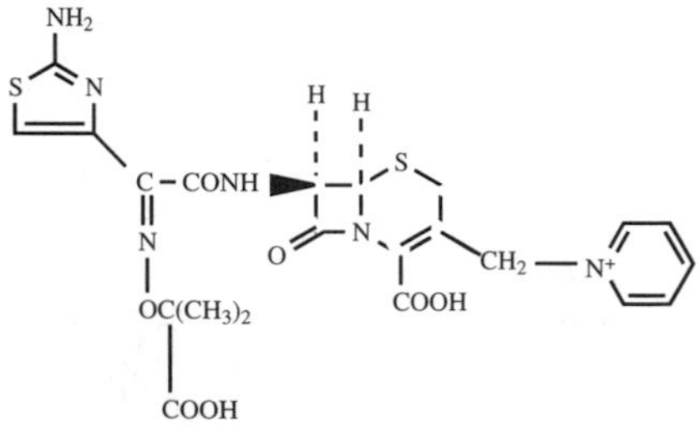

Cefteram (mol wt 479) An orally active cephalosporin antibiotic that inhibits bacterial cell wall synthesis.

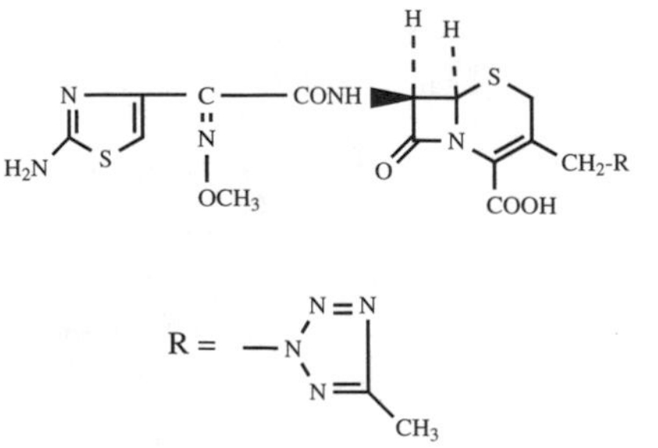

Ceftibuten (mol wt 410) A third-generation cephalosporin antibiotic that inhibits bacterial cell wall synthesis.

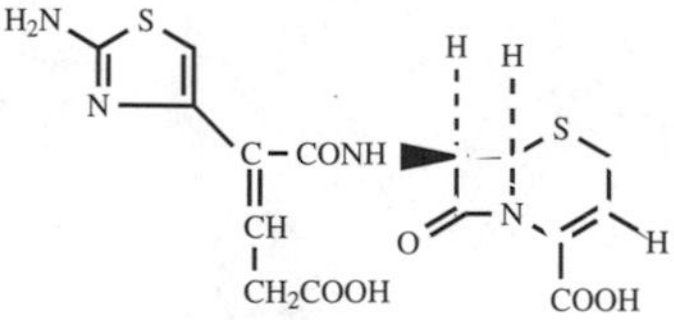

Ceftin A trade name for cefuroxime, a cephalosporin-type antibiotic.

Ceftiofur (mol wt 524) A third-generation cephalosporin antibiotic that inhibits bacterial cell wall synthesis.

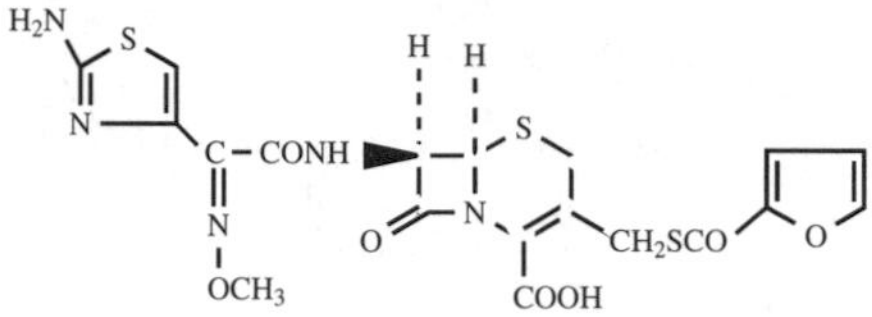

Ceftizoxime (mol wt 383) A third-generation cephalosporin antibiotic, that inhibits bacterial cell wall synthesis.

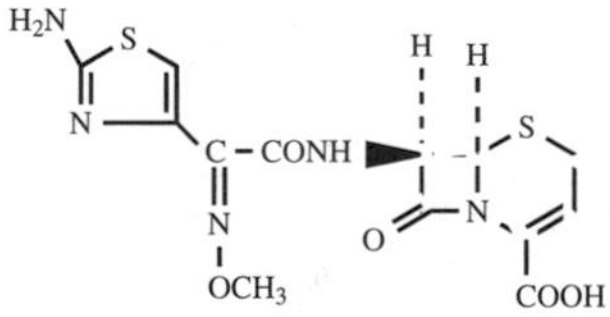

Ceftriaxone (mol wt 555) A third-generation cephalosporin antibiotic that inhibits bacterial cell wall synthesis.

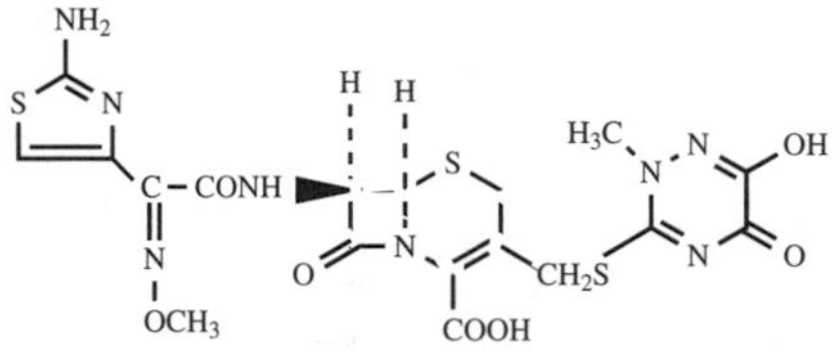

Cefuroxime (mol wt 424) An antibacterial agent.

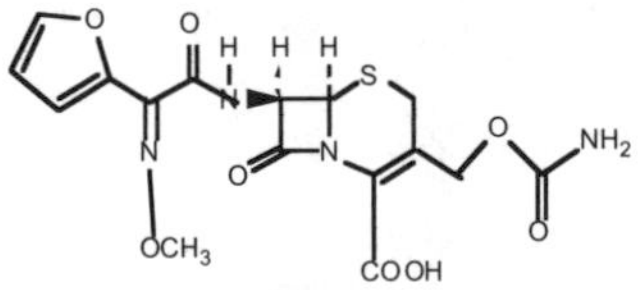

Cefuzonam (mol wt 514) A third-generation cephalosporin antibiotic that inhibits bacterial cell wall synthesis.

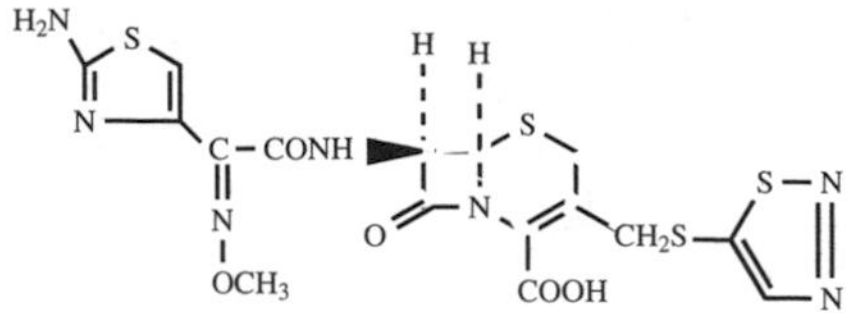

Cefzil A trade name for cefprozil monohydrate, a cephalosporin-type antibiotic.

CEI Abbreviation for converting enzyme inhibitors.

CelII (EspI) A restriction endonuclease from *Coccochloris elabbens* with the following specificity:

5′....GC↓TNAGC....3′
3′....CGANT↑CG....5′

Celebrex A trade name for celecoxib, a nonsteroidal anti-inflammatory and analgesic agent.

Celecoxib (mol wt 381) A nonsteroidal anti-inflammatory and analgesic agent.

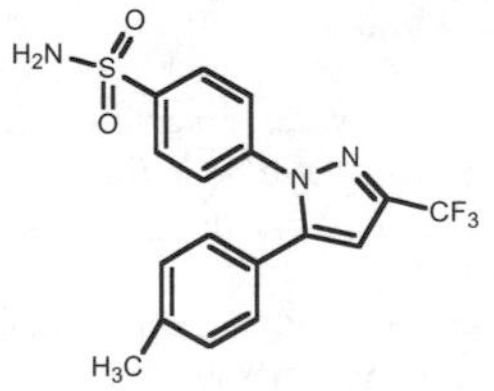

Celestoderm A trade name for betamethasone valerate, a corticosteroid.

Celestone A trade name for betamethasone, a glucocorticoid used as an anti-inflammatory agent.

Celexa A trade name for citalopram used as an antidepressant.

Celiac Disease A violent intestinal upset resulting from eating gluten.

Celioma A tumor of the belly.

Celiprolol (mol wt 380) An antihypertensive and antianginal agent.

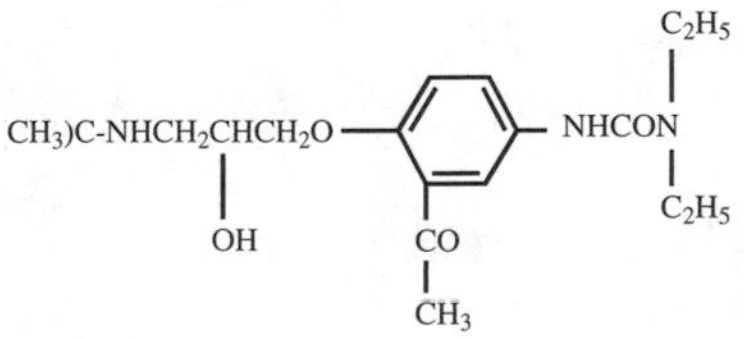

Cell (BamHI) A restriction endonuclease from *Coccochloris ellabens* with the same specificity as BamHI.

Cell The basic structure and functional unit of life.

Cell Adhesion Molecules Molecules on the cell surface that mediate the cell-to-cell binding or cohesive interaction between cells.

Cell Affinity Chromatography A method for obtaining a functionally homogenous population of cells from a mixed culture using affinity chromatographic techniques in which affinity adsorbents are prepared by linking cell-surface-specific protein (e.g., antibody and lectin) for adsorption of specific cell-types. The adsorbed cell type is then eluted from the affinity absorbent.

Cell Biology Science that deals with the structures and functions of cell.

Cell Blotting A technique to blot cells onto nitrocellulose paper for subsequent detection, identification, and characterization.

Cell Body The portion of a nerve cell that contains the nucleus.

Cell Coat The outer layer of eukaryotic cells that is rich in glycoprotein and mucopolysaccharide.

Cell Cortex Specialized layer of cytoplasm on the inner face of plasma membrane. It is an actin-rich layer in animal cells responsible for cell surface movement.

Cell Counting The method used to enumerate cells in a culture or in a sample employing a hemocytometer or other appropriate device.

Cell Culture The *in vitro* growth of cells.

Cell Cycle The reproductive cycle of the cell; the sequence of events by which the genetic information in the nucleus is duplicated and parceled out into two daughter cells. The cell cycle is divided into four phases: M (mitosis), S (DNA synthesis), and G1 and G2 (gap phases).

Cell Determination An event in embryogenesis that specifies the developmental pathway that a cell will follow.

Cell Differentiation A process of cell specialization through the selective expression of genes.

Cell Disruption Any procedure that breaks the cell and releases its cellular content, e.g., homogenization and enzymatic, or ultrsonic treatment.

Cell Division The process by which a cell divides into two cells (see mitosis and meiosis).

Cell Entrapment The entrapment of free, mobile cells into a gel matrix (e.g., polyacrylamide gel). The entrapped cells can be used for production of specific bioactive compounds or to perform a specific biochemical reaction.

Cell Extract A preparation that contains a large number of broken cells and their released cellular content.

Cell Fractionation A technique (e.g., centrifugation, chromatography, electrophoresis) for separation of cell content into functional subcellular and noncellular components following cell disruption.

Cell Free Extract The fluid resulting from centrifugation or filtration of disrupted cell suspension; it contains subcellular components (e.g., ribosomes) and soluble biomolecules (e.g., DNA, RNA, proteins, and carbohydrates).

Cell Free Protein Synthesis Protein synthesis carried out in the laboratory with cell-free extracts without the presence of living cells.

Cell Free System A biosynthetic system without the presence of cells, e.g., a system for cell-free protein synthesis.

Cell Free Translation System A cell-free extract that contains all components required for protein synthesis (e.g., ribosomes, tRNAs, amino acids, enzymes and cofactors) and is able to translate the added mRNA into protein.

Cell Fusion Fusion of two different cells to form a hybrid with the aid of polyethylene glycol, electroporation, or viruses.

Cell Homogenate A preparation that contains broken cells and their released contents.

Cell Hybrid A somatic cell containing chromosomes derived from parental cells of different spceies (e.g., man-mouse somatic cell hybrid).

Cell Hybridization Production of viable somatic cell hybrids with cell fusion techniques.

Cell Immobilization Conversion of cells from a free, mobile state to a fixed state either by binding onto an insoluble substrate or by entrapment into a gel matrix (e.g., polyacrylamide gel). The immobilized cells can be used for production of a specific bioactive compound or to perform a specific reaction or activity.

Cell Junction Specialized region of connection on the plasma membrane where contact between two adjacent animal cells occurs.

Cell Line Population of cells of plant or animal origin capable of dividing indefinitely in culture.

Cell Locomotion Active movement of a cell from one location to another, e.g., migration of a cell over the surface.

Cell Lysate A mixture of cellular components obtained from cell lysis.

Cell Mediated Immunity Immunity that is mediated by the effector T cells, e.g., destruction of foreign cells by cytotoxic T cells in an allograft rejection reaction or delayed hypersensitivity reaction mediated by T_D lymphocytes.

Cell Membrane The structure surrounding a cell that consists of a lipid bilayer and proteins, it regulates the flow of material into and out of the cell (also called plasma membrane).

Cell Plate Flattened membrane-bound structure that forms fusing vesicles in the cytoplasm of a dividing plant cell. It is the precursor of a new cell wall.

Cell Proliferation The process of increasing the number of cells through mitotic division.

Cell Respiration The cellular chemical reactions that release energy from fuel molecules, e.g., glycolysis, citric acid cycle, and electron transport system.

Cell Sap 1. Fluid in the vacuole of a plant cell. 2. Cytoplasm without cell organelles, e.g., cytosol.

Cell Sorter A device to separate different cell types in tissues that have been treated with trypsin or collegenase to destroy the intercellular matrix.

Cell Surface Receptor A protein in, on, or transversing the membrane that recognizes and binds specific bioactive molecules.

Cell Theory The theory that the cell is the basic structural and functional unit for all organisms.

Cell Wall Extracellular matrix deposited by a cell outside its plasma membrane. It is a prominent structure in cells of bacteria, plants, algae, and fungi, but not in animal cells.

Cellcept A trade name for mycophenolate mofetil used as an immunosuppressive agent.

Cellifugal In a direction away from the cell body.

Cellipetal In a direction toward the cell body.

Cellobiase (β-glucosidase) The enzyme that catalyzes the hydrolysis of glucoside to glucose and alcohol.

Cellobiose (mol wt 343) A β-1,4-linked disaccharide of glucose.

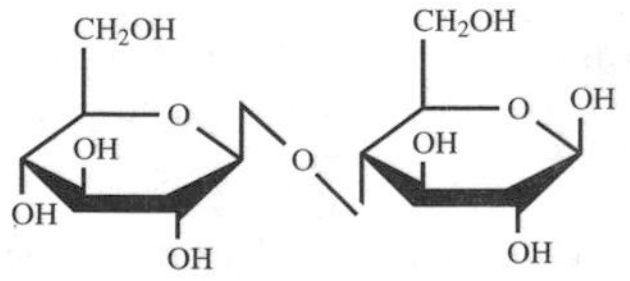

Cellobiose Dehydrogenase The enzyme that catalyzes the following reaction:

Cellobiose + acceptor

⇌

Cellobiono-1,5-lactone + reduced acceptor

Cellobiose Epimerase The enzyme that catalyzes the following reaction:

$$\text{Cellobiose} \rightleftharpoons \text{D-glucosyl-D-mannose}$$

Cellobiose Oxidase The enzyme that catalyzes the following reaction:

$$\text{Cellobiose} + O_2 \rightleftharpoons \text{Cellobiono-1,5-lactone} + H_2O_2$$

Cellobiose Phosphorylase The enzyme that catalyzes the following reaction:

$$\text{Cellobiose} + \text{Pi} \rightleftharpoons \text{Glucose 1-phosphate} + \text{glucose}$$

Cellocidin (mol wt 112) An antibiotic produced by *Streptomyces chibaensis.*

$$H_2NCOC \equiv CCONH_2$$

Cellodextrin Phosphorylase The enzyme that catalyzes the following reaction:

$$(1,4\text{-}\beta\text{-D-Glucosyl})_n + \text{Pi} \rightleftharpoons 1,4\text{-}\beta\text{-D-glucosyl})_{n-1} + \alpha\text{-D-glucose 1-phosphate}$$

Cellogel The gelatinized cellulose acetate used as electrophoretic medium.

Cellophane A trade name for a flexible, transparent cellulose acetate sheet used in dialysis.

Cellose Synonym of cellobiose.

Cellosolve A trade name for ethylene glycol monoethyl ether.

Cellotriose A trisaccharide consisting of β-1,4-glucosidic-linked glucose.

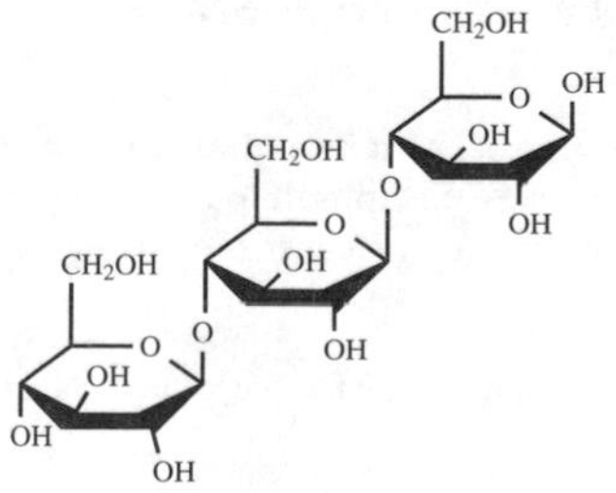

Cellular Pertaining to or derived from a cell.

Cellular Immunity See cell-mediated immunity.

Cellular Plasminogen Activator A protease that catalyzes the conversion of plasminogen to plasmin by cleaving the peptide bond between arginine and valine in plasminogen.

Cellular Respiration See cell respiration.

Cellular Slime Mold The slime mold in which the vegetative stage consists of uninucleate amoeboid cells that aggregate to form multicellular pseudoplasmodium.

Cellulase The enzyme that catalyzes the endohydrolysis of 1,4-β-glucosidic linkage in cellulose.

Cellulite Referring to fat deposited beneath the skin.

Cellulitis An infection of skin, e.g., infection of subcutaneous skin tissue by *Streptococci* or *Staphylococci.*

Cellulomonas A genus of aerobic or facultative anaerobic, asporogenous bacteria.

Cellulose A polymer of glucose consisting of 1,4-β-glucosidic linkages of glucose with a molecular weight ranging from 20,000 to 40,000; the major constituent of the plant cell wall.

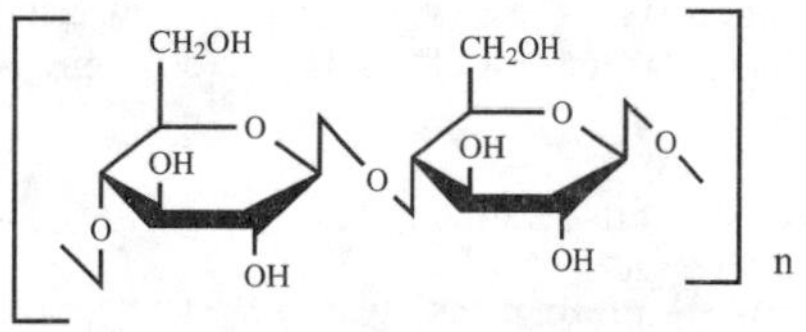

Cellulose 1,4-β-Cellobiosidase The enzyme that catalyzes the hydrolysis of 1,4-β-D-glucosidic linkages in cellulose and cellotetraose, releasing cellobiose from the nonreducing ends of the chain.

Cellulose Polysulfatase The enzyme that catalyzes the hydrolysis of the 2- and 3-sulfate groups of polysulfate of cellulose.

Cellulose Synthetase (GDP-forming) The enzyme that catalyzes the following reaction:

$$\text{GDP-glucose} + (1,4\text{-}\beta\text{-D-glucosyl})_n \rightleftharpoons \text{GDP} + (1,4\text{-}\beta\text{-D-glucosyl})_{n+1}$$

Cellulose Synthetase (UDP-forming) The enzyme that catalyzes the following reaction:

$$\text{UDP-glucose} + (1,4\text{-}\beta\text{-D-glucosyl})_n \rightleftharpoons \text{UDP} + (1,4\text{-}\beta\text{-D-glucosyl})_{n+1}$$

Cellulosic Substance made from cellulose or a derivative of cellulose.

Cellulosome A cellulose-binding structure that consists of cellulase.

Celontin A trade name for methsuximide, an anticonvulsant.

Celsius Temperature Scale The temperature scale in which the freezing and boiling point of water at one atm pressure are zero degrees and 100 degrees, respectively (also known as centigrade temperature scale).

CEM Abbreviation for 1. a combination drug containing CCNU, etoposide, and methotrexate; 2. conventional electron microscope.

Cementification The process of cementum formation.

Cementoblast Cells that are active in the formation of cementum.

Cementum The external calcified bony layer of the tooth within the gum.

Cenafed A trade name for pseudoephedrine hydrochloride, a nasal decongestant.

Cena-K A trade name for potassium, an electrolyte.

Cenogenesis The development of new genetic characters as a result of adaptation to the environment.

Centesis Surgical puncture.

Centi- A prefix meaning 1/100.

Centibar A unit of atmospheric pressure; 1/100 of a bar.

Centigrade Temperature Scale See Celsius temperature scale.

Centigram 1/100 gram.

Centiliter 1/100 liter.

Centimeter 1/100 meter.

Centimeter-Gram-Second System (CGS System) A metric unit system that expresses length, mass, and time in units of centimeter, gram, and second, respectively.

Centimorgan A unit of physical distance on a chromosome equivalent to a 1% frequency of recombination between closely linked genes.

Centipeda A genus of Gram-negative, anaerobic, rod-shaped bacteria.

Central Dogma The principle of flow of genetic information from DNA to RNA to protein or from RNA to DNA to RNA to protein (in retroviruses).

Central Lymphoid Organs See primary lymphoid organs.

Central Nervous System The part of the nervous system in vertebrates that contains the brain and spinal cord.

Central Vacuole A membrane-bound structure in plant cells that is responsible for maintaining turgor pressure and intracellular digestion.

Centrax A trade name for prazepam, an antianxiety agent.

Centrifugal 1. Away from the center or axis. 2. Efferent.

Centrifugal Force The force exerted on a rotating particle outward from the center of rotation.

Centrifugation The process of sedimentation or separation of cellular components or cellular organelles through the use of a centrifuge. The force generated during centrifugation can be calculated by the equation:

$$F = \omega^2 x \text{ where}$$

F = force generated by centrifugation.
ω = angular velocity.
x = distance from the center of rotation.

Centrifuge An instrument used for generating centrifugal force for sedimentation and separation of cell organelles or bioactive molecules.

Centrifuge Rotor A device that holds centrifuge tubes during centrifugation.

Centriole Short cylindrical array of microtubules that functions as a mitotic spindle organizer. A pair of centrioles is usually found in the center of a centrosome in an animal cell.

Centripetal 1. Moving toward the center. 2. Afferent.

Centromere The constricted region of a mitotic chromosome that holds the two sister chromatides together during mitosis and meiosis.

Centromeric Index The percentage of the total length of a chromsome encompassed by its short arm.

Centrosome The microtubule-organizing center that acts as the spindle pole during mitosis. It contains a pair of centrioles in most animal cells.

CEOH Abbreviation for cholesteryl ester hydroxide.

CEOOH Abbreviation for cholesteryl ester hydroperoxide.

CEP Abbreviation for a combination drug containing CCNU, etoposide, and prednimustine.

CEPA Abbreviation for chloroethane phosphoric acid.

Cephacetrile Sodium (mol wt 361) A semisynthetic cephalosporin antibiotic that inhibits bacterial cell wall synthesis.

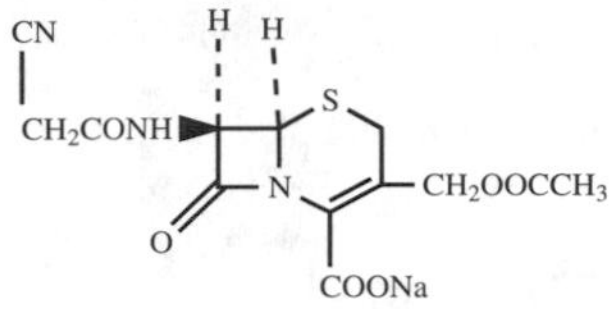

Cephalalgia Headache.

Cephalexin (mol wt 347) A semisynthetic cephalosporin antibiotic that inhibits bacterial cell wall synthesis.

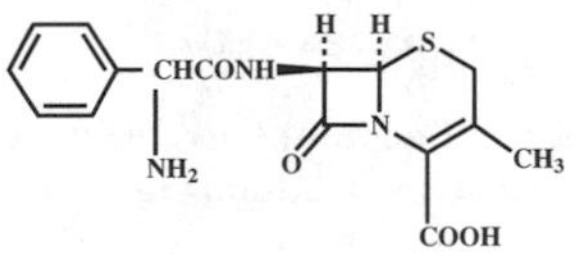

Cephalic Pertaining to the head.

Cephalin A phospholipid (phosphatidyl ethanolamine) found in animal tissues (see phosphatidylethanolamine for structure).

Cephalin-Cholesterol Flocculation Test A flocculation test for hepatitis in which serum from individuals with hepatitis forms a flocculant upon treatment with a cephalin-cholesterol suspension. It is a liver function test.

Cephaloglycin (mol wt 405) A semisynthetic cephalosporin antibiotic that inhibits bacterial cell wall synthesis.

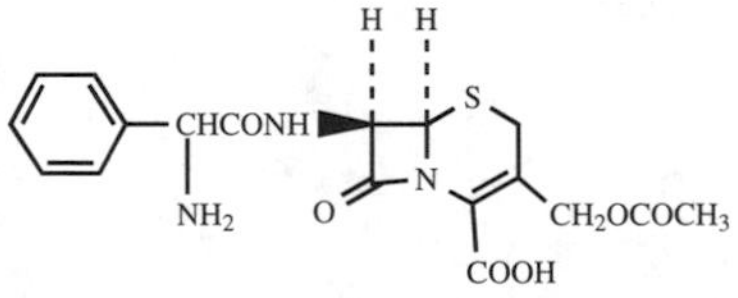

Cephaloridine (mol wt 416) A semisynthetic cephalosporin antibiotic that inhibits bacterial cell wall synthesis.

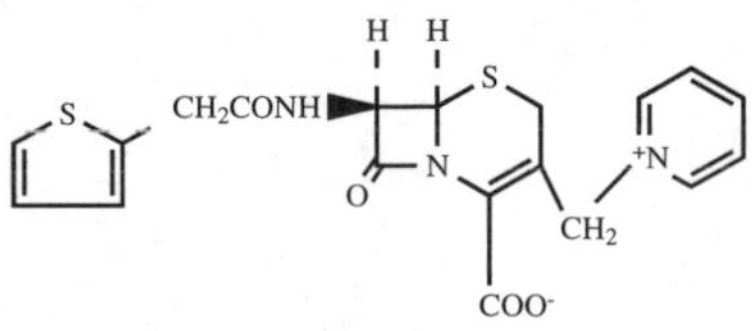

Cephalosporin A heterogeneous group of natural and semisynthetic antibiotics that inhibit bacterial cell wall synthesis in a range of Gram-positive and Gram-negative bacteria by inhibiting the formation of cross-links in peptidoglycan. Cephalosporins consist of a dihydrothiazine rings instead of thiazolidine and β-lactam rings and are less sensitive than penicillins to β-lactamase.

Cephalosporin C (mol wt 415) A type of cephalosporin antibiotic produced by *Cephalosporium* species that inhibits bacterial cell wall synthesis.

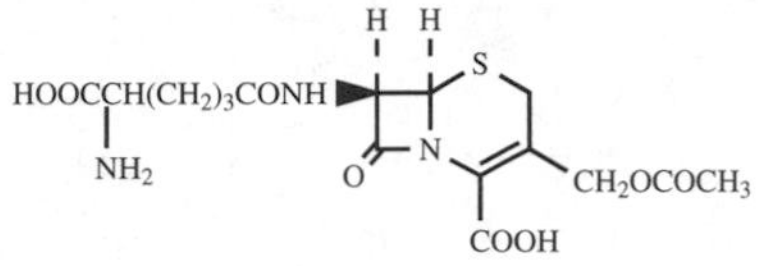

Cephalosporin C Deacetylase The enzyme that catalyzes the following reaction:

Cephalosporin C + H_2O

⇌

deacetylcephalosporin c + acetate

Cephalosporin C Transaminase The enzyme that catalyzes the following reaction:

Cephalosporin C + pyruvate

⇌

7-(5-Carboxyl-5-oxopentanyl)-
aminocephalosporinate + L-alanine

Cephalosporin P1 (mol wt 575) A steroid antibiotic produced by *Cephalosporium* species.

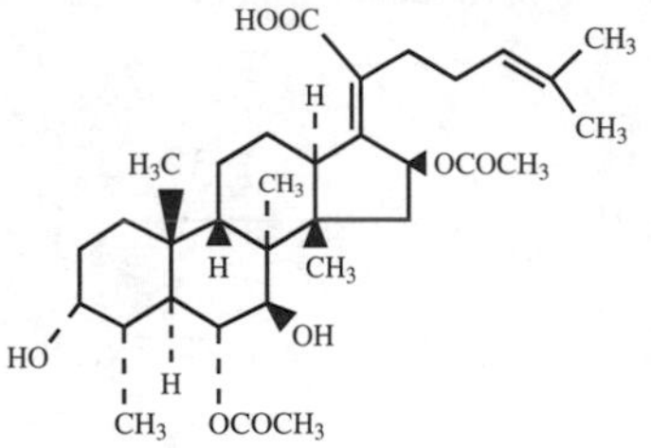

Cephalosporinase Synonym of b-lactamase.

Cephalosporium A genus of fungi that produce cephalosporins.

Cephalothin (mol wt 396) A semisynthetic cephalosporin antibiotic that inhibits bacterial cell wall synthesis.

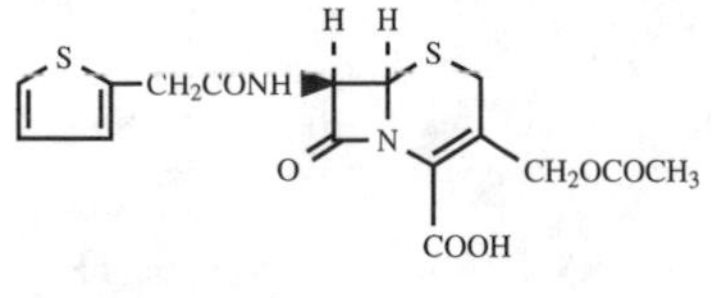

Cephamycin A family of β-lactam antibiotics produced by various species of *Streptomyces* that inhibits bacterial cell wall synthesis.

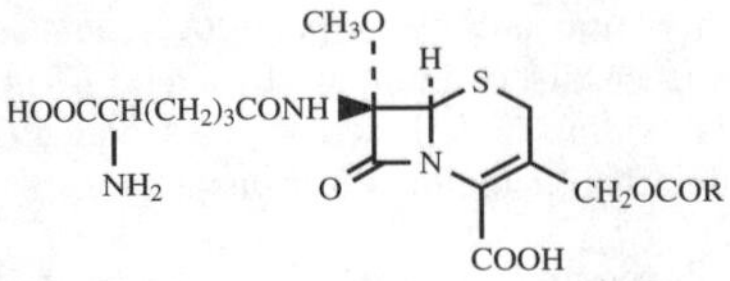

Cephamycin A:

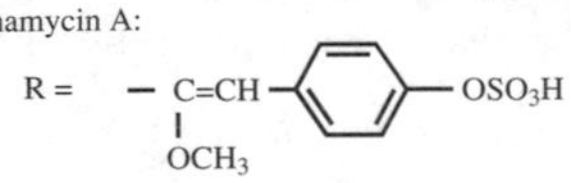

Cephamycin B:

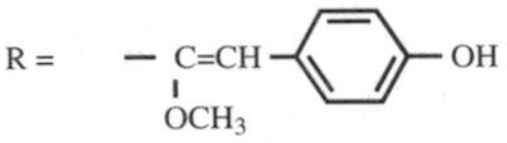

Cephamycin C: R = $-NH_2$

Cephapirin Sodium (mol wt 445) A broad spectrum cephalosporin antibiotic that inhibits bacterial cell wall synthesis.

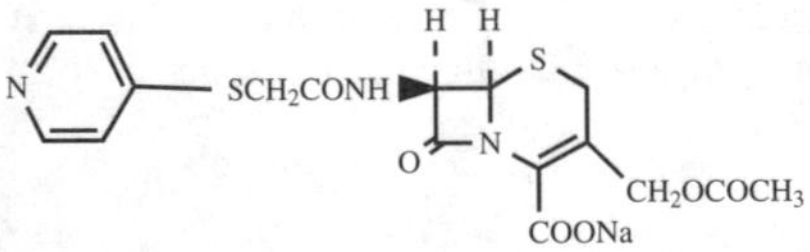

Ceph-Flo Abbreviation for cephalin flocculation test.

Cephradine (mol wt 349) A semisynthetic cephalosporin antibiotic that inhibits bacterial cell wall synthesis.

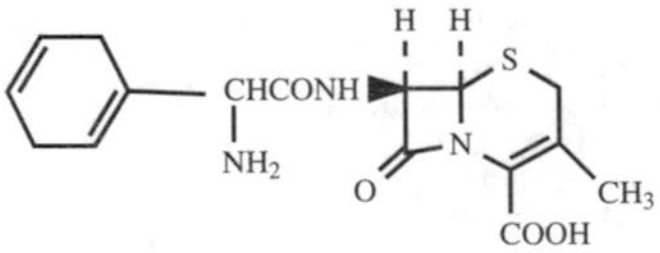

Cephulac A trade name for lactulose, a laxative.

Cephulax A trade name for lactulose, used as a laxative.

Ceporacin A trade name for cephalothin, an antibiotic that inhibits bacterial cell wall synthesis.

Ceporex A trade name for cephalexin monohydrate, an antibiotic that inhibits bacterial cell wall synthesis.

Ceptaz A trade name for ceftazidime, a third generation cephalosporin antibiotic.

CeqI (EcoRV) A restriction endonuclease from *Corynebacterium equii* with the following specificity:

↓
5′.........GATATC.........3′
3′.........CTATAG.........5′
↑

Cer Abbreviation for ceramide.

Ceramidase The enzyme that catalyzes the following reaction:

N-Acylsphingosine + H_2O
⇅
A carboxylate + sphingosine

Ceramide A sphingolipid and a major membrane component.

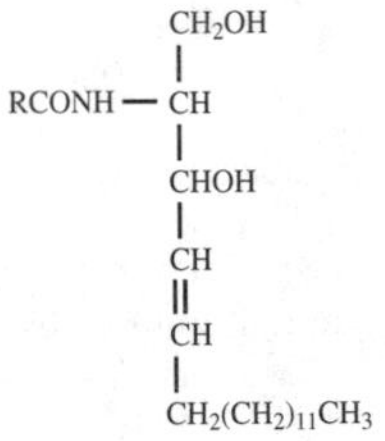

R = acyl group

Ceramide Cholinephosphotransferase The enzyme that catalyzes the following reaction:

CDP-choline + N-acylsphingosine
⇅
CMP + sphingomyelin

Ceramide Glucosyltransferase The enzyme that catalyzes the following reaction:

UDP-glucose + N-acylsphingosine
⇅
UDP + D-glucosyl-N-acylsphingosine

Ceramide Glycanase Synonym of glucosylceramidase.

Ceramide Kinase The enzyme that catalyzes the following reaction:

ATP + ceramide
⇅
ADP + ceramide 1-phosphate

CERBC Abbreviation for chicken embryo red blood cell.

Cerberoside (mol wt 859) A cardiotonic agent.

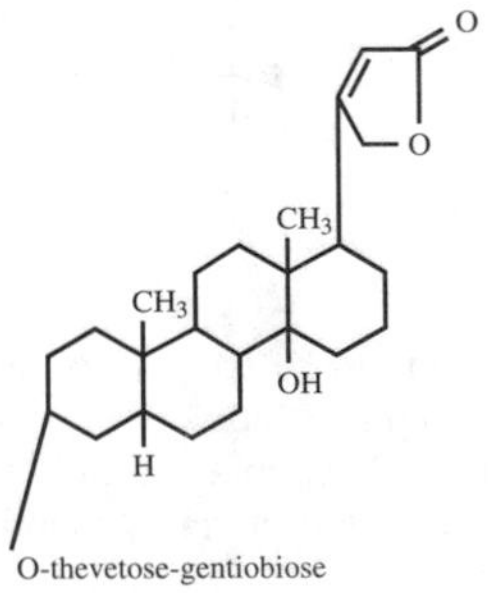

Cercaria The final free-swimming stage of a trematode.

Cerebellar Pertaining to cerebellum.

Cerebellar Angioblastoma A tumor of the brain.

Cerebellitis Inflammation of the cerebellum.

Cerebellospinal Tract A nerve tract carrying impulses from the cerebellum to the spinal cord.

Cerebellum A part of the central nervous system located below and posterior to the cerebrum and above the pons and medulla.

Cerebral Pertaining to the brain.

Cerebral Angiography X-ray examination of the blood vessels of the brain using radiopaque contrast medium.

Cerebral Palsey A motor nerve disorder resulting from the damage to the brain before or during birth.

Cerebral Thrombosis Blood clot in blood vessels of the brain.

Cerebrocuprien A copper-containing protein found in the brain.

Cerebroside A sphingoglycolipid associated with myelin sheath of the nerve; it contains sphingosine, fatty acid, and sugar (glucose or galactose).

CH_3-$(CH_2)_{12}$-CH=CHCH(OH)CHCH$_2$O — X

NH — CO — R

R = fatty acid
X= glucose (glucocerebroside)
= galactose (galactocerebroside)

Cerebroside Sulfatase The enzyme that catalyzes the following reaction:

A cerebroside 3-sulfate + H_2O

⇅

Cerebroside + sulfate

Cerebrospinal Pertaining to the brain and spinal cord.

Cerebrospinal Fluid The lymphlike fluid in the central cavity of the brain and spinal cord.

Cerebrospinal Meningitis The inflammation of the meninges of the brain and spinal cord.

Cerebrospinal Nerve The nerves that begin in the brain and spinal cord.

Cerebrospinal Tract A nerve tract that carries impulses from the cerebrum to the spinal cord.

Cerebrum The large uppermost part of the brain excluding medulla, pons, and cerebellum.

Cerebyx A trade name for fosphenytoin, an antiepileptic agent.

Ceredase A trade name for enzyme alglucerase produced by DNA recombinant technology.

Cerespan A trade name for papaverine, a vasodilator that relaxes smooth muscle by inhibiting phosphodiesterase leading to the increase of cAMP concentration.

Cerevisin A proteolytic enzyme from baker's yeast.

Cerezyme A trade name for human glucosylceramidase produced by recombination DNA technology.

Cer(Hex) Abbreviation for ceramide monohexoside.

Cer(Hex)$_2$ Abbreviation for ceramide dihexoside.

Cer(Hex)$_3$ Abbreviation for ceramide trihexoside.

Cerium A chemical element with atomic weight 140, valence 2.

Cerivastatin Sodium (mol wt 482) An HMG CoA inhibitor used as an antihypertensive agent.

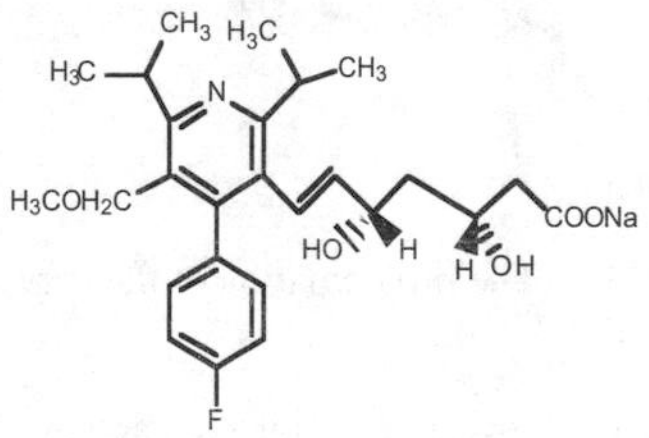

Ceroid A lipid granule found in cirrhosis of the liver.

Ceroid Pigment Referring to yellowish aging pigment.

Ceronapril (mol wt 440) An antihypertensive agent.

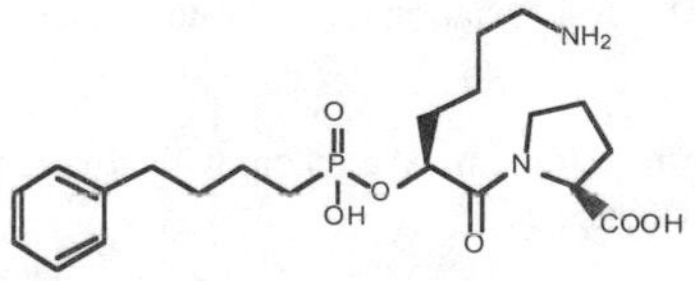

Cerotic Acid (mol wt 397) A 26-carbon saturated fatty acid.

$CH_3(CH_2)_{24}COOH$

Cerotoyl Referring to the group of:

CH_3-$[CH_2]_{24}$-CO-

Cerubidin A trade name for daunorubicin hydrochloride, an antibiotic that interferes with the activity of DNA-dependent RNA polymerase.

Cerulenin (mol wt 223) An antigfungal antibiotic and an inhibitor of fatty acid biosynthesis.

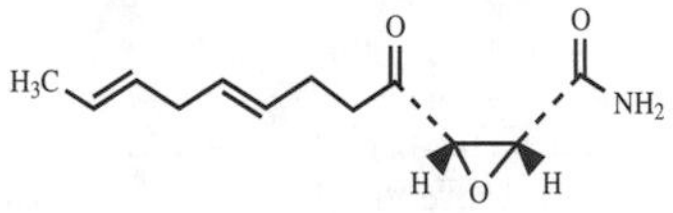

Ceruletide A decapeptide capable of stimulating gastric, pancreatic, and biliary secretion.

Ceruloplasmin A copper-containing α_2-globulin in mammalian blood responsible for copper transport in the blood.

Cerumen Earwax.

Cerumenex A trade name for triethanolamine polypeptide oleate condensate, a ceruminolytic agent used to emulsify and disperse accumulated cerumen.

Ceruminolytic Any agent capable of dissolving earwax.

Cervicarcin (mol wt 392) An antibiotic.

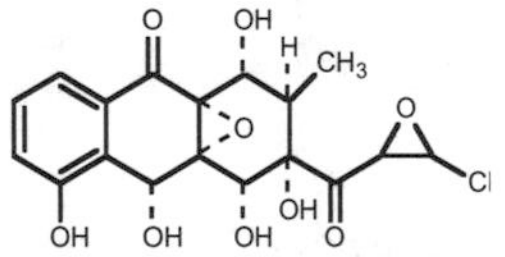

Cervicitis The inflammation of the cervix of the uterus.

Cervidil A trade name for prostagladin E2.

Cervix 1. Any necklike portion of an organ. 2. The narrowed portion of the uterus.

Ceryl Referring to the group of:

$$CH_3\text{-}[CH_2]_{24}\text{-}CH_2\text{-}$$

CESD Abbreviation for cholesteryl ester storage disease.

Cesium A chemical element with atomic weight 133, valence 1.

Cesium 137 A radioisotope of cesium with half-life of about 30 years.

Cesium Chloride (CsCl, mol wt 168) A standard density-gradient centrifugation medium used for fractionation of macromolecules, e.g., nucleic acids.

Cesium Chloride Gradient Centrifugation A type of ultracentrifugation for fractionation of macromolecules in which the concentrated cesium chloride containing the macromolecules to be separated is centrifuged until the cesium chloride reaches equilibrium distribution and produces a linear density gradient in the centrifuge tube. Macromolecules with different densities sediment or float to the positions where their densities equal those of CsCl in the gradient.

C-Esterase See acetylesterase.

Cestodes Parasitic flatworms, e.g., tapeworms.

Cetacort A trade name for hydrocortisone, used an anti-inflammatory agent.

Cetalkonium Chloride (mol wt 396) A topical anti-infective agent.

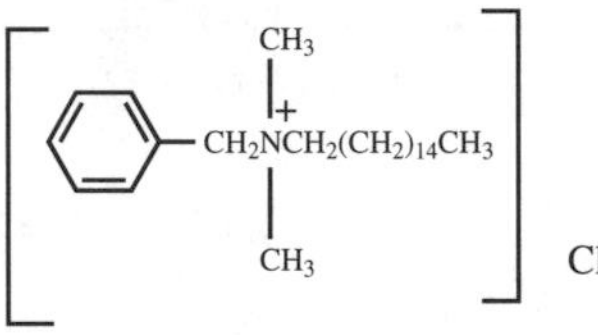

Cetamolol (mol wt 310) An antihypertensive agent.

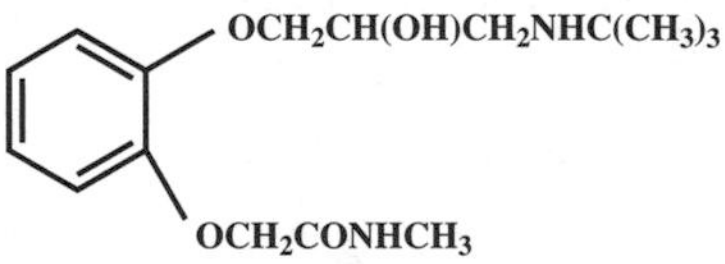

Cetane A trade name for vitamin C (ascorbic acid).

Cetapred Ointment A trade name for a combination drug containing sulfacetamide sodium and prednisolone acetate, used as an anti-infective agent.

Cetavlon A trade name for cetyltrimethyl ammonium bromide, an antiseptic agent.

Cethexonium Bromide (mol wt 449) An antiseptic agent.

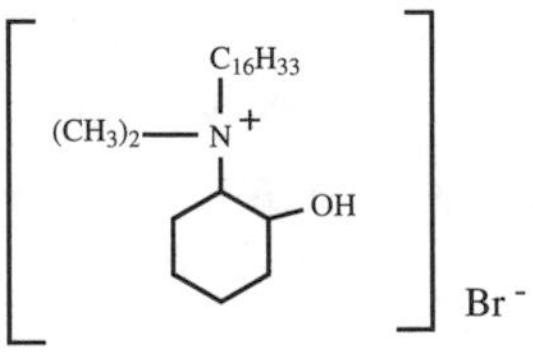

Cetiedil (mol wt 350) A vasodilator.

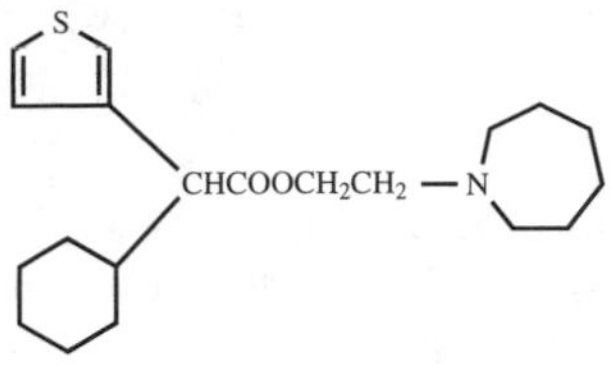

Cetirizine (mol wt 389) An antihistaminic agent.

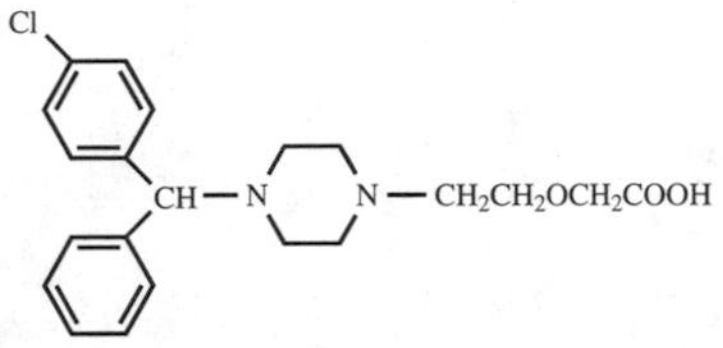

Cetixime (mol wt 255) An antihistaminic agent.

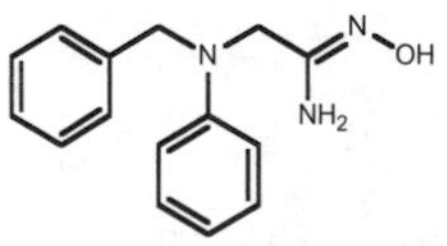

Cetotiamine (mol wt 427) A vitamin B_1 source.

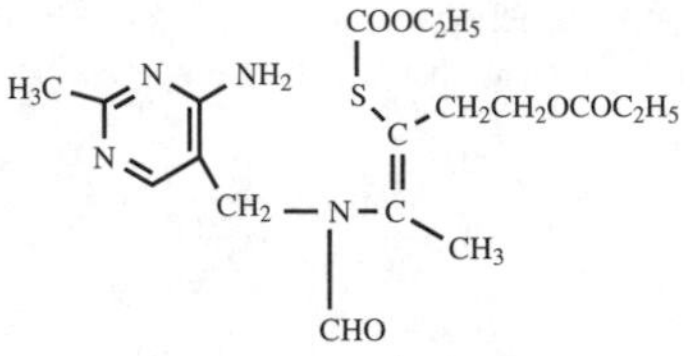

CETP Abbreviation for cholesteryl ester transfer protein.

Cetraxate (mol wt 305) An antiulcerative agent.

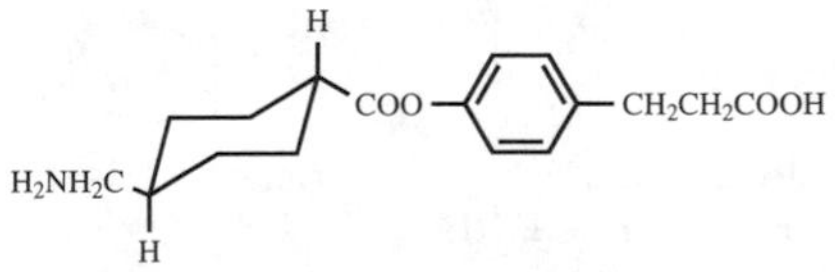

Cetrimonium Bromide (mol wt 364) An antiseptic agent.

$$[CH_3\text{-}(CH_2)_{15}N(CH_3)_3]Br$$

Cetyl Referring to $CH_3\text{-}[CH_2]_{14}\text{-}CH_2\text{-}$ group.

Cetyl Alcohol (mol wt 242) An alcohol obtained from spermaceti by saponification and used as an emulsifying and softening agent.

$$CH_3-(CH_2)_{14}CH_2OH$$

Cetyldimethylethylammonium Bromide (mol wt 378) A cationic detergent used for the disruption of the plasma membrane and precipitation of nucleic acids and mucopolysaccharide.

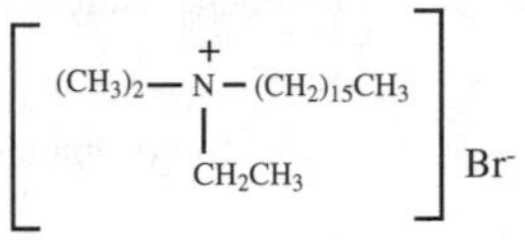

Cetylpyridinium Chloride (mol wt 340) An antiseptic agent.

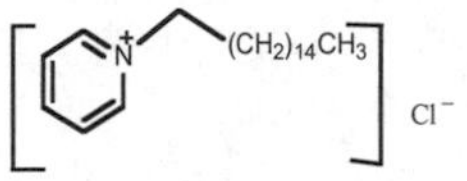

CEV Abbreviation for a combination drug containing cyclophosphamide, etoposide, and vincristine.

CEVD Abbreviation for a combination drug containing CCNU, etoposide, vincristine, and dexamethasone.

CEX Abbreviation for cation exchanger.

Cf Abbreviation for confer (compare).

CF Abbreviation for 1. carbol fuchsin; 2. chemotactic factor; 3. chick fibroblast; 4. Christmas factor; 5. citrovorum factor; 6. colicin factor; 7. complement fixation; 8. cystic fibrosis.

CF Test Abbreviation for complement fixation test. A serological test used for clinical diagnosis and identification of specific antigen. It is a two antigen-antibody system involving the use of hemolysin, erythrocytes, complement, testing antigen, and antibody.

C3f Abbreviation for a subcomponent resulting from cleavage of C3 (third component of complement).

CF_0 A component of ATP synthetase (ATPase) complex of the chloroplast.

CF_0-CF_1 Complex Referring to chloroplast ATPase complex that has properties similar to mitochondrial F_0F_1-ATPase.

CF_1 A component of ATP synthetase (ATPase) complex with chloroplast (also called chloroplast coupling factor).

CFA Abbreviation for 1. complement fixing antibody; 2. complete Freund's adjuvant.

CFC Abbreviation for chlorofluorocarbons.

CFDA Abbreviation for 5,6-carboxyfluorescein diacetate.

CFI Abbreviation for chemotactic factor inhibition.

CFI Test Abbreviation for chemotactic factor inhibition test.

CflI (PstI) A *Cellulomonas flavigena* restriction endonuclease with the following specificity:

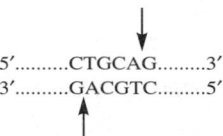

5′..........CTGCAG.........3′
3′..........GACGTC.........5′

CFM Abbreviation for chlorofluoromethane.

c-*fos* Abbreviation for cellular oncogen *fos*.

CfoI (HhaI) A restriction endonuclease from *Clostridium formicoaceticum* with the same specificity as HhaI.

CFP Abbreviation for 1. a combination drug containing cytoxan, fluorouracil, and prednisone; 2. cerebrospinal fluid protein.

CfrI A restriction endonuclease from *Citrobacter freundii* RFL 2 with the following specificity:

5′..........PyGGCCPu.........3′
3′...........PuCCGGPy........5′

Cfr4I (AsuI) A restriction endonuclease from *Citrobacter freundii* RFL 4 with the same specificity as AsuI.

Cfr5I (EcoRII) A restriction endonuclease from *Citrobacter freundii* RFL5 with the same specificity as EcoRII.

Cfr6I (PvuII) A restriction endonuclease from *Citrobacter freundii* RFL 6 with the following specificity:

5′..........CAGCTG.........3′
3′..........GTCGAC.........5′

Cfr7I (BstEII) A restriction endonuclease from *Citrobacter freundii* RFL 7 with the following specificity:

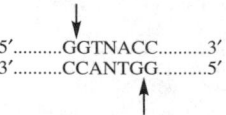

5′..........GGTNACC..........3′
3′..........CCANTGG..........5′

Cfr8I(AsuI) A restriction endonuclease from *Citrobacter freundiin* RFL 8 with the same specificity as AsuI.

Cfr9I A restriction endonuclease from *Citrobacter freundiin* with the following specificity:

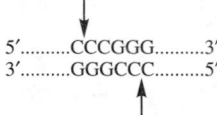

5′..........CCCGGG..........3′
3′..........GGGCCC..........5′

Cfr10I A restriction endonuclease from *Citrobacter freundii* RFL 10 with the following specificity:

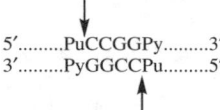

5′.........PuCCGGPy.........3′
3′.........PyGGCCPu.........5′

Cfr11I (EcoRII) A restriction endonuclease from *Citrobacter freundii* RFL 11 with the same specificity as EcoRII.

Cfr13I A restriction endonuclease from *Citrobacter freundii RFL 13* with the following specificity:

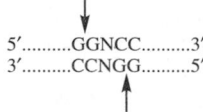

5′..........GGNCC..........3′
3′..........CCNGG..........5′

Cfr14I (CfrI) A restriction endonuclease from *Citrobacter freundii* RFL 14 with the same specificity as CfrI.

Cfr19I (BstEII) A restriction endonuclease from *Citrobacter freundii* RFL 19 with the same specificity as Bst EII.

Cfr20I (EcoRII) A restriction endonuclease from *Citrobacter freundii* RFL 20 with the same specificity as EcoRII.

Cfr22I(EcoRII) A restriction endonuclease from *Citrobacter freundii* RFL 22 with the same specificity as EcoRII.

Cfr23I (AsuI) A restriction endonuclease from *Citrobacter freundii* RFL 23 with the same specificity as AsuI.

Cfr24I (EcoRII) A restriction endonuclease from *Citrobacter freundii* RFL 24 with the same specificity as Eco RII.

Cfr25I (EcoRII) A restriction endonuclease from *Citrobacter freundii* RFL 25 with the same specificity as EcoRII.

Cfr27I (EcoRII) A restriction endonuclease from *Citrobacter freundii* RFL 27 with the same specificity as EcoRII.

Cfr28I (EcoRII) A restriction endonuclease from *Citrobacter freundii* RFL 28 with the same specificity as EcoRII.

Cfr29I (EcoRII) A restriction endonuclease *from Citrobacter freundii* RFL 29 with the same specificity as EcoRII.

Cfr30I (EcoRII) A restriction endonuclease from *Citrobacter freundii* RFL30 with the same specificity as EcoRII.

Cfr31I (EcoRII) A restriction endonuclease from *Citrobacter freundii* RFL31 with the same specificity as EcoRII.

Cfr32I (HindIII) A restriction endonuclease from *Citrobacter freundii* RFL32 with the same specificity as HindIII.

Cfr33I (AsuI) A restriction endonuclease from *Citrobacter freundii* RFL 23 with the same specificity as AsuI.

Cfr35I (EcoRII) A restriction endonuclease from *Citrobacter freundii* RFL 35 with the same specificity as EcoRII.

Cfr37I (SacII) A restriction endonuclease from *Citrobacter freundii* RFL 37 with the same specificity as SacII.

Cfr38I (CfrI) A restriction endonuclease from *Citrobacter freundii* RFL 38 with the same specificity as CfrI.

Cfr39I (CfrI) A restriction endonuclease from *Citrobacter freundii* RFL 39 with the same specificity as CfrI.

Cfr40I (CfrI) A restriction endonuclease from *Citrobacter freundii* RFL 40 with the same specificity as CrfI.

Cfr41I (SacII) A restriction endonuclease from *Citrobacter freundii* RFL 41 with the same specificity as SacII.

Cfr42I (SacII) A restriction endonuclease from *Citrobacter freundii* RFL 42 with the same specificity as SacII.

Cfr43I (SacII) A restriction endonuclease from *Citrobacter freundii* RFL 43 with the same specificity as SacII.

Cfr45I (AsuI) A restriction endonuclease from *Citrobacter freundii* RFL 45 with the same specificity as AsuI.

Cfr45II (SacII) A restriction endonuclease from *Citrobacter freundii* RFL 45 with the same specificity as SacII.

Cfr46I (AsuI) A restriction endonuclease from *Citrobacter freundii* RFL 46 with the same specificity as AsuI.

Cfr47I (AsuI) A restriction endonuclease from *Citrobacter freundii* RFL 47 with the same specificity as AsuI.

Cfr481 (AsuI) A restriction endonuclease from *Citrobacter freundii* RFL 48 with the same specificity as HgiJII.

Cfr51I (PvuI) A restriction endonuclease from *Citrobacter freundii* RFL 51 with the same specificity as PvuI.

Cfr52I (AsuI) A restriction endonuclease from *Citrobacter freundii* RFL 52 with the same specificity as AsuI.

Cfr54I (AsuI) A restriction endonuclease from *Citrobacter freundii* RFL 54 with the same specificity as AsuI.

CfrA4I (PstI) A restriction endonuclease from *Citrobacter freundii* RFL A4 with the same specificity as PstI.

CfrNI (AsuI) A restriction endonuclease from *Citrobacter freundii* with the same specificity as AsuI.

CfrS37I (EcoRII) A restriction endonuclease from *Citrobacter freundii* S 39 with the same specificity as EcoRII.

CFS Abbreviation for chronic fatigue syndrome.

CFT Abbreviation for complement fixation test.

CFTR Abbreviation for cystic fibrosis transmembrane regulator. The product of the gene associated with cystic fibrosis.

CFU Abbreviation for colony-forming unit.

CfuI (DpnI) A *Caulobacter fusiformis* restriction endonuclease with the following specificity:

```
           ↓
5′..........GATC..........3′
3′..........CTAG..........5′
           ↑
```

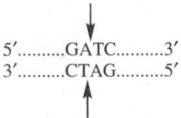

CFU-C Abbreviation for colony-forming unit in culture.

CFU-E Abbreviation for colony-forming unit in erythroid.

CFU-EOS Abbreviation for colony-forming unit in eosinophil.

CFU-F Abbreviation for colony-forming unit in fibroblast.

CFU-GEMM Abbreviation for colony-forming unit in granulocyte, erythrocyte, monocyte, and megakaryocyte.

CFU-GM Abbreviation for colony-forming unit in granulocyte and macrophage.

CFU-L Abbreviation for colony-forming unit in lymphoid.

CFU-M Abbreviation for colony-forming unit in megakaryocyte.

CFU-NM Abbreviation for colony-forming unit in neutrophil and monocyte.

CFU-S Abbreviation for colony-forming unit in spleen.

CG Abbreviation for 1. chorionic gonadotropin, a hormone produced by placenta and 2. cytosine-guanine content in DNA.

Cγ Abbreviation for constant region of immunoglobulin chain.

CGA 1. A genetic code or codon for the amino acid arginine. 2. Abbreviation for catabolite gene activator protein.

CGB Agar Abbreviation for canavanine-glycine-bromthymol blue agar used to distinguish strains of *Crptococcus.*

CGC A genetic code or codon for the amino acid arginine.

CGD Abbreviation for chronic granulomatous disease.

CGG A genetic code or codon for the amino acid arginine.

cGI-PDE Abbreviation cGMP-inhibited PDE (phosphodiesterase).

cGK Abbreviation for cyclic GMP-dependent protein kinase.

cGMP (mol wt 345) Abbreviation for cyclic guanosine monophosphate, a metabolic regulator.

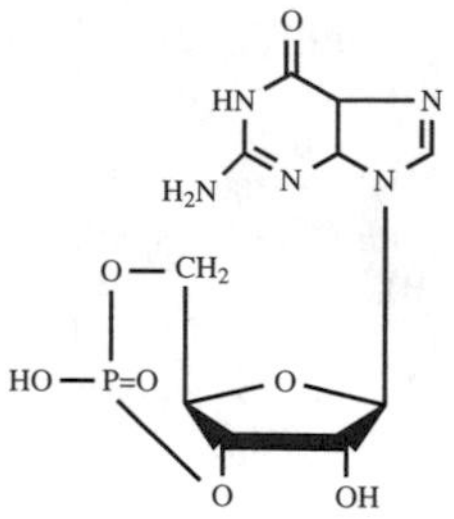

CGP Abbreviation for 1. N-carbobenzoyl-glycyl-phenylalanine; 2. choline glycerophosphatide.

cGPDH Abbreviation for cytosolic GPDH (glycerate 3-phosphate dehydrogenase).

cGPx (cGPX) Abbreviation for cytosolic glutathione peroxidase.

CGR Abbreviation for crystal growth rate.

CGRP Abbreviation for calcitonin-gene-related peptide.

cgs Abbreviation for centimeter-gram-second system.

cGSH Abbreviation for cytosolic glutathione.

cGSH-Px Abbreviation for cytosolic glutathione peroxidase.

CGT Abbreviation for N-carbobenzoyl-γ-glutamyl-L-tyrosine.

CGTT Abbreviation for cortisol glucose tolerance test.

CGU A genetic code or codon for the amino acid arginine.

CH Abbreviation for 1. collagen homology; 2. calponin homology.

C_H Abbreviation for the constant region of the heavy chains of an immunoglobulin.

Ch Abbreviation for choline.

$CH_{18:2}$ Abbreviation for cholesteryl linoleate.

CHAD Abbreviation for cold hemagglutinin disease.

Chagas Disease A disease caused by the protozoan parasite (*Trypanosoma cruzi*), transmitted by blood-sucking bugs, and characterized by prolonged high fever, edema, and enlargement of spleen, liver, and lymph nodes.

Chain Terminator 1. Substance that terminates the extension of a DNA chain during replication (e.g., dideoxynucleotides). 2. Substance that stops the growth of a chain polymerization.

Chair Conformation The arrangement of atoms in a molecule that resembles the outline of a chair.

Chalcomycin (mol wt 701) An antibiotic produced by™ *Streptomyces bikiniensis.*

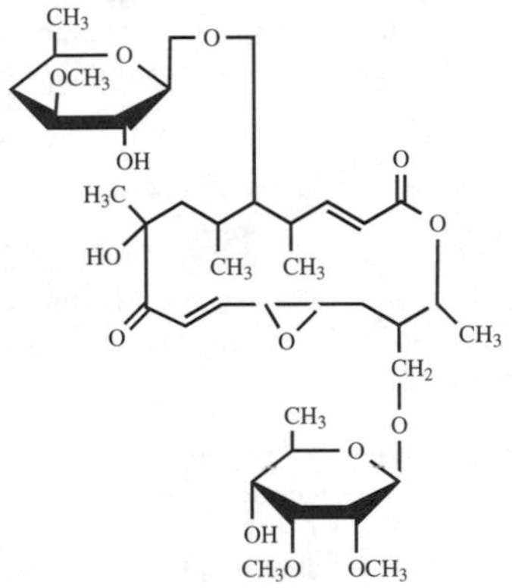

Chalconase The enzyme that catalyzes the following reaction:

A chalcone ⇌ a flavanone

Chalcone A group of bioflavonoids (pigments).

D-Chalcose (mol wt 162) A sugar and component in chalcomycin.

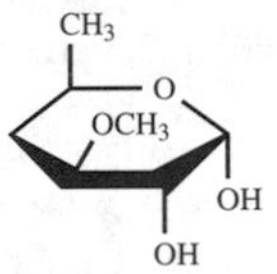

Chalicosis A lung disease caused by inhalation of calcium dust or stone dust.

Chalone Substances produced by the mature cell for inhibition of cell division.

Chambon's Rule The rule states that the nucleotide sequences of all introns start with GT and end with AG except introns for tRNA.

CHAMOA Abbreviation for a combination drug containing cytoxan, hydroxyurea, actinomycin D, methotrexate, oncovin, and adriamycin.

Chancre The primary sore of syphilis.

Channel Former Ionophore that functions by forming a hydrophobic channel within the plasma membrane for transport of ions across the membrane.

Channel-Forming Integral Membrane Protein 28 An integral membrane protein that forms a water-specific channel that provides the plasma membranes of red cells and kidney proximal tubules with high permeability to water.

Channel Protein Proteins that form water-filled pores or channels across the membrane and are responsible for transporting solutes across the membrane, e.g., porin protein in bacterial cell membrane.

Chaoptin An extracellular membrane glycoprotein required for *Drosophila melanogaster* photoreceptor cell morphogenesis, it contains a leucine-rich repeat.

Chaotropic Agents Ions or substances that denature proteins, e.g., thiocyanate(SCN^-) and perchlorate (CLO^-), used for solubilization of membrane proteins.

Chaperone Proteins that help other proteins to avoid misfolding pathways that may lead to the production of an inactive protein or protein aggretate.

CHAPS (mol wt 615) Abbreviation for 3-[(cholamidopropyl)-dimethylammonio]-1-propane sulfonate, a nondenaturing biological detergent used in the study of membrane biochemistry.

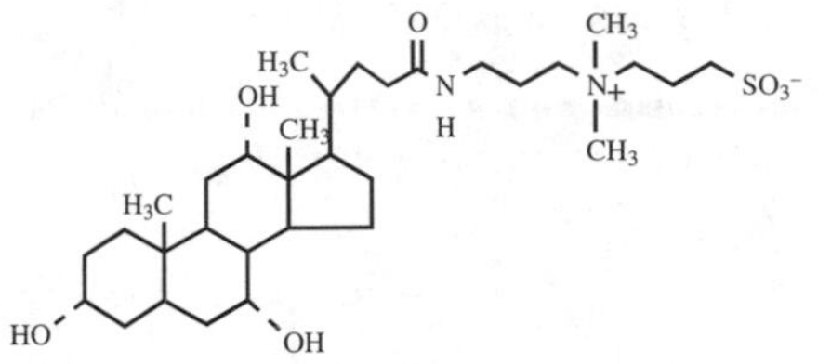

Charcoaide Referring to activated charcoal used to inhibit absorption of drugs or chemicals from the GI tract.

Charcocaps Referring to activated charcoal.

Chargaff's Rule The rule for the description of base composition of the double-stranded DNA. It states that the relationship of the mole percent concentrations of adenine, guanine, thymine, and cytosine are 1. A = T; G = C; 2. A/T = G/C; and 3. sum of purines = sum of pyrimidines.

Charge Density The charge per unit of area.

Charge Relay System A phenomenum observed in the serine proteinase in which the partial negative charge on the oxygen atom of the CH_2-OH group increases the nucleophilicity of the serine side chain at the active site.

Charge Repulsion The separating force due to the presence of the same electric charges on two ions, two ionic molecules, or two ionic regions of a molecule.

Charge Shielding The effect of metal ions on enzymatic function in which metal ion and substrate form a complex and electrostatically shield the negative charge of the substrate.

Charge Shift Immunoelectrophoresis A type of immunoelectrophoresis used to distinguish amphiphilic proteins from hydrophobic proteins in the presence of nonionic detergents. The amphiphilic proteins bind nonionic detergents and change their electrophoretic mobilities while the hydrophobic proteins are unaffected.

Charge Transfer Complex A noncovalent interaction in which an electron pair is partially transferred from a donor to an acceptor so that the two are held together by an electrostatic attraction.

Charged Polar Amino Acid Polar amino acid that carries either negative or positive charges, e.g., aspartic acid.

Charged tRNA The tRNA that carries a covalently linked amino acid at its 3′ end (also called aminoacyl-tRNA).

Charles's Law The law states that at constant volume the pressure of a confined gas is proportional to its absolute temperature.

Charon Bacteriophage A bacteriophage derived from λ phage and used as a cloning tool.

Chartreusin (mol wt 641) An antibiotic produced by *Streptomyces chartreusis*.

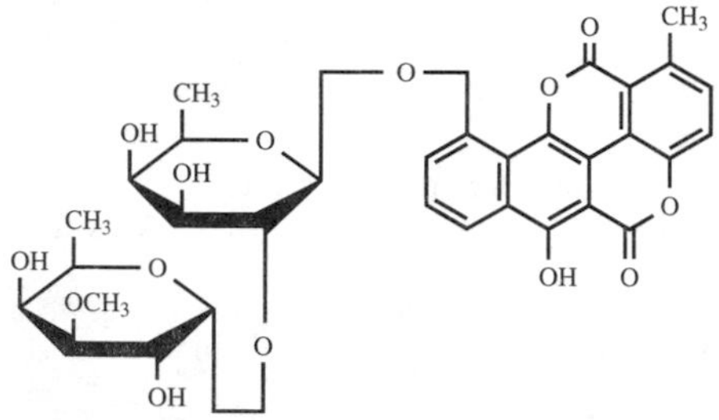

Charybdotoxin A 37-residue peptide inhibitor for calcium-activated potassium channel and voltage-dependent potassium channel, isolated from venom of the scorpion *Leiurus quinquestriatus herbraeus*.

Chase To follow the metabolic fate of a radioactively labeled compound in a pulse-chase experiment after terminating the brief exposure of a radioactivelly labeled compound (see pulse-chase experiment).

Chaulmoogric Acid (mol wt 280) A cyclic fatty acid.

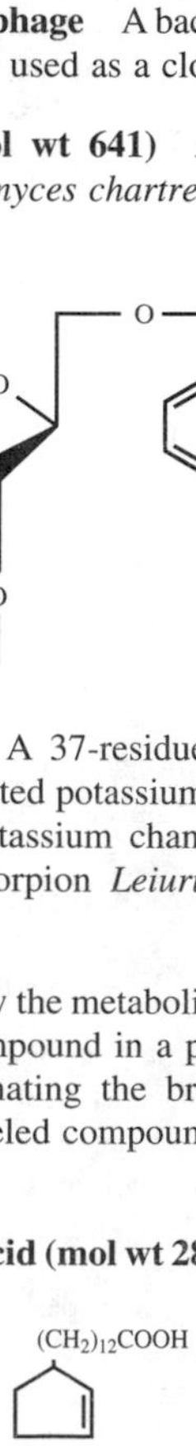

CH_3-Cbl Abbreviation for methyl-cobalamin.

CHD Abbreviation for coronary heart disease.

ChDNA Abbreviation for chloroplast DNA.

ChE Abbreviation for cholinesterase.

Checkmate A trade name for sodium fluoride, used for bone remineralization.

Chediak-Higashi Syndrome A genetic disorder involving leukocytes and melanocytes; characterized by partial albinism and susceptibility to pyogenic infection.

CHEF Abbreviation for Chinese hamster embryo fibroblast.

Chelate A ring structure formed by the reaction of a metal ion with one or more groups on an organic ligand molecule.

Chelating Agent A compound capable of forming a chelate with metal ion, e.g., EDTA.

Chelation The binding of a metal ion by two or more atoms on a chelating agent.

Chelex 100 A trade name for a synthetic ion-exchange resin used for binding metal ions.

Chemet A trade name for succimer, a chelating agent and antidote for heavy metal poisoning. It forms water-soluble complexes with heavy metals (e.g., lead).

Chemical Referring to 1. A compound produced by a chemical reaction. 2. Pertaining to chemistry.

Chemical Antidote Any compound capable of reacting with a poison to form a harmless or less harmful complex.

Chemical Bond The linkage between two atoms in a molecule or linkage between two groups of ions or molecules.

Chemical Coupling Hypothesis The hypothesis states that electron transport yields reactive intermediates whose subsequent breakdown drives oxidative phosphorylation (ATP formation).

Chemical Energy The energy resulting from a chemical reaction.

Chemical Equation The representation of a chemical reaction using symbols to show the stochiometric relationship between reactants and products.

Chemical Equilibrium The condition in a chemical reaction in which the rate of reaction in one direction equals the rate in the reverse direction.

Chemical Equivalent 1. Atomic weight of an element divided by its valence. 2. The number of parts of weight of an element or radical that will combine with or displace eight parts by weight of oxygen or one part by weight of hydrogen. 3. The molecular weight of a salt divided by the valence of the particular element considered.

Chemical Fusogen Any substance capable of fusing two cells together, e.g., polyethylene glycol.

Chemical Kinetics The science that deals with the rate and mechanism of reaction.

Chemical Mutagens Substances that cause mutation.

Chemical Potential The partial molar free energy of a substance.

Chemical Quenching Absorption of radiation energy either by the sample or by the other substance present in the solution as observed in scintillation counting.

Chemical Reaction The reaction between two or more chemicals whereby new bonds are formed and energy is subsequently exchanged.

Chemical Synapse The junction between two neurons at which a specific substance (neurotransmitter) transmits a nerve impulse across the junction.

Chemical Taxonomy Classification of organisms based upon the chemical characteristics of the organisms, e.g., DNA sequence and amino acid composition of the proteins (also called chemotaxonomy).

Chemical Transmitter A substance capable of transmitting nerve impulses, e.g., acetylcholine.

Chemiluminescence Emission of the absorbed energy as light resulting from a chemical reaction, e.g., bioluminescence.

Chemiluminescence Labeling A method for labeling DNA probes in which two labeled DNA probes for the adjacent sequences of a gene hybridize with the complementary segments and emit light that can be detected by a photomultiplier.

Chemiosmotic Hypothesis The hypothesis states that the free energy of electron transport is conserved by creation of an electrochemical H^+ gradient (pH gradient) across the inner mitochondrial membrane. The electrochemical potential of the pH gradient is the driving force for oxidative phosphorylation.

Chemisorption The formation of bonds between surface molecules and other substances in contact with the surface.

Chemistry The science that deals with the composition, structure, properties, and transformation of substances.

Chemoautotroph An organism that uses carbon dioxide as a carbon source and obtains energy for growth by oxidizing inorganic substances (e.g., sulfur, hydrogen, and nitrite).

Chemoheterotroph An organism that obtains energy from oxidation of organic compounds and carbon from preformed organic compounds.

Chemokinesis Chemically stimulated random movement or cellular activity.

Chemolithotroph See chemoautotroph.

Chemometrics Application of computer data analysis for classification, assimilation, and interpretation of chemical information.

Chemoorganotroph See heterotroph.

Chemoprophylaxis The use of chemicals, e.g., antibodies, to prevent microbial infection.

Chemoreceptor The receptor on the cell that senses and interacts with specific substances or stimulants.

Chemosis Swelling of the conjunctival tissue around the cornea.

Chemosmosis A chemical reaction, activity, or process that takes place through a semipermeable membrane.

Chemosorption Adsorption in which chemical energy causes an accumulation of the dispersed substances.

Chemostat An apparatus used to maintain bacterial cultures in the log phase of growth by the continuous supply of fresh medium.

Chemosterilant Substance or process that sterilizes insects.

Chemotactic Pertaining to chemotaxis.

Chemotactic Hormone A hormone capable of exerting a chemotactic effect.

Chemotaxin A substance capable of tropism.

Chemotaxis The movement of cells or organisms toward or away from a chemical stimulus.

Chemotaxonomy See chemical taxonomy.

Chemotherapeutic Agent Any agent that is used for treatment of disease.

Chemotherapeutic Index The ratio of the maximum tolerated dose of a chemical agent or drug used in chemotherapy to its minimum effective dose.

Chemotherapy The treatment or prevention of disease with chemicals or antibiotics.

Chemotropism The response of an organism to a chemical stimulant.

Chemovar The difference in chemical composition of plants grown in different geographic locations.

Chenix A trade name for chenodiol, used as an inhibitor for synthesis of cholesterol and cholic acid. It is capable of dissolution of gallstones.

Chenodeoxycholic Acid (mol wt 393) A major bile acid.

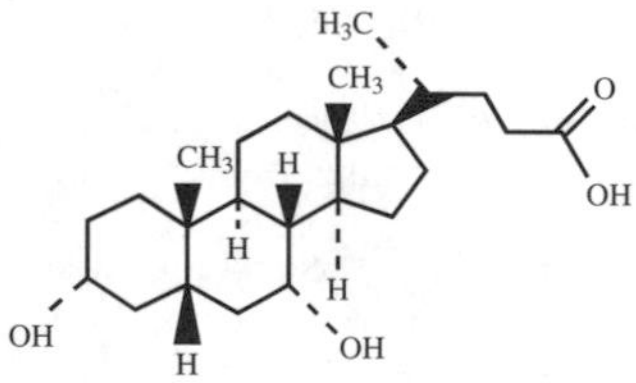

Chenodiol (mol wt 393) An agent that suppresses hepatic synthesis of cholesterol and cholic acid.

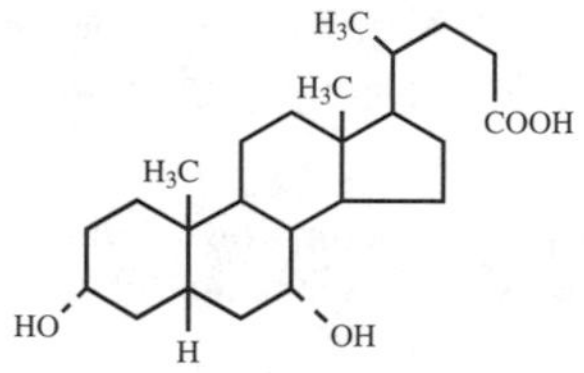

Che-W A cytoplasmic protein involved in the intracellular signaling process of chemotaxis.

Che-Y A cytoplasmic protein involved in the intracellular signaling process of chemotaxis.

Che-Z A cytoplasmic protein involved in the intracellular signaling process of chemotaxis.

CHF Abbreviation for 1. chick heart fibroblast; 2. congestive heart failure.

CHFP Abbreviation for a combination drug containing cytoxan, hexamethylmelamine, fluorouracil, and platinol.

***chi* Sequence** The sequence of GCTGGTGG on the DNA in *E. coli* that provides a hot spot for RecA-mediated genetic recombination.

***chi* Square Method** A statistical method that enables one to determine how closely an experimentally obtained set of data fits a given theoretical expectation.

***chi* Structure** 1. The structure formed at the point of cross-over between two double-stranded DNA genomes. 2. A structure formed by cleavage of a dimeric circular DNA at a single site on each of the circular DNAs by the restriction endonuclease.

Chiasma 1. Site of DNA exchange between two chromatids. 2. Chromosomal sites where crossing over produces an exchange of homologous parts between nonsister chromatides. 3. The crossing of the optic nerves.

Chiasmata Plural of chiasma.

Chicken Pox An infectious disease caused by varicella zoster, a virus in the family Herpesviridae.

Chilomastix A genus of protozoa parasitic in the intestine. It may cause diarrhea.

Chimera An organism composed of a mixture of genetically dissimilar cells.

Chimerin Referring to the GTPase-activating proteins.

Chimeric DNA A recombinant DNA molecule carrying unrelated foreign DNA.

Chimeric Vector A cloning vector or a plasmid that carries foreign DNA fragment.

Chinine A trade name for quinine bisulfate, used as an antimicrobial agent.

CHIP 28 Abbreviation for channel-forming integral membrane protein 28.

Chiral Asymmetrical molecules that are mirror images of each other.

Chiral Atom The asymmetric atom of a chiral compound.

Chiral Center Synonym for chiral atom.

Chiral Compound A compound that contains an asymmetric center (chiral atom or chiral center) and thus can occur in two nonsuperimposible mirror-image forms.

Chirocaine A trade name for levobupivacaine, an anesthetic agent.

Chitan Chitin in which all the glucosamine residues are N-acetylated.

Chitin A mucopolysaccharide occurring in anthropod exoskeletons. It consists of β-(1,4)-N-acetylglucosamine residues.

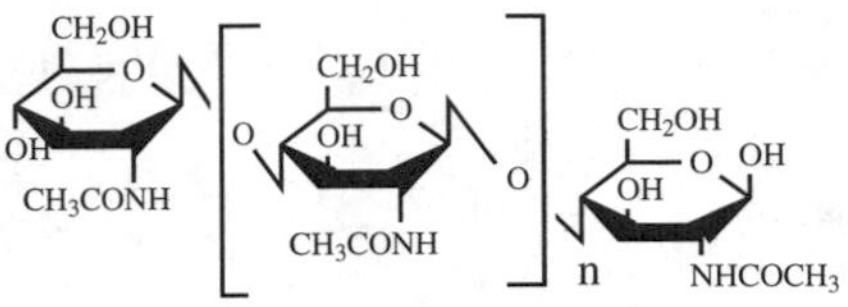

Chitin Deacetylase The enzyme that catalyzes the following reaction:

$$\text{Chitin} + H_2O \rightleftharpoons \text{Chitosan} + \text{acetate}$$

Chitin-UDP N-Acetylglucosaminyltransferase See chitinase.

Chitinase The enzyme that hydrolyzes chitin and chitodextrin.

Chitin Synthetase The enzyme that catalyzes the synthesis of chitin.

Chitoamidohydrolase See chitin deacetylase.

Chitodextrinase See chitinase.

Chitosan The deacylated derivative of chitin.

Chitosan N-Acetylglucosaminohydrolase The systematic name for chitosanase.

Chitosanase The enzyme that catalyzes the hydrolysis of β-1,4-linkages between N-acetyl-D-glucosamine and D-glucosamine residues in partially acetylated chitosan.

Chitosome A membrane-bound structure or cell organelle found in fungi containing enzyme chitin synthetase.

Chl Abbreviation for chlorophyll.

Chl a Abbreviation for chlorophyll a.

Chl b Abbreviation for chlorophyll b.

CHLA Abbreviation for cyclohexyl linoleic acid.

Chlamydia A genus of Gram-negative bacteria (family Chlamydiaceae).

Chlamydomonas A genus of green algae.

Chlamydospores The thick-walled resting spores produced asexually by certain types of fungi from somatic hyphae.

Chlophedianol (mol wt 290) An antitussive agent.

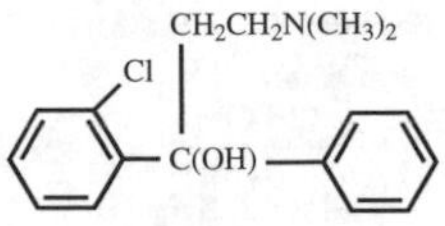

Chlor-3 A solution containing 50% sodium chloride, 30% potassium chloride, and 20% magnesium chloride.

Chloracizine (mol wt 361) A coronary vasodilator.

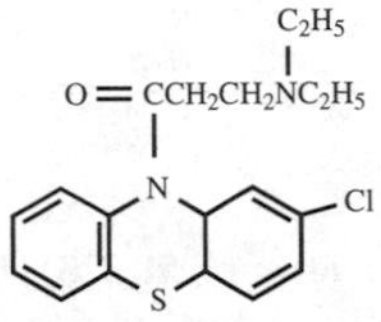

Chloral Formamide (mol wt 192) A sedative and hypnotic agent.

$$HCONHCH(OH)CCl_3$$

Chloral Hydrate (mol wt 165) A sedative and hypnotic agent and a reagent used for manufacturing DDT.

$$Cl_3CH(OH)_2$$

Chloralantipyrine (mol wt 354) A hypnotic and analgesic agent.

$$CCl_3CH(OH)_2C_{11}H_{12}N_2O$$

α-Chloralose (mol wt 310) A sedative and hypnotic agent.

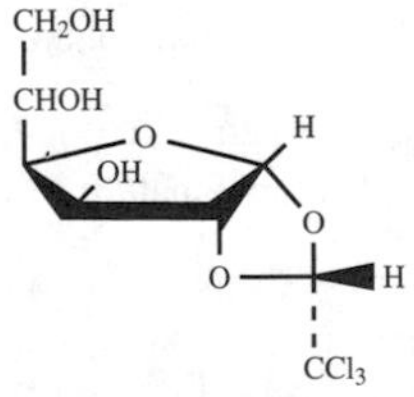

Chloramben (mol wt 206) An herbicide.

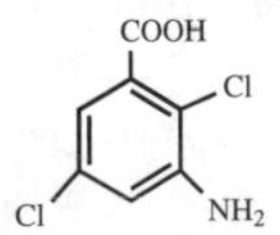

Chlorambucil (mol wt 304) An alkylating and antineoplastic agent that cross-links strands of cellular DNA interfering with transcription and translation.

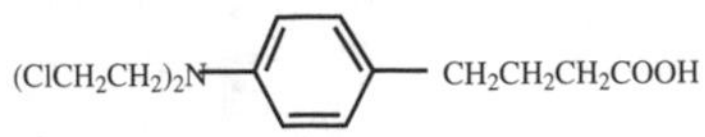

Chloramine-B (mol wt 214) An antibacterial agent.

$C_6H_5ClNNaO_2S$

Chloramine-T (mol wt 228) An antibacterial agent.

$C_7H_7ClNNaO_2S$

Chloraminophenamide (mol wt 286) A diuretic agent.

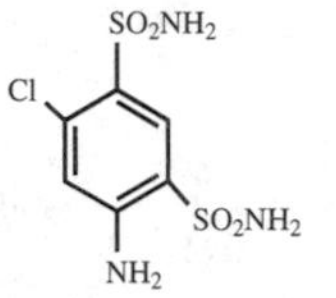

Chloramphenicol (mol wt 323) A broad spectrum antibiotic produced by *Streptomyces venezuelae* that inhibits bacterial protein synthesis by binding to 50S ribosomes to prevent peptide formation.

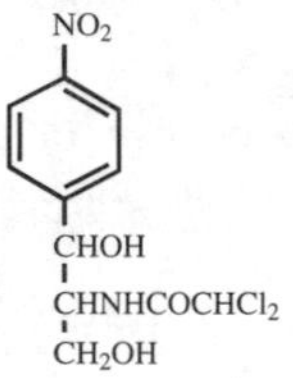

Chloramphenicol O-Acetyltransferase The enzyme that catalyzes the following reaction:

$$\text{Acetyl-CoA} + \text{chloramphenicol} \rightleftharpoons \text{CoA} + \text{chloramphenicol 3-acetate}$$

Chlorangiopancreatography An X-ray examination of the bile ducts and pancreas.

Chloranil (mol wt 246) A fungicide and a reagent used for dehydrogenation of hydroaromatic compounds.

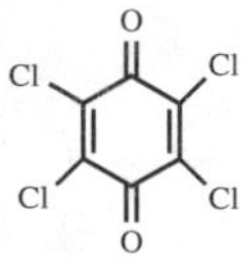

Chloranilic Acid (mol wt 209) A reagent used for paper-chromatographic detection of metals.

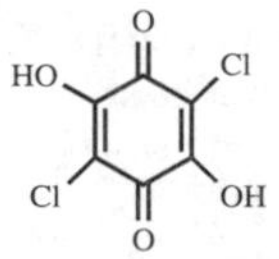

Chlorate A salt that contains radical ClO_3.

Chlorate Reductase The enzyme that catalyzes the following reaction:

$$\text{AH} + \text{chlorate} \rightleftharpoons \text{A} + H_2O + \text{chlorite}$$

Chlorazanil (mol wt 222) A diuretic agent.

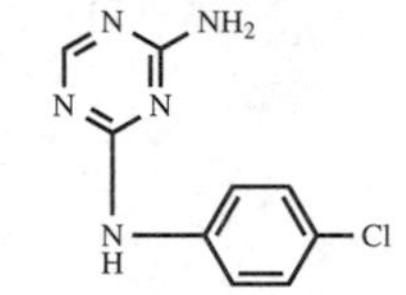

Chlorbenzoxamine (mol wt 435) An anticholinergic agent.

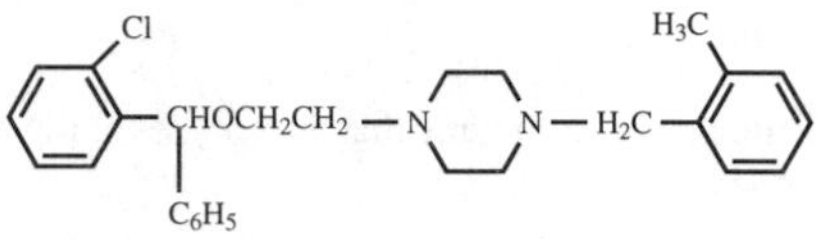

Chlorbetamide (mol wt 331) An antiamebic agent.

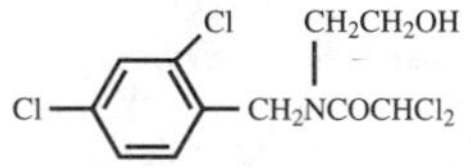

Chlorcyclizine (mol wt 301) An antihistaminic agent.

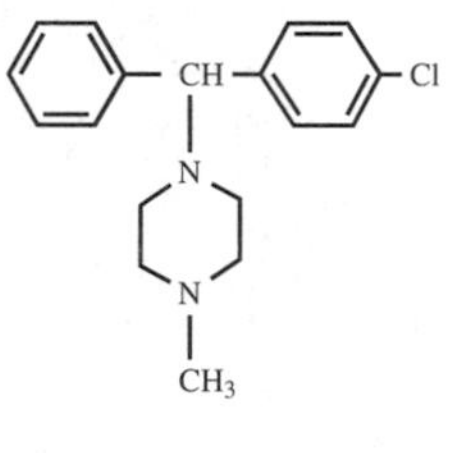

Chlordantoin (mol wt 348) A fungicide.

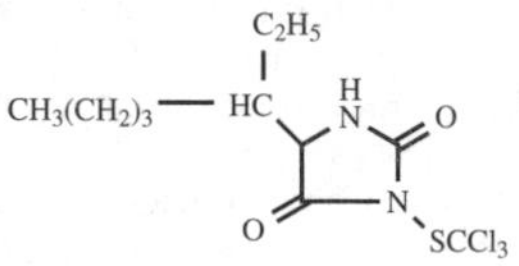

Chlordiazepoxide (mol wt 300) An anxiolytic agent and a tranquilizer. It is an inhibitory neurotransmitter.

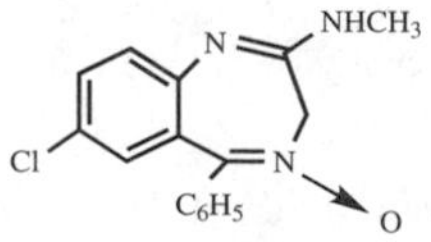

Chlorella A genus of nonmotile, unicellular green algae.

Chloremia The presence of a large quantity of chlorine in the blood.

Chlorenchyma Plant tissue that contains chloroplasts.

Chlorfenac (mol wt 240) An herbicide.

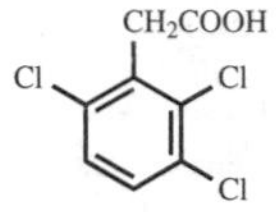

Chlorhexidine (mol wt 505) An antibacterial agent.

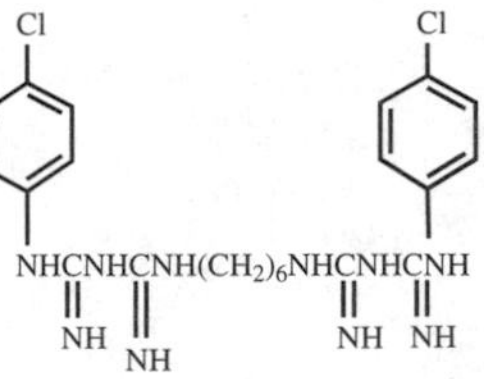

Chlorhydria A high concentration of hydrochloric acid in the stomach.

Chloride Compound of chlorine.

Chloride Channel A channel responsible for transport of chloride ions across a membrane. γ-Aminobutyrate (GABA) and glycine open chloride channels and increase the chloride conductance and lead to membrane hyperpolarization that triggers an action potential.

Chloride Hydrogen Peroxide Oxidoreductase Systematic name for chloride peroxidase.

Chloride Ionophore I An ionophore for chloride.

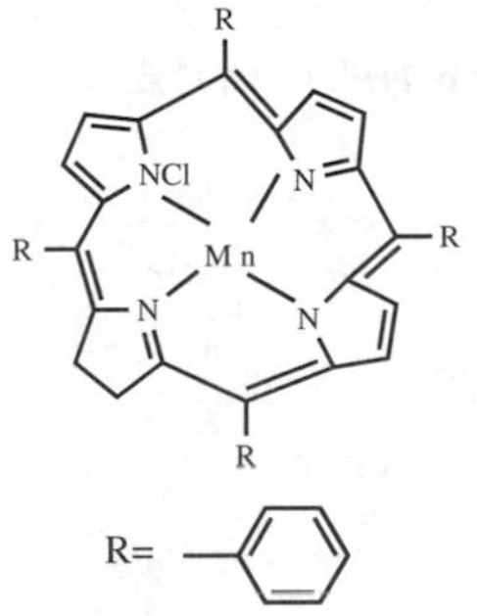

Chloride Peroxidase The enzyme that catalyzes the following reaction:

$$2RH + 2Cl^- + H_2O_2 \rightleftharpoons 2RCl + 2H_2O$$

Chloridometer An instrument used for analysis of chloride.

Chloriduria The presence of a large quantity of chlorine in the urine.

Chlorinated Hydrocarbons Synthetic organic molecules in which one or more hydrogen atoms have been replaced by chlorine atoms.

Chlorinated Organic Pesticides Pesticides of chlorinated hydrocarbons, e.g., DDT.

Chlorination The process of disinfecting water with chlorine to kill microorganisms.

Chlorindanol (mol wt 169) A spermaticide.

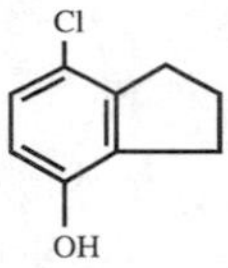

Chlorine A chemical element with atomic weight 35, valence 1.

Chlorine-36 Radioactive chlorine with half-life of 440,000 years.

Chlorine Dioxide (mol wt 67) A bleaching and antimicrobial agent.

ClO_2

Chlorine Number The number of chlorines of bleaching powder absorbed by 100 grams of oven-dried cellulose pulp.

Chlorine Water A decoloring agent and disinfectant consisting of approximately 0.4% chlorine.

Chlorinolysis The chlorination of organic compounds that ruptures the carbon-to-carbon bond yielding chlorocompounds with fewer carbons than in the original.

Chloriodized Oil Referring to chlorinated and iodinated vegetable oil.

Chlorisondamine Chloride (mol wt 429) An antihypertensive agent.

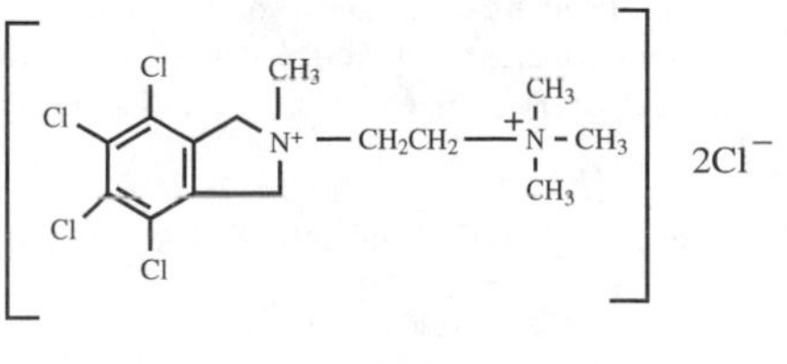

Chlorite The salt of H_2ClO_2.

Chlormadinone Acetate (mol wt 405) An orally active progestogen with antiandrogenic activity used as an oral contraceptive.

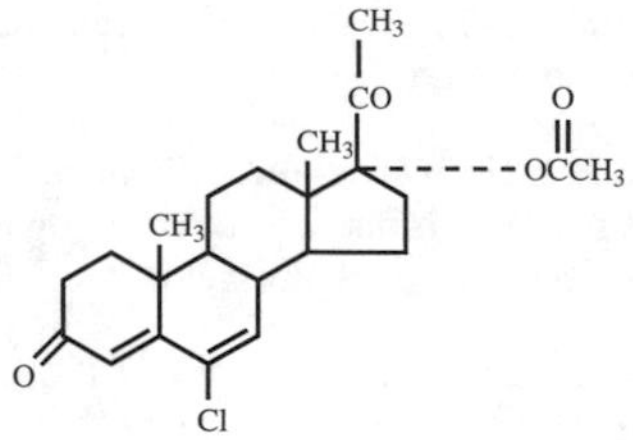

Chlormequat Chloride (mol wt 158) A plant growth regulator.

$[ClCH_2CH_2N^+(CH_3)_3]Cl^-$

Chlormerodrin (mol wt 367) A mercurial diuretic agent.

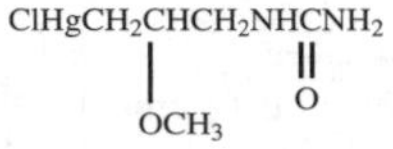

Chlormezanone (mol wt 274) An anxiolytic agent and muscle relaxant.

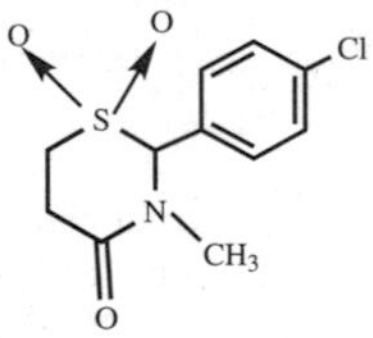

Chlormycetin Otic A trade name for chloramphenicol, an antibiotic.

Chlormycetin Ophthalmic A trade name for chloramphenicol, an antibiotic.

Chlornaphazine (mol wt 268) An antineoplastic agent.

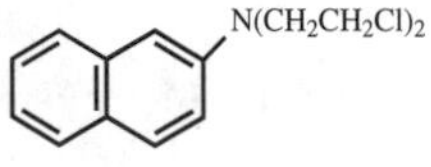

Chlor-Niramine A trade name for chlorpheniramine maleate, an antihistaminic agent that competes for H-1 receptor sites on the effector cells.

Chloroacetaldehyde (mol wt 79) A fungicide.

$ClCH_2CHO$

Chloroacetic Acid (mol wt 95) A monochloracetic acid and an herbicide.

$CH_2ClCOOH$

2-Chloro-Adenosine (mol wt 302) An adenosine receptor agonist.

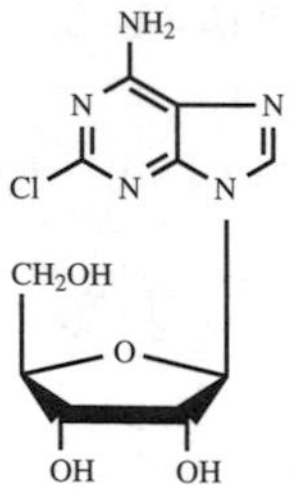

b-Chloroalanine (mol wt 160) A synthetic antibiotic that functions as an inhibitor of alanine racemase.

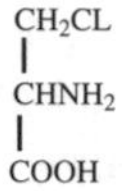

Chloroazodin (mol wt 183) A topical anesthetic agent.

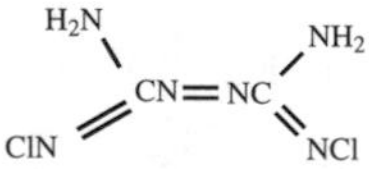

Chlorobenzilate (mol wt 325) A pesticide for control of spider mites.

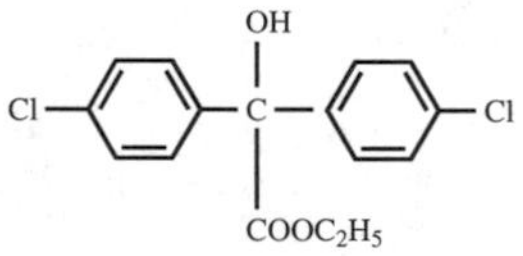

Chlorobium A genus of phototrophic green sulfur bacteria (family Chlorobiaceae).

Chlorobutanol (mol wt 177) A dental analgesic agent.

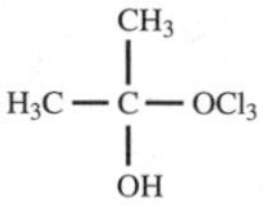

Chlorocystis A genus of unicellular green algae.

2-Chlorodeoxyadenosine (mol wt 286) An antineoplastic agent.

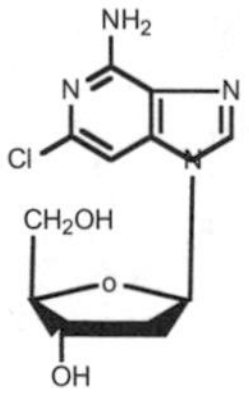

Chloroethane A trade name for ethyl chloride.

Chlorofair A trade name for chloramphenicol, an antibiotic that inhibits bacterial protein synthesis.

Chloroform (mol wt 120) An anesthetic agent.

$CHCl_3$

Chlorogenic Acid (mol wt 354) An antifungal metabolite found in plants.

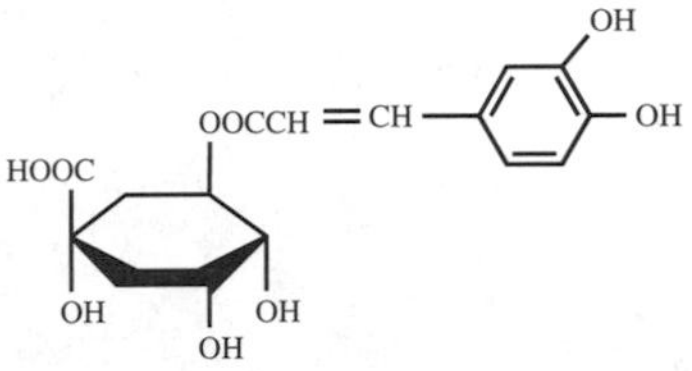

Chlorogloeopsis A genus of filamentous cyanobacteria.

Chlorogonium A genus of unicellular biflagellate green algae closely related to *Chlamydomonas* species.

Chloroma A tumor arising from myeloid tissue containing a pale green pigment.

Chlorometry An instrument used for quantitation and measurement of chlorine.

Chloromycetin A trade name for chloramphenicol, an antibiotic that inhibits bacterial protein synthesis.

Chloronema A genus of filamentous mesophilic bacteria (family Chlorflexaceae).

Chloropeptide A hepatotoxic mycotoxin produced by *Penicilllium islandicum*. It binds to actin and modifies cytoskeleton.

Chloroperoxidase Synonym of chloride peroxidase.

Chlorophenol Red (mol wt 423) A pH indicator.

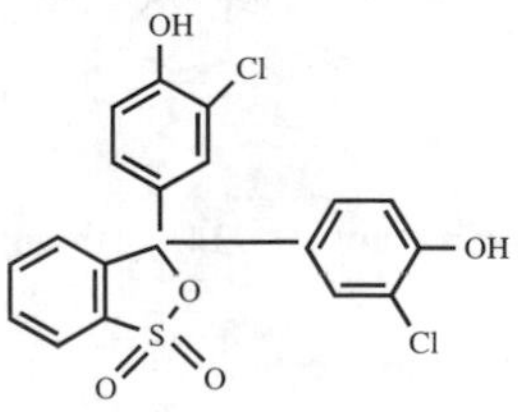

9-Chlorophenyl-9-Phenylxanthene (mol wt 292) A deoxynucleotide 5′-O-protecting agent used in oligonucleotide synthesis.

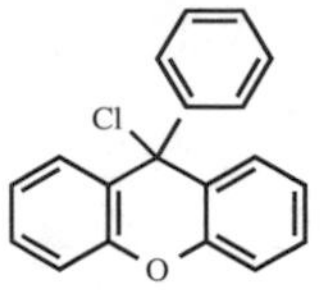

Chlorophyll Major photosynthetic pigment in photosynthetic organisms.

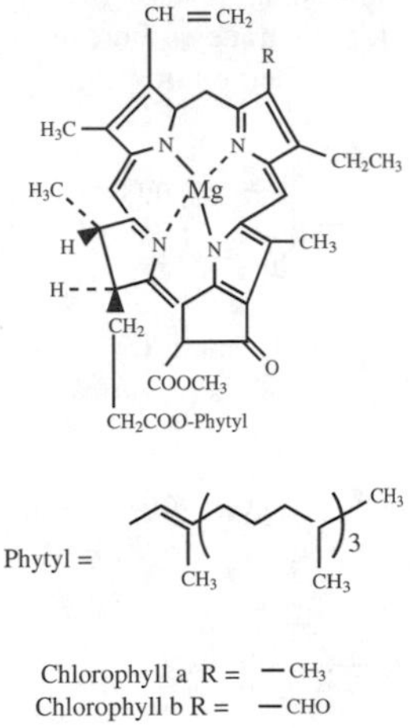

Chlorophyllide Chlorophyll without phytyl side chain.

Chlorophyllin Copper Complex Sodium A drug used to promote healing and relieve itching, discomfort, and skin irritation.

Chloroplast Membrane-bound cell organelle of photosynthetic eukaryotes (e.g., green algae and plants) that contains chlorophyll and is responsible for the biochemical conversion of light energy to ATP and synthesis of carbohydrate. Chloroplast that contains its own DNA and ribosomes.

Chloroplast ATPase A membrane-bound, multisubunit complex of ATPase in chloroplast that has the property similar to the ATPase in mitochondria. It catalyzes the reaction in which the synthesis of ATP is coupled to the discharge of proton gradients.

Chloroplast Genome Referring to the DNA in chloroplasts.

Chloroplast Ribosome The 70S ribosome found in the chloroplast. The protein synthesis on the 70S ribosome can be inhibited by chloramphenicol but not cycloheximide.

Chloroplast rRNA Referring to 23S, 16S, 5S, and 4.5S rRNA found in the chloroplast.

Chloroplast tRNA Referring to tRNAs found in the chloroplast. They are encoded by chloroplast DNA and differ in structure from the cytoplasmic tRNAs.

Chloroprocaine Hydrochloride (mol wt 307) A local anesthetic agent that blocks depolarization by interfering with sodium-potassium exchange across the nerve cell membrane and prevents generation and conduction of nerve impulses.

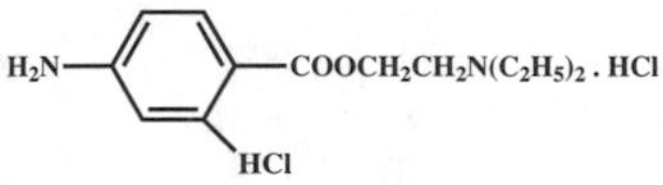

Chloroptic A trade name for chloramphenicol, an antibiotic that inhibits bacterial protein synthesis.

Chloropyramine (mol wt 290) An antihistaminic agent.

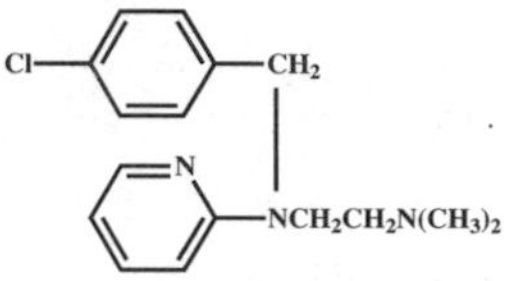

Chloroquine (mol wt 320) An antimalarial, antiamebic, and antirheumatic agent.

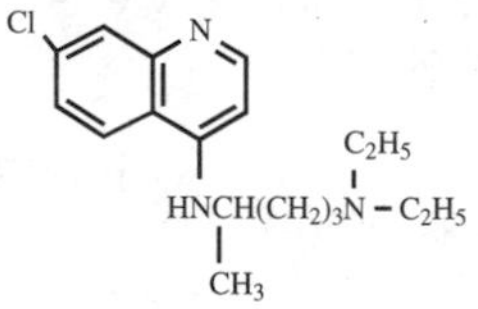

Chlorosis 1. The yellowing of leaves due to loss of chlorophyll in plants. 2. A form of anemia due to a diminution of red blood cells and hemoglobin.

Chlorosome A cylindrically shaped, intracellular photosynthetic structure in green bacteria (Chlorobiaceae).

Chlorotab A trade name for chlorpheniramine, used as an antihistaminic agent that competes with histamine for H-1 receptors.

Chlorothiazide (mol wt 296) A diuretic agent that increases urine excretion of sodium and water by inhibiting sodium reabsorption in the cortical diluting site of the nephron.

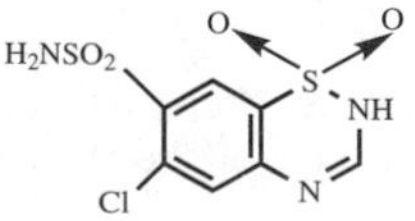

Chlorothricin (mol wt 956) A macrolide antibiotic produced by *Streptomyces antibioticus* that is active against Gram-positive bacteria.

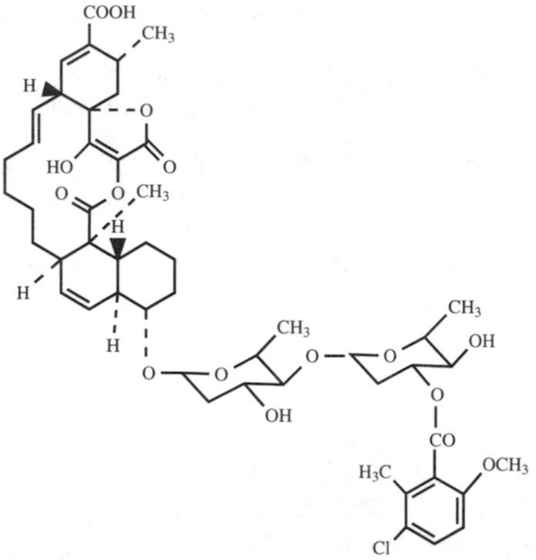

Chlorotrianisene (mol wt 381) An estrogen used to treat prostate cancer and to dry up breast milk after birth. It inhibits the release of FSH and LH from the pituitary.

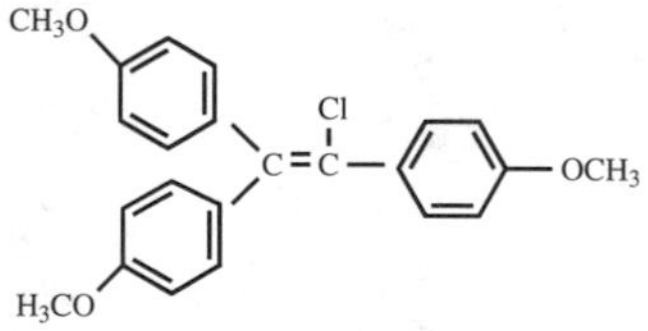

Chloroxylenol (mol wt 157) A topical antiseptic agent.

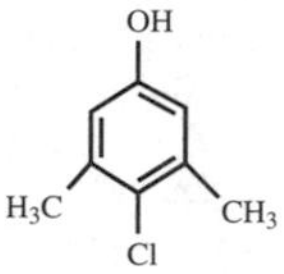

Chloroxine (mol wt 214) An antiseborrheic agent.

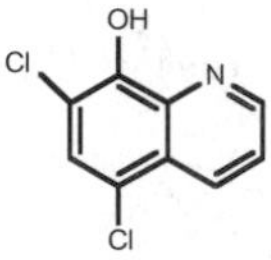

Chlorozotocin (mol wt 314) An antineoplastic agent.

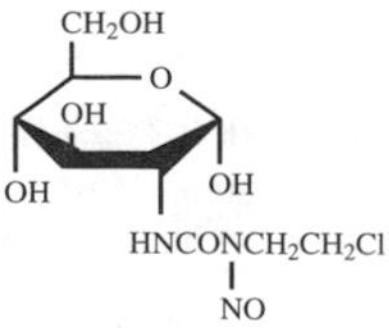

Chlorpazine A trade name for prochlorperazine maleate, an antiemetic agent.

Chlorphed A trade name for brompheniramine maleate, an antihistaminic agent that competes with histamine for the H-1 receptors.

Chlorphenesin (mol wt 203) A tropical antifungal agent.

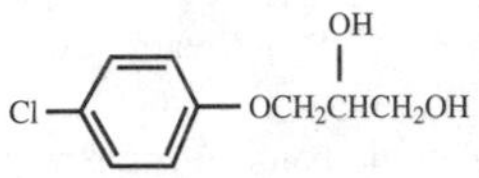

Chlorpheniramine (mol wt 275) An antihistaminic agent that competes with histamine for H-1 receptors.

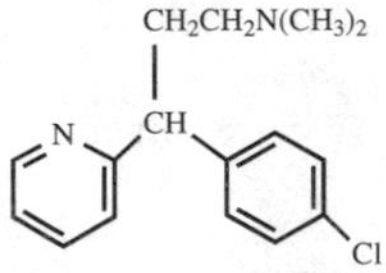

Chlorphenoxamide (mol wt 399) An antiamebic agent.

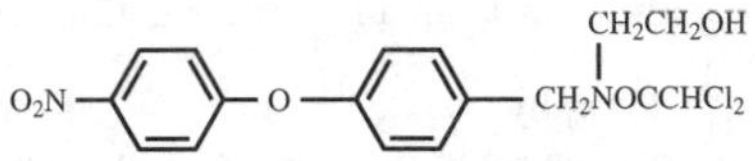

Chlorphenoxamine (mol wt 304) An anticholinergic agent.

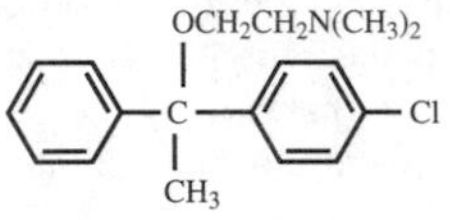

Chlor-Pro A trade name for chlorpheniramine maleate, an antihistaminic agent that competes with histamine for H-1 receptors.

Chlorproethazine (mol wt 347) A skeletal muscle relaxant.

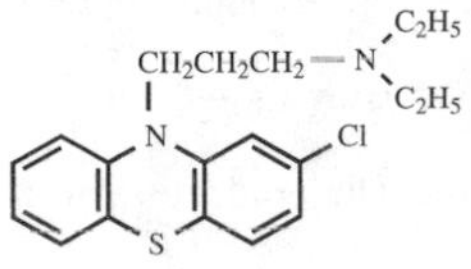

Chlorpromanyl A trade name for chlorpromazine, an antiemetic and antipsychotic agent that blocks the postsynaptic dopamine receptors in the brain.

Chlorpromazine (mol wt 319) A sedative and antiemetic agent that blocks the postsynaptic dopamine receptor in the brain.

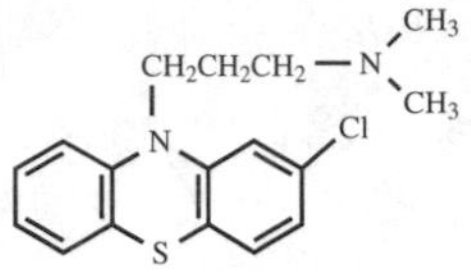

Chlorpropamide (mol wt 277) An antidiabetic agent that stimulates insulin release from the pancreatic beta cell and reduces glucose output by the liver.

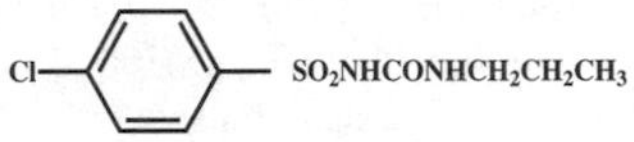

Chlorpropham (mol wt 214) A plant growth regulator and herbicide.

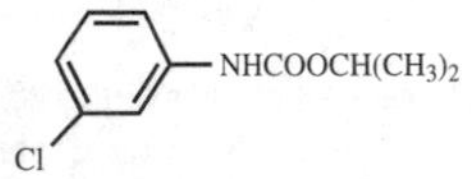

Chlorprothixene (mol wt 316) An antipsychotic agent that blocks postsynaptic dopamine receptors in the brain.

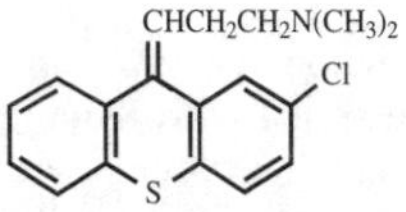

Chlorquin A trade name for chloroquine hydrochloride, an antimalarial agent.

Chlorsig A trade name for chloramphenicol, an antibiotic that inhibits bacterial protein synthesis.

Chlorspan-12 A trade name for chlorpheniramine maleate, an antihistaminic agent that competes with histamine for H-1 receptors.

Chlortab A trade name for chlorpheniramine maleate, an antihistaminic agent that competes with histamine for H-1 receptors.

Chlortetracycline (mol wt 479) An broad spectrum antibiotic produced by *Streptomyces aureofaciens* that inhibits bacterial protein synthesis.

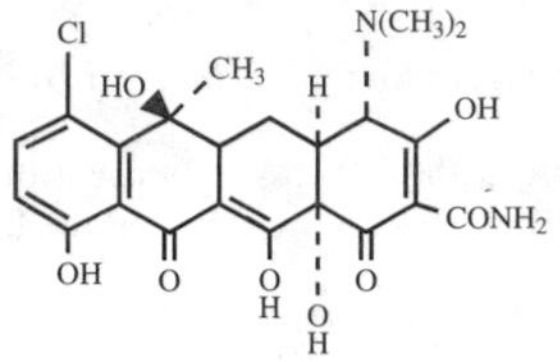

Chlorthalidone (mol wt 339) A diuretic agent that increases urine excretion of sodium and water by inhibiting sodium reabsorption in the cortical diluting site of the nephron.

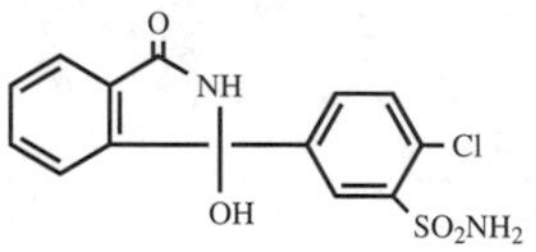

Chlor-trimeton A trade name for chlorpheniramine maleate, an antihistaminic agent that competes with histamine for H-1 receptors.

Chlorzoxazone (mol wt 170) A skeletal muscle relaxant that reduces the transmission of impulses from spinal cord to skeletal muscle.

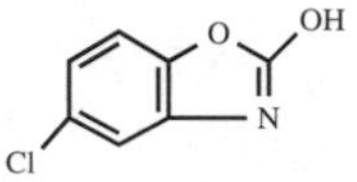

Chlotride A trade name for chlorothiazide, a diuretic agent that increases urine excretion of sodium and water by inhibiting sodium reabsorption in the cortical diluting site of the nephron.

CHM Abbreviation for chicken heart mesenchymal cells.

CHO Abbreviation for 1. a combination drug containing cytoxan, hydroxydaunomycin and oncovin; 2. carbohydrate; 3. chinese hamster ovary.

ChO Abbreviation for cholesterol oxidase.

CHO Cell Line A somatic cell line derived from Chinese hamsters. The cell contains many chromosomes with deletions, translocations, and aberrations.

CHOB Abbreviation for a combination drug containing cytoxan, hydroxydaunomycin, oncovin, and bleomycin.

$CH_{18:2}$-OH Abbreviation for cholesteryl linoleate hydroxide.

Chol Abbreviation for cholesterol.

Cholac A trade name for lactulose, used as a laxative that produces osmotic effects and decreases pH in the colon and promotes peristalsis.

Cholagogue Any agent that promotes flow of bile.

Cholangiocarcinoma A cancer of liver.

Cholangiography X-ray examination of bile ducts after ingestion of radiopaque medium.

Cholangiole One of the terminal branches of the bile duct.

Cholangiolitis Inflammation of the fine tubules (capillaries) of the bile duct system.

Cholangitis Inflammation of bile duct.

Cholate-CoA Ligase The enzyme that catalyzes the following reaction:

$$\text{ATP} + \text{cholate} + \text{CoA} \rightleftharpoons \text{AMP} + \text{PPi} + \text{Choloyl-CoA}$$

Cholaxin A trade name for dextrothyroxine sodium, an antihyperlipidemic agent.

Cholecalciferol Synonym for vitamin D_3.

Cholecalciferol 25-Hydroxylase The enzyme that hydroxylates vitamin D_3 to 25-hydroxycholecalciferol.

Cholecalcin An intracellular calcium-binding protein.

Cholecyst The gallbladder.

Cholecystagogue An agent that causes discharge of the bile from the gallbladder.

Cholecystitis Inflammation of the gallbladder.

Cholecystography X-ray examination of the gallbladder after ingestion of radiopaque medium.

Cholecystokinen Variant spelling of cholecystokinin.

Cholecystokinin A polypeptide hormone secreted by the duodenum that stimulates the secretion of digestive enzymes by the pancreas. Also stimulates contraction of the gallbladder.

Choledocholithiasis A disorder characterized by the presence of calculi in the gallbladder and bile duct.

Choledyl A trade name for choline theophyllinate, a bronchodilator that inhibits phosphodiesterase and increases cAMP concentration.

Choleglobin A green pigment present in the bile.

Cholelith Referring to gallstones.

Cholelithiasis The production of gallstones.

Cholemia The presence of excessive quantities of bile in the blood (an indication of liver disease).

Cholera An acute infectious disease caused by *Vibrio cholerae* and characterized by diarrhea, vomiting, and dehydration.

Choleragen Synonym of cholera toxin.

Cholera Toxin A protein toxin produced by *Vibrio cholerae.* It catalyzes the transfer of ADP-ribose from NAD^+ to Arg side chains of the G protein causing stimulation of cyclic adenylate cyclase activity and increasing cAMP concentration.

Choleretic Promoting bile secretion; an agent capable of promoting bile secretion.

Cholestanol (mol wt 389) A minor sterol occurs in human feces, gallstones, and eggs.

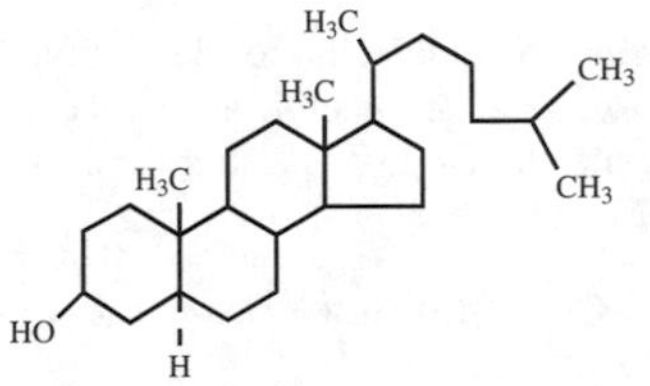

Cholestasis Interruption of the flow of bile through the biliary system.

Cholestatic Hepatitis Inflammation of the liver caused by an infection.

Cholesterase See cholesterol esterase.

Cholesterol (mol wt 387) The principal sterol of vertebrates present in all body tissues and in animal fat and oil.

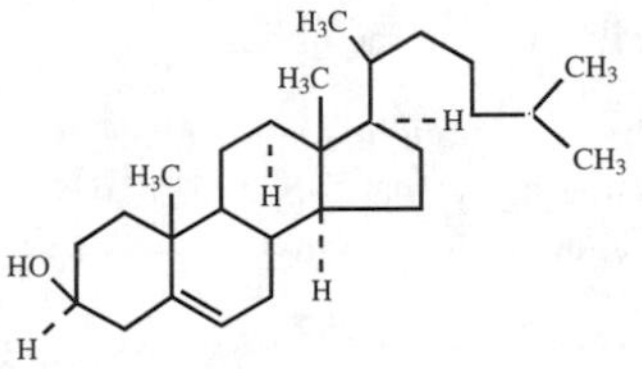

Cholesterol Acyltransferase The enzyme that catalyzes the following reaction:

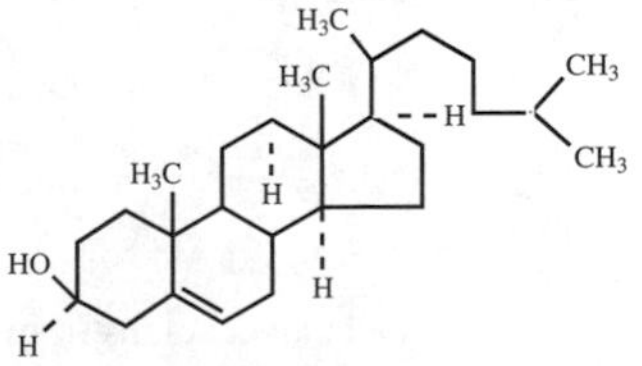

Cholesterol Binding Proteinase Synonym of pancreatic endopeptidase E.

Cholesterol Desmolase See cholesterol monooxygenase.

Cholesterol Ester An ester formed from the combination of cholesterol and fatty acid.

Cholesterol Ester Synthetase Synonym of sterol esterase.

Cholesterol Esterase The enzyme that catalyzes the following reaction:

A cholesterol ester + water

⇅

Cholesterol + fatty acid

Cholesterol Monooxygenase The enzyme that catalyzes the following reaction:

Cholesterol + reduced adrenal ferredoxin + O_2

⇅

pregnenolone + 4-methylpentanal +
oxidized adrenal ferredoxin + H_2O

Cholesterol Oxidase The enzyme that catalyzes the following reaction:

Cholesterol + O_2

⇅

Cholest-4-en-3-one + H_2O_2

Cholesterol Oxygen Oxidoreductase The systematic name for cholesterol oxidase.

Cholesterolemia The presence of excessive quantities of cholesterol in the blood.

Cholesterosis The excessive cholesterol deposits in tissues and organs.

Cholestyramine Resin A synthetic, basic cation-exchange resin.

Cholic Acid (mol wt 409) A bile acid and reagent for solubilization of membrane-bound proteins.

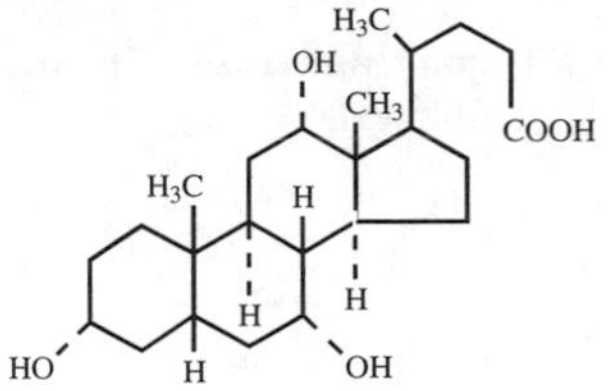

Choline (mol wt 104) A component of phospholipid (lecithin).

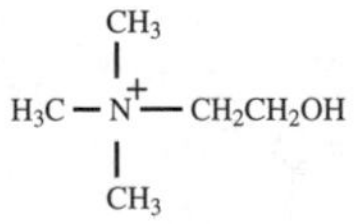

Choline Acetyltransferase The enzyme that catalyzes the following reaction:

Acetyl-CoA + choline

⇅

Acetylcholine + CoA

Choline Dehydrogenase The enzyme that catalyzes the following reaction:

$$\text{Choline} + \text{acceptor} \rightleftharpoons \text{Betaine aldehyde} + \text{reduced acceptor}$$

Choline Esterase The enzyme that catalyzes the following reaction:

$$\text{An acylcholine} + H_2O \rightleftharpoons \text{Choline} + \text{carboxylate}$$

Choline Kinase The enzyme that catalyzes the following reaction:

$$\text{Choline} + \text{ATP} \rightleftharpoons \text{Phosphocholine} + \text{ADP}$$

Choline Oxidase The enzyme that catalyzes the following reaction:

$$\text{Choline} + O_2 \rightleftharpoons \text{Betaine aldehyde} + H_2O$$

Choline Phosphatase The enzyme that catalyzes the following reaction:

$$\text{A phosphatidylcholine} + H_2O \rightleftharpoons \text{Choline} + \text{phosphatidate}$$

Choline Phosphotransferase The enzyme that catalyzes the following reaction:

$$\text{CDP-choline} + \text{1,2-diacylglycerol} \rightleftharpoons \text{CMP} + \text{phosphatidylcholine}$$

Choline Salicylate (mol wt 241) An analgesic and antipyretic agent.

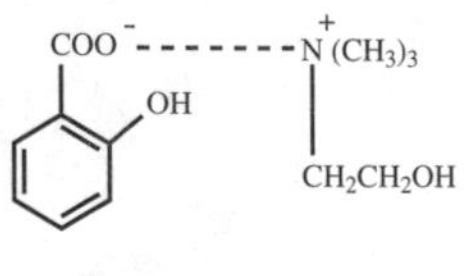

Cholinergic Referring to the nerve fibers that release nerve-signal carrier acetylcholine at the nerve endings.

Cholinergic Blocking Agent Substance capable of blocking the action of acetylcholine.

Cholinergic Nerve A nerve capable of releasing nerve-signal carrier acetylcholine so that the signal is carried to another nerve.

Cholinergic Receptor See acetylcholine receptor.

Cholinergic Synapses The synapses that use acetylcholine as neurotransmitter.

Cholinesterase See choline esterase.

Cholinolytic Agent or action capable of blocking the action of acetylcholine.

Cholinomimetic Substance capable of initiating the action of acetylcholine.

Choloxin A trade name for dextrothyroxine sodium, used as an antilipemic agent that accelerates hepatic metabolism of cholesterol and lowers cholesterol level.

Choloyl-CoA Synthetase See cholate-CoA ligase.

Choloylglycine Hydrolase The enzyme that catalyzes the following reaction:

$$3\alpha, 7\alpha, 12\alpha\text{-Trihydroxy-}5\beta\text{-cholan-24-oylglycine} + H_2O \rightleftharpoons 3\alpha, 7\alpha, 12\alpha\text{-Trihydroxy-}5\beta\text{-cholanate} + \text{glycine}$$

Choloyltaurine Hydrolase Synonym of choloylglycine hydrolase.

Choluria Bile in the urine.

Cholybar A trade name for cholestyramine, an antilipemic agent that forms insoluble complexes with bile acid for excretion.

Chondral Pertaining to cartilage.

Condrin Materials that resemble gelatin produced when cartilage is boiled.

Chondriosome Referring to mitochondria.

Chondrocalcin The C-terminal peptide found in human procollagen.

Chondroblast A cell that develops from the connective tissue of the embryo and forms cartilage (also called chondroplast).

Chondroclast Cells capable of absorbing cartilage.

Chondrocytes Cell of cartilage that secretes special collagens and glucosaminoglycan.

Chondrofibroma A fiberlike tumor containing cartilage elements.

Chondrogenesis The formation of cartilage.

Chondroid Resembling cartilage.

Chondroitin ABC Eliminase Synonym of chondroitin ABC lyase.

Chondroitin ABC Lyase The enzyme that catalyzes the eliminative degradation of polysaccharides containing 1,4-β-D-hexaminyl and 1,3-β-D-glucuronosyl or 1,3-α-L-iduronosyl linkages to disaccharides containing 4-deoxy-β-D-glucose-4-enuronosyl groups.

Chondroitin AC Eliminase Synonym of chondroitin AC lyase.

Chondroitin AC Lyase The enzyme that catalyzes the eliminative degradation of polysaccharides containing 1,4-β-D-hexaminyl and 1,3-β-D-glucuronosyl linkages to disaccharides containing 4-deoxy-β-D-glucose-4-enuronosyl groups.

Chondroitin Sulfate A mucopolysaccharide consisting of N-acetylchondrosine repeating units that occurs in animal tissue.

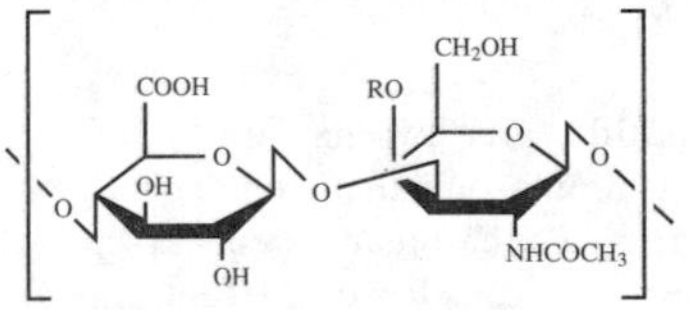

R = SO_3H

Chondroitin Sulfate Lyase Synonym of chondroitin AC lyase.

Chondroitin 4-Sulfotransferase The enzyme that catalyzes the following reaction:

3′-Phosphoadenylylsulfate + chondroitin

⇌

Adenosine 3′,5′-bisphosphate + chondroitin 4-phosphate

Chondroitin 6-Sulfotransferase The enzyme that catalyzes the following reaction:

3′-Phosphoadenylylsulfate + chondroitin

⇌

Adenosine 3′,5′-bisphosphate + chondroitin 6′-sulfate

Chondroitinase Synonym of chondroitin AC lyase.

Chondrolipoma A nonmalignant tumor found in connective tissue of the embryo.

Chondroma A harmless cartilage tumor.

Chondronectin A factor that mediates the attachment of chondrocytes to collagen.

Chondrosamine Synonym for galactosamine.

Chondrosarcoma A malignant bone tumor derived from the cartilage cell that erodes the bone and invades the adjacent soft tissue.

Chondrosine (mol wt 355) A disaccharide.

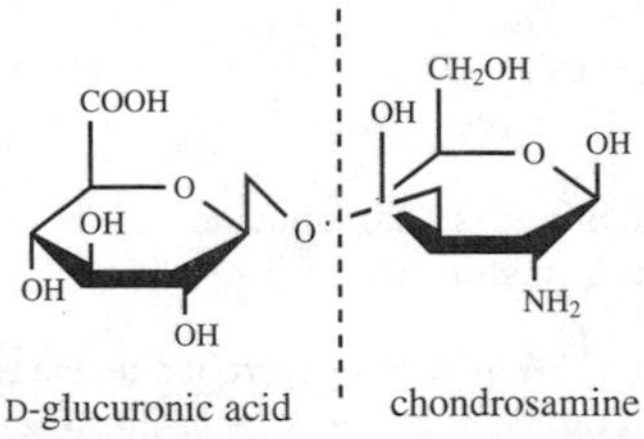

Chondro-4-Sulfatase The enzyme that catalyzes the following reaction:

4-Deoxy-β-D-gluc-4-enuronosyl-(1,3)-N-acetyl-D-galactosamine 4-sulfate + H_2O

⇌

4-Deoxy-β-D-gluc-4-enuronosyl-(1,3)-N-acetyl-D-galactosamine + sulfate

Chondro-6-Sulfatase The enzyme that catalyzes the following reaction:

4-Deoxy-β-D-gluc-4-enuronosyl-(1,3)-N-acetyl-D-galactosamine 6-sulfate + H_2O

⇌

4-Deoxy-β-D-gluc-4-enuronosyl-(1,3)-N-acetyl-D-galactosamine + sulfate

$Ch_{18:2}$-OOH Abbreviation for cholesteryl linoleate hydroperoxide.

CHOP Abbreviation for a combination drug containing cycloheximide, hydroxydaunorubicin, oncovin, and prednisone.

CHOPB Abbreviation for a combination drug containing cytoxan, hydroxydaunomycin, oncovin, prednisone, and bleomycin.

Chorditis Inflammation of spermatic cord or vocal cord.

CHORE Abbreviation for carbohydrate-response element.

Chorex A trade name for gonadotropin, a hormone capable of stimulating ovulation and promoting secretion of gonadal steroid hormones.

Chorio- A prefix meaning the membrane that protects the fetus.

Chorioallantoic Membrane The membrane surrounding the embryo of the chicken used for cultivation of viruses, e.g., vaccinia and influenza.

Chorioamnionitis Inflammation of fetal membranes.

Chorioangioma A benign vascular tumor of the chorion.

Choriocarcinoma A malignant tumor found in testicles in males and the uterus or ovaries in females.

Choriogenesis Growth of the membrane surrounding the fetus.

Choriomeningitis Inflammation of the cerebral membranes (meninges).

Chorion The membrane surrounding the embryo or fetus, it forms the placenta connecting the mother and the fetus.

Chorionic Pertaining to the chorion.

Chorionic Gonadotropin A polypeptide hormone from the placenta that stimulates the release of progesterone and estrogen.

Chorioretinitis Inflammation of the choroid and retina of the eye.

Chorismate Mutase The enzyme that catalyzes the following reaction:

Chorismate ⇌ Prephenate

Chorismate Pyruvate Lyase The enzyme that catalyzes the following reaction:

Chorismate + L-glutamine

⇅

Anthranilate + pyruvate + L-glutamate

Chorismate Pyruvate Mutase See chorismate mutase.

Chorismate Synthetase The enzyme that catalyzes the following reaction:

3-Phospho-5-enolpyruvoylshikimate

⇅

Chorismate + Pi

Chorismic Acid (mol wt 226) An intermediate for the biosynthesis of aromatic amino acids.

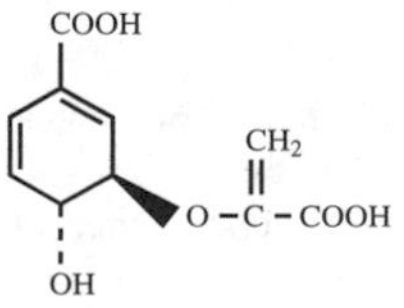

Choroid The middle vascular layer of the eye.

Choroidermia A genetic disorder characterized by the pregressive degeneration of the vascular layer of the eye (chorid).

Choroiditis Inflammation of the vascular coat of the eye.

Choron 10 A trade name for chorionic gonadotropin, a hormone.

Chou-Fasman Scheme An empirical scheme to predict the native three-dimensional structure from the known amino acid sequence of the protein.

CHPX Abbreviation for chicken pox.

Christmas Disease Referring to hemophilia B, a hereditary sex-linked hemorrhagic disease due to the absence of a coagulation factor.

Christmas Factor A blood clotting factor (factor IX) that activates clotting factor X.

Chromaffin Cell Cell that stores adrenaline in the secretory vesicle and secretes it at time of stress when stimulated by the nervous system.

Chromagen A trade name for a nutritional gelatin capsule containing multi-vitamins and minerals.

Chromatids The two daughter strands of a duplicated chromosome that are joined together by a centromere. The chromatids become separate chromosomes after division of the centromere.

Chromatin A complex of DNA and histone proteins or nonhistone proteins found in the eukaryotic nuclei.

Chromatin-Negative A male (usually) whose nuclei lack sex chromatin.

Chromatin-Positive A female (usually) whose nuclei contain sex chromatin.

Chromatogram The profile of a chromatographic separation.

Chromatography A technique used for the separation of a mixture of substances based upon the differences in their electric charges, particle sizes, and chemical properties.

Chromatophore The photosynthetic, submicroscopic structure from the photosynthetic prokaryotic cells containing photosynthetic pigment.

Chromatosome A structure of DNA-protein complex consisting of linker DNA and H1 histone protein.

Chromium A chemical element with atomic weight 52, valence 1 – 6.

Chromium-51 An artificial radioisotope of chromium used as chromate for labeling blood cells.

Chromium-51 Assay See ^{51}Cr Assay.

Chromium Picolinate (mol wt 418) A nutritional supplement.

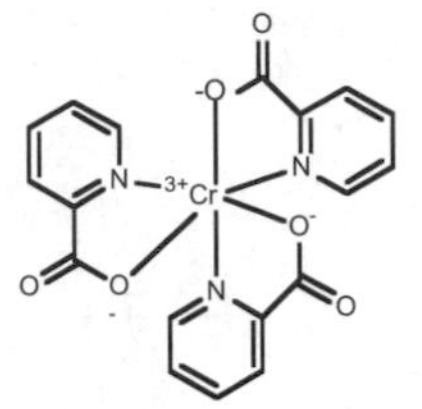

Chromobacterium A genus of chemoorganotrophic Gram-negative bacteria.

Chromoblastomycosis A skin disorder caused by black molds, e.g., *Caldosporium carrionii.*

Chromocenter An aggregate of heterochromatins from different chromosomes.

Chromogen Substance capable of producing color.

Chromogenic Label Any chemical label that generates a color compound as a means of visualizing the location and quantity of bound probe.

Chromomere One of the darkly stained beads on a eukaryotic chromosome due to the local coiling of the chromosome thread.

Chromonar (mol wt 361) A coronary vasodilator.

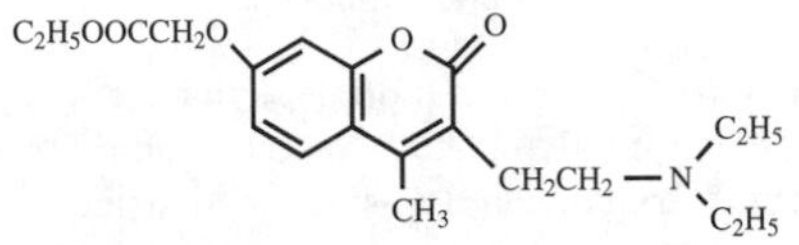

Chromonema Referring to the chromosome thread of eukaryotic cells.

Chromoneme Referring to the DNA thread of bacteria and bacteriophages.

Chromophilic Cells or organisms that can be easily stained.

Chromophobic Cells or organisms that can not be easily stained.

Chromophore 1. The light-absorbing group of a substance. 2. The group within a dye molecule that is responsible for the color of the dye.

Chromoplast Carotenoid-containing plastid responsible for color appearance in ripe fruits and flowers.

Chromosomal Pertaining to chromosomes.

Chromosomal Aberration Alteration or rearrangement of genetic material in the chromosome.

Chromosomal Fiber A bundle of microtubules that extends outward from kinetochores of a chromosome toward one of the two poles of the mitotic spindle (also called kinetochore fiber).

Chromosomal Map See chromosome map.

Chromosomal Polymorphism The existence of two or more different structural rearrangements of the chromosomal material within a population.

Chromosomal Puff The local uncoiling of a polytene chromosome that is undergoing transcription.

Chromosomal Substitution Replacement of one or more chromosomes by a chromosome from another source.

Chromosomal Tubules Referring to the microtubules of the spindle apparatus originating at the kinetochores of the centromeres.

Chromosomal RNA The RNA molecules associated with chromosomes, e.g., primer RNA.

Chromosome 1. The thread-like structure in the nucleus of a eukaryotic cell consisting of DNA and histone proteins. 2. The circular DNA of a prokaryotic cell that carries genetic information.

Chromosome Alteration See chromosomal aberration.

Chromosome Banding Technique Technique for staining human chromosomes, e.g., staining chromosomes with Giemsa (G-banding) or with flurochrome (Q-banding).

Chromosome Condensation The shortening and thickening of the eukaryotic chromosome during prophase.

Chromosome Congression The migration of chromosomes to the spindle equator during mitosis.

Chromosome Fiber See chromosomal fiber.

Chromosome Jumping See chromosome walking.

Chromosome Map The map showing the locations of genes on a chromosome.

Chromosome Mapping Determination of the order of the genes on a chromosome (also known as genetic map or cytogenetic map).

Chromosome Polymorphism See chromosomal polymorphism.

Chromosome Puff See chromosomal puff.

Chromosome Substitution See chromosomal substitution.

Chromosome Walking The sequential isolation of clones carrying overlapping restriction sequences from large regions of the chromosome in order to reach a particular locus of interest. The technique is used to isolate a locus of interest that is known to be linked to a gene that has been identified and cloned.

Chromotrope Substances capable of altering the color; a metachromatic dye.

Chronic A disease or infection that persists for a relatively long period of time.

Chronic Exposure Radiation exposure of long duration at a low dose.

Chronulac A trade name for lactulose, a laxative.

Chroococcidiopsis A genus of unicellular cyanobacteria.

Chrysanthemaxanthin (mol wt 585) A carotenoid pigment.

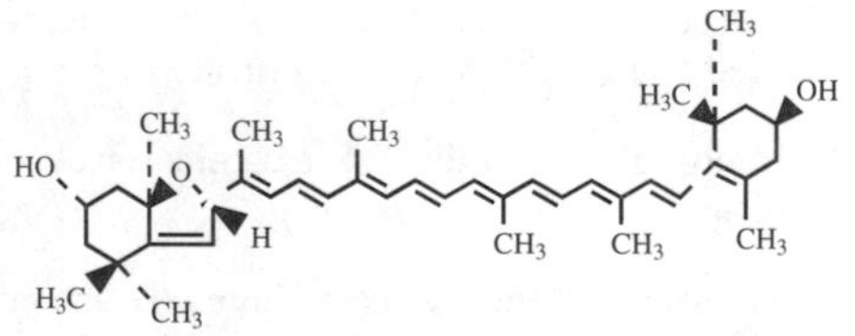

Chryseomonas A genus of yellow-pigmented, aerobic, catalase-positive, Gram-negative bacteria.

Chrysolaminarin A polysaccharide consisting of 1,3-β-linkages of glucan with some 1,6-β branching linkages.

Chrysops A genus of bloodsucking flies, commonly called deer flies. In the U.S., the *C. discalis* is a vector of tularemia.

Chrysose Synonym of chrysolaminarin.

Chrysotherapy The use of gold-containing substance for the treatment of disease.

ChTX Abbreviation for charybdotoxin.

ChuI (HindIII) A restriction endonuclease from *Corynebacterium humiferum* with the same specificity as HindIII.

ChuII (HindII) A restriction endonuclease from *Corynebacterium humiferum* with the following specificity:

↓
5′..........GTPyPuAC..........3′
3′..........CAPuPyTG..........5′
↑

CHX Abbreviation for cycloheximide.

ChyI (StuI) A restriction endonuclease from *Corynebacterium hydrocarboclastum* with the following specificity:

↓
5′..........AGGCCT..........3′
3′..........TCCGGA..........5′
↑

Chyle A milky lymphatic fluid containing lymph and the products of digestion.

Chylomicron Lipoprotein fat globule that functions to deliver dietary triacylglycerols to the muscle and adipose tissue and cholesterol to the liver.

Chylomicronemia The presence of a large number of microscopic particle of fat in the blood.

Chylomicron Remnant The chylomicron without triacylglycerol.

Chyloperitoneum The accumulation of milky liquid in the peritoneal cavity.

Chylopoiesis The formation of chyle.

Chyluria The presence of chyle in the urine giving a white turbid appearance.

Chymase A mast cell protease that cleaves peptide bonds involving the carboxyl groups of phenylalanine, tryptophan, tyrosine, and leucine.

Chyme The partially digested, semifluid mass passing from stomach into the intestine.

Chymodenin A basic peptide that stimulates the pancreatic secretion of chymotrypsinogen.

Chymopapain A protease from papaya having an activity similar to papain but differing in electrophoretic mobility, solubility, and stability.

Chymosin A protease from the stomach of calves.

Chymotropic Pigment Pigment present in the vacuole of the plant cell.

Chymotrypsin A protease that preferentially hydrolyzes peptide bonds involving carboxyl groups of aromatic amino acid residues.

Chymotrypsinogen The precursor of chymotrypsin.

Chymotryptic Peptides The peptides obtained from chymotrypsin digestion.

CI Abbreviation for crystalline insulin.

Ci Abbreviation for curie, a unit of radioactivity.

cI Gene The gene in λ phage that encodes repressor for lysogeny.

cI Protein A lambda phage repressor protein encoded by lambda phage DNA.

CIA Abbreviation for 1. a combination drug containing CCNU, isophosphamide, and adriamycin; 2. chymotrypsin inhibitor activity.

CIB Abbreviation for cytomegalic inclusion body.

Cibacalcin A trade name for calcitonin, a calcium regulator.

Cibalith-S A trade name for lithium carbonate, a psychotic agent.

CIC Abbreviation for circulating immune complex.

Cicletanine (mol wt 262) An antihypertensitive agent.

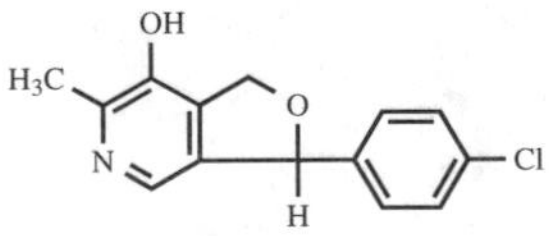

Ciclonicate (mol wt 247) A vasodilator.

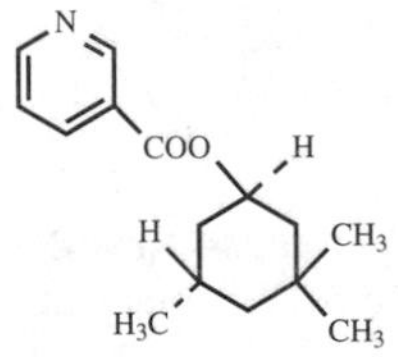

Ciclopirox (mol wt 207) An antifungal agent.

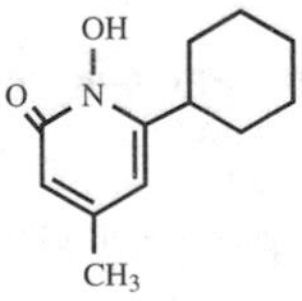

Ciclosidomine (mol wt 280) An antihypertensive agent.

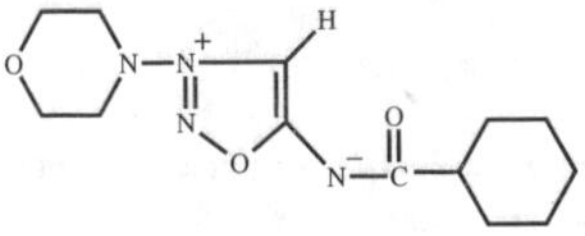

CICR Abbreviation for calcium-induced calcium release.

Cicrotoic Acid (mol wt 168) A choleretic agent.

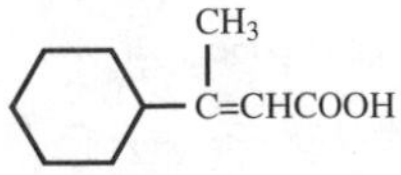

CID Abbreviation for 1. cytomegalic inclusion disease; 2. collision-induced dissociation.

CIDNP Abbreviation for chemically induced dynamic nuclear polarization.

Cidofovir (mol wt 279) An antiviral agent that inhibits viral DNA replication.

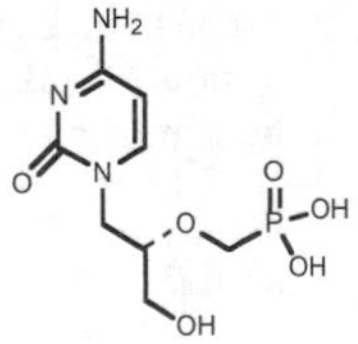

Cidomycin A trade name for gentamicin, an antibiotic that inhibits protein synthesis.

CIDS Abbreviation for cellular immunity deficiency syndrome.

CIE Abbreviation for 1. countercurrent immunoelectrophoresis; 2. crossed immunoelectrophoresis.

CIEP Abbreviation for countercurrent immunoelectrophoresis.

CIF Abbreviation for 1. calcium influx factor; 2. clone inhibiting factor.

Cifenline (mol wt 262) An antiarrhythmic agent.

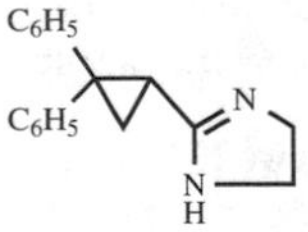

CIG Abbreviation for 1. cold insoluble immunoglobulin; 2. cytoplasmic immunoglobulin.

cIgM Abbreviation for cytoplasmic IgM.

Cilamox A trade name for amoxicillin trihydrate, an antibiotic that inhibits bacterial cell wall synthesis.

Cilastatin (mol wt 358) A dipeptidase inhibitor.

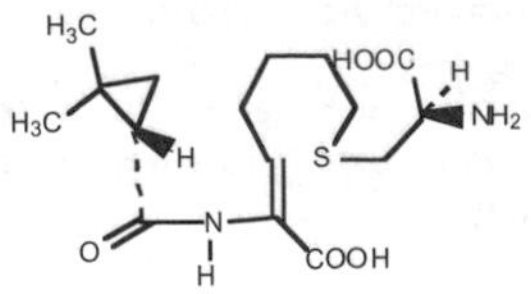

Cilazapril (mol wt 436) An antihypertensive agent.

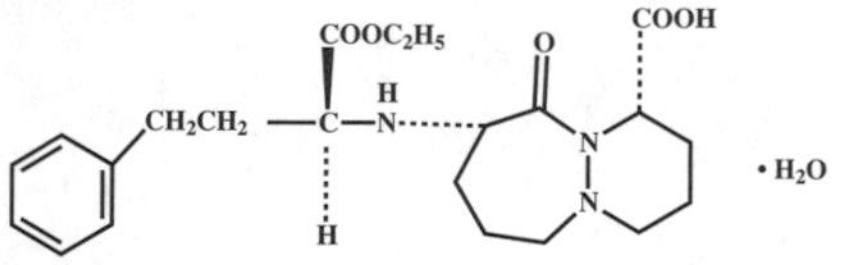

Cilia Plural of cilium.

Ciliary Movement Cilium- or microtubule-mediated movement.

Cilicane VK A trade name for penicillin VK, an antibiotic that inhibits bacterial cell wall synthesis.

Cilium Membrane-bound appendage or hair-like extension on the surface of eukaryotic cells, it consists of a core bundle of microtubules and is responsible for cell motility.

Cilnidipine (mol wt 493) An antihypertensive agent.

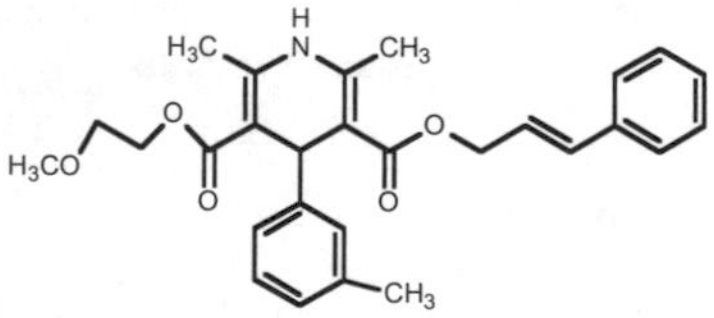

Cilostazol (mol wt 370) An antithrombotic agent.

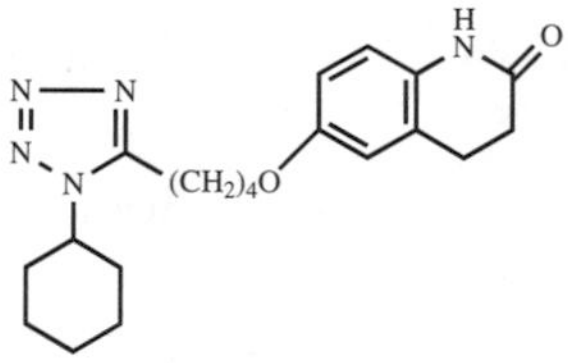

Ciloxin A trade name for ciprofloxcin hydrochloride, an antibacterial agent.

Cimetidine (mol wt 252) An antiulcerative agent that inhibits acid secretion in the stomach.

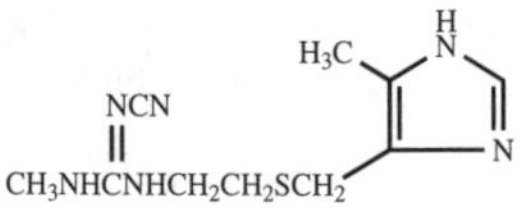

cIMP Abbreviation for cyclic inosine monophosphate.

CIM6PR Abbreviation for cation-independent mannose 6-phosphate receptor.

Cinalone A trade name for triamcinolone diacetate, a corticosteroid used as an anti-inflammatory agent.

Cinchona The dried bark of Cinchona trees used to treat malaria, to stimulate appetite, and to prevent hemorrhage and diarrhea.

Cinchonism Poisoning caused by overdose of cinchona.

Cineangiocardiography A form of angiocardiography in which X-ray pictures are recorded on the film. This allows the dynamic movements of the heart to be studied when the film is projected.

Cinepazet Maleate (mol wt 509) An antianginal agent.

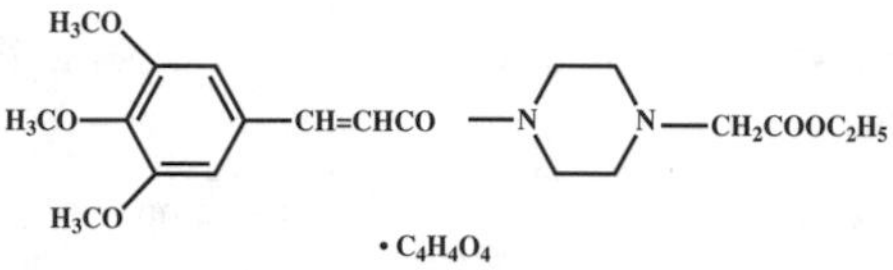

Cinepazide (mol wt 418) A vasodilator.

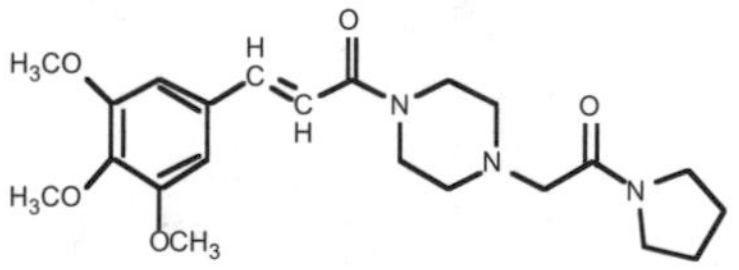

Cinnamaldehyde (mol wt 132) Substance responsibe for the flavor of cinnamon.

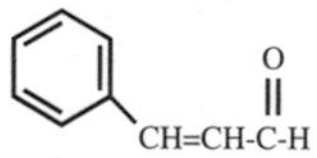

Cinnamate β-D-glucosyltransferase The enzyme that catalyzes the following reaction:

UDP-glucose + trans-cinnamate

⇅

UDP + trans-cinnamoyl-β-D-glucoside

***trans*-Cinnamate 4-monooxygenase** The enzyme that catalyzes the following reaction:

trans-Cinnamate + NADPH + O_2

⇅

4-Hydroxycinnamate + $NADP^+$ + H_2O

Cinnamic Acid (mol wt 148) An unsaturated acid.

$C_6H_5CH{=}CHCOOH$

Cinnamic Acid 2-Hydroxylase The enzyme that catalyzes the following reaction:

trans-Cinnamate + NADPH + O_2

⇅

2-Hydroxycinnamate + $NADP^+$

Cinnamic Acid 4-Hydroxylase The enzyme that catalyzes the following reaction:

trans-Cinnamate + NADPH + O_2

⇅

4-Hydroxycinnamate + $NADP^+$

Cinnamoyl-CoA Reductase See cinnamoyl reductase.

Cinnamoyl Reductase The enzyme that catalyzes the following reaction:

Cinnamaldehyde + CoA + $NADP^+$

⇅

Cinnamoyl-CoA + NADPH

Cinnamyl-Alcohol Dehydrogenase The enzyme that catalyzes the following reaction:

Cinnamyl alcohol + $NADP^+$

⇅

Cinnamaldehyde + NADPH

Cinnarizine (mol wt 369) A vasodilator and antihistaminic agent.

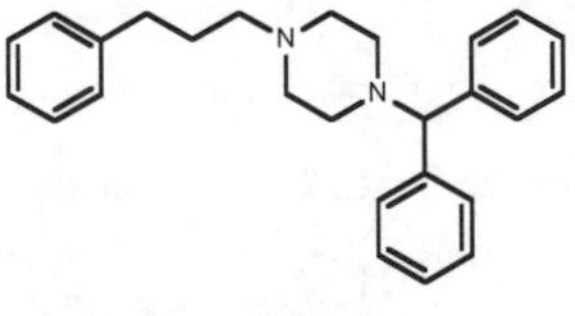

Cinobac A trade name for cinoxacin, an antimicrobial agent that inhibits microbial DNA synthesis.

Cinolazepam (mol wt 358) A sedative and hypnotic agent.

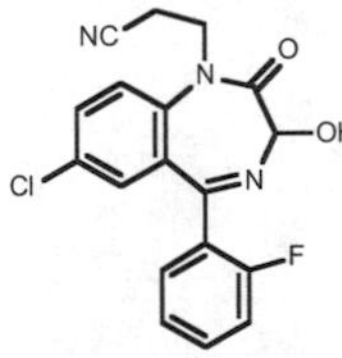

Cinonide A trade name for triamcinolone acetonide, used as an anti-inflammatory agent.

Cinoxacin (mol wt 262) An antibacterial agent that inhibits bacterial DNA synthesis.

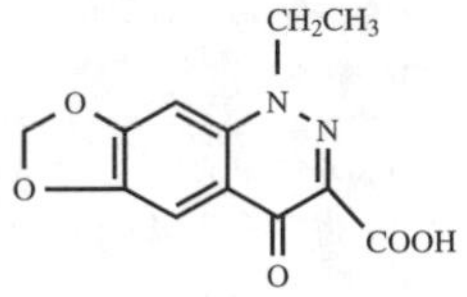

Cin-Quin A trade name for quinidine sulfate, an antiarrhythmic agent that prolongs action potential.

CiNU Abbreviation for lomustine.

Cioteronel (mol wt 252) An antiacne agent.

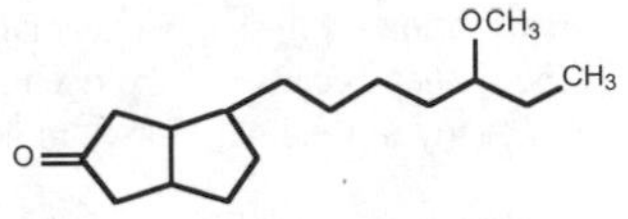

CIP Abbreviation for 1. CDK-dependent interacting protein; 2. calf intestinal phosphatase; 3. cold insoluble protein.

Cipro A trade name for ciprofloxacin, an antbiotic that inhibits bacterial DNA gyrase.

Ciprofibrate (mol wt 289) An antihyperlipoproteinemic agent.

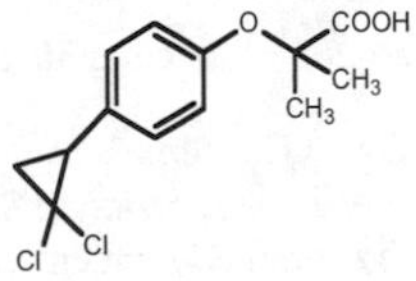

Ciprofloxacin (mol wt 331) An antibacterial agent that inhibits bacterial DNA synthesis (an DNA gyrase inhibitor).

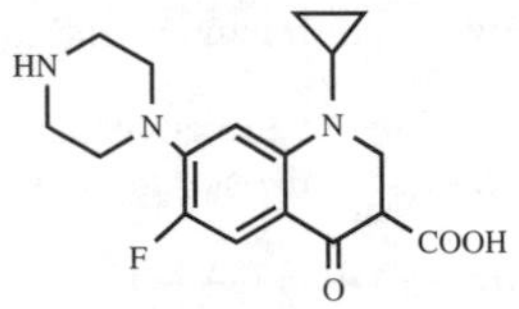

Ciramadol (mol wt 249) An analgesic agent.

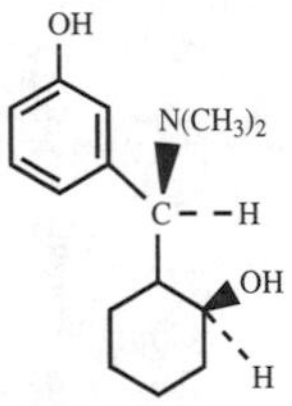

Circadian Describing biological activity that exhibits an endogenous periodicity of approximately 24 hours independently of any daily variation in the environment.

Circular Birefringence The birefringence resulting from the effect of left and right circularly polarized light.

Circular Dichroism The property of molecules that shows differences between extinction coefficients of the left and right circularly polarized light at a given wavelength, e.g., the phenomenon observed in helical proteins.

Circular DNA DNA with a closed ring structure, it may be a double-stranded circular or single-stranded circular structure.

Circular Genetic Map The genetic map of a circular DNA genome, e.g., genetic map of *E. coli.*

Circular Permutation Physically linear DNA with a circular genetic map because of the occurrence of terminal redundancy in the linear DNA molecule.

Circulins Polypeptide antibiotics produced by *Bacillus circulans* that increase the membrane permeability of bacterial cells.

Cirrhosis A chronic disorder of the liver characterized by the loss of normal lobular architecture.

cis Referring to 1. configuration of geometrical isomers, 2. two mutations that lie on the same chromosome, and 3. genes inherited together on the same chromosome.

***cis*-Acting Element** See cis-acting locus.

***cis*-Acting Locus** The genetic region that affects the activity of genes on the same DNA molecule. *cis*-Acting loci generally do not encode protein but serve as attachment sites for DNA-binding protein.

***cis*-Acting Mutation** A mutation that alters the nucleotide sequence of a DNA-binding site for a transcriptional regulatory protein that affect only the physically linked structural genes.

***cis*-Acting Protein** A protein capable of acting on gene expression at the starting point.

Cisapride (mol wt 466) A peristaltic agent.

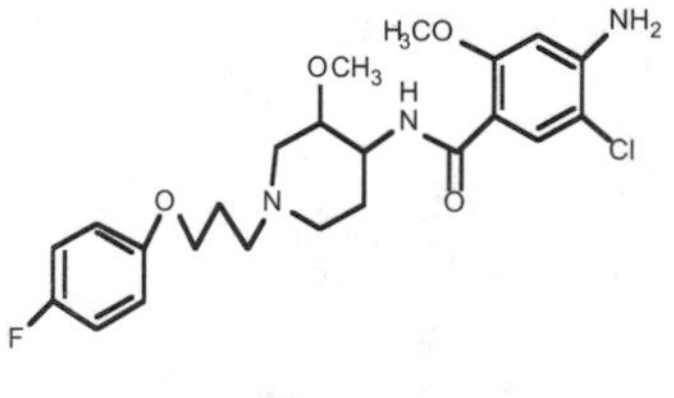

***cis*-Confirguration** Configuration of geometrical isomers in which two chemical groups lie on the same side of the plane of the double bond.

***cis*-Dominance** Genetic element that affects the expression of one or more adjacent loci on the same chromosome.

***cis*-Effect** The effect of one gene on the expression of another gene on the same chromosome.

CISM Cells Abbreviation for cat iris sphincter smooth-muscle cells.

CISP Abbreviation for corticotropin-induced secreted protein.

Cisplatin (mol wt 300) An antineoplastic agent that binds with DNA and forms two platinum-nitrogen bonds with N7 atoms of the adjacent guanine.

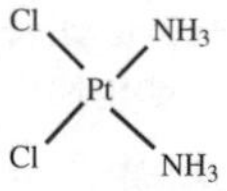

Cisterna The membrane-bound flattened sac in the cell, e.g., endoplasmic reticulum, Golgi apparatus.

Cisternal Space The interior of the cisternae, e.g., space in the endoplasmic reticulum or Golgi complex.

***cis-trans* Isomerase** The enzyme that catalyzes a cis-trans isomerization in the metabolism of unsaturated fatty acids.

***cis-trans* Test** A complementation test to determine whether two mutations affecting the same character lie within the same or different cistrons, it yields a mutant phenotype if the two mutants are in the *trans* position (on the same cistron), it yields a wild type phenotype if two mutants are in the *cis* position (on different cistrons).

Cistron A unit of genetic function or a structural gene on a chromosome that encodes a single polypeptide.

Cit Abbreviation for citrolline.

Citalopram (mol wt 324) An antidepressant.

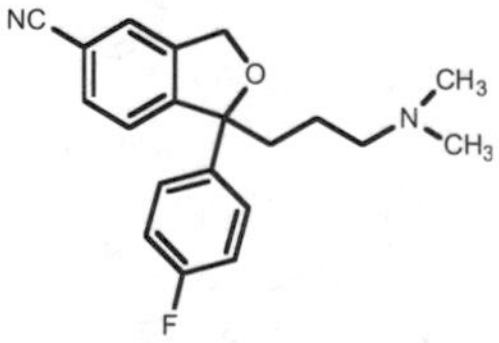

Citicoline (mol wt 488) A naturally occurring nucleotide coenzyme.

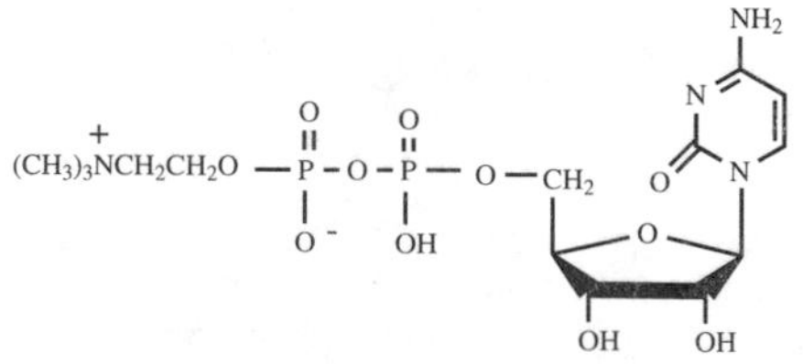

***Cit*-Plasmid** An enterobacterial plasmid that encodes a citrate transport system.

Citracal A trade name for calcium-citrate dietary supplement.

Citraderm A trade name for an ascorbic acid solution (10%) for dermatological use.

Citramalate-CoA Transferase The enzyme that catalyzes the following reaction:

Acetyl-CoA + citramalate

$\rightleftharpoons$

Acetate + citramalyl-CoA

Citrase The enzyme catalyzes the following reaction:

$$\text{Citrate} \rightleftharpoons \text{Acetate} + \text{oxaloacetate}$$

Citratase Synonym for citrase.

Citrate A salt of citric acid.

Citrate Aldolase Synonym of citrate lyase.

Citrate CoA Transferase The enzyme that catalyzes the following reaction:

$$\text{Acetyl-CoA} + \text{citrate} \rightleftharpoons \text{Acetate} + \text{citryl-CoA}$$

Citrate CoA Ligase The enzyme that catalyzes the following reaction:

$$\text{ATP} + \text{citrate} + \text{CoA} \rightleftharpoons \text{ADP} + \text{Pi} + \text{citryl-CoA}$$

Citrate CoA-transferase The enzyme that catalyzes the following reaction:

$$\text{Acetyl-CoA} + \text{citrate} \rightleftharpoons \text{Acetate} + \text{citryl-CoA}$$

Citrate Condensing Enzyme See citrate synthetase.

Citrate Cycle See tricarboxylic acid cycle.

Citrate Dehydratase The enzyme that catalyzes the following reaction:

$$\text{Citrate} \rightleftharpoons \textit{cis}\text{-Aconitate} + H_2O$$

Citrate Lyase The enzyme that catalyzes the following reaction:

$$\text{Citrate} + \text{ATP} + \text{CoA} \rightleftharpoons \text{Acetyl-CoA} + \text{ADP} + \text{Pi} + \text{oxaloacetate}$$

Citrate of Magnesia A trade name for magnesium citrate, an antacid and laxative agent.

Citrate Oxaloacetate Lyase The systematic name for citrate lyase.

Citrate Synthetase The enzyme that catalyzes the following reaction:

$$\text{Oxaloacetate} + \text{acetyl- CoA} + H_2O \rightleftharpoons \text{Citrate} + \text{CoA}$$

Citrate Test A test to determine the ability of an organism to use citrate as sole source of carbon.

Citreoviridin A yellow pigment and neurotoxic mycotoxin produced by *Penicillium citreoviride*.

Citric Acid (mol wt 192) A tricarboxylic acid and an intermediate in the Krebs cycle.

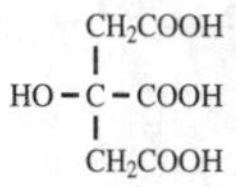

Citric Acid Cycle A central metabolic pathway in all aerobic organisms for the production of energy using acetyl-CoA as starting material (also called Krebs cycle, tricarboxylic acid cycle).

Citridesmolase Synonym of citrate lyase.

Citrinin (mol wt 250) An antibiotic produced by species of *Aspergillus* and *Penicillium*.

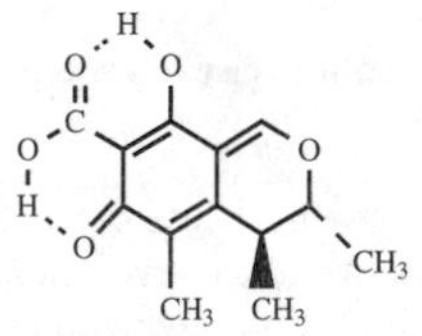

Citritase Synonym of citrate lyase.

Citrobacter A genus of Gram-negative bacteria of the family Enterobacteriaceae.

Citrocarbonate A trade name for sodium carbonate, an antacid.

Citrogenase Synonym of citrate synthetase.

Citroma A trade name for magnesium citrate, used as a laxative; it draws water into the lumen and causes peristalsis.

Citromag A trade name for magnesium citrate, used as a laxative; it draws water into the lumen and causes peristalsis.

Citromycetin (mol wt 290) An antibiotic produced by *Penicillium frequentans*.

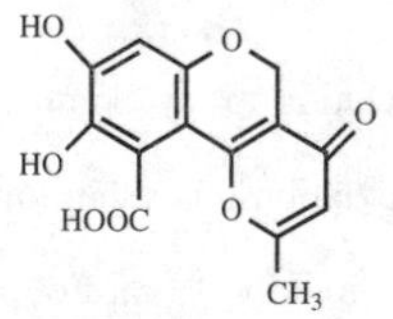

Citro-Nesia A trade name for magnesium citrate, used as a laxative; it draws water into the lumen and causes peristalsis.

Citrovorum Factor Referring to the reduced form of folic acid.

Citrucel A trade name for methylcellulose, used as a laxative.

Citrullinase The enzyme that catalyzes the following reaction:

L-Citrolline + H_2O

⇅

L-Ornithine + CO_2 + NH_3

Citrulline (mol wt 175) A nonprotein amino acid and an intermediate in the urea cycle and in the pathway for the synthesis of arginine.

$NH_2CH_2CH_2CH_2CHNH_2COOH$

Citrulline Aspartate Ligase The enzyme that catalyzes the following reaction:

ATP + L-citrolline + aspartate

⇅

AMP + PPi + argininosuccinate

Citrulline Phosphorylase See ornithine carbamoyltransferase.

Citrullinemia A genetic disorder characterized by the presence of an excessive amount of citruline in the blood, urine, and cerebrospinal fluid due to a defect in the metabolism of citrulline.

Citryl-CoA Lyase The enzyme that catalyzes the following reaction:

Citryl-CoA ⇌ Acetyl-CoA + oxaloacetate

c-jun Abbreviation for cellular oncogen *jun*.

CK Abbreviation for 1. casein kinase; 2. choline kinase; 3. creatine kinase; 4. cytokinin.

Cκ Abbreviation for constant region of κ light chain of an immunoglobulin.

CK1 Abbreviation for creatine kinase 1 isozyme.

CK2 Abbreviation for 1. casein kinase 2; 2. creatine kinase-2 isozyme.

CK3 Abbreviation for creatine kinase 3 isozyme or isoenzyme.

CK8 Abbreviation for cytokeratin-8.

CK1a Abbreviation for casein kinase 1a.

CL Abbreviation for 1. cardiolipin; 2. cholesterol-lecithin.

Cl Symbol for chlorine.

C_L Abbreviation for the constant region of the light chains of an immunoglobulin.

CL Test Abbreviation for cholesterol-lecithin test.

Cλ Abbreviation for constant region of λ light chain of an immunoglobulin.

CLA Abbreviation for cyclic lysine anhydride.

ClaI A restriction endonuclease from *Coryophanon latum* L with the following specificity:

↓

5′ ....ATGCAT..........3′
3′...........TACGTA..........5′

↑

Cladistics The science that deals with a classification system based on the phylogenetic relationships of organisms.

Cladogenesis The splitting of an evolutionary lineage into two or more lineages.

Cladogram A phylogenetic tree displaying relationships between taxa.

Cladosporiosis Any infection or disease caused by the species *Cladosporium.*

Cladosporium A genus of fungi (*Hyphomycetes*).

Cladribine (mol wt 286) An antineoplastic agent that blocks repair and synthesis of DNA.

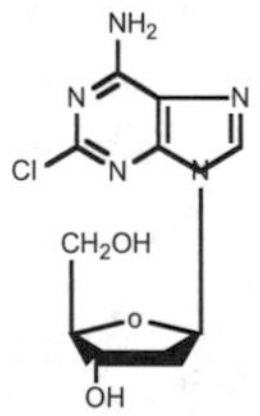

Claforan A trade name for cefotaxime sodium, an antibiotic that inhibits bacterial cell wall synthesis.

Clanobutin (mol wt 348) A choleretic agent.

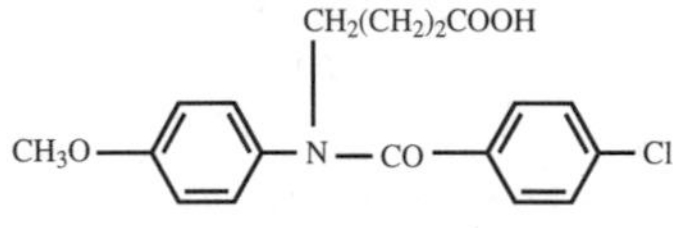

Claripex A trade name for clofibrate, an antilipemic agent.

Clarithromycin (mol wt 747) A semisynthetic antibiotic derived from erythromycin that inhibits bacterial protein synthesis.

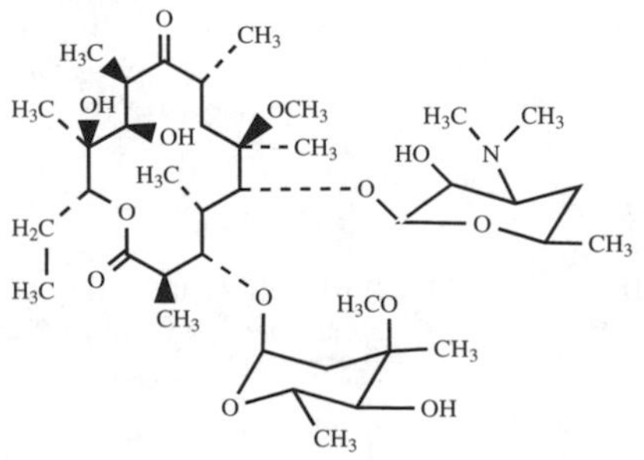

Claritin A trade name for loratadine, an antihistaminic agent.

Clark Electrode A type of oxygen electrode.

CLAS Abbreviation for circulating lupus anticoagulant syndrome.

Class A taxonomic group of organisms with greater similarity than a phylum or subphylum but less than an order.

Class I MHC Antigens Gene products encoded by HLA-A, HLA-B, and HLA-C loci in humans, or H2-K and H2-D loci in mice. Class I antigens are responsible for allograft rejection and cytotoxic T cell action.

Class II MHC Antigens Gene products encoded by HLA-D loci in humans or the H2-I region in mice. Class II antigens act as peptide receptors on the surface of antigen-presenting cells.

Class III MHC Antigens Products encoded by genes located on the MHC complex. Class III antigens are involved in the regulation and synthesis of complement.

Class Switching Change in expression of the different constant regions of an immunoglobulin during lymphocyte differentiation.

Classical Complement Pathway An antibody-dependent, complement-mediated cytolytic pathway that is initiated by antigen-antibody-mediated activation of complement and formation of membrane attack complexes leading to cell lysis.

Clastogen Substance that causes chromosomal abnormalities.

Clathrate Compound capable of trapping other compounds.

Clathrate Complex A complex formed through molecular entrapment, e.g., trapping of nonpolar substances by water.

Clathrin Protein that forms a polyhedral cage on the cytoplasmic side of a membrane so as to produce a clathrin-coated pit, which buds off to form a clathrin-coated vesicle.

Clavulanic Acid (mol wt 199) An antibacterial agent.

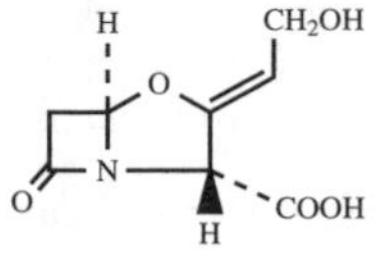

Clavulin A trade name for a combination drug containing amoxicillin and calvulanate, used as an antibacterial agent.

Clazuril (mol wt 373) An anticoccidial agent.

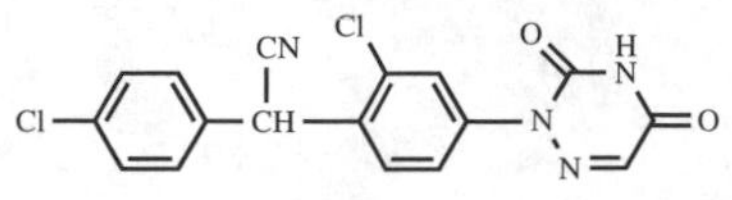

ClcI (PstI) A restriction endonuclease from *Cylindrospermum lichenforme* with the same specificity as PstI.

ClcII (Mst I) A restriction endonuclease from *Cylindrospermum lichenforme* with the following specificity:

↓
5′..........TGCGCA..........3′
3′..........ACGCGT..........5′
↑

Clear Lysate The clear supernatant resulting from centrifugation of suspensions of lysed cells.

Clear Plaque A plaque in which all the cells within the plaque are lysed.

Clearance Test A method to test the kidney function for removing a substance from the blood by measuring the amount of the substance excreted in the urine within a given period of time.

Cleavage 1. Physical splitting of a cell into two. 2. Specialized type of cell division seen in many early embryos whereby a large cell becomes subdivided into many smaller cells without growth. 3. Splitting a compound into two or more simpler small molecules.

Cleavage Furrow The indentation of the plasma membrane of an animal cell leading to the separation of daughter nuclei.

Cleavage Map A profile produced by electrophoresis of a DNA sample treated with restriction endonuclease (also known as restriction map).

Clebopride (mol wt 390) An antiemetic agent.

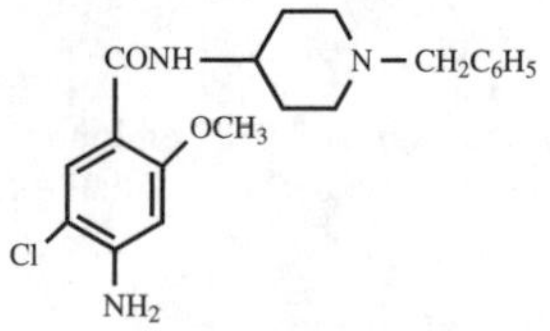

Cleland's Reagents Referring to reducing agents, e.g., DTE (dithioerythritol) and DTT (dithiothreitol), used for the protection of sulfhydryl groups.

Clemastine (mol wt 344) An antihistaminic agent that competes with histamine for H-1 receptors on target cells.

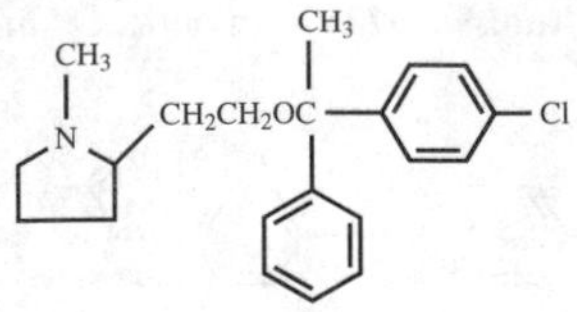

Clemizole (mol wt 326) An antihistaminic agent.

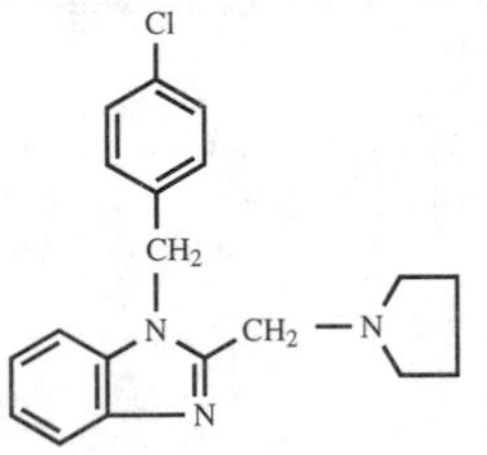

Clenbuterol (mol wt 277) An antiasthmatic agent.

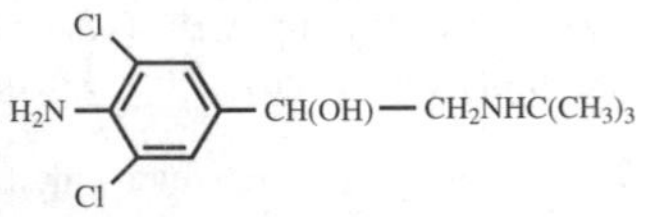

Clentiazem (mol wt 449) An antihypertensive agent.

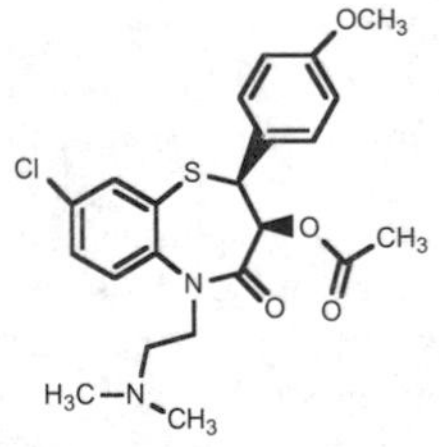

Cleocin A trade name for clindamycin, an antibiotic that inhibits bacterial protein synthesis by binding to 50S ribosome subunit.

CliI (AvaII) A restriction endonuclease from *Cylindrospermum lichenforme* with the following specificity:

↓
5′..........GG(A/T)CC..........3
3′..........CC(T/A)GG..........5
↑

CliII (MstI) A restriction endonuclease from *Cylindrospermum lichenforme* with the same specificity as MstI.

Clidanac (mol wt 279) An anti-inflammatory and antipyretic agent.

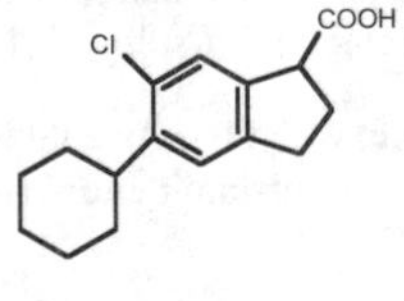

Clidinium Bromide (mol wt 432) An anticholinergic agent that competitively blocks acetylcholine, decreases GI mobility, and inhibits gastric acid secretion.

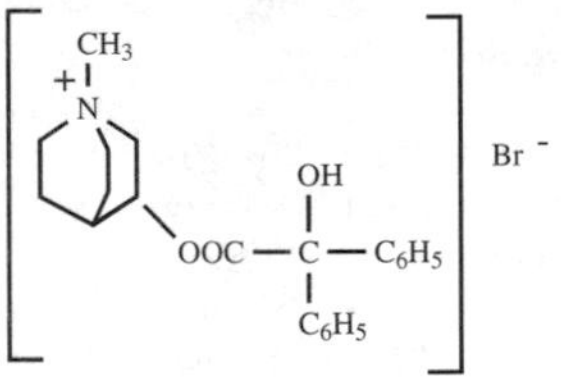

Climara Transdermal System A trade name for a system of transdermal delivery of estradiol. There are different systems, each system has a contact surface area.

Clinafloxacin (mol wt 366) An antibacterial agent.

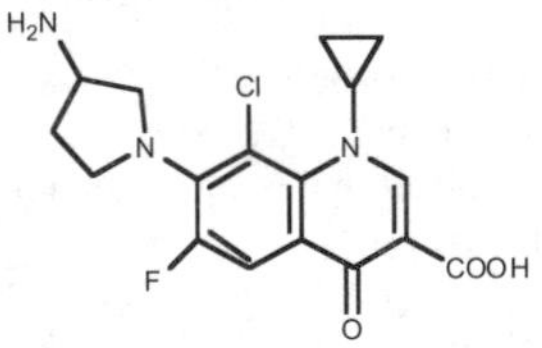

Clinda-Derm A trade name for clindamycin, an antibiotic.

Clindamycin (mol wt 425) A semisynthetic antibiotic derived from linomycin that inhibits bacterial protein synthesis by binding to 50S ribosomal subunit.

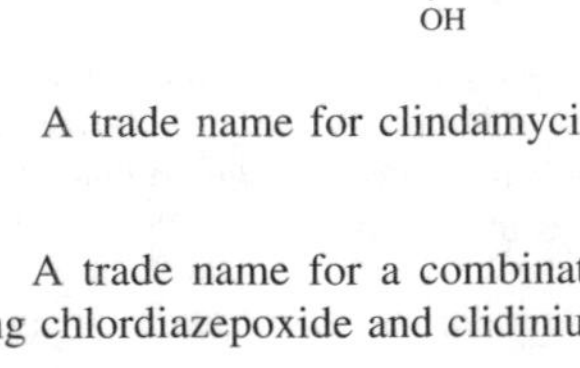

Clindets A trade name for clindamycin, an antibiotic.

Clindex A trade name for a combination drug containing chlordiazepoxide and clidinium.

Clinical Chemistry Medical science that deals with the qualitative and quantitative determination of chemical substances related to medicine.

Clinistil A trade name for a group of chemically treated paper strips, used for semiquantitive determinations of chemical components in urine and/or in the blood.

Clinofibrate (mol wt 469) An antiathersclerosis agent.

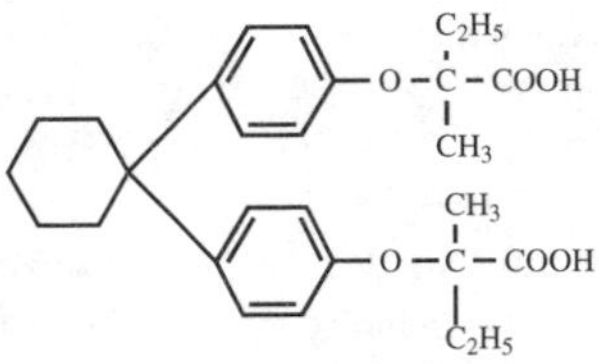

Clinoril A trade name for sulindac, a nonsteroidal anti-inflammatory agent.

Clinoscope A device for the measurement of cyclophobia (tendency of one eye to deviate).

Clioquinol A local anti-infective agent (see also iodochlorhydroxyquin).

Clioxanide (mol wt 542) An anthelmintic agent.

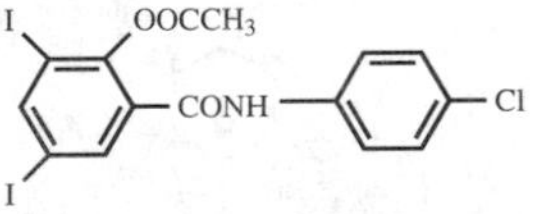

Clinoxide A trade name for a combination drug containing chlordiazepoxide and clidinium.

Clipoxide A trade name for a combination drug containing chlordiazepoxide and clidinium.

CLIP-170 Abbreviation for cytoplasmic linker protein 170, a phosphorylation-regulated microtubule-binding protein.

CLL Abbreviation for chronic lymphocytic leukemia.

CLM Abbreviation for caveolin-rich light membrane.

ClmI (HaeIII) A restriction endonuclease from *Caryophanon latum* with the same specificity as HaeIII.

ClmII (AvaII) A restriction endonuclease from *Caryophanon latum* with the same specificity as AvaII.

CLN (Cln) Abbreviation for 1. calnexin; 2. cyclin.

Cloacin DF13 A bacteriocin produced by strains of *Enterobacter cloacae* and *E. coli* that possess the CloDF13 plasmid.

Clobenfurol (mol wt 259) A coronary vasodilator.

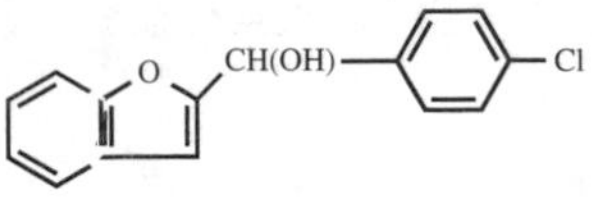

Clobenztropine (mol wt 342) An antihistaminic agent.

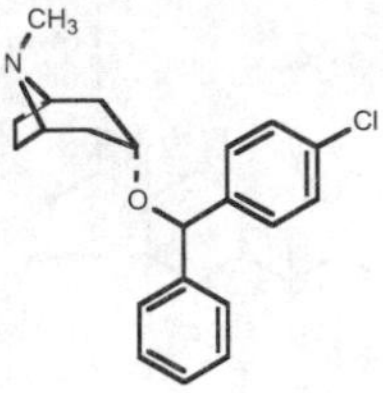

Clobenzepam (mol wt 316) An antihistaminic agent.

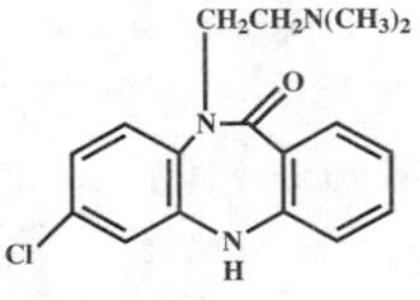

Clobetasol (mol wt 411) A glucocorticoid and anti-inflammatory agent.

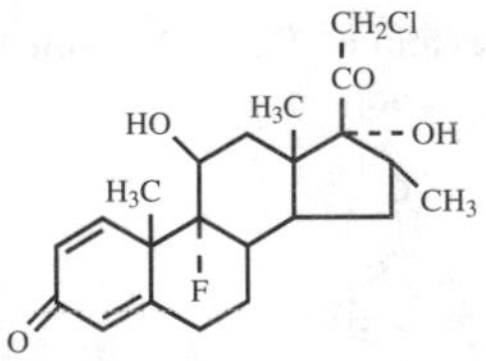

Clobetasone (mol wt 409) A glucocorticoid with anti-inflammatory activity.

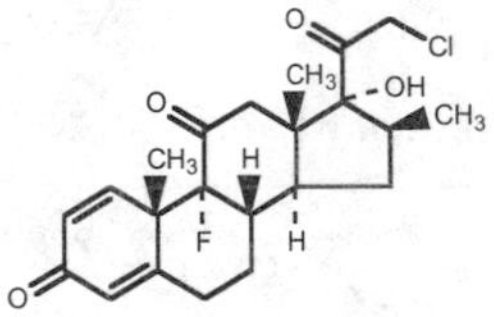

Clobutinol (mol wt 256) An antitussive agent.

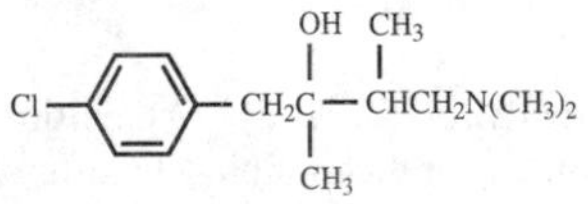

Clobuzarit (mol wt 305) An antirheumatic agent.

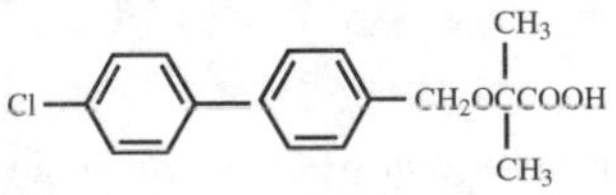

Clocinizine (mol wt 403) An antihistaminic agent.

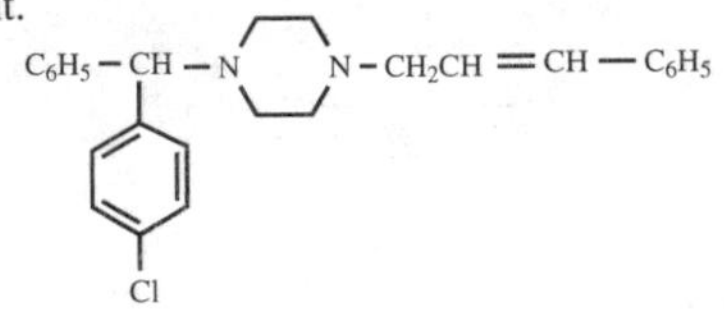

Clocortolone (mol wt 411) A glucocorticoid.

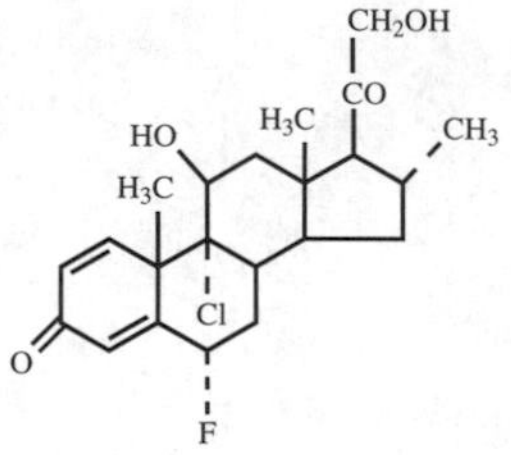

CloDF13 A plasmid capable of producing bacteriocin cloacin DF13 in *E. coli* and *Enterobacter cloacae*.

Clodronic Acid (mol wt 245) A calcium regulator.

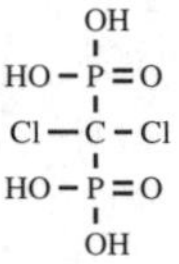

Clofazimine (mol wt 473) An antituberculostatic and antileprostastic agent.

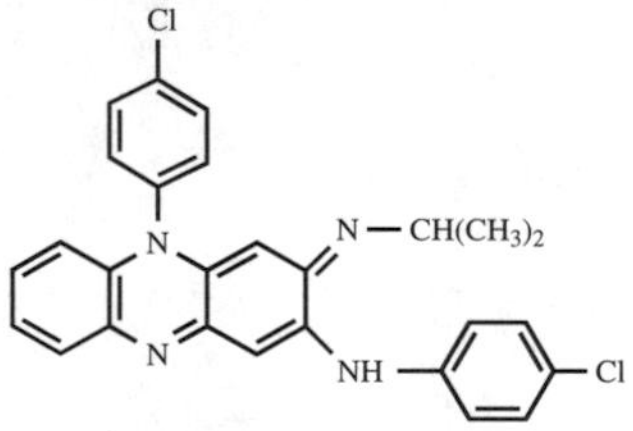

Clofenciclan (mol wt 310) A CNS stimulant.

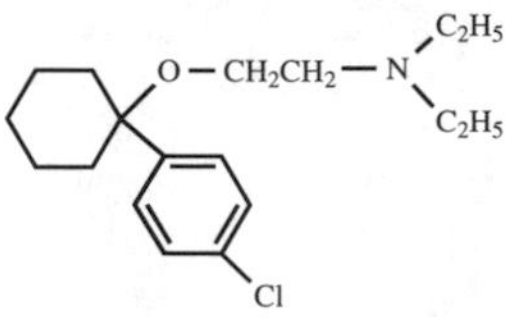

Clofibrate (mol wt 243) An antihyperlipoproteinemic agent that inhibits cholesterol biosynthesis.

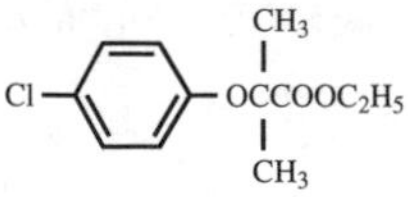

Clofibric Acid (mol wt 215) An antihyperlipoproteinemic agent and an inhibitor of cholesterol biosynthesis.

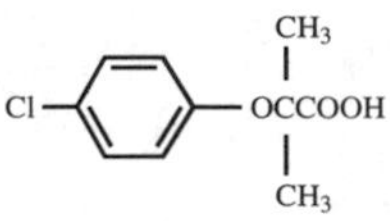

Cloflucarban (mol wt 349) A disinfectant.

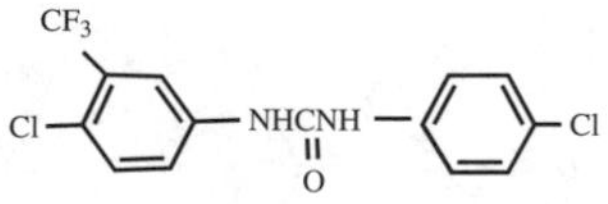

Clofoctol (mol wt 365) An antibacterial agent that is effective against both Gram-positive and Gram-negative bacteria.

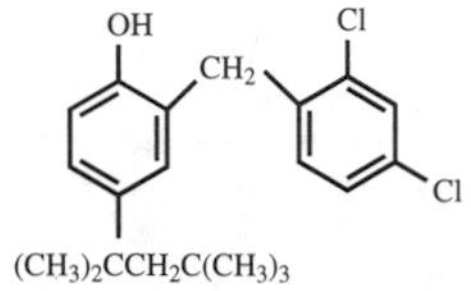

Clomestrone (mol wt 358) An analgesic agent.

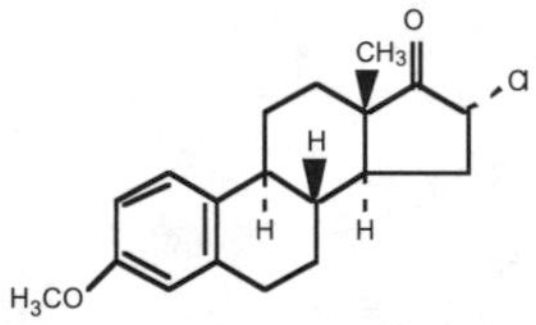

Clometacin (mol wt 358) An analgesic agent.

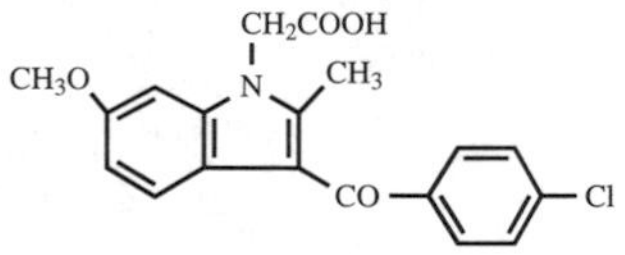

Clometocillin (mol wt 433) A semisynthetic antibiotic related to penicillin that inhibits bacterial cell wall synthesis.

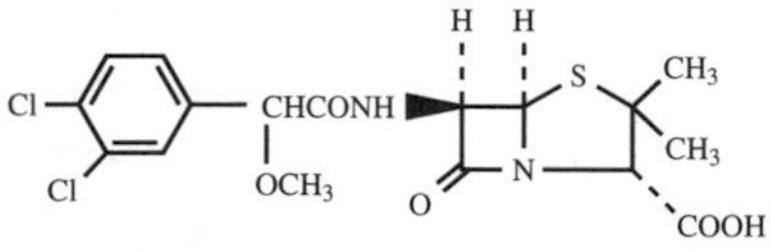

Clomid A trade mane for clomiphene, a drug that stimulates the release of pituitary gonadotropins, folicle-stimulating hormone, and luteinizing hormone.

Clomiphene (mol wt 406) A gonad-stimulating agent that stimulates the release of pituitary gonadotropins, folicle-stimulating hormone, and luteinizing hormone.

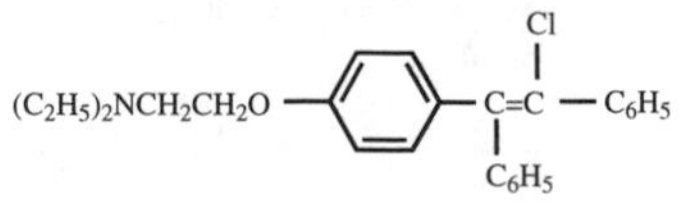

Clomipramine (mol wt 315) An antidepressant.

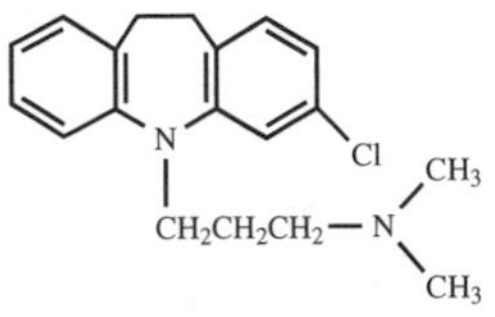

Clomocycline (mol wt 509) A semisynthetic broad-spectrum antibiotic related to tetracycline.

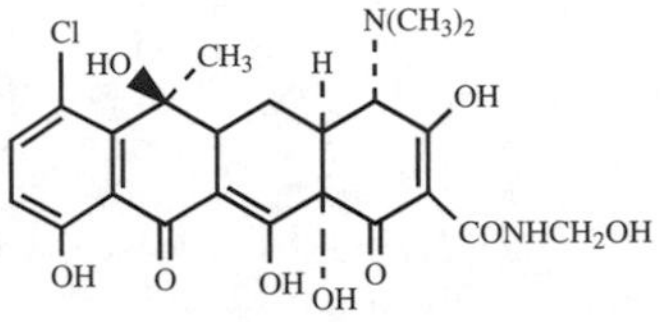

Clonal Pertaining to a clone.

Clonal Deletion Creation of tolerance by deletion of cells that possess specificity for a given antigen and, therefore are no longer available to respond upon subsequent exposure to that antigen.

Clonal Selection Theory The theory states that antibodies are produced by clones of proliferated lymphocytes that are specifically selected by antigen epitopes among a large number of lymphoid clones. The cells in each clone can react with only one epitope. The clones responsive to selfepitopes are destroyed during fetal life.

Clonapam A trade name for clonazepam, an antiepileptic agent.

Clonazepam (mol wt 316) An anticonvulsant.

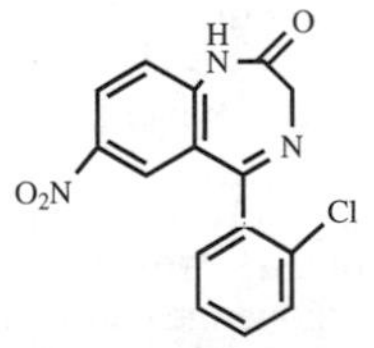

Clone 1. Population of cells or organisms formed by repeated asexual division from a single common ancestral cell or organism. 2. Genetically engineered replicas of DNA sequences.

Cloned DNA Any DNA sequence or fragment that has been inserted into a cloning vector and replicated in the host organism.

Cloned Library A collection of cloned DNA sequences representative of the genome of the organism under study.

Clonidine (mol wt 230) A blood pressure regulator and antihypertensive agent that inhibits the central vasomotor center thereby decreasing sympathetic outflow.

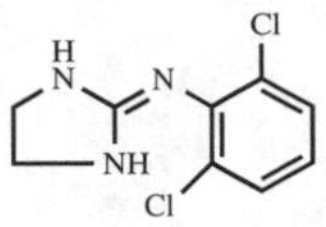

Cloning A process of propagation of a clone in a host, e.g., propagation of a gene or a fragment of DNA with a cloning vector.

Cloning Vector A self-replicating plasmid, bacteriophage, or virus that is used to transfer a segment of foreign DNA into a host cell for propagation.

Clonitazene (mol wt 387) A narcotic analgesic agent.

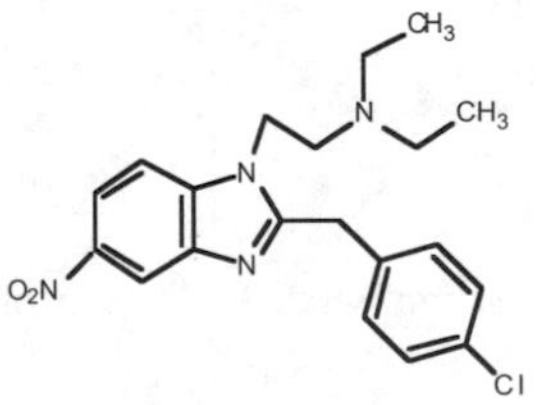

Clonixin (mol wt 263) An analgesic agent.

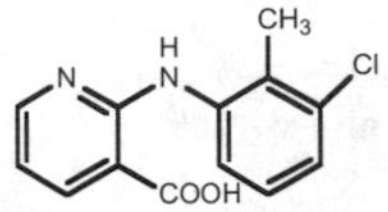

Clonorchiasis A condition caused by the presence of the fluke *Clonorchis sinensis* in the bile duct.

Clonotype 1. The homogeneous cell type of a clone of cells. 2. The phenotype of a clone of cells.

Clopamide (mol wt 346) An antihypertensive agent.

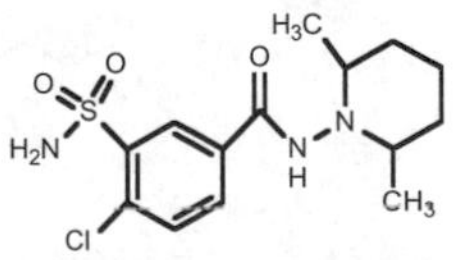

Cloperastine (mol wt 330) An antitussive agent.

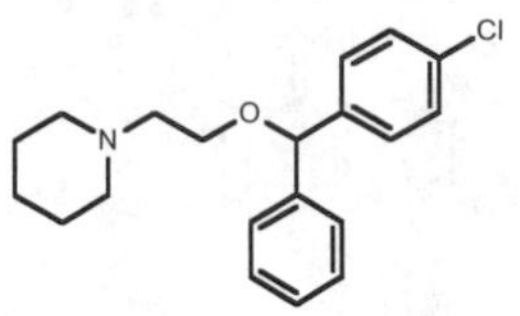

Clopidogrel (mol wt 322) An ADP receptor antagonist used as an antiplatelet agent.

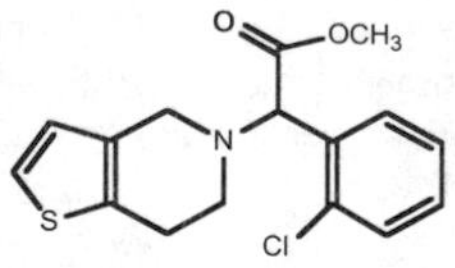

Clopirac (mol wt 264) An anti-inflammatory agent.

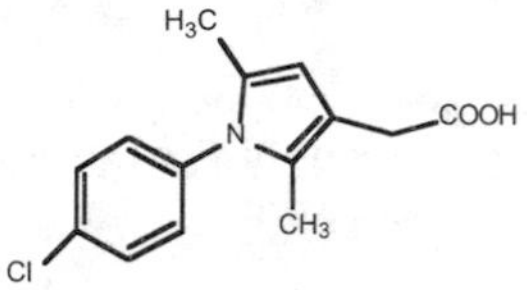

Clopra A trade name for metoclopramide, an antiemetic and GI stimulant.

Cloranolol (mol wt 292) An antiarrhythmic agent.

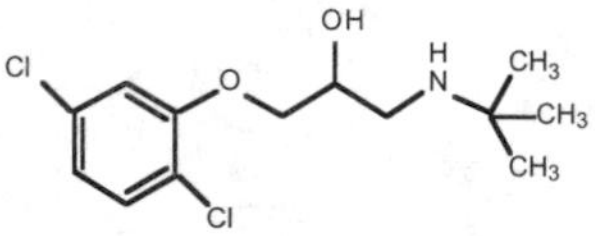

Clorazepate (mol wt 333) An anxiolytic agent.

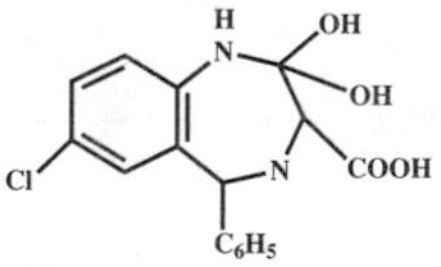

Clorexolone (mol wt 239) A diuretic agent.

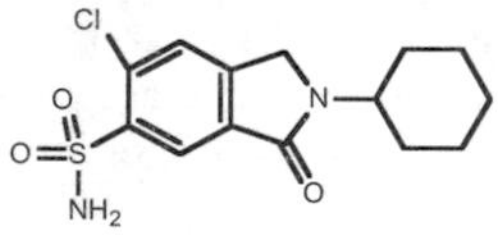

Cloricromen (mol wt 257) A vasodilator and antithrombotic agent.

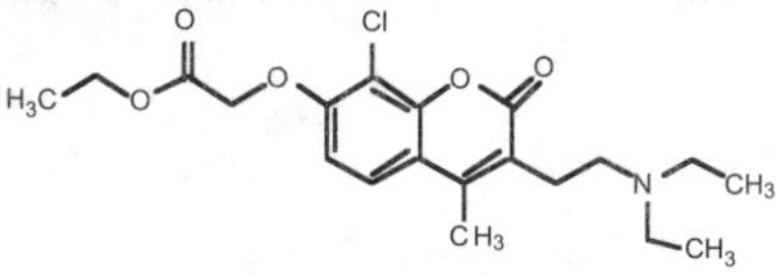

Clorindione (mol wt 257) An anticoagulating agent.

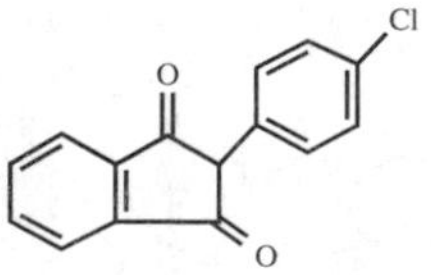

Clorophene (mol wt 219) A disinfectant.

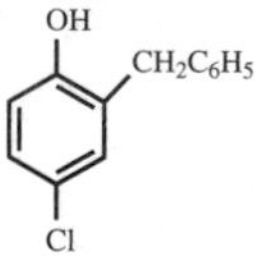

Clorprenaline (mol wt 214) A bronchodilator.

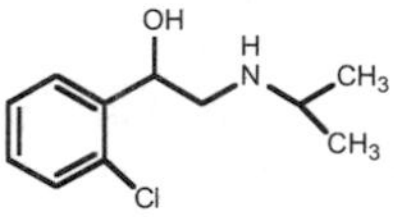

Clorpres A trade name for a combination drug containing clonidine hydrochloride and chlorthalidone.

Closed Reading Frame A sequence in mRNA that cannot be translated.

Closed System A thermodynamic system that neither exchanges energy nor matter with its surroundings.

Clostridial Myonecrosis The death of individual muscle cells caused by *Clostridia*.

Clostridiopeptidase A The microbial collagenase that cleaves peptide bonds involving the amino group of glycine in native collagen.

Clostridiopeptidase B See clostripain.

Clostridium A genus of endospore-forming, chemoorganotrophic, obligate anaerobic bacteria.

Clostripain A microbial protease that preferentially cleaves peptide bonds involving the carboxyl group of arginine (including the bond between arginine and proline).

Clot-Promoting Factor See Hageman factor.

Clotrimadern A trade name for clotrimazole, an antifungal agent.

Clotrimazole (mol wt 345) An antifungal agent that alters cell wall permeability of fungi.

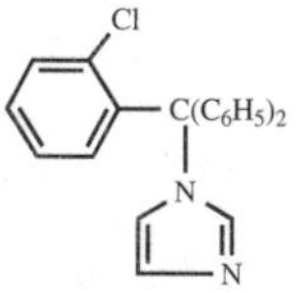

Clotting Factor Proteins that promote blood clotting.

Cloverleaf Structure Referring to the characteristic secondary structure of tRNA; it consists of 4 loops and a CCA-terminus.

Cloxacillin (mol wt 436) A semisynthetic antibiotic related to penicillin that inhibits bacterial cell wall synthesis.

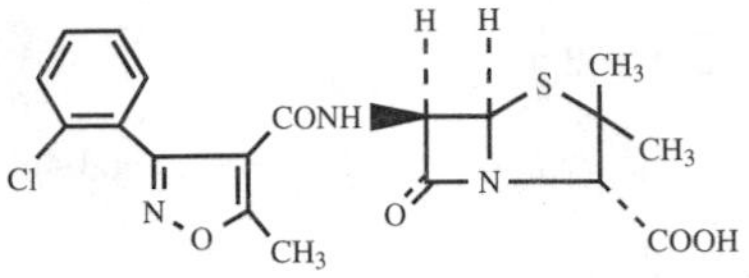

Cloxpen A trade name for cloxacillin sodium, an antibiotic that inhibits bacterial cell wall synthesis.

Cloxyquin (mol wt 180) An antifungal agent.

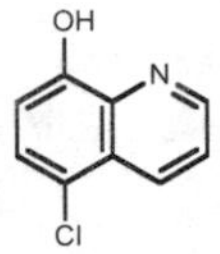

Clozapine (mol wt 327) An antipsychotic agent that binds to the dopamine receptor and interferes with adrenergic, cholinergic, and serotoninergic receptors.

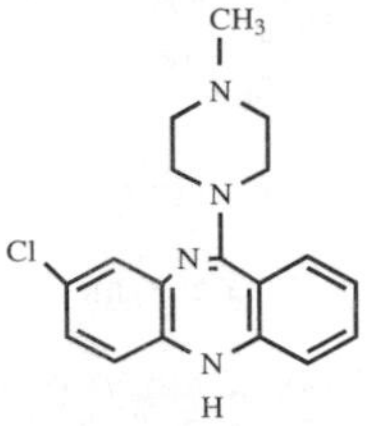

Clozaril A trade name for clozapine, an antipsychotic agent.

C-LPL Abbreviation for cellular LPL.

CLSH Abbreviation for chronic lympho-sarcoma leukemia.

CLSM Abbreviation for confocal laser scanning electron microscope.

CltI (HaeIII) A restriction endonuclease from *Caryophanon latum* with the following specificity:

↓
5′..........GGCC..........3′
3′..........CCGG..........5′
↑

Clupein A protamine isolated from herring, it consists of 30 amino acid residues.

Cluster Gene A gene that encodes multifunctional proteins.

Cluster of Differentiation Cell surface molecules, detectable by monoclonal antibodies that define a particular cell line or state of cellular differentiation (also known as cluster designation or CD marker).

Clysis Infusion of fluid into the body.

Clysma Referring to enema.

Clysodrast A trade name for a combination drug containing bisacodyl and tannic acid.

Clyster Referring to enema.

cm Abbreviation for centimeter.

cM Abbreviation for centimorgan.

CM Abbreviation for 1. carboxymethyl; 2. complete medium; 3. conditioned medium; 4. contrast medium.

Cμ Abbreviation for the constant region of μ chain of an IgM.

CM Cellulose Carboxymethyl cellulose, a cation exchanger used in ion exchange chromatography.

CM Dextran Abbreviation for carboxymethyl dextran, a cation exchanger used in ion exchange chromatography.

CM Sephadex Abbreviation for carboxymethyl-Sephadex, a cation exchanger used in ion exchange chromatography.

CM Sepharose Abbreviation for carboxymethyl-Sepharose, a cation exchanger used in ion exchange chromatography.

CMC Abbreviation for 1. carboxymethyl- cellulose; 2. critical micelle concentration; 3. cell-mediated cytotoxicity; 4. a combination drug containing cyclophosphamide, methotrexate, and CCNU.

C-Meiosis Colchicine-blocked meiosis.

C-Metaphase Colchicine-blocked metaphase.

CMF Abbreviation for a combination drug containing cyclophosphamide, methotrexate, and fluorouracil.

CMFDA Abbreviation for chloromethyl fluorescein diacetate.

CMFH Abbreviation for a combination drug containing cyclophosphamide, methotrexate, fluorouracil, and hydroxyurea.

CMFP Abbreviation for a combination drug containing cyclophosphamide, methotrexate, fluorouracil, and prednisone.

CMFT Abbreviation for a combination drug containing cyclophosphamide, methotrexate, fluorouracil, and tamoxifin.

CMFVP Abbreviation for a combination drug containing cytoxan, methotrexate, fluorouracil, vincristine, and prednisone.

CMH Abbreviation for ceramide monohexoside.

CMI Abbreviation for 1. cell-mediated immunity; 2. carbohydrate metabolism index.

C-Mitosis Colchicine-blocked mitosis.

CML Abbreviation for 1. cell-mediated lympholysis; 2. cell-mediated lymphocytotoxicity.

CMM Abbreviation for complete minimal medium.

CMO Abbreviation for corticosterone methyl oxidase.

CMo Abbreviation for centimorgan.

CMOAT Abbreviation for canalicular multi-specific organic anion transporter.

CMOPP Abbreviation for a combination drug containing cyclophosphamide, nitrogen mustard, oncovin, procarbazine, and prednisone.

CMP Abbreviation for 1. cytidine monophosphate (cytidylic acid). 2. cytidine 5′-monophosphate (5′-cytidylic acid).

CMPF Abbreviation for a combination drug containing CCNU, methotrexate, prednisone, and 5-fluorouracil.

CMP-N-Acylneuraminate Phosphodiesterase The enzyme that catalyzes the following reaction:

$$\text{CMP-N-acylneuraminate} + H_2O \rightleftharpoons \text{CMP-} + \text{N-acylneuraminate}$$

CMP-Sialate Hydrolase See CMP-N-acylneuraminate phosphodiesterase.

CMP-Sialate Synthetase The enzyme that catalyzes the following reaction:

$$\text{CTP} + \text{N-acylneuraminate} \rightleftharpoons \text{PPi} + \text{CMP-N-acylneuraminate}$$

CMR Abbreviation for ^{13}C-nuclear magnetic resonance.

CMU Abbreviation for chlorophenyl dimethylurea.

CMV Abbreviation for cytomegalovirus.

c-myc A cellular oncogen.

CnA Abbreviation for calcineurin catalytic subunit.

CnB Abbreviation for calcineurin regulatory subunit.

CNBr Abbreviation for cyanogen bromide.

CN[^{51}Co]Cbl Abbreviation for ^{51}Co-labeled cyanocobalamin.

CN[^{57}Co]Cbl Abbreviation for ^{57}Co-labeled cyanocobalamin.

CNS Abbreviation for central nervous system.

CNTF Abbreviation for ciliary neurotrophic factor.

CNX Abbreviation for Na^+/Ca^{2+} exchanger.

Co Symbol for cobalt.

Co-I Referring to coenzyme I (NAD).

Co-II Referring to coenzyme II (NADP).

CoA (mol wt 768) Coenzyme A involved in a variety of biochemical reactions.

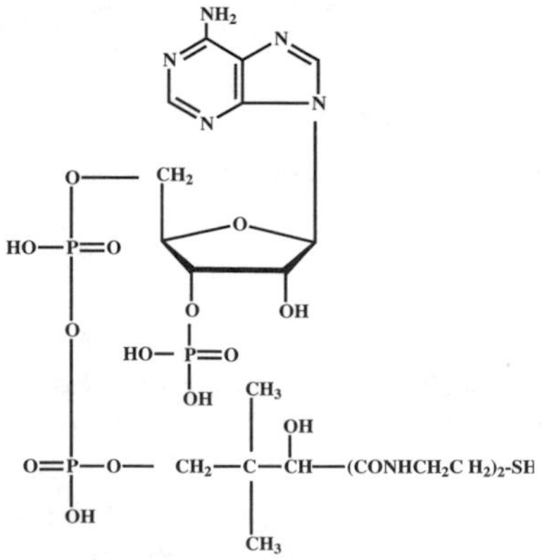

CoA-Disulfide Reductase The enzyme that catalyzes the following reaction:

$$\text{NADH} + \text{CoA-disulfide} \rightleftharpoons NAD^+ + 2\ \text{CoA}$$

Coagglutination Agglutination of different strains of a microorganism by a given antiserum.

CoA-Glutathione Reductase The enzyme that catalyzes the following reaction:

$$\text{NADPH} + \text{CoA-glutathione} \rightleftharpoons \text{CoA} + \text{glutathione} + \text{NADP}^+$$

Coagula Plural of coagulum.

Coagulant Any agent capable of causing coagulation.

Coagulase The enzyme that coagulates blood plasma, e.g., the enzyme produced by pathogenic *Staphylococci* causing blood coagulation.

Coagulase Test A test to determine if a given bacterial strain produces a coagulase.

Coagulation The formation of a clot, e.g., blood coagulation.

Coagulation Factor I Referring to fibrinogen.

Coagulation Factor II Referring to prothrombin.

Coagulation Factor III Referring to thromboplastin (tissue factor).

Coagulation Factor IV Referring to calcium ion required in coagulation.

Coagulation Factor V Referring to proaccelerin.

Coagulation Factor VII Referring to proconvertin.

Coagulation Factor VIIa A protease that catalyzes the selective cleavage of the peptide bond between Arg-Ile in factor X to form factor Xa.

Coagulation Factor VIII Referring to antihemophilic factor.

Coagulation Factor IX Referring to Christmas Factor.

Coagulation Factor X Referring to Stuart Factor.

Coagulation Factor Xa A protease that catalyzes the selective cleavage of peptide bonds between Arg-Thr and then Arg-Ile in prothrombin to form thrombin.

Coagulation Factor XI Referring to plasma thromboplastin antecedent.

Coagulation Factor XIa A protease that catalyzes the selective cleavage of the peptide bond between Arg-Ala and Arg-Val in factor IX to form factor IXa.

Coagulation Factor XII Referring to Hageman Factor.

Coagulation Factor XIIa A protease that catalyzes the selective cleavage of peptide bond between Arg-Ile in factor VII to factor VIIa.

Coagulation Factor XIII Referring to fibrin-stabilizing factor.

Coagulation Factor XIIIa Referring to protein-glutamine γ-glutamyltransferase.

Coagulator Any agent that causes coagulation.

Coagulogen A 16-kDa fibrinogen-like protein found in the hemocyte of horseshoe crab (*Limulus polyphemus*). It participates in hemostasis.

Coagulometer A device used for measuring the time required for a sample to coagulate.

Coagulum A clot or curd.

Coal Tar A by-product in the destructive distillation of coal, it consists of benzene, toluene, naphthalene, anthracene, xylene, phenol, and cresol and is used as an antieczematic agent.

COAP Abbreviation for a combination drug containing cytoxan, oncovin, ara-C, and prednisone.

COAPB Abbreviation for a combination drug containing cytoxan, oncovin, ara-C, prednisone, and bleomycin.

CoA-S-Ac Abbreviation for Acetyl coenzyme A or acetyl-CoA.

CoASH Abbreviation for coenzyme A.

Coat Protein 1. Proteins that make up the outer layer of a structure. 2. Viral capsid protein.

Coated Pit A specialized region of plasma membranes with clustered receptors and clathrin backing; it mediates endocytosis through invagination to form coated vesicles.

Coated Vesicle See coated pit.

COB Abbreviation for a combination drug containing cisplatin, oncovin, and bleomycin.

$^{60}CoB_{12}$ Abbreviation for ^{60}Co-labeled cyanocobalamin.

Cobalamin A cobalt-containing coenzyme derived from vitamin B_{12} and involved in a variety of enzymatic reactions.

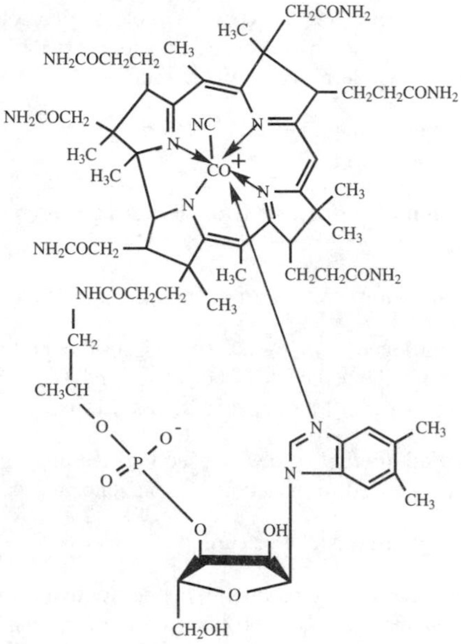

Cob(I)alamin Adenosyltransferase The enzyme that catalyzes the following reaction:

$$\text{ATP} + \text{cob(I)alamin} + H_2O \rightleftharpoons \text{Pi} + \text{PPi} + \text{adenosylcobalamin}$$

Cob(II)alamin Reductase The enzyme that catalyzes the following reaction:

$$\text{NADH} + 2\ \text{cob(II)alamin} \rightleftharpoons \text{NAD}^+ + 2\ \text{cob(I)alamin}$$

Cobalt A chemical element with atomic weight of 39 and valences of 2 or 3.

Cobalt-60 A radioactive cobalt.

Cobamamide (mol wt 1580) A cobalt-containing coenzyme derived from vitamin B12, and hematopoietic vitamin.

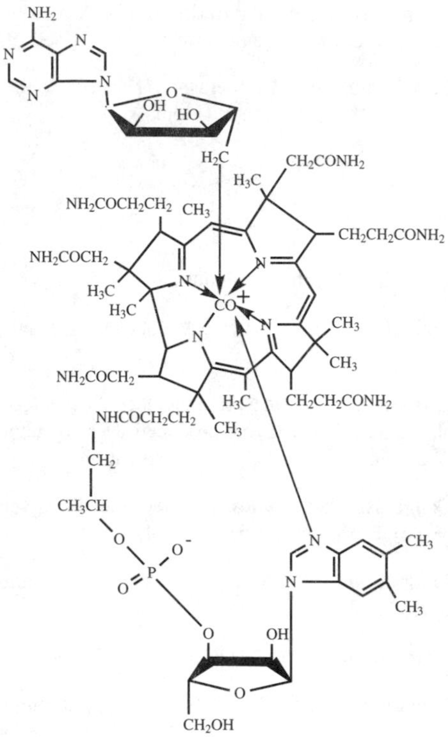

Cobamide A cobalt-containing coenzyme derived from vitamin B_{12} and involved in a variety of enzymatic reactions.

COBMAM Abbreviation for a combination drug containing cytoxan, oncovin, bleomycin, methotrexate, adriamycin, and MeCCNU.

Coboglobulin A hemoglobin-like molecule in which iron is replaced by cobalt.

Cobra Venom Factor Either of two protein factors from cobra venom that affects the alternative pathway of complement activation. It catalyzes the formation of C3 convertase.

Cobrotoxin A protein toxin from cobra venom consisting of 62 amino acid residues.

Cocaethylene (mol wt 317) An anesthetic agent.

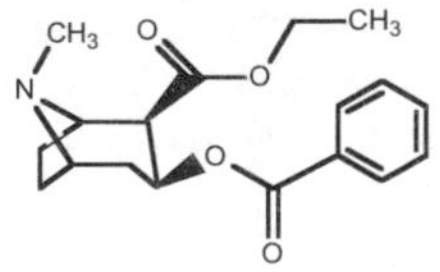

Cocaine (mol wt 303) 1. A topical anesthetic agent, 2. plus its other characteristics.

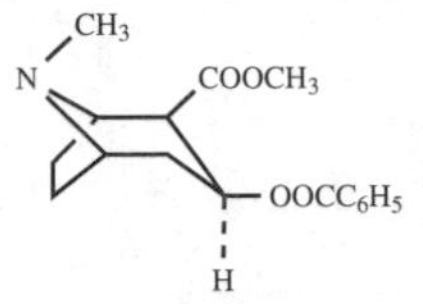

Cocaine Esterase The enzyme that catalyzes the following reaction:

A carboxylic ester + H_2O

⇅

An alcohol + carboxylate

Cocarboxylase Referring to thiamin pyrophosphate.

Cocarcinogen Any substance that increases the activity of a carcinogen.

Cocci Plural of coccus.

Coccidioides A genus of fungi and the causal agent of coccidiodomycosis.

Coccidioidin Any antigenic preparation derived from *Coccidioides immitis* used in a diagnostic skin test for coccidiodomycosis.

Coccidioidomycosis Infection or disease caused by *Coccidioides immitis.*

Coccidiosis Any infection or disease caused by protozoa of the order Coccidia.

Coccobacillus A bacterial cell intermediate in the morphology between coccus and bacillus.

Coccolysin A protease that catalyzes the preferential cleavage of peptide bonds involving amino groups of leucine, phenylalanine, tyrosine, and alanine.

Coccomyxa A genus of unicellular, nonmotile green algae.

Coccus A spherical or near spherical bacterium. Cocci may occur singly, in pairs, or in groups of four or more.

Cocktail The solution of fluorophores used for liquid scintillation counting.

Coconut Cadang Cadang Viroid An RNA viroid that infects coconut plants.

Cocoonase A protease capable of digesting silk protein.

Codalan A trade name for a combination drug containing acetaminophen, codeine phosphate, and caffeine, used as an analgesic agent.

Codaminaphen A trade for a combination drug containing acetaminophen, codeine phosphate, and caffeine, used as an analgesic agent.

Codecarboxylase Referring to pyridoxal phosphate.

Codeine (mol wt 299) A narcotic, analgesic, and antitussive agent that binds with opiate receptors at many sites in the CNS.

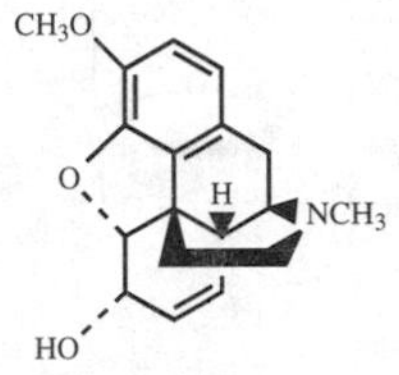

Codesol A trade name for prednisolone sodium phosphate, used as an anti-inflammatory agent.

Codiclear DH Syrup A trade name for a combination drug containing hydrocodone bitartrate and guaifenesin used as an expectorant and antitussive agent.

Codimal DH A trade name for a combination drug containing hydrocodone bitartrate, phenylephrine hydrochloride, pyrilamine maleate, potassium guaiacolsulfonate, sodium citrate, and citric acid.

Coding Triplet Referring to codon or genetic code on mRNA.

Codon Sequence of three adjacent nucleotides (triplet) on mRNA that specifies a given amino acid.

Codon Dictionary See genetic dictionary.

Codroxocobalamin Referring to vitamin B_{12a} (hydroxocobalamin).

Coefficient A numerical parameter for determining the effect or change produced by variations of specified conditions or of the ratio between two quantities.

Coelomocyte A phagocytic cell found in the coelomic cavity of invertebrates.

Coenobium A colony of unicellular eukaryotes surrounded by a common membrane.

Coenocyte A multinucleate cell or organism.

Coenzyme A nonprotein factor required for an enzymatic reaction.

Coenzyme I Referring to the coenzyme NAD.

Coenzyme II Referring to the coenzyme NADP.

Coenzyme A (mol wt 768) A coenzyme involved in a variety of biochemical reactions.

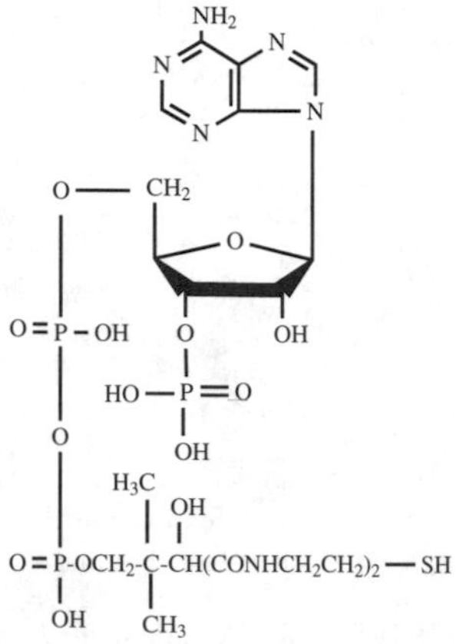

Coenzyme B_{12} (mol wt 1579) 5′-deoxyadenosylcobalamin involved in reactions catalyzed by methylmalonyl-CoA mutase.

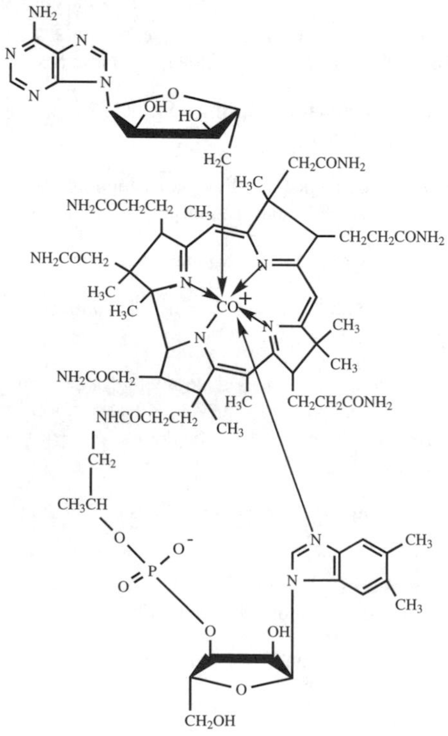

Coenzyme F_{420} A coenzyme derived from flavin.

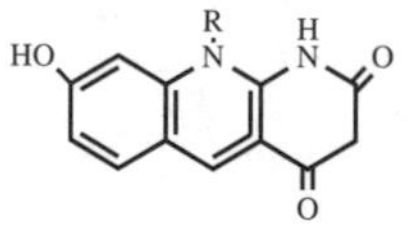

Coenzyme F_{420} Hydrogenase The enzyme that catalyzes the following reaction:

$$H_2 + \text{coenzyme } F_{420} \rightleftharpoons \text{Reduced coenzyme } F_{420}$$

Coenzyme M Referring to the coenzyme, 2-mercaptoethanesulfonic acid, involved in methane formation.

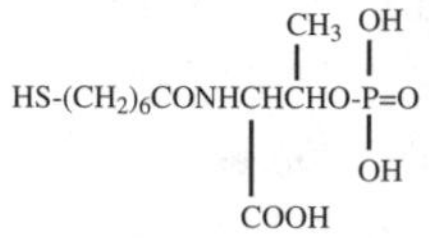

Coenzyme Q (ubiquinone) A coenzyme that functions as an electron carrier in the electron transport system.

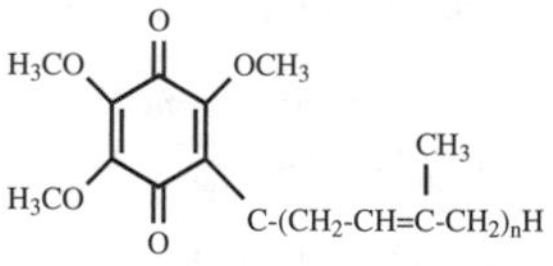

coenzyme Q_4 : n = 4
coenzyme Q_9 : n = 9
coenzyme Q_{10} : n = 10

Coenzyme Q_4 See coenzyme Q.

Coenzyme Q_9 See coenzyme Q.

Coenzyme Q_{10} See coenzyme Q

Coenzyme R Referring to biotin.

CoF Abbreviation for cobra factor.

Cofactor Ion or molecule that serves as a factor in an enzymatic reaction.

Cogentin A trade name for benztropine mesylate, an antiparkinsonian agent that blocks central cholinergic receptors.

Co-Gesic A trade name for a combination drug containing acetaminophen and hydrocodone bitartrate, used as an analgesic agent.

Cognate Referring to two biomolecules that normally interact with each other, e.g., enzyme with substrate and ligand with receptor.

Cognate tRNA Referring to tRNA that can be recognized by aminoacyl-tRNA synthetase.

Cognex A trade name for tacrine hydrochloride, a cholinesterase inhibitor used as an Alzheimer's drug.

COHb Abbreviation for carboxyhemoglobin or carbon monooxyhemoglobin.

Coherin A peptide cofactor involved in regulation of intestinal motility in mammals.

Cohesive End The ends of a double-stranded DNA molecule having a single-stranded terminus at each end.

Cohesive End Ligation Joining of two DNA fragments that have cohesive ends with the enzyme DNA ligase.

Cointegrate The circular product resulting from fusion of two circular replicons (e.g., two plasmids or a plasmid and a bacterial chromosome) mediated by a transposable element.

Coisogenic Individuals that are genetically identical at all loci except one.

Colace A trade name for docusate sodium, a laxative that reduces surface tension of interfacing liquid contents of the bowel.

Colaspase A trade name for asparaginase, an antineoplastic agent.

Colbenemid A trade name for a combination drug containing probenecid and colchicine, used as a uricosuric and renal tubular blocking agent that inhibits the tubular re-absorption of urate and increases urinary excretion of uric acid.

Colcemide (mol wt 371) A tubulin-binding substance that interferes with microtubule-dependent function.

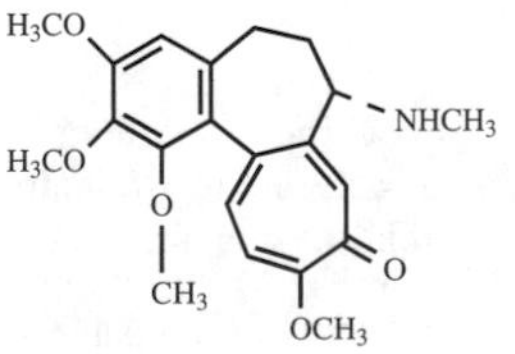

Colchicine (mol wt 385) An alkaloid from *Colchicum autumnale* (Liliaceae) used as an antigout agent. It reduces leukocyte mobility, phagocytosis, and lactic acid production and leads to a decrease of urate deposition and inflammation. It is also used to produce polyploid varieties of species important in horticulture.

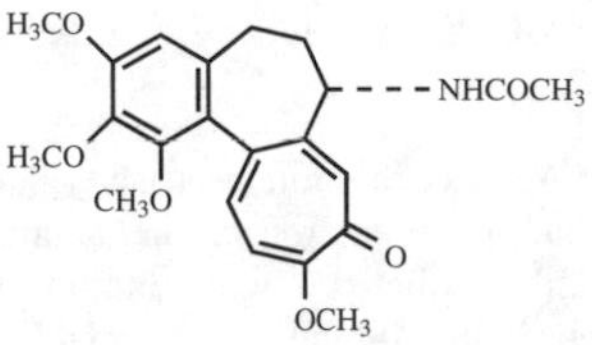

Cold Agglutinins Antibodies or substances that agglutinate bacteria or erythrocytes more efficiently at temperatures below 37° C.

Cold Hemagglutinin The hemagglutinin that causes hemagglutination at low temperature but disperses cells at higher temperature.

Cold Insoluble Globulin Globulins (e.g., fibronectin) that are insoluble at low temperature because the low temperature weakens the hydrophobic interactions.

Cold Sensitive Enzyme The enzyme that has an unusual sensitivity to low temperature owing to the weakening of hydrophobic interactions and formation of inactive subunits at low temperature.

Cold Sensitive Mutant A mutant that is defective at low temperature but functional at normal temperature.

Cold Sore A clustered sore of the lips caused by herpes simplex virus (also called fever blister).

Colectomy Surgical removal of part or all of the large intestine (colon).

Colestid A trade name for colestipol hydrochloride, an antihyperlipidemic agent.

Colestipol A basic anion exchange resin and a copolymer of diethylenetriamine and 1-chloro-2,3-epoxypropane. It is an antilipemic agent that combines with bile acid to form insoluble molecules for excretion; it triggers the synthesis of new bile acid from cholesterol and leads to the reduction of low-density lipoprotein and cholesterol level.

Col-Factor The genetic element that controls the production of colicin.

Colfosceril Palmitate (mol wt 734) A lung surfactant.

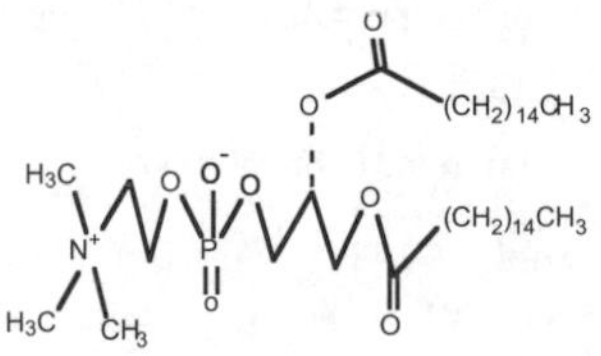

Colgout A trade name for colchicine, an antigout agent.

Colibacillosis Any infection or disease caused by *E. coli*.

Colicin A class of proteins (bacteriocins) produced by strains of enterobacteria containing colicin plasmids. Colicins are proteins produced by some bacteria that inhibit closely related bacteria.

Colicin A A colicin produced by strains of enterobacteria containing Col-A plasmids. It binds to the outer membrane protein F (OmpF) leading to the formation of pores in the cytoplasmic membrane.

Colicin B A colicin produced by strains of enterobacteria containing Col-B plasmids. It binds to the outer membrane protein leading to the formation of pores in the cytoplasmic membrane.

Colicin D A colicin produced by strains of enterobacteria containing Col-D plasmids. It binds to outer membrane protein leading to the formation of pores in the cytoplasmic membrane.

Colicinogen See Col factor.

Colicinogenic Capable of producing colicin.

Colicinogenic Factor See Col factor.

Colicinogenic Plasmids The plasmids that encode colicins.

Coliforms Any Gram-negative, rod-shaped, asporogenous, facultative anaerobic bacteria capable of fermenting lactose.

Coliform Test A test used to detect the presence of coliform bacteria in a sample (e.g., water sample).

Colinear Having the corresponding part arranged in the same linear order, e.g., DNA and RNA and mRNA and amino acids in protein.

Colipase A pancreatic protein that forms complexes with lipase; it inhibits the surface denaturation of lipase.

Coliphage A virus that infects *Eschericia coli.*

Colistimethate Sodium A polypeptide antibiotic.

Colistin A cyclopeptide antibiotic produced by *Bacillus colistinus*.

Colitis Inflammation of the colon.

Colitose (mol wt 148) A monosaccharide (3,6-dideoxygalactose) isolated from *E. coli.*

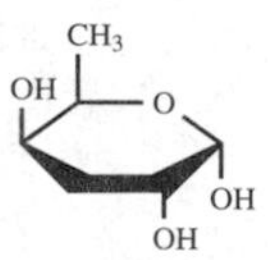

Collagen The major fibrous protein of connective tissue such as bone, teeth, tendon, skin, and blood vessels and components of the extracellular matrix.

Collagen Disease Diseases characterized by the swelling and breakdown of fibers in connective tissue.

Collagenase Enzyme that catalyzes the hydrolysis of collagen.

Collagenase A A microbial collagenase that cleaves the peptide bonds involving the amino group of glycine in the triple helical region of native collagen.

Collagenase Ointment A drug used for treatment of bed sores, burns, and skin disorders.

Collectin One of a family of plasma lectins that has a trimeric structure similar to complement C1q.

Colligation The formation of covalent bond by means of two combining groups each donating one electron.

Colligative Property A property of a solution that depends on the number of solute particles per unit volume of the solution, e.g., freezing point depression.

Collimate To make light rays parallel.

Collimator A device that converts incident radiation into a narrow beam of parallel rays.

Collinomycin (mol wt 536) An antibiotic produced by *Streptomyces collinus.*

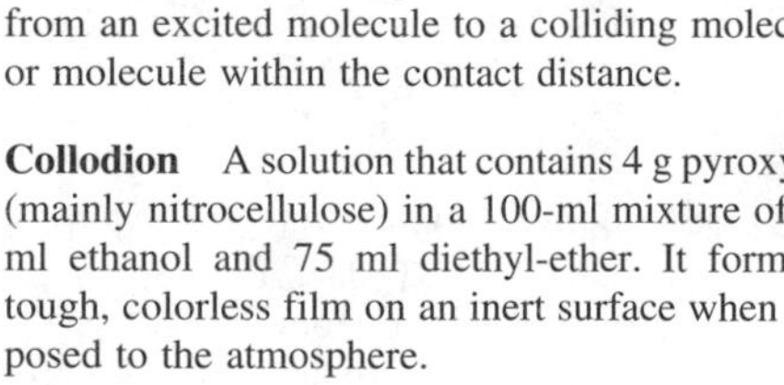

Collisional Quenching The transfer of energy from an excited molecule to a colliding molecule or molecule within the contact distance.

Collodion A solution that contains 4 g pyroxylin (mainly nitrocellulose) in a 100-ml mixture of 25 ml ethanol and 75 ml diethyl-ether. It forms a tough, colorless film on an inert surface when exposed to the atmosphere.

Colloid A macromolecule or a particle with one dimension having a length at least 10^{-9} to 10^{-6} m and incapable of passing through a semipermeable membrane.

Colloid Solution A thermodynamically stable solution consisting of colloidal macromolecules and solvent; it can be readily reconstituted after separation from the solvent.

Collyrium A trade name for boric acid.

Colocynthin (mol wt 719) A glucoside from fruit of *Citrullus colocynthis* used as a cathartic agent.

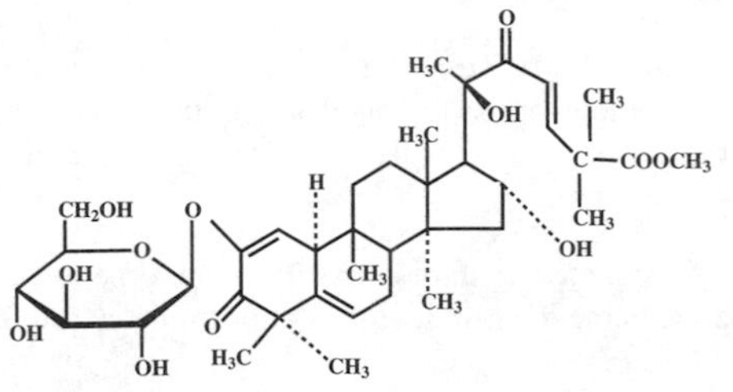

Cologel A trade name for methylcellulose, a laxative that absorbs water and increases bulk and moisture content of the stools.

Colominic Acid The capsular polysaccharide from strains of *E. coli* K12 consisting of a linear polymer of N-acetylneuraminic acid.

Colon The large intestine.

Colonoscope A device or fiberscope used for visual examination of the large intestine.

Colonoscopy The visual examination of the large intestine with a colonoscope.

Colony A population of cells growing on a solid medium that arose from a single cell.

Colony Counter A device for counting the number of bacterial colonies on a solid medium.

Colony Forming Unit An unit for quantifying cells capable of forming a colony from a single cell; it is expressed as number of colony forming units per volume of sample.

Colony Hybridization A procedure for the direct detection of a particular DNA sequence within an array of bacterial cells (clones) using a radiolabeled DNA or RNA probe that is complementary to the sequence being investigated. The cells from a colony are transferred to nitrocellulose paper and then lysed, the resultant DNA from the lysed cells are denatured, bound to the nitrocellulose paper, and identified with labeled probe.

Colony Stimulation Factor Any of the cytokines that controls the differentiation of hemopoietic stem cells.

Coloproctitis Inflammation of both the large intestine and the rectum.

Color Quenching The absorption of light emitted by the fluorophlores in a scintillation is counted by the color component in the sample.

Color Vision The capacity to perceive color.

Colorado Tick Fever A tick-borne viral disease caused by orbivirus.

Colorimeter A photoelectric instrument used for the quantitative determination of colored compounds in a solution.

Colorimetry A method for quantitative determination of color compounds in a solution with a colorimeter.

Colosigmoidoscopy The visual examination of the large intestine with a sigmodoscope (a flexible tube with lighted device).

Colostrum The fluid released from the mammory glands before milk production begins (also known as first milk).

Coloxyl A trade name for docusate sodium, a laxative that reduces the surface tension of the interfacing liquid contents of the bowel and promotes incorporation of additional liquid into the stool.

Colpectomy The surgical removal of the vagina.

Colpitis Inflammation of the vagina and bladder.

Colposcope A lighted device for direct visual examination of the vagina and cervix.

Colposcopy Visual examination of the vagina and cervix with a colposcope.

Colsalide A trade name for colchicine, an antigout agent.

Column Chromatography A type of chromatography in which the stationary phase is packed in a column and the mobile phase percolates through the column thereby separating the components, e.g., gel filtration, ion exchange, and absorption chromatography.

Coly-mycin M A trade name for colistimethate sodium, an antibacterial agent

Colyte A trade name for polyethylene glycol electrolyte solution, used as a laxative agent.

Coma A state of profound unconsciousness.

COMA Abbreviation for a combination drug containing cyclophosphamide, oncovin, methotrexate, and adriamycin.

COMAA Abbreviation for a combination drug containing cytoxan, oncovin, methotrexate, adriamycin, and ara-C.

COMB Abbreviation for a combination drug containing cyclophosphamide, oncovin, methotrexate, and bleomycin.

Combantrin A trade name for pyrantel embonate, an anthelintic agent that blocks neuromuscular activity.

Combinatorial Any process that is governed by specific combination factors. The different combinations have different effects.

Combinatorial Association The association of immunoglobulin molecules from any one class of heavy chain with molecules from any one type of light chain.

Combinatorial Translocation The association of a gene of any variable region with a gene of any constant region of an immunoglobulin in the same multigene family. The two genes are brought together through deletion of intervening DNA sequences.

Combined Oxygen The oxygen that is attached to hemoglobin.

Combipres A trade name for a combination drug containing chlorthalidone and clondine hydrochloride, used as an antihypertensive.

Combivent Inhalation Aerosol An inhalation aerosol containing a micro-crystalline suspension of ipratropium bromide and albuterol sulfate used as a bronchodilator.

Combivir A trade name for a combination drug containing lamivudine and zidovudine used as an antiviral agent for treatment of HIV infection.

COMF Abbreviation for a combination drug containing cyclophosphamide, oncovin, methotrexate, and 5-fluorouracil.

COMLA Abbreviation for a combination drug containing cytoxan, oncovin, methotrexate, leucogen, and ara-C.

COMLEC Abbreviation for a combination drug containing cytoxan, oncovin, methotrexate, leucogen, etoposide, and cytarabine.

Commaless Genetic Code Successive codons that are contiguous without noncoding bases (e.g., genes without introns).

Commensalism The symbiosis in which one symbiont receives benefits from the association and the other symbiont receives neither benefit nor harm.

Common Cold An upper respiratory disease caused by a number of viruses, e.g., rhinovirus, coronavirus, influenzavirus, parainfluenza virus, and reovirus.

Comoviruses A group of plant viruses that contain positive-stranded ssRNA.

COMP Abbreviation for 1. a combination drug containing cyclophosphamide, oncovin, methotrexate, and prednisone; 2. cartilage oligomeric matrix protein.

CompA Abbreviation for compound A (11-dehydrocorticosterone).

Compatibility Test Serological test used to detect if blood or tissue from a prospective donor can be transfused or transplanted without immunological rejection.

Compazine A trade name for prochlorperazine, an antiemetic agent that alters dopamine action in the CNS.

CompB Abbreviation for compound B (corticosterone).

CompE Abbreviation for compound E (cortisone).

Competence Referring to 1. a bacterial cell capable of undergoing transformation, 2. a lymphocyte capable of recognizing an antigen and synthesizing antibody, and 3. the ability of a group of embryonic cells to react to a morphogenic stimulus and initiate differentiation.

Competitive Inhibition A type of inhibition in which the inhibitor competes with substrate or ligand for the binding site on the enzyme. The competitive inhibition can be reversed by increasing substrate concentration.

Competitive Inhibitor A substance capable of competing with substrate or ligand for the binding site on the enzyme.

CompF Abbreviation for compound F (cortisol or hydrocortisone).

Complement A group of proteins and proenzymes in the plasma of vertebrates that can be activated by an antigen-antibody reaction (classical pathway) leading to cell lysis.

Complement Activation Conversion of complement components to their active forms through the classical pathway of activation or through the alternative pathway of activation leading to cell lysis and the generation of pharmacologically active fragmentation products.

Complement Cascade Referring to the cascading action of the pathways of complement activation. This leads to the conversion of components of complement to active subcomponents and fragments leading to the lysis of cells.

Complement Factor B A complement factor involved in the alternative pathway of complement activation.

Complement Factor D A complement factor involved in the alternative pathway of complement activation. It catalyzes the cleavage of complement factor B.

Complement Factor I A complement factor involved in the alternative pathway of complement activation. It catalyzes the cleavage of complement C3b.

Complement Fixation Test A sensitive serological test for detection of the presence of specific antigen or the presence of complement-fixing antibody. It consists of two serological systems: a hemolytic indicator system and an antigen-antibody testing system.

Complement Fixing Antibody The antibody capable of activation or fixation of complement, e.g., IgG and IgM.

Complement Subcomponents The subcomponent of the complement, e.g., C1q, C1r and C1s are subcomponents of C1.

Complementary Base Pair Referring to the A:T, A:U, and G:C pairs in DNA or RNA.

Complementary DNA The DNA molecule synthesized from mRNA template using reverse transcriptase; it consist of no intron.

Complementary Genes Two genes that produce similar phenotypes when they are present separately but produce different phenotypes when present together.

Complementary Strand The DNA strand that has a complementary base sequence to another strand.

Complementation The ability of a gene to compensate for a functional defect in a homologous gene when present in the same organism.

Complementation Test A test used to determine if two mutants are located on the same cistron or on a different cistron.

Complete Blood Count The quantitation of the number of white and red blood cells per cubic millimeter.

Complete Freund's Adjuvant The Freund's adjuvant that contains dead cells of mycobacteria.

Complete Medium A type of culture medium that contains nutrients necessary for the growth of auxotrophs.

Complete Transduction Transduction in which the DNA from the donor bacterium is completely integrated into the chromosome of the recipient cell.

Complex I Referring to NADH-dehydrogenase complexes in the electron transport system. It catalyzes the oxidation of NADH by coenzyme Q.

Complex II Referring to succinate dehydrogenase complexes in the electron transport chain. It catalyzes the oxidation of succinate by coenzyme Q.

Complex III Referring to Cytochrome b_{c1} complexes in the electron transport chain. It catalyzes the oxidation of reduced coenzyme Q by cytochrome c.

Complex IV Referring to cytochrome oxidase complexes in the electron transport chain. It catalyzes oxidation of reduced cytochrome c by molecular oxygen.

Complex-Mediated Hypersensitivity A type of hypersensitivity that is mediated by soluble antigen-antibody complexes, e.g., conditions like serum sickness or glomerulonephritis.

Complex Virion A complete virus particle with a complex symmetry (not a simple cubic nor a simple helical symmetry). It may be a combination of cubic and helical symmetry.

Complexone Referring to ionophore.

Compound W A trade name for salicylic acid.

Compoz Diahist A trade name for diphenhydramine hydrochloride, an antihistaminic agent that competes with histamine for H-1 receptors on the target cells.

CompS Abbreviation for compound S (11-deoxycortisol).

Computed Tomography Scan A specialized form of X-ray examination in which the X-ray source and detector rotate around the object to be scanned and the information obtained can be used to produce cross-section images by computer.

COMT Abbreviation for catechol-O-methyltransferase.

Co-mutagen Any non-mutagenic agent that enhances the effect of a mutagen.

Comvax A trade name for *Hemophilus* b conjugate and hepatitis B recombinant vaccine.

ConA Abbreviation of concanavalin A.

Conalbumin An iron-binding protein found in egg white and avian blood.

Conantokin G A toxic peptide (also called sleeper peptide) from the venom of *Conus geographicus* (fish-hunting cone snail) that is an antagonist of brain NMDA receptor.

Conarachin A minor globulin occurring in the seeds of peanut.

Concanavalin A A mitogenic lectin from jack bean *(Canavalia ensiformis)*. It stimulates the proliferation of T lymphocytes.

Concanavalin B A minor crystallizable protein obtained from the pipe seeds of the jack bean.

Concatemer Two or more DNA or RNA molecules that are covalently joined (end to end) in the same orientation.

Concentraid A trade name for desmopressin, an antidiuretic hormone.

Concentration Gradient A condition in which the concentration of a solute changes with distance, e.g., sucrose density gradient used in density gradient centrifugation and pH gradient generated in mitochondria by the electron transport system.

Concerted Acid-Base Catalysis A catalytic reaction that consists of simultaneous actions of both acidic and basic catalytic groups.

Concerted Catalysis A catalytic reaction in which there are more than one catalytic grouping in the active site of an enzyme.

Concerted Feedback Inhibition See concerted inhibition.

Concerted Inhibition Feedback inhibition in which two or more products are simultaneously involved, e.g., inhibition of glutamine synthetase (an allosteric enzyme) by the six products of glutamine metabolism.

Concerted Model of Allosteric Enzyme A model describing the action of an allosteric enzyme in which the enzyme exists in two conformational forms (R-form: relaxed form and T-form, tensed form), the two forms differ in their capacity to bind substrate, positive effector, and negative effector. The binding of an effector or substrate shifts the equilibrium from one form to the other leading to regulation of enzyme activity.

Concussion A condition caused by injury to the head.

Condensation 1. A type of chemical reaction in which two or more molecules combine with the release of water, alcohol, or simple substance. 2. The change of state of a substance (e.g., from vapor to liquid or to solid form).

Condensing Enzyme Referring to citrate synthetase.

Conditional Mutant A mutant that grows as a normal organism under permissive conditions and expresses a lethal mutation phenotype under the nonpermissive condition (e.g., conditional lethal mutation).

Condrin-LA A trade name for a combination drug containing phenylpropanolamine hydrochloride and chlorpheniramine maleate, used as an antihistaminic agent.

Conductance The electrical property of a solution defined as the reciprocal of the resistance.

Conductiometric Method A method of chemical analysis in which the end point of a reaction is determined by the measurement of conductance.

Conductivity The ability of a substance or a mixture of substances to transfer heat or electricity. It is the reciprocal of resistivity.

Cone A light sensitive cell present in the retina.

Cone Pigment Gene Genes that encode pigments in the cone cells of the retina and are responsible for color vision.

Configuration The spatial arrangement of an organic molecule that is conferred by the presence of either a double bond or a chiral center around which substitution groups are arranged in a specific sequence. Configuration isomers cannot be interconverted without breaking a covalent bond or bonds.

Confluence A state in cell culture where cells have multiplied to cover the surface of the growth vessel.

Conformation Isomer Any one or more isomers that differ only in their stereochemical conformation.

Conformational Isozyme The multiple forms of a single gene product that differ in secondary or tertiary structure.

Congenic Strain The strains of organisms that are different from one another in a small chromosomal segment.

Congenital Present at the time of birth.

Congeric Of the same species.

Congespirin A trade name for a combination drug containing phenylephrine hydrochloride and acetaminophen, used as a bronchodilator.

Congest A trade name for estrogen.

Conglutination Agglutination of particulate entities bearing C3b by conglutinin.

Conglutinin A protein from bovine serum capable of binding with complement C3b leading to agglutination of particles bearing C3b.

Congo Red (mol wt 697) A dye and pH indicator.

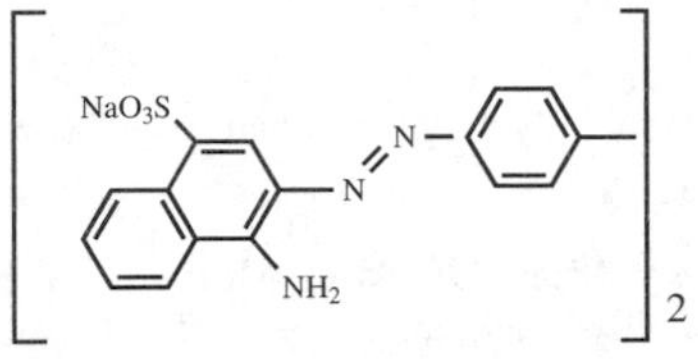

Conidia Plural of conidium.

Conidium An asexual haploid spore borne on an aerial hypae.

Coniferin (mol wt 342) The principal glucoside of conifers.

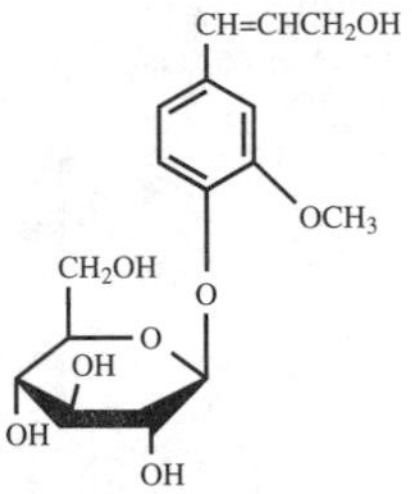

Coniferyl Alcohol (mol wt 180) An alcohol obtained from hydrolysis of coniferin.

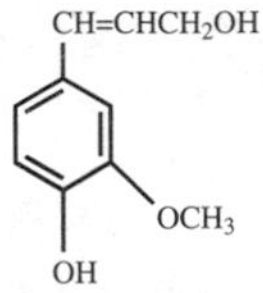

Coniine (mol wt 127) The toxic principle of poison hemlock *(Conium maculatum).*

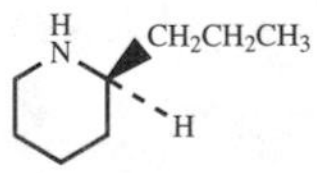

Conjec-B A trade name for brompheniamine maleate, an antihistaminic drug that competes with histamine for H-1 receptor on effector cells.

Conjugase Synonym of γ-glu-X carboxypeptidase.

Conjugate Acid-Base Pair An proton donor and its corresponding deprotonated species (e.g., acetic acid, donor and acetate, acceptor).

Conjugate Redox Pair An electron donor and its corresponding electron acceptor (e.g., NADH, donor and NAD^+, acceptor).

Conjugated Double Bond Any of the two or more double bonds in a molecule where each double bond is separated from the next by a single bond.

Conjugated Protein A protein that is linked to a nonprotein group (prosthetic group).

Conjugation 1. Union of gametes from opposite sexes. 2. Transfer of genetic material from a male strain (F^+) to the female strain (F^-) in bacteria.

Conjugation Labeling A covalently coupling procedure for introducing a label into a large molecule.

Conjugative Plamid Any plasmid that mediates the transfer of DNA by conjugation.

Conjugon Referring to that genetic element essential for bacterial conjugation.

Connective Tissue Any supporting tissue that lies between other tissue and consists of cells embedded in the extracellular matrix, e.g., bone, cartilage, and loose connective tissue.

Connexin The main protein component of connexon.

Connexon Water-filled pores in the cytoplasmic membrane formed by a ring of six protein subunits. The connexons from two adjoining cells join to form a continuous channel between the two cells.

Conotoxin Any of the several peptides of the family of ω-conotoxins isolated from the venoms of marine snails. ω-Conotoxins are neurotoxins that inhibit the voltage-gated calcium channel.

Consensus Sequence Average or most typical form of a sequence that is reproduced with minor variations in a group of related DNA, RNA, or protein sequences. The consensus sequence shows the nucleotide or amino acid most often found at each position.

Conservative Substitution Replacement of one amino acid residue in a protein by another amino acid with similar properties, e.g., substitution of glutamate with aspartate.

Conserved Sequence A sequence of nucleotides that has not changed noticeably over a long evolutionary period.

Conspecific Within the same species.

Constant Region The portion of an immunoglobulin molecule that is encoded by the constant region gene. The amino acid sequence in the constant region remains constant in various types of immunoglobulins.

Constilac A trade name for lactulose, a laxative that increases water content and softens the stool.

Constitutive Produced in constant amount.

Constitutive Enzyme An enzyme that is required in constant amounts by a cell (also known as a house-keeping enzyme).

Constitutive Genes Genes that are expressed as a function of the interaction of RNA polymerase with a promoter without additional regulation.

Constulose A trade name for lactulose, a laxative that increases water content and softens the stool.

Contac Capsules A trade name for a combination drug containing phenylpropanolamine and chlorpheniramine, used as an antihistaminic agent.

Contac Cough Formula A trade name for dextromethorphan hydrobromide, an antitussive agent.

Contact Inhibition Cessation of cell growth and cell division upon contact with other cells. Contact inhibition is responsible for the formation of a cell monolayer on the surface of a culture vessel.

Contact Insecticide A substance that kills insects by penetrating the body surface.

Continuous Cell Line A population of cells derived from either a tumor or tissue culture following transformation (also known as an established cell line).

Continuous Culture The maintenance of a microorganism in exponential growth by continuous inflow of fresh medium to the culture vessel.

Continuous Fiber The microtubules that connect the two poles of the mitotic apparatus.

Continuous Flow Centrifugation A type of centrifugation in which the suspension to be centrifuged continuously flows into the centrifuge and the supernatant is continuously removed.

Contrapsin A glycoprotein and a plasma protease inhibitor.

Contrast Medium Substance used to improve the visibility of structure during radiography.

Controlling Gene The gene that switches the cistron on and off.

Convallatoxin (mol wt 551) A cardiotonic agent from the blossoms of the lily (*Convallaria majalis*).

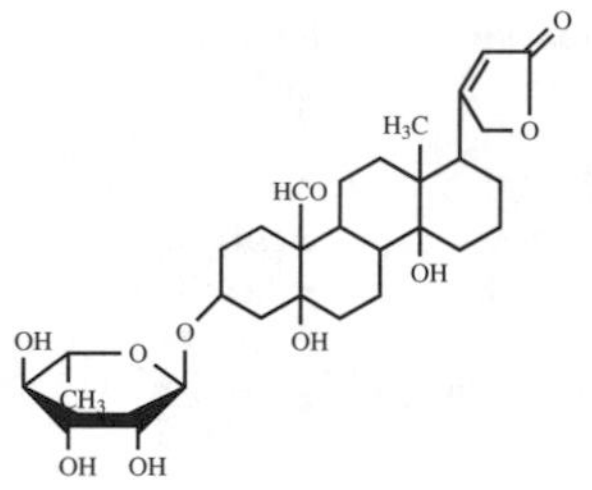

Convertin Synonym of blood coagulation factor VIIa.

Conway Microdiffusion Method A microanalytical method for analysis of substance that liberates ammonia.

COOH-Terminal Abbreviation for carboxyl terminal of protein.

Coomassie Brilliant Blue A trade name for a group of dyes used for staining proteins.

Coomassie Brilliant Blue G 250 A protein stain used in gel electrophoresis.

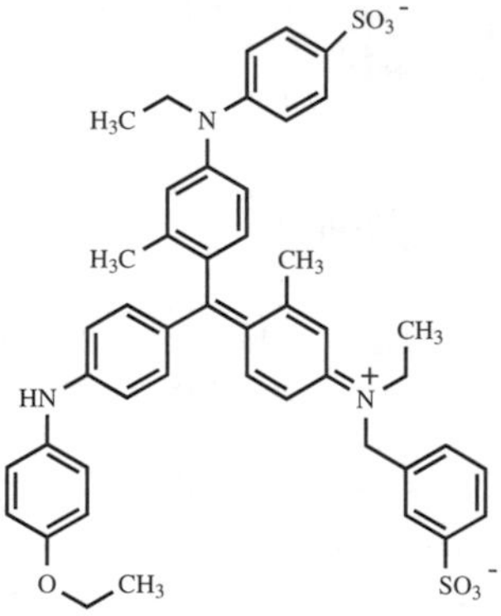

Coomassie Brilliant Blue R 250 A protein staining dye for gel electrophoresis.

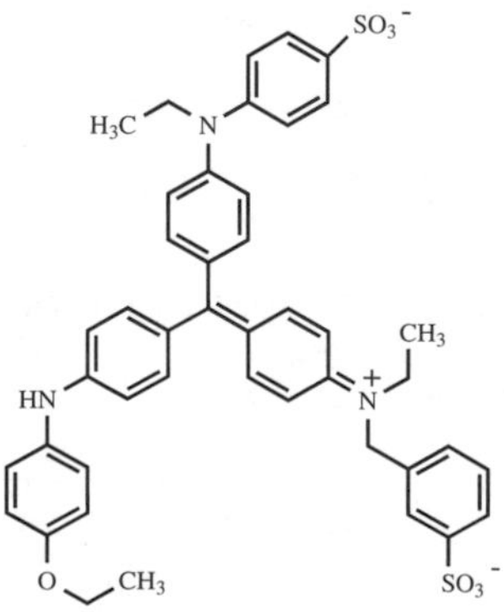

Coombs' Test A test used to identify the presence of antierythrocyte antibodies (also called antiglobulin test).

Cooperative Binding The binding of one ligand to its binding site affects the binding of subsequent ligands to the other sites on the same molecule.

Cooperative Hydrogen Bonding The interaction among adjacent hydrogen bonds in a molecule in which the energy required to form them is less than the sum of the energies required to form the individual bonds but the energy required to break these bonds is greater than the sum of the energies required to break individual bonds.

Cooperative Ligand Binding The binding of one ligand to one site on the macromolecule increases or decreases the affinity of ligand to the other site.

Coordinate Covalent Bonding A covalent bond in which two electrons are donated to only one of the bonded atoms.

Coordinate Enzymes The enzymes that are under the control of an operon system.

Coordinate Induction Induction of a series of enzyme syntheses in an operon system by a single inducer.

Coordinate Regulation The regulation of induction or repression of structural genes in an operon system by a single inducer or repressor.

Coordinate Repression The repression of a series of enzyme syntheses in an operon system by a single repressor.

COP Abbreviation for a combination drug containing cyclophosphamide, oncovin, and prednisone.

COPA Abbreviation for a combination drug containing cyclophosphamide, oncovin, prednisone, and adriamycin.

COPAB Abbreviation for a combination drug containing cyclophosphamide, oncovin, prednisone, adriamycin, and bleomycin.

COPAC Abbreviation for a combination drug containing CCNU, oncovin, prednisone, adriamycin, and cyclophosphamide.

Copaxone A trade name for glatiramer acetate, an acetate salt of synthetic polypeptide used for the treatment of multiple sclerosis.

COPB Abbreviation for a combination drug containing cyclophosphamide, oncovin, prednisone, and bleomycin.

Cope A trade name for a combination drug containing aspirin, caffeine, magnesium hydroxide, and aluminum hydroxide, used as an antipyretic and analgesic agent.

Cophene-B A trade name for brompheniramine maleate, an antihistaminic drug that competes with histamine for H-1 receptors on effector cells.

Copolymer A polymeric molecule containing more then one type of monomer unit.

COPP Abbreviation for a combination drug containing CCNU, oncovin, procarbazine, and prednisone.

Copper A chemical element with an atomic weight of 64, valence 1 or 2.

Coproantibody An antibody in the lumen of the gastrointestinal tract or feces.

Coprococcus A genus of Gram-positive, asporogenous, anaerobic bacteria.

Coprophilic Referring to organisms that grow preferentially in animal feces.

Coprozoic Referring to protozoa that are coprophilic.

Copy Choice Hypothesis The hypothesis states that the new strand of DNA alternates between the paternal and maternal strands of DNA during its replication.

Copy DNA Referring to complementary DNA (cDNA).

Copy Error An error made during DNA replication.

Copy Mutant A mutant plasmid whose copy number differs from the wild-type plasmid.

Copy Number The number of copies of any gene or plasmid in a given cell.

CoQ Abbreviation for Coenzyme Q.

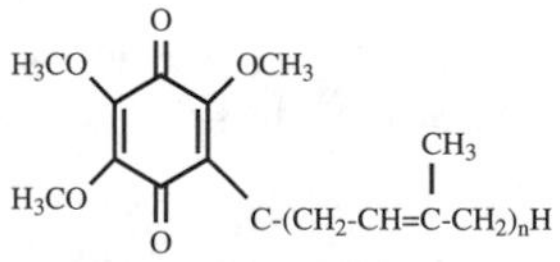

coenzyme Q_4 : n = 4
coenzyme Q_9 : n = 9
coenzyme Q_{10} : n = 10

$CoQH_2$ Abbreviation for reduced form of coenzyme Q.

Coquinone A combination drug containing coenzyme Q_{10} and alpha lipoic acid used to increase ATP production in mitochondria.

CoR Abbreviation for Congo red.

Corax A trade name for chlordiazepoxide hydrochloride, an anti-anxiety agent.

Cord Factor A toxic glycolipid from *Mycobacterium.*

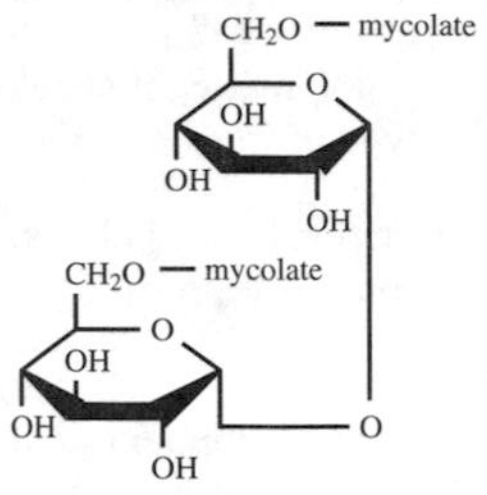

Cordycepin (mol wt 251) A nucleoside antibiotic from *Cordyceps militaris* that inhibits RNA biosynthesis.

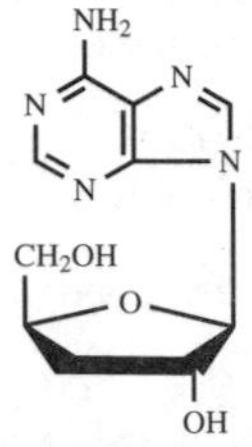

Core DNA The DNA segment in a nucleosome that wraps around a histone octamer.

Core Enzyme RNA polymerase complex without a sigma subunit.

Core Particle The product resulting from nucleosome digestion that consists of histone octamer and core DNA.

Core Polysaccharide A component of lipopolysaccharide in bacterial cell walls consisting of ketodeoxyoctonate, heptoses, glucose, and N-acetylglucosamine.

Core Protein The protein in proteoglycan to which glycosaminoglycans attach.

Coreg A trade name for carvedilol, an alpha/beta adrenergic blocker used as an antihypertensive agent.

Corepressor A substance capable of combining with an inactive repressor to form an active repressor.

Corgard A trade name for nadodol, an antianginal drug that blocks the stimulation of beta-1 and beta-2 receptors.

Cori Cycle A cycle for transfer of lactate produced by glycolysis in the muscle to the bloodstream and then liver where it is converted to glucose and then transported from the liver back to the muscle.

Coricidin A trade name for a combination drug containing chlorpheniramine maleate and acetaminophen, used as an antihistaminic agent.

Cori's Disease A disorder characterized by a large deposition of glycogen in the liver.

Corium A layer of skin underneath the epidermis.

Cork Tissue produced by cork cambium in the outer part of the cortex of some stems and roots.

Cork Cambium The layer of cells capable of undergoing repeated division in the production of cork cells.

Corlopam A trade name for fenoldopam mesylate, a rapid-acting vasodilator.

Cormax A trade name for clobetasol propionate used as an anti-inflammatory and immunosuppressive agent.

Corn Steep Water A concentrated liquid obtained by steeping corn grains in water containing 0.2% SO_2 at temperature 46 to 50°C.

Coronary Angioplasty A procedure in which a segment of coronary artery narrowed by atheroma is stretched by the inflation of a balloon introduced into it by means of a cardiac catheterization under X-ray screening.

Coronary Bypass Graft Coronary revascularization in which a segment of a coronary artery narrowed by atheroma is bypassed by an autologous section of healthy saphenous vein.

Coronary Revascularization A surgical procedure of improving the blood flow through coronary arteries narrowed by atheroma.

Coronary Thrombosis The blockage of the coronary artery of the heart by a thrombus.

Coronavirus A virus of the family Coronaviridae capable of causing breathing disorders.

Coroxon (mol wt 347) A cholinesterase inhibitor and an anthelmintic agent.

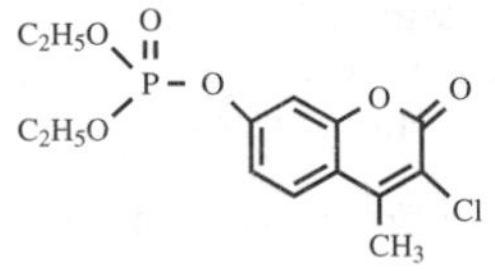

Corpuscle 1. Any cell of the body not forming continuous tissue. 2. A cell capable of moving freely.

Correlation Coefficient A measure of the degree of correlation, a value of +1, –1, and 0 indicating perfect positive, perfect negative, and lack of correlation, respectively.

Correndonuclease An endonuclease involved in repairing damaged DNA.

Correndonuclease II A deoxyribonuclease from *E. coli*.

Corrin Ring System The ring structure of vitamin B_{12} in which a cobalt atom is chelated.

Corrinoid Any compound containing the corrin ring system.

Cortaid A trade name for hydrocortisone.

Cortalone A trade name for prednisolone, used as an anti-inflammatory agent.

Cortate A trade name for hydrocortisone cypionate, a hormone.

Cortef A trade name for hydrocortisone.

Cortenema A trade name for hydrocortisone retention enema used as an anti-inflammatory agent for treatment of ulcerative colitis.

Cortex 1. A structural layer inside the spore coat of a bacterial spore. 2. The outer primary tissue of the stem or root extending from the primary phloem to the epidermis. 3. The outer layer of an organ, e.g., cerebral cortex, adrenal cortex, etc.

Cortic Ear Drop A trade name for an ear drop solution containing chloroxylenol, pramoxine HCl, and hydrocortisone.

Corticaine A trade name for hydrocortisone.

Corticorelin Ovine Triflutate A trifluoroacetate salt of synthetic peptide used for the determination of pituitary corticotrophic responsiveness.

Corticostatin A bioactive peptide that inhibits ACTH-stimulated corticosteroid synthesis.

Corticosteroid Any of the adrenal cortex-derived steroid hormones, e.g., glucocorticoids.

Corticosterone (mol wt 346) A glucocorticoid derived from progesterone.

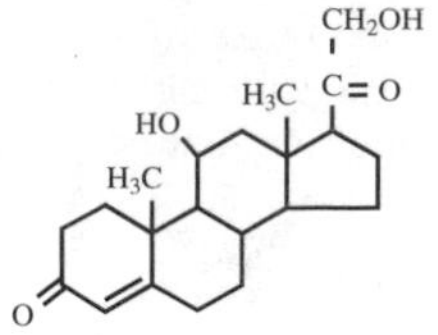

Corticotrophin Variant spelling of corticotropin.

Corticotropin See adrenocorticotropic hormone.

Corticotropin Receptor One of a number of membrane proteins that binds corticotropin.

Corticotropin Releasing Hormone The hypothalamic hormone that controls the secretion of corticotropin.

Corticoviridae A family of icosahydral, lipid-containing, nonenveloped bacteriophages.

Cortisol See hydrocortisone.

Cortisone (mol wt 360) A glucocortoid that regulates carbohydrate metabolism and suppresses normal immune responses; used as an anti-inflammatory agent.

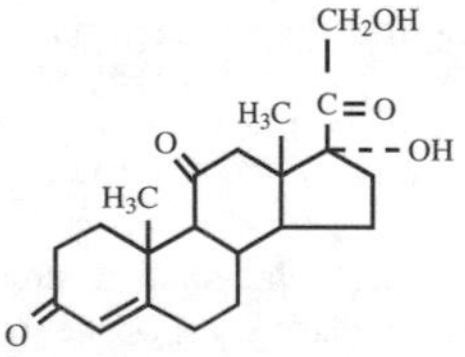

Cortisone α-Reductase The enzyme that catalyzes the following reaction:

$$4,5\text{-}\alpha\text{-dihydrocortisone} + NADP^{+} \rightleftharpoons \text{Cortisone} + NADPH$$

Cortisone β-Reductase The enzyme that catalyzes the following reaction:

$$4,5\text{-}\beta\text{-hydrocortisone} + NADP^{+} \rightleftharpoons \text{cortisone} + NADPH$$

Cortisporin Cream A trade name for a combination drug containing neomycin, polymyxin B sulfate, and hydrocortisone acetate used as an anti-infective and anti-inflammatory agent.

Cortivazol (mol wt 531) A glucocorticoid.

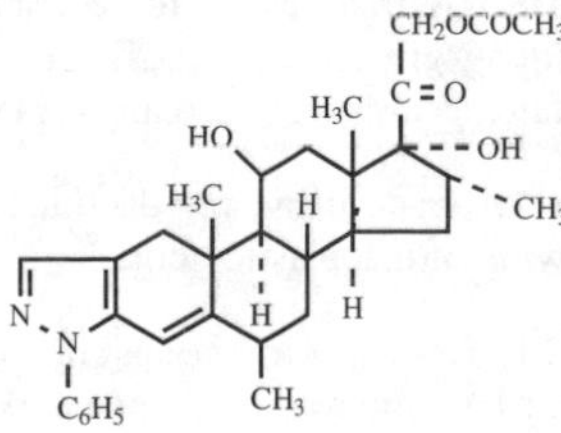

Cortizone A trade name for hydrocortisone, a hormone.

Cortone A trade name for cortisone.

Cortrosyn A trade name for cosyntropin, a synthetic peptide having similar biological activity as ATCH.

Corvert A trade name for ibutilide fumarate, an antiarrhythmic agent.

Corydaline (mol wt 369) An alkaloid from species of *Corydalis* (e.g., squirrel corn).

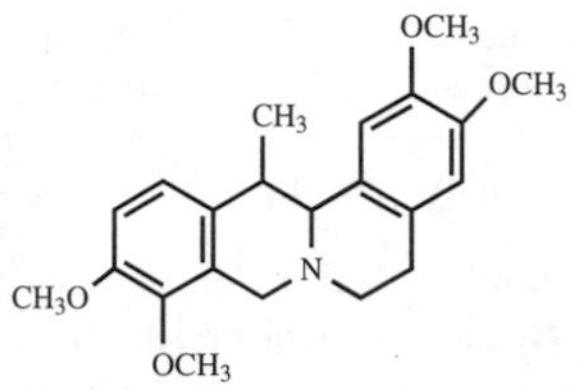

Corynebacterium A genus of gram positive, aerobic, nonmotile rod-shaped bacteria.

Corynecin Any antibiotic that is produced by the *Corynebacterium*.

Coryneform 1. A club-shaped structure. 2. Referring to Gram-positive, asporogenous, pleomorphic, rod-shaped bacteria.

Corynephage A bacteriophage that infects *Corynebacterium*.

Corynephage-B A bacteriophage of *Corynebacterium diphtheriae* responsible for the production of diptheriatoxin.

Coryza An inflammation of the mucous membranes of the nose.

Corzide A trade name for a combination drug containing nadolol and bendroflumethiazide.

COS Cell Any of a number of cell lines derived from monkey cells that contain an integrated segment of SV40 DNA coding for T antigen.

COS-1 Abbreviation for a type of cell line.

COS-7 Abbreviation for a type of cell line.

Cosamin DS A trade name for a combination drug containing glucosamine HCl, sodium chondroitin sulfate, ascorbate, and manganese.

Cosmegan A trade name for dactinomycin, an antibiotic with antineoplastic activity.

Cosmetic Referring to any preparation or procedure intended to improve the appearance of an individual.

Cosmic Rays The high-energy ionizing radiation that originates outside the earth's atmosphere; consists primarily of protons and other nuclei.

Cosmid A cloning vector derived from bacteriophage lambda used to carry large DNA fragments in and out of cells.

Cosopt A trade name for an ophthalmic solution containing dorzolamide hydrochloride and timolol maleate.

Cos-Site The single-stranded cohesive end of a linear double-stranded DNA, it is responsible for conversion of double-stranded linear DNA to double-stranded circular DNA.

Cosyntropin A peptide adrenocorticotropic hormone.

C_{ot} The concentration (mole/L) of single-stranded DNA in a DNA reassociation reaction multiplied by the time of incubation in seconds.

$C_{ot1/2}$ The C_{ot} required to proceed to half completion of the DNA reassociation reaction.

C_{ot} Curve The curve obtained by plotting log of C_{ot} versus percentage of reassociation.

C_{ot} Method A graphical method for evaluating the renaturation kinetics of denatured and fragmented double-stranded DNA.

C_{ot} Plot A plot of percentage of reassociation versus log C_{ot} in a DNA reassociation experiment.

Cotinine (mol wt 176) An oxidation product of nicotine containing antidepressant activity.

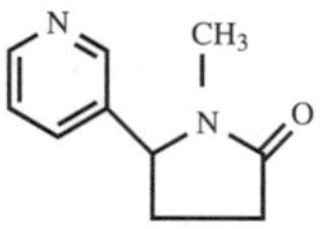

Cotransduction The transduction of two or more identifiable genes in a single event.

Cotransitional Transport Coupling of protein transport across a biological membrane with the synthesis of a peptide signal on a peptide to be transported.

Cotransport The simultaneous transport of two solutes by a single transporter.

Cotyledon A seed leaf; a food-digesting and storing part of a plant embryo.

Coulomb A unit of electrical quantity equal to the amount of charge transferred in one second by a steady current of one ampere.

Coulomb's Law The law states that the attraction or repulsion between two charged particles, bodies, or magnetic poles is proportional to the magnitude of their charges or pole lengths and inversely proportional to the sequence of their distance from each other.

Coulometer A voltmeter for measuring a quantity of electricity.

Coulometry A method of chemical analysis based on measurements of the quantity of electricity produced in a quantitative electrode reaction.

Coulter Counter An electric device that identifies, sorts, and counts cells in cell suspensions.

Coumadin A trade name for warfarin sodium, which inhibits the vitamin K-dependent activity of clotting factors (factor II, III, IX and X).

Coumaphos (mol wt 363) An insecticide.

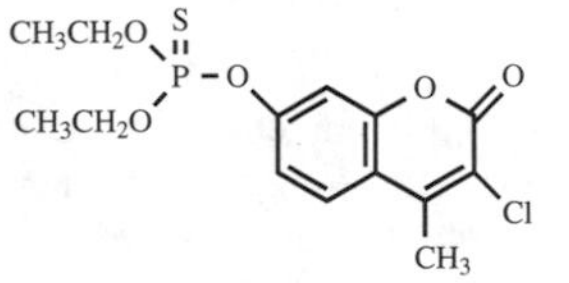

Coumetarol (mol wt 380) An anticoagulant.

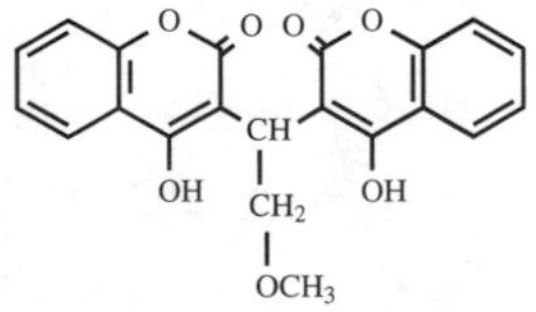

Coumithoate (mol wt 368) A cholinesterase inhibitor and insecticide.

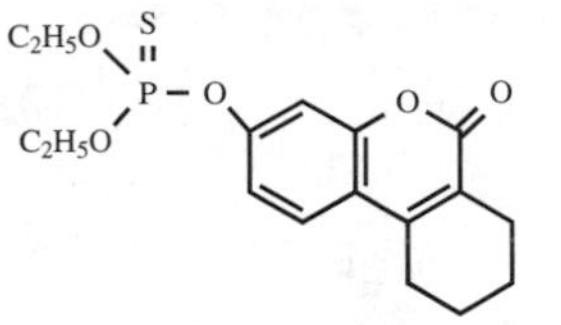

Counter Current Distribution A separation procedure based on solubility differences of compounds in two immiscible liquid phases. The compounds are partitioned repeatedly between the two immiscible phases as they move along a large number of partition tubes.

Counter Current Immunoelectrophoresis A technique based on the movement of an antibody and an antigen toward each other in an electric field resulting in the rapid formation of a detectable antigen-antibody precipitate.

Counter Stain The staining of a stained specimen with a second dye to highlight the background or reveal another type of cellular constituent.

Counting Efficiency The ratio of the number of registered radioactive counts to the number of actual radioactive disintegrations that occurred during the same time (generally multiplied by 100 to give percent efficiency).

Coupled Neutral Pump A coupled pump in which the movement of one ion across the membrane must be linked to the movement of another ion with opposite charge and equal valence.

Coupled Pump A transport mechanism that pumps one solute across a membrane that is coupled to the transport of a second solute in the opposite direction across the same membrane.

Coupled Reactions 1. Two chemical reactions that have a common intermediate through which energy can be transferred from one reaction to the other. 2. The linking of an exogonic and an endogonic reaction in which there is only a small net change in free energy.

Coupled Transport The transport of one solute across the membrane that is linked to the transport of a second solute across the same membrane in the opposite direction.

Coupling Factor Referring to the F_1 factor of the mitochondrial ATPase (the nontransmembrane head piece of the membrane-bound mitochondrial ATPase).

COUP-TF Abbreviation for chicken ovalbumin upstream promoter transcription factor.

COV Abbreviation for cross-over value.

Covalent Bond Chemical bond between two atoms formed by the sharing of a pair of electrons.

Covalent Catalysis Catalysis that requires the transient formation of catalyst-substrate covalent bonds.

Covalent Intermediate A transient substance formed during covalent catalysis.

Covera-HS A trade name for verapamil hydrochloride, a calcium channel blocker used as an antianginal, antihypertensive, and antiarrhythmic agent.

Cover-Glass Synonym for coverslip.

Coverslip A very thin piece of glass or plastic that is placed on top of a specimen for viewing of the specimen under a microscope.

Cowan I, II, and III Referring to the strains of *Staphylococcus aureus* used for producing standard typing antisera.

Cowdria A genus of Gram-negative bacteria of the tribe *Ehrlichieae*.

Cowper's Gland Either of the two pea-sized glands at the end of the male urinary canal.

Cowpox An avirulent strain of smallpox used for vaccination against small pox.

COX Abbreviation for cyclo-oxygenase.

COX-1 Abbreviation for constitutive cyclo-oxygenase.

COX-2 Abbreviation for inducible cyclo-oxygenase.

Coxackievirus A virus in the family of picornaviridae.

Coxiella A genus of Gram-negative bacteria of Rickettsiae.

Cozaar A trade name for losartan potassium, an angiotensin II receptor antagonist used as an antihypertensive agent.

Cozymase Referring to a heat stable fraction obtained from cell-free extracts from yeast; it is a mixture of cofactors of NAD, NADP, ATP, ADP, as well as a metal.

CP Abbreviation for 1. capillary pressure; 2. carbamoyl phosphate; 3. carboxypeptidase; 4. complement protein; 5. crude protein; 6. cyclophosphamide.

C1P Abbreviation for ceramide 1-phosphate.

C_p Symbol for excess heat capacity.

CP Buffer Abbreviation for citrate/phosphate buffer.

CPA Abbreviation for 1. carboxypeptidase A; 2. cyclopiazonic acid; 3. cyclopentyladenosine.

CpaI (MboI) A restriction endonuclease from *Clostridium pasteurianium* with the same specificity as MboI.

CPAE Cell Abbreviation for calf pulmonary artery endothelial cells.

CpaseY Synonym of serine-type carboxypeptidase.

C_3-Pathway A pathway for conversion of CO_2 to carbohydrate via the ribulose-1-5-bisphosphate carboxylase system (so called because CO_2 is first converted to a stable 3-carbon phosphoglycerate).

C_4-Pathway A pathway for conversion of CO_2 to organic compounds using phosphoenol-pyruvate carboxylase system (so called because CO_2 is first converted to a 4-carbon oxaloacetate).

CPB Abbreviation for 1. cetyl pyridinium bromide; 2. competitive protein binding; 3. a combination drug containing cytoxan, platinol, and BCNU.

CPBA Abbreviation for competitive protein binding assay.

CPC Abbreviation for cetylpyridinium chloride.

CPD Abbreviation for citrate/phosphate/dextrose.

cpDNA Abbreviation for chloroplast DNA.

CPE Abbreviation for cytopathic effects, i.e., histological evidence resulting from infection of a cell monolayer by a virus.

CpeI (BclI) A restriction endonuclease from *Corynebacterium Petrophilum* with the same specificity as BclI.

CpfI (MboI) A restriction endonuclease from *Clostridium perfringens* with the same specificity as MboI.

CpG Island An unmethylated genome region about 0.5 to 1.0 kbp long at the 5′ end of genes in which CpG is frequently observed.

CP-H Abbreviation for carboxypeptidase H.

CP-HPLC Abbreviation for chiral phase HPLC.

CPK Abbreviation for creatine phosphokinase.

CPK BB Abbreviation for BB isoform of creatine phosphokinase.

CPK Isoenzymes Multiple molecular forms of creatine phosphokinae. The increase of a specific form of CPK isoenzyme in the blood indicates a clinical disorder (e.g., heart disease).

cPKC Abbreviation for conventional or classical PKC.

CPK-MB Abbreviation for MB isoform of creatine phosphokinase.

$cPLA_2$ Abbreviation for cytosolic phospholipase A_2.

C_3-Plant Plants that possess C_3 pathway but no C_4 pathway.

C_4-Plant Plants that possess vascular bundle sheath cells in both C_3 and C_4 pathways.

CPLIM Abbreviation for cysteine-peptone-liver infusion medium.

cpm Abbreviation for counts per minute, e.g., the number of radioactive counts per minute.

CPM Abbreviation for a combination drug containing CCNU, procarbazine, and methotrexate.

CPO Abbreviation for chloro-peroxidase.

CpoI A restriction endonuclease from *Caseobacter polymorphus* with the following specificity:

5'........CG↓G(A/T)CCG........3'
'........GCC(T/A)GG↑C........5'

CPOB Abbreviation for a combination drug containing cytoxan, prednisone, oncovin, and bleomycin.

CPPD Abbreviation for calcium pyrophosphate dihydrate.

CPR Abbreviation for 1. chlorophenyl red; 2. cardiopulmonary resuscitation.

cPR (CPR) Abbreviation for chicken progesterone receptor.

C-Propeptide Abbreviation for C-terminal of propeptide.

CPS Abbreviation for carbamoyl phosphate synthetase.

CPS-1 Abbreviation for carbamoyl-phosphate synthetase-1.

CPSase Abbreviation for carbamoyl phosphate synthetase.

CPS-cAMP Abbreviation for 8-chlorophenylthio-cAMP.

CPT Abbreviation for 1. carnitine palmitoyl transferase; 2. choline phosphotransferase.

CPT-1 Abbreviation for carnitine palmitoyl transferase-1.

CPT-II Abbreviation for carnitine palmitoyltransferase-II.

CPT-cAMP Abbreviation for chlorophenylthio-cAMP.

CPY Abbreviation for carboxypeptidase Y.

CQ Abbreviation for chloroquine.

C1q Abbreviation for a subcomponent of C_1 (first component of complement).

CR Abbreviation for complement receptor on the cell surface, e.g., CR1, CR2, CR3, CR4, CR5 and CR6.

Cr-51 Assay A method for assaying cytotoxicity using radioactive chromium-51.

C1r Abbreviation for a subcomponent of C_1 (first component of complement).

CRABP Abbreviation for cellular retinoic acid binding protein.

Crabtree Effect The inhibition of oxygen consumption in cellular respiration and the occurrence of high levels of fermentative metabolism by increasing the concentration of glucose (see also Pasteur effect).

CRAC Abbreviation for calcium-release-activated calcium channel.

Crack A purified form of cocaine.

Craig Apparatus An apparatus used in countercurrent distribution experiment.

Crambin A protein from the plant seeds of *Crambe abyssinica*.

Craniopharyngioma A tumor on the pituitary gland.

Craniostenosis A birth defect of the skull.

Crasulacean Acid Metabolism A mechanism for storing CO_2 absorbed at night in C_4 plants via the C_4 pathway of photosynthesis during the daytime.

CRBP Abbreviation for cellular retinol-binding protein.

CRD Abbreviation for carbohydrate recognition domain.

9cRDH Abbreviation for 9-cis-retinol dehydrogenase.

CRE Abbreviation for 1. cAMP-response element; 2. Ca^{2+}/cAMP response element.

C-Reactive Protein (CRP) A beta-globulin found in the serum of patients with diverse inflammatory diseases; it reacts with pneumococcal Type C polysaccharide.

CRE/AP-1 Abbreviation for cAMP-response element/activator protein-1.

Creatinase The enzyme that catalyzes the following reaction:

$$\text{Creatine} + H_2O \rightleftharpoons \text{Urea} + \text{sarcosine}$$

Creatine (mol wt 131) A substance present in muscular tissue of vertebrates serving as a free energy storage compound.

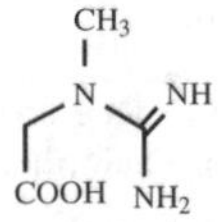

Creatine Amidinohydrolase See creatinase.

Creatine Kinase The enzyme that catalyzes the following reaction:

ATP + creatine ⇌ ADP + phosphocreatine

Creatine Phosphokinase (CPK) See creatine kinase.

Creatininase The enzyme that catalyzes the following reaction:

Creatinine + H_2O ⇌ Creatine

Creatinine (mol wt 113) The end product in creatine metabolism and a constituent of urine.

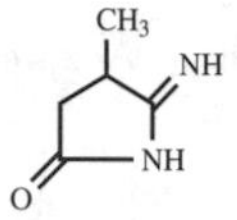

Creatinine Amidohydrolase The systematic name for creatinase.

Creatinine Clearance The rate of removal of endogenous or exogenous creatinine from the blood by the kidney.

Creatinine Coefficient The number of milligrams of creatinine excreted in 24 hours per kilogram of body weight.

Creatinine Deaminase The enzyme that catalyzes the following reaction:

Creatinine + H_2O
⇅
N-methylhydantoin + NH_3

Creatinine Deiminase Synonym of creatinine deaminase.

Creatinine Iminohydrolase The systematic name for creatinine deaminase.

Creatinura The presence of excessive amounts of creatine in the urine.

Creatorrhea The passage of excessive nitrogen in the feces due to the failure of digestion or absorption in the small intestine.

CREBP Abbreviation for cAMP-response- element-binding protein.

C-Region (Constant Region) The C-terminal portion of the H or L chain that is identical within immunoglobulin molecules of a given class and subclass apart from genetic polymorphisms.

Crenation The shrinking of a cell due to loss of water, e.g., red blood cells in a hypertonic solution (also called plasmolysis).

Crenothrix A genus of iron bacteria.

Creon A trade name for pancrelipase, a digestive enzyme.

Crescent Referring to the mature crescent-shaped gametes formed by some species of *Plasmodium.*

Cresol (mol wt 108) A phenolic compound with disinfectant activity. It consists of three isomers.

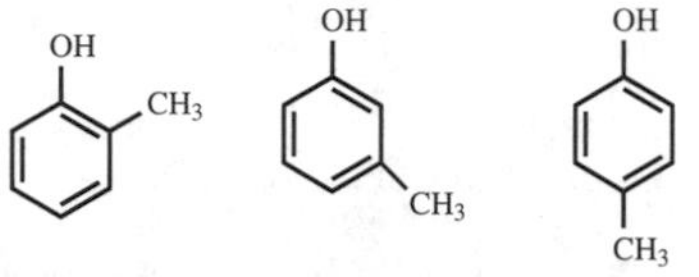

o-cresol *m*-cresol *p*-cresol

4-Cresol Dehydrogenase The enzyme that catalyzes the following reaction:

4-Cresol + acceptor
⇅
4-Hydroxybenzaldehyde + reduced acceptor

Cresol Dye (mol wt 382) An indicator dye.

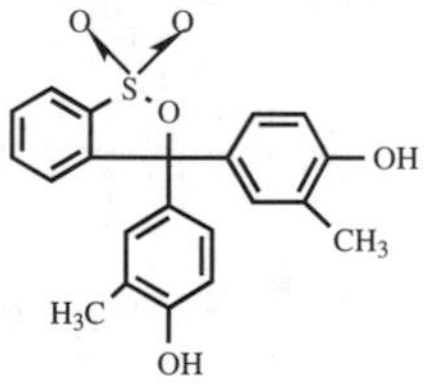

Cresol Red (mol wt 382) A dye and pH indicator.

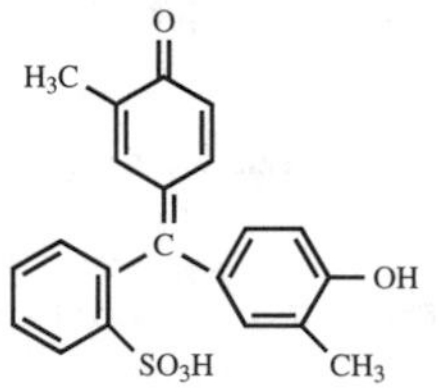

Cresolase See monophenol monooxygnase.

Cretinism A congenital condition characterized by the arrested mental and physical development due to the lack of thyroid hormone.

Creutzfeldt-Jakob Disease A human spongiform encephalopathy disorder caused by prion.

CRF Abbreviation for corticotropin releasing factor (see also corticotropin releasing hormone).

CRH Abbreviation for corticotropin releasing hormone.

Crigler-Najjar Syndrome An inherited metabolic defect characterized by jaundice and CNS disorders, due to a deficiency of the enzyme glucuronyl transferase in the liver.

Crinone A trade name for progesterone, a hormone.

Crista 1. The infolding structure of the inner membrane of the mitochondrion. 2. A sensory structure in the inner ear.

Cristae Plural of crista.

Crit Abbreviation for hemacrit.

Critical Micelle Concentration (CMC) The concentration of a surface-active substance that determines the formation of micelles by this substance. At a concentration above CMC, the added substance forms micelles.

Critical Point Drying A method for removing liquids from a specimen in scanning electron microscopy by adjusting the temperature and pressure so that the liquid and gas phases of the liquid are in equilibrium with each other for the purpose of minimizing the disruption of biological structures.

Crixivan A trade name for indinavir sulfate, an antiviral agent.

Crk A protooncogen of avian sarcoma viruses CT10. The name derives from CT10 regulator of kinase.

CRLR Abbreviation for calcitonin-receptor-like receptor.

CRM 197 A nontoxic mutant protein related to diphtheria toxin.

CRM-A Abbreviation for cytokine-response modifier-A.

cRNA Abbreviation for complementary RNA.

Cro Protein A regulatory protein in lambda phage life cycle that blocks the synthesis of lambda repressor.

CRO Abbreviation for cathode ray oscillograph.

Cromoglicic A trade name for cromolyn sodium, an anti-allergic agent.

Cromolyn (mol wt 468) A prophylactic antiasthmatic and antiallergic agent.

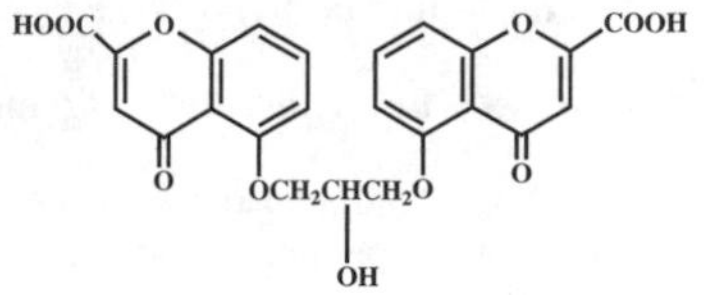

Cropropamide (mol wt 240) An analgesic agent.

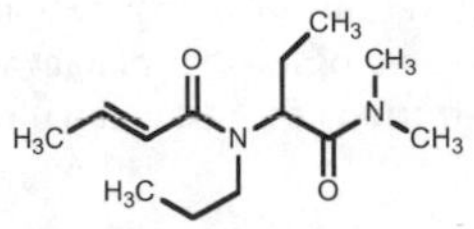

Cross Agglutination Test A serological test in which erythrocytes from an unknown individual are mixed with serum of a known blood type.

Cross-Bridge One of the many protein projections extending from each end of the thick myofilaments in the fibrils within a muscle cell.

Crossed-Immunoelectrophoresis A two dimensional electrophoresis in which antigens are first separated by gel electrophoresis, and the separated antigens are then electrophoresed in gel containing antibodies at right angle to the first separation.

Cross-Hybridization The hybridization of a polynucleotide probe to another polynucleotide molecule.

Cross-Infection The infection of a bacterium by two or more different phage mutants.

Cross-Linking The formation of covalent bonds among polymeric chains.

Cross-Matching of Blood Matching of blood types of donor and recipient through hemagglutination testing by mixing erythrocytes of the donor with serum of the recipient.

Cross-Over The exchange of genetic material between homologous chromosomes.

Cross-Over Junction Endonuclease The enzyme that catalyzes the endonucleolytic cleavage at the junction between two homologous DNA duplexes.

Crossover-Nuclease A deoxyribonuclease involved in the process of DNA recombination.

Cross-Over Unit The distance in terms of the probability of crossing over occurring between two linked loci on a chromosome. One crossing over unit is the crossing over value of 1% between a pair of linked genes.

Cross-Reacting Antibody An antibody that can combine with antigens that are specific for different antibodies.

Cross-Reacting Antigen An antigen that can combine with antibodies that are produced against different antigens.

Cross-Reaction The reaction of an antibody with an antigen other than the one that induced its formation.

Cross-Reactivation The reappearance of activity in the progeny of a lethal mutant virus following mixed infection of a host cell with one or more active viruses.

Cross-Reactive Antibodies Antibodies capable of combining with nonhomologous antigens or antigens which are not involved in elicitation of antibody production.

Cross-Resistance The resistance of a bacterium to an antibiotic that is associated with its resistance to one or more antibiotics.

Cross-Sensitization The sensitization of an organism with an antigen that is different from the antigen that will be used subsequently to trigger an anaphylactic response.

Cross-Tolerance 1. The tolerance toward one antigen resulting from the administration of a cross-reacting antigen. 2. The decrease of effectiveness of a drug due to the exposure to another drug.

Crotalase A protease from the diamond-back rattlesnake that hydrolyzes peptide bonds involving COOH groups of arginine in fibrinogen to form fibrin.

Crotamine A polypeptide from the venom of the Brazilian rattlesnake.

Crotamiton (mol wt 203) An antipruritic agent and insecticide (scabicide).

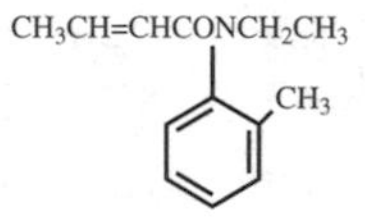

Crotethamide (mol wt 226) An analgesic agent.

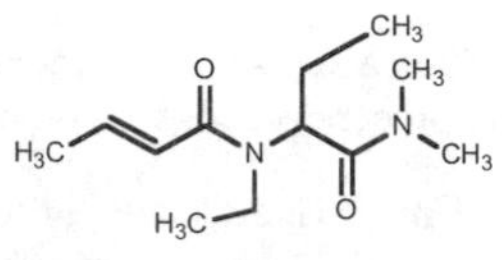

Croton Oil Oil expressed from seeds of *Croton tiglium* (Euphorbiaceae).

Crotonase Synonym of enoyl-CoA hydratase.

Crotonic Acid (mol wt 86) An unsaturated alphatic acid found in croton oil.

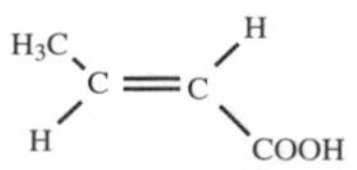

Crotoxin A polypeptide neurotoxin found in rattlesnake venom.

Crotoxyphos (mol wt 314) An insecticide that functions as a cholinesterase inhibitor.

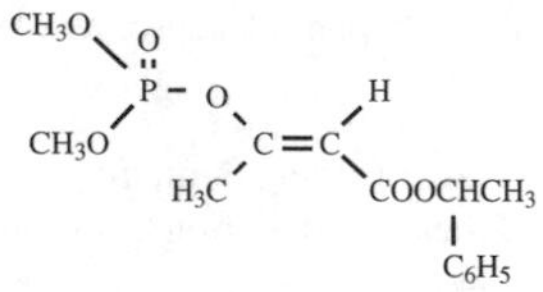

Crown Gall A neoplastic plant tumor caused by the infection of *Agrobacterium tumefaciens.*

CRP Abbreviation for 1. C-reactive protein. 2. Cyclic AMP receptor protein.

CRPA Abbreviation for C-reactive protein antiserum.

CRPF Abbreviation for chloroquine-resistant *Plasmodium falciparum.*

CRP-XL Abbreviation for collagen-related- peptide-cross-linked.

CRS Abbreviation for codon-recognizing site.

CRT Abbreviation for calreticulin.

Crt Abbreviation for cerotoyl group.

Crucible A vessel used for melting and calcining a substance that requires a high degree of heat.

Cruciform 1. A cross-shaped structure. 2. The cross-shaped structure formed in DNA due to the presence and pairing of the inverted repeats.

Crude Extract An extract resulting from centrifugation of homogenates of living cells and tissues.

Cruex A trade name for a combination drug containing undecylenic acid and zinc undecylenate, used as a local anti-infective agent.

Cruzipain A cysteine proteinase that hydrolyses chromogenic peptides at the carboxyl arginine or lysine residue.

Cry-Ab Abbreviation for cryptococcal antibody.

Cry-Ag Abbreviation for cryptococcal antigen.

Cryo- A prefix meaning cold or low temperature.

Cryobiology The science that deals with the effect of low temperature on living systems.

Cryoenzymology Low temperature enzymology.

Cryogen 1. A substance capable of causing freezing or creating low temperature. 2. A freezing mixture.

Cryogenic 1. Pertaining to cold or low temperatures. 2. Capable of producing low temperature.

Cryoglobinemia A disorder characterized by the presence of large amounts of cryoglobulins in the blood (e.g., in multiple myeloma patient).

Cryoglobulin A protein with the property of forming a precipitate or gel at cold temperatures.

Cryophilic Thriving at low temperatures.

Cryoprecipitagogue A substance capable of inducing the formation of cryoprecipitates.

Cryoprecipitate Precipitate formed from a solution at low temperature.

Cryopreservation A method of storage of living organisms at low temperature, e.g., use of liquid nitrogen.

Cryoprotectant Substance added to the suspensions of cells prior to cryopreservation to enhance the survival of cells.

Cryoprotein A protein that can be precipitated from a solution by lowering the temperature.

Cryoscope An instrument used for the determination of freezing points.

Cryoscopic Method The determination of either molecular weight or osmotic pressure from the depression of the freezing point of a solvent resulting from the addition of a solute.

Cryosolvent Any mixture of water and organic polar solvent used to maintain the proteins at temperature below 0°C.

Cryostat A device for maintaining constant low temperature for sectioning of frozen tissue by a microtome.

Cryosublimation The process whereby water is sublimed and removed from a frozen sample.

Cryotome A modified microtome that maintains the specimen in a frozen state during the process of sectioning.

Cryoxicide A sterilant that contains ethylene oxide (11%), trichlorofluoromethane (79%), and dichlorodifluoromethane (10%).

Cryptic DNA DNA with unknown function.

Cryptic Gene A gene that has been silenced by a single nucleotide substitution.

Cryptic Mutant A mutant that lacks one or more components of a transport system and is incapable of transporting substrate across a membrane.

Cryptic Plasmid A plasmid that has no apparent effect on the phenotypic expression of the host cell.

Cryptic Species Phenotypically similar species that do not form hybrids in nature.

Crypto- A prefix denoting secret, hidden, concealed, or unrecognized.

Cryptochrome A protein involved in circadian photoreception in *Drosophila*.

Cryptococcosis An infection caused by the fungus *Cryptococcus neoformans*.

Cryptococcus A genus of nonfermentative imperfect yeast (Hyphomycetes).

Cryptoendomitosis Somatic polyploidization that occurs within an intact nuclear envelope.

Cryptogam A plant that does not produce a true flower or seed.

Cryptogram A viral classification scheme using physical characteristics of viral nucleic acid, virion morphology, host, and vector as classification criteria.

Cryptosporidiosis An infection caused by a species of *Crytosporidium*.

Cryptosporidium A genus of parasitic protozoa.

Cryptotope A hidden immunological determinant.

Cryptoxanthin (mol wt 553) A carrotenoid pigment with vitamin A activity.

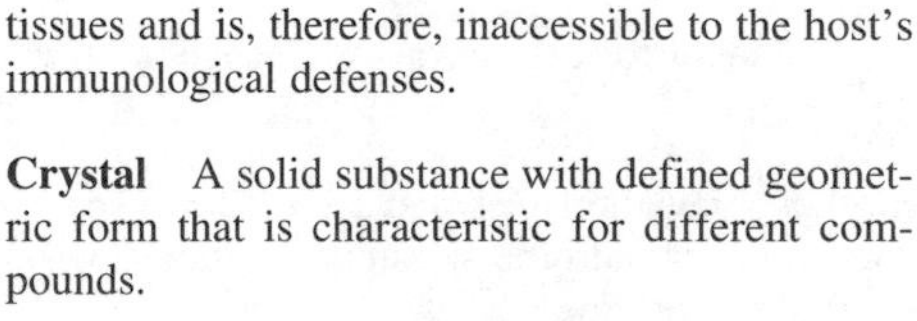

Cryptozoite Stage in the life cycle of *Plasmodium* in which the parasite is within the host's tissues and is, therefore, inaccessible to the host's immunological defenses.

Crystal A solid substance with defined geometric form that is characteristic for different compounds.

Crystal Lattice The three-dimensional arrangement of the atoms in a crystal.

Crystal Scarlet (mol wt 502) A dye.

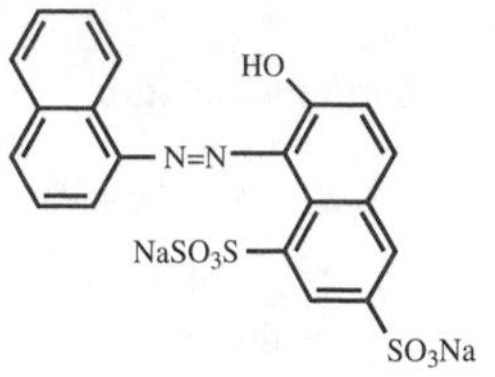

Crystal Violet (mol wt 408) A dye.

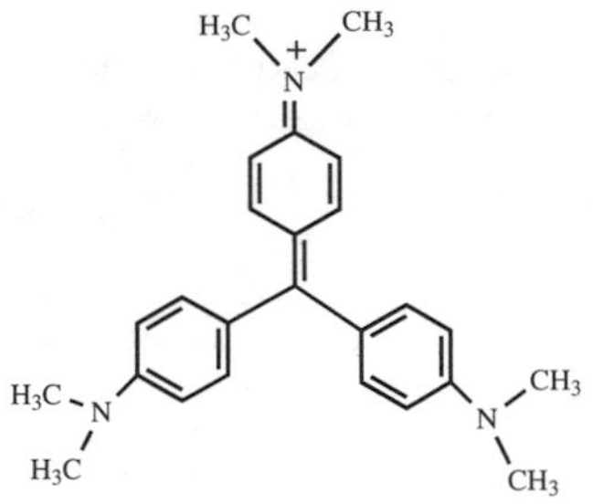

Crystallin Structural proteins of the lens of the vertebrate eye.

Crystallizable Fraction Referring to the Fc fragment of an immunoglobulin.

Crystallization The joining of molecules or ions from a liquid state to form a solid state.

Crystallographic Model A molecular model in which the bond lengths and the bond angles are specified.

Crystallography The study of the geometric forms of crystals.

Crystalloid A substance that forms a true solution and is capable of being crystallized.

Crystalluria The presence of crystals in the urine.

Crystamine A trade name for vitamin B_{12}.

Crystapen A trade name for penicillin G potassium, an antibiotic that inhibits bacterial cell wall synthesis.

Crysticillin AS A trade name for penicillin G procaine, an antibiotic that inhibits bacterial cell wall synthesis.

Crystodigin A trade name for digitoxin which inhibits sodium-potassium activated ATPase.

Cs Symbol for cesium (atomic weight 133, valence 1).

CS Abbreviation for 1. calf serum; 2. citrate synthetase; 3. chondroitin sulfate; 4. corticoid sensitive; 5. cycloserine.

^{137}Cs Abbreviation for radioactive cesium.

C_{1s} Abbreviation for a subcomponent of C_1 complement.

C4S Abbreviation for chondro-4-sulfatase.

C6S Abbreviation for chondro-6-sulfatase.

C3S Mutation Abbreviation for a mutation in which cysteine in position 3 is replaced by serine.

CS-I Abbreviation for citrate synthetase isoenzyme-I.

CS-II Abbreviation for citrate synthetase isoenzyme-II.

Cs-A Abbreviation for cyclosporin-A.

CSA Abbreviation for 1. Cockayne syndrome group A; 2. colon-specific antigen; 3. cyclosporin-A; 3. chondroitin sulfate A.

CSAID Abbreviation for cytokine suppressive anti-inflammatory drug.

CSAP Abbreviation for colon-specific antigen protein.

CSB Abbreviation for 1. caffeine/sodium/benzoate; 2. Cockayne syndrome group B.

CscI (SacII) A restriction endonuclease from *Calothrix scopulorum* with the following specificity:

CSD Abbreviation for cold-shock domain.

CSF Abbreviation for 1. cerebrospinal fluid. 2. colony-stimulating factor.

CSF-1 Abbreviation for colony-stimulating factor-1.

CsG Abbreviation for cyclosporin-G.

CSGE Abbreviation for conformation-sensitive gel electrophoresis.

CsH Abbreviation for cyclosporin-H.

CSIF Abbreviation for cytokine synthesis inhibitory factor.

CSL Abbreviation for cardiolipin synthetic lecithin.

CSM Abbreviation for 1. chorionic somatomammotropin; 2. crude synaptic membrane.

CSP Abbreviation for 1. cell surface protein; 2. cold-shock protein; 3. cysteine-string protein.

CspI A restriction endonuclease from *Clostridium* species with the following specificity:

Csp2I (HaeIII) A restriction endonuclease from *Corynebacterium* species with the same specificity as Hae III.

Csp6I A restriction endonuclease from *Corynebacterium* species RFL6 with the following specificity:

```
      ↓
5'........GTAC........3'
3'........CATG........5'
           ↑
```

Csp45I A restriction endonuclease from *Clostridium sporogenes* with the following specificity:

```
        ↓
5'........TTCGAA........3'
3'........AAGCTT........5'
            ↑
```

CspAI A restriction endonuclease from *Chromatium vinosum* with the following specificity:

```
        ↓
5'........ACCGGT........3'
3'......  TGGCCA........5'
              ↑
```

c-src A cellular oncogen present in various vertebrae. It hybridizes with *src* oncogens of Rous sarcoma viruses.

CSV Abbreviation for crude synaptic vesicle.

CT Abbreviation for 1. calcitonin; 2. cholera toxin; 3. chymotrypsin; 4. computed tomography or computerized tomography; 5. connective tissue; 6. Coombs' test.

CT Scan Abbreviation for computerized tomographic scan.

CT1 Abbreviation for cardiac triadin-1.

CTA Abbreviation for 1. cyproterone acetate; 2. cytotoxicity assay.

CTAB Abbreviation for cetyltrimethylammonium bromide, a precipitant for nucleic acids and mucopolysaccharides.

CTAP Abbreviation for connective tissue activating peptides.

CTC Abbreviation for chlorotetracycline.

CTC Solution A solution of copper tartrate/carbonate.

ctDNA Referring to chloroplast DNA.

C-Telopeptide Abbreviation for C-terminal telopeptide.

CTEM Abbreviation for conventional transmission electron microscope.

C-Terminal Abbreviation for carboxyl terminal of a protein or peptide.

C-Terminal Amino Acid The amino acid on the C-terminus of a peptide that carries a free carboxyl group. It is the last amino acid added to a growing peptide chain.

C-Terminus The carboxyl-terminal end of a protein molecule.

CTF Abbreviation for 1. Colorado tick fever; 2. cytotoxic factor.

CTFE Abbreviation for chlorotrifluoroethylene.

CTGF Abbreviation for connective tissue growth factor.

CTH Abbreviation for ceramide trihexoside.

CthI (BclI) A restriction endonuclease from *Clostridium thermocellum* with the following specificity:

```
       ↓
5′..........TGATCA..........3′
3′..........ACTAGT..........5′
                ↑
```

CthII (EcoRII) A restriction endonuclease from *Clostridium thermocellum* with the same specificity as EcorRII.

CTL Abbreviation for cytotoxic lymphocytes.

CTP (mol wt 325) Abbreviation for 1. cytidine triphosphate. 2. cytidine-5′-triphosphate.

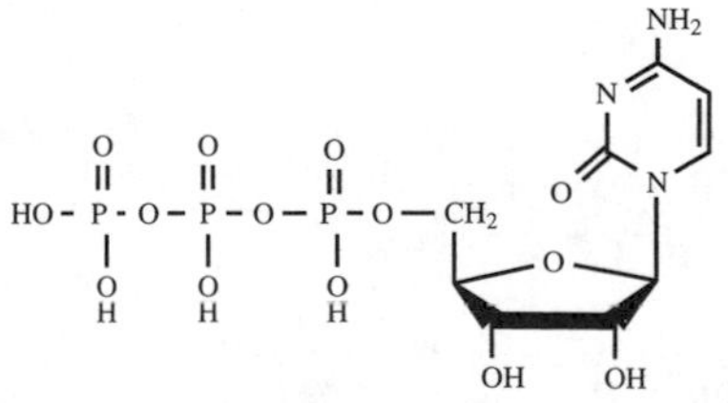

cytidine 5′ triphosphate

CTP Synthetase The enzyme that catalyzes the following reaction:

$$UTP + ATP + NH_3 \rightleftharpoons ADP + CTP + Pi$$

CTP-³H Abbreviation for tritium-labeled CTP (cytidine triphosphate).

CTS Abbreviation for computerized tomographic scanner.

CT-Scan Abbreviation for computed tomographic scan or computerized tomographic scan.

CTSH Abbreviation for chorionic thyroid-stimulating hormone.

CTU Abbreviation for centigrade thermal unit.

C-Type Particle A type of retroviral particle with a centrally located, spherical RNA-containing nucleoid. C-type viral particles are associated with many sarcomas and leukemias.

CTX Abbreviation for 1. cefotaxime; 2. cytoxan (cyclophosphamide); 3. cholera toxin.

CTZ Abbreviation for chemo-receptor trigger zone.

Cu Symbol for copper (atomic weight 64, valence 1).

^{61}Cu Abbreviation for radioactive copper.

^{64}Cu Abbreviation for radioactive copper.

CUA A genetic code or codon for the amino acid leucine.

Cubic Symmetry A rotational symmetry used for the description of spherical-shaped virions.

CUC A genetic code or codon for the amino acid leucine.

^{14}C-UCBR Abbreviation for radioactive carbon-labeled, unconjugated bilirubin.

Cucumisin A protease with broad specificity derived from musk melon (*Cucumis melo*).

Cucurbitine (mol wt 116) A naturally occurring nonprotein amino acid.

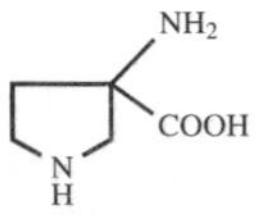

CUG A genetic code or codon for the amino acid leucine.

Culdocentesis Aspiration of pus or fluid from the rectouterine pouch through a transvaginal puncture.

Culdoscope A lighted device used for visual examination of the pelvic cavity and its contents.

Culex A genus of mosquito.

Culicide Any substance that kills mosquitoes.

Culicifuge Agent or substance that expels mosquitoes.

Cultivar A variety of plant produced through selective breeding by humans and maintained by cultivation.

Culture Cultivation of microbial cells or tissue cells in or on nutrient medium.

Cuminaldehyde Thiosemicarbazone (mol wt 221) An antiviral agent.

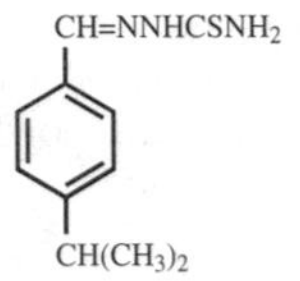

cUMP Abbreviation for cyclic UMP.

Cumulative Product Feedback Inhibition The feedback inhibition for an enzyme reaction resulting from the actions of two or more products produced separately and independently from branched pathways.

Cup Fungi Fungi that form cuplike fruiting bodies (e.g., some species of *Ascomycetes*).

Cuprammonium Rayon Cellulose that has been regenerated from a solution of cuprammonium hydroxide.

Cuprid A trade name for trientine hydrochloride, which chelates copper and increases urinary secretion.

Cuprimine A trade name for penicillamine, an anti-inflammatory agent.

Cupro-protein A copper-containing protein.

Cuproxoline (mol wt 965) An antirheumatic agent.

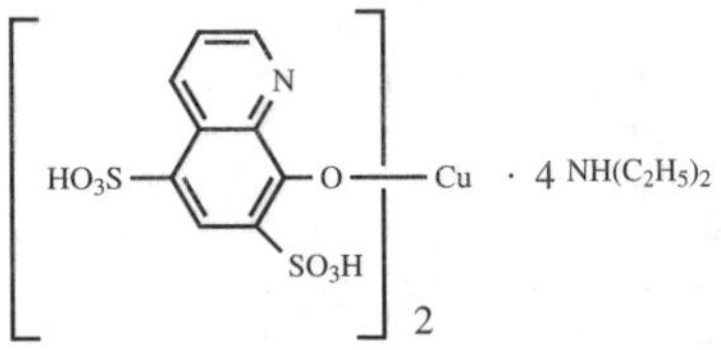

Curare A mixture of alkaloids from plants (*Strychnos toxifera*) that function as neurotoxins that block the transmission of nerve impulses.

Curdlan A polymer of glucose (α,1,3-glucan) found in *Alicaligenes faecalis* and *Agrobacterium* species. It forms a nonreversable gel upon heating.

Curie (Ci) A unit of radioactivity (1Ci = 3.7 × 10^{10} dps).

Curing 1. A method of food preservation by permeation with solution containing NaCl, $NaNO_2$, and $NaNO_3$. 2. Elimination of a plasmid without loss of bacterial viability.

Curium (Cm) A chemical element with atomic weight 247, valence 3 or 4.

Current Movement of electric charges form one point to another through a conductor.

Curtobacterium A genus of aerobic, asporogenous bacteria (Actinomycetales).

Cushing's Disease A metabolic disorder owing to the overproduction of adrenocorticotropin and characterized by the puffiness of the face, neck, trunk, and muscular weakness with impaired carbohydrate tolerance.

Cut and Patch A mechanism for repair of DNA by enzymatically removing incorrect nucleotides and substituting correct ones.

Cutaneous Pertaining to the skin.

Cutaneous Anaphylaxis A skin reaction produced by intradermal injection of allergen into allergic individuals.

Cutaneous Basophil Hypersensitivity An immunologically mediated inflammation of the skin with a prominent basophil infiltration after injection of the sensitizing antigen.

Cuticle A waxy protective layer on the aerial surface of plant tissues.

Cutln A heterogeneous polymer of fatty acids that forms cuticles covering the aerial parts of higher plants. It consists of two main fatty acids (16 and 18-carbon fatty acids).

Cutinase The enzyme that digests cutin.

CUU A genetic code or codon for the amino acid leucine.

Cuvette A small glass, plastic, or quartz vessel of specific dimension used to hold a sample for measurement of absorbance by spectrophotometry.

CuZnSOD Abbreviation for copper-zinc superoxide dismutase.

CV Abbreviation for 1. cardiovascular; 2. cell volume; 3. cerebrovascular; 4. cresyl violet; 5. crystal violet.

CV-1 An animal cell line derived from the kidney of a male adult of the African green monkey.

CVA Abbreviation for a combination drug containing cyclohexamide, vincristine, and adriamycin.

C-Value The amount of DNA (in picograms) per cell in the haploid genome of a species.

CVB Abbreviation for a combination drug containing CCNU, vinblastine, and bleomycin.

CVF Abbreviation for 1. a combination drug containing cytoxan, vincristine, and fluorouracil; 2. cobra venom factor.

CviAI (MboI) A restriction endonuclease from strain of *Chlorella* NC64A (PBCV-1) with the same specificity as MboI.

CviBI (HinfI) A restriction endonuclease from strain of *Chlorella* NC64A (NC-1A) with the following specificity:

↓
5′..........GANTC..........3′
3′..........CTNAG..........5′
↑

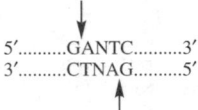

CViBII (MboI) A restriction endonuclease from strain of *Chlorella* NC64A (NC-1A) with the same specificity as MboI.

CviBIII (TaqI) A restriction endonuclease from strain of *Chlorella* NC64A (NC-1A) with the same specificity as TaqI.

CviCI (HinfI) A restriction endonuclease from strain of *Chlorella* NC64A (NE-8A) with the same specificity as HinfI.

CVID Abbreviation for common variable immunodeficiency.

CviDI (HinfI) A restriction endonuclease from strain of *Chlorella* NC64A (A1-2C) with the same specificity as HinfI.

CviEI (HinfI) A restriction endonuclease from the strain of *Chlorella* NC64A (MA-1E) with the same specificity as HinfI.

CviFI (HinfI) A restriction endonuclease from the strain of *Chlorella* NC64A (NY-2F) with the same specificity as HinfI.

CviHI (MboI) A restriction endonuclease from strain of *Chlorella* NC64A (NC-1C) with the same specificity as MboI.

CviJI A restriction endonuclease from strain of *Chlorella* NC64A (IL-3A) with the following specificity:

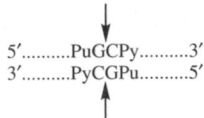

↓
5′..........PuGCPy..........3′
3′..........PyCGPu..........5′
↑

CviKI (CviJI) A restriction endonuclease from strain of *Chlorella* NC64A (CA-1A) with the same specificity as CviJI.

CviLI (CviJI) A restriction endonuclease from strain of *Chlorella* NC64A (CA-2A) with the same specificity as CviJI.

CviMI (CviJI) A restriction endonuclease from strain of *Chlorella* NC64A (IL-2A) with the same specificity as CviJI.

CviNI (CviJI) A restriction endonuclease from strain of *Chlorella* NC64A (Il-2B) with the same specificity as CviJI.

CviOI (CviJI) A restriction endonuclease from strain of *Chlorella* NC64A (IL-3D) with the same specificity as CviJI.

CviQI (RsaI) A restriction endonuclease from strain of *Chlorella* NC64A (NY-2A) with the following specificity:

```
       ↓
5′..........GTAC..........3′
3′..........CATG..........5′
         ↑
```

C-Virus Abbreviation for Coxsackie-virus.

CvnI (SauI) A restriction endonuclease from *Chromatium vinosum* with the following specificity:

```
          ↓
5′..........CCTNAGG..........3′
3′..........GGANTCC..........5′
               ↑
```

CVP Abbreviation for a combination drug containing cyclohexamide, vincristine, and prednisone.

CVPP Abbreviation for a combination drug containing CCNU, vinblastine, procarbazine, and prednisone.

CW Abbreviation for cell wall.

CWDB Abbreviation for cell wall deficiency bacteria.

CWP Abbreviation for cell wall-associated protein.

CX Abbreviation for 1. connexin; 2. cyclohexImide.

Cyacetacide (mol wt 99) An antibacterial agent.

$$NCCH_2CONHNH_2$$

Cyanabin A trade name for cyanocobalamin or vitamin B_{12}.

Cyanamide Hydratase The enzyme that catalyzes the following reaction:

$$\text{Urea} \rightleftharpoons \text{Cyanamide} + H_2O$$

Cyanate Hydratase The enzyme that catalyzes the following reaction:

$$\text{Formamide} \rightleftharpoons \text{Cyanate} + H_2O$$

Cyanate Hydrolase See cyanate lyase.

Cyanate Lyase The enzyme that catalyzes the following reaction:

$$\text{Cyanate} + \text{bicarbonate} \rightleftharpoons C_2O + \text{carbamate}$$

Cyanazine (mol wt 241) An herbicide.

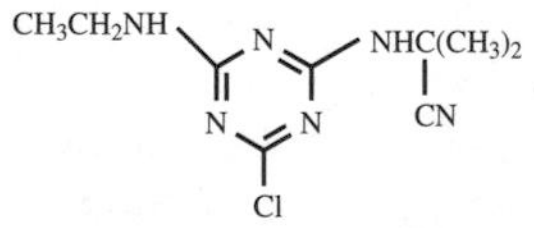

Cyanelles Endosymbiontic cyanobacteria found in various eukaryotes.

Cyanide Any compound that contains the CN^- group. It inhibits both oxidized and reduced forms of cytochrome oxidase at the final step of the electron transport chain.

Cyanide Test A test used to determine the ability of an organism to grow in the presence of cyanide. It is used to identify certain species of bacteria.

Cyanidin Chloride (mol wt 323) A substance used for treatment of night blindness.

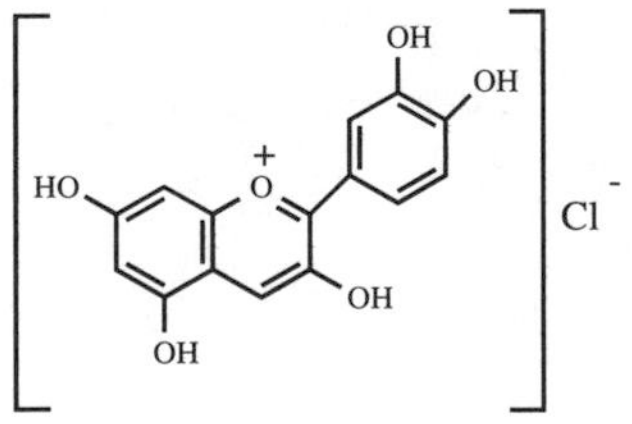

Cyano- A prefix denoting 1. the cyanide ion or hydrocyanic acid; 2. the –CN group in covalent linkage; 3. blue.

β-Cyanoalanine (mol wt 114) A naturally occurring nonprotein amino acid.

$$N{\equiv}CCH_2CH(NH_2)COOH$$

Cyanoalanine Hydratase The enzyme that catalyzes the following reaction:

$$\text{L-Asparagine} \rightleftharpoons \text{3-Cyanoalanine} + H_2O$$

Cyanoalanine Nitrilase The enzyme that catalyzes the following reaction:

$$\text{3-Cyano-L-alanine} + 2\,H_2O \rightleftharpoons \text{L-Asparate} + NH_3$$

Cyanobacteria (Blue-Green Algae) A large group of prokaryotic, photosynthetic organisms that contain chlorophyll (not bacterial chlorophyll) and carry out oxygenic photosynthesis.

Cyanocobalamin Synonym of vitamin B_{12}. A hematopoietic compound used to treat anemia.

Cyanocobalamin Reductase The enzyme that catalyzes the following reaction:

$$\text{NADPH} + 2\ \text{cob(III)alamin} \rightleftharpoons \text{NADP}^+ + 2\ \text{Cob(II)balamin}$$

Cyanofenphos (mol wt 303) An insecticide.

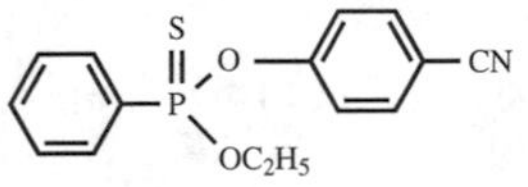

Cyano-Gel A trade name for cyanocobalamin or vitamin B_{12}.

Cyanogen (mol wt 52) A highly poisonous gas.

$$N{\equiv}C{-}C{\equiv}N$$

Cyanogen Bromide (mol wt 106) A reagent used in protein sequencing and immobilization of protein onto a support surface.

CNBr

Cyanogenesis The production or yield of cyanide ions or hydrocyanic acid.

Cyanoginosin A toxic cyclic heptapeptide.

Cyanohemoglobin A derivative of hemoglobin in which the sixth coordination position of iron is occupied by cyanide anion.

Cyanoject A trade name for cyanocobalamin or vitamin B_{12}.

Cyanophage Viruses that infect cyanobacteria.

Cyanophil Any cell or tissue element that can be readily stained with blue dye.

Cyanophilic Having an affinity for blue dye, e.g., lactophenol cotton blue.

Cyanophycean Starch A glycogen-like storage polysaccharide found in Cyanobacteria.

Cyanophycin A high molecular-weight linear polymer consisting of equal amounts of L-aspartic acid and L-arginine.

Cyanopsia A vision disorder in which all objects appear to be tinted blue.

Cyanopsin A photoreceptor protein found in the retinal cone cells of fresh-water and migratory fish and certain amphibians.

Cyanosis A bluish discoloration of the skin and mucous membranes due to insufficient oxygenation of the blood.

Cyanosome A phycobilisome of Cyanobacteria.

Cybrid A hybrid resulting from the fusion of a cell with the cytoplasm of another cell.

CyCAP Abbreviation for cyclophilin-C- associated protein.

Cycasin (mol wt 252) A toxic compound found in the seeds of *Cycas revoluta*.

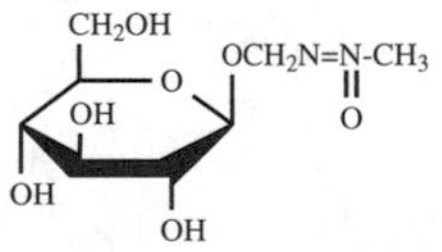

Cyclacillin (mol wt 341) A semisynthetic antibiotic related to penicillin.

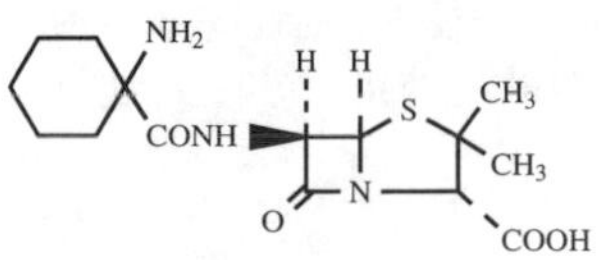

Cyclamic Acid (mol wt 179) A nonnutritive sweetener.

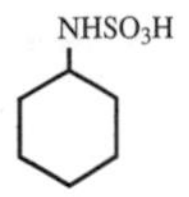

Cyclan A trade name for cyclandelate, a vasodilator that inhibits phosphodiesterase and increases cAMP concentration.

Cyclandelate (mol wt 276) A vasodilator that inhibits phosphodiesterase and increases cAMP concentration.

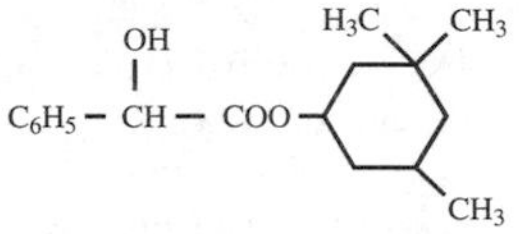

Cyclarbamate (mol wt 368) An anxiolytic agent and a muscle relaxant.

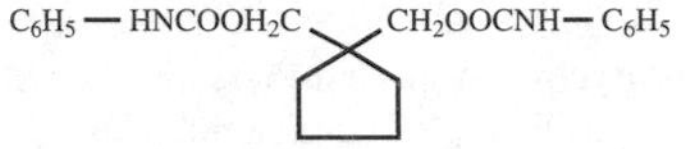

Cycle A time interval in which a regularly repeated sequence occurs.

Cyclethrin (mol wt 328) An insecticide for flies, roaches, and grain pests.

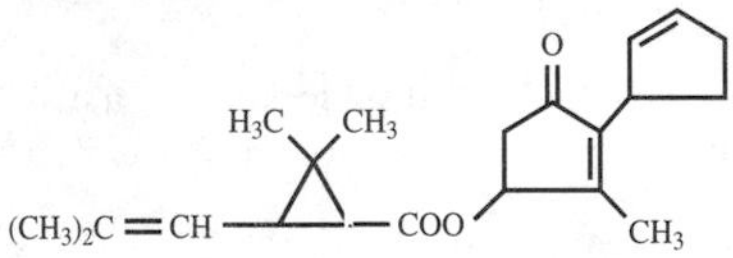

Cyclexanone (mol wt 263) An antitussive agent.

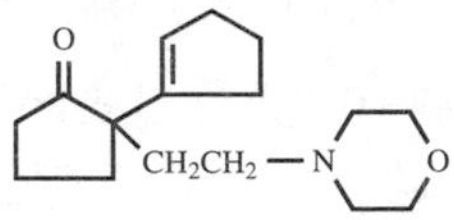

Cyclic Adenylic Acid See cyclic AMP.

Cyclical Transmission A mode of transmission of a parasite by a vector in which the parasite undergoes one or more essential stages of its life cycle in the vector.

Cyclic AMP (mol wt 329) An abbreviation for adenosine 3′,5′-monophosphate or adenosine 2′,3′-monophosphate. An important intracellular regulator or second messenger for a number of cellular processes in animals, bacteria, fungi, and plants.

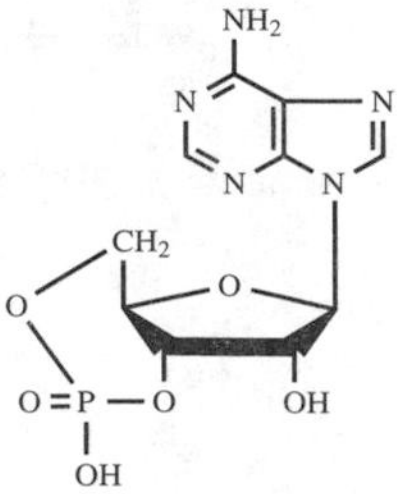

adenosine 3,5′-cyclic monophosphate

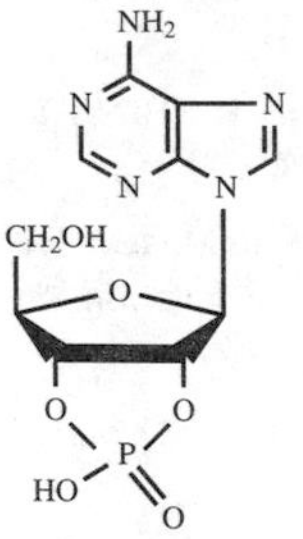

adenosine 2′,3′-cyclic monophosphate

3′,5′-Cyclic AMP Synthetase The enzyme that catalyzes the following reaction:

$$\text{ATP} \rightleftharpoons 3',5'\text{-Cyclic AMP} + \text{PP}$$

Cyclic CMP Abbreviation for cyclic cytidine monophosphate (either 2′,3′-cyclic cytidine monophosphate or 3′,5′-cyclic cytidine monophosphate).

2′,3′-Cyclic CMP (cCMP, mol wt 305) Abbreviation for 2′,3′-cyclic cytidine monophosphate, a cyclic nucleotide.

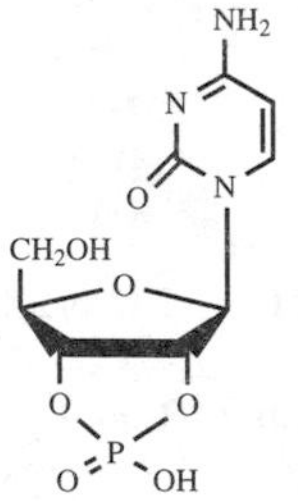

3′,5′-Cyclic CMP (cCMP, mol wt 305) Abbreviation for 3′,5′-cyclic cytidine monophosphate, a cyclic nucleotide.

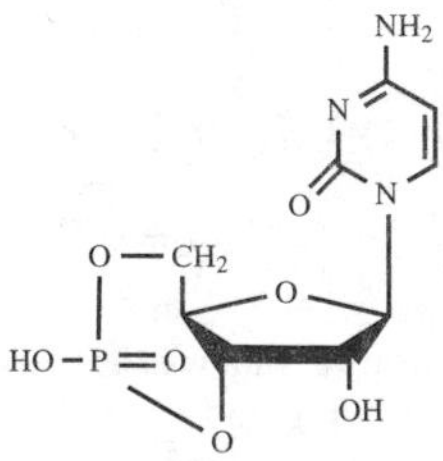

Cyclic CMP Synthetase The enzyme that catalyzes the following reaction:

$$\text{CTP} \rightleftharpoons \text{Cyclic CMP} + \text{PPi}$$

Cyclic Electron Flow The cyclic flow of electrons within photosystem I for generation of ATP (the electrons ejected from P_{700} returns to P_{700}).

Cyclic Fatty Acid Any of a class of fatty acids that contain a carbocylic unit such as manoic acid.

Cyclic GMP (cGMP, 345) Abbreviation for either 3′,5′-cyclic guanosine monophosphate or 2′,3′-cyclic guanosine monophosphate, an intracellular regulator.

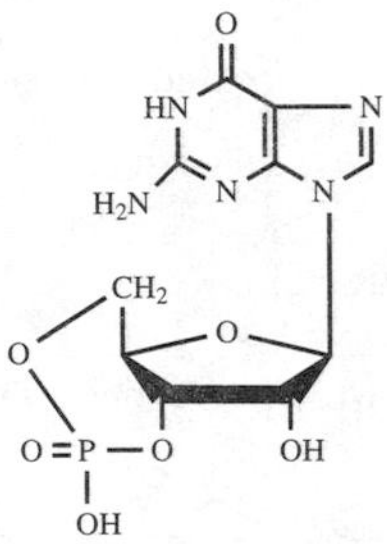

guanosine 3′,5′-cyclic monophosphate

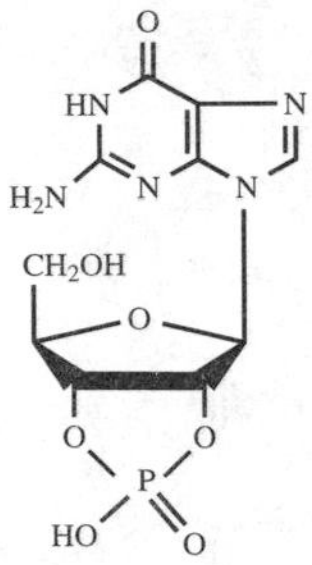

guanosine 2′,3′-cyclic monophosphate

3′,5′-Cyclic GMP Phosphodiesterase The enzyme that catalyzes the following reaction:

Guanosine 3′,5′-cyclic monophosphate + H_2O

$\rightleftharpoons$

Guanosine 5′-monophosphate

Cyclic GMP Synthetase The enzyme that catalyzes the following reaction:

GTP $\rightleftharpoons$ Cyclic GMP + PPi

3′,5′-Cyclic GMP Synthetase The enzyme that catalyzes the following reaction:

GTP $\rightleftharpoons$ 3′,5′-Cyclic GMP + PPi

2′,3′-Cyclic GMP Synthetase The enzyme that catalyzes the following reaction:

GTP $\rightleftharpoons$ 2′,3′-Cyclic GMP + PPi

2′,3′-Cyclic Nucleotide Phosphodiesterase The enzyme that catalyzes the conversion of 2′,3′-cyclic nucleotide to nucleotide monophosphate.

3′,5′-Cyclic Nucleotide Phosphodiesterase The enzyme that catalyzes the conversion of 3′,5′-cyclic nucleotide to nucleotide monophosphate.

Cyclic Oxidative Photophosphorylation See photophosphorylation or cyclic flow of electrons.

Cyclic Peptide A covalently linked circular polypeptide.

Cyclic Photophosphorylation The synthesis of ATP from cyclic flow of electrons in photosystem I without production of NADPH.

Cyclic UMP (mol wt 306) A cyclic form of uridylic acid.

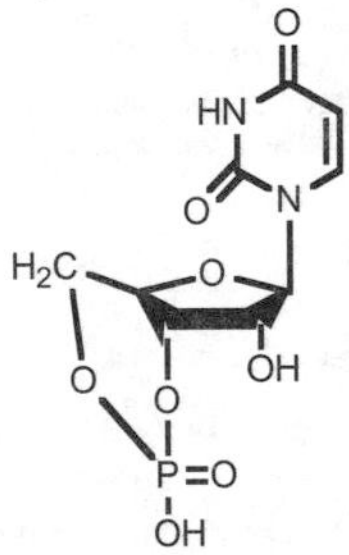

Cyclidox A trade name for doxycycline hydrochloride, an antibiotic that inhibits microbial protein synthesis by binding with 30S ribosome subunits.

Cyclin A protein that appears at various stages of the eukaryotic cell cycle; it activates protein kinase and controls the progression from one stage of the life cycle to another. Its concentration rises and falls in various stages of the eukaryotic cell cycle.

Cyclin-Dependent Protein Kinase The protein kinase that has to be complexed with cyclin in order to function. It involves triggering different steps in the cell-division cycle by phosphorylating specific target proteins.

Cylindroma A benign epithelium tumor.

Cyclindruria The presence of casts in the urine.

Cyclitol Antibiotic Any antibiotic that contains a cyclic alcohol.

Cyclizine (mol wt 266) An antiemetic agent.

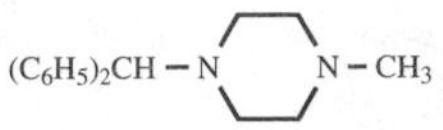

Cyclizine Lactate A derivative of cyclizine.

Cycloallin (mol wt 177) A naturally occurring nonprotein amino acid.

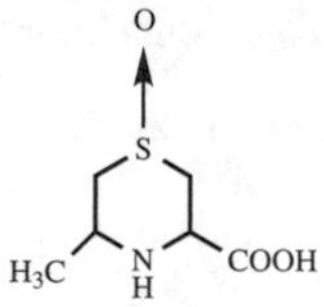

Cyclobarbital (mol wt 236) A sedative-hypnotic agent.

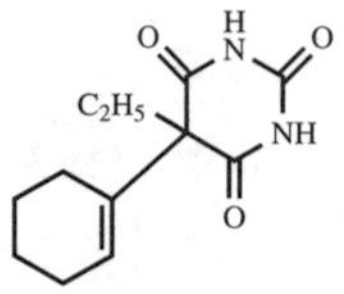

Cyclobenzaprine (mol wt 275) A skeletal muscle relaxant that reduces transmission of impulses from the spinal cord to skeletal muscle.

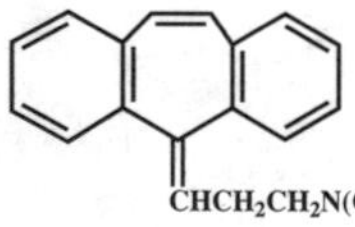

Cycloblastin A trade name for cyclophosphamide, an alkylating agent that cross-links strands of cellular DNA.

Cyclocort A trade name for amcinonide, a corticosteroid used as an anti-inflammatory agent.

Cyclocumarol (mol wt 322) An anticoaggulant.

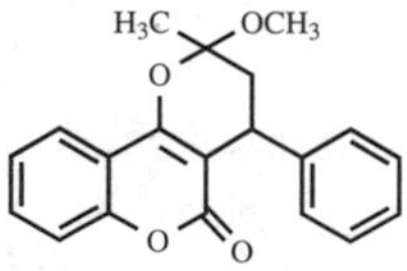

Cyclodextrins (Cycloamylose) A group of naturally occurring clathrates and products by the action of *Bacillus macerans* amylase on starch, e.g., α, β, γ-cyclodextrin.

α-Cyclodextrin (mol wt 972) A cylomaltohexaose (consisting of six glucose residues).

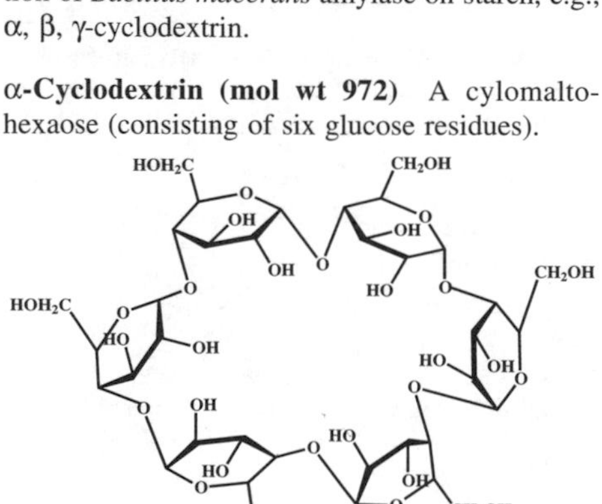

β-Cyclodextrin (mol wt 1135) A cyclomaltoheptaose (consisting of seven glucose residues).

γ-Cyclodextrin (mol wt 1297) A cyclomaltooctaose (consisting of eight glucose residues).

Cyclodextrin Glucanotransferase The enzyme that cyclizes part of a 1,4-α-D-glucan chain by formation of a 1,4-α-D-glucosidic linkage.

Cyclodextrin Hydrolase See cyclodextrinase.

Cyclodrine (mol wt 319) An anticholinergic agent.

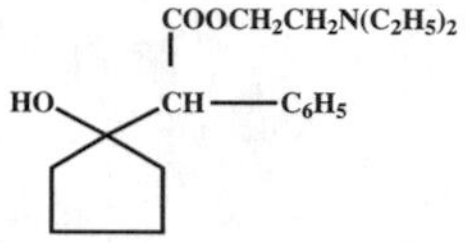

Cyclofenil (mol wt 364) A gonad-stimulating agent.

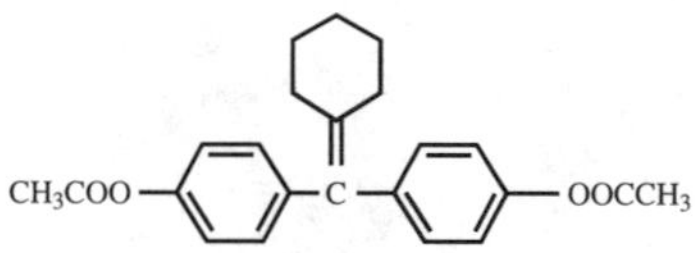

Cycloguanil (mol wt 252) An antimalarial agent.

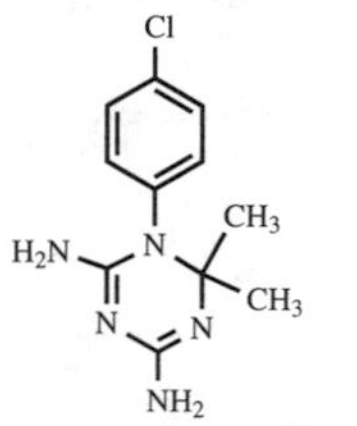

Cyclohexanone Monooxygenase The enzyme that catalyzes the following reaction:

$$\text{Cyclohexanone} + \text{NADPH} + O_2 \rightleftharpoons \text{6-hexanolide } \text{NADP}^+ + H_2O$$

Cyclohexanone NADPHS Oxygen Oxidoreductase The systematic name for cyclohexanone monooxygenase.

Cycloheximide (mol wt 281) An antibiotic from streptomycin-producing strains of *Streptomyces griseus.* It inhibits protein synthesis by preventing peptide bond formation on 80S ribosomes. It is also a plant growth regulator and a fungicide.

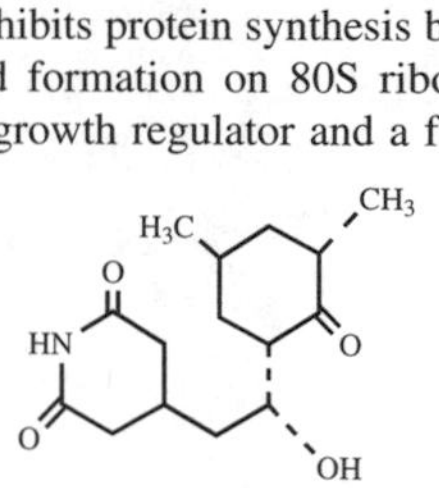

3-(Cyclohexylamino)-2-Hydroxypropanesulfonic Acid (mol wt 237) A reagent used for blotting of strongly basic proteins for transfer from SDS-polyacrylamide gels to nitrocellulose paper.

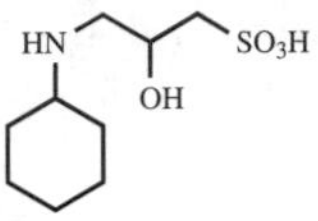

Cycloleucine (mol wt 129) A synthetic amino acid and valine antagonist.

Cyclomaltodextrinase The enzyme that catalyzes the following reaction:

Cyclodextrin + H_2O

⇅

Linear maltodextrin

Cyclomen A trade name for danazol, a steroid that suppresses ovulation.

Cyclomethycaine (mol wt 360) A topical anesthetic agent.

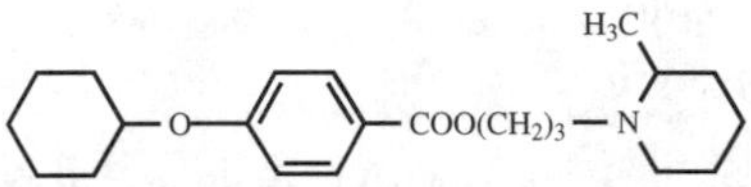

Cyclomydril Ophthalmic A trade name for a combination drug containing cyclopentolate and phenylephrine, used as an anticholinergic agent.

Cyclonium Iodide (mol wt 471) An antispasmodic agent.

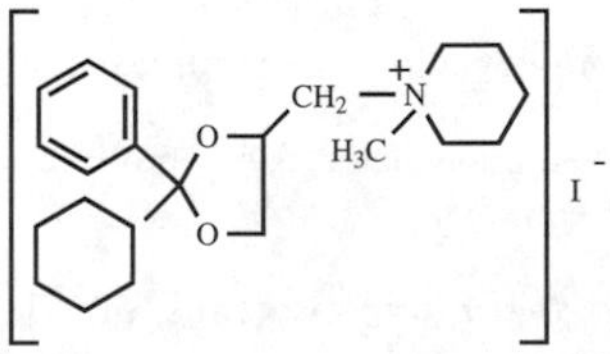

Cyclooxygenase Synonym of prostaglandin-endoperoxide synthetase.

Cyclooxygenase Pathway An enzymatic pathway for the metabolism of cell membrane-derived arachidonic acid for the production of prostaglandins.

Cyclooxygenation The process that introduces one molecule of dioxygen at C-9 and one at C-15 of a molecule of arachidonic acid.

Cyclopentamine (mol wt 141) A vasoconstrictor and nasal decongestant.

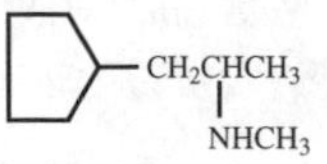

Cyclopentanone Monooxygenase The enzyme that catalyzes the following reaction:

Cyclopentanone + NADPH + O_2

⇅

5-Valerolactone + $NADP^+$ + H_2O

Cyclopentanone NADPH Oxygen Oxidoreductase The systematic name for cyclopentanone monooxygenase.

Cyclopentanoperhydrophenanthrene The parent ring structure of steroids.

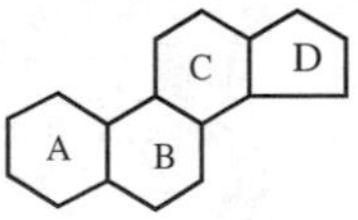

Cyclopenthiazide (mol wt 380) An antihypertensive agent.

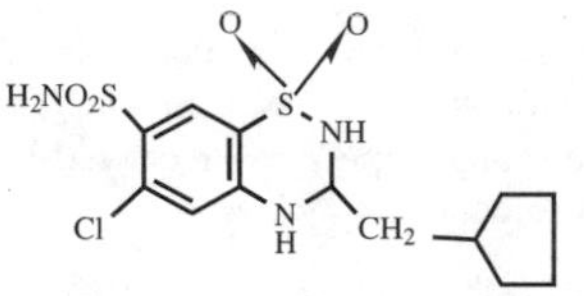

Cyclopentobarbital (mol wt 234) A sedative-hypnotic agent.

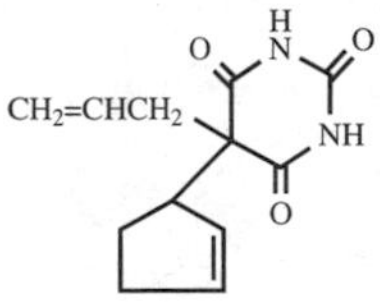

Cyclopentolate (mol wt 291) A mydriatic agent that possesses anticholinergic activity.

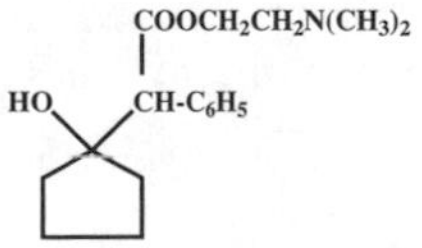

Cyclophilin A protein found in the cytoplasm of T cells with strong binding affinity for cyclosporin.

Cyclophosphamide (mol wt 261) An antineoplastic and immunosuppressive agent that interferes with DNA synthesis and transcription.

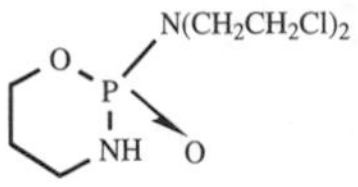

Cyclopiazonic Acid (mol wt 336) An inhibitor for calcium-mediated ATPase.

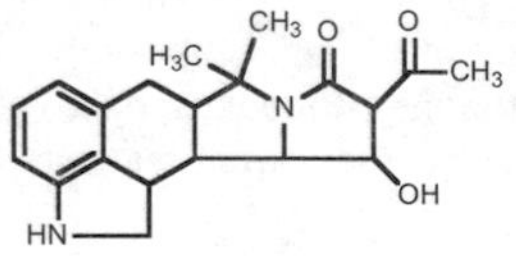

Cyclopropane (mol wt 42) A colorless, inflammable, explosive gas used as a general anesthetic agent.

C_3H_6

Cycloserine (mol wt 102) An antibiotic produced by *Streptomyces garyphalus* that binds to D-alanyl-D-alanine synthetase and inhibits bacterial cell wall synthesis.

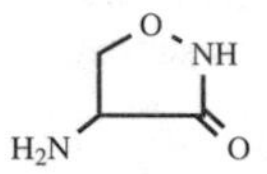

Cyclosis Circular flow of cell content around a central vacuole.

Cyclospasmol A trade name for cyclandelate, a vasodilator that inhibits phosphodiesterase and increases cAMP concentration.

Cyclosporin A A cyclic undecapeptide isolated from the fungi *Cylindrocarpon lucidum*, it is a potent immunosuppressive agent used to prevent allograft rejection.

Cyclosporins A group of nonpolar, cyclic oligopeptide antibiotics with immunosuppressive activity produced by fungi, e.g., *Tolypocladium* and *Trichodera*.

Cyclothiazide (mol wt 390) A diuretic agent.

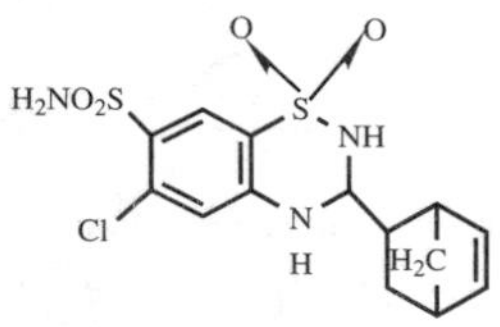

Cyclouridine (mol wt 226) An antiviral and anticancer agent.

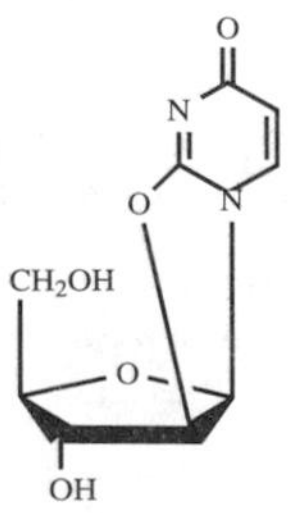

Cyclovalone (mol wt 366) A choleretic agent.

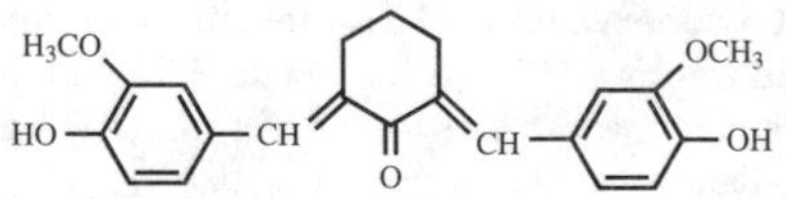

Cycrimine Hydrochloride (mol wt 324) An anticholinergic agent.

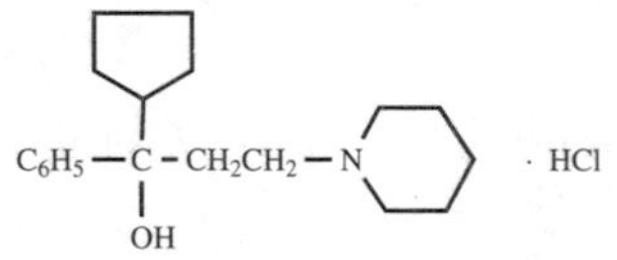

Cycrin A trade name for medroxyprogesterone acetate, a hormone that suppresses ovulation.

CyD Abbreviation for cyclodextrin.

Cyd Abbreviation for cytidine.

CydP Abbreviation for cytidine phosphate.

Cyd2'P Abbreviation for cytidine 2'-phosphate.

Cyd3'P Abbreviation for cytidine 3'-monophosphate.

Cyd2'3'P Abbreviation for cytidine 2',3'-phosphate (cyclic).

Cyd3'5'P Abbreviation for cytidine 3',5'-phosphate (cyclic).

Cyd5'P Abbreviation for cytidine 5'-monophosphate.

Cyd5'PP Abbreviation for cytidine 5'-diphosphate.

Cyd5'PPP Abbreviation for cytidine 5'-triphosphate.

CYE Abbreviation for charcoal yeast extract.

CYL Abbreviation for casein yeast lactate.

Cylert A trade name for pemoline, used as a cerebral stimulant.

Cylindrospermum A genus of filamentous Cyanobacteria.

Cyn Abbreviation for cyanide.

Cynarin(e) (mol wt 516) A choleretic agent.

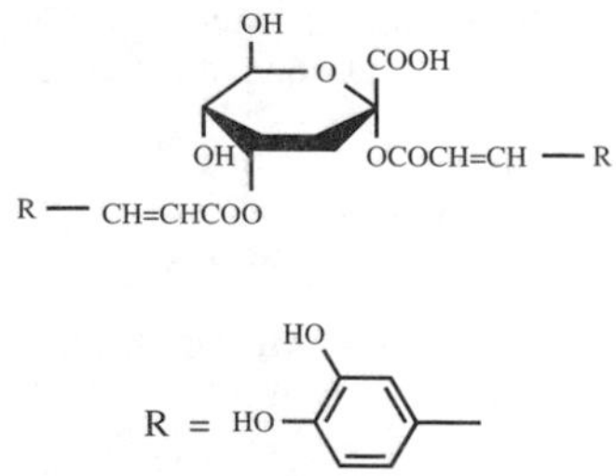

Cyoctol (mol wt 252) An antiacne and anti-alopecia agent.

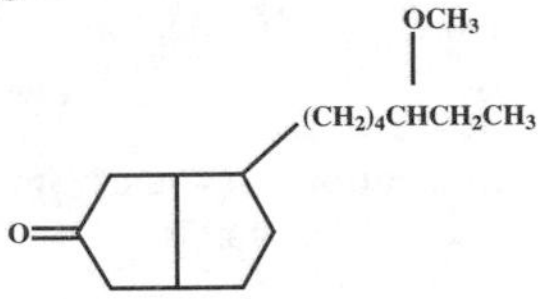

Cyomin A trade name for vitamin B_{12}.

CyP Abbreviation for cyclophilin.

CyPA Abbreviation for cyclophilin-A.

CyPB Abbreviation for cyclophilin-B.

CyPC Abbreviation for cyclophilin-C.

Cypermethrin (mol wt 416) An ectoparasiticide.

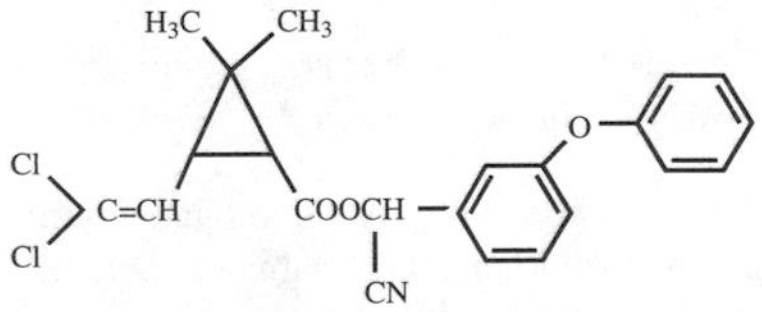

Cyprenorphine (mol wt 424) An etorphine antagonist.

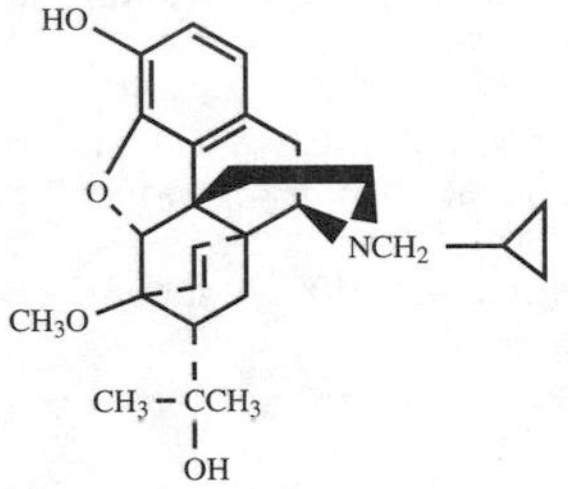

Cypridopathy Referring to venereal disease.

Cyproheptadine (mol wt 287) An antihistaminic and antipruritic agent that competes with histamine for H-1 receptors on effector cells.

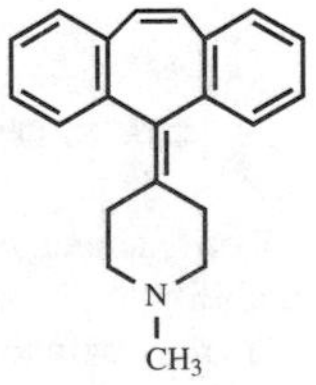

Cyproterone (mol wt 375) A progesterone derivative with anti-androgen activity.

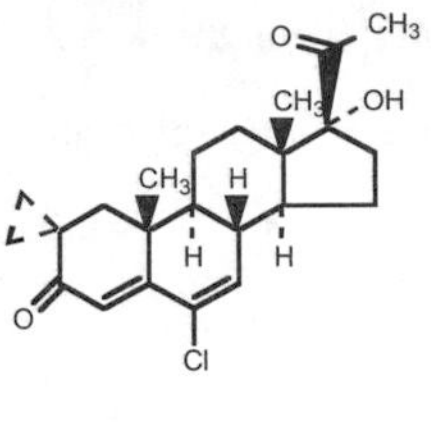

Cyr61 A growth factor binding protein expressed in the cell cycle.

Cyromazine (mol wt 166) An ectoparasiticide.

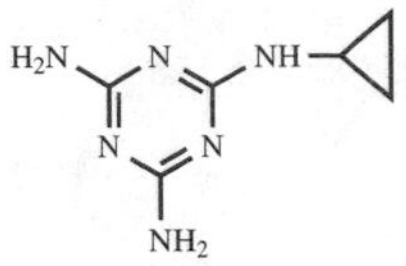

Cyronine A trade name for the hormone liothyronine (T_3).

Cys Abbreviation for cysteine.

CysNO Abbreviation for S-nitrosocysteine.

Cyst 1. A dormant form or resting structure in certain organisms. 2. A closed abnormal sac containing fluid or semisolid material in the body. 3. A bladder.

Cystadenoma A cystic neoplasm lined with epithelial cells and filled with retained materials.

Cystalgia Pain in the bladder.

Cystamine (mol wt 152) The decarboxylation product of cystine.

$$H_2N\text{-}CH_2\text{-}CH_2\text{-}S\text{-}S\text{-}CH_2\text{-}CH_2\text{-}NH_2$$

Cystathionine (mol wt 222) A substance that consists of L-homocysteine and L-serine residues.

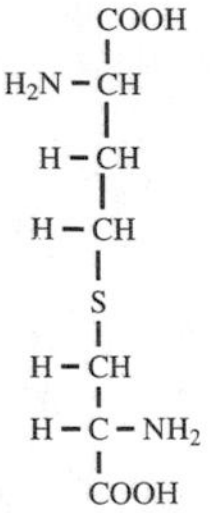

Cystathionine β-Lyase The enzyme that catalyzes the following reaction:

$$\text{L-Cystathionine} + H_2O \rightleftharpoons \text{Homocysteine} + NH_3 + \text{pyruvate}$$

Cystathionine β-Synthetase The enzyme that catalyzes the following reaction:

$$\text{L-Serine} + \text{L-homocysteine} \rightleftharpoons \text{Cystathionine} + H_2O$$

Cystathionine γ-Lyase The enzyme that catalyzes the following reaction:

L-Cystathionine $+ H_2O$

$\updownarrow$

L-cysteine $+ NH_3 \ + \ \alpha$-ketobutanoate

Cystathionine γ-Synthetase The enzyme that catalyzes the following reaction:

L-Cystathionine $+ H_2O$

$\updownarrow$

L-Cysteine $+ NH_3 \ + \ \alpha$-ketobutanoate

Cystathioninuria An inherited metabolic disorder characterized by the presence of a large amount of cystathionine in the urine due to a defect in amino acid metabolism.

Cystatin A group of proteins present in tissues and body fluids that inhibit cysteine proteinase.

Cysteamine (mol wt 77) A sulfydryl compound with a variety of biological functions.

$$HSCH_2CH_2\text{-}NH_2$$

Cysteic Acid (mol wt 169) An oxidation product of cysteine and a naturally occurring nonprotein amino acid.

$$HOOC - \overset{\overset{\displaystyle NH_2}{|}}{CH} - CH_2 - \overset{\overset{\displaystyle O}{\|}}{\underset{\underset{\displaystyle O}{\|}}{S}} - OH$$

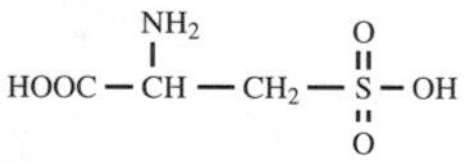

Cysteine (mol wt 121) A protein amino acid containing sulfur.

$$
\begin{array}{c}
H \\
S \\
| \\
CH_2 \\
| \\
CHNH_2 \\
| \\
COOH
\end{array}
$$

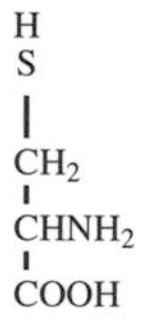

Cysteine Desulfhydrase The enzyme that catalyzes the following reaction:

D--Cysteine $+ H_2O$

$\updownarrow$

Sulfide $+ NH_3 +$ pyruvate

Cysteine Endopeptidase Any thiol proteinase or endopeptidase that has cysteine residue at the active center.

Cysteine Lyase The enzyme that catalyzes the following reaction:

L-Cysteine $+$ sulfite $\rightleftharpoons$ L-Cysteate $+$ sulfide

Cysteine Peptidase A protease whose active site contains the sulfhydryl group of cysteine.

Cysteine Synthetase The enzyme that catalyzes the following reaction:

Acetyl-L-serine $+$ hydrogen sulfide

$\updownarrow$

L-Cysteine $+$ acetate

Cysteine Transaminase The enzyme that catalyzes the following reaction:

L-Cysteine $+ \ \alpha$-ketoglutarate

$\updownarrow$

Mercaptopyruvate $+$ glutamate

Cysteine-tRNA Ligase The enzyme that catalyzes the following reaction:

ATP $+$ cysteine $+ $ tRNA^{cys}

$\updownarrow$

AMP $+$ PPi $+$ L-cysteine-tRNA^{cys}

Cysteinyl-tRNA Synthetase Synonym of cysteine-tRNA Ligase.

Cystex A trade name for a combination drug containing methenamine, salicylamide, sodium salicylate, and benzoic acid, used as an anti-infective agent.

Cystic Fibrosis An inherited disorder characterized by the functional failure of the mucus-secreting glands.

Cysticercosis Infestation with the larval form of pork tapeworm (*Taenia solium*).

Cystine (mol wt 240) A sulfur-containing protein amino acid.

$$HOOC - \overset{\overset{\displaystyle NH_2}{|}}{CH} - CH_2 - S - S - CH_2 - \overset{\overset{\displaystyle NH_2}{|}}{CH} - COOH$$

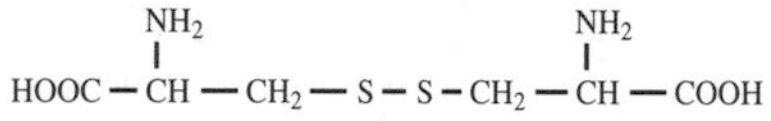

Cystine Knot A knot-like structure in a protein formed from three disulfide bonds among six cysteine residues.

Cystine Reductase The enzyme that catalyzes the following reaction:

NADH $+$ L-cystine $\rightleftharpoons$ NAD$^+$ $+$ 2 L-cysteine

Cystinosis An inherited metabolic disorder characterized by the inability to utilize cystine leading to the accumulation and precipitation of cystine in tissue.

Cystinuria A metabolic disorder characterized by the presence of excessive amounts of cystine, lysine, and arginine in the urine.

Cystitis An inflammation of the bladder.

Cystography X-ray examination of the bladder following ingestion of radiopaque medium.

Cystolithiasis The presence of stones in the urinary bladder.

Cystoma A tumor that contains cysts.

Cystoscope An instrument used for examination of disorders of the bladder and kidney.

Cystoscopy Examination of the urinary tract with a cystoscope.

Cystospaz A trade name for hyoscyamine sulfate, an anticholinergic agent that blocks acetylcholine, decreases GI mobility, and inhibits gastric acid secretion.

Cyt Abbreviation for 1. cytochrome; 2. cytosine.

Cyt a Abbreviation for cytochrome a.

Cyt-b_5 Abbreviation for cytochrome b_5.

Cyt b_{245} A component of the phagocytic oxidase complex.

Cyt b_{599} A cytochrome of photosystem II of photosynthesis.

Cyt c Abbreviation for cytochrome c.

Cyt c_1 A protein component of respiratory chain in the inner mitochondrial membrane.

Cyt cd Synonym of *Pseudomonas* cytochrome oxidase.

Cyt P_{450} Abbreviation for cytochrome P450.

Cyt P_{450}-OR Abbreviation for cytochrome P450 oxidoreductase.

Cyt-A Abbreviation for cytolytic toxin-A from *Bacillus thuringiensis.*

Cytadren A trade name for aminoglutethimide, an antineoplastic agent and a hormone antagonist.

Cytarabine (mol wt 243) A nucleoside that acts as an antiviral and antineoplastic agent.

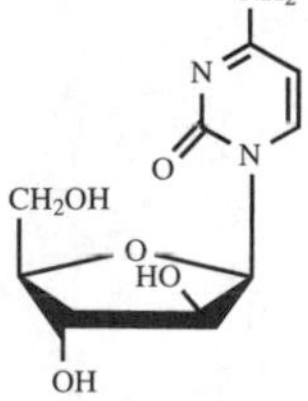

Cyt-B Abbreviation for cytolytic toxin-B from *Bacillus thuringiensis.*

Cytidine (mol wt 243) A ribonucleoside of cytosine and a constituent of RNA.

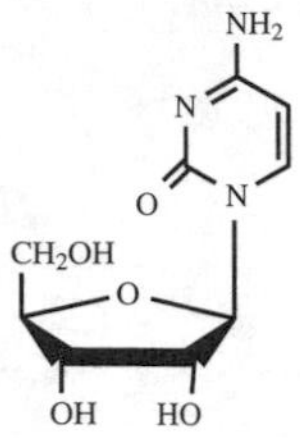

Cytidine Cyclease See cytidylate cyclase.

Cytidine 2′-3′-Cyclic Monophosphate (mol wt 305) A cyclic ribonucleotide of cytosine (cCMP).

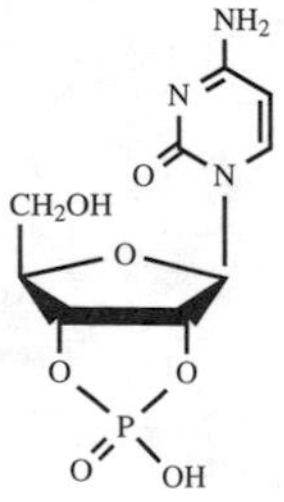

Cytidine 3′-5′-Cyclic Monophosphate (mol wt 305) A cyclic ribonucleotide of cytosine (cCMP).

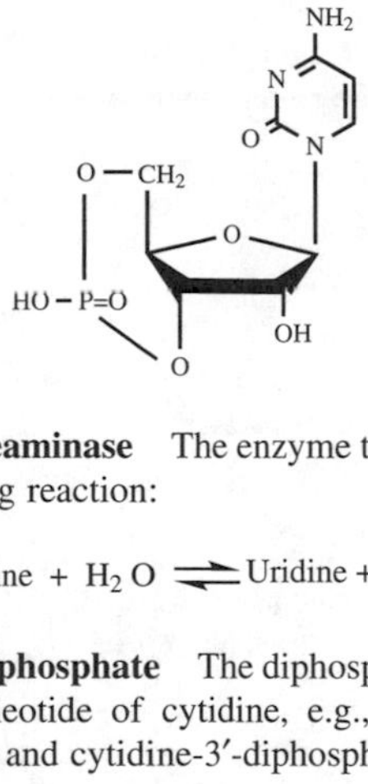

Cytidine Deaminase The enzyme that catalyzes the following reaction:

$$\text{Cytidine} + H_2O \rightleftharpoons \text{Uridine} + NH_3$$

Cytidine Diphosphate The diphosphate form of the ribonucleotide of cytidine, e.g., cytidine-5′-diphosphate and cytidine-3′-diphosphate.

Cytidine 3′-Diphosphate (mol wt 403) The diphosphate form of the ribonucleotide of cytosine.

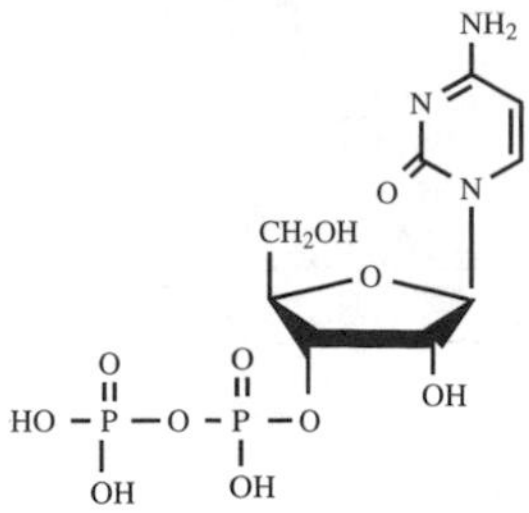

Cytidine 5′-Diphosphate (mol wt 403) The diphosphate form of the ribonucleotide of cytosine.

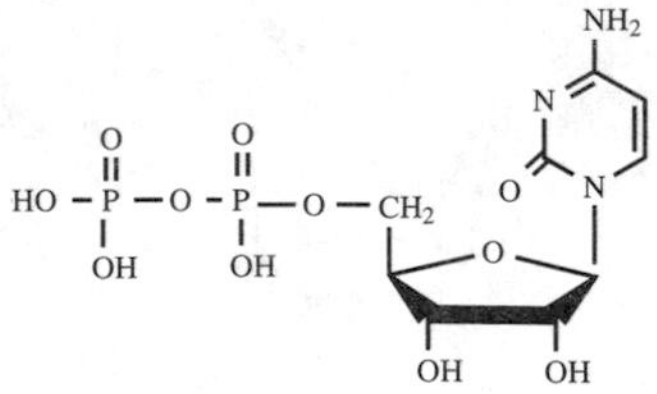

Cytidine Diphosphocholine (mol wt 488) A choline donor molecule.

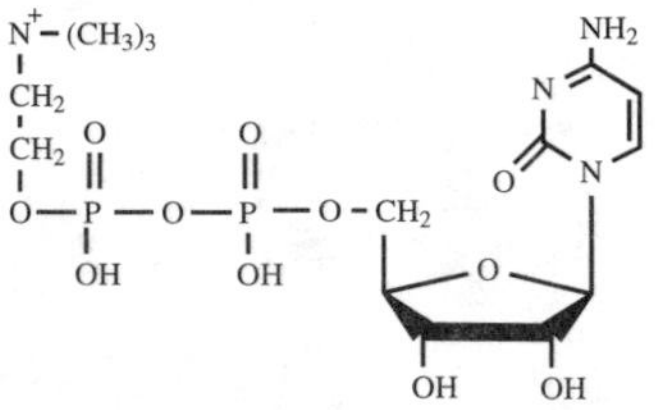

Cytidine Monophosphate (mol wt 325) The monophosphate ribonulceotide of cytosine (see cytidine 3′ or 5′-monophosphate).

Cytidine 2′-Monophosphate (mol wt 325) A monophosphate ribonucleotide of cytosine.

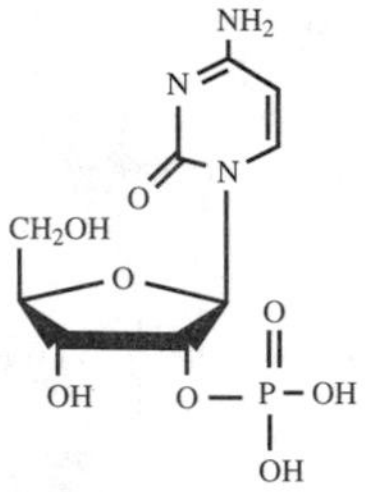

Cytidine 3′-Monophosphate (mol wt 325) A monophosphate ribonucleotide of cytosine.

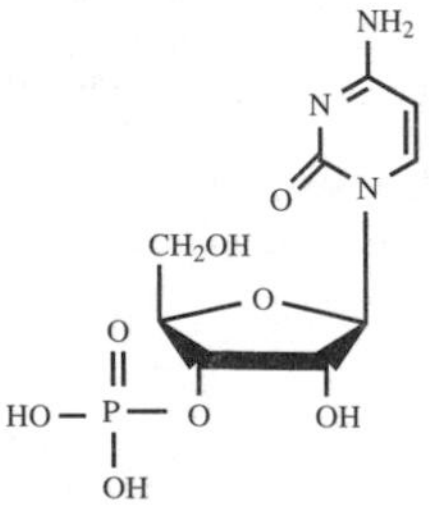

Cytidine 5′-Monophosphate (mol wt 325) A monophosphate ribonucleotide of cytosine.

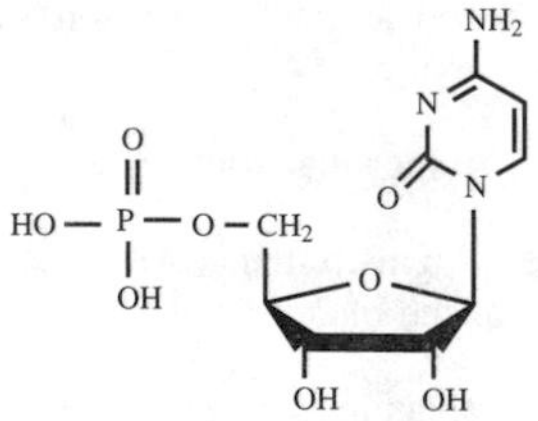

Cytidine 5′-Triphosphate (mol wt 483) The triphosphate ribonucleotide of cytosine, a component used for synthesis of RNA.

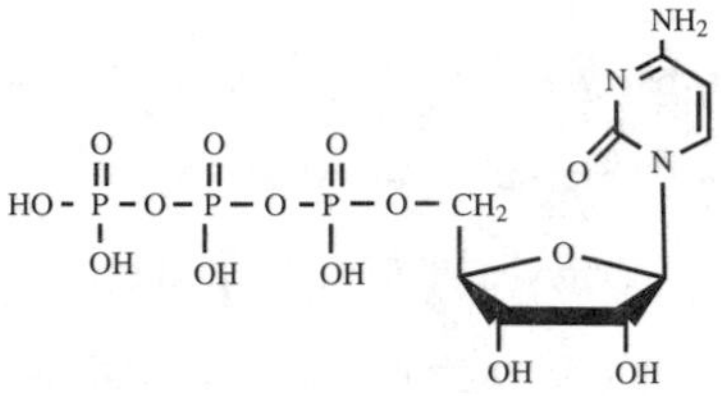

Cytidyl Cyclase See cytidylate cyclase.

Cytidylate Cyclase The enzyme that catalyzes the following reaction:

$$\text{CTP} \rightleftharpoons \text{3',5'-Cyclic CMP} + \text{PPi}$$

Cytidylate Kinase The enzyme that catalyzes the following reaction:

$$\text{ATP} + \text{dCMP} \rightleftharpoons \text{ADP} + \text{dCDP}$$

Cytidylic Acid (mol wt 323) The monophosphate form of the cytidine nucleotide (see cytidine-5′-monophosphate or monophosphate).

Cytidylyl Cyclase See cytidylate cyclase.

Cytoanalyzer A device used for screening smears that contain cells suspected of malignancy.

Cytochalasin A (mol wt 478) A fungal metabolite that inhibits the polymerization of tubulin.

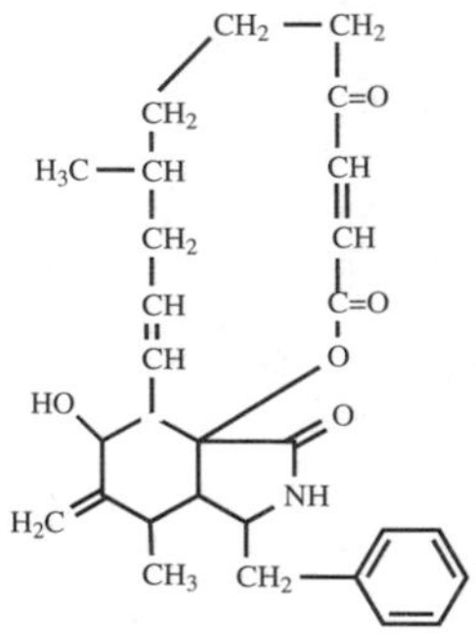

Cytochalasin B (mol wt 480) One of a group of fungal metabolites that inhibits glucose transport across the basal membrane of the human placental syncytiotrophoblast.

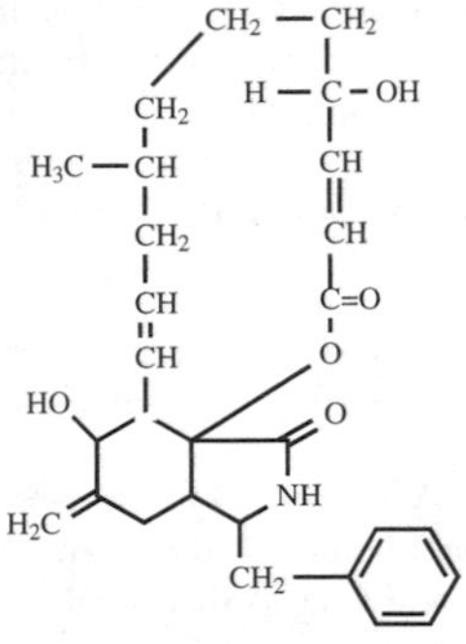

Cytochalasin D (mol wt 508) One of a group of fungal metabolites.

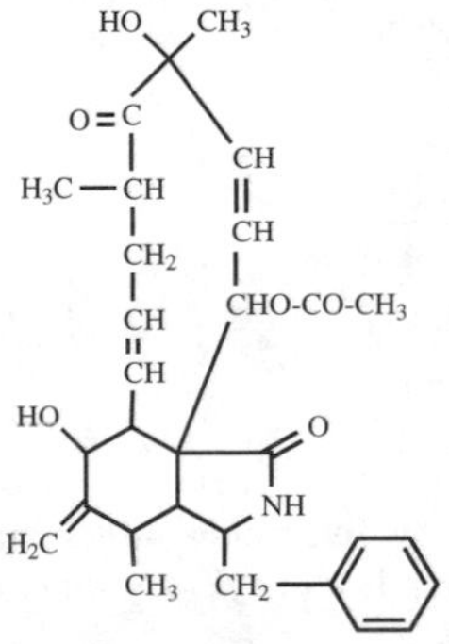

Cytochalasin E (mol wt 496) One of a group of fungal metabolites that inhibits actin polymerization in the blood platelets.

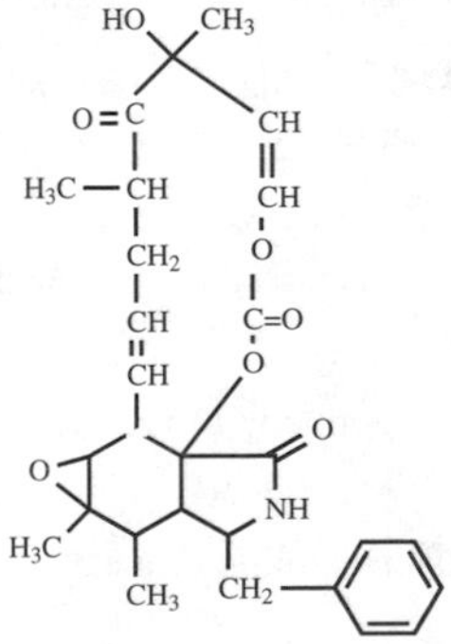

Cytochalasin H (mol wt 494) One of a group of fungal metabolites.

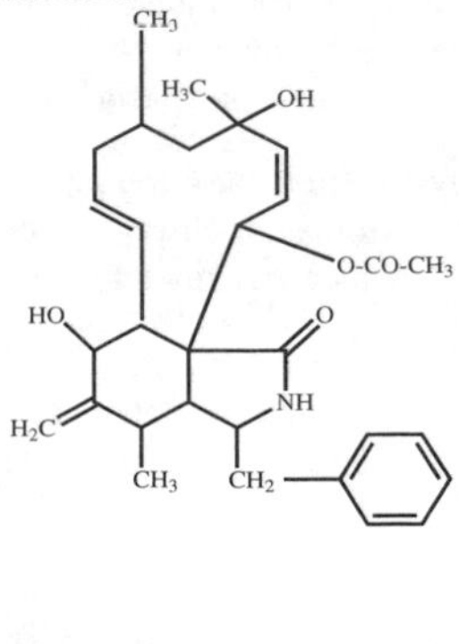

Cytochalasin J (mol wt 452) One of a group of fungal metabolites.

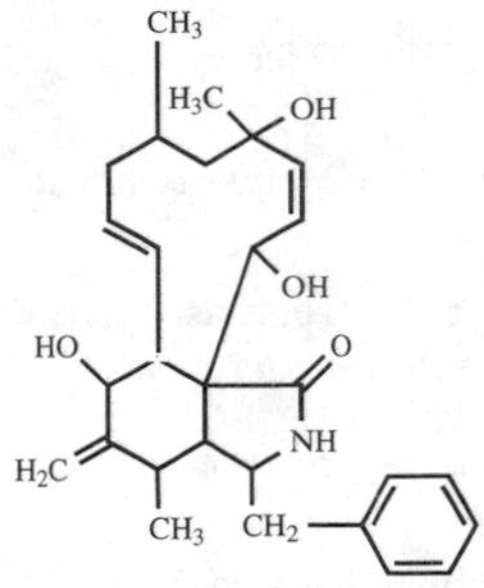

Cytochrome a/a$_3$ A cytochrome complex associated with cytochrome oxidase activity at the terminal step of the electron transport chain or the respiratory chain. The heme prosthetic group contains a formyl group.

Cytochrome b A cytochrome whose heme group contains no formyl group.

Cytochrome b$_5$ A cytochrome involved in fatty acid desaturation (production of unsaturated fatty acid).

Cytochrome b$_5$ Reductase The enzyme that catalyzed the following reaction:

$$\text{NADH} + 2\ \text{ferricytochrome b}_5 \rightleftharpoons \text{NAD}^+ + 2\ \text{ferrocytochrome b}_5$$

Cytochrome b$_6$ A cytochrome involved in the formation of b$_6$-f complexes for the generation of a proton gradient in photosystem I.

Cytochrome b$_6$-f Complex A cytochrome complex involved in the generation of a proton gradient in photosystem I.

Cytochrome c A cytochrome that carries electrons from the cytochrome reductase complex to cytochrome oxidase.

Cytochrome c Oxidase The enzyme that catalyzes the following reaction:

$$4\ \text{Ferrocytochrome c} + O_2 \rightleftharpoons 4\ \text{Ferricytochrome c} + H_2O$$

Cytochrome c Peroxidase The enzyme that catalyzes the following reaction:

$$4\ \text{Ferrocytochrome c} + H_2O_2 \rightleftharpoons 2\ \text{Ferricytochrome c} + 2H_2O$$

Cytochrome c Reductase The enzyme that catalyzes the following reaction:

$$\text{NADH} + 2 \text{ ferricytochrome c} \rightleftharpoons \text{NAD}^+ + 2 \text{ ferrocytochrome c}$$

Cytochrome c_3 Hydrogenase The enzyme that catalyzes the following reaction:

$$2 H_2 + \text{Ferricytochrome } c_3 \rightleftharpoons 4 H^+ + \text{ferrocytochrome } c_3$$

Cytochrome f A type of cytochrome in the electron transport chain that connects photosystems I and II.

Cytochrome P_{450} A cytochrome that catalyzes the monooxygenation of a variety of hydrophobic substances. It has a maximum absorption peak at 450 nm.

Cytochromes A group of heme-containing proteins involved in the electron transport chain for oxidative phosphorylation and photophosphorylation.

Cytocidal Capable of killing cells.

Cytocide Substance that kill cells.

Cytocuprein Synonym of superoxide dismutase.

Cytofluorometry A flow cytometry that involves the detection of specific fluorescence of the fluorochrome markers.

Cytogenetics The science that deals with cellular heredity.

CytoGram A trade name for CMV immune globulin IV produced by DNA recombinant technology and used for the prevention of cytomegalovirus (CMV).

Cytohemin (mol wt 894) A substance found in heart muscle.

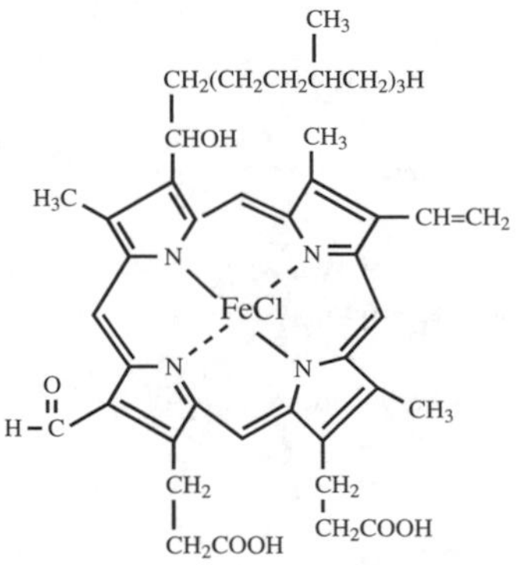

Cytokeratin Member of a family of intermediate filament proteins characteristic of epithelial cells.

Cytokine The biologically active substances produced by cells that mediate cell-to-cell communication, e.g., lymphokines from lymphocytes and monokines from monocytes.

Cytokinesin A plant growth regulator that affects cell division.

Cytokinesis The division and separation of the cytoplasm during mitosis and meiosis.

Cytokinins A group of N-substituted derivatives of adenine that function as plant growth regulators (e.g., kinetin and zeatin) that simulates cell division.

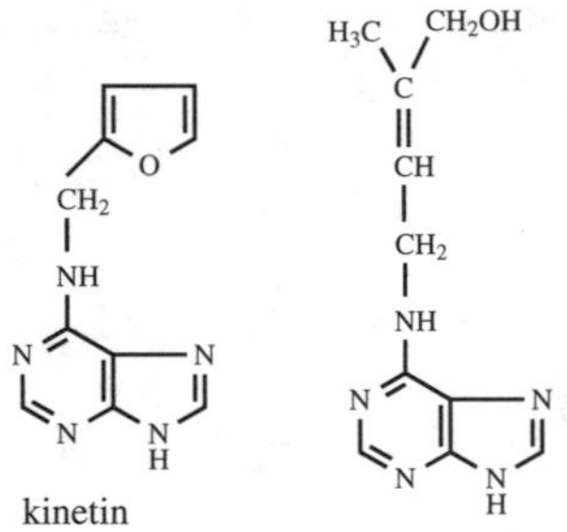

kinetin

zeatin

Cytolipin H A glycosphingolipid containing lactose and ceramide (N-acyl fatty acid derivative of a sphingosine).

Cystolith Referring to bladder stones.

Cystolithiasis The presence of stones in the bladder.

Cytological Map A diagrammatic representation showing the locations of genes on a particular chromosome.

Cytology The science that deals with structure, function, and life cycles of cells.

Cytolysin An antibody or substance capable of lysing cells.

Cytolysis Disintegration of a cell.

Cytomegalovirus A virus of the Herpesviridae. It consists of a dsDNA genome and causes cellular enlargement and the formation of eosinophilic inclusion bodies in the nucleus. It may also cause congenital deformities if present during pregnancy.

Cytomegalovirus Immune Globulin An IgG antibody against cytomegalovirus used for control or treatment of cytomegalovirus infection.

Cytomel A trade name for liothyronine sodium, a hormone that stimulates metabolism of body tissue and accelerating cellular oxidation (also known as T_3).

Cytometer A device used for counting and measuring blood cells.

Cytometry A method for counting and measuring cells.

Cytomorphosis The change undergone by a cell in the course of its life cycle.

Cyton The cell body of a neuron.

Cytopathic Producing cytopathological changes in cells.

Cytopathic Effect The degenerative cytopathological change on a cell monolayer caused by virus.

Cytopenia The diminution of the cellular elements in the blood.

Cytophagy The ingestion of cells by another cell, e.g., phagocytes.

Cytophilic Having an affinity for cells.

Cytophilic Antibody Antibody that binds to the surface of cells bearing appropriate Fc receptors.

Cytophotometry A technique employing both microscope and spectrophotometer for determination of cells or cellular components.

Cytoplasm The protoplasmic contents of a cell excluding cell organelles.

Cytoplasmic Gene Referring to the genes present in mitochondria and chloroplasts.

Cytoplasmic Inheritance The inheritance that is not associated with nuclear DNA but depends upon the replication of cytoplasmic organelles, e.g., mitochondria and chloroplasts.

Cytoplasmic Membrane The membrane that forms the outer limit of the protoplast, it consists of lipid bilayers and proteins.

Cytoplasmic Microtubules The loosely organized, dynamic network of microtubules in the cytoplasm of eukaryotic cells.

Cytoplasmic Polyhedrosis Virus An insect virus of the family Baculoviridae. It has been used as a cloning vehicle for eukaryotic genes and for biological control of caterpillars.

Cytoplasmic Streaming Flow of cytoplasm within a cell or between adjacent cells.

Cytoplast 1. A unified structure that provides the rigidity needed to hold the various structures of the eukaryotic cell in their appropriate locations. 2. The intact cytoplasmic content of a cell excluding the nucleus.

Cytopoiesis Production of cells.

Cytosar-U A trade name for cytarabine (cytosine arabinoside), an antitumor agent that inhibits cellular DNA synthesis.

Cytosegresome An intracellular membrane vacuole that contains cellular constituents formed during autophagy.

Cytosine (mol wt 110) A nitrogenous pyrimidine base in nucleic acids.

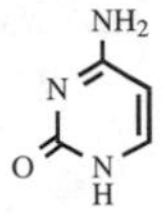

Cytosine Aminohydrolase Synonym of cytosine deaminase.

Cytosine Arabinoside (mol wt 244) An antimetabolite and antitumor agent that inhibits DNA synthesis.

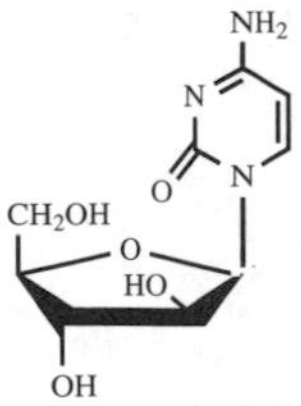

Cytosine Deaminase The enzyme that catalyzes the following reaction:

$$\text{Cytosine} + H_2O \rightleftharpoons \text{Uracil} + NH_3$$

-cytosis A suffix meaning an increase in the number of cells.

Cytoskeleton The complex network of microtubules, microfilaments in the cytoplasm that provide structure to the cytoplasm of the eukaryotic cell and plays an important role in cell movement and maintaining the characteristic shape of the cells (also called cytosketetal network).

Cytosol The soluble portion of the cytoplasm excluding of organelles.

Cytosome The cell body without a nucleus.

Cytostatic Capable of suppressing growth and multiplication of cells.

Cytotaxigen Any substance that indirectly induces cytotaxis or the formation of cytotaxin.

Cytotaxin Any substance that acts to induce cytotaxis.

Cytotaxis Synonym of chemotaxis.

Cytotec A trade name for misoprostol, an antiulcer agent.

Cytotoxic Capable of causing cell death.

Cytotoxic Anaphylaxis An anaphylactic reaction occurs when cytotropic antibody (e.g., IgE) is present or injected into an individual or animal.

Cytotoxic Antibody An antibody capable of causing cell death in the presence of complement, e.g., complement fixing antibody.

Cytotoxic Cell A cell capable of killing other cells, e.g., cytotoxic T cell or activated macrophage.

Cytotoxic Drug Any drug that kills cells.

Cytotoxic Hypersensitivity Type II hypersensitivity; an antibody-mediated cytotoxic reaction, e.g., blood transfusion reaction.

Cytotoxic T Cells A subpopulation of T cells that recognize and kill histoincompatible target cells in the absence of antibody.

Cytotoxins Substances capable of causing death of cells or inhibiting growth of the cells.

Cytotropic Having an affinity for living cells.

Cytotropic Antibodies Antibodies capable of sensitizing target cells for subsequent anaphylaxis, e.g., with IgE antibody.

Cytovene A trade name for ganciclovir, an antiviral agent that inhibits viral DNA synthesis.

Cytovillin Synonym of ezrin.

Cytoxan A trade name for cyclophosphamide, an alkalating agent.

Cytozoic Living within a cell.

Cytozoon A protozoan parasite within a cell.

Cz Abbreviation for cefazolin, an antibacterial agent.

CZE Abbreviation for capillary zone electrophoresis.

CZI Abbreviation for crystalline zinc insulin.

D

d Abbreviation for 1. deoxyribose; 2. dextrorotary.

d Abbreviation for dextro(rotary), the opposite of *l*.

D Abbreviation for 1. aspartic acid; 2. Dalton; 3. dextro in chemical configuration, opposite of L, e.g., D or L amino acid; 4. diffusion coefficient; 5. Deuterium.

δ The fourth letter of the Greek alphabet.

D Symbols for 1. configuration of a chemical structure, e.g., α-D-glucose, 2. diffusion coefficient, and 3. deuterium.

D_{10} See D_{10} value.

D°_{20w} The value of the standard diffusion coefficient calculated from the data that have been extrapolated to zero concentration in water at 20° C.

D_2 A trade name for dihydrotachysterol, a calcium regulator.

1D Abbreviation for one dimensional.

2D Abbreviation for two dimensional.

2,4-D (mol wt 221) Abbreviation for 2,4-dichlorophenoxy acetic acid, a synthetic plant growth hormone.

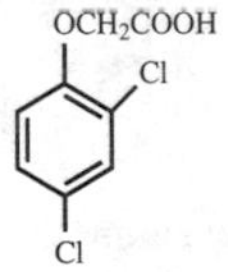

3D Abbreviation for three dimensional.

$[D]_{50}$ Abbreviation for concentration of denaturant that gives 50% unfolding.

1252-D3 A trade name for calcitriol, a vitamin and calcium regulator.

D Loop A single-stranded DNA loop formed when one strand of dsDNA is displaced by a single strand of partially homologous invading DNA.

D Type Particle A type of ring-shaped retrovirus particle located near the plasma membrane.

D_{10} Value The time required to reduce the number of viable cells to 10% of the original number of a given organism at a given temperature.

DA Abbreviation for 1. direct agglutination; 2. dopamine.

Da Abbreviation for dalton, a molecular weight or atomic mass unit.

dA Abbreviation for deoxyadenosine.

DAB Abbreviation for 1. diaminobutyric acid; 2. 4-dimethyl aminoazobenzene; 3. 3,3'-diaminobenzidene.

DABA Abbreviation for diaminobenzoic acid.

DABITC Abbreviation for dimethylamino benzene isothiocyanate.

Dac Abbreviation for dactinomycin.

Dacarbazine (mol wt 182) An antineoplastic agent that disrupts the synthesis of DNA, RNA, and protein.

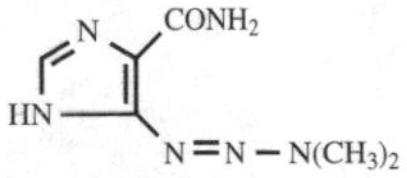

Daclizumab A trade name for an immunosupressive monoclonal antibody produced by recombinant-DNA technology that reacts specifically with exposed receptor sites on the activated T-lymphocytes to block immune response associated with the allograft rejection.

Dacodyl A trade name for bisacodyl, a laxative.

Dacryoadenitis Inflammation of the tear gland of the eye.

Dacryocyst A tear sac at the inner corner of the eye.

Dacryocystitis Inflammation of the tear sac of the eye.

Dacryorrhea Excessive tear flow.

Dact Abbreviation for dactinomycin.

Dactinomyces A genus of fungi.

Dactinomycin Synonym for actinomycin D.

Dactylitis Inflammation of the fingers and toes.

Dactylococcopsis A genus of unicellular Cyanobacteria.

Dactylosporangium A genus of mycelial bacteria (Actinomycetales).

dAdo Abbreviation deoxyadenosine.

dAdo5'P Abbreviation for deoxyadenosine 5'-phosphate.

dADP Abbreviation for deoxyadenosine diphosphate.

DAF Abbreviation for 1. decay-accelerating factor; 2. decay-activating factor.

DaFd Abbreviation for ferredoxin from *Desulfovibrio africanus*.

DAG Abbreviation for diacylglycerol.

DAGT Abbreviation for direct anti-gamma globulin test.

Dakin's Solution An antiseptic solution containing hypochlorite and sodium bicarbonate.

Dalacaine A trade name for lidocaine hydrochloride, an antiarrhythmic agent that shortens the action potential.

Dalacin C A trade name for clindamycin hydrochloride, an antibiotic that inhibits bacterial protein synthesis.

Dalalone A trade name for daxamethasone sodium, a corticosteroid used as an anti-inflammatory agent.

Dalmane A trade name for flurazepam, a sedative agent.

Dalpro A trade name for valproic acid, an anticonvulsant.

Dalteparin (mol wt 4000-6000 daltons) A low molecular weight fragment of porcine mucosal heparin with antithrombotic activity.

Dalton 1. An atomic mass unit that is defined as 1/12th the mass of a ^{12}C atom. 2. The weight of a single hydrogen atom (1.66×10^{-24} gram).

Daltroban (mol wt 353) An antithrombotic agent.

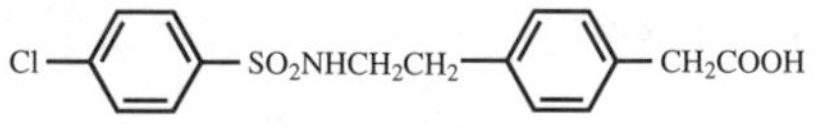

DAM Abbreviation for 1. diacetyl monoxime; 2. diacetyl morphine.

***dam* Gene** A gene in *E. coli* that encodes DNA methylase for methylation of adenine in DNA.

dam Methylase A DNA methylation enzyme encoded by the *dam* gene in *E. coli*. It methylates the adenine base of the GATC sequence.

Daminozide (mol wt 160) A plant growth regulator.

$HOOCCH_2CH_2CONHN(CH3)_2$

dAMP Abbreviation for deoxyadenosine monophosphate.

Danaparoid Sodium A low molecular heparin from porcine intestinal mucosa that inhibits thrombus and clot formation.

Danazol (mol wt 337) An antigonadotropin.

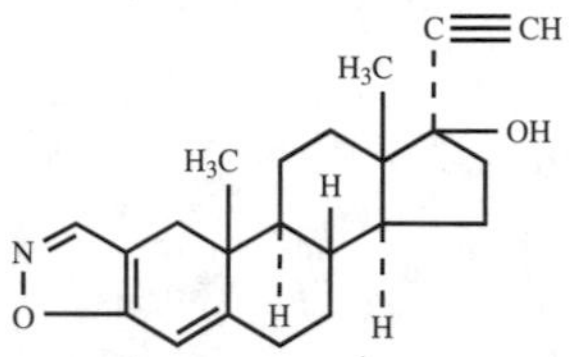

Dandruff A condition in which the scalp is covered with small flakes of dead skin.

Danielli-Davson Model A model of unit membrane structure describing the arrangement of membrane proteins with the lipid bilayer.

Danielli-Davson-Robertson Model A model of unit membrane structure describing the arrangement of membrane proteins with the lipid bilayer.

Danofloxacin (mol wt 357) An antibacterial agent.

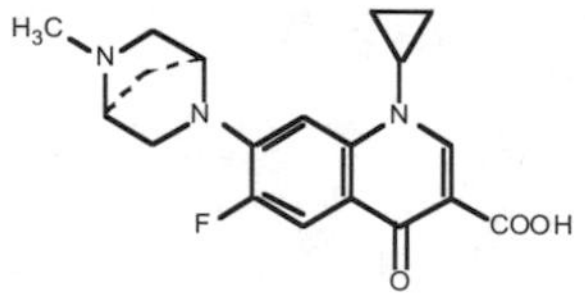

dansCaM Abbreviation for dansylated calmodulin.

dansPCaM Abbreviation for dansylated and phosphorylated calmodulin.

Dansyl Amino Acid A fluorescent amino acid derivative produced by reaction of an amino acid with dimethylaminonaphthalenesulfonyl chloride.

Dansyl Chloride (mol wt 270) Dimethylaminonaphthalenesulfonyl chloride, a compound used for labeling the N-terminal amino acid residue in proteins.

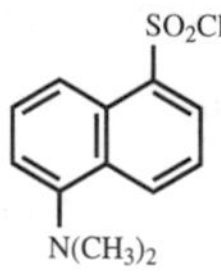

Dansylaminophenylboronic Acid (mol wt 370) A carbohydrate ligand and an inhibitor of boronic acid serine protease.

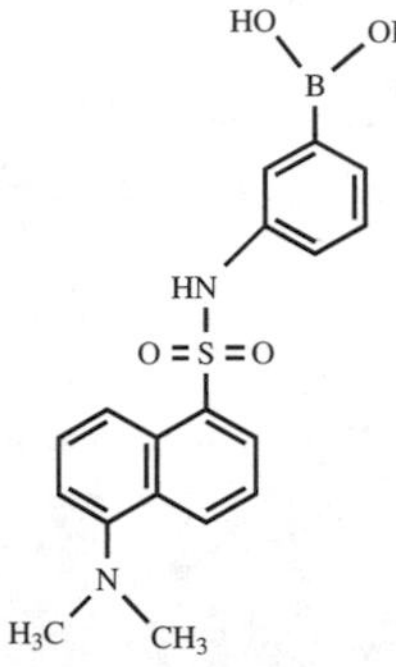

Dansylation The reaction of the dansyl group with an organic compound.

Dansylcadaverine (mol wt 335) An inhibitor of fibrin-stabilizing factor and an epidermal growth factor.

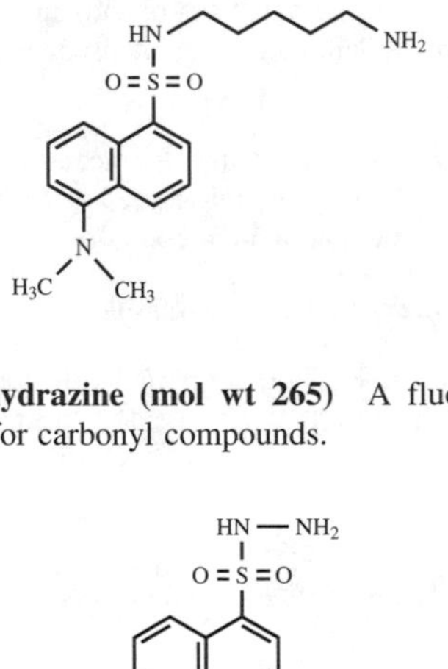

Dansylhydrazine (mol wt 265) A fluorescent marker for carbonyl compounds.

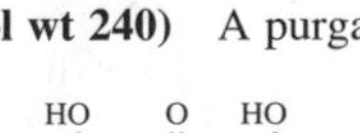

Danthron (mol wt 240) A purgative agent.

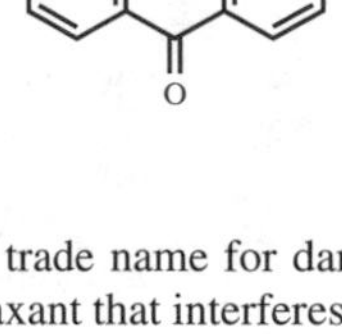

Dantrium A trade name for dantrolene, a skeletal muscle relaxant that interferes with intracellular calcium movement.

Dantrolene (mol wt 314) A skeletal muscle relaxant that interferes with intracellular calcium movement.

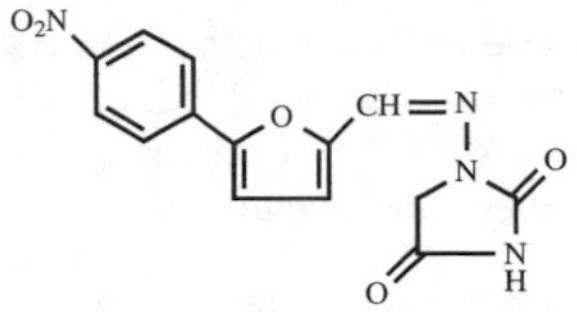

DAO Abbreviation for diamine oxidase.

DAP Abbreviation for 1. diaminopimelic acid; 2. dihydroxyacetone phosphate.

Dapa A trade name for acetaminophen.

DAP-DC Abbreviation for diaminopimelate decarboxylase.

Dapex A trade name for phentermine hydrochloride. It promotes transmission of nerve impulses by releasing stored norepinephrine from nerve terminals in the brain.

Dapiprazole (mol wt 325) An antiglaucoma agent that blocks α-adrenergic receptors in smooth muscle.

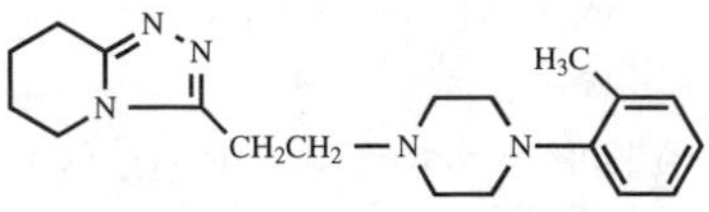

Dapsone (mol wt 248) An antibacterial and antiprotozoan agent that inhibits folic acid biosynthesis in susceptible organisms.

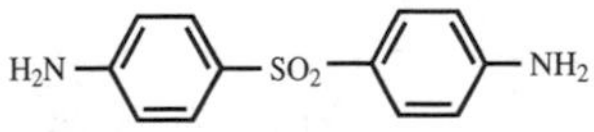

DAPT Abbreviation for direct agglutination pregnancy test.

Daranide A trade name for dichlorophenamide, used to reduce aqueous humor secretion.

Daraprim A trade name for pyrimethamine, an antimalarial agent and an inhibitor of dihydrofolic acid reductase.

Darbid A trade name for isopropamide, an anticholinergic agent that inhibits gastric acid secretion.

Daricon A trade name for oxyphencyclimine hydrochloride, used as an anesthetic agent.

Dark Field Microscope A type of microscope that uses only light rays diffracted by the specimen to produce a bright image on a dark background.

Dark Reaction The reactions for synthesis of carbohydrate from carbon dioxide via C_3 and C_4 pathways using ATP and NADPH generated in the light reaction of photosynthesis.

Dark Repair A light-independent DNA repair mechanism (e.g., removing thymine dimer and replacing it with a normal base sequence).

Darvocet-N 100 A trade name for a combination drug containing propoxyphene napsylate and acetaminophen used as an analgesic agent.

Darvon A trade name for propoxyphene hydrochloride, an analgesic agent that binds with opiate receptors in the CNS, thus altering perception and emotional response to pain.

Darwinism The theory of evolution proposed by Charles Darwin that states biological evolution is due to the natural selection of organisms best suited for survival in their environment.

DAT Abbreviation for 1. a combination drug containing daunomycin, ara-C, and thioguanine; 2. differential agglutination test; 3. diphtheria antitoxin; 4. direct agglutination test; 5. direct antiglobulin test.

d(A-T) Abbreviation for deoxyadenylate-deoxythymidylate.

Dative Bond Semipolar linkage between two atoms where one atom contributes both electrons (also called coordinate covalent bond).

dATP Abbreviation for deoxyadenosine triphosphate.

Datril A trade name for acetaminophen, an analgesic agent.

Daughter-Strand Gap Repair See recombination repair or postreplication repair.

Daun Abbreviation for daunorubicin.

Daunomycin Hydrochloride (mol wt 564) A potent anticancer agent that inhibits synthesis of DNA and RNA.

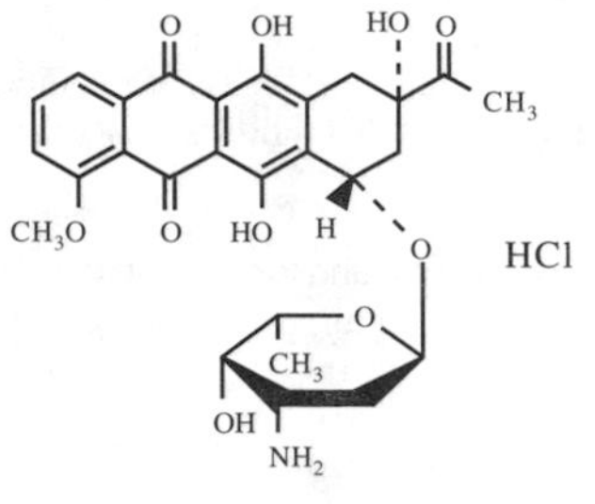

Daunorubicin (mol wt 528) An anthracycline antibiotic isolated from *Streptomyces pencetius* with antineoplastic activity. It inhibits the synthesis of DNA and RNA.

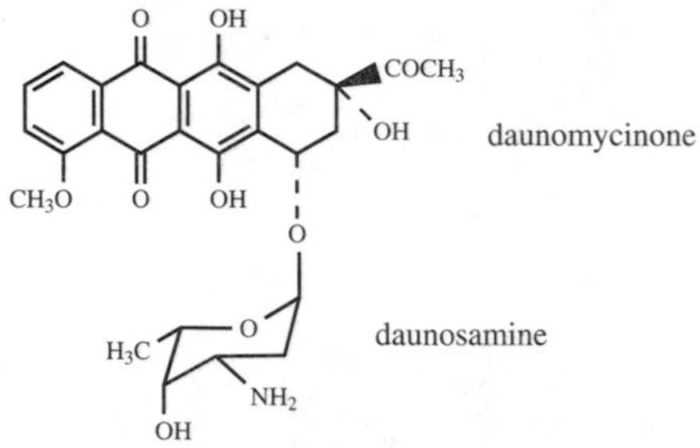

DaunoXome A trade name for a liposomal form of chemotherapeutic daunorubicin, used for first line treatment of HIV-related Kaposi's sarcoma.

Davson-Danielli Model Synonym of Danielli-Davson Model (membrane model).

Day-Barb A trade name for butabarbital sodium. It inhibits the transmission of nerve impulses.

Daypro A trade name for oxaprozin, a nonsteroidal anti-inflammatory agent used as an antipyretic and analgesic drug.

Dazamide A trade name for acetazolamide, a carbonic anhydrase inhibitor used as an antiglaucoma, diruetic, and antiepileptic drug.

DB Abbreviation for dextran blue.

2,4-DB (mol wt 249) Abbreviation for 4-(2,4-dichlorophenoxy)butanoic acid, a postemergent herbicide.

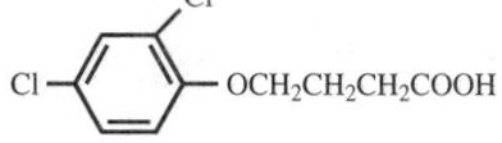

DBD Abbreviation for DNA-binding domain.

DβH Abbreviation for dopamine β-hydroxylase.

DBM Paper See diazobenzyloxymethyl paper.

DBP Abbreviation for dibutylphthalate.

DBV Abbreviation for a combination drug containing dacarbazine, BCNU, and vincristine.

DC Abbreviation for 1. dansylcadaverine; 2. decarboxylase; 3. deoxycholate; 4. dendritic cell; 5. direct current; 6. donor's cells.

dC Abbreviation for deoxycytidine.

DCA Abbreviation for 1. deoxycholate citrate agar; 2. deoxycholic acid; 3. deoxycorticosterone acetate; 4. dichloroacetate or dichloroacetic acid.

DCC Abbreviation for dextran-coated charcoal.

DCCD (N,N′-Dicyclohexylcarbodiimide, mol wt 206) An inhibitor for (F_0F_1)-type ATPase.

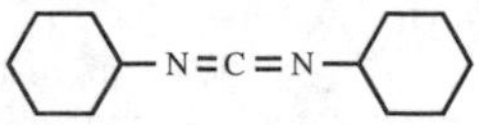

DCCD-Binding Proteolipid The component of F_0 in mammalian F_0F_1 ATPase.

dCDP Abbreviation for deoxycytidine diphosphate.

DCF Abbreviation for 1. direct centrifugal floatation; 2. dichlorofluorescein; 3. a trade name for pentostatin, an antibiotic with antineoplastic activity.

DCH Abbreviation for delayed cutaneous hypersensitivity.

DCI Abbreviation for direct chemical ionization.

DCIP Abbreviation for dichlorophenolindo-phenol.

dCK Abbreviation for deoxycytidine kinase.

DCLS Abbreviation for deoxycholate-citrate-lactose-saccharose agar.

***dcm* Gene** A gene in *E. coli* that encodes a methylase for methylation of DNA.

***dcm* Methylase** The enzyme encoded by *dcm* gene in *E. coli* that methylates cytosine in the CC(A/T)GG sequence of DNA.

DCMP Abbreviation for a combination drug containing daunomycin, cytosine arabinoside, 6-mercaptopurine, and prednisone.

dCMP Abbreviation for deoxycytidine monophosphate.

dCMP Aminohydrolase See dCMP deaminase.

dCMP Deaminase The enzyme that catalyzes the following reaction:

$$\text{dCMP} + H_2O \rightleftharpoons \text{dUMP} + NH_3$$

DCMU Abbreviation for dichlorophenyldimethylurea, an herbicide and inhibitor for flow of electrons between photosystem I and II.

DCNB Abbreviation for 1,2-dichloro-4-nitrobenzene.

Dco Abbreviation for decanoyl group.

$$CH_3\text{-}[CH_2]_8\text{-}CO\text{-}$$

DCP Abbreviation for dibasic calcium phosphate.

DCPIP Abbreviation for 2,6-dichlorophenol-indolphenol.

DCR Abbreviation for 2,4-dienoyl-CoA reductase.

DCT Abbreviation for direct Coombs' test.

DCTMA Abbreviation for deoxy-corticosterone trimethyl acetate.

dCTP Abbreviation for deoxycytidine triphosphate.

dCTP Aminohydrolase See dCTP deaminase.

DCTPA Abbreviation for deoxycorticosterone triphenyl acetate.

dCTPase The enzyme that catalyzes the following reaction:

$$\text{dCTP} + H_2O \rightleftharpoons \text{dCMP} + \text{PPi}$$

dCTP Deaminase The enzyme that catalyzes the following reaction:

$$\text{dCTP} + H_2O \rightleftharpoons \text{dUTP} + NH_3$$

dCTP Pyrophosphatase See dCTPase.

DCV Abbreviation for a combination drug containing dacarbazine, CCNU, and vincristine.

dd Abbreviation for dideoxy.

DD Abbreviation for 1. death domain; 2. DNA-dependent.

ddA Abbreviation for dideoxyadenosine.

ddATP (mol wt 475) Abbreviation for dideoxyadenosine triphosphate, used in Sanger's dideoxy method of DNA sequencing.

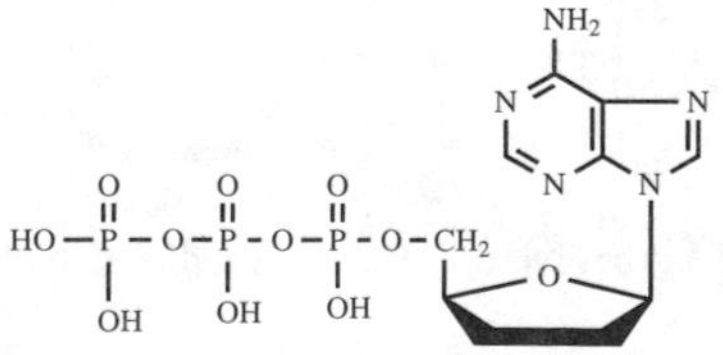

DDAVP A trade name for desmopresin acetate, a hormone.

DDC Abbreviation for diethyldithiocarbamate.

ddC Abbreviation for 1. dideoxy-cytidine; 2. a trade name for zalcitabine, an antiviral agent.

ddCTP (mol wt 451) Abbreviation for dideoxycytidine triphosphate, used in Sanger's dideoxy method of DNA sequencing.

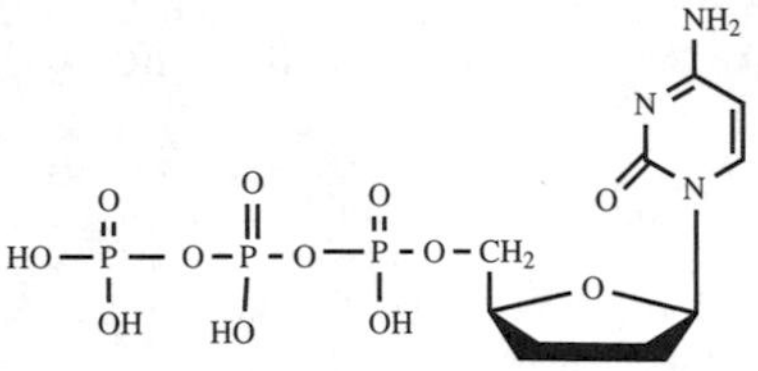

DDD (mol wt 350) Abbreviation for dihydroxydinaphthyldisulfide, a reagent used for determination of protein-bound sulfhydryl groups; also an inhibitor for adrenocortical function.

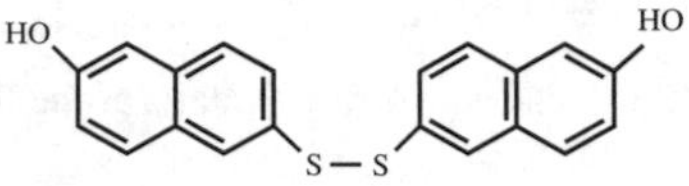

DDDE Abbreviation for dichloro-diphenyl-dichloro-ethylene.

DDDP Abbreviation for DNA-dependent DNA polymerase.

DdeII (XhoI) A restriction endonuclease from *Desulfovibrio desulfuricans* Norway with the same specificity as XhoI.

DDG Abbreviation for deoxy-D-glucose.

ddG Abbreviation for dideoxyguanosine.

ddGTP (mol wt 491) Abbreviation for dideoxyguanosine triphosphate, used in Sanger's dideoxy method of DNA sequencing.

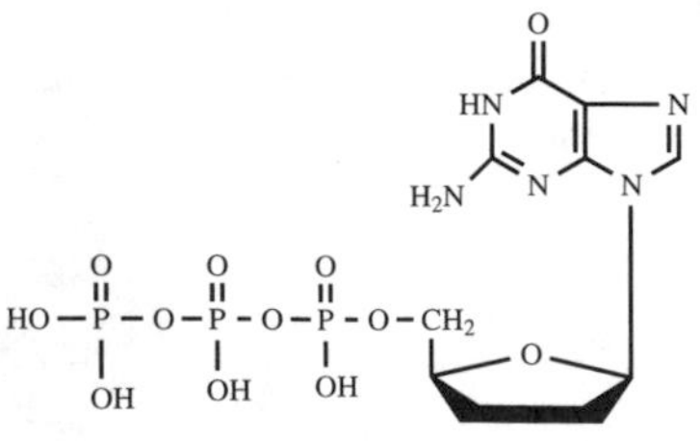

DDH Abbreviation for dihydrodiol dehydrogenase.

ddI 1. A trade name for didanosine, a synthetic nucleotide that inhibits HIV infection. 2. An abbreviation for dideoxyinosine.

ddNTP Abbreviation for dideoxynucleoside triphosphate, e.g., ddATP, ddCTP, ddGTP, and ddTTP.

DD-PCR Abbreviation for differential display polymerase chain reaction.

DDS Abbreviation for diaminodiphenylsulfone. See also dapsone.

DdsI (BamHI) A restriction endonuclease from *Desulfovibrio desulguricans* with the same specificity as BamHI.

ddT Abbreviation for dideoxythymidine.

DDT (mol wt 355) Abbreviation for dichlorodiphenyltrichloroethane, a polychlorinated and nondegradable pesticide.

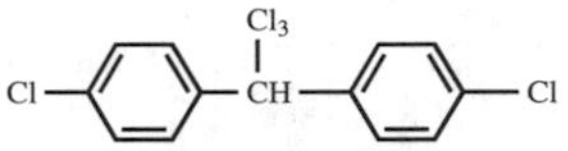

DDTC Abbreviation for diethyl-dithio-carbamate.

DDT-Dehydrochlorinase The enzyme that catalyzes the following reaction:

1,1,1-Trichloro-2,2-bis(4)-chlorophenyl-ethane

⇌

1,1-Dichloro-2,2-bis(4)-chlorophenyl-ethylene
+ chloride

ddTTP (mol wt 466) Abbreviation for 2′,3′-dideoxythymidine 5′-triphosphate, an analog of thymidine triphosphate used in Sanger's dideoxy method for DNA sequencing. ddTTP is also known as 3′-deoxythymidine-5′-triphosphate since thymidine itself is a 2′-deoxynucleoside.

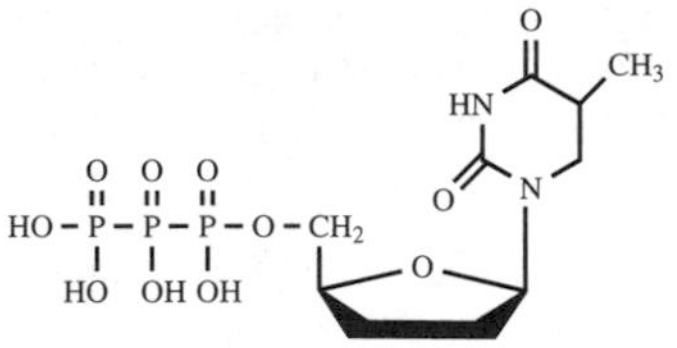

De- A prefix meaning out or away.

DEA Abbreviation for diethylamine.

Deacylated tRNA The tRNA whose attached aminoacyl group or peptidyl group has been removed.

Deacylation The removal of an acyl group from a compound.

DEAD-Box Helicases Abbreviation for a family of ATP-dependent DNA or RNA helicases with D E A D amino acid consensus (A = Ala; D = Asp; E = glu).

DEAE Abbreviation for diethylaminoethyl.

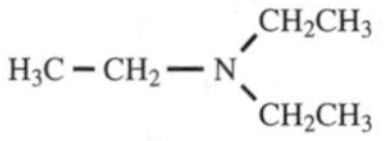

DEAE-Agarose An anion exchanger used in ion exchange chromatography.

DEAE-Cellulose An anion exchanger used in ion exchange chromatography.

DEAE-Dextran An anion exchanger used in ion exchange chromatography.

DEAE-Sephacel An anion exchanger used in ion exchange chromatography.

DEAE-Sephadex An anion exchanger used in ion exchange chromatography.

DEAE-Sepharose An anion exchanger used in ion exchange chromatography.

Deamidase The enzyme that catalyzes the hydrolytic cleavage of the C-N bond in a carboxamide.

Deamido-NAD$^+$ L-Glutamine Amido-Ligase (AMP-forming) The systematic name for NAD$^+$ Synthetase.

Deaminases The enzymes that catalyze the removal of an amino group from an organic compound.

Deamination The process of removal of an amino group from an organic compound.

Deaminooxytocin (mol wt 992) A potent analog of the posterior pituitary hormone oxytocin.

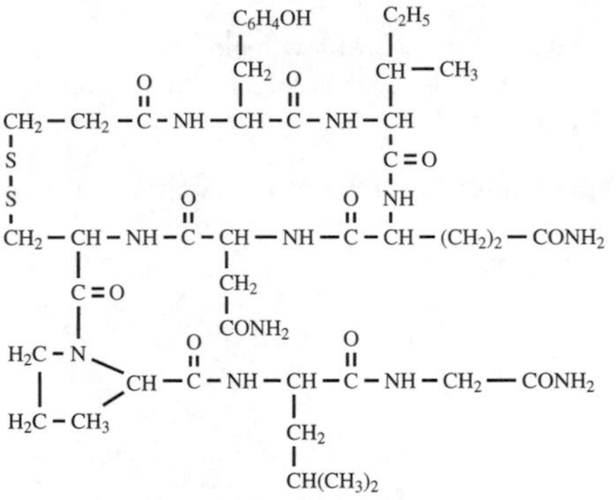

Dean and Webb Titration A serological technique to determine the equivalence zone of precipitin in which a constant volume of a given antiserum is mixed with varying dilutions of the homologous antigen.

DEA/NO Abbreviation for diethylamine/nitric oxide.

Deanol Acetamidobenzoate (mol wt 268) An antidepressant.

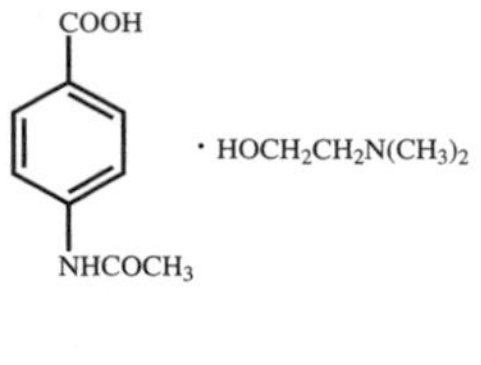

DEBA Abbreviation for diethylbarbituric acid.

Deblocking Antibody Antibody that overcomes the inhibitory effect of a blocking factor.

Debranching Enzyme The enzyme that hydrolyzes the α-1,6-glycosidic linkage in amylopectin, glycogen, and related polysaccharides.

Debrisoquin (mol wt 175) An antihypertensive agent.

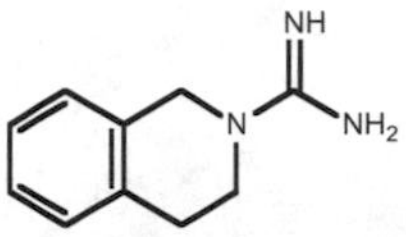

Debrox A trade name for carbamide peroxide, a cerumenolytic agent that emulsifies and disperses accumulated cerumen.

DEBS Abbreviation for 6-deoxyerythronolide B synthetase.

DEC Abbreviation for dendritic epidermal cell.

Deca- A prefix meaning 10.

Decaderm A trade name for dexamethasone, a corticosteroid used as an anti-inflammatory agent.

Decadrol A trade name for dexamethasone sodium phosphate, a corticosteroid used as an anti-inflammatory agent.

Decadron A trade name for dexamethasone, a corticosteroid used as an anti-inflammatory agent.

Deca-Durabolin A trade name for nandrolone decanoate, a male hormone that promotes tissue building.

Decaject A trade name for dexamethasone sodium phosphate, a corticosteroid used as an anti-inflammatory agent.

Decalcification The loss of calcium from bone or teeth.

Decameth A trade name for dexamethasone sodium phosphate, a corticosteroid used as an anti-inflammatory agent.

Decamethonium Bromide (mol wt 418) A synthetic skeletal muscle relaxant.

$$\left[(CH_3)_3\overset{+}{N}(CH_2)_{10}\overset{+}{N}(CH_3)_3\right] \cdot 2Br^-$$

Decanoic Acid See capric acid.

Decanoyl Group Referring to the univalent acyl group of CH_3-$[CH_2]_8$-CO-.

Decarboxylase The enzyme that liberates carbon dioxide from the carboxyl group of an organic compound.

Decarboxylase Test A test used to determine the ability of bacterial strains to decarboxylate arginine, lysine, or ornithine.

Decarboxylation The process of removal of carbon dioxide from the COOH group of an organic molecule.

Decatenation The separation of catenane into two circular molecules.

Deci- A prefix meaning 1/10.

Decidua The mucous membrane of the inner lining of the uterus.

Deciduitis An inflammation of the lining membrane of the uterus.

Decimemide (mol wt 337) An anticonvulsant.

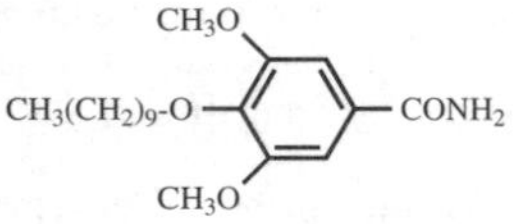

Declomycin A trade name for demeclocycline, an antibiotic that inhibits bacterial protein synthesis.

Decofed A trade name for pseudoephedrine hydrochloride, a nasal decongestant.

Decolone A trade name for nandrolone decanoate, an anabolic hormone used to promote tissue building processes.

Decongestant Substance capable of clearing nasal and bronchial passages by shrinking the mucous membrane of these regions.

Decorin A collagen-binding protein of the extracellular matrix.

Decose Any aldose having a chain of 10 carbon atoms.

Deculose Any ketose having 10 carbon atoms.

Decyl Group Referring to $CH_3[CH_2]_8$-CH_2- group.

Decyl-β-D-Glucopyranoside (mol wt 320) A nonionic detergent used for solubilization of membrane protein.

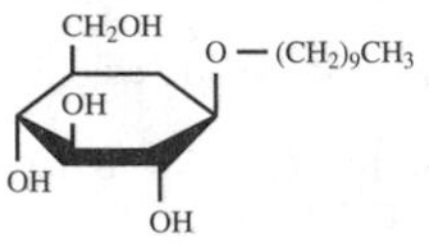

Decyl-β-D-Maltoside (mol wt 483) A nonionic detergent.

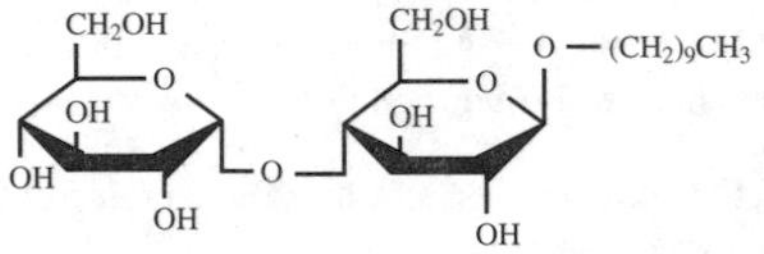

DED Abbreviation for death effector domain.

Deep Etching A freeze etching technique in which a volatile cryoprotectant is used to extend the etching period and to remove a deep layer of ice, thereby producing a large exposed area of the specimen surface for the examination of the interior structure of a cell.

DEF Abbreviation for differentiation-enhancing factor.

Defecation The elimination of fecal material from the digestive tract.

Defective Prophage A prophage that is incapable of replication upon induction.

Defective Virus A virus that lacks one or more genetic functions necessary for its replication and maturation.

Defen-LA A trade name for a combination drug containing pseudoephedrine hydrochloride and guaifenesin.

Defensins A class of low molecular weight cationic peptides from neutrophils with antibacterial, antifungal, and antiviral activity.

Deferoxamine (mol wt 561) A chelating agent and an antidote to iron poisoning.

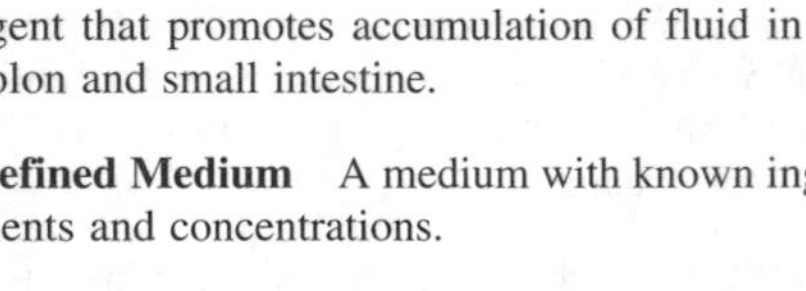

Defibrinated Blood Blood from which fibrin is removed.

Deficol A trade name for bisacodyl, a laxative agent that promotes accumulation of fluid in the colon and small intestine.

Defined Medium A medium with known ingredients and concentrations.

Deflazacort (mol wt 442) A glucocorticoid used as an anti-inflammatory agent.

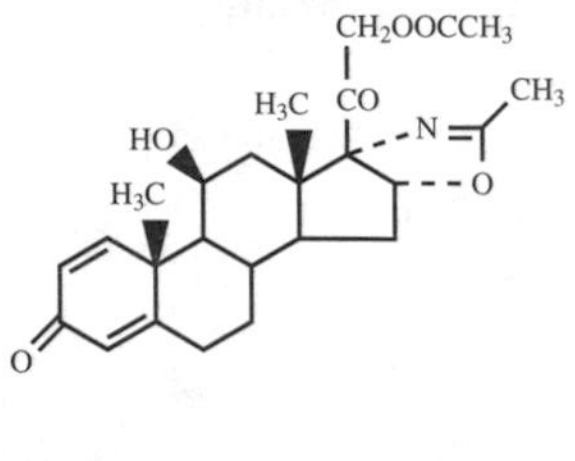

Deformylase The enzyme that catalyzes the hydrolysis or removal of the formyl group at the amino terminus of newly synthesized protein.

Defosfamide (mol wt 342) An antineoplastic agent.

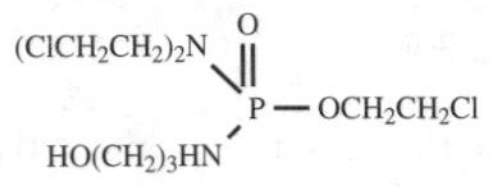

Degenerate Codon Two or more codons that encode the same amino acid.

Degranulation 1. A process whereby cytoplasmic organelles of phagocytic cells fuse with phagosomes and discharge their contents in the phagosome for digestion of the phagocytized materials. 2. A process whereby mast cell releases vasoactive compounds, e.g., histamine.

Dehiscence The process of splitting open.

Dehist A trade name for brompheniramine maleate, an antihistaminic agent that competes with histamine for H_1 receptors on effector cells.

Dehydral A trade name for methenamine, a urinary tract anti-infective agent.

Dehydrase The enzyme that catalyzes the dehydration reaction.

Dehydration Reaction A water-removing chemical reaction.

Dehydration Synthesis A process of polymer formation in which monomers are linked together by removing a molecule of water.

Dehydroascorbic Acid (mol wt 174) The oxidized form of ascorbic acid.

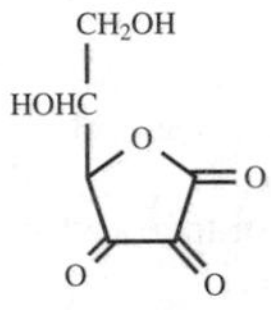

Dehydrocholesterol (mol wt 385) Precursor of vitamin D_3; it becomes vitamin D_3 upon irradiation with UV light.

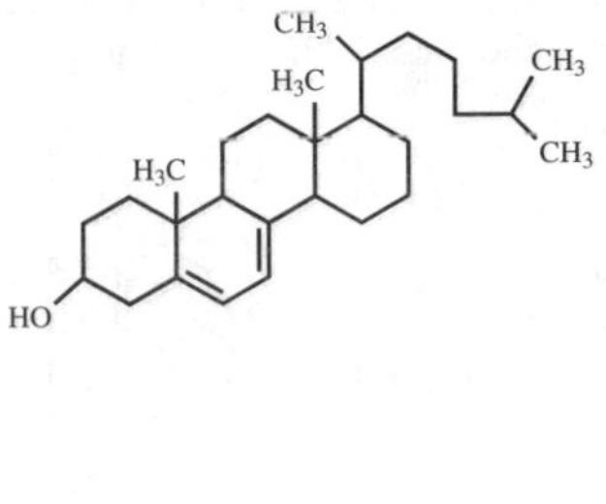

Dehydrocholic Acid (mol wt 403) A choleretic agent.

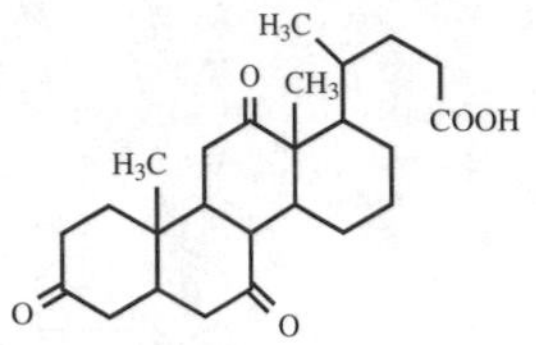

Dehydrogenase Any enzyme that catalyzes the transfer of hydrogen from one compound to another.

3,4-Dehydro-L-Proline (mol wt 113) An analog of proline and hydroxyproline.

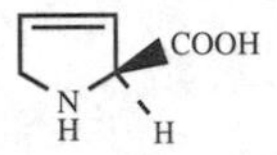

Dehydroquinate Dehydratase The enzyme that catalyzes the following reaction:

$$\text{3-Dehydroquinate} \rightleftharpoons \text{3-Dehydroshikimate} + H_2O$$

Dehydroquinate Synthetase The enzyme that catalyzes the following reaction:

$$\text{3-Deoxyarabinoheptulosonate 7-phosphate} \rightleftharpoons \text{3-Dehydroquinate} + \text{Pi}$$

Deinococcus A genus of Gram-negative bacteria.

Deionization The process of removal of ions from a solution by ion-removing substance, e.g., ion exchange resin.

Deionized Water Water from which ions are removed. Deionized water has low conductivity.

Deka (Deca) A prefix meaning 10.

Dekkera A genus of yeast (Saccharomycetaceae).

Delacort A trade name for hydrocortisone used as an anti-inflammatory agent.

Delalone A trade name for dexamethasone sodium phosphate, a hormone.

Delalutin A trade name for hydroxyprogesterone caproate used to suppress ovulation.

Delapril (mol wt 453) An antihypertensive agent.

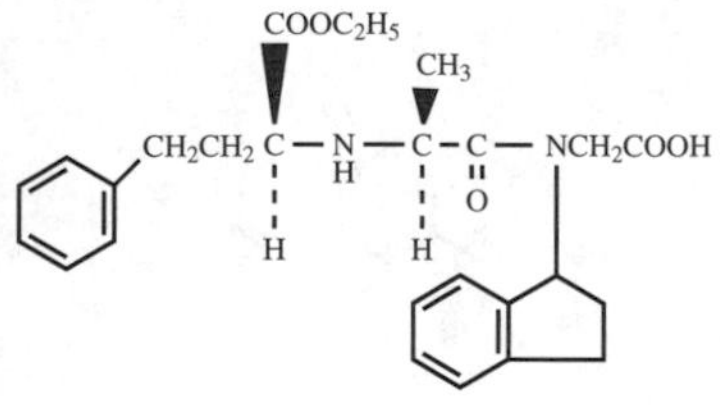

Delatestryl A trade name for testosterone, a hormone.

Delavirdine (mol wt 457) A viral inhibitor that inhibits HIV reverse transcriptase.

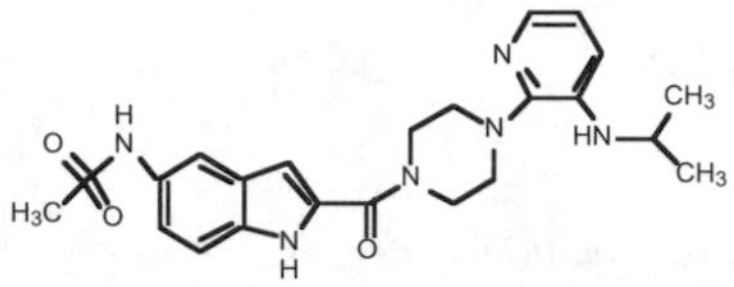

Delaxin A trade name for methocarbamol, a skeletal muscle relaxant that reduces transmission of nerve impulses from spinal cord to muscle.

Delayed Hypersensitivity See delayed-type hypersensitivity.

Delayed-Type Hypersensitivity A T-lymphocyte-mediated immunity that produces a cellular infiltrate and an edema between 24 to 48 hours after injection of antigen (e.g., tuberculin vaccination).

Delestrogen A trade name for estradiol valerate, a hormone.

Deletion A type of mutation in which a nucleotide or section of nucleotide is lost in the DNA genome of an organism.

Deletion Mapping The use of overlapping deletions to locate the position of a gene on a chromosome.

Delipidation Removal of lipid from a sample.

Delsym A trade name for dextromethorphan hydrobromide, a non-narcotic antitussive agent.

Delta (δ) A Greek letter that denotes 1. difference between two values and 2. the fourth carbon atom from the carbon atom that carries functional groups.

Delta Agent See delta hepatitis virus.

Delta Chain The heavy chain of IgD.

Delta Endotoxin A glycoprotein toxin produced by *Bacillus thuringiensis* and used as a microbial insecticide for biological control.

Delta Hemolysin A hemolysin produced by *Staphylococcus* with approximate molecular weight of 103,000 daltons.

Delta Hepatitis Virus A defective single-stranded circular RNA virus that replicates only in the presence of helper hepatitis B virus (also called delta agent, delta virus, or hepatitis B virus).

Delta Herpesvirus A herpesvirus related to human varicella-zooster virus.

Delta T50H The difference between the temperature at which DNA of homoduplexes and heteroduplexes undergo 50% dissociation.

Delta Virus See delta hepatitis virus.

Delta-Cortef A trade name for prednisolone, a hormone.

Deltakephalin A synthetic hexapeptide and a potent δ-opiate receptor agonist.

Tyr-Thr-Gly-Phe-Leu-Thr

Delta-Sleep-Inducing Peptide A nonapeptide (Trp-Ala-Gly-Gly-Asp-Ala-Ser-Gly-Glu) present in the plasma of rats and rabbits that induces low-wave sleep on injection into awake animals.

Deltasone A trade name for prednisone, a hormone.

DEM Abbreviation for diethylmaleate.

Demadex A trade name for torsemide, a diuretic agent.

DE-MALDI Abbreviation for delayed extraction matrix-assisted laser desorption ionization.

Dematin A protein involved in the formation of actin bundles.

Demazin A trade name for chlorpheniramine, an antihistaminic agent.

Demecarium Bromide (mol wt 717) A cholinergic agent and anticholinesterase drug.

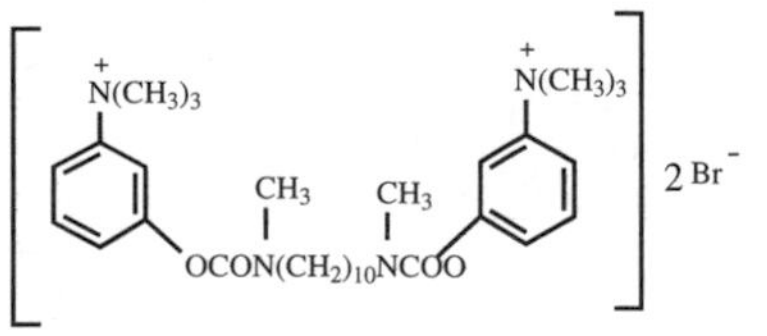

Demeclocycline (mol wt 465) An antibiotic related to tetracycline produced by *Streptomyces aureofaciens* that inhibits bacterial protein synthesis by binding 30S ribosomes.

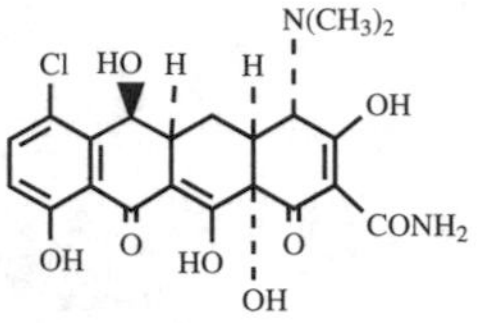

Demecolcine (mol wt 371) An antineoplastic agent.

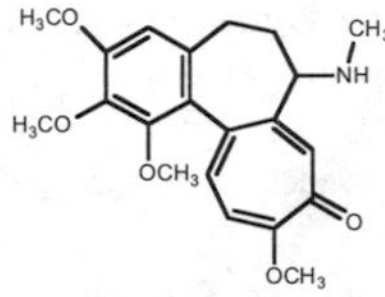

Demer-Idine A trade name for meperidine, an analgesic agent.

Demerol A trade name for meperidine hydrochloride, used as an analgesic agent.

demi- A prefix denoting half.

Demonecrotic Capable of causing necrosis.

Demser A trade name for metyrosine, an antihypertensive agent.

Demulen A trade name for a combination drug containing ethynodiol diacetate and ethinyl estradiol used as an oral contraceptive agent.

Demyelination Destruction or removal of the myelin sheet from nerve fibers.

Denaturant Any substance capable of changing the three-dimensional structure and physical or biological property of a macromolecule.

Denaturation Loss of the natural three-dimensional structure of a macromolecule, e.g., protein or nucleic acid structural integrity loss caused by disruption of hydrogen bonds and altering the secondary or tertiary structure.

Denatured Alcohol Ethyl alcohol made unsuitable for human consumption by addition of a poisonous substance.

Denatured DNA Single-stranded DNA molecules resulting from unwinding of double-stranded DNA duplex by denaturant (e.g., heat) leading to change of physical properties (e.g., viscosity and optical density).

Denatured Protein Unfolding of a polypeptide chain through the disruption of hydrogen bonds leading to the loss of biological activity and change in physical property.

Denavir Cream A trade name for penciclovir, an antiviral and antiseptic agent.

Dendrite A fiber of a nerve cell that receives impulses and transmits them inward toward the cell body.

Dendritic Cells Mononuclear cells that possess long cytoplasmic processes and function as antigen-trapping and antigen-presenting cells.

Dengue (Breakbone) Fever An acute, tropical, and subtropical human disease caused by any of the four serotypes of dengue virus transmitted by mosquitoes.

Denhardts Solution A buffer solution containing Ficoll, polyvinylpyrrolidine and bovine serum, used as a blocking reagent for hybridization with nitrocellulose filters.

Denhart's Solution A solution of ficoll (polyvinylpyrrolidone) and bovine serum employed for treatment of filters containing bound nucleic acids to prevent the binding of single-stranded DNA probes.

Denileukin Diftitox A trade name for a cytotoxic fusion protein designed to direct the cytocidal action of diphtheria toxin to the target cells and used for treatment of lymphoma.

Denitrification A process of energy-yielding metabolism in which nitrate or nitrite is converted to molecular nitrogen or nitrous oxide by nitrifying bacteria.

Denitrifying Bacteria The bacteria capable of nitrification, e.g., *Bacillus licheniformis.*

Denopterin (mol wt 469) An antineoplastic agent.

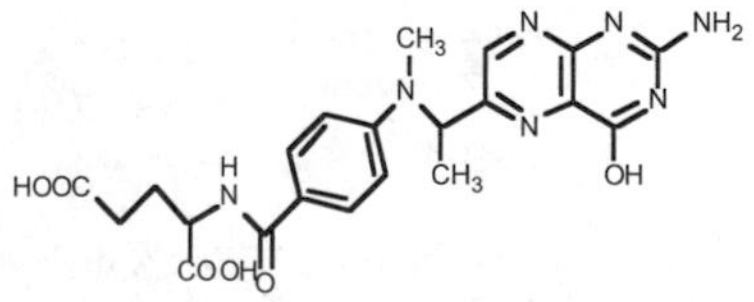

Densitometer A device used for measuring the absorbance in areas of interest, e.g., scanning a chromatogram or an electrophoregram with a densitometer.

Density Mass per unit volume, e.g., g/ml.

Density Gradient Solution in which the concentration of a solute changes with distance.

Density Gradient Centrifugation A technique for fractionation of macromolecules in which a mixture of macromolecules to be separated are centrifuged in a density gradient tube. Different macromolecules sediment at different positions in the density gradient tube after centrifugation.

Density Gradient Equilibrium Centrifugation A type of density gradient centrifugation in which a concentrated salt solution (e.g., cesium chloride) containing macromolecules to be separated is ultracentrifuged until the salt achieves its equilibrium and forms a linear salt gradient. The macromolecules in the salt density gradient migrate to the positions where their densities equal those of the gradient (also called isopycnic density gradient).

Density Marker Bead A colored bead with known density used for calibrating density gradients.

Dental Plaque The matrix of microbial cells and microbiologically produced extracellular polysaccharides that forms on the tooth surface.

Dentin The chief calcified structure of the teeth surrounding the inner part of the tooth. It is covered by enamel on the crown or cementum on the roots.

Dentinogenesis The formation of dentin of the teeth.

Deossification Removal of the mineral elements of bones.

2′-Deoxyadenosine (mol wt 251) A deoxyribonucleoside of adenine.

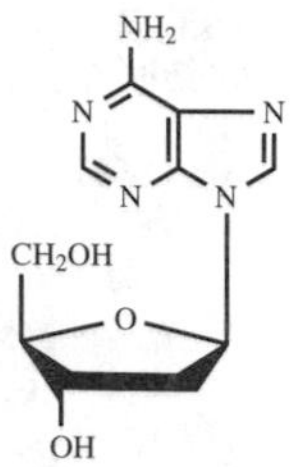

3′-Deoxyadenosine (mol wt 251) A deoxyribonucleoside of adenine.

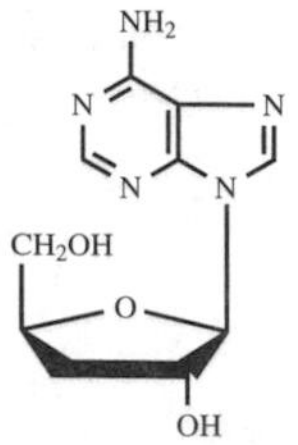

2′-Deoxyadenosine 5′-Diphosphate (dADP, mol wt 411) A diphosphate form of the nucleotide of adenine.

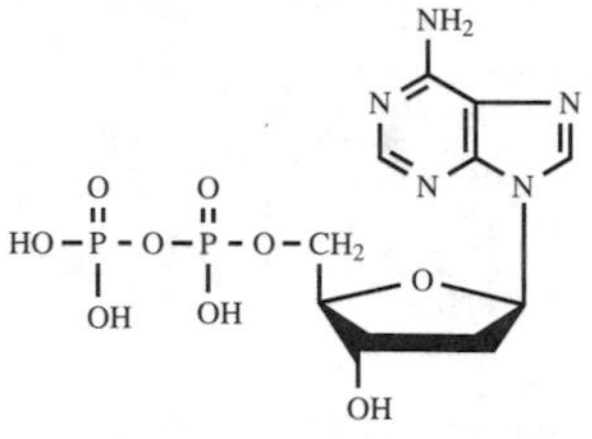

Deoxyadenosine Kinase The enzyme that catalyzes the following reaction:

$$\text{ATP} + \text{deoxyadenosine} \rightleftharpoons \text{ADP} + \text{dAMP}$$

2′-Deoxyadenosine 5′-Monophosphate (dAMP, mol wt 331) A monophosphate form of the nucleotide of adenine.

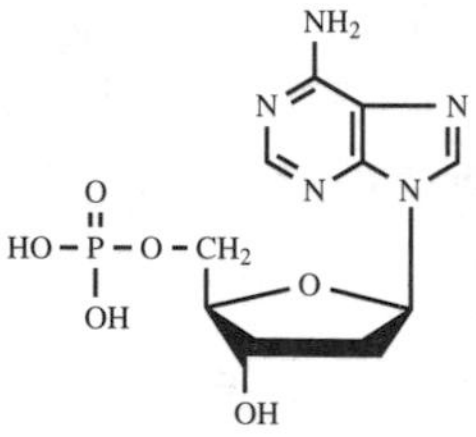

2′-Deoxyadenosine 5′-Triphosphate (dADP, mol wt 491) A triphosphate form of the deoxynucleotide of adenine.

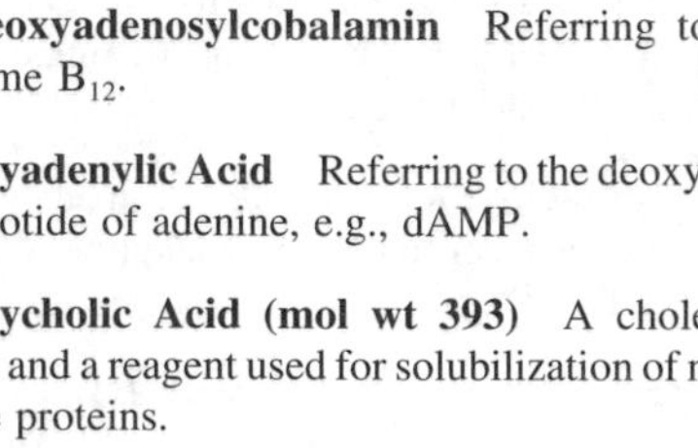

5′-Deoxyadenosylcobalamin Referring to coenzyme B_{12}.

Deoxyadenylic Acid Referring to the deoxyribonucleotide of adenine, e.g., dAMP.

Deoxycholic Acid (mol wt 393) A choleretic agent and a reagent used for solubilization of membrane proteins.

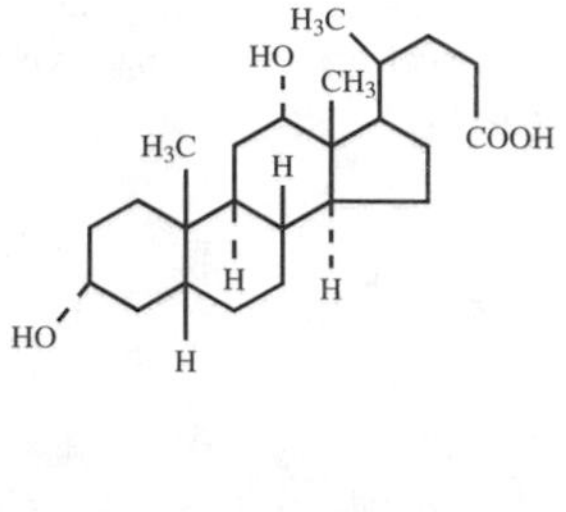

Deoxycorticosterone (mol wt 330) A mineral corticoid.

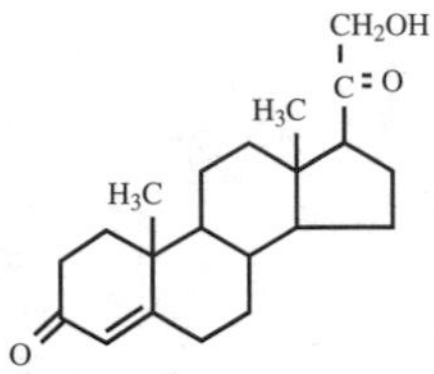

2′-Deoxycytidine (mol wt 227) A deoxyribonucleoside of cytosine.

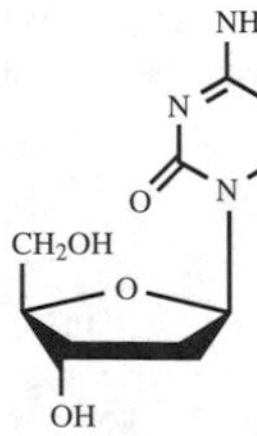

Deoxycytidine Aminohydrolase See deoxycytidine deaminase.

Deoxycytidine Deaminase The enzyme that catalyzes the following reaction:

$$\text{dCMP} \rightleftharpoons \text{dUMP} + NH_3$$

2′-Deoxycytidine 5′-Diphosphate (dCDP, mol wt 387) A diphosphate form of the deoxyribonucleotide of cytosine.

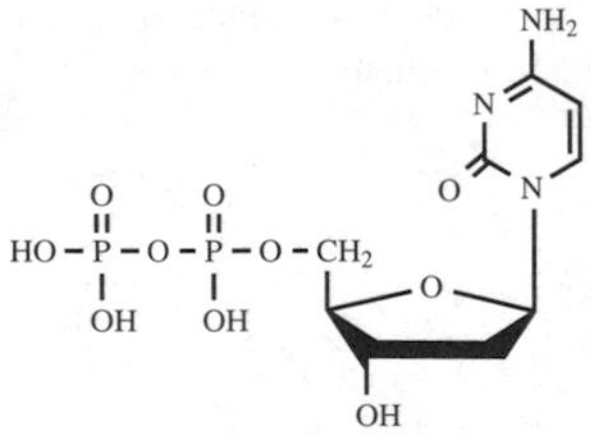

2′-Deoxycytidine 5′-Monophosphate (dCMP, mol wt 307) A monophosphate form of the deoxynucleotide of cytosine.

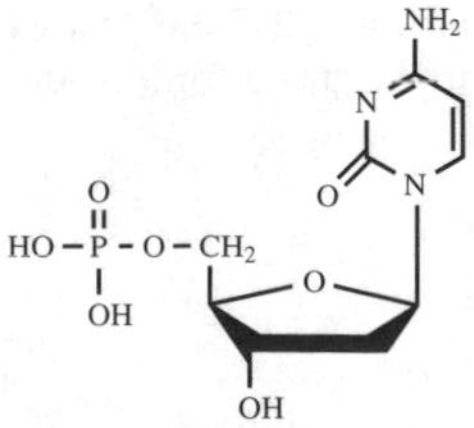

Deoxycytidine Triphosphatase The enzyme that catalyzes the following reaction:

$$\text{dCTP} + H_2O \rightleftharpoons \text{dCMP} + \text{PPi}$$

2′-Deoxycytidine 5′-Triphosphate (dCTP, mol wt 467) A triphosphate form of the nucleotide of cytosine.

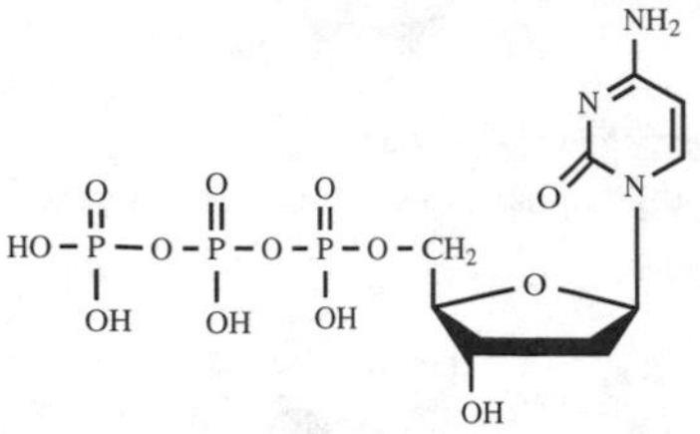

Deoxycytidylate 5-Hydroxymethyl Transferase The enzyme that catalyzes the following reaction:

$$\text{5-Hydroxymethyl dCMP} + \text{tetrahydrofolate} \rightleftharpoons \text{dCMP} + \text{5,10-methylenetetrahydrofolate} + H_2O$$

Deoxycytidylate Kinase The enzyme that catalyzes the following reaction:

$$\text{ATP} + \text{dCMP} \longrightarrow \text{ATP} + \text{dCDP}$$

2′-Deoxycytidylic Acid Referring to deoxyribonucleotide of cytosine, e.g., dCMP.

Deoxydihydrostreptomycin (mol wt 568) An antibiotic derived from streptomycin, used for the treatment of tuberculosis.

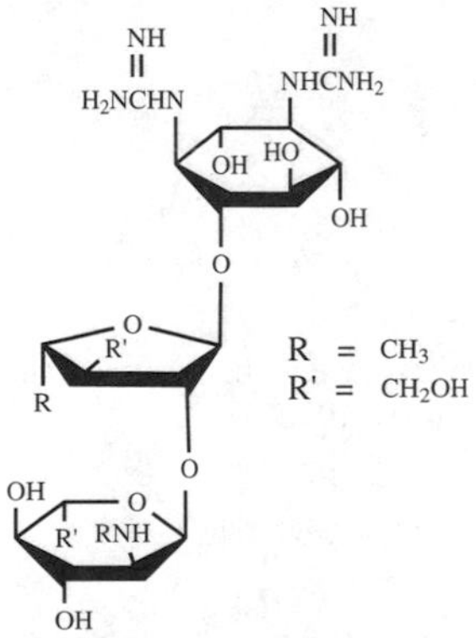

3′-Deoxy-3′-Fluorothymidine (mol wt 244) An antiviral agent.

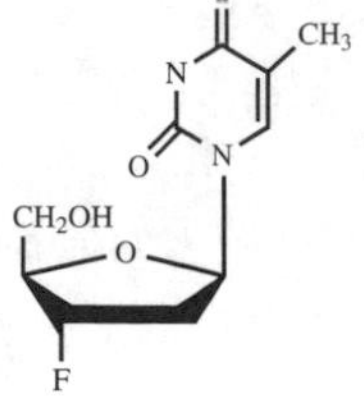

2-Deoxy-D-Galactose (mol wt 164) A deoxy sugar.

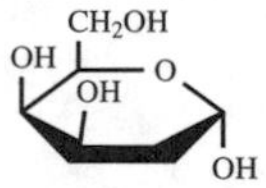

6-Deoxygalactose (mol wt 164) A deoxysugar found in blood group substances (also known as fucose).

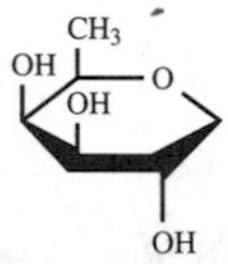

2-Deoxygalactosidase The enzyme that catalyzes the following reaction:

2-Deoxygalactoside + H_2O ⇌ An alcohol + 2-deoxy-D-galactose

2-Deoxygluconate Dehydrogenase The enzyme that catalyzes the following reaction:

2-Deoxygluconate + NAD^+ ⇌ 2-deoxy-3-keto-gluconate + NADH

6-Deoxy-D-Glucose (mol wt 164) A derivative of glucose.

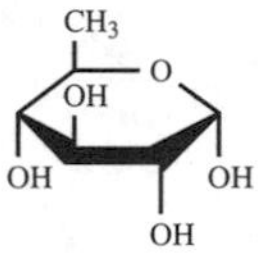

6-Deoxyglucosidase The enzyme that catalyzes the following reaction:

6-Deoxyglucoside + H_2O ⇌ An alcohol + 6-deoxy-D-glucose

Deoxy-GTPase The enzyme that catalyzes the following reaction:

dGTP + H_2O ⇌ Deoxyguanosine + triphosphate

2′-Deoxyguanosine (mol wt 267) A deoxyribonucleoside of guanine.

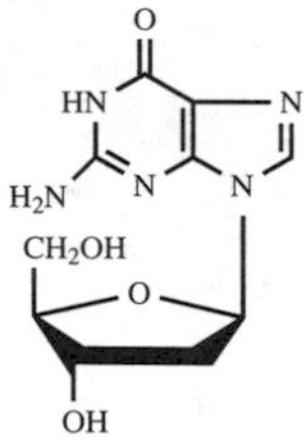

2′-Deoxyguanosine 5′-Diphosphate (dGDP, mol wt 427) The diphosphate form of deoxyribonucleotide of guanine.

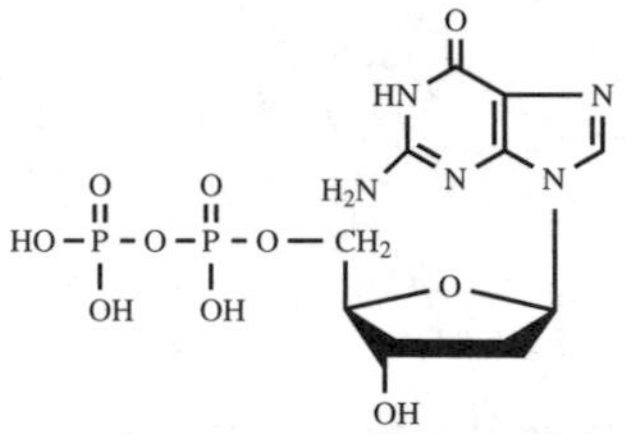

Deoxyguanosine Kinase The enzyme that catalyzes the following reaction:

ATP + deoxyguanosine ⇌ ADP + dGMP

2′-Deoxyguanosine 5′-Monophosphate (dGMP, mol wt 347) A monophosphate form of the deoxyribonucleotide of guanine.

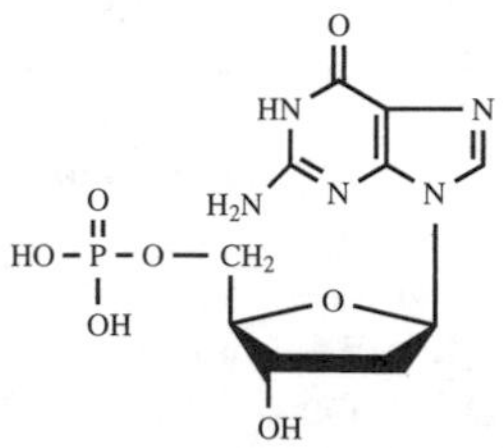

2′-Deoxyguanosine 5′-Triphosphate (dGTP, mol wt 507) A triphosphate form of the deoxyribonucleotide of guanine.

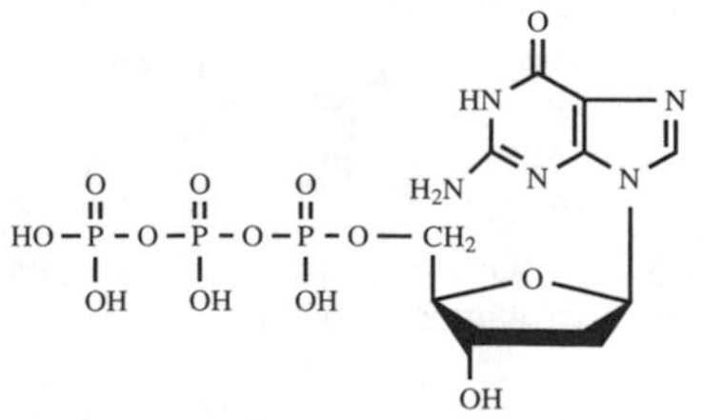

Deoxyguanylate Kinase The enzyme that catalyzes the following reaction:

$$ATP + GMP \rightleftharpoons ADP + GDP$$

Deoxyguanylic Acid Referring to deoxyribonucleotide of guanine.

Deoxyhemoglobin Hemoglobin without molecular oxygen attached.

2′-Deoxyinosine (mol wt 252) A deoxyribonucleoside of hypoxanthine.

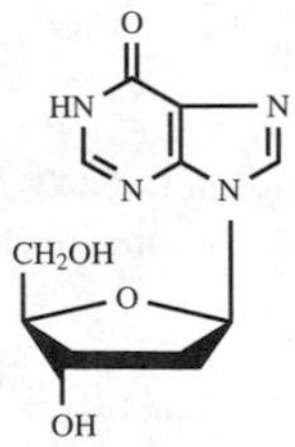

2′-Deoxyinosine 5′-Diphosphate (mol wt 412) A diphosphate form of hypoxanthine.

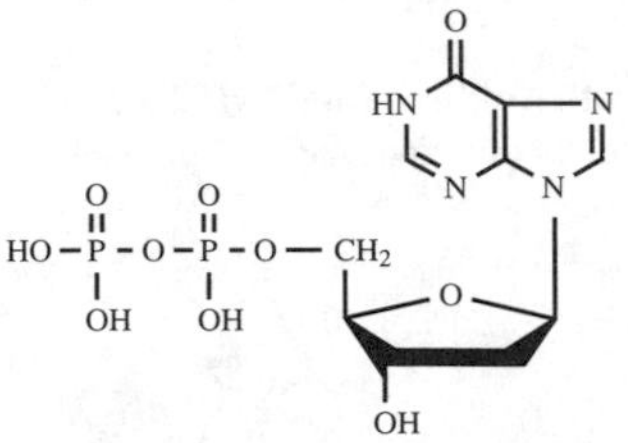

2′-Deoxyinosine 5′-Monophosphate (mol wt 332) A diphosphate form of hypoxanthine.

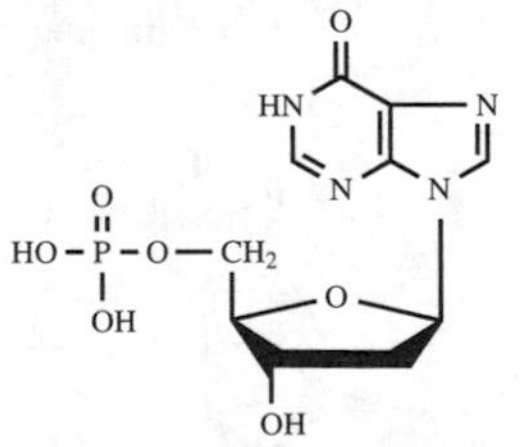

2′-Deoxyinosine 5′-Triphosphate (dITP, mol wt 492) A triphosphate form of the deoxyribonucleotide of hypoxanthine.

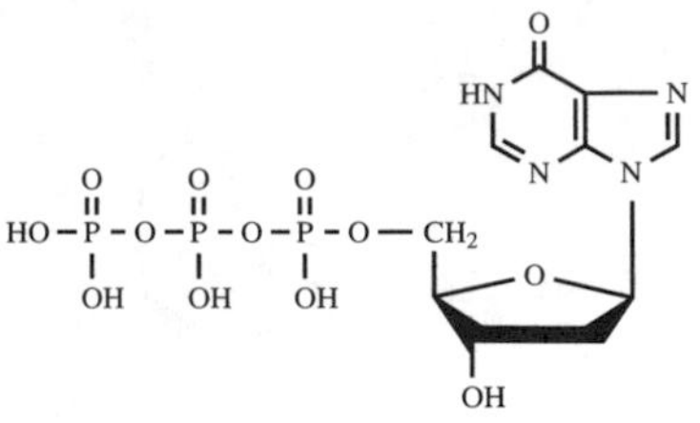

deoxyMb Abbreviation for deoxymyoglobin.

Deoxymyoglobin Myoglobin without molecular oxygen attached.

1-Deoxynojirimycin (mol wt 163) An antibiotic derived from nojirimycin that is an α-glucosidase inhibitor.

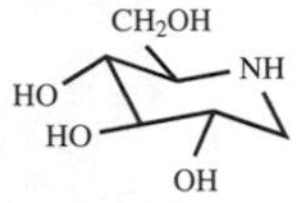

3′-Deoxynucleosidase The enzyme that catalyzes the following reaction:

$$\text{A nucleoside 3'-monophosphate} + H_2O \rightleftharpoons \text{Deoxynucleoside} + \text{Pi}$$

Deoxynucleotides Referring to the deoxyribonucleotide of adenine, guanine, thymine, or cytosine.

Deoxyriboaldolase The enzyme that catalyzes the following reaction:

$$\text{Deoxyribose 5'-phosphate} \rightleftharpoons \text{Glyceraldehyde 3-phosphate} + \text{acetaldehyde}$$

Deoxyribonuclease (DNase or DNAase) The enzyme that catalyzes the internal hydrolysis of DNA.

Deoxyribonuclease I An endonuclease that hydrolyzes the phosphodiester linkage of DNA yielding 5′-phosphate terminated mono- or oligonucleotides with a free hydroxyl group at the 3′ position.

Deoxyribonuclease II An endonuclease that hydrolyzes the phosphodiester linkage of DNA yielding 3′-phosphate terminated mono- or oligonucleotides with a free hydroxyl group at the 5′ position.

Deoxyribonuclease III An exonuclease that catalyzes exonucleolytic degradation of double-stranded DNA progressively in the 3’ to 5’ direction releasing 5’-phosphomononucleotides.

Deoxyribonuclease IV A T_4-induced deoxyribonuclease that catalyzes the preferential endonucleolytic cleavage of single-stranded DNA yielding 5′-phosphooligonucleotides.

Deoxyribonuclease V The enzyme that catalyzes the endonucleolytic cleavage of DNA near pyrimidine dimers yielding products with 5′-phosphate.

Deoxyribonuclease S_1 The enzyme that catalyzes the endonucleolytic cleavage of DNA yielding 5′-phosphomononucleotides and 5′-phosphooligonucleotides.

Deoxyribonuclease X The enzyme that catalyzes the endonucleolytic cleavage of supercoiled DNA to a linear duplex.

Deoxyribonucleic Acid (DNA) Carrier molecule of the genetic information in cells, composed of chains of phosphate, sugar, deoxyribose, adenine, guanine, thymine and cytosine. The DNA molecules can be either linear, circular, single stranded, or double stranded.

Deoxyribonucleoside A deoxynucleoside consisting of a base and a deoxyribose.

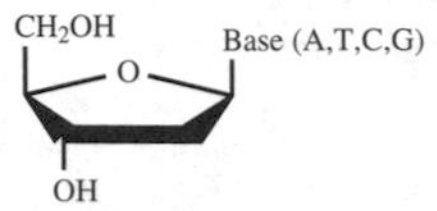

Deoxyribophage Referring to a bacteriophage containing DNA as genetic material.

Deoxyribopyrimidine Endonucleosidase The enzyme that cleaves the N-glycosidic bond between the 5'-pyrimidine residue in cyclobutadipyrimidine (in DNA) and the corresponding deoxy-D-ribose residue.

2′-Deoxyribose (mol wt 134) A 5-carbon sugar and component of DNA.

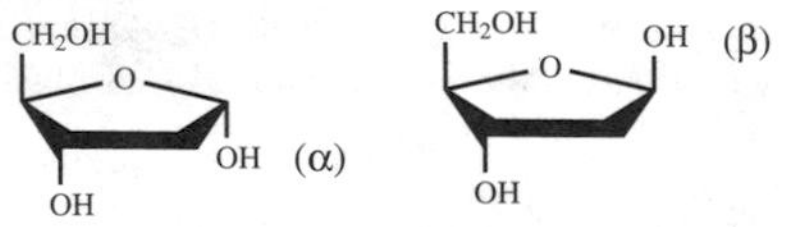

3′-Deoxyribose (mol wt 134) A deoxyribose.

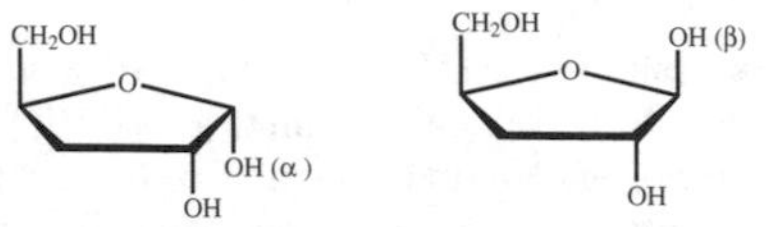

2′-Deoxyribose 5′-Monophosphate (mol wt 214) A phosphate form of deoxyribose.

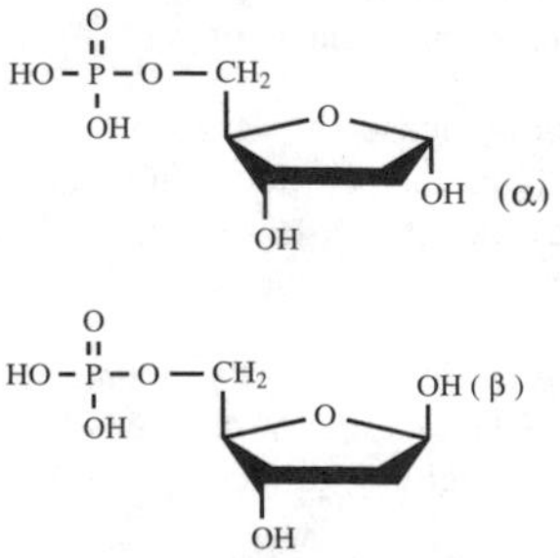

Deoxyribose 5-Phosphate Aldolase See deoxyriboaldolase.

Deoxyribotide Synonym of deoxyribonucleotide.

Deoxyribovirus Referring to viruses containing DNA as genetic material.

2-Deoxystreptamine (mol wt 162) A component of many aminoglycoside antibiotics.

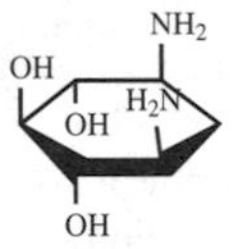

Deoxy-Sugar A sugar with one of its hydroxyl groups reduced to a hydrogen (e.g., 6-deoxygalactose).

2′-Deoxythymidine (mol wt 242) Referring to thymidine because thymidine itself is a 2′-deoxynucleoside.

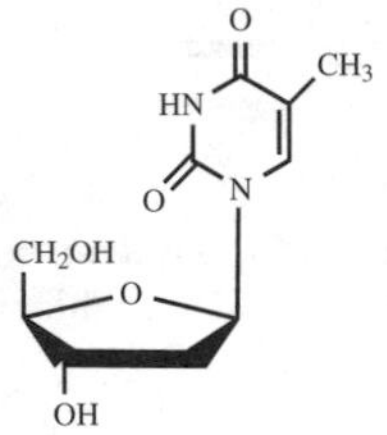

3′-Deoxythymidine (mol wt 226) A dideoxynucleoside of thymine since thymidine is a 2′-deoxynucleoside.

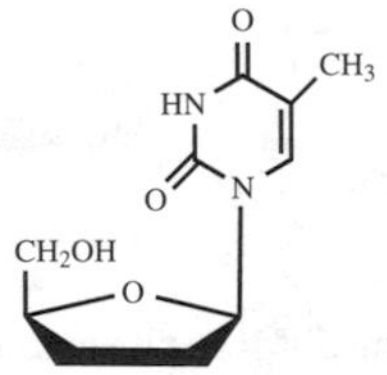

5′-Deoxythymidine (mol wt 226) A dideoxynucleoside of thymine since thymidine is a 2′-deoxynucleoside.

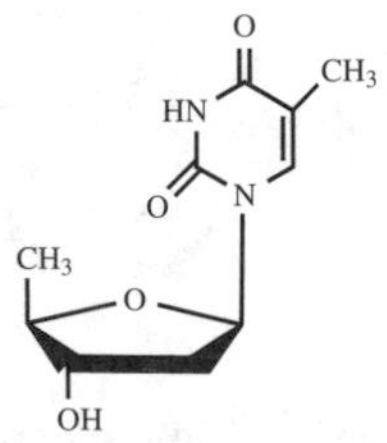

Deoxythymidine 5′-Diphosphate (mol wt 402) A diphosphate form of the deoxynucleotide of thymine.

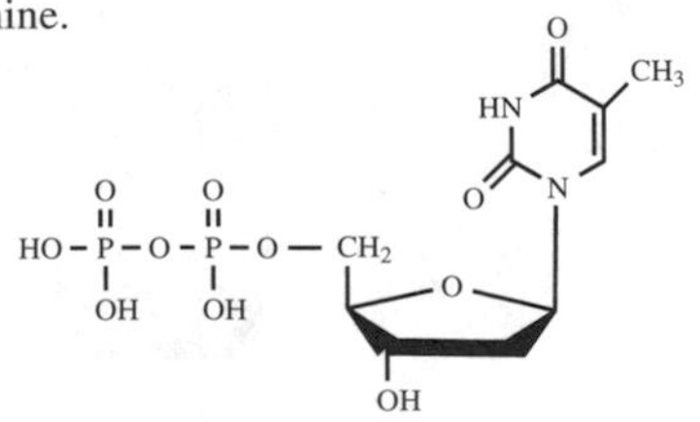

3′-Deoxythymidine 5′-Diphosphate (mol wt 386) A diphosphate form of the dideoxynucleotide of thymine since thymidine itself is a 2′-deoxynucleoside.

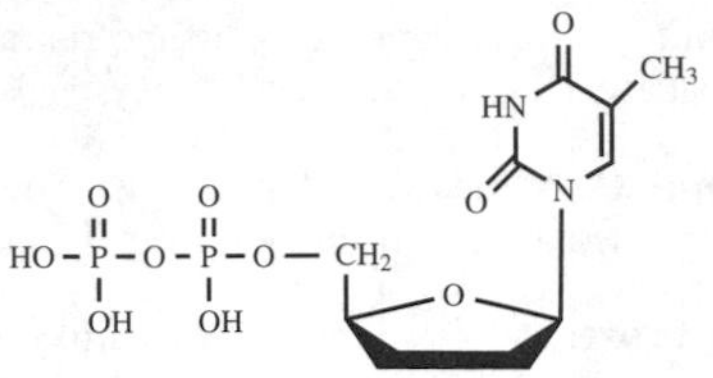

Deoxythymidine 5′-Monophosphate (mol wt 322) A monophosphate form of the deoxyribonucleotide of thymine.

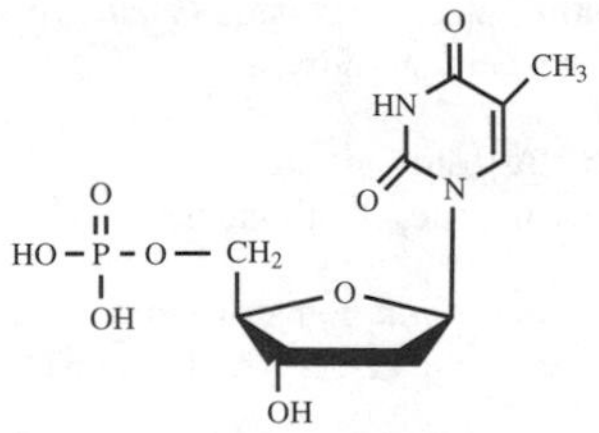

3′-Deoxythymidine 5′-Monophosphate (mol wt 306) A monophosphate form of the dideoxyribonucleotide of thymine (thymidine itself is a 2′-deoxynucleoside).

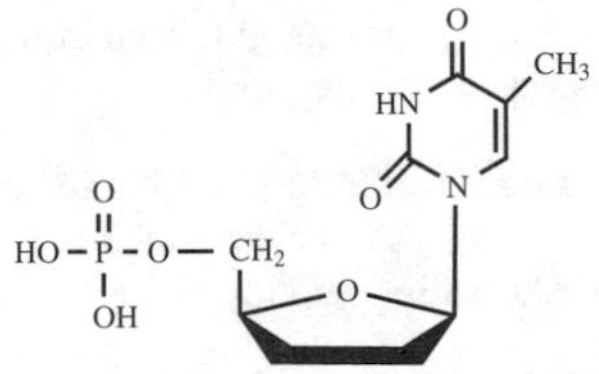

Deoxythymidine 5′-Triphosphate (mol wt 482) A triphosphate form of the deoxynucleotide of thymine.

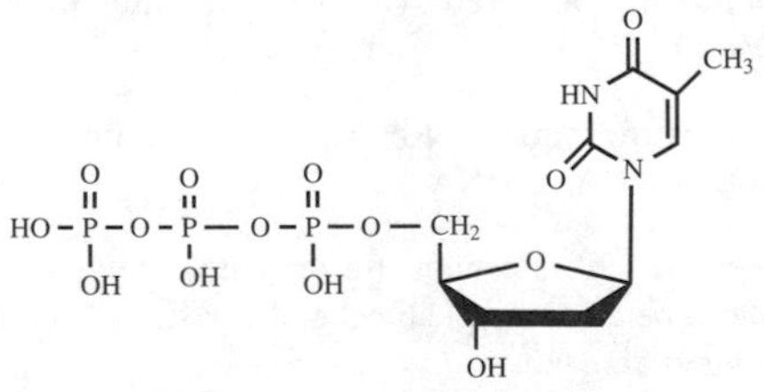

3′-Deoxythymidine 5′-Triphosphate (ddTTP, mol wt 466) A triphosphate form of the dideoxynucleotide of thymine (thymidine itself is a 2′-deoxynucleoside).

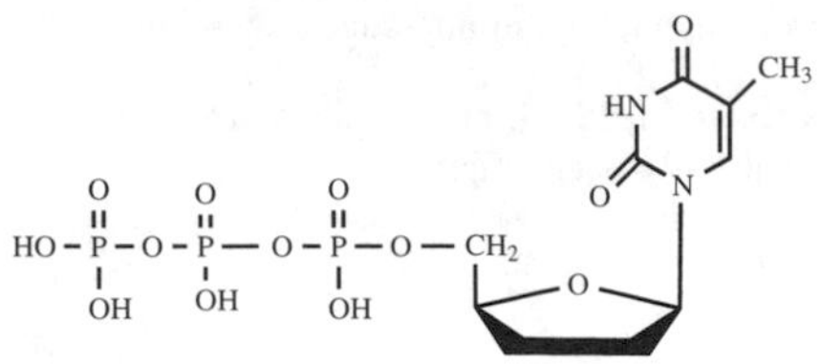

Deoxythymidylic Acid Referring to deoxynucleotide of thymine.

2′-Deoxyuridine (mol wt 228) A deoxynucleoside of uracil.

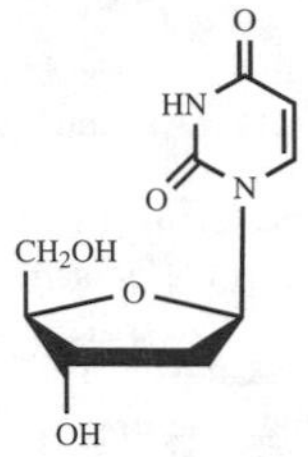

2′-Deoxyuridine 5′-Diphosphate (dUDP, mol wt 388) A diphosphate form of the deoxynucleotide of uracil.

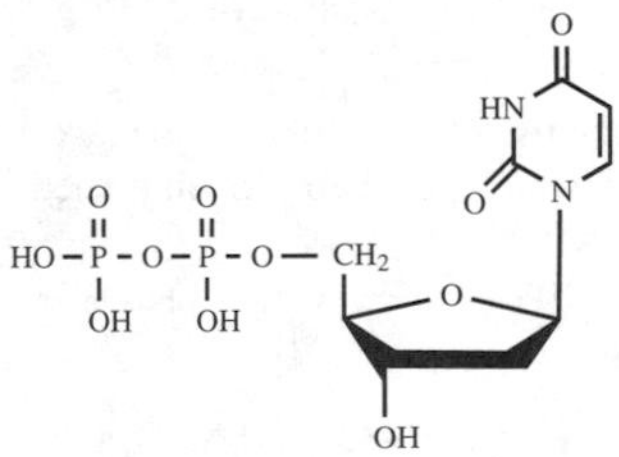

2′-Deoxyuridine 5′-Monophosphate (dUMP, mol wt 308) A monophosphate form of the deoxynucleotide of uracil.

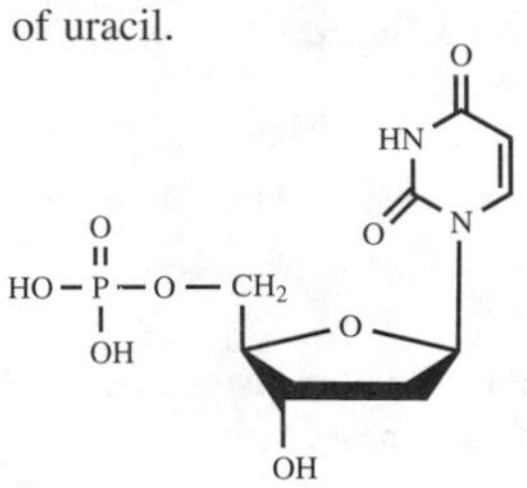

2′-Deoxyuridine Triphosphate (dUTP, mol wt 468) A triphosphate form of the deoxynucleotide of uracil.

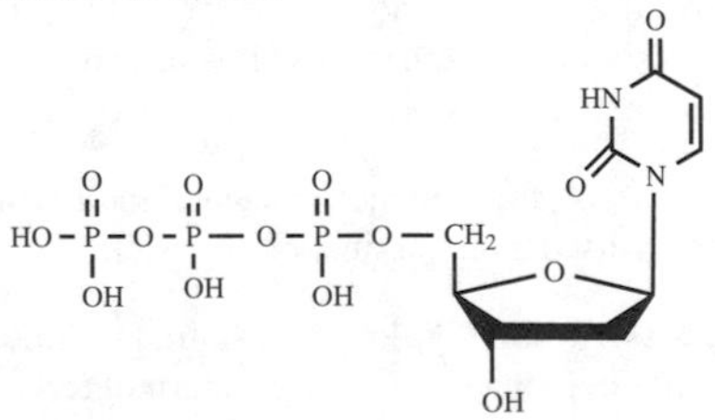

Deoxyuridylic Acid Referring to the deoxynucleotide of uracil.

Depa A trade name for valproic acid, an anticonvulsant.

Depacon A trade name for valproic acid or valproate, an anti-epileptic agent.

Depactin A protein isolated from starfish eggs that depolymerizes the F-actin and inhibits the actin polymerization.

Depakene A trade name for valproate sodium, used as an anticonvulsant.

DepAndro A trade name for testosterone, an androgen hormone.

Depen A trade name for penicillamine, a chelating agent used as an anti-rheumatic drug.

Dependovirus Referring to viruses whose replication depends upon the presence of helper virus.

Dephospho-CoA Kinase The enzyme that catalyzes the following reaction:

$$\text{ATP} + \text{dephospho-CoA} \rightleftharpoons \text{ADP} + \text{CoA}$$

Dephospho-CoA Pyrophosphorylase The enzyme that catalyzes the following reaction:

$$\text{ATP} + \text{pantetheine 4}'\text{-phosphate} \rightleftharpoons \text{PPi} + \text{dephospho-CoA}$$

Dephosphorylase Kinase The enzyme that catalyzes the following reaction:

$$4\ \text{ATP} + 2\ \text{phosphorylase b} \rightleftharpoons 4\ \text{ADP} + \text{phosphorylase a}$$

Dephosphophosphorylase Kinase Synonym of phosphorylase kinase.

Dephosphorylation Removal of a phosphate group from an organic compound.

DepMedalone A trade name for methylprednisolone sodium succinate, used as an anti-inflammatory agent.

Depocyt A trade name for cytarabine-liposome complex used as an antineoplastic agent.

Depo-Estradiol Cypionate A trade name for estradiol cypionate, an estrogen hormone.

Depoject A trade name for methylprednisolone acetate, used as an anti-inflammatory agent.

Depolarization Elimination of a polarized state or an electric potential across the membrane of a nerve cell or a muscle cell.

Depolymerizing Enzyme The enzyme that catalyzes the degrading of polymer to monomers or oligomers.

Depo-Medrol A trade name for methyprednisolone acetate, a corticosteroid used as an anti-inflammatory agent.

Deponit A trade name for nitroglycerin, an antianginal agent that decreases cardiac oxygen demand.

Depopred A trade name for methylprednisolone acetate, used as an anti-inflammatory agent.

Depo-Provera A trade name for medroxyprogesterone acetate, a hormone with antineoplastic activity.

Depot-Fat The stored fat in the body.

Depressant Any substance capable of reducing bodily functional activity.

Deproic A trade name for valproic acid or valproate, an antiepileptic agent.

Deprol A combination drug that contains meprobamate and benactyzine hydrochloride.

Deproteinization The process of removing proteins from a biological sample.

Depside An ester formed by condensation of two or more phenols and carboxylic acids.

Depsipeptide A polypeptide consisting of alternating peptide and ester bonds.

Depsipeptide Antibiotic Antibiotics consisting of amino and hydroxy acids, which are linked together by alternating peptide and ester bonds.

Deptran A trade name for doxepin hydrochloride, an antidepressant that increases the amounts of norepinephrine and serotonin in the CNS.

Depurination Removal of purine bases from a DNA or RNA.

Depyrimidination Removal of pyrimidine bases from a DNA or RNA.

Deralin A trade name for propanolol hydrochloride, a beta adrenergic blocker that reduces cardiac oxygen demand.

DeRib Abbreviation for deoxyribose.

Derivative A substance derived from a parent compound through chemical modification.

Derived Carbohydrate Compounds derived from sugars, e.g., amino sugars or sugar acids.

Derived Lipid Lipid obtained from hydrolysis of naturally occurring lipid.

Derived Protein Proteins resulting from treatment by heat, acid, base, enzyme, or other agent.

DermaCort A trade name for hydrocortisone, a steroid used as an anti-inflammatory agent.

Dermal Pertaining to the skin.

Dermaseptin An antimicrobial peptide derived from amphibian skin.

Dermatan Sulfate A glycosaminoglycan or mucopolysaccharide consisting of iduronate, *N*-acetylgalactosamine, and glucuronic acid.

Dermatitis The inflammation of skin.

Dermatobia A genus of flies. The parasitic maggots of *D. hominis* that can cause a serious skin disease in humans.

Dermatomycosis Fungal infection of the skin.

Dermatomyositis An inflammation disorder of the skin and underlying tissues including muscles.

Dermatop A trade name for prednicarbate, an anti-inflammatory agent.

Dermatophilus A genus of aerobic or facultative anaerobic, catalase-positive bacteria (*Actinomycetales*).

Dermatophytes A group of fungi capable of degrading keratinized tissue, e.g., skin, nails, and hair.

Dermatophytosis Any fungus infection of the skin.

Dermatosclerosis A skin disorder characterized by the appearance of patches of thick, leathery skin.

Dermatotropic Having a selective affinity for the skin.

Dermocarpella A genus of unicellular *Cyanobacteria.*

Dermolate A trade name for hydrocortisone, used as an anti-inflammatory agent.

Dermorphin Any of a group of heptapeptides amides with opiate-like activity, isolated from the skin of frogs with the following general structure:

H-Tyr-Ala-Phe-Gly-Tyr-Xaa-Ser-

Dermovate A trade name for clobetasol propionate, a corticosteroid.

Dermtex A trade name for hydrocortisone, used as an anti-inflammatory agent.

Deronil A trade name for dexamethasone, used as an anti-inflammatory agent.

DERV Abbreviation for duck embryo rabies vaccine.

Derxia A genus of Gram-negative, aerobic, catalase-negative, chemoorganotrophic bacteria.

Des- A prefix meaning specific lack.

DES Abbreviation for diethylstilbestrol.

Desalanine Insulin Insulin from which the alanine at the C-terminus of the β-chain is removed.

Desalting A procedure in which salts are removed from a solution by techniques such as gel filtration or dialysis.

Desaspidin (mol wt 446) An uncoupler for oxidative and photophosphorylation reactions.

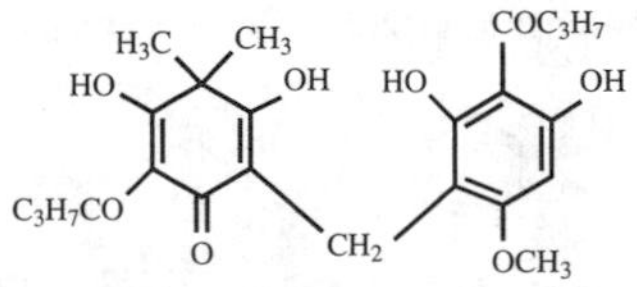

Desaturase The enzyme that catalyzes the conversion of a saturated fatty acid to an unsaturated fatty acid.

Descarboxy-Clotting Factor Any of the abnormal blood clotting factors that contains glutamic acid residues instead of γ-carboxyglutamic acid residue.

Descending Chromatography A chromatographic method in which the mobile eluting phase moves downward along the support medium.

Desensitization 1. Reduction of receptor response of a target cell to signaling ligand. 2. Modification of an allosteric enzyme through mutation or chemical modification leading to the loss of allosteric response to effector without loss of catalytic activity. 3. Reduction or elimination of allergic response to allergen through the repeated injection of small doses of allergen.

Desensitized Enzyme Modification of an allosteric enzyme so that it loses its allosteric response to an effector.

Deseril A trade name for methysergide maleate, a serotonin agonist and adrenergic blocker.

Deserpidine (mol wt 579) An antihypertensive agent that inhibits norepinephrine release.

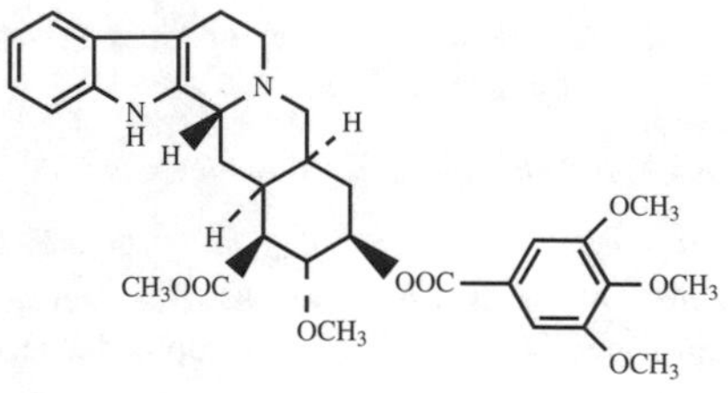

Desferal A trade name for deferoxamide mesylate, a substance that chelates iron.

Desflurane (mol wt 268) An anesthetic agent.

$CHF_2OCHFCF_3$

Desiccant An agent capable of absorbing moisture from a substance (e.g., calcium chloride).

Desiccater A device that holds desiccant for the purpose of drying material.

Desipramine (mol wt 266) An antidepressant agent.

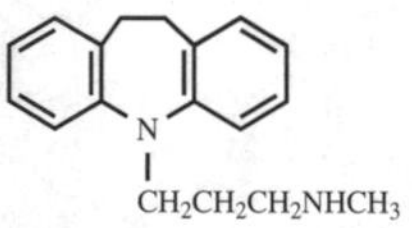

Desmids A group of freshwater unicellular green algae.

Desmin A protein in striated muscle that binds microfilaments together.

Desmo- A prefix denoting ligament or bind.

Desmofibrin The stabilized fibrin polymer formed from soluble fibrin polymer in the final stages of the blood clotting process.

Desmoglein A membrane protein of the mature desmosomal junction involved in the mediating cell-cell adhesion.

Desmolase An enzyme complex that catalyzes the removal of side chains from cholesterol.

Desmopressin An analog peptide of vasopressin with antidiuretic activity.

Desmosine (mol wt 879) A covalent cross-linker for polypeptide chains in elastin, a naturally occurring amino acid.

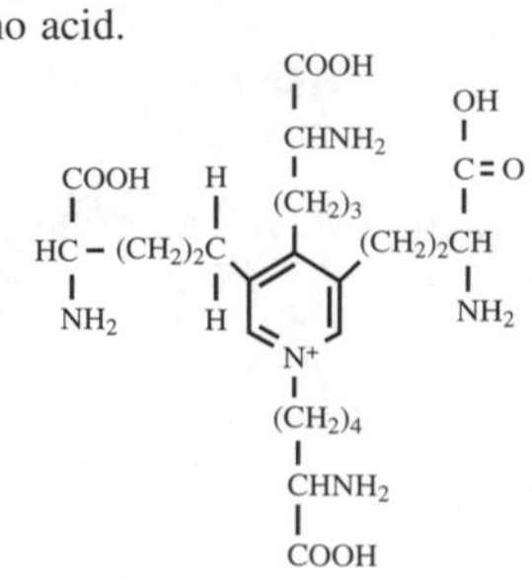

Desmosome An intercellular junction in animal cells that functions in cell adhesion.

Desmotubule The tubular structure that lies in the central channel of plasmodesma between two plant cells.

Desogen A trade name for a combination drug containing desogestrel and ethinyl estradiol used as an oral contraceptive agent.

Desogestrel (mol wt 310) A contraceptive agent.

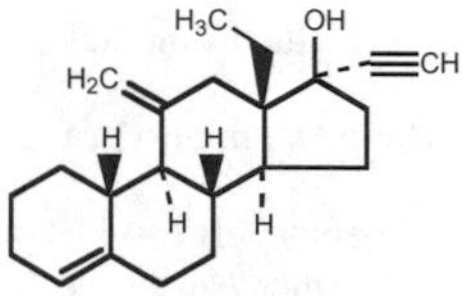

Desomorphine (mol wt 271) A narcotic analgesic agent.

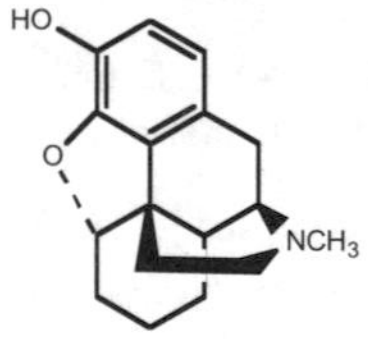

Desonide (mol wt 417) An anti-inflammatory agent.

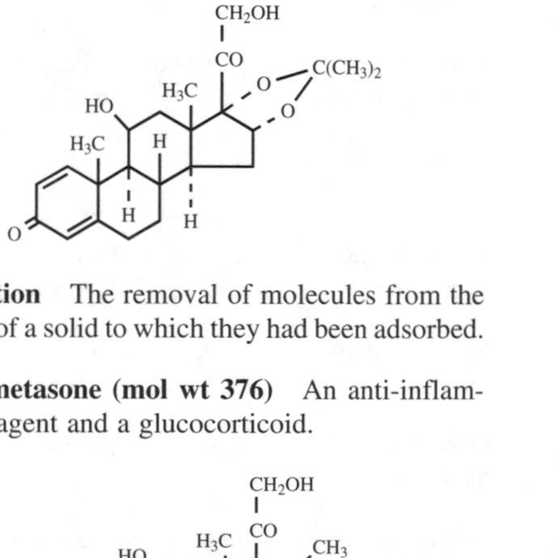

Desorption The removal of molecules from the surface of a solid to which they had been adsorbed.

Desoximetasone (mol wt 376) An anti-inflammatory agent and a glucocorticoid.

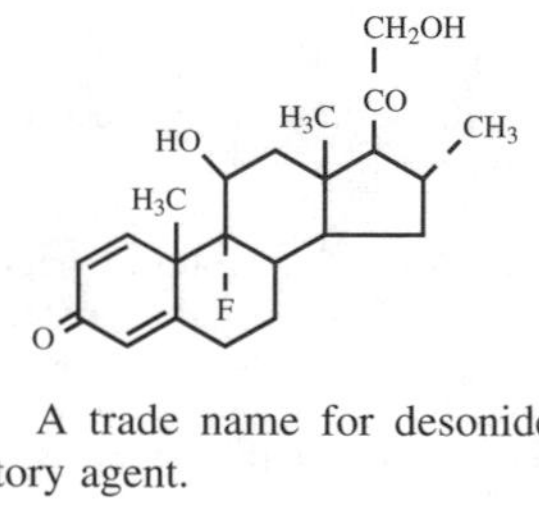

Desowen A trade name for desonide, an anti-inflammatory agent.

Desoximetasone (mol wt 376) An anti-inflammatory agent.

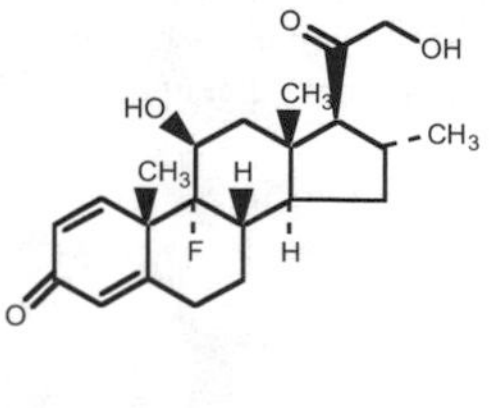

Desoxyn A trade name for methamphetamine hydrochloride, a cerebral stimulant that promotes transmission of nerve impulses by releasing stored norepinephrine.

Desquan A trade name for benzoyl peroxide gel, used as an antimicrobial and comedolytic agent.

Desthio- A prefix denoting the replacement of a sulfur atom by two hydrogen atoms.

Destomycin A (mol wt 527) A broad spectrum antimicrobial and anthelmintic agent.

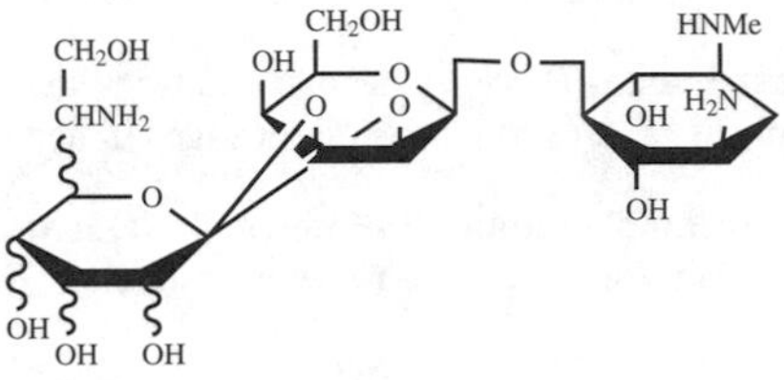

Destrin An actin-binding protein.

Desulfobacter A genus of Gram-negative sulfate-reducing bacteria.

Desulfobulus A genus of Gram-negative sulfate-reducing bacteria.

Desulfococcus A genus of Gram-negative sulfate-reducing bacteria.

Desulfomonas A genus of Gram-negative sulfate-reducing bacteria.

Desulfonema A genus of sulfate-reducing bacteria.

Desulforubidin A sirohaem-containing sulfite reductase from *Desulfovibrio.*

Desulfosarcina A genus of Gram-negative sulfate-reducing bacteria.

Desulfotomaculum A genus of Gram-negative endospore-forming bacteria.

Desulfovibrio A genus of Gram-negative sulfate-reducing bacteria.

Desulfoviridin A sulfite reductase occurring in cytoplasm in species of *Desulfovibrio.*

Desulfurococcus A genus of chemolithotrophic archaebacteria.

Desulfuromonas A genus of Gram-negative, obligate anaerobic bacteria.

Desxasone-LA A trade name for dexamethasone acetate, a corticosteroid hormone.

Desyrel A trade name for trazodone hydrochloride, an antidepressant that inhibits serotonin uptake in the brain.

DET Abbreviation for diethyltryptamine.

Detergent Gel Electrophoresis Synonym of SDS gel electrophoresis.

Determinant The region on an antigen or a protein that determines the antigenic specificity of the antigen or protein.

Deterpenation The elimination of terpenes from essential oils.

Detoxin Complex A group of antagonists of blasticidin produced by *Streptomyces caespitosus* that counteracts the inhibitory action of blasticidin S against *Bacillus cereus.*

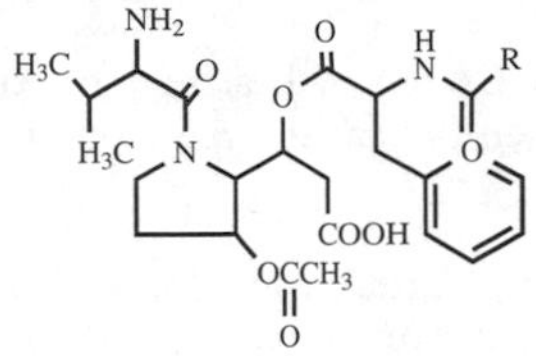

Detroit-6 An established cell line derived from human sternal bone marrow.

Detrol A trade name for tolterodine tartrate, an antimuscarinic agent that competitively blocks muscarinic receptor sites and decreases bladder contraction.

Dettol Chelate A disinfectant containing chloroxylenol and EDTA.

Deuterated Labeled with deuterium.

Deuterium Heavy hydrogen; its nucleus contains one proton and one neutron.

Deuteromycetes Fungi without a known sexual stage (also called fungi imperfecti).

Deuteron A deuterium nucleus containing one proton and one neutron.

DEV Abbreviation for 1. duck egg vaccine; 2. duck egg virus; 3. duck embryo vaccine; 4. duck embryo virus.

Devoret Test A test for detection of carcinogenic substance based upon the ability of the substance on induction of λ phage in lysogenic *E. coli* cells.

Devrom A trade name for bismuth subgallate, an antidiarrheal agent.

Dex Abbreviation for 1. dexamethasone; 2. dextran.

Dex A trade name for dexamethasone sodium phosphate, used as an anti-inflammatory agent.

DEXA Abbreviation for dual energy X-ray absorptiometry, a method for measuring bone density.

Dexamethasone (mol wt 392) A glucocorticoid and an anti-inflammatory agent.

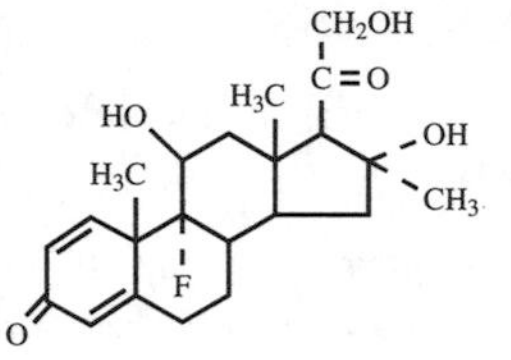

Dexasone A trade name for dexamethasone sodium phosphate, a corticosteroid.

Dexedrine LA A trade name for dextroamphetamine, an anorexiant and central nervous system stimulant.

Dexferrum A trade name for iron dextran used to elevate the serum iron concentration.

Dexon A trade name for dexamethasone sodium phosphate, used as an anti-inflammatory agent.

Dexpanthenol (mol wt 205) A GI stimulant.

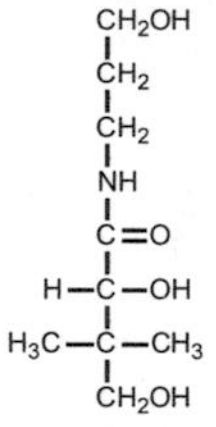

Dextran A branched polymer of glucose produced by bacteria growing on sucrose; it consists mainly of α-(1,6)-glucosidic linkages.

Dextran Bead Beads made by cross-linking dextran and used as a gel filtration medium.

Dextran Blue A high molecular weight dextran dye used as a marker in gel filtration chromatography.

Dextran α-D-(1,2)-Debranching Enzyme The enzyme that catalyzes the hydrolysis of α-D-(1,2) glucosidic linkages at the branched point of dextran.

Dextran-1-6-Glucosidase The enzyme that catalyzes the hydrolysis of α-(1,6)-glucan links in dextrin containing short 1,6-linked side-chains.

Dextran 1,6-α-Isomatotriosidase The enzyme that catalyzes the hydrolysis of 1,6-α-D-glucosidic linkages in dextran to remove successive isomaltose units from the nonreducing ends of the dextran chain.

Dextranase The enzyme that catalyzes the hydrolysis of 1,6-α-D-glucosidic linkages in dextran.

Dextransucrase The enzyme that catalyzes the following reaction:

$$(\alpha\text{-}1,6\text{-glucosyl})n + \text{D-fructose} \rightleftharpoons (\alpha\text{-}1,6\text{-glucosyl})n\text{-}1 + \text{sucrose}$$

Dextrin A polymer of glucose with intermediate chain length produced from partial degradation of starch by heat, acid, or enzyme.

Dextrinase The enzyme that catalyzes the hydrolysis of α-(1,6) glucosidic linkages in dextrin.

Dextroamphetamine Sulfate (mol wt 368) A stimulant for the central nervous system.

$C_{18}H_{28}N_2O_4S$

Dextromethorphan Hydrobromide (mol wt 370) An antitussive agent.

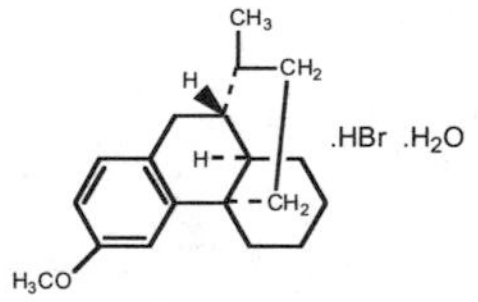

Dextromoramide (mol wt 393) A narcotic and analgesic agent.

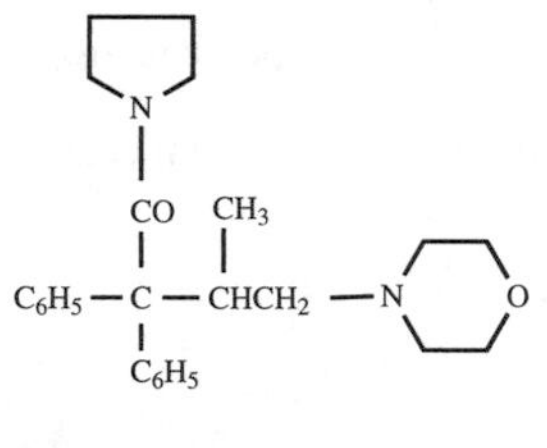

Dextropantothenyl Alcohol A trade name for dexpanthenol, a GI stimulant.

Dextropropoxyphene A trade name for propoxyphene, a narcotic agonist analgesic agent.

Dextrorotatory Substance capable of rotating plane-polarized light to the right or clockwise.

Dextrose Synonym for glucose.

Dextrostat A trade name for dextroamphetamine sulfate, a CNS stimulant.

Dezocine (mol wt 245) An analgesic agent.

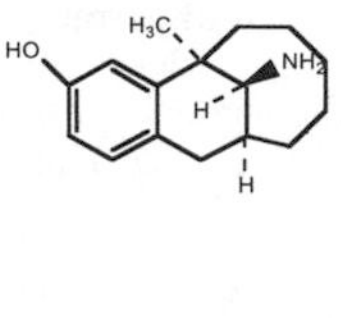

DFA Abbreviation for direct fluorescent antibody.

DFMO Abbreviation for a-difluoro-methyl-ornithine.

DFP Abbreviation for diisopropyl fluorophosphate, a reagent used for analysis of the active site of the proteases.

DFPase Abbreviation for diisopropyl fluorophosphatase, the enzyme that catalyzes the following reaction:

$$\text{Diisopropyl phosphofluoridate} + H_2O \rightleftharpoons \text{Diisopropyl phosphate} + \text{fluoride}$$

DFP Peptide Peptide containing diisopropyl fluorophosphate.

DFU Abbreviation for dideoxyfluorouridine.

DG Abbreviation for 1. deoxyglucose; 2. deoxyguanosine; 3. diglyceride.

dG Abbreviation for 1. deoxyglucose; 2. deoxyguanosine or deoxyguanylate.

2dG Abbreviation for 2-deoxy-D-glucose.

DGAT Abbreviation for diacylglycerol acyltransferase.

DGC Abbreviation for density gradient centrifugation.

DGDG Abbreviation for digalactosyldiacylglycerol.

dGDP Abbreviation for deoxyguanosine diphosphate.

DGE Abbreviation for density gradient electrophoresis.

DGK Abbreviation for diacylglycerol kinase.

dGMP Abbreviation for deoxyguanosine monophosphate.

DGPP Abbreviation for diacylglycerol pyrophosphate.

dGTP Abbreviation for deoxyguanosine triphosphate.

DGV Abbreviation for dextrose-gelatin-veronal solution.

DH Abbreviation for 1. dehydrogenase; 2. delayed hypersensitivity.

DHA Abbreviation for 1. dehydroacetic acid; 2. dehydroascorbic acid; 3. dihydroalanine; 4. dihydroxyacetone; 5. docosahexaenoic acid.

DHAP Abbreviation for dihydroxyacetone phosphate.

DHAPAT Abbreviation for dihydroxyacetone phosphate acyltransferase.

DHB Abbreviation for dihydrobutyrine.

DHBV Abbreviation for duck hepatitis B virus.

DHC Abbreviation for 1. dehydrocholesterol; 2. dehydrocholic acid; 3. dynein heavy chain.

DHE Abbreviation for dihydroergotamine.

DHE 45 A trade name for dihydroergotamine mesylate, an anti-migraine agent.

DHEA Abbreviation for dehydro-epi-androsterone.

DHF Abbreviation for 1. dengue hemorrhagic fever; 2. dihydrofolate.

DHFR Abbreviation for dihydrofolate reductase.

DHIA Abbreviation for dehydroisoandrosterone.

DHM Abbreviation for dihydromorphine.

DHO-DH Abbreviation for dihydro-orotate dehydrogenase.

DHP Abbreviation for dehyrogenated polymer.

DHPA Abbreviation for dihydroxypropyl-adenine, an antiviral agent and analog of adenosine.

DHPG A trade name for ganciclovir sodium, an antiviral agent that inhibits viral DNA replication.

DHPR Abbreviation for dihydropyridine receptor.

DHQase Abbreviation for dehydroquinase.

DHQS Abbreviation for dehydroquinate synthetase.

DHR Abbreviation for delayed hypersensitivity reaction.

DHT Abbreviation for dihydrotestosterone.

DHU Abbreviation for dihydrouracil or dihydrouridine.

DHU Arm The base-paired stem on the tRNA (also called DHU stem).

DHU Loop The loop on the DHU arm or DHU stem of tRNA.

DiaBeta A trade name for glyburide, used for treatment of diabetes.

Diabetes A condition characterized by excessive urine production.

Diabetes Insipidus A metabolic disorder characterized by thirst and heavy urination without excess excretion of sugar.

Diabetes Mellitus A disorder characterized by deficiency of the hormone insulin and consequent elevation of glucose concentrations in the blood and urine.

Diabetogenic Capable of producing diabetes.

Diabetogenic Hormone Hormone that stimulates gluconeogenesis.

Diabinese A trade name for chlorpropamide, an antidiabetic agent that stimulates the release of insulin from pancreatic beta cells and reduces glucose output.

Diacetinase An enzyme that hydrolyzes diacetyl glycerol ester to glycerol and two fatty acids.

Diacetyl Reductase Synonym of acetoin dehydrogenase.

Diachlor A trade name for chlorothiazide, a diuretic agent that increases urine secretion.

Diacylglycerol (DAG) An ester of two fatty acids and a glycerol (also called diglyceride).

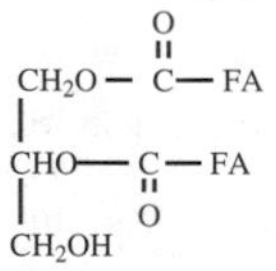

FA = fatty acid

Diacylglycerol Acyltransferase The enzyme that catalyzes the following reaction:

Diacylglycerol + acyl-CoA

$\rightleftharpoons$

Triacylglycerol + CoA

Diacylglycerol Choline-Phosphotransferase The enzyme that catalyzes the following reaction:

CDP-choline + 1,2-diacylglycerol

$\rightleftharpoons$

CMP + phosphatidylcholine

1,2-Diacylglycerol 3-β-Galactosyltransferase The enzyme that catalyzes the following reaction:

UDP-galactose + 1,2-diacylglycerol

$\rightleftharpoons$

UDP 3-β-D-galactosyl-1,2-diacylglycerol

Diacylglycerol Kinase The enzyme that catalyzes the following reaction:

Diacylglycerol + ATP

$\rightleftharpoons$

Phosphatidic acid + ADP

Diacylglycerol Lipase The enzyme that catalyzes the following reaction:

1,2-Diacylglycerol

$\rightleftharpoons$

Monoacylglycerol + fatty acid

Diacylglycerol Sterol O-Acyltransferase The enzyme that catalyzes the following reaction:

1,2-Diacylglycerol + sterol

$\rightleftharpoons$

Monoacylglycerol + sterol ester

Diacytosis The discharge of empty pinocytotic vesicles from cells after releasing their content.

Diagonal Electrophoresis Two-dimensional electrophoresis for the identification of chemically modified peptides in which the electrophoretically separated peptides in paper or gel are chemically modified and reelectrophoresed; the nonmodified peptides form a diagonal on the electrophoretogram while the modified peptides lie off the diagonal of the electrophoretogram.

Diakinesis The stage of late prophase I of meiosis when the chromosomes are well separated.

Dialifor (mol wt 394) An insecticide.

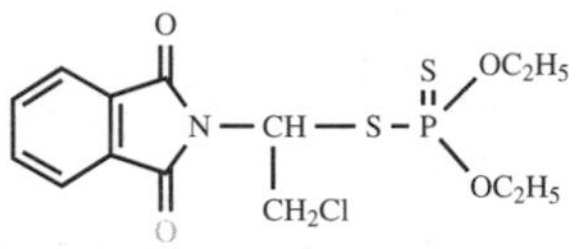

Diallate (mol wt 270) An herbicide.

Dialose A trade name for docusate potassium, used as a stool softener.

Dialume A trade name for aluminum hydroxide, an antacid and antiflatulent.

Dialysate The fluid outside the dialysis membrane sac after a dialysis procedure.

Dialysis 1. A technique for separation of micromolecules from macromolecules in colloid solution through a semipermeable membrane sac in which the micromolecules in the membrane sac diffuse out and the macromolecules remain inside the membrane sac. 2. A procedure for removal of undesired materials from the blood.

Dialysis Fermentation A type of fermentation in which cells are retained within the membrane-separated fermentation vessel while the products are allowed to diffuse out through the membrane.

Diamine Oxidase The enzyme that catalyzes the following reaction:

$$RCH_2NH_2 + H_2O \rightleftharpoons RCHO + NH_3 + H_2O_2$$

Diamine Transaminase The enzyme that catalyzes the following reaction:

An α–ω-Diamine + a-keto-glutarate

⇅

A ω-aminoaldehyde + L-glutamate

3,5-Diaminobenzoic Acid (mol wt 152) A compound used in microfluorimetric determination of DNA.

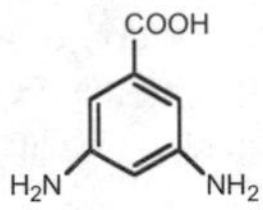

Diaminobutyrate Pyruvate Transaminase The enzyme that catalyzes the following reaction:

2,4-Diaminobutanoate + pyruvate

⇅

L-Aspartate 4-semialdehyde + L-alanine

Diaminopimelate Decarboxylase The enzyme that catalyzes the following reaction:

$$\text{Diaminopimelate} \rightleftharpoons \text{Lysine} + CO_2$$

Diamino-Pimelic Acid (mol wt 190) An intermediate in the biosynthesis of lysine.

$$HOOC—CHNH_2—(CH_2)_3CHNH_2—COOH$$

Diaminopimelic Acid Pathway A pathway for biosynthesis of lysine in bacteria, plant, and certain fungi.

2,6-Diaminopurine (mol wt 150) A naturally occurring purine antagonist.

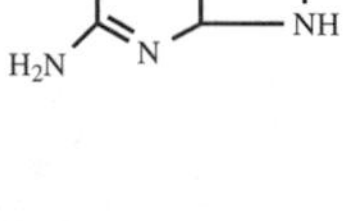

Diamox A trade name for acetazolamide, a diuretic agent that promotes renal excretion of sodium, potassium, bicarbonate, and water.

Diampromide (mol wt 324) An analgesic agent.

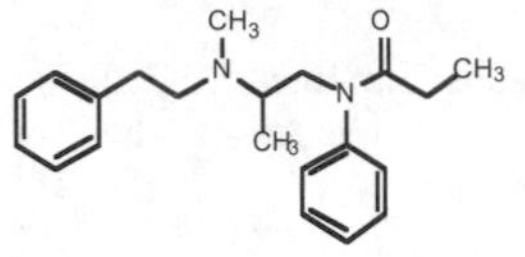

Diapause A period of arrested growth and development in the life cycle of insects and certain other animals.

Diapedesis The outward passage of white blood cells (e.g., neutrophils) through intact blood vessel walls.

Diaphenylsulfone A trade name for dapsone, a leprostatic agent.

Diaphorase The enzyme that catalyzes the transfer of electrons from NADH or NADPH to an artificial electron acceptor, e.g., dye, ferricyanide.

Diaphoresis Artificially induced sweat or perspiration.

Diapid A trade name for the hormone lypressin, an antidiuretic agent that promotes reabsorption of water and produces concentrated urine.

Diarrhea A common symptom of gastrointestinal disease characterized by the abnormally frequent passage of soft or loose stools.

Diasolysis A process of diffusion of organophilic compounds in organic solvent through a gum, plastic, or rubber membrane for separation of hydrophilic and colloidal substances that do not dialyze.

Diastase Referring to a crude preparation of amylase.

Diastase Malt A commercial preparation containing amylolytic enzymes.

Diastatic Index A parameter for determination of amylase activity. It is the number of milliliters of 0.1% starch solution hydrolyzed by the enzyme present in 1 ml of sample at 37° C in 30 minutes.

Diastereomers The non-mirror-image optical stereoisomers, e.g., two D and two L isomers.

Diastole The stage of the heart cycle in which the heart muscle is relaxed, allowing the chambers to fill with blood.

Diastolic Pressure The lowest blood pressure measured between the contractions of the heart.

Diathymosulfone (mol wt 571) An antibacterial agent (Leprostatic).

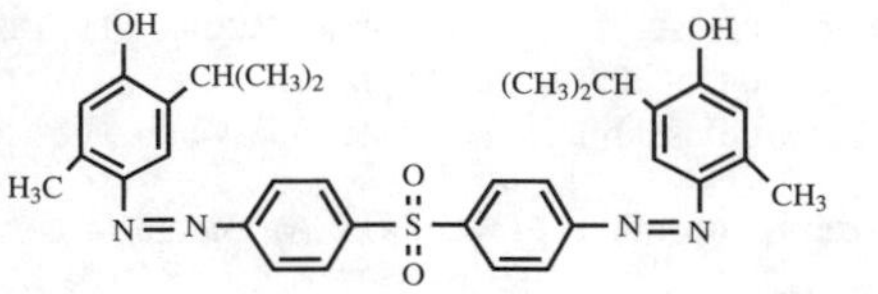

Diatomaceous Earth A fine siliceous material from diatoms used as an adsorbent in adsorption chromatography.

Diatrizoate Sodium (mol wt 636) A density gradient reagent for blood cell separation.

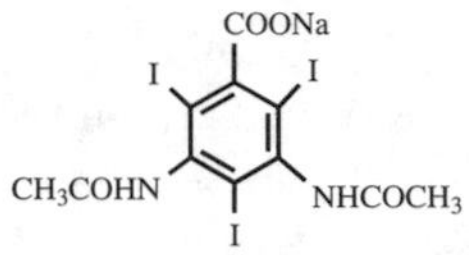

Diauxie The phenomenon for preferential utilization of one of two sources of carbon when an organism metabolizes one carbon source completely before utilizing the other.

Diazemuls A trade name for diazepam, an antianxiety and anticonvulsant agent.

Diazepam (mol wt 285) A muscle relexant and anxiolytic agent.

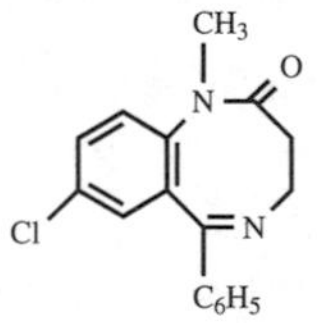

Diazinon (mol wt 304) An insecticide.

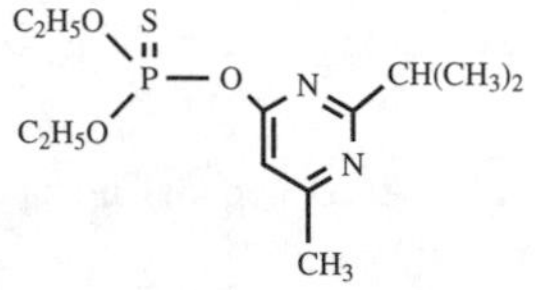

Diaziquone (mol wt 364) An antineoplastic agent.

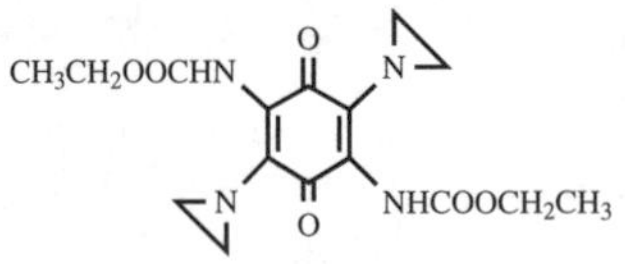

Di-Azo A trade name for phenazopyridine hydrochloride, an antipyretic and nonnarcotic analgesic agent.

Diazoate The ion of R - N = N - O⁻ derived from diazonium salt.

Diazobenzloxymethyl Paper Paper treated with diazobenzyloxymethyl compounds and used for blotting experiments; it binds single-stranded DNA, RNA, and protein.

Diazo-Compound A compound with the general formula Ar-N=N-X where Ar = aromatic and X = organic.

Diazonium Group The following chemical group:

$$—N \equiv N$$

Diazonium Salt The salt that contains a diazonium group.

6-Diazo-5-Oxo-L-Norleucine (mol wt 171) An antibiotic and antitumor agent from species of *Streptomyces* that inhibits purine synthesis.

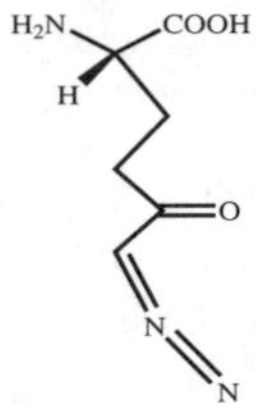

Diazo-Paper Paper coated with diazo light-sensitive dye.

Diazotization The reaction of a primary aromatic amine with nitrous acid in the presence of an excess amount of mineral acid to produce a diazo (-N=N-) compound.

Diazotroph An organism capable of nitrogen fixation.

Diazoxide (mol wt 231) An antihypertensive agent that relaxes smooth muscle, releases insulin from the pancreas, and decreases peripheral utilization of glucose.

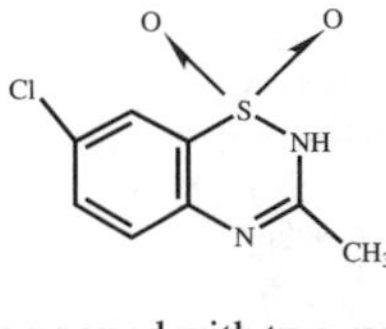

Dibasic A compound with two metal replaceable hydrogen atoms, or an acid capable of yielding two hydrogen ions.

DiZ Éacin (mol wt 452) A semisynthetic analog of kanamycin active against kanamycin-resistant bacteria.

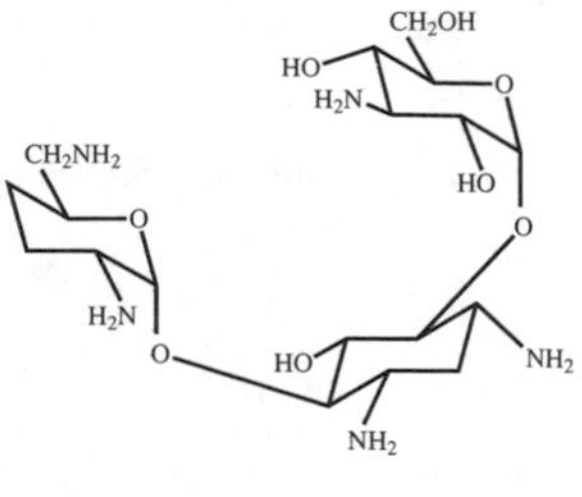

Dibent A trade name for dicyclomine hydrochloride, an anticholinergic agent.

Dibenzepin (mol wt 295) An antidepressant agent.

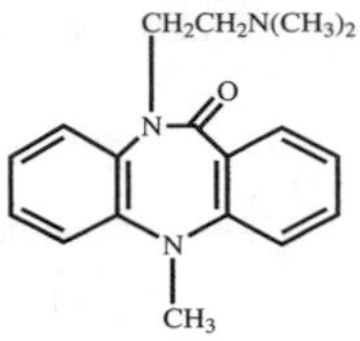

Dibenzyline A trade name for phenoxybenzamide hydrochloride, an antihypertensive agent that noncompetitively blocks the effect of catecholamine on alpha adrenergic receptors.

3,5-Dibromo-L-Tyrosine (mol wt 339) A thyroid inhibitor.

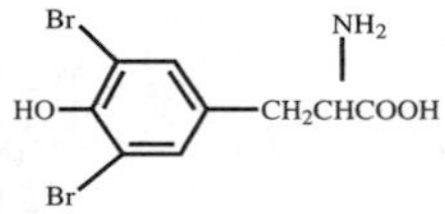

Dibromsalicil (mol wt 400) An antiseptic agent.

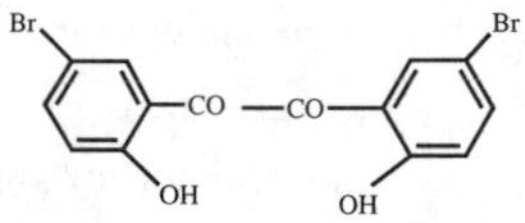

3,5-Dibromosalicylaldehyde (mol wt 280) An antibacterial agent.

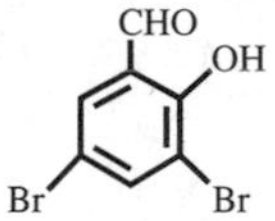

Dibucaine Hydrochloride (mol wt 380) A local anesthetic agent.

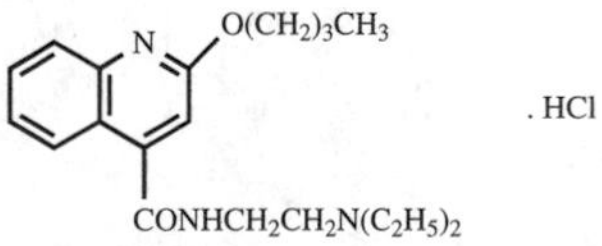

Dibutoline Sulfate (mol wt 642) An anticholinergic agent.

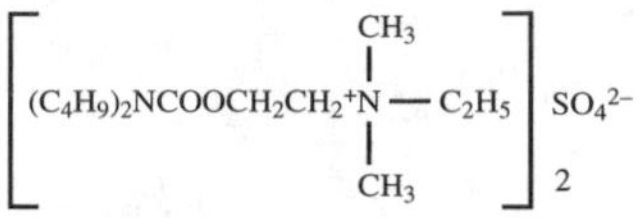

DIC Abbreviation for 1. disseminated intravascular coagulation; 2. dexamethasone-induced complex; 3. dynein intermediate chain.

Dicapthon (mol wt 298) A cholinesterase inhibitor.

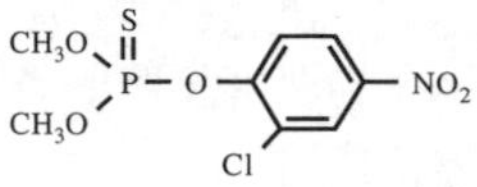

Dicarbosil A trade name for calcium carbonate, an antacid.

Dicarboxylate-CoA Ligase The enzyme that catalyzes the following reaction:

$$\text{ATP} + \alpha,\omega\text{-dicarboxylic acid} \rightleftharpoons \text{AMP} + \omega\text{-carboxyacyl-CoA} + \text{PPi}$$

Dicarboxylic Acid Cycle See glyoxylate cycle.

Dicentric A chromosome with two centromeres.

Dichlofluanide (mol wt 333) An antifungal agent.

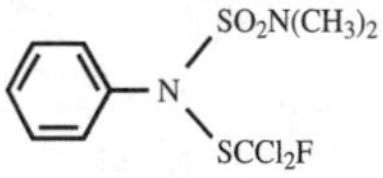

Dichloride A compound that contains two chloride atoms per molecule.

Dichlorisone (mol wt 413) An antipuritic agent.

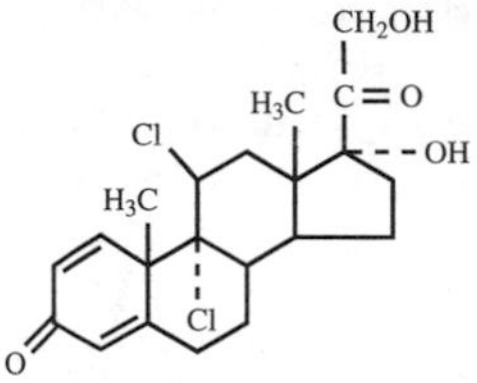

Dichlorobenzyl Alcohol (mol wt 177) An antiseptic agent.

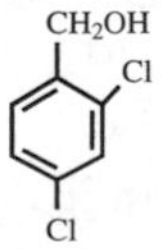

Dichlorofluorescein (mol wt 401) A fluorescent indicator.

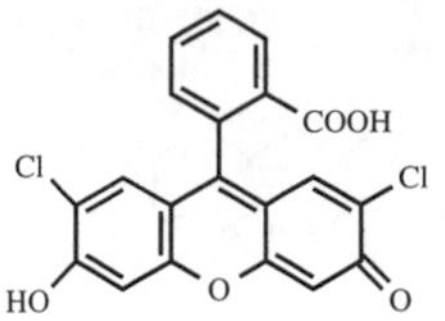

Dichloroisoproterenol (mol wt 248) An adrenergic agent.

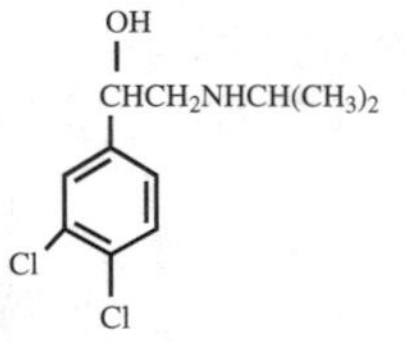

Dichloromethane Dehalogenase The enzyme that catalyzes the following reaction:

Dichloromethane + H_2O

⇅

Formaldehyde + 2 chloride

Dichlorophenamide (mol wt 305) A diuretic agent and an inhibitor for carbonic anhydrase.

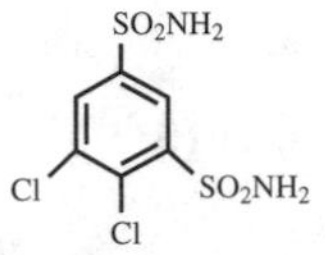

2,4-Dichlorophenol Hydroxylase See 2,4-Dichlorophenol 6-monooxygenase.

2,4-Dichlorophenol 6-Monooxygenase The enzyme that catalyzes the following reaction:

2,4-Dichlorophenol + NADPH + O_2

⇅

3,5-Dichlorocatechol + $NADP^+$ + H_2O

Dichlorophenoxyacetic Acid (2,4-D, mol wt 221) An herbicide.

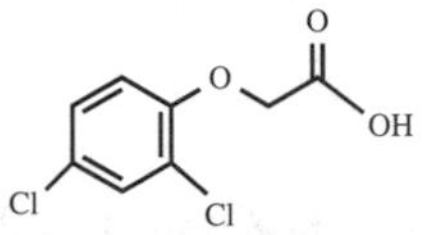

Dichlorophenyltrichloroethane (DDT, mol wt 282) An insecticide (see DDT for structure). It is also known as 1,1-bis(*p*-chlorophenyl)-2,2,2-trichloroethane.

Dichlororiboflavin (mol wt 417) A riboflavin antagonist.

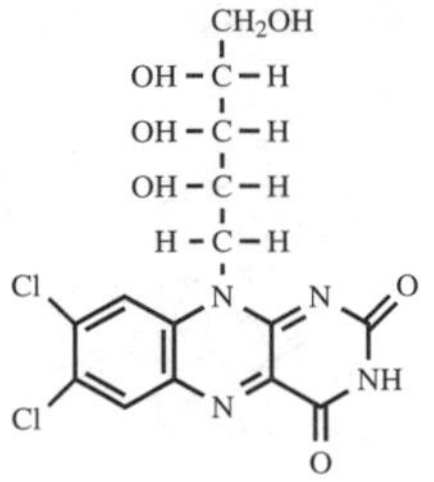

Dichlortride A trade name for hydrochlorothiazide, a diuretic agent that increases urine secretion of sodium and water by inhibiting sodium reabsorption.

Dichlorvos (mol wt 221) A cholinesterase inhibitor and insecticide.

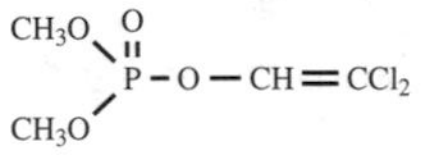

Dichromatism A disorder in color perception, the spectrum is seen as composed of only two colors separated by an achromatic or colorless band.

Dichromophil Denoting tissue that takes both acidic and basic stains for different areas.

Dichrostachinic Acid (mol wt 252) A naturally occurring nonprotein amino acid.

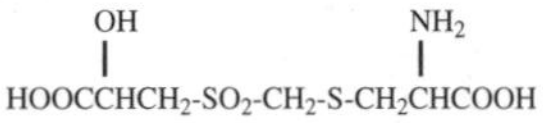

DICI Abbreviation for direct intracytoplasmic injection.

Dick Test A test for susceptibility to scarlet fever.

Diclofenac Sodium (mol wt 318) An anti-inflammatory agent and inhibitor for prostaglandin synthesis.

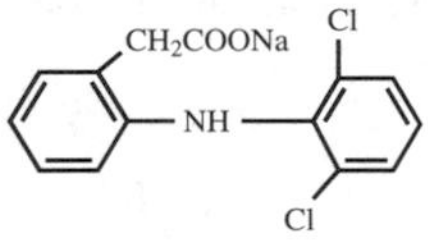

Dicloxacillin (mol wt 470) A semisynthetic antibiotic related to penicillin. It is a penicillinase-resistant antibiotic.

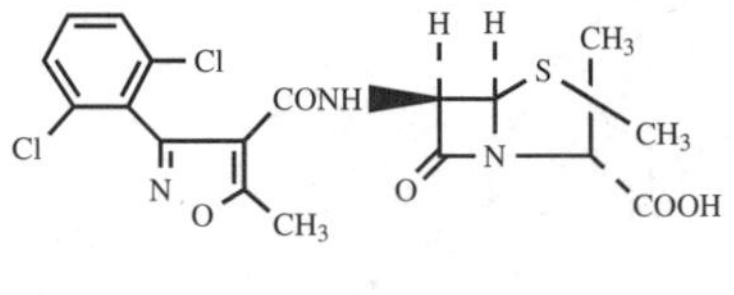

Dicotyledon A plant whose embryo has two cotyledons.

Dicrotophos (mol wt 237) A cholinesterase inhibitor.

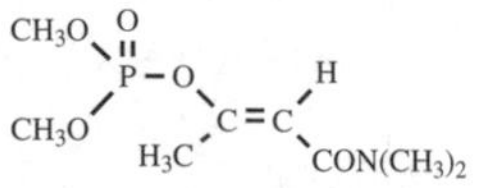

Dictyoglomus A genus of anaerobic, asporogenous, chemoorganotrophic bacteria.

Dictyoma A cancer of the retina.

Dictyostelium A genus of slime mold.

Dictysomes The individual stack of membrane vesicles that forms a Golgi apparatus.

Dicumarol (mol wt 336) An anticoagulant that inhibits vitamin K-dependent activation of clotting factors II, VII, IX, and X.

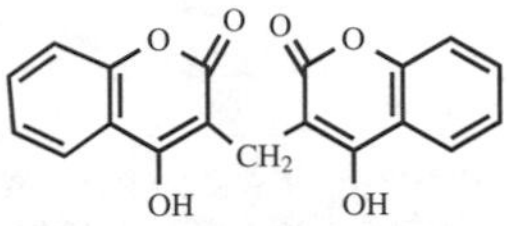

Dicyanodiamide (mol wt 84) A reagent used as a condensing agent for amino acids.

Dicyclohexylcarbodiimide (mol wt 206) A reagent used for coupling of amino acids.

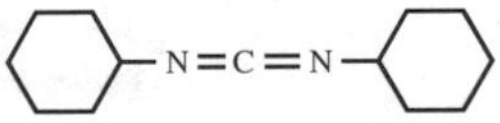

Dicyclomine Hydrochloride (mol wt 346) An anticholinergic agent that exerts a spasmolytic action on smooth muscle.

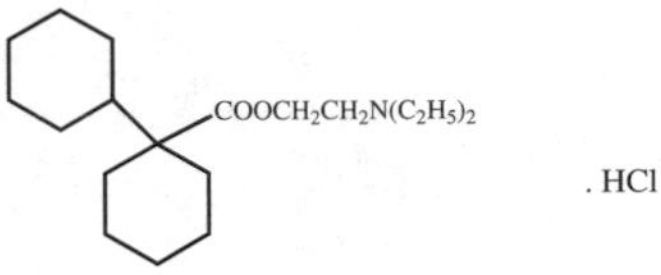

Di-Cyclonex A trade name for dicyclomine hydrochloride, an anticholinergic drug.

Didanosine Synonym for dideoxyinosine, it inhibits RNA-dependent DNA polymerase (reverse transcriptase).

Didemnins The biologically active depsipeptides from Caribbean tunicate (e.g., sea squirt) with antiviral and antitumor activity.

Dideoxy Method of DNA Sequencing A method of DNA sequencing in which DNA synthesis is carried out by the Klenow fragment of DNA-polymerase I using radioactive dNTP, primer, and template-DNA in the presence of ddNTP (dideoxynucleotide triphosphate). The incorporation of ddNTP in the newly synthesized DNA terminates the synthesis of DNA resulting in a series of new DNA fragments that can be separated in polyacrylamide gel and the DNA sequence can be read directly from the X-ray film.

2′,3′-Dideoxyadenosine (mol wt 235) The dideoxynucleoside of adenine and an antiviral agent.

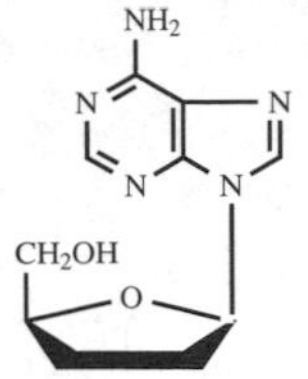

2′,3′-Dideoxyadenosine 5′-Diphosphate (mol wt 395) A diphosphate form of the dideoxynucleotide of adenine.

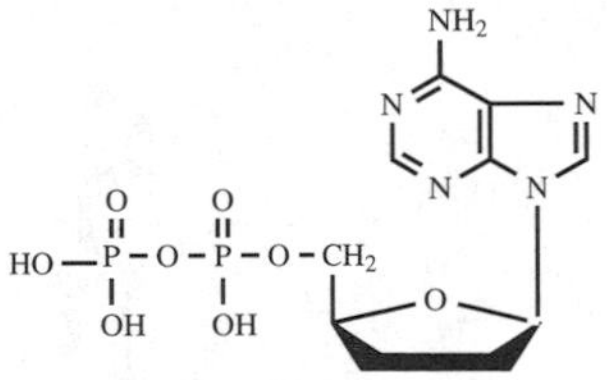

2′,3′-Dideoxyadenosine 5′-Monophosphate (mol wt 315) A monophosphate form of the dideoxynucleotide of adenine.

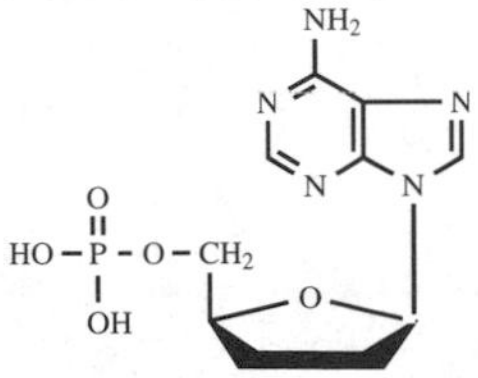

2′,3′-Dideoxyadenosine 5′-Triphosphate (mol wt 475) The triphosphate form of the dideoxyadenosine nucleotide, a reagent used for the dideoxy method of DNA sequencing.

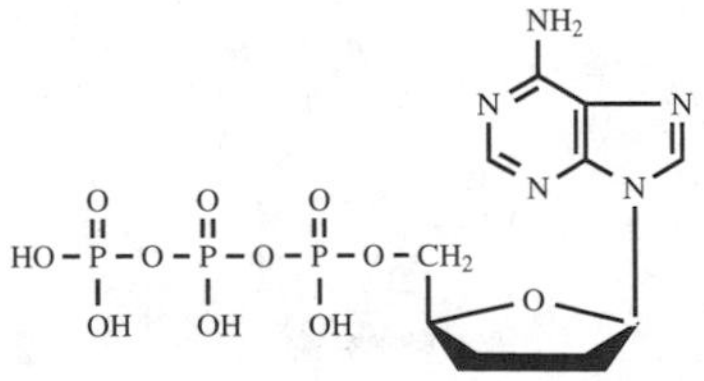

2′,3′-Dideoxycytidine (mol wt 211) A dideoxynucleoside of cytosine and an antiviral agent.

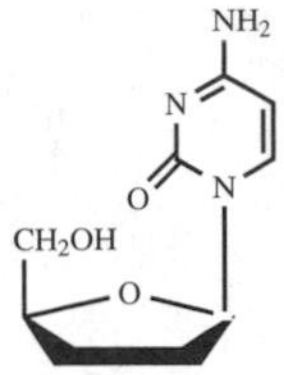

2′,3′-Dideoxycytidine 5′-Diphosphate (mol wt 371) A diphosphate form of the dideoxynucleotide of cytosine.

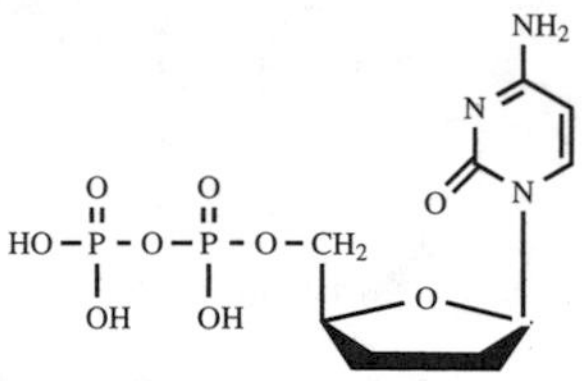

2′,3′-Dideoxycytidine 5′-Monophosphate (mol wt 291) A monophosphate form of the dideoxynucleotide of cytosine.

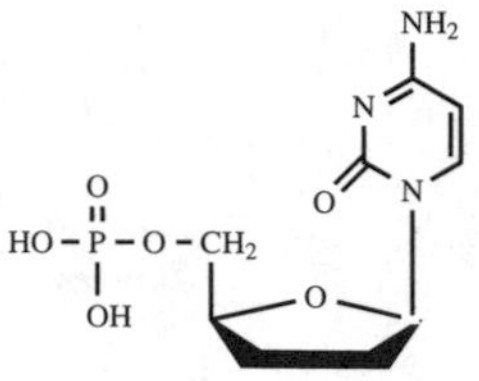

2′,3′-Dideoxycytidine 5′-Triphosphate (mol wt 451) A triphosphate form of dideoxycytidine, a reagent used in the dideoxy method of DNA sequencing.

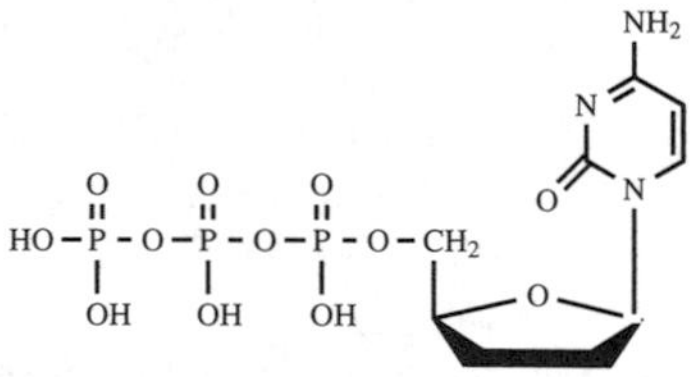

2′,3′-Dideoxyguanosine (mol wt 226) A dideoxynucleoside of guanine.

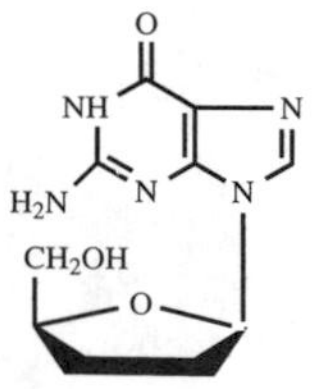

2′,3′-Dideoxyguanosine 5′-Diphosphate (mol wt 396) A diphosphate form of the dideoxynucleotide of guanine.

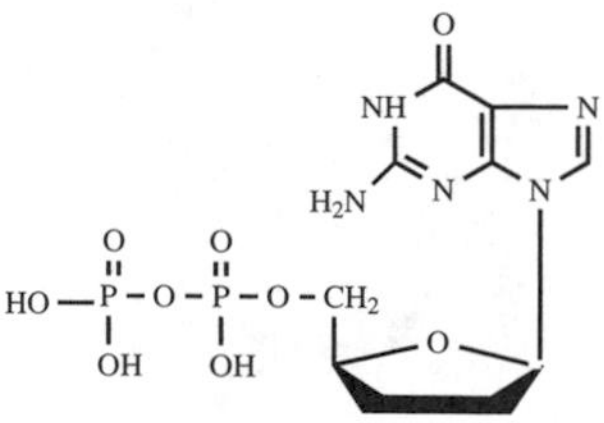

2′,3′-Dideoxyguanosine 5′-Monophosphate (mol wt 316) A monophosphate form of the dideoxynucleotide of guanine.

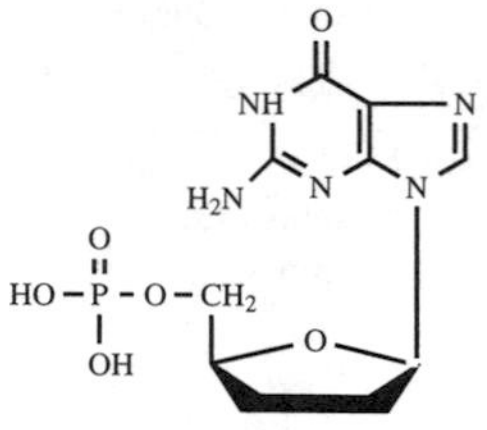

2′,3′-Dideoxyguanosine 5′-Triphosphate (mol wt 491) A triphosphate form of dideoxyguanosine and a reagent used in the dideoxy method of DNA sequencing.

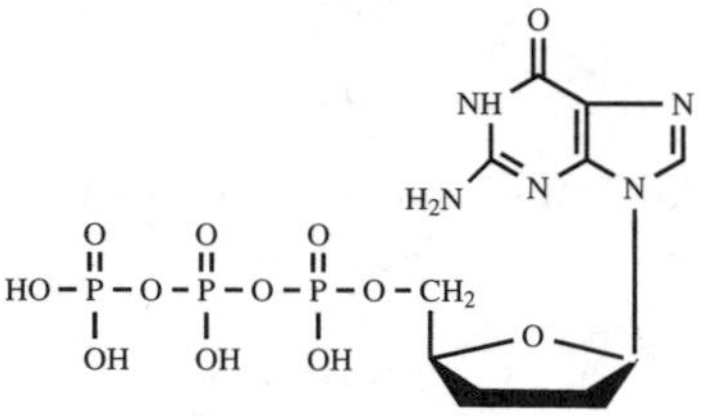

2′,3′-Dideoxyinosine (mol wt 236) A dideoxynucleoside of hypoxanthine.

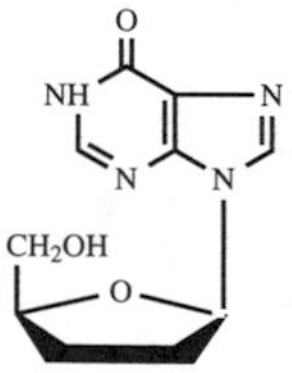

2′,3′-Dideoxyinosine 5′-Diphosphate (mol wt 396) A diphosphate form of the dideoxynucleotide of hypoxanthine.

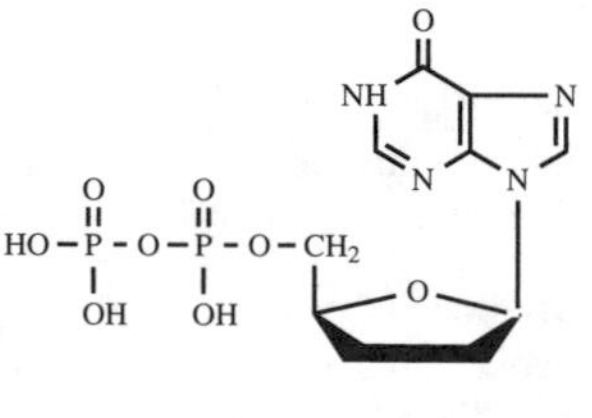

2′,3′-Dideoxyinosine 5′-Monophosphate (mol wt 316) A monophosphate form of the dideoxynucleotide of hypoxanthine.

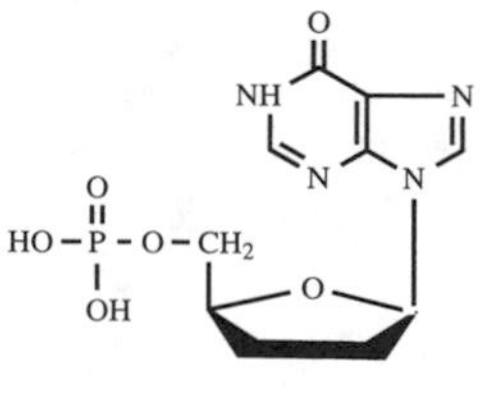

2′,3′-Dideoxyinosine 5′-Triphosphate (mol wt 476) A triphosphate form of the dideoxynucleotide of hypoxanthine.

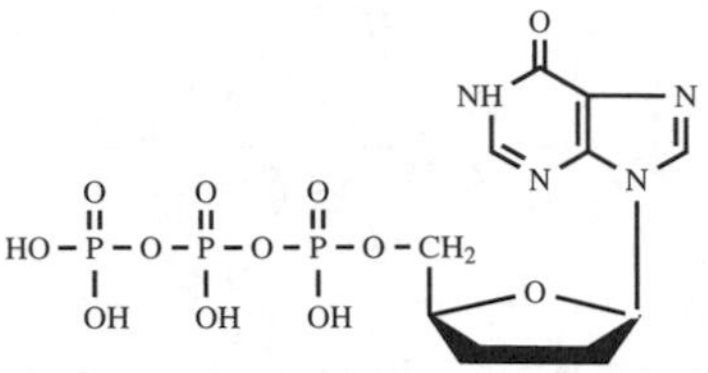

2′,3′-Dideoxythymidine 5′-Diphosphate (mol wt 386) A dideoxynucleotide of thymine (also known as 3′-deoxythymidine 5′-diphosphate since thymidine itself is a 2′-deoxynucleoside).

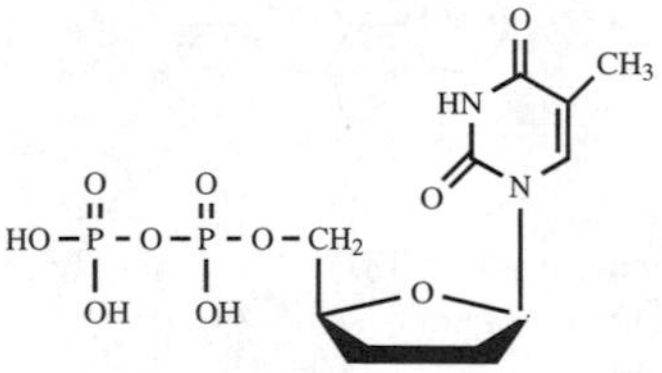

2′,3′-Dideoxythymidine Triphosphate (mol wt 466) A triphosphate form of the dideoxynucleotide of thymine used in the dideoxy method of DNA sequencing (also known as 3′-deoxythymidine 5′-triphosphate since thymidine itself is a 2′-deoxynucleoside).

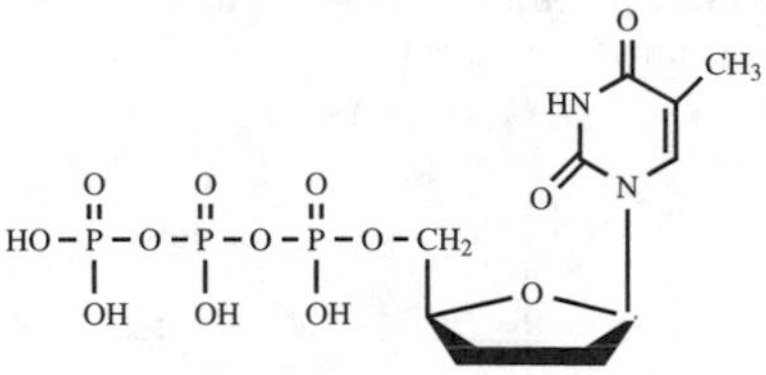

2′,3′-Dideoxythymine 5′-Monophosphate (mol wt 306) A monophosphate form of the dideoxynucleotide of thymine (also known as 3′-deoxythymidine 5′-monophosphate since thymidine itself is a 2′-deoxynucleoside).

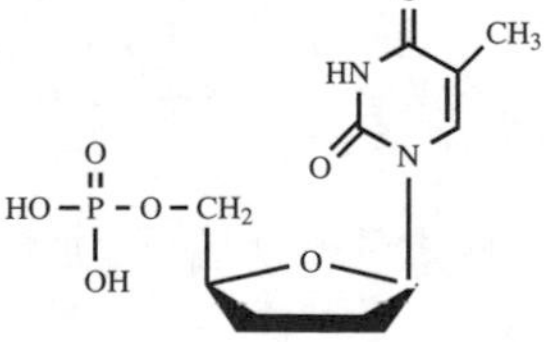

2′,3′-Dideoxyuridine (mol wt 212) A dideoxynucleoside of uracil.

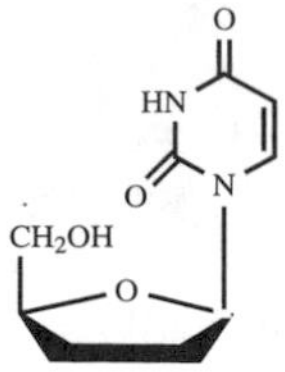

2′,3′-Dideoxyuridine 5′-Diphosphate (mol wt 372) A diphosphate form of the dideoxynucleotide of uracil.

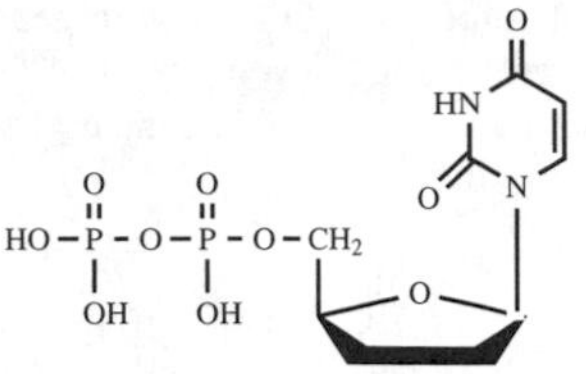

2′,3′-Dideoxyuridine 5′-Monophosphate (mol wt 292) A monophosphate form of the dideoxynucleotide of uracil.

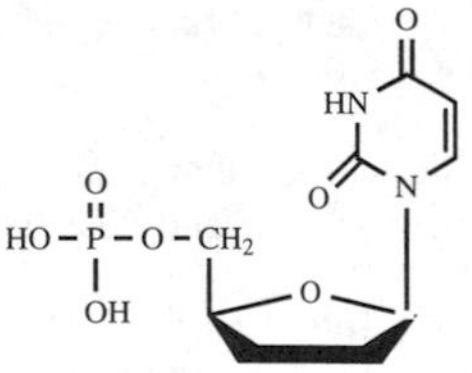

2′,3′-Dideoxyuridine 5′-Triphosphate (mol wt 452) A triphosphate form of the dideoxynucleotide of uracil.

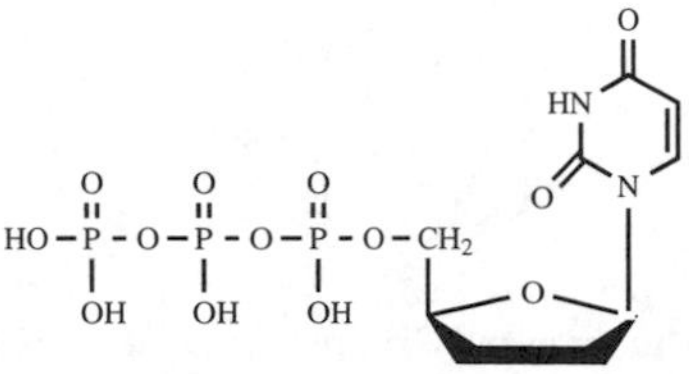

Didrex A trade name for benzphetamine hydrochloride, a cerebral stimulant that promotes transmission of nerve impulses by releasing stored norepinephrine from the nerve terminals in the brain.

Didronel A trade name for etidronate disodium, a calcium regulator.

Didymus Male sex organ, a testis.

Dieldrin (mol wt 381) A pesticide.

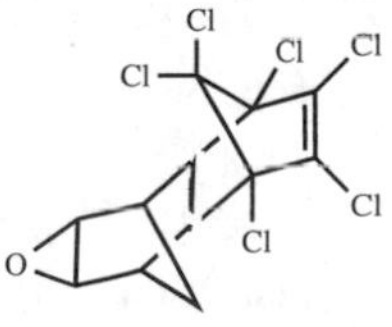

Dielectric A nonconductor of electric current.

Dielectric Constants The ratio of the electric force in a vacuum to that in the medium.

-diene A suffix denoting the presence of an unsaturated aliphatic hydrocarbon chain containing two double bonds.

Dienestrol (mol wt 266) An estrogen that promotes the growth and development of female sex organs and maintenance of secondary sex characteristics in women.

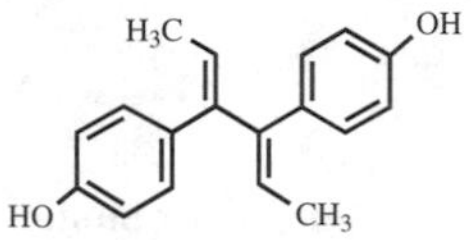

Diesterase See phosphodiesterase.

Dietary Fiber Plant substances that are indigestible in the digestive system, e.g., cellulose.

Dietetics The science that deals with the relationship of diet to health and disease.

Diethanolamine (mol wt 105) An amine used in manufacturing cationic detergents.

$NH(CH_2CH_2OH)_2$

Diethazine (mol wt 298) An anticholinergic agent.

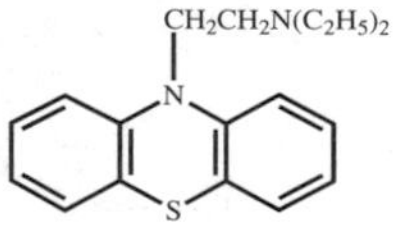

Diethylaminoethyl Group A chemical group that can be incorporated into polymeric carrier for production of ion exchange resins.

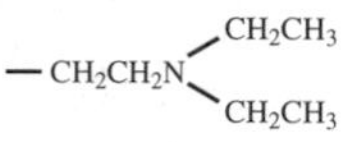

Diethylaminoethyl-Cellulose An ion exchanger.

Diethylaminoethyl-Sephadex An anion exchanger.

Diethylaminoethyl-Sepharose An anion exchanger.

Diethylpropion (mol wt 205) An anorexic agent.

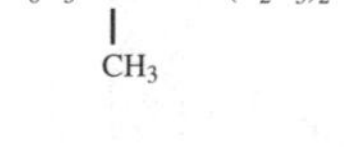

Diethylpyrocarbonate (mol wt 162) A reagent used for modification of histidine residues in proteins.

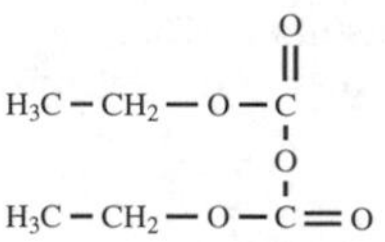

Diethylstilbestrol (mol wt 268) An estrogen that increases DNA, RNA, and protein synthesis in responsive tissues; it also reduces FSH and LH release from the pituitary.

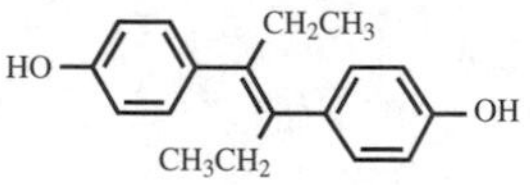

Diethylstilbestrol Dipropionate (mol wt 380) An estrogen that promotes growth and development of female sex organs and maintenance of secondary sex characteristics in women.

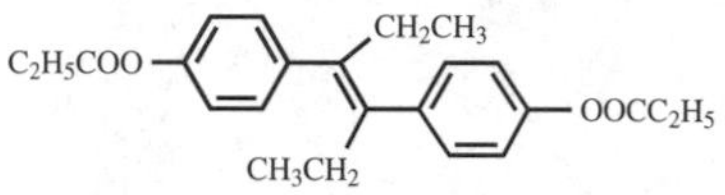

Difemerine (mol wt 327) An anticholinergic and antispasmodic agent.

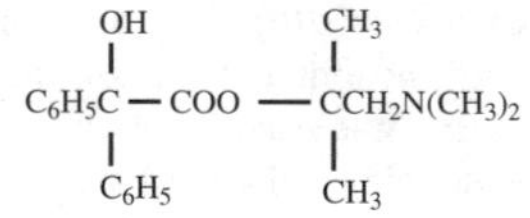

Difenamizole (mol wt 334) An analgesic and anti-inflammatory agent.

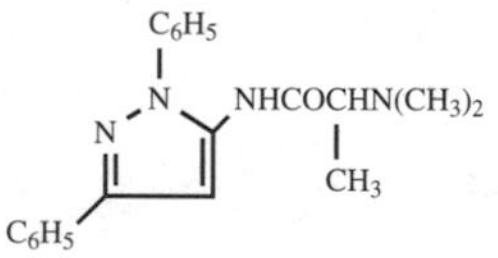

Difenoxin (mol wt 425) An antiperistaltic and antidiarrheal agent that inhibits excess GI mobility.

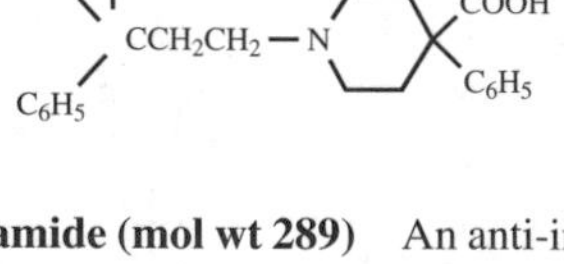

Difenpiramide (mol wt 289) An anti-inflammatory agent.

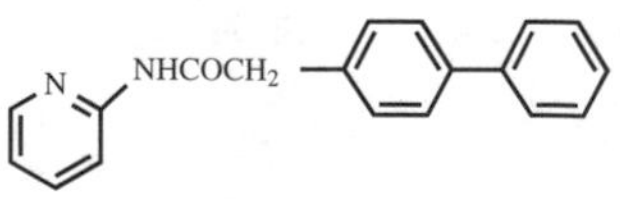

Diferric-Transferrin Reductase The enzyme that catalyzes the following reaction:

$$\text{Transferrin}[\text{Fe(II)}]_2 + \text{NAD}^+ \rightleftharpoons \text{Transferrin }[\text{Fe(III)}]_2 + \text{NADH}$$

Differential Blood Count Procedure for determining the ratios of various types of blood cells.

Differential Centrifugation Separation of particles or cell organelles of different sizes by the different speeds of centrifugation.

Differential Display PCR A variation of regular polymerase chain reaction (PCR) used to identify differentially expressed genes.

Differential Leukocyte Count A determination for the proportions of the different types of white cells present in a sample of blood.

Differential Medium Growth medium used for differentiation and identification of bacteria.

Differential Transcription Selective activation of unique subsets of genes in different cell types to produce distinct populations of cellular mRNAs.

Differentiation The process of development in which cells become specialized in function and structure.

Differentiation Antigen Any cell surface antigen whose expression varies during successive developmental stage.

Differentiation Marker Specific gene product that is expressed on the surface of a particular cell type or expressed at a characteristic stage during development.

Difflugia A genus of testate amoebae.

Diffraction Grating A device to separate light into its component wavelengths.

Diffusate The substance that passes through a sealed dialysis bag.

Diffuse Cortex Region in the peripheral lymphoid tissue where T lymphocytes reside.

Diffusion Free, nonmediated movement of a solute with direction and rate dictated by the difference in solute concentration between two different regions.

Diffusion Test A method of serological test involving diffusion of antigen, antibody, or both through agar.

Diflorasone (mol wt 410) A glucocorticoid used as an anti-inflammatory agent.

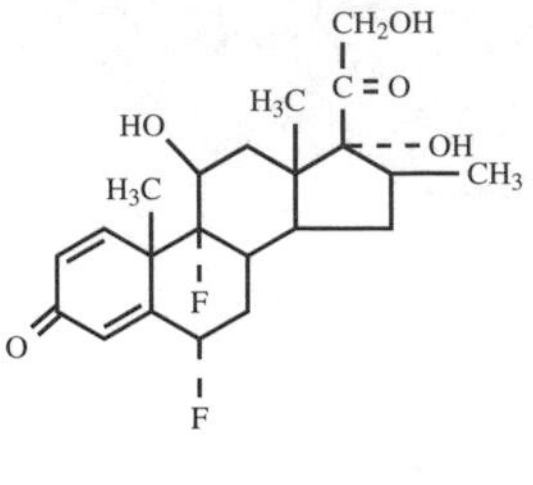

Diflucan A trade name for fluconazole, an antifungal agent that inhibits fungal sterol synthesis.

Diflucortolone (mol wt 394) An anti-inflammatory agent.

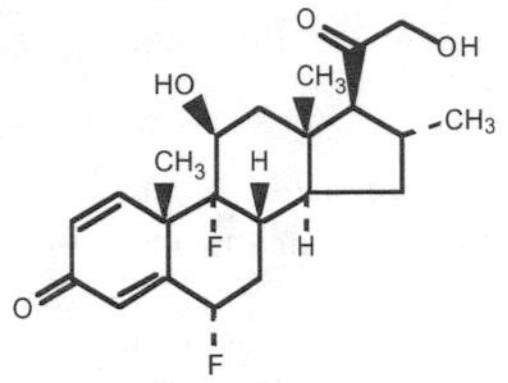

Diflunisal (mol wt 250) An analgesic and an anti-inflammatory agent.

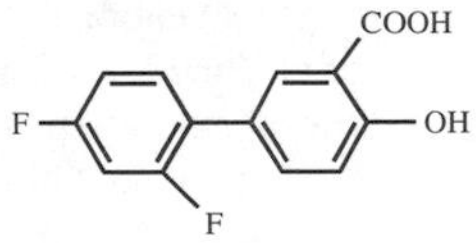

Difluprednate (mol wt 508) An anti-inflammatory agent.

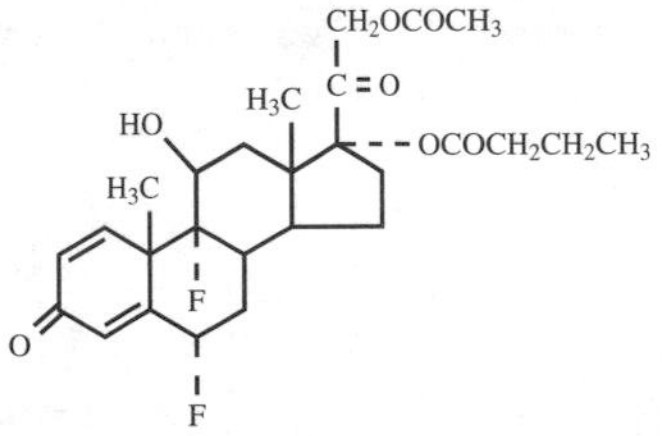

DIFP Abbreviation for diisopropyl fluorophosphate.

DIG Abbreviation for 1. digoxigenin; 2. detergent-insoluble glycolipids.

dIgA (DIgA) Abbreviation for dimer IgA.

Digbind A trade name for digoxin immune Fab that binds molecules of digoxin and digitoxin.

DIGD Abbreviation for detergent insoluble glycolipid-rich domain.

Di-Gel A trade name for a combination drug containing aluminum hydroxide, magnesium hydroxide, and simethicone.

Digenetic A parasite that carries out part of its life cycle in two different host species.

DiGeorge Syndrome An immune disorder due to the deficiency of T lymphocytes stemming from incomplete fetal development of the thymus.

Digess 8000 A trade name for pancrelipase, a digestive enzyme.

Digestion The process of breaking down food into molecules small enough for the body to absorb.

Digestive System The organs, structures, and glands involved in digestion, e.g., mouth, stomach, and small and large intestines.

Digestive Vacuoles The membrane-bound organelles formed from the fusion of lysosomes with phagocytic vesicles for digestion of engulfed foreign materials.

Digibind A trade name for an antigen-binding fragment of anti-digoxin antibody used as an antidote.

Digital Angiography Computerized equipment for the enhancement of X-ray photographs of the heart.

Digital Radiography The application of digitization technique to the conventional radiography, allowing the storage of images on hard disk and their subsequent retrieval and interpretation using TV monitors rather than photographic film.

Digital Spot Imaging The X-ray screening in which the images produced are stored digitally and viewed on a TV monitor.

Digitalin (mol wt 713) A cardiotonic agent from seeds of *Digitalis purpuirea.*

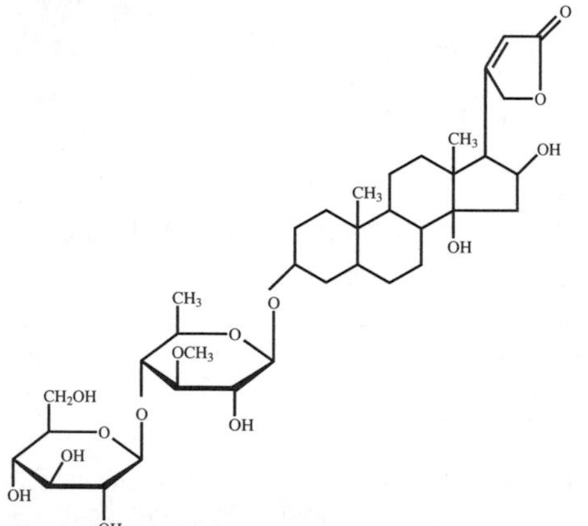

Digitalis 1. The extract from leaves of foxglove (*Digitalis)* containing a mixture of cardiac glycosides, e.g., digitonin, digitalin, and digitalosamin. 2. A genus of plant (Scrophulariaceae).

Digitalization Treatment of heart disorders with digitalis glucosides to achieve a desired therapeutic effect.

Digitalose (mol wt 178) A methylated sugar.

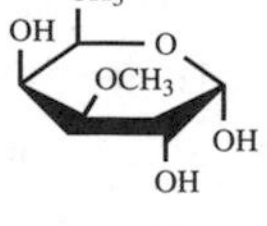

Digitization The assignment of a numerical value to each pixel of a gray-scale image to allow storage, electronic manipulation, and transfer via computer links of images produced by devices such as X-ray machines, CT scanners, MRI scanners, or ultrasound probes.

Digitogenin (mol wt 449) The aglucon of digitonin.

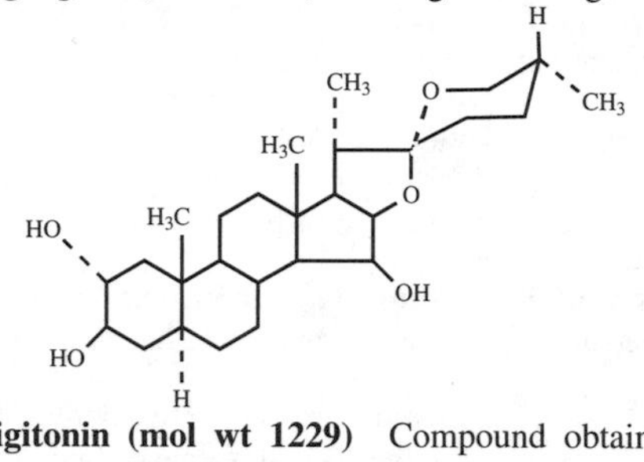

Digitonin (mol wt 1229) Compound obtained from the seeds of *Digitalis purpurea* used as clinical reagent for determination of cholesterol.

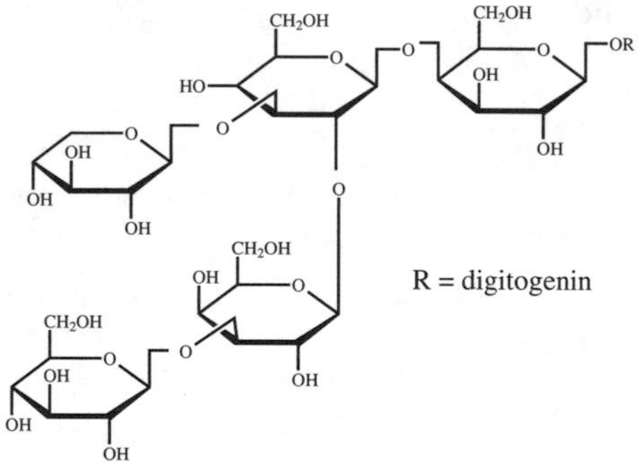

Digitoxigenin (mol wt 374) A component of digitoxin.

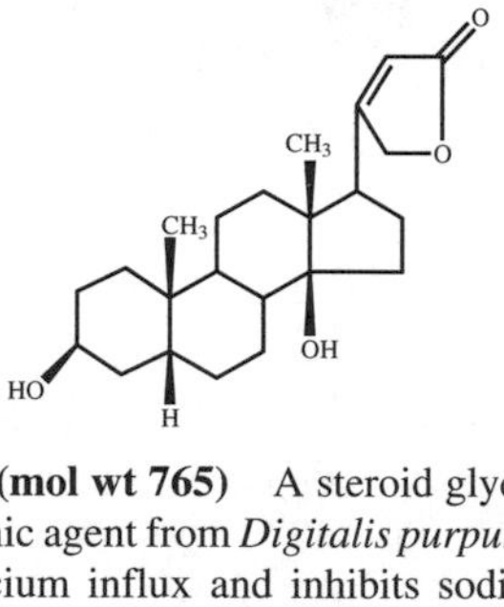

Digitoxin (mol wt 765) A steroid glycoside and a cardiotonic agent from *Digitalis purpurea*; it promotes calcium influx and inhibits sodium-potassium activated ATPase.

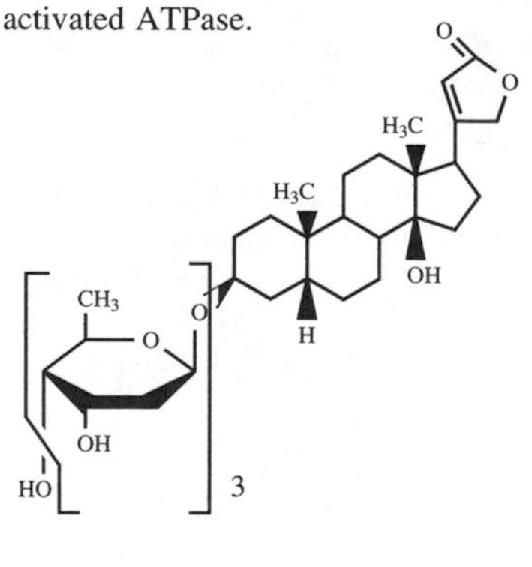

Digitoxose (mol wt 148) An aldosugar derivative obtained from the hydrolysis of digitoxin.

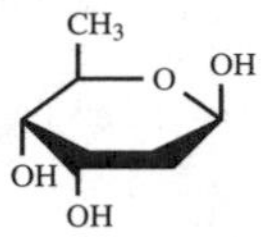

Diglyceride See diacylglycerol.

Diglyceride Kinase See diacylglycerol kinase.

Diglyceride Lipase Synonym of lipoprotein lipase.

Digoxin (mol wt 781) A glycoside obtained from *Digitalis lanata* and used for treatment of congestive heart failure.

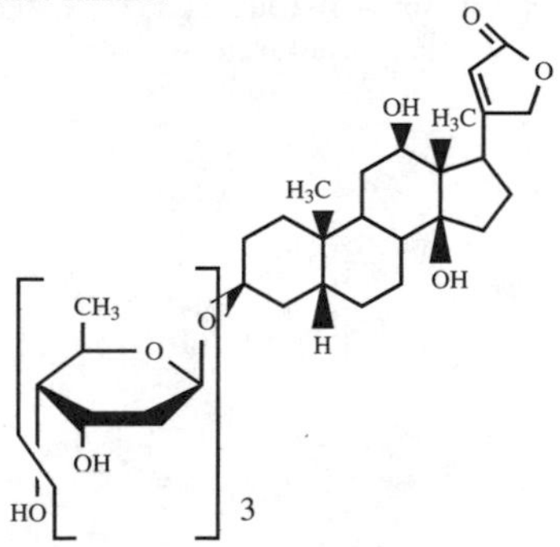

Digoxin Immune Fab An antigen-binding fragment of antidigoxin antibody used as an antidote.

Diguanidinobutanase The enzyme that catalyzes the following reaction:

1,4-Diguanidinobutane + H_2O ⇌ Agmatine + urea

DIHPPA Abbreviation for diiodohydroxyphenyl pyruvic acid.

Dihybrid A hybrid that is heterogeneous at two genetic loci.

Dihydergot A trade name for mesylate dihydroergotamine, an adrenergic blocker that inhibits the effect of epinephrine and norepinephrine.

Dihydrate A compound that has two molecules of water.

Dihydrex A trade name for diphenhydramine hydrochloride, an antihistamine that competes with histamine for H_1 receptors on effector cells.

Dihydric Describing a chemical compound containing two hydroxyl groups per molecule.

Dihydro- A prefix denoting the presence of two additional hydrogen atoms.

Dihydroallin (mol wt 179) A naturally occurring nonprotein amino acid.

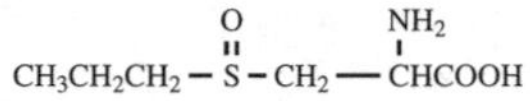

Dihydrocodeine (mol wt 301) A narcotic analgesic agent.

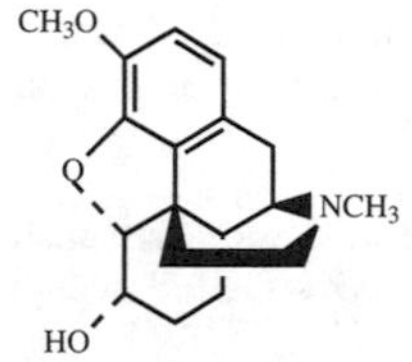

Dihydrocodeinone Enol Acetate (mol wt 341) A narcotic analgesic and antitussive agent.

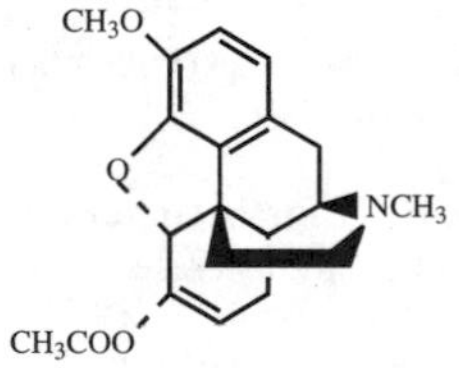

Dihydrocoumarin Hydrolase The enzyme that catalyzes the following reaction:

Dihydrocoumarin + H_2O ⇌ Melilotate

Dihydrofolate Reductase The enzyme that catalyzes the following reaction:

5,6,7,8-Tetrahydrofolate + $NADP^+$ ⇌ 7,8-Dihydrofolate + NADPH

Dihydrofolate Synthetase The enzyme that catalyzes the following reaction:

ATP + dihydropteroate + L-glutamate ⇌ ADP + Pi + dihydrofolate

Dihydrofolic Acid (mol wt 443) A coenzyme and a substrate for dihydrofolate reductase.

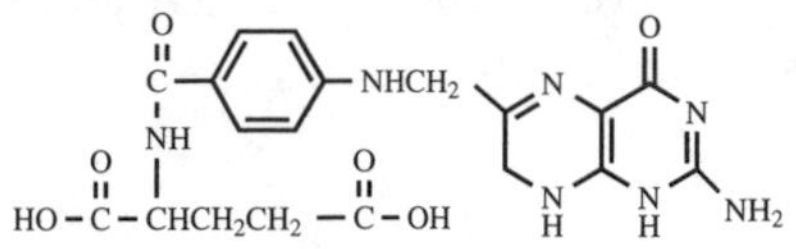

Dihydrolipoamide Acetyltransferase The enzyme that catalyzes the following reaction:

Acetyl-CoA + dihydrolipoamide

⇌

CoA + *S*-acetylhydrolipoamide

Dihydrolipoamide Dehydrogenase The enzyme that catalyzes the following reaction:

Dihydrolipoamide + NAD^+

⇌

Lipoamide + NADH

Dihydrolipoamide NAD^+ Oxidoreductase The systematic name for dihydrolipoamide dehydrogenase.

Dihydrolipoamide Succinyltransferase The enzyme that catalyzes the following reaction:

Succinyl-CoA + dihydrolipoamide

⇌

CoA + S-succinyl-dihydrolipoamide

Dihydrolipoyl Transacetylase See dihydrolipoamide acetyltransferase.

Dihydrolipoyl Transsuccinylase See dihydrolipoamide succinyltransferase.

Dihydroorotase The enzyme that catalyzes the following reaction:

4,5-Dihydroorotate + H_2O

⇌

N-Carbamoyl aspartate

Dihydroorotate Dehydrogenase The enzyme that catalyzes the following reaction:

4,5-Dihydroorotate + acceptor

⇌

Orotate + reduced acceptor

Dihydroorotate NAD^+ Oxidoreductase The systematic name for dihydroorotate dehydrogenase.

Dihydroorotate Oxidase The enzyme that catalyzes the following reaction:

Dihydroorotate + O_2 ⇌ Orotate + H_2O_2

Dihydroorotic Acid (mol wt 158) An intermediate in the biosynthesis of uridine monophosphate.

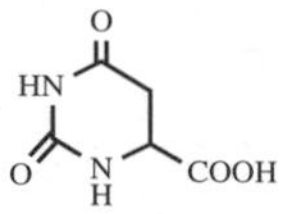

Dihydropterine Reductase The enzyme that catalyzes the following reaction:

NADPH + 6,7-dihydropterine

⇌

5,6,7,8-tetrahydropterine + $NADP^+$

Dihydropyrimidinase See dihydrouracil dehydrogenase.

Dihydropyrimidine Dehydrogenase The enzyme that catalyzes the following reaction:

5,6-Dihydrouracil + $NADP^+$

⇌

Uracil + NADPH

Dihydrosphingosine Kinase See sphingosine kinase.

Dihydrostreptomycin (mol wt 584) A semisynthetic antibiotic derived from streptomycin.

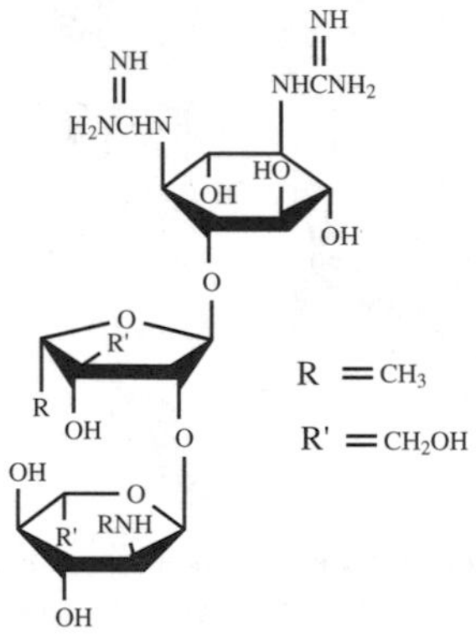

Dihydrostreptomycin 6-Phosphate 3′-α-Kinase The enzyme that catalyzes the following reaction:

ATP + dihydrostreptomycin 6-phosphate

⇌

ADP + dihydrostreptomycin α-3,6-bisphophate

Dihydrotachysterol (mol wt 399) A calcium regulator that stimulates calcium absorption from the GI tract and promotes secretion of calcium from the bone to the blood.

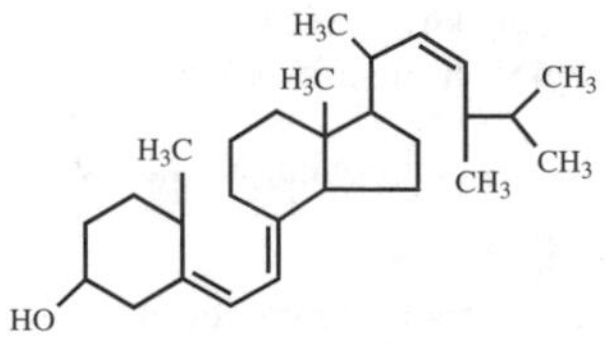

Dihydrothymidine Dehydrogenase See dihydropyrimidine dehydrogenase.

Dihydro-U Abbreviation for dihydrouracil.

Dihydro-U Arm One of the arms or stems in tRNA.

Dihydro-U Stem See dihydro-U arm.

Dihydrouracil (mol wt 114) A base found in tRNA.

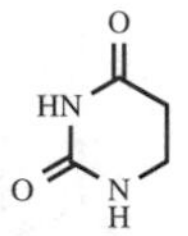

Dihydrouracil Dehydrogenase The enzyme that catalyzes the following reaction:

4,5-Dihydrouracil + NAD^+ ⇌ Uracil + NADH

Dihydrouracil Oxidase The enzyme that catalyzes the following reaction:

5,6-Dihydrouracil + O_2
⇅
Uracil + H_2O_2

Dihydrouridine (mol wt 246) A ribonucleoside of dihydrouracil found in tRNA.

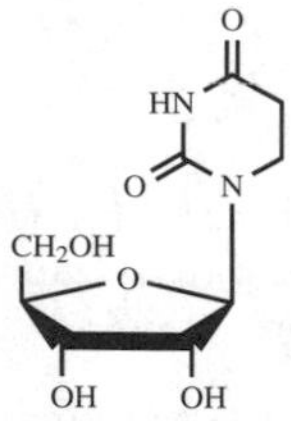

Dihydroxyacetone (mol wt 90) A product in dihydroxyacetone fermentation.

CH_2OH
|
C=O
|
CH_2OH

Dihydroxyacetone Fermentation An aerobic fermentation by *Gluconobacter oxydans* for conversion of glycerol to dihydroxyacetone.

Dihydroxyacetone Kinase The enzyme that catalyzes the following reaction:

ATP + dihydroxyacetone
⇅
ADP + dihydroxyacetone 3-phosphate

Dihydroxyacetone 3-Phosphate (mol wt 170) An intermediate in glycolysis.

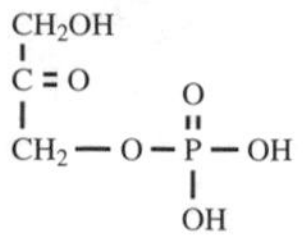

Dihydroxyacetone Phosphate Shuttle See glycerol phosphate shuttle.

Dihydroxyacetone Reductase The enzyme that catalyzes the following reaction:

Glycerol + $NADP^+$
⇅
Dihydroxyacetone + NADPH

Dihydroxyacetone Transferase See transaldolase.

Dihydroxyaluminum Acetylsalicylate (mol wt 240) An analgesic and antipyretic agent.

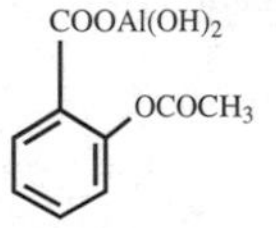

Dihydroxyaluminum Aminoacetate (mol wt 135) An antacid.

$NH_2CH_2COOAl(OH)_2$

Dihydroxyaluminium Sodium Carbonate (mol wt 144) An antacid.

$(OH)_2AlOCO_2Na$

Dihydroxyfumarate Decarboxylase The enzyme that catalyzes the following reaction:

Dihydroxyfumarate
⇅
Tartronate semialdehyde + CO_2

Dihydroxyphenylalanine (mol wt 197) A precursor for synthesis of melanin.

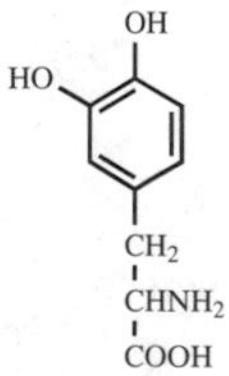

Dihydroxyphenylalanine Amonia-Lyase The enzyme that catalyzes the following reaction:

3,4-Dihydro-L-phenylalanine
⇅
trans-Caffeate + NH_3

Dihydroxyphenylalanine Transaminase The enzyme that catalyzes the following reaction:

3,4-Dihydroxyphenylalanine + α-ketoglutarate
⇅
3,4-Dihydroxyphenylpyruvate + L-glutamate

Diimine (mol wt 30) An intermediate in nitrogen fixation.

H-N=N-H

2,4-Diiodohistidine (mol wt 406) A naturally occurring nonprotein amino acid.

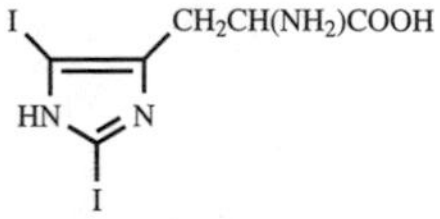

Diiodohydroxyquin See iodoquinol.

Diiodothyronine (mol wt 525) A naturally occurring nonprotein amino acid from bovine thyroid gland.

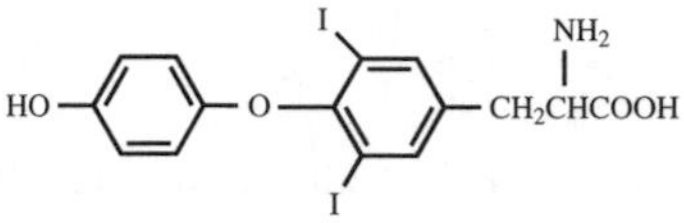

Diiodotyrosine (mol wt 433) A naturally occurring nonprotein amino acid.

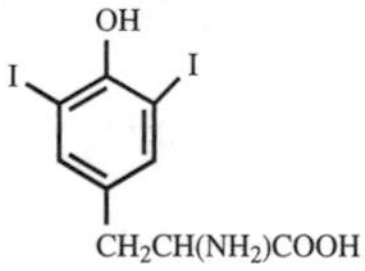

Diiodotyrosine Transaminase The enzyme that catalyzes the following reaction:

3,5-Diiodotyrosine + α-keto-glutarate

⇅

3,5-Diiodo-4-hydroxyphenylpyruvate + L-glutamate

Diiosopromine (mol wt 295) An antispasmodic agent.

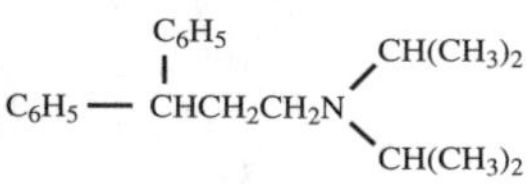

Diisopropylamine Dichloroacetate (mol wt 230) A vasodilator and hypotensive agent.

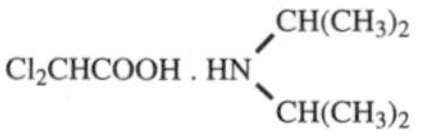

Diisopropyl Paraoxon (mol wt 303) A cholinergic agent.

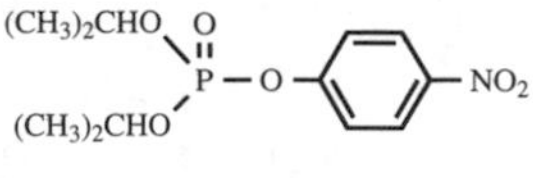

Diisothiocyanatostilbene Disulfonic Acid (mol wt 454) A fluorescent reagent for the modification of proteins.

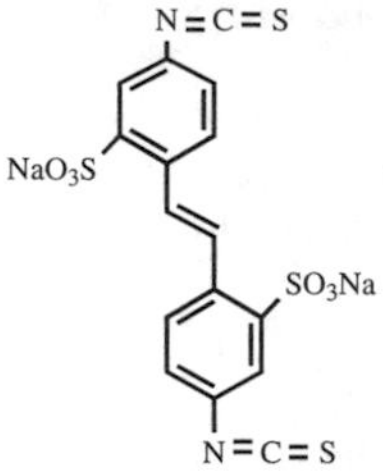

Dikaryon Cell with two different nuclei.

Dikaryotes Cells with two different nuclei.

Dikegulac (mol wt 274) A plant growth regulator.

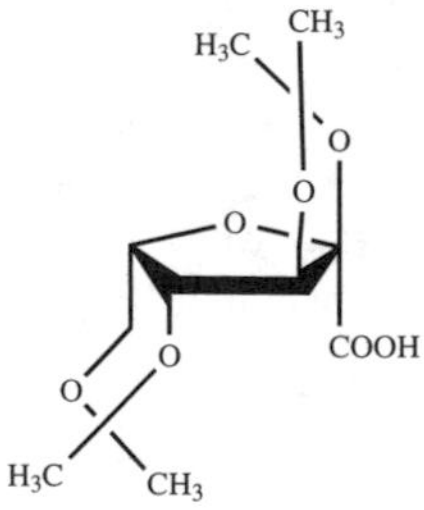

Diketose Any monosaccharide derivative containing two ketonic carbonyl groups or two potential ketonic carbonyl groups.

Dilacor XR A trade name for diltiazem hydrochloride, an antianginal agent that inhibits calcium ion influx and decreases cardiac oxygen demand.

Dilantin A trade name for phenytoin, an anticonvulsant.

Dilation The enlargement of, or expansion of, a hollow organ such as a blood vessel.

Dilatrate-SR A trade name for isosorbide dinitrate, an antianginal agent that reduces cardiac oxygen demand.

Dilaudid A trade name for hydromorphone hydrochloride, an analgesic and antitussive agent.

Dilazep (mol wt 605) A coronary vasodilator.

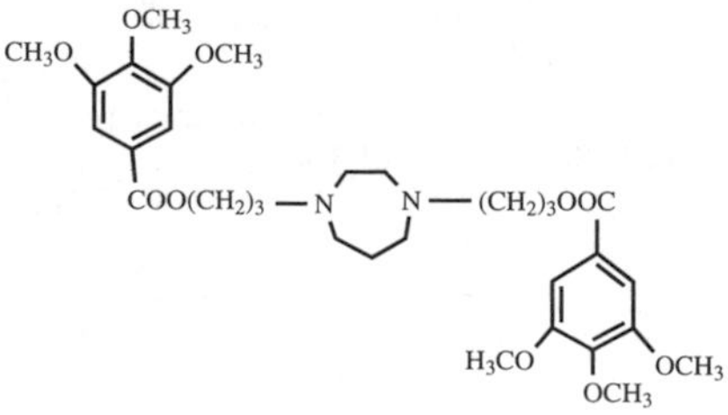

Dilevalol (mol wt 328) An antihypertensive agent.

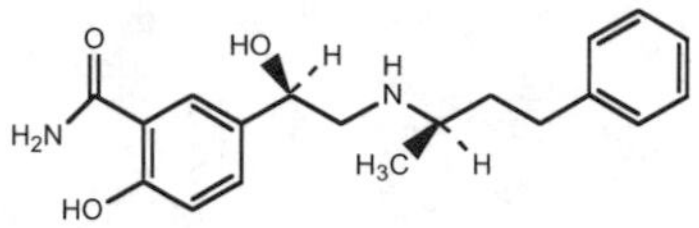

Dilin A trade name for dyphyline, a bronchodilator that inhibits phosphodiesterase and increases cAMP concentration.

Dilocaine A trade name for lidocaine hydrochloride, a local anesthetic agent that blocks depolarization by interfering with sodium-potassium exchange.

Dilomine A trade name for dicyclomine hydrochloride, an anticholinergic agent

Dilor A trade name for dyphylline, a bronchodilator that inhibits phosphodiesterase and increases cAMP concentration.

Dilosyn A trade name for methdilazine hydrochloride, an antihistaminic agent that competes with histamine for H_1 receptors on target cells.

Diltiazem (mol wt 415) An antianginal agent that inhibits calcium influx and decreases cardiac oxygen demand.

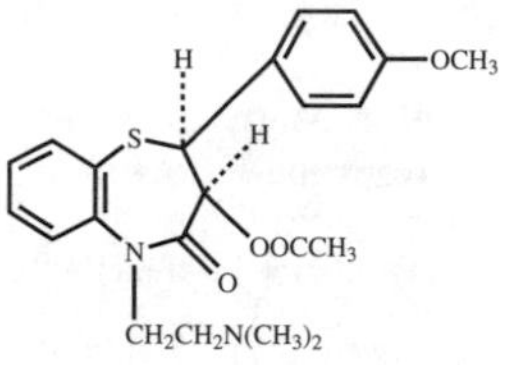

Diluent Solution used for making dilutions.

Dimecrotic Acid (mol wt 222) A choleretic agent.

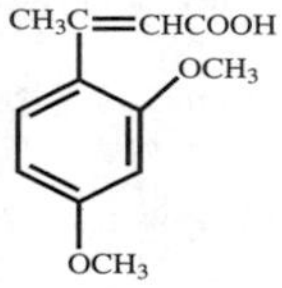

Dimefox (mol wt 154) An insecticide.

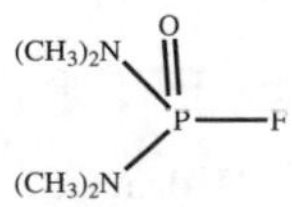

Dimelor A trade name for acetohexamide, an antidiabetic agent that stimulates the release of insulin from beta cells.

Dimemorfan (mol wt 255) An antitussive agent.

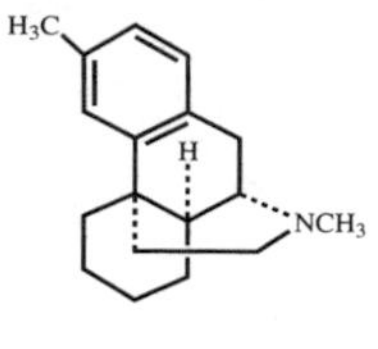

Dimenhydrinate (mol wt 470) An antiemetic agent that inhibits nausea and vomiting, it is also an antihistaminic agent.

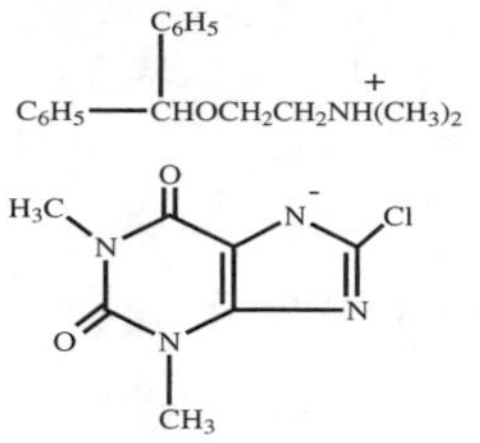

Dimenoxadol (mol wt 327) A narcotic analgesic agent.

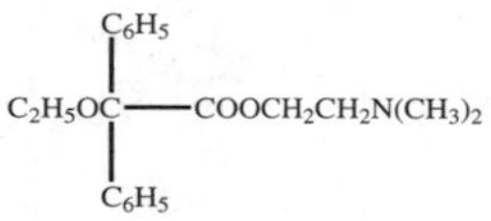

Dimentabs A trade name for dimenhydrinate, an antihistaminic and antiemetic agent.

Dimepheptanol (mol wt 311) A narcotic analgesic agent.

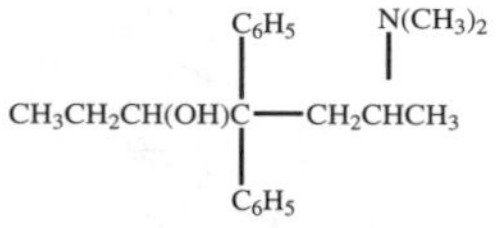

Dimercaprol (mol wt 124) An antidote to mercury poisoning and a chelating agent that forms complexes with heavy metal.

$$HOCH_2CH(SH)CH_2SH$$

Dimestrol (mol wt 296) An estrogen.

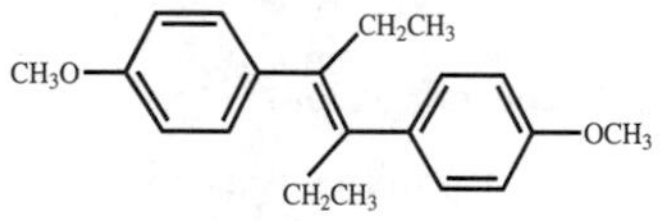

Dimetabs A trade name for dimenhydrinate, an anticholinergic and antihistaminic agent.

Dimetane A trade name for bromopheniramine maleate, an antihistaminic agent that competes with histamine for H_1 receptors on effector cells.

Dimetapp Allergy A trade name for brompheniramine maleate, an antihistaminic agent.

Dimethadione (mol wt 129) An anticonvulsant.

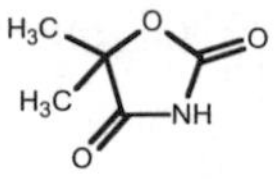

Dimethazan (mol wt 251) An antidepressant.

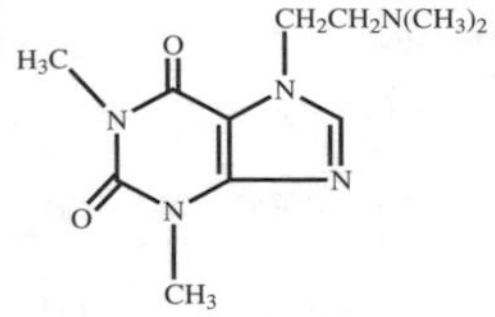

Dimethindene (mol wt 292) An antihistaminic agent.

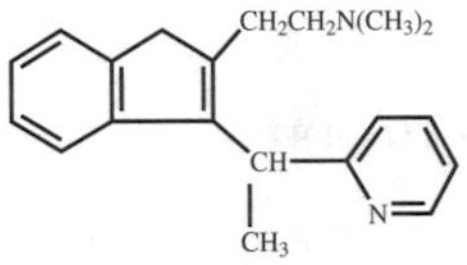

Dimethisoquin (mol wt 272) A topical anesthetic agent.

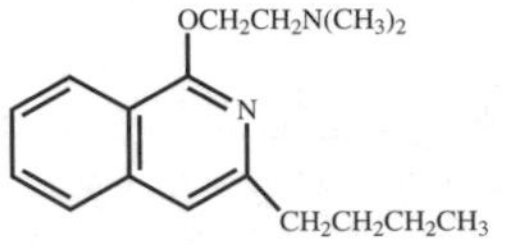

Dimethisterone (mol wt 340) A pregestogen.

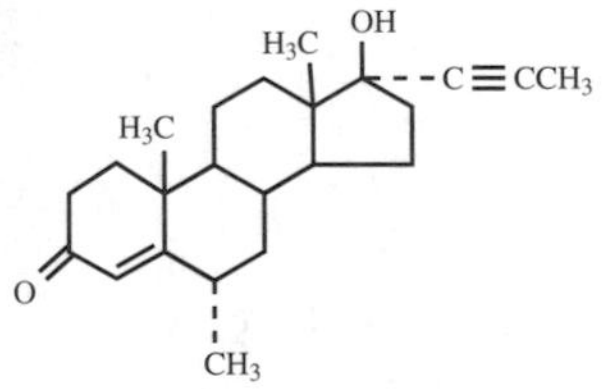

Dimethocaine (mol wt 278) An anesthetic agent.

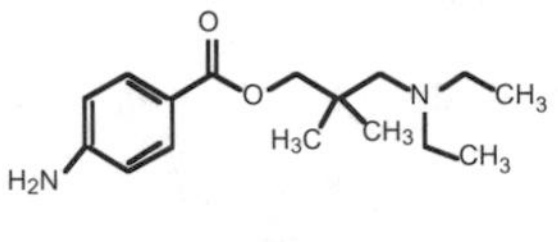

Dimethoxanate (mol wt 358) An antitussive agent.

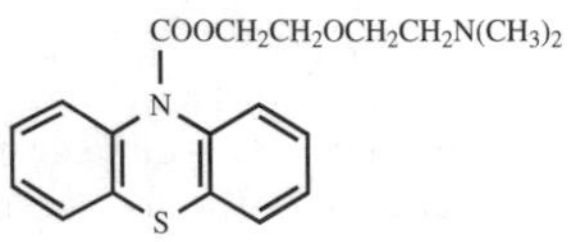

Dimethoxymethyl Benzoquinone (mol wt 182) A coenzyme (coenzyme Q_0).

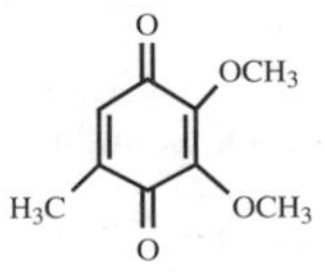

Dimethylargininase The enzyme that catalyzes the following reaction:

Dimethylarginine + H_2O

⇅

Methylamine + urea

Dimethyldithiopropionimidate Di-hydrochloride (mol wt 309) A reagent used for cross-linking of membrane proteins and hemoglobins.

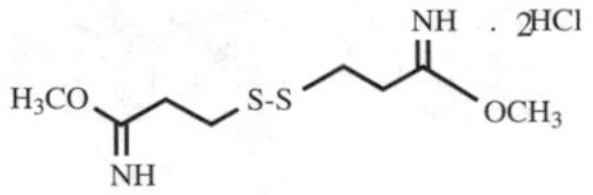

Dimethylformamidase The enzyme that catalyzes the following reaction:

N,N-Dimethylformamide + H_2O

⇅

N,N-Dimethylamine + formate

Dimethylglycine Dehydrogenase The enzyme that catalyzes the following reaction:

Dimethylglycine + acceptor + H_2O

⇅

Sarcosine + formaldehyde + reduced acceptor

Dimethylglycine Oxidase The enzyme that catalyzes the following reaction:

Dimethylglycine + H_2O + O_2

⇅

Sarcosine + formaldehyde + H_2O_2

Dimethylhistidine Methyltransferase The enzyme that catalyzes the following reaction:

S-Adenosyl-methionine + dimethylhistidine

⇅

S-Adenosylhomocysteine + trimethylhistidine

Dimethylmalate Dehydrogenase The enzyme that catalyzes the following reaction:

Dimethylmalate + NAD^+

⇅

3-Methyl-2-keto-butanoate + NADH + CO_2

Dimethylmalate Lyase The enzyme that catalyzes the following reaction:

2-4-Dimethylmalate ⇌ Propanoate + pyruvate

Dimethylmaleate Hydratase The enzyme that catalyzes the following reaction:

Dimethylmalate ⇌ Dimethylmaleate + H_2O

Dimethylmaleic Anhydride (mol wt 126) A reagent used for dissociation of ribosomal proteins.

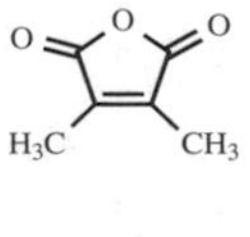

Dimethyl-Pyruvate Lyase The enzyme that catalyzes the following reaction:

Dimethylmalate ⇌ Pyruvate + propanoate

Dimethyl-Sulfate (mol wt 126) A methylating reagent used in DNA sequencing.

$(CH_3O)_2SO_2$

Dimethyl-Sulfate Protection Protection from methylation of adenine and guanine in DNA by dimethyl sulfate upon the attachment of enzyme (e.g., RNA-polymerase) or protein onto the DNA molecule.

Dimethylthiambutene (mol wt 263) A narcotic analgesic agent.

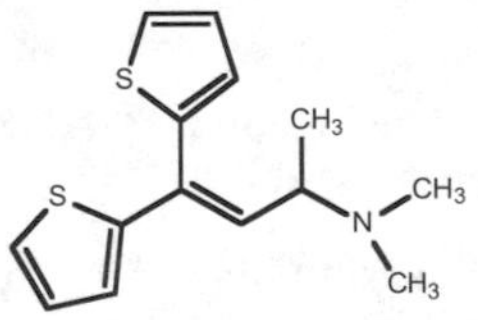

Dimethyltryptamine (mol wt 188) A metabolite of 5-hydroxytryptamine and a hallucinogenic agent.

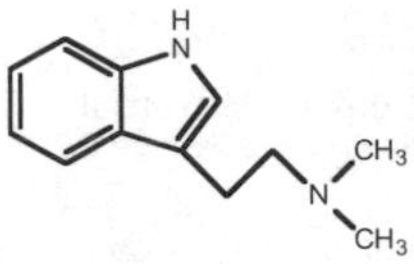

Dimetofrine (mol wt 227) An antihypotensive agent.

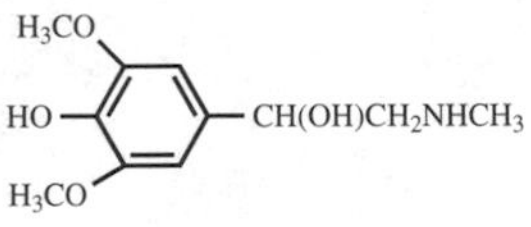

Dimorphism The existence within a species of two distinct forms.

Dimoxyline (mol wt 367) A vasodilator.

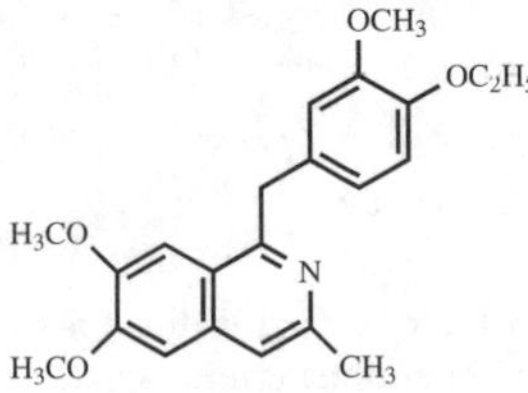

Dimycor A trade name for a combination drug containing pentaerythritol tetranitrate and phenobarbital sodium, used as an antianginal agent.

Dinate A trade name for dimenhydrinate, an antiemetic agent.

2,4-Dinitro-5-Fluoroaniline (mol wt 201) A reagent used for labeling amino groups in proteins.

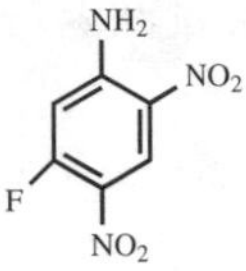

2,4-Dinitro-1-Fluorobenzene (mol wt 186) A reagent used in modification and determination of amino acids, peptides, and phenols.

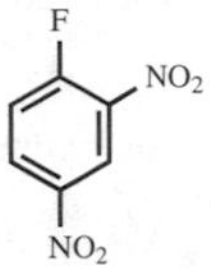

2,4-Dinitrophenol (mol wt 184) An oxidative phosphorylation uncoupler and a reagent used as a hapten for the study of antigen-antibody interaction.

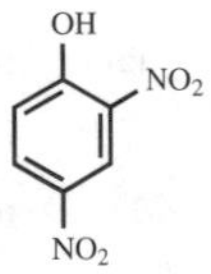

Dinitrophenyl Amino Acid A colored derivative of an amino acid resulting from treatment of an amino acid with 1-fluoro-2,4-dinitrobenzene.

2,4-Dinitrophenylhydrazine (mol wt 198) A reagent used for determination of aldehydes and ketones.

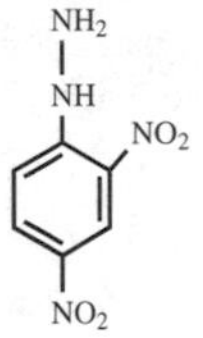

Dinoprostone (mol wt 353) A prostaglandin (prostaglandin E_2) used as an abortifacient.

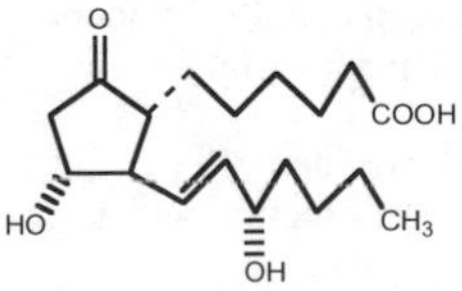

Dintzis Procedure A procedure for investigating the direction and rate of biosynthesis of a polypeptide.

Dinucleotide Nucleotidohydrolase The systematic name for nucleotide pyrophosphatase.

Diocto A trade name for docusate sodium, a laxative that promotes incorporation of water into the stool.

Dioctyl Sodium Sulfosuccinate See docusate.

Diode Laser A portable laser used for treating disease of the retina of the eye by producing small burns in the retina.

Diodoquin A trade name for iodoquinol, an antiprotozoal agent.

Dioecious Having the male and female organs in different plants.

Dioeze A trade name for docusate sodium, a laxative that promotes incorporation of water into the stool.

Diol Any organic compound that contains two hydroxy groups.

Diolax A trade name for a combination drug containing docusate sodium and casanthranol, used as a laxative.

Diolein Referring to glycerol 1,2- and 1,3-dioleate.

Diomycin A trade name for erythromycin, an antibiotic.

Diopterin (mol wt 571) A folic acid analog.

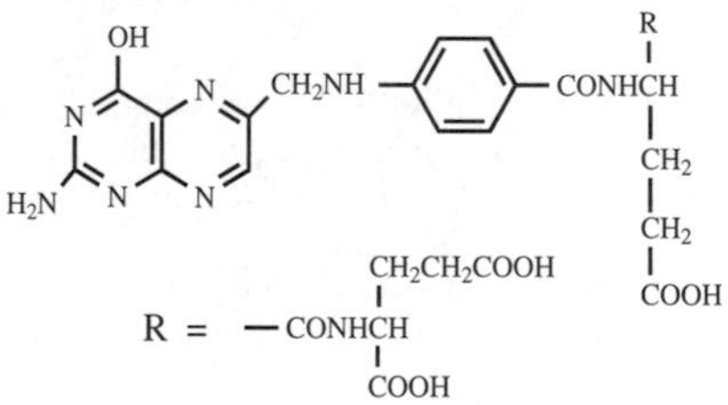

Diosuccin A trade name for docusate sodium, a laxative that promotes incorporation of water into the stool.

Dio-Sul A trade name for docusate sodium, a laxative that promotes incorporation of water into the stool.

Dioval A trade name for estradiol valerate, an estrogen that promotes growth and development of female sex organs and maintenance of secondary sex characteristics in women.

Diovan A trade name for valsartan, an angiotensin II receptor blocker used as an antihypertensive agent.

Diovol A trade name for a combination drug containing aluminum hydroxide and magnesium hydroxide, used as an antacid.

Dioxathion (mol wt 457) An insecticide.

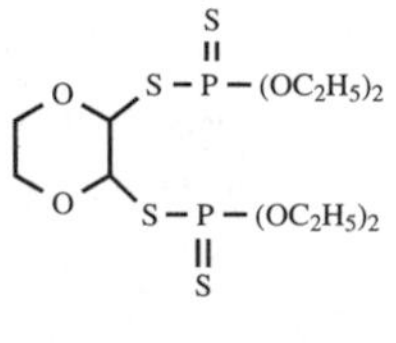

Dioxethedrine (mol wt 211) A bronchodilator.

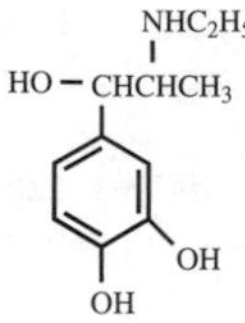

Dipeptidase Enzyme that hydrolyzes dipeptides.

Dipeptidase M The enzyme that catalyzes the hydrolysis of dipeptides involving COOH groups of methionine.

Dipeptide A peptide containing two amino acid residues.

Dipeptide Hydrolase Synonym of peptidyl dipeptidase A.

Dipeptidyl Aminopeptidase II Synonym of dipeptidyl-peptidase II.

Dipeptidyl Arylamidase I Synonym of dipeptidyl-peptidase.

Dipeptidyl Carbosypeptidase The enzyme that catalyzes the release of dipeptide from the C-terminal of a polypeptide chain.

Dipeptidyl Peptidase The enzyme that catalyzes the release of dipeptide from the N-terminal of a polypeptide chain.

Diperodon (mol wt 398) A local anesthetic agent.

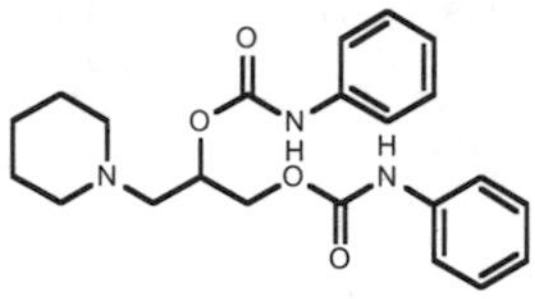

DIPF Abbreviation for diisopropyl phosphofluoridate.

Diphemanil Methylsulfate (mol wt 390) An anticholinergic agent.

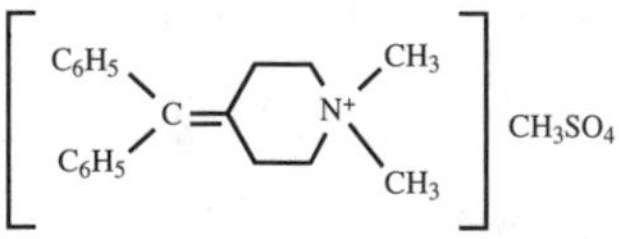

Diphen A trade name for diphenhydramine hydrochloride, an antihistaminic agent.

Diphen Cough A trade name for diphenhydramine hydrochloride, an antihistaminic agent.

Diphenacen-50 A trade name for diphenhydramine hydrochloride, an antihistaminic agent.

Diphenadione (mol wt 340) An anticoagulant.

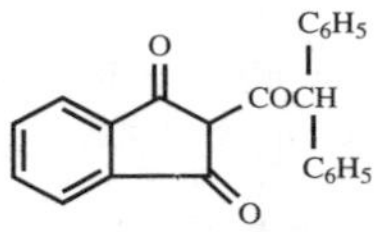

Diphenadryl A trade name for diphenhydramine, used as an antihistaminic agent.

Diphenatol A trade name for diphenoxylate hydrochloride, an antidiarrheal agent.

Diphenazoline (mol wt 266) An antihistaminic agent.

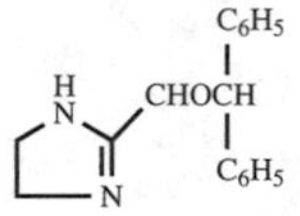

Diphenhist A trade name for diphenhydramine, an antihistaminic agent.

Diphenhydramine (mol wt 255) An antihistaminic agent that competes with histamine for H_1 receptors on effector cells.

Diphenicillin Sodium (mol wt 418) A semisynthetic antibiotic related to penicillin.

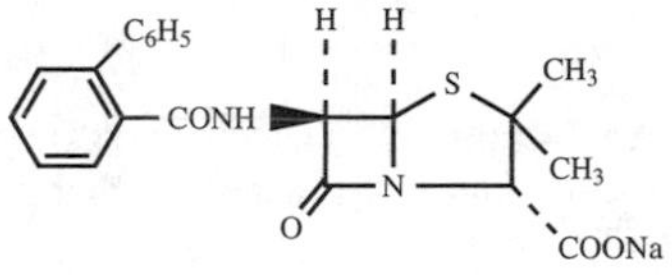

Diphenidol (mol wt 309) An antiemetic agent for control of nausea and vomiting.

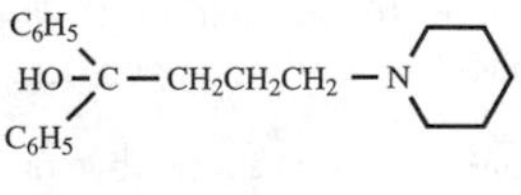

Diphenol Oxidase The enzyme that catalyzes the following reaction:

$$2\ \text{Catechol} + O_2 \rightleftharpoons 2\ \text{Benzoquinone} + 2\ H_2O$$

Diphenoxylate (mol wt 453) An antidiarrheal agent that inhibits mobility and propulsion and diminishes secretion.

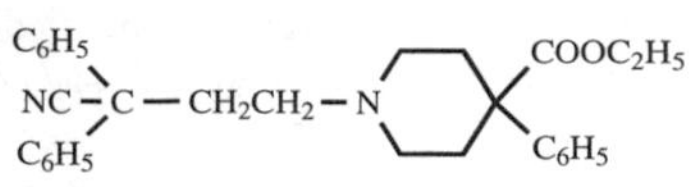

Diphenylamine (mol wt 169) A reagent used for colorimetric determination of DNA. It reacts with deoxyribose of DNA.

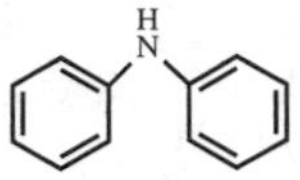

Diphenylan A trade name for phenytoin sodium, an anticonvulsant.

Diphenylhydantoin A synonym of phenytoin.

Diphenylpyraline (mol wt 281) An antihistaminic agent.

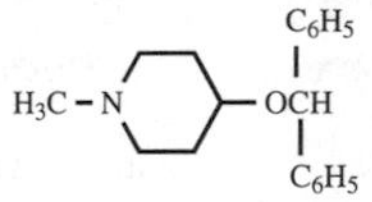

Diphosphoglycerate Phosphatase The enzyme that catalyzes the following reaction:

$$\text{2,3-Diphospho-D-glycerate} + H_2O \rightleftharpoons \text{3-Phospho-D-glycerate} + \text{Pi}$$

1,3-Diphosphoglyceric Acid (mol wt 266) An intermediate in glycolysis (also known as 1,3-bisphosphoglyceric acid).

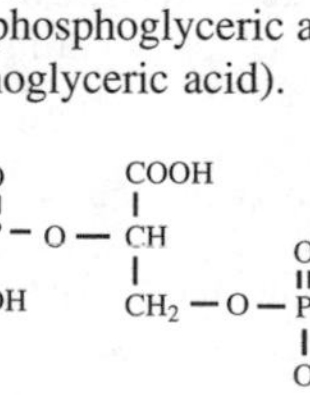

2,3-Diphosphoglyceric Acid (mol wt 266) An isomer of 1,3-diphosphoglyceric acid (also known as 2,3-bisphosphoglyceric acid).

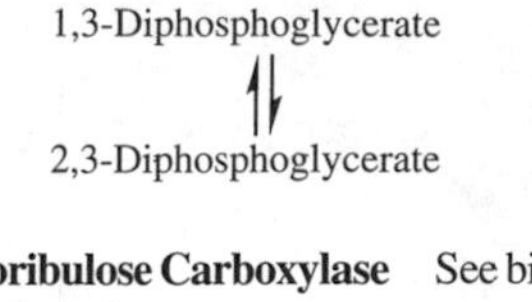

Diphosphoglyceromutase The enzyme that catalyzes the following reaction:

$$\text{1,3-Diphosphoglycerate} \rightleftharpoons \text{2,3-Diphosphoglycerate}$$

Diphosphoribulose Carboxylase See bisphosphoribulose carboxylase.

Diph-Tet Abbreviation for *Diphtheria tetanus.*

Diphtheria An acute, communicable human disease caused by strains of *Corynebacterium diphtheriae.*

Diphtheria Antitoxin Referring to the antibody against diphtheria toxin.

Diphtheria Toxin A protein toxin produced by strains of *Corynebacterium diphtheriae* that inhibits protein synthesis in mammalian cells.

Diphtheria Toxoid The chemically modified avirulent diphtheria toxin.

Diphtheroid Referring to the nonpathogenic strain of *Corynebacterium.*

Diph-Tox Abbreviation for *Diphtheria* toxin.

Diplobiont Referring to an organism that has both haploid and diploid somatic stages in the life cycle.

Diplochromosome A chromosome that consists of four chromatides resulting from the abnormal duplication in which the centromere fails to divide and the daughter chromosomes fails to move apart.

Diplococci Cocci occurring in pairs.

Diplo-haplont An organism in which the products of meiosis form haploid gametophyte producing gametes. The fertilization of gametes generates diploid sporophytes in which meiosis takes place.

Diploid Cell A cell containing two sets of chromosomes (2N) that are inherited from both parents.

Diplont An organism that has a diploid somatic stage.

Diplopia With double vision.

Diplornaviridae A family of animal viruses that contain double-stranded RNA.

Diplotene A stage in prophase of meiosis in which two chromatides of each chromosome become visible.

Dipolar ion See zwitterion.

Dipole A molecule that carries opposite charges at opposite poles.

Diponium Bromide (mol wt 404) An antispasmodic agent.

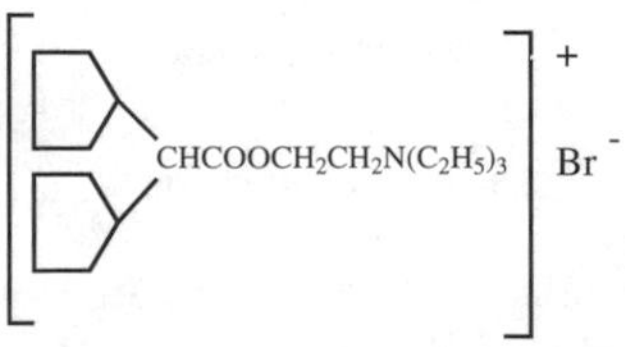

DIPP Abbreviation for diphosphoinositol polyphosphate phosphohydrolase.

Dipridacot A trade name for dipyridamole, an antianginal and antiplatelet agent.

Diprivan A trade name for propofol, an anesthetic agent.

Diprolene A trade name for beta-methasone dipropionate, a glucocorticoid used as an anti-inflammatory agent.

Dipropalin (mol wt 281) An herbicide.

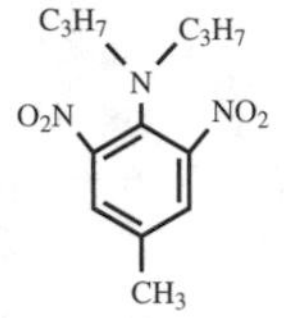

Diprosone A trade name for beta-methasone dipropionate, a glucocorticoid used as an anti-inflammatory agent.

Diprotic Acid An acid that contains two dissociable protons.

Dipyrocetyl (mol wt 238) An antipyretic and analgesic agent.

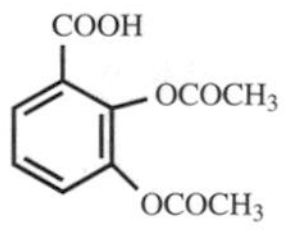

Direct Agglutination The direct agglutination of erythrocytes, microorganisms, or other particles by antibody.

Direct Coomb's Test A test utilizing antibody-coated erythrocytes.

Direct Immunofluorescence The detection of antigens by the direct reaction of fluorescently labeled antibody with the antigen.

Direct Repeat The presence of two or more identical or closely related DNA sequences in the same orientation of the same DNA molecule.

DIS Abbreviation for death-inducing signal.

Disalcid A trade name for salsalate, an analgesic and antipyretic activity.

DISC Abbreviation for death-inducing signaling complex.

Disc Diffusion Test An antibiotic-sensitivity test in which the antibiotic-impregnated paper disk is placed onto an agar plate seeded with bacteria.

Disc Gel Electrophoresis A type of zonal electrophoresis in which discontinuities of pH, ionic strength, buffer composition, and gel concentration are incorporated into the gel system to obtain a high resolution of protein separation. The most commonly used gel system is polyacrylamide disc gel electrophoresis.

Dische Reaction A reaction used for colorimetric determination of DNA using diphenylamine as a reagent that combines with deoxyribose from DNA forming a color complex.

Discoidin A group of endogenous carbohydrate-binding protein (lectins) produced by cells of the slime mould *Dictyostelium discoideum* during differentiation.

Discontinuous Density Gradient A nonlinear density gradient in which the density changes are in a stepwise fashion.

Discontinuous Replication The process of synthesis of a lagging strand during DNA replication. The lagging strand consists of a series of Okazaki fragments before ligation (see also discontinuous strand of DNA).

Discontinuous Strand of DNA The strand of DNA that grows discontinuously and lags behind the replication of the continuous strand. The lagging strand consists of a series of Okazaki fragments before ligation (see also discontinuous replication).

Disinfectants Substance capable of disinfection.

Disinfection The destruction of microorganisms.

Disjunction Referring to normal separation of chromosomes in mitosis and meiosis.

Dislodgement The phenomenon of elimination of resident plasmid by the introduction of a second compatible plasmid into the cell.

Disonate A trade name for docusate sodium, a laxative.

Disophenol (mol wt 391) An anthelmintic agent for hookworm.

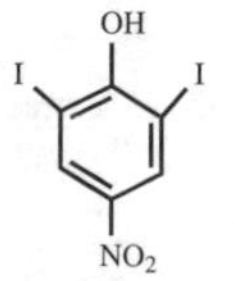

Disophrol Chronotabs A trade name for a combination drug containing dexbrompheniramine maleate and pseudoephedrine, used as an antihistaminic agent.

Disoprofol A trade name for propofol, a general anesthetic agent.

Disopyramide (mol wt 339) A cardiac depressant that prolongs the action potential and also has a membrane stabilizing effects.

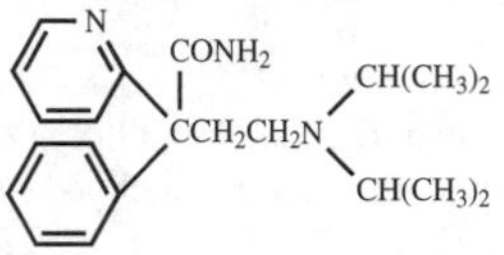

Di-Sosul A trade name for docusate sodium, a laxative.

Disotate A trade name for edetate sodium, a chelating agent.

Disparlure (mol wt 283) An insect sex attractant.

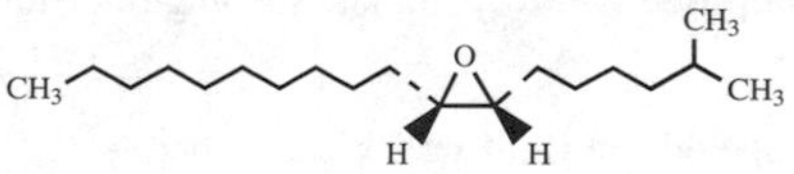

Dispase Synonym of leucolysin.

Di-Spaz A trade name for dicyclomine hydrochloride, an anticholinergic agent.

Disproportionating Enzyme Synonym of 4-α-glucanotransferase.

Disseminated Intravascular Coagulation A disorder resulting from over-stimulation of the blood clotting mechanisms in response to disease or injury.

Dissemination The dispersion of microorganisms or disease.

Dissimitory Nitrate Reduction An energy-yielding nitrate reduction reaction (see denitrification).

Dissociation Separation of a complex molecule or a structural complex into subcomponents.

Dissociation Constant The equilibrium constant of dissociation.

Distigmine Bromide (mol wt 576) A cholinesterase inhibitor.

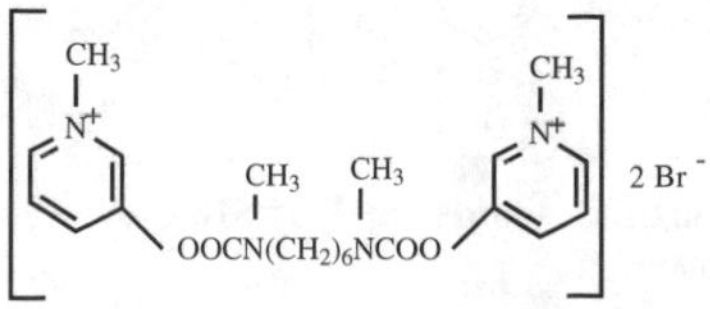

Disulfamide (mol wt 285) A diuretic agent.

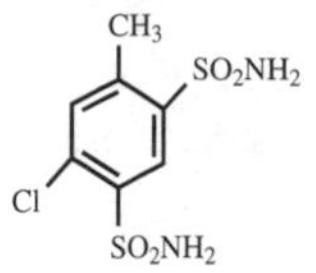

Disulfide Bonds The S-S bond formed between two sulfhydryl-containing cysteine residues in a protein.

Disulfide Knot A region at the center of a fibrinogen where the six polypeptide chains are linked together by eight disulfide bonds.

Disulfiram (mol wt 297) An alcohol deterrent that blocks the oxidation of alcohol at the aldehyde stage to give an unpleasant smell as a deterrent for alcoholism.

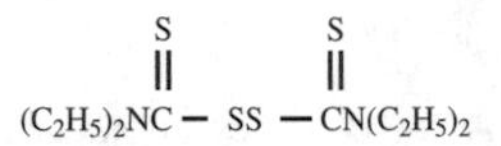

Disulphide Knot Alternate spelling of disulfide knot.

Disul-Sodium (mol wt 309) An herbicide.

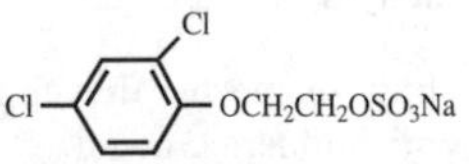

DIT Abbreviation for diiodotyrosine.

Dital A trade name for phendimetrazine, an anorexic agent.

Ditate-DS A trade name for a combination drug containing estradiol valerate and testosterone enanthate.

Ditazol (mol wt 324) An anti-inflammatory agent.

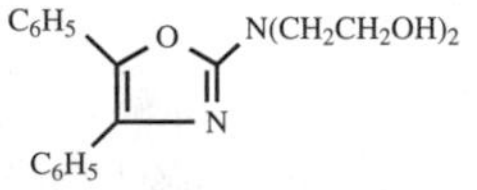

Dithianone (mol wt 296) A fungicide.

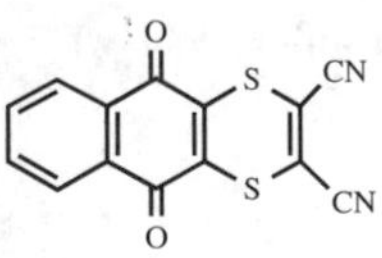

Dithiazanine Iodide (mol wt 518) An anthelmintic agent.

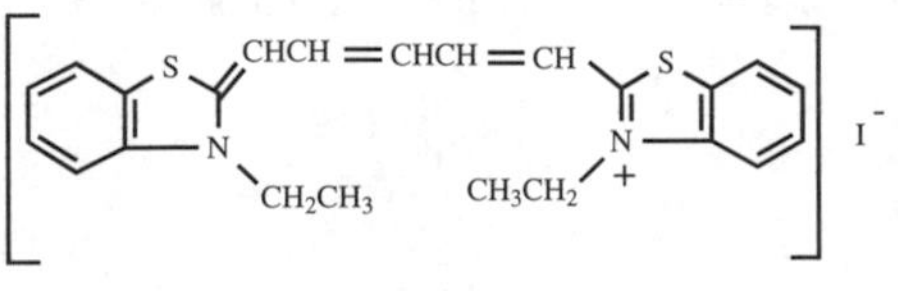

Dithiodipyridine (mol wt 220) A reagent used for determination of sulfhydryl groups in biological material.

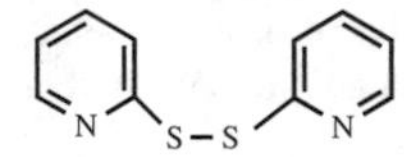

Dithiothreitol (mol wt 154) A protective agent for SH groups.

$$HSCH_2(CHOH)_2CH_2SH$$

dITP Abbreviation for deoxyinosine 5′-triphosphate.

Ditropan A trade name for oxybutynin chloride, a spasmolytic agent.

Diucardin A trade name for hydroflumethiazide, a diuretic agent that increases urine secretion of sodium and water by inhibiting sodium reabsorption.

Diuchlor H A trade name for hydrochlorothiazide, a diuretic agent that increases urine secretion of sodium and water by inhibiting sodium reabsorption.

Diupres A trade name for a combination drug containing chlorothiazide and reserpine.

Diurese A trade name for trichlormethiazide, a diuretic agent that increases urine secretion of sodium and water by inhibiting sodium reabsorption.

Diuresis The increased formation and release of urine.

Diuret A trade name for chlorothiazide, a diuretic agent.

Diuretics Agents that promotes urine secretion.

Diurigen A trade name for chlorothiazide, a diuretic agent that increases urine secretion of sodium and water by inhibiting sodium reabsorption.

Diuril A trade name for chlorothiazide, a diuretic agent that increases urine secretion of sodium and water by inhibiting sodium reabsorption.

Diurnal Rhythm A cycle based upon a daily periodicity.

Diuron (mol wt 233) A preemergent herbicide.

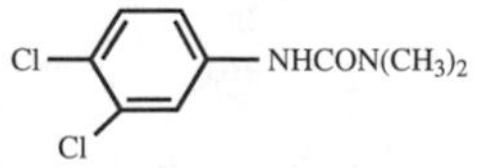

Divalent Referring to 1. a substance having a chemical valence of two or 2. a substance or protein having two binding sites (e.g., bivalent or divalent antibody).

Divergence The percent difference of nucleotide sequences between two strands of DNA or the percent difference of amino acid sequences between two peptide chains.

Diversity Gene The genes for supervariable regions of immunoglobulins.

Dixon Plot A plot for determination of the inhibition constant of an enzymatic reaction inhibitor.

Dizac A trade name for diazepam, an antianxiety, antiepileptic agent.

Dizmiss A trade name for meclizine, an antiemetic agent that inhibits nausea and vomiting.

Dizygotic Twins Twins arising from two fertilized eggs.

Djenkolic Acid (mol wt 254) A naturally occurring nonprotein amino acid.

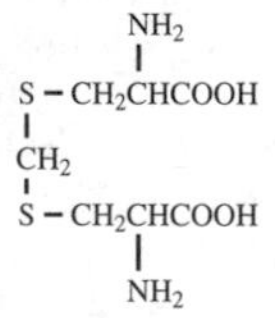

DKTC Abbreviation for dog kidney tissue culture.

DL A prefix in the chemical nomenclature denoting an equimolar mixture of D and L enantiomers.

DLC Abbreviation for 1. differential leukocyte count; 2. dynein light chain.

DLD Abbreviation for dihydrolipoamide dehydrogenase.

DLDH Abbreviation for dihydrolipoamide dehydrogenase.

DLIC Abbreviation for dynein light-intermediate chain.

D-Loop (Displacement Loop) A displaced single-stranded loop formed during early DNA replication through the displacement of one of the parental strands by a newly synthesized leading strand.

dm Abbreviation for decimeter(s).

DM Abbreviation for diabetes mellitus.

DMA Abbreviation for 1. dimethylacetamide; 2. dimethyl adenine; 3. dimethyl arginine.

DMARD Abbreviation for disease modifying antirheumatic drug.

dMb Abbreviation for deoxymyoglobin.

DMCC Abbreviation for direct microscope clump count.

DMCTC Abbreviation for dimethylchloro-tetracycline.

DMEM Abbreviation for Dulbecco's modified Eagle's medium.

DMF Abbreviation for dimethylformamide.

DMG Abbreviation for dimyristoyl glycerol.

1,2-DMG Abbreviation for 1,2-dimyristoylglycerol.

1,3-DMG Abbreviation for 1,3-dimyristoylglycerol.

DMMP Abbreviation for dimethylmethylene blue.

DMP Abbreviation for dimethyl phthalate.

DMPC Abbreviation for dimyristoyl-α-phosphatidylcholine.

DMPD Abbreviation for N,N-dimethyl-1,4-phenylene diamine.

DMPO Abbreviation for 5,5-dimethyl-1-pyrroline-N-oxide.

DMPP Abbreviation for dimethylphenylpiperazinium.

DMS Abbreviation for dimethyl sulfate.

DMSO Abbreviation for dimethyl sulfoxide.

DMT Abbreviation for 1. dimethyltryptamine; 2. divalent metal transporter.

DMTU Abbreviation for dimethylthiourea.

DMU Abbreviation for dimethyluracil.

DNA Abbreviation for deoxyribonucleic acid, the molecule of genetic information containing deoxynucleotides of adenine, guanine, cytosine, and thymine.

DNA Base Composition The molar ratio of adenine, guanine, cytosine, and thymine in DNA.

DNA Binding Protein Referring to single-strand binding protein.

DNA Blotting Transfer of separated DNA fragments from an electrophoretic gel to a nitrocellulose sheet for subsequent detection.

DNA Chimera DNA molecule with an inserted foreign DNA sequence.

DNA Clone A DNA sequence that is inserted into a cloning vector or plasmid.

DNA Cloning The process of generating multiple copies of a specific DNA sequence by replication of a recombinant plasmid within bacterial cells or by use of the polymerase chain reaction.

DNA Deoxyinosine Glycosidase The enzyme that catalyzes the release of free hypoxanthine.

DNA Dependent DNA Polymerase See DNA polymerase.

DNA Directed DNA Polymerase See DNA polymerase.

DNA Dependent RNA Polymerase See RNA polymerase.

DNA Directed RNA Polymerase See RNA polymerize.

DNA Double Helix The helical structure formed by coiling of two antiparallel polydeoxyribonucleotide chains around the same axis.

DNA Fingerprinting A technique to reveal the patterns of fragments of DNA resulting from restriction endonuclease treatment of DNA from different individuals. The great variability existing from one human chromosome to the next in the number of times a sequence is repeated produces an enormous degree of fragment-size variation.

DNA Footprinting A method of identifying regions of the DNA to which regulatory proteins bind.

***dna* Genes** The genes in *E. coli* whose products are involved in DNA replication.

DNA Glycosidase Enzyme that recognizes a deaminated base and removes it from the DNA molecule by cleaving the glycosidic bond between the base and the deoxyribose sugar to which it is attached.

DNA α-Glucosyl Transferase The enzyme that catalyzes the transfer of α-D-glucose from UDP-glucose to the hydroxymethyl cytosine of T-even phage DNA.

DNA β-Glucosyl Transferase The enzyme that catalyzes the transfer of β-D-glucose from UDP-glucose to the hydroxymethyl cytosine of T-even phage DNA.

DNA Groove Referring to the two grooves (major and minor) in DNA that run its length.

DNA Gyrase The enzyme that catalyzes the breakage, passage, and rejoining of the DNA (see also DNA topoisomerase).

DNA Helicase The DNA unwinding enzyme that catalyzes the energy-dependent unwinding of the double helix during DNA replication.

DNA Homology The degree of similarity of base sequences in DNA from different organisms.

DNA Hybridization A technique for assessing the relationships between DNA molecules by measuring the ability of a single-stranded DNA from one source to anneal with single-stranded DNA or RNA from different sources.

DNA Joinase See DNA ligase.

DNA Library The collection of cloned DNA molecules representing either an entire genome (genomic library) or DNA copies of mRNA produced by a cell (cDNA library).

DNA Ligase The enzyme that catalyzes the formation of phosphodiester bonds between the 3′-OH end and 5′-phosphate end of the same strand. There are two types of DNA ligase (ATP-dependent or NAD-dependent).

DNA Ligase (ATP-dependent) The enzyme that catalyzes the following reaction:

$$\text{ATP} + (\text{deoxyribonucleotide})_m + (\text{deoxyribonucleotide})_n \rightleftharpoons \text{AMP} + \text{PPi} + (\text{deoxyribonucleotide})_{m+n}$$

DNA Ligase (NAD-dependent) The enzyme that catalyzes the following reaction:

$$\text{NAD} + (\text{deoxyribonucleotide})_m + (\text{deoxyribonucleotide})_n \rightleftharpoons \text{AMP} + (\text{deoxyribonucleotide})_{m+n} + \text{nicotinamide nucleotide}$$

DNA Looping The interaction of proteins bounded at the distant sites on a DNA molecule so that it intervenes to form a loop.

DNA 3-Methyladenine Glycosidase I The enzyme that catalyzes the hydrolysis of alkylated DNA releasing 3-methyladenine.

DNA 3-Methyladenine Glycosidase II The enzyme that catalyzes the hydrolysis of alkylated DNA releasing 3-methyladenine and 3-methylguanine.

DNA Methylase The enzyme that catalyzes the methylation of the bases in DNA.

DNA Methylation The addition of methyl groups ($-CH_3$) to bases of DNA after DNA synthesis is completed.

DNA Methyltransferase The enzyme that catalyzes the following reaction:

$$S\text{-Adenosyl-methionine} + \text{DNA} \rightleftharpoons S\text{-Adenosyl-homocysteine} + \text{methylated DNA}$$

DNA Modification A process in which bases in DNA are substituted with methyl or other groups.

DNA Nucleotidyl Exotransferase The enzyme that catalyzes the following reaction:

$$\text{Deoxynucleoside triphosphate} + \text{DNA}_n \rightleftharpoons \text{Pyrophosphate} + \text{DNA}_{n+1}$$

DNA Nucleotidyltransferase See DNA polymerase.

DNA Phage Bacteriophage containing DNA.

DNA Photolase The photoreacting enzyme that splits thymine dimers (in DNA repair).

DNA Polymerases A family of enzymes that catalyze the synthesis of a new DNA strand on a DNA template by the successive addition of nucleotides to 3′-OH ends of a primer RNA or a growing DNA strand. There are a number of DNA polymerases. *E. coli* has polymerase I, II, and III and all of them have polymerase and exonuclease activity; polymerase I and II appear to function in DNA repair and editing whereas polymerase III functions mainly on DNA replication. DNA polymerase α, β, and γ are eukaryotic DNA polymerases; polymerase α seems to be responsible for chromosome replication whereas polymerase γ appears to carry out mitochondrial DNA replication.

DNA Polymerase Chain Reaction See PCR or polymerase chain reaction.

DNA Primase The enzyme that catalyzes the formation of primer DNA on a single-stranded DNA template.

DNA Primer A single-stranded DNA segment to which deoxyribonucleotides are added by DNA polymerase during the replication of DNA.

DNA Print The pattern formed by denaturing the DNA in the bacterial colonies grown on a support and fixing it to the support.

DNA Probe A radioactively labeled segment or an enzyme-linked segment of nucleic acid used to locate or detect a gene with a complementary sequence.

DNA Repair A process that repairs damage in DNA.

DNA Replication A process in which the DNA duplex unwinds and serves as a template for the synthesis of progeny DNA.

DNA RNA Hybrid A double helix that consists of one chain of DNA hydrogen bonded to a chain of RNA by means of complementary base pairs.

DNA Sequencing Determination of the order of nucleotides in a DNA molecule or DNA fragment by DNA sequencing techniques, e.g., Maxam-Gilbert's method or Sanger's method.

DNA Supercoiling The coiling of DNA upon itself as a result of bending, underwinding, or overwinding of the DNA duplex.

DNA Swivelase Synonym of DNA topoisomerase.

DNA Topoismerase A group of enzymes that catalyzes the breakage, passage, and joining of DNA leading to interconversion of topological isomers of DNA.

DNA Vector A replicon (e.g., plasmid or phage) that transfers foreign DNA into a host organism.

DNase Abbreviation for deoxyribonuclease.

DNase I Hypersensitivity Site Specific location near an active gene that shows extreme sensitivity to digestion by DNase I.

DNAse Protection A method for estimating the size of a DNA sequence that is protected from DNAse hydrolysis by binding of an enzyme or a protein onto the DNA.

DNBP Abbreviation for dinitrobutylphenol.

DNCB Abbreviation for 2,4-dinitrochlorobenzene.

DNFB Abbreviation for dinitrofluorobenzene.

dNM Abbreviation for deoxynojirimycin.

2D-NMR Abbreviation for two-dimensional NMR spectroscopy.

DNP Abbreviation for dinitrophenyl.

DNPC Abbreviation for 2,3-dinitrophenyl-β-D-cellobiosidc.

DNPH Abbreviation for dinitrophenyl hydrazine.

DNPM Abbreviation for dinitrophenyl morphine.

DNP-SG Abbreviation for 2,4-dinitrophenyl- S-glutathione.

DNR A trade name for daunorubicin citrate liposomal complex, an antibiotic with antineoplastic activity.

dNTP Abbreviation for deoxynucleotide triphosphate.

DO Abbreviation for diamine oxidase.

DOAP Abbreviation for a combination drug containing daunorubicin, oncovin, adriamycin, and prednisone.

Dobutamine (mol wt 301) A cardiotonic agent and an analog of isoproterenol that stimulates beta-1 receptors in the heart to increase myocardial contraction.

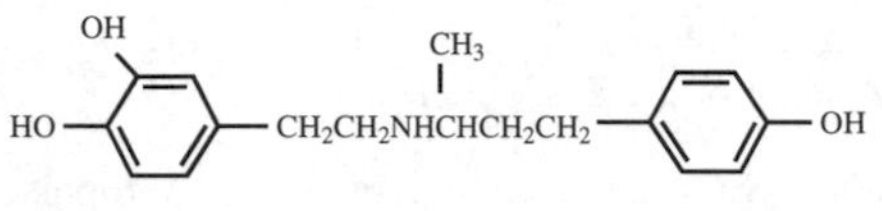

Dobutrex A trade name for dobutamine hydrochloride, a cardiotonic agent and an analog of isoproterenol that stimulates beta-1 receptors in the heart to increase myocardial contraction.

DOCA Abbreviation for deoxycorticosterone acetate.

Docetaxel (mol wt 808) An antineoplastic agent.

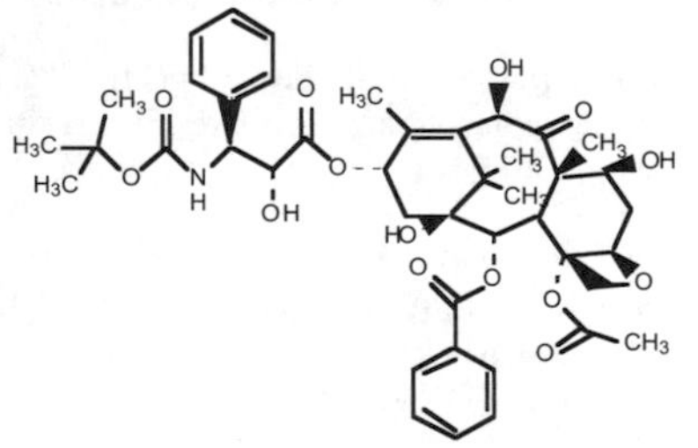

Docking The binding of any macromolecule to its specific harboring site on another molecule.

Docking Protein Membrane protein in the rough endoplasmic reticulum that serves as the receptor for the SRP-blocked ribosomal complex, thereby overcoming the translation block (SRP: signal recognition protein).

Docking Site A location on a protein where another protein may fit with near perfect conformity.

Docosa- A prefix denoting 22 or 22 times.

Docosadienoic Acid Any straight-chain fatty acid having 22 carbon atoms and 2 double bonds.

Docosahexaenoic Acid Any straight-chain fatty acid having 22 carbon atom and 6 double bonds.

Docosanoic Acid A straight-chain saturated fatty acid with 22 carbon atoms.

Docosapentaenoic Acid Any straight-chain fatty acid having 22 carbon atoms and 5 double bonds.

Docosatetraenoic Acid Any straight-chain fatty acid having 22 carbon atoms and 4 double bonds.

Docosatrienoic Acid Any straight-chain fatty acid having 22 carbon atoms and 3 double bond.

Docosenoic Acid Any straight-chain fatty acid having 22 carbon atoms and 1 double bond.

Docusate Calcium (mol wt 883) A stool softener.

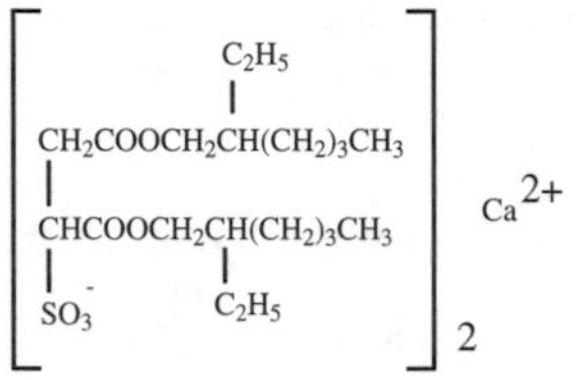

Docusate Sodium (mol wt 445) A stool softener.

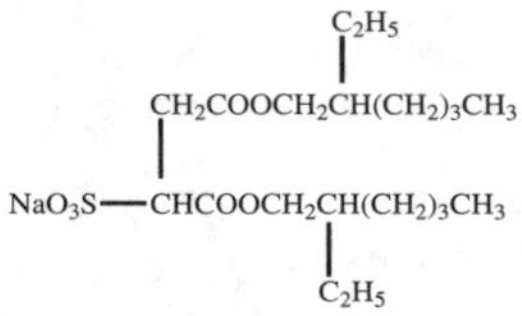

Dod Referring to dodecyl group.

Dodeca- A prefix denoting 12.

Dodecandrin A ribosome-inactivating protein.

Dodecanoic Acid A straight-chain saturated fatty acid having 12 carbon atoms.

Dodecarbonium Chloride (mol wt 397) An anti-infective agent.

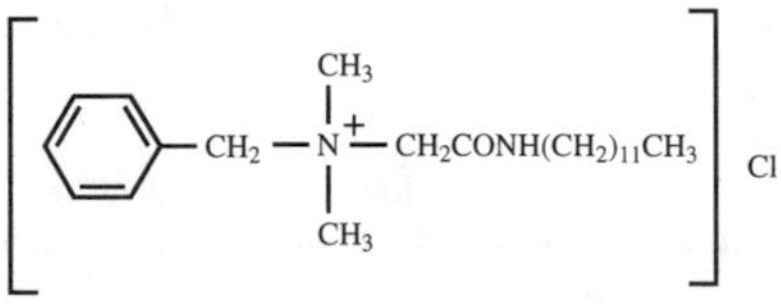

Dodecenoic Acid A straight-chain fatty acid having 12 carbons and 1 double bond.

Dodecyl-β-D-Glucopyranoside (mol wt 348) A nonionic detergent used for solubilization of membrane proteins.

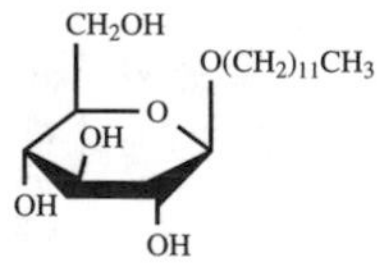

Dodecyl Referring to the group of

$$CH_3\text{-}[CH_2]_{10}\text{-}CH_2\text{-}$$

Dodecyl-D-Maltoside (mol wt 511) A nonionic detergent used for stabilization and activation of enzymes.

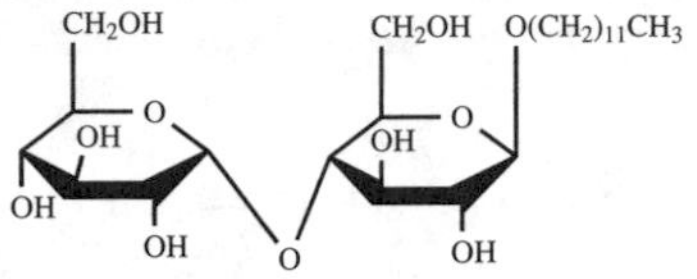

Dodecyltriethylamonium Bromide (mol wt 308) A detergent used for disruption of the plasma membrane.

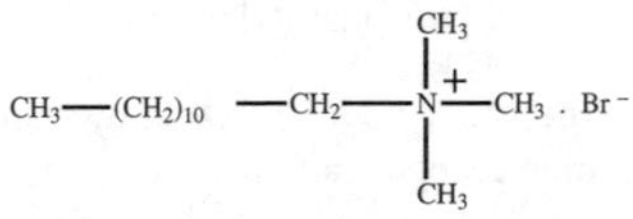

DOG Abbreviation for dioleoylglycerol.

1,2-DOG Abbreviation for 1,2-dioleoyl glycerol.

1,3-DOG Abbreviation for 1,3-dioleoyl glycerol.

DOK Liquid A trade name for docusate sodium, a laxative.

Doktors A trade name for phenylephrine hydrochloride, a vasoconstrictor.

Dol Abbreviation for dolichol.

Dolacet A trade name for a combination drug containing acetaminophen and hydrocodone bibartrate.

Dolanex A trade name for acetaminophen, an analgesic and antipyretic agent.

Dolene A trade name for propoxyphene, an analgesic agent.

Dolichol Kinase The enzyme that catalyzes the following reaction:

CTP + dolichol

⇅

CDP + dolichol phosphate

Dolichol O-Acetyltransferase The enzyme that catalyzes the following reaction:

Palmitoyl-CoA + dolichol

⇅

CoA-SH + dolichol palmitate

Dolichol Phosphate Activated lipid carrier to which monosaccharides are added stepwise to form the core oligosaccharide of glycoproteins.

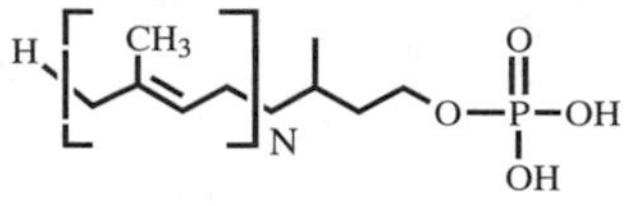

Dolobid A trade name for diflunical, an analgesic and antipyretic agent.

Dolophine A trade name for methadone hydrochloride, an analgesic agent.

Doloxene A trade name for propoxyphene napsylate, an analgesic agent.

Dol-P Abbreviation for dolichyl phosphate or dolichol phosphate.

Domain 1. The structural and functional portion of a polypeptide that folds independently of the other portion of the protein; a region of a protein with tertiary structure. 2. Regions of H or L chains of the immumoglobulin molecule that are folded into functional three-dimensional structures (e.g., CH domains and VH domains).

Dominant 1. A gene that produces a phenotypic effect in a heterozygote. 2. A trait that is shown phenotypically by both homozygotes and heterozygotes.

Dominant Allele The allele that is phenotypically expressed despite the presence of the other alleles of the same gene.

Dominant Gene A gene that can suppress the expression of an allele.

Domiodol (mol wt 244) A mucolytic agent.

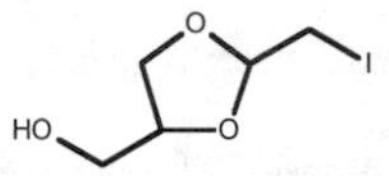

Domiphen Bromide (mol wt 414) A topical anti-infectant.

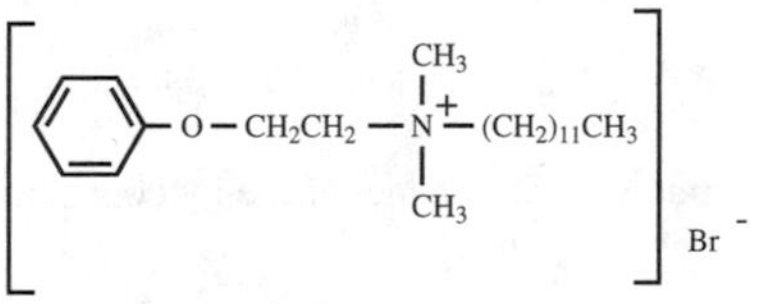

Dommanate A trade name for the antiemetic agent dimenhydrinate.

Domperidone (mol wt 426) An antiemetic agent.

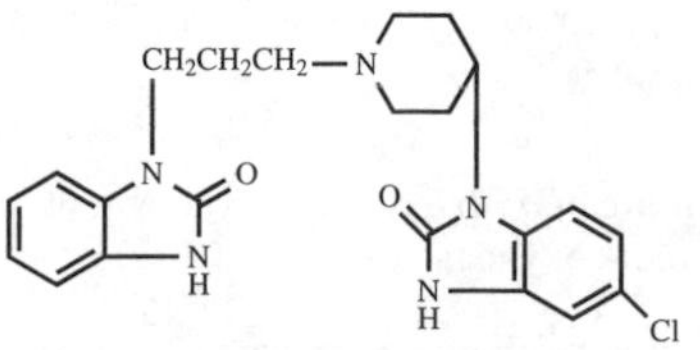

DON Abbreviation for diazo-oxo-norleucine.

Donepezil Hydrochloride (mol wt 416) A cholinesterase inhibitor used as an Alzheimer's drug.

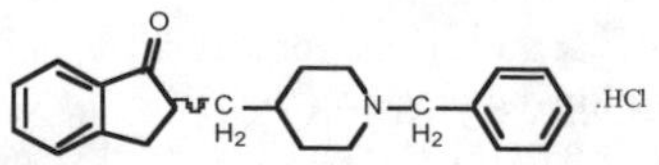

Donnagel-MB A trade name for a mixture of Kaolin and pectin, an antidiarrheal agent.

Donnan-Gibbs Equilibrium The unequal equilibrium distribution of diffusible ions established at equilibrium on both sides of the membrane if one side contains nondiffusible ions.

Donnazyme A trade name for a combination digestant containing pancreatin, pepsin, bile salt, hyposcyamine sulfate, atropine sulfate, scopolamine hydrobromide, and phenobarbital.

Donor Hydrogen Peroxide Oxidoreductase The systematic name for peroxidase.

DOP (Dop) Abbreviation for dopamine.

Dopa (mol wt 197) Abbreviation for dihydroxyphenylalanine, a naturally occurring amino acid and precursor for synthesis of melanin.

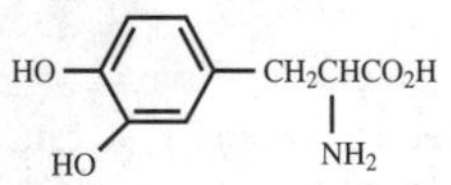

Dopa Decarboxylase The enzyme that catalyzes the following reaction:

L-tryptophan ⇌ Tryptamine + CO_2

Dopa Transaminase The enzyme that catalyzes the following reaction:

3,4-Dihydroxyphenylalanine + α-ketoglutarate

⇅

3,4-Dihydroxyphenylpyruvate + glutamate

Dopamet A trade name for methyldopa, an antihypertensive agent.

Dopamine (mol wt 153) A cardiotonic and antihypertensive agent that stimulates dopaminergic, beta-adrenergic, and α-adrenergic receptors of the sympathetic nervous system.

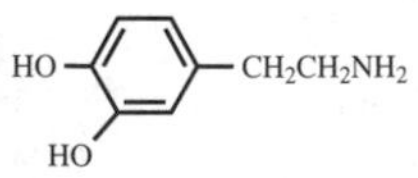

Dopamine β-Hydroxylase Synonym of dopamine β-monooxygenase.

Dopamine β-Monooxygenase The enzyme that catalyzes the following reaction:

3,4-Dihydroxyphenethylamine + ascorbate + CO_2

⇅

Noradrenaline + dehydroascorbate + H_2O

Dopar A trade name for levopoda, an antiparkinsonian agent.

Dopastin (mol wt 215) An inhibitor for dopamine β-hydroxylase from *Pseudomonas.*

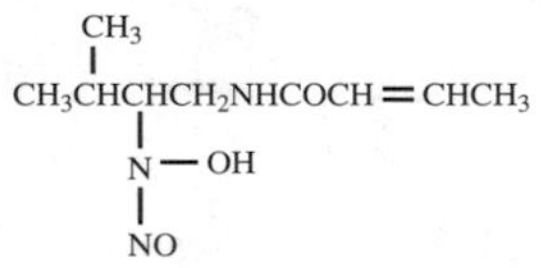

DOPC Abbreviation for 1. dodecyl phosphocholine; 2. dioleoyl phosphatidylcholine.

Dopexamine (mol wt 357) A cardiotonic agent and dopamine receptor agonist.

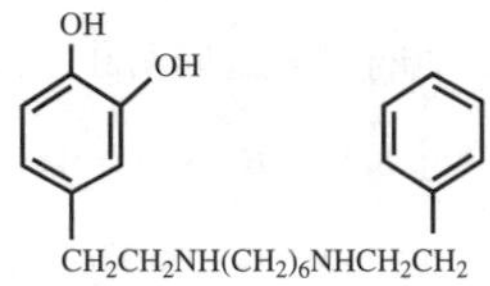

Dopram A trade name for doxapram hydrochloride, a respiratory and cerebral stimulant.

Doractin A trade name for cimetidine, an antiulcer agent that decreases gastric acid secretion.

Doral A trade name for quazepam, a sedative agent.

Doraphen A trade name for propoxyphene hydrochloride.

Doriden A trade name for gultethimide, a sedative-hypnotic agent.

Doriglute A trade name for glutethimide, a sedative-hypnotic agent.

Dormarex A trade name for diphenhydramine hydrochloride, an antihistaminic agent.

Dornase Alfa A trade name for recombinant human deoxyribonuclease (DNase) that breaks thick, sticky mucus that clogs the airways by the destruction of DNA molecules in the sticky mucus.

Dorsiventral Having distinctly different upper and lower surfaces.

Doryx A trade name for doxycycline hyclate, an anti-infective agent.

Dorzolamide (mol wt 324) An antiglaucoma agent.

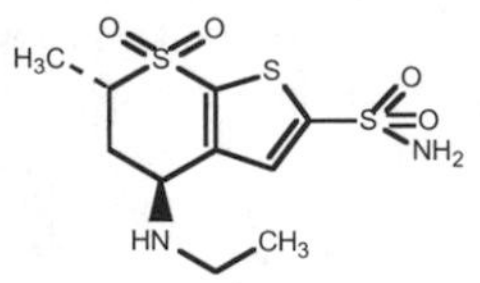

Dosage Effect Genetic effect whereby the phenotype of a cell is altered by an increased amount of a particular gene product.

Dosimeter A device to measure the cumulative dose of radiation.

Dostinex A trade name for cabergoline, a dopamine receptor agonist.

Dot Blot A modified Southern blotting technique to quantify a given sequence of nucleic acid in which denatured sample is applied as a dot and followed by the addition of radioactively labeled probe.

Dot Hybridization A semiquantitative method for evaluating the relative abundance of nucleic acid sequences in a sample by applying it as a spot on the nitrocellulose film and followed by treatment with radioactively labeled probe. The degree of hybridization is determined by measuring the radioactivity of the bound probe.

Dothiepin (mol wt 295) An antidepressant.

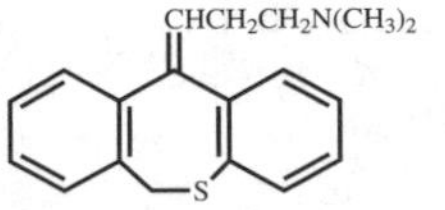

Double Diffusion Method A serological precipitin reaction in which an antigen and an antibody diffuse toward each other in the gel.

Double Helix See DNA double helix.

Double-Reciprocal Plot Graphic method for analyzing enzyme kinetic data by plotting 1/*v* versus 1/[S], used in the determination of Vmax and Km.

Double-stranded RNA The duplex formed from two single-stranded RNA through complementary base pairing.

Dovonex A trade name for calcipotriene, an antiseptic agent.

Dowex A trade name for a group of ion-exchange resins.

Down's Syndrome A birth defect characterized by gross mental retardation and assorted physical changes resulting from the presence of three copies of chromosome 21 in body cells.

Downstream The direction in which a nucleic acid chain or a polypeptide chain is synthesized.

DOX (Dox) Abbreviation for doxorubicin, an antibiotic with antineoplastic activity.

Doxaphene A trade name for propoxyphene hydrochloride.

Doxapram (mol wt 379) A respiratory stimulant that acts either directly on the respiratory center in the medulla or indirectly on the chemoreceptors.

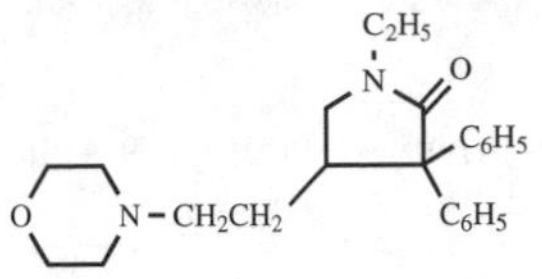

Doxazosin (mol wt 451) An antihypertensive agent that acts on peripheral vasculature to produce vasodilation.

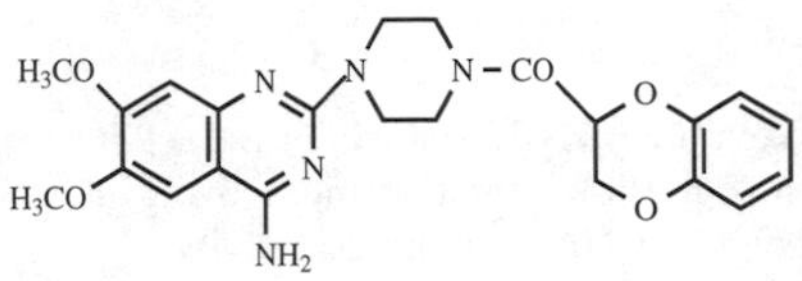

Doxefazepam (mol wt 359) A sedative and hypnotic agent.

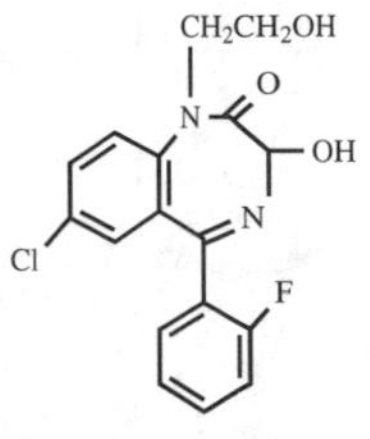

Doxenitoin (mol wt 238) An anticonvulsant.

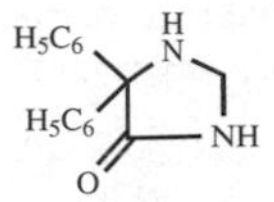

Doxepin (mol wt 279) An antihypertensive agent that increases the amount of norepinephrine or serotonin in the CNS by blocking uptake by presynaptic neurons. It allows neurotransmitter to accumulate.

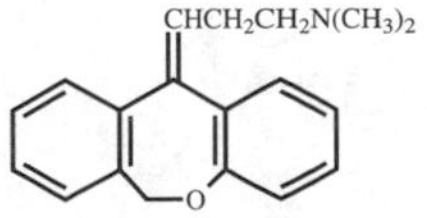

Doxidan A trade name for a combination laxative containing docusate calcium and phenophthalein.

Doxifluridine (mol wt 246) An antineoplastic agent.

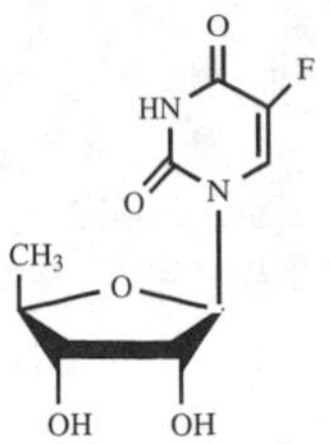

Doxil A trade name for doxorubicin hydrochloride.

Doxinate A trade name for docusate sodium, a laxative.

Doxofylline (mol wt 266) A bronchodilator.

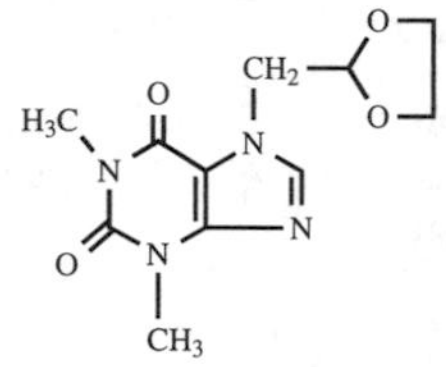

Doxorubicin Hydrochloride (mol wt 580) An antineoplastic agent that interferes with DNA-dependent RNA polymerase activity.

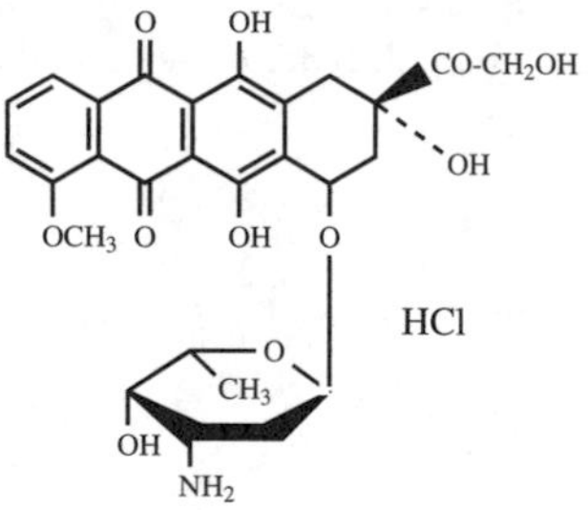

Doxy-Caps A trade name for doxycycline, an antibiotic.

Doxychel A trade name for doxycycline hyclate that binds to 30S ribosome inhibiting bacterial protein synthesis.

Doxycycline Hydrochloride (mol wt 481) An antibiotic that binds to 30S ribosome subunits and inhibits protein synthesis.

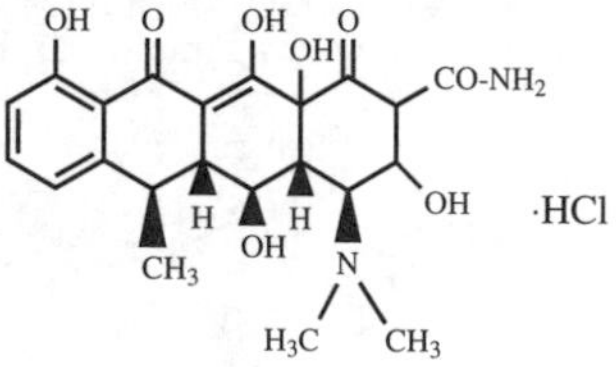

Doxylamine (mol wt 270) An antihistaminic agent.

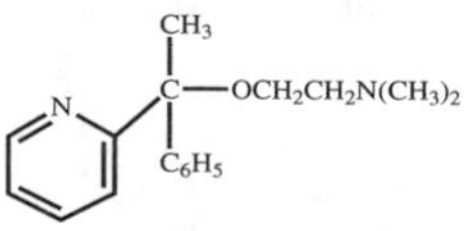

Doxylin A trade name for doxycycline, an antibiotic that inhibits bacterial protein synthesis.

Doxytec A trade name for doxycycline, an antibiotic.

DP Abbreviation for 1. degree of polymerization (number of monomeric units in the polymer); 2. dipropionate.

DPA Abbreviation for docosapentaenoic acid.

DPBS Abbreviation for Dulbecco's phosphate buffered saline.

DPG Abbreviation for 1. dipalmitoyl glycerol; 2. 2,3-diphosphoglycerate.

1,2-DPG Abbreviation for 1,2-dipalmitoyl glycerol.

1,3-DPG Abbreviation for 1,3-dipalmitoyl glycerol.

2,3-DPG Abbreviation for 2,3-diphosphoglycerate.

DPI Abbreviation for 1. diphenylene iodonium; 2. days post infection.

dpm Abbreviation for disintegrations per minute.

DPN Abbreviation for diphosphopyrimdine nucleotide, currently called NAD (nicotinamide adenine dinucleotide).

DPNase Synonym of NAD^+ nucleosidase.

DPNH Reduced DPN, currently called NADH.

DpnI A restriction endonuclease from *Diplococcus pneumoniae* with the following specificity:

```
        ↓
5′...........GATC..........3′
3′...........CTAG..........5′
        ↑
```

DpnII (MobI) A restriction endonuclease from *Diplococcus pneumoniae* with the same specificity as MboI.

DPP Abbreviation for dipeptidyl peptidase.

DPPA Abbreviation for diphenylphosphoryl azide.

DPPC Abbreviation for D-α-dipalmitoylphosphatidylcholine.

DPPD Abbreviation for diphenylphenylenediamine.

DPPS Abbreviation for dipalmitoylphosphatidylserine.

DPRA Abbreviation for dot-plot re-association assay.

dps Abbreviation for disintegrations per second.

DPT Abbreviation for 1. a vaccine mixture of diphtheria, pertussis, and tetanus; 2. dipropyltryptamine.

DPTV Abbreviation for a combination of diphtheria, pertussis, and tetanus vaccine

DQ A human class II MHC allele.

DQF Abbreviation for double-quantum filtered.

DR A human class II MHC allele.

DRA Abbreviation for dextran-reactive antibody.

DraI (AhaIII) A restriction endonuclease from *Deinococcus radiophilus* with the following specificity:

5′..........TTTAAA..........3′
3′..........AAATTT..........5′

DraII A restriction endonuclease from *Deinococcus radiophilus* with the following specificity:

5′..........PuGGNCCPy..........3′
3′..........PyCCNGGPu..........5′

DraIII A restriction endonuclease from *Deinococcis radiophilus* with the following specificity:

5′..........CACNNNGTG..........3′
3′..........GTGNNNCAC..........5′

Dracontiasis A tropical disease caused by the parasitic nematode *Dracunculus medinensis*.

Dramamine A trade name for dimenhydrinate, an antiemetic agent.

Dramarate A trade name for dimenhydrinate, an antiemetic agent.

Dramilin A trade name for dimenhydrinate, an antiemetic agent.

Dramocen A trade name for dimenhydrinate, an antiemetic agent.

Dramoject A trade name for dimenhydrinate, an antiemetic agent.

DrdI A restriction endonuclease from *Deinococcus radiodurans* with the following specificity:

5'........GACNNNNNNGTC........3'
3,........CTGNNNNNNCAG........5'

Drenison A trade name for flurandrenolide, a corticosteroid.

Dressing Material or preparation applied to a wound or lesion to prevent external infection.

D-Rex-65 A trade name for a combination drug containing acetaminophen and propoxyphene hydrochloride, used as an analgesic agent.

dRib Abbreviation for deoxyribose.

Drisdol A trade name for vitamin D_2.

Dristan A trade name for a combination bronchodilator containing phenylephrine hydrochloride, chlorpheniramine maleate, and acetaminophen.

Drixoral A trade name for a combination drug containing dexbrompheniramine maleate and pseudoephedrine sulfate, used as an antihistaminic agent.

Drize A trade name for a combination drug containing phenylpropanolamine hydrochloride and chlorpheniramine maleate, used as an antihistaminic agent.

DRM Abbreviation for detergent resistant membrane.

dRNA Abbreviation for double-stranded RNA.

Drocarbil (mol wt 430) An anthelmintic agent. A mixture of:

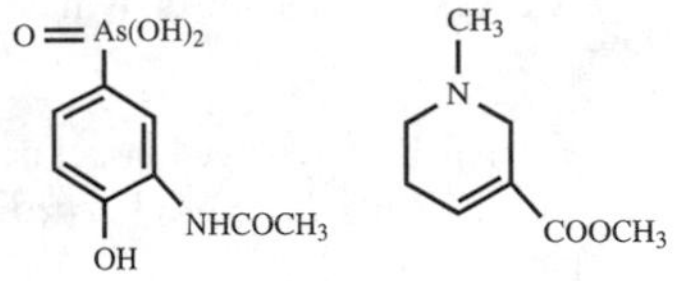

36% 64%

Drofenine (mol wt 317) An antispasmodic agent.

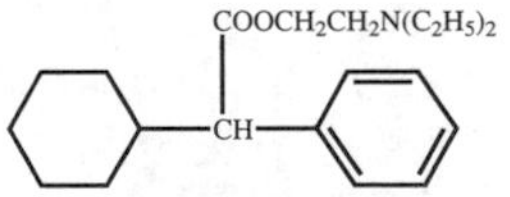

Droleptan A trade name for droperidol, an anesthetic agent.

Droloxifene (mol wt 388) A hormone with antineoplastic activity.

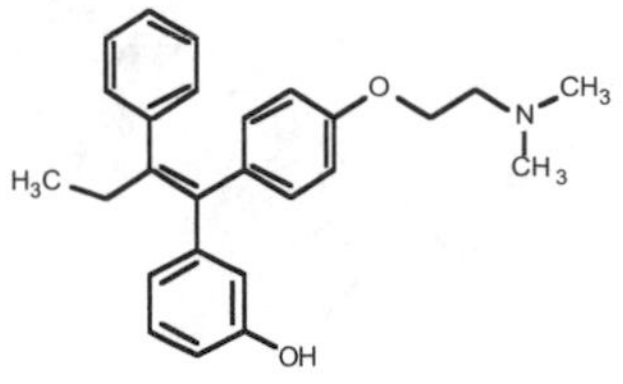

Dromject A trade name for dimenhydrinate, an antiemetic agent.

Dromostanolone (mol wt 304) An antineoplastic agent.

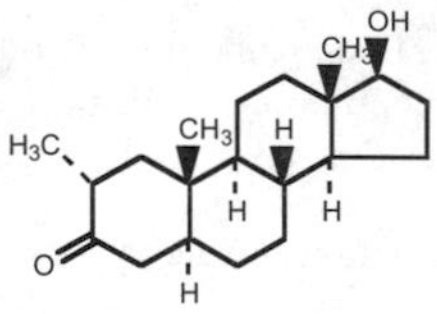

Dronabinol A trade name for delta-9 tetrahydrocannabinol, an anti-emetic agent.

Droperidol (mol wt 379) An anesthetic agent that acts at the subcortical level to produce sedation.

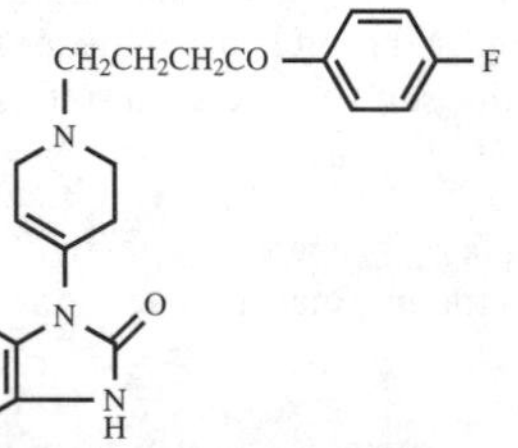

Droprenilamine (mol wt 336) A vasodilator.

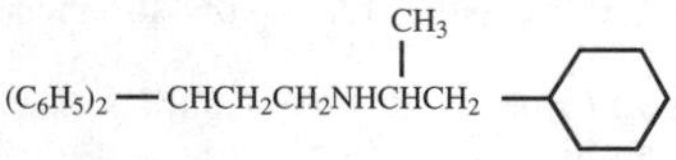

Dropsy An excessive amount of lymph in the tissue (edema).

Drosophila A small two-winged insect used extensively in genetic and biochemical experiments. Also known as fruitfly.

Drotebanol (mol wt 333) An antitussive agent.

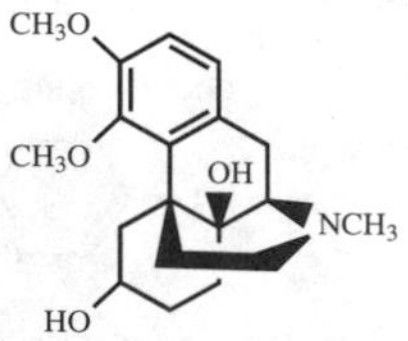

Droxia A trade name for hydroxyurea, an antineoplastic agent.

Droxicam (mol wt 357) An anti-inflammatory agent.

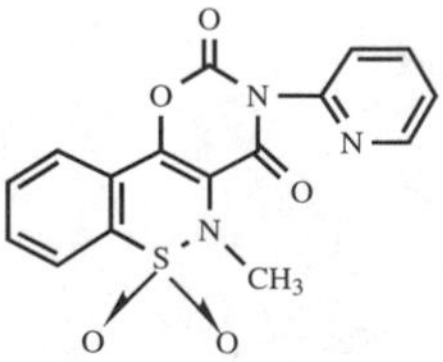

Droxidopa (mol wt 213) An antiparkinsonian agent.

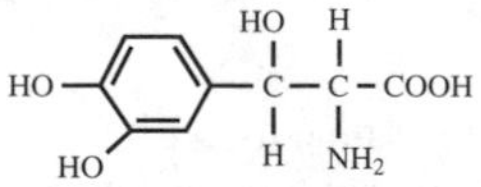

Drug-Fast Referring to microorganisms that resist the action of a drug.

Drug Resistant Gene A gene that encodes a product (e.g., an enzyme) that enables the host to resist the effect or action of a drug.

Dry Ice Solid carbon dioxide.

DS Abbreviation for 1. dermatan sulfate; 2. detergent soluble; 3. dextrose/saline; 4. Disaccharide; 5. Down's syndrome.

ds Abbreviation for double stranded.

DSA Abbreviation for *Datura stramonium* agglutinin.

DsaI A restriction endonuclease from *Dactylococcopsis salina* with the following specificity:

↓
5′..........CCPuPyGG..........3′
3′..........GGPyPuCC..........5′
↑

DsaII (HaeIII) A restriction endonuclease from *Dactylococcopsis salina* with the following specificity:

↓
5′..........GGCC..........3′
3′..........CCGG..........5′
↑

DSC Abbreviation for differential scanning calorimetry.

dsDNA Abbreviation for double-stranded DNA.

DseDI A restriction endonuclease from *Deinococcus* species Dx with the following specificity:

↓
5'........GACNNNNNNGTC........3
3'........CTGNNNNNNCAG........5'
↑

DSG Abbreviation for distearoyl glycerol.

DSI Abbreviation for digital spot imaging.

DSP Abbreviation for dithiobispropionate.

DspII (SacII) A restriction endonuclease from *Deinococcus species* RFLI with the same specificity as SacII.

DSPC Abbreviation for desaturated phosphatidylcholine.

dsRNA Abbreviation for double stranded RNA.

DSS A trade name for docusate sodium, a laxative.

DSS Plus A trade name for a combination laxative containing docusate sodium and casanthranol.

DST Abbreviation for dexamethasone suppression test.

***dsx* Gene** Abbreviation for double-sex gene in *Drosophila melanogaster.* Alternative splicing of the same transcript yields the male-specific and the female-specific protein.

DT Abbreviation for 1. deoxythymidine; 2. Diphtheria tetanus; 3. Diphtheria toxin; 4. Diphtheria toxoids.

dT Abbreviation for deoxythymidine.

DTAB Abbreviation for dodecylethylammonium bromide.

DTAF Abbreviation for dichlorotriazinyl aminofluorescein.

dTDP Abbreviation for deoxythymidine diphosphate.

DTE Abbreviation for dithioerythritol.

DTH Abbreviation for delayed-type hypersensitivity, a type of cell-mediated hypersensitivity.

dThd Abbreviation for deoxythymidine.

dThd5'P Abbreviation for deoxythymidine 5'-phosphate.

dThd5'PP Abbreviation for deoxythymidine 5'-diphosphate.

dThd5'PPP Abbreviation for deoxythymidine or thymidine 5'-triphosphate.

d-Thyroxine Sodium See dextrothyroxine sodium.

DTIC A trade name for decarbazine, an alkylating agent that cross-links strands of cellular DNA interfering with RNA transcription.

dTMP Abbreviation for deoxythymidine monophosphate.

dTMP Kinase The enzyme that catalyzes the following reaction:

$$\text{ATP} + \text{dTMP} \rightleftharpoons \text{ADP} + \text{dTDP}$$

DTNB Abbreviation for 5,5'-dithiobis-(2-nitrobenzoic acid).

DTNBA Abbreviation for 5,5'-dithiobis-(2-nitrobenzoic acid).

D-topoII Abbreviation for *Drosophila* topoisomerase-II.

DTP Abbreviation for diphtheria-tetanus- pertussis.

DTP Vaccine Abbreviation for vaccine against diphtheria, tetanus, and pertussis.

DTPA Abbreviation for diethylene-triamine penta-acetic acid.

DTPV Abbreviation for diphtheria-tetanus- pertussis vaccine.

DTRP Abbreviation for *Drosophila* transient receptor protein.

DTT Abbreviation for a reducing agent dithiothreitol.

dTTP Abbreviation for deoxythymidine triphosphate.

DTTP A trade name for diphtheria and tetanus toxoids and pertussis vaccine.

DTX Abbreviation for detoxification.

Dubin-Johnson Syndrome A congenital defect of excretory function of the liver, characterized by jaundice, i.e., the accumulation of large quantities of bilirubin in the blood.

dU Abbreviation for deoxyuridine.

Duchenne's Muscular Dystrophy A birth defect characterized by the progressive wasting of the muscles in the legs and pelvis due to a gene on the distal portion of the short arm of human X chromosome.

Duck Embryo Vaccine A killed rabies virus vaccine prepared from duck embryo.

Duct A tubular passageway conveying the products of glands to another part of the body.

Ductless Gland A structure that secretes hormone or hormones into the body fluids (also known as endocrine gland).

dUDP Abbreviation for deoxyuridine diphosphate.

Duffy Blood Group System A type of blood classification system. Duffy substances are located on the surface of erythrocyte and serve as receptors for *Plasmodium vivax.* An individual homozygous for the inactive allele (Fy^-) is immune to tertian malaria.

Dulcin (mol wt 180) A nonnutritive sweetener.

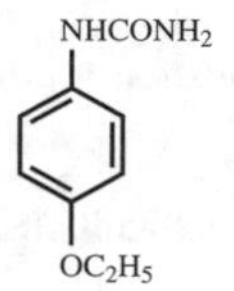

Dulcitol (mol wt 18) Sugar alcohol of galactose.

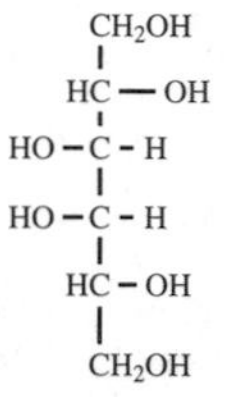

Dulcolax A trade name for bisacotyl, a laxative that promotes accumulation of fluid in the colon and small intestine.

Dull-C A trade name for vitamin C.

Duloxetine (mol wt 297) An antidepressant.

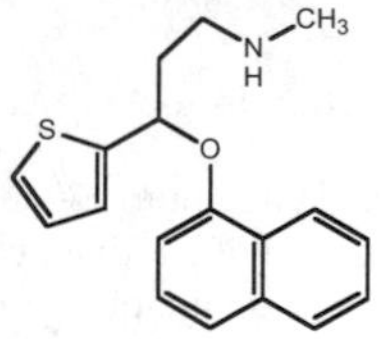

dUMP Abbreviation for deoxyuridine monophosphate.

Duocet A trade name for a combination drug containing acetaminopen and hydrocodone bitartrate, used as an analgesic agent.

Duo-Medihaler A trade name for a combination drug containing isoproterenol hydrochloride and phenylephrine bitartrate, used as a bronchodilator.

Duosol A trade name for ducusate sodium, a laxative.

Duotrate A trade name for pentaerythritol tetranitrate, an analgesic agent.

Duphalac A trade name for lactulose, a laxative.

Duplex DNA See DNA.

Dura Mater The outermost of the three meninges covering the brain and spinal cord.

Durabolin A trade name for nandrolone phenpropionate, used as an anabolic steroid.

Duraclon A trade name for clonidine hydrochloride, an antihypertensive and analgesic agent.

Duradyne A trade name for a combination drug containing acetaminophen, aspirin, and caffeine, used as an analgesic and antipyretic agent.

Duradyne DHC A trade name for a combination drug containing acetaminophen and hydrocodone bitartrate, used as an analgesic agent.

Duragen A trade name for estradiol valerate, an estrogen that increases the synthesis of DNA, RNA, and protein.

Duragesic A trade name for fentanyl, an analgesic agent.

Duralith A trade name for lithium carbonate, an anti-manic agent.

Duralone A trade name for methylprednisolone acetate, a corticosteroid used as an anti-inflammatory agent.

Duralutin A trade name for hydroxyprogesterone caproate, an estrogen used to suppresses ovulation.

Durameth A trade name for methylprednisolone acetate, a corticosteroid used as an anti-inflammatory agent.

Duramist Plus A trade name for oxymetazoline hydrochloride, a nasal agent that produces local vasoconstriction of dilated arterioles to reduce blood flow and nasal congestion.

Duramorph P F A trade name for morphine sulfate, used as an analgesic agent.

Duranest A trade name for etidocaine, a local anesthetic agent.

Durapam A trade name for flurazepam hydrochloride, a sedative.

Duraphyl A trade name for theophylline, a bronchodilator.

Duraquin A trade name for quinidine gluconate, an antiarrhythmic agent.

Duratest A trade name for testosterone cypionate, an anabolic hormone.

Durathate A trade name for testosterone enanthate, an anabolic hormone.

Duratuss A trade name for a combination drug containing guaifenesin, hydrocodone, and pseudoephedrine.

Dura-Vent A trade name for a combination drug containing phenylpropanolamine and guaifenesin.

dUrd Abbreviation for deoxyuridine.

dUrd5'P Abbreviation for deoxyuridine 5'-phosphate.

dUrd5'PP Abbreviation for deoxyuridine 5'-diphosphate.

DUrd5'PPP Abbreviation for deoxyuridine 5'-triphosphate.

Duretic A trade name for methyclothiazide, a diuretic agent.

Duricef A trade name for cefadroxil monohydrate, an antimicrobial agent that inhibits bacterial cell wall synthesis.

Duromine A trade name for phentermine hydrochloride, a cerebral stimulant.

Durraz A trade name for hydroxyzine hydrochloride, an antianxiety agent.

dUTP Abbreviation for deoxyuridine triphosphate.

dUTP Pyrophosphatase The enzyme that catalyzes the following reaction:

$$\text{dUTP} + H_2O \rightleftharpoons \text{dUMP} + \text{PPi}$$

dUTPase The enzyme that catalyzes the following reaction:

$$\text{dUTP} + H_2O \rightleftharpoons \text{dUMP} + \text{PPi}$$

Duvadilan A trade name for isoxyprine hydrochloride, a vasodilator.

Duvoid A trade name for bethanechol chloride, a cholinergic agent.

D-Vert A trade name for meclizine hydrochloride, an anti-emetic and antihistaminic agent.

DVP-Asp Abbreviation for a combination drug containing daunorubicin, vincristine, prednisone, and asparaginase.

Dyad A pair of sister chromatides joined at the centromere, as in the first division of meiosis.

Dyad Symmetry A macromolecule that produces an identical structure when it is rotated 180°.

Dyazide A trade name for a combination drug containing triamterene and hydrochlorothiazide, used as a diuretic agent.

Dycill A trade name for dicloxacillin sodium, a penicillinase-resistant antibiotic.

Dyclone A trade name for dyclonine hydrochloride.

Dyclonine (mol wt 289) A topical anesthetic agent.

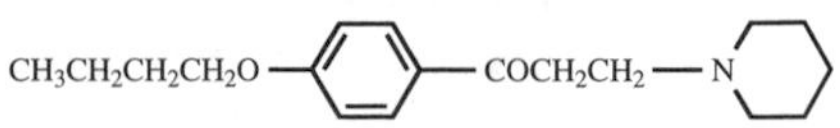

Dydrogesterone (mol wt 312) A progestogen.

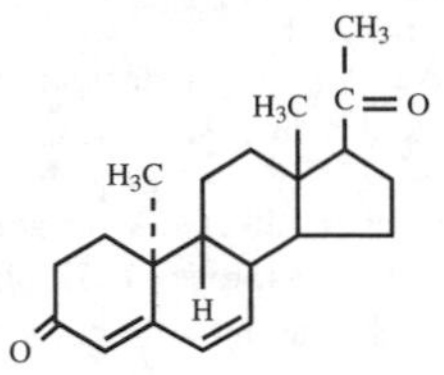

Dye Exclusion Test A cell viability test based on the fact that living cells exclude some dyes.

Dyflex A trade name for dyphylline, a bronchodilator.

Dyllin A trade name for dyphylline, a bronchodilator.

Dymadon A trade name for acetaminophen, an analgesic and antipyretic agent.

Dymelor A trade name for acetohexamide, an anti-diabetic agent.

Dymelorm A trade name for acetohexamide, an antidiabetic agent that stimulates the release of insulin from pancreatic beta cells.

Dymenate A trade name for the antiemetic agent dimenhydrinate.

dyn Abbreviation for dynes.

Dynabac A trade name for dirithromycin, an antibiotic.

DynaCirc A trade name for isradipine, an antihypertensive agent.

Dynamin A microtubule-associated protein that is capable of binding and hydrolyses GTP.

Dynamometer An instrument used for measuring the amount of energy in the contraction of muscle.

Dynapan A trade name for the antibiotic dicloxacillin sodium, which inhibits bacterial cell wall synthesis.

Dyne A unit of energy that is defined as the force that acts on a mass of 1 gram and imparts to it an acceleration of 1 cm/sec^2.

Dynein A protein with ATPase activity found in the armlike extension of the nine peripheral doublets of axonemes.

Dynein ATPase The enzyme that catalyzes the formation of dynein arms that link the microtubules of the cilia and induces their relative mobility.

Dynorphin A group of potent neuropeptides.

Dynorphin A A potent neuropeptide consisting of 17 amino acid residues.

Dynorphin B A potent neuropeptide consisting of 13 amino acid residues.

Dyphylline (mol wt 254) A bronchodilator that inhibits phosphodiesterase and relaxes smooth muscle of both bronchial airways and pulmonary blood vessels.

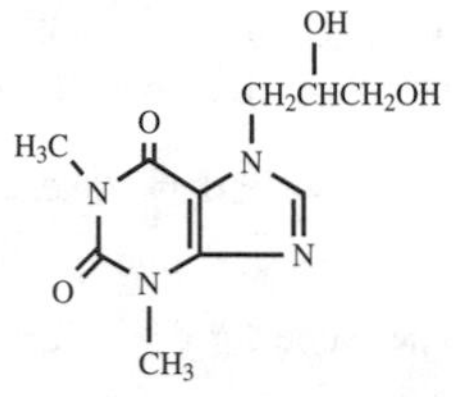

Dyrenium A trade name for triamterene, a diuretic agent.

Dyrexan OD A trade name for phendimetrazine tartrate, a cerebral stimulant.

Dys- A prefix denoting abnormal, diseased, impaired, painful, or difficult.

Dysarthrosis Deformity of joints.

Dysautonomia A dysfunction of the autonomic nervous system.

Dyschromia The abnormal pigmentation of the skin.

Dyscrasia An imbalance of the components of the blood.

Dysentery Inflammation of intestine characterized by abnormal pain and diarrhea.

Dysfunctional Immunity 1. An immune response that produces an undesirable physiological state, e.g., an allergic reaction. 2. Lack of proper immune response.

Dysgammaglobulinemia The abnormal production of gamma globulins in the blood.

Dysgenesis 1. Defective formation of an organ. 2. Lacking the ability to reproduce.

Dysgerminoma A malignant tumor of the ovary.

Dyshidrosis Abnormal sweating

Dyspepsia Indigestion.

Dysphagia Difficulty in swallowing.

Dysplasia An abnormal growth or development of the tissues or organs.

Dysploidy A species in a genus that has different diploid numbers but does not represent a polyploid series.

Dyspnea Difficulty in breathing.

Dysprosium A chemical element with atomic weight 162, valence 3.

Dysproteinemia Abnormal protein content in the blood.

Dystonia An impairment of muscle tone.

Dystrophin A protein product of the human Duchenne muscular dystrophy gene with molecular weight of about 400,000 daltons.

Dystrophy An abnormal condition caused by faulty nutrition and characterized by increased muscle size and decreased muscle strength (also called dystrophia).

Dysuria Painful urination.

Dytac A trade name for triamterene, a diuretic agent.

DZ Abbreviation for dizygote.

E

e Symbol for the base of natural logarithm (a constant equal to 2.7183).

e$^-$ Abbreviation for the negatively charged electron.

E Abbreviation for 1. amino acid glutamic acid; 2. electric field strength; 3. electric potential difference; 4. electromotive force; 5. energy; 6. enzyme.

ε The letter epsilon in the Greek alphabet.

E32 Abbreviation for a protein produced by B lymphocyte that has a role in immunoglobulin heavy chain transcription.

E74 A protein expressed during the development of *Drosophila*.

E_1, E_2, E_3 Referring to the first, second, and third generation of organisms following some experimental manipulation (e.g., irradiation).

Eo′ Symbol for standard redox potential of a redox couple at pH 7 and concentration of 1 *M*.

Eo Redox potential of a redox couple at pH 0 and concentration of 1 *M*.

$E^{1\%}_{1cm}$ The absorption extinction coefficient of a 1% solution in a 1-cm cuvette.

E Face The hydrophobic interior face of the external half of the membrane lipid bilayer obtained from freeze-fracture electron microscopy.

E Protein An F-plasmid-encoded protein acting as a positive initiator for F-plasmid replication.

EA Abbreviation for 1. erythrocyte-antibody (erythrocyte-amboceptor) and 2. antibody-sensitized erythrocytes.

E_a (E_A) Symbol for Arrhenius activation energy.

EAA Abbreviation for 1. electrothermal atomic absorption; 2. essential amino acid; 3. excitatory amino acid.

EAC Abbreviation for erythrocyte-antibody-complement or erythrocyte sensitized with amboceptor (antibody) and complement.

εACA Abbreviation for epsilon amino caproic acid.

EAC-Rosette A cluster of antibody-complement-sensitized erythrocytes around a B lymphocyte.

Eadie-Hofstee Plot A plot of *v*/[S] versus *v* in an enzymatic reaction where *v* is velocity of the reaction and the [S] is substrate concentration in *M*. The intercept on the ordinate is the V_{max} and the slope is -Km.

EaeI (cfrI) A restriction endonuclease from *Enterobacter aerogenes* with the following specificity:

5′..........PyGGCCPu..........3′
3′..........PuCCGGPy..........5′

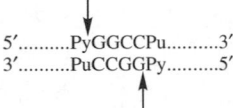

EaePI (PstI) A restriction endonuclease from *Enterobacter aerogenes* with the same specificity as PstI.

EagI (Xma II) A restriction endonuclease from *Enterobacter agglomerans* with the following specificity:

5′..........CGGCCG..........3′
3′..........GCCGGC..........5′

EagKI (EcoRII) A restriction endonuclease from *Enterobacter agglomerans* with the same specificity as EcoRII.

Eagle's Medium A medium for growth and maintenance of cell cultures.

EagMI (AvaII) A restriction endonuclease from *Enterobacter agglomerans* with the following specificity:

5′..........GG(A/T)CC..........3′
3′..........CC(T/A)GG..........5′

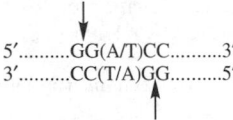

EAHLG Abbreviation for equine anti-human lymphoblast globulin.

EAHLS Abbreviation for equine anti-human lymphoblast serum.

Eam1104I A restriction endonuclease from *Enterobacter amnigenus* RFL 1104 with the following specificity:

5'........CTCTTCNNNN........3'
3'....... GAGAAGNNNN........5'

Eam1105I A restriction endonuclease from *Enterobacter amnigenus* RFL 1105 with the following specificity:

5'........GACNNN↓NNGTC........3'
3'........CTGNN↑NNNCAG........5'

EarI A restriction endonuclease from *Enterobacter aerogenes* with the following specificity:

5'........CTCTTC$(N)_1$↓........3'
3'........GAGAAG$(N)_4$↑........5'

Earle's Balanced Salt Solution A salt solution used for tissue culture medium.

Earle's BSS Abbreviation for Earle's balanced salt solution.

Early Antigen The viral encoded protein antigen that appears soon after infection of a cell by a virus.

Early Enzymes The viral specific enzyme resulting from transcription of early genes.

Early Genes Genes that are transcribed early in a viral infection.

Early Proteins Virus-specific proteins synthesized in the early stage of infection.

Early Quitter Any incomplete polypeptide formed in an *in vitro* translation system.

Early RNA The specific viral RNA transcribed in the early stage of viral infection.

EA-Rosettes Rosettes formed by antibody-coated erythrocytes.

Easprin A trade name for aspirin, an analgesic and antipyretic agent.

Eastern Equine Encephalomyelitis (EEE) An acute encephalitis (encephalomyelitis) of humans and horses, caused by an alphavirus.

Eaton's Agent Referring to *Mycoplasma pneurmoniae.*

EB Abbreviation for 1. Epstein-Barr virus; 2. ethidium bromide; 3. extraction buffer.

EB Virus Abbreviation for Epstein-Barr virus, a virus of the family Herpesviridae. It is the causal agent of Burkitt's lymphoma.

Ebastine (mol wt 470) An antihistaminic agent.

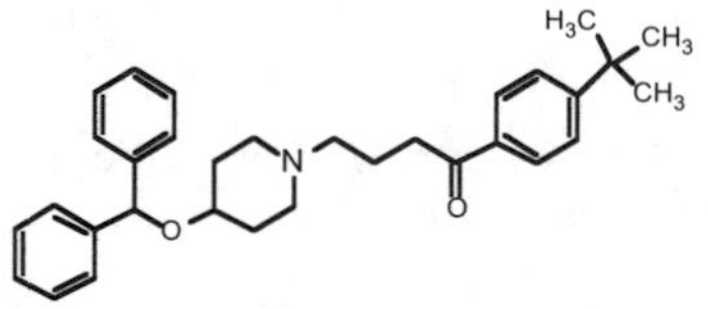

EBNA Abbreviation for Epstein-Barr virus nuclear antigen.

Ebola Virus An enveloped helical ssRNA virus that causes clinically similar hemorrhagic fevers in humans.

4E-BP Abbreviation for eIF4E-binding protein (eukaryotic initiation factor E4-binding protein).

Ebrotidine (mol wt 477) An antiulcerative agent.

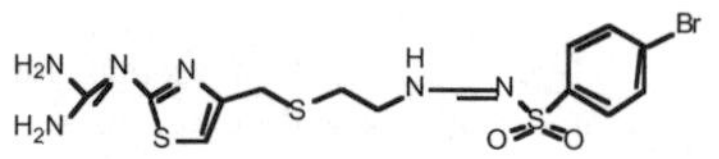

EBS Abbreviation for Earle's balance salt solution.

Eburnamonine (mol wt 294) A vasodilator.

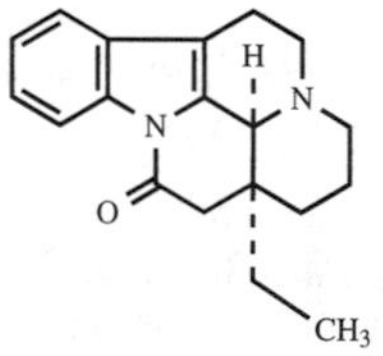

EBV Abbreviation for Epstein-Barr virus, a virus in the family of Herpesviridae. It is the causal agent of Burkitt's lymphoma.

EBVB Abbreviation for Epstein-Barr virus-transformed B-cell.

EBV-LPD Abbreviation for Epstein-Barr virus-induced B-cell lymphoproliferative diseases.

EBVNA Abbreviation for Epstein-Barr virus nuclear antigen.

EC Abbreviation for 1. electron capture; 2. effective concentration; 3. endothelial cell; 4. enzyme commission; 5. ether-chloroform mixture.

EC_{50} Abbreviation for half effective concentration.

ECA Abbreviation for endothelial cell antibodies.

EcaI (BstEII) A restriction endonuclease from *Enterobacter cloacae* with the following specificity:

↓
5′..........GGTNACC..........3′
3′..........CCANTGG..........5′
↑

EcaII (EcoRII) A restriction endonuclease from *Enterobacter cloacae* with the same specificity as EcoRII.

Ecabet (mol wt 381) An antiulcerative agent.

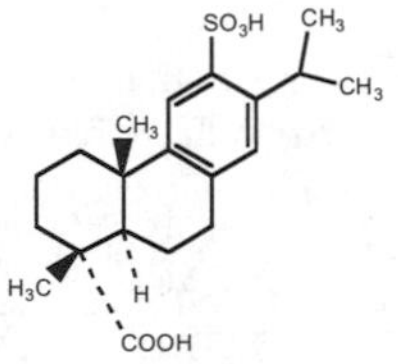

E-CAT (ECAT) Abbreviation for emission computer-assisted tomography.

ECBO Virus Abbreviation for enteric cytopathogenic bovine orphan virus.

Ecbolic Agent An agent that induces childbirth by stimulating contractions of the uterus.

EccI (SacII) A restriction endonuclease from *Enterobacter cloacae* with the same specificity as SacII.

Eccentric 1. Located away from the center. 2. Deviating from the established norm.

Ecchondroma A benign cartilaginous tumor.

Eccrine Glands The sweat glands.

Eccrinology The science that deals with secretion and secretory organs.

Eccrisis The excretion of a waste product.

ECD Abbreviation for 1. electron capture detector; 2. electrochemical detection;

Ecdemic Referring to a disease that is brought into the area from the outside.

Ecdysis The shedding of an outer layer.

Ecdysones Insect molting hormones that control pupation.

ECE Abbreviation for endothelin-converting enzyme.

ECF Abbreviation for 1. electron transfer flavoprotein; 2 eosinophil chemotactic factor; 3. extracellular fluid.

ECFA (ECF-A) Abbreviation for eosinophil chemotactic factor of anaphylaxis.

ECG Abbreviation for 1. Echocardiogram; 2. echocardiography; 3. electrocardiogram.

ECGF Abbreviation for endothelial cell growth factor.

ECH Abbreviation for enoyl-CoA hydratase.

Echinacea The extract of *Echinacea pallida*, which has been used experimentally as a hyaluronidase antagonist.

Echino- A prefix meaning spine or spiny.

Echinochrome A (mol wt 266) The red pigment in eggs of the sea urchin.

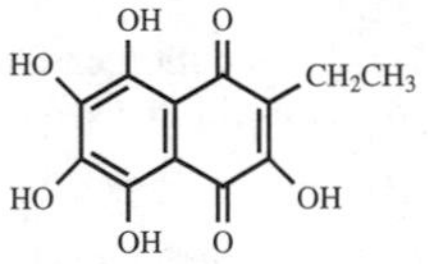

Echinococcosis Infection by larvae of *Echinococcus* tapeworm.

Echinococcus A genus of tapeworm.

Echinocyte Erythrocytes that have shrunk in hypertonic solution so that the surface is spiky.

Echinomycin (mol wt 1101) A potent inhibitor for nucleic synthesis.

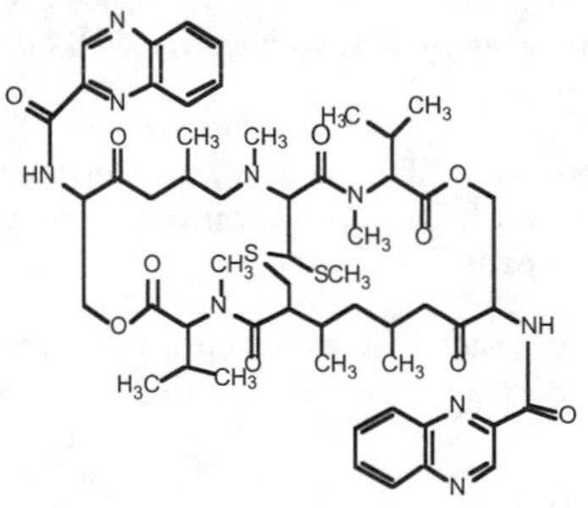

Echinosphaerium A genus of protozoa that is multinucleate.

Echinostelium A genus of slime mold.

Echinulate Covered with small spines.

Echonazole (mol wt 382) An antifungal agent.

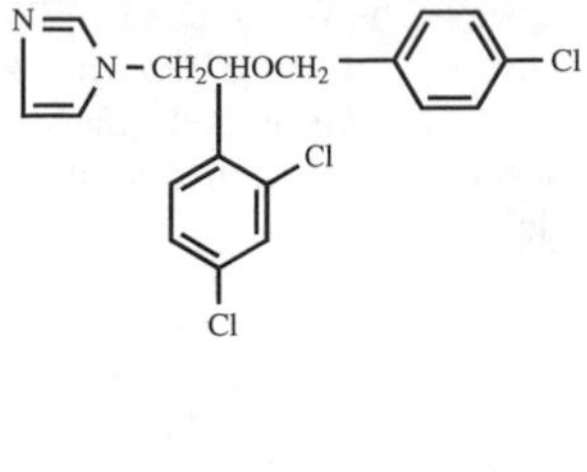

Echothiophate Iodide (mol wt 383) A cholinergic, mitotic agent and inhibitor of cholinesterase.

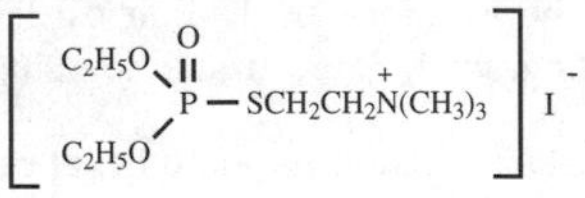

Echovirus A virus of Picornaviridae (also called *Enterovirus*).

ECI Abbreviation for enoyl-CoA isomerase.

ECL Abbreviation for enhanced chemiluminescence.

EclII (EcoRII) A restriction endonuclease from *Enterobacter cloacae* with the same specificity as EcoRII.

Ecl28I (SacII) A restriction endonuclease from *Enterobacter cloacae* with the same specificity as SacII.

Ecl37I (SacII) A restriction endonuclease from *Enterobacter cloacae* with the same specificity as Sac II.

Ecl66I (EcoRII) A restriction endonuclease from *Enterobacter cloacae* with the same specificity as EcoRII.

Ecl77I (PstI) A restriction endonuclease from *Enterobacter cloacae* with the same specificity as PstI.

Ecl133I (PstI) A restriction endonuclease from *Enterobacter cloacae* with the same specificity as PstI.

Ecl136I (EcoRII) A restriction endonuclease from *Enterobacter cloacae* with the same specificity as EcoRII.

Ecl136II (SacI) A restriction endonuclease from *Enterobacter cloacae* with the same specificity as SacI.

Ecl593I (PstI) A restriction endonuclease from *Enterobacter cloacae* with the same specificity as PtsI.

ECLAM Abbreviation for endothelial cell leukocyte adhesion molecule.

E-Classification The classification of enzymes based upon the recommendation of the committee on enzyme nomenclature of the International Union of Biochemistry. The first number following the letter EC indicates the type of enzyme, the second and third numbers indicate subsidiary grouping, and the last number is assigned arbitrarily in numerical order by the commission.

EclHKI A restriction endonuclease from *Enterobacter cloacae* with the following specificity:

↓
5'........GACNNNNNGTC........3'
3'........CTGNNNNNCAG........5'
↑

Eclipse Period The period in the early stage of a viral infection during which the virus exists as free nucleic acid within the host cell.

EclJI (PvuI) A restriction endonuclease from *Enterobacter cloacae* with the following specificity:

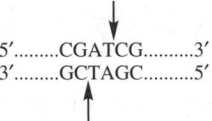

↓
5'.........CGATCG..........3'
3'.........GCTAGC..........5'
↑

EC-Loop Abbreviation for extracellular loop.

Eclosion Hormone Any peptide hormone that programs the death of certain muscles and neurons during metamorphosis of insects.

EclS39I (EcoRII) A restriction endonuclease from *Enterobacter cloacae* with the same specificity as EcoRII.

ECLT Abbreviation for euglobulin clot lysis time.

EclXI A restriction endonuclease from *Enterobacter cloacae* 590 with the following specificity:

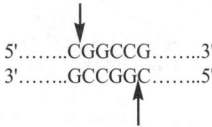

↓
5'........CGGCCG........3'
3'........GCCGGC........5'
↑

ECM Abbreviation for extracellular matrix.

ECMO Virus Abbreviation for enteric cytopathogenic monkey orphan virus.

EC-Naprosyn A trade name for naproxen, a nonsteroidal anti-inflammatory agent.

EcoVIII (HindIII) A restriction endonuclease from *E. coli* EI 858-68 with the following specificity:

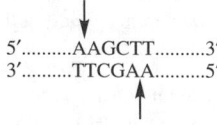

↓
5'..........AAGCTT..........3'
3'..........TTCGAA..........5'
↑

Eco24I (HgiJII) A restriction endonuclease from *E. coli* RFL 24 with the following specificity:

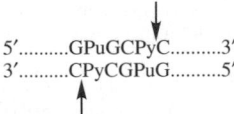

5′..........GPuGCPyC..........3′
3′..........CPyCGPuG..........5′

Eco25I (HgiJII) A restriction endonuclease from *E. coli* RFL 25 with the same specificity as HgiJII.

Eco26I (HgiJII) A restriction endonuclease from *E. coli* RFL 26 with the same specificity as HgiJII.

Eco31I A restriction endonuclease from *E. coli* RFL 31 with the following specificity:

5′...........GGTCTC(N)$_1$...........3′
3′...........CCAGAG(N)$_5$..........5′

Eco32I (EcoRV) A restriction endonuclease from *E. coli* RFL 32 with the same specificity as EcoRV.

Eco35I (HgiJII) A restriction endonuclease from *E. coli* RFL 35 with the same specificity as HgiJII.

Eco38I (EcoRII) A restriction endonuclease from *E. coli* RFL 38 with the same specificity as EcoRII.

Eco39I (AsuI) A restriction endonuclease from *E. coli* RFL 39 with the same specificity as AsuI.

Eco40I (EcoRII) A restriction endonuclease from *E. coli* RFL 40 with the same specificity as EcoRII.

Eco41I (EcoRII) A restriction endonuclease from *E. coli* RFL 41 with the same specificity as EcoRII.

Eco42I (Eco31I) A restriction endonuclease from *E. coli* RFL 42 with the same specificity as Eco31I.

Eco43I (ScrFI) A restriction endonuclease from *E. coli* RFL 43 with the following specificity:

5′..........CCNGG..........3′
3′..........GGNCC..........5′

Eco47I (AvaII) A restriction endonuclease from *E. coli* RFL 47 with the following specificity:

5′..........GG(A/T)CC..........3′
3′..........CC(T/A)GG..........5′

Eco47II (AsuI) A restriction endonuclease from *E. coli* RFL 47 with the same specificity as AsuI.

Eco47III A restriction endonuclease from *E. coli* RFL 47 with the following specificity:

5′..........AGCGCT..........3′
3′..........TCGCGA..........5′

Eco48I (PstI) A restriction endonuclease from *E. coli* RFL 48 with the same specificity as PstI.

Eco49I (PstI) A restriction endonuclease from *E. coli* RFL 49 with the same specificity as PstI.

Eco50I (HgiCI) A restriction endonuclease from *E. coli* RFL 50 with the following specificity:

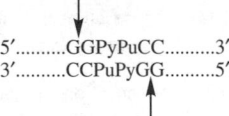

5′..........GGPyPuCC..........3′
3′..........CCPuPyGG..........5′

Eco51I (Eco31I) A restriction endonuclease from *E. coli* RFL 51 with the same specificity as Eco31I.

Eco51II (ScrFI) A restriction endonuclease from *E. coli* RFL 51 with the following specificity:

5′..........CCNGG..........3′
3′..........GGNCC..........5′

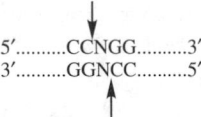

Eco52I (XmaIII) A restriction endonuclease from *E. coli* RFL 52 with the following specificity:

5′..........CGGCCG..........3′
3′..........GCCGGC..........5′

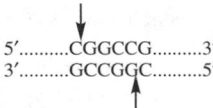

Eco55I (SacII) A restriction endonuclease from *E. coli* RFL 55 with the same specificity as SacII.

Eco56I (NaeI) A restriction endonuclease from *E. coli* RFL 56 with the following specificity:

5′..........CGGCCG..........3′
3′..........GCCGGC..........5′

Eco57I A restriction endonuclease from *E. coli* RFL 57 with the following specificity:

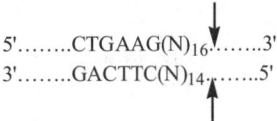

5'........CTGAAG(N)$_{16}$........3'
3'........GACTTC(N)$_{14}$........5'

Eco60I (EcoRII) A restriction endonuclease from *E. coli* RFL 60 with the same specificity as EcoRII.

Eco61I (EcoRII) A restriction endonuclease from *E. coli* RFL 61 with the same specificity as EcoRII.

Eco64I (HgiCI) A restriction endonuclease from *E. coli* RFL 64 with the following specificity:

```
              ↓
5′..........CCPuPyGG..........3′
3′..........GGPyPuCC..........5′
             ↑
```

Eco65I (HindIII) A restriction endonuclease from *E. coli* RFL 65 with the same specificity as HindIII.

Eco67I (EcoRII) A restriction endonuclease from *E. coli* RFL 67 with the same specificity as EcoRII.

Eco68I (HgiJII) A restriction endonuclease from *E. coli* RFL 68 with the same specificity as HgiJII.

Eco70I (EcoRII) A restriction endonuclease from *E. coli* RFL 70 with the same specificity as EcoRII.

Eco71I (EcoRII) A restriction endonuclease from *E. coli* RFL 71 with the same specificity as EcoRII.

Eco72I (PmaCI) A restriction endonuclease from *E. coli* RFL 72 with the following specificity:

```
              ↓
5′..........GTGCAC..........3′
3′..........CACGTG..........5′
              ↑
```

Eco76I (SauI) A restriction endonuclease from *E. coli* RFL 76 with the same specificity as SauI.

Eco78I (NarI) A restriction endonuclease from *E. coli* RFL 78 with the following specificity:

```
              ↓
5′.........CCGCGG.........3′
3′.........GGCGCC.........5′
              ↑
```

Eco80I (ScrFI) A restriction endonuclease from *E. coli* RFL 80 with the following specificity:

```
              ↓
5′..........CCNGG..........3′
3′..........GGNCC..........5′
               ↑
```

Eco81I A restriction endonuclease from *E. coli* RFL 81 with the following specificity:

```
              ↓
5′..........CCTNAGG..........3′
3′..........GGANTCC..........5′
                 ↑
```

Eco82I (EcoRI) A restriction endonuclease from *E. coli* RFL 82 with the same specificity as EcoRI.

Eco83I (PstI) A restriction endonuclease from *E. coli* RFL 83 with the same specificity as PstI.

Eco85I (ScrFI) A restriction endonuclease from *E. coli* RFL 85 with the same specificity as ScrFI.

Eco88I (AvaI) A restriction endonuclease from *E. coli* RFL 88 with the same specificity as AvaI.

Eco90I (CfrI) A restriction endonuclease from *E. coli* RFL 90 with the following specificity:

```
             ↓
5′..........PyGGCCPu..........3′
3′..........PuCCGGPy..........5′
                 ↑
```

Eco91I (BstEII) A restriction endonuclease from *E. coli* RFL 91 with the following specificity:

```
                ↓
5′..........CCANTGG..........3′
3′..........GGTNACC..........5′
             ↑
```

Eco92I (SacII) A restriction endonuclease from *E. coli* RFL 92 with the same specificity as SacII.

Eco93I (ScrFI) A restriction endonuclease from *E. coli* RFL 93 with the same specificity as ScrFI.

Eco95I (Eco3II) A restriction endonuclease from *E. coli* RFL 95 with the same specificity as Eco3II.

Eco96I (SacII) A restriction endonuclease from *E. coli* RFL 96 with the same specificity as SacII.

Eco97I (Eco31I) A restriction endonuclease from *E. coli* RFL 97 with the same specificity as Eco31I.

Eco98I (HindIII) A restriction endonuclease from *E. coli* RFL 98 with the same specificity as HindIII.

Eco99I (SacII) A restriction endonuclease from *E. coli* RFL 99 with the same specificity as SacII.

Eco100I (SacII) A restriction endonuclease from *E. coli* RFL 100 with the same specificity as SacII.

Eco101I (Eco31I) A restriction endonuclease from *E. coli* RFL 101 with the same specificity as Eco31I.

Eco104I (SacII) A restriction endonuclease from *E. coli* RFL 104 with the same specificity as SacII.

Eco105I (SnaBI) A restriction endonuclease from *E. coli* RFL 105 with the following specificity:

5′..........TACGTA..........3′
3′..........ATGCAT..........5′

Eco113I (HgiJII) A restriction endonuclease from *E. coli* RFL 113 with the same specificity as HgiJII.

Eco115I (SauII) A restriction endonuclease from *E. coli* RFL 115 with the same specificity as SauII.

Eco118I (SauI) A restriction endonuclease from *E. coli* RFL 118 with the same specificity as SauI.

Eco120I (Eco31I) A restriction endonuclease from *E. coli* RFL 120 with the same specificity as Eco31I.

Eco121I (CauI) A restriction endonuclease from *E. coli* RFL 121 with the same specificity as CauI.

Eco125I (Eco57I) A restriction endonuclease from *E. coli* RFL 125 with the same specificity as Eco57I.

Eco128I (EcoRI) A restriction endonuclease from *E. coli* RFL 128 with the same specificity as EcoRI.

Eco129I (Eco31I) A restriction endonuclease from *E. coli* RFL 129 with the same specificity as Eco31I.

Eco130I (StyI) A restriction endonuclease from *E. coli* RFL 130 with the following specificity:

5′..........CC(C/T)(A/T)GG..........3′
3′..........GG(T/A)(T/A)CC..........5′

Eco133I (PstI) A restriction endonuclease from *E. coli* RFL 133 with the same specificity as the PstI.

Eco134I (SacII) A restriction endonuclease from *E. coli* RFL 134 with the same specificity as the SacII.

Eco135I (SacII) A restriction endonuclease from *E. coli* RFL 135 with the same specificity as SacII.

Eco136I (EcoRII) A restriction endonuclease from *E. coli* RFL 136 with the same specificity as EcoRII.

Eco136II (SacI) A restriction endonuclease from *E. coli* RFL 136 with the same specificity SacI.

Eco137II (EcoRII) A restriction endonuclease from*E. coli* RFL 136 with the same specificity as EcoRII.

Eco141I (PstI) A restriction endonuclease from *E. coli* RFL 141 with the same specificity as PstI.

Eco143I (BssHII) A restriction endonuclease from *E. coli* RFL 143 with the following specificity:

5′.........GCGCGC.........3′
3′.........CGCGCG.........5′

Eco147I (StuI) A restriction endonuclease from *E. coli* RFL 147 with the following specificity:

5′..........AGGCCT..........3′
3′..........TCCGGA..........5′

Eco149I (KpnI) A restriction endonuclease from *E. coli* RFL 149 with the same specificity as KpnI.

Eco153I (ScrFI) A restriction endonuclease from *E. coli* RFL 153 with the same specificity as ScrFI.

Eco155I (Eco31I) A restriction endonuclease from *E. coli* RFL 155 with the same specificity as Eco31I.

Eco158I (SacII) A restriction endonuclease from *E. coli* RFL 158 with the same specificity as SacII.

Eco158II (SnaBI) A restriction endonuclease from *E. coli* RFL 158 with the same specificity as SnaBI.

Eco159I (EcoRI) A restriction endonuclease from *E. coli* RFL 159 with the same specificity as EcoRI.

Eco161I (PstI) A restriction endonuclease from *E. coli* RFL 161 with the same specificity as PstI.

Eco162I Eco31I) A restriction endonuclease from *E. coli* RFL 162 with the same specificity as Eco31I.

Eco164I (CfrI) A restriction endonuclease from *E. coli* RFL 164 with the following specificity:

5′..........PyGGCCPu..........3′
3′..........PuCCGGPy..........5′

Eco165I (EcoRII) A restriction endonuclease from *E. coli* RFL 165 with the same specificity as EcoRII.

Eco167I (PstI) A restriction endonuclease from *E. coli* RFL 167 with the same specificity as PstI.

Eco168I (HgiCI) A restriction endonuclease from *E. coli* RFL 168 with the same specificity as HgiCI.

Eco169I (HgiCI) A restriction endonuclease from *E. coli* RFL 169 with the same specificity as HgiCI.

Eco170I (EcoRII) A restriction endonuclease from *E. coli* RFL 170 with the same specificity as EcoRII.

Eco171I (HgiCI) A restriction endonuclease from *E. coli* RFL 171 with the same specificity as HgiCI.

Eco173I (HgiCI) A restriction endonuclease from *E. coli* RFL 173 with the same specificity as HgiCI.

Eco178I (EcoRV) A restriction endonuclease from *E. coli* RFL 178 with the same specificity as EcoRV.

Eco179I (CauII) A restriction endonuclease from *E. coli* RFL 179 with the same specificity as CauII.

Eco180I (HgiJII) A restriction endonuclease from *E. coli* RFL 180 with the same specificity as HgiJII.

Eco182I (SacII) A restriction endonuclease from *E. coli* RFL 182 with the same specificity as SacII.

Eco185I (Eco31I) A restriction endonuclease from *E. coli* RFL 185 with the same specificity as Eco31I.

Eco188I (HindIII) A restriction endonuclease from *E. coli* RFL 188 with the same specificity as HindII.

Eco190I (CauI) A restriction endonuclease from *E. coli* RFL 190 with the same specificity as CauI.

Eco191I (Eco31I) A restriction endonuclease from *E. coli* RFL 191 with the same specificity as Eco31I.

Eco193I (EcoRII) A restriction endonuclease from *E. coli* RFL 193 with the same specificity as EcoRII.

Eco195I (HgiCI) A restriction endonuclease from *E. coli* RFL 195 with the same specificity as HgiCI.

Eco196I (SacII) A restriction endonuclease from *E. coli* RFL 196 with the same specificity as SacII.

Eco196II (AsuI) A restriction endonuclease from *E. coli* RFL 196 with the same specificity as AsuI.

Eco200I (ScrFI) A restriction endonuclease from *E. coli* RFL 200 with the same specificity as ScrFI.

Eco201I (AsuI) A restriction endonuclease from *E. coli* RFL 201 with the same specificity as AsuI.

Eco203I (Eco31I) A restriction endonuclease from *E. coli* RFL 203 with the same specificity as Eco31I.

Eco204I (Eco31I) A restriction endonuclease from *E. coli* RFL 204 with the same specificity as Eco31I.

Eco205I (Eco31I) A restriction endonuclease from *E. coli* RFL 205 with the same specificity as Eco31I.

Eco255I A restriction endonuclease from *E. coli* RFL 255 with the following specificity:

```
          ↓
5'.........AGTACT........3'
3'.........TCATGA........5'
             ↑
```

Ecoblast The prospective ectoderm before separation of the germ layers.

EcoHI (CrfI) A restriction endonuclease from *E. coli* with the following specificity:

```
            ↓
5'..........PyGGCCPu..........3'
3'..........PuCCGGPy..........5'
                  ↑
```

EcoICRI (SacI) A restriction endonuclease from *E. coli* 2bT with the following specificity:

```
                ↓
5'..........GAGCTC..........3'
3'..........CTCGAG..........5'
            ↑
```

E. coli An abbreviation used for *Escherichia coli.*

Ecology The science that deals with the relationship between the environment and its associated organisms.

EcoNI A restriction endonuclease from *E. coli* with the following specificity:

```
               ↓
5'.........CCTNNNNNNAGG.........3'
3'.........GGANNNNNNTCC.........5'
                ↑
```

Econazole (mol wt 382) An antifungal agent that alters fungal cell wall permeability.

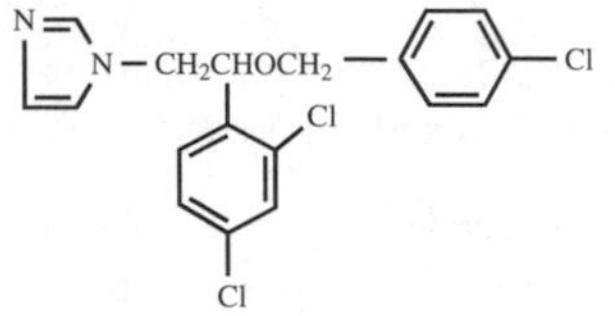

Econazole Nitrate A derivative of econazole and an antifungal agent.

Econochlor A trade name for chloramphenicol, an antibiotic that inhibits bacterial protein synthesis.

Econopred Ophthalmic A trade name for prednisolone acetate, an ophthalmic anti-inflammatory agent.

EcoO65I (BstEII) A restriction endonuclease from *E. coli* K11a with the following specificity:

↓
5′..........GGTNACC..........3′
3′..........CCANTGG..........5′
↑

EcoO109I (DraII) A restriction endonuclease from *E. coli* H709c with the following specificity:

↓
5′..........PuGGNCCPy..........3′
3′..........PyCCNGGPu..........5′
↑

EcoRI A restriction endonuclease from *E. coli* RY13 with the following specificity:

↓
5′..........GAATTC..........3′
3′..........CTTAAG..........5′
↑

EcoRII A restriction endonuclease from *E. coli* R245 with the following specificity:

↓
5′..........CC(A/T)GG..........3′
3′..........GG(T/A)CC..........5′
↑

EcoRV A restriction endonuclease from *E. coli* pLG74 with the following specificity:

↓
5′..........GATATC..........3′
3′..........CTATAG..........5′
↑

Ecosone A trade name for hydrocortisone.

Ecostatin A trade name for econazole nitrate, an anti-infective agent.

EcoT22I (AvaIII) A restriction endonuclease from *E. coli* TB22 with the following specificity:

↓
5′..........ATCGAT..........3′
3′..........TAGCTA..........5′
↑

EcoT38I A restriction endonuclease from *E. coli* TH 38 with the following specificity:

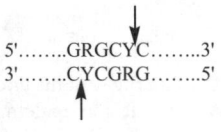

↓
5'........GRGCYC........3'
3'........CYCGRG........5'
↑

R= A or G Y= C or T

EcoT104I (StyI) A restriction endonuclease from *E. coli* TB104 with the same specificity as StyI.

EcoT14I (StyI) A restriction endonuclease from *E. coli* TB14 with the following specificity:

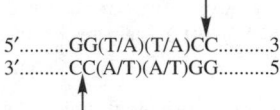

↓
5′..........GG(T/A)(T/A)CC..........3′
3′..........CC(A/T)(A/T)GG..........5′
↑

EcoT22I (AvaIII) A restriction endonuclease from *E. coli* TB22 with the following specificity:

↓
5′..........TACGTA..........3′
3′..........ATGCAT..........5′
↑

EcoT38I (HgiJII) A restriction endonuclease from *E. coli* TH38 with the same specificity as HgiJII.

Ecotin A monomeric periplasmic protein of *E. coli* that inhibits pancreatic chymotrypsin, trypsin, and elastase.

Ecotrin A trade name for aspirin, an analgesic and antipyretic agent.

Ecotropic Virus 1. A virus that replicates only or mainly in a single species. 2. A virus that can replicate only in the host of the species in which it originated.

Ecotype A distinct population of a given species that has adapted genetically to its environment.

ECP Abbreviation for eosinophil cationic protein.

ECPO Virus Abbreviation for enteric cytopathogenic porcine orphan virus.

ECSO Virus Abbreviation for enteric cytopathogenic swine orphan virus.

EC-SOD Abbreviation for extracellular superoxide dismutase.

ECT Abbreviation for euglobulin clot time.

Ectasis The dilation of a tube, duct, or hollow organ.

ECTEOLA-Cellulose Abbreviation for epichlorohydrin triethanolamine cellulose, an anion exchanger for ion exchange chromatography.

Ecto- A prefix meaning outside.

Ectoderm The outermost of the three embryonic layers that gives rise to the epidermis, sense organs, and nervous system.

Ectodermosis A disorder due to maldevelopment of any organ or tissue derived from ectoderm.

Ectoenzyme Any enzyme that is attached to the external surface of the plasma membrane of a cell.

Ectomere Any of the cells formed by division of the fertilized egg that participates in the formation of ectoderm.

-ectomy A suffix meaning surgical removal.

Ectoparasite A parasite that lives or feeds on the outer surface of the host's body, e.g., a louse.

Ectopic Expression The occurrence of gene expression in a tissue in which it is normally not expressed. Ectopic expression can be caused by the juxtaposition of novel enhancer elements to a gene.

Ectopic Integration The insertion of an introduced gene at a site other than its usual locus.

Ectoplasm The outer stiffer portion or region of the cytoplasm of a cell, which may be differentiated in texture from the inner portion (endoplasm).

Ectosymbiont An organism that lives in a symbiotic association on the surface of the host.

Ectosymbiosis A type of symbiosis in which one organism remains outside of the other organism.

Ectotherm An animal unable to control its body temperature (cold-blooded animal).

Ectothiorhodospira A genus of photosynthetic bacteria (Ectothiorhodospiraceae).

Ectothrix Referring to a type of fungal infection in which the hyphae grow both within and on the surface of the hair shaft.

Ectotoxin Synonym for exotoxin.

Ectylurea (mol wt 156) A sedative and hypnotic agent.

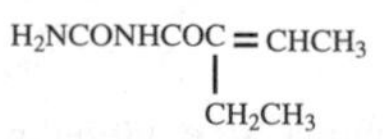

ECV Abbreviation for endosomal carrier vesicle.

E-Cypionate A trade name for estradiol cypionate, a hormone that increases the synthesis of DNA, RNA, and protein.

Eczema A skin disorder characterized by redness, thickening, swelling, and formation of papules, vesicles or crusts.

ED Abbreviation for 1. electrodialysis; 2. Entner-Doudoroff pathway; 3. effective dose.

ED_{50} The dose of a given agent that causes a specific effect in 50% of the tested subjects (also known as effective dose ED_{50}).

Edatrexate (mol wt 467) An antineoplastic agent.

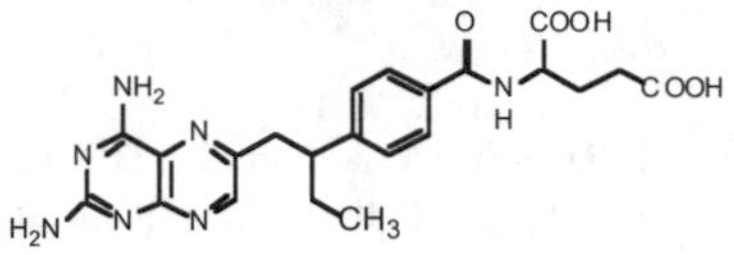

Edecril A trade name for ethacrynic acid, a diuretic agent.

Edecrin A trade name for ethacrynic acid, a diuretic agent.

Edeines A group of basic peptide antibiotics produced by *Bacillus brevis* effective against both Gram-negative and Gram-positive bacteria.

Edema Swelling of any part of the body due to the excessive accumulation of fluid.

Edestin One of the major storage plant proteins from hemp seed.

Edetate Calcium Disodium (mol wt 374) An agent that forms stable soluble complexes with metals (e.g., lead).

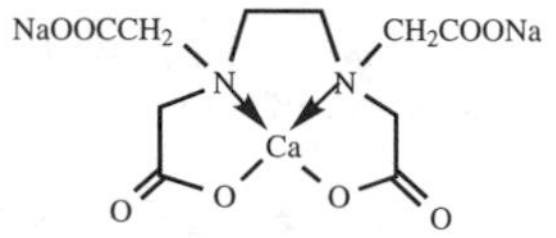

Edetate Disodium (mol wt 336) A chelating agent.

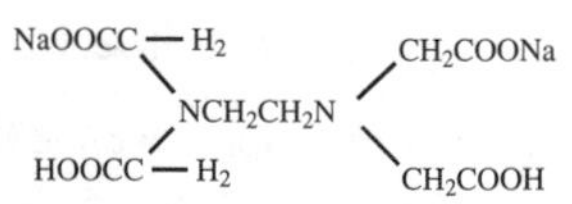

Edex A trade name for alprostadil, a prostaglandin used to relax vascular smooth muscle.

EDG Abbreviation for endothelial differentiation gene.

Editing Enzyme Referring to DNA-polymerase-I, which removes incorrectly incorporated nucleotides during DNA synthesis.

Editosome A macromolecular complex involved in the editing of RNA transcript.

Edman Degradation A method for determining the amino acid sequence by a stepwise removal of amino acids from the N-terminal of a peptide or a protein.

Edman's Reagent Referring to phenylisothiocyanate, which reacts with free alpha amino groups in a peptide or protein.

Edoxudine (mol wt 256) An antiviral agent (antiherpes).

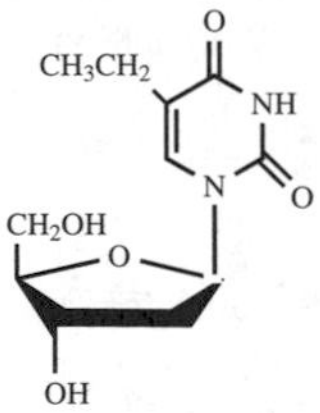

EDP Abbreviation for electron dense particle.

ED-Pathway Abbreviation for Entner-Doudoroff pathway, a pathway that converts glucose 6-phosphate to 3-phosphoglyceraldehyde and pyruvate, in some bacteria.

EDRF Abbreviation for endothelium-dependent releasing factor.

Edrophonium Chloride (mol wt 202) A cholinergic agent that inhibits the destruction of acetylcholine released from parasympathetic and somatic efferent nerves.

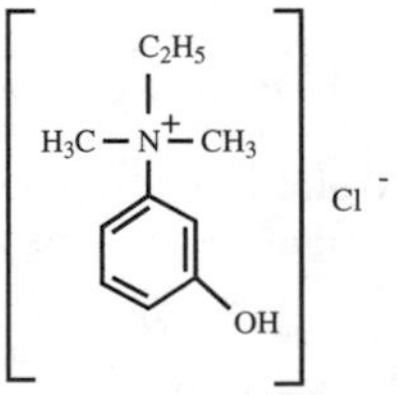

EDTA Abbreviation for ethylenediaminetetraacetic acid, a chelator that forms a complex with metallic ions; used to remove metals from a sample.

EDTA-RA Abbreviation for EDTA-releasing assay.

Edwardsiella A genus of Gram-negative bacteria (Enterobacteriaceae).

Edward Syndrome A congenital defect of humans caused by the presence of an extra chromosome number 18.

EDXA Abbreviation for energy dispersive X-ray analysis.

EE Abbreviation for 1. embryo extract; 2. equine encephalitis.

EEA-1 Abbreviation for early endosomal autoantigen-1.

E_1E_2-ATPase A type of ATPase occurring in the membrane of certain fungi for creation of ion-motive pumps to transport ions. The enzyme has two conformational states, namely E_1 and E_2.

EEE Abbreviation for 1. Eastern equine encephalitis; 2. Eastern equine encephalomyelitis.

EEDQ (mol wt 247) Abbreviation for N-ethoxycarbonyl-2-ethoxy-1,2-dihydroquinoline, an agent for peptide condensation and an irreversible inhibitor for mitochondrial ATPase.

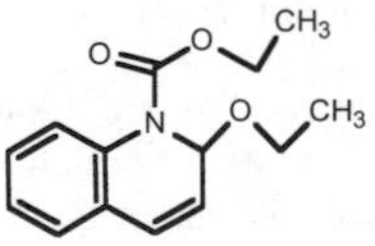

EEE Virus Abbreviation for 1. Eastern equine encephalitis virus; 2. Eastern equine encephalomyelitis virus.

eEF Abbreviation for eukaryotic elongation factor in protein synthesis.

EEG Abbreviation for electroencephalograph, a method for recording the rhythmical waves of electrical potential occurring in the brain.

EES A trade name for erythromycin ethylsuccinate, an antibiotic that inhibits bacterial protein synthesis.

EF Abbreviation for 1. electric field; 2. elongation factor; 3. enhancing factor; 4. extrinsic factor.

EFA Abbreviation for essential fatty acid.

E-FABP Abbreviation for epidermal-type fatty acid binding protein.

E-Face The hydrophobic interior face of the external half of the membrane lipid bilayer obtained from freeze-fracturing for electron microscopy.

EFAD Abbreviation for essential fatty acid deficiency.

Efavirenz (mol wt 316) A non-nucleoside antiviral drug that inhibits reverse transcriptase.

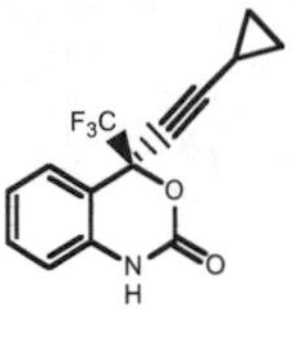

EFE Abbreviation for ethylene-forming enzyme.

Effective Dose The dose of a given agent that gives a specific effect.

Effective Half Life The time required for a radioactive isotope to diminish to 50% in an animal body.

Effector 1. Substance capable of regulating allosteric enzyme activity or changing the property of an allosteric protein. 2. A metabolite capable of engaging in feedback control in a metabolic pathway.

Effector Cell 1. Cell that directly attacks or effects specific target cells. 2. Lymphocyte capable of engaging in cell-mediated or humoral immunity, e.g., effector T cell (capable of mediating cytotoxicity) or effector B cell (capable of synthesizing antibody). 3. Cell that produces specific effects in the living organism.

Effector Molecule Small, biologically active molecule that acts as a regulator to control the activity of a protein or an enzyme by binding to a specific region on the protein or enzyme (e.g., binding of allolactose with the *lac* repressor to initiate *lac* operon action).

Effector Site Region of an allosteric protein at which the effector binds and affects the functional properties of the protein.

Efferent Conveying away from a cell or an organ, e.g., efferent nerves carrying impulses away from the central nervous system.

Effexor XR A trade name for venlafaxine, an antidepressant.

Effluent The fluid that flows out.

Efflux The outward flow of fluid.

Effuse Spreading out widely and thinly on a surface.

Effusion The outflowing of the fluid from an anatomical vessel or structure.

EF-G Elongation factor G, a protein that promotes the process of translocation in bacterial protein synthesis.

Eflornithine (mol wt 182) An antineoplastic, antipneumocystic, and antiprotozoal agent and an inhibitor of ornithine carboxylase.

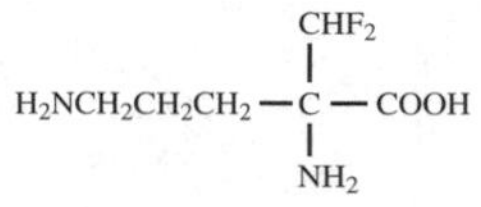

Efloxate (mol wt 324) A coronary vasodilator.

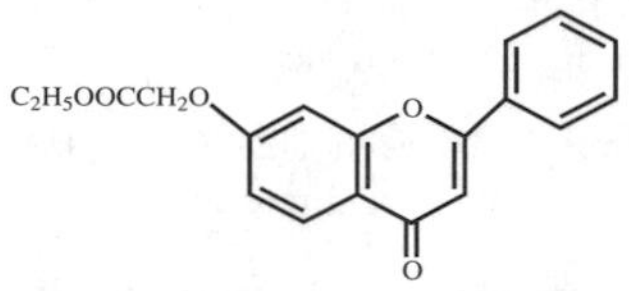

Efrapeptin A lipophilic polypeptide that inhibits F_0F_1-ATPase.

EF-T The elongation factor T in protein synthesis that consists of two components, namely Tu and Ts.

EF-Ts A heat stable protein component of the elongation factor T in bacterial protein synthesis.

EF-Tu A heat unstable protein component of elongation factor T in bacterial protein synthesis.

EF-TuTs Referring to elongation factor T in protein synthesis.

Efudex A trade name for fluorouracil, an anticancer agent.

EGDF Abbreviation for embryonic growth and development factor.

EgeI A restriction endonuclease from *Enterobacter gergoviae* NA with the following specificity:

5'........GGC↓GCC........3'
3'........CCG↑CGG........5'

Egest The release of a substance or waste from the cell or body.

EGF Abbreviation for epidermal growth factor.

EGFR Abbreviation for EGF-receptor.

EGF-Urogastrone A polypeptide epidermal growth factor and an inhibitor of gastric acid secretion.

Eglin C A protease inhibitor consisting of 70 amino acid residues.

EGME Abbreviation for ethylene glycol monoethyl ether.

EGOT Abbreviation for erythrocyte glutamate oxaloacetate transaminase.

EGS Abbreviation for ethylene glycol succinate.

EGTA Abbreviation for ethyleneglycol-bis-(β-aminoethyl)-*N,N,N′,N′*-tetra-acetic acid, a chelating agent.

EH Abbreviation for ethidium homodimer.

EheI A restriction endonuclease from *Erwinia herbicola* strain 915 with the following specificity:

Ehringhaus Compensator A device used in interference or polarization microscopy to reduce the brightness of the object to zero in order to measure the phase retardation.

Ehrlich Ascites A mouse ascites tumor cell line maintained by passage in animals.

Ehrlich Reagent 1. A reagent containing *p*-dimethylaminobenzaldehyde and used for determination of porphobilinogen concentrations, δ-aminolevulinic acid. 2. A reagent containing diazotized sulfanilic acid for determination of bilirubin concentrations.

Ehrlichia A genus of Gram-negative bacteria (Ehrlichieae).

Ehrlichiosis A disease caused by a species of *Ehrlichia*.

EI Abbreviation for enzyme inhibitor.

EIA Abbreviation for 1. electro-immuno-assay; 2. enzyme immuno-assay; 3. exercise-induced anaphylaxis; 4. exercise-induced asthma.

EIAV Abbreviation for equine infectious anemia virus.

EIB Abbreviation for exercise-induced bronchospasm.

Eicosadienoic Acid Any straight-chain fatty acid having 20 carbons and 2 double bonds.

Eicosanoic Acid (mol wt 319) A 20-carbon saturated fatty acid found in peanut oil and butter (see also arachidonic acid).

$$CH_3 — (CH_2)_{18}COOH$$

Eicosanoids Collective term for substances derived from arachidonic acid such as prostaglandins, thromboxanes, leukotrienes, and lipoxins.

5,8,11,14,17-Eicosapentaenoic Acid (mol wt 302) An antihyperlipoproteinemic agent.

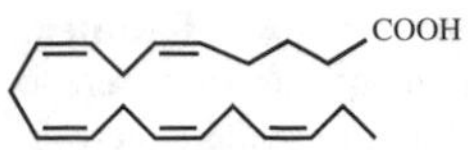

Eicosatetraenoic Acid Any straight-chain fatty acid having 20 carbons and 4 double bonds.

Eicosatetraynoic Acid Any straight-chain fatty acid having 20 carbons and 4 triple bonds.

Eicosatrienoic Acid Any straight-chain fatty acid having 20 carbons and 3 double bonds.

Eicosatriynoic Acid Any straight-chain fatty acid having 20 carbons and 3 triple bonds.

Eicosenoic Acid Any straight-chain fatty acid having 20 carbons and 1 double bond.

EID Abbreviation for 1. egg-infective dose; 2. electro-immuno-diffusion.

eIF Abbreviation for eukaryotic initiation factor in eukaryotic protein synthesis.

EIF Abbreviation for erythrocyte initiation factor.

eIF1 Abbreviation for eukaryotic initiation factor 1.

eIF2 Abbreviation for eukaryotic initiation factor 2.

eIF2α Abbreviation for a subunit of eukaryotic initiation factor 2.

eIF3 Abbreviation for eukaryotic initiation factor 3.

eIF5A Abbreviation for eukaryotic initiation factor 5A.

eIF2B Abbreviation for eukaryotic initiation factor 2B.

eIF4B Abbreviation for eukaryotic initiation factor 4B.

eIF4E Abbreviation for eukaryotic initiation factor 4E.

eIF4E-BP Abbreviation for eukaryotic initiation factor 4E-binding protein.

EIF4F Abbreviation for eukaryotic initiation factor 4F.

Eijkman Test A test to determine the ability of enterobacteria to produce gas from lactose.

Eikenella A genus of anaerobic, facultative aerobic, oxidase-positive, catalase-negative, chemoorganotrophic Gram-negative bacteria.

Eimeria A genus of parasitic protozoa.

Einstein Referring to one mole of photons.

Einsteinium A stable isotope with atomic weight 252, valence 3.

Ejectisome A type of extrusome found in *Cryptophyta* and *Chlorophyta,* which can be discharged from the organisms upon proper stimulation.

EK Abbreviation for erythrokinase.

EKG Abbreviation for electrocardiography.

EKTR Abbreviation for enterokinase-responsive α-thrombin receptor.

Elaidic Acid (mol wt 282) An organic acid.

$$CH_3(CH_2)_7CH{=}CH(CH_2)_7COOH$$

Elaidinization The *cis-trans* isomerization of saturated fatty acids.

Elaiomycin (mol wt 258) An antibiotic produced by *Streptomyces hepaticus* that inhibits growth of *Mycobacterium tuberculosis.*

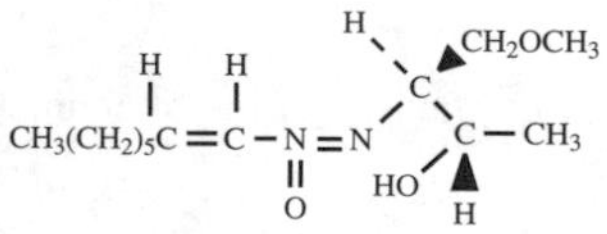

Elaioplast A lipid-storing organelle.

ELAM Abbreviation for endothelial leukocyte adhesion molecule.

ELAM-1 Abbreviation for endothelial leukocyte adhesion molecule 1.

ELA-Max A trade name for lidocaine, a topical anesthetic agent.

Elase A trade name for fibrinolysin, which breaks fibrin of a blood clot.

Elastase A protease obtained from pancreas that is capable of hydrolyzing elastin and proteins of elastic fibers.

Elastic Fiber Yellowish elastic fiber found in the intercellular substance of connective tissue.

Elastin Elastic, load-bearing protein fibers of animal connective tissue, particularly the ligaments of vertebrate and the walls of large arteries.

Elastatinal A peptide produce by *Actinomycetes* that inhibits serine proteinase and elastase.

Elastomer A polymer that has elastic properties (either natural or synthetic).

Elastonectin Elastin-binding protein found in extracellular matrix and produced by skin fibroblasts.

ELAT Abbreviation for enzyme-linked antiglobulin test.

Elavil A trade name for amitriptyline hydrochloride, an antidepressant that increases the level of norepinephrine or serotonin in the CNS.

Elcatonin (mol wt 3364) A synthetic polypeptide hormone (analog of eel calcitonin) that regulates calcium and is effective in reducing plasma calcium.

Eldepryl A trade name for selegiline hydrochloride, an antiparkinsonian agent and inhibitor of monoamino oxidase.

Eldisine A trade name for vindesine sulfate, an antimitotic agent that arrests mitosis in metaphase and blocking cell division.

Eldofe A trade name for ferrous fumarate, a hematinic agent that provides iron for the formation of hemoglobin.

Eldopaque A trade name for hydroquinone.

Eldoquin A trade name for hydroquinone.

Electrical Synapse Junction between two nerve cells that allows a nerve impulse to be transmitted by direct electrical connection without the involvement of chemical neurotransmitters.

Electric Birefringence The birefringence produced by molecules resulting from the application of an electric field.

Electric Dichroism The dichroism resulting from the absorption of polarized light by molecules.

Electrobiology The biological science that deals with the electric phenomena in living organisms.

Electroblotting Transfer of DNA, RNA, or protein from an electrophoretic gel to nitrocellulose paper by an electric current.

Electrocardiography A technique for recording changes in electrical potential associated with heart activity.

Electrochemical Gradient Transmembrane gradient of an ionic substance resulting from differences of charge and chemical concentration across a membrane.

Electrochemical Potential The electrochemical potential resulting from the differences in charges and concentrations of ionic substances across a membrane.

Electrochemical Proton Gradient The transmembrane gradient due to differences in charges and concentration of protons across a membrane.

Electrochemical Sensor An analytical device for measurement of chemical activity employing either a potentiometric or an amperometric electrode.

Electrochemistry The science that deals with relationships of electricity and chemical energy.

Electrocoagulation The hardening of a diseased tissue by high frequency current.

Electrode 1. A device by which a current leaves and enters an electrolytic cell. 2. A conductor of electricity through which current leaves and enters a medium.

Electrodeposition The deposition of materials onto an electrode by the application of an electric field.

Electrodialysis A technique for removal of small ionic molecules from a dialysis sac by applying electrical current.

Electrodiffusion Diffusion of charged molecules that is facilitated by an electric current.

Electroencephalography A technique for recording brain waves or activity.

Electroendosmosis The movement of a charged fluid, relative to a fixed medium carrying the opposite charge, under the influence of an electrical field.

Electrofocusing See isoelectric focusing.

Electrofusion Fusion of cells by high frequency electric current.

Electrogastrograph An instrument used for recording the bioelectric potential associated with GI activity.

Electrogenic Generating electrochemical potential.

Electrogenic Pump The pump that generates a transmembrane electrochemical potential.

Electrogenic Transport Transport of ions or ionic substances across a membrane resulting in the generation of a transmembrane electrical potential difference.

Electroimmunoassay Synonym for immunoelectrophoresis.

Electrolysis A process of decomposition of a substance by means of an electric current.

Electrolyte Substance capable of dissociating into ions and conducting electric current.

Electrolyte Balance The maintenance of a constant internal environment in the body in respect to the distribution of electrolytes between various fluid compartments.

Electrolytic Desalt The removal of salts from a dialysis sac by electric current.

Electromagnetic Spectrum A spectrum of the entire wavelengths or frequency of electromagnetic radiation.

Electromorph Allozymes that can be separated or distinguished by electrophoresis.

Electromyography A technique for recording electric impulses in the muscle.

Electron A negatively charged particle present in the atom with a mass approximately 0.00055 amu and a charge of –1.

Electron Acceptor Substance capable of accepting an electron or capable of being reduced.

Electron Affinity The tendency to accept electrons.

Electron Capture The orbital electron of an atom attracted to the nucleus and combining with the proton in that nucleus.

Electron Carrier Substance capable of serving as electron donor and acceptor.

Electron Donor Substance capable of donating electrons or capable of being oxidized.

Electron Dense Substance capable of absorbing electrons or capable of preventing passage of electrons; appears dark in electron micrographs.

Electron Dense Label The use of electron dense material as a marker in the electron microscopy.

Electron Gun A device that emits a controlled beam of accelerated electrons.

Electron Ionization Mass Spectrometry A mass spectrometric technique that requires volatilization of sample prior to ionization.

Electron Magnetic Resonance See Electron Paramagnetic Resonance.

Electron Micrograph A photographic record of a specimen produced with the electron microscope.

Electron Microscope An instrument that uses a beam of electrons to visualize cellular structures and examine cellular architecture.

Electron Paramagnetic Resonance A technique to characterize a substance (e.g., organic molecule, free radical) with unpaired electrons. A molecule with unpaired electrons has a characteristic electron paramagnetic resonance spectrum because its unpaired electrons interact with the magnetic field generated by the nucleus and other electrons of the molecule.

Electron Spin Resonance See electron paramagnetic resonance.

Electron Transfer Flavoprotein A FAD-containing protein that serves as electron carrier.

Electron Transfer Protein Protein capable of serving as an electron carrier in an oxidation-reduction reaction, e.g., cytochrome.

Electron Transparent A substance that is incapable of absorbing or preventing passage of electrons; appears light in electron micrographs.

Electron Transport Chain A series of electron carriers arranged in order of increasing electron affinity along which electrons are transferred from donor to acceptor with an accompanying release of energy at each transfer step.

Electron Transport Phosphorylation The phosphorylation of ADP to ATP driven by the energy of a transmembrane gradient of protons generated by transfer of electrons along the electron transport chain.

Electron Transport System See electron transport chain.

Electron Volt The energy imparted to an electron by a potential of one volt (equal to 1.6×10^{12} erg).

Electronarcosis The induction of sleep by passing weak electrical current.

Electronegative The term describes 1. an atom or a group of atoms that tend to attract electrons; 2. any chemical or entity that carries a negative charge and tend to move to the anode.

Electroneutral Symport A type of symport that transports two different molecules in the same direction without net movement of charge across the membrane.

Electroneutral Transport A transport that results in no net transfer of charges across the membrane.

Electronic Transition The passage of an electron in a chemical entity from one energy level to another.

Electroosmosis See electroendosmosis.

Electrophile An electron-deficient compound that tends to react with negatively charged ions or substance, e.g., H^+ or metal ion.

Electrophilic Catalysis A catalysis in which the catalyst attracts a pair of electrons from a reactant.

Electrophilic Substitution Reaction in which an electron-deficient (electropositive) molecule attacks an electron-rich (electronegative) molecule and results in the formation of a covalent bond.

Electrophoregram The profile of an electrophoretic pattern.

Electrophoresis A method of separating charged macromolecules or particles according to their charge, size, and shape as they migrate through a medium in an electrical field.

Electrophoretic Effect The phenomenon of decreased electrophoretic mobility of a charged macromolecule caused by the movement of counter ions or solvent molecules in the opposite direction.

Electrophoretic Mobility The velocity of a charged molecule in electrophoresis; it is expressed in units of cm^2/sec◊volt.

Electrophoretic Transfer Synonym of electroblotting.

Electrophoretic Velocity The velocity of a charged particle or molecule during electrophoresis. It is generally proportional to the electric field strength.

Electroporation A technique for alteration of plasma membrane permeability of cells facilitating the introduction of foreign DNA into the cell by means of an electric current.

Electropositive The term describing 1. an atom or a group of atoms that tend to give up electrons; 2. any chemical or entity that carries a positive charge and tends to move the cathode.

Electrostatic Catalysis The catalysis in which charge distribution around the active sites of an enzyme is arranged so as to stabilize the transition states of the catalyzed reaction.

Eledoisin (mol wt 1188) A bioactive peptide from the octopus; it is a potent vasodilator and hypotensive agent.

Eleidin A protein found in the skin and mucous membrane.

Element A substance that contains only one type of atom.

Elgodipine (mol wt 525) An antianginal agent.

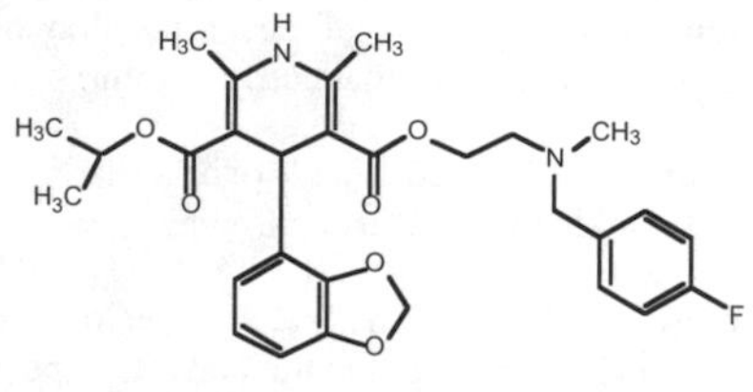

Elicitor Substance or agent capable of inducing the formation of phytoalexins in higher plants.

ELIEDA Abbreviation for enzyme-linked immuno-electro-diffusion assay.

Elimite A trade name for permethrin, an antiparasitic agent.

ELISA Abbreviation for enzyme-linked immunosorbent assay; it involves the reaction of specific antibody with an antigen. The antigen-antibody complex is detected by the use of an enzyme-labeled second antibody against the first antibody through an enzyme-mediated color reaction.

Elixicon A trade name for theophylline, a bronchodilator that inhibits phosphodiesterase and increases cAMP concentration.

Elixomin A trade name for theophylline, a bronchodilator that inhibits phosphodiesterase and increases cAMP concentration.

Elixophyllin A trade name for theophylline, a bronchodilator that inhibits phosphodiesterase and increases cAMP concentration.

Ellagic Acid (mol wt 338) A plant phenol and an inhibitor of glutathione S-transferase.

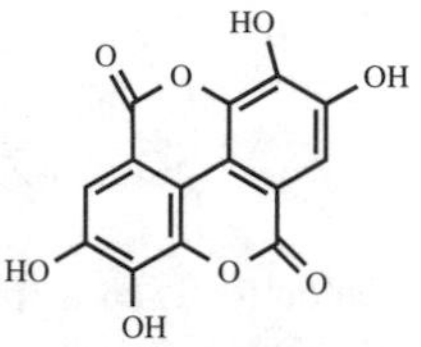

Ellipsosome Membrane-bound compartment containing cytochrome-like pigment, found in the retinal cones of some fish.

Elliptinium Acetate (mol wt 336) An antineoplastic agent.

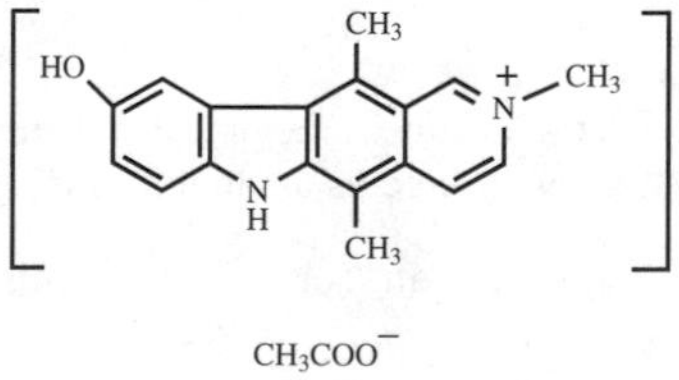

Elliptocyte An oval-shaped red blood cell.

Elliptocytosis The presence of a large number of elliptocytes.

Ellman's Reagent A reagent containing 5,5′-dithio-bis(2-nitrobenzoic acid), used for determination of the sulfhydryl group in protein.

Elocon A trade name for mometasone furoate, a topical corticosteroid.

Elongation Factors The factor for elongation of a polypeptide chain during protein synthesis, e.g., EF-Tu, EF-Ts, and EF-G.

Elson-Morgan Reaction A colorimetric reaction for the estimation of combined and free hexosamines.

Elspar A trade name for asparaginase, used as an antineoplastic agent.

ELT Abbreviation for euglobulin lysis time.

Eltor A trade name for pseudoephedrine hydrochloride, which stimulates alpha-adrenergic receptors in the respiratory tract.

Eltroxin A trade name for levothyroxine sodium, a hormone that accelerates cellular oxidation.

Eluate The solute or substance that is being separated and collected from a chromatographic column.

Eluent The solution or buffer that is used for the elution of solute from a chromatographic column.

Elution A process by which a substance or solute is separated, eluted, and collected from a chromatographic column.

Elution Profile A plot of eluate property versus elution volume, e.g., absorbance at 280 nm of proteins versus elution volume.

Elution Volume The volume of eluate collected from a chromatographic column.

EM Abbreviation for electron microscope.

E_m Symbol for molar extinction coefficient (concentration in g-mole/liter).

EMA Abbreviation for epithelial membrane antigen.

EMAC Abbreviation for a combination drug containing etoposide, methotrexate, adriamycin, and citrovorum.

EMB Abbreviation for eosin-methylene blue.

EMB Agar Abbreviation for eosin-methylene blue agar, an agar-base medium used for the isolation and differentiation of enterobacteria.

Embden-Meyerhof Pathway A pathway in which glucose is converted to pyruvate through a series of biochemical reactions (see glycolysis).

Embden-Meyerhof-Parnas Pathway A pathway in which glucose is converted to pyruvate through a series of biochemical reactions (see glycolysis).

Embedding Embedding tissue in wax or plastic in the preparation of sections for microscopic examination.

Embolectomy Surgical removal of a clot.

Embolism The obstruction of a blood vessel by a clot or other object.

Embolus A piece of a circulating blood clot or a foreign object that travels through the blood stream until it becomes lodged in a blood vessel.

Embramine (mol wt 348) An antihistaminic agent.

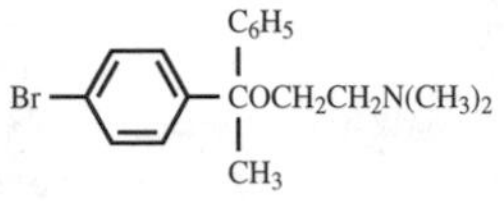

Embryo 1. The early development of an organism. 2. The stage prior to birth. 3. The early state of growth and differentiation characterized by cleavage; laying down of fundamental tissue, and formation of organ systems.

Embryo Transfer A method by which an early embryo is collected from a donor female and introduced into a recipient female who serves as a surrogate mother.

Embryogenesis The formation and development of embryo.

Embryoid A mass of plant or animal tissue that resembles an embryo.

Embryology The biological science that deals with embryos and their development.

Embryonated Egg An egg that contains a live, developing embryo.

Embryonic Stem Cells Cells in the blastocyst that give rise to the embryo itself, not the embryonic membranes.

Embryonin Synonym of bovine α_2-macroglobulin.

EMC Virus Abbreviation for encephalomyocarditis virus.

Emcodeine A trade name for a combination drug containing aspirin and codeine phosphate, used as an analgesic and antipyretic agent.

Emcyt A trade name for estramustine phosphate, a hormone with antineoplastic activity.

Emedastine (mol wt 302) An antihistaminic agent.

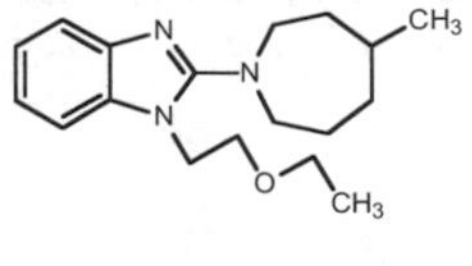

Emepronium Bromide (mol wt 362) An antispasmodic agent.

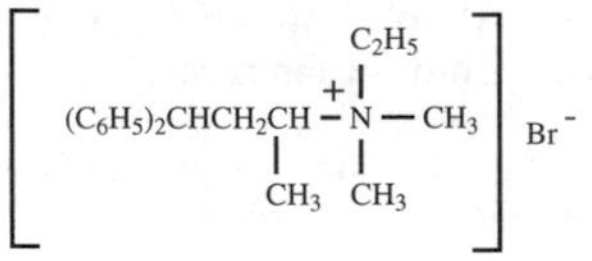

Emerson Enhancement Effect The enhancement of photosynthetic activity or efficiency by supplement of 700 nm wavelength with a short 680 nm wavelength.

Emesis Vomiting.

Emete-Con A trade name for benzquinamide hydrochloride, an antiemetic agent.

Emetic Agent Substance or agent that produces vomiting.

Emetine (mol wt 481) An alkaloid from the roots of *Uragoga ipecacuanba*; it inhibits eukaryotic protein synthesis.

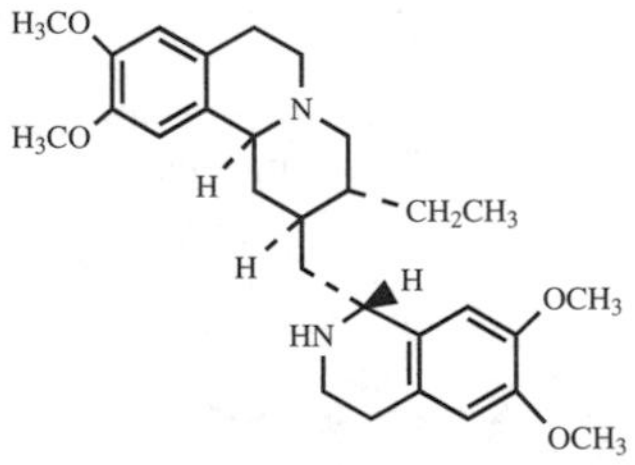

EMF Abbreviation for 1. eosinophil maturation factor; 2. erythrocyte maturation factor.

emf Abbreviation for electromotive force.

Emfabid A trade name for dyphylline, a bronchodilator that inhibits phosphodiesterase and increases cAMP concentration.

EMG Abbreviation for electromyogram or electromyography.

Emgel A trade name for a topical gel containing 2% erythromycin used as an antibacterial agent.

EMI Abbreviation for enzyme-multiplied immuno-assay.

EMIA Abbreviation for enzyme-multiplied immuno-assay.

EMIAT Abbreviation for enzyme-multiplied immunoassay technique.

Eminase A trade name for a fibrinolytic enzyme.

Emiocytosis A form of exocytosis (release of secretory substance from cell).

Emission Spectrum A plot of the emission of electromagnetic radiation by a molecule versus wavelength.

EMIT Abbreviation for enzyme-multiplied immunoassay technique.

Emitrip A trade name for amitriptyline hydrochloride, an antidepressant that increases norepinephrine and serotonin concentrations in the CNS.

Emla Cream A trade name for a combination drug containing 2.5% lidocaine and 2.5% prilocaine.

Emodin (mol wt 270) A cathartic agent from rhubarb root.

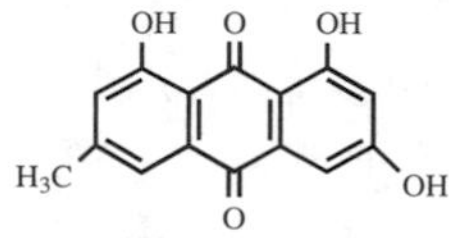

Emollient Any agent that softens and soothes the skin or mucous membrane.

Emorfazone (mol wt 239) An anti-inflammatory and analgesic agent.

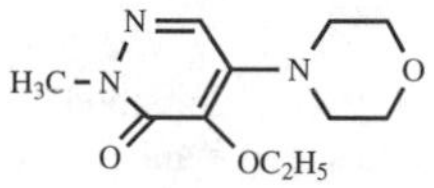

EMP Pathway See Embden-Meyerhof-Parnas pathway.

Emphore Nonenzyme protein that is capable of binding ligand.

Emphysema 1. Swelling due to the abnormal presence of air in the tissue or cavity of the body. 2. A lung disorder associated with chronic bronchitis.

Empirical Formula Chemical formula that describes the number and types of atoms in a compound.

Empirin A trade name for aspirin, an analgesic and antipyretic agent.

Empracet A trade name for a combination drug containing acetaminophen and codeine phosphate, used as an analgesic and antipyretic agent.

EMS (mol wt 124) Abbreviation for ethylmethane sulfonate, an alkylating agent.

$$CH_3SO_3C_2H_5$$

EMSA Abbreviation for electrophoretic mobility-shift assay.

EMU Abbreviation for electromagnetic unit

Emulsan A polymer bioemulsifier produced by *Acinetobacter calcoaceticus*, consisting of *N*-acetylgalactosamine and *N*-acetylhexosaminuronic acid and fatty acid.

Emulsification The formation of an emulsion; a suspension of two immiscible liquids.

Emulsifying Agent Substance capable of promoting the formation and stabilization of an emulsion.

Emulsoil A trade name for castor oil, which promotes accumulation of fluid in the colon and small intestine.

EMU-V A trade name for erythromycin base, an antibiotic that inhibits bacterial protein synthesis by binding to 50S ribosomes.

E-Mycin A trade name for erythromycin, an antibiotic that inhibits bacterial protein synthesis by binding to 50S ribosomes.

Emylcamate (mol wt 145) An anxiolytic agent.

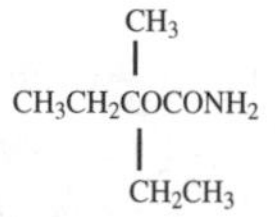

ENA Abbreviation for 1. extractable nuclear antibody; 2. extractable nuclear antigen.

ENaC Abbreviation for epithelial Na^+ channel.

Enalapril (mol wt 376) An inhibitor for angiotensin converting enzyme.

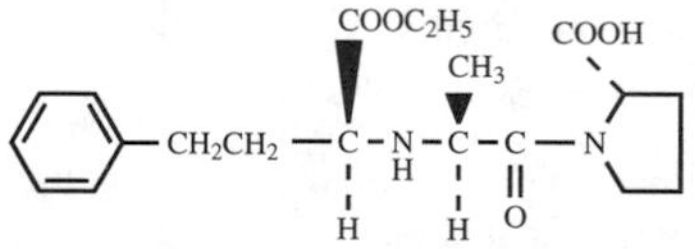

Enalaprilat (mol wt 384) An inhibitor for nonsulfhydryl dipeptide angiotensin converting enzyme and an antihypertensive agent.

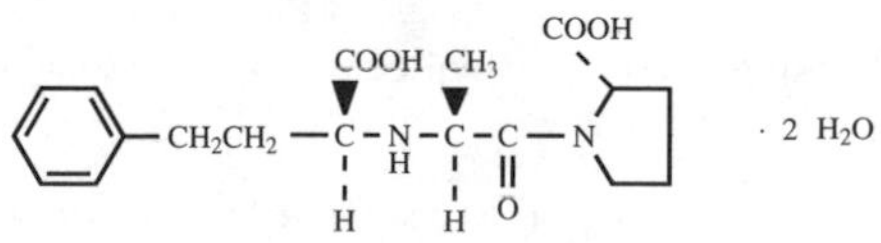

Enallypropymal (mol wt 224) A sedative-hypnotic agent.

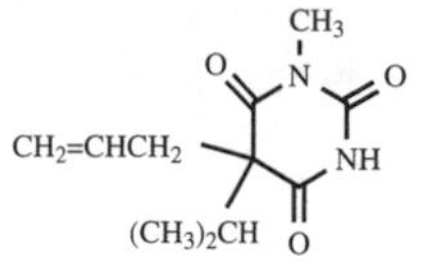

Enamel The hard, calcified substance that covers a tooth.

Enanthem A rash or eruption on a mucous membrane.

Enanthotoxin (mol wt 258) A toxin from a toxic plant *Oenanthe crocata* (Umbelliferae).

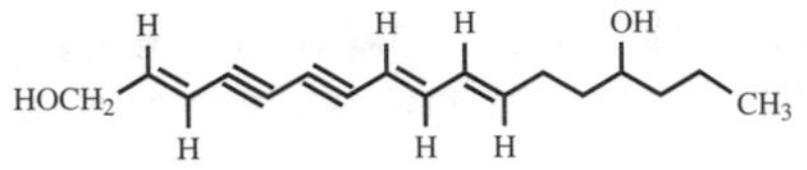

Enantiomers Mirror-image, nonsuperimposable isomers.

Enbrel A trade name for etanercept, an antiarthritis drug.

Encainide (mol wt 352) An antiarrhythmic agent.

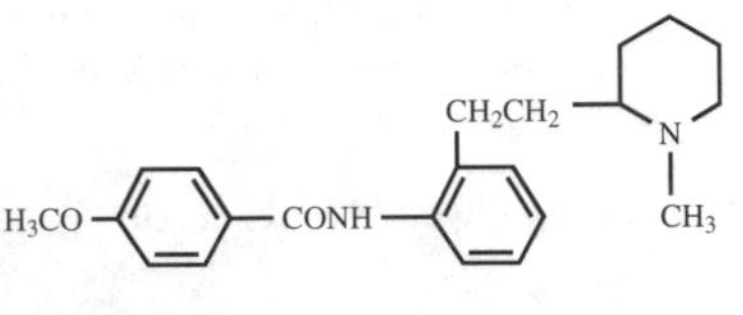

Encapsidation Incorporation of viral nucleic acid into the viral capsid.

Encephalic Pertaining to the brain.

Encephalitis Inflammation of the brain.

Encephalitozoon A genus of protozoa (class Microsporea).

Encephalography X-ray examination of the brain.

Encephalomeningitis Inflammation of the brain and its coverings (meninges).

Encephalomyelitis Inflammation or disease of the brain and spinal cord.

Encephalomyocarditis Diseases or inflammation of the brain, spinal cord, and heart.

Encephalon The brain.

Enchondroma A tumor consisting of cartilaginous tissue.

Encyst To form or to become enclosed in a cyst.

Encystation A process of forming a cyst or becoming enclosed in a capsule.

3' End Referring to the end of a linear polynucleotide chain at which the 3'-hydroxyl group of the nucleoside residue is normally free and not linked to other nucleoside residue.

5' End Referring to the end of a linear polynucleotide chain at which the 5'-hydroxyl group of the nucleoside residue is not linked to other nucleoside residue but is normally phosphorylated.

End Filling The conversion of the sticky ends of a DNA fragment to blunt ends by synthesis of a new single strand complementary to the single-stranded extension in the sticky ends.

End Group Analysis Analysis of the characteristics of the end groups of a polymer, e.g., N-terminal analysis of proteins.

End Labeling The attachment of a radioactive group to the end of a polymer, e.g., labeling of 5¢-end of a DNA or an RNA with ^{32}P.

End Plate The terminal portion of a motor nerve fiber that transmits nerve impulses to the muscle.

End Plate Potential A potential formed at the neuromuscular junction following release of the neurotransmitter from the nerve ending.

End Point Dilution Assay A method of quantifying a virus sample by diluting the virus sample to a point where it produces 50% infection in the test subject.

End Point Titration A method of assaying the concentration of a specific antibody or antigen in a given sample by determining the highest dilution of the antibody that gives a positive reaction under the specified test conditions.

End Product The final product of a chain metabolic reaction.

End Product Efflux The outflow of the end product of a chain of metabolic reactions from a growing cell.

End Product Inhibition The inhibition of the initial enzymatic reaction in a chain of metabolic reactions by the end product of the chain reactions (also known as feed-back inhibition).

Endal HD A trade name for a combination drug containing hydrocodone bitartrate, phenylephrine hydrochloride, and chlorpheniramine.

Endamoeba A genus of cyst-forming amoebae, which is parasitic in the gut of invertebrates.

Endantadine Synonym of amantadine hydrochloride, an antiviral agent.

Endarteritis An inflammation of the arterial intima.

Endemic A disease that is present in a community at all times but only in small numbers of cases.

Endemic Goiter Enlargement or swelling of the thyroid gland due to the lack of iodine in the diet.

Endemosarca A genus of endoparasitic, plasmodium-forming eukaryotic organisms that are parasitic in ciliates.

Endep A trade name for amitriptyline hydrochloride, an antidepressant.

Endergonic Reaction An energy-requiring reaction characterized by a positive free energy change.

Endo- A prefix meaning inside.

Endoamylase Any amylase that hydrolyzes nonterminal glycosidic linkages of polysaccharide.

Endobenzyline Bromide (mol wt 410) An anticholinergic agent.

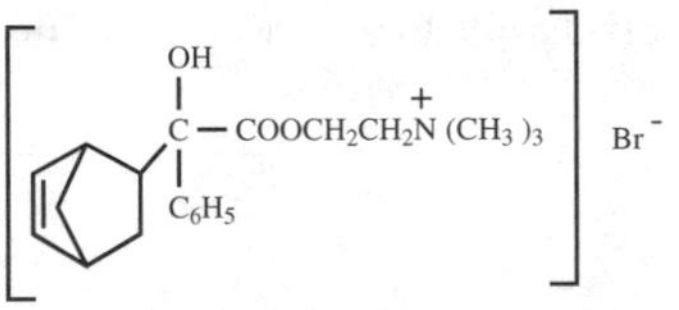

Endobiotic Living within a host organism.

Endocan A trade name for a combination drug containing aspirin and oxycodone hydrochloride, used as an analgesic agent.

Endocarditis Inflammation of the inner membrane of the heart due to a viral or bacterial infection.

Endocardium The membrane lining of the heart chamber.

Endocet A trade name for a combination drug containing acetaminophen and oxycodone hydrochloride, used as an analgesic agent.

Endocrine Gland Ductless, hormone-secreting gland, e.g., pituitary, thyroid, parathyroid, and adrenal gland.

Endocrine Pancreas The portion of pancreas that releases hormones.

Endocrine System The network of ductless glands that release hormones into the bloodstream.

Endocrinology The science that deals with structure and function of the endocrine system.

Endocytosis The uptake of material by a cell through the formation of a membrane-bound endocytotic vesicle, e.g., pinocytosis and phagocytosis.

Endodan A trade name for a combination drug containing hydrocodone bitartrate, hematropine methylbromide used as an antitussive agent.

Endodeoxyribonuclease See restriction endonuclease.

Endoderm A germ layer remote from the surface of the embryo that gives rise to the internal tissues.

Endodermis A sheath of cells surrounding the vascular bundle of the root of a plant.

Endoenzyme 1. Enzymes that are formed within the cell and not excreted into the medium (also called intracellular enzyme). 2 Any enzyme that cleaves the internal linkages in a polymer chain.

EndoF Abbreviation for endoglycosidase F.

Endogenote 1. The segment of chromosome in a partially diploid bacterial cells that is homologous to the chromosome transmitted by the donor cell. 2. The recipient bacterial cell's own genetic material into which the donor DNA can integrate.

Endogenous Originating from within.

Endogenous Infection Infection caused by an organism that is part of the normal body microflora.

Endogenous Pyrogen A fever-inducing substance released by the cells, e.g., interleukine-1 produced by mononuclear phagocytes.

Endogenous Respiration The use of stored reserves to maintain an organism in the absence of growth.

Endogenous Retrovirus The provirus form of a retrovirus.

Endogenous Virus The provirus form of a virus.

Endoglucanase An endoglucosidase that catalyzes the cleavage of internal glucosidic bonds in a glucan.

Endo-1,3-β-D-glucosidase The enzyme that catalyzes the hydrolysis of 1,3-β-D-glucosidic linkages in a 1,3-β-D-glucan.

Endoglycoceramidase See Endoglycosylceramidase.

Endoglycosidase An enzyme that hydrolyzes the internal glycosidic linkages in a polysaccharide.

Endoglycosylceramidase The enzyme that catalyzes the following reaction:

$$\text{Oligoglycosylglucosylceramide} + H_2O \rightleftharpoons \text{Oligoglycosylglucose} + \text{ceramide}$$

EndoH Abbreviation for endoglycosidase H.

EndoHf Abbreviation for endoglycosidase Hf.

Endolimax A genus of parasitic protozoa.

Endolithic Living within a rock.

Endolor Esgic A trade name for a combination drug containing acetaminophen, caffeine, and butalbital.

Endolyn-78 A glycoprotein present in substantial amounts in the membranes of endosomes and lysosomes.

Endolysin A substance that breaks bacterial cell walls, e.g., lysozyme.

Endometritis Inflammation of the endometrium.

Endometrium The inner mucous membrane lining of the uterus.

Endomitosis Chromosomal replication that is not accompanied by either nuclear or cytoplasmic division.

Endomyocarditis An inflammatory disorder of the muscle and lining membrane of the heart.

Endone A trade name for oxycodone hydrochloride, an analgesic agent.

Endonexin A calcium-dependent membrane-binding protein located on the endoplasmic reticulum of fibroblasts.

Endonuclease An enzyme that cleaves phosphodiester bonds within a nucleic acid strand (see also restriction endonuclease).

Endonucleobiosis Symbiont that occurs within the nucleus of the host.

Endoparasite A parasite that lives inside its host.

Endopectin Lyase The enzyme that catalyzes the eliminative cleavage of polysaccharides containing β-mannuronate residues.

Endopeptidase The enzyme that catalyzes the internal cleavage of peptide bonds in a polypeptide.

Endopeptidase K A protease from *Tritirachium album* (mold) that catalyzes the hydrolysis of keratin with subtilisin-like specificity.

Endopeptidase La A protease from *E. coli* that catalyzes the hydrolysis of protein in the presence of ATP.

Endoperoxide Isomerase Synonym of prostaglandin-E synthetase.

Endophthalmitis Inflammation of the interior of the eye.

Endophyte An organism that is parasitic partly or wholly within a plant.

Endoplasm The inner, granule-rich cytoplasm of amoeba.

Endoplasmic Reticulum (ER) A network of membranous tubules and flattened sacs in the cytoplasm of eukaryotic cells that is composed of ribosome-studded (rough ER) and ribosome-free (smooth ER) regions.

Endoplasmin Proteins in mammalian microsomal preparations (100-fold more concentrated in the endoplasmic reticulum than elsewhere.

Endopolygeny Asexual reproduction in which many daughter cells form within the parent cell.

Endopolyphosphatase The enzyme that catalyzes the following reaction:

$$\text{Polyphosphate} + nH_2O \rightleftharpoons (n+1)\text{-Oligophosphates}$$

Endoprotease Glu-C A serine protease with preferential cleavage of peptide bonds involving COOH groups of glutamate or aspartate

Endoribonuclease I See pancreatic ribonuclease.

Endoribonuclease IV The enzyme that catalyzes the endonucleolytic cleavage of poly(A) to fragments terminated by 3′-hydroxyl and 5′-phosphate groups.

Endoribonuclease H The enzyme that catalyzes the endonucleolytic cleavage of RNA yielding 5′-phosphomonoester.

β-Endorphin Acetyltransferase The enzyme that catalyzes the following reaction:

$$\text{Acetyl-CoA} + \text{peptide} \rightleftharpoons \text{CoA} + \text{N-acetylpeptide}$$

Endorphins A group of neuropeptides that are endogenous ligands of the opiate receptors. Endorphins prevent pain and invoke a feeling of euphoria.

Endoscope A lighted instrument used for interior examination of a body cavity or organ.

Endoscopy The interior examination of a body cavity or organ, e.g., stomach.

Endoskeleton A skeleton buried within the soft tissues of an animal, e.g., the spicules of sponges, the plates of exhinoderms, and the bony skeletons of vertebrates.

Endosome A membrane vesicle formed by endocytosis, e.g., phagosome or pinosome.

Endosmosis The osmatic passage of fluid into a cell or vessel.

Endosperm A nutrient-rich structure in seed plants formed within the embryo sac.

Endospore Spore that is formed intracellularly by the parent cell or formed under the condition of nutrient limitation (e.g., endospore formed within the bacterial cell).

Endosporium See endospore.

Endosymbiont An organism that establishes a symbiotic relationship within a eukaryotic cell or tissue of another organism.

Endosymbiosis A type of symbiosis in which one organism is found within another organism.

Endosymbiotic Hypothesis The theory that self-replicating eukaryotic organelles (e.g., mitochondria or chloroplast) arose when a prokaryote established an endosymbiotic relationship with the eukaryotic ancestor and then evolved into eukaryotic organelles.

Endosymbiotic Infection An infection of cells by viruses in which viral replication without cytopathic effect to the cell occurs.

Endosymbiotic Theory See endosymbiotic hypothesis.

Endothelins Peptides produced by endothelium cells.

Endothelium A layer of flattened cells derived from endoderm that lines the internal surfaces of the blood vessels and lymph vessels in vertebrates.

Endothelium Cell Cells derived from endoderm.

Endothermic Reaction A reaction that requires energy. The endothermic reaction has a positive enthalpy change.

Endotherms Animals that use metabolic energy to maintain a constant body temperature.

Endothiapepsin A protease with the similar specificity as pepsin from *Ascomycete Endothia parasitica.*

Endotoxin B Referring to a protein toxin that is the cause of staphytococcal food poisoning.

Endotoxins The heat-stable lipopolysaccharides of the outer membranes of certain Gram-negative bacteria that are released after lysis of bacterial cells.

Endotrypanum A genus of protozoa (family Trypanosomatidae) parasitic in the erythrocytes of sloths.

Endoxan-Asta A trade name for cyclophosphamide, an alkylating agent that cross-links strands of cellular DNA interfering with transcription.

Endoxylanase Synonym of endo-1,4-β-xylanase.

Endralazine (mol wt 269) An antihypertensive agent.

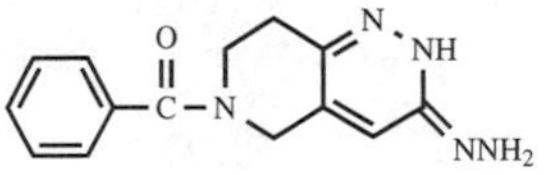

Endrate A trade name for edetate disodium, a chelating agent.

Enduracididine (mol wt 172) A naturally occurring nonprotein amino acid.

Enduracidin A peptide antibiotic produced by *Streptomyces fungicidicus.*

Enduracin A cyclodepsipeptide antibiotic produced by *Streptomyces fungicidicus.*

Enduron A trade name for methyclothiazide, a diuretic agent.

Enduronyl A trade name for a combination drug containing methyclothiazide and deserpidine, used as an antihypertensive agent.

ENE Abbreviation for ethyl-nor-epinephrine.

-ene A suffix denoting the presence of one or more carbon to carbon double bonds in an organic compound.

Enediol Any acyclic organic compound in which there is a hydroxyl group attached to each of the two carbons that are linked by a double bond.

Enema An injection of fluid into the rectum for the purpose of cleaning or treatment.

Ener-B A trade name for vitamin B_{12} (cyanocobalamin).

Energy Capacity to do work or ability to cause specific changes.

Energy Charge The fractional degree to which the ATP/ADP/AMP system is filled with high energy phosphate groups.

Energy Coupling The transfer of energy from one process to another.

Energy-Rich Bond A chemical bond that releases a large quantity of energy upon cleavage, e.g., energy-rich phosphate bonds in ATP.

Enfenamic Acid (mol wt 241) An anti-inflammatory and analgesic agent.

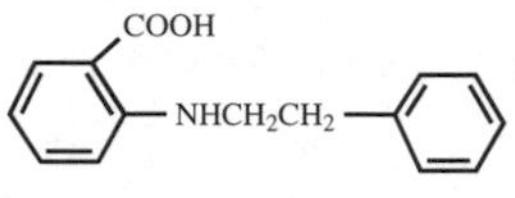

Enflurane (mol wt 184) An anesthetic agent.

$$C_3H_2ClF_5O$$

Engerix-B A trade name for a recombinant hepatitis B vaccine.

Engram The location where a memory is stored in the brain.

Enhancer A regulatory sequence in DNA that increases the transcription of a gene. It is the binding site for gene-regulatory proteins.

Enhancer Element A specific sequence on DNA that contains a binding site(s) for transcription factors and confers transcriptional efficiency and specificity for the associated structural gene. The enhancer sequence can be upstream or downstream from the transcription unit.

Enhancer Trapping A technique that employs a vector containing a transposable element, which can be used to find tissue-specific enhancer elements.

Enilconazole (mol wt 297) An antifungal agent.

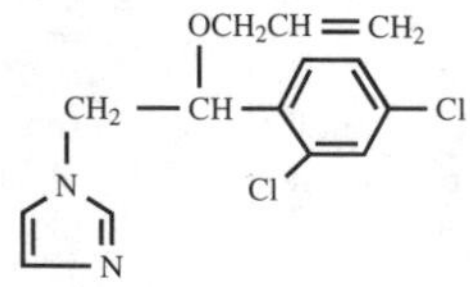

Enkephalins The pentapeptide pain-relieving hormones produced in the brain and anterior pituitary.

Enlon A trade name for edrophonium chloride, a cholinergic agent.

Enniatins Cyclic depsipeptide antibiotics and monocationic ionophores produced by species of *Fusarium*.

2-Enoate Reductase The enzyme that catalyzes the following reaction:

$$\text{Butanoate} + \text{NAD}^+ \rightleftharpoons \text{2-Butenoate} + \text{NADH}$$

Enocitabine (mol wt 566) An antineoplastic agent.

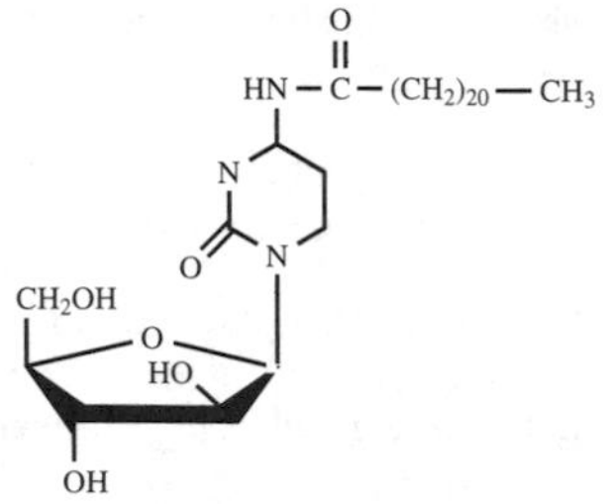

Enol Any acyclic organic compound having a hydroxyl group attached to either of the two carbons that are linked by a double bond.

Enolase The enzyme that catalyzes the following reaction:

$$\text{2-Phospho-D-glycerate} \rightleftharpoons \text{Phosphoenol pyruvate} + H_2O$$

Enology The science that deals with wine and wine making.

eNOS Abbreviation for endothelial NOS.

Enovid A trade name for a birth control pill containing estrogen, mestranol, and progestin.

Enovil A trade name for amitriptyline hydrochloride, an antidepressant.

Enoxacin (mol wt 320) A flurorinated quinolone antibiotic that inhibits bacterial DNA gyrase and prevents DNA replication.

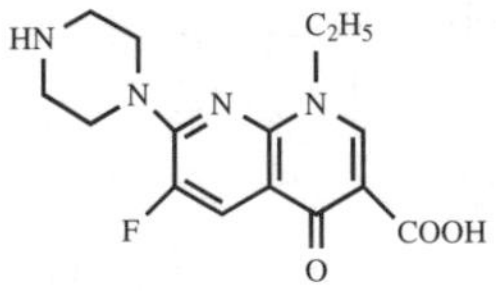

Enoxaparin A fragment of heparin possessing antithrombotic activity.

Enoximone (mol wt 248) A cardiotonic agent and phosphodiesterase inhibitor with vasodilating and iontropic activity.

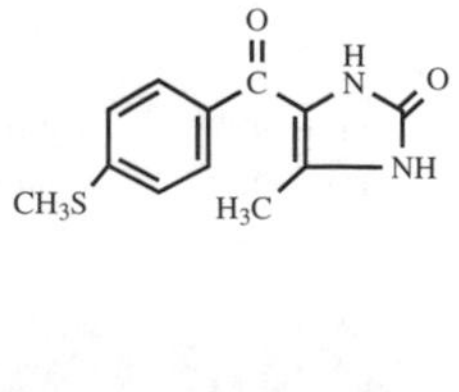

Enoxolone (mol wt 471) An anti-inflammatory agent.

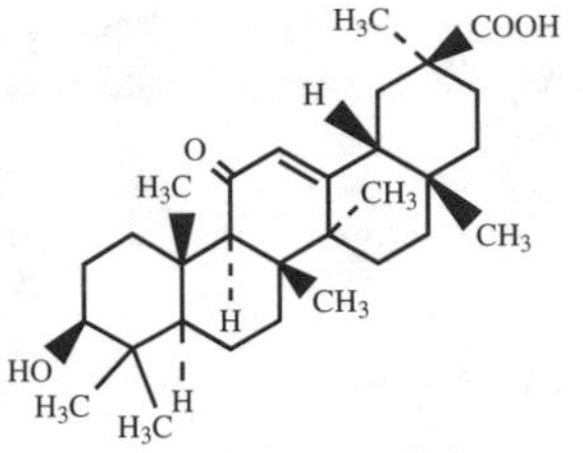

Enoyl-ACP Reductase (NAD$^+$-specific) The enzyme that catalyzes the following reaction:

$$\text{Acyl-ACP} + \text{NAD}^+ \rightleftharpoons \textit{trans}\text{-2,3-dehydroacylacyl-ACP} + \text{NADH}$$

Enoyl-CoA Hydratase The enzyme that catalyzes the following reaction:

$$\text{3-Hydroxyacyl-CoA} \rightleftharpoons \textit{trans}\text{-2 or 3-enoyl-CoA} + H_2O$$

Enprofylline (mol wt 194) A bronchodilator and an antiasthmatic agent.

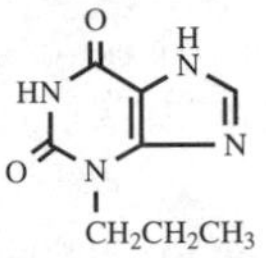

Enprostil (mol wt 400) An antiulcerative and an antisecretary agent.

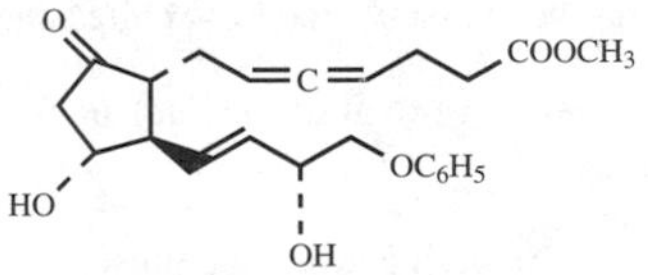

Enrichment Culture A technique to increase the number of organisms in a culture relative to the number of other types of organisms by providing favorable growth conditions for the organism or providing conditions unfavorable for contaminating organisms.

Enrofloxacin (mol wt 359) A fluorinated quinolone antibiotic.

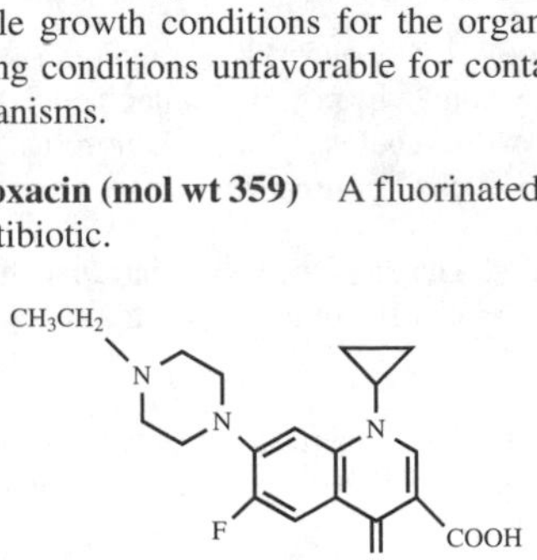

Ensifer A genus of Gram-negative, chemoorganotrophic bacteria.

Entactin A sulfated calcium-binding protein found in the basement membrane and involved in cell adhesion.

Entacyl A trade name for piperazine adipate, an anthelmintic agent.

Entamoeba A genus of parasitic intestinal protozoa.

Enterectomy Surgical removal of part of the intestine.

Enteric Pertaining to the intestine.

Enteric Coated Tablet Drug tablets that are coated with a substance that enables them to pass through the stomach to the intestine unchanged.

Enteric Fever Any typhoidlike human disease caused by *Salmonella paratyphi* or *S. typhimurium*.

Enteritis Inflammation of the lining of the intestine.

Enterobacter A genus of Gram-negative bacteria of Enterobacteriaceae.

Enterobacteria 1. Bacteria of the family Enterobacteriaceae. 2. Bacteria found in the intestine.

Enterobacterial Common Antigen A heteropolysaccharide occurring in the outer membrane in wild type bacteria of the family Enterobacteriaceae.

Enterobactin (mol wt 670) A physiologically active, macrocyclic, iron-sequestering agent involved in microbial transport and metabolism of iron.

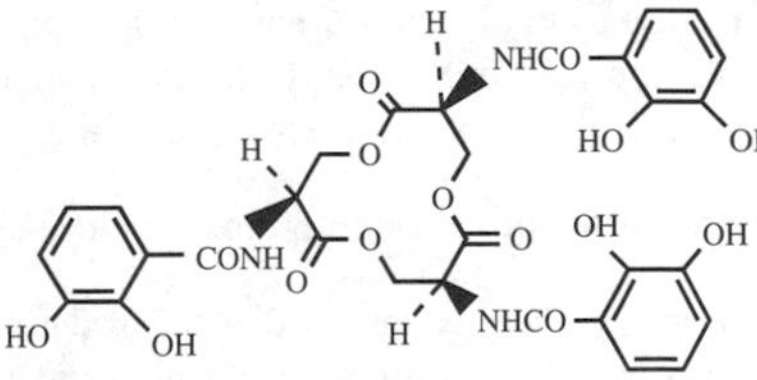

Enterobiasis An infection by *Enterobius vermicularis* (pinworm).

Enterobius vermicularis A parasitic worm (e.g., pinworm, threadworm).

Enterochelin See enterobactin.

Enterochromaffin Cell Any gut endocrine cell containing biogenic monoamines that gives a positive chromaffin reaction.

Enterocolitis Swelling of the large and small intestines.

Enterocrinin A peptide hormone that stimulates duodenal and jejunal secretion of digestive enzyme.

Enterocytes Cells of the intestinal epithelium.

Enterocytozoon A genus of protozoa.

Enterogastrone A hormone from the duodenum that inhibits gastric activity.

Enterokinase See enteropeptidase.

Enterolith A stone within the intestine.

Enteromorpha A genus of macroscopic green algae.

Enteromycin (mol wt 188) An antibiotic from *Streptomyces albireticulies*.

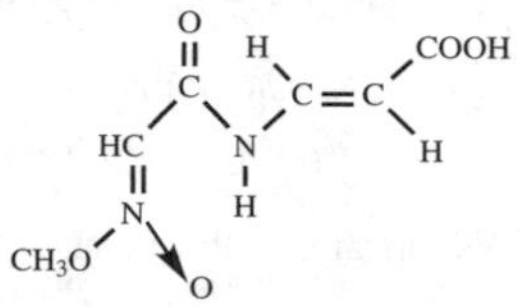

Enteropathogenic Pathogenic for the intestine.

Enteropeptidase Protease secreted by the intestine that catalyzes the activation of trypsinogen by selective cleavage of peptide bonds between cysteine and isoleucine.

Enteroscope An illuminated optical instrument used to inspect the interior of the small intestine.

Enterotoxigenic An organism that produces enterotoxins.

Enterotoxin A group of bacterial exotoxins that act on the intestinal mucosa and perturb ion and water transport systems causing vomiting and diarrhea, e.g., cholera toxin and food poisoning.

Enterotoxinogenic See enterotoxigenic.

Enterovirus Viruses (family Picornaviridae) that replicate primarily in the mammalian intestinal tract, e.g., coxsackie virus.

Entex A trade name for a combination drug containing phenylephrine hydrochloride, phenylpropanolamine hydrochloride, and guaifenesin, used as an adrenergic agent.

Entex-LA A trade name for a combination drug containing phenlpropanolamine hydrochloride and guaifenesin, used as a bronchodilator.

Enthalpy Heat content of a system.

Enthalpy Change (ΔH) The difference between the energy used to break bonds and the energy gained by the formation of new bonds.

Entner-Doudoroff Pathway (ED Pathway) An alternative pathway for the breakdown of glucose to pyruvate and glyceraldehyde 3-phosphate via 2-keto-3-deoxy-6-phosphogluconate (occurring in some bacteria).

Entocort A trade name for budenoside, a corticosteroid.

Entomogenous Growing in or on insects.

Entomology The science that deals with insects.

Entomopoxviruses A pox virus that infects insects.

Entomotoxin A toxin against insects.

Entozyme Tablets A trade name for a combination drug containing pancreatic pepsin and bile salt, used for promotion of digestion.

Entrapment The immobilization of enzyme, proteins, or cells by trapping them within polymeric meshes during the process of polymerization, e.g., entrapment of enzymes or cells in polyacrylamide gel.

Entrophen A trade name for aspirin, an antipyretic, analgesic, anti-inflammatory, antiplatelet agent.

Entropy The randomness or disorder in a system. The higher the entrophy, the more disorder in the system.

ENU Abbreviation for ethylnitrosourea.

Enucleate To remove the nucleus from a cell.

Enucleated Cell A eukaryotic cell in which the nucleus has been inactivated (also called cytoplast).

Enucleation Removal of the nucleus from a eukaryotic cell.

Enulose A trade name for lactulose, a laxative.

***Env* Glycoprotein** The glycoprotein found in the lipoprotein envelope of enveloped viruses, e.g., retroviruses.

Envelope 1. The outer lipid-containing layer possessed by some virions (also called peplos). 2. The envelope that encloses the chromosome and defines the nuclear compartment.

Enveloped Viruses Any virus in which the nucleoprotein core is surrounded by a lipoprotein envelope.

Envelysin The enzyme that catalyzes the hydrolysis of proteins of the fertilization envelope and dimethylcasein.

Enviomycin (mol wt 686) A polypeptide antibiotic from *Streptomyces griseoverticillatus*.

Enzootic The moderate prevalence of a disease in a given animal population.

Enzyme Bioactive protein that catalyzes the biochemical reactions in the living cell. Occasionally, nucleic acid may also act as a biological catalyst, e.g., catalytic RNA.

Enzyme I A soluble enzyme involved in the transport of sugar across bacterial membrane.

Enzyme II A membrane-bound enzyme involved in the transport of sugar across the bacterial membrane.

Enzyme III An enzyme involved in the transport of sugar across the plasma membrane.

Enzyme Blotting A technique for transfer or blotting of enzyme to nitrocellulase paper and for subsequent detection by specific reaction with substrate.

Enzyme Detergent Any detergent that incorporates an enzyme to assist its cleansing action.

Enzyme Electrode A combination of an immobilized enzyme with an ion-selective electrode sensor, used for measuring the concentration of reactants or the product from an enzymatic reaction.

Enzyme Immunoassay (EIA) An immunoassay in which an antibody-enzyme complex (e.g., antibody-peroxidase) is used as a marker to assay the presence of specific antigens or antibodies.

Enzyme Induction The synthesis of an enzyme in response to the presence of an enzyme inducer.

Enzyme Kinetics Quantitative analysis of the rate of enzyme reactions and the manner in which they are influenced by a variety of factors.

Enzyme Linked Immunosorbent Assay See ELISA.

Enzyme Paper Graft An enzyme immobilized on filter paper which is frequently impregnated with indicators for making analytical devices.

Enzyme pH Electrode An enzyme electrode that measures pH changes.

Enzyme Reactor A device employing immobilized enzyme for synthetic or processing reactions.

Enzyme Unit A unit for measurement of enzyme concentration. One unit is defined as amount of enzyme that catalyzes the transformation of 1 μmole of substrate per minute.

Enzymoblotting See enzyme blotting.

Enzymology The science that deals with enzymes.

Eoff Process The conversion of dihydroxyacetone-3-phosphate to glycerol under alkaline condition.

Eosin B (mol wt 624) A dye for staining tissue.

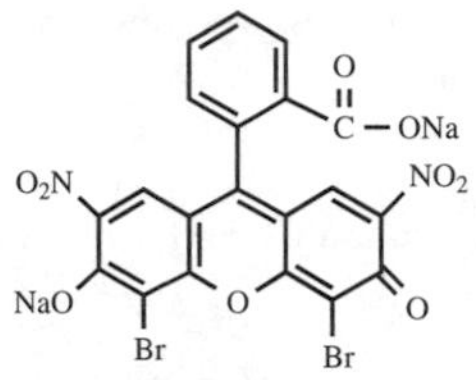

Eosin Y (mol wt 692) A dye used for the spectrophotometric determination of silver.

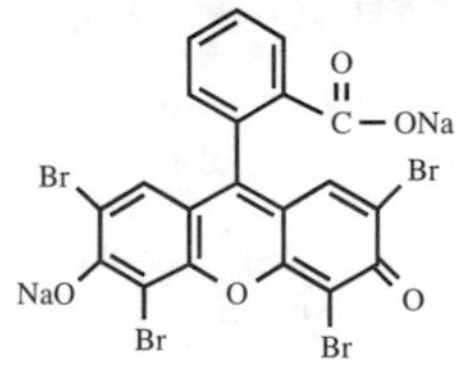

Eosine Variant spelling of eosin.

Eosinophil A two-lobed white blood cell that can be easily stained with neutral dye.

Eosinophil Cationic Protein An arginine-rich protein in the granules of eosinophils that possesses a destructive effect on *schistosomula in vitro*.

Eosinophil Chemotactic Factor See eosinophil chemotactic peptide.

Eosinophil Chemotactic Peptide Tetrapeptides released by mast cells to attract and activate eosinophil such as:

Val-Gly-Ser-Glu

Ala-Gly-Ser-Glu

Eosinophilia The presence of a large number of eosinophils in the circulation due to parasitism by helminths or allergy.

Eosinophilic Readily stained with eosin dye.

Eosinophilopoietin A peptide released by T lymphocytes that regulates eosinophil development in the bone marrow.

EP Abbreviation for electrophoresis.

E4P Abbreviation for erythrose 4-phosphate.

EPA Abbreviation for eicosapentaenoic acid with 20 carbons and 5 double bonds.

E-Pam A trade name for diazepam, an antianxiety agent.

Epanolol (mol wt 369) An antihypertensive and antianginal agent.

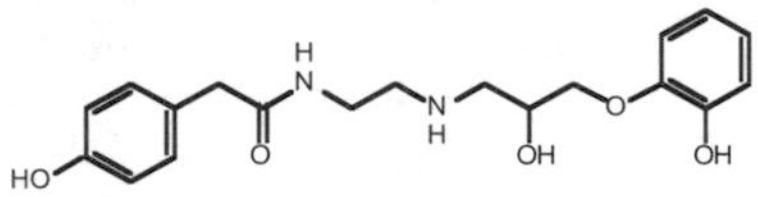

EPE Abbreviation for erythropoietin-producing enzyme.

Ependyma The inner lining of the brain and spinal cord cavities.

Ependymal Cells Cells that line the cavities in the CNS.

Ependymoma A type of brain tumor.

Eperisone (mol wt 259) A skeletal muscle relaxant.

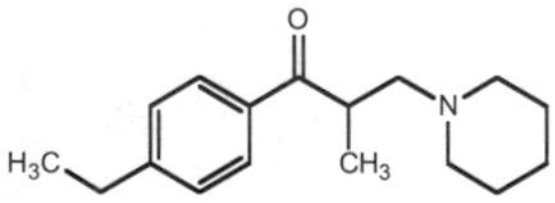

Eperythrozoon A genus of Gram-negative bacteria of the family Anaplasmataceae.

Eperythrozoonosis Any disease caused by an *Eperythrozoon* species.

EPF Abbreviation for endothelial proliferating factor.

Ephed II A trade name for ephedrine sulfate, a bronchodilator.

Ephedrine (mol wt 165) A bronchodilator that stimulates alpha- and beta-adrenergic receptors.

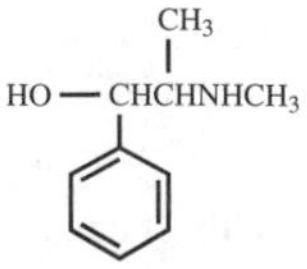

Ephedrine Sulfate A derivative of ephedrine and a bronchodilator.

EPI Abbreviation for echo planar imaging.

Epiblast The outer layer of cells in an embryo.

Epicillin (mol wt 351) A semisynthetic antibiotic related to penicillin.

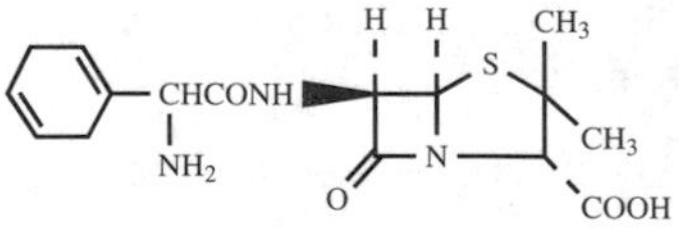

Epidemic The unusual prevalence or sudden appearance of a disease in a given geographical region for a limited period of time.

Epidemic Keratoconjunctivitis A highly infectious disease of the eye caused by an adenovirus.

Epidemiology The science that deals with the interrelationship among pathogens, environments, and host populations.

Epidermal Growth Factor Peptide mitogen that stimulates epidermal cells and a variety of other cell types to divide and proliferate. It binds to the plasma membrane and stimulates the activity of protein kinases.

Epidermis 1. The outer layer of the skin. 2. The dermal tissue system of plants.

Epidermomycosis The fungal infection of the skin.

Epideroblast A type of animal cell that gives rise to the epidermis.

Epifoam A trade name for hydrocortisone acetate, an anti-inflammatory agent.

Epifrim A trade name for epinephrine, a hormone and a bronchodilator.

Epifrin A trade name for epinephrine hydrochloride, a hormone.

Epigenetic All processes relating to the expression and interaction of genes.

Epiglycanin Mucinlike molecule on the surface of murine mammary carcinoma cells that may mask histocompatibility antigens.

Epilepsy Any one of a group of brain disorders characterized by the recurrent seizures that have sudden onset.

Epiligrin An extracellular matrix protein secreted by the cultured epidermal keratinocytes.

Epilim A trade name for valproate sodium, an anticonvulsant.

E-Pilo A trade name for a combination drug containing epinephrine bitartrate and pilocarpine hydrochloride, used as a cholinergic agent.

Epimeric Carbon The asymmetric carbon of epimers.

Epimerase The enzyme that catalyzes the interconversion between two epimers.

Epimers The isomers that differ in configuration at one asymmetric carbon atom (the epimeric carbon).

Epimestrol (mol wt 302) An anterior pituitary activator.

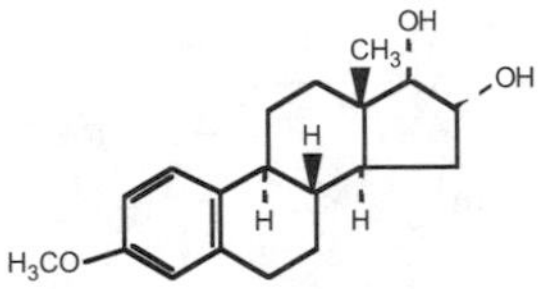

Epimorph A trade name for morphine hydrochloride, an analgesic agent that binds to opiate receptors.

Epinal A trade name for epinephryl borate, a mydriatic agent.

Epinastine (mol wt 249) An antihistaminic agent.

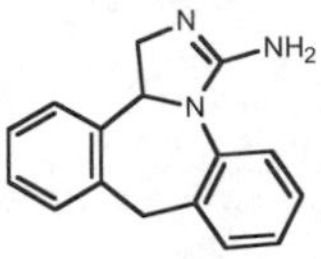

Epinephrine (mol wt 183) Hormone secreted by the adrenal medulla that is capable of regulating carbohydrate metabolism and exerting various effects on the cardiovascular system and muscular tissue. It stimulates alpha- and beta-adrenergic receptors.

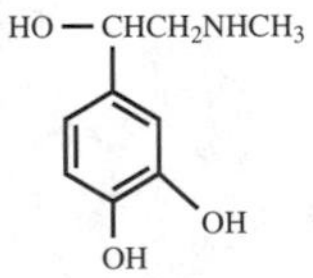

Epinephrine Bitartrate An epinephrine derivative used as a bronchodilator.

Epinephrine Hydrochloride An epinephrine derivative used as a bronchodilator.

Epinephryl Borate An epinephrine derivative used as a mydriatic agent.

Epi-Pen A trade name for epinephrine hydrochloride, a hormone and bronchodilator.

Epirizole (mol wt 234) An analgesic, antipyretic, and anti-inflammatory agent.

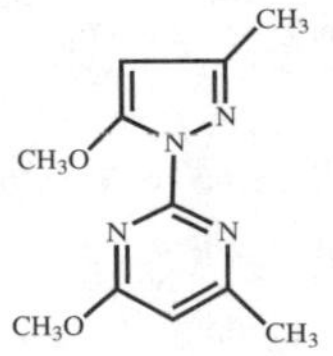

Epirubicin (mol wt 544) An analog of the anthracycline antibiotic doxorubicin.

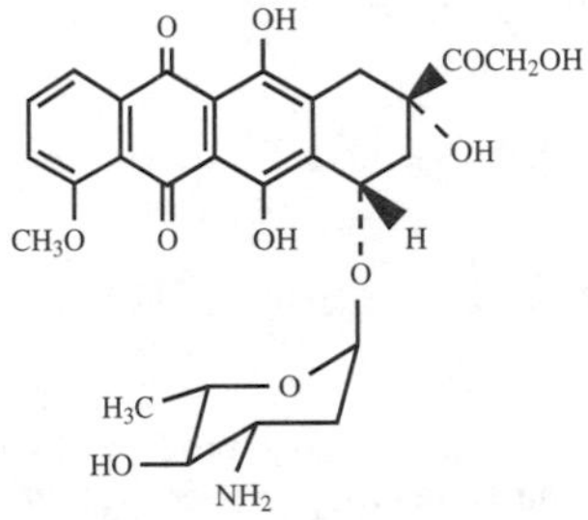

Episome A plasmid that can exist either independently or reversibly integrated into a bacterial chromosome.

Epistasis A phenomenon in which one gene alters the expression of another independently inherited gene.

Epistatic Gene A gene whose expression suppresses or reduces the expression of another nonallelic gene.

Epitectin A glycoprotein found on the surface of human tumor cells.

Epithelial Cells Cells derived from the ectoderm lining the surfaces of the body.

Epithelioma A tumor of the skin.

Epithelium A layer of cells covering or lining the surface of the organs of the body, e.g., lining of the blood vessels.

Epithiazide (mol wt 426) An antineoplastic agent.

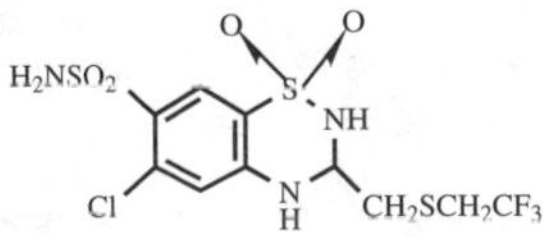

Epitol A trade name for carbamazepine, an anticonvulsant.

Epitope An antigenic determinant present on a complex antigenic molecule and capable of combining with antibody or T cell receptors (also known as an antigenic determinant).

Epitrate A trade name for epinephrine bitartrate, a hormone.

Epival A trade name for divalproex sodium, an anticonvulsant.

Epivir A trade name for lamivudine, an antiviral agent.

Epizoic Living on an animal.

Epizootic A sudden outbreak of a disease in an animal population.

EPMR Abbreviation for electron para-magnetic resonance.

EPMSA Abbreviation for electrophoretic mobility shift assay.

Epoetin Alfa A hormone produced by DNA recombination technology that controls the rate of cell production and functions as a growth and differentiation factor.

Epogen A trade name for epoetin alfa., a DNA recombinant hormone.

Epoprostenol A naturally occurring prostacyclin that inhibits platelet aggregation and has anti-proliferative effects.

Epoxide (mol wt 120) A chromosome-breaking and alkylating agent.

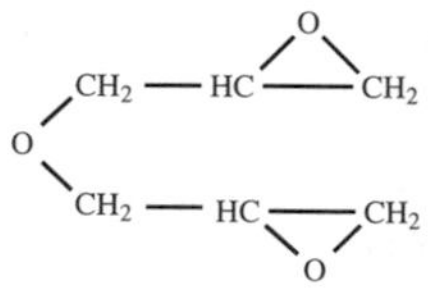

Epoxide Hydratase See epoxide hydrolase.

Epoxide Hydrolase The enzyme that catalyzes the following reaction:

$$\text{An epoxide} + H_2O \rightleftharpoons \text{A glycol}$$

EPR Abbreviation for electron paramagnetic resonance.

Eprazinone (mol wt 381) An antitussive agent.

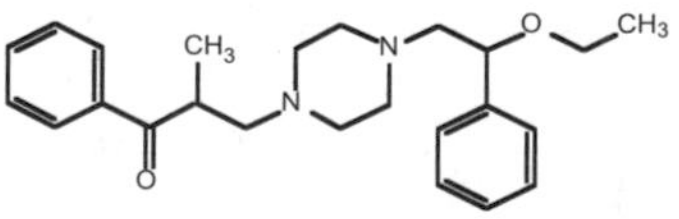

Eprex A trade name for epoetin, a recombinant human erythropoietin (a hormone).

Eprolin A trade name for vitamin E.

Eprosartan (mol wt 425) An antihypertensive agent.

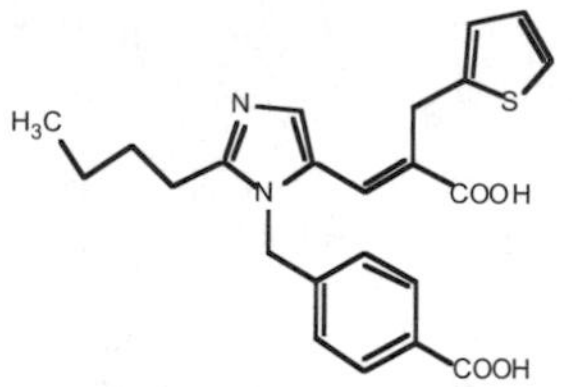

Eprozinol (mol wt 354) A bronchodilator.

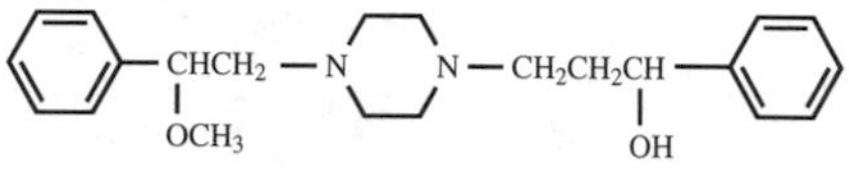

EPRS Abbreviation for electron paramagnetic resonance spectroscopy.

Epsilon The fifth letter of the Greek alphabet.

Epsilon Chain The heavy chain of immunoglobulin E (IgE).

Epsom Salt Referring to magnesium sulfate, used as a laxative.

EPSP Abbreviation for excitatory postsynaptic potential.

Epstein Barr Virus A herpesvirus and a causal agent for mononucleosis and Burkitt's lymophoma.

Eptifibatide (mol wt 832) An antiplatelet drug.

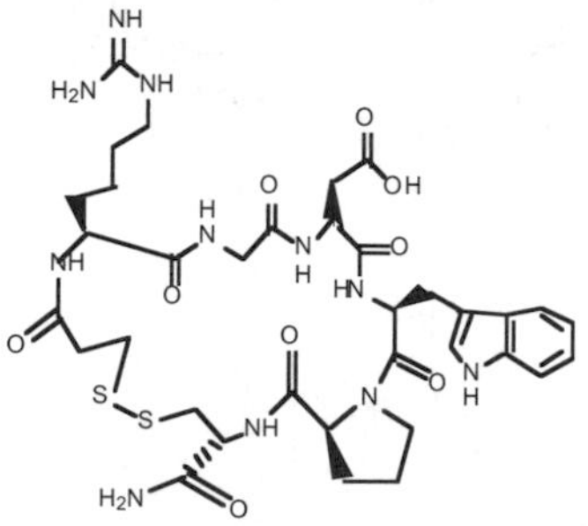

Eq Abbreviation for equation.

Equagesic A trade name for a combination drug containing aspirin and meprobamate.

Equal A trade name for aspartame, a non-nutritious sweetener.

Equanil A trade name for meprobamate, a sedative and antianxiety drug.

Equalactin A trade name for calcium polycarbophil, a laxative.

Equilib Abbreviation for equilibrium.

Equilet A trade name for calcium carbonate, an antiflatulent and antacid agent.

Equilibrium A state in which the forward process and the reverse process occur at the same rate and the free energy is at the minimum at equilibrium.

Equilibrium Constant (K_{eq}) A constant characteristic for each chemical reaction. It is the ratio of product activities to reactant activities for a given chemical reaction when the reaction has reached equilibrium.

Equilibrium Density Centrifugation A ultracentrifugation technique that separates macromolecules on the basis of differences in densities. Banding of the macromolecules with different densities in a linear density gradient medium occurs in a region where the density of medium is equal to the density of the macromolecule (also called isopyknic density gradient centrifugation).

Equilibrium Dialysis A technique for measuring the extent of binding of ligands with macromolecules. Dialysis of known concentrations of ligand and protein until equilibrium is established. At the equilibrium the free unbound ligand on both sides of the dialysis bag are the same; the bound ligand inside the dialysis bag can be calculated, and the number of binding sites on the macromolecule can be determined.

Equilibrium Potential Membrane potential that will offset the effect of the concentration gradient for a given ionic species.

Equilin (mol wt 268) An estrogenic steroid hormone isolated from the urine of pregnant mares.

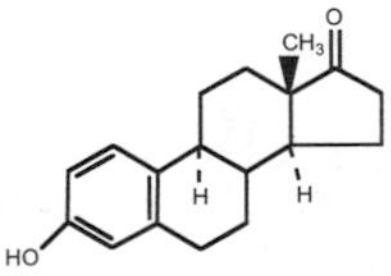

Equine Pertaining to horses.

Equine Encephalitis An insect-transmitted viral disease of horses. Humans are secondary hosts.

Equiv Abbreviation for equivalent.

Equivalence Zone A zone in a serological assay where maximal antigen-antibody precipitation occurs.

ER Abbreviation for 1. endoplasmic reticulum; 2. estrogen receptor.

Er Symbol for the chemical element erbium.

ERA Abbreviation for estrogen receptor assay.

ERα Abbreviation for estrogen receptor alpha.

erA An oncogene originally found in avian erythroblastosis virus that encodes thyroid hormone receptor.

Erabutoxins Neurotoxic peptides isolated from the venom of sea snakes.

Eramycin A trade name for erythromycin, an antibiotic.

***erb*-B** An oncogene in chicken erythro-leukemia and fibrosarcoma that encodes protein kinase (tyrosine) which acts as an epidermal growth factor receptor.

Erbium (Er) A rare chemical element with atomic weight 167, valence 3.

ERC Abbreviation for erythropoietin responsive cells.

Ercaf A trade name for a combination drug containing ergotamine tartrate and caffeine, used as an adrenergic blocker.

ERCC1 Abbreviation for excision repair cross-complementary group 1.

Erdin (mol wt 385) An antibiotic produced by *Aspergillus terreus*.

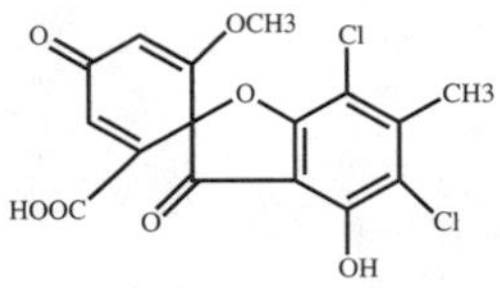

Erdosteine (mol wt 249) A mucolytic agent.

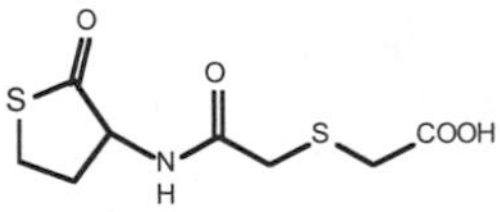

ERE Abbreviation for estrogen-responsive element.

eRF Abbreviation for eukaryotic releasing factor (in protein synthesis).

ERFC Abbreviation for erythrocyte rosette forming cells.

erg A unit of energy, the equivalent of the work done by one dyne of force acting through one cm of distance; 1 erg = 10^{-7} joule.

ERG Abbreviation for electroretinogram.

Ergamisol A trade name for levamisol hydrochloride, an immunosuppressant.

Ergo-caff PB A trade name for a combination drug containing ergotamine tartrate, belladonna alkaloids, and pentobarbital sodium, used as an adrenergic blocker.

Ergocalciferol Referring to Vitamin D_2.

Ergoloid Mesylates The α-adrenergic blockers used for treatment of impaired mental function of the elderly.

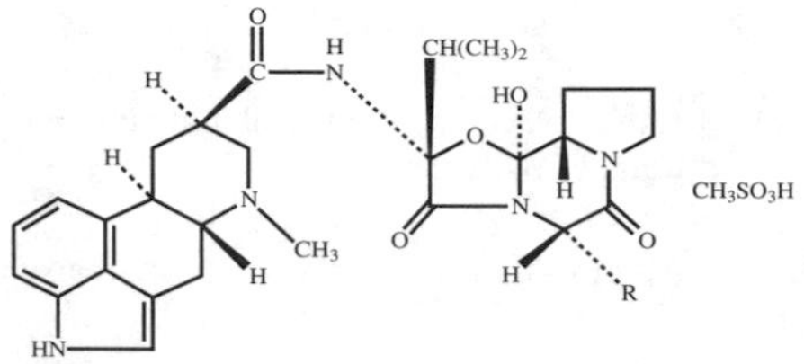

Dihydroergocornine	R = —CH(CH$_3$)$_2$
Dihydroergocristine	R = —CH$_2$—C$_6$H$_5$
Dihydro-α-ergocryptine	R = —CH$_2$CH(CH$_3$)$_2$
Dihydro-β-ergocryptine	R = —CH(CH$_3$)CH$_2$CH$_3$

Ergomar A trade name for ergotamine tratrate, an adrenergic blocker.

Ergonovine (mol wt 325) An oxytocic agent.

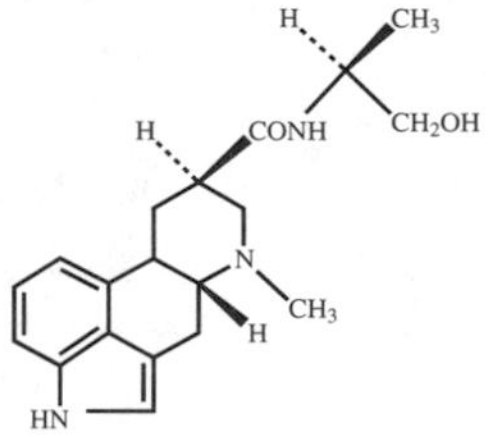

Ergostat A trade name for ergotamine tartrate, an adrenergic blocker.

Ergosterol (mol wt 397) A provitamin D.

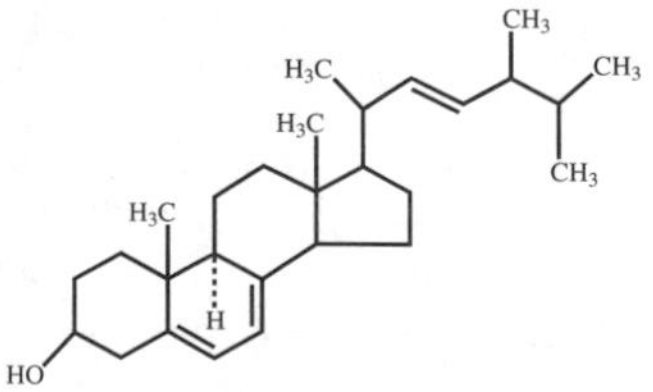

Ergot The dried sclerotium of *Claviceps purpurea* (fungus of Ascomycetes).

Ergot Alkaloids A group of indole alkaloids isolated from fungi (genus *Claviceps).*

Ergotamine (mol wt 582) An ergot alkaloid with antimigraine activity.

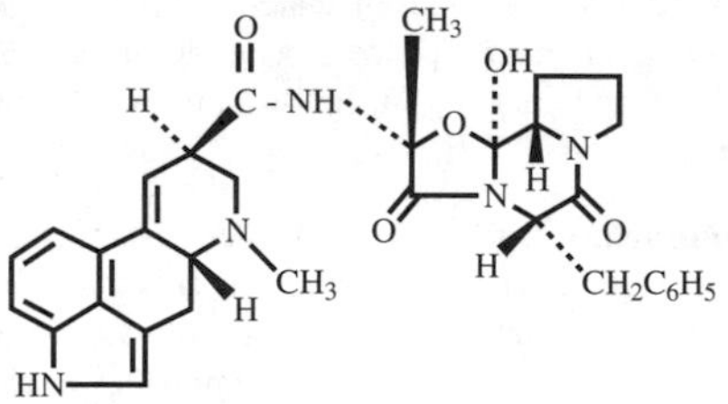

Ergotamine Tartrate A derivative of ergotamine used as an adrenergic blocker, that inhibits the effects of epinephrine and norepinephrine.

Ergothionine (mol wt 229) A naturally occurring nonprotein amino acid found in the ergot of fungi (*Claviceps*).

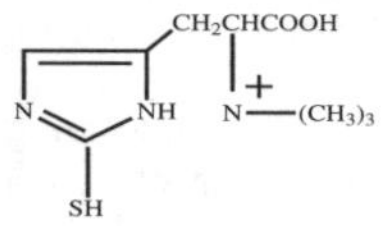

Ergotism A condition of intoxication resulting from the ingestion of grains contaminated by ergot alkaloids produced by *Claviceps purpurea*.

Ergotrate Maleate A trade name for ergonovine, an oxytocic agent that increases the strength, duration, and frequency of uterine contractions.

ErhI A restriction endonuclease from *Erwinia rhaponici* B9 with the following specificity:

W= A or T

ERIA Abbreviation for electro-radio-immunoassay.

ERICA Abbreviation for estrogen receptor immuno-cytochemical assay.

Eridium A trade name for phenazopyridine hydrochloride, an analgesic and antipyretic agent.

Eritadenine (mol wt 253) An anticholesteremic agent from mushrooms (*Lentinus edodes*).

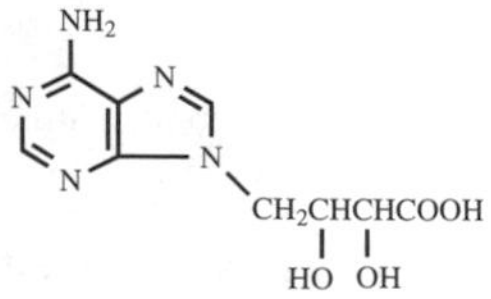

ERK Abbreviation for extracellular-signal-regulated kinase or extracellular regulated kinase.

ERK-1 Abbreviation for extracellular-signal-regulated kinase-1.

ERK-2 Abbreviation for extracellular-signal-regulated kinase-2.

E-rosette Attachment of a cluster of erythrocytes (e.g., sheep red blood cell) around T lymphocytes (e.g., mouse or human T lymphocytes).

ERP Abbreviation for estrogen receptor protein.

ErpI (AvaII) A restriction endonuclease from *Erwinia rhaponici* with the following specificity:

```
      ↓
5′.........GG(A/T)CC.........3′
3′.........CC(T/A)GG.........5′
                   ↑
```

ER-Resident Protein Any protein that is retained by the endoplasmic reticulum.

Erucic Acid (mol wt 339) A fatty acid.

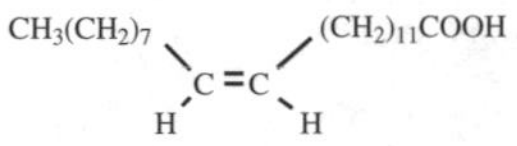

Erwinia A genus of bacteria (family Enterobacteriaceae).

Erybid A trade name for erythromycin, an antibiotic.

Eryc A trade name for erythromycin base, an antibiotic.

Eryc Sprinkle A trade name for erythromycin base, an antibiotic.

Erycette A trade name for erythromycin, an antibiotic.

EryDerm A trade name for erythromycin, an antibiotic.

EryGel A trade name for erythromycin, an antibiotic.

Erymax A trade name for erythromycin, an antibiotic.

Erypar A trade name for erythromycin stearate, an antibiotic.

EryPed A trade name for erythromycin ethylsuccinate, a derivative of erythromycin.

Erysipelas An acute inflammation of the dermal layer of the skin.

Ery-Tab A trade name for erythromycin, an antibiotic.

Erythema Redness or swelling of the skin or mucous membrane.

Erythrasma A chronic skin infection caused by *Corynebacterium minutissimum.*

Erythremia An abnormal increase in the number of erythrocytes in the blood.

Erythritol (mol wt 122) A major photosynthetic product in certain unicellular and filamentous green algae.

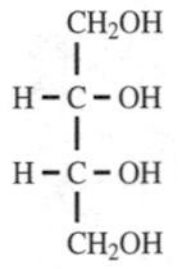

Erythritol Kinase The enzyme that catalyzes the following reaction:

$$\text{ATP} + \text{erythritol} \rightleftharpoons \text{ADP} + \text{D-erythritol 4-phosphate}$$

Erythrobacter A genus of Gram-negative aerobic halophilic, ovoid to rod-shaped, orange- or pink-pigmented bacteria.

Erythroblast The nucleated cell from which an erythrocyte is formed.

Erythroblastoma Tumor derived from erythroblast cells.

Erythroblastosis The presence of erythroblasts (nucleated precursors of erythrocytes) in the blood.

Erythrocin A trade name for erythromycin ethylsuccinate, an antibiotic.

Erythrocruorin Any of a group of respiratory pigments of invertebrate that contain 30 to 400 heme groups per molecule.

Erythrocuprein Synonym of superoxide dismutase.

Erythrocyanosis The mottled purplish discoloration on the legs and thighs.

Erythrocyte A red blood cell.

Erythrocyte Ghost Purified erythrocyte plasma membrane obtained by removal of cytoplasm from the erythrocyte.

Erythrocyte Sedimentation Rate The rate at which red blood cells settle out of the suspension in the blood plasma, measured under standard conditions.

Erythrogenesis See erythropoiesis.

Erythroimmunoassay (EIA) A sensitive immunoassay that employs erythrocytes for detecting and quantifying specific antigens or antibodies.

Erythromid A trade name for erythromycin, an antibiotic.

Erythromycin (mol wt 734) An antibiotic produced by *Streptomyces erythreus* that inhibits protein synthesis. There are three major erythromycins, namely erythromycin A, B, and C.

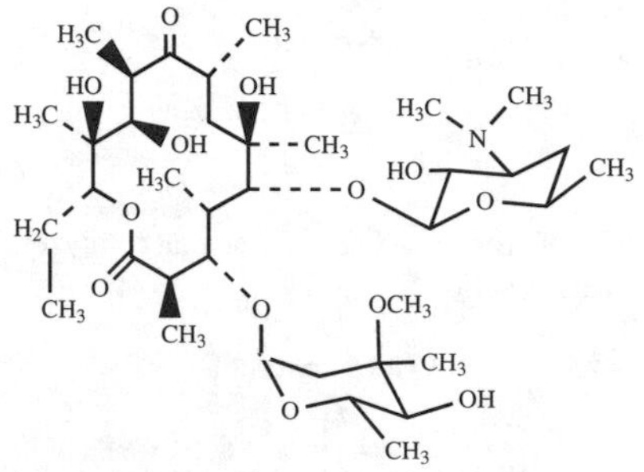

erythromycin A

Erythromycin Acistrate (mol wt 1060) An erythromycin derivative.

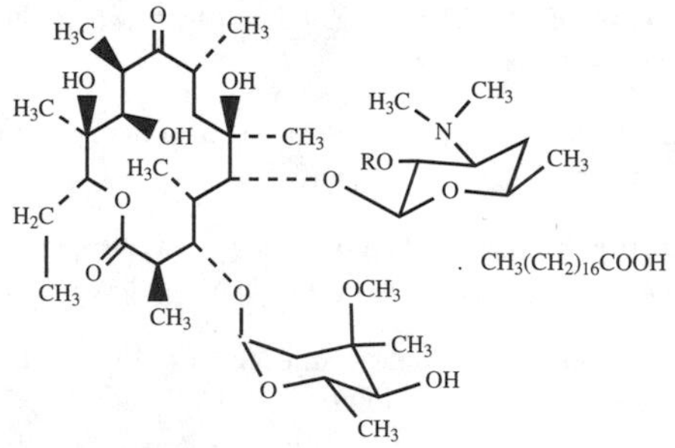

R = $COCH_3$

Erythromycin Base An erythromycin derivative.

Erythromycin Estolate (mol wt 1056) An erythromycin derivative.

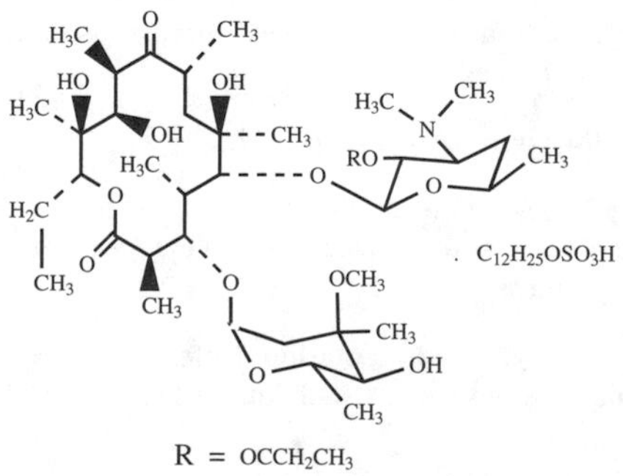

R = $OCCH_2CH_3$

Erythromycin Ethylsuccinate An erythromycin derivative.

Erythromycin Glucoheptonate A semisynthetic antibiotic related to erythromycin.

Erythromycin Lactobionate A semisynthetic antibiotic related to erythromycin.

Erythromycin Stearate An erythromycin derivative.

Erythropoiesis The process of erythrocyte production.

Erythropoietin A protein hormone produced by recombinant DNA technology that stimulates erythrocyte formation.

Erythrose (mol wt 120) An aldotetrose.

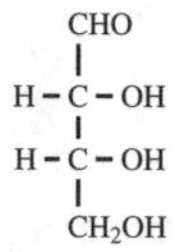

Erythrose 4-Phosphate (mol wt 200) A phosphorylated four-carbon aldosugar and intermediate in the C_3 pathway.

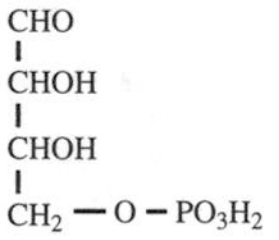

Erythrosine (mol wt 880) A biological stain.

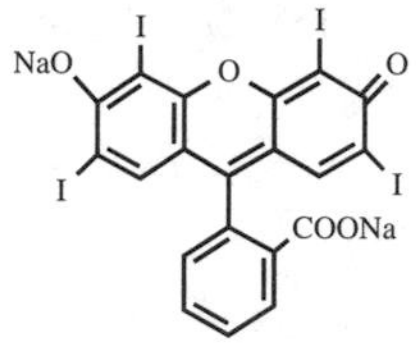

Erythrulose (mol wt 120) A four-carbon ketosugar.

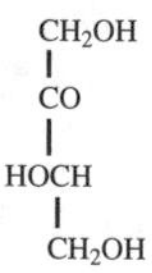

L-erythrulose

Erythrulose Reductase The enzyme that catalyzes the following reaction:

Erythritol + $NADP^+$

⇅

D-Erythrulose + NADPH

Eryzole A trade name for a combination drug containing erythromycin and sulfisoxazole.

ES Abbreviation for 1. embryonic stem cell; 2. enzyme-substrate complex.

ESAF Abbreviation for endothelial cell-stimulating angiogenesis factor.

Esaprazole (mol wt 225) An antiulcerative agent.

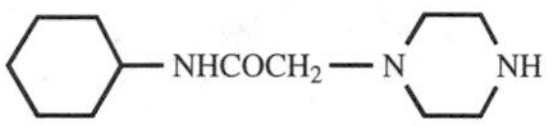

ESC Abbreviation for embryonic stem cell.

Escherichia A genus of Gram-negative bacteria of the enterobacteriaceae, e.g., *Escherichia coli.*

Esclim A trade name for estradiol, a hormone.

Esculetin (mol wt 178) An inhibitor for 5- and 12-lipoxygenase.

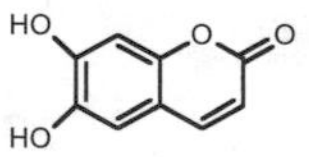

Esculin (mol wt 340) A skin protectant from horse chestnut.

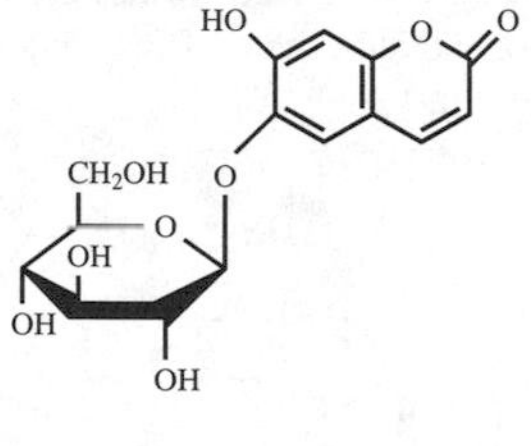

Eseridine (mol wt 291) A cholinergic agent from calabar bean (*Physostigma venenosum).*

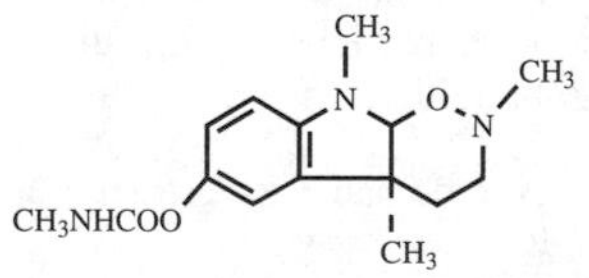

ESF Abbreviation for erythropoietic-stimulating factor.

Esgic A trade name for a combination drug containing butalbital, acetaminophen, and caffeine.

ESI Abbreviation for 1. elastase-specific inhibitor; 2. electro-spray ionization.

Esidrix A trade name for hydrochlorothiazide, a diuretic agent.

ESI-MS Abbreviation for electrospray ionization mass spectrometry.

E-site A site on ribosomes that transiently binds the outgoing tRNA.

Eskalith A trade name for lithium carbonate, which causes alteration of chemical transmitters in the CNS.

Esmolol (mol wt 295) An antiarrhythmic agent that decreases myocardial contractility and blood pressure.

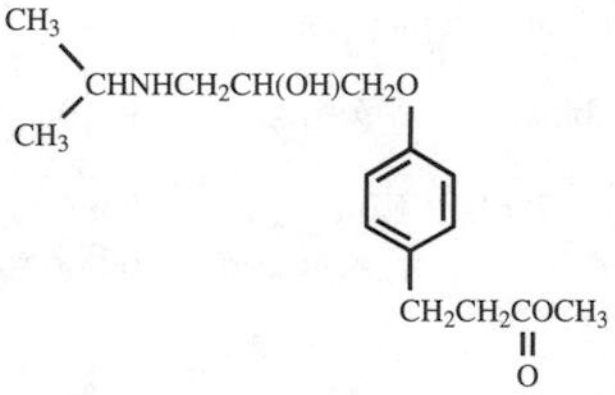

Esophagitis Inflammation of the esophagus.

Esophagogastroduodenoscopy An endoscopic examination of the upper alimentary tract using a fiberoptic or video instrument.

Esophagoscope An optical instrument used to inspect the interior of the esophagus.

EspI A restriction endonuclease from *Eucapsis* species with the following specificity:

↓
5′..........GCTNAGC...........3′
3′..........CGANTCG...........5′
↑

Esp3I A restriction endonuclease from *Erwinia* species RFL3 with the following specificity:

↓
5'........CGTCTC(N)1........3'
3'........GCAGAG(N)5........5'
↑

Esp141I (PstI) A restriction endonuclease from *Enterobacter* species RFL 141 with the same specificity as PstI.

Esp1396I A restriction endonuclease from *Erwinia* species RFL 1396 with the following specificity:

↓
5'........CCANNNNNTGG........3'
3'........GGTNNNNNACC........5'
↑

ESR Abbreviation for 1. electron spin resonance; 2. erythrocyte sedimentation rate.

Essential Amino Acids Amino acids that cannot be synthesized by the organism and must be obtained in the diet.

Essential Elements Referring to the elements carbon (C), hydrogen (H), nitrogen (N), potassium (K), phosphorus (P), magnesium (Mg), sulfur (S), calcium (Ca), and iron (Fe).

Essential Fatty Acids The fatty acids that cannot be synthesized by the organism and must be obtained from dietary sources.

EST Abbreviation for expressed sequence tag.

Established Cell Line A cell line capable of unlimited *in vitro* propagation (also known as continuous cell line).

Estazolam (mol wt 295) A sedative and hypnotic agent that binds to specific benzodiazepine receptors in the CNS.

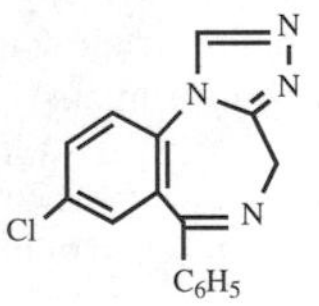

Ester A compound formed from an acid and an alcohol by losing 1 molecule of water.

Esterase Enzyme that catalyzes the hydrolysis of esters.

Esterification Formation of an ester.

Esterolytic Protease Protease capable of hydrolyzing ester bonds.

Estinyl A trade name for ethinyl estradiol, an estrogen that increases the synthesis of DNA, RNA, and protein.

Estivation A physiological state characterized by slow metabolism and inactivity.

Estivin II A trade name for naphazoline hydrochloride, a vasoconstriction agent.

Estolide An intermolecular lactone formed from hydroxy fatty acids.

Estrace A trade name for estradiol, an estrogen that increases the synthesis of DNA, RNA, and protein.

Estracyst A trade name for estramustine phosphate sodium, a hormone with antineoplastic activity.

Estraderm A trade name for estradiol, a hormone that increases the synthesis of DNA, RNA, and protein.

Estradiol (mol wt 272) A steroid sex hormone for regulating feminine characteristics.

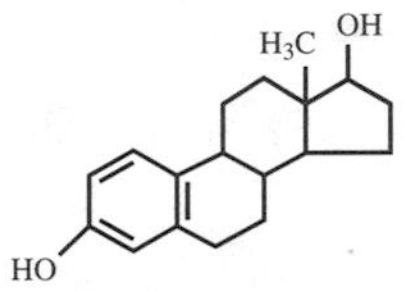

Estradiol 17α-Dehydrogenase The enzyme that catalyzes the following reaction:

$$\text{Estradiol-17-}\alpha + \text{NADP}^+ \rightleftharpoons \text{estrone} + \text{NADPH}$$

Estradiol 17β-Dehydrogenase The enzyme that catalyzes the following reaction:

$$\text{Estradiol-17-}\beta + \text{NADP}^+ \rightleftharpoons \text{estrone} + \text{NADPH}$$

Estradiol Benzoate (mol wt 377) An estrogen.

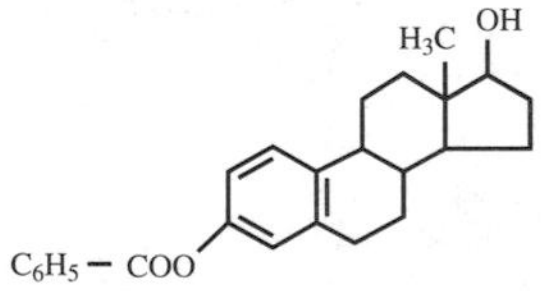

Estradiol 17β-Cypionate (mol wt 397) A steroid hormone that reduces FSH and LH release from the pituitary.

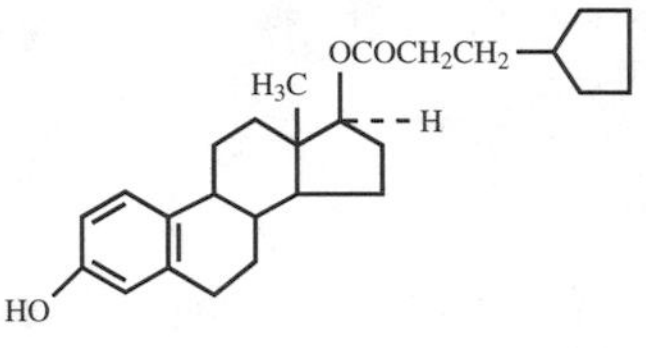

Estradiol LA A trade name for estradiol valerate, a hormone that increases synthesis of DNA, RNA, and protein.

Estra-L A trade name for estradiol, a hormone.

Estramustine (mol wt 440) An antineoplastic agent.

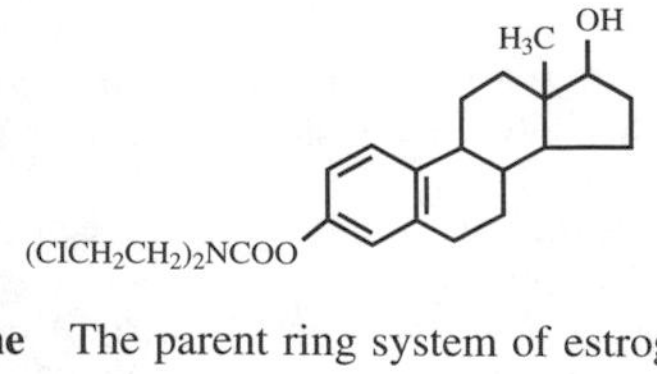

Estrane The parent ring system of estrogens.

Estratest H.S. A trade name for a combination steroid drug containing esterified estrogen and methyltestosterone.

Estraval A trade name for estradiol valerate, an estrogen that increases the synthesis of DNA, RNA, and protein.

Estring A trade name for estradiol, a hormone.

Estriol (mol wt 288) An estrogen.

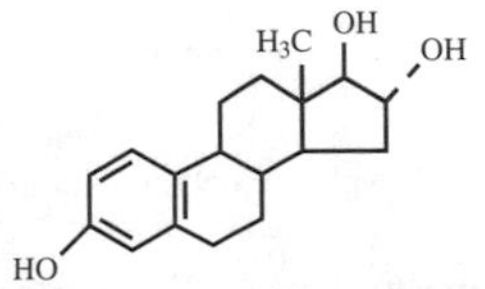

Estrogen Receptor A specific site on the surface of a cell that binds to estrogen.

Estrogens A group of female steroid hormones.

Estrone (mol wt 270) An estrogen.

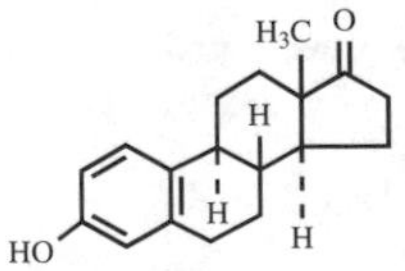

Estronol A trade name for the hormone estrone, which increases the synthesis of DNA, RNA, and protein.

Estrophilin A protein receptor for estrogen.

Estrostep 21 A trade name for a combination drug containing norethindrone acetate and ethinyl estradiol used as a contraceptive agent.

Estrovis A trade name for quinestrol, a hormone that increases the synthesis of DNA, RNA, and protein.

esu (e.s.u) Abbreviation for electrostatic unit.

eT Abbreviation for electron transfer.

Et Abbreviation for ethyl or ethyl group C_2H_5-.

ET Abbreviation for 1. endothelin; 2. endotoxin.

η (eta) Symbol of viscosity.

ETAF Abbreviation for epithelial thymic-activating factor.

Etafedrine (mol wt 193) A bronchodilator.

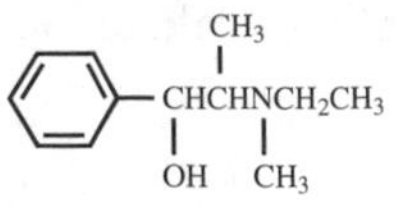

Etafenone (mol wt 325) A vasodilator.

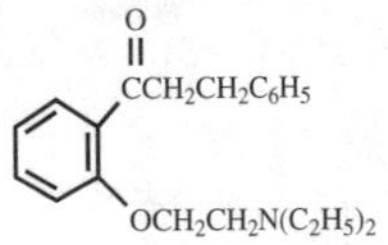

Etamiphyllin (mol wt 279) A bronchodilator.

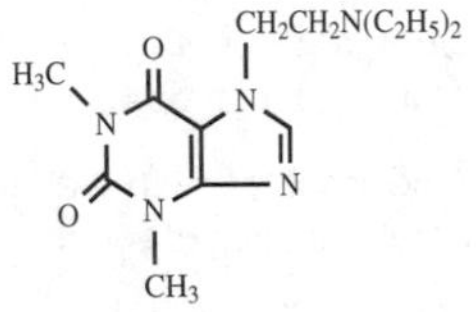

Etamycin (mol wt 879) A lactone antibiotic produced by *Streptomyces grisens.*

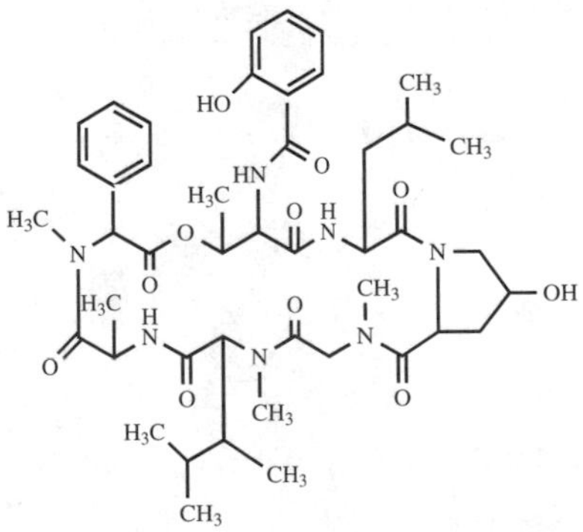

Etanercept A genetically engineered tumor necrosis factor receptor from Chinese hamster ovary cells.

Etaqualone (mol wt 264) A sedative and hypnotic agent.

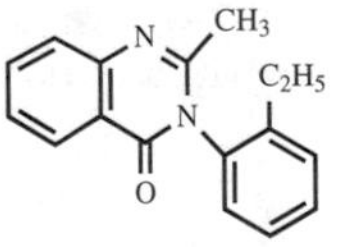

EtBr Abbreviation for ethidium bromide.

ETC Abbreviation for electron transport chain.

Eterobarb (mol wt 320) An anticonvulsant.

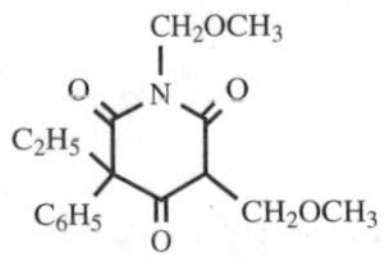

Etersalate (mol wt 357) An antipyretic agent.

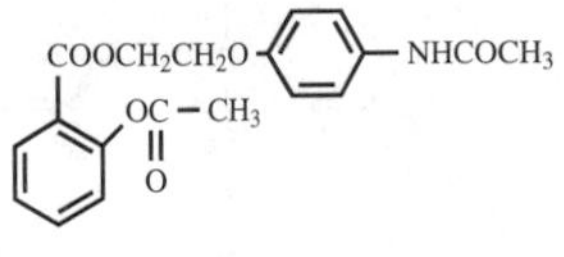

eTF Abbreviation for electron transfer flavoprotein.

Ethacridine (mol wt 253) An antiseptic agent.

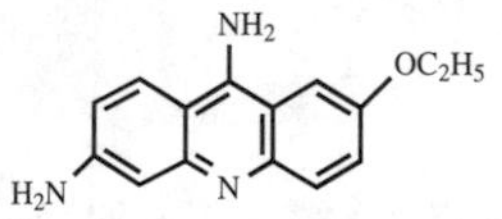

Ethacrynate Sodium The sodium form of ethacrynic acid and a diuretic agent that inhibits the reabsorption of sodium and chloride.

Ethacrynic Acid (mol wt 303) A diuretic agent that inhibits the reabsorption of sodium and chloride.

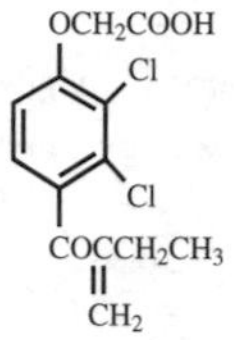

Ethadione (mol wt 157) An anticonvulsant.

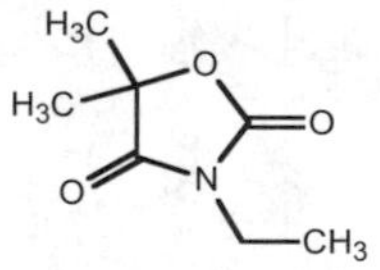

Ethalfluralin (mol wt 333) A preemergent herbicide.

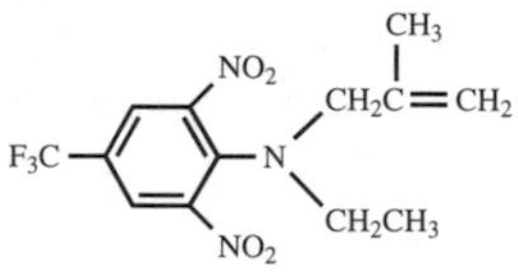

Ethambutol (mol wt 204) An antituberculostatic agent that interferes with bacterial cellular metabolism.

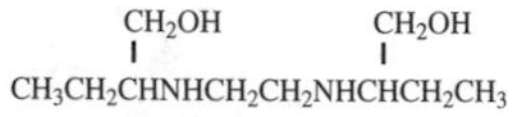

Ethamivan (mol wt 223) A respiratory stimulant.

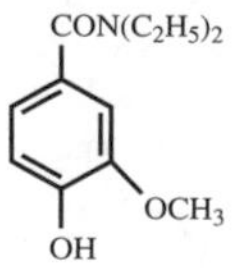

Ethamoxytriphetol (mol wt 420) An estrogen antagonist.

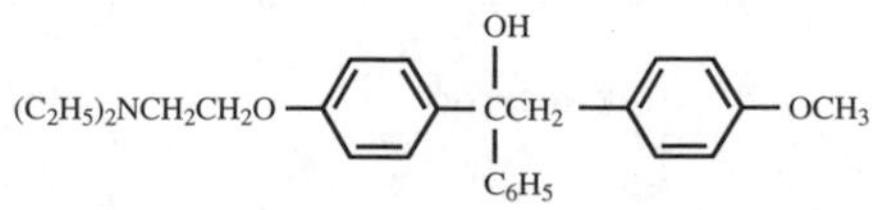

Ethanol (mol wt 46) A two-carbon alcohol.

$$CH_3CH_2OH$$

Ethanolamine (mol wt 61) A component of phospholipids.

$$NH_2CH_2CH_2NH_2$$

Ethanolamine Kinase The enzyme that catalyzes the following reaction:

$$\text{ATP} + \text{ethanolamine} \rightleftharpoons \text{ADP} + \text{phosphoethanolamine}$$

Ethanolamine Oxidase The enzyme that catalyzes the following reaction:

$$\text{Ethanolamine} + H_2O + O_2 \rightleftharpoons \text{Glycoaldehyde} + NH_3 + H_2O_2$$

Ethanolamine Phosphate Phospholyase The enzyme that catalyzes the following reaction:

$$\text{Ethanolamine phosphate} + H_2O \rightleftharpoons \text{Acetaldehyde} + NH_3 + \text{Pi}$$

Ethanolamine Phosphotransferase The enzyme that catalyzes the following reaction:

$$\text{CDP-ethanolamine} + \text{1,2-diacylglycerol} \rightleftharpoons \text{CMP} + \text{phosphatidyl ethanolamine}$$

Ethanolic Fermentation A process by which glucose is converted to ethanol and carbon dioxide.

Ethanolysed Cellulose A powdered and highly purified cellulose used as a supporting medium in zone electrophoresis.

Ethaquin A trade name for ethaverine hydrochloride, a vasodilator that inhibits phosphodiesterase, increases concentrations of cAMP, and relaxes smooth muscle.

Ethatab A trade name for ethaverine hydrochloride, a vasodilator that inhibits phosphodiesterase, increases concentrations of cAMP, and relaxes smooth muscle.

Ethaverine (mol wt 395) An antispasmodic agent that inhibits phosphodiesterase, increases concentrations of cAMP, and relaxes smooth muscle.

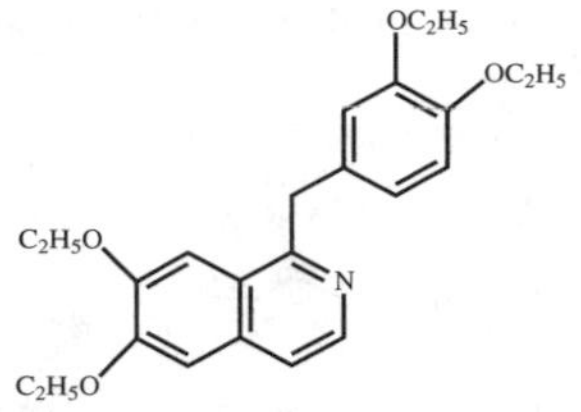

Ethavex-100 A trade name for ethaverine hydrochloride, a vasodilator that inhibits phosphodiesterase, increases concentrations of cAMP, and relaxes smooth muscle.

Ethchlorvynol (mol wt 145) A sedative and hypnotic agent.

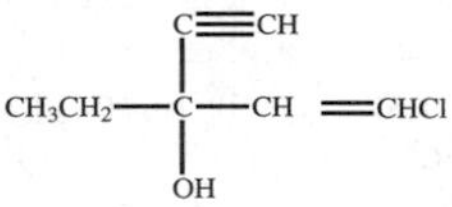

Ethenzamide (mol wt 165) An analgesic agent.

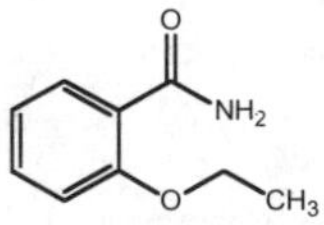

Ether An organic compound derived from an alcohol by replacing the hydrogen of the hydroxyl group with an organic radical.

Ethiazide (mol wt 326) A diuretic agent.

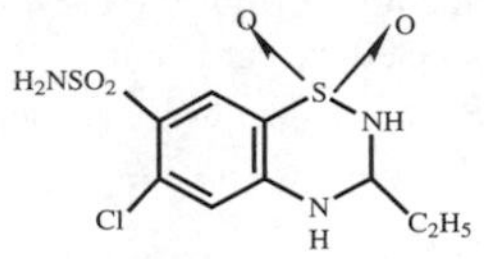

Ethidium Bromide (mol wt 394) A dye and mutagen that intercalates the stacked bases of nucleic acids (double stranded most efficiently) and causes the molecules to fluoresce under UV light.

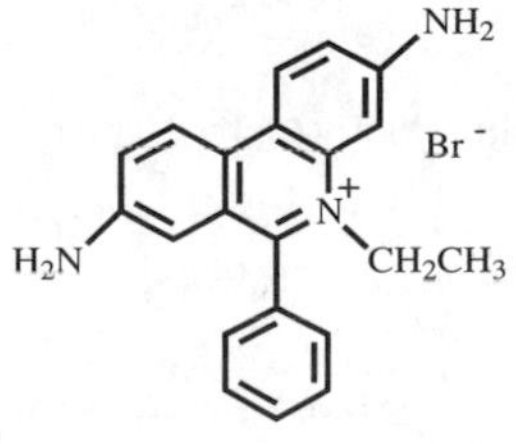

Ethinamate (mol wt 167) A sedative and hypnotic agent.

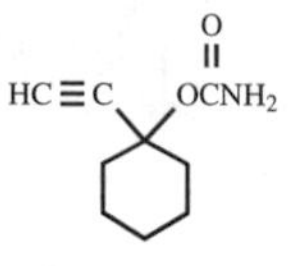

Ethinyl Estradiol (mol wt 296) An estrogen that increase the synthesis of DNA, RNA, and protein.

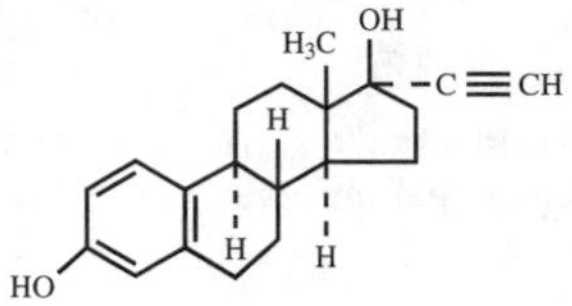

Ethion (mol wt 166) A cholinesterase inhibitor.

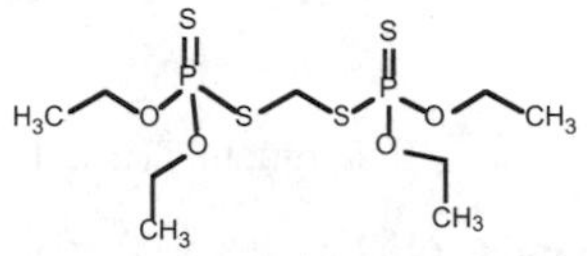

Ethionamide (mol wt 166) An antituberculostatic agent.

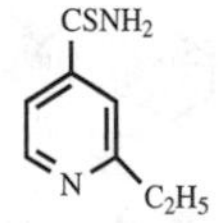

Ethionine (mol wt 163) An amino acid analog that can be incorporated into protein during protein synthesis.

$C_2H_5SCH_2CH_2CH(NH_2)COOH$

Ethirimol (mol wt 209) A fungicide used as a seed dressing and as a foliar spray for the control of powdery mildews in cereals.

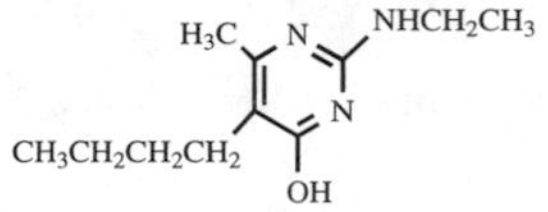

Ethmozine A trade name for moricizine hydrochloride, an antiarrhythmic agent.

Ethoheptazine (mol wt 261) An analgesic agent.

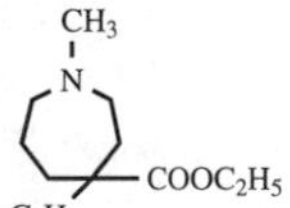

Ethopropazine (mol wt 312) An anticholinergic and antiparkinsonian agent.

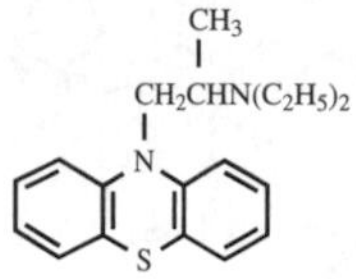

Ethosuximide (mol wt 141) An anticonvulsant that depresses nerve transmission in the motor cortex.

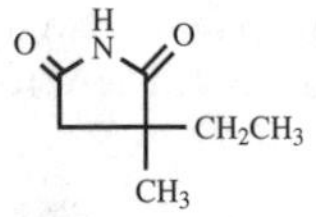

Ethotoin (mol wt 204) An anticonvulsant that stabilizes neuronal membranes and limits seizure activity.

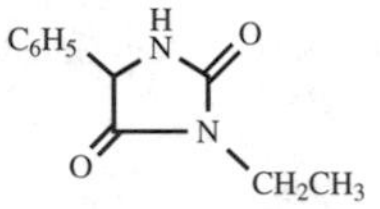

Ethoxy-Group Referring to CH_3-CH_2-O-.

Ethoxzolamide (mol wt 258) A carbonic anhydrase inhibitor.

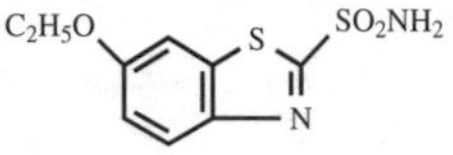

Ethril A trade name for erythromycin stearate, an antibiotic.

Ethyl acetate (mol wt 88) An acetic acid ethyl ester.

$$CH_3COOC_2H_5$$

Ethyl Acetoacetate (mol wt 130) An acetoacetic acid ethyl ester.

$$CH_3COCH_2COOC_2H_5$$

Ethyl Alcohol (mol wt 46) Ethanol.

$$CH_3CH_2OH$$

Ethyl Aminobenzoate (mol wt 165) A topical anesthetic agent.

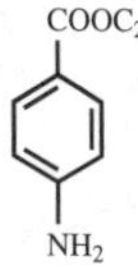

Ethyl Biscoumacetate (mol wt 408) An anticoagulant.

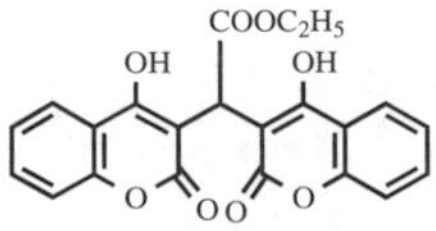

Ethyl Chloride (mol wt 65) A topical anesthetic agent.

$$C_2H_5Cl$$

Ethyl Dibunate (mol wt 348) An antitussive agent.

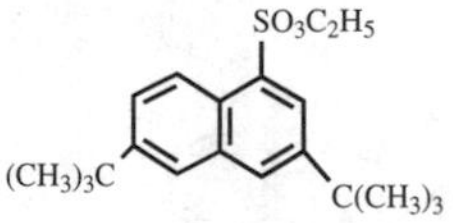

Ethyl Methane Sulfonate (mol wt 124) A reagent capable of combining with guanine to cause mutation.

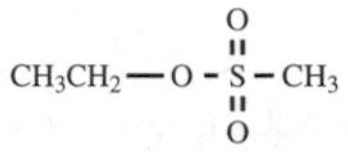

Ethylene (mol wt 28) An unsaturated hydrocarbon gas that acts as a plant hormone responsible for fruit ripening, growth inhibition, leaf abscission, and aging.

$$CH_2 = CH_2$$

Ethylenediamine (mol wt 60) A urinary acidifier and emulsifier.

$$H_2N — CH_2 — CH_2 — NH_2$$

Ethylenediaminetetraacetic Acid (mol wt 292) A chelating agent (abbreviated EDTA).

Ethylene Dibromide (mol wt 188) An insecticide and suspected cancer agent.

$$Br - CH_2 — CH_2 — Br$$

Ethylene Glycol (mol wt 62) A reagent useful in denaturation of ribosomes and DNA.

$$HOCH_2CH_2OH$$

Ethylene Oxide (mol wt 44) A disinfectant.

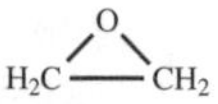

Ethylenimine (mol wt 33) A mutagen.

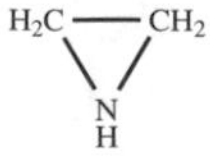

Ethylestrenol (mol wt 288) An anabolic steroid.

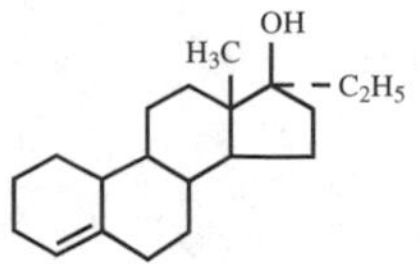

Ethylidene Dicoumarol (mol wt 350) An anticoagulant.

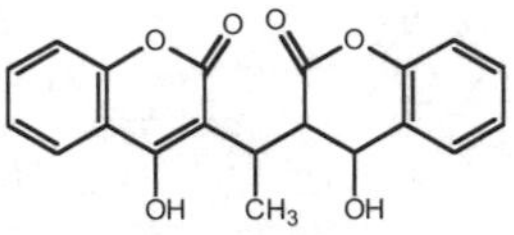

Ethylmalate Synthetase The enzyme that catalyzes the following reaction:

3-Ethylmalate + CoA

⇅

Butanoyl-CoA + H_2O + glyoxylate

N-Ethylmaleimide (mol wt 125) A reagent used to modify amino acid residues in protein.

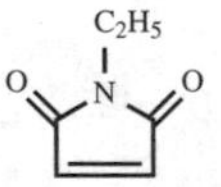

N-Ethylmaleimide-Sensitive Fusion Protein A cytoplasmic protein required for vesicle-mediated transport from the endoplasmic reticulum to the Golgi stack.

Ethylmethylthiambutene (mol wt 277) A narcotic analgesic agent.

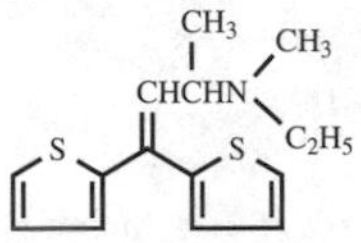

Ethylmorphine (mol wt 313) A narcotic analgesic and antitussive agent.

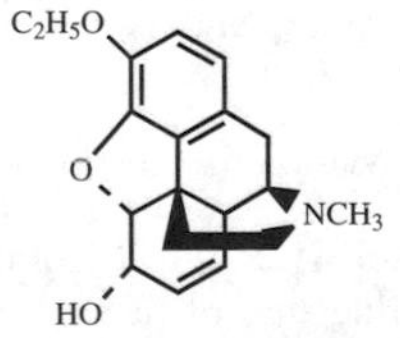

Ethylnorepinephrine (mol wt 197) A bronchodilator that relaxes bronchial smooth muscle by acting on beta-adrenergic receptors.

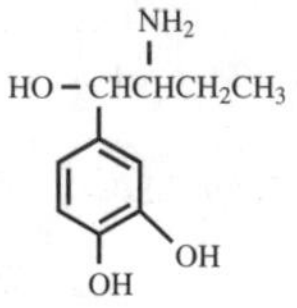

Ethyloestrenol See ethylestrenol.

5-Ethyl-5-(1-piperidyl)-barbituric Acid (mol wt 239) A sedative and hypnotic agent.

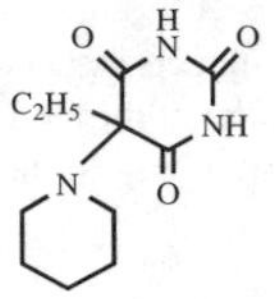

Ethylthiotrifluoroacetate (mol wt 120) A reagent used for modification of N-terminal amino acid residues in protein.

$F_3C\text{-}CO\text{-}S\text{-}CH_2\text{-}CH_3$

Ethynodiol (mol wt 300) An oral contraceptive agent.

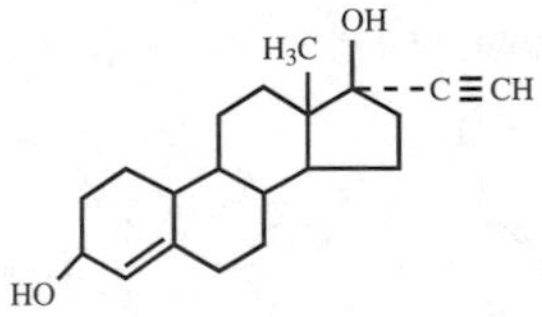

ETI Abbreviation for 1. eicosatriynoic acid; 2. egg trypsin inhibitor.

Etibi A trade name for ethambutol hydrochloride, a tuberculostatic agent.

Etidocaine (mol wt 276) A local anesthetic agent that blocks depolarization by interfering with sodium potassium exchange across the nerve cell membrane.

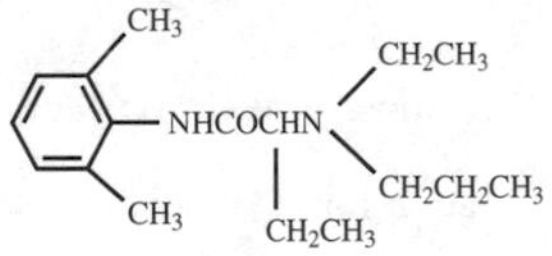

Etidronic Acid (mol wt 206) A calcium regulator that decreases mineral release and collagen breakdown in the bone.

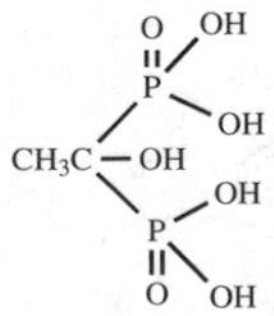

Etifelmin (mol wt 237) An antihypotensive agent.

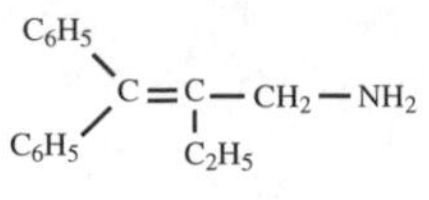

Etiocobalamin (mol wt 1042) The vitamin B_{12} factor obtained by removal of the nucleotide from cyanocobalamin.

Etiolation The characteristic pattern or syndrome of plants grown under continuous darkness. It consists of small, yellow leaves and abnormally long internodes.

Etiological Agent Agent capable of causing disease.

Etiology The science that deals with causal agents of disease.

Etioplast A form of plastid containing precursors of chloroplasts present in plants grown in the dark. It develops into a chloroplast upon exposure to light.

Etiroxate (mol wt 819) An antihyperlipoproteinemic agent.

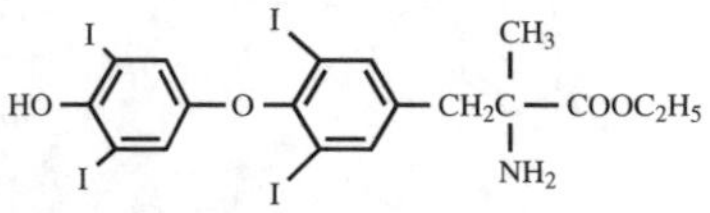

Etizolam (mol wt 343) An anxiolytic agent.

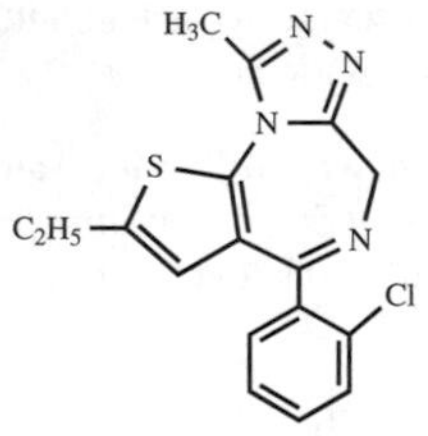

ETM Abbreviation for erythromycin.

Etn Abbreviation for ethanolamine.

Et_2O Abbreviation for ether.

Etodolac (mol wt 287) An anti-inflammatory and analgesic agent that inhibits prostaglandin biosynthesis.

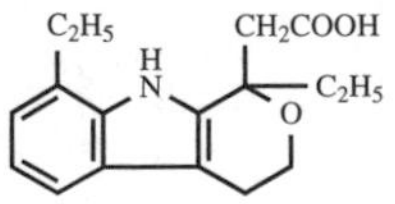

Etodroxizine (mol wt 419) An hypnotic agent.

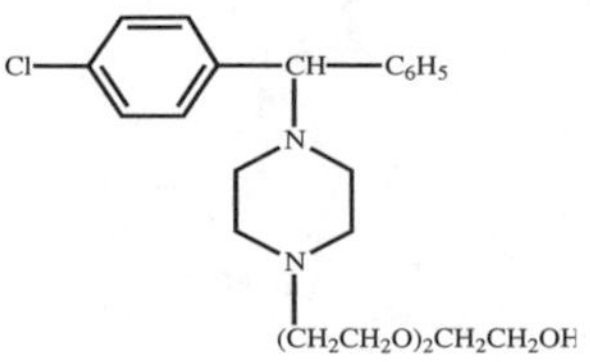

Etofenamate (mol wt 369) An anti-inflammatory agent.

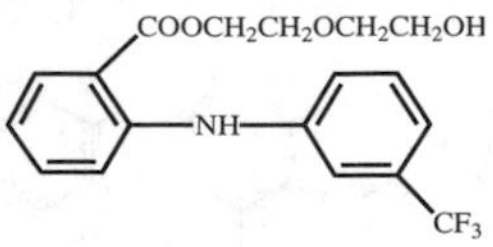

Etofibrate (mol wt 364) An antihyperlipoproteinemic agent.

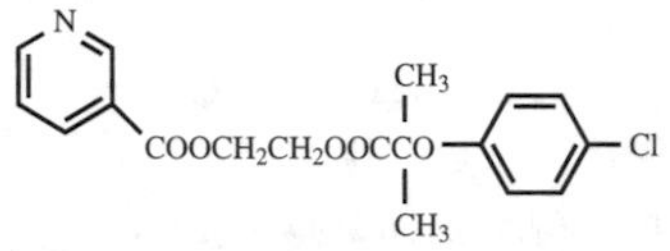

Etofylline (mol wt 224) A bronchodilator.

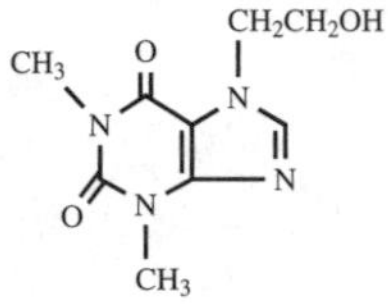

Etofylline Nicotinate (mol wt 329) A vasodilator.

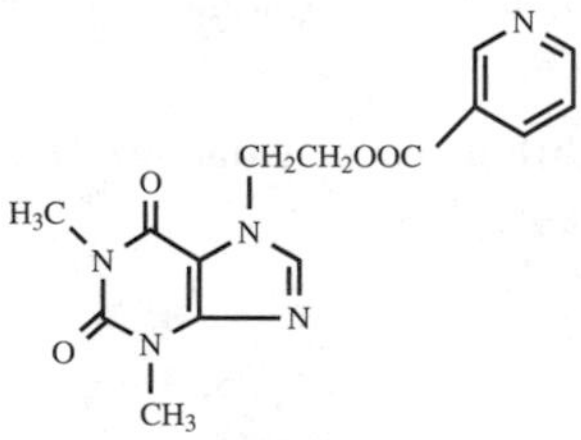

Etoglucid (mol wt 262) An antineoplastic agent.

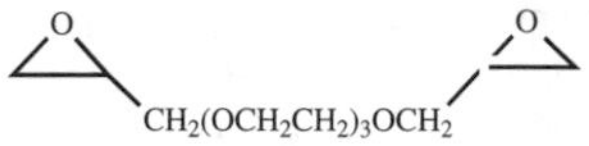

EtOH Abbreviation for ethanol or ethyl alcohol.

Etomidate (mol wt 244) A hypnotic agent that inhibits the firing rate of neurons within the ascending reticular-activating system.

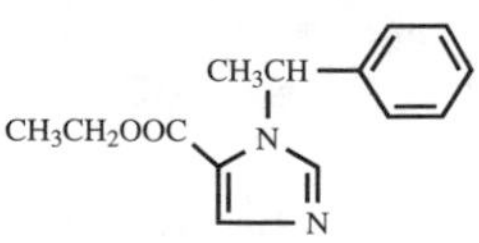

Etomidoline (mol wt 380) An anticholinergic agent.

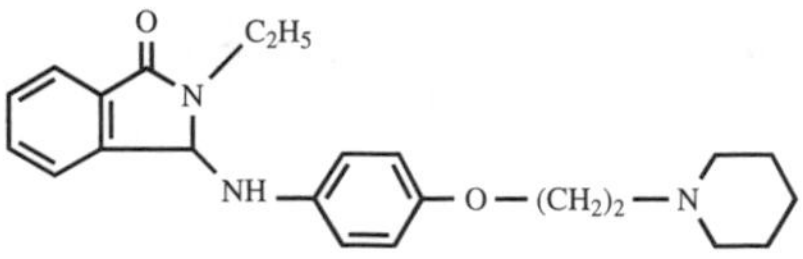

Etoperidone (mol wt 378) An antidepressant.

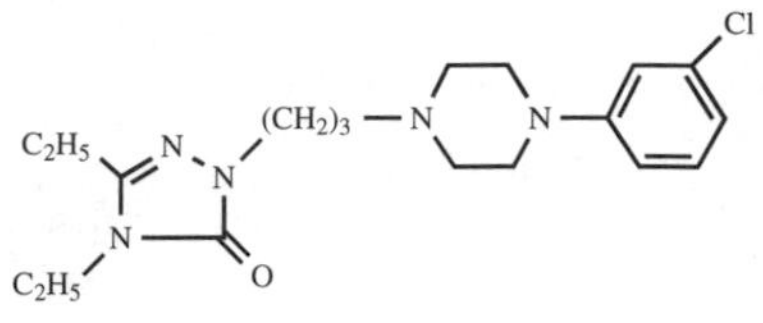

Etopophos A trade name for etoposide phosphate, an antineoplastic agent.

Etoposide (mol wt 589) A semisynthetic derivative of podophylloxin and antineoplastic agent that arrests cell mitosis.

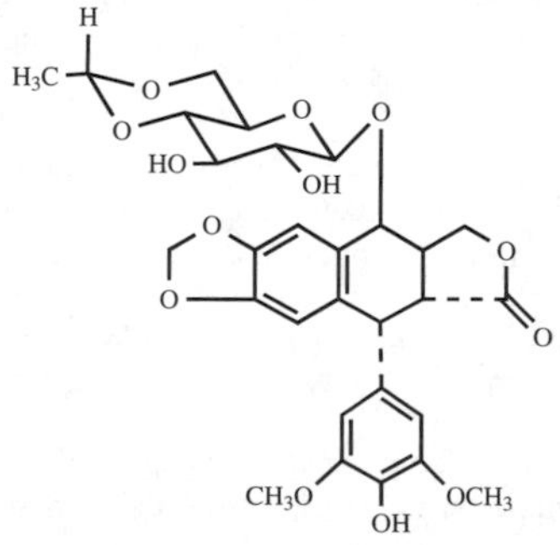

Etoxadrol (mol wt 261) An intravenous anesthetic agent.

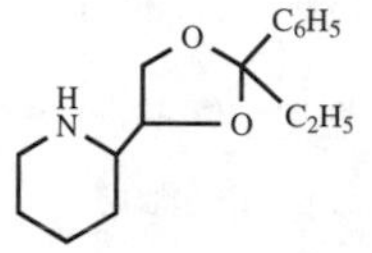

Etozolin (mol wt 284) A diuretic agent.

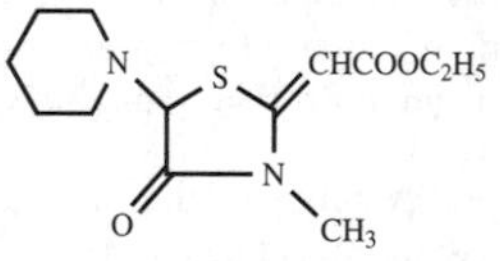

ETP Abbreviation for electron transport particle.

ET-R (ETR) Abbreviation for endothelin receptor.

Etrafon A trade name for a combination drug containing perphenazine (tranquilizer) and amitriptyline hydrochloride (antidepressant).

Etretinate (mol wt 354) An antipsoriatic agent.

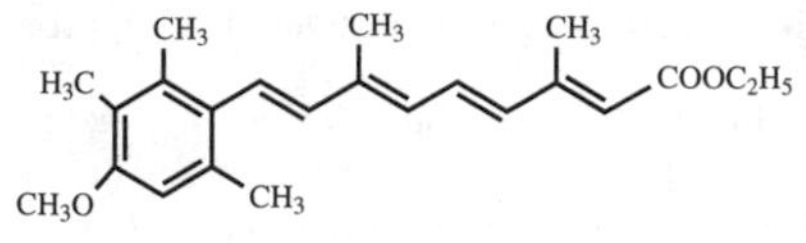

Etrimfos (mol wt 292) An agricultural insecticide.

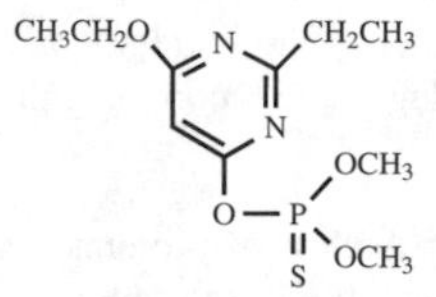

ETS Abbreviation for 1. electron transfer system; 2. expression tagged site.

ETS A proto-oncogen in transforming avian erythroblastosis virus E26, which encodes a family of transcription factors.

Etymemazine (mol et 327) A tranquilizer.

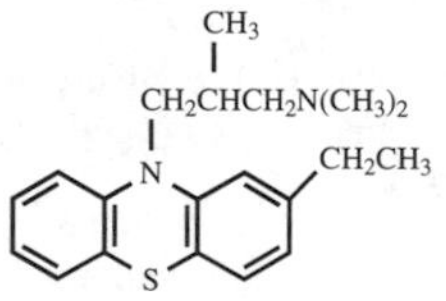

EU Abbreviation for 1. endotoxin unit; 2. enzyme unit.

Eubacteria A kingdom of bacteria including all procaryotic microorganisms not classified as an Archaebacteria.

β-Eucaine (mol wt 247) A local anesthetic agent.

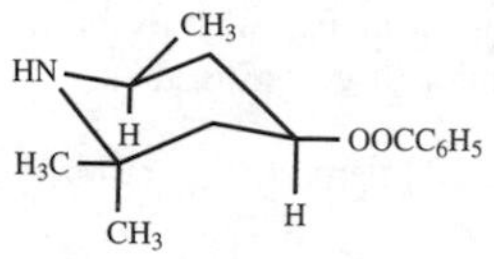

Eucaryon The nucleus of a eucaryotic cell (see also eukaryon).

Eucaryote See eukaryote.

Eucaryotic Cell See eukaryotic cell.

Euchromatin Diffuse, uncondensed chromatin prominent during interphase.

Euchromomatic With euchromatin.

Eucollagen A highly purified form of collagen that can be transformed easily into gelatin.

Eugenics The science that deals with the improvement of genetic quality of a species or race.

Eugenol (mol wt 164) An analgesic agent.

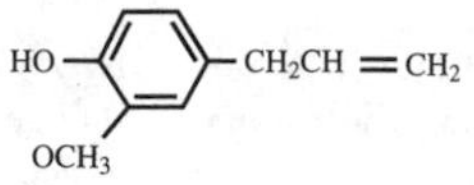

Euglena A genus of photosynthetic euglenoid flagellates.

Euglenoids A group of algae or protozoa with stigma and flagella that contain chlorophyll *a* and *b*.

Euglobulin A class of proteins that is insoluble in water but soluble in salt solutions.

Euglucon A trade name for glyburide, an antidiabetic agent.

Euglypha A genus of testate amoebae.

EuISA Abbreviation for europium-linked immunosorbent assay.

Eukaryon The nucleus of an eukaryotic cell.

Eukaryote Cell or cells of an organism that have a membrane-bound nucleus, cell organelles, and 80S ribosomes.

Eukaryotic Pertaining to eukaryote.

Eukaryotic Cells Cells of higher organisms that have nuclear membranes, membrane-bound cell organelles, and 80S ribosomes.

Eulexin A trade name for flutamide, an antineoplastic agent that inhibits androgen uptake or prevents binding of androgen in the nucleus of cells within target tissues.

Euploid An organism with exact multiple of the haploid number of chromosomes.

Eurax A trade name for crotamiton, a laxative.

Eurithermophile An organism capable of tolerating a wide range of temperature.

Europium (Eu) A chemical element with atomic weight 152, valence 2 and 3.

Euroxic An organism exhibiting a wide range of oxygen tolerance.

Eurysaline Tolerant of a wide salinity range.

Eurythermic Pertaining to organisms that exhibit a wide range of temperature tolerance.

Euryxenous With a broad host range.

Euthenics The science that deals with the improvement of human living conditions.

Euthroid A trade name for liotrix, a hormone that stimulates cellular oxidation.

Eutrophication 1. A process by which a body of water become deficient in oxygen due to the addition to water of chemical substances and subsequent excessive growth of organisms. 2. Enrichment of an aquatic environment with nutrient.

eV Abbreviation for eltron volt.

EVA Abbreviation for 1. a combination drug containing etoposide, vincristine, and adriamycin; 2. ethyl violet azide.

Evan's Blue (mol wt 961) A dye.

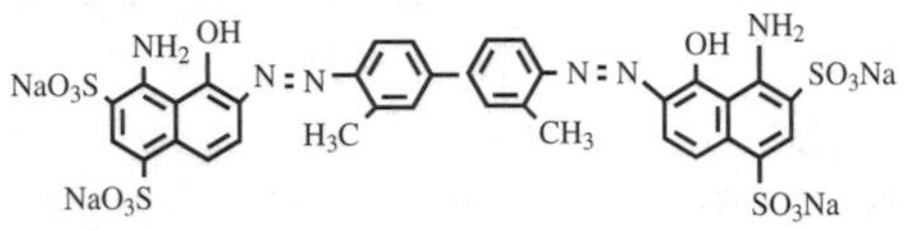

EVAP Abbreviation for a combination drug containing etoposide, vincristine, ara-C, and prednisone.

Everone A trade name for testosterone enanthate, a steroid that stimulates target tissue to develop normally in the adrogen-deficient male.

E-Vista A trade name for hydroxyzine hydrochloride, an antianxiety agent.

Evolution The process of changes by which descendents become distinct from their ancestors.

Ewingella A genus of bacteria of the Enterobacteriaceae.

Ewing's Sacroma Cancer of bone marrow.

EWL Abbreviation for egg white lysozyme.

Exa- A prefix meaning 10^{18}.

Exanthema A skin rash.

Excedrin PM A trade name for a combination drug containing acetaminophen and diphenhydramine, used as an antipyretic and analgesic agent.

Excellospora A genus of thermophilic soil bacteria (order Achnomycetales).

Exchange Diffusion The equal molar passive transport of two solutes across a biological membrane in opposite directions.

Exchange Pairing The paring of homologous chromosomes to allow genetic crossing over to occur.

Excision The enzymatic removal of a DNA fragment or a provirus DNA from a chromosome (e.g., excision of prophage from bacterial chromosome).

Excision Repair Repair of DNA by enzymatic removal of incorrect nucleotides from the DNA and replacing it with the correct nucleotides.

Excisionase The enzyme that catalyzes the excision of the prophage or a DNA fragment from the host chromosome.

Excitable Cells Cells, such as neurons and muscle cells, capable of utilizing changes in membrane potential to conduct signals.

Excitation 1. The transition of an atom or molecule from a low to high energy level. 2. A change in living matter resulting from an external stimulus.

Excitatory Amino Acid Amino acids such as L-glutamate and L-aspartate or their synthetic analogs that have the properties of excitatory neurotransmitters in the CNS and act as excitotoxins.

Excitatory Postsynaptic Potential (EPSP) The change of membrane potential on a postsynaptic caused by the binding of a neurotransmitter to its receptors.

Excitatory Synapses The change of membrane potential on a postsynaptic membrane caused by the binding of a neurotransmitter to its receptors.

Excited State The energy-enhanced state of an atom or molecule after an electron has been moved from its normal stable orbital to an outer orbital having a higher energy level.

Exciton Transfer An energy transfer in which an excited molecule directly transfers its excitation energy to a nearby unexcited molecule with a similar electronic property (also called resonance energy transfer).

Excitotoxin The toxins that damage neurons and produce lesions in the CNS, e.g., excitatory amino acids.

Exclusion Chromatography A method of chromatographic separation based on the principle that large molecules cannot enter the pores of the solid support granules and will be "excluded" and hence elute rapidly, whereas smaller molecules can enter the pores of gel granules and will be eluted more slowly and require a larger elution volume (also called gel filtration).

Exclusion Limit The molecular weight of a macromolecule that is too large to penetrate the pores of the gel particle in gel filtration chromatography.

Exclusion Volume The volume eluted from a gel filtration column that contains the macromolecules too large to pass through the pores of the gel particles.

Exconjugant A female bacterial cell (F^-) that has separated from the donor cell (Hfr) after conjugation.

Exdol A trade name for acetaminophen, an antipyretic agent.

Exelderm A trade name for sulconazole nitrate, an anti-infective.

Exergonic Pertaining to an energy-rcleasing reaction characterized by a negative free energy change.

Exergonic Reaction An energy-releasing reaction characterized by a negative standard free energy change.

Exflagellation The formation and release of mature flagellated male sex cells.

Exfoliatin An epidermolytic toxin produced by some strains of *Staphylococcus aureus* that causes detachment of the outer layer of the skin and produces scalded skin syndrome.

Exfoliation The shedding or peeling of tissue.

Exfoliative Toxin A toxin that causes loss of the surface layer of the skin.

Exgest LA A trade name for a combination drug containing phenylpropanolamine hydrochloride and guaifenesin.

Exite Site See e-site.

Ex-Lax A trade name for phenolphthale, a laxative.

Exna A trade name for benzthiazide, a diuretic agent.

Exo- A prefix meaning outside or external.

Exo-Cellobiohydrolase Synonym of cellulose 1,4-β-cellobiosidase.

Exochelins Iron-solubilizing peptides secreted by *Mysobacterium* species.

Exocrine Glands Glands such as sweat and salivary glands that excrete their products into tubes or ducts that empty onto an epithelial surface.

Exocytosis A process of discharge of cellular material by fusion of vesicular membranes with the plasma membrane so that contents of the vesicle can be secreted into the extracellular environment. It is a process of reverse endocytosis.

Exocytotic Vesicle Vesicle that fuses with the plasma membrane to release its contents.

Exodeoxyribonuclease I The enzyme that catalyzes the exonucleolytic cleavage in the 3′ to 5′ direction to yield 5′-phosphomononucleotides.

Exodeoxyribonuclease III The enzyme that catalyzes the exonucleolytic cleavage in the 3′ to 5′ direction to yield 5′-phosphomononucleotides (preference for double-stranded DNA).

Exodeoxyribonuclease V The enzyme that catalyzes the exonucleolytic cleavage (in the presence of ATP) in either the 5′ to 3′ or 3′ to 5′ direction to yield 5′-phosphomononucleotides.

Exodeoxyribonuclease VII The enzyme that catalyzes exonucleolytic cleavage in either the 5′ to 3′ or 3′ to 5′ direction to yield 5′-phosphomononucleotides (preference for single-stranded DNA).

Exoenzyme 1. Enzyme that is synthesized by the living cell and released into the surrounding environment. 2. Enzyme that catalyzes the stepwise removal of monomer from the end of a polymer.

Exogenote Genetic element received from the donor cell through the process of transformation.

Exogenous Not arising within the organism.

Exogenous Virus A virus that infects a cell and replicates vegetatively and is not vertically transmitted in a gametic genome.

Exoglucanase The enzyme that catalyzes the hydrolysis of glucan externally from the end of the glucan chain.

Exoglycosidase An enzyme that hydrolyzes glycosidic linkages externally from the end of a polysaccharide chain.

Exoglycosylase The enzyme that removes monosaccharide from the ends of a polysaccharide or oligopolysaccharide.

Exon The coding region of the eucaryotic DNA genome that can be transcribed into mRNA and protein.

Exon Shuffling The formation of a new gene by the rearrangement of several coding sequences that have previously specified different proteins or different domains of the same protein through intron-mediated recombination.

Exonuclease An enzyme that catalyzes the sequential removal of nucleotide from the end of a polynucleotide chain.

3′-Exonuclease The enzyme that catalyzes the exonucleolytic cleavage in the 5′ to 3′ direction to yield 3′-phosphomononucleotides (also known as spleen exonuclease).

5′-Exonuclease The enzyme that catalyzes the successive removal of 5′-nucleotides from the 3′-hydroxy termini of 3′-hydroxy-terminated oligonucleotides (also known as phosphodiesterase I).

Exopeptidase The enzyme that catalyzes the sequential removal of amino acids from the end of a polypeptide chain.

Exopoly-α-galacturonosidase The enzyme that catalyzes the hydrolysis of pectic acid from its nonreducing end, releasing digalacturonate.

Exopolyphosphatase The enzyme that catalyzes the following reaction:

$$(\text{Polyphosphate})_n + H_2O \rightleftharpoons (\text{Polyphosphate})_{n-1} + \text{Pi}$$

Exoribonuclease II The enzyme that catalyzes the exonucleolytic cleavage in the 3′ to 5′ direction yielding 3-phosphomononucleotides.

Exoribonuclease H The enzyme that catalyzes the exonucleolytic cleavage to yield 5′-phosphomonoester oligonucleotides in both the 5′ to 3′ and 3′ to 5′ directions (attack RNA in duplex with DNA strand).

Exorphin Any peptide with morphine-like activity.

Exo-α-sialidase The enzyme that catalyzes the hydrolysis of α-2,3-, α-2,6-, and α-2,8-glycosidic linkages at the terminal sialic residue in oligopolysaccharides, glycoproteins and colominic acid.

Exoskeleton The hard-surface structure of an animal, such as the shell of a mollusk or the suticle of an arthropod.

Exosome A fragment of exogenous DNA that is not readily integrated into the chromosome but can replicate and be expressed.

Exosmosis The outward osmotic flow.

Exosurf Neonatal A trade name for colfosceril palmitate, used for treatment of respiratory distress syndrome in neonates.

Exothermic Process An energy-releasing process or reaction characterized by a negative free energy change.

Exotoxin Heat-liable toxin secreted into the surrounding environment by the living microorganism, e.g., Gram-negative bacteria (also called extracellular toxin).

Expansin Any of a group of proteins located within the plant cell wall that play an essential role in loosening cell walls during cell growth.

Expectorant Substance capable of promoting discharge or expulsion of mucous.

Explant Tissue or organ fragment used for starting *in vitro* tissue culture experiments.

Exponential Phase The phase of the growth curve in which the rate of cell multiplication is maximum and constant.

Expression Vector A vector that is designed to permit transcription and translation of the inserted DNA sequence. It contains a regulatory sequence that controls the expression of the cloned gene.

Extendryl A trade name for a combination drug containing phenylephrine hydrochloride and chlorpheniramine nitrate, used for treatment of nasal congestion.

Extensin Glycoproteins found in the cell walls of plants and fungi.

Extinction Coefficient See molar absorptivity.

Extracellular External to the cells of an organism.

Extracellular Matrix Cellular materials secreted into the surrounding environment, e.g., linking proteins (fibronectin) and space-filling molecules (glycosaminoglycans).

Extrachromosomal Element An inheritable genetic element that is not associated with nuclear chromosomes (e.g., plasmid).

Extrachromosomal Inheritance The nonmendelian inheritance due to the DNA in cell organelles (e.g., DNA in mitochondria and chloroplasts).

Extractive Fermentation A type of fermentation in which the fermentation product is continuously removed by an extraction process that is not harmful to the fermentation system.

Extragenic Reversion A mutational change that eliminates or suppresses the mutant phenotype of the original gene.

Extranuclear Genes Referring to cytoplasmic genes.

Extrapolation The extension of a graph from a point of experimental data to a region devoid of data.

Extravascular Not contained in the body vessels.

Extrinsic Blood Coagulation A cascade of blood clotting reactions that involve the sequential interactions of circulation factors and tissue factors (see extrinsic pathway).

Extrinsic Factor See vitamin B_{12}.

Extrinsic Fluorescence The fluorescence due to the activity of the ligand bound to the protein.

Extrinsic Muscle A muscle that has its origin some distance from the part on which it acts.

Extrinsic Pathway An alternative blood clotting pathway. It is initiated by the proteolytic activation of proconvertin (factor VII). The activated proconvertin in turn activates factor X in the presence of accessory tissue factor.

Extrinsic Proteins See peripheral proteins.

Extrusome A vacuole, excreted by parasitic protists containing substances for penetration of the host cell.

Exudate Fluid accumulated at the site of inflammation or lesion.

Exudate Cells Leukocytes that exudate or enter the tissue from blood vessels.

EY Abbreviation for egg yolk.

Eye-Piece Micrometer A scale in the visual field of the eye piece of a microscope, used for measuring the size of a specimen.

Eye-Sed Ophthalmic A trade name for zinc sulfate, an ophthalmic vasoconstrictor.

Eye-Structure The structure formed by the two replication forks on the double-stranded DNA during the process of bidirectional replication.

Eyesine A trade name for tetrahydrozoline hydrochloride, a nasal decongestant.

EYPC Abbreviation for egg yolk phosphatidyl choline.

Ezrin A protein involved in connections of major cytoskeletal structures to the plasma membrane.

F

F Abbreviation for 1. amino acid phenylalanine; 2. Fahrenheit; 3. Faraday constant; 4. fluorine; 5. force.

°F Degrees Fahrenheit.

FI, FII, FIII, FIV, FV, FVI, VII, VIII, IX, X, XI Abbreviation for blood coagulating factor I, II, III, IV, V, VI. VII, VIII, IX, X, XI.

F_0 A component of the mitochondrial ATPase complex embedded in the inner mitochondrial membrane.

F_1 Abbreviation for 1. component of the mitochondrial ATPase complex and 2. first-generation offspring resulting from a cross between parental strains.

F_2 Second-generation offspring resulting from crosses between first-generation members.

F^+ Bacterial cell that possesses fertility factor or F plasmid (donor bacteria or male bacteria).

F^- Recipient bacteria or female bacteria.

F_1 Antigen A heat-labile, water-soluble capsular antigen from virulent strains of *Yersinia pestis.*

F^+ Cell Abbreviation for *E. coli* cell that contains F^+ factor.

F Episome Synonym of F plasmid.

F Factor The fertility factor. Bacterial cell that possesses F factor is designated as F^+ (see also F plasmid).

F_1 Fragment The separated globular head portion of a myosin molecule.

F_2 Fragment The separated fibrous part of the heavy meromyosin.

F Pili Hairlike projections produced in the F^+ strains of bacterial cells. It provides the adsorption sites for male-specific phages.

F plasmid A fertility factor that gives rise to F^+ bacterial cells. Bacterial cell that possesses F plasmid is designated as F^+, which enables bacterial cells to conjugate.

F′ Plasmid The F plasmid that has undergone aberrant excision from the bacterial chromosome. It carries segments of the bacterial chromosome.

FA Abbreviation for 1. fatty acid; 2. fluorescent antibody; 3. a combination drug containing 5-fluorouracil and adriamycin; 4. folic acid; 5. free acid.

FAA Abbreviation for formalin-acetate-alcohol solution.

Fab The antigen-binding fragment produced by enzymatic digestion of an immunoglobulin molecule with papain.

FAB Abbreviation for fast action bombardment or fast atom bombardment.

$F(ab)_2$ A fragment obtained by pepsin digestion of immunoglobulin molecules containing two Fab fragments linked by disulfide bonds.

FAB-MS Abbreviation for fast-atom bombardment mass spectrometry.

FABP Abbreviation for fatty acid-binding protein.

Fabry's Disease A kidney disease due to a deficiency of the enzyme trihexosyl ceramide α-galactosylhydrolase (α-galactosidase A).

FAC Abbreviation for a combination anticancer drug containing fluorouracil, adriamycin (doxorubicin), and cyclophosphamide.

FACE Abbreviation for fluorophore-assisted carbohydrate electrophoresis.

Faciliated Diffusion Carrier-mediated passive transport.

FACS Abbreviation for 1. a combination drug containing 5-fluorouracil, adriamycin, cyclophosphamide, and streptozocin; 2. fluorescence-activated cell sorter.

F-actin A long fibrous protein microfilament consisting of polymerized G-actin monomers.

Factor I Referring to fibrinogen, a blood clotting factor.

Factor II Referring to to prothrombin, a blood clotting factor.

Factor III Referring to thromoplastin, a blood clotting factor.

Factor IV Referring to calcium involved in the blood clotting system.

Factor V Referring to proaccelerin, a blood clotting factor.

Factor VI Factor V was once called factor VI; therefore, no factor VI is designated.

Factor VII Referring to proconvertin, a blood clotting factor.

Factor VIII Referring to antihemophilic factor.

Factor IX Referring to Christmas factor in the blood clotting system.

Factor X Referring to Stuart factor in the blood clotting system.

Factor XI Referring to plasma thromboplastin antecedent in the blood clotting system.

Factor XII Referring to Hageman factor in the blood clotting system.

Factor XIII Referring to fibrin stabilizing factor in the blood clotting system.

Factor XIV Referring to protein C whose activated form can inactivate factor V and factor VIII.

Factor B A factor present in the plasma and involved in the alternative pathway of complement activation.

Factor D A glycoprotein factor present in the plasma and involved in the alternative pathway of complement activation.

Factor F Referring to the initiation factor in protein synthesis.

Factor G Referring to translocase (in protein synthesis).

Factor H A protein factor present in the plasma and involved in the alternative pathway of complement activation.

Factor I A glycoprotein factor present in the plasma and involved in the alternative pathway of complement activation.

Factor R Referring to the releasing factor in protein synthesis.

Factorial Referring to the continuous product of all integers to a given number, e.g., factorial 5 = 5 × 4 × 3 × 2 × 1.

Factrel A trade name for gonadorelin hydrochloride, a luteinizing hormone.

Facultative Capable of adapting to more than one condition.

Facultative Anaerobe An anaerobe capable of growing in either the presence or absence of oxygen.

Facultative Heterochromatin A chromatin that may behave as a heterochromatin in some cells and as a euchromatin in other cells, e.g., one of the X chromosomes of a mammalian female.

FACV Abbreviation for a combination drug containing fluorouracil, adriamycin, cyclophosphamide, and VP-16.

FAD Abbreviation for flavine adenine dinucleotide.

FAD Nucleotide Hydrolase See FAD pyrophosphorylase.

FAD Pyrophosphorylase The enzyme that catalyzes the following reaction:

$$\text{Flavin mononucleotide} + \text{ATP} \rightleftharpoons \text{FAD} + \text{PPi}$$

FADH$_2$ The reduced form of FAD.

FAE Abbreviation for feruloyl esterase.

Fagarine (mol wt 229) An antiarrhythmic agent from *Fagara coco* (Rutaceae).

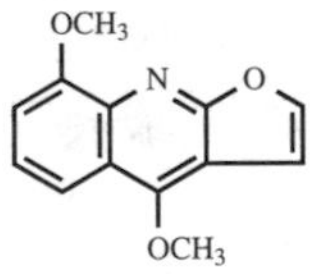

FAH Abbreviation for fumaryl acetoacetate hydrolase.

Fahrenheit Temperature Scale A temperature scale on which the freezing point of water is 32° and the boiling point is 212°.

FAK Abbreviation for focal adhesion kinase.

Falapen A trade name for penicillin G potassium, an antibiotic.

Fallout The radioactive substance generated from a nuclear explosion.

False Negative A test that is erroneously recorded as negative because of an imperfect test method or procedure.

False Positive A test that is erroneously recorded as positive because of an imperfect test method or procedure.

FAM Abbreviation for an anticancer combination drug containing fluorouracil, adriamycin (doxorubin), and mitomycin.

FAMA Abbreviation for fluorescent antibody to membrane antigen.

Famciclovir (mol wt 321) An antiviral agent.

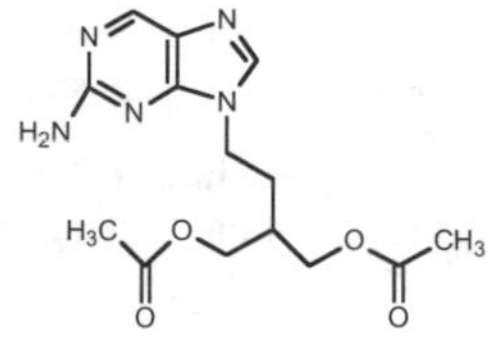

FAME Abbreviation for 1. a combination drug containing fluorouracil, adriamycin, and MeCCNU; 2. fatty acid methyl ester.

Familial Goiter A genetic disorder characterized by an excessive loss of iodinated tyrosine from the thyroid gland due to a deficiency of iodotyrosine dehalogenase.

Familial Hypercholesterolemia A genetic disorder characterized by the presence of high levels of low density lipoprotein. It is inherited as an autosomal dominant gene at a locus on chromosome number 14.

Familial Polyposis A genetic disorder characterized by the formation of polyps in the colon and rectum.

Familial Trait A trait shared by members of a family.

Famir A trade namc for famciclovir sodium, an antiviral agent.

Famotidine (mol wt 337) An antiulcerative agent that decreases gastric acid secretion.

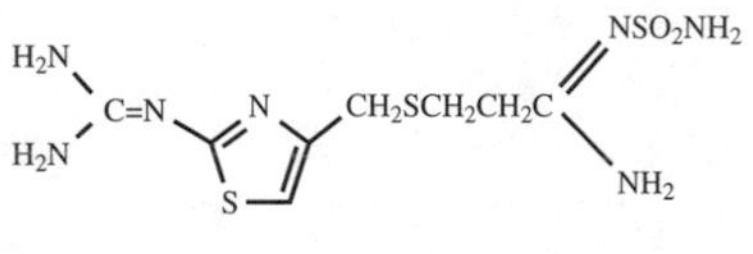

FAMS Abbreviation for a combination drug containing fluorouracil, adriamycin, mitomycin C, and streptozocin.

Famvir A trade name for famciclovir, an antiviral agent.

FAN Abbreviation for a dye mixture containing fuchsin, amidoblack, and naphthol yellow.

FANA Abbreviation for fluorescent antinuclear antibody.

Fanconi's Anemia A genetic disorder characterized by the reduction in red and white blood cells due to a deficiency in DNA repair.

Fansidar A trade name for a combination drug containing pyrimethamide and sulfadoxine, used as an antimalarial agent.

Fantofarone (mol wt 550) An antihypertensive agent.

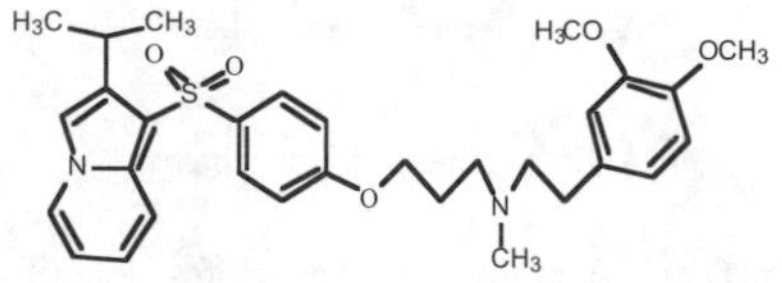

FAP Abbreviation for a combination drug containing fluorouracil, adriamycin, and platinol.

Faraday Constant One Faraday is the quantity of electricity that will deposit or dissolve one weight-equilivent of any substance (one Faraday = 96490 coulombs).

Farber's Disease A genetic disorder characterized by the difficulty in respiration due to deficiency of the enzyme ceramidase (also known as Farber's lipogranulomatosis).

Fareston A trade name for toremifene, an estrogen receptor modulator used as an antineoplastic agent.

Farital A trade name for a combination drug containing aspirin, caffeine, and butalbital, used as an antipyretic and analgesic agent.

Farmer's Lung A lung disorder due to the inhalation of dusts from moldy hay.

Farnesol (mol wt 222) A compound present in many essential oils.

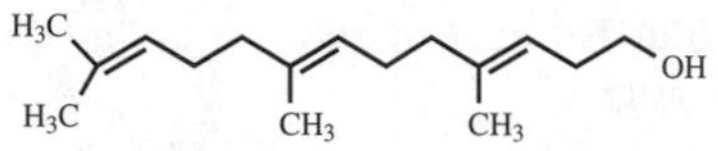

Farnesol Dehydrogenase The enzyme that catalyzes the following reaction:

2-*trans*, 6-*trans* Farnesol + $NADP^+$

⇅

2-*trans*, 6-*trans* Farnesal + NADPH

Farnesol 2-Isomerase The enzyme that catalyzes the following reaction:

2-trans, 6-trans Farnesol

⇅

2-cis, 6-trans Farnesol

Farnesyl Diphosphate Farnesyltransferase The enzyme that catalyzes the following reaction:

2-Farnesyl diphosphate

⇅

PPi + presqualine diphosphate

Farnesyl Diphosphate Kinase The enzyme that catalyzes the following reaction:

$$\text{ATP} + \text{farnesyl diphosphate} \rightleftharpoons \text{ADP} + \text{farnesyl triphosphate}$$

Farr Test A radioimmunoassay for quantifying antibody and determining the antigen-binding capacity of an antiserum.

FAS Abbreviation for fatty acid synthetase.

Fasciculus A small bundle of muscle fibers or a bundle of nerve fibers.

Fasciitis Inflammation of connective tissue.

Fasciola A genus of *Tematoda* (liver flukes).

Fascioliasis Infection caused by *Fasciola* (liver flukes).

Fascioloposis A genus of intestinal Trematoda.

FAST Abbreviation for 1. fluorescent allergosorbent test; 2. fluorescent antibody staining technique.

Fast Blue B Salt (mol wt 475) Reagent used for TLC-detection of aflotoxins.

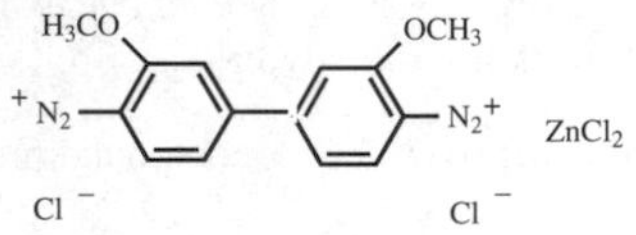

Fast Blue BB (mol wt 300) A dye used in cytochemistry.

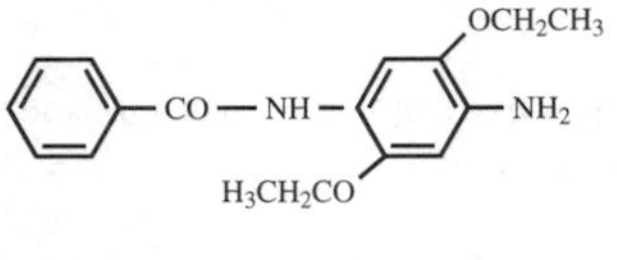

Fast Blue RR (mol wt 272) A reagent used in cytochemistry.

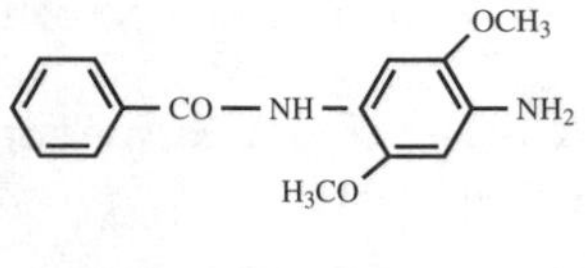

Fast Component 1. The fragments of DNA that consist of a highly repetitive sequence and renature first in a renaturation experiment. 2. The component that migrates farthest from the origin during electrophoresis.

Fast Green FCF (mol wt 809) A dye for staining proteins.

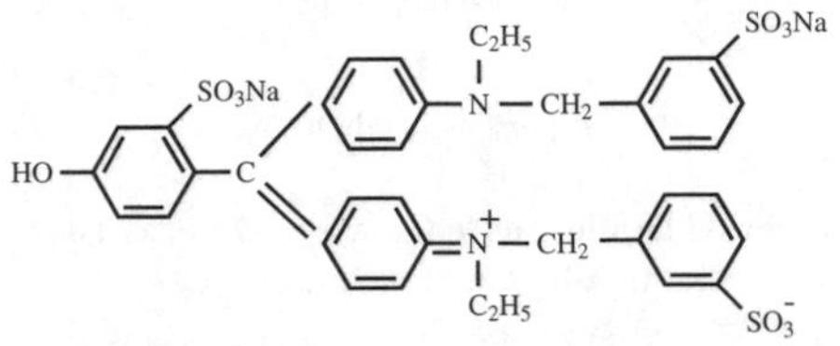

Fast Reaction A reaction or a step in a reaction sequence that has a large rate constant.

Fast Red Violet LB Salt (mol wt 376) A reagent used for the determination of acid and alkaline phsophatase.

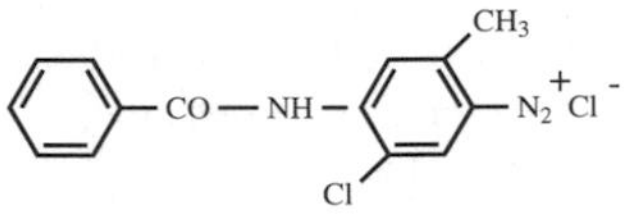

Fastidious Organism An organism that is difficult to isolate or cultivate on ordinary culture media.

Fastin A trade name for phentermine hydrochloride, a cerebral stimulant that promotes transmission of nerve impulses by releasing stored norepinephrine from the nerve terminals in the brain.

Fast-Twitch Muscle Fiber A muscle fiber capable of rapid development of high tension.

Fasudil (mol wt 291) A vasodilator.

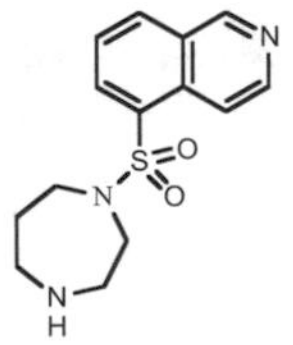

Fat Ester of fatty acids with glycerol.

CH_2O-CO-FA
|
CH_2-O-CO-FA
|
CH_2-O-CO-FA

triacylglycerol or triglyceride

FAT Abbreviation for 1. fatty acid translocase; 2. fluorescent antibody test.

Fat Embolism The blockage of an artery by a blob (embolus) of fat in the circulation.

Fat Soluble Vitamin Referring to vitamins A, D, E, and K that are soluble in organic solvent.

Fat Solvent Referring to nonpolar solvent, e.g., chloroform.

Fate Map A map of the developmental fate of a zygote or an early embryo showing the areas that are destined to develop into specific adult tissues or organs.

F_o-ATPase Abbreviation for F_o component of ATPase.

F_1ATPase Abbreviation for F_1 component of ATPase.

Fatty Acid A long chain of carboxylic acid that may either be saturated (without double bond) or nonsaturated (with double bond).

Fatty Acid Activating Enzyme Referring to fatty acid thiokinase.

Fatty Acid Activation The process of converting fatty acid to fatty acyl-CoA.

Fatty Acid Desaturase The enzyme that catalyzes the desaturation of saturated fatty acids to unsaturated fatty acids.

Fatty Acid O-Methyltransferase The enzyme that catalyzes the following reaction:

S-Adenosyl-L-methioine + fatty acid

$\rightleftharpoons$

S-Adenosyl-L- homocysteine + a fatty acid methyl ester

Fatty Acid Oxidation The pathway that converts fatty acid to acetyl-CoA for generation of energy, e.g., beta oxidation.

Fatty Acid Peroxidase The enzyme that catalyzes the following reaction:

Palmitate + $2H_2O_2$

$\rightleftharpoons$

Pentadecanal + CO_2 + $3H_2O$

Fatty Acid Synthetase The enzyme that catalyzes the synthesis of fatty acid from acetyl-CoA and malonyl-CoA.

Fatty Acid Thiokinase The enzyme that catalyzes the following reaction:

Fatty acid + CoA + ATP

$\rightleftharpoons$

Fatty acyl-CoA + AMP + PPi

Fatty Acyl Carnitine A complex of fatty acid and carnitine involved in the transport of fatty acid to mitochondria for metabolism.

carnitine

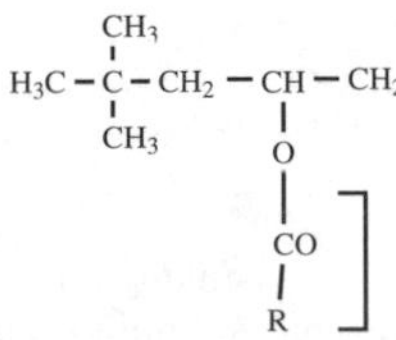

fatty acid

Fatty Acyl-ACP See fatty acyl-S-ACP.

Fatty Acyl-CoA A fatty acid and CoA complex formed during fatty acid oxidation.

CH_3CH_2......CO-S-CoA

Fatty Acyl-CoA Desaturase The enzyme that catalyzes the following reaction:

Saturated fatty acyl-CoA + NADH + O_2

$\rightleftharpoons$

Unsaturated fatty acyl-CoA + NAD^+ + 2 H_2O

Fatty Acyl-CoA Reductase The enzyme that catalyzes the following reaction:

Hexadecanal + CoA + NAD^+

$\rightleftharpoons$

Hexadecanoyl-CoA + NADH

Fatty Acyl-ethyl Ester Synthetase The enzyme that catalyzes the following reaction:

A long chain fatty acyl ethyl ester + H_2O

$\rightleftharpoons$

A long chain fatty acid + ethanol

Fatty Acyl-S-ACP An intermediate in the biosynthesis of fatty acid. It is a complex of fatty acid and ACP.

CH_3CH_2.......CO-ACP

Fatty Degeneration The degeneration of an organ with replacement of its normal structure by fat.

Fatty Liver A liver disorder characterized by fatty degeneration.

FauNDI A restriction endonuclease from *Flavobacterium aquatile* ND with the following specificity:

5'........CA↓TATG........3'
3'........GTAT↑AC........5'

Favism A type of acute anemia caused by eating beans or breathing pollen from the fava plant. The disorder is due to the lack of glucose 6-phosphate dehydrogenase.

Favus Infection of the scalp by fungus *Microsporum gypseum* and characterized by the formation of yellow crusts.

Fazadinium Bromide (mol wt 604) A skeletal muscle relaxant.

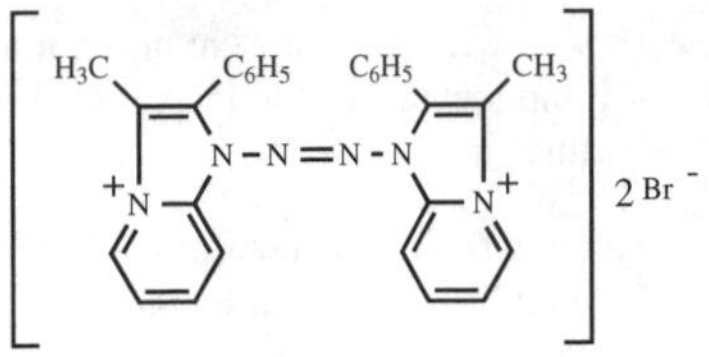

FB Abbreviation for feed-back.

FB1 Abbreviation for fumonisin B-1.

FbaI (BclI) A restriction endonuclease from *Flavobacterium balustinum* with the following specificity:

↓
5′............TGATCA............3′
3′............ACTAGT............5′
↑

FBHE Abbreviation for fetal bovine heart endothelial.

FBP Abbreviation for 1. fibrin-binding protein or fibrin-binding peptide; 2. fibrinogen breakdown product; 3. folate-binding protein; 4. fructose 1,6-bisphosphate.

F1,6BP Abbreviation for 1. fructose 1,6-bisphosphatase; 2. fructose 1,6-bisphsphate.

F2,6BP Abbreviation for 1. fructose 2,6-bisphosphatase; 2. fructose 2,6-bisphsphate.

FBPase Abbreviation for the enzyme fructose 1,6-bisphosphatase.

FbrI (Fnu4H I) A restriction endonuclease from *Flavobacterium breve* with the following specificity:

↓
5′..........GCNGC..........3′
3′..........CGNCG..........5′
↑

FBS Abbreviation for fetal bovine serum.

FBS-AChE Abbreviation for fetal bovine serum acetylcholinesterase.

FC Abbreviation for 1. 5-fluorocytosine; 2. free cholesterol.

Fc Abbreviation for crystallizing fragment of the antibody molecule.

5FC A trade name for 5-fluorocytosine, an antifungal agent.

Fc Fragment A crystallizable fragment obtained by papain digestion of immunoglobulin that consists of the C-terminal half of two H (heavy) chains linked by the disulfide bonds.

Fc Receptor A receptor present on various subclasses of lymphocytes for the Fc fragment of immunoglobulins.

FCA Abbreviation for 1. ferritin-conjugated antibody; 2. freund's complete adjuvant.

FCP Abbreviation for a combination drug containing fluorouracil, cytoxan, and prednisone.

FcR Abbreviation for receptor of Fc fragment (crystallizing fragment of the antibody molecule).

FcαR Referring to Fc receptor for IgA.

FcδR Referring to Fc receptor for IgD.

FcεR Referring to Fc receptor for IgE.

FcγR Referring to Fc receptor for IgG.

FcμR Referring to Fc recepor for IgM.

FCRase Abbreviation for ferricyanide reductase.

FCS Abbreviation for fetal calf serum.

Fd Abbreviation for ferredoxin.

FD Abbreviation for freeze-dried.

Fd_{540} Symbol for a membrane-bound ferredoxin with redox potential of –0.59V.

Fd_{590} Symbol for a membrane-bound ferredoxin with redox potential of –0.54V.

Fd Fragment A protein fragment obtained from papain hydrolysis of an immunoglobulin.

FDC Abbreviation for follicular dendritic cell.

FDDC Abbreviation for ferric dimethyl dithiocarbonate.

FDG Abbreviation for fluorodeoxyglucose.

FdiI (AvaII) A restriction endonuclease from *Fremyella displosiphon* with the following specificity:

↓
5′..........GG(A/T)CC..........3′
3′..........CC(T/A)GG..........5′
↑

FdiII (MstI) A restriction endonuclease from *Fremyella diplosiphon* with the following specificity:

↓
5′..........TGCGCA..........3′
3′..........ACGCGT..........5′
↑

FDNB Abbreviation for 1-fluoro-2,4-dinitrobenzene, the Sanger's reagent for amino acid residue modification.

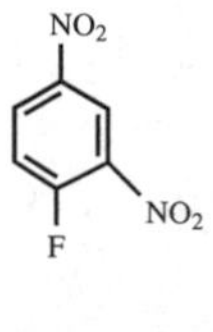

FDP Abbreviation for fructose 1,6-diphosphate (new designation: fructose 1,6-biphosphate).

F-Duction A process by which genetic material is transferred from male bacteria to female bacteria.

FdU Abbreviation for fluorodeoxyuridine.

FdUMP Abbreviation for 5-fluorodeoxyuridine monophosphate.

FdUrd Abbreviation for fluorodeoxyuridine.

FDV Abbreviation for Friend Disease Virus.

Fe Symbol for element iron.

^{59}Fe Abbreviation for radioactive iron.

Fe Protein Referring to an Fe-containing component of nitrogenase.

Febantel (mol wt 446) An anthelmintic agent.

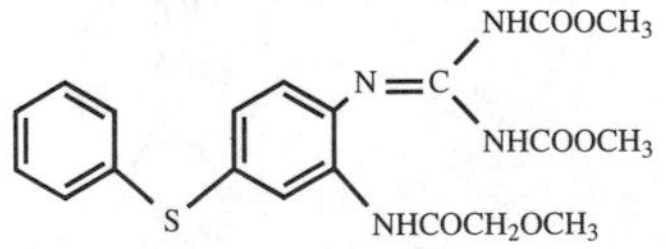

Febarbamate (mol wt 405) An antidepressant and thymoanaleptic agent.

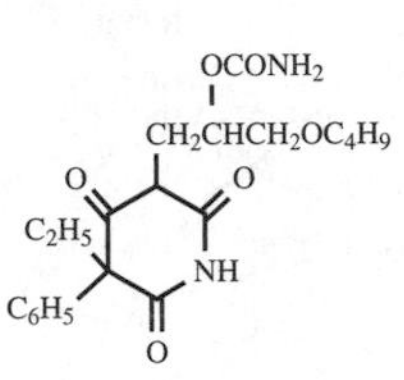

FEBP Abbreviation for fetoneonatal estrogen-binding protein.

Febridyne A trade name for a combination drug containing acetaminophen, caffeine, and butalbital, used as an antipyretic and analgesic agent.

Febuprol (mol wt 224) A choleretic agent.

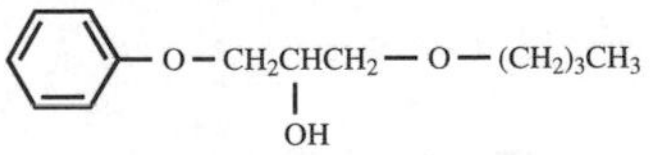

Feclemine (mol wt 359) An antispasmodic agent.

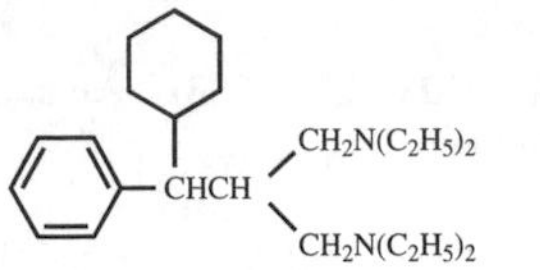

Fe(III)-Cyt-c Abbreviation for ferricytochrome c.

Fedahist A trade name for a combination drug containing pseudoephedrine hydrochloride and chlorpheniramine maleate, used as an antihistaminic agent.

Fedrine A trade name for ephedrine hydrochloride, a bronchodilator that stimulates both alpha- and beta-adrenergic receptors.

Feedback A mechanism by which the output of a system acts to modify that reaction system.

Feedback Activation The binding of the reaction product in a pathway onto the allosteric site of an allosteric enzyme to activate the enzyme or to increase the enzyme activity.

Feedback Inhibition The inhibition of the activity of the first enzyme of a biosynthetic pathway by the end product of that pathway.

Feeder Cell An irradiated cell that is capable of metabolizing but not dividing.

Feeder Pathway A pathway that supplies metabolites to another metabolic pathway.

Fehling's Test A test that determines reducing sugar based upon the reduction of cupric ions by a reducing sugar.

Felbamate (mol wt 238) An antiepileptic agent.

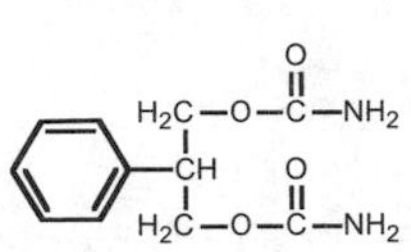

Felbatol A trade name for felbamate, an antiepileptic agent.

Felbinac (mol wt 212) An anti-inflammatory and analgesic agent.

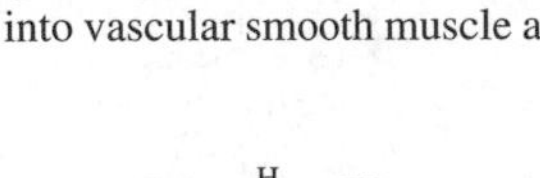

Feldene A trade name for piroxicam, an anti-inflammatory, analgesic, and antipyretic agent.

Felodipine (mol wt 384) An antihypertensive and antianginal agent that blocks the entry of calcium ions into vascular smooth muscle and cardiac cells.

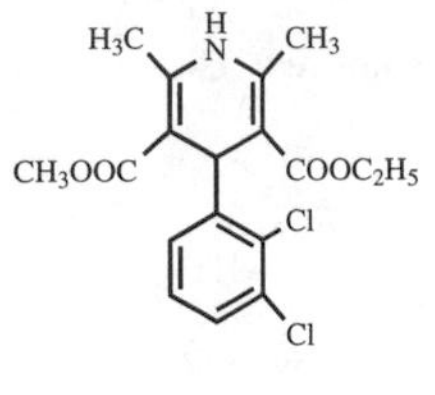

FeLV Abbreviation for feline leukemia virus.

Felypressin (mol wt 1040) A vasoconstrictor and a bioactive peptide.

Femara A trade name for letrozole, an anti-estrogen.

Femcap A trade name for a combination drug containing acetaminophen, caffeine, and butalbital, used as an analgesic and antipyretic agent.

Femcet A trade name for a combination drug containing acetaminophen, caffeine, and butalbital, used as an analgesic and antipyretic agent.

Feminate A trade name for estradiol valerate, a hormone that increases the synthesis of DNA, RNA, and protein.

Feminone A trade name for ethinyl estradiol, a hormone that increases the synthesis of DNA, RNA, and protein.

Femogex A trade name for estradiol valerate, a hormone that increases the synthesis of DNA, RNA, and protein.

FeMo Protein Referring to the nitrogenase complex that catalyzes the conversion of N_2 to NH_3.

Femoxetine (mol wt 311) An antidepressant.

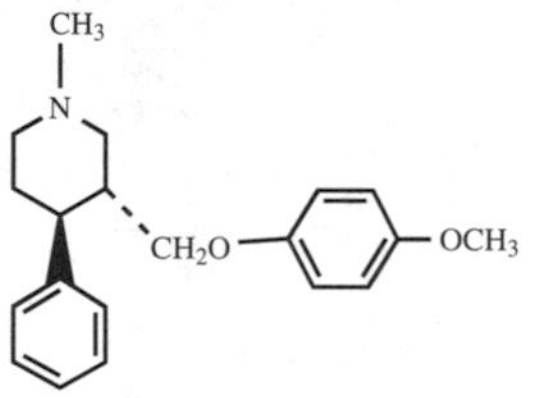

FemPatch A trade name for estradiol, a hormone.

Femstat A trade name for butoconazole nitrate, an anti-infective agent that alters the permeability fungus membrane.

Femto- A prefix denoting 10^{-15} part of.

Fenadiazole (mol wt 162) A hypnotic agent.

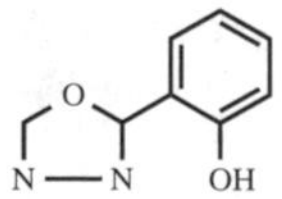

Fenalamide (mol wt 334) An antispasmodic agent.

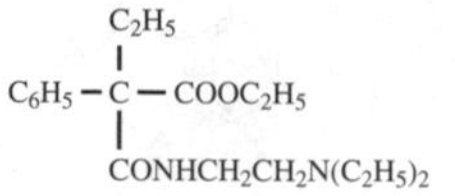

Fenalcomine (mol wt 313) A cardiac stimulant and a local anethetic agent.

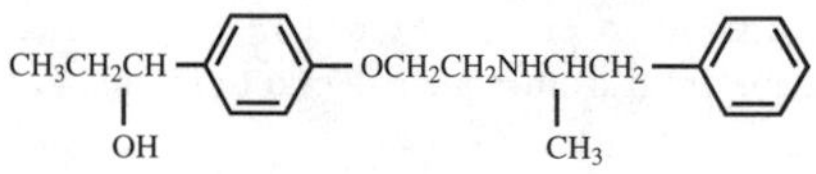

Fenbenicillin (mol wt 426) A semisynthetic antibiotic related to penicillin.

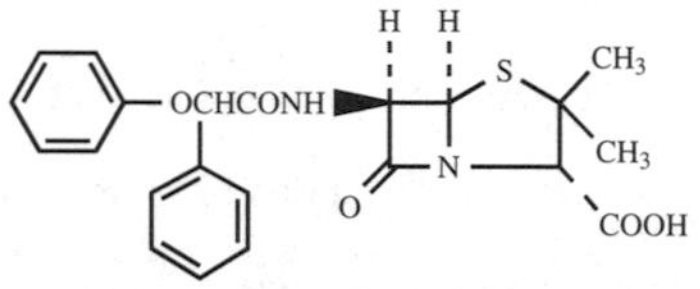

Fenbufen (mol wt 254) An anti-inflammatory agent.

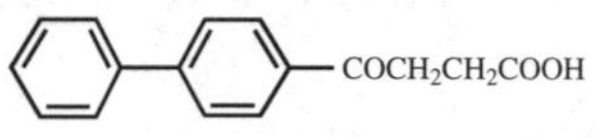

Fenbutrazate (mol wt 367) An anorexic agent.

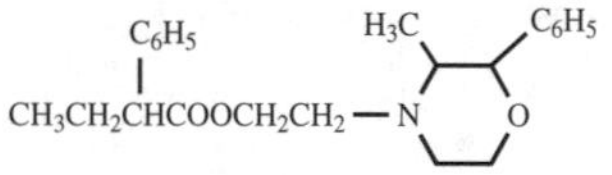

Fencibutirol (mol wt 262) A choleretic agent.

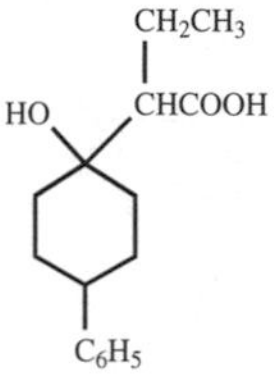

Fenclofenac (mol wt 297) An anti-inflammatory agent.

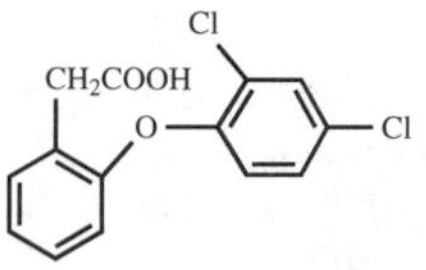

Fenclorac (mol wt 278) An anti-inflammatory agent.

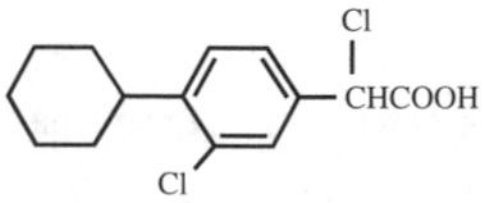

Fenclozic Acid (mol wt 253) An anti-inflammatory agent.

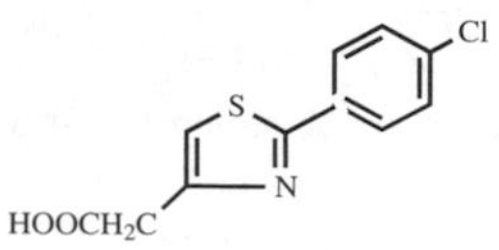

Fendiline (mol wt 315) A coronary vasodilator.

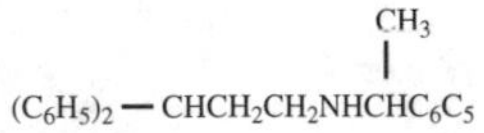

Fendosal (mol wt 381) An anti-inflammatory agent.

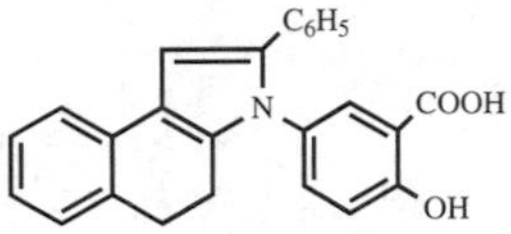

Fenesin DM A trade name for a combination drug containing dextromethorphan hydrobromide and guaifenesin used as an expectorant and antitussive agent.

Fenethazine (mol wt 270) An antihistaminic agent.

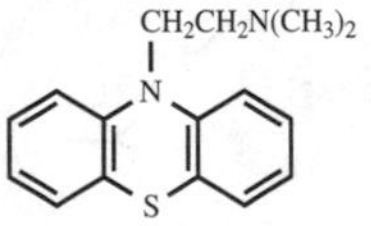

Fenethylline (mol wt 341) A CNS stimulant.

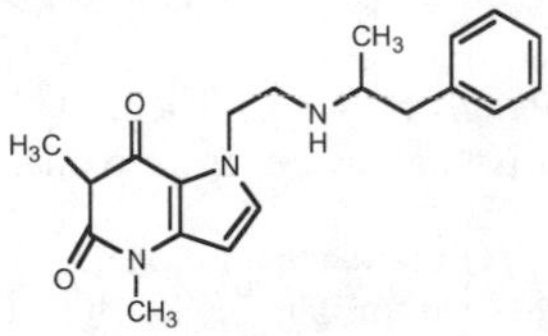

Fenfluramine (mol wt 231) A cerebral stimulant that acts on the ventromedial nucleus of the hypothalamus.

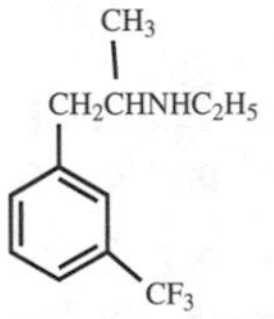

Fenicol A trade name for chloramphenicol, an antibiotic that inhibits bacterial protein synthesis.

Fenitrothion (mol wt 277) A cholinesterase inhibitor.

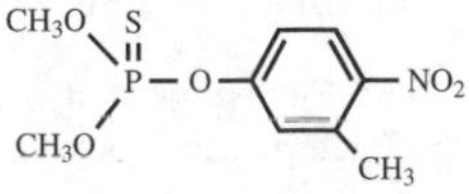

Fenofibrate (mol wt 361) An antihyperlipoproteinemic agent.

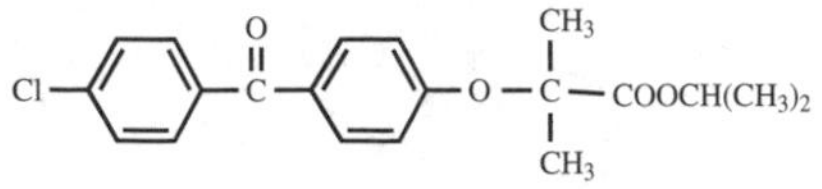

Fenoldopam (mol wt 306) An antihypertensive agent.

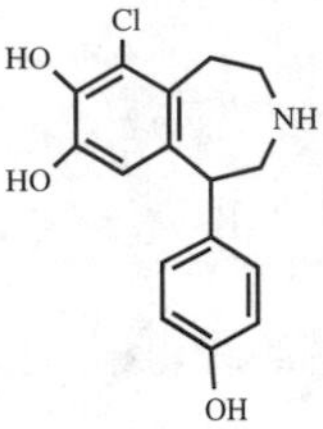

Fenoprofen (mol wt 242) An anti-inflammatory and analgesic agent.

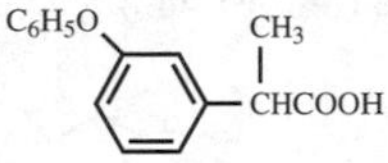

Fenoterol (mol wt 303) A bronchodilator and tocolytic agent.

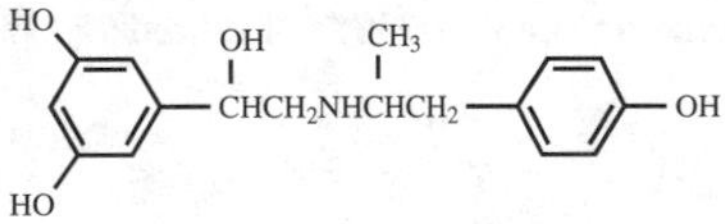

Fenoverine (mol wt 460) An antispasmodic agent.

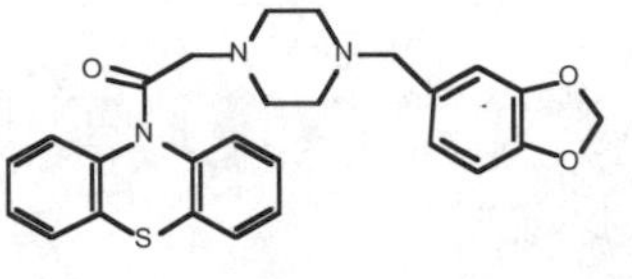

Fenoxedil (mol wt 487) A vasodilator.

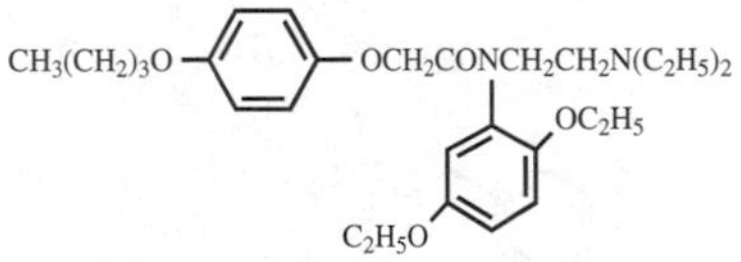

Fenpentadiol (mol wt 229) An antidepressant.

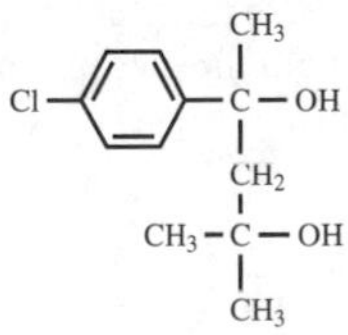

Fenpiprane (mol wt 279) An antiallergic and antispasmodic agent.

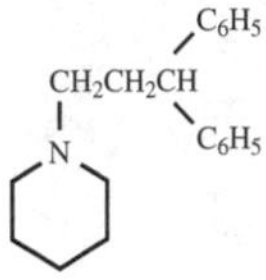

Fenpiverinium Bromide (mol wt 417) An antispasmodic agent.

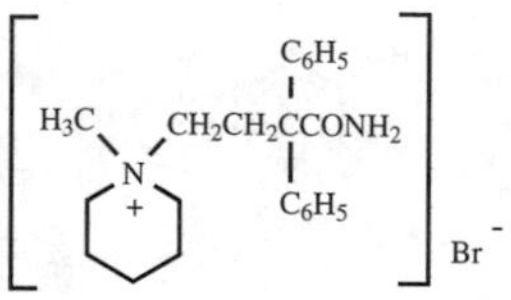

Fenproporex (mol wt 188) An anorexic agent.

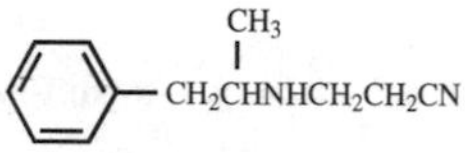

Fenprostalene (mol wt 402) A luteolysin.

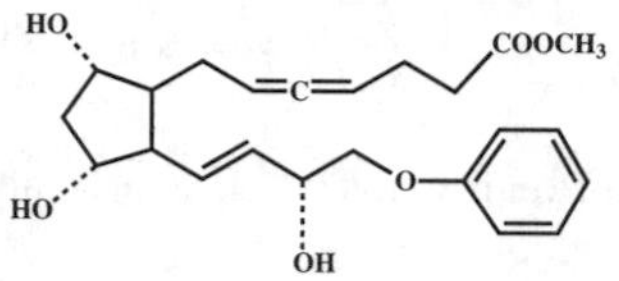

Fenquizone (mol wt 338) A diuretic agent.

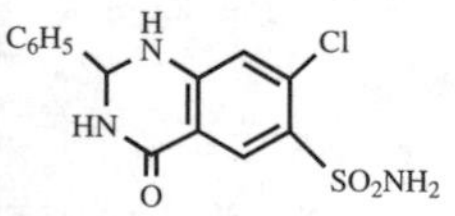

Fenretinide (mol wt 392) An antineoplastic agent.

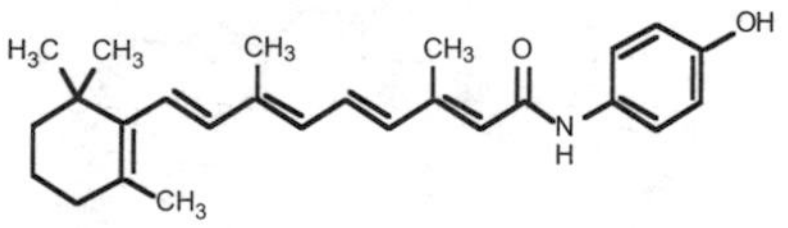

Fenspiride (mol wt 260) A bronchodilator.

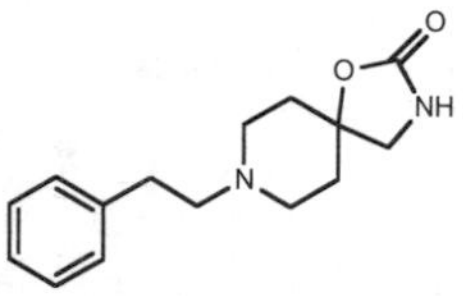

Fentanyl (mol wt 336) A narcotic analgesic agent that binds with opiate receptors at many sites in the CNS and alters both the perception and emotional response to pain.

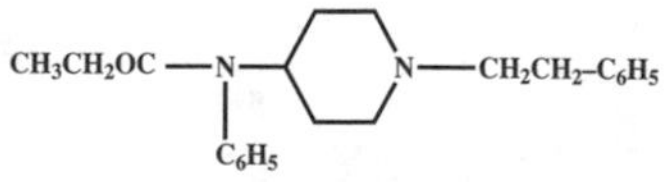

Fentiazac (mol wt 330) An inflammatory agent.

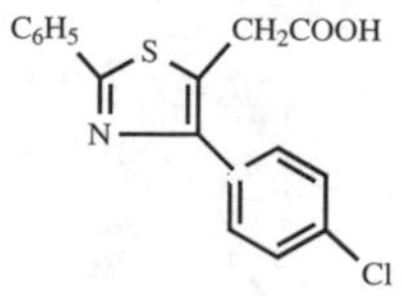

Fenticlor (mol wt 287) An anti-infective agent.

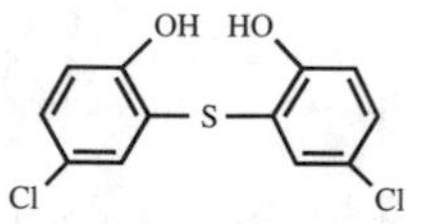

Fenticonazole (mol wt 455) An antifungal agent.

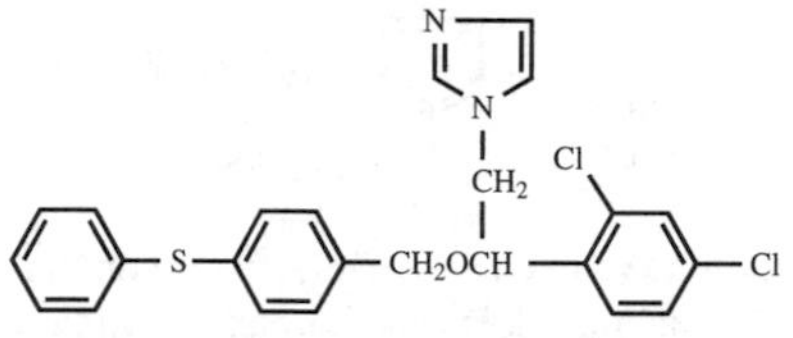

Fentonium Bromide (mol wt 564) An anticholinergic and antispasmodic agent.

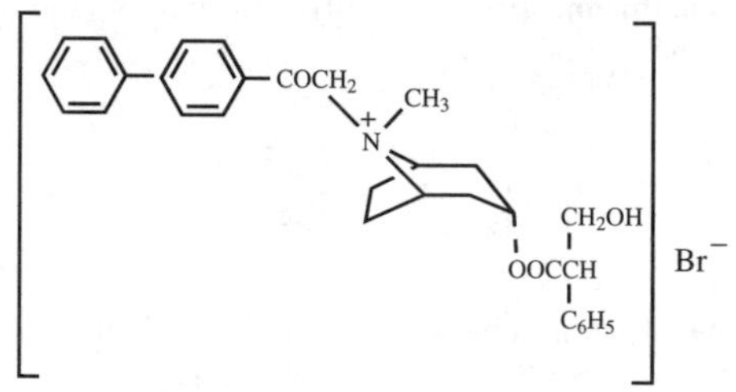

Fenton's Reagent An oxidizing reagent containing ferrous ion and hydrogen peroxide.

Fenylhist A trade name for diphenhydramine hydrochloride, an antihistaminic agent that competes with histamine for H_1 receptors on target cells.

Feosol A trade name for ferrous sulfate, a hematinic agent that provides iron for the synthesis of hemoglobin.

Feostat A trade name for ferrous fumarate, a hematinic agent that provides iron for the synthesis of hemoglobin.

Fe(II)-Oxygen Oxidoreductase The systematic name for ferroxidase.

***FepA* Protein** An outer membrane protein encoded by the *fepA* gene in *E. coli* that acts as a receptor for colicins.

F-Episome Referring to fertility factor.

F′-Episome The F-episome that carries genetic elements of the bacterial chromosome.

Fepradinol (mol wt 209) An anti-inflammatory agent.

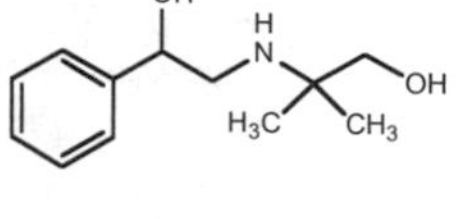

Feprazone (mol wt 320) An anti-inflammatory agent.

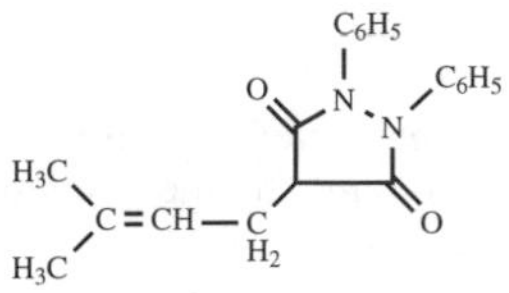

Fe-Protein A component of the enzyme nitrogenase that contains iron.

Fergon A trade name for ferrous gluconate, a hematinic agent that provides iron for the synthesis of hemoglobin.

Ferguson Equation An equation describing the relationship between the electrophoretic mobility of a protein and the concentration of polyacrylamide.

Fer-In-So A trade name for ferrous sulfate used to elevate the serum iron concentration.

Fer-Iron A trade name for ferrous sulfate used to elevate the serum iron concentration.

Fergon A trade name for ferrous gluconate used to elevate serum iron concentration.

Feritard A trade name for ferrous sulfate, a hematinic agent that provides iron for the synthesis of hemoglobin.

Fermalox A trade name for a combination drug containing ferrous sulfate and magnesium hydroxide, used as a hematinic agent.

Fermentation Breakdown of carbohydrate or fuel molecule via non-oxygen-requiring pathways. The fermentation products may be ethanol, lactic acid, or butanol, etc.

Fermentor A device in which fermentation is carried out.

Fermocolase A trade name for catalase.

Ferndex A trade name for dextroamphetamine sulfate, a cerebral stimulant that promotes transmission of nerve impulses by releasing the stored norepinephrine from the nerve terminals in the brain.

Ferocyl A trade name for a combination drug containing iron and docusate sodium, used as a hematinic agent.

Ferodan A trade name for ferrous salt used to elevate serum iron concentration.

Fero-Folic-500 A trade name for iron with vitamin C and folic acid for enhancement of iron absorption.

Fero-Grad A trade name for ferrous sulfate, a hematinic agent that provides iron for the synthesis of hemoglobin.

Fero-Gradumet A trade name for ferrous sulfate, a hematinic agent that provides iron for the synthesis of hemoglobin.

Ferolix A trade name for ferrous sulfate, a hematinic agent that provides iron for the synthesis of hemoglobin.

Ferospace A trade name for ferrous sulfate, a hematinic agent that provides iron for the synthesis of hemoglobin.

Ferralet A trade name for ferrous gluconate, a hematinic agent that provides iron for the synthesis of hemoglobin.

Ferralyn A trade name for ferrous sulfate, a hematinic agent that provides iron for the synthesis of hemoglobin.

Ferranol A trade name for ferrous fumarate, a hematinic agent that provides iron for the synthesis of hemoglobin.

Ferredoxin An iron-containing protein that acts as an electron carrier in photosynthesis and nitrogen fixation.

Ferredoxin NAD Reductase The enzyme that catalyzes the noncyclic transfer of electrons from the reduced form of ferredoxin to NAD^+.

$$\text{Reduced ferredoxin} + NAD^+ \rightleftharpoons \text{NADH} + \text{ferredoxin}$$

Ferredoxin NADP Reductase The enzyme that catalyzes the noncyclic transfer of electrons from the reduced form of ferredoxin to $NADP^+$.

$$\text{Reduced ferredoxin} + NADP^+ \rightleftharpoons \text{NADPH} + \text{ferredoxin}$$

Ferredoxin NADP⁺ Oxidoreductase The systematic name for ferredoxin-$NADP^+$ reductase.

Ferredoxin Nitrate Reductase The enzyme that catalyzes the following reaction:

$$\text{Nitrite} + H_2O + 2\ \text{oxidized ferredoxin} \rightleftharpoons \text{Nitrate} + 2\ \text{reduced ferredoxin}$$

Ferredoxin Nitrite Reductase The enzyme that catalyzes the following reaction:

$$\text{Ammonia} + 3\ \text{oxidized ferredoxin} \rightleftharpoons \text{Nitrite} + 3\ \text{reduced ferredoxin}$$

Ferric Iron that has a valence of 3 (Fe^{+3}).

Ferric Sodium Edetate (mol wt 367) An iron source.

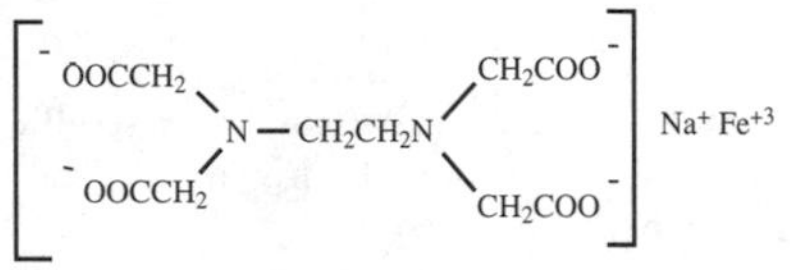

Ferrichromes A group of iron cyclic hexapeptides containing 3 glycine residues and 3 hydroxyornithine residues.

Ferricytochrome A cytochrome containing Fe^{+3}.

Ferrihemoprotein P-450 Reductase Synonym of NADPH-ferrihemoprotein reductase.

Ferrihemoprotein Reductase The enzyme that catalyzes the following reaction:

$$2\ \text{Ferrihemoprotein} + \text{NADPH} \rightleftharpoons \text{NADP}^+ + 2\ \text{ferrocytochrome}$$

Ferritin A protein that contains about 20% iron and serves as an electron-dense marker.

Ferritin-Labeled Antibody A ferritin-antibody complex used for locating the antigen-antibody complex in specimens for electron microscopy.

Ferrlecit A trade name for sodium ferric gluconate complex used to replete the total body content of iron.

Ferro- A prefix denoting containing ferrous.

Ferrocheletase The enzyme that catalyzes the following reaction:

$$Fe^{+2} + \text{protoporphyrin IX} \rightleftharpoons \text{Protoheme} + 2\ H^+$$

Ferrocytochrome A cytochrome containing Fe^{+2}.

Ferrocytochrome Nitrate Oxidoreductase The systematic name for nitrate reductase (cytochrome).

Ferrocytochrome Oxygen Oxidoreductase The systematic name for cytochrome-c oxidase.

Ferrous Iron element that has a valence of 2 (Fe^{+2}).

Ferrous Fumarate (mol wt 170) A hematinic agent that provides iron for the synthesis of hemoglobin.

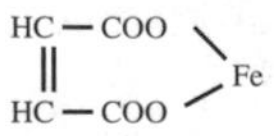

Ferrous Gluconate (mol wt. 446) A hematinic agent that provides iron for the synthesis of hemoglobin.

$$Fe[HOCH_2(CHOH)_4CO_2]_2$$

Ferrous Succinate (mol wt 172) A hematinic agent that provides iron for the synthesis of hemoglobin.

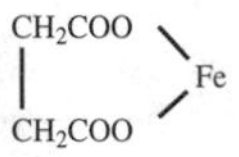

Ferrous Sulfate (mol wt 152) A hematinic agent that provides iron for the synthesis of hemoglobin.

$$FeSO_4$$

Ferroxidase The enzyme that catalyzes the following reaction:

$$4\ Fe(II) + 4\ H^+ + O_2 \rightleftharpoons 4\ Fe(III)\ 2\ H_2O$$

Fersamal A trade name for ferrous fumarate, a hematinic agent that provides iron for the synthesis of hemoglobin.

Fertility Factor A plasmid present in some bacterial cells that confers maleness and sexual conjugation (see also F factor).

Fertilization Membrane The membrane that grows outward from the point of contact between the egg and sperm.

Fertilizin Referring to the substance secreted by the ovum to attract sperm.

Fertinex A trade name for urofollitropin, a hormone.

Fertinorm HP A trade name for urofollitropin, a hormone.

Ferulic Acid (mol wt 194) A food preservative.

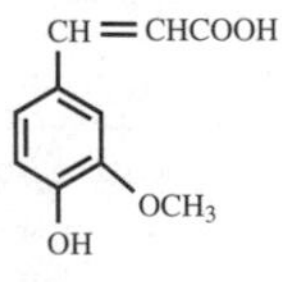

Fertinic A trade name for ferrous gluconate, a hematinic agent that provides iron for the synthesis of hemoglobin.

fes An oncogen present in feline sarcoma virus.

Fe-SOD Abbreviation for iron-containing superoxide dismutase.

Fesofor A trade name for ferrous sulfate used to elevate iron concentration.

Fespan A trade name for ferrous sulfate, a hematinic agent that provides iron for the synthesis of hemoglobin.

Festal II Tablets A trade name for pancrelipase, a digestant.

Fetal Antigen Tumor-associated antigen that is present in the embryonic stage but not in adult tissues. It reappears during the development of certain types of cancer.

Fetal Hemoglobin A hemoglobin present in the developing fetus, it rapidly diminishes after birth. The reappearance of these proteins in the adult tissue may serve as a cancer indicator.

Feticide An agent that destroys a fetus in the uterus.

Feto- A prefix denoting fetus.

α-Fetoprotein A tumor marker protein. It occurs in adult tissue in association with certain types of cancer.

Fetrin A trade name for a nutritional combination of ferrous fumarate, ascorbic acid, and cyanocobalamin.

Feulgen Reaction A reaction for staining DNA with Feulgen reagent.

Feulgen Reagent A Schiff reagent used for staining DNA.

FEUO Abbreviation for external use only.

Fexofenadine (mol wt 502) An antihistaminic agent.

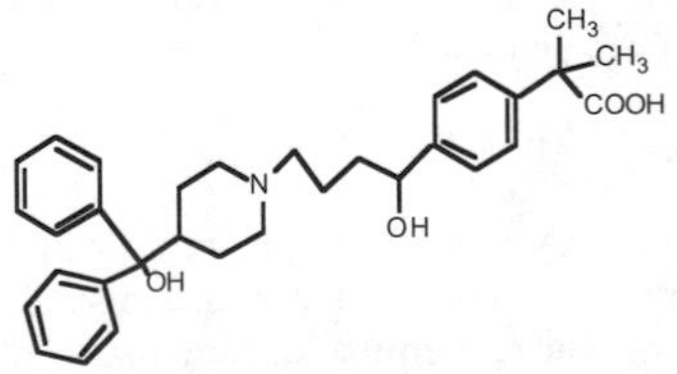

FF Abbreviation for 1. fat-free; 2. fertilization factor; 3. filtration fraction.

FFA Abbreviation for free fatty acid.

F_oF_1-ATPase The proton-translocating ATP synthetase present in mitochondria, chloroplast, and bacteria.

FFT Abbreviation for fast Fourier transform.

FGAR Abbreviation for formylglycinamide ribotide, an intermediate in the biosynthesis of inosine monophosphate.

FGF Abbreviation for fibroblast growth factor.

FGF-2 Abbreviation for fibroblast growth factor-2.

FGFR Abbreviation for fibroblast growth factor receptor.

FGR A gene encoding a non-receptor tyrosine kinase that transforms most mammalian fibroblasts but not epithelial cells.

FH Abbreviation for familial hypercholesterolemia.

FH_2 Abbreviation for dihydrofolic acid.

FH_4 Abbreviation for tetrahydrofolic acid.

FIA Abbreviation for 1. fluoro-immuno-assay; 2. freund's incomplete adjuvant.

Fialuridine (mol wt 372) An antiviral agent.

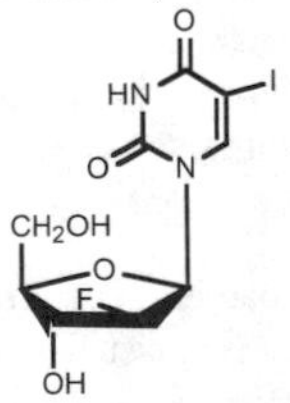

Fiber Autoradiography A type of autoradiography in which the tritium-labeled DNA fibers are stretched out on a slide.

Fiberall A trade name for polycarbophil, a laxative.

FiberCon A trade name for calcium polycarbophil, a laxative.

Fiberoptics The use of synthetic fibers with special optical properties for transmission of light images and relay of pictures from inside the body.

Fiberscope An endoscope that uses fiberoptics for transmission of images from the interior of the body.

Fibrillin A glycoprotein and the calcium-binding component of connective tissue microfibrils isolated from fibroblast cells.

Fibrin An elastic, fibrous, insoluble protein in blood clots.

Fibrin Stabilizing Factor See factor XIII.

Fibrinase A protease that catalyzes the preferential cleavage of peptide bonds involving carboxyl groups of arginine and lysine and converts fibrin to soluble products (also known as plasmin, fibrinolysin).

Fibrinogen A soluble glycoprotein and precursor of fibrin.

Fibrinogenase A protease that catalyzes the selective cleavage of peptide bonds between arginine and glycine in fibrinogen to form fibrin-releasing fibrinopeptides A and B.

Fibrinoligase Synonym of protein glutamine γ-glutamyltransferase.

Fibrinolysin See fibrinase or plasmin.

Fibrinolysis The degradation of fibrin.

Fibrinopeptide The nonfibrin peptide resulting form the conversion of fibrinogen to fibrin by the enzyme thrombin or fibrinogenase.

Fibroadenoma A tumor consisting of a large quantity of fibrous tissue.

Fibroblast A type of cell derived from mesoderm that gives rise to interstitial tissue and is capable of synthesizing fibrous proteins (e.g., collagen).

Fibroblast Interferon Interferon produced by the fibroblast in mammalian connective tissue (also known as beta interferon).

Fibrocyte An inactive cell present in fully differentiated connective tissue. It is derived from a fibroblast.

Fibrogenopenia Lack of fibrinogen in the blood.

Fibroin Protein filament of silk fiber.

Fibrolysin A proteolytic enzyme capable of dissolving a fibrin.

Fibroma A benign tumor of connective tissue.

Fibromyositis Inflammation of the fibro-muscular tissue.

Fibronectin A dimeric protein produced by mast cells and macrophages. It is a chemotactic factor and plays an important role in tissue repair and cell-to-cell adhesion.

Fibroplasia The production of the fibrous tissue during the healing of wounds.

Fibrosarcoma A tumor derived from collagen-producing fibroblasts.

Fibrosis The thickening and scarring of connective tissue.

Fibrositis Inflammation fibrous connective tissue.

Fibrous Actin See F-actin.

Fibrous Protein Protein with an extensive a-helix or b-pleated sheet structure that confers a highly ordered fibrous or sheet structure.

Fibulin A calcium-binding glycoprotein of the extracellular matrix.

FICA Abbreviation for fluoro-immuno-cyto-adherence.

Ficain A protease with activity similar to papain.

Ficin See ficain.

Fick's First Law of Diffusion The law states that the rate of diffusion of a substance in the direction to eliminate concentration gradient is proportional to the magnitude of the gradient.

Ficoll A synthetic, water soluble, nonionic copolymer of sucrose and epichlorhydrin used to prepare density gradients for centrifugation.

Ficoll-Paque A trade name for an aqueous solution of ficoll and diatrizoate sodium used for the one-step separation of lymphocytes from anticoagulant-treated blood.

FID Abbreviation for 1. flame ionization detector; 2. free induction decay.

Field's Stain A staining procedure for detecting species of *plasmodium* and trypanosomes in thick blood smears. It is carried out by staining first with azure and then with eosin.

FIF Abbreviation for 1. feedback inhibition factor; 2. fibroblast inferferon.

FIGE Abbreviation for field inverted gel electrophoresis.

FIGEP Abbreviation for field inverted gel electrophoresis.

FIH Abbreviation for fat-induced hyperglycemia.

Fijivirus A plant virus (Reoviridae) causing Fijπi disease of sugarcane.

Filaform Threadlike form.

Filaggrin A protein that aggregates keratin intermediate filaments and promotes disulfide bond formation during terminal differentiation of mammalian epidermis.

Filament A threadlike structure found in tissues and cells of the body.

Filamentous Charactarized by a threadlike structure.

Filamentous Phage Male-specific, single-stranded DNA bacteriophages (Inoviridae) that infect bacterial pilli (e.g., M13, fd).

Filamin A fibrous protein that links actin filaments into a random, fibrous, three-dimensional network.

Filaria A genus of parasitic, threadlike nematode.

Filariasis A disorder or infection caused by the *Filaria* nematode.

Fildes' Enrichment Agar An agar medium that contains factor X and V used to isolate species of *Haemophilus*.

Filial Generation Any progeny after the parental generation.

Filibacter A genus of Gram-negative, strictly aerobic, filamentous gliding bacteria.

Filipin An antifungal antibiotic complex from *Streptomyces filipenesis*, e.g., filipin II, III, and IV.

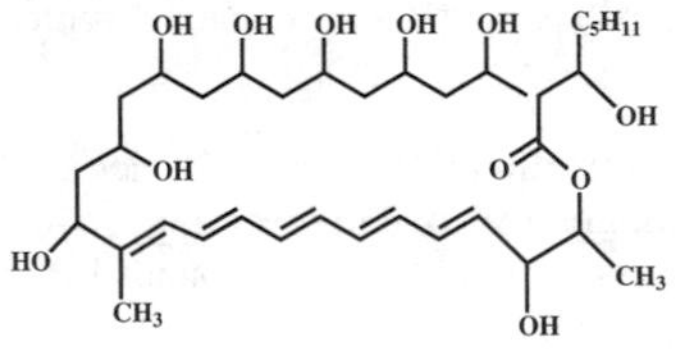

pilipin III

Filter A porous sheet through which filterable and nonfilterable materials can be separated.

Filter Binding Assay A technique to study how ligands bind onto a filter that has bonded receptors, e.g., proteins or nucleic acids.

Filter Enrichment A technique for recovering auxotrophic mutants of filamentous fungi.

Filter Hybridization A method of hybridization in which denatured DNA fragments immobilized on nitrocellulose paper are incubated with radioactively labeled RNA or DNA probe for assaying hybridization.

Filtrate Substance that has passed through a filter.

Filtration The passage of a fluid under pressure through a membrane or filter.

Fimbriae Thin, short, numerous, proteinaceous, plasmid-encoded appendages on the surface of certain bacterial cells. They are involved in cell-to-cell adhesion.

Fimbrin An actin-binding protein.

FinII (HpaII) A restriction endonuclease from *Flavobacterium indologenes* with the same specificity as HpaII.

Finasteride (mol wt 373) An androgen hormone inhibitor used for treatment of prostatic hypertrophy.

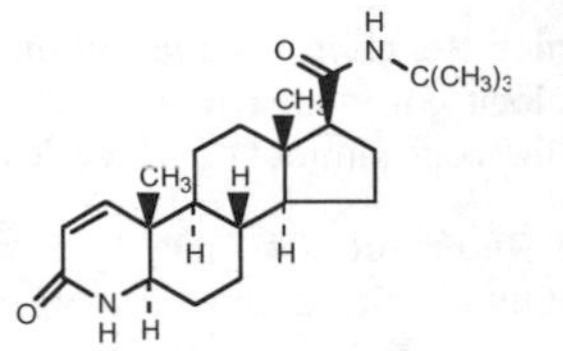

Fingerprinting A two-dimensional separation technique to analyze different relationships of proteins or nucleic acids employing chromatography (first dimension) and electrophoresis (second dimension).

FinSI (HaeIII) A restriction endonuclease from *Flavobacterium indoltheticum* with the same specificity as HaeIII.

Fioricet A trade name for a combination drug containing acetaminophen, caffeine, and butalbital, used as an antipyretic and analgesic agent.

Fiorinal A trade name for a combination drug containing butalbital, aspirin, phenacetin, and caffeine, used as an antipyretic and analgesic agent.

Fiorpap A trade name for a combination drug containing butalbital, APAP, and caffeine.

Fiortal A trade name for a combination drug containing butalbital, aspirin, and caffeine.

FIR Abbreviation for far infrared.

Firefly Luciferase The enzyme that catalyzes the following reaction:

$$\text{Luciferin} + O_2 + \text{ATP} \rightleftharpoons \text{Oxidized luciferin} + CO_2 + H_2O + \text{AMP} + \text{PPi} + h\nu$$

Firefly Luciferin (mol wt 280) A light emission substance from the firefly (*photinus pyralis*).

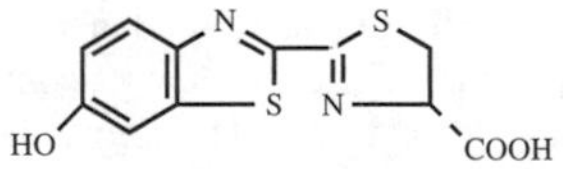

First Law of Photochemistry The law states that energy must be absorbed by the molecule before a photochemical reaction can take place.

First Law of Thermodynamics The law states that energy can be converted from one form to another but cannot be created or destroyed.

First Messenger The external signals that initiate a multicyclic cascade reaction system, e.g., binding of epinephrine (first messenger) on the cell surface for the activation of adenylate cyclase in the cascade metabolism of carbohydrate.

First Order Reaction A reaction in which the rate or velocity of the reaction is directly proportional to the concentration(s) of the reactant(s).

First Set Rejection The initial rejection of the transplant by a recipient in an incompatible transplant.

Fischerella A genus of filamentous, thermophilic Cyanobacteria.

FISH Abbreviation for fluorescence *in situ* hybridization.

Fiske-SubbaRow Method A colorimetric method for the determination of phosphate with ammonium molybdate and aminonaphthol sulfonic acid.

Fission 1. A type of asexual reproduction in which a cell divides in half to form two identical daughter cells. 2. The splitting of a large atom into two atoms of lighter elements with the release of a large quantity of energy.

Fissure A split or groove on the surface of an organ.

FITC (mol wt 389) Abbreviation for fluorescein isothiocyanate. A reactive derivative of fluorescein used for labeling proteins.

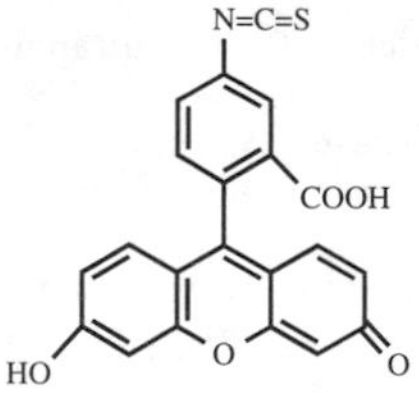

Fite-Faraco Stain An acid-fast stain for detecting *Mycobacterium leprae* in tissue section.

FIUO Abbreviation for internal use only.

Fixation 1. The conversion of a raw material into a biologically useful form, e.g., conversion of CO_2 into carbohydrate. 2. The preparation of tissues for cytological and histological examination.

Fixative Substance used for fixation of tissue for histological examination.

Fixed Allele An allele for which all members in the population are homozygous and no other alleles for this locus exist in the population.

Fixed Breaking Point The point in dsDNA from which unwinding starts.

Fixed Virus A strain of rabies virus attenuated by serial passage through rabbits.

FK Abbreviation for forskolin.

FL Abbreviation for firefly luciferase.

***fla* Gene** A gene that encodes flagella.

Flagella Plural of flagellum.

Flagellates Organisms having flagella.

Flagellin Protein subunit that constitutes the bacterial flagellum.

Flagellum A long hairlike locomotory appendage on the surface of a cell.

Flagyl A trade name for metronidazole, an antiprotozoa drug.

Flamazine A trade name for silver sulfadiazine, an anti-infective agent that is active against both Gram-positive and Gram-negative bacteria and yeast.

Flame Ionization Detector A device used in gas chromatography to detect organic compounds through ionization of the compounds by a flame.

Flame Photometry A technique for determination of an element in solution by measuring the emission spectrum produced by spraying the element into a flame.

Flash Chromatography A type of column chromatography in which the separation process is accelerated by the use of external pressure on the eluting solvent.

Flash Evaporator A device for removal of solvent from a sample by rotating evaporation under vacuum.

Flash Fermentation A type of fermentation in which the volatile product resulting from fermentation is removed by circulating the broth from the fermentor into a vacuum where volatile product is removed.

Flatulex A trade name for a combination drug containing simethicone and activated charcoal, used as an antacid.

Flavanone (mol wt 224) A bioflavonoid.

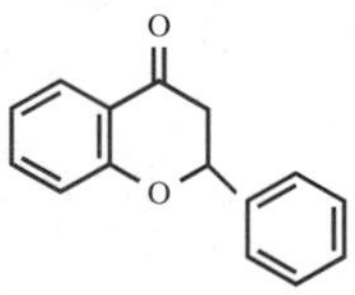

Flavettes A trade name for vitamin C.

Flavianic Acid (mol wt 314) A reagent used for precipitating arginine and tyrosine.

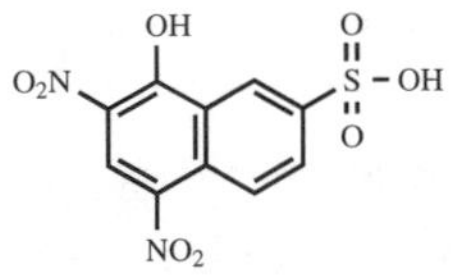

Flavin Adenine Dinucleotide (mol wt 786) An electron carrier molecule that acts as a coenzyme in energy-transfer reactions (abbreviated as FAD).

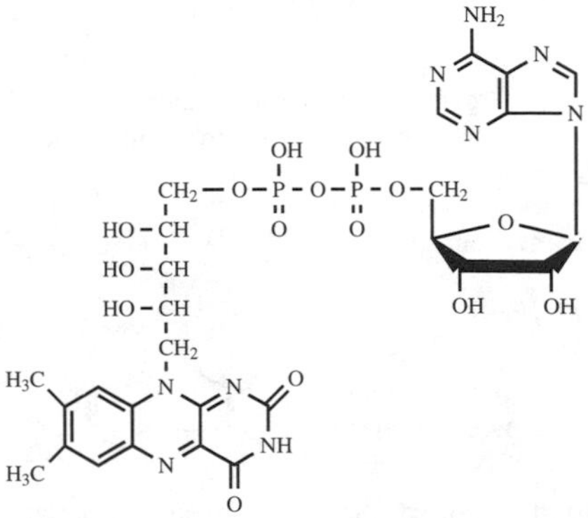

Flavin Enzyme Any enzyme that is a flavoprotein.

Flavin Mononucleotide (FMN) A flavin nucleotide that acts as a coenzyme in flavin-protein mediated dehydrogenation reactions (also called riboflavin phosphate).

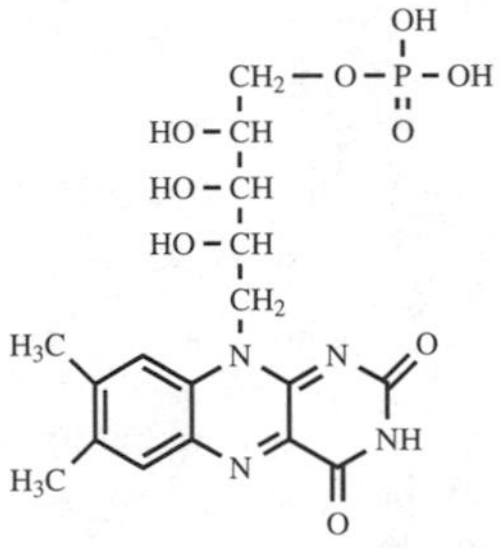

Flavin Nucleotide A general term for FAD or FMN.

Flaviridae A family of enveloped ssRNA-containing viruses, e.g., yellow fever virus.

Flavo- A prefix denoting 1. yellow; 2. containing flavin nucleotide.

Flavobacterium A genus of aerobic, chemoorganotrophic, Gram-negative bacteria.

Flavodoxin A flavoprotein capable of replacing ferredoxin as the electron carrier.

Flavoenzyme An enzyme that requires flavin nucleotide as coenzyme or prosthetic group.

Flavokinase See riboflavin kinase.

Flavone (mol wt 222) A compound isolated from *Primula malacoides* (Primulaceae).

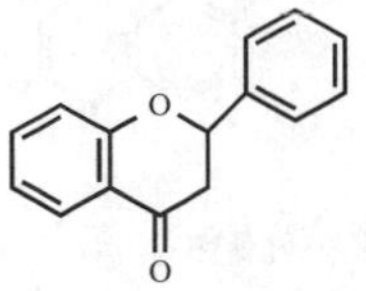

Flavonoid A group of aromatic, colored compounds that are related to the parent compound flavone, e.g., flavanone and anthocyanine.

Flavoprotein Protein that has a bound flavin nucleotide and serves as a biological electron carrier in the electron transport chain of mitochondrion.

Flavoxate (mol wt 391) An antispasmodic and local anesthetic agent.

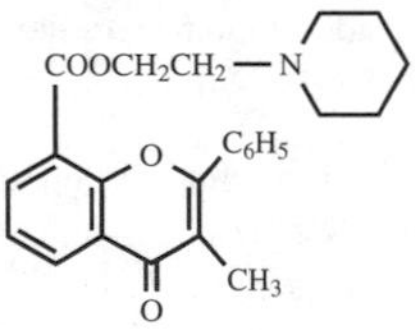

Flaxedil A trade name for gallamine triethiodide, a neuromuscular blocker that prevents acetylcholine from binding to receptors on muscle cells.

FLC Abbreviation for Friend leukemia cell.

Flecainide (mol wt 414) An antiarrhythmic agent (cardiac depressant).

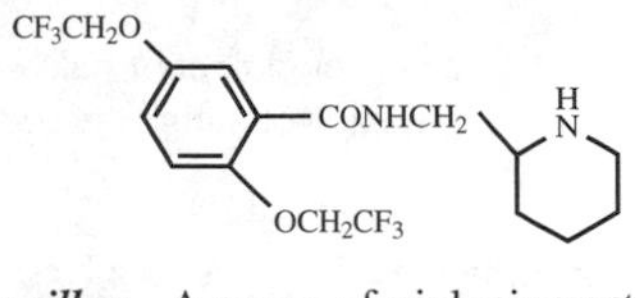

Flectobacillus A genus of pink-pigmented bacteria (Spirosomaceae).

Fleet Babylax A trade name for glycerin, a laxative that draws water from the tissues into the feces and stimulates evacuation.

Fleet Bisacodyl A trade name for bisacodyl, a laxative that promotes fluid accumulation in the colon.

Fleet Enema A trade name for a saline laxative containing monobasic sodium phosphate monohydrate and dibasic sodium phosphate heptahydrate.

Fleet Glycerin Laxatives A trade name for laxatives of glycerin suppositories or glycerin liquid suppositories.

Fleroxacin (mol wt 369) A fluorinated quinolone and antibacterial agent.

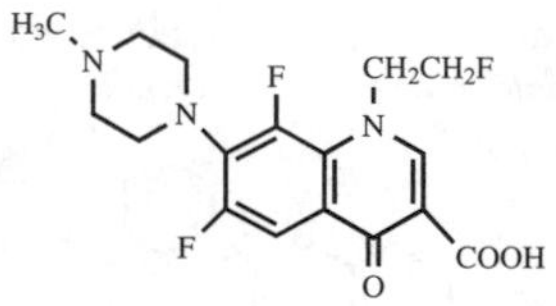

Flesinoxan (mol wt 415) An anxiolytic agent.

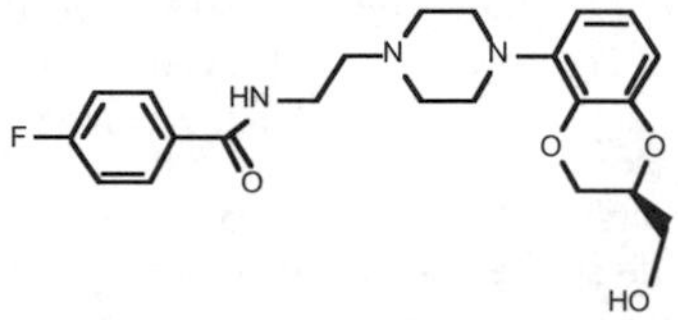

Flexeril A trade name for cyclobenzaprine, a skeletal muscle relaxant that reduces transmission of impulses from spinal cord to skeletal muscle.

Flexoject A trade name for orphenadrine citrate, a skeletal muscle relaxant.

Flexon A trade name for orphenadrine citrate, a muscle relaxant that reduces transmission of impulses from spinal cord to skeletal muscle.

Flipase The enzyme that catalyzes flip-flop movement.

Flip-Flop A mechanism for transverse movement of a molecule across a lipid bilayer by rotating the molecule through the plane of the bilayer so that the molecule can move from one monolayer to another.

FLK Abbreviation for fetal liver kinase.

Flocculation The development of flaky or fluffy precipitate, e.g., formation of antigen-antibody precipitin.

Floctafenine (mol wt 406) An analgesic agent.

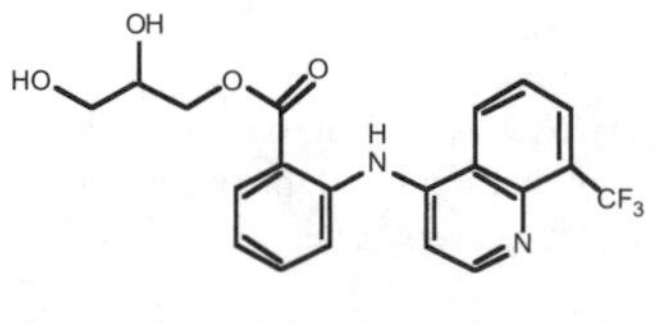

Flolan A trade name for epoprostenol, a prostaglandin that inhibits platelet aggregation.

Flomax A trade name for tamsulosin, an alpha adrenergic blocker.

Flomoxef (mol wt 496) A semisynthetic oxacephalosporin antibiotic.

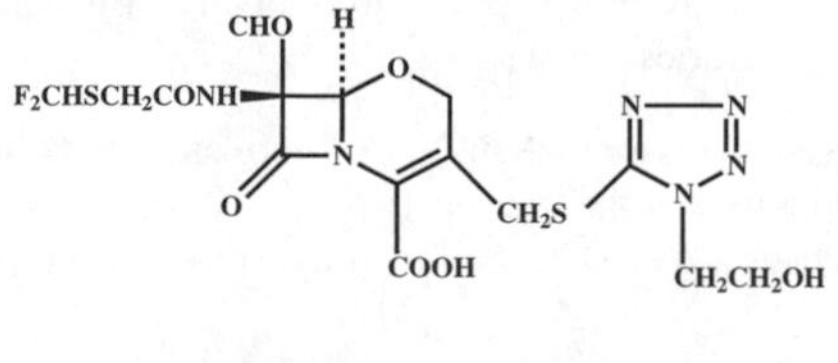

Flonase A trade name for fluticasone, an anti-allergic and anti-inflammatory agent.

Flopropione (mol wt 182) An antispasmodic agent.

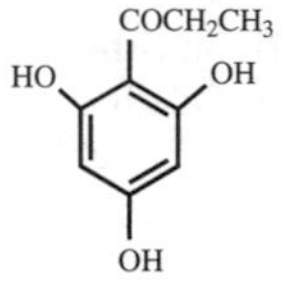

Flora Organisms or microorganisms present in a given location or organ, e.g., intestinal flora and flora of soil.

Florantyrone (mol wt 302) A choleretic agent.

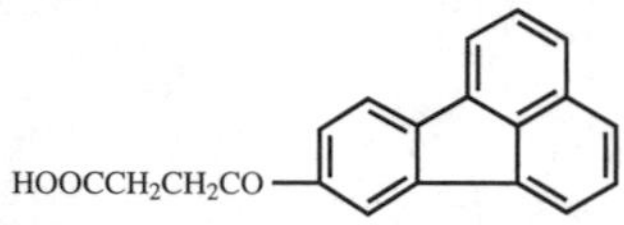

Floredil (mol wt 295) A coronary vasodilator.

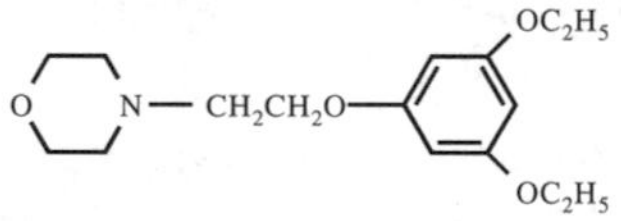

Florfenicol (mol wt 358) An antibacterial agent.

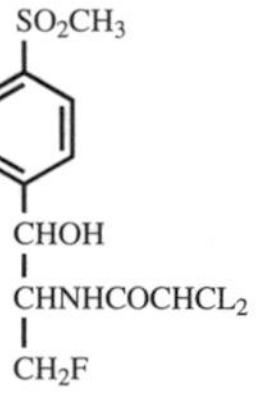

Florical A trade name for a fluoride and calcium supplement in the form of sodium fluoride and calcium carbonate.

Floridean Starch An amylopectin-like glycan in red algae.

Florinef A trade name for fludrocortisone acetate, a corticosteroid that increases sodium reabsorption and secretion of potassium and hydrogen.

Florisil A trade name for magnesium silicate, an adsorbent used in column chromatography.

Florone A trade name for diflorasone diacetate, an anti-inflammatory drug.

Floropryl A trade name for isoflurophate, an eye ointment that inhibits the destruction of acetylcholine by acetylcholinesterase.

Flosequinan (mol wt 239) An antihypertensive agent.

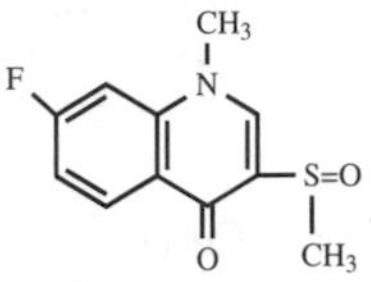

Flovent A trade name for fluticasone, an anti-allergic and anti-inflammatory agent.

Flovent Rotadisk A trade name for fluticasone, an anti-allergic and anti-inflammatory agent.

Flow Cytofluorimeter A flow cytometer for counting and monitoring fluorescence-labeled object.

Flow Cytometer A device used to analyze and to sort a mixed cell population by passing fluorescence-labeled cells through a narrow laser beam (see also flow cytometry).

Flow Cytometry A technique in which a mixed population of fluorescence-labeled cells are counted, sorted, and analyzed by passing through a narrow laser beam and by the analysis of the fluorescent signal emitted by the cells upon excitation by a laser beam.

Flow Dichroism The dichroism due to orientation of asymmetrical macromolecules when the solution containing the macromolecules is subjected to shear.

Flow Meter A device for measurement of volume of liquid or gas passing through a system.

Floxacillin (mol wt 454) A semisynthetic antibiotic active against penicillin-resistant *Staphylococci.*

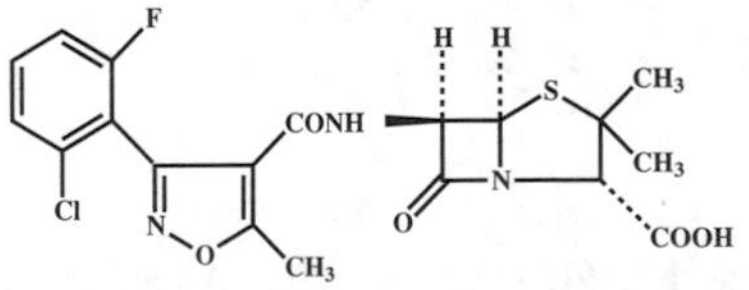

Floxin A trade name for ofloxacin, an antibiotic that inhibits bacterial DNA gyrase and prevents DNA replication.

Floxuridine (mol wt 246) An antifungal and antineoplastic agent that inhibits DNA synthesis.

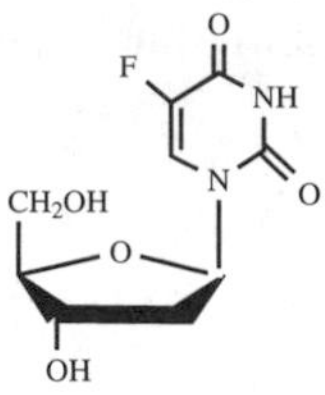

FLSP Abbreviation for fluorescein-labeled serum protein.

FLTM Abbreviation for fluorescence lifetime imaging microscopy.

Flu See influenza.

Fluacizine (mol wt 394) An antidepressant.

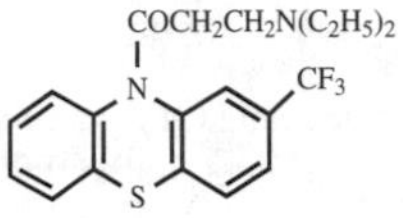

Fluanisone (mol wt 356) A neuroleptic agent.

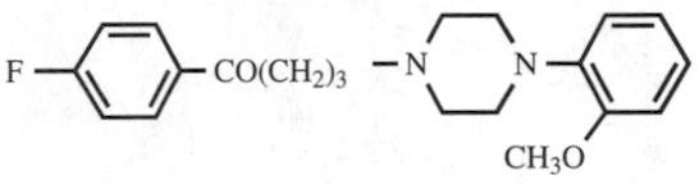

Fluazacort (mol wt 460) An anti-inflammatory agent.

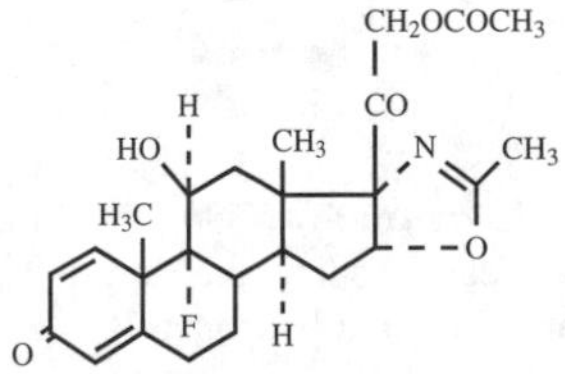

Flubendazole (mol wt 313) An anthelmintic agent.

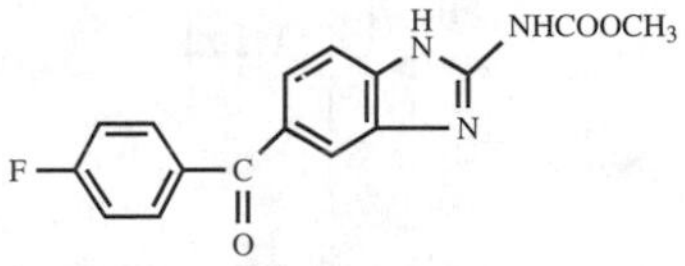

Flucloronide (mol wt 487) A glucocortioid.

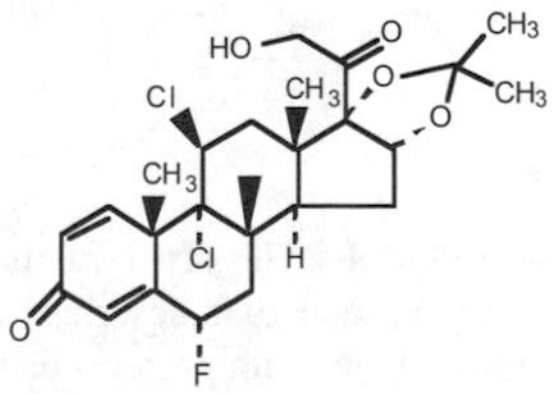

Fluconazole (mol wt 306) An antifungal agent.

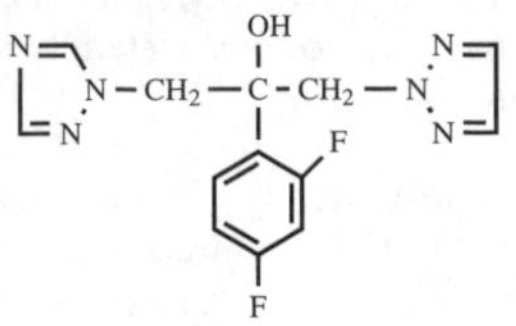

Fluctuation Test A statistical analysis designed by Luria and Delbruck to prove that bacteriophage-resistant mutants of bacteria arose prior to the exposure of the selective agent.

Flucytosine (mol wt 129) 5-Fluorocytosine, an antifungal agent that inhibits DNA synthesis.

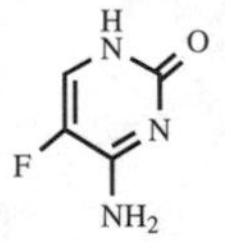

Fludara A trade name for fludarabine phosphate, an antineoplastic antimetabolite that inhibits DNA polymerase.

Fludarabine (mol wt 285) An antineoplastic agent.

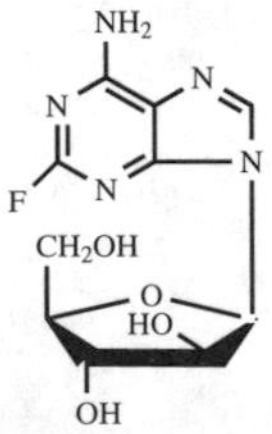

Fludrocortisone (mol wt 380) A mineral corticoid that increases sodium reabsorption and secretion of potassium and hydrogen.

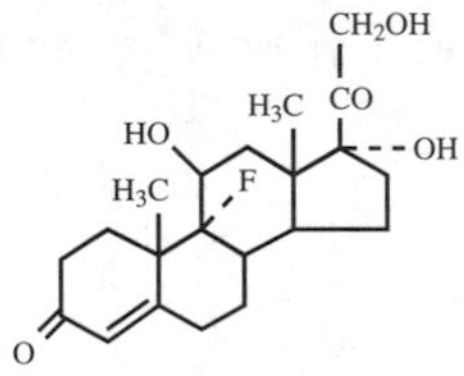

Flufenamic Acid (mol wt 281) An anti-inflammatory and analgesic agent.

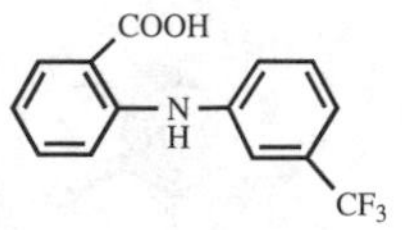

Fluid Mosaic Model The structural model of the cytoplasmic membrane that describes the membrane as a lipid bilayer with proteins distributed in a mosaic-like pattern both on the surface and in the interior of the membrane.

Fluidity The property of membranes indicating the ability of lipids to move laterally within their particular monolayers.

Fluidized Bed Reactor A fermenter in which the nutrient solution is pumped upward through a column of small particles coated with cells. The upward flow of liquid causing rapid interaction between cell-coated particles and medium.

Fluindione (mol wt 240) An anticoagulant.

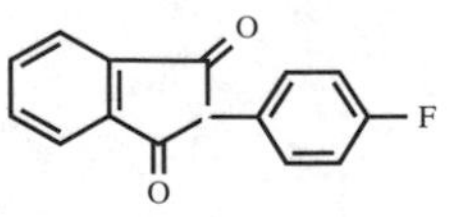

Fluke A parasitic flatworm (Trematoda).

Flumadine A trade name for rimantadine hydrochloride, an antiviral agent.

Flumazenil (mol wt 303) A benzodiazepine antagonist that competitively inhibits the action of benzodiazepine.

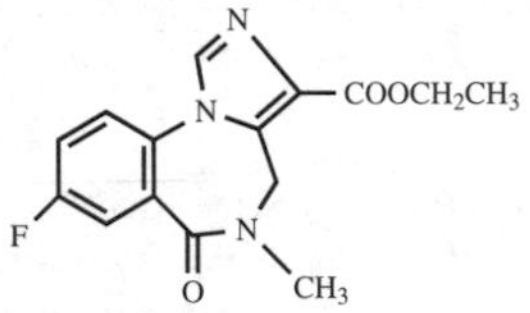

Flumedroxone Acetate (mol wt 441) An antimigraine agent.

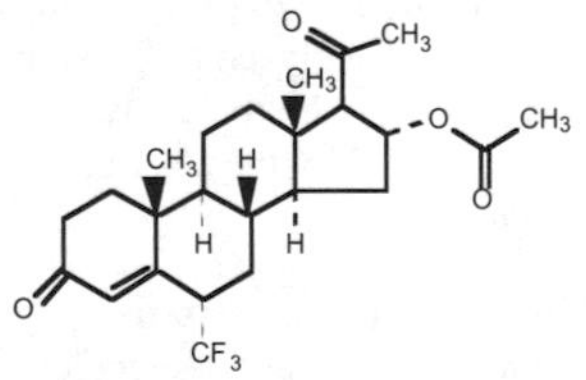

Flumethasone (mol wt 410) An anti-inflammatory agent and a glucocorticoid.

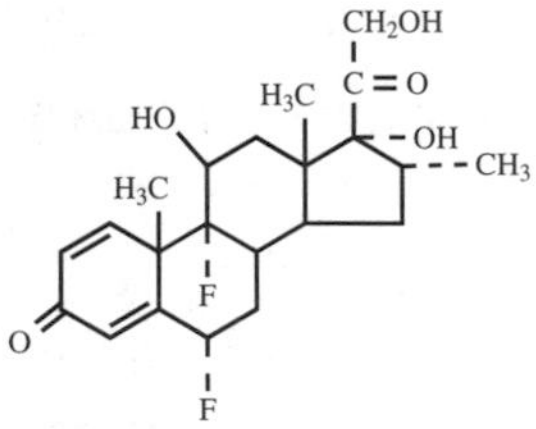

Flumethiazide (mol wt 329) A carbonic anhydrase inhibitor.

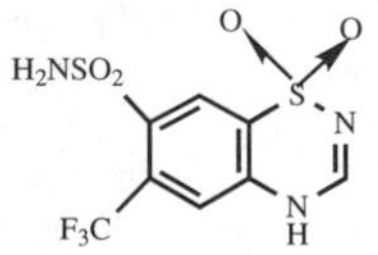

Flumethrin (mol wt 510) An insecticide.

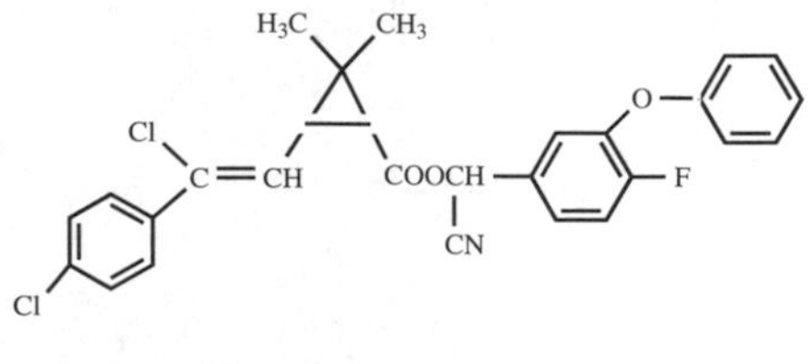

Flumetramide (mol wt 245) A skeletal muscle relaxant.

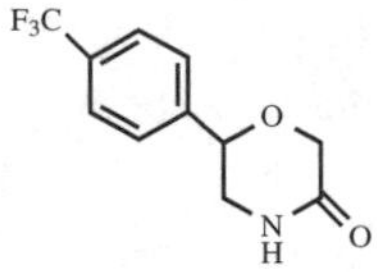

Flunarizine (mol wt 405) A calcium channel blocker and vasodilator.

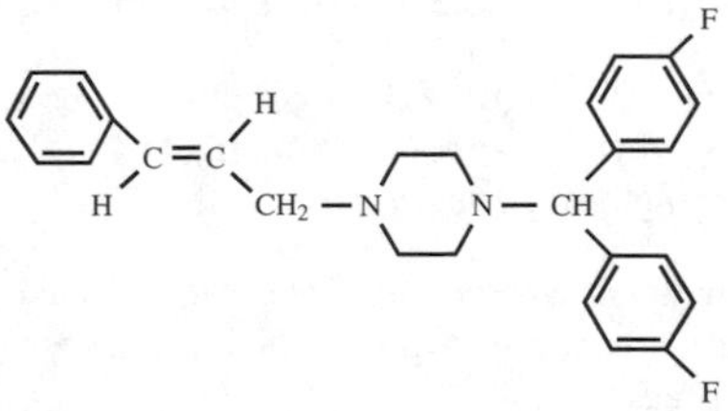

Flunisolide (mol wt 435) A synthetic fluorinated corticosteroid related to prednisolone that decreases inflammation and suppresses the immune response.

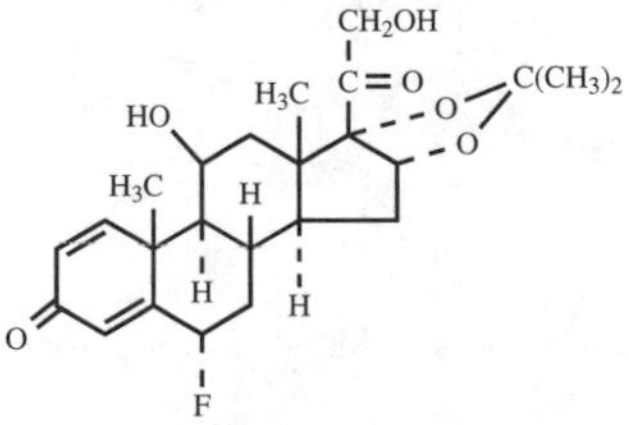

Flunitrazepam (mol wt 313) A hypnotic agent.

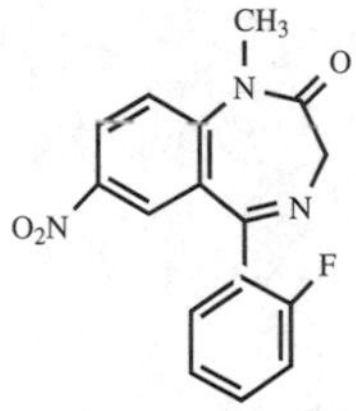

Flunixin (mol wt 296) An anti-inflammatory and analgesic agent.

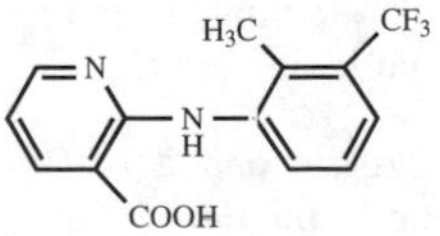

Fluocinolone Acetonide (mol wt 453) An adrenocortical steroid and anti-inflammatory agent.

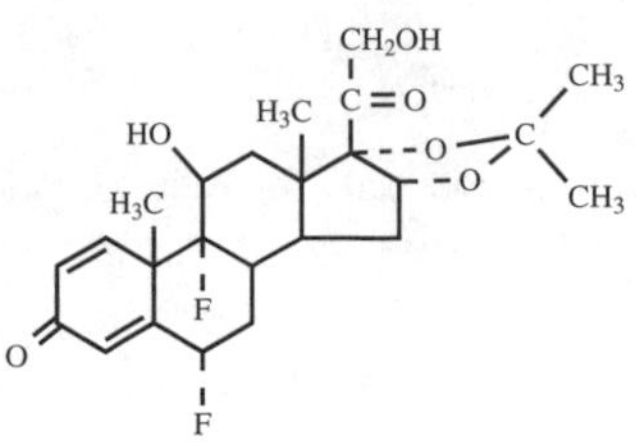

Flunoxaprofen (mol wt 285) An anti-inflammatory agent.

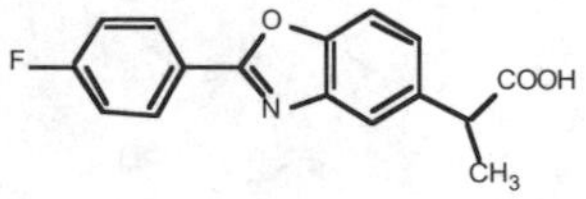

Fluocinonide (mol wt 495) A glucocorticoid and an anti-inflammatory agent.

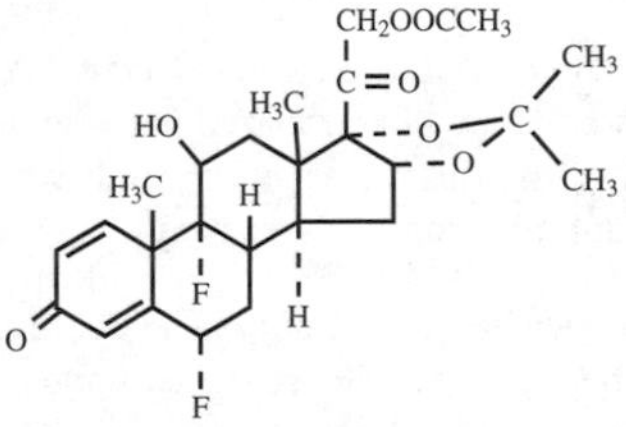

Fluocortin Butyl (mol wt 447) An anti-inflammatory agent.

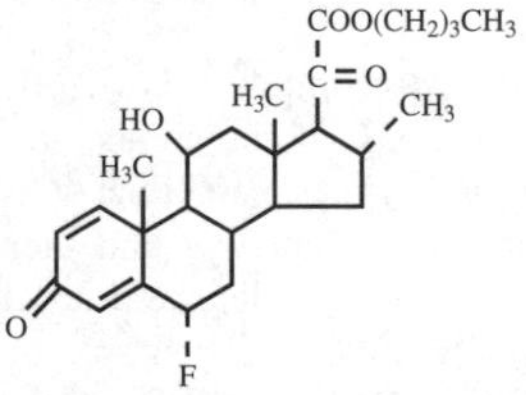

Fluogen A trade name for influenza virus vaccine, trivalent, type A and B.

Fluogen Split A trade name for a split virion antigen of influenza virus, used as a vaccine against influenza (referring to the 1992-1993 trivalent types A and B).

Fluonex A trade name for fluocinonide, an anti-inflammatory agent.

Fluonid A trade name for fluocinolone acetonide, a corticosteroid.

Fluor Substance used in scintillation counters that emits a flash of light upon excitation by radioactive radiation.

Fluorescamine (mol wt 278) A reagent used for labeling proteins in the electrophoretic gel that reacts with amines in protein to form a highly fluorescent derivative.

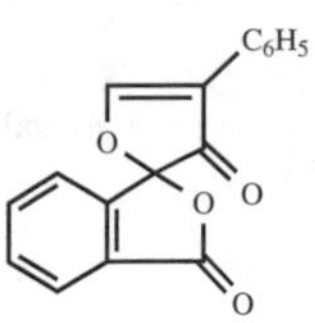

Fluorescein (mol wt 332) A reagent used as label for immunoassay.

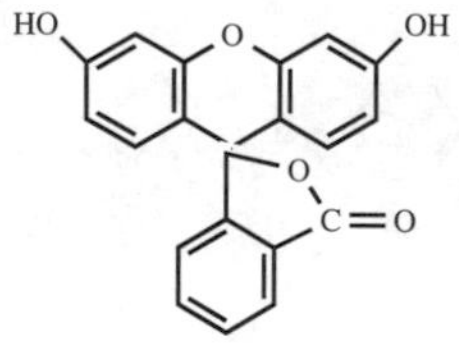

Fluorescence Property of molecules that absorb ultraviolet light (short wave length) and reemit the energy as visible light (longer wavelength).

Fluorescence Activated Cell Sorter A device in which cells are characterized and sorted by the intensity of the fluorescence they emit when passing through an exciting laser beam.

Fluorescence Enhancement The fluorescent difference between free fluorescence ligand and the bound fluorescence ligand.

Fluorescence *in situ* Hybridization A method that uses fluorescein-labeled DNA probe to locate specific regions of DNA in chromosomes.

Fluorescence Microphotolysis A fluorescence technique for study of membrane transport and diffusion in which intramembrane substances are labeled with fluorescent dye and then bleached with a laser beam to render them nonfluorescent. The rate at which the bleached area of the membrane recovers its fluorescence due to the exchange of unbleached and bleached fluorescent molecule across the membrane and can be monitored by a fluorescent photometric technique.

Fluorescence Microscopy A microscopic technique that forms a bright image on a dark background from fluorescent light produced by a specimen stained with fluorescent dye.

Fluorescence Photobleaching Recovery See fluorescence microphotolysis.

Fluorescence Probe Referring to any small molecule that undergoes changes of its fluorescence properties as a result of non-covalent interaction with a protein or other macromolecule.

Fluorescence Quenching A technique to investigate the binding property between ligand and receptor (e.g., hapten and antibody) in which aromatic amino acids in the receptor protein are excited to fluoresce. The binding of ligand to the fluorescing protein (e.g., antibody) causes the decrease in fluorescence.

Fluorescent Antibody Antibody that is labeled with fluorescent dye.

Fluorescent Antibody Technique A technique for the detection of antigen by a fluorescence-labeled antibody.

Fluorescent Antigen Antigen that is labeled with fluorescent dye.

Fluorescent Chromophore A molecule that emits visible light upon irradiation with UV light.

Fluorescent Dye A substance that emits visible light when ultraviolet light is absorbed.

Fluorescent Light The light emitted by a substance when it is irradiated with light of a different wavelength (e.g., short wavelength).

Fluorescent Screen A plate that is coated with material, e.g., zinc sulfate, which fluoresces upon UV irradiation.

Fluorescin (mol wt 334) A substance that can be readily oxidized to fluorescein.

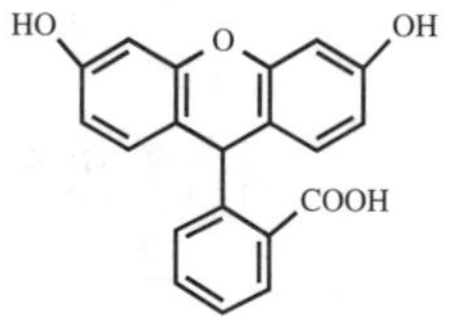

Fluoresone (mol wt 188) An anticonvulsant and anxiolytic agent.

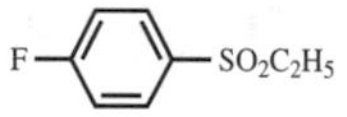

Fluoridamid (mol wt 296) A plant growth retardant.

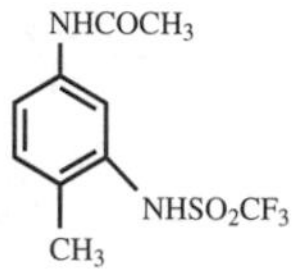

Fluoridation Addition of fluoride to drinking water to prevent tooth decay.

Fluoride 1. A compound of fluorine. 2. The monovalent ion of fluoride.

Fluorigard A trade name for sodium fluoride, used for bone remineralization.

Fluorine (F) A chemical element with atomic weight 19, valence 1.

Fluorinse A trade name for sodium fluoride, used for bone remineralization.

Fluoritab A trade name for sodium fluoride, used for bone remineralization.

Fluoroacetate Hydrolase The enzyme that catalyzes the following reaction:

$$\text{Fluoroacetate} + H_2O \rightleftharpoons \text{Acetate} + \text{fluoride}$$

Fluoroacetic Acid (mol wt 178) A toxic substance found in some plants.

$$FCH_2COOH$$

Fluorochrome A dye that absorbs light and then emits light of different color (always of a longer wavelength).

5-Fluorocytosine (mol wt 129) An inhibitor for DNA synthesis.

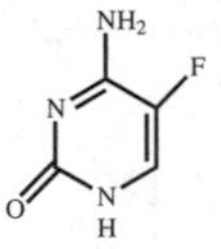

5-Fluoro-2-deoxyuridine (mol wt 246) An inhibitor for DNA synthesis.

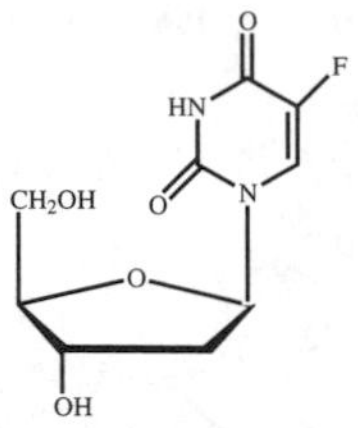

1-Fluoro-2,4-Dinitrobenzene (mol wt 186) Sanger's reagent for protein analysis.

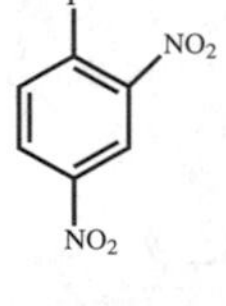

Fluoroimmunoassay An immunoassay employing fluorescence-labeled antigen or antibody.

Fluorometer A device for measuring fluorescence.

Fluorometholone (mol wt 376) A glucocorticoid and anti-inflammatory agent.

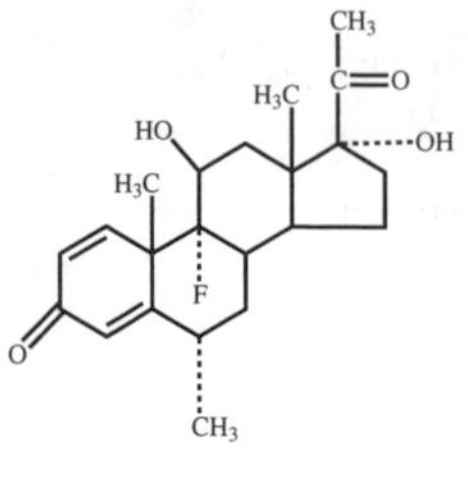

Fluorometry A technique for measuring fluorescence.

4-Fluoro-3-Nitrophenyl Azide (mol wt 182) A photoreactive reagent used to determine the active site of antibody, immobilize asparaginase, and modify trypsin and chymotrypsin.

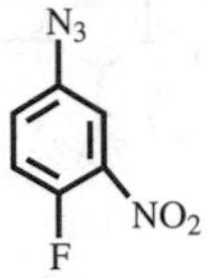

Fluoro-Op A trade name for fluorometholone, an anti-inflammatory agent.

Fluorophore The fluorescent group in a molecule.

Fluoroplex A trade name for fluorouracil, an anticancer drug that inhibits DNA synthesis.

Fluorosalan (mol wt 439) A disinfectant.

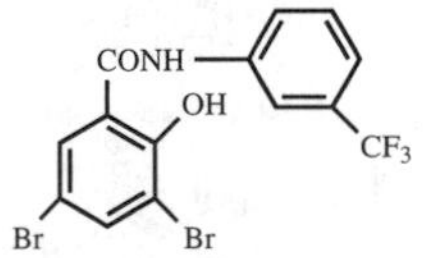

Fluoroscope A device used to show an X-ray image on a fluorescent screen.

Fluorosis A disorder caused by the excessive intake of fluorine.

Fluorouracil (mol wt 130) An antineoplastic agent that inhibits DNA synthesis.

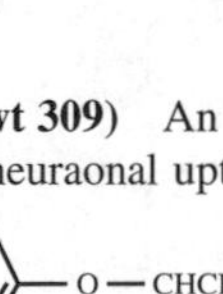

Fluorsyn A trade name for fluocinolone, a glucocorticoid and an anti-inflammatory agent.

Fluothane A trade name for halothane, an anesthetic agent.

Fluoxetine (mol wt 309) An antidepressant that inhibits the CNS neuraonal uptake of serotonin.

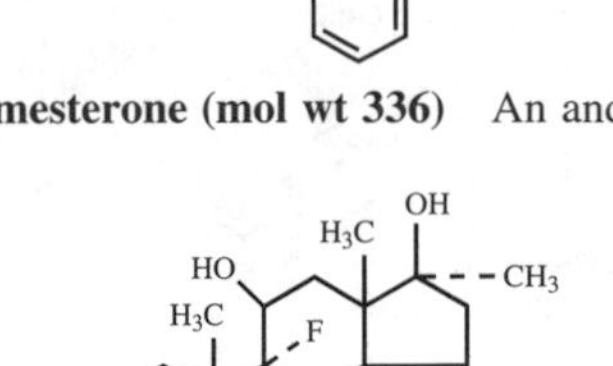

Fluoxymesterone (mol wt 336) An androgen.

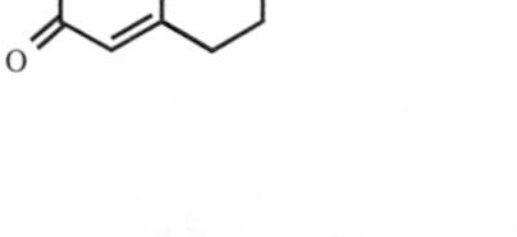

Fluperolone Acetate (mol wt 435) An anti-inflammatory agent and a glucocorticoid.

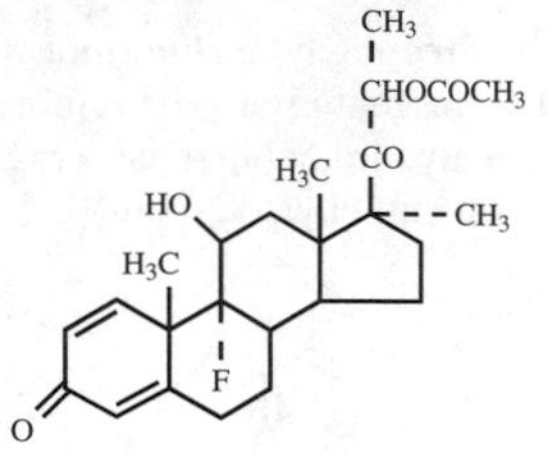

Fluphenazine (mol wt 438) An antipsychotic agent and a tranquilizer that blocks postsynatptic dopamine receptors in the brain.

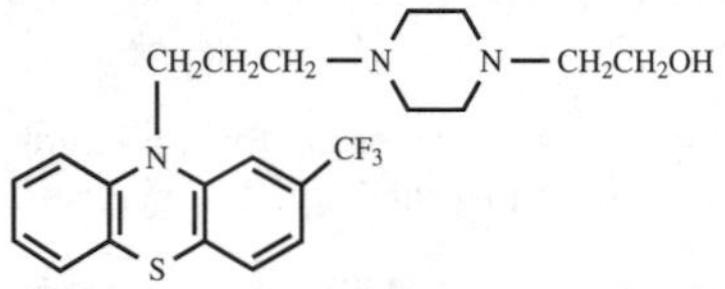

Flupirtine (mol wt 304) An analgesic agent.

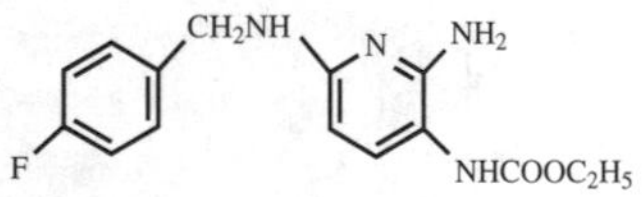

Fluprednidene Acetate (mol wt 432) An anti-inflammatory agent.

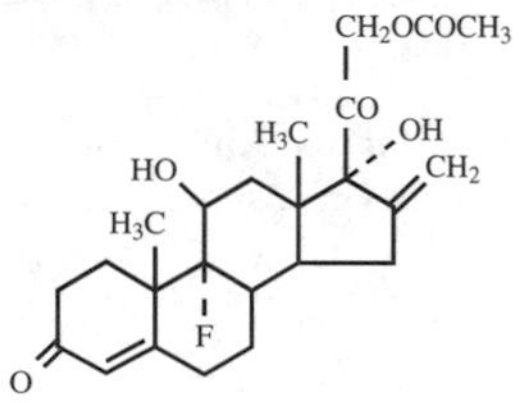

Fluprednisolone (mol wt 378) An anti-inflammatory agent.

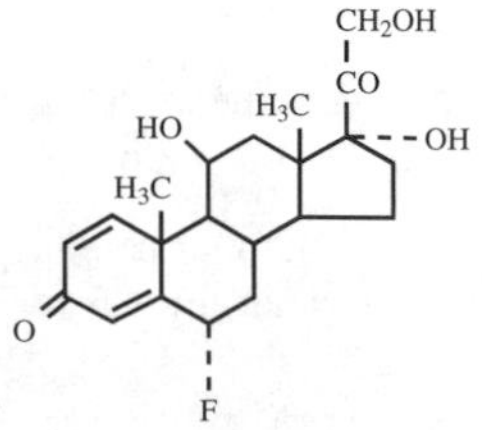

Fluproquazone (mol wt 296) An analgesic agent.

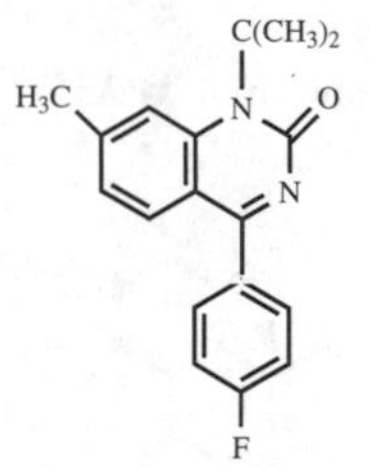

Flura A trade name for sodium fluoride, used for bone remineralization.

Fluram (mol wt 278) A nonfluorescent reagent that reacts with amino groups in amino acids and peptides to form fluorescent compounds.

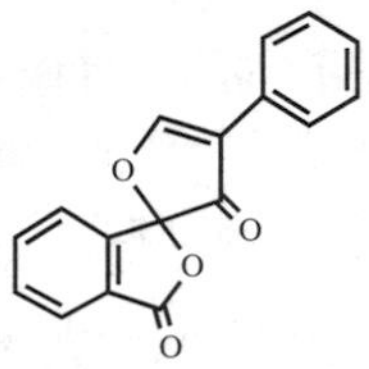

Flurandrenolide (mol wt 437) A glucocorticoid and anti-inflammatory agent.

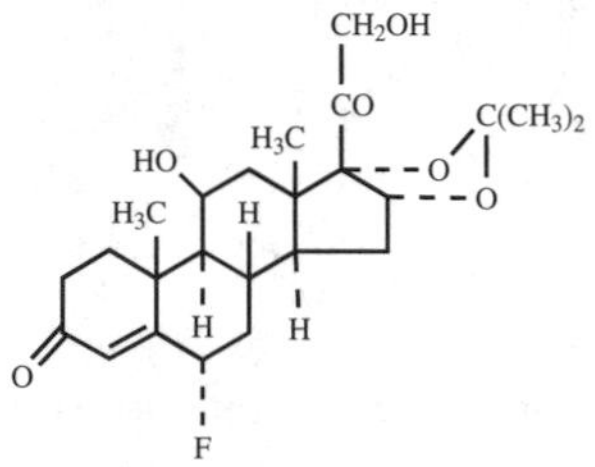

Flurazepam (mol wt 388) A sedative-hypnotic agent.

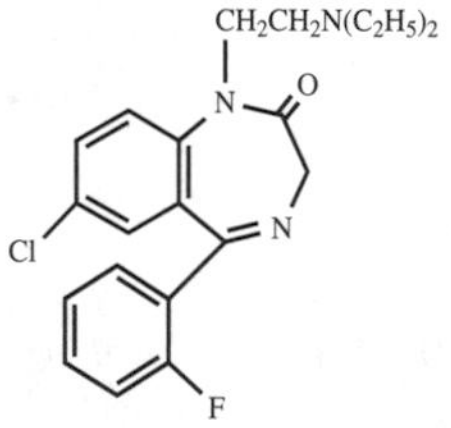

Flurbiprofen (mol wt 244) An anti-inflammatory and analgesic agent that interferes with prostaglandin biosynthesis.

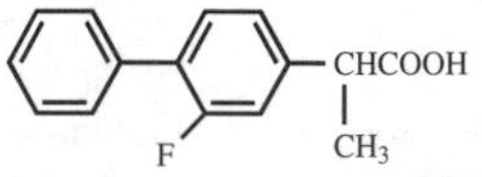

Fluress A trade name for a combination drug containing sodium fluorescein and benoxinate hydrochloride, used as an ophthalmic agent.

Flurosyn A trade name for fluocinolone acetonide, a corticosteroid.

Flury Virus An avianized strain of rabies virus used for rabies vaccination.

Flush End The ends of a double-stranded DNA without the single-stranded appearance.

Flu-Shield A trade name for influenza virus vaccine.

Flutamide (mol wt 276) An antiandrogen that inhibits androgen uptake or prevents binding of androgens in the nucleus within the target tissue.

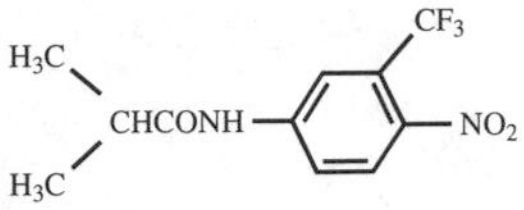

Flutex A trade name for triamcinolone acetonide, a dermatomucosal agent used for the control of skin infection.

Fluticasone Propionate (mol wt 501) An anti-allergic and anti-inflammatory agent.

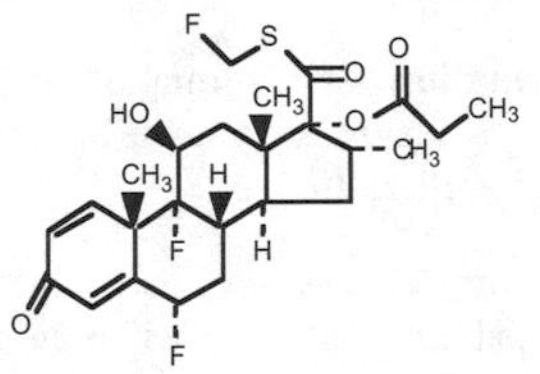

Flutone A trade name for diflorasone diacetate, a dermatomucosal agent used for the control of skin infection.

Flutropium Bromide (mol wt 478) A broncho-dilator.

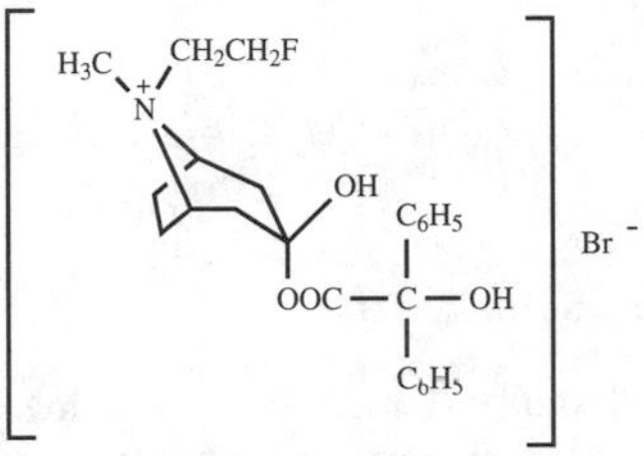

Fluvastatin (mol wt 411) An antihyper-lipoproteinemic agent that inhibits the enzyme catalyzing the first step in the cholesterol synthesis.

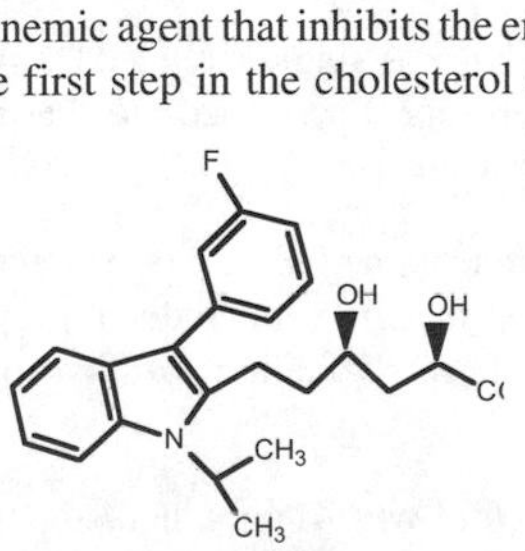

Fluvirin A trade name for influenza virus vaccine.

Fluvoxamine (mol wt 318) An antidepressant.

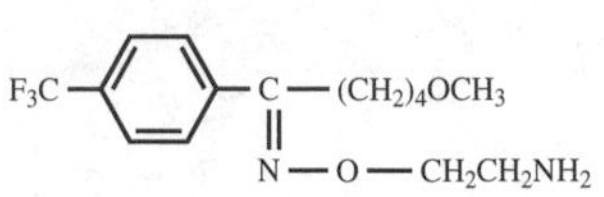

Flux 1. The flow rate of transfer in terms of number of particles passing through a unit area per unit time. 2. Movement of solute or molecule across a membrane. 3. Excessive discharge of bodily secretion.

Fluzone A trade name for influenza virus vaccine, a trivalent, type A and B.

FLV Abbreviation for Friend leukemia virus.

FM Abbreviation for 1. feed-back; 2. fibrin monomer; 3. flavin mononucleotide; 4. fluorescent microscope; 5. forensic medicine.

FMDV Abbreviation for foot and mouth disease virus.

fMet Abbreviation for formylmethionine.

fMet-releasing Enzyme See *N*-formylmethionyl peptidase.

fMet-tRNA Abbreviation for *N*-formylmethionyl tRNA.

FML Liquifilm Ophthalmic A trade name for fluorometholone, an ophthalmic drug that decreases the infiltration of leukocytes at the site of inflammation.

FML S.O.P. A trade name for fluorometholone, an ophthalmic drug that decreases the infiltration of leukocytes at the site of inflammation.

fMLP Abbreviation for formylmethioninyl leucylphenylalanine.

FMN Abbreviation for flavin mononucleotide.

FMN Adenylyl Transferase The enzyme that catalyzes the following reaction:

$$\text{ATP} + \text{FMN} \rightleftharpoons \text{Pyrophosphate} + \text{FAD}$$

FMN Reductase The enzyme that catalyzes the following reaction:

$$\text{NAD(P)H} + \text{FMN} \rightleftharpoons \text{FMN-H}_2 + \text{NAD(P)}^+$$

FMN-H$_2$ The reduced form of FMN (an electron donor).

fmol Abbreviation for femtomole (10^{-15} mole).

fms An oncogen present in McDonough strain of feline sarcoma virus.

FMS Abbreviation for fat-mobilizing substance.

FN Abbreviation for false negative.

Fn Abbreviation for fibronectin.

Fn-1 Abbreviation for fibronectin-1.

Fn-2 Abbreviation for fibronectin-2.

FN40 Abbreviation for 40 kD chymotryptic fragment of fibronectin.

FN120 Abbreviation for 120 kD chymotryptic fragment of fibronectin.

FNPA Abbreviation for 4-fluoro-3-nitrophenyl azide.

FnuAI (HinfI) A restriction endonuclease from *Fusobacterium nucleatum* A with the following specificity:

```
         ↓
5′...........GANTC...........3′
3′...........CTNAG...........5′
                ↑
```

Fnu4HI A restriction endonuclease from *Fusobacterium nucleatum* 4H with the following specificity:

```
           ↓
5′...........GCNGC...........3′
3′...........CGNCG...........5′
              ↑
```

FnuAII (MboI) A restriction endonuclease from *Fusobacterium nucleatum* A with the same specificity as MboI.

FnuCI (MboI) A restriction endonuclease from *Fusobacterium nucleatum* C with the same specificity as MboI.

FnuDI (HaeIII) A restriction endonuclease from *Fusobacterium nucleatum* D with the following specificity:

```
            ↓
5′...........GGCC..........3′
3′...........CCGG..........5′
            ↑
```

FnuDII A restriction endonuclease from *Fusobacterium nucleatum* D with the following specificity:

```
            ↓
5′...........CGCG..........3′
3′...........GCGC..........5′
            ↑
```

FnuDIII (HhaI) A restriction endonuclease from *Fusobacterium nucleatum* D with the following specificity:

```
               ↓
5′...........GCGC..........3′
3′...........CGCG..........5′
           ↑
```

FnuEI (MboI) A restriction endonuclease from *Fusobacterium nucleatum* E with the same specificity as MboI.

FOA Abbreviation for 5-fluoro-orotic acid.

FOAM Abbreviation for a combination drug containing fluorouracil, oncovin, adriamycin, and mitomycin C.

Foamy Virus A virus of the Spumavirinae (subfamily of retroviruses).

Focus Cluster of morphologically transformed cells in a monolayer cell culture that was initiated by a transforming agent, e.g., oncogenic virus.

Focus Assay A screening procedure for the detection of certain transforming oncogens on a monolayer cell culture.

Focus Formation The formation of dense clusters of cells by the transformed eukaryotic cells on a monolayer cell culture.

Focus-forming Viruses Viruses (e.g., oncogenic viruses) capable of forming foci on the monolayer cell culture.

Fodrin A protein from bovine brain that is immunologically related to spectrin.

FokI A restriction endonuclease from *E. coli* strain carrying plasmid encoding Fok I gene with the following specificity:

```
                     ↓
5'........GGAATG(N)9........3'
3'........CCTAAC(N)13.......5'
                     ↑
```

Folacin See folic acid.

Folch Method A method for isolation of lipid from tissue by chloroform/methanol/water extraction.

Foldback DNA A single-stranded DNA that contains inverted repeat sequences so that it can hybridize with itself.

Folex A trade name for methotrexate, an antineoplastic agent that prevents reduction of folate to tetrtahydrofolate by binding to dihydrofolate reductase.

Folic Acid (mol wt 441) A hematopoietic vitamin (vitamin B_9) that is necessary for normal erythropoiesis and nuceloprotein synthesis.

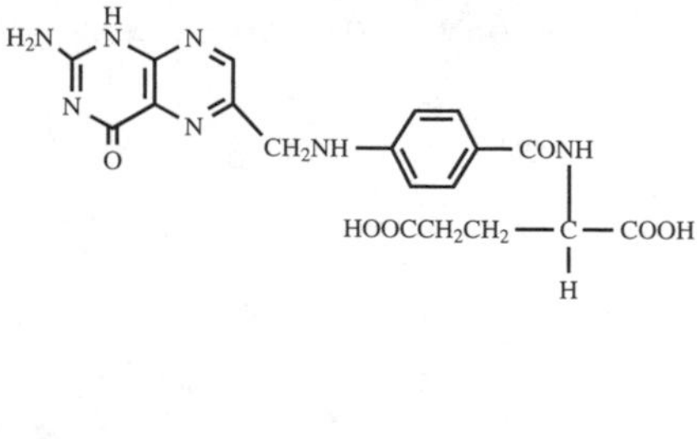

Folic Acid Conjugate Referring to folic acid derivatives that contain different numbers of glutamyl residues.

Folic Acid Reductase The enzyme that catalyzes the following reaction:

$$NADP^+ + \text{5,6,7,8-tetrahydrofolate} \rightleftharpoons NADPH + \text{7,8-dihydrofolate}$$

Folin Reaction A colorimetric reaction for determination of amino acids based upon the production of blue color by treatment of sample with 1,2-naphthoquinone 4-sulfonate.

Folin-Ciocalteau Reaction A colorimetric reaction for determination of protein based upon the production of blue color by treatment of sample with Folin-Ciocalteau's phenol reagent.

Folin-Ciocalteau's Phenol Reagent A reagent containing phosphomolybdotungstic acid and used for determination of protein.

Folin-Wu Method A method for the determination of glucose in the blood with alkaline copper tartrate and phosphomolybdic acid following deproteinization of the blood with tungstic acid.

Folinic Acid (mol wt 512) The reduced and formylated derivative of folic acid.

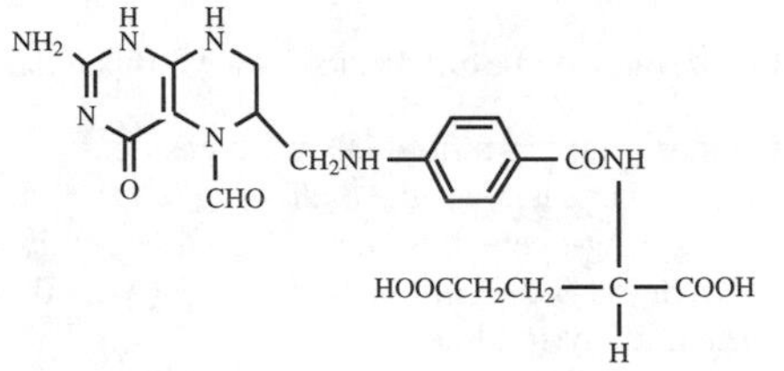

Follicle Spherical mass of cells usually contained in the skin from which the hair emerges.

Follicle-Stimulating Hormone A gonadotrophic hormone that stimulates ovulation, estrogen synthesis, and growth of the ovarian follicles in females. It also causes spermatogenesis in males.

Follipsin A serine proteinase from pig ovarian fluid.

Follistatin An activin-binding glycoprotein that regulates biosynthesis and secretion of follicle-stimulating hormone.

Follistin A trade name for follitropin used for treatment of infertility.

Follistin Alfa (Alpha) A trade name for follitropin alpha used for treatment of infertility.

Follistin Beta A trade name for follitropin beta used for treatment of infertility.

Follitropin Synonym of follicle-stimulating hormone.

Follutein A trade name for gonadotropin, a hormone that promotes secretion of gonadal steroids.

Folvit A trade name for vitamin B_9 (folic acid).

FOM Abbreviation for a combination drug containing 5-fluorouracil, oncovin, and mitomycin C.

Fomecins A group of antibacterial substances produced by *Fomes juniperinus*.

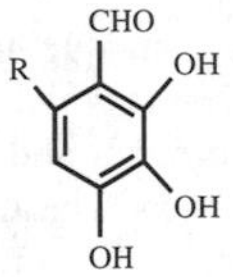

Fomepizole (mol wt 82) An antidote to methanol and ethylene glycol poisoning.

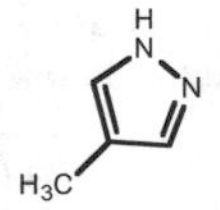

Fomes 1. Inanimate objects that can act as carriers of an infectious agent. 2. A genus of fungi.

Fominoben (mol wt 401) An antitussive and respiratory stimulant.

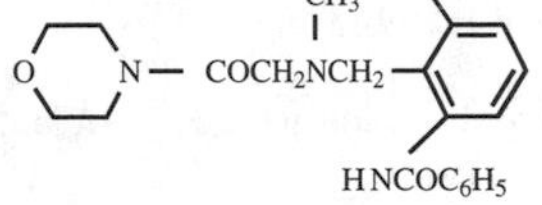

Fomites Objects or materials that have been associated with infected persons or animals and are potentially dangerous for harboring pathogenic microorganisms.

Fomocaine (mol wt 311) An anesthetic agent.

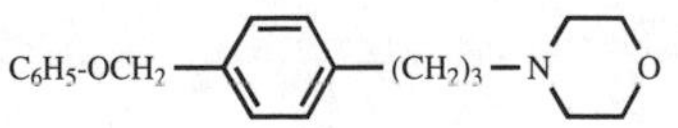

Fonazine (mol wt 392) A serotonin inhibitor.

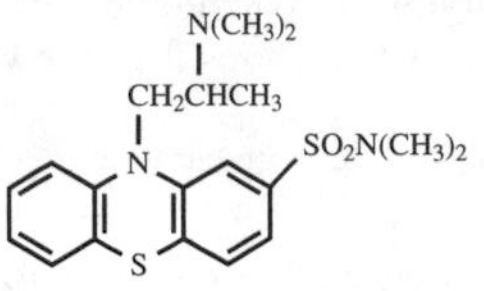

Fonofos (mol wt 246) A cholinesterase inhibitor.

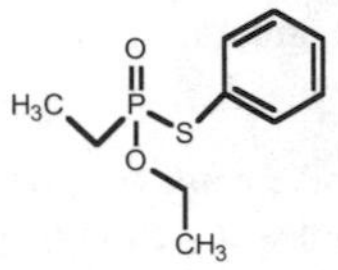

Food Additive Substance added to food to improve its color, flavor, or shelf life.

Food Poisoning Referring to the acute gastroenteritis caused by ingestion of food contaminated with pathogenic microorganisms or their toxic products.

Footprinting A technique to identify protein-binding sites on DNA in which DNA is allowed to bind with specific proteins and then is subjected to endonuclease digestion. The regions of the DNA with bounded proteins are protected from enzymatic digestion. The fragments of DNA obtained after digestion are electrophoresed, characterized, and compared with the control.

Forane A trade name for isoflurane, a general inhalation anesthetic agent.

Forbes' Disease A disorder caused by excessive deposits of glycogen in the liver (glycogen storage disease).

Forensic Medicine Medical science dealing with scientific investigation of causes of injury and death of the unexplained circumstances.

Forewater The amniotic fluid between the fetus and the fetal membrane.

Formaldehyde (mol wt 30) An antimicrobial agent and a fixative.

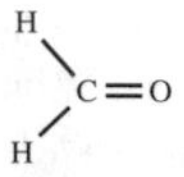

Formaldehyde Dehydrogenase The enzyme that catalyzes the following reaction:

Formaldehyde + NAD$^+$ + H_2O

⇅

Formate + NADH

Formaldehyde Dehydrogenase (Glutathione) The enzyme that catalyzes the following reaction:

Formaldehyde + glutathione + NAD$^+$

⇅

Formylglutathione + DADH

Formaldehyde Dismutase The enzyme that catalyzes the following reaction:

Formaldehyde + formaldehyde

⇅

Formate + methanol

Formaldehyde NAD$^+$ Oxidoreductase (Glutathione-forming) The systematic name for formaldehyde dehydrogenase.

Formaldehyde Transketolase The enzyme that catalyzes the following reaction:

D-Xylulose 5-phosphate + formaldehyde

⇅

Glyceraldehyde 3-phosphate + glycerone

Formalin 37% (w/v) solution of formaldehyde.

Formamidase The enzyme that catalyzes the following reaction:

Formamide + H_2O ⇌ Formate + NH_3

Formamide (mol wt 45) An organic substance used for denaturation of double-stranded DNA. It combines with the NH_2 groups of adenine and prevents the formation of A-T base pairs.

$$H-\overset{O}{\overset{\|}{C}}-NH_2$$

Formamide Amidohydrolase See formamidase.

Formamidopyrimidine-DNA Glycosidase The enzyme that catalyzes the hydrolysis of DNA containing ring-opened N^7-methylguanine residues, releasing 2,6-diamino-4-hydroxy-5-(*N*-methyl) formamidopyrimidine.

Formate C-Acetyltransferase The enzyme that catalyzes the following reaction:

Acetyl-CoA + formate ⇌ CoA + pyruvate

Formate Dehydrogenase (Cytochrome) The enzyme that catalyzes the following reaction:

Formate + ferricytochrome b_1

⇅

Ferrocytochrome b_1 + CO_2

Formate Dehydrogenase (Cytochrome-c-553) The enzyme that catalyzes the following reaction:

Formate + ferricytochrome c-553

⇅

Ferrocytochrome c-553 + CO_2

Formate Dehydrogenase ($NADP^+$) The enzyme that catalyzes the following reaction:

$$\text{Formate} + NADP^+ \rightleftharpoons CO_2 + NADPH$$

Formate Dihydrofolate Ligase The enzyme that catalyzes the following reaction:

$$\text{ATP} + \text{formate} + \text{dihydrofolate} \rightleftharpoons \text{ADP} + \text{Pi} + \text{10-formyldihydrofolate}$$

Formate Kinase The enzyme that catalyzes the following reaction:

$$\text{ATP} + \text{formate} \rightleftharpoons \text{ADP} + \text{formylphosphate}$$

Formate Tetrahydrofolate Ligase The enzyme that catalyzes the following reaction:

$$\text{ATP} + \text{formate} + \text{tetrahydrofolate} \rightleftharpoons \text{ADP} + \text{Pi} + \text{10-formyltetrahydrofolate}$$

Formate Tetrahydrofolate Synthetase See F formate tetrahydrofolate ligase.

Formazan A sugar derivative formed from the reaction of a carbohydrate phenylosazone with diazo compound.

Formebolone (mol wt 344) An anabolic hormone.

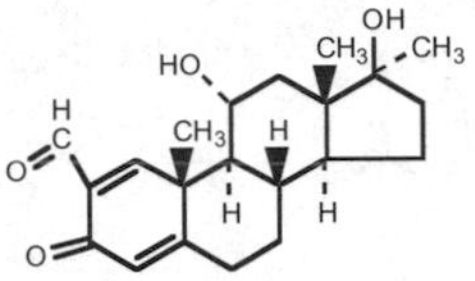

Formed Element A collective term for 1. cellular components of the blood and 2. components present in the urine, e.g., crystals, bacteria, ova, or parasites.

Formestane (mol wt 302) A hormone with antineoplastic activity.

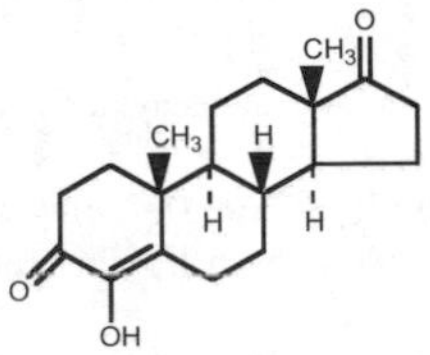

Formic Acid (mol wt 46) A counterirritant and an astringent agent.

HCOOH

Formic Dehydrogenase See formaldehyde dehydrogenase.

Formicin (mol wt 89) An antiseptic agent.

$CH_3CO—NHCH_2OH$

Formimino Group The group - CH=NH.

Formiminoaspartate Deformylase The enzyme that catalyzes the following reaction:

$$N\text{-Formyl-L-aspartate} + H_2O \rightleftharpoons \text{Formate} + \text{L-asparate}$$

Formiminoaspartate Deiminase The enzyme that catalyzes the following reaction:

$$N\text{-Formimino-L-aspartate} + H_2O \rightleftharpoons N\text{-Formyl-L-aspartate} + NH_3$$

Formiminoglutamase The enzyme that catalyzes the following reaction:

$$N\text{-Formimino-L-glutamate} + H_2O \rightleftharpoons \text{L-Glutamate} + \text{formamide}$$

Formiminoglutamate Deiminase The enzyme that catalyzes the following reaction:

$$N\text{-Formimino-L-glutamate} + H_2O \rightleftharpoons N\text{-Formyl-L-glutamate} + NH_3$$

Formimino-L-Glutamic Acid Transferase Synonym of glutamate formiminotransferase.

Formiminotetrahydrofolate Cyclodeaminase The enzyme that catalyzes the following reaction:

$$\text{5-Formiminotetrahydrofolate} \rightleftharpoons \text{5,10-methenyltetrahydrofolate} + NH_3$$

5-Formimino-Tetrahydrofolate L-Glutamate N-Formiminotransferase The systematic name for glutamate formiminotransferase.

Forminitrazole (mol wt 173) An antiprotozoal agent (*Trichomonas*).

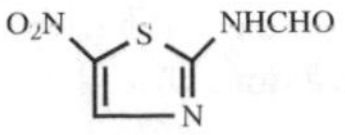

Formoterol (mol wt 344) An antiasthmatic agent.

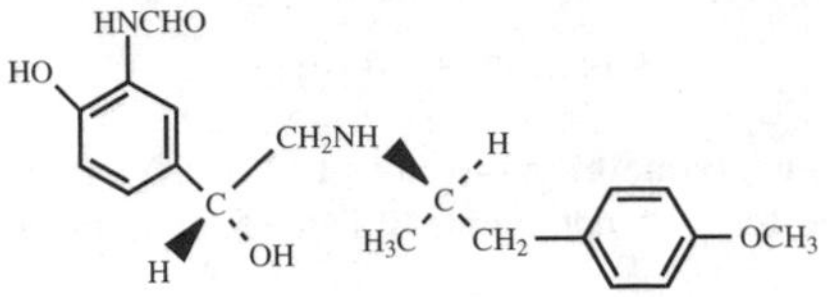

Formothion (mol wt 257) A systematic insecticide.

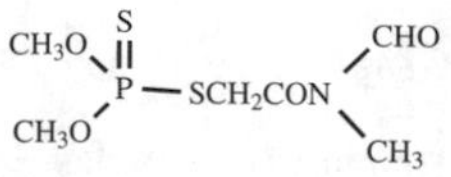

Formula Weight A value obtained by summing the quantity for all the elements in a compound.

Formulex A trade name for dicylomine hydrochloride, an antispasmodic, anticholinergic, antimuscarinic agent.

Fomvar Polyvinylformyl substance used for preparation of supporting film for electron microscopy.

Formycins (mol wt 267) A group of antibiotics produced by *Norcardia interforma.*

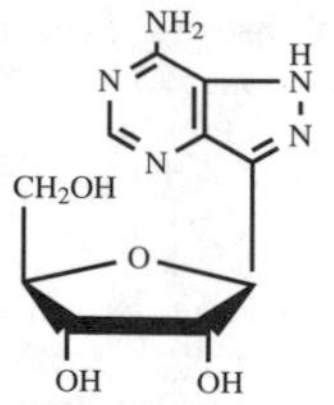

formycin A

Formyl The group - CH=O.

Formylaspartate Deformylase The enzyme that catalyzes the following reaction:

N-Formyl-L-aspartate + H_2O
⇌
Formate + L-aspartate

Formyl-CoA Hydrolase The enzyme that catalyzes the following reaction:

Formyl-CoA + H_2O ⇌ CoA + formate

Formylglutamate Formyltransferase The enzyme that catalyzes the following reaction:

N-Formyl-L-glutamate + tetrahydrofolate
⇌
L-glutamate + 5-formyltetrahydrofolate

Formylkynureninase The enzyme that catalyzes the following reaction:

N-Flormyl-L-kynurenine + H_2O
⇌
Formate + L-kynurenine

***N*-Formylmethionine (mol wt 177)** The formylated form of methionine. It is the first amino acid incorporated into the peptide chain during protein synthesis.

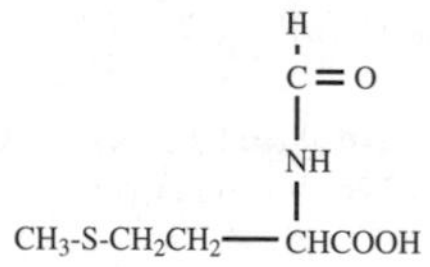

***N*-Formylmethionyl Aminoacyl-tRNA Deformylase** The enzyme that catalyzes the following reaction:

N-Formylaminoacyl-tRNA + H_2O
⇌
Formate + L-methionylaminoacyl-tRNA

***N*-Formylmethionyl Aminoacyl-tRNA Amidohydrolase** See *N*-formylmethionyl aminoacyl-tRNA deformylase.

***N*-Formylmethionyl Peptidase** The enzyme that catalyzes the release of an *N*-formylmethionyl residue from the *N*-terminal of a polypeptide.

Formylmethionyl-tRNA (fmet-tRNA) The aminoacyl-tRNA complex of formylmethionine and tRNA, it serves as an initiator for protein synthesis.

N^2-Formylsulfisomidine (mol wt 306) An antibacterial agent.

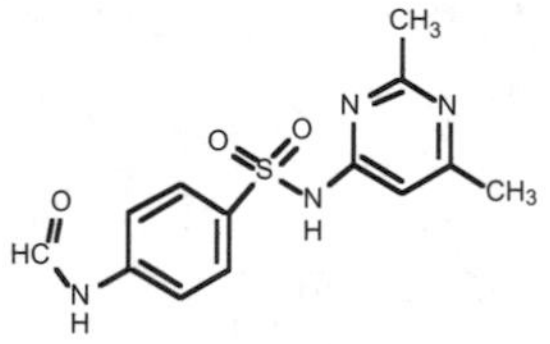

10-Formyltetrahydrofolate Amidohydrolase See formyltetrahydrofolate deformase.

5-Formyltetrahydrofolate Cycloligase The enzyme that catalyzes the following reaction:

ATP + N^5-Formyltetrahydrofolate
⇌
ADP + Pi + N^5,N^{10}-methenyltetrahydrofolate

Formyltetrahydrofolate Deformylase The enzyme that catalyzes the following reaction:

10-Formyltetrahydrofolate + H_2O
⇌
Formate + tetrahydrofolate

Formyltetrahydrofolate Ligase See formate tetrahydrofolate synthetase.

Forskolin (mol wt 410) A substance isolated from plants that stimulates adenylate cyclase activity.

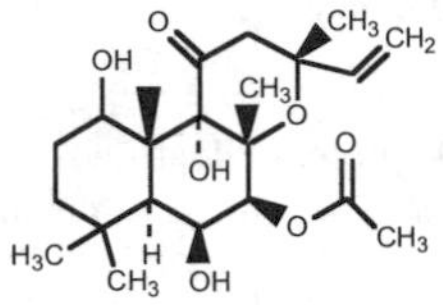

Forssman Antigen Any heterophil antigen that can induce the production of an antibody capable of combining with sheep red blood cells.

Fort Bragg Fever A human leptospirosis caused by *Leptospira interrogans* and characterized by pretibial rash, malaise, and fever.

Fortaz A trade name for ceftazidime, an antibiotic that inhibits bacterial cell wall synthesis.

Fortovase A trade name for saquinavir, an antiviral agent.

Fortral A trade name for pentazoxine hydrochloride, an analgesic agent.

Fortran A computer language used for scientific and mathematical computations.

Forward Mutation A mutation that converts a wild type allele to a mutant allele.

fos An oncogene whose products form AP-1 gene regulatory protein in mouse esteosarcoma.

Fosamax A trade name for alendronate sodium, a calcium regulator.

Foscarnet Sodium (mol wt 192) An inhibitor for viral DNA polymerase and reverse transcriptase.

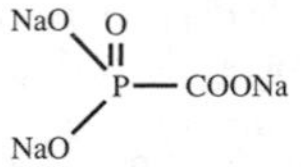

Foscavir A trade name for foscarnet sodium, an antiviral drug for herpesviruses.

Fosfestrol (mol wt 428) An estrogen used for treatment of prostatic carcinoma. It inhibits angiotensin-converting enzyme and prevents the conversion of angiotensin I to angiotensin II.

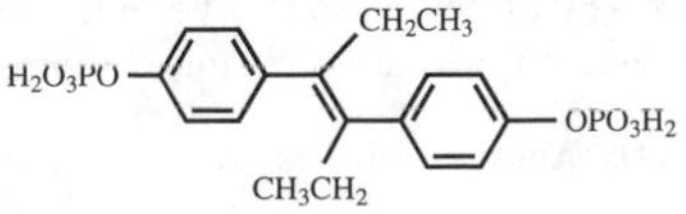

Fosfomycin (mol wt 138) An antibiotic produced by *Streptomyces* and *Pseudomonas syringae.*

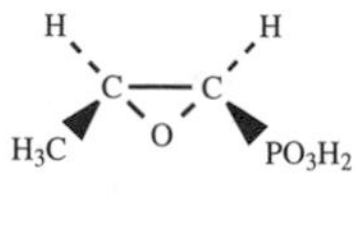

Fosfosal (mol wt 218) An analgesic agent.

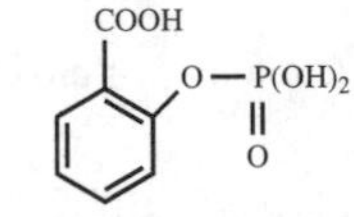

Fosfree A trade name for nutritional supplement of calcium, iron, and multivitamin.

Foshay's Vaccine An antitularaemia vaccine prepared from killed *Francisella tularensis.*

Fosinopril (mol wt 564) An antihypertensive agent.

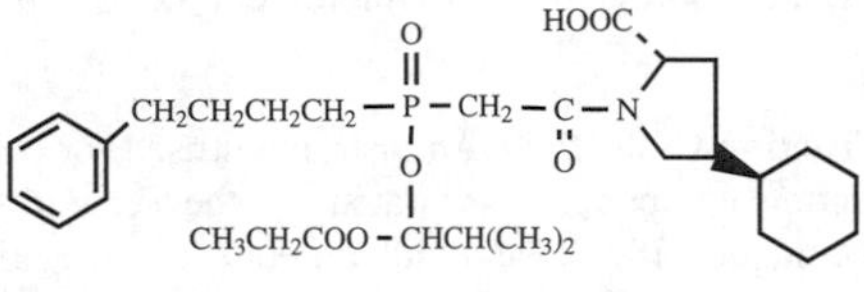

Fospirate (mol wt 306) An anthelmintic agent.

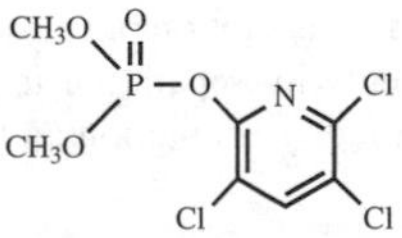

Fostex A trade name for benzoyl peroxide gel or lotion, an antimicrobial and comedolytic agent.

Fotemustine (mol wt 316) An antineoplastic agent.

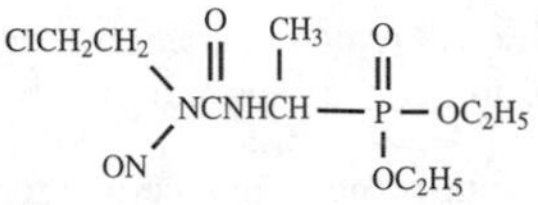

Fouchet's Test A colorimetric test for determination of bilirubin in the urine based upon the production of green color on treatment of the urine with ferric chloride and trichloroacetic acid.

Fourier Transformations The mathematical equation used to convert raw X-ray data into an electron density map.

FP Abbreviation for 1. false positive; 2. Fibrinopeptide; 3. flavin phosphate; 4. freezing point.

F1P Abbreviation for fructose 1-phosphate.

F6P Abbreviation for fructose 6-phosphate.

F1,6-P Abbreviation for fructose-1,6-bisphosphate.

FPH_2 Abbreviation for flavin phosphate (reduced form).

FPIA Abbreviation for fluorescent polarization immunoassay.

F-pilus (plural F-pili) A male-specific appendage.

FPLC Abbreviation for fast protein liquid chromatography.

F-plasmid See fertility factor.

F′-plasmid The F-plasmid that contains host genetic element.

FPR Abbreviation for fluorescent photobleaching recovery.

FR Abbreviation for floacculation reaction.

FRA Abbreviation for fluorescent rabies antibody.

Fraction Collector An instrument used for collecting components separated by the separation techniques. It collects fluid components emerging from continuous separation at intervals of volume or time.

Fractionation Separation of substances into discrete components by separation techniques, e.g., ammonium sulfate precipitation and chromatography.

Fraction-Crystalline Referring to the Fc fragment of an immunoglobulin that can be readily crystallized.

Fracture Face The exposed surface of a specimen obtained by freeze fracture technique.

Fragile Site A region in a chromosome that can be induced to generate chromosome breaks.

Fragile X Syndrome An inherited form of mental retardation in males associated with fragile sites on the X chromosome.

Fragmentation Map The electrophoretic patterns of DNA fragments resulting from restriction endonuclease digestion.

Fragmin A protein that binds monomeric actin (G-protein) and prevents the formation of actin filament (F-protein) from G-protein.

Frameshift Mutation Insertion or deletion of one of more nucleotide base pairs into (or from) a DNA molecule, resulting in a change in the reading frame of the mRNA molecule and leading usually to a garbled message.

Francis Skin Test An immediate type hypersensitivity test for the presence of antibody to pneumococci in which pneumococcal capsular polysaccharide is injected into the skin and produces a wheal-and-flare response.

Frankia A genus of bacteria (order Actinomycetales).

FRAT Abbreviation for free radical assay technique.

Frataxin A protein identified from a mutant found in *Friedreich's ataxia*, an inherited disease.

Frateuria A genus of Gram negative, obligately aerobic, chemoorganotrophic bacteria (Pseudomonadaceae).

Fraudulant DNA A DNA that contains purine or pyrimidine analogs.

Fraxiparine The low molecular weight fraction (4500 daltons) of heparin with antithrombic activity.

Flazier's Medium An agar medium that contains 0.4% gelatin, used for detection of microbial hydrolysis of gelatin.

Fredericamycin A (mol wt 540) An antitumor antibiotic from *Streptomyces griseus.*

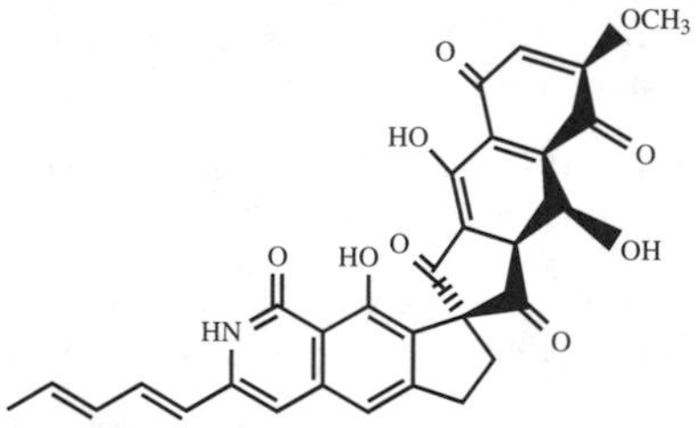

Free Diffusion Referring to the passive diffusion of solutes across a membrane without the involvement of carrier molecules.

Free Electrophoresis Referring to moving boundary electrophoresis.

Free Energy That component of the total energy of a system that can do work under constant temperature and pressure.

Free Energy Change Thermodynamic parameter used to quantify the net free energy liberated or required by a reaction. It is abbreviated as ΔG (also called Gibbs free energy).

Free Energy of Activation The initial input of energy necessary to start a chemical reaction.

Free Fatty Acid A non-esterified fatty acid.

Free Radical A molecule that contains one or more unpaired electrons.

Freeze Drying Removal of solvent from a frozen sample under vacuum and low temperature.

Freeze Etching A freeze fracturing technique used in transmission electron microscopy to examine the topography of a surface exposed by fracturing. The surfaces exposed by cleavage are etched by sublimation at –100 °C and a thin replica of the etched surface is then prepared for EM- examination.

Freeze Fracturing An electron microscopic technique in the preparation of specimens for electron microscopy by fracturing the frozen specimen with a knife edge to yield a cleavage surface. The exposed surface is then replicated by metal casting for EM examination.

Freezing-Point Osmometer An osmometer that is based upon the lowering of the freezing point of a solvent by the addition of a solute.

French Pressure Cell A device used to disrupt chloroplasts, bacteria, yeasts. and other cells.

Frenolicin (mol wt 346) An antibiotic produced by *Streptomyces fradiae.*

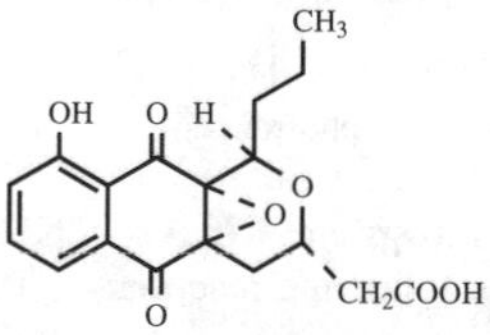

Freon A group of halogenated hydrocarbons used as cryogen or as intermediate fluid in critical point drying.

Frequency The number of vibrations or cycles per unit time.

Frequency Histogram A step curve in which the frequencies of various arbitrarily bounded classes are graphed.

FRET Abbreviation for fluorescence resonance energy transfer.

FRET Probe Abbreviation for fluorescence resonance energy transfer probe.

Freund's Complete Adjuvant An oil-water emulsion that contains killed mycobacteria.

Freund's Incomplete Adjuvant An oil-water emulsion that contains all of the elements of Freund's complete adjuvant with the exception of killed mycobacteria.

FRF Abbreviation for follicle-stimulating hormone releasing factor.

FRH Abbreviation for follicle-stimulating hormone-release hormone.

Friedreich's Ataxia An autosomal recessive inherited disease that involves the central and peripheral nervous systems and the heart.

Friend Cell A cultured cell line of leukemic mouse erythroblasts obtained by transformation with Friend leukemia virus.

FriOI A restriction endonuclease from *Flavobacterium* species 09 with the following specificity:

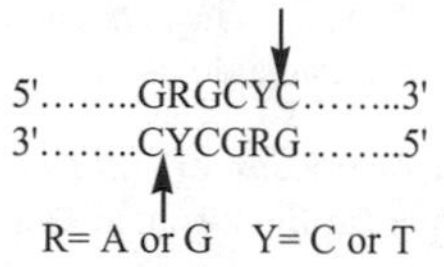

Froben A trade name for flurbiprofen, a nonsteroidal anti-inflammatory, and analgesic agent.

Fru Abbreviation for fructose.

Fructan A polymer of fructose occurring in plants.

Fructan b-Fructosidase The enzyme that catalyzes the hydrolysis of terminal nonreducing 2,1- and 2,6- linked b-D-fructofuranose residues in fructans.

β-Fructofuranosidase The enzyme that catalyzes the hydrolysis of terminal nonreducing β-D-fructopfuranoside residues in β-D-fructofuranosides.

Fructofuranoside Fructohydrolase The systematic name for β-D-fructofuranosidase.

Fructokinase The enzyme that catalyzes the following reaction:

$$\text{ATP} + \text{fructose} \rightleftharpoons \text{ADP} + \text{fructose 6-phosphate}$$

Fructose (mol wt 180) A six-carbon ketosugar (also known as fruit sugar).

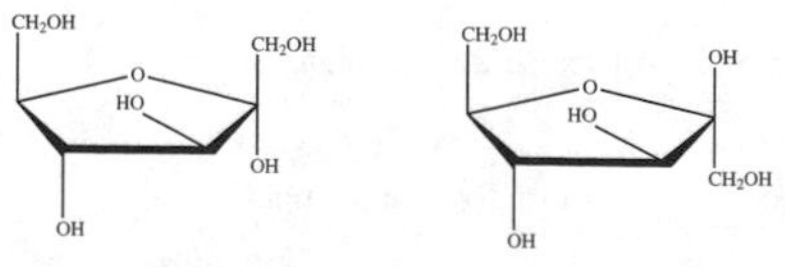

a-D-fructofuranose b-D-fructofuranose

Fructose Bisphphophatase The enzyme that catalyzes the following reaction:

$$\text{D-Fructose 1,6-bisphosphate} + H_2O \rightleftharpoons \text{D-Fructose 6-phosphate} + \text{Pi}$$

Fructose 1-6-Bisphosphate (mol wt 340) An intermediate in glycolysis.

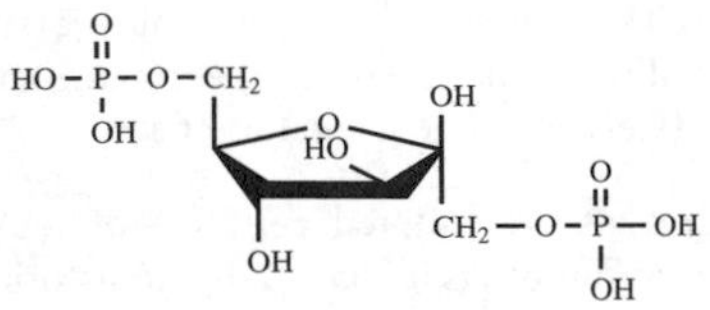

Fructose 1,6-Bisphosphate 1-Phosphohydrolase The systematic name for fructose bisphosphatase.

Fructose 1,6-Bisphosphate D-Glyceraldehyde-3-Phosphate Lyase The systematic name for fructose bisphosphate aldolase.

Fructose-2-6-Bisphosphate (mol wt 340) An inhibitor for fructose-6-phosphate kinase.

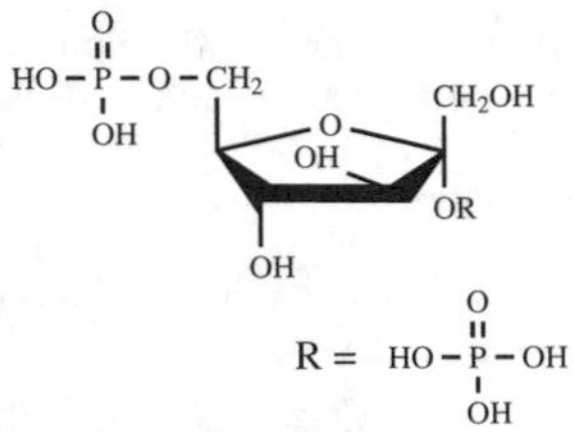

Fructose Bisphosphate Aldolase The enzyme that catalyzes the following reaction:

Fructose 1,6-bisphosphate

⇌

Dihydroacetone phosphate +
glyceraldehyde 3-phosphate

Fructose 2,6-Bisphosphate 2-Phosphatase The enzyme that catalyzes the following reaction:

D-Fructose 2,6-bisphosphate + H_2O

⇌

D-Fructose 6-phosphate + Pi

Fructose 2,6-Bisphosphate 6-Phosphatase The enzyme that catalyzes the following reaction:

D-Fructose 2,6-bisphosphate + H_2O

⇌

D-Fructose 2-phosphate + Pi

Fructose 5-Dehydrogenase The enzyme that catalyzes the following reaction:

D-Fructose + acceptor

⇌

5-Dehydro-D-fructose + reduced acceptor

Fructose 5-Dehydrogenase ($NADP^+$) The enzyme that catalyzes the following reaction:

D-Fructose + $NADP^+$

⇌

5-Dehydro-D-fructose + NADPH

Fructose 1-6-Diphosphate See fructose-1-6-bisphosphate.

Fructose 1,6-Diphosphatase Synonym of fructose bisphosphatase.

Fructose Intolerance A genetic disorder due to a deficiency in the enzyme fructose 1-phosphate-aldolase. Individuals with fructose intolerance develop a strong distaste for anything sweet.

Fructose 1-Phosphate (mol wt 260) A phosphorylated form of fructose.

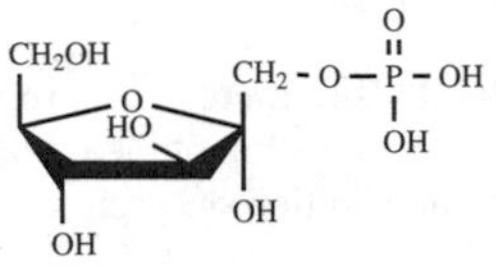

α-D-fructose 1-phosphate

Fructose 1-Phosphate Aldolase The enzyme that catalyzes the following reaction:

Fructose 1-phosphate

⇌

Dihyroxyacetone phosphate + glyceraldehyde

Fructose 1-Phosphate Kinase The enzyme that catalyzes the following reaction:

ATP + fructose 1-phosphate

⇌

ADP + fructose 1,6-bisphosphate

Fructose 1-Phosphate Pathway A pathway for metabolism of fructose via fructose 1-phosphate aldolase to form glyceraldehyde and dihydroxyacetone phosphate. The glyceraldehyde thus formed is phosphorylated to become glyceraldehyde 3-phosphate for entering glycolysis.

Fructose 6-Phosphate (mol wt 260) The monophosphate form of fructose and an intermediate in glycolysis.

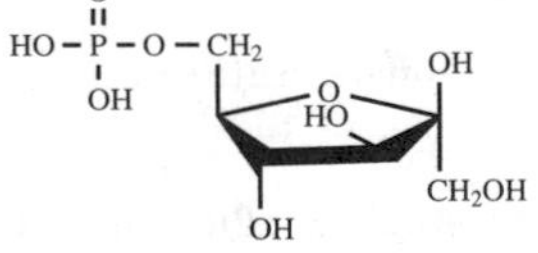

Fructose 6-Phosphoketolase The enzyme that catalyzes the following reaction:

Fructose 6-phosphate + Pi

⇌

Acetylphosphate + erythrose 4-phosphate

Fructosuria A genetic disorder characterized by the presence of excessive amounts of fructose in the urine due to a deficiency of the enzyme fructokinase.

Fructuronate Reductase The enzyme that catalyzes the following reaction:

D-mannonate + NAD^+

⇅

D-fructuronate + NADH

Fruit Bromelain A protease from pineapple with broad specificity for hydrolysis of peptide bonds.

Fruiting Body A specialized microbial structure that bears sexually or asexually derived spores.

Frusemide See furosemide.

FseI A restriction endonuclease from *Frankia* species Eul IB with the following specificity:

```
            ↓
5'........GGCCGGCC........3'
3'........CCGGCCGG........5'
            ↑
```

F-Sequence A transposable element in the genome of *Drosophila.*

FSF Abbreviation for fibrin-stabilizing factor.

FSH Abbreviation for follicle-stimulating hormone.

FSHRF Abbreviation for follicle-stimulating hormone releasing factor.

FSH-RH Abbreviation for follicle-stimulating hormone releasing hormone.

FSP Abbreviation for fibrin-split product.

FspI (MstI) A restriction endonuclease from *Fischerella* species with the following specificity:

```
              ↓
5'.........TGCGCA.........3'
3'.........ACGCGT.........5'
              ↑
```

FspII (AsuII) A restriction endonuclease from *Fischerella* species with the following specificity:

```
            ↓
5'.........TTCGGA.........3'
3'.........AAGCTT.........5'
                ↑
```

Fsp4HI A restriction endonuclease from *Flavobacterium* species 4H with the following specificity:

```
            ↓
5'........GCNGC........3'
3'........CGNCG........5'
            ↑
```

FspMI (FnuDII) A restriction endonuclease from *Flavobacterium* species with the same specificity as FnuDII.

FspMSI (AvaII) A restriction endonuclease from *Fischerella* species with the following specificity:

```
            ↓
5'.........GG(A/T)CC.........3'
3'.........CC(T/A)GG.........5'
                  ↑
```

***fstI* Gene** A gene in *E. coli* that encodes penicillin-binding protein.

FTA-ABS Abbreviation for fluorescence *Treponema* antibody adsorption or fluorescent-absorbed treponemal antibody.

FTA-AT Abbreviation for fluorescence *treponema* antibody adsorption test.

FTase Abbreviation for farnesyltransferase.

fTHF Abbreviation for N^{10}-formyltetrahydrofolate.

FT-IR (FTIR) Abbreviation for Fourier-transform infrared.

FT-IRS (FTIRS) Abbreviation for Fourier-transform infrared spectroscopy.

FsuI(Tth111I) A restriction endonuclease from *Flavobacterium suaveolens* with the following specificity:

```
                 ↓
5'.........GACNNNGTC.........3'
3'.........CTGNNNCAG.........5'
                 ↑
```

5-FU Abbreviation for 5-fluorouracil.

Fuc Abbreviation for fucose.

Fuchsin A reagent for determination of sulfite in food.

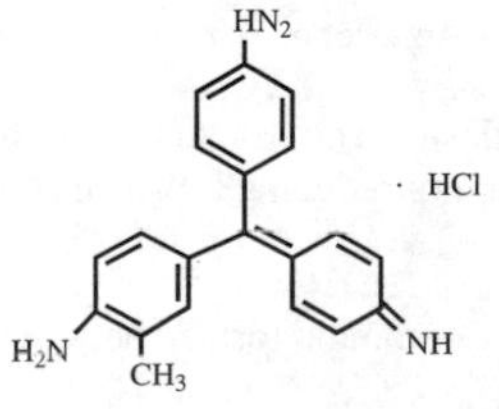

Fucoidan A polysaccharide composed predominantly of sulfated fucose.

Fucoidanase The enzyme that catalyzes the hydrolysis of 1,2-α-L-fucoside linkages in fucoidan without releasing sulfate.

L-Fucokinase The enzyme that catalyzes the following reaction:

ATP + 6-deoxy-L-galactose
⇅
ADP + 6-deoxygalactose 1-phosphate

Fucolipid A lipid that contains fucose (glycolipid).

Fucosamine (mol wt 163) An amino sugar derived from galactose (2-amino-2,6-dideoxygalactose).

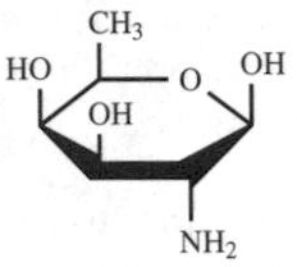

Fucosan A polymer of fucose.

L-Fucose (mol wt 164) A sugar found in blood group substances.

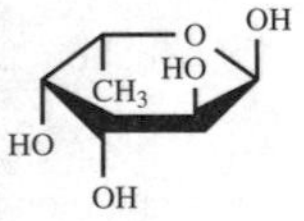

Fucose (L-fucose) Dehydrogenase Synonym of D-threo-aldose 1-dehydrogenase.

Fucose 1-Phosphate Guanylyltransferase The enzyme that catalyzes the following reaction:

GTP + L-fucose 1-phosphate
⇅
PPi + GDP-L-fucose

α-L-Fucosidase The enzyme that catalyzes the following reaction:

An α-L-fucoside + H_2O
⇅
An alcohol + L-fucose

b-D-Fucosidase The enzyme that catalyzes the hydrolysis of nonreducing b-D-fucose residues in a b-D-fucoside.

Fucoside Compound formed between fucose and alcohol.

Fucosidosis A genetic disorder characterized by cerebral degeneration, muscle spasticity, and accumulation of H-isoantigen due to deficiency of a-fucodase.

Fucosyl N-Acetyl-Glucosaminyl Glycoprotein Fucohydrolase The systematic name for 1,3-α-L-fucosidase.

5FUDP Abbreviation for 5-fluorouridine diphosphate.

FUDR Abbreviation for fluorodeoxyuridine.

Fulcin A trade name for griseofulvin microsize, an antifungal agent that disrupts membrane permeability of fungi.

Fuligo A genus of slime mold.

Fulvicin A trade name for griseofulvin, an antifungal drug that disrupts membrane permeability of fungi.

Fumagillin (mol wt 459) An antiprotozoal agent.

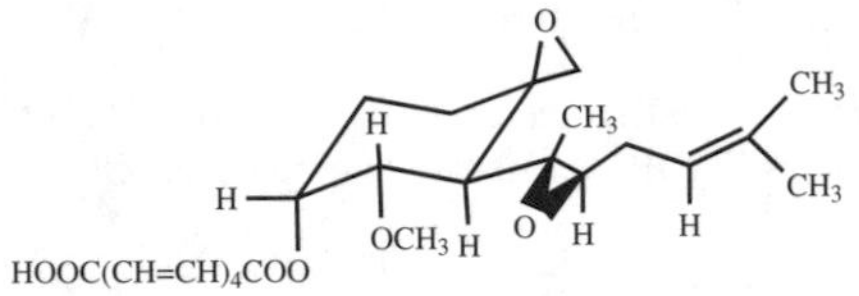

Fumarase The enzyme that catalyzes the following reaction:

Malate ⇌ Fumarate + H_2O

Fumarate Hydrase See fumarase.

Fumarate Hydratase The enzyme that catalyzes the following reaction:

(S)-Malate
⇅
Fumarate + H_2O

Fumarate Reductase See succinate dehydrogenase.

Fumaric Acid (mol wt 116) An intermediate in the Krebs cycle.

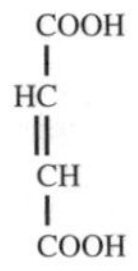

Fumaric Pathway The pathway that converts phenylalanine or tryosine to fumarate and acetoacetate.

Fumarylacetoacetatase The enzyme that catalyzes the following reaction:

4-Fumarylacetoacetate + H_2O
⇅
Acetoacetate + fumarate

Fumarylacetoacetate Fumarylhydrolase See fumarylacetoacetatase.

Fumasorb A trade name for ferrous fumarate, a hematinic agent that provides iron for the synthesis of hemoglobin.

Fumerin A trade name for ferrous fumarate, a hematinic agent that provides iron for the synthesis of hemoglobin.

Fumonisin B1 (mol wt 722) A mycotoxin produced by *Fusarium moniliforme.*

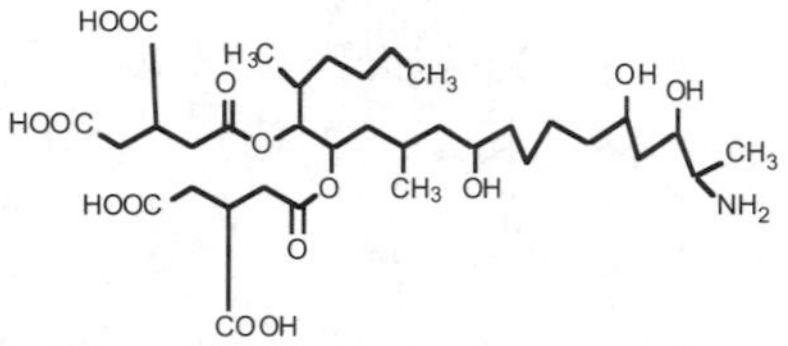

FUMP Abbreviation for 5-fluorouridine monophosphate.

Functional Group Atom or group of atoms in a molecule that confers the characteristic chemical properties to the molecule.

Fungaemia The presence of fungi in the blood.

Fungatin A trade name for folnaftate, an antifungal agent.

Fungi Plural of fungus.

Fungichromin (mol wt 671) An antifungal antibiotic related to filipin.

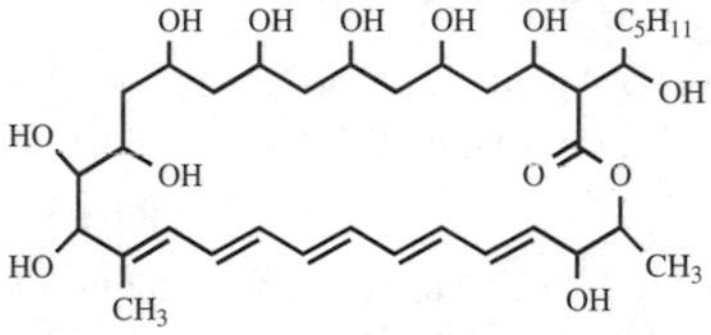

Fungicide Substance capable of killing or inhibiting growth of fungi.

Fungi Imperfecti Fungi without a sexual stage that reproduce by means of asexual formation of conidia (also called Deuteromycete).

Fungilin Oral A trade name for amphotericin B, an antifungal agent that alters membrane permeability of fungi.

Fungistasis The prevention or inhibition of fungal growth.

Fungizone A trade name for amphotericin B, an antifungal drug that alters fungus membrane permeability.

Fungoid Resembling a fungus or fungus-like growth.

Fungus A kingdom of organisms including eukaryotic, unicellular (yeasts), or multicellular heterotrophs.

Funis The umbilical cord.

FUra Abbreviation for fluorouracil.

Furacin A trade name for nitrofurazone, an antimicrobial skin medication.

Furadantin A trade name for nitrofurantoin, an anti-infective agent.

Furalan A trade name for nitrofurantoin, an anti-infective agent.

Furaltadone (mol wt 324) An antibacterial agent.

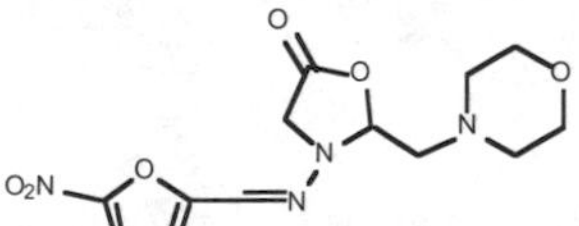

Furan A five-membered heterocyclic structure.

Furanite A trade name for nitrofurantoin, an anti-infective agent.

Furanomycin (mol wt 157) A naturally occurring amino acid and an antibiotic.

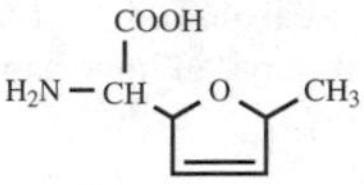

Furanose A monosaccharide with a five-membered ring structure, e.g., ribose and fructose.

Furazolidone (mol wt 225) An antimicrobial and antiprotozoal agent.

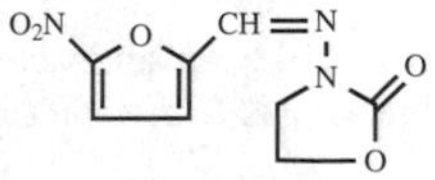

FUrd Abbreviation for fluorouridine.

Furfural (mol wt 96) A heterocyclic aldehyde.

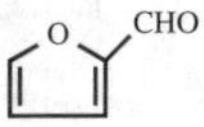

5-Furfuryl-5-isopropylbarbituric Acid (mol wt 250) A sedative and hypnotic agent.

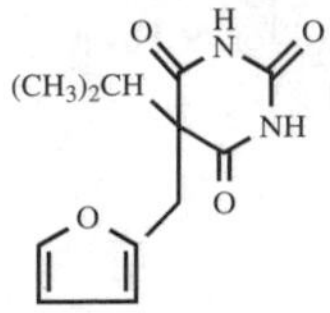

Furomide M.D. A trade name for furosemide, a diuretic agent that inhibits reabsorption of sodium and chloride.

Furonazide (mol wt 229) An antituberculostatic agent.

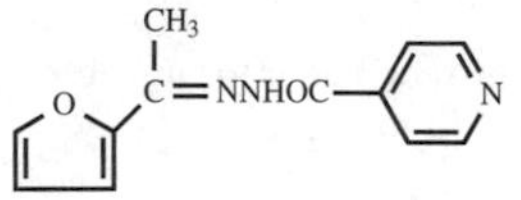

Furosemide (mol wt 331) A diuretic and antihypertensive agent that inhibits reabsorption of sodium and chloride.

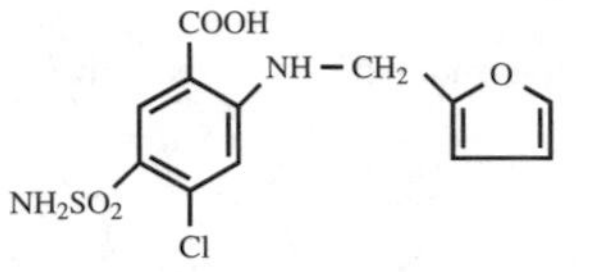

Furoside A trade name for the diuretic agent furosemide which inhibits reabsorption of sodium and chloride.

Furosine (mol wt 254) A naturally occurring nonprotein amino acid.

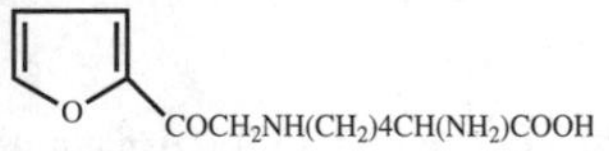

Fusaric Acid (mol wt 179) An inhibitor of plant polyphenol oxidase isolated from fungus *Fusarium heterosporium*. It plays an important role in pathogenesis of plants.

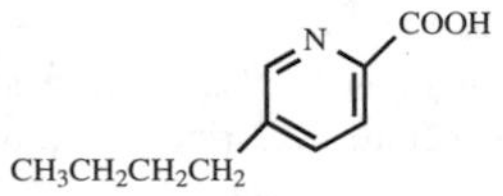

Fusaritoxicosis A mycotoxicosis due to the toxin produced by *Fusarium* species.

Fuscin (mol wt 276) An antibacterial pigment produced by the fungus *Oidiodendron fuscom*.

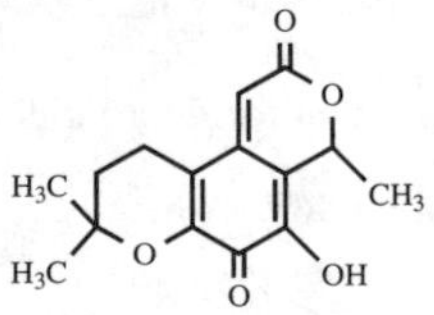

Fused Gene A hybrid gene resulting from recombination of DNA of interest with another gene or plasmid.

Fusel Oil Referring to a mixture of amyl, isoamyl, and propyl alcohol formed as a by-product in alcohol fermentation.

Fused Protein A protein produced by a fused gene (also called hybrid protein).

Fusidic Acid (mol wt 517) An antibiotic from the fungus *Fusidium coccineum* that prevents translation by interfering with the activity of elongation factor G.

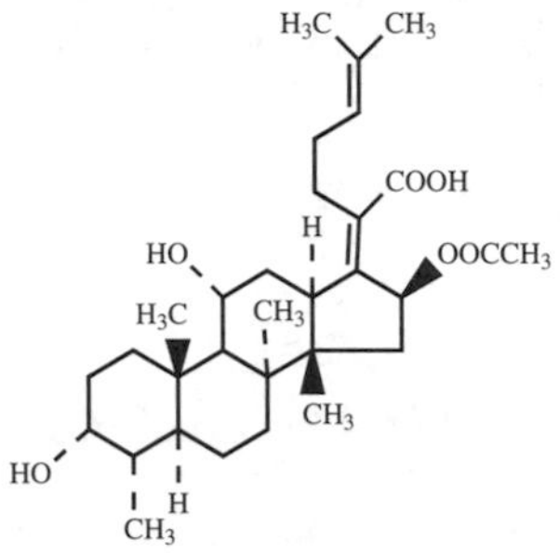

Fusiform Spindle shaped (tapered at both ends).

Fusion Formation of 1. a heavier atomic complex by nuclear fusion, 2. a hybridoma by fusion of plasma cell with myeloma cell, 3. a hybrid gene by fusion of different genes, and 4. a fused protein from a fused gene.

Fusion Gene A hybrid gene that consists of parts of two other genes resulting from deletion of chromosome segments between two linked genes.

Fusion Protein A protein that promotes fusion between two host cells or between the viral envelope and plasma membrane.

Fusobacterium A genus of Gram-negative bacteria (Bacteroidaceae).

Fusogenic Agent Substance capable of inducing cell fusion, e.g., polyethylene glycol or Sendai virus.

Futile Cycle The useless consumption of ATP or energy by two opposing reactions that convert a substrate to an intermediate and then back to the same substrate at a comparable rate such as:

ATP + Fructose-6-phosphate

↓ fructose phosphokinase

Fructose-1-6-bisphosphate

↓ fructose-1-6-bisphosphatase

Fructose-6-phosphate + Pi

5FUTP Abbreviation for 5-fluorouridine triphosphate.

Fuzzy Layer Layer of material exterior to the cell coat of animal cells consisting of collagen and glycosaminoglycans (see also glycocalyx).

FV Abbreviation for Friend virus.

Fv Fragment Referring to the variable regions in the Fab fragment of an immunoglobulin.

FVAC Abbreviation for a combination drug containing fluorouracil, vinblastine, adriamycin, and cyclophosphamide.

FW Abbreviation for fresh weight.

-fylline A suffix meaning a drug capable of relaxing or improving air passage or breathing.

FYN A gene encoding non-receptor tyrosine kinase.

Fynex A trade name for diphenhydramine hydrochloride, used as an antihistaminic agent.

G

g Gravity used in description of relative centrifugal force, e.g., 5000 × g (referring to sedimentation 5000 times that of earth's gravity).

G Abbreviation for 1. gram, 2. Gibbs free energy, 3. guanine, and 4. glycine.

γ Abbreviation for gamma, a letter in the Greek alphabet.

g% Abbreviation for gram percent (e.g., grams/100 ml).

G_0 1. A type of G protein of unknown function that occurs in the brain, heart, and other tissues. 2. A stage in the cell cycle in which cell division is arrested (it constitutes an exit from the cell cycle).

G1 A phase in the cell cycle that constitutes the gap between the end of mitosis and the onset of S phase.

G2 A phase in the cell cycle that constitutes the gap between the S phase and the onset of mitosis.

ΔG Symbol for free energy change.

ΔG_0 Symbol for standard free energy change. It is defined as the value of free energy change in Kcal/mole under standard conditions of 25° C, 1 atm pressure, and concentrations of reactants and products at 1 mole and pH 0.

$\Delta G'_0$ Referring to ΔG_0 at pH 7.

ΔG′ The symbol for steady state free energy change at nonstandard conditions when the concentrations of reactants and products are not at 1 mole.

GA Abbreviation for 1. glucuronic acid; 2. gramcidin A.

Ga Abbreviation for gallium, a chemical element.

GAA A genetic code or codon for glutamic acid.

Gα Abbreviation for G protein alpha subunit.

G2A Mutation Abbreviation for mutation in which glycine in position 2 is replaced by alanine (A).

GABA Abbreviation for gamma aminobutyric acid (see γABA).

γABA (mol wt 103) Abbreviation for gamma amino butyric acid, a primary inhibitory neurotransmitter.

$$H_2NCH_2CH_2CH_2COOH$$

GABA Receptor The binding site for GABA (γ-aminobutyric acid) on the surface of most nerve cell bodies.

GABA Shunt A pathway occurring in bacteria for conversion of α-ketoglutaric acid to succinic acid that differs from the normal Krebs cycle.

Gabase The enzyme that catalyzes the conversion of γ-aminobutyric acid to succinic aldehyde and then to succinic acid in the presence of $NADP^+$.

Gabapentin (mol wt 171) An anticonvulsant and antiepileptic agent.

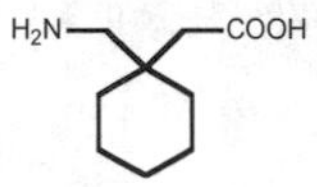

Gabexate (mol wt 321) A nonpeptide protease inhibitor.

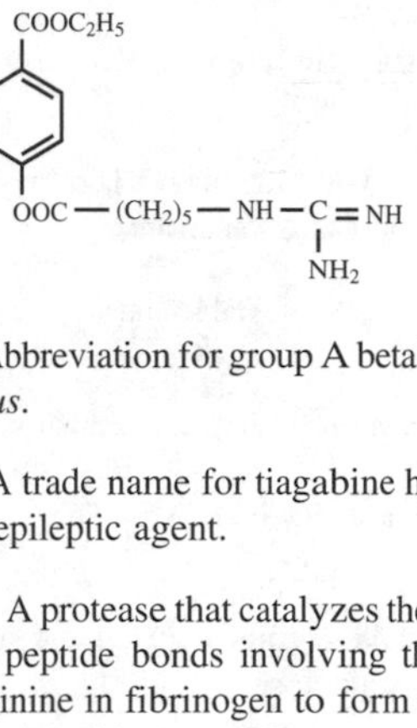

GABHS Abbreviation for group A beta hemolytic *Streptococcus.*

Gabitrol A trade name for tiagabine hydrochloride, an antiepileptic agent.

Gabonase A protease that catalyzes the selective cleavage of peptide bonds involving the COOH group of arginine in fibrinogen to form fibrin and release fibrinopeptide A and B.

GAC A genetic code or codon for aspartic acid.

G-actin The globular monomeric form of actin that polymerizes to form F-actin filaments.

GAD Abbreviation for glutamic acid decarboxylase.

GADDP Abbreviation for growth arrest and DNA damage-inducible protein.

GADH Abbreviation for glutamic acid dehydrogenase.

Gadolinium (Gd) A chemical element with atomic weight 64, valence 3.

GAG A genetic code or codon for glutamic acid.

***gag* Gene** A gene in the retrovirus genome that encodes viral capsid protein.

gag Protein Protein encoded by the *gag* gene in the retrovirus.

GAIP (GαIP) Abbreviation for G-alpha interacting protein.

Gal Abbreviation for 1. galactose and 2. gallon.

Gal Operon An operon in *E. coli* that encodes for galactokinase, galactose transferase, and galactose epimerase for metabolism of galactose.

GAL4 A gene involved in the regulation of galactose utilization in yeast.

GalI (SacII) A restriction endonuclease from *Gluconobacter albidus* with the following specificity:

5′..........CCGC↓GG..........3′
3′..........GG↑CGCC..........5′

Galactan Polymer of galactose.

Galactanase See arabinogalactan endo-1,3-β-galactosidase.

Galactarate Dehydratase The enzyme that catalyzes the following reaction:

D-Galactarate ⇌ 5-dehydro-4-deoxy-D-glucarate + H_2O

D-Galactarate Hydrolyase See galactarate dehydratase.

Galactaric Acid (mol wt 210) A sugar acid derived from galactose.

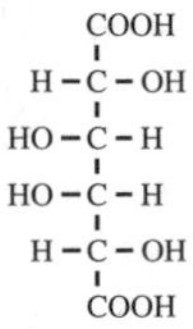

-galactia A suffix meaning secretion of milk.

Galactinol Raffinose Galactosyltransferase The enzyme that catalyzes the following reaction:

1-α-D-Galactosyl-*myo*-inositol + raffinose ⇌ *myo*-Inositol + stachose

Galactinol Sucrose Galactosyltransferase The enzyme that catalyzes the following reaction:

1-α-D-Galactosyl-*myo*-inositol + sucrose ⇌ *myo*-Inositol + raffinose

Galactinol Synthetase The enzyme that catalyzes the following reaction:

UDP-galactose + *myo*-inositol ⇌ UDP + α-D-galactosyl-*myo*-inositol

Galactitol (mol wt 182) A sugar alcohol derived from galactose (also known as dulcitol).

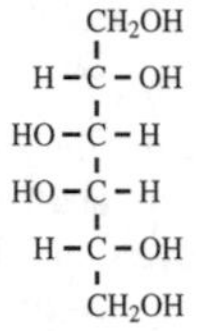

Galactitol 2-Dehydrogenase The enzyme that catalyzes the following reaction:

Galactitol + NAD^+ ⇌ D-tagatose + NADH

Galactobiose (mol wt 342) A disaccharide that consists of β-D-(1,6) linked galactose.

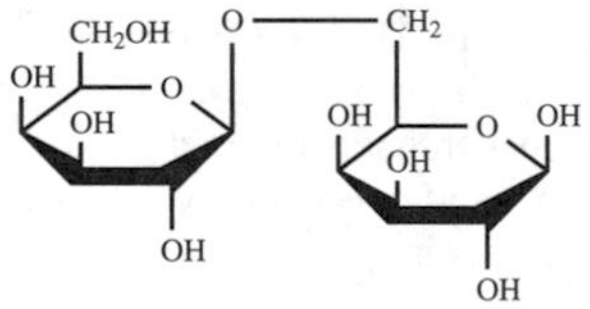

Galactocarolose An extracellular polysaccharide from *Penicillium charlesii.* It consists of 1,5-linked galactofuranosyl residues.

Galactocele A fluid-filled sac formed from blockage of milk ducts in the breast.

Galactocerebrosidase The enzyme that catalyzes the following reaction:

Galactocerebroside ⇌ Ceramide + galactose

Galactocerebroside A cerebroside containing galactoses.

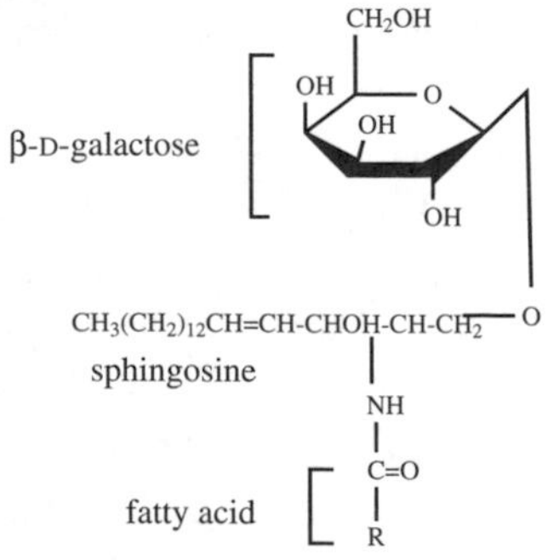

Galactoflavin (mol wt 406) A riboflavin antagonist.

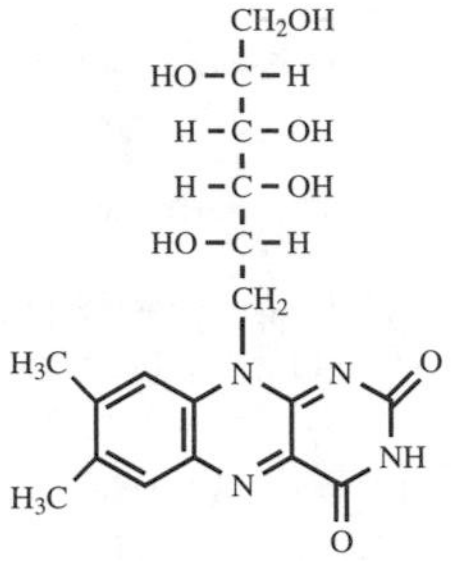

Galactokinase The enzyme that catalyzes the following reaction:

ATP + α–D-galactose

⇅

ADP + α-D-galactose-1-phosphate

Galactokinase Deficiency A genetic disorder characterized by the accumulation of galactose in the blood due to the deficiency in galactokinase.

Galactolipid Any glycolipid containing residues of galactose or N-acetylgalactosamine.

Galactomannan Any heteroglycan containing residues of galactose and D-mannose.

Galactonate Dehydratase The enzyme that catalyzes the following reaction:

D-Galactonate

⇅

2-Dehydro-3-deoxy-D-galactonate + H_2O

Galactonic Acid (mol wt 196) An organic acid derived from galactose.

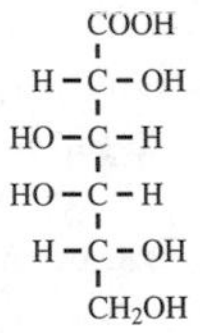

Galactonic Acid γ-Lactone (mol wt 178) A derivative of galactose.

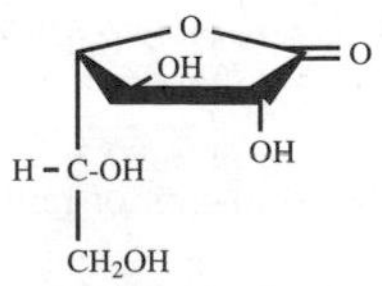

Galactonolactone Dehydrogenase The enzyme that catalyzes the following reaction:

L-Galactono-γ–lactone + 2 ferricytochrome C

⇅

L-ascorbate + 2 ferrocytochrome C

L-Galactonolactone Oxidase The enzyme that catalyzes the following reaction:

L-Galactono-1,4-lactone + O_2

⇅

L-Ascorbate + H_2O_2

Galactosamine (mol wt 179) An amino sugar derived from galactose.

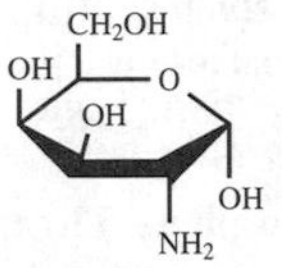

α-D-galactosamine

Galactose (mol wt 180) A six-carbon aldose sugar.

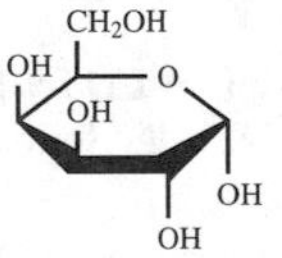

α–D-Galactose

Galactose Binding Protein A bacterial periplasmic protein in *E. coli* that acts both as a sensory element in the detection of galactose and as a chemotactic signal in the uptake of sugar.

Galactose Dehydrogenase The enzyme that catalyzes the following reaction:

D-Galactose + NAD^+

⇅

D-galactono-1,4-γ–lactone + NADH

Galactose Dehydrogenase ($NADP^+$) The enzyme that catalyzes the following reaction:

D-Galactose + $NADP^+$

⇅

D-Galactonolactone + NADPH

Galactose NAD^+ 1-Oxidoreductase The systematic name for galactose 1-dehydrogenase.

Galactose Operon See gal operon.

Galactose Oxidase The enzyme that catalyzes the following reaction:

D-Galactose + O_2

⇅

D-Galacto-hexodialdose + H_2O_2

Galactose Oxygen 6-Oxidoreductase The systematic name for galactose oxidase.

Galactose 1-Phosphate (mol wt 260) A phosphate form of galactose.

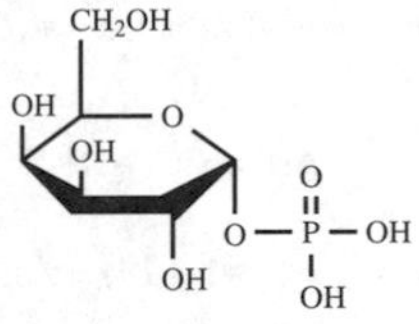

Galactose 1-Phosphate Pathway A pathway for conversion of galactose to glucose via glucose 1-phosphate by the UDP-glucose 4-epimerase.

Galactose 1-Phosphate Thymidyl Transferase The enzyme that catalyzes the following reaction:

TTP + α-D-galactose 1-phosphate

⇌

pyrophosphate + TDP-galactose

Galactose 1-Phosphate Uridyl Transferase The enzyme that catalyzes the following reaction:

Galactose 1-phosphate + UDP-glucose

⇌

UDP-galactose + glucose 1-phosphate

Galactose 6-Phosphate (mol wt 260) A phosphate form of galactose.

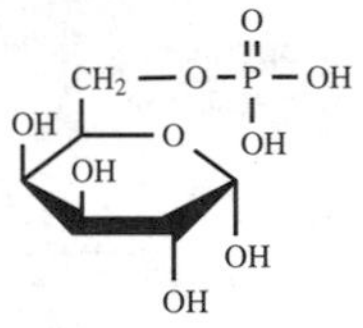

Galactose 6-Sulfurylase The enzyme that catalyzes the elimination of sulfate from β-D-galactose-6-sulfate residues of porphyran to produce 3,6-anhydrogalactose residues.

Galactose Tolerance Test A test for the ability of the liver to remove galactose from the blood and to convert it to glycogen.

Galactosemia A genetic disorder characterized by the inability to metabolize galactose in milk due to a deficiency in galactose 1-phosphate uridyl transferase.

α-Galactosidase The enzyme that catalyzes the following reaction:

An α-D-galactoside + H_2O

⇌

An alcohol + α–D-galactose

β-Galactosidase The enzyme that catalyzes the following reaction:

A β–D-galactoside + H_2O

⇌

An alcohol + β–D-galactose

Galactoside Acetyltransferase The enzyme that catalyzes the following reaction:

Acetyl-CoA + β-D-galactoside

⇌

CoA + 6-acetyl-β-D-galactoside

Galactoside 2-α-L-Fucosyltransferase The enzyme that catalyzes the following reaction:

GDP-L-fucose + β-D-galactosyl-R

⇌

GDP + α-L-fucosyl-1,2-β-D-galactosyl-R

Galactoside Galactohydrolase Synonym of galactosidase.

Galactoside Permease The enzyme that catalyzes the transport of lactose into the cell (see also lactose permease).

Galactoside α-2,3-Sialyltransferase The enzyme that catalyzes the following reaction:

CMP-N-acetylneuraminate
+
β-D-galactosyl-1,3-N-acetyl-α-D-galactosaminyl-R

⇌

CMP + α-N-acetylneuraminyl-2,3-β-D-galactosyl-1,3-N-acetyl-α-D-galactosaminyl-R

Galactoside α-2,6-Sialyltransferase The enzyme that catalyzes the following reaction:

CMP-N-acetylneuraminate
+
β-D-galactosyl-1,4-N-acetyl-α-D-galactosamine

⇌

CMP + α-N-acetylneuraminyl-2,6-β-D-galactosyl-1,4-N-acetyl-β-D-galactosamine

Galactosis The formation of milk.

Galactosphingolipid See galactocerebroside.

Galactostat A trade name for the galactose oxidase reagent used to test for galactose.

Galactosuria The presence of an excessive amount of galactose in the urine due to galactosemia.

Galactosylceramidase The enzyme that catalyzes the following reaction:

D-Galactosyl *N*-acylsphingosine + H_2O

⇌

D-Galactose + *N*-acylsphingosine

Galactosyl Sphingosine Transferase See UDP-galactose-sphingosine galactosyl transferase.

Galactosyl Transferase The enzyme that catalyzes the following reaction:

UDP-galactose + *N*-acetyl glucosamine

⇌

N-acetyl lactosamine

Galactouronokinase The enzyme that catalyzes the following reaction:

ATP + D-galactouronate

⇌

ADP + α-D-Galacturonate 1-phosphate

Galactozymase Referring to the starch-digesting enzyme.

Galacturan 1,4-α-Galacturonidase The enzyme that catalyzes the following reaction:

$(1,4\text{-}\alpha\text{-D-Galacturonide})_n$ + H_2O

⇌

$(1,4\text{-}\alpha\text{-D-Galactouronide})_{n-1}$ + D-galacturonide

Galacturonan A polysaccharide that consists of galacturonic acid residues.

D-Galacturonic Acid (mol wt 194) The uronic acid derived from galactose.

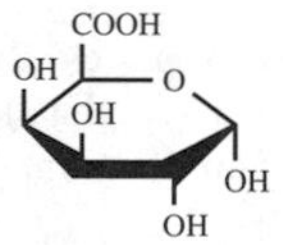

Galanin A neuropeptide from the small intestine and the central and peripheral nervous systems that regulates gut motility and activity of endocrine pancreas.

Galanthamine (mol wt 287) An inhibitor for cholinesterase.

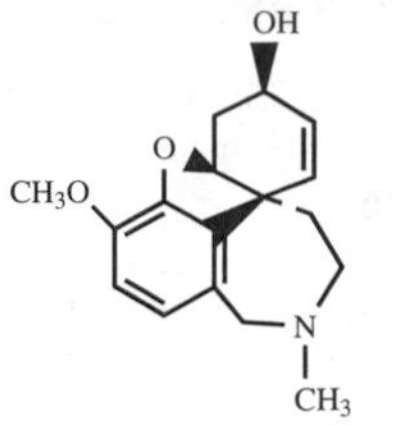

Galbiose (mol wt 342) A disaccharide consisting of two galactoses in a,1,4-linkage (see also galactobiose).

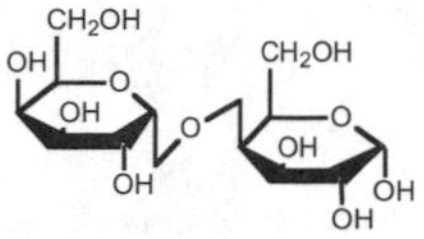

GalCer Abbreviation for galactosylceramide.

Gall An abnormal plant structure produced in response to parasitic infections, e.g., crown gall produced by the infection of *Agrobacterium tumeficiens*.

Gallacetophenone (mol wt 168) An antiseptic agent.

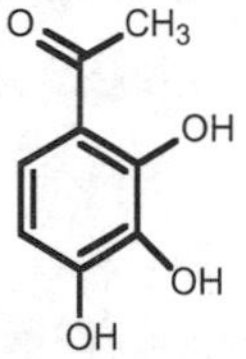

Gallamine Triethiodide (mol wt 892) A skeletal muscle relaxant that prevents acetylcholine from binding to its receptor.

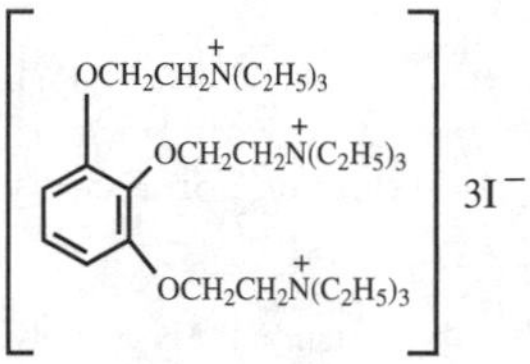

Gallate Carboxylase The enzyme that catalyzes the following reaction:

3,4,5-Trihydroxybenzoate

⇌

Pyrogallol + CO_2

Gallate 1-β-Glucosyltransferase The enzyme that catalyzes the following reaction:

UDP-glucose + gallate

⇌

UDP + 1-galloyl-β-D-glucose

Gallein (mol wt 364) A biological stain and an acid and base indicator.

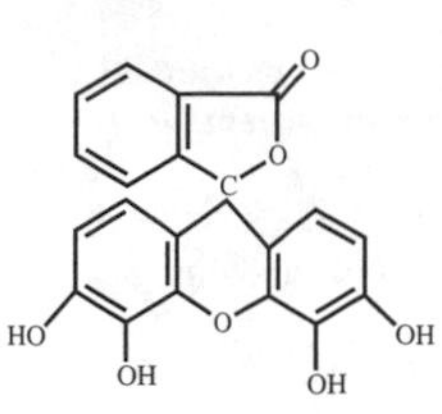

Gallic Acid (mol wt 170) An astringent agent.

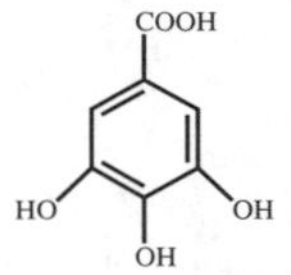

Gallionella A genus of Gram-negative iron bacteria.

Gallium (Ga) A chemical element with atomic weight 70, valence 3, 2, and 1.

Gallium Nitrate (mol wt 256) A substance that reduces hypercalcemia by inhibiting resorption of bone.

$$Ga(NO_3)_3$$

Gallopamil (mol wt 485) A calcium channel blocking agent.

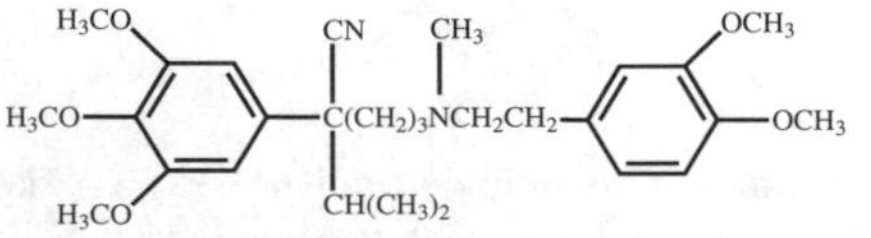

Gallstone A solid body formed in the gall bladder or bile duct.

GalNAc Abbreviation for *N*-acetylgalactosamine.

GalNAc Abbreviation for N-acetyl-D-galactosamine.

GalNAcT Abbreviation for N-acetyl-D-galactosaminyl transferase.

GalNH$_2$ Abbreviation for galactosamine.

GalO Abbreviation for galactose oxidase.

Gal-1-P Abbreviation for galactose 1-phosphate.

Gal-6-P Abbreviation for galactose 6-phosphate.

GalT Abbreviation for galactosyl transferase.

GALT Abbreviation for gut-associated lymphoid tissue.

GalU Abbreviation for galacturonic acid.

Galvanometer An instrument for measuring quantity of electric current.

Galvanoscope An instrument used to detect the presence of direct current.

Galvanotaxis A taxis in which the stimulus is an electrical potential gradient.

Gamastan A trade name for intramuscular immune globulin.

Gamete Either of the two germ cells that fuse to form a zygote.

Gametic Number Referring to the haploid number.

Gametocide An agent that kills gametocytes.

Gametoclone Plants generated from cell cultures derived from meiospores, gametes, or gametophytes.

Gametocytaemia The presence of a gametocyte of a parasite in the blood.

Gametocyte Any cell that gives rise to gametes.

Gametogenesis The process leading to the formation of gametes (see also gametogony).

Gametogony The formation of gametes from gametocytes.

Gametophyte Haploid generation in the life cycle of an organism that alternates between haploid and diploid forms.

Gamimune-N A trade name for a 10% human immunoglobulin solution.

Gamma (γ) A Greek letter used to denote 1. a unit of weight equal to one microgram or 2. the third carbon from the carbon atom that carries a functional group of a molecule.

Gamma Amino Butyric Acid (mol wt 103) A nonprotein amino acid and inhibitory neurotransmiter in the brain, heart, lung, and kidney.

$$NH_2-CH_2\text{-}CH_2\text{-}CH_2\text{-}COOH$$

Gamma Amino Butyric Acid Bypass The pathway for converting a-ketoglutaric acid to succinic acid that differs from the normal Krebs cycle.

Gamma Benzene A trade name for lindane, an insecticide that inhibits neuronal membrane function in the arthropod.

Gamma Chain Referring to the heavy chain of IgG.

Gamma Counter An instrument used to measure the amount of gamma irradiation emitted by a radioactive substance.

Gamma Efferent Fibers The nerve fibers that carry impulses from the central nervous system to the fibers of the muscle.

Gamma Globulin Serum proteins with low electrophoretic mobility (gamma mobility) that make up the majority of immunoglobulins (abbreviated as γ-globulin).

Gamma Interferon Interferon produced by T lymphocytes.

Gamma Radiation A high-frequency electromagnetic radiation from certain radioisotopes that consists of protons originating from the nucleus of the atom.

Gamma Ray High-energy, short-wave electromagnetic radiation emitted by a radioactive substance.

Gamma Toxin A protein toxin produced by *Staphylococcus aureus* that has high hemolytic activity against rabbit erythrocytes.

Gammagard A trade name for immune globulin.

Gammagard S/D A trade name for detergent treated human immunoglobulin.

Gammopathy A disorder that causes an abnormal increase of gamma globulin in the blood.

Gamone Referring to a substance released by an egg or sperm that acts as a chemotactic agent to attract gametes of the opposite sex.

Gamont Referring to the haploid adult in protoctists that has both haploid and diploid phases in the life cycle.

Gamulin Rh A trade name for Rh_o (D) immune globulin.

Ganciclovir (mol wt 255) An antiviral agent used for treatment of cytomegalovirus infection.

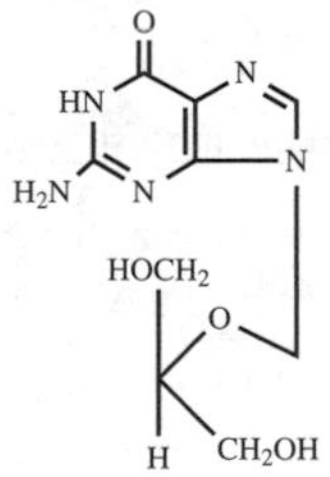

Ganglefene (mol wt 335) A coronary vasodilator.

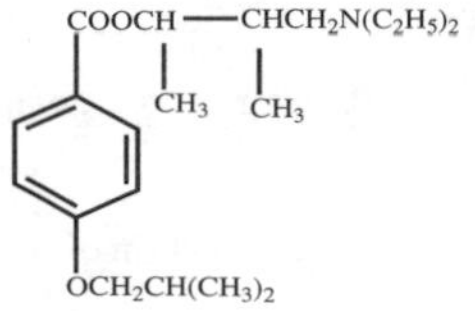

Ganglioblast An embryonic cell from which ganglion cells arise.

Ganglioma A tumor of nerve cells of ganglion.

Ganglion A collection of nerve cell bodies outside the brain.

Ganglionitis Inflammation of nerve cells.

Ganglioplegic Substance that blocks the transmission of impulses through autonomic ganglion.

Ganglioside A group of sphingoglycolipids found in the CNS and other tissues that contain galactose, *N*-acetylgalactosamine, *N*-acetylneuraminidate (sialic acid), stearic acid, and sphingosine.

Gangliosidosis A genetic disorder due to a deficiency in the enzymes that metabolize gangliosides. It is characterized by the accumulation of specific gangliosides in the CNS.

Gangrene The necrosis or death of tissue resulting from the loss of blood supply.

Ganite A trade name for gallium nitrite, an antihypercalcemic agent that inhibits calcium resorption from bone.

Gantanol A trade name for sulfamethoxazole, an antibiotic that inhibits the synthesis of bacterial folic acid.

Gantrisin A trade name for sulfisoxazole, an antibiotic that inhibits the synthesis of folic acid in bacteria.

GAP Abbreviation for 1. glutamate aminopeptidase; 2. glyceraldehyde 3-phosphate; 3. growth-associated protein; 4. GTP-activated protein; 5. GTPase-activating protein.

Gap The region where one or more nucleotides are missing in the dsDNA leading to the formation of broken chains.

GAP-43 Abbreviation for growth-associated protein of 43 kDa.

Gap Filament A thin filament observed between the A band and the I band of striated muscle.

Gap Junction The region with clusters of transmembrane protein channels (connexon) between plasma membranes of two animal cells where exchanges of chemical substances and transmission of signals take place.

GAPDH Abbreviation for glyceraldehyde-3-phosphate dehydrogenase.

GAR Abbreviation for glycinamide ribotide, an intermediate in the biosynthesis of purine nucleotides.

GAR Synthetase The enzyme that catalyzes the following reaction:

$$\text{5-Phosphoribosylamine + ATP + glycine} \rightleftharpoons \text{Glycinamide ribotide + ADP + Pi}$$

GAR Transformylase The enzyme that catalyzes the following reaction:

$$\text{Glycinamide ribotide} + N^{10}\text{-formyl-THFA} \rightleftharpoons \text{Formylglycinamide ribotide + THFA}$$

Garamycin A trade name for gentamicin sulfate, an antibiotic that binds to 30S ribosomes and inhibits bacterial protein synthesis.

Gardenal A trade name for phenobarbital, an anticonvulsant that depresses monosynaptic and polysynaptic transmission in the CNS.

Gardol (mol wt 293) A detergent and foaming agent.

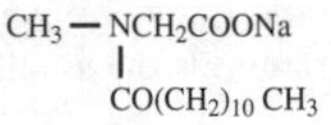

GARFP Abbreviation for glycinamide ribonucleotide formyltransferase of G protein.

Gas Chromatogram The profile of a gas chromatographic separation pattern for a mixture of substances.

Gas Chromatography A column partition chromatography, e.g., gas solid chromatography in which the mobile phase is a gas and the stationary phase is a solid, or gas liquid chromatography in which a carrier coated with nonvolatile liquid acts as the stationary phase and an inert gas acts as the mobile phase.

Gas Constant Physical constant that is derived from a gas law and used in thermodynamic calculations.

Gas Embolism Blockage of one or more small blood vessels by bubbles of gases.

Gas Ionization Formation of ion pairs from a gas that is subjected to ionizing radiation.

Gas Liquid Chromatography Column-partition chromatography in which the stationary phase is a carrier coated with nonvolatile liquid and the mobile phase is an inert gas.

Gas Relief A trade name for simethicone, an anti-flatulent.

Gas Solid Chromatography Column-partition chromatography in which the stationary phase is a solid and the mobile phase is an inert gas.

Gasogenic Gas producing.

Gasohol A mixture of gasoline and ethanol.

GasPak An anaerobic jar used for growing anaerobic bacteria.

Gastrectomy Surgical removal of all or part of the stomach.

Gastric Pertaining to the stomach.

Gastric Glands The glands in the stomach that produce hydrochloric acid and enzymes.

Gastric Inhibitory Peptide Peptide hormone in the mucous membrane of the small intestine; it causes the pancreas to release insulin and prevents the secretion of gastric acid and pepsin.

Gastric Juice Digestive juice from the stomach consisting of hydrochloric acid and enzymes.

Gastricsin A protease in gastric juice that catalyzes the preferential cleavage of peptide bonds involving the COOH group of tyrosine. It has high activity for hemoglobin.

Gastrin A peptide hormone secreted by the stomach that stimulates secretion of hydrochloric acid and enzymes in the stomach.

Gastritis Inflammation of the lining of the stomach.

Gastro- A prefix denoting stomach.

Gastroblenorrhea Excessive secretion of mucus by the stomach.

Gastrocoli Pertaining to stomach and colon.

Gastrocrom A trade name for cromolyn sodium, which inhibits mast cell degranulation.

Gastrodermis The lining of the digestive tract of an animal.

Gastroduodenoscopy The technique for viewing the inside of the stomach and duodenum with an endoscope.

Gastroenteric Pertaining to stomach and intestine.

Gastroenteritis Inflammation of stomach and intestine.

Gastrogenic Originating in the stomach.

Gastrohepatic Pertaining to stomach and liver.

Gastrohydrorrhea Excessive secretion of watery fluid by the stomach.

Gastrointestinal Pertaining to the stomach and the intestine.

Gastrolith A calculus in the stomach.

Gastrolithiasis The presence of one or more calculi in the stomach.

Gastrology The science that deals with the stomach.

Gastroscope A device used for examination of the interior of the stomach.

Gastrostomy Surgical creation of an opening from the abdominal wall to the stomach.

Gas-X A trade name for simethicone, which prevents the formation of mucus-surrounded pockets in the GI tract.

Gated Channels Protein channels in the membrane that permit transient passage of external substances or ions into a cell when properly triggered.

Gating The switching on or off of a gated channel in the membrane.

Gating Current The current resulting from movement of ions through gated channels.

GAU A genetic code or codon for aspartic acid.

Gaucher Cell The large, lipid-filled cells found in the reticuloendothelial system of the patient with Coucher's disease.

Gaucher's Disease A genetic disorder characterized by the enlargement of liver, spleen, and lymph nodes due to the accumulation of cerebrosides in the tissues and deficiency of the enzyme glucocerebrosidase.

Gavage Tube feeding of a patient.

Gaviscon A trade name for a combination drug containing aluminum hydroxide and magnesium carbonate, used as an antacid.

Gb Abbreviation for G-protein beta subunit.

GBA Abbreviation for ganglionic-blocking agent.

G-Band The bands that appear on metaphase chromosomes upon staining with giemsa stain.

G-Banding Technique for staining chromosomes with Giemsa stain that reveals patterns of deeply stained bands.

GBBHS Abbreviation for group B beta hemolytic *Streptococcus*.

GBG Abbreviation for gonadal steroid-binding globulin.

GBM Abbreviation for glomerular basement membrane.

GBP Abbreviation for 1. galactose-binding protein; 2. glucose-binding protein; 3. glucose bisphosphate.

G-1,6BP Abbreviation for glucose 1,6-bisphosphate.

GBR1 A seven-transmembrane domain protein with high affinity for γ-aminobutyric acid (GABA).

GBS Abbreviation for group B beta hemolytic *Streptococcus*.

GBSS Abbreviation for granule-bound starch synthetase.

GC Abbreviation for 1. galactocerebroside; 2. ganglion cells; 3. gas chromatography; 4. glucocorticoid; 5. guanylate cyclase; 6. guanine/cytosine.

GC Box A guanine- and cytosine-rich sequence in the promoter of many eukaryotic housekeeping genes.

G–C Content Percentage of guanine and cytosine in a nucleic acid (also called G–C value).

G–C Pair The guanine-cytosine pair in the double-stranded DNA molecule.

G–C Value See G–C content.

GCA A genetic code or codon for the amino acid alanine.

G-Cell A gastrin-producing cell found in the stomach and to a lesser extent in the mucosa of the duodenum.

GC/AT Ratio The ratio of guanine + cytosine to adenine + thymine in double-stranded DNA.

GCC A genetic code or codon for the amino acid alanine.

GceI (SacII) A restriction endonuclease from *Gluconobacter cerinus* with the following specificity:

```
          ↓
5′..........CCGCGG..........3′
3′..........GGCGCC..........5′
           ↑
```

GceGLI (SacII) A restriction endonuclease from *Gluconobacter cerinus* with the same specificity as SacII.

GCFT Abbreviation for gonorrhea complement fixation test.

GCG A genetic code or codon for the amino acid alanine.

GC-MS (GCMS) Abbreviation for gas chromatographic mass spectrometry.

GCP Abbreviation for 1. glutamate carboxypeptidase; 2. glycine carboxypeptidase.

GCR Abbreviation for glucocorticoid receptor.

GCRE Abbreviation for goblet-cell response element.

GCS Abbreviation for 1. gamma-glutamylcysteine synthetase; 2. glucosylceramide synthetase.

GCSF Abbreviation for granulocyte colony stimulating factor, a glycoprotein that stimulates proliferation and differentiation of hemopoietic cells.

GCS_h Abbreviation for gamma-glutamylcysteine synthetase heavy chain.

GCS-HS Abbreviation for GCS heavy subunit.

GCS-LS Abbreviation for GCS light subunit.

GCU A genetic code or codon for the amino acid alanine.

Gd Abbreviation for gadolinium, a chemical element.

GDC Abbreviation for 1. glutamate decarboxylase; 2. glycine decarboxylase.

GDF Abbreviation for growth and development factor.

GDH Abbreviation for 1. glutamate dehydrogenase; 2. glyerophosphate dehydrogenase; 3. growth and differential hormone.

GDI Abbreviation for GDP-dissociation inhibitor.

GdiI (StuI) A restriction endonuclease from *Gluconobacter dioxyacetonicus* with the following specificity:

```
          ↓
5′..........AGGCCT..........3′
3′..........TCCGGA..........5′
          ↑
```

GdmCl Abbreviation for guanidinium chloride.

G4-DNA Referring to DNA that carries G-quartet.

GdnCl Abbreviation for guanidinium chloride.

GDNF Abbreviation for glial cell line-derived neurotropic factor.

GdoI (HamHI) A restriction endonuclease from *Gluconobacter dioxyacetonicus* with the same specificity as HamHI.

GDP Abbreviation for 1. gel diffusion precipitation; 2. guanosine diphosphate; 3. glucose diphosphate.

G-1,6-DP Abbreviation for glucose 1,6-diphosphate.

GDPase See guanosine diphosphatase.

GDP-4-Dehydro-D-Rhamnose Reductase The enzyme that catalyzes the following reaction:

$$\text{GDP-6-deoxy-D-mannose} + NADP^+ \rightleftharpoons \text{GDP-4-dehydro-6-deoxy-D-mannose} + NADPH$$

GDP-Glucosidase The enzyme that catalyzes the following reaction:

$$\text{GDP-glucose} + H_2O \rightleftharpoons \text{GDP} + \text{D-glucose}$$

GDPM Abbreviation for GDP-mannose.

GDP-mannose Dehydrogenase The enzyme that catalyzes the following reaction:

$$\text{GDP-mannose} + 2NAD^+ + H_2O \rightleftharpoons \text{GDP-mannuronate} + 2NADH$$

GDP-mannose 3,5-Epimerase The enzyme that catalyzes the following reaction:

$$\text{GDP-mannose} \rightleftharpoons \text{GDP-L-galactose}$$

GDP-Phosphohydrolase See guanosine diphosphatase.

GE Abbreviation for gel electrophoresis.

GEE Abbreviation for glycine ethyl ester.

Gee-Gee A trade name for guaifenesin, an antitussive agent that increases the production of respiratory tract fluid to liquefy the viscosity of tenacious secretions.

GEF Abbreviation for 1. gonadotropin-enhancing factor; 2. guanine-nucleotide-exchange factor.

Geiger-Mueller Counter A radioactive radiation counter.

Gefarnate (mol wt 401) An antiulcerative agent.

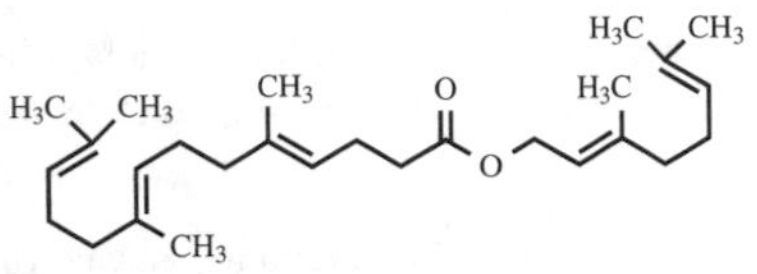

Gel A firm colloidal mass consisting of a network of molecular sieves or pores, e.g., starch gel, polyacrylamide gel, or agarose gel.

Gel Chromatography See gel filtration.

Gel Diffusion A serological procedure performed in agarose gel for identification of antigen or antibody.

Gel Electrofocusing Electrofocusing procedure performed in a gel.

Gel Electrophoresis Electrophoretic separation of macromolecules, e.g., DNA, RNA, or protein in a gel system (e.g., polyacrylamide gel, agarose gel).

Gel Exclusion Chromatography See gel filtration.

Gel Filtration A type of column chromatography in which the column consists of gel particles (e.g., Sephadex) of controlled size and porosity (molecular sieve). The molecules are separated through the column on the basis of their sizes and molecular weights.

Gel Permeation Chromatography See gel filtration.

Gel Retardation Assay A technique used to identify the binding of cellular protein to a specific DNA sequence.

Gelarose A trade name for a group of agarose gels, used in gel filtration.

Gelatin A heterogeneous mixture of water-soluble proteins obtained from the hydrolysis of collagens.

Gelatinase The enzyme that catalyzes the cleavage of gelatin.

Gelsemine (mol wt 322) A CNS stimulant isolated from root and rhizome of *Gelsemium sempervirens*.

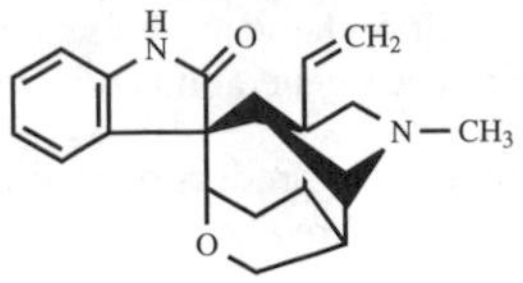

Gelsolin An actin severing protein found in mammalian cells and blood plasma. It decreases the viscosity of systic fibrosis sputum samples *in vitro*.

Gel-Tin A trade name for sodium fluoride, used for bone remineralization.

Gelusil A trade name for a combination drug containing aluminum hydroxide, magnesium hydroxide, and simethicone, used as an antiflatulent.

GEM Abbreviation for glycolipid enriched membrane.

Gemcitabine (mol wt 263) An antineoplastic agent.

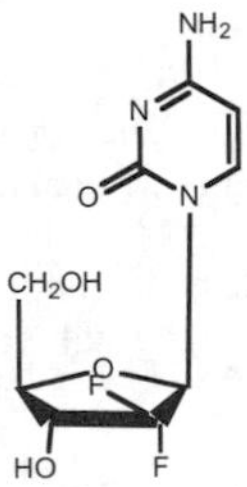

Gemella A genus of Gram-positive bacteria (Streptococcaceae).

Gemeprost (mol wt 384) An analog of prostaglandin E_1 used as an abortifacient and oxytocic agent.

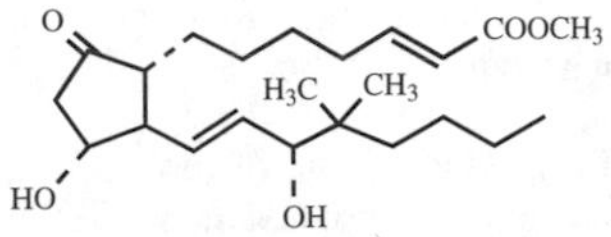

Gemfibrozil (mol wt 250) An antihyperlipoproteinemic agent.

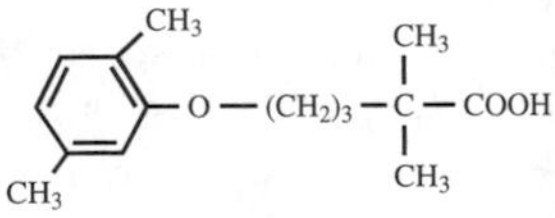

Gemini Viruses A group of small icosahedral, single-stranded DNA-containing plant viruses in which the virions usually occur in pairs.

Gemmata A genus of fresh-water, budding bacteria.

Gemmiger A genus of Gram-variable, anaerobic, budding bacteria.

Gemnisyn A trade name for a combination drug containing aspirin and acetaminophen, used as an analgesic and antipyretic agent.

Gemzar A trade name for gemcitabine hydrochloride, an antineoplastic agent.

Genabid A trade name for papaverine hydrochloride, a vasodilator that inhibits phosphodiesterase and increases the concentration of cAMP.

Genalac A trade name for calcium carbonate, which reduces total acid load in the GI tract.

Genallerate A trade name for chlorpheniramine maleate, an antihistamine that competes with histamine for H_1 receptors on target cells.

Gen-Amantadine A trade name for amantadine hydrochloride, an antiviral and anti-Parkinsonism drug.

Genapap A trade name for acetaminophen, an antipyretic and analgesic agent.

Genapax A trade name for gentian violet (methylrosaniline chloride, crystal violet), an anti-infective agent.

Genapax Tampon A trade name for a tampon treated with gentian violet to prevent infection.

Genapol A series of nonionic detergents with the following structure:

$$CH_3\text{-}(CH_2)_x\text{-}O\text{-}(CH_2CH_2O)_y\text{-}H$$

Genasal A trade name for oxymetazoline, a nasal decongestant.

Genasoft A trade name for docusate sodium, a laxative agent that promotes incorporation of fluid into the stool.

Genaspor A trade name for tolnaftate, an antifungal agent.

Gen-Atenolol A trade name for atenolol, a beta adrenergic blocking agent used as an antihypertensive agent.

Genatuss A trade name for guaifenesin, an expectorant.

Gen-Baclofen A trade name for baclofen, a skeletal muscle relaxant.

Gene A segment of the DNA molecule that encodes a functional product, e.g., polypeptide chain or RNA molecule.

Gene Activator Protein A protein that facilitates RNA polymerase binding to a particular promoter sequence.

Gene Amplification Selective creation of multiple copies of a particular gene permitting the production of vast quantities of the gene products.

Gene Cloning The replication of a cloned foreign DNA gene inserted by recombinant DNA technology into a cloning vector for the production of a large number of individual genes.

Gene Cluster A cluster of functionally different genes each of which encodes a separate gene product.

Gene Conversion The alteration of one of two near-homologous DNA sequences to the sequence of the other.

Gene Disruption The targeted inactivation of a gene by the homologous recombination.

Gene Dose The number of copies of a particular gene in a cell or nucleus.

Gene Expression The flow of the genetic information from DNA to RNA and the production of proteins.

Gene Flow The exchange of genes between different species or between different populations of the same species.

Gene Frequency The proportion of a given allele in a population.

Gene Fusion The use of recombinant DNA technology to join (fuse) two or more genes that encode different products.

Gene Insertion The introduction of a foreign gene or genes into a cell.

Gene Library A collection of randomly cloned fragments that encompasses the entire genome of a given species (also called genomic library or bank).

Gene Locus The chromosomal position occupied by a particular gene.

Gene Manipulation The *in vitro* joining of DNA fragments of interest to a vector so as to allow their incorporation into the host organism for propagation.

Gene Mapping Depiction of the linear order of genes along a chromosome.

Gene Pair The two alleles present in a diploid organism at a specific gene locus on two homologous chromosomes.

Gene Pool All of the genes present in a population during a given generation or period.

Gene Product The product of gene action, e.g., protein or RNA (abbreviated as gp).

Gene Recombination See genetic recombination.

Gene Redundancy The presence of multiple copies of a gene in a chromosome.

Gene Regulatory Protein Protein that regulates gene expression and binding of RNA polymerase to promoter.

Gene Reiteration The presence of multiple copies of a particular gene.

Gene Splicing 1. The enzymatic manipulations by which one DNA fragment is attached to another. 2. The process by which the introns are removed and exons joined during mRNA synthesis.

Gene Substitution The replacement of one allele by another allele of the same gene.

Gene Superfamily The evolutionarily related genes or gene products with divergent functions, e.g., immunoglobulin superfamily.

Gene Therapy The introduction of a functional gene or genes into a recipient to correct a genetic defect.

Genera Plural of genus.

General Acid-Base Catalysis A form of catalysis that depends on transfer of protons.

General Anaphylaxis An IgE-mediated allergic reaction characterized by itching, swelling, or edema and wheezing respiration due to release of vasoactive amines (e.g., histamine) from mast cells.

General Transduction A phage-mediated transfer of host DNA from a donor cell to a recipient cell.

Generation Time The length of the cell cycle or time period needed for a cell population to double its numbers.

Generic Pertaining to genus or a substance not protected by patent.

Generic Name A technical, unsystematic type of name used in describing a drug.

-genesis A suffix meaning 1. origin and 2. formation or production.

Genetic Burden See genetic load.

Genetic Code Referring to the nucleotide triplets on mRNA that specify different amino acids in the process of translation.

Genetic Code Dictionary Referring to the 64 nucleotide triplets or codons resulting from the combination of four ribonucleotides of adenine, guanine, cytosine, and uracil.

Genetic Complementation The gene products of two mutant genes that can combine to give rise to a wild phenotype.

Genetic Cross 1. Mating of two organisms to produce genetic recombinants. 2. The progeny that contains genotypes of two or more parents, e.g., simultaneous infection of a bacterial cell with several types of phages. 3. The progeny derived from mating.

Genetic Disease 1. A disease due to changes in the genetic material. 2. A disease that is inherited in a mendelian fashion.

Genetic Dissection The use of recombination and mutation to piece together the various components of a given biological function.

Genetic Drift Changes in genotype or gene frequencies from generation to generation in a population as a result of random processes.

Genetic Engineering The in vitro manipulation of DNA to generate new, desirable recombinant sequences, genes, or organisms.

Genetic Equilibrium The frequency of a gene remains constant from generation to generation.

Genetic Expression See gene expression.

Genetic Homeostasis The self-regulating capacity of populations to adopt to the changing environment.

Genetic Load The average number of recessive lethal genes carried in the heterozygous condition by an individual in a population (also known as genetic burden).

Genetic Mapping A depiction of the linear order of genes along a chromosome.

Genetic Marker A detectable and genetically controlled marker on a chromosome of an organism.

Genetic Mosaic An organism that contains cells of different genotypes.

Genetic Polymorphism The presence of two or more alleles at a gene locus over a succession of generations.

Genetic Profiling A technique for providing profiles of DNA fragments resulting from digestion with restriction enzymes.

Genetic Recombination The combining of two different DNA molecules to produce a third molecule that is different from either of the original two.

Genetic Transformation The genetic change or expression caused by the introduction of exogenous DNA into the cell.

Genetic Variance The phenotypic variance resulting from the presence of different genotypes in the population.

Geneticist A scientist who specializes in genetics.

Genetics The science that deals with inheritance.

Genetotrophic Disease A genetic disease whose genetic insufficiency can be restored by the supply of nutrients.

Gen-Fibro A trade name for gemfibrozil, an antihyperlipidemic agent.

Gen-Glybe A trade name for glyburide, an antidiabetic agent.

Genic Pertaining to genes.

Genistein (mol wt 270) A competitive inhibitor at the ATP-binding site.

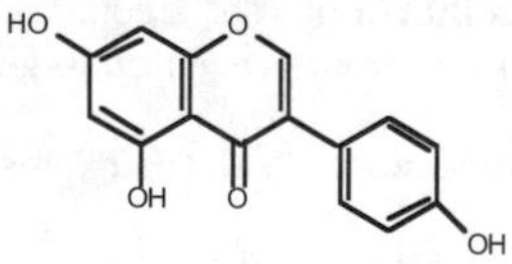

Genital Pertaining to the reproductive system.

Genital Herpes A sexually transmitted disease caused by herpes simplex virus.

Genitalia The reproductive organs (also called genitals).

Genital Warts A disease characterized by the formation of benign tumors caused by human *Papillomavirus* and transmitted by sexual contact.

Genitourinary Tract Referring to the urinary and reproductive systems.

Genome The complete set of genes of an organism.

Genomic Blotting A DNA blotting in which the fragments from a restriction endonuclease digest of total DNA from a particular organism's cell are visualized by annealing to a radioactive probe.

Genomic Control Regulatory mechanism in eukaryotes for the selective suppression or amplification of specific genetic information of the genome.

Genomic Imprinting A process by which genes are differentially expressed from the maternal or paternal genome.

Genomic Library A collection of randomly cloned fragments that encompass the entire genome of a given species (also called genomic library or bank).

Genophore Referring to the genetic material of viruses and prokaryotes.

Genoptic A trade name for gentamicin sulfate, an anti-infective agent that inhibits bacterial protein synthesis..

Genotoxin Any agent that damages DNA.

Genotropin A trade name for human growth hormone somatropin produced by DNA recombinant technology and used for the treatment of children with growth hormone deficiency.

Genotype The genetic constitution of a cell or organism.

Genpril A trade name for ibuprofen, an anti-inflammatory, antipyretic, and analgesic agent.

Gentabs A trade name for acetaminophen, an antipyretic and analgesic agent.

Gentamicin An antibiotic complex produced by *Micromonospora purpurea* that consists of three closely related components, gentamicin C_1, C_2, and C_{1a}. It inhibits bacterial protein synthesis.

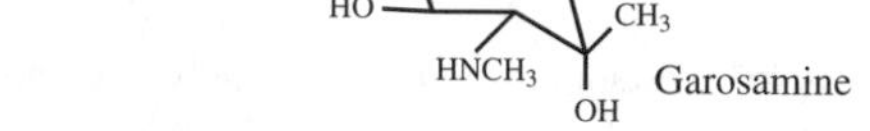

Gentamicin C_1 : $R_1 = R_2 = CH_3$
Gentamicin C_2 : $R_1 = CH_3$; $R_2 = H$
Gentamicin C_{1a} : $R_1 = R_2 = H$

Gentian Violet (mol wt 408) A dye and a topical anti-infective agent that possesses fungistatic and bactericidal activity.

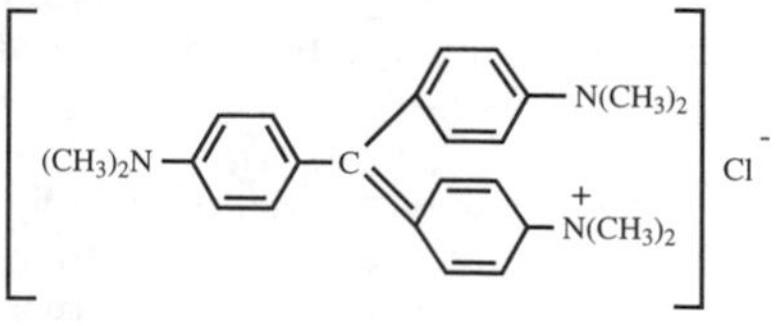

Gentiobiase See β-glucosidase.

Gentiobiose (mol wt 342) A disaccharide consisting of β-1,6-linked glucose.

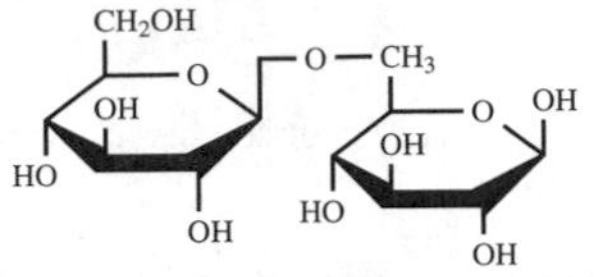

β-gentiobiose

Gentiopicrin (mol wt 356) An antimalarial agent.

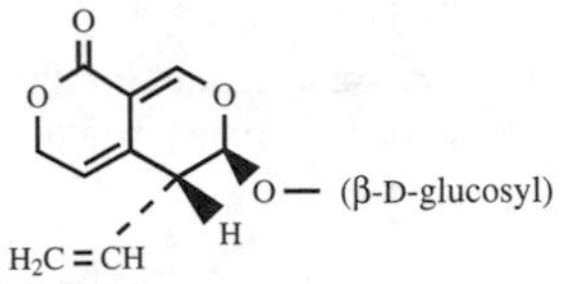

Gentisate Decarboxylase The enzyme that catalyzes the following reaction:

2,3-Dihydroxybenzoate ⇅ Hydroquinone + CO_2

Gentisic Acid (mol wt 154) A product from *Penicillium patulum* with analgesic and anti-inflammatory activity.

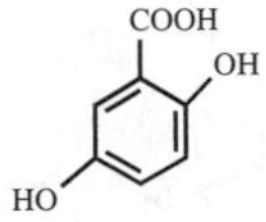

Gentran 40 A trade name for low molecular weight dextran, used for expansion of plasma volume and providing fluid replacement.

Gentran 75 A trade name for high molecular weight dextran, used for expansion of plasma volume and providing fluid replacement.

Genus A taxonomic group directly above the species level.

Gen-Xene A trade name for clorazepate dipotassium, an anticonvulsant.

Geocillin A trade name for carbenicillin indanyl sodium, an antibiotic that inhibits bacterial cell wall synthesis.

Geodermatophilus A genus of aerobic, catalase-positive bacteria (Actinomycetales).

Geometric Isomer Isomers that differ from each other in the configuration of groups attached to the two carbon atoms that are linked by a double bond.

Geopen A trade name for carbenicillin disodium, an antibiotic that inhibits bacterial cell wall synthesis.

Geosmin (mol wt 182) A compound produced by actinomycetes and cyanobacteria that causes the characteristic odor of soil and water.

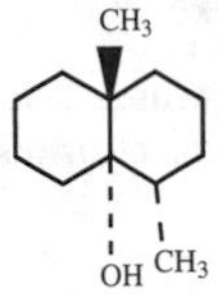

Geotaxis A type of taxis in which the stimulus is gravity.

Geotrichosis An upper-respiratory tract disorder caused by fungus *Geotrichum candidum.*

Geotrichum A genus of fungi.

Gepefrine (mol wt 151) An antihypotensive agent.

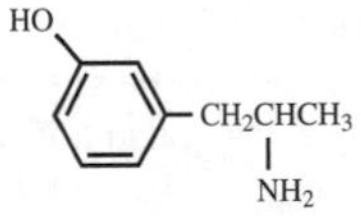

Gephryin A cytoplasmic protein associated with microtutules. It is required for clustering of glycine receptors in the spinal cord.

Gephyrotoxin (mol wt 287) A neurotoxin from the skin secretion of the Colombian poison-dart frog *Dendrobates histrionicus.*

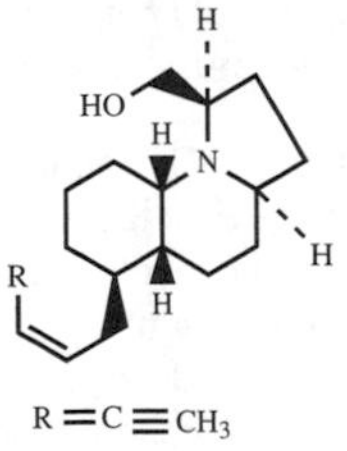

Gepirone (mol wt 359) An anxiolytic agent.

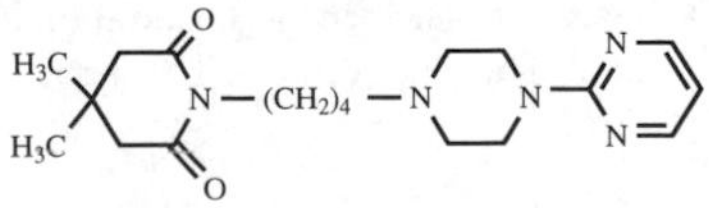

Geraniol (mol wt 154) An insect attractant.

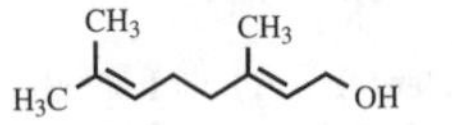

Geraniol Dehydrogenase The enzyme that catalyzes the following reaction:

Geraniol + $NADP^+$ ⇌ Geranial + NADPH

Gerbich System A blood group system whose antigenic determinants are located on the extracellular glycosylated domain of glycophorin C and D.

Geref A trade name for human growth hormone sermorelin produced by recombinant technology and used for the treatment of children with growth hormone deficiency.

Geriatrics The medical science that deals with aging and the treatment of diseases affecting the aged.

Geridium A trade name for phenazopyridine hydrochloride, an antipyretic drug.

GERL Abbreviation for Golgi-endoplasmic reticulum lysosome. The region in the Golgi apparatus from which lysosomes are derived.

Germ Pathogenic microorganism.

Germanium (Ge) A chemical element with atomic weight 73, valence 2 and 4.

German Measles A systemic infectious disease of humans caused by rubella viruses (Togaviridae) characterized by a rash (also known as rubella).

Germ Cell Reproductive cell responsible for creating the next generation, e.g., egg cell or sperm cell.

Germ Layer One of the three original cell layers formed during the early development of the embryo, e.g., ectoderm, endoderm, and mesoderm.

Germ Line The cell lineage consisting of germ cells or cells capable of contributing genetic material to subsequent generations.

Germ Line Theory The theory that explains the production of diverse antibodies. It states that all cells possess the same set of genes for immunoglobulins as those in the germ cells. The diversity of antibodies results from the rearrangement of genes for variable regions of immunoglobulins in the germ cells.

Germ Plasm 1. The genetic material (genes) of germ cells. 2. The germ cells. 3. The germ cells of an organism.

Germicide An agent capable of killing germs.

Germinal Cell The cell capable of developing into a gamete upon meiosis.

Germinal Center A histologically discernible region of lymph nodes and spleens populated mostly by B lymphocytes.

Germinal Mutation Mutations occurring in the cells that are destined to develop into gametes.

Germination 1. Growth of seed or other reproductive organs after dormancy. 2. Growth and development of an organism from the time of fertilization to the formation of an embryo. 3. The process from which a spore is converted to a vegetative cell or hyphae.

Gerontology The science that deals with processes of aging.

GES Abbreviation for glucose electrolyte solution.

Gesterol 50 A trade name for the hormone progesterone, which suppresses ovulation.

Gesterol LA A trade name for the hormone hydroxyprogesterone caproate, which suppresses ovulation.

Gestodene (mol wt 310) An oral contraceptive.

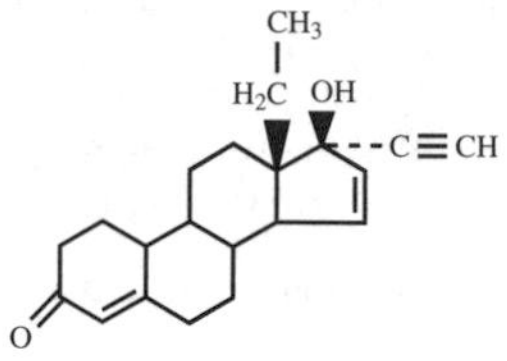

Gestonoron Caproate (mol wt 415) A progestogen used in the treatment of prostate hypertrophy.

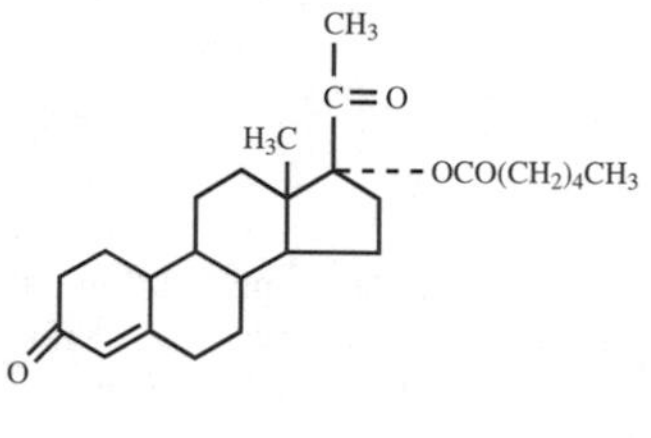

Gestrinone (mol wt 308) An antigonadotropin.

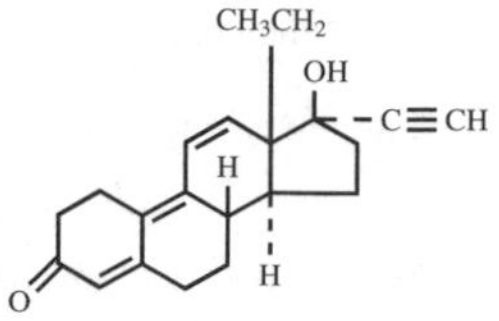

GeV Abbreviation for giga electron volt (10^9 volt).

GF Abbreviation for 1. growth factor; 2. germfree.

G-Factor Referring to the elongation factor G in protein synthesis (also known as translocase).

GFAP Abbreviation for glial fibrillary acidic protein.

GFP Abbreviation for green fluorescent protein.

GFR Abbreviation for glomerular filtration rate.

GG Abbreviation for 1. gamma globulin; 2. geranylgeranyl moiety; 3. glycylglycine.

γG Abbreviation for gamma globulin.

Gγ Abbreviation for gamma subunit of G protein.

GG-CEN A trade name for an antitussive drug guaifenesin (glyceryl guaiacolate), which increases production of respiratory tract fluid to liquefy and reduce the viscosity of tenacious secretions.

GGE Abbreviation for gradient gel electrophoresis.

GGPP Abbreviation for geranylgeranyl pyrophosphate.

GGT Abbreviation for 1. γ–glutamyl transpeptidase; 2. gamma glutamyl transferase.

GGTase-I Abbreviation for geranylgeranyl transferase type-I.

GGTase-II Abbreviation for geranylgeranyl transferase type-II.

GGTP Abbreviation for γ–glutamyl transpeptidase.

GH Abbreviation for growth hormone.

GHA Abbreviation for glucoheptanoic acid.

GHb Abbreviation for glycosylated hemoglobin.

GHB Abbreviation for gamma hydrobutyrate.

GHBP Abbreviation for glial hyaluronate-binding protein.

GHIF Abbreviation for growth hormone inhibiting factor.

GHIH Abbreviation for growth hormone inhibiting hormone.

Ghost Cell Erythrocyte without normal cytoplasm.

GHRF Abbreviation for growth hormone-releasing factor.

GHRH Abbreviation for growth hormone-releasing hormone.

GHRIH Abbreviation for growth hormone-release-inhibiting hormone.

GI Abbreviation for gastrointestinal tract.

Giardia A genus of parasitic flagellate protozoa.

Giardiasis An infection of human intestine caused by *Giardia lamblia* (protozoa).

Gibberellic Acid (mol wt 346) A plant growth regulator produced by *Gibberella fujikuroi.*

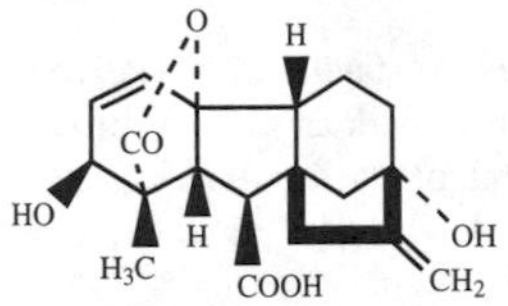

Gibberellins A class of related plant hormones that stimulate growth of the stem and leaves, trigger germination of seeds, and break bud dormancy.

Gibbs Free Energy See Free energy.

Gibbs-Donnon Equilibrium The unequal distribution of diffusible ions across the membrane at equilibrium in the presence of nondiffusible ions.

Gibbs-Duhen Equation An equation describing the chemical potentials of the different components in a system.

Gibbs Reagent (mol wt 210) A reagent used for determination of phenols.

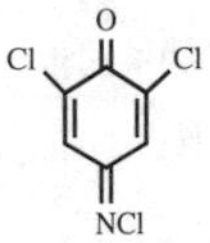

Giberalla A genus of fungi.

Giemsa's Stain A biological stain that contains both basic and acidic dyes, used to stain blood smears that are believed to contain protozoan parasites. It also distinguishes basic and acidic granules in granulocytes.

GIF Abbreviation for growth hormone-inhibitory factor.

Giga A prefix meaning one billion (10^9).

GIH Abbreviation for gastrointestinal hormone.

GIK Abbreviation for glucose-insulin-potassium.

Gilbert-Maxam Method See Maxam-Gilbert's method.

Gilbert's Syndrome A disorder of the liver characterized by excessive bile color in the blood.

GinI (BamHI) A restriction endonuclease from *Gluconobacter industricus* with the same specificity as Bam HI.

Gingiva The gum of the mouth.

Gingivitis Inflammation of the gums.

Ginkgo Biloba Extract from leaves of *Ginkgo biloba,* used as a dietary supplement to increase blood circulation to the brain and extremities. It contains ginkgolides.

Ginkgolides A family of bioactive terpenes isolated from root bark and leaves of *Ginkgo biloba.* It is a potent platelet activator factor antagonist. Ginkgolide A, B, and C have been identified.

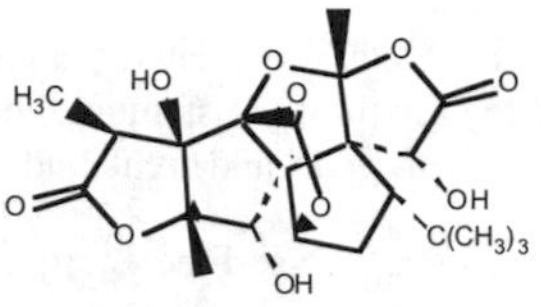

Ginkgolide B

GIP Abbreviation for 1. gastrointestinal peptide or gastrointestinal polypeptide; 2. gastric inhibitory peptide.

GIPL Abbreviation for glycoinositol phospholipid.

Giractide An adrenocorticotropic peptide hormone that consists of 18 amino acid residues.

GIT Abbreviation for gastrointestinal tract.

Gitoxin (mol wt 781) A cardiotonic agent.

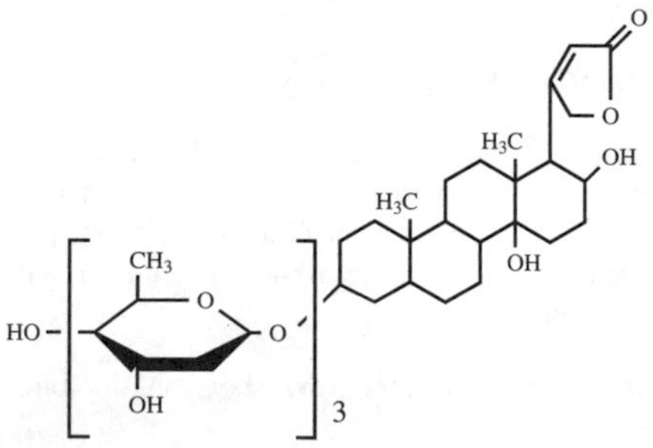

GITT Abbreviation for glucose-insulin tolerance test.

GK Abbreviation for 1. glycerol kinase; 2. glucokinase; 3. glycocyamine kinase.

G_k A type of G protein isolated from human red blood cells that activates K^+ channels in the heart muscle.

g/L Abbreviation for grams per liter.

GL Abbreviation for 1. gastric lipase; 2. glycolipid.

Gladiolic Acid (mol wt 222) A fungistatic antibiotic produced by *Penicillium gladioli.*

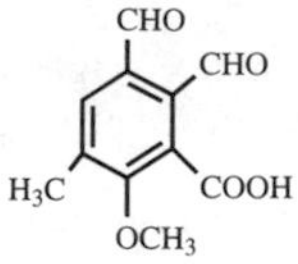

Glafenine (mol wt 373) An analgesic agent.

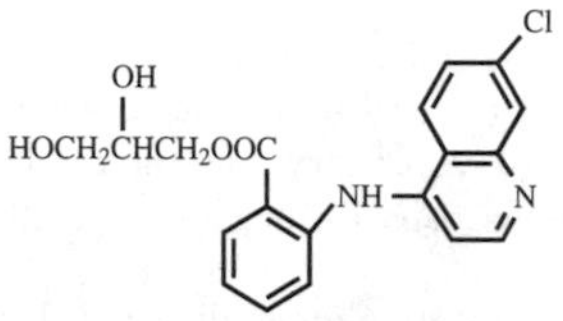

Gland An organ that produces and releases substances for regulating functions of other organs or tissues.

Glandular Pertaining to a gland.

Glandular Fever A self-limiting disease caused by Epstein-Barr virus (infectious mononucleosis) and characterized by the appearance of many large lymphoblasts in the circulation.

Glass Electrode An electrode used for determination of pH.

Glaucoma A disorder of the eye due to the blockage of the normal flow of watery fluid in the space between the cornea and lens.

Glaucon A trade name for epinephrine hydrochloride, a hormone that dilates the pupil by contracting the dilator muscle, used as a mydriatic agent.

Glc Abbreviation for glucose.

GLC Abbreviation for gas liquid chromatography.

GlcA Abbreviation for gluconic acid.

GlcCer Abbreviation for glucosylceramide or glucoceramide.

GlcK Abbreviation for glucokinase or glucose kinase.

GlcN Abbreviation for glucosamine.

GlcNAc Abbreviation for *N*-acetylglucosamine.

GlcNAc-PI Abbreviation for N-acetyl glucosaminyl phosphatidyl inositol.

GlcNAc-T Abbreviation for N-acetyl glucosaminyl transferase.

GlcNS Abbreviation for N-sulfated glucosamine.

Glc-6-P Abbreviation for glucose 6-phosphate.

Glc6PDH Abbreviation for glucose 6-phosphate dehydrogenase.

GlcU Abbreviation for glucuronic acid.

GlcUA Abbreviation for glucuronic acid.

GLDH Abbreviation for glutamate dehydrogenase.

GLI Abbreviation for glucagon-like immunoreactant.

Glia The nonneuronal tissue of the brain.

Gliadel A trade name for carmustine, an alkylating agent with antineoplastic activity.

Gliadin One of the prolamine proteins from wheat that is soluble in 30 to 80% alcohol and insoluble in 100% alcohol.

Glibornuride (mol wt 366) An antidiabetic agent.

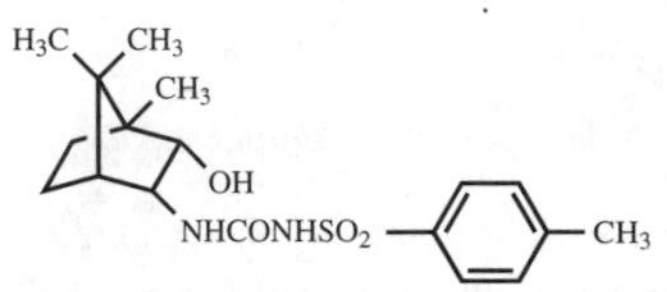

Gliclazide (mol wt 323) An antidiabetic agent.

Gliding Bacteria A nontaxonomic group of bacteria and cyanobacteria that exhibit gliding mobility.

Gliding Mobility An active movement on a contact surface by an organism that has neither a visible locomotory organelle nor a distinct change in the shape of the organism. It is a surface-associated locomotion.

Glimepiride (mol wt 491) An anti-diabetic agent.

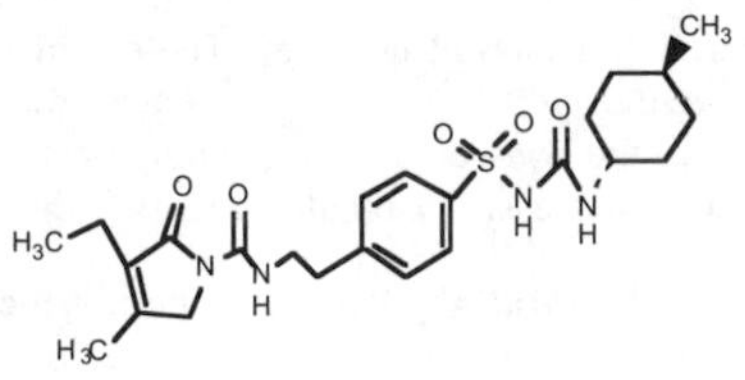

Glioblastoma A brain tumor.

Glioma A tumor derived from the tissue surrounding the brain and spinal cord.

Gliosarcoma A cancer of the nerve cells.

Gliotoxin An antiviral agent produced by various species of *Trichoderma, Gladiocladium fimbriatum*, and *Aspergillus fumigatus*.

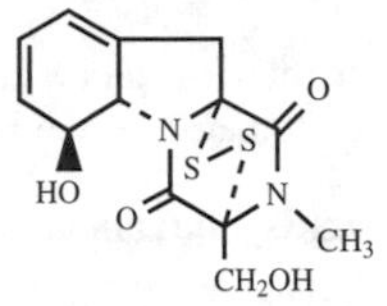

GLIP Abbreviation for glucagon-like insulinotropic peptide.

Glipizide (mol wt 446) An antidiabetic agent that stimulates insulin release from the pancreatic beta cells and reduces glucose output by the liver.

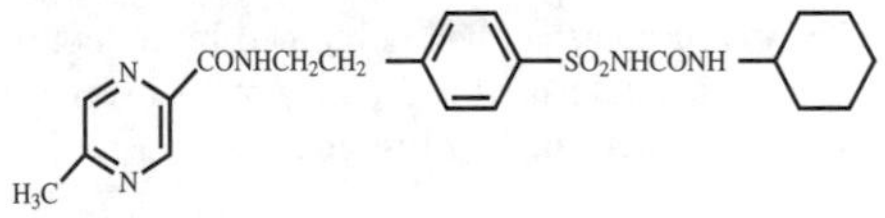

Gliquidone (mol wt 528) An antidiabetic agent.

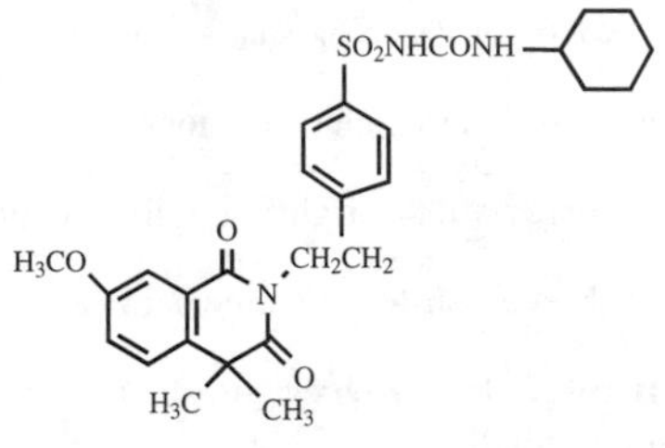

Glisoxepid (mol wt 450) An antidiabetic agent.

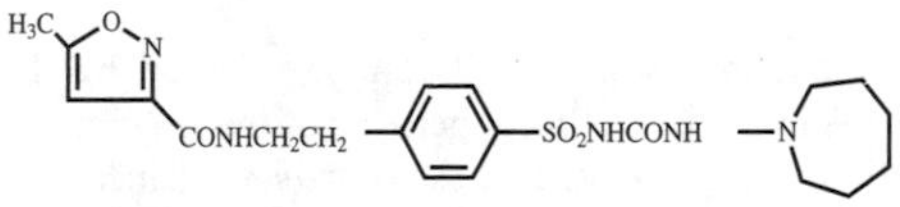

Gln Abbreviation for glutamine.

GlN Abbreviation for glutamine.

Glo Abbreviation for glyoxalate.

Globin A single protein with an iron-porphyrin group, e.g., myoglobin.

Globinometer An instrument used for determination of oxyhemoglobin in the blood.

Globoside A cerebroside that contains two or more sugars or sugar derivatives.

Globular Protein The general name for a group of water soluble proteins in which the polypeptide chains are coiled into a more or less globular shape.

Globulins Proteins that are insoluble in water but soluble in dilute salts of strong acids or bases.

Globulinuria The presence of globulin in the urine.

Glomerular Filtration A process by which blood is filtered by the glomeruli of the kidney.

Glomerulonephritis Inflammation of the glomeruli of the kidney.

Glomerulosclerosis Fibrosis and degeneration of the structures within the glomeruli of the kidney.

Glomerulus Tuftlike structures in the kidney where blood filtration takes place.

G Loop The unpaired G bubble or loop in the dsDNA of a bacteriophage obtained by induction of lysogenic bacteria. It is formed by the inversion of G the segment during lysogeny.

-glossia A suffix meaning condition of the tongue.

Glossina A genus of blood-sucking flies that transmits African sleeping sickness in humans.

Glossitis Inflammation of the tongue.

GLP Abbreviation for glucagon-like peptide.

Glu Abbreviation for glutamic acid.

Glu Microtubules Abbreviation for microtubules rich in glu-tubulin.

GluA Abbreviation for glucuronic acid.

Glucagon A peptide hormone released by the cells in the islets of Langerhans of the pancreas. It stimulates glycogeneolysis in the liver and increases the concentration of glucose in the blood. It is used as an antidiabetic agent.

Glucagon-Like Immunoreactant Any of a group of peptides found in the extract of mammalian gastrointestinal tract that reacts with N-terminal-specific anti-glucagon antibodies.

Glucagon-Like Peptide Receptor Any of the membrane proteins that bind glucagon-like peptides and mediate their intracellular effects.

Glucagonoma A neoplasm (benign or malignant) originating from the A cells of pancreatic islets.

Glucametacin (mol wt 519) An anti-inflammatory agent.

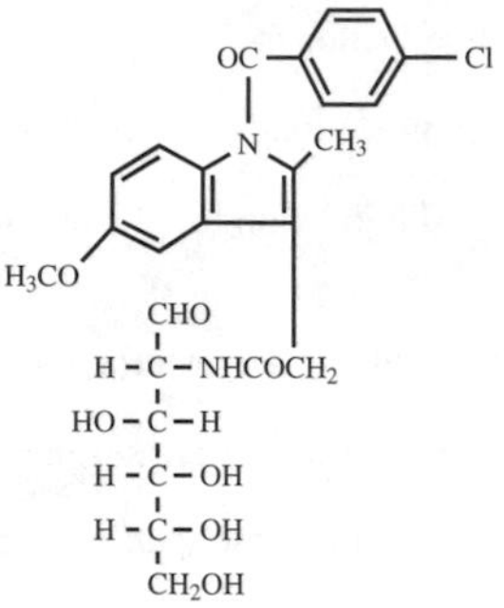

Glucamide A trade name for chlorpropamide, which stimulates insulin release from the pancreatic beta cells and reduces glucose output by the liver.

Glucamine (mol wt 181) A derivative of glucose.

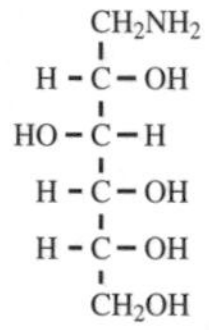

Glucan The polymer of glucose.

Glucanase The enzyme that catalyzes the hydrolysis of glucan.

Glucan Endo-α-Glucosidase The enzyme that catalyzes the random hydrolysis of α-glucosidic linkages in α-D-linked glucans.

Glucan Endo-β-Glucosidase The enzyme that catalyzes the random hydrolysis of β-D-glucosidic linkages in β-D-linked glucans.

Glucan α-Glucosidase The enzyme that catalyzes the hydrolysis of terminal α-D-glucosidic linkages in α-D-linked glucan to remove successive glucose units.

Glucan β-Glucosidase The enzyme that catalyzes the hydrolysis of terminal β-D-glucosidic linkages in β-D-glucan to remove successive glucose units.

Glucan 1,6-α-Isomaltosidase The enzyme that catalyzes the hydrolysis of 1,6-α-D-glucosidic linkages from the nonreducing end in polysaccharides to remove successive isomaltose units.

Glucan 1,4-α-Maltohydrolase The enzyme that catalyzes the hydrolysis of 1,4-α-D-glucosidic linkages from the nonreducing end in polysaccharides to remove successive maltose units.

α-Glucan Phosphorylase The enzyme that catalyzes the following reaction:

$$(\alpha\text{-1,4-glucosyl})_n + \text{Pi} \rightleftharpoons (\alpha\text{-1,4-glucosyl})_{n-1} + \alpha\text{-D-glucose 1-phosphate}$$

1,3-α-Glucan Synthetase The enzyme that catalyzes the following reaction:

$$\text{UDP-glucose} + (1,3\text{-}\alpha\text{-D-glucosyl})_n \rightleftharpoons \text{UDP} + (1,3\text{-}\alpha\text{-D-glucosyl})_{n+1}$$

1,3-β-Glucan Synthetase The enzyme that catalyzes the following reaction:

$$\text{UDP-glucose} + (1,3\text{-}\beta\text{-D-glucosyl})_n \rightleftharpoons \text{UDP} + (1,3\text{-}\beta\text{-D-glucosyl})_{n+1}$$

Glucarate Dehydratase The enzyme that catalyzes the following reaction:

$$\text{D-Glucarate} \rightleftharpoons \text{5-Dehydro-4-deoxy-D-glucarate} + H_2O$$

D-Glucaric Acid (mol wt 210) A sugar acid derived from glucose.

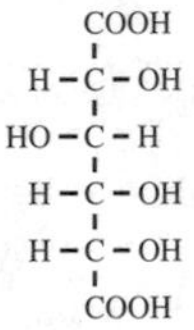

Glucitol See sorbitol.

Glucoamylase See glucan 1,4-α-glucosidase.

Glucocerebrosidase The enzyme that catalyzes the hydrolysis of glucocerebroside to yield glucose and ceramide.

Glucocerebroside A glucose-containing cerebroside.

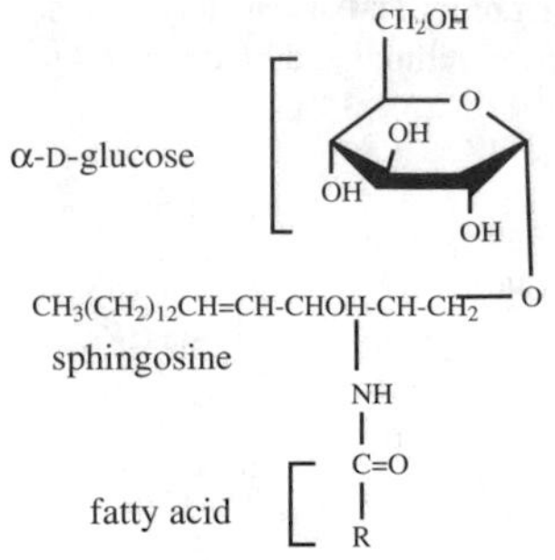

Glucocorticoid A group of 21 carbon steroid hormones, e.g., corticosterone, cortisome, and cortisol produced by the adrenal cortex; they are involved in metabolism of carbohydrates (e.g., gluconeogenesis), lipids, and proteins and also possess anti-inflammatory and antiallergic activity.

Glucocorticoid Response Element A specific DNA sequence that mediates the effects of glucocorticoids.

Glucocortin A protein induced by the glucocorticoids in adipocyte, liver, fibroblast, and thymus.

Glucodextrinase See glucan 1,6-α-glucosidase.

Glucogenic Pertaining to gluconeogenesis.

Glucogenic Amino Acid Amino acids whose carbon skeletons can be metabolized to pyruvate, α-ketoglutaric acid, succinyl-CoA, fumarate, or oxaloacetate to serve as glucose precursors.

Glucoheptonic γ-Lactone (mol wt 208) A seven-carbon compound derived from glucoheptonic acid.

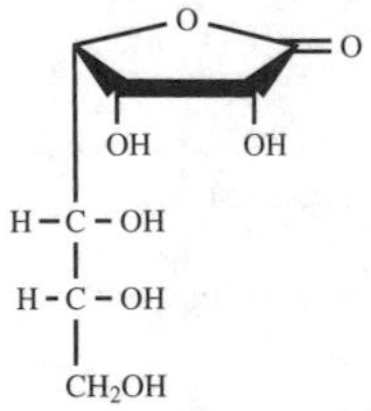

Glucoinvertase Synonym of α-glucosidase.

Glucokinase The enzyme that catalyzes the following reaction:

$$\text{ATP} + \text{D-glucose} \rightleftharpoons \text{D-glucose 6-phosphate} + \text{ADP}$$

Glucolipid Lipid that contains covalently linked glucose.

Glucomannan Polymer of glucose and mannose (hemicellulose).

Gluconate Acceptor 2-Oxidoreductase The systematic name for gluconate 2-dehydrogenase.

Gluconate Dehydratase The enzyme that catalyzes the following reaction:

$$\text{D-Gluconate} \rightleftharpoons \text{2-Dehydro-3-deoxy-D-gluconate} + H_2O$$

Gluconate Dehydrogenase The enzyme that catalyzes the following reaction:

D-Glucose + acceptor
⇅
α-keto-gluconate + reduced acceptor

Gluconate Kinase Synonym of gluconokinase.

Gluconeogenesis Synthesis of glucose in the liver from noncarbohydrate precursors such as amino acids, glycerol, or lactate.

Gluconeogenic Pathway The pathway for conversion of non-carbohydrate precursors to glucose 6-phospahte.

Gluconic Acid (mol wt 196) A glucose derivative.

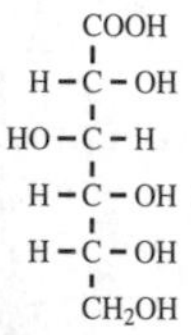

Gluconobacter A genus of Gram-negative bacteria (Acetobacteraceae).

Gluconokinase The enzyme that catalyzes the following reaction:

ATP + D-gluconate
⇅
ADP + 6-phosphogluconate

Gluconolactonase The enzyme that catalyzes the following reaction:

D-glucon-δ–lactone + H_2O
⇅
D-gluconate

Gluconolactone (mol wt 178) An intermediate in the pentose phosphate pathway.

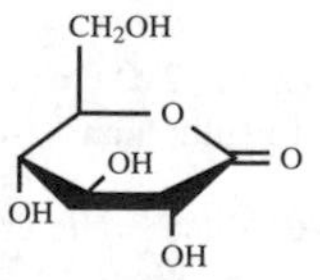

Glucophage A trade name for an antidiabetic agent metformin that increases the number of insulin receptors.

Glucoplastic Amino Acid Amino acids capable of contributing to gluconeogenesis.

Glucopyranose See glucose.

Glucortrol A trade name for glipizide, an antidiabetic agent used to lower the blood glucose concentration.

Glucosaminate Ammonia-lyase The enzyme that catalyzed the following reaction:

D-Glucosaminate
⇅
2-Dehydro-3-deoxy-D-gluconate + NH_3

Glucosamine (mol wt 179) An amino sugar of glucose and an anticarthritic agent.

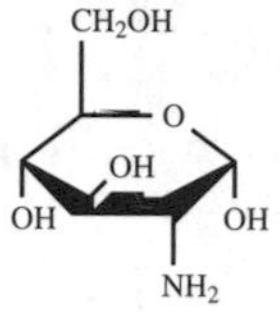

Glucosamine Acetylase See glucosamine *N*-acetyltransferase.

Glucosamine *N*-Acetyltransferase The enzyme that catalyzes the following reaction:

Acetyl-CoA + D-glucosamine
⇅
N-Acetylglucosamine + CoA

Glucosamine Kinase The enzyme that catalyzes the following reaction:

ADP + D-glucosamine
⇅
D-Glucosamine phosphate + ADP

Glucosamine Phosphate Acetyltransferase The enzyme that catalyzes the following reaction:

Glucosamine 6-phosphate + acetyl-CoA
⇅
CoA + *N*-acetylglucosamine 6-phosphate

Glucosamine Phosphate Isomerase The enzyme that catalyzes the following reaction:

D-glucosamine 6-phosphate + H_2O
⇅
D-fructose-6-phosphate + NH_3

Glucose (mol wt 180) An aldose sugar and a key compound in cellular and biochemical metabolism (also called dextrose).

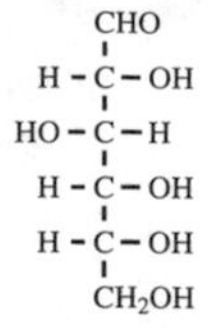

D-glucose

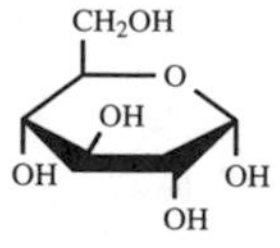

α-D-glucose

Glucose Aerodehydrogenase See glucose oxidase.

Glucose Alanine Cycle An aminotransferase-mediated pathway in which alanine in the blood is transported to the liver where it undergoes transamination to yield pyruvate for gluconeogenesis.

Glucose 1,6-Bisphosphate (mol wt 340) The diphosphate form of fructose.

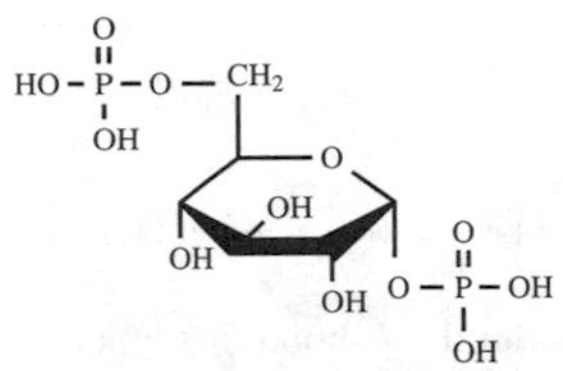

α-D-glucose 1,6-bisphosphate

Glucose 1,6-Bisphosphate Synthetase The enzyme that catalyzes the following reaction:

3-Phosphoglyceroyl phosphate + D-glucose 1-phosphate

⇅

3-Phosphoglycerate + glucose1,6-bisphosphate

Glucose Dehydrogenase ($NADP^+$) The enzyme that catalyzes the following reaction:

D-Glucose + $NADP^+$

⇅

D-glucono-δ–lactone + NADPH

Glucose Electrode An electrode that contains incorporated enzyme for the determination of glucose.

Glucose $NADP^+$ 1-Oxidoreductase ($NADP^+$) The systematic name for glucose 1-dehydrogenase.

Glucose $NADP^+$ Oxidoreductase The systematic name for glucose dehydrogenase.

β-Glucose Oxidase The enzyme that catalyzes the following reaction:

β-D-glucose + O_2

⇅

D-glucono-1,5–lactone + H_2O_2

Glucose Oxyhydrase Synonym of glucose oxidase.

Glucose 1-Phosphatase The enzyme that catalyzes the following reaction:

D-glucose 1-phosphate + H_2O

⇅

D-glucose + Pi

Glucose 6-Phosphatase The enzyme that catalyzes the following reaction:

D-Glucose 6-phosphate + H_2O

⇅

D-glucose + orthophosphate

α-D-Glucose 1-Phosphate (mol wt 260) The monophosphate form of glucose.

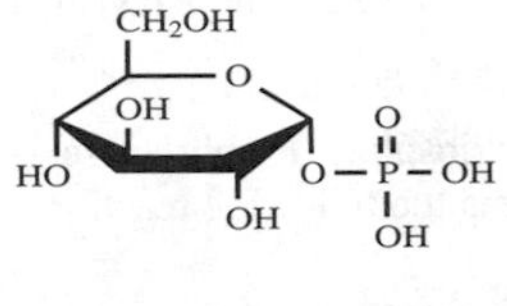

α-D-Glucose 6-Phosphate (mol wt 260) The phosphate form of glucose and key component in carbohydrate metabolism.

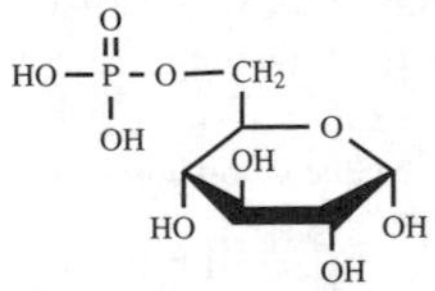

Glucose 1-Phosphate Adenylyltransferase The enzyme that catalyzes the following reaction:

ATP + α-D-glucose 1-phosphate

⇅

PPi + ADP-glucose

Glucose 1-Phosphate Cytidylyltransferase The enzyme that catalyzes the following reaction:

CTP + D-glucose 1-phosphate

⇅

PPi + CDP-glucose

Glucose 1-Phosphate Guanylyltransferase The enzyme that catalyzes the following reaction:

GTP + α-D-glucose 1-phosphate

⇅

PPi + GDP-glucose

Glucose 1-Phosphate Thymidyltransferase The enzyme that catalyzes the following reaction:

TTP + α-D-glucose 1-phosphate

⇅

PPi + TDP-glucose

Glucose 1-Phosphate Uridyltransferase The enzyme that catalyzes the following reaction:

Glucose 6-Phosphate Dehydrogenase The enzyme that catalyzes the following reaction:

D-Glucose 6-phosphate + $NADP^+$
⇅
D-glucono-δ–lactone + NADPH

Glucose 6-Phosphate Dehydrogenase Deficiency A genetic disorder due to a deficiency of the enzyme glucose-6-phosphate dehydrogenase in the red blood cell for metabolism of carbohydrate. This disorder develops into favism upon eating beans.

Glucose 6-Phosphate 1-Epimerase The enzyme that catalyzes the following reaction:

α-D-Glucose 6-phosphate
⇅
β-D-Glucose 6-phosphate

Glucose 6-Phosphate Isomerase The enzyme that catalyzes the following reaction:

D-Glucose 6-phosphate
⇅
D-fructose 6- phosphate

Glucose 6-Phosphate Ketol-Isomerase Synonym of glucose 6-phosphate isomerase.

Glucose 6-Phosphate Phosphohydrolase The systematic name for glucose 6-phosphatase.

UTP + α-D-glucose 1-phosphate
⇅
UDP-glucose + PPi

Glucose Phosphomutase See phosphoglucomutasse.

Glucose Sensitive Operon A bacterial operon whose function is inactivated by the presence of glucose.

Glucose Tolerance Test A test used for diagnosis of diabetes mellitus.

Glucose Transport Protein responsible for transporting glucose.

α-D-Glucosidase The enzyme that catalyzes the hydrolysis of the terminal, nonreducing 1,4-linked α-D-glucose residues releasing α-D-glucose in glucan.

β-D-Glucosidase The enzyme that catalyzes the hydrolysis of the terminal, nonreducing 1,4-linked β-D-glucose residues in a glucan releasing of β-D-glucose.

Glucoside A compound that contains a glucose linked to another molecule by the glucosidic bond.

Glucoside (α-D) Glucohydrolase The systematic name for α-glucosidase.

Glucoside (β-D) Glucohydrolase The systematic name for β-glucosidase.

β-Glucoside Kinase The enzyme that catalyzes the following reaction:

ATP + cellobiose
⇅
ADP + 6-phospho-β-D-glucosyl-(1,4)-D-glucose

Glucosidic Bond A bond that links a glucose to another molecule through an intervening oxygen (O-glucosidic) or nitrogen (N-glucosidic).

Glucosidosucrase Synonym of α-glucosidase.

Glucosphingolipid Sphingolipid containing glucose.

Glucosphingosine Glucosylhydrolase Synonym of glucosylceramidase.

Glucostat A trade name for a glucose determination reagent containing glucose oxidase and peroxidase.

Glucosulfatase The enzyme that catalyzes the following reaction:

D-Glucose 6-sulfate + H_2O
⇅
D-Glucose + sulfate

Glucosulfone Sodium (mol wt 781) An antibacterial agent.

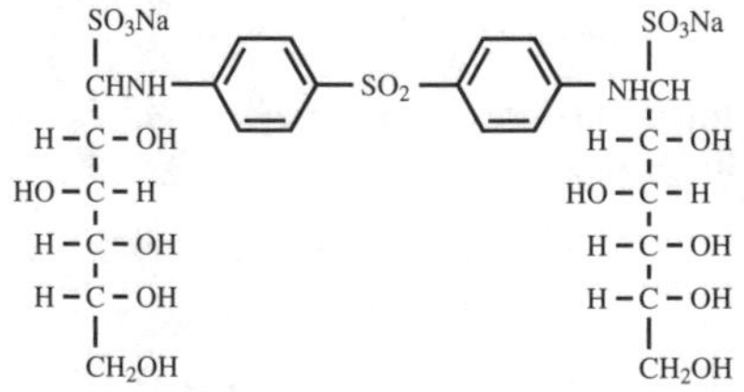

Glucosuria The presence of excessive amounts of glucose in the urine due to diseases such as diabetes mellitus.

Glucosyl Pertaining to glucose.

Glucosylation Formation of a glucosylated compound by the introduction of glucose into an organic compound (e.g., glucosylated DNA in T-even phage).

Glucosyl-DNA α-Glucosyltransferase The enzyme that catalyzes the transfer of α-D-glucosyl residue from UDP-glucose to the hydroxymethylcytosine residue in DNA.

Glucosyl-DNA β-Glucosyltransferase The enzyme that catalyzes the transfer of β-D-glucosyl residues from UDP-glucose to hydroxymethylcytosine residues in DNA.

Glucosylceramidase The enzyme that catalyzes the hydrolysis of glucocerebroside to glucose and ceramide.

Glucosylceramide Lipidosis A genetic disorder characterized by the accumulation of cerebrosides in the tissue and enlargement of spleen and liver due to deficiency of the enzyme glucocerebrosidase (also called Gaucher's disease).

Glucotrol A trade name for glipizide that stimulates insulin release from the pancreatic beta cells and reduces glucose output by the liver.

Glucozyme Synonym of glucan 1,4-α-glucosidase.

Glucuronate Isomerase The enzyme that catalyzes the following reaction:

D-glucuronate ⇌ D-fructuronate

Glucuronate Pathway The pathway in which the glucuronate is converted to xylulose 5-phosphate, an intermediate in pentose phosphate pathway.

Glucuronate 1-Phosphate Uridylyltransferase The enzyme that catalyzes the following reaction:

UTP + α-D-Glucuronate 1-phosphate

⇅

PPi + UDP-glucuronate

Glucuronate Reductase The enzyme that catalyzes the following reaction:

L-gulonate + $NADP^+$

⇅

D-glucuronate + NADPH

Glucuronic Acid (mol wt 194) A sugar acid derived from glucose.

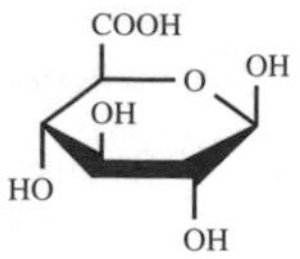

β-D-glucuronic acid

β-D-Glucuronidase The enzyme that catalyzes the following reaction:

β–D-Glucuronoside + H_2O

⇅

An alcohol + D-glucuronate

Glucuronide A compound formed by combining glucuronate with substances such as alcohol.

Glucuronokinase The enzyme that catalyzes the following reaction:

ATP + D-glucuronate

⇅

ADP + α-D-glucuronate 1-phosphate

Glucuronolactone (mol wt 176) A compound found in the plant gums and a structural constituent of fibrous and connective tissue.

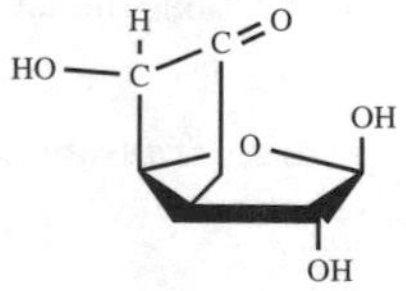

Glucuronolactone Dehydrogenase The enzyme that catalyzes the following reaction:

D-Glucurono-γ–lactone + NAD^+ + H_2O

⇅

D-glucarate + NADH

Glucuronolactone Reductase The enzyme that catalyzes the following reaction:

L-Gulono-1,4-lactone + $NADP^+$

⇅

D-Glucurono-3,6-lactone + NADPH

Glucuronoside Glucuronosohydrolase The systematic name for b-glucuronidase.

Glucuronosyltransferase The enzyme that catalyzes the following reaction:

UDP-glucuronate + acceptor

⇅

UDP + acceptor β-D-glucuronoside

GluK A trade name for potassium gluconate, which regulates the body's potassium.

$GluNH_2$ Symbol for glutamine.

GLUT1 Abbreviation for glucose transport 1 (basal growth-regulated glucose transporter).

GLUT4 Abbreviation for glucose transport 4 (insulin-responsive glucose transporter).

Glutamate Alanine Transaminase The enzyme that catalyzes the following reaction:

L-Alanine + α–ketoglutarate ⇌ Pyruvate + L-glutamate

Glutamate Aminopeptidase See glutamyl aminopeptidase.

Glutamate Ammonia Ligase The enzyme that catalyzes the following reaction:

ATP + L-glutamate + NH_3 ⇌ ADP + Pi + L-glutamine

Glutamate Aspartate Carrier A transport protein that facilitates the transport of aspartate and glutamate between the mitochondrial matrix and cytosol.

Glutamate Aspartate Transaminase The enzyme that catalyzes the following reaction:

L-Aspartate + 2-oxoglutarate ⇌ Oxaloacetate + L-glutamate

Glutamate Carboxypeptidase The enzyme that catalyzes the release of the C-terminal glutamate residue from a polypeptide.

Glutamate Cyclase The enzyme that catalyzes the following reaction:

D-Glutamate ⇌ 5-oxo-D-proline + H_2O

Glutamate Cysteine Ligase The enzyme that catalyzes the following reaction:

ATP + L-glutamate + L-cysteine ⇌ ADP + Pi + γ-L-glutamyl-L-cysteine

Glutamate Decarboxylase The enzyme that catalyzes the following reaction:

L-Glutamate ⇌ 4-Aminobutyrate + CO_2

Glutamate Dehydrogenase (NAD⁺) The enzyme that catalyzes the following reaction:

L-Glutamate + NAD^+ + H_2O ⇌ α-Ketoglutarate + NADH + NH_3

Glutamate Dehydrogenase (NADP⁺) The enzyme that catalyzes the following reaction:

L-Glutamate + $NADP^+$ + H_2O ⇌ α-Ketoglutarate + NADPH + NH_3

Glutamate Endopeptidase See glutamyl endopeptidase.

Glutamate Ethylamine Ligase The enzyme that catalyzes the following reaction:

ATP + L-glutamate + ethylamine ⇌ ADP + Pi + N^5-ethyl-L-glutamine

Glutamate Formiminotransferase The enzyme that catalyzes the following reaction:

N-Formimino-L-glutamate + tetrahydrofolate ⇌ L-glutamate + 5-formiminotetrahydrofolate

Glutamate Formyltranferase Synonym of glutamate formino-transferase.

Glutamate Histidine Synthetase See glutamyl-histidine synthetase.

Glutamate Kinase The enzyme that catalyzes the following reaction:

ATP + L-glutamate ⇌ L-Glutamate phosphate

Glutamate NADP⁺ Oxidoreductase The systematic name for glutamate dehydrogenase.

Glutamate Oxaloacetate Transaminase The enzyme that catalyzes the following reaction:

Glutamate + Oxaloacetate ⇌ Aspartate + α-keto-glutarate

Glutamate Oxidase The enzyme that catalyzes the following reaction:

D-Glutamate + H_2O + O_2 ⇌ α-Keto-glutarate + NH_3 + H_2O_2

Glutamate Oxygen Oxidoreductase The systematic name for glutamate oxidase.

Glutamate Pyruvate Transaminase The enzyme that catalyzes the following reaction:

Glutamate + pyruvate
⇅
α-keto-glutarate + alanine

Glutamate Racemase The enzyme that catalyzes the following reaction:

L-Glutamate ⇌ D-glutamate

Glutamate Semialdehyde An intermediate in the metabolism of arginine.

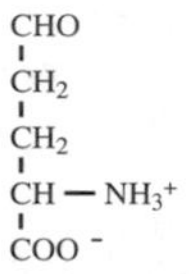

Glutamate Semialdehyde Dehydrogenase The enzyme that catalyzes the following reaction:

Glutamate semialdehyde + Pi + $NADP^+$
⇅
Glutamyl 5-phosphate + NADPH

Glutamate tRNA Ligase The enzyme that catalyzes the following reaction:

ATP + L-glutamate + $tRNA^{glu}$
⇅
AMP + PPi + L-glutamyl-$tRNA^{glu}$

Glutamate tRNA Synthetase See glutamate tRNA ligase.

Glutamic Acid (mol wt 147) An amino acid found in protein.

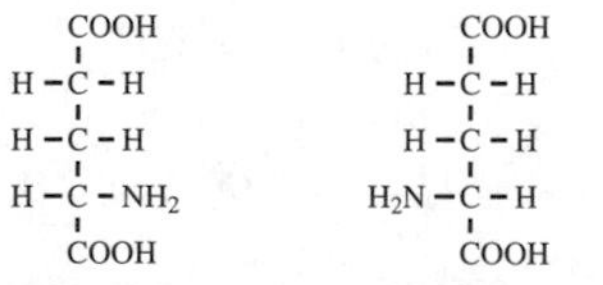

D-glutamic acid L-glutamic acid

Glutamicacidemia A genetic disorder characterized by the accumulation of glutamate in the blood due to deficiency of the enzyme that metabolizes glutamate and marked by mental and physical retardation, convulsions, and fragile hair growth.

Glutamic Semialdehyde See glutamate semialdehyde.

Glutaminase The enzyme that catalyzes the following reaction:

L-Glutamine + H_2O ⇌ L- glutamate + NH_3

Glutamine (mol wt 146) A basic amino acid found in protein.

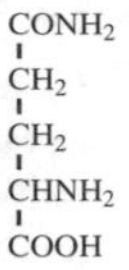

Glutamine Amidohydrolase The systematic name for glutaminase.

Glutamine Fructose 6-Phosphate Amino Transferase The enzyme that catalyzes the following reaction:

Glutamine + fructose 6-phosphate
⇅
Glutamate + 2-amino-2-deoxy-D-glucose 6-phosphate

Glutamine N-Phenylacetyltransferase The enzyme that catalyzes the following reaction:

Phenylacetyl-CoA + L-glutamine
⇅
CoA + α-*N*-phenylacetyl-L-glutamine

Glutamine Phenylpyruvate Transaminase The enzyme that catalyzes the following reaction:

L-glutamine + phenylpyruvate
⇅
α-Ketoglutamate + L-phenylalanine

Glutamine Phosphoribosyl Pyrophosphate Aminotransferase The enzyme that catalyzes the following reaction:

5-Phospho-β-D-ribosylamine+ pyrophosphate + L-glutamate
⇅
L-glutamine + 5-phospho-α–D-ribose-1-diphosphate

Glutamine Synthetase The enzyme that catalyzes the following reaction:

ATP + L-glutamate + NH_3
⇅
ADP + Pi + glutamine

Glutamine tRNA Ligase The enzyme that catalyzes the following reaction:

ATP + L-glutamine + $tRNA^{gln}$
⇅
AMP + PPi + L-glutaminyl-$tRNA^{gln}$

Glutaminyl Pertaining to glutamine.

Glutaminyl Cyclase See glutaminyl peptide cyclotransferase.

γ-Glutaminyl Cycle An energy-driven transport system coupled to the synthesis and breakdown of glutathione for the transfer of amino acids across a membrane.

Glutaminyl Peptide Cyclotransferase The enzyme that catalyzes the following reaction:

L-Glutaminyl-peptide
⇅
5-Oxoprolyl-peptide + NH_3

Glutaminyl tRNA Ligase See glutamine tRNA ligase.

Glutaminyl tRNA Synthetase See glutamine tRNA ligase.

Glutamyl Pertaining to glutamate.

Glutamyl Aminopeptidase The enzyme that catalyzes the release of the N-terminal glutamate from a peptide.

Glutamyl Carboxypeptidase The enzyme that cleaves γ-glutamyl bond to release an un-substituted C-terminal amino acid.

Glutamyl Endopeptidase The enzyme that catalyzes the preferential cleavage of peptide bonds involving COOH groups of glutamate and aspartate.

Glutamyl Histamine Synthetase The enzyme that catalyzes the following reaction:

ATP + L-glutamate + histamine
⇅
ADP + L-Glutamylhistamine + Pi

γ-Glutamyl Hydrolase Synonym of glutamyl carboxypeptidase.

Glutamyl Transferase The enzyme that catalyzes the following reaction:

5-L-Glutamylpeptide + an amino acid
⇅
Peptide + 5-L-glutamyl amino acid

Glutamyl Transpeptidase Synonym of γ-glutamyl transferase.

Glutamyl tRNA Ligase See glutamate tRNA ligase.

Glutamyl-tRNA Synthetase See glutamate tRNA ligase.

Glutaraldehyde (mol wt 100) A disinfectant and a reagent used as a fixative for electron microscopy.

CHO
|
CH_2
|
CH_2
|
CH_2
|
CHO

Glutaredoxin A monomeric disulfide-containing protein involved in reducing ribonucleotide to deoxyribonucleotide.

Glutaric Acid (mol wt 132) A dicarboxylic acid and an inhibitor for succinate dehydrogenase.

COOH
|
CH_2
|
CH_2
|
CH_2
|
COOH

Glutaryl-CoA (mol wt 881) An intermediate in lysine metabolism.

COOH
|
CH_2
|
CH_2
|
CH_2
|
O = C – S – CoA

Glutaryl-CoA Dehydrogenase The enzyme that catalyzes the following reaction:

Glutaryl-CoA + FAD
⇅
Glotonoyl-CoA + + CO + $FADH_2$

Glutaryl-CoA Synthetase The enzyme that catalyzes the following reaction:

ATP + glutarate + CoA
⇅
ADP + Pi + glutaryl-CoA

Glutathione (mol wt 307) A tripeptide hormone consisting of glutamyl-cysteinyl-glycine that serves as a coenzyme and antioxidant (abbreviated as GSH).

$H_2NCH(CH_2)_2CONHCHCONHCH_2COOH$ (with COOH on the first CH and CH_2SH on the second CH)

Glutathione CoA Glutathione Transhydrogenase The enzyme that catalyzes the following reaction:

CoA + oxidized glutathione ⇅ CoA-glutathione + glutathione

Glutathione Cysteine Transhydrogenase The enzyme that catalyzes the following reaction:

2 Glutathione + cystine ⇅ Oxidized glutathione + 2 cysteine

Glutathione Dehydrogenase The enzyme that catalyzes the following reaction:

2 Glutathione + dehydroascorbate ⇅ Oxidized gluthionine + ascorbate

Glutathione Homocysteine Transhydrogenase The enzyme that catalyzes the following reaction:

2 Glutathione + homocystine ⇅ Oxidized glutathione + 2 homocysteine

Glutathione Hydrogen-Peroxide Oxidoreductase The systematic name for glutathione peroxidase.

Glutathione Insulin Transhydrogenase The enzyme that catalyzes the following reaction:

2 Glutathione + protein-disulfide ⇅ Oxidized glutathione + protein-dithiol

Glutathione Oxidase The enzyme that catalyzes the following reaction:

2 Glutathione + O_2 ⇅ Oxidized glutathione + H_2O_2

Glutathione Oxygen Oxidoreductase The systematic name for glutathione oxidase.

Glutathione Peroxidase The enzyme that catalyzes the following reaction:

2 GSH + organic hydroperperoxide ⇅ GS-SH + ROH + H_2O

Glutathione Reductase The enzyme that catalyzes the following reaction:

NADPH + oxidized glutathione ⇅ $NADP^+$ + 2 glutathione

Glutathione Synthetase The enzyme that catalyzes the following reaction:

ATP + γ-L-glutamyl-L-cysteine + glycine ⇅ ADP + Pi + glutathione

Glutathione Thioesterase The enzyme that catalyzes the following reaction:

S-Acylglutathione + H_2O ⇅ Glutathione + a carboxylate

Glutathione Transferase The enzyme that catalyzes the following reaction:

RX + glutathione ⇌ HX + R-S-G

Glutelins A group of basic proteins that consist of basic amino acids and are soluble in dilute acids or alkaline.

Gluten Reserve protein from wheat that contains a mixture of glutelin and gliadin.

Glutethimide (mol wt 217) A sedative and hypnotic agent.

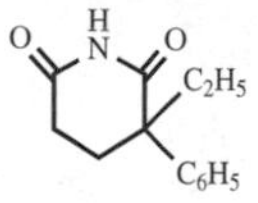

Glutose-15 A trade name for a lemon-flavored, dye-free glucose oral gel used for treatment of insulin reaction or hypoglycemia.

Glu-Tubulin Abbreviation for tubulin dimer containing a tubulin lacking C-terminal tyrosine, exposing a glutamic residue at the end.

Glyphosine (mol wt 263) A chemical ripener.

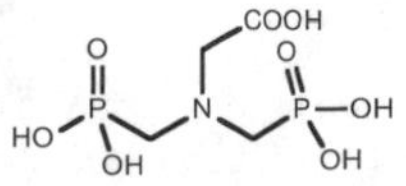

Glx Abbreviation for amino acid glutamate or glutamine.

Gly Abbreviation for amino acid glycine.

Glyate A trade name for guaifenesin, which stimulates the production of respiratory-tract fluids to reduce the viscosity of tenacious secretions.

Glyburide (mol wt 494) An antidiabetic agent that stimulates insulin release from the pancreatic beta cells and reduces glucose output by the liver.

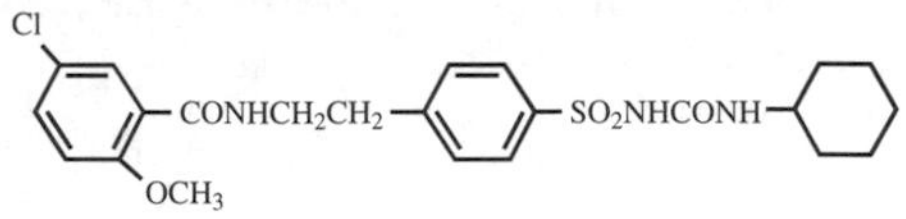

Glybuthiazole (mol wt 312) An antidiabetic agent.

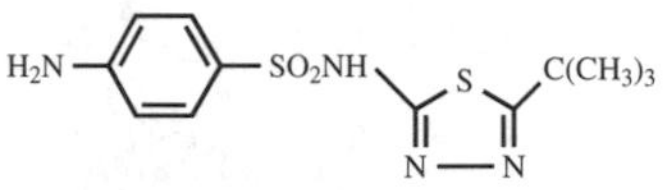

Glybuzole (mol wt 297) An antidiabetic agent.

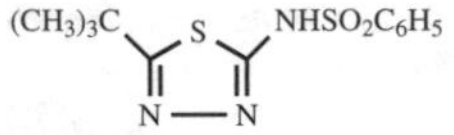

GlyCAM Abbreviation for glycosylation-dependent cell-adhesion molecule.

Glycan Polysaccharide.

Glycarsamide (mol wt 275) An anthelmintic agent.

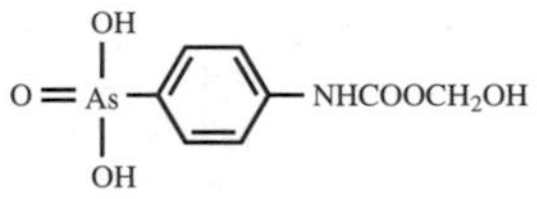

Glycaric Acid A compound formed from monosaccharide by oxidation of -CHO groups and $-CH_2OH$ groups to COOH.

Glycate A trade name for calcium carbonate, which is used to reduce the acid load in the GI tract and to elevate gastric pH.

-glycemia A suffix meaning condition of sugar in the blood.

Glyceraldehyde (mol wt 90) A three-carbon compound containing an aldehyde group.

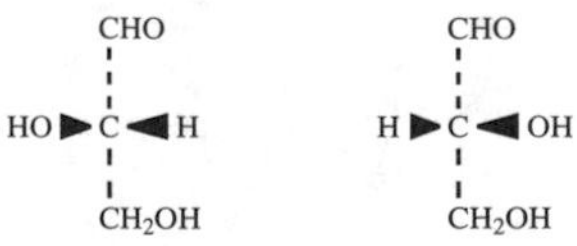

L-form D-form

Glyceraldehyde Kinase The enzyme that catalyzes the following reaction:

$$\text{Glyceraldehyde + ATP} \rightleftharpoons \text{Glyceraldehyde 3-phosphate + ADP}$$

Glyceraldehyde 3-Phosphate (mol wt 170) An intermediate in glycolysis.

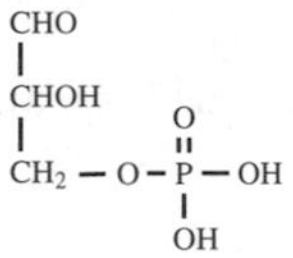

Glyceraldehyde 3-Phosphate Dehydrogenase The enzyme that catalyzes the following reaction:

$$\text{D-Glyceraldehuyde 3-phosphate + Pi + NAD}^+ \rightleftharpoons \text{3-phospho-D-glyceroyl phosphate + NADH}$$

Glyceraldehyde 3-Phosphate Ketol Isomerase Synonym of triose phosphate isomerase.

Glyceraldehyde 3-Phosphate NAD⁺ Oxidoreductase The systematic name for glyceraldehyde 3-phosphate dehydrogenase.

Glycerate Dehydrogenase The enzyme that catalyzes the following reaction:

$$\text{Glycerate + NADP}^+ \rightleftharpoons \text{Hydroxypyruvate + NADPH}$$

Glycerate Kinase The enzyme that catalyzes the following reaction:

$$\text{ATP + Glycerate} \rightleftharpoons \text{ADP + phosphoglycerate}$$

Glyceric Acid (mol wt 106) A three-carbon organic acid.

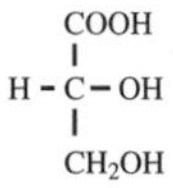

Glyceride An ester derived from glycerol and fatty acids.

Glycerin A pharmaceutical preparation containing glycerol and used as a moistener and laxative (e.g., suppositories for constipation).

Glyceroglycolipid Glycerol lipid that contains carbohydrate.

Glycerol (mol wt 92) A three-carbon alcohol and essential component of fat.

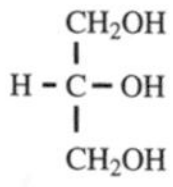

Glycerol Dehydrogenase The enzyme that catalyzes the following reaction:

$$\text{Glycerol} + \text{NAD}^+ \rightleftharpoons \text{Dihydroxyacetone} + \text{NADH}$$

Glycerol Dehydrogenase (NADP⁺) The enzyme that catalyzes the following reaction:

$$\text{Glycerol} + \text{NADP}^+ \rightleftharpoons \text{Dihydroxyacetone} + \text{NADPH}$$

Glycerol Hydratase The enzyme that catalyzes the following reaction:

$$\text{Glycerol} \rightleftharpoons \text{3-Hydroxypropanal} + H_2O$$

Glycerol Kinase The enzyme that catalyzes the following reaction:

$$\text{ATP} + \text{glycerol} \rightleftharpoons \text{ADP} + \text{glycerol 3-phosphate}$$

Glycerol NAD⁺ Oxidoreductase The systematic name for glycerol dehydrogenase (NAD⁺).

Glycerol NADP⁺ Oxidoreductase The systematic name for glycerol dehydrogenase (NADP⁺).

Glycerol Oxygen 2-Oxidoreductase The systematic name for glycerol 3-phosphate oxidase.

Glycerol 3-Phosphate NAD⁺ 2-Oxidoreductase The systematic name for glycerol 3-phosphate dehydrogenase.

Glycerol 1-Phosphatase The enzyme that catalyzes the following reaction:

$$\text{Glycerol 1-phosphate} + H_2O \rightleftharpoons \text{Glycerol} + \text{P}$$

Glycerol 2-Phosphatase The enzyme that catalyzes the following reaction:

$$\text{Glycerol 2-phosphate} + H_2O \rightleftharpoons \text{Glycerol} + \text{Pi}$$

Glycerol 3-Phosphate Acyltransferase The enzyme that catalyzes the following reaction:

$$\text{Acyl-CoA} + \text{glycerol 3-phosphate} \rightleftharpoons \text{CoA} + \text{acylglycerol 3-phosphate}$$

Glycerol 3-Phosphate Cytidylyltransferase The enzyme that catalyzes the following reaction:

$$\text{CTP} + \text{glycerol 3-phosphate} \rightleftharpoons \text{PPi} + \text{CDP-glycerol}$$

Glycerol 3-Phosphate Dehydrogenase The enzyme that catalyzes the following reaction:

$$\text{Glycerol 3-phosphate} + \text{NAD}^+ \rightleftharpoons \text{Dihydroxyacetone 3-phosphate} + \text{NADH}$$

Glycerol 3-Phosphate Dehydrogenase (NADP⁺) The enzyme that catalyzes the following reaction:

$$\text{Glycerol 3-phosphate} + \text{NADP}^+ \rightleftharpoons \text{Dihydroxyacetone 3-phosphate} + \text{NADPH}$$

Glycerol 3-Phosphate Glucose Phosphotransferase The enzyme that catalyzes the following reaction:

$$\text{Glycerol 3-phosphate} + \text{glucose} \rightleftharpoons \text{Glycerol} + \text{glucose 6-phosphate}$$

Glycerol 3-Phosphate Oxidase The enzyme that catalyzes the following reaction:

$$\text{Glycerol 3-phosphate} + O_2 \rightleftharpoons \text{Dihydroxyacetone phosphate} + H_2O_2$$

Glycerol Phosphate Shuttle A mechanism for transferring electrons from NADH in the cytosol to FAD in the mitochondrion for the formation of mitochondrial $FADH_2$.

Glycerol Phosphatide Referring to phospholipid.

Glycerophosphocholine Phosphodiesterase The enzyme that catalyzes the following reaction:

$$\textit{Sn}\text{-Glycero-3-phosphocholine} + H_2O \rightleftharpoons \text{Choline} + \text{sn-glycerol 3-phosphate}$$

Glycerophospholipid Referring to phospholipid, e.g., phosphatidyl choline and phosphatidyl serine.

Glycerophosphoric Acid (mol wt 172) The phosphoester of glycerol.

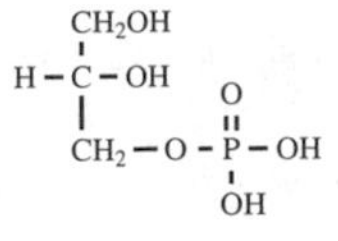

Glyceryl Guaiacolate See guaifenesin.

Glycidol (mol wt 74) An inhibitor for triose phosphate isomerase.

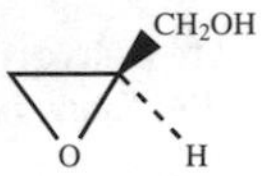

Glycidol Phosphate (mol wt 154) An inhibitor for triose phosphate isomerase.

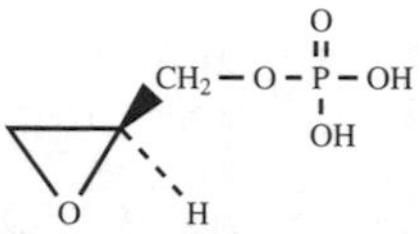

Glycinamide Ribonucleotide Synthetase The enzyme that catalyzes the following reaction:

ATP + phospho-ribosylamine + glycine ⇌ ADP + Pi +5-Phosphoribosyl glycinamide

Glycinamide Ribotide Transformylase The enzyme that catalyzes the following reaction:

5-Phosphoribosyl-*N*-formylglycinamide + tetrahydrofolate ⇌ 5'-Phosphoribosylglycinamide + 5-10-methenyltetrahydrofolate

Glycine (mol wt 75) The simplest protein amino acid.

$CHNH_2$
|
$COOH$

Glycine Acetyltransferase The enzyme that catalyzes the following reaction:

Acetyl-CoA + glycine ⇌ CoA + 2-amino-3-ketobutanoate

Glycine Acyltransferase The enzyme that catalyzes the following reaction:

Acyl-CoA + glycine ⇌ CoA + *N*-acylglycine

Glycine Alantoin Cycle A pathway for synthesis of urea that occurs in some urea-accumulating organisms.

Glycine Amidinotransferase The enzyme that catalyzes the following reaction:

Arginine + glycine ⇌ Ornithine + guanidinoacetate

Glycine Aminotransferase The enzyme that catalyzes the following reaction:

Glycine + α–ketoglutarte ⇌ Glyoxylate + glutamate

Glycine *N*-Benzoyltransferase The enzyme that catalyzes the following reaction:

Benzoyl-CoA + glycine ⇌ CoA + *N*-benzoylglycine

Glycine Cleavage System A multiple enzyme complex for metabolism of glycine to N^5-N^{10}-methylenetetrahydrofolic acid and CO_2.

Glycine Cresol Red (mol wt 579) A biological dye.

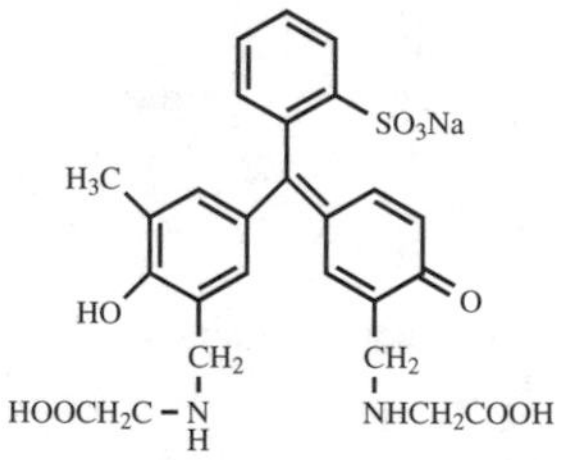

Glycine Dehydrogenase The enzyme that catalyzes the following reaction:

Glycine + H_2O + NAD^+ ⇌ Glyoxylate + NH_3 + NADH

Glycine Formiminotransferase The enzyme that catalyzes the following reaction:

Formiminotetrahydrofolate + glycine ⇌ Tetrahydrofolate + *N*-formiminoglycine

Glycine Hydroxymethyltransferase The enzyme that catalyzes the following reaction:

5,10-Methylenetetrahydrofolate + glycine + H_2O ⇌ Tetrahydrofolate + L-serine

Glycine *N*-Methyltransferase The enzyme that catalyzes the following reaction:

S-Adenysyl-L-methionine + glycine ⇌ *S*-Adenosyl-homocysteine + sacrosine

Glycine Oxaloacetate Transaminase The enzyme that catalyzes the following reaction:

Glycine + oxaloacetate
⇅
Gloxylate + L-aspartate

Glycine Transaminase The enzyme that catalyzes the following reaction:

Glycine + α-keto-glutarate
⇅
Glyoxylate + L-glutamate

Glycine tRNA Ligase The enzyme that catalyzes the following reaction:

ATP + glycine = $tRNA^{gly}$
⇅
AMP + PPi + glycyl-$tRNA^{gly}$

Glycinin Major protein of soybean.

Glycocalyx A cell coat structure consisting of a fuzzy layer of polysaccharide or glycoprotein surrounding certain animal cells.

Glycocholic Acid (mol wt 466) A glycine-cholic acid conjugate and bile salt.

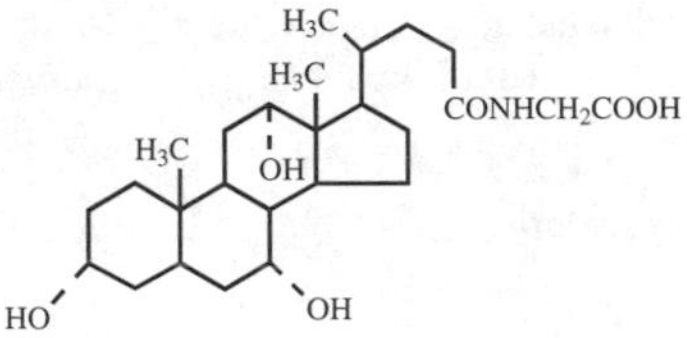

Glycocyamine (mol wt 117) A guanidine acetic acid complex.

H_2N — $CNHCH_2COOH$
‖
NH

Glycocyamine Kinase The enzyme that catalyzes the following reaction:

ATP + guanidinoacetate
⇅
ADP + phosphoguanidinoacetate

Glycogen Highly branched, storage polysaccharide in animal cells consisting of α-D-glucose repeating units linked by α-1,4- and α-1,6-glucosidic bonds (also known as animal starch).

Glycogen Branching Enzyme The enzyme that catalyzes the formation of 1,6-glucosidic branching in glycogen.

Glycogen Debranching Enzyme The enzyme that catalyzes the hydrolysis of 1,6-linked glucosidic bonds in glycogen.

Clycogen 6-Glucanohydrolase Synonym of isoamylase.

Glycogen Phosphorylase The enzyme that catalyzes the following reaction:

$(\alpha\text{–}1,4\text{–glucosyl})_n$ + Pi
⇅
$(\alpha\text{–}1,4\text{–glucosyl})_{n-1}$ + glucose 1-phosphate

Glycogen Storage Disease A group of genetic disorders characterized by abnormal accumulation of glycogen in the liver and other body tissues due to deficiency in the enzymes for metabolism of glycogen.

Glycogen Synthetase The enzyme that catalyzes the synthesis of glycogen:

UDP-glucose + $(\alpha\text{-}1,4\text{-glucosyl})_n$
⇅
UDP + $(\alpha\text{-}1,4\text{-glucosyl})_{n+1}$

Glycogen Synthetase α-Kinase Synonym of protein kinase.

Glycogenase Synonym of α-amylase.

Glycogenesis The formation of glycogen from glucose.

Glycogenic Amino Acid Amino acids that are capable of serving as precursors for synthesis of glucose and glycogen in carbohydrate metabolism.

Glycogenin A protein involved in glycogen synthesis.

Glycogenin Glucosyltransferase The enzyme that catalyzes the following reaction:

UDP-glucose + glycogenin
⇅
UDP + glucosylglycogenin

Glycogenolysis The breakdown of glycogen to glucose.

Glycogenosis The development of glycogen storage disease.

Glycogen-UDP Glucosyltransferase The enzyme that catalyzes the following reaction:

UDP-glucose + glycogen
⇅
UDP + $(glycogen)_{n+1}$

Glycoglycerolipid Referring to glycolipid.

Glycol An alcohol that contains two hydroxyl groups.

Glycolaldehyde Transferase Synonym of transketolase.

Glycolate Dehydrogenase The enzyme that catalyzes the following reaction:

$$\text{Glycolate} + \text{acceptor} \rightleftharpoons \text{Glyoxylate} + \text{reduced acceptor}$$

Glycolate NAD$^+$ Oxidoreductase The systematic name for glycolate reductase

Glycolate Oxidase The enzyme that catalyzes the following reaction:

$$\text{Glycolate} + O_2 \rightleftharpoons \text{Glyoxylate} + H_2O_2$$

Glycolate Oxygenase See glycolate oxidase.

Glycolate Reductase The enzyme that catalyzes the following reaction:

$$\text{Glycolate} + \text{NAD}^+ \rightleftharpoons \text{Glyoxylate} + \text{NADH}$$

Glycolate Pathway A pathway occurring in photosynthetic cells by which the phosphoglycolate generated in the chloroplast by the oxygenase activity of ribulose-1-5-bisphosphate carboxylase is converted to glyoxylate, glycine, and glycerate in the peroxisome and mitochondrion (see also photorespiration).

Glycolic Acid (mol wt 76) A two-carbon organic acid and constiuent of sugarcane juice.

COOH
|
CH_2OH

Glycolipids Lipids that contain polar, hydrophilic carbohydrate groups.

Glycolysis A pathway by which glucose is catabolized to pyruvate without the involvement of oxygen; it generates ATP and NADH (also known as glycolytic pathway or Embden-Meyerhof pathway).

Glycolytic Pertaining to glycolysis.

Glycolytic Pathway See glycolysis.

Glycone The carbohydrate moiety of a glycoside.

Glyconeogenesis The formation of glycogen from noncarbohydrate substances.

Glyconiazide (mol wt 295) An antibacterial (tuberculostatic) agent.

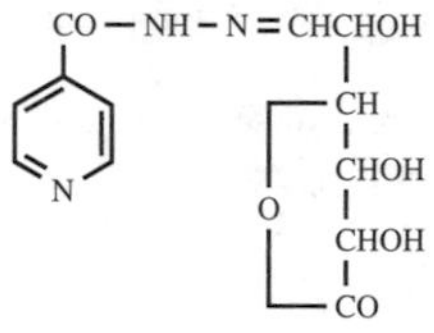

Glyconic Acid An acid formed from oxidation of the CHO group of an aldose to COOH.

Glycopeptidase Synonym of peptide-N4-(N-Acetyl-β-glucosaminyl)asparagine amidase.

Glycopeptide A peptide that contains a carbohydrate moiety.

Glycopeptide α-N-Acetylgalactosaminidase The enzyme that catalyzes the hydrolysis of terminal D-galactosyl-N-acetyl-α-D-galactosaminidic residues from a variety of glycopeptides and glycoprotreins.

Glycophorin An erythrocyte transmembrane glycoprotein that consists of sialic acid.

Glycophorin A A membrane glycoprotein on red blood cells that acts as a receptor site for influenza virus and the malarial parasite (*Plasmodium falciparum*).

Glycopyrrolate (mol wt 398) An anticholinergic agent.

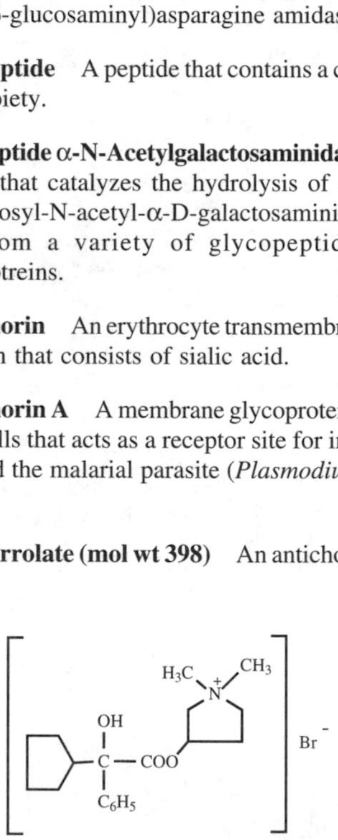

Glycosaminoglycan An unbranched polysaccharide consisting of alternating uronic acid and hexosamine residues (also known as mucopolysaccharide).

Glycosaminoglycan Galactosyltransferase The enzyme that catalyzes the following reaction:

$$\text{UDP-galactose} + \text{glycosaminoglycan} \rightleftharpoons \text{UDP} + \text{galactosylglycosaminoglycan}$$

Glycosidase The enzyme that catalyzes the hydrolysis of glycosidic bonds in a polysaccharide.

Glycoside Sugar derivative that contains a nonsugar moiety linked by glycosidic bonds.

Glycosidic Bond The bond formed between a sugar and another molecule through intervening oxygen (*O*-glycosidic) or nitrogen (*N*-glycosidic).

Glycosome A microbody occurring in *Trypanosoma* that contains enzymes involved in glycolysis starting from 3-phosphoglyceraldehyde.

Glycosphingolipid Sphingolipid that contains carbohydrate moiety.

Glycosphingolipid Deacylase The enzyme that catalyzes the hydrolysis of gangliosides and neutral glycosphingolipid by releasing fatty acid to form fatty-acid derivatives.

Glycosulfatase The enzyme that catalyzes the following reaction:

$$\text{Glucose 6-sulfate} + H_2O \rightleftharpoons \text{Glucose} + \text{sulfate}$$

Glycosuria A disorder characterized by the presence of a large amount of reducing sugars in the urine.

Glycosylase The enzyme that catalyzes the attachment of a sugar moiety to DNA or RNA or protein.

Glycosylation Covalent attachment of a carbohydrate molecule to polypeptide or polynucleotide (e.g., glycosylated DNA in T- even phage).

Glycosyl Ceramidase The enzyme that catalyzes the following reaction:

$$\text{Glycosyl-}N\text{-acylsphingosine} + H_2O \rightleftharpoons \text{A sugar} + N\text{-acylsphingosine}$$

Glycosyl Ceramide A ceramide that contains carbohydrate.

Glycosyl Glyceride Referring to glyceride that contains carbohydrate.

Glycosyl Group Referring to the sugar moiety that is linked to another molecule by a glycosidic bond.

Glycosyl Transferase The enzyme that catalyzes the transfer of a sugar moiety from the UDP-sugar to an acceptor.

Glycotropic Increasing concentration of glucose in the blood.

Glycotuss A trade name for guaifenesin, which increases the production of respiratory-tract fluid to reduce the viscosity of the tenacious secretion.

Glycuronic Acid A sugar acid formed by oxidation of terminal CH_2OH groups to COOH.

Glycyl tRNA Ligase See glycine tRNA ligase.

Glycyl tRNA Synthetase See glycine tRNA ligase.

Glycryrrhizic Acid (mol wt 823) A sweet compound from licorice root. It causes retention of sodium and water and excessive secretion of potassium.

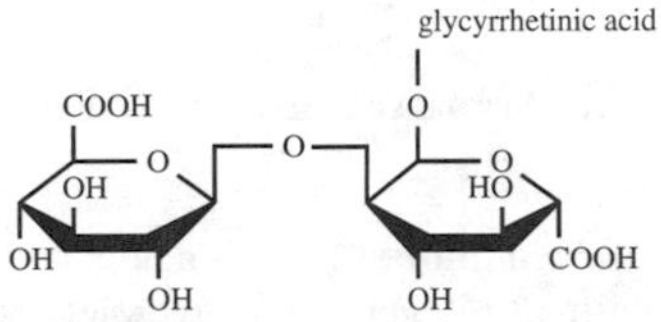

Glyhexamide (mol wt 322) An antidiabetic agent.

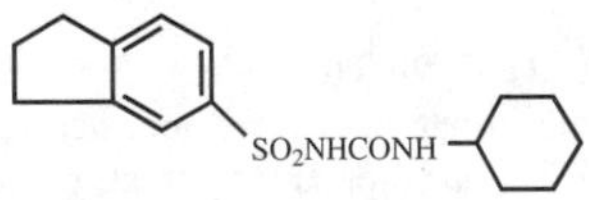

Glymidine (mol wt 309) An antidiabetic agent.

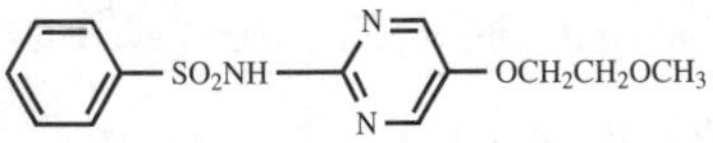

Glynase Prestab A trade name for glyburide, an antidiabetic agent that stimulates insulin release from the pancreatic beta cell and reduces glucose output by the liver.

Glyoxal (mol wt 58) A reagent used to denature RNA and DNA that reacts specifically with guanosine residues and prevents nucleotide pairing.

CHO
|
CHO

Glyoxylate Bypass See glyoxylate cycle.

Glyoxylate Cycle A metabolic pathway occurring in plants and some microorganisms. It involves conversion of isocitrate to succinate and glyoxylate. The glyoxylate thus formed is combined with acetyl-CoA to form malate and bypass the normal Krebs cycle sequence.

Glyoxylate Dehydrogenase The enzyme that catalyzes the following reaction:

$$\text{Glyoxylate} + \text{CoA} + \text{NADP} \rightleftharpoons \text{Oxalyl-CoA} + \text{NADPH}$$

Glyoxylate NAD+ Oxidoreductase The systematic name for glyoxylate reductase.

Glyoxylate Oxidase The enzyme that catalyzes the following reaction:

$$\text{Glyoxylate} + H_2O + O_2 \rightleftharpoons \text{Oxalate} + H_2O_2$$

Glyoxylate Pathway See glyoxylate cycle.

Glyoxylate Reductase The enzyme that catalyses the following reaction:

$$\text{Glycolate} + \text{NAD} \rightleftharpoons \text{Glyoxylate} + \text{NADH}$$

Glyoxylate Shunt See glyoxylate cycle.

Glyoxylate Transacetylase See malate synthetase.

Glyoxylic Acid (mol wt 74) A compound occurring in unripe fruit and young green leaves.

Glyoxylic Acid Reaction A colorimetric reaction for determination of compounds containing an indol ring (e.g., tryptophan) by treating the sample with glyoxylic acid and sulfuric acid.

Glyoxysomes The membrane-bound organelles that contain the enzymes for the glyoxylate cycle.

Glyphosate (mol wt 169) An herbicide.

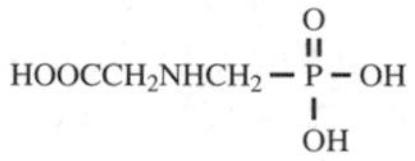

Glyphosine (mol wt 263) A chemical ripener.

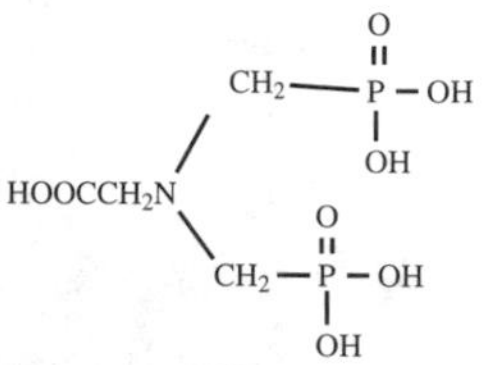

Glypinamide (mol wt 332) An oral hypoglycemic agent.

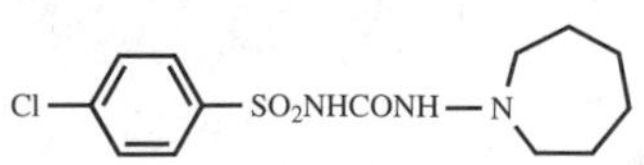

Gly-Pro-Naphthylamidase Synonym of dipeptidyl peptidase IV.

Glyset A trade name for miglitol, an anti-diabetic agent.

Glytuss A trade name for guaifenesin, an antitussive agent that increases the production of respiratory fluid to reduce the viscosity of tenacious secretions.

Gm Abbreviation for 1. gram. 2. An allotype marker on immunoglobulins.

GM Abbreviation for Geiger-Mueller counter for counting radioactivity.

Gm% Abbreviation for gram percent.

GMCL Abbreviation for goldfish macrophage cell line.

GMCSF Abbreviation for granulocyte macrophage colony stimulating factor, a glycoprotein.

GMK Abbreviation for green monkey kidney.

GMP Abbreviation for guanosine 5′-monophosphate.

cGMP Abbreviation for cyclic guanosine monophosphate.

GMP Reductase The enzyme that catalyzes the following reaction:

$$\text{NADPH} + \text{GMP} \rightleftharpoons \text{NADP}^+ + \text{IMP} + NH_3$$

GMP Synthetase The enzyme that catalyzes the following reaction:

$$NH_3 + \text{ATP} + \text{Xanthoine 5'-phosphate} \rightleftharpoons \text{AMP} + \text{PPi} + \text{GMP}$$

GMP Synthetase (Glutamine-Hydrolyzing) The enzyme that catalyzes the following reaction:

$$\text{ATP} + \text{xanthosine 5'-phosphate} + \text{L-glutamine} \rightleftharpoons \text{AMP} + \text{PPi} + \text{GMP} + \text{L-glutamate}$$

GMS Abbreviation for gel mobility shift.

GMSF Abbreviation for granulocyte/macrophage stimulating factor.

G-Myticin A trade name for gentamicin sulfate, an antibiotic that inhibits bacterial protein synthesis.

GN Abbreviation for glomerulonephritis.

Gn Abbreviation for gonadotropin.

GNA Abbreviation for *Galanthus nivalis* agglutinin.

GNB Abbreviation for gram negative bacteria.

GnHCl Abbreviation for guanidine hydrochloride.

GndHCl Abbreviation for guanidinium hydrochloride.

Gnotobiotic Referring to 1. the microbiologically monitored environment or animal in which the identities of microorganisms present are identified and 2. germ-free animal.

Gnotobiotic Animal A germ-free animal that has known microbial flora.

GnRF Abbreviation for gonadotropin-releasing hormone.

GnRH Abbreviation for gonadotropin-releasing hormone.

GO Abbreviation for glucose oxidase.

Goblet Cell A special cell that releases mucus.

Goblin A protein from avian erythrocyte plasma membrane.

Goiter An overgrown or enlarged thyroid gland.

Goitrin (mol wt 129) An antithyroid compound from seeds of *Brassica*.

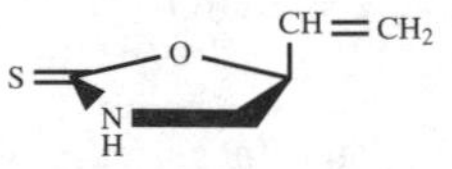

Goitrogen Substances that cause goiter.

Goitrogenic Capable of causing goiter.

Goitrogenic Glycoside A glycoside capable of causing hyperthyroidism or goiter.

Gold A chemical element with atomic weight 197, valence 1 and 3.

Gold Sodium Thiomalate (mol wt 490) An antirheumatic agent.

$AuNa_3O_6S_4$

Gold-198 A radioactive isotope of gold.

Goldberg-Hogness Box The TATA sequence on eukaryotic DNA located about 25 bp upstream from the site where transcription starts.

Goldman Equation The equation relating electrical potential across membrane to the distribution and permeability constants of the ions separating the membrane.

Golgi Apparatus The eukaryotic intracellular organelle consisting of a flattened, parallel membrane related to the endoplasmic reticulum; serves as a collecting and packaging center for secretary products.

Gomori Stain Referring to methenamine silver.

Gonad The male or female sex organ or gamete-producing organ in animals.

Gonadal Hormones Sex hormones.

Gonadotrophic Capable of influencing gonads.

Gonadotropin-Releasing Factor A factor consisting of 10 amino acid residues that stimulates the adenohypophysis to release luteinizing hormone.

Gonadotropins Hormones that stimulate the activities of the testes and ovaries, e.g., follicle stimulating and luteinizing hormone.

Gonal-F A trade name for follitropin alpha, a gonadotropic releasing hormone, and a human follicle stimulating hormone produced by recombinant DNA technology.

Gonidia Plural of gonidium.

Gonidium The reproductive cell of unicelluar green algae.

Gonococcus A bacterium of the species *Neisseria gonorrheoae*.

Gonorrhea A sexually transmitted disease caused by *Neisseria gonorrhoeae*.

G-Orange (mol wt 480) A biological dye.

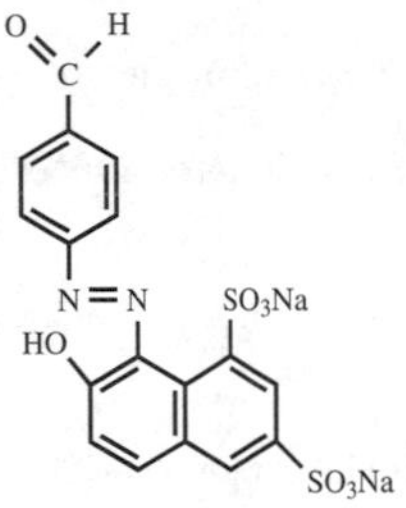

Goserelin A peptide hormone with antineoplastic activity.

Gossypol (mol wt 519) A poisonous pigment from cotton seed.

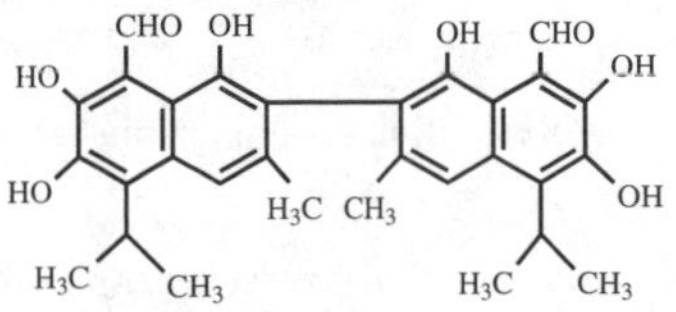

GOT Abbreviation for 1. glucose oxidase test; 2. glutamate-oxaloacetate transaminase; 3. glutamine-oxaloacetate transaminase.

Gougerotin (mol wt 443) An antibiotic from *Streptomyces aougerotii* that inhibits protein synthesis in both prokaryotes and eukaryotes.

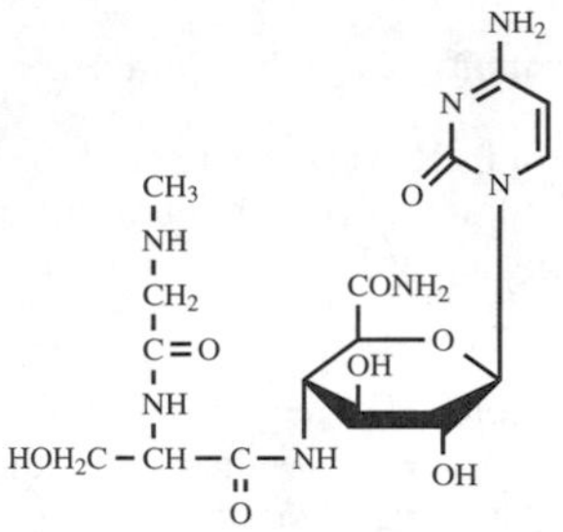

Gout A disease characterized by the painful swelling of the joints due to the deposition of uric acid in the joints.

GoxI (BamHI) A restriction endonuclease from *Gluconobacter oxydans* sub *melonogenes* with the same specificity as BamHI.

GP Abbreviation for 1. gene product; 2. glycogen phosphorylase; 3. Glycoprotein; 4. gram positive.

gp Abbreviation for gene product and glycoprotein.

G-1-P Abbreviation for glucose 1-phosphate.

G-3-P Abbreviation for gyceraldehyde 3-phosphate.

G-6-P Abbreviation for glucose 6-phosphate.

G1,6-P Abbreviation for glucose-1,6-bisphosphate.

gp32 An envelope glycoprotein from HIV-related viruses.

gp120 Abbreviation for a surface 120-kD glycoprotein of human immuno-deficiency virus type 1.

gp130 Abbreviation for glycoprotein of 130 kDa.

gp160 Abbreviation for a vaccine that contains a cloned segment of the envelope protein of 160 kD from HIV-1.

GPA Abbreviation for 1. guinea pig albumin; 2. glycogen phosphorylase a.

GPAIS Abbreviation for guinea pig anti-insulin serum.

G6Pase Abbreviation for glucose 6-phosphatase.

GPAT Abbreviation for glycerol 3-phosphate acyltransferase.

GPB Abbreviation for Gram positive bacteria.

GPC Abbreviation for gel permeation chromatography.

GPCP Abbreviation for glycerophosphocholine phosphodiesterase.

GPCR Abbreviation for G-protein coupled receptor.

G3PD Abbreviation for glyceraldehyde 3-phosphate dehydrogenase.

G-6-PD Abbreviation for glucose 6-phosphate dehydrogenase.

GPDH Abbreviation for glyceraldehyde phosphate dehydrogenase.

G3PDH Abbreviation for glyceraldehyde 3-phosphate dehydrogenase.

G6PDH Abbreviation for glucose 6-phosphate dehydrogenase.

GPF Abbreviation for granulocytosis-promoting factor.

GPGG Abbreviation for guinea pig gamma globulin.

GP?G Abbreviation for guinea pig gamma globulin.

GPI Abbreviation for 1. glucose phosphate isomerase; 2. glucose-potassium-insulin; 3. glycosylphosphatidylinositol.

GPI-PLC Abbreviation for glycophosphatidyl inositol-specific phospholipase C.

GPI-PLD Abbreviation for glycophosphatidyl inositol-specific phospholipase D.

GPK Abbreviation for guinea pig kidney.

GPKC Abbreviation for guinea pig kidney cells.

GPLA Abbreviation for guinea pig lymphocyte antigen (major histocompatibility complex).

GPN Abbreviation for glycyl-L-phenylalanine naphthylamide.

GPO Abbreviation for glycerol 3-phosphate oxidase.

G-Protein A GTP-binding membrane protein that is capable of hydrolyzing GTP, activating membrane-bound cAMP, and mediating a variety of signal transducing systems.

GPT Abbreviation for glutamate pyruvate transaminase.

GPUT Abbreviation for galactose phosphate uridinyl transferase.

GPVI Abbreviation for platelet glycoprotein VI.

GPX Abbreviation for glutathione peroxidase.

G-Quartet A quadruple helix structure formed in both DNA and RNA that contains a cyclic H-bonded array of guanines.

GR Abbreviation for 1. glucocorticoid receptor; 2. glutathione reductase.

Graafian Follicle A fluid-filled vesicle in the mammalian ovary containing an oocyte attached to its wall.

Gradient Elution The removal of solute molecules from a chromatographic system (e.g., column) by means of a linear gradient salt solution or buffer solution.

Gradient Former A device used for preparation of linear density gradient medium in a column or tube for density gradient centrifugation or gradient electrophoresis.

Gradient Gel Electrophoresis Gel electrophoresis performed in a concentration gradient gel with progressively decreasing pore size.

Gradient Plate Technique A technique for isolating antibiotic-resistant bacterial mutants in which an agar plate containing concentration gradient of antibiotic is inoculated with testing bacteria.

Gradostat An apparatus used for continuous culture in which two different solutions flow simultaneously in opposite directions.

Graft vs. Host Reaction The disorder due to the transfer of immunocompetent cells from the donor into a nonhistoincompatible and immunodeficient recipient.

Grahamella A genus of Gram-negative bacteria (Bartonellaceae).

Graiacol (mol wt 124) An expectorant and intestinal disinfectant.

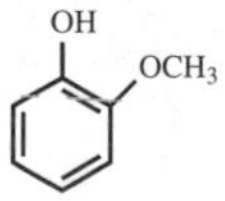

Gram A unit of weight equal to the mass of 1 cm^3 of water at 4° C.

Gram Atomic Weight The atomic weight expressed in gram units.

Gram Equivalent Weight The weight of a substance in grams that can release or combine with 1 gram of hydrogen or 8 grams of oxygen.

Gram Molecular Weight The molecular weight expressed in gram units.

Gram Negative See Gram stain.

Gram Positive See Gram stain.

Gram Stain A differential stain by which bacteria are classified as Gram (+) or Gram (–) depending upon whether they retain or lose the primary stain (e.g., crystal violet) when subjected to decolorization treatment. Bacteria that can be decolorized easily after primary staining are classified as Gram negative while those that can not be decolorized by decolorizing treatment are classified as Gram positive.

Gram Variable Bacteria that are Gram positive but tend to become Gram negative after culturing.

Gramicidin(s) A group of linear polypeptide antibiotics, e.g., gramicidin A, B, C, D, produced by *Bacillus brevis*, and active against Gram-positive bacteria by changing the ionic permeability of the bacterial membrane.

Gramicidin S A cyclic peptide ionophorous antibiotic produced by *Bacillus brevis* and an oxidative phosphorylation uncoupler.

Grana Plural of granum.

Grancalcin A calcium-binding protein found in neutrophils and monocytes.

Granisetron (mol wt 312) An anti-emetic agent.

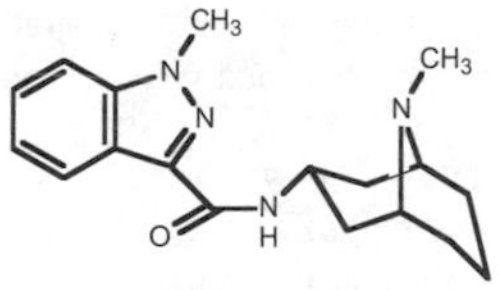

Granules The small intracellular particles that can be stained selectively.

Granuloblast An embryonic cell capable of developing into a granulocyte.

Granulocytes Leukocytes with distinct cytoplasmic granules, e.g., eosinophils, basophils and neutrophils.

Granulocytopenia The reduction in the numbers of circulating granulocytes in the blood.

Granulocytosis The presence of a large number of circulating granulocytes in the blood.

Granuloma 1. A tumor composed of granulation tissue. 2. An organized structure in mononuclear cells that is the hallmark of cell-mediated immunity.

Granulopoietin A glycoprotein derived from monocytes that controls the production of granulocytes by the bone marrow (colony-stimulating factor).

Granulosis An insect disease caused by granulosis virus.

Granulosis Virus A virus of the family Baculoviridae.

Granum A stack of chlorophyll-containing thylakoid disks within the chloroplast.

GRAS Abbreviation of *generally recognized as safe*.

Gratuitous Inducer 1. A substance that induces the synthesis of specific enzymes but itself is not a substrate. 2. An analog inducer in an operon system that acts as an inducer but can not be metabolized by the cell.

Gratuitous Induction The induction of inducible enzyme production by a nonnatural substrate or inducer.

Grave's Disease A disorder characterized by the enlargement of thyroid glands due to the production of excessive amounts of thyroid hormone.

Gravimeter A device used to measure specific gravity.

Gravol A trade name for dimenhydrinate, which inhibits nausea and vomiting.

Gray (Gy) A unit of absorbed radiation equal to 1 joule of energy absorbed by 1 kg of material.

Gray Crescent Region on the surface of the zygote at which invagination occurs at the onset of gastrulation in amphibian embryogenesis.

Gray Matter Tissue of vertebrate CNS containing numerous cell bodies, dendrites of nerve cells, and terminations of nerve fibers.

Grb2 A growth factor receptor-bound protein that binds to the activated epidermal growth factor and platelet-derived growth factor receptor.

GRE Abbreviation for glucocorticoid response element.

Green Bacteria Bacteria of the family Chlorobineae (also known as green sulfur bacteria).

Green Fluorescent Protein A protein that contains a highly fluorescent fluorophore.

Green House Effect The increase of global temperature resulting from the increase in the concentration of atmospheric carbon dioxide.

Grepafloxacin (mol wt 359) An antibiotic.

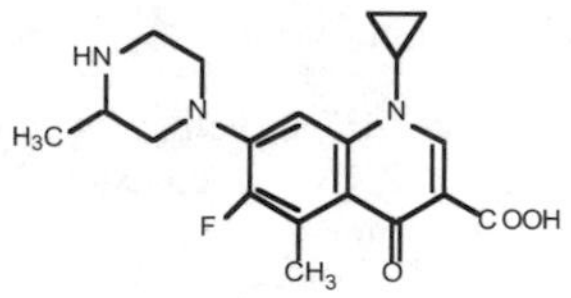

GRF Abbreviation for growth hormone-releasing factor, e.g., somatostatin.

GRH Abbreviation for growth regulatory hormone.

Grid A metal screen used for mounting specimens for electromicroscopy.

GRIF Abbreviation for growth hormone release-inhibiting factor.

Grifulvin A trade name for the antifungal agent griseofulvin.

Grippe Refering to influenza.

Grisactin A trade name for the antifungal drug griseofulvin.

Grisein A cyclic peptide antibiotic produced by *Streptomyces griseus* and active against Gram-negative bacteria.

Griseofulvin (mol wt 353) An antifungal antibiotic produced by *Penicillium griseofulvum* that disrupts mitotic spindles of fungal cells.

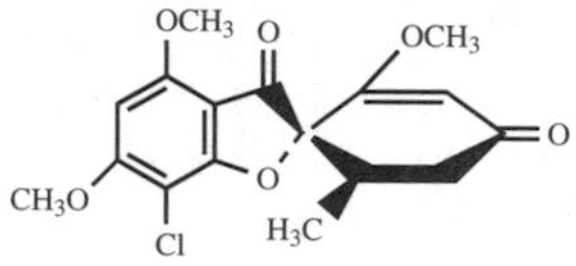

Griseoviridin (mol wt 478) An antibiotic produced by *Streptomyces griseus.*

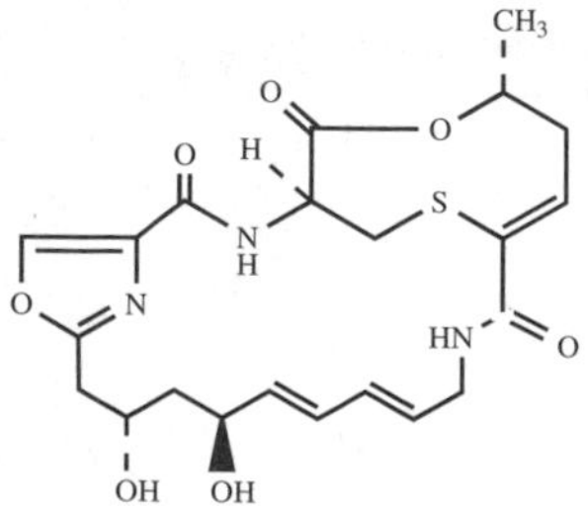

Gris-Peg A trade name for griseofulvin, an antifungal agent.

GRK Abbreviation for G-protein coupled receptor kinase.

GroEL Gene A gene in *E. coli* that encodes heat shock protein.

GroES Gene A gene in *E. coli* that encodes heat shock protein.

Groove A narrow, elongated depression, e.g., groove occurring in double-stranded DNA.

Ground State 1. The normal, stable form of an atom or a molecule. 2. The stage of a cell in the absence of activation of a developmental regulating signal.

Ground Substance The mucopolysaccharide-containing matrix of connective tissue.

Group Activation The transfer of a high-energy group, e.g., pyrophosphate group, from one compound to another.

Group Transfer Potential The ability of a compound to donate an activated group (e.g., phosphate group or acyl group).

Growth Curve A plot of cell number in a culture as a function of time.

Growth Factor Any of a number of highly specific proteins that stimulate cell division in particular types of mammalian cells, e.g., platelet-derived growth factor (PDGF), epidermal growth factor (EGF), nerve growth factor (NGF), and interleukin-2 (IL-2).

Growth Factor Receptor A membrane-spanning protein that selectively binds with its growth factor and transduces a cellular signal (e.g., signal for cell division).

Growth Hormone The hormones secreted by the anterior lobe of the pituitary gland that stimulate body growth.

Growth Substance Referring to plant growth hormone, e.g., auxin.

Growth Vitamin Referring to vitamin A.

GRP Abbreviation for gastrin-releasing protein.

GRP78 Abbreviation for glucose-regulated protein of 78 kDa.

GRPS Abbreviation for glucose-ringer phosphate solution.

GS Abbreviation for 1. galactosialidosis; 2. guanidino specificity; 2. glycogen synthetase.

GSA Abbreviation for 1. glutamate 1-semialdehyde; 2. guanidinosuccinic acid.

GSA-AT Abbreviation for glutamate 1-semialdehyde aminotransferase.

GSBG Abbreviation for gonadal steroid-binding globulin.

GSBP Abbreviation for gonadal steroid-binding protein.

GSC Abbreviation for gas-solid chromatography.

GSD Abbreviation for glycogen storage disease.

GseI (AsuI) A restriction endonuclease from *Gloeocapsa* species with the same specificity as AsuI.

GseII (PstI) A restriction endonuclease from *Gloeocapsa* species with the same specificity as PstI.

GseIII (BamHI) A restriction endonuclease from *Gloeocapsa* species with the same specificity as HamHI.

GSH Abbreviation for glutathione.

GSK Abbreviation for glycogen synthetase kinase.

GSK3b Abbreviation for glycogen synthetase kinase 3b.

GSL Abbreviation for glycosphingolipid.

GSP Abbreviation for gene-specific primer.

GspI (PvuII) A restriction endonuclease from *Gleothece* species with the same specificity as PvuII.

GspAI (AvaII) A restriction endonuclease from *Gloeotricia* species with the same specificity as AvaII.

GspAII (MstI) A restriction endonuclease from *Gloeotricia* species with the same specificity as MstI.

Gs Protein A type of G protein that regulates the activity of adenylate cyclease.

GSSG Abbreviation for oxidized form of glutathione.

GST Abbreviation for glutathione S-transferase.

GST-RAP Abbreviation for glutathione S-transferase receptor-associated protein.

GsuI A restriction endonuclease from *Gluconobacter suboxydans* H-15T with the following specificity:

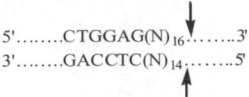

$$\begin{array}{l} 5'\ldots\ldots\text{CTGGAG(N)}_{16}\downarrow\ldots\ldots 3' \\ 3'\ldots\ldots\text{GACCTC(N)}_{14}\uparrow\ldots\ldots 5' \end{array}$$

GT Abbreviation for 1. galactosyl transferase; 2. glucose tolerance; 2. glutamyl transpeptidase.

GTA Abbreviation for gene transfer agent.

GT-AG Rule The rule states that the intron junctions start with GT and end with AG.

GTBP Abbreviation for G/T mismatch-binding protein.

GTE Abbreviation for glucose-tris-EDTA buffer.

GTF Abbreviation for 1. glucose tolerance factor; 2. gonadotropic factor.

GTH Abbreviation for gonadotropic hormone.

GTP Abbreviation for guanosine 5′-triphosphate.

dGTP Abbreviation for deoxyriboguanosine triphosphate, a nucleotide for the synthesis of DNA.

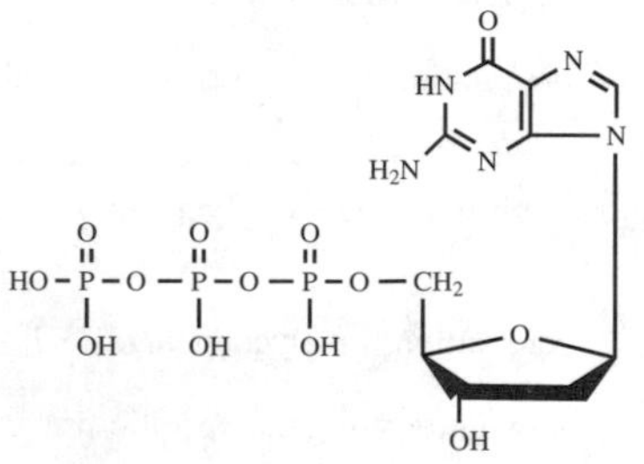

GTP mRNA Guanylyltransferase The systematic name for mRMA guanylyltransferase.

GTP Pyrophosphokinase The enzyme that catalyzes the following reaction:

ATP + GTP
⇅
AMP + guanosine 3′-diphosphate 5′-triphosphate

GTPase Abbreviation for guanosine triphosphatase, the enzyme that catalyzes the following reaction:

$$GTP + H_2O \rightleftharpoons \text{Guanosine} + \text{triphosphate}$$

dGTPase Abbreviation for deoxyguanosine triphosphatase, the enzyme that catalyzes the following reaction:

$$dGTP + H_2O \rightleftharpoons \text{Deoxyguanosine} + \text{triphosphate}$$

GTP[S] Abbreviation for guanosine 5'-(γ-thio)triphosphate.

GTP[βS] Abbreviation for guanosine 5'-[β-thio]triphosphate.

GTP[γS] Abbreviation for guanosine 5'-[γ-thio]triphosphate.

GTRH Abbreviation for gonadotropin-releasing hormone.

GTT Abbreviation for glucose tolerance test.

GUA A genetic code or codon for the amino acid valine.

Gua Abbreviation for guanine.

GU-AG Rule The rule states that an intron starts with GU at the 5′-terminal and ends with AG at the 3′-terminal.

Guaiacol (mol wt 124) An expectorant.

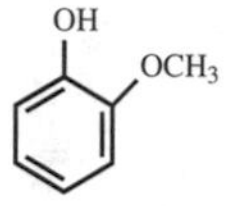

Guaiacol Carbonate (mol wt 274) An expectorant.

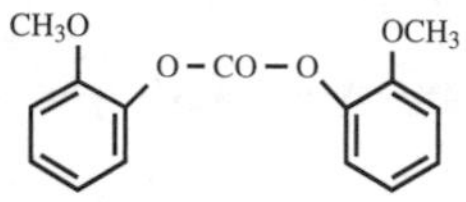

Guaiactamine (mol wt 223) A spasmolytic agent.

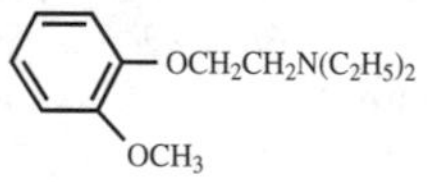

Guaiapate (mol wt 323) An antitussive agent.

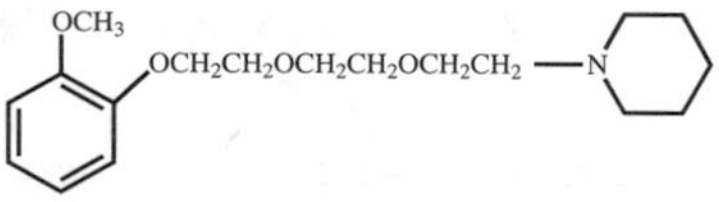

Guaiazulene (mol wt 198) An anti-inflammatory agent.

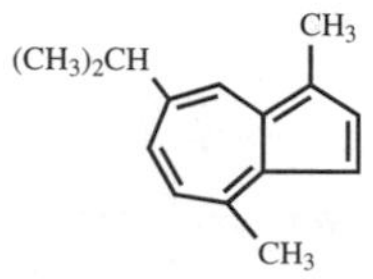

Guaifed A trade name for a combination drug containing pseudoephedrine hydrochloride and guaifenesin.

Guaifenesin (mol wt 198) An expectorant that increases production of respiratory-tract fluid to reduce viscosity of tenacious secretions.

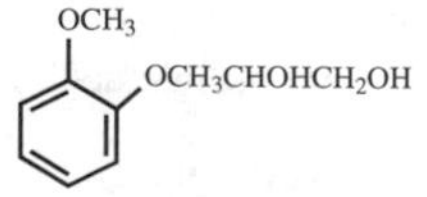

Guamecycline (mol wt 627) An antibacterial agent.

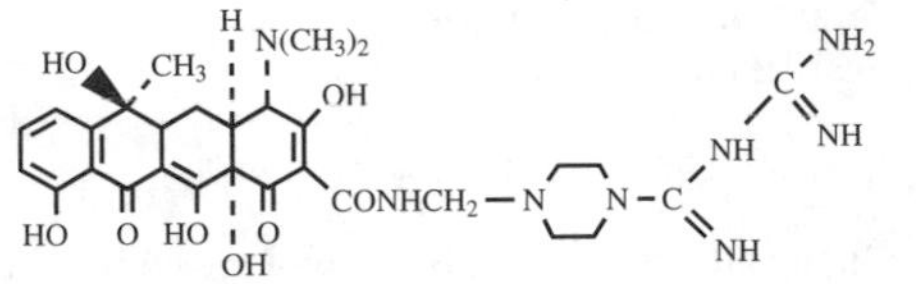

Guanabenz (mol wt 231) An antihypertensive agent that inhibits the central vasomotor center, thereby decreasing sympathetic outflow.

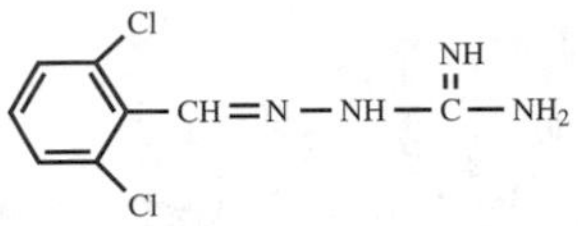

Guanacline (mol wt 182) An antihypertensive agent.

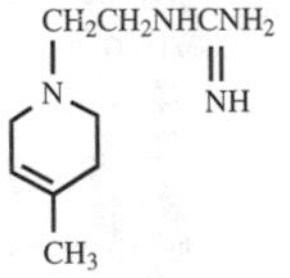

Guanadrel (mol wt 213) An antihypertensive agent that prevents release of norepinephrine from the adrenergic nerve ending.

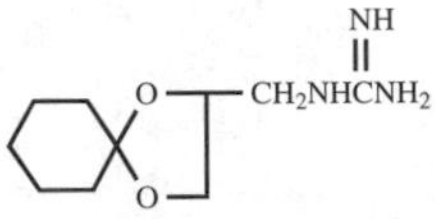

Guanase See guanine deaminase.

Guanazodine (mol wt 184) An antihypertensive agent.

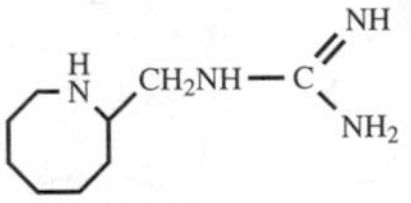

Guanethidine (mol wt 198) An antihypertensive agent.

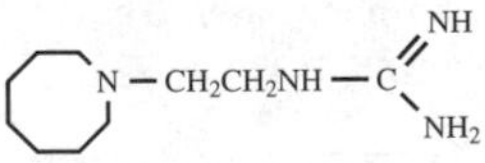

Guanfacine (mol wt 246) An antihypertensive agent that stimulates CNS a_2 adrenoreceptors.

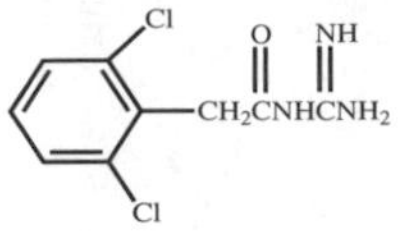

Guanidine (mol wt 59) A compound from turnip, mushroom, corn germ, rice hulls, and earthworms with antiviral, antipyretic, and antifungal activity. It also possesses muscle stimulatory activity.

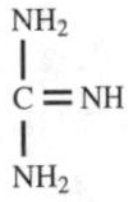

Guanidine Hydrochloride (mol wt 96) An ionic denaturant that rapidly and effectively denatures most proteins.

NH_2-C=NH-NH_2. HCl

Guanidine Thiocyanate (mol wt 118) A protein denaturant.

H_2-C=NH-NH_2. HSCN

Guanidinium Referring to $[C(NH_2)_3]^+$, a cation.

Guanidino Referring to –HN-(C=NH)-NH_2 group.

Guanidinoacetase The enzyme that catalyzes the following reaction:

Guanidinoacetate + H_2O $\rightleftharpoons$ Glycine + urea

Guanidinoacetate Kinase The enzyme that catalyzes the following reaction:

Guanidinoacetate + ATP $\rightleftharpoons$ ADP + phosphoguanidinoacetate

Guanidinoacetate *N*-Methyltransferase The enzyme that catalyzes the following reaction:

S-Adenosyl-L-methionine + guanidinoacetate $\rightleftharpoons$ *S*-Adenosyl-L-homocysteine + creatine

Guanidinobutyrase The enzyme that catalyzes the following reaction:

4-Guanidinobutanoate + H_2O $\rightleftharpoons$ 4-Aminobutanoate + urea

Guanidinium Group See guanido group.

Guanidino Group See guanido group.

Guanidinopropionase The enzyme that catalyzes the following reaction:

3-Guanidinopropanoate + H_2O $\rightleftharpoons$ β-Alanine + urea

Guanido Group The group that is present in the amino acid arginine.

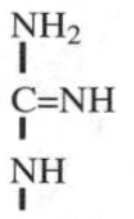

Guanine (mol wt 15) A constituent base in nucleic acids.

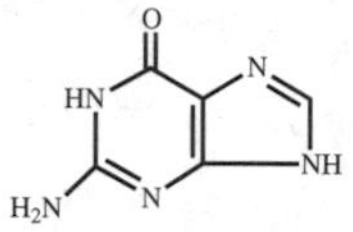

Guanine Aminase See guanine deaminase.

Guanine Nucleotide Binding Protein See G protein.

Guanochlor (mol wt 263) An antihypertensive agent.

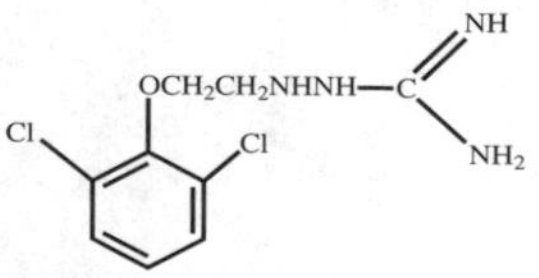

Guanosine (mol wt 283) A nucleoside and constituent of nucleotides.

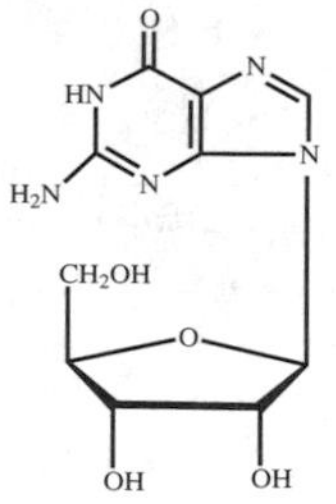

Guanosine 3′,5′-Cyclic Monophosphate (mol wt 345) A cyclic ribonucleotide and intracellular messenger participating in a variety of intracellular signaling processes (abbreviated as cGMP).

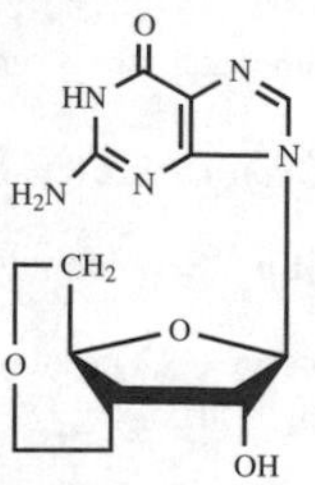

Guanosine Deaminase The enzyme that catalyzes the following reaction:

$$\text{Guanine} + H_2O \rightleftharpoons \text{Xanthine} + NH_3$$

Guanosine Diphosphatase The enzyme that catalyzes the following reaction:

$$\text{GDP} + H_2O \rightleftharpoons \text{GMP} + \text{Pi}$$

Guanosine 5¢-Diphosphate (mol wt 443) The diphosphate form of guanosine nucleotide.

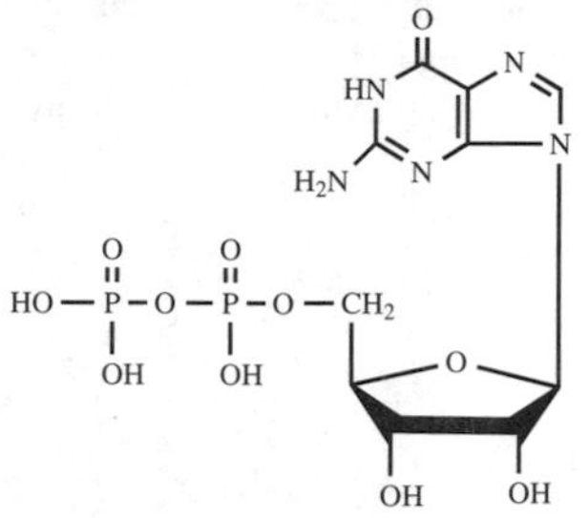

Guanosine 3′-Monophosphate (mol wt 363) The monophosphate form of guanosine nucleotide.

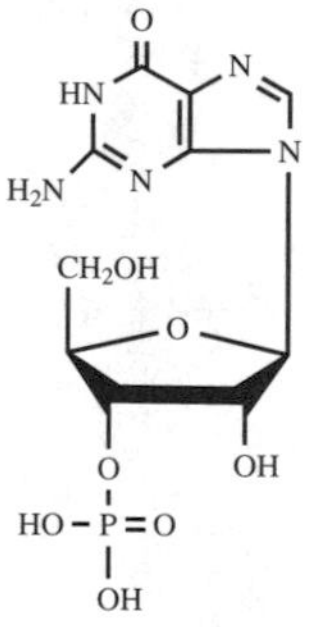

Guanosine 5′-Monophosphate (mol wt 363) The monophosphate form of guanosine nucleotide.

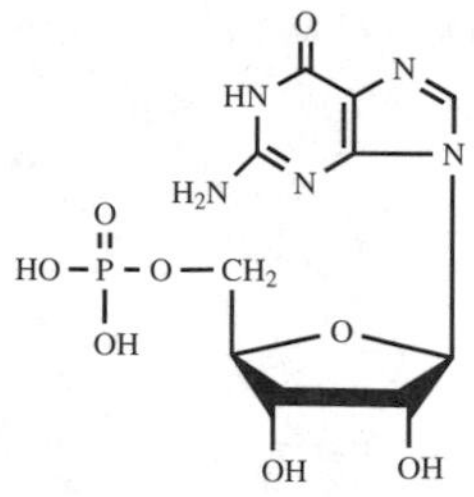

Guanosine Phosphorylase The enzyme that catalyzes the following reaction:

$$\text{Guanosine} + \text{Pi} \rightleftharpoons \text{Guanine} + \text{ribose 1-phosphate}$$

Guanosine 5′-Triphosphate (GTP, mol wt 672) The triphosphate form guanosine nucleotide.

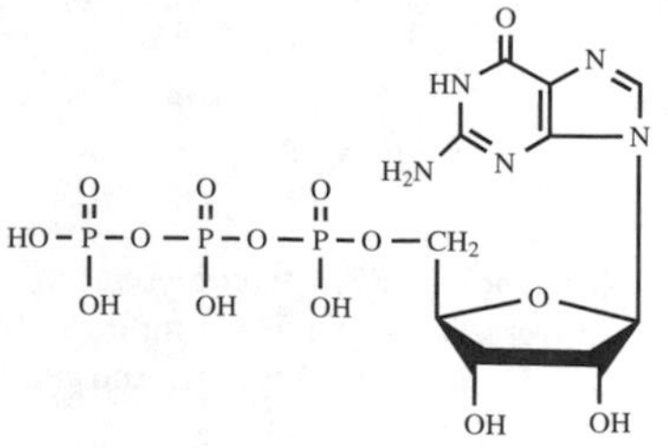

Guanoxabenz (mol wt 212) An antihypertensive agent.

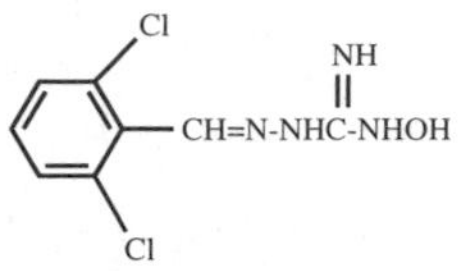

Guanoxan (mol wt 207) An antihypertensive agent.

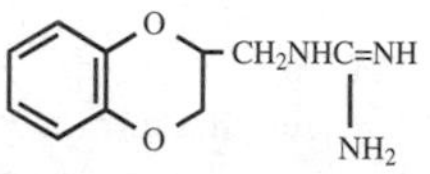

Guanylate Cyclase The enzyme that catalyzes the following reaction:

$$GTP + H_2O \rightleftharpoons \text{Cyclic GMP} + PPi$$

Guanylate Kinase The enzyme that catalyzes the following reaction:

$$ATP + GMP \rightleftharpoons ADP + GDP$$

Guanylic Acid (mol wt 363) The monophosphate form of guanosine nucleotide; may be in either guanosine 3′-monosphate or guanosine 5′-monophosphate form.

Guanylin A pentapeptide hormone from rat jejunum that activates the guanylate cyclase.

Guanyloribonuclease Synonym of ribonuclease T2.

Guaran A polysaccharide from endosperm of guar seeds.

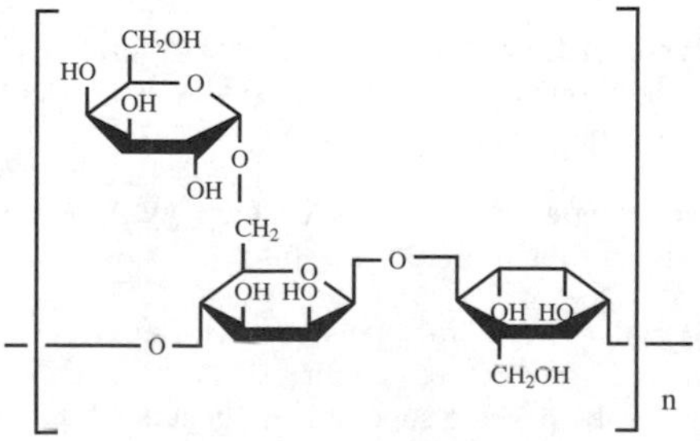

Guard Cells Two cells surrounding a stoma in the epidermis of a plant leaf to regulate the opening and closing of the stoma.

Guarnieri Bodies A type of inclusion bodies that are formed in the cytoplasm of cells infected with vaccinia or cowpox virus.

GUC A genetic code or codon for the amino acid valine.

GUG A genetic code or codon for the amino acid valine.

GuHCl Abbreviation for guanidine hydrochloride.

Guiatus A trade name for guaifenesin, an antitussive agent that increases the production of respiratory-tract fluids to reduce tenacious secretions.

Guillain-Barre Syndrome An acute disease caused by cytomegalovirus infection.

Guinea Green B (mol wt 691) A dye.

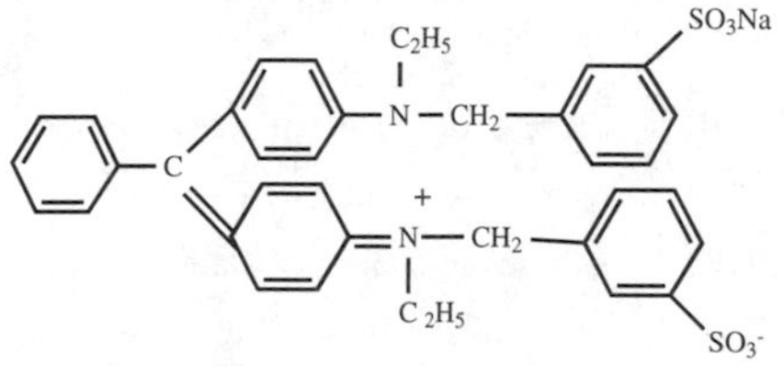

Gulonate 3-Dehydrogenase The enzyme that catalyzes the following reaction:

$$\text{L-Gulonate} + NAD^+ \rightleftharpoons \text{3-Dehydro-L-gulonate} + NADH$$

Gulonic Acid (mol wt 196) A sugar acid derived from gulose.

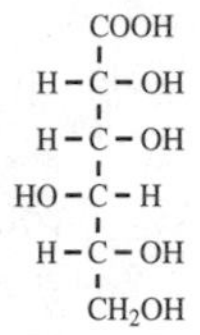

Gulono-γ-Lactone (mol wt 178) A derivative from gulose.

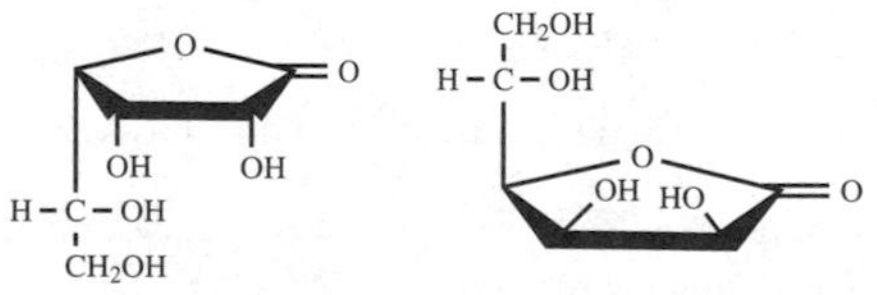

D-gulono-γ-lactone L-gulonic-γ-lactone

Gulose (mol wt 180) A six-carbon aldose.

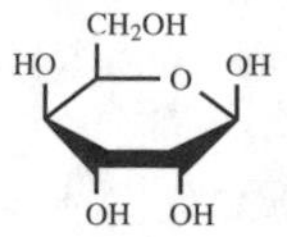

Gum 1. Viscous sap exuded by some plants. 2. Substance that forms a sticky gel or mucilage.

Gum Arabic A branched polymer of galactose, rhamnose, arabinose,and glucuronic acid in the form of calcium, magnesium, and potassium salts.

Gum Xanthan A hydrophilic colloid produced by the fermentation of dextrose with *Xanthomonas campestris.* It is a polymer of glucose, mannose, potassium glucuronate, acetate, and pyruvate.

Guo Abbreviation for guanosine.

Guo2'P Abbreviation for guanosine 2'-phosphate.

Guo2'3'P Abbreviation for guanosine 2',3'-phosphate.

Guo3'P Abbreviation for guanosine 3'-phosphate.

Guo3'5'P Abbreviation for guanosine 2',3'-phosphate.

Guo5'P Abbreviation for guanosine 5'-phosphate.

Guo5'PP Abbreviation for guanosine 5'-diphosphate.

Guo5'PPP Abbreviation for guanosine 5'-triphosphate.

Guo5'PPPS Abbreviation for guanosine 5'-γ-thiotriphosphate.

Guo5'PPS Abbreviation for guanosine 5'-γ-thiodiphosphate.

GuoPPMan Abbreviation for guanosine diphosphomannose.

GuoPP[NH$_2$]P Abbreviation for guanosine 5'-[β,γ-methylene] triphosphate.

GuoPP[NH]P Abbreviation for guanosine 5'-[β,γ-imido] triphosphate.

GuSCN Abbreviation for guanidine thiocyanate.

Gustalac A trade name for calcium carbonate, used as an antacid.

Guthrie Test A test for phenylketouria employing strains of bacteria that cannot grow in the absence of phenylalanine.

GUU A genetic code or codon for the amino acid valine.

Guvacine (mol wt 127) A growth factor for *Staphylococcus aureus* and *Proteus vulgaris* from betel nut, the seed of *Areca catechu.*

GV Abbreviation for gentian violet.

GVBD Abbreviation for germinal vesicle breakdown.

GVH Abbreviation for graft vs. host reaction.

GVHD Abbreviation for graft-vs.-host disease.

GVHR Abbreviation for graft-vs.-host reaction.

G-Well A trade name for lindane, an antiparasitic agent.

GYC Agar A type of medium containing glucose, yeast extract, and calcium carbonate for growing bacteria.

Gymnosperm A plant whose seeds do not develop within an ovary (e.g., conifers).

Gynandromorph An organism that exhibits a mosaic of tissues of male and female genotype (also known as sex mosaic).

Gynecology The science that deals with the diseases and hygiene of women.

Gynecort A trade name for hydrocortisone acetate, a corticosteroid.

Gyne-Lotrimin A trade name for clotrimazole, an antifungal agent that alters fungal cell wall synthesis.

Gynergen A trade name for ergotamine tartrate, an a-adrenergic blocking agent that inhibits the effects of epinephrine and norepinephrine.

Gynodioey A sexual dimorphism in plants.

Gynodiol A trade name for estradiol, a hormone.

Gynogen LA A trade name for estradiol valerate, a hormone that increases synthesis of DNA, RNA, and protein in responsive tissues.

gyrA Gene A gene that encodes gyrA protein that is a subunit of DNA gyrase.

Gyrase A type of topoisomerase that promotes unwinding of closed circular DNA helix and removes the positive superhelicity generated during the process of replication by introducing negative supercoils ahead of the replication fork.

gyrB Gene A gene that encodes gyrB protein that is a subunit of DNA gyrase.

H

h Symbol for Planck's constant equal to 6.626×10^{-27} erg-sec.

H Abbreviation for 1. amino acid histidine; 2. hydrogen.

^{2}H Abbreviation for deuterium.

^{3}H Abbreviation for tritium, an isotope of hydrogen, a weak beta emitter used for labeling substances.

H Antigen Referring to 1. the precursor antigen of A, B, O blood group system; 2. the bacterial flagellar antigen.

H-2 Antigen Referring to an antigen encoded by the H-2 region of the major histocompatibility complex in the mouse.

H Chain Referring to the heavy chain of immunoglobulin (see heavy chain).

H-2 Complex Referring to the major histocompatibility complex in house that encodes class I, II, and III histocompatibility antigens.

H1 Receptor Referring to histamine receptors on the mammalian cells.

H2 Receptor Referring to histamine receptors on the mammalian cells.

H Strand 1. The strand in the double-stranded DNA that has a high buoyant density due to its high G-C content. 2. The strand of a polynucleotide that has been labeled with a heavy isotope.

H Substance The precursor substance for blood group antigens A and B.

H Zone The central, less-dense portion of the A band of the myofibrils of striated muscle.

HA Abbreviation for 1. hemadsorbent; 2. hemagglutinating activity; 3. hemaglutinin; 3. hemolytic anemia; 4. hepatitis A; 5. hyaluronic acid.

H2A Abbreviation for H2A histone.

HAA Abbreviation for 1. hemolytic anemia antigen; 2. hepatitis-associated antigen.

HA-Ab Abbreviation for hepatitis A antibody.

HA-Ag Abbreviation for hepatitis A antigen.

Habitrol A trade name for nicotine, a smoking deterrent.

Habituation The acquired ability of a population of cells to grow and divide independently without an exogenous supply of growth regulators.

HAC Abbreviation for a combination drug containing hexamethylmelamine, adriamycin, and cyclophosphamide.

HAc Abbreviation for acetic acid.

HacI (MboI) A restriction endonuclease from *Halococcus acetoinfaciens* with the following specificity:

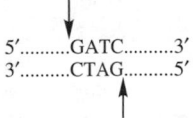

```
       ↓
5′.........GATC.........3′
3′.........CTAG.........5′
               ↑
```

HAc-Ab Abbreviation for hepatitis A core antibody.

HAc-Ag Abbreviation for hepatitis A core antigen.

hAChE (HAChE) Abbreviation for human acetylcholine esterase.

HAChT Abbreviation for high affinity choline transport.

HAD Abbreviation for 1. a combination drug containing hexamethylmelamine, adriamycin, and DDP; 2. haloacid dehydrogenase; 3. 3-hydroxyacyl-CoA dehydrogenase.

HAd Abbreviation for 8-hydroxyadenine.

Hadacidin (mol wt 119) An antitumor antibiotic and analog of aspartic acid, produced by *Penicillium frequentans.*

```
       OH
       |
OHC — NCH2COOH
```

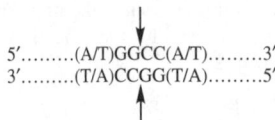

HAE Abbreviation for hereditary angioedema.

HaeI A restriction endonuclease from *Haemophilus aegyptius* with the following specificity:

```
                 ↓
5′.........(A/T)GGCC(A/T).........3′
3′.........(T/A)CCGG(T/A).........5′
                 ↑
```

HaeII A restriction endonuclease from *Haemophilus aegyptius* with the following specificity:

5′..........PuGCGCPy..........3′
3′..........PyCGCGPu..........5′

HaeIII A restriction endonuclease from *Haemophilus aegyptius* with the following specificity:

5′..........GGCC..........3′
3′..........CCGG..........5′

HaeIII Methylase A restriction methylase that methylates the internal cytosine residues of the following sequence:

CH_3

5′..........GGCC..........3′
3′..........CCGG..........5′

CH_3

Haem Variant spelling for heme.

Haemadin An anticoagulant peptide from Indian leech *Haemadipsa sylvestris*.

Haemlbartonella A genus of Gram-negative bacteria of the family Anaplasmataceae.

Haemophilus A genus of pathogenic Gram-negative bacteria (Pasteurellaceae).

Haemophilus influenzae Bacterium that causes pneumonia and meningitis.

Hafnia A genus of Gram-negative bacteria (Enterobacteriaceae).

Hafnium (Hf) A chemical element with atomic weight 178, valence 4.

Hageman Factor An intrinsic blood clotting factor (factor XII) that activates plasma thromboplastin antecedent.

HAGG Abbreviation for hyperimmune anti-variola gamma globulin.

HAHTG Abbreviation for horse anti-human thymus globulin.

HAI Abbreviation for 1. hemagglutinating inhibition; 2. hemagglutinin inhibition.

HAI Test Abbreviation for hemgglutination inhibition test, a method for identification of hemagglutinating viruses.

Hair Cells 1. Cells that are found in the epithelial lining of the labyrinth of the inner ear. 2. Hair-like cells that cover a plant surface.

Hairpin Structure A self-pairing double helical region on a single-stranded DNA or RNA molecule.

Halazepam (mol wt 353) An anxiolytic agent.

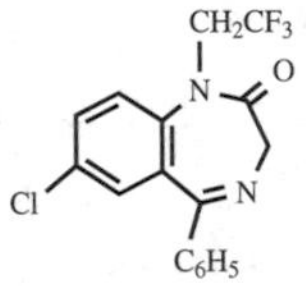

Halciderm A trade name for halcinonide, an anti-inflammatory agent

Halcinonide (mol wt 455) An anti-inflammatory agent.

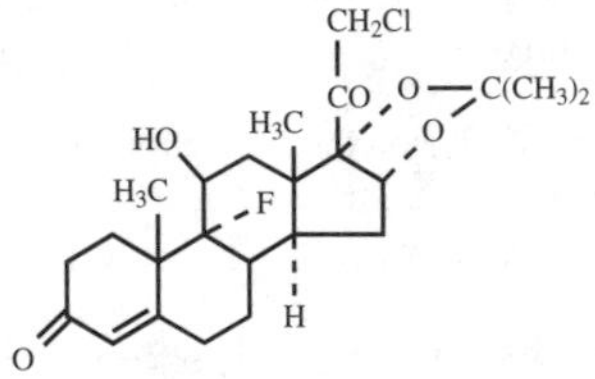

Halcion A trade name for triazolam, an hypnotic agent.

Haldane Effect The release of protons resulting from the oxygenation of hemoglobin.

Haldol A trade name for haloperidol, a tranquilizer.

Haldrone A trade name for paramethasone acetate, an anti-inflammatory agent.

Halenol A trade name for acetaminophen, an antipyretic and analgesic agent.

Halethazole (mol wt 361) An antiseptic and antifungal agent.

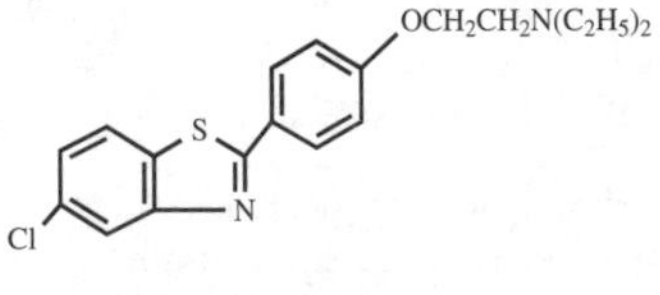

Half-Cell Experiment An experimental device for determination of redox potential of the redox couple. It consists of two half-cells (sample half-cell and standard hydrogen half-cell).

Half-Ester Any monoester of a dibasic acid.

Half-Life The time required for completion of half of a defined activity, reaction, or decay.

Halfprin A trade name for aspirin, an antipyretic and analgesic agent.

Half Reaction See half-cell.

Halides Derivatives of halogens or salts derived from halogens.

Halimide (mol wt 354) A cationic surface active agent.

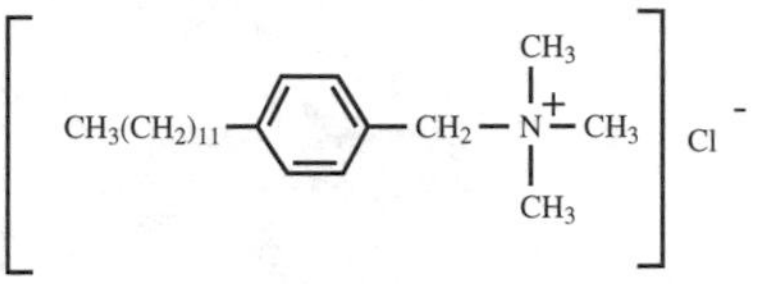

Haliscomenobacter A genus of Gram-negative, rod-shaped bacteria.

Hallucination Perception of objects or events that do not exist.

Hallucinogen Any substance that is capable of inducing hallucination.

Halo- A prefix denoting 1. containing halogen; 2. related to the sea or salt.

Haloacetate Dehalogenase The enzyme that catalyzes the following reaction:

Haloacetate + H_2O ⇌ Glycolate + halide

Haloacid Dehalogenase The enzyme that catalyzes the following reaction:

Haloacid + H_2O ⇌ Formaldehyde + bromide + chloride

Haloamine Halogen derivative of ammonia, e.g., chloroamine ($ClNH_2$).

Halobacteria Bacteria that live in conditions of high salinity (e.g., *Halobacterium halobium*).

Halobacterium halobium Photosynthetic bacterium that has patches of purple membrane containing the pigment bacteriorhodopsin.

Halobetasol Proprionate (mol wt 485) An anti-inflammatory agent.

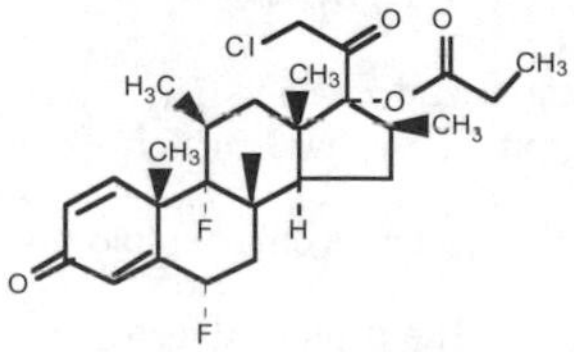

Halofantrine (mol wt 500) An antimalarial agent.

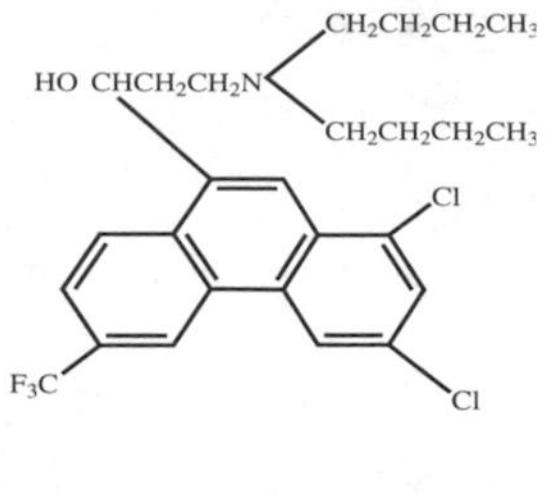

Halofed A trade name for pseudoephedrine hydrochloride, an adrenergic agent that stimulates alpha receptors in the respiratory tract producing vasoconstriction.

Halofuginone (mol wt 415) An antiprotozoal agent.

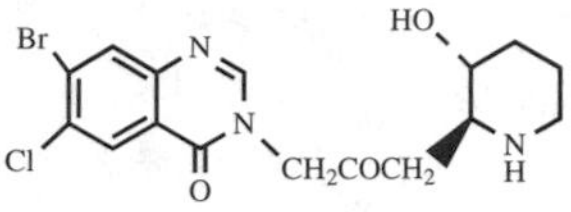

Halog A trade name for halcinonide, an anti-inflammatory agent.

Halogen Referring to any of the elements of fluorine, chlorine, bromine, and iodine.

Halogenated Hydrocarbon Synthetic organic compound containing one or more atoms of the halogen elements (e.g., chlorine, fluorine, and bromine).

Halogenation 1. Combination of a halogen molecule with microbial cell wall to inhibit or destroy bacteria. 2. Introduction of halogen atom(s) into an organic molecule.

Halometasone (mol wt 445) An anti-inflammatory and antipuritic agent.

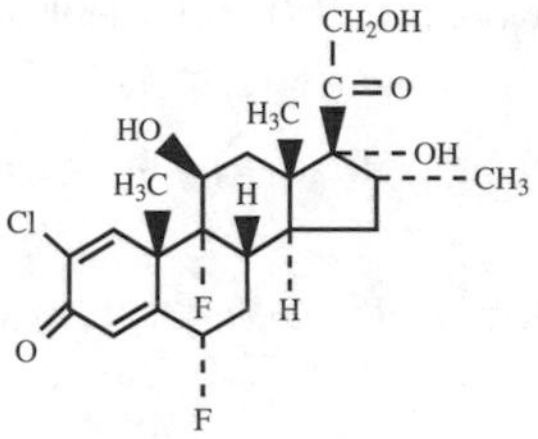

Halomethanes The chlorinated organic compounds formed during the process of chlorination of water.

Halomonas A genus of aerobic, facultative anaerobic, chemoorganotrophic, Gram-negative bacteria.

Haloperidol (mol wt 376) A tranquilizer that blocks postsynaptic receptors in the brain.

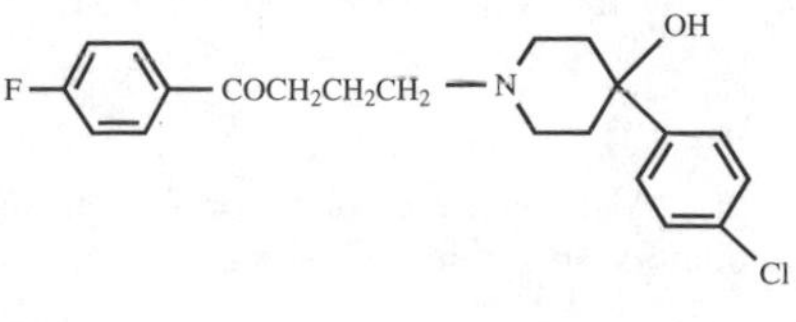

Haloperon A trade name for haloperidol, a tranquilizer.

Halophile An organism that requires a high concentration of NaCl for growth.

Halophyte A plant that can withstand high salinity.

Halopredone Acetate (mol wt 559) An anti-inflammatory agent.

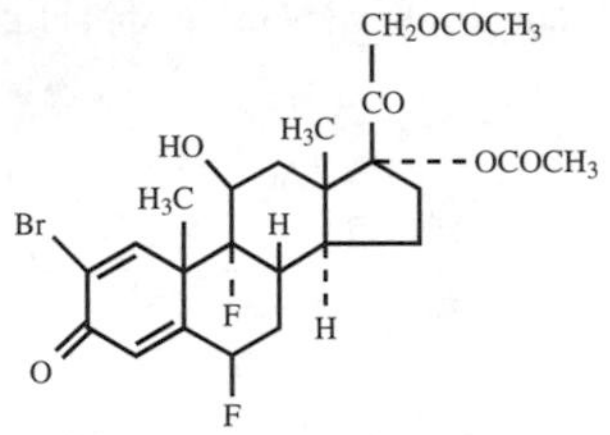

Haloprogin (mol wt 361) An antifungal and antibacterial agent.

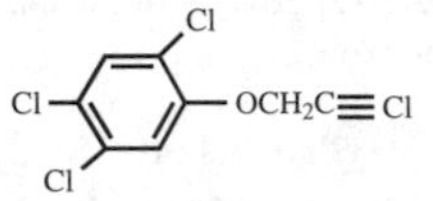

Halorhodopsin A protein pigment occurring in the purple membrane of *Halobacterium solanarium* that is involved in ion translocation across membranes.

Halotestin A trade name for fluoxymesterone, an androgen hormone (an analog of testosterone).

Halotex A trade name for haloprogin, an antibacterial and antifungal agent.

Halothane (mol wt 197) An anesthetic agent.

$$C_2HBrClF_3$$

Halotussin A trade name for guaifenesin, an expectorant.

Haloxazolam (mol wt 377) A sedative and hypnotic agent.

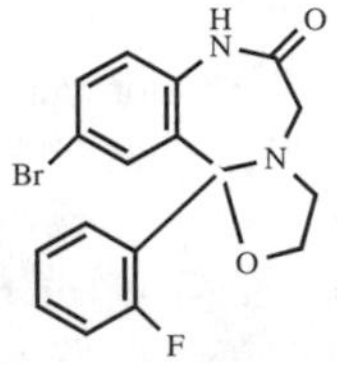

Haloxon (mol wt 416) An anthelmintic agent.

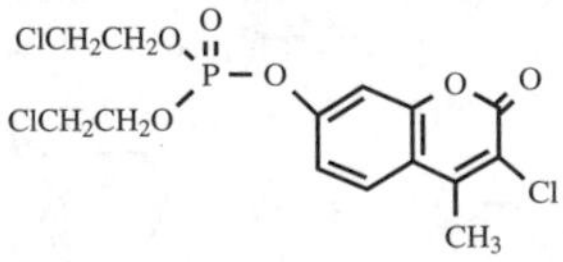

Haltran A trade name for ibuprofen, an anti-inflammatory and antipyretic drug.

HALV Abbreviation for human AIDS-lymphotrophic virus.

HAM Abbreviation for a combination drug containing hexamethylmelamine, adriamycin, and methotrexate.

HAMA Abbreviation for human anti-mouse antibody.

Hamamelose (mol wt 180) A six-carbon aldosugar.

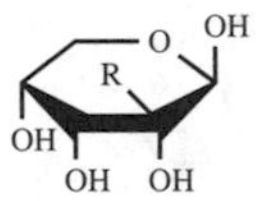

R = CH_2OH

Hamamelose Kinase The enzyme that catalyzes the following reaction:

$$\text{ATP} + \text{D-hamamelose} \rightleftharpoons \text{ADP} + \text{D-hamamelose 2-phosphate}$$

Hamartoma Tumorlike but nonneoplastic overgrowth of tissue.

Hanes Plot A plot of [S]/*v* vs. [S] in analysis of enzyme kinetic data in which [S] is substrate concentration and *v* is the velocity of the enzyme reaction.

Hanks' BSS Abbreviation for Hanks' balanced salt solution, used in tissue culture for maintenance of cell viability.

HapII (HpaII) A restriction endonuclease from *Haemophilus aphrophilus* with the following specificity:

Haploid A cell that has only one copy of each type of chromosome.

Haploidization Production of haploid from a diploid through chromosome loss.

Haploid Number The gametic chromosome number (symbolized by N).

Haploid Parthenogenesis The development of a hapoid from an egg without fertilization.

Haplont A plant with only haploid somatic cells.

Haplophase The haploid stage of the life cycle.

Haplosis The reduction of the chromosome number to half during meiosis.

Haplotype Set of closely linked alleles on a specific chromosome carried by an individual and inherited as a unit, e.g., alleles of the major histocompatibility complex on chromosome number 6 in humans.

Haptenic Determinant The hapten on a carrier molecule that is responsible for its antigenicity.

Haptens Small nonantigenic molecules that are capable of stimulating an immune response (e.g., production of antibody) when chemically coupled to a large protein carrier.

Hapto- A prefix denoting ability to bind or combine.

Haptocorrin A blood cobalamin-binding glycoprotein.

Haptoglobin The α_2-plasma glycoprotein that binds oxyhemoglobin.

Haptomer Any substance that can interact with the cell membrane or bind to the effectomer.

Harborage Transmission A mode of transmission in which the infectious microorganism does not undergo morphological and physiological changes in the vector.

Hard Soap The sodium salt of the long-chain fatty acid.

Hard Water Water that contains high concentrations of calcium, magnesium, and iron.

Harlequin Chromosome A condition in which a pair of sister chromatids that stain differently so that one appears dark and the other light.

Harmine (mol wt 212) An inhibitor of monoamine and diamine oxidase.

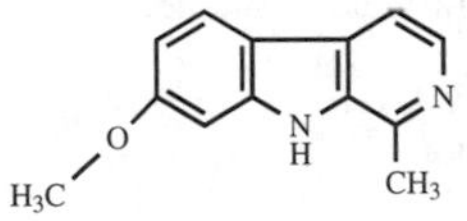

Harnonyl A trade name for deserpidine, an antihypersensitive agent.

Harris-Ray Test A test for vitamin C that uses 2,6-dichlorophenol indophenol.

Hartnup Disease A genetic disorder of amino acid transport characterized by the excessive loss of monoamino monocarboxylic acids in the urine and marked by a pellagra-like rash upon exposure to sunlight.

Harvey Murine Sarcoma Virus A replication-defective, v-onc$^+$, fibroblast-transforming murine sarcoma virus.

HAsAb Abbreviation for hepatitis A surface antibody.

HAsAg Abbreviation for hepatitis A surface antigen.

Hashimoto's Disease An immune disorder characterized by the development of goiter in the thyroid gland.

HAST or hAST Abbreviation for human arylsulphotransferase.

hAT1 Abbreviation for human angiotension II type 1 receptor.

hAT2 Abbreviation for human angiotension II type 2 receptor.

HAT Medium Abbreviation for hypoxanthine-aminopterin-thymidine medium, used for selection of hybrid cells in monoclonal antibody production.

Hatch-Slack-Kortshak Pathway See Hatch-Slack pathway.

Hatch-Slack Pathway A pathway in C_4 plants in which carbon dioxide is fixed into oxaloacetate (a four carbon compound) in the mesophyll cells and then transported into vascular boundle sheath cells where the CO_2 is released and refixed into sugar by the C_3 pathway.

H^+-ATPase A type of ATPase (ATP synthetase) that requires no other cations for activation.

HAV Abbreviation for hepatitis A virus.

Haverhill Fever An infection caused by *Streptobacillus moniliformis*.

Havrix A trade name for inactivated hepatitis A vaccine.

HA-VSMC Abbreviation for human aortic vascular smooth muscle cell.

Haworth Projection A model that depicts the perspective representation of the cyclic forms of sugars (e.g., pyran and furan form).

Hay Fever A type of allergy involving the upper respiratory tract. It is an IgE-mediated type I hypersensitivity.

Hb Abbreviation for hemoglobin.

HB Abbreviation for 1. homogenization buffer; 2. hepatitis B.

H2B Abbreviation for H2B histone.

HbA Abbreviation for adult hemoglobin.

HbA_{1c} See hemoglobin A_{1c}.

HbA_2 A minor hemoglobin in adults (about 2% of the total hemoglobin).

hBABP Abbreviation for human bile acid binding protein.

HB-Ag Abbreviation for hepatitis B antigen.

HbC Abbreviation for hemoglobin C. A type of hemoglobin in which glutamate at position 6 has been replaced by lysine in the beta chain.

HBc Abbreviation for hepatitis B core antigen or core antigen of hepatitis B.

HBC Abbreviation for high blood cholesterol.

HBcAb Abbreviation for hepatitis B core antibody.

HBcAg Abbreviation for hepatitis B core antigen.

HbCO Abbreviation for carbon monoxide hemoglobin.

HBD Abbreviation for hydroxybutyric dehydrogenase.

HBDH Abbreviation for hydroxybutyric dehydrogenase.

HbE An abnormal hemoglobin.

HBe-Ag An antigen resulting from the cleavage of HBcAg by a protease.

HbF See hemoglobin F.

HbH Abbreviation for a hemoglobin consisting of four identical tetramers of beta chains.

HBIg Abbreviation for hepatitis B immunoglobulin.

H2-Blocker Referring to drugs that bind to the H2 receptors, preventing action of histamine.

HBLV Abbreviation for human B lymphotropic virus.

HbM Abbreviation for hemoglobin M in which heme groups are oxidized or partially oxidized to the ferric state (Fe^{3+}).

HbO_2 Abbreviation for oxygenated hemoglobin.

HBP Abbreviation for heparin-binding protein.

HBS Abbreviation for Hepes-buffered saline.

HbS Abbreviation for sickle cell hemoglobin (see hemoglobin S).

HBs-Ag Abbreviation for hepatitis B surface antigen.

HbSC Abbreviation for sickle cell hemoglobin.

HBSS Abbreviation for Hanks balanced salt solution.

HBV Abbreviation for 1. hepatitis B vaccine; 2. hepatitis B virus; 3. honey bee venom.

Hc Abbreviation for hematocrit.

HC Abbreviation for 1. heparin cofactor; 2. hepatic catalase; 3. hepatitis C; 4. hydro-carbon; 5. hydro-cortisone; 6. hydroxycorticoid.

HC Toxin A toxin produced by *Helminthosporium carbonum* (fungus), which is toxic to certain plants.

HCA Abbreviation for 1. hydrocortisone acetate; 2. hydrophobic cluster analysis.

HCAF Abbreviation for a combination drug containing hexamethylmelamine, cytoxan, adriamycin, and fluorouracil.

HCAP Abbreviation for a combination drug containing hexamethylmelamine, cytoxan, adriamycin, and platinol.

HCC Abbreviation for 1. hepatocellular carcinoma; 2. hydroxycholecalciferol.

hCD36 Abbreviation for a human membrane protein involved in a variety of membrane processes.

HCG Abbreviation for human chorionic gonadotropin, a glycoprotein hormone synthesized by chorionic tissue of the placenta and found in the urine during pregnancy.

hCGH (HCGH) Abbreviation for human chorionic gonadotrophic hormone.

hCGR (HCGR) Abbreviation for human choriogonadotropin receptor.

HCH Abbreviation for hexachloro-hexane.

H-Chain Abbreviation for heavy chain.

hChE Abbreviation for human cholinesterase.

HCHWA-D Abbreviation for hereditary cerebral hemorrhage with amyloidosis, Dutch type.

HCI Abbreviation for heme-controlled inhibitor.

HCMV Abbreviation for human cytomegalovirus.

HCO Referring to a formyl group.

H2-Complex Mouse major histocompatibility complex encoding class I, II, and III MHC antigens.

HCP Abbreviation for histidine-rich Ca^{2+}-binding protein.

hCS Abbreviation for human chorionic somatomammotropin.

HCS Abbreviation for hydroxycorticosteroids.

17HCS Abbreviation for 17-dehydrocortico-steroids.

hCSM Abbreviation for human chorionic somatomammotropin.

Hct Abbreviation for hematocrit.

HCT Abbreviation for hydrochlorothiazide.

HCTZ Abbreviation for hydrochlorothiazide.

HCU Abbreviation for homocystinuria.

HCV Abbreviation for hepatitis C virus.

Hcy Abbreviation for homocysteine.

hCyP Abbreviation for human cyclophilin.

HD Abbreviation for 1. hemodialysis; 2. hemolyzing dose; 3. hepatitis D; 4. high density; 5. Hodgkin's disease; 6. Huntington's disease.

5HD Abbreviation for 5-hydroxydecanoic acid.

HD_{50} Abbreviation for hemolytic dose 50, the amount of complement required for lysis of 50% of a standard suspension of sensitized erythrocytes.

HDA Abbreviation for hydrodopamine.

HDC Abbreviation for 1. L-histidine decarboxylase; 2. human diploid cell.

hDCRV Abbreviation for human diploid cell rabies vaccine.

HDCV Abbreviation for human diploid cell vaccine, used against rabies.

HDEL Motif A tetrapeptide motif consisting of His-Asp-Glu-Leu.

HDF (hDF) Abbreviation for human diploid fibroblast.

HdG Abbreviation for 8-hydroxydeoxyguanosine.

HDH Abbreviation for histidine dehydrogenase.

HDL Abbreviation for high-density lipoprotein.

$HDL_{(3)}$ Abbreviation for high density lipoprotein subfraction 3.

HDLC Abbreviation for high density lipoprotein cholesterol.

HDM Abbreviation for high density microsomal fraction.

HDN Abbreviation for hemolytic disease of the newborn.

h-DNA Abbreviation for hybrid DNA.

HDP Abbreviation for 1. helix-destabilizing protein; 2. hydroxy-dimethyl pyrimidine.

hdpDNA Abbreviation for human deproteinized DNA.

HD-Protein Abbreviation for helix-destabilizing protein.

HDQase Abbreviation for dehydroquinase.

HDRV Abbreviation for human diploid-cell rabies vaccine.

HDV Abbreviation for hepatitis D virus or hepatitis delta virus.

HE Abbreviation for 1. hemagglutinating encephalomyelitis; 2. holoenzyme; 3. hemoglobin electrophoresis.

He Abbreviation for helium, a chemical element with atomic weight of 4.

He Antigen An MNS blood group antigen.

$H/2e^-$ Ratio The protons translocated across a membrane by the passage of 2 electrons along the electron transport system.

Headful Mechanism The mechanism that determines the amount of bacteriophage DNA to be packaged into the phage head by the quantity of DNA that can fit into a head.

Heaf Test A skin test to determine whether or not an individual is immune to tuberculosis.

HEAT Abbreviation for human erythrocyte agglutination test.

Heat Shock Proteins Proteins that are synthesized in an organism in response to a sudden rise in temperature or to certain other type of stresses.

Heat Shock Transcription Factor A factor expressed in *Drosophila* following abrupt increase in temperature that binds to the promoter of heat shock gene and regulates transcription of heat shock protein.

Heavy Atom An isotopic form of an atom that contains more than the common number of neutrons.

Heavy Chain The large polypeptide component of the immunoglobulin molecule (also called the H chain). Immunoglobulins are classified on the basis of the types of heavy chains they possess (e.g., α-heavy chain in IgA, δ-heavy chain in IgD, ε-heavy chain in IgE, γ-heavy chain in IgG, and μ-heavy chain in IgM).

Heavy Chain Diseases A group of disorders characterized by the presence of incomplete immunoglobulin heavy chains in serum or urine.

Heavy Hydrogen Referring to deuterium or tritium.

Heavy Isotope An isotope that contains a greater number of neutrons than the commonly observed isotope form (e.g., N^{15}).

Heavy Meromyosin Fragment of myosin molecule that contains a globular head with ATPase activity.

Heavy Nitrogen Referring to ^{15}N.

Heavy Ribosome Ribosome that has been labeled with a heavy isotope.

Heavy Strand 1. The strand in double-stranded DNA that has a high buoyant density due to the high G-C content. 2. The polynucleotide that has been labeled with a heavy isotope.

Heavy Water Water molecule in which the hydrogens are replaced by deuterium (D_2O) or tritium.

HEC Abbreviation for 1. human endothelial cell; 2. hamster embryo cell; 3. hydroxyergocaliferol.

HE-cellulose Abbreviation for hydroxyethyl-cellulose (an ion exchanger).

Hecto- A prefix meaning 100.

Hectogram 100 grams.

Hectometer 100 meters.

Hedaquinium Chloride (mol wt 554) An antiseptic agent.

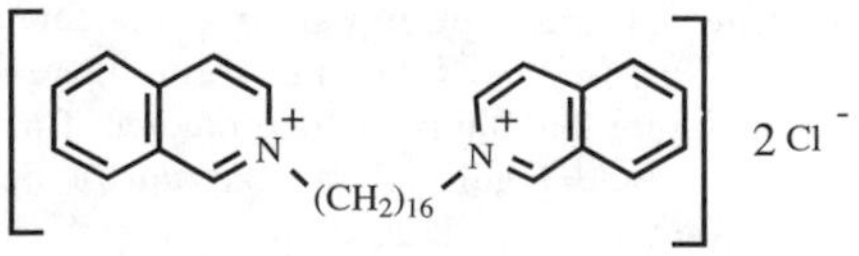

Hedgehog Protein A transmembrane protein involved in segment polarity and cell to cell signaling during embryogenesis and metamorphosis in *Drosophila*.

HEF Abbreviation for hamster embryo fibroblast.

Heinz Body A type of inclusion body in red blood cells of patients with hemolytic anemia that contains altered hemoglobin.

HEIR Abbreviation for high energy ionization radiation.

HEIS Abbreviation for high energy ion scattering.

HEK Abbreviation for human embryonic kidney or human embryo kidney.

HEK-293 A type of cell line of human embryonic kidney.

HEL Abbreviation for human embryonic lung.

HeLa Cell An established cell line of human cervical carcinoma cells originally derived from a Ms. Helen Lane. It is used for culture of a wide range of viruses.

Helenynolic Acid (mol wt 294) A nonsaturated fatty acid from seed of *Helichrysum bracteatum*.

$$CH_3(CH_2)_4C \equiv CCH = CHCH(OH)(CH_2)_7COOH$$

Helical Symmetry Referring to a viral capsid structure that consists of protein units in a coiled helical arrangement.

Helical Viruses Any virus in which the protein capsid exhibits a helical arrangement (e.g., TMV and influenza virus).

Helicase Enzymes capable of unwinding the DNA double helix beginning at the replication fork.

Helicin (mol wt 284) A β-D-glucoside that forms Schiff bases with amino acids.

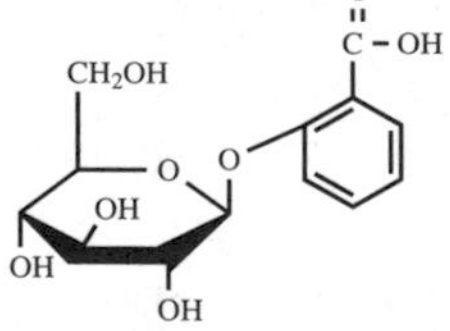

Helicobacter pylori A genus of flagellated Gram negative bacteria which is found in the stomach within the mucous layer and gastric ulceration.

Heliobacterium A genus of photosynthetic bacteria.

Helium (He) A chemical element with atomic weight 4.

Helix A coiled, spiral biological polymer with repeat patterns (e.g., protein and nucleic acid), formed through the interaction of noncovalent bonds among the monomeric units.

α-Helix The protein helix that has 3.6 peptide units per turn and a pitch of 5.4 Angstroms.

π-Helix The protein helix that has 4.4 peptide units per turn and a pitch of 5.2 Angstroms, it is wider and shorter than the α-helix.

Helix-Breaker Any amino acid residue in a protein whose occurrence interrupts helical structure of the protein.

Helix-Breaking Amino Acid An amino acid such as proline that interrupts the α-helical structure of the protein and forms a bend in the polypeptide.

Helix-Coil Transition The transition from an ordered, helical conformation to a disordered random coil conformation in nucleic acid or protein.

Helix-Destabilizing Proteins Single-stranded DNA binding proteins involved in DNA replication.

Helix-Former Any amino acid residue in a protein whose occurrence promotes helical structure of the protein.

Helix-Loop-Helix Protein A type of secondary structure for binding protein (e.g., DNA-binding protein) in which two alpha helix regions (recognizing helix and stabilizing helix) are separated by a loop.

Helixate A trade name for antihemophilic factor VIII.

Hellebrin (mol wt 725) A cardiac glycoside isolated from the rhizome of *Helleborus niger* (Ranunculaceae).

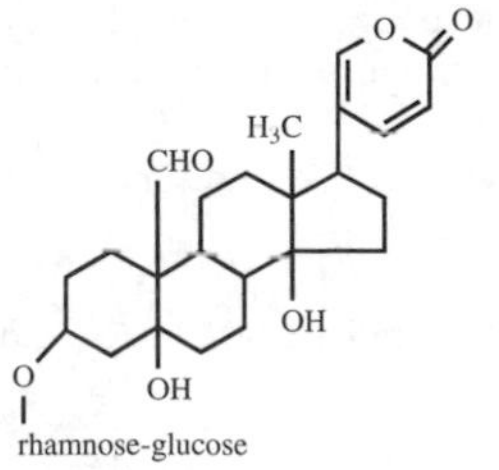

Heller's Test A test for protein in the urine based upon the formation of a white precipitate by treatment of the urine sample with nitric acid.

Helly's Fluid or Solution A mixture of potassium dichromate, sodium sulfate, mercuric chloride, formaldehyde, and distilled water used for the preservation of bone marrow.

Helmholtz Free Energy The maximum amount of energy available to do work resulting from changes in a system at constant volume.

Helminth A disease-causing parasitic worm.

Helminthiasis A disease caused by parasitic worms.

Helminthology The study of parasitic worms.

Helminthosporol (mol wt 236) A plant growth regulator.

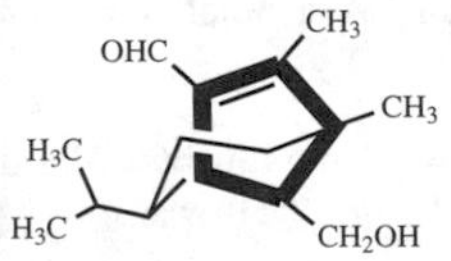

Helper Factor A group of factors (e.g., interleukins) from helper T-lymphocytes that act specifically or nonspecifically to help other classes of lymphocytes by assisting in proliferation and antibody production.

Helper T Cell A type of T-lymphocyte that promotes the immune response of other lymphocytes by releasing soluble helper factors (e.g., interleukins).

Helper Virus Any virus that can provide function necessary for the replication of a defective virus.

Helvolic Acid (mol et 569) An antibiotic produced by *Aspergillus fumigatus*.

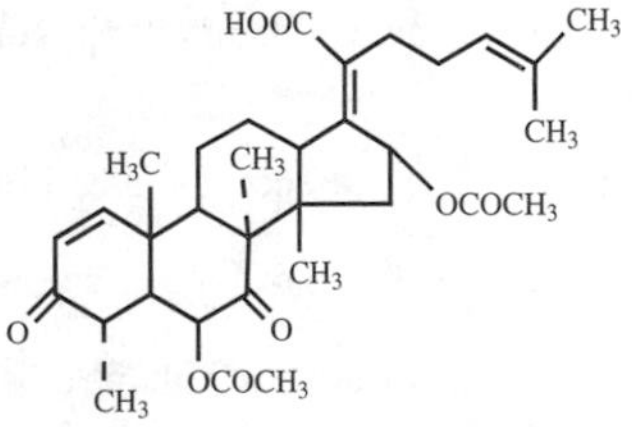

Hem- A prefix meaning blood.

Hema- A prefix meaning blood.

Hemabate A trade name for carboprost, an abortifacient.

Hemadsorption 1. The attachment of red blood cells on another cell surface or particle. 2. The attachment of substances or viral particles onto the surface of erythrocytes.

Hemagglutination The agglutination or clumping of red blood cells.

Hemagglutination Inhibition (HI) The inhibition of virus-induced hemagglutination, a technique used for the identification of hemagglutinating viruses.

Hemagglutinin Antibody or substance that causes agglutination of erythrocytes.

Hemangiectasis Dilation of blood vessels.

Hemangioblast An embryonic cell that gives rise to endothelium of blood vessels and all types of blood-forming elements.

Hemangioblastoma A brain tumor.

Hemangioma A nonmalignant tumor of blood vessels.

Hematemesis Vomiting of red blood cells due to bleeding of the upper GI tract.

Hematherm Warm-blooded animal.

Hematin (mol wt 633) A heme in which the iron is in the ferric form.

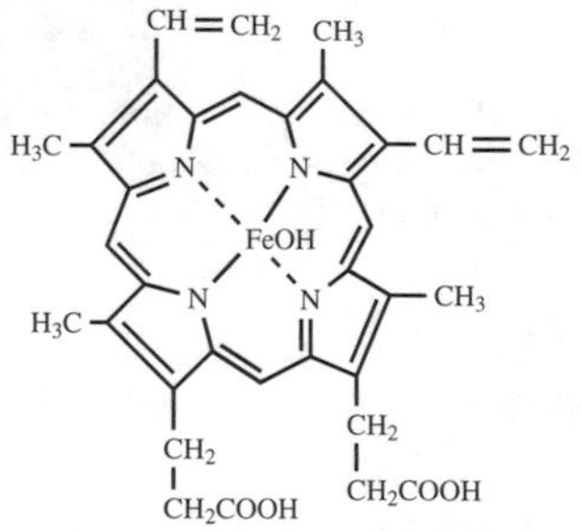

Hematinics Agents or drugs that stimulate blood cell formation or increase hemoglobin concentration in the blood.

Hematinuria The presence of heme in urine.

Hematochezia Passage of bloody stools.

Hematocrit 1. A measure of volume of red blood cells to the volume of whole blood. 2. A device for measuring the volume of cells and plasma in blood.

Hematocyanin See hemocyanin.

Hematocytometer See hemocytometer.

Hematocyturia The presence of erythrocytes in the urine.

Hematogenesis 1. Derived from the blood. 2. Circulated by the blood.

Hematoid Pertaining to the blood or resembling blood.

Hematology The science that deals with blood or blood forming tissues.

Hematolysis See hemolysis.

Hematomanometer A device used for measuring blood pressure.

Hematometra The collection and retention of menstrual fluid within the uterus.

Hematometry The determination of number, types, and proportion of blood cells or hemoglobin content in the blood.

Hematophagus Feeding on blood.

Hematopoiesis The development and formation of blood cells in the bone marrow.

Hematopoietic Pertaining to blood forming.

Hematopoietic Stem Cell Actively dividing cell that gives rise to all types of blood cells.

Hematopoietic Tissue Blood-forming tissue.

Hematoporphyrin (mol wt 599) A compound derived from hemin and used as an antidepressant.

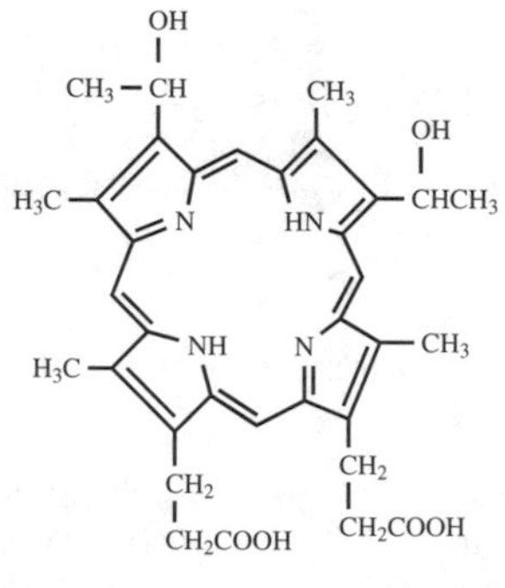

Hematoporphyrinuria The appearance of blackish urine due to the presence of porphyrin resulting from destruction or lysis of red blood cells.

Hematospermia The presence of blood in the semen.

Hematoxylin (mol wt 302) A biological stain.

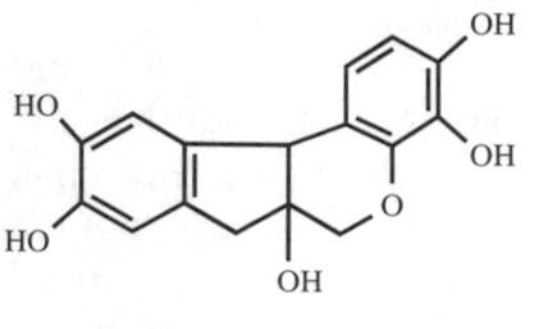

Hematuria The presence of erythrocytes in the urine.

Heme (mol wt 616) An iron-porphyrin that forms the oxygen-binding portion of hemoglobin and also serves as a prosthetic group in cytochrome b and c.

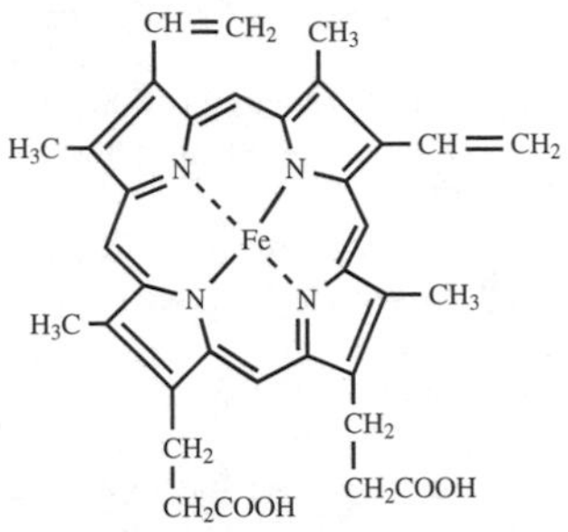

Heme-Controlled Inhibitor A protein kinase that phosphorylates the α subunit of eIF2, thus blocking the recycling of eIF2.

Heme-Heme Interaction The cooperative interaction of heme groups in hemoglobin for binding and unloading of oxygen.

Hemerythrin A nonheme, oxygen-carrying protein in various invertebrates.

Hemi- A prefix meaning 1. half and 2. one side of the body.

Hemic Pertaining to the blood.

Hemicellulose A general term for plant polysaccharides that are not classified as pectic substances which can be extracted by aqueous alkaline solution.

Hemichannel The cell-to-cell transmembrane channel formed from the protein connexin.

Hemicholinium-3 (mol wt 547) A choline analog and an inhibitor of choline kinase.

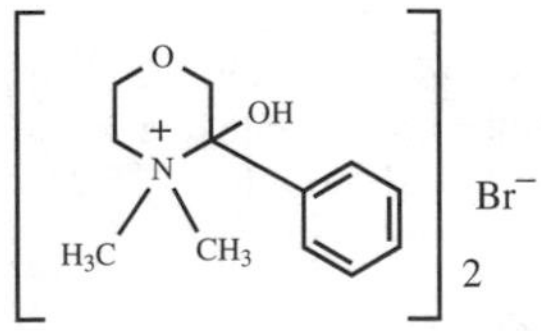

Hemidesmosome A desmosome-like structure that serves to join the basal surface of the epithelial cells to the underlying basal lamina.

Hemihypoplasia Incomplete development of one side or one half of an organ or body.

Hemiketal A compound formed by reaction of a ketone with an alcohol group.

Hemin An ion-containing heme group derived from hemoglobin.

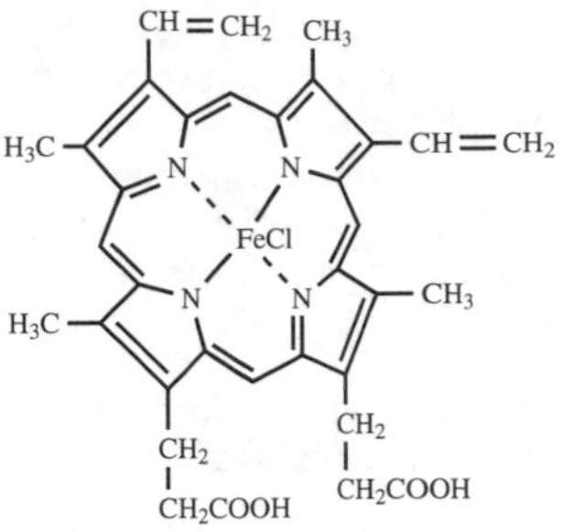

Hemin-Controlled Repressor See heme-controlled inhibitor.

Hemizygous Gene The gene that is present in only one copy in a diploid cell (e.g., X chromosome in an XY male).

Hemo- A prefix meaning blood.

Hemoagglutination Variant spelling of hemagglutination.

Hemochromatosis A disease caused by excessive absorption and deposition of iron in the body leading to liver enlargement and skin discoloration.

Hemocoel A body cavity (e.g., in arthropods) that contains fluid and functions as part of the circulatory system.

Hemocuprein Synonym of superoxide dismutase.

Hemocyanin A nonheme, oxygen-carrying protein found in various invertebrates, e.g., keyhole-limpet.

Hemocytes Blood cells of invertebrate animals.

Hemocytometer A device used for estimating the number of blood cells in the blood.

Hemodialysis Removal of impurities from the blood by means of a hemodialyzer.

Hemodialyzer An apparatus used for blood dialysis by circulating blood through a series of dialysis membranes to remove waste products or poisonous substances.

Hemodilution The decrease in the proportion of red blood cells relative to the plasma.

Hemodynamics Science that deals with the circulation of the blood and the mechanisms involved in blood circulation.

Hemofil-M A trade name for antihemophilic factor VIII.

Hemoflagellate A flagellate protozoan parasite found in the blood.

Hemoglobin Oxygen-carrying protein found in red blood cells; it consists of two pairs of polypeptide chains (alpha and beta chains) and an iron-containing heme group. There are a number of hemoglobin variants that differ in the type of peptide chain and amino acid composition in the peptide chains.

Hemoglobin A Adult hemoglobin that consists of two alpha and two beta chains.

Hemoglobin A_{1c} A minor hemoglobin type whose formation is proportional to the blood sugar level and thus diabetics have a higher proportion of hemoglobin A_{1c} than normal individuals.

Hemoglobin A_2 See HbA_2.

Hemoglobin C See HbC.

Hemoglobin C Disease A genetic blood disorder due to the presence of hemoglobin C.

Hemoglobin E An abnormal hemoglobin.

Hemoglobin F Normal fetal hemoglobin.

Hemoglobin H A hemoglobin consisting of four identical tetramers of beta chains.

Hemoglobin M See HbM.

Hemoglobin S An abnormal hemoglobin that causes sickle cell anemia; it differs from normal hemoglobin by a single amino acid residue in the beta chain (valine for glutamate at position 6).

Hemoglobin Variant Any type of hemoglobin except hemoglobin A.

Hemoglobinemia The presence of free hemoglobin in the plasma caused by lysis of erythrocytes in the blood vessels.

Hemoglobinometer A device for measuring the amount of hemoglobin in the blood.

Hemoglobinopathy A group of genetic disorders characterized by changes in the molecular structure of hemoglobin.

Hemoglobinuria The presence of hemoglobin in the urine.

Hemogram A profile of number, proportion, and morphology of cellular elements in the blood.

Hemolin An insect hemolymph protein, it belongs to the immunoglobulin superfamily.

Hemolymph The circulatory fluid of invertebrate animals that is functionally comparable to blood and lymph of vertebrates.

Hemolysate The soluble fraction derived from lysed red blood cells.

Hemolysin An antibody or substance that can lyse red blood cells in the presence of complement.

Hemolysis The lytic destruction of red blood cells with release of intracellular hemoglobin.

Hemolytic Anemia A disorder characterized by the premature destruction of red blood cells.

Hemolytic Antibody Antibody capable of lysing red blood cells in the presence of complement.

Hemolytic Disease of the Newborn A disease associated with Rh-factor incompatibility between Rh-positive fetal and Rh-negative mother, resulting in maternal antibody activity against fetal red blood cells (also known as erythroblastosis fetalis).

Hemolytic System An assay system that contains red blood cells, complement, and hemolysin used in the complement fixation test as an indicator system.

Hemopath Any disorder of the blood or blood-forming tissue.

Hemoperfusion Passage of blood through an absorbent for removal of toxic substances.

Hemopericardium The buildup of blood in the sac surrounding the heart.

Hemoperitoneum The passage or presence of blood in the peritoneal cavity.

Hemopexin A β-glycoprotein in the plasma that binds heme.

Hemophagocyte Phagocytes that engulf red blood cells.

Hemophagocytosis The process of engulfment of red blood cells by phagocytes.

Hemophilia A genetic disease characterized by uncontrollable bleeding due to a sex-linked recessive deficiency of blood-clotting factor (usually of Factor VIII).

Hemopoiesis The formation or production of various types of blood cells.

Hemoprotein A conjugated protein that has a heme as a prosthetic group.

Hemopsonin Antibody or substance that is capable of combining with erythrocytes, rendering them susceptible to phagocytosis.

Hemoptysis The expectoration of blood.

Hemorphin Any of a group of hemoglobin-derived peptides with affinity for opioid receptors.

Hemorrhage Loss of blood from the blood vessels.

Hemorrhage Disease of Newborn A bleeding disorder of newborns caused by vitamin K deficiency. It can be corrected by uptake of vitamin K.

Hemorrhagin Referring to substance or toxin that destroys endothelial cells in the capillaries and causes hemorrhage.

Hemosiderin A mammalian iron-storage protein related to ferritin but less abundant.

Hemosiderosis Excessive deposit of iron or hemosiderin in the tissues.

Hemospermia Blood in the semen.

Hemostasis Stoppage of bleeding through blood clotting and contraction of blood vessels.

Hemostatic Any agent capable of arresting hemorrhage.

Hemothorax An accumulation of blood in the chest cavity.

Hemotoxin Toxin capable of destroying red blood cells.

Hempa (mol wt 161) An aziridine mutagen.

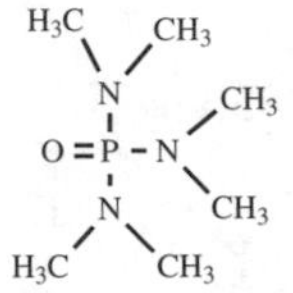

Henderson-Hasselbalch Equation A mathematical relationship between the pKa of an acid and the pH of a solution containing the acid and its conjugate base.

$$pH = PKa + \log \frac{[A^-]}{[HA]}$$

Henna An extract from leaves of *Lawsonia* species used for dyeing hair and fingernails.

Hepadnaviridae A family of enveloped DNA-containing animal viruses that can cause hepatitis B in human.

Hepalean A trade name for heparin sodium, an anticoagulant.

Heparan Any polysaccharide derived by desulfation of heparan sulfate or heparin.

Heparan α-Glucosamine *N*-Acetyltransferase The enzyme that catalyzes the following reaction:

Acetyl-CoA + heparan sulfate α-D-glucosaminide

⇅

CoA + heparan sulfate *N*-acetyl α-D-glucosaminide

Heparan Sulfate Compound closely related to heparin and containing iduronate, glucosamine, and *N*-acetylglucosamine, it differs from heparin in being smaller and less sulfated.

Heparin A sulfated mucopolysaccharide and anticoagulant found in the granules of mast cells. It contains repeating units of D-glucosamine, D-glucuronic acid, or L-iduronic acid.

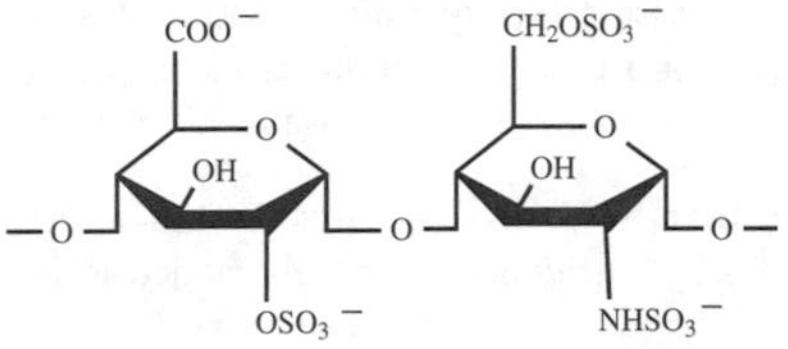

Heparin Eliminase Synonym of heparin lyase.

Heparin Lyase The enzyme that catalyzes the selective cleavage of heparin to yield oligosaccharides.

Heparin Sulfate Eliminase Synonym of heparin sulfate lyase.

Heparin Sulfate Lyase The enzyme that catalyzes the elimination of sulfate from heparin.

Heparinase See heparin lyase.

Heparitinase Synonym of heparin sulfate lyase.

Hepatectomy Surgical removal of part of the liver.

Hepatic Pertaining to the liver.

Hepatitis Inflammation of the liver and characterized by yellowing skin, enlargement of the liver, and abnormal liver function.

Hepatitis A An infectious hepatitis caused by an enterovirus (a single-stranded RNA virus). It is also known as infectious hepatitis.

Hepatitis Associated Antigen See hepatitis surface antigen.

Hepatitis B A viral hepatitis caused by a DNA-containing hepatitis B virus (Hepadnaviridae). It is also called serum hepatitis.

Hepatitis B Core Antigen Antigen from the core protein of the capsid of the Hepatitis B virus.

Hepatitis B Immune Globulin (HBIG) An immune globulin to hepatitis B virus used for protection against hepatitis B virus.

Hepatitis B Surface Antigen The hepatitis B viral envelope antigen that appears in the circulation as noninfectious, DNA-free protein particles (also called Australia antigen).

Hepatitis B Virus A DNA-containing virus of the family Hepadnaviridae that causes hepatitis B in human and is apparently also the causal agent for human hepatocellular carcinoma.

Hepatitis C Virus A virus of the family Flaviviridae that contains single-stranded RNA with positive polarity.

Hepatitis D Virus A satelite virus (also known as delta virus) that requires hepatitis B virus as helper for replication. It contains a covalently closed, circular, negative-sense RNA genome.

Hepatitis E Virus A virus in the family of Togaviridae. It contains single-stranded RNA.

Hepatocellular Carcinoma A carcinoma of liver cells associated with hepatitis B.

Hepatocyte Epithelial cell of the liver.

Hepatogram A radioisotope scan of the liver.

Hepatolith Calculi in the liver.

Hepatolithiasis Production of stones in the liver.

Hepatoma A malignant tumor of the liver.

Hepatotoxin A toxin that damages the liver.

Hep-B-Gammagee A trade name for hepatitis B immune globulin (human).

HEPES (mol wt 238) Abbreviation for 2-hydroxyethylpiperazineethanesulphonic acid, a widely used reagent for preparation of bilogical buffer and tissue culture media.

Hep-forte A trade name for a balanced formulation of vitamins, minerals, lipotropic factor, and vitamin-protein supplements.

HepG2 Abbreviation for a type of cell line.

Hep-Lock A trade name for heparin sodium, an anticoagulant.

Hepronicate (mol wt 506) A vasodilator.

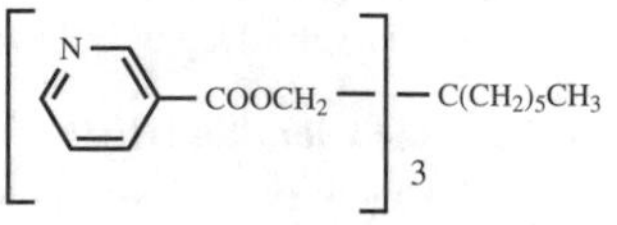

Hepsin A serine protease.

Hepta- A prefix meaning seven.

Heptabarbital (mol wt 250) A sedative and hypnotic agent.

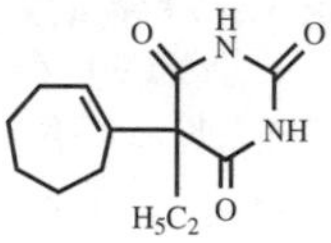

Heptagon A polygon with seven angles and seven sides.

Heptalac A trade name for lactulose, a laxative.

Heptaminol (mol wt 145) A cardiotonic agent.

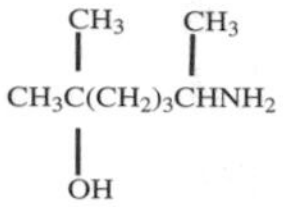

Heptanoic Acid (mol wt 130) A fatty acid found in various fuel oils.

$$CH_3-(CH_2)_5COOH$$

Heptanoyl An acyl group, $CH_3(CH_2)_5CO$-, derived from heptanoic acid.

Heptenophos (mol wt 251) An insecticide.

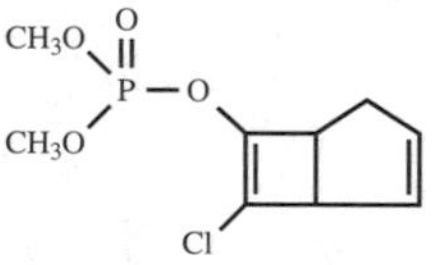

Heptitol Any alditol with seven carbon.

Heptonic Acid (mol wt 229) A sugar acid derived from heptose.

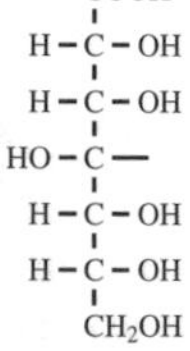

Heptonic γ-Lactone (mol wt 208) A derivative of heptonic acid.

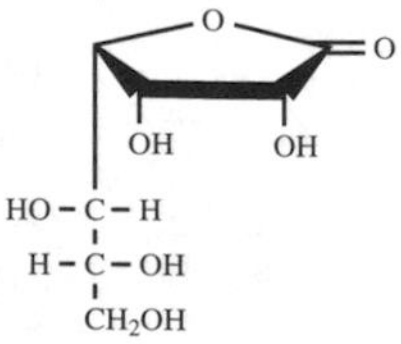

Heptose (mol wt 210) A seven-carbon aldosugar.

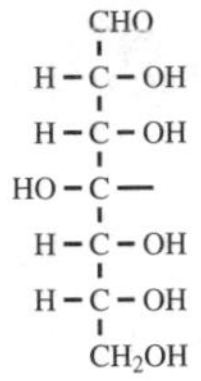

Heptulose (mol wt 210) A seven-carbon ketosugar.

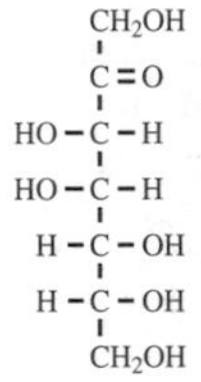

Heptulose Kinase See sedoheptulose kinase.

hER Abbreviation for human estrogen receptor.

Herbicide A chemical that selectively kills plants.

Herbicolin A An acyl peptide antibiotic that inhibits the growth of yeast and filamentous fungi.

Herbivore A plant-eating animal.

Hereditary Angioedema A disorder characterized by the uncontrolled production of C2-kinin due to a deficiency in C1 inhibitor.

Heregulin One of several glycoproteins that bind to the transmembrane tyrosine kinase and stimulate its activity.

Heroin (mol wt 369) An addictive narcotic agent (also known as diacetyl morphine).

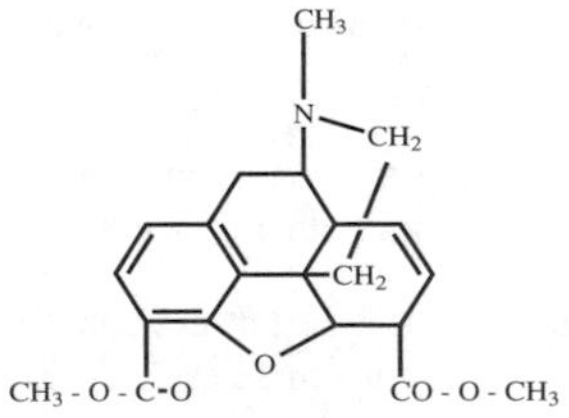

Herpangina A disorder caused by coxsackie virus (*Picornaviridae*) characterized by sore throat, headache, and pain in stomach, neck, arms and legs.

Herpes Any disease caused by herpes simplex virus.

Herpes Simplex Virus A virus of the family Herpesviridae that attacks the skin and nervous system.

Herpes Zoster An acute human skin disorder caused by Varicella zoster virus (Herpesviridae). It is characterized by the formation of painful blisters on the skin.

Herpesviridae A family of enveloped, dsDNA-containing, icosahydral viruses.

Herpetosiphon A genus of gliding bacteria (Cytophagales) that occurs in aquatic habitats.

Herplex A trade name for idoxuridine, an antiviral drug.

Hers Disease A glycogen-storage disease caused by a deficiency of liver phosphorylase, which utilizes glycogen, leading to the accumulation of glycogen in the liver (also called glycogen storage disease type IV).

Hershey-Chase Experiment The experiment demonstrates that DNA is the genetic material in bacteriophage and responsible for the production of bacteriophage progeny.

Hershey-Circle Referring to double-stranded DNA formed from linear DNA with cohesive ends (e.g., λ-DNA circularization).

Hertone A nonhistone chromosomal protein.

Hertz A unit of measurement of frequency equal to one cycle per second.

HES 1. Abbreviation for hydroxyethyl starch. 2. A trade name for hetastarch hydroxyethyl starch, a plasma expander.

Hespan A trade name for hetastarch hydroxyethyl starch, a plasma expander.

Hesperidin (mol wt 611) A bioflavonoid from citrus fruits.

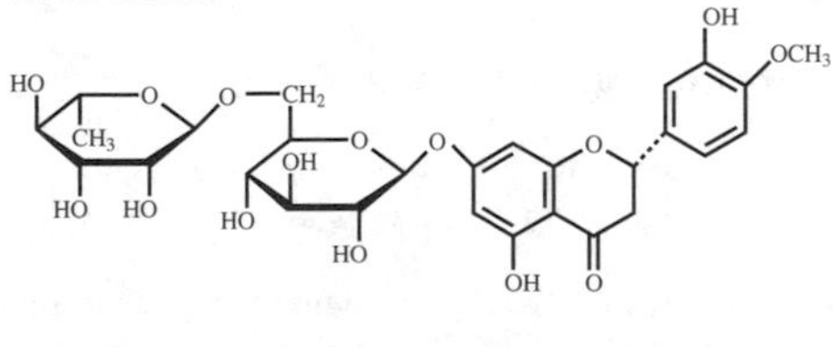

hET Abbreviation for human endothelin.

hETA (hET-A) Abbreviation for human endothelin-A.

Hetacillin (mol wt 389) A semisynthetic antibiotic related to penicillin.

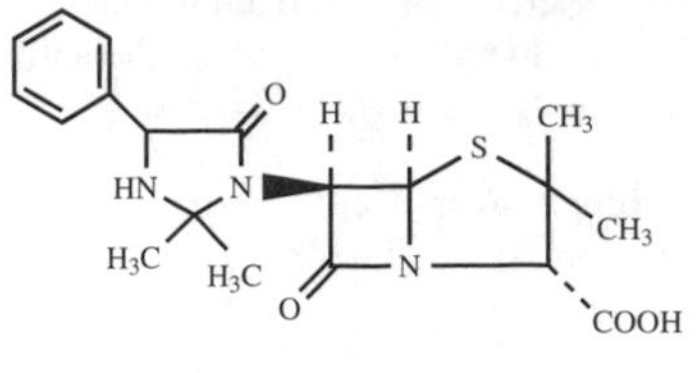

hETAR (hET-R) Abbreviation for human endothelin-A receptor.

Hetastarch A starch derivative capable of increasing the volume of blood plasma, used as a plasma volume extender.

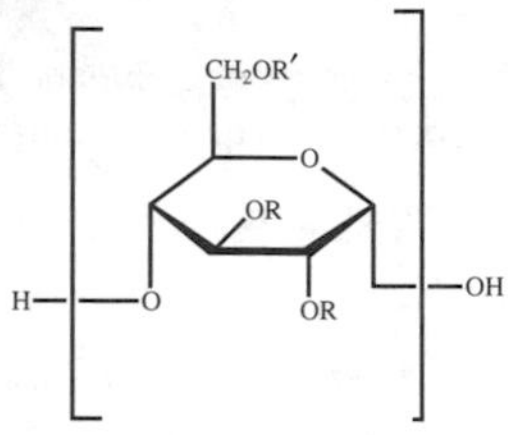

R or R′ = H or CH_2CH_2OH

hETB (hET-B) Abbreviation for human endothelin-B.

hETBR (hET-BR) Abbreviation for human endothelin-B receptor.

5-HETE (mol wt 320) Abbreviation for hydroxyeicosatetraenoic acid, an intermediate in the biosynthesis of immunoactive compounds from arachidonic acid.

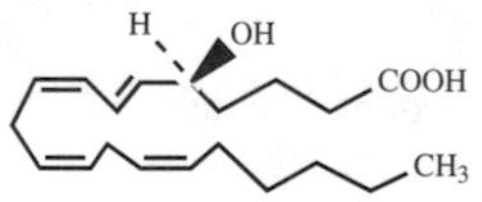

Hetero- A prefix meaning different.

Heteroantibody Any antibody that reacts with antigens from another species.

Heteroantigen See heterologous antigen.

Heteroatom A noncarbon atom in a ring structure.

Heterocaryon See heterokaryon.

Heterochromatin Chromosomal regions that remain condensed during interphase.

Heterochronic Mutation Mutation that causes particular cell divisions to be skipped or repeated, thereby accelerating or delaying terminal differentiation.

Heterocyclic Pertaining to an organic ring structure that consists of one or more noncarbon atoms.

Heterocytotropic Antibodies An antibody that has a greater affinity when fixed to mast cells of a species other than the one in which the antibody is produced.

Heterodimer A molecule consisting of two different monomers or subunits.

Heteroduplex 1. A double-stranded DNA molecule in which the two strands originated from different sources and therefore do not have completely complementary base sequences. 2. RNA:DNA hybrid.

Heteroduplex Mapping A technique that uses heteroduplex analysis to determine the location of various inserts, deletions, or heterogeneities in a DNA.

Heteroenzyme Functionally identical enzyme from different species.

Heterofermentation Fermentation that produces nonidentical products (e.g., fermentation of glucose to lactic acid and acetic acid).

Heterogamous Sexual reproduction in which the gametes are morphologically different.

Heterogamy Union of two dissimilar gametes.

Heterogeneity A mixture of macromolecules of different sizes, charges, structures and properties.

Heterogeneous Nuclear RNA A class of RNA found in the nucleus with a wide range of sizes (abbreviated as hnRNA).

Heterogenote A partial diploid cell or a merozygot in which a donor chromosome (exogenote) segment carries a different allele in comparison with a chromosome segment of the recipient (endogenote).

Heterogenotic Merozygote A partially heterozygous bacterial cell that contains a different exogenote from the donor cell.

Heteroglycan Any polysaccharide that contains more than one type of monosaccharide.

Heterograft Tissue graft from one species to another.

Heteroimmunization Immunization with an antigen derived from another species.

Heterokaryon A cell that contains two nuclei, each from a different source and generated by cell fusion.

Heterokont Referring to a pair of flagella on a diflagellate that are different from each other.

Heterolactic Acid Fermentation Fermentation that produces a mixture of lactic acid and other products (e.g., fermentation of glucose to lactic acid and acetate).

Heterologous Antigen An antigen that participates in a cross-reaction.

Heterolytic Bond Cleavage The breakage of a covalent bond in which the electron pair in the covalent bond remains with one of the atoms.

Heteromorphic Morphologically different.

Heteromultimeric Protein A protein that consists of nonidentical subunits encoded by different genes.

Heteronium Bromide (mol wt 412) An anticholinergic agent.

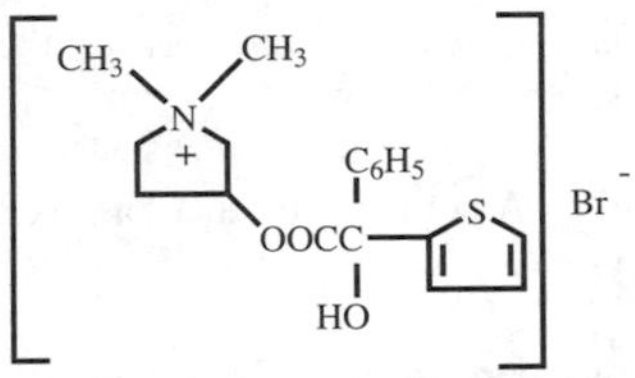

Heterooligomer Any oligomer that consists of two or more types of repeating units.

Heterophile Antibodies Antibodies that react with antigens from different species.

Heterophile Antigens Immunologically related antigens found in unrelated species.

Heterophyiasis Infestation of the small intestine with the parasitic fluke *Heterophyes*.

Heteroplasmon A cell whose cytoplasm contains a mixture of genetically unlike mitochondria or chloroplasts.

Heteroplastic Transplantation Transplantation between individuals of different species in the same genus.

Heteroploid Having a chromosome number that is not a simple multiple of the haploid chromosome number.

Heteropolymer A polymer that contains different types of monomers.

Heteropolysaccharide Polysaccharide that contains more than one type of sugar monomer.

Heterosis The superiority of the hybrid over the parent.

Heterosporous Capable of production of more than one type of spore.

Heterotroph An organism that cannot use inorganic CO_2 for synthesis of organic compounds; it requires preformed organic molecules as a source of carbon and energy.

Heterotrophic Effect Interaction between nonidentical ligands (e.g., activator or inhibitor) on binding of a substrate by the enzyme.

Heterotrophic Nitrification The nitrification carried out by a chemoheterotrophic microorganism.

Heterotropic Regulatory Enzyme An enzyme-mediated reaction that can be stimulated or inhibited by a specific effector or modulator other than substrate.

Heteroxenous Requiring more than one host to complete a life cycle.

Heterozygosity Having one or more pairs of dissimilar alleles at one or more loci.

Heterozygote A diploid cell that contains two different alleles.

Heterozygous Pertaining to a herterozygote.

Heterozygous Probing The use of a labeled nucleic acid probe to identify related molecules by hybridization.

Hetolin (mol wt 488) An anthelmintic agent.

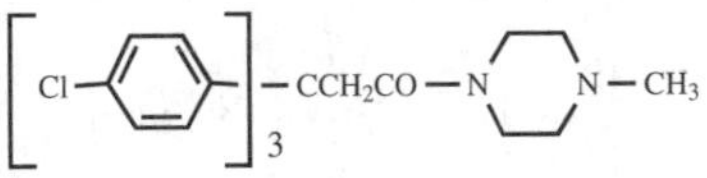

HETP Abbreviation for hexa-ethyl-tetra-phosphate.

hETR (hET-R) Abbreviation for human endothelin receptor.

HEV Abbreviation for 1. hepatitis E virus; 2. high endothelial venules.

Hevein An N-acetyl-D-glucosamine and N-acetyl D-neuraminic acid binding lectin.

HEWL Abbreviation for hen-egg-white lysozyme.

Hex Abbreviation for hexanoyl.

Hexa- A prefix meaning six.

Hexabrachion A six-armed structure in which regions of protein radiate outwards like spokes of a wheel.

Hexacarbacholine Bromide (mol wt 536) A muscle relaxant.

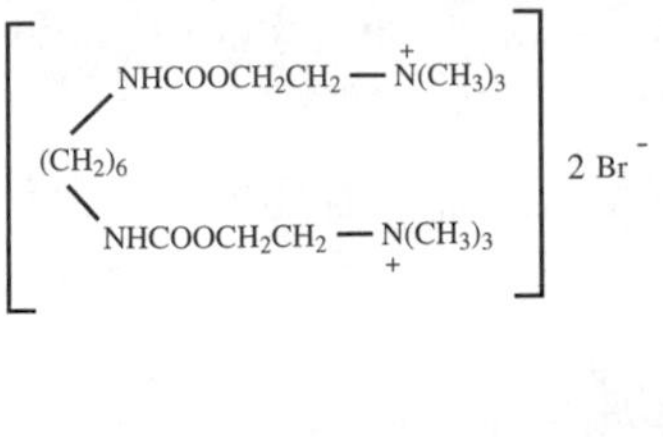

Hexachlorophene (mol wt 407) A topical anti-infective agent.

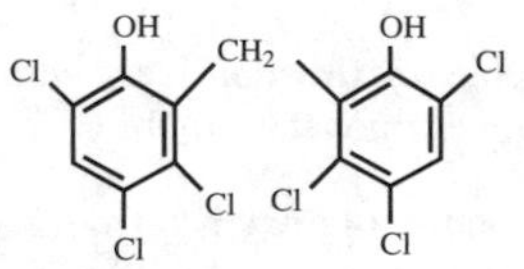

Hexacyclonate Sodium (mol wt 194) A CNS stimulant.

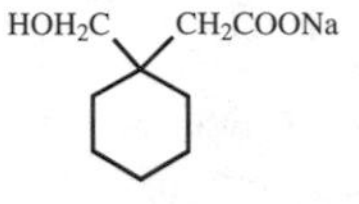

Hexadecadienoic Acid Any straight chain fatty acid having 16 carbons and 2 double bonds.

Hexadecanoic Acid Synonym of palmitic acid.

Hexadecenoic Acid Any straight chain fatty acid having 16 carbons and 1 double bond.

Hexadimethrine Bromide An antiheparin agent.

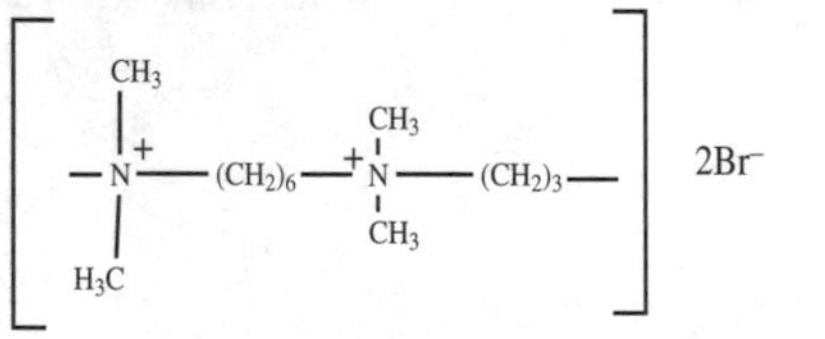

Hexadrol A trade name for dexamethasone sodium, used as an anti-inflammatory agent.

Hexafluorenium Bromide (mol wt 663) A skeletal muscle relaxant and succinylcholine synergist.

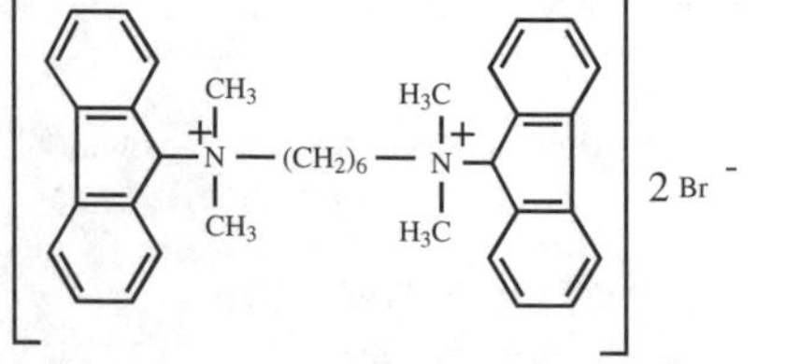

Hexagon A structure with six angles and six sides.

Hexalen A trade name for altretamine, an antitumor drug.

Hexalol A combination drug containing methenamine, phenyl salicylate, atropine sulfate, hyoscyamine, benzoic acid, and methylene blue; it is used as an anti-infective agent.

Hexamer An oligomer consisting of six monomers.

Hexamethylomelamine (mol wt 306) An antitumor agent.

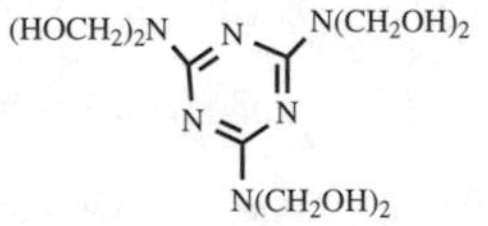

Hexamidine (mol wt 354) An antiseptic agent.

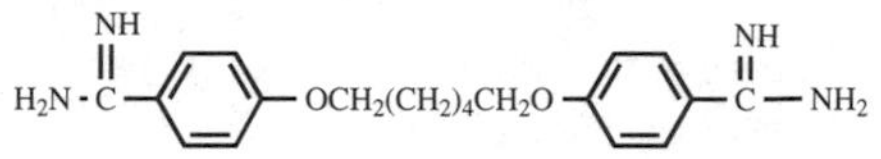

Hexamita A genus of protozoa of the Diplomonadida.

Hexaploid A somatic cell that has six chromosome sets.

Hexapropymate (mol wt 181) A sedative and hypnotic agent.

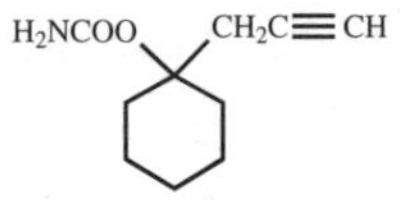

Hexaric Acid An aldaric acid derived from hexose by oxidation at both C-1 and C-6.

Hexazinone (mol wt 252) An herbicide.

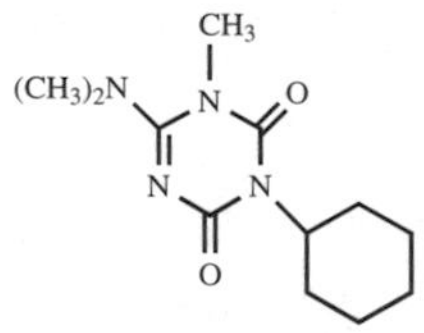

Hexedine (mol wt 352) An antibacterial agent.

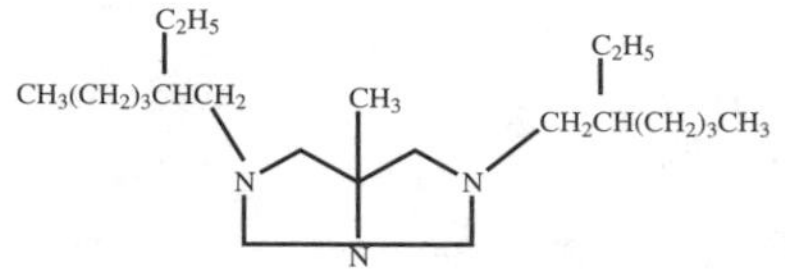

Hexestrol (mol wt 270) An estrogen and antineoplastic agent.

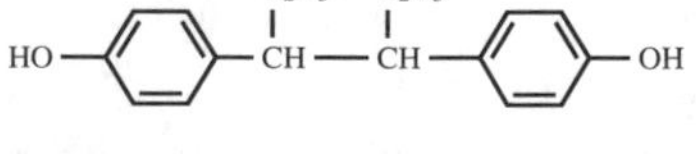

Hexestrol Bis(*b*-diethylaminoethyl ether) (mol wt 469) A coronary vasodilator.

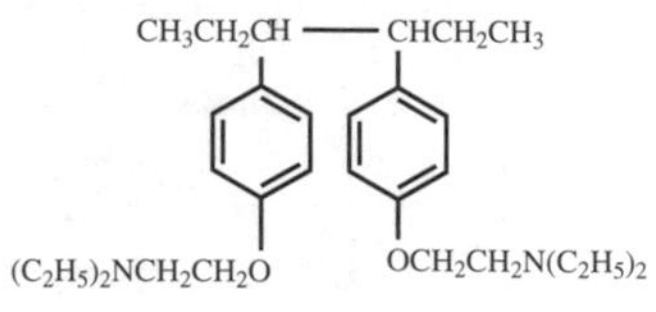

Hexethal Sodium (mol wt 262) A sedative and hypnotic agent.

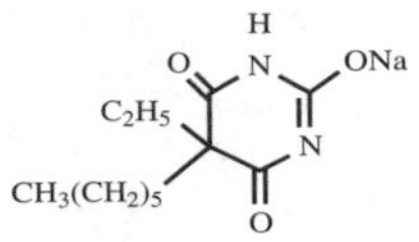

Hexetidine (mol wt 340) A fungicide.

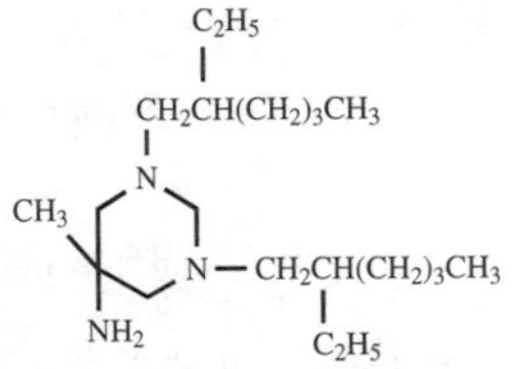

Hexitol A sugar alcohol derived from a six-carbon sugar, e.g., sorbitol and mannitol.

HexN Abbreviation for hexosamine.

Hexobarbital (mol wt 236) A sedative and hypnotic agent.

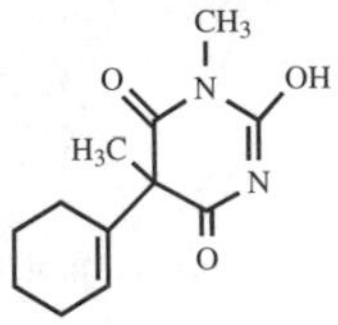

Hexocyclium Methyl Sulfate (mol wt 429) An anticholinergic agent that competitively blocks acetylcholine and inhibits gastric acid secretion.

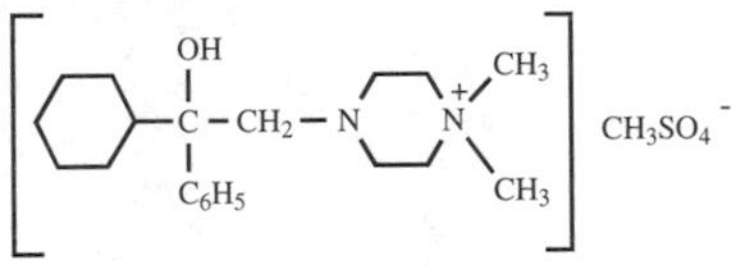

Hexokinase The enzyme that catalyzes the following reaction:

ATP + D-hexose

⇅

ADP + D-hexose 6-phosphate

Hexon A type of capsomer with six angles and six sides, found in the capsid of icosahedral virions.

Hexonic Acid Any nonocarboxylic acid derived from a hexose by oxidation of C-1.

Hexoprenaline (mol wt 421) A bronchodilator.

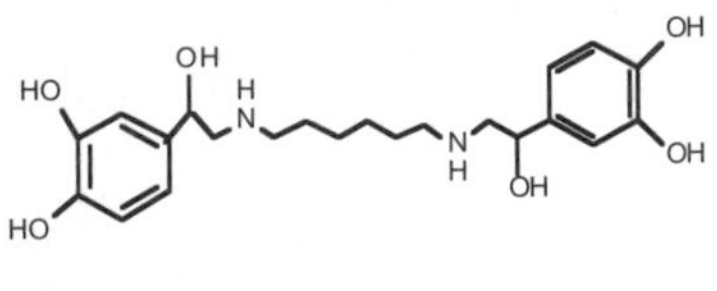

Hexosamine Aminosugar derived from a six-carbon sugar, e.g., glucosamine and galactosamine.

Hexosaminidase An enzyme that catalyzes the degradation of sphingoglycolipid. A deficiency in this enzyme leads to Tay-Sachs and Sandhoff disease.

Hexosaminidase A The enzyme that catalyzes the following reaction:

Ganglioside G_{m2}

⇅

Ganglioside G_{m3} + *N*-acetylgalactosamine

Hexosaminidase B The enzyme that catalyzes the following reaction:

Globoside

⇅

Trihexosylceramide + *N*- acetylgalactosamine

Hexosan Polysaccharide that consists of hexose.

Hexose Monosaccharide that contains six carbon atoms, e.g., glucose, galactose and mannose.

Hexose Bisphosphatase The enzyme that catalyzes the following reaction:

Fructose 1-6-bisphosphate + H_2O

⇅

Fructose 6-phosphate + Pi

Hexose Bisphosphate Pathway See Embden-Meyerhof-Parnas pathway.

Hexose Diphosphatase See hexose bisphosphatase.

Hexose Monophosphate Shunt A pathway in which glucose is converted to five-carbon sugars with concomitant production of NADPH and carbon dioxide (also called pentose phosphate pathway).

Hexose Oxidase The enzyme that catalyzes the following reaction:

D-Glucose + O_2

⇅

D-Glucose 1,5-lactone + H_2O_2

Hexose Phosphate Isomerase See glucose phosphate isomerase.

Hexose 1-Phosphate Uridylyl Transferase The enzyme that catalyzes the following reaction:

UDP-glucose + α-D-galactose 1-phosphate

⇅

α-D-glucose 1-phosphate + UDP-galactose

Hexulose Referring to a six-carbon ketosugar, e.g., fructose.

Hexulose Phosphate Pathway A cyclic pathway used by some methylotrophic bacteria for assimilation of formaldehyde (also known as allulose phosphate pathway and ribulose monophosphate pathway.

Hexuronic Acid A six-carbon sugar acid formed from the oxidation of CH_2OH groups to COOH.

Hexylresorcinol (mol 194) An anthelmintic agent.

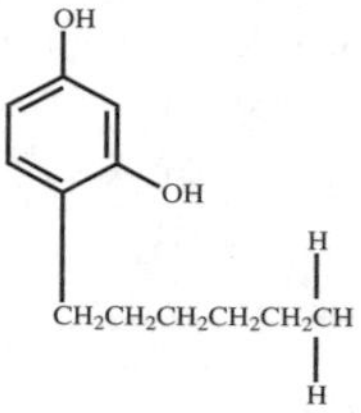

HF Abbreviation for 1. Hageman factor; 2. hemorrhagic factor; 3. human ferritin; 4. human fibroblast.

HFABP Abbreviation for heart type fatty acid binding protein.

HFBA (mol wt 214) Abbreviation for heptafluorobutyric acid, a reagent used for protein sequence analysis.

$$CF_3CF_2CF_2COOH$$

HFDK Abbreviation for human fetal diploid kidney cell.

HFI Abbreviation for hereditary fructose intolerance.

HFIF Abbreviation for human fibroblast interferon.

Hflu (HFLU) Abbreviation for *Hemophilus influenza.*

Hfr Abbreviation for high frequency of recombination. A bacterium in which the F factor has been integrated into the chromosome so that it is capable of transferring its genome DNA to an F cell at a very high frequency.

HFSH Abbreviation for human follicle stimulatory hormone.

HFT Abbreviation for high frequency transduction.

Hg Symbol for mercury with atomic weight 201, valences 1, 2.

HgaI A restriction endonuclease from *Haemophilus gallinarum* with the following specificity:

5′..........GACG(N)$_5$↓............3′
3′..........CTGC(N)$_{10}$↑..........5′

HGF Abbreviation for 1. hepatocyte growth factor; 2. hyperglycemic-glycogenolytic factor.

HGG (hGG) Abbreviation for human gamma globulin.

hγG Abbreviation for human gamma globulin.

HGH Abbreviation for human growth hormone.

HgiAI A restriction endonuclease from *Herpetosiphon giganteus* HP 1023 with the following specificity:

5′..........G(A/T)GC(A/T)↓C..........3′
3′..........C↑(T/A)CG(T/A)G..........5′

HgiBI (AvaII) A restriction endonuclease from *Herpetosiphon giganteus* with the same specificity as AvaII.

HgiCI A restriction endonuclease from *Herpetosiphon giganteus* with the following specificity:

5′..........G↓GPyPuCC..........3′
3′..........CCPuPyG↑G..........5′

HgiCII (AvaII) A restriction endonuclease from *Herpetosiphon giganteus* with the same specificity as AvaII.

HgiCIII (SalI) A restriction endonuclease from *Herpetosiphon giganteus* with the following specificity:

5′............G↓TCGAC............3′
3′............CAGCT↑G............5′

HgiDI (AcyI) A restriction endonuclease from *Herpetosiphon giganteus* with the following specificity:

5′............GPu↓CGPyC............3′
3′............CPyGC↑PuG............5′

HgiDII A restriction endonuclease from *Herpetosiphon giganteus* Hpa2 with thc following specificity:

5′..........GTCGAC..........3′
3′..........CAGCTG..........5′

HgiEI (AvaII) A restriction endonuclease from *Herpetosiphon giganteus* Hpg24 with the same specificity as AvaII.

HgiGI (AcyI) A restriction endonuclease from *Herpetosiphon giganteus* Hpa1 with the following specificity:

5′..........GPuCGPyC..........3′
3′..........CPyGCPuG..........5′

HgiHI (HgiCI) A restriction endonuclease from *Herpetosiphon giganteus* Hp1049 with the following specificity:

5′..........GGPyPuCC..........3′
3′..........CCPuPyGG..........5′

HgiHII (AcyI) A restriction endonuclease from *Herpetosiphon giganteus* with the following specificity:

5′..........GPuCGPyC..........3′
3′..........CPyGCPuG..........5′

HgiHIII (AvaII) A restriction endonuclease from *Herpetosiphon giganteus* with the following specificity:

5′..........GG(A/T)CC..........3′
3′..........CC(T/A)GG..........5′

HgiJI (AvaII) A restriction endonuclease from *Herpetosiphon giganteus* with the same specificity as AvaII.

HgiJII A restriction endonuclease from *Herpetosiphon giganteus* with the following specificity:

5′..........GPuGCPyC..........3′
3′..........CPyCGPuG..........5′

HgiS21I (CauII) A restriction endonuclease from *Herpetosiphon giganteus* with the same specificity as CauII.

HGPRT Abbreviation for enzyme hypoxanthine-guanine phosphoribosyl transferase. The enzyme that catalyzes the following reaction:

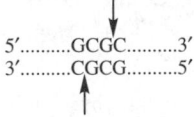

IMP + pyrophosphate
⇅
Hypoxanthine + 5-phosphoribosylpyrophosphate

hGST (HGST) Abbreviation for human GST (glutathione transferase).

hGSTA-4 (HGSTA-4) Abbreviation for human glutathione transferase-A4.

HHA Abbreviation for hereditary hemolytic anemia.

HhaI A restriction endonuclease from *Haemophilus haemolyticus* with the following specificity:

5′..........GCGC..........3′
3′..........CGCG..........5′

HhaII (HinfI) A restriction endonuclease from *Haemophilus haemolyticus* with the following specificity:

5′..........GANTC..........3′
3′..........CTNAG..........5′

HhgI (HaeIII) A restriction endonuclease from *Haemophilus haemoglobinophilus* with the same specificity as HaeIII.

HHL Abbreviation for histidino-hydroxy-lysinonorleucine.

HHR A combination drug containing hydrochlorothiazide, hydrazine hydrochloride, and reserpine.

HHV Abbreviation for human herpes virus.

HI Abbreviation for 1. hemagglutination inhibition; 2. hydroxyindole.

HIA Abbreviation for hemagglutination inhibition activity.

5HIAA Abbreviation for 5-hydroxy-indole acetic acid.

HIB Vaccine A vaccine against *Haemophilus influenzae* type B.

Hibernation A condition of reduced metabolic activity in which some animals pass the winter.

Hibiciens A trade name for chlorhexidine gluconate, an antiseptic agent.

HIC Abbreviation for hydrophobic interaction chromatography.

Hidden Immunological Determinant Any antigenic determinant so positioned that it is not accessible for recognition by lymphocytes or antibodies.

Hidr- A prefix denoting sweat.

Hidradenitis The inflammation of the sweat gland.

Hidrosis The excretion of sweat or excessive sweat.

HIE Abbreviation for hyper-immunoglobulin E syndrome.

Hierarchical Assembly Synthesis of biological structure from simple starting molecules to a more complex structure.

HIF Abbreviation for hypoxia inducible factor.

HIFBS Abbreviation for heat-inactivated fetal bovine serum.

hIg (HIg) Abbreviation for human immunoglobulin.

hIgGFc Abbreviation for Fc fragment of human IgG.

High Copy Number A large number of repetitive copies of a gene that are produced in gene cloning.

High Density Lipoprotein (HDL) A plasma lipoprotein with an approximate density of 1.063 to 1.210 g/ml that is involved in the transport of cholesterol and fat from the blood to the tissue.

High Dose Tolerance An immunological unresponsiveness produced by repeated injections of large amounts of antigen.

High Energy Bond A bond that yields a large amount of free energy upon breakage (5 kcal/mole or more).

High Energy Phosphate Compound A phosphorylated compound having a high negative standard free energy change upon hydrolysis.

High Frequency Recombinant A bacterial strain in which F factor is integrated into the bacterial chromosome; it exhibits a high frequency of gene transfer and recombination during conjugation.

High Frequency Transduction A transduction in which the transducing phages constitute a large proportion of the total phage population.

High Mobility Group A group of nonhistone, high electrophoretic mobility chromosomal proteins that are capable of binding to nucleosomes and regulating gene transcriptional activity.

High Molecular Weight Kininogen Referring to fibrin-stabilizing factor.

High Performance Liquid Chromatography A type of column chromatographic method employing high resolution column, pressure, and gradient elution systems for separation of solutes or macromolecules.

High Pressure Liquid Chromatography See high performance liquid chromatography.

Hill Equation An equation that describes the kinetics of binding of ligands to a protein.

Hill Reaction The reaction demonstrates that CO_2 does not participate directly in the O_2-producing photosynthetic reaction since the illumination of chloroplast in the absence of CO_2 and in the presence of artificial electron acceptor produces oxygen.

Hill Reagent The artificial electron acceptor in the Hill reaction, e.g., ferric oxalate.

Hilum 1. A scar on a seed. 2. The nucleus of a starch grain 3. A depression or pit of an organ.

HinII (AcyI) A restriction endonuclease from *Haemophilus influenzae* RFL1 with the following specificity:

```
            ↓
5′..........GPuCGPyC..........3′
3′..........CPyGCPuG..........5′
                ↑
```

HinIII (NlaIII) A restriction endonuclease from *Haemophilus influenzae* RFL1 with the following specificity:

```
                ↓
5′..........CATG..........3′
3′..........GTAC..........5′
            ↑
```

Hin2I (HpaII) A restriction endonuclease from *Haemophilus influenzae* RFL2 with the same specificity as HpaII.

Hin3I (CauII) A restriction endonuclease from *Haemophilus influenzae* RFL3 with the same specificity as CauII.

Hin5I (HpaII) A restriction endonuclease from *Haemophilus influenzae* RFL5 with the same specificity as HpaII.

Hin5II (AsuI) A restriction endonuclease from *Haemophilus influenzae* RFL5 with the same specificity as AusI.

Hin5III (HindIII) A restriction endonuclease from *Haemophilus influenzae* RFL5 with the same specificity as HindIII.

Hin6I (HhaI) A restriction endonuclease from *Haemophilus influenzae* RFL6 with the following specificity:

↓
5′..........GCGC..........3′
3′..........CGCG..........5′
↑

Hin7I (HhaI) A restriction endonuclease from *Haemophilus influenzae* RFL7 with the same specificity as Hin6I.

Hin8I (AcyI) A restriction endonuclease from *Haemophilus influenzae* RFL8 with the following specificity:

↓
5′..........GPuCGPyC..........3′
3′..........CPyGCPuG..........5′
↑

Hin8II (NlaIII) A restriction endonuclease from *Haemophilus influenzae* RFL8 with the same specificity as NalII.

Hin173I (HindIII) A restriction endonuclease from *Haemophilus influenzae* 173 with the same specificity as HindIII.

Hin1056I (FnuDII) A restriction endonuclease from *Haemophilus influenzae* 1056 with the same specificity as FunDII.

Hin1160II (HindII) A restriction endonuclease from *Haemophilus influenzae* 1160 with the same specificity as HindII.

Hin1161II (HindII) A restriction endonuclease from *Haemophilus influenzae* 1161 with the same specificity as HindII.

HinbIII (HindIII) A restriction endonuclease from *Haemophilus influenzae* Rb with the same specificity as HindIII.

HincII (HindII) A restriction endonuclease from *Haemophilus influenzae* Rc with the same specificity as HindII.

HindI A specific methylase from *Haemophilus influenzae* that methylates the adenine nucleotide with the following sequence:

CH_3
↓
5′..........CAC..........3′
3′..........GTG..........5′

HindII A restriction endonuclease from *Haemophilus influenzae* Rd with the following specificity:

↓
5′..........GTPyPuAC..........3′
3′..........CAPuPyTG..........5′
↑

HindIII A restriction endonuclease from *Haemophilus influenzae* Rd with the following specificity:

↓
5′..........AAGCTT..........3′
3′..........TTCGAA..........5′
↑

HindIV A specific methylase from *Haemophilus influenzae* Rd that methylates the nucleotide with the following sequence:

CH_3
↓
5′..........GAT...........3′
3′..........CGA..........5′

Hinderin (mol wt 622) A thyroid inhibitor.

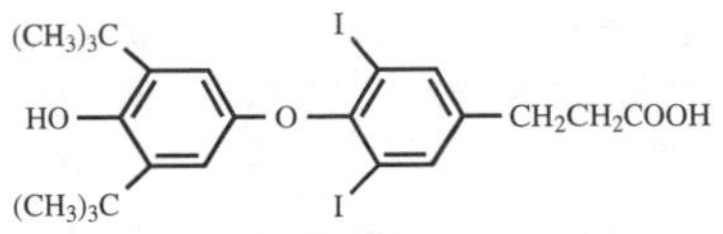

HinfI A restriction endonuclease from *Haemophilus influenzae* Rf with the following specificity:

↓
5′..........GANTC..........3′
3′..........CTNAG..........5′
↑

HinfII (HindIII) A restriction endonuclease from *Haemophilus influenzae* Rf with the same specificity as HindIII.

Hinge Region The portion of the constant region in the heavy chains of immunoglobulins located near the Fc fragment.

HinGUI (HhaI) A restriction endonuclease from *Haemophilus influenzae* GU with the same as specificity HhaI.

HinGUII (FokI) A restriction endonuclease from *Haemophilus influenzae* GU with the following specificity:

↓
5′..........GGATG$(N)_9$..........3′
3′..........CCTAG$(N)_{13}$.........5′
↑

HinHI (HaeII) A restriction endonuclease from *Haemophilus influenzae* H-I with the same as specificity HaeII.

HinJCI (HindII) A restriction endonuclease from *Haemophilus influenzae* JC9 with the following specificity:

5′..........GTPy↓PuAC..........3′
3′..........CAPu↑PyTG..........5′

HinJCII (HindIII) A restriction endonuclease from *Haemophilus influenzae* JC9 with the same specificity as HindIII.

HinP1I (HhaI) A restriction endonuclease from *Haemophilus influenzae* P1 with the following specificity:

5′..........G↓CGC..........3′
3′..........CGC↑G..........5′

HinS1I (HhaI) A restriction endonuclease from *Haemophilus influenzae* S1 with the same specificity as HhaI.

HinS2I (HhaI) A restriction endonuclease from *Haemophilus influenzae* S2 with the same specificity as HhaI.

HIOMT Abbreviation for hydroxyindol-O-methyltransferase.

Hippocalcin A neuron-specific calcium-binding protein.

Hippurate Hydrolase The enzyme that catalyzes the following reaction:

Hippurate + H_2O ⇌ Benzoate + glycine

Hippuric Acid (mol wt 179) A compound found in the urine resulting from reaction of benzoic acid with glycine.

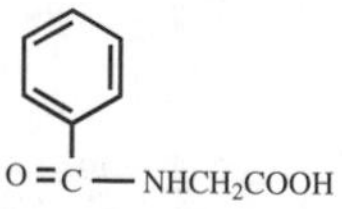

Hiprex A trade name for methenamine hippurate, an antibacterial agent.

Hirsutism Excessive growth of body hair due to a hormonal imbalance.

Hirudin A peptide inhibitor for thrombin.

His Abbreviation for histidine.

***his* Operon** An operon system encoding all the enzymes necessary for the biosynthesis of histidine.

Hismanal A trade name for astemizole, an antihistamine agent.

Histaminase See amine oxidase.

Histamine (mol wt 111) A potent vasodilator involved in allergic reactions.

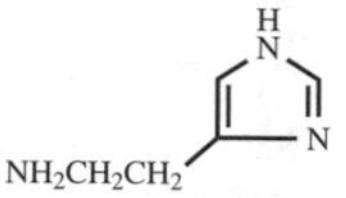

Histamine *N*-Methyltransferase The enzyme that catalyzes the following reaction:

S-Adenosyl-L-methionine + histamine

⇅

S-Adenosyl-L-homocysteine + *N*-methylhistamine

Histanil A trade name for promethazine hydrochloride, an antihistamine agent.

Histapyrrodine (mol wt 280) An antihistaminic agent.

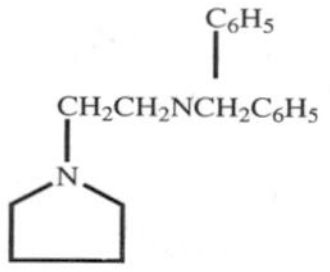

Histerone A trade name for testosterone, an anabolic hormone.

Histidase See histidine ammonia lyase.

Histidinase See histidine ammonia lyase.

Histidine (mol wt 155) A protein amino acid.

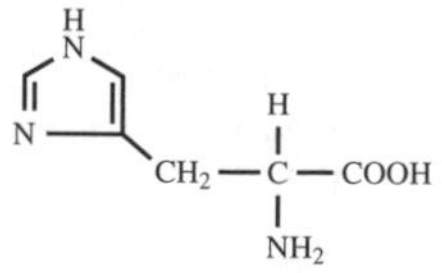

Histidine Ammonia-lyase The enzyme that catalyzes the following reaction:

Histidine ⇌ Urocanate + NH_3

Histidine Carboxy Lyase Synonym of histidine decarboxylase.

Histidine Deaminase See histidine ammonia-lyase.

Histidine Decarboxylase The enzyme that catalyzes the following reaction:

$$\text{Histidine} \rightleftharpoons \text{Histamine} + CO_2$$

Histidine Operon See *his* operon.

Histidine tRNA-Ligase The enzyme that catalyzes the following reaction:

$$\text{ATP} + \text{L-histidine} + \text{tRNA}^{his} \rightleftharpoons \text{AMP} + \text{PPi} + \text{L-histidyl-tRNA}^{his}$$

Histidinemia A genetic disorder characterized by the elevation of histidine due to a deficiency in the enzyme histidase.

Histidinol (mol wt 144) A histidine derivative.

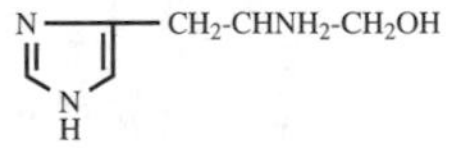

Histidinol Dehydrogenase The enzyme that catalyzes the following reaction:

$$\text{Histidinol} + 2\ NAD^+ \rightleftharpoons \text{Histidine} + 2\ \text{NADH}$$

Histidyl-tRNA Synthetase See histidine tRNA ligase.

Histiocyte Macrophage of connective tissue.

Histiocytosis Proliferation of histiocytes.

Histo- A prefix denoting tissue.

Histochemistry The science that deals with the chemistry of tissues and cells.

Histocompatibility Ability to accept transplants from another member of the same species.

Histocompatibility Antigen Cell surface glycoproteins that are crucial in tissue transplantation. There are three classes of histcompatibility antigens (class I, class II, and class III).

Histocompatibility Complex Referring to the major histocompatibility complex that encodes class I, II, and III antigens.

Histocompatibility Gene The genes on the major histocompatibility complex on chromosome number 6 in humans and chromosome number 17 in the mouse that encode compatibility antigens.

Histoelectrofocusing A technique in which an unfixed frozen section of tissue (25 to 40 μm thick) is applied to support gel and then subjected to electrofocusing.

Histogenesis The formation and development of tissues.

Histogeny See histogenesis.

Histogram A graphical representation of statistical data by means of rectangle bars to show the relationship between two factors.

Histology The science that deals with tissue structure and organization.

Histolysain The enzyme that catalyzes the hydrolysis of basement membrane collagen and azocasein.

Histolysis The breakdown of bodily tissues.

Histones Arginine-lysine-rich proteins that are in close association with the nuclear DNA of most eukaryotic organisms; they are classified on the basis of the lysine/arginine ratio present in the proteins.

Histone Acetyltransferase The enzyme that catalyzes the following reaction:

$$\text{Acetyl-CoA} + \text{histone} \rightleftharpoons \text{CoA} + \text{acetylhistone}$$

Histone Kinase See protamine kinase.

Histone Lysine *N*-Methyltransferase The enzyme that catalyzes the following reaction:

$$\textit{S}\text{-Adenosyl-L-methionine} + \text{histone-lysine} \rightleftharpoons \textit{S}\text{-Adenosyl-L-homocysteine} + \text{histone } \textit{N}\text{-methyl-lysine}$$

Histopathology The science that deals with cellular and tissue changes due to disease.

Histoplasmin An extract from fungus *Histoplasma capsulatum* used for an intradermal injection test for histoplasmosis.

Histoplasmin Test A skin test analogous to the tuberculin test used for the diagnosis of histoplasmosis.

Histoplasmosis An infection caused by fungus *Hisplasma capsulatum*.

Histussin D Liquid A trade name for a combination drug containing hydrocodone bitartrate and pseudoephedrine hydrochloride used as an antitussive and decongestant.

Histussin HC Syrup A trade name for a combination drug containing hydrocodone bitartrate, phenylephrine hydrochloride, and chlorpheniramine maleate, used as an antitussive and decongestant.

His_6-Ub Abbreviation for hexahistidine-tagged ubiquilin.

HIT Abbreviation for 1. hemagglutination inhibition test; 2. histamine inhalation test; 3. histamine-induced thrombocytopenia.

HI-Test Abbreviation for hemagglutination inhibition test.

HIV Abbreviation for human immunodeficiency virus, a virus in the family of Retroviridae that causes AIDS; it consists of two molecules of RNA genome and possesses reverse transcriptase.

HIV-1 Protease An aspartyl protease that is essential for the life of HIV.

Hives An anaphylactic skin reaction mediated by histamine that is released from the degranulation of mast cells.

Hivid Atrade name for zalcitabine or dideoxycytidine, an antiviral agent for treatment of HIV.

HIV-Ig Abbreviation for human immunodeficiency virus immunoglobulin.

HIVRT Abbreviation for human immunodeficiency virus reverse transcriptase.

HjaI (EcoRV) A restriction endonuclease from *Hyphomonas Jannaschiana* with the same specificity as EcoRV.

HK Abbreviation for 1. heat-killed; 2. hexokinase.

(H^+-K^+)-ATPase A type of ATPase in mammalian mucosa involved in transport of H^+ and K^+ ions.

H^+/K^+ Exchange Enzyme Referring to H^+-K^+-activated ATPase.

HLA Abbreviation for human lymphocyte antigens, the major histocompatibility antigens of humans responsible for graft rejection (also called human leukocyte antigen).

HLA-A Abbreviation for human lymphocyte antigen A, a type of Class I histocompatibility antigen.

HLA-B Abbreviation for human lymphocyte antigen B, a type of Class I histocompatibility antigen.

HLA-C Abbreviation for human lymphocyte antigen C, a type of Class I histocompatibility antigen.

HLA Complex Abbreviation for human lymphocyte antigen complex, the major histocompatibility antigen complex of human located on chromosome number 6.

HLA-D Abbreviation for human lymphocyte antigen D, a type of Class II histocompatibility antigen.

HLA-G Abbreviation for a polymorphic class-I HLA antigen.

HLA-H Abbreviation for a nonfunctional HLA-A class I antigen that lacks cysteine residue at position 164.

HLA-LD Abbreviation for human lymphocyte antigen lymphocyte defined.

HLA-Proteins Abbreviation for human lymphocyte antigen proteins encoded by genes in the HLA complex.

HLA-SD Abbreviation for human lymphocyte antigen serologically defined.

HLB Abbreviation for hypotonic lysis buffer.

hLF Abbreviation for human lactoferrin.

HLF Abbreviation for heat-labile factor.

hLF^{-2N} Abbreviation for human lactoferrin lacking two N-terminal residues.

hLF^{-3N} Abbreviation for human lactoferrin lacking three N-terminal residues.

hLF^{-5N} Abbreviation for human lactoferrin lacking five N-terminal residues.

HLH Abbreviation for human luteinizing hormone.

HLHP Abbreviation for helix-loop-helix protein.

hLig-I Abbreviation for human DNA ligase I.

HLP Abbreviation for hyperlipoproteinemia.

hLPP Abbreviation for human LPP (lipid phosphate phosphohydrolase).

HLT Abbreviation for human lymphocyte transformation.

hLZ Abbreviation for human lysozyme.

hMA Abbreviation for human monoclonal antibody.

HMB Abbreviation for hematropine methyl bromide.

HMBA Abbreviation for hexamethylene bisacetamide.

HMC Abbreviation for 5-hydroxymethylcytosine, a modified base of cytosine found in the DNA of T-even phage.

H-meromyosin Abbreviation for heavy meromyosin, a fragment of myosin containing ATPase activity.

HMG Abbreviation for 1. high-mobility group and 2. hydroxymethyl glutarate.

HMG-CoA (mol wt 919) Abbreviation for hydroxymethyl glutaryl CoA, an intermediate in the biosynthesis of acetoacetate from acetyl CoA.

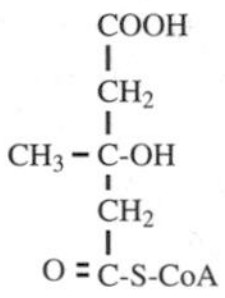

HMG-CoA Lyase The enzyme that catalyzes the following reaction:

Hydroxymethyl glutaryl-CoA

⇅

Acetoacetate + acetyl-CoA

HMG-CoA Reductase The enzyme that catalyzes the following reaction:

HMG-CoA + 2 NADPH

⇅

2 $NADP^+$ + CoA + mevalonate

HMG-CoA Reductase Kinase The enzyme that catalyzes the following reaction:

Active nonphosphorylated HMG-CoA reductase + ATP

⇅

Less active phosphorylated HMG-CoA reductase + ADP

HMG-CoA Synthetase The enzyme that catalyzes the following reaction:

Acetoacetyl-CoA + acetyl-CoA

⇅

HMG-CoA + CoA

HMG-Proteins Abbreviation for high-mobility group proteins, a group of nonhistone, high electrophoretic mobility chromosomal proteins capable of regulating transcriptional activity.

HMK Abbreviation for high molecular weight kininogen.

HMM Abbreviation for heavy meromyosin, a fragment of myosin containing ATPase activity.

HMMA Abbreviation for 4-hydroxy-3-methoxymandelic acid.

HMP Shunt Abbreviation for hexose monophosphate shunt.

HMQC Abbreviation for heteronuclear multiple quantum correlation.

HMQCE Abbreviation for heteronuclear multiple quantum coherence effect.

hMR-1 Abbreviation for human muscarinic receptor type –1.

HMS Abbreviation for hexose mnophosphate shunt, a pathway for the metabolism of glucose and production of NADPH.

HMU Abbreviation for hydroxymethyl uracil.

HMW Abbreviation for high molecular weight.

HNB Abbreviation for 2-hydroxy-5-nitrobenzyl bromide.

HNBB Abbreviation for hydroxynitrobenzyl bromide.

HNF Abbreviation for hepatocyte nuclear factor.

HNF-1, 2, 3, 4 Abbreviation for hepatocyte nuclear factor 1, 2, 3, and 4.

HNMT Abbreviation for histamine N-methyltransferase.

hnRNA Abbreviation for heterogenous nuclear RNA. A group of primary transcripts from DNA from which introns are removed.

hnRNP (NRNP) Abbreviation for heterogeneous nuclear ribonucleoprotein.

Ho Abbreviation for chemical element holmium with atomic weight 164, valence 3.

HO-1 Abbreviation for hem-oxygenase-1.

H^+/O Abbreviation for the number of protons pumped to the external medium by the electron transport complexes per oxygen consumed.

Hodgkin's Disease A human lymphoma affecting spleen, lymph nodes, and occasionally the bone marrow. It is characterized by the appearance of fever, weight loss, and lassitude.

Hoechst 33258 (mol wt 624) A DNA-binding fluorescent dye used in staining chromosomes.

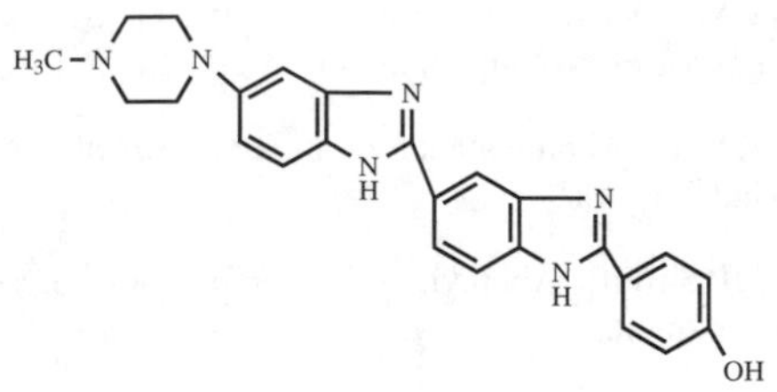

Hofmeister Series The order of the effectiveness of various ions in stabilizing a protein or their capacity to salt out proteins.

Hogness Box A nearly universal TATA sequence in the promoter region of eukaryotic DNA located about 27 bp upstream from the transcription starting point. It involves the selection of a transcriptional starting site.

HOHAHA Abbreviation for homonuclear Hartman-Hahn spectroscopy.

Holandric A type of inheritance controlled by genes linked to the Y chromosome (e.g., a trait transmitted from male to male).

Holandric Gene A gene that is linked to the Y chromosome.

Holiday Structure A four-stranded branch of a DNA intermediate that arises from genetic recombination by exchange of single strands between two homologous DNA duplexes.

Holmium (Ho) A chemical element with atomic weight 165, valence 3.

Holocellulose A complex of cellulose and hemicellulose.

Holocentric A chromosome with a diffuse centromere.

Holocytochrome-c Synthetase The enzyme that catalyzes the following reaction:

$$\text{Cytochrome-c} \rightleftharpoons \text{Apocytochrome-c} + \text{heme}$$

Holoenzyme The complete form of an enzyme (complex of all protein subunits, prosthetic group and coenzyme).

Hologynic A sex-linked character that appears only in the female.

Holokinetic See holocentric.

Holomycin (mol wt 214) An antibiotic produced by *Streptomyces griseus.*

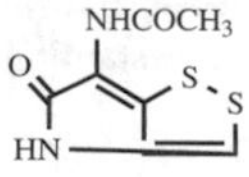

Holoprotein A functional protein that contains a protein part (apoprotein) and ligand or ligands.

Holorepressor A functional repressor that contains an apoprotein and a co-repressor.

Holospora A genus of Gram-negative bacteria that are obligate endosymbionts.

Holo-T Abbreviation for holotransducin.

HOM Abbreviation for a combination drug containing hexamethylmelamine, oncovin, and methotrexate.

Homarine (mol wt 137) A naturally occurring amino acid found in *Arenicola marina.*

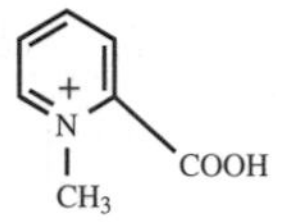

Homatrine A trade name for homatropine hydrobromide, an anticholinergic agent.

Homatropine (mol wt 275) An anticholinergic agent.

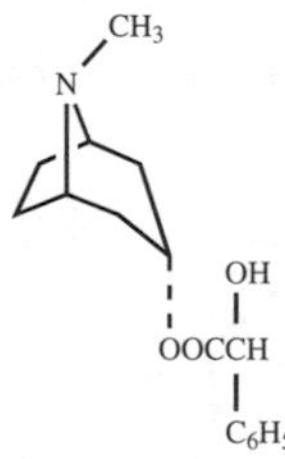

Homeo- A prefix meaning similar.

Homeobox A highly conserved DNA sequence of about 180 base pairs that encodes the site-specific domain of a DNA-binding protein involved in the regulation of gene expression during development.

Homeodomain The protein domain encoded by homeobox involved in DNA binding and recognition.

Homeostasis The maintenance of a dynamic steady-state physiological condition of the body by self-regulation.

Homeothermic Capable of self-regulation of body temperature (warm blooded).

Homeotic Gene The genes that specify the segmental identity in *Drosophila.* Similar genes are found in vertebrates.

Homeotic Mutation 1. A mutation that causes tissues to alter their normal differentiation pattern. 2. Mutation that causes cells in one region of the body to behave as though they were located in another.

Homidium (mol wt 314) An inhibitor of DNA synthesis through intercalation of double-stranded DNA.

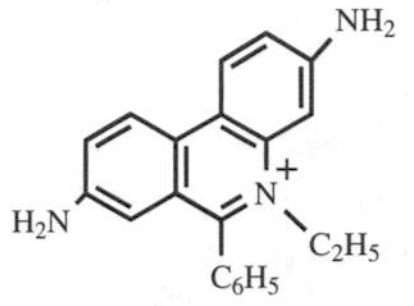

Homing Receptors Cell surface molecules of lymphocytes that enable lymphocytes to bind to ligands on endothelial cells as an initial step toward leaving the blood stream.

Homo- A combining form meaning like, common, or same.

Homoamino Acid An amino acid that contains an additional CH_2 group, e.g., homoserine and homocysteine.

Homoarginine (mol wt 189) A nonprotein amino acid and inhibitor of alkaline phosphatase.

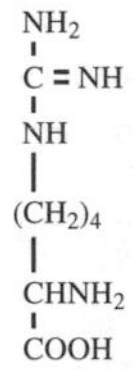

Homobetaine (mol wt 131) A naturally occurring amino acid.

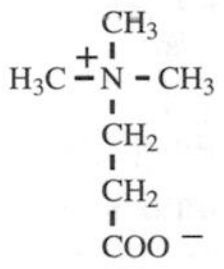

Homobifunctional Describing a chemical reagent that carries two identical reactive groups.

Homocamfin (mol wt 152) A CNS stimulant.

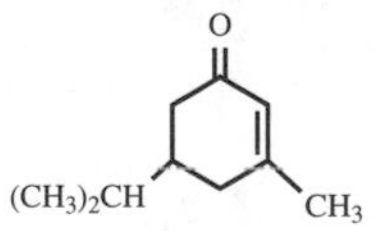

Homocaryon See homokaryon.

Homochlorcyclizine (mol wt 315) A serotonin antagonist.

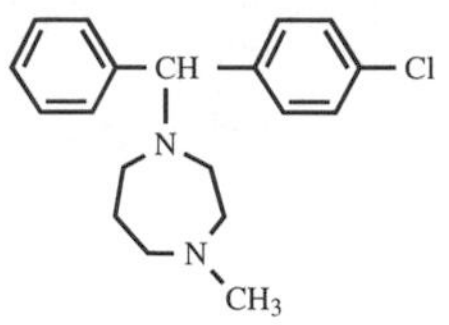

Homocitrate Synthetase The enzyme that catalyzes the following reaction:

$$\text{Hydroxybutane tricarboxylate} + \text{CoA} \rightleftharpoons \text{Acetyl-CoA} + \alpha\text{-ketoglutarate} + H_2O$$

Homocodonic Referring to genetic codons with three identical bases, e.g., AAA, UUU, CCC, and GGG.

Homocopolymer A polymer that consists of a single monomer.

Homocyclic Compounds that contain a closed-ring system in which all the atoms are the same.

Homocysteine (mol wt 135) A nonprotein amino acid and intermediate in methionine metabolism.

$$HS-CH_2CH_2-CHNH_2-COOH$$

Homocysteine Desulfhydrase The enzyme that catalyzes the following reaction:

$$\text{L-Homocysteine} + H_2O \rightleftharpoons \text{Sulfide} + NH_3 + \text{2-ketobutanoate}$$

Homocysteine *S*-Methyltransferase The enzyme that catalyzes the following reaction:

$$S\text{-Adenosyl-L-methionine} + \text{L-homocysteine} \rightleftharpoons S\text{-Adenosyl-homocysteine} + \text{L-methionine}$$

Homocystine (mol wt 268) A naturally occurring amino acid found in human urine.

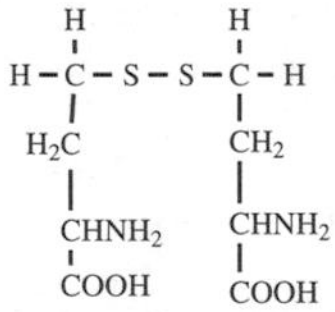

Homocysteinuria A genetic disorder due to the deficiency in cystathione synthetase and characterized by the elevation of concentrations of methionine and homocysteine in the blood.

Homodimer Any macromolecular structure that consists of two identical subunits.

Homofenazine (mol wt 452) A sedative agent.

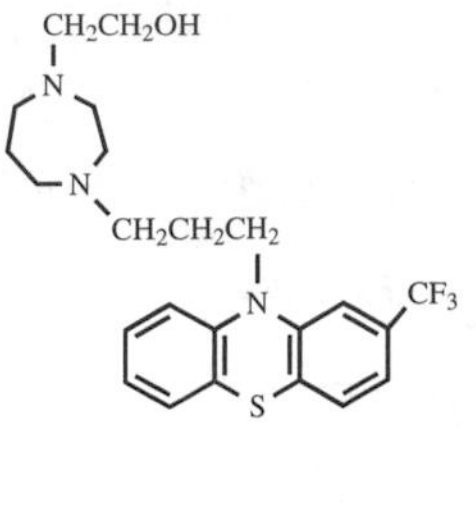

Homogenate The uniform biological fluid obtained by homogenation or disruption of tissues or cells that contains a uniform suspension of broken tissue or lysed cells.

Homogeneous Uniform in nature.

Homogenizer An apparatus for producing homogenate.

Homogenote A bacterial merozygote in which the donor chromosome (exogenote) carries the same alleles as the recipient (endogenote).

Homogentisate Dioxygenase The enzyme that catalyzes the following reaction:

$$\text{Homogentisate} + O_2 \rightleftharpoons \text{4-maleylacetoacetate}$$

Homogentisate Oxygenase See homogentisate dioxygenase.

Homogentisic Acid (mol wt 168) An intermediate in the metabolism of tyrosine and phenylalanine.

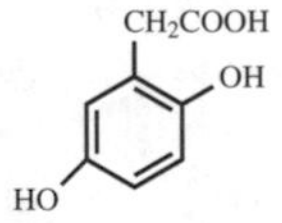

Homogentisicase See homogentisate oxygenase.

Homoglutathione Synthetase The enzyme that catalyzes the following reaction:

ATP + γ-L-glutamyl-L-cysteine + β-alanine
⇅
ADP + Pi + γ-glutamyl-L-cysteinyl-β-alanine

Homograft Tissue graft between two members of the same species.

Homokaryon A multinucleate cell that contains nuclei of only one genotype.

Homolactic Fermentation The fermentation of glucose that produces lactic acid as the sole product.

Homolanthionine (mol wt 236) A naturally occurring amino acid found in *E coli.*

$$HOOCCH(NH_2)(CH_2)_2S(CH_2)CH(NH_2)COOH$$

Homolo- A prefix meaning similar.

Homologous Chromosomes One of the two copies of a particular chromosome in the diploid cell (each copy derived from a different parent).

Homologous Genetic Recombination Recombination of DNA molecules between two similar sequences occurring during meiosis and mitosis.

Homologous Proteins The evolution-related proteins that have sequences and functions similar in different species.

Homologous Sequences Regions in different macromolecules that have a similar order of monomers.

Homology 1. The presence of a similar phenotypic or genotypic feature in different species or groups. 2. The degree of sequence similarity between two genes. 3. Similarity in structure.

Homolytic Bond Cleavage The breakage of a covalent bond between two atoms in which the electrons from the electron pair are shared equally between the two atoms.

Homomultimeric Protein A protein that consists of identical subunits.

Homonicotinic Acid (mol wt 137) A vasodilator.

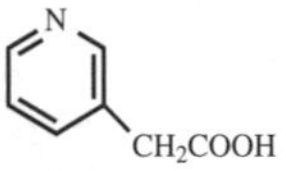

Homooligomer An oligomer that contains only one type of constituent repeating unit.

Homophilic Bonding The attraction force between similar chemical groups.

Homoplasmy The existence of only one type of chloroplast or mitochondria in an organism.

Homopolymer A polymer that consists of repeating units of a single monomer.

Homopolymer Tailing A procedure for joining DNA fragments using deoxynucleotidyl terminal transferase to add a homopolymer extension (e.g., poly-dA or poly- dT) to the 3′-end of the double-stranded DNA.

Homopolysaccharide Polysaccharide that contains only one type of sugar monomer.

Homoserine (119) A nonprotein amino acid.

Homoserine Acetyltransferase The enzyme that catalyzes the following reaction:

Homoserine + acetyl-CoA
⇅
O- Acetyl-L-homoserine + CoA

Homoserine Dehydratase The enzyme that catalyzes the following reaction:

Homoserine + H_2O

⇅

2-ketobutyrate + NH_2 + H_2O

Homoserine Dehydrogenase The enzyme that catalyzes the following reaction:

L-Homoserine + NADP

⇅

L-Aspartate-semealdehyde + NADPH

Homoserine Kinase The enzyme that catalyzes the following reaction:

ADP + Phosphohomoserine ⇌ Homoserine + ATP

Homoserine Succinyltransferase The enzyme that catalyzes the following reaction:

Succinyl-CoA + homoserine

⇅

CoA + *O*-succinyl-L-homoserine

Homostachydrine (mol wt 156) A naturally occurring amino acid found in alfalfa.

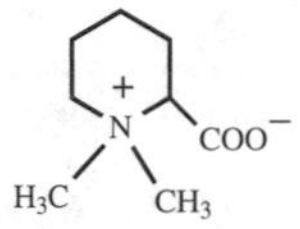

Homo-Tet A trade name for tetanus immune globulin (human).

Homotransplant Synonym of homograft.

Homotrophic Effect Allosteric interaction between binding sites and identical ligands, e.g., binding of four O_2 by a molecule of hemoglobin.

Homotropic Enzyme An allosteric enzyme that uses its substrate as a modulator.

Homovanillic Acid (mol wt 182) A metabolite of adrenalin found in urine.

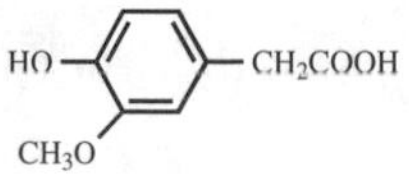

Homozygote A diploid cell or an organism that has two identical alleles of a specified gene.

Homozygous Having identical alleles at one or more loci on a pair of homologous chromosomes.

Homozygous Recessive Having two copies of a recessive allele in a given gene pair.

Honvol A trade name for diethylstilbestrol diphosphate, a hormone.

HOP Abbreviation for a combination drug containing hydroxydaunomycin, oncovin, and prednisone.

Hopantenic Acid (mol wt 233) A cerebral activator.

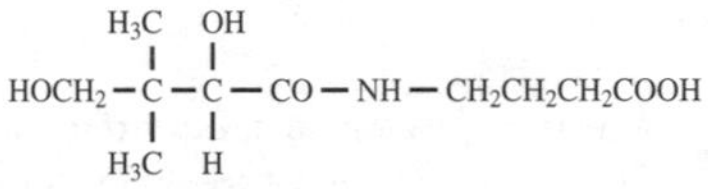

Hopkins-Cole Reaction A colorimetric reaction for determination of tryptophan and other compounds containing an indol ring.

Hordein A type of glutelin found in barley.

Hordeiviruses A group of rod-shaped plant viruses containing divided RNA gnome (e.g., barley-stripe mosaic).

Hormones Substance that is produced in a small quantity in a gland of an animal or one part of a plant and exerts its specific effects on other parts of the body or plant.

Horseradish Peroxidae A hemin-containing peroxidase commonly used in ELISA.

Hostacycline P A trade name for tetracycline hydrochloride, an antibiotic that binds 30S ribosomes and thus inhibiting bacterial protein synthesis.

Host-Cell-Mediated Repair Referring to cut-and-patch repair of a UV-induced lesion on bacteriophage DNA by the host enzymes after phage enter the host cell.

Host-Controlled Modification The modification of bacteriophage DNA by the host modification enzyme.

Host-Dependent Mutant A conditional lethal mutant of bacteriophage that contains an amber mutation in an essential gene and thus can only replicate in a permissive host that contains amber suppresser mutation.

Host-Range Mutant A mutant virus in which the mutation changes the range of hosts of the virus.

Hot-Antigen Suicide A technique in which an antigen is labeled with a high-specific-activity radioisotope and used *in vivo* or *in vitro* to inhibit specific lymphocyte function by attaching the labeled antigen to an antigen-binding lymphocyte and killing it by radiolysis.

Hot-Spots Sites in genes at which mutations occur with exceptionally high frequency.

Housekeeping Genes Genes that provide general functions for the cell, e.g., genes encoding the enzymes for glycolysis and the citric acid cycle.

HPA Abbreviation for 1. hydrophobic affinity; 2. hydroxyphenylacetate or hydroxyphenyl acetic acid.

HpaI A restriction endonuclease from *Haemophilus parainfluenzae* with the following specificity:

↓
5′..........GTTAAC..........3′
3′..........CAATTG..........5′
↑

HpaII A restriction endonuclease from *Haemophilus parainfluenzae* with the following specificity:

↓
5′..........CCGG..........3′
3′..........GGCC..........5′
↑

HPAA Abbreviation for hydroxyphenyl acetic acid.

HPAEC Abbreviation for high performance anion-exchange chromatography or high pH anion-exchange chromatography.

HPAEC-PAD Abbreviation for high performance anion-exchange chromatography with pulsed amperometric detection.

HPCE Abbreviation for high performance capillary electrophoresis.

HPETE Abbreviation for hydroperoxyeicosatetraenoic acid.

HPF Abbreviation for heparin-precipitable fraction.

hPFSH Abbreviation for human pituitary follicle-stimulating hormone.

hPG Abbreviation for human pituitary gonadotropin.

HphI A restriction endonuclease from *Haemophilus parainfluenzae* with the following specificity:

↓
5′..........GGTGA(N)$_8$..........3′
3′..........CCACT(N)$_7$...........5′
↑

HPK1 Abbreviation for hemopoietic progenitor kinase 1.

hPL Abbreviation for 1. human parotide lysozyme; 2. human peripheral lymphocytes; 3. human placental lactogen.

HPLA Abbreviation for hydroxyphenyl lactic acid.

HPLC Abbreviation for high-performance liquid chromatography.

HPLC-ECD Abbreviation for HPLC with electrochemical detection.

hPLD Abbreviation for human PLD (phospholipase D).

HPPA Abbreviation for hydroxyphenyl pyruvic acid.

Hpr A membrane protein responsible for transport of glucose across the bacterial membrane.

HPRP Abbreviation for hypoxanthine-guanine phosphoribosyl transferase.

HPSEC Abbreviation for high-performance size exclusion chromatography.

HPTLC Abbreviation for high performance thin layer chromatography.

H^+-Pump See hydrogen pump.

HPV Abbreviation for 1. *Hemophilus pertussis* vaccine; 2. human papilloma virus; 3. high passage virus.

HR Abbreviation for hormone receptor.

HRA Abbreviation for histamine releasing activity.

HRBC Abbreviation for 1. horse red blood cells; 2. human red blood cells.

HRE Abbreviation for hormone response element.

H1-Receptor Abbreviation for histamine receptor.

H2-Receptor Abbreviation for histamine receptor.

H3-Receptor Abbreviation for histamine receptor.

HRF Abbreviation for histamine releasing factor.

HRG Abbreviation for heregulin, a glycoprotein that binds transmembrane tyrosine kinase.

HRGP Abbreviation for hydroxyproline-rich glycoprotein, a class of plant glycoproteins and proteoglycans found in the plant cell wall and produced in response to injury.

hRIG Abbreviation for human rabies immune globulin.

HRIK Abbreviation for hem-regulated inhibitory kinase.

HRI-Kinase Abbreviation for hem-regulated inhibitory kinase.

HR-LPL Abbreviation for heparin-releasable lipoprotein lipase.

HRM Abbreviation for hem-regulatory motif.

hRNA (HRNA) Abbreviation for heterogeneous RNA.

HRP Abbreviation for horseradish peroxidase.

HRP-C Abbreviation for horseradish peroxidase isoenzyme C.

HRPO Abbreviation for horseradish peroxidase.

HRV Abbreviation for 1. human rhinovirus; 2. human rotavirus; 3. human retrovirus.

hRyR Abbreviation for human ryanodine receptor.

HS Abbreviation for 1. heat stable; 2. heme synthetase; 3. heparan sulfate; 4. horse serum.

HSA Abbreviation for 1. horse serum albumin; 2. human serum albumin.

HSAG Abbreviation for Hepes-saline-albumin-gelation.

HSC Abbreviation for 1. heat shock cognate protein; 2. hematopoietic stem cell; 3. hepatic satellite cell.

hsc 70 (Hsc 70) Abbreviation for heat shock cognate protein of 70 kD.

HSD Abbreviation for hydroxysteroid dehydrogenase.

HSDH Abbreviation for hydroxysteroid dehydrogenase.

HSE Abbreviation for heat-shock element.

Hse Abbreviation for homoserine.

HSF Abbreviation for 1. human skin fibroblast; 2. heat-shock factor.

hSGF (HSGF) Abbreviation for human skeletal growth factor.

HSK Pathway Abbreviation for Hatch-Slack-Kortschak pathway (see also C_4 pathway).

HSL Abbreviation for hormone sensitive lipase.

HSP Abbreviation for heat shock protein whose synthesis is induced by rising temperature or by environmental stress.

Hsp 60 (hsp 60) Abbreviation for 60-kD heat-shock protein.

HSP-70 A family of 70-kD heat shock proteins whose synthesis is induced by environmental stress or heat.

Hsp 90 (hsp 90) Abbreviation for 90-kD heat-shock protein.

Hsp2I (AvaII) A restriction endonuclease from *Hafnia* species with the same specificity as AvaII.

Hsp92I A restriction endonuclease from *Haemophilus influenzae* with the following specificity:

5'........G(A/G)↓CG(T/C)C........3'
3'........C(T/C)GC↑(A/G)G........5'

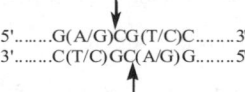

Hsp92II A restriction endonuclease from *Haemophilus influenzae* 92 with the following specificity:

5'........CATG↓........3'
3'........↑GTAC........5'

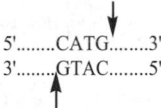

HspAI A restriction endonuclease from *Haemophilus* species A1 with the following specificity:

5'........G↓CGC........3'
3'........CGC↑G........5'

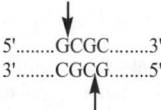

HSPG Abbreviation for heparan sulfate proteoglycan.

HSS Abbreviation for hypersensitive site.

HSSBP (hSSBP) Abbreviation for human single-strand-binding protein.

hST (HST) Abbreviation for human serum transferrin.

HSTF Abbreviation for heat shock transcription factor, a 110-kD protein responsible for regulation of synthesis of heat shock protein in eukaryotes.

H-Strand Abbreviation for heavy strand.

HsuI (HindIII) A restriction endonuclease from *Haemophilus suis* with the same specificity as HindIII.

H-substance A precursor for synthesis of the blood group substances A and B.

H-subunit A subunit of lactate dehydrogenase.

HSV Abbreviation for herpes simplex virus.

HSV-1 Abbreviation for herpes simplex virus type 1.

HSV-2 Abbreviation for herpes simplex virus 2.

HSV-TK Abbreviation for herpes simplex virus thymidine kinase.

Ht Abbreviation for hematocrit.

HT Abbreviation for 1. hemagglutination titer; 2. human thrombin; 3. hydroxytryptamine.

3HT Abbreviation for 3-hydroxytryptamine.

5HT Abbreviation for 5-hydroxytryptamine.

^{3}HT Abbreviation for tritiated thymidine.

HTAC Abbreviation for hexadecyltrimethylammonium chloride.

hTACS (HTACS) Abbreviation for human thyroid adenylcyclase stimulator.

hTAT (HTAT) Abbreviation for human tetanus antitoxin.

HTC Abbreviation for hepatoma tissue culture cell.

HTH Abbreviation for 1. helix-turn-helix; 2. homeostatic thymic hormone.

HTL Abbreviation for human thymic leukemia.

HTLA Abbreviation for 1. high-titer, low-avidity antibody; 2. human T lymphocyte antigen.

HTLV Abbreviation for human T cell leukemia virus.

HTLV-1 Abbreviation for human T-lymphotrophic virus type 1, a retrovirus causing leukemia.

HTLV-II Abbreviation for human T-lymphotrophic virus type II, related to HTLV-1.

hTopo-II Abbreviation for human topoisomerase-II.

HTP Abbreviation for 5-hydroxytryptophan.

5HTP Abbreviation for 5-hydroxytryptophan.

HTRP Abbreviation for human TRP (transient receptor potential) protein.

hTS Abbreviation for human thyroid stimulator.

hTSH Abbreviation for human thyroid stimulating hormone.

HU Abbreviation for 1. hemagglutination unit; 2. hemolytic unit; 2. 5-hydroxyuracil.

Humalog A trade name for insulin, an anti-diabetic agent.

Human Chorionic Gonadotropin A gonadotropic hormone produced by the placenta.

Human Immunodeficiency Virus A retrovirus that causes human acquired immune deficiency syndrome (AIDS) in humans. It possesses two genomic RNA molecules and the enzyme reverse transcriptase.

Human Papilloma Virus A member of papovavirus.

Human Prostate Specific Antigen A prostate cancer indicator.

Humate-P A trade name for antihemophilic factor VIII.

Humatin A trade name for paromycin sulfate, an amebicide.

Humatrope A trade name for somatropin, a hormone.

Humegon A trade name for menotropins, a hormone used as a fertility drug.

Humibid LA A trade name for guaifenesin, an expectorant and antitussive agent.

Humibid Sprinkle A trade name for guaifenesin, an expectorant.

Humor Fluid that occurs normally in the body.

Humoral Pertaining to body fluid, e.g., plasma, lymph, and tissue fluid.

Humoral Immune Response Immune response that is mediated by antibody.

Humoral Immunity Immunity that is mediated by circulating antibodies.

Humorsol A trade name for demecarium bromide, an anticholinesterase drug.

Humulin A synthetic insulin manufactured through the use of DNA recombination technology.

Humulon (mol wt 362) An antibiotic constituent of hops (*Humulus lupulus*).

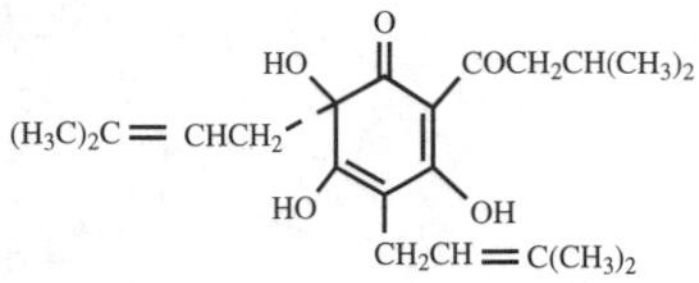

Hunter Syndrome An X-linked genetic disorder of connective tissue characterized by the storage of mucopolysaccharide due to deficiency of iduronate sulfatase and heparan sulfate sulfatase.

HU-Protein A histonelike protein that facilitates the binding of dnaA protein to DNA.

HUR Abbreviation for hydroxyurea.

H_2Urd Abbreviation for dihydrouridine.

Hurler's Disease An autosomal recessive disorder of connective tissue characterized by the accumulation of dermatan and heparan sulfate due to deficiency of the enzyme iduronidase.

Hurricaine A trade name for 25% solution of benzocaine, used as a topical anesthetic agent.

Hu-Tet A trade name for tetanus immune globulin (human).

HUVEC Abbreviation for human umbilical-vein endothelial cell.

HVA Abbreviation for homovanillic acid.

HVE Abbreviation for high voltage electrophoresis.

hwt-MCADH Abbreviation for human wild type medium chain acyl-CoA dehydrogenase.

HX Abbreviation for hypoxanthine.

Hyalgan A trade name for sodium hyaluronate used for the treatment of pain in osteoarthritis.

Hyaloplasm 1. Cytoplasm without cell organelles. 2. Cytosol.

Hyalouronoglucuronidase The enzyme that catalyzes the random hydrolysis of 1,3-linkages between b-D-glucuronate and *N*-acetylglucosamine residues in hyaluronate.

Hyaluronic Acid A mucopolysaccharide consisting of alternating β-(1,3) glucuronic and β-(1,4) glucosaminidic linkages. It serves as a lubricating fluid of the joints and as a blocking agent for spreading of microorganisms in tissue.

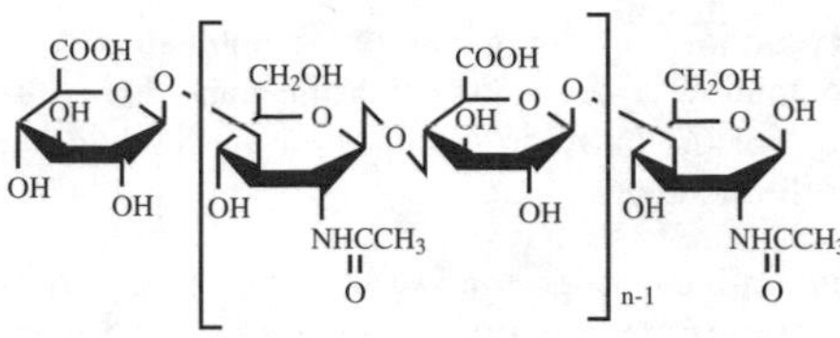

Hyaluronidase See hyalouronoglucosaminase.

Hyaluronoglycosaminidase The enzyme that catalyzes the random hydrolysis of 1,4-linkages between b-D-*N*-acetylglucosamine and D-glucuronate residues in hyaluronate.

HY-Antigen The transplantation antigen that is present on the tissue of the heterogametic sex of mammals. The gene that encodes the HY antigen in humans that is located on the Y chromosome.

Hybolin A trade name for nandrolone decanoate, an anabolic hormone that promotes tissue-building processes.

Hybrid Offspring of a cross between genetically different parents.

Hybrid Arrest Translation A technique used for identification of cDNA corresponding to a mRNA by the ability of cDNA to pair with mRNA *in vitro* to inhibit translation.

Hybrid Vector A vector that contains inserted foreign DNA.

Hybrid Vigor Increased vitality of the hybrid offspring.

Hybridization 1. Formation of a hybrid from genetically different parents. 2. Formation of a double-stranded structure from two single-stranded molecules, e.g., DNA:DNA, DNA:RNA, and RNA:RNA. 3. Formation of oligomeric protein from protein subunits of different sources.

Hybridoma Hybrid cell line used for the production of monoclonal antibodies, generated by fusion of an antibody-producing B lymphocyte with a myeloma cancer cell.

Hycamtin A trade name for topotecan hydrochloride, an anti-tumor agent with topoisomerase-inhibitory activity.

Hycanthone (mol wt 356) An anthelmintic agent.

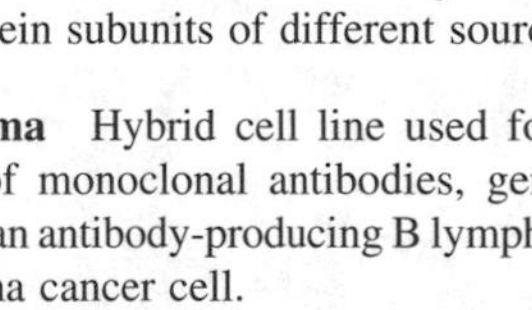

Hycodon A trade name for a combination drug containing hydrocodone bitartrate and homatropine methylbromide, used as an antitussive and analgesic agent.

Hycomine Compound A trade name for a combination drug containing hydrocodone bitartrate, chlorpheniramine maleate, phenylephrine hydrochloride, acetaminophen, and caffeine, used as an antitussive, anti-allergic agent.

Hycort A trade name for hydrocortisone, used as an anti-inflammatory agent.

Hycortole A trade name for hydrocortisone, used as an anti-inflammatory agent.

Hycotuss A trade name for a combination drug containing hydrocodone bitartrate and guaifenesin, used as an antitussive and expectorant agent.

Hydeltrasol A trade name for prednisolone sodium phosphate, used as an anti-inflammatory agent.

Hydergine A combination of an adrenergic blocker drug containing dihydroergocornine mesylate, dihydroergocristine mesylate, and dihydroergocryptine mesylate.

Hydextran A trade name for iron dextran, an iron provider.

Hydnocarpic Acid (mol wt 252) An antibacterial agent from seeds of *Hydnocarpus wightiana.*

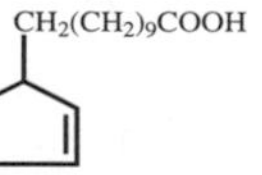

Hydopa A trade name for methyldopa, an antihypertensive agent that decreases sympathetic outflow.

Hydracarbazine (mol wt 153) An antihypertensive and diuretic agent.

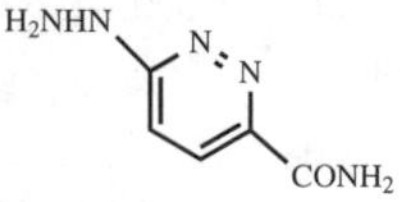

Hydralazine Hydrochloride (mol wt 160) An antihypertensive agent capable of relaxing arteriolar smooth muscle.

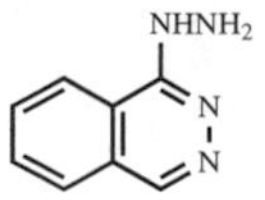

Hydramethylnon (mol wt 495) An insecticide.

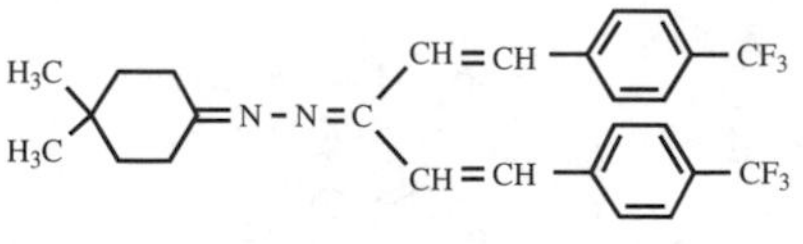

Hydramine A trade name for diphenhydramine hydrochloride, an antihistaminic agent.

Hydramitrazine (mol wt 253) An antispasmodic agent.

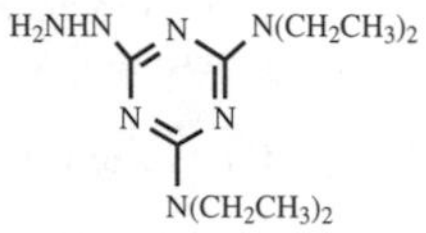

Hydramyn A trade name for diphenhydramine hydrochloride, an antihistaminic agent.

Hydrargaphen (mol wt 982) An anti-infective agent.

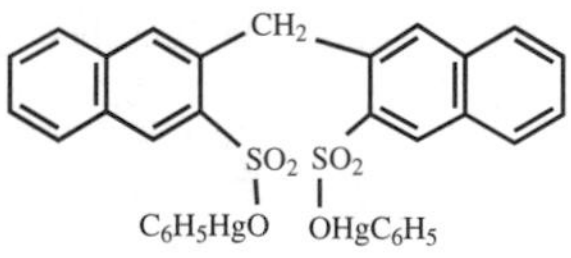

Hydratase The enzyme that catalyzes the reversible hydration of a double bond.

Hydrazine (mol wt 32) A reagent used in DNA and protein sequencing experiments.

$$H_2NNH_2$$

Hydrazinolysis of Protein Cleavage of peptide bonds in protein by treatment of protein with hydrazine.

Hydrea A trade name for hydroxyurea, an antitumor agent.

Hydride Ion A proton with an associated pair of electrons, an anion ($H:^-$).

Hydril A trade name for diphenhydramine hydrochloride, an antihistaminic agent.

Hydrindantin (mol wt 322) A reagent used for photometric determination of amino acid.

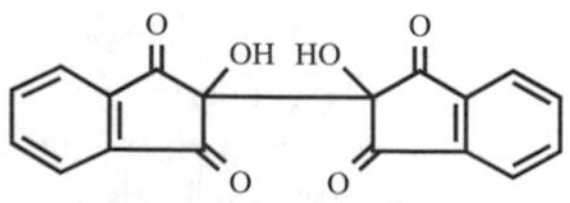

Hydrisalic A trade name for salicylic acid, a topical antibacterial and antifungal agent.

Hydrobexan A trade name for vitamin B_{12a} (hydroxocobalamin).

Hydrocarbon Compounds that consist of only hydrogen and carbon.

Hydrocet A trade name for a combination drug containing hydrocodone hydrochloride and acetaminophen, used as an antitussive and analgesic agent.

Hydrochloric Acid (mol wt 37) An acid secreted by the stomach.

Hydrochlorothiazide (mol wt 298) A diuretic agent capable of inhibiting sodium reabsorption.

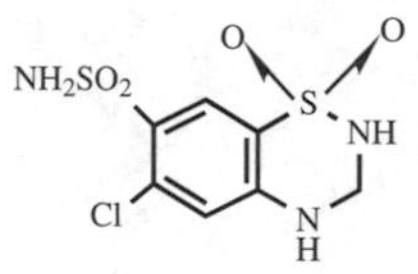

Hydrocodone (mol wt 299) A narcotic analgesic and antitussive agent.

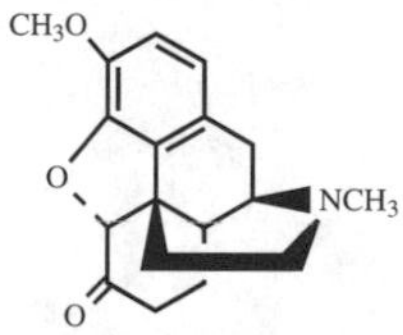

Hydrocortamate (mol wt 476) A glucocorticoid.

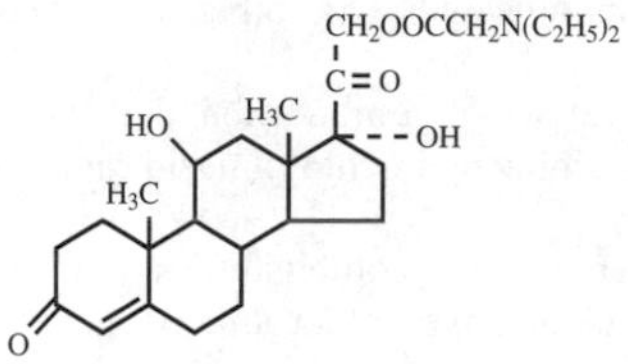

Hydrocortex A trade name for hydrocortisone.

Hydrocortisone (mol wt 362) An anti-inflammatory hormone capable of suppressing immune response, stimulating bone marrow, and regulating metabolism of lipid and carbohydrate (also called compound F).

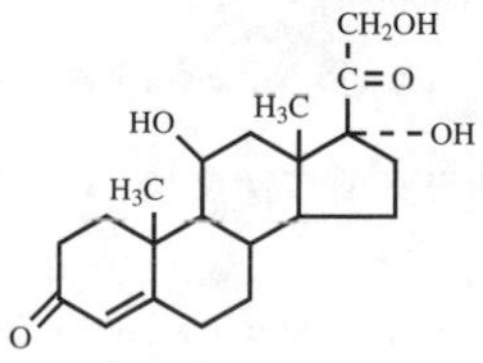

Hydrocortone A trade name for hydrocortisone, used an anti-inflammatory agent.

HydroDiueil A trade name for hydrochlorothiazide, a diuretic agent.

Hydrodynamics The science that deals with fluids in motion and the forces affecting the motion.

Hydrogel Any gel in which water is the liquid component.

Hydrogen A chemical element with atomic weight 1, valence 1.

Hydrogenase The enzyme that catalyzes the following reaction:

$$2 \text{ Reduced ferrodoxin} + 2 H^+ \rightleftharpoons 2 \text{ Oxidized ferredoxin} + H_2$$

Hydrogen Bond A weak, noncovalent bond between a hydrogen atom and an electronegative atom such as oxygen or nitrogen.

Hydrogen Carrier An electron carrier.

Hydrogen Chloride (mol wt 36) An inorganic acid.

$$HCl$$

Hydrogen Dehydrogenase The enzyme that catalyzes the following reaction:

$$H_2 + NAD^+ \rightleftharpoons H^+ + NADH$$

Hydrogen Half Cell The standard reference of half-cell that consists of an electrode immersed in a 1 *M* H^+ solution (pH 0) that is in equilibrium with H_2 gas at 1 atmosphere pressure.

Hydrogen Ion Pump An ATPase-mediated hydrogen transfer that generates H^+-gradient for the transport of nutrients.

Hydrogen Ionophore II (mol wt 345) A hydrogen ion ionophore used in microelectrodes and assay of hydrogen ion activity in a single cell.

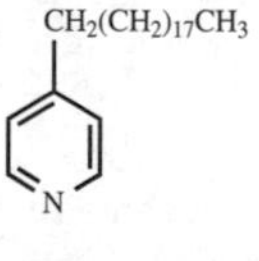

Hydrogen Ionophore IV (mol wt 375) A hydrogen ion ionophore used in a pH sensor.

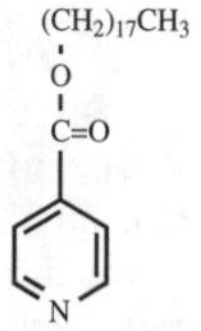

Hydrogen Oxidizing Bacteria Bacteria that are capable of growing chemolithoautotrophically by obtaining energy from oxidation of hydrogen gas with oxygen via an electron transport system.

Hydrogen Peroxide (mol wt 34) An anti-infective agent.

H_2O_2

Hydrolase The enzyme that catalyzes the reaction of hydrolysis.

Hydrolysate Substance or substances produced by hydrolysis.

Hydrolysis Splitting of one molecule into two by incorporation of one water molecule.

Hydrolytic Enzymes Enzymes that catalyze the reaction of hydrolysis.

Hydrometer A device used for the measurement of specific gravity of a liquid.

Hydrometry The determination of specific gravity of a liquid.

Hydromorphone (mol wt 285) A narcotic analgesic agent capable of binding with opiate receptors in the CNS.

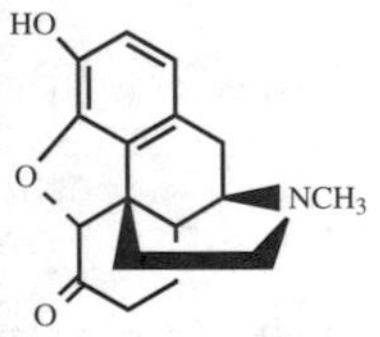

Hydromox A trade name for quinethazone, a diuretic agent that increases urine secretion of sodium and water.

Hydronium Ion Hydrated hydrogen or proton (H_3O^+).

Hydroorotic Acid (mol wt 158) A precursor for thymine, uracil, and cytosine.

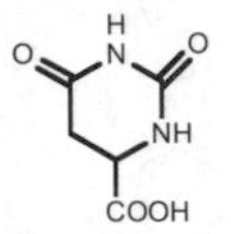

Hydro-Par A trade name for hydrochlorothiazide, a diuretic agent.

Hydrophilic Polar molecules or polar parts of molecules that readily associate with water (water loving).

Hydrophobic Nonpolar molecules or nonpolar parts of molecules that do not readily associate with water (also known as lipophilic or water hating). Hydrophobic molecules do not dissolve in water.

Hydropine A combination antihypertensive agent containing hydroflumethiazide and reserpine.

Hydropoiesis The production of sweat.

Hydroponics The culture of plants without soil.

Hydropres-25 A combination antihypertensive drug containing hydrochlorothiazide and reserpine.

Hydrops Fetalis The massive edema in the fetus or newborn with severe erythroblastosis fetalis due to the lack of the α-globin gene.

Hydroquinidine (mol wt 326) An antiarrhythmic agent.

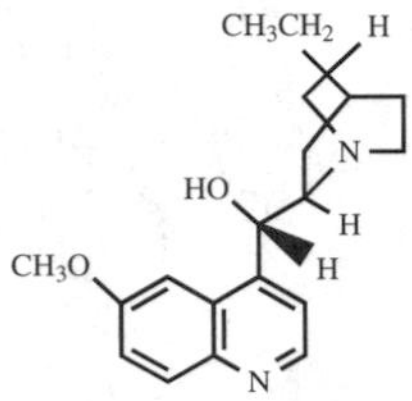

Hydroquinone (mol wt 110) A depigmentor used for treatment of hyperpigmentation.

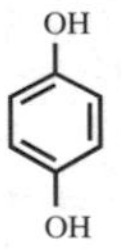

Hydro-Reserp A combination antihypertensive drug containing hydrochlorothiazide and reserpine.

Hydro-Serp A combination antihypertensive drug containing hydrochlorothiazide and reserpine.

Hydroserpine A combination antihypertensive drug containing hydrochlorothiazide and reserpine.

Hydrosis 1. Excessive perspiration. 2. Disorder of the sweat gland.

Hydrostatic Pertaining to the pressures exerted by a liquid at rest.

Hydrotensin-25 A combination antihypertensive drug containing hydrochlorothiazide and reserpine.

HydroTex A trade name for hydrocortisone.

Hydrotropism A tropism in which water is the stimulus.

Hydroxacen A trade name for hydroxyzine hydrochloride, an antianxiety agent.

Hydroxocobalamin An analog of Vitamin B_{12} (Vitamin B_{12a}) and a coenzyme.

Hydroxy- A prefix denoting the presence of one or more hydroxyl groups.

Hydroxyacid Oxidase The enzyme that catalyzes the following reaction:

Hydroxyacid + O_2 $\rightleftharpoons$ Ketoacid + H_2O_2

Hydroxyacyl-ACP An intermediate in fatty acid biosynthesis.

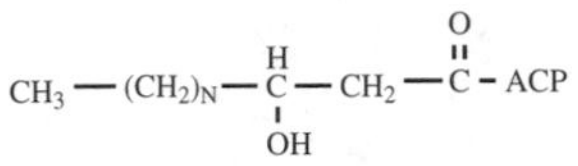

β-Hydroxyacyl-ACP Dehydrase The enzyme that catalyzes the following reaction:

β-Hydroxybutyryl-ACP

$\rightleftharpoons$

α,β-Trans- butenoyl-ACP

Hydroxyacyl-CoA An intermediate in fatty acid metabolism.

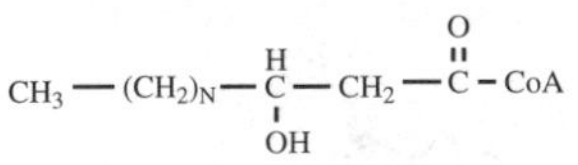

3-Hydroxyacyl-CoA Dehydrogenase The enzyme that catalyzes the following reaction:

3-Hydroxyacyl-CoA + NAD^+

$\rightleftharpoons$

3-Ketoacyl- CoA + NADH

3-Hydroxyacyl-CoA NAD⁺ Oxidoreductase The systematic name for 3-hydroxyacyl-CoA dehydrogenase.

Hydroxyacylglutathione Hydrolase The enzyme that catalyzes the following reaction:

S-Hydroxyacyl glutathione + H_2O

$\rightleftharpoons$

Glutathione + 2-hydroxyacid anion

Hydroxyalkyl-Protein Kinase Synonym of protein kinase.

Hydroxyamphetamine (mol wt 151) An adrenergic agent.

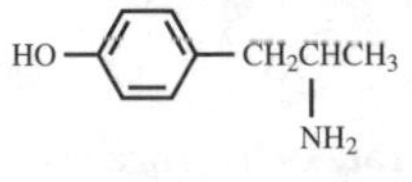

3-Hydroxyanthranilate (mol wt 152) An intermediate in metabolism of tryptophan.

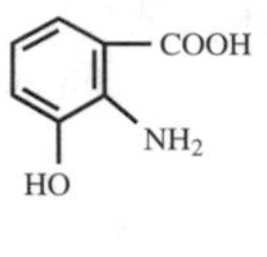

3-Hydroxyanthranilate-3,4-Dioxygenase The enzyme that catalyzes the following reaction:

3-Hydroxyanthranilate + O_2

$\rightleftharpoons$

2-Amino 3-carboxymuconate 6-semialdehyde

Hydroxyapatite A form of calcium phosphate that binds dsDNA and used as column packing material for chromatographic separation of double-stranded DNA from a mixture containing both single-stranded and double-stranded DNA.

Hydroxyaspartate Aldolase The enzyme that catalyzes the following reaction:

Hydroxy-L-asparate

$\rightleftharpoons$

Glycine + glyoxylate

Hydroxyaspartic Acid (mol wt 149) A derivative of aspartic acid.

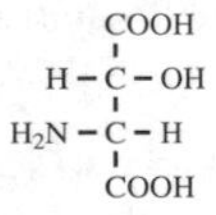

Hydroxybenzoate-CoA Ligase The enzyme that catalyzes the following reaction:

ATP + hydroxybenzoate + CoA

$\rightleftharpoons$

AMP + PPi + hydroxybenzoate-CoA

Hydroxybenzoate Decarboxylase The enzyme that catalyzes the following reaction:

Hydroxybenzoate $\rightleftharpoons$ Phenol + CO_2

4-Hydroxybenzoate 3-Monooxygenase The enzyme that catalyzes the following reaction:

4-Hydroxybenzoate + NADPH + O_2

$\rightleftharpoons$

Protocatechuate + NADP+ + H_2O

4-Hydroxybenzoate NADPH Oxygen Oxidoreductase The systematic name for 4-hydroxybenzoate 3-monooxygenase.

Hydroxybenzylpenicillin Sodium (mol wt 372) An antibiotic produced by mutant of *Penicillium chrysogenum*.

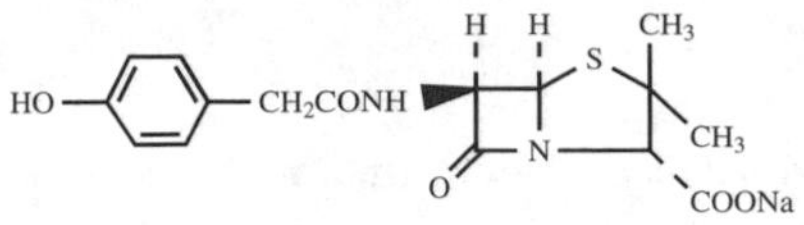

Hydroxybutyl-CoA (mol wt 853) An active hydroxybutyric acid.

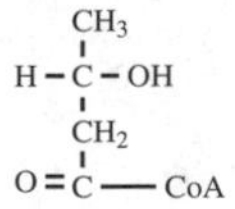

3-Hydroxybutyl-CoA Dehydratase The enzyme that catalyzes the following reaction:

3-Hydroxybutanoyl-CoA

⇅

Crotonoyl-CoA + H_2O

3-Hydroxybutyrate Dehydrogenase The enzyme that catalyzes the following reaction:

3-Hydroxybutyrate + NAD^+

⇅

Acetoacetate + NADH

4-Hydroxybutyrate Dehydrogenase The enzyme that catalyzes the following reaction:

4 Hydroxybutanoate + NAD^+

⇅

Succinate semialdehyde + NADH

Hydroxybutyric Acid (mol wt 104) A four-carbon organic acid.

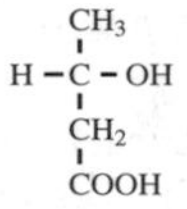

3-Hydroxycamphor (mol wt 168) A topical antipruritic agent.

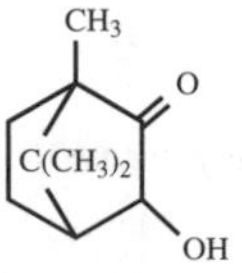

Hydroxychloroquine (mol wt 336) An antimalarial, antirheumatic, and antilupus erythematosus agent.

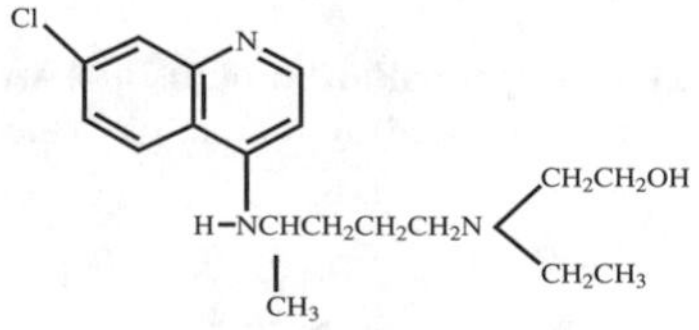

Hydroxydione Sodium (mol wt 455) An anesthetic agent.

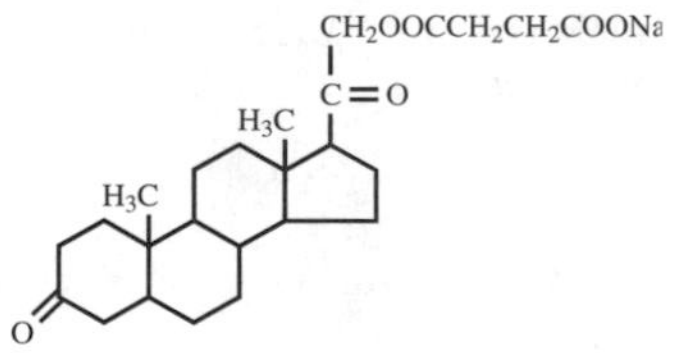

***p*-Hydroxyephedrine (mol wt 181)** An adrenergic agent.

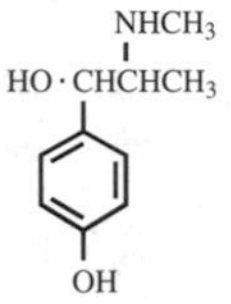

***N*-Hydroxyethylpromethazine Chloride (mol wt 365)** An antihistaminic agent.

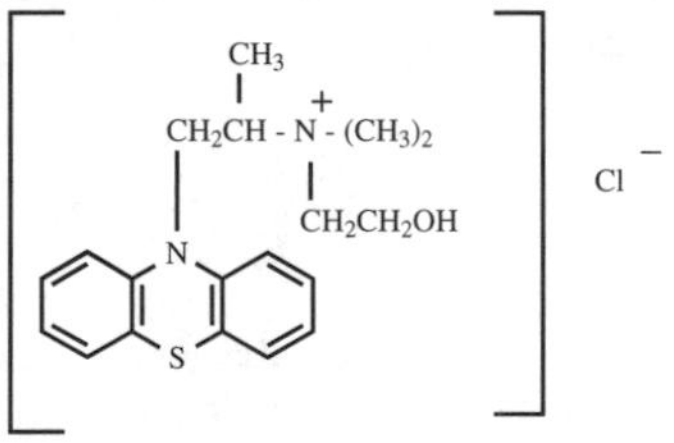

Hydroxyethyl-Starch Synonym of hetastarch, a plasma expander.

Hydroxyglutamate Decarboxylase The enzyme that catalyzes the following reaction:

3-Hydroxy-L-glutamate

⇅

4-Amino-3-hydroxybutanoate + CO_2

Hydroxyglutamate Transaminase The enzyme that catalyzes the following reaction:

Hydroxy-L-glutamate + α-ketoglutarate

⇅

α-Keto-4-hydroxyglutarate + L-glutamate

Hydroxyglutamic Acid (mol wt 163) A nonprotein amino acid.

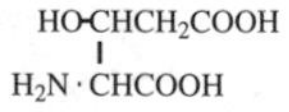

Hydroxyglutarate Dehydrogenase The enzyme that catalyzes the following reaction:

Hydroxyglutarate + acceptor

⇅

α-Keto-glutarate + reduced acceptor

Hydroxyglutarate Synthetase The enzyme that catalyzes the following reaction:

Propanoyl-CoA + glyoxylate + H_2O
⇌
Hydroxyglutarate + CoA

Hydroxyisobutyric Acid (mol wt 104) An intermediate in the metabolism of branched-chain amino acids.

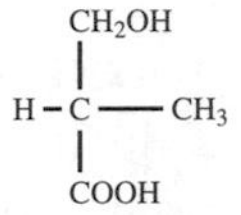

β-Hydroxyisobutyryl CoA Hydrolase The enzyme that catalyzes the following reaction:

β-Hydroxybutyryl CoA + H_2O
⇌
β-Hydroxyisobutyrate + CoASH

Hydroxykynurenine An intermediate in tryptophan metabolism.

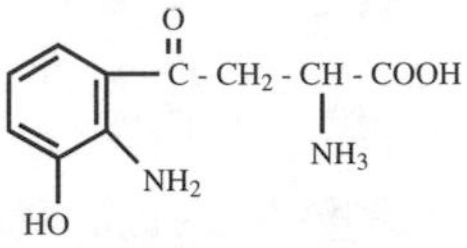

Hydroxyl Group A polar, functional OH group.

Hydroxyl Ion A negatively charged OH group (OH^-).

Hydroxylamine (mol wt 33) A reducing agent and antioxidant for fatty acids.

NH_2OH

Hydroxylamine Oxidase The enzyme that catalyzes the following reaction:

Hydroxylamine + O_2 ⇌ Nitrite + H_2O

Hydroxylamine Reductase The enzyme that catalyzes the following reaction:

Ammonia + acceptor
⇌
Hydroxylamine + reduced acceptor

Hydroxylase An enzyme that introduces a hydroxyl group into a molecule.

Hydroxylation Reaction Chemical reaction that introduces a hydroxyl group into an organic compound.

Hydroxylysine (mol wt 163) A lysine derivative found in collagen, it is formed through the process of posttranscription.

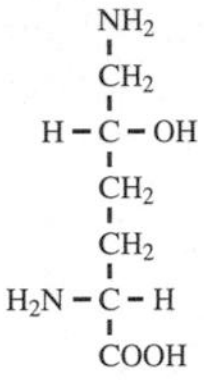

Hydroxylysine Kinase The enzyme that catalyzes the following reaction:

GTP + 5-hydroxylysine
⇌
GDP + 5-phosphohydoxy-L-lysine

Hydroxymagnesium Aluminate Synonym of magaldrate, an antacid.

Hydroxymandelonitrile Hydroxy-benzaldehyde Lyase The enzyme that catalyzes the following reaction:

4-Hydroxymandelonitrile
⇌
Cyanide + 4-hydroxybenzaldehyde

Hydroxymandelonitrile Lyase Synonym of hydroxymandelonitrile hydroxybenzaldehyde lyase.

5-Hydroxymethyl Cytosine (mol wt 141) An unusual base found in the DNA of T-even phage.

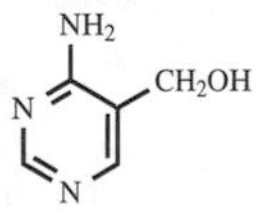

β-Hydroxy-β-methylglutaric acid (mol wt 162) A derivative of glutaric acid.

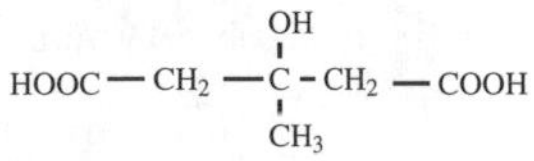

β-Hydroxy-β-methylglutaryl-CoA (mol wt 919) An intermediate in the biosynthesis of cholesterol.

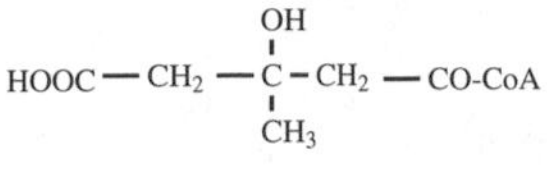

Hydroxymethylglutaryl-CoA Reductase The enzyme that catalyzes the following reaction:

Meralonate + CoA + NAD^+

⇌

3-Hydroxy-3-methylglutaryl-CoA + NADH

Hydroxymethylphenylalanine (mol wt 195) A naturally occurring amino acid found in *E. coli*.

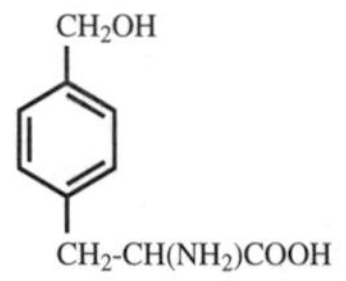

Hydroxyminaline (mol wt 127) A naturally occurring amino acid found in *Penicilium* and *Aspergillus*.

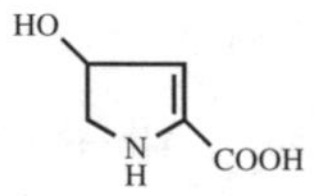

Hydroxynitrile Lyase Synonym of mandelonitrile lyase.

Hydroxypethidine (mol wt 263) A narcotic analgesic agent.

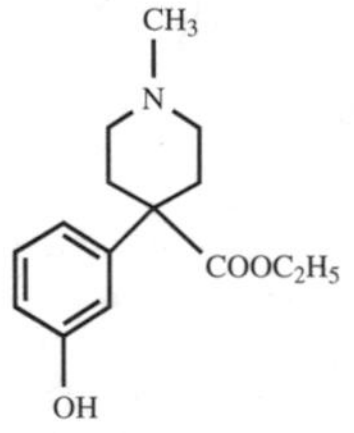

Hydroxyphenamate (mol wt 209) An anxiolytic agent.

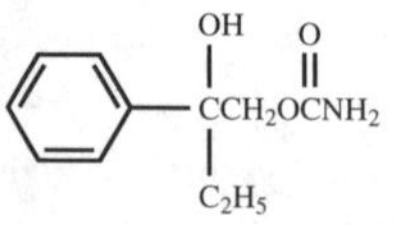

Hydroxyphenylacetic Acid (mol wt 152) A reagent used for fluorometric determination of uric acid.

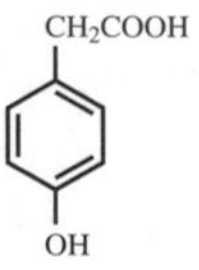

Hydroxyphenylpyruvate (mol wt 179) An intermediate in the metabolism of phenylalanine.

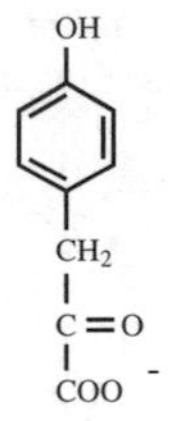

Hydroxyphenylpyruate Dioxygenase The enzyme that catalyzes the following reaction:

p-Hydroxyphenylpyruvate + ascorbate + O_2

⇌

Homogentisate + dihydroascorbate + H_2O + CO_2

Hydroxyprocaine (mol wt 252) An anesthetic agent.

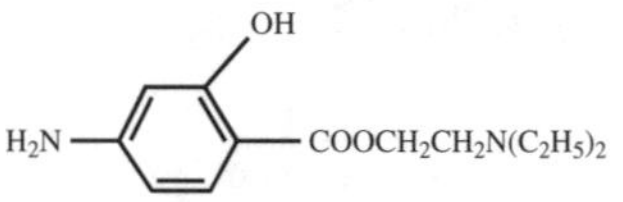

17α-Hydroxyprogesterone (mol wt 330) A hormone that is capable of suppressing ovulation.

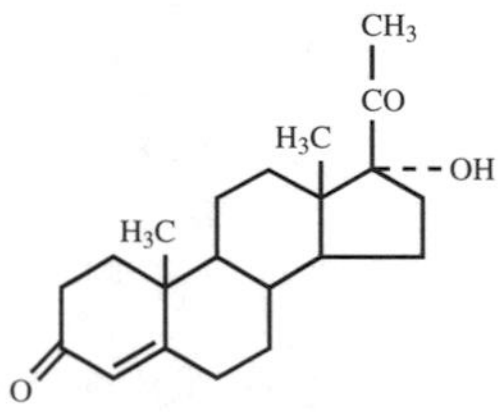

17α-Hydroxyprogesterone Caproate (mol wt 429) A progestogen derivative capable of suppressing ovulation.

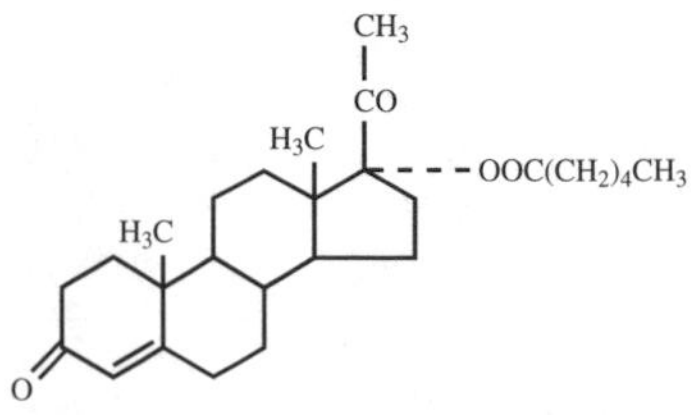

Hydroxyproline (mol wt 131) A nonessential amino acid.

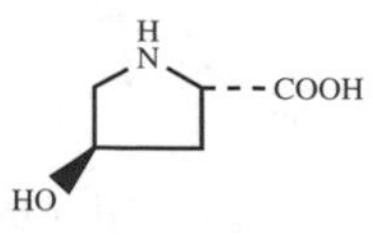

Hydroxyproline Epimerase The enzyme that catalyzes the following reaction:

L-Hydroxyproline ⇌ D-Allohydroxyproline

Hydroxyprolinemia A metabolic disorder characterized by the high levels of free 4-hydroxyproline in the blood plasma and urine.

3-Hydroxypropionate Dehydrogenase The enzyme that catalyzes the following reaction:

$$\text{3-Hydroxypropionate} + NAD^+ \rightleftharpoons \text{Malonate semialdehyde} + NADH$$

Hydroxypropyl Cellulose A non-ionic, water soluble ether of cellulose that produces a solution having a wide range of viscosity.

Hydroxypropyl Methylcellulose A non-ionic, water soluble ether of cellulose that produces solution having a wide range of viscosity.

Hydroxyprostaglandin Dehydrogenase The synonym of 3-α-hydroxysteroid dehydrogenase.

8-**Hydroxyquinoline (mol wt 145)** An antiseptic agent.

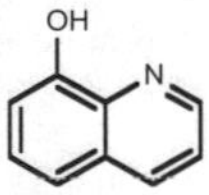

Hydroxypyruvate Reductase The enzyme that catalyzes the following reaction:

$$\text{Hydroxypyruvate} + NADH \rightleftharpoons \text{glycerate} + NAD^+$$

Hydroxypyruvic Acid (mol wt 104) An intermediate in photorespiration for the metabolism of phosphoglycolate.

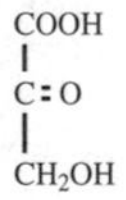

3-Hydroxystachydrine (mol wt 159) A naturally occurring nonprotein amino acid found in *Courbonia virgata*.

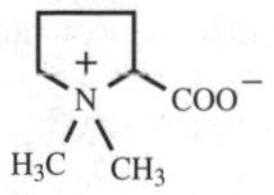

21-Hydroxysteroid Dehydrogenase The enzyme that catalyzes the following reaction:

$$\text{Steroid-21-alcohol} + NAD^+ \rightleftharpoons \text{Steroid-21- aldehyde} + NADH$$

Hydroxystreptomycin (mol wt 598) An antibiotic produced by *Streptomyces reticuli.*

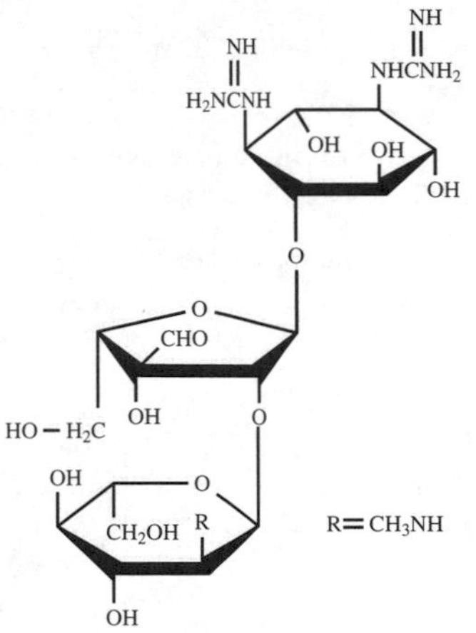

Hydroxytetracaine (mol wt 280) A topical anesthetic agent.

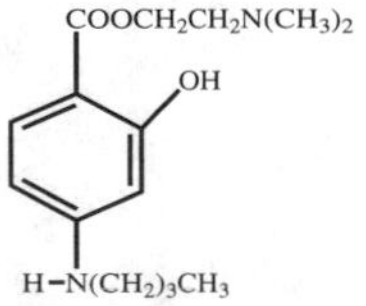

5-Hydroxytryptamine See serotonin.

Hydroxytryptophan (mol wt 220) A tryptophan derivative used as an antidepressant.

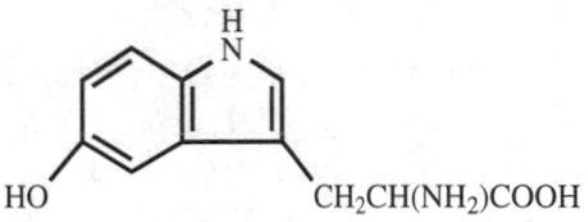

Hydroxyurea (mol wt 76) An antineoplastic agent that inhibits DNA synthesis.

$$H_2NCONHOH$$

Hydroxyzine (mol wt 375) An antianxiety and antihistaminic agent. It is a H_1 receptor antagonist.

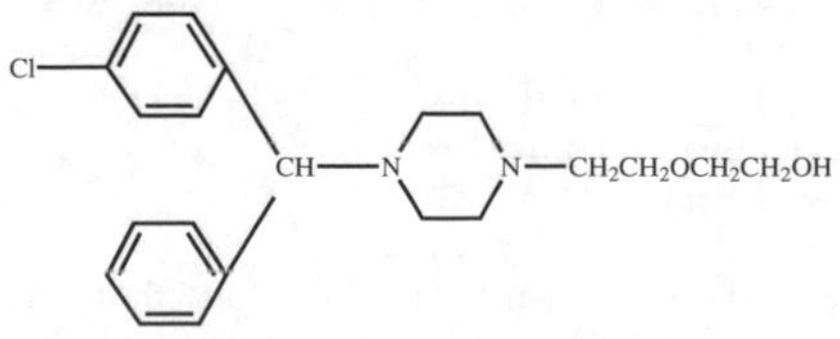

Hydroxyzine Embonate An antihistaminic and antianxiety agent.

Hydroxyzine Hydrochloride An antihistaminic and antianxiety agent.

Hydroxyzine Pamoate An antihistaminic and antianxiety agent.

H262Y-GDH Abbreviation for GDH (glucose dehydrogenase) in which His^{262} has been altered to tyrosine (Y).

Hy-Gesterone A trade name for hydroxyprogesterone.

Hygrometer An instrument used for measuring relative humidity of the atmosphere.

Hygromycin (mol wt 511) An antibiotic produced by *Streptomyces hygroscopicus.*

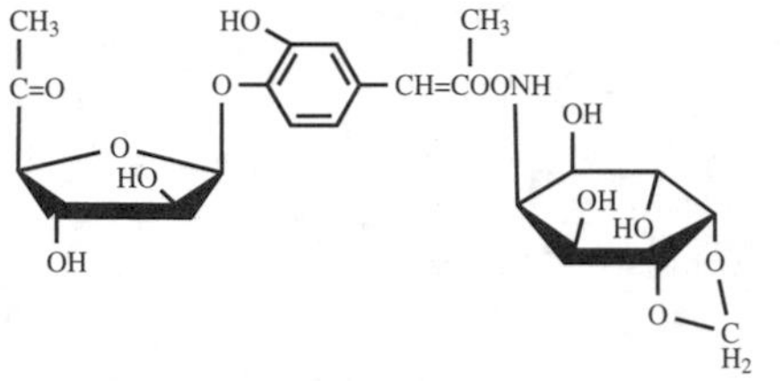

Hygromycin B An antibiotic produced by *Streptomyces hygroscopicus.*

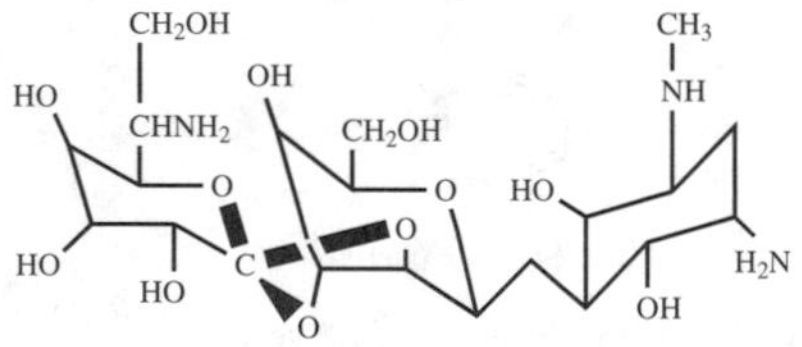

Hygroton A trade name for chlorthaliodone, a diuretic agent that increases excretion of sodium and water by inhibiting sodium reabsorption.

Hyl Abbreviation for hydroxylysine.

Hylorel A trade name for guanadrel sulfate, an antihypertensive agent that inhibits release of norepinephrine.

Hylutin A trade name for hydroxyprogesterone caproate, a hormonal agent that suppresses ovulation.

Hyoscine See scopolamine.

Hyoscine Butylbromide See scopolamine butylbromide.

Hyoscine Hydrobromide See scopolamine hydrobromide.

Hyoscyamine (mol wt 289) Anticholinergic agent that competitively blocks acetylcholine.

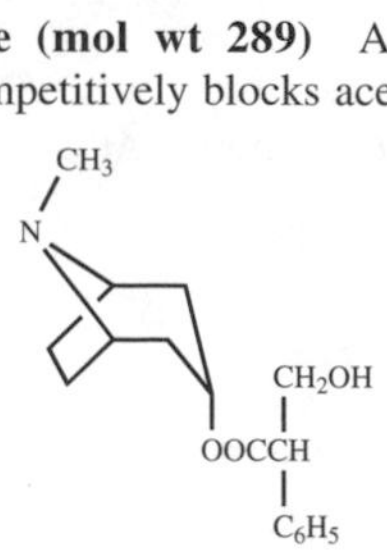

Hy-Pam A trade name for hydroxyzine pamoate, an antianxiety agent.

Hypaphorine (mol wt 246) A convulsive poison from seed of *Erythrina americana.*

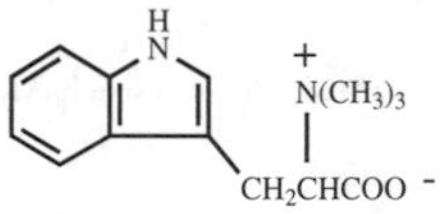

Hypelcin A A 20-residue peptide of microbial origin that modifies the permeability of phospholipid bilayers.

Hyper- A prefix meaning above or over.

Hyperab A trade name for rabies immune globulin.

Hyperadrenalism Over-activity of the adrenal gland.

Hyperadrenocorticism The excessive secretion of adrenocortical hormone.

Hyperaldosteronism The excessive secretion of aldosterone by the adrenal cortex (also known as aldosteronism).

Hyperammonemia The presence of a high concentration of ammonia in the blood.

Hyperandrogenism The excessive secretion of androgen in women.

Hyperbetalipoproteinemia A disorder in fat metabolism characterized by the presence of a high concentration of low-density lipoprotein (beta-lipoprotein), cholesterol, and phospholipid.

Hyperbilirubinemia The presence of a large quantity of bilirubin in the blood.

Hyperbolic A convex upward curve on a graph that rises quickly and then levels off.

Hypercalcemia The presence of a large quantity of calcium in the blood.

Hypercalciuria The presence of a large quantity of calcium in the urine due to hypercalcemia.

Hypercapnia Abnormally high blood CO_2 concentration.

Hyperchloremia The presence of a high concentration of chloride in the blood.

Hyperchlorhydria The excessive secretion of gastric juice due to abnormal function of the stomach.

Hypercholesterolemia The presence of a high concentration of cholesterol in the blood.

Hypercholia The excessive secretion of bile by the liver.

Hyperchromia The abnormal increase of hemoglobin in red blood cells.

Hyperchromic Shift An increase in absorbance at 260 nm by denaturation of double-stranded DNA (separation of dsDNA into ssDNA).

Hypercytosis The increase in number of blood cells.

Hyperdynamia The excessive activity of muscle.

Hyperemesis Excessive vomiting.

Hyperemia Excessive blood accumulation in one area.

Hypergammaglobulinemia The presence of a high concentration of gamma globulin in the blood.

Hyperglycemia The presence of an excessive amount of glucose in the blood.

Hyperhydrosis Excessive perspiration.

Hypericin (mol wt 504) An antidepressant.

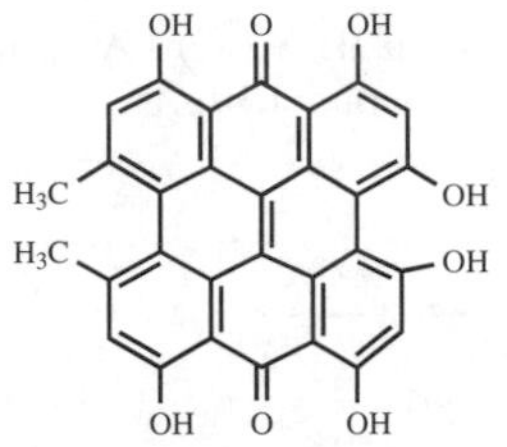

Hyperimmune Serum Serum from an animal that has received repeated antigen injections that contains a high concentration of polyclonal antibodies against that antigen.

Hyperinsulinism The excess secretion of insulin by the islet cells of the pancreas.

Hyperkalemia The presence of an abnormally high concentration of potassium in the blood (also known as hyperpotassemia).

Hyperkeratosis The thickening of the outer horny layer of the skin.

Hyperlactation Excessive secretion of milk.

Hyperlipidemia The presence of an elevated level of low-density lipoprotein in the blood.

Hyperlipoproteinemia Disorders of fat metabolism characterized by the presence of high concentrations of lipoproteins in the blood. Hyperlipoproteinemias are classified into five main types: type I (presence of a large amount of chylomicrons in the blood); type II (presence of a large amount of betalipoprotein, cholesterol, and phospholipid); type III (presence of a high concentration of very-low density lipoprotein); type IV (presence of a large amount of glyceride and very low density lipoprotein); type V (presence of a large amount of chylomicron and very-low density lipoprotein and triglyceride).

Hyperlysinemia A genetic disorder characterized by the presence of a large amount of lysine in the blood leading to mental retardation, anemia, hypotonia, and other symptoms.

Hyperlysinuria The presence of an abnormally high concentration of lysine in the urine due to hyperlysinemia.

Hypermagnesemia The presence of a large amount of magnesium in the blood.

Hypermyotrophy Excessive development of muscular tissue.

Hypernatremia The presence of an abnormally high concentration of sodium in the blood.

Hyperosomotic A liquid with a higher osmotic pressure.

Hyperosmotic Solution A solution with a greater osmotic pressure than the physiological level.

Hyperostosis Abnormal growth of bone tissue.

Hyperoxaluria The presence of an excessive amount of oxalic acid in the urine due to a defect in glyoxylic acid metabolism.

Hyperparasite An organism that parasitizes another parasite.

Hyperparathyroidism An overfunctioning of the parathyroid gland that causes muscular weakness, bone decalcification, and an increase in calcium in the blood and urine.

Hyperpepsinia The excessive secretion of pepsin (an acidic protease).

Hyperphenylalaninemia The presence of an abnormal high concentration of phenylalanine in the blood.

Hyperplasia An increase in the size of an organ or part of an organ due to increased cellular proliferation.

Hyperploid Having a chromosome number greater than diploid.

Hyperpolarization A negative shift in a cell's resting potential which is normally negative, or an increase in positive charge.

Hypersensitivity A state of excessive and potentially damaging immune responsiveness mediated by either antibody or T cells due to previous exposure or sensitization to antigen (see also allergy). Hypersensitivities are classified into five main types: Type I (IgE-mediated); type II (ADCC-mediated); type III (Ag-Ab-complex-mediated); type IV (delayed-type hypersensitivity).

Hypersplenism The decrease in the number of red cells, white cells, and platelets in the blood due to the enlargement of the spleen.

Hyperstat A trade name for diazoxide, an antihypertensive agent that relaxes arteriolar smooth muscle.

Hypertension Abnormally high blood pressure.

Hypertensive Pertaining to high blood pressure.

Hyperthermia Body temperature above 37° C.

Hyperthermophilic Enzyme The enzymes that are produced by the hyperthermophilic microorganisms such as *Thermotoga* species and are capable of functioning at a temperature between 80 to 100°C.

Hyperthyroidism Overactivity of the thyroid gland resulting in an increase in metabolism.

Hypertonic Solution See hyperosmotic solution.

Hypertrophy Enlargement of an organ or part of an organ.

Hyperuricemia The presence of a large amount of uric acid in the blood.

Hyperuricosuria See hyperuricuria.

Hyperuricuria The presence of a large amount of uric acid in the urine.

Hypervariable Region A small area in the variable region of an immunoglobulin molecule where the greatest variations in amino acid sequence occur.

Hypervolemia Larger volume of circulating blood than normal.

Hypha The threadlike structures that make up the body of a fungus.

Hyphen A combination drug containing acetaminophen and hydrocodone bitartrate, used as an analgesic agent.

Hypnomidate A trade name for the anesthetic agent etomidate.

Hypnotic Any drug that produces sleep.

Hypo- A prefix meaning under or less.

Hypocalcemia The presence of lower than normal concentration of calcium in the blood.

Hypochlorhydria The presence of abnormally low concentration of hydrochloric acid in the gastric juice.

Hypochloride See hypochlorous acid.

Hypochlorous Acid (mol wt 52) An antioxidizing agent used as a disinfectant and discoloring agent.

$$HClO$$

Hypocholesterolemia The presence of abnormally low concentrations of cholesterol in the blood.

Hypochromic Shift A decrease in absorbance at 260 nm by the renaturation of single-stranded DNA (formation of dsDNA from ssDNA).

Hypoderic Beneath the skin.

Hypogammaglobulinemia A condition characterized by the underproduction of one or more classes of immunoglobulin.

Hypoglycemia An abnormally low concentration of glucose in the blood.

Hypoglycin A (mol wt 141) An unusual amino acid found in unripe ackee fruit that can cause Jamaican vomiting sickness.

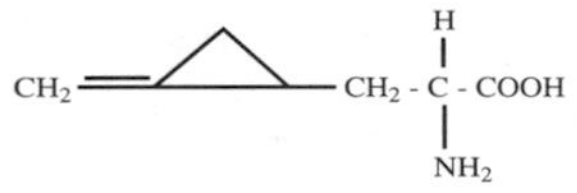

Hypoinsulinism A deficiency of insulin.

Hypokalemia An abnormally low concentration of potassium in the blood.

Hypolipoproteinemia The presence of abnormally low concentrations or the absence of lipoprotein.

Hypomagnesemia An abnormally low concentration of magnesium in the blood.

Hyponatremia An abnormally low concentration of sodium in the blood.

Hypoosmotic A solution with a lesser osmotic pressure than the physiological level.

Hypoparathyroidism Low activity of the parathyroid gland.

Hypophosphatasia A genetic disorder in humans characterized by the deficiency of the enzyme alkaline phosphatase.

Hypophosphatemia The presence of abnormally low concentrations of phosphate in the blood.

Hypophysectomy Surgical removal of the pituitary gland.

Hypopituitarism The deficiency in production of growth hormones by the pituitary gland.

Hypoplasia Reduction in size of an organ or part of an organ.

Hypoploid Referring to a cell or an organism that contains one or more fewer chromosomes or chromosome segments than the normal chromosome number.

Hypopotassemia The presence of a low concentration of potassium in the blood (also known as hypokalemia).

Hypoproteinemia The decrease in the quantity of protein in the blood.

Hypotension Having lower than normal blood pressure.

Hypothalmic Hormone Referring to the hypothalamic stimulatory and inhibitory hormones produced by the hypothalamus. The hypothalamic stimulatory hormones stimulate the release of hormones such as corticotropic hormone, luteinizing hormone, and growth hormone. The hypothalmic inhibitory hormones inhibit the release of hormones such as prolactin and malanocyte-stimulatory hormone.

Hypothermia A subnormal body temperature (below 37° C).

Hypothrombinemia A decreased level of thrombin that may result in a tendency to bleed.

Hypothyroidism A disorder due to the deficiency in production of thyroid hormone and characterized by obesity, cold dry skin, and low metabolic rate.

Hypotonic Solution See hypoosmotic.

Hypotrophy The underdevelopment of an organ or part of an organ.

Hypovolemia Abnormal decrease in the volume of circulating blood.

Hypoxanthine (mol wt 136) A purine base and the product of deamination of adenine.

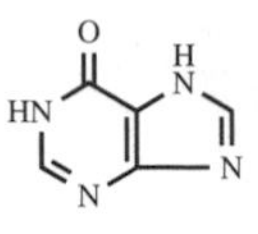

Hypoxanthine Guanine Phosphoribosyl Transferase (HGPRT) The enzyme that catalyzes the following reaction:

$$\text{IMP + PPi} \rightleftharpoons \text{Hypoxanthine + 5-phosphoribosyl pyrophosphate}$$

Hypoxanthine Oxidase Synonym of xanthine oxidase.

Hypoxanthine Phosphoribosyltransferase See hypoxanthine-guanine phosphoribosyl transferase (HGPRT).

Hypoxia A condition characterized by the reduced level of oxygen.

HypRho-D A trade name for Rh_o(D) immune globulin (human).

Hyprogest A trade name for hydroxyprogesterone caproate, a hormone that suppresses ovulation.

Hyproval P.A. A trade name for hydroxyprogesterone caproate, a hormone that suppresses ovulation.

Hypurin Neutral A trade name for regular zinc insulin.

Hyrexin-50 A trade name for diphenhydramine hydrochloride, an antihistaminic agent.

Hyroxon A trade name for hydroxyprogesterone, a hormone that suppresses ovulation.

Hysteroscope An optical instrument used for observing the interior of the uterus.

Hytakerol A trade name for dihydrotachysterol, which stimulates calcium absorption and promotes secretion of calcium from bone to the blood.

Hytone A trade name for hydrocortisone.

Hytrin A trade name for terazosine hydrochloride, an antihypertensive agent.

Hytus A trade name for guaifenesin, an antitussive agent.

Hyzaar A trade name for losartan potassium hydrochlorothiazide, an anti-vasoconstrictor and antihypertensive agent.

Hyzine-50 A trade name for hydroxyzine hydrochloride, an antianxiety and antihistaminic agent.

HZ Abbreviation for herpes zoster.

H-Zone (H Band) The central, lighter zone of A band in the myofibrils of striated muscle.

I

I Abbreviation for 1. inosine, 2. isoleucine, 3. iodine, and 4. electric current.

^{131}I The radioactive iodine used for clinical diagnosis and labeling proteins.

I_{50} Abbreviation for concentration of an inhibitor that causes 50% inhibition.

I Band The lighter band of the myofibril in striated muscle seen under the electron microscope and composed of thin actin-containing filaments (also known as isotropic band).

***i* Gene** The regulatory gene in the *lac* operon of *E. coli* that encodes the repressor for the *lac* operon.

Ia Antigen Class II histocompatibility molecules encoded by the I region of the MHC of mice.

IAA Abbreviation for indole acetic acid, a plant growth hormone.

IAGT Abbreviation for indirect anti-globulin test.

IAHA Abbreviation for immune adherence hemagglutination.

IAHAA Abbreviation for immune adherence hemagglutination assay.

IAM Abbreviation for iodoacetamide.

IAPP Abbreviation for islet amyloid polypeptide.

-iasis A suffix denoting a diseased condition.

IAT Abbreviation for 1. indirect antiglobulin test; 2. iodine azide test.

Iatrogenic Infection or disorder resulting from a medical procedures or medical intervention.

IB Abbreviation for inclusion body.

Ibafloxacin (mol wt 275) An antibacterial agent.

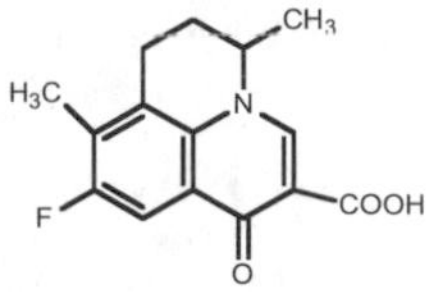

IBD Abbreviation for inflammatory bowel disease.

IBMX Abbreviation for isobutylmethyl xanthine.

Ibiamax A trade name for amoxicillin trihydrate, an antibiotic that inhibits bacterial cell wall biosynthesis.

Ibogaine (mol wt 310) An indole alkaloid isolated from plant *Tabernanthe iboga* used for treatment of heroin addiction.

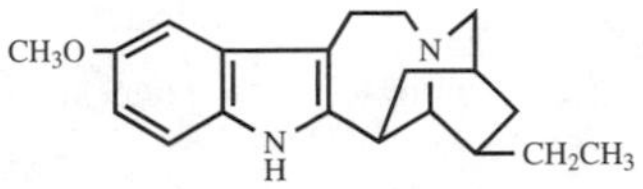

Ibopamine (mol wt 307) A cardiotonic agent with dopaminergic and adrenergic agonist activity.

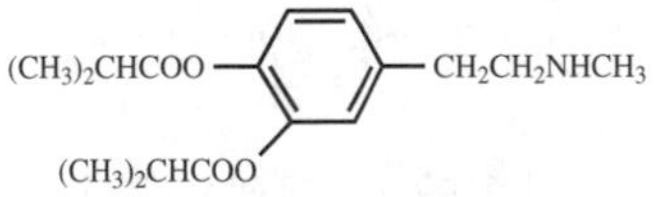

Ibotenic Acid (mol wt 158) A cardiotonic agent found in mushrooms.

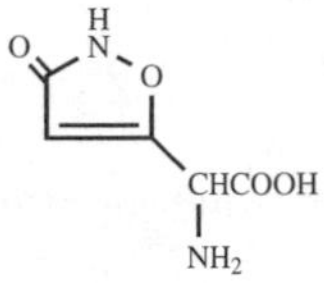

Ibrotamide (mol wt 208) A sedative and hypnotic agent.

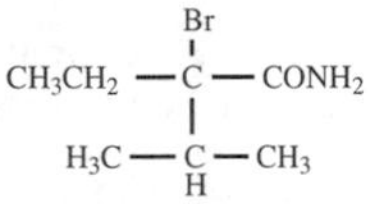

iBu Abbreviation for isobutyryl.

IBU A trade name for ibuprofen, an anti-inflammatory, analgesic and anitpyretic agent that inhibits the prostaglandin biosynthesis and suppresses pain and reduces fever.

Ibudilast (mol wt 230) An antihistaminic agent and cerebral vasodilator.

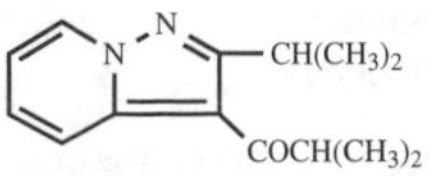

Ibufenac (mol wt 192) An analgesic and anti-inflammatory agent.

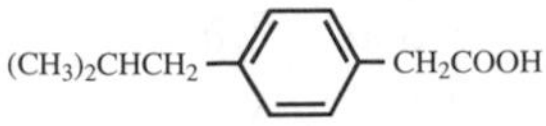

Ibuprin A trade name for ibuprofen, an anti-inflammatory agent that inhibits prostaglandin biosynthesis, suppresses pain, and reduces fever.

Ibuprobm A trade name for ibuprofen, an anti-inflammatory, analgesic and anitpyretic agent that inhibits the prostaglandin biosynthesis and suppresses pain and reduces fever.

Ibuprofen (mol wt 206) An anti-inflammatory agent that inhibits prostaglandin biosynthesis, suppresses pain, and reduces fever.

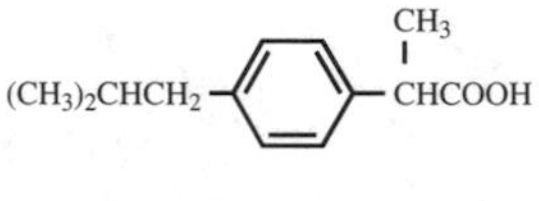

Ibuproxam (mol wt 221) An anti-inflammatory agent.

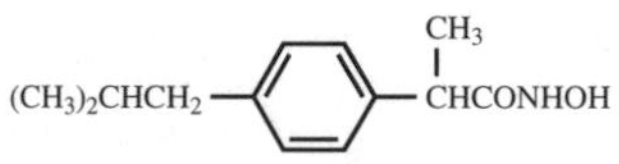

Ibutilide (mol wt 385) An anti-arrhythmic agent.

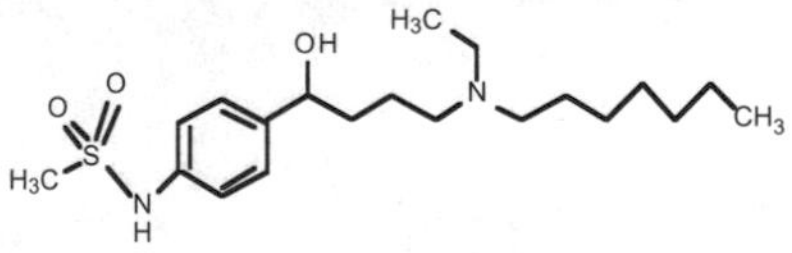

IBV Abbreviation for infectious bronchitis virus.

IC Abbreviation for 1. immune complex; 2. immune cytotoxicity; 3. immunochemistry; 4. intracellular.

IC_{50} Abbreviation for concentration of an inhibitory substance that causes 50% inhibition.

ICa Abbreviation for calcium current.

ICAM Abbreviation for intercellular adhesion molecule, which is expressed on fibroblasts and endothelial cells.

ICAM-1, 2, 3 Abbreviation for intercellular adhesion molecule 1, 2, and 3.

ICBP Abbreviation for intracellular-binding protein.

ICC Abbreviation for 1. immunocompetent cell; 2. immunocytochemistry.

ICD Abbreviation for 1. immune complex disease; 2. isocitric dehydrogenase.

ICDH Abbreviation for isocitrate dehydrogenase.

ICE Abbreviation for interleukin-1 beta-converting enzyme.

Ice Nucleation Bacteria Bacteria that promote water-to-ice transition.

Ice Nucleation Proteins Proteins found in certain pathogenic plant bacteria that induce frost damage and help bacteria to invade plants.

I-Cell Disease A genetic disorder of connective tissue characterized by progressive psychomotor retardation, skeletal deformities, and early death due to the deficiency of hydrolases in the lysosome. The lysosome from the victims of I-cell disease contains large inclusions (after which the disease is named) of glycosaminoglycan and glycolipid.

ICER Abbreviation for inducible cAMP early repressor.

I-CeuI A restriction endonuclease from the chloroplast large rRNA of *Chlamydomonas*, cloned and over-expressed in *E. coli* with the following specificity:

5'........TAACTATAACGGTCCTAA↓GGTAGCGA........3'
3'........ATTGATATTCGCA↑GGATTCCATCGCT..........5'

ICF Abbreviation for intracellular fluid.

ICFA Abbreviation for incomplete Freund's adjuvant.

ICG Abbreviation for indocyanine green.

Ichthyo- A prefix meaning fish.

Ichthyocin A peptide hormone consisting of nine amino acid residues secreted by the posterior lobe of the pituitary gland of the bony fishes.

Ichthyology The science that deals with fishes.

Ichthyosis An inherited skin disorder characterized by a dry, cracked and fish-scalelike skin condition.

Ichthyotoxin Toxin that kills gill-bearing animals.

ICL Abbreviation for isocitrate lyase.

I-ClaI A restriction endonuclease from *Bacillus sphaericus* with the following specificity:

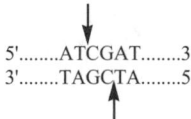

5'........AT↓CGAT........3'
3'........TAGC↑TA........5'

IC-Loop Abbreviation for intracellular loop.

Icosahedral Symmetry The symmetry of an icosahedron that has 12 vertices, 20 faces, and 30 edges. It is used to describe the structure of spherical virions (isometric virions).

Icosahedral Virus A virus that has an icosahedral symmetry.

Icosahedron A geometric structure composed of 12 vertices, 20 faces, and 30 edges.

ICP-AES Abbreviation for inductively coupled plasma atomic emission spectrometry.

I_{CRAC} Abbreviation for current carried by ions moving through CRAC (calcium-release-activated calcium channel).

I/CRAC Abbreviation for current carried by ion moving through CRAC (calcium-release-activated calcium channel).

ICSBP Abbreviation for interferon consensus-sequence binding protein.

ICSH Abbreviation for interstitial cell stimulating hormone.

ICSI Abbreviation for intracytoplasmic sperm injection.

ICT Abbreviation for 1. indirect Coomb's test; 2. insulin clearance test.

Icterus A condition characterized by the excess of bile pigments in the blood and tissues leading to a yellow pigmentation of the skin.

Icterus Index An index in a liver function test for the determination of bilirubin in serum or plasma by comparing the color of the plasma or serum with a standard solution of potassium dichromate.

Ictotest A colorimetric test for the presence of bilirubin in the urine based on the production of blue-purple color by diazotization of bilirubin.

ID Abbreviation for 1. iditol dehydrogenase; 2. immunodeficiency; 3. infectious disease; 4. infectious dose; 5. internal diameter.

ID_{50} Abbreviation for the infectious dose 50, the dose of an infectious agent that produces a 50% infection in experimental animals or a testing system.

IDA Abbreviation for iminodiacetic acid.

Idamycin A trade name for idarubicin, an antibiotic with antineoplastic activity.

Idarubicin (mol wt 498) An antibiotic with antineoplastic activity.

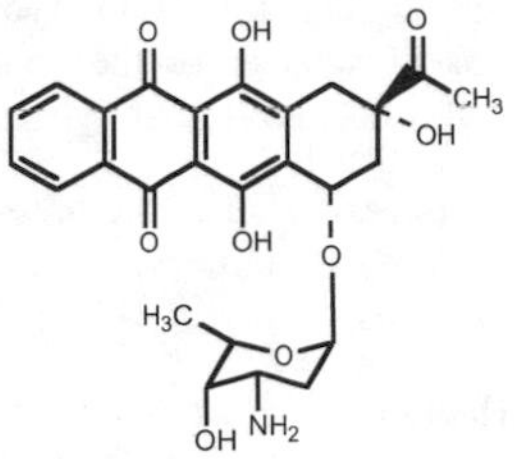

IDAV Abbreviation for immune deficiency-associated virus.

IDD Abbreviation for insulin-dependent diabetes.

IDDM Abbreviation for insulin dependent diabetes mellitis.

IDE Abbreviation for insulin degrading enzyme.

Ideal Gas A gas that obeys gas laws.

Ideal Solution A solution in which the thermodynamic activity of each component is proportional to its mole fraction.

IDH Abbreviation for isocitric dehydrogenase.

Idio- A prefix denoting peculiarity to the individual.

Idiogram A diagrammatic representation of a chromosome complement of a cell or individual based on a karyotype of the entire chromosome complement present.

Idiolite A secondary metabolite that is not required for growth and vital function in the organism, e.g., penicillin in *Penicilium*.

Idiopathic A pathological condition without a known cause.

Idiopathic Thrombocytopenic Purpura An autoimmune disease in which platelets are destroyed, leading to spontaneous bruising.

Idiophase A phase of metabolism in which secondary metabolism is dominant over the primary growth-directed metabolism, e.g., the production of secondary product of penicillin in the microorganism.

Idiotope One of the antigenic determinants that makes up the idiotype of the variable regions of an immunoglobulin.

Idiotype The antigenic determinants of a particular variable domain on the variable region of a specific immunoglobulin. It is a collective term for the idiotopes on the variable regions of an immunoglobulin.

Idiotype Network Series of idiotype-anti-idiotype reactions postulated for the control of humoral and cell-mediated immunity.

Idiotype Suppression The suppression of the synthesis of idiotype antibodies caused by anti-idiotype antibodies.

Idiotypic Marker The antigenic determinant or epitope in the antigen-binding site of an idiotypic antibody (also known as idiotope).

Iditol (mol wt 182) A sugar alcohol and an inhibitor for L-iduronidase.

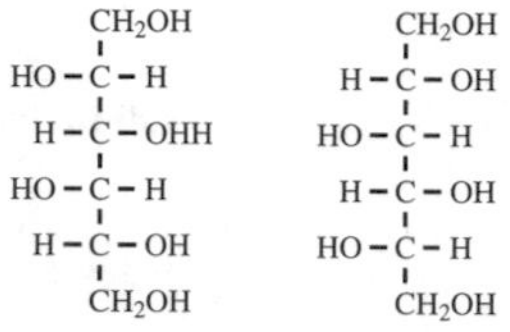

D-iditol L-iditol

D-Iditol Dehydrogenase The enzyme that catalyzes the following reaction:

$$\text{D-Iditol} + \text{NAD}^+ \rightleftharpoons \text{D-Sorbose} + \text{NADH}$$

L-Iditol Dehydrogenase The enzyme that catalyzes the following reaction:

$$\text{L-Iditol} + \text{NAD}^+ \rightleftharpoons \text{L-Sorbose} + \text{NADH}$$

IDL Abbreviation for intermediate density lipoprotein with density between VLDL (very low density lipoprotein) and LDL (low density lipoprotein).

Idling Reaction The reaction in the production of pppGpp and ppGpp by ribosomes when uncharged tRNA is present at the A site of the ribosome.

I-DmoI A restriction endonuclease from *Desulfurococcus mobilis* with the following specificity:

ID-MS Abbreviation for isotope dilution mass spectrometry.

iDNA Abbreviation for initiator DNA, which is formed by extension of primer RNA by DNA polymerase.

IdoA Abbreviation for iduronic acid.

IdoA(2S) Abbreviation for iduronic acid 2-sulphate.

Idose (mol wt 180) A six-carbon aldosugar.

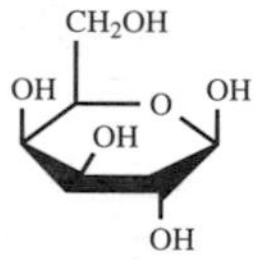

β-D-idose

Idoxuridine (mol wt 354) An antiviral drug that interferes with viral DNA synthesis.

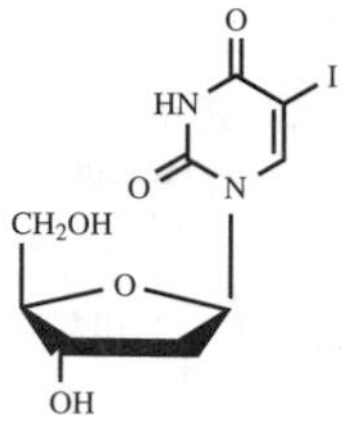

5-Iodo-2′-deoxyuridine

IDP Abbreviation for inosine diphosphate or inosine-5′-diphosphate.

Idrocilamide (mol wt 191) A skeletal muscle relaxant.

IDU Abbreviation for idoxuridine.

Iduronate Sulfatase The enzyme that catalyzes the hydrolysis of the sulfate group of iduronate sulfate unit of dermatan sulfate, heparan sulfate, and heparin.

Iduronic Acid (mol wt 164) A uronic acid derived from the sugar idose.

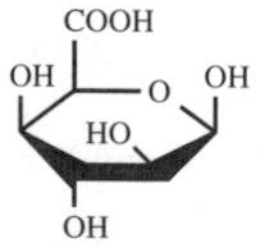

β-D-Iduronic acid

Iduronidase The enzyme that catalyzes the hydrolysis of α-L-iduronisidic linkages in dermatan sulfate.

IE Abbreviation for immunoelectrophoresis.

IEA Abbreviation for 1. immuno-electroadsorption; 2. immuno-enzyme assay; 3. intravascular erythrocyte aggregation.

IEC Abbreviation for ion exchange chromatography.

IEF Abbreviation for isoelectric focusing.

IEM Abbreviation for immunoelectron microscopy.

IEP Abbreviation for 1. immuno-electrophoresis; 2. isoelectric point.

IEX Abbreviation for ion-exchange chromatography.

IF Abbreviation for 1. immuno-fluorescence; 2. initiation factor; 3. intrinsic factor.

IF1, IF2, IF3 Abbreviation for initiation factor 1, 2, and 3 in protein biosynthesis.

IFA Abbreviation for 1. incomplete Freund's adjuvant; 2. indirect fluorescent antibody.

I-FABP Abbreviation for intestinal fatty acid-binding protein.

IFE Abbreviation for immuno-fixation electrophoresis.

Ifenprodil (mol wt 325) A cerebral and peripheral vasodilator.

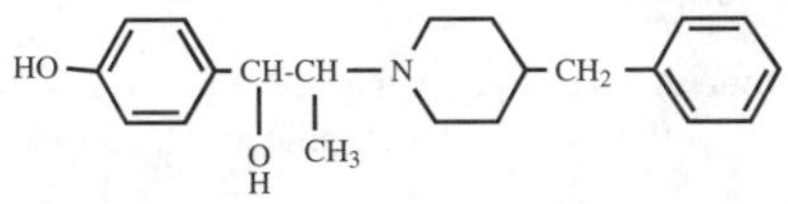

Ifex A trade name for ifosfamide, an antineoplastic agent.

IFN Abbreviation for interferon.

Ifosfamide (mol wt 261) An antineoplastic agent that interferes with RNA transcription.

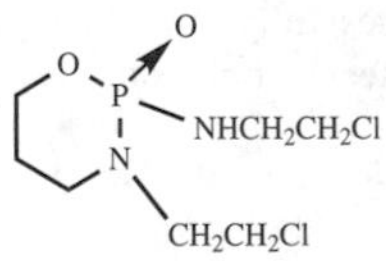

IFRA Abbreviation for indirect fluorescent rabies antibody.

IFT Abbreviation for immuno-fluorescence technique or test.

Ig Abbreviation for immunoglobulin.

IgA Abbreviation for immunoglobulin A, an immunoglobulin that occurs primarily in mucus, semen, and secretions such as saliva, tears, and sweat; it consists of two alpha (a) heavy chains and plays a major role in protecting mucous membrane surface against infection.

IgA1 A subclass of IgA.

IgA1 Protease A bacterial extracellular protease that cleaves the hinge region of human IgA1 heavy chains but has no effect on IgA_2.

IgA_2 A subclass of IgA.

Igα Abbreviation for immunoglobulin alpha, a protein on the B cell surface.

IgA Protease Synonym of IgA-specific serine endopeptidase.

Igβ Abbreviation for immunoglobulin beta, a protein on the B cell surface.

IG-Cell The intestinal gastrin cell, a small granulated gastrin-producing cell.

IGD Abbreviation for inter-globular domain.

IgD Abbreviation for immunoglobulin D, an immunoglobulin that occurs on the surface of some lymphocytes, consists of two delta (d) heavy chains and appears to play a regulatory role in lymphocyte activity.

IgE Abbreviation for immunoglobulin E, an immunoglobulin that is normally present in the blood serum in a very low concentration but becomes elevated in an allergic individual. It consists of two epsilon (ε) heavy chains and binds to mast cells and basophils causing mast-cell degranulation and release of histamine leading to type I hypersensitivity.

i-Gene The regulator gene in the *lac* operon of *E. coli* that encodes the repressor for *lac* operon.

IGF Abbreviation for insulin-like growth factor.

IGF-BP Abbreviation for insulin-like growth factor binding protein.

IGG Abbreviation for indocyanine green.

IgG Abbreviation for immunoglobulin G, a major serum immunoglobulin that plays a major role in protecting the body against infection. It consists of two identical gamma (γ) heavy chains. It is also known as 7S antibody or 7S immunoglobulin.

IgG1, IgG2, IgG3, IgG4 Subclasses of IgG.

IGH Abbreviation for idiopathic growth hormone.

iGluR Abbreviation for inotropic glutamate receptor.

IgM Abbreviation for immunoglobulin M, a high-molecular-weight immunoglobulin occurring as a pentamer that is formed prior to IgG in the primary immune response. It plays an important role in the early immune response.

IgN An IgG-like immunoglobulin that occurs in some reptiles.

IGSS Abbreviation for immuno-gold silver staining.

IgX An IgG-like immunoglobulin that occurs in amphibia.

IgY An IgG-like immunoglobulin that occurs in birds.

IH Abbreviation for infectious hepatitis.

IHA Abbreviation for indirect hemagglutination.

IHF Abbreviation for integration host factor, a factor in the host that is required for integration of phage DNA into host chromosomes.

IHP Abbreviation for inositol hexaphosphate.

IHSA Abbreviation for iodinated human serum albumin.

IICR Abbreviation for $InsP_3$-induced calcium release.

IIF Abbreviation for indirect immuno-fluorescence.

Ikaros A protein that binds and activates the enhancer of gene CD-3-δ involved in T cell specification and maturation.

IL Abbreviation for interleukin.

IL-1 ... IL-15 Abbreviation for interleukin -1 to interleukin -15.

ILA Abbreviation for insulin-like activity.

Ile Abbreviation for the amino acid isoleucine.

Iletin PZI A trade name for insulin, an antidiabetic hormone.

Ileum The posterior portion of the small intestine, the preferred site for vitamin B_{12} absorption.

Ileus The blockage of the passage of intestinal content.

Illegitimate Recombination The insertion of genes by a transposome into the recipient DNA that requires no homology between the donor and the recipient DNA.

Illicit Transport The transport of a substance across the membrane by a transport system that is designed for another substance.

Illiozyme A trade name for enzyme pancrelipase.

Illudins A group of antitumor antibiotic substances produced by the poisonous fungus *Clitocybe illudens* (Basidiomycetes).

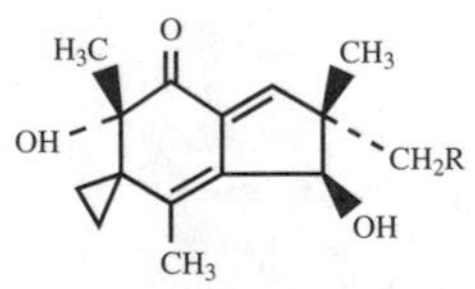

Illudin M: R = H

Illudin S: R = OH

Ilopan A trade name for dexpanthenol, a GI stimulant and cofactor in the synthesis of neurotransmitter acetylcholine.

Iloprost (mol wt 360) A vasodilator and antithrombotic agent.

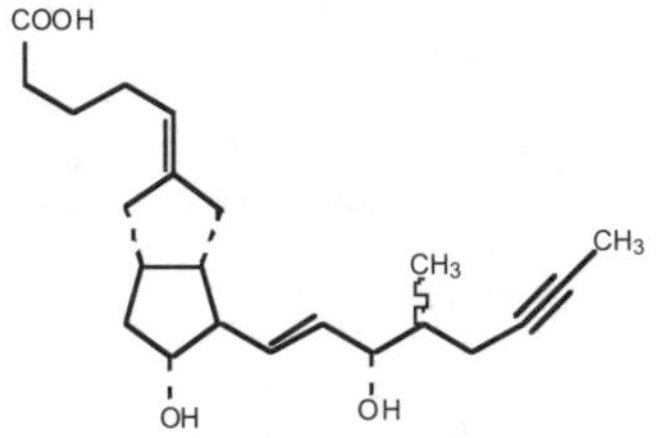

IL-R Abbreviation for interleukin receptor.

Ilsone A trade name for erythromycin estolate, an antibiotic that binds to 30S ribosomes inhibiting bacterial protein synthesis.

***ilv*-Operon** The operon that specifies the enzymes for biosynthesis of isoleucine, leucine, and valine.

IMAC Abbreviation for immobilized metal-ion affinity chromatography.

Imbibition The absorption of water and the swelling of colloidal materials, e.g., uptake of water by a dry gel.

ImD_{50} Symbol for a dose of antigen or vaccine capable of successfully immunizing 50% of a particular animal test population.

Imdur A trade name for isosorbide mononitrate, a vasodilator.

Imferon A trade name for iron dextran, a hematologic agent that provides iron for the synthesis of hemoglobin.

IMHT Abbreviation for indirect micro-hemagglutination test.

Imidapril (mol wt 405) An antihypertensive agent.

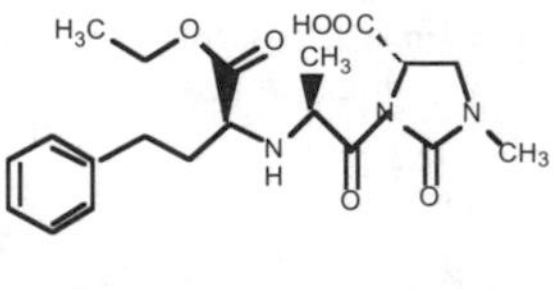

Imidazole (mol wt 68) The heterocyclic ring system of histidine and an inhibitor of histamine.

Imidazole Acetic Acid (mol wt 126) A complex of acetic acid and imidazole.

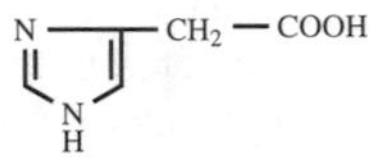

Imidazole Aceto Phosphate (mol wt 206) The phosphate form of imidazole acetic acid.

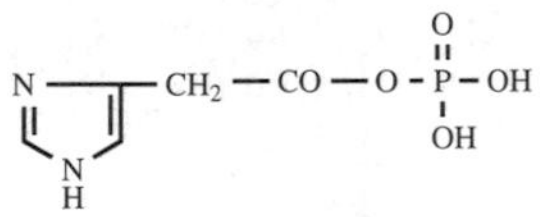

Imidazole Group Referring to the heterocyclic ring of the amino acid histidine.

Imidazole *N*-Acetyltransferase The enzyme that catalyzes the following reaction:

Acetyl-CoA + imidazole

⇅

CoA + *N*-acetylimidazole

Imidazole Glycerol Phosphate (mol wt 239) An intermediate in the biosynthesis of histidine.

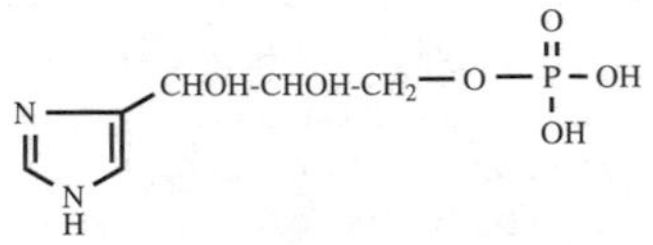

Imidazole Glycerol Phosphate Dehydratase The enzyme that catalyzes the following reaction:

Imidazole glycerol phosphate

⇅

Imidazole acetol phsphate + H_2O

4-Imidazolone 5-Propionase The enzyme that catalyzes the following reaction:

4-Imidazolone 5-propionate + H_2O

⇅

N-formimino-L-glutamate

4-Imidazolone 5-Propionic Acid (mol wt 108) An intermediate in the conversion of histidine to glutamate.

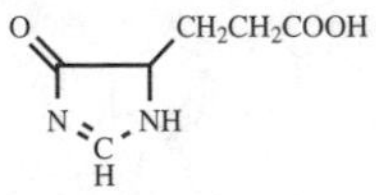

Imidodipeptidase Synonym of X-pro dipeptidase.

Imiglucerase A human β-glucocerebrosidase produced by recombinant DNA technology. It catalyzes the breakdown glucocerebroside to glucose and ceramide.

Imino Referring to the -NH- group where the nitrogen is part of a ring system or is united by both bonds to a single atom.

Imino Acid An acid in which the nitrogen atom is part of the ring system, e.g., proline and hydroxyproline.

Imino Group The -NH- group.

Iminoglycinuria The abnormal excretion of glycine, proline, and hydroxproline in the urine due to a defect in amino transport system.

Iminothiolane (mol wt 101) A cross-linking reagent that cross-links the epsilon amino groups of lysine residues in a protein.

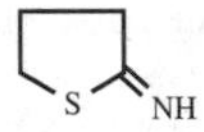

Imipenem (mol wt 317) A broad spectrum semisynthetic antibiotic that inhibits bacterial cell wall biosynthesis.

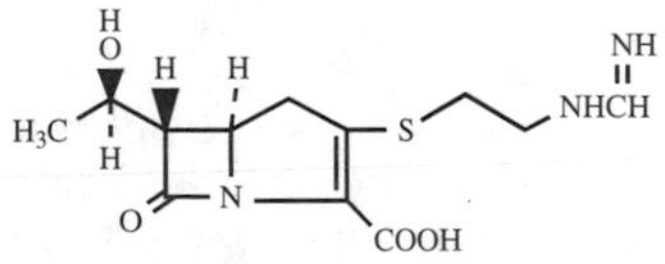

Imipramine (mol wt 280) An antidepressant that causes accumulation of norepinephrine or serotonin allowing neurotransmitter to accumulate.

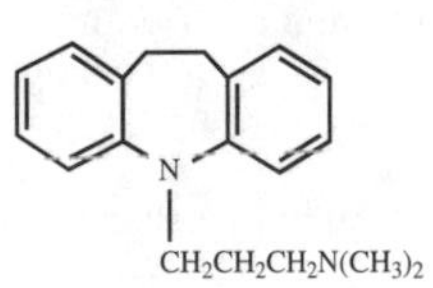

Imipramine *N*-Oxide (mol wt 296) An Imipramine derivative and an antidepressant that causes accumulation of norepinephrine or serotonin allowing neurotransmitter to accumulate.

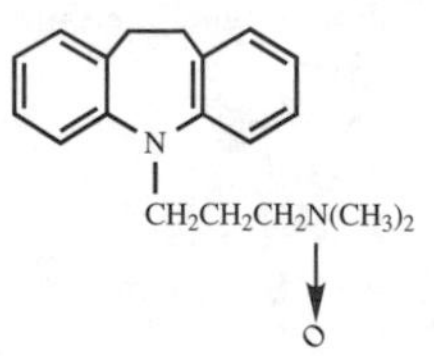

Imiprin A trade name for imipramine hydrochloride, an antidepressant that causes accumulation of norepinephrine or serotonin allowing neurotransmitter to accumulate.

Imiquimod (mol wt 240) An antiviral agent and immunomodulator.

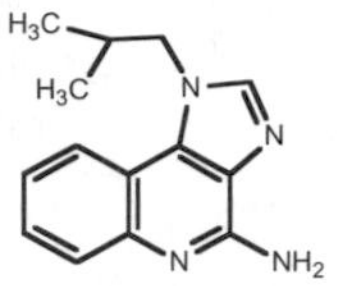

Imitrex A trade name for sumatriptan succinate, an anti-migraine agent.

Immaturin A protein produced by *Paramecium caudatum* that represses its mating activity.

Immediate Type Hypersensitivity An IgE-mediated allergic reaction that occurs in sensitized individuals within minutes after re-exposure to an allergen. It involves binding of allergen to the IgE on the surface of the mast cell leading to mast cell degranulation and release of vasoactive compounds e.g., histamine and serotonin (also known as type-I hypersensitivity).

Immersion Oil An oil with the refractive index about 1.5, used in an oil immersion objective of the light microscope.

Immobilized Enzyme An enzyme that is covalently linked to an insoluble carrier, e.g., agarose or acrylamide for continuous use and recovery.

Immortal Cell Culture See continuous cell culture.

Immortalization A change in the cells to escape from the normal limitation on growth of a finite number of division cycles leading to indefinite growth in culture.

Immune Resistant to a pathogen due to previous exposure.

Immune Adsorbent The insoluble substance that is capable of adsorbing antibody, e.g., protein A from bacteria.

Immune Competent Cells See immunocompetent cells.

Immune Complex Multimolecular antigen-antibody complexes that may be soluble or insoluble depending on their size.

Immune Complex Diseases Disorders due to the presence of soluble immune complexes in the body, e.g., Arthus reaction and serum sickness (also known as type-III hypersensitivity or complex-mediated hypersensitivity).

Immune Cytolysis The lysis of cells caused by antibody against cell surface antigens on erythrocytes in the presence of complement.

Immune Deficiency Disease Diseases in which immune responses are suppressed or reduced (e.g., AIDS).

Immune Exclusion The prevention of entry of antigens across mucosal membranes by secretary IgA.

Immune Globulin Referring to immunoglobulin.

Immune Hemolysis The lysis of erythrocytes by hemolysin (antibody) in the presence of complement.

Immune Opsonin See opsonin.

Immune Prophylaxis Protection from disease by active immunization or passive transfer of antibody or immune cells.

Immune Reaction Specific reaction between antigen and antibody or between antigen and T cell antigen receptors.

Immune Response The responses of an individual or animal to an antigenic stimulus such as antigen presentation by macrophage, lymphocyte activation, antibody production, and antigen elimination.

Immune Response Genes Genes that control the ability of an individual to respond to an antigen.

Immune Serum Serum obtained from an immunized animal (see also antiserum).

Immune Surveillance The survey or patrol of the body by the immune lymphocytes for detection and destruction of altered-self cells or transformed cells.

Immune System The defense system in the vertebrate to recognize and eliminate antigens, reject incompatible grafts, and kill pathogenic microorganisms. The immune system in humans consists of primary lymphoid organs (e.g., bone marrow and thymus) and secondary lymphoid organs (e.g., spleen, lymph node, etc.).

Immunity Resistance to disease acquired as a result of previous exposure to an immunogen or vaccination.

Immunization Process of creating a state of immunity to an antigen through the administration of the appropriate avirulent immunogen (e.g., vaccination).

Immuno- A prefix meaning immunogenic, immunology, or immunity.

Immunoadsorption Removal of antibody from a sample by use of an antigen or substance capable of adsorbing antibody or removing antigen by antibody.

Immunoaffinity Chromatography A type of chromatography in which either an antigen or antibody is coupled to an insoluble matrix and used for the separation and purification of a desired substance.

Immunoassay The techniques for detection of antigen-antibody reaction.

Immunobeads Small solid inert spheres coated with antibody or antigen that are used for various immunoassays.

Immunoblast Precursor of immunocyte.

Immunoblotting Techniques e.g., western blotting, in which proteins separated in an electrophoretic gel are transferred onto nitrocellulose sheets and then detected by enzyme-linked IgG or radioactively labeled IgG.

Immunochemistry The science that deals with the chemical aspects of immunology and immunological techniques.

Immunocompetent Cells See immunocytes.

Immunoconglutinin Antibodies that react with fixed complement component, e.g., C3b and C4b.

Immunocyte Any cell that participates in the immune response, e.g., lymphocyte and antigen-presenting cells.

Immunocytoadherence A technique for identifying immunoglobulin-bearing cells by formation of rosettes with cells (e.g., erythrocytes) containing homologous antigen.

Immunocytochemistry The science that deals with the chemistry of staining cells or tissues using radioactively labeled, enzyme-linked antibodies, or fluorescently conjugated antibodies.

Immunodeficiency See immune deficiency diseases.

Immunodiffusion A technique in which antigen and antibody are allowed to diffuse in agar toward each other to form a visible precipitin in the agar. The patterns of the precipitin lines in the agar are used for identification of antigens.

Immunoelectrofocusing A separation technique that combines gel electrofocusing with immunodiffusion or immunoelectrophoresis.

Immunoelectron Microscopy An electron microscopic technique that employs immunochemical method, e.g., ferritin-labeled antibody, for identification of cellular components.

Immunoelectrophoresis A technique that combines electrophoresis and immunodiffusion for identification of electrophoretically separated antigens (e.g., crossed immunoelectrophoresis and rocket immunoelectrophoresis).

Immunoferritin Technique A ferritin-labeled antibody technique used in immunoelectron microscopy (the ferritin serving as an electron-dense marker for electromicroscopic observation.

Immunofixation Electrophoresis A technique that uses antibodies for identification of electrophoretically separated proteins.

Immunofluorescence A cytochemical technique for the detection of antigens on cells or tissue through the use of fluorescently conjugated specific antibody.

Immunogen A substance that is capable of provoking an immune response.

Immunogenicity The ability of a substance to elicit an immune response.

Immunoglobulin Antibodies that are capable of binding to specific antigenic determinant. The monomeric form of immunoglobulin consists of 4 peptide chains (two identical heavy chains and two identical light chains. There are five main classes of immunoglobulin, namely immunoglobulin G (consisting of two γ heavy chains), immunoglobulin A (consisting of two α heavy chains), Immunoglobulin E (consisting of two ε heavy chains), immunoglobulin M (consisting of two μ heavy chains), and immunoglobulin D (consisting of two δ heavy chains).

Immunoglobulin Chains Referring to the heavy chains and light chains of immunoglobulin. There are five main types of heavy chains (α, δ, ε, γ, μ) and two types of light chains (κ and λ).

Immunoglobulin Class Referring to the five classes of immunoglobulin (e.g., IgG, IgM, IgA, IgD, IgE) that are classified on the basis of types of heavy chains they possess. IgG possesses γ heavy chains, IgM possesses μ heavy chains, IgE possesses ε heavy chains, IgD possesses δ heavy chains, and IgA possesses α heavy chains.

Immunoglobulin Class Switch A process whereby an IgM-producing B cell switches isotypes leading to the production of IgG or IgA.

Immunoglobulin Heavy Chains Referring to the five types of heavy chains (α, γ, δ, ε, and μ) present in the different classes of immunoglobulin.

Immunoglobulin Light Chains Referring to the two types of light chains (κ and λ) present in immunoglobulin.

Immunoglobulin Subclasses Referring to the subclasses of immunoglobulin that are classified on the basis of structural and antigenic differences in the heavy chains in each of the classes. IgG has four main subclasses (IgG1, IgG2, IgG3, and IgG4). IgA has two subclasses (IgA1 and IgA2).

Immunoglobulin Superfamily A structurally related group of genes that encode immunoglobulins, T cell receptors, β_2-microglobulin, and other proteins.

Immunogold Technique A form of immunoelectron microscopy in which antibodies or protein-A molecules are conjugated with gold to serves as electron-dense marker.

Immunohemotology The science that deals with antigen, antibody, and their reactions and effects on blood.

Immunohistochemical Histochemical and immunological method for analysis of living cells and tissues.

Immunoliposome A liposome that bears a chemically coupled monoclonal antibody.

Immunologic Pertaining to immunology or immune response.

Immunological Pertaining to immunology or immune response.

Immunological Memory The rapid and extensive immunological responses to a subsequent exposure of an antigen that had been previously encountered by the immune system.

Immunological Paralysis Referring to immunological unresponsiveness.

Immunological Surveillance See immune surveillance.

Immunological Test A test based on the principles of antigen-antibody reactions.

Immunological Tolerance Specific unresponsiveness to an antigen due to genetic factors or exposure to antigen during fetal life. Tolerance can also be induced experimentally in animals.

Immunologist A scientist who specializes in immunology.

Immunology The science that deals with immunity.

Immunometry The measurement of amounts of substances by the use of a specific antigen-antibody reaction.

Immunomodulation Manipulation of the immune response by substances that are capable of increasing or decreasing the ability of the immune system to respond to antigens or invading microorganisms.

Immunopathic Damage to cells, tissues, or organs caused by an immune response.

Immunopathogenesis Pathological effects resulting from immune responses.

Immunoperoxidase Method A method that employs peroxidase-labeled antibody for detection of specific antigen-antibody reactions.

Immunopotency The ability of antigen to provoke an immune response.

Immunoprecipitation The precipitation resulting from the formation of large antigen-antibody complexes from soluble antigen and soluble antibody.

Immunoradiometry Determination of antigen-antibody reactions by employing radioactively labeled antibody or antigen.

Immunoselection A method for isolation of cell variants that lack specific antigen from a population of cells by treatment of cells with antibody against specific cell surface antigen in the presence of complement. The treatment leads to the lysis of antigen bearing cells the salvage of the cell variants that lack corresponding antigen.

Immunosorbent Assay An immunoassay in which antigen or antibody is immobilized onto a solid surface or carrier.

Immunostaining A staining technique in which specific antigen on a structure is stained by a stain-generating immunochemical reaction, e.g., a peroxidase-linked antibody detection system.

Immunosuppressant A drug or substance that suppresses the immune response.

Immunosuppression The reduction of T- and/or B-cell clones by activation of specific or nonspecific T-suppresser lymphocytes or by drugs that suppress effector T- or B-lymphocytes.

Immunotherapy Treatment of diseases with immunostimulants or immunosuppressants or by transfer of immunocompetent cells from compatible donor to the patient.

Immunotoxins Toxins that are conjugated to an immuoglobulin directed against a specified cell surface antigen on a target cell. The toxin-conjugated antibody combines with antigen on the target cell leading to the killing of the target cell.

Immunotyping The typing of cells with immunological markers such as monoclonal antibodies.

IMN Abbreviation for infectious mononucleosis.

Imodium A trade name for loperamide, an antidiarrheal agent.

Imogam A trade name for rabies immunoglobulin.

Imolamine (mol wt 260) An antianginal agent.

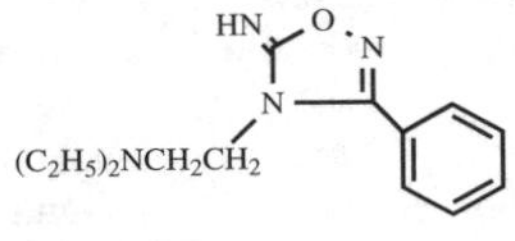

Imovax A trade name for rabies vaccine prepared from human diploid cells.

IMP Abbreviation for inosine monophosphate or inosine 5′-monophosphate.

IMP Cyclohydrolase The enzyme that catalyzes the following reaction:

IMP + H_2O

⇅

5-Formamido-1-(5-phosphoribosyl) imidazole-4-carboxamide

IMP Dehydrogenase The enzyme that catalyzes the following reaction:

Inosine 5-phosphate + NAD^+ + H_2O

⇅

Xanthosine 5-phosphate + NADH

IMP Pyrophosphate Phospho-D-ribosyltransferase The systematic name for hypoxanthine phosphoribosyl transferase.

IMP Pyrophosphorylase The enzyme that catalyzes the following reaction:

IMP + PPi

⇅

Hypoxanthine + 5-phospho-α-D-ribose 1-diphosphate

Impedence The resistance of flow of an alternating electric current.

Impedimetry A microbiological technique for measurement of microbial growth by observing changes of electrical impedence of the growth medium.

Impeller A rotating, multibladed device in a fermenter that regulates oxygen transfer and keeps cells in suspension.

Imperfect Fungi Fungi that reproduce only by asexual means.

Imperfect Stage The stage in the life cycle of a fungus in which only asexual reproduction occurs.

Impetigo An acute, inflammatory skin disease caused by bacteria and characterized by the appearance of small blisters, weeping fluid, and crusts.

Implant Material artificially placed into an organism.

Implantation 1. The attachment of the embryo to the uterine wall. 2. Addition of tissue to an organism (e.g., tissue graft).

Impril A trade name for imipramine, an antidepressant.

Improsulfan (mol wt 409) An antineoplastic agent.

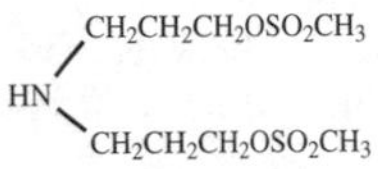

Impulse The wave of excitation or signal that passes along a nerve fiber. It is due to the electrochemical gradient across the membrane.

Imuran A trade name for azathioprine, an immunosuppressant that inhibits purine biosynthesis.

In Symbol for the chemical element Indium with atomic weight 114, valence 1, 2, and 3.

IN Abbreviation for integrase.

In situ In the normal, natural, original, or appropriate position.

***in situ* Hybridization** A type of hybridization performed by denaturing the DNA of cells smashed on a microscope slide so that the radioactively labeled ssDNA or RNA is allowed to hybridize.

in vitro Outside the organism, e.g., an experiment performed in the test tube.

in vivo Occurring within the living organism.

Inactivated Serum Serum that has been heated at 56° C for 30 minutes to inactivate complement.

Inactivated Vaccine A vaccine containing pathogenic microorganisms that has been treated with inactivating agent so that it is incapable of causing disease.

INAH Abbreviation for iso-nicotinic acid hydrazide.

Inanition An exhausted condition resulting from the lacking of food and water or defect in assimilation.

Inaperisone (mol wt 245) A muscle relaxant.

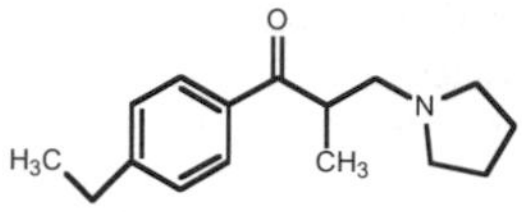

Inapsine A trade name for droperidol, an anesthetic agent.

Inborn Error of Metabolism An inherited metabolic disorder due to a defect in an enzyme or protein leading to a metabolic block with pathological consequence.

Inbred Strain Any strain of animal or plant resulting from mating between genetically related organisms that results in high homozygosity.

Inbreeding Mating between genetically related individuals leading to increased homozygosity.

Inbreeding Depression A decrease in vigor and fitness resulting from inbreeding.

Incipient Species Organisms that are too distinct to be considered as subspecies but not sufficiently different to be regarded as different species.

Incision 1. Surgical creation of an opening into an organ in the body. 2. Creation of a nick in DNA or RNA.

Inclusion A discrete heterogeneous mass or structure contained in a cell.

Inclusion Bodies Nuclear or cytoplasm structures with characteristic staining properties that may be a site of viral multiplication and accumulation.

Inclusion Conjunctivitis An acute, pus-filled infection caused by *Chlamydia trachomatis.*

Incompatibility Physical interaction between tissues of two organisms leading to a rejection, e.g., graft rejection or incompatible blood transfusion.

Incompatibility Plasmids The inability of two plasmids to coexist.

Incomplete Antibody An antibody that can bind to an antigen but cannot cause a visible serological reaction (also known as blocking antibody).

Incomplete Antigen See hapten.

Incomplete Freund's Adjuvant An oil-in-water emulsion that is capable of enhancing the immune response when administered together with antigen. It contains no killed *Mycobacterium.* (Compare to complete Freund's adjuvant.)

Incubation The maintenance of microorganisms or a reaction mixture under the favorable conditions, e.g., constant ambient temperature and humidity.

Incubation Period 1. The time period between exposure to an infectious agent and the appearance of disease symptoms. 2. The time period that inoculated microorganisms are allowed to grow or the reaction mixture is allowed to proceed.

Incubator 1. An apparatus used to provide a controlled environment for maintaining growth of cells and microorganisms. 2. A device used to preserve the life of a premature baby.

Indalpine (mol wt 228) An antidepressant and an inhibitor of serotonin uptake.

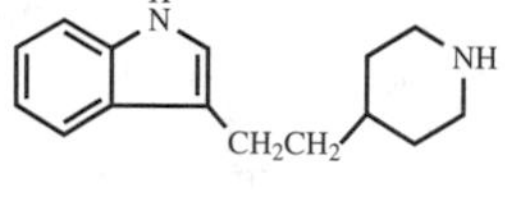

Indameth A trade name for indomethacin sodium trihydrate, an anti-inflammatory agent.

Indanazoline (mol wt 201) A vasodilator and nasal decongestant.

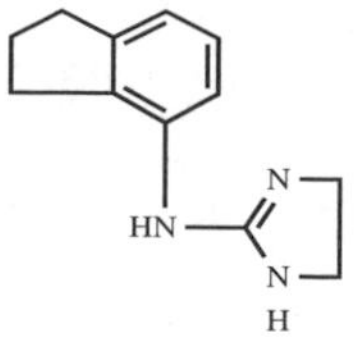

Indapamide (mol wt 366) A diuretic agent that inhibits sodium reabsorption.

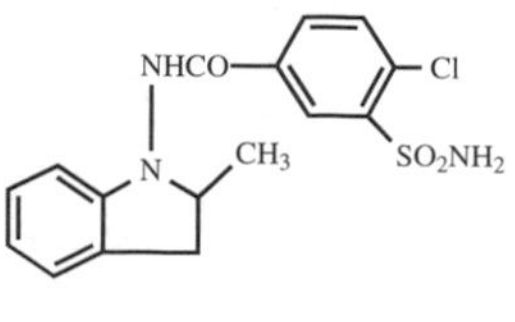

Indecainide (mol wt 308) A cardiac depressant.

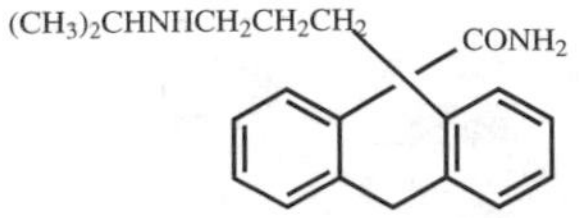

Indeloxazine Hydrochloride (mol wt 267) An antidepressant and an inhibitor of serotonin uptake.

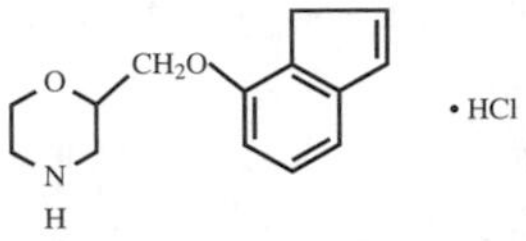

Indenolol (mol wt 247) An antihypertensive, antiarrhythmic, antianginal agent, and a beta-adrenergic blocker.

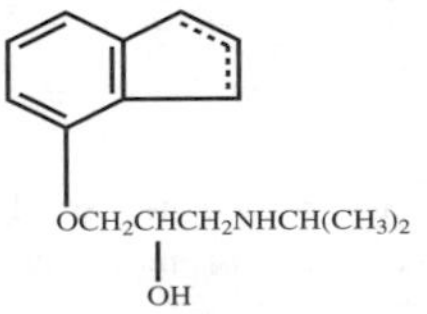

Independent Assortment The segregation of two or more pairs of alleles that lie on different chromosome.

Independent Binding The binding of one ligand to one binding site on a macromolecule that exerts no effect on the binding of subsequent ligands to other binding sites.

Inderal A trade name for propranolol hydrochloride, a beta-adrenergic blocker that reduces cardiac oxygen demand.

Inderide A trade name for a combination antihypertensive drug containing propranolol hydrochloride and hydrochlorothiazide.

Indican An anionic polysaccharide produced by *Beijerinckia indicus* that contains glucose, mannose, rhaminose, uronic acid, and acetyl groups. It has a high viscosity and forms a gel upon heating.

Indicator Organism An organism used to monitor a particular condition, e.g., the presence of *E. coli and Streptococcus faecalis*, thus indicating the degree of water pollution due to fecal contamination.

Indigenous Referring to organisms that are native to a particular habitat.

Indigo (mol wt 262) A stain.

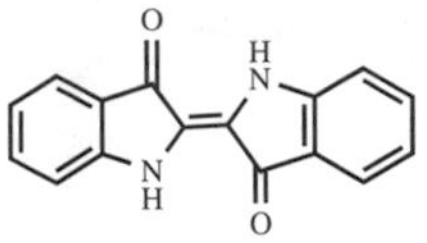

Indigo Carmine (mol wt 466) A reagent used for detection of nitrate, chlorate, and also for testing of milk.

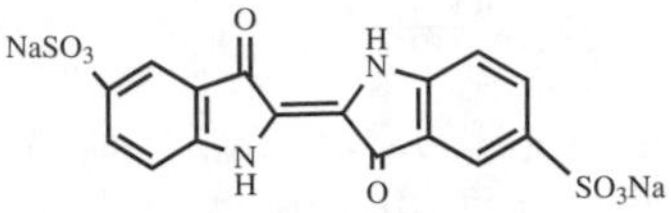

Indigo 5,5′,7-Trisulfonic Acid Tripotassium (mol wt 617) A reagent used for the determination of ozone in water.

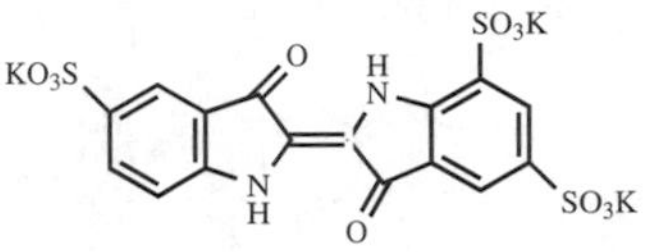

Indinavir (mol wt 614) An antiviral agent used for the treatment of AIDS.

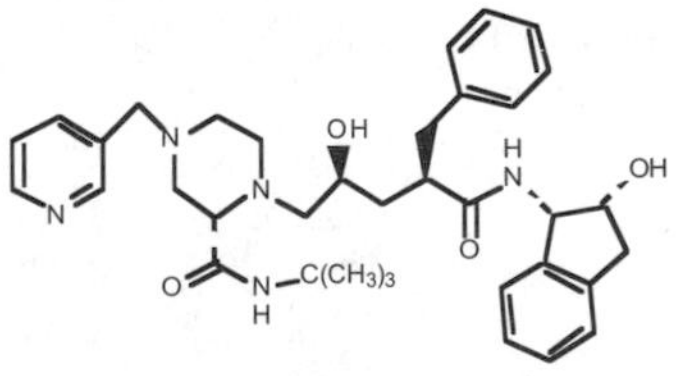

Indirect Agglutination Agglutination of particles or erythrocytes with antibodies directed against antigens that have been coupled to the surface of a second source of particles or erythrocytes.

Indirect Fluorescent Antibody Technique See indirect immunofluorescence.

Indirect Immunofluorescence A method of fluorescent staining whereby the unlabeled antibody is incubated with the antigen to be identified and then reacted with fluorescently labeled anti-IgG for subsequent detection by fluorescent microscopy.

Indium (In) A chemical element with atomic weight 114, valence 1, 2, and 3.

Indo I (mol wt 840) A fluorescent calcium chelator that shifts the fluorescence emission from 480 nm to 400 nm upon binding of calcium.

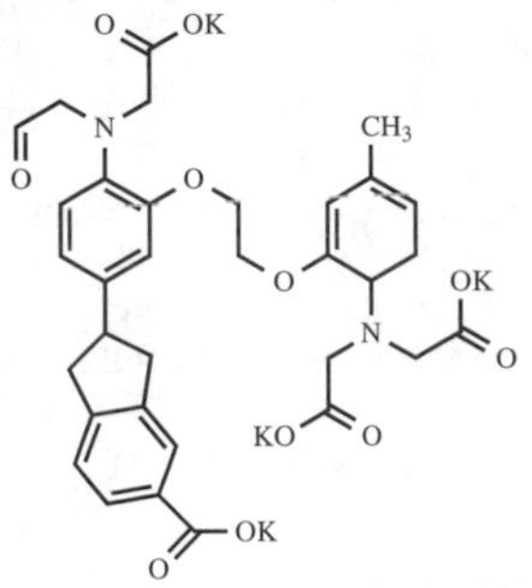

Indobufen (mol wt 295) An antithrombotic agent.

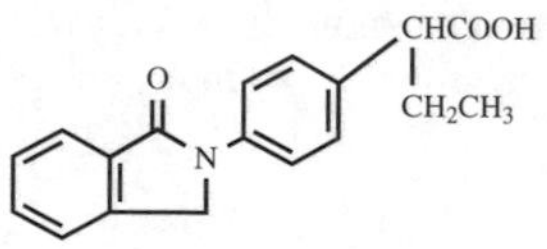

Indocid A trade name for indomethacin sodium trihydrate, an anti-inflammatory agent.

Indocin A trade name for indomethacin, an anti-inflammatory agent.

Indoine Blue A biological dye.

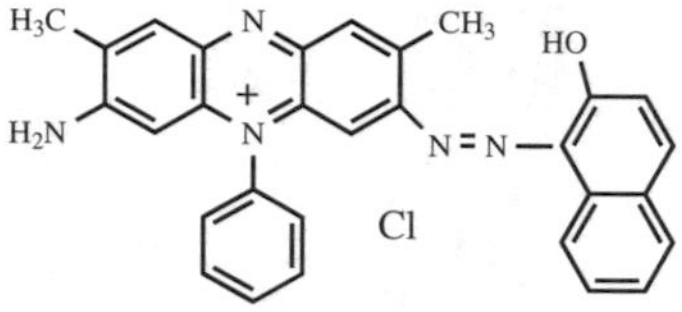

Indole (mol wt 117) The ring system of the amino acid tryptophan.

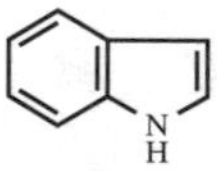

Indole Oxidase See indole-2,3-dioxygenase.

Indole 3-Acetaldehyde (mol wt 263) An indole-containing acetaldehyde.

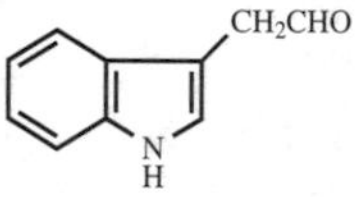

Indole 3-Acetaldehyde Oxidase The enzyme that catalyzes the following reaction:

2-Indole-3-acetaldehyde + O_2

⇌

2-Indole-3-acetate + 2 H_2O

Indole 3-Acetaldehyde Reductase (NADH) The enzyme that catalyzes the following reaction:

Indole-3-ethanol + NAD^+

⇌

Indole-3-acetaldehyde + NADH

Indole 3-Acetaldehyde Reductase (NADPH) The enzyme that catalyzes the following reaction:

Indole-3-ethanol + $NADP^+$

⇌

Indole-3-acetaldehyde + NADPH

Indole 3-Acetamide (mol wt 174) A compound containing indole.

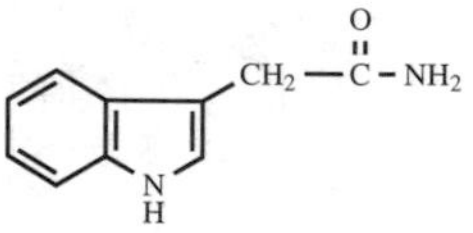

Indole 3-Acetate β-Glucosyltransferase The enzyme that catalyzes the following reaction:

UDP-glucose + indole 3-acetate

⇌

UDP + indole 3-acetyl-β-D-glucose

Indole Acetic Acid (IAA, mol wt 175) A plant hormone or growth regulator that stimulates the synthesis of RNA and proteins.

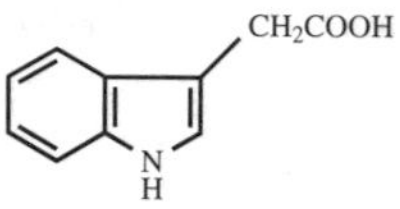

3-Indole Acetonitrile (mol wt 156) A plant growth regulator that promotes callus growth in tobacco.

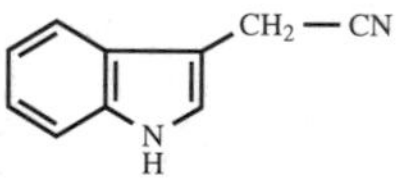

Indole Acetylglucose Inositol *O*-Acyltransferase The enzyme that catalyzes the following reaction:

Indole 3-acetylglucose + *myo*-inositol

⇌

D-Glucose + indole 3-acetyl-*myo*-inositol

3-Indole Acrylic Acid (mol wt 187) An inhibitor for conversion of indole to tryptophan by tryptophan synthetase.

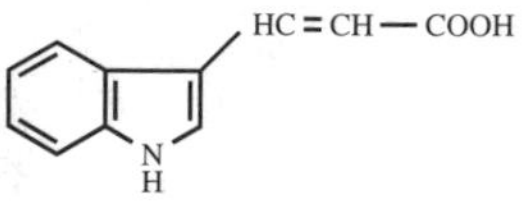

Indole Butyric Acid (mol wt 203) A cancer causing agent.

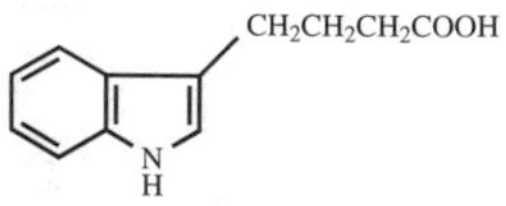

Indole 2,3-Dioxygenase The enzyme that catalyzes the following reaction:

Indole + O_2

⇌

2-Formylaminobenzaldehyde

Indole Lactate Dehydrogenase The enzyme that catalyzes the following reaction:

Indole-lactate +NAD^+

⇌

Indole-pyruvate NADH

Indole Lactic Acid (mol wt 205) An indole-containing compound.

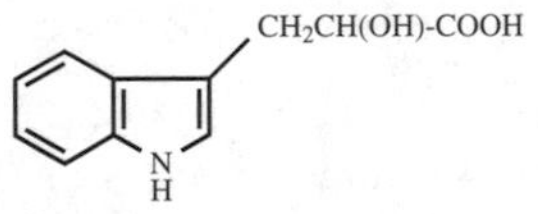

Indole 3-Pyruvic Acid (mol wt 203) An indole derivative containing pyruvate.

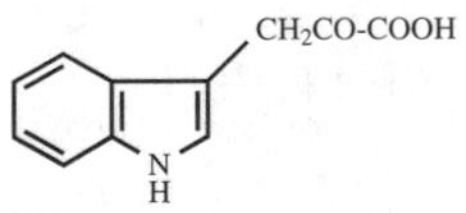

Indolicidin A bactericidal and fungicidal peptide from cytoplasmic granules of bovine neutrophils.

Indolmycin (mol wt 257) An antibiotic produced by *Streptomyces albus.*

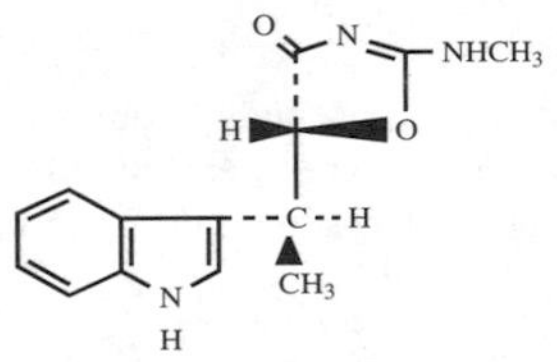

Indomethacin (mol wt 357) An anti-inflammatory agent that inhibits prostaglandin synthesis.

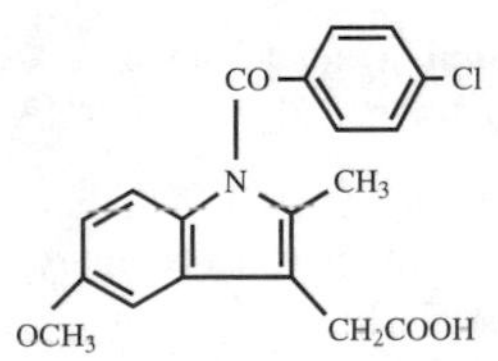

Indoprofen (mol wt 281) An analgesic and anti-inflammatory agent.

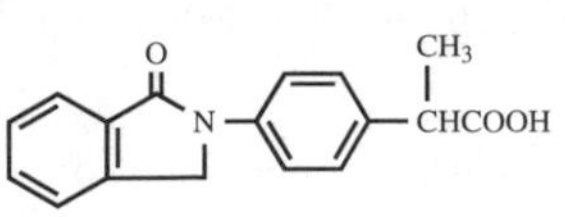

Indoramin (mol wt 348) An antihypertensive agent.

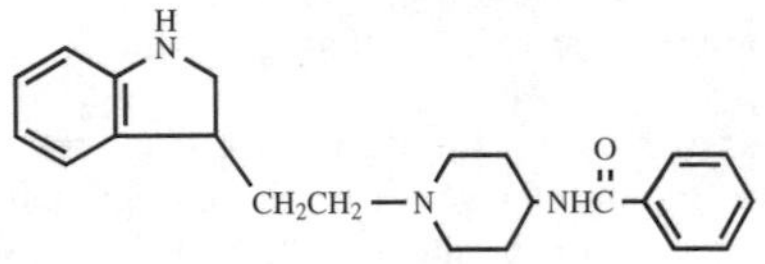

Indospicine (mol wt 173) A naturally occurring amino acid found in *Endigofera spicata.*

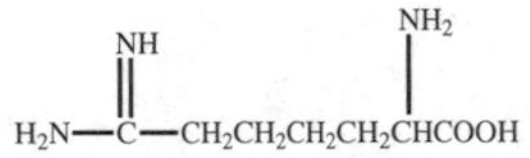

Indotec A trade name for indomethacin, a nonsteroidal anti-inflammatory agent.

Indoxyl Acetate (mol wt 175) A fluorogenic substrate for assay of cholinesterase, lipase, and acid phosphatase.

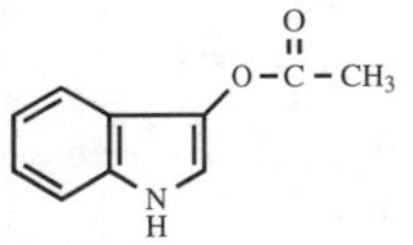

3-Indoxy Phosphate (mol wt 213) A reagent used for histochemical demonstration of alkaline phosphatase.

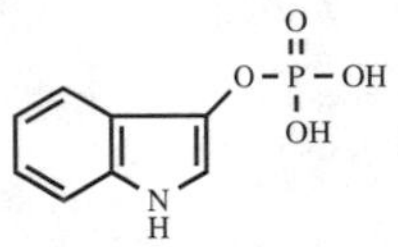

Induced Enzyme See inducible enzyme.

Induced Fit Model The model states that substrate induces a conformation change on the enzyme for the binding of substrate to the active site for conversion of the substrate to product.

Inducer 1. A substance that stimulates differentiation of cells or development of a particular structure. 2. A substance that activates a particular gene or triggers transcription. 3. A substance that induces the production of a specific enzyme.

Inducer Cells Cells that induce other nearby cells to differentiate into a specified pathway.

Inducible Enzyme The enzyme that is synthesized only in the presence of the substrate or inducer.

Induction 1. A process of initiation of virus production in a lysogenic bacterial cell. 2. An increase in the rate of enzyme synthesis due to the presence of substrate or inducer.

Inductive Interaction A regulatory mechanism in which the activity of cells of a tissue is influenced by the surrounding cells or tissues to differentiate into a specified pathway or pattern.

Inductive Phase 1. The time period between the administration of an antigen into the animal and the appearance of antibody. 2. The time period during which a normal cell changes into a cancerous cell.

Industrial Mycology The science that deals with fungi of importance in industry.

Inert Chemically, biologically, and physiologically inactive.

Inert Gas Referring to argon, helium, and krypton.

Inert Support A nonactive matrix used for enzyme immobilization or affinity chromatography by attachment of enzyme or ligand.

INF Abbreviation for interferon.

INF-α Abbreviation for interferon alpha (leukocyte interferon).

INF-β Abbreviation for interferon beta (fibroblast interferon).

INF-γ Abbreviation for interferon gamma (interferon produced by lymphocytes, e.g., T cell).

Infasurf A trade name for calfactant, a lung surfactant that reduces surface tension.

Infection A condition in which pathogenic microorganisms have become established in the host or the tissues of a host organism.

Infectious Capable of producing disease in a susceptible host.

Infectious DNA DNA isolated from a bacteriophage or virus that can successfully enter the host and initiate development of viral particles.

Infectious Dose The dose of pathogen that is needed to overcome the host defense mechanisms and establish an infection.

Infectious Hepatitis See hepatitis A.

Infectious Mononucleosis A disease caused by Epstein-Barr virus and characterized by fever, sore throat, and swollen lymph glands.

Infectious Nucleic Acid See infectious DNA or RNA.

Infectious RNA RNA from plus-stranded virus that can enter the host and produce infectious virus particles.

Infective Center The bacteriophage that produces a plaque on a bacterial lawn in a plaque assay.

Infectivity The ability of a pathogen, e.g., bacteria or viruses, to establish infection in a host.

InFed A trade name for iron dextran, an iron supplement and an antianemic agent.

Infergen A trade name for interferon produced by DNA recombinant technology and used for the treatment of hepatitis C.

Infertile Not being able to produce offspring.

Inflamase Forte A trade name for prednisolone sodium phosphate, an anti-inflammatory ophthalmic agent.

Inflamase Ophthalmic A trade name for prednisolone sodium phosphate solution, an anti-inflammatory ophthalmic agent.

Inflammation The response of the animal to injury, microbial infection, or presence of foreign matter that is characterized by swelling, redness, pain, and accumulation of phagocytes at the site of inflammation.

Inflammatory Exudate Puslike material deposited in the tissues or on the tissue surfaces resulting from a defensive response of the host to injury.

Inflammatory Response See inflammation.

Influent The fluid that enters into a process or system.

Influenza An acute, highly communicable disease that tends to occur in epidemic form, caused by an orthomyxovirus and characterized by malaise, headache, and fever.

Influenza Virus A virus of the family Orthomyxoviridae that causes influenza in humans.

Influx Inward flow.

Informational Macromolecule Referring to polynucleotides, e.g., DNA or RNA, that carry genetic information.

Informofers Proteins that complex with hnRNA (heterogeneous RNA).

Informosome The heterogeneous complex of RNA and nonribosomal proteins.

Infra- A prefix meaning below or under.

Infrared Pertaining to electromagnetic radiation with wavelengths longer than those of visible light.

Infrared Radiation Dry-heat sterilization.

Infrared Spectrophotometer An instrument used for the measurement of absorption of infrared radiation by a sample that provides information on the interatomic bonds that have a characteristic frequency within the infrared range.

Infrared Spectrum The wavelength range of 7.5 $\times 10^{-5}$ to 4.2×10^{-2} cm.

Infufer A trade name for iron dextran, a source of iron.

Infusate A fluid that is given intravenously.

Infusion 1. The introduction of fluid (e.g., nutrient, drug) via the vein. 2. The extract obtained by a steeping process.

INH Abbreviation for isoniazide, an anti-tubercular and antibacterial agent that inhibits *Mycobacterium* cell wall synthesis.

Inheritance The transmission of genetic information from parents to offspring.

Inhibin A gonadal polypeptide hormone that selectively inhibits the secretion of FSH (follicle stimulating hormone).

Inhibition 1. Prevention of growth or multiplication of microorganisms. 2. Reduction or prevention in the rate of enzymatic activity. 3. Repression of chemical or physical activity.

Inhibition Zone The zone of excess antigen or antibody in an antigen-antibody precipitin reaction curve.

Inhibitors Substances that repress a chemical or biological action.

Inhibitory Post Synaptic Potential Change in the potential of the postsynaptic membrane so that the membrane is hyperpolarized thereby reducing the amplitude of an excitatory postsynaptic potential.

Inhibitory Synapse A synapse in which an action potential in the presynaptic cell reduces the probability of an action potential occurring in the postsynaptic cell.

Inhibitory Transmitter A substance released by a neuron that inhibits the firing of another neuron (e.g., GABA).

Initiation Codon The codon 5′-AUG in messenger RNA that recognizes formyl-methionyl-tRNA in bacteria and methionyl-tRNA in eukaryotes.

Initiation Complex The complex of mRNA, N-formylmethionine-tRNA, and ribosome formed through a series of processes involving initiation factors and at the expense of the hydrolysis of GTP. The initiation complex is needed for starting protein synthesis.

Initiation Factor Protein factors involved in the formation of the initiation complex required for translation (abbreviated as IF). There are three initiation factors in prokaryotes, namely IF_1, IF_2 and IF_3. The initiation factor in eukaryotes is abbreviated eIF.

Initiator RNA Referring to RNA primer or primer RNA.

Initiator Transfer RNA Referring to tRNA that is responsible for initiation of protein synthesis (e.g., N-formylmethionyl-tRNA in prokaryotes, methionyl-tRNA in eukaryotes).

Innate Immunity The defenses acquired at birth that are independent of the exposure to specific antigens. The efficiency of innate immunity cannot be improved by repeated infection or vaccination.

Inner Volume The volume of water or solvent within gel particle, e.g., water within a Sephadex gel particle.

Innovar A trade name for fentanyl/droperidol, an anesthetic agent.

Ino Abbreviation for inosine.

Inocid A trade name for indomethacin, a nonsteroidal anti-inflammatory agent.

Inocor A trade name for amrinone lactate, which increases myocardial contractility.

Inoculation The artificial introduction of microorganisms into the body or into a culture medium.

Inoculum The material containing viable microorganisms used for inoculation.

Inorganic Chemical compounds that do not contain carbon atoms as an integral part of their molecular structure.

Inorganic Phosphate Ions derived from orthophosphoric acid.

Inorganic Pyrophosphatase The enzyme that catalyzes the following reaction:

$$\text{Pyrophosphate} + H_2O \rightleftharpoons 2\ \text{Orthophosphate}$$

iNOS Abbreviation for inducible NOS (nitric oxide synthetase).

Inosinase See inosine nucleosidase.

Inosinate Nucleosidase The enzyme that catalyzes the following reaction:

$$5'\text{-Inosinate} + H_2O \rightleftharpoons \text{Hypoxanthine} + \text{D-ribose 5-phosphate}$$

Inosine (mol wt 268) A nucleoside found in the wobble position of some tRNAs.

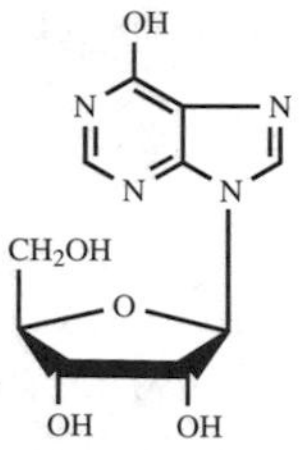

Inosine 3′,5′-Cyclic Monophosphate (mol wt 330) A cyclic form of inosine monophosphate.

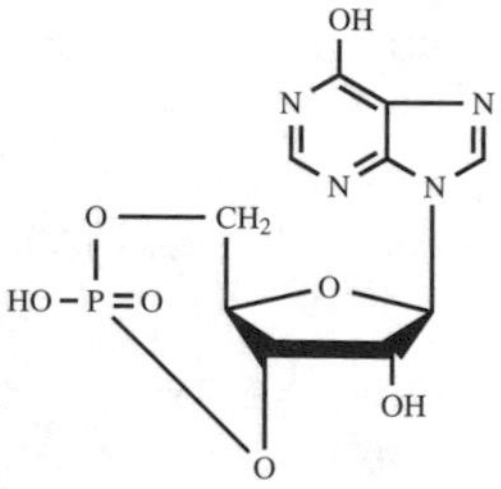

Inosine 5′-Diphosphate (mol wt 426) A diphosphate form of inosine nucleotide.

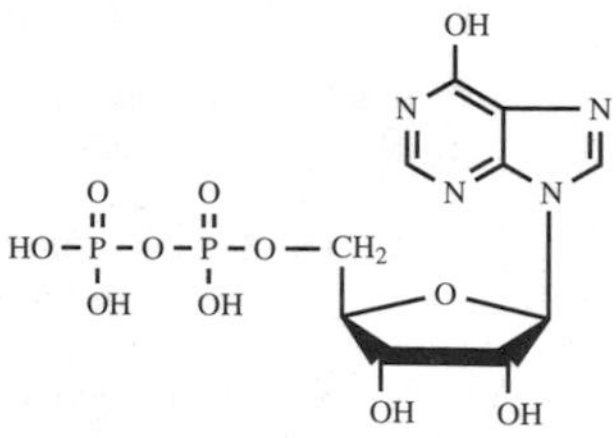

Inosine Kinase The enzyme that catalyzes the following reaction:

$$\text{ATP} + \text{inosine} \rightleftharpoons \text{ADP} + \text{IMP}$$

Inosine 5′-Monophosphate (mol wt 346) A nucleotide found in tRNA and an intermediate in biosynthesis of purine nucleotide.

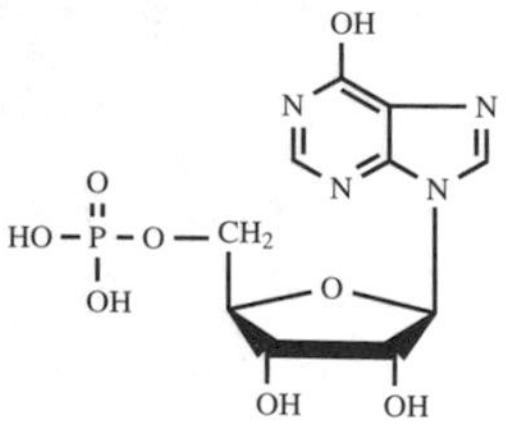

Inosine Nucleosidase The enzyme that catalyzes the following reaction:

$$\text{Inosine} + H_2O \rightleftharpoons \text{Hypoxanthine} + \text{D-ribose}$$

Inosine Phosphorylase Synonym of purine nucleoside phosphorylase.

Inosine Pranobex (mol wt 1115) An immunomodulator and antiviral agent.

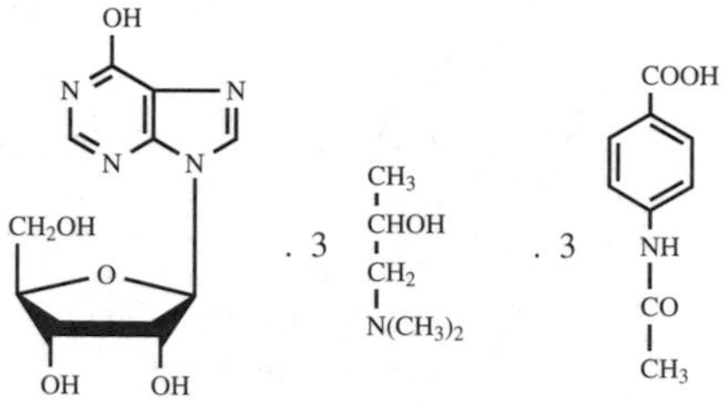

Inosine 5′-Triphosphate (mol wt 506) The triphosphate form of inosinic acid.

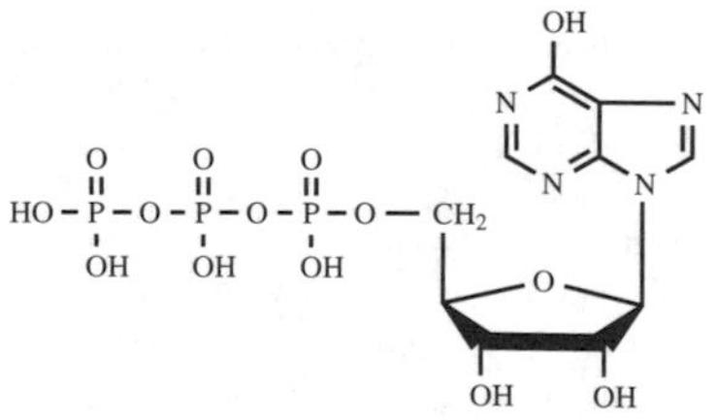

Inosinic Acid See Inosine 5′-monophosphate.

Inositol (mol wt 180) A sugar alcohol, an important component of membrane lipid and a growth factor for animals and microorganisms. It has been classified as part of the vitamin B complex.

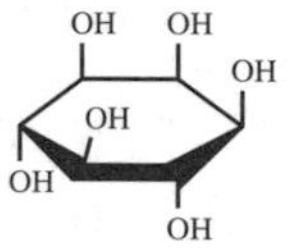

Inositol Bisphosphate (mol wt 338) The diphosphate form of inositol.

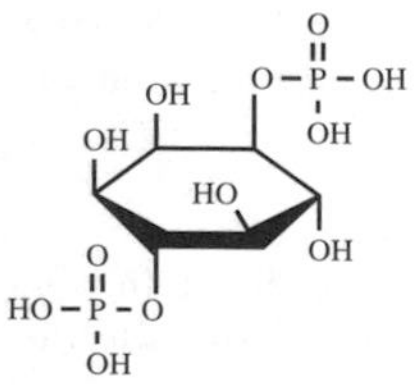

inositol 1,4-bisphosphate

Inositol 1,3-Bisphosphate 3-Phosphatase The enzyme that catalyzes the following reaction:

Inositol 1,3-bisphosphate + H_2O ⇌ Inositol 1-phosphate + Pi

Inositol 3,4-Bisphosphate 4-Phosphatase The enzyme that catalyzes the following reaction:

Inositol 3,4-bisphosphate + H_2O ⇌ Inositol 3-phosphate + Pi

Inositol 2-Dehydrogenase The enzyme that catalyzes the following reaction:

Inositol + NAD^+ ⇌ Pentahydroxycyclohexanone + NADH

Inositol Diphosphate See inositol bisphosphate.

Inositol 1-α-Galactosyl Transferase The enzyme that catalyzes the following reaction:

UDP-galactose + inositol ⇌ UDP + α-D-galactosyl-inositol

Inositol Hexaphosphate (mol wt 660) An inositol that contains six phosphates (also known as phytic acid).

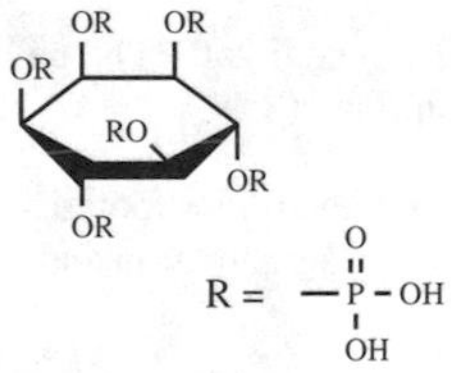

Inositol Kinase The enzyme that catalyzes the phosphorylation of inositol at the expense of hydrolysis of ATP.

Inositol Methyltransferase The enzyme that catalyzes the following reaction:

S-Adenosyl-L-methionine + inositol ⇌ *S*-Adenosyl-L-homocysteine + *O*-methyl-inositol

Inositol Monophosphate (mol wt 260) The monophosphate form of inositol.

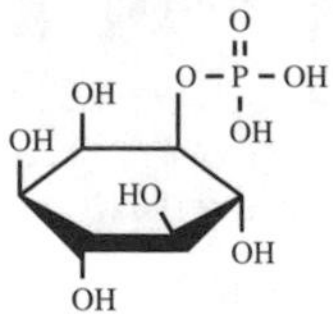

Inositol NAD⁺ 2-Oxidoreductase The systematic name for *myo*-inositaol 2-dehydrogenase.

Inositol Niacinate (mol wt 810) A vasodilator.

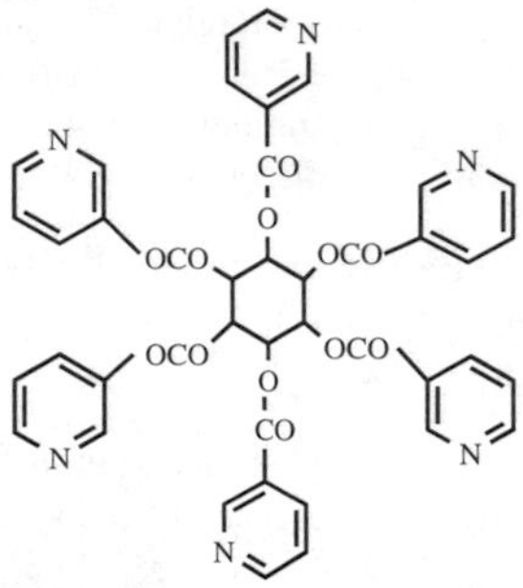

Inositol Oxygenase The enzyme that catalyzes the following reaction:

Inositol + O_2 ⇌ D-Glucuronate + H_2O

Inositol 1-Phosphate Synthetase The enzyme that catalyzes the following reaction:

D-Glucose 6-phosphate ⇌ Inositol 1-phosphate

Inositol Phospholipid Phospholipid derived from inositol that is important in signal transduction in eukaryotes.

Inositol 1,4,5-Triphosphate (mol wt 420) A compound that acts as a second messenger in cellular signal transduction.

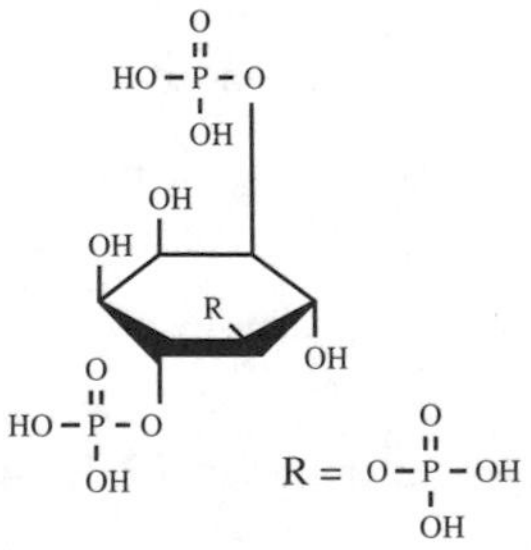

Inotropic Capable of influencing muscular contractibility.

Inoviridae A family of single-stranded DNA bacteriophages with filamentous or rod-shaped virion.

Inovirus A virus in the family of Inoviridae with filamentous virion containing single-stranded DNA.

Insecticides Substances that kill insects or chemicals used to control insect populations.

Insectivore An animal that feeds primarily on insects.

Insertion Hot Spot A region in a genome at which transposible genetic elements (transposome) preferentially insert.

Insertion Inactivation Inactivation of a functional gene by inserting a foreign DNA and rendering the gene nonfunctional, e.g., insertion of a foreign DNA into an antibiotic-resistant site causing the loss of antibiotic resistance.

Insertion Mutagenesis Production of a mutation or alteration of a DNA sequence by the insertion of a foreign DNA.

Insertion Mutation A mutation produced by insertion of one or more nucleotides into a gene.

Insertion Sequences A transposable element of DNA that carries only genes for transposition and is capable of inserting itself into a number of sites in a genome.

Insertion Site A restriction site in a cloning vector at which foreign DNA inserts.

Insertion Vector A cloning vector that has a single site at which an exogenous DNA can be inserted.

Insertosome A DNA element (800–1400 bp) capable of inserting itself randomly into the *E. coli* chromosome and producing mutation.

Insoluble Enzyme The enzyme that has been covalently linked onto an insoluble particle such as agarose or polyacrylamide without destruction of catalytic activity.

Insomnia Inability to sleep or to remain asleep throughout the night.

$InsP_3$ Abbreviation for inositol triphosphate.

$Ins(1,3,4)P_3$ Abbreviation for inositol 1,3,4-triphosphate.

$Ins(1,4,5)P_3$ Abbreviation for inositol 1,4,5-triphosphate.

$InsP_4$ Abbreviation for inositol 1,3,4,5-tetrakisphosphate.

$InsP_3R$ Abbreviation for inositol 1,4,5-triphosphate receptor.

$InsP_4R$ Abbreviation for inositol 1,3,4,5-tetrakisphosphate receptor.

Insulin A polypeptide hormone produced by beta cells of the pancreas. It regulates metabolism of carbohydrate and lipid and influences the biosynthesis of protein and RNA.

Insulin Kinase The enzyme that inactivates insulin.

Insulinase An enzyme that catalyzes the hydrolysis of A and B chains of insulin but has no effect on intact insulin.

Insulin-Dependent Diabetes Mellitus The inability of the body to metabolize carbohydrate that is caused by an insulin deficiency. It is characterized by loss of weight and diminished strength.

Insulin-Independent Diabetes Mellitus A form of diabetes that occurs in adults whose insulin is near the normal level. The defect may be due to a low level of insulin receptors.

Insulin-Like Growth Factor Polypeptides with considerable sequence similarity to insulin. They are capable of eliciting some of the same biological responses as insulin.

Insulinogenic Capable of promoting the production and release of insulin by the pancreas.

Insulinoma A benign tumor of the insulin-secreting cells of the pancreas.

Insulinopenia A low level of insulin in the blood.

Insulinotropin A potent insulin stimulator released from the perfused rat pancreas.

Insulin-Shock A hypoglycemia shock caused by overdose of insulin and characterized by sweating, trembling, chilliness, hunger, and hallucination.

Insulin-Zinc Suspension Sterile suspension of insulin prepared with zinc chloride.

***Int* Gene** The gene involved in the integration of phage DNA into host DNA.

Intal A trade name for cromolyn sodium, which inhibits mast cell degranulation and release of histamine.

Integral Membrane Protein Hydrophobic protein located within the interior of the membrane but usually having hydrophilic regions protruding from the membrane on one or both sides.

Integral Proteins Referring to integral membrane proteins.

Integrase The enzyme that integrates a viral genome into the genome of the host cell, e.g., integration of λ phage DNA into bacterial chromosome or integration of retroviral cDNA into the eukaryote genome.

Integration The process of integration of viral or foreign DNA into genomic DNA of a host cell.

Integration Host Factor A bacterial protein that acts in concert with λ phage-encoded integrase for integration of λ phage DNA into host chromosome.

Integrilin A trade name for eptifibatide, an antiplatelet drug.

Integrins A superfamily of adhesion proteins involved in the adhesion of cells to extracellular matrix. Integrins consist of noncovalently linked heterodimers and share a common β chain but have different α chains.

Integument An external covering or coat of an organ.

Intein An internal peptide sequence of a protein precursor that is spliced out during the processing to form the mature protein.

Intercalary Deletion Deletion of genetic material that occurs in the internal part of a chromosome but not the terminal part.

Intercalating Agent Any substance (e.g., acridine) capable of inserting or intercalating between adjacent base pairs and thus disrupting the alignment and pairing of the bases in double-stranded DNA or in base pair regions of ssDNA. This process leads to the formation of mutation.

Intercellular Between cells.

Intercistronic Region 1. The DNA sequence that separates two genes. 2. The nucleotide sequence between the terminal codon of one gene and the initiation codon of another gene in a polycistronic transcription unit.

Interconvertible Enzyme An enzyme that can undergo covalent modification and interconversion by another enzyme, e.g., muscle phosphorylase *a* can be converted to phosphorylase *b* by phosphorylase kinase.

Interdigitating Cells Cells found in thymus-dependent regions of lymph nodes that have dendritic morphology and accessory cell function. They can act as antigen-presenting cells.

Interdoublet Link The link between adjacent doublets in the axoneme of a cilium or flagellum.

Interesterification A process in which fatty acyl residues are interchanged with various triglycerides to form a new ester.

Interface The boundary between two phases, e.g., interface between phenol and water.

Interface Centrifugation A process of centrifugation for transfer of solutes into different phases, e.g., centrifugation of a phenol-water mixture separates water soluble compounds within the water phase, phenol-soluble compounds in the phenol phase, and insoluble compounds in the interface or in the bottom of the centrifuge tube.

Interference Microscope A microscope that relies on the destructive and/or additive interference of light waves to achieve contrast.

Interferometer A device used to measure the velocity and absorption of sonic or ultrasonic waves in a gas or a liquid.

Interferon α A family of glycoproteins produced by peripheral blood leukocytes upon exposure to interferon inducers, e.g., virus, double-stranded RNA, or bacterial products.

Interferon-α-$_{2a}$ A sterile interferon protein produced by DNA recombination technology.

Interferon-α-$_{2b}$ A sterile interferon protein produced by DNA recombination technology.

Interferon-α-$_{n3}$ A naturally occurring antiviral agent derived from human leukocytes.

Interferon β (Beta Interferon) Interferons produced by fibroblasts.

Interferon γ (Gamma Interferon) Interferons produced by mitogen-stimulated T-cells that possess antiviral and immunomodulatory activity.

Interferons (INF) A family of species-specific glycoproteins produced by vertebrates in response to inducers, e.g., virus or double-stranded RNA that confers a broad range of activities such as resistance to viral infection, modulation of the immune response, and regulation of cell proliferation.

Intergenic Complementation The complementation produced by two mutants that carry mutations at different genes in which the unmutated gene on one chromosome complements the mutated gene on the other.

Intergenic Mutation Mutation that involves more than one gene.

Intergenic Suppression The suppression of a mutant phenotype due to a mutation in a different gene.

Intergenote A partial zygot in which the donor DNA is different from that of the recipient.

Interkinesis A short resting stage between the first and second meiotic division. No DNA replication occurs during interkinesis.

Interleukin-1 (IL-1) Lymphokines secreted by macrophages that stimulate the resting helper T-cell to secrete interleukin-2.

Interleukin-2 (IL-2) Lymphokines secreted by helper T cells that bind to cell-surface interleukin-2 receptors of activated T-cells and causes them to proliferate (also called T-cell growth factor).

Interleukin-3 (IL-3) A T-cell product that stimulates proliferation and differentiation of other lymphocytes and some hematopoietic cells.

Interleukin-4 (IL-4) A factor produced by helper T-cells that stimulates the growth of both T and B cells. It is also a switch factor for synthesis of IgE.

Interleukin-5 (IL-5) A factor produced by helper T cells that stimulates B cells and eosinophils. It also facilitates differentiation of B cells that secretes IgA.

Interleukin-6 (IL-6) A factor that is co-induced with interferon in fibroblasts that functions as a B-cell differentiation factor, a hybridoma growth factor, and an inducer of acute phase proteins.

Interleukin-7 (IL-7) A factor produced by stromal cells that causes lymphoid stem cell differentiation into progenitor B cell and T cells.

Interleukin-8 (IL-8) A factor produced by macrophages that possesses chemotactic activity for T-cells and neutrophils.

Interleukin-9 (IL-9) A glycoprotein factor that facilitates growth of some T helper cell clones but not cytotoxic T cell clones.

Interleukin-10 (IL-10) A protein factor expressed by $CD4^+$ and $CD8^+$ T cells, monocytes, macrophages and activated B cells. It acts as a cytokine synthesis inhibitory factor.

Interleukin-11 (IL-11) A protein factor that induces IL-6-dependent murine plasmacytoma cells to proliferate.

Interleukin-12 (IL-12) A protein factor that acts on T cells as a cytotoxic lymphocyte maturation factor.

Interleukin-13 (IL-13) A protein factor produced by activated T cells that inhibits inflammatory cytokine production by lipopolysaccharide in human peripheral blood monocytes.

Interleukin-14 (IL-14) A protein factor produced by follicular dendritic cells, germinal T cells and some malignant B cells, it enhances the proliferation of B cells and induces memory B cell production and maintenance.

Interleukin-15 (IL-15) A T cell growth factor that enhances peripheral blood T cell proliferation.

Interleukins A group of bioactive proteins produced by leukocytes, monocytes and other cells that regulate the immune response (also called lymphokines).

Intermediary Metabolism The chemical pathways or reactions in a cell that transform food or nutrients into energy and molecules needed for cell growth.

Intermediate Compound generated in a reaction pathway or metabolic cycle.

Intermediate Filaments Fibrous protein filaments that play a structural or tension-bearing role in the cytoskeleton of eukaryotic cells. These filaments are intermediate in size between microtubules and microfilaments.

Intermediate Filament Typing A technique for identification of cell type by determining the type of intermediate filaments present in the cell. This technique is useful for tracing cell lineages during development and for pathological classification of tumor origin.

Intermedin A peptide hormone from the intermediate lobe of the pituitary gland that causes dispersion of melanin in melanophores (also known as melanocyte stimulating hormone).

Intermittent Fever A fever that occurs in cycles (as in malaria).

Intermolecular Recombination The recombination or transposition between two separate DNA molecules.

Internal Compensation A type of optical activity in which a molecule contains two asymmetric centers that produce equal and opposite rotation of the plane of polarization with the result that the molecule is optically inactive.

Internal Medicine The medical science that deals with physiopathology of the internal organs, medical diagnosis, and treatment of diseases.

Internal Membrane Eukaryotic cell membrane other than the plasma membrane, e.g., membrane of the Golgi apparatus or endoplasmic reticulum.

Internal Radiation The radioactive radiation from the substances that are deposited in the tissue.

Internal Resolution Site Referring to the site-specific recombination that occurs at an A-T rich region of the DNA molecule.

Internal Standard A standard agent or chemical that is added to a sample.

Interneuron An excitatory or inhibitory neuron in the CNS situated between the primary afferent neuron and the final motor neuron. It is involved in the intermediate processing of signals.

Interoceptor Any sensory nerve ending in the cells of the viscera that responds to stimuli from within the body regarding the function of the internal organs (e.g., digestion, excretion, and blood pressure).

Interphase A stage of the cell that is not undergoing mitosis. The interval between nuclear divisions.

Interpolate Estimation of a value of a function between data values already obtained or known either graphically or by calculation.

Interrupted Gene A gene that contains introns.

Interrupted Mating A technique used to map the gene order of the bacterial chromosome by interrupting the process of DNA transfer from the donor to the recipient at different time intervals during conjugation.

Intersome The precursor of ribosome or subribosomal particles produced by the stepwise removal of proteins.

Interspecific Heterokaryons Cells that contain nuclei from two different species that are produced by cell fusion.

Interstitial Pertaining to the interspaces of tissues or structures.

Interstitial Cystitis Inflammation of the bladder.

Interstitial Fluid The fluid that fills the spaces between the cells.

Interstitial Nephritis The inflammation of interstitial tissue of the kidney.

Interstitial Pneumonia A chronic inflammation of the lung.

Interstitial Volume The volume of mobile solvent within a chromatographic column.

Intervening Sequence Referring to the noncoding sequence between genes (see also intron).

Interventricular Between the two ventricles of the heart.

Interwound Helix A DNA helix structure in which the duplex axis is twisted around itself.

Intestinal Epithelium The endodermally derived epithelium of the intestine with microvilli on the surface.

Intestinal Ischemia An impaired blood supply to the intestine due to arteriosclerosis of the vessels that supply blood to the intestine.

Intima The inner layer of the blood vessel.

Intimin An actin-polymerization-inducing protein produced by enteropathogenic and enterohemorrhagic strains of enteric bacteria.

Intoxicant Any intoxicating agent, e.g., alcohol.

Intoxication A poisoned state caused by intoxicating substances, e.g., drugs or alcohol.

Intra- A prefix meaning within or inside.

Intrabutazone A trade name for phenylbutazone, an anti-inflammatory agent.

Intracardiac Located within the heart.

Intracather A thin, flexible plastic tube inserted into the blood vessel for the purpose of supplying blood, fluid, or medication.

Intracellular Within the cell.

Intracellular Receptor Receptors located within the cytoplasm of a cell.

Intracellular Transport Movement of substances across membranes of organelles inside the cell.

Intracerebral Within the cerebrum.

Intracranial Within the skull.

Intracristal Space The space between the inner and outer membranes of the mitochondria.

Intracutaneous Within the skin.

Intradermal Within the dermal layer of the skin.

Intradermal Test A procedure used to identify allergens by injecting the individual with extracted allergens into the skin for observation of sign of wheal surround by redness.

Intraepidermal Within the outermost nonvascular layer of the skin (without blood vessels).

Intraepithelial Within epithelial cells.

Intragenic Complementation The complementation between two mutants that occurs at different sites in the same gene.

Intragenic Molecular Recombination Recombination or transposition between two different sites in the same DNA molecule.

Intragenic Mutation Mutation that involves only a single gene.

Intragenic Recombination The recombination between mutation sites of the same gene.

Intragenic Suppresser Mutation A second mutation that negates the effect of the first mutation within a single gene.

Intragenic Suppression The suppression of a mutant phenotype by a second mutation in the same gene.

Intrahepatic Within the liver.

Intramedullary Within the bone marrow.

Intramural Within the wall of an organ.

Intramuscular Within the muscle.

Intranasal Within the nose.

Intraneural Within the nerve.

Intraocular Within the eyeball.

Intraocular Implant A plastic lens that is inserted into the eye to replace the natural lens, which was removed because of a cataract.

Intraperitoneal Within the abdominal cavity.

Intrapleural Within the chest cavity.

Intrapulmonary Within the lung.

Intrarenal Within the kidney.

Intraspecific Relation Relations between members of the same species.

Intrathecal Within a sheet.

Intrathoracic Within the chest cavity.

Intrathoracic Goiter An enlargement of the thyroid gland that protrudes into the thoracic cavity.

Intrathylakoid Space Space within the membranes of the thylakoids and the stroma lamellae.

Intrauterine Within the cavity of the uterus.

Intrauterine Device (IUD) A plastic or metal device inserted in the uterus for use as a contraceptive.

Intrauterine Transfusion A process in which blood is directly injected into the fetus through the mother's abdominal wall for the uterus to combat Rh factor disease.

Intraval A trade name for thiopentone sodium, an anesthetic agent.

Intravascular Within the blood or lymphatic vessels.

Intravenous Within or into a vein.

Intravenous Cholangiography An X-ray procedure for detecting stones within the bile duct after intravenous injection of contrast medium.

Intravenous Injection A hypodermic injection of an agent or substance directly into the vein.

Intravenous Pyelography An X-ray procedure for visualization of the kidney and uterus after intravenous injection of contrast medium.

Intraventricular Space in a ventricle of the heart or brain.

Intravital Stain Staining cells without killing them.

Intrinsic 1. Natural or inherent. 2. Within an organ or tissue. 3. Not dependent upon external factors.

Intrinsic Asthma A nonseasonal and nonallergic form of asthma.

Intrinsic Blood Coagulation Blood coagulation without involvement of external factors (see intrinsic pathway).

Intrinsic Factor A glycoprotein factor secreted by the stomach that is responsible for the absorption of vitamin B_{12}.

Intrinsic Fluorescence The fluorescence due to the aromatic amino acids of the protein, not the fluorescence due to the labeled ligand.

Intrinsic Pathway A series of cascade reactions in blood coagulation that involves the factors present in the circulation.

Intrinsic Proteins Referring to integral membrane proteins.

Intron A segment of DNA that is present between exons and does not contain expressed genes. Introns are transcribed into RNA but are excised out after transcription.

Intron A A trade name for interferon $alfa_{2b}$.

Intron Intrusion The disruption of a functional gene by the insertion of an intron into the gene.

Intron Splicing The splicing out of intron transcript before RNA is translated.

Intropin A trade name for dopamine hydrochloride that stimulates dopaminergic, beta-adrenergic, and alpha-adrenergioc receptors of the sympathetic nervous system.

Intubation Introduction of a tube into a cavity, e.g., the passage of a tube into the larynx to keep the air passage open.

Intumescence An enlargement or swelling of an organ or part of an organ.

Inulin (mol wt approx. 5000) A polymer of fructose and a reagent used in kidney function tests.

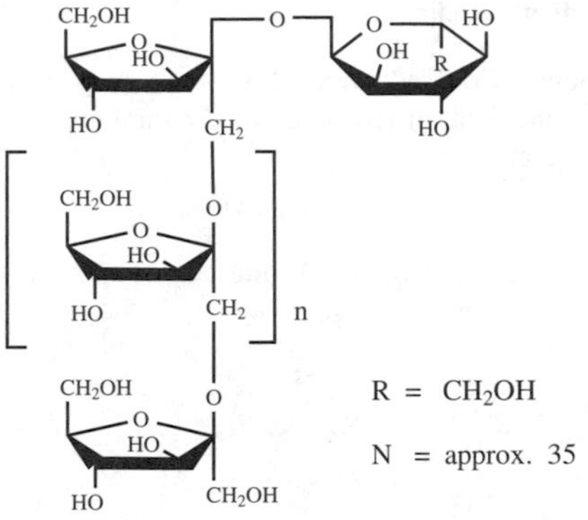

Inulin Clearance Test A test of the rate of filtration of inulin in the glomerulus of the kidney.

Inulin Fructotransferase The enzyme that catalyzes the successive removal of the terminal D-fructosyl-D-fructofuranosyl group from inulin.

Inulinase The enzyme that catalyzes the hydrolysis of 2,1-β-D-fructosidic linkages in inulin.

Inulosucrase The enzyme that catalyzes the following reaction:

$$\text{Sucrose} + (2,1\text{-}\beta\text{-D-fructosyl})_n \rightleftharpoons \text{Glucose} + (2,1\text{-}\beta\text{-D-fructosyl})_{n+1}$$

***inv* Marker** An allotypic marker on the constant region of the kappa light chains of human immunoglobulin.

Invagination Local infolding or inpocketing of a membrane leading to the formation of a pouch or a saclike structure.

Invariant Identical amino acid sequence in a protein isolated from various species.

Invariant Residues Amino acid residues that exhibit homology in evolutionary related proteins.

Invasin Referring to hyaluronidase.

Invasiveness The spread of pathogens or cancer cells from the infection sites or tumor sites.

Inversine A trade name for mecamylamine hydrochloride, an antihypertensive agent and ganglionic blocker that competes with acetylcholine for ganglionic cholinergic receptors.

Inversion Chromosome segment that has been rearranged so that sequences in the rearranged segment are in inverse order.

Inversion Heterozygote Individual in which one chromosome contains an inversion whereas the homologous chromosome does not.

Invert Sugar A mixture of glucose and fructose obtained from hydrolysis of sucrose.

Invertase The enzyme that catalyzes the hydrolysis of sucrose to glucose and fructose (also know as fructofuranosidase).

Invertebrate An animal that lacks a backbone, e.g., earthworm and snail.

Inverted Repeat Two copies of DNA sequences that have the same sequence but in opposite orientation.

Inverted Terminal Repeat The short identical DNA sequences that have the reverse orientation present at each end of a DNA genome.

Invirase A trade name for saquinavir, an antiviral agent used for the treatment of AIDS.

Iocetamic Acid (mol wt 614) A substance used as radiopaque medium in diagnosis.

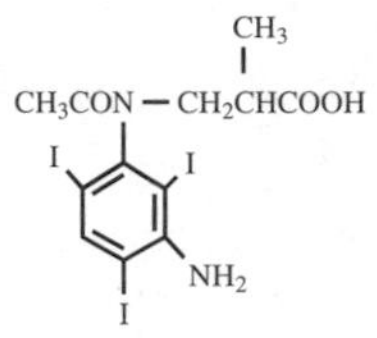

IOD Abbreviation for integrated optical density.

Iodamide (mol wt 628) A substance used as radiopaque medium in diagnostic procedures.

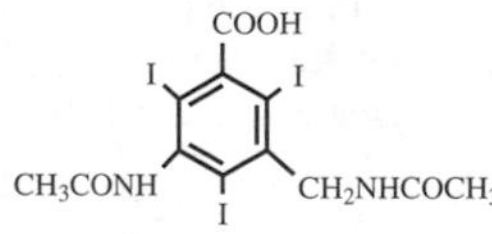

Iodic Acid (mol wt 176) An astringent and disinfectant.

$$HIO_3$$

Iodide Hydrogen-Peroxide Oxidoreductase Systematic name for iodide peroxidase.

Iodide Peroxidase The enzyme that catalyzes the following reaction:

$$\text{Iodide} + H_2O_2 \rightleftharpoons \text{Iodine} + 2\,H_2O$$

Iodide Pump An active transport mechanism that concentrates iodide in the thyroid gland.

Iodide Tyrosine Deiodase See iodide peroxidase.

Iodinase See iodide peroxidase.

Iodinated Density Gradient Medium Medium used for isopycnic ultracentrifugation for separation of macromolecules, e.g., tri-iodinated benzamide derivative of glucose (metrizamide) or tri-iodinated derivative of benzoic acid.

Iodinated Glycerol (mol wt 258) An expectorant that increases production of respiratory tract fluid.

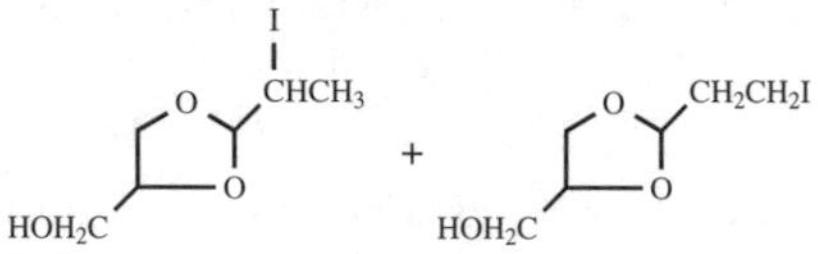

Iodinated Trichloride (mol wt 233) A topical anti-infective agent.

$$Cl_3I$$

Iodination The replacement of a hydrogen atom by an atom of iodine in a molecule, a common procedure for labeling protein.

Iodine (I) A chemical element with atomic weight 127, valence 1 to 7.

Iodine–125 (^{125}I) An artificial radioactive nuclide emitting low-energy gamma radiation (half life 59 days).

Iodine–131 (^{131}I) An artificial radioactive nuclide emitting beta radiation (half life 8 days).

Iodine Number A measurement of the degree of unsaturation in a fat. It is the number of grams of iodine taken up by 100 grams of fat. The greater the number, the greater the unsaturation of the fat sample.

Iodine Value See iodine number.

Iodinin (mol wt 244) An antibiotic pigment from *Chromobacterium iodinum.*

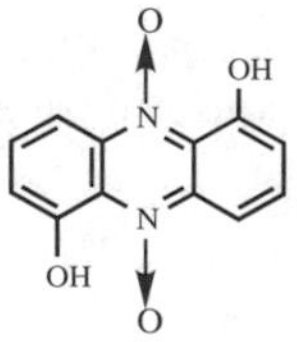

Iodipamide (mol wt 1140) A substance used as radiopaque medium in X-ray diagnoses.

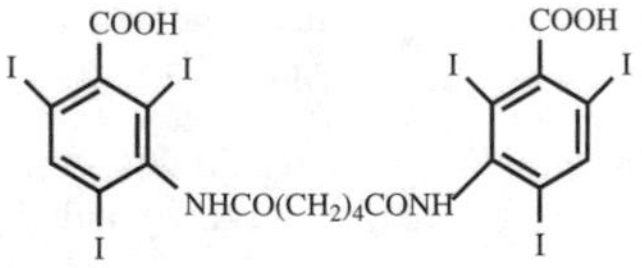

Iodism Iodine poisoning.

Iodoacetic Acid (mol wt 186) A reagent used for chemical modification of the thiol group in a protein.

$$ICH_2COOH$$

Iodochlorhydroxyquin (mol wt 306) An intestinal anti-infective agent.

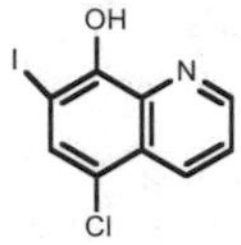

5-Iodo-2′-Deoxyuridine (mol wt 354) A potent inhibitor of thymidine kinase and thymidylate synthetase.

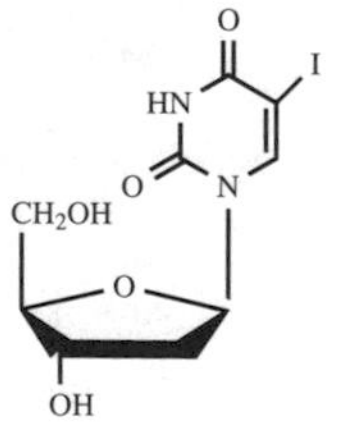

Iodohippurate Sodium (mol wt 327) A radiopaque compound used as a contrast medium in radiology.

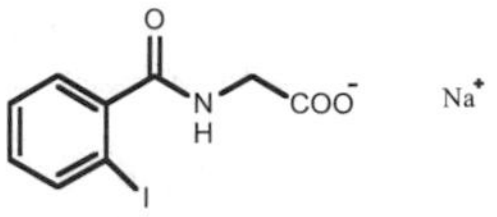

Iodonitrotetrazolium Chloride (mol wt 506) An electron acceptor for the colorimetric assay of lactate dehydrogeanse.

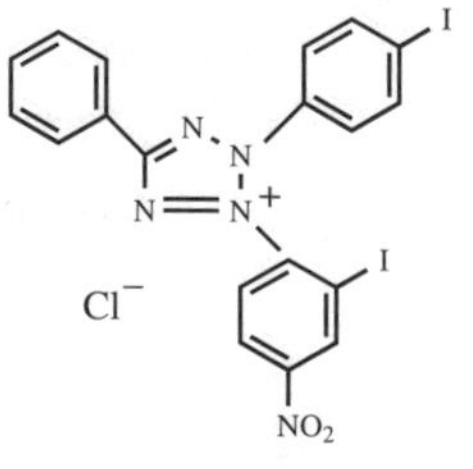

Iodopsin A protein pigment found in the retinal cone cell.

Iodopyrrole (mol wt 570) A topical antiseptic agent.

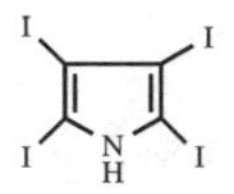

Iodoquinol (mol wt 397) An anti-amebic agent.

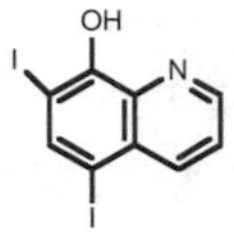

Iodosobenzoic Acid (mol wt 264) A reagent for chemical cleavage of peptide bonds in a protein between tryptophan and tyrosine.

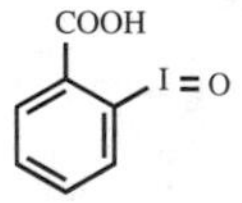

Iodo-L-tyrosine (mol wt 307) An inhibitor for tyrosine hydroxylase.

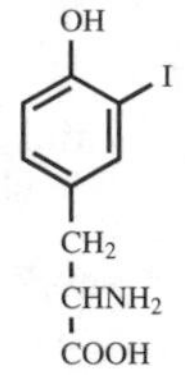

Iodo-uridine (mol wt 307) A uridine derivative.

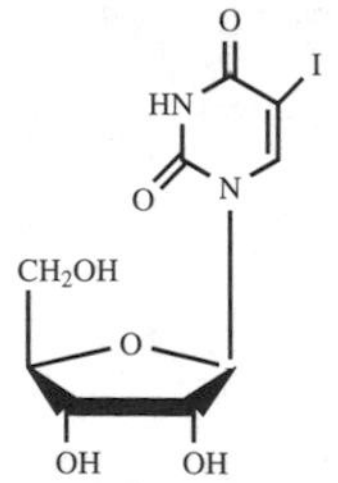

Ioglycamic acid (mol wt 1127) A radiopaque medium used as a diagnostic aid.

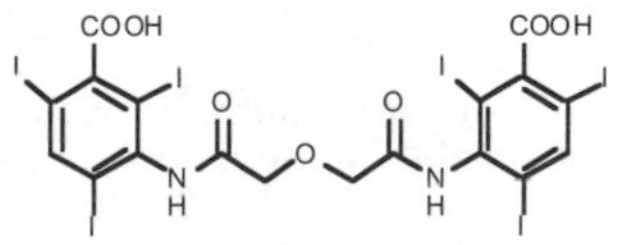

Ion A charged atom or molecule.

Ion Antagonism The inactivation or inhibition of a reaction or a set of reactions by one or more ions.

Ion Carrier Substance (e.g., ionophore) that carries ions in its hydrophobic cavity, shielding the ions from the hydrophobic region of the lipid bilayer of the membrane to facilitate its transport across a biological membrane.

Ion Channel 1. An integral membrane protein channel that regulates and transports ions across a membrane. 2. Ionophore (e.g., gramicidin) that provides a channel through which ions can pass the membrane (known as channel ionophore).

Ion Dipole Interaction The attractive or repulsive force between an ion and a dipole.

Ion Etch A technique for treatment of a specimen with a beam of inert ions prior to examination by scanning electron microscopy.

Ion Exchange Chromatography A type of column chromatography that employs ion exchange resin in a column for the separation of a mixture of solutes on the basis of their charges and affinity for the exchange resin.

Ion Exchanger Referring to both cation or anion exchange resin used in ion exchange chromatography.

Ion Exchanger Conditioning A procedure used for cleaning an ion exchange resin or to convert it to the free base or free acid form by washing with NaOH or HCl.

Ion Exchange Resin An insoluble polymeric resin (e.g., polystyrene) that contains fixed charge groups and used in ion exchange chromatography for separation of solutes.

Ion Pair The association of two ionic groups of opposite charge.

Ion Pump Transmembrane-protein mediated active transport of ions across a biological membrane.

Ionamin A trade name for phentermine, an anorexic agent.

Ionic Pertaining to ions.

Ionic Bond Bond formed by the force between ions of the opposite charges, a type of noncovalent bond.

Ionic Detergent A surface-active substance that carries charges.

Ionic Strength The summation of the molar concentration of each ionic species in a solution multiplied by the square of their valences.

Ionizating Radiation A type of radiation, e.g., X-ray, which causes loss of electrons from an organic molecule, thus making them more reactive.

Ionization 1. Split of a molecule into ions. 2. Formation of a charged group in a molecule.

Ionization Constant The dissociation constant of a reaction that converts a molecule to an ion.

Ionogen Any compounds or atoms or group of atoms that are capable of becoming ionized.

Ionomycin (mol wt 747) An ionophore antibiotic with a greater selective activity for calcium than magnesium.

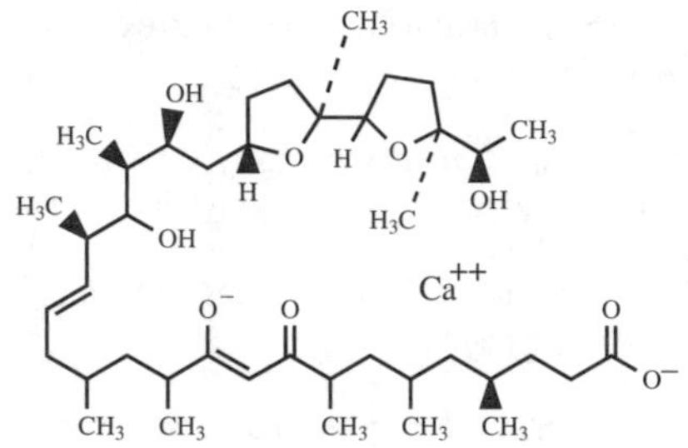

Ionophore A compound that is capable of binding metal ion(s) and diffusing across a membrane.

Ionophorous Antibiotics Antibiotic that functions as an ionophore.

Ionotropic Describing a type of receptor that mediates its effects by regulating ion channels.

Iopanoic Acid (mol wt 571) An iodine-containing radiopaque medium used as a diagnostic aid in radiology.

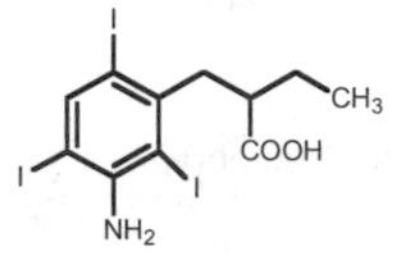

Iophendylate (mol wt 416) An iodine-containing radiopaque medium used as a diagnostic aid in radiology.

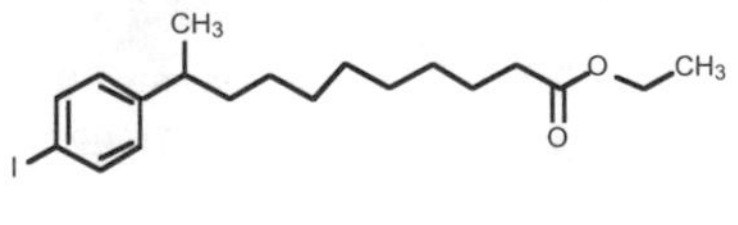

Iopidine A trade name for apraclonidine, an adrenergaic agonist.

Iopol A trade name for inactivated polio virus vaccine.

Iosopan A trade name for magaldrate (hydroxymagnesium aluminate), an antacid.

Iothiouracil (mol wt 254) A thyroid inhibitor.

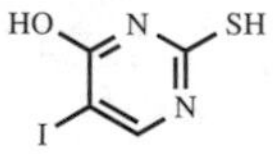

IP Abbreviation for 1. immuno-precipitation; 2. induced protein; 3. inosine phosphorylase; 4. intraperitoneal; 5. isoelectric point.

IP_1 Abbreviation of inositol monophosphate.

IP_1 Phosphatase The enzyme that catalyzes the hydrolysis of inositol-1-phosphate to inositol and inorganic phosphate.

IP_2 Abbreviation of inositol bisphosphate.

IP_2 Phosphatase The enzyme that catalyzes the hydrolysis of inositol-diphosphate to inositol phosphate and inorganic phosphate.

IP_3 Abbreviation of inositol triphosphate, a sugar alcohol that causes release of calcium from the endoplasmic reticulum.

IP_3 Phosphatase The enzyme that catalyzes the hydrolysis of inositol triphosphate to inositol bisphosphate and inorganic phosphate.

IP_4 Abbreviation for inositol 1,3,4,5- tetraphosphate.

IP_5 Abbreviation for inositol pentaphosphate.

IP_6 Abbreviation for inositol hexaphosphate.

IPA Abbreviation for isopropyl alcohol.

IPB Abbreviation for immuno-precipitation buffer.

iPCR Abbreviation for inverse PCR.

Ipecac An emetic agent from rhizome and roots of *Cephaelis ipecacuanha* (known as Panama Ipecac).

IPOL A trade name for inactivated poliovirus vaccine.

Ipodate (mol wt 598) Substance used as radiopaque medium in diagnosis.

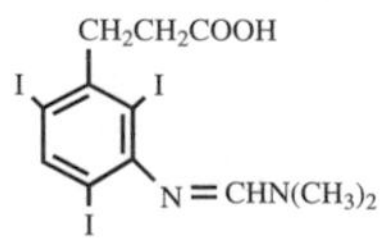

Ipomea A resin extracted from dry root of *Impomoea orizabensis* used as a cathartic agent.

I-PpoI A restriction endonuclease from nuclear extrachromosomal ribosomal DNA of *Physarum polycephalum,* cloned and over-expressed in *E. coli* with the following specificity:

5'.....CTCTCTTAA↓GGTAGC.....3'
3'......GAGAG↑AATTCCATCG.....5'

Ipratropium Bromide (mol wt 412) A bronchodilator and antiarrhythmic agent.

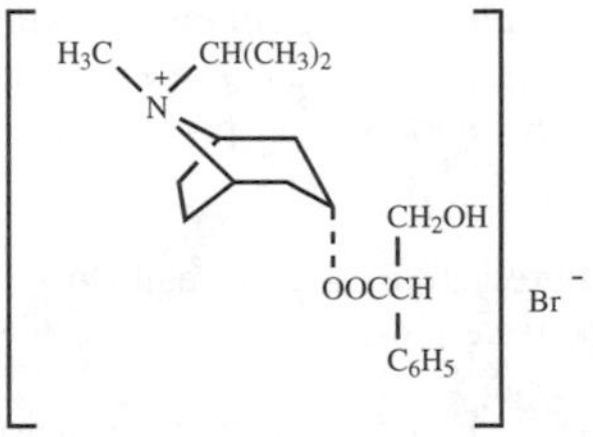

Ipriflavone (mol wt 280) A calcium regulator.

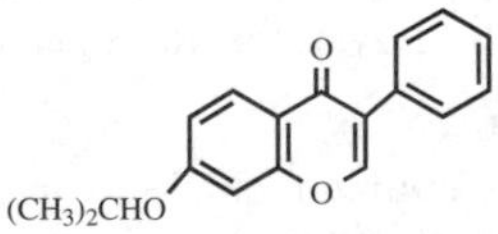

Iprindole (mol wt 284) An antidepressant.

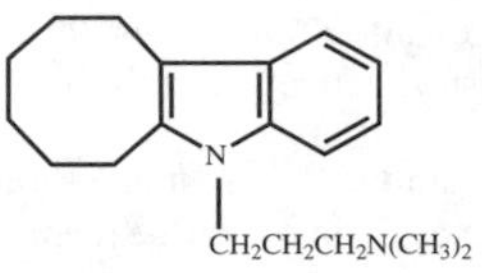

Iproclozide (mol wt 243) An antidepressant.

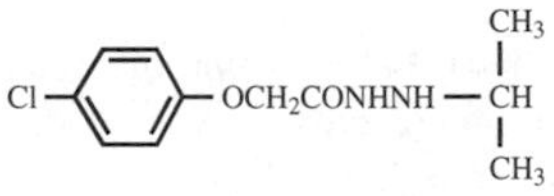

Iprodione (mol wt 330) A fungicide.

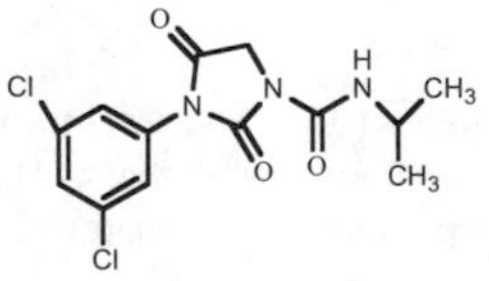

Ipronidazole (mol wt 169) An antiprotozoal and antimicrobial agent.

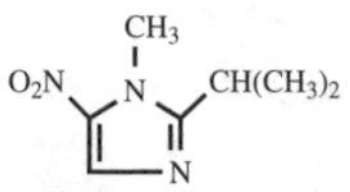

Ipronizid (mol wt 179) An antidepressant and inhibitor for monoamine oxidase.

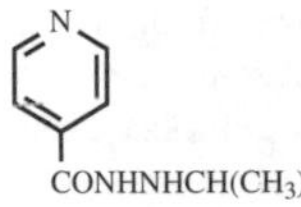

IPT Abbreviation for idiopathic thrombocytopenic purpura.

IPTG Abbreviation for isopropylthiogalactoside, a gratuitous inducer for *lac* operon.

IPTH Abbreviation for immuno-reactive parathyroid hormone.

IPV A trade name for inactivated poliovirus vaccine.

IR Abbreviation for 1. infrared and 2. inverted repeat.

Ir Symbol of iridium, a chemical element with atomic weight 192, valence 1,3,4.

IRA Abbreviation for immunoradiometric assay.

IRABP Abbreviation for intracellular retinoic acid-binding protein.

IRAK Abbreviation for interleukin-1 receptor-associated kinase.

IRAK-Wt Abbreviation for wild type interleukin-1 receptor-associated kinase.

IRAP Abbreviation for interleukin-1 receptor antagonist protein.

Irbesartan (mol wt 429) An antihypertensive agent.

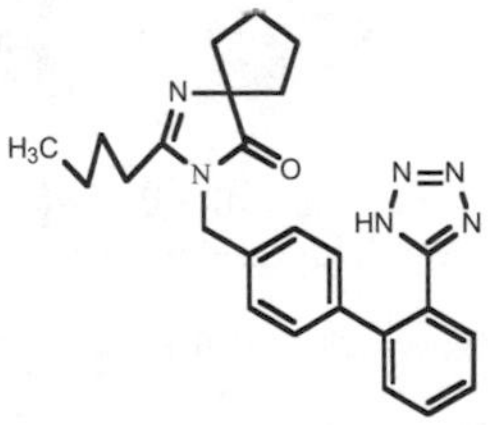

Ircon A trade name for ferrous fumarate, an iron source.

IRE Abbreviation for 1. iron regulatory element or iron responsive element; 2. interferon regulatory element or interferon-gene regulatory element.

I-region The region in the H-2 major histocompatibility complex of mouse that contains genes that encode class II antigens, antigens responsible for regulating immune responses.

IRF Abbreviation for interferon regulatory factor.

IRG Abbreviation for immuno-reactive glucagon.

Ir-genes Immune regulatory genes or immune response genes located within the I-region of the histocompatibility complex of the mouse.

IRIA Abbreviation for indirect radio-immunoassay.

Iridium (Ir) A chemical element with atomic weight 199, valence 1, 3, and 4.

Iridochoroiditis Inflammation of both iris and vascular coat of the eyeball.

Iridocyclitis Inflammation of iris and ciliary body of the eye.

Iridomyrmecin (mol wt 168) An antibacterial agent.

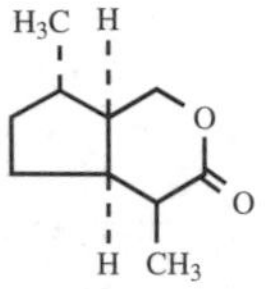

Iridoviridae A family of double-stranded DNA isometric viruses that infect mammals, fishes, and insects.

Irinotecan (mol wt 587) An antineoplastic agent.

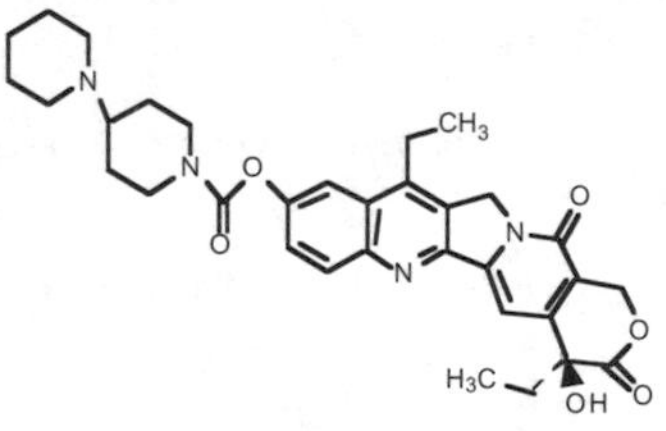

Iris The circular contracting disc situated between the cornea and the crystalline lens of the eye.

Irititis Inflammation of the iris.

IRMA Abbreviation for immuno-radiometric assay.

Iron (Fe) A chemical element with atomic weight 56, valence 2, and 3.

Iron-59 (^{59}Fe) The artificial radioactive nuclide emitting beta radiation (half life 44 days)

Iron Bacteria Bacteria capable of depositing oxides or hydroxides of iron or magnesium.

Iron Binding Protein Protein capable of binding iron, e.g., transferrin.

Iron Deficiency Anemia Anemia caused by lack of iron.

Iron Dextran A substance used for treatment of iron-deficiency anemia.

Iron Protoporphyrin IX A form of heme found in hemoglobin and cytochromes b, c, and c_1.

Iron Response Element A mRNA nucleotide sequence involved in mediating iron-dependent translation of ferritin mRNA and iron-dependent destablization of transferrin receptor mRNA.

Iron Sulfur Cluster The prosthetic groups of iron-sulfur protein. The two common types of iron-sulfur clusters are 2Fe-2S and 4Fe-4S.

Iron Sulfur Protein Protein that contains iron and sulfur atoms that are complexed with cysteine groups of the protein and serves as an electron carrier in the electron transport chain, e.g., ferredoxin.

IRP Abbreviation for iron regulatory protein or iron responsive protein.

Irradiation Exposure to a beam of ionizing or electromagnetic radiation.

Irreversible Inhibition Covalent binding of an inhibitor to an enzyme that cause permanent inactivation.

Irreversible Inhibitor Molecule that covalently binds to an enzyme and causes irreversible loss of catalytic activity.

IRS Abbreviation for 1. insulin-receptor substrate; 2. insulin receptor synthetase.

IS Abbreviation for 1. immune serum; 2. immunosuppressive; 3. insertion sequence; 4. internal standard.

ISA Abbreviation for iodinated serum albumin.

Isatin (mol wt 147) A chromatographic spray reagent for amino acids.

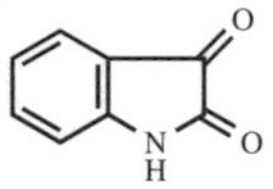

Isaxonine (mol wt 137) A neurotropic agent that promotes neurite out-growth that has been used for treatment of peripheral neuropathies.

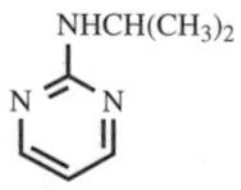

Isazofos (mol wt 314) A cholinesterase inhibitor.

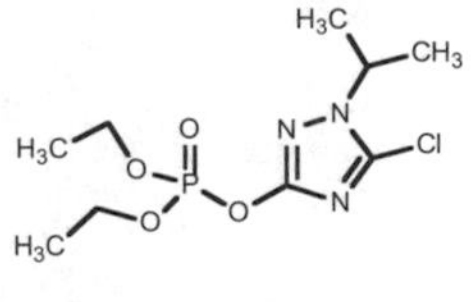

Isbogrel (mol wt 281) An antithrombotic agent.

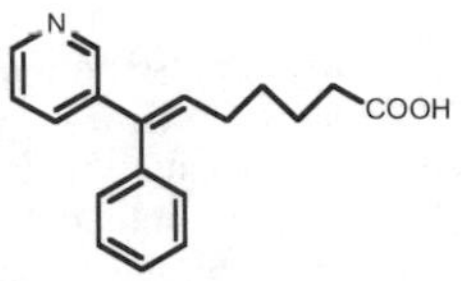

I-SceI A restriction endonuclease from *E. coli* TG1 with the following specificity:

5'.......TAGGGATAA↓CAGGGTAAT.......3'

Ischemia (Ischaemia) Inadequate blood flow in the tissue characterized by pain and organ dysfunction.

Ischuria Retention or suppression of the urine.

ISDN Abbreviation for isosorbide dinitrate.

Isepamicin (mol wt 570) A semisynthetic derivative of antibiotic gentamicin.

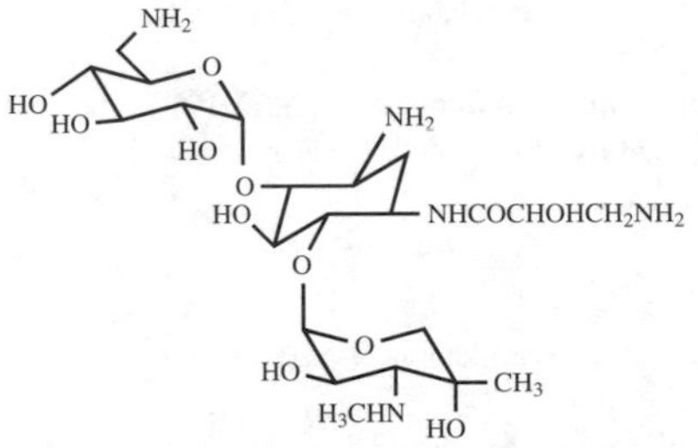

ISG Abbreviation for immune serum globulin.

IS1 A transposable element in the genome of *E. coli.*

Islet Cells Cells of the Islets of Langerhans in the pancreas.

Islets of Langerhans Cluster of endocrine cells within the pancreas that secret insulin (beta cells) and glucagon (alpha cells).

Ismelin A trade name for guanethidine sulfate, an antihypertensive agent that inhibits norepinephrine release and depletes norepinephrine in the nerve endings.

Ismo A trade name for the antianginal agent isosorbide mononitrate, which reduces cardiac oxygen demand.

Ismotic A trade name for isosorbide, which promotes redistribution of water-producing diuresis.

Iso- A prefix meaning same or like or identical.

Isoaccepting tRNA Different tRNAs that specify the same amino acid.

Isoagglutination Agglutination caused by isoagglutinin, e.g., agglutination of isoagglutinogen of one individual by the antibody from another member of the same species.

Isoagglutinin An agglutinating antibody that agglutinates cells of other individuals of the same species.

Isoagglutinogen The antigen that induces the production of isoagglutinin.

Isoaminile (mol wt 244) An antitussive agent.

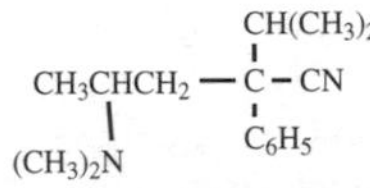

Isoamyl Acetate (mol wt 130) The sting pheromone of the honeybee.

$$CH_3CO_2CH_2CH_2CH(CH_3)_2$$

Isoamylase The enzyme that catalyzes the hydrolysis of 1,6-α-D-glucosidic linkages in glycogen and amylopectin.

Isoantibody Antibody to antigen derived from another member of the same species (e.g., anti-A of the blood group antigen A).

Isoantigen Antigen from one individual that is immunogenic in another individual of the same species.

Isoascorbic Acid (mol wt 176) An antioxidant and antimicrobial agent.

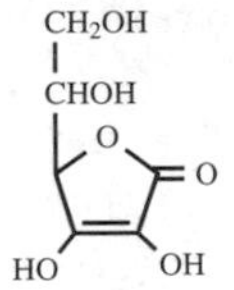

Iso-Bid A trade name for isosorbide dinitrate, an antianginal agent that reduces cardiac oxygen demand.

Isobutal A combination drug containing aspirin, caffeine, and butalbital.

Isobutylamine (mol wt 73) A topical anesthetic agent.

$$(CH_3)_2CH—CH_2NH_2$$

Isobutyl-*p*-aminobenzoate (mol wt 193) A topical anesthetic agent.

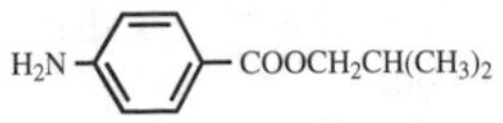

3-Isobutyl-1-Methylxanthine (mol wt 222) A potent inhibitor of cyclic nucleotide phosphodiesterase.

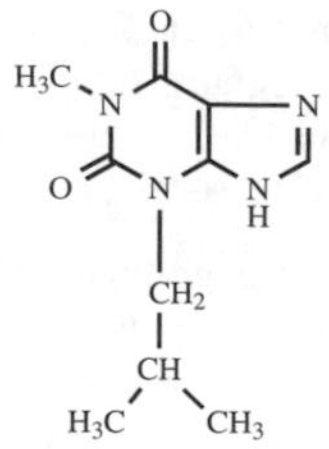

Isobutyryl-CoA Mutase The enzyme that catalyzes the following reaction:

2-Methylpropanoyl-CoA $\rightleftharpoons$ Butanoyl-CoA

***N*²-Isobytyryl-2′-Deoxyguanosine (mol wt 337)** A deoxyguanosine derivative.

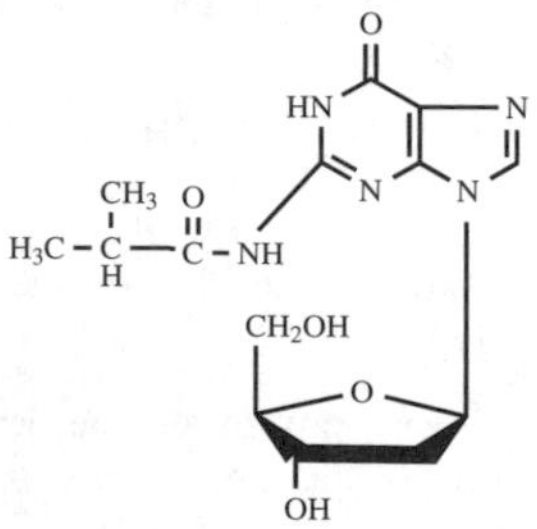

Isocaine A trade name for mepivacaine hydrochloride, a local anesthetic agent that interferes with sodium-potassium exchange across the nerve cell membrane preventing generation and conduction of nerve impulses.

Isocaproic Acid (mol wt 116) An organic acid.

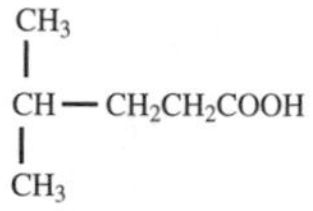

Isocarboxazid (mol wt 231) An antidepressant that promotes accumulation of neurotransmitters by inhibition of monoamine oxidase.

Isocaudamers Different restriction endonucleases that produce identical cohesive ends.

Isochore Referring to the changes in temperature and pressure at constant volume.

Isochorismatase The enzyme that catalyzes the following reaction:

Isochorismate + H_2O $\rightleftharpoons$ Dihydroxy-dihydrobenzoate + pyruvate

Isochorismate Synthetase The enzyme that catalyzes the following reaction:

Chorismate $\rightleftharpoons$ Isochorismate

Isochromosome A metameric chromosome resulting from a transverse split of the centromer instead of a longitude split; the chromosome arms are equal in length.

Isocil (mol wt 247) An herbicide.

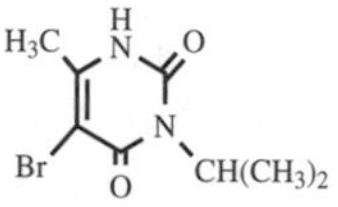

Isocitrase See isocitrate lyase.

Isocitratase See isocitrate lyase.

Isocitrate Dehydrogenase (NAD⁺) The enzyme that catalyzes the following reaction:

Isocitrate + NAD^+ $\rightleftharpoons$ α-Ketoglutarate + NADH + CO_2

Isocitrate Dehydrogenase (NADP⁺) The enzyme that catalyzes the following reaction:

Isocitrate + $NADP^+$ $\rightleftharpoons$ α-Ketoglutarate + NADPH + CO_2

Isocitrate Glyoxylate Lyase See isocitrate lyase.

Isocitrate Lyase The enzyme that catalyses the following reaction:

Isocitrate $\rightleftharpoons$ Succinate + glyoxylate

Isocitric Acid (mol wt 192) An intermediate in the tricarboxylic acid cycle.

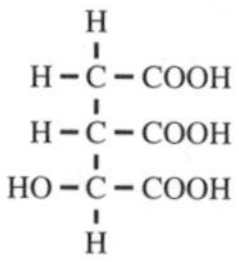

Isocitritase See isocitrate lyase.

Isoconazole (mol wt 416) An antibacterial and antifungal agent.

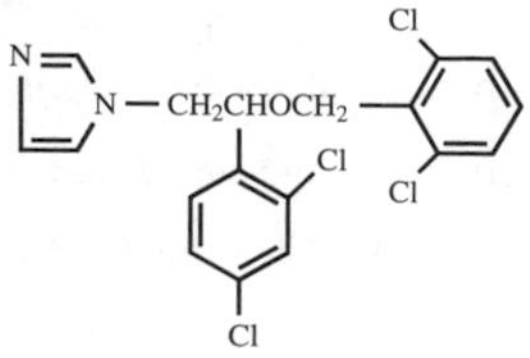

Isocratic Elution A nongradient chromatographic elution.

Isoelectric Focusing Electrophoresis performed in a pH gradient, a high-resolution method for separation of proteins based on the isoelectric points of the proteins.

Isoelectric pH The pH at which a protein possesses no net charge (see also isoelectric point).

Isoelectric Point The pH at which a molecule has no net charge.

Isoelectric Precipitation Precipitation of a protein at its isoelectric pH.

Isoelectronic Molecules or ions that have the same number of electrons.

Isoenzymes Multiple molecular forms of an enzyme that catalyze the same reaction but differ in amino acid sequence, substrate affinity, and electrophoretic mobility.

Isoetharine (mol wt 239) A bronchodilator that relaxes bronchial smooth muscle by acting on B_2-adrenergic receptors.

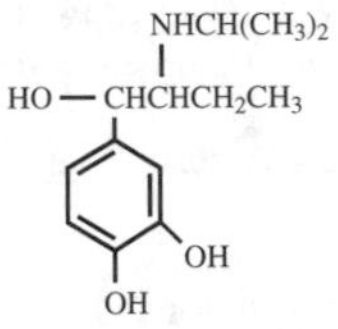

Isofenphos (mol wt 345) An insecticide.

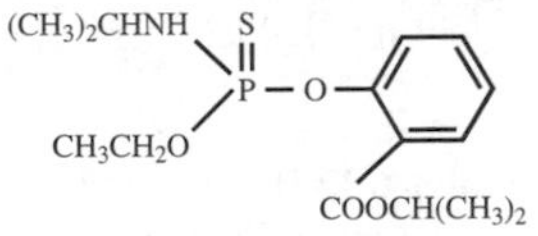

Isofezolac (mol wt 354) An anti-inflammatory, antipyretic and analgesic agent.

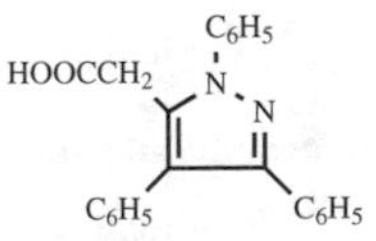

Isoflupredone (mol wt 378) An anti-inflammatory agent.

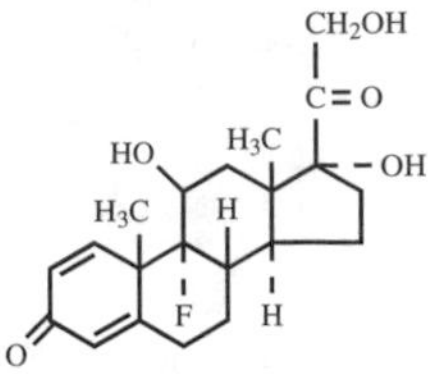

Isoflurane (mol wt 179) A general inhalation anesthetic agent.

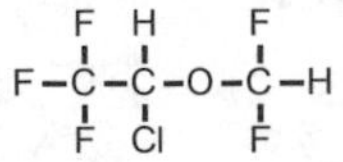

Isoflurophate (mol wt 184) A miotic agent that inhibits the enzymatic destruction of acetylcholine by inactivating cholinesterase.

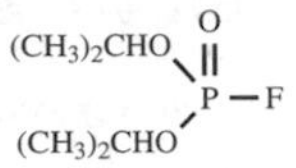

Isoform Multiforms of a functional protein that are different in amino acid sequence and electrophoretic mobilities. They may be produced by different genes or by alternative splicing of RNA transcripts from the same gene.

Isofunctional Enzymes Different enzymes that catalyze the same reaction.

Isogamy Fertilization in which the gametes are similar in size, morphology, and behavior.

Isogeneic Referring to identical genetic background.

Isogenic Organisms or cells that have the same genotypes.

Isoglutamine (mol wt 146) An amino acid with the following structure:

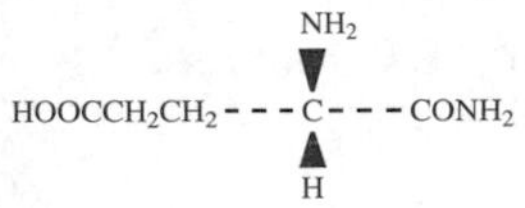

Isograft A graft between two genetically identical individuals.

Isohemagglutinins Antibodies directed against antigenic determinants on erythrocytes from other members of the same species.

Isohydric Shift Reactions in red blood cells in which oxygenation and deoxygenation of hemoglobin is coupled to the reversible ionization of carbonic acid so that no changes of the intracellular pH of the red blood cell will occur.

Isoimmunization Stimulation of an immune response by antigen from another member of the same species. (e.g., development of Rh antibody in Rh negative individuals).

Isoionic Protein Protein that has an equal number of protonated NH_3^+ groups and deprotonated COO^- groups.

Isokont Cell with flagella of the same length.

Isoladol (mol wt 273) An analgesic agent.

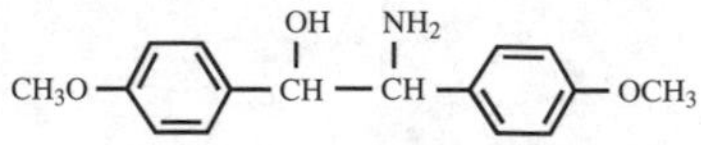

Isolectin Any of the two or more molecular forms of lectins with apparently the same biological properties from the same origin.

Isoleucine (mol wt 131) An essential amino acid.

$CH_3—CH_2—CH(CH_3)CH(NH_2)COOH$

Isoleucine Aminotransferase The enzyme that catalyzes the following reaction:

$$\text{Isoleucine} + \alpha\text{-keto-glutarate} \rightleftharpoons \alpha\text{-Keto-methylvalerate} + \text{glutamate}$$

Isoleucinium Referring to isoleucine cation.

Isoleucine tRNA Ligase See isoleucine-tRNA synthetase.

Isoleucine-tRNA Synthetase The enzyme that catalyzes the following reaction:

$$\text{ATP} + \text{isoleucine} + \text{tRNA}^{\text{ile}} \rightleftharpoons \text{AMP} + \text{PPi} + \text{isoleucyl-tRNA}^{\text{ile}}$$

Isoleucinyl-tRNA Ligase See isoleucine-tRNA synthetase.

Isoleucinyl-tRNA Synthetase See isoleucine-tRNA synthetase.

Isollyl A combination drug containing aspirin, caffeine, and butalbital used as an antipyretic and analgesic agent.

Isologous Cell Line Cell lines derived from identical twins or from highly inbred animals.

Isomaltase The enzyme that catalyzes the hydrolysis of 1,6-α-D-glucosidic linkage in isomaltose and glycogen or dextrin produced from starch.

Isomalto-Dextranase The enzyme that catalyzes the hydrolysis of 1,6-α-D-glucosidic linkages in a polysaccharide to remove successive isomaltose units from the nonreducing end of the chain.

Isomaltose (mol wt 342) A reducing disaccharide of glucose with a 1,6-glucosidic linkage.

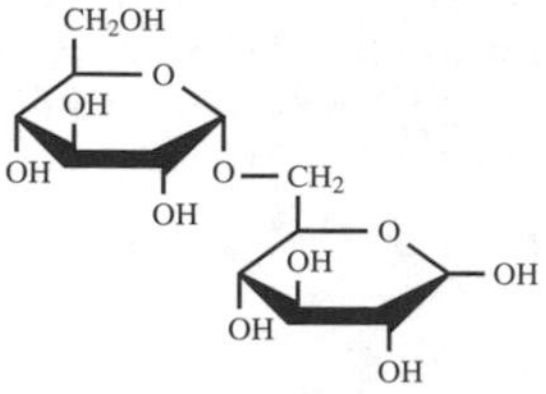

Isomeprobamate Synonym of carisoprodol.

Isomerase Enzyme that catalyzes the rearrangement of atoms within a molecule.

Isomerism The phenomenon of compounds possessing the same molecular weight but differing in chemical structures and properties.

Isomerization The interconversion of different isomers.

Isomers Molecules containing identical numbers and types of atoms but having a different 3-dimensional structure and properties.

Isomethadone (mol wt 309) A narcotic analgesic agent.

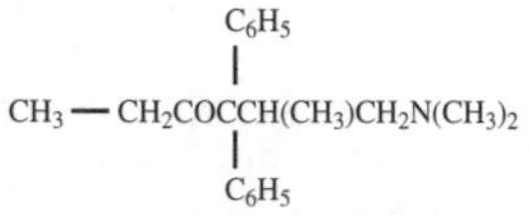

Isometheptene (mol wt 141) A sympathomimetic drug used in the treatment of migraine.

$(CH_3)_2C{=}CH_2CH_2CH(NHCH_3)CH_3$

Isometric Growth The growth of different organs or parts of an organism at the same rate.

Isometric Virus A virus that exhibits icosahedral symmetry.

Isomorphic Morphologically similar or identical.

Isonate A trade name for isosorbide dinitrate, an antianginal agent that reduces cardiac oxygen demand.

Isoniazid (mol wt 137) An antibacterial agent that inhibits bacterial cell wall synthesis.

Isoniazid Methanesulfonate (mol wt 231) An antibacterial agent.

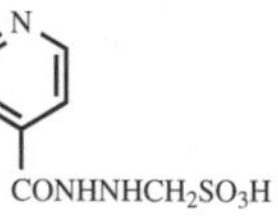

Isonicotinamide (mol wt 122) A pyridine carboxylic acid amide.

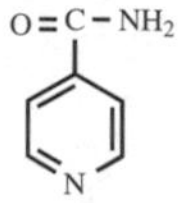

Isonicotinic Acid (mol wt 123) A pyridine carboxylic acid.

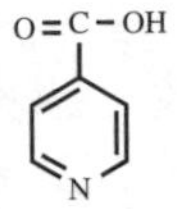

Isonixin (mol wt 274) An analgesic and anti-inflammatory agent.

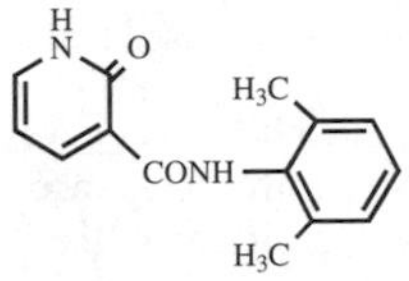

Isopentenyl Adenosine (mol wt 335) An inhibitor of cAMP phosphodiesterase.

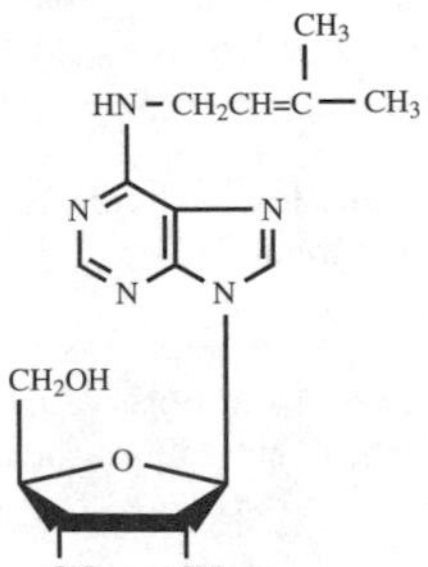

Isopeptide Bond Peptide bond formed from nonalpha amino groups and nonalpha carboxyl groups, e.g., peptide bond formed between the epsilon NH_2 group of lysine and side chain COOH group of aspartic acid.

Isophane Insulin Suspension Sterile suspension in a buffered medium made from zinc-insulin crystals.

Isoprene (mol wt 68) An unsaturated five-carbon unit and the parent compound of isoprenoids.

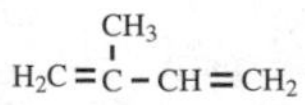

Isoprenoids Lipid molecules with the carbon skeleton based on the multiple five-carbon isoprene units.

Isopromethazine (mol wt 284) An antihistaminic agent.

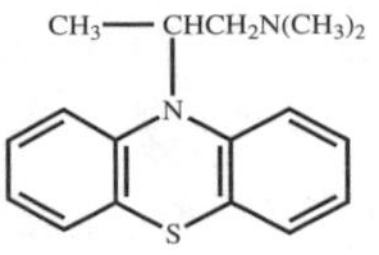

Isopropamide Iodide (mol wt 480) An anticholinergic agent that competitively blocks acetylcholine.

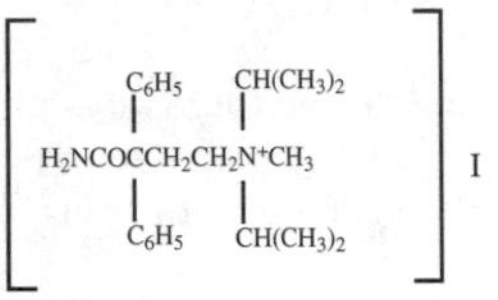

Isopropenyl Chloroformate (mol wt 120) Reagent used in the activation of carboxylic acid for esterification and peptide coupling.

$$Cl\text{-}CO_2C(CH_3){:}CH_2$$

Isopropyl Alcohol (mol wt 60) An antiseptic alcohol.

$$CH_3 - CHOH - CH_3$$

***N*⁵-Isopropylglutamine (mol wt 188)** A naturally occurring amino acid found in *Lunaria annua.*

Isopropylmalate Dehydrase The enzyme that catalyzes the following reaction:

$$\text{Isopropylmalate} \rightleftharpoons \text{Isopropylmaleate} + H_2O$$

Isopropylmalic Acid (mol wt 174) An intermediate in the biosynthesis of the amino acid leucine.

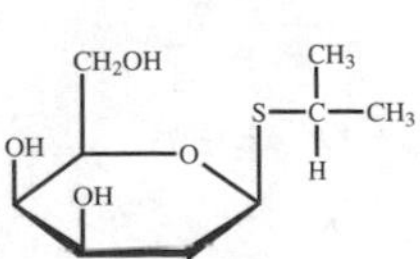

Isopropylthiogalactoside (mol wt 238) A gratuitous inducer of beta-galactosidase in the *lac* operon system.

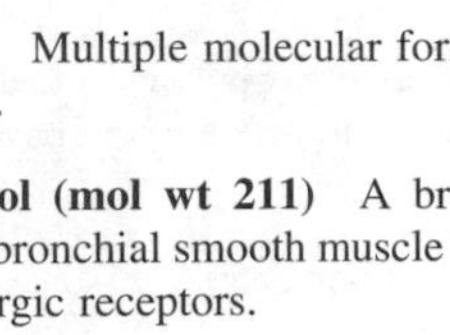

Isoproteins Multiple molecular forms of a specific protein.

Isoproterenol (mol wt 211) A bronchodilator that relaxes bronchial smooth muscle by acting on $beta_2$ adrenergic receptors.

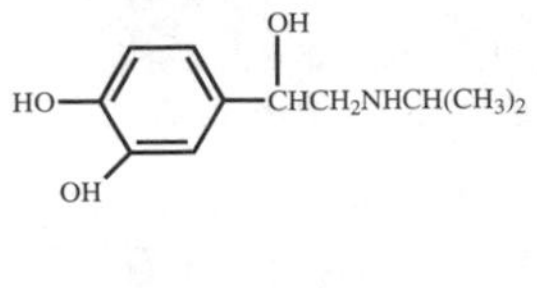

Isoproturon (mol wt 206) An herbicide.

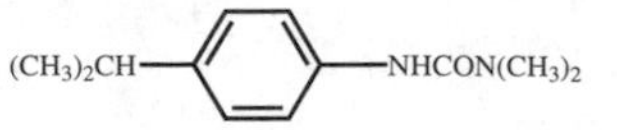

Isoptin A trade name for verapamil hydrochloride, an antianginal agent that inhibits calcium influx across the membrane of cardiac and smooth muscle cells.

Isopycnic Having the same density.

Isopycnic Density Gradient Centrifugation A method used for the separation of macromolecules on the basis of their different densities in a density gradient column that has been centrifuged until equilibrium is reached (e.g., cesium chloride equilibrium density gradient centrifugation).

Isorbid A trade name for isosorbide dinitrate, an antianginal agent that reduces cardiac oxygen demand.

Isordil A trade name for isosorbide dinitrate, an antianginal agent that reduces cardiac oxygen demand.

Isoschizomer Referring to a restriction endonuclease that recognizes the same sequence as another restriction endonuclease.

Isosorbide (mol wt 146) A diuretic drug that promotes redistribution of water and thus produces diuresis.

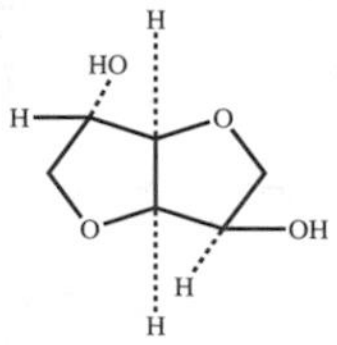

Isosorbide Dinitrate (mol wt 236) An antianginal agent that reduces cardiac oxygen demand.

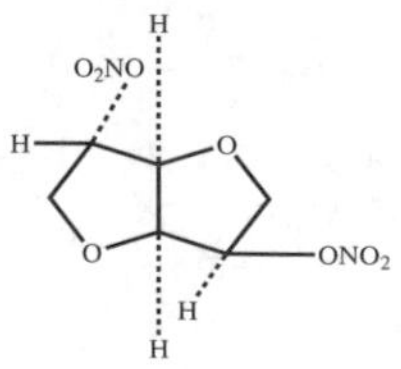

Isotactic Polymer A polymer in which all the R groups of the monomers are arranged on one side of the plane.

Isotamine A trade name for isoniazide, an antitubercular agent.

Isotherm Referring to changes in volume or pressure at a constant temperature.

Isothermal Having a constant temperature.

Isothipendyl (mol wt 285) An antihistaminic agent.

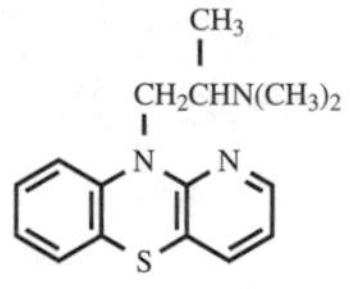

Isotocin (mol wt 966) A peptide hormone secreted by the posterior lobe of the pituitary gland of bony fish that consists of 9 amino acid residues.

Isotones Atomic nuclei that have the same number of neutrons but differ in number of protons and therefore possess a different atomic number.

Isotonic Solution A solution that has the same osmotic pressure to the one under comparison.

Isotope A chemical element that has the same number of protons and electrons but differs in the number of neutrons contained in the nucleus.

Isotope Incorporation The introduction of one isotope into a compound.

Isotope Tracer An isotope used to label a compound and to permit the observation of the compound through the chemical, physical and biological process.

Isotrex A trade name for isotretinoin used for treatment severe cystic acne.

Isotropic Band See I band.

Isotype An antigenic variant that exists in all individuals of the same species.

Isotype Switching The process of a change in synthesis of immunoglobulin heavy chain, e.g., change in synthesis of m heavy chain to g heavy chain in B cells.

Isotopic Variation The antigenic differences of isotypic antibodies.

Isotrate A trade name for isosorbide dinitrate, an antianginal agent that reduces cardiac oxygen demand.

Isovaleric Acid (mol wt 102) An intermediate in amino acid metabolism.

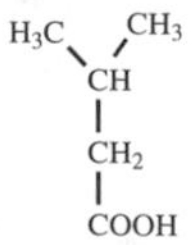

Isovaleric Acidemia A genetic disorder in humans characterized by elevated levels of α-ketoisovaleric acid in the blood and urine due to the deficiency of the enzyme isovaleryl-CoA dehydrogenase.

Isovaleryl-CoA (mol wt 851) An intermediate in amino acid metabolism.

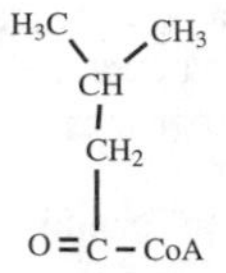

Isovaleryl-CoA Dehydrogenase The enzyme that catalyzes the following reaction:

FAD + isovaleryl-CoA

⇅

β-Methylcrotonyl-CoA + $FADH_2$

Isovaleryl-Diethylamide (mol wt 157) A sedative.

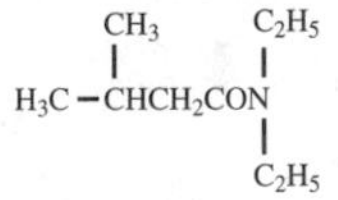

Isovaline (mol wt 117) A nonprotein amino acid.

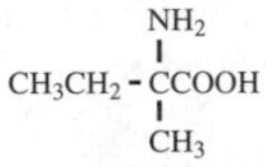

Isovalthine (mol wt 221) A naturally occurring amino acid found in human urine.

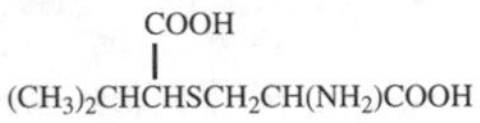

Isovex A trade name for ethaverine hydrochloride, a vasodilator that inhibits phosphodiesterase and increases cAMP concentration.

Isoxepac (mol wt 268) An anti-inflammatory agent.

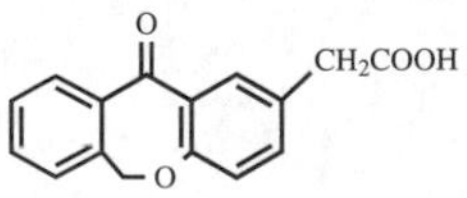

Isoxicam (mol wt 335) An anti-inflammatory agent.

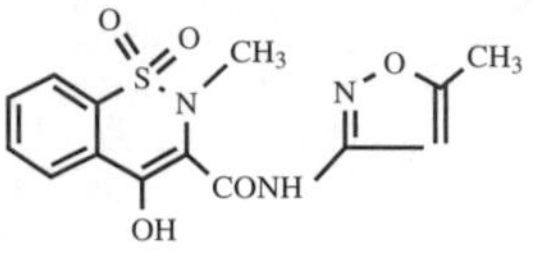

Isoxsuprine A vasodilator that stimulates beta receptors.

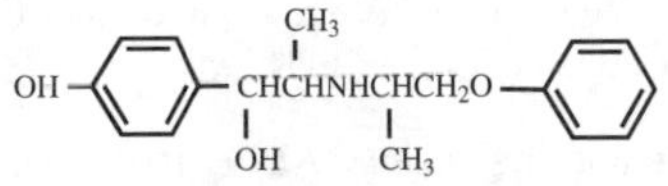

Isozymes See isoenzymes.

Isradipine (mol wt 371) An antihypertensive and antianginal agent that inhibits calcium influx in cardiac and smooth muscle cells.

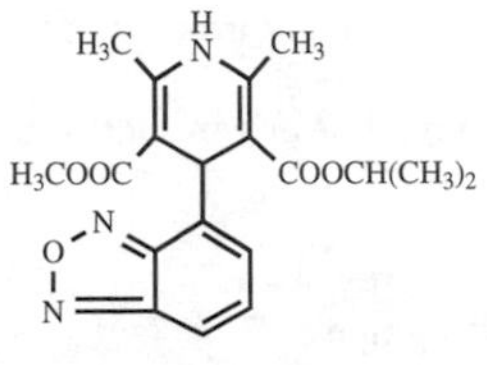

ISRE Abbreviation for INF-stimulated (interferon-stimulated) response element.

IST Abbreviation for insulin sensitivity test.

Isuprel A trade name for isoproterenol, a bronchodilator that acts on alpha- and beta-adrenergic receptors.

ISV Abbreviation for ion-source voltage.

ItaI A restriction endonuclease from *Ilyobacter tartaricus* with the following specificity:

↓
5'........GCNGC........3'
3'........CGNCG........5'
↑

Itaconate CoA-Transferase The enzyme that catalyzes the following reaction:

Succinyl-CoA + citramalate

⇅

Succinate + citramalyl-CoA

Itaconic Acid (mol wt 130) A methylsuccinic acid produced by *Aspergillus terreus* and used as a resin in detergents.

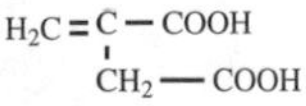

Itaconyl-CoA Hydratase The enzyme that catalyzes the following reaction:

Citramalyl-CoA ⇌ Itaconyl-CoA + H_2O

ITAM Abbreviation for immuno-receptor tyrosine activation motif.

ItdU Abbreviation for 5'-iodo-4'-thio-2'-deoxyuridine.

Iteron The repeated sequence in or near the replication origin in certain plasmids.

Iteroparous Organism An organism that reproduces more than once in its lifetime.

ITF Abbreviation for intestinal trefoil factor.

ITFBP Abbreviation for intestinal trefoil factor binding protein.

-itis A suffix meaning inflammation.

ITLC Abbreviation for instant thin layer chromatography.

ITLC-SG Abbreviation for silica gel instant thin layer chromatography.

ITP Abbreviation for I 1. idiopathic thrombocytopenic purpura; 2. inosine triphosphate.

Itraconazole (mol wt 706) A broad spectrum antifungal agent.

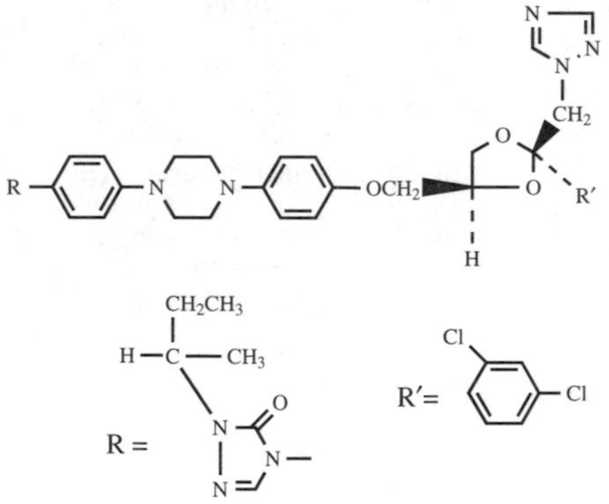

Itramin Tosylate (mol wt 278) A vasodilator.

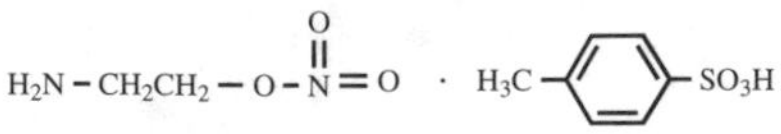

IU International unit.

IUD A birth control device that consists of a strip of plastic (or other material) that is inserted into the uterus to prevent pregnancy.

IUdR Abbreviation for idoxuridine.

IUI Abbreviation for intrauterine insemination.

IV Abbreviation for 1. intravenous and 2. intraventicular.

Iveegam A trade name for immune globulin that provides passive immunity by increasing antibody titer.

IVF Abbreviation for *in vitro* fertilization.

Ivocort A trade name for hydrocortisone.

IVS Abbreviation for intervening sequence (see also intron).

IVTT (IVT/T) Abbreviation for *in vitro* transcription and translation.

Ixodes Parasitic ticks that transmit disease, e.g., rocky mountain spotted fever.

Ixodiasis A disorder characterized by a skin lesion and fever, caused by ticks of the family Ixodidae.

Ixodicide Any substance that kills ticks.

Ixodid Pertaining to ticks of the genus Ixodes.

J

J Abbreviation for joule unit of energy equal to 10^7 ergs.

J Chain A cysteine-rich polypeptide chain involved in the formation of dimeric and pentameric forms of IgA and IgM, respectively.

J Genes 1. A gene of the λ phage genome that encodes J protein for attachment of λ phage to bacterial cells. 2. A short exon that encodes the J segments involved in joining V and C genes in heavy and light chains of the immunoglobulins.

J Protease The protease that catalyzes the hydrolysis of peptide bonds between glutamine-histidine, serine-histidine, and alanine-leucine in the B chain of insulin and the peptide bond between proline-phenylalanine in angiotensin I.

J Protein The protein encoded by J gene in λ phage genome responsible for the attachment of phage to the bacterial cells.

Jack Bean Lectin Referring to the mitogen concanavalin A.

Jacob-Monod Model A model for genetic regulation of protein synthesis in prokaryotes in which the structural genes that determine the primary structures of the proteins are controlled by other regions of DNA upstream from the structural genes, such as promoter and regulator.

Jail Fever Referring to typhus fever.

Jamaican Vomiting Sickness A violent vomiting disorder due to consumption of unripe ackee fruit that causes a deficiency of acyl-CoA dehydrogenase.

Janiemycin A peptide antibiotic that inhibits bacterial cell wall synthesis.

Janimine A trade name for imipramine hydrochloride, an antidepressant that increases levels of norepinephrine and serotonin in the CNS allowing neurotransmitter to accumulate.

Janthinobacterium A genus of Gram-negative, strictly aerobic, catalase-positive, chemoorganotrophic bacteria.

Janus Green B (mol wt 511) A basic dye used as a vital stain.

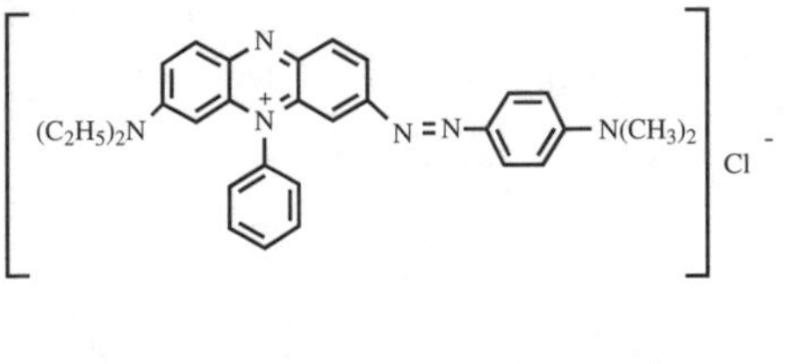

Japanese B Encephalitis A viral encephalitis in humans caused by a flavivirus (Flaviviridae). It is also known as Japanese encephalitis and Russian autumn encephalitis.

Japanese River Fever Referring to scrub typhus.

Jarvik-7 An artificial heart designed by R. K. Jarvik.

Jaundice Yellow discoloration of the skin and eyes due to the presence of bile pigment in the blood.

Javanicin (mol wt 290) An antibiotic produced by *Fusarium javanicum.*

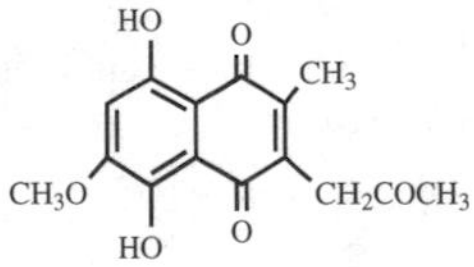

JBAM Abbreviation for jack-bean α-mannosidase.

JBE Abbreviation for Japanese B encephalitis.

JC Virus A human papovavirus (Papovaviridae).

Jejunitis The inflammation of the jejunum.

Jejunum The middle portion of the small intestine between the duodenum and ileum.

Jelly Fungi The fungi that form gelatinous fruiting bodies, e.g., *Exidia glandulosa.*

Jenamicin A trade name for gentamicin sulfate, an antibiotic that binds to 30S ribosomes and inhibiting bacterial protein synthesis.

Jenner's Stain A dye for staining peripheral blood smears that consists of eosin and methylene blue.

Jet Lag A feeling of lassitude or desynchronization of biological rhythms occurring among individuals who cross a number of time zones.

Jet Loop Fermenter A type of fermenter in which the culture medium is continuously withdrawn and recirculated back to the fermenter to promote liquid circulation and gas absorption.

JE-VAX A trade name for Japanese encephalitis vaccine.

Jeyes Fluid A disinfectant obtained by solubilization of coal tar acid with a soap prepared from pine resin and alkali.

JH Abbreviation for juvenile hormone.

J_L Abbreviation for rate of proton leak.

JNK Abbreviation for 1. *c-jun* N-terminal kinase. 2. *c-jun* kinase.

J_o Abbreviation for rate of oxygen consumption.

Job's Syndrome A disorder due to a defect in neutrophil chemotaxis leading to predisposition to infection by *Staphylococci* and an elevated level of plasma IgE.

Johnston-Ogston Effect The changes in values of the sedimentation coefficients resulting from cocentrifugation with other molecules.

Joinase Synonym of DNA ligase.

Joint A contact surface between two individual bones (also called articulation).

Josamycin (mol wt 828) An antibiotic produced by *Streptomyces narbonesis*

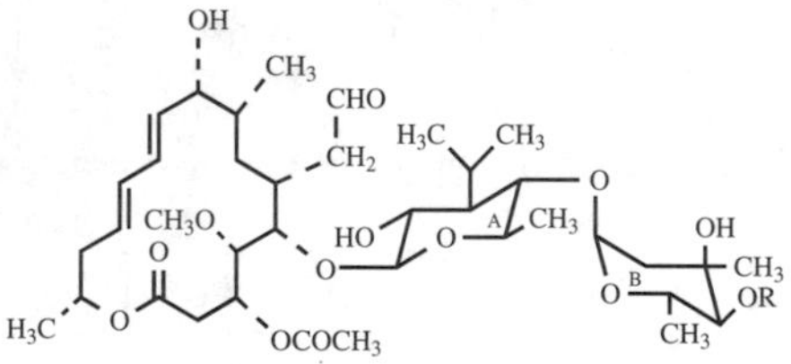

R = $COCH_2CH(CH_3)_2$

Joule A unit of energy equal to 10^7 ergs.

J_p Abbreviation for rate of phosphorylation.

Jugate Structures that are joined.

Juglone (mol wt 174) A naphthoquinone derivative derived from walnuts.

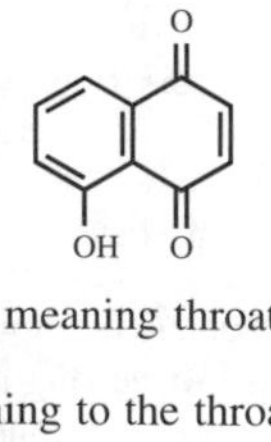

Jugu- A prefix meaning throat or neck.

Jugular Pertaining to the throat or neck.

Jugular Vein The veins in the neck that drain the blood from the head down to the large veins emptying into the heart.

Jumping Gene Referring to transposon or insertion element.

jun A gene family encoding nuclear transcription factors; *v-jun* is an oncogen from avian sarcoma.

Junction The specialized regions of the plasma membranes of adjacent cells responsible for cell-to-cell communication, e.g., desmosome and gap junction.

Junctional Basal Lamina Specialized region of the extracellular matrix surrounding a muscle cell at the neuromuscular junction.

Junctional Complex Referring to a specialized region of intercellular adhesion observed under the electron microscope.

Junctional Receptor Acetylcholine receptors that occur in a cluster at the nerve-muscle junction.

Junctional Sliding The variation in location of the intron-exon junction within members of a gene family.

JUNK Abbreviation for *c-jun* N-terminal kinase.

Junk DNA Referring to 1. the repetitive DNA that serves no useful purpose in the host (also called selfish DNA), and 2. branched DNA.

***jun* Protein** A transcription protein factor encoded by an oncogen.

Justicidins Lignins from various species of *Justicia* (Acanthaceae).

Juvenile Diabetes See insulin-dependent diabetes.

Juvenile Hormones (JH) A group of hormones in the arthropods that promotes retention of larval characteristics.

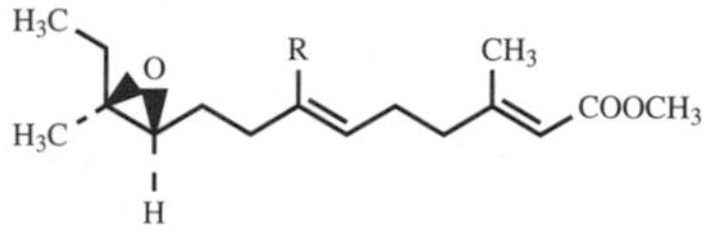

different JHs differ in the R group

Juvenile Hormone C_{16} or C_{16} Juvenile Hormone (mol wt 266) A juvenile hormone that controls larval morphosis in insects.

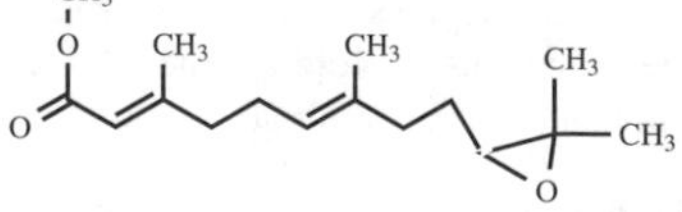

Juxtaglomerular Adjacent to a kidney glomerulus.

Juxtaposition A close anatomical relationship, e.g., two side-by-side structures.

JY A B-lymphoblastoid cell line.

JYF Abbreviation for jungle yellow fever.

K

k Symbol of 1. rate constant; 2. Boltzmann constant.

k A Greek letter used to denote 1. one of the two types of light chains in immunoglobulins; 2. a type of killer particle (kappa particle) present in certain strains of *Paramecium*.

K Abbreviation for 1. amino acid lysine; 2. kilo (10^3).

K' Symbol for apparent (concentration) equilibrium constant.

K° Symbol for standard equilibrium constant.

K_1 Abbreviation for 1. vitamin K_1; 2. symbol for first equilibrium (dissociation or association) constant.

K_2 Abbreviation for second equilibrium (dissociation or association) constant.

K_3 Abbreviation for 1. vitamin K_3; 2. third equilibrium (dissociation or association) constant.

K_4 Abbreviation for 1. vitamin K_4; 2. fourth equilibrium (dissociation or association) constant.

K88 A fimbrial antigen in certain pathogenic strains of *E. coli.*

17K Abbreviation for 17-ketosteroid.

K^+ Symbol for positively charged potassium ion.

K99 A fimbrial antigen in certain pathogenic strains of *E. coli.*

K Antigen Referring to the capsule antigen of bacteria.

K Cell Referring killer cell (non-T, non-B lymphocyte).

K Gene A gene that is carried by certain strains of *Paramecium aurelia* reponsible for the production of kappa particles that kill other paramecia.

K^+ATPase Referring to the K^+-activated ATPase involved in the active transport of potassium across a biological membrane.

K^+ Pump Referring to the mechanism that involves energy-dependent pumping of potassium or the active transport of K^+ ion across a biological membrane by K^+-activated ATPase.

K^+ Transport Referring to K^+-ATPase-mediated active transport of potassium.

KA Abbreviation for keto-acidosis.

KAAD Abbreviation for a mixture of kerosene, alcohol, acetic acid, and dioxane.

Kabikinase A trade name for streptokinase, which catalyzes the conversion of plasminogen to plasmin.

Kabolin A trade name for nandrolone decanoate, an anabolic steroid that promotes tissue-building processes.

Kadian A trade name for morphine sulfate, a narcotic agonist and an analgesic agent.

Kaempferol (mol wt 286) A plant flavonoid that functions as an enzyme cofactor and causes growth inhibition.

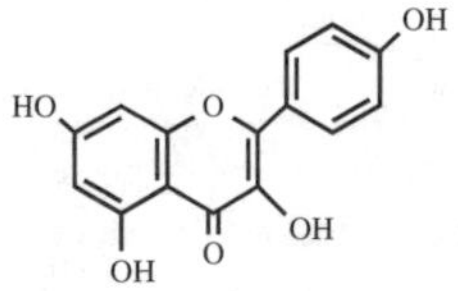

KAF Abbreviation for conglutinogen activating factor, a factor involved in the alternative pathway of complement activation (also called factor I).

Kahn Test A test for syphilis.

Kainate A salt of kainic acid.

Kainic Acid (mol wt 213) A neurotoxin and anthelmintic agent.

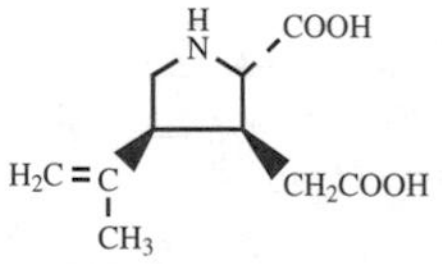

Kalcinate A trade name for calcium gluconate, which replaces and maintains calcium levels in the body.

Kalemia The presence of potassium in the blood.

Kaliopenia Deficiency of potassium in the body.

Kaliuresis The excretion of potassium in the urine.

Kalium Latin for potassium.

Kaliuretic Pertaining to kaliuresis.

Kallidin (mol wt 1188) A decapeptide and a vasodilator with smooth-muscle stimulating activity.

Lys-Arg-Pro-Gly-Phe-Ser-Pro-Phe-Arg

Kallikrein A plasma serine protease that catalyzes the conversion of kininogen to kinin. It consists of kallikrein A and kallikrein B.

Kallistatin An acidic protein in human tissues that inhibits kallikrein.

Kaluril A trade name for amiloride hydrochloride, a diuretic agent that inhibits sodium reabsorption and potassium excretion.

Kanamycin A group of antibiotics produced by *Streptomyces kanamyceticus* that binds to 30S ribosomes inhibiting bacterial protein synthesis.

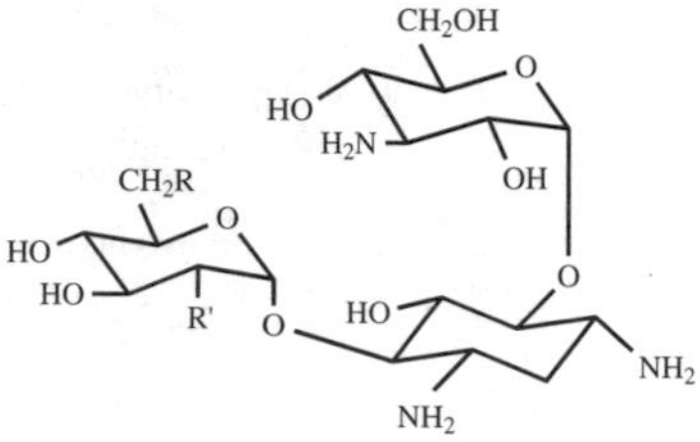

different Kanamycins differ in R and R′ group

Kanamycin 6′-*N*-Acetyltransferase The enzyme that catalyzes the following reaction:

$$\text{Acetyl-CoA} + \text{kanamycin} \rightleftharpoons \text{CoA} + N^{6'}\text{-acetylkanamycin}$$

Kanamycin Kinase The enzyme that catalyzes the following reaction:

$$\text{ATP} + \text{kanamycin} \rightleftharpoons \text{ADP} + \text{kanamycin phosphate}$$

Kanamycin Sulfate A derivative of kanamycin that binds to 30S ribosomes causing inhibition of bacterial protein synthesis.

Kanasig A trade name for kanamycin sulfate, an antimicrobial agent that binds to 30S ribosomes causing inhibition of bacterial protein synthesis.

Kantrex A trade name for kanamycin sulfate, an antimicrobial agent that binds to 30S ribosomes causing inhibition of bacterial protein synthesis.

KAOBP Abbreviation for lysine/arginine/ornithine-binding protein.

Kaochlor-Eff A trade name for a combination of electrolytes containing potassium chloride, potassium citrate, potassium bicarbonate, and betaine hydrochloride.

Kao-Con A trade name for an antidiarrheal agent that consists of kaolin and pectin, and decreases fluid content in the stool.

Kaolin The hydrated form of aluminum silicate used as an absorbent.

Kaolin and Pectin Mixture An antidiarrheal mixture containing koalin and pectin that decreases the fluid content of the stool.

Kaon-Cl A trade name for potassium chloride, used to replace and maintain the level of potassium in the body.

Kaopectate A trade name for the antidiarrheal mixture of kaolin and pectin that decreases the fluid content in the stool.

Kao-Tin A trade name for an antidiarrheal mixture of kaolin and pectin, which decreases the fluid content in the stool.

KAP Abbreviation for kinase anchor protein.

Kapectolin A trade name for the antidiarrheal mixture of kaolin and pectin, which decreases the fluid content in the stool.

Kaposi's Sarcoma A malignant tumor involving the skin, lymph nodes, and gastrointestinal tract. It is a common feature of AIDS.

K_{app} Symbol for the apparent value of the rate constant.

Kappa Light Chain One of the two types of light chains of immunoglobulin (also known as κ light chain).

Kappa Particle A self-duplicating nucleoprotein particle found in various strains of *Paramecium* and capable of killing other sensitive strains of *Paramecium*. It is an obligate endosymbiont.

Karatanase Synonym of karatan sulfate endo-1,4-β-galactosidase.

Karatan Sulfate Endo-1,4-β-Galactanohydrolase The systematic name for karatan sulfate endo-1,4-β-galactosidase.

Karatan Sulfate Endo-1,4-β-Galactosidase The enzyme that catalyzes the endohydrolysis of 1,4-β-D-galactosidic linkages in keratan sulfate.

Karidium A trade name for sodium fluoride, used to catalyze bone remineralization.

Karigel A trade name for sodium fluoride, used to catalyze bone remineralization.

Karsil (mol wt 260) An herbicide.

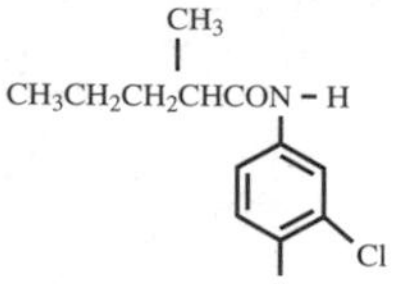

Kary- A prefix denoting a cell nucleus.

Karyogamy Fusion of two gamete nuclei.

Karyokinesis The characteristic nuclear division of mitosis.

Karyolymph Referring to the clear ground substance in the nucleus of a cell (also known as nucleoplasm).

Karyolysis Dissolution of a cell nucleus.

Karyon The nucleus of a cell.

Karyoplasm Referring to nucleoplasm.

Karyoplast A nucleus obtained from a eukaryotic cell that is surrounded by a narrow rim of cytoplasm and a plasma membrane.

Karyorrhexis Degeneration of the nucleus of a cell.

Karyosome Referring to a nucleolus or a nucleolus-like structure or Feulgen-positive body.

Karyostasis The resting stage of the nucleus between cell divisions.

Karyotheca The nuclear membrane.

Karyotic Pertaining to nucleus.

Karyotype The chromosomal profile of a given cell, e.g., the number, size, and morphology of the chromosome set of a cell.

Karyotyping Technique A method for identification and classification of cells, organisms, or tissues on the basis of the karyotype (e.g., the number, size and morphology of the chromosome set of a cell).

Kasof A trade name for docusate potassium, a laxative that promotes incorporation of fluid into the stool.

Kasugamycin (mol wt 378) An aminoglycoside antibiotic produced by *Streptomyces kasugaensis* that inhibits protein synthesis by altering the methylation of 16S RNA leading to the formation of an altered 30S ribosomal subunit.

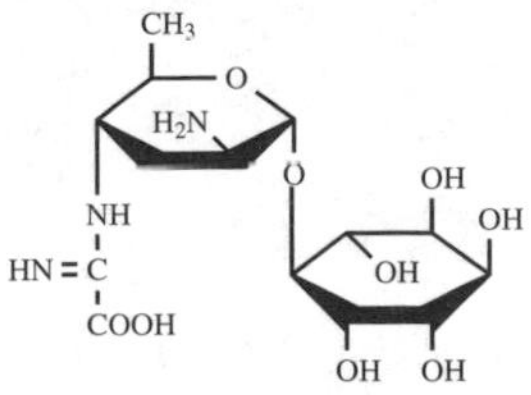

Kat See katal.

Katacalcin A 21-residue peptide derived from the protein precursor of calcitonin.

Katal A unit of enzyme activity, the amount of enzyme that catalyzes the conversion of one mole of substrate to product per second.

Katanin A microtubule-stimulated ATPase that severs and disassembles microtubules to tubulin dimer.

Katharometer A thermal conductivity detector used in gas chromatography for detection of a gas mixture.

Kathode Variant spelling of cathode.

Kation Variant spelling of cation.

Kato powder A trade name for potassium chloride, used to replace and maintain the level of potassium in the body.

K_{ATP} Channel Abbreviation for ATP-sensitive K^+ channel.

K-ATPase K^+-activated ATPase involved in active transport of potassium across a biological membrane.

Kay Ciel A trade name for potassium chloride, used to replace and maintain the level of potassium in the body.

Kaybovite A trade name for vitamin B_{12}.

Kayexalate A trade name for a potassium-removing resin, which exchanges sodium ions for potassium ions in the intestine, e.g., sodium polystyrene sulfonate.

Kaylixer A trade name for potassium gluconate, used to replace and maintain the level of potassium in the body.

KB Abbreviation for 1. ketone bodies; 2. kilo base; 3. kilobyte.

KB Cells An established continuous cell line derived from a human carcinoma.

KBGS Abbreviation for Kell blood group system.

KBP Abbreviation for kilobase pair, a unit of length of double-stranded DNA or double-stranded RNA.

KC Abbreviation for Kupffer cell.

K_c Symbol for equilibrium (concentration equilibrium) constant.

Kcal Abbreviation for kilocalories.

K_{cat} Abbreviation for catalytic constant (used for measuring catalytic efficiency of the enzyme-catalyzed reaction).

KCCT Abbreviation for kaolin-cephalin clotting time.

K-Cell Non-T and non-B lymphoid cell that mediates ADCC (antibody-dependent cell-mediated cytotoxicity).

kCi Abbreviation for kilo Curie.

KD Abbreviation for kilodalton, a molecular weight unit.

K_d Dissociation constant or equilibrium constant for a dissociation reaction.

$$A \rightarrow B + C \quad K_d = \frac{[B][C]}{[A]}$$

KDa Abbreviation for kilodalton, a molecular weight unit.

kDNA Abbreviation for kinetoplast DNA.

KDO Abbreviation for 2-keto-3-deoxyoctanoate.

KDPG Abbreviation For 2-keto-3-deoxy-6-phosphogluconate.

KDPG Aldolase The enzyme that catalyzes the following reaction:

2-Keto-3-deoxy-6-phosphogluconate

⇅

Pyruvate + 3-phosphoglyceraldehyde

K-Dur A trade name for potassium chloride, used to replace and maintain the level of potassium in the body.

Kebuzone (mol wt 322) An antirheumatic agent.

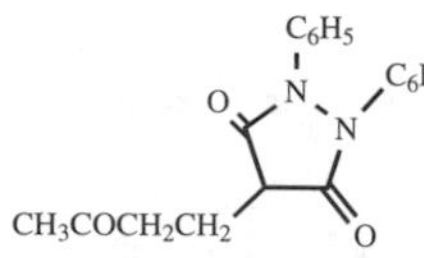

Keflet A trade name for cephalexin monohydrate, an antibacterial agent that inhibits bacterial cell wall synthesis.

Keflex A trade name for cephalexin monohydrate, an antibacterial agent that inhibits bacterial cell wall synthesis.

Keflin A trade name for cephalothin sodium, an antibacterial agent that inhibits bacterial cell wall synthesis.

Keftab A trade name for cephalexin monohydrate, an antibacterial agent that inhibits bacterial cell wall synthesis.

Kefurox A trade name for cefuroxime sodium, an antibacterial agent that inhibits bacterial cell wall synthesis.

Kefzol A trade name for cefazolin sodium, an antibacterial agent that inhibits bacterial cell wall synthesis.

Keilin-Hartree Particle A submitochondrial particle that is capable of electron transport but incapable of conducting oxidative phosphorylation.

Kellogg's Castor Oil A trade name for castor oil, used as a laxative.

Kelvin Temperature Scale Absolute temperature scale in which absolute zero is –273.15°C.

Kemadrin A trade name for procyclidine hydrochloride, which blocks cholinergic receptors.

Kemptide (mol wt 772) A peptide and substrate for assaying cAMP-dependent kinase.

Leu-Arg-Arg-Ala-Ser-Leu-Gly

Kenacort A trade name for triamcinolone, an anti-inflammatory agent.

Kenaject A trade name for triamcinolone acetonide, an anti-inflammatory agent.

Kenalog A trade name for triamcinolone acetonide, an anti-inflammatory agent.

Kenalone A trade name for triamcinolone acetonide, an anti-inflammatory agent.

K-enzyme An allosteric enzyme that changes the apparent K_m in response to binding of either negative or positive modulators without a change in V_{max}.

K_{eq} The symbol for equilibrium constant.

$$A + B \rightarrow C + D \qquad K_{eq} = \frac{[B][C]}{[A][B]}$$

Keratanase An endo-β-galactosidase.

Keratan Sulfate A mucopolysaccharide (glysosaminoglycan) consisting of D-galactose and *N*-acetyl-D-glucosamine 6-sulfate.

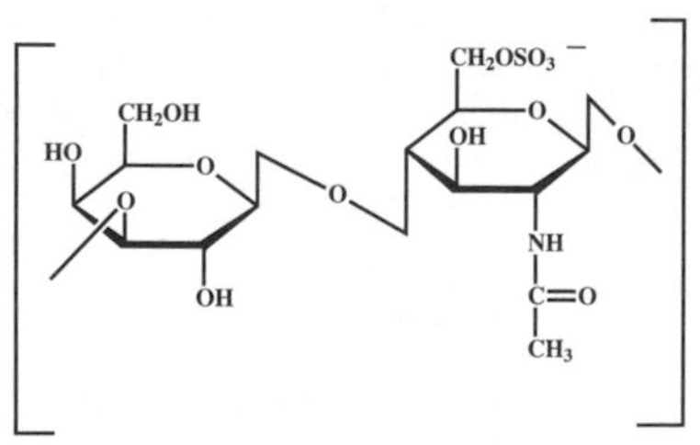

Keratan Sulfate Endo-1,4-β-Galactosidase The enzyme that catalyzes the hydrolysis of 1,4-β-D-galactosidic linkages in keratan sulfate.

Keratan Sulfotransferase The enzyme that catalyzes the following reaction:

3′-Phosphoadenylylsulfate + keratan

⇅

Adenosine 3′5′-bisphosphate + keratan 6′-sulfate

Keratic Pertaining to the cornea.

Keratin Insoluble protective or structural proteins such as finger nails and feathers (also called albuminoids). There are two classes of keratins, namely α and β keratins. The α keratins are rich in cysteine in nails, horns, and wools. The β keratins contain little or no cysteine and are rich in amino acids with small side chains in silk fiber spun by spiders.

Keratinase A protease that catalyzes the hydrolysis of peptide bonds in keratin, e.g., wool.

Keratinization A process by which skin cells become horny tissue.

Keratinocyte A specialized epidermal cell that produces keratin.

Keratinous Pertaining to keratin or containing keratin.

Kerotinophilic With affinity for keratin, e.g., fungus capable of growing on keratin.

Keratitis Inflammation of the cornea.

Keratoacanthoma A tumor of the skin composed largely of a hard keratin substance.

Keratoconjunctivitis Inflammation of the cornea and the conjunctiva (mucous membrane lining the eyelid).

Keratoderma A bony condition of the skin.

Kerotogeneiss Formation of bony skin.

Keratogenic Capable of inducing the formation of bony skin.

Keratohyalin A protein associated with keratin fiber which consists mainly of a protein known as filaggrin.

Keratoiritis Inflammation of the cornea and iris.

Keratoma A tumor of the outer horny skin (callus).

Keratomalacia A disorder of the eye characterized by dryness and ulceration of the cornea due to a severe deficiency of vitamin A.

Keratomycosis Fungus infection of the cornea.

Keratosis Outgrowth and thickening of outer skin layer.

Kerlone A trade name for betaxolol hydrochloride, an antihypertensive agent that decreases blood pressure.

Kernicterus A disorder characterized by the degeneration of brain tissue and elevation of unconjugated bilirubin in the serum.

KERV Abbreviation for Kentucky equine respiratory virus.

Kestrone A trade name for the hormone estrone, which increases the synthesis of DNA, RNA, and protein.

Ketalar A trade name for ketamine hydrochloride, an anesthetic agent.

Ketamine (mol wt 238) An anesthetic agent.

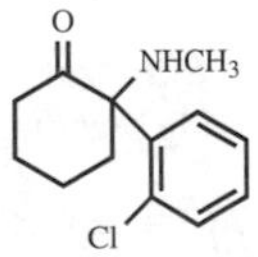

Ketanserin (mol wt 395) An antihypertensive agent and serotonin S_2-receptor antagonist.

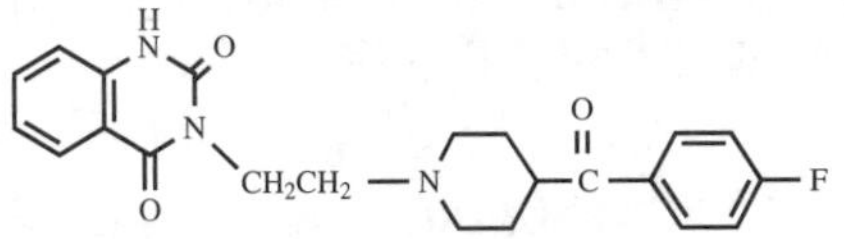

Ketazolam (mol wt 369) An anxiolytic agent.

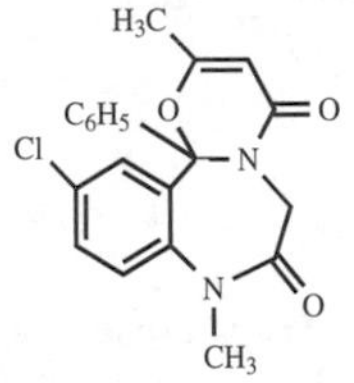

Kethoxal (mol wt 148) An antiviral agent.

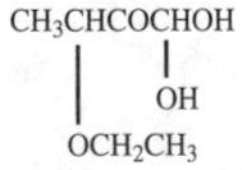

Keto- A prefix meaning ketose or ketone.

Ketoacid An organic acid that has a keto group.

α-Ketoacid Carboxylase The enzyme that catalyzes the following reaction:

An α-ketoacid ⇌ An aldehyde + CO_2

α-Ketoacid Decarboxylase See α-ketoacid carboxylase.

Ketoacidosis The presence of an excessive amount of ketone bodies in the blood and tissue.

Ketoaciduria The presence of an excessive amount of ketone bodies in the urine.

β-Ketoacyl-ACP An intermediate in the biosynthesis of fatty acid.

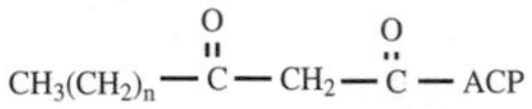

ACP = acyl carrier protein

β-Ketoacyl-ACP Reductase The enzyme that catalyzes the following reaction:

NADPH + acetoacetyl-ACP
⇌
β-Hydroxybutyryl-ACP + $NADP^+$

β-Ketoacyl-ACP Synthetase The enzyme that catalyzes the following reaction:

Acetyl-ACP + malonyl-ACP
⇌
ACP + acetoacetyl-ACP + CO_2

β-Ketoacyl-CoA An intermediate in the metabolism of fatty acid.

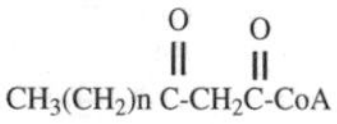

β-Ketoacyl-CoA Thiolase The enzyme that catalyzes the following reaction:

β–Ketoacyl-CoA + CoA
⇌
Fatty acyl-CoA + acetyl-CoA

Ketoacyl-CoA Transferase The enzyme that catalyzes the following reaction:

Acetoacetate + succinyl-CoA
⇌
Acetoacetyl-CoA + succinate

Ketoadipic Acid (mol wt 160) An intermediate in the metabolism of lysine.

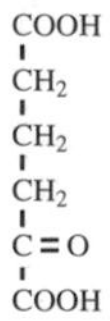

Ketoaldose Any monosaccharide that contains both keto- and aldo-groups.

Ketobemidone (mol wt 247) A narcotic analgesic agent.

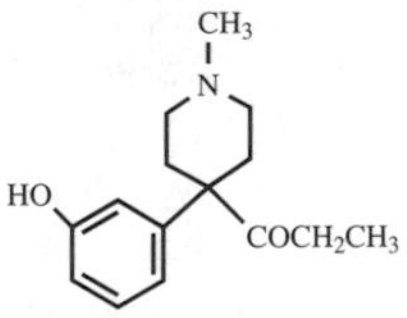

Ketobutyrate Dehydrogenase The enzyme that catalyzes the following reaction:

α–Ketobutryate + CoA + NAD^+
⇌
Propionyl-CoA + NADH + CO_2

α-Ketobutyric Acid (mol wt 102) An intermediate in the metabolism of amino acids.

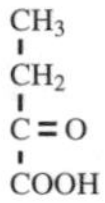

Ketoconazole (mol wt 531) An antifungal agent and an inhibitor for nucleic acid and protein synthesis.

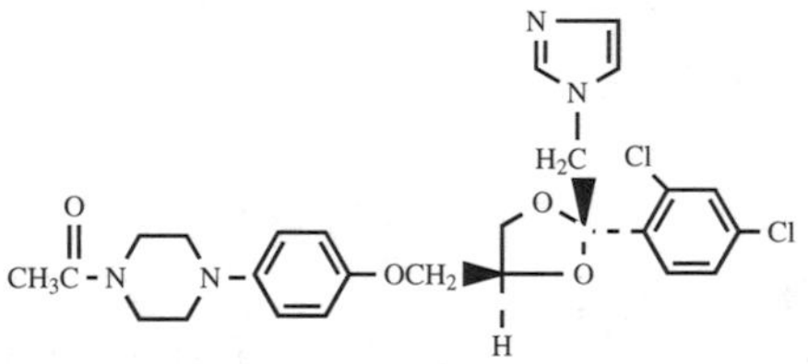

2-Keto-3-Deoxyarabinoheptulosonic Acid 7-Phosphate (mol wt 289) An intermediate in the biosynthesis of aromatic amino acids.

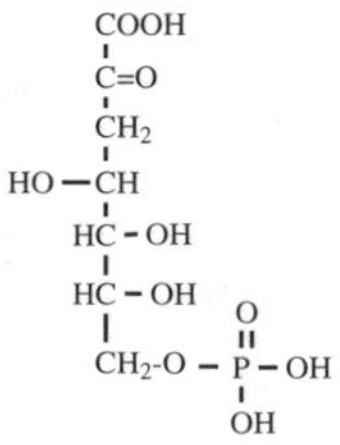

2-Keto-3-deoxyarabinoheptulosonate 7-phosphate Synthetase The enzyme that catalyzes the following reaction:

Phosphoenol pyruvate + erythrose 4-phosphate
⇌
2-Keto-3-deoxyarabinoheptulosonate 7-phosphate

2-Keto-3-Deoxy 6-Phosphogluconic Acid (mol wt 258) An intermediate in the metabolism of glucose via the Entner-Doudoroff pathway.

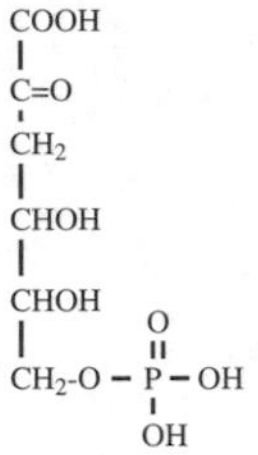

Keto-Enol Tautomerism The shift of a hydrogen atom in a molecule resulting in the formation of either keto- or enol-type isomer.

Ketofuranose A ketomonosaccharide that is in the furanose form.

Ketogenesis A process of formation of ketone bodies by the conversion of acetyl-CoA to acetoacetate, acetone, and beta-hydroxybutyrate.

Ketogenic Amino Acid Amino acids whose carbon skeletons serve as precursors for ketone bodies.

Ketogenic Fermentation A fermentation process in which polyhydric alcohols are converted to ketones.

Ketogenic Hormone Hormone that stimulates fatty acid metabolism.

Ketogenic Substance Compound that provides a source of ketone bodies.

Ketogluconate 2-Dehydrogenase The enzyme that catalyzes the following reaction:

D-Gluconate + $NADP^+$

⇅

Keto-D-gluconate + NADPH

α-Ketogluconate Kinase The enzyme that catalyzes the following reaction:

α–Ketogluconate + ATP

⇅

α-Ketogluconate 6-phosphate + ADP

α-Ketogluconic Acid (mol wt 193) An intermediate in the metabolism of glucose.

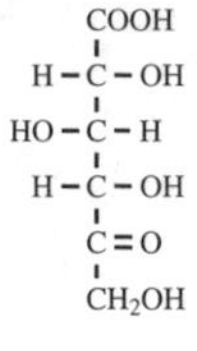

α-Ketoglutarate Dehydrogenase The enzyme that catalyzes the following reaction:

NADH + succinyl-CoA + CO_2

⇅

α–Ketoglutarate + CoA + NAD^+

α-Ketoglutarate Pathway A pathway by which products derived from amino acids (e.g., lysine, glutamate, glutamine, arginine, proline and histidine) enter the Krebs cycle by way of α-ketoglutaric acid.

α-Ketoglutaric Acid (mol wt 146) An intermediate in the Krebs cycle.

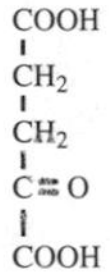

α-Ketoglutaric dehydrogenase Synonym α-ketoglutarate dehydrogenase.

β-Ketoglutaric Isocitric Carboxylase Synonym of isocitrate dehydrogenase (NAD^+).

2-Keto-L-Gulonic Acid (mol wt 194) An intermediate in the biosynthesis of vitamin C.

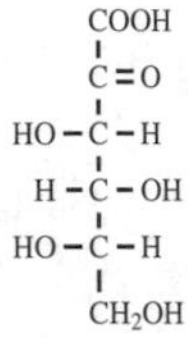

Ketohexokinase The enzyme that catalyzed the following reaction:

ATP + D-fructose

⇅

ADP + D-fructose 6-phosphate

Ketohexose A six-carbon monosaccharide containing one ketone group, e.g., fructose.

α-Ketoisocaproic Acid (mol wt 130) An intermediate in the biosynthesis of leucine.

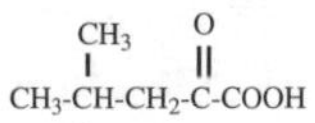

α-Ketoisovalerate Dehydrogenase The enzyme that catalyzes the following reaction:

α-Ketoisovalerate + CoA + NAD^+

⇅

Isovaleryl-CoA + NADH

α-Ketoisovaleric Acid (mol wt 116) An intermediate in the biosynthesis of amino acid valine.

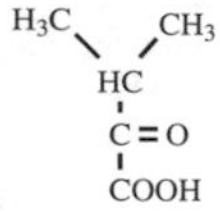

Ketolase The enzyme that catalyzes the cleavage of sugars at the carbonyl carbon position.

Ketolysis Dissolution of ketone bodies.

α-Keto-β-Methylvaleric Acid (mol wt 130) An intermediate in isoleucine biosynthesis.

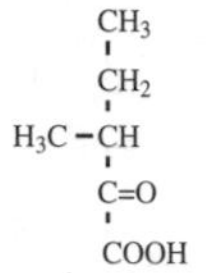

Ketone Organic compound that contains a ketone group.

Ketone Aldehyde Mutase Synonym of lactoylglutathione lyase.

Ketone Bodies Referring to acetoacetic acid, acetone, and beta-hydroxybutyric acid in the blood and urine.

Ketone Group The -CO- group.

Ketonemia The presence of an excessive amount of ketone bodies in the blood that is also detectable in the individual's breath as a distinctive odor.

Ketonuria The presence of an excessive amount of ketone bodies in the urine.

Ketoprofen (mol wt 254) An anti-inflammatory and analgesic agent that inhibits prostaglandin biosynthesis.

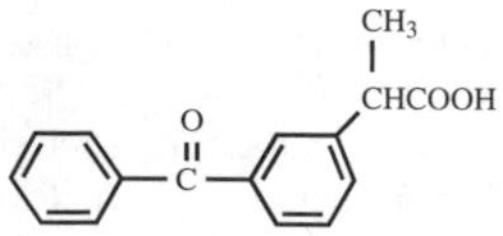

Ketopyranose A ketose sugar that is in the pyranose form.

Ketorolac (mol wt 255) An anti-inflammatory and analgesic agent that blocks prostaglandin biosynthesis.

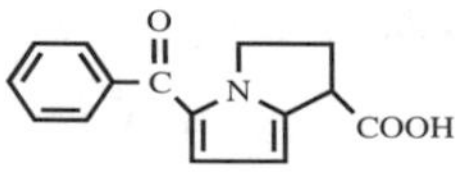

Ketose A monosaccharide with a ketone group.

Ketose Sugar A monosaccharide with a ketone group.

Ketosis A condition in which the level of ketone bodies in the blood, tissue, and urine is abnormally high.

3-Ketosphingamine (mol wt 299) An intermediate in the biosynthesis of ceramide.

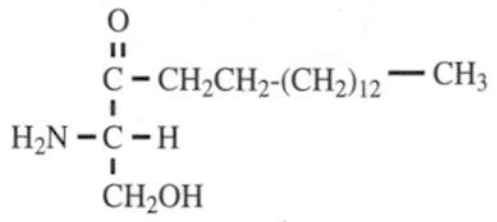

3-Ketosphingamine Reductase The enzyme that catalyzes the following reaction:

3-Ketosphingamine + NADPH

⇅

Sphingamine + $NADP^+$

3-Ketosphingamine Synthetase The enzyme that catalyzes the following reaction:

Palmitoyl-CoA + serine

⇅

3-Ketosphingamine + CO_2 + CoA

Ketosteroid A group of neutral steroids that has keto groups attached to carbon 17 of the steroid ring (also known as 17-ketosteroid). They represent the degradation products of steroids.

Ketostix Test A test of the presence of ketone bodies in urine or serum.

Ketosuccinic acid See oxaloacetic acid.

Ketosugar See ketose sugar.

Ketothiolase The enzyme that catalyzes the following reaction:

3-Ketoacyl-CoA + CoA

⇅

Acetyl-CoA + acyl-CoA $_{(n-2)}$

Ketotic Pertaining to ketosis.

Ketotic Hyperglycinemia A disorder characterized by the combination of hyperglycinemia, hyperglycinuria, ketoacidosis, and hyperammonemia.

Ketotifen (mol wt 309) An antiasthmatic agent with antihistaminic activity.

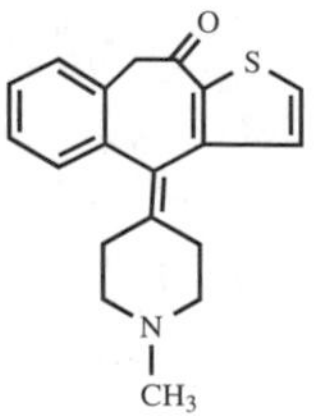

keV (KEV) Abbreviation for kilo electron volt.

Kexin A protease that catalyzes the cleavage of peptide bonds involving the COOH group in the lys-arg-X and arg-arg-X sequence.

K-Exit A trade name for sodium polystyrene sulfonate, a potassium-removing resin.

Key-Pred A trade name for prednisolone acetate, an anti-inflammatory agent.

Key-Pred-SP A trade name for prednisolone sodium phosphate, an anti-inflammatory agent.

K-FeRON A trade name for iron dextran, which provides essential iron for the synthesis of hemoglobin.

K-Flex A trade name for orphenadrine citrate, a skeletal muscle relaxant that reduces transmission of the nerve impulses.

kg Abbreviation for kilogram.

K-G Elixir A trade name for potassium gluconate, used to replace and maintain the potassium level in the body.

K-Gen ET A trade name for potassium bicarbonate used to replace and maintain the potassium level in the body.

KGF Abbreviation for keratinocyte growth factor.

KGS Abbreviation for ketogenic steroid.

KHB Abbreviation for Krebs-Henseleit bicarbonate buffer.

Khellin (mol wt 260) A vasodilator.

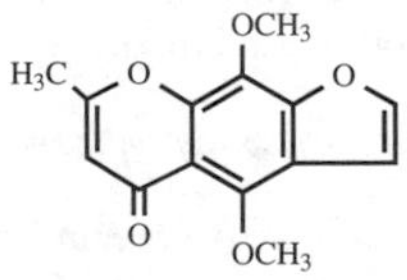

KHF Abbreviation for Korea hemorrhagic fever.

KH/hst A human oncogene with no known viral counterpart, its product is related to fibroblast growth factor and is expressed in Kaposi's sarcoma.

K_i Abbreviation for the dissociation constant of an enzyme-inhibitor complex.

$[K^+]_i$ Abbreviation for intracellular K^+ concentration.

KIC Abbreviation for ketoisocaproate.

K-Ide A trade name for potassium bicarbonate, used to replace and maintain the potassium level in the body.

Kidney Stone The hard aggregate present in the kidney due to abnormal calcification or aggregation of organic or inorganic materials.

Kidrolase A trade name for L-asparaginase used for the treatment of leukemia.

Killed Vaccine Vaccines that are produced by chemical or physical inactivation of a virulent pathogenic microorganism, e.g., killed bacterial vaccine or inactivated viral vaccine.

Killer Cells Referring to non-T, non-B lymphocytes that lyse antibody-coated target cells and mediate antibody-dependent cellular cytotoxicty (ADCC).

Killer Factor Any of the toxic proteins produced by the killer strains of *Saccharomyces cerevisiae* due to the presence of killer plasmid.

Killer Paramecia Strain of *Paramecium* capable of killing other sensitive strains of Paramecia.

Killer Plasmid A double-stranded RNA plasmid in yeast, e.g., *Saccharomyces cerevisiae* that secretes toxic protein capable of killing other sensitive yeast strains.

Killer T Cell Referring to antigen-stimulated T lymphocytes or cytotoxic T cells that are capable of destroying incompatible or foreign target cells.

Killer Toxin The toxic protein encoded by a killer plasmid in yeast.

Killer Yeast Yeast strains that produce toxic protein that kill other yeast strains due to the presence of killer plasmid in the yeast cells.

Kilo- A prefix meaning 1000.

Kilobase A unit of length for nucleic acids equal to 1000 bases or nucleotides.

Kilobase Pairs A unit of length of double-stranded DNA or double-stranded RNA equal to 1000 basepairs.

Kilocalories A unit of energy equal to 1000 calories.

Kilogram A unit of weight equal to 1000 grams.

Kilojoule A unit of energy equal to 1000 joules.

Kilometer A unit of length equal to 1000 meters.

Kinase The enzyme that catalyzes the transfer of a phosphate group from one compound to another.

Kineosporia A genus of bacteria (Actinomycetales) that form a substrate mycelium but not aerial hyphae.

Kinesed Tablets A trade name for a combination drug containing atropine sulfate, scopolamine hydrochloride, hyoscyamine hydrochloride, and phenobarbital, used as an anticholinergic agent.

Kinesin A type of motor protein that uses energy from ATP hydrolysis to move along a microtuble.

Kinesis The change of activity rate in response to a stimulus.

-kinesis a suffix denoting movement.

Kinete A motile zygote or a motile form derived from a zygote.

Kinetensin A bioactive peptide containing nine amino acid residues.

Ile-Ala-Arg-Arg-His-Pro-Try-Phe-Leu

Kinetic Analysis An analysis that employs the kinetic data of reactions measured under specified conditions.

Kinetic Constant See rate constant.

Kinetic Energy The energy of motion.

Kinetic Proofreading 1. A mechanism by which the ribosome selects correct codon-anticodon interactions on the basis of kinetic considerations. 2. A mechanism that permits an enzyme to discriminate between correct and incorrect substrate on the basis of kinetic considerations.

Kinetics Science that deals with the rate of reaction.

Kinetin (mol wt 215) A plant cytokinin that stimulates cell division.

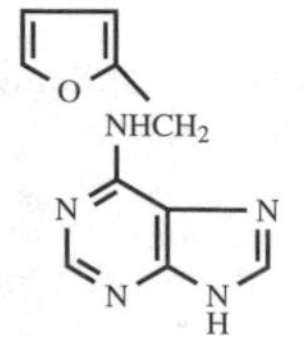

Kinetin Riboside (mol wt 347) An anticancer and antiviral agent.

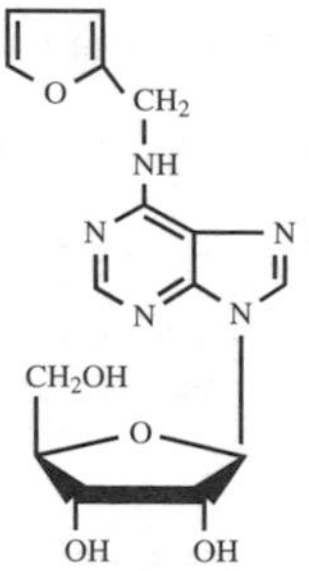

Kineto- A prefix pertaining to the relationship to motion.

Kinetochore A complex, electron-dense protein structure on the centromere to which microtubules attach.

Kinetodesma The cytoplasmic fibrils associated with kinetosomes of the ciliate protozoa. Each kinetodesma arises from a kinetosome.

Kinetography The technique used for recording muscular or organ movement.

Kinetoplasm 1. The chromophil substance of the nerve cell. 2. The most contractile portion of a cell.

Kinetoplast Autonomous, membrane-bound organelle associated with the basal body at the base of the flagella in certain flagellates such as trypanosomes.

Kinetosome The cytoplasmic structure or basal body at the base of the flagellum or cilium.

Kinety A row of kinetosomes with the associated kinetodesmata in the ciliate protozoa.

Kingella A genus of catalase-negative bacteria (Neisseriaceae).

Kinidin Durules A trade name for quinidine bisulfate, an antiarrhythmic agent that prolongs the action potential.

Kinin A group of vasoactive peptides produced from kininogen by the action of kallikrein. These peptides cause dilation of blood vessels and alteration of vascular permeability.

Kinin System An amplification system initiated by the activation of coagulation factor XII leading to the formation of kallikrein.

Kininase A protease that acts on kinin.

Kininogen The inactive precursor of kinin.

Kininogenase See kallikrein.

Kininogenin Synonym of plasma kallikrein.

Kinollium A type of cilium that contains one central pair of microfibrils and nine peripheral pairs.

Kinomer Synonym of centromer.

Kinosinogenin See kallikrein.

Kinyoun Stain An acid-fast stain.

Kirby-Bauer Method A disc diffusion method for the determination of the suceptibility of microorganisms to a chemotherapeutic agent.

Kirsten Murine Sarcoma Virus A replication-defective murine sarcoma virus carrying oncogene *v-ras*, causing erythro-leukemia and sarcoma in newborn mice.

Kissing Disease Referring to infectious mononucleosis.

Kistrin A protein and potent platelet aggregation inhibitor isolated from the venom of the Malayan pit viper *Agkistrodon rhodostoma.*

kit An oncogen that encodes a protein kinase.

Kitasatoa A genus of bacteria that resembles *Streptomyces*.

Kitol (mol wt 573) A provitamin A obtained from mammalian liver oil.

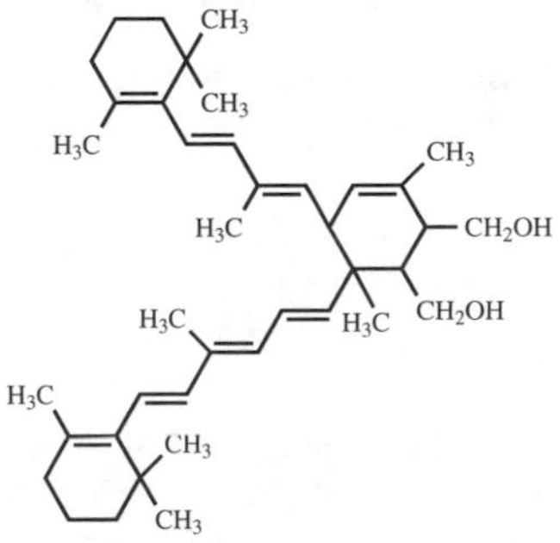

Kjeldahl Method A method for the determination of total nitrogen (organic or inorganic) by digesting the sample with concentrated sulfuric acid in the presence of sodium selenate, potassium sulfate, and copper sulfate.

Klavikordal A trade name for nitroglycerin, an antianginal agent that reduces cardiac oxygen demand.

Klebcil A trade name for kanamycin sulfate, an antibacterial agent that binds to 30S ribosomes inhibiting bacterial protein synthesis.

Klebsiella A genus of Gram-negative, nonmotile, rodlike bacteria.

Kleinschmidt Monolayer Technique A method for electron microscopic examination of nucleic acid in which the DNA sample is mixed with globular protein (e.g., cytochrome c) to form a monomolecular film.

Klenow Fragment A polypeptide fragment obtained by partial proteolytic digestion of DNA polymerase I. It possesses DNA polymerase and 3′ to 5′ exonuclease activity.

KLH Abbreviation for keyhole limpet hemocyanin.

Klinefelter's Syndrome A human genetic disorder in which the male has an extra X chromosome (XXY).

Kline Test A microscopic flocculation test for the diagnosis of syphilis.

Klonopin A trade name for clonazepam, an anticonvulsant.

K-Lor A trade name for potassium chloride, used to replace and maintain potassium levels in the body.

Klor-Con A trade name for potassium chloride, used to replace and maintain the level of potassium in the body.

Klorvess A trade name for potassium chloride, used to replace and maintain the level of potassium in the body.

Klotrix A trade name for the potassium chloride, used to replace and maintain the level of potassium in the body.

Kluyvera A genus of motile bacteria of the family Enterobactericeae.

K-Lyte/Cl A trade name for potassium chloride, used to replace and maintain the level of potassium in the body.

K_m Symbol for the Michaelis constant, the substrate concentration at the half-maximal velocity of an enzyme and used to describe the affinity of an enzyme for a substrate.

Km Marker An allotypic marker on the kappa light chain of human immunoglobulin.

KMV Abbreviation for killed measles virus vaccine.

Knallgas Bacteria Referring to the genus of bacteria *Hydrogemonas* that obtains energy from the reaction of hydrogen and oxygen to form water.

Knallgas Reaction An energy yielding reaction.

$$2H_2 + O_2 \rightleftharpoons 2H_2O + \text{energy}$$

Knop's Solution A solution used for growing plants that contains calcium nitrate, KCl, magnesium sulfate, and potassium dihydrogen phosphate.

Koate-KS A trade name for an antihemophilic factor.

Koate-HT A trade name for an antihemophilic factor.

Koettstorfer Number Referring to saponification number.

Koffex A trade name for the antitussive agent dextromethorphan hydrobromide.

Kogenate A trade name for antihemophilic factor VIII.

Kojibiose A reducing disaccharide consisting of glucose.

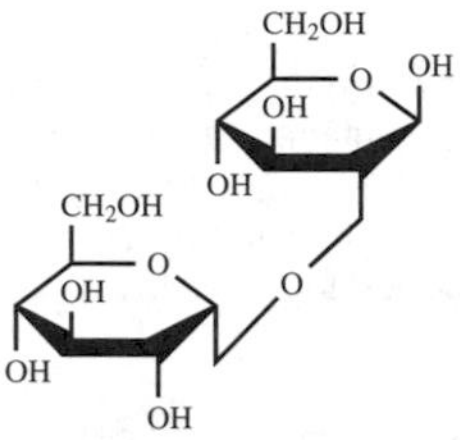

Kojic Acid (mol wt 142) An antibiotic produced by a variety of aerobic microorganisms.

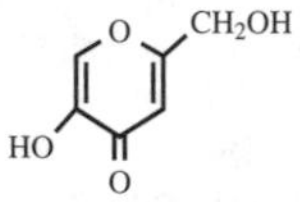

Koji Fermentation A type of fermentation for the production of soy sauce in which a mixture of moistened soybeans and wheat is inoculated with spores of *Aspergillus oryzae* so that fungi can grow on the surface.

Kolmer CFT A standard complement fixation test for syphilis.

Kolyum A trade name for a combination drug containing potassium chloride and potassium gluconate, used to replace and maintain the level of potassium in the body.

Kondremul A trade name for mineral oil, used as a laxative.

Konsyl A trade name for a laxative from psyllium that increases bulk and moisture content of the stool.

Konyne-HT A trade name for blood clotting factor IX.

Koplil's Spots Lesions or spots occurring on the mucous membranes of the mouth during early stages of measles. The lesions are characterized by a blue whitish center in each spot.

Korean Haemorrhagic Fever A viral haemorrhagic fever characterized by renal dysfunction, proteinuria, and oliguria caused by a virus of the family Bunyaviridae.

Kornberg Enzyme Referring to DNA polymerase I.

Koserella A genus of Gram-negative bacteria of Enterobacteriaceae.

Kosin An antihelmintic agent isolated from flowers of *Hagenia abyssinica*.

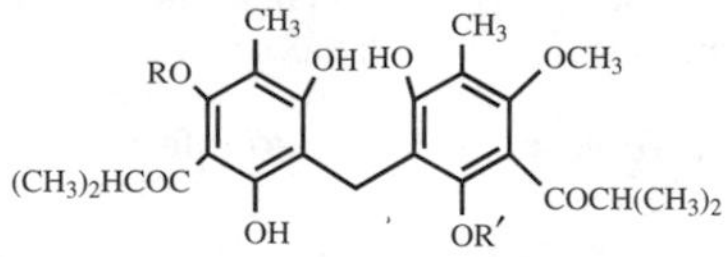

α-kosin : R = CH_3 ; R′ = H
β-kosin : R = H ; R′ = CH_3

Koster's Stain A stain containing safranin and KOH for detecting *Brucella* species in mammalian tissue.

Kotonkan Virus A virus in the family of Rhabdoviridae.

Kovacs' Indole Reagent A reagent used for testing for indole that contains dimethylaminobenzaldehyde in amyl alcohol and HCl.

KoxI (BstEII) A restriction endonuclease from *Klebsiella oxytoca* with the following specificity:

```
           ↓
5′ .......... GGTNACC .......... 3′
3′ .......... CCANTGG .......... 5′
                    ↑
```

KoxII (HgiJ III) A restriction endonuclease from *Klebsiella oxytoca* with the following specificity:

```
                    ↓
5′ .......... GPuGCPyC .......... 3′
3′ .......... CPyCGPuG .......... 5′
              ↑
```

Kox165I (EcoRII) A restriction endonuclease from *Klebsiella oxytoca* RFL 165 with the same specificity as EcoRII.

K-P A trade name for an antidiarrheal mixture containing koalin and pectin, which decreases fluid content in the stool.

K-Pek A trade name for an antidiarrheal mixture containing koalin and pectin, which decreases fluid content in the stool.

K-Phen A trade name for promethazine hydrochloride, an antihistaminic agent that competes with histamine for H_1 receptors on the target cells.

KpnI A restriction endonuclease from *Klebsiella pneumoniae* OK8 with the following specificity:

```
                   ↓
5′ .......... GGTACC .......... 3′
3′ .......... CCATGG .......... 5′
               ↑
```

Kpn2I (BspMII) A restriction endonuclease from *Klebsiella pneumoniae* with the following specificity:

5′ .... TCCGGA .... 3′
3′ .... AGGCCT .... 5′

Kpn10I (EcoRII) A restriction endonuclease from *Klebsiella pneumoniae* RFL10 with the same specificity as EcoRII.

Kpn12I (Pst I) A restriction endonuclease from *Klebseilla pnemoniae* RFL 12 with the same specificity as Pst I.

Kpn13 (EcoRII) A restriction endonuclease from *Klebsiella pneumoniae* RFL 13 with the same specificity as EcoRII.

Kpn14I (EcoRII) A restriction endonuclease from *Klebsiella pneumoniae* RFL 14 with the same specificity as EcoRII.

Kpn16I (EcoRII) A restriction endonuclease from *Klebsiella pneumoniae* RFL 16 with the same specificity as EcoR II.

Kpn30I (BssHII) A restriction endonuclease from *Klebsiella pneumoniae* RFL 30 with the same specificity as BssH II.

KpnK14I (KpnI) A restriction endonuclease from *Klebsiella pneumonia* K14 with the same specificity as KpnI.

K*p*NPPase Abbreviation for K^+-stimulated *p*-nitrophenol phosphatase.

Kr Abbreviation for the chemical element krypton with atomic weight 83.

Krabbe's Disease A genetic disorder characterized by mental retardation and accumulation of galactocerebrosides due to a deficiency of the enzyme galactosyl ceramide β-galactosidae (galactocerobrosidase).

KRB Abbreviation for Krebs/Ringer buffer.

Krebs Cycle A cycle of reactions that oxidize pyruvic acid to hydrogen ions, electrons, and carbon dioxide. The electrons are passed along the electron-carriers in the electron transport chain for the production of ATP (oxidative phosphorylation).

Krebs Henseleit Cycle Referring to urea cycle.

Krebs-Kornberg Cycle Referring to glyoxylate cycle.

KRH Abbreviation for Krebs-Ringer-Hepes buffer.

Kristalose A trade name for lactulose, a synthetic disaccharide used for the treatment of constipation.

Kronofed-A A trade name for a combination drug containing pseudoephedrine hydrochloride and chlorpheniramine maleate used for the relief of upper respiratory and nasal congestion associated with the common cold.

KRP Abbreviation for Krebs-Ringer phosphate.

KRRS Abbreviation for kinetic resonance Raman spectroscopy.

Krypton A chemical element, atomic weight 83.

K_s Dissociation constant of an enzyme-substrate complex.

KS Abbreviation for 1. kaposi's sarcoma; 2. keratan sulfate; 3. ketosteroid.

KSP Abbreviation for kidney-specific protein.

K-Tab A trade name for potassium chloride, used to replace and maintain the level of potassium in the body.

KTX Abbreviation for kaliotoxin.

Kunin Antigen The common antigen of enterobacteria.

Kunitz Inhibitor A trypsin inhibitor from soybean.

Kunjin Virus A virus in the family of Flaviviridae.

Kupffer Cells Macrophages of the liver.

Kura Disease A disease of the central nervous system caused by a prion occurring among Fore people of Papua in New Guinea.

Kurthia A genus of Gram-positive, asporogenous, catalase-positive, obligatory aerobic bacteria.

Kusnezovia A genus of manganese-depositing bacteria.

Ku-Zyme HP A trade name for the enzyme pancrelipase, used to aid digestion.

kV Abbreviation for kilo volt.

Kveim Test A delayed-hypersensitivity test for sarcoidosis in which potent antigenic extracts from sarcoid tissue are injected intradermally and observed for the appearance of a papule, nodule, or superficial necrosis around the site of injection.

K_w Referring to the ion products of water.

Kwashiokor Disease A disorder due to dietary protein deficiency, characterized by retarded growth, anemia, liver failure, and depigmentation of the skin and hair.

Kwell A trade name for lindane, an insecticide.

Kwellada A trade name for lindane, an insecticide.

Kymograph An instrument used for recording pressure vibration, e.g., muscle contraction or nerve conduction.

Kynurenic Acid (mol wt 189) A metabolic product from tryptophan.

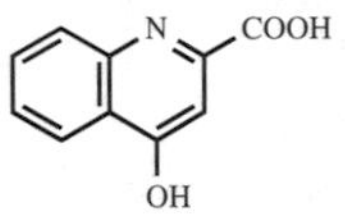

Kynureninase The enzyme that catalyzes the following reaction:

$$\text{L-kynurenine} + H_2O \rightleftharpoons \text{Anthranilate} + \text{alanine}$$

Kynurenine (mol wt 208) A nonprotein amino acid derived from tryptophan.

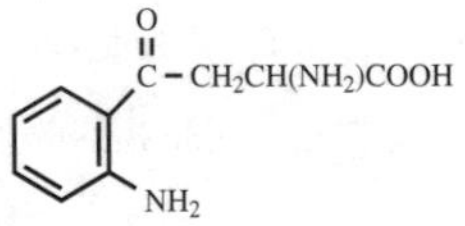

Kynurenine α-Ketoglutarate Transaminase The enzyme that catalyzes the following reaction:

$$\text{L-Kynurenine} + \alpha\text{-ketoglutarate} \rightleftharpoons \text{Aminophenyldioxobutanoate} + \text{glutamate}$$

Kynurenine 3-Monooxygenase The enzyme that catalyzes the following reaction:

$$\text{L-Kynurenine} + \text{NADPH} + O_2 \rightleftharpoons \text{3-Hydroxy-L-kynurenine} + \text{NADP}^+ + H_2O$$

Kyotorphin An analgesic dipeptide (L-Tyr-L-Arg) isolated from bovine brain.

Kytril A trade name for granisetron hydrochloride, an antiemetic agent.

L

L Abbreviation for 1. levorotatory, optical activity of a molecule that rotates polarized light to the left; 2. liter; 3. leucine; 4. lysidine; and 5. lysine.

L1, L2, L3,… L34 Referring to the ribosomal proteins from large ribosomal subunits. These proteins are numbered from L1 to L34.

L Form Bacteria A bacterial cell in which the cell wall is defective or absent and the cell becomes spherical.

La Abbreviation for lanthanum.

LA Abbreviation for 1. lactic acid; 2. late antigen; 3. latex agglutination; 4. leucine aminopeptidase; 4. leukemia antigen; 5. linoleic acid; 6. linolenic acid.

LA-12 A trade name for vitamin B_{12a}.

La Cross Virus A virus of the family Bunyaviridae.

LAA Abbreviation for 1. leukemia-associated antigen; 2. leukocyte ascorbic acid.

Label A radioactive atom used to label a compound and to facilitate observation of its metabolic transformations or identification of target structure.

Labeled Compound Substance that carries a radioactive or an enzyme label.

Labetalol (mol wt 328) An antihypertensive agent that blocks the response to alpha and beta stimulation and decreases renin secretion.

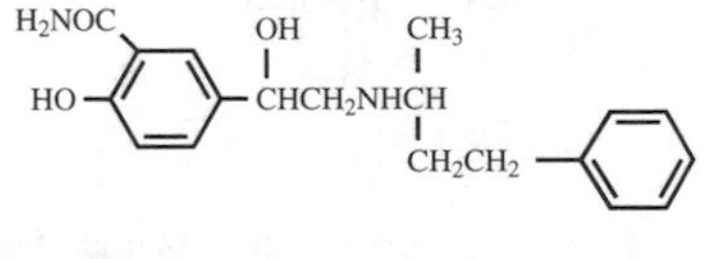

Labile 1. Substances that are readily undergoing chemical, physical, or biological changes. 2. Unstable.

Labile Methyl Group The methyl group that is transferred from one compound to another.

Labile Phosphate Group The phosphate group that can be readily liberated from a compound by hydrolysis at 100°C in 1*N* HCl.

Labile Sulfur Sulfur that can be readily liberated from proteins as H_2S under acid pH.

Labyrinth 1. A group of intercommunicating channels. 2. The internal ear.

Labyrinthitis Inflammation of the labyrinth of the inner ear.

Lac Abbreviation for lactose.

Lac Operator A segment of DNA in the *lac* operon that binds with repressor and inhibits the operation of the *lac* operon.

***lac* Operon** A cluster of genes in *E. coli* consisting of structural genes, regulatory gene, promoter, and operator. This gene cluster functions coordinately in the induction of enzymes for the metabolism of lactose.

Lac Plasmid A plasmid that specifies the uptake and metabolism of lactose occurring in some *Streptococci.*

Lac Repressor A protein encoded by *lac*I gene in the *lac* operon capable of binding with *lac* operator and repressing the *lac* operon.

LacA Gene A structural gene in the *lac* operon that encodes the enzyme transacetylase.

Laccase The enzyme that catalyzes the following reaction:

$$4 \text{ Benzenediol} + O_2 \rightleftharpoons 4 \text{ Benzosemiquinone} + 2\,H_2O$$

Lachnospira A genus of Gram-positive bacteria of the family Bacteroidaceae that ferment glucose to acetate, ethanol, formate, lactate, CO_2, and H_2.

Lachrymal Gland Tear gland of the eye beneath the upper eyelid that secrets tears.

Lac-Hydrin A trade name for ammonium lactate used as a skin lotion.

LacI Gene The gene in the *lac* operon that encodes *lac* repressor protein.

Lacidipine (mol wt 456) An antihypertensive agent.

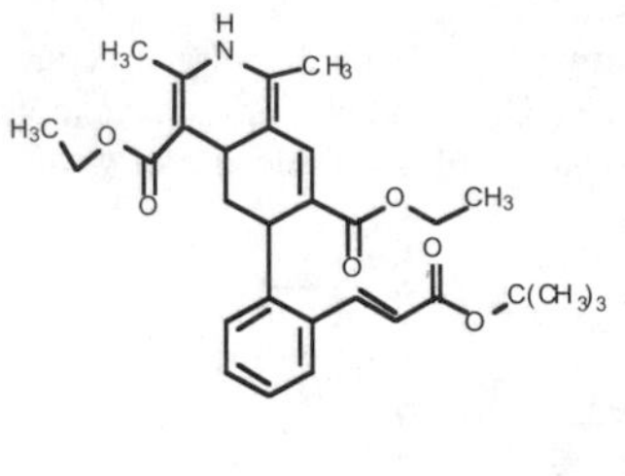

Lacinate Lobed or notched.

LacNAc Abbreviation for N-acetyl-lactosamine.

Lacrimal Pertaining to tears, tear ducts, or tear-secreting organs.

Lacrimal Apparatus The functional and structural network for secreting and draining tears.

Lacrimal Gland The tear gland that secretes tears.

Lacrimation Secretion of tears.

Lacrimator Any agent that causes secretion of tears.

LacT Abbreviation for lactose tolerance.

Lactalbumin Water soluble proteins found in milk.

Lactaldehyde Dehydrogenase The enzyme that catalyzes the following reaction:

$$\text{Lactaldehyde} + NAD^+ + H_2O \rightleftharpoons \text{Lactate} + NADH$$

Lactaldehyde Reductase The enzyme that catalyzes the following reaction:

$$\text{1,2-Propanediol} + NADP^+ \rightleftharpoons \text{D-lactaldehyde} + NADPH$$

Lactam The keto form of an organic cyclic ring structure that is formed by linkage of a NH_2 and COOH group, e.g., the β-lactam ring in penicillin.

β-Lactam Referring to the lactam ring in the penicillins and cephalosporins.

β-Lactam Antibiotics A family of antibiotics that contain a β-lactam ring, e.g., cephalosporins and penicillins.

β-Lactam Hydrolase The systematic name for β-lactamase, it catalyzes the following reaction:

$$\beta\text{-Lactam} + H_2O \rightleftharpoons \text{A substituted } \beta\text{-amino acid}$$

β-Lactam Ring A cyclic ring structure formed between a COOH and NH_2 group in a molecule by the elimination of one molecule of water.

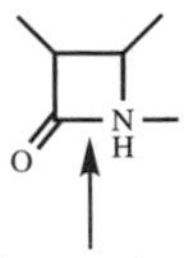

β-lactam ring

β-Lactamase The enzyme encoded by the bacterial genome that cleaves the lactam ring and inactivates the lactam antibiotics.

β-Lactamase Inhibitor Substances that inhibit the destruction of β-lactam antibiotics. They can be combined with β-lactamase-sensitive antibiotics to form clinically useful drugs.

Lactamhydrolase See β-lactamase.

Lactase The enzyme that catalyzes the hydrolysis of lactose (see β-galactosidase).

Lactate 1. Salts of lactic acid. 2. To secrete milk.

Lactate Acidosis A disorder characterized by the accumulation of lactic acid in the blood leading to the reduction of the pH in the blood.

Lactate Aldolase The enzyme that catalyzes the following reaction:

$$\text{Lactate} \rightleftharpoons \text{Formate} + \text{acetaldehyde}$$

Lactate Dehydrogenase The enzyme that catalyzes the following reaction:

$$\text{Lactate} + NAD^+ \rightleftharpoons \text{Pyruvate} + NADH$$

Lactate Dehydrogenase Virus An enveloped, isomeric, RNA-containing virus of the family Togaviridae that causes elevation of plasma lactate dehydrogenase in an infected animal.

Lactate Fermentation An anaerobic fermentation of carbohydrates with lactate as the end product, e.g., homolactate fermentation.

Lactate Malate Transhydrogenase The enzyme that catalyzes the following reaction:

$$\text{Lactate} + \text{oxaloacetate} \rightleftharpoons \text{Pyruvate} + \text{malate}$$

Lactate Monooxygenase The enzyme that catalyzes the following reaction:

$$\text{Lactate} + O_2 \rightleftharpoons \text{Acetate} + CO_2 + H_2O$$

Lactate NAD⁺ Oxidoreductase The systematic name for lactate dehydrogenase.

Lactate Oxidase The enzyme that catalyzes the following reaction:

$$\text{L-lactate} + O_2 \rightleftharpoons \text{Acetate} + CO_2 + H_2O_2$$

Lactate Racemase The enzyme that catalyzes the interconversion of D and L forms of lactate.

Lactation Secretion of milk from mammary glands.

Lacteal 1. The intestinal lymphatics that absorb chyl. 2. Pertaining to or resembling milk.

Lactic Pertaining to milk products, lactic acid, or lactose.

Lactic Acid (mol wt 90) An organic acid produced from lactate fermentation.

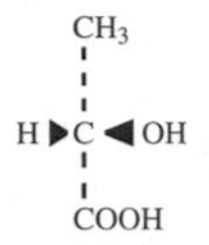

D-lactic acid

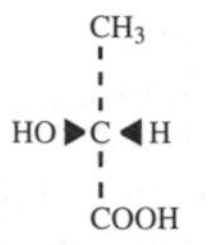

L-lactic acid

Lactic Acidosis See lactate acidosis.

Lactic Acid Bacteria A group of Gram-positive, nonsporing bacteria that carry out lactic acid fermentation.

Lactic Acid Fermentation See lactate fermentation.

Lactic Acid Starter The starter culture used for initiation of lactic acid fermentation.

Lactic Dehydrogenase See lactate dehydrogenase.

Lactic Dehydrogenase Virus See lactate dehydrogenase virus.

Lactiferous Secreting or conveying milk.

Lactifuge An agent that reduces the secretion of milk.

Lactim The enol form of an organic cyclic ring structure, e.g., enol form of uracil.

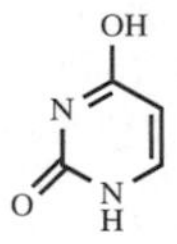

Lactinex A trade name for a drug containing the bacteria *Lactobacillus* for treatment of gastrointestinal disorders.

Lactinol A trade name for a skin lotion containing lactic acid.

Lactitol (mol wt 304) A sugar alcohol containing galactose.

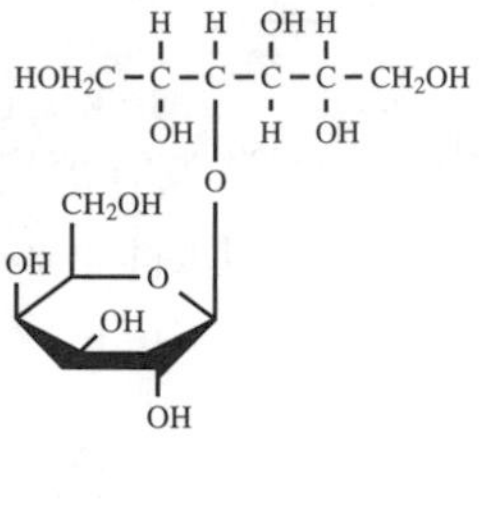

Lactivorous Feeding on milk.

Lactobacillic Acid (mol wt 296) A lipid constituent of various microorganisms.

Lactobacillus A genus of Gram-positive, asporogenous bacteria that produce lactic acid either homofermentatively or heterofermentatively.

Lactobionic Acid (mol wt 358) A sugar acid obtained by the oxidation of lactose.

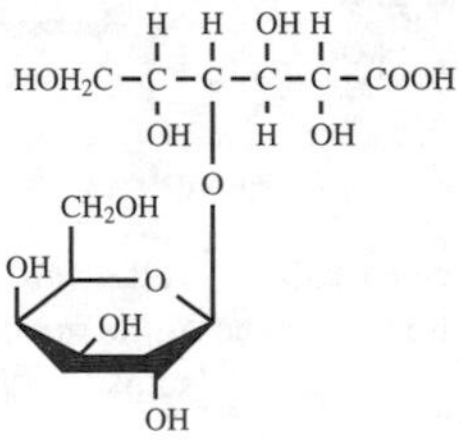

Lactochrome See lactoflavin.

Lactoferrin An antimicrobial iron-containing protein occurring in milk, tears, and neutrophil granules that binds the iron necessary for microbial growth.

Lactoflavin Riboflavin from milk.

Lactogen Substance or agent that stimulates the production and secretion of milk.

Lactogenic Hormone Hormones secreted by the anterior lobe of the pituitary gland that initiates lactation in mammals.

Lactoglobulin Protein found in milk that is soluble in 50% saturated ammonium sulfate solution and insoluble in water.

Lactonase The enzyme that catalyzes the hydrolysis of glucono-1,5-lactone to gluconic acid.

Lactone An intramolecular ester formed between OH and COOH groups by the elimination of one molecule of water, e.g., glucono-1,5-lactone.

Lactoperoxidase Peroxidase from milk used to catalyze the iodination of tyrosine-containing protein.

r-Lactophenetide (mol wt 209) An analgesic and antipyretic agent.

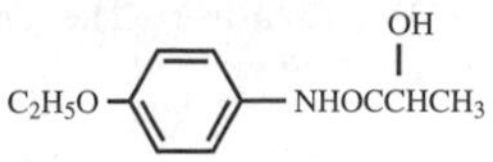

Lactose (mol wt 342) A disaccharide in human and bovine milk and a by-product from the cheese industry that consists of galactose and glucose.

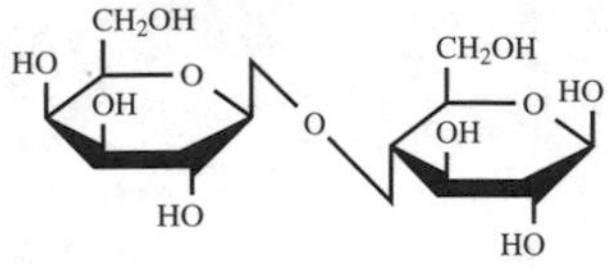

β-D-lactose

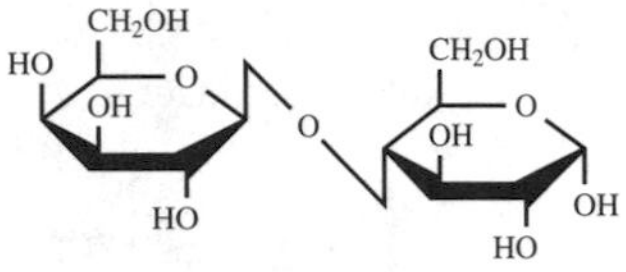

α-D-lactose

Lactose Carrier Protein Referring to a permease encoded by the *lac*Y gene of the *lac* operon that is responsible for uptake of lactose by *E. coli.*

Lactose Galactohydrolase The systematic name for lactase.

Lactose Intolerance A disorder in the digestion of lactose caused by a deficiency in β-galactosidase. The disorder is characterized by bloating, intestinal gas, nausea, diarrhea, and abdominal cramps.

Lactose Operon See *lac* operon.

Lactose Permease Referring to the β-galactoside permease encoded by *lac*Y gene that utilizes the proton gradient across the bacterial cell membrane for cotransport of H^+ and lactose.

Lactose Plasmid See *lac* plasmid.

Lactose Repressors See *lac* repressor.

Lactose Synthetase The enzyme that catalyzes the following reaction:

UDPG-galactose + D-glucose

⇅

UDP + lactose

Lactose Tolerance Test A test for the presence of intestinal lactase activity by a measurement of the glucose concentration as a function of time in the blood following administration of lactose.

Lactosuria The presence of lactose in urine.

Lactoyl-CoA Dehydratase The enzyme that catalyzes the following reaction:

$$\text{Lactoyl-CoA} \rightleftharpoons \text{Acryloyl-CoA} + H_2O$$

Lactoyl-Glutathione Lyase The enzyme that catalyzes the following reaction:

L-Lactoyl-glutathione

⇅

Glutathione + methylglyoxal

Lactoylglutathione Methylglyoxal Lyase The systematic name for lactoglutathione lyase.

Lactulax A trade name for the laxative lactulose.

Lactulose (mol wt 342) A disaccharide used as a laxative that consists of fructose and galactose.

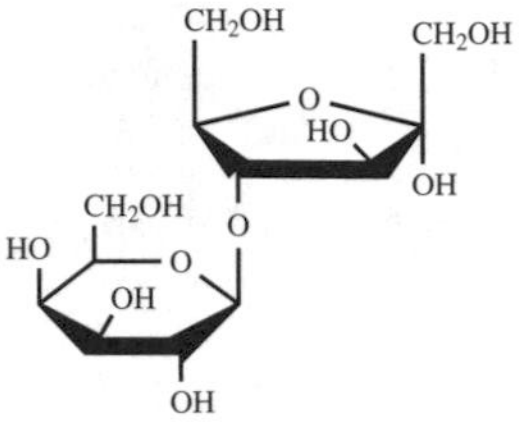

Lacunose Referring to a structure covered with pits or indentation.

LacY Gene A structure gene in the *lac* operon that encodes β-galactoside permease.

LacZ Gene A structural gene in the *lac* operon that encodes β-galactosidase.

LAD Abbreviation for 1. leukocyte adhesion deficiency; 2. lipoamide dehydrogenase; 3. liver alcohol dehydrogenase; 4. lymphocyte-activating determinant.

Ladder Sequencing Technique Referring to the DNA sequencing technique in which the sequence patterns in the electrophoretic gel has the appearance of a ladder, such as Maxam-Gilbert's method or Sanger's dideoxy method of DNA sequencing.

LADH Abbreviation for 1. lipoamide dehydrogenase; 2. liver alcohol dehydrogenase.

Laemmili Gel Electrophoretic Technique Referring to SDS-polyacrylamide gel electrophoresis developed by Laemmili.

Laevorotatory Description of a compound with optical activity that rotates polarized light to the left.

Laevulic Acid See levulic acid.

Laevulose See levulose.

LAF Abbreviation for 1. leukocyte-activating factor; 2. lymphocyte-activating factor.

LAG Abbreviation for lymph-angiogram.

Lag Phase A period following the introduction of microorganisms into a culture medium during which the cell number does not increase.

Lagging Strand of DNA A daughter strand of DNA that is synthesized discontinuously during DNA replication that consists of ligated Okazaki fragments.

LAH Abbreviation for lactalbumin hydrolysate.

LAHV Abbreviation for leukocyte-associated herpes virus.

LAI Abbreviation for 1. latex agglutination inhibition; 2. leukocyte adherence inhibition.

LAIA Abbreviation for leukemia-associated inhibitory activity.

Laidlomycin (mol wt 699) A polyether ionophore antibiotic.

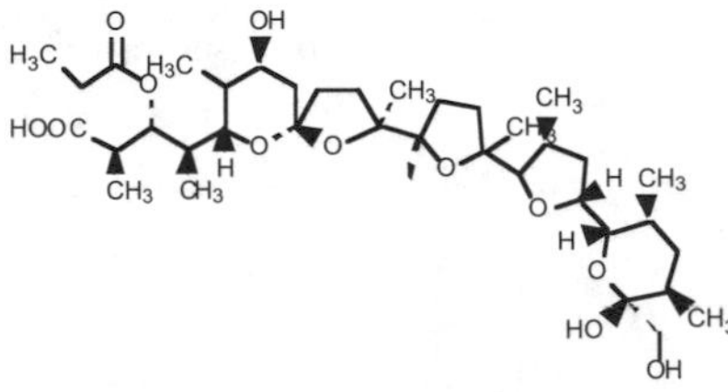

LAIF Abbreviation for leukocyte adherence inhibition factor.

LAIT Abbreviation for latex agglutination inhibition test.

LAK Cell Abbreviation for 1. lymphocyte-associated killer cell; 2. lymphokine-activated killer cell.

Laked Blood The blood in which degeneration of red blood cells has occurred.

Laki-Lorand Factor Referring to fibrin stabilizing factor or blood clotting factor XIII.

Laking Lysis of erythrocytes with release of hemoglobin.

LAM Abbreviation for leukocyte adhesion molecule.

***lam*B Gene** The gene in *E. coli* that encodes *lam*B protein.

***lam*B Protein** An *E. coli* outer membrane protein encoded by *lam*B gene that functions as receptor for λ phage.

Lambda (λ) A Greek letter used to denote 1. a bacteriophage and 2. a unit of volume equal to 1 microliter.

Lambda Bacteriophage A double-stranded DNA bacteriophage of the family Styloviridae with long noncontractile tail that can undergo either a lytic or lysogenic cycle (abbreviated as λ bacteriophage).

Lambda Chain Referring to the one of the two types of light chain in immunoglobulin (abbreviated as λ chain).

Lambda Cloning Vector Referring to the lambda bacteriophage that is genetically engineered so that it can accept foreign DNA and serve as a cloning tool in DNA recombination experiments.

Lambda DNA The DNA genome of λ bacteriophage. It possesses cohesive sites and is capable of integrating into the bacterial chromosome.

Lambda Particle An old name for *Lyticum*, a genus of Gram-negative bacteria that occurs as endosymbionts in the cytoplasm of certain strains of *Paramecium aurelia.*

Lambda Phage See lambda bacteriophage.

Lambdoid Phage A group of temperate phages whose genomes can undergo recombination with each other, e.g., phage λ, 21, 82, 434, and P 22.

Lambert's Law A law of spectrophotometry that states that the absorbance of a light-absorbing compound is directly proportional to the length of the light path passing through the substance.

Lamella 1. A thin platelike structure, e.g., thylakoid membrane in the chloroplast. 2. A layer of tissue or saclike structure.

Lamellae Plural of lamella.

Lamellar Pertaining to lamellae.

Lamellipodium Flattened projection from the surface of a cell.

Lamictal A trade name for lamotrigine, an anticonvulsant and antiepileptic agent.

Lamifiban (mol wt 469) An antithrombotic agent, a specific nonpeptide platelet fibroinogen receptor antagonist.

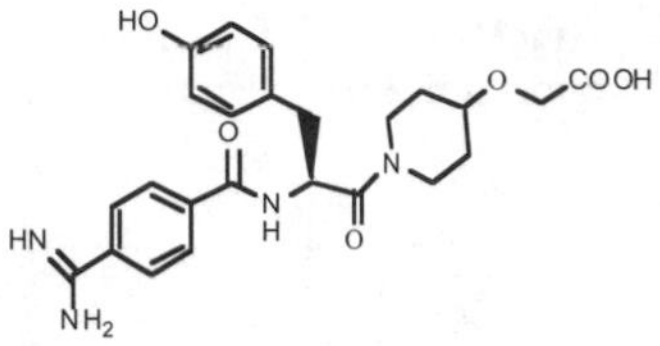

Lamina A thin flat sheet or a layer of tissue or a scalelike structure.

Lamina Densa The microscopically dense region of the glomerular basement membrane.

Lamina Rara The microscopically light region of the glomerular basement membrane.

Laminae Plural of lamina.

Laminar Flow 1. The flow of air currents along parallel flow lines that is used in a laminar flow hood to provide air free from microbes over a work area. 2. The undisturbed flow of a liquid.

Laminaribiose (mol wt 342) A reducing disaccharide consisting of β-(1,3)-glucosidic linkage.

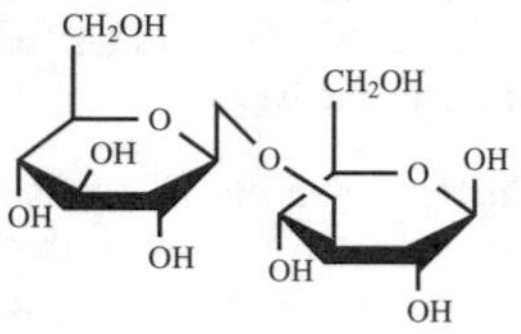

Laminaribiose Phosphorylase The enzyme that catalyzes the following reaction:

3-β-D-Glucosyl-D-glucose + Pi

⇅

D-Glucose + α-D-glucose 1-phosphate

Laminarin An antithrombotic polysaccharide that consists of mainly β-1,3-linked glucose residues and some β-1,4-links. It is the principle storage polysaccharide in brown algae.

Larminarin Phosphotransferase The enzyme that catalyzes the following reaction:

$(1,3\text{-}\beta\text{-D-glucosyl})_n$ + Pi

⇅

$(1,3\text{-}\beta\text{-D-glucosyl})_{n-1}$ +
α-D-glucose 1-phosphate

Laminarinase The enzyme that catalyzes the hydrolysis of 1,3-β-D-glucosidic linkages in a 1,3-β-D-glucan.

Laminin Protein component of the basement membrane.

Laminine (mol wt 220) A naturally occurring amino acid found in *Laminaria angustata*.

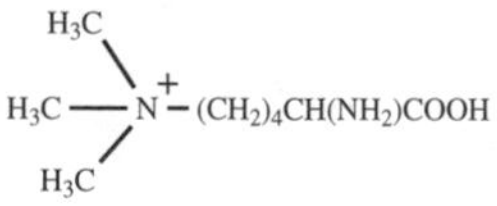

L-amino Acid The L-form stereoisomer of those amino acids occurring in proteins.

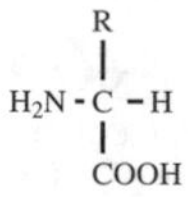

L amino acid

Lamisil A trade name for terbinafine, an antifungal agent.

Lamivudine (mol wt 229) An antiviral agent used for the treatment of HIV.

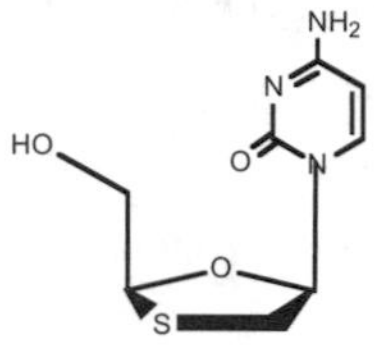

LAMMA Abbreviation for laser microprobe mass analyzer.

Lamotrigine (mol wt 256) An anticonvulsant.

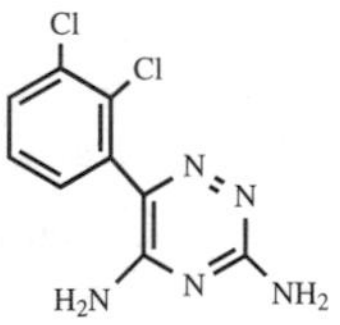

LAMP Abbreviation for lysosomal-associated membrane protein.

Lampbrush Chromosome A large chromosome found in amphibian eggs, characterized by the lateral DNA loops that produce a brushlike appearance.

Lamprene A trade name for clofazimine, an antituberculotic agent.

Lamprocystis A genus of photosynthetic bacteria in the family Chromatiaceae.

Lampropedia A genus of aerobic Gram-negative, chemoorganotrophic bacteria.

Lanabiotic A trade name for a combination drug containing polymyxin B sulfate, neomycin sulfate, bacitracin, and lidocaine, used as a local anesthetic agent.

Lanacane A trade name for benzocaine, an anesthetic agent.

Lanacort A trade name for hydrocortisone acetate.

Lanatosides A family of four cardiotonic glycosides A, B, C, D isolated from species of *Digitalis*.

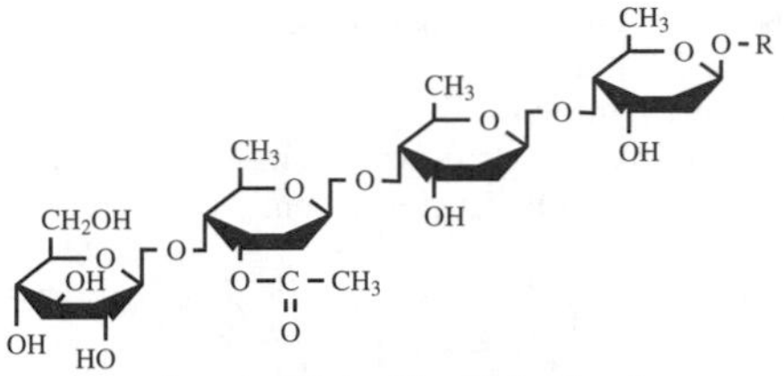

Lanatoside A R = digitoxigenin
Lanatoside B R = ditoxigenin
Lanatoside C R = digoxigenin
Lanatoside D R = diginatigenin

Lanatrate A trade name for a combination drug containing ergotamine tartrate and caffeine.

Lancefield's Grouping Test A test procedure for identifying and classifying Streptococci.

Landsteiner's Classification The classification of human blood into A, B, AB and O group based on the presence of blood group antigen A and B.

Landsteiner's Rule The rule states that blood group antigen and antibody do not coexist in the same individual.

Langerhan's Cell Nonphagocytic cells (e.g., dendritic cells) found in the skin and related to macrophages and possessing MHC class II antigen. It is an effective antigen-presenting cell.

Laniazide-CF A trade name for isoniazid, an antituberculotic agent that inhibits bacterial cell wall synthesis.

Laniroif A trade name for a combination drug containing aspirin, caffeine, and butalbital.

Lankamycin (mol wt 833) An antibiotic produced by *Streptomyces violaceoniger*.

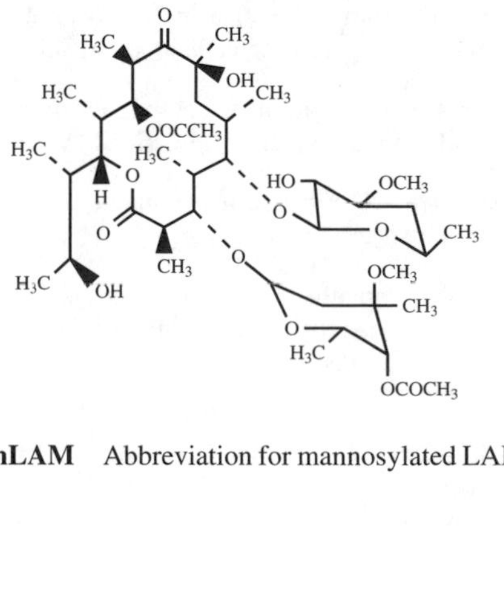

LanLAM Abbreviation for mannosylated LAM.

Lannorinal A trade name for a combination drug containing aspirin, caffeine, and butalbital, used as an analgesic and antipyretic agent.

Lanolin A wool fat or wax that consists of a complex mixture of esters and polyesters of alcohols, fatty acids, and steroids.

Lanophyllin A trade name for theophylline, a bronchodilator that inhibits phosphodiesterase and increases cAMP concentration.

Lanosterol (mol wt 427) An unsaturated sterol occurring in wool fat and yeast.

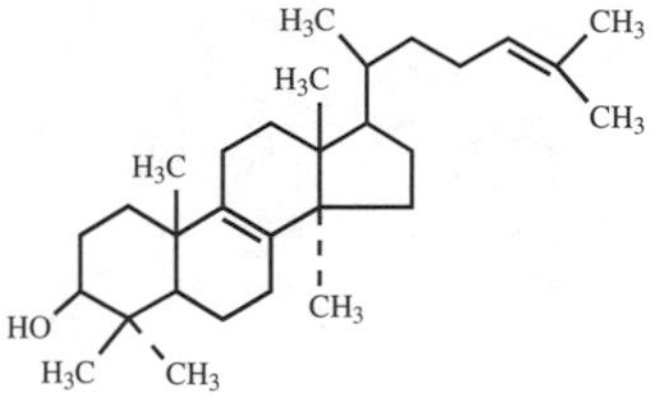

Lanosterol Synthetase The enzyme that catalyzes the following reaction:

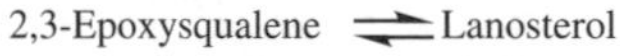

Lanoxicaps A trade name for digoxin, which promotes movement of calcium from the extracellular to intracellular cytoplasm and inhibits sodium-potassium activated ATPase.

Lanoxin A trade name for digoxin, which promotes movement of calcium from the extracellular to intracellular cytoplasm and inhibits sodium-potassium activated ATPase.

Lansoprazole (mol wt 369) An antiulcerative agent and gastric acid-pump inhibitor.

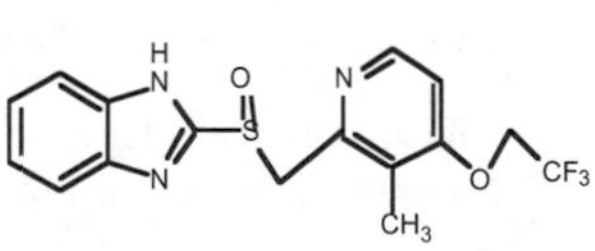

Lanthanum A chemical element with atomic weight 139, valence 3. The salt of lanthanum can be used as a negative stain in electron microscopy and as a calcium channel blocker.

Lanthionine (mol wt 208) An unusual amino acid found in wool protein.

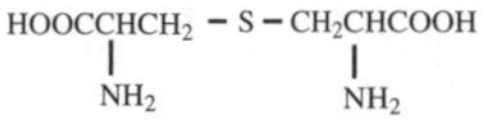

Lantibiotic An antimicrobial polypeptide isolated from *Lactococcus lactis*.

Lanvis A trade name for 6-thioguanine, an antineoplastic agent.

LAO Binding Protein Abbreviation for lysine-arginine-ornithine binding protein.

LAP Abbreviation for 1. liver activator protein or liver-activating protein; 2. leucine aminopeptidase; 3. leukocyte alkaline phosphatase.

Lapachol (mol wt 242) A substance in the hardwood.

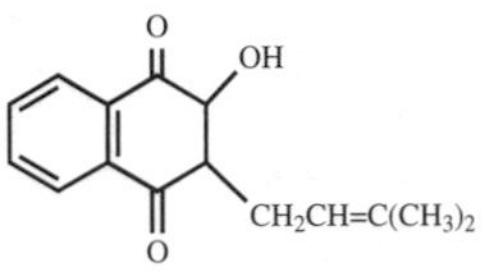

Laparoscope An instrument with a lighted tube and magnifying lense used for examination of the abdominal cavity by insertion through the abdominal wall.

Laparotomy A surgical incision into the abdomen cavity.

LA-PCR Abbreviation for ligation-mediated-anchor-PCR.

Lapinized Vaccine A vaccine that has been attenuated by serial passage through rabbits.

LAPOCA Abbreviation for a combination drug containing L-asparaginase, prednisone, oncovin, cytosine arabinoside, and adriamycin.

La-Protein A protein that is transiently bound to the unprocessed precursor RNAs in the nucleus.

Lapyrium Chloride (mol wt 399) A cationic emulsifier, deodorant, detergent-germicide, and antistatic agent.

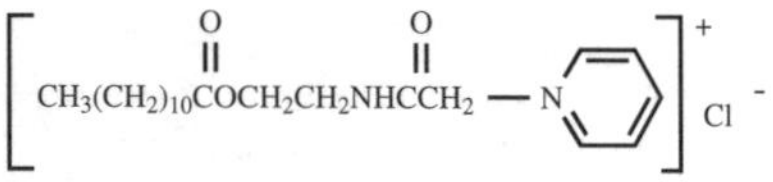

Lard Congealed fat obtained from boiled animal fat.

Lard Fard Referring to vitamin A.

Largactil A trade name for chlorpromazine hydrochloride, an antipsychotic agent that blocks the postsynaptic receptors in the brain.

Large Calorie Referring to kilocalorie (1000 calories).

Large Ribosomal Subunit The component of a 80S or 70S ribosome with a sedimentation coefficient of 60S (60×10^{-13} sec) in eukaryotes and 50S (50×10^{-13} sec) in prokaryotes.

Large T Antigen The product of one of the three genes of the murine polyoma virus involved in the induction of cell transformation.

Lariam A trade name for mefloquine hydrochloride, an antimalarial agent.

Lariat Structure A structure of an intron formed immediately after excision in which the 5′ end loops back and forms a 5′ to 2′ linkage with another nucleotide.

Larodopa A trade name for levodopa, an antiparkinsonian agent.

Laroxyl A trade name for amitriptyline hydrochloride, an antidepressant that blocks the reuptake of norepinephrine and serotonin by the presynaptic neurons.

Larygopharyx The lower portion of the pharynx, lying adjacent to the larynx. Also known as hypopharynx.

Laryngectomy Surgical removal of the larynx.

Laryngitis Inflammation of the larynx. It is characterized by the temporary loss of the ability to speak and by pain in the throat.

Larynx The voicebox or sound-producing organ located between the base of the tongue and the windpipe.

LAS Abbreviation for leucine acetylsalicylate.

Laser A device for producing an intense, monochromatic light beam with an amplified vibration, used in surgical and physiological procedures.

Laser Ablation Use of a highly focused laser beam to destroy specific cells in an organism.

Laser Beam A highly focused nonspreading beam of monochromatic light emitted by a laser.

Laser Dye A dye used for production of a laser beam at a specific wavelength.

Laser Microprobe A microscopic device that uses a laser beam to vaporize a small area of tissue and the vapor thus obtained is analyzed by spectrophotometry.

Lasing Droplet A droplet that emits laser radiation when irradiated with a laser beam. The emitted radiation highlights the liquid-air interface permitting observation on changes in droplet size, shape, and orientation.

Lasix A trade name for furosemide, a diuretic agent that inhibits the reabsorption of sodium and chloride.

Lassa Fever A viral hemorrhagic fever caused by lassa virus (family Arenaviridae).

Latamoxef See moxalactam.

Lassa Virus A virulent and highly infectious arenavirus (Arenaviridae) whose normal host is a rodent.

Latanoprost (mol wt 433) An antiglaucoma agent.

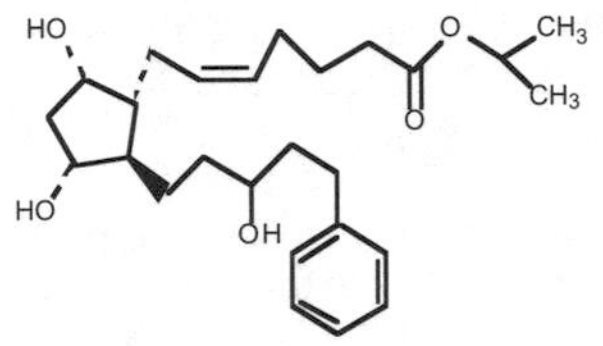

Late Enzyme The viral-encoded enzyme that is synthesized during the late period of the infection cycle.

Late Gene The genes that function during the late period of the viral infection cycle. These genes encode mainly structural proteins of the virus.

Late mRNA The mRNA that is transcribed from late genes, e.g., genes in bacteriophages that encode mRNA responsible for synthesis of capsid proteins.

Late Protein The viral proteins that are translated from the late genes.

Latency 1. The period between the exposure to a stimulus and the response to the stimulus 2. The persistence of a virus within a host cell in a concealed state.

Latent Infection A state of asymptomatic, persistent infection in which the infectious agents are not being produced.

Latent Period 1. The period between the exposure to a stimulus and the response to the stimulus. 2. The time interval between the entrance of a virus into the cell and appearance of intracellular virion. 3. The length of time required for a reaction or response to develop.

Latent Virus The asymptomatic persistence of a virus within a host cell without the active production of infective viral particles and manifestation of disease.

Lateral Located on the side.

Lateral Diffusion Movement of lipid or protein molecules within the plane of a membrane.

Lateral Inhibition The signal produced by one cell that prevents the adjacent cell from achieving its ultimate destination.

Lateral Transmission Transmission of a disease from one individual to another contemporary individual (also known as horizontal transmission).

Latex A milky fluid found in certain plants such as opium poppy, milkweed, and rubber tree.

Latex Cell Cell that contains latex.

Latex Fixation Test A passive agglutination test in which synthetic latex particles coated with antibody or antigen are used in an agglutination test.

Lathyrism The abnormalities of bones, joints, and blood vessels resulting from the ingestion of seeds of sweet pea (*Lathyrus odoratus*) containing β-aminopropionitrile that inhibit collagen cross-linking by the enzyme lysyl oxidase.

α-Latrotoxin A neurotoxic protein from black widow spider that causes massive release of acetylcholine at the neuromascular junction.

LATS Abbreviation for long-acting thyroid stimulator.

Lattice A three-dimensional geometric network formed by atoms or molecules.

Lattice Hypothesis The hypothesis dealing with the formation of a three-dimensional lattice network when antigens and antibodies are combined in optional proportions.

Laurell's Rocket Immunoelectrophoresis See rocket immunoelectrophoresis.

Lauric Acid (mol wt 200) A 12-carbon saturated fatty acid from coconut.

$$CH_3(CH_2)_{10}COOH$$

Lauroguadine (mol wt 377) An antiprotozoal (*Trichomonas*) agent.

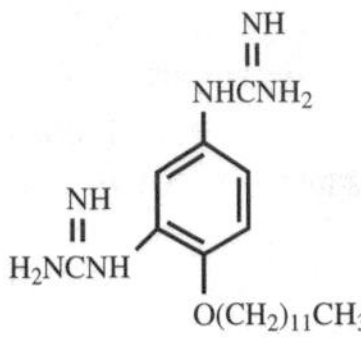

Laurolinium Acetate (mol wt 387) An antiseptic agent.

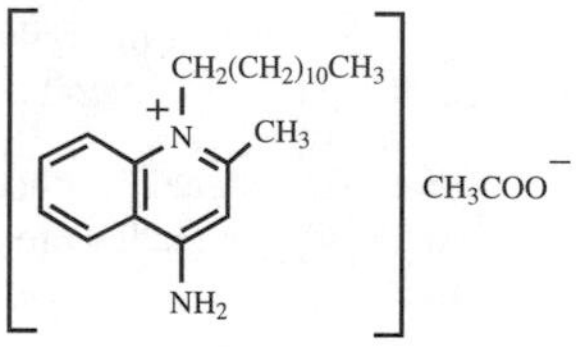

Lauryl Gallate (mol wt 338) An antioxidant.

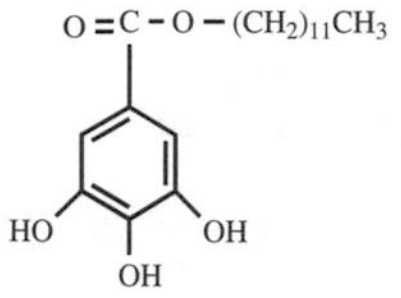

LAV Abbreviation for lymphadenopathy-associated viruses that infect lymphoid cells and give rise to lymphomas.

Lawn Plate A supporting medium plate whose surface bears the confluent growth of bacteria.

Lawrencium (Lr) A chemical element with the atomic number 103, valence 3.

LAX (Lax) Abbreviation for laxative.

Laxative Substance that promotes the evacuation of the bowels.

Laxilose A trade name for lactulose, a laxative and ammonia reduction agent.

Laxinate A trade name for docusate sodium, a laxative that promotes the incorporation of fluid into the stool.

Laxit A trade name for bisacodyl, a laxative that promotes accumulation of fluid in the colon and the small intestine.

Layered Metabolic Pathway The pathway in which the product of one reaction initiates a second set of reactions.

Lazabemide (mol wt 200) An antiparkinsonian agent.

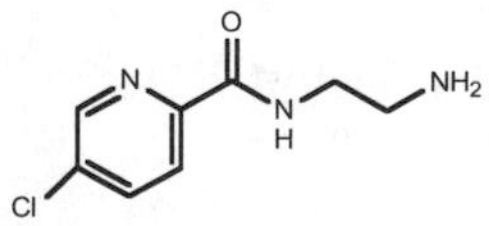

Lazy Leucocyte Syndrome A condition or disorder in which neutrophils display poor locomotion activity toward sites of infection.

Lb Abbreviation for leghemoglobin.

LB Medium Abbreviation for 1. Luria broth; 2. Luria-Bertani medium.

LBD Abbreviation for ligand-binding domain.

LBF Abbreviation for *Lactobacillus bulgaricus* factor.

LBGS Abbreviation for Lewis blood group system.

LBL Abbreviation for lymphoblastic lymphoma.

LBP Abbreviation for 1. laminin-binding protein; 2. lipopolysaccharide-binding protein.

LBS Abbreviation for lipoprotein-binding sites.

LBTI Abbreviation for lima bean trypsin inhibitor.

LC Abbreviation for 1. langerhans cell; 2. lethal concentration; 3. light chain; 4. lipid cytosome; 5. liquid chromatography.

Lc Protein An outer membrane protein (porin) that forms a water-filled transmembrane channel in *E. coli.*

LCA Abbreviation for leukocyte common antigen.

LCAD Abbreviation for long chain acyl-CoA dehydrogenase.

LCADH Abbreviation for long chain acyl-CoA dehydrogenase.

LCAM Abbreviation for leukocyte cell adhesion molecule.

LCAT Abbreviation for lecithin-cholesterol acyltransferase.

LCDG Abbreviation for long-chain diglyceride.

L-Cell An established cell line derived from mouse fibroblast cells.

LCFA Abbreviation for long chain fatty acid.

LCHAD Abbreviation for long chain 3-hydroxyacyl-CoA dehydrogenase.

L-Chain See light chain.

LCL Abbreviation for 1. lympho-cytic leukemia; 2. lympho-cytic lymphosarcoma.

LCLC Abbreviation for large cell lung carcinoma.

LCM Abbreviation for lymphocytic choriomeningitis.

LCM Virus Abbreviation for lymphocytic choriomeningitis virus (Arenaviridae).

LCMG Abbreviation for long-chain mono-glyceride.

LCMV Abbreviation for lymphocytic choriomeningitis virus.

L-Configuration The stereochemical configuration of a molecule that is based upon its stereochemical relation to L-glyceraldehyde.

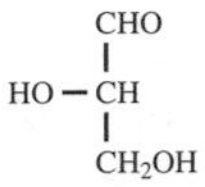

L-glyceraldehyde

LCOX Abbreviation for long chain acyl-CoA oxidase.

LCPT-I Abbreviation for liver CPT-I (carnitine palmitoyltransferase 1).

LCT Abbreviation for 1. Legionnaires' disease; 2. long chain triglyceride; 3. lympho-cytotoxicity test.

LD Abbreviation for 1. lactate dehydrogenase; 2. lethal dose (the dose that causes the death in test subjects or species); 3. levo-dopa; 4. Lyme disease; 5. lymphocyte-defined.

LD_{50} Referring to a lethal dose of a substance that causes death in 50% of the tested animals.

LD_{100} Abbreviation for lethal dose 100 (the concentration or dose that kills 100% of the test subjects).

LD Antigens Abbreviation for lymphocyte defined antigens.

LDC Abbreviation for lysine decarboxylase.

LDCC Abbreviation for 1. lecithin-dependent cellular cytotoxicity; 2. lectin-dependent cell-mediated cytotoxicity.

LDCF Abbreviation for lymphocyte-derived chemotactic factor.

L-dCyd Abbreviation for b-L-2'-deoxycytidine.

LDH Abbreviation for 1. lactate dehydrogenase; 2. leucine dehydrogenase; 3. lysine dehydrogenase.

LDHI Abbreviation for lactate dehydrogenase isoenzymes.

LDL Abbreviation for low-density lipoprotein.

LDL Receptor Membrane protein that binds low-density lipoprotein (LDL).

LDLC Abbreviation for low density lipoprotein cholesterol.

LDLR Abbreviation for low density lipoprotein receptor.

LDM Abbreviation for low density microsomal fraction.

LDMS Abbreviation for laser desorption mass spectrometry.

L-DOPA (L-dopa) Abbreviation for L-dihydroxyphenylalanine.

LDS Abbreviation for lithium dodecyl sulfate.

LDV Abbreviation for lactic dehydrogenase virus.

LE Abbreviation for lupus erythematosus.

Le Antigen Abbreviation for Lewis antigen, a blood group antigen in the Lewis blood group system.

LE Body A globular mass of nuclear material associated with lesions of systemic lupus erythematosus that are stained with haematoxylin.

LE Factor Antinuclear antibodies present in the blood sera of patients with systemic lupus erythematosus.

Le^a Abbreviation for Lewis a, a blood group antigen.

Leaching Bacteria Referring to bacteria that are capable of assisting in solubilization of metals from their ores.

Lead A chemical element with atomic weight 207, valence 2 and 4.

Leader Peptidase An endopeptidase that removes the signal peptide from the secretory protein after transporting across the membrane (see also signal peptidase).

Leader Peptide 1. A regulatory peptide produced by the leader sequence of a messenger RNA. 2. Signal peptide in a secretary protein.

Leader Sequence 1. The nucleotide sequence at the 5′ end of a mRNA (or a polycistronic mRNA) extending to the initiation codon that plays an important role in regulation of transcription. However, the leader sequence is usually not translated. 2. A sequence of amino acids at the N-terminal end of a newly synthesized protein that determines the ultimate destination of a protein.

Leading Strand The strand of a dsDNA that is synthesized by the continuous addition of deoxyribonucleotide in the 5′ to 3′ direction.

Leaf Peroxisome Membrane-bounded organelle that contains enzymes involved in photorespiration.

Leaky Mutant Gene An allele with reduced activity relative to that of the normal allele.

Le^b Abbreviation for Lewis b, a blood group antigen.

Lecithin A phospholipid found in all living organisms of plants and animals (also known as phosphatidylcholine).

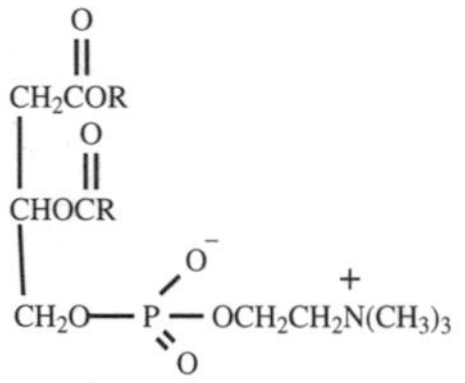

Lecithinase The extracellular phospholipid-splitting enzymes that catalyze the breakdown of a lecithin into its constituents.

Lecithinase A The enzyme that catalyzes the following reaction:

Phosphatidylcholine + H_2O ⇌ 1-Acylglycerophosphocholine + fatty acid

Lecithinase B The enzyme that catalyzes the following reaction:

2-Lysophosphatidylcholine + H_2O ⇌ Glycerophosphocholine + fatty acid

Lecithinase C The enzyme that catalyzes the following reaction:

Phosphatidylcholine + H_2O ⇌ 1,2-Diacylglycerol + choline phosphate

Lecithinase D The enzyme that catalyzes the following reaction:

Phosphatidylcholine + H_2O ⇌ Choline + phosphatidate

Lecithin-Cholesterol Acyltransferase The enzyme that catalyzes the following reaction:

Phosphatidylcholine + cholesterol ⇌ Lysolecithin + cholesterol ester

Lectin A group of proteins derived chiefly from plants that are capable of agglutinating erythrocytes, binding sugars, and stimulating mitosis, e.g., concanavalin A and phytohemagglutinins.

Lederberg Technique A method for rapid isolation of individual bacterial cells for demonstration of the spontaneous origin of bacterial mutants.

Ledercillin VK A trade name for *Penicillin* V that inhibits bacterial cell wall synthesis.

LEF Abbreviation for LE factor.

Lefetamine (mol wt 225) An analgesic agent.

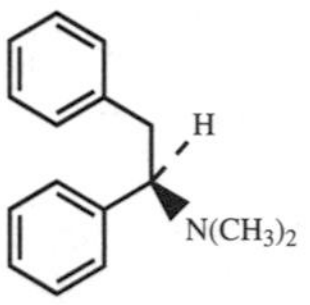

Leflunomide (mol wt 270) An immuno-modulator and inhibitor for dihydroorotate dehydrogenase.

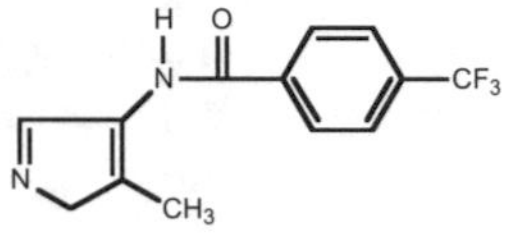

Left Splicing Junction The region between the right end of an exon and left end of an intron.

Legatrin A trade name for quinine sulfate, an antimalarial agent.

Leghemoglobin A form of hemoglobin found in the nitrogen-fixing root nodules of legumes that binds oxygen, and thus protects the nitrogen-fixing enzyme nitrogenase from inactivation by oxygen.

Leghemoglobin Reductase The enzyme that catalyzes the following reaction:

NADPH + 2 ferrileghemoglobin ⇌ $NADP^+$ + 2 ferroleghemoglobin

Legionella A genus of Gram-negative, asporogenous bacteria that causes Legionnaire's disease.

Legionellosis Any disease caused by the species of *Legionella.*

Legionnaire's Disease A form of pneumonia caused by *Legionella pneumophila.*

Legumin Major storage protein of the seeds of peas and other legumes.

Leiomyosarcoma A malignant tumor of smooth muscle.

Leiopyrrole (mol wt 348) An antispasmodic agent.

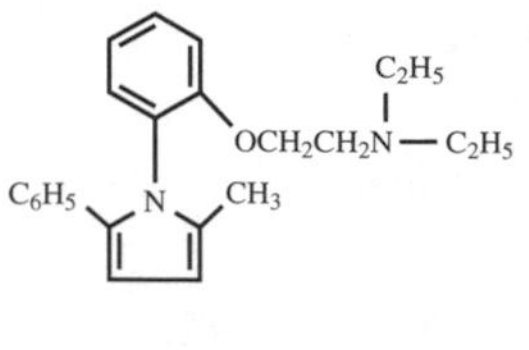

Leiotonin A protein involved in smooth contraction.

Leishmania A genus of parasitic protozoas (family Trypanosomatidae).

Leishmaniasis A disease caused by the species *Leishmania.*

Leishmanolysin A protease from *Leishmania.*

Leishman's Stain A mixture of basic and acidic dyes used to stain blood smears and to distinguish various classes of leucocytes.

Leminorella A genus of nonmotile, Gram-negative bacteria of the family Enterobacteriaceae.

Lenampicillin (mol wt 461) An orally active antibiotic related to penicillin that inhibits bacterial cell wall synthesis.

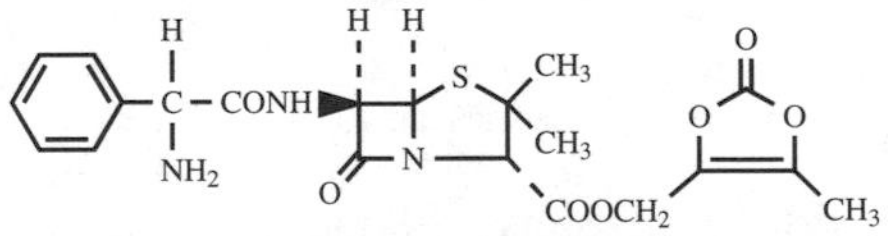

Lenitive 1. A soothing agent. 2. Soothing.

Lens A transparent structure located behind the pupil of the vertebrate eyes that focuses light rays on the retina.

Lente Humulin L A trade name for human insulin.

Lente Iletin I A trade name for human insulin.

Lente Insulin A trade name for insulin zinc suspension, a protein hormone that regulates carbohydrate metabolism.

Lentigo Small, tan, brownish spots on the skin caused by exposure to sun or other agents.

Lentinan A neutral polysaccharide from edible mushrooms with antitumor and immunostimulatory activity.

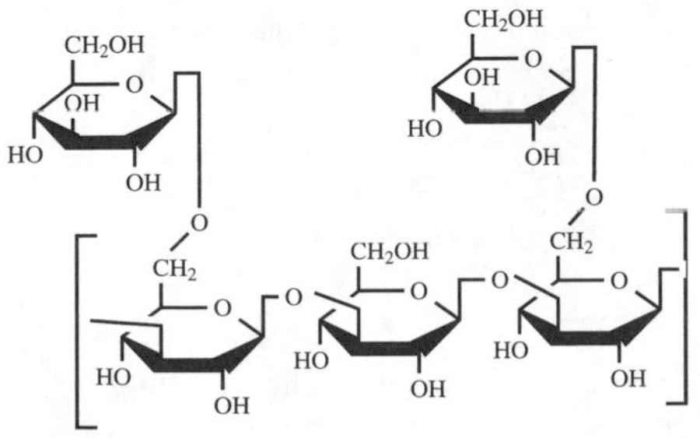

Lentivirinae A subfamily of nononcogenic retroviruses that cause *slow diseases* and characterized by a long incubation period and a chronic progressive phase.

Leovora A trade name for a combination drug containing levonorgestrel and ethinyl estradiol, used as an oral contraceptive agent.

LEP Abbreviation for low-egg passage, referring to a viral strain that has undergone several passages in chicken embryo.

Lepidosis Any scaly eruption of the skin.

Lepirudin A polypeptide and an inhibitor of thrombin produced by recombinant DNA technology.

Leprid The early skin lesion of leprosy.

Leprology The science that deals with leprosy.

Leprostatic Any agent that inhibits the growth of leprosy bacteria.

Leprosy A chronic, contagious human skin disease caused by *Mycobacterium leprae.*

Leptin A weight-controlling protein hormone produced by the adipocytes.

Lepto- A prefix meaning fine, narrow, or thin.

Leptocyte An abnormal blood cell that has a pigmented border surrounding the clear area with a pigmented center.

Leptocytosis The presence of leptocytes in the blood.

Leptodactyline A neuromuscular blocker.

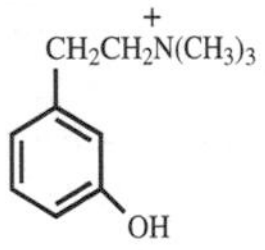

Leptospira A genus of obligate, aerobic bacteria with bent or hooked flagella.

Leptospirosis Disease caused by various serotypes of *Leptospira.*

Leptotene The first stage of meiotic prophase, referring to the chromosomes that appear as thin threads having well-defined chromomeres.

Leptothrix A genus of sheathed Gram-negative iron bacteria.

Leptotrichia A genus of Gram-negative bacteria in the family Bacteroidaceae.

Lercanidipine (mol wt 612) An antihypertensive agent.

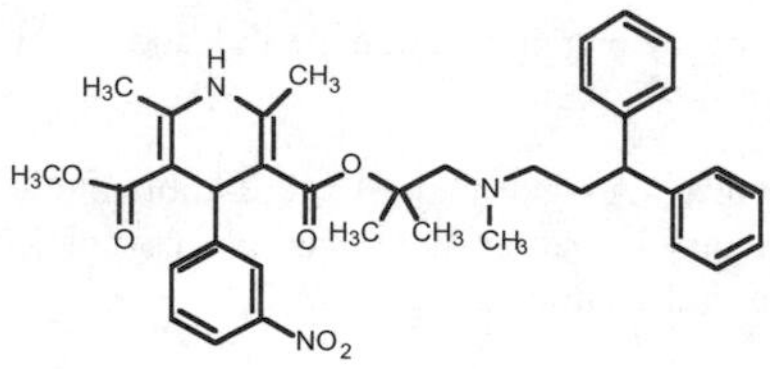

Lesch-Nyhan Syndrome A sex-linked metabolic disease characterized by severe mental retardation and a tendency to self-mutilation, caused by a deficiency in the enzyme HGPRT (hypoxanthine-guanine phosphoribosyl transferase).

Lescol A trade name for fluvastatin, an antihyperlipidemic agent.

Lesion A region of destructive or abnormal change in the tissue.

LET Abbreviation for lupus erythematosus test.

Lethal Allele An allele whose expression causes premature death of an organism.

Lethal Dose The amount of a toxin that causes the death of an organism.

Lethal Gene A gene whose expression causes premature death of an organism (also known as lethal mutation).

Lethal Mutation Mutation that causes premature death of an organism.

Letosteine (mol wt 279) A mucolytic agent.

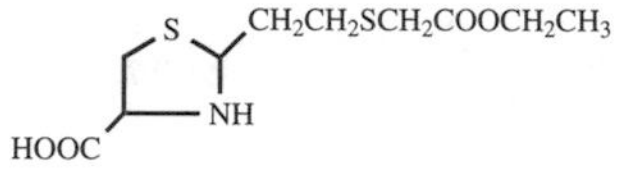

Letrozole (mol wt 285) An antineoplastic and anti-estrogen agent.

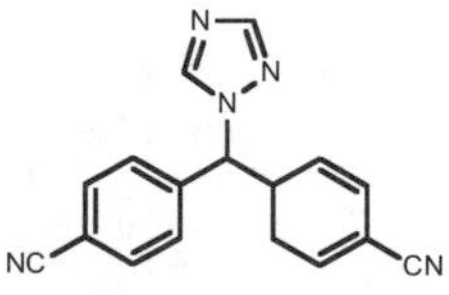

LETS Abbreviation for large extracellular transformation-sensitive protein. Also known as fibronectin.

LeTx Abbreviation for anthrax lethal toxin.

Leu Abbreviation for the amino acid leucine.

Leucine (mol wt 131) An essential protein amino acid.

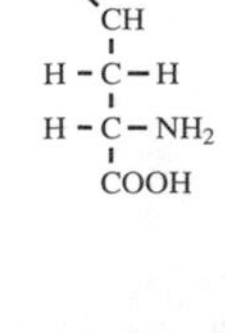

Leucine N-Acetyltransferase The enzyme that catalyzes the following reaction:

Acetyl-CoA + L-leucine ⇌ CoA + *N*-acetyl-L-leucine

Leucine 2,3-Aminomutase The enzyme that catalyzes the following reaction:

α-Leucine ⇌ β-Leucine

Leucine Amino Peptidase An enzyme that catalyzes the release of leucine from the N-terminal of a polypeptide.

Leucine Dehydrogenase The enzyme that catalyzes the following reaction:

L-Leucine + H_2O + NAD ⇌ 4-methyl-2-oxopentanoate + NH_3 + NADH

Leucine Enkaphalin The N-terminal pentapeptide of hormone β-endorphin that binds to opiate receptors and acts as an opiate.

Tyr-Gly-Gly-Phe-Leu

Leucine NAD⁺ Oxidoreductase The systematic name for leucine dehydrogenase.

Leucine 2-Naphthylamide (mol wt 256) A substrate used for assaying leucine aminopeptidase.

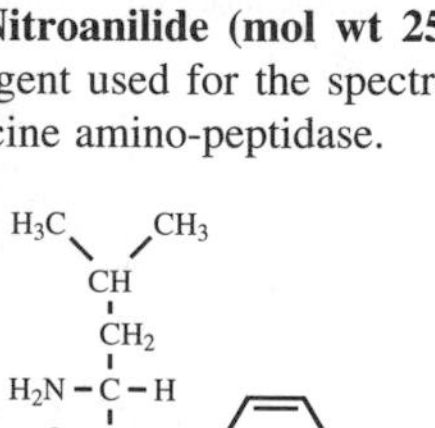

Leucine 4-Nitroanilide (mol wt 251) A chromogenic reagent used for the spectrophotometric assay of leucine amino-peptidase.

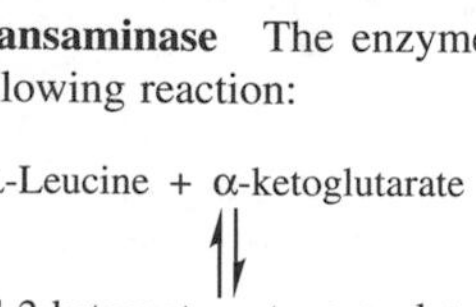

Leucine Transaminase The enzyme that catalyzes the following reaction:

L-Leucine + α-ketoglutarate ⇌ 4-Methyl-2-ketopentanoate + L-glutamate

Leucine tRNA Ligase The enzyme that catalyzes the following reaction:

$$\text{L-Leucine} + \text{ATP} + \text{tRNA}^{leu} \rightleftharpoons \text{AMP} + \text{PPi} + \text{L-leucinyl-tRNA}^{leu}$$

Leucine tRNA Synthetase See leucine tRNA ligase.

Leucine Zipper The hydrophobic strips formed by interdigitating leucine residues in the DNA-binding domain of DNA-binding proteins. The strips resemble the teeth of a zipper and are involved in the process of transcriptional activation and regulation.

Leucinol (mol wt 117) A derivative of leucine.

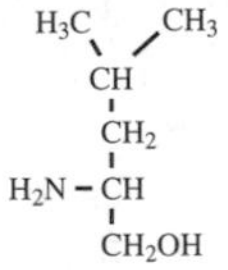

Leucinosis The presence of excessive amounts of leucine in the tissue.

Leucinuria The presence of leucine in the urine.

Leuco- A prefix meaning white or colorless.

Leucocidin Any substance that destroys leukocytes.

Leucocyte See leukocyte (preferred).

Leucolysin Substance or toxin that lyses leukocytes.

Leucomalachite Green (mol wt 330) A reagent used for the colorimetric determination of hemoglobbin and other heme-containing compounds.

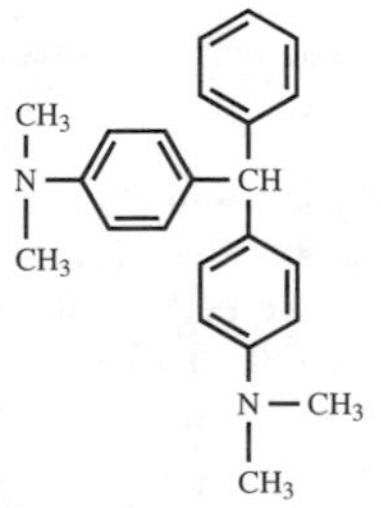

Leuco-Methylene Blue The colorless, reduced form of methylene blue.

Leucomycins An antibiotic complex produced by *Streptomyces kitasatoensis* that possesses at least six components.

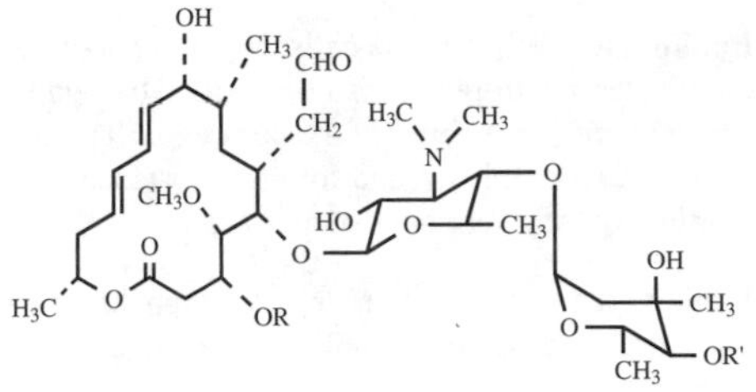

Leuconostoc A genus of bacteria in the family Streptococcaceae that ferments glucose heterofermentatively to lactic acid, ethanol, and carbon dioxide.

Leucopenia A decrease in the number of leukocytes in the blood.

Leucoplast A colorless plastid, the storage center, e.g., amyloplast (starch storage center) and proteinoplast (protein storage center) in plants.

Leucosin The storage polysaccharide of golden-brown algae composed mostly of b-1,3 linked glucose residues.

Leucothrix A genus of gliding bacteria of the family Leucotrichaceae.

Leucovorin Referring to folic acid.

Leucyl Aminopeptidase Synonym of leucine aminopeptidase.

Leucyl Dehydrogenase Synonym of leucine dehydrogenase.

Leucyl tRNA Synthetase Synonym of leucine tRNA synthetase.

LeuDH Abbreviation for leucine dehydrogenase or leucyl dehydrogenase.

Leukaemia See leukemia.

Leukemia Cancer of leukocytes or blood-forming organs, characterized by the presence of large numbers of the immature white blood cells in the bone marrow, thymus, lymph node, spleen, and circulating blood.

Leukeran A trade name for chlorambucil, an alkylating agent.

Leukine A trade name for granulocyte-macrophage colony stimulating factor.

Leuko- A prefix meaning white or colorless.

Leukoblast The precursor of leukocytes.

Leukocidin A toxic substance that destroys leukocytes.

Leukocyte White blood cells, e.g., lymphocytes, monocytes, neutrophils, eosinophils, and basophils. They function in defense of the body against pathogenic microorganisms, and foreign substances and elimination of altered-self materials.

Leukocyte Inhibitory Factor A lymphokine that inhibits the migration of polymorphonuclear neutrophils.

Leukocytosis An increased number of leukocytes in the blood.

Leukoderma The loss of skin pigment.

Leukokinin A cytophilic immunoglobulin (IgG) that binds to autologous polymorphonuclear leukocytes.

Leukomethylene Blue The colorless, reduced form of methylene blue.

Leukopenia A decrease in number of leukocytes (below normal 5×10^3 cells/mm^3) in the blood.

Leukophoresis Fractionation or separation of white blood cells for identification, quantitation and purification.

Leukoplast See leucoplast.

Leukopoiesis The process of the formation of white blood cells.

Leukorrhea A white discharge from the vagina.

Leukosis The proliferation of leucocyte-forming tissue.

Leukosulfakinin A sulfated myotropic, 11-residue peptide isolated from cockroach. It exhibits sequence homology with human gastrin II and cholecystokinin.

Leukotaxis The migration of white blood cells toward the site of inflammation or source of chemotactic factor.

Leukotoxin Toxin that destroys leukocytes.

Leukotriene A_4 Hydrolase The enzyme that catalyzes the conversion of leukotriene A_4 to leukotriene B_4.

Leukotriene A_4 Synthetase Synonym of arachidonate 5-lipoxygenase.

Leukotrienes A family of immunoactive substances that are generated from the metabolism of arachidonic acid by the action of lipoxygenases.

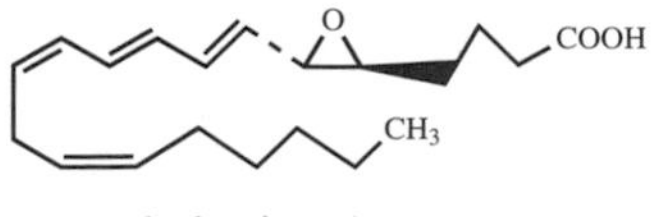

leukotriene A_4

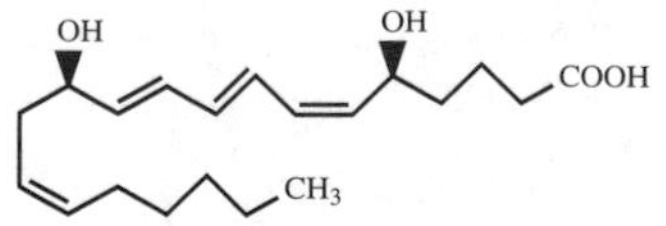

leukotriene B_4

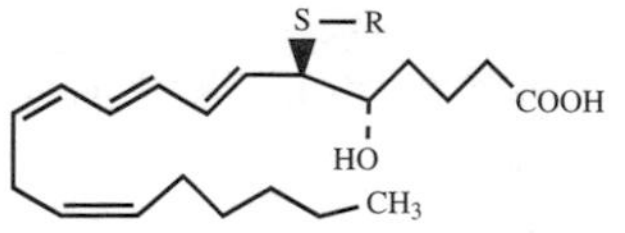

leukotriene C_4

R= — $CH_2CHCONHCH_2COOH$ (on CH: $NHCOCH_2CH_2CHCOOH$, on CH: NH_2)

leukotriene D_4

R = — $CH_2CHCONHCH_2COOH$ (on CH: NH_2)

leukotriene E_4

R = — $CH_2CHCOOH$ (on CH: NH_2)

leukotriene F_4

R= — $CH_2CHCOOH$ (on CH: $NHCOCH_2CH_2CHCOOH$, on CH: NH_2)

Leukoviruses Former name of retroviruses.

Leupeptin A group of modified tripeptides that act as protease inhibitor that are produced by various species of *Actinomycetes.*

Leuprolide A peptide antineoplastic agent that inhibits the release of folicle-stimulating hormone and luteinizing hormone.

Leustatin A trade name for cladribine, an antineoplastic agent.

Levalbuterol Hydrochloride (mol wt 276) An agent that relaxes the smooth muscles of the airways. It activates the beta-2 adrenergic receptors on the airway smooth muscles.

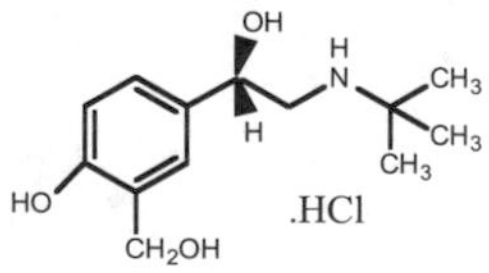

Levallorphan (mol wt 283) A narcotic antagonist.

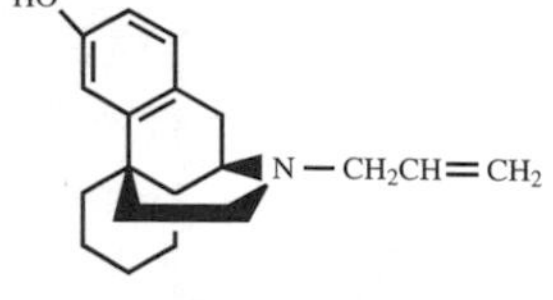

Levamisole (mol wt 203) An anthelmintic drug with immunostimulatory activity and a potent inhibitor of alkaline phosphatase.

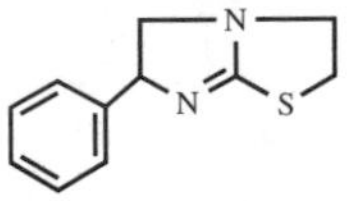

Levan A polysaccharrde consisting of 2,6-β-linked repeating units of D-fructose that is produced by a range of microorganisms and higher plants.

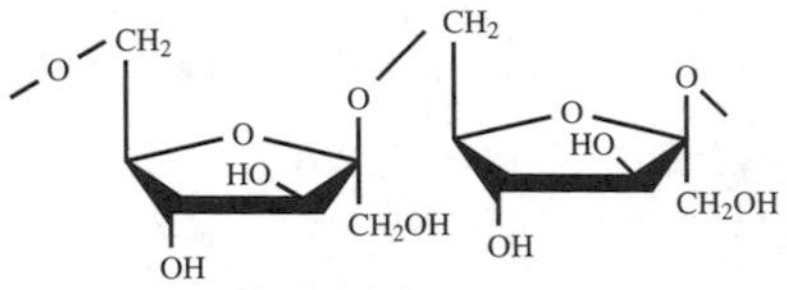

Levaquin A trade name for levofloxacin, an antibiotic.

Levarterenol A trade name for norepinephrine bitartrate, a hormone.

Levate A trade name for amitriptyline hydrochloride, an antidepressant that increases the level of norepinephrine and serotonin in the CNS.

Levatol A trade name for penbutolol sulfate, an antihypertensive agent that blocks the $alpha_1$ and $alpha_2$ adrenergic receptors.

Levbid A trade name for hyoscyamine sulfate, an anticholinergic agent used as an antispasmodic drug.

Leviviridae A family of small, icosahedral-shaped, plus-stranded RNA phages, e.g., MS2, f2, R17, and QB.

Levlen A trade name for a combination drug containing ethinyl estradiol and levonorgestrel, used as an oral contraceptive that inhibits ovulation.

Levlite A trade name for a combination drug containing levonorgestrel and ethinyl estradiol, used as an oral contraceptive agent.

Levo- A prefix meaning left.

Levobunolol (mol wt 291) An antiglaucoma agent and a beta blocker.

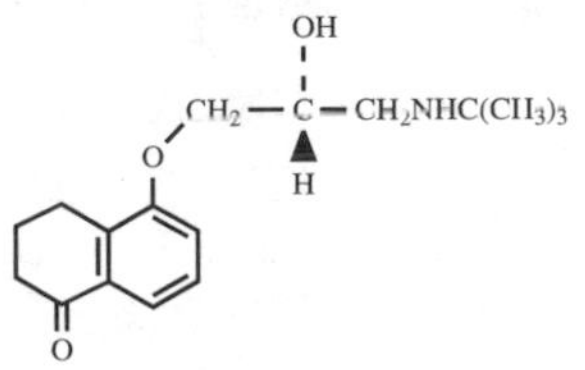

Levobupivacaine Hydrochloride (mol wt 325) A local anesthetic agent.

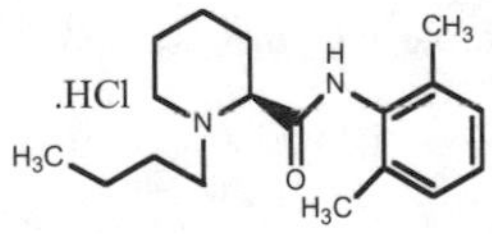

Levocabastine (mol wt 421) An antihistaminic agent.

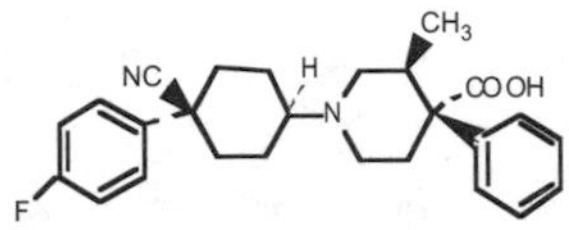

Levocarnitine Referring to L-carnitine that facilitates the transport of fatty acid from cytoplasm to mitochondria.

Levodopa (mol wt 197) An antiparkinsonian agent.

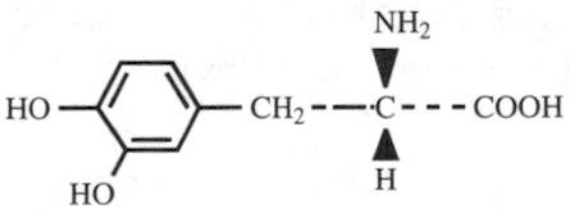

Levo-Dromoran A trade name for levorphanol tartrate, a narcotic and opioid analgesic agent that binds to the opiate receptors in the CNS.

Levoid A trade name for levothyroxine sodium, a thyroid hormone that accelerates cellular oxidation (also known as T_4).

Levomepate (mol wt 303) An anticholinergic agent.

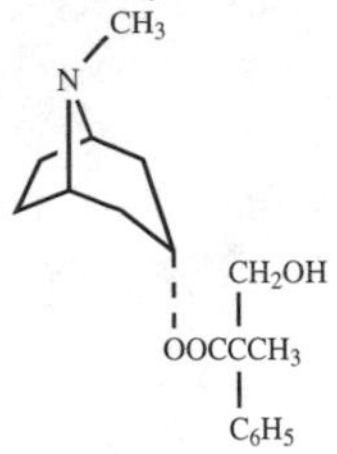

Levomethadyl Acetate (mol wt 354) A narcotic agonist and analgesic agent.

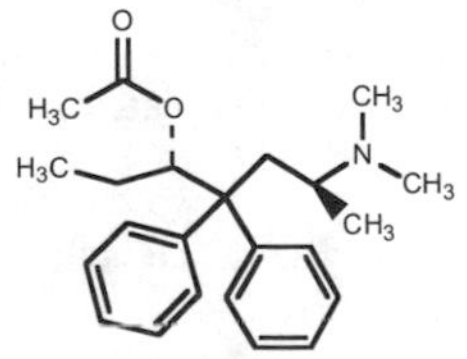

Levonorgestrel (mol wt 313) A hormone and component of a contraceptive agent.

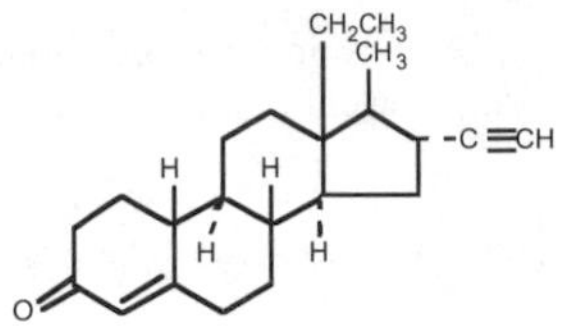

Levophacetoperane (mol wt 233) An antidepressant and anorexic agent.

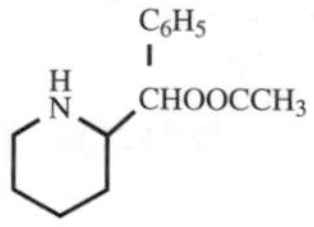

Levophed A trade name for norepinephrine, a hormone.

Levoprome A trade name for methofrimeprazine hydrochloride, a sedative-hypnotic agent.

Levorotatory The optical activity of a molecule that rotates the plane of the plane-polarized light to the left or counterclockwise.

Levorphanol (mol wt 257) A narcotic analgesic agent that binds to opiate receptors.

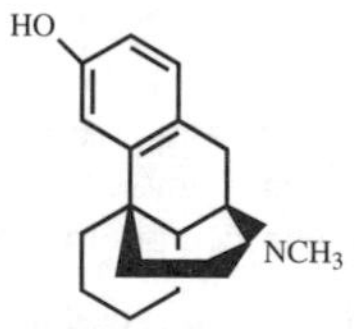

Levo-T A trade name for levothyroxine sodium, a thyroid hormone.

Levothroid A trade name for levothyroxine sodium, a thyroid hormone that accelerates cellular oxidation.

Levothyroxine Sodium (mol wt 799) A thyroid hormone that accelerates the rate of cellular oxidation.

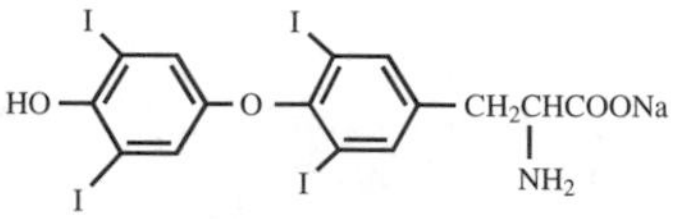

Levoxine A trade name for levothyroxine sodium, a thyroid hormone accelerates cellular oxidation.

Levoxyl A trade name for levothyroxine, a thyroid hormone.

Levsin A trade name for hyoscyamine sulfate, an anticholinergic agent that blocks acetylcholine action.

Levsinex A trade name for hyoscyamine sulfate, an anticholinergic agent that blocks acetylcholine action.

Levulinic Acid (mol wt 116) An organic acid.

$CH_3COCH_2CH_2COOH$

Levulose Referring to D-fructose.

Levulosemia The presence of fructose in the blood.

Levulosuria The presence of fructose in the urine.

Lewis Acid An atom, ion, or molecule that acts as an electron pair acceptor.

Lewis Antigen Blood group substance used for classification or typing of red blood cells.

Lewis Acid-Base Catalysis A form of catalysis in which the catalyst is a Lewis acid (electron pair acceptor) or Lewis base (electron pair donor).

Lewis Base An atom, ion, or molecule that acts as an electron pair donor.

Lewis Blood System The use of Le^a and Le^b antigens for blood group classification. It was named after a Mrs. Lewis in whom the antibodies were discovered.

LexA Gene The gene whose product represses the SOS system for repairing DNA.

LexA Protein The protein produced by LexA gene that represses the SOS system for repairing DNA.

Lexxel A trade name for a combination drug containing enalapril maleate and felodipine, used as an antihypertensive agent.

Leydig Cell 1. Cells found in the epidermis with characteristics of macrophages. 2. The interstitial cell of the mammalian testis involved in the synthesis of testosterone.

LF Abbreviation for 1. lactoferrin; 2. lassa fever; 3. lipotropic factor.

LFA Abbreviation for lymphocyte function-associated antigen.

LFA-1 Abbreviation for lymphocyte function-related antigen-1, a heterodimeric lymphocyte plasma membrane protein and one of the integrin superfamily of adhesion molecules.

LFA-3 Abbreviation for lymphocyte function-related antigen-3, a ligand for the CD2 adhesion receptor that is expressed on cytolytic T-cells.

L-FABP Abbreviation for liver fatty acid-binding protein.

LFN Abbreviation for lactoferrin, a protein that combines with iron and competes with microorganisms for it.

L-Form Bacterium Referring to a defective, spherical or irregular shaped bacterial cell whose cell wall is partially or totally absent as observed in *Proteus*, *Streptococcus*, and *Vibrio*.

LFT Abbreviation for 1. latex fixation test; 2. latex flocculation test; 3. liver function test.

LFV Abbreviation for lassa fever virus.

LGH Abbreviation for lactogenic hormone.

LGL Abbreviation for large granular lymphocytes.

LGMD Abbreviation for limb-girdle muscular dystrophy.

Lgp120 Abbreviation for lysosomal glycoprotein of 120 kDa.

LH Abbreviation for luteinizing hormone, a gonadotropin produced by the anterior pituitary. It stimulates the testes to produce testosterone in males and stimulates the corpus luteum to produce progesterone in females.

LHC Abbreviation for light-harvesting complex.

LHCP Abbreviation for light-harvesting chlorophyll protein.

LHFSH Abbreviation for luteinizing hormone follicle-stimulating hormone.

LHRF Abbreviation for luteinizing hormone releasing factor.

LHRH Abbreviation for lutenizing hormone releasing hormone.

Li Abbreviation for lithium.

LIA Abbreviation for leukemia-associated inhibitory activity.

Librax A trade name for a combination drug containing chlordiazepoxide hydrochloride and clidinium bromide, used as an antianxiety agent.

Libritabs A trade name for chlordiazepoxide, an antianxiety agent.

Librium A trade name for chlordiazepoxide hydrochloride, an antianxiety agent.

LIBS Abbreviation for 1. Lendrum's inclusion body stain; 2. ligand-induced binding site.

Lichen A composite organism formed from symbiotic association of a true fungus and a cyanobacterium or a unicellular alga.

Lichenase See licheninase.

Lichenification The process by which skin becomes thick and hard.

Lichenin A linear polysaccharide consisting of β-D-1,3-linked cellotriose units (three β-1,4-linked glucose unit).

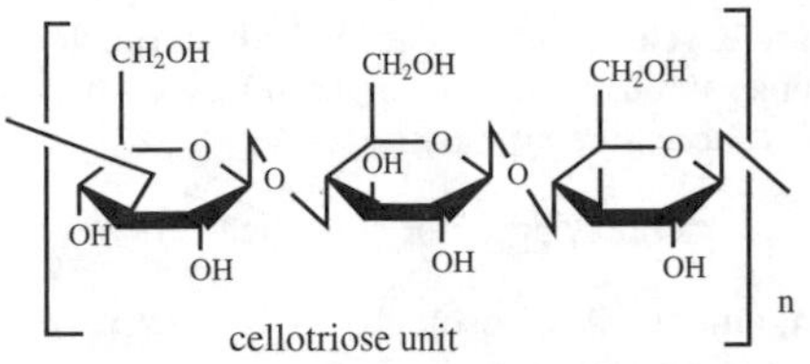

Licheninase The enzyme that catalyzes the hydrolysis of 1,4-β-D-glucosidic linkages in β-D-glucan containing only 1,3 and 1,4 β-glucosidic linkages.

Lichenology The science that deals with lichens.

Lichenophagous Feeding on lichens.

Lidamidine (mol wt 220) A antiperistaltic and anti-diarrheal agent.

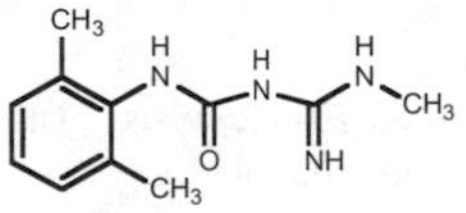

Lidemol A trade name for fluocinonide, a corticosteroid.

Lidex A trade name for fluocinonide, a corticosteroid.

Lidocaine (mol wt 234) An antiarrhythmic agent that shortens the action potential.

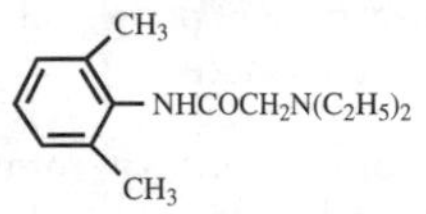

Lidoflazine (mol wt 492) A coronary vasodilator.

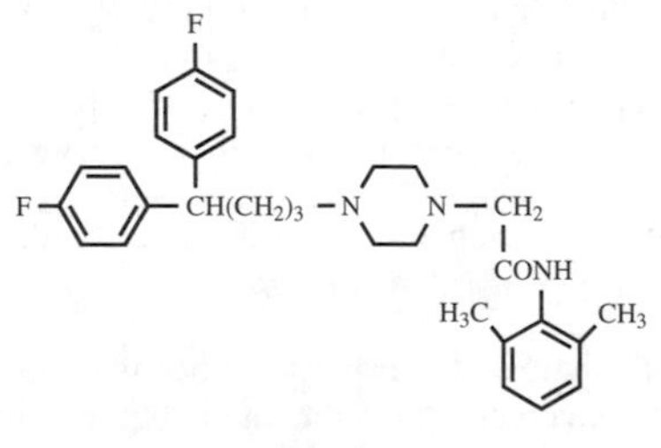

Lidoject A trade name for lidocaine hydrochloride, an antiarrhythmic agent.

Liebermann Burchard Reaction A colorimetric reaction for the determination of chlolesterol based on the color production upon treatment of the sample with acetic anhydride and sulfuric acid.

Lienography The X-ray examination of the spleen following injection of a dye.

LIF Abbreviation for 1. leukemia inhibitory factor; 2. leukocyte-inducing factor; 3. leukocyte inhibitory factor.

Life Cycle The functional and morphological stages through which an organism passes between two successive primary stages.

Lig-I Abbreviation for DNA-ligase I.

Ligament A type of fibrous connective tissue that joins bones together at the joints.

Ligand Substance that bonds to a specific receptor, initiating a particular event or a series of events.

Ligand-Exchange Chromatography A form of column chromatography in which one ligand in the mobile phase exchanges or replaces the ligand that bind to the column.

Ligand-Gated Ion Channel A transmembrane ion channel that opens transiently on the binding of a specific ligand.

Ligand-Induced Endocytosis The endocytotic uptake of ligands that bind to the cell surface receptor via coated pits.

Ligase The enzyme that catalyzes the linking of two molecules e.g., linking of two DNA fragments by DNA ligase.

Ligate 1. To bind a ligand. 2. To join two fragments. 3. To tie off a blood vessel to prevent bleeding.

Ligatin A filamentous plasma membrane protein for the attachment of peripheral glycoproteins to the external cell surface.

Ligatin A polypeptide that forms polymeric fibrils on the outside of the chick neural retina cells.

Ligation 1. The enzymatic joining of two fragments of DNA or RNA by formation of 3′-5′-phosphodiester linkages. 2. Tying off a blood vessel or duct to prevent bleeding (also known ligature).

Ligature Tying off a blood vessel or duct to prevent bleeding during surgery.

Light Chains Referring to either the kappa or lambda chain present in the immunoglobulin molecule.

Light Green SF Yellow (mol wt 793) A biological dye.

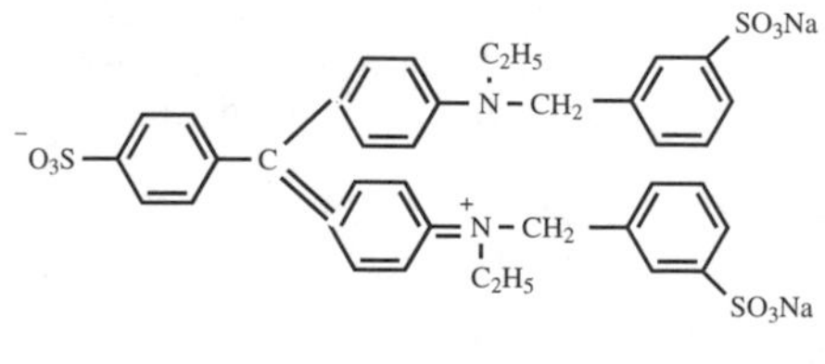

Light Harvesting Chlorophyll Protein A chlorophyll-binding protein capable of transfer of the absorbed photon energy to both photosystem I and II.

Light Harvesting Complex The organized protein-pigment complexes that harvest and transfer radiant energy to photosynthetic reaction centers for photosynthetic reactions.

Light Meromyosin The terminal fragment of the myosin molecule obtained by treatment of myosin with trypsin.

Light Microscope An instrument used for the enlargement of an image in a specimen to be viewed microscopically.

Light Reactions 1. The reactions of photosynthesis in the thylakoid membranes of the chloroplast for conversion of light energy into ATP, NADPH, and evolving oxygen. 2. A light-dependent reaction.

Light Repair See photoreactivation.

Light Scattering The scattering of light by the suspended particles in a solution. The degree of scattering is related to the size and shape of the particles and is used for the quantitation of suspended particles.

Light Strand 1. The strand of a dsDNA that has a lower buoyant density than the complementary strand due to its low G-C content. 2. A polynucleotide strand that is not labeled with a heavy isotope.

Lignans A class of dibenzylbutane derivatives found in higher plants and in the fluid of humans and animals.

Lignin Polymeric substance deposited with cellulose in the cell walls of plants and fungi to provide added strength. Hydrolysis of lignin yields syringaldehyde, hydroxybenzaldehyde, and vanillin.

Ligninase An enzyme that catalyzes the breakdown of lignin.

Ligninolytic Capable of degrading lignin.

Lignocellulose Referring to lignin-cellulose complex.

Lignoceric Acid (mol wt 369) A 24-carbon saturated fatty acid.

$$CH_3 - (CH_2)_{22}COOH$$

Lignotic Capable of degrading wood.

Limaprost (mol wt 381) An antianginal agent.

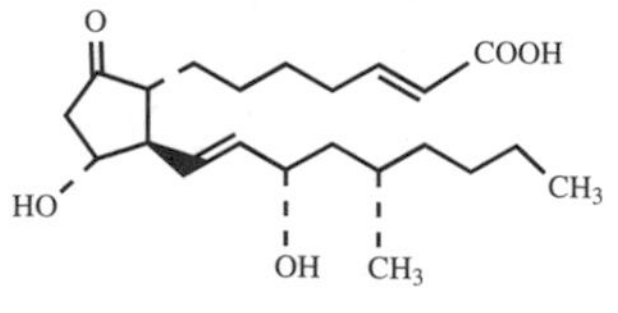

Limbitrol A trade name for a combination drug containing chlordiazepoxide and amitriptyline hydrochloride, used as an antidepressant.

Limit Dextrin The branched core of amylopectin resulting from the exhaustive digestion of amylopectin by β-amylase.

Limited Chromosome The chromosome that occurs only in the nucleus of germ cells and not in somatic cells.

Limiting Factor A factor that restricts the growth and development of an organism or a culture or a population.

Limnology The science that deals with freshwater biology.

Limosis Abnormal hunger or an excessive desire for food.

Limulin A lectin found in horseshoe crab, capable of agglutinating *E. coli* and many other Gram-negative bacteria.

Linamarin (mol wt (247) A compound found in the seed or embryo of flax.

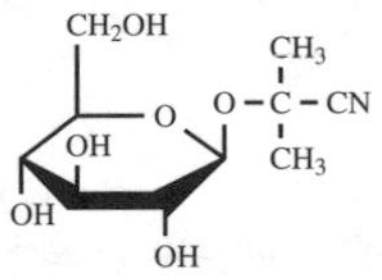

Linamarin Synthetase The enzyme that catalyzes the following reaction:

UDP-glucose +
2-hydroxy-2-methylpropanenitrile

⇅

UDP + linamarin

Lincocin A trade name for lincomycin, an antibiotic that binds to 50S ribosomes, inhibiting bacterial protein synthesis.

Lincomycin (mol wt 407) An antibiotic produced by *Streptomyces lincolnensis* that inhibits bacterial protein synthesis by binding to the 50S ribosome subunit.

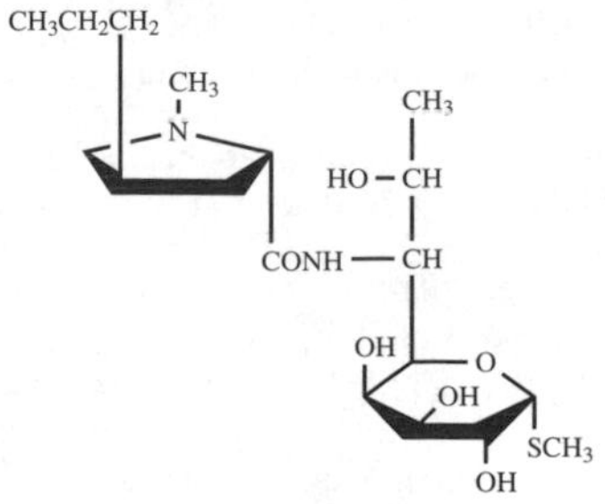

Lincorex A trade name for lincomycin, an antibiotic.

Lindane (mol wt 291) An insecticide that inhibits neuronal membrane function in arthropods.

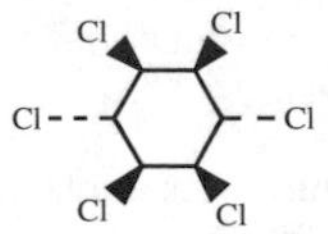

Lineage 1. The descent from a common progenitor. 2. The ancestry of a cell during development or evolution.

Linear Absorption Coefficient The reduction in the intensity of a beam of radiation per unit of thickness of an absorber.

Linear Density Gradient A gradient in which the concentation of the substance varies linearly with distance, e.g., sucrose density gradient used for density gradient centrifugation.

Linear Elution A type of elution system in column chromatography in which the concentration of the eluant solution varies linearly with the volume of solution used.

Linear Energy Transfer The energy dissipation per micron of tissue; used in the description of a particular type of ionizing radiation passing through the tissue or other matter.

Linear Phosphate A linear polyphosphate consisting only of phosphate.

$$H_2PO_3\text{-}(H_2PO_3)\text{-}H_2PO_3$$

Linear Polymer An unbranched chain or unbranched polymer.

Linear Velocity Distance moved in a given direction per unit time.

Lineweaver-Burk Equation A linear equation obtained by the inversion of the Michaelis-Menten equation used in the determination of V_{max} and K_m of an enzyme-mediated reaction.

$$\frac{1}{v} = \frac{1}{V} + \frac{Km}{V} \cdot \frac{1}{[S]}$$

where
v = velocity
V = maximum velocity
[S] = substrate concentration
Km = Michaelis constant

Lineweaver-Burk Plot A straight line obtained by plotting 1/*v* versus 1/{S}. It is used for the analysis of the kinetics of an enzyme-catalyzed reaction and is useful in determining the V_{max} and K_m of an enzyme-mediated reaction.

Lingual Gland Glands that deliver their secretions onto the surface of the tongue.

Liniment An oily medicinal liquid applied onto the skin as a counterirritant.

Linin The fine, threadlike, nonstaining substance of the cell nucleus.

Lining Epithelium An epithelium that lines the duct, cavity, or vessel.

Linitis Inflammation of the cellular tissue of the stomach.

Linkage 1. The tendency of alleles for different characters to remain associated from one generation to the next. 2. The location of genes on the same chromosome.

Linkage Equilibrium Condition in a population in which all possible combinations of linked genes are present at equal frequency.

Linkage Map The linear sequence of known genes on a chromosome.

Linked Genes Genes that are located on the same chromosome.

Linked Recognition A condition in which lymphocytes need to receive two stimulating signals in order to be activated.

Linked Transduction Bacterial transduction in which two or more linked genes are cotransduced.

Linked Transformation A bacterial transformation in which two or more linked genes are cotransduced.

Linker DNA 1. Chemically synthesized oligonucleotides of the defined sequence containing the recognition cleavage site that is used to join two DNA fragments. 2. The DNA segment in a chromosome that connects or links adjacent nucleosomes.

Linking Number The number of times that one closed circular DNA strand winds around the other or the number of times the two strands of a closed DNA duplex cross over each other.

Linoleate Oxygen 13-Oxidoreductase The systematic name for lipoxygenase.

Linoleic Acid (mol wt 280) An 18-carbon essential fatty acid.

Linolenate A salt or ester of linolenic acid.

Linolenic Acid (mol wt 278) An 18-carbon essential fatty acid with three double bonds.

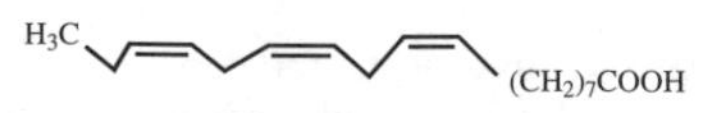

γ-Linolenic Acid (mol wt 278) A derivative of linoleic acid.

Linuron (mol wt 249) An herbicide.

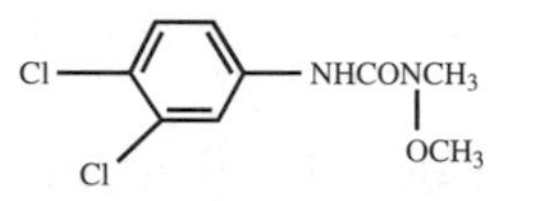

Lioresal A trade name for baclofen, a muscle relaxant that reduces the transmission of nerve impulses from spinal cord to skeletal muscle.

Liothyronine (mol wt 651) A thyroid hormone that stimulates tissue metabolism by accelerating the rate of cellular oxidation (also known as T_3).

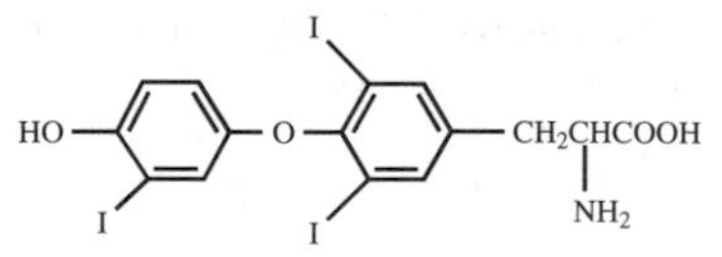

Liotrix A mixture of T_3 (liothyronine sodium) and T_4 (levothyroxine sodium), the thyroid hormones that stimulate metabolism by accelerating cellular oxidation.

LIP Abbreviation for 1. liver inhibitory protein; 2. lymphocytic interstitial pneumonia.

Lipaemia See lipemia.

Lipase The enzyme that catalyzes the hydrolysis of triglycerides yielding fatty acid and diacylglycerol.

Lipectomy Surgical removal of fat beneath the skin or the abdominal wall.

Lipemia The presence of excessive amounts of lipid in the blood.

Lipid Referring to hydrophobic organic molecules such as steroids, fats, fatty acids, phospholipids, and water-insoluble vitamins.

Lipid A A lipid occurring in the lipopolysaccharide of Gram-negative bacterial cell walls.

Lipid Bilayer The membrane structure consisting of two layers of phospholipids arranged so that their polar hydrophilic heads face the outer surface of the membrane while the hydrophobic tails cluster within the interior of the bilayer.

Lipid Crimson (mol wt 365) A lipoprotein stain.

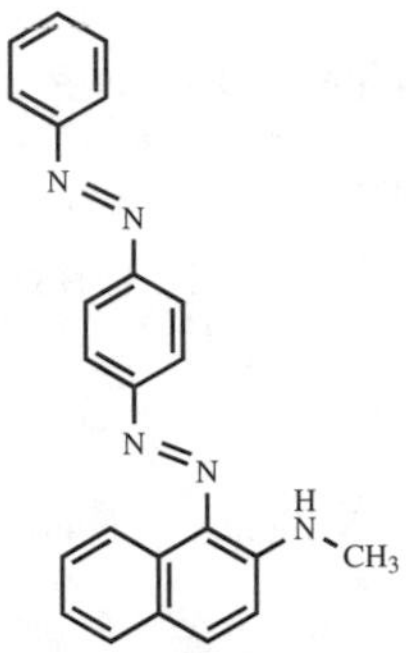

Lipid Metabolism Processes involved in the synthesis and degradation of lipids.

Lipid Monolayer Single layer of lipid molecules oriented so that the hydrophilic heads are on one side and the hydrophobic tails are on the other side.

Lipid Peroxidation The nonenzymatic oxidation of fatty acids to hydroperoxides by strong oxidizing agents such as hydrogen peroxide or superoxide.

Lipid Soluble Vitamins Referring to vitamins A, E, D, K.

Lipid Stain A dye that stains lipid.

Lipid Storage Diseases See lipidosis.

Lipidosis A disorder characterized by the accumulation of an abnormal amount of sphingolipids in the body due to a deficiency in enzymes for degradation of sphingolipid.

Lipiduria The presence of fat bodies (lipid) in the urine.

Lipin An old name for lipid.

Lipitor A trade name for atorvastatin, an antilipidemic drug used for lowering cholesterol.

Lipo- A prefix meaning lipid.

Lipoamide The functional form of lipoic acid in which the carboxyl group of lipoic acid is attached to the protein by an amide linkage to a lysine residue.

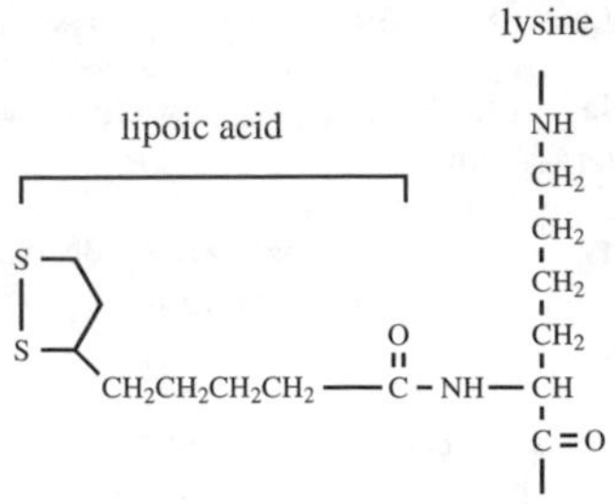

Lipoamide Dehydrogenase The enzyme that catalyzes the following reaction:

FAD + reduced form dihydrolipoamide

⇌

$FADH_2$ + lipoamide

Lipoamide Reductase The enzyme that catalyzes the following reaction:

Dihydrolipoamide + NAD^+

⇌

Lipoamide + NADH

Lipoamino Acid 1. A compound formed by linking a fatty acid or a long chain alcohol to an amino acid by an ester or by an amide bond. 2. An ester of an amino acid with phosphatidylglycerol.

Lipoate Acetyltransferase The enzyme that catalyzes the following reaction:

Acetyl-CoA + dihydrolipoate

⇌

CoA + 6-*S*-acetylhydrolipoate

Lipocaic A lipotropic preparation from the pancreas that stimulates the oxidation of fatty acid.

Lipocalin Any of a large group of ligand-binding proteins that binds hydrophobic ligands.

Lipochrome Referring to various fat-soluble pigments such as carotenoid.

Lipocortins A group of proteins that inhibit phospholipase.

Lipocyte Referring to fat cell.

Lipofection A lipid-mediated DNA-transfection technique.

Lipofuscin A group of lipid pigments found in cardiac and smooth muscle cells, macrophages, parenchyma, and interstitial cells that are involved in the process of cell aging (also known as age pigment).

Lipogenesis The biosynthesis of fatty acid by acetyl-CoA.

Lipoglycan Polysaccharide that contains a covalently linked lipid moiety.

Lipoic Acid (mol wt 206) A vitamin B and a growth factor for some microorganisms; it is a cofactor for decarboxylation of keto-acids.

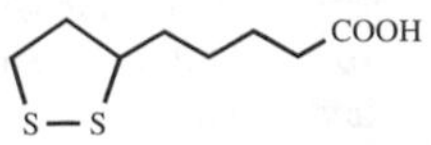

Lipoid A fatlike substance or resembling fat.

Lipoidosis The deposition of fat in the tissue or the replacement of tissue with fat.

Lipolysis The breakdown or hydrolysis of lipid.

Lipolytic Pertaining to lipolysis.

Lipolytic Enzymes Enzymes that hydrolyze lipid.

Lipoma A tumor that consists of fat cells or a tumor of adipose tissue.

Lipomatosis An abnormal tumorlike accumulation of lipid in the tissue.

Lipomodulin Referring to lipocortin.

Lipomul A trade name for corn oil, used as a source of fat.

Lipophilic 1. Nonpolar or fat soluble. 2. With affinity for lipids.

Lipophilic Stain A stain for lipid.

Lipophilicity The physical property of a molecule that renders it readily soluble in a nonpolar solvent.

Lipophorin A group of high-density lipoproteins from insect hemolymph that transport diacylglycerols.

Lipophosphodiesterase I See phospholipase C.

Lipophosphodiesterase II See phospholipase D.

Lipophylin Membrane lipoprotein occurring in brain myelin.

Lipopolysaccharide Any polysaccharide that contains lipid. It is an important component of the outer membrane of the Gram-negative bacteria.

Lipopolysaccharide *N*-Acetylglucosaminyl Transferase The enzyme that catalyzes the following reaction:

UDP-*N*-acetylglucosamine + lipopolysaccharide

⇅

UDP + *N*-acetylglucosaminyl polysaccharide

Lipopolysaccharide Glucosyl Transferase The enzyme that catalyzes the following reaction:

UDP-glucose + lipopolysaccharide

⇅

UDP + glucosyl lipopolysaccharide

Lipoprotein A class of conjugated serum proteins consisting of a lipid linked to lipid. Lipoproteins are classified according to their densities, e.g., very low density (VLDL), low density (LDL), and high density (HDL).

Lipoprotein Lipase The enzyme that catalyzes the hydrolysis of glycerides in the presence of a lipoprotein complex.

Triacylglycerol + H_2O

⇅

Diacylglycerol + a fatty acid

Liposarcoma Malignant tumor of adipose tissue.

Liposome A closed vesicle of lipid bilayer or lipid monolayer. The interior of the liposome may be used to encapsulate exogenous materials or drugs for ultimate delivery into the cells by fusion with the cell.

Liposyn A trade name for fat emulsions, used as a source of fat to prevent fatty acid deficiency.

Lipoteichoic Acid The complex of teichoic acid and glycolipid found in the walls of Gram-positive bacteria.

Lipotropic Having affinity for lipid compounds.

Lipotropic Agent Substance capable of preventing fatty acid infiltration and the formation of fatty liver.

Lipotropic Hormone Any hormone with lipolytic activity on adipose tissue.

Lipotropin A lipotropic polypeptide hormone produced by the pituitary gland that is capable of stimulating lipolytic activity.

β-Lipotropin A lipotropic peptide hormone produced by the cleavage of pro-opiomelanocortin (POMC). The cleavage of POMC yields ATCH and β-lipotropin.

γ-Lipotropin A lipotropic peptide hormone produced by cleavage of β-lipotropin. The cleavage of β-lipotropin yields γ-lipotropin and β-endorphin.

Lipovitellenin A low density lipoprotein found in egg yolk.

Lipovitellin A high-density lipoprotein.

Lipoxidase Synonym of lipoxygenase.

Lipoxide A trade name for chlordiazepoxide, an antianxiety agent.

Lipoxins A family of metabolites derived from arachidonic acid by the action of the lipoxygenase pathway.

Lipoxygenase Enzyme that initiates the conversion of arachidonic acid to hydroperoxyeicosatetraenoic acid (HETE) in the process of synthesis of leukotrienes.

Lipoxygenase Pathway Enzymatic metabolism of arachidonic acid and synthesis of leukotrienes.

Lipoyl Dehydrogenase Synonym of dihydrolipoamide dehydrogenase.

Lipuria The presence of fat in the urine.

Liquaemin Sodium A trade name for heparin sodium, an anticoagulant.

Liquefacient Agent that promotes liquefaction.

Liquefaction Conversion of a solid or a gas to liquid.

Liqui-Char A trade name for activated charcoal.

Liquid Chromatography Referring to liquid-liquid, liquid-solid, paper, thin layer, ion exchange, and gel chromatography.

Liquid Liquid Chromatography A type of partition chromatography in which the mobile phase is a liquid and the stationary phase is solid that is coated with a liquid.

Liquid Gel Chromatography A type of chromatography in which the stationary phase is a gel.

Liquid Nitrogen Nitrogen in liquid state, it has a boiling point of –196°C and is used as a freezing and cooling agent.

Liquid Oxygen Oxygen in liquid state, it has a boiling point of –1823°C.

Liquid Pred A trade name for prednisone, a glucocorticoid hormone.

Liquid Scintillation Counter An instrument used for the detection of radioactivity in which a radioactive sample is dissolved or suspended in a solution containing fluorescent substances that emit a pulse of light when struck by radiation and the number of light pulses are detected and counted electronically.

Liquid Solid Chromatography A chromatographic technique in which the mobile phase is a liquid and the stationary phase is a solid (also known as adsorption chromatography).

Liquid Tumor Referring to any tumor of the circulating cells of the blood.

Liquiprin A trade name for the antipyretic and analgesic agent acetaminophen, which blocks generation of pain impulses.

Lisinopril (mol wt 442) An antihypertensive agent that inhibits angiotensin-converting enzyme, preventing conversion of angiotensin I to angiotensin II. It reduces sodium and water retention and blood pressure.

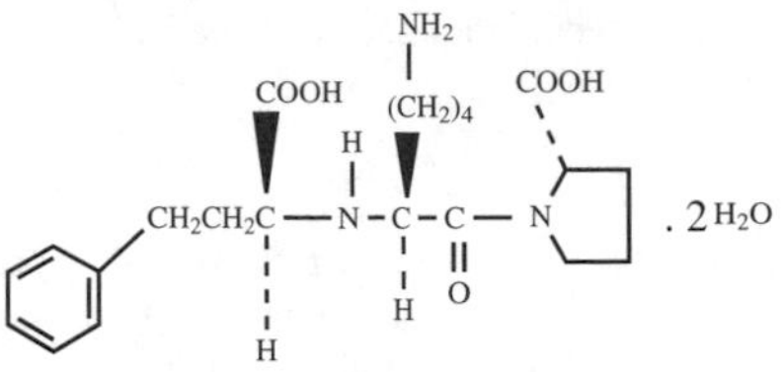

Lissamine Green B (mol wt 577) A biological dye.

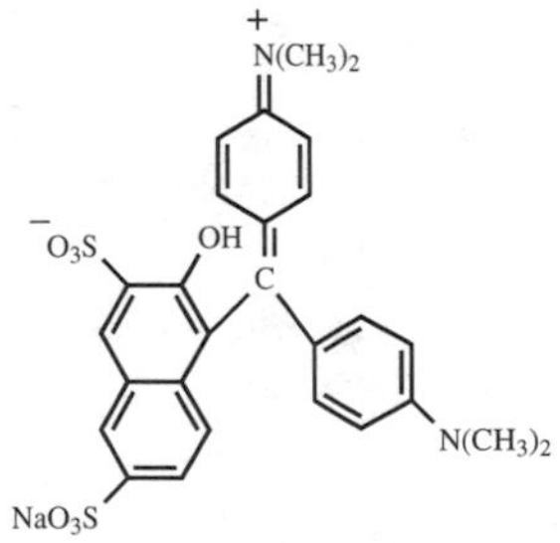

Listeria A genus of small, Gram-positive, motile, asporogenous, aerobic, facultative anaerobic bacteria.

Listeriosis A disease of humans and animals caused by *Listeria monocytogenes*.

Lisuride (mol wt 338) An antimigraine and a prolactin inhibitor.

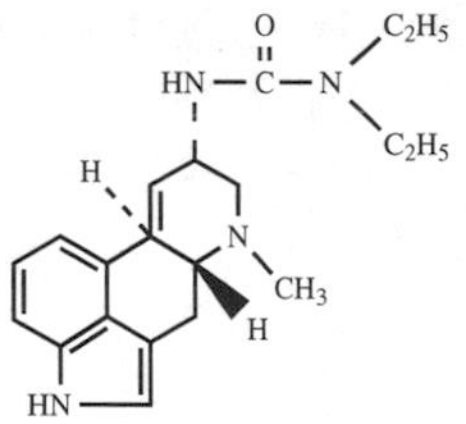

Liter A unit of volume equal to 1000 cm^3 or 1000 ml.

Lith- A prefix denoting stone or calculus.

Lithagogue An agent that promotes the removal of stones or calculi.

Lithague An agent that promotes dislodging of a calculus or stone.

Lithane A trade name for lithium carbonate, which alters chemical transmitters in the CNS.

Lithiasis Formation of stones or calculi from mineral salts in the hollow organs or ducts of the body, e.g., biliary or urinary duct.

Lithicarb A trade name for lithium carbonate, which alters chemical transmitters in the CNS.

Lithium A chemical element with atomic weight 7, valence 1.

Lithium Carbonate (mol wt 74) A psychotherapeutic agent that alters the chemical transmitter in the CNS.

Li_2CO_3

Lithium Citrate (mol wt 210) A psychotherapeutic agent that alters chemical transmitters in the CNS that is used for the treatment of depressive disorders.

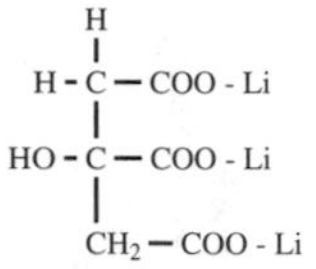

Lithizine A trade name for lithium carbonate, a psychotherapeutic agent that alters chemical transmitters in the CNS.

Litho- A prefix denoting stone or calculus.

Lithobid A trade name for lithium carbonate, a psychotherapeutic agent that alters chemical transmitters in the CNS.

Lithocholic Acid (mol wt 377) A bile acid that causes calcium release from the endoplasmic reticulum.

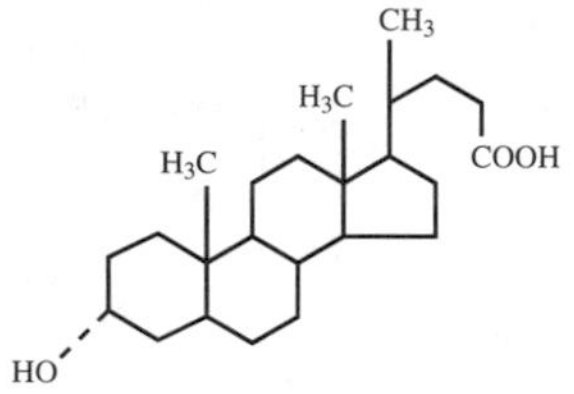

Lithodialysis The dissolution of stones or calculus of the bladder.

Lithogenic Capable of promoting the formation of stones or calculus.

Lithonate A trade name for lithium carbonate, a psychotherapeutic agent that alters chemical transmitters in the CNS.

Lithostat A trade name for acetohydroxamic acid, which prevents formation of renal stones by inhibiting bacterial urease activity.

Lithotabs A trade name for lithium carbonate, a psychotherapeutic agent that alters chemical transmitters in the CNS.

Lithotomy Surgical removal of stones or calculus from the urinary tract.

Lithotrophs An organism that obtains energy from the oxidation of inorganic matter, e.g., use of an inorganic compound as electron donor.

Litmus A substance obtained from lichens that is used as a pH indicator.

Litmus Paper A paper impregnated with litmus that is used for the test of pH.

Little Drop Technique A method for isolating single cells in which a drop of cellular suspension containing a single cell is transferred with a capillary pipette to an appropriate culture medium.

Little t Antigen or Small t Antigen One of the three gene products in polyomavirus that acts coordinately for transformation of rat embryo fibroblasts. The three antigens are large T antigen (mol wt 105,000), middle T antigen (mol wt 56,000) and small t antigen (mol wt 22,000).

LIVBP Abbreviation for leucine/isoleucine/valine-binding protein.

Live Vaccine A vaccine that contains attenuated pathogenic microorganisms or viruses.

Liver Fluke Any of the trematodes that infects liver, e.g., *Clonorchis sinensis.*

Liver Function Test A test to evaluate various functions of the liver, e.g., functions of liver metabolism, excretion, galactose tolerance, levels of alkaline phosphatase, and bilirubin.

Livetins Proteins found in the egg yolk, e.g., α-livetin, β-livetin, and γ-livetin.

Livostin A trade name for levocabastine, an antihistaminic agent.

Lixolin A trade name for theophylline, a bronchodilator that inhibits phosphodiesterase and increases the concentration of cAMP.

LK Abbreviation for lombricine kinase.

LL Abbreviation for lymphoblastic lymphoma.

LLC Abbreviation for 1. Lewis lung carcinoma; 2. liquid liquid chromatography.

LLD Factor Referring to vitamin B_{12}.

LLDH Abbreviation for liver lactate dehydrogenase.

LM Abbreviation for lipomannan.

LMA Abbreviation for liver membrane auto-antibody.

LMCT Abbreviation for ligand to metal charge transfer.

LMF Abbreviation for leukocyte mitogenic factor.

LMM Abbreviation for light meromyosin, the rodlike portion of the myosin heavy chain that is involved in the lateral interactions with other LMM to form a thick filament of striated muscle.

LM-PCR Abbreviation for ligation-mediated PCR (polymerase chain reaction).

LMWD Abbreviation for low molecular weight dextran.

LMWDX (LMWDx) Abbreviation for low molecular weight dextran.

lmw-UPA Abbreviation for low molecular- mass urokinase plasminogen activator.

Ln Natural logarithm.

LNA Abbreviation for linolenic acid.

LNPF Abbreviation for lymph node permeability factor.

Lobenzarit (mol wt 292) An antirheumatic agent.

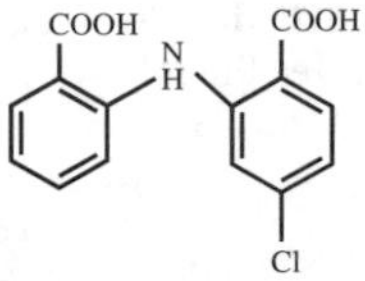

Lobomycosis Infection caused by fungus *Loboa loboi.*

Lock and Key Model A model describing the binding of substrate to an enzyme such that the active site on the enzyme and the substrate match each other as a key fits into a lock.

Locke's Solution A balanced salt solution containing NaCl, KCl, $CaCl_2$, $NaHCO_3$ and glucose.

Locoid A trade name for hydrocortisone.

Locus The site of a given gene on a chromosome.

Lodine A trade name for etodolac (ultradol), an anti-inflammatory agent.

Lodoxamide (mol wt 312) An anti-allergic agent.

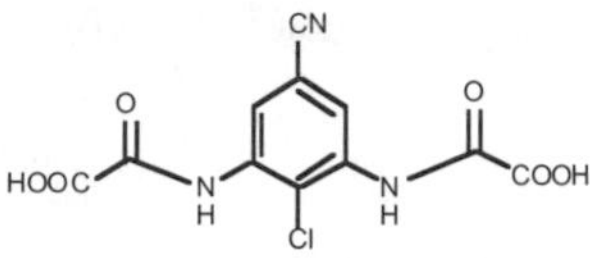

Lodrane A trade name for theophylline, a bronchodilator that inhibits phosphodiesterase and increases the concentration of cAMP.

Loeffler's Medium A medium that contains sterile serum and nutrient broth used to culture *Corynebacterium diphtheriae* in the diagnosis of disease.

Loestrin A trade name for a combination drug containing norethindrone acetate and ethinyl estradiol, used as an oral contraceptive agent.

Lofene A trade name for diphenoxylate hydrochloride, an antidiarrheal agent that inhibits mobility and propulsion and diminishes secretion in the GI tract.

Lofentanil (mol wt 409) A narcotic analgesic agent.

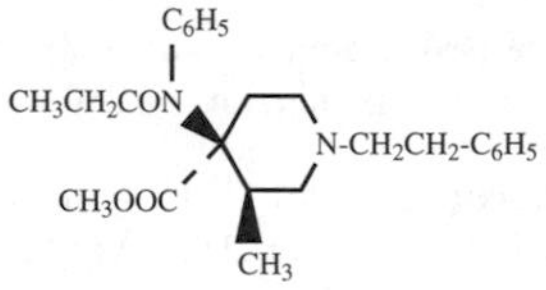

Lofepramine (mol wt 419) A psychotropic agent and antidepressant.

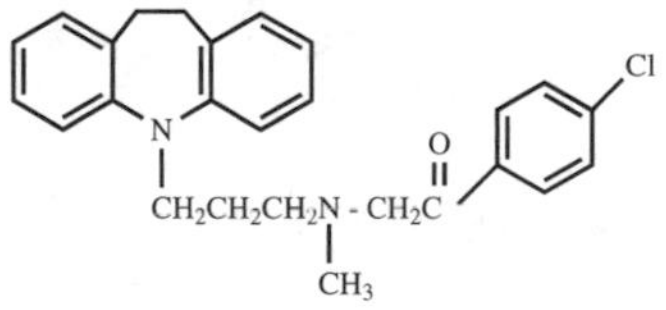

Lofexidine (mol wt 259) An antihypertensive and vasoactive agent.

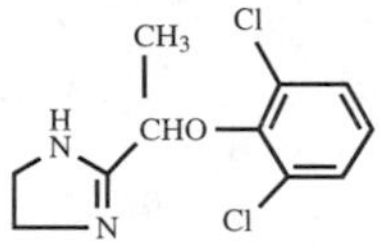

Loflucarban (mol wt 315) An antifungal agent.

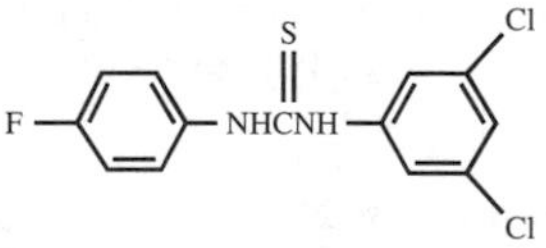

Log Phase of Growth The exponential cell growth characterized by rapid dividing cells.

Loganin (mol wt 390) An intermediate in the biosynthesis of indole alkaloids.

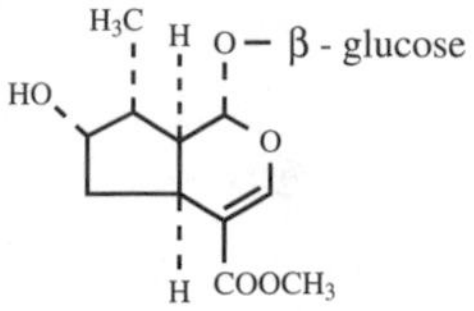

Logen A trade name for diphenoxylate hydrochloride, an antidiarrheal agent that inhibits motility and propulsion and diminishes secretion in the GI tract.

Logoderm A trade name for alclometasone diproprionate, a corticosteroid.

-logy A suffix denoting field of study.

Lohmann Reaction Referring to the reaction catalyzed by creatine kinase.

Lohmann's Enzyme Referring to creatine kinase.

Loiasis Infection by nematodes of the genus *Loa*.

Lomasome Membranous structure located between the plasma membrane and cell wall of fungi.

Lomefloxacin (mol wt 351) An antibacterial agent that inhibits bacterial DNA gyrase.

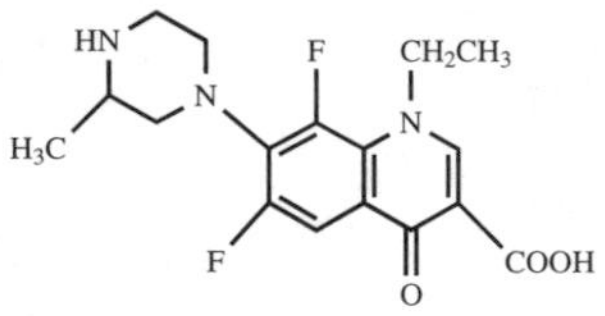

Lomerizine (mol wt 469) An antimigraine.

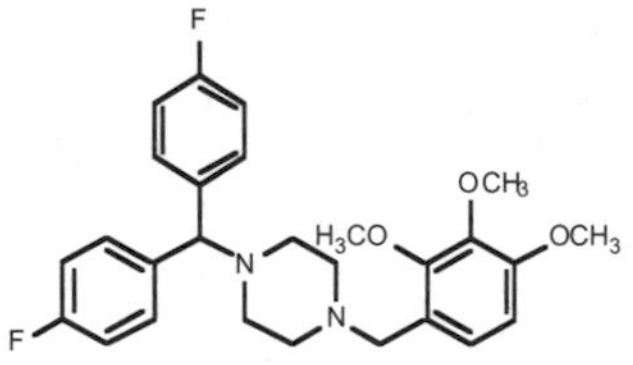

Lomine A trade name for dicyclomine hydrochloride, an anticholinergic agent.

Lomotil A trade name for diphenoxylate hydrochloride, an antidiarrheal agent that inhibits motility and propulsion and diminishes secretion in the GI tract.

Lomustine (mol wt 234) An antineoplastic and alkylating agent that cross-links strands of DNA and interferes with RNA transcription, causing an imbalance of growth and leading to cell death.

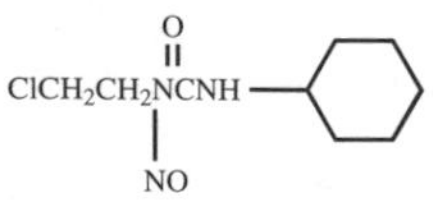

Lonavar A trade name for oxandrolone, an anabolic steroid that promotes tissue-building processes.

Lonazolac (mol wt 313) An anti-inflammatory agent.

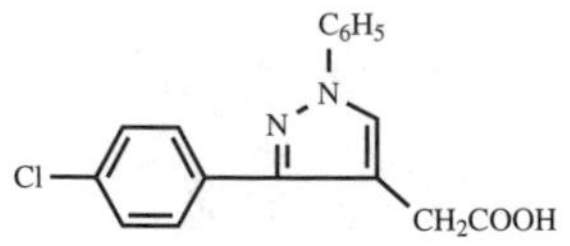

Long Chain Acyl-CoA Dehydrogenase The enzyme that catalyzes the following reaction:

Acyl-CoA + acceptor
⇅
Dehydroacyl-CoA + reduced acceptor

Long Chain Alcohol Dehydrogenase The enzyme that catalyzes the following reaction:

A long chain alcohol + 2 NAD^+ + H_2O
⇅
A long chain carboxylate + 2 NADH

Long Chain Alcohol Oxidase The enzyme that catalyzes the following reaction:

2 Long chain alcohol + O_2
⇅
2 Long chain aldehyde + 2 H_2O

Long Chain Aldehyde Dehydrogenase The enzyme that catalyzes the following reaction:

A long chain aldehyde + NAD^+
⇅
A long chain carboxylate + NADH

Long Chain Enoyl-CoA Hydratase The enzyme that catalyzes the following reaction:

3(S)-3-Hydroxyacyl-CoA
⇅
trans-2-enoyl-CoA + H_2O

Long Chain Fatty Acid-CoA Ligase The enzyme that catalyzes the following reaction:

ATP + long chain carboxylate + CoA
⇅
AMP + PPi + acyl-CoA

Long Chain Fatty Acid-CoA Reductase The enzyme that catalyzes the following reaction:

A long chain aldehyde + $NADP^+$
⇅
A long chain carboxylate + NADPH

Long Chain Fatty Acid Thiokinase The enzyme that catalyzes the formation of acyl-CoA from fatty acids with more than 12 carbons.

Long-Terminal Repeats The identical DNA sequences (several hundred nucleotides long) found at either ends of transposons and the proviral DNA, formed by reverse transcription of retroviral RNA. The long terminal repeats play an essential role in integration of the transposon or provirus into the host DNA.

Lonidamine (mol wt 321) An antineoplastic agent.

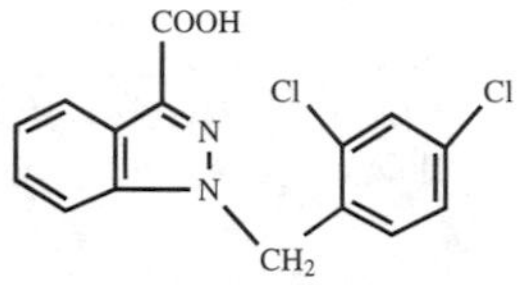

Loniten A trade name for minoxidil, an antihypertensive agent that promotes arteriolar vasodilation.

Lonox A trade name for diphenoxylate hydrochloride, an antidiarrheal agent that inhibits motility and propulsion and diminishes secretion in the GI tract.

Loop-Back DNA Synonym of hairpin DNA.

Lo/Ovral A trade name for a combination drug containing ethinyl estradiol and norethindrone acetate, used as an oral contraceptive to inhibit ovulation.

Lophotoxin (mol wt 416) A neuromuscular toxin.

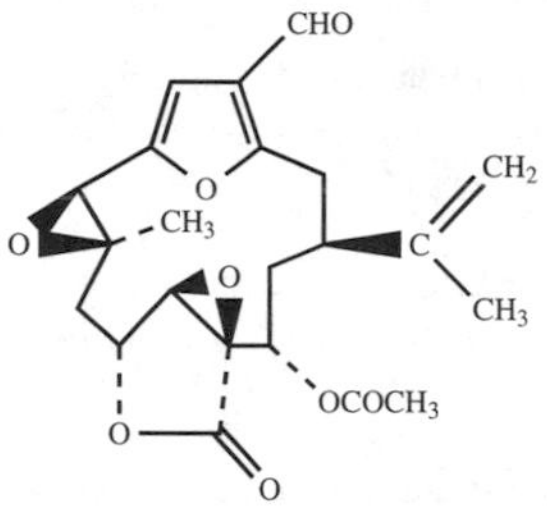

Lophotrichous A cell that has a cluster of flagella at one or both ends.

Lopid A trade name for gemfibrozil, an antilipemic agent that lowers glyceride levels and increases the concentration of high-density lipoprotein.

LOPP Abbreviation for a combination drug containing leukeran, oncovin, procarbazine, and prednisone.

Lopresor A trade name for metoprolol tartrate, an antihypertensive agent that blocks cardiac beta receptors and decreases renin secretion.

Loprox A trade name for ciclopirox olamine, a local anti-infective agent (antifungal).

Lopurin A trade name for allopurinol, an antigout agent.

Lorabid A trade name for loracarbef, which inhibits bacterial cell wall synthesis.

Loracarbef (mol wt 350) An antibiotic that inhibits bacterial cell wall synthesis.

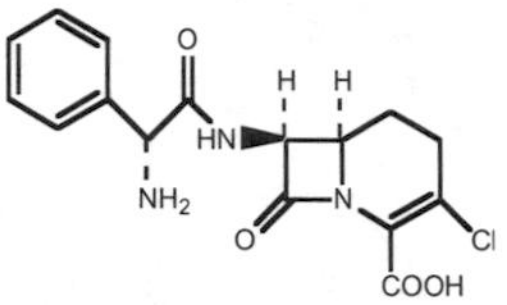

Lorajmine (mol wt 413) A cardiac depressant.

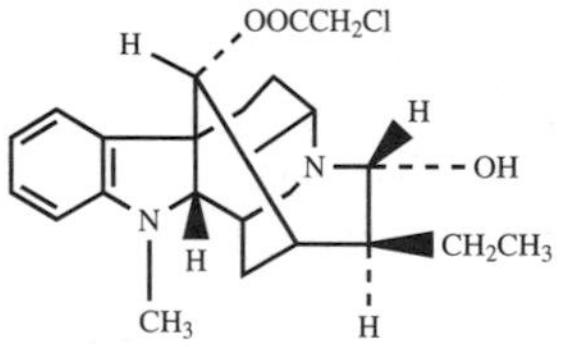

Loratadine (mol wt 383) An antihistaminic agent that competes with histamine receptors.

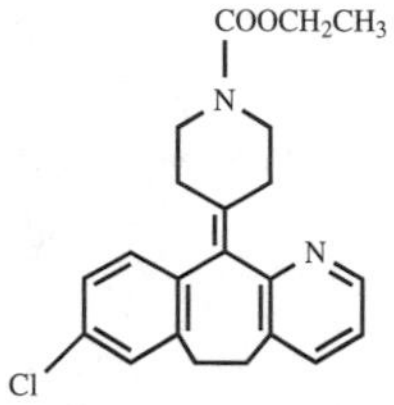

Loraz A trade name for lorazepam, an antianxiety agent.

Lorazepam (mol wt 321) An anxiolytic agent that depresses the CNS.

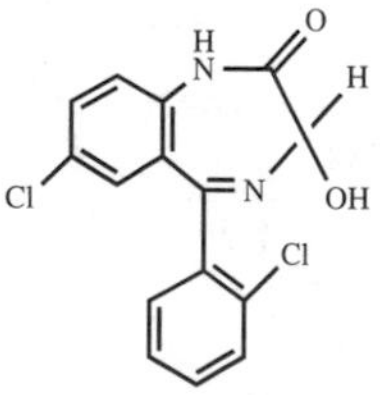

Lorcet-HD A trade name for a combination drug containing acetaminophen and hydrocodone bitartrate, used as an analgesic agent.

Lorelco A trade name for probucol, which inhibits cholesterol transport from the intestine and prevents oxidation of low-density lipoprotein.

Loricrin The major protein of the cornified cell envelope of the terminally differentiated epidermal keratinocytes.

Lormetazepam (mol wt 335) A sedative and hypnotic agent.

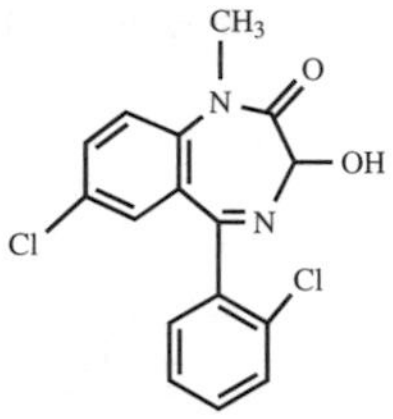

Loroxide A trade name for benzoyl peroxide lotion, which possesses antimicrobial and comedolytic activity.

Lortab A trade name for a combination drug containing acetaminophen and hydrocodone bitartrate, used as an analgesic agent.

Losartan (mol wt 423) An antihypertensive agent.

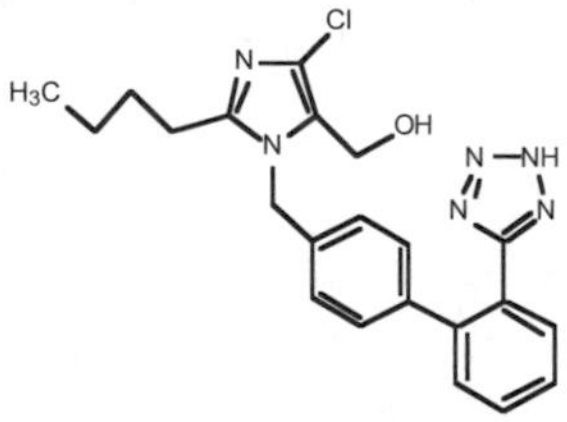

Lotensin A trade name for benazepril hydrochloride, an antihypertensive agent that inhibits angiotensin-converting enzyme and thus preventing the conversion of angiotensin I to angiotensin II.

Loteprednol Etabonate (mol wt 467) An anti-inflammatory agent.

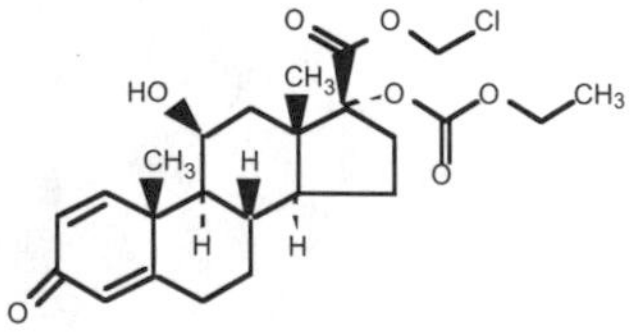

Lotrifen (mol wt 280) An abortifacient.

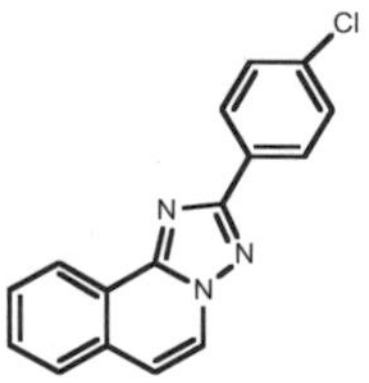

Lotrimin A trade name for clotrimazole, which alters fungal cell wall permeability.

Lotrisone Cream An anti-infective cream containing clotrimazole and beta-methasone dipropionate.

Lo-Trol A trade name for diphenoxylate hydrochloride, an antidiarrheal agent that inhibits motility and propulsion and diminishes secretion in the GI tract.

Lovastatin (mol wt 405) An antilipemic agent that inhibits 3-hydroxy-3-methylglutaryl-CoA (HMG-CoA) reductase, preventing cholesterol biosynthesis.

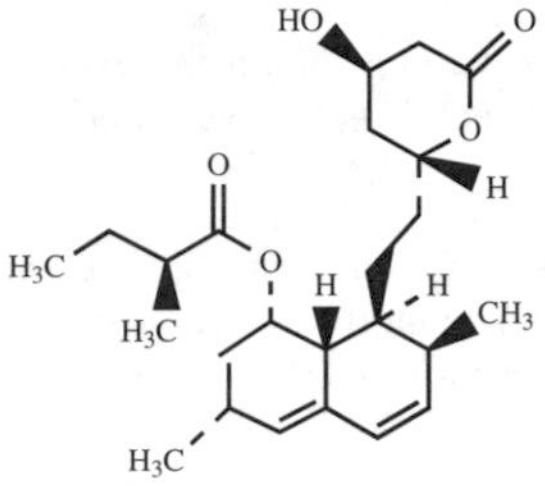

Low-Density Lipoprotein The lipoprotein that has a density of 1.006-1.063 g/ml.

Low-Density Lipoprotein Receptor A cell-surface protein that mediates endocytosis of LDL by cells. Genetic defects in LDL-receptors lead to abnormal levels of LDL in the serum (hypercholesterolemia).

Low-Dose Tolerance A transient and incomplete state of tolerance induced by subimmunogenic doses of soluble antigen.

Low-Energy Phosphate Compound A phosphorylated compound that yields a relatively low standard free energy on hydrolysis.

Low-Quel A trade name for diphenoxylate hydrochloride, an antidiarrheal agent that inhibits motility and propulsion and diminishes secretion in the GI tract.

Lowry's Method A sensitive colorimetric method for the determination of protein based upon Folin-Ciocateau reaction, which produces a blue color upon treatment of protein with complex phosphomolybdotungstic acid reagent.

Lowsium A trade name for magaldrate (aluminum-magnesium complex), which reduces total acid load in the GI tract.

Loxapac A trade name for loxapine hydrochloride, an antipsychotic agent that blocks postsynaptic dopamine receptors in the brain.

Loxapine (mol wt 328) An anxiolytic agent that blocks postsynaptic dopamine receptors in the brain.

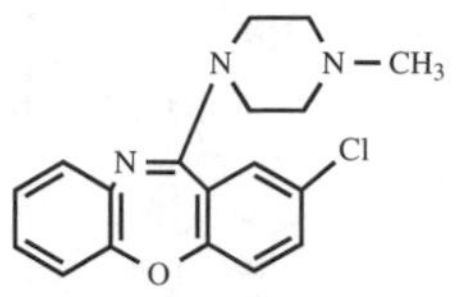

Loxitane A trade name for the antipsychotic agent loxapine succinate, which blocks the dopamine receptors in the brain.

Loxoprofen (mol wt 246) An anti-inflammatory and analgesic agent.

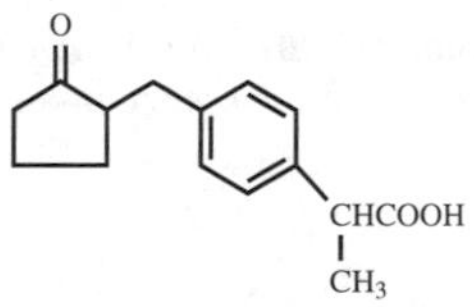

Lozide A trade name for indapamide, a diuretic drug that inhibits sodium reabsorption.

LPF Abbreviation for 1. leukocytosis-promoting factor; 2. lymphocytosis-promoting factor.

LPG Abbreviation for lipophosphoglycan.

LPH Abbreviation for lipotropic hormone.

L-PK Abbreviation for L-type pyruvate kinase.

LPL Abbreviation for lipoprotein lipase.

LPP Abbreviation for lipid phosphate phosphohydrolase.

LP-PLA$_2$ Abbreviation for lipoprotein-associated phospholipase A_2

LPR Abbreviation for late-phase reaction.

LPS Abbreviation for lipopolysaccharide.

LP(S$_2$) Symbol for an oxidized form of lipoic acid.

LP(SH)$_2$ Symbol for a reduced form of lipoic acid.

Lr Abbreviation for lawrencium, a chemical element.

LRF Abbreviation for luteinizing hormone releasing factor.

LRH Abbreviation for luteinizing hormone releasing hormone.

LRP Abbreviation for LDL receptor protein or LDL-related protein.

LRR Abbreviation for leucine-rich repeat.

LRS Abbreviation for lactated Ringer's solution.

LSA Abbreviation for 1. lipid-bound sialic acid; 2. lymphosarcoma.

LSB Abbreviation for Laemmli sample buffer used in gel electrophoresis.

LSC Abbreviation for 1. liquid scintillation counting; 2. liquid-solid chromatography.

LSCG Abbreviation for liquid-solid chromatography.

LSD Abbreviation for lysergic acid diethylamide, a hallucinogenic agent.

LSF Abbreviation for 1. leukocyte stimulating factor; 2. lymphocyte-stimulating factor.

LSGP Abbreviation for leukocyte sialoglycoprotein.

LSH Abbreviation for 1. leukocyte stimulating hormone; 2. lymphocyte-stimulating hormone; 3. lutein-stimulating hormone.

L-Shaped Structure Referring to the three-dimensional structure of tRNA.

LSM Abbreviation for lymphocyte separation medium.

LSSA Abbreviation for lipid-soluble secondary antioxidant.

L-Strand Abbreviation for light strand.

LT Abbreviation for 1. leukotriene; 2. levothyroxine; 3. lymphotoxin.

LTA Abbreviation for 1. lipoteichoic acid and 2. leukotriene A.

LTA$_4$ Abbreviation for leukotriene A_4.

LTB Abbreviation leukotriene B.

LTB$_4$ Abbreviation for leukotriene B_4.

LTBP Abbreviation for latent transforming growth factor B-binding protein.

LTC Abbreviation for leukotriene C.

LTC$_4$ Abbreviation for leukotriene C_4.

LTD Abbreviation for leukotriene D.

LTD$_4$ Abbreviation for leukotriene D_4.

LTE Abbreviation for leukotriene E.

LTF Abbreviation for 1. lymphocyte-transforming factor; 2. lipotropic factor.

LTPP Abbreviation for lipothiamide pyrophosphate.

LTQ Abbreviation for lysine-tyrosine quinone.

LTR Abbreviation for long terminal repeats in retrovirus.

LTT Abbreviation for 1. lymphoblast transformation test; 2. lymphocyte transformation test.

L-Type Structure Referring to the three-dimensional structure of tRNA.

Lu Abbreviation for lutetium, a chemical element.

Luc Abbreviation for luciferase.

Lucanthone Hydrochloride (mol wt 377) An anthelmintic (*Schistosoma*) agent.

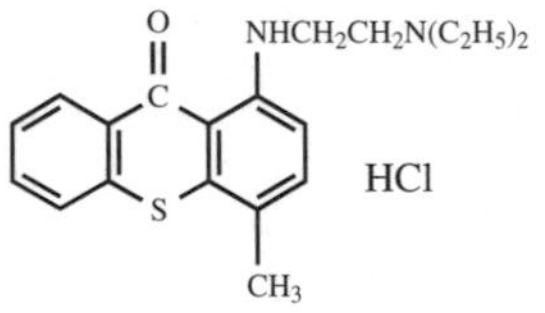

Lucensomycin (mol wt 708) An antifungal antibiotic produced by *Streptomyces lucensis*.

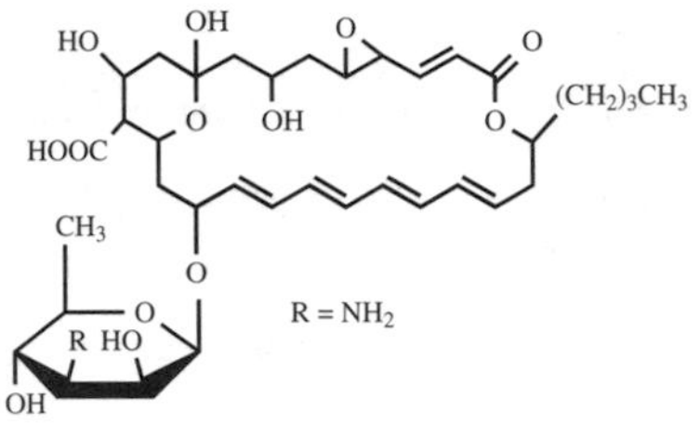

Lucifer Yellow CH Dilithium (mol wt 457) A highly fluorescent dye used for labeling protein.

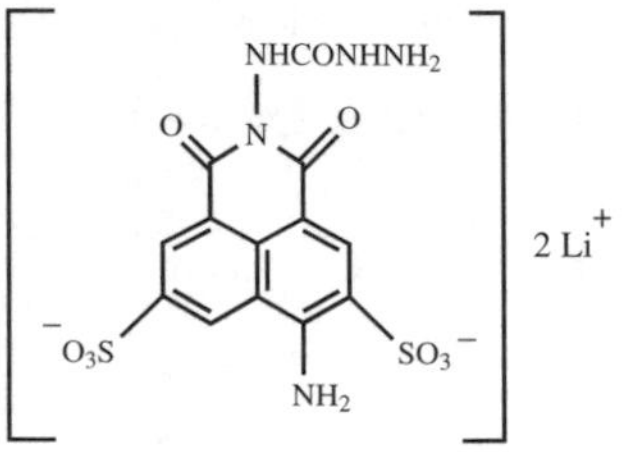

Lucifer Yellow VS Dilithium (mol wt 550) A highly fluorescent dye used for labeling protein.

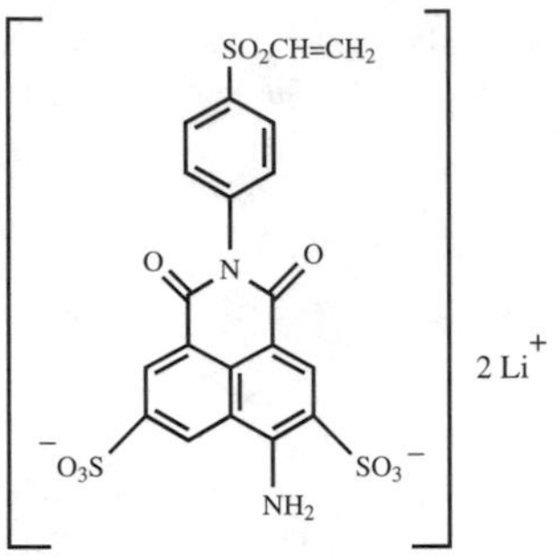

Luciferase An enzyme from firefly tails that catalyzes bioluminescence by oxidation of luciferin and production of visible light.

Luciferin (mol wt 280) A substrate for luciferase that produces bioluminescence upon oxidation by luciferase.

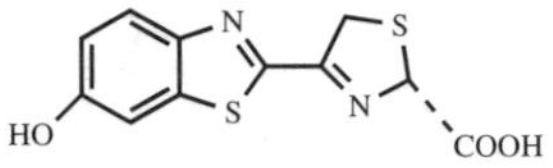

D-luciferin

Lucigenin (mol wt 511) A substrate for assaying the metabolic activation of leukocytes by chemiluminescence. It emits light upon oxidation by superoxide.

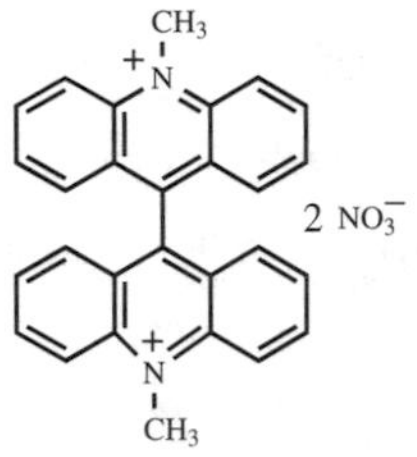

Lucrin A trade name for leuprolide acetate, an antineoplastic agent.

Ludiomil A trade name for maprotiline hydrochloride, an antidepressant that increases the concentrations of norepinephrine and serotonin in the CNS by blocking their reuptake.

Luffin A ribosome-inactivating protein.

Lufylin A trade name for dyphylline, a bronchodilator that inhibits phosphodiesterase and increases the concentration of cAMP.

Lugol's Solution A strong iodine solution (5% or 10% w/v).

Lumen The internal cavity within a structure.

Lumichrome (mol wt 242) A fluorescent compound formed by photolysis of riboflavin in acidic solution.

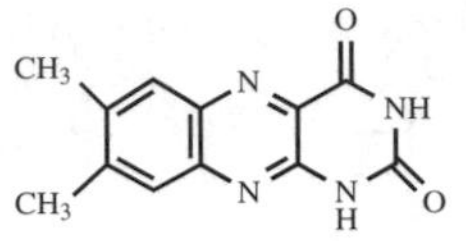

Lumicolchicine (mol wt 399) A fluorescent derivative of colchicine.

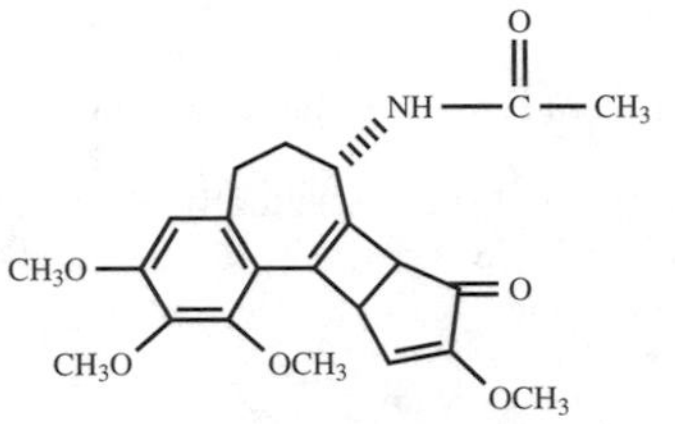

Lumiflavine (mol wt 256) A fluorescent compound formed from photolysis of riboflavin in basic solution.

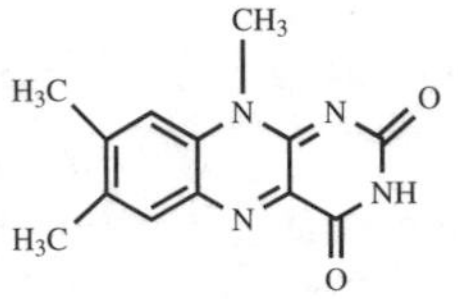

Luminal A trade name for phenobarbital, an anticonvulsant.

Luminal Protein Proteins that are retained within the cisternae of the rough endoplasmic reticulum.

Luminescence The emission of light from a chemical reaction or a physiological process.

Luminol (mol wt 177) A compound used as substrate in assaying the metabolic activation of leukocytes by chemiluminescence.

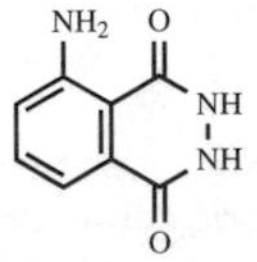

Luminophore 1. Any substance that emits light at room temperature. 2. Agent that promotes luminescence of the other organic compound.

Lumpectomy The surgical removal of a hard mass of tissue, e.g., removal of a breast tumor without removing large amounts of surrounding tissue.

Lung Surfactants Substances secreted by lung tissue that are essential for maintaining proper surface tension of alveoli, e.g., phosphatidyl choline.

Lunularic Acid (mol wt 258) A growth regulator for lower plants.

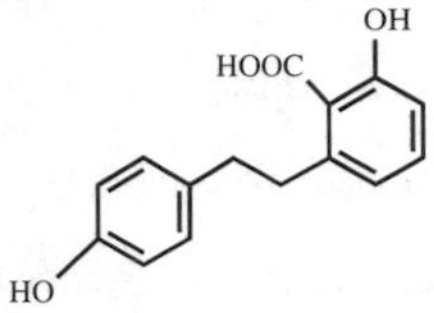

Lupinosis A mycotoxicosis caused by *Phomopsis leptostromiformis* and characterized by the appearance of jaundice and liver damage.

Lupron A trade name for leuprolide acetate, an antineoplastic agent.

Lupron Depot A trade name for leuprolide acetate, an antineoplastic agent.

Luprostiol (mol wt 445) A luteolytic agent.

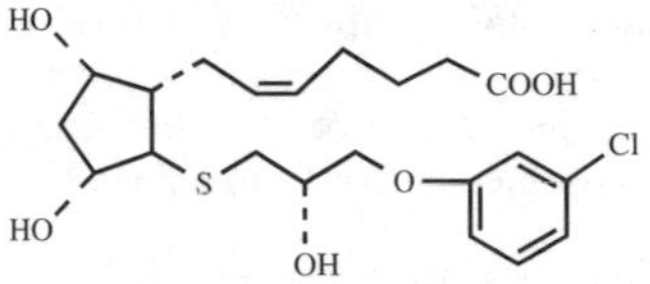

Lupulon (mol wt 415) An antimicrobial substance.

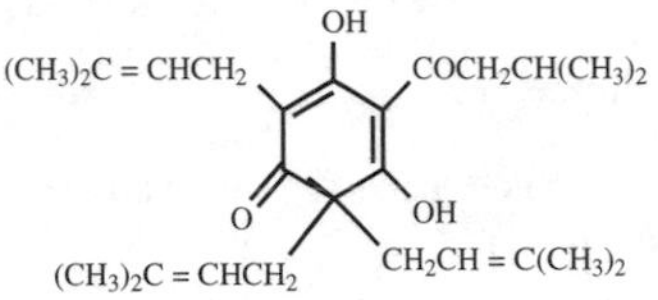

Lupus Erythematosus A disease caused by autoimmune reactions against a variety of self-components including DNA, RNA, and nuclear proteins, leading to the development of arthritis, glomerulonephritis, and other disorders.

Lurselle A trade name for probucol, which inhibits cholesterol transport from the intestine and prevents oxidation of low-density lipoproteins.

Luteal Pertaining to the corpus luteum of the female ovary.

Lutein The yellow colored material contained in egg yolk.

Luteinization 1. The changes that occur in an ovary after discharging a mature egg. 2. The process of the formation of corpus luteum.

Luteinizing Hormone A gonadotropin produced by the anterior pituitary. It stimulates the gonads to produce sex hormones.

Luteinizing Hormone Releasing Factor A peptide hormone released from the hypothalamus to stimulate the release of luteinizing hormone.

Luteochrome Referring to progesterone.

Luteolytic Inhibition or repression of corpus luteum.

Luteoma An ovarian tumor derived from corpus luteum.

Luteotropic Hormone Referring to prolactin, a hormone produced by the pituitary gland for the stimulation of milk production.

Luteotropin Referring to prolactin, a hormone produced by the pituitary gland for the stimulation of milk production.

Lutetium A chemical element with atomic weight 175, valence 3.

Lutheran Blood Group One of the blood group classification systems based on the presence of *Lu* antigens, which are encoded by the genes located on chromosome number 19 in humans.

Lutrepulse A trade name for the hormone gonadorelin acetate.

Lutropin Referring to luteinizing hormone.

LUV Abbreviation for large unilamellar vesicle.

Luvox A trade name for fluvoxamine maleate, an antidepressant.

Lux Proteins Proteins that mediate bioluminesence in marine bacteria luciferase system.

Luxury Genes The genes that encode luxury proteins.

Luxury Protein Proteins that are produced for the specific function of the cells and are not required for general cell maintenance, e.g., immunoglobulins of plasma cells.

LV Abbreviation for 1. lipovitellin; 2. leukemia virus.

LVB Abbreviation for a combination drug containing lomustine, vinblastine, and bleomycin.

LVP Abbreviation for lysine vasopressin.

Lvr Abbreviation for leucovorin.

LW Antigen Abbreviation for Landsteiner-Weiner antigen, the blood group substances used for classification of blood groups.

LX Abbreviation for local irradiation.

LY Agar Abbreviation for egg-yolk agar.

Lyapolate Sodium An anticoagulant.

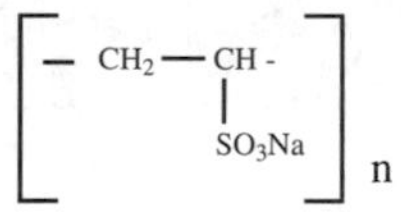

Lyases The enzymes that catalyze the nonhydrolytic cleavage of substrate, e.g., aldolases and decarboxylases.

Lyb Antigen Surface antigens of mouse B cells.

Lycin (mol wt 118) A naturally occurring amino acid found in *Lycium barbarum* (also known as glycine betaine).

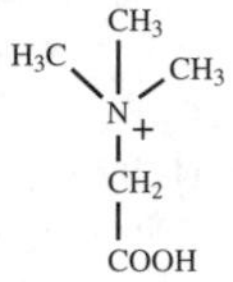

Lycogala A genus of slime mold (Myxomycete).

Lycomarasmin A peptide antibiotic produced by *Fussarium lycopersici.*

Lycopene (mol wt 537) Carotenoid occurring in ripe fruit, e.g., tomato.

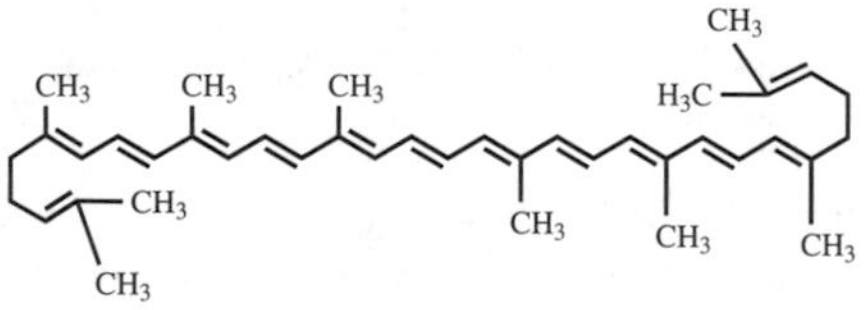

Lycophyll (mol wt 569) Carotenoid pigment from *Lycopersicum esculentum.*

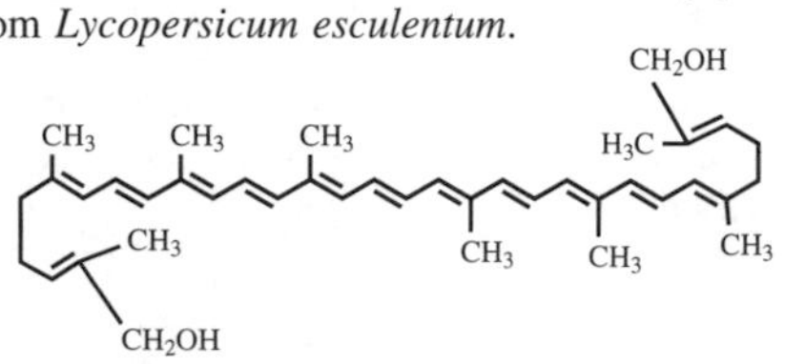

Lyme Disease A human disease caused by *Borelia burgdorferi,* transmitted by the bites of ticks and characterized by fever, swelling lymph nodes, muscular pain, and raised red spots on the skin.

Lymecycline (mol wt 602) A semisynthetic antibiotic related to tetracycline.

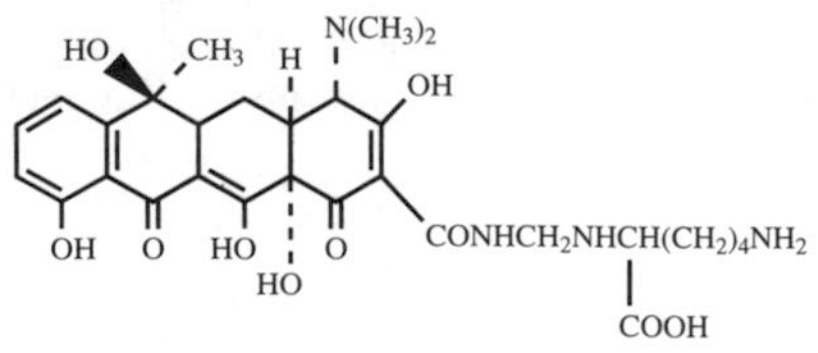

Lymerix A trade name for Lyme disease vaccine.

Lymph The colorless fluid that circulates through the vessels of the lymphatic system.

Lymph Gland See lymph node.

Lymph Nodes Small, secondary lymphoid organs consisting of lymphocytes, macrophages, and dendritic cells. Lymph nodes serve as filters through which foreign antigens or pathogens are trapped and lymphocytes activated.

Lymphadenectomy Surgical removal of lymph nodes.

Lymphadenitis Inflammation of lymph nodes.

Lymphadenopathy Pathological disorders of the lymph nodes.

Lymphagogue Agent that promotes the formation and flow of the lymph.

Lymphangiectasis Dilation of the lymphatic vessels.

Lymphangitis Inflammation of a lymphatic vessel.

Lymphatic Pertaining to lymph node or lymph vessel.

Lymphatic System A system of vessels and nodes that convey lymph in the vertebrate.

Lymphatic Tissue Tissues that are rich in lymphocytes.

Lymphectasia Dilatation of the lymphatic vessels.

Lympho- A prefix denoting lymph or lymphatic system.

Lymphoblast 1. Dividing lymphocytes. 2. Immature lymphocytes. 3. Antigen-stimulated lymphocytes.

Lymphoblastoid Cell Line A lymphocyte cell line that grows indefinitely in culture and derived from lymphocytes that are immortalized by fusion with a cancer cell.

Lymphocele A cystic mass that contains lymph.

Lymphocyte White blood cells derived from stem cells of the primary lymphoid organs, (e.g., bone marrow and thymus) and are responsible for mediation of immune responses. Two main classes of lymphocytes are T lymphocytes and B lymphocytes.

Lymphocyte Activation The change that occurs upon exposure of lymphocytes to antigen or mitogen and leads to the differentiation and production of lymphoblasts and effector cells.

Lymphocyte-Defined Antigens The histocompatibility antigens that are identified primarily by the mixed lymphocyte reaction. They are class-II histocompatibility antigens.

Lymphocyte Immune Globulin The antilymphocyte globulin that inhibits cell-mediated immunity by either altering T cell function or eliminating antigen-reactive T cells.

Lymphocyte Transformation See lymphocyte activation.

Lymphocytopenia The abnormally low numbers of lymphocytes in the blood due to a blood disorder (e.g., infectious mononucleosis).

Lymphocytosis The production of a large number of lymphocytes in the blood or in the tissue fluid.

Lymphogranuloma Venereum A sexually transmitted disease caused by *Chlamydia*.

Lymphoid Pertaining to lymph or lymphatic tissue.

Lymphoid Cell Cells of lymphoid lineage, e.g., lymphocytes.

Lymphoid Organs Organs of vertebrates that contain large numbers of lymphocytes, e.g., primary lymphoid organs (bone marrow and thymus) and secondary lymphoid organs (lymph nodes and spleen).

Lymphoid Tissue See lymphatic tissue.

Lymphokine Substances produced by lymphocytes (chiefly T cells) following antigen or mitogen stimulation that are responsible for mediation of immune responses.

Lymphokine-Activated Killer Cells Killer cells or natural killer cells (NK cells) that exhibit enhanced killing capability toward the target cells following activation by interleukin.

Lympholysis Destruction or lysis of lymphocytes.

Lymphoma A malignant neoplastic disorder of lymphoid tissue that produces a distinct tumor mass, e.g., Burkitt's lymphoma.

Lymphon An entire immune system.

Lymphopenia Abnormally low numbers of lymphocytes in the blood.

Lymphopoiesis Generation or production of lymphocytes.

Lymphotoxins Cytotoxic products produced by lymphocytes, e.g., tumor necrosis factor.

LYN A proto-oncogene encoding a tyrosine kinase, it belongs to the *src* family.

Lynestrenol (mol wt 284) An oral contraceptive.

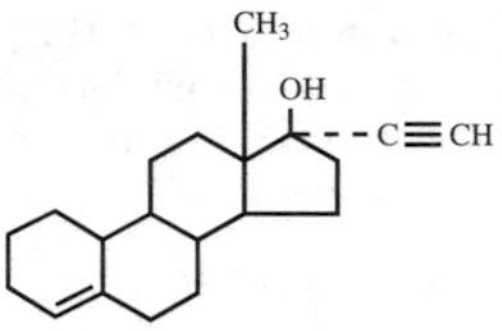

Lyophilization The process of drying a sample by rapidly freezing and then dehydrating under low temperature and a high vacuum in which water is removed by sublimation (also known as freeze drying).

Lyophilized Culture A desiccated culture prepared by freeze-drying procedure.

Lyophilizer Synonym of freeze-drier.

Lyophobic Solvent rejecting.

LYP Agar Abbreviation for lactose-yeast-peptone agar.

Lypressin (mol wt 1056) An anti-tussive peptide and a vasopressor that increases the permeability of renal tubular epithelium to adenosine monophosphate and water.

LYPS Abbreviation for lysophospholipase.

Lys Abbreviation for the amino acid lysine.

Lysate The product of cell lysis.

Lyseric Acid Diethylamide (mol wt 321) A potent hallucinogenic substance that induces schizophrenic-like state in human.

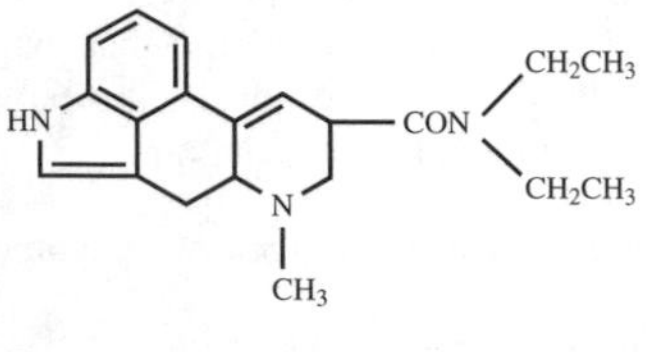

Lysin Antibody or substance that causes lysis of cells under appropriate conditions.

Lysine (mol wt 146) A basic amino acid.

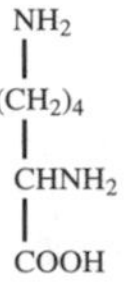

Lysine Acetylsalicylic Acid (mol wt 326) An analgesic, antipyretic, and anti-inflammatory agent.

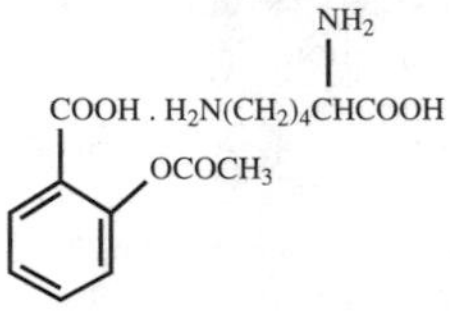

Lysine *N*-Acetyltransferase The enzyme that catalyzes the following reaction:

Acetyl phosphate + L-lysine
⇅
Pi + acetyllysine

Lysine 5,6-Aminomutase The enzyme that catalyzes the following reaction:

D-Lysine ⇌ 5,6-Diaminohexanoate

Lysine Carbamoyl Transferase The enzyme that catalyzes the following reaction:

Carbamoyl phosphate + lysine
⇅
Pi + homocitruline

Lysine Carboxylase The systematic name for lysine decarboxylase.

Lysine Decarboxylase The enzyme that catalyzes the following reaction:

L-Lysine ⇌ Cadaverine + CO_2

Lysine Dehydrogenase The enzyme that catalyzes the following reaction:

Lysine + NAD^+
⇅
Didehydropiperidine carboxylate + NH_3 + NADH

Lysine Lactamase The enzyme that catalyzes the following reaction:

L-Lysine 1,6-lactam + H_2O ⇌ L-Lysine

Lysine 2-Monooxygenase The enzyme that catalyzes the following reaction:

Lysine + O_2
⇅
5-Aminopentanamide + CO_2 + H_2O

Lysine 6-Monooxygenase The enzyme that catalyzes the following reaction:

Lysine + O_2
$\rightleftharpoons$
Hydroxy-L-lysine + H_2O

Lysine Oxidase The enzyme that catalyzes the following reaction:

Lysine + O_2 + H_2O
$\rightleftharpoons$
6-Amino-2-ketohexanoate

Lysine Oxoglutarate Reductase Synonym of saccharopine dehydrogenase (NAD+, L-lysine-forming).

Lysine Oxygen 2-Oxidoreductase The systematic name for L-lysine oxidase.

Lysine Pyruvate Transaminase The enzyme that catalyzes the following reaction:

Lysine + pyruvate
$\rightleftharpoons$
Alanine + aminoadipate semialdehyde

Lysine Racemase The enzyme that catalyzes the following reaction:

L-Lysine $\rightleftharpoons$ D-Lysine

Lysine tRNA Ligase The enzyme that catalyzes the following reaction:

ATP + L-lysine + $tRNA^{lys}$
$\rightleftharpoons$
AMP + PPi + L-lysyl-$tRNA^{lys}$

Lysine Vasopressin A peptide vasopressin found in hogs in which the eighth amino acid residue is replaced by lysine.

Lysine Vasotocin A vasotocin found in hogs in which the eighth amino acid residue is lysine.

Lysis Rupture of cell membranes and loss of cytoplasm.

-lysis A prefix meaning to break or to dissolve.

Lysodren A trade name for mitotane, an antineoplastic agent.

Lysogen A bacterial cell that carries a prophage.

Lysogenic Referring to a bacterium that carries a prophage.

Lysogenic Cell A bacterial cell that carries a prophage in its chromosome.

Lysogenic Conversion The change in phenotypic character of bacteria resulting from the integration of a prophage, e.g., the conversion of nontoxin-producing bacteria to toxin-producing bacteria by the action of a prophage.

Lysogenic Cycle A process in which the viral genome becomes incorporated into the bacterial host chromosome as a prophage. In the lysogenic cycle, no viral replication occurs.

Lysogenic Viruses Referring to viruses that are capable of establishing lysogeny in bacterial hosts.

Lysogeny A state in which phage genome becomes integrated into the host chromosome. The integrated phage DNA or genome is termed a prophage.

Lysol Trade name for a disinfectant consisting of a mixture of *o*-, *m*-, and *p*-cresols solubilized with an excess of a potassium soap.

Lysolecithin A lecithin that lacks one fatty acid and an active hemolytic agent.

Lysolecithinase Synonym of lysophospholipase.

Lysolecithin Acylmutase The enzyme that catalyzes the following reaction:

2-Lysolecithin $\rightleftharpoons$ 3-Lysolecithin

Lysophosphatidylcholine Acylhydrolase Synonym of lysophospholipase.

Lysophospholipase The enzyme that catalyzes the following reaction:

2-Lysophosphatylcholine + H_2O
$\rightleftharpoons$
Glycerolphosphocholine + a carboxylate

Lysopine (mol wt 204) A naturally occurring amino acid found in calf thymus histone.

$H_2N(CH_2)_4CHCOOH$
|
NH
|
$H_3CCHCOOH$

Lyso-PtdCho Abbreviation for lysophosphatidylcholine.

Lysosomal Carboxypeptidase A synonym of serine-type carboxypeptidase.

Lysosomal Enzymes A group of degradative enzymes present in the lysosome, e.g., acid phosphatase, protease, and DNAse.

Lysosomal Storage Disease Diseases resulting from deficiency of one or more specific lysosomal enzymes and characterized by the undesirable accumulation of excessive amounts of specific substances. This condition may lead to neurological disorder.

Lysosome A cell organelle that contains hydrolytic enzymes involved in autolysis and degradation of foreign materials engulfed by phagocytes.

Lysosome-Phagosome A structure formed by fusion of phagosome with lysosome for the release of lysosomal enzymes into the phagosome and digestion of phagocytized materials. Also known as a phagolysosome.

Lysostaphin A protein antibiotic complex produced by *Staphylococcus staphyloliticus* with highly specific lytic activity against other *Staphylococcus species*.

Lysozyme An enzyme that catalyzes the hydrolysis of the 1,4-β-linkage between *N*-acetylmuramic acid and *N*-acetylglyosamine in the mucopeptide or mucopolysaccharide of bacterial cell wall.

Lyspafen A trade name for difenoxin hydrochloride, an antidiarrheal agent.

Lysopeptide The N-terminal sequence of a protein involved in transport of synthesized protein through the membrane.

Lyssa Virus A virus of the family Rhabdoviridae, a type of rabies.

Lysyl Endopeptidase The enzyme that catalyzes the preferential cleavage of internal peptide bonds involving the carboxyl group of lysine.

Lysyl-tRNA Synthetase See lysine tRNA ligase.

Lyt Antigen Cell surface glycoproteins on mouse lymphocytes.

Lytic Pertaining to lysis or a lysin.

Lytic Bacteriophage A bacteriophage that infects and replicates in the host cell leading to the release of virion and lysis of the host cell.

Lytic Complex The cytolytic complex formed from polymerization of components of complement (also known as membrane attack complex).

Lytic Cycle A type of viral replication cycle in which the production and release of new virions leads to the lysis of host cells.

Lytic Growth A mode of viral growth in which the production and release of new virions leads to the lysis of host cells.

Lytic Infection Viral infection leading to the lysis of host cells.

Lytic Phage Bacteriophages that cause lysis of host cells.

Lytic Viruses Viruses that cause lysis of host cells.

Lyxoflavine (mol wt 376) A growth-promoting agent and analog of riboflavin.

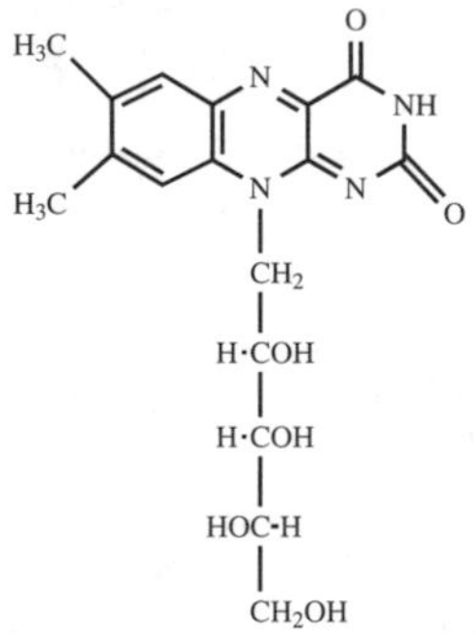

D-Lyxose (mol wt 150) A five-carbon aldose.

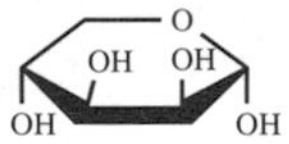

α-D-lyxose

LZ Abbreviation for 1. leucine zipper; 2. lysozyme.

LZM Abbreviation for lysozyme.

M

m 1. Abbreviation for meter. 2. A prefix meaning modified form of interconvertable enzyme, e.g., m-glycogen phosphorylase *a*.

M Abbreviation for 1. mole or molar concentration 2. molecular weight 3. methionine.

M12 An icosahydron RNA phage in the family of Leviviridae.

M13 A filamentous bacteriophage containing single-stranded DNA (Inoviridae).

M Antigen 1. A protein antigen from *Streptococci* that cross-reacts with muscle antigen. 2. A LPS-protein surface antigen from a species of *Brucella.* 3. A capsular antigen in certain enterobacteria. 4. Myeloma protein.

M Cells Cells associated with the surface of lymphoid tissue.

M Chromosome Referring to human mitochondrial chromosome.

M Fimbriae The fimbriae that are found on the pyelonephritic strains of *E. coli.*

M9 Medium A defined minimal medium for culturing bacteria.

M Phase A stage in the eukaryotic cell cycle in which both nucleus and cytoplasm divide.

M1 Protein A protein encoded by one of the RNA segments of the influenza viral genome that binds to the underside of the viral envelope.

M2 Protein A nonstructural protein of the influenza virion encoded by one of the RNA segments of the influenza viral genome.

M Subunit Referring to one of the two types of subunits of lactate dehydrogenase.

MA Abbreviation for muramic acid.

mA Abbreviation for milli-ampere.

3MA Abbreviation for 3-methyladenine.

m^1A Abbreviation for 1-methyladenosine.

MAA Abbreviation for melanoma-associated antigens.

Maalox No1 A trade name for a combination drug containing aluminum hydroxide and magnesium hydroxidase, used as an antacid.

Maalox Plus Tablets A trade name for a combination drug containing aluminum hydroxide, Simethicone, and magnesium hydroxide, used as an antacid.

Maalox TC Tabelets Same as Maalox No1 but concentrations of magnesium hydroxide and aluminum hydroxide are higher than Maalox No1.

mAb Abbreviation for monoclonal antibody.

MABOP Abbreviation for a combination drug containing mechlorethamine, adriamycin, bleomycin, oncovin, and prednisone.

Mabuterol (mol wt 311) A bronchodilator and an antiasthmatic agent.

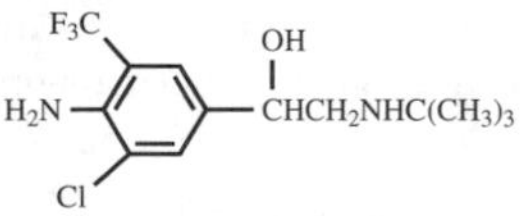

MAC Abbreviation for 1. a combination drug containing methotrexate, adriamycin, and cytoxan; 2. membrane attack complex.

MACC Abbreviation for a combination drug containing methotrexate, adriamycin, cyclophosphamide, and CCNU.

MacConkey's Agar An agar-based medium containing peptone, lactose, sodium glycocholate, sodium chloride, and neutral red, used for isolation of *Salmonella* and *Shigella* from feces.

Machupo Virus A virus of the family Arenaviridae that causes severe hemorrhagic fever in humans.

MACI Abbreviation for membrane attack complex inhibitor.

MAC-IP Abbreviation for membrane attack complex inhibitory protein.

MACOB Abbreviation for a combination drug containing methotrexate, adriamycin, cytoxan, oncovin, prednisone, and bleomycin.

Macro- A prefix meaning large.

Macrobid A trade name for a combination drug containing nitrofurantoin macrocrystal and nitrofurantoin monohydrate, used as an anti-infective agent.

Macrocyte An erythrocyte with diameter or mean corpuscular volume exceeding the normal mean value.

Macrocytic Anemia A disorder characterized by the presence of abnormally large and fragile red blood cells.

Macrocytosis A condition in which red blood cells are larger than normal.

Macrodantin A trade name for nitrofurantoin microcrystal, an anti-infective agent.

Macrodex A trade name for a high molecular weight dextran, used for expansion of plasma volume.

Macrofibril An aggregation of microfibrils that is visible under the light microscope.

Macroglobulin Referring to IgM (immunoglobulin M).

α_2-Macroglobulin Referring to fibrinolysin inhibitor secreted by macrophages.

Macroglobulinemia A disorder marked by the excessive production of IgM and abnormally high levels in the blood.

Macroions Charged macromolecules.

Macrolide Antibiotics A large group of antibiotics that consist of a large lactone ring structure (12 to 14 carbon atoms), e.g., erythromycin, angolamycin, and chalcomycin.

Macromolecule Polymeric molecule consisting of repeating monomer units, with molecular weights ranging from a few thousand to hundreds of millions, e.g., proteins, DNA, RNA, and polysaccharide.

Macronucleus 1. The large nucleus of the two nuclei in ciliate protozoa. 2. A nucleus that occupies a large area in a cell.

Macronutrient 1. The nutrients that are needed in appreciable quantity by an organism, e.g., carbohydrate, protein, and lipid. 2. The element required in large quantity for growth of plants.

Macrophage Mononuclear phagocytes of mammalian tissue derived from monocytes. Macrophages from different tissues have different properties and bear different names, e.g., Kupffer cells in the liver; histocyte in the connective tissue; and microglia in the brain.

Macrophage Activating Factor A factor that enhances the phagocytic and cytotoxic capacity of macrophages.

Macrophage Colony Stimulating Factor A factor that stimulates the committed stem cells of bone marrow to differentiate and proliferate into monocytes (mononuclear phagocytes).

Macrophage Derived Mitogenic Factor Referring to interleukin I.

Macrophage Inhibition Factor A group of lymphokines produced by activated T-lymphocytes that reduce macrophage mobility and increase the adhesion property of macrophages.

Macropinocytosis An up-take mechanism of cells in which the plasma membrane invaginates around the substances or particles to be ingested, forming a small pinocytotic vesicle.

Macroporous Resin A type of ion exchange resin that has an open porous structure, capable of adsorbing large ions.

Macroscopic Visible to the unaided eye.

Macrosorb A trade name for a number of macroporous particles, used in the chromatographic separation of substances, e.g., macrosorb SPR.

Macrosorb SPR A type of gel granule prepared by copolymerization of dimethylacrylamide, ethylenebisacrylamide, and sarcosine methyl ester, used in continuous flow peptide synthesis.

Macrotetralide Antibiotics A group of antibiotics that contain four similar tetrahydrofuran rings, which act as ionophores and inhibit Gram-positive bacteria.

Macula Spot of discoloration on the skin.

MAD Abbreviation for multiple autoimmune disorders.

MAdCAM Abbreviation for mucosal addressin cell adhesion molecule.

MADU (mol wt 257) Abbreviation for 5-methylamino-2-deoxyuridine, an antirival agent.

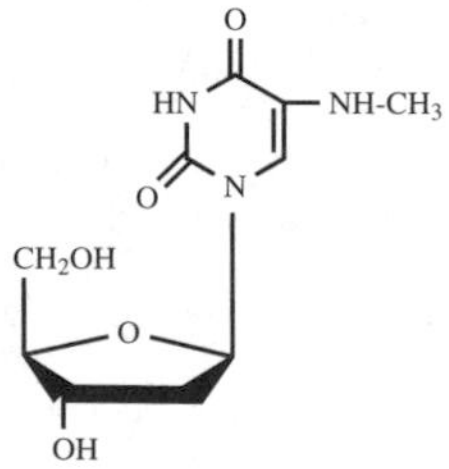

Maduromycosis A subcutaneous fungal infection caused by *Madurella mycetoma.*

Madurose (mol wt 194) Methyl-galactose, a derivative of galactose.

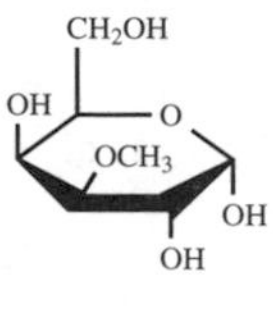

MaeI A restriction endonuclease from *Methanococcus aeolicus* with the following specificity:

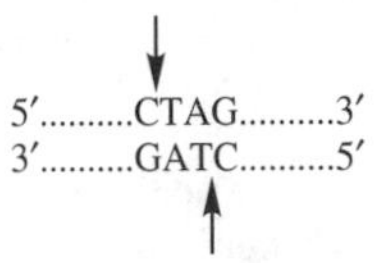

MaeII A restriction endonuclease from *Methanococcus aeolicus* with the following specificity:

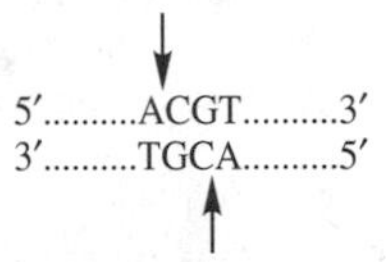

MaeIII A restriction endonuclease from *Methanococcus aeolicus* with the following specificity:

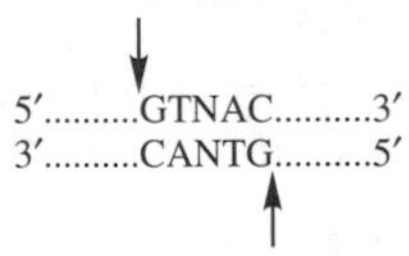

MAF Abbreviation for 1. macrophage-activating factor; 2. macrophage-agglutinating factor.

Mafenide (mol wt 186) An antibacterial agent that interferes with bacterial cellular metabolism.

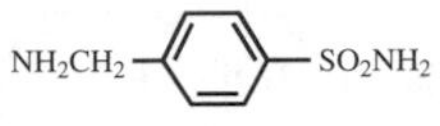

MAG Abbreviation for monoacylglycerol.

Magainins A group of potent antimicrobial peptides from glands of clawed toads (*Xenopus laevis*).

Magaldrate Aluminum magnesium hydroxide used as an antacid to reduce total acid load in the GI tract.

$$Al\ Mg\ (OH)_7$$

Magan A trade name for magnesium salicylate, an analgesic agent.

Magenta I (mol wt 338) A fungicide.

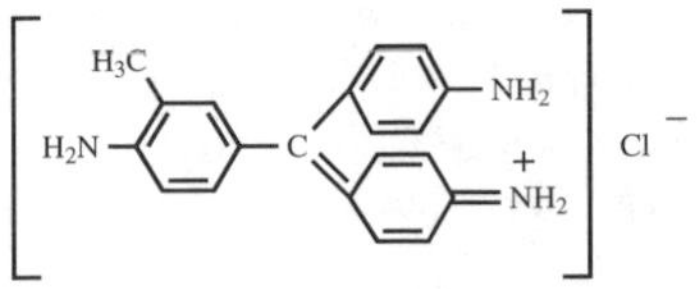

MAggF Abbreviation for macrophage agglutinating factor.

MAGL Abbreviation for monoacylglycerol lipase.

Magnacef A trade name for ceftazidine, an antibiotic that inhibits bacterial cell wall synthesis.

Magnatril A trade name for a combination drug containing aluminum hydroxide, magnesium hydroxide, and magnesium trisilicate, used as an antacid to reduce the total acid load in the GI tract.

Magnesemia The presence of magnesium in the blood.

Magnesia A trade name for magnesium salt used as an antacid and laxative.

Magnesium A chemical element with atomic weight 24, valence 2.

Magnesium-28 The artificial radioactive nuclide of magnesium emitting gamma radiation with a half life of 21 hours.

Magnesium Carbonate (mol wt 114) A weak antacid.

$$MgCO_3$$

Magnesium Carbonate Hydroxide (mol wt 485) An antacid.

$$(MgCO_3)_4\ Mg(OH)_2.5H_2O$$

Magnesium Chloride (mol wt 95) An electrolyte used for maintaining magnesium levels in the body and as an anticonvulsant.

$$MgCl_2$$

Magnesium Hydroxide (mol wt 58) A laxative.

$$Mg\ (OH)_2$$

Magnesium Ionophore I (mol wt 341) An ionophore for magnesium.

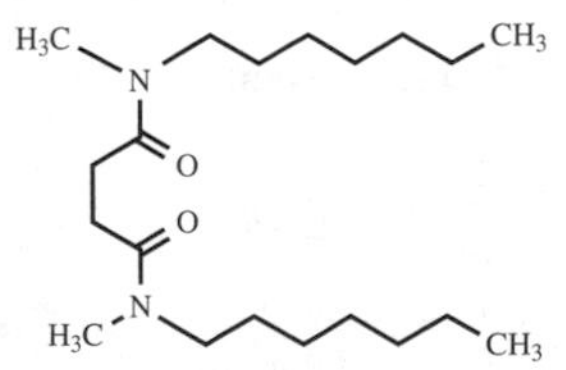

Magnesium Ionophore II (mol wt 567) An ionophore for magnesium.

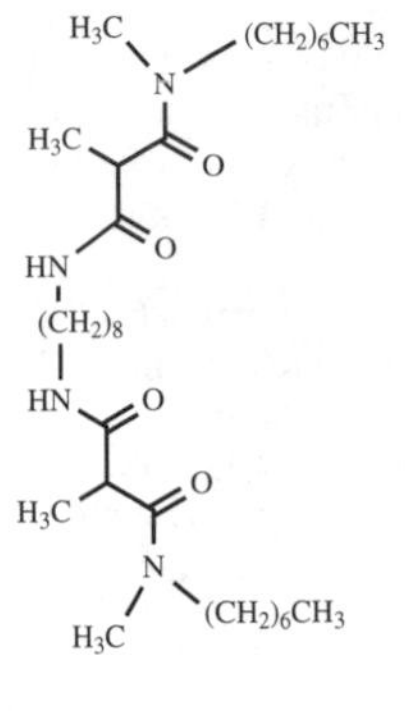

Magnesium Ionophore III (mol wt 539) An ionophore for magnesium.

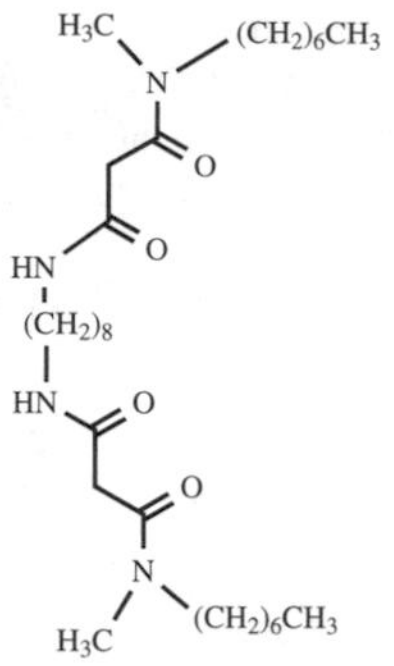

Magnesium Ionophore IV (mol wt 863) An ionophore for magnesium.

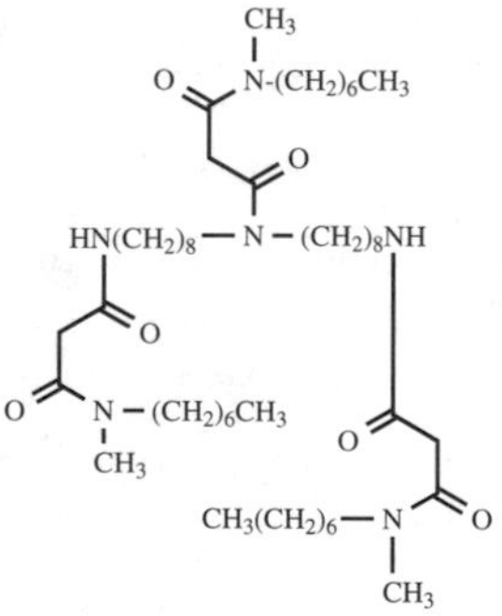

Magnesium Oxide (mol wt 40) An antacid.

MgO

Magnesium Salicylate (mol wt 299) An analgesic agent that blocks the generation of pain impulses.

$Mg\ [C_6H_4\ (OH)\ COO]_2$

Magnesium Sulfate (mol wt 120) An anticonvulsant.

$Mg\ SO_4$

Magnet A substance that attracts particles of iron and effects an electric current.

Magnetic Dipole A substance with two magnetic poles.

Magnetic Dipole Movement The tendency of a substance to become oriented in a magnetic field; equal to the product of the strength of the magnetic pole and the length of the magnet.

Magnetic Field The space in the vicinity of a magnet through which the magnetic forces act.

Magnetic Immunoassay A type of immuno-assay in which an antibody is bound to a particulate, inert, magnetic support material.

Magnetic Resonance Imaging (MRI) A diagnostic technique for the computerized analysis of the absorption and transmission of high frequency radio waves by the molecules in tissues under the strong magnetic field.

Magnetic Resonance Spectroscopy A technique that utilizes the phenomenon of nuclear magnetic resonance to obtain a biochemical profile of tissue.

Magnetic Separation The use of magnetic beads or particles for separation of substances, e.g., use of a complex of protein A and magnetic particles for separation of immunoglobulin from the blood.

Magnetic Stirrer A plastic- or glass-coated magnet used as a stirring tool (stirring bar).

Magnetism 1. The property of mutual attraction or repulsion produced by a magnet or by an electric current. 2. The force exhibited by a magnetic field.

Magnetoliposome An artificially prepared liposome containing ferromagnetic particles.

Magnetophoresis The movement of magnetizable particles through a fluid under the action of a magnetic field.

Magnetosomes Dense inclusion bodies within bacterial cells that contain iron granules and act as magnetes, allowing bacteria to orient themselves in a magnetic field.

Magnetotactic Bacteria Referring to Gram-negative bacteria that form intracellular, enveloped magnetic structures (magnetosomes).

Magnetotaxis A type of taxis in which the stimulus is a magnetic field.

Magon (mol wt 411) A reagent used for the determination of magnesium in the blood.

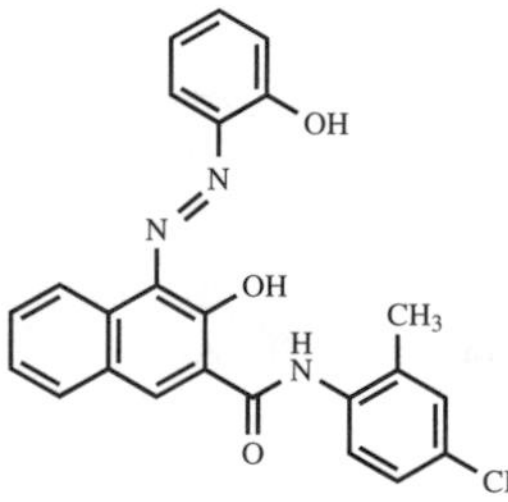

Magonate A trade name for magnesium gluconate, a magnesium source.

Mag-Ox 400 A trade name for magnesium oxide, used as an antacid.

MAHA Abbreviation for macro-angiopathic hemolytic anemia.

MAIDS Abbreviation for murine acquired immuno-deficiency syndrome.

Maillard Reaction The reaction in which the amino groups of the amino acids react with aldehydes, ketones or reducing sugars in the absence of an enzyme.

Maize Streak Virus A plant virus of the group Geminiviruses.

Major Groove The larger of the two grooves in the double-stranded helical structure of DNA, resulting from twist of the two strands around each other.

Major Histocompatibility Antigens Antigens that are encoded by the genes on the histocopatibility complex on chromosome number 6 in human and chromosome number 17 in mice.

Major Histocompatibility Complex A set of genes on a chromosome (number 6 in humans; number 17 in mice) that encode histocompatibility antigens and factors influencing immune responses (e.g., graft rejection).

Mal- A prefix denoting disease, disorder, or abnormality.

Malachite Green (mol wt 365) A dye with antifungal and antiseptic activity.

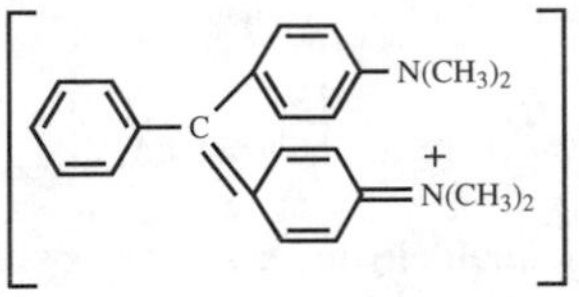

Malaise A general feeling of illness, accompanied by weakness and discomfort.

Malaria An infectious chronic disease caused by *Plasmodium oritixia,* transmitted by mosquitoes and characterized by intermittent fever, chills, and sweating.

Malate The ionic form of malic acid, or a salt of malic acid.

Malate Aspartate Shuttle A pathway that shuttles the electrons from cytoplasmic NADH into the mitochondrial electron transport chain for oxidation.

Malate Dehydrogenase The enzyme that catalyzes the following reaction:

$$\text{Malate} + \text{NAD}^+ \rightleftharpoons \text{Oxaloacetate} + \text{NADH}$$

Malate Isomerase The enzyme that catalyzes the following reaction:

$$\text{Malate} \rightleftharpoons \text{Fumarate}$$

Malate α-Ketoglutarate Carrier A carrier protein that transports aspartate from mitochondrial matrix to cytosol in exchange for cytosolic glutamate.

Malate Oxidase The enzyme that catalyzes the following reaction:

$$\text{Malate} + O_2 \rightleftharpoons \text{Maleate} + H_2O_2$$

Malate Synthetase The enzyme that catalyzes the following reaction:

$$\text{L-malate} + \text{CoA} \rightleftharpoons \text{Acetyl-CoA} + \text{glyoxylate} + H_2O$$

Malathion (mol wt 330) An insecticide that inactivates acetylcholinesterase.

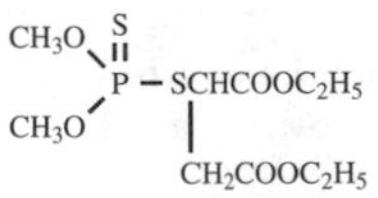

MalBSA Abbreviation for maleated bovine serum albumin.

MALDI Abbreviation for matrix-assisted-laser-desorption ionization.

MALDI-TOFMS Abbreviation for matrix-assisted laser desorption/ionization-time-of-flight mass spectrometry.

Male Specific Phage Bacteriophages that infect male bacteria, e.g., filamentous phages.

Maleate The ion form of maleic acid.

Maleate Hydratase The enzyme that catalyzes the following reaction:

$$\text{Malate} \rightleftharpoons \text{Maleate} + H_2O$$

Maleate Isomerase The enzyme that catalyzes the following reaction:

$$\text{Maleate} \rightleftharpoons \text{Fumarate}$$

Maleic Acid (mol wt 116) An unsaturated organic acid used for preparation of biological buffers.

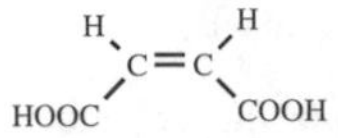

Maleic Anhydride (mol wt 98) A reagent used for dissociation of protein into subunits.

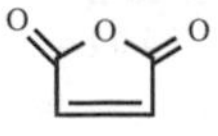

Maleoyl Group A bivalent acyl group derived from maleic acid.

–CO-CH=CH-CO-

Maleyl Group A univalent acyl group derived from maleic acid.

HOOC-CH=CH-CO-

Maleylacetate Reductase The enzyme that catalyzes the following reaction:

3-Ketoadipate + $NADP^+$
⇅
2-Maleylacetate + NADPH

Maleylacetoacetate Isomerase The enzyme that catalyzes the following reaction:

4-Maleylacetoacetate ⇌ 4-fumarylacetoacetate

Maleylpyruvate Isomerase The enzyme that catalyzes the following reaction:

3-Maleylpyruvate ⇌ 3-Fumarylpyruvate

Malic Acid (mol wt 134) An organic acid and an intermediate in the Krebs cycle.

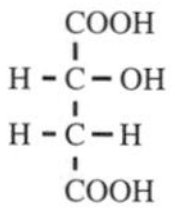

Malic Dehydrogenase See malate dehydrogenase.

Malic Enzyme (NAD^+) The enzymes that catalyze the following reactions:

Malate + NAD^+ ⇌ pyruvate + NADH + CO_2

Malic Enzyme ($NADP^+$) The enzymes that catalyze the following reactions:

Malate + $NADP^+$ ⇌ pyruvate + NADPH + CO_2

Malignant The primary tumor that has the property of uncontrollable growth and dissemination.

Malignant Hepatoma A malignant tumor of the liver.

Malignant Melanoma A malignant tumor of melanin-forming cells.

Malignant Neoplasm A tumor that tends to grow, invade nearby tissue, and spread through the blood stream.

Malignant Transformation The conversion of noncancerous cells to cancerous cells by an oncogen or a carcinogen.

Malignant Tumor An invasive or metastatic tumor.

Mallamint A trade name for calcium carbonate, used as an antacid.

Mallergan A trade name for promethazine hydrochloride, an antihistaminic agent that competes with histamine for H_1 receptors on target cells.

Mallory Bodies Large irregular masses located in the hepatocytes of the liver.

Mallory's Triple Stain A histological stain containing aniline blue or methyl blue, orange G, and oxalic acid.

Malogan A trade name for testosterone, an anabolic steroid that promotes tissue building processes.

Malogex A trade name for testosterone enanthate, an anabolic hormone that promotes tissue building processes.

Malolactic Fermentation A type of fermentation that yields a mixture of malic acid and lactic acid.

Malonate The ionic form of malonic acid or a salt of malonic acid.

Malonate CoA-transferase The enzyme that catalyzes the following reaction:

Acetyl-CoA + malonate
⇅
Acetate + malonyl-CoA

Malonate Semialdehyde Dehydratase The enzyme that catalyzes the following reaction:

Malonate semialdehyde
⇅
Acetylene monocarboxylate + H_2O

Malonate Semialdehyde Dehydrogenase The enzyme that catalyzes the following reaction:

Malonate semialdehyde + CoA + NAD^+
⇅
Acetyl-CoA + CO_2 + NADH

Malonic Acid (mol wt 104) An organic acid and substance for synthesis of fatty acid.

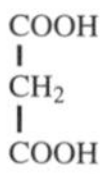

Malonic Semialdehyde An intermediate in metabolism of pyrimidine (see also methylmalonic semialdehyde).

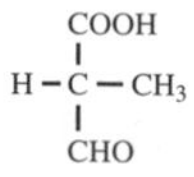

Malonyl-ACP Abbreviation for malonyl acyl carrier protein, an essential intermediate in the synthesis of fatty acid.

Malonyl-CoA (mol wt 854) An intermediate in the biosynthesis of fatty acid.

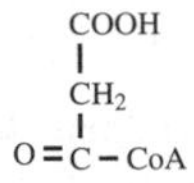

Malonyl-CoA ACP-transacylase The enzyme that catalyzes the following reaction:

$$\text{ACP + malonyl-CoA} \rightleftharpoons \text{Malonyl-ACP + CoA}$$

Malonyl-CoA Decarboxylase The enzyme that catalyzes the following reaction:

$$\text{Malonyl-CoA} \rightleftharpoons \text{Acetyl-CoA} + CO_2$$

Malotuss A trade name for guaifenesin, an antitussive agent.

Maloyl Group A bivalent acyl group derived from malic acid.

$$\text{-OC-CH(OH)CH}_2\text{-CO-}$$

Malt A preparation obtained from germinated barley seeds that contains partially degraded starch and protein.

MALT Abbreviation for mucosa-associated lymphoid tissue.

Malta Extract An extract from malt used for culturing yeasts and molds.

Maltaner Antigen A cardiolipin-lecithin antigen used for the testing for *Syphilis*.

Maltase The enzyme that converts maltose to glucose.

Malting The process of enzymatic conversion of barley by plant amylases and proteases, used to prepare grain for microbial alcoholic fermentation.

Maltitol (mol wt 344) A sugar alcohol consisting of glucose and sorbitol.

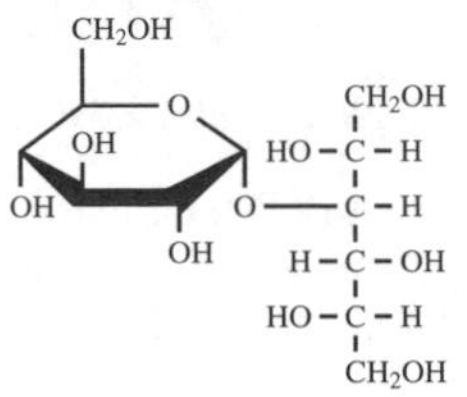

Maltobiose Synonym for maltose.

Maltogenic Amylase Synonym of 4-α-glucanotransferase.

Maltohexaose (mol wt 991) An oligosaccharide consisting of glucose.

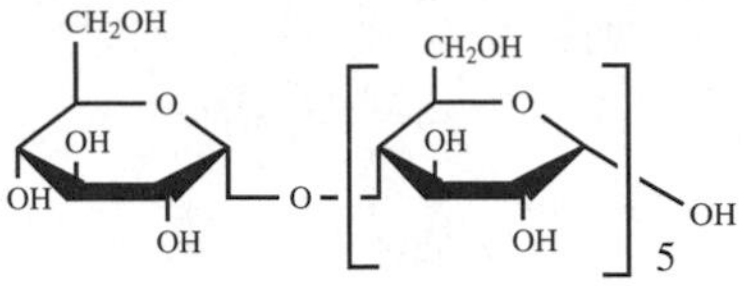

Maltopentose (mol wt 829) An oligosaccharide consisting of glucose.

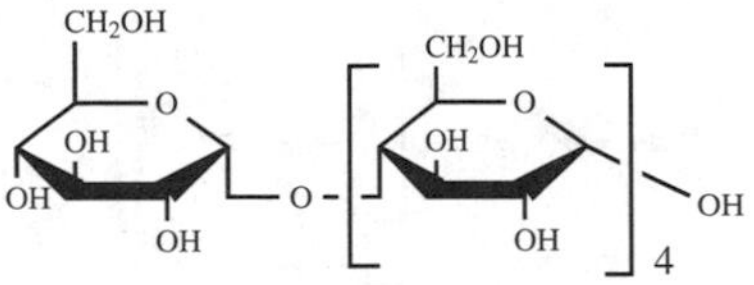

Maltoporin A transmembrane protein in *E. coli* that transports maltodextrins into the cell and acts as a receptor for lambda and some phages.

Maltose (mol wt 342) A reducing disaccharide consisting of α-1,4-linked glucose residues.

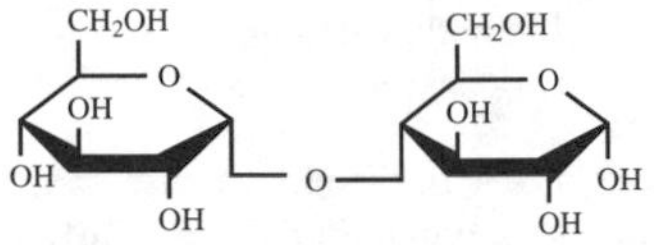

Maltose Acetyltransferase The enzyme that catalyzes the following reaction:

$$\text{Acetyl-CoA + maltose} \rightleftharpoons \text{CoA + acetyl-maltose}$$

Maltose Binding Protein A protein involved in the uptake of maltose in *E. coli.*

Maltose Glucosyltransferase The enzyme that catalyzes the following reaction:

$$(\alpha\text{-1,4-glucosyl})_n + \text{D-glucose} \rightleftharpoons (\alpha\text{-1,4-glucosyl})_{n-1} + \text{maltose}$$

Maltose Orthophosphate 1-β-D-Glucosyltransferase The systematic name for maltose phosphorylase.

Maltose 6-Phosphate Glucosidase The enzyme that catalyzes the following reaction:

$$\text{Maltose 6-phosphate} + H_2O \rightleftharpoons \text{Glucose 6-phosphate + Glucose}$$

Maltose Phosphorylase The enzyme that catalyzes the following reaction:

Maltose + Pi ⇌ Glucose 1-phosphate + glucose

Maltose Synthetase The enzyme that catalyzes the following reaction:

2 α-D-Glucose 1-phosphate
⇅
Maltose + 2 Pi

Maltotetraose (mol wt 667) An oligosaccharide consisting of glucose.

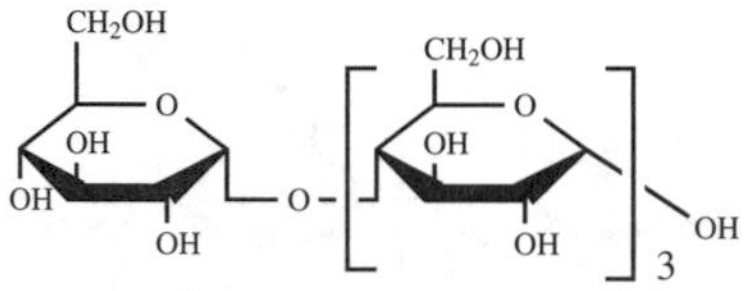

Maltotriose (mol wt 504) A trisaccharide that contains three α-1,4-linked glucose residues.

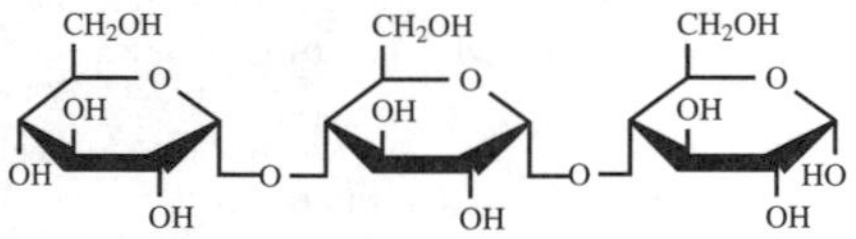

MAM Abbreviation for 1. methyl-azo-methanol; 2. mitochondria-associated membrane.

MamI A restriction endonuclease from *Microbacterium ammoniaphilum* with the following specificity:

5.........GATNN↓NNATC........3'
3'.........CTANN↑NNTAG........5'

Mammalian Expression Vector A vector capable of genetic expression in eukaryotic mammalian cells, e.g., shuttle vector.

Mammary Gland A gland responsible for secretion of milk.

Mammary Tumor Virus Referring to the mouse mammary tumor virus (MMTV), a member of Retroviridae that is transmitted through milk and causes mammary tumors in mice.

Mammogram The X-ray examination of the soft tissues of the breast for identification of cysts or tumors.

mAmp Abbreviation for milliampere.

Man Abbreviation for mannose.

Man-BP Abbreviation for mannan-binding protein.

Man-BSA Abbreviation for mannose-conjugated bovine serum albumin.

Mandameth A trade name for methenamine mandelate, an anti-infective agent.

Mandelamine A trade name for methenamine mandelate, an anti-infective agent.

Mandelonitrile Benzaldehyde Lyase The systematic name for mandelonitrile lyase.

Mandelonitrile Lyase The enzyme that catalyzes the following reaction:

Mandelonitrile
⇅
Cyanide + benzaldehyde

Mandol A trade name for cefamandol, an antibiotic that inhibits bacterial cell wall synthesis.

Manganese A chemical element with atomic weight 55, valence 2, 4, and 7.

Manganese Nodules Nodules produced by microbial oxidation of manganese oxides.

Manganese Peroxidase The enzyme that catalyzes the following reaction:

2 Mn (II) + 2 H^+ + H_2O_2
⇅
2 Mn (III) + H_2O

Mangatrace A trade name for manganese chloride.

Manidipine (mol wt 611) An antihypertensive agent.

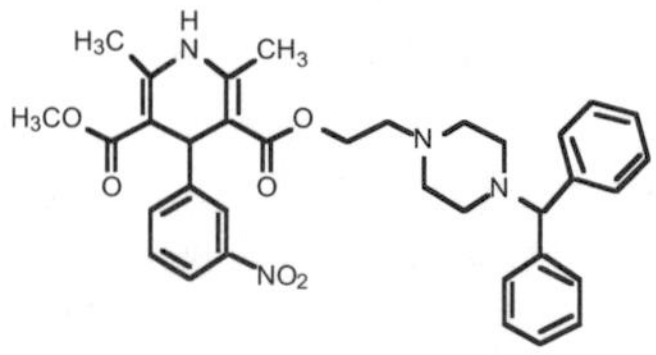

Maniron A trade name for ferrous fumarate, a hematinic agent that provides iron for the synthesis of hemoglobin.

Mannan A polymer that consists of mannose.

Mannan Endo-1,4-β-Mannosidase The enzyme that catalyzes the random hydrolysis of 1,4-β-D-mannosidic linkage in mannans, galactomannans, and glucomannans.

Mannan(1,4-β-D) Mannanohydrolase Synonym of mannan endo-1,4-β-mannosidase.

Mannase The enzyme that catalyzes the hydrolysis of nonreducing β-D-mannose residues in β-D-mannoside.

Mannitol (mol wt 182) A sugar alcohol and a diuretic agent derived from mannose.

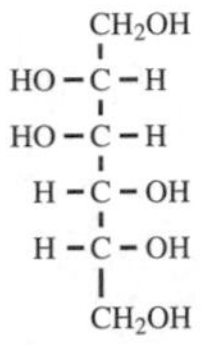

D-mannitol

Mannitol Dehydrogenase (NAD^+) The enzyme that catalyzes the following reaction:

Mannitol + NAD^+ ⇌ Fructose + NADH

Mannitol Dehydrogenase ($NADP^+$) The enzyme that catalyzes the following reaction:

Mannitol + $NADP^+$ ⇌ Fructose + NADPH

Mannitol NAD^+ 2-Oxidoreductase The systematic name for mannitol 2-dehydrogenase.

Mannitol 1-Phosphate Dehydrogenase The enzyme that catalyzes the following reaction:

Mannitol 1-phosphate + NAD^+
⇌
D-fructose 6-phosphate + NADH

Mannitol 6-Phosphate Isomerase The enzyme that catalyzes the following reaction:

D-Mannose 6-phosphate
⇌
D-Fructose 6-phosphate

Mannoheptose (mol wt 210) A 7-carbon aldosugar.

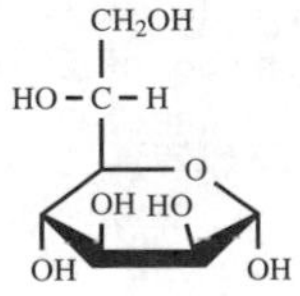

Mannoheptulose (mol wt 210) A 7-carbon ketosugar.

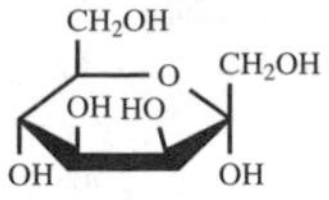

Mannokinase The enzyme that catalyzes the following reaction:

ATP + mannose
⇌
Mannose 6-phosphate + ADP

Mannomustine (mol wt 378) An antineoplastic agent.

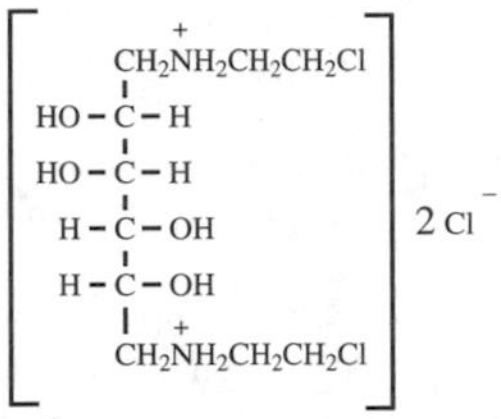

Mannonat Dehydratase The enzyme that catalyzes the following reaction:

D-mannonate
⇌
2-Keto-3-deoxy-D-gluconate + H_2O

Mannonic γ-Lactone (mol wt 178) A derivative of mannose.

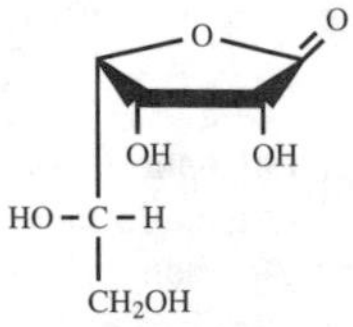

Mannosamine (mol wt 180) An aminosugar derived from mannose.

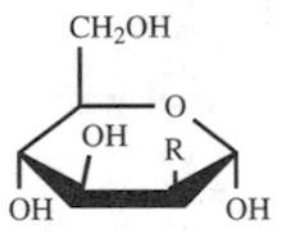

$R = NH_2$

Mannose (mol wt 180) A six-carbon aldo-sugar.

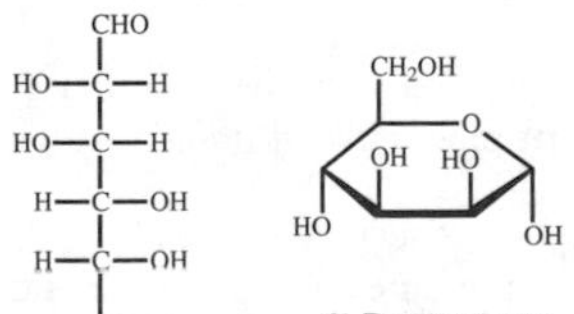

α-D-mannose

Mannose Isomerase The enzyme that catalyzes the following reaction:

D-Mannose ⇌ D-fructose

Mannose 1-Phosphate Guanylyltransferase The enzyme that catalyzes the following reaction:

GTP + manose 1-phosphate
⇌
PPi + GDP-mannose

α-Mannosidase The enzyme that catalyzes the following reaction:

$$\alpha\text{-D-mannoside} + H_2O \rightleftharpoons \text{Alcohol} + \text{D-mannose}$$

β-Mannosidase The enzyme that catalyzes the following reaction:

$$\beta\text{-D-mannoside} + H_2O \rightleftharpoons \text{Alcohol} + \text{D-mannose}$$

Mannoside α-D-Mannohydrolase The systematic name for α-mannosidase.

Mannosyl Glycoprotein Endo-β-N—Acetylglucosaminidase The enzyme that catalyzes the endohydrolysis of the N-N'-diacetylchitobiosyl unit in high-mannose glycopeptides and glycoproteins.

Mannosyl Oligosaccharide 1,2-α-Mannosidase The enzyme that catalyzes the hydrolysis of terminal 1,2-linked α-D-mannose residues in the mannosyl-oligosaccharide.

Mannosyl Oligosaccharide 1,3-1,6-α-Mannosidase The enzyme that catalyzes the hydrolysis of terminal 1,3 and 1,6-linked α-D-mannose residues in the mannosyl-oligosaccharide $Man_5(GlcNAc)_3$.

1,2-α-Mannosyl Oligosaccharide α-D-Mannohydrolase The systematic name for mannosyl-oligosaccharide 1,2-α-mannosidase.

1,3-(1,6)-Mannosyl Oligosaccharide α-D-Mannohydrolase The systematic name for 1,3-(1,6)-mannosyl-oligosaccharide α-D-mannohydrolase.

Mannuronic Acid (mol wt 194) A sugar acid.

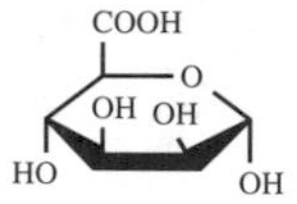

α-D-mannuronic acid

Manometer An instrument used for the measurement of pressure of a liquid or a gas.

Manometry A technique used for measuring the changes in pressure of a gas or liquid resulting from a chemical reaction or biological action.

Manox A trade name for magnesium oxide, used as an antacid.

Man6P Abbreviation for mannose 6-phosphate.

Mantadil A trade name for a skin cream containing hydrocortisone acetate and chlorcyclizine hydrochloride (antihistamine).

Mantoux Test An intracutaneous test for detection of infection by *Mycobacterium tuberculosis*.

MAO Abbreviation for monoamine oxidase.

MAO Inhibitor Any drug that prevents the activity of the monoamine oxidase in the brain tissue.

MAOI Abbreviation for monoamine oxidase inhibitor.

Maolate A trade name for chlorphenesin carbamate, a muscle relaxant that reduces the transmission of nerve impulses from the spinal cord to skeletal muscle.

Maox-420 A trade name for magnesium oxide, an antacid and laxative.

MAP Abbreviation for 1. methylaminopurine; 2. mercapturic acid pathway; 3. methyl acceptor protein; 4. microtubule-associated protein; 5. mitogen-activated protein.

Map Unit A unit for measuring the distance between two linked genes that equals 1% recombination frequency between the two linked genes.

MAPK Abbreviation for mitogen-activated protein kinase.

MAPK-AP Abbreviation for MAPK-activated protein.

MAP-Kinase Abbreviation for mitogen-activated protein kinase.

MAPKK Kinase Abbreviation for mitogen-activated protein kinase kinase kinase.

MAPKKK Abbreviation for mitogen-activated protein kinase kinase kinase.

Maple Bark Disease A lung disease caused by *Cryptostroma corticale* (a mold) found in the bark of maple trees.

Maple Syrup Urine Disease A disorder caused by a deficiency in the enzyme for degrading the branched-chain amino acids. It is so-called because the buildup of branched-chain amino acids render the urine an odor of maple syrup.

Maprotiline (mol wt 277) An antidepressant that increases the amount of norepinephrine or serotonin in the CNS by blocking their uptake by presynaptic neurons.

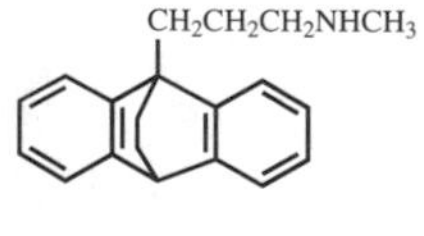

MAR Abbreviation for mouse aldose reductase.

Marax A trade name for a combination drug containing theophylline, ephedrine sulfate, and hydroxyzine hydrochloride, used as a bronchodilator.

Marbaxin-750 A trade name for methocarbamol, a skeletal relaxant that reduces the transmission of nerve impulses from the spinal cord to skeletal muscle.

Marburg Fever A viral hemorrhagic fever caused by the Marburg virus. It was first discovered in Marburg, Germany, in 1967.

Marburg Virus A single-stranded RNA virus.

Marcaine A trade name for bupivacaine hydrochloride, a local anesthetic agent that interferes with sodium-potassium exchange across the nerve membrane.

Marcillin A trade name for ampicillin, an antibiotic.

MARCK Abbreviation for myristoylated alanine-rich C-kinase.

MARCKS Abbreviation for myristoylated alanine-rich C-kinase substrate.

Marek's Disease A cancer of the lymphoid system (lymphomatosis) in chickens, caused by a contagious herpesvirus.

Marezine A trade name for cyclizine hydrochloride, an antiemetic agent.

Marfan's Syndrome A disorder caused by the abnormal condition of the type I collagen and characterized by excess bone length.

Marflex A trade name for orphenadrine citrate, a skeletal muscle relaxant that reduces the transmission of nerve impulses from the spinal cord to skeletal muscle.

Margaric Acid (mol wt 270) An organic acid.

$$CH_3(CH_2)_{15}COOH$$

Margesic A-C A combination drug containing aspirin, propoxyphene hydrochloride, and caffeine and used as an analgesic agent.

Margination Attachment of leukocytes (e.g., neutrophil) to the endothelium of the blood vessel.

MARIA Abbreviation for macroaggregated radioiodinated albumin.

Marinol A trade name for dronabinol, an antiemetic agent.

Marker 1. A genetic locus that is associated with a particular phenotypic characteristic. 2. A molecule that is linked to another substance or biological structure to serve as a detectable marker. 3. A molecule with known physical characteristics and used as a reference, e.g., molecular weight markers used in SDS-PAGE (sodium dedecylsulfate polyacrylamide gel electrophoresis).

Marker Genes Genes whose effect and position are known.

Marker Rescue A condition in which a defective virus produces infectious progeny virion due to the presence of another virus (defective or nondefective) as a result of recombination between the two viruses.

Marmine A trade name for dimenhydrinate, an antiemetic agent.

Marnal A trade name for a combination drug containing aspirin, caffeine, and butalbital, used as an antipyretic and analgesic agent.

Marplan A trade name for isocarboxazid, an antidepressant drug and a MAO (monoamine oxidase) inhibitor that promotes accumulation of neurotransmitter.

Marrow The soft tissue mass inside the long bone.

Marrow Transplant Referring to a bone marrow transplant from a donor to a recipient.

Marseilles Fever A disease caused by *Rickettsia conorii,* carried by a brown tick and characterized by chills, fever, and a black crust at the site of the tick bite.

Marzine A trade name for cyclizine lactate, an antiemetic agent.

***MAS* Gene** A human transforming gene that encodes angiotensin III receptor on the membrane.

Masculining Tumor of the Ovary A tumor of the ovary that causes the appearance of male characteristics in a female, e.g., hair on the face and a deepening of the voice.

Maser A device used for conversion of electromagnetic radiation into a beam of highly amplified monochromatic radiation at a frequency within the microwave region.

MASER Abbreviation for 1. microwave amplification by stimulated emission of radiation; 2. molecular application by stimulated emission of radiation.

Masked mRNA Long-lived and stable mRNAs found in oocytes that cannot be translated until specific regulatory substances become available.

Masked Residue The amino acid residue in a protein that is not accessible for action or reaction.

Maso-Cort A trade name for hydrocortisone.

Mass Number The sum of the number of protons and neutrons per atom of a given nuclide.

Mass Spectrogram The graphic record or profile obtained from mass spectrometer.

Mass Spectrometer A device in which the molecules are ionized, separated according to their mass to charge ratio, and detected by electronic amplification.

Mass Spectrum The characteristic patterns obtained from a mass spectrometer.

Mass Unit See atomic mass unit.

Mast Cell A type of granulocyte that bears Fc receptors for IgE. Mast cells are rich in cytoplasmic granules that contain vasoactive substances such as histamine and serotonin that mediate for type-I hypersensitivity reactions.

Mast Cell Degranulating Peptide A neurotoxic peptide from bee venom, it consists of 22-residue peptides and stimulates mast cell degranulation.

Mastectomy Surgical removal of a breast.

Master Plate A culture plate that contains the original microbial colonies from which replica plates are made.

Mastigonemes The lateral projections from a eukaryotic flagella.

Mastitis Inflammation of the breast.

Mastocytogenesis The formation of mast cells.

Mastocytoma Tumorlike nodule or mass that contains mast cells.

Mastoparan A peptide toxin from wasp venom that activates GTP-binding protein.

Mastopathy Any disease of the breast.

MATB Abbreviation for *Mycobacterium avium* tuberculosis.

Mate Killer The endosymbiont-containing strain of *Paramecium aurelia* (ciliate protozoa) that kills the conjugal partner upon conjugation.

Maternal Antibody The antibody transferred transplacentally from mother to fetus.

Maternal Effect A condition in which the genotype of the mother influences the phenotype of the offspring by a substance present in the cytoplasm of the egg.

Maternal Effect Gene The gene whose product produces maternal effect.

Maternal Immunity The immunity of the fetus obtained from the mother via the placenta or colostrum.

Maternal Inheritance Inheritance of a trait through cytoplasmic factors or orgnelles (e.g., mitochondria and chloroplast) derived from the female gamete.

Mating Bridge Structure necessary for the transfer of DNA from a male bacterial cell to a female cell.

Mating Type A stain of organism capable of interacting sexually with other genetically distinct strains.

Matrilysin A metalloproteinase.

Matrin A zinc finger DNA-binding protein.

Matrix 1. Ground substance or gellike substance that fills a space, e.g., mitochondrial matrix. 2. A loose meshwork within which cells are embedded, e.g., extracellular matrix. 3. Homogeneous intercellular substance of tissue.

Matrix Gla Protein A protein that mediates the association of organic matrix of bone and cartilage.

Matrix Protein 1. A protein of the outer membrane of Gram-negative bacteria that forms a water-filled transmembrane channel (pore) permitting the passage of ions or molecules (also known as porin). 2. A protein encoded by one of the segment of the influenza viral genome.

Matroclinous Inheritance A type of inheritance in which all offspring have the nucleus-determined phenotype of the mother.

Matulane A trade name for procarbazine hydrochloride, an antineoplastic agent that inhibits the synthesis of DNA, RNA, and protein.

Maturation of Germ Cells Process of development of mature sperm and ova from cells in the testis and ovary, respectively.

Maturation Protein A protein encoded by a RNA-phage that is required for production of a complete infectious virion.

Maturing Face The concave face of the Golgi complex, usually oriented toward the cell surface (also called the trans face).

MAV Abbreviation for transmembrane activation voltage.

Mavik A trade name for trandolapril, an antihypertensive agent and an angiotensin-converting enzyme inhibitor.

MAX The symbol for a gene family encoding DNA-binding proteins that function in the control of cell growth.

Maxair A trade name for pirbuterol, a bronchodilator.

Maxalt A trade name for rizatriptan, an antimigraine agent.

Maxam-Gilbert's Method A chemical method for DNA sequencing in which the double-stranded DNA is labeled with ^{32}P at its 5′ or 3′ ends, treated with dimethyl sulfate, cleaved under appropriate conditions, and electrophoresed in polyacrylamide gel and detected by autoradiography.

Maxaquin A trade name for lomefloxacin hydrochloride, an antibacterial agent that inhibits bacterial DNA gyrase.

Maxeran A trade name for metoclopramide hydrochloride, an antiemetic agent that stimulates the motility of the upper GI tract.

Maxibolin A trade name for ethylestrenol, an anabolic steroid that promotes tissue building processes.

Maxicell A bacterial cell that is capable of synthesizing plasmid-encoded gene products. Maxicell is obtained by UV-irradiation of a bacterial cell containing multicopies of a plasmid to render the bacterial chromosome incapable of replication. The plasmid surviving radiation continues the synthesis of plasmid-encoded gene product.

Maxidex A trade name for dexamethasone, an ophthalmic anti-inflammatory agent.

Maxiflor A trade name for diflorasone diacetate, a corticosteroid.

Maximum Velocity The maximum reaction rate of an enzyme-catalyzed reaction at maximum substrate concentration.

Maxipime A trade name for cefepime hydrochloride, a third generation cephalosporin antibiotic.

Maxivate A trade name for betamethasone dipropionate, a corticosteroid.

Maxzide A trade name for a combination drug containing hydrochlorothiazide and triamterene, used as an antihypertensive agent.

Mayaro Fever A disease of humans caused by an alphavirus, transmitted by mosquitoes.

Mazanor A trade name for mazindol, a cerebral stimulant that inhibits the neural uptake of norepinephrine and dopamine.

Mazepine A trade name for carbamazepine, an anticonvulsant.

Mazicon A trade name for flumazenil, a benzodiazepine antagonist that inhibits the action of benzodiazepine on gamma-aminobutyric acid and benzodiazepine receptors.

Mazindol (mol wt 285) A CNS stimulant that inhibits neuronal uptake of norepinephrine and dopamine.

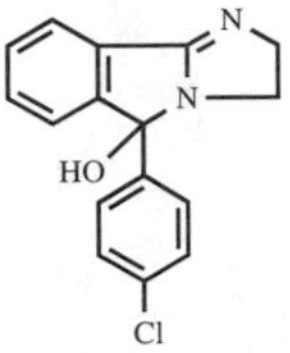

Mazipredone (mol wt 479) An anti-inflammatory agent.

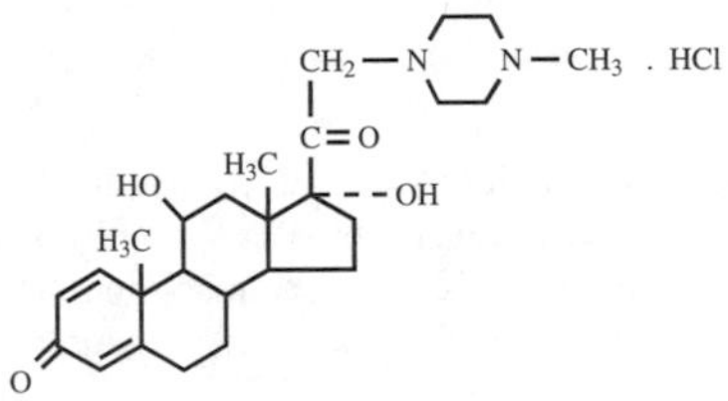

Mazzini Test A test for syphilis.

Mb Abbreviation for myoglobin.

MB Abbreviation for 1. methylene blue; 2. megabase; 3. mega-byte.

MBA Abbreviation for methyl bovine albumin.

MBACOD Abbreviation for a combination drug containing methotrexate, bleomycin, adriamycin, cytoxan, oncovin, and dexamethasone.

M-band Central region of the A-band of the sarcomere in striated muscle.

MBC Abbreviation for a combination drug containing methotrexate, bleomycin, and cisplatin.

MBCD Abbreviation for methyl-β-cyclodextrin.

MbCO Symbol for the complex of carbon monoxide and myoglobin.

MBD Abbreviation for a combination drug containing methotrexate, bleomycin, and DDP.

MBH_2 Abbreviation for reduced form of methylene blue.

MBK Abbreviation for methyl butyl ketone.

MBL Abbreviation for mannan-binding lectin.

MBLA Abbreviation for methyl benzyl linoleic acid.

MBM Abbreviation for mineral basal medium.

MbO_2 Symbol for oxymyoglobin.

MboI A restriction endonuclease from *Moraxella bovis* with the following specificity:

↓
5′..........GATC..........3′
3′..........CTAG..........5′
↑

MboII A restriction endonuclease from *Moraxella bovis* with the following specificity:

↓
5′..........GAAGA(8N)..........3′
3′..........CT T CT(7N)..........5′
↑

MBP Abbreviation for 1. a combination drug containing methotrexate, bleomycin and platinol; 2. myelin basic protein; 2. maltose-binding protein; 3. methylene bisphosphate; 4. major basic protein.

MBR Abbreviation for methylene blue reduced.

MBSA Abbreviation for methylated bovine serum albumin.

MC Abbreviation for 1. mast cell; 2. mitotic cycle.

mC Abbreviation for millicurie.

μC Abbreviation for microcurie.

m^3C Abbreviation for 3-methylcytosine.

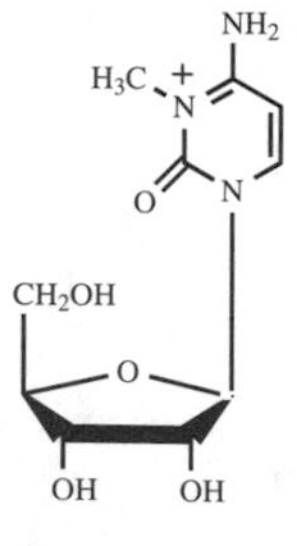

MC29 An acute avian leukemia virus.

MCA Abbreviation for 1. a combination drug containing megestrol, cytoxan, and adriamycin; 2. monoclonal antibody.

MCAb Abbreviation for monoclonal antibody.

MCAC Abbreviation for metal-chelate-affinity chromatography.

MCAD Abbreviation for medium-chain acyl-CoA dehydrogenase.

MCADH Abbreviation for medium chain acyl-CoA dehydrogenase.

MCAF Abbreviation for macrophage chemotactic and activating factor.

McArdle's Disease An inherited glycogen storage disorder characterized by the accumulation of glycogen in skeletal muscle and painful muscle cramps due to the deficiency of glycogen phosphorylase in the muscle tissue.

MCBM Abbreviation for muscle capillary basement membrane.

MCBP Abbreviation for a combination drug containing melphalan, cyclophosphamide, BCNU, and prednisone.

MCCase Abbreviation for methylcrotonoyl-CoA carboxylase.

McClung Toabe Egg-Yolk Agar An agar-based medium for isolation and identification of clostridia and other anaerobes. It contains peptone, yeast extract, glucose, eggyolk, Na_2HPO_4, NaCl, $MgSO_4$, and agar.

MCCNU Abbreviation for methyl CCNU (methyl-1-(2-chloroethyl)-3-cyclohexyl-1-nitrosourea).

MCD Abbreviation for 1. magnetic circular dichroism; 2. mast cell degranulation.

McDonough Feline Sarcoma Virus A retrovirus that causes various neoplastic and degenerative diseases of the hematopoietic system.

MCDT Abbreviation for mast cell degranulation test.

MCF Abbreviation for 1. a combination drug containing mitoxantrone, cytoxan, and fluorouracil; 2. macrophage chemotactic factor.

MCF Virus Abbreviation for mink cell focus-forming virus, a variant of murine leukemia virus (Retroviridae). It can be detected and assayed by its ability to form foci on the monolayer of mink lung cells.

MCFA Abbreviation for medium-chain fatty acid.

McFarlane's Method A method for radioiodination of proteins in which proteins are reacted with iodine monochloride which has been previously equilibrated with radioiodide.

mcg Abbreviation for microgram.

MCH Abbreviation for mean corpuscular hemoglobin.

mCi Abbreviation for millicurie.

MCK Abbreviation for muscle creatine kinase.

M-Components An abnormal immunoglobulin that appears in an increased concentration in the serum or urine in patients with macroglobulinaemia (Waldenstrom's syndrome) and multiple myeloma.

mCoul Abbreviation for millicoulomb.

MCP Abbreviation for 1. a combination drug containing melphalan, cytoxan, and prednisone; 2. membrane cofactor protein; 3. mast cell protease; 4. metaclopramide; 5. methyl-accepting chemotaxis protein, a protein that can be methylated in response to signal transmission; 6. monocyte chemoattractant protein.

MCP Abbreviation for methyl-accepting chemotaxis protein, a protein of the inner face of the cytoplasmic membrane that can be methylated in response to signal transmission.

MCP-1 Abbreviation for monocyte chemo-attractant protein-1.

MCPA (mol wt 201) Abbreviation for 2-methyl-4-chlorophenoxy-acetic acid, a selective weed killer.

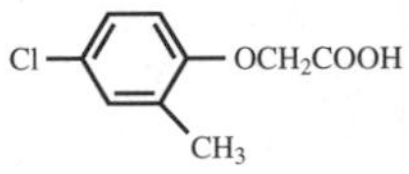

MCPA-CoA Abbreviation for methylenecyclopropylacetyl-CoA.

McrBCI A restriction endonuclease from *E. coli* K-12 with the following specificity:

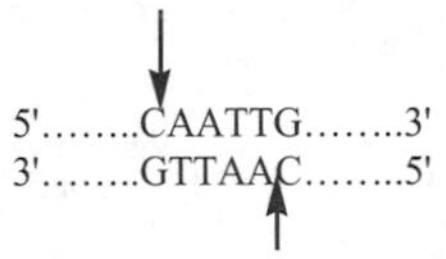

mCRP Abbreviation for monomeric CRP.

MCS Abbreviation for malonyl-CoA synthetase.

MCSF Abbreviation for macrophage colony-stimulating factor.

MCT Abbreviation for 1. medium-chain triglyceride; 2. mono-carboxylate transporter.

MCTD Abbreviation for mixed-connective tissue disease.

mcU Abbreviation for micro-unit.

MCV Abbreviation for mean cell volume.

MD Abbreviation for 1. maximum dose; 2. minimum dose; 3. molecular dynamics; 4. muscular dystrophy.

MDA Abbreviation for monodehydroascorbate.

Mda Abbreviation for megadalton (10^6 daltons).

MDC Abbreviation for 1. malonyl-CoA decarboxylase; 2. methionine decarboxylase; 3. Monodansylcadaverine; 4. minimum detectable concentration.

MDC Protein Abbreviation for metallo-proteinase-like, disintegrin-like, cysteine-rich protein.

MDCK Cell Abbreviation for Madin-Darby canine kidney cells, a heteroploid cell line derived from kidney of a dog.

MDH Abbreviation for 1. malate dehydrogenase; 2. methanol dehydrogenase.

MDHAR Abbreviation for monodehydroascorbate reductase.

M-Disk The central region of the A band in striated muscle (also known M-line).

MDMA Abbreviation for 3,4-methylene dioxymethamphetamine.

mDNA Abbreviation for mitochondrial DNA.

MDP Abbreviation for muramyl dipeptide (*N*-acetylmuramyl-L-alanyl-D-isoglutamine, a component of bacterial cell wall and an immunoregulator.

MDR Abbreviation for 1. minimum daily requirement; 2. multi-drug resistant.

MDR Protein Abbreviation for multi-drug resistant protein, an integral transmembrane glycoprotein involved in multiple drug resistance.

Me Abbreviation for methyl group.

2ME Abbreviation for 2-mercaptoethanol.

MEA Abbreviation for mercapto-ethylamine.

Mean The average of a series of values divided by the number of values.

Mean Corpuscular Hemoglobin The number of grams of hemoglobin per 100 ml of packed red blood cells.

Mean Corpuscular Volume The volume of the average red blood cell in a given blood sample.

Measles An acute, infectious childhood disease caused by a paramyxovirus and characterized by coryza, cough, fever, and appearance of rash on the body.

Measurin A trade name for aspirin, an antipyretic and analgesic agent.

Mebaral A trade mane for mephobarbital, an anticonvulsant.

Mebhydroline (mol wt 276) An antihistaminic agent.

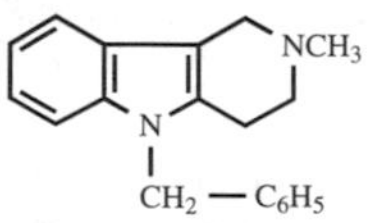

Mebiquine (mol wt 401) An antidiarrheal agent.

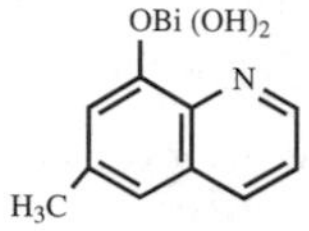

Mebutamate (mol wt 232) An antihypertensive agent.

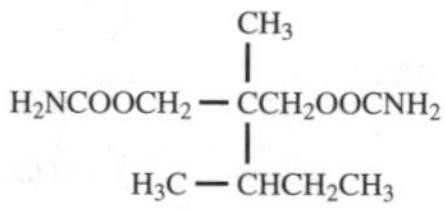

MEC Abbreviation for minimum effective concentration.

Mecamylamine (mol wt 167) An antihypertensive agent and a ganglionic blocker that competes with acetylcholine for ganglionic choline receptor.

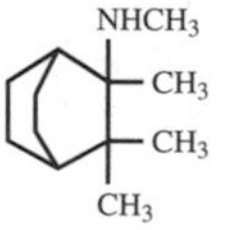

MeCCNU Abbreviation for methyl-CCNU.

Mechanism The fundamental physical and chemical processes involved in an action or a reaction.

Mechanochemical Coupling Hypothesis The coupling of ATP synthesis to the operation of the electron transport system.

Mechanoreceptors Sensory receptors that respond to a mechanical stimulation such as pressure, touch, stretch notion, and sound.

Mechlorethamine (mol wt 156) An antineoplastic and alkylating agent that cross-links strands of cellular DNA and interferes with RNA transcription.

$$CH_3N\ (CH_2CH_2Cl)_2$$

Mechlorethamine Oxide Hydrochloride (mol wt 209) An antineoplastic agent.

Meclizine (mol wt 391) An antiemetic agent.

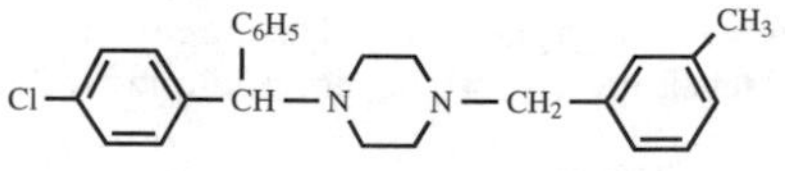

Meclocycline (mol wt 477) An antibacterial agent.

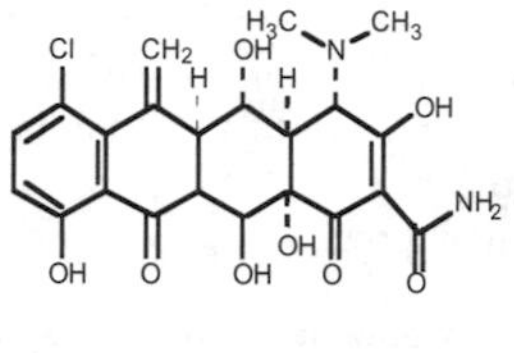

Meclofenamic Acid (mol wt 296) An anti-inflammatory and antipyretic agent.

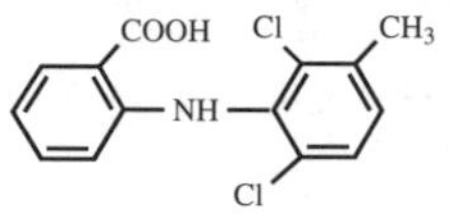

Meclofenoxate (mol wt 258) A CNS stimulant.

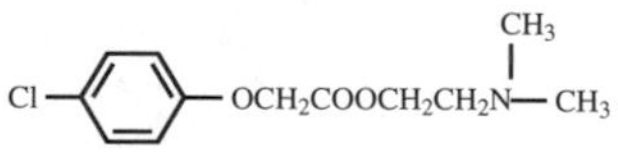

Meclomen A trade name for meclofenamate, an anti-inflammatory and antipyretic agent.

Mecloxamine (mol wt 318) An anticholinergic and sedative agent.

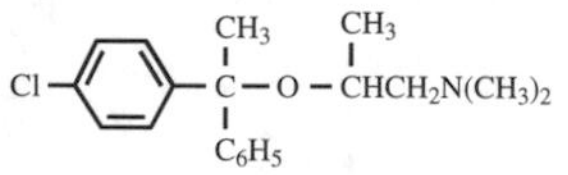

Mecoqualone (mol wt 270) A sedative and hypnotic agent.

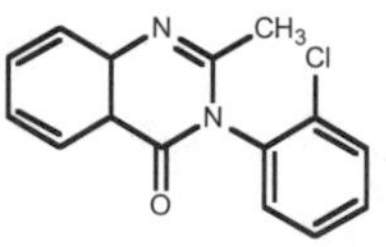

mEC-SOD Abbreviation for mutant extracellular superoxide dismutase.

Meda Cap A trade name for acetaminophen, an analgesic and antipyretic agent.

Meda Tab A trade name for acetaminophen, an analgesic and antipyretic agent.

Medetomidine (mol wt 200) A sedative and analgesic agent.

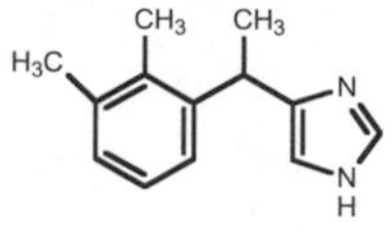

Medial Cisternae of Golgi Apparatus One of three different types of flattened membranous sacs (e.g., cis, medial, and trans cisternae) of the Goligi apparatus.

Mediated Transport The transport of solute across a biological membrane that requires the participation of transport protein or a nonprotein transporting agent.

Medibazine (mol wt 386) A coronary vasodilator and a bronchodilator.

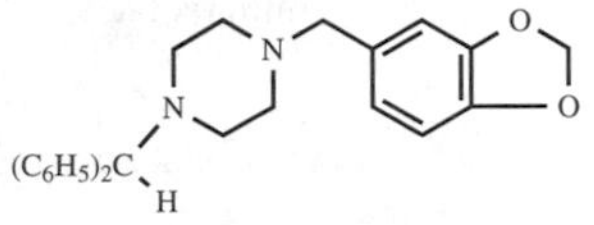

Medical Microbiology The science that deals with the pathogenicity of microorganisms and their effects on humans.

Medical Mycology The science that deals with human diseases caused by fungi.

Medicyline A trade name for tetracycline, an antibiotic.

Medifoxamine (mol wt 257) An antidepressant.

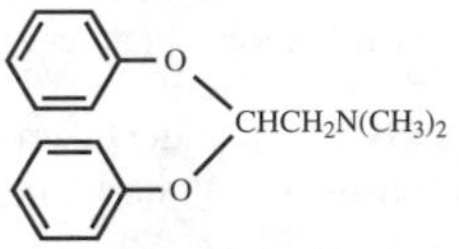

Medigesic A trade name for a combination drug containing acetaminophen, caffeine, and butalbital, used as an analgesic and antipyretic agent.

Medihaler-Epi A trade name for epinephrine bitartrate, a bronchodilator that stimulates both alpha and beta adrenergic receptors.

Medihaler-Ergotamine A trade name for ergotamine tartrate, an adrenergic blocker that acts on beta-adrenergic receptors.

Medihaler-Iso A trade name for isoproterenol hydrochloride, an adrenergic agonist used as a bronchodilator and antiasthmatic agent.

Medilax A trade name for phenolphthalein, a laxative that promotes fluid accumulation in the colon and small intestine.

Medilium A trade name for chlordiazepoxide hydrochloride, an antianxiety agent that depresses the CNS.

Medipren A trade name for ibuprofen, an analgesic and antipyretic agent.

Mediquell A trade name for dextromethorphan hydrobromide, an antitussive agent.

Meditran A trade name for meprobamate, an antianxiety agent that depresses the CNS.

Medium A lipid or solid nutrient preparation used for culture of microorganisms, cells, and tissues.

Medium-Chain Acyl-CoA Dehydrogenase The enzyme involved in the metabolism of medium-chain-length fatty acids. It has been implicated in the infant sudden death syndrome.

Medium-Chain Fatty Acid Referring to fatty acids with 4 to 12 carbon atoms.

Medium-Chain Fatty Acid Thiokinase The enzyme that catalyzes the formation of acyl-CoA from fatty acids with 4 to 12 carbon atoms.

Medmain (mol wt 202) A serotonin inhibitor.

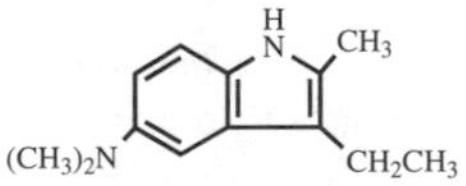

Medralone A trade name for methylprednisolone acetate, a corticosteroid used as an anti-inflammatory agent.

Medrol A trade name for corticosteroid methylprednisolone acetate, used as an anti-inflammatory agent.

Medrone A trade name for methylprednisolone acetate, a corticosteroid used as an anti-inflammatory agent.

Medroxyprogesterone (mol wt 344) A progesterone and an estrus regulator that suppresses ovulation.

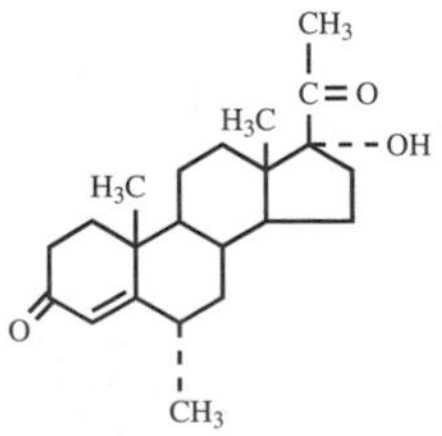

Medrylamine (mol wt 285) An antihistaminic agent.

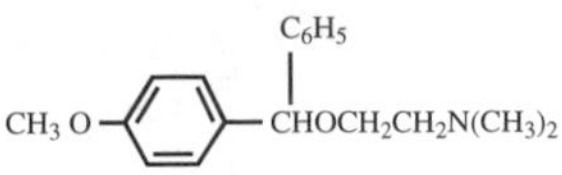

Medrysone (mol wt 344) A glucocorticoid that decreases the infiltration of leukocytes at the site of inflammation.

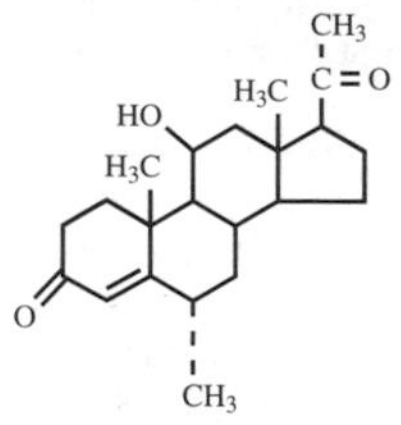

Med-Seltzer A trade name for a combination drug containing sodium bicarbonate, aspirin, and citric acid, used as an antacid and an adsorbent.

Medulla The central portion of an organ or structure.

Medulloblastoma A cancer of the brain.

MEE Abbreviation for methylethyl ester.

MEF Abbreviation for mouse embryonic fibroblast or mouse embryo fibroblast.

Mefenamic Acid (mol wt 241) An anti-inflammatory agent.

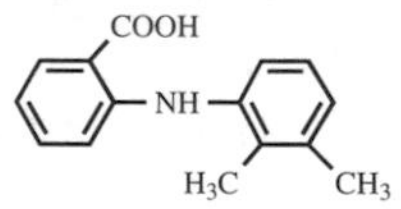

Mefloquine Hydrochloride (mol wt 415) An antimalarial agent.

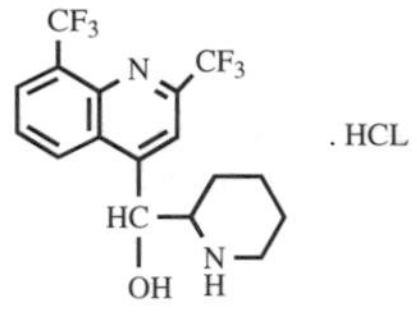

Mefluidide (mol wt 310) An herbicide and a plant growth regulator.

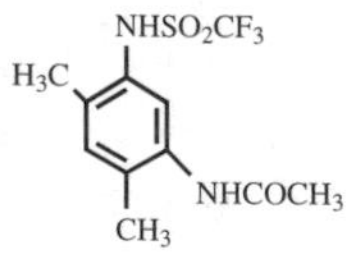

Mefoxin A trade name for cefoxitin sodium, an antibiotic that inhibits bacterial cell wall synthesis.

Mefruside (mol wt 383) A diuretic agent.

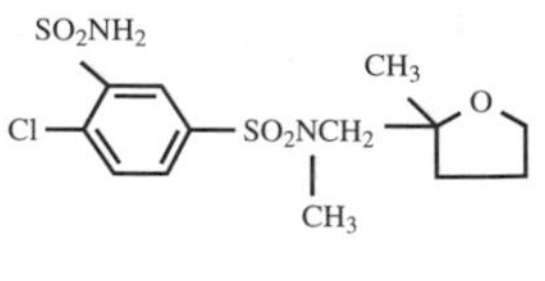

MEG Abbreviation for mercaptoethyl guanidine.

1-MeG Abbreviation for 1-methylguanine.

7-MeG Abbreviation for 7-methylguanine.

Mega- A prefix meaning 1. 1 million (10^6) and 2. large.

Megace A trade name for megestrol acetate, an antineoplastic agent.

Megacillin A trade name for penicillin G potassium, an antibiotic that inhibits bacterial cell wall synthesis.

Megacins Bacteriocins produced by strains of *Bacillus megaterium.*

Megadalton 10^6 daltons.

Megadyne An unit of force equal to 1 million dynes.

Megagesic A trade name for a combination drug containing acetaminophen and hydrocodone bitartrate, used as an analgesic agent.

Megakaryoblast A cell that gives rise to the platelet-forming megakaryocyte.

Megakaryocyte A large cell with multilobed nucleus from which blood platelets are derived.

Megalencephaly A disorder characterized by an abnormal overgrowth of brain tissue.

Megaloblast A large, immature red blood cell.

Megaloblastic Anemia A blood disease caused by the deficiency of vitamin B_{12} and characterized by the appearance and spreading of abnormally large red blood cells (megaloblasts).

Megalocyte A large primitive red blood cell.

Megasphaera A genus of Gram-negative bacteria (Vellonellaceae).

Megaspore Haploid cell produced by the meiotic division of megasporocytes in flowering plants.

Megasporocyte Primordial germ cell in the ovary of flowering plants.

Megestrol Acetate (mol wt 385) A progestogen antineoplastic agent and estrus regulator.

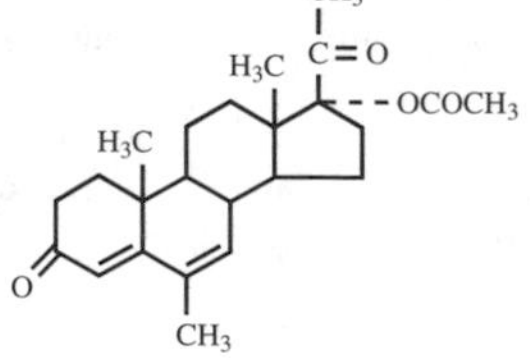

MEGF (mEGF) Abbreviation for mouse epidermal growth factor.

Meglutol (mol wt 162) An antihyperlipoproteinemic agent.

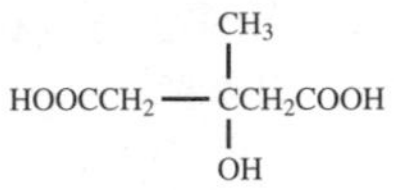

Megostat A trade name for megestrol acetate, an antineoplastic agent.

MEH Abbreviation for microsomal epoxide hydrolase.

Mehler Reaction A photosynthetic reaction that produces peroxide from water and molecular oxygen.

Meiocyte Any cell capable of undergoing meiosis.

Meiosis A process in which two successive cell divisions produce one duplication of chromosomes and four haploid cells (also called reduction division).

Meiospore Spore formed by meiosis.

Meiotic Pertaining to meiosis.

Meiotic Spindle The meiotic equivalent of the mitotic spindle.

Meister Cycle A cyclic metabolic pathway for transport of amino acids across a cell membrane through the synthesis and breakdown of glutathione.

MEK Abbreviation for 1. mitogen extracellular-regulated protein kinase; 2. methylethyl ketone.

MEL Abbreviation for murine erythroleukemia.

Melamine (mol wt 126) A cancer-causing agent.

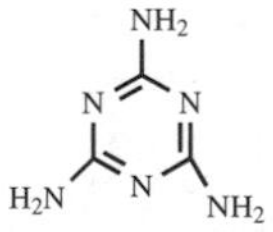

Melanex A trade name for hydroquinone, a depigmentation agent used for the treatment of hyperpigmented skin condition.

Melanins A pigment of animal origin and polymer of indole quinone.

Melanism The abnormal pigmentation of the skin or tissue due to the accumulation of melanin.

Melanocyte The pigment cells that produce melanin.

Melanocyte Stimulating Hormone A peptide hormone produced by the pituitary gland that stimulates the production of melanocytes and malanins.

α-Melanocyte Stimulating Hormone A peptide hormone resulting from cleavage of ACTH (adrenocorticotropic hormone).

γ-Melanocyte Stimulating Hormone A peptide hormone resulting from cleavage of proopiomelanocortin (POMC).

Melanoderm The dark pigmentation of the skin resulting from the accumulation of melanin.

Melanogen Substance which can be transformed into melanin.

Melanoma A tumor derived from melanocytes.

Melanophore Cells found in the skin of lower vertebrates (e.g., amphibian skin, fish scales) that consist of melanin-containing granules.

Melanosis The development of pigment (e.g., melanin) in the skin.

Melanosome A tyrosinase-containing intracellular organelle of the melanocyte.

Melanotrophin Variant spelling of melanotropin.

Melanotropin Referring to melanocyte-stimulating hormone.

Melanuria The presence of melanin in the urine giving the urine a blackish color.

Melatonin (mol wt 232) A hormone produced by the pineal gland. It has been used to relieve jet lag and in the treatment of insomnia.

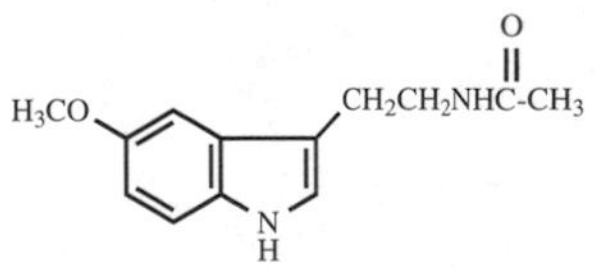

Melengestrol (mol wt 354) A progestogen and an antineoplastic agent.

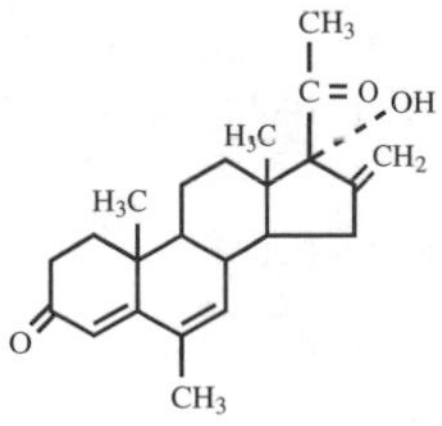

Melezitose (mol wt 504) A trisaccharide containing 2 glucoses and one fructose.

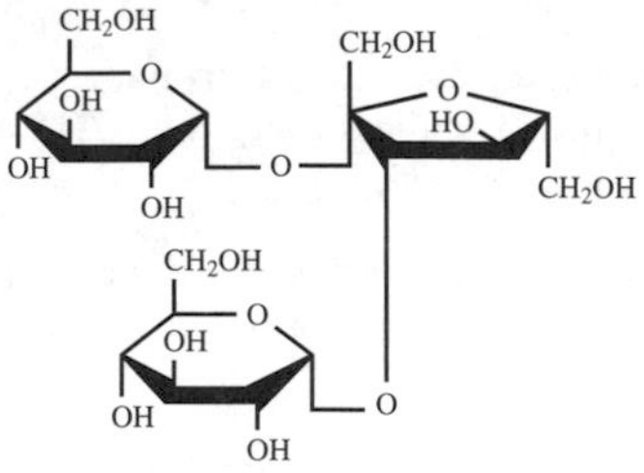

Melfiat A trade name for phendimetrazine tartrate, a CNS stimulant that promotes transmission of nerve impulses.

Melibiose (mol wt 342) A disaccharide containing galactose and glucose.

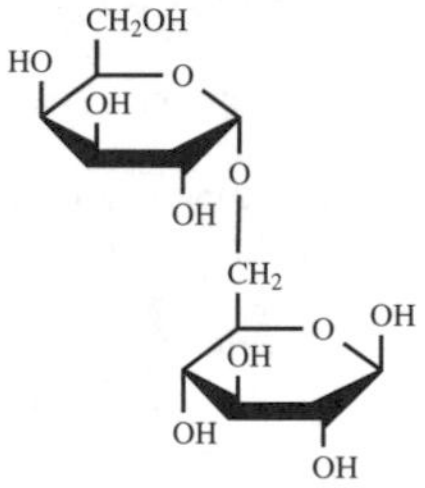

Melissic Acid (mol wt 453) An organic acid.

$CH_3(CH_2)_{28}COOH$

Melitracen (mol wt 291) An antidepressant.

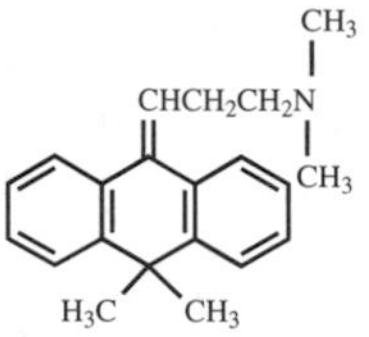

Melittin A basic polypeptide from venom of the honeybee.

Mellaril A trade name for thioridazine hydrochloride, an antipsychotic agent that blocks the postsynaptic dopamine receptors in the brain.

Mellaril-S A trade name for thioridazine hydrochloride, an antipsychotic agent that blocks the postsynaptic dopamine receptors in the brain.

Melphalan (mol wt 305) An alkylating and antineoplastic agent that cross-links strands of cellular DNA and interferes with RNA transcription causing cell death.

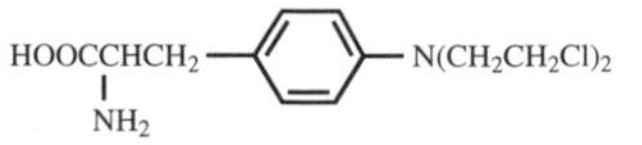

Melting 1. An increase in fluidity with increased temperature of a membrane. 2. Thermal denaturation of double-stranded DNA into two component strands.

Melting Curve A graph obtained by plotting changes in absorbance at 260 nm as a function of temperature for the thermal transition of double-stranded DNA to single-stranded DNA when a sample of double-stranded DNA is heated.

Melting Temperature A temperature at which the transition of double-stranded DNA to single-stranded DNA is halfway complete during thermal denaturation (abbreviated as T_m).

MEM Abbreviation for 1. macrophage electrophoretic mobility; 2. minimal essential medium; 3. modified Eagle's medium.

Memantine (mol wt 179) A skeletal muscle relaxant.

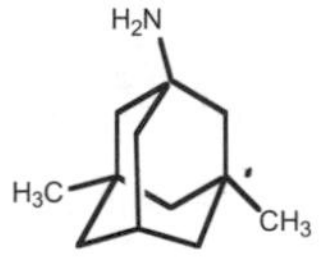

Membrane Permeability barrier surrounding delineating cells or organelles that consists of a bilayer of phospholipids and its associated proteins.

Membrane Anchorage Sequence The N-terminal sequence of a protein that is essential for initiation of transfer of protein across the membrane.

Membrane Asymmetry Property of a membrane based on the differences between the two monolayers and the proteins associated with the membrane.

Membrane Attack Complex Complex formed by components of complement that creates cytolytic pores in the membrane of a cell leading to the lysis of the cell.

Membrane Channel A transmembrane complex that allows the small solutes, ions, or molecules to diffuse passively across a membrane.

Membrane Fluidity The viscous property of the interior of the biological membrane.

Membrane Lipids Lipids associated with the biological membrane, e.g., phospholipids, and cholesterol.

Membrane Permeability Relative ability of a membrane to allow a specific solute molecule to traverse across a membrane.

Membrane Potential The voltage that exists across a cell membrane (usually, the inside of a cell is negatively charged with respect to the outer surface). A typical potential of a eukaryotic cell is about –60mV.

Membrane Proteins The proteins associated with the membrane. There are two types of membrane proteins, namely peripheral and integral membrane proteins.

Membrane Receptor Integral membrane protein that has a binding site on the membrane surface for binding of a specific ligand. The binding of a specific ligand with receptor initiates a particular intracellular event or series of events.

Membrane Recycling A process by which a membrane is inserted, fused with an internal membranous compartment, and then re-incorporated into the plasma membrane.

Membrane Transport The transfer of solutes across a membrane.

Membrane Turnover Changes in membrane composition, e.g., continual removal and replacement of the lipids or proteins in a membrane.

Membrane Vesicle An enclosed membrane structure formed from actions, phagocytosis, or endocytosis.

Memory B Cells The reserved population of B cells directed against a specific antigen that proliferate into antibody-secreting plasma cells upon subsequent exposure to the same antigen.

Memory Cells Referring to memory B and T cells that mediate rapid, efficient secondary immunological responses upon exposure to the same antigen.

Memory T Cells Reserved population of T cells directed against a specific antigen that proliferate into killer T cells upon subsequent exposure to the same antigen.

Menadione (mol wt 172) A synthetic naphthoquinone derivative having vitamin K properties.

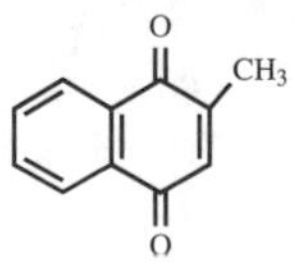

Menadione Reductase Synonym of NADPH dehydrogenase.

Menadol A trade name for ibuprofen, an anti-inflammatory, antipyretic, and analgesic agent.

Menaval A trade name for estradiol cypionate, which increases the synthesis of DNA, RNA, and protein in responsive tissue.

Menazon (mol wt 281) An insecticide (acaricide).

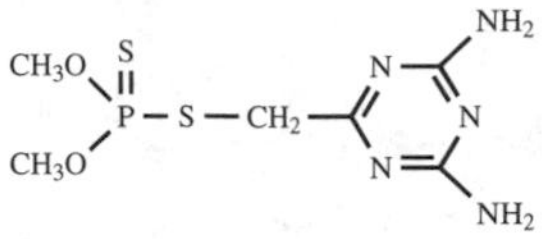

Menbutone (mol wt 258) A choleretic agent.

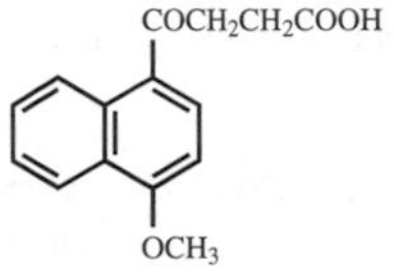

Mendelian Inheritance Laws The laws that govern inheritance and explain the chromosomal segregation (law of segregation) and independent assortment (law of assortment).

Menest A trade name for esterified estrogens that increase synthesis of DNA, RNA, and protein in responsive tissues. It also reduces the release of FSH and LH from the pituitary.

Mengovirus A virus in the family of Picornaviridae that causes encephalitis in rodents.

Meni-D A trade name for mecliqine hydrochloride, an antiemetic agent.

Meniere's Disease A disorder caused by the disruption of sodium metabolism and characterized by intense dizziness or vertigo.

Meninges The special layer of tissue that protects the brain and the spinal cord.

Meningitis Inflammation of meninges (outer covering) of the brain and spinal cord.

Meningitis Vaccine A killed bacterial vaccine that promotes active immunity against meningitis caused by bacterial infection.

Meningococcal Meningitis An epidemic meningitis caused by *Neisseria meningitidis.*

Meningoencephalitis Inflammation of the brain and its meninges.

Meningomyelitis Inflammation of the spinal cord and its meninges.

Meniscus The curved upper surface of a liquid column.

Meniscus A genus of Gram-negative, anaerobic bacteria.

Menogaril (mol wt 542) An antineoplastic agent.

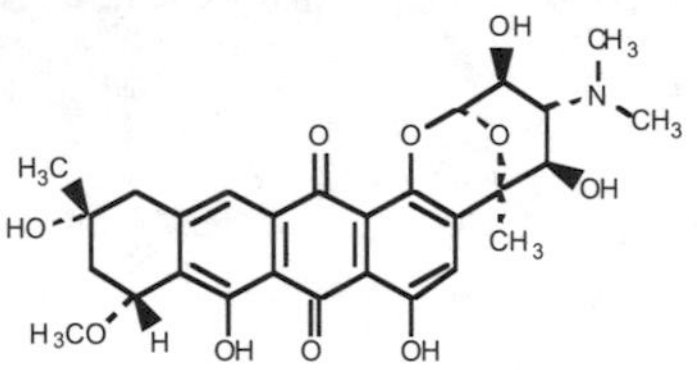

Menomune-A/C A trade name for meningitis vaccine, a vaccine active against bacterial infections.

Menomune-A/C/Y/W-135 A trade name for meningitis vaccine, a vaccine active against bacterial infections.

Menstrual Cycle A monthly repeating cycle of the uterus to receive the fertilized egg and to discharge the uterus lining.

Menstruation Monthly discharge of a bloody mass from the uterus.

Menthol (mol wt 156) A phenol derivative (2-methyl-5-isopropylphenol) obtained from peppermint oil that gives a sensation of coolness by selective stimulation of nerve endings sensitive to cold, also an anti-itching agent.

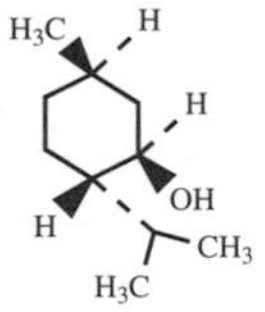

MeOH Abbreviation for methyl alcohol.

MEOS Abbreviation for microsomal ethanol oxidizing system.

Meparfynol (mol wt 98) A hypnotic and sedative agent.

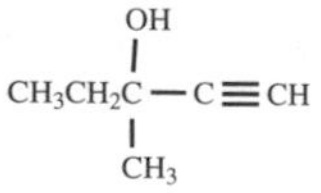

Meparfynol Carbamate (mol wt 141) A hypnotic and sedative agent.

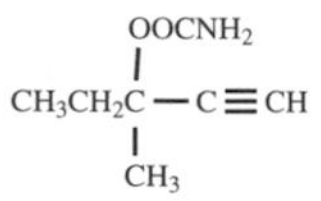

Mepazine (mol wt 310) A tranquilizer.

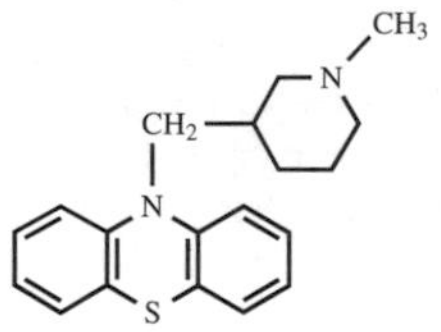

Mepenzolate Bromide (mol wt 420) An anticholinergic agent.

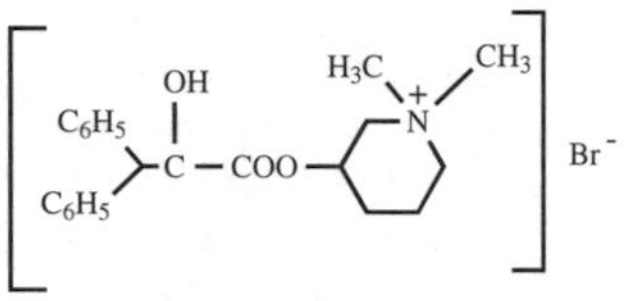

Mepergan A trade name for a combination drug containing meperidine and promethazine HCl used as an analgesic and sedative agent.

Meperidine (mol wt 247) A narcotic analgesic agent that binds opiate receptors in the CNS.

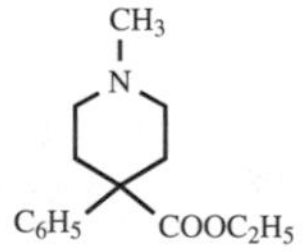

Mephaquin A trade name for mefloquine hydrochloride, an antimalarial agent.

Mephenhydramine (mol wt 269) An antihistaminic agent.

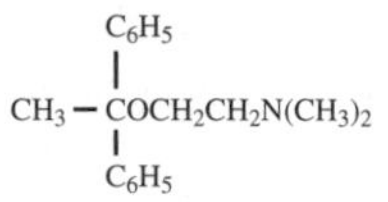

Mephentermine (mol wt 163) A vasopressor that stimulates alpha- and beta-adrenergic receptors.

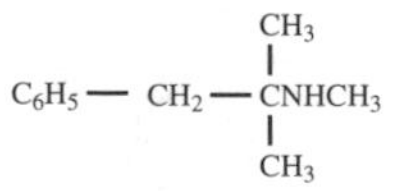

Mephenytoin (mol wt 218) An anticonvulsant that either increases efflux or decreases influx of sodium ions across cell membranes in the motor cortex during generation of nerve impulses.

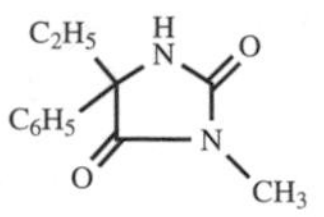

Mephobarbital (mol wt 246) An anticonvulsant and sedative agent that depresses monosynaptic and polysynaptic transmission in the CNS.

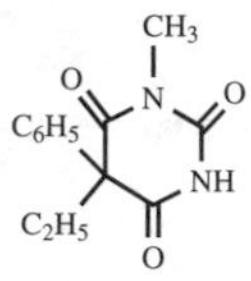

Mephyton A trade name for vitamin K_1 (phytonadione).

Mepindolol (mol wt 262) An antihypertensive and an antianginal agent.

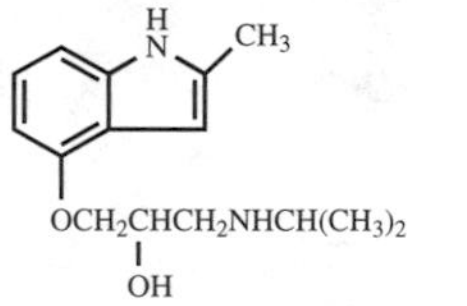

Mepiprazole (mol wt 305) A tranquilizer.

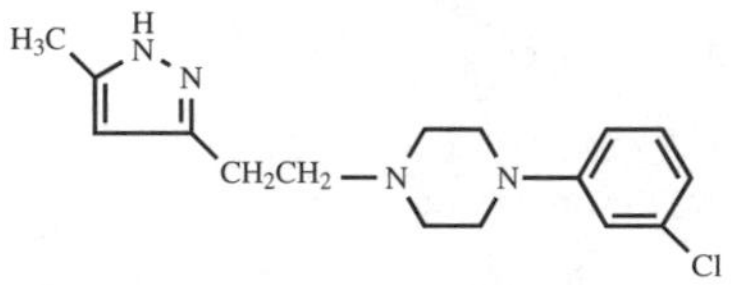

Mepitiostane (mol wt 405) An antineoplastic agent.

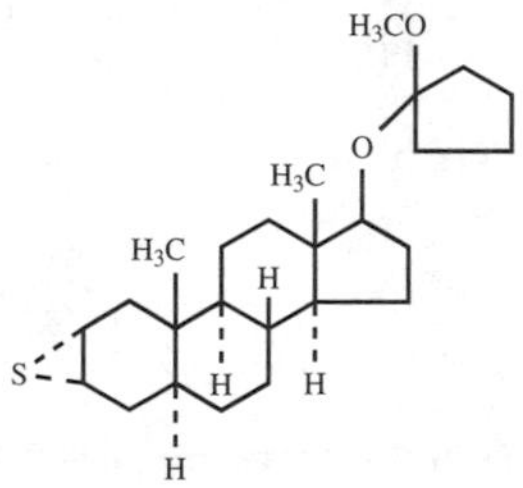

Mepivacaine (mol wt 246) A local anesthetic agent that blocks depolarization by interfering with sodium-potassium exchange across the nerve cell membrane preventing generation of nerve impulses.

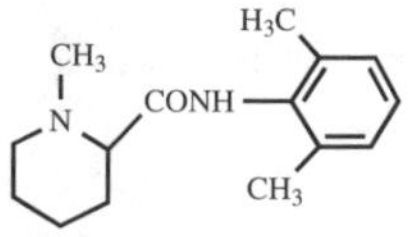

Meprin A A protease.

Meprobamate (mol wt 218) An antianxiety agent that depresses the CNS.

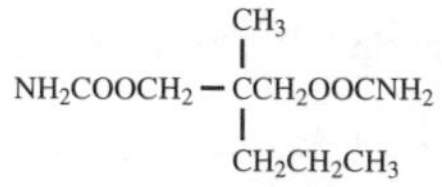

Meprolone A trade name for methylprednisolone acetate, a corticosteroid.

Mepron A trade name for atovaquone, an antiprotozoal agent.

Meprospan A trade name for meprobamate, an antianxiety agent.

Meptazinol (mol wt 233) A narcotic analgesic agent.

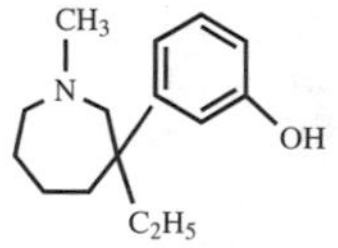

Meq Abbreviation for milliequivalent.

Mequitazine (mol wt 322) An antihistaminic agent.

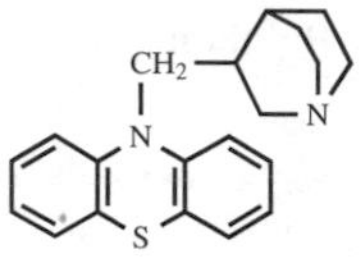

Meralluride (mol wt 611) A diuretic agent.

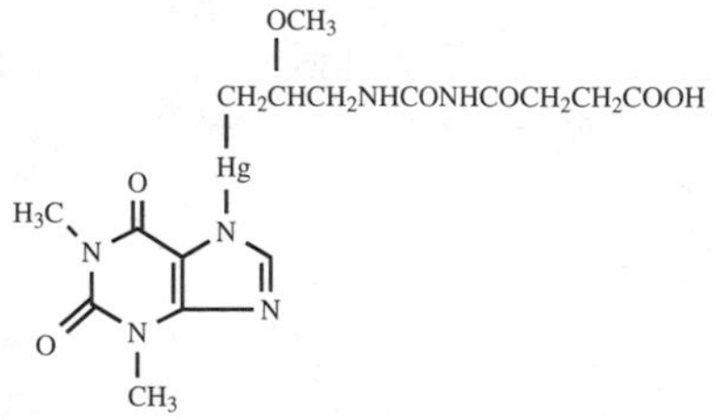

Meravil A trade name for amitriptyline hydrochloride, an antidepressant agent that increases the level of norepinephrine and serotonin in the CNS.

Merbentyl A trade name for dicyclomine hydrochloride, an anticholinergic agent.

Mercamphamide (mol wt 488) A diuretic agent.

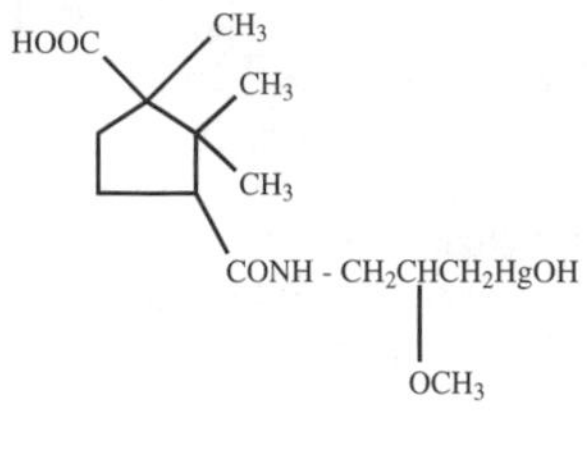

Mercaptan An organic compound that has a -SH group directly connected to a carbon atom (also known as thiol).

Mercapto Group Referring to the -SH group.

Mercaptoacetic Acid (mol wt 92) A protein protector.

$$HS\text{-}CH_2COOH$$

Mercaptoethanol (mol wt 78) A water-soluble thiokol used to protect sulfhydryl groups of enzymes or proteins against oxidation.

$$HS\text{-}CH_2CH_2OH$$

β-Mercaptoethylamine (mol wt 76) A component of acetyl-CoA or CoA.

$$NH_2\text{-}CH_2CH_2SH$$

β-Mercaptolactate Cysteine Disulfide (mol wt 242) A naturally occurring amino acid found in human urine.

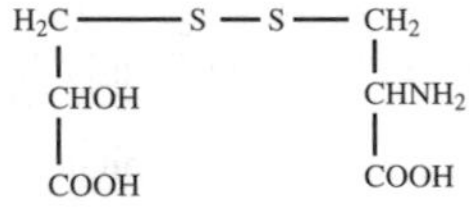

Mercaptomerin Sodium (mol wt 606) A diuretic agent.

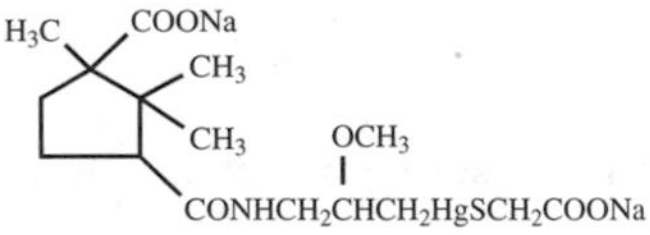

6-Mecaptopurine (mol wt 152) A substance that inhibits biosynthesis of AMP and GMP.

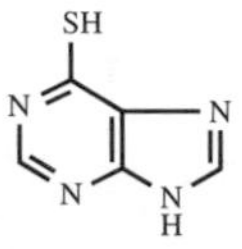

Mercaptopurine Riboside (mol wt 284) An analog of purine nucleoside.

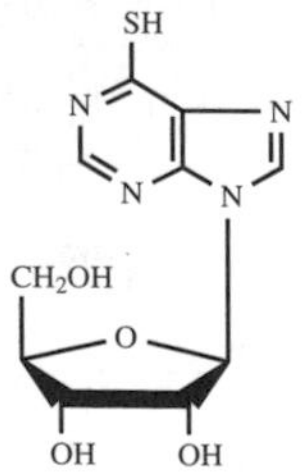

Mercaptopyruvate Sulfurtransferase The enzyme that catalyzes the following reaction:

3-Mercaptopyruvate + cyanide

⇅

Pyruvate + thiocyanate

Mercerization Treatment of cellulose with 20% sodium hydroxide to increase the affinity of cellulose for dyes and for greater tensile strength.

Mercumallylic Acid (mol wt 479) A diuretic agent.

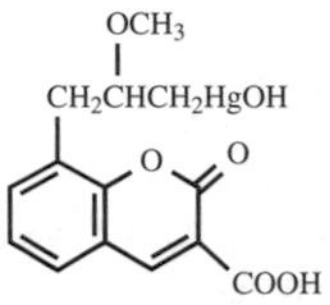

Mercurial Organic substance containing mercury.

Mercuric Succinimide (mol wt 397) An antibacterial agent.

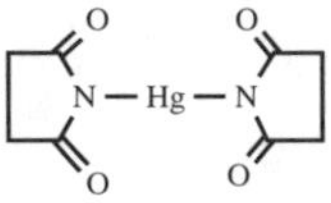

Mercury (Hg) A chemical element with atomic weight 201, valence 1 and 2.

Mercury (II) Reductase The enzyme that catalyzes the following reaction:

$$Hg + NADP^+ \rightleftharpoons Hg^{2+} + NADPH$$

Meridia A trade name for sibutramine hydrochloride monohydrate, an anorexic agent.

Meristem Rapid dividing, undifferentiated cells that give rise to different cell types of plant tissue.

Merlin A membrane-stabilizing protein from fetal brain and other tissues such as kidney, lung, and breast.

Mero- A prefix meaning part.

Merocrine Gland A type of gland that discharges secretary products without loss of cytoplasm, e.g., salivary gland.

Meromyosin Fragment of the myosin molecule obtained by trypsin digestion that contains ATPase activity and calcium-binding properties (also known as heavy meromyosin).

Meropenem (mol wt 437) An antibacterial agent.

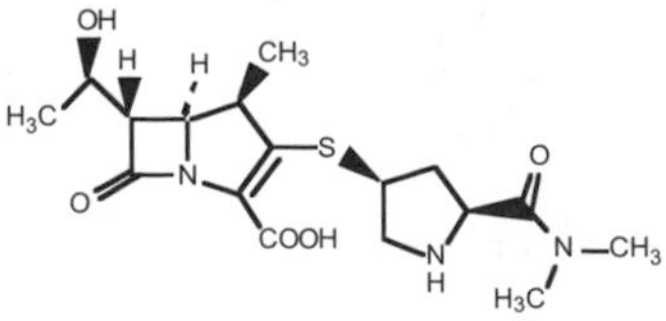

Merosin A tissue-specific basement-membrane protein and the M chain of laminin.

Merozygote A bacterium that is part diploid, part haploid.

Merrem A trade name for meropenem, an antibacterial agent.

mers A term used to denote number of monomers in an oligomer, e.g., an oligonucleotide consisting of 16 nucleotides is called 16 mers.

Mersalyl (mol wt 506) A diuretic agent and an inhibitor for exchange of phosphate and hydroxyl ions across the inner mitochondrial membrane.

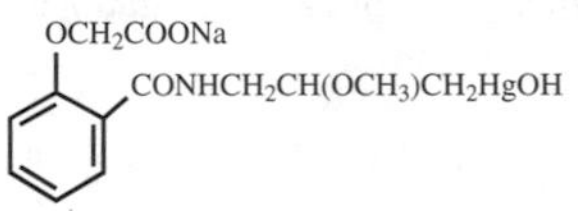

Meruvax II A trade name for an attenuated live vaccine of rubella virus.

MES (mol wt 194) Abbreviation for 4-morpholinoline-ethanesulfonic acid used for preparation of buffer.

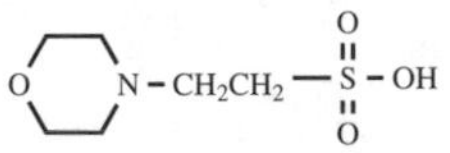

Mesalamine (mol wt 153) An anti-inflammatory agent used for treatment of ulcerative colitis.

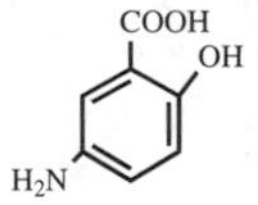

Mesantoin A trade name for mephenytoin, an anticonvulsant.

MeSATP Abbreviation for 2-methylthio-ATP.

Mescaline (mol wt 211) A narcotic agent found in cactus.

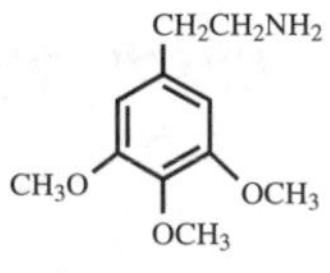

Meselson-Stahl Experiment Experiment to demonstrate the mode of semiconservative replication of dsDNA employing heavy isotope and normal isotope through density gradient centrifugation.

Mesenchyme The immature, unspecialized form of connective tissue in animals, consisting of cells embedded in a tenuous extracellular matrix.

Mesentery Membranes that secure the stomach and intestines of the vertebrate to the body wall and contain blood vessels, nerves, and lymph vessels serving the gut.

Meslon A trade name for morphine sulfate, an analgesic agent.

Mesna (mol wt 164) A mucolytic and antineoplastic agent.

$$[HSCH_2CH_2SO_3]^- Na^+$$

Mesnex A trade name for mesna, an antineoplastic agent.

Meso- A prefix used to denote a structure located in the middle or a stage that appears at some intermediate time.

Meso-Carbon The carbon atom that has two identical or two nonidentical substituentes attached.

Mesoderm The middle of the three embryonic tissue layers that gives rise to supporting tissues, e.g., skeleton, muscles, bones, blood, and connective tissue.

Meso-Inositol One of the stereoisomers of inositol.

Mesophiles Organisms whose optimum growth temperature ranges from 20 to 45°C.

Mesophyll The tissue of leaf that is sandwiched between thc upper and lower epidermis in which photosynthetic reactions occur.

Mesophyll Cell Cells found in the interior of plant leaves in which the C_4 pathway is located.

Mesoridazine (mol wt 387) An antipsychotic agent that blocks postsynaptic dopamine receptors in the brain.

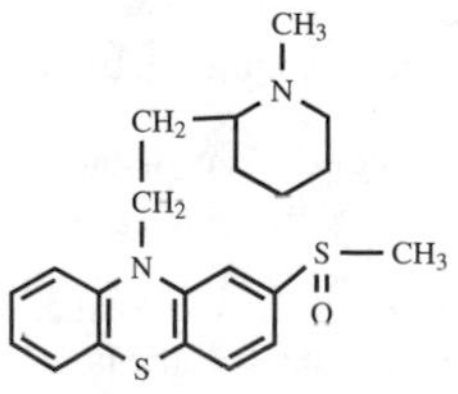

Mesosome An extensively infolded portion of the prokaryotic plasma membrane.

Mesotendon The connective tissue membrane that surrounds a tendon.

Messenger RNA (mRNA) The RNA that specifies the amino acid sequence for a polypeptide.

Mestinon A trade name for pyridostigmine bromide, which inhibits acetylcholinesterase preventing the destruction of acetylcholine.

Mestranol (mol wt 310) An estrogen used in combination with progestogen as an oral contraceptive.

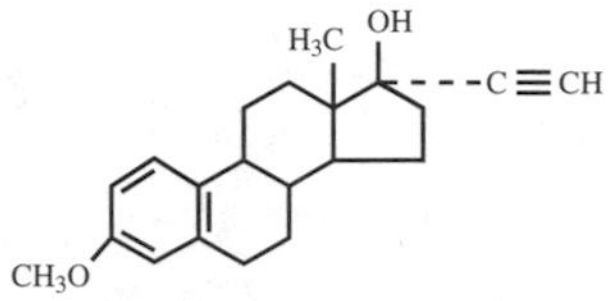

Met Abbreviation for methionine or methionyl.

Meta- A prefix used in biology to denote a change or a shift to a new form or level, e.g., metamorphosis.

Metabiosis A phenonmenon in which the growth and metabolism of one organism alters the environmental conditions allowing the growth of another organism.

Metabolic Pertaining to metabolism.

Metabolic Acidosis An acidosis resulting from metabolic changes leading to increase in concentration of acid or decrease of concentration of base in the body.

Metabolic Alkalosis An alkalosis resulting from a metabolic change leading to an abnormal loss in acid or increase in base.

Metabolic Antagonist A substance that inhibits a specific metabolic reaction due to its similarity in structure to the natural metabolite.

Metabolic Burst Biochemical response of phagocytes (e.g., neutrophils) for the elimination of foreign substances, antigens, or microorganisms, leading to the production of antimicrobial substances, e.g., superoxide, singlet oxygen and hydroxyl radical (also known as respiratory burst).

Metabolic Disorder Any disorder resulting from abnormality in metabolism.

Metabolic Inhibitor A substance that blocks a metabolic pathway at a specific point, thereby causing accumalation of the preceeding intermediate.

Metabolic Pathway A sequence of enzyme-mediated reactions that transform one compound to another and provide intermediates and energy for cellular functions. The metabolic pathway can be linear (e.g., glycolysis) or cyclic (e.g., Krebs cycle).

Metabolic Quotient A parameter used for measuring the rate of uptake or discharge of a metabolite by an organism.

Metabolic Rate A measure of the rate of a chemical reaction in an organism, e.g., rate of oxygen consumption.

Metabolic Shunt A pathway that uses some reactions of one pathway and bypasses the others.

Metabolism The overall enzymatic reactions that take place in an organism that include anabolic reactions (e.g., building of complex molecules) and catabolic reactions (e.g., breakdown of molecules to provide energy).

Metabolites The products or intemediates from any metabolic pathway.

Metabolize To transform by means of metabolism.

Metabutoxycaine Hydrochloride (mol wt 345) A local anesthetic agent.

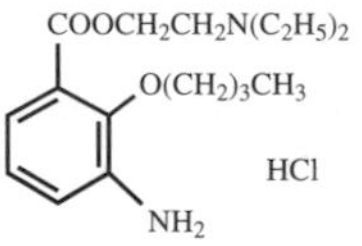

Metacentric Chromosome A chromosome whose centromere is located near the center so that the two arms are equal.

Metachromatic Dye A dye that stains cells or tissues with a color different from the color of the dye.

Metachromatic Granules The cytoplasmic granules of polyphosphate occurring in the cells of certain bacteria that appears different in color when stained with a basic dye.

Metachromasia The property by which a cell or tissue stains in a color different from the dye.

Metachromatic Leukodystrophy A disorder due to the deficiency of the enzyme arylsulfatase-A, which converts sulfatide to galactocerebroside.

Metachrosis The ability to change color.

Metaclazepam (mol wt 394) An anxiolytic agent.

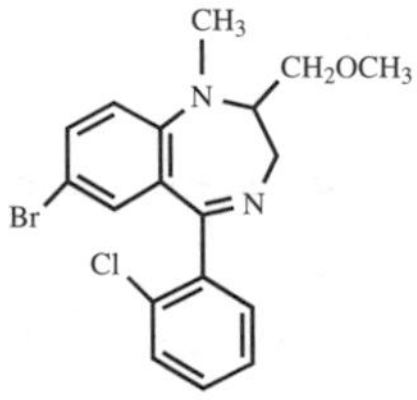

Metahydrin A trade name for trichlormethiazide, a diuretic agent that inhibits sodium reabsorption and increase the urine excretion of sodium and water.

Metal Ion Catalysis An enzyme-mediated catalytic reaction that requires the participation of metal ion.

Metallochromic Describing a dye, indicator, or stain that exhibits a distinctive color change when complexed with metal ions.

Metalloenzyme Enzymes that contain tightly bound metal ion or ions.

Metallogenium A genus of iron bacteria.

Metalloprotease A type of protease that requires metal ion for the catalytic reaction. The metalloprotease is sensitive to chelating agent (e.g., EDTA).

Metalloprotein A protein that contains a bound metal ion as part of its structure.

Metalloproteinase Synonym for metalloprotease.

Metallothionein A cysteine-rich ion-binding protein that binds heavy metal ion.

Metalone-TBA A trade name for prednisolone tebutate, a corticosteroid used as an anti-inflammatory agent.

Metamere A repeated unit or segment of a structure.

Metamerism Division of the body into segments, (e.g., in insects).

Metaminodiazepoxide Hydrochloride Synonym of chlordiazepoxide, an anti-anxiety agent.

Metamivam (mol wt 237) A cardiac and respiratory stimulant.

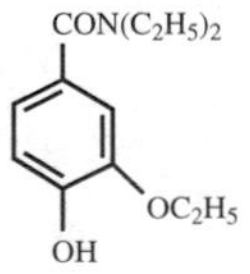

Metamorphosis Morphological and physiological changes that transform an organism from one stage to another (e.g., from lava to adult).

Metampicillin (mol wt 361) A semisynthetic antibiotic related to penicillin.

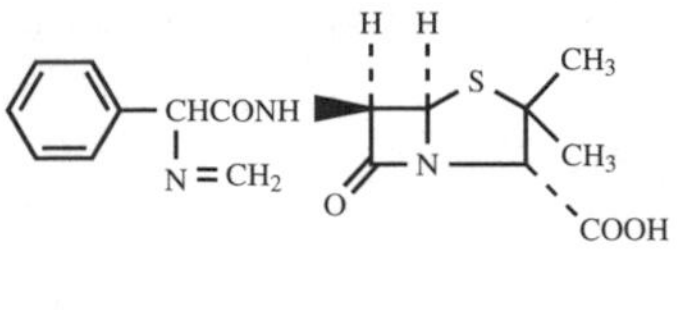

Metamucil A trade name for *Psyllium*, a laxative from seeds of *Psyllium* that absorbs water and expands to increase bulk and moisture content of the stool.

Metanil Yellow (mol wt 375) A dye and pH indicator.

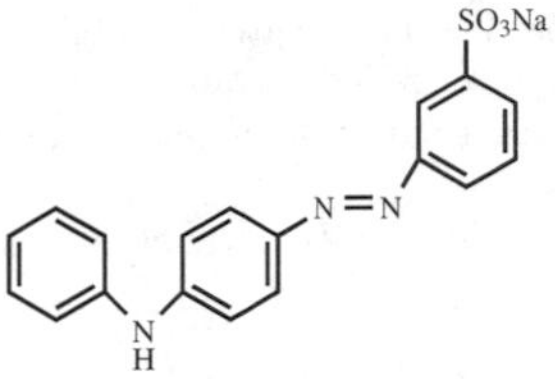

Metaphase A stage in mitosis in which chromosomes become condensed and attached to the spindle and migrate to the equator.

Metaphase Plate Plane between the two poles in which chromosomes are positioned at metaphase.

Metaplasia The change from one differentiated phenotype to another.

Metapramine (mol wt 238) An antidepressant.

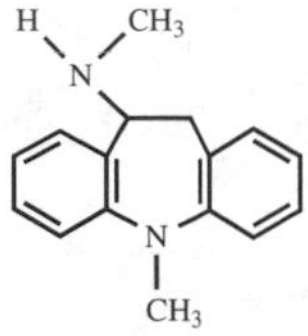

Metaprel A trade name for metaproterenol sulfate, a bronchodilator that acts on $beta_2$ adrenergic receptors.

Metaproteins Derived or denatured proteins that are soluble in acid and base but insoluble in neutral aqueous solvent.

Metaproterenol (mol wt 211) A bronchodilator that relaxes bronchial smooth muscle by acting on $beta_2$ adrenergic receptors.

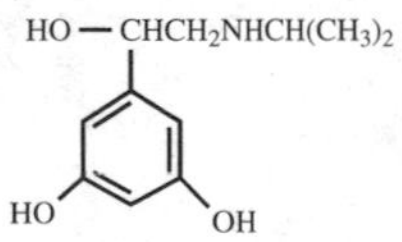

Metaraminol (mol wt 167) An adrenergic agent that stimulates alpha-adrenergic receptors within the synapathetic nervous system.

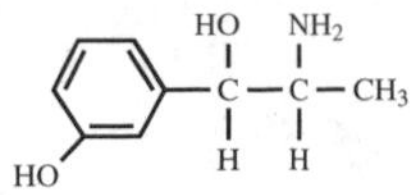

Metarhodopsin The structurally altered form of rhodopsin resulting from exposure of rhodopsin to light.

Metastasis The ability of a cancer cell to invade surrounding tissues, to enter the circulatory system, and to establish malignancy at a new site.

Metatensin Tablets A trade name for a combination drug containing trichlormethiazide and reserpine, used as an antihypertensive agent.

Metaxalone (mol wt 221) A skeletal muscle relaxant.

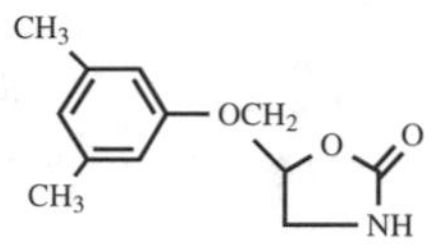

Metazocine (mol wt 231) A narcotic analgesic agent.

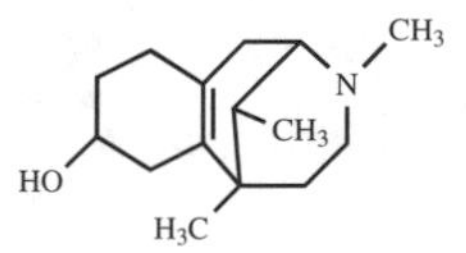

Metcaraphen (mol wt 317) An anticholinergic agent.

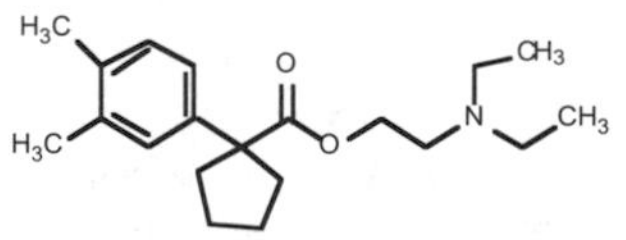

Meter An unit of length equal to 100 cm.

Metergoline (mol wt 404) A prolactin inhibitor.

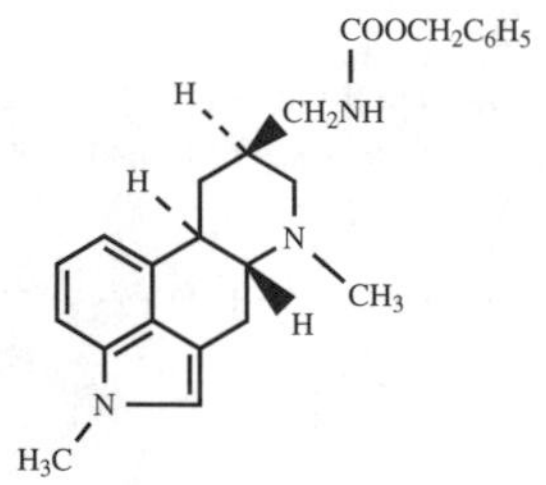

Meterosim The distention of the abdomen or intestine because of gas.

Metformin (mol wt 129) An antidiabetic agent.

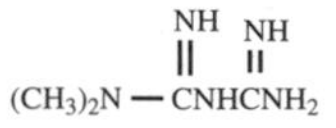

Methadone Hydrochloride (mol wt 346) A narcotic analgesic agent that binds with opiate receptors in the CNS.

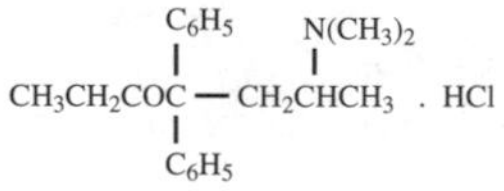

Methadose A trade name for methadone, an analgesic agent that binds opiate receptors in the CNS.

Methafurylene (mol wt 245) An antihistaminic agent.

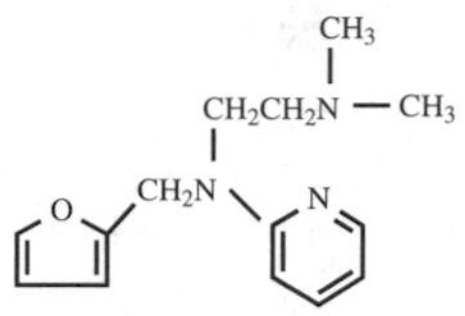

Methallatal (mol wt 226) An antiemetic agent.

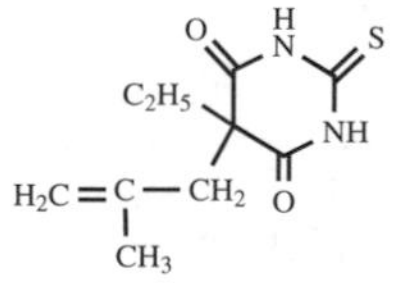

Methamphetamine (mol wt 149) A cerebral stimulant that promotes transmission of nerve impulses by releasing stored norepinephrine from the nerve terminals in the brain.

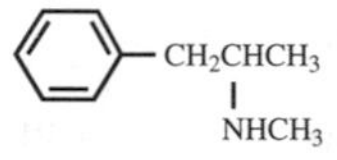

Methane (mol wt 16) A flammable gas.

$$CH_4$$

Methane Hydroxylase See methane monooxygenase.

Methane Monooxygenase The enzyme that catalyzes the following reaction:

$$\text{Methane} + \text{NADPH} \rightleftharpoons \text{Methanol} + \text{NADP}^+ + H_2O$$

Methano- A prefix denoting methane or the presence of $-CH_2-$ bridge in a polycyclic hydrocarbon.

Methanobacillus A genus of methanogenic bacteria in the family of Methanobacteriaceae.

Methanobacterium A genus of methanogenic bacteria in the family of Methanobacteriaceae.

Methanobrevibacter A genus of methanogenic bacteria in the family of Methanobacteriaceae.

Methanococcoides A genus of methanogenic bacteria in the family of Methanosarcinaceae.

Methanogenic Bacteria Anaerobic bacteria that derive energy by converting carbon dioxide H_2, formate, and acetate to methane.

Methanogenesis The energy yielding formation of methane by methanogenic bacteria.

Methanogenium A genus of methanogenic bacteria in the family of Methanomicrobiaceae.

Methanogens Referring to methanogenic bacteria.

Methanol (mol wt 32) A pungent alcohol.

$$CH_3OH$$

Methanol Dehydrogenase The enzyme that catalyzes the following reaction:

$$\text{Methanol} + NAD^+ \rightleftharpoons \text{Formaldehyde} + NADH$$

Methanol Oxidase The enzyme that catalyzes the following reaction:

$$\text{Methanol} + O_2 \rightleftharpoons \text{Formaldehyde} + H_2O_2$$

Methanomicrobium A genus of methanogenic bacteria in the family of Methanomicrobiaceae.

Methanoplanus A genus of methanogenic bacteria in the family of Methanoplanaceae.

Methanosarcina A genus of methanogenic bacteria of the family Methanosarcinaceae.

Methanospirillum A genus of methanogenic bacteria in the family of Methanomicrobiaceae.

Methanothermus A genus of methanogenic bacteria in the family of Methanothermaceae.

Methanothrix A genus of methanogenic bacteria in the family of Methanosarcinaceae.

Methanotroph Organism capable of using methane as sole source of carbon and energy.

Methantheline Bromide (mol wt 420) An anticholinergic agent that blocks acetylcholine and inhibits gastric acid secretion.

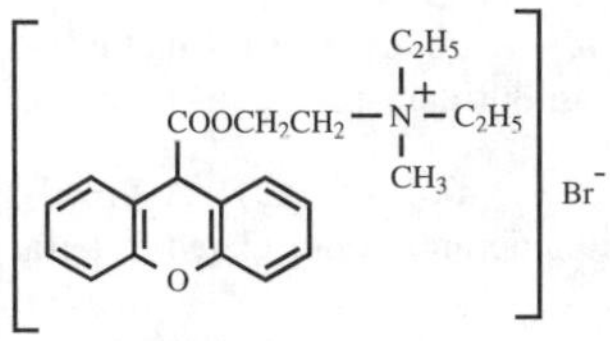

Methaphenilene (mol wt 260) An antihistaminic agent.

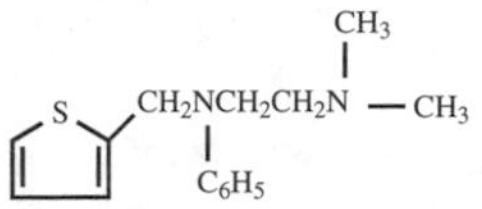

Methapyrilene (mol wt 261) An antihistaminic agent.

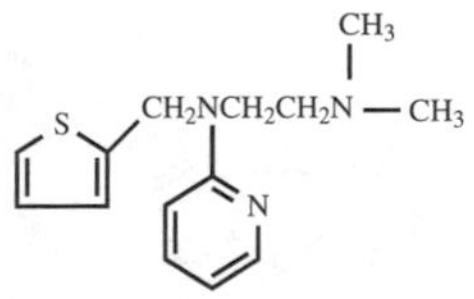

Methaqualone (mol wt 250) An hypnotic and sedative agent.

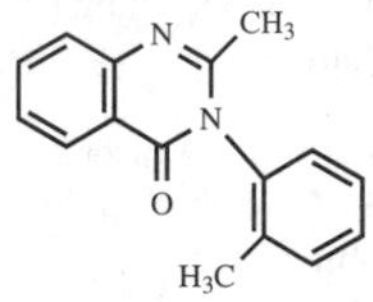

Metharbital (mol wt 198) An anticonvulsant.

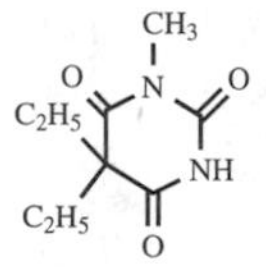

Methazolamide (mol wt 236) A diuretic agent that increases urine excretion of sodium and water by inhibiting sodium reabsorption.

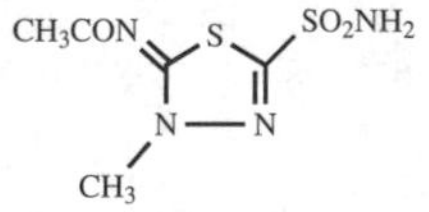

Methazotrophic Organism capable of using methylamines as the sole source of nitrogen, e.g., *Candida utilis*.

MetHb Abbreviation for methemoglobin.

Methblue A trade name for methylene blue, an antidote and urinary tract anti-infective.

Methcycline (mol wt 442) A broad spectrum, semi-synthetic antibiotic related to tetracycline.

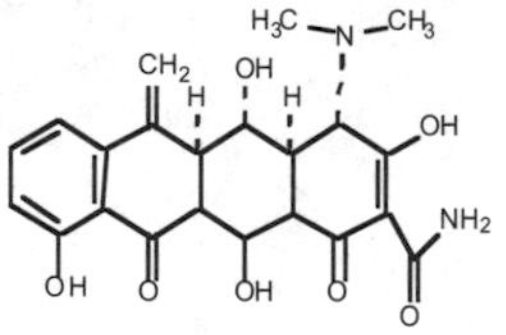

Methdilazine (mol wt 296) An antipruritic and antihistáminic agent that competes with histamine for H_1 receptor sites on effector cells.

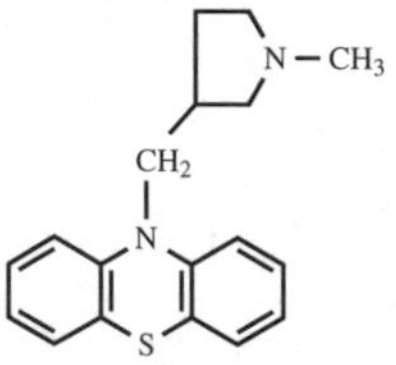

Methemoglobin An altered hemoglobin whose iron is in the ferric state. Methemoglobin does not have the capacity to carry oxygen and is found in circulating blood after poisoning with certain chemicals such as cyanide.

Methemoglobinemia A disorder characterized by the appearance of bluish skin due to the presence of methemoglobin.

Methemoglobin Reductase The enzyme that converts methmoglobin (FeIII) to the FeII form.

Methenamine (mol wt 140) An antibacterial agent used for the treatment of urinary infection.

Methenamine-Silver Stain A dye used for staining and detecting actinomycetes in tissue sections.

Methene Referring to the trivalent diatomic =CH- group.

Methenyl Group Referring to the -CH= group.

Methenyltetrahydrofolate Cyclohydrolase The enzyme that catalyzes the following reaction:

5,10-Methenyltetrahydrofolate + H_2O

⇅

10-Formyltetrahydrofolate

5,10-Methenyltetrahydrofolate Synthetase The enzyme that catalyzes the following reaction:

ATP + 5-Formyltetrahydrofolate

⇅

ADP + Pi + 5,10-methenyltetrahydrofolate

Methergine A trade name for methylergonovine maleate, used to increase the motor activity of the uterus.

Methetoin (mol wt 218) An anticonvulsant.

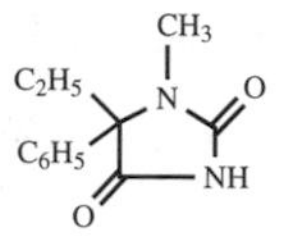

MeTHFA Abbreviation for methyl tetrahydrofolate or methyl tetrahydrofolic acid.

Methicillin Sodium (mol wt 402) A semisynthetic penicillinase-resistant antibiotic related to penicillin that inhibits bacterial cell wall synthesis.

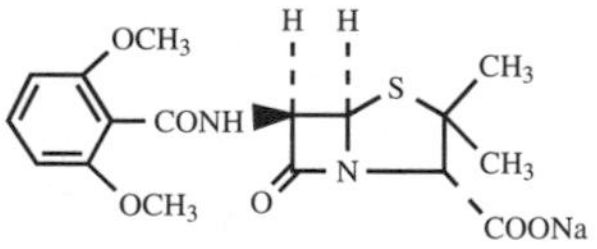

Methimazole (mol wt 114) An antihyperthyroid agent that inhibits oxidation of iodine in the thyroid gland and blocks the ability of iodine to combine with tyrosine to form thyroxine and triiodothyronine.

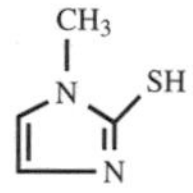

Methionase Synonym of methionine γ-lyase.

Methionine (mol wt 149) A sulfur-containing protein amino acid.

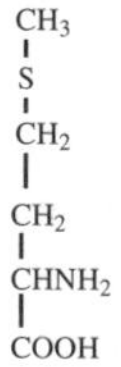

Methionine Adenosyl Transferase The enzyme that catalyzes the following reaction:

Methionine + ATP + H_2O

⇅

S-adenosylmethionine + Pi + PPi

Methionine Aminopeptidase The enzyme that catalyzes the removal of N-terminal methionine from a polypeptide.

Methionine Decarboxylase The enzyme that catalyzes the following reaction:

Methionine

⇅

3-Methylthiopropanamine + CO_2

Methionine Glyoxylate Transaminase The enzyme that catalyzes the following reaction:

Methionine + glyoxylate

⇅

4-Methylthio-2-ketobutanoate + glycine

Methionine Methanethiol-Lyase Systematic name for methionine γ-lyase.

Methionine Pyruvate Transaminase The enzyme that catalyzes the following reaction:

D-Methionine + pyruvate

⇅

4-Methylthio-2-ketobutanoate + L-alanine

Methionine Racemase The enzyme that catalyzes the following reaction:

L-Methionine ⇌ D-Methionine

Methionine tRNA Ligase The enzyme that catalyzes the following reaction:

Methionine + $tRNA^{met}$ + ATP

⇅

Methionine-$tRNA^{met}$ + AMP + PPi

Methionine tRNA Synthetase See methionine-tRNA ligase.

Methioninium The methionine cation.

$$CH_3\text{-}S\text{-}[CH_2]_2CH(NH_3^+)\text{-}COOH$$

Methionyl Transfer RNA The tRNA that is responsible for carrying the amino acid methionine.

Methionyl tRNA Synthetase See methionine-tRNA ligase.

Methioprim (mol wt 171) A tumor antagonist in mice.

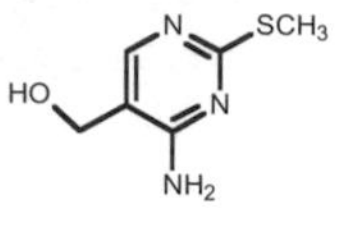

Methisazone (mol wt 234) An antiviral agent.

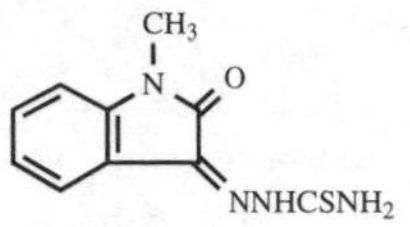

Methitural (mol wt 288) A sedative and hypnotic agent.

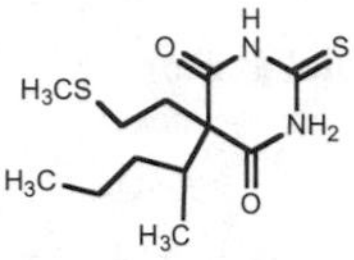

Methixene (mol wt 309) An anticholinergic agent.

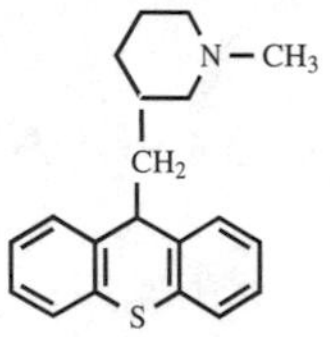

Methocarbamol (mol wt 241) A skeletal muscle relaxant that reduces transmission of nerve impulses from the spinal cord to skeletal muscle.

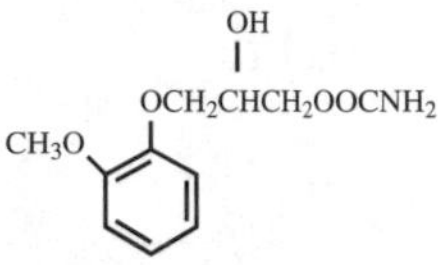

Methohexital Sodium (mol wt 284) An anesthetic agent.

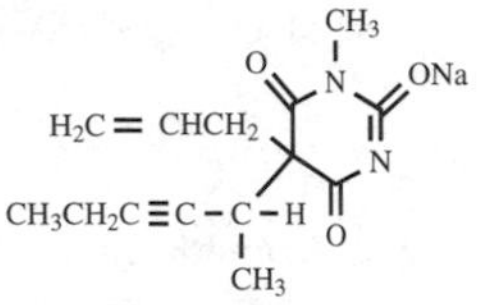

Methohexitone Sodium See methohexital sodium.

Methotrexate (mol wt 454) An analog of dihydrofolate and an antineoplastic agent that inhibits the action of dihydrofolate reductase.

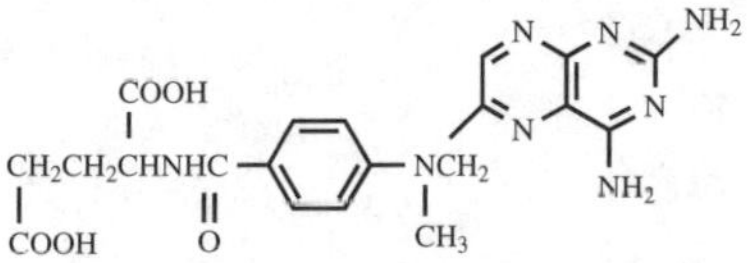

Methotrimeprazine (mol wt 328) An analgesic agent that acts on CNS to produce a hypnotic effect.

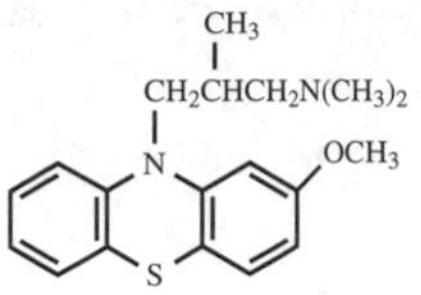

Methoxamine Hydrochloride (mol wt 248) An adrenergic agent.

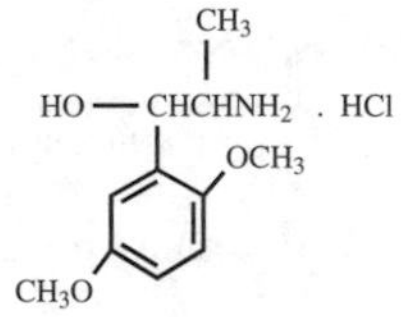

Methoxsalen (mol wt 216) A pigmentation agent that enhances melanogenesis.

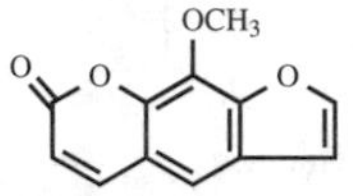

Methoxy Group Referring to CH_3-O- group.

Methoxyflurane (mol wt 165) An anesthetic agent.

$$C_3H_4Cl_2F_2O$$

Methoxyphenamine (mol wt 179) A bronchodilator.

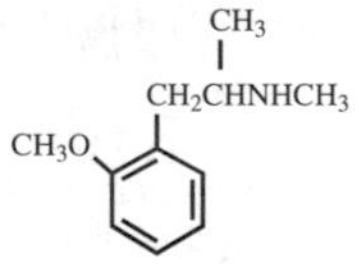

Methoxypromazine (mol wt 314) A neuroleptic agent.

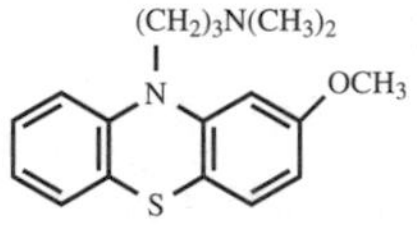

Methscopolamine Bromide (mol wt 398) An anticholinergic agent that blocks acetylcholine and inhibits gastric acid secretion.

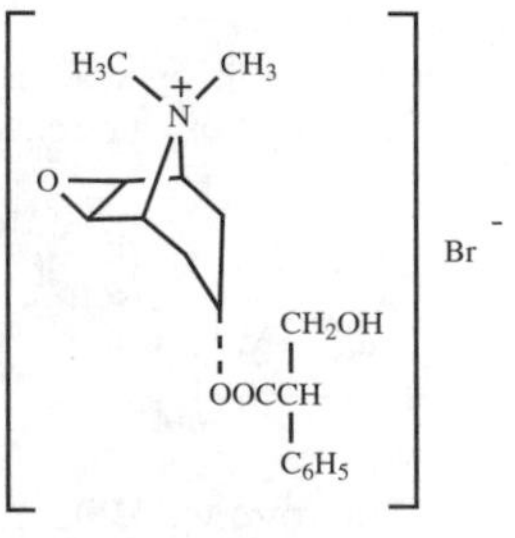

Methsuximide (mol wt 203) An anticonvulsant.

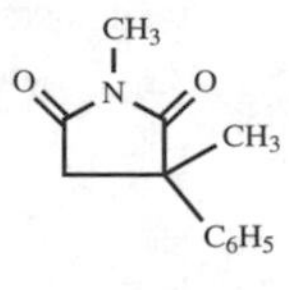

Methyclothiazide (mol wt 360) A diuretic agent that increases urine secretion of sodium and water by inhibiting sodium reabsorption.

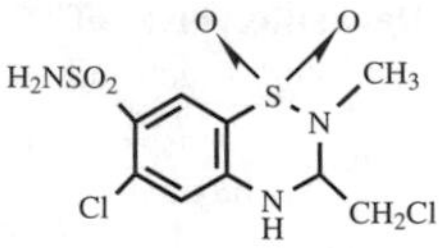

Methyl Referring to methyl group (CH_3-).

Methyl Accepting Chemotaxis Protein A group of bacterial transmembrane proteins responsible for transmission of chemotactic signals across the cytoplasmic membrane.

Methyl Adenine (mol wt 149) A methylated form of adenine.

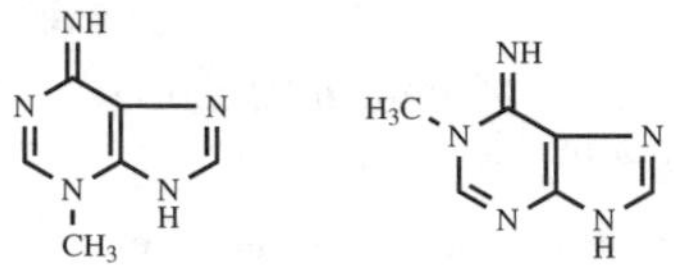

3-methyl adenine 1-methyl adenine

Methyl Adenosine (mol wt 281) A methylated nucleoside of adenine.

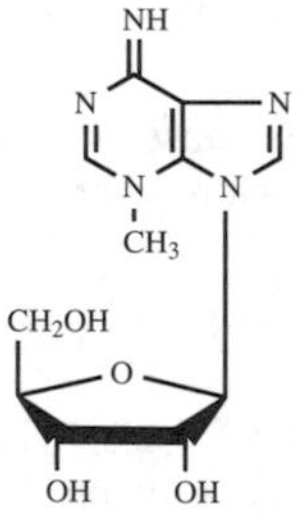

3-methyl adenosine

Methyl Adenosine Nucleosidase The enzyme that catalyzes the hydrolysis of methyl adenosine to methyl adenine and ribose.

Methyl Agarose Bead A methylated agarose granule used as a hydrophobic resin in chromatography.

N-Methyl Alanine (mol wt 103) A naturally occurring amino acid found in the gifblaar.

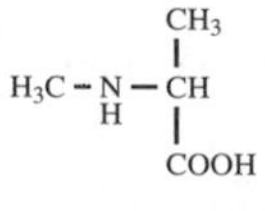

Methyl Albumin Methylated albumin used as an adsorbent for the chromatographic fractionation of nucleic acids.

Methyl Amine (mol wt 31) A flammable liquid.

CH_3NH_2

N-Methyl Arginine (mol wt 188) An arginine derivative and a nonprotein amino acid.

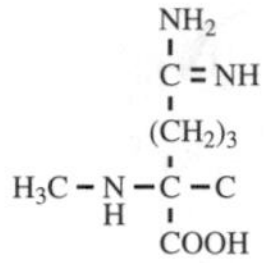

3-Methyl Arsacetin (mol wt 273) An antimalarial agent.

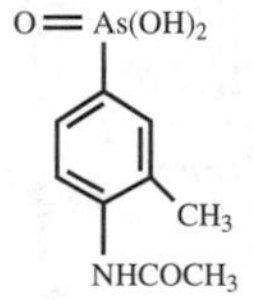

***N*-Methyl Asparagine (mol wt 146)** A naturally occurring amino acid.

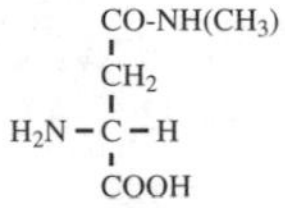

***N*-Methyl Aspartic Acid (mol wt 147)** A ligand for a class of receptors in the CNS.

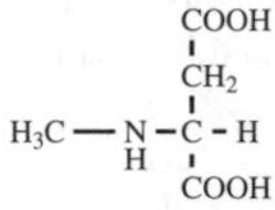

Methyl Blue (mol wt 800) A dye and an antiseptic agent.

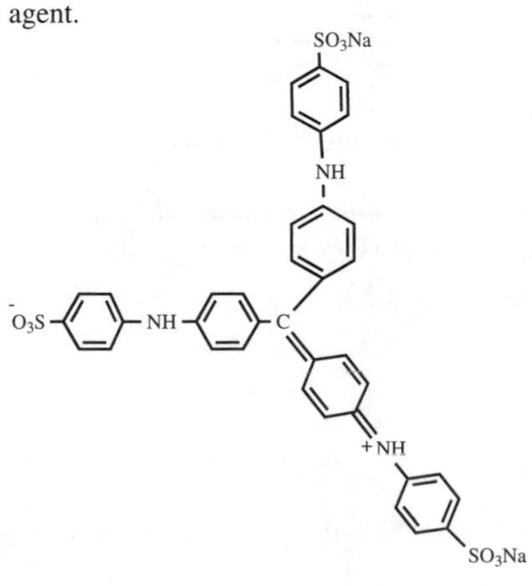

Methyl Butyrase Synonym of carboxyesterase.

Methyl Calcein (mol wt 277) A dye.

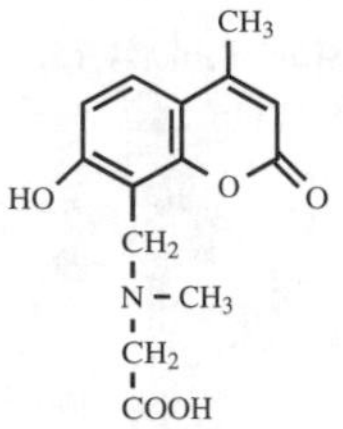

Methyl CCNU See semustine.

Methyl Cellulose A cellulose methyl ester used as a laxative.

Methyl Crotonyl-CoA (mol wt 849) An intermediate in the metabolism of leucine.

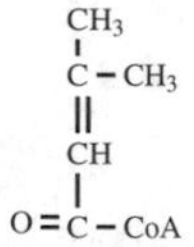

Methyl Crotonyl-CoA Carboxylase The enzyme that catalyzes the following reaction:

Methyl crotonyl-CoA + CO_2

$\rightleftharpoons$

Methyl glutaconyl-CoA

Methyl Crotonyl-CoA Hydratase The enzyme that catalyzes the following reaction:

Methyl crotonyl-CoA + H_2O

$\rightleftharpoons$

Hydroxymethyl glutaryl-CoA

Methyl L-Cysteine (mol wt 135) A naturally occurring nonprotein amino acid.

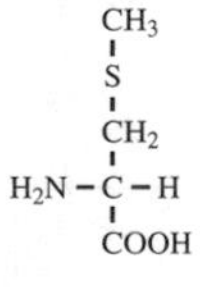

Methyl Cytidine (mol wt 257) A modified nucleoside of cytidine.

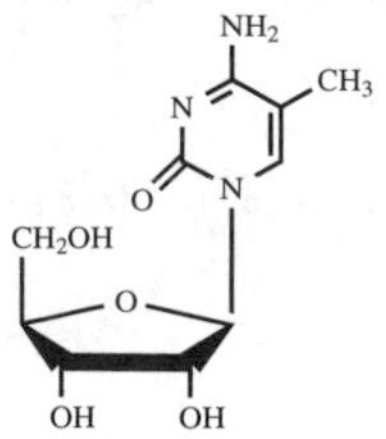

5-Methyl Cytosine (mol wt 125) A methylated form of cytosine.

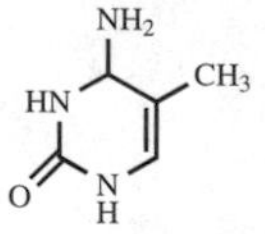

5-Methyl 2′-Deoxycytidine (mol wt 241) A derivative of deoxynucleoside of cytosine.

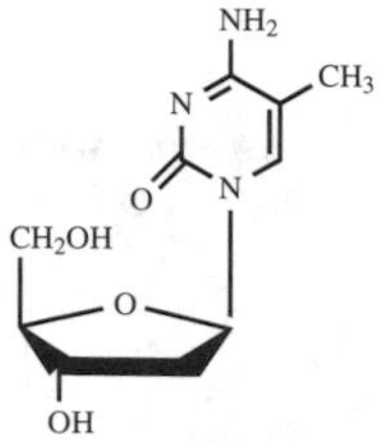

5-Methyl 2′,3′-Dideoxycytidine (mol wt 225) A derivative of didexoynucleoside of cytosine.

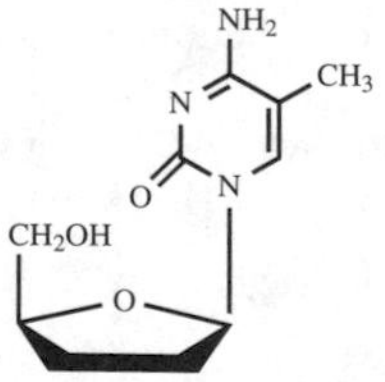

Methyl Dopa (mol wt 211) An antihypertensive agent that inhibits the central vasomotor center and decreases sympathetic outflow.

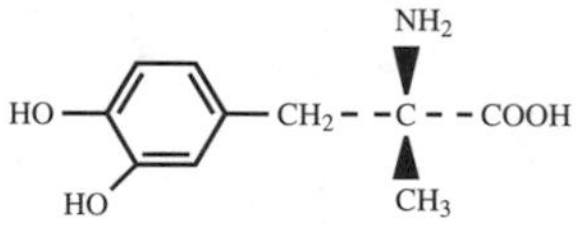

Methyl E Eosin (mol wt 684) A biological dye.

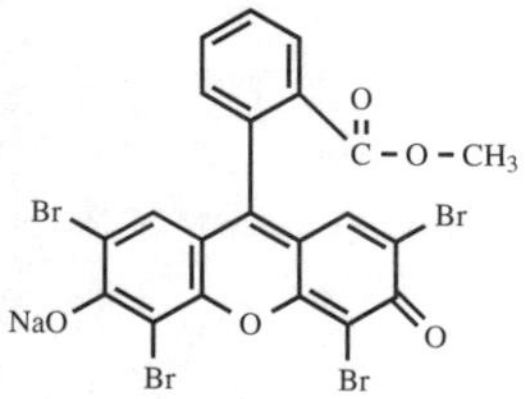

***N*-Methyl Epinephrine (mol wt 197)** An adrenergic agent.

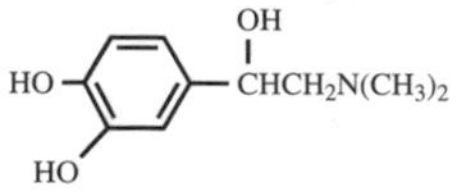

Methyl Ergonovine (mol wt 339) An oxytocic agent.

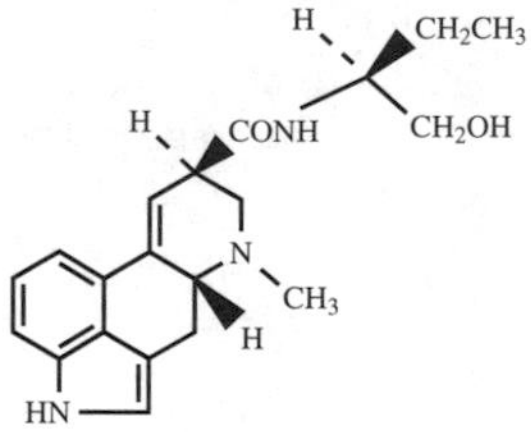

Methyl β-D-Galactoside (mol wt 194) A *lac* operon inducer.

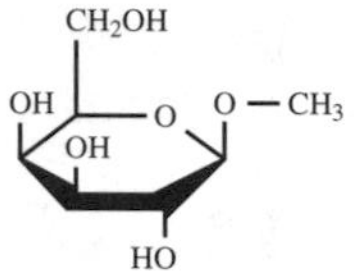

Methyl Gallate (mol wt 184) A substrate for tannase.

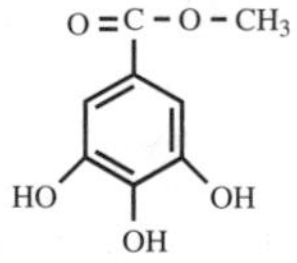

Methyl Glucoside (mol wt 194) A glucose derivative that is formed by a glucosidic linkage with a methyl alcohol.

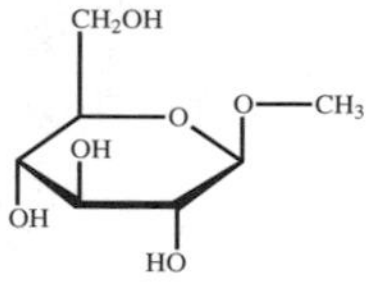

methyl β-D-glucoside

γ-Methyl Glutamic Acid (mol wt 141) A naturally occurring nonprotein amino acid.

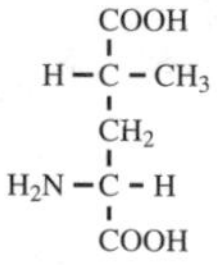

Methyl Green (mol wt 517) A biological stain.

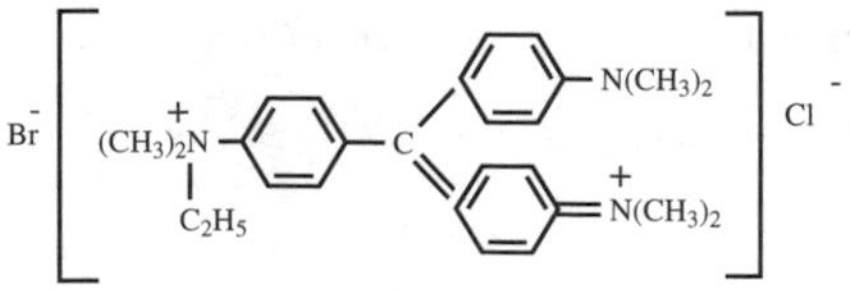

Methyl Green Pyronin Stain A stain used to distinguish DNA from RNA that stains DNA green and RNA red.

Methyl Group Referring to $-CH_3$ group.

Methyl Guanine (mol wt 165) A methylated form of guanine.

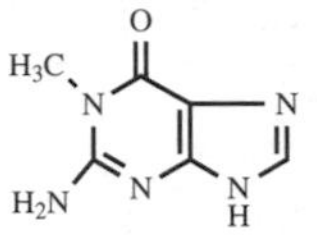

1-methyl guanine

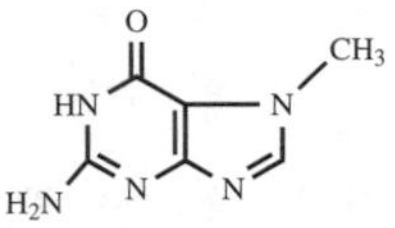

7-methyl guanine

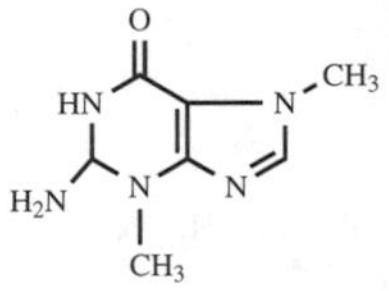

3-methyl guanine

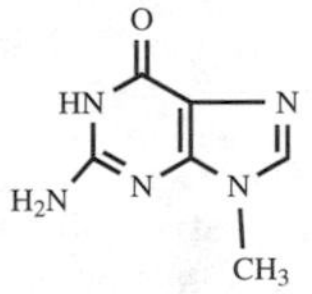

9-methyl guanine

Methyl Hexaneamine (mol wt 115) An adrenergic agent.

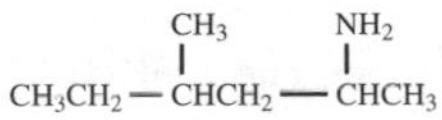

Methyl Histidine (mol wt 169) A naturally occurring cyclic amino acid found in urine.

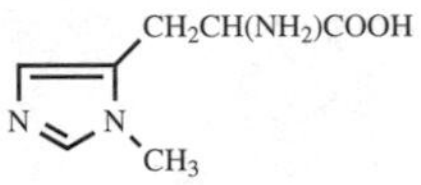

1-methyl histidine

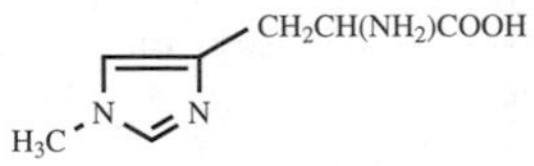

3-methyl histidine

***N*-Methyl Isoleucine (mol wt 145)** A naturally occurring amino acid found in enniatin A.

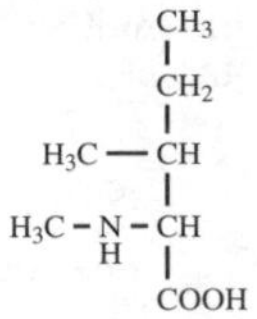

Methyl Isothiocyanate (mol wt 73) A pesticide.

$CH_3N{=}CS$

***N*-Methyl Leucine (mol wt 145)** A naturally occurring amino acid found in enniatin A.

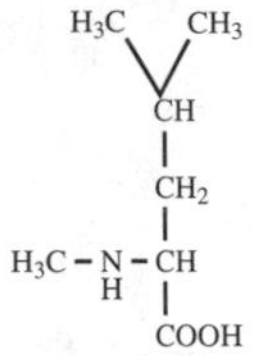

ε-*N*-Methyl Lysine (mol wt 161) A rare amino acid found in actin and histone.

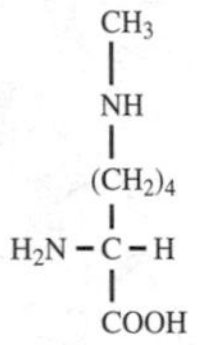

Methyl Malonate Semialdehyde Dehydrogenase The enzyme that catalyzes the following reaction:

Methylmalonate semialdehyde + NAD^+ + CoA

⇅

Propionyl-CoA + CO_2 + NADH

Methyl Malonic Acid (mol wt 118) A methylated form of malonic acid.

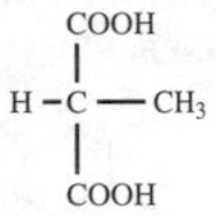

Methyl Malonyl Acidemia A genetic disorder characterized by massive ketosis due to the deficiency of malonyl-CoA carboxymutase.

Methyl Malonyl-CoA (mol wt 868) An intermediate in the metabolism of methionine, valine, and isoleucine succinyl-CoA.

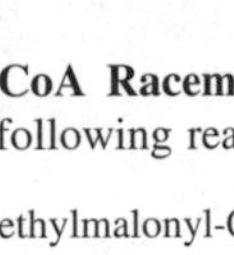

Methyl Malonyl-CoA Mutase The enzyme that catalyzes the following reaction:

Methylmalonyl-CoA ⇌ Succinyl-CoA

Methyl Malonyl-CoA Racemase The enzyme that catalyzes the following reaction:

(*S*)-Methylmalonyl-CoA

⇅

(*R*)-methylmalonyl-CoA

***N*-Methyl *O*-Methyl Serine (mol wt 133)** A naturally occurring amino acid found in *Mycobacterium butyricum*.

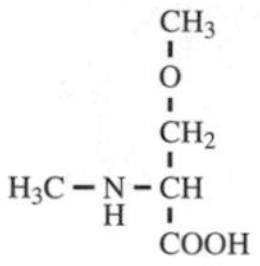

Methyl Orange (mol wt 327) A dye and pH indicator.

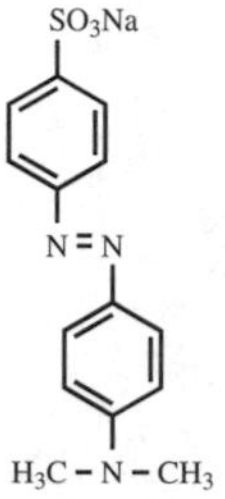

Methyl Phenidate (mol wt 233) A CNS stimulant that promotes transmission of nerve impulses by releasing stored norepinephrine from nerve terminals in the brain.

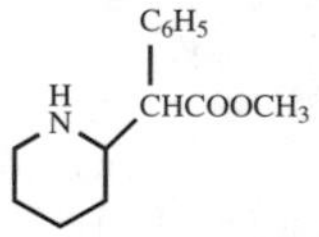

***N*-Methyl Phenylglycine (mol wt 166)** A naturally occurring amino acid found in etamycin.

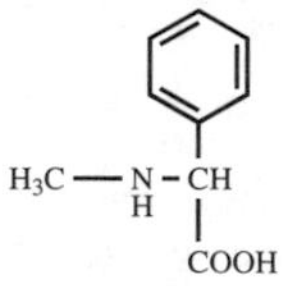

Methyl Prednisolone (mol wt 374) A corticosteroid used as an anti-inflammatory agent.

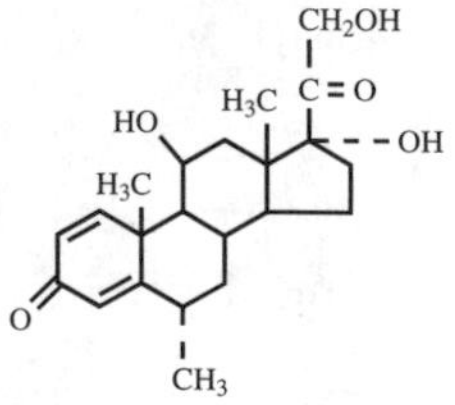

Methyl Proline (mol wt 129) A naturally occurring amino acid found in apple.

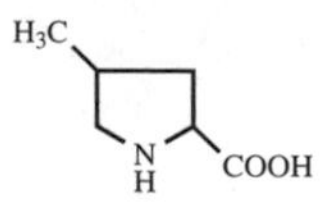

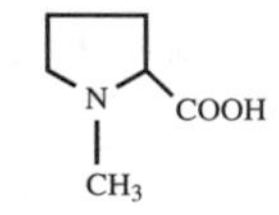

4-methyl proline N-methyl proline

Methyl Red (mol wt 269) A dye and pH indicator.

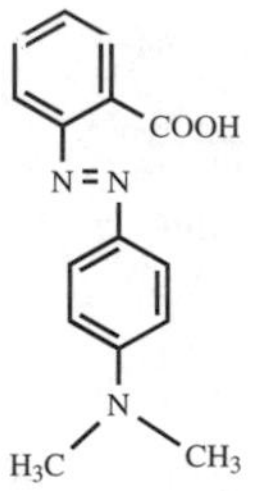

Methyl Salicylate (mol wt 152) A counterirritant that temporarily blocks pain.

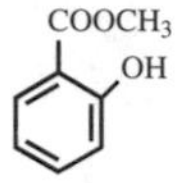

***N*-Methyl Streptolidine (mol wt 202)** An amino acid found in streptothricin.

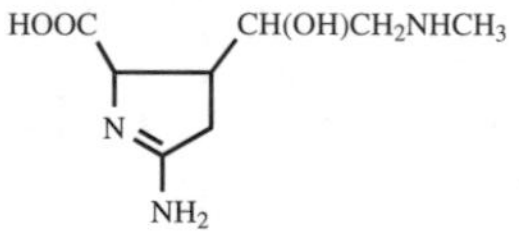

17-Methyl Testosterone (mol wt 302) An androgen.

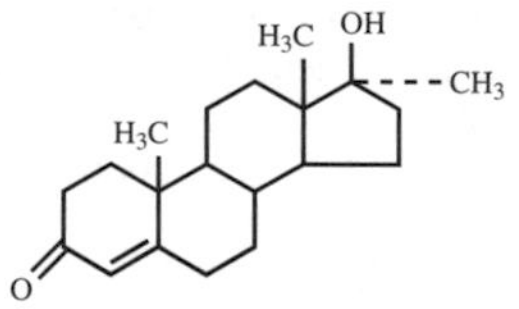

5-Methyl THF Abbreviation for 5-methyltetrahydrofolate.

Methyl β-D-Thiogalactoside (mol wt 210) A *lac* operon inducer.

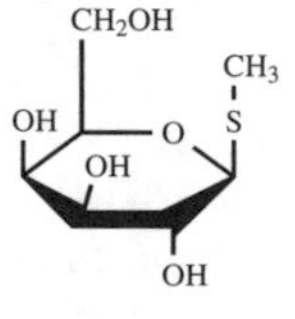

***N*-Methyl Threonine (mol wt 133)** A naturally occurring amino acid.

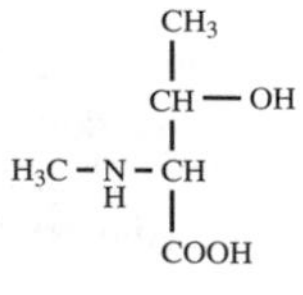

Methyl Transferase The enzyme that catalyzes the transfer of a methyl group from a methyl group donor, e.g., S-adenosyl methionine, to an organic acceptor.

β-Methyl Tryptophan (mol wt 218) A naturally occurring amino acid found in telomycin.

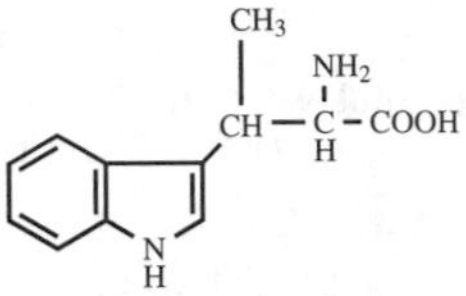

***N*-Methyl Tyrosine (mol wt 195)** A naturally occurring amino acid.

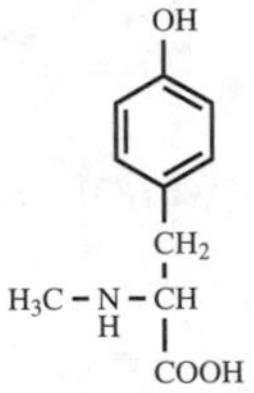

Methyl Valine (mol wt 131) A naturally occurring amino acid found in actinomycin.

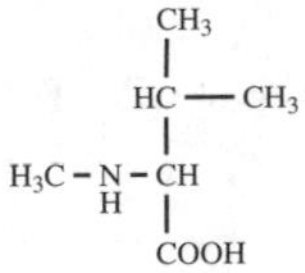

Methyl Violet (mol wt 394) A dye and pH indicator.

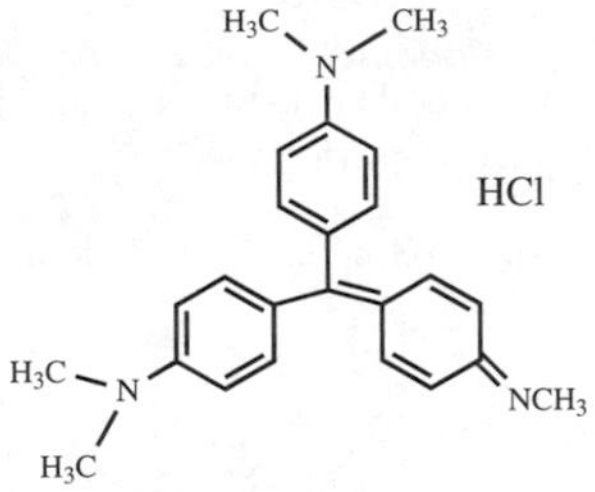

Methyl Xanthine (mol wt 166) A methylated xanthine, a base present in tRNA.

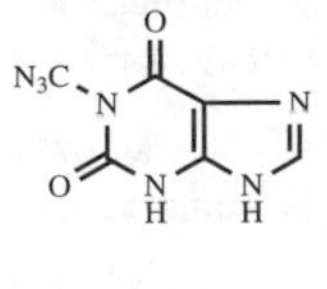

1-methyl xanthine

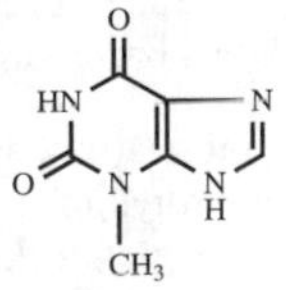

3-methyl xanthine

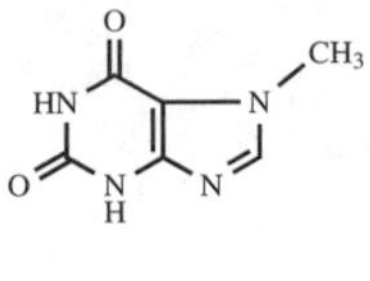

7-methyl xanthine

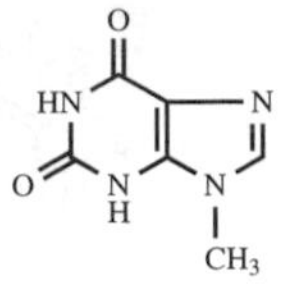

9-methyl xanthine

Methylase The enzyme that catalyzes the methylation reaction.

Methylene Blue (mol wt 320) A dye and an antimethemoglobinemic agent.

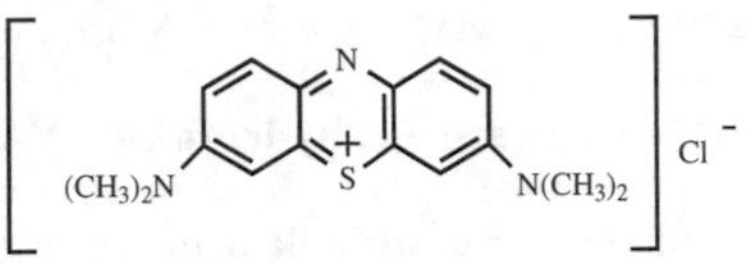

Methylene Blue Test A test employing methylene blue used to determine the number of microorganisms in milk.

γ-Methylene Glutamic Acid (mol wt 159) A naturally occurring amino acid.

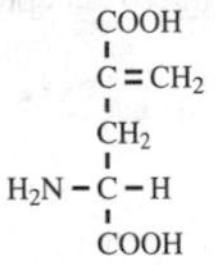

γ-Methylene Glutamine (mol wt 158) A naturally occurring amino acid.

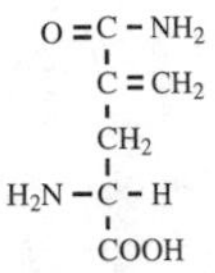

Methylene Green (mol wt 365) A histological stain.

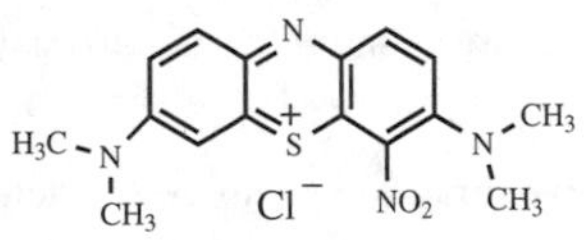

Methylene Group Referring to $=CH_2$ group.

4-Methylene Proline (mol wt 127) A naturally occurring nonprotein amino acid.

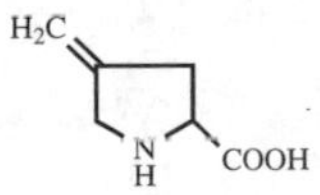

Methylene Tetrahydrofolate Dehydrogenase (NAD^+) The enzyme that catalyzes the following reaction:

$$\text{5,10-Methylene tetrahydrofolate} + NAD^+ \rightleftharpoons \text{5,10-Methenyl tetrahydrofolate} + NADH$$

Methylene Tetrahydrofolate Dehydrogenase ($NADP^+$) The enzyme that catalyzes the following reaction:

5,10-Methylene tetrahydrofolate + $NADP^+$
⇌
5,10-Methenyl tetrahydrofolate + NADPH

5,10-Methylene Tetrahydrofolate $NADP^+$ Oxidoreductase The systematic name for methylenetetrahydrofolate dehydrogenase ($NADP^+$).

Methylene Tetrahydrofolate Reductase The enzyme that catalyzes the following reaction:

5-methyltetrahydrofolate + acceptor
⇌
5,10-Methylene tetrahydrofolate + reduced acceptor

Methylene Violet (mol wt 256) A biological dye.

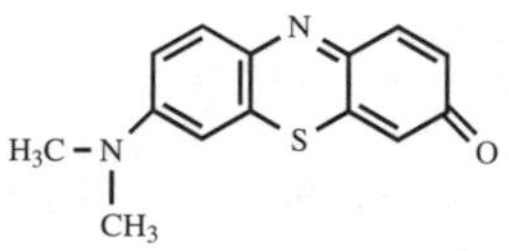

Methylenomycins A group of antibiotics related to sarkomycins from *Streptomyces violaceoruber.*

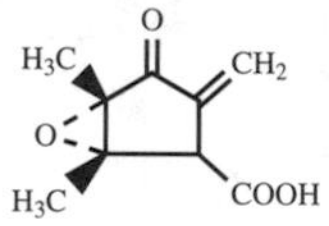

Methylenomycin A

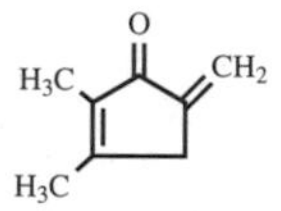

Methylenomycin B

Methylesterase Any of the various enzymes that catalyze the hydrolysis of methyl esters.

Methylglyoxalase Synonym of lactoylglutathione lyase.

Methylneogenesis The formation of a methyl group.

Methylol Riboflavine Derivative of riboflavine and enzyme cofactor and vitamin source.

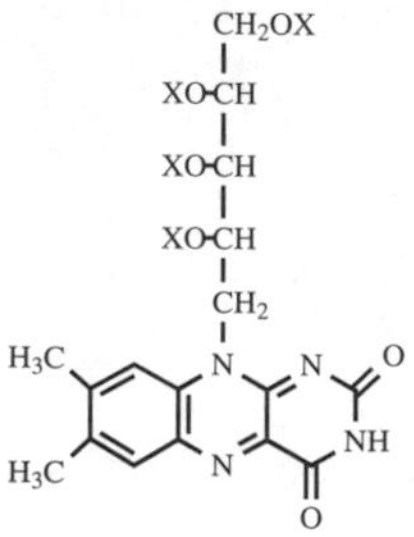

X = H or CH_2OH

Methylone A trade name for methylprednisolone, a corticosteroid.

Methylotroph Organism that use methanol as an energy source, e.g., yeast (*Hansenula polymorpha*).

Methymycin (mol wt 470) An antibiotic from *Streptomyces venezuelae.*

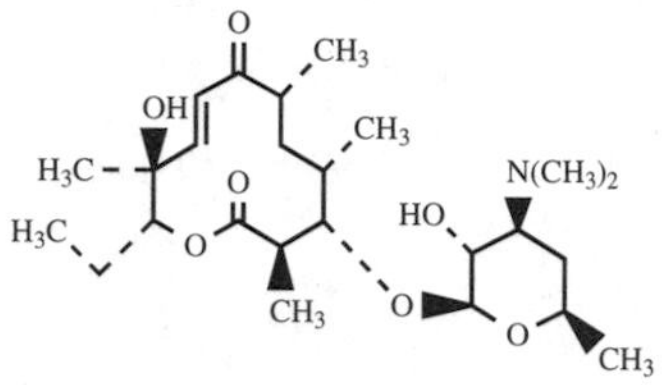

Methyprylon (mol wt 183) A sedative and hypnotic agent.

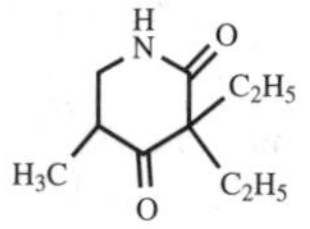

Metiazinic Acid (mol wt 271) An anti-inflammatory agent.

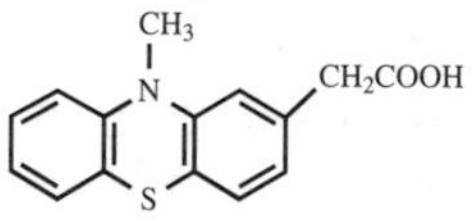

Meticorten A trade name for an anti-inflammatory corticosteroid prednisone.

Meticrane (mol wt 275) A diuretic agent.

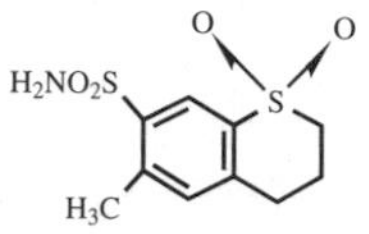

Metin A trade name for methicillin, an antibiotic that inhibits bacterial cell wall synthesis.

Metipranolol (mol wt 309) An anti-hypertensive, antiarrhythmic, antiglaucoma agent, and a beta-adrenergic blocker.

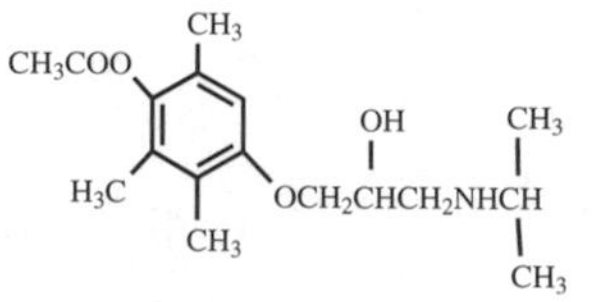

Metizol A trade name for metronidazole, an antiprotozoal agent.

Metizoline (mol wt 230) An adrenergic agent and nasal decongestant.

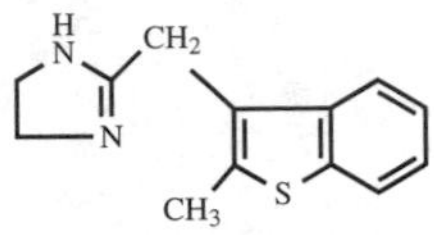

MetMb Abbreviation for metmyoglobin.

Metmyoglobin The Fe^{+++} form of myoglobin.

Metoclopramide (mol wt 300) An antiemetic agent.

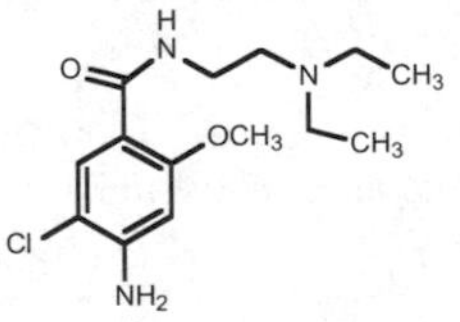

Metocurine Iodide (mol wt 907) An antiemetic agent and a neuromuscular blocker.

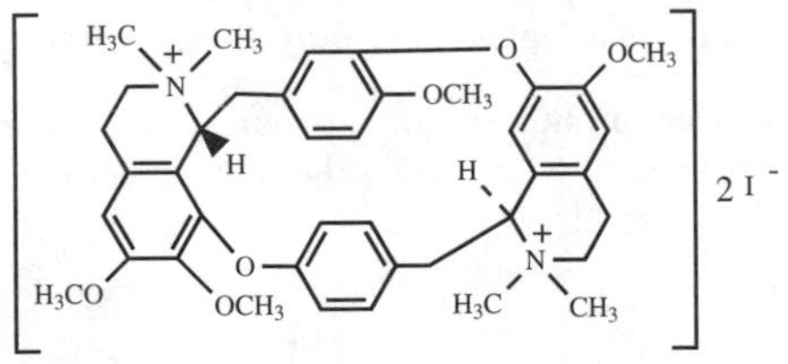

Metolazone (mol wt 366) A diuretic and an antihypertensive agent that increases urine secretion of sodium and water by inhibiting sodium reabsorption.

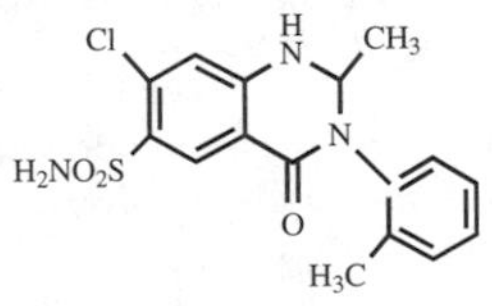

Metomidate (mol wt 230) A hypnotic agent.

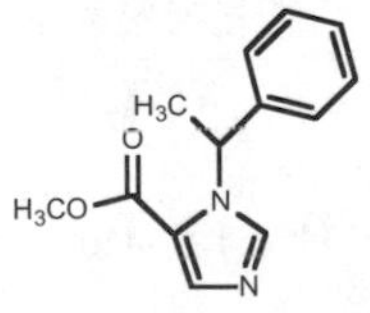

Metopon (mol wt 299) A narcotic analgesic agent.

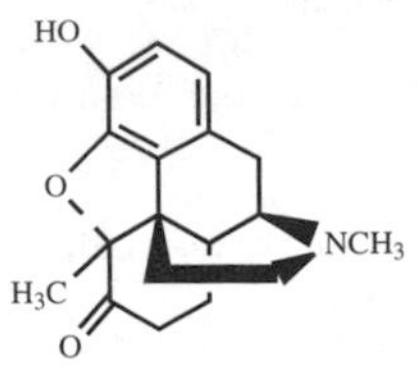

Metoprolol (mol wt 267) An antihypertensive, antianginal, and antiarrhythmic agent.

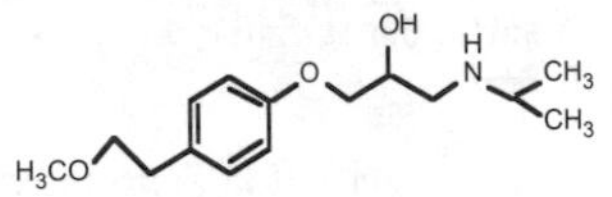

Metoserpate (mol wt 429) A sedative agent.

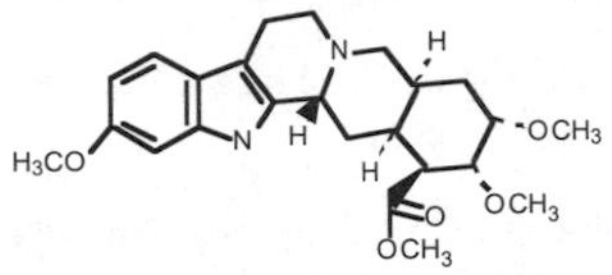

Metr- A prefix denoting the uterus.

Metra A trade name for the cerebral stimulant phendimetrazine tartrate, which promotes the transmission of nerve impulses by releasing stored norepinephrine from nerve terminals in the brain.

Metric System A decimal system of measurement based on the meter as the unit of length, the liter as the unit of volume, and the gram as the unit of weight.

Metritis Inflammation of the uterus.

Metrizamide (mol wt 789) A nonionic substance used in density gradient centrifugation.

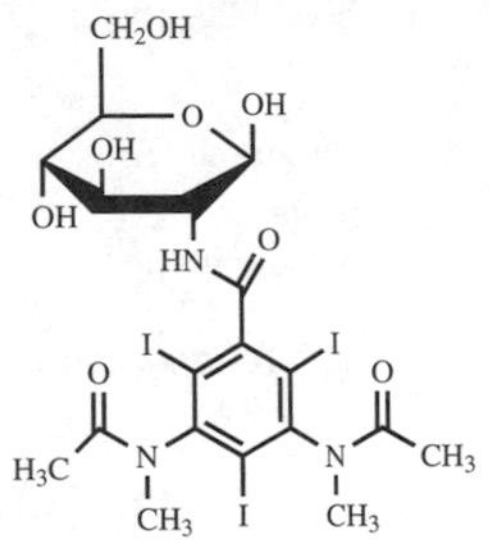

MetRNAS Abbreviation for methionyl-tRNA synthetase.

Metrodin A trade name for urofollitropin, a hormone used as a fertility drug.

MetroGel A trade name for metronidaqole, a local anti-infective agent.

Metrogyl A trade name for metronidazole, an antiprotozoal agent.

Metronidazole (mol wt 171) An antiprotozoal and antibacterial agent.

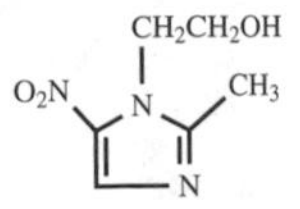

Metrozine A trade name for the antiprotozoal agent metronidazole.

-metry A suffix denoting the process or science of measurement.

Metryl A trade name for metronidazole, an antiprotozoal agent.

Met-tRNA Referring to methionyl-tRNA.

Metubine A trade name for metocurine iodide, a neuromuscular blocker.

Metyrosine (mol wt 195) An antihypertensive agent that inhibits tyrosine hydroxylase and endogenous catecholamine synthesis.

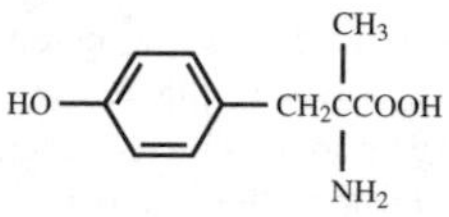

MeuI (MboI) A restriction endonuclease from *Micrococcus euryhalis* with the following specificity:

↓
5′..........GATC..........3′
3′..........CTAG..........5′
↑

MeV Abbreviation for mega-electronvolt (1 MeV equal 10^6 electronvolts).

Mevacor A trade name for iovastatin, an antilipemic agent that inhibits 3-hydroxy-3-methylglutaryl-CoA reductase.

Meval A trade name for diazepam, an antianxiety agent.

Mevalolactone (mol wt 130) An ester of mevalonate involved in enzyme modification.

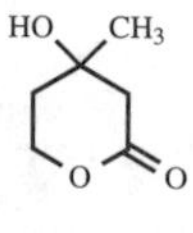

Mevalonate 5-Phosphotransferase The enzyme that catalyzes the following reaction:

Mevalonate + ATP ⇌ Phosphomevalonate + ADP

Mevalonic Acid (mol wt 148) An organic acid.

$$HOOC \cdot CH_2 - \underset{CH_3}{\overset{OH}{C}} - CH_2CH_2OH$$

Mevalonic Acid Lactone See mevalolactone.

Mevastatin (mol wt 391) An inhibitor for HMG-CoA reductase.

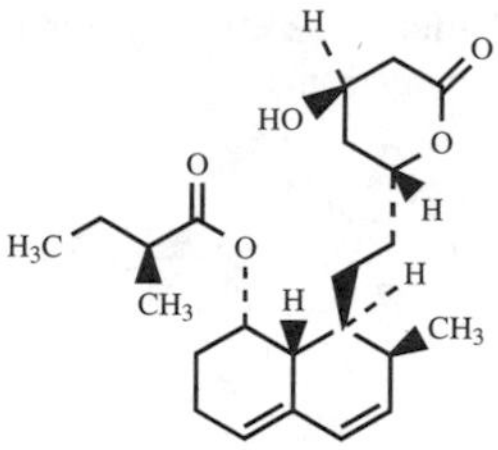

Mevinolin A trade name for lovastatin, an HMG CoA inhibitor used as an antihypertensive agent.

Mexate A trade name for methotrexate sodium, which prevents reduction of folic acid to tetrahydrofolate.

Mexicanain A proteinase from fruit of *Pileus mexicanus* with an activity similar to papain.

Mexitil A trade name for mexiletine hydrochloride, an antiarrhythmic agent that shortens the action potential.

Mezlin A trade name for mezlocillin sodium, an antibiotic that inhibits bacterial cell wall synthesis.

Mezlocillin (mol wt 540) A semisynthetic antibiotic related to penicillin that inhibits bacterial cell wall synthesis.

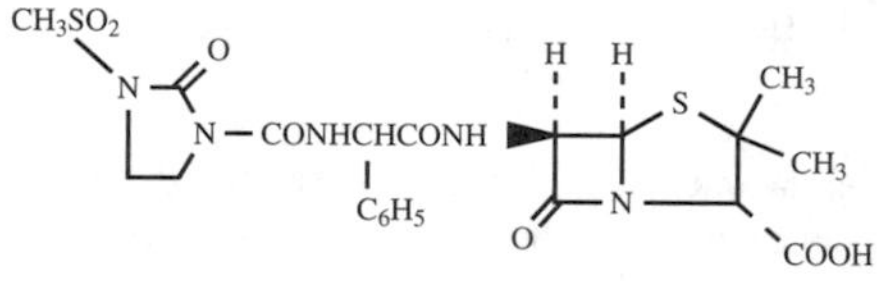

MF Abbreviation for 1. mitogenic factor and 2. microfilament.

MflI (XhoII) A restriction endonuclease from *Microbacterium flavum* with the following specificity:

↓
5′..........PuGATCPy..........3′
3′..........PyCTAGPu..........5′
↑

MF Solution Abbreviation for merthiolate-formaldehyde solution.

MFA Abbreviation for 1. methyl fluoroacetate; 2. monofluoroacetate.

M-FABP Abbreviation for myelin fatty acid-binding protein.

MFID Abbreviation for multielectrode flame ionization detector.

MFP Abbreviation for monofluorophosphate.

mg Abbreviation for milligram (1/1000 of a gram).

mg Percent Referring to a solution expressed in number of milligrams per 100 ml.

Mg Abbreviation for magnesium, atomic weight 24, valence 2.

MG Abbreviation for 1. methyl glucose; 2. methyl glucoside; 3. methyl guanine; 4. monoglyceride.

μg Abbreviation for microgram.

3MG Abbreviation for 3-O-methyl glucose.

m^{7}G Abbreviation for N^7-methylguanosine.

MGBG Abbreviation for methyl-glyoxal bisguanylhydrozone.

MGBGH Abbreviation for methyl-glyoxal bisguanylhydrozone.

MGD Abbreviation for molybdopterin guanine dinucleotide.

MGDG Abbreviation for mono-galacctosyl-diacylglycerol.

M-Gesic A trade name for a combination drug containing acetaminophen and codeine phosphate, used as an analgesic and antipyretic agent.

MGM Abbreviation for 2-methylene glutarate mutase.

mgm Abbreviation for milligram.

MgP Abbreviation for magnesium protoporphyrin IX.

m^{7}GTP Abbreviation for 7-methyl GTP (guanosine triphosphate).

MH Abbreviation for 1. malignant histiocytosis; 2. mammotropic hormone; 3. melanophore hormone.

mH Abbreviation for millihenry.

MH Virus Abbreviation for 1. Marek's herpes virus; 2. murine hepatitis virus.

MHA Abbreviation for micro-hemagglutination.

MHb Abbreviation for methemoglobin.

MHC Abbreviation for 1. major histo- compatibility complex; 2. myosin heavy chain.

MHC Antigens Proteins encoded by genes located in the major histocompatibility complex. There are three major classes of MHC antigens (Class I, II, and III).

MHC Associated Recognition See MHC restriction.

MHC Molecules Referring to MHC antigens.

MHC Proteins Proteins encoded by genes in the MHC.

MHC Restriction A phenomenon in which the recognition of foreign antigen by T lymphocytes is associated with the MHC antigens. For example, cytotoxic T cells kill viral-infected cells that have the same class I antigen as cytotoxic T cells.

MHCK Abbreviation for myosin heavy chain kinase.

MHD Abbreviation for 1. minimum hemolytic dose, the smallest quantity of complement needed to lyse a standardized suspension of sensitized erythrocytes. 2. minimum hemagglutinating dose, the smallest quantity of hemagglutinating agent capable of causing hemagglutination in a standardized suspension of erythrocytes.

MHR Abbreviation for major histo-compatibility region.

MHS Abbreviation for major histo-compatibility system.

MHV Abbreviation for mouse hepatitis virus.

MHVD Abbreviation for Marek's herpes virus disease.

MI Abbreviation for 1. mercapto-imidazole; 2. myocardial infarction.

Miacalcin A trade name for calcitonin, which decreases osteoclastic activity.

Mibefradil (mol wt 496) An antibacterial.

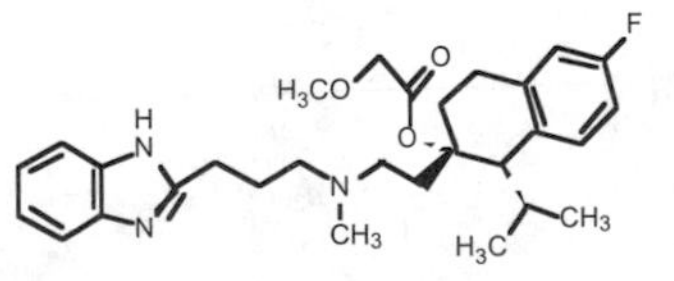

Miboplatin (mol wt 437) An antineoplastic agent.

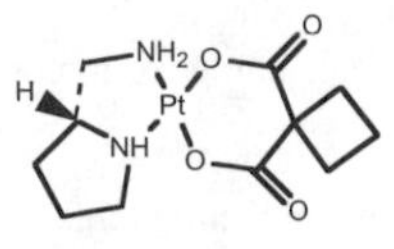

MIC Abbreviation for minimum inhibitory concentration, the lowest concentration of an antibiotic that inhibits a given type of microorganism under standard test conditions.

Micardis A trade name for telmisartan, an angiotensin II receptor antagonist used as an antihypertensive agent.

Micatin A trade name for miconazole, a local anti-infective agent that disrupts cell membrane permeability of fungi.

Micelle An aggregate formed by amphipathic molecules in water such that their polar ends are in contact with water and their nonpolar portions are in the interior of the aggregate.

Michaelis Complex Referring to an enzyme-substrate complex.

Michaelis Constant Referring to Km, the rate constants for a given substrate of an enzyme mediated reaction. It is expressed as follows:

$$[E] + [S] \underset{k_2}{\overset{k_1}{\rightleftharpoons}} [ES] \xrightarrow{k_3} P + E$$

$$Km = (k_2 + k_3)/k_1$$

Km is also defined as substrate concentration at which the enzyme-catalyzed reaction is at one-half of the maximum velocity.

Michaelis-Menten Equation The mathematical description of the relationship between the rate of an enzymatic reaction and the substrate concentration.

$$\upsilon = V[S] / (K_m + [S])$$

υ = initial velocity
V = maximum velocity
S = substrate concentration
K_m = Michaelis constant

Miconazole (mol wt 416) An antifungal agent that disrupts permeability of the fungal cell membrane.

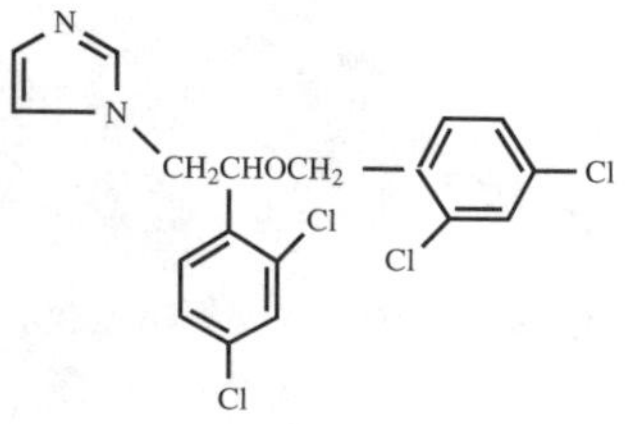

Micozole A trade name for miconazole, an antifungal agent.

Micro- A prefix meaning small.

Microautophagy A process by which lysosomes take up and degrade cytosolic protein by invagination of the lysosomal membrane.

Microbacterium A genus of catalase-positive, asporogenous bacteria (Actinomycetales).

Microbes Referring to microscopic organisms or microorganisms.

Microbial Ecology The science that deals with the interactions of microorganisms with their biotic and abiotic environments.

Microbial Genetics The science that deals with genetics of microorganisms.

Microbial Pesticides Pathogenic or predatory microorganisms that are toxic or antagonistic toward a particular pest population.

Microbicidal Agent capable of destroying, killing, or inactivating microorganisms.

Microbiological Assay The employment of microorganisms for assaying activity of bioactive compounds.

Microbiology The science that deals with microorganisms and their effects on humans and other organisms.

Microbispora A genus of bacteria that form branched aerial and substrate mycelium (Actinomycetales).

Microbistatic Capable of inhibiting growth and reproduction of microorganisms (microcidal).

Microbodies Referring to a variety of membrane-enclosed structures in eukaryotic cells containing enzymes for specific metabolic pathways (e.g., glyoxylate cycle in peroxysomes).

Microcarrier Microscopic beads or spheres, made of dextrans or agarose, that are used in tissue culture for attachment of anchorage-dependent cells.

Microcentrifuge A small table-top centrifuge used for centrifugation of small sample volume.

Microcins Low molecular weight bactericidal antibiotics produced by the bacteria of Enterobacteriaceae (usually plasmid encoded).

Micrococcus A genus of Gram-positive, aerobic, chemoorganotrophic, asporogenous, catalase-positive bacteria (Micrococcaceae).

Microcort A trade name for hydrocortisone.

Microcurie 10^6 curies.

Microcytes 1. Abnormally small red blood cells observed in patients with anemia. 2. A type of resting bacterial cell.

Microcythemia A disorder characterized by the presence of abnormally small red blood cells (also known as microcystic anemia).

Microcytic Anemia A blood disorder characterized by the presence of abnormally small red blood cells.

Microcytosis The presence of abnormally small red cells in the blood.

Microdrop Technique A method for assaying antibody synthesis by an individual lymphocyte in a microdrop medium containing antigen.

Microelectrode An electrode with an extremely small, fine tip, capable of nondestructive puncturing of the plasma membrane for injection of ionic solution or study of cellular activity (e.g., recording cellular resting and action potentials).

Microequivalent The equivalent weight expressed in micrograms.

Microfibrils Threadlike structures found in the cell walls of filamentous fungi and plants.

Microfilament Polymer of actin that is an integral part of the cytoskeleton and involved in the support, maintaining shape, and mobility of the eukaryotic cell.

Microfilament Based Movement Movement mediated by microfilaments, e.g., muscle contraction, amoeboid movement, and cytoplasmic streaming.

Microfilament Cross-Linking Protein Protein that binds microfilaments together to form a stable network with gellike properties.

Microfilament Severing Protein Protein that breaks actin microfilaments leading to the disruption of microfilament networks, causing gel-to-sol transition during the process of cytoplasmic streaming.

Microfuge Referring to microcentrifuge.

Microglia Macrophages of the central nervous system.

Microglobulin Any plasma globulin or globulin fragment with molecular weight less than 40 kDa.

β_2-Microglobulin A protein associated with MHC class I antigen but not encoded by MHC genes.

Microgram One-millionth of a gram or 1/1000 of a milligram (abbreviated as μg).

Microinjection Injection of molecules into a single cell with a microelectrode.

Micro-K Extencaps A trade name for potassium chloride, used to replace and to maintain the level of potassium in the body.

Micro-Kjeldahl Method A modified Kjeldahl method used for the determination of small amounts of nitrogen in the range of 0.1 to 1.0 mg.

Microliter 10^{-6} liter or 10^{-3} ml.

Micromanipulator A device used for dissection, injection, retraction of microscopic specimens, or for the isolation of a single cell.

Micrometer 1. A device used in conjunction with a microscope or other optical instrument for measuring minute lengths or distances. 2. One millionth of a meter or one-thousandth of a millimeter.

Micromineral Referring to an element that is required in only minute quantities, e.g., Zn, Cu, I, F, Cr, Se, and Mo.

Micromole 10^{-6} mole.

Micromonospora A genus of bacteria (Actinomycetales).

Micron A unit of length equal to 1/1000 mm.

Micronase A trade name for glyburide, which stimulates insulin release from pancreatic beta cells.

MicroNefrin A trade name for epinephrine, a hormone.

Micronomicin (mol wt 464) An antibiotic produced by *Micromonospora sagamiensis* var *nonreducans*.

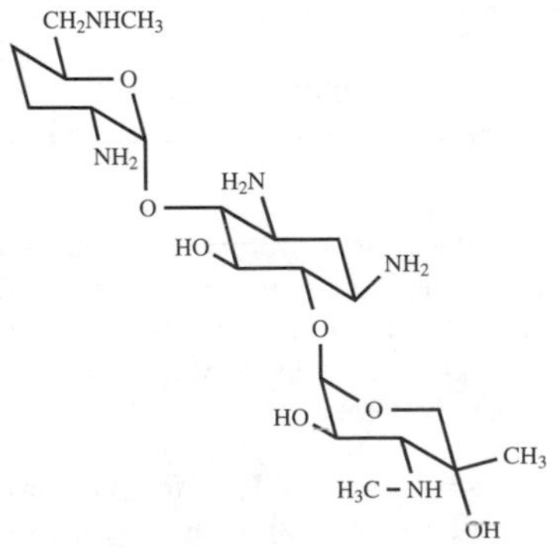

Micronor A trade name for norethindrone, a hormone that suppresses ovulation.

Micronucleus One of the two types of nuclei observed in the ciliate protozoa that is characterized by the absence of nucleoli and presence of inactive DNA that does not undergo transcription.

Micronutrient Nutrient needed by an organism in relatively minute amounts, e.g., vitamins and minerals.

Microorganisms Microscopic organisms, including algae, bacteria, fungi, protozoa, and viruses.

Micropinocytosis A type of uptake mechanism in which the plasma membrane invaginates around the substance to form a very small pinocytotic vesicle.

Micropolyspora A genus of bacteria (Actinomycetales).

Micropyle 1. A channel in the outer coat of a seed through which a pollen tube passes during fertilization. 2. An opening in the egg membrane that allows the entry of the sperm.

Microsomal Fraction One of the fractions obtained by differential centrifugation of cell homogenate during subcellular fractionation, which contains microsomes and ribosomes.

Microsome Vesicle formed by fragments of endoplasmic reticulum when tissue is homogenized.

Microspore A spore that develops into a male gametophyte.

Microspore Mother Cell A cell in seed plants in which meiosis occurs leading to the production of four microspores.

Microsulfon A trade name for sulfadiazine, an antimicrobial agent that decreases bacterial folic acid synthesis.

Microtetraspora A genus of bacteria (Actinomycetales).

Microtome Instrument used to slice an embedded biological specimen into thin sections for light microscopy.

Microtubule Polymer of the protein tubulin that is an integral part of the cytoskeleton. It is involved in maintaining shape and motility of the eukaryotic cells. It is also found in the cilia and flagella of many eukaryotic cells.

Microtubule Associated Protein Proteins associated with microtubules, influencing the stability and organization of the microtubules.

Microtubule-Associated Protein 2 Kinase Synonym of Ca^{2+}/calmodulin-dependent protein kinase.

Microvilli Plural of microvillus.

Microvillus Finger-like projection from the outer surface of a cell that increases the effective surface area of the membrane and plays an important role in cells that have an absorption function.

Microviridae A family of icosahydral, lytic, ssDNA bacteriophage, e.g., ϕx174.

Microzide Capsules A trade name for hydrochlorothiazide, a diuretic agent.

Micrugy Microsurgery that is carried out under microscopic magnification, e.g., micromanipulation of a single cell with a micromanipulator.

***Micrurus fulvius* Antivenin** An anti-coral snake venom agent that binds and neutralizes coral venom.

Mictrin A trade name for hydrochlorothiazide, a diuretic agent that increases urine excretion of sodium and water by inhibiting sodium reabsorption.

MID Abbreviation for 1. minimum inhibitory dose; 2. minimum infective dose.

Midamor A trade name for amiloride hydrochloride, a diuretic agent that inhibits sodium reabsorption and potassium excretion.

Midazolam (mol wt 326) An anesthetic agent that depresses the CNS.

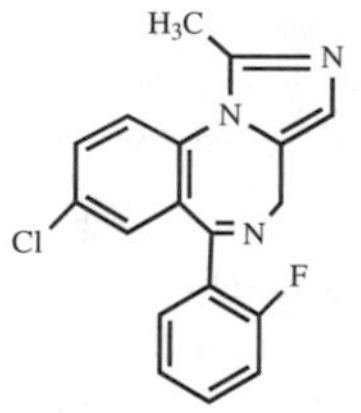

Middle Lamella The outer layer of extracellular substance of the plant cell wall that connects the cell wall between two adjacent cells.

Middle Point Potential 1. The electrode potential at which the redoxant and oxidant are present at equal concentrations 2. The middle point of an oxidation-reduction titration curve.

Middle T Antigen A polyomavirus-encoded antigen involved in the formation of tumors in animals.

Midecamycins A macrolide antibiotic complex from *Streptomyces mycarofaciens.*

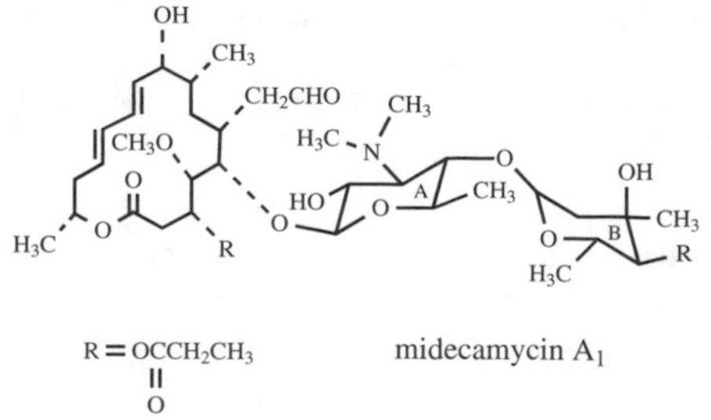

midecamycin A_1

Midodrine (mol wt 254) An antihypotensive and alpha-adrenergic agent.

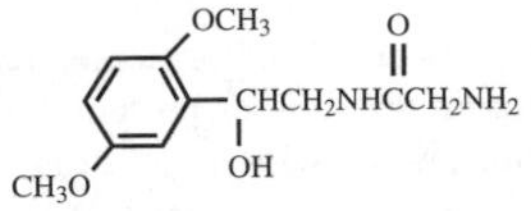

Midol A trade name for a combination drug containing aspirin, caffeine, and cinnamedrine hydrochloride.

Midol-200 A trade name for ibuprofen, an anti-inflammatory, analgesic, and antipyretic drug.

Midol PMS A trade name for a combination drug containing acetaminophen, pamabrom, and pyrilamine maleate, used as an antipyretic, analgesic, and anti-inflammatory agent.

MIF Abbreviation for 1. macrophage-inhibiting factor; 2. merthiolate-iodine-formaldehyde solution; 3. micro-immuno-fluorescence; 4. migration-inhibiting factor or migration inhibition factor.

Mifegynel A trade name for mifepristone, an abortifacient.

Mifentidine (mol wt 228) A histamine H_2-receptor antagonist.

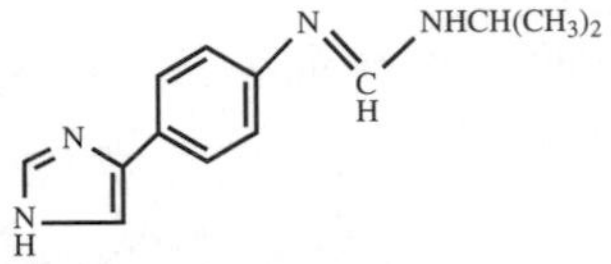

Mifepristone (mol wt 430) An abortifacient.

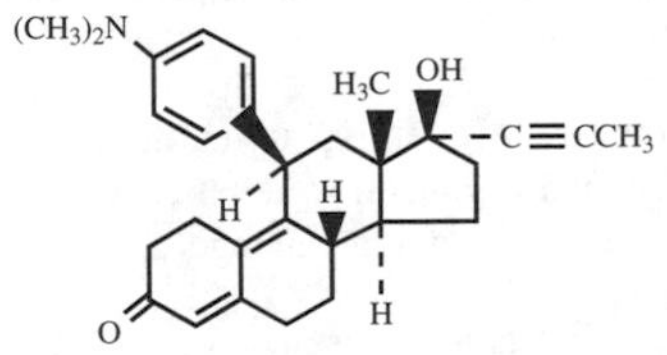

MIg Abbreviation for 1. malaria immunoglobulin; 2. membrane immunoglobulin; 3. measles immunoglobulin.

Miglitol (mol wt 207) An antidiabetic agent.

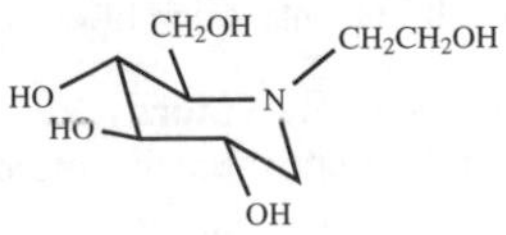

Migral A trade name for a combination drug containing ergotamine tartrate, caffeine, and cyclizine hydrochloride, used as an adrenergic blocker.

Migraine A condition marked by a throbbing headache, severe pain, and sensitivity to light.

Migranol A trade name for dihydroergotamine mesylate, an anti-migraine agent.

Migration-Inhibition Factor Protein factors that inhibit the movement of macrophages.

MIH A trade name for procarbazine hydrochloride, an antineoplastic agent.

Mikamycin An antibiotic complex from *Streptomyces mitakaensis*.

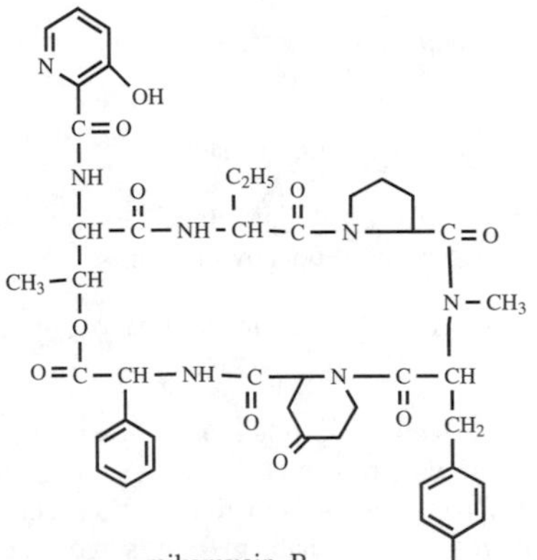

mikamycin B

Milbemycins A family of macrolide antibiotics with insecticidal and acaricidal activity from *Streptomyces hygroscopicus*.

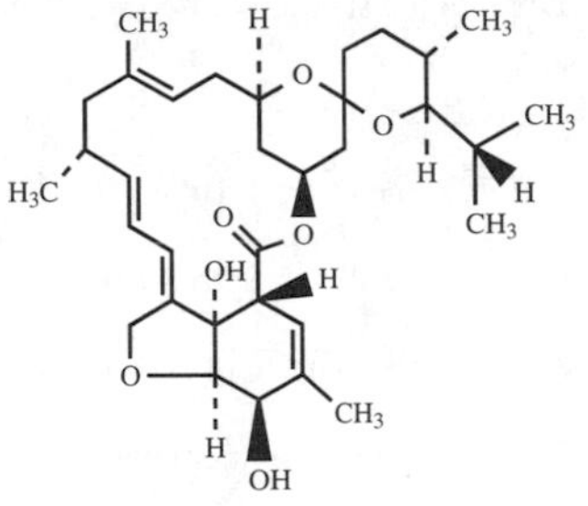

milbemycin D

Mildew A variety of plant diseases in which the mycelium of the parasitic fungus is visible on the surface of the affected plant.

Mildiomycin (mol wt 514) A nucleoside antibiotic with antimildew activity from *Streptoverticilium rimofaciens*.

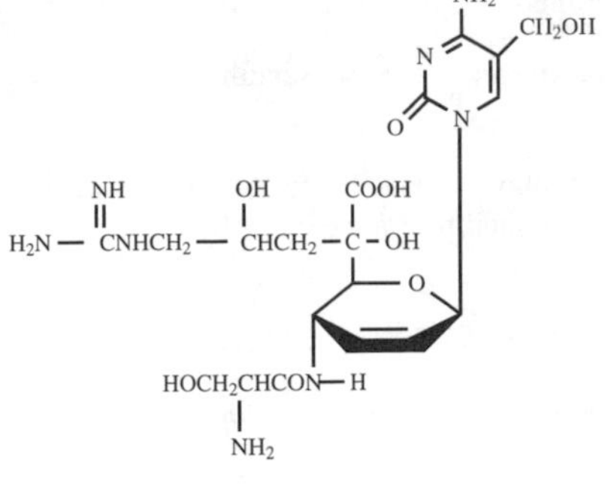

Milk Acidophilus Milk that contains beneficial bacteria, e.g., *Lactobacillus acidophilus* used for treatment of intestinal disorders.

Milk Intolerance A disorder resulting from the inability of an individual to digest milk due to a deficiency in enzyme β-D-galactosidase leading to a painful digestive upset.

Milk of Magnesia Synonym of milk of magnesium.

Milk of Magnesium Referring to magnesium hydroxide used as a laxative.

Milk Sugar Referring to lactose.

Milkers' Nodule A mild, localized viral infection caused by a pseudocowpox virus.

Milkinol A trade name for mineral oil, used as a laxative.

Miller's Spread A technique to mount chromosomes for electron microscopic examination in which chromosomes from the smashed cells are centrifuged in 1% formalin in 0.1 M sucrose, spread onto a membrane-coated grid, stained with phosphotungstic acid, and examined under the electron microscope.

Miller's Tree The pattern of rRNA transcribed from the DNA in the salamander oocyte. The rRNA molecules attached to the chromatin fiber resemble a tree.

Milli- A prefix meaning 1/1000 (10^{-3}).

Milliampere One-thousandth of an ampere.

Millicurie One-thousandth of a curie (10^{-3} curie).

Milliequivalent The equivalent weight expressed in milligrams.

Milligram One-thousandth of a gram (10^{-3} g).

Milligram % A solute concentration that is expressed in number of milligrams per 100 ml.

Milliliter A unit of volume equal to 1/1000 of a liter (abbreviated as ml).

Millimeter 10^{-3} meter.

Millimicron One-thousandth of a micron or one nanometer.

Millimolar Pertaining to a solution that contains one-thousandth of a mole or 10^{-3} mole of a solute per liter.

Millimole One-thousandth of a mole.

Milliosmole One-thousandth of an osmole.

Millipore Filler A trade name for a type of synthetic bacterial filter with specified pore size.

Millon's Reaction A colorimetric reaction for determination of tyrosine or protein, based on the treatment of sample with a solution containing mercurous and mercuric nitrates in the presence of concentrated nitric acid.

Milontin A trade name for phensuximide, an anticonvulsant.

Miloxacin (mol wt 263) An antibacterial agent.

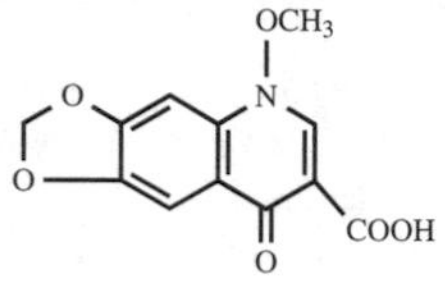

Milprem A trade name for a combination drug containing meprobamate and conjugated estrogen, used as an antianxiety agent.

Milrinone (mol wt 211) A cardiotonic agent.

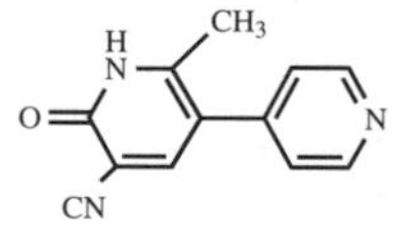

Miltefosine (mol wt 405) An antineoplastic agent.

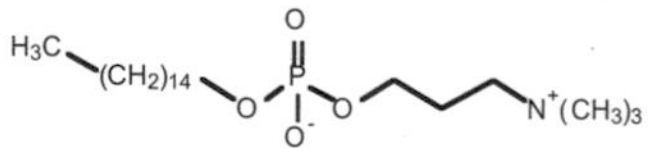

Miltown A trade name for meprobamate, an antianxiety agent.

Mimicry 1. Mimicking something. 2. Resemblance of one organism to another to provide an offensive or defensive advantage.

Mineral Inorganic substance other than water.

Mineral Oil Any oil derived from a nonliving source, used as a laxative.

Mineralocorticoid A hormone released by the adrenal glands that regulates metabolism of water and mineral salt and maintains blood volume.

Mineralocorticoid Receptors Receptors that bind ligand and mediate the action of mineralocorticoid.

Mineralocorticosteroid See mineralocorticoid.

Mini- A prefix meaning small.

Minidiab A trade name for glipizide, which stimulates insulin release from the pancreatic beta cells and reduces glucose output by the liver.

Minicell A bacterial cell produced by a cell division that generates a cytoplasm without nuclear material.

Mini-F Plasmid Any small plasmid constructed from a fragment of the F plasmid.

Minimal Medium A defined growth medium for growth of wild type microorganisms in which all components other than the carbon source are inorganic compounds.

Minims A trade name for atropine sulfate, an anticholinergic, antimuscarinic, and antiparkinsonism agent.

Minimum Hemagglutinating Dose The smallest quantity of hemagglutinating agent that causes a complete hemagglutinating reaction in a standard volume of red blood cells.

Minimum Hemolytic Dose The smallest quantity of complement that causes complete hemolysis of a standard volume of sensitized red blood cells.

Minimum Inhibitory Concentration The concentration of an antimicrobial agent necessary to inhibit the growth of a particular strain of microorganism.

Minimum Lethal Dose The smallest quantity of a toxic substance, bacterium, or virus that causes death of 100% of the test subjects.

Minimum Molecular Weight The molecular weight of a substance that is determined by assaying one of its structural elements.

Minimyosin A type of myosin that binds actin.

Mini-Prep A rapid, small-scale procedure for isolation and purification of plasmid DNA from a biological source.

Minipress A trade name for prazosin hydrochloride, an antihypertensive agent that blocks postsynatptic alpha receptors.

Minirin A trade name for the hormone desmopressin acetate, which promotes reabsorption of water and produces concentrated urine.

Mini-T Plasmid A tumor-inducing plasmid in plants isolated from *Agrobacterium tumefaciens* in which a portion of the DNA in the plasmid is nonessential for replication and is excised. It is used as a clonal vehicle in plant genetic engineering.

Minitran A trade name for nitroglycerin, an antianginal agent.

Minizide A trade name for a combination drug containing polythiazide and prazosin hydrochloride, used as an antihypertensive agent.

Mink Cell Focus-Forming Virus A variant of murine leukemia virus (Retroviridae) that forms foci on the mink lung cell monolayer.

Minocin A trade name for minocycline hydrochloride, an antibacterial agent that binds to 30S ribosomes, inhibiting bacterial protein synthesis.

Minocycline (mol wt 457) A semisynthetic antibiotic against tetracycline-resistant *Staphylococci.* It binds to 30S ribosomes and inhibits bacterial protein synthesis.

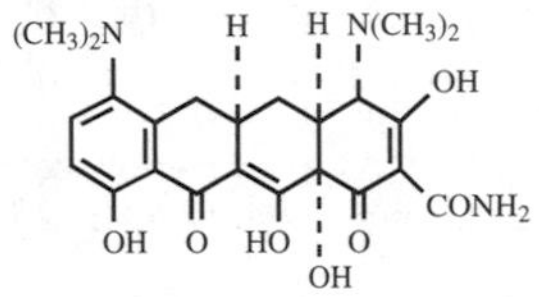

Minodyl A trade name for minoxidil, an antihypertensive agent that produces arteriolar vasodilation.

Minomycin A trade name for minocycline hydrochloride, an antibiotic that binds 30S ribosomes inhibiting bacterial protein synthesis.

Minor Bases Referring to a group of uncommon purine and pyrimidine bases that occur in tRNA and viral DNA, e.g., pseudouracil, methylated guanine, ribosylthymidine, and hydroxymethylcytosine.

Minor Groove The shallow and narrow grove of the two grooves in double-stranded DNA, resulting from the twisting of the two strands around each other.

Minox A trade name for minoxidil, a vasodilator and antihypertensive agent.

Minoxidil (mol wt 209) An antihypertensive agent that produces direct arteriolar vasodilation.

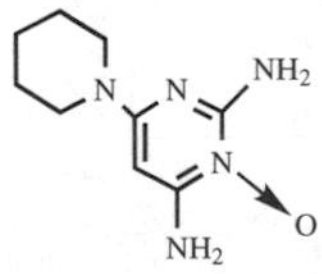

Minoxigaine A trade name for minoxidil, a vasodilator and antihypertensive agent.

Mintezol A trade name for thiabendazole, an anthelmintic agent.

Minus End The end of a microtubule or actin filament at which the addition of monomers occurs less readily.

Minus Strand DNA The strand of a dsDNA that does not serve as template for the synthesis of mRNA.

Minus Strand RNA Viral RNA that does not serve as mRNA upon infection.

Minute Gel A trade name for sodium fluoride, used for bone mineralization.

Miocarpine A trade name for pilocarpine hydrochloride, which causes contraction of the sphincter muscles of the iris, resulting in miosis.

Miochol A trade for acetylcholine chloride used as a cholinergic drug.

Miostat A trade name for carbachol, used as a cholinergic drug.

MIP Abbreviation for 1. macrophage inflammatory protein; 2. mitochondrial intermediate peptidase.

MIP-1α Abbreviation for macrophage inflammatory protein 1α.

MIPP Abbreviation for multiple inositol polyphosphate phosphatase.

Miraculin A taste-modifying protein and sweetening agent with molecular weight of about 44,000 daltons.

Miralax A trade name for pramipexole, a dopamine agonist used for the treatment of Parkinson's disease.

Mirapex A trade name for pramipexole, a dopamine receptor agonist used as an antiparkinsonism agent.

Mircette A trade name for a combination drug containing desogestrel and ethinyl estradiol used as a contraceptive agent.

Mireze A trade name for nedocromil sodium, an anti-allergic and anti-asthmatic agent.

Mirtazepine (mol wt 265) An antidepressant.

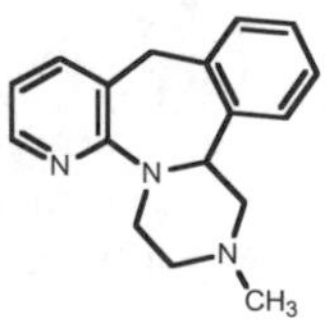

MisI (NaeI) A restriction endonuclease from *Micrococcus* species with the same specificity as NaeI.

Mischarge The incorrect matching and linkage between a tRNA and an amino acid.

Mismatch Repair A DNA repair system that detects, removes, and replaces defective or mismatched bases in the DNA.

Misoprostol (mol wt 383) An antiulcerative agent that stimulates gastric acid secretion.

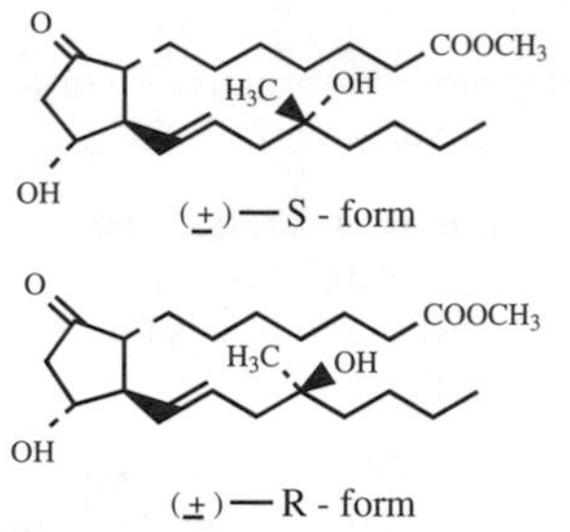

Missense Codon An altered codon that encodes a different amino acid.

Missense Mutation A point mutation in which a codon is changed so that it encodes a different amino acid.

Mistranslation A translation process that incorporates an incorrect amino acid into the peptide.

MIT Abbreviation for mono-iodotyrosine.

mITF Abbreviation for mouse intestinal trefoil factor.

Mithracin A trade name for plicamycin, an antibiotic that binds DNA and interferes with transcription.

Mithramycin See plicamycin.

Mitobronitol (mol wt 308) An antineoplastic agent.

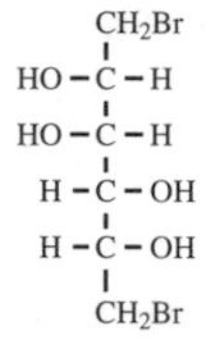

Mitochondria Plural of mitochondrion.

Mitochondrial ATPase Referring to the ATPase that consists of two components designated as F_0 and F_1. The F_0 component is embedded in the lipid-bilayer serving as a proton channel, while the F_1 component is on the surface of the membrane and responsible for synthesis of ATP.

Mitochondrial DNA The DNA or genetic material of the mitochondrion, which is a circular, histone-free, double-stranded DNA and encodes proteins and enzymes for the mitochondrion.

Mitochondrial Matrix The fluid interior of the mitochondrion enclosed by the inner mitochondrial membrane.

Mitochondrial RNA Any RNA that is complementary to mitochondrial DNA.

Mitochondrion A cell organelle and site of electron transport system for generation of ATP in eukaryotic cells. It contains enzymes for the Krebs cycle, oxidative phosphorylation, and electron transport system.

Mitogen Any substance capable of inducing a cell to begin DNA synthesis and cell division.

Mitogen Receptor Transmembrane protein that binds with specific mitogen at the surface of the cell, thereby initiating mitosis and cell division.

Mitogillin A purine-specific ribonuclease that attacks 28S rRNA. It is also an IgE-binding protein and a powerful allergen.

Mitoguazone (mol wt 184) An antineoplastic agent.

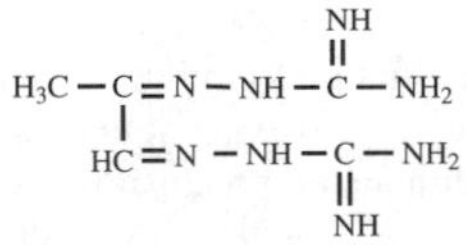

Mitolactol (mol wt 308) An antineoplastic agent.

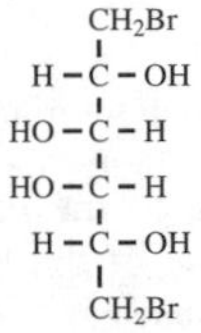

Mitomycins A group of antibiotics from *Streptomyces caespitosus* that cross-links strands of double-stranded DNA, inhibiting DNA replication and transcription.

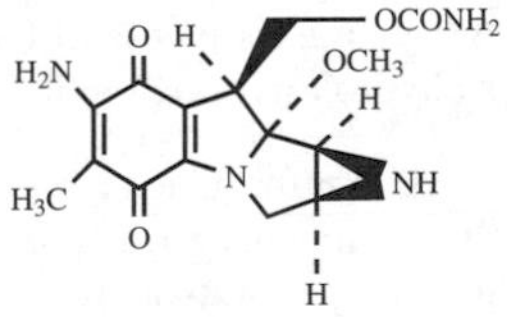

mitomycin C

Mitoplast A mitochondrion without an outer membrane.

Mitoribosomes Referring to mitochondrial ribosomes that resemble prokaryotic ribosomes.

Mitosis A process in which two genetically identical daughter nuclei are produced from one nucleus through chromosome duplication and segregation. Each daughter cell has the same genetic material as the parental cell. The mitotic process consists of four phases: prophase, metaphase, anaphase and telophase.

Mitotane (mol wt 320) An antineoplastic agent.

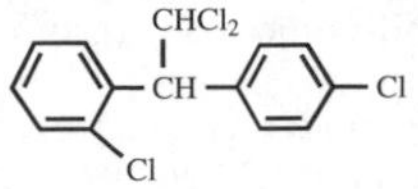

Mitotic Apparatus See mitotic spindle.

Mitotic Center Cellular region that organizes the microtubules for mitosis.

Mitotic Index The proportion of cells present in a culture that are undergoing mitosis in a given sample.

Mitotic Recombination The somatic crossing over and recombination between homologous chromosomes during mitosis.

Mitotic Shake-off Method A method for collecting cells from a cell culture during mitosis since cultured cells become less firmly attached to the culture substratum during mitosis.

Mitotic Spindle Microtubular structure responsible for separating chromosomes during mitosis.

Mitoxantrone (mol wt 444) An antineoplastic agent.

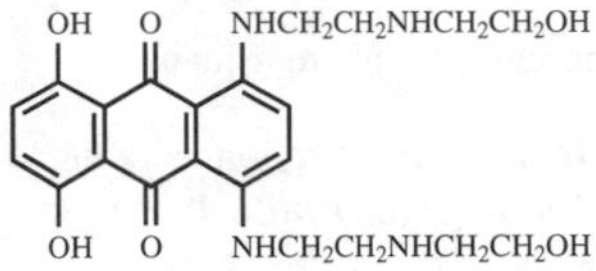

Mitran A trade name for chlordiazepoxide hydrochloride, an anti anxiety agent.

Mitrolan A trade name for calcium polycarbophil, a laxative agent that absorbs water and expands stool bulk.

Mivacron A trade name for mivacurium chloride, a skeletal muscle relaxant.

Mixed Acid Fermentation A type of fermentation carried out by bacteria of the family Enterobacteriaceae that ferments glucose to a mixture of different acids, e.g., acetic and lactic acid.

Mixed Bed Demineralizer A demineralizer that contains both cationic and anionic exchangers and used for removal of ions from water or a solution.

Mixed Infection The concurrent infection of a cell or an individual with more than one pathogenic microorganism.

Mixed Lymphocyte Reaction An *in vitro* test for identification of class-II histocompatibility antigens based on the proliferation of lymphocytes in the presence of cells with nonhistocompatibility antigens.

Mixed Order Reaction The rate of a chemical reaction that cannot be described by a simple first-, second-, or third-order rate equation.

Mixed Triglyceride A triacylglycerol that contains different fatty acids.

Mixed Type Inhibition Inhibition of an enzymatic reaction by an inhibitor that causes the alterations or changes of V_{max} and K_m.

Mixed Vaccine A vaccine capable of giving protection against more than one pathogen.

Mixotrophs Organisms capable of utilizing both autotrophic and heterotrophic metabolic processes.

Mixtard A trade name for isophane-insulin suspension.

Mizoribine (mol wt 259) An immunosuppressant.

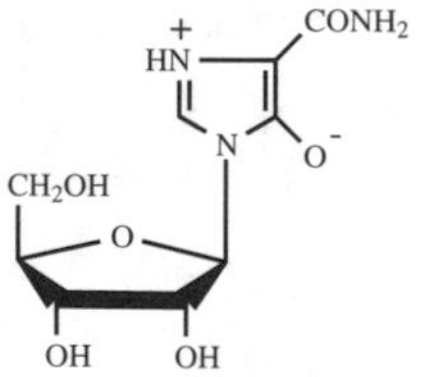

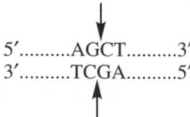

MJ Abbreviation for marijuana.

MjaI (MaeI) A restriction endonuclease from *Methanococcus japannshii* with the same specificity as MaeI.

MjaII (AsuI) A restriction endonuclease from *Methanococcus japannshii* with the same specificity as AsuI.

MK Cell Abbreviation for monkey kidney cell.

MKC-CSA Abbreviation for megakaryocytic colony stimulatory activity.

MkiI (HindIII) A restriction endonuclease from *Moraxella kingae* with the same specificity as HindIII.

MKK Abbreviation for MAP kinase kinase.

MKKK Abbreviation for MAP kinase kinase kinase.

MkrI (PstI) A restriction endonuclease from *Micrococcus kristinae* with the same specificity as PstI.

MKS System Abbreviation for meter-kilogram-second system.

M-kya A trade name for a combination drug containing quinine sulfate, vitamin E, and lecithin.

ml Abbreviation for milliliter.

MlaI (AsuII) A restriction endonuclease from *Mastigocladus laminosus* with the following specificity:

↓
5′..........TTCGAA..........3′
3′..........AAGCTT..........5′
↑

MLC Abbreviation for 1. minimum lethal concentration; 2. mixed leukocyte culture; 3. mixed lymphocyte culture; 4. multilamellar cytosome; 5. myosin light chain.

MLC_{20} Abbreviation for 20-kDa myosin light chain.

MLCK Abbreviation for myosin light-chain kinase, an enzyme that catalyzes the phosphorylation of the myosin light-chain.

MLCP Abbreviation for myosin light chain phosphatase.

MLD Abbreviation for minimum lethal dose, the smallest dose of a toxic agent that causes death of the testing animals or organisms.

MleI (BamHI) A restriction endonuclease from *Micrococcus luteus* with the same specificity as BamHI.

MLEC Abbreviation for mink lung epithelial cell.

M-Line Referring to the dark line that is observed in the middle of the H zone of the myofibrils of striated muscle.

MLK Abbreviation for mixed-lineage kinase.

MLO Abbreviation for *Mycoplasma*-like organisms.

mLPP Abbreviation for mouse LPP (lipid phosphate phosphohydrolase).

MLR Abbreviation for mixed lymphocyte reaction, a reaction commonly used for identification of class-II histocompatibility antigens.

MltI (AluI) A restriction endonuclease from *Micrococcus luteus* with the following specificity:

↓
5′..........AGCT..........3′
3′..........TCGA..........5′
↑

MluI A restriction endonuclease from *Micrococcus luteus* with the following specificity:

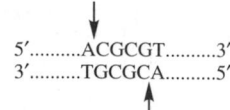

↓
5′..........ACGCGT..........3′
3′..........TGCGCA..........5′
↑

MluNI A restriction endonuclease from *Micrococcus luteus* N1 with the following specificity:

```
         ↓
5'.......TGGCCA........3'
3'........ACCGGT........5'
           ↑
```

MLV Abbreviation for 1. Moloney leukemia virus; 2. mouse leukemia virus; 4. multilamellar vesicle; 3. murine leukemia virus.

MM Abbreviation for 1. malignant melanoma; 2. multiple myeloma; 3. minimal medium used for the culture of wild-type microorganisms.

mM Abbreviation for millimolar or millimole/liter.

mm Abbreviation for millimeter.

μM Abbreviation for micromolar.

MM21 Abbreviation for a monoclonal antibody to myosin heavy chain.

MM Blood Group A type of blood group classified on the basis of the presence of M substance.

MMA Abbreviation for methylmalonic acid.

M-macroglobulin An abnormal IgM occurring in patient with Waldenstrom's macroglobulinemia.

MMC Abbreviation for mitomycin C.

MMCM Abbreviation for methylmalonyl-CoA mutase.

MmeII (MboI) A restriction endonuclease from *Methylophilus methylotrophus* with the same specificity as MboI.

mmHg A unit of pressure equal to 1 millimeter of mercury (1 mmHg = 1333.3224 pascals).

MMLV Abbreviation for Moloney murine leukemia virus.

MMO Abbreviation for methane monooxygenase.

mMole Abbreviation for millimole.

μMole Abbreviation for micromole.

MMP Abbreviation for matrix metallo-proteinase.

MMPI Abbreviation for matrix metallo-proteinase inhibitor.

MMPR Abbreviation for methyl-mercapto-purine riboside.

MMR Abbreviation for measles-mumps-rubella vaccine.

MMR-II A trade name for measles, mumps, and rubella virus vaccine (live).

MMTV Abbreviation for mouse mammary tumor virus.

MMU Abbreviation for mercaptomethyl uracil.

MMUP Abbreviation for mouse major urinary protein.

MN Blood Group System Transmembrane glycoproteins of the erythrocyte used for classification of blood groups. The combination of MN substances produces three types of blood groups, namely MM, MN, and NN.

MNCF Abbreviation for mononuclear cell factor.

MNF Abbreviation for monocyte nuclear factor.

MniI (HaeIII) A restriction endonuclease from *Moraxella nonliquefaciens* with the same specificity as HaeIII.

MniII (HpaII) A restriction endonuclease from *Moraxella nonliquefaciens* with the same specificity as HpaII.

MNL Abbreviation for mononuclear leukocyte.

MnlI A restriction endonuclease from *Moraxella nonliquefaciens* with the following specificity:

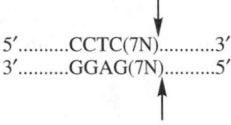

```
                   ↓
5'..........CCTC(7N)...........3'
3'..........GGAG(7N)..........5'
                   ↑
```

MnnI (HindII) A restriction endonuclease from *Moraxella nonliquefaciens* with the same specificity as HindII.

MnnII (HaeIII) A restriction endonuclease from *Moraxella nonliquefaciens* with the same specificity as HaeIII.

MnoI (HpaII) A restriction endonuclease from *Moraxella nonliquefaciens* with the following specificity:

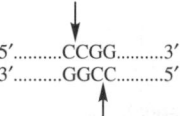

```
           ↓
5'.........CCGG..........3'
3'..........GGCC..........5'
              ↑
```

MnoIII (MboI) A restriction endonuclease from *Moraxella nonliquefaciens* with the same specificity as MboI.

MnSOD Abbreviation for manganese superoxide dismutase.

Mo Abbreviation for the element molybdenum, with atomic weight 96, valence 2, 3, 4, 5, 6.

MO_2 Abbreviation for myocardial oxygen.

MoAb Abbreviation for monoclonal antibody.

MOAT Abbreviation for multiple organic-anion transporter.

Moban A trade name for molindone hydrochloride, an antipsychotic agent that blocks postsynaptic dopamine receptors in the brain.

Mobenol A trade name for tolbutamide, which stimulates the release of insulin from pancreatic beta cells and reduces glucose output by the liver.

Mobidin A trade name for magnesium salicylate, which produces analgesia.

Mobile Genetic Element See transposome.

Mobile Ion Carrier See ionophore.

Mobile Phase The mobile phase of a carrier gas or a carrier liquid in a chromatographic system for separation of solute mixtures.

Mobilization Lipase The enzyme that catalyzes the release of lipids or fatty acids from adipose tissue.

Mobiluncus A genus of anaerobic, asporogenous, Gram-negative and Gram-variable bacteria.

MOCA Abbreviation for a combination drug containing methotrexate, oncovin, cytoxan, and adriamycin.

Moclobemide (mol wt 269) An antidepressant.

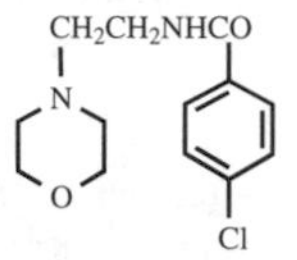

Moctanin A trade name for monoctanoin, which dissolves or solubilizes gallstones.

Modane A trade name for phenolphthalein, a laxative.

mODC Abbreviation for mouse ornithine decarboxylase.

Modecate Decanoate A trade name for fluphenazine decanoate, an antipsychotic drug that blocks postsynaptic dopamine receptors in the brain.

Modeccin A 57-kDa lectin from a turnip-like plant in southern Africa, *Modecca digitata*.

Moderil A trade name for rescinnamine, an antihypertensive agent that inhibits the release of norepinephrine.

Modicon A trade name for a combination drug containing ethinyl estradiol and norethindrone, used as an oral contraceptive.

Modification A process by which a substance is transformed from one form to another, e.g., modification of newly synthesized DNA by methylation, or modification of proteins or enzyme by phosphorylation.

Modification and Demodification Reaction The reaction of covalent modification and demodification for conversion of an enzyme from an inactive form to an active form or vice versa.

Modification Methylase The enzyme that catalyzes the methylation of DNA for protection against endogenous restriction endonucleases.

Modifier Gene A gene whose function alters the phenotypic expression of one or more genes at loci other than its own.

Moditen A trade name for fluphenazine, an antipsychotic drug that blocks postsynaptic dopamine receptors in the brain.

Modrastane A trade name for trilostane, an antineoplastic agent that alters hormone balance in the body.

Modulon A trade name for trimebutine maleate, which regulates intestinal motility.

Moduretic A trade name for a combination drug containing amiloride hydrochloride and hydrochlorothiazide, used as a diuretic agent.

Moebiquin A trade name for iodoquinol, an antiprotozoal agent.

Moellerella A genus of bacteria of the family Enterobacteriaceae.

Moexipril Hydrochloride (mol wt 535) An angiotensin-converting enzyme inhibitor.

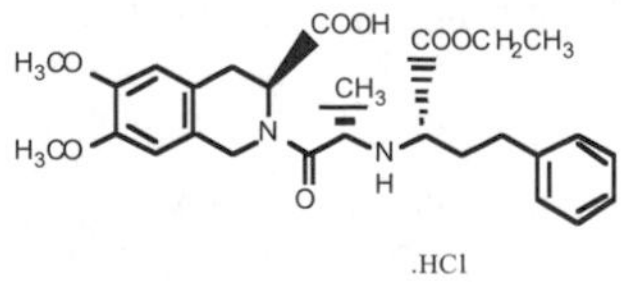

MoFd Abbreviation for molybdoferredoxin (a component of nitrogenase).

Mofebutazone (mol wt 232) An anti-inflammatory agent.

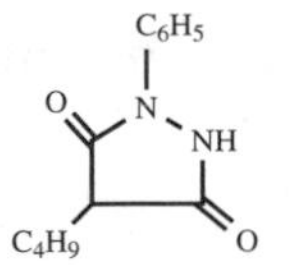

MoFe Protein Referring to nitrogenase for nitrogen fixation.

Mofegiline (mol wt 197) An antiparkinsonian agent.

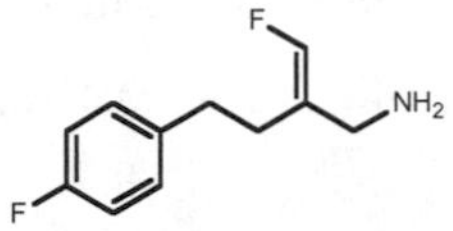

Mofezolac (mol wt 339) An anti-inflammatory and analgesic agent.

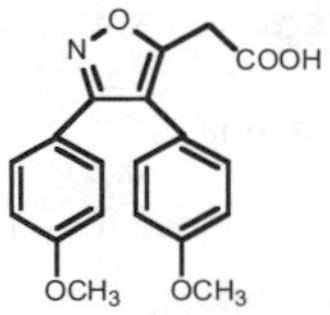

MOG Abbreviation for mono-oleoylglycerol.

MOI Abbreviation for multiplicity of infection.

Moiety A structural component of a molecule that imports a characteristic chemical property.

Molal Referring to one liter of solvent to which one mole of a solute is added.

Molality The concentration of a solute that is expressed as number of mole or moles dissolved in one liter of solvent.

Molar Pertaining to mole.

Molar Absorptivity The absorbance of one molar solution of a given substance measured at a specific wavelength in a cuvette with a diameter of one cm.

Molar Activity of Enzyme Referring to the number of moles of substrate transformed to product per mole of enzyme under optimal substrate concentration.

Molar Concentration The number of moles of a solute per liter of solution.

Molar Extinction Coefficient See molar absorptivity.

Molar Solution A solution whose concentration is expressed in the number of moles or moles per liter.

Molarity The concentration of a solute that is expressed in number of moles of solute per liter of solution.

Molasses A syrupy liquid and a by-product from the sugar industry that contains sucrose and non-sugar substances.

Mold Fungi with filamentous mycelia that form a visible mycelial layer on the surface of the infected material.

Mole The gram molecular weight of a substance, which contains 6.02×10^{23} molecules.

Mole% G + C Referring to the molar proportion of guanine and cytosine in a dsDNA macromolecule.

Molecular Pertaining to molecule.

Molecular Biology The science that deals with biological processes at the molecular level.

Molecular Chaperone A protein that assists in the folding of a second protein and preventing the formation of an inactive or incorrect structure.

Molecular Cloning Referring to genetic engineering and DNA cloning.

Molecular Formula A formula that shows the number of atoms of each element present in a molecule.

Molecular Genetics The science that deals with the study of genetics at the molecular level.

Molecular Imprinting A condition in which the differential expressions of genes occur between maternally and paternally inherited genes.

Molecular Mimicry The antigenic determinants on a pathogenic organism that resemble that of the host, thereby eliciting no immunological response.

Molecular Sieve Referring to the cross-linked polymers that form porous sieves and are used as supporting medium for chromatographic separation of mixtures of solutes.

Molecular Sieve Chromatography A method for separating molecules on the basis of size and mass (see gel filtration).

Molecular Sieve Coefficient Referring to the ratio of the equilibrium concentration of the solute within the gel matrix to its concentration in the mobile phase.

Molecular Taxonomy The classification of organisms on the basis of the distribution and composition of chemical substances in the organisms.

Molecular Weight The sum of the atomic weights of the atoms in a molecule.

Molecule The smallest unit of a compound consisting of covalently linked atoms.

Molindone (mol wt 276) An antipsychotic agent that blocks postsynaptic dopamine receptors in the brain.

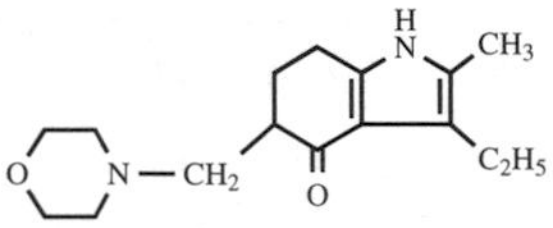

Molisch Test A colorimetric test for the determination of carbohydrate based on the color production upon treatment of the sample with α-naphthol in the presence of concentrated sulfuric acid.

Mollicutes Referring to a class of prokaryotic organisms that do not form cell walls, e.g., *Mycoplasma.*

Molluscum Contagiosum A disease caused by a pox virus (Poxviridae) and characterized by the formation of multiple, firm, rounded, whitish, transparent nodules on the skin.

Moloney Murine Leukemia Virus A replication competent v-*onc*$^-$ murine leukemia virus (Retroviridae) causing thymic leukemia in mice.

Moloney Murine Sarcoma Virus A replication-defective retrovirus (v-*onc*$^+$) capable of inducing fibrosarcomas *in vitro* and transforming cells in culture.

Moloney Test A skin test for immunity to diphtheria by the intradermal injection of diphtheria toxoid.

Molsidomine (mol wt 242) An antianginal agent.

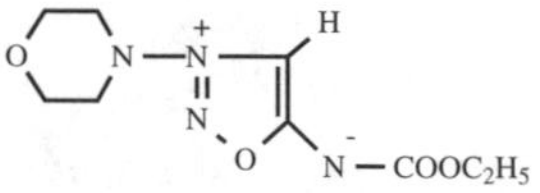

Molybdenum (Mo) A chemical element with atomic weight 96, valence 2, 3, 4, 5, and 6.

Molybdoenzymes Enzymes that contain molybdenum, e.g., aldehyde oxidase, nitrogenase, sulfite oxidase, and xanthine oxidase.

Molybdoferrodoxin Referring to nitrogenase that contains Mo and Fe.

M.O.M. Abbreviation for milk of magnesia (magnesium hydroxide), which is a laxative.

Mometasone Furoate (mol wt 521) A topical anti-inflammatory agent.

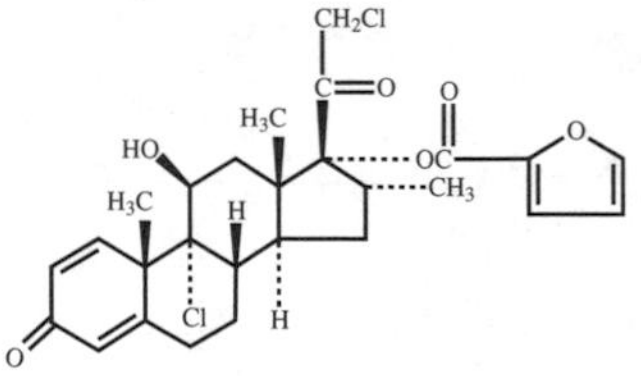

Momordin A ribosome inactiviating protein.

MoMSV Abbreviation for Moloney murine sarcoma virus.

MoMuLV Abbreviation for Moloney murine leukemia virus.

Monad 1. Referring to the unicellular, free-living flagellate stage of a single organism. 2. The haploid set of chromosomes found in the nucleus of an ootide or spermatide.

Monazole A trade name for miconazole nitrate, an antifungal agent.

Monckeberg's Arteriosclerosis A type of hardening of the arteries in which calcium deposits are found in the lining of the artery.

Monellins A low-calorie peptide sweetener from tropical plant *Discoreophyllum cumminsii.*

Monensin (mol wt 671) A sodium-binding ionophore.

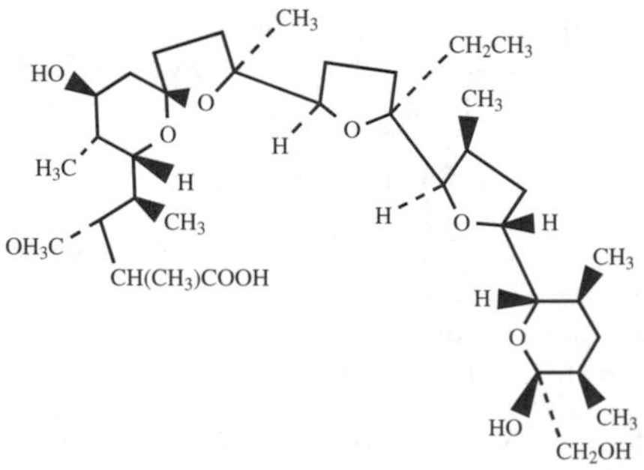

Monera Referring to unicellular prokaryotic protists that contain no true nuclei nor cell organelles, e.g., bacteria or blue green bacteria.

Mongolism A genetic disorder due to the abnormality of chromosome number 21.

Moniliform A form that resembles a string of beads.

Monine A trade name for factor IX, an antihemophilic agent.

Monistat A trade name for miconazole, an antifungal agent.

Monitan A trade name for acebutolol, an antihypertensive agent and a beta blocker that decreases myocardial contraction and heart rate.

Monkey Pox A disease of humans caused by monkey poxvirus (Orthomyxoviridae).

Monkey Poxvirus A virus of the family Orthomyxoviridae.

Mono- A prefix meaning single, one, or alone.

Monoacylglycerol An acylglycerol that contains one fatty acid.

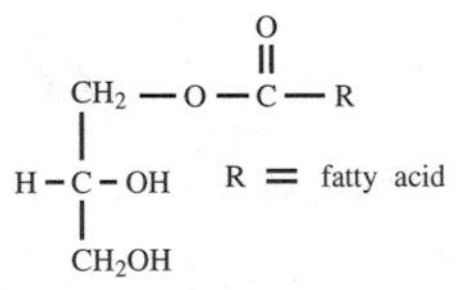

1-monoacylglycerol

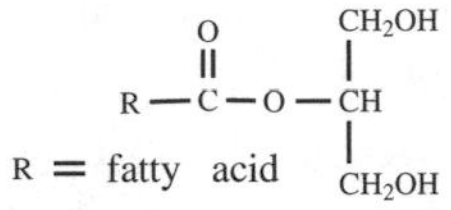

2-monoacylyglycerol

2-Monoacylglycerol Acyltransferase The enzyme that catalyzes the following reaction:

CoA + Diacylglycerol ⇌ 2-Monoacyl-glycerol + acyl-CoA

Monoacylglycerol Kinase The enzyme that catalyzes the following reaction:

ATP + acylglycerol ⇌ ADP + acylglycerol 3-phosphate

Monoacylglycerol Lipase The enzyme that catalyzes the hydrolysis of glycerol monoesters of long chain fatty acids.

Monoacylglycerol Pathway A pathway in which diacylglycerol or triacylglycerol are formed through the sequential acylation of monoacylglycerol by acyl-CoA

Monoamine Referring to an organic compound that contains only one amine group.

Monoamine Oxidase The enzyme that catalyzes the oxidative deamination of a variety of biogenic amines (e.g., serotonin, norepinephrine, epinephrine, or dopamine) to NH_3, hydrogen peroxide, and aldehyde.

Monobasic A compound that has one hydrogen atom that is replaceable by a metal ion.

Monobasic Acid An acid which has only one replaceable hydrogen atom.

Monobasic Salt Salt that has only one of its dissociable hydrogens replaced with another cation, e.g., NaH_2PO_4.

Monobenzone (mol wt 200) A depigmentation agent used for treatment of abnormal skin coloration.

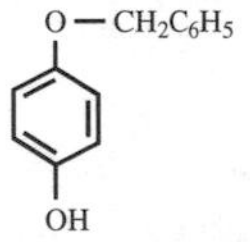

Monoblast Immature monocyte.

Monocentric Chromosome Chromosome that has a single centromer.

Monochromatic Consisting of a single wavelength.

Monochromatic Radiation Radiation of a single wavelength.

Monochromator A device used for selecting a single wavelength from a wide range of wavelengths of radiation.

Monocid A trade name for cefonicid sodium, an antibiotic that inhibits bacterial cell wall synthesis.

Monocins Bacteriocins produced by *Listeria monocytogenes.*

Monocistronic mRNA Messenger RNA molecule that yields only one polypeptide chain upon translation.

Monoclate A trade name for an antihemophilic factor VIII, used to replace the deficient clotting factor.

Monoclonal Originating from a single precursor cell.

Monoclonal Antibody Homogeneous immunoglobulin derived from a single clone of cells.

Monoclonal Hypergammaglobulinemia An increase in serum immunoglobulins produced by a single clone of cells.

Monoclonal Protein A protein produced by the progeny of a single clone.

Monocot Short for monocotyledon.

Monocotyledons Referring to flowering plants (angiosperms) in which the embryo bears one seed leaf (cotyledon).

Monocyclic Cascades Covalent modification and demodification of an enzyme for the regulation of enzyme activity, e.g., cyclic interconversion of an enzyme from the nonactive to the active form through covalent modification and demodification.

Monocytes Circulating granular white blood cells with lobulated nucleus that can differentiate into macrophages.

Monocytopenia The reduction of monocytes in the blood.

Monocytosis The increase in number of monocytes in the blood.

Monodehydroascorbate Reductase The enzyme that catalyzes the following reaction:

$$\text{NADH} + 2\ \text{monodehydroascorbat} \rightleftharpoons \text{NAD}^+ + 2\ \text{ascorbate}$$

Monodentate Ligand A ligand that is chelated to a metal ion through one donor atom.

Monodox A trade name for doxycyline monohydrate, an antibacterial agent.

Monoecious The condition in which male and female structures are present in the same organism.

Monoenergetic Radiation A type of radiation in which all particles or photons have the same quantity of energy.

Monoenoic Referring to alkenyl carboxylic acid containing only one double bond.

Monogalactosyl Diacylglycerol A glycolipid found in the thylakoid membrane.

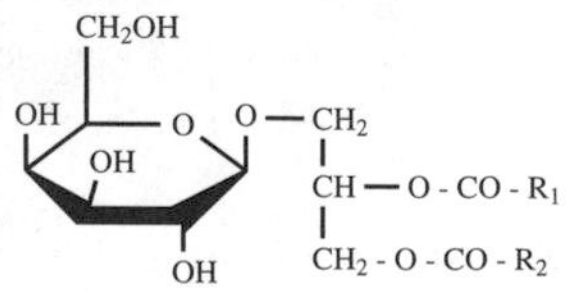

R_1, R_2 = different fatty acids

Monogeneric Referring to a classification system that has only one genus.

Mono-Gesic A trade name for salsalate, an anti-inflammatory and analgesic drug.

Monoglyceride See monoacylglycerol.

Monoglyceride Acyltransferase See monoacylglycerol transferase.

Monohybrid Cross A genetic cross between two organisms that differ in only a single gene.

Monohydric Referring to any chemical compound containing only one hydroxyl group.

Monokaryotic Referring to a cell that has only one nucleus (one nucleus per cell).

Monoket A trade name for isosorbide, an osmotic diuretic agent.

Monokine Regulatory proteins released by monocytes.

Monolayer Referring to 1. a single layer of cells formed on the surface of a culture vessel and 2. a single layer of molecules or particles.

Monolayer Cells A single layer of cells formed on the surface of a culture vessel in tissue culture or cell culture. The formation of a single monolayer of cells in culture due to the property of contact inhibition of the cells.

Monomer Basic unit of polymer, e.g., amino acids in proteins or monosaccharides in polysaccharides.

Monomeric IgM A single unit of the normal pentameric IgM.

Monomeric Protein A protein that consists of a single polypeptide chain.

Monomethyl Sulfatase The enzyme that catalyzes the following reaction:

$$\text{Monomethyl sulfate} + H_2O \rightleftharpoons \text{Methanol} + \text{sulfate}$$

Monomorphic An organism that exists in only one form.

Mononuclear Having only one nucleus.

Mononuclear Leukocyte Referring to mononuclear phagocytes, e.g., macrophages.

Mononuclear Phagocyte Referring to macrophages and monocytes.

Mononuclear Phagocytic System The cells of the macrophage family and their precursors.

Mononucleoside A single nucleoside that consists of a nitrogenous base and a sugar, e.g., adenosine, cytidine, uridine, thymidine, and guanosine.

Mononucleosis A condition in which large numbers of monocytes are present in the blood.

Mononucleotide A single nucleotide that consists of a nitrogenous base, a sugar, and a phosphate, e.g., adenosine monophosphate, guanosine monophosphate, cytidine monophosphate, and uridine monophosphate.

Monooxygenase The enzyme that catalyzes the incorporation of one atom of molecular oxygen into a substrate molecule.

Monophen (mol wt 486) A radiopaque medium used as a diagnostic aid.

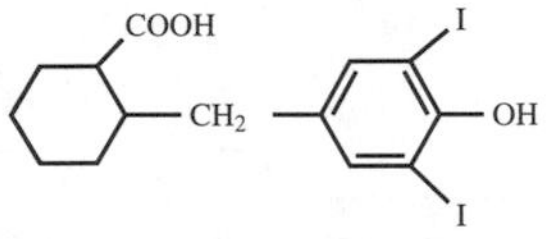

Monophenol, L-Dopa Oxygen Oxidoreductase The systematic name for monophenol monooxygenase.

Monophenol Monooxygenase The enzyme that catalyzes the following reaction:

$$\text{L-Tyrosine} + \text{L-dopa} + O_2 \rightleftharpoons \text{L-Dopa} + \text{dopaquinone} + H_2O$$

Monophenol Oxidase See monophenol monooxygenase.

Monophobia An extreme fear of being alone.

Monophosphatidylinositol Phosphodiesterase Synonym of 1-phosphatidylinossitol phosphodiesterase.

Monophyletic Referring to a taxonomic group that has derived from a single ancestral lineage.

Monopril A trade name for fosinopril sodium, an antihypertensive agent that inhibits angiotensin converting enzyme and preventing conversion of angiotensin I to angiotensin II.

Monoprotic Capable of donating only one proton.

Monosaccharide Simple sugar, the repeating unit of polysaccharides.

Monose Referring to monosaccharide.

Monosodium Glutamate (mol wt 169) A derivative of glutamate used as a flavor enhancer.

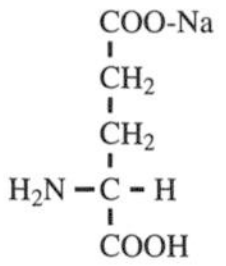

Monosome 1. A single ribosome dissociated from polysome 2. A single ribosome attached to a mRNA.

Monosomic Referring to a diploid cell that has one less chromosome than the normal diploid.

Monosomy A condition in which a cell or an organism has only one copy of a particular chromosome instead of the expected two copies of a diploid state.

Monospecific Antiserum An antiserum that reacts with only one specific antigenic determinant.

Monotard HM A trade name for insulin zinc suspension.

Monotard MC A trade name for insulin zinc suspension.

Monotrichous Possessing only one flagellum.

Monovalent 1. A chemical element with a valence of one. 2. A molecule that has only one binding site.

Monovalent Chromosome Referring to a single unpaired chromosome in meiosis.

Montelukast (mol wt 586) A leukotriene receptor antagonist used as an anti-asthmatic agent.

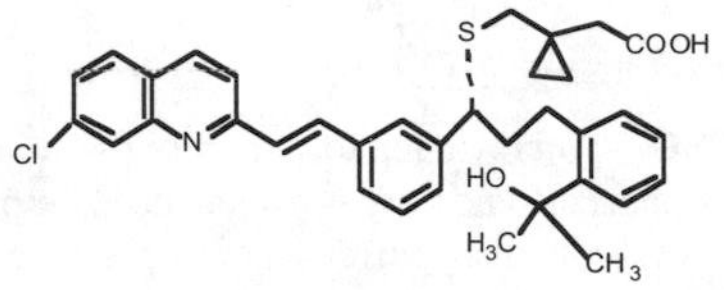

Montenegro Test A test used for diagnosis of cutaneous leishmaniasis using killed *Leishmania*.

Montoux Test A test for tuberculosis by intradermal injection of purified protein derivative.

Monozygotic Twins Twins originating from the same fertilized egg.

Monurol A trade name for fosfomycin tromethamine, a urinary tract anti-infective agent.

MOP Abbreviation for a combination drug containing mechlorethamine, oncovin, and procarbazine.

8-MOP A trade name for methoxsalen, an agent that produces an increased synthesis of melanin and an increased number of melanocytes.

Moperone (mol wt 355) An antipsychotic agent.

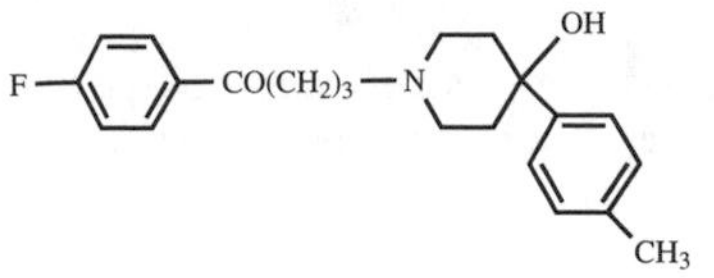

Mopidamol (mol wt 422) An antineoplastic agent.

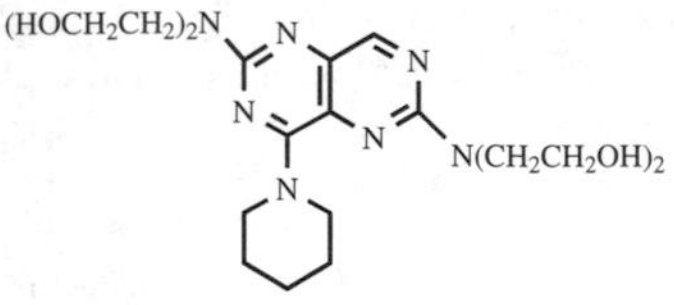

MOPP Abbreviation for a combination drug containing mechlorethamine, oncovin, prednisone, and procarbazine.

Moprolol (mol wt 239) A β-adrenergic blocker and antihypertensive agent.

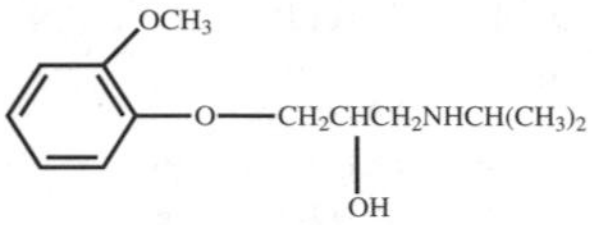

Mo-Protein Abbreviation for a protein component of nitrogenase from *Klebsiella pneumoniae*.

MOPS (mol wt 209) Abbreviation for 4-morpholinepropanesulfonic acid. A regent used for preparation of biological buffers at a pH range between 6.5 and 7.9.

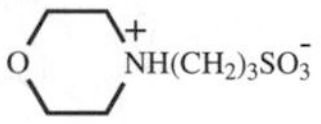

MOPSO Abbreviation for β-hydroxy-4-morpholinepropanesulfonic acid. A reagent used for preparation of biological buffers in the pH range between 6.2 and 7.9.

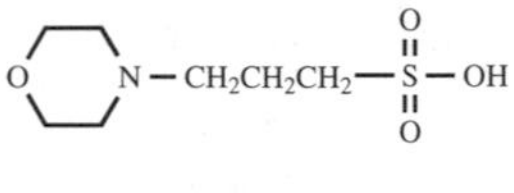

Moquizone (mol wt 351) A choleretic agent.

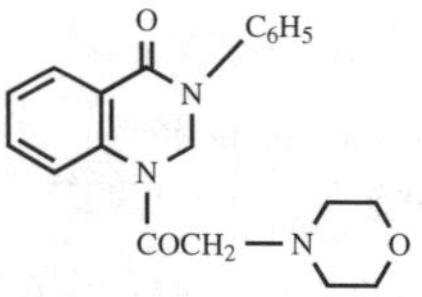

Morantel (mol wt 220) An anthelmintic agent.

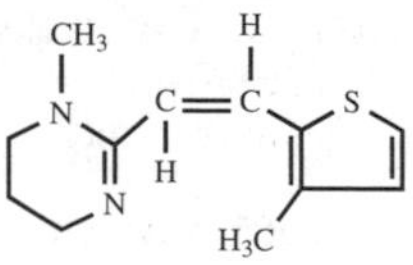

Moraxella A genus of oxidase-positive, aerobic, Gram-negative bacteria (Neisseriaceae).

Morazone (mol wt 377) An analgesic, antipyretic, and anti-inflammatory agent.

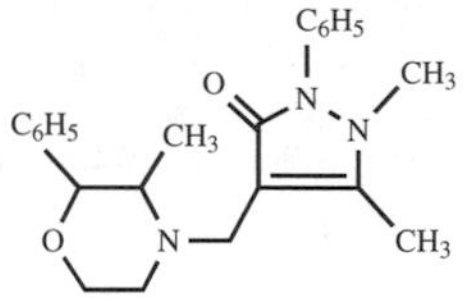

Morbillivirus A virus in the family of Paramyxoviridae.

Morclofone (mol wt 406) An antitussive agent.

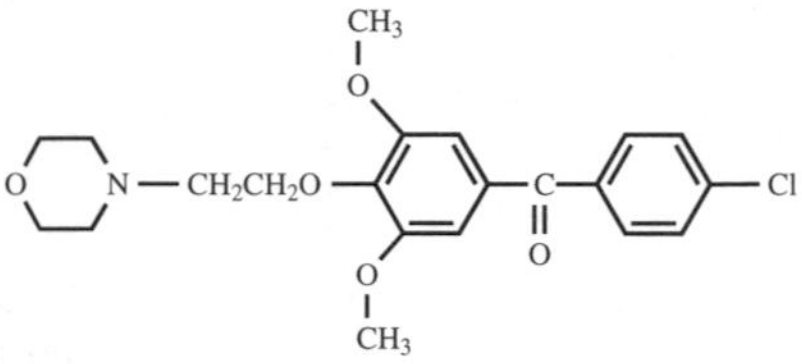

Mordant Any substance that increases the affinity of a stain for a biological specimen.

Morgan Unit A unit for measuring the distance between two genes. One Morgan unit equals a crossover value of 100%, a crossover value of 10% and 1% are called decimorgan and centimorgan, respectively.

Moricizine (mol wt 428) An antiarrhythmic agent that reduces the fast inward current carried by sodium ions.

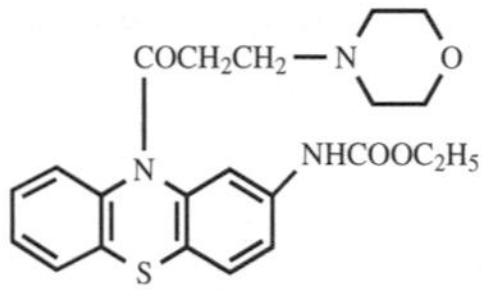

Moroxydine (mol wt 171) An antiviral agent.

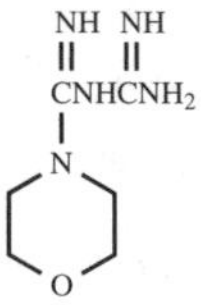

Morphazinamide (mol wt 222) An antituberculostatic agent.

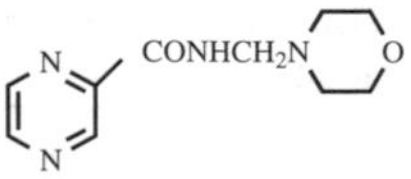

Morphiceptin (mol wt 397) A bioactive opioid dipeptide consisting of tyrosine and arginine.

Morphine (mol wt 285) An alkaloid of opium used as a narcotic analgesic agent that binds with opiate receptors altering both perćeption and emotional responses to pain.

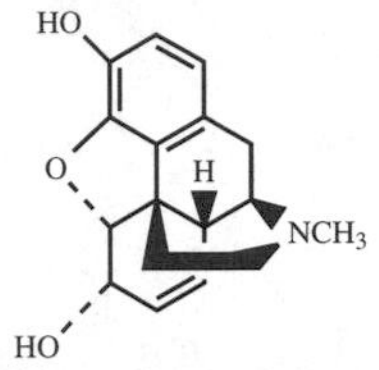

Morphine 6-Dehydrogenase The enzyme that catalyzes the following reaction:

$$\text{Morphine} + \text{NADP}^+ \rightleftharpoons \text{Morphinone} + \text{NADPH}$$

Morphinism Addiction to morphine.

Morphitec A trade name for morphine hydrochloride, an alkaloid of opium used as a narcotic analgesic agent that binds with opiate receptors altering both perception and emotional responses to pain.

Morphogen 1. Substance secreted by one group of cells that causes specific changes in the cellular destination and morphogenesis of another group of cells. 2. Diffusible substances produced in the embryo at specific locations that create a gradient to allow cells to locate their position in the embryo.

Morphogenesis The growth and differentiation of cells to form tissues and organs.

7-Morpholinomethyltheophylline (mol wt 279) A diuretic agent.

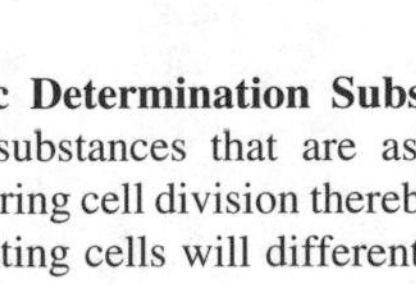

Morphogenetic Gene The gene that encodes the morphogen.

Morphogenic Determination Substances The cytoplasmic substances that are asymmetrically distributed during cell division thereby influencing how the resulting cells will differentiate.

Morpholine Salicylate (mol wt 225) An analgesic, antipyretic, and anti-inflammatory agent.

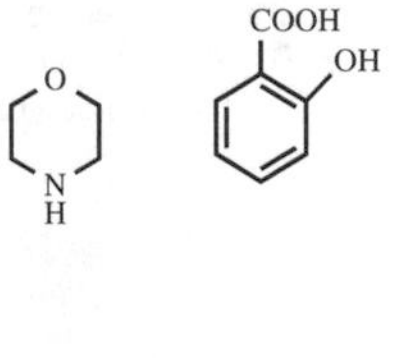

2-Morpholinoethyl Isocyanide (mol wt 140) A coupling reagent used for peptide synthesis.

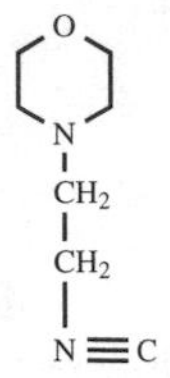

Morphology The science that deals with the structures and forms of living organisms.

MOS A gene family encoding serine/threonine kinases, the *v-mos* is the oncogene of transforming murine Moloney sarcoma virus.

Mosaic 1. The mottled or variegated color appearance of a leaf caused by a viral infection (e.g., tobacco mosiac caused by TMV). 2. An individual composed of two or more cell lines from the same species.

Mosaic Evolution The evolution of different features of an organism at different rates.

mOsm Abbreviation for milliosomole.

Most Probable Number The statistical estimate of the viable bacterial population in a sample through the use of dilution and multiple tube inoculations.

Motif Region of the secondary structure in a protein that consists of segments of α helix and/or a β sheet connected by looped regions of varying length.

Motilin A peptide hormone that stimulates intestinal motility that consists of 22 amino acid residues.

Motilium A trade name for domperidone, which blocks peripheral dopamine receptors and enhances motility in the stomach and intestine.

Motion-Aid A trade name for dimenhydrinate, an antiemetic agent that inhibits nausea and vomiting.

Motofen A trade name for difenoxin hydrochloride, an antidiarrheal agent that slows intestinal motility.

Motor Fiber A fiber in the spinal nerves that carries the movement of signal to muscle fiber.

Motor Neuron Nerve cell that transmits impulses from the central nervous system to an effector, e.g., muscle fiber.

Motor Proteins Protein molecules that generate movement by advancing along a surface using energy derived from the hydrolysis of ATP (e.g., myosin, dynein, and kinesin).

Motretinide (mol wt 354) An antiacne agent.

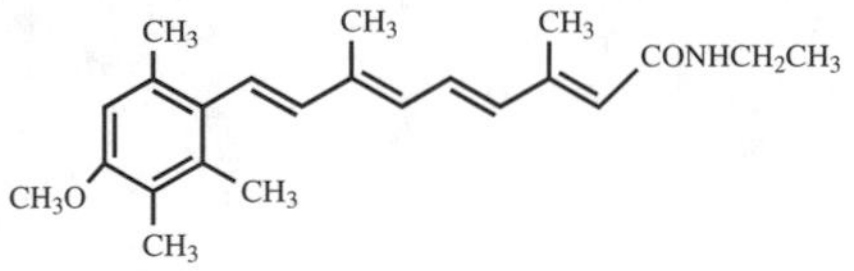

Motrin A trade name for ibuprofen, an anti-inflammatory, analgesic, and antipyretic agent.

Mould Variant spelling of mold.

Mouse L Cells A strain of continuous mouse fibroblast cell line.

Mouse Leukemia Virus An oncogenic virus in the family of Retrovividae.

Mouse Mammary Tumor Virus A virus of Retroviridae that is transmitted through milk and causes mammary tumors in mice.

Moveltipril (mol wt 399) An antihypertensive agent.

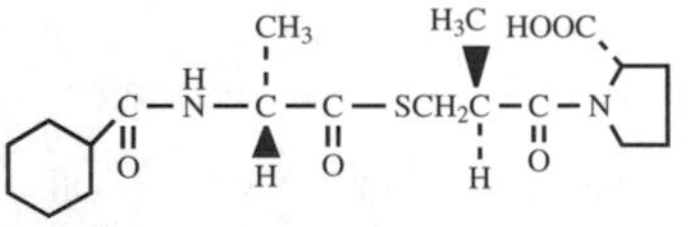

Moving Boundary Centrifugation A type of analytical ultracentrifugation in which a uniform solution containing macromolecules is centrifuged and a boundary is formed. The velocity of boundary sedimentation can be followed and measured with an optical device. It is used for analysis of sedimentation coefficient of macromolecules.

Moving Boundary Electrophoresis A type of electrophoretic technique in which the protein boundary formed during electrophoresis can be followed and measured. It is suited for the determination of electrophoretic mobility of macromolecules.

MOX Abbreviation for methanol oxidase.

Moxacin A trade name for amoxicillin trihydrate, an antibiotic that inhibits bacterial cell wall synthesis.

Moxalactam (mol wt 520) An antibacterial agent that inhibits bacterial cell wall synthesis.

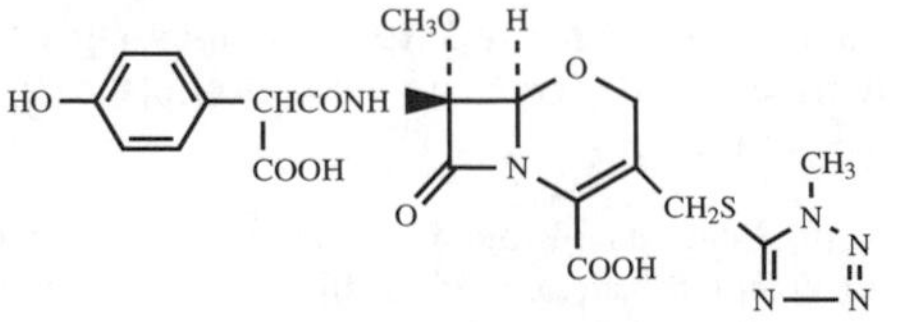

Moxam A trade name for maxalactam disodium, an antibiotic that inhibits bacterial cell wall synthesis.

Moxaverine (mol wt 307) An antispasmodic agent.

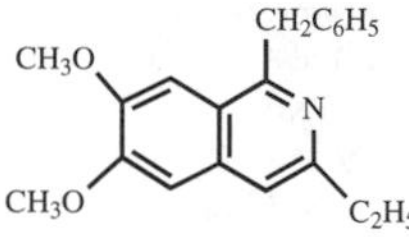

Moxisylyte (mol wt 279) A vasosilator.

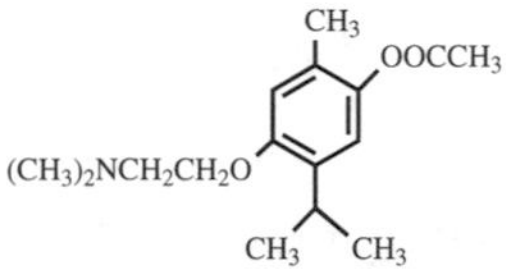

mp Abbreviation for melting point.

MP Abbreviation for 1. melphalan (L-phenylalanine mustard); 2. methyl prednisone; 3. mitogenic protein; 4. monophosphate; 5. mucopolysaccharide; 6. myeloma protein; 7. plasma membrane.

m.p. Abbreviation for melting point.

6-MP Abbreviation for 6-mercaptopurine, an antimetabolite that inhibits DNA and RNA synthesis.

M6P Abbreviation for mannose 6-phosphate.

MPA Abbreviation for methyl prednisone acetate.

MPEH Abbreviation for methyl-phenyl-ethyl hydantoin.

MPF Abbreviation for maturation-promoting factor, a protein kinase that controls the transition from G_2 phase to M phase during mitosis.

MPGM Abbreviation for monophosphoglycerate mutase.

MPH Abbreviation for milk protein hydrolysate.

MphI (EcoRII) A restriction endonulcease from *Moraxella phenylpyruvica* with the same specificity as EcoRII.

Mph1103I A restriction endonuclease from *Moraxella phenylpyruvica* RFL 1103 with the following specificity:

M-Phase Abbreviation for M phase of mitosis.

MPI Abbreviation for mucus proteinase inhibitor.

mPKC Abbreviation for murine PKC.

MPM Abbreviation for mouse peritoneal macrophages.

MPN Abbreviation for most probable number.

MPO Abbreviation for myeloperoxidase.

MPP Abbreviation for 1. mitochondrial processing peptidase; 2. mitochondrial processing peptide.

mPPARα Abbreviation for mouse peroxisome proliferator-activated receptor α.

M6PR Abbreviation for mannose 6-phosphate receptor.

M-Prednisol A trade name for methylprednisolone acetate, a corticosteroid.

M-Protein 1. A protein found in the M line of myofibrils of striated muscle. 2. A cell surface protein of *Brucella*.

MPT Abbreviation for mitochondrial permeability transition.

MPTP (mol wt 173) 1-Methyl-4-phenyl-1,2,3,6-tetrahydropyridine, a piperidine derivative that causes irreversible symptoms of Parkinsonism in humans and monkeys.

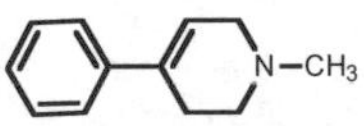

MpuI (XhoI) A restriction endonuclease from *Micromonospora purpurea* with q`% same specificity as XhoI.

MR Abbreviation for 1. magnetic resonance; 2. methyl red; 3. mineralocorticoid receptor.

MRA Abbreviation for magnetic resonance angiography.

MraI (SacII) A restriction endonuclease from *Micrococcus radiodurans* with the same specificity as SacII.

mRBC Abbreviation for 1. monkey red blood cell; 2. mouse red blood cell; 3. murine red blood cell.

MRD Abbreviation for minimum reacting dose.

MRE Abbreviation for metal regulatory element.

MRF Abbreviation for 1. melanotropin-releasing factor; 2. myogenic regulatory factor.

MRH Abbreviation for melanotropin-releasing hormone.

MRHA Abbreviation for mannose-resistant hemagglutination.

MRI Abbreviation for magnetic resonance image.

MRIH Abbreviation for melanotropin-releasing inhibiting hormone.

mRNA Abbreviation for messenger RNA, which is capable of protein translation.

mRNA Capping Enzyme The enzyme that catalyzes the addition of guanosine nucleotide to the 5′-end of a mRNA.

mRNA Coding Triplet Referring to codon on mRNA.

mRNP Abbreviation for messenger ribonucleoprotein.

MroI (BspMII) A restriction endonuclease from *Micrococcus roseus* with the following specificity:

5′..........TCCGGA..........3′
3′..........AGGCCT..........5′

MroNI A restriction endonuclease from *Micrococcus roseus* N with the following specificity:

5'........GCCGGC........3'
3'........CGGCCG........5'

MroXI A restriction endonuclease from *Micrococcus roseus* X with the following specificity:

5'........GAANNNNTTC........3'
3'........CTTNNNNAAG........5'

MRP Abbreviation for metal regulatory protein.

MRS Abbreviation for magnetic resonance spectroscopy.

MRSA Abbreviation for methicillin-resistant *Staphylococcus aureus* or multiple-resistant *Staphylococcus aureus*.

M-R-VaxII A trade name for a combined live vaccine against measles and rubella virus.

MS Abbreviation for 1. mass spectrometry; 2. methionine synthetase; 3. multiple sclerosis.

MS-2 A single-stranded RNA phage in the family of Leviviridae.

MSA Abbreviation for mitotic spindle apparatus.

MSAFP Abbreviation for maternal serum alpha fetoprotein.

MscI A restriction endonuclease from *Micrococcus* species or E. coli strain carrying a clone gene from the *Micrococcus* species, with the following specificity:

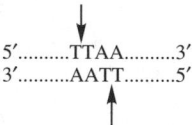

5'........TGGCCA........3'
3'........ACCGGT........5'

MS-Contin A trade name for morphine sulfate, an alkaloid of opium used as a narcotic analgesic agent that binds with opiate receptors altering both perception and emotional responses to pain.

msDNA Abbreviation for multicopy single-stranded DNA.

MseI A restriction endonuclease from *Micrococcus* species with the following specificity:

5′..........TTAA..........3′
3′..........AATT..........5′

MSF Abbreviation for macrophage spreading factor.

MSG Abbreviation for monosodium glutamate.

MSH Abbreviation for melanocye stimulating hormone, a peptide hormone with melanocyte stimulating activity and secreted by the pituitary gland. Three MSH-peptides have been identified, namely α-MSH, β-MSH, and γ-MSH (also called melanotropin).

α-MSH See MSH.

β-MSH See MSH.

γ-MSH See MSH.

MSH-IF Abbreviation for melanocyte-stimulating hormone-inhibiting factor.

MSH-RF Abbreviation for melanocyte-stimulating hormone-releasing factor.

MS/L A trade name for morphine sulfate, a narcotic analgesic agent.

MslI A restriction endonuclease from *Moraxella osloensis* with the following specificity:

5'........CAPyNNNNPuTG.........3'
3'........GTPuNNNNPyAC.........5'

MSLA Abbreviation for mouse specific lymphocytic antigen.

MSP Abbreviation for mouse serum protein.

MspI (HpaII) A restriction endonuclease from *Moraxella* species with the following specificity:

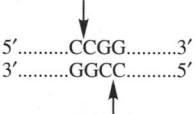

5′..........CCGG..........3′
3′..........GGCC..........5′

Msp17I A restriction endonuclease from *Moraxella* species 17 with the following specificity:

5'........GRCGYC........3'
3'........CYCGRG........5'

R=A or G Y= C or T

Msp67I (ScrFI) A restriction endonuclease from *Moraxella* species with the following specificity:

5′..........CCNGG..........3′
3′..........GGNCC..........5′

Msp67II (MboI) A restriction endonuclease from *Moraxella* species with the same specificity as MboI.

MspA1I A restriction endonuclease from *Moraxella* species with the following specificity:

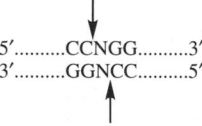

5'........C(A/C)GC(G/T)G........3'
3'........G(T/G)CG(C/A)C........5'

MspCI A restriction endonuclease from *Micrococcus* species with the following specificity:

5'........CTTAAG........3'
3'........GAATTC........5'

MspR9I A restriction endonuclease from *Micrococcus* species R9 with the following specificity:

5'........CCNGG........3'
3'........GGNCC........5'

MstI A restriction endonuclease from *Microcoleus* species with the following specificity:

5′..........TGCGCA..........3′
3′..........ACGCGT..........5′

MstII (SauI) A restriction endonuclease from *Microcoleus* species with the following specificity:

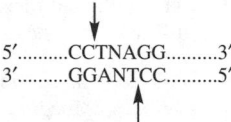

5′..........CCTNAGG..........3′
3′..........GGANTCC..........5′

MSU Abbreviation for 1. maple syrup urine; 2. middle stream of urine; 3. monosodium urate.

MSV Abbreviation for 1. Moloney's sarcoma virus; 2. murine sarcoma virus.

MT Abbreviation for 1. metallothionein; 2. methyltyrosine; 3. microtubule.

MTase Abbreviation for methyltransferase.

MTB Abbreviation for methylthymol blue.

mtCK Abbreviation for mitochondrial creatine kinase.

mtDNA Abbreviation for mitochondrial DNA.

MTdU Abbreviation for 5'-methyl-4'-thio-2-deoxyuridine.

MTF Abbreviation for 1. melanotransferrin; 2. methionyl-tRNA formyltransferase.

MTG Abbreviation for methylthiogalactoside.

MTGuo Abbreviation for 7-methyl-6-thioguanosine.

MTH-amino Acid Abbreviation for methylthiohydantoin amino acid.

MTHF Abbreviation for methyltetrahydrofolate.

MTHFA Abbreviation for methyltetrahydrofolic acid.

mt-mRNA Referring to mitochondrial mRNA.

MTOC Abbreviation for microtubule organizing center.

MTP Abbreviation for microsomal triglyceride transfer protein.

MTR Abbreviation for 5-methylthioribose.

mt-rRNA Referring to mitochondrial ribosomal RNA.

MtTK Abbreviation for mitochondrial thymidine kinase.

mt-tRNA Referring to mitochondrial tRNA.

MTU Abbreviation for methythiouracil.

MTV Abbreviation for mammary tumor virus.

MTX Abbreviation for methotrexate.

mu (μ) A Greek letter used to denote: 1. micron, a unit of length (one-thousandth of a mm), 2. a temperate bacteriophage that contains double-stranded DNA and infects enterobacteria, and 3. the heavy chain of IgM.

Mu Abbreviation for murine.

mu (μ) A letter in the Greek alphabet.

4MU Abbreviation for 4-methylumbelliferone.

μ Chain The heavy chain of IgM.

2μ Circle See 2μ DNA plasmid or 2m plasmid.

2μ DNA Plasmide See 2μ plasmid.

2μ Plasmid A cccDNA plasmid (6318 bp) present in certain strains of *Saccharomyces cerevisiae.*

3μ Plasmid A cccDNA plasmid of unknown function present in *Saccharomyces.*

Mucate A salt of mucic acid.

Mucic Acid (mol wt 210) The product from the oxidation of galactose.

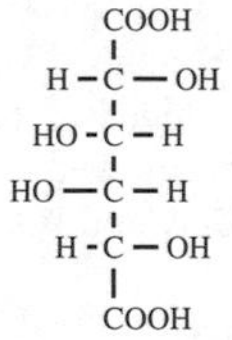

Mucigel A mucilaginous layer that covers the root tips and root hairs of plants.

Mucilage A sticky mixture of carbohydrates from plants.

Mucins Glycoproteins and major constituents of mucus.

Muco- A prefix meaning amino sugar.

Mucocyst A small, membrane-bound vesicular organelle in the ciliate protozoans responsible for discharge of mucuslike material.

Mucoid Mucuslike material.

Mucolipid Synonym of ganglioside.

Mucolipidosis A metabolic disorder characterized by the accumulation of both mucopolysaccharide and lipid in the tissues.

Mucolytic Any agent that dissolves or breaks down mucus.

Mucomyst A trade name for acetylcysteine, which reduces the viscosity of tenacious secretions, used as an antitussive agent.

Mucopepsin A protease isolated from fungi *Mucor pusillus.*

Mucopeptide Referring to peptidoglycan or any peptide that contains mucopolysaccharide or amino sugar (e.g., *N*-acetylglucosamine).

Mucopolysaccharide Polysaccharide that contains amino sugars.

Mucopolysaccharidosis A metabolic disorder characterized by the excessive accumulation and secretion of oligomucpolyosaccharide due to the deficiency of enzymes that break down glycosaminoglycan.

Mucoprotein Protein that contains a mucopolysaccharide moiety, e.g., proteoglycan.

Mucor A genus of fungus.

Mucosa An epithelial membrane containing cells that secrete mucus.

Mucosal Homing The ability of immunologically competent cells derived from mucosal follicles to traffic back to mucosal areas.

Mucosal Immune System The lymphoid tissues associated with the mucosal surfaces of the gastrointestinal, respiratory, and urogenital tracts that produce secretory IgA and immunity for the mucosal surfaces.

Mucosil-10 A trade name for acetylcysteine, an antitussive agent.

Mucositis Inflammation of mucous membranes.

Mucosol A trade name for acetylcysteine, which reduces the viscosity of tenacious secretions.

Mucous Pertaining to mucus.

Mucous Membrane The epithelium membrane or a sheet of tissue cells that lines the surface of the various parts of body cavities.

Mucrolysin A protease isolated from Chinese habu snake venom.

Mucus The viscous fluid secreted by mucous glands that contains mucin and inorganic salts.

Multicellular Composed of more than one cell.

Multiciliate Having many cilia present on a sperm or a spore or other type of ciliated cell.

Multicistronic mRNA See polycistronic mRNA.

Multicomponent Virus A virus that exists in two or more different particles and each particle contains only part of the viral genome.

Multicopy Plasmid A plasmid whose copy number in a given cell is high, e.g., pBR322.

Multienzyme Complex The enzymes that form complexes to catalyze a specific biochemical process or pathway, e.g., enzyme complex for synthesis of fatty acid in prokaryotic cells.

Multifunctional Enzyme A single enzyme that catalyzes several reactions, e.g., DNA polymerase I that has polymerase activity and 5′ to 3′ or 3′ to 5′ exonuclease activity.

Multimeric Protein Protein that consists of more than one polypeptide chains.

Multipartite A virus that exists in more than one form of particle (see also multicomponent virus).

Multipax A trade name for hydroxyzine hydrochloride, an antianxiety agent.

Multiple Antibiotic Resistance The ability of a microorganism or a strain of one microorganism to resist the effects of two or more unrelated antibiotics.

Multiple Codon Recognition The recognition by a single tRNA of more than one type of codons in mRNA.

Multiple Myeloma Cancer of the B cell characterized by the presence of a high level of Bence Jones protein (monoclonal immunoglobulin light chain) in the serum and urine.

Multiple Sclerosis A disorder in which the lipid sheath of nerve cells is progressively degenerated.

Multiple Tube Method Referring to the most probable number method used to estimate the number of viable microorganisms in an aqueous sample.

Multiplicity of Infection Referring to the number of infectious particles per cell in an infectious mixture (e.g., the number of bacteriophages and bacterial cells in an infection mixture used to initiate the infection).

Multipotent Capable of giving rise to several kinds of cells, tissues, or structures.

Multiseriate With more than one row of cells.

Multispecificity The ability of a given immunoglobulin to bind with more than one type of antigen molecule.

Multivesicular Body Structure formed by inward budding of an uncoated vesicle membrane.

MuLV Abbreviation for murine leukemia virus.

Mumps A contagious human disease caused by a paramyxovirus (Paramyxoviridae) and characterized by fever, pain, and swelling of salivary glands.

Mumpsvax A trade name for a live vaccine of mumps virus.

MunI A restriction endonuclease from *E. coli* carrying the cloned MunIR gene from *Mycoplasma*, with the following specificity:

```
          ↓
5'........CAATTG........3'
3'........GTTAAC........5'
               ↑
```

MUP Abbreviation for mouse major urinary protein.

Mupirocin (mol wt 501) An antibiotic produced by *Pseudomonas fluorescens* that inhibits bacterial RNA and protein synthesis.

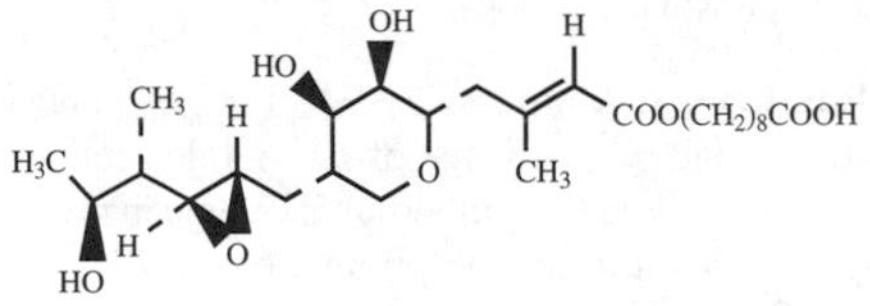

Mur Abbreviation for muramic acid.

Muramic Acid (mol wt 251) A derivative of an amino sugar occurring in bacterial cell wall.

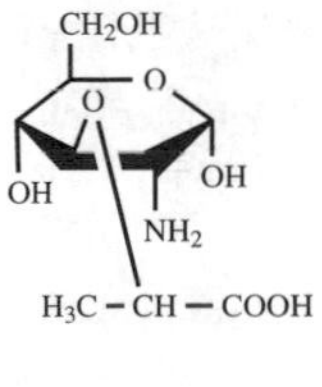

Muramidase See lysozyme.

Muramoylpentapeptide Carboxypeptidase The enzyme that catalyzes the cleavage of bonds of UDP-*N*-acetylmuramoyl-L-alanyl-D-γ-glutamyl 6-carboxy-L-lysinyl-D-alanyl-D-alanine.

Muramoyltetrapeptide Carboxypeptidase The enzyme that catalyzes the cleavage of bonds of *N*-acetyl D-glucosaminyl *N*-acetylmuramoyl-L-alanyl-D-glutamyl 6-carboxy-L-lysinyl-D-alanine.

Muramyl Dipeptide (mol wt 492) An immunologically active glycopeptide subunit of bacterial cell wall.

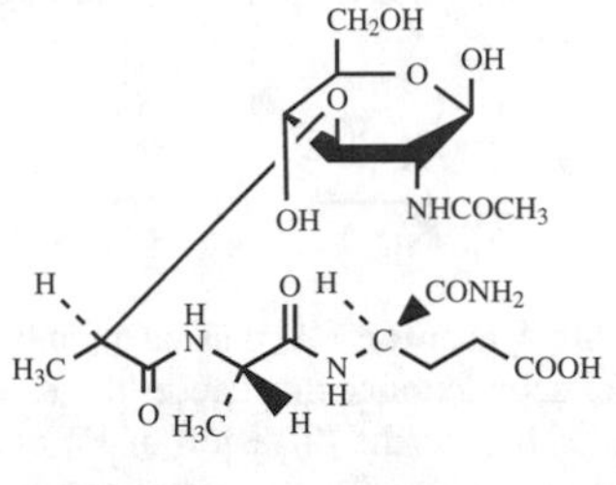

Murein A peptidoglycan that forms the backbone of bacterial cell wall that consists of alternating *N*-acetylglucosamine and *N*-acetylmuramic acid residues crossed-linked by a peptide bridge.

Murexide Test A colorimetric test for the determination of purines based upon the treatment of sample with concentrated nitric acid and ammonium hydroxide.

Murexine (mol wt 224) A neuromuscular blocker from *Murex trunculus* and other related species of *Mollusks*.

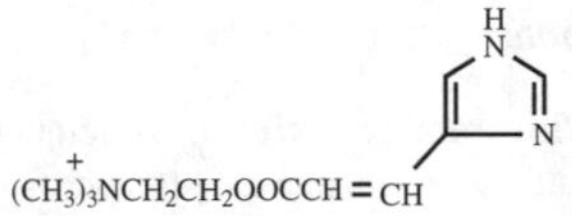

Muriatic Acid An old name for hydrochloric acid.

Murine Referring to mice and rats.

Murine Sarcoma Viruses A group of oncogenic, replicative-defective retroviruses.

MurNAc Abbreviation for N-acetyl muramic acid or N-acetyl muramate.

Murocoll-2 A trade name for a combination drug containing scopolamine hydrochloride and phenylephrine hydrochloride, used as a mydriatic agent.

Muroctasin (mol wt 887) An immunostimulant.

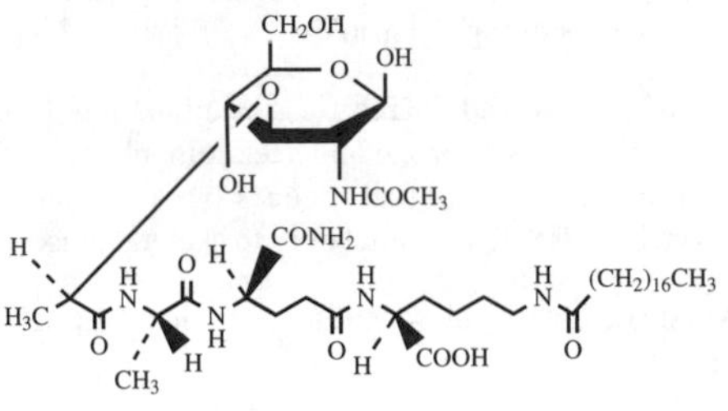

Muromonab-CD3 A murine monoclonal antibody to the antigen of human T cells used as an immuno-suppressant.

Murray Valley Fever A human disease caused by a flavivirus (Flaviviridae).

Muscarine (mol wt 174) An alkaloid produced by the poisonous mushroom (*Amanita mascara*).

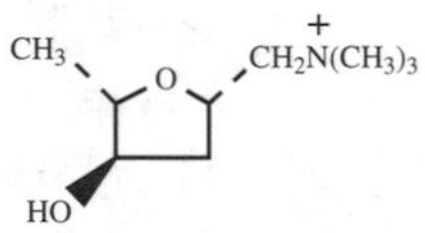

Muscarine Receptor A plasma membrane receptor for acetylcholine that causes the K^+ channel to open leading to the diffusion of K^+ out of the cell, hyperpolarizing the plasma membrane.

Muscarinic Synapse A synapse that contains muscarinic receptors.

Muscimol (mol wt 114) A GABA-agonist affecting the CNS.

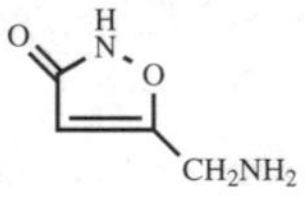

Muscle Fiber Long, thin, multinucleate cells in the skeletal muscle that is specialized for contractile function.

Muscle Phosphofructokinase Deficiency A disorder due to the deficiency of phosphofructokinase leading to the abnormal buildup of glycolytic metabolites, e.g., glucose-6-phosphate and fructose-6-phosphate, causing glycogen accumulation in the muscle and type-VII glycogen storage disease.

Muscle Phosphorylase Deficiency A disorder due to the deficiency of muscle phosphorylase and characterized by painful muscle cramps caused by the inability of the muscle to breakdown glycogen (also known as McArdle's disease or type-V glycogen storage disease).

Muscle Sugar Referring to inositol.

Muscular Dystrophy A disease characterized by the degeneration of skeletal cells due to a defect in the gene encoding dystrophin.

Mus-Lax A trade name for a combination drug containing chloroxazone and acetaminophen, used as a muscle relaxant that reduces transmission of nerve impulses from spinal cord to skeletal muscle.

Mustard Gas (mol wt 159) A toxic agent used in chemical warfare.

$$(ClCH_2CH_2)_2S$$

Mustargen A trade name for mechlorethamine hydrochloride, an alkylating agent that cross-links strands of DNA and interferes with transcription.

Mutagen A chemical or physical agent that interacts with DNA and causes a mutation.

Mutagenesis The production of mutations.

Mutagenic Capable of causing mutagenesis.

Mutamycin A trade name for mitomycin, an antibiotic that cross-links strands of DNA and interferes with RNA transcription and alters cell growth causing death of the cells.

Mutan A water soluble glycan consisting of α-1,3- and α-1,6-glycosidic linkages.

Mutanolysin A proteolytic enzyme used for cell lysis.

Mutant An organism that is genetically different from its parent.

Mutarotase The enzyme that catalyze the interconversion of α-D-glucose to β-D-glucose.

Mutarotation The change in the optical activity of an optically active substance when it is dissolved in water or some other solvent.

Mutase Referring to the enzyme that catalyzes intramolecular changes, e.g., intramolecular transfer of phosphate groups.

Mutation 1. A process by which a gene undergoes change (e.g., change of nucleotide sequence in DNA) leading to an inheritable change in phenotype or alteration in the product encoded by the gene. 2. Modification of a gene leading to a change in phenotypic expression.

Mutational Spectrum A genetic map of point mutations produced by exposure to a mutagen.

Mutator Gene A gene that promotes the spontaneous mutation rate for other genes.

Mutator Phage A bacteriophage which causes an increase in rate of mutation in the host cell, e.g., phage mu.

Mutator Strain A strain of organism that carries mutator gene.

Muton The smallest genetic element within which a change causes mutation.

Mutualism A symbiotic relationship in which both the host and the symbiont benefit.

Muzolimine (mol wt 272) A diuretic and an antihypertensive agent.

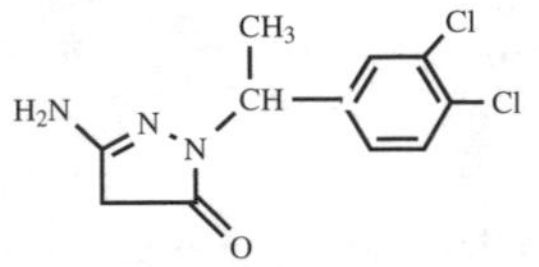

MV A trade name for a combination drug containing mitoxantone, an etoposide used for treatment of leukemia.

mV Abbreviation for millivolt.

MV-L3 A group of mycoplasma viruses of the family Podoviridae.

MvaI A restriction endonuclease from *E. coli* carrying cloned MvaIR gene from *Micrococcus varians* RFl19, with the following specificity:

5'........CC↓(A/T)GG........3'
3'........GG(T/A)↑CC........5'

Mva1269I A restriction endonuclease from *Micrococcus varians* RFL1269 with the following specificity:

5'........GAATGCN↓........3'
3'........CTTAC↑GN........5'

MVAC A trade name for a combination drug containing methotrexate, vinblastine, adriamycin, and cisplatin, used for treatment of genitourinary cancer.

MVM Abbreviation for minute virus of mice.

MvnI (FnuDII) A restriction endonuclease from *Methanococcus vannielii* with the following specificity:

5′..........CG↓CG..........3′
3′..........GC↑GC..........5′

MVPP A trade name for a combination drug containing mechlorethamine, vinblastine, procarbazine, and prednisone, used for treatment of Hodgkin's lymphoma.

MwoI A restriction endonuclease from *E. coli* strain carrying cloned genes from *Methanobacterium wolfelii,* with the following specificity:

5'........GC(N)$_5$↓(N)$_2$GC........3'
3'........CG(N)$_2$↑(N)$_5$CG........5'

Myalgia Referring to muscle pain.

Myambutol A trade name for ethambutol hydrochloride, an antibacterial agent.

Myapap A trade name for acetaminophen, an antipyretic and analgesic agent.

Myasthenia Gravis An autoimmune disease due to the production of antibodies against self-acetylcholine receptors leading to muscle weakness.

myb An oncogene in avian myelolastosis virus (Retroviridae).

myc An oncogene in avian myelocytomatosis virus (Retroviridae).

Mycelex A trade name for clotrimazole, an anti-infective agent that alters the permeability of the fungal cell.

Mycelia Plural of mycelium.

Mycelium Hyphae of fungus (threadlike cellular structure).

Mycifradin A trade name for neomycin sulfate, an antibiotic that binds to 30S ribosomes inhibiting bacterial protein synthesis.

Mycil A trade name for chlorphenesin carbamate, a skeletal muscle relaxant.

-mycin A suffix meaning antibiotics produced by *Streptomyces*.

Myciguent A trade name for neomycin sulfate, an antibiotic that binds to 30S ribosomes inhibiting bacterial protein synthesis.

Mycitracin Ointment A trade name for a combination drug containing polymyxin B sulfate, bacitracin, and neomycin sulfate, used as an anti-infective agent.

Mycitracin Ophthalmic Ointment A trade name for a combination drug containing polymyxin B sulfate, neomycin sulfate, and bacitracin, used as an ophthalmic anti-infective agent.

Myclo A trade name for clotrimazole, an antifungal agent.

Myco- A prefix meaning fungus.

Mycobacidin (mol wt 217) An antibiotic from *Streptomyces lavendulae*.

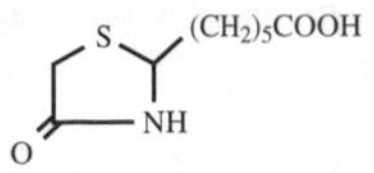

Mycobacillin A cyclic peptide antibiotic produced by *Bacillus subtilis*.

Mycobacteriophage Any bacteriophage that infects species of *Mycobacterium.*

Mycobacteriosis A tuberculosislike disorder caused by mycobacteria other than *Mycobacterium tuberculosis.*

Mycobacterium A genus of Gram-positive, aerobic, nonmotile, asporogenous bacteria.

Mycobacterium tuberculosis A bacterium that causes tuberculosis.

Mycobactins A group of iron-chelating factors derived from *Mycobacterium paratuberculosis.* There are many different types of mycobactins, e.g., mycobactin A, F, H, N, P, R, S, and T.

Mycobiont The fungal partner in a lichen.

Mycobutin A trade name for rifabutin, an antituberculous agent.

Mycocerosate Synthetase The enzyme that catalyzes the following reaction:

Acyl-CoA + n methylmalonyl-CoA + 2n NADPH

⇅

Multi-methyl-branched acyl-CoA + n CoA + n CO_2 + 2n $NADP^+$

Mycocide Fungicide.

Mycolic Acids High molecular weight, branched, hydroxy fatty acid and component of the cell envelope of *Mycobacteria.*

Mycology The science that deals with fungi.

Mycomycin (mol wt 198) An antibiotic from *Nocardia acidophilus.*

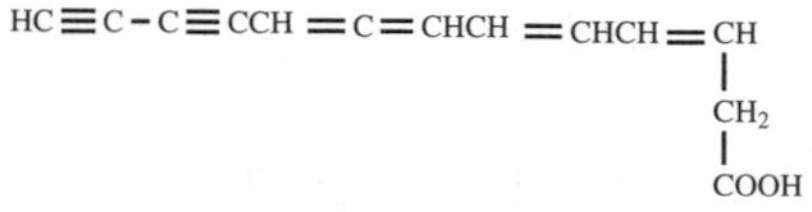

Mycophagous Fungus-eating.

Mycophenolic Acid (mol wt 320) An antineoplastic agent.

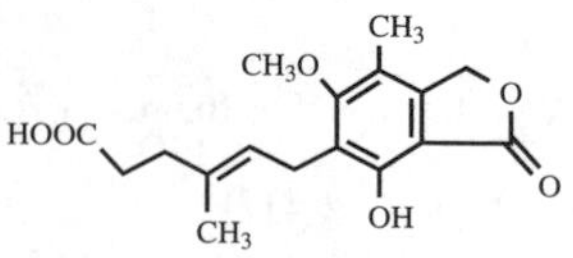

Mycoplana A genus of Gram-negative, aerobic bacteria.

Mycoplasma A genus of cell wall-less, sterol requiring, catalase-negative bacteria (family of Mycoplasmataceae).

Mycoplasmavirus Bacteriophages that infect bacteria of Mycoplasmatales.

Mycorrhiza Symbiotic association between the mycelium of a fungus and the roots of a higher plant.

MyCort Lotion A trade name for a lotion containing hydrocortisone acetate.

Mycosamine (mol wt 163) An amino sugar and an antifungal agent.

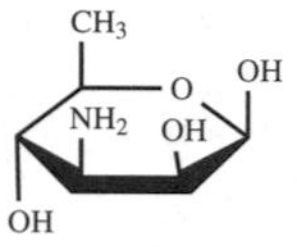

Mycosis An infection caused by a fungus.

Mycostatin A trade name for nystatin, an antifungal agent that alters the permeability of fungal cells.

Mycotoxicosis Poisoning resulting from ingestion of mycotoxin.

Mycotoxins Toxic substances produced by fungi, e.g., aflatoxin, amatoxin, and ergot alkaloids.

Mycovirus A virus that infects fungi.

Mycovore Fungus-eating.

Mydfrin A trade name for phenylephrine hydrochloride, a mydriatic agent that causes dilation of the muscle of the pupil.

Mydriacyl A trade name for tropicamide, a mydriatic agent that causes dilation of the muscle of the pupil.

Mydriasis The widening of the pupil of the eye.

Mydriatic Any agent that causes dilation of the pupil of the eye.

Myectomy The surgical removal of a muscle or part of a muscle.

Myelin Sheath of stacked membranes that surround the axons. It serves to insulate the axons electrically and to increase the rate of transmission of nerve impulses.

Myelin Basic Protein Referring to the major protein in the mammalian CNS.

Myelination The formation of myelin sheath surrounding the axons.

Myelitis Inflammation of spinal cord.

Myeloblast Precursor of myelocytes.

Myelocyte An immature white blood cell (e.g., granulocyte).

Myeloencephalitis Inflammation of the spinal cord and the brain.

Myeloid System All the granulocytes and their precursors.

Myeloid Tissue Referring to tissue in which white blood cells are formed.

Myeloma A tumor of plasma cells.

Myeloma Protein A monoclonal immunoglobulin chain produced in patients with myeloma.

Myeloperoxidase The enzyme that catalyzes the oxidation of halide ions to hypohalite during the respiratory burst of neutrophils accompanying the destruction of injested microorganisms.

Myelopoiesis The formation and growth of bone marrow.

Myelosome An organelle formed by fusion of myelinated nerve axons during the process of homogenization.

Myidone A trade name for primidone, an anticonvulsant.

Myidyl A trade name for triprolidine hydrochloride, an antihistaminic agent that competes with histamine for H_1 receptors on effector cells.

Mykrox A trade name for metolazone, a diuretic agent that increases urine excretion of sodium and water by inhibiting sodium reabsorption.

Mylanta Tablets A combination drug containing aluminum hydroxide, magnesium hydroxide and simethicone used as an antacid.

Myleran A trade name for busulfan, an alkylating agent that cross-links strands of double-stranded DNA interfering with transcription.

Mylicon A trade name for simethicone, which depresses or prevents formation of mucus-surrounded gas pockets in the GI tract.

Mymethasone A trade name for dexamethasone sodium phosphate, a corticosteroid that decreases inflammation and suppresses the immune response.

Myo- A prefix denoting muscle.

Myoadenylate Deaminase Deficiency A disorder characterized by fatigue and muscle cramps due to deficiency of myoadenylate deaminase.

Myoblast Stem cell that gives rise to muscle tissue.

Myocardial Pertaining to heart muscle.

Myocardial Infarction Referring to heart attack or blood flow stoppage.

Myocardiograph An instrument for determining the action of the heart muscle.

Myocarditis Inflammation of the heart muscle.

Myocardium The muscular tissue of the heart wall.

Myocyte Muscle cell.

Myodine A trade name for iodinated glycerol, used for reduction of viscosity of thick, tenacious secretion.

Myodynia Pain in the muscle.

Myoepithelioma A sweat-gland tumor occurring in the skin.

Myofibril Functional unit of contraction, consisting of highly organized arrays of thin actin filaments and thick myosin filaments found in the cytoplasm of skeletal muscle.

Myofibroma A benign tumor composed of muscle and fibrous tissue.

Myogenesis The formation of muscle.

Myogenin A protein involved in muscle differentiation that induces fibroblasts to differentiate into myoblasts.

Myoglobin Monomeric heme-containing, oxygen-binding protein that occurs in muscle.

Myograph A device for recording the muscle movement.

Myokinase See adenylate kinase.

Myolin A trade name for orphenadrine citrate, which reduces the transmission of nerve impulses from the spinal cord to skeletal muscle.

Myolipoma A tumor composed of muscle tissue, fat tissue, and fibrous tissue.

Myology The study of structure, function, and diseases of muscle.

Myoma Benign tumor of muscle.

Myomalacia Degeneration of muscle.

Myomectomy Surgical removal of a myoma, e.g., surgical removal of myoma from the uterus.

Myopathy Diseases of muscle.

Myoquin A trade name for quinine bisulfate, an antimalarial agent.

Myosarcoma A malignant tumor of muscle.

Myosin The main protein in muscle that consists of two heavy chains and two distinct pairs of light chains. Myosin is a motor protein that advances along actin filaments using energy derived from the hydrolysis of ATP. The intact myosin molecules are rod-shaped with two globular heads with a flexible tail. The ATPase activity and the actin-binding capability are located in the globular heads.

Myosin Heavy-Chain Kinase The enzyme that catalyzes the phosphorylation of myosin heavy chain leading to the inhibition of the actin-activated ATPase activity of myosin.

Myosin Light-Chain Kinase The enzyme that catalyzes the phosphorylation of myosin light chain leading to smooth muscle contraction.

Myosin Light-Chain Phosphatase The enzyme that catalyzes the removal of phosphate groups from myosin light chain leading to smooth muscle relaxation.

Myositis Inflammation of muscle.

Myotonachol A trade name for bethanechol chloride, a cholinergic agent that binds cholinergic receptors.

Myotonia Any genetically distinct hereditary disease characterized by specific muscle malformations.

Myotrophin A 12-kDa protein of hypertrophied myocardium of hypertensive rats that stimulates protein synthesis.

Myoviridae A family of bacteriophages characterized by having an icosahedral head and a long, contractile tail, e.g., T_2 bacteriophage.

Myricetin (mol wt 318) An inhibitor for α-glucosidase.

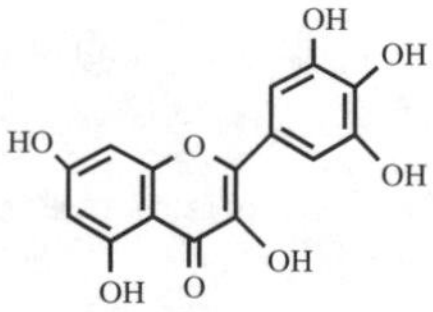

Myristica Dried ripe seed of *Myristica fragrans* used as carmintive agent that promotes expulsion of gas from the alimentary canal.

Myristic Acid (mol wt 228) A 14-carbon saturated fatty acid.

$$CH_3(CH_2)_{12}COOH$$

Myristoleic Acid (mol wt 226) A 14-carbon unsaturated fatty acid with a single double bond.

$$CH_3(CH_2)_3CH{=}CH(CH_2)_7COOH$$

Myristyltrimethylammonium Bromide (mol wt 336) A disinfectant and deodorant.

$$[CH_3(CH_2)_{12}N(CH_3)_3]^+Br^-$$

Myrmecia A genus of unicellular green algae.

Myrophine (mol wt 586) A narcotic analgesic agent.

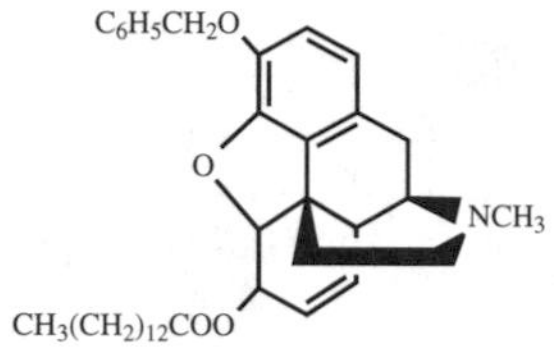

Myrosemide A trade name for a diuretic agent furosemide that inhibits the reabsorption of sodium and chloride.

Myrosinase Synonym of thioglucosidase.

Myrtecaine (mol wt 265) A local anesthetic agent.

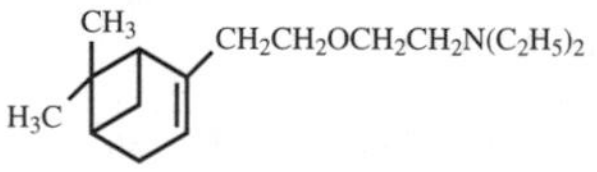

Mysoline A trade name for primidone, an anticonvulsant.

Mytatrienediol (mol wt 316) A hormone and antilipemic and hypocalciuric agent.

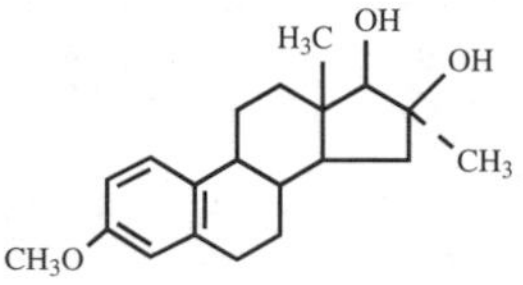

Mytelase A trade name for ambenonium chloride, an acetylcholinesterase inhibitor that inhibits the destruction of acetylcholine released from the efferent nerves.

Mytrate A trade name for epinephrine bitartrate, a mydriatic agent that dilates the pupil.

Mytussin A trade name for guaifenesin, an expectorant.

Myxedema A metabolic disorder caused by the insufficient function of the thyroid gland that is characterized by a puffy appearance of tissues, dryness of skin, and gaining weight.

Myxidium A genus of protozoa (Myxosporea).

Myxin (mol wt 258) A broad-spectrum antibiotic from *Sorangium* species.

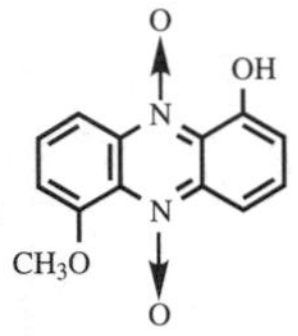

Myxo- A prefix meaning mucus or slime.

Myxobacter A gliding bacteria (Myxobacterales).

Myxobacterales An order of small, rod-shaped, glidding myxobacteria normally embedded in a slime layer.

Myxobolus A genus of protozoa (Myxosporea).

Myxoma A tumor of connective tissue consisting of a large amount of mucous tissue.

Myxomycete A class of slime mold.

Myxophage Bacteriophage that infects gliding bacteria of Myxobacterales.

Myxosarcina A genus of unicellular bacteria of Myxobacterales.

Myxosoma A genus of protozoa.

Myxothiazol (mol wt 448) A potent inhibitor of the mitochondrial respiratory chain.

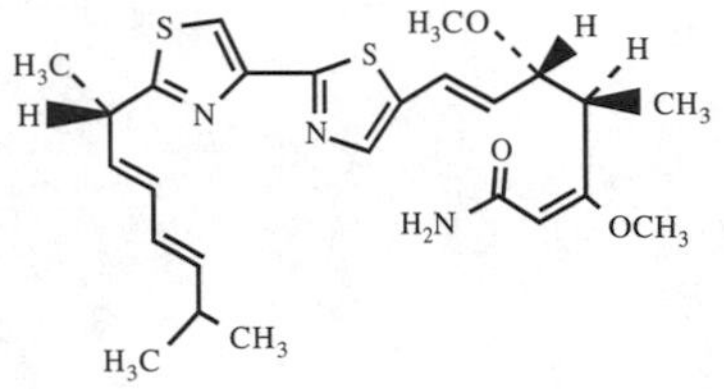

Myxoviruses A group of nontaxonomic viruses of RNA-containing viruses including viruses in the families of Orthomyxoviridae and Paramysoviridae.

N

N Abbreviation for 1. asparagine; 2. normality of an acid or base; 3. nitrogen; 4. Avogadro's number; 5. haploid number; 6. neutron number.

N- A prefix in chemical nomenclature denoting the substitution on nitrogen.

^{15}N A nonradioactive heavy isotope of nitrogen used as a marker for labeling nitrogen compounds.

N Cell Cells found in the upper small intestine in humans that contain homogeneous granules in which neurotensin is stored.

N Terminal Referring to the amino terminus of a protein or polypeptide that is the terminus where the first amino acid is incorporated during mRNA translation.

NA Abbreviation for 1. naphthylamide; 2. neuraminidase; 3. neutralizing antibody; 4. nicotinic acid; 5. noradrenaline; 6. numerical aperture.

Na^+ Symbol for positively charged sodium ion.

NAA Abbreviation for 1. naphthalene acetic acid; 2. nicotinic acid amide.

NAAP Abbreviation for N-acetyl-4-aminophenazone.

Na^+-ATPase A sodium-dependent ATPase involved in sodium extrusion in *Streptococcus faecalis*.

NAB Hepatitis Abbreviation for non-A, non-B hepatitis.

Nabam (mol wt 256) An agricultural fungicide that causes violent vomiting.

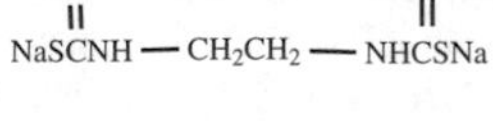

Nabi-HB A trade name for human hepatitis B immune globulin.

Nabilone (mol wt 373) An antiemetic agent.

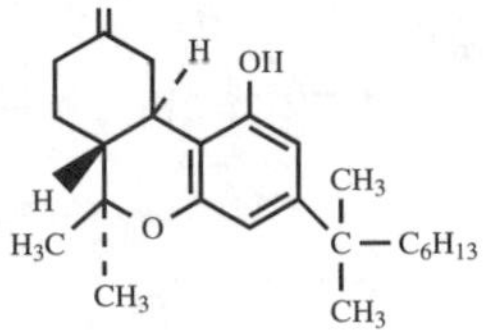

Nabumetone (mol wt 228) An anti-inflammatory agent that inhibits the synthesis of prostaglandin.

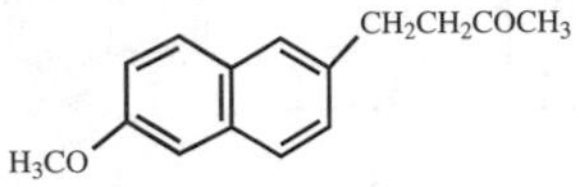

$[Na^+]_c$ Abbreviation for cytoplasmic-free sodium concentration.

NAC Abbreviation for N-acetylcysteine.

NAc Abbreviation for N-acetyl.

Na^+-Ca^{++}-Antiport A system that pumps Na^+ out of the cell and Ca^{2+} into the cell resulting in increase of intracellular Ca^{2+} concentration and creation of the force for cardiac muscle contraction.

Na^+-Channel The transaxonal membrane proteins that function as a voltage-sensitive channel for Na^+ and K^+ transport. The opening and closing of the channel is regulated by the membrane potential (also known as voltage-gated channel).

NAChR Abbreviation for nicotinic acetylcholine receptor.

NACM Abbreviation for neural cell adhesion molecule, a transmembrane protein that mediates cell-to-cell adhesion in neurons.

NAD (mol wt 665) Abbreviation for nicotinamide adenine dinucleotide, a compound found in all living cells that exists in two interconvertible forms (oxidized form NAD^+ and reduced form NADH) and serves as a coenzyme for a variety of biochemical reactions.

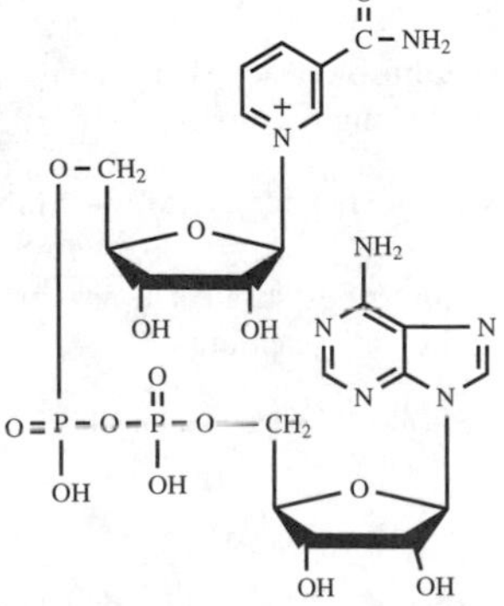

oxidized form

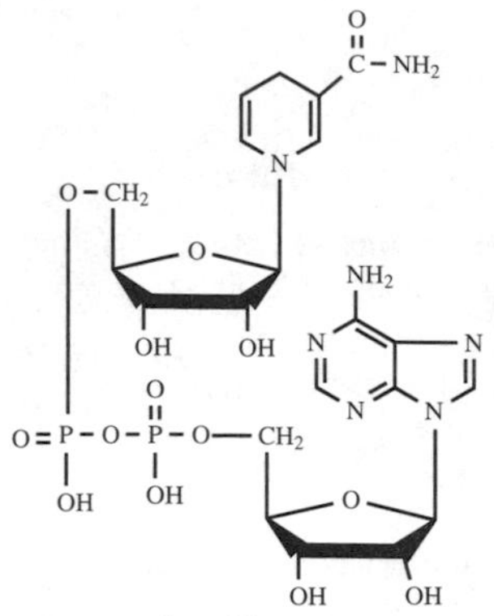

reduced form

NAD+ The oxidized form of NAD and an electron acceptor or an oxidizing agent (see NAD for structure.

NAD+ ADP-Ribosyltransferase The enzyme that catalyzes the following reaction:

$$\text{NAD}^+ + (\text{ADP-ribosyl})_n\text{-acceptor} \rightleftharpoons \text{Nicotinamide} + (\text{ADP-ribosyl})_{n+1}\text{-acceptor}$$

NAD+ Glycohydrolase Systematic name for NAD+ nucleosidase.

NAD+ Kinase The enzyme that catalyzes the following reaction:

$$\text{ATP} + \text{NAD}^+ \rightleftharpoons \text{ADP} + \text{NADP}^+$$

NAD+ Lactate Dehydrogenase Synonym of lactate dehydrogenase.

NAD+ Nucleosidase The enzyme that catalyzes the following reaction:

$$\text{NAD}^+ + H_2O \rightleftharpoons \text{Nicotinamide} + \text{ADP-ribose}$$

NAD+ Pyrophosphatase The enzyme that catalyzes the following reaction:

$$\text{NAD} + H_2O \rightleftharpoons \text{AMP} + \text{NMN}$$

NAD+ Pyrophosphorylase The enzyme that catalyzes the following reaction:

$$\text{ATP} + \text{nicotinamide mononucleotide} \rightleftharpoons \text{PPi} + \text{NAD}^+$$

NAD+ Synthetase The enzyme that catalyzes the following reaction:

$$\text{ATP} + \text{glutamine} + H_2O + \text{nicotinate adenine dinucleotide} \rightleftharpoons \text{NAD}^+ + \text{AMP} + \text{PPi} + \text{glutamate}$$

NADase The enzyme that catalyzes the following reaction:

$$\text{NAD}^+ + H_2O \rightleftharpoons \text{Nicotinamide} + \text{ADP-ribose}$$

NADH The reduced form of NAD and an electron donor or reducing agent (see NAD for structure).

NADH Coenzyme Q Reductase Complex (Complex 1) The enzyme complex that catalyzes the transfer of electrons from NADH to coenzyme Q.

NADH Cytochrome b_5 Reductase An electron transfer system for mammalian terminal desaturase that occurs on the inner surface of endoplasmic reticulum and is not associated with oxidative phosphorylation.

NADH Dehydrogenase The enzyme complex of the electron transport chain involved in transfer of electrons from NADH to coenzyme Q.

NADH FMN Oxidoreductase Synonym of NADH dehydrogenase.

NADH Hydrogen Peroxide Oxidoreductase The systematic name for NADH oxidase.

NADH Kinase The enzyme that catalyzes the following reaction:

$$\text{NADH} + \text{ATP} \rightleftharpoons \text{ADP} + \text{NADPH}$$

NADH Oxidase The enzyme that catalyzes the following reaction:

$$\text{NADH} + H_2O_2 \rightleftharpoons \text{NAD}^+ + 2H_2O$$

NADH Peroxidase The enzyme that catalyzes the following reaction:

$$\text{NADH} + H_2O_2 \rightleftharpoons \text{NAD}^+ + H_2O$$

NADH UQ Oxidoreductase Referring to complex I of the mitochondrial electron transport chain (see also NADH-coenzyme Q reductase complex).

Nadide (mol wt 663) A coenzyme necessary for the alcoholic fermentation of glucose and an alcohol consumption deterrent.

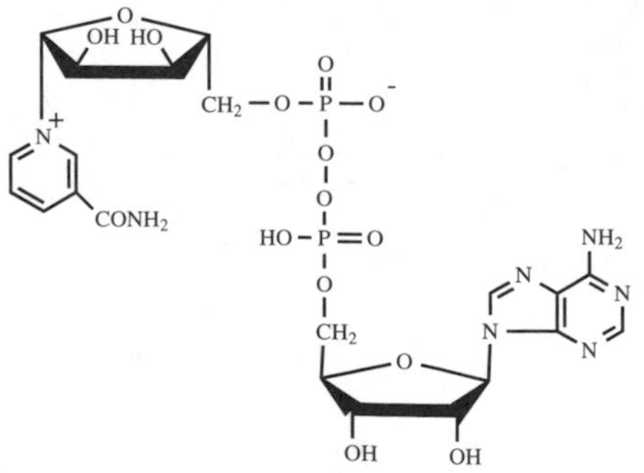

Nadifloxacin (mol wt 360) An antibacterial agent.

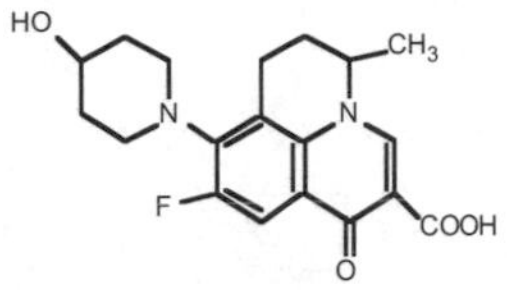

NADK Abbreviation for NAD^+ kinase.

Nadolol (mol wt 309) An antihypertensive and antianginal agent, and a beta-adrenergic blocker that reduces cardiac oxygen demand.

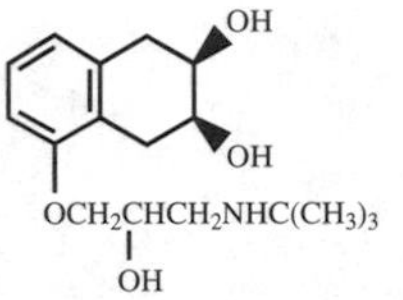

Nadopen-V A trade name for penicillin-V, an antibiotic that inhibits bacterial cell wall synthesis.

Nadostine A trade name for nystatin, an antifungal agent that alters the permeability of fungal cells.

Nadoxolol (mol wt 260) An antiarrhythmic agent.

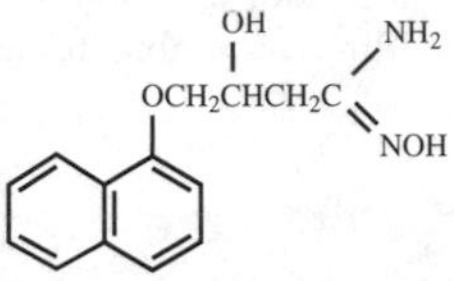

$NADP^+$ (mol wt 743) Abbreviation for nicotinamide adenine dinucleotide phosphate, which exists in two forms (reduced form and oxidized form) and serves as coenzyme in many biochemical reactions.

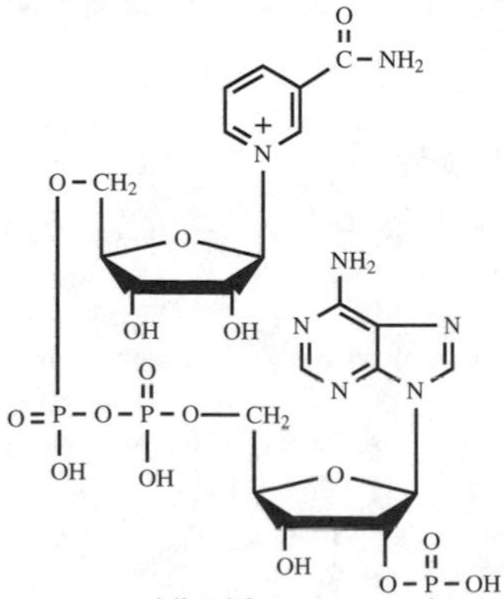

oxidized form

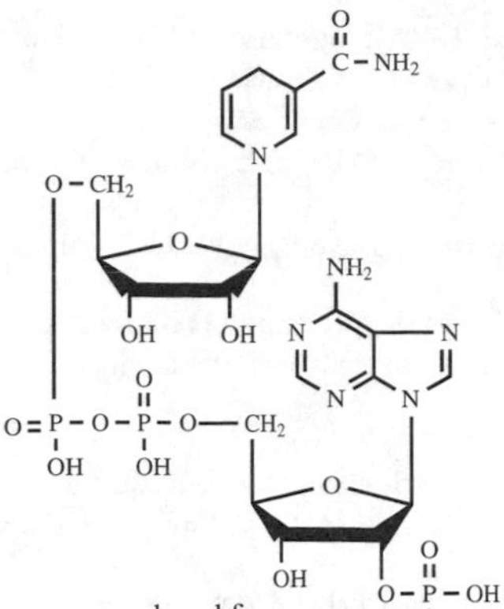

reduced form

$NADP^+$ The abbreviation for the oxidized form of NADP (see NADP for structure).

$NADP^+$ Arginine ADP-Ribosyltransferase The enzyme that catalyzes the following reaction:

$$NAD^+ + \text{L-arginine} \rightleftharpoons \text{Nicotinamide} + N^2\text{-(ADP-D-ribosyl)-L-arginine}$$

$NADP^+$ Nucleosidase The enzyme that catalyzes the following reaction:

$$\text{NADPH} + H_2O \rightleftharpoons \text{Nicotinamide} + \text{ADP-phosphoribose}$$

$NADP^+$ Transhydrogenase The enzyme that catalyzes the following reaction:

$$\text{NADPH} + NAD^+ \rightleftharpoons NADP^+ + \text{NADH}$$

NADPH The abbreviation for the reduced form of NADP (see NADP for structure).

NADPH Cytochrome c_2 Reductase The enzyme that catalyzes the following reaction:

$$\text{NADPH} + 2\ \text{ferricytochrome}\ c_2 \rightleftharpoons NADP^+ + 2\ \text{ferrocytochrome}\ c_2$$

NADPH Dehydrogenase The enzyme that catalyzes the following reaction:

$$\text{NADPH} + \text{acceptor} \rightleftharpoons NADP^+ + \text{reduced acceptor}$$

NADPH Dehydrogenase (FMN) The enzyme that catalyzes the following reaction:

$$NADPH + FMN \rightleftharpoons NADP^+ + FMNH_2$$

NADPH Diaphorase See NADPH dehydrogenase.

NADPH Ferricytochrome Oxidoreductase The systematic name for NADPH ferrihemoprotein reductase.

NADPH Ferrihemoprotein Reductase The enzyme that catalyzes the following reaction:

$$NADPH + 2\ \text{ferricytochrome} \rightleftharpoons NADP^+ + 2\ \text{Ferrocytochrome}$$

NADPH FMN Oxidoreductase The systematic name for NADPH reductase (FMN).

NADPH Nitrate Oxidoreductase The systematic name for nitrate reductase (NADPH).

NADPH Oxidized Glutathione Oxido-reductase The systematic name of glutathione reductase (NADPH).

NADPH Peroxidase The enzyme that catalyzes the following reaction:

$$NADPH + H_2O_2 \rightleftharpoons NADP^+ + 2\ H_2O$$

NADPH P-450 Reductase Synonym of NADPH-ferrihemoprotein reductase.

Nadroparin (mol wt 4500 daltons) A low molecular weight fraction of heparin.

NaeI A restriction endonuclease from *Nocardia aerocolonigenes* with the following specificity:

5′..........GCC↓GGC..........3′
3′..........CGG↑CCG..........5′

Naepaine (mol wt 250) A topical anesthetic agent.

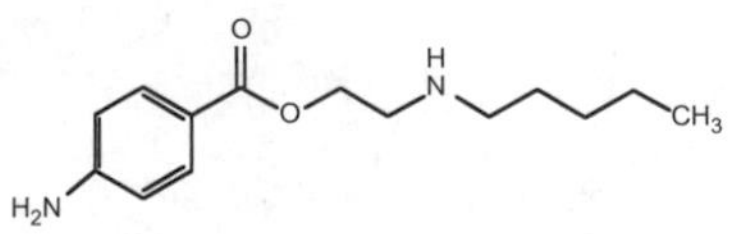

Nafamostat (mol wt 347) An inhibitor for trypsin, thrombin, kallikrein, and plasmin.

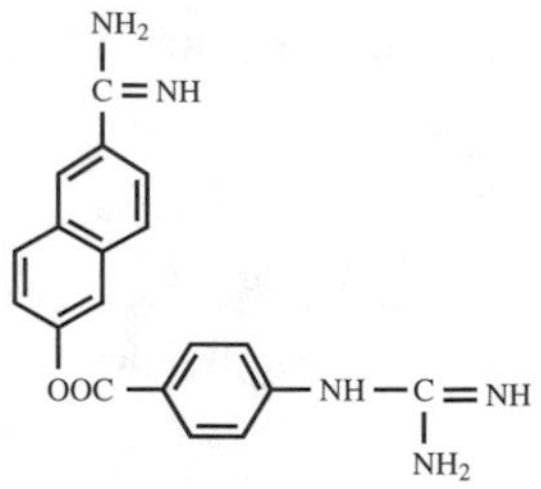

Nafarelin (mol wt 1322) A synthetic peptide agonist analog of gonadotropin-releasing hormone that decreases the release of follicle-stimulating hormone and luteinizing hormone.

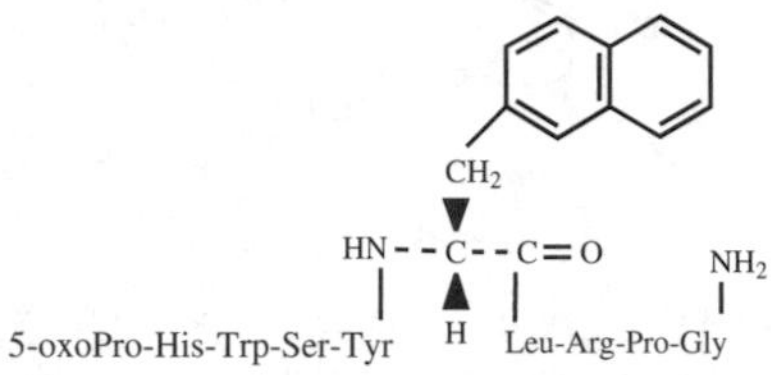

Nafcil A trade name for nafcillin sodium, a penicillinase-resistant penicillin that inhibits bacterial cell wall synthesis.

Nafcillin Sodium (mol wt 436) A penicillinase-resistant penicillin that inhibits bacterial cell wall synthesis.

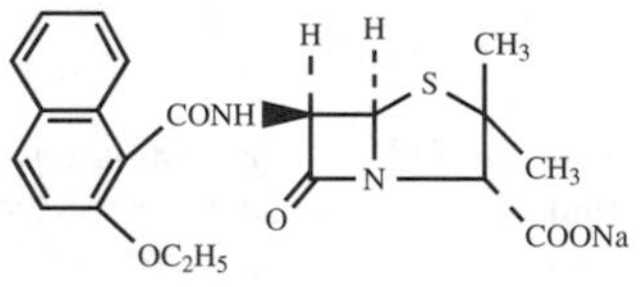

Nafiverine (mol wt 539) An antispasmodic agent.

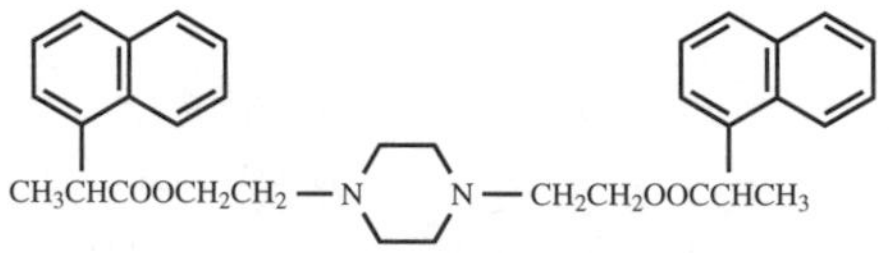

Nafronyl (mol wt 384) A vasodilator.

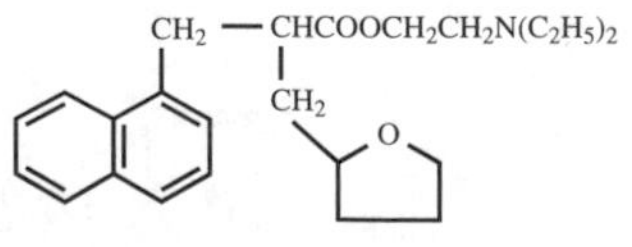

Naftalofos (mol wt 349) An anthelmintic agent.

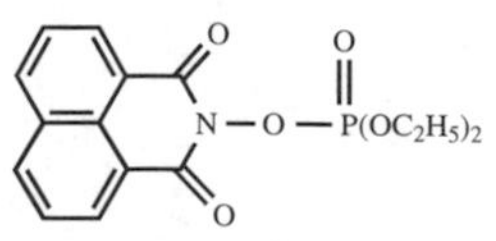

Naftifine (mol wt 287) An antifungal agent that inhibits sterol biosynthesis and alters the permeability of fungal cells.

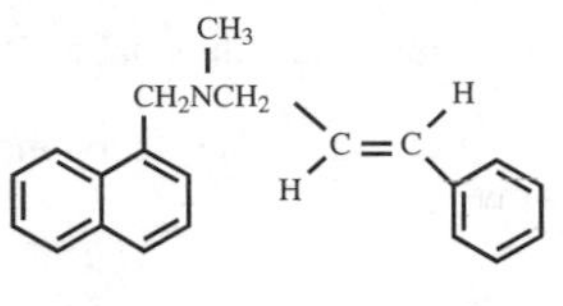

Naftin A trade name for naftifine, an antifungal agent that alters the permeability of fungal cells.

NAG Abbreviation for 1. N-acetyl-glucosamine; 2. N-acetylglucosaminidase.

Nagler's Test A test for detection of *Clostridium perfringens* based on the production of colonies surrounded by a zone of opacity on egg-yolk agar medium due to the production of lecithinase by the bacteria.

Na^+-Glucose Symport A system for transporting and concentrating glucose in brush border cells of the intestinal epithelium by a Na^+-dependent symport. The energy needed for active glucose transport is derived from a Na^+ gradient generated by Na^+K^+-AtPase.

NAH Abbreviation for N-acetylhexosaminidase.

$[Na^+]_i$ Abbreviation for intracellular sodium concentration.

NAIBS Abbreviation for nonadrenergic imidazoline binding site.

Na^+K^+-ATPase A transport system mediated by the Na^+K^+-dependent, membrane-bound ATPase. The Na^+K^+-ATPase is responsible for active transport of sodium ions out of the cell and active transport of potassium ions into the cell with the hydrolysis of ATP.

Na^+K^+-Pump Referring to the Na^+K^+-ATPase system that mediates the active transport of Na^+ out of the cell and K^+ into the cell (see Na^+K^+-ATPase).

Naked Virion A virion without envelope.

***nalA* Gene** A gene that encodes subunit protein of DNA gyrase.

Nalbuphine (mol wt 357) A narcotic analgesic agent that binds to opiate receptors in the CNS and alters both perception and emotional response to pain.

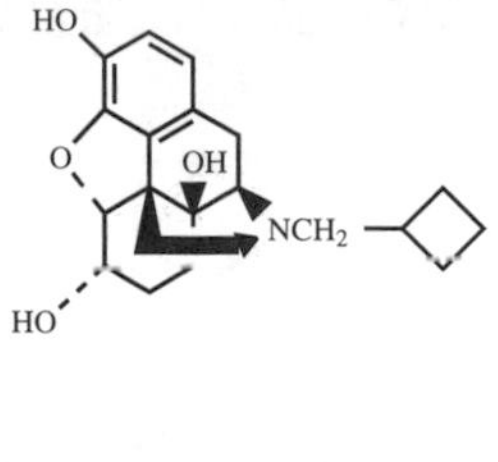

Nalcrom A trade name for cromolyn sodium, which inhibits degranulation of sensitized mast cells.

Naldecon A trade name for a combination drug containing phenylephrine hydrochloride, phenylpropanolamine hydrochloride, phenyltoloxamine citrate, and chloropheniramine maleate, used as an antihistaminic agent.

Nalfon A trade name for fenoprofen, a nonsteroidal anti-inflammatory and analgesic agent.

Nalidixic Acid (mol wt 232) An antibacterial agent that blocks DNA synthesis by inhibiting DNA gyrase.

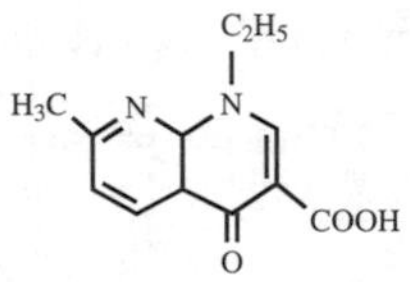

Nallpen A trade name for nafcillin sodium, a penicillinase-resistant antibiotic that inhibits bacterial cell wall synthesis.

Nalmefene (mol wt 339) A narcotic antagonist used to block the effects of opioids.

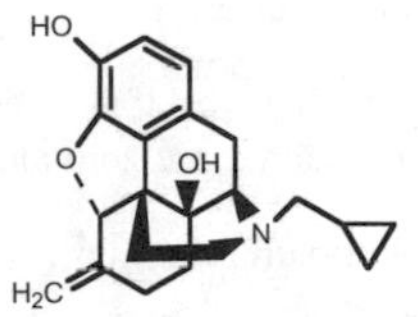

Nalorphine (mol wt 311) A narcotic antagonist.

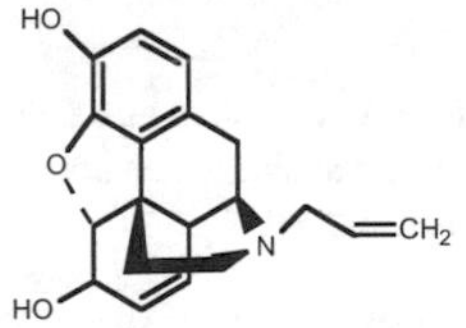

Nalorphine Dinicotinate (mol wt 522) A narcotic antagonist.

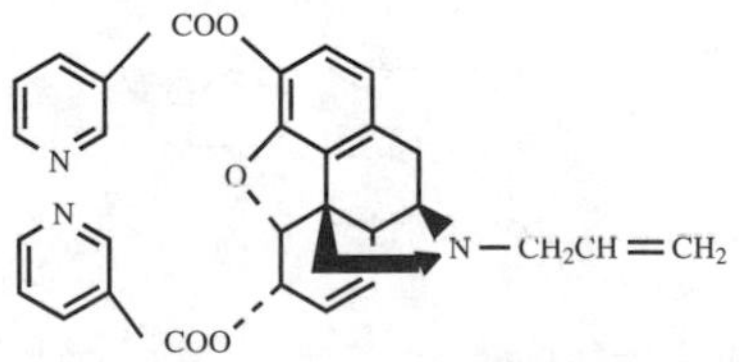

Naloxone (mol wt 327) A narcotic antagonist that displaces previously administered narcotic analgesics from their receptors.

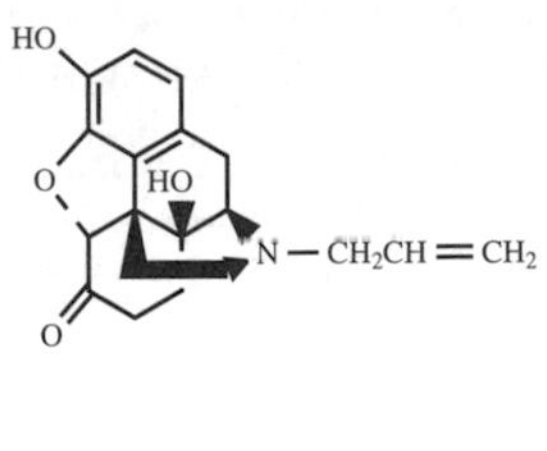

Naloxone Reductase See morphine 6-dehydrogenase.

Naltrexone (mol wt 341) A narcotic antagonist that blocks the effects of the previously administered opioids by occupying opiate receptors in the brain.

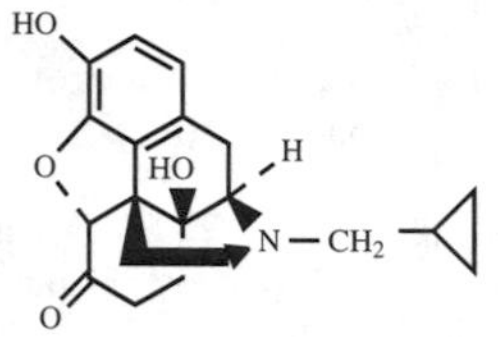

NAM Abbreviation for *N*-acetylmuramic acid.

NamI (NarI) A restriction endonuclease from *Nocardia argentinesis* with the following specificity:

5′..........GG↓CGCC..........3′
3′..........CCGC↑GG..........5′

Namalwa Cells A lymphoblastoid cell line used for production of interferon.

NANA Abbreviation for 1. N-acetylneuraminic acid; 2. N-acetylneuraminic acid aldolase.

NANBH Abbreviation for non-A, non-B hepatitis.

NANB Hepatitis Referring to non-A-non-B hepatitis.

Nandrobolic A trade name for nandrolone phenpropionate, an anabolic hormone that promotes tissue building processes.

Nandrolone (mol wt 274) An anabolic hormone that promotes tissue building and stimulates erythropoiesis.

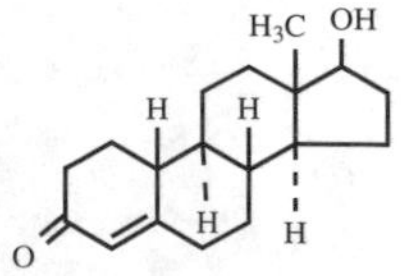

Nandrolone Decanoate A derivative of nandrolone, an anabolic hormone that promotes tissue building processes and stimulates erythropoiesis.

Nandrolone Phenpropionate An anabolic hormone derived from nandrolone that promotes tissue building processes and stimulates erythropoiesis.

Nandrolone Propionate A nandrolone-derived anabolic hormone that promotes tissue building processes and stimulates erythropoiesis.

Nano- A prefix meaning one-billionth (10^{-9}).

Nanogram 10^{-9} gram.

Nanometer Unit of length equal to 10^{-9} m.

Nanomolar Referring to a solution that contains 1 nanomole per liter.

Nanomole Referring to 10^{-9} mole.

NAP Abbreviation for 1. neutrophil activating peptide; 2. neutrophil alkaline phosphatase.

Napamide A trade name for disopyramide phosphate, an antiarrhythmic agent that prolongs the action potential.

Naphazoline (mol wt 210) A vasoconstrictor and decongestant that produces vasoconstriction by local adrenergic action on the blood vessels of the conjunctiva.

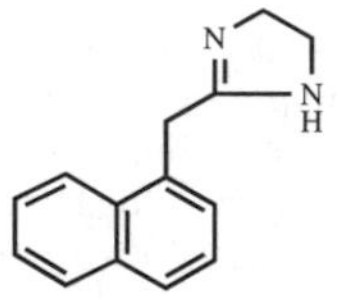

Naphcon A trade name for naphazoline hydrochloride, a vasoconstrictor that produces vasoconstriction by local adrenergic action on the blood vessels of the conjunctiva.

Naphcon Forte A trade name for naphazoline hydrochloride, a vasoconstrictor that produces vasoconstriction by local adrenergic action on the blood vessels of the conjunctiva.

Naphthalene Acetic Acid See naphthyl acetic acid.

Naphthoic Acid (mol wt 172) An aromatic organic acid.

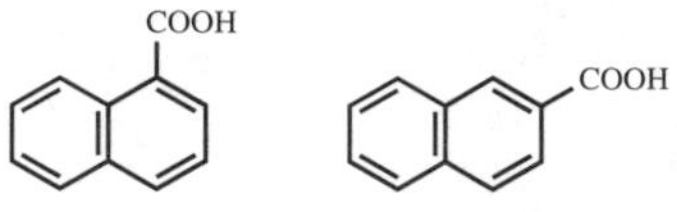

1-naphthoic acid   2-naphthoic acid

α-Naphthoic Acid See 1-naphthoic acid.

β-Naphthoic Acid See 2-naphthoic acid.

1-Naphthol (mol wt 144) A substrate for histochemical detection of peroxidase and an agent that induces oxidant stress in erythrocytes.

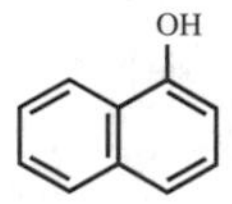

2-Naphthol (mol wt 144) A substrate for fluorometric assay of phenolsulfotransferase.

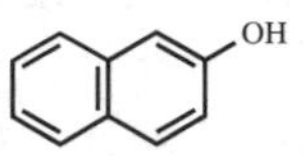

α-Naphthol See 1-naphthol.

β-Naphthol See 2-naphthol.

Naphthol AS (mol wt 263) A derivative of naphthol.

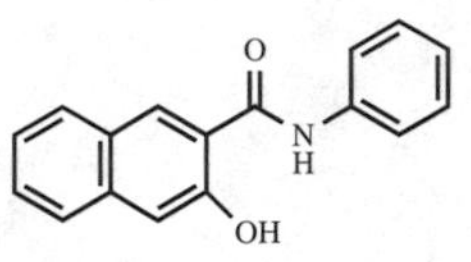

Naphthol AS Acetate (mol wt 305) A substrate for histochemical detection of esterase.

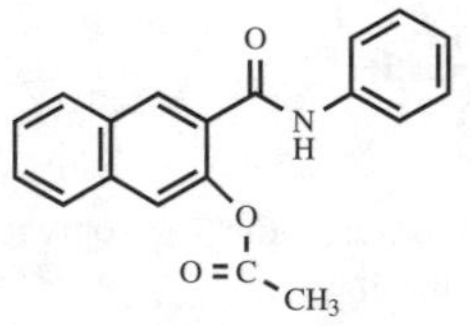

Naphthol AS Chloroacetate (mol wt 340) A substrate for histochemical detection of esterase.

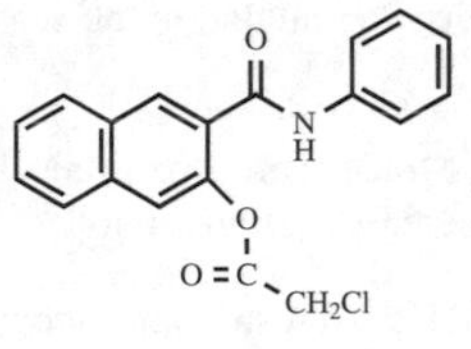

Naphthol Blue Black (mol wt 617) A dye used for staining protein.

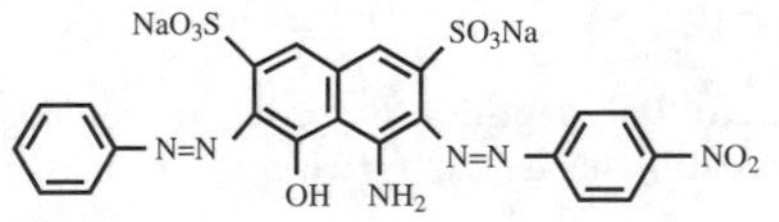

Naphthol Green B (mol wt 878) A dye.

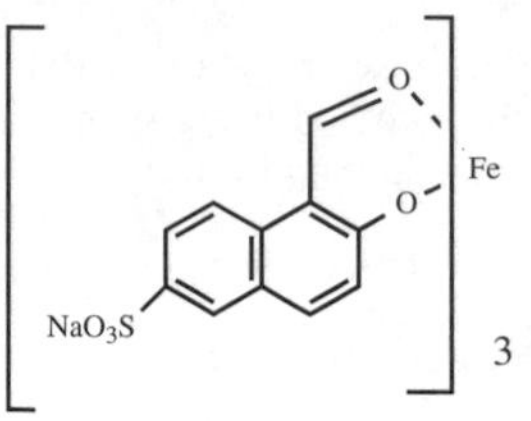

Naphthol Phthalein (mol wt 418) A dye and a pH indicator.

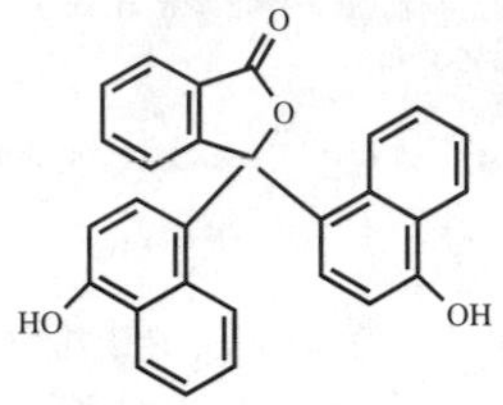

Naphthol Yellow S (mol wt 358) A dye and a reagent used for precipitation of amino acids and peptides.

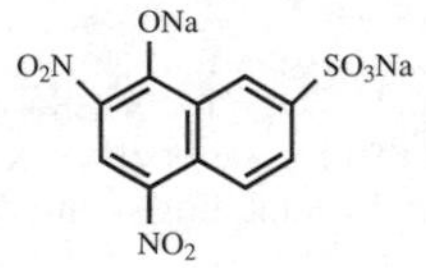

1-Naphthyl Acetate (mol wt 186) A substrate for assaying esterase activity.

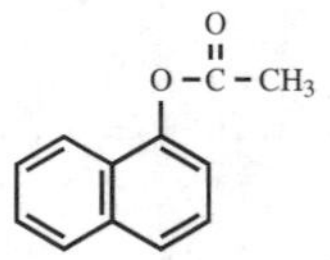

2-Naphthyl β-D-Galactosidase (mol wt 306) A substrate used for assaying β-galactosidase activity.

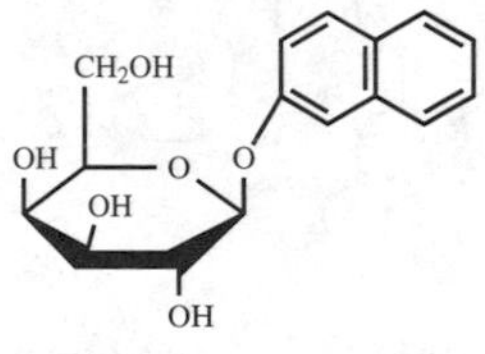

Naphthyl Laurate (mol wt 326) A reagent used for determination of lipase and esterase activity.

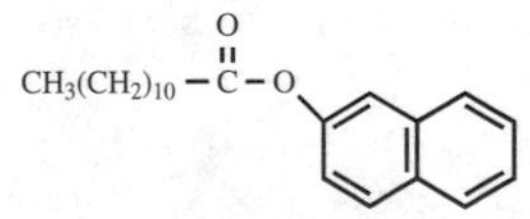

Naphthyl Red (mol wt 248) A dye and a pH indicator.

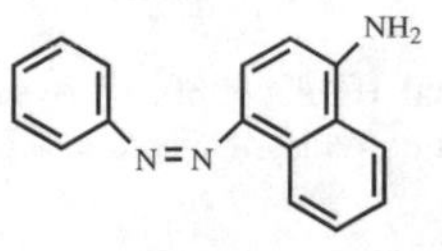

Naprelan A trade name for naproxen, a nonsteroidal, anti-inflammatory, and analgesic agent.

Naprogesic A trade name for naproxen sodium, an anti-inflammatory agent.

Napron X A trade name for naproxen, a nonsteroidal, anti-inflammatory, and analgesic agent.

Naprosyn A trade name for naproxen, an anti-inflammatory agent.

Naproxen (mol wt 230) An anti-inflammatory agent.

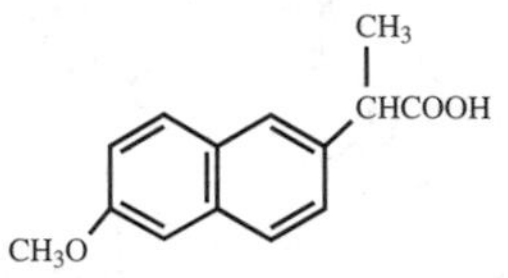

Naptrate A trade name for pentaerythritol tetranitrate, an antianginal agent that reduces cardiac oxygen demand.

Naqua A trade name for trichlormethiazide, a diuretic agent that increases urine excretion of sodium and water by inhibiting sodium reabsorption.

NarI A restriction endonuclease from *Nocardia argentinensis* with the following specificity:

5'........GG↓CGCC........3'
3'........CCGC↑GG........5'

Narbomycin (mol wt 510) An antibiotic.

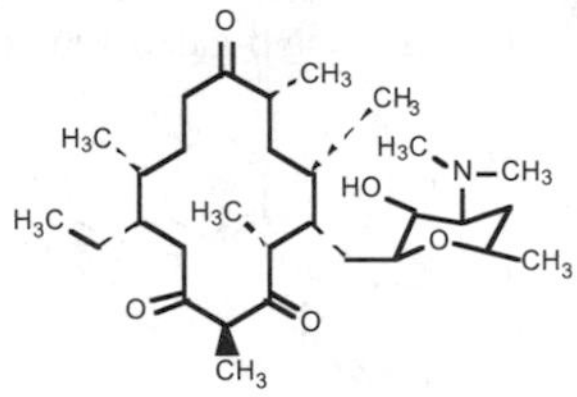

Narcan A trade name for naloxone hydrochloride, which displaces previously administered narcotic analgesics from their receptors.

Narcein (mol wt 445) An antitussive agent.

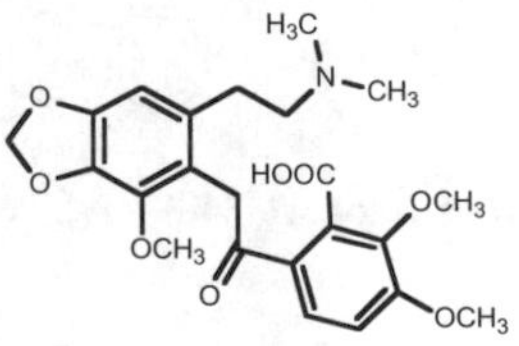

Narcobarbital (mol wt 303) A sedative, hypnotic and anticonvulsant agent.

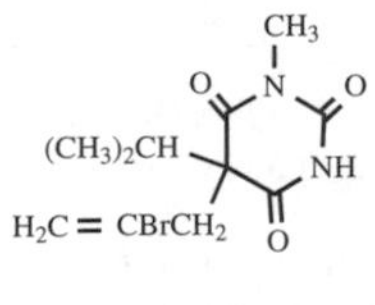

Narcolepsy A disorder characterized by sudden deep-sleep attack.

Narcosis A state of unconsciousness (stupor) caused by a narcotic drug.

Narcotic 1. Any drug that dulls the senses, relieves pain, and induces narcosis. 2. An addictive drug or presumed addictive drug. 3. Substance which produces addiction.

Nardil A trade name for phenelzine sulfate, a MAO inhibitor and an antidepressant.

Naringin (mol wt 581) A bitter principle in grapefruit.

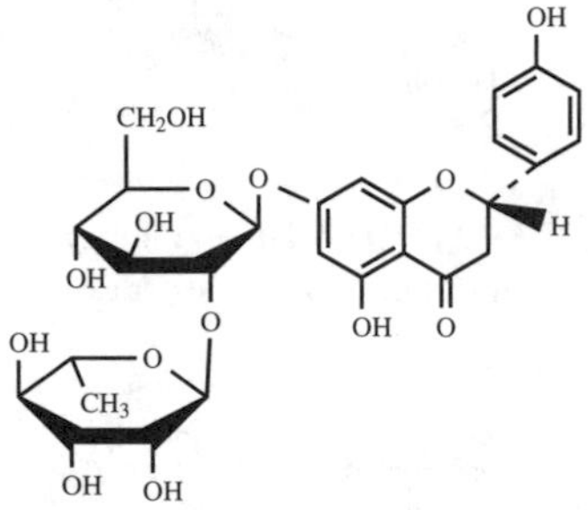

Naropin A trade name for ropivacaine hydrochloride, an anesthetic agent.

Nasacort A trade name for triamcinolone acetonide, a glucocorticosteroid.

Nasalcrom A trade name for cromolyn sodium, which inhibits degranulation of sensitized mast cells.

Nasalide A trade name for flunisolide, a nasal agent that decreases inflammation.

NasBI (BamHI) A restriction endonuclease from *Nocardia asteroides* with the same specificity as BamHI.

Nascent Referring to the macromolecule that is in the process of being synthesized.

Nascent Polypeptide Referring to polypeptide that is still attached to a tRNA.

Nascent Protein A protein or polypeptide chain in the process of being synthesized but not completed.

Nascent RNA Referring to an RNA molecule that is in the process of being synthesized (before posttranscriptional modification).

Nascobal A trade name for cyanocobalamin, a vitamin.

Nasonex A trade name for mometasone furoate monohydrate, an anti-inflammatory corticosteroid.

Nasopharyngitis Inflammation of the nose and throat.

NasSI (SacI) A restriction endonuclease from *Nocardia asteroides* with the following specificity:

5′..........GAGCTC..........3′
3′..........CTCGAG..........5′

NasWI (NaeI) A restriction endonuclease from *Nocardia asteroides* with the same specificity as NaeI.

NAT Abbreviation for N-acetyl-L-tyrosine.

NATA Abbreviation for N-acetyltryptophanamide.

Natacyn A trade name for natamycin, an anti-infective drug that alters the permeability of fungal cells.

Natamycin (mol wt 666) An antifungal agent produced by *Streptomyces natalensis* that increases fungal membrane permeability.

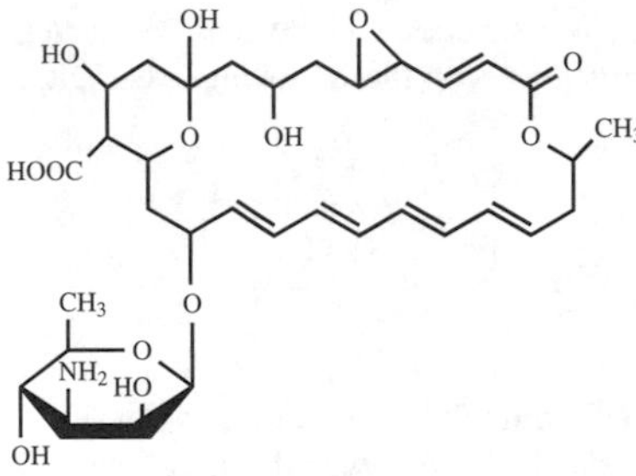

Native Conformation A macromolecule, e.g., protein or DNA, that is in its natural three-dimensional shape with biological activity.

Natremia The presence of large amounts of sodium in the blood.

Natrilix A trade name for indapamide, a diuretic agent that inhibits sodium reabsorption in the cortical diluting site of the nephron.

Natrimax A trade name for hydrochlorothiazide, a diuretic agent that increases urine excretion of sodium and water by inhibiting sodium reabsorption.

Natrium Latin for sodium.

Natriuresis The excess excretion of sodium in the urine.

Natriuretic Any agent that promotes excretion of sodium.

Natriuretic Hormone Hormones that increase the excretion of sodium in the urine by inhibiting reabsorption of sodium by the kidney.

Natulan A trade name for procarbazine hydrochloride, an antineoplastic agent that inhibits DNA, RNA, and protein synthesis.

Naturacil A trade name for psyllium, a laxative.

Natural Amino Acids Referring to the L-isomer of the amino acids used in protein biosynthesis.

Natural Antibody Antibody present in the serum in the absence of apparent specific antigenic contact in the host.

Natural Immunity Nonspecific resistance against infection that is not dependent upon immunization by infection or vaccination.

Natural Killer Cell A type of leukocyte that kills certain tumor cells and virus-infected cells.

Natural Selection The differential success in the reproduction of different phenotypes due to the interaction of organisms with their environment. The members of a species that are better adapted to their environment tend to eliminate those that are not well adapted.

Naturetin A trade name for bendroflumethiazide, a diuretic agent that increases urine excretion of sodium and water by inhibiting sodium reabsorption.

Nauseant Any agent that induces nausea.

Nauseatol A trade name for dimenhydrinate, an antiemetic agent.

Navane A trade name for thiothixene, an antipsychotic agent that blocks the postsynaptic dopamine receptors in the brain.

Navelbine A trade name for vinorelbine tartrate, a mitotic inhibitor used as an antineoplastic agent.

Navonaprox A trade name for naproxen, an anti-inflammatory agent.

Naxen A trade name for naproxen, an anti-inflammatory agent.

NB Abbreviation for Northern blot.

Nb Abbreviation for niobium, a chemical element.

NbaI (NaeI) A restriction endonuclease from *Nocardia brasiliensis* with the same specificity as NaeI.

NBAT Abbreviation for neutral and basic acid transporter.

4-NBC Abbreviation for 4-nitrobenzylchloride.

NBC Medium Abbreviation for nitrogen base-casein medium.

NBD Abbreviation for nucleotide-binding domain.

NBF Abbreviation for nucleotide-binding factor.

NblI (PvuI) A restriction endonuclease from *Nocardia blackwellii* with the following specificity:

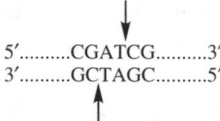

↓
5′..........CGATCG..........3′
3′..........GCTAGC..........5′
↑

NBMPR Abbreviation for nitrobenzyl-mercapto purine ribonucleoside.

NBP Abbreviation for nucleotide-binding protein.

NbrI (NaeI) A restriction endonuclease from *Nocardia brasiliensis* with the same specificity as NaeI.

NBS Abbreviation for N-bromosuccinimide.

NBT Abbreviation for nitrobluetetrazolium.

NBT Test A metabolic assay for the reduction of nitrobluetetrazolium dye during activation of the hexose monophosphate shunt in phagocytic cells.

NBTGR Abbreviation for nitrobenzylguanosine ribonucleoside.

NBTI Abbreviation for nitrobenzylthioinosine.

NcaI (HinfI) A restriction endonuclease from *Neisseria caviae* with the following specificity:

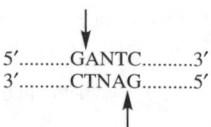

↓
5′..........GANTC..........3′
3′..........CTNAG..........5′
↑

NCAM Abbreviation for neural cell-adhesion molecule.

NCC Abbreviation for natural cytotoxic cell.

NCF Abbreviation for neutrophil chemotactic factor.

NCF-A Abbreviation for neutrophil chemotactic factor of anaphylaxis.

NciI (CauII) A restriction endonuclease from *Neisseria cinerea* with the following specificity:

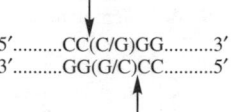

↓
5′..........CC(C/G)GG..........3′
3′..........GG(G/C)CC..........5′
↑

NcoI A restriction endonuclease from *Nocardia corallina* with the following specificity:

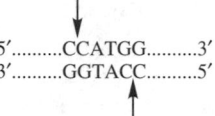

↓
5′..........CCATGG..........3′
3′..........GGTACC..........5′
↑

NCP Abbreviation for nucleosomal core particle.

NCTC Abbreviation for natural cytotoxic T cell.

NcuI (MboII) A restriction endonuclease from *Nocardia cuniculi* with the same specificity as MboII.

NCV Abbreviation for Newcastle disease virus.

ND Abbreviation for Newcastle disease.

Nd Abbreviation for neodymium, a chemical element.

NdeI A restriction endonuclease from *Neisseria denitrificians* with the following specificity:

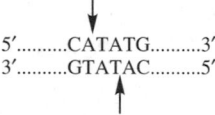

↓
5′..........CATATG..........3′
3′..........GTATAC..........5′
↑

NdeII (MboI) A restriction endonuclease from *Neisseria denitrificians* with the following specificity:

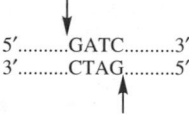

↓
5′..........GATC..........3′
3′..........CTAG..........5′
↑

NDGA Abbreviation for nordihydroguaiaretic acid.

NDIRA Abbreviation for non-dispersive infrared analysis.

nDNA Abbreviation for nuclear DNA.

NDP Abbreviation for nucleotide diphosphate.

NDP Kinase Abbreviation for nucleoside diphosphokinase.

NDV Abbreviation for newcastle disease virus (Paramyxoviridae).

NE Abbreviation for norepinephrine.

Ne Abbreviation for neon.

Nealbarbital (mol wt 238) A sedative and hypnotic agent.

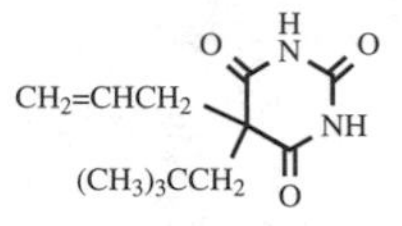

Neamine (mol wt 322) A component of the antibiotic complex neomycin.

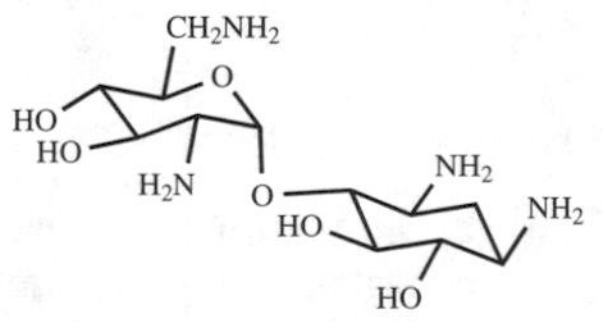

Nearest Neighbor Analysis An experimental analysis used for determination of the polarity of the complementary strands in double-stranded DNA. The analysis is based on the fact that when a ^{32}P-labeled NTP is added to a growing chain of polynucleotide during DNA synthesis, the labeled alpha phosphate of the NTP is transferred to the nearest neighbor upon hydrolysis of newly synthesized DNA.

Nebcin A trade name for tobramycin sulfate, an antibiotic that inhibits bacterial protein synthesis by binding to 30S ribosomes.

Nebivolol (mol wt 405) An antihypertensive agent.

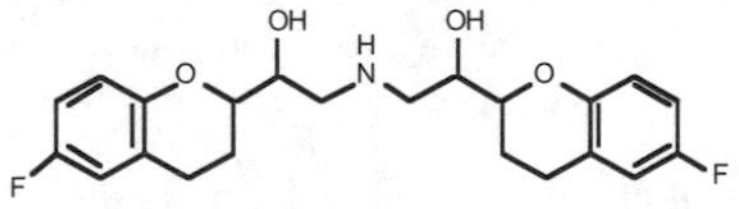

Nebulin Protein in the N-line of sarcomere of striated muscle.

Nebulization A process of converting a liquid into a fine spray.

Nebulizer A device capable of converting a solution to a fine spray.

NebuPent A trade name for pentamidine isethionate, an antiprotozoal agent that interferes with synthesis of DNA, RNA, phospholipid, and protein.

Necatoriasis An infestation of the small intestine by the parasitic hookworm *Necator americanus.*

Necon A trade name for a combination drug containing norethindrone and ethinyl estradiol used as a contraceptive agent.

Necrosis The death of cell or tissue caused by injury or infection.

Nectin The protein found in the stalk region of the mitochondrial proton ATPase.

Nedocromil (mol wt 271) An antiallergic and antiasthmatic agent that prevents the release of histamine from mast cells.

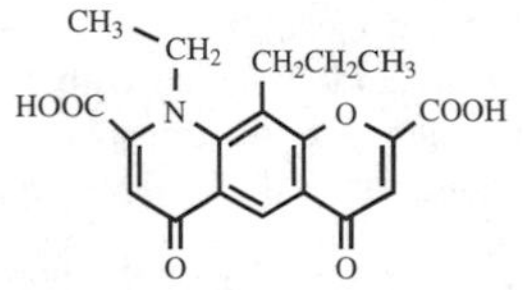

NEFA Abbreviation for non-esterified fatty acid.

Nefazodone (mol wt 470) An antidepressant.

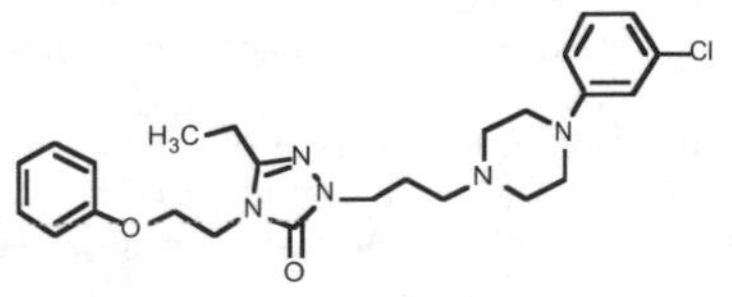

Nefopam (mol wt 253) An analgesic and an antidepressant agent.

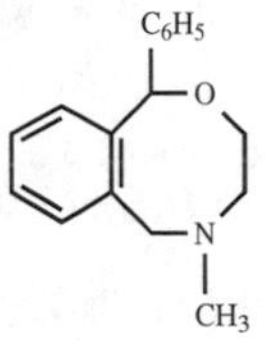

Negamycin (mol wt 248) An antibiotic isolated from *Streptomyces purpeoeidem*.

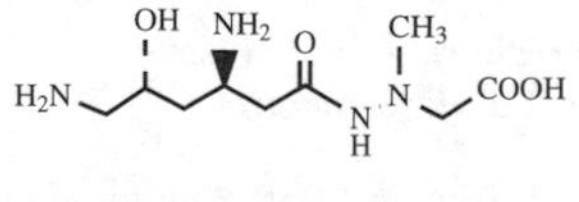

Negative Complementation The suppression of the wild type activity of one subunit of a multimeric protein by a mutant allelic subunit.

Negative Contrast Optical system in which a lightly stained specimen is seen against a dark, heavily stained background.

Negative Control The mechanism for regulation of gene activity in which a regulatory macromolecule (e.g., repressor) functions to turn off transcription, e.g., binding of regulatory protein on the DNA inhibiting transcription.

Negative Cooperativity The binding of one ligand to one site on the macromolecule decreases the affinity for the binding of the subsequent ligand to the other site on the same molecule.

Negative Feedback 1. A control mechanism in which the product (usually the end product) of a process or a pathway inhibits the early reaction in the process or pathway. 2. A decrease in body function in response to stimulation.

Negative Regulation See negative control.

Negative Staining An electron-microscopic technique in which an electron-dense substance, e.g., phosphotungstic acid, is mixed with specimen so that the specimen appears transparent against a dark background under the electron microscope.

Negative Strand RNA The RNA that has a base sequence complementary to the corresponding mRNA. Negative-stranded RNA cannot directly act as mRNA.

Negative-Stranded Virus A virus that contains negative-stranded RNA as genetic material.

Negative Supercoil A form of double-stranded circular DNA formed by winding of the duplex DNA in a direction opposite to the turns of the strands of the double helix.

Negative Superhelix See negative supercoil.

Negatron Referring to a negatively charged electron.

NegGram A trade name for nalidixic acid, which inhibits microbial DNA synthesis.

Negri Bodies Intracytoplasmic inclusion bodies in nerve cells associated with rabies virus infection.

Neisseria A genus of aerobic, oxidase-positive, Gram-negative bacteria (Neisseriaceae).

Neisseria Gonorrhoeae A bacterium that causes the venereal disease gonorrhea.

Neisserial Infection Referring to gonorrhea.

Nekton Referring to the active motile organisms in the water.

Nelova A trade name for a combination drug containing ethinyl estradiol and norethindrone, used as an oral contraceptive.

NEM Abbreviation for N-ethylmaleimide.

Nematocide Any substance that kills nematode.

Nematodes Roundworms of the phylum Nematoda; some members are pathogenic.

Nematodiasis Infection or disease caused by nematodes.

Nematophagous Fungi Fungi that derive nutrient from nematode worms.

Nematosome A cytoplasmic inclusion observed in certain neurons.

Nembutal A trade name for pentobarbital, used as a sedative-hypnotic agent.

Nemonapride (mol wt 388) An anti-psychotic agent.

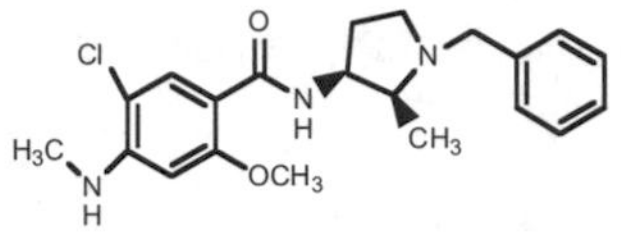

Neo- A prefix meaning new.

Neoadjuvant Chemotherapy A chemotheraphy given before treatment of the primary tumor for the purpose of improving the results of other treatment and preventing the development of metastases.

Neoantigens 1. Nonself antigens that arise spontaneously on cell surfaces (usually during the neoplasia). 2. An antigen formed *in vivo* by the combination of a self protein with an exogenous substance or hapten.

Neoblastic Pertaining to new tissue or new growth.

Neocin A trade name for neomycin, an antibiotic.

Neo-Codema A trade name for hydrochlorothiazide, a diuretic agent.

Neo-Cortef Ointment A trade name for a combination drug containing hydrocortisone acetate and neomycin sulfate, used as an anti-infective agent.

Neo-Cultol A trade name for liquid petrolatum (mineral oil), used as a laxative agent.

Neocyten A trade name for orphenadrine citrate, which reduces the transmission of nerve impulses from the spinal cord to the skeletal muscle.

NeoDecadron Cream A trade name for a combination drug containing dexamethasone phosphate and neomycin sulfate, used as a local anti-infective agent.

NeoDecadron Ophthalmic Ointment A trade name for a combination drug containing dexamethasone phosphate and neomycin sulfate, used as an ophthalmic anti-infective agent.

Neo-DM A trade name for dextromethorphan hydrobromide, used as an antitussive agent.

Neo-Durabolic A trade name for nandrolone decanoate, an anabolic hormone that promotes tissue building processes.

Neodymium (Nd) A chemical element with atomic weight 144, valence 2, 3, and 4.

Neo-Estrone A trade name for esterified estrogen.

Neo-Fed A trade name for pseudoephedrine hydrochloride, which stimulates alpha-adrenergic receptors and promotes vasoconstriction.

Neo-Flo A trade name for boric acid, used as an ophthalmic anti-infective.

Neoloid A trade name for castor oil, used as a laxative agent.

Neomethymycin (mol wt 470) A macrolide antibiotic from *Streptomyces venezuelae*.

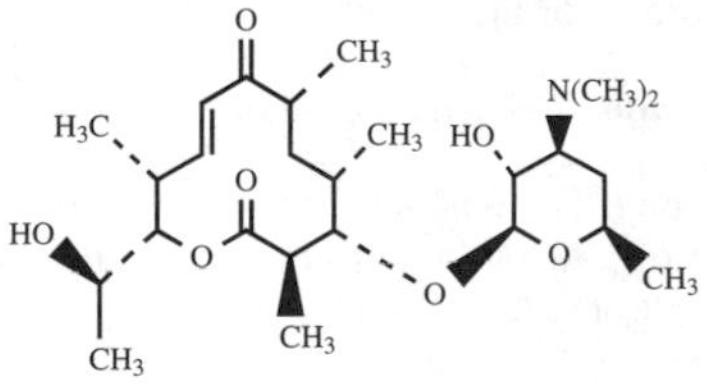

Neo-Metric A trade name for metronidazole, used as an antiprotozoal agent.

Neomycin A complex antibiotic composed of neomycin A, B, and C, and isolated from *Streptomyces fradiae* that binds to 30S ribosomes and inhibits bacterial protein synthesis.

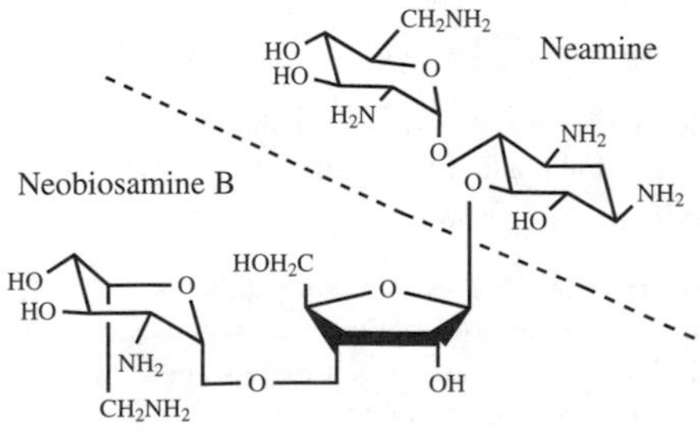

Neomycin B

Neomycin Kanamycin Phosphotransferase Synonym of kanamycin kinase.

Neomycin Phosphotransferase Synonym of kanamycin kinase.

Neon A chemical element with atomic weight 20.

Neonatal The period shortly after birth.

Neonate A newborn.

Neonatology The science that deals with the newborn.

Neopap A trade name for acetaminophen, used as an analgesic and antipyretic agent.

Neoplasic Pertaining to neoplasm.

Neoplasm The formation of a growing tissue mass resulting from the uncontrolled proliferation of cells due to the loss of normal growth control. Neoplasms may be benign or malignant (invasive).

Neoplastic Transformation Conversion of a tissue that shows normal growth regulation into a tumor that grows in a progressive, uncontrolled manner.

Neo-Polycin Ointment A trade name for a combination drug containing polymyxin B sulfate, zinc bacitracin, and neomycin sulfate, used as a local anti-infective agent.

Neopterin (mol wt 253) Precursor of biopterin.

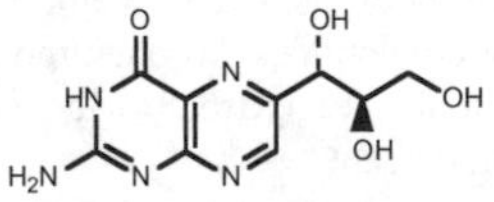

Neoquess A trade name for hyoscyamine sulfate, used as an antichlolinergic agent that decreases the motility of the GI tract and inhibits gastric acid secretion.

Neoral A trade name for cyclosporine, an immuno-suppressant.

Neorickettsia A genus of Gram-negative bacteria (Ehrlichieae).

Neo-Rx A trade name for neomycin sulfate, an antibiotics used as a local anti-infective agent that inhibits bacterial protein synthesis.

Neosar A trade name for cyclophosphamide, used as an alkylating agent that cross-links cellular DNA strands interfering with transcription.

Neo-Spec A trade name for guaifencsin, used as an antitussive agent that increases production of respiratory fluids and reduces viscosity of tenacious secretions.

Neosporin Cream A trade name for a combination drug containing polymyxin B sulfate and neomycin sulfate, used as a local anti-infective agent.

Neosporin Ointment A trade name for a combination drug containing polymyxin B sulfate, zinc bacitracin, and neomycin sulfate, used as a local anti-infective agent.

Neosporin Ophthalmic Ointment A trade name for a combination drug containing polymyxin B sulfate, neomycin sulfate, and zinc bacitracin, used as an anti-infective agent.

Neostigmine (mol wt 209) A cholinergic agent that inhibits the destruction of acetylcholine released from parasympathetic nerves.

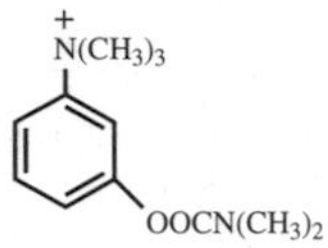

Neosulf A trade name for neomycin sulfate, an antibiotic that binds to 30S ribosomes inhibiting bacterial protein synthesis.

Neo-Synephrine A trade name for phenylephrine hydrochloride, which stimulates alpha-adrenergic receptors in the sympathetic nervous system.

Neotal A trade name for a combination drug containing polymycin B sulfate, neomycin sulfate, and bacitracin zinc, used as a local anti-infective agent.

Neotep-Granucaps A trade name for a combination drug containing chlorpheniramine maleate and phenylephrine hydrochloride, used as an antihistaminic agent.

Neothylline-GG Tablets A trade name for a combination drug containing dyphylline and guaifenesin, used as a bronchodilator.

Neotrace-4 A trade name for a mixture of minerals containing zinc sulfate, copper sulfate, manganese sulfate, and chromium chloride, used as a mineral supplement.

Neo-Tric A trade name for metronidazole, an antibiotic.

NEP Abbreviation for neutral endopeptidase.

Nepenthesin A protease from the insectivorous plant (*Nepenthes* species) with properties similar to pepsin.

Nephelometer An instrument used for measuring the turbidity of a sample containing suspended particles or cells.

NEpHGE Abbreviation for non-equilibrium pH gradient electrophoresis.

Nephelometry The determination of concentrations of particles or cells in a suspension by the measurement of the light scattered by the suspended particles or cells.

Nephr- A prefix meaning kidney.

NephrAmine A trade name for amino acid infusions, used for treatment of renal failure.

Nephrectomy The surgical removal of a kidney.

Nephric Pertaining to renal.

Nephritic Factor Factors occurring in the serum of patients with proliferative glomerulonephritis that activate the alternative complement pathway.

Nephritis Kidney inflammation.

Nephritogenic Capable of causing nephritis.

Nephrocalcinosis A condition marked by the calcification of the tubules of the kidney.

Nephrolithiasis The presence of stones in the kidney.

Nephrolithotomy The surgical removal of calculi or stones from the kidney.

Nephrology The science that deals with diseases or disorders of the kidney.

Nephroma A malignant tumor of the kidney.

Nephron The functional unit of the kidney consisting of glomerular capsule, convoluted tubule, and nephronic loop.

Nephronex A trade name for nitrofurantoin microcrystals, used as an antibacterial agent.

Nephrosclerosis A kidney disorder associated with high blood pressure and characterized by hardening of the arteries and impaired kidney function.

Nephroscopy A technique for the examination of the interior of the kidney.

Nephrotoxic Toxic to the kidneys.

Nepoviruses A group of nematode-transmitted, polyhydral plant viruses containing single-stranded RNA.

Neprilysin A protease that catalyzes the preferential cleavage of peptide bonds between hydrophobic amino acids. It is a membrane-bound glycoprotein found in animal tissue.

Neptazane A trade name for methyclothiazide, used as a diuretic agent that increases urine excretion of sodium and water by inhibiting sodium reabsorption.

NER Abbreviation for nucleotide excision repair.

Neriifolin (mol wt 535) A cardiotonic agent.

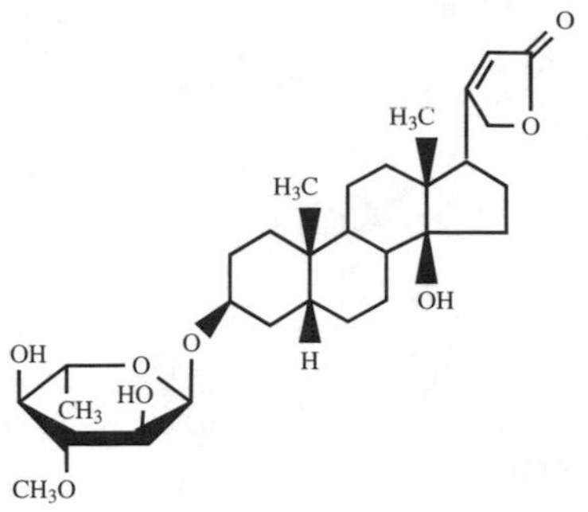

Nernst Equation An equation for calculation of equilibrium potential or membrane potential across a membrane. It is a quantitative expression that relates the equilibrium ratio of an ion on either side of the membrane to the voltage difference across the membrane.

Nerve A cord-like anatomical structure consisting of bundles of signal-carrying fibers that transmit impulses and stimuli from the brain and spinal cord to the other parts of the body.

Nerve Cell See neuron.

Nerve Cell Body The portion of a neuron that contains the cell nucleus.

Nerve Ending The end portion of a neuron where a nerve transmits or receives impulses.

Nerve Fiber Any of the processes or filaments of a neuron.

Nerve Growth Factor (NGF) A protein growth factor found in a variety of peripheral tissues that maintains the viability of neurons and stimulates the growth of nerve cell processes.

Nerve Impulse Signal transmitted along nerve cells by a wave or depolarization-repolarization event.

Nerve Motor A nerve that causes muscle contraction.

Nervine Nighttime Sleep-Aid A trade name for diphenhydramine hydrochloride, which is used as an antihistaminic agent.

Nervocaine A trade name for lidocaine hydrochloride (lignocaine hydrochloride), used as a local anesthetic agent that blocks depolarization by interfering with sodium-potassium exchange across the nerve membrane.

Nervon A galactocerebroside in which the acyl group is nervonoyl.

Nervonic Acid (mol wt 267) A nonsaturated fatty acid with a single double bond.

$$CH_3(CH_2)_7CH = CH(CH_2)_{13}COOH$$

Nervonoyl The acyl group derived from nervonic acid.

$$CH_3\text{-}[CH_2]_7\text{-}CH{=}CH\text{-}[CH_2]_{13}\text{-}CO\text{-}$$

Nervous System A system of interconnecting cells (neurons) in the vertebrate. It consists of the brain, spinal cord, nerves, ganglia and organs that collect, process, and respond to information from the environment and from within the organism by the transmission of electrical impulses and exchange of chemical signals.

NES Abbreviation for nuclear export signal.

Nesacaine A trade name for chloroprocaine hydrochloride, used as a local anesthetic agent that blocks depolarization by interfering with sodium-potassium exchange across the nerve membrane.

Nessler's Reagent A reagent used for colorimetic determination of nitrogen that contains mercuric iodide, potassium iodide, and potassium hydroxide.

Net Charge The sum of positive and negative charges on a molecule; a molecule with five negative and two positive charges has three negative net charges.

Net Radiation The arithmetic difference between incoming solar radiation and outgoing terrestrial radiation.

Netilmicin (mol wt 476) A semisynthetic antibiotic.

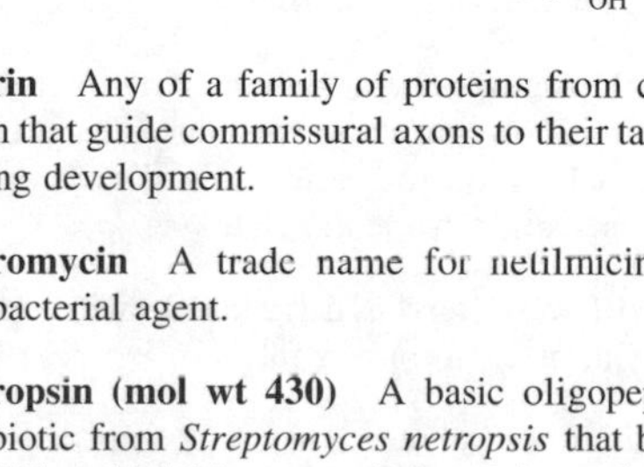

Netrin Any of a family of proteins from chick brain that guide commissural axons to their targets during development.

Netromycin A trade name for netilmicin, an antibacterial agent.

Netropsin (mol wt 430) A basic oligopeptide antibiotic from *Streptomyces netropsis* that binds to DNA and inhibits transcription. It is used as an antibacterial and antineoplastic agent.

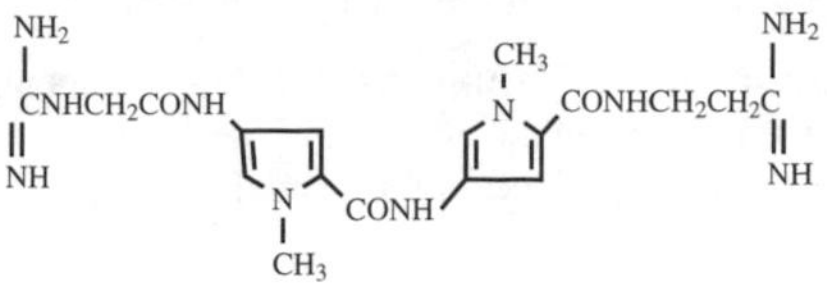

Network Hypothesis A theory of immunoregulation by a cascade of idiotype-antiidiotype reactions involving T-cell receptors and antibodies.

Neu Abbreviation for neuraminic acid.

neu An oncogene associated with a neuroblastoma in rat that encodes a transmembrane glycoprotein with intrinsic tyrosine kinase activity.

NeuAc Abbreviation for *N*-acetylneuramic acid.

Neumega A trade name for recombinant human interleukin-11 that stimulates proliferation of hematopoietic stem cells and megakaryocyte progenitor cells.

NeuNAc Abbreviation for N-acetylneuraminic acid.

Neupogen A trade name for filgrastim produced by DNA recombinant technology used for the treatment of chemotherapy-induced neutropenia, bone marrow transplant accompanied neutropenia, and severe chronic neutropenia.

Neur- A prefix denoting nerve or nervous system.

Neural Pertaining to nerve or nervous system.

Neural Cell-Adhesion Molecule Referring to molecules in the immunoglobulin superfamily that function as molecular recognition molecules.

Neural Cleft See neural groove.

Neural Crest A group of embryonic cells derived from the roof of the neural tube that migrate to different locations and form various types of adult cells, e.g., nerve cells, ganglia, melanocytes, and Schwann cells.

Neural Crest Cells Cells that migrate away from the neural tube to specific locations along the periphery of the embryo, where they give rise to pigment cells, cartilage, and numerous other tissues.

Neural Groove Invagination of the ectoderm along the dorsal surface of the embryo, induced by the dorsal mesoderm during animal embryogenesis; also called the neural cleft.

Neural Tube Nervous tissue initially formed from ectoderm in the form of a tube.

Neuramidase Spikes The projections from surfaces of influenza viruses containing neuraminidase that are involved in the release of viruses from infected cells following viral replication.

Neuraminic Acid (mol wt 267) A compound derived from mannosamine and pyruvate and a major building block in the animal cell coat.

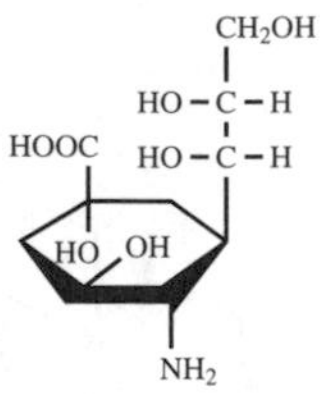

Neuraminidase The enzyme that catalyzes the cleavage of *N*-acetyl neuraminic acid from mucopolysaccharide.

Neuregulin Any of a family of closely related proteins involved in the regulation of neural and muscle development.

Neurilemma The covering or sheath of a nerve fiber.

Neurites The thin cytoplamic extensions that emerge from the immature neurons and develop into axons and dendrites.

Neuritis An inflammation or degeneration of a nerve.

Neuro- A prefix denoting nerve or nervous system.

Neuroblast Cells that give rise to nerve cells.

Neuroblastoma A malignant tumor of immature nerve cells that can be maintained in culture and induced to proliferate.

Neurocalcin A calcium-binding protein involved in the calcium-dependent regulation of rhodopsin phosphorylation and found in retina and brain neurons.

Neurocytolysin Toxic substance capable of lysis of nerve cells.

Neurocrinology The science that deals with the interaction of the nervous system with endocrine glands.

Neuroendocrine Pertaining to the nervous and endocrine systems.

Neurofascin A chick glycoprotein implicated in axon extension.

Neurofibril A threadlike structure found in the neurons.

Neurofibroma A nonmalignant tumor composed of nerve and fibrous tissues.

Neurofilament A class of intermediate filaments found in the axons of the nerve cells. It consists of three distinct protein subunits.

Neurogenic Pertaining to the nerve tissue.

Neuroglia The nonneural connective tissue of the central nervous system of higher animals.

Neurohormone A hormone secreted by the specialized neurons that acts upon cells located at some distance from its releasing point.

Neurohumor Chemical substances synthesized and carried by a neuron that effects the passage of nerve impulses from one cell to another at the synapse.

Neuroid Resembling a nerve.

Neurolemmoma A tumor of the nerve sheath.

Neurolepsis An altered state of consciousness marked by reduced physical movement and anxiety.

Neuroleptic Any substance capable of acting on the nervous system and causing neurolepsis.

Neuroleukin A lymphokine produced by lectin-activated T cells that promotes the survival of some neurons in tissue culture.

Neurologist Scientist who specializes in structure, function, and diseases of the nervous systems.

Neurology The science that deals with structure, function, and disease of the nervous system.

Neurolysin A protease that catalyzes the cleavage in neurotensin.

Neuromuscular Pertaining to nerves and muscles.

Neuromuscular Junction The junction or synapse between a nerve cell axon and skeletal muscle cell.

Neuromyelitis Inflammation of nerves and spinal cord.

Neuron A nerve cell consisting of a cell body, axons, and dendrites that receives, conducts, and transmits signals in the nervous system.

Neuron Specific Enolase Synonym of phosphopyruvate hydratase.

Neuronal Pertaining to a nerve cell.

Neuronitis An inflammation of a nerve or nerve cell.

Neurontin A trade name for gabapentin, an antiepileptic agent.

Neuropathy Any abnormal condition marked by swelling and wasting of the nerve.

Neuropeptide Peptide secreted by neurons as a signal molecule, e.g., peptide neurotransmitter and peptide neuromodulator.

Neurophysin The carrier protein that transports neurohypophysial hormone along axons from hypothalamus to the posterior lobe of the pituitary.

Neuroplasm The protoplasm of a nerve cell.

Neuroretinitis Inflammation of the optic nerve and retina.

Neurosecretory Cells Cells that are capable of carrying impulses and secreting hormones.

Neurosecretory Granule Granule found in the posterior lobe of the pituitary gland that contains hormones, oxytocin, vasopressin, and the neurophysin.

Neurosome Mitochondrion of a nerve cell.

Neurospora A genus of fungi, e.g., *Neurospora crassa* (Ascomycetes) that is a classical organism for genetic research.

Neurotensin A peptide hormone found in the mammalian brain and gut that possesses neuroendocrine activity.

Neurotoxin A toxin capable of destroying nerve tissue or interfering with neural transmission.

Neurotransmitter The chemical messenger released from the presynaptic nerve cell at a chemical synapse to relay signal to the postsynaptic cell, e.g., acetylcholine, GABA, noradrenaline, serotonin, and dopamine.

Neurotropic Having an affinity for neural tissue or capability of growing toward the neural tissue, e.g., neurotropic rabies virus.

Neurotubules Microtubules in the neuron.

Neurovesicles Microscopic sacs in the axon terminals that contain neurotransmitter.

Neut A trade name for sodium bicarbonate, an electrolyte, urinary alkalinizer, and an antacid.

Neutral 1. Neither acidic or basic (pH 7). 2. Neither positively or negatively charged. 3. An atom with an equal number of protons and electrons.

Neutral Alleles Alleles that encode functionally identical products.

Neutral Amino Acid Amino acids with one NH_2 group and one COOH group.

Neutral Fat An ester of glycerol and fatty acid (see triacylglycerol or triglyceride).

Neutral Glycolipid A glycolipid whose polar moiety consists of only neutral sugars.

Neutralization 1. The formation of water between an acid and a base. 2. Inactivation of the infectivity of a virus or microorganism by a homologous antibody.

Neutralization Test An antigen-antibody test used for identification of antigen or microorganism.

Neutral Mutation 1. A mutation that results in no measurable phenotypic change or effect. 2. A mutation that does not affect the fitness of an organism in a particular environment.

Neutral Protease Protease with optimal activity at neutral pH.

Neutral Red (mol wt 289) A biological dye and a pH indicator.

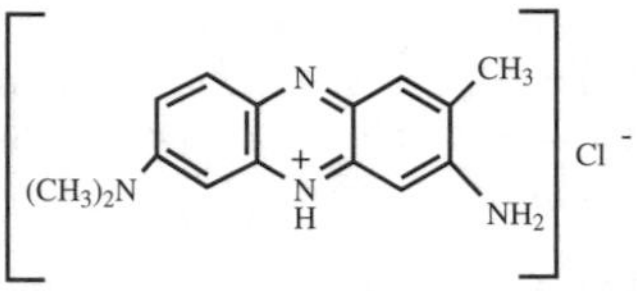

Neutral Sugar Sugars without a charged group.

Neutra-phos A trade name for a combination drug containing phosphorus and sodium, used as an electrolyte replacement solution.

Neutrexin A trade name for trimetrexate glucuronate, a synthetic inhibitor of dihydrofolate reductase used as an antineoplastic agent.

Neutrino An uncharged subatomic particle emitted from a radioactive nucleus with zero mass.

Neutrocyte Synonym for neutrophil.

Neutrocytopenia See neutropenia.

Neutron An electrically neutral subatomic particle with approximately the same mass as a proton.

Neutropen A preparation that contains penicillinase used to destroy penicillin for treatment of penicillin allergy.

Neutropenia The presence of an abnormally low number of neutrophils in the blood.

Neutrophil A large granular leukocyte with lobed nucleus and cytoplasmic granules. Neutrophils are phagocytic cells and can be stained with neutral dyes such as eosin (see also polymorphonuclear neutrophil).

Neutrophil Microbicidal Assay A test for the ability of neutrophils to kill intracellular bacteria.

Neutrophilia An increase in number of neutrophils in the blood.

Neutrophilin A neutrophil-derived platelet activator.

Nevirapine (mol wt 266) An antiviral agent.

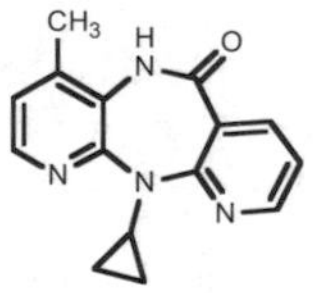

Nevskia A genus of aquatic bacteria.

Nexin Protein that connects and maintains the spatial relationship of adjacent microtubule doublets in the axoneme of flagella.

NEY Agar Abbreviation for neomycin egg yolk agar.

NF Abbreviation for 1. nephritic factor; 2. nuclear factor.

NF-1 Abbreviation for nuclear factor 1.

NFAT Abbreviation for nuclear factor of activated T-cell.

NF-*k*B Abbreviation for nuclear factor *k*B.

NflI (MboI) A restriction endonuclease from *Neisseria flavescens* with the same specificity as MboI.

NflAI (EcoRV) A restriction endonuclease from *Neisseria flavescens* with the same specificity as EcoRV.

NflAII (MboI) A restriction endonuclease from *Neisseria flavescens* with the same specificity as MboI.

NflBI (MboI) A restriction endonuclease from *Neisseria flavescens* with the same specificity as MboI.

***n*-Fold Helix** A helical molecule that has n residues per helical turn.

NF-Y Abbreviation for nuclear factor Y.

ng Abbreviation for nanogram (10^{-9} gram).

NgbI (PstI) A restriction endonuclease from *Nocardia globerula* with the same specificity as PstI.

NGD⁺ Abbreviation for nicotinamide guanine dinucleotide (oxidized form).

NGF Abbreviation for nerve growth factor.

NgoI (HaeII) A restriction endonuclease from *Neisseria gonorrhoea* with the following specificity:

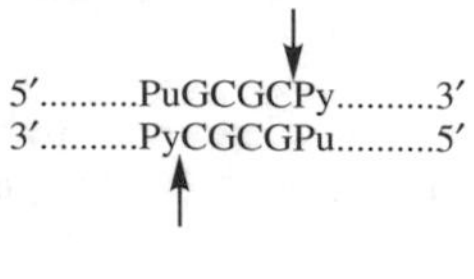

NgoII (HaeIII) A restriction endonuclease from *Neisseria gonorrhoea* with the following specificity:

5′..........GG↓CC.........3′
3′..........CC↑GG.........5′

NgoIII (SaeII) A restriction endonuclease from *Neisseria gonorrhoea* KH7764-45 with the following specificity:

5′..........CCGC↓GG..........3′
3′..........GG↑CGCC..........5′

NgoAIV A restriction endonuclease from *Neisseria gonorrhea* with the following specificity:

5'........G↓CCGGC........3'
3'........CGGCC↑G........5'

NgoDI (SacII) A restriction endonuclease from *Neisseria gonorrhoea* JKD211 with same specificity as Sac II.

NgoDIII (DpnI) A restriction endonuclease from *Neisseria gonorrhoea* JKD211 with the following specificity:

5′..........GA↓TC..........3′
3′..........CT↑AG..........5′

NgoMI (NaeI) A restriction endonuclease from *Neisseria gonorrhoea* MS11 with the following specificity:

5′..........GCC↓GGC..........3′
3′..........CGG↑CCG..........5′

NgoPII (HaeIII) A restriction endonuclease from *Neisseria gonorrhoea* P9-2 with the following specificity:

5′..........GG↓CC..........3′
3′..........CC↑GG..........5′

NgoPIII (SaeII) A restriction endonuclease from *Neisseria gonorrhoea* P9-2 with the following specificity:

5′..........CCGC↓GG..........3′
3′..........GG↑CGCC..........5′

NGS Abbreviation for normal goat serum.

NHCP Abbreviation for non-histone chromosomal protein.

NheI A restriction endonuclease from *Neisseria mucosa* with the following specificity:

5′..........G↓CTAGC..........3′
3′..........CGATC↑G..........5′

NHS Abbreviation for 1. normal horse serum; 2. normal human serum.

NHTBE Cell Abbreviation for normal human tracheobronchial epithelial cell.

NH_2-Terminal Abbreviation for the amino terminal of protein.

Ni Abbreviation for nickel, a chemical element.

Niac A trade name for niacin (vitamin B_3).

Niacin Vitamin B_3 (see nicotinic acid).

Niacinamide (mol wt 122) A vitamin and an enzyme cofactor that is essential for lipid metabolism, glycogenolysis, and tissue respiration.

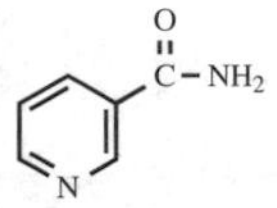

Niacor A trade name for niacin (see nicotinic acid).

Nialamide (mol wt 298) An antidepressant.

Niaprazine (mol wt 356) A sedative and hypnotic agent.

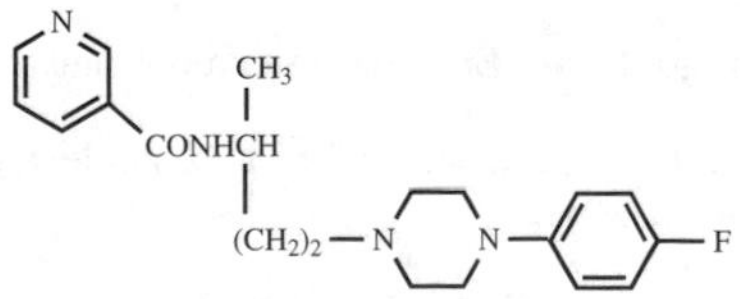

Niaspan A trade name for niacin, vitamin B_3.

Nicametate (mol wt 222) A vasodilator.

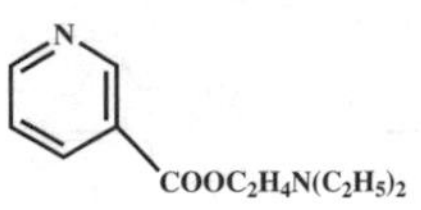

Nicardipine (mol wt 480) An antihypertensive and antianginal agent that inhibits calcium ion influx across cardiac and smooth muscle cells, thus decreasing myocardial contractility and oxygen demand.

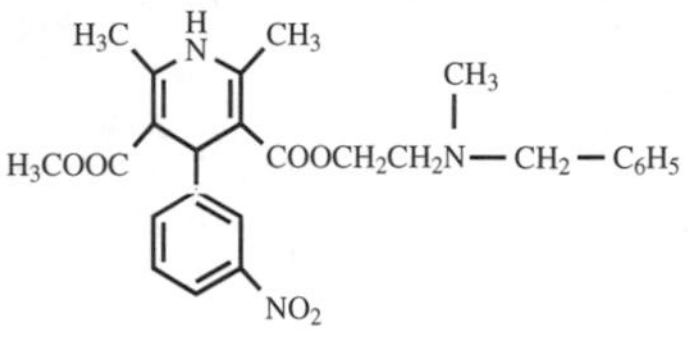

Nicergoline (mol wt 484) A vasodilator.

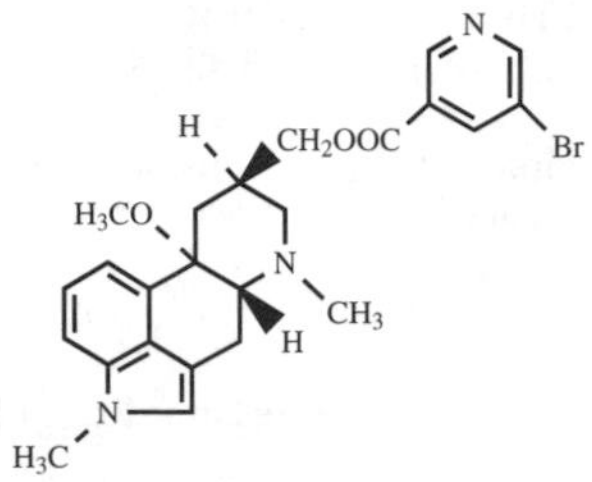

Niceritrol (mol wt 557) An antihyperlipoproteinemic agent.

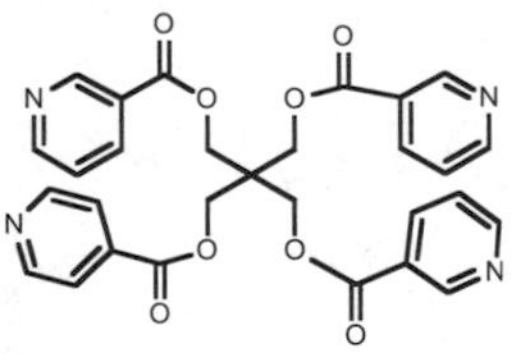

Nick A break in a strand of DNA or RNA caused by an enzyme or radiation (e.g., topoisomerase or UV radiation).

Nickase The enzyme that catalyzes the introduction of nicks into DNA or RNA.

Nickel (Ni) A chemical element with atomic weight 59, valence 2.

Nicking-Closing Enzyme See topoisomerase I.

Niclocide A trade name for niclosamide, an anthelmintic agent.

Niclosamide (mol wt 327) An anthelmintic agent that inhibits the metabolic process of oxidative phosphorylation in tapeworms.

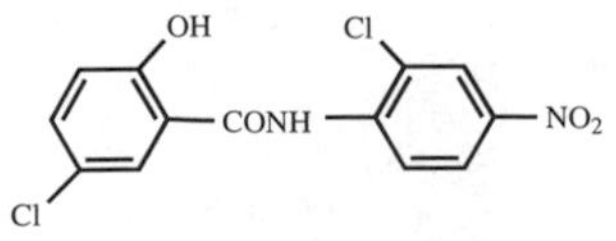

Nicoclonate (mol wt 290) An antilipemic agent.

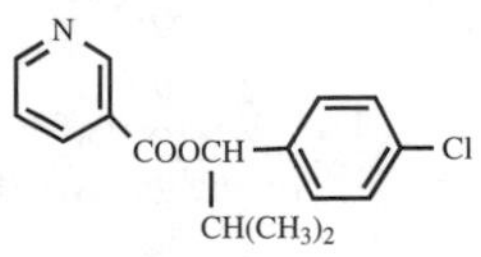

Nicoderm A trade name for a nicotine transdermal system, used for relief of nicotine withdrawal symptoms in patients undergoing smoking cessation. It provides nicotine and stimulates nicotinic acetylcholine receptors in the CNS, neuromuscular junction, autonomic ganglia, and adrenal medulla.

Nicofibrate (mol wt 306) An antihyperlipoproteinemic agent.

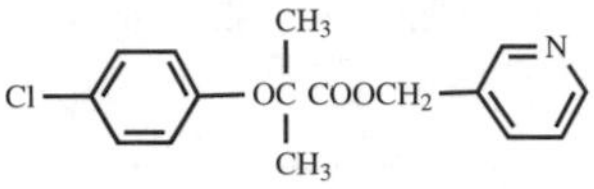

Nicofuranose (mol wt 601) A vasodilator.

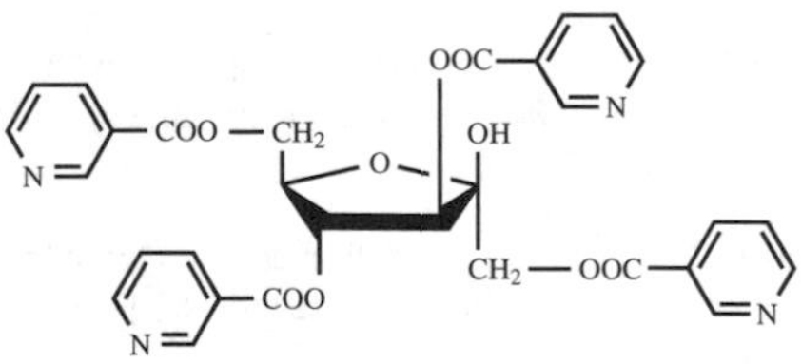

Nicomol (mol wt 641) An anticholesteremic agent.

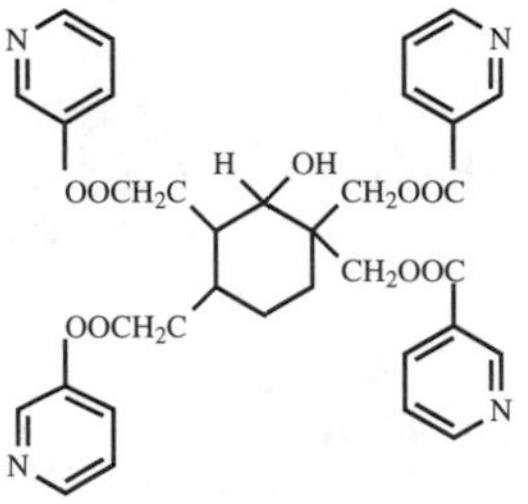

Nicomorphine (mol wt 496) A narcotic-analgesic agent.

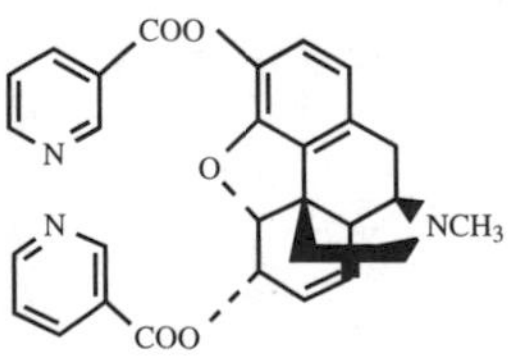

Nicorandil (mol wt 211) A coronary vasodilator.

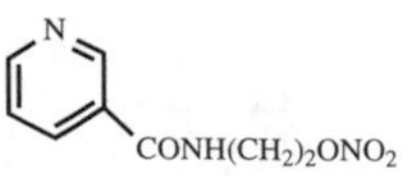

Nicorette A trade name for a nicotine-resin complex used for relief of nicotine withdrawal symptoms in patient undergoing smoking cessation. It provides nicotine and stimulates nicotinic acetylcholine receptors in the CNS, neuromuscular junction, autonomic ganglia, and adrenal medulla.

Nicotinamide (mol wt 122) An active component of NAD and a factor for lipid metabolism, tissue respiration, and glycogenolysis (see also niacinamid).

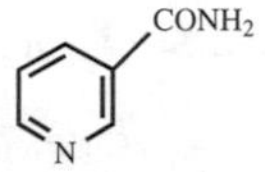

Nicotinamide Adenine Dinucleotide (mol wt 654) A compound found in all living cells that exists in two interconvertible forms (oxidized form NAD^+ and reduced form NADH) and serves as a coenzyme for a variety of biochemical reactions.

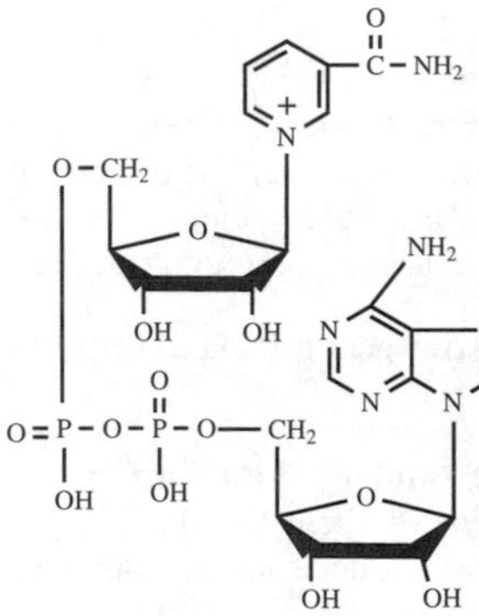

oxidized form

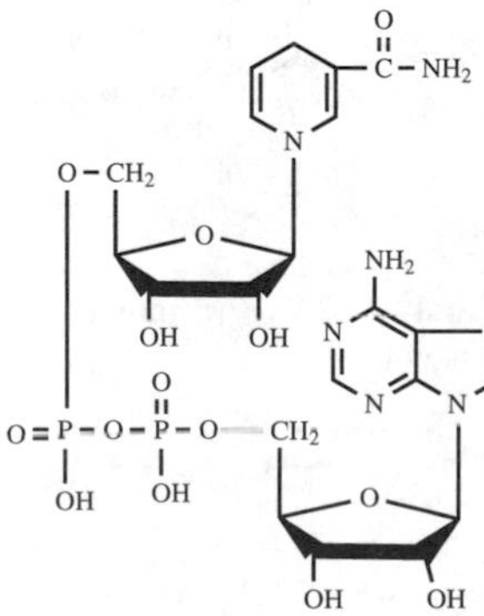

reduced form

Nicotinamide Adenine Dinucleotide Phosphate (mol wt 744) An electron carrier that exists in two forms (reduced form and oxidized form) and serves as coenzyme in many biochemical reactions.

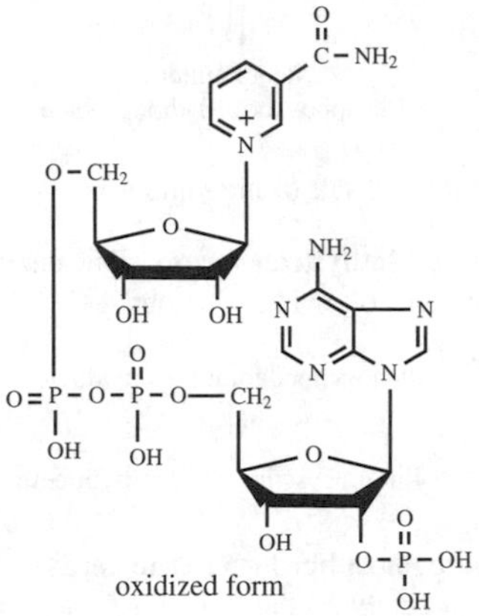

oxidized form

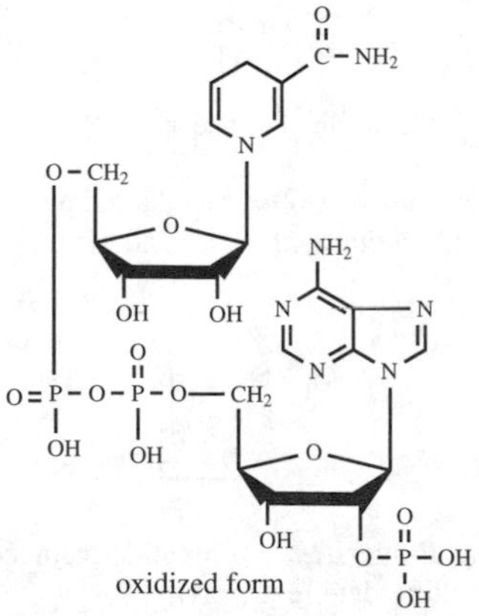

oxidized form

Nicotinamide Deaminase The enzyme that catalyzes the following reaction:

$$\text{Nicotinamide} + H_2O \rightleftharpoons \text{Nicotinate} + NH_3$$

Nicotinamide Mononucleotide (mol wt 334) A compound involved in the DNA ligase reaction.

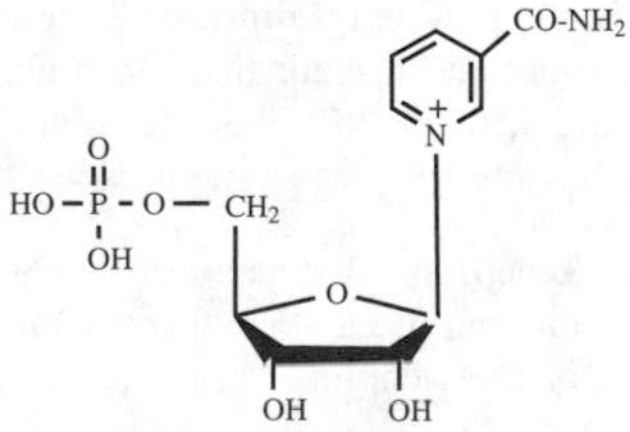

Nicotinamide Nucleotide Adenylyltransferase The enzyme that catalyzes the following reaction:

$$\text{ATP} + \text{nicotinamide ribonucleotide} \rightleftharpoons \text{Pyrophosphate} + NAD^+$$

Nicotinamide Phosphoribosyl Transferase The enzyme that catalyzes the following reaction:

Nicotinamide ribonucleotide + PPi

⇌

Nicotinamide +
5-Phospho-ribose 1-diphosphate

Nicotinate A salt of nicotinic acid.

Nicotinate Methyltransferase The enzyme that catalyzes the following reaction:

S-adenosylmethionine + nicotinate

⇌

S-adenosylhomocysteine + *N*-methylnicotinate

Nicotinate Phosphoribosyl Transferase The enzyme that catalyzes the following reaction:

Nicotinate + PRPP

⇌

Nicotinate ribonucleotide + PPi

Nicotine (mol wt 162) An alkaloid derived from tobacco (*Nicotiana tabacum)* and other species of *Nicotiana.*

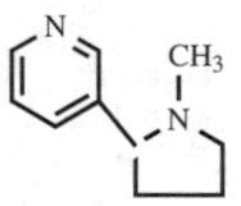

Nicotine Polacrilex A nicotine-resin complex with smoking deterrent activity.

Nicotinex A trade name for vitamin B_3 (niacin).

Nicotinic Acid (mol wt 123) An enzyme cofactor.

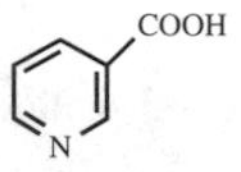

Nicotinic Acid Acetylcholine Receptor A plasma membrane receptor that binds acetylcholine causing diffusion of sodium ions into the cell and depolarizing the plasma membrane.

Nicotinic Receptor A synaptic acetylcholine receptor for binding of nicotine, thereby mimicking the action of acetylcholine.

Nicotinic Synapse A synapse that contains nicotinic receptor.

Nicotinyl Alcohol (mol wt 109) A vasodilator.

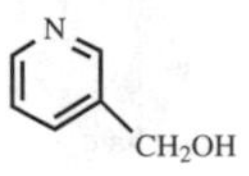

Nicotrol A trade name for nicotine, a smoking deterrent.

NidaGel A trade name for metronidazole, an antibacterial agent.

NIDD Abbreviation for non-insulin dependent diabetes.

NIDDM Abbreviation for non-insulin dependent diabetes mellitus.

Nidogen A sulfated, laminin-associated glycoprotein in the basement membrane.

Nidroxyzone (mol wt 242) An anti-infective agent.

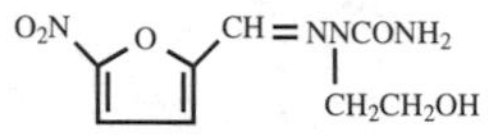

Nidulin (mol wt 444) An antibiotic.

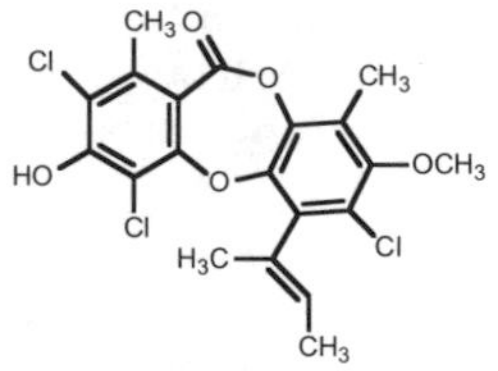

Niemann-Pick Disease A genetic disorder characterized by the accumulation of sphingomyelin in the tissue due to deficiency in the enzyme sphingomyelinase leading to mental retardation.

***nif* Gene** Any gene associated with nitrogen fixation.

Nifedipine (mol wt 346) An antianginal and antihypertensive agent that inhibits calcium ion influx across cardiac and smooth muscle cells, decreases myocardial contractility and oxygen demand, and dilates coronary arteries and arterioles.

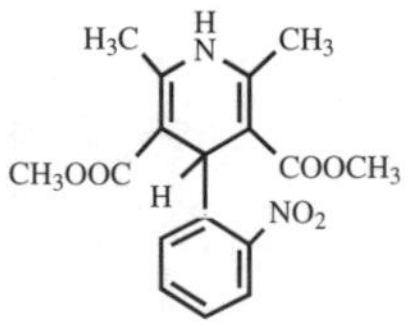

Nifenalol (mol wt 224) An antianginal and antiarrhythmic agent.

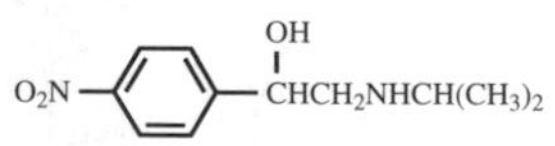

Nifenazone (mol wt 308) An analgesic and antipyretic agent.

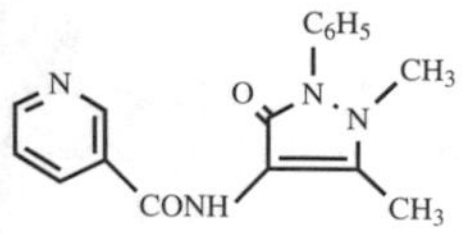

Niflumic Acid (mol wt 282) An anti-inflammatory agent.

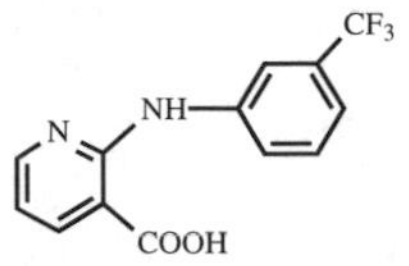

Nifuradene (mol wt 224) An antibacterial agent.

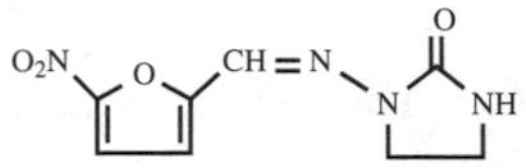

Nifuraldezone (mol wt 226) An antibacterial agent.

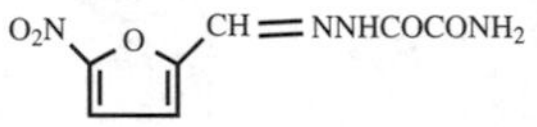

Nifuratel (mol wt 285) An antibacterial and an antifungal agent.

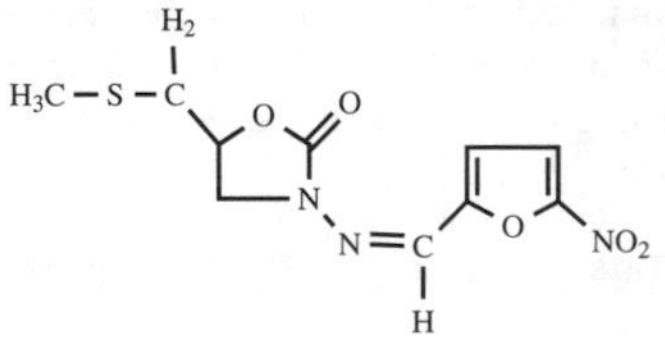

Nifurfoline (mol wt 337) An antibacterial agent.

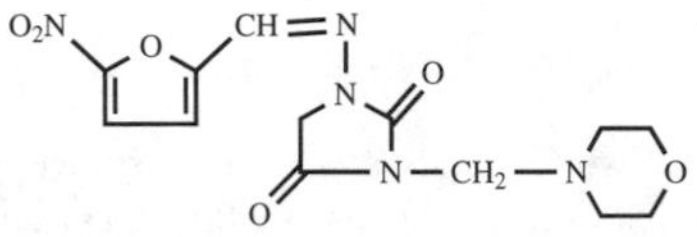

Nifuroquine (mol wt 300) An antibacterial agent.

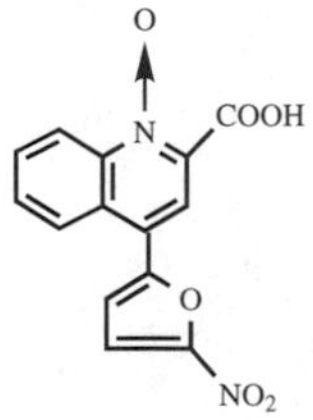

Nifuroxazide (mol wt 275) An intestinal antiseptic agent.

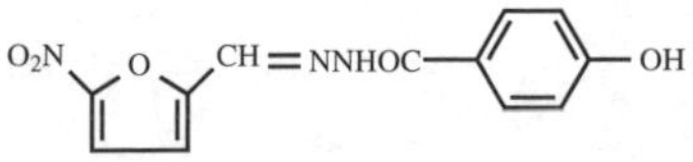

Nifuroxime (mol wt 156) A topical anti-infective and antiprotozoal agent.

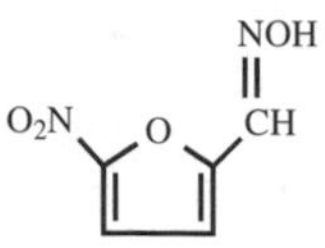

Nifurprazine (mol wt 232) A topical antibacterial agent.

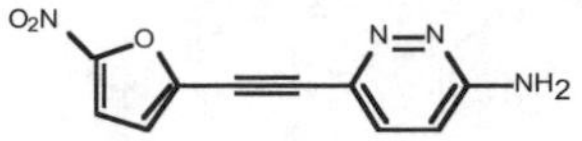

Nifurtimox (mol wt 287) An antiprotozoal agent (*Trypanosoma*).

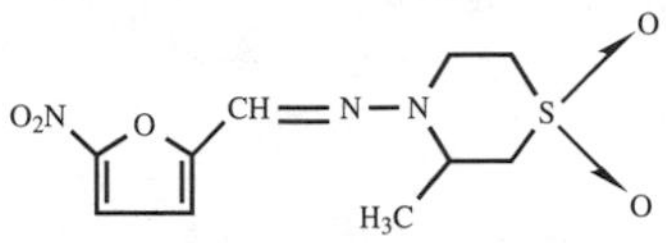

Nifurtoinol (mol wt 268) An antibacterial agent.

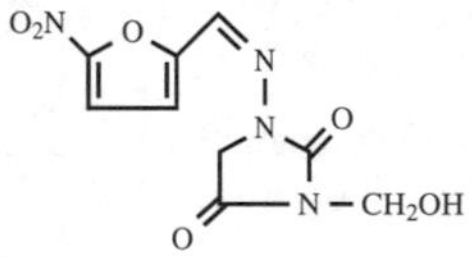

Nifurzide (mol wt 336) An anti-infective agent.

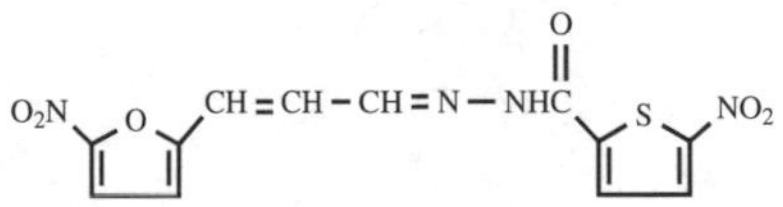

Nigeran A linear α-D-glucan with alternating α-1, 3- and α-1, 4-glycosidic linkages.

Nigericin (mol wt 724) A polyester antibiotic and an ionophore that exchanges K^+ for H^+ across the membrane.

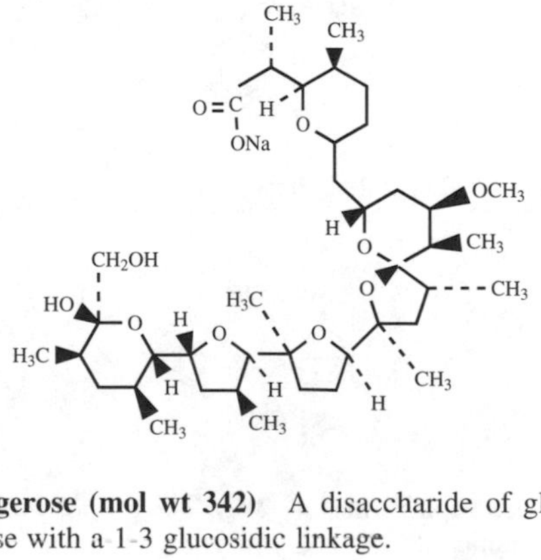

Nigerose (mol wt 342) A disaccharide of glucose with a 1-3 glucosidic linkage.

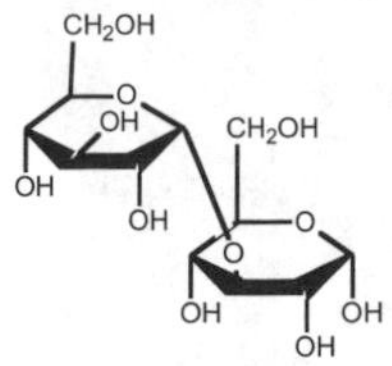

Nihydrazone (mol wt 197) An antibacterial and antiprotozoal agent.

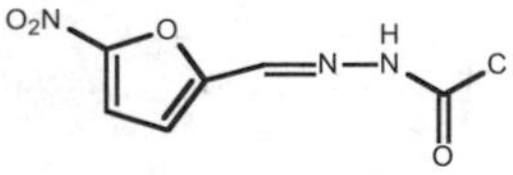

Ni-IDA Abbreviation for Ni^{2+}-imidodiacetic acid.

Nikethamide (mol wt 178) A respiratory stimulant.

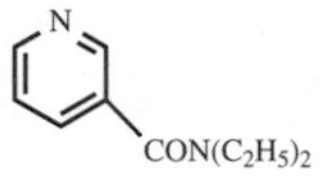

Nilandron A trade name for nilutamide, an antiandrogen.

Nilstat A trade name for nystatin, an antifungal agent that alters the permeability of fungal cells.

Nilutamide (mol wt 317) An antiandrogen.

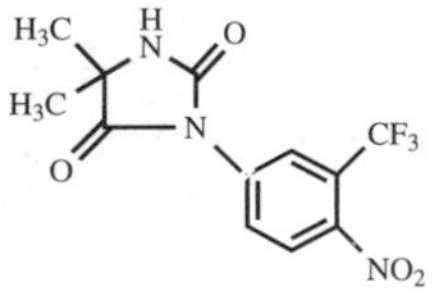

Nilvadipine (mol wt 385) An antihypertensive and antianginal agent.

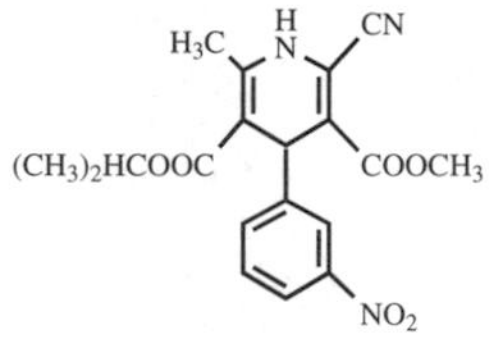

Nimesulide (mol wt 308) An anti-inflammatory agent.

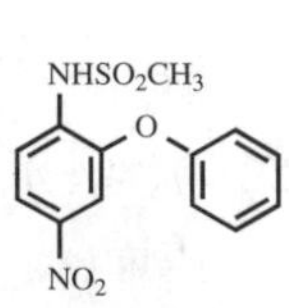

Nimetazepam (mol wt 295) An anticonvulsant and a muscle relaxant.

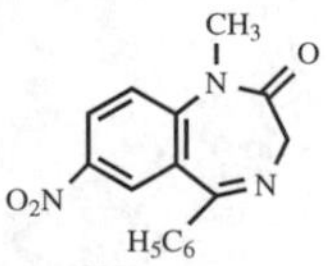

Nimodipine (mol wt 418) A vasodilator that inhibits calcium ion influx across cardiac and smooth muscle cells, thus decreasing myocardial contractility and oxygen demand and dilating coronary and cerebral arteries and arterioles.

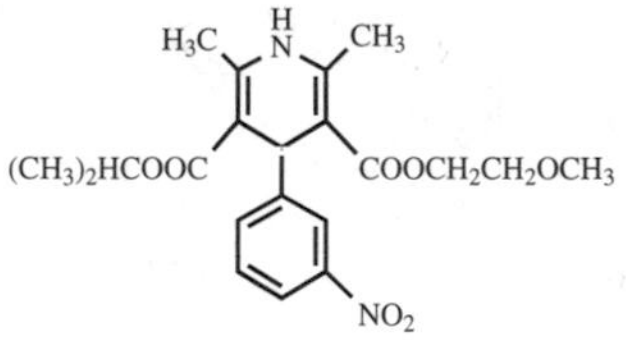

Nimorazole (mol wt 226) An antiprotozoal agent for *Trichomonas*.

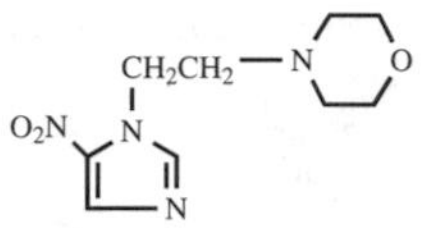

Nimotop A trade name for nimodipine, used as a vasodilator that inhibits calcium ion influx across the cardiac and smooth muscle cells, thus decreasing myocardial contractility and oxygen demand and dilating coronary and cerebral arteries and arterioles.

Nimustine (mol wt 273) An antineoplastic agent.

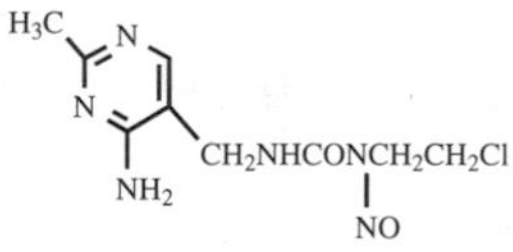

Nine + Two (9 + 2) System The arrangement of microtubules in the eukaryotic flagella or cilia that consists of nine peripheral pairs of microtubules surrounding two single central microtubules.

Ninhydrin (mol wt 178) A reagent used for colorimetric determination of amino acids.

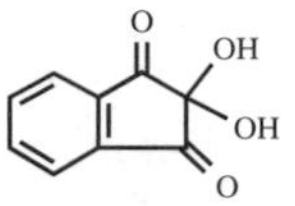

Ninhydrin Reaction A colorimetric reaction for detection and determination of amino acids.

Ninopterin (mol wt 455) An antineoplastic agent.

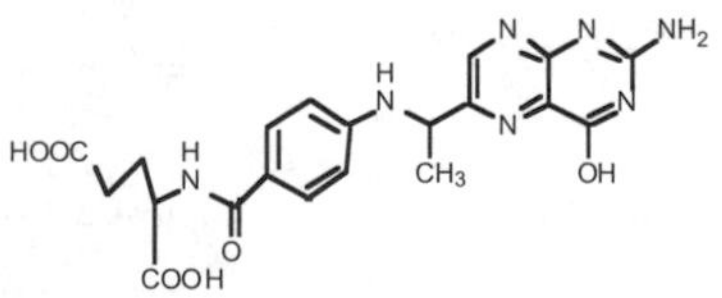

Ni-NTA-Agarose Abbreviation for Ni-nitrilotriacetic acid agarose.

Niobium A chemical element with atomic 93, valences 2, 3, 4.

Niong A trade name for nitroglycerin, used to reduce cardiac oxygen demand and increase blood flow.

Nipent A trade name for pentostatin, used as an antineoplastic agent that inhibits adenosine deaminase.

Nipradilol (mol wt 326) An antianginal and antihypertensive agent.

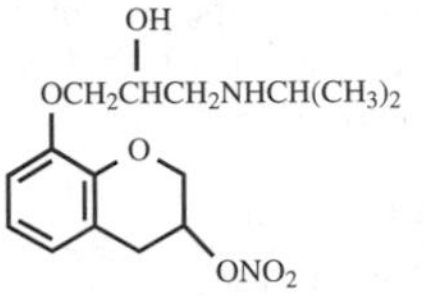

Nipride A trade name for nitroprusside sodium, used as an antihypertensive agent that relaxes both arteriolar and venous smooth muscle.

NIR Abbreviation for nitrite reductase.

Niridazole (mol wt 214) An anthelmintic agent.

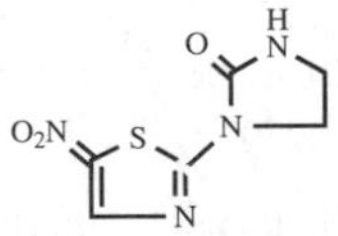

Niscort A trade name for prednisolone acetate, used as an anti-inflammatory agent.

Nisin A polypeptide antibiotic produced by *Streptomyces lactis* that inhibits peptidoglycan synthesis.

NiSOD Abbreviation for nickel-containing superoxide dismutase.

Nisoldipine (mol wt 388) An antihypertensive and antianginal agent.

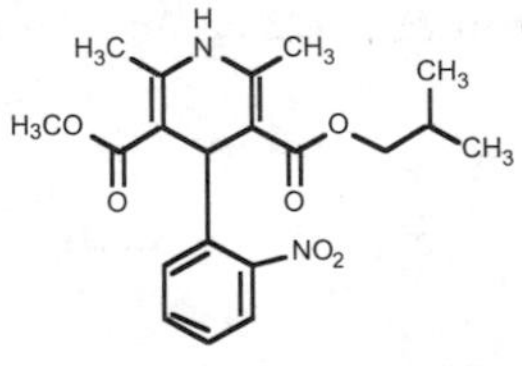

Ni-Span A trade name for vitamin B_3 (niacin).

Nithiazide (mol wt 216) An antiprotozoal agent.

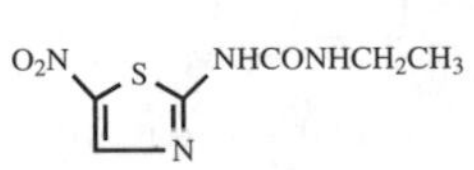

Nitracrine (mol wt 324) A derivative of acridine and an antineoplastic agent.

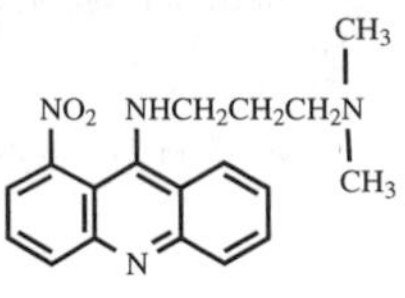

Nitradisc A trade name for nitroglycerin, which reduces cardiac oxygen demand and increases blood flow.

Nitratase See nitrate reductase.

Nitrate A salt of nitric acid and the major source of nitrogen for higher plants.

Nitrate Reductase (NAD$^+$) The enzyme that catalyzes the following reaction:

$$\text{NADH} + \text{nitrate} \rightleftharpoons \text{NAD}^+ + \text{nitrite} + H_2O$$

Nitrate Reductase (NADP$^+$) The enzyme that catalyzes the following reaction:

$$\text{NADPH} + \text{nitrate} \rightleftharpoons \text{Nitrite} + \text{NADP}^+ + H_2O$$

Nitrate Reduction The reduction of nitrate to nitrite or ammonia.

Nitrate Respiration An aerobic respiration for reduction of nitrate to nitrite catalyzed by the enzyme nitrate reductase.

Nitrazepam (mol wt 281) An anticonvulsant and a hypnotic agent.

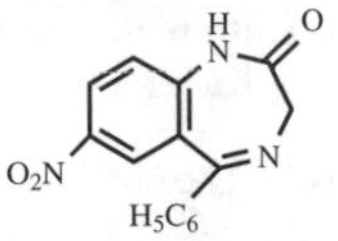

Nitrazine Yellow (mol wt 542) A dye and a pH indicator.

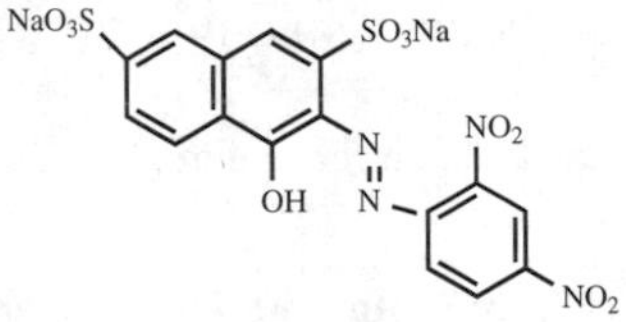

Nitrefazole (mol wt 248) An alcohol deterrent.

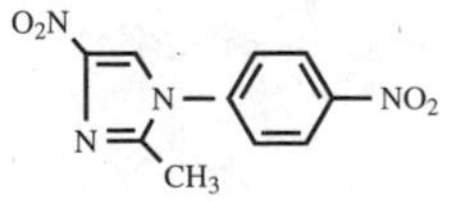

Nitrendipine (mol wt 360) An antihypertensive agent.

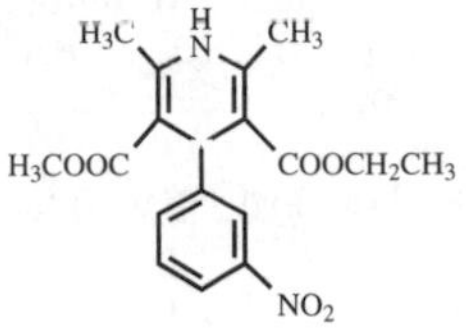

Nitric Acid (mol wt 63) An inorganic acid.

$$HNO_3$$

Nitric Oxide Synthetase The enzyme that catalyzes the following reaction:

$$\text{L-Arginine} + n\text{NADH} + m\text{O}_2 \rightleftharpoons \text{Citruline} + \text{nitric oxide} + n\text{NADP}^+$$

Nitrification The process in which ammonia is oxidized to nitrite and to nitrate by aerobic, chemolithotrophic bacteria.

Nitrifying Bacteria Bacteria that oxidize ammonia to nitrite and to nitrate, e.g., Gram-negative bacteria in the family of Nitrobacteraceae.

Nitrilase The enzyme that catalyzes the following reaction:

$$\text{Nitrile} + H_2O \rightleftharpoons \text{Carboxylate} + NH_3$$

Nitrile An organic cyanide containing the CN group that yields carboxylate and ammonia upon hydrolysis.

Nitrite A salt of nitrous acid used as a food preservative.

Nitrite Ammonification Reduction of nitrite to ammonium ions by bacteria.

Nitrite Reductase The enzyme that catalyzes the following reaction:

$$3\ \text{NADPH} + \text{nitrite} \rightleftharpoons 3\ \text{NADP}^+ + NH_4OH + H_2O$$

Nitro Abbreviation for 1. nitroglycerin; 2. NO_2 group.

Nitroakridin 3582 (mol wt 428) An antiseptic agent.

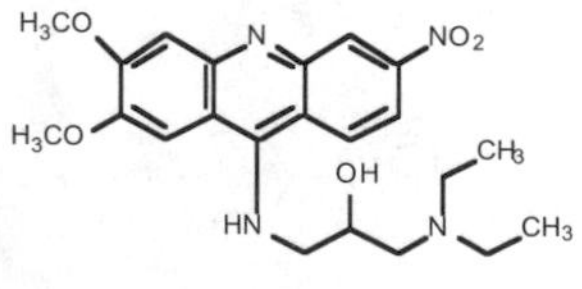

Nitrobacter A genus of Gram-negative bacteria of the family Nitrobacteraceae.

Nitro-Bid A trade name for nitroglycerin, used for reduction of cardiac oxygen demand and increase of blood flow.

4-Nitrobiphenyl (mol wt 199) A human carcinogen.

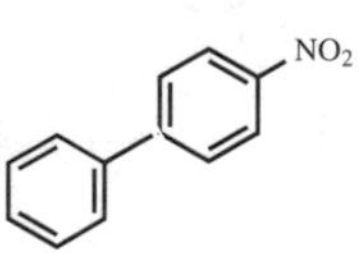

Nitrocap A trade name for nitroglycerin, used for reduction of cardiac oxygen demand and increase of blood flow.

Nitrocellulose Paper Paper or membrane with a high absorbing power for biological molecules (e.g., DNA, RNA, protein), used in blotting experiments (e.g., Southern, northern blotting).

Nitrocine A trade name for nitroglycerin, used for reduction of cardiac oxygen demand and increase of blood flow.

Nitrodan (mol wt 296) An anthelmintic agent.

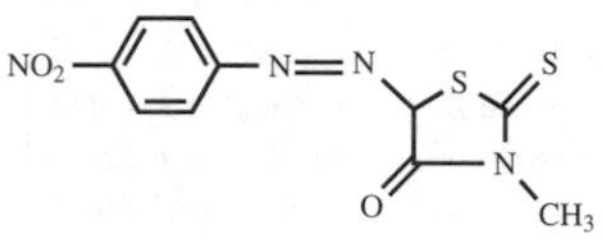

Nitro-Derm A trade name for nitroglycerin, an antianginal agent.

Nitrodisc A trade name for nitroglycerin, used for reduction of cardiac oxygen demand and increase of blood flow.

Nitro-Dur A trade name for nitroglycerin, used for reduction of cardiac oxygen demand and increase of blood flow.

Nitrofan A trade name for nitrofurantoin microcrystals, an antibacterial agent.

Nitrofen (mol wt 284) An herbicide.

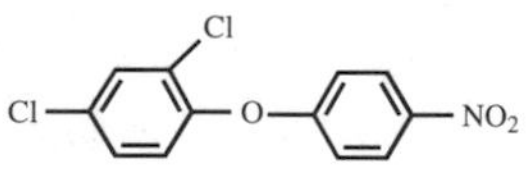

Nitrofurantoin (mol wt 238) An antibacterial agent.

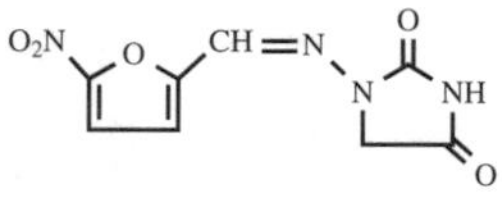

Nitrofurazone (mol wt 198) A topical anti-infective agent.

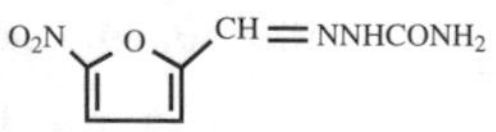

Nitrogard A trade name for nitroglycerin, used for reduction of cardiac oxygen demand and increase of blood flow.

Nitrogen (N) A chemical element with atomic weight 14, valence 3 and 5.

Nitrogen-13 An artificial radioactive nitrogen nuclide (^{13}N).

Nitrogen-15 A heavy nitrogen, the stable naturally occurring nitrogen nuclide (^{15}N).

Nitrogen Cycle The cyclic interconversion of organic nitrogen and inorganic compounds in the soil by both living organisms and nonbiological processes.

Nitrogen Fixation Conversion of atmospheric N_2 to NH_3 by nitrogen fixation bacteria (e.g., *Clostridium, Klebsiella, Rhizobium*) catalyzed by nitrogenase.

Nitrogen Fixation Bacteria Bacteria (*Clostritridium, Klebsiella, Rhizobium*) that are capable of transforming atmospheric nitrogen into NH_3 and organic nitrogen compounds.

Nitrogen Mustard (mol wt 157) An alkylating agent that causes mutation and depurination in DNA.

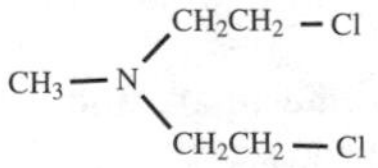

Nitrogenase The enzyme that catalyzes the reduction of atmospheric N_2 to NH_3. It is a FeMo-protein complex.

Nitrogenous Relating to or containing nitrogen.

Nitrogenous Bases Referring to purine and pyrimidine bases in DNA and RNA.

Nitroglycerin (mol wt 227) An antianginal agent used for reduction of cardiac oxygen demand and increase of blood flow.

CH_2 — ONO_2
|
CH — ONO_2
|
CH_2 — ONO_2

Nitroglyn A trade name for nitroglycerin, an antianginal agent.

Nitrol A trade name for nitroglycerin, used for reduction of cardiac oxygen demand and increase of blood flow.

Nitrolate Ointment A trade name for nitroglycerin, which reduces the cardiac oxygen demand and increases blood flow.

Nitrolin A trade name for nitroglycerin, used for reduction of cardiac oxygen demand and increase of blood flow.

Nitrolingual A trade name for nitroglycerin, used for reduction of cardiac oxygen demand and increase of blood flow.

Nitromersol (mol wt 352) A topical disinfectant.

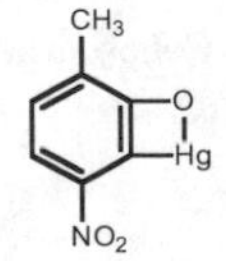

Nitromide (mol wt 211) An antibacterial agent.

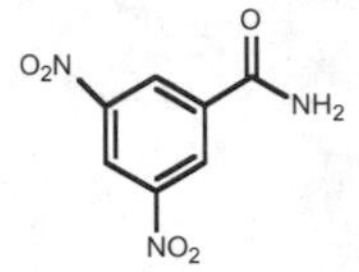

Nitronet A trade name for nitroglycerin, used for reduction of cardiac oxygen demand and increase of blood flow.

Nitrong A trade name for nitroglycerin, used for reduction of cardiac oxygen demand and increase of blood flow.

2-Nitrophenol (mol wt 139) Substrate used for assaying β-galactosidase.

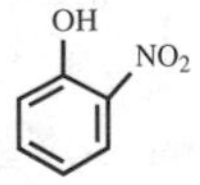

2-Nitrophenol 2-Monooxygenase The enzyme that catalyzes the following reaction:

$$\text{2-Nitrophenol} + \text{NADPH} + O_2 \rightleftharpoons \text{Catechol} + \text{nitrite} + \text{NADP}^+ + H_2O$$

4-Nitrophenol (mol wt 139) A substance used for assaying phosphatase.

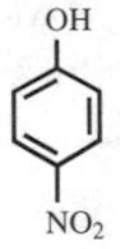

4-Nitrophenol 2-Monooxygenase The enzyme that catalyzes the following reaction:

$$\text{2-Nitrophenol} + \text{NADH} + O_2 \rightleftharpoons \text{Nitrocatechol} + \text{NAD}^+ + H_2O$$

4-Nitrophenyl α-L-Arabinofuranoside See 4-phenyl α-L-arabinoside.

4-Nitrophenyl α-L-Arabinoside (mol wt 271) A chromogenic substrate for α-L-arabinosidase.

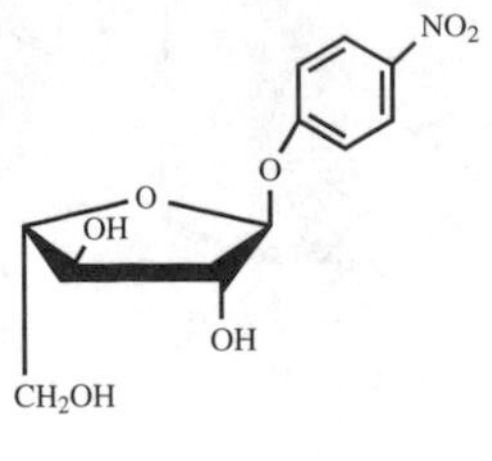

4-Nitrophenyl β-D-Cellobioside (mol wt 463) A substrate for assaying exoglucanase activity.

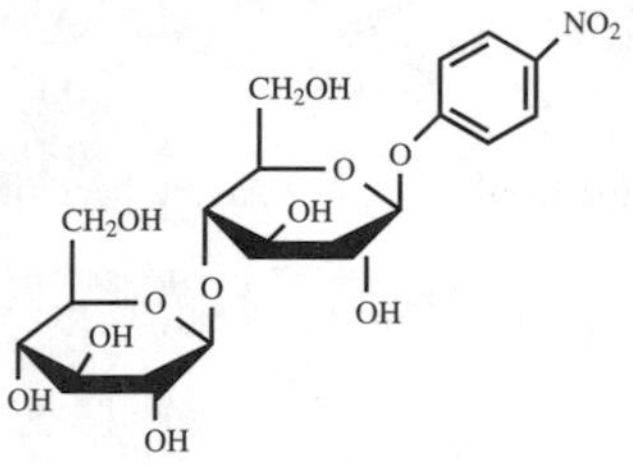

4-Nitrophenyl β-D-Galactopyranoside See 4-nitrophenyl β-D-galactoside.

4-Nitrophenyl β-D-Galactoside (mol wt 301) A substrate for β-D-galactosidase.

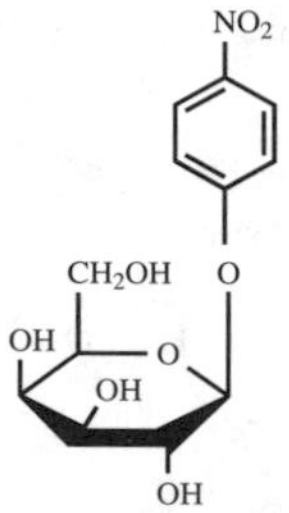

2-Nitrophenyl β-D-Glucopyranoside See 2-nitropheyl β-D-glucoside.

2-Nitrophenyl β-D-Glucoside (mol wt 301) A substrate for assaying β-glucosidase.

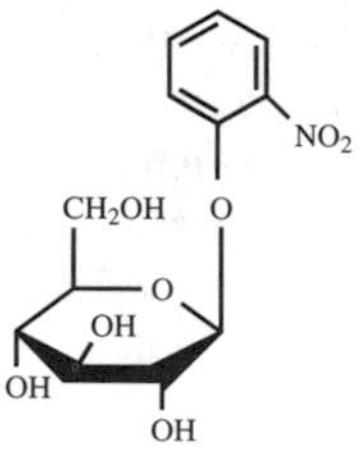

4-Nitrophenyl β-D-Glucopyranoside See 4-nitrophenyl β-D-glucoside.

4-Nitrophenyl β-D-Gluconide (mol wt 315) A substrate for β-gluconidase.

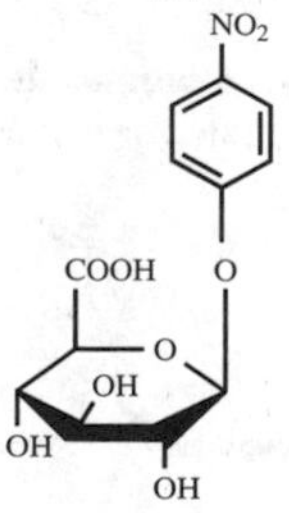

4-Nitrophenyl β-D-Glucoside (mol wt 301) A substrate for β-glucosidase.

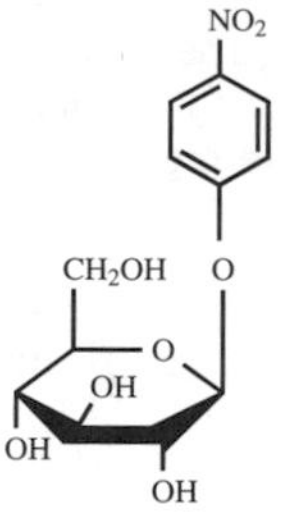

4-Nitrophenyl Hydrazine (mol wt 153) A reagent used for assaying aldehydes and ketones.

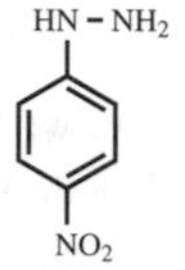

4-Nitrophenyl Phosphate (mol wt 219) A substrate for phosphatase.

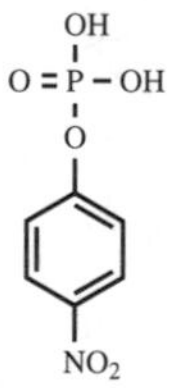

2-Nitrophenyl β-D-Thiogalactoside (mol wt 317) A galactoside.

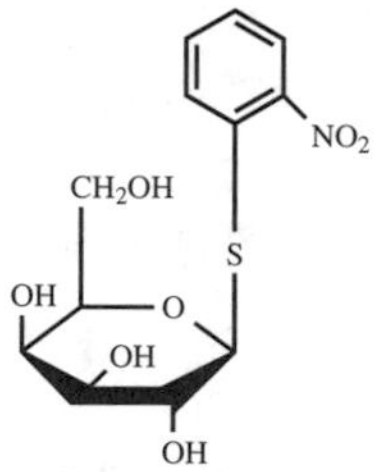

Nitropress A trade name for nitroprusside sodium, which relaxes both arteriolor and venous smooth muscle.

5'-Nitro-2'-Propoxyacetanilide (mol wt 238) An antipyretic and analgesic agent.

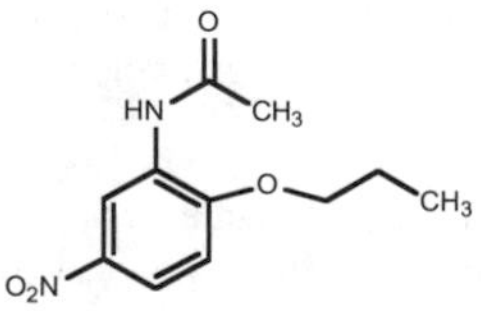

Nitroprusside Test A colorimetric test for determination of cysteine or sulfhydryl groups in proteins based on the production of red color following treatment of sample with sodium nitroprusside and ammonia.

Nitro-Red (mol wt 512) A dye.

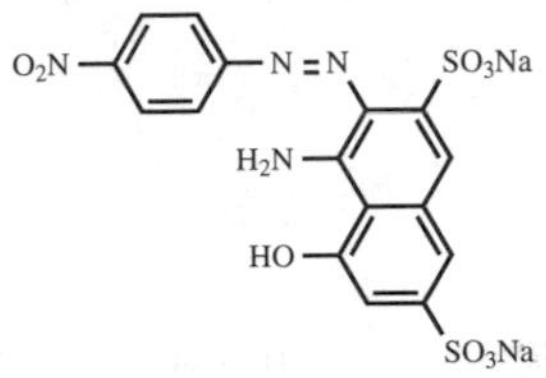

Nitrosamine A group of mutagenic molecules that contain nitroso groups causing mutation in dsDNA by transition of A:T to G:C or G:C to A:T in the dsDNA.

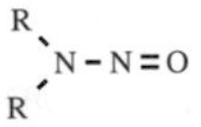

nitroso group

Nitrosation The introduction of a nitroso group into a molecule.

Nitroso Compounds Compounds that contain a nitroso group.

Nitroso Group The group of - N = O.

Nitrosococcus A genus of Gram-negative bacteria (family Bacteraceae).

Nitrosolobus A genus of Gram-negative bacteria (family Nitrobacteraceae).

Nitrosomonas A genus of Gram-negative bacteria (family Nitrobacteraceae).

Nitrosospira A genus of Gram-negative bacteria (family Nitrobacteraceae).

Nitrosovibrio A genus of Gram-negative bacteria (family Nitrobacteraceae).

Nitrospan A trade name for nitroglycerin, used for reduction of cardiac oxygen demand and increase of blood flow.

Nitrospina A genus of Gram-negative bacteria (family Nitrobacteraceae).

Nitrostat A trade name for nitroglycerin, used for reduction of cardiac oxygen demand and increase of blood flow.

Nitro-Time A trade name for nitroglycerin, an antianginal agent.

Nitrotym-plus A trade name for a combination drug containing nitroglycerin and butabarbital sodium, used as an antianginal agent and a calcium blocker.

Nitrous Acid (mol wt 47) A reagent that causes deamination of cytosine to uracil or adenine to hypoxanthine leading to mutations.

$$HNO_2$$

Nitrovin (mol wt 360) An antibacterial agent.

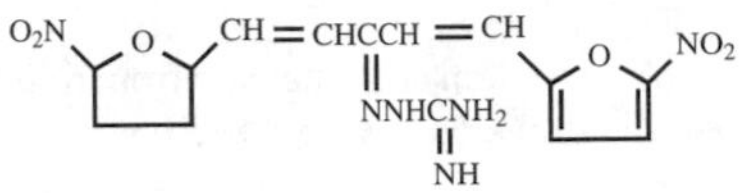

Nitroxoline (mol wt 190) An antibacterial agent.

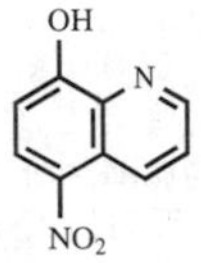

Nitroxynil (mol wt 290) An anthelmintic agent.

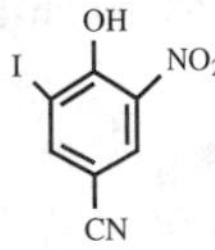

Nivaquine A trade name for chloroquine sulfate, used as an anti-malarial agent.

Nix A trade name for permethrin, used as a pesticide.

Nizoral A trade name for ketoconazole, used as an antifungal agent that inhibits synthesis of DNA, RNA, and protein.

NK Cells Abbreviation for natural killer cell, non-T, non-B lymphocytes capable of killing some tumor cells and some virus-infected cells.

NKA Abbreviation for neurokinin A.

NKCSF Abbreviation for natural killer cell stimulating factor.

NKHS Abbreviation for nonketotic hyperosmotic syndrome.

NKSF Abbreviation for natural killer stimulating factor.

NlaI (HaeIII) A restriction endonuclease from *Neisseria lactamica* with the same specificity as HaeIII.

NlaII (MboI) A restriction endonuclease from *Neisseria lactamica* with the following specificity:

5′..........↓GATC..........3′
3′..........CTAG↑..........5′

NlaIII A restriction endonuclease from *Neisseria lactamica* with the following specificity:

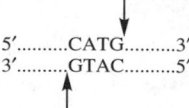

5′..........CATG..........3′
3′..........GTAC..........5′

NlaIV A restriction endonuclease from *Neisseria lactamica* with the following specificity:

5′..........GGNNCC..........3′
3′..........CCNNGG..........5′

NlaDI (MboI) A restriction endonuclease from *Neisseria lactamica* 5841 with the same specificity as MboI.

NlaDII (AsuI) A restriction endonuclease from *Neisseria lactamica* 5841 with the following specificity:

5′..........GGNCC..........3′
3′..........CCNGG..........5′

NlaDIII (SacII) A restriction endonuclease from *Neisseria lactamica* 5841 with the same specificity as SacII.

NlaSI (SacII) A restriction endonuclease from *Neisseria lactamica* with the same specificity as SacII.

NlaSII (AcyI) A restriction endonuclease from *Neisseria lactamica* with the following specificity:

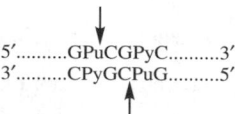

5′..........GPuCGPyC..........3′
3′..........CPyGCPuG..........5′

Nle Abbreviation for norleucine.

NLF Abbreviation for nonlactose-fermenting bacteria.

NliI (AvaI) A restriction endonuclease from *Nostoc linckia* with the same specificity as AvaI.

NliII (AvaII) A restriction endonuclease from *Nostoc linckia* with the same specificity as AvaII

NLS Abbreviation for nuclear localization signal.

nm Abbreviation for nanometer (equals $10^{-3}\mu$ or 10^{-9} m).

nM Abbreviation for 1. nanomole; 2. nanometer.

NM Abbreviation for nitrogen mustard.

NMDA (mol wt 147) Abbreviation for *N*-methyl-D-aspartic acid, a potent agonist for NMDA-receptors found in some vertebrate nerve cells.

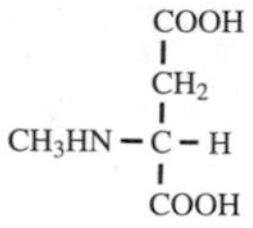

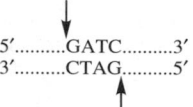

NmeCI (MboI) A restriction endonuclease from *Neisseria meningitidis* C114 with the following specificity:

5′..........GATC..........3′
3′..........CTAG..........5′

NMF Abbreviation for non-magnetic fraction.

NmiI (KpnI) A restriction endonuclease from *Nocardia minima* with the same specificity as KpnI.

NMN Abbreviation for nicotinamide mononucleotide.

NMP Abbreviation for nucleoside monophosphate.

NMR Abbreviation for nuclear magnetic resonance, the resonant absorption of electromagnetic radiation at a specific frequency by the atomic nuclei in a magnetic field. The NMR so}ãårum provides information for the chemical environment of the T›clei. The two-dimensional NMR is used for the determination of the three-dimensional structure of small proteins.

NmuI (NaeI) A restriction endonuclease from *Neisseria mucosa* with the same specificity as NaeI.

NmuAI (AvaI) A restriction endonuclease from *Nostoc muscorum* with the same specificity as AvaI.

NmuAII (AvaII) A restriction endonuclease from *Nostoc muscorum* with the same specificity as AvaII.

NmuEI (DpnI) A restriction endonuclease from *Neisseria mucosa* with the same specificity as DpnI.

NmuEII (AsuI) A restriction endonuclease from *Neisseria mucosa* with the same specificity as AsuI.

NmuFI (NaeI) A restriction endonuclease from *Neisseria mucosa* with the same specificity as NaeI.

nNOS Abbreviation for neuronal NOS (nitric oxide synthetase).

NO Abbreviation for nitric oxide.

No Abbreviation for nobelium, a chemical element.

Nobelium (No) A chemical element with atomic weight 259, valence 2 and 3.

Nobesine A trade name for diethylpropion hydrochloride, a cerebral stimulant that promotes transmission of nerve impulses.

Noble Agar Referring to purified agar.

Noble Gas Referring to the gas of helium, neon, argon, krypton, xenon, and radon that occur as a minor constituent of the atmosphere.

NocI (PstI) A restriction endonuclease from *Nocardia otitidis-caviarum* with the same specificity as PstI.

Nocardia A genus of Gram-positive, aerobic, chemoorganotrophic, nonmotile bacteria.

Nocardicins A class of b-lactam antibiotics produced by species of *Nocardia.*

Nocardioides A genus of aerobic soil bacteria (Actinomycetales).

Nocardiopsis A genus of bacteria (Actinomycetales).

Nocardiosis Any disorder of humans and animals caused by a species of *Nocardia.*

Noci- A prefix denoting pain or injury.

Nociceptor A receptor that responds to the stimuli responsible for sensation of pain.

Noctec A trade name for chloral hydrate, used as a sedative drug.

Nodaviridae A family of insect-infecting viruses that contain ssRNA.

NodB A *Rhizobium*-derived protein that is involved in generating plant-specific nodulation signal.

Nodes of Ranvier The periodic interruptions in the myelin sheath that expose the plasma membrane to the underlying axon.

NOD-Mouse Abbreviation for nonobese diabetic mouse, a strain of mouse that develops diabetes mellitus due to the absence or destruction of pancreatic beta cells.

Nodularia A genus of filamentous fresh-water and marine cyanobacteria.

Nodules A structure resulting from the enlargement or swelling on roots of legumes and certain other plants in response to infection by symbiotic nitrogen-fixing bacteria within which the infecting bacteria fix atmospheric nitrogen.

Nodulins Root nodule-specific proteins encoded by genes of host plant that are involved in symbiotic nitrogen fixation and nitrogen metabolism.

NOE Abbreviation for 1. nuclear Overhauser effect; 2. nuclear Overhauser enhancement.

NOES Abbreviation for 1. nuclear Overhauser effect spectroscopy; 2. nuclear Overhauser enhancement spectroscopy.

Noformicin (mol wt 197) An antiviral agent from *Nocardia formica.*

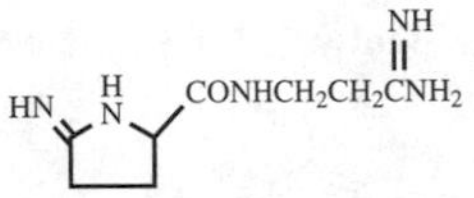

Nogalamycin (mol wt 788) An antineoplastic agent.

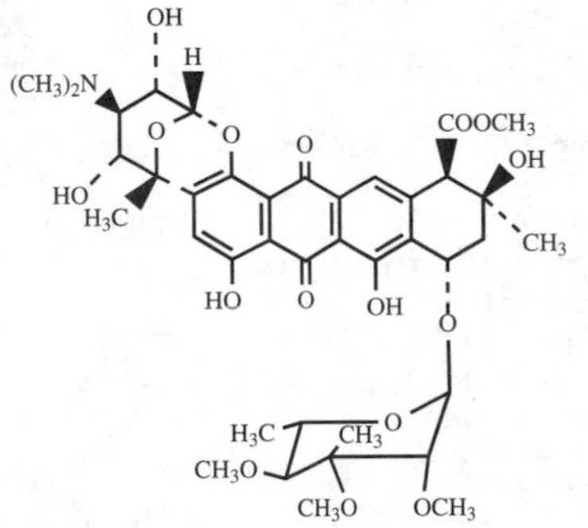

Nogalose (mol wt 204) A mannose derivative and a carbohydrate component of nogalamycin.

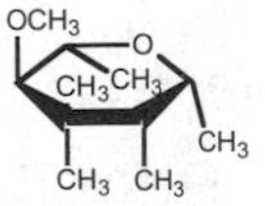

Nojirimycin (mol wt 179) An antibiotic isolated from *Streptomyces*. It inhibits intestinal α-glucosidase and pancreatic α-amylase.

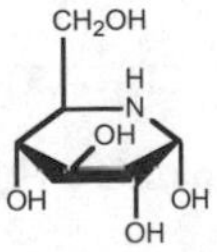

Nolahist A trade name for phenindamine tartrate used as an antihistaminic agent.

Nolamine A trade name for a combination drug containing chlorpheniramine maleate, phenindamine tartrate, and phenylpropanolamine hydrochloride, used as an antihistaminic drug.

Noludar A trade name for methyprylon, used as a sedative drug.

Nolvadex A trade name for tamoxifen citrate, an antineoplastic agent that acts as an estrogen antagonist.

Nomarski Differential Interference Microscope A specialized type of interference microscope employing a polarizing filter, an interference contrast condenser, and a prism analyzer plate to produce high-contrast images of the unstained specimens with a three-dimensional appearance.

Nomenclature The method of assigning names in the classification of organisms.

Nomifensine (mol wt 238) An antidepressant.

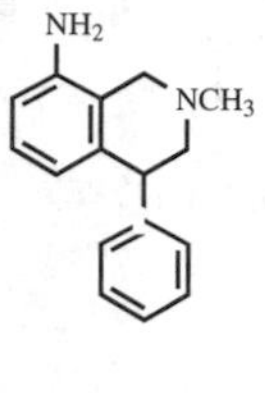

Nona- A prefix meaning 9.

Nonacosa- A prefix denoting 29.

Nonactin (mol wt 737) An antibiotic produced by species of *Streptomyces.*

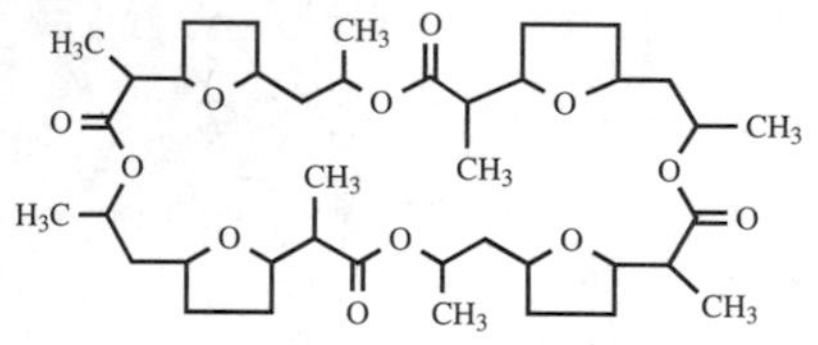

Nonadeca- A prefix denoting 19.

Nonaethylene Glycol Monododecyl Ether (mol wt 582) A nonionic detergent.

$$HO(CH_2CH_2O)_9(CH_2)_{11}CH_3$$

Nonamer A compound that consists of nine monomers or nine subunits.

Nonanoic Acid (mol wt 159) A 9-carbon fatty acid.

$$CH_3[CH_2]_7\text{-}COOH$$

Non-A-non-B Hepatitis Hepatitis that is not caused by hepatitis A virus or hepatitis B virus.

***N*-Nonanoyl-*N*-Methylglucamine (mol wt 335)** A water soluble detergent with high solubilizing power and nondenaturing properties.

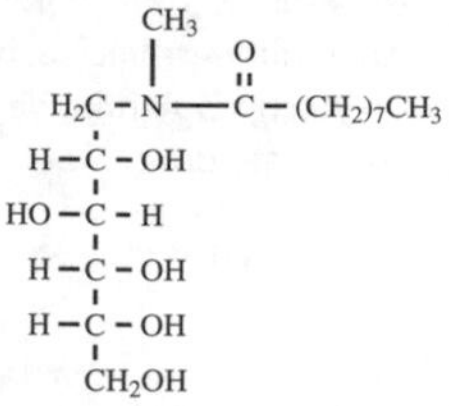

Non-Asp-49-PLA$_2$ Abbreviation for PLA$_2$ having an amino acid other than aspartate at position 49.

Nonbasic Chromosomal Proteins Referring to nonhistone proteins associated with chromosomes.

Noncellulosic Matrix Component of plant and fungal cell walls that consists of hemicellulose, pectin, and lignin, and the protein extensin.

Noncoding Strand The DNA strand that has the complementary sequence to mRNA.

Noncompetitive Inhibition Inhibition of enzyme activity by a noncompetitive inhibitor that cannot be revealed by increasing the substrate concentration.

Noncompetitive Inhibitor An inhibitor that binds the enzyme at a site other than the active site.

Non-Conjugative Plasmid A plasmid that does not encode the function(s) for its own intercellular transmission.

Noncovalent Bond The bond not resulting from the sharing of electrons, e.g., hydrogen bond or hydrophobic bond.

Noncyclic Electron Flow Continuous, unidirectional flow of electrons, e.g., light-induced flow of electrons from water to $NADP^+$ in oxygen-evolving photosynthetic reactions.

Noncyclic Photophosphorylation The production of ATP by noncyclic flow of electrons.

Nondisjunction Failure of separation of homologous chromosomes or of sister chromatids during nuclear division, producing cells or organisms having an aneuploid number of chromosomes.

Nonelectrolyte A compound that does not dissociate into ions in water and thereby does not conduct electric current.

Nonessential Amino Acids Amino acids that can be synthesized by the organism, e.g., alanine, asparagine, aspartate, cysteine, glutamate, glutamine, glycine, proline, serine and tyrosine in humans.

Nonheme Protein Protein that contains iron and sulfur but no heme group.

Nonhistone Chromosomal Protein Nonbasic proteins associated with DNA in chromatin, but is not a part of the nucleosome structure.

Nonhomologous Recombination Recombination involving little or no homology between the donor DNA and the region of the DNA in the recipient where insertion occurs.

Nonidet P40 A nonionic detergent of nonylphenylpolyethylene glycol used for solubilization of membrane protein.

Nonionic Detergent A surface-active agent that contains polar and nonpolar groups but no charge.

Nonketotic Hyperglycemia An inherited human disease characterized by the accumulation of large amounts of glycine in the body fluid due to the deficiency in glycine cleavage systems leading to mental retardation.

Nonmediated Transport The transport of solute across a membrane without involvement of transport protein. The force for flow of the substance is a simple concentration gradient.

Non-Mendelian Inheritance The inheritance that fails to follow Mendel's law of segregation, independent assortment, and linkage (also called cytoplasmic inheritance).

Nonose An aldose having 9-carbons.

Nonpermissive Host 1. A host cell that does not allow successful replication of an infecting virus. 2. A host in which a conditional lethal mutant fails to survive.

Nonpolar Without dipole movement or polarity.

Nonpolar Amino Acid An amino acid that has a nonpolar side chain, e.g., leucine, valine, and isoleucine and is generally not soluble in water.

Nonpolar Bond 1. A covalent bond in which two atoms share electrons evenly 2. Hydrophobic bond.

Nonpolar Covalent Bond Covalent bond in which electrons are shared equally between two bonded atoms.

Nonpolar Molecule 1. A molecule that lacks any asymmetrical accumulation of either positive or negative charge and is generally insoluble in water. 2. Water insoluble molecule.

Nonpolar Solvent Solvent without significant concentration of charged groups and/or dipoles.

Nonprotein Amino Acid An amino acid that is not used for biosynthesis of protein, e.g., D-form amino acids and other naturally occurring amino acids.

Nonradioactive Tracer A labeled substance that can be monitored by a method other than radioactivity, e.g., fluorescence, or color.

Nonreducing End The terminal of an oligopolysaccharide or polysaccharide that does not contain a hemiacetal or hemiketal group, thereby incapable of carrying out a reducing reaction.

Nonreducing Sugar A sugar that does not contain an aldehyde group or potential aldehyde group, thereby incapable of carrying out a reducing reaction.

Nonsaponifiable Lipid A lipid that cannot be hydrolyzed with alkali to form soap.

Nonsecretor An individual who does not secrete blood group antigen (e.g., antigen A or B) into the body fluid.

Nonsense Codon Referring to the three termination codons, UAA (Ochre codon), UAG (Amber codon), and UGA (Opal codon) that do not encode any amino acid, leading to the premature termination of a peptide chain (also called termination codons).

Nonsense Mutations A mutation that alters an encoding triplet leading to the formation of one of the three termination codons, resulting in premature termination of the polypeptide chain and formation of nonfunctional protein.

Nonsense Suppresser A tRNA that has been altered so that its anticodon can recognize a nonsense codon, thereby suppressing the function of the termination codon leading to the extension of the polypeptide chain and formation of functional protein.

Nonspecific Immunity The constitutive resistance of the body produced by nonimmunological mechanisms that involve phagocytes, interferon, lysozyme, and acute phase protein.

Non-Spiking The capability of a neuron to convey information without generating action potential.

Non-Treponemal Test A test for syphylis.

Nonyl β-D-Glucopyranoside (mol wt 306) A nonionic detergent for solubilization of membrane proteins.

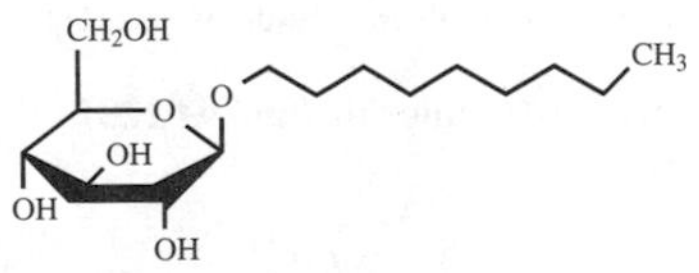

Nopaline (mol wt 304) A plant tumor metabolite and a rare amino acid derivative.

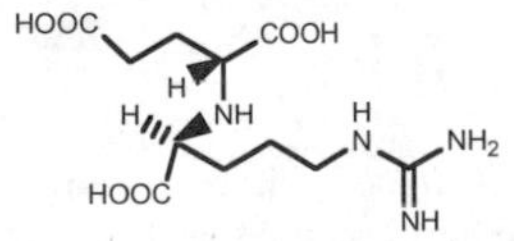

NopI (SalI) A restriction endonuclease from *Nocardia opaca* with the following specificity:

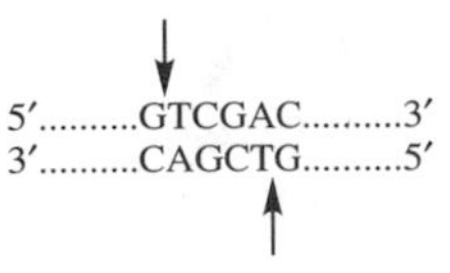

Noprylsulfamide (mol wt 494) An antibacterial agent.

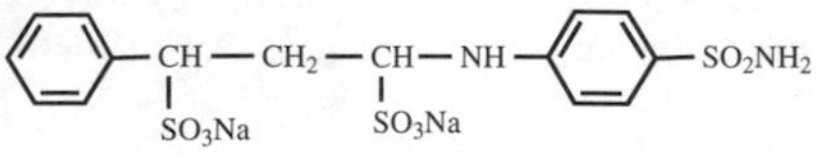

NOR Abbreviation for nucleolar-organizing region.

Noradex A trade name for orphenadrine citrate, used to reduce the transmission of nerve impulses from spinal cord to skeletal muscle.

Noradrenaline See norepinephrine.

Noradryl A trade name for diphenhydramine hydrochloride, an antihistaminic agent that competes with histamine for H_1 receptor on effector cells.

Norbolethone (mol wt 316) An anabolic hormone that promotes tissue-building processes.

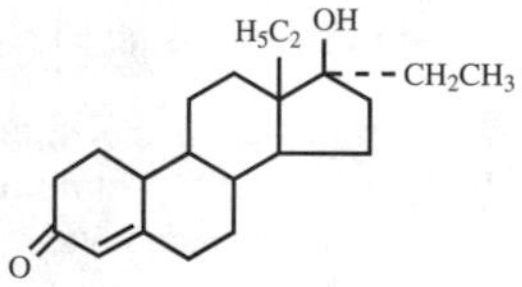

Norcet A trade name for a combination drug containing acetaminophen and hydrocodone bitartrate, used as a narcotic analgesic agent.

Norco A trade name for a combination drug containing hydrocodone bitartrate and acetaminophen used as an antitussive and analgesic agent.

Norcuron A trade name for vecuronium bromide, used as a neuromuscular blocker.

Nordefrin Hydrochloride (mol wt 220) A vasoconstrictor.

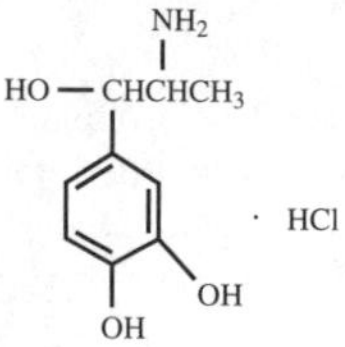

Nordette A trade name for a combination drug containing ethinyl estradiol and levonorgestrel, used as an oral contraceptive for inhibition of ovulation.

Nordihydroguaiaretic Acid (mol wt 302) An antioxidant in fat and oil.

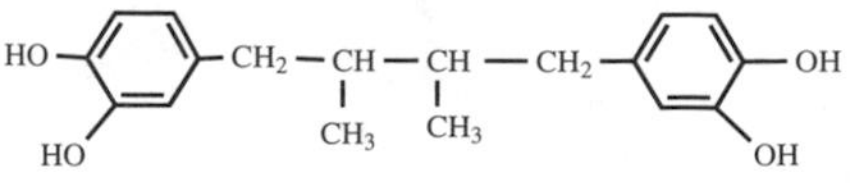

Norditropin A trade name for human growth hormone produced by DNA recombinant technology and used for the treatment of growth hormone deficiency in children.

Nordryl A trade name for diphenhydramine chloride, an antihistaminic agent that competes with histamine for H_1 receptors on effector cells.

Norea (mol wt 222) An herbicide.

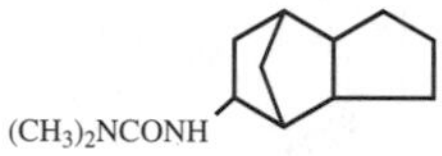

Norepinephrine (mol wt 169) A hormone produced by the adrenal medulla that stimulates both alpha- and beta-adrenergic receptors within the sympathetic nervous system.

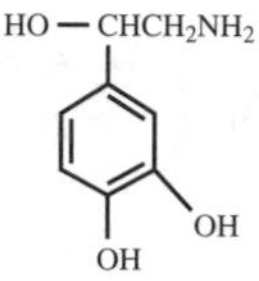

Norethandrolone (mol wt 302) An androgen.

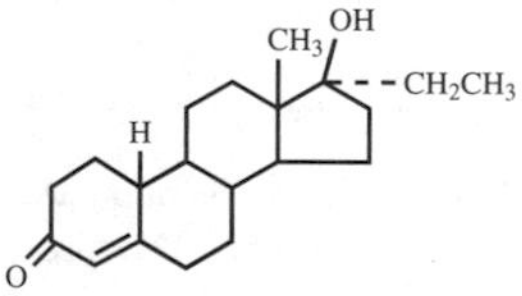

Norethindrone (mol wt 298) A progestogen that suppresses ovulation.

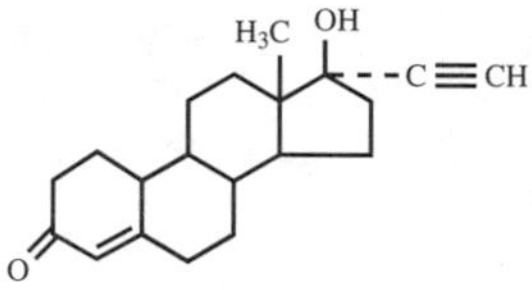

Norethynodrel (mol wt 298) A progestogen.

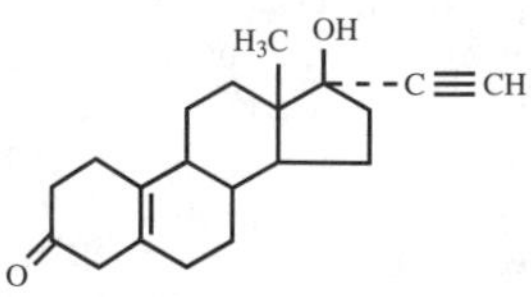

Norfenefrine (mol wt 153) An adrenergic agent.

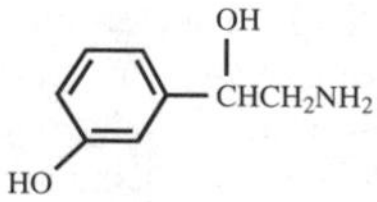

Norflex A trade name for orphenadrine citrate, used as a skeletal muscle relaxant to reduce the transmission of nerve impulses.

Norfloxacin (mol wt 319) An antibacterial agent that inhibits bacterial DNA synthesis by inhibiting DNA gyrase.

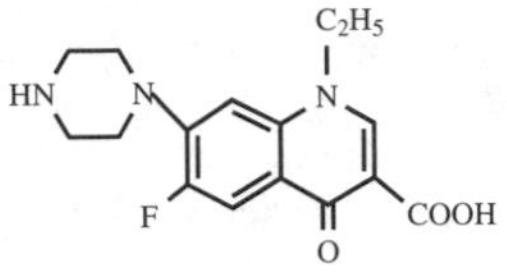

Norflurazon (mol wt 304) An herbicide.

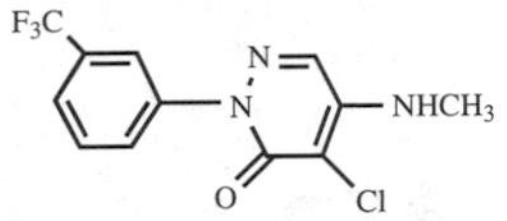

Norgesic Forte A trade name for a combination drug containing orphenadrine citrate, aspirin, and caffeine, used as a skeletal muscle relaxant.

Norgesterone (mol wt 300) A progestogen.

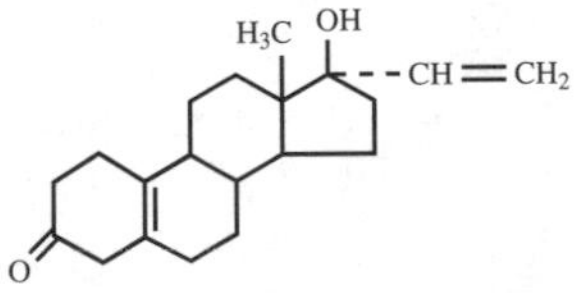

Norgestimate (mol wt 370) A progestogen used as an oral contraceptive in combination with estrogen.

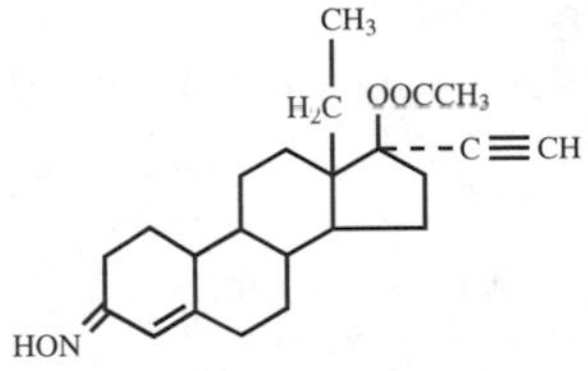

Norgestrel (mol wt 312) A hormone that suppresses ovulation.

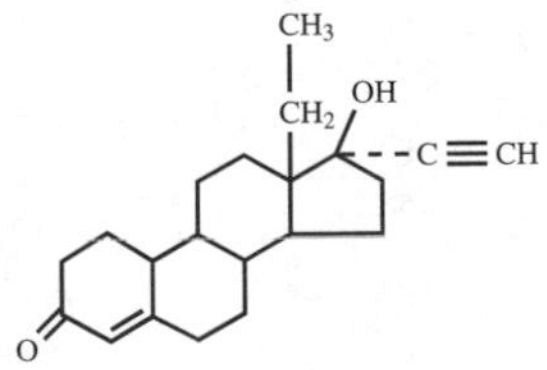

Norgestrienone (mol wt 294) A progestogen.

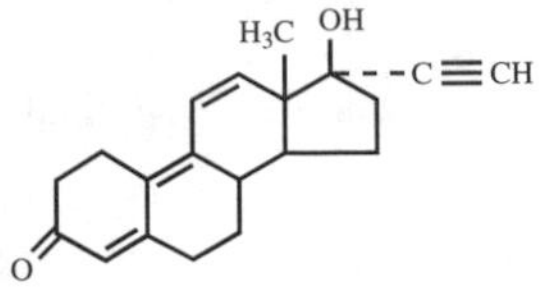

Norisodrine Aerotrol A trade name for isoproterenol hydrochloride, used as a bronchodilator.

Norlestrin 21 A trade name for an oral contraceptive containing eithinyl estradiol and norethindrone acetate.

Norleu Abbreviation for norleucine.

Norleucine (mol wt 131) A straight chain isomer of leucine that can be incorporated into protein during protein synthesis.

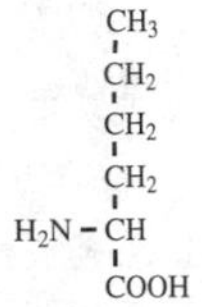

Norlevorphanol (mol wt 243) A narcotic analgesic agent.

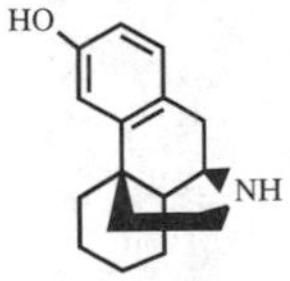

Norlutate A trade name for norethindrone acetate, used to suppress ovulation.

Norlutin A trade name for norethindrone, used to suppress ovulation.

Normal Solution A solution that contains 1 gram equivalent of a substance in one liter of solution.

Normality The number of equivalents of a solute per liter of solution.

Normethadone (mol wt 295) A narcotic analgesic and antitussive agent.

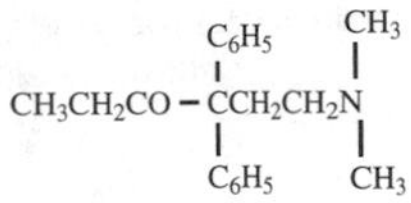

Normethandrone (mol wt 288) An androgen.

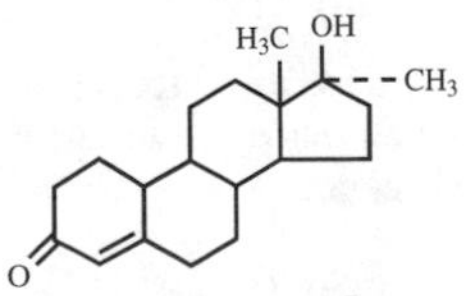

Normiflo A trade name for ardeparin sodium, an anti-thrombotic agent.

Nor-Mil A trade name for diphenoxylate hydrochloride, an antidiarrheal agent that inhibits GI tract mobility and propulsion and diminishes secretion.

Normoblast A nucleated cell of the myeloid series found in the bone marrow that gives rise to red blood cells.

Normodyne A trade name for labetalol hydrochloride, used as an antihypertensive agent that decreases renin secretion.

Normorphine (mol wt 271) A narcotic analgesic agent

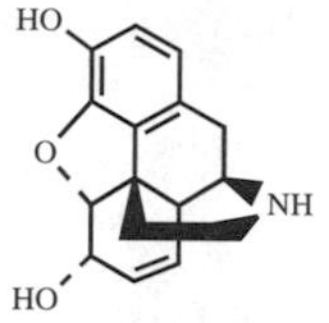

Normozide A trade name for a combination drug containing labetalol hydrochloride and hydrochlorothiazide, used as an antihypertensive agent.

Nornicotine (mol wt 148) An agricultural insecticide.

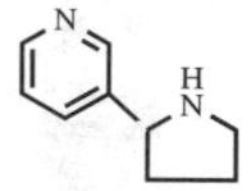

Nornidulin (mol wt 430) An antibiotic.

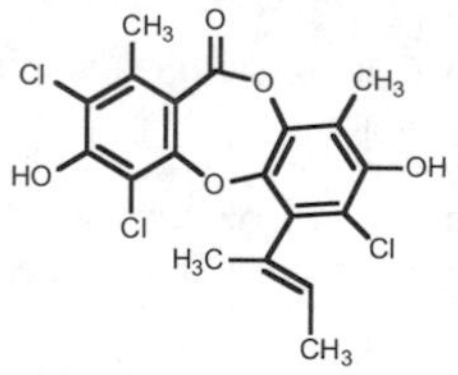

Noroxin A trade name for norfloxacin, an antibiotic that inhibits bacterial DNA synthesis by inhibiting DNA gyrase.

Norpace A trade name for disopyramide phosphate, used as an antiarrhythmic agent that prolongs action potential.

Norpanth A trade name for propantheline bromide, used as an anticholinergic agent that competitively blocks acetylcholine.

Norpramin A trade name for desipramine hydrochloride, used as an antidepressant agent that increases levels of norepinephrine and serotonin in the CNS.

Norpseudoephedrine (mol wt 151) An anorexic agent.

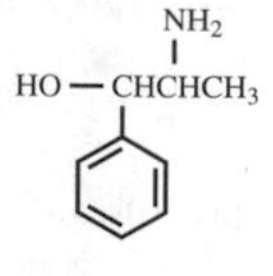

Norsteroid A steroid-like molecule or a modified steroid in which the ring has been contracted.

Nor-Tet A trade name for tetracycline hydrochloride, an antibiotic that binds to 30S ribosomes inhibiting bacterial protein synthesis.

Northern Blotting A method for detecting RNA fragments that are separated by gel electrophoresis, blotted on a nitrocellulose film or paper, and probed with labeled DNA or RNA.

Northern Transfer Technique See northern blotting.

Nortriptyline (mol wt 263) An antidepressant agent that increases the amount of norepinephrine or serotonin in the CNS.

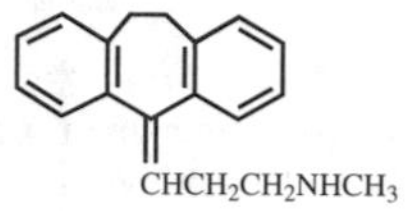

Nortussin A trade name for guaifenesin, used as an antitussive agent that increases the production of respiratory tract fluid to help liquefy and reduce the viscosity of the tenacious secretions.

Norvaline (mol wt 117) A straight chain isomer of valine and an inhibitor of ornithine carbamyltransferase.

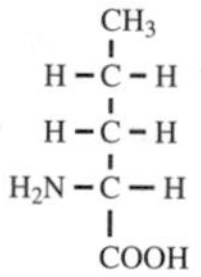

Norvasc A trade name for amlodipine, a calcium channel blocker used as an antianginal and antihypertensive agent.

Norventyl A trade name for nortriptyline hydrochloride, an antidepressant.

Norvinisterone (mol wt 300) A progestogen.

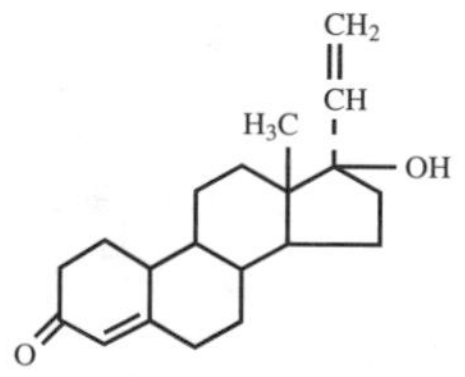

Norvir A trade name for ritonavir, an inhibitor for HIV protease.

Norwich A trade name for aspirin, an antipyretic, analgesic, anti-inflammatory, and antiplatelet agent.

Norzine A trade name for thiethylperazine maleate, an anti-emetic agent.

NOS Abbreviation for nitric oxide synthetase.

NOS2 Abbreviation for type-2 nitric acid synthetase.

Noscapine (mol wt 413) An antitussive agent.

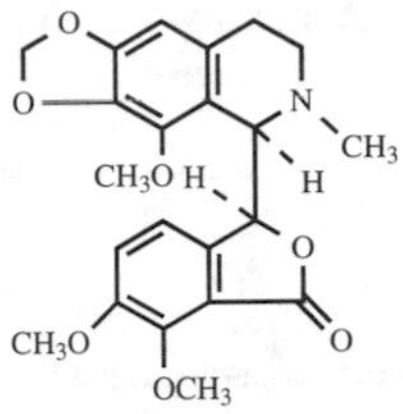

Nosocomial Infection An infection acquired while in the hospital.

Nosology The naming and classification of diseases.

Nostoc A genus of filamentous cyanobacteria.

Nostril A trade name for phenylephrine hydrochloride, used to produce local vasoconstriction of dilated arterioles to reduce blood flow and nasal congestion.

Nastrilla A trade name for oxymetazoline hydrochloride, used to produce local vasoconstriction and reduce blood flow and nasal congestion.

Notatin A flavoprotein glucose oxidase produced by *Penicillium notatum* that possesses antimicrobial activity due to its ability to produce hydrogen peroxide.

Noten A trade name for atenolol, an antihypertensive agent that blocks the response to beta stimulation and decreases renin secretion.

Notexin A phospholipase A_2 from snake venom that acts as a presynaptic neurotoxin.

NotI A restriction endonuclease from *Nocardia otitidis-caviarum* with the following specificity:

```
          ↓
5′..........GCGGCCGC..........3′
3′..........CGCCGGCG..........5′
                ↑
```

Nov II (HinfI) A restriction endonuclease from *Neisseria ovis* with the following specificity:

```
          ↓
5′..........GANTC..........3′
3′..........CTNAG..........5′
               ↑
```

Novafed A A trade name for a combination drug containing pseudoephedrine hydrochloride and chlorpheniramine maleate, used as an antihistaminic agent.

Novahistine A trade name for dextromethorphan hydrochloride, used as an antitussive agent.

Novamobarb A trade name for amobarbital, a sedative and hypnotic agent.

Novamoxin A trade name for amoxicillin trihydrate, an antibiotic that inhibits bacterial cell wall synthesis.

Novantrone A trade name for mitoxantrone hydrochloride, used as an antineoplastic agent.

Novasen A trade name for aspirin, used as an antipyretic and analgesic agent.

Novel Antigen A weak or nonimmunogenic molecule that becomes immunogenic upon conjugation to an immunogenic molecule or carrier.

Novembichin (mol wt 255) An antineoplastic agent.

$$CH_3-\underset{}{\overset{Cl}{\overset{|}{C}}}HCH_2\overset{CH_2CH_2Cl}{\overset{|}{N}}-CH_2CH_2Cl$$

Noviose (mol wt 224) A methylated hexose.

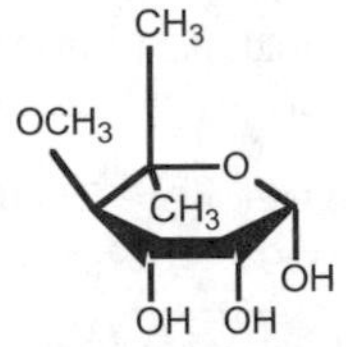

Novo Ampicillin A trade name for ampicillin, an antibiotic that inhibits bacterial cell wall synthesis.

Novobiocin (mol wt 613) An antibiotic produced by *Streptomyces spheroides* that inhibits prokaryotic DNA gyrase.

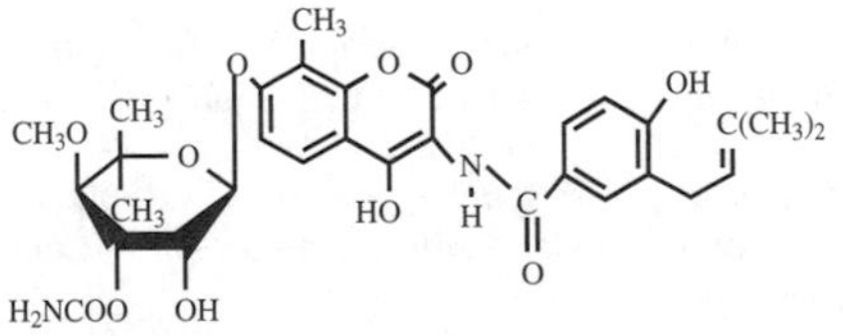

Novobutamide A trade name for tolbutamide, used to stimulate insulin release from pancreatic beta cell.

Novobutazone A trade name for phenylbutazone, used as an anti-inflammatory and antipyretic agent.

Novocain A trade name for procaine hydrochloride, used as a local anesthetic agent that blocks depolarization by interfering with sodium potassium exchange across the nerve cell membrane.

Novochlorhydrate A trade name for chloral hydrate, used as a sedative agent.

Novochlorocap A trade name for chloramphenicol, an antibiotic that binds to 50S ribosomes inhibiting bacterial protein synthesis.

Novoclopate A trade name for clorazepate dipotassium, used as an antianxiety agent.

Novocloxin A trade name for cloxacillin sodium, an antibiotic that inhibits bacterial cell wall synthesis.

Novodigoxin A trade name for digoxin, which promotes movement of calcium from extracellular to intracellular cytoplasm and inhibits sodium-potassium activated adenosine triphosphatase (ATPase).

Novodimenate A trade name for dimenhydrinate, used as an antiemetic agent.

Novodipam A trade name for diazepam, used as an antianxiety agent.

Novoferrogluc A trade name for ferrous gluconate, used to provide iron.

Novoferrosulfa A trade name for ferrous sulfate, used to provide iron in the body.

Novofibrate A trade name for clofibrate, used as an antiemetic agent.

Novoflupam A trade name for flurazepam hydrochloride, used as a sedative agent.

Novo-Flurazine A trade name for trifluoperazine hydrochloride, used as an antipsychotic agent.

Novofolacid A trade name for vitamin B_9 (folic acid).

Novofumar A trade name for ferrous fumarate, used to provide iron for the body.

Novofuran A trade name for nitrofurantoin microcrystals, used as an antimicrobial agent.

Novohexidyl A trade name for trihexyphenidyl hydrochloride, used for treatment of Parkinsonian disease.

Novohydrazide A trade name for hydrochlorothiazide, used as a diuretic agent that increases urine excretion of sodium and water by inhibiting sodium reabsorption.

Novohydroxyzin A trade name for hydroxyzine hydrochloride, used as an antianxiety agent.

Novo-Hylazin A trade name for hydralazine hydrochloride, used as an antihypertensive agent.

Novolexin A trade name for cephalexin monohydrate, an antibiotic that inhibits bacterial cell wall synthesis.

Novolin A trade name for insulin.

Novolorazem A trade name for lorazepam, used as an antianxiety agent.

Novomedopa A trade name for methyldopa, used as an antihypertensive agent.

Novo-Mepro A trade name for oxazepam, used as an antianxiety agent.

Novomethacin A trade name for indomethacin sodium trihydrate, used as an anti-inflammatory agent.

Novometoprol A trade name for metoprolol tartrate, used as an antihypertensive agent that blocks cardiac beta receptors and decreases renin secretion.

Novonal (mol wt 155) An hypnotic agent.

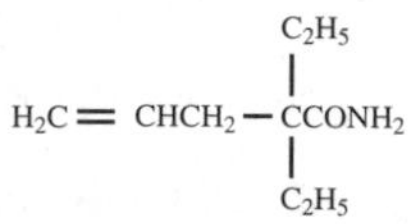

Novonaprox A trade name for naproxen, used as an anti-inflammatory agent.

Novonidazol A trade name for metronidazole, used as an antiprotozoal agent.

Novo-Nifedin A trade name for nifedipine, used as an antianginal agent that inhibits calcium ion influx across cardiac and smooth muscle cells and reduces cardiac oxygen demand.

NovoPen G A trade name for penicillin G potassium, an antibiotic that inhibits bacterial cell wall synthesis.

NovoPen VK A trade name for penicillin V potassium, an antibiotic that inhibits bacterial cell wall synthesis.

Novopentobarb A trade name for pentobarbital sodium, used as a sedative and hypnotic agent that interferes with transmission of impulses from the thalamus to the cortex of the brain.

Novoperidol A trade name for haloperidol, used as an antipsychotic agent that blocks postsynaptic dopamine receptors in the brain.

Novopheniram A trade name for chlorpheniramine maleate, used as an antihistaminic agent that competes with histamine for H_1 receptors on effector cells.

Novopirocam A trade name for piroxicam, used as an anti-inflammatory agent.

Novopoxide A trade name for chlordiazepoxide hydrochloride, used as an antianxiety agent.

Novo-Pramine A trade name for imipramine hydrochloride, used as an antidepressant.

Novopranol A trade name for propranolol hydrochloride, an antianginal agent and a beta blocker that reduces cardiac oxygen demand.

Novo-Prednisolone A trade name for prednisone, used as an anti-inflammatory agent.

Novoprofen A trade name for ibuprofen, used as an analgesic and antipyretic agent.

Novo-Propamide A trade name for chlorpropamide, used as an antidiabetic agent that stimulates the release of insulin from pancreatic beta cells.

Novopropoxyn A trade name for propoxyphene hydrochloride, an opioid analgesic agent that binds to opiate receptors altering both perception and emotional response to pain.

Novoquindin A trade name for quinidine sulfate, used as an antiarrhythmic agent that prolongs action potential.

Novoquinine A trade name for quinine sulfate, used as an antimalarial agent.

Novoreserpine A trade name for reserpine, an antihypertensive agent that inhibits norepinephrine release.

Novoridazine A trade name for thioridazine hydrochloride, an antipsychotic agent that blocks the postsynaptic dopamine receptors in the brain.

Novorythro A trade name for erythromycin base, an antibiotic that binds to 50S ribosomes inhibiting bacterial protein synthesis.

Novosecobarb A trade name for secobarbital sodium, a sedative-hypnotic agent that interferes with the transmission of impulses from thalamus to the cortex of the brain.

Novosemide A trade name for furosemide, used as a diuretic agent that inhibits the reabsorption of sodium and chloride.

Novoseven A trade name for recombinant human coagulation factor VIIa.

Novosorbide A trade name for isosorbide dinitrate, an antianginal agent that reduces cardiac oxygen demand and increases blood flow.

Novosoxazole A trade name for sulfasoxazole (sulfafuraxole, sulphafuraxole), an antimicrobial agent that inhibits microbial synthesis of dihydrofolic acid.

Novospiroton A trade name for spironolactone, a diuretic agent that increases excretion of sodium and water by inhibiting sodium reabsorption.

Novotetra A trade name for tetracycline hydrochloride, an antibiotic that binds to 30S ribosomes inhibiting bacterial protein synthesis.

Novothalidone A trade name for chlorthalidone, a diuretic agent that increases excretion of sodium and water by inhibiting sodium reabsorption.

Novotrimel A trade name for cotrimoxazole (sulfamethoxazole-trimethoprim), an antimicrobial agent that inhibits microbial synthesis of dihydrofolic acid.

Novo-Triptyn A trade name for amitriptyline hydrochloride, an antidepressant drug that increases levels of norepinephrine and serotonin in the CNS.

Novoxapam A trade name for oxazepam, used as an antianxiety agent.

Noxiptilin (mol wt 294) An antidepressant.

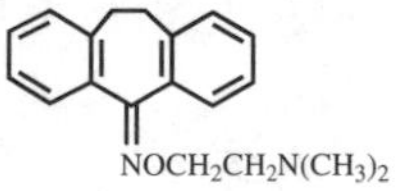

Noxythiolin (mol wt 120) An antiseptic agent.

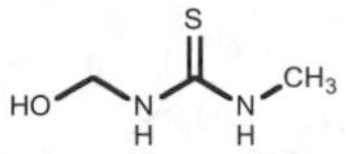

Nozinan A trade name for methotrimeprazine hydrochloride (levomepromazine hydrochloride), used as a sedative-hypnotic agent.

NP Abbreviation for 1. normal plasma; 2. nucleoprotein; 3. nucleocapsid protein; 4. nucleoside phosphorylase.

NP-27 A trade name for tolnaftate, used as an antifungal agent.

NP Detector Abbreviation for nitrogen-phosphorus detector.

NP-40 Abbreviation for Nonidet P40, a detergent.

4NPA Abbreviation for 4-nitrophenylacetate.

NPH A trade name for neutral protamine hagedorn insulin, an isophane insulin suspension.

NphI A restriction endonuclease from *Neisseria pharyngis* C245 with the following specificity:

```
      ↓
5′.........GATC..........3′
3′.........CTAG..........5′
               ↑
```

NPN Abbreviation for non-protein nitrogen.

NPOE Abbreviation for N-phenyloctyl ester.

NPP Abbreviation for 4-nitrophenyl phosphate.

N-Propeptide Abbreviation for N-terminal of the propeptide.

NPV Abbreviation for nucleopolyhydrosis virus.

NPY Abbreviation for neuropeptide Y.

NR Abbreviation for 1. nitrate reductase; 2. nuclear receptor; 3. neutral red.

NRBC Abbreviation for nucleated red blood cell.

NRE Abbreviation for negative regulatory element.

nRNA Abbreviation for nuclear RNA.

NRP Abbreviation for negative regulatory protein.

NRPS Abbreviation for non-ribosomal peptide synthetase.

NRS Abbreviation for normal rabbit serum.

NruGI A restriction endonuclease from *Nocardia rugosa G* with the following specificity:

↓
5'........GACNNNNNGTC........3'
3'........CTGNNNNNCAG........5'
↑

NSAIA Abbreviation for nonsteroidal anti-inflammatory agent.

NSAID Abbreviation for nonsteroidal anti-inflammatory drug.

NSE Abbreviation for neuron-specific enolase.

NSFP Abbreviation for N-ethylmaleimide-sensitive fusion protein.

NsiI (AvaIII) A restriction endonuclease from *Neisseria sicca* C351 with the following specificity:

↓
5′..........ATGCAT..........3′
3′..........TACGTA..........5′
↑

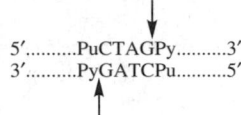

NsiAI (MboI) A restriction endonuclease from *Neisseria sicca* with the same specificity as MboI.

NsiCI (EcoRV) A restriction endonuclease from *Neisseria sicca* C351 with the following specificity:

↓
5′..........GATATC..........3′
3′..........CTATAG..........5′
↑

NsiHI (HinfI) A restriction endonuclease from *Neisseria sicca* with the same specificity as NinfI.

NSLTP Abbreviation for non-specific lipid-transfer protein.

NSO Abbreviation for neosporin ointment.

NSP Abbreviation for non-structural protein.

NspI A restriction endonuclease from *Nostoc* species with the following specificity:

↓
5′..........PuCTAGPy..........3′
3′..........PyGATCPu..........5′
↑

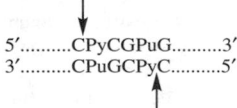

NspII (SduI) A restriction endonuclease from *Nostoc* species with the following specificity:

↓
5′..........G(A/G/T)GC(A/C/T)C..........3′
3′..........C(T/C/A)CG(T/G/A)G..........5′
↑

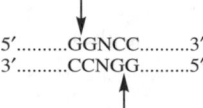

NspIII A restriction endonuclease from *Nostoc* species with the following specificity:

↓
5′..........CPyCGPuG..........3′
3′..........CPuGCPyC..........5′
↑

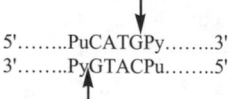

NspIV A restriction endonuclease from *Nostoc* species with the following specificity:

↓
5′..........GGNCC..........3′
3′..........CCNGG..........5′
↑

NspV (AsuII) A restriction endonuclease from *Nostoc* species with the same specificity as AsuII.

Nsp7524I A restriction endonuclease from *Nostoc* species with the following specificity:

↓
5'........PuCATGPy........3'
3'........PyGTACPu........5'
↑

Nsp7524V A restriction endonuclease from *Nostoc* species with the following specificity:

5'........TTCGAA........3'
3'........AAGCTT........5'

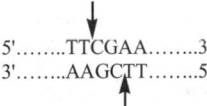

NspAI (MboI) A restriction endonuclease from *Nocardia* species with the same specificity as MboI.

NspBI (AsuII) A restriction endonuclease from *Nostoc* species with the same specificity as AsuII.

NspBII A restriction endonuclease from *Nostoc* species with the following specificity:

5′..........C(A/C)GC(G/T)G..........3′
3′..........G(T/G)CG(C/A)C..........5′

NspDI (AvaI) A restriction endonuclease from *Nostoc* species with the same specificity as AvaI.

NspDII (AvaII) A restriction endonuclease from *Nostoc* species with the same specificity as AvaII.

NspEI (AvaI) A restriction endonuclease from *Nostoc* species with the same specificity as AvaI.

NspFI (AsuII) A restriction endonuclease from *Nostoc* species with the same specificity as AsuII.

NspGI (AvaII) A restriction endonuclease from *Nostoc* species with the same specificity as AvaII.

NspHI A restriction endonuclease from *Nostoc* species with the following specificity:

5′..........PuCATGPy..........3′
3′..........PyGTACPu..........5′

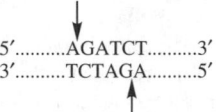

NspHII (AvaII) A restriction endonuclease from *Nostoc* species with the same specificity as AvaII.

NspHIII (MstI) A restriction endonuclease from *Nostoc* species with the same specificity as MstI.

NspJI (AsuII) A restriction endonuclease from *Nostoc* species with the same specificity as AsuII.

NspKI (AvaII) A restriction endonuclease from *Nostoc* species with the same specificity as AvaII.

NspLI (MstI) A restriction endonuclease from *Nostoc* species with the same specificity as MstI.

NspLII (AsuI) A restriction endonuclease from *Nostoc* species with the same specificity as AsuI.

NspMI (MstI) A restriction endonuclease from *Nostoc* species with the same specificity as MstI.

NspMACI A restriction endonuclease from *Nostoc* species with the following specificity:

5′..........AGATCT..........3′
3′..........TCTAGA..........5′

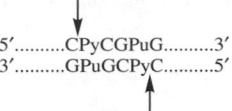

NspSAI A restriction endonuclease from *Nostoc* species with the following specificity:

5′..........CPyCGPuG..........3′
3′..........GPuGCPyC..........5′

NspSAII A restriction endonuclease from *Nostoc* species with the following specificity:

5′..........GGTNACC..........3′
3′..........CCANTGG..........5′

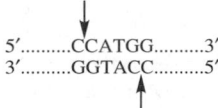

NspSAIII (NcoI) A restriction endonuclease from *Nostoc* species with the following specificity:

5′..........CCATGG..........3′
3′..........GGTACC..........5′

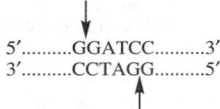

NspSAIV (BamHI) A restriction endonuclease from *Nostoc* species with the following specificity:

5′..........GGATCC..........3′
3′..........CCTAGG..........5′

NspWI (NaeI) A restriction endonuclease from *Nocardia* species with the same specificity as NaeI.

NT Abbreviation for 1. neurotensin; 2. neurotrophin.

NTA Abbreviation for nitrilotriacetate.

NtaI (Tth111I) A restriction endonuclease from *Nocardia tartaricans* with the following specificity:

5′..........GACNNNGTC..........3′
3′..........CTGNNNCAG..........5′

NtaSI (StuI) A restriction endonuclease from *Nocardia tartaricans* with the same specificity as StuI.

NtaSII (NaeI) A restriction endonuclease from *Nocardia tartaricans* with the same specificity as NaeI.

N-Terminal The amino terminal of a protein that carries a free NH_2 group that is the terminal where the first amino acid is incorporated into the protein during translation.

N-Terminus See also N-terminal.

NTG Abbreviation for 1. nitroglycerin; 2. nitrosoguanidine.

NTP Abbreviation for nucleoside triphosphate.

NTR Abbreviation for neurotensin receptor.

NTS A trade name for nitroglycerin, used for reduction of cardiac oxygen demand and increase of blood flow.

Nu-Alpraz A trade name for alprazolam, an antianxiety agent.

Nu-Amoxi A trade name for amoxicillin, an antibiotic.

Nu-Ampi A trade name for ampicillin, an antibiotic.

Nubain A trade name for nalbuphine hydrochloride, used as an analgesic agent.

Nu-Capto A trade name for captopril, an ACE (angiotensin-converting enzyme) inhibitor used as an antihypertensive agent.

Nu-Cephalex A trade name for cephalexin, an antibiotic.

Nu-Cimet A trade name for cimetidine, a histamine H2 antagonist.

Nuclear Area See nucleoid.

Nuclear Body See nucleoid nuclear zone or nuclear region.

Nuclear cortex See nuclear lamina.

Nuclear Envelope The membrane of the eukaryotes that encloses the nucleus, separating it from the cytoplasm.

Nuclear Fast Red (mol wt 357) A reagent used for determination of calcium.

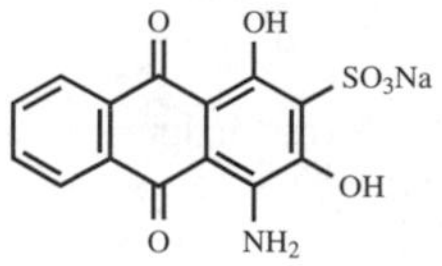

Nuclear Fission The splitting of a heavier atomic nucleus and releasing a large amount of energy.

Nuclear Fusion The union of two atomic nuclei to form a heavier nucleus and releasing a large amount of energy.

Nuclear Hypersensitivity The sensitivity of the specific region of the chromosome to DNAase due to the action or attachment of nonhistone protein in the region.

Nuclear Isomers Atoms with the same atomic number and mass but having different rates of radioactive decay.

Nuclear Lamina Electron-dense layer of fibrous material on the inner surface of the inner nuclear membrane, composed of protein subunits called lamins (also called the nuclear cortex).

Nuclear Lamins Proteins found in the nuclear lamina.

Nuclear Localization Sequence A short amino acid sequence that targets proteins for uptake into the nuceus.

Nuclear Magnetic Resonance A nondestructive, noninvasive method for studying structures, conformations, and interactions of molecules both *in vitro* and *in vivo* based on the absorption and consequent emission of electromagnetic radiation by the nuclei of certain types of atoms in the presence of a strong magnetic field.

Nuclear Matrix Filament Structural filament within the nucleus.

Nuclear Matrix Filament network that provides supporting framework for the nucleus.

Nuclear Membrane The membrane that surrounds the nucleus of a eukaryotic cell (see also nuclear envelope).

Nuclear Medicine The science that employs radioactive materials for diagnosis and treatment of disease.

Nuclear Polyhydrosis Virus An insect virus of the Baculoviridae. It is used as a viral pesticide and a clonal vehicle.

Nuclear Pore Channels in the nuclear envelope that allow certain molecules to pass between the nucleus and cytoplasm.

Nuclear Region The region of a cytoplasmic mass of a prokaryotic cell that contains DNA (see also nucleoid).

Nuclear RNA A group of heterogenous RNAs present in the nucleus, some of which are associated with protein to form small nuclear ribonucleoproteins.

Nuclear Transplantation A process by which a nucleus is transferred from one cell to the cytoplasm of another for study of nucleo-cytoplasmic interaction.

Nuclear Transport Passage of molecules in and out of the nucleus.

Nuclear Zone See nucleoid or nuclear region.

Nuclearia A genus of amoebae (class Flosea).

Nuclease The enzyme that catalyzes the cleavage of phosphodiester bonds.

Nucleic Acid Unbranched polymer, e.g., DNA and RNA, composed of pentose (ribose or deoxyribose), phosphate, and nitrogenous base (guanine, cytosine, adenosine, and thymine or uracil).

Nucleic Acid Base Referring to nitrogenous, aromatic bases found in nucleic acid e.g., adenine, guanine, cytosine, thymine, and uracil.

Nucleobindin A DNA-binding protein derived from lupus erythematosus-proned mouse that contains a signal peptide, leucine zipper, and basic amino acid-rich regions.

Nucleocapsid Referring to the protein-nucleic acid complex in the virion.

Nucleocapsid Protein The proteins found in the nucleocapsid.

Nucleoid The region of a cytoplasmic mass of a prokarryotic cell that contains genetic material (DNA).

Nucleodisome Chromosome fragment that consists of 2 nucleosomes connected by a linker DNA.

Nucleohistone Complex of histone protein and nucleic acid.

Nucleolar Organizer A region on a chromosome that contains a cluster of ribosomal RNA genes that gives rise to the nucleolus following nuclear division.

Nucleolar Organizer Region See nuclear organizer.

Nucleolus Large, discrete structure present in the nucleus of a eukaryotic cell, the site of ribosomal RNA synthesis and processing, and assembly of ribosomal subunits.

Nucleolus Organizer See nucleolar organizer.

Nucleon A collective term for the constituents in the atomic nucleus, e.g., proton or neutron.

Nucleophilic Atom An atom that has excess electrons.

Nucleophilic Catalysis A type of catalysis in which the catalyst donates a pair of electrons to a reactant.

Nucleophilic Reagent Chemical groups that act by donating or sharing their electrons.

Nucleophilic Substitution Reaction in which an electron-rich (electronegative) molecule or region of an electron-rich molecule donates electrons to an electron-deficient (electropositive) molecule or region of an electron-deficient molecule, resulting in the formation of a covalent bond.

Nucleoplasm The fluidlike material that fills the interior of the nucleus.

Nucleoplasmin An acidic protein found in the nucleus that binds to histone and participates in nucleosome assembly by acting as a molecular chaperone to bring DNA and histone protein together.

Nucleoprotein A complex of protein and nucleic acid.

Nucleosidase The enzyme that catalyzes the hydrolysis of nucleoside to pentose and nitrogenous base.

Nucleoside A component of a nucleotide that consists of a nitrogenous base (purine or pyrimidine) linked to a pentose sugar (ribose or deoxyribose).

Nucleoside Antibiotic An antibiotic that contains a nucleoside structure.

Nucleoside Cyclic Monophosphate A monophosphate nucleotide in which the phosphate group is esterified with two hydroxyl groups on the sugar, e.g., cyclic AMP or cyclic GMP.

Nucleoside 2',3'-Cyclic Phosphate 2'-Nucleotidohydrolase The systematic name for 2',3'-cyclic-nucleotide 3'-phosphodiesterase.

Nucleoside Deoxyribosyltransferase The enzyme that catalyzes the following reaction:

$$\text{2-Deoxyribosyl-base}^1 + \text{base}^2 \rightleftharpoons \text{2-Deoxyribosyl-base}^2 + \text{base}^1$$

Nucleoside Diphosphatase The enzyme that catalyzes the following reaction:

$$\text{A dinucleoside diphosphate} + H_2O \rightleftharpoons \text{A nucleotide} + \text{Pi}$$

Nucleoside Diphosphate A nucleoside that contains two phosphate groups, e.g., ADP and GDP.

Nucleoside Diphosphate Kinase The enzyme that catalyzes the following reaction:

$$\text{ATP + nucleoside diphosphate} \rightleftharpoons \text{ADP + nucleoside triphosphate}$$

Nucleoside Diphosphate Sugar A nucleoside diphosphate with sugar moiety attached, e.g., UDP-G (uridine diphosphate glucose). It is a coenzyme-like carrier of a sugar molecule, functioning in biosynthesis of polysaccharide or sugar derivative.

Nucleoside Monophosphate A nucleoside with a phosphate group (see nucleotide).

Nucleoside Phosphate Kinase The enzyme that catalyzes the following reaction:

$$\text{ATP + nucleoside phosphate} \rightleftharpoons \text{ADP + nucleoside diphosphate}$$

Nucleoside Phosphoacylhydrolase The enzyme that catalyzes the hydrolysis of mixed phosphoanhydride bonds.

Nucleoside Phosphorylase Synonym of purine nucleoside phosphorylase.

Nucleoside Phosphotransferase The enzyme that catalyzes the following reaction:

$$\text{A nucleotide + a 2}'\text{-deoxynucleoside} \rightleftharpoons \text{A 2}'\text{-deoxynucleoside monophosphate + a nucleoside}$$

Nucleoside Triphosphatase The enzyme that catalyzes the following reaction:

$$\text{NTP} + H_2O \rightleftharpoons \text{NDP} + \text{Pi}$$

Nucleoside Triphosphate A nucleoside that contains three phosphate groups, e.g., ATP, GTP, and CTP.

Nucleoside Triphosphate Pyrophosphatase The enzyme that catalyzes the following reaction:

$$\text{A nucleoside triphosphate} + H_2O \rightleftharpoons \text{A nucleotide + PPi}$$

Nucleoside Triphosphate RNA Nucleotidyltransferase The systematic name for DNA-directed RNA polymerase.

Nucleosome Basic structural unit of chromosomes consisting of about 200 base pairs of DNA wrapped around a group of histone proteins. It is the structural unit for packing chromatin.

Nucleotidase The enzyme that catalyzes the hydrolysis of nucleotide to nucleoside and inorganic phosphate.

3′-Nucleotidase The enzyme that catalyzes the following reaction:

$$\text{A 3}'\text{-Nucleotide} + H_2O \rightleftharpoons \text{A ribonucleoside + Pi}$$

5′-Nucleotidase The enzyme that catalyzes the following reaction:

$$\text{A 5}'\text{-Nucleotide} + H_2O \rightleftharpoons \text{A ribonucleoside + Pi}$$

Nucleotide The basic building block of nucleic acids, it consists of a nitrogenous base (a purine or a pyrimidine), a pentose (a ribose or a deoxyribase) and a phosphate.

Nucleotide Pyrophosphatase The enzyme that catalyzes the following reaction:

$$\text{A dinucleotide} + H_2O \rightleftharpoons \text{2 Mononucleotides}$$

Nucleotide Pyrophosphokinase The enzyme that catalyzes the following reaction:

$$\text{ATP + nucleoside 5}'\text{-phosphate} \rightleftharpoons \text{AMP + 5}'\text{-phosphonuccleoside 3}'\text{-diphosphate}$$

Nucleus 1. The membrane-bounded cell organelle of the eukaryotic cell that houses chromosomes and nucleoli. 2. The center core of an atom consisting of protons and neutrons. 3. The ring structure of an organic compound.

Nuclide An atomic species characterized by the constitution of its nuclear composition, e.g., number of protons and neutrons.

Nuclidic Mass See atomic mass.

Nu-Clonidine A trade name for clonidine hydrochloride, an antihypertensive and analgesic agent.

Nu-Cloxi A trade name for cloxacillin sodium, an antibiotic.

Nucofed A trade name for a combination drug containing codeine phosphate and pseudoephedrine hydrochloride, used as an antitussive and decongestant.

Nude Mouse A hairless mouse that congenitally lacks a thymus, characterized by the deficiency of thymus derived T lymphocytes.

Nu-Diclo A trade name for diclofenac sodium, an analgesic and anti-inflammatory agent.

Nu-Diflunisal A trade name for diflunisal, an analgesic, antipyretic, and anti-inflammatory agent.

Nu-Diltiaz A trade name for diltiazem hydrochloride, a calcium channel blocker used as an antianginal and antihypertensive agent.

Nu-Doxycycline A trade name for doxycycline, an antibiotic.

NuDP-Sugar Referring to nucleoside diphosphate sugar, e.g., UDPG.

Nu-Erythromycin-S A trade name for erythromycin, an antibiotic.

Nu-Hydral A trade name for hydralazine hydrochloride, a vasodilator and an antihypertensive agent.

Nu-Ketoprofen A trade name for ketoprofen, a nonsteroidal anti-inflammatory and analgesic agent.

Null Allele An allele that does not encode a functional gene product.

Null Cells Lymphoid cells that lack the characteristic T and B cell markers; they are non-T and non-B lymphocytes.

Nu-Loraz A trade name for lorazepam, a sedative, hypnotic and anti-anxiety agent.

Nu-Medopa A trade name for methyldopa an antihypertensive and sympatholytic agent.

Numerical Taxonomy A statistical method for classifying organisms based on the number of similarities of the measurable phenotypic characters among organisms.

Nu-Metoclopramide A trade name for metoclopramide, a GI stimulant and an anti-emetic agent.

Nu-Metop A trade name for metoprolol tartrate, an antihypertensive agent.

Numorphan A trade name for oxymorphone hydrochloride, an analgesic agent that binds with opiate receptors in the CNS.

NunII (NarI) A restriction endonuclease from *Nocardia uniformis* with the following specificity:

```
            ↓
5′.........GGCGCC..........3′
3′.........CCGCGG..........5′
              ↑
```

Nu Particle Abbreviation for nucleosome.

Nu-Pindol A trade name for pindolol, an antihypertensive agent.

Nu-Pirox A trade name for piroxicam, a nonsteroidal anti-inflammatory agent.

Nu-Prazo A trade name for prazosin hydrochloride, an alpha adrenergic blocker used as an antihypertensive agent.

Nu-Ranit A trade name for ranitidine, histamine H2 antagonist, which competitively inhibits the action of histamine.

Nuprin A trade name for ibuprofen, an analgesic, antipyretic agent.

Nuromax A trade name for doxacurium chloride, a neuromuscular blocker that competes for acetylcholine receptors.

Nurse Cell Cell that is connected to an oocyte by cytoplasmic bridges allowing transfer of nutrient into the growing oocyte.

nus Any of the three genes (*nusA, nusB*, and *nusG*) in *E coli* that are involved in the transcriptional termination.

Nu-Tetra A trade name for tetracycline, an antibiotic.

Nutracort A trade name for the hormone hydrocortisone.

Nutrasweet See aspartame.

Nu-Trazodone A trade name for trazodone hydrochloride, an antidepressant.

Nutrient A growth-supporting substance.

Nutrient Agar A liquid medium gelified with agar.

Nutrient Broth A liquid growth medium used for culture of microorganisms.

Nutritional Mutations Mutations that alter the nutritional requirement of a microorganism.

Nutropin A trade name for human growth hormone somatropin, produced by DNA recombinant technology and used for the treatment of growth hormone deficiency in children.

Nva Abbreviation for norvaline.

NVOC Abbreviation for 6-nitroveratryloxycarbonyl.

Nybomycin (mol wt 298) An antibiotic produced by *Streptomyces*.

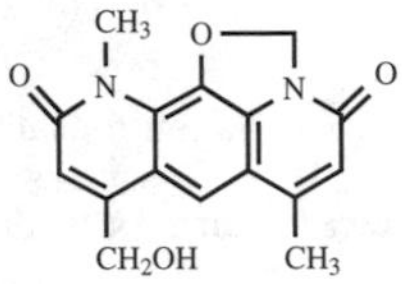

Nyctalopia Night blindness caused by a deficiency of vitamin A.

Nyctophobia The extreme fear of darkness.

Nydrazid A trade name for isoniazid, an antimicrobial agent that inhibits bacterial cell wall synthesis.

Nylander's Reagent A reagent for the determination of reducing sugars.

Nylon Membrane A synthetic polymer of polyhexamethylene adipamide used for blotting DNA or protein.

Nyquil Nighttime A trade name for a combination drug containing pseudoephedrine, dextromethorphan, and doxylamine.

Nystatin An antibiotic used in the treatment of *Candida* infections.

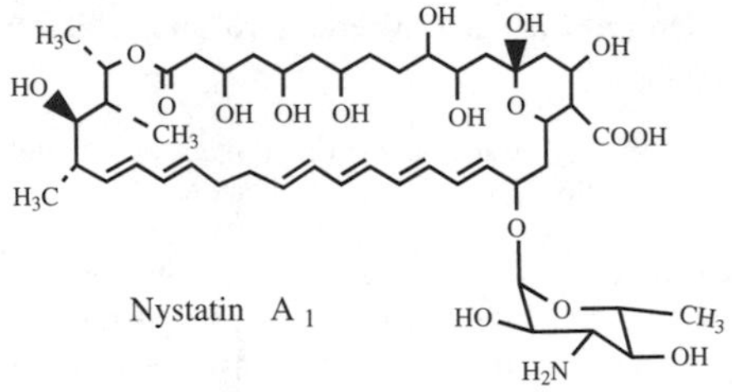

Nystel A trade name for nystatin, an antibiotic.

Nystop A trade name for nystatin, an antifungal agent.

Nystex A trade name for nystatin, used as an antifungal agent that alters the permeability of the fungal cells.

Nytol with DPH A trade name for diphenhydramine hydrochloride, an antihistaminic agent that competes with histamine for H_1 receptors on effector cells.

NZB Mouse A genetically inbred strain of mice in which autoimmune disease resembling systemic lupus erythematosus develops spontaneously.

NZBM Abbreviation for New Zealand black mouse.

NZO Mouse A New Zealand obese mouse.

NZWM Abbreviation for New Zealand white mouse.

O

O_2 Symbol for molecular oxygen.

$^1O_2$ Abbreviation for singlet oxygen.

o- A prefix in chemical nomenclature denoting ortho.

O_2^- Symbol for superoxide anion, an oxygen metabolite toxic to microorganisms.

O Antigen Referring to heat-stable, alcohol-resistant lipopolysaccharide-protein complex (somatic antigen) from the Gram-negative bacteria of the family Enterobacteriaceae. The lipopolysaccharide component of the O antigen consists of the repeating trisaccharide units, 2-keto-3-deoxyoctonate and lipid A.

OA Abbreviation for 1. ovalbumin; 2. oleic acid; 3. okadaic acid; 4. oxalic acid.

OAA Abbreviation for oxaloacetic acid, an intermediate in the Krebs cycle.

OAF Abbreviation for osteoclast-activating factor.

OAG Abbreviation for 1-oleoyl-2-acetyl glycerol.

OAP Abbreviation for a combination drug containing oncovin, ara-C, and prednisone.

OAT Abbreviation for ornithine amino-transferase.

Obalan A trade name for phendimetrazine tartrate, used as a cerebral stimulant to promote transmission of nerve impulses.

Obelin A luminescent protein obtained from the jellyfish *Obelia geniculata.*

Obe-Nix A trade name for phentermine hydrochloride, used as a cerebral stimulant to promote transmission of nerve impulses.

Obephen A trade name for phentermine hydrochloride, used as a cerebral stimulant to promote transmission of nerve impulses.

Obermine A trade name for phentermine, used as a cerebral stimulant to promote transmission of nerve impulses.

Obesity A condition characterized by an abnormal increase of fat in the stomach, intestine, and in the tissue beneath the skin.

Obestin A trade name for phentermine hydrochloride, used as a cerebral stimulant to promote transmission of nerve impulses.

Obesumbacterium A genus of nonmotile, slow-growing bacteria of the family Enterobacteriaceae.

Obeval A trade name for phendimetrazine tartrate, used as a cerebral stimulant to promote transmission of nerve impulses.

Obidoxime Chloride (mol wt 359) A cholinesterase reactivator.

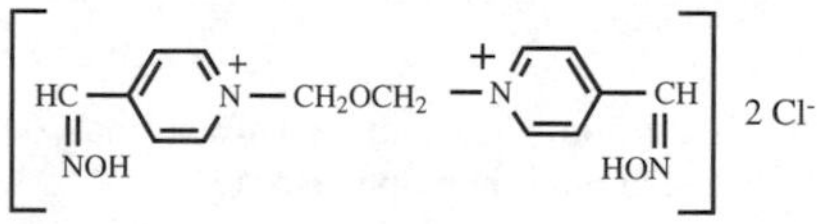

Objective Lens The microscope lens closest to the object under view.

Obligate 1. Necessary or essential for life. 2. Restricted to a specified condition of life.

Obligate Aerobes Organisms that require oxygen for cellular growth and cellular respiration.

Obligate Anaerobes Organisms that cannot use molecular oxygen and grow only under anaerobic conditions.

Obligate Intracellular Parasites Organisms that can live and reproduce only within the cells of other organisms, e.g., viruses.

Obligate Thermophiles Organisms that grow only at high temperatures.

OBP Abbreviation for odorant-binding protein.

Obstetrics Medical science that deals with the care of women during pregnancy.

Oby-Trim A trade name for phentermine hydrochloride, used as a cerebral stimulant to promote transmission of nerve impulse.

OC Abbreviation for osteocalcin.

OCA Abbreviation for a combination drug containing oncovin, cyclophosphamide, and adriamycin.

Occlucort A trade name for betamethasone dipropionate, a hormone.

Occludin An integral membrane protein that localizes in tight junctions in chick liver.

Oceanospirillum A genus of Gram-negative, asporogenous marine bacteria.

Ochratoxins Toxic metabolites and inhibitors for the phosphorylase and mitochondrial respiratory chain from *Aspergillus ochraceus,* e.g., ochratoxin A and ochratoxin B.

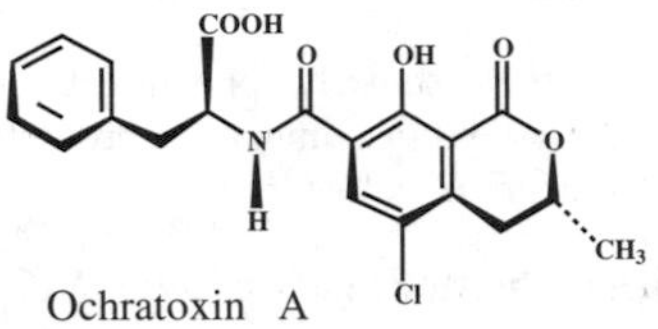

Ochratoxin A

Ochre Codon Referring to the UAA termination codon.

Ochre Mutation The mutation of a normal codon to Ochre termination codon (UAA).

Ochre Suppression Suppression of Ochre termination codon (UAA) allowing continuation of translation and preventing premature termination of protein synthesis.

Ochre Suppressor A gene that encodes an altered tRNA whose anticodon can recognize the Ochre termination codon (UAA), allowing continuation of translation, preventing premature termination of protein synthesis.

Ochrobium A genus of iron bacteria.

OCT Abbreviation for ornithine carbamoyl transferase.

***oct* Plasmid** A plasmid of *Pseudomonas* that encodes the activity for metabolism of octane and decane.

Octa- A prefix meaning eight.

Octacaine (mol wt 234) A local anesthetic agent.

$$CH_3-\underset{\displaystyle N(C_2H_5)_2}{CH}CH_2CONHC_6H_5$$

Octadecadienoic Acid (mol wt 281) A straight chain fatty acid having 18 carbons and 2 double bonds.

Octadecapentaenoic Acid (mol wt 274) A straight chain fatty acid having 18 carbons and 5 double bonds.

Octadecatetraenoic Acid (mol wt 276) A straight chain fatty acid having 18 carbons and 4 double bonds.

Octadecatrienoic Acid (mol wt 278) A straight chain fatty acid having 18 carbons and 3 double bonds.

Octadecenoic Acid (mol wt 282) A straight chain fatty acid having 18 carbons and 1 double bond.

Octahydron Any solid geometrical figure having 8 plane triangular faces, 12 edges, and 6 tetrameric vertices.

Octamer An oligomer that consists of eight monomers.

Octamide A trade name for metoclopramide that stimulates GI tract motility.

Octamoxin (mol wt 144) An antidepressant.

$$CH_3(CH_2)_5-\underset{\displaystyle NHNH_2}{CH}CH_3$$

Octanal (mol wt 128) An aldehyde derived from octanol.

$$CH_3(CH_2)_7CHO$$

Octanol (mol wt 130) An alcohol.

$$CH_3(CH_2)_7OH$$

Octanol Dehydrogenase The enzyme that catalyzes the following reaction:

$$\text{1-Octanol} + NAD^+ \rightleftharpoons \text{1-Octanal} + NADH$$

Octapeptins A group of antibiotics related to polymyxins.

Octaverine (mol wt 397) An antispasmodic agent.

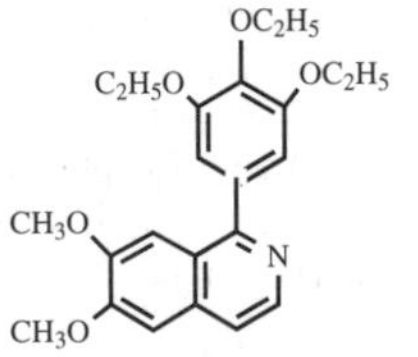

Octocaine A trade name for lidocaine hydrochloride, used as a local anesthetic agent that blocks depolarization by interfering with sodium-potassium exchange across the nerve cell membrane.

Octodrine (mol wt 129) A decongestant.

$$H_3C-\underset{\displaystyle CH_3}{CH}-(CH_2)_3-\underset{\displaystyle NH_2}{CH}-CH_3$$

Octonic Acid (mol wt 144) An organic acid.

$CH_3(CH_2)_6COOH$

***N*-Octonoyl-*N*-Methylglucamine (mol wt 321)** A water-soluble detergent with high solubilization power and nondenaturing property.

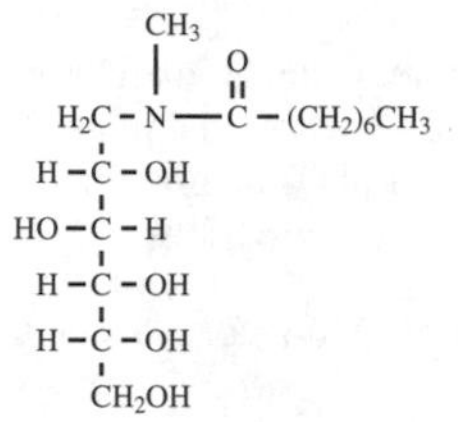

Octopamine (mol wt 153) A biogenic amine and a neurotransmitter that acts as an adrenergic agonist.

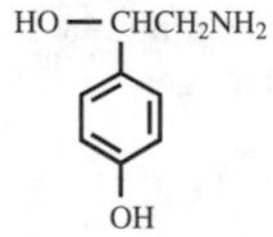

Octopine (mol wt 246) A compound found in the tumors of crown-gall disease of plants and in the muscle of certain invertebrates.

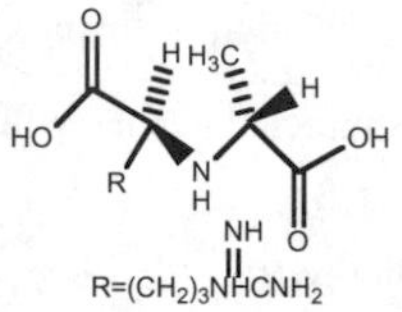

D-Octopine Dehydrogenase The enzyme that catalyzes the following reaction:

N^2-(D-1-Carboxyethyl)-L-Arginine

$NAD^+ + H_2O$

⇅

L-Arginine + Pyruvate + NADH

D-Octopine Synthetase Synonym of D-octopine dehydrogenase.

Octose A monosaccharide that contains eight carbon atoms.

Octostim A trade name for desmopressin acetate, a hormone.

Octreotide (mol wt 1019) A peptide and an analog of somatostatin.

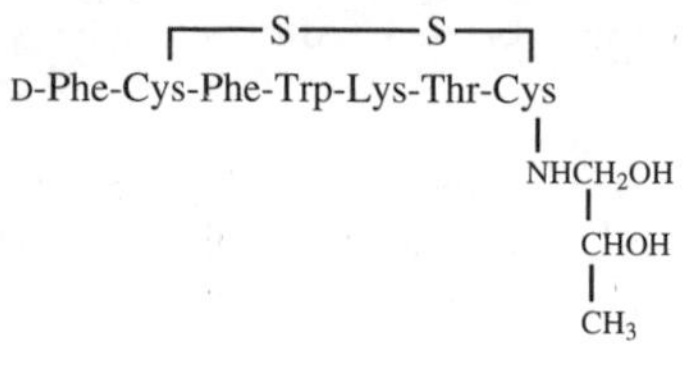

Octulose Any ketose having a chain of eight carbons.

Octyl-β-D-Glucopyranoside (mol wt 292) A nonionic detergent with absorbance at 228 nm and a reagent used for solubilization of membrane protein.

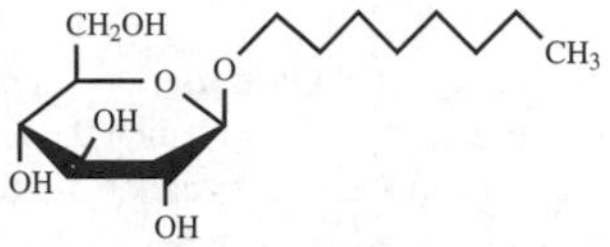

Octyl-β-D-1-Thioglucopyranoside (mol wt 308) A nonionic detergent with absorbance at 228 nm and a reagent used for solubilization of membrane protein.

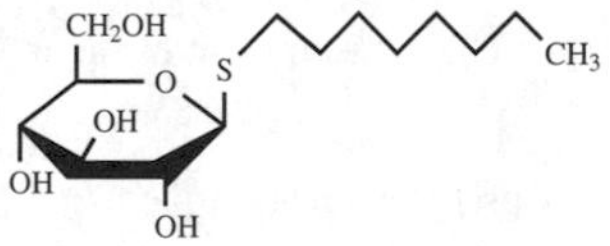

Octyl-thioglucoside (mol wt 308) A mild nonionic detergent.

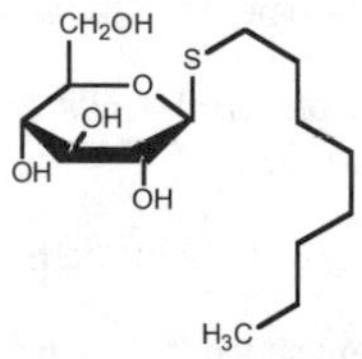

Ocufen A trade name for flurbiprofen sodium, used as an ophthalmic anti-inflammatory agent.

Ocuflox A trade name for ofloxacin, an antibiotic.

Ocular Pertaining to the eye or the eyepiece of a microscope or an optical instrument.

Ocular Lens The eyepiece of a microscope or an optical instrument.

Oculinum A trade name for botulinum toxin type A that produces a neuromuscular paralysis by binding to the acetylcholine receptors on the motor end-plate.

Oculo- A prefix denoting the eye.

Ocu-Pred A trade name for prednisolone sodium phosphate, used as an ophthalmic anti-inflammatory agent that reduces the infiltration of leukocytes at the site of inflammation.

Ocupress A trade name for carteolol hydrochloride, a beta adrenergic blocker used as an antihypertensive agent.

Ocusert Pilo A trade name for pilocarpine hydrochloride, used as a cholinergic drug. It also causes miosis.

OD Abbreviation for 1. optical density; 2. outside diameter.

ODC Abbreviation for ornithine decarboxylase.

Odd-Chain Fatty Acid Oxidation The oxidation of odd-numbered fatty acids in which the propionyl-CoA resulting from the final cycle of β-oxidation is converted to succinyl-CoA to enter the Krebs cycle.

Odd-Numbered Chain Fatty Acid Any fatty acid that has an odd number of carbon atoms.

ODN Abbreviation for oligodeoxyribonucleotide.

Odometry The measurement of the intensity of odors.

Odont- A prefix denoting a tooth.

Odontoblast Specialized cells in the dental papilla that secret dentin.

Odontogenesis The development of teeth.

Odontogenic Epithelial Cells Epithelial layer that gives rise to teeth.

Odontology The study of teeth.

Odorant Agent added to the dangerous substance to warn its presence.

ODP Abbreviation for orotidine diphosphate or orotidine 5′-diphosphate.

OE Abbreviation for oleoylethanol.

OEC Abbreviation for oxygen-evolving complex.

Oerskovia A genus of asporogenous bacteria (Actinomycetales).

Oestradiol Variant spelling for estradiol.

Oestrogen Variant spelling for estrogen.

OFA Abbreviation for onco-fetal antigen.

OFAGE Abbreviation for orthogonal field agarose gel electrophoresis.

O-Flex A trade name for orphenadrine citrate, a skeletal muscle relaxant that reduces the transmission of nerve impulses from the spinal cord to the skeletal muscle.

Ofloxacin (mol wt 361) A broad spectrum antibiotic that inhibits DNA gyrase preventing DNA replication.

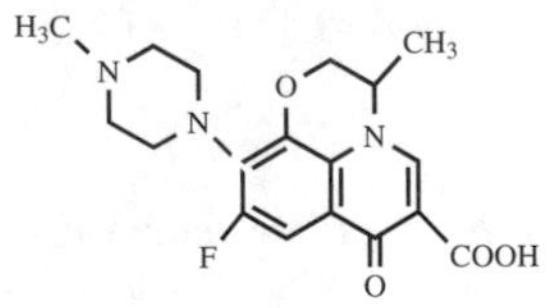

2-OG Abbreviation for 2-oxo-glutarate.

OGCP Abbreviation for oxoglutarate/malate carrier protein, a mitochondrial inner membrane integral protein that plays an important role in a number of biochemical processes.

OGDH Abbreviation for 2-oxoglutarate dehydrogenase.

Ogen A trade name for estropipate (piperazine ostrone sulfate) that increases synthesis of DNA, RNA, and proteins.

-ogen A suffix meaning inactive precursor of an enzyme, e.g., pepsinogen.

OGS Abbreviation for oxogenic steroid.

3′-OH End The unbounded hydroxyl group at the 3′-posistion of ribose or deoxyribose at the end of a nucleic acid molecule.

OH-Cbl Abbreviation for hydroxycobalamin.

17-OHCS Abbreviation for 17-hydroxycorticosteroids.

8-OHdG Abbreviation for 8-hydoxydeoxyguanosine.

Ohm A unit of electric resistance equal to the resistance of a circuit in which a potential difference of one volt produces a current of one ampere.

Ohm's Law The law states that the strength of the direct current is directly proportional to the potential difference and inversely proportional to the resistance.

17-OHP Abbreviation for 17-hydroxy-progesterone.

-oic A suffix in chemical nomenclature denoting the carboxyl group.

-oid A suffix meaning resembling.

OIH Abbreviation for ortho-iodo-hippurate.

Oil Immersion Objective An objective lens used to increase resolving power by filling the space between the specimen and the objective with an oil.

Oil Overlay A thin film of oil placed on the top of a liquid or solid culture medium to reduce the rate of dehydration from the surface.

Oil Red O (mol wt 409) A lipoprotein stain.

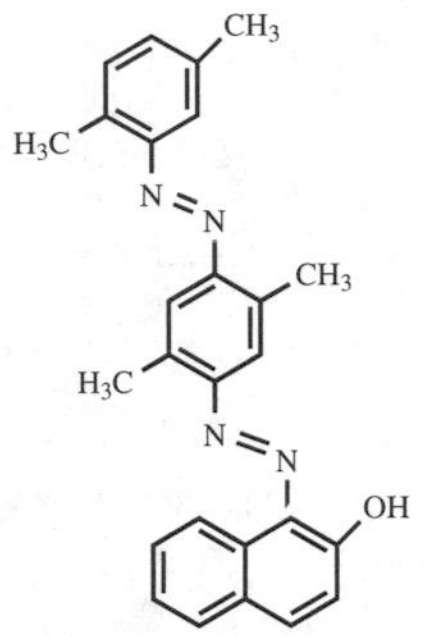

Ointment Any of the soft, bland, high viscous preparations used as a vehicle for external medication as an emollient or as a cosmetic.

Okazaki Fragments Short ssDNA fragments produced on the DNA template during the early stage of discontinuous replication of DNA. They represent the precursor fragments of the lagging strand and are subsequently linked together by DNA-ligase to form a linear daughter strand.

-ol A suffix denoting the presence of a hydroxyl group attached to a carbon.

Olanzapine (mol wt 312) A dopaminergic blocking agent used as an anti-psychotic agent.

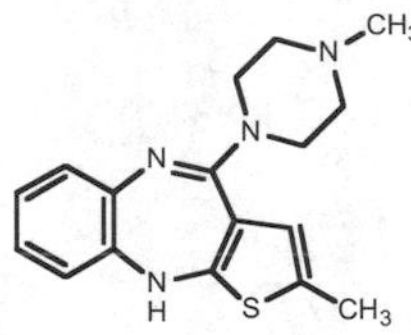

Old Yellow Enzyme Referring to the NADPH dehydrogenase.

Oleaginous Referring to a cell that accumulates large amount of lipid.

Oleandomycin (mol wt 688) An antibiotic related to erythromycin from *Streptomyces antibioticus*. It inhibits bacterial protein synthesis.

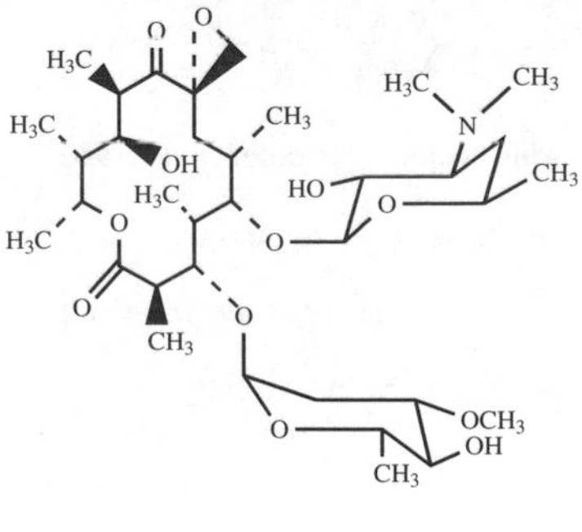

Oleandrin (mol wt 577) A cardiotonic and diuretic agent.

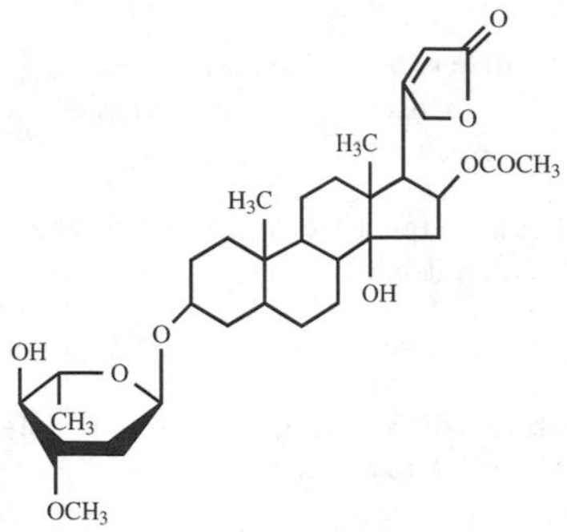

Oleandrose (mol wt 178) 2,6-Dideoxy-3-O-methyl-arabino-hexose, a component of some cardiac glycosides.

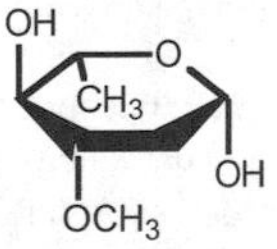

Oleate A salt of oleic acid.

Olein Ester of glycerol and oleic acid.

Oleometer A device for measuring the specific gravity of an oil.

Oleophilic Having an affinity for oil.

Oleophobic Lacking an affinity for oil.

Oleoresin A natural mixture of essential oils and a resin.

Oleosome A plant cell structure that is rich in lipid that serves as a storage granule in seeds and fruits.

Olfactometer A device for measuring and testing the sense of smell.

Olfactory Organ An organ of smell.

OLH Abbreviation for 1. ovine lactogenic hormone; 2. ovine luteinizing hormone.

Oligaemia Variant spelling of oligemia.

Oligemia The deficiency in the volume of blood.

Oligo- A prefix meaning scant or little.

2′,5′ Oligoadenylate A 2′,5′ linked oligomer of adenine nucleotide that is capable of converting inactive nuclease to active nuclease.

2′,5′-Oligoadenylate Synthetase The enzyme that catalyzes the formation of oligoadenylate joined by 2′,5′-phosphodiester linkage rather than by the usual 2′,3′-phosphodiester linkage (abbreviated as 2,5-A).

Oligodendrocyte Cell in the central nervous system that is responsible for forming the myelin sheath around a nerve axon.

Oligodendrocytoma A slow-growing brain tumor (see also oligodendroglioma).

Oligodendroglioma A slow-growing brain tumor.

Oligoglucan-Branching Glycosyltransferase Synonym of 1,4-α-glucan 6-α-glucosyltransferase.

Oligo-1,6-Glucosidase The enzyme that catalyzes the hydrolysis of 1,6-glucosidic linkage in isomaltose and dextran.

Oligoglycosylglucosyl Ceramide Glycohydrolase The systematic name for endoglycosylceramidase.

Oligomer A polymer that consists of only a small number of monomeric units, e.g., short chain nucleotide (oligonucleotide), short chain polysaccharide (oligopolysaccharide), or short chain peptide (oligopeptide).

Oligomycins Antibiotics produced by a species of *Streptomyces,* e.g., oligomycin A, B, C, and D, that bind to mitochondrial F_0F_1-ATPase and inhibit the transfer of electrons and oxidative phosphorylation.

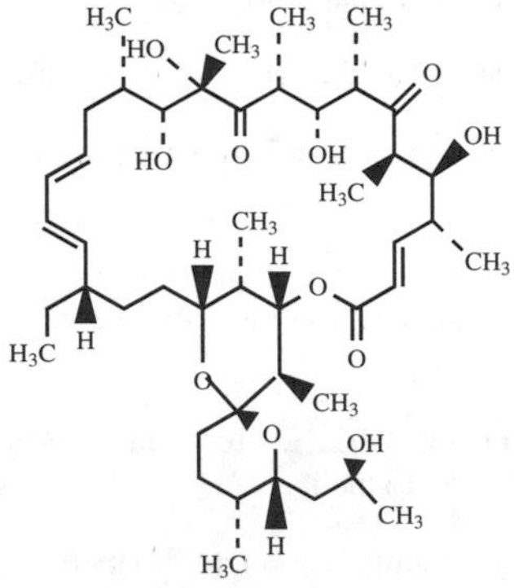

Oligomycin A

Oligomycin Sensitivity-Conferring Protein A protein found in the stalk region of the mitochondrial F_0F_1-ATPase.

Oligonucleate 5'-Nucleotidohydrolase The systematic name for phosphodiesterase I.

Oligonucleotide A short polynucleotide consisting of up to about 20 nucleotides.

Oligopeptide A short polypeptide consisting of up to about 40 amino acid residues.

Oligosaccharide Linear or branched carbohydrate that consists of up to about 20 monosaccharide units linked by glycosidic bonds.

Oligosaccharin An oligosaccharide derived from plant cell wall that induces physiological responses and acts as a molecular signal.

Oligospermia Insufficient number of sperm in the semen.

Oligotroph Organisms that can grow in a nutrient-poor environment.

Oligotrophic Pertaining to a body of water that is low in nutrients.

Oliguria The passage of insufficient amount of urine or inability to secret urine.

Olivomycins A mixture of antibiotics produced by *Streptomyces olivoreticuli*, e.g., olivomycin A, B, C, and D.

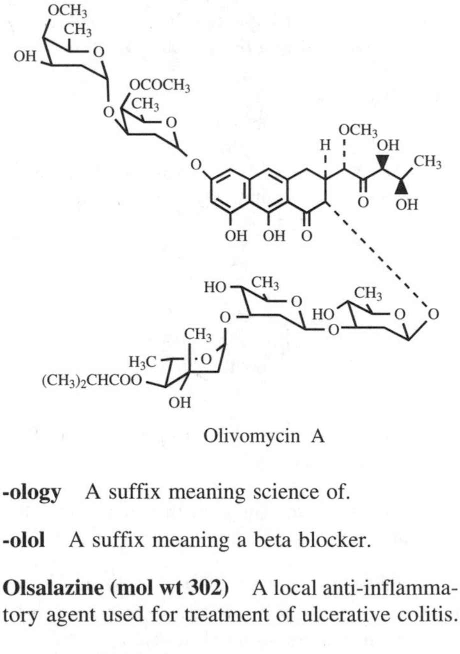

Olivomycin A

-ology A suffix meaning science of.

-olol A suffix meaning a beta blocker.

Olsalazine (mol wt 302) A local anti-inflammatory agent used for treatment of ulcerative colitis.

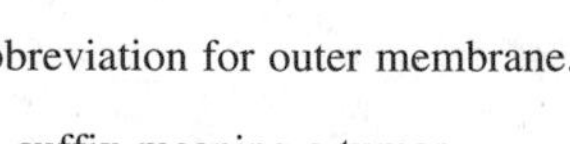

OM Abbreviation for outer membrane.

-oma A suffix meaning a tumor.

Omega The last letter of the Greek alphabet.

Omega-3 Fatty Acid A fatty acid that has a double bond at the number 3 carbon from the omega end (CH_3 end).

Omega Oxidation A pathway for oxidation of fatty acid in which the CH_3 group of a fatty acid at the nonpolar end is oxidized to a COOH group leading to the formation of a dicarboxylic acid.

Omega Protein Referring to type-I topoisomerase that relaxes negatively supercoiled DNA.

Omeprazole (mol wt 345) An antiulcerative agent that inhibits the activity of the acid pump (H^+K^+-ATPase) and blocks the formation of gastric acid.

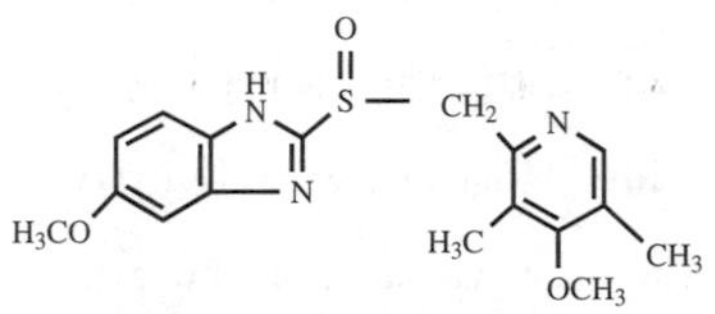

3OMGlc Abbreviation for 3-O-methylglucose.

Omicron An endosymbiotic, Gram-negative, rod-shape bacteria occurring in the cytoplasm of protozoa, e.g., *Euplotes*.

Omnicef A trade name for cefdinir, an antibiotic.

OmniHIB A trade name for hemophilus b conjugated vaccine with tetanus toxoid.

Omnipen A trade name for the antibiotic ampicilin, which inhibits bacterial cell wall synthesis.

Omnivorous Referring to 1. protozoa that feed on microscopic animals or plants, and 2. hetrotrophic animals that consume both meat and plant materials.

Omoconazole (mol wt 424) An antifungal agent.

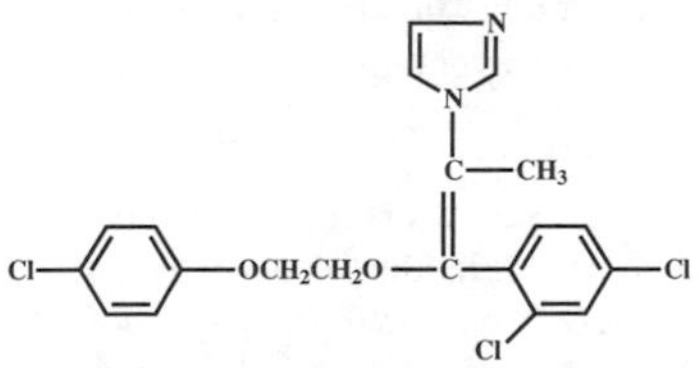

OMP Abbreviation for 1. orotidine monophosphate and 2. outer membrane protein in Gram-negative bacteria.

OMP Decarboxylase Abbreviation for orotidine monophosphate decarboxylase, the enzyme that catalyzes the following reaction:

Orotidine monophosphate

$\rightleftharpoons$

Uridine monophosphate + CO_2

OmPA Abbreviation for outer membrane protein A.

OMS-Concentrate A trade name for morphine sulfate, a narcotic analgesic agent.

ON Abbreviation for osteonectin.

Oncaspar A trade name for pegasparagase, an antineoplastic agent.

Onco- A prefix meaning tumor.

Oncofetal Antigens Antigens expressed during normal fetal development that disppear during the adult life and reappear during cancer development, e.g., carcinoembryonic antigen (CEA).

Oncogene One of a large number of genes whose presence can cause a cell to become malignant. The oncogen in a retrovirus *(v-onc)* that causes tumor formation when introduced into a nontumor cell. The oncogen of the eukaryotic cell (proto-oncogen or cellular oncogen c-onc) arises by mutation of normal genes involved in control of cell growth and division.

Oncogenesis The process of producing neoplasia or malignancy.

Oncogenic Capable of causing cancer.

Oncogenic Viruses Viruses that are capable of causing cell transformation or tumors, e.g., oncogenic DNA viruses and oncogenic RNA viruses (retroviruses).

Oncology The science that deals with tumors.

Oncolysis The destruction of tumor or tumor cell.

Oncolytic Capable of destroying tumor cells.

Oncomouse A laboratory mouse that carries activated human cancer genes.

Oncornavirus See retroviruses.

Oncostatin M A polypeptide produced by a monocytelike cell line that inhibits the replication of A375 melanoma cells and certain human tumor cells.

Oncovin A trade name for vincristine sulfate, an antineoplastic drug that blocks cell division.

Oncovirinae A subfamily of viruses of the retroviridae; there are three types (type B, type C and type D).

Ondansetron (mol wt 293) An antiemetic agent and an antagonist of serotonin receptors.

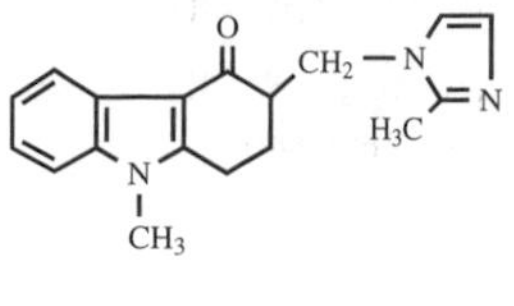

One Carbon Transfers The enzymatic reaction that mediates the transfer of a one-carbon fragment from one compound to another, e.g., transfer of a methyl group or a formyl group.

One Gene One Enzyme The original hypothesis for gene translation stating that each gene controls the synthesis of one enzyme. The hypothesis is only partially true since many genes specify proteins that are not enzymes and many proteins or enzymes consisting subunits that are specified by different genes.

One Gene One Polypeptide The hypothesis stating that each gene specifies one polypeptide. The principle is not completely accurate since many genes specify only functional RNA not polypeptides, e.g., genes for tRNA, rRNA.

One Step Growth Curve The graphic representation that describes the lytic reproduction cycle of both bacterial and animal viruses.

One Step Growth Experiment A procedure for quantitative study of the reproduction cycle of a lytic bacteriophage or animal virus.

$ONOO^-$ Symbol for peroxynitrite ion.

ONPG (mol wt 301) Abbreviation for *O*-nitrophenyl β-D-galactoside, a substrate for determination of β-D-glactosidase activity.

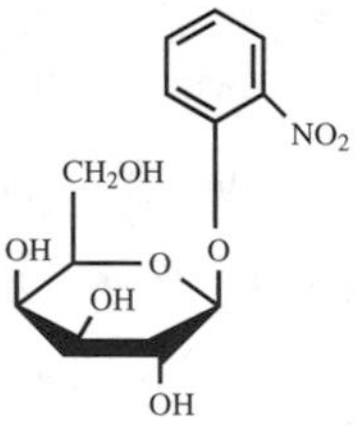

Ontianil (mol wt 282) An antifungal agent.

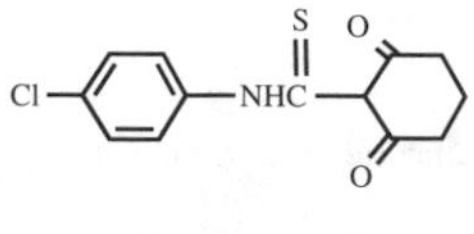

Ontogeny The developmental history of an individual organism.

Onychomycosis Mycosis that affects the finger- or toe-nails.

O'nyong-nyong Fever A mosquito-transmitted disease caused by alphavirus and characterized by fever, headache, and rash.

Oo- A prefix meaning egg.

Oocyte The female sex cell that undergoes meiosis while producing ovum.

Oogamy The fertilization that involves morphologically distinguishable male and female gametes, e.g., a motile male gamete and a relatively large female gamete.

Oogenesis The process of egg formation.

Oogonium 1. The female gametangium of algae and fungi. 2. Cell that serves as source of oocytes.

Ookinete The motile, elongated zygote of the malarial parasite (*Plasmodium*) formed after fertilization of the macrogamete.

Oolemma The plasma membrane of the ovum.

Oophoritis Inflammation of the ovary.

Ooplasm The cytoplasm of an oocyte.

Oosphere A female gamete that develops inside an oogonium.

Oospore A thick-walled resting spore produced from a fertilized oosphere.

Ootide Nucleus One of the four nuclei formed by meiotic division of a primary oocyte.

OP Abbreviation for 1. organophosphate; 2. ovine prolactin; 3. *o*-phenanthroline.

OPA Abbreviation for *o*-phthalaldehyde.

***opal* Codon** The termination codon UGA that causes termination of protein synthesis.

Opal Mutation A mutation that results in change of a normal codon to the UGA *opal* termination codon.

***opal* Suppression** The suppression of the *opal* termination codon allowing continuation of protein synthesis and preventing premature termination of translation.

Opal Suppresser A gene that encodes an altered tRNA whose anticodon can recognize the *opal* termination codon allowing continuation of protein synthesis and preventing premature termination of translation.

Opalina A genus of parasitic protozoa.

Open Chain A noncyclic polymer chain in which the two ends are not covalently linked together, e.g., nucleic acid or noncyclic polypeptide.

Open Culture Synonym for continuous culture.

Open Hemoprotein A hemoprotein in which the 5th and/or 6th coordination position of the heme are not occupied.

Open Reading Frame A nucleotide sequence between initiation codon and termination codon.

Open System A system that exchanges matter as well as energy with its surroundings.

Open System Culture Synonym for continuous culture.

Operator A site in the operon that interacts with specific repressor and thereby exerts control over transcription of its adjacent structural genes.

Operator Constitutive Mutant Mutant of an operator gene that is incapable of binding with repressor so that a previously inducible enzyme becomes constitutive. The mutant is characterized by the production of gene product regardless of the presence or absence of the inducer.

Operon A cluster of functionally related genes whose expression or operation is regulated by the interaction of repressor protein with operator gene. The operon consists of a promoter gene, an operator gene, and closely linked structure genes.

Operon Fusion The fusion of two operons so that the structural genes of the two operons are under the control of the same operator gene.

Operon Model See operon.

Operon Network A collection of operons that are capable of interacting with each other so that gene products of one operon act as effectors or repressors for the regulation of the other operon.

Ophthaine A trade name for proparacaine hydrochloride, used as a topical ophthalmic anesthetic agent that prevents the initiation and transmission of nerve impulses at the nerve cell membrane.

Ophthalgan A trade name for glycerin (anhydrous), used to remove excess fluid from the cornea.

Ophthalmia Inflammation of the eye or eyeball.

Ophthalmic Pertaining to the eye.

Ophthalmitis Inflammation of the eye.

Ophthalmodiaphanoscope A device for examination of the interior of the eye.

Ophthalmodynamoter A device used for measuring blood pressure of the retinal vessels and the power of convergence of the eye.

Ophthalmology The science that deals the with eye.

Ophthoclor Ophthalmic A trade name for chloramphenicol, an antibiotic that inhibits bacterial protein synthesis and used for treatment of bacterial infections of the eye.

Ophthocort A trade name for a combination drug containing chloramphenicol, polymyxin B sulfate, and hydrocortisone acetate, used as an ophthalmic anti-infective agent.

Opiate Receptor Synonym for opioid receptor.

Opiates Alkaloids derived from opium poppy with high pharmacological activity.

Opines The Ti-plasmid encoded substances produced by cells of crown gall of plants, e.g., octopine and nopaline.

Opiniazide (mol wt 329) An antituberculostatic agent.

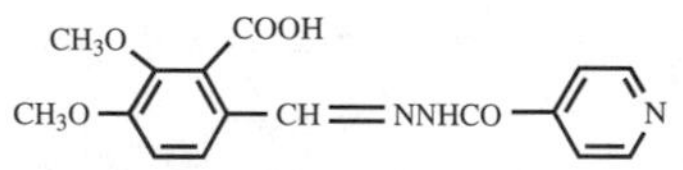

Opiniazide (mol wt 329) An antibacterial agent.

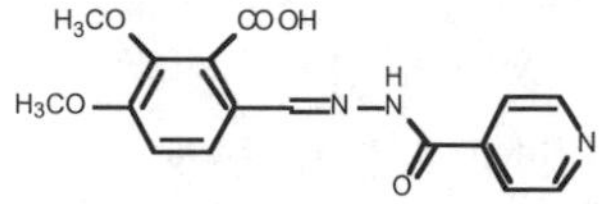

Opioid Natural and synthetic substances that possess properties and characteristics of opiate narcotics but not derived from opium.

Opioid Peptide A group of peptides that are capable of binding with opioid receptors, e.g., endorphin.

Opioid Receptor Membrane proteins on animal cells, e.g., brain cells, that bind opiate peptide neurotransmitters. The name is given because opiates are potent agonists that bind to the receptors and mimic the action of the natural transmitter.

Opiomelanocortin A polypeptide present in the intermediate lobe of the pituitary gland.

Opipramol (mol wt 363) An antidepressant and antipsychotic agent.

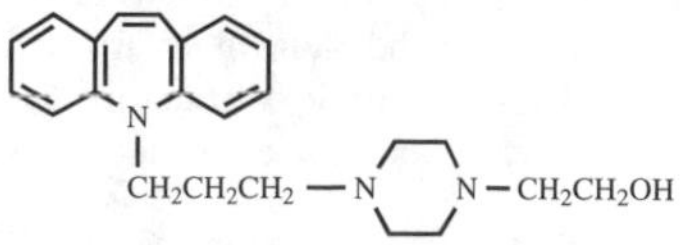

Opium Air-dried, milky exudation from unripe pods of opium poppy, e.g., *Papaver somniferum* or *P. album* (Papaveraceae). It contains a number of

Opium Poppy An annual Eurasian poppy, e.g., *Papaver somniferum* or *P. album* (Papaveraceae).

OPL Abbreviation for a combination drug containing oncovin, prednisone, and leucogen.

Opportunistic Infection Infections caused by opportunistic pathogens of relatively low virulence in individuals with an altered immunity.

Opportunistic Pathogens The organisms that exist as part of the normal body microbiota but may become pathogenic when the normal body defense mechanisms have been impaired.

OPPP Abbreviation for oxidative pentose phosphate pathway.

Oprelvekin A recombinant human interleukin-11 produced by *E. coli* used as a hematopoietic growth factor.

OPRT Abbreviation for orotate phosphoribosyl transferase.

OPRTase Abbreviation for orotate phosphoribosyl transferase.

Opsin The apoprotein (protein moiety) of visual pigment of vertebrate or of bacteriorhodopsin.

Opsonic Effect See opsonization.

Opsonic Index The ratio of phagocytic activity of a patient's blood phagocytes to that of a normal individual in respect to a specific opsonin.

Opsonin Substance capable of binding to the surface of foreign cells or particles and enhancing phagocytosis by the phagocytes. Antibody and complement are two major opsonins occurring in the body.

Opsonization The process of coating a foreign cell or particle with opsonin to enhance phagocytosis by the phagocytes.

Optazine A trade name for naphazoline hydrochloride, used as an ophthalmic vasoconstrictor.

Optical Activity The property possessed by certain substances in rotating the plane of polarized light due to the presence of asymmetric atoms; the substances that rotate the plane of polarized light to the right are designated as dextrorotatory and those to the left are designated as levorotatory.

Optical Brightner 1. Compound that absorbs ultraviolet light and emits visible light. 2. A bleaching agent.

Optical Density (OD) Synonym for absorbance.

Optical Isomers Isomers that differ from each other in their configurations or spatial arrangement due to the presence of asymmetric carbon atoms. They are mirror image of one another but are not superimposable. It should be noted that not all optical isomers exhibit this optical rotation property.

Optical Rotation The rotation of the plane of polarized light by an optically active substance due to the presence of asymmetric atoms; the substances that rotate the plane of polarized light to the right are designated as dextrorotatory and those to the left are designated as levorotatory.

Opticrom A trade name for cromolyn sodium, which inhibits mast cell degranulation.

Optigene A trade name for tetrahydrozoline hydrochloride, an ophthalmic vasoconstrictor that produces vasoconstriction.

Optimal Growth Temperature The temperature at which microbes exhibit the maximal growth rate and maximal product yield.

Optimal Oxygen Concentration The oxygen concentration at which microbes exhibit the maximal growth rate and maximal product yield.

Optimal pH The pH at which an enzyme exhibits its maximal activity.

Optimine A trade name for azatadine maleate, an antihistaminic agent that competes with histamine for H_1 receptors on effector cells.

Optimum pH See optimal pH.

Optimyd A trade name for a combination drug containing prednisolone phosphate and sulfacetamide sodium, used as an ophthalmic anti-infective agent.

OptiPranolol A trade name for metipranolol hydrochloride, used as a beta-adrenergic blocker.

Optometrist A specialist licensed to practice optometry.

Optometry Medical science that deals with the examination and detection of faults of refraction of the eye and prescribes correctional lenses.

OPV A trade name for the polio virus vaccine

Orabase HCA A trade name for hydrocortisone.

Orabolin A trade name for ethylestrenol, an anabolic hormone that promotes tissue-building processes.

Oracet Blue B (mol wt 328) A dye.

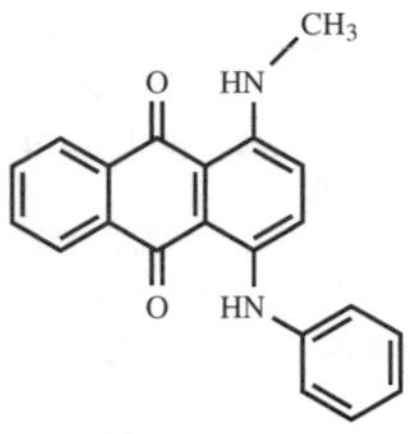

Orafen A trade name for ketoprofen, a non-narcotic analgesic agent.

Orahist A trade name for a combination drug containing phenylpropanolamine hydrochloride and chlorpheniramine maleate, used as an antihistaminic agent.

Oramide A trade name for tolbutamide, which stimulates the release of insulin from pancreatic beta cells.

Oraminic II A trade name for brompheniramine, an antihistaminic agent that competes with histamine for H_1 receptors on effector cells.

Oramorph A trade name for morphine sulfate, a narcotic analgesic agent.

Orange I (mol wt 350) A dye used for staining proteins.

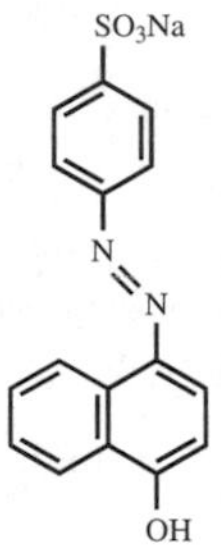

Orange II (mol wt 350) A dye and pH indicator used for spectrophotometric determination of cationic surfactants.

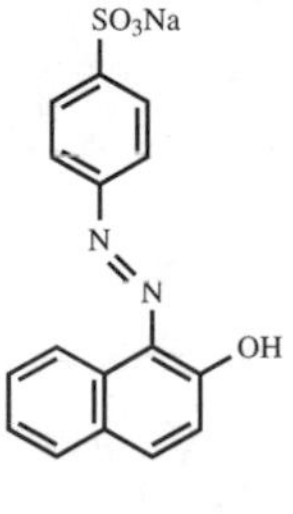

Orange G (mol wt 452) A dye.

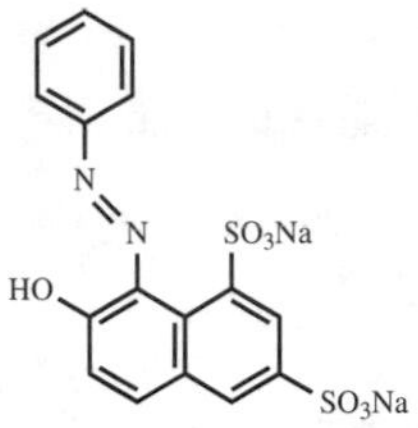

Orap A trade name for pimozide, an antipsychotic agent that blocks dopaminergic receptors.

Oraphen-PD A trade name for acetaminophen, an analgesic and antipyretic agent that blocks the generation of pain impulses.

Orasone A trade name for prednisone, an anti-inflammatory agent.

Ora-Testryl A trade name for fluoxymesterone, an anabolic steroid that promotes tissue building processes.

Orazinc A trade name for zinc, a source of zinc.

Orbenin A trade name for cloxacillin sodium, a penicillinase-resistant antibiotic that inhibits bacterial cell wall synthesis.

Orbivirus A virus of the family Reoviridae that contains double-stranded RNA, e.g., Colorado tick fever.

Orchitis Inflammation of the testis.

Orcinol (mol wt 124) A reagent used for the determination ribose and indirectly for the determination of RNA.

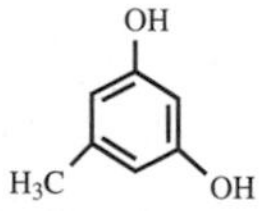

Orcinol Hydroxylase See orcinol 2-monooxygenase.

Orcinol 2-Monooxygenase The enzyme that catalyzes the following reaction:

$$\text{Orcinol} + \text{NADPH} + O_2 \rightleftharpoons \text{Catechol} + H_2O + \text{NAD}^+$$

Orcinol Reaction A colorimetric method for the determination of ribose or pentose based on the production of green color upon treatment of the sample with orcinol and ferric chloride.

Orcipren A trade name for metaproterenol, a beta-2 adrenergic agonist used as a bronchodilator and an antiasthmatic agent.

Ord Abbreviation for orotidine.

Order A taxonomic category ranked below class and above family.

Ordinate The vertical axis in a plane and a rectangular coordinate system.

Oretic A trade name for hydrochlorothiazide, used as a diuretic agent.

Oreticyl Forte A trade name for a combination drug containing hydrochlorothiazide and deserpidine, used as an antihypertensive agent.

Oreton Methyl A trade name for methyltesterone, an anabolic hormone that promotes tissue building processes.

ORF Abbreviation for open reading frame.

Orfenace A trade name for orphenadrine citrate, a skeletal muscle relaxant.

Orflagen A trade name for orphenadrine citrate, a muscle relaxant that reduces transmission of impulses from the spinal cord to the skeletal muscle.

Organ Perfusion A method for study of tissue metabolism in which a specific organ is surgically removed from the animal and connected to an artificial circulation system. The composition of the material entering the organ or metabolic products formed are controlled and monitored.

Organelles Membrane-surrounded structures found in eukaryotic cells that posses specific functions.

Organic Pertaining to carbon-containing compounds.

Organic Acid An acid that contains carbon.

Organic Chemistry The science that deals with compounds of carbon and their derivatives.

Organic Compound See organic molecule.

Organic Molecule Molecule containing two or more covalently linked carbon atoms in addition to hydrogen and other atoms.

Organic Peroxide Compounds that contain the following structure:

R-O-O-R′

Organidine A trade name for iodinated glycerol, used to increase production of respiratory tract fluid to help liquefy and reduce the viscosity of thick, tenacious secretion.

Organism A living entity that is either unicellular or multicellular.

Organogenesis The process of the formation of specific organs in plants and animals.

Organophosphates A large group of phosphate-containing organic compounds used as pesticides, e.g., parathion and malathion.

Organotroph An organism that uses organic compounds as electron donor for production of energy.

Orgaran A trade name for danaparoid sodium, an antithrombotic agent.

Orgotein Water-soluble protein isolated from red blood cells and other tissue, used as an anti-inflammatory agent.

***ori*C Locus** The location where DNA replication in *E. coli* is initiated (replication origin).

Oriental Sore A skin disorder caused by *Leishmania tropica* (protozoa).

Origin of Replication A point DNA on a chromosome at which DNA replication is initiated.

Origin of Transfer A point on the DNA of F-plasmid at which the transfer of the plasmid from an F^+ (male) bacterial cell to an F^- (female) recipient cell begins during conjugation.

Orimune A trade name for an oral poliovirus vaccine.

Orinase A trade name for tolbutamide, which stimulates insulin release from pancreatic beta cells and reduces glucose output by the liver.

Orlaam A trade name for levomethadyl acetate, a narcotic analgesic agent.

Orlistat (mol wt 496) A lipid inhibitor used as a weight loss agent.

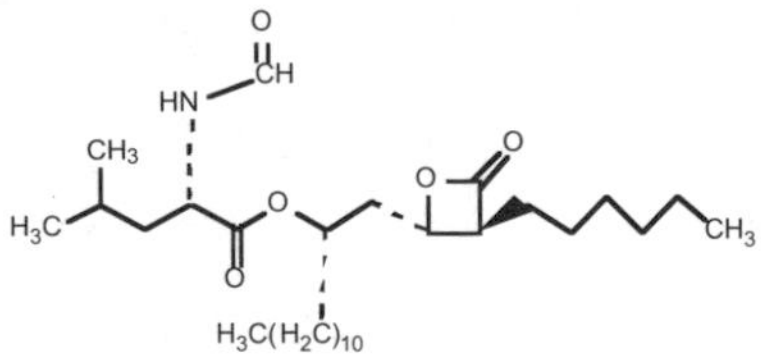

Orn Abbreviation for the amino acid ornithine.

Ornade Spansule A trade name for a combination drug containing phenylpropanol hydrochloride and chlorpheniramine maleate used as a nasal congestant and antihistaminic agent.

Ornidazole (mol wt 220) An anti-infective agent.

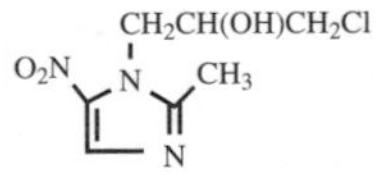

Ornidyl A trade name for eflornithine hydrochloride, used as an antiprotozoal agent.

Ornipressin (mol wt 1042) A synthetic peptide and an analog of vasopressin, used as a vasoconstrictor.

Cys-Tyr -Ph e-Gln-Asn-Cys-Pro-Orn-GlyNH_2
└——S————S——┘

Ornithine (mol wt 132) A nonprotein amino acid and an intermediate in the urea cycle.

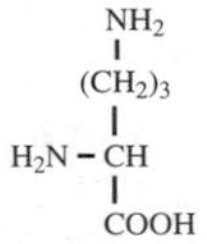

Ornithine Aminotransferase The enzyme that catalyzes the following reaction:

Ornithine + α-ketoglutarate

⇅

Glutamate + glutamate-5-semialdehyde

Ornithine Ammonia-lyase See ornithine cyclodeaminase.

Ornithine Carbamoyltransferase The enzyme that catalyzes the following reaction:

Carbamoyl phosphate + L-ornithine

⇅

Pi + L-citrulline

Ornithine Carboxylyase See ornithine decarboxylase.

Ornithine-Citrulline Cycle See urea cycle.

Ornithine Cycle Referring to urea cycle.

Ornithine Cyclodeaminase The enzyme that catalyzes the following reaction:

L-Ornithine ⇌ L-Proline + NH_3

Ornithine Decarboxylase The enzyme that catalyzes the following reaction:

L-Ornithine ⇌ Putrescine + CO_2

Ornithine Ketoacid Aminotransferase The enzyme that catalyzes the following reaction:

L-Ornithine + a ketoacid

⇅

L-Glutamate 5-semialdehyde + an L-aminoacid

Ornithine Racemase The enzyme that catalyzes the interconversion of D-ornithine and L-ornithine.

Ornithine Transcarbamoylase The enzyme that catalyzes the following reaction:

Carbamoyl Phosphate + ornithine

⇅

Citrulline + Pi

Ornoprostil (mol wt 411) An antiulcerative agent.

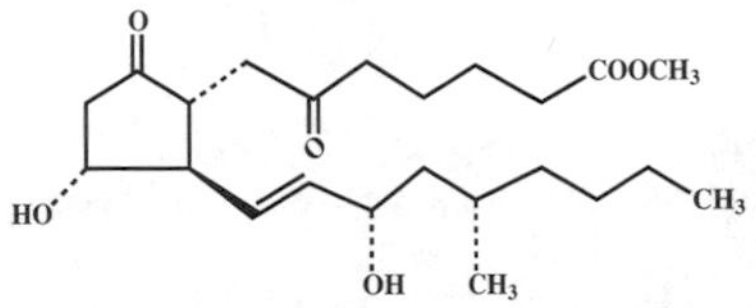

Orosomucoid Plasma glycoprotein of mammals and birds that contains about 38% carbohydrate. The level of orosomucoid increases with inflammation, pregnancy, and certain diseases.

Orotate Phosphoribosyl Transferase The enzyme that catalyzes the following reaction:

Orotate + PRPP

⇅

Orotidine monophosphate + PPi

Orotate Reductase (NAD^+) The enzyme that catalyzes the following reaction:

Dihydroorotate NAD^+ ⇌ Orotate + NADH

Orotate Reductase ($NADP^+$) The enzyme that catalyzes the following reaction:

Dihydroorotate $NADP^+$ ⇌ Orotate + NADPH

Orotic Acid (mol wt 156) A component of the orotidine.

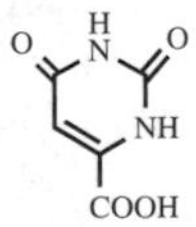

Orotic Aciduria An inherited disorder characterized by the secretion of a large amount of orotic acid in the urine, retarded growth, and severe anemia.

Orotidine (mol wt 288) A nucleoside of orotic acid.

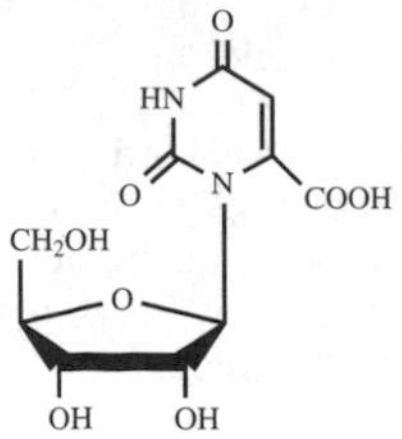

Orotidine Monophosphate (mol wt 368) An intermediate in the biosynthesis of pyrimidine nucleotide.

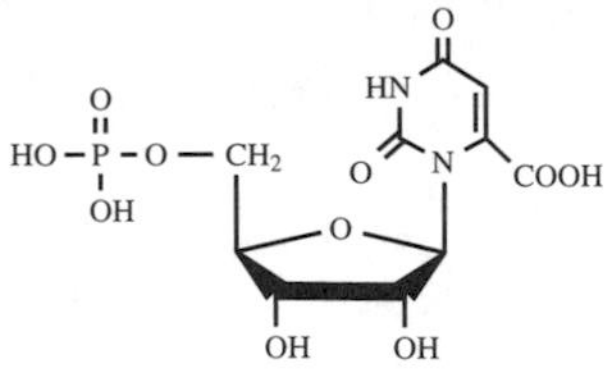

orotidine 5′-monophosphate

Orotidine 5'-Phosphate Carboxylase The systematic name for orotidine 5'-phosphate decarboxylase.

Orotidine 5′-Phosphate Decarboxylase The enzyme that catalyzes the following reaction:

$$\text{Orotidine 5-phosphate} \rightleftharpoons \text{UMP} + CO_2$$

Orotidine 5'-Phosphate Pyrophosphate Phospho-α-D-Ribosyltransferase The systematic name for orotate phosphoribosyltransferase.

Orotidine 5'-Phosphate Pyrophosphorylase Synonym of orotate phosphoribosyltransferase.

Orotidylic Acid A nucleotide of orotic acid (see orotidine monophosphate).

Oroxine A trade name for levothyroxin sodium (T_4 or L-thyroxin sodium) that accelerates the rate of cellular oxidation.

Orphan Drug A drug used for the treatment of rare diseases.

Orphan Virus A virus that is not known to cause disease.

Orphenadrine (mol wt 269) A skeletal muscle relaxant that reduces transmission of nerve impulses from the spinal cord to skeletal muscle.

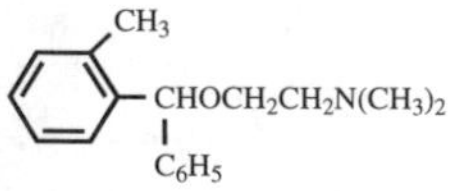

Orphenate A trade name for orphenadrine citrate, used as a skeletal muscle relaxant that reduces transmission of nerve impulses from the spinal cord to skeletal muscle.

Ortho- A prefix meaning 1. two substituents of adjacent atoms of the ring in an aromatic compound, 2. something straight, normal, or correct, and 3. hydrated or hydroxylated to the highest degree, e.g., orthophosphoric acid.

Ortho Dienestrol A trade name for dienestrol, a hormone capable of increasing DNA, RNA, and protein synthesis.

Orthocaine (mol wt 167) An anesthetic agent.

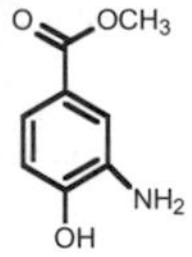

Ortho-Cept A trade name for a combination drug containing desogestrel and ethinyl estradiol used as a contraceptive agent.

Orthochromatic Dye Any dye that stains cells with a single color.

Orthoclone OKT A trade name for a murine monoclonal antibody to the antigen of human T cells used as an immuno-suppressant.

Orthodontics Medical science that deals with irregularities of the teeth and their correction.

Orthodontist A dental specialist licensed to practice orthodontics.

Ortho-Est A trade name for estropipate, a hormone.

Orthologous Genes Gene loci in different species that are sufficiently similar in their nucleotide sequences to suggest that they originated from a common ancestral gene.

Orthomyxoviridae A family of enveloped, minus-stranded RNA animal viruses with divided genome (e.g., influenza).

Ortho-Novum A trade name for a combination drug containing norethindrone and ethinyl estradiol used as an oral contraceptive agent.

Orthophosphate Cleavage Referring to the removal of orthophosphate group from ATP or ADP.

Orthoreovirus Synonym for reovirus.

Ortho-Tri-Cyclen A trade name for a combination drug containing norgestimate and ethinyl estradiol used as an oral contraceptive agent.

Orudis A trade name for ketoprofen, used as an anti-inflammatory agent.

Oruvail SR A trade name for ketoprofen, a nonnarcotic analgesic agent.

Oryzalin (mol wt 346) An herbicide.

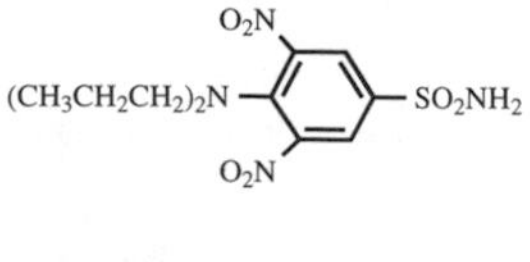

Oryzamin Referring to vitamin B1.

Oryzenin A glutelin found in the seeds of rice.

OS Abbreviation for osteosclerosis.

Os Abbreviation for osmium, a chemical element.

Osalmid (mol wt 229) A choleretic agent.

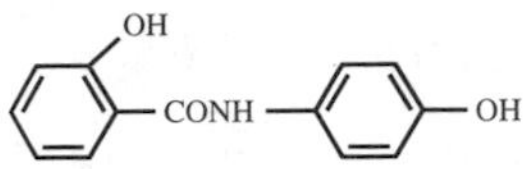

Osaterone (mol wt 365) An anti-androgen used for treatment of benign prostatic hypertrophy.

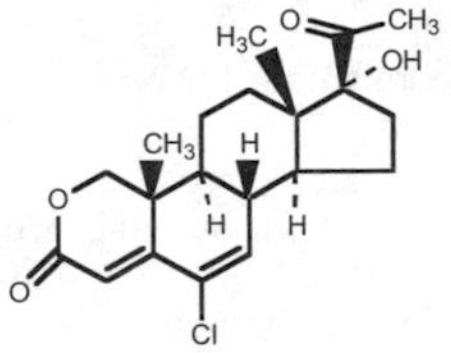

OSCF Abbreviation for oligomycin-sensitivity-conferring factor.

Oscillatoria A genus of cyanobacteria.

OSCP Abbreviation for oligomycin-sensitivity-conferring protein.

-ose A prefix denoting a sugar.

-oside A suffix denoting glycoside.

-osis A suffix denoting 1. a process or state e.g. necrosis; 2. a diseased condition, e.g., tuberculosis; 3. an increase or excess, e.g., leukocytosis.

OsM Abbreviation for osmole.

OSM Abbreviation for oncostatin M, a polypeptide cytokine.

Osmiophilic With an affinity for osmium.

Osmiophilic Globules Small lipid bodies that stain with osmium tetroxide.

Osmitrol A trade name for mannitol, used as a diuretic agent.

Osmium (O_s) A chemical element with atomic weight 190, valence 1 through 8.

Osmoglyn A trade name for glycerin, a hyperosmolar laxative and an osmotic diuretic agent.

Osmolal Solution Solute concentration expressed in number of osmoles in 1000 grams of solvent.

Osmolality Solute concentration expressed in number of osmoles in 1000 grams of solvent.

Osmolar Solution Solute concentration expressed in number of osmoles per liter of solution

Osmolarity Solute concentration expressed in number of osmoles per liter of solution.

Osmole Number of the dissociated particles or ions per mole of solute, e.g., one mole NaCl equals to two osmoles (one osmole Na^+ and one osmole Cl^-).

Osmolute An osmotically active solute or solute particle.

Osmometer A device used to measure the rate of osmosis or osmotic pressure.

Osmophiles Organisms that grow best or only in or on media of relatively high osmotic pressure.

Osmophilic Pertaining to osmophile.

Osmoreceptor Receptors in the CNS that respond to changes in osmotic pressure of the blood.

Osmoregulation Regulation of the salt and water content in the body.

Osmosis The movement of water molecules across a semipermeable membrane driven by a difference in solute concentration.

Osmotic Pertaining to osmosis.

Osmotic Lysis The lysis of cells in a hypotonic solution (cells are in a solution whose molar concentration of the solute is less than that of the cell).

Osmotic Potential Tendency of a solution to gain water when separated from water by a semipermeable membrane, a negative osmotic pressure tends to cause water to move into the solution.

Osmotic Pressure Pressure that must be exerted on the high solute concentration side of the semipermeable membrane to prevent flow of water across the membrane due to osmosis.

Osmotic Shock Lysis of cells in a hypotonic medium caused by the movement of water into the cells leading to the rupture or lysis of the cells.

Osmotic Work The energy-requiring process by which cells transport and concentrate solutes from the environment.

Osmotolerant Organisms that can withstand high osmotic pressures or grow in solutions of high solute concentrations.

Ossification The hardening of the bone (calcification).

Ost- A prefix denoting bone.

Oste- A prefix denoting bone.

Osteitis Inflammation of the bone.

Osteo- A prefix meaning bone or relationship to bone.

Osteoarthritis A disorder associated with the degeneration of the bone and cartilage.

Osteoblast Bone-forming cell.

Osteocalcin A protein found in the extracellular matrix of the bone involved in regulating calcium in the bone and teeth.

Osteochondritis Inflammation of bone and its cartilage.

Osteoclast A large multinucleated cell that participates in bone resorption.

Osteoclast Activating Factor A lymphokine that promotes the resorption of bone.

Osteocyte A bone-forming cell derived from osteoblasts.

Osteodentin Substance that partially fills the pulp cavity of teeth of the aged.

Osteodynia Pain in the bone.

Osteogenesis Bone formation.

Osteogenesis Imperfecta A disorder associated with fragile bones due to reduced stability and decrease in type-1 collagen.

Osteogenin A bone-inducing protein associated with extracellular matrix that binds heparin.

Oteoid 1. The products of osteoblasts consisting mainly of collagen. 2. Resembling bone.

Osteology Science that deals with bones.

Osteolysis Dissolution of the bone.

Osteoma A benign tumor of the bone.

Osteomalacia A vitamin D deficiency disorder characterized by the formation of deformed and soft bones.

Osteomyelitis Inflammation or infection of bone.

Osteonectin A bone-specific protein that binds both to collagen and to hydroxyapatite.

Osteopetrosis A disorder characterized by the abnormal thickening and hardening of the bones.

Osteopontin A bone-specific protein found in calcified bone that links cells and hydroxyapatite.

Osteoporosis A disorder characterized by the progressive loss of both the organic matrix and mineral content of bone leading to the enlargement of bone spaces and disturbance of nutrition and mineral metabolism.

Osteosclerosis The abnormal hardening of the bones.

Ot- A prefix denoting the ear.

OT Abbreviation for oxytocin.

OTA Abbreviation for orthotoluidine arsenite.

Otalgia Pain in the ear.

OTC Abbreviation for 1. ornithine carbamoylase; 2. oxytetracycline.

OTC Drug Abbreviation for over-the-counter drug.

OTCase Abbreviation for ornithine carbamoylase.

OTF Abbreviation for ovotransferrin.

Otic- A prefix pertaining to the ear.

Otitis Inflammation or infection of the ear.

Otivin A trade name for xylometazoline hydrochloride, a nasal agent used to produce local vasoconstriction of dilated arterioles and to reduce blood flow and nasal congestion.

OTMS Abbreviation for orthotoluidine manganese sulfate.

Otocort A trade name for a combination drug containing hydrocortisone, neomycin, and polymyxin B used as an anti-infective agent.

Otology The science that deals with the ear.

Otomycosis A mycosis of the ear.

Otorhinolarynogology The science that deals with ear, nose, and throat diseases.

Otorrhagia Bleeding from the ear.

Otorrhea Discharge from the ear.

Otosclerosis A disorder characterized by the progressive deafness due to aging.

Otoscope An instrument used for examination of the ear.

OTP Abbreviation for orotidine triphosphate.

OtuI (AluI) A restriction endonuclease from *Oerskovia turbota* with the following specificity:

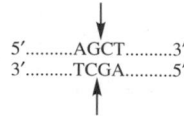

```
      ↓
5′..........AGCT..........3′
3′..........TCGA..........5′
              ↑
```

OtuNI (AluI) A restriction endonuclease from *Oerskovia turbota* with the same specificity as AluI.

Ouabain (mol wt 585) A cardiac glucoside from seeds of *Strophanthus gratus* and an inhibitor of sodium and potassium transport across the cell membrane.

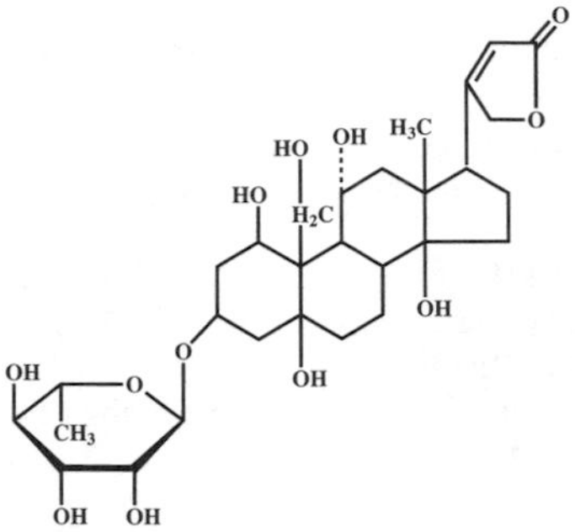

Ouchterlony Double Diffusion An immunodiffusion technique in which antigen and antibody are allowed to diffuse toward each other in agar and form visible immune precipitates. It is a method used for identification of soluble antigens.

Oudin's Diffusion Technique A one dimensional immunodiffusion technique in which antibodies are immobilized in agar and the antigens are allowed to diffuse into the antibody-containing agar to form antigen-antibody precipitin line.

Outbreeding The breeding between two different species for the purpose of producing hybrids.

Outcross Mating of one species with another for purposes of producing hybrids.

Outer Membrane The outer lipopolysaccharide-protein structure of the cell wall of Gram-negative bacteria that controls entrance of the molecules into the bacterial cell.

Ov- A prefix meaning egg.

Ovalbumin The major water-soluble protein of egg white.

Ovary Female organ that produces eggs in both animals and flowering plants.

Ovcon 50 A trade name for a combination drug containing norethindrone and ethinyl estradiol used as a contraceptive agent.

Overlapping Gene Genes whose nucleotide sequences overlap to some degree, the common nucleotide sequence is read in two or more different reading frames, thus producing different peptides.

Overlapping Method A method to determine the sequence of amino acids in a protein or sequence of nucleotides in a nucleic acid by comparing the overlapping sequences of different fragments to deduce the sequence of the original molecule.

Overproducer A genetically engineered organism that produces a large quantity of a given gene product.

Ovex (mol wt 303) A miticide.

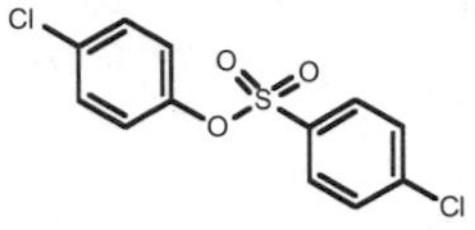

Ovi- A prefix meaning egg.

Ovicide Any substance or agent that destroys eggs.

Ovide A trade name for malathion, an insecticide that inhibits cholinesterase.

Oviduct The tube that serves to transport eggs from the ovary to the uterus.

Oviductin A high molecular glycoprotein secreted by the non-ciliated secretory cells of the oviduct.

Oviparous A type of development in which young hatch from eggs laid outside the mother's body.

Ovoflavin A riboflavin preparation from egg white.

Ovomucoid A mucoprotein found in egg white.

Ovoviviparous A type of development in which young hatch from the egg retained in the mother's uterus.

Ovral A trade name for a combination drug containing norgestrel and ethinyl estradiol used as an oral contraceptive agent.

Ovrette A trade name for norgestrel, a hormone used to suppress ovulation.

Ovulation The release of an egg from an ovary.

Ovum A haploid, unfertilized egg cell or female gamete.

Owl-Eye Cells The enlarged cells that have been infected by cytomegalovirus and contain large inclusion bodies.

Ox Abbreviation for oxacillin.

Oxa- A prefix in the chemical nomenclature denoting replacement of methylene group in a specified acyclic or monocyclic hydrocarbon by an oxygen.

OxaI (AluI) A restriction endonuclease from *Oerskovia xanthineolytica* with the same specificity as AluI.

Oxaceprol (mol wt 173) An anti-inflammatory agent.

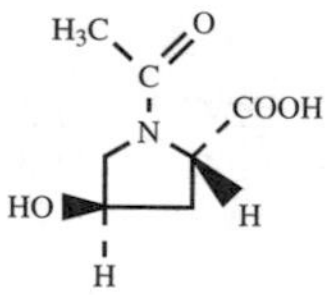

Oxacillin (mol wt 401) A semisynthetic, penicillinase-resistant antibiotic that inhibits bacterial cell wall synthesis.

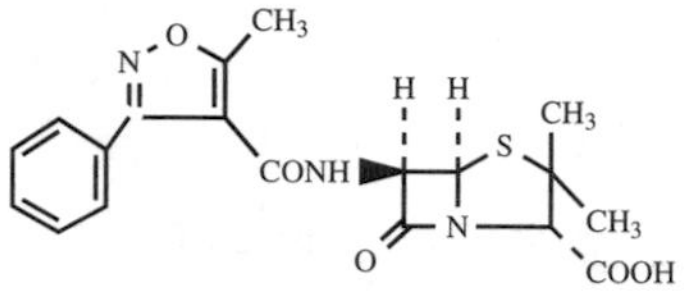

Oxadiazon (mol wt 345) An herbicide.

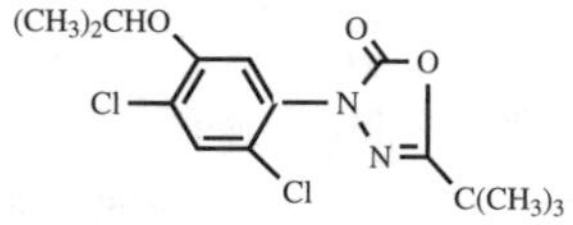

Oxaflozane (mol wt 273) An antidepressant.

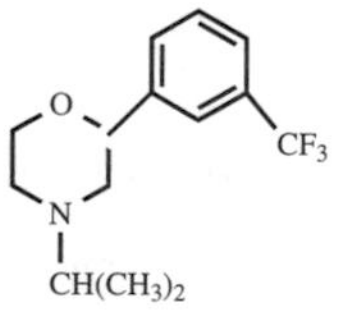

Oxaflumazine (mol wt 508) An antipsychotic agent.

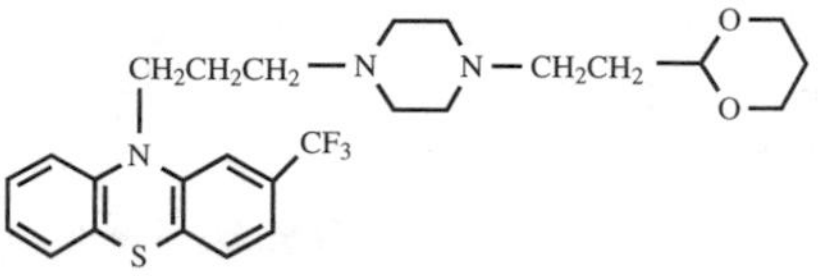

Oxalate Carboxy-Lyase The systematic name for oxalate decarboxylase.

Oxalate CoA Ligase The enzyme that catalyzes the following reaction:

ATP + oxalate + CoA ⇌ PPi + oxalyl-CoA

Oxalate CoA Transferase The enzyme that catalyzes the following reaction:

Succinyl-CoA + oxalate ⇌ Succinate + oxalyl-CoA

Oxalate Decarboxylase The enzyme that catalyzes the following reaction:

Oxalate ⇌ Formate + CO_2

Oxalate Oxidase The enzyme that catalyzes the following reaction:

Oxalate + O_2 ⇌ 2 CO_2 + H_2O_2

Oxalated Blood Blood to which a soluble oxalate has been added to prevent coagulation.

Oxalic Acid (mol wt 90) A toxic substance produced by plants that binds with calcium leading to the precipitation of calcium oxalate in the kidneys and preventing calcium uptake in the gut.

COOH
|
COOH

Oxaliplatin (mol wt 397) An antineoplastic agent.

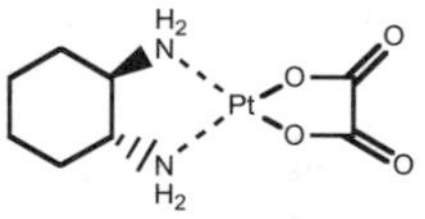

Oxalo- A prefix denoting the presence of acyl group, HOOC-CO-.

Oxaloacetase The enzyme that catalyzes the following reaction:

Oxaloacetate + H_2O ⇌ Oxalate + acetate

Oxaloacetate Acetylhydrolase See oxaloactase.

Oxaloacetate Carboxy-Lyase The systematic name for oxaloacetate decarboxylase.

Oxaloacetate Decarboxylase The enzyme that catalyzes the following reaction:

Oxaloacetate ⇌ Pyruvate + CO_2

Oxaloacetate Keto-enol Isomerase See oxaloacetate tautomerase.

Oxaloacetate Tautomerase The enzyme that catalyzes the following reaction:

Keto-oxaloacetate ⇌ Enol-oxaloacetate

Oxaloacetate Transacetase See citrate synthetase.

Oxaloacetic Acid (mol wt 132) An intermediate in the Krebs cycle.

COOH
|
CH_2
|
C=O
|
COOH

Oxalomalate Lyase The enzyme that catalyzes the following reaction:

3-Oxalomalate

⇅

Oxaloacetate + glyoxylate

Oxalosis A metabolism disorder that causes the deposition of oxalate in the kidney.

Oxalosuccinate Decarboxylase See isocitrate dehydrogenase.

Oxalosuccinic Acid (mol wt 256) An intermediate in the Krebs cycle.

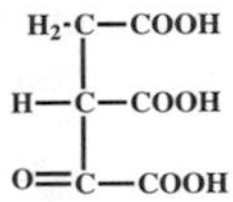

Oxaluria The presence of oxalic acid or oxalates in the urine.

Oxalyl-CoA Decarboxylase The enzyme that catalyzes the following reaction:

Oxalyl-CoA ⇌ Formyl-CoA + CO_2

Oxalyl-CoA Synthetase See oxalate-CoA ligase.

Oxamarin (mol wt 391) A hemostatic and anti-inflammatory agent.

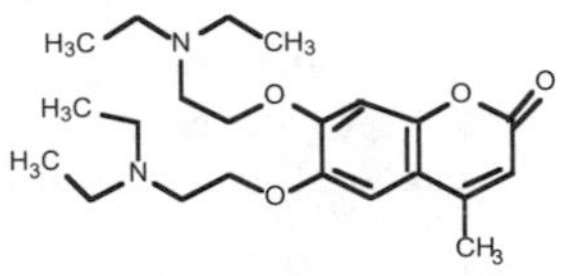

Oxametacine (mol wt 373) An anti-inflammatory agent.

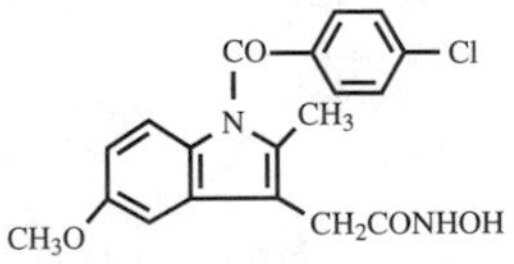

Oxamniquine (mol wt 279) An anthelmintic agent that reduces the egg load of *Schistosoma mansoni*.

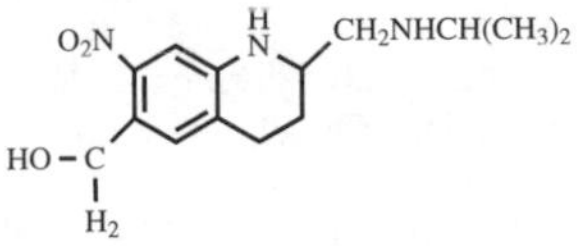

OxaNI A restriction endonuclease from *Oerskovia xanthineolytic* N with the following specificity:

↓

5′..........CCTNAGG..........3

3′..........GGANTCC..........5

↑

Oxanamide (mol wt 157) An anxiolytic agent.

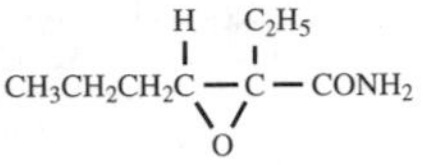

Oxandrin A trade name for an anabolic steroid hormone.

Oxandrolone (mol wt 306) An anabolic steroid that promotes tissue building processes.

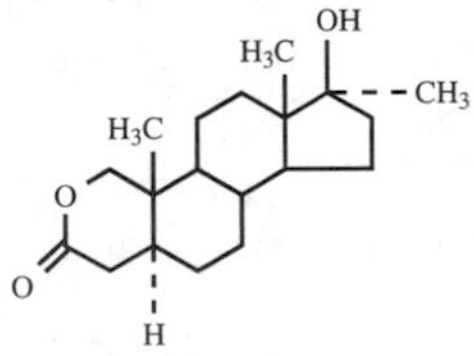

Oxantel (mol wt 216) An anthelmintic agent.

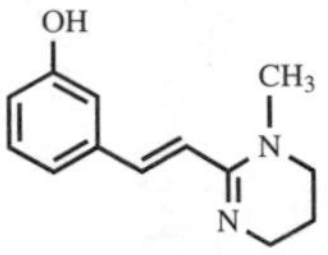

Oxapropanium Iodide (mol wt 273) A cholinergic agent.

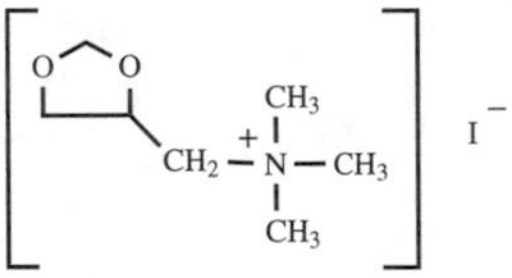

Oxaprotiline (mol wt 293) An antidepressant.

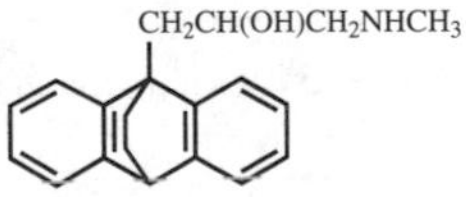

Oxaprozin (mol wt 293) An anti-inflammatory agent.

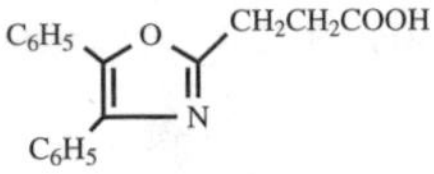

Oxatomide (mol wt 427) An antiallergic and antiasthmatic agent.

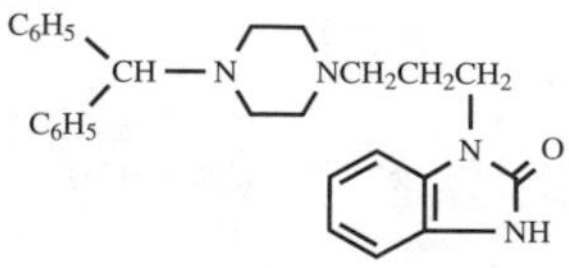

Oxazepam (mol wt 287) An antianxiety agent that depresses the CNS activity.

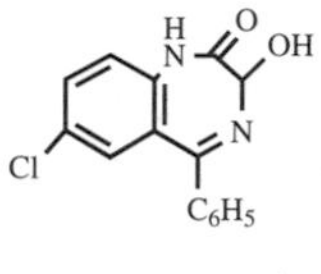

Oxazidione (mol wt 321) An anticoagulant.

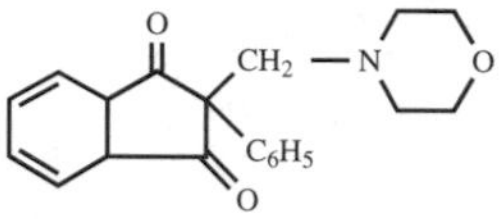

Oxazolam (mol wt 329) An anxiolytic agent.

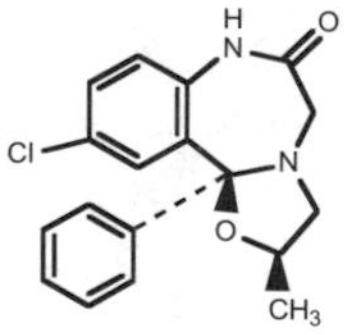

Oxcarbazepine (mol wt 252) An anticonvulsant.

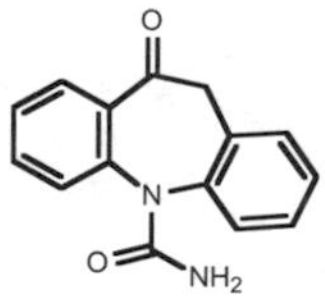

Oxeladin (mol wt 335) An antitussive agent.

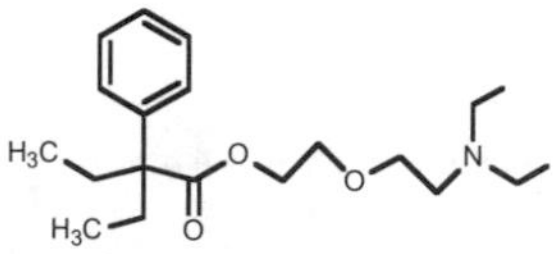

Oxetorone (mol wt 319) An analgesic agent.

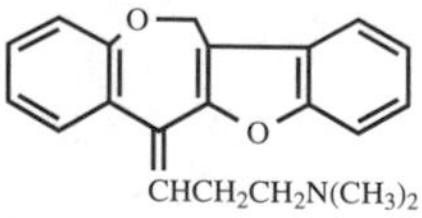

Oxfendazole (mol wt 315) An anthelmintic agent.

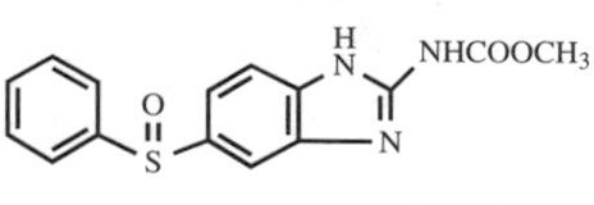

Oxibendazole (mol wt 249) An anthelmintic agent.

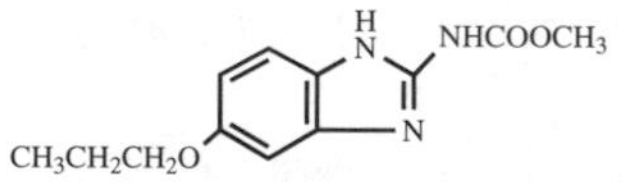

Oxiconazole Nitrate (mol wt 492) An antifungal agent.

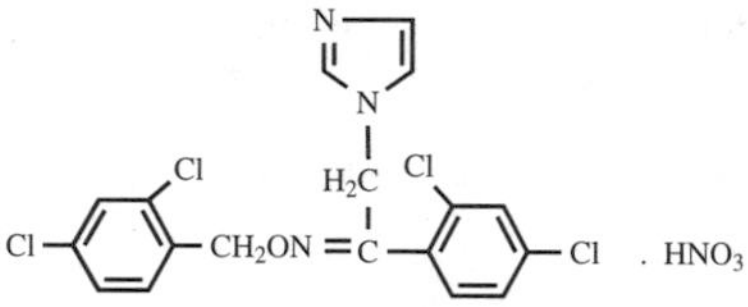

Oxidant An oxidizing agent that accepts electrons, or hydrogens, from a reducing agent or reductant.

Oxidase An enzyme that catalyzes a reaction in which electrons removed from a substrate are donated directly to molecular oxygen.

Oxidation Removal of hydrogens or electrons from a compound or an element.

β-Oxidation Oxidative degradation of fatty acid to acetyl-CoA by successive cycles of oxidations at the β-carbon atom of an activated form of the fatty acid (fatty acyl-CoA).

Oxidation Reduction Couple A pair of molecules between which electrons are transferred (also called a redox couple).

Oxidation Reduction Potential A quantitative expression of the tendency of a given redox couple to donate or accept elections from another redox couple.

Oxidation Reduction Reaction A reaction in which electrons are transferred from a donor to an acceptor molecule (also known as redox reaction).

Oxidative Deamination The process of release of free ammonia from an amino acid with concomitant oxidation of the molecule to an α-keto acid.

Oxidative Decarboxylation The process of removal of carbon dioxide from an organic molecule by oxidation.

Oxidative Phosphorylation The synthesis of ATP from ADP and inorganic phosphate within the mitochondrion mediated by an electrochemical proton gradient generated during the electron transfer along the electron transport chain.

Oxidative Photophosphorylation The synthesis of ATP from ADP and inorganic phosphate within the chloroplast by an electrochemical gradient generated by light-initiated electron transfer along the electron transport system.

Oxidimethiin (mol wt 210) A plant growth regulator.

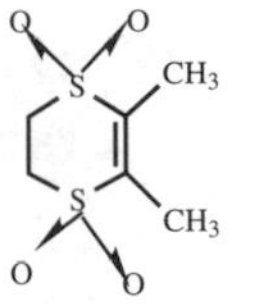

Oxidizing Agent A substance capable of accepting electrons from other substances.

Oxidoreductase Any enzyme that catalyzes the oxidoreduction reaction.

Oxidoreduction Reaction See oxidation-reduction reaction.

Oximeter A device used for determining the degree of oxygenation of blood.

Oxiniacic Acid (mol wt 139) An antihyperlipoproteinemic agent.

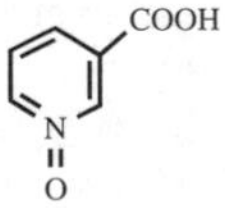

Oxistat A trade name for oxiconazole, an antifungal agent.

OxLDL Abbreviation for oxidized LDL (low density lipoprotein).

2-Oxo-Acid Carboxy-Lyase The systematic name for pyruvate decarboxylase.

3-Oxo-Acid CoA-Transferase The enzyme that catalyzes the following reaction:

Succinyl-CoA + a 3-oxo acid

⇌

Succinate + a 3-oxoacyl-CoA

Oxoglutarate Decarboxylase Synonym of oxoglutarate dehydrogenase (lipoamide).

Oxoglutarate Dehydrogenase (lipoamide) The enzyme that catalyzes the following reaction:

Oxoglutarate + lipoamide

⇌

S-Succinyldihydrolipoamide + CO_2

2-Oxoglutarate Lipoamide 2-Oxidoreductase Synonym of oxoglutarate dehydrogenase.

Oxolamine (mol wt 245) An anti-inflammatory agent.

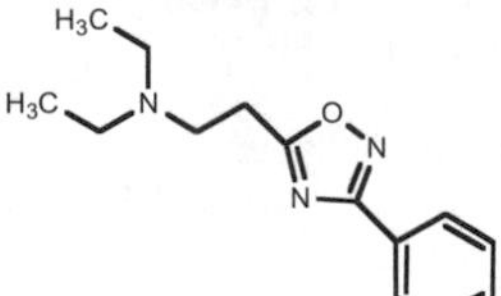

Oxolaurate Decarboxylase The enzyme that catalyzes the following reaction:

3-Oxododecanoate

⇌

2-undecanone + CO_2

Oxolinic Acid (mol wt 261) A prokaryotic DNA gyrase inhibitor and an antibacterial agent.

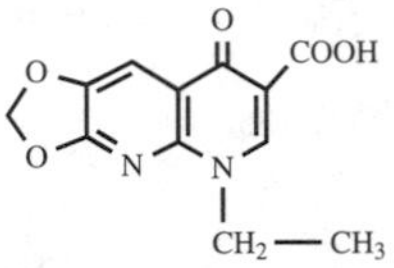

Oxonium Ion The ion of

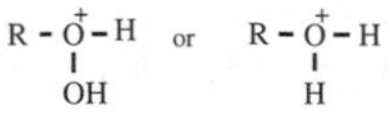

5-Oxoprolinase (ATP-hydrolyzing) The enzyme that catalyzes the following reaction:

ATP + 5-oxo-L-proline + $2H_2O$

⇌

ADP + orthophosphate + L-glutamate

5-Oxoproline Amidohydrolyase (ATP-hydrolyzing) The systematic name for 5-oxoprolinase.

4-Oxoproline Reductase The enzyme that catalyzes the following reaction:

4-Hydroxyproline + NAD^+

⇌

4-Oxoproline + NADH

5-Oxoprolyl Peptidase Synonym of pyroglutamyl peptidase I.

3-Oxopropanoate Hydro-Lyase The systematic name for malonate semialdehyde dehydrogenase.

3-Oxopropanoate NADP⁺ Oxidoreductase The systematic name for malonate semialdehyde dehydrogenase.

3-Oxopropanoate NADP⁺ Oxidoreductase (Decarboxylating, CoA-acetylating) The systematic name for malonate semialdehyde dehydrogenase (acetylating).

3-Oxosteroid 1-Dehydrogenase The enzyme that catalyzes the following reaction:

A 3-oxosteroid + acceptor

⇌

A 3-oxo-Δ^1-steroid + reduced

3-Oxo-5a-Steroid Acceptor D4-Oxidoreductase The systematic name for 3-oxo-5α-steroid 4-dehydrogenase.

3-Oxo-5α-Steroid 4-Dehydrogenase The enzyme that catalyzes the following reaction:

$$\text{A 3-oxo-5}\alpha\text{-steroid + acceptor} \rightleftharpoons \text{A 3-oxo-}\Delta^4\text{-steroid + reduced acceptor}$$

3-Oxosteroid D5-D4-Isomerase Systematic name for steroid D-isomerase.

6-Oxotetrahydronicotinate Dehydrogenase Synonym of 6-hydroxynicotinate reductase.

Oxprenolol (mol wt 265) An antihypertensive, antianginal, and antiarrhythmic agent.

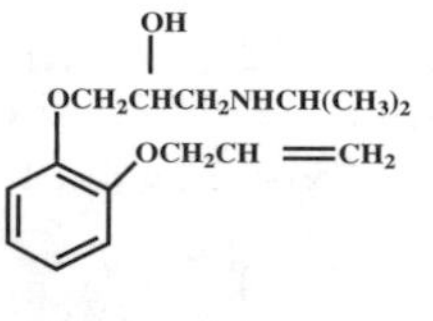

Oxsoralen A trade name for methoxalen, used to enhance melanogenesis.

Oxy A trade name for benzoyl peroxide lotion (e.g., oxy 10, oxy 5) that has antimicrobial and comedolytic activity.

Oxyanion The negatively charged carbonyl oxygen.

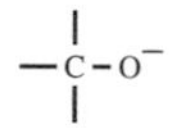

Oxyanion Hole The region or pocket on an enzyme to which carbonyl oxygen (oxyanion) binds.

Oxybiontic Organism capable of utilizing molecular oxygen for growth.

Oxybiotin A synthetic analog of biotin in which the sulfur is replaced by oxygen.

Oxybutynin Chloride (mol wt 394) An anticholinergic agent used for treatment of neurogenic bladder.

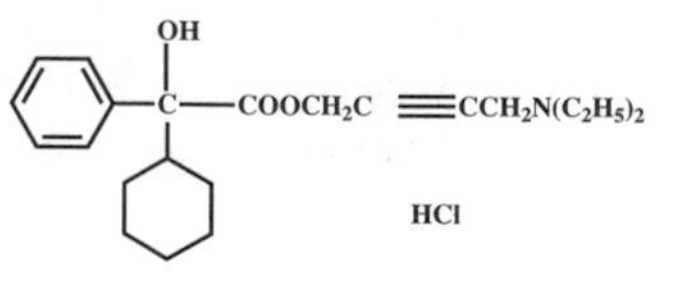

Oxycellulose The oxidized form of cellulose used as cation exchanger.

Oxychlorosene (mol wt 407) An antiseptic agent.

$C_{20}H_{34}O_3SHOCl$

Oxycinchophen (mol wt 265) An antidiuretic and uricosuric agent.

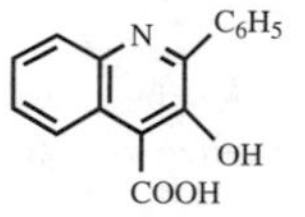

Oxycocet A trade name for a combination drug containing acetainophen and oxycodone hydrochloride, used as an opioid analgesic agent.

Oxycodan A trade name for a combination drug containing aspirin and oxycodone hydrochloride, used as an opioid analgesic agent.

Oxycodone (mol wt 315) An opioid analgesic agent that binds with opioid receptors in the CNS altering the perception and emotional response to pain.

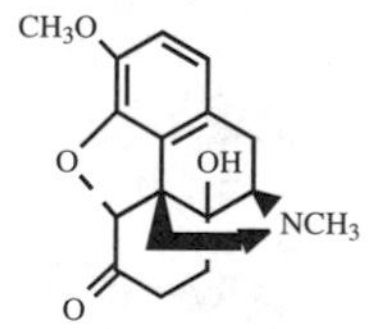

OxyContin A trade name for oxycodone, an analgesic agent that binds with opioid receptors in the CNS altering the perception and emotional response to pain.

Oxydess II A trade name for dextroamphetamine sulfate, a cerebral stimulant that stimulates the transmission of nerve impulses by releasing stored norepinephrine from nerve terminals in the brain.

Oxyfast A trade name for oxycodone, an analgesic agent that binds with opioid receptors in the CNS altering the perception and emotional response to pain.

Oxyfedrine (mol wt 313) An antianginal agent.

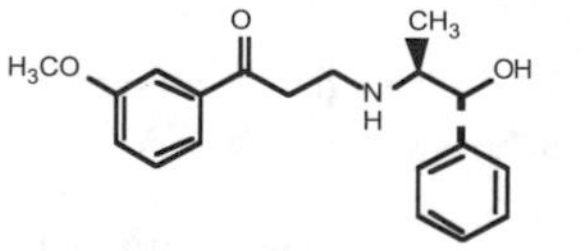

Oxygen (O) A chemical element with atomic weight 16, valence 2.

Oxygen Capacity of the Blood The oxygen-combining power of the blood.

Oxygen Carrier Substance capable of carrying and transporting oxygen, e.g., hemoglobin and myoglobin.

Oxygen Debt The amount of oxygen needed to oxidize the lactic acid buildup in muscle cells during heavy exercise.

Oxygen Electrode A device for measuring the rate of oxygen consumption that consists of an Ag/AgCl reference electrode and Pt electrode with an oxygen-permeable Teflon membrane.

Oxygen Evolving Complex The complex in the photosystem II that binds H_2O molecules producing O_2.

Oxygen Saturation Cure A plot of amount oxygen bound to a given concentration of hemoglobin or myoglobin as a function of oxygen partial pressure.

Oxygenase The enzyme that catalyzes the introduction of oxygen to an acceptor molecule, e.g., monooxygenase (introduction of one oxygen atom into an acceptor molecule) and dioxygenase (introduction of two oxygen atoms into an acceptor molecule).

Oxygenation 1. Introduction of oxygen into a compound. 2. Saturation of blood with oxygen.

Oxygenic Photoautotroph Organism that utilizes water as the electron donor in photosynthesis and release of molecular oxygen by splitting water molecules.

Oxyhemoglobin Hemoglobin that carries oxygen (oxygenated hemoglobin or a complex of oxygen and hemoglobin).

OxyIR A trade name for oxycodone, an analgesic agent that binds with opioid receptors in the CNS altering the perception and emotional response to pain.

Oxymesterone (mol wt 318) An androgen and an anabolic hormone that promotes tissue building processes.

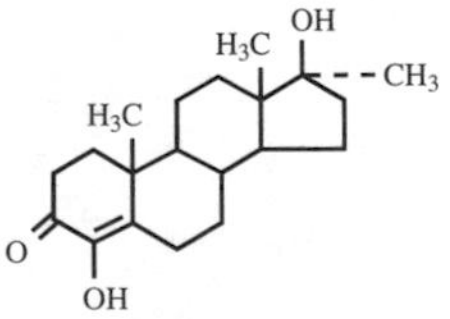

Oxymetazoline (mol wt 260) An agent that produces local nasal vasoconstriction of dilated arterioles to reduce blood flow and nasal congestion.

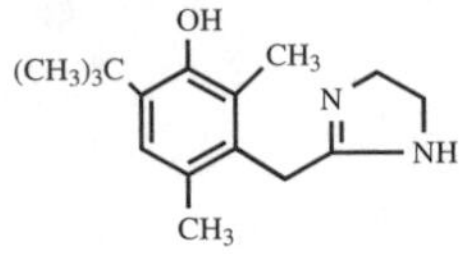

Oxymetholone (mol wt 332) An anabolic steroid that promotes tissue-building processes and stimulates erythropoiesis.

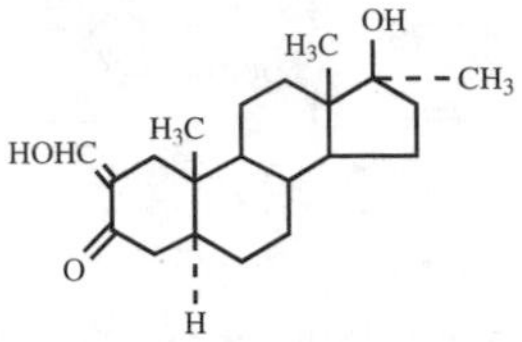

Oxymethurea (mol wt 120) An antiseptic agent.

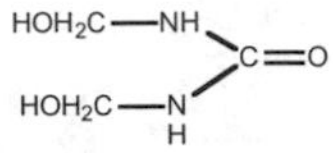

Oxymorphone (mol wt 301) An opioid analgesic agent that binds to opioid receptors in the CNS altering both the perception and emotional response to pain.

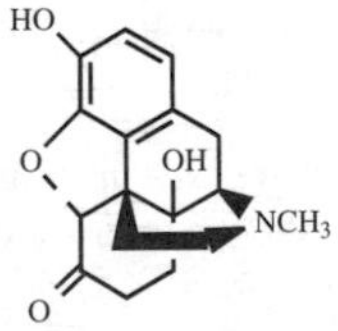

Oxyntic Cells The acid-secreting cells of the stomach.

Oxypendyl (mol wt 371) An antiemetic agent.

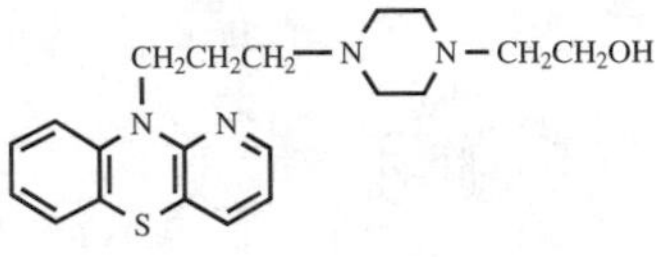

Oxypertine (mol wt 380) An antidepressant agent.

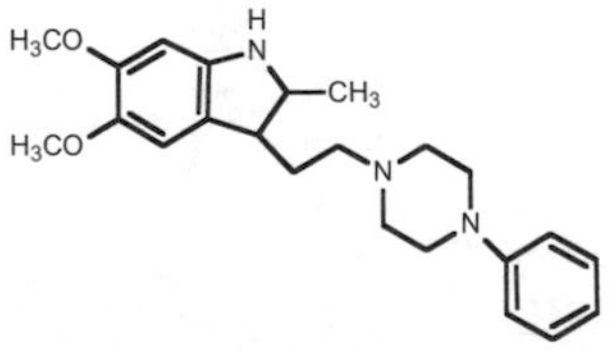

Oxyphenbutazone (mol wt 324) An anti-inflammatory agent that produces anti-inflammatory and antipyretic effects by inhibition of prostaglandin synthesis.

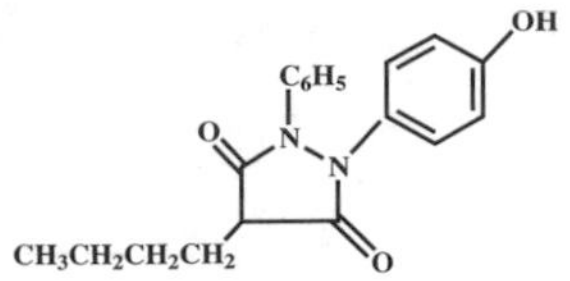

Oxyphencyclimine (mol wt 344) An anticholinergic agent.

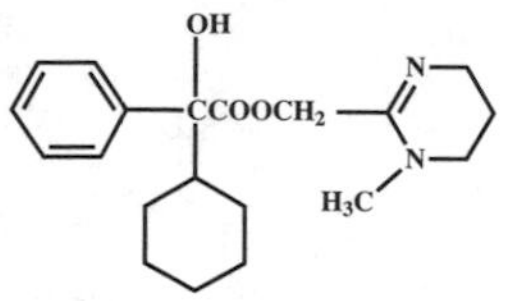

Oxyphenisatin Acetate (mol wt 401) A cathartic agent.

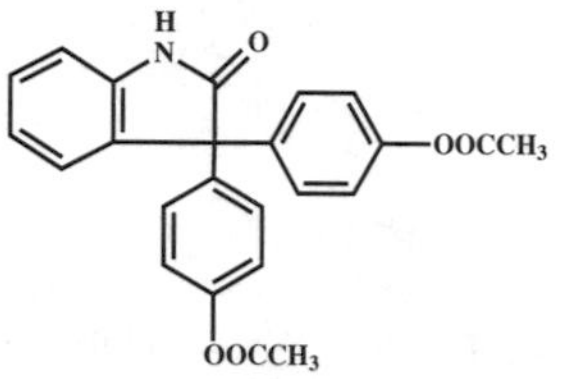

Oxyphotobacteria Bacteria capable of using water as electron donor in photosynthesis by splitting water to form oxygen.

Oxysome Referring to a macromolecular aggregate capable of functioning as a unit in oxidative phosphorylation.

Oxytetracycline (mol wt 460) An antibiotic that binds to 30S ribosomes and inhibits bacterial protein synthesis.

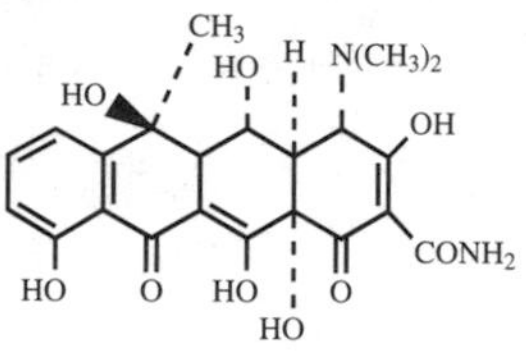

Oxytocic Capable of inducing contraction of uterine smooth muscle.

Oxytocin A peptide hormone secreted by the pituitary gland and used as an oxytocic agent to induce contraction of uterine smooth muscle.

Cys-Tyr-Ile-Gin-Asn-Cys-Pro-Leu-GlyNH$_2$
└─ S ──────── S ─┘

Oxytocinase An aminopeptidase.

Oxyzin An endopeptidase from *Aspergillus oryzae.*

Oz Marker An allotypic marker in the constant region of lambda light chain of immunoglobulin.

Ozagrel (mol wt 228) An antithrombotic and antianginal agent.

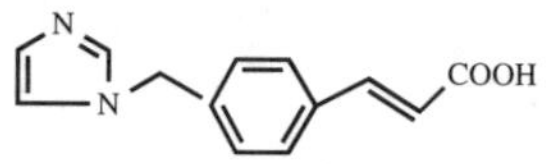

Ozone Triatomic oxygen with molecular weight 48.

Ozone Shield The layer of ozone gas (O_3) in the upper atmosphere that shields the earth from ultraviolet radiation.

Ozonolysis The cleavage of a double bond by the addition of ozone to an unsaturated organic compound.

P

p- A prefix meaning para position

p Abbreviation for plasmid, e.g., pBR322.

p20 Abbreviation for 20 kDa peptide or protein.

p120 Abbreviation for proliferation-associated nucleolar protein antigen of 120 kDa that is expressed in many human cancers.

p174 A shuttle vector for vertebrate cells and *E. coli* that contains Amp^r (ampicillin resistant) marker and BclI, EcoRI, ClaI, HindIII, EcoRV, BamHI, SalI, BstEI, SmaI, SphI, and SstII cleavage sites.

p304 A shuttle vector for vertebrate cells and *E. coli* that contains Amp^r (ampicillin resistant) marker and HindIII, SphI, PstI, SalI, AccI, BamHI, KpnI, and SstI cleavage sites.

p305 A shuttle vector for vertebrate cells and *E. coli* that contains Amp^r (ampicillin resistant) marker and HindIII, SphI, PstI, SalI, AccI, XbaIIBamHI, KpnI, SstI, and EcoRI cleavage sites.

p308 A shuttle vector for vertebrates cell and *E. coli* that contains Amp^r (ampicillin resistant) marker and HindIII, SphI, PstI, SalI, HincII, AccI, BamHI, and KpnI cleavage sites.

p53 A gene that mutates to form an important oncogene for human cancer.

P Abbreviation for 1. phosphorus or phosphate, 2. product, 3. proline, 4. pressure, and 5. pico, e.g., pg for picogram (10^{-12} gram).

~P The symbol for phosphate group in a high energy compound.

^{32}P Radioactive isotope of phosphorus that emits strong β particles with a half-life of 14.3 days.

P_o Abbreviation for open channel probability.

P1 Bacteriophage A temperate bacteriophage (family Myoviridae) that infects *E. coli.*

P2 Bacteriophage A temperate bacteriophage of the family Myoviridase that infects *E. coli.*

P4 Bacteriophage A satellite bacteriophage that infects *E. coli* and requires all of the late genes of a helper phage to complete its replication cycle.

P7 Bacteriophage A bacteriophage related to P1.

P22 Bacteriophage A temperate bacteriophage of the family P odoviridae that infects *Salmonella.*

P-50 A trade name for penicillin G potassium, an antibiotic that inhibits bacterial cell wall synthesis.

P_{50} Abbreviation for oxygen pressure at 50% saturation.

P450 Abbreviation for cytochrome P450.

P_{582} A siroheme-containing sulfite reductase from *Desulfotomaculum nigrificans.*

P_{680} Photosynthetic pigment of the photosystem II in the eukaryotic photosynthetic cell that functions in conjuction with photolysis of water and production of oxygen.

P_{700} Photosynthetic pigment in the photosystem I in both prokaryotic and eukaryotic photosynthetic cells, it is responsible for cyclic photophosphorylation in the photosynthesis.

P_{870} A bacterial chlorophyll with absorption band at 870 nm.

P_{960} A bacteriochlorophyll.

P_I Abbreviation for photosystem I (see also P_{700}).

P_{II} Abbreviation for photosystem II (see also P_{680}).

P Agar An agar medium containing peptone (1%), yeast extract (0.5%), NaCl (0.5%), glucose (0.1%), and agar (1.5%).

P Element A group of transposable elements from *Drosophila melanogaster.*

P Face The fracture face derived from the half of a membrane adjacent to the cytoplasm during freeze-fracture.

P Sequence A transposable element in the genome of *Drosophila.*

P Site Referring to the peptidyl site on the ribosome for the attachment of peptidyl-tRNA complex during protein synthesis.

P System A blood group system in which blood group antigens are detected on erythrocytes, erythroblasts, platelets, megakaryocytes, and fibroblast. They are designated as P^k, P, and P1.

PA Abbreviation for 1. phenyl alcohol; 2. phosphatidic acid; 3. phosphoarginine; 4. plasminogen activator; 5. pernicious anemia; 6. perchloric acid; 7. pre-albumin; 8. pyridoxic acid.

4PA Abbreviation for 4-pyridoxic acid.

PA2 Bacteriophage A bacteriophage of the family Styloviridae.

PAA Abbreviation for 1. phenylacetic acid; 2. phosphoamino acid; 3. pyridine acetic acid.

pAA31 A plasmid in *E. coli* that contains Amp^r (ampicillin resistant), Gal^s (galactose sensitive) markers and EcoRI, HindIII, SalI cleavage sites.

pAA31P A plasmid in *E. coli* that contains Cml^r (chloramphenicol resistant), Gal^s (galactose sensitive) markers and PvuI, HpaI, SalI, HindIII cleavage sites.

PAb Abbreviation for polyclonal antibody.

PABA Abbreviation for *p*-aminobenzoic acid.

***p*ABA** Abbreviation for *para*-amino benzoic acid.

Pabanol A trade name for methyl salicylate, used as a pain killer.

PAbs Abbreviation for polyclonal antibodies.

PacI A restriction endonuclease from *Nostoc* species B with the following specificity:

↓
5'........TTAATTAA........3'
3'........AATTAATT........5'
↑

***pac* Site** The packaging initiation site on DNA of bacteriophage.

PACAP Abbreviation for pituitary adenylyl-cyclase-activating peptide.

Pacaps A trade name for a combination drug containing acetaminophen, caffeine, and butalbital, used as an analgesic and antipyretic agent.

PACE Abbreviation for paired amino acid cleaving enzyme.

Pacemaker 1. A bodily structure of the heart that activates the contraction of cardiac muscle and heart beat. 2. An artificial electronic device that may be implanted into the body to perform the same function as a pacemaker.

Pacerone A trade name for amiodarone hydrochloride, an antiarrhythmic agent.

Pachy- A prefix meaning thickened.

Pachydactyly The abnormal thickening of the fingers or toes.

Pachyderma Abnormal thickening of the skin.

Pachymeningitis Inflammation and thickening of the dura (the outer membrane covering the brain).

Pachynema Synonym for pachytene.

Pachytene A stage of prophase I of meiosis characterized by synapsis of homologous chromosomes.

Packed Bed Reactor A tubular bioreactor in which the interior of the tube is filled or coated with cells or enzymes so that the substrate percolated through the tube is converted to product that emerges in the effluent.

Packed Cell Volume Referring to the volume (%) of packed red blood cells in whole blood.

Paclitaxel (mol wt 854) An antineoplastic agent.

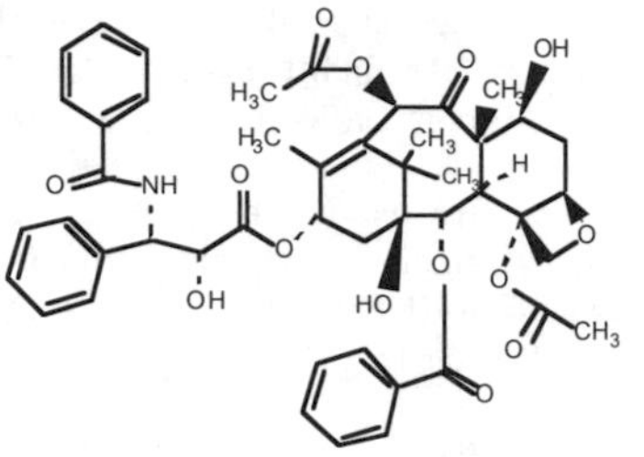

$PACO_2$ Abbreviation for 1. alveolar carbon dioxide partial pressure; 2. arterial carbon dioxide partial pressure.

Pactamycin (mol wt 559) An antitumor antibiotic produced by *Streptomyces pactum var pactum* that inhibits protein synthesis in both prokaryotic and eukaryotic cells.

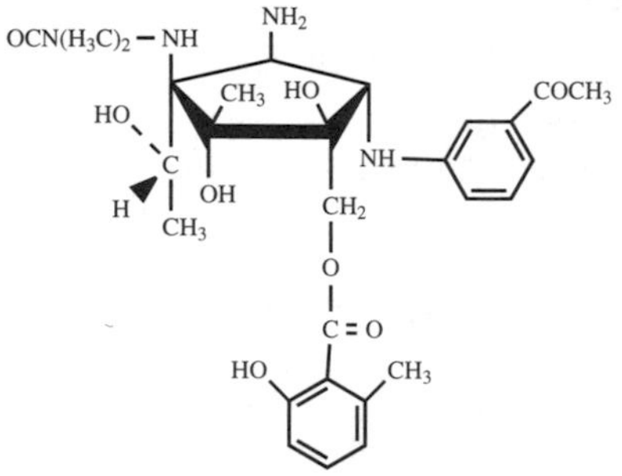

pACYC177 A plasmid of *E. coli* that contains Cml^r (chloramphenicol resistant) and Tet^r (tetracycline resistant) markers and HindII, PstI, XhoI, NruI, SmaI, and HindIII cleavage sites.

PAD Abbreviation for pulsed amperometric detector.

PaeI (SphI) A *Pseudomonas aeruginosa* restriction endonuclease with the following specificity:

↓
5′..........GCATGC..........3′
3′..........CGTACG..........5′
↑

Pae177I (BamHI) A restriction endonuclease from *Pseudomonas aeruginosa* RFL177 with the same specificity as BamHI.

Pae181I (CauII) A restriction endonuclease from *Pseudomonas aeruginosa* RFL181 with the same specificity as CauII.

PAE Cell Abbreviation for porcine aortic endothelial cell.

PaeAI (SacII) A restriction endonuclease from *Pseudomona aeruginosa* with the following specificity:

5′..........CCGC↓GG..........3′
3′..........GG↑CGCC..........5′

PaeR7I (XhoI) A restriction endonuclease from *Pseudomonas aeruginosa* with the following specificity:

5′..........C↓TCGAG..........3′
3′..........GAGCT↑C..........5′

Paedomorphosis The retention of juvenile characters in an adult organism.

PAF Abbreviation for 1. platelet-activating factor; 2. platelet-aggregation factor.

PAGE Abbreviation for polyacrylamide gel electrophoresis.

Paget's Disease A bone disorder characterized by bone deformity and loss of calcium.

PAH Abbreviation for 1. *para*-aminohippuric acid; 2. phenylalanine hydroxylase; 3. polycyclic aromatic hydrocarbon.

PAHA Abbreviation for *para*-aminohippuric acid, a reagent used in the PAHA sodium clearance test for kidney damage.

PAHA Sodium Clearance Test A test for the determination of the ability of the kidney to remove *para*-aminohippuric acid from the blood.

PAI Abbreviation for plasminogen activator inhibitor.

PaiI (HaeIII) A restriction endonuclease from *Pseudomonas aeruginosa* with the same specificity as HaeIII.

PAIgG Abbreviation for platelet-associated IgG (immunoglobulin G).

PAK Abbreviation for p21-activated kinase.

PAK1 Abbreviation for p21-activated kinase 1.

PAL Abbreviation for phenylalanine ammonia lyase.

pAL4 A plasmid of *E. coli* that contains Kan[r] (kanamycin resistant) and Neo[r] (neomycin resistant) markers and EcoRI, PstI, KpnI cleavage sites.

pAL181 A plasmid of *E. coli* that contains Amp[r] (ampicillin resistant) markers.

PalI A restriction endonuclease from *Providencia alcalifaciens* with the following specificity:

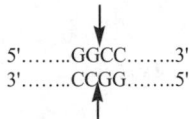

5'........GG↓CC........3'
3'........CC↑GG........5'

Palade Pathway A pathway for transport of proteins from their sites of synthesis to cellular and extracellular destinations.

Palafer A trade name for ferrous fumarate, used to provide iron for the body.

Palatinose (mol wt 342) A disaccharide (α-1,6-glucopyranosyl-D-fructose).

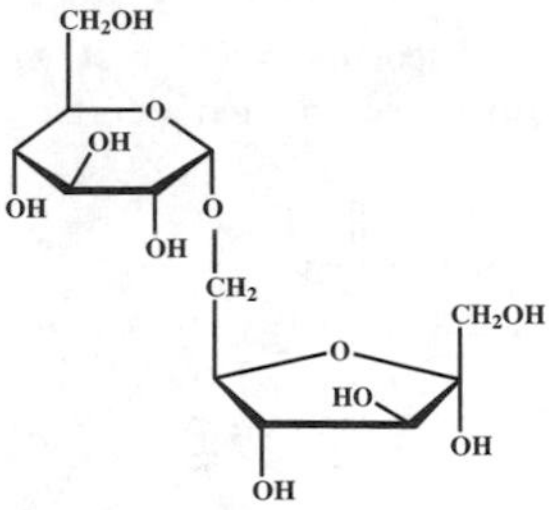

Paleo- A prefix denoting 1. ancient; 2. primitive.

Palindrome A fragment of double-stranded DNA that displays identical sequences when read either from the 5′ to 3′ or 3′ to 5′ direction such as:

5′..........ACTAGT..........3′
3′..........TGATCA..........5′

Palindromia The recurrence of a pathological condition.

Palindromic Referring to a fragment of double-stranded DNA that displays identical base sequences when read either from the 5′ to 3′ or 3′ to 5′ direction.

Palindromic Sequence A fragment of double-stranded DNA that displays identical sequences when read from either the 5′ to 3′ or 3′ to 5′ direction. The palindromic sequences are often the recognition sites for restriction endonucleases.

Palindromic Symmetry The symmetry of DNA that has the same backward and forward sequence.

Palisade Cells Elongated cells of the mesophyll located beneath the upper epidermis of a leaf.

Palisade Mesophyll Tissue below the upper epidermis consisting of chloroplast-bearing parenchyma cells oriented perpendicular to the leaf surface.

Palivizumab A murine and human monoclonal antibody produced by recombinant DNA technology specific to an antigenic site of the respiratory syncytial virus.

Palladium (Pd) A chemical element with atomic weight 106, valence 2 and 4.

Palmitate A salt of palmitic acid.

Palmitic Acid (mol wt 256) A saturated fatty acid.

$$CH_3(CH_2)_{14}COOH$$

Palmitoleic Acid (mol wt 254) An unsaturated fatty acid with one double bond.

$$CH_3(CH_2)_5CH = CH(CH_2)_7COOH$$

Palmitoyl-ACP A complex of palmitic acid and ACP, and an intermediate in the biosynthesis of plamitic acid.

$$CH_3 - (CH_2)_{14}CO - S - ACP$$

Palimitoyl-L-Ascorbic Acid (mol wt 415) A complex of palmitic acid and ascorbic acid.

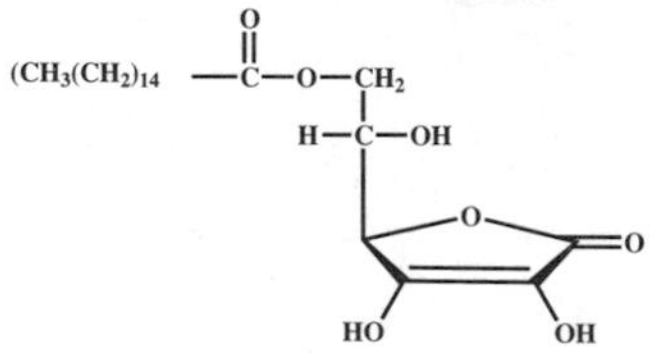

Palmitoyl-Carnitine (mol wt 401) A complex of palmitic acid and carnitine and an intermediate in fatty acid metabolism.

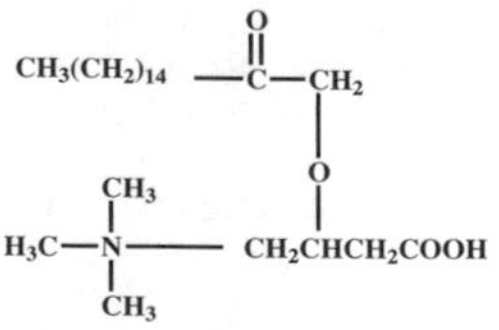

Palmitoyl-CoA (mol wt 1006) A complex of CoA and palmitic acid and an intermediate in the metabolism of palmitic acid.

$$CH_3(CH_2)_{14}CO- S - CoA$$

Palmitoyl-CoA Hydrolase The enzyme that catalyzes the following reaction:

$$\text{Palmitoyl-CoA} + H_2O \rightleftharpoons \text{CoA} + \text{palmitate}$$

Palmitoyl-CoA Synthetase The enzyme that catalyzes the following reaction:

$$\text{ATP} + \text{palmitate} + \text{CoA} \rightleftharpoons \text{AMP} + \text{PPi} + \text{palmitoyl-CoA}$$

Palmitoyl-Thioesterase The enzyme that catalyzes the following reaction:

$$\text{Palmitoyl-ACP} + H_2O \rightleftharpoons \text{ACP} + \text{palmitate}$$

Palsy Referring to nerve paralysis or degeneration

PAM Abbreviation for 1. penicillin aluminum monosterate; 2. peptidylglycine α-amidating monooxygenase; 3. phenylalanine mustard; 4. percentage of accepted point mutations.

Pama No1 A trade name for calcium carbonate, used to reduce total acid in the GI tract.

Pamabrom (mol wt 347) A diuretic agent.

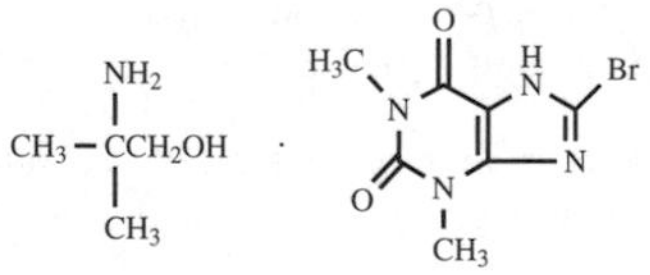

Pamaquine (mol wt 315) An antimalarial agent.

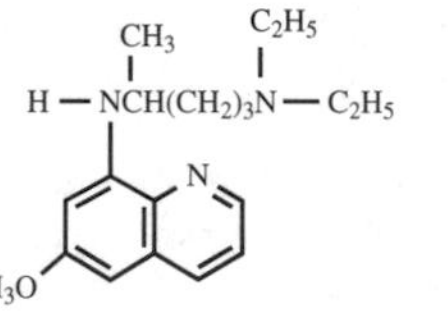

Pamelor A trade name for nortriptyline hydrochooride, an antidepressant that increases the level of norepinephrine and serotonin in the CNS.

Pamidronic Acid (mol wt 235) An antihypercalcemia agent that inhibits resorption of bone.

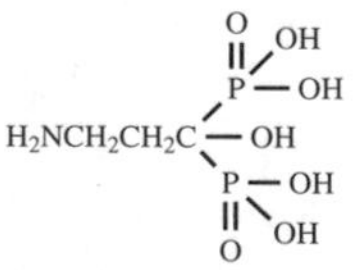

Pamine A trade name for methscopolamine bromide, which blocks acetylcholine, decreases GI motility, and inhibits gastric acid secretion.

Pamprin-1B A trade name for ibuprofen, used as an anti-inflammatory agent.

PAN Abbreviation for peroxyacetylnitrates, a compound present in photochemical smog that is toxic to plants and animals (irritating to eyes, nose, and throat of humans).

PanI (XhoI) A restriction endonuclease from *Pseudomonas alkanolytica* with the following specificity:

5′..........C↓TCGAG..........3′
3′..........GAGCT↑C..........5′

Panadol A trade name for acetaminophen, an analgesic and antipyretic agent that blocks the generation of pain impulses.

Panafcort A trade name for prednisone, a corticosteroid that decreases inflammation.

Panafil A trade name for an enzymatic healing ointment that contains papain, urea, and chlorophyllin copper complex used for the treatment of acute or chronic lesions, burns, or infected wounds.

Panamax A trade name for acetaminophen, an analgesic and antipyretic agent that blocks the generation of pain impulses.

Panasol A trade name for prednisone, a corticosteroid that decreases inflammation.

Pancreas A gland that secrets enzymes into the intestine for digestion of food and manufacture of insulin.

Pancrease Capsules A trade name for pancrelipase used for the assistance of digestion and absorption of fat.

Pancreatectomy The surgical removal of the pancreas.

Pancreatic Acinar Cells Cells in the pancreas that secret enzymes for digestion of food.

Pancreatic Hormones Hormones secreted by the pancreas, e.g., insulin and glucagon.

Pancreatic Islet Synonym of islet Langerhans.

Pancreatic Secretory Trypsin Inhibitor A protein trypsin inhibitor isolated from bovine pancreas.

Pancreatic Trypsin Inhibitor Synonym of pancreatic secretory trypsin inhibitor.

Pancreatin A pancreatic enzyme preparation from swine or cattle used as a digestive aid and for treatment of various intestinal disorders.

Pancreatitis Inflammation of the pancreas.

Pancreatoduodenectomy The surgical removal of both pancreas and duodenum (first portion of the intestine).

Pancrelipase Pancreatic enzyme preparation containing lipase, amylase, and proteases that is used as a digestive aid for starch, lipid, and protein.

Pancreozymin A peptide hormone secreted by the mucosal cells of the duodeum that stimulates contraction of the gallbladder and secretion of digestive enzymes by the pancreas.

Pancuronium Bromide (mol wt 733) A skeletal muscle relaxant that promotes acetylcholine from binding to receptors.

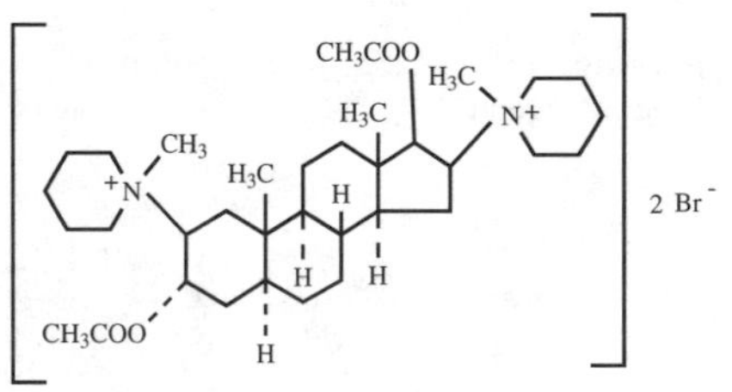

Pancytopenia The decrease of red blood cells and white blood cells in the blood.

Pandel Cream A trade name for a hydrocortisone cream used as an anti-inflammatory and antipruritic agent.

Pandemic A disease occurring in epidemic proportion over a large geographic area.

Panectyl A trade name for trimeprazine tartrate, an antihistaminic agent that competes with histamine for H_1 receptors on target cell.

Panhematin A trade name for hemin derived from processed red blood cells used for the treatment of acute porphyrias

Panhysterectomy Surgical removal of the entire uterus, including the cervix.

Panipenem (mol wt 339) An antibacterial agent.

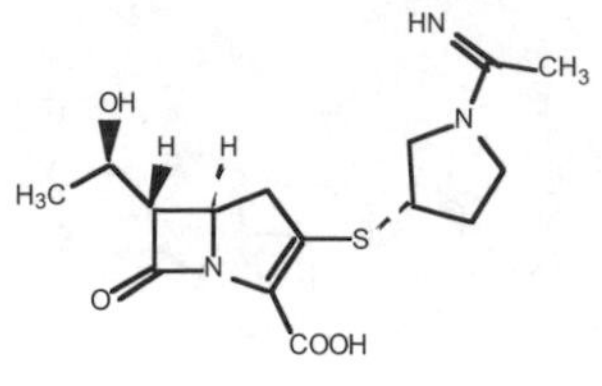

Pankrin An endopeptidase from hog or beef pancreas.

Panmycin P A trade name for tetracycline hydrochloride that binds to 30S ribosomes inhibiting protein synthesis.

Pantetheine (mol wt 278) A growth factor for *Lactobacterium bulgaricus* and an intermediate in the biosynthesis of CoA in mammalian liver.

Pantetheine Kinase The enzyme that catalyzes the following reaction:

ATP + pantetheine

⇌

ADP + pantetheine-4-phosphate

Pantethine (mol wt 555) An antihyperlipoproteinemic agent.

Panthoderm A trade name for dexpanthenol, an agent used to relieve itching and mild skin irritation.

Pantoate Activating Enzyme The enzyme that catalyzes the following reaction:

ATP + L-pantoate + β-alanine

⇌

AMP + PPi + L-pantothenate

Pantoate β-Alanine Ligase See pantoate activating enzyme.

Pantoate 4-Dehydrogenase The enzyme that catalyzes the following reaction:

Pantoate + NAD^+

⇌

4-Hydropantoate + NADH

Pantoprazole (mol wt 383) An antiulcerative agent.

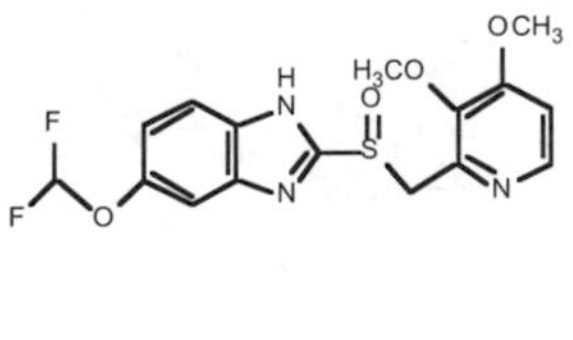

Pantothenase The enzyme that catalyzes the following reaction:

Pantothenate + H_2O

⇌

Pantoate + β-alanine

Pantothenate Amidohydrolase See pantothenase.

Pantothenate Kinase The enzyme that catalyzes the following reaction:

Pantothenate + ATP

⇌

ADP + 4-phosphopantothenate

Pantothenate Synthetase See pantoate activating enzyme.

Pantothenic Acid (mol wt 219) A water soluble vitamin required as a growth factor by yeast, fungi, and bacteria (also called vitamin B_3). It is also a component of coenzyme A.

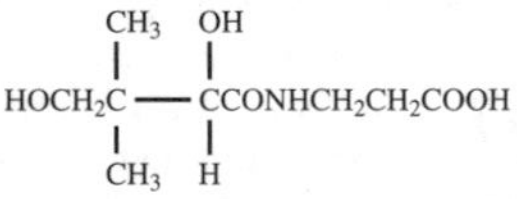

Pantropic Referring to microorganisms that infect a wide range of cell types.

Panwarfin A trade name for warfarin sodium, used as an anticoagulant.

PAO Abbreviation for 1. polyamine oxidase; 2. phenylarsine oxide.

PAO_2 Abbreviation for 1. alveolar oxygen partial pressure; 2. arterial oxygen partial pressure.

PAP Abbreviation for 1. peroxidase-antiperoxidase; 2. pokeweed antiviral protein; 3. potato acid phosphatase; 4. proline aminopeptidase; 5. prostatic acid phosphatase.

Papain An enzyme with broad specificity that catalyzes the hydrolysis of proteins with preference for residues bearing a large hydrophobic side chain.

Papaverine (mol wt 339) A smooth muscle relaxant.

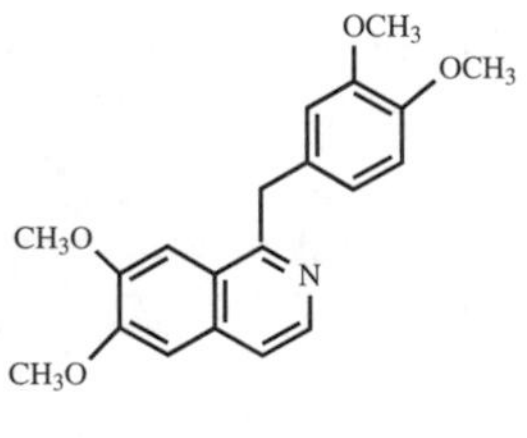

Paper Chromatogram The developed patterns or profiles on paper of separated compounds following paper chromatography.

Paper Chromatography A separation technique in which a moistened filter paper sheet acts as stationary phase for the separation of a mixture of substances according to their partition coefficients as the mobile phase (usually organic solvent) flows along the paper.

Paper Electrophoresis A type of electrophoresis in which filter paper is employed as a supporting medium for differential migration of charged substances in an electric field.

Papilloma A small neoplasm or benign growth on the skin (usually called a wart).

Papillomavirus A virus of the family Papovaviridae, e.g., rabbit papillomavirus or human papillomavirus.

***p*APMA** Abbreviation for *p*-aminophenylmercuric acetate.

PAPOVA Abbreviation for papillomavirus, polyomavirus, and vacuolating agent.

Papovaviridae A family of small icosahedral viruses containing double-stranded circular DNA (e.g., polyomavirus).

Papovaviruses Viruses of the family Papovaviridae.

PAPP Abbreviation for pregnancy-associated plasma protein.

PAPS Abbreviation for 3'-phosphoadenosine 5'-phosphosulfate.

Pap Smear A screening procedure for uterine cancer in which cells present in the vaginal secretions are examined under the microscope.

PAR Abbreviation for protease-activated receptor.

Para- A prefix meaning 1. beside or closely related and 2. two substituents on opposite carbon atoms in an aromatic ring compound.

***Para*-Aminobenzoic Acid (mol wt 137)** A water-soluble vitamin essential for the metabolism of many bacteria. It is also a sunburn protection agent.

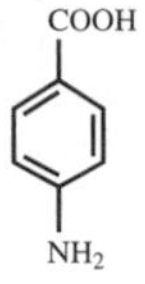

Parabiont Either of the organisms joined in parabiosis.

Parabiosis The anatomical and physiological union of two organisms either naturally or artificially produced.

Parabromdylamine Maleate Synonym of brompheniramine maleate, an antihistaminic agent.

Paracasein One of the digestion products of casein by rennin.

Paracentric Inversion A chromosome aberration in which the inverted segment does not contain the centromere.

Paracet Forte A trade name for a combination drug containing chlorzoxazone and acetaminophen, used as a skeletal muscle relaxant.

Paracoccus A genus of aerobic, oxidase-positive, Gram-negative bacteria.

Paracortex The zone located between the cortex and medulla of lymph nodes in which T cells predominate.

Paracrine Hormone Hormone that is synthesized and released by cells and acts only on the cells in the immediate environment, rather than on cells of target tissues in distant locations.

Paradione A trade name for paramethadione, used as an anticonvulsant.

Paraffin A mixture of solid hydrocarbons derived from petroleum and used for raising the melting temperature of ointment and for embedding tissue for histological sectioning.

Paraffin Method See paraffin section.

Paraffinoma A tumor caused by injection of paraffin oil beneath the skin.

Paraffin Section A method for histological sectioning of tissue for microscopic examination that employs paraffin as the embedding medium.

Parafilm A flexible, moisture-proof synthetic sheet used for sealing containers in the laboratory.

Paraflex A trade name for chlorzoxazone which is used as a skeletal muscle relaxant.

Paraflutizide (mol wt 406) A diuretic agent.

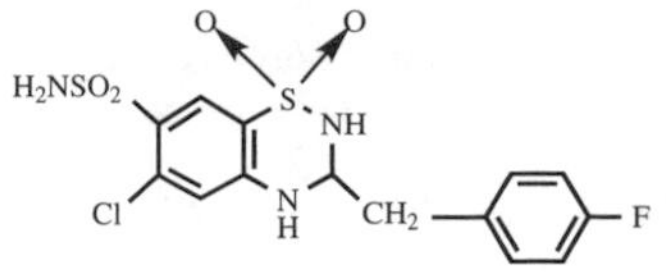

Parafon Forte DSC A trade name for chlorzoxazone, used as a skeletal muscle relaxant that reduces the transmission of nerve impulses from spinal cord to skeletal muscle.

Paraformaldehyde A solid polymeric form of formaldehyde that yields formaldehyde gas upon heating.

Parainfluenza Virus A virus of the family Paramyxoviridae, e.g., sendai virus. The human type parainfluenza viruses cause mild respiratory infection.

Paral A trade name for paraldehyde, a sedative and hypnotic agent.

Paraldehyde (mol wt 132) A sedative-hyphotic agent.

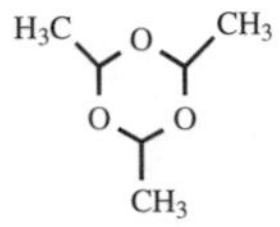

Parallel Strands A double-stranded DNA whose two 3′-terminals or two 5′-terminals run in the same direction.

Paralogous Gene A gene that is originated by duplication and then diverged from the parent copy by mutation or selection.

Paramecium A genus of protozoa of the order Hymenostomatida. It feeds on bacteria and reproduces asexually by binary fission and sexually by conjugation.

Paramethadione (mol wt 157) An anticonvulsant.

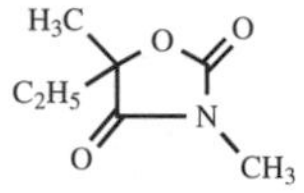

Paramethasone (mol wt 392) A glucocorticoid used as an anti-inflammatory agent.

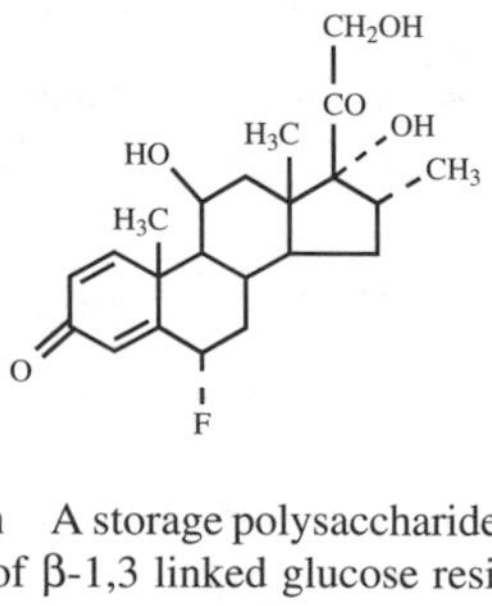

Paramylon A storage polysaccharide in *Euglena* composed of β-1,3 linked glucose residues.

Paramyosin A protein of the thick filament of the invertebrate muscle.

Paramyxoviridae A family of enveloped, minus-stranded RNA animal viruses (e.g., measles, mumps, rinderpest).

Paranaplasma A genus of bacteria of the family Anaplasmataceae.

Paranuclear Located near the nucleus.

Paranucleus An accessory nucleus or a small chromatin body resembling a nucleus.

Paranyline (mol wt 296) An anti-inflammatory agent.

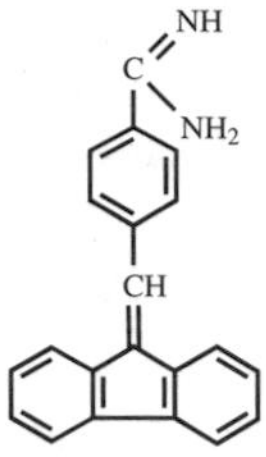

Paraoxon (mol wt 275) A cholinesterase inhibitor and an insecticide.

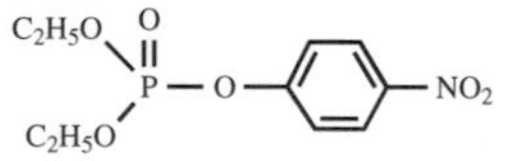

Paraoxonase A serum esterase exclusively associated with high density lipoprotein.

Paraphen A trade name for acetaminophen, used as an antipyretic and analgesic agent that blocks the generation of pain impulses.

Paraplatin A trade name for carboplatin, an antineoplastic agent.

Paraprotein An abnormal monoclonal immunoglobulin formed in patients with multiple myeloma or Waldenström's macroglobulinemia that is characterized by a well-defined peak on electrophoresis.

Paraproteinemia A condition occurring in a heterogeneous group of diseases characterized by the presence in the serum or urine of a monoclonal immunoglobulin.

Parapyruvate A dimer of pyruvate.

Parasexual Cycle A series of nonsexual events leading to genetic recombination in vegetative or somatic cells. It involves formation of heterokaryon by fusion of unlike haploid nuclei to yield a heterozyous diploid nuclei and followed by recombination and segregation at mitosis by mitotic crossing over and loss of chromosomes.

Parasites Organisms that live on or in another living organism (the host) from which they derive their nutrients.

Parasitism An association between species in which organisms of one species (the parasite) obtain their nutrients by living on or in another species (the host), often with harmful effects to the host.

Parasitology The science that deals with parasites.

Parastatin A 73-residue peptide hormone that inhibits parathyrin secretion.

Parasympathetic Referring to parasympathetic nerve or parasympathetic nervous system.

Parasympathetic Nerves The involuntary or autonomic nervous system that supplies nerves to the eyes, glands, heart, lungs, abdominal organs, genitals, etc.

Parasympathetic Nervous System A portion of the autonomic (involuntary) nervous system that produces effects such as decreased blood pressure and decelerated heart beat. The neurotransmitter for this system is acetylcholine.

Parasympatholytics Referring to drugs that block the actions of parasympathetic nerves.

Parasympathomimetic Drug Drugs that stimulate the parasympathetic nerves.

Parathion (mol wt 291) An insecticide that inhibits the activity of acetylcholine esterase and prevents hydrolysis of acetylcholine to choline and acetic acid.

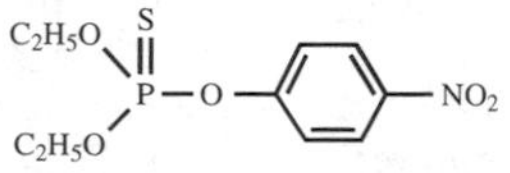

Parathormone See parathyroid hormone.

Parathyrin See parathyroid hormone.

Parathyroid Glands Small endocrine glands of vertebrates located near the thyroid that secret parathyroid hormone controlling the metabolism of calcium and phosphate.

Parathyroid Hormone A peptide hormone secreted by parathyroids that controls metabolism of calcium and phosphate.

Paratopes 1. The antigen-combining sites on immunoglobulin. 2. The site at which a cell receptor binds with an exogenous molecule.

Paratose (3,6-dideoxy-D-glucose) A simple sugar from *Salmonella paratyphi* and a component of O antigen.

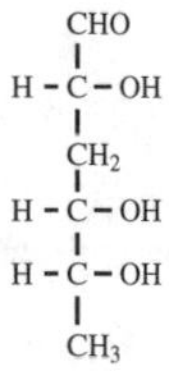

Paratyphoid Fever An acute infectious disease caused by *Salmonella paratyphi.*

Parbendazole (mol wt 247) An anthelmintic agent.

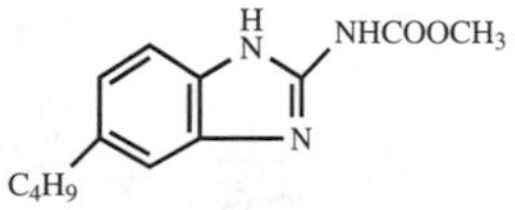

pARC5 A plasmid of *E. coli* that contains Amp^r (ampicillin resistant) marker and EcoRI, BamHI, HindIII, PstI, SalI, XbaI, SmaI, SstI cleavage sites.

Parenchyma Ground tissue of plant composed of relatively unspecialized cells.

Parenteral Medication Introduction of medicine through a route other than by mouth, e.g., injection.

Parepectolin A trade name for a combination drug containing opium, kaolin, pectin, and alcohol, used as an antidiarrheal agent.

Parethoxycaine (mol wt 265) A local anesthetic agent.

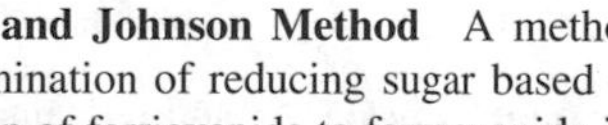

Paricalcitol A synthetically produced vitamin D analog used to reduce parathyroid hormone levels.

Parietal Cells Cells in the stomach lining that secrete hydrochloric acid.

Park and Johnson Method A method for the determination of reducing sugar based on the reduction of ferricyanide to ferrocyanide in alkaline solution.

Parkinsonism A group of nervous disorders characterized by muscular rigidity and tremor due to the loss of dopamine, norepinephrine, and serotonin.

Parlodel A trade name for bromocriptin mesylate, an antiparkinsonian agent that acts as a dopamine receptor agonist.

Par-Mag A trade name for magnesium oxide, used as an antacid.

Parmine A trade name for phentermine hydrochloride, used as a cerebral stimulant that promotes transmission of impulses by releasing stored norepinephrine from the nerve terminals in the CNS.

Parmol A trade name for acetaminophen, used as an analgesic and an antipyretic agent.

Parnate A trade name for tranylcypromine sulfate, an antidepressant that promotes accumulation of neurotransmitter by inhibiting MAO.

Paromomycin (mol wt 616) A broad spectrum aminoglycoside antibiotic produced by species of *Streptomyces.*

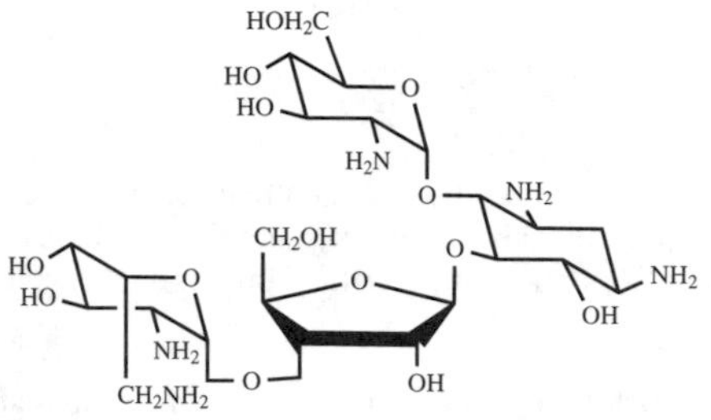

Parotitis Inflammation of parotid salivary gland.

Paroxetine (mol wt 329) An antidepressant.

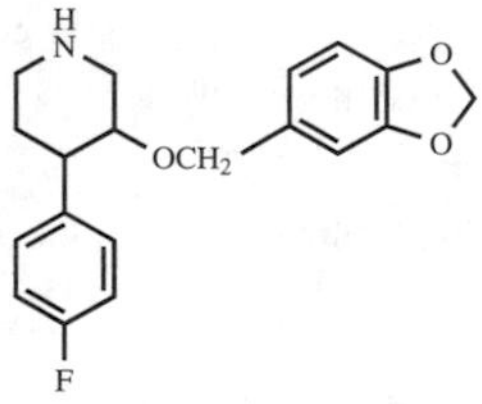

Paroxypropione (mol wt 150) A pituitary gonadotropic hormone inhibitor.

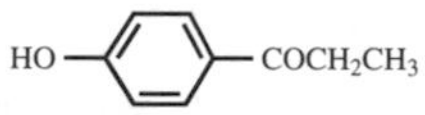

PARP Abbreviation for 1. poly-ADP-ribose polymerase; 2. proline-arginine-rich protein.

Parsalmide (mol wt 246) An anti-inflammatory and an analgesic agent.

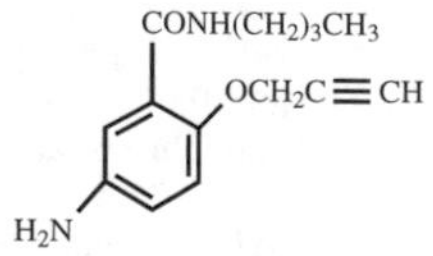

Parser Granule A trade name for *p*-aminosalicylic acid used as an antibacterial agent.

Parthenocarp Formation of fruit without fertilization leading to the production of seedless fruits.

Parthenogenesis The development of an organism from an unfertilized egg.

Partial Denaturation Referring to a process by which regions of a dsDNA molecule are converted to single-stranded loops.

Partial Diploid A bacterium that contains an exogenotic chromosome fragment donated by the F^+ cell.

Partial Pressure The pressure exerted by a specific gas in a gas mixture.

Particle Specific Volume The reciprocal of density (volume/mass or ml/g).

Particle-Bound Aminopeptidase Synonym of membrane alanyl aminopeptidase.

Particulate Antigen Antigen that is part of an insoluble structure (insoluble antigen).

Particulate Enzyme The enzyme that is linked to an insoluble particle or bead, e.g., agarose bead.

Partition The distribution of solutes between stationary and mobile phases.

Partition Chromatography A type of chromatographic method for separation of solutes based on the solubilities or partition coefficients of the solutes in the stationary phase and mobile phase.

Partition Coefficient The equilibrium constant defined as:

$$K = \frac{\text{concentration of solute in stationary phase}}{\text{concentration of solute in mobile phase}}$$

Partition Isotherm Referring to partition coefficient.

Parturition Giving birth.

Parvalbumins The calcium-binding proteins in the muscle of lower vertebrates.

Parvolex A trade name for acetylcysteine, used as a mucolytic agent.

Parvoviridae A family of nonenveloped isometric, ssDNA viruses.

Parvovirus A virus in the family Parvoviridae containing ssDNA that infects vertebrates.

PAS Abbreviation for 1. periodic acid-Schiff's reagent for the determination of carbohydrate and 2. *para*-aminosalicylic acid.

pASI A plasmid of *E. coli* that contains Ampr (ampicillin resistant) marker and BamHI cloning site and l promoter.

PAS Band The protein bands in SDS-PAGE gels stained with PAS reagent.

PAS Stain Abbreviation for *p*-aminosalicylic stain.

PASA Abbreviation for para-aminosalicylic acid.

Pasiniazide (mol wt 290) A sedative-hypnotic agent.

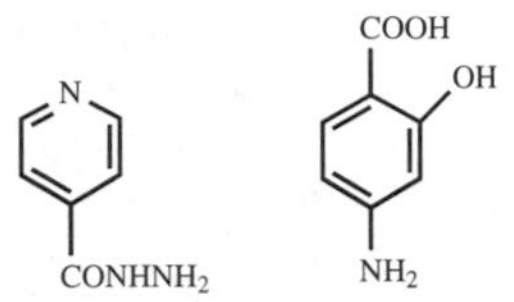

PASR Abbreviation for periodic acid-Schiff reaction.

Passage Number The number of times a culture has been subcultured.

Passive Cutaneous Anaphylaxis A test for detecting cytotropic antibody responsible for immediate-type hypersensitivity.

Passive Agglutination A serological procedure in which the soluble antigen is made readily detectable by prior adsorption onto erythrocytes or insoluble particles.

Passive Diffusion The passive (nonmediated) movement of molecules from an area of high concentration to an area of low concentration without expenditure of energy.

Passive Immunity Immunity conferred by the transfer of preformed immune cells or immune products, such as antibodies or sensitized T cells, into a nonimmune individual.

Passive Mediated Transport Mediated transport or flow of specific molecule from high concentration to low concentration.

Passive Transfer of Immunity See passive immunity.

Passive Transport See passive mediated transport.

Pasteur Effect The inhibition by oxygen on the production of ethanol due to the inhibition of phosphofructoinase by ATP produced from oxidative phosphorylation in the presence of oxygen. The yeast cells consume more glucose and produce more ethanol under anaerobic than aerobic conditions.

Pasteur Pipette An open-ended glass tube with one end drawn out to an internal diameter of about 1 mm.

Pasteurella A genus of Gram-negative bacteria of the family Pasteurellaceae.

Pasteuria A genus of budding bacteria.

Pasteurization A form of heat treatment for the destruction of pathogenic microorganisms in the milk as well as for agents that may cause deterioration of the milk.

PAT Abbreviation for palmitoyl acyltransferase.

Patanol A trade name for olopatadine hydrochloride, a mast cell stabilizer and an antihistaminic agent.

Patatin Any of a family of soluble glycoprotein from potato.

Patch Clamp Technique A technique in which a tiny micropipette is placed on the surface of a cell for measuring the movement of ions through the individual ion channel.

Patching The reorganization of a cell surface membrane component into discrete patches.

PATCO Abbreviation for a combination drug containing prednisone, ara-C, thioguanine, cytoxan, and oncovin.

PATH Abbreviation for pituitary adrenotropic hormone.

pATH1, pATH2, pATH3, pATH10, pATH11, pATH20, pATH21, pATH22, pATH23 Plasmids of *E. coli* that contain Amp^r marker and *try* promoter. They are different in restriction endonuclease cleavage sites.

Pathocil A trade name for dicloxacillin sodium, a penicillinase-resistant penicillin that inhibits bacterial cell wall synthesis.

Pathogens Organisms capable of causing disease in animals, plants, or microorganisms.

Pathogenic Capable of causing disease.

Pathogenic RNA Referring to viroid.

Pathogenicity The ability of an organism to cause disease.

Pathogenesis Related Protein Proteins that accumulate in plant tissues as part of the hypersensitivity response to viral or fungal infection.

Pathology The science that deals with the nature of disease.

Patulin (mol wt 154) An antibiotic from the species of *Aspergillus*.

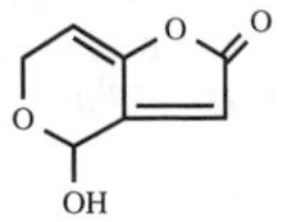

PauI A restriction endonuclease from *Paracoccus alcaliphilus* ZVK3-3 with the following specificity:

5'........G↓CGCGC........3'
3'........CGCGC↑G........5'

Paul-Bunnell Test A hemagglutination test used for diagnosis of infectious mononucleosis. It measures the titer of the heterophilic antibodies that react with antigens on the surface of sheep red blood cells.

Pavabid A trade name for papaverine hydrochloride, a vasodilator that inhibits phosphodiesterase and increases the level of cAMP.

Pavagen TD A trade name for papaverine, a smooth muscle relaxant.

Pavarine Spancaps A trade name for papaverine hydrochloride, a vasodilator that inhibits phosphodiesterase and increases the level of cAMP.

Pavasule A trade name for papaverine hydrochloride, a vasodilator that inhibits phosphodiesterase and increases the level of cAMP.

Pavatine A trade name for papaverine hydrochloride, a vasodilator that inhibits phosphodiesterase and increases the level of cAMP.

Pavatym A trade name for papaverine hydrochloride, a vasodilator that inhibits phosphodiesterase and increases the level of cAMP.

Pavementing The margination of leukocytes from the blood stream to the endothelium near a site of tissue damage or inflammation.

Paveral A trade name for codeine phosphate, an opioid analgesic agent that binds with opioid receptors in the CNS, altering both perception and emotional response to pain.

Pavulon A trade name for pancuronium bromide, a neuromuscular blocker that prevents acetylcholine from binding to its receptors.

pAW101 A plasmid that contains Amp^r (ampicillin resistant) and Tet^r (tetracycline resistant) markers.

Pawly Reaction A colorimetric reaction for the determination of histidine that uses diazotized sulfanilic acid in alkaline solution.

Paxil A trade name for paroxetine, an antidepressant.

Paxillin An adhesion protein associated with vinculin.

Paxipam A trade name for halazepam, used as an antianxiety agent.

Pazufloxacin (mol wt 318) An antibacterial agent.

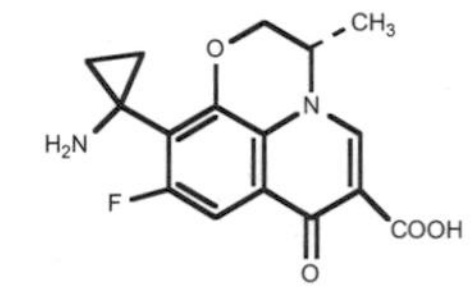

Pb Symbol for lead.

PB Abbreviation for phenol barbital.

PBA Abbreviation for polyclonal B-cell activator.

PBB Abbreviation for polybromated biphenyl.

PBC Abbreviation for 1. peripheral blood cells; 2. primary biliary cirrhosis.

pBC16 A plasmid of *Bacillus* that contains Str^r (streptomycin resistant) and Tet^r (tetracycline resistant) markers and BamHI and XbaI cleavage sites.

pBD6 A plasmid of *Bacillus* and *Staphylococcus aureus* that contains Str^r (streptomycin resistant) and Kan^r (kanamycin resistant) markers and BamHI, TacI, BglII, and HindIII cleavage sites.

pBD8 A plasmid of *Bacillus* and *Staphylococcus aureus* that contains Cml^r (chloramphenicol resistant), Str^r (streptomycin resistant) and Kan^r (kanamycin resistant) markers and BamHI, XbaI, EcoRI, HindIII, and BglII cleavage sites.

pBD9 A plasmid of *Baccillus* and *Staphylococcus aureus* that contains Ery^r (erythromycin resistant) and Kan^r (kanamycin resistant) markers and EcoRI, BamHI, TacI, PstI, BglII, BclI, and HpaI cleavage sites.

pBD10 A plasmid of *Baccillus* and *Staphylococcus aureus* that contains Cml^r (chloramphenicol resistant), Ery^r (erythromycin resistant), and Kan^r (kanamycin resistant) markers and BamHI, XbaI, BglII, HpaI, and BclI cleavage sites.

pBD11 A plasmid of *Baccillus* and *Staphylococcus aureus* that contains Ery^r (erythromycin resistant) and Kan^r (kanamycin resistant) markers and XbaI, BamHI, BglII, HpaI, and BclI cleavage sites.

pBD12 A plasmid of *Baccillus* and *Staphylococcus aureus* that contains Cml^r (chloramphenicol resistant) and Kan^r (kanamycin resistant) markers and EcoRI, XbaI, BamHI, TacI, HindIII, and BglII cleavage sites.

pBD20 A plasmid of *Baccillus* and *Staphylococcus aureus* that contains Cml^r (chloramphenicol resistant) and Ery^r (erythromycin resistant) markers and BclI, HpaI, PstI, XbaI, and HindIII cleavage sites.

PB-Fe Abbreviation for protein-bound iron.

pBGP120 A plasmid of *E. coli* that contains the Amp^r (ampicillin resistant) marker and *lac* promoter.

pBGS8 A plasmid of *E. coli* that contains Kan^r (kanamycin resistant) marker and *lac* promoter and HindIII, PstI, SalI, AccI, HincII, BamHI, SmaI, and EcoRI cleavage sites.

PBI Abbreviation for protein-bound iodine.

PBK Abbreviation for phosphorylase b kinase.

PBL Abbreviation for peripheral blood lymphocytes (e.g., lymphocytes found in the circulating blood).

pBLA11 A plasmid of *E. coli* that contains the Tet^r (tetracycline resistant) marker and ribosomal binding site of the β-lactamase gene.

PBMC Abbreviation for peripheral blood mononuclear cell.

PBMNC Abbreviation for peripheral blood mononuclear cell.

PBO Abbreviation for penicillin in beeswax ointment.

PBP Abbreviation for penicillin-binding protein.

***pbp*A** A gene in *E. coli* that encodes penicillin-binding protein.

pBR322 A general purpose plasmid vector of *E. coli* that contains Tet^r (tetracycline resistant) and Amp^r (ampicillin resistant) markers and EcoRI, ClaI, HindIII, EcoRV, BamHI, SphI, SalI, XmaI, NruI, AvaI, BalI, PvuII, Tth111I, NdeI, PstI, PvuI, ScaI, and AatII cleavage sites.

pBR328 A plasmid of *E. coli* that contains Amp^r (ampicillin resistant), Cml^r (chloramphenicol resistant), and Tet^r (tetracycline resistant) markers and AatII, AsuII, AvaI, ClaI, HgiEII, HindIII, Tth111I, BamHI, EcoRI, EcoRV, BalI, BamHI, NcoI, NruI, PstI, PvuI, PvuII, SalI, SphI, and XmaIII cleavage sites.

pBRH1 A plasmid of *E. coli* that contains the Amp^r (ampicillin resistant) marker.

PBS Abbreviation for phosphate-buffered saline.

PBS1 Bacteriophage A bacteriophage of the family Myoviridae.

PBT$_4$ Abbreviation for protein-bound thyroxine.

PBV Abbreviation for a combination drug containing platinol, bleomycin, and velban.

PBZ A trade name for tripelennamine citrate, used as an antihistaminic agent that competes with histamine for H$_1$ receptors on effector cells.

PC Abbreviation for 1. packed cells; 2. palmitoyl carnitine; 3. phosphatidylcholine; 4. phosphocreatine; 5. plastocyanin; 6. platelet count; 7. prostacyclin; 8. pyruvate carboxylase.

pC194 A plasmid of *Bacillus* and *Staphylococcus aureus* that contains the chloramphenicol resistant marker (Cml^r) and HaeIII, HindIII, and BglI cleavage sites.

PCA Abbreviation for 1. passive cutaneous anaphylaxis; 2. perchloric acid; 3. phenylcarboxylic acid.

PCB Abbreviation for polychlorinated biphenyl. A group of widely used chlorinated hydrocarbons that are extremely resistant to breakdown and known to be carcinogenic.

pCB1 A plasmid of *E. coli* that contains the ampicillin resistant (Amp^r) marker.

pCB101 A plasmid of *Corynebacterium glutamicum, bacillus subtilis* that contains Kan^r (kanamycin resistant) and Spc^r (spectinomycin resistant) markers and PstI and BglII cleavage sites.

PCBP Abbreviation polychlorinated biphenyl.

PCC Abbreviation for propionyl-CoA carboxylase.

PCCase Abbreviation for propionyl-CoA carboxylase.

PCD Abbreviation for 1. phosphate/citrate/dextrose; 2. programmed cell death.

PCE Abbreviation for 1. pseudocholinesterase; 2. polymer-coated erythromycin.

PCE Disperstabs A trade name for erythromycin base, which binds to 50S ribosomes inhibiting bacterial protein synthesis.

PCF Abbreviation for prothrombin conversion factor.

pCFM526 A plasmid of *E. coli* that contains ampicillin resistant (Amp^r) and kanamycin resistant (Kan^r) markers.

PCH Abbreviation for paroxysmal cold hemaglobinuria.

pCHR82 A plasmid of *E. coli* that contains Kan^r (tetracycline resistant) and Tmp^r (temperature resistant) markers and ClaI, XbaI, EcoRI, and HindIII cleavage sites.

pCHR83 A plasmid of *E. coli* that contains Cml^r (chloramphenicol resistant) and Tmp^r (temperature resistant) markers and PvuII and EcoRI cleavage sites.

PCI Abbreviation for protein-C inhibitor.

PCM Abbreviation for protein carboxyl methylase.

PCMB Abbreviation for *para*-chloromercuribenzoic acid.

***p*-CMBS** Abbreviation for *p*-chloromercuric benzenesulfonate.

PCNA Abbreviation for proliferating cell nuclear antigen.

PCNB Abbreviation for pentachloronitrobenzene.

PCoA Abbreviation for palmitoyl-CoA.

PCOOH Abbreviation for phosphatidylcholine hydroperoxide.

PCOR Abbreviation for protochlorophyllide oxidoreductase (NADPH-dependent).

PCP Abbreviation for 1. parachlorophenol; 2. Pentachlorophenol; 3. *Pneumocystis carinii* pneumonia.

PCPA Abbreviation for *para*-chlorophenylalanine.

PCr Abbreviation for phosphocreatine.

PCR Abbreviation for polymerase chain reaction, a technique for amplifying specific regions of DNA by multiple cycles of polymerization, each followed by a brief heat treatment for separation of complementary strand.

pCR1 A plasmid of *E. coli* that contains kanamycin resistant (Kan^r) marker and HindIII cleavage site.

PCS Abbreviation for paradoxical concanavalin-A staining.

pCS3 A plasmid of *E. coli* that contains Ampr (ampicillin resistant) and Tet^r (tetracycline resistant markers and EcoRI, PvuII, PstI, PvuI, ClaI, HindIII, SpHI, SalI, and HamHI cleavage sites.

PCV Abbreviation for 1. a combination drug containing procarbazine, cytoxan, and velban; 2. packed cell volume; 3. polychlorinated vinyl.

PCVP Abbreviation for a combination drug containing procarbazine, cytoxan, vinblastine, and prednisone.

PCZ Abbreviation for 1. procarbazine; 2. prochlorperazine.

PD Abbreviation for 1. Parkinson's disease; 2. porphobilinogen deaminase.

Pd Abbreviation for palladium, a chemical element.

pD553 A plasmid of *E. coli* that contains ampicillin resistant (Amp^r) marker and EcoRI, HindIII, and HpaI cleavage sites and antiterminating λ gene N.

PDAB Abbreviation for *para*-dimethylaminobenzaldehyde.

PDB Abbreviation for 1. phorbol 12, 13-dibutyrate; 2. protein database.

PDBu Abbreviation for phorbol 12, 13-dibutyrate.

PDC Abbreviation for 1. phenylalanine decarboxylase; 2. pyruvate decarboxylase complex.

PDD Abbreviation for phorbol 12,13-didecanoate.

PDE Abbreviation for phosphodiesterase.

PD-ECGF Abbreviation for platelet-derived epithelial cell growth factor.

pDG141 A plasmid of *E. coli* that contains ampicillin resistant marker and SacI cleavage site.

PDGF Abbreviation for platelet-derived growth factor.

PDGFR Abbreviation for platelet-derived growth factor receptor.

PDH Abbreviation for 1. phenylalanine dehydrogenase; 2. proline dehydrogenase; 3. pyruvate dehydrogenase.

PDI Abbreviation for protein disulfide isomerase.

PDK Abbreviation for 1. pyruvate dehydrogenase kinase; 2. 3-phosphoinositide dependent protein kinase.

PDK1 Abbreviation for 3-phosphoinositide-dependent protein kinase-1.

pDM1 A plasmid of *E. coli* that contains ampicillin resistant marker and SalI, AccI, and HincII cleavage sites.

pDR42 A plasmid of *E. coli* that contains ampicillin resistant (Amp^r) and tetracycline resistant (Tet^r) markers and EcoRI, HindIII, BamHI, and SalI cleavage sites.

PDTC Abbreviation for pyrollidine dithiocarbamate.

PE Abbreviation for 1. phosphatidylethanolamine; 2. phycoerythrin; 3. paper electrophoresis; 4. protein electrophoresis.

PEA Abbreviation for 1. phosphatidylethanolamine; 2. *Pseudomonas* exotoxin A.

pEA300 A plasmid of *E. coli* that contains ampicillin resistant (Amp^r) marker, tryptophan promoter, and ClaI cleavage site.

Pebulate (mol wt 203) An herbicide.

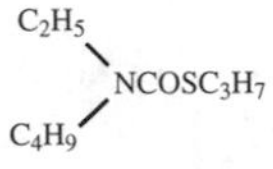

PEC-60 A 60-residue endogenous regulatory polypeptide in the intestinal tissue and central nervous system that inhibits the formation of cyclic AMP and activates Na^+- K^+-ATPase.

PECAM Abbreviation for platelet endothelial cell adhesion molecule, a member of the immunoglobulin superfamily.

Pecilocin (mol wt 291) An antifungal agent.

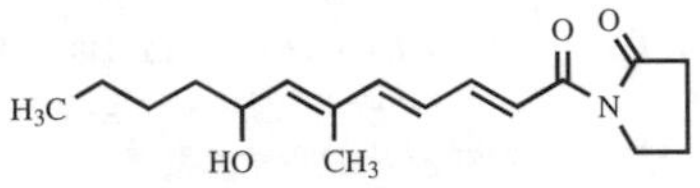

Pectase See pectinesterase.

Pectate A salt of pectic acid.

Pectate Lyase The enzyme that catalyzes the depolymerization of pectins (also known as pectic depolymerase).

Pectate Transeliminase Synonym of pectate lyase.

Pectic Acid A polymer of galacturonic acid.

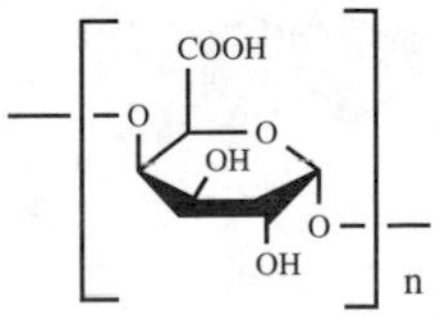

Pectic Enzymes The enzymes that hydrolyzes pectins.

Pectic Polysaccharides Polysaccharides from plant cell walls that contain pectins, pectic acid, and neutral polysaccharide.

Pectic Substance Referring to pectic acids, pectins, and related compounds.

Pectin A polymer of galacturonic acid deposited with cellulose in the cell wall of plant tissue that functions as a intercellular cementing material. It contains partial methyl ester of a α-(1,4)-galacturonate sequence interrupted with (1,2)-linked sugar residues.

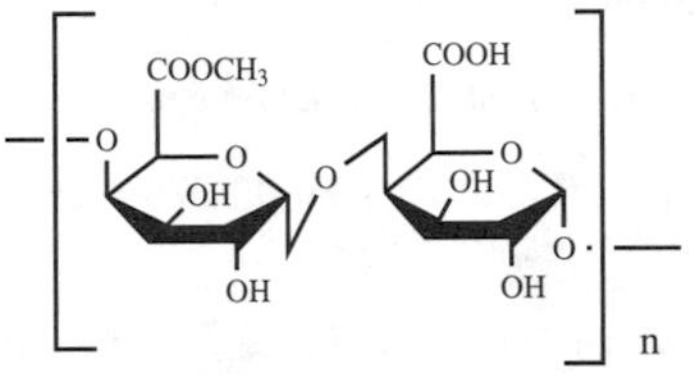

Pectin Demethoxylase Synonym of pectinesterase.

Pectin Lyase The enzyme that catalyzes the eliminative cleavage of pectin producing oligosaccharides with terminal 4-deoxy-6-methyl-α-D-galact-4-enuronosyl groups.

Pectin Methoxylase Synonym of pectinesterase.

Pectin Methylesterase Synonym of pectinesterase.

Pectin Pectylhydrolase The systematic name for pectinesterase.

Pectinase The enzyme that catalyzes the hydrolysis of a-1,4-D-galacturonide links in pectate (see also polygalacturonase).

Pectinatus A genus of Gram-negative bacteria (family Bacteroidaceae).

Pectinesterase The enzyme that catalyzes the following reaction:

$$\text{Pectin} + n\ H_2O \rightleftharpoons n\ \text{Methanol} + \text{pectate}$$

Pectinolytic Capable of degradating pectin.

Pectolyase Synonym of pectin lyase.

Pediacare-1 A trade name for dextromethorpan hydrobromide, used as an antitussive agent.

Pediaflor A trade name for sodium fluoride, used as a nutrition mineral.

Pediamycin A trade name for erythromycin ethylsuccinate, an antibiotic that binds to 50S ribosomes inhibiting bacterial protein synthesis.

Pediapred A trade name for prednisolone sodium phosphate, used as an anti-inflammatory hormone.

Pediaprophen A trade name for ibuprofen, an analgesic and anti-inflammatory agent.

Pediatrics The medical science that deals with the health of children.

Pediazole A trade name for a combination drug containing sulfisoxazole and erythromycin ethylsuccinate, used as an antibacterial agent.

Pediculicide Any substance that kills lice.

Pediculosis A skin infestation caused by lice.

Pedigree The record or profile of the inheritance of an individual.

Pediococcus A genus of Gram-positive bacteria of the family Streptococcaceae.

Pedomicrobium A genus of budding prosthecate bacteria found in the soil.

Pedric A trade name for acetaminophen, used as an antipyretic and analgesic agent that blocks the generation of pain impulses.

Pedtrace-4 A trade name for a combination drug containing copper sulfate, magnesium sulfate, and chromium chloride, used as nutritional minerals.

Pedvax HIB A trade name for hemophilus (haemophilus) b vaccine.

Pefloxacin (mol wt 333) A quinolone antibiotic.

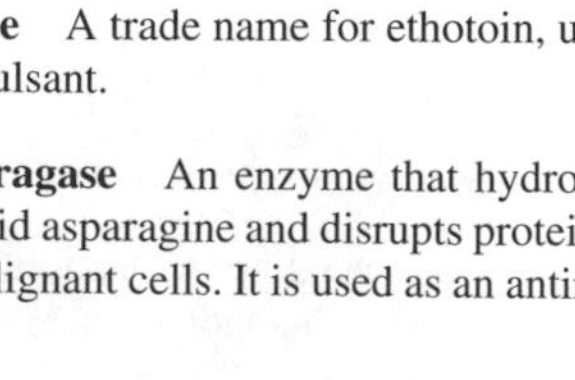

PEG Abbreviation for polyethylene glycol.

Peganone A trade name for ethotoin, used as an anticonvulsant.

Pegasparagase An enzyme that hydrolyzes the amino acid asparagine and disrupts protein synthesis in malignant cells. It is used as an antineoplastc agent.

PEG-L A trade name for pegasparagase.

PEG-PS Abbreviation for polyethylene glycol polystyrene.

PEI-Cellulose Abbreviation for polyethyleneimine cellulose, an anion exchanger used in ion exchange chromatography.

Pelargonic Acid (mol wt 158) A saturated organic acid.

$CH_3(CH_2)_7COOH$

P-Element A transposable element in the genome of *Drosophila.*

Pellagra A vitamin-B deficient disorder characterized by rash on the body surface, diarrhea, and ulceration in the mouth.

Pellagra-Preventing Factor Referring to nicotinic acid (niacin or vitamin B_3).

Pellet Referring to sediment or precipitate resulting from centrifugation.

Pelletierine (mol wt 141) An anthelmintic agent.

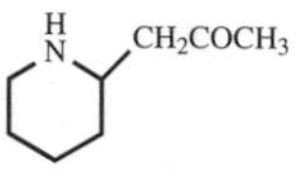

Pellicle 1. A thin protective membrane occurring around some protozoa (also known as a periplast). 2. A continuous or fragmentary film formed at the surface of a liquid culture.

Pelobacter A genus of Gram-negative, rod-shaped anaerobic bacteria.

Pelodictyon A genus of photosynthetic bacteria (family Chlorobiaceae).

Pelvic Inflammatory Disease An inflammation or infection of the pelvic organs (e.g., ovaries) caused by *Neisseria gonorrhoeae*.

pEMBL8$^+$ A plasmid of *E. coli* that contains ampicillin-resistant (Ampr) marker and EcoRI, AvaI, SmaI, XmaI, BamHI, SalI, AccI, PstI, HindII, HindIII, ClaI, and NaeI cleavage sites.

pEMBL9 A plasmid of *E. coli* that contains ampicillin-resistant (Ampr) marker and EcoRI, SmaI, XmaI, BamHI, SalI, AvaI, PstI, and HindIII cleavage sites.

Pemirolast (mol wt 228) An antiallergic agent.

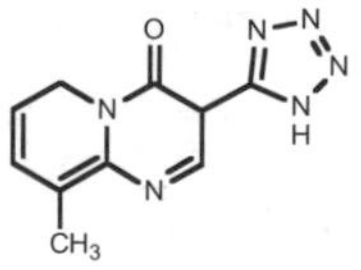

Pemoline (mol wt 176) A CNS stimulant that promotes transmission of nerve impulses by releasing stored norepinephrine from the nerve terminals in the brain.

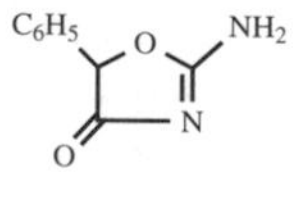

Pempidine (mol wt 155) An antihypertensive agent.

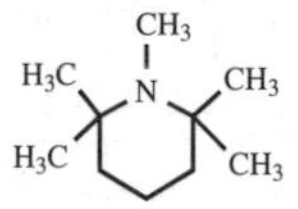

PEMT Abbreviation for phosphatidyl-ethanolamine N-methyltransferase.

Penamecillin (mol wt 406) A semisynthetic antibiotic related to penicillin that inhibits bacterial cell wall synthesis.

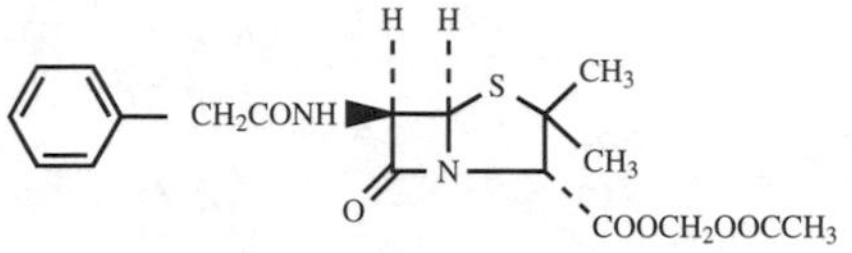

Penamp A trade name for ampicillin trihydrate, which inhibits bacterial cell wall synthesis.

Penapar VK A trade name for penicillin V potassium, an antibiotic that inhibits bacterial cell wall synthesis.

Penbec V A trade name for penicillin V, an antibiotic.

Penbritin A trade name for ampicillin trihydrate that inhibits bacterial cell wall synthesis.

Penbutolol (mol wt 291) An antihypertensive, antianginal, and antiarrhythmic agent that blocks both beta-1 and beta-2 adrenergic receptors.

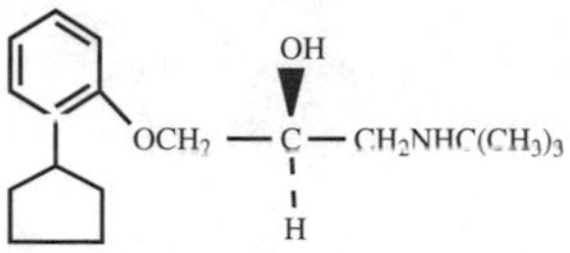

Penciclovir (mol wt 253) An antiviral agent.

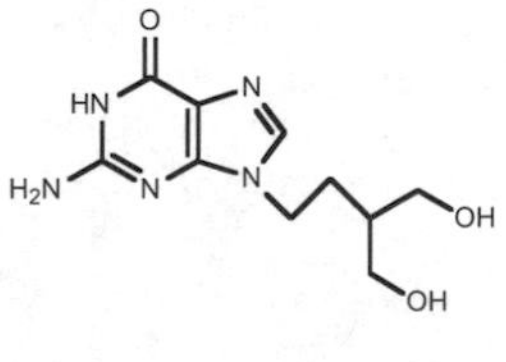

Pendred's Syndrome A genetic disorder characterized by the low level incorporation of iodine into thyroglobulin.

Penecort A trade name for hydrocortisone.

Penethamate Hydriodide (mol wt 562) An antibiotic related to penicillin that inhibits bacterial cell wall synthesis.

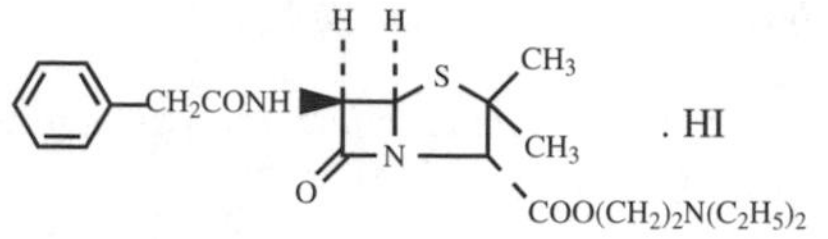

Penetration The entrance of the virion or virus nucleic acid into a host cell.

Penetrex A trade name for enoxacin, which inhibits bacterial DNA gyrase preventing bacterial replication.

Penfluridol (mol wt 524) An antipsychotic agent.

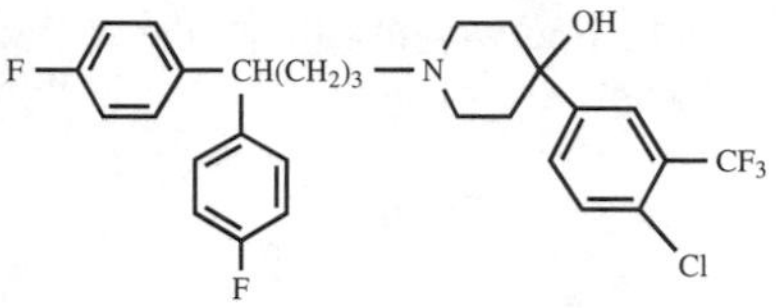

PenG Abbreviation for penicillin G.

Penglobe A trade name for bacampicillin hydrochloride, an aminopenicillin that inhibits bacterial cell wall synthesis

-penia A suffix denoting lack or deficiency.

Penicillamine (mol wt 149) A chelating and anthelmintic agent.

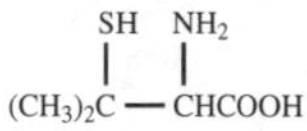

Penicillanic Acid (mol wt 200) A component of penicillin.

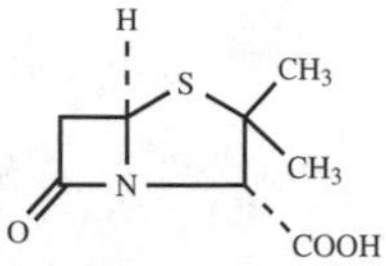

Penicillic Acid (mol wt 170) An antibiotic produced by *Penicillium puberulum.*

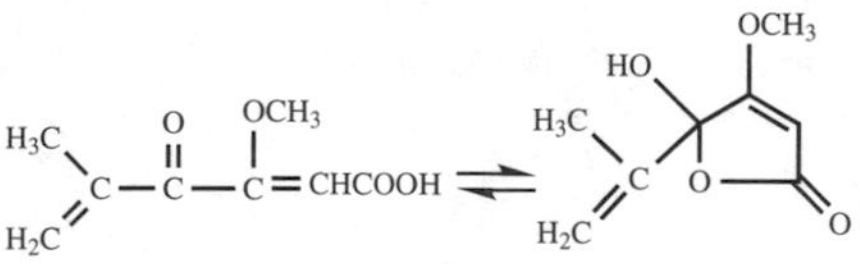

Penicillin A group of natural and semisynthetic antibiotics with a β-lactam ring that are active against Gram-positive bacteria, inhibiting the formation of cross-links in the peptidoglycan of growing bacteria.

Penicillin Acylase Synonym of penicillin amidase.

Penicillin Amidase The enzyme that catalyzes the following reaction:

$$\text{Penicillin} + H_2O \rightleftharpoons \text{A carboxylate} + \text{6-aminopenicillanate}$$

Penicillin Amidohydrolase Synonym of penicillin amidase.

Penicillin Binding Proteins Proteins in the cell envelope of Gram-positive bacteria that bind covalently to penicillin and penicillin-related antibiotics.

Penicillin BT (mol wt 346) Butylthiomethyl penicillin, an antibiotic that inhibits bacterial cell wall synthesis.

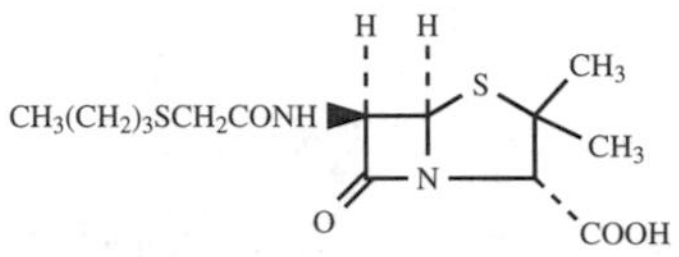

Penicillin G (mol wt 334) Benzylpenicillin, one of the natural penicillins produced by the species *Penicillium.*

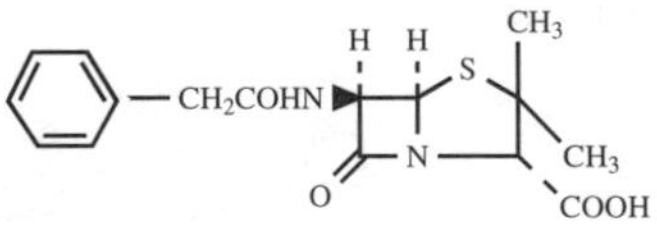

Penicillin G Benethamine (mol wt 546) A semisynthetic penicillin prepared from penicillin G.

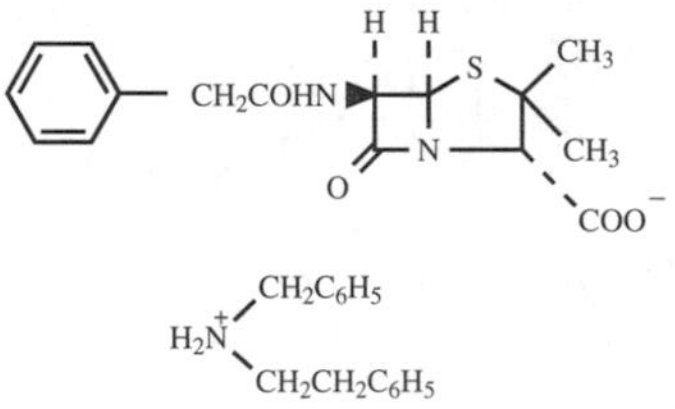

Penicillin G Calcium (mol wt 707) The calcium form of penicillin that inhibits bacterial cell wall synthesis.

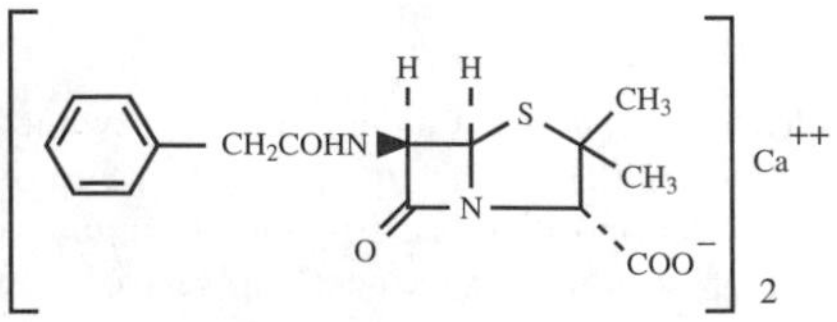

Penicillin G Potassium (mol wt 372) The potassium form of penicillin G that inhibits bacterial cell wall synthesis.

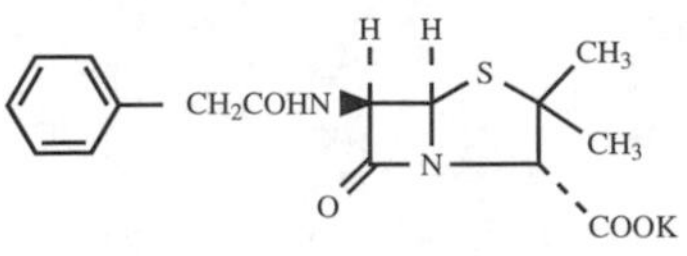

Penicillin N (mol wt 359) An antibiotic produced by the species *Cephalosporium.*

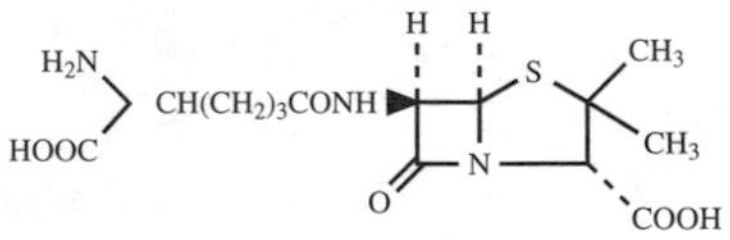

Penicillin O (mol wt 330) An antibiotic produced by *Penicillium chrysogenum.*

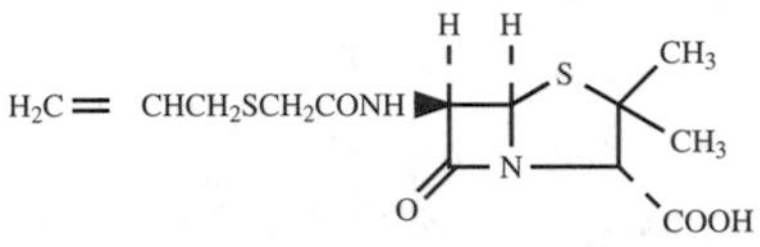

Penicillin S Potassium (mol wt 417) A β-lactam antibiotic that inhibits bacterial cell synthesis.

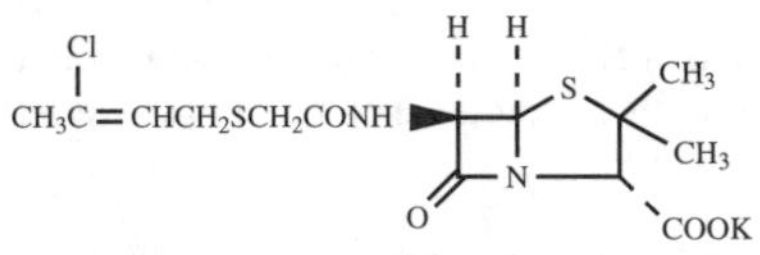

Penicillin V (mol wt 350) A β-lactam antibiotic that inhibits bacterial cell wall synthesis.

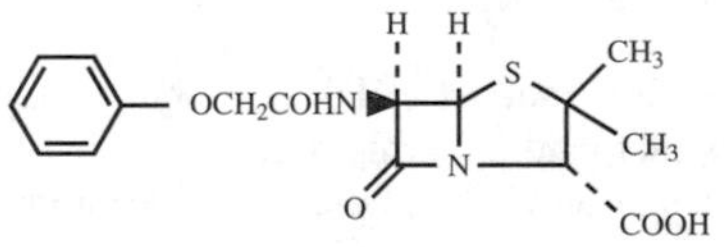

Penicillinase The enzyme that catalyzes the following reaction:

$$\text{Penicillin} + H_2O \rightleftharpoons \text{Penicilloate}$$

Penicilloic Acids The product obtained from the cleavage of penicillin.

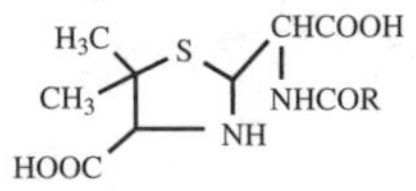

Penicillopepsin An acid aspartyl protease from *Penicillium janthinellum* with an aspartyl residue in the active site.

PENMT Abbreviation for phosphoethanolamine N-methyl transferase.

Pennoxide A sterilant containing ethylene oxide (12%) and dichlorodifluoromethane (88%).

Penta- A prefix meaning five.

Pentacarinat A trade name for pentamidine isothionate, an antiprotozoal agent.

Pentachlorophenol (mol wt 266) A fungicide and a preservative for leather, paper, and textiles. It is also used as an insecticide for termite control.

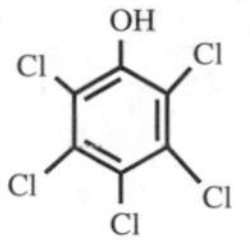

Pentachlorophenol Monooxygenase The enzyme that catalyzes the following reaction:

$$\text{Pentachlorophenol} + \text{NADPH} + O_2 \rightleftharpoons \text{Tetrachlorohydroquinone} + \text{NADP}^+ + \text{chloride}$$

Pentacynium Bis(methyl sulfate) (mol wt 644) An antihypertensive agent.

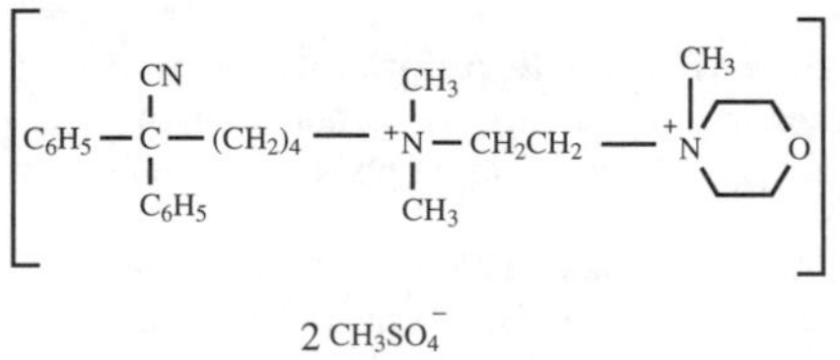

Pentaerythritol Chloral (mol wt 726) A sedative-hypnotic agent.

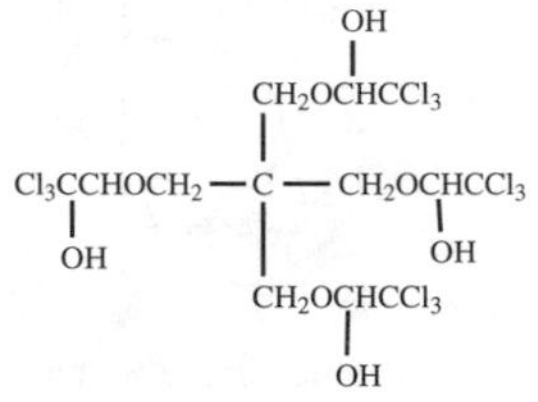

Pentaerythritol Tetraacetate (mol wt 304) An antilipemic agent.

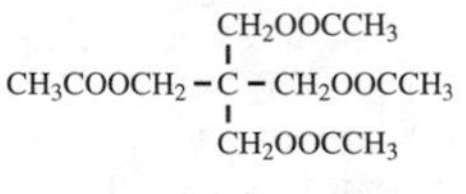

Pentaerythritol Tetranitrate (mol wt 316) A vasodilator that reduces cardiac oxygen demand.

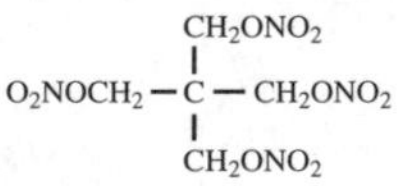

Pentagonal Capsomer A capsomer that consists of five promoters.

Pentam 300 A trade name for pentamidine isethionate, used as an antiprotozoal agent.

Pentamer An oligomer that consists of five monomers, e.g., a protein consisting of five subunits.

Pentamethonium Bromide (mol wt 348) An antihypertensive agent.

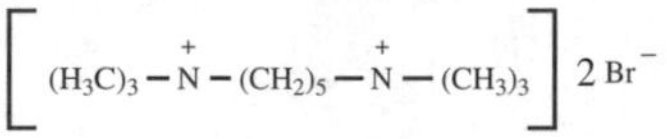

Pentamidine (mol wt 340) An antiprotozoal agent that interferes with biosynthesis of DNA, RNA, phospholipid, and proteins in susceptible organisms.

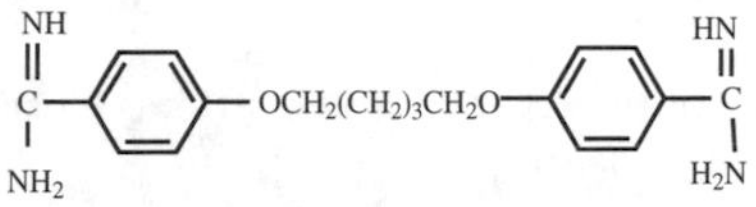

Pentamycetin A trade name for chloramphenicol sodium succinate, which binds to 50S ribosomes, inhibiting bacterial protein synthesis.

Pentaric Acid Any aldaric acid obtained by oxidation of a pentose.

Pentasa A trade name for mesalamine, an anti-inflammatory agent.

Pentasaccharide Any oligosaccharide that contains five monosaccharide units.

Pentazine A trade name for promethazine hydrochloride, used as an antihistaminic agent that competes with histamine for H_1 receptors on effector cells.

Pentazocine (mol wt 285) A narcotic analgesic agent that binds to opiate receptors in the CNS.

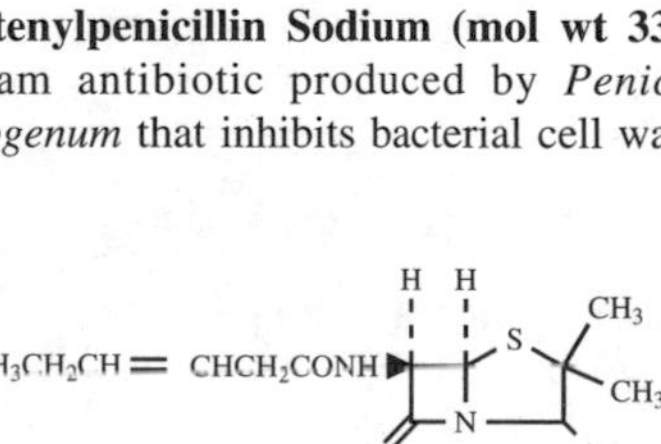

2-Pentenylpenicillin Sodium (mol wt 334) A β-lactam antibiotic produced by *Penicillium chrysogenum* that inhibits bacterial cell wall synthesis.

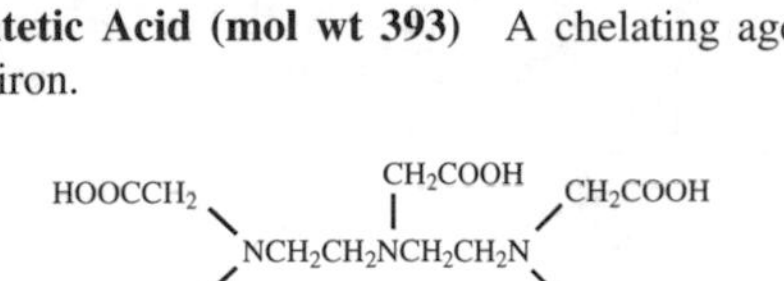

Pentetic Acid (mol wt 393) A chelating agent for iron.

Penthienate Bromide (mol wt 420) An anticholinergic agent.

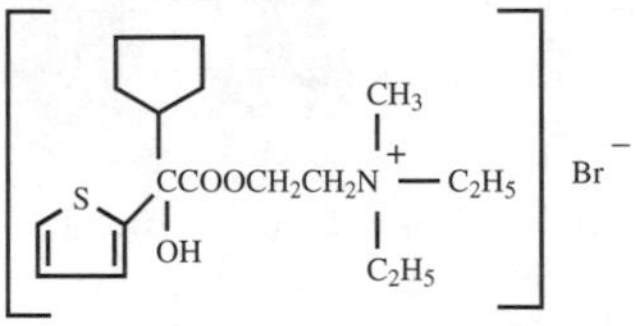

Penthothal A trade name for thiopental, an anesthetic agent.

Penthrane A trade name for methoxyflurane, an anesthetic agent.

Pentids A trade name for penicillin G potassium, which inhibits bacterial cell wall synthesis.

Pentifylline (mol wt 264) A vasodilator.

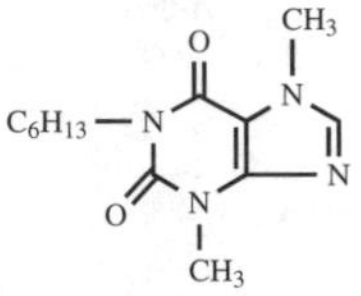

Pentigetide (mol wt 589) A pentapeptide and an antiallergic agent that inhibits IgE-mediated hypersensitivity.

Asp-Ser-Asp-Pro-Arg

Pentitol A five-carbon alcohol, e.g., ribotol.

Pentobarbital Sodium (mol wt 248) A sedative-hypnotic agent that interferes with transmission of nerve impulses from the thalamus to the cortex of the brain.

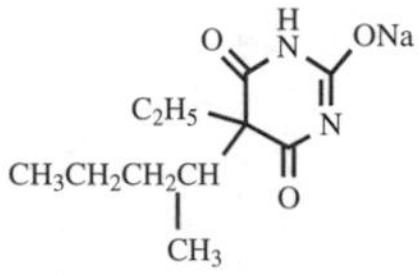

1-Pentol (mol wt 96) An intermediate in the biosynthesis of vitamin A.

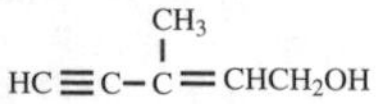

Pentolinium Tartrate (mol wt 539) An antihypertensive agent.

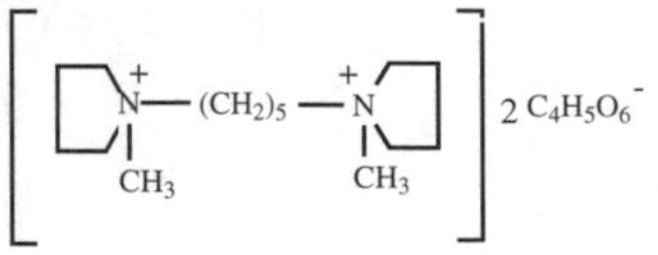

Pentosan Polymer of pentoses.

Pentosan Polysulfate An anticoagulant.

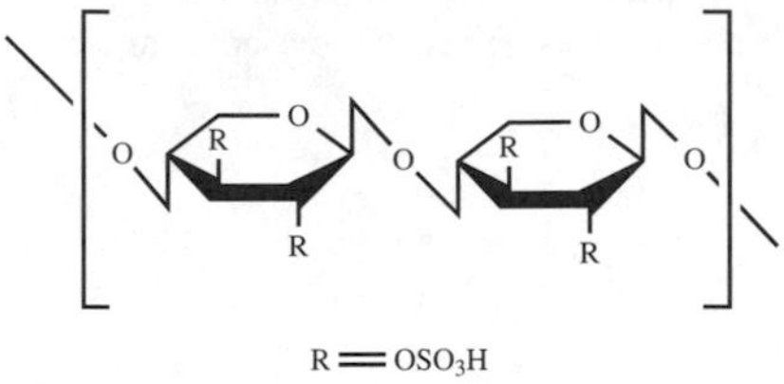

Pentosanase Synonym of endo-1,4-β-xylanase.

Penton A capsomer and a morphological unit that consists of five promoters.

Pentose A sugar that contains five carbon atoms, e.g., ribose, ribulose.

Pentose Oxidation Cycle See pentose phosphate pathway or hexose monophosphate shunt.

Pentose Phosphate Pathway A pathway for metabolism of glucose and production of NADPH (also called hexose monophosphate shunt).

Pentostatin (mol wt 268) An antineoplastic agent and an inhibitor for adenosine deaminase from *Streptomyces antibioticus*.

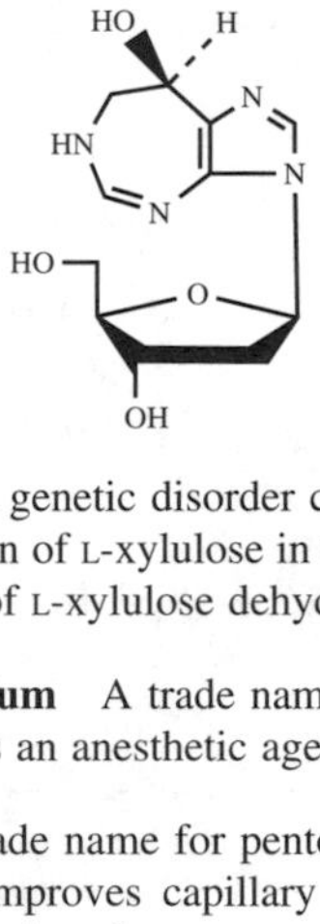

Pentosuria A genetic disorder characterized by the accumulation of L-xylulose in the urine due to the deficiency of L-xylulose dehydrogenase.

Pentothal Sodium A trade name for thiopental sodium, used as an anesthetic agent.

Pentoxil A trade name for pentoxifylline, a vasodilator that improves capillary blood flow by lowering blood viscosity.

Pentoxyl (mol wt 156) A leukopoietic stimulant.

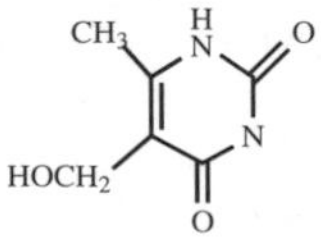

Pentrinitrol (mol wt 271) A vasodilator.

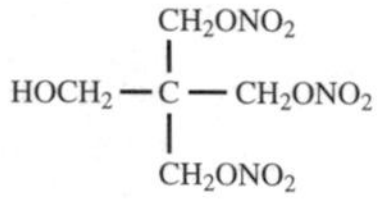

Pentritol A trade name for pentaerythritol tetranitrate, used as an antianginal agent to reduce cardiac oxygen demand.

Pentulose Any ketose having five carbon atoms.

Pentylan A trade name for pentaerthritol tetranitrate, used as an antianginal agent to reduce cardiac oxygen demand.

Pentylenetetrazole (mol wt 138) A CNS stimulant.

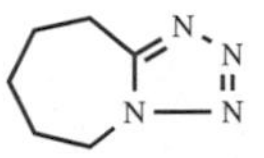

PenV Abbreviation for penicillin V.

Pen-Vee A trade name for penicillin V.

Pen-Vee K A trade name for penicillin V potassium, an antibiotic that inhibits bacterial cell wall synthesis.

Pen-VK A trade name for penicillin V potassium.

PEP Abbreviation for 1. phosphoenolpyruvate; 2. polyestradiol phosphate; 3. protein electrophoresis; 4. positron emission tomography.

PEP Carboxykinase The enzyme that cayalyzes the following reaction:

$$\text{Oxaloacetate} + \text{ATP} \rightleftharpoons \text{Phosphoenol pyruvate} + \text{ADP} + CO_2$$

PEP Carboxylase The enzyme that catalyzes the following reaction:

$$\text{PEP} + CO_2 \rightleftharpoons \text{Oxaloacetate} + \text{Pi}$$

PEP Carboxytransphosphorylase The enzyme that catalyzes the following reaction:

$$\text{PEP} + CO_2 + \text{Pi} \rightleftharpoons \text{Oxaloacetate} + \text{PPi}$$

PEP Dependent Phosphotransferase System An active transport system present in both Gram-negative and Gram-positive bacteria in which sugars (e.g., hexose) are first phosphorylated and then transported across the membrane using phosphoenol pyruvate as energy source.

PEP Synthetase The enzyme that catalyzes the following reaction:

$$\text{Pyruvate} + \text{ATP} \rightleftharpoons \text{PEP} + \text{AMP} + \text{Pi}$$

Pepcid A trade name for famotidine, used as an antiulcer agent that decreases gastric acid secretion.

Pepcidine A trade name for famotidine, used as an antiulcer agent that decreases gastric acid secretion.

PEPCK Abbreviation for phosphoenol pyruvate carboxykinase.

Peplomers The protein projects on the outer surface of a virus envelope.

Peplos Synonym for viral envelope.

Pepsin A protease that catalyzes the preferential hydrolysis of peptide bonds involving aromatic or dicarboxylic amino acid residues.

Pepsinogen The precursor of pepsin (the inactive form of pepsin).

Pepsinostrepin A bioactive peptide that binds with pepsin and inhibits its activity.

Pepstatin A peptide inhibitor for pepsin.

Pepstatin A A peptide inhibitor for pepsin and renin.

Peptic Peptides The peptides resulting from pepsin digestion of protein.

Peptidase Enzyme that hydrolyzes peptide bonds in a small peptide chain.

Peptidase A Synonym of penicillopepsin.

Peptidase D Synonym of X-pro dipeptidase.

Peptidase E Synonym of membrane alanyl aminopeptidase.

Peptidase S Synonym of leucyl aminopeptidase.

Peptide A short peptide chain that contains a few amino acid residues linked through peptide bonds.

Peptide N^4-(N-Acetyl-β-Glucosaminyl)-Asparagine Amidase The enzyme that catalyzes the hydrolysis of an N^4-(acetyl-β-glucosaminyl)-asparagine residue in which the glucosamine residue may be further glycosylated to yield an N-acetyl-β-D-glucosaminylamine and a peptide containing an aspartic residue.

Peptide Bond A covalent bond between two amino acids in which the alpha-amino group of one amino acid is bonded to the alpha-carboxyl group of the other amino acid.

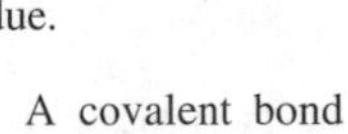

R and R′ = Amino acid

Peptide Histidine Valine 42 A 42-residue peptide with N-terminal histidine and C-terminal valine.

Peptide Linkage See peptide bond.

Peptide Mapping A technique for determination of relationships of various proteins through analysis by combined chromatography and electrophoresis of partial peptide digests (also called fingerprinting).

Peptide Nucleic Acid The nucleic acid analogy in which α-aminoethyl glycine-glycine linkages replace the normal phosphodiester backbone.

Peptide Synthetase See peptidyl transferase.

Peptide T An octapeptide segment of the human HIV envelope glycoprotein, named peptide T because of its high threonine content. It blocks the binding of HIV envelope to human leukocyte receptor CD4.

Peptide Tryptophan 2,3-Dioxygenase The enzyme that catalyzes the following reaction:

Peptide tryptophan + O_2
⇅
Peptide formylkynurenine

Peptide YY A 36-residue peptide amide with N-terminal tyrosine (single letter code Y), it is a gut hormone that inhibits exocrine pancreatic secretion.

Peptidoglycan Component of bacterial cell walls that consists of polysaccharide chains cross-linked by small peptides.

Peptidoglycan *N*-Acetylmuramoylhydrolase See lysozyme.

Peptidoglycoid Referring to a compound that contains peptide, carbohydrate, and lipid.

Peptidyl Arginine Deiminase Synonym of protein-arginine deiminase.

Peptidyl Dipeptidase A The enzyme that catalyzes the release of a C-terminal dipeptide, Xaa/Xbb-Xcc when neither Xaa nor Xbb is proline.

Peptidyl Glutaminase The enzyme that catalyzes the following reaction:

Peptidyl-L-glutamine + H_2O
⇅
Peptidyl-glutamate + NH_3

Peptidyl Glycinamidase The enzyme that catalyzes the cleavage of C-terminal glycinamide from a polypeptide.

Peptidyl Proline *cis-trans* Isomerase The systematic name for peptidylprolyl isomerase.

Peptidyl Site Site on a ribosome at which the complex of tRNA and growing polypeptide chain are attached (also known as P site).

Peptidyltransferase Enzyme on the 50S ribosomes that catalyze the formation of a peptide bond during the process of protein synthesis.

Peptidyl-tRNA A tRNA that carries a peptide or a polypeptide at its 3′-CCA terminal.

Peptidyl-tRNA Binding Site See peptidyl site.

Pepto-Bismol A trade name for bismuth subsalicylate, used as an antidiarrheal agent.

Peptococcus A genus of Gram-positive, asporogenous, anaerobic bacteria.

Peptol A trade name for cimetidine, an antihistaminic agent.

Peptolipid Referring to a complex that contains lipid and peptide or lipid and amino acid.

Peptone A partially hydrolyzed protein used as microbiological culture medium. It contains water-soluble low-molecular weight proteins and amino acids. Peptone cannot be coagulated by heat, cannot be precipitated by neutral salt, but can be precipitated with phosphotungstic acid.

Peptone Water A medium containing 1-2% peptone and 0.5% NaCl used for culture of microorganisms.

Peptonization The enzymatic production of peptone from protein.

Peptostreptoccocus A genus of Gram-positive, asporogenous, anaerobic bacteria.

Perazine (mol wt 339) An antipsychotic agent.

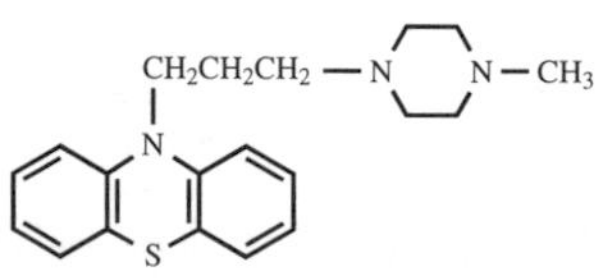

Perbuzem A trade name for a combination drug containing pentaerythritol tetranitrate and butabarbital sodium, used as an antianginal agent.

Percent Saturation The salt concentration of a solution to that of the saturated solution of the same salt in percent.

Percent Solution Referring to the concentration of a solute that is expressed in either gram(s) per 100 ml solution (w/v%) or ml per 100 ml solution (v/v%).

Percent Transmittance The expression of the amount of light that passes through a sample as measured in a spectrophotometer.

Percocet A trade name for a combination drug containing acetaminophen and oxycodone hydrochloride, used as an analgesic agent.

Percodan A trade name for a combination drug containing aspirin, oxycodone hydrochloride, and oxycodone terephthalate, used as an analgesic agent.

Percodan-Demi A trade name for a combination drug containing aspirin, oxycodone hydrochloride, and oxycodone terephthalate, used as an analgesic agent.

Percolation The slow movement of a liquid through a bed or column of solid particles.

Percoll A trade name for a colloidal silica coated with polyvinylpyrrolidone, used for density gradient centrifugation for the separation of cells, viruses, and subcellular organelles.

Percolone A trade name for oxycodone, an analgesic agent.

Perdiem Plain A trade name for psyllium, used as a laxative that increases the bulk and moisture content of the stool.

Perforin Cytolytic molecules produced by T_c (cytotoxic T cell) cells that form transmembrane plugs and cause lysis of target cells.

Performic Acid (mol wt 62) A reagent used for cleavage of disulfide bonds.

HCOOOH

Perfringens Food Poisoning A self-limiting disorder characterized by abdominal pain and diarrhea caused by the ingestion of *Clostridium perfringens* type A.

Perfusate A liquid that has been pumped through an organ or perfused through an organ.

Perfusion The flow of a liquid through an organ.

Pergolide (mol wt 314) An antiparkinsonian agent that stimulates dopamine receptors.

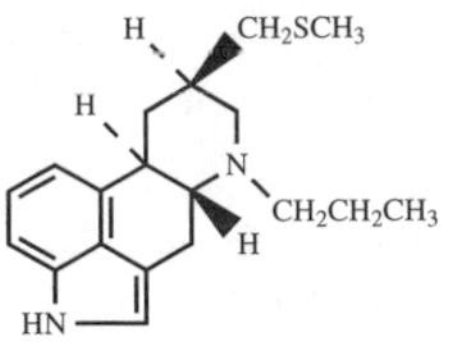

Pergonal A trade name for menotropins which mimics folicle-stimulating hormone in inducing follicular growth and lutenizing hormone in aiding follicular maturation.

Perhexiline (mol wt 278) A vasodilator and a diuretic agent.

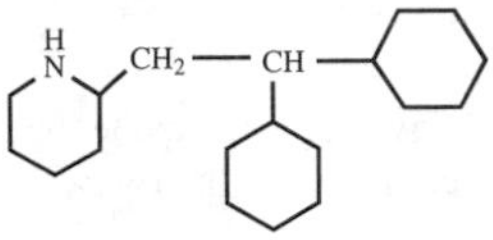

Perhydrocyclopentanophenanthrene The parent ring structure of the steroids (also known as cyclopentanoperhydrophenanthrene).

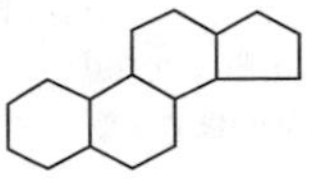

Peri- A prefix meaning surrounding.

Periactin A trade name for cyproheptadine hydrochloride, used as an antihistaminic agent that competes with histamine for H_1 receptors on effector cells.

Periadenitis Inflammation of tissues surrounding a gland.

Periarthritis Inflammation of the tissue surrounding a joint.

Peribacteroid Membrane The membrane in the nodules of legume plants that surrounds the nitrogen-fixing bacteroid in the nodule for nitrogen fixation.

Pericarditis The inflammation of the sheath surrounding the heart.

Pericardium The membranous sac enclosing the heart.

Pericentric Involving the centromere of a chromosome.

Pericentric Inversion The chromosome inversion in the region including a centromere.

Perichondrial Cell Cells of the perichondrium surrounding the cartilage.

Perichondritis Inflammation of perichondrium.

Perichondrium The fibrous sheath encasing cartilage.

Peri-Colace A trade name for a combination drug containing docusate sodium and casanthranol, a laxative that promotes fluid accumulation in the colon and small intestine.

Pericyazine (mol wt 366) An antipsychotic agent.

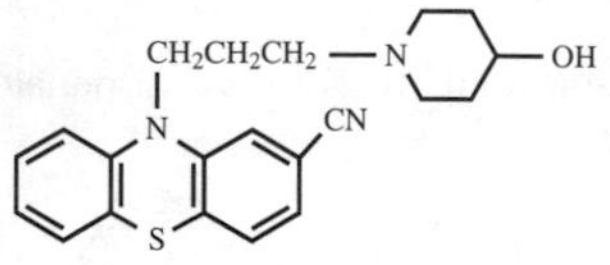

Pericycle A layer of cells inside the endodermis but outside the phloem of the roots and the stems.

Pericyte Cells associated with the walls of small blood vessels.

Periderm The outer cork layer of the plant.

Peridex A trade name for chlorhexidine gluconate, used as an antibacterial agent (both Gram-positive and Gram-negative).

Peridin-C A trade name for a combination drug containing two popular antioxidants of vitamin C and bioflavonoids.

Peridol A trade name for haloperidol, used as an antipsychotic agent.

Peridontium The supporting structure of the teeth or tissue surrounding the teeth.

Perihepatitis Inflammation of the peritoneal coat of the liver.

Perikaryon The cell body of the neuron.

Perilipin An adipocyte-specific phosphoprotein.

Perimethazine (mol wt 385) An antipsychotic agent.

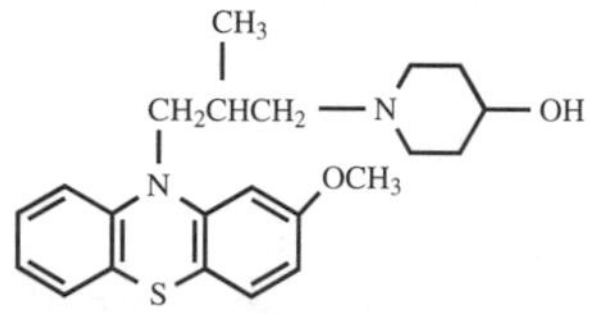

Perinatology The medical science that deals with the fetus.

Perindopril (mol wt 368) An antihypertensive agent.

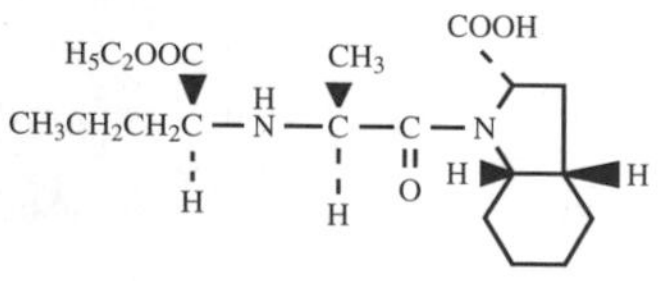

Perineuritis The inflammation of the sheath surrounding a nerve.

Perinuclear Space Fluid-filled compartment between the two nuclear membranes.

Periodate Oxidation Cleavage of bonds between two adjacent carbon atoms by the periodate oxidation reaction in which the periodate ion is reduced to the iodate ion ($IO_4^- \longrightarrow IO_3^-$).

Periodic Acid (mol wt 228) A reagent used in periodate oxidation.

$$H_5IO_6$$

Periodic Acid-Schiff's Reagent The reagent used for staining of carbohydrate or the carbohydrate moiety of glycoprotein.

Periodic Table A table of the elements arranged in the order of atomic numbers. The elements with similar properties are placed together to form groups of elements.

Periodontal Disease The disease of the tissues surrounding the teeth.

Periodontitis The inflammation of the periodontium (the tissues surrounding a tooth).

Peripheral Pertaining to the external surface.

Peripheral Lymphoid Organs Lymphoid organs that are not essential to the ontogeny of the immune system, e.g., spleen, lymph nodes, tonsils, and Peyer's patches (also called secondary lymphoid organs).

Peripheral Lymphoid Tissue Referring to the lymphoid tissues of secondary lymphoid organs, e.g., lymph node and spleen.

Peripheral Membrane Proteins The proteins that are not integrated into the membrane and can be washed off or dissociated from the membrane with mild treatment.

Peripheral Nervous System Sensory and motor neurons of the nervous system that control skeletal voluntary movement and involuntary activities of muscles and glands.

Peripherin A glycoprotein that is essential for eye-disk morphogenesis.

Periplasm The region and its content between the cytoplasmic membrane and the outer cell wall membrane of Gram-negative bacteria.

Periplasmic Binding Proteins The transport proteins in the region of the periplasmic space.

Periplasmic Space The area between the outer cell wall membrane and the cytoplasmic membrane in Gram-negative bacteria.

Periseptal Annulus A structure associated with cell division in Gram-negative bacteria.

Perisoxal (mol wt 272) An anti-inflammatory and an analgesic agent.

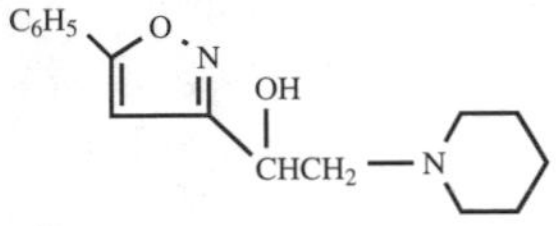

Peristalsis Waves of contraction of digestive smooth muscle that push food along the digestive tract.

Peritoneal Exudate Cells Inflammatory cells present in the peritoneum of animals injected with an inflammatory agent.

Peritoneum Membrane lining the abdominal wall.

Peritonitis The inflammation of the peritoneum (the membrane that lines the abdominal cavity).

Peritrate A trade name for pentaerythritol tetranitrate, an antianginal agent that reduces cardiac oxygen demand.

Peritrate Forte A trade name for pentaerythritol tetranitrate, an antianginal agent that reduces cardiac oxygen demand.

Peritrate SA A trade name for pentaerythritol tetranitrate, an antianginal agent that reduces cardiac oxygen demand.

Peritrichous Referring to the uniform distribution of bacterial flagella over the surface of the cell.

Perkinsus A genus of protozoa.

Perlapine (mol wt 291) A hypnotic agent.

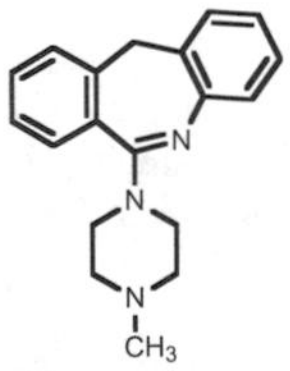

Permapen A trade name for penicillin G benzathine, an antibiotic.

Permax A trade name for pergolide mesylate, an antiparkinsonian agent that stimulates dopamine receptors.

Permeaplast A lysozyme-EDTA-treated cyanobacterial cell used for transformation of certain cyanobacteria.

Permease Integral membrane protein that mediates the passage of specific substances across the membrane.

Permeability The property of cell membranes that permit transport of molecules and ions in solution across the membrane.

Pemeability Factor Referring to vitamin P (rioflavonoids).

Permeaphore Substance within the membrane that facilitates the transport of solutes across the membrane.

Permeation Chromatography See gel filtration.

Permissive Cell A host cell enabling certain phages or viruses or mutants to grow and produce progeny.

Permethrin (mol wt 391) An antiparasitic agent that disrupts the sodium channel current causing paralysis of the parasite.

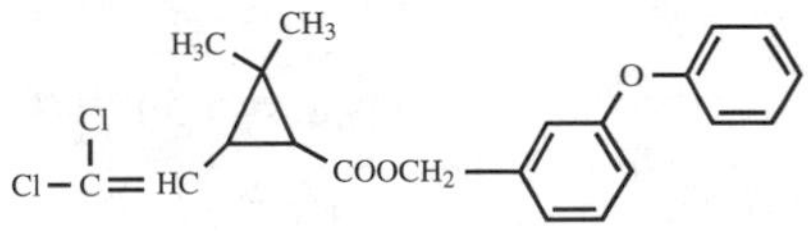

Permitil A trade name for fluphenazine hydrochloride, used as an antipsychotic agent.

Permutite Process The process of softening hard water by passing water through metal silicate or earth aluminum silicate.

Pernicious Anemia A disorder caused by a deficiency in vitamin B_{12} and characterized by decreased numbers of red blood cells, low hemoglobin level, and progressive neurological deterioration.

Peroxidase The enzyme that catalyzes the reaction in which the oxidation is coupled with the reduction of hydrogen peroxide.

$$\text{Donor} + H_2O_2 \rightleftharpoons \text{Oxidized donor} + 2\ H_2O$$

Peroxidase-Antibody Conjugate An antibody-peroxidase complex used in ELISA (enzyme-linked immunosorbent assay).

Peroxidation Oxidation of fatty acid to hydroperoxide.

Peroxide The anion of O_2^- or HO_2^-, or a compound that contains O_2^- or HO_2^-.

Peroxide Number The amount of iodine liberated from KI by a fat, used for measurement of the fat content.

Peroxisomes Intracellular organelles that contain enzymes involved in hydrogen peroxide metabolism, purine degradation, photorespiration, and the glyoxylate cycle.

Peroxy Containing a —O—O— group.

Peroxy Acid Any acid that contains –CO-O-OH

Perphenazine (mol wt 404) An antipsychotic agent that blocks postsynaptic dopamine receptors in the brain.

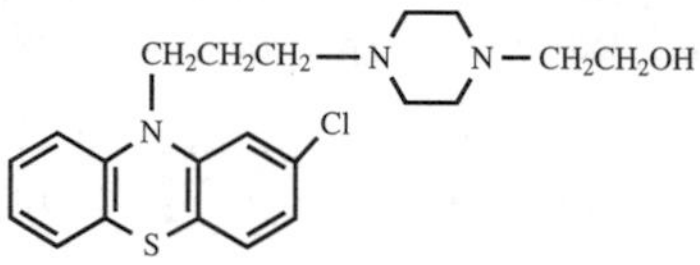

Persa-Gel A trade name for benzoyl peroxide gel, an antimicrobial and comedolytic agent.

Persantin A trade name for dipyridamole, a vasodilator that inhibits platelet adhesion and inhibits adenosine deaminase and phosphodiesterase.

Persantine See persantin.

Perspiration A process of sweating or releasing of fluid from sweat glands.

Pertofran A trade name for desipramine hydrochloride, an antidepressant that increases levels of norepinephrine and serotonin.

Pertofrane See pertofran.

Pertropin A trade name for vitamin E, an antioxidant and a cofactor.

Pertussin A trade name for dextromethorphan hydrobromide, an antitussive agent.

Pertussis Referring to whooping cough caused by *Bordetella pertussis* toxin.

Pertussis Toxin An exoprotein toxin produced by *Bordetella pertussis*. It possesses ADP-ribosyl transferase activity and interacts with G protein causing an increase in intracellular cAMP.

Pervaporation The evaporation of solvent through a dialysis membrane used for concentrating macromolecules that are incapable of passing through the membrane.

PES Abbreviation for phenazine ethosulfate.

PEST Referring to the tetrapeptide with amino acid sequence of proline(P)-glutamate(E)-serine(S)-threonine(T).

PEST Hypothesis A hypothesis that the presence of one or more PEST sequence in protein confers susceptibility to rapid intracellular proteolysis.

Pesticide Substance that kills undesirable organisms.

Pesticin-I A bacteriocin produced by strains of *Yersinia pestis* that is active against *Yersinia pseudotuberculosis* and strains of *E. coli.*

Pestivirus A virus in the family of Togaviridae that infects pigs.

Pestle A club-shaped implement used for grinding material in a mortar.

PEt Abbreviation for phosphatidylethanol.

pET1 A plasmid of *E. coli* that contains ampicilin-, chloramphenicol-, and tetracycline-resistant markers and AvaI, BclI, EcoRI, PstI, XhoI, XmaI, SalI, SphI, BamHI, and EcoRV cleavage sites.

PET Genes A group of nuclear genes whose products are necessary for mitochondrial morphogenesis, respiration, and oxidative phosphorylation.

Peta- A prefix denoting 10^{15}.

Pethidine A trade name for meperidine hydrochloride, a non-narcotic analgesic agent.

Petri Dish A round, shallow, flat-bottomed dish with a vertical edge together with a loosely fitting lid used for culture of microorganisms.

Petrochemical Chemicals derived from petroleum base.

PETN A trade name for pentaerythritol tetranitrate, an antianginal agent that reduces cardiac oxygen demand.

PETT Abbreviation for positron emission transaxial tomography or positron emission transaxial tomography.

Peyer's Patches Nodules of lymphoid tissue in the submucosa of the small intestines that contain follicles and diffuse areas of abundant lymphocytes and plasma cells.

PF Abbreviation for platelet factor.

Pf 155 A protein encoded by *Plasmodium falciparum* found in the erythrocyte infected with the *P. falciparum.*

pfaI (MboI) A restriction endonuclease from *Pseudomonas facilis* with the same specificity as MboI.

PFC Abbreviation for 1. perfluorocarbon; 2. plaque-forming cell.

PFD Abbreviation for prion-forming domain.

pfdA2 A plasmid of *E. coli* that contains Kan^r (kanamycin resistant) marker and EcoRI and HaeII cleavage sites.

pfdA3 A plasmid of *E. coli* that contains Kan^r (kanamycin resistant) and Amp^r (ampicillin resistant) markers and HindIII, PstI, SmaI, and XhoI cleavage sites.

pfdA4 A plasmid of *E. coli* that contains Kan^r (kanamycin resistant) and Cml^r (chloramphenicol resistant) markers and EcoRI, HindIII, PvuI, SmaI, and XhoI cleavage sites.

pfdA4 A plasmid of *E. coli* that contains Kan^r (kanamycin resistant) and BglII, EcoRI, HaeII, PstI, SalI, and XhoI cleavage sites.

pfdB2 A plasmid of *E. coli* that contains Kan^r (kanamycin resistant) and the EcoRI cleavage site.

Pfeiffer Phenomenon Referring to the rapid lysis of *Vibrio cholerae* by specific antibody and complement.

Pfeiffer's *Bacillus* Referring to *Haemophilus influenzae*.

Pfeiffer's Disease Referring to infectious mononucleosis.

PFGE Abbreviation for pulsed-field gel electrophoresis.

Pfizerpen-AS A trade name for penicillin G potassium, an antibiotic that inhibits bacterial cell wall synthesis.

PFK Abbreviation for phosphofructokinase.

Pfl23II A restriction endonuclease from *Pseudomonas fluorescens* RFL23 with the following specificity:

5'........↓CGTACG........3'
3'........GCATGC↑........5'

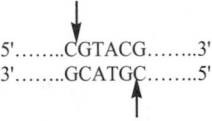

PflAI (FnuDII) A restriction endonuclease from *Pseudomonas fluorescens* type A with the same specificity as FnuDII.

PflMI A restriction endonuclease from *Pseudomonas fluorescens* with the following specificity:

5'..........CCANNNN↓NTGG..........3'
3'..........GGTN↑NNNNACC..........5'

PflNI (XhoI) A restriction endonuclease from *Pseudomonas fluorescens* with the same specificity as XhoI.

PflWI (XhoI) A restriction endonuclease from *Pseudomonas fluorescens* with the same specificity as XhoI.

PFK Abbreviation for phosphofructokinase.

pFN (PFN) Abbreviation for plasma fibronectin.

PFO Abbreviation for perfluoro-octanoic acid.

PFP Abbreviation for 1. platelet-free plasma; 2. pore-forming protein.

PFU Abbreviation for 1. plaque forming unit; 2. pock forming unit.

PfuI (SplI) A restriction endonuclease from *Pseudomonas fluorescens* with the following specificity:

5'..........C↓GTACG..........3'
3'..........GCATG↑C..........5'

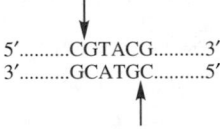

PG Abbreviation for 1. phosphatidylglycerol; 2. phosphogluconate; 3. pituitary gonadotropin; 4. plasma glucose; 5. polygalacturonate; 6. prostaglandin; 7. proteoglycan.

pg Abbreviation for picogram (10^{-12} gram).

Pg Abbreviation for plasminogen.

2PG Abbreviation for 2-phosphoglycerate.

3PG Abbreviation for 3-phosphoglycerate.

6PG Abbreviation for 1. 6-phosphogalactose; 2. 6-phosphoglucose; 3. 6-phosphogluconate.

pG1 A plasmid of *E. coli* that contains an Amp^r (ampicillin resistant) marker and BamHII and PstI cleavage sites.

PGA Abbreviation for 1. phosphoglyceric acid and 2. prostaglandin A.

3PGA Abbreviation for 3-phosphoglyceric acid.

PGA_1 Abbreviation for prostaglandin A_1.

PGA_2 Abbreviation for prostaglandin A_2.

PGA_3 Abbreviation for prostaglandin A_3.

PGAL Abbreviation for phosphoglyceraldehyde, an intermediate in glycolysis and the C_3 pathway.

Pgal Abbreviation for phenyl-β-D-galactoside.

PGAM Abbreviation for phosphoglyceric acid mutase.

PGB Abbreviation for prostaglandin B.

PGB$_1$ Abbreviation for prostaglandin B_1.

PGB$_2$ Abbreviation for prostaglandin B_2.

PGB$_3$ Abbreviation for prostaglandin B_3.

PGC Abbreviation for prostaglandin C.

PGC$_1$ Abbreviation for prostaglandin C_1.

PGC$_2$ Abbreviation for prostaglandin C_2.

PGC$_3$ Abbreviation for prostaglandin C_3.

PGD Abbreviation for prostaglandin D.

PGD$_1$ Abbreviation for prostaglandin D_1.

PGD$_2$ Abbreviation for prostaglandin D_2.

PGD$_3$ Abbreviation for prostaglandin D_3.

PGDH Abbreviation for 1. phosphogluconate dehydrogenase; 2. phosphoglyceraldehyde dehydrogenase.

PGE Abbreviation for prostaglandin E.

PGE$_1$ Abbreviation for prostaglandin E_1.

PGE$_2$ Abbreviation for prostaglandin E_2.

PGE$_3$ Abbreviation for prostaglandin E_3.

PGE$_4$ Abbreviation for prostaglandin E_4.

PGF Abbreviation for prostaglandin F.

PGF$_1$ Abbreviation for prostaglandin F_1.

PGF$_{1\alpha}$ Abbreviation for prostaglandin $F_{1\alpha}$.

PGF$_2$ Abbreviation for prostaglandin F_2.

PGF$_{2\alpha}$ Abbreviation for prostaglandin $F_{2\alpha}$.

PGF$_3$ Abbreviation for prostaglandin F_3.

PGG Abbreviation for prostaglandin G.

PGG$_1$ Abbreviation for prostaglandin G_1.

PGG$_2$ Abbreviation for prostaglandin G_2.

PGH Abbreviation for prostaglandin H

PGH$_1$ Abbreviation for prostaglandin H_1.

PGH$_2$ Abbreviation for prostaglandin H_2.

pGH6 A plasmid of *E. coli* that contains Amp^r (ampicillin resistant) and Tet^r (tetracycline resistant) markers and the pBR322 unique cleavage site.

PGI Abbreviation for 1. phosphoglucose isomerase; 2. prostaglandin I.

PGI$_1$ Abbreviation for prostaglandin I_1.

PGI$_2$ Abbreviation for prostaglandin I_2.

PGK Abbreviation for phosphoglycerate kinase.

PGL Abbreviation for phospho-glyco-lipid.

PglI (NaeI) A restriction endonuclease from *Pseudomonas glycinae* with the same specificity as NaeI.

pGL101 A plasmid of *E. coli* that contains the Amp^r (ampicillin resistant) marker, *lac* promoter, and EcoRI and PvuII cleavage sites.

PGlcI Abbreviation for glucose 6-phosphate isomerase or phosphoglucose isomerase.

6PglcDH Abbreviation for 6-phosphoglucose dehydrogenase.

PGlcM Abbreviation for phosphoglucomutase.

6PgluDH Abbreviation for 6-phosphogluconate dehydrogenase.

PGlyM Abbreviation for phosphoglycerate mutase.

PGM Abbreviation for 1. phospho-gluco-mutase; 2. phosphoglycerate mutase.

PGP Abbreviation for phosphatidylglycerol phosphate.

PGS Abbreviation for prostaglandin synthetase.

PGUT Abbreviation for phosphogalactose uridyltransferase.

PGYE Medium Abbreviation for peptone/glucose/yeast extract medium.

PH Abbreviation for pleckstrin homology.

pH An expression of hydrogen ion concentration of an aqueous solution, defined as the negative logarithm of the hydrogen ion concentration in moles per liter of an aqueous solution. A solution with a pH of 7 is said to be neutral; a pH value higher than 7 is increasingly basic and less than 7 is increasingly acidic.

PH Domain Abbreviation for pleckstrin homology domain.

pH Electrode A platinum electrode used for measuring pH of a solution.

pH 5 Enzyme Referring to aminoacyl-tRNA synthetase.

pH 5 Fraction A subcellular fraction obtained from the disruption of cells and precipitation at pH 5 that contains aminoacyl-tRNA synthetase.

pH Gradient Electrophoresis See isoelectric focusing.

pH Meter An instrument used to measure the pH value or concentration of H^+ of a solution.

PHA Abbreviation for phytohemagglutinin, a plant lectin from kidney beans, which activates principally T lymphocytes.

pHA10 A plasmid of *E. coli* that contains the Amp^r (ampicillin resistant) marker and SalI and BamHI, HpaI cleavage sites.

PHAA Abbreviation for phytohemagglutinin antigen.

Phacitis Inflammation of the lens of the eye.

Phacolysis Dissolution of the lens of the eye.

Phaeophytin A type of chlorophyll in which magnesium is replaced by two hydrogen atoms.

Phage Short for bacteriophage.

Phage Cross Production of recombinant phage by infection of a bacterium with several different bacteriophage mutants, the resulting progeny of phages carry genes from both parental types.

Phage Induction A process in which the lysogenic bacteria change from a lysogenic condition to a lytic cycle (production of infective bacteriophage).

Phage Lysate A suspension containing newly released bacteriophage resulting from lysis of phage-infected bacteria.

Phage Typing A method of typing in which strains of bacteria are distinguished on the basis of their differences in susceptibilities to various types of bacteriophages.

Phagocytes A general term for phagocytic cells such as macrophages and neutrophils that are capable of engulfing microorganisms or foreign particles.

Phagocytic Vacuole Membrane-bounded structure containing phagocytized particles.

Phagocytosis The process by which phagocytes, e.g., neutrophils and macrophages, ingest or engulf particulate matter into the cells via membrane vesicles that pinch off from the plasma membrane.

Phagolysosome A membrane-bound intracellular vesicle produced by the fusion of a phagosome with a lysosome following phagocytosis.

Phagosome A structure in the eukaryotic cell resulting from ingestion of particulate materials by phogocytosis.

Phalloidin (mol wt 789) A toxin isolated from the poisonous fungus *Amanita phalloides.*

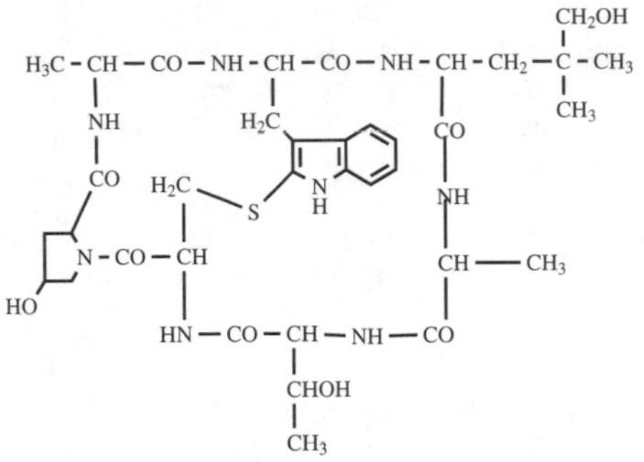

Phallotoxin A group of cyclic heptapeptide toxins produced by *Amanita phalloides*, e.g., phalloidin.

PHA-M Abbreviation for phytohemagglutinin M.

Phanquinone (mol wt 210) An antiamebic agent.

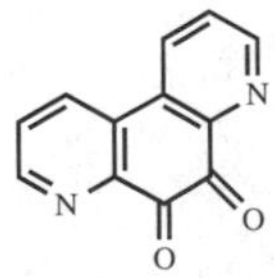

Pharmaceutical Chemistry Science that deals with preparation, composition, and testing of drugs.

Pharmacodynamics The science that deals with drugs and their effects on the body.

Pharmacology Science that deals with the chemistry, composition, and identification of drugs and their effects on the living system.

Pharmacy Science that deals with the chemistry, composition, preparation, and dispensing of drugs.

Pharmadine A trade name for povidone-iodine, used an antimicrobial and antiviral agent.

Pharyngitis Inflammation of the pharynx caused by *Streptococcus pyogenes* or virus, characterized by sore throat, fever, and pharyngeal exudate.

Pharyngorhinitis Inflammation of the nose and throat.

Pharynx The part of the gut between the mouth and the esophagus.

PHAS Protein Abbreviation for phosphorylated, heat and acid-stable (protein).

Phase Contrast Microscope A type of microscope that achieves enhanced contrast of the specimen by altering the phase of light that passes through the specimen for viewing of live specimens.

Phase Partition A separation technique in which substances are allowed to partition between two or more immiscible or partially miscible phases.

Phase Variation A phenomenon of alteration in the expression of surface antigens in bacteria by the spontaneous switch in synthesis of surface antigen from one given type to another.

Phaseolin (mol wt 322) An antifungal agent isolated from French bean plants (*Phaseolus vulgaris*).

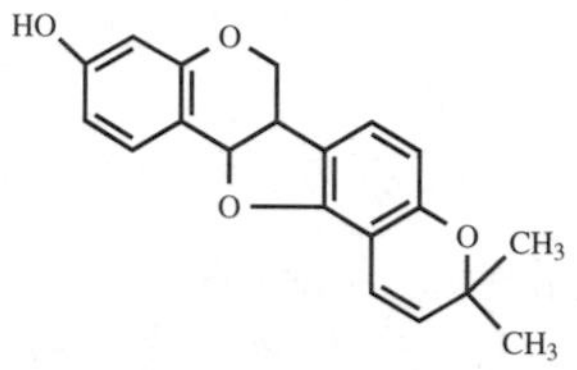

Phaseolotoxin A chlorosis-inducing toxin in plants produced by *Pseudomonas syringae* that inactivates ornithine carbamoyltransferase.

Phasin A poisonous agglutinin from beans.

Phasmid A hybrid plasmid obtained by recombination between l genome and a plasmid containing the phage l attaching site.

Phazyme A trade name for simethicone, used to prevent the formation of mucus-surrounding gas pockets in the GI tract.

PHBB Abbreviation for propyl-hydroxy-benzyl benzimidazole.

pHC79 A plasmid of *E. coli* that contains Amp^r (ampicillin resistant) and Tet^r (tetracycline resistant) markers and EcoRI, ClaI, BamHI, SalI, EcaI, and PstI cleavage sites.

PHCP Abbreviation for phospholipid hydroperoxide cysteine peroxidase.

PHCPP Abbreviation for perhydrocyclopentanophenanthrene.

Phe Abbreviation for phenylalanine or phenylalanyl.

pHE3 A plasmid of *E. coli* that contains the Cml^r (chloramphenicol resistant) marker and PstI, HincII, BamHI, and HindIII cleavage sites.

Phelloderm The tissue in the bark of some plants derived from cork cambium.

Phellogen The meristematic tissue in plant that gives rise to cork phellem and phelloderm cells.

pH$_{en}$ Abbreviation for pH in the endosomal compartment.

Phenacaine Hydrochloride (mol wt 335) A topical anesthetic agent.

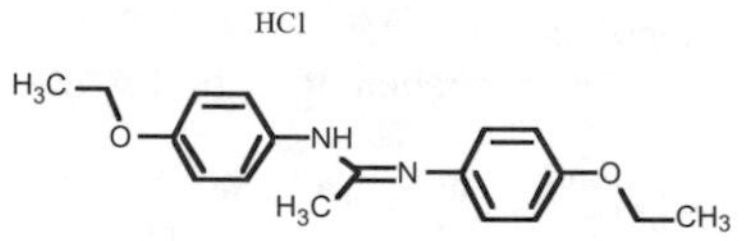

Phenacetin (mol wt 179) An analgesic-antipyretic agent.

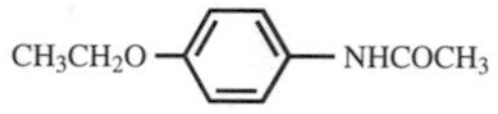

Phenactropinium Chloride (mol wt 430) An antihypertensive agent.

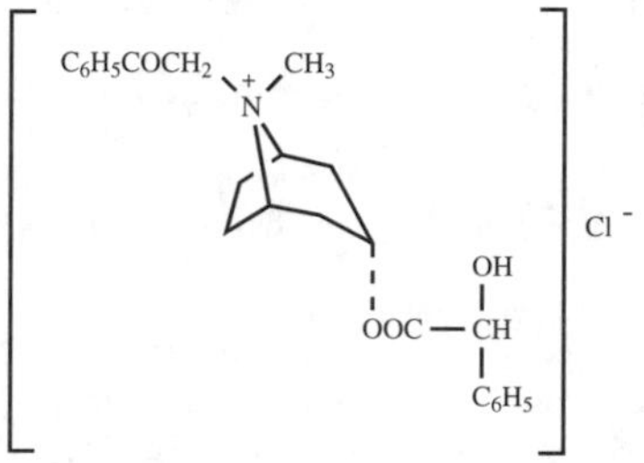

Phenaglycodol (mol wt 215) An anxiolytic agent.

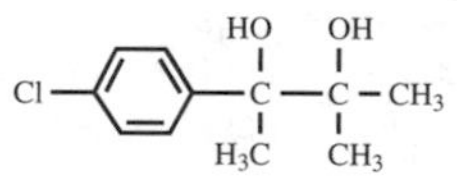

Phenallymal (mol wt 244) A sedative-hypnotic agent.

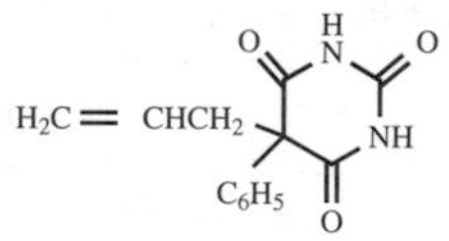

Phenamacide Hydrochloride (mol wt 258) An antispasmodic agent.

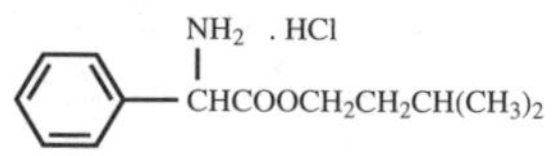

Phenamet (mol wt 435) An antineoplastic agent.

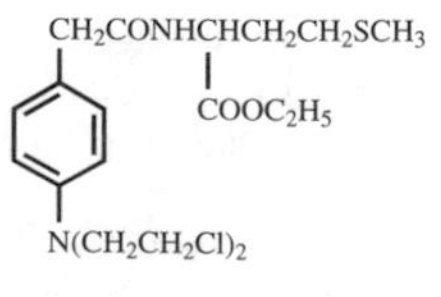

Phenameth A trade name for promethazine hydrochloride, an antihistaminic agent that competes with histamine for H_1 receptors on effector cells.

Phenamidine (mol wt 254) An antiprotozoal agent.

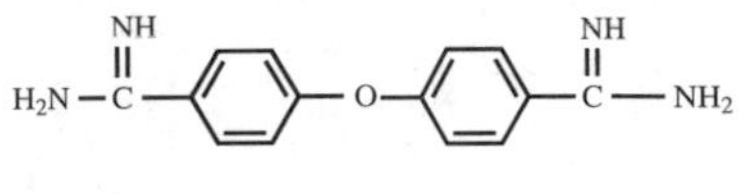

Phenaphen A trade name for acetaminophen, an analgesic-antipyretic agent that prevents generation of pain impulses.

Phenarsone Sulfoxylate (mol wt 355) An antiamebic agent.

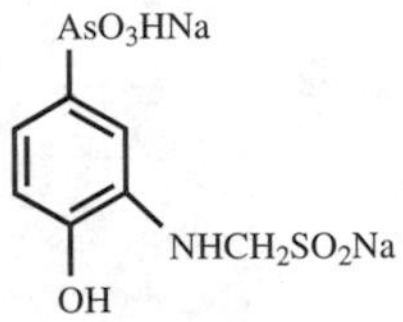

Phenazine Methosulfate (mol wt 306) An artificial electron acceptor used for detection of hydrogenase.

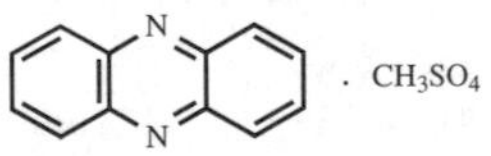

Phenazo A trade name for phenazopyridine hydrochloride, used as an analgesic-antipyretic agent.

Phenazocine (mol wt 321) A narcotic analgesic agent.

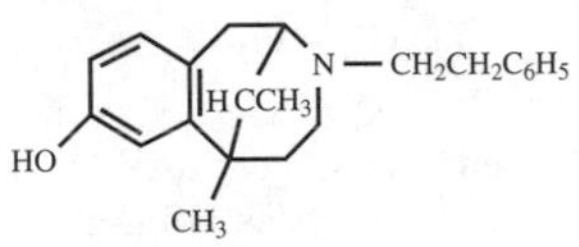

Phenazopyridine Hydrochloride (mol wt 250) An analgesic agent for the urinary tract.

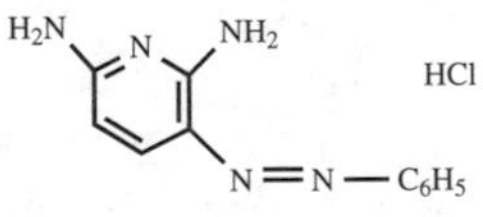

Phenbenzamine (mol wt 254) An antihistaminic agent.

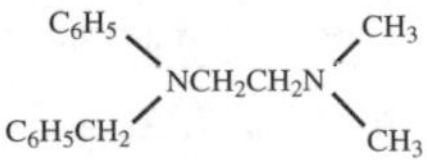

Phencarbamide (mol wt 328) An anticholinergic agent.

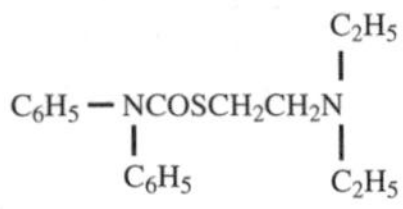

Phencen-50 A trade name for promethazine hydrochloride, an antihistaminic agent that competes with histamine for H_1 receptors on effector cells.

Phencyclidine (mol wt 243) An analgesic and anesthetic agent.

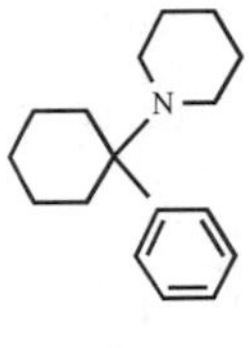

Phendimetrazine (mol wt 191) An anorexic agent.

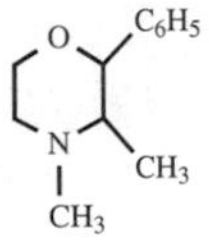

Phenelzine (mol wt 136) An antidepressant.

$C_6H_5CH_2CH_2NHNH_2$

Phenerbel-S A trade name for a combination drug containing ergotamine tartrate, levorotatory belladonna alkaloids, and phenobarbital, used as an adrenergic blocker.

Phenergan A trade name for promethazine hydrochloride, an antihistaminic agent that competes with histamine for H_1 receptors on effector cells.

Phenesterine (mol wt 645) An antineoplastic agent.

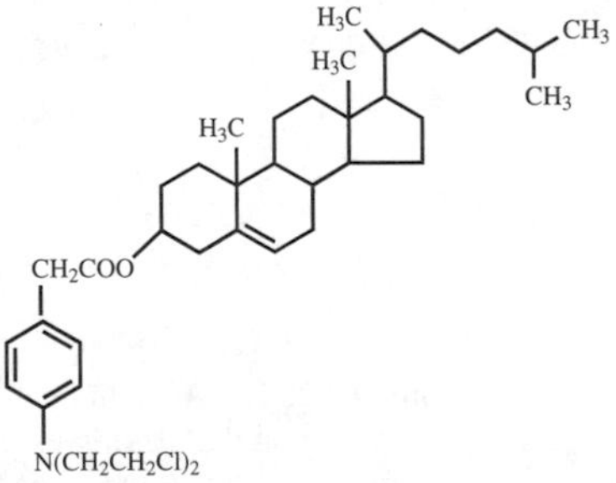

Phenetharbital (mol wt 260) An anticonvulsant.

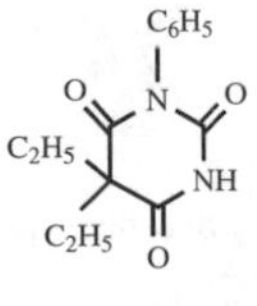

Phenethicillin Potassium (mol wt 403) A semisynthetic antibiotic related to penicillin that inhibits bacterial cell wall synthesis.

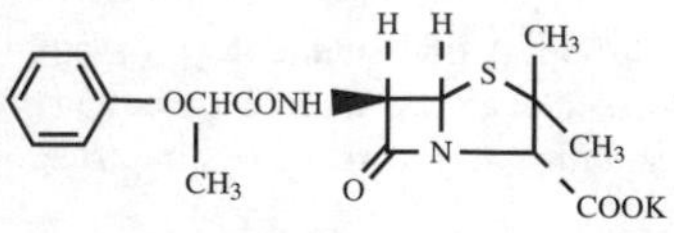

Phenetron A trade name for chlorpheniramine maleate, an antihistaminic agent that competes with histamine for H_1 receptors on effector cells.

Pheneturide (mol wt 206) An anticonvulsant.

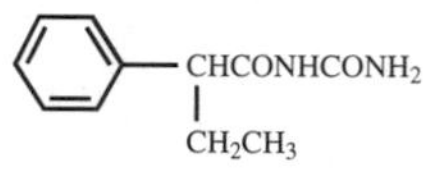

Phenformin (mol wt 205) An antidiabetic agent.

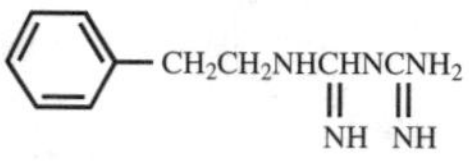

Phenglutarimide (mol wt 288) An anticholinergic agent.

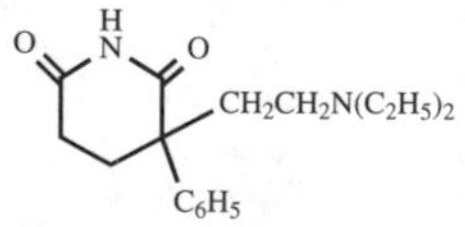

Phenindione (mol wt 222) An anticoagulant.

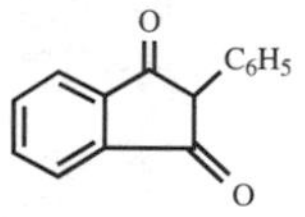

Pheniramine (mol wt 240) An antihistaminic agent.

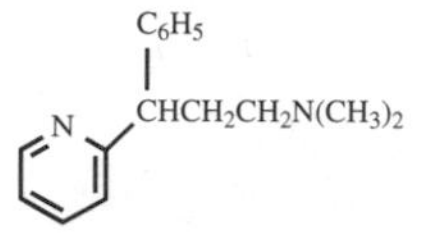

Phenobarbital (mol wt 232) An anticonvulsant, sedative, and hypnotic agent that depresses monosynaptic and polysynaptic transmission in the CNS.

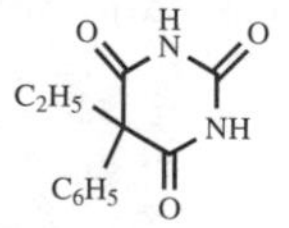

Phenobarbitone See phenobarbital.

Phenobarbitone Sodium The sodium form of phenobarbital.

Phenocoll (mol wt 194) An analgesic-antipyretic agent.

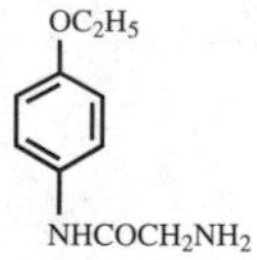

Phenocopy An environmentally produced phenotype, e.g., F^+-donor cells of *E. coli* that lose their donor characteristics when grown to maximum density in aerated broth.

Phenoject-50 A trade name for promethazine hydrochloride, an antihistaminic agent that competes with histamine for H_1 receptors on effector cells.

Phenol (mol wt 94) An aromatic alcohol.

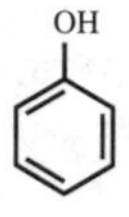

Phenol β-Glucosyltransferase The enzyme that catalyzes the following reaction:

$$\text{UDP-glucose} + \text{a phenol} \rightleftharpoons \text{UDP} + \text{an aryl } \beta\text{-D-glucoside}$$

Phenol Hydroxylase See phenol 2-monooxygenase.

Phenol 2-Monooxygenase The enzyme that catalyzes the following reaction:

$$\text{Phenol} + \text{NADPH} + O_2 \rightleftharpoons \text{Catechol} + \text{NADP}^+ + H_2O$$

Phenol Reagent A reagent used for protein determination (also known as Folin-Ciocalteu's reagent).

Phenol Red (mol wt 354) A pH indicator dye.

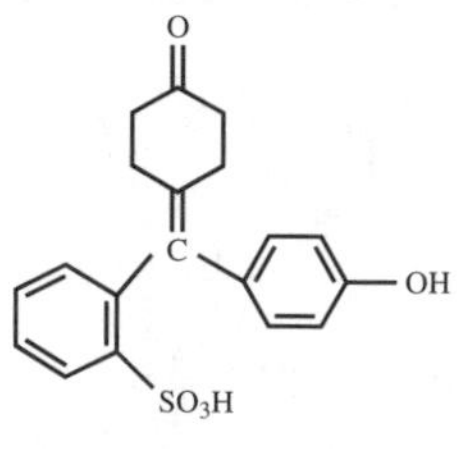

Phenolase Synonym of catechol oxidase.

Phenolphthalein (mol wt 318) A pH indicator (pH 8.3 – 10).

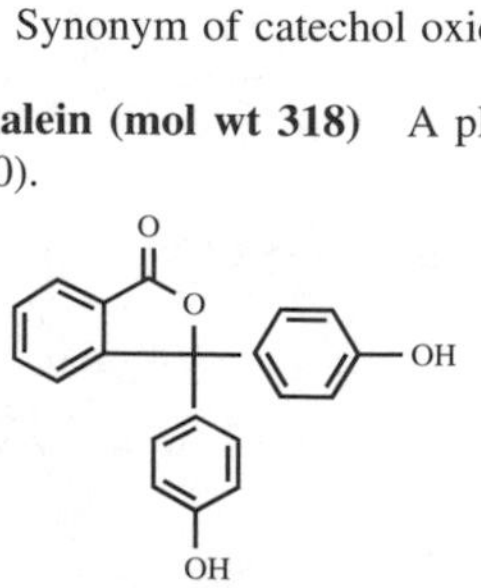

Phenolphthalol (mol wt 306) A cathartic agent.

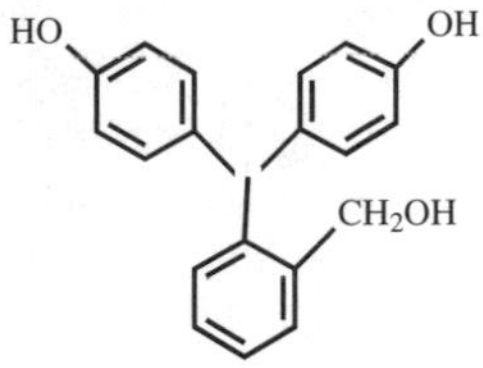

Phenolsulfonphthalein (mol wt 354) A reagent used in a test for renal function.

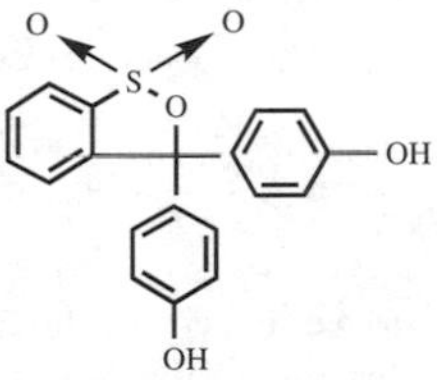

Phenolsulfonphthalein Test A test used for testing kidney function in which the rate of urine excretion is measured following injection of phenolsulfonphthalein.

Phenoltetrachlorophthalein (mol wt 456) A cathartic agent.

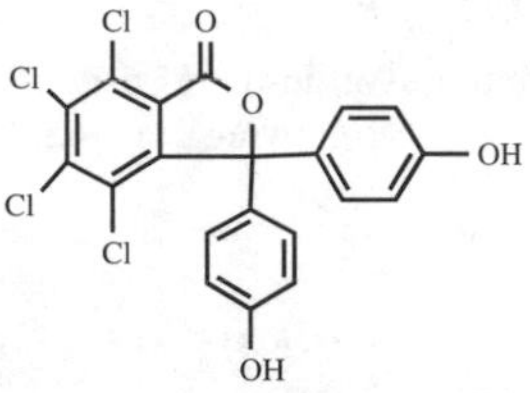

Phenon A type of classification in which organisms are grouped on the basis of their similar phenotypes.

Phenoperidine (mol wt 367) A narcotic analgesic agent.

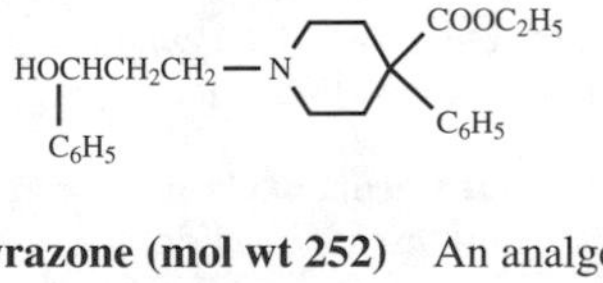

Phenopyrazone (mol wt 252) An analgesic and an antipyretic agent.

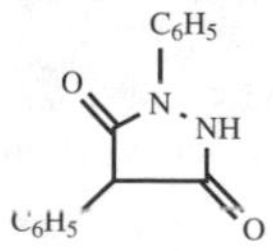

Phenosafranin (mol wt 323) A biological stain.

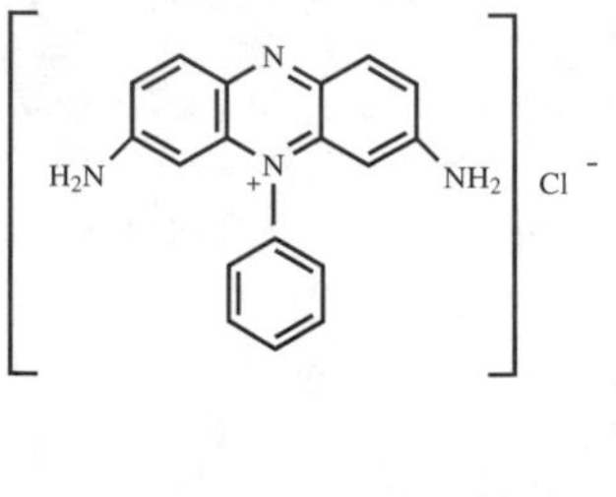

Phenothiazine (mol wt 199) An anthelmintic agent.

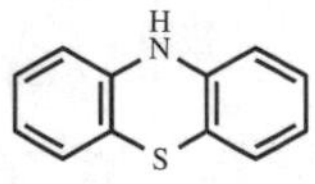

Phenotype The observable properties of an organism that have developed under the combined influences of the genetic constitution of the organism and the effects of environmental factors.

Phenotypic Lag The delay in expression of an acquired phenotype in a mutation or in gene transfer.

Phenotypic Mixing The production of progeny virions resulting from a mixed infection in which the phenotype of the progeny does not match its genotype.

Phenotypic Suppression The suppression of a mutant phenotype by a nongenetic factor, e.g., suppression of phenotypic expression by the presence of streptomycin or 5-fluorouracil.

Phenoxyacetyl Cellulose A cellulose derivative used for enzyme immobilization.

Phenoxybenzamine (mol wt 304) An antihypertensive agent that blocks the effect of catecholamines on alpha-adrenergic receptors.

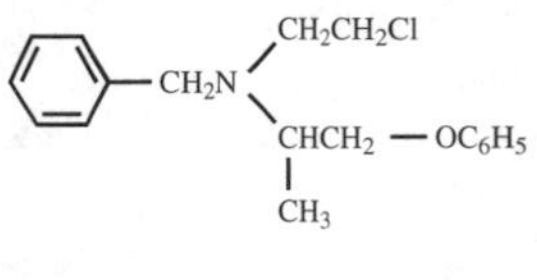

Phenoxymethyl Penicillin Synonym of penicillin V.

Phenoxymethyl Penicillin Potassium See penicillin V potassium.

Phenpentermine (mol wt 163) An anorexic agent.

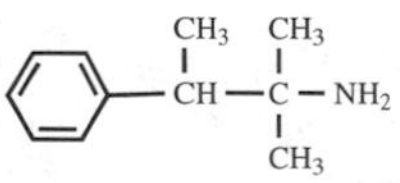

Phensuximide (mol wt 189) An anticonvulsant and antiepileptic agent.

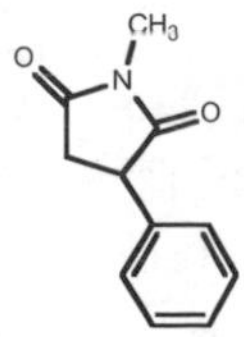

Phentolamine (mol wt 281) An antihypertensive agent that blocks the effect of catecholamine on alpha-adrenergic receptors.

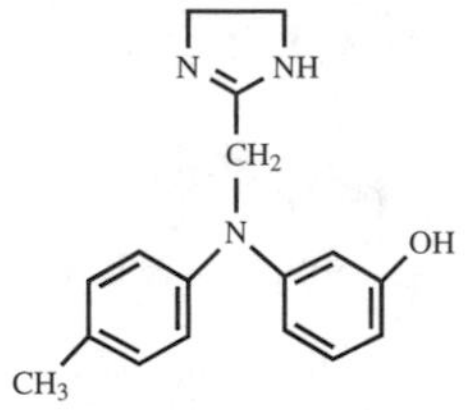

Phenurone A trade name for phenacemide, used as an anticonvulsant.

Phenyl-Aminosalicylate (mol wt 229) An antituberculostatic agent.

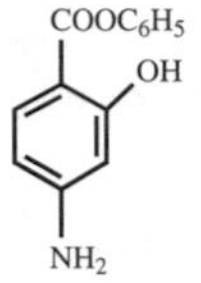

Phenylacetaldehyde Dehydrogenase The enzyme that catalyzes the following reaction:

Phenylacetaldehyde + NAD^+ + H_2O ⇌ Phenylacetate + NADH

Phenylacetate-CoA Ligase The enzyme that catalyzes the following reaction:

ATP + phenylacetate + CoA ⇌ Phenylacetyl-CoA + ADP + Pi

Phenylacetic Acid (mol wt 136) A phenyl derivative of acetic acid.

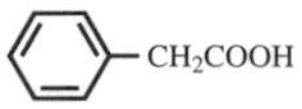

Phenylaceturic Acid A conjugate of phenylacetic acid with glycine.

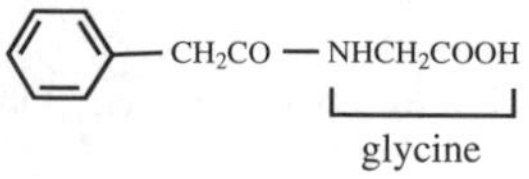

Phenyl-Agarosen A hydrophobic agarose with a phenyl group attached and used as an affinity chromatographic medium.

Phenylalaninase The enzyme that catalyzes the following reaction:

L-phenylalanine + dihydrobiopterine + O_2 ⇌ L-tyrosine + biopterin + H_2O

Phenylalanine (mol wt 165) An essential protein amino acid.

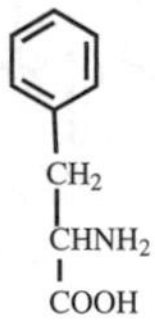

Phenylalanine *N*-Acetyltransferase The enzyme that catalyzes the following reaction:

Acetyl-CoA + phenylalanine ⇌ CoA + N-acetyl-phenylalanine

Phenylalanine Adenylyltransferase The enzyme that catalyzes the following reaction:

ATP + L-phenylalanine ⇌ PPi + *N*-adenylyl-phenylalanine

Phenylalanine Agar An agar medium containing 0.3% phenylalanine, 0.1% Na_2HPO_4, 0.5% NaCl, and 1.2% agar.

Phenylalanine Ammonia Lyase The enzyme that catalyzes the following reaction:

L-phenylalanine ⇌ *trans*-cinnamate + NH_3

Phenyl-Alanine Carboxy-Lyase Synonym of phenylalanine decarboxylase.

Phenylalanine Deaminase The enzyme that catalyzes the following reaction:

L-phenylalanine + H_2O ⇌ *enol*-pyruvate + NH_3

Phenylalanine Deaminase Test A method for testing the ability of an organism to deaminate phenylalanine to enolpyruvate.

Phenylalainine Decarboxylase The enzyme that catalyzes the following reaction:

Phenylalanine ⇌ Phenylethylamine + CO_2

Phenylalanine Dehydrogenase The enzyme that catalyzes the following reaction:

L-Phenylalanine + NAD^+ + H_2O ⇌ Phenylpyruvate + NH_3 + NADH

Phenylalanine Hydroxylase The enzyme that catalyzes the following reaction:

O_2 + phenylalanine + tetrahydrobiopterin ⇌ Tyrosine + dihydrobiopterin + H_2O

Phenylalanine 2-Monooxygenase The enzyme that catalyzes the following reaction:

$$\text{Phenylalanine} + O_2 \rightleftharpoons \text{Phenylacetamide} + CO_2 + H_2O$$

Phenylalanine 4-Monooxygenase The enzyme that catalyzes the following reaction:

$$\text{Phenylalanine} + \text{tetrahydrobiopterin} + O_2 \rightleftharpoons \text{Tyrosine} + \text{dihydrobiopterin} + H_2O$$

Phenyl-Alanine NAD+ Oxidoreductase The systematic name for phenylalanine dehydrogenase.

Phenylalanine Oxidase See phenylalanine 2-monooxygenase.

Phenylalanine Racemase (ATP-hydrolyzing) The enzyme that catalyzes the following reaction:

$$\text{L-Phenylalanine} + \text{ATP} + H_2O \rightleftharpoons \text{D-Phenylalanine} + \text{AMP} + \text{PPi}$$

Phenylalanine tRNA Referring to tRNA that carries amino acid phenylalanine.

Phenylalaninemia The presence of phenylalanine in the blood.

Phenylalanine tRNA Ligase The enzyme that catalyzes the following reaction:

$$\text{ATP} + \text{L-phenylalanine} + \text{tRNA}^{phe} \rightleftharpoons \text{AMP} + \text{PPi} + \text{L-phenylalanyl-tRNA}^{phe}$$

Phenylalanyl tRNA Ligase See phenylalanine tRNA-ligase.

Phenylalanyl tRNA Synthetase See phenylanine tRNA-ligase.

Phenyl-Aminosalicylate (mol wt 229) An antibacterial agent.

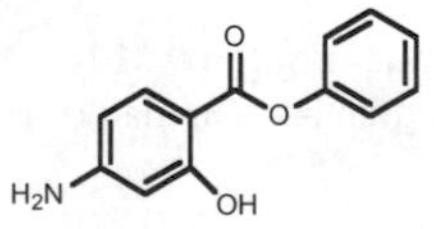

Phenylbutazone (mol wt 308) A nonsteroidal anti-inflammatory agent.

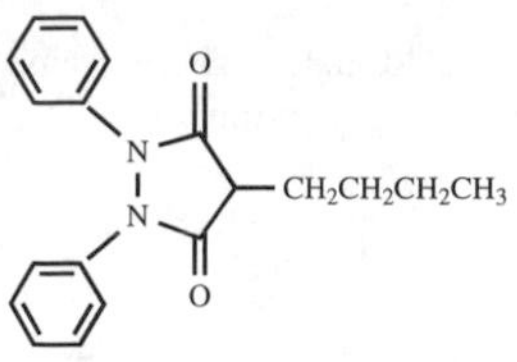

Phenylbutyric Acid (mol wt 164) A phenyl derivative of butyric acid.

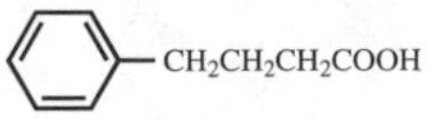

Phenylcarboxylase See phenylalanine decarboxylase.

***o*-Phenylenediamine (mol wt 108)** A substrate used for assaying peroxidase activity in the ELISA procedure.

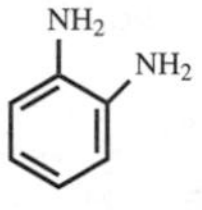

***p*-Phenylenediamine (mol wt 108)** A substrate used for assaying ceruloplasmin activity.

Phenylephrine Hydrochloride (mol wt 204) A decongestant.

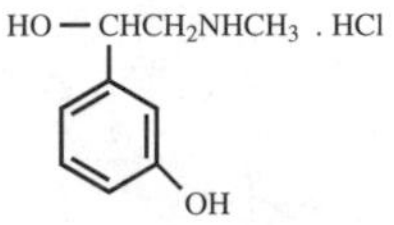

Phenylethanolamine (mol wt 137) A topical vasoconstrictor.

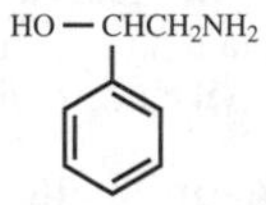

Phenylethanolamine *N*-Methyltransferase The enzyme that catalyzes the following reaction:

$$\text{Phenylethanolamine} + \text{adenosyl-L-methionine} \rightleftharpoons \textit{S}\text{-Adenosyl-L-homocysteine} + \textit{N}\text{-methylphenylethanolamine}$$

Phenyl-β-D-Galactopyranoside (mol wt 256) An acceptor substrate for detection of fucosyltransferase activity.

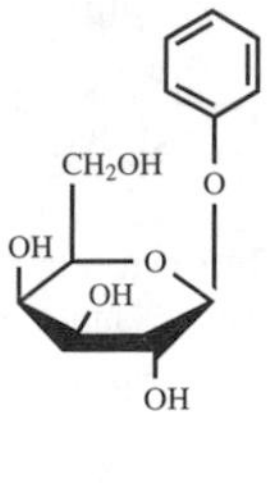

Phenyl-β-D-Glucopyranoside (mol wt 256) A substrate used for the detection of β-D-glucosidase.

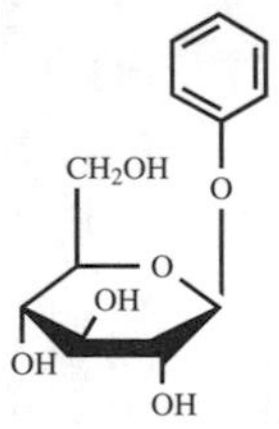

Phenylglyoxal (mol wt 134) A reagent used for the modification of arginine residues.

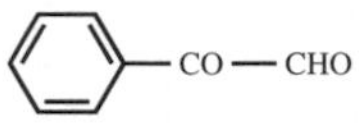

Phenylhydrazine (mol wt 108) A hemolytic agent.

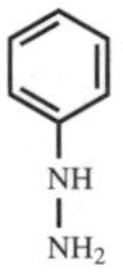

Phenylisothiocyanate (mol wt 135) A reagent used for the analysis of N-terminal amino acids of proteins.

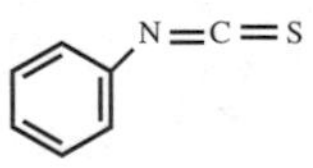

Phenylketonuria A genetic disorder caused by the absence or deficiency of phenylalanine hydroxylase, a disease characterized by mental retardation and accumulation of phenylalanine and the appearance of phenylketone in the urine.

Phenylmercury Borate (mol wt 339) A topical antiseptic agent.

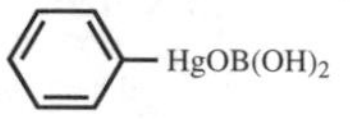

Phenylmethanesulfonyl Fluoride (mol wt 174) An inhibitor for trypsin, chymotrypsin, and serine proteases.

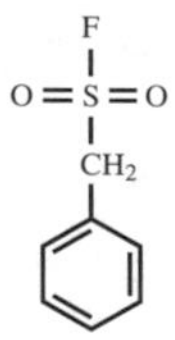

Phenylmethylbarbituric Acid (mol wt 218) An anticonvulsant, sedative, and hypnotic agent.

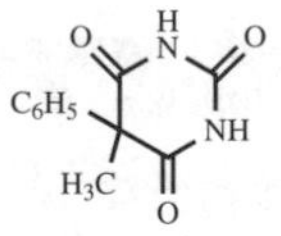

Phenylpropanolamine Hydrochloride (mol wt 188) A decongestant and an anorexic agent.

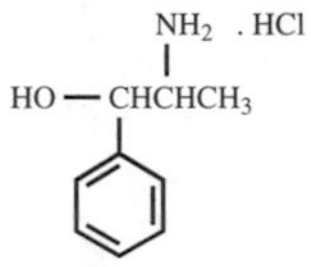

Phenylpropylmethylamine (mol wt 149) An adrenergic agent.

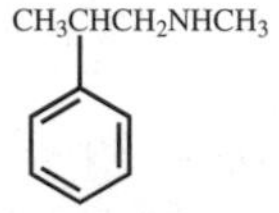

Phenylpyruvate Decarboxylase The enzyme that catalyzes the following reaction:

Phenylpyruvate $\rightleftharpoons$ Phenylacetaldehyde + CO_2

Phenylpyruvate Keto-Enol-Isomerase The systematic name for phenylpyruvate tautomerase.

Phenylpyruvate Tautomerase The enzyme that catalyzes the following reaction:

keto-Phenylpyruvate

$\rightleftharpoons$

enol-Phenylpyruvate

Phenylpyruvic Acid (mol wt 164) An intermediate in phenylalanine degradation.

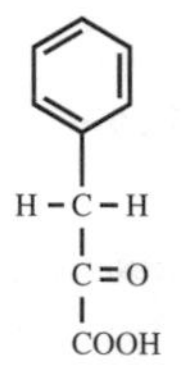

Phenylsalicylate (mol wt 214) An analgesic, antipyretic, and anti-inflammatory agent.

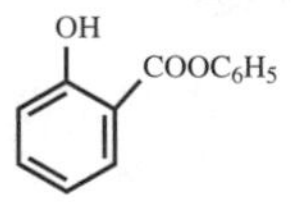

Phenylserine Aldolase The enzyme that catalyzes the following reaction:

L-*threo*-3-Phenylserine

$\rightleftharpoons$

Glycine + benzylaldehyde

Phenytex A trade name for phenytoin, an anticonvulsant and antiepileptic agent.

Phenytoin (mol wt 252) An anticonvulsant and an antiepileptic agent.

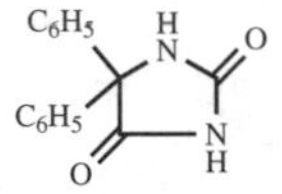

Phenzine A trade name for a cerebral stimulant that promotes transmission of nerve impulses.

Pheo Abbreviation for pheophytin.

Pheo- A prefix denoting dark colored.

Pheo α Abbreviation for pheophytin α.

Pheophytin A type of chlorophyll in which the central magnesium is replaced by two hydrogen atoms.

Pheophytin α An electron carrier pigment in the photosynthetic system. It is a derivative of chlorophyll *a* in which the Mg^{++} in the chlorophyll *a* is replaced by two hydrogen atoms.

Pheromone Substance released by one member of a species that influences the behavior of another member of the same species.

Phethenylate Sodium (mol wt 280) An anticonvulsant.

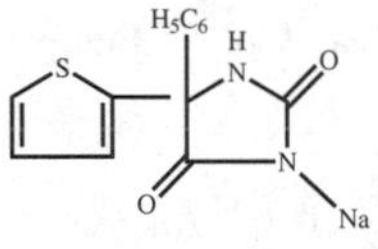

pHG165 A plasmid of *E. coli* that contains an Amp[r] (ampicillin resistant) marker and HindIII, PstI, SalI, BamHI, SmaI, and EcoRI cleavage sites.

pHG276 A plasmid of *E. coli* that contains an Amp[r] (ampicillin resistant) marker and SalI, BamHI, SmaI, and EcoRI cleavage sites.

PHGPX Abbreviation for phospholipid hydroperoxide glutathione peroxidase.

pH_i Abbreviation for intracellular pH.

PHI Abbreviation for phosphohexose isomerase.

phi (ϕ) A Greek letter.

ϕ6 A bacteriophage in the family Cystoviridae containing double-stranded RNA.

ϕX 174 A bacteriophage of the family Microviridae containing single-stranded DNA genome.

Philadelphia Chromosome A chromosomal disorder (chromosome number 22 in human) associated with chronic myelogenous leukemia. It is a translocational aberration in which approximately one half the long arm of chromosome number 22 is removed to a terminal position of another chromosome.

-phile A suffix meaning similar to or having an affinity for.

Philopodia Filamentous projections composed entirely of ectoplasm.

pHiso Hex A trade name for hexachlorophene, used as a bacteriostaic agent.

PHK Abbreviation for phosphohexokinase.

Phleb- A prefix denoting a vein or veins.

Phlebo- A prefix denoting a vein or veins.

Phlebitis Inflammation of the wall of a vein.

Phlebothrombosis The development of a blood clot in a vein.

Phlebovirus A virus in the family Bunyaviridae that contains single-stranded RNA as the genetic material.

Phlegm Thick mucus secreted from the respiratory tract, e.g., lung or bronchial tubes.

Phleom Plant tissue in the vascular system that is responsible for transport of substances to various parts of the plant.

PHLLA Abbreviation for post-heparin lipolytic activity.

Phlogistic Capable of causing inflammation.

Phloretin (mol wt 274) An inhibitor of glucose transport across the membrane of erythrocytes.

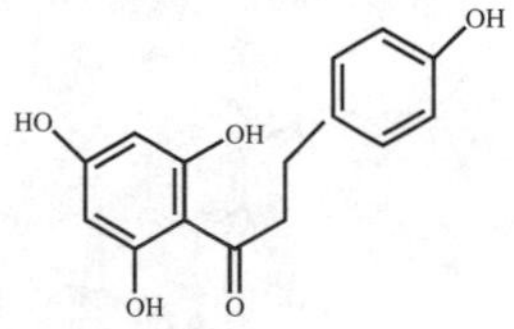

Phloretin Hydrolase The enzyme that catalyzes the following reaction:

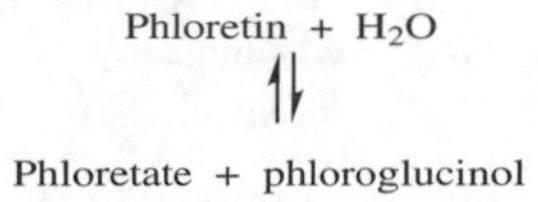

Phloridzin (mol wt 436) An inhibitor for Na^+-dependent glucose transport.

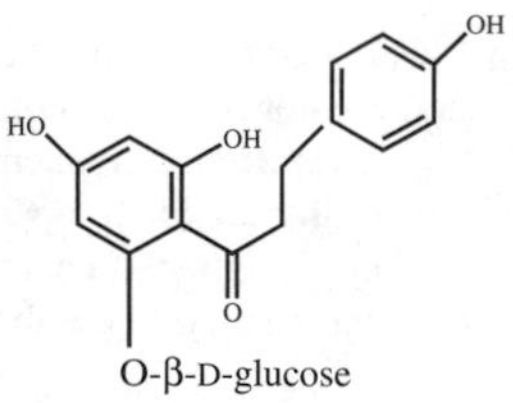

Phloroglucinol (mol wt 126) An antispasmodic agent.

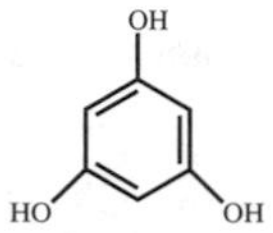

PHM Abbreviation for peptidylglycine α-hydroxylating mono-oxygenase.

PHMB (*p*HMB) Abbreviation for *p*-hydroxymecuribenzoate.

***pho* A Gene** A gene in the regulon of *E. coli* involved in the response to phosphate starvation that encodes a periplasmic alkaline phosphatase.

***pho* B Gene** A gene in the regulon of *E. coli* involved in the response to phosphate starvation whose product controls the operation of *pho* regulon.

***pho* E Gene** A gene in the regulon of *E. coli* involved in the response to phosphate starvation that encodes a porin.

***pho* Regulon** Regulatory genes that encode proteins responsible for the response to phosphate starvation in *E. coli.*

***pho* S Gene** A gene in the regulon of *E. coli* involved in the response to phosphate starvation that encodes a phosphate-binding protein.

-phobic A suffix meaning having an aversion for or lacking affinity for.

Pholcodine (mol wt 398) An antitussive agent.

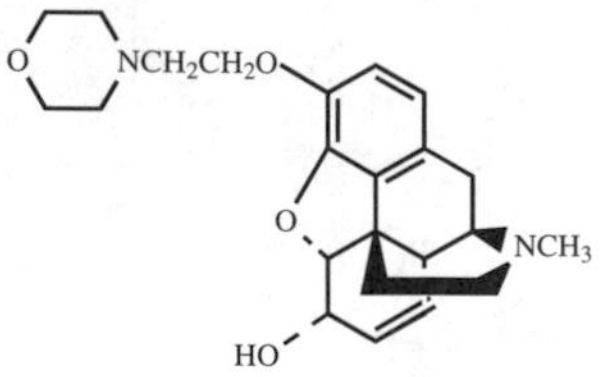

Pholedrine (mol wt 165) A sympathomimetic agent and a circulatory stimulant.

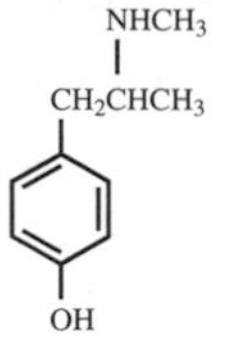

Phorate (mol wt 260) An insecticide and an inhibitor for cholinesterase.

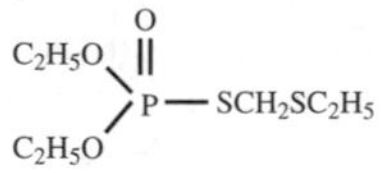

Phorbol (mol wt 364) An alcohol found in croton oil.

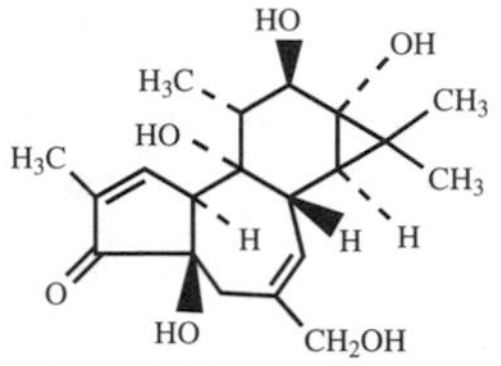

Phorbol Esters Derivatives of phorbol and a tumor-promoting agent found in croton oil.

-phore A suffix meaning carrier or bearer.

Phoresis A process of transporting ionic compounds into the tissues by means of an electric current.

Phormidium A genus of cyanobacteria.

Phoschol A trade name for phosphatidylcholine.

Phos-Ex A trade name for calcium acetate, used to maintain calcium levels in the body.

Phos-Flur A trade name for sodium fluoride, used to catalyze bone remineralization.

Phosgene (mol wt 99) A highly toxic gas.

$$Cl_2C = O$$

Phosmet (mol wt 317) An insecticide (acaricide).

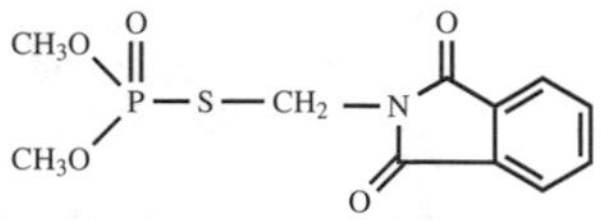

Phosphagen Referring to substances that are capable of storing high-potential phosphoryl groups in muscle, e.g., arginine phosphate or creatine phosphate.

Phosphamidase See phosphoamidase.

Phosphamidon (mol wt 300) A cholinesterase inhibitor.

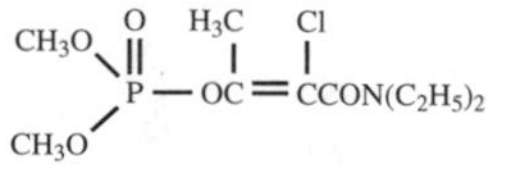

Phosphatase The enzyme that catalyzes the hydrolysis of the ester of phosphoric acid. Phosphatases are classified according to their pH optima, e.g., acid phosphatase and alkaline phosphatase.

Phosphatase (acid)   The enzyme that catalyzes the hydrolysis of a number of phosphomonoesters at acid pH but not phosphodiesters.

$$\text{Phosphoric monoester} + H_2O$$
$$\Updownarrow$$
$$\text{Alcohol} + \text{phosphoric acid}$$

Phosphatase (alkaline)   The enzyme that catalyzes the hydrolysis of phosphomonoester at alkaline pH.

Phosphatase Test   A test for detecting the presence of alkaline phosphatase in milk. Alkaline phosphatase is present in untreated raw milk but disappears after pasteurization.

Phosphate   1. A salt of phosphoric acid. 2. An anionic radical of phosphoric acid.

Phosphate Acetyltransferase   The enzyme that catalyzes the following reaction:

$$\text{Acetyl-CoA} + Pi \rightleftharpoons \text{CoA} + \text{acetylphosphate}$$

Phosphate Bond Energy   Referring to the free energy change when a phosphorylated compound undergoes hydrolysis to equilibrium at pH 7, 25° C in a 1.0 M solution.

Phosphate Butyryltransferase   The enzyme that catalyzes the following reaction:

$$\text{Butanoyl-CoA} + Pi$$
$$\Updownarrow$$
$$\text{CoA} + \text{butanoylphosphate}$$

Phosphate Group Transfer Potential   Referring to the negative values of the free energy change in the phosphate transfer reactions.

Phosphate Potential   Referring to the ratio of the concentrations of ATP to ADP and inorganic phosphate in a biological system.

Phosphatemia   The presence of excessive quantities of phosphate in the blood.

Phosphatidase   See phospholipase.

Phosphatidate   Referring to diacylglycerol phosphate.

$$\begin{array}{l} H_2C-O-CO-R_1 \\ \quad | \\ HC-O-CO-R_2 \\ \quad | \qquad\quad O \\ \quad | \qquad\quad \| \\ H_2C-O-P-OH \\ \qquad\qquad\quad | \\ \qquad\qquad\quad O^- \end{array}$$

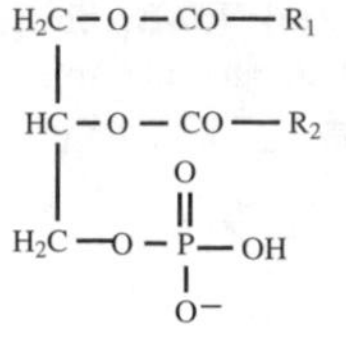

$$R_1 \text{ and } R_2 = \text{fatty acid}$$

Phosphatidate Cytidylyltransferase   The enzyme that catalyzes the following reaction:

$$\text{CTP} + \text{phosphatidate}$$
$$\Updownarrow$$
$$\text{PPi} + \text{CDP-diacylglycerol}$$

Phosphatidate Phosphatase   The enzyme that catalyzes the following reaction:

$$\text{A phosphatidate} + H_2O$$
$$\Updownarrow$$
$$\text{A diacylglycerol} + Pi$$

Phosphatide   Referring to phosphatidate or phosphatidic acid.

Phosphatidic Acid   Referring to phosphoglyceride that consists of two fatty acids and a phosphate group linked by ester bonds to glycerol, it is a key intermediate in the synthesis of other phosphoglycerides.

$$\begin{array}{l} R_1-CO-O-CH_2 \\ \qquad\qquad\quad | \\ R_2-CO-O-CH \\ \qquad\qquad\quad | \qquad\quad O \\ \qquad\qquad\quad | \qquad\quad \| \\ \qquad\qquad CH_2-O-P-OH \\ \qquad\qquad\qquad\qquad\quad | \\ \qquad\qquad\qquad\qquad\quad OH \end{array}$$

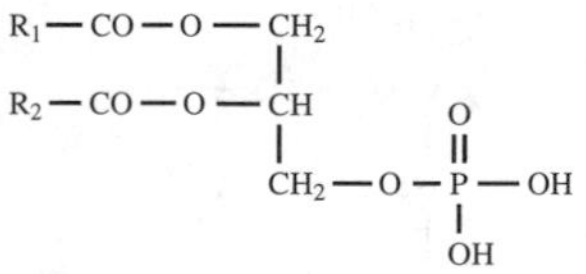

$$R_1 \text{ and } R_2 = \text{fatty acid}$$

Phosphatidic Acid Phosphatase   The enzyme that catalyzes the following reaction:

$$\text{Phosphatidic acid} + H_2O \rightleftharpoons \text{Diacylglycerol} + Pi$$

Phosphatidylcholine   A phospholipid consisting of glycerol, fatty acids, phosphate and choline.

$$\begin{array}{l} H_2C-O-CO-R_1 \\ \quad | \\ HC-O-CO-R_2 \\ \quad | \qquad\quad O \\ \quad | \qquad\quad \| \qquad\qquad\qquad\qquad CH_3 \\ H_2C-O-P-O-CH_2CH_2-N^+-CH_3 \\ \qquad\qquad\quad | \qquad\qquad\qquad\qquad\qquad | \\ \qquad\qquad\quad OH \qquad\qquad\qquad\qquad\quad CH_3 \end{array}$$

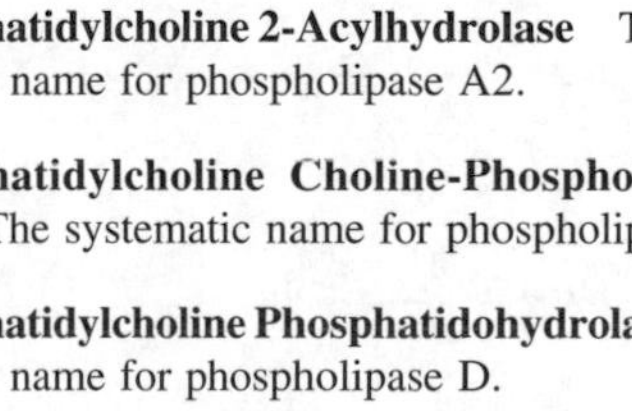

$$R_1 \text{ and } R_2 = \text{fatty acid}$$

Phosphatidylcholine 2-Acylhydrolase   The systematic name for phospholipase A2.

Phosphatidylcholine Choline-Phosphohydrolase   The systematic name for phospholipase C.

Phosphatidylcholine Phosphatidohydrolase   Systematic name for phospholipase D.

Phosphatidylcholine Sterol O-Acyltransferase   The enzyme that catalyzes the following reaction:

$$\text{Phosphatidylcholine} + \text{sterol}$$
$$\Updownarrow$$
$$\text{1-Acylglycerolphosohocholine} + \text{sterol ester}$$

Phosphatidylethanolamine A phospholipid consisting of glycerol, fatty acids, phosphate, and ethanolamine.

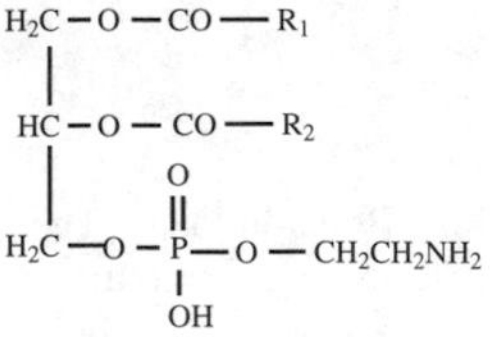

R_1 and R_2 = fatty acid

Phosphatidylethanolamine Methyltransferase The enzyme that catalyzes the following reaction:

S-Adenosylmethionine + phosphatidylethanolamine

⇅

S-adenosylhomocysteine + *N*-methylphosphatidylethanolamine

Phosphatidylglycerol Referring to the phosphatidate with an additional glycerol molecule attached to the phosphate group.

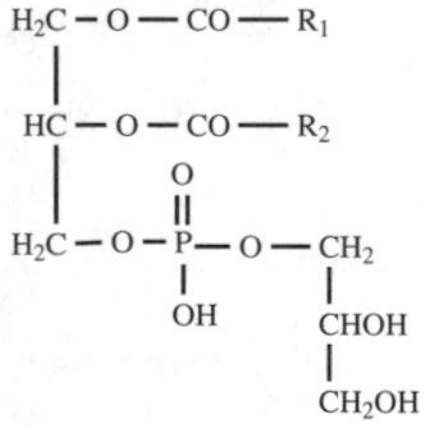

R_1 and R_2 = fatty acid

Phosphatidyl-Glycerophosphatase The enzyme that catalyzes the following reaction:

Phosphatidyl glycerophosphate + H_2O

⇅

Phosphatidylglycerol + Pi

Phosphatidylglycerol Phosphate An intermediate in the biosynthesis of phosphatidylglycerol.

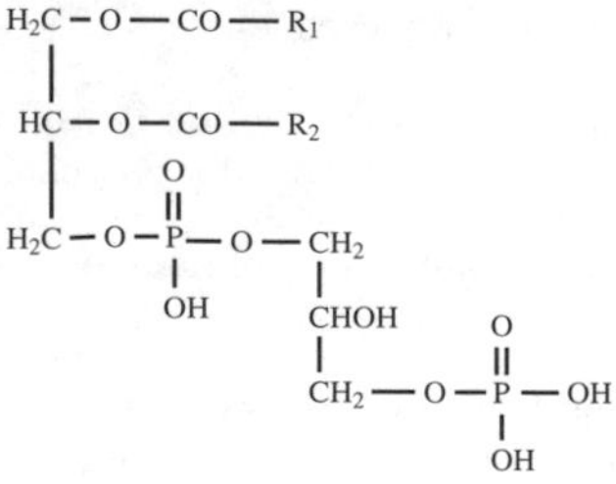

R_1 and R_2 = fatty acid

Phosphatidylinositol (PI) A phospholipid.

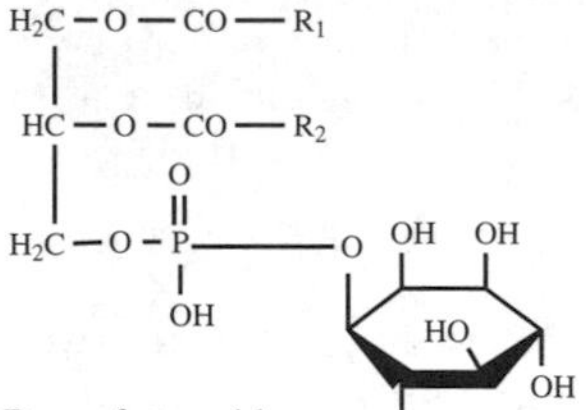

R_1 and R_2 = fatty acid

Phosphatidylinositol *N*-Acetylglucosaminyl Transferase The enzyme that catalyzes the following reaction:

UDP-*N*-acetylglucosamine + phosphatidylinositol

⇅

UDP + *N*-acetyl-D-glucosaminyl phosphatidylinositol

Phosphatidylinositol 4,5-Bisphosphatase The enzyme that catalyzes the following reaction:

Phosphatidylinositol 4,5-bisphosphate + H_2O

⇅

Phosphatidylinositol 4-phosphate + Pi

Phosphatidylinositol 4,5-Bisphosphate (PIP_2) A phospholipid and an intermediate in the insositol phosphate cycle. It mediates the action of calcium as a second messenger.

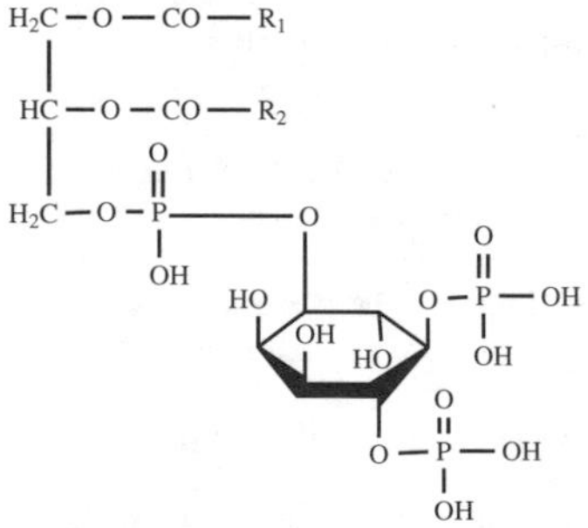

R_1 and R_2 = fatty acid

Phosphatidylinositol 4,5-Bisphosphate Phosphodiesterase The enzyme that catalyzes the following reaction:

Phosphatidylinositol 4,5-bisphosphate + H_2O

⇅

Inositol 1,4,5-triphosphate + diacylglycerol

Phosphatidylinositol Deacylase The enzyme that catalyzes the following reaction:

Phosphatidylinositol + H_2O
⇌
1-Acylglycerophosphoinositol + a carboxylate

Phosphatidylinositol 3-Phospshatase The enzyme that catalyzes the following reaction:

Phosphatidyl 3-phosphate + H_2O
⇌
Phosphatidylinositol + Pi

Phosphatidylinositol Phosphate (PIP) A phospholipid and an intermediate in the inositol phosphate cycle.

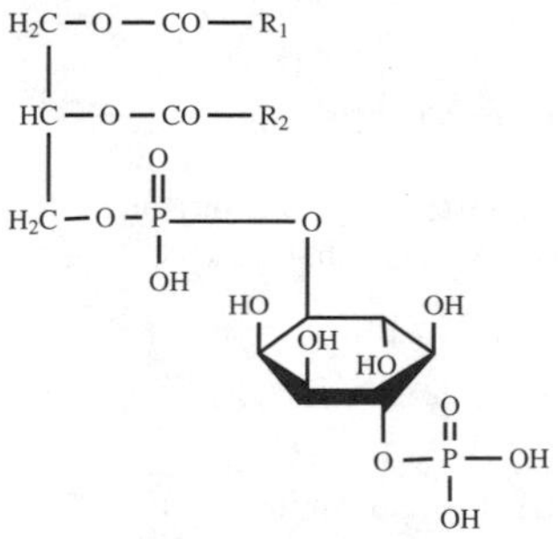

R_1 and R_2 = fatty acid

1-Phosphatidylinositol Phosphodiesterase The enzyme that catalyzes the following reaction:

Phosphatidyl-D-*myo*-inositol
⇌
D-*myo*-Inositol 1,2-cyclic phosphate + diacylglycerol

Phosphatidylinositol Phospholipase C Synonym of 1-phosphatidylinositol phosphodiesterase.

Phosphatidylinositol Synthetase The enzyme that catalyzes the following reaction:

CDP-diacylglycerol + inositol
⇌
CMP + phosphatidylinositol

Phosphatidylserine A phospholipid whose phosphate group is linked to a serine molecule.

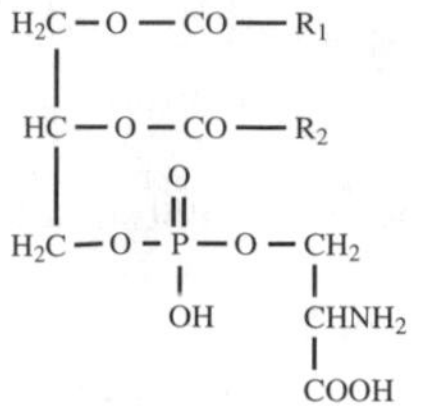

R_1 and R_2 = fatty acid

Phosphatidylserine Carboxy-lyase See phosphatidylserine decarboxylase.

Phosphatidyl-Serine Decarboxylase The enzyme that catalyzes the following reaction:

Phosphatidylserine
⇌
Phosphatidyl ethanolamine + CO_2

Phosphatidylserine Synthetase The enzyme that catalyzes the following reaction:

CDP-diacylglycerol + serine
⇌
CMP + phosphatidylserine

Phosphaturia The presence of large quantities of phosphate in the urine.

Phosphine Oxide Referring to substance that contains the group:

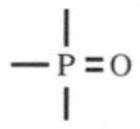

Phosphinothricin (mol wt 181) A naturally occurring amino acid and a glutamate analog.

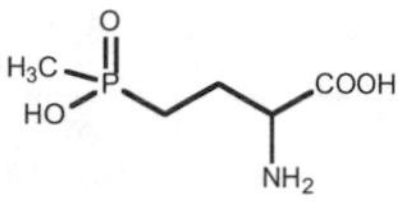

Phosphite A salt or an ester of phosphoric acid.

Phosphite Triester Method A method used for synthesis of oligonucleotide using a dimethoxytrityl group to protect the 5′-end of the growing polynucleotide chain (3′-end is anchored onto a solid support).

Phosphoacetyl Glucosamine Mutase The enzyme that catalyzes the following reaction:

N-Acetylglucosamine 1-phosphate
⇌
N-Acetylglucosamine 6-phosphate

Phosphoadenylylsulfatase The enzyme that catalyzes the following reaction:

3-Phosphoadenylylsulfate + H_2O
⇌
Adenosine 3′,5′-bisphosphate + sulfate

Phosphoamidase The enzyme that catalyzes the following reaction:

N-Phosphocreatine + H_2O
⇌
Creatine + Pi

Phosphoamide Hydrolase See phosphoamidase.

Phosphoanhydride Bond The high energy bond formed between two phosphate groups, e.g., in ADP or ATP.

Phosphoarginine (mol wt 254) A compound with high phosphate group transfer potential.

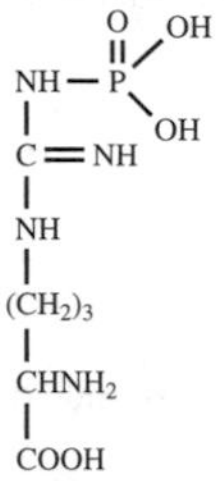

Phosphocellulose A cation exchanger used in ion exchange chromatography.

Phosphocholine (mol wt 184) A compound involved in the biosynthesis of phosphatidylcholine.

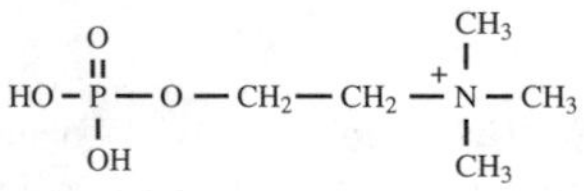

Phosphocolamine (mol wt 141) A potent inhibitor for ornithine decarboxylase.

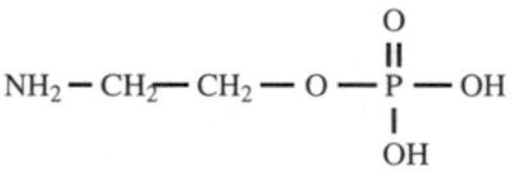

Phosphocreatine (mol wt 211) A compound with high phosphate group transfer potential.

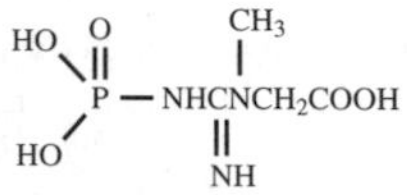

Phosphodeoxyriboaldolase The enzyme that catalyzes the following reaction:

2-Deoxyribose 5-phosphate

⇌

Glyceraldehyde 3-phosphate + acetaldehyde

Phosphodiester The compound whose two hydroxyl groups are esterified to a phosphate group, e.g., two adjacent nucleotides in DNA and RNA.

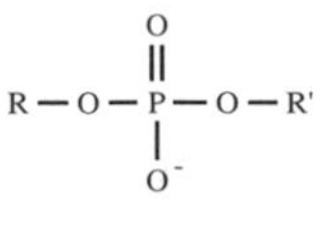

Phosphodiesterase The enzyme that catalyzes the hydrolysis of phosphodiester bond in the polynucleotides or cyclic nucleotides.

Phosphodiester Bond A covalent bond formed by ester linkage of two hydroxyl groups to the same phosphate group, e.g., linkage of adjacent nucleotides in DNA and RNA.

3′,5′-Phosphodiester Bond A covalent ester linkage in which a phosphoric acid is esterified to the 3′ hydroxyl of one nucleoside and the 5′ hydroxyl of another nucleoside.

Phosphoenolpyruvate Carboxykinase (ATP-specific) The enzyme that catalyzes the following reaction:

ATP + oxaloacetate

⇌

ADP + phosphoenolpyruvate + CO_2

Phosphoenolpyruvate Carboxykinase (GTP-specific) The enzyme that catalyzes the following reaction:

Oxaloacetate + GTP

⇌

Phosphoenolpyruvate + GDP + CO_2

Phosphoenolpyruvate Carboxylase The enzyme that catalyzes the following reaction:

Phosphoenolpyruvate + CO_2

⇌

Oxaloacetate + Pi

Phosphoenolpyruvate Glycerone Phosphotransferase The enzyme that catalyzes the following reaction:

Phosphoenolpyruvate + glycerone

⇌

Pyruvate + glycerone phosphate

Phosphoenolpyruvate Kinase See pyruvate kinase.

Phosphoenolpyruvate Mutase The enzyme that catalyzes the following reaction:

Phosphoenolpyruvate

⇌

3-Phosphonopyruvate

Phosphoenolpyruvate Phosphatase The enzyme that catalyzes the following reaction:

Phosphoenolpyruvate + H_2O

⇌

Pyruvate + Pi

Phosphoenolpyruvate Synthetase The enzyme that catalyzes the following reaction:

ATP + pyruvate

⇅

AMP + Pi + phosphoenolpyruvate

Phosphoenolpyruvic Acid (mol wt 168) A compound for initiation of the C_4 pathway.

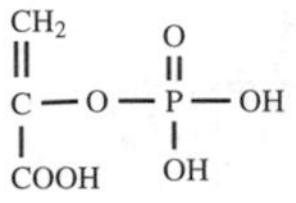

Phosphoester Bond Ester bond formed by removing a hydroxide ion from a phosphate group and a hydrogen ion from an alcohol group.

Phosphoethanolamine (mol wt 111) An intermediate in the biosynthesis of phosphatidylethanolamine.

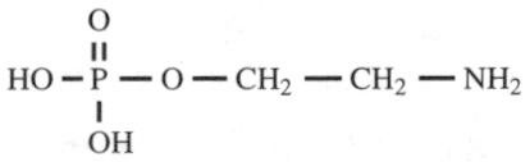

Phosphoethanolamine *N*-Methyltransferase The enzyme that catalyzes the following reaction:

S-Adenosyl-methionine +
ethanolamine phosphate

⇅

S-Adenosyl-homocysteine +
N-methylethanolamine phosphate

Phosphofructokinase The enzyme that catalyzes the following reaction:

Fructose 6-phosphate + ATP

⇅

Fructose 1-6-bisphosphate

Phosphofructokinase-2 The enzyme that catalyzes the following reaction:

Fructose 6-phosphate + ATP

⇅

Fructose 2-6-bisphosphate

6-Phosphofructose See fructose-6-phosphate.

6-Phosphogalactosidase The enzyme that catalyzes the following reaction:

6-Phospho-β-D-galactoside + H_2O

⇅

An alcohol + 6-phospho-β-D-galactose

Phosphoglucokinase The enzyme that catalyzes the following reaction:

Glucose 1-phosphate + ATP

⇅

Glucose 1-6-bisphosphate + ADP

Phosphoglucomutase The enzyme that catalyzes the following reaction:

α-D-Glucose 1-phosphate

⇅

α-D-glucose 6-phosphate

Phosphogluconate Dehydratase The enzyme that catalyzes the following reaction:

6-Phosphogluconate

⇅

2-Keto-3-deoxygluconate + H_2O

Phosphogluconate Dehydrogenase The enzyme that catalyzes the following reaction:

6-Phosphogluconate + $NADP^+$

⇅

Ribulose-5-phosphate + NADPH + CO_2

Phosphogluconate Oxidative Pathway Referring to pentose phosphate pathway or hexose monophosphate shunt.

6-Phospho-D-Gluconate $NADP^+$ 2-Oxidoreductase The systematic name for phosphogluconate dehydrogenase (decarboxylating).

6-Phosphogluconic Acid (mol wt 276) An intermediate in pentose phosphate pathway.

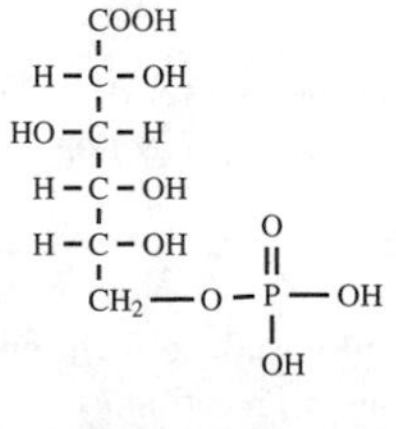

Phosphogluconic Acid Dehydrogenase Synonym of phosphogluconate dehydrogenase.

6-Phosphogluconolactonase The enzyme that catalyzes the following reaction:

6-Phosphoglucono-1,5-lactone + H_2O

⇅

6-Phosphogluconate

6-Phosphogluconolactone (mol wt 258) An intermediate in the pentose phosphate pathway.

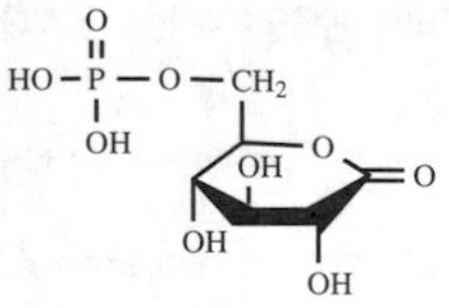

6-Phosphoglucosamine (mol wt 259) A phosphorylated amino sugar derived from glucose.

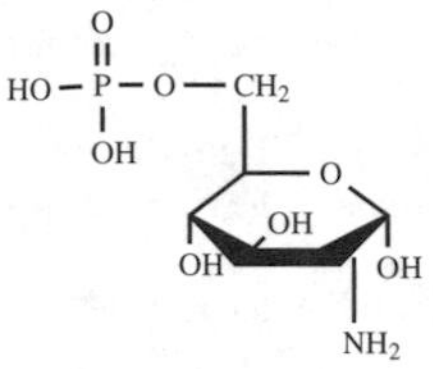

Phosphoglucosamine Acetylase The enzyme that catalyzes the following reaction:

Acetyl-CoA + glucosamine 6-phosphate

$\rightleftharpoons$

CoA + *N*-acetyl-glucosamine 6-phosphate

Phosphoglucosamine Transacetylase See phosphoglucosamine acetylase

6-Phosphoglucose (mol wt 260) An active glucose involved in the metabolism of glucose through a number biochemical pathways.

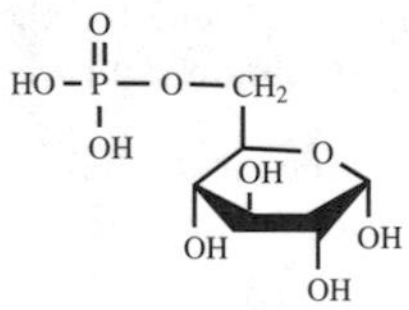

6-Phosphoglucose Isomerase The enzyme that catalyzes the following reaction:

Glucose 6-phosphate $\rightleftharpoons$ Fructose 6-phosphate

6-Phospho-β-glucosidase The enzyme that catalyzes the following reaction:

6-Phospho-β-D-glucosyl-(1,4)-D-glucose + H_2O

$\rightleftharpoons$

D-Glucose 6-phosphate + D-glucose

3-Phosphoglycerate A salt of phosphoglyceric acid (see 3-phosphoglyceric acid).

Phosphoglycerate Carboxy-Lyase The systematic name for ribulose bisphosphate carboxylase.

Phosphoglycerate Dehydrase The enzyme that catalyzes the following reaction:

2-Phospho-D-glycerate

$\rightleftharpoons$

Phosphoenolpyruvate + H_2O

3-Phosphoglycerate Dehydrogenase The enzyme that catalyzes the following reaction:

3-Phosphoglycerate + NAD^+

$\rightleftharpoons$

3-Phosphohydroxypyruvate + NADH

Phosphoglycerate Hydro-Lyase Systematic name for phosphopyruvate hydratase.

3-Phosphoglycerate Kinase The enzyme that catalyzes the following reaction:

3-Phosphoglycerate + ATP

$\rightleftharpoons$

1,3-Bisphosphoglycerate

3-Phosphoglycerate Mutase The enzyme that catalyzes the following reaction:

3-Phosphoglycerate $\rightleftharpoons$ 2-Phosphoglycerate

Phosphoglycerate NAD^+ 2-Oxidoreductase The systematic name for glycerate dehydrogenase.

Phosphoglycerate Phosphatase The enzyme that catalyzes the following reaction:

Glycerate phosphate + H_2O

$\rightleftharpoons$

Glycerate + Pi

Phosphoglycerate 2,3-Phosphomutase The systematic name for phosphoglycerate mutase.

Phosphoglyceric Acid (mol wt 186) An intermediate in glycolysis.

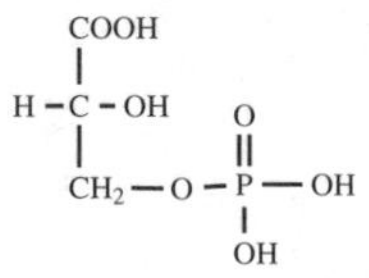

3-phosphoglyceric acid

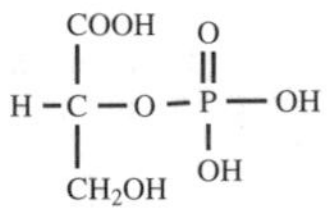

2-phosphoglyceric acid

Phosphoglyceride An ester of glycerol with fatty acids.

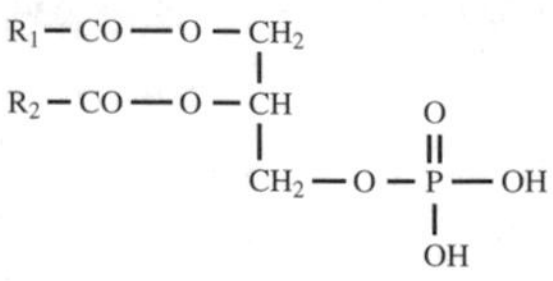

R_1 and R_2 = fatty acid

3-Phosphoglycerol (mol wt 173) An alcohol involved in transfer of electrons from cytoplasmic NADH to the mitochondrial electron transport chain (see also glycerophosphate shuttle).

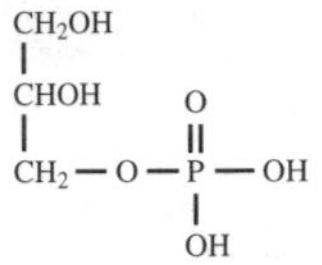

3-Phosphoglycerol Dehydrogenase The enzyme that catalyzes the following reaction:

Dihydroxyacetone phosphate + NADH

⇅

3-phosphoglycerol + NAD^+

Phosphoglyceromutase Synonym of phosphoglycerate mutase.

Phosphoglycolate Phosphatase The enzyme that catalyzes the following reaction:

Phosphoglycolate + H_2O

⇅

Glycolate + Pi

Phosphoglycolic Acid (mol wt 156) An intermediate in photorespiration pathway.

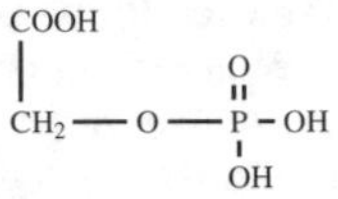

Phosphohexokinase Synonym of 6-phosphofructokinase.

Phosphohexomutase Synonym of mannose 6-phosphate isomerase.

Phosphohexose Isomerase Synonym of mannose 6-phosphate isomerase.

Phosphohistidine (mol wt 220) The phosphate form of histidine involved in the transport of sugar across the membrane in bacteria.

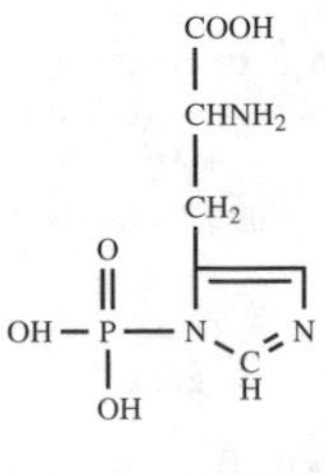

Phosphohomoserine (mol wt 199) An intermediate in biosynthesis of methionine from threonine.

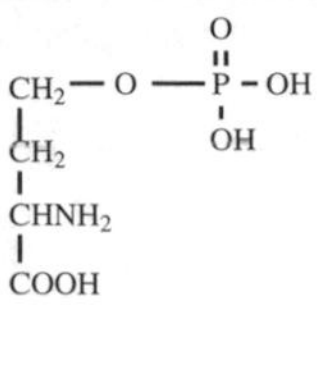

3-Phosphohydroxypyruvic Acid (mol wt 184) An intermediate in biosynthesis of serine from 3-phosphoglycerate.

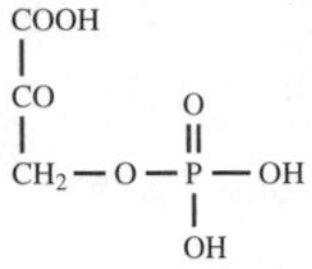

Phosphoinositide Referring to inositol phospholipid.

Phosphoinositide Signaling Pathway A pathway by which extracellular signaling molecules activate phospholipase C producing the second messengers inositol triphosphate and diacylglycerol.

Phospho-2-Keto-3-Deoxy Galactonate Aldolase The enzyme that catalyzes the following reaction:

6-Phospho-2-keto-3-deoxygalactonate

⇅

Pyruvate + glyceraldehyde 3-phosphate

Phospho-2-Keto-3-Deoxygluconate Aldolase The enzyme that catalyzes the following reaction:

6-Phospho-2-keto-3-deoxygluconate

⇅

Pyruvate + glyceraldehyde 3-phosphate

Phosphoketolase The enzyme that catalyzes the following reaction:

Xyluose 5-phosphate + Pi

⇅

Acetylphosphate + glyceraldehyde 3-phosphate + H_2O

Phosphoketolase Pathway A pathway for heterolactic fermentation of pentose and hexose employing the enzyme phosphoketolase.

Phosphoketotetrose Aldolase The enzyme that catalyzes the following reaction:

Erythrose 1-phosphate

⇅

Glycerone phosphate + formaldehyde

Phospholine Iodide A trade name for echothiophate iodide, used as a miotic agent.

Phospholipase The enzymes that catalyze the breakdown of phospholipids and release of fatty acids or other compounds from phosphoacylglycerol or phosphoglyceride.

Phospholipase A Enzyme that catalyzes the hydrolysis of ester bonds that link fatty acids to glycerol in phospholipids thereby releasing free fatty acids.

Phospholipase A_1 The enzyme that catalyzes the removal of fatty acyl groups from phosphoglyceride at carbon-1 position.

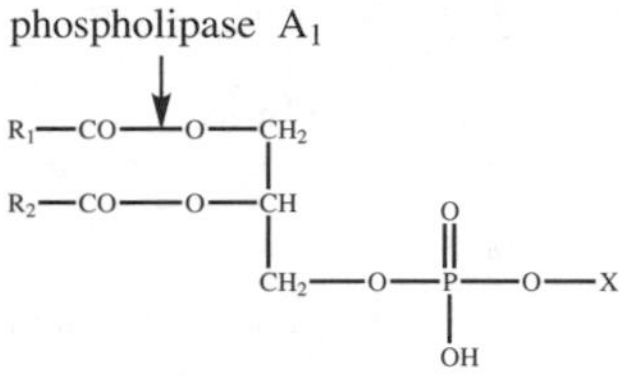

R_1 and R_2 = fatty acid

X = An organic molecule

Phospholipase A_2 The enzyme that catalyzes the removal of fatty acyl groups from phosphoglyceride at the carbon-2 position.

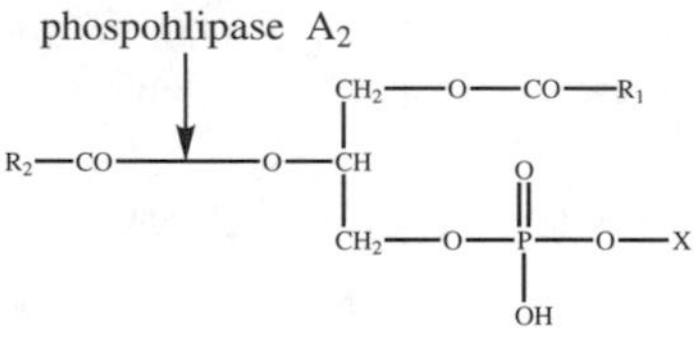

R_1 and R_2 = fatty acid

X = An organic molecule

Phospholipase C The enzyme that catalyzes the cleavage of bond between phosphate group and glycerol in phosphoglyceride.

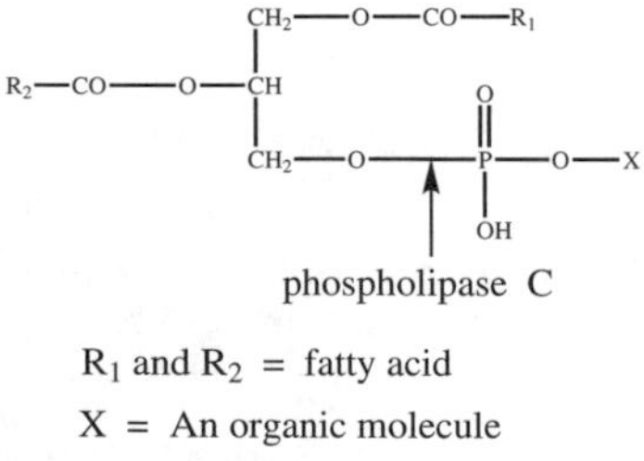

R_1 and R_2 = fatty acid

X = An organic molecule

Phospholipase D The enzyme that catalyzes the removal of components attached to the phosphate in phosphoacylglycerol.

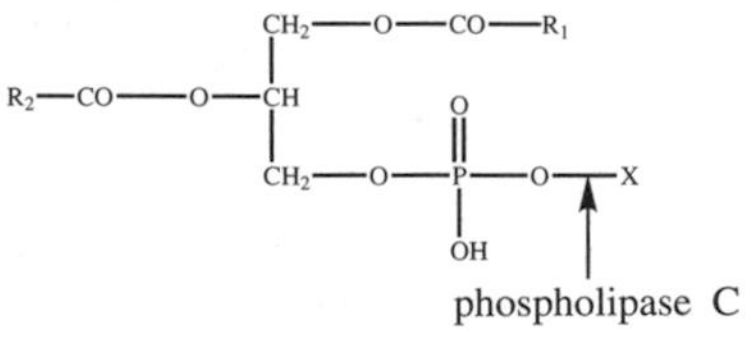

R_1 and R_2 = fatty acid

X = an organic molecule

Phospholipase L_1 The enzyme that catalyzes the removal of phosphate groups from the product formed by the action of phospholipase A_1.

Phospholipase L_2 The enzyme that catalyzes the removal of phosphate groups from the product formed by the action of phospholipase A_2.

Phospholipid The major lipid component of the biological membrane. It consists of glycerol, fatty acid, phosphate, and an organic component, e.g., choline, ethanolamine, inositol, of sphingosine.

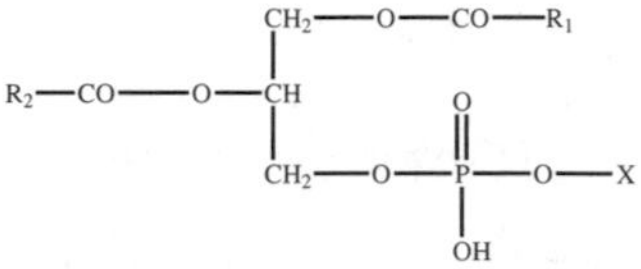

R_1 and R_2 = fatty acid

X = an organic molecule

Phospholipid Bilayer A basic structure of the biological membrane formed from phospholipids by the hydrophobic interaction.

Phospholipid Cholesterol Acyltransferase Synonym of phosphatidylcholine sterol O-acyltransferase.

Phospholipid Transfer Protein The cytoplasmic protein that recognizes a specific kind of phospholipid and mediates the transfer of that phospholipid from one membrane to another.

Phospholipid Translocator The membrane protein that catalyzes the flip-flop of the membrane lipids from one side of the lipid bilayer to another.

Phosphomannan Mannosephospho-Transferase The enzyme that catalyzes the following reaction:

$$\text{GDP-mannose} + (\text{phosphomannan})_n \rightleftharpoons \text{GMP} + (\text{phosphomannan})_{n+1}$$

Phosphomannose Isomerase The enzyme that catalyzes the following reaction:

Mannose 6-phosphate ⇌ Fructose 6-phosphate

Phosphomannose Mutase The enzyme that catalyzes the following reaction:

D-Mannose 1-phosphate ⇌ D-Mannose 6-phosphate

Phosphomonoesterase Synonym of alkaline phosphatase.

Phosphonecrosis Destruction of tissues caused by an excessive amount of phosphorus in the system.

Phosphonoformic Acid (mol wt 126) An antiviral agent that inhibits DNA polymerase activity.

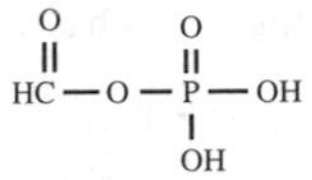

Phosphonolipid A phospholipid in which the phosphate group is linked to a carbon atom.

4′-Phosphopantothenic Acid (mol wt 298) A compound used for synthesis of CoA.

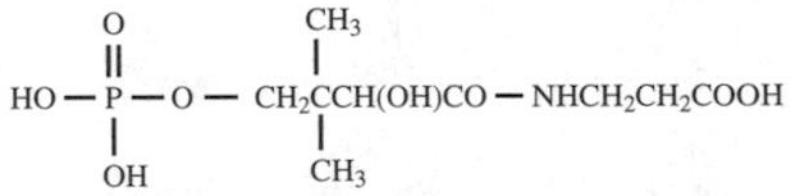

Phosphopantothenoylcysteine (mol wt 401) An intermediate in biosynthesis CoA from pantothenate.

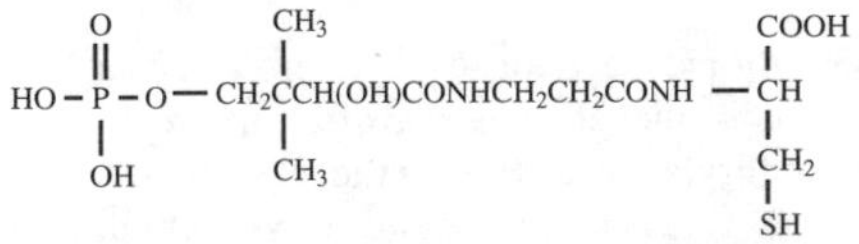

Phosphopantothenoylcysteine Decarboxylase The enzyme that catalyzes the following reaction:

Phosphopantothenolycysteine

⇅

4′-Phosphopantetheine + CO_2

Phosphopantothenolycysteine Synthetase The enzyme that catalyzes the following reaction:

4′-phosphopantothenate + ATP + cysteine

⇅

4′-Phosphopantothenoylcysteine + ADP + pi

Phosphopentokinase Synonym of phosphoribulokinase.

Phosphopentose Epimerase The enzyme that catalyzes the following reaction:

Xylulose-5-phosphate ⇌ Ribulose-5-phosphate

Phosphoprotein Proteins that contain phosphate groups.

Phosphoprotein Phosphatase The enzyme that catalyzes the following reaction:

A phosphoprotein + H_2O

⇅

A protein + orthophosphate

Phosphoprotein Phosphohydrolase The systematic name for phosphoprotein phosphatase.

Phosphopyruvate Hydratase The enzyme that catalyzes the following reaction:

2-Phospho-D-glycerate

⇅

Phosphoenolpyruvate + H_2O

Phosphor A substance used in the scintillation counter that emits a flash of light when excited by a radioactive radiation.

Phosphorescence The emission of light by a substance that has absorbed excited radiation, the light emitted is longer wavelength than the excited radiation and it continues for a noticeable time after the source of excited radiation has stopped.

Phosphoriboisomerase See phosphoribose isomerase.

Phosphoribose (mol wt 230) The phosphate form of ribose involved in the C_3 pathway of carbon dioxide assimilation.

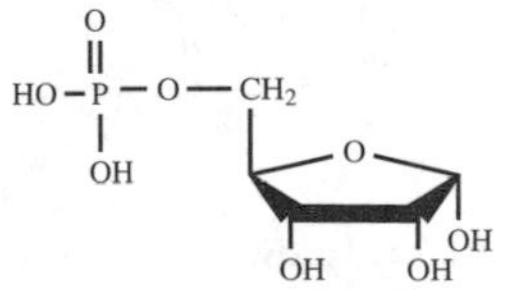

5-phospho-α,D-ribose

Phosphoribose Isomerase The enzyme that catalyzes the following reaction:

Ribose 5-phosphate ⇌ Ribulose 5-phosphate

Phosphoribose Kinase The enzyme that catalyzes the following reaction:

ATP + ribose 5-phosphate

⇅

ADP + ribose 1,5,-bisphosphate

Phosphoribosylamine (mol wt 229) An intermediate in the biosynthesis of purine nucleotides.

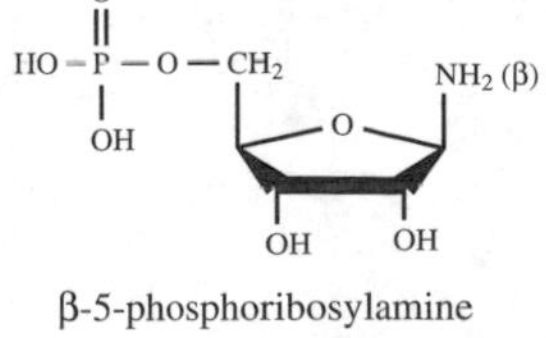

β-5-phosphoribosylamine

Phosphoribosylamine Glycine Ligase The enzyme that catalyzes the following reaction:

ATP + 5-phosphoribosylamine + glycine

⇅

5-Phosphoribosyl glycineamide + Pi + ADP

Phosphoribosylamine Synthetase The enzyme that catalyzes the following reaction:

$$ATP + \text{ribose 5-phosphate} + NH_3 \rightleftharpoons ADP + Pi + \text{5-phosphoribosylamine}$$

Phosphoribosyl-Pyrophosphate (PRPP) An intermediate in the biosynthesis of purine nucleotide.

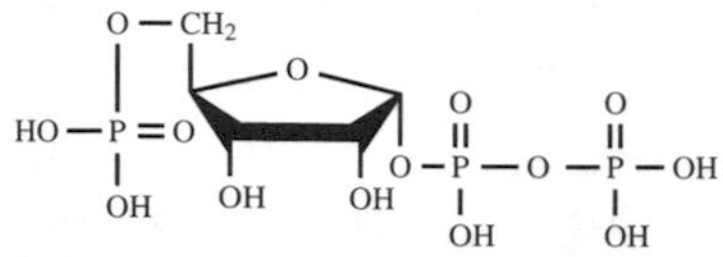

Phosphoribulokinase See phosphoribulose kinase.

Phosphoribulose (mol wt 230) The phosphate form of ribulose and an intermediate in the C_3 pathway of carbon dioxide assimilation.

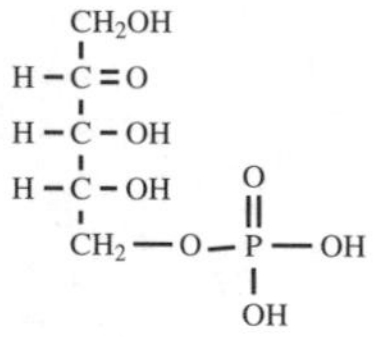

Phosphoribulose Epimerase The enzyme that catalyzes the following reaction:

$$\text{Ribulose 5-phosphate} \rightleftharpoons \text{Xylulose 5-phosphate}$$

Phosphoribulose Kinase The enzyme that catalyzes the following reaction:

$$\text{Ribulose 5-phosphate} + ATP \rightleftharpoons \text{Ribulose 1-5-bisphosphate} + ADP$$

Phosphoric Acid (mol wt 98) A tribasic acid (also known as orthophosphoric acid).

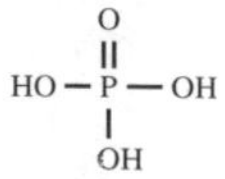

Phosphorimetry The use of phosphorescence in chemical analysis.

Phosphoroclastic Split The cleavage of a molecule by the addition of an inorganic phosphate such as:

$$Pi + \text{pyruvate} \rightleftharpoons \text{Formate} + \text{acetyl phosphate}$$

Phosphorus (P) A chemical element with atomic weight 31, valence 3 and 5.

Phosphorus-31 The stable nuclide of phosphorus.

Phosphorus-32 An artificial radioactive nuclide (^{32}P), it emits β radiation with a half life of 14 days.

Phosphorus-33 An artificial radioactive nuclide (^{33}P), it emits β radiation with a half life of 125 days.

Phosphorylase The enzyme that catalyzes the following reaction:.

$$Pi + (\text{glycogen})_n \rightleftharpoons \text{Glucose 1-phosphate} + (\text{glycogen})_{n-1}$$

Phosphorylase *a* The phosphorylated form of phosphorylase, a more active form of phosphorylase.

Phosphorylase *b* The dephosphorylated form of phosphorylase, a less active form of phosphorylase.

Phosphorylase Kinase The enzyme that catalyzes the phosphorylation of serine residues of phosphorylase, converting the less active form of phosphorylase *b* to the more active form of phosphorylase *a*.

Phosphorylase Phosphatase The enzyme that catalyzes the dephosphorylation of phosphorylase *a*.

Phosphorylating Transport A process of inward transfer of a sugar into a bacterial cell in which phosphorylation of sugar to be transported is an integrating part of the uptake mechanism (see phosphotransferase system).

Phosphorylation A reaction in which phosphate is added to a compound, e.g., the formation of ATP from ADP and inorganic phosphate.

Phosphorylation Potential The membrane potential generated from transfer of electrons or other mechanisms for phosphorylation reactions.

Phosphorylation/Dephosphorylation The activation or inactivation of an enzyme by the addition of a phosphate group to the enzyme or removal of a phosphate group from the enzyme.

Phosphorylcholine (mol wt 220) A substance used for treatment of hepatobillary dysfunction.

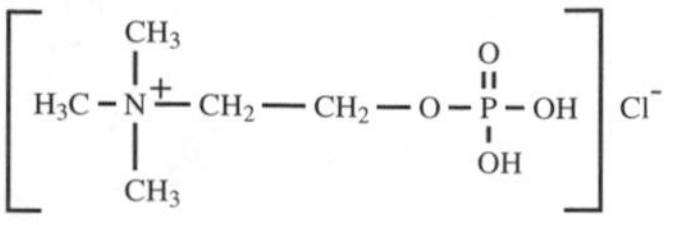

Phosphoryn Major phosphate-rich protein in dentin.

Phosphoserine (mol wt 185) A substance used in combination with glutamate and vitamin B_{12} as a roborant.

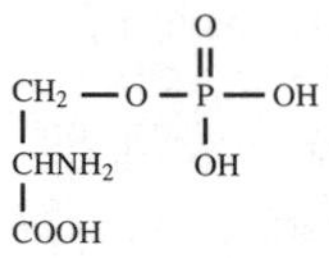

Phosphoserine Phosphatase The enzyme that catalyzes the following reaction:

$$\text{Phosphoserine} + H_2O \rightleftharpoons \text{Serine} + \text{Pi}$$

Phosphoserine Transaminase The enzyme that catalyzes the following reaction:

Phosphoserine + α-keto-glutarate

⇅

3-Phosphonoxypyruvate + glutamate

Phosphosphingolipid A sphingolipid that contains a phosphate group.

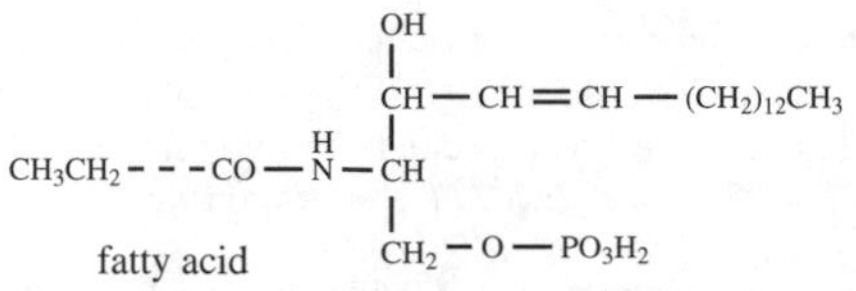

Phosphotransacetylase The enzyme that catalyzes the following reaction:

Acetyl-CoA + Pi

⇅

CoA + acetyl phosphate

Phosphotransferase System A type of sugar transport system in which phosphorylation of sugar to be transported occurs as it passes through the membrane of a bacterium (see also phosphorylating transport).

Phosphotriose Isomerase The enzyme that catalyzes the following reaction:

Glyceraldehyde 3-phosphate

⇅

Glycerone phosphate

Phosphotyrosine Phosphatase Synonym of protein-tyrosine phosphatase.

Phosphovitin A phosphoprotein of egg-yolk in oviparous vertebrates.

Phosvitin A phosphoglycoprotein from egg yolk and an anticoagulant.

Photinus-Luciferin 4-Monooxygenase (ATP-Hydrolyzing) The enzyme that catalyzes the following reaction:

Photinus luciferin + O_2 + ATP

⇅

Oxidized Photinus luciferin + CO_2 + H_2O + AMP + pyrophosphate + hv

Photinus-Luciferin Oxygen 4-Oxidoreductase Synonym of photinus-luciferin 4-monooxygenase (ATP-hydrolyzing).

Photoactivated Cross-Linking A light-mediated cross-linking between two compounds or components, e.g., photoaffinity labeling or light-induced cross-linking of aminoacyl-tRNA synthetase to its cognate tRNA.

Photoactivating Enzyme The enzyme that catalyzes the removal of thymine dimer in DNA. It splits the thymine dimer upon absorption of 300 to 600 nm wavelength of light.

Photoaffinity Labeling A light-induced covalent linking of a photoreactive ligand to a compound, e.g., linking of a substrate that carries a photoreactive label to an enzyme by exposing photoactively labeled substrate and the enzyme to light to form a covalently linked substrate-enzyme complex for investigation of the active site on the enzyme.

Photoautotroph An organism that is capable of utilizing energy from light and carbon dioxide as carbon source.

Photobacterium A genus of Gram-negative bacteria of the family Vibrionaceae.

Photobiont A photosynthetic symbiont.

Photobleaching The loss of color or fluorescence through the action of visible or near UV-radiation.

Photocell A photoelectric device that responds to changes in the intensity of light.

Photochemical Action Spectrum A plot of photochemical action as a function of wavelength.

Photochemical Reaction Chemical reaction caused or initiated by light.

Photochemical Reaction Center The region of a photosynthetic complex in phototrophic cells where the energy of the absorbed photon causes charge separation and initiates transfer of electrons.

Photochemical Reduction Transfer of photoexcited electrons from one molecule to another.

Photochemistry The chemical science that deals with the interaction of radiant energy and chemical processes.

Photochrome A molecule or a group of atoms in a molecule that is capable of producing photochromism.

Photochromism The change of color by a molecule upon absorption of light (e.g., infrared or UV light).

Photochromogen Referring to microorganisms or strains of a microorganism that require exposure to light for development of pigment.

Photodissociation Dissociation of a complex molecule caused by the absorption of light energy.

Photodynesis Light-induced cytoplasmic streaming.

Photoexcitation Excitation of an electron to a higher energy level caused by the absorption of a photon.

Photofootprinting Technique A technique used for investigating interactions between protein and DNA. The method is based on the principle that interaction of protein with DNA causes distortion of the configuration of the double-helix DNA and initiates a photoreaction at the site of contact.

Photoheterotroph An organism that is capable of utilizing energy from light but must obtain its carbon source from organic compounds.

Photolithotrophs Phototrophic organisms that use inorganic substances as electron donors in photosynthesis.

Photolyase Referring to the enzyme that catalyzes the split of thymine dimers in DNA (see also photoreactivating enzyme).

Photolysis Light-induced cleavage of a molecule, e.g., light-dependent oxidative splitting of water into oxygen and hydrogen ions and electrons.

Photomonas A genus of protozoa (family of Trypanosomatidae), a parasitic in certain plants.

Photomorphogenesis The light-induced or light-dependent development or formation of new cells, tissues, or organs.

Photon A unit of light energy. The energy of a photon is given by $E=hv$, where E is photon energy, v is frequency of the radiation, and h is Planck's constant (6.625×10^{-27} ergs).

Photoorganotroph Organisms that are capable of utilizing energy from light but must use organic compounds as carbon source.

Photophore A light-emitting organ in animals.

Photophosphorylation The enzymatic process for generation of ATP through a protonmotive force generated from electron transport in the thylakoid membrane of the chloroplast during photosynthesis.

Photopolymerization A process of light-induced polymerization, e.g., light-induced formation of polyacrylamide gel from the monomer solutions.

Photoreactivating Enzymes The enzymes that binds to the thymine dimer on DNA causing cleavage of thymine dimer upon exposure to light of wavelength 300 to 600 nm.

Photoreactivation A mechanism of DNA repair in which the thymine dimers induced by UV-radiation are cleaved and restored to their normal monomeric forms by photoreactivating enzyme upon exposure to radiation of wavelength 300 to 600 nm (see also photoreactivating enzyme).

Photoreceptor A sense structure or region of a structure that responds to light stimulus.

Photoreduction Light-dependent process for the generation of NADPH by the transfer of energized electrons from photo-excited chlorophyll molecules to $NADP^+$ via a series of electron carriers.

Photorespiration A pathway in which ribulose 1,5-bisphosphate is oxidized to phosphoglycolate in the chloroplast and then transported to peroxisomes where it is reoxidized to glyoxylate and H_2O.

Photorestoration Synonym for photoreactivation.

Photosensitization A process in which a photochemical reaction is induced to occur by the presence of a photosensitizing reagent.

Photosensitizer 1. A substance that is capable of absorbing and transferring absorbed light energy to facilitate a photochemical reaction. 2. Any substance that causes photosensitization.

Photosensor A photoreceptor that initiates a reaction upon receiving light stimuli.

Photosynthate Referring to carbohydrate produced from photosynthesis.

Photosynthesis The process in which light energy is absorbed by the photosynthetic pigments (e.g., chlorophyll *a* and accessory pigments) and converted to chemical energy, ATP. The ATP resulting from the light reaction is subsequently used for conversion of CO_2 to carbohydrate via the C_3 pathway.

Photosynthetic Bacteria Bacteria that possess photosynthetic systems and are capable of carrying out photosynthesis.

Photosynthetic Phosphorylation The process by which light energy is utilized to produce ATP (phosphorylation).

Photosynthetic Quotient Referring to the ratio of number of moles of oxygen evolved to the number of moles of CO_2 taken up in photosynthesis.

Photosystem The light-harvesting unit in photosynthesis that consists of chlorophyll, accessory pigment, proteins, and electron carriers for absorption of light energy and generation of ATP and NADPH.

Photosystem I Photosynthetic system present in both prokaryotic and eukaryotic organisms in which light of longer wavelength is required for function. It is responsible for the cyclic photophosphorylation and generation of NADPH (also known as P_{700}).

Photosystem II One of the photosynthetic systems present in the eukaryotic organism in which light of shorter wavelength is required for function. It is involved in splitting of water, production of molecular oxygen, and supply of electrons for photosystem II (also known as P_{680}).

Phototaxis A type of taxis in which an organism moves toward (positive) or away from (negative) the stimulus of light.

Phototrophs Organisms that are capable of converting light energy to chemical energy.

Phototropism A tendency for an organism to migrate toward (positive) or away from (negative) the stimulus of light.

PhOx Abbreviation for phagocyte oxidase.

Phoxim (mol wt 298) An insecticide.

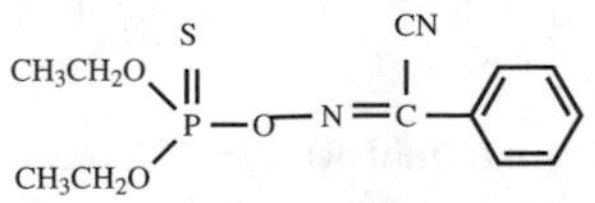

pHP34 A plasmid of *E. coli* that contains Amp^r (ampicillin resistant) and Tet^r (tetracycline resistant) markers and ClaI, HindIII, BamHI, SalI, PvuII, PstI, PvuI, and SphI cleavage sites.

Phragmoplast A cylindrical structure in the central region of a dividing plant cell formed by a parallel array of microtubules and involved in cell plate formation.

Phragmosome A region in the cytoplasm in the plant cell where the nucleus is located during nuclear division.

pHSG250 A plasmid of *E. coli* that contains colicin E-1 immunity and EcoRI and BamHI, cleavage sites.

pHSG262 A plasmid of *E. coli* that contains Kan^r (kanamycin resistant) marker and EcoRI, BamHI, and HincII cleavage sites.

pHSG429 A plasmid of *E. coli* that contains Kan^r (kanamycin resistant), Cml^r (chloramphenicol resistant) and Amp^r (ampicillin resistant) markers and PvuII, and HincII, cleavage sites.

pHSG439 A plasmid of *E. coli* that contains Cml^r (chloramphenicol resistant) marker and AvaI, BamHI, HincII, and EcoRI cleavage sites.

pHSREM1 A plasmid of *E. coli* that contains Amp^r (ampicillin resistant) marker and M13 promoter and PstI, HindIII, EcoRV, EcoRI, SmaI, HamHI, ClaI, and SalI cleavage sites.

Phthalofyne (mol wt 246) An anthelmintic agent.

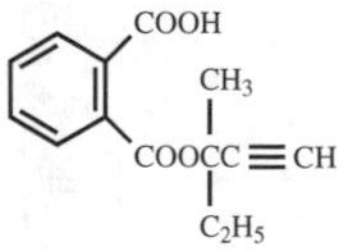

Phthalylsulfacetamide (mol wt 362) An antibacterial agent.

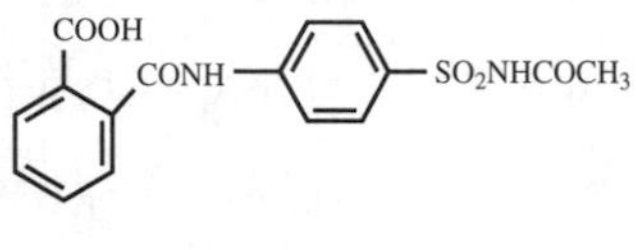

Phthalylsulfathiazole (mol wt 403) An antibacterial agent.

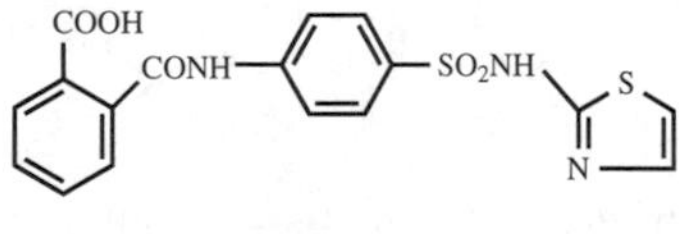

Phthiocol (mol wt 188) An antibiotic produced by *Mycobacterium tuberculosis*.

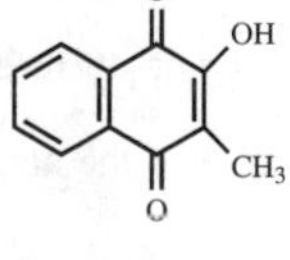

pHUB2 A plasmid of *E. coli* that contains Kan^r (kanamycin resistant), Tet^r (tetracycline resistant) markers and HpaI, EcoRI, BamHI, and SalI cleavage sites.

pHUB4 A plasmid of *E. coli* that contains Kan^r (kanamycin resistant) marker and BamHI and SalI cleavage sites.

Phyco- A prefix denoting seaweed or algae.

Phycobiliprotein Protein found in the phycobilisome that consists of many covalently attached billins as prosthetic groups.

Phycobilisomes Large protein assemblies on the outer face of the thylakoid membrane found in cyanobacteria and red algae that harvest green and yellow light.

Phycobillins Photosynthetic pigments found in certain algae and cyanobacteria.

Phycobiont An algae symbiont (e.g., algae partner of a lichen).

Phycocyanin A phycobiliprotein found in cyanobacteria and blue-green algae.

Phycocyanobilin A blue photosynthetic pigment in cyanobacteria.

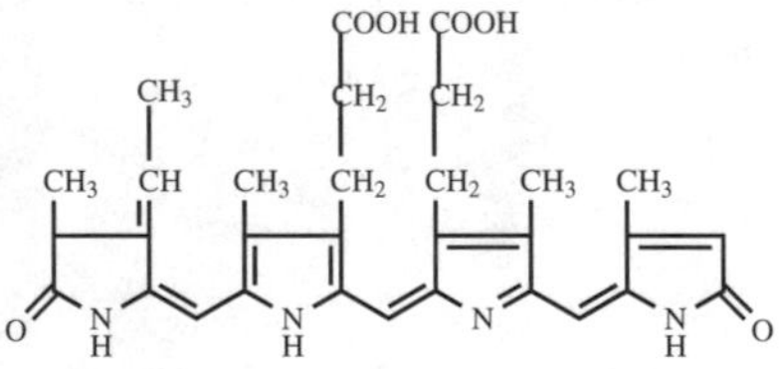

Phycoerythrin Type of pigment in cyanobacteria and red algae that confers red color.

Phycoerythrobilin Red photosynthetic pigment in red algae.

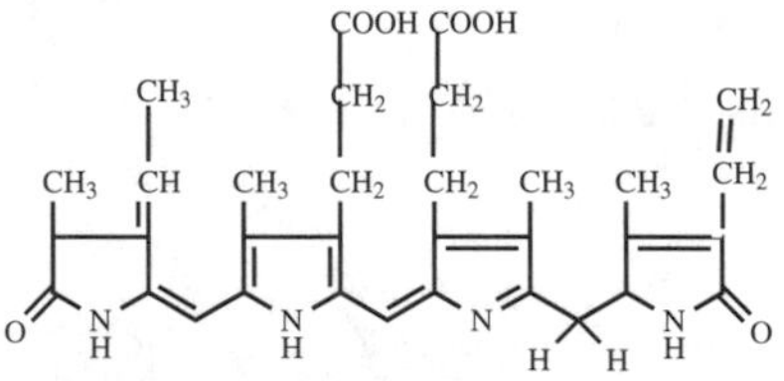

Phycoflour Any artificial conjugate of a phycobiliprotein with a molecule such as avidin or protein A to confer biological specificity.

Phycology The science that deals with algae.

Phycotoxin Any toxin produced by algae.

Phycovirus Viruses of the algae or viruses that infect algae.

Phylaxis Protection against infection.

Phyllobacterium A genus of Gram-negative bacteria of Rhizobiaceae.

Phyllocontin A trade name for aminophylline, used as a bronchodilator.

Phylloquinone (mol wt 451) An electron carrier in the photosystem I (also known as vitamin K_1).

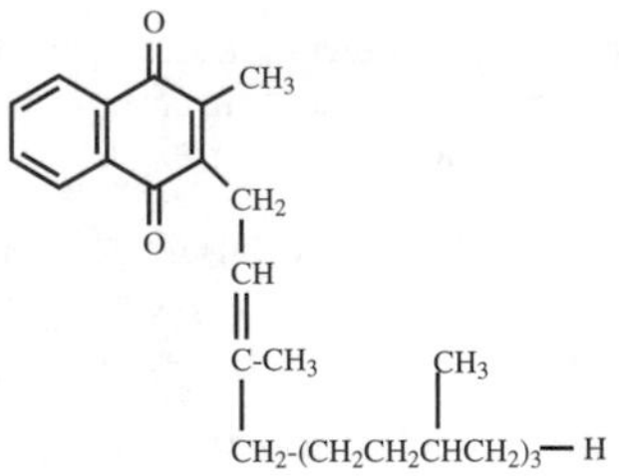

Phylloquinone Reductase Synonym of NADPH dehydrogenase (quinone).

Phylogenetic Referring to the classification system based on the evolutionary relationships of organisms.

Phylogeny The evolutionary history of a given species of an organism.

Phylum A major taxonomic category above class in classification systems of animals and plants.

Physalaemin A peptide with a capability of stimulating salivary secretion, intestinal contraction, and vasodilation.

Physaropepsin The enzyme that catalyzes the preferential cleavage of peptide bonds in the B chain of insulin: Gly^8-Ser; Leu^{11}-Val; Phe^{24}-Phe. The enzyme also possesses milk clotting activity.

Physeptone A trade name for methadone hydrochloride, used as an opioid analgesic agent.

Physical Biochemistry The science that deals with physical aspects of cellular activity and biochemical reactions.

Physiological Biophysics The science that deals with the use of physical mechanisms to explain the behavior and the functioning of living systems.

Physiological Saline Referring to 0.85% of NaCl (w/v).

Physiology The science that deals with life activities of living systems.

Physostigmine (mol wt 275) An acetylcholine inhibitor derived from Calabar bean.

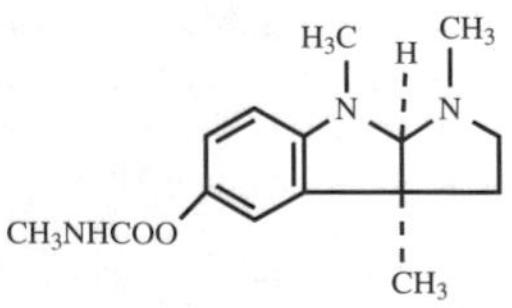

Phytanate-CoA Ligase The enzyme that catalyzes the following reaction:

$$\text{ATP} + \text{phytanate} + \text{CoA} \rightleftharpoons \text{AMP} + \text{PPi} + \text{phytanoyl-CoA}$$

Phytanic Acid (mol wt 292) A degradation product of phytol.

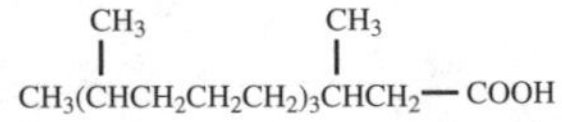

Phytanic Acid Storage Disease A genetic disorder characterized by progressive neurological difficulties (e.g., tremors and poor night vision) due to the accumulation of phytanic acid.

Phytase The enzyme that catalyzes the hydrolysis of phytic acid to inositol and inorganic phosphate.

Phytic Acid (mol wt 660) An inositolhexaphosphoric acid and an important compound for phosphate storage in the plant.

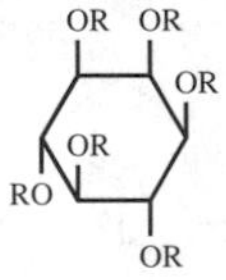

$R=PO_3H_2$

Phytin A mixture of calcium–magnesium salts of phytic acid.

Phyto- A prefix meaning plant.

Phytoalexins Substances (e.g., stress proteins, phenolic compounds) produced by higher plants in response to infection or other stresses (also known as plant antibiotics). Phytoalexins possess nonspecific antifungal and antibacterial activity.

Phytochromes Protein pigments from plants that mediate physiological responses and govern light-sensitive processes, e.g., photoperiodic control of flowering.

Phytoestrogen Substances from plants with estrogenic activity.

Phytohemagglutinin A carbohydrate-binding protein from red kidney bean that is capable of agglutinating red blood cells. It is also a T-cell mitogen.

Phytol (mol wt 297) A highly hydrophobic 20-carbon alcohol present in chlorophyll.

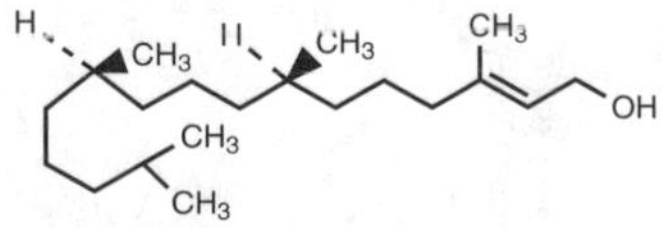

Phytomitogens Mitogens derived from plants (e.g., lectins) that stimulate DNA synthesis and the proliferation of lymphocytes.

Phytoncide Substance produced by a plant that confers resistance to disease or infection.

Phytone Products resulting from papain digestion of plant materials, e.g., soybean meal used as a medium.

Phytoreovirus Plant viruses that contain double-stranded RNA.

Phytosterol Referring to sterols obtained from higher plants.

pI Abbreviation for isoelectric point of an amphoteric substance, e.g., protein or amino acid.

Pi Abbreviation for inorganic phosphate.

PI Abbreviation for 1. phosphatidyl inositol; 2. phospho-inositide; 3. propidium iodide; 4. protamine insulin.

pi (π) 1. A Greek letter. 2. A mathematical constant to express the ratio of circumference to diameter of a circle (equal to 3.1416).

pi Bond A chemical bond formed by electrons from the pi orbital.

pi Electron An electron in the pi orbital.

PI Kinase Abbreviation for phosphatidylinositol kinase, the enzyme that catalyzes the formation of phosphatidylinositol phosphate from phosphatidylinositol.

***pi* Protein (π Protein)** A protein required for the initiation of DNA replication in an antibiotic-resistance plasmid.

PIA Abbreviation for plasma insulin activity.

Piberaline (mol wt 281) An antidepressant.

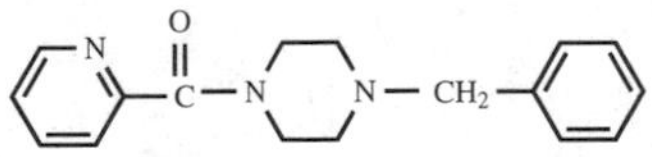

PIC Abbreviation for 1. pre-initiation complex; 2. phosphoinositidase C.

pIC7 A plasmid of *E. coli* that contains Amp^r (ampicillin resistant) marker and EcoRI, ClaI, EcoRV, XbaI, BglII XhoI, SacI, NruI, and HindIII cleavage sites.

pIC19H A plasmid of *E. coli* that contains Amp^r (ampicillin resistant) marker and HindIII, PstI, SalI, BamHII, SmaI, EcoRI, ClaI, EcoRV, XbaI, BglII, XhoI, SacI, NruI, and NarI cleavage sites.

pIC20H A plasmid of *E. coli* that contains Amp^r (ampicillin resistant) marker and ClaI, EcoRV, BglII, XhoI, SacI, NruI, HindIII, SphI, PstI, SalI, BamHI, SmaI, and KpnI cleavage sites.

Picadex (mol wt 162) An anthelmintic agent.

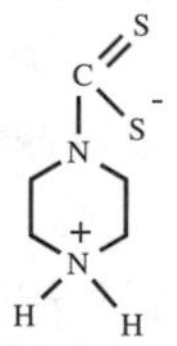

pICEM19H⁺ A plasmid of *E. coli* that contains Amp^r^ (ampicillin resistant) marker and HindIII, PstI, SalI, BamHI, SmaI, EcoRI, EcoRV, XbaI, BglII, XhoI, SacI, NruI, and NaeI cleavage sites.

pICEM19H⁻ A plasmid of *E. coli* that contains Amp^r^ (ampicillin resistant) marker and NaeI, HindIII, PstI, SalI, BamHI, SmaI, EcoRI, EcoRV, XbaI, BglII, XhoI, SacI, and NruI cleavage sites.

pICEM19R⁺ A plasmid of *E. coli* that contains Amp^r^ (ampicillin resistant) marker and EcoRI, EcoRV, XbaI, BglII, XhoI, SacI, NruI, HindIII, PstI, SalI, BamHI, SmaI, and NaeI cleavage sites.

pICEM19R⁻ A plasmid of *E. coli* that contains Amp^r^ (ampicillin resistant) marker and EcoRI, EcoRV, XbaI, BglII, XhoI, SacI, NruI, HindIII, PstI, SalI, BamHI, SmaI, and NaeI cleavage sites.

Picloram (mol wt 241) A herbicide.

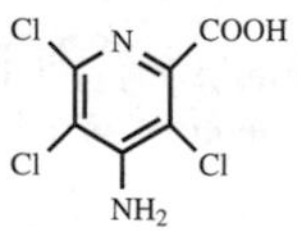

Picilorex (mol wt 236) An anorexic agent.

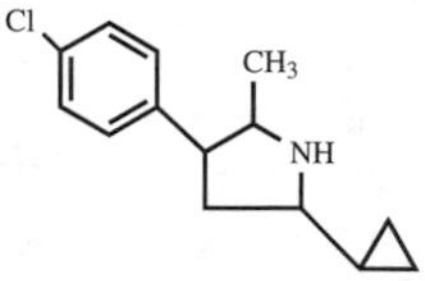

Picloxydine (mol wt 475) A topical antibacterial agent.

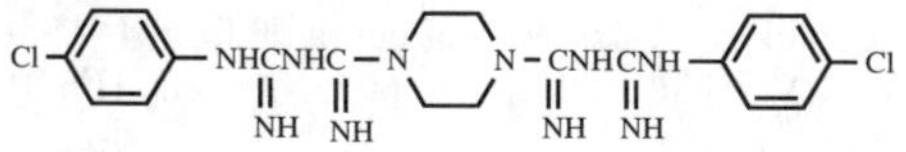

Pico- A prefix meaning 10^{-12} or small.

Picogram 10^{-12} gram.

Picolinic Acid (mol wt 123) An isomer of nicotinic acid.

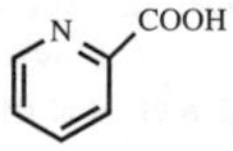

Picomole 10^{-12} mole.

Picoperine (mol wt 295) An antitussive agent.

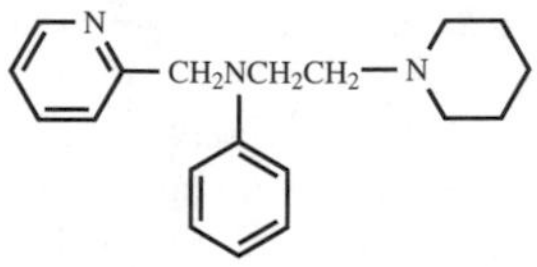

Picornain A protease from picornaviruses.

Picornaviruses Small, naked, icosahydral, RNA-containing animal viruses of the family Picornaviridae, e.g., poliovirus.

Picosulfate Sodium (mol wt 481) A cathartic agent.

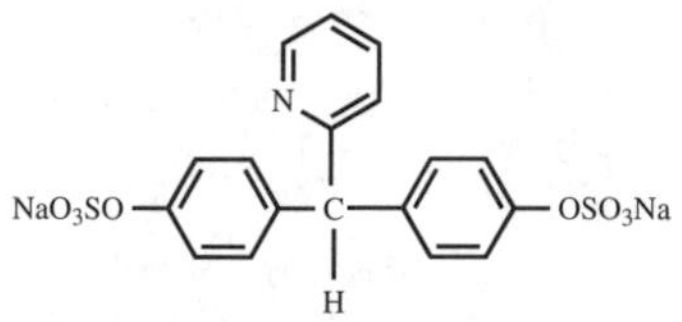

Picotamide (mol wt 376) An anticoagulant, antithrombotic, and fibrinolytic agent.

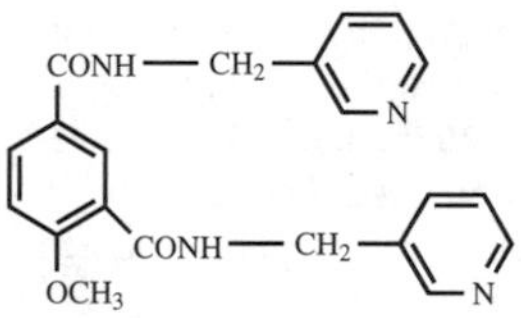

Picric Acid (mol wt 229) An antimicrobial agent.

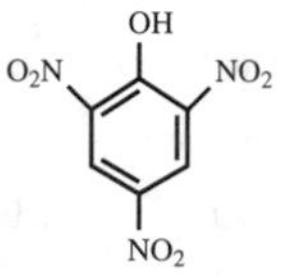

Picromycin (mol wt 526) An antibiotic isolated from species of *Actinomyces*.

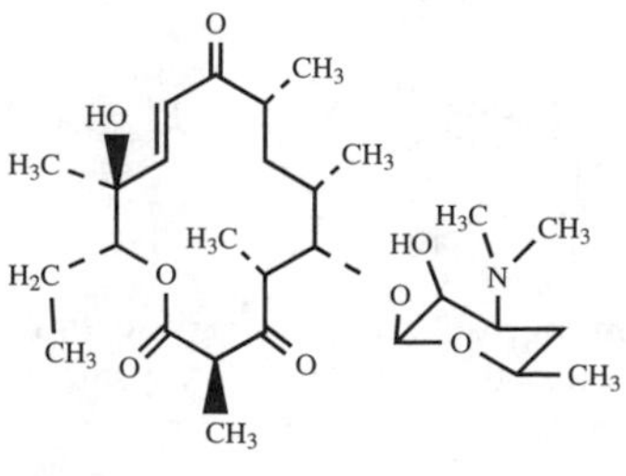

Picrotoxin (mol wt 602) A CNS stimulant from the seed of *Anamirta cocculus* that consists of picrotoxinin and picrotin.

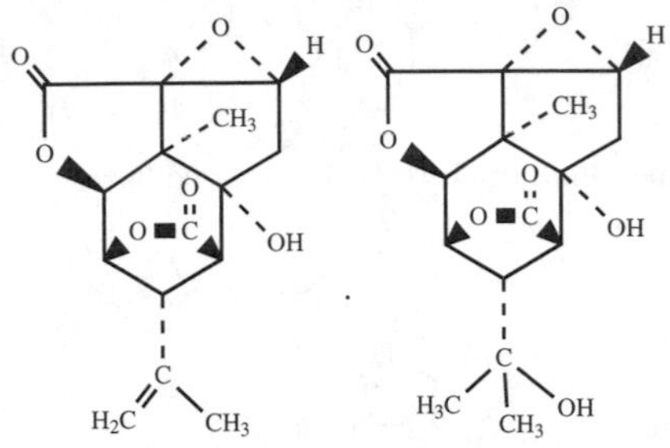

PICROTOXININ PICROTIN

Picrotoxinin (mol wt 292) A component of picrotoxin.

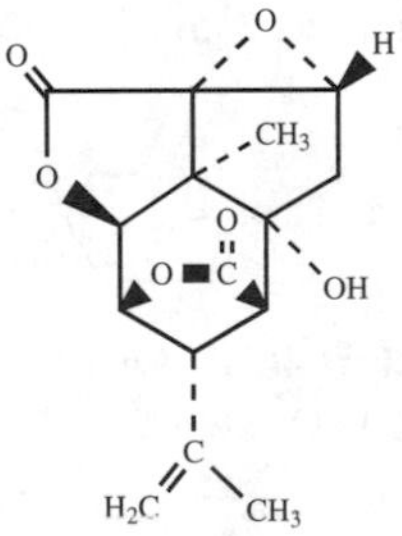

Picumast (mol wt 441) An antiallergic agent.

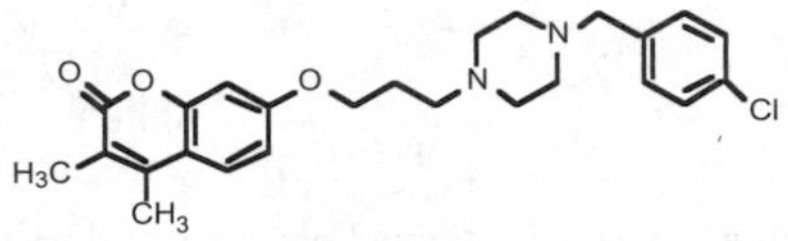

PIF Abbreviation for 1. prolactin-inhibiting factor; 2. proliferation-inhibiting factor.

Pifarnine (mol wt 425) An antiulcerative agent.

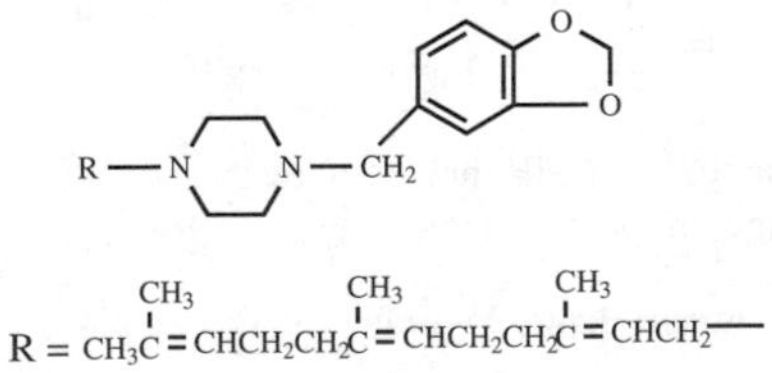

Pifoxime (mol wt 276) An anti-inflammatory agent.

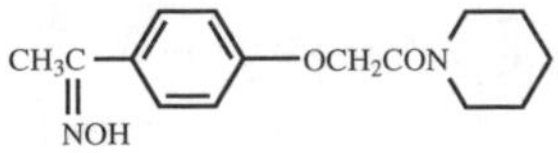

PIFT Abbreviation for platelet immunofluorescence test.

PIG Abbreviation for phosphatidylinositol glycan.

Pigmented Retinal Epithelium A layer of phagocytic epithelium cells lying below the photoreceptor of the vertebrate eye.

PIG-Tailed Protein Proteins that are anchored to a membrane by the linkage to phosphatidylinositol glycan.

PIH Abbreviation for prolactin inhibitory hormone.

PII Abbreviation for plasma inorganic iodine.

PIIF Abbreviation for proteinase-inhibitor inducing factor. A factor produced by a plant in response to insect attack.

Piketoprofen (mol wt 344) A topical anti-inflammatory agent.

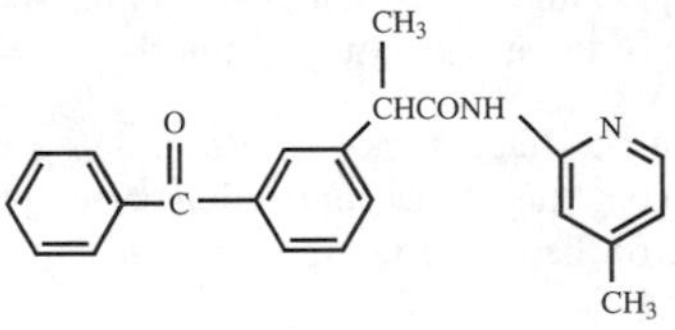

Pildralazine (mol wt 197) An antihypertensive agent.

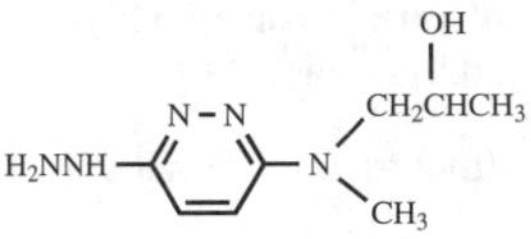

Pili Plural of pilus.

Pilimelia A genus of keratinophilic bacteria (Actinomycetales).

Pilin Referring to a protein subunit of pilus.

Pilocar A trade name for pilocarpine hydrochloride, a miotic agent that causes contraction of the sphincter muscle of the iris resulting in miosis.

Pilocarpine Hydrochloride (mol wt 245) A cholinergic agent that causes contraction of the sphincter muscle of the iris, resulting in miosis.

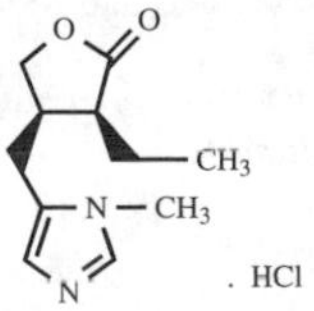

Pilocarpine Nitrate (mol wt 271) A cholinergic agent that causes contraction of the sphincter muscle of the iris resulting in miosis.

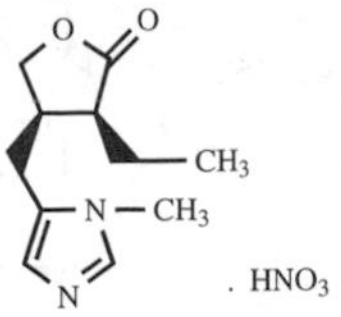

Pilocel A trade name for pilocarpine hydrochloride, a miotic agent that causes contraction of the sphincter muscle of the iris resulting in miosis.

Pilomiotin A trade name for pilocarpine hydrochloride, a miotic agent that causes contraction of the sphincter muscle of the iris resulting in miosis.

Pilopine HS A trade name for pilocarpine hydrochloride, a miotic agent that causes contraction of the sphincter muscle of the iris resulting in miosis.

Pilopt A trade name for pilocarpine hydrochloride, a miotic agent that causes contraction of the sphincter muscle of the iris resulting in miosis.

Piloptic A trade name for pilocarpine, a cholinergic agent that causes contraction of the sphincter muscle of the iris resulting in miosis.

Pilostat A trade name for pilocarpine, a cholinergic agent that causes contraction of the sphincter muscle of the iris resulting in miosis.

Pilot Protein A protein that plays an important role for 1. transfer of DNA from donor cell to recipient cell during bacterial conjugation, and 2. attachment of certain bacteriophage virions onto the host cell during infection.

Pilsicainide (mol wt 72) An antiarrhythmic agent.

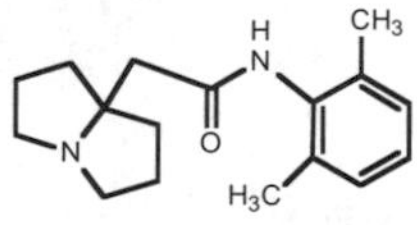

Pilus Long, hairlike projection of an F^+ (male) bacterial cell that facilitates the transfer of DNA during conjugation between the F^+ cell and an F^- (female) cell.

PIM Abbreviation for phosphatidyl-*myo*-inositol mannoside.

Pima A trade name for potassium iodide, used to inhibit thyroid hormone formation by blocking the synthesis of iodotyrosine and iodohydronine.

Pimaricin (mol wt 666) An antibiotic produced by *Streptomyces natalensis*.

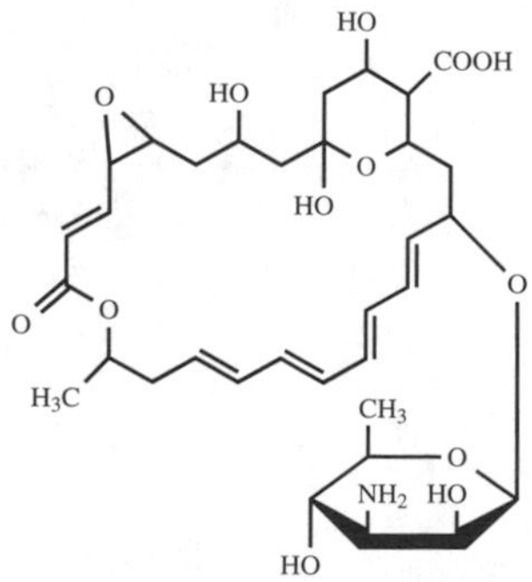

Pimeclone (mol wt 195) A respiratory stimulant.

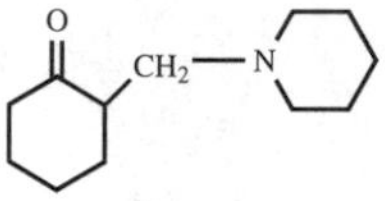

Pimefylline (mol wt 314) A vasodilator.

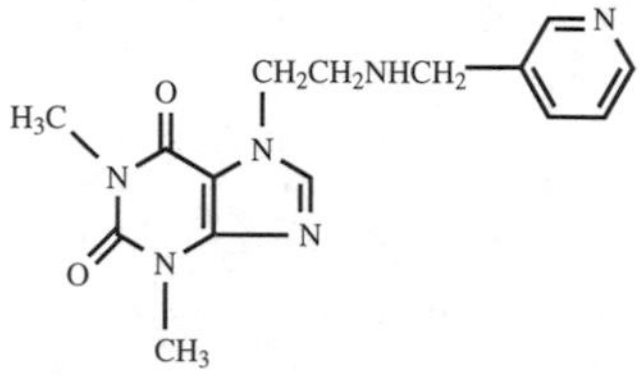

Pimeluria The presence of fat in the urine.

Piminodine (mol wt 366) A narcotic analgesic agent.

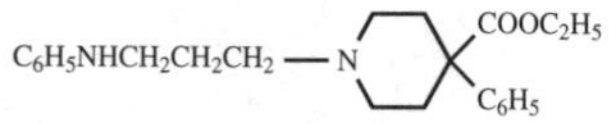

Pimozide (mol wt 462) An antipsychotic agent that blocks dopaminergic receptors.

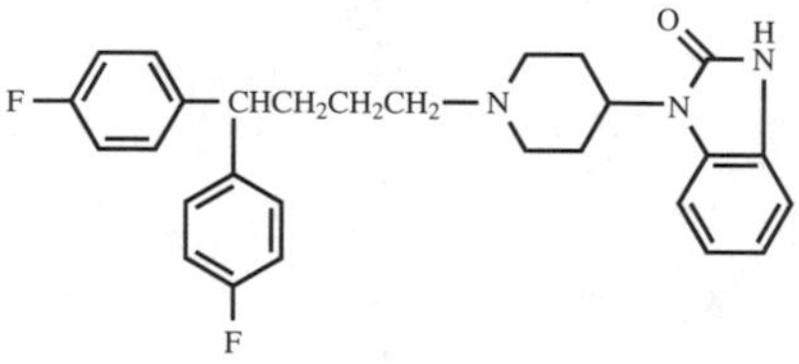

Pimple A small inflamed swelling on the skin that contains pus.

Pinacidil (mol wt 263) An antihypertensive agent.

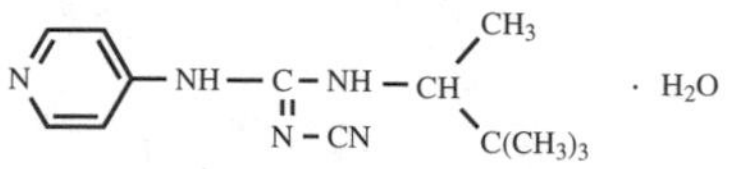

Pinacocyte Cells that form the surface layer of a sponge and synthesize collagen.

Pinaverium Bromide (mol wt 591) A spasmolytic agent.

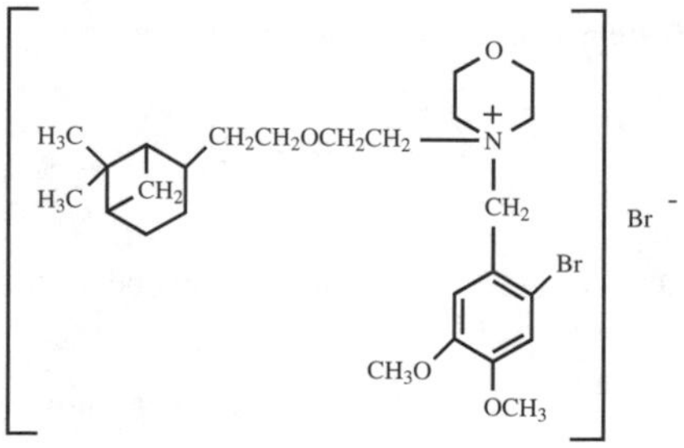

Pindolol (mol wt 248) An antihypertensive, antianginal, antiarrhythmic, and antiglaucoma agent.

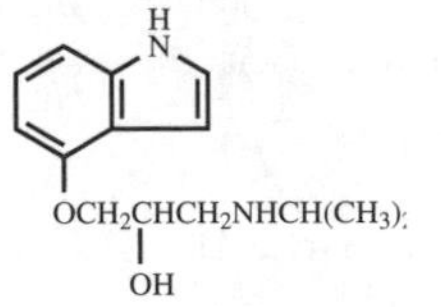

Pineal Gland A small gland in the brain that secrets melatonin and vasotocin.

Pinealoma A tumor of the pineal gland.

Ping-Pong Mechanism A mechanism for enzymatic reactions in which the first substrate binds to the enzyme resulting in the production of the first product and modified form of the enzyme. The binding of second substrate to the modified form of enzyme produces a second product and conversion of the modified form of the enzyme back to the original form.

Pinocytosis A type of endocytosis in which droplets of soluble materials are taken up from the environment and incorporated into pinocytotic vesicles for digestion, also known as cell drinking.

Pinocytotic Vesicle Fluid-filled endocytotic vesicle.

Pinosome See pinocytotic vesicle.

Pinosylvin (mol wt 212) An antifungal agent.

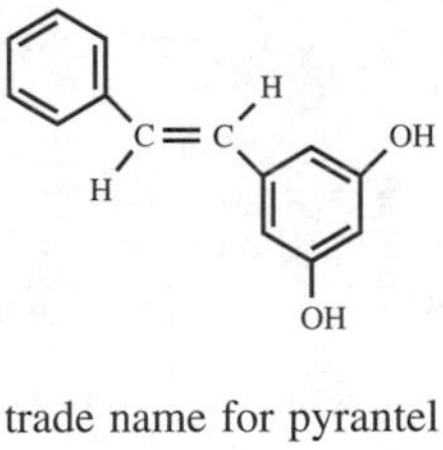

Pin-Rid A trade name for pyrantel pamoate, an antihelmintic agent.

Pinworm A parasitic nematode worm of the genus *Enterobius* that lives in the upper part of the large intestine.

Pin-X A trade name for pyrantel pamoate, an antihelmintic agent.

Pioglitazone (mol wt 356) An antidiabetic agent.

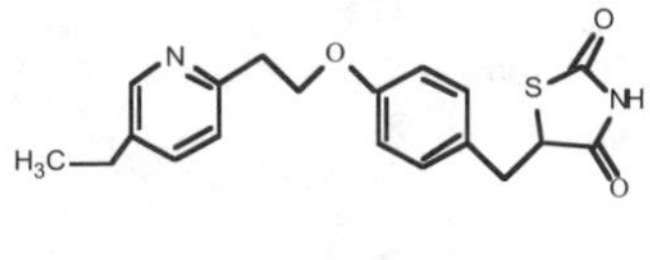

PIP Abbreviation for phosphatidylinositol phosphate.

PIP_2 Abbreviation for phosphatidylinositol 4,5-bisphosphate.

PI(3)P Abbreviation for phosphatidylinositol 3-phosphate.

PI(4)P Abbreviation for phosphatidylinositol 4-phosphate.

PI(3,4)P_2 Abbreviation for phosphatidylinositol 3,4-bisphosphate.

PI(3,4,5)P_3 Abbreviation for phosphatidylinositol 3,4,5-triphosphate.

Pipamazine (mol wt 402) An antiemetic agent.

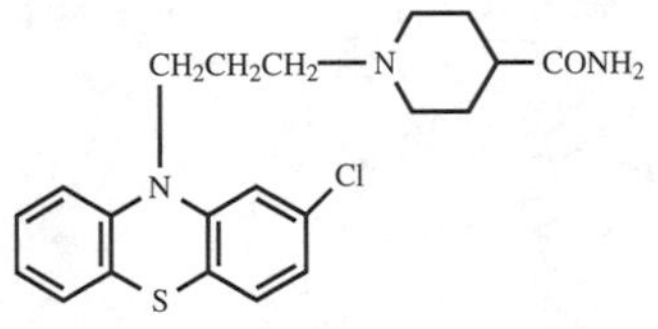

Pipazethate (mol wt 400) An antitussive agent.

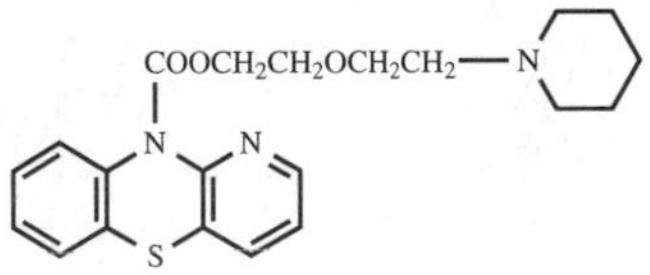

Pipebuzone (mol wt 421) An anti-inflammatory, antipyretic, and analgesic agent.

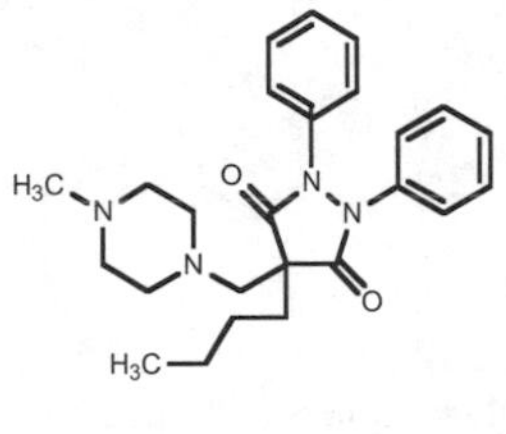

Pipecurium Bromide (mol wt 788) A skeletal muscle relaxant that competes with acetylcholine for receptor sites at the motor end plate.

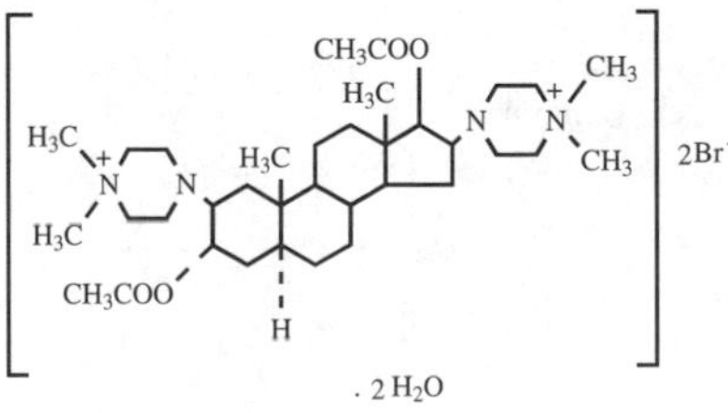

Pipemidic Acid (mol wt 303) An antibacterial agent.

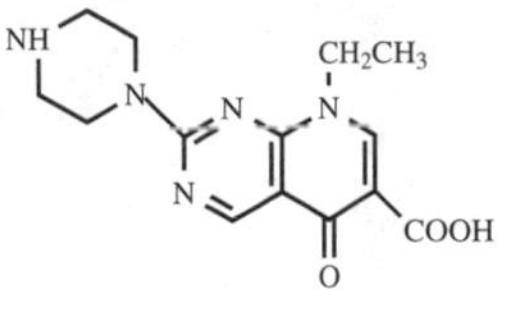

Pipenzolate Bromide (mol wt 434) An anticholinergic agent.

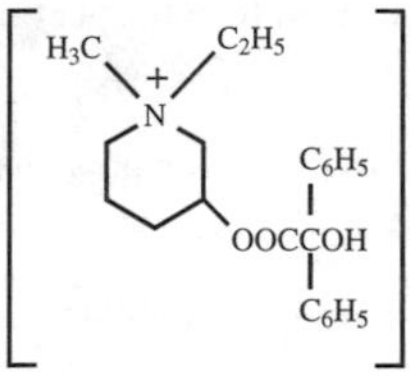

Piperacetazine (mol wt 411) A tranquilizer.

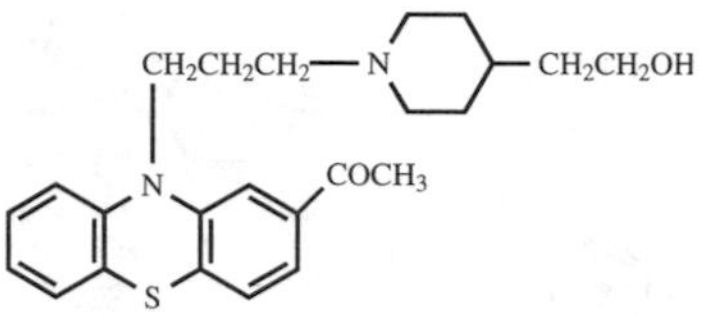

Piperacillin (mol wt 518) A broad spectrum semisynthetic antibiotic that inhibits bacterial cell wall synthesis.

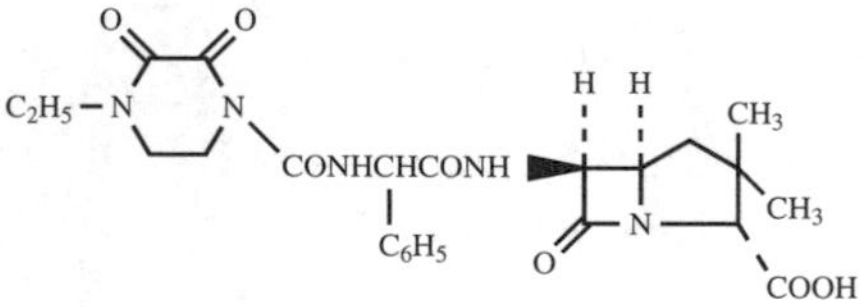

Piperazine (mol wt 86) An anthelmintic agent that blocks neuromuscular action.

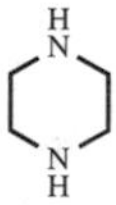

Piperidine (mol wt 85) A reagent used in Maxam Gilbert's DNA sequencing procedure.

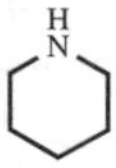

Piperidione (mol wt 169) A sedative and an antitussive agent.

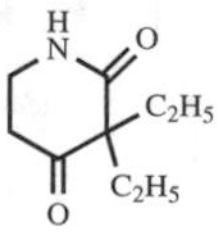

Piperidolate (mol wt 323) An anticholinergic agent.

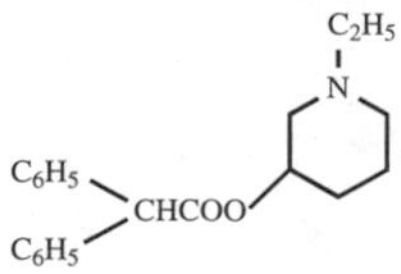

Piperilate (mol wt 339) An anticholinergic and antispasmodic agent.

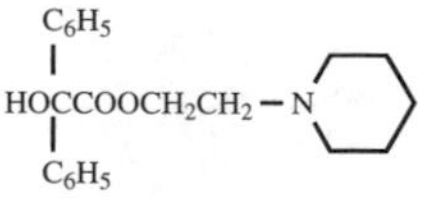

Piperocaine (mol wt 261) A local anesthetic agent.

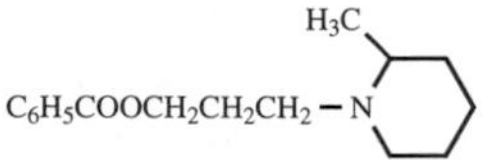

Piperoxan (mol wt 233) An antihypertensive agent.

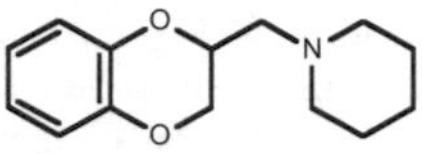

Piperylone (mol wt 285) An analgesic agent.

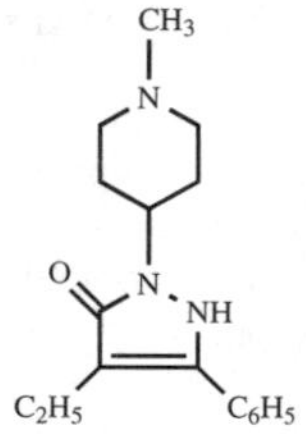

PIPES (mol wt 302) Abbreviation For 1,4-piperazinediethanesulfonic acid. A reagent used for preparation of biological buffers.

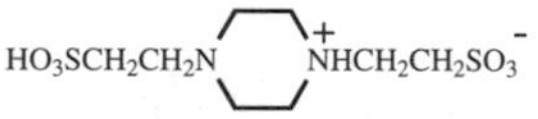

PIP-Kinase The enzyme that catalyzes the formation of phosphatidylinositol 4,5-triphosphate from phosphatidylinositol phosphate.

PI-PLC Abbreviation for phosphatidylinositol specific phospholipase C.

Piposulfan (mol wt 386) An antineoplastic agent.

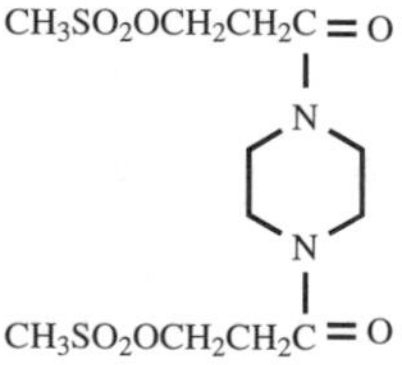

Pipotiazine (mol wt 476) An antipsychotic agent.

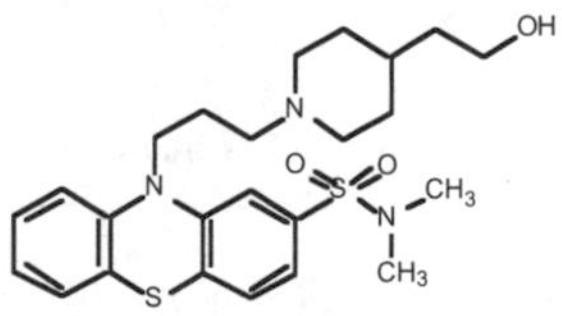

Pipoxolan Hydrochloride (mol wt 388) An antispasmodic agent.

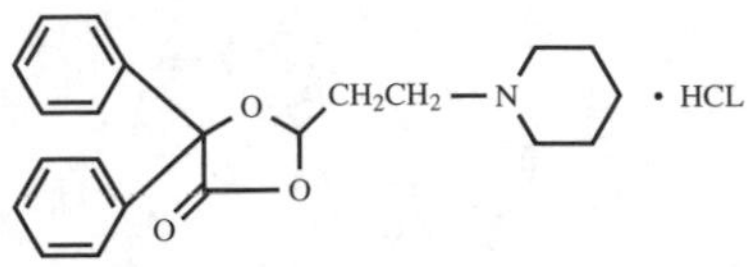

PIP-Phosphomonoesterase The enzyme that catalyzes the dephosphorylation of PIP.

PIP_2-Phosphomonoesterase The enzyme that catalyzes the dephosphorylation of PIP_2.

Pipracil A trade name for piperacillin sodium, an antibiotic that inhibits bacterial cell wall synthesis.

Pipradrol (mol wt 267) A CNS stimulant.

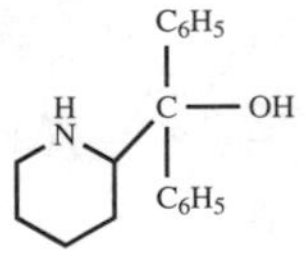

Pipril A trade name for piperazine citrate, used as an anthelmintic agent.

Piprinhydrinate (mol wt 496) An antihistaminic and antiemetic agent.

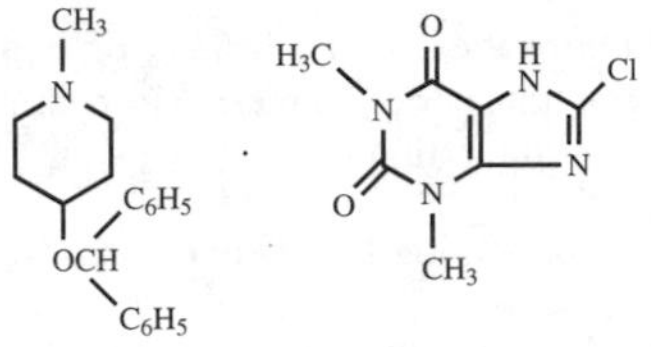

Piprozolin (mol wt 298) A choloretic agent.

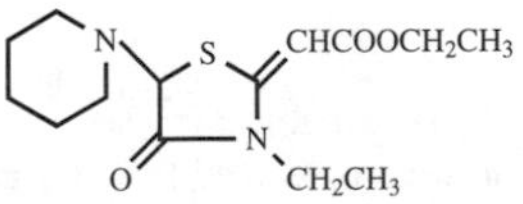

Pirarubicin (mol wt 628) An antineoplastic agent.

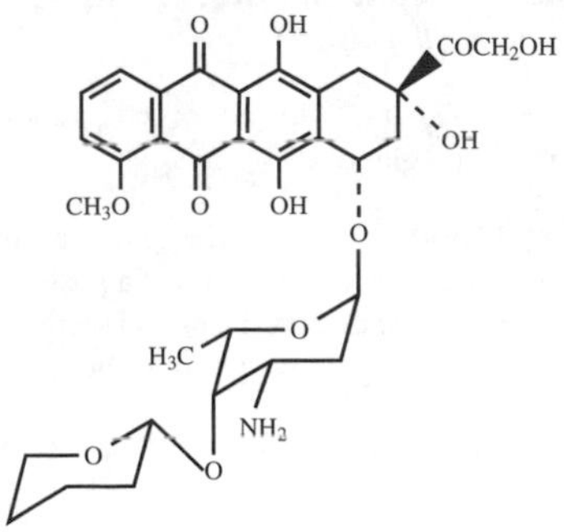

Pirenzepine (mol wt 351) A gastric acid inhibitor.

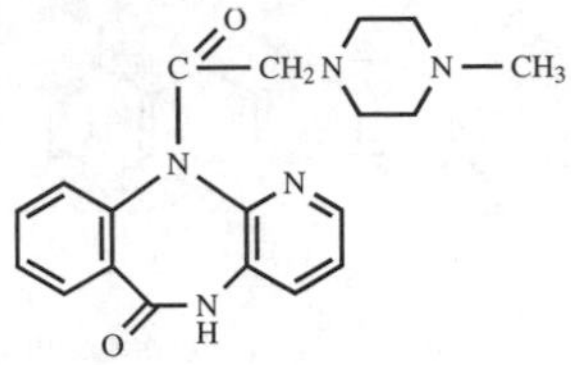

Piretanide (mol wt 362) A diuretic agent.

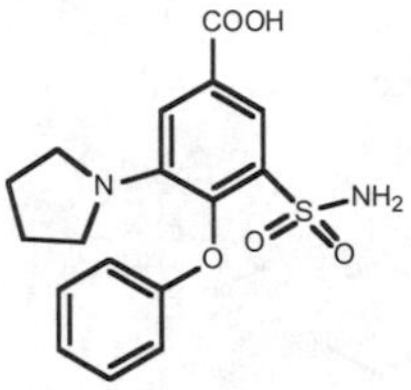

Piribedil (mol wt 298) A vasodilator.

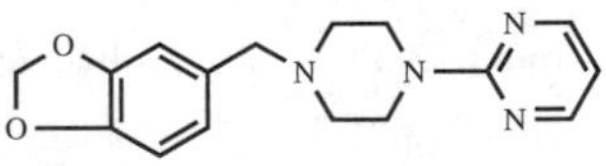

Piricularin A toxic copper-containing protein produced by *Pyricularia oryzae* that is toxic to fungus.

Piridocaine (mol wt 248) A local anesthetic agent.

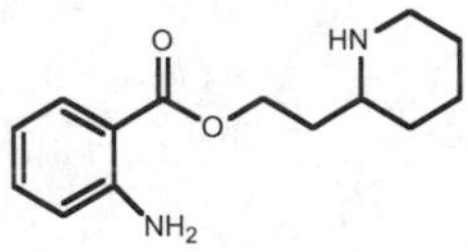

Pirifibrate (mol wt 336) An antihyperlipoproteinemic agent.

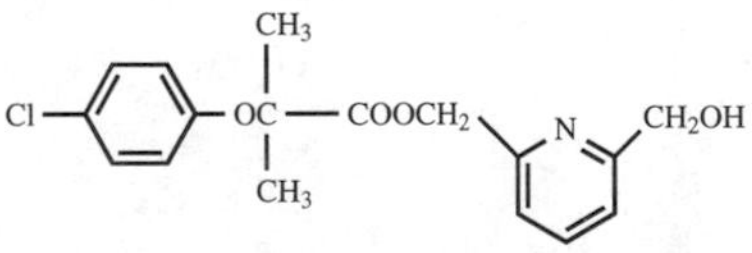

Piriton A trade name for chlorpheniramine maleate, an antihistaminic agent that competes with histamine for H_1 receptors on effector cells.

Piritramide (mol wt 431) A narcotic analgesic agent.

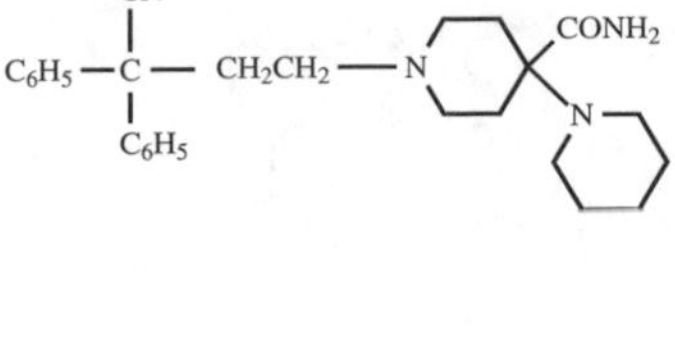

Piritrexim (mol wt 325) An antineoplastic agent.

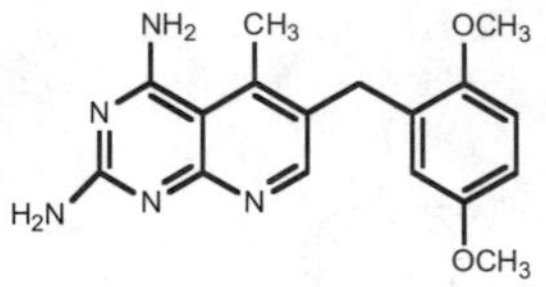

pIRL19 A plasmid of *E. coli* that contains Amp[r] (ampicillin resistant) marker and EcoRI, EcoRV, HindIII, ClaI, HgiEII, NspcI, HaeIII, and FnuDII cleavage sites.

Pirlimycin (mol wt 411) An antibacterial agent.

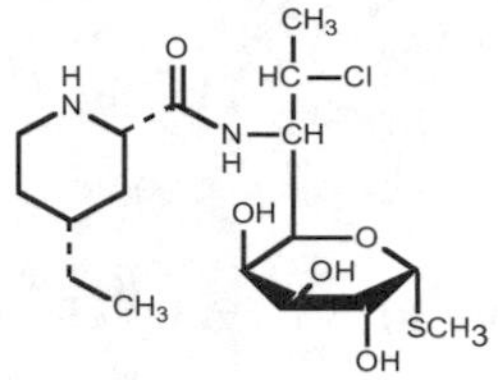

Pirmenol (mol wt 338) An antiarrhythmic agent.

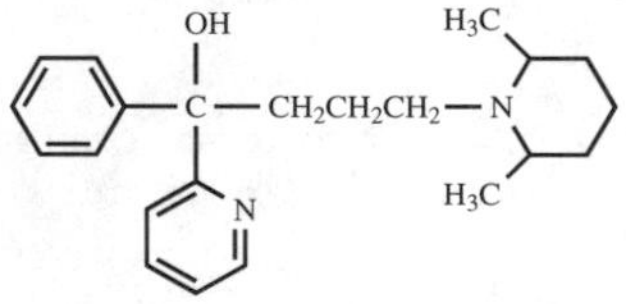

Piroctone (mol wt 237) An antiseborrheic agent.

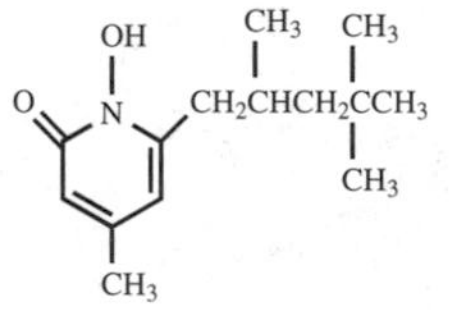

Piroheptine (mol wt 303) An antiparkinsonian agent.

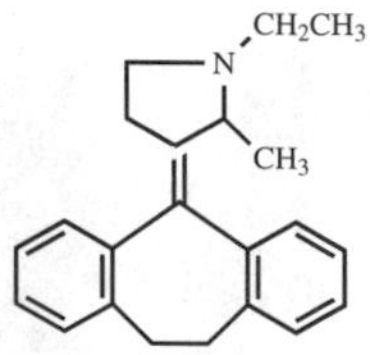

Piromidic Acid (mol wt 288) An antibacterial agent.

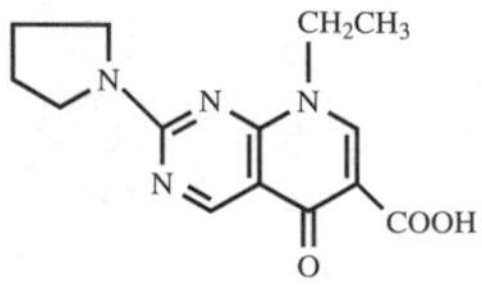

Piroxicam (mol wt 331) An anti-inflammatory and antipyretic agent that inhibits prostaglandin synthesis.

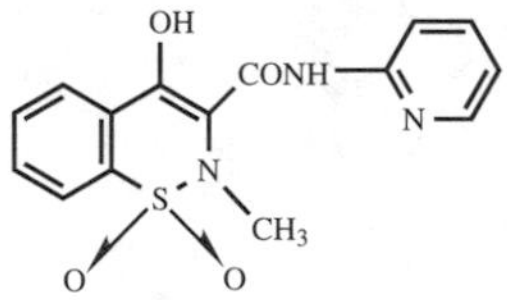

Pirozadil (mol wt 528) An antihyperlipoproteinemic agent.

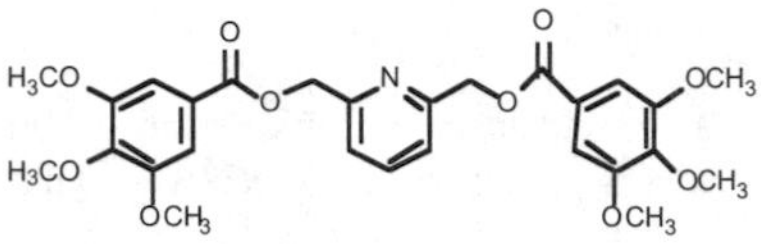

Pirprofen (mol wt 252) An anti-inflammatory agent.

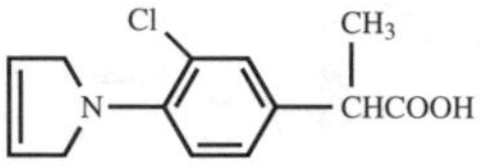

Pisatin An isoflavonoid phytoalexin produced by the pea.

Pit 1. A region in plant cell wall in which the secondary wall is interrupted exposing the underlying primary cell wall. 2. An indented or depressed area of the body.

PIT Abbreviation for 1. plasma iron transfer; 2. plasma iron transporter; 3. plasma iron turnover; 4. prothrombin inhibition test.

PITC Abbreviation for phenylisothiocyanine.

Pitocin A trade name for oxytocin, an oxytocic hormone.

PITP Abbreviation for phosphatidylinositol transfer protein.

Pitressin A trade name for vasopressin, an antidiuretic hormone produced by pituitary glands.

Pitrilysin A protease from *E. coli*.

Pittsburg Pneumonia Agent Referring to *Legionella micdadei*.

Pituicytes The intrinsic cells of the posterior lobe of the pituitary gland.

Pituitary Gland An endocrine gland at the base of the hypothalamus that consists of a posterior lobe and an anterior lobe that produces and secretes many hormones regulating diverse body functions.

PIV Abbreviation for para-influenza virus.

Pivampicillin (mol wt 464) A semisynthetic antibiotic related to penicillin that inhibits bacterial cell wall synthesis.

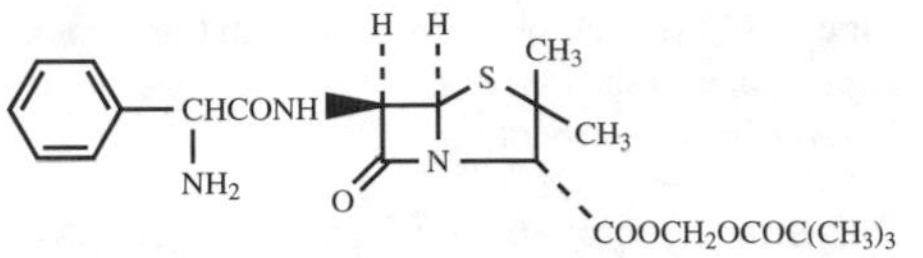

Pivcefalexin (mol wt 462) An antibacterial agent.

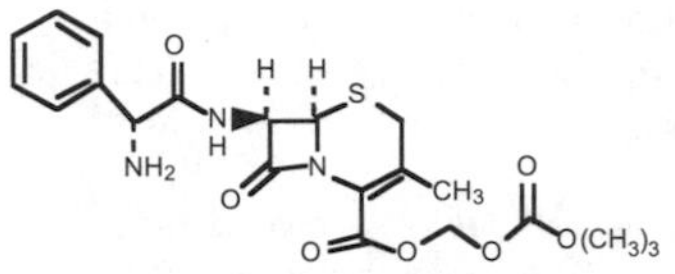

Pizotyline (mol wt 294) An antidepressant and a serotonin inhibitor.

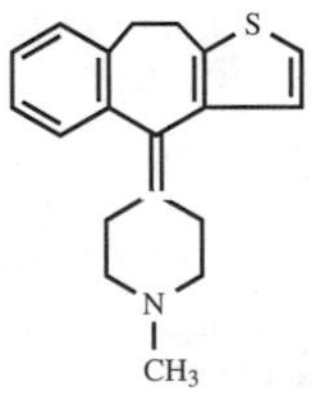

pJB8 A plasmid of *E. coli* that contains Amp^r (ampicillin resistant) marker and HindIII and SalI cleavage sites.

pJB61 A plasmid of *E. coli* that contains Amp^r (ampicillin resistant) marker and SalI, PvuII, and HincII cleavage sites.

pJEL144 A plasmid of *E. coli* that contains Amp^r (ampicillin resistant) marker and EcoRI and BamHI cleavage sites.

pJH200 A plasmid of *E. coli* that contains Amp^r (ampicillin resistant) marker and EcoRI, BglII, SmaI, HindIII, and ClaI cleavage sites.

pJS133 A plasmid of *E. coli* that contains Amp^r (ampicillin resistant) and Cmlr (chloramphenicol resistant) markers and an EcoRI cleavage site.

pJSC73 A plasmid of *E. coli* that contain Amp^r (ampicillin resistant and Cml^r (chloramphenicol resistant) markers and EcoRI, PstI, HincII, and PvuII cleavage sites.

pK The negative log of an equilibrium constant.

PK Abbreviation for 1. protein kinase; 2. pyruvate kinase.

PK Test Abbreviation for Prausnitz-Kustner test, a test used to detect the presence of IgE that is homologous to a given allergen.

pKa The negative log of the equilibrium constant for an acid.

PKA Abbreviation for protein kinase-A.

PKA-C Abbreviation for catalytic subunit of PKA (protein kinase A).

PKA-R Abbreviation for regulatory subunit of PKA (protein kinase A).

PKB Abbreviation for protein kinase B.

pKB111 A plasmid of *E. coli* that contains Amp^r (ampicillin resistant) marker and KpnI, SmaI, BamHI, HindIII, and PstI cleavage sites.

pKB686 A plasmid of *E. coli* that contains Amp^r (ampicillin resistant) marker and KpnI and ClaI cleavage sites.

pKB706 A plasmid of *E. coli* that contains Amp^r (ampicillin resistant) marker and KpnI and XhoI cleavage sites.

PKC Abbreviation for protein kinase C.

pKC7 A plasmid of *E. coli* that contains Amp^r (ampicillin resistant) and Kan^r (kanamycin resistant) markers and HindIII, BglII, and BamHI cleavage sites.

pKC16 A plasmid of *E. coli* that contains Amp^r (ampicillin resistant) marker and a BamHI cleavage site.

pKC30 A plasmid of *E. coli* that contains Amp^r (ampicillin resistant) marker and HpaI, BamHI, HindIII, and EcoRI cleavage sites.

PKCα Abbreviation for protein kinase Cα, an isotype of PKC.

PKCβ Abbreviation for protein kinase Cβ, an isotype of PKC.

PKCγ Abbreviation for protein kinase Cγ, an isotype of PKC.

pKG1800 A plasmid of *E. coli* that contains Amp^r (ampicillin resistant) marker and SmaI and HindIII cleavage sites.

pKG1901 A plasmid of *E. coli* that contains Amp^r (ampicillin resistant) marker and SmaI, BamHI, and EcoRI cleavage sites.

pKH4 A plasmid of *E. coli* that contains Amp^r (ampicillin resistant) marker and ClaI, EcoRI, and BamHI cleavage sites.

pKH47 A plasmid of *E. coli* that contains Amp^r (ampicillin resistant) and Tet^r (tetracycline resistant) markers and EcoRI, PstI, ClaI, HindIII, BamHI, SalI, and AvaI cleavage sites.

pKH502 A plasmid of *E. coli* that contains Amp^r (ampicillin resistant) marker and EcoRI and BamHI cleavage sites.

***p*KM1** A plasmid of *E. coli* that contains Amp^r (ampicillin resistant) marker and EcoRI and HindIII cleavage sites.

PKN Abbreviation for protein kinase N.

pKN001 A plasmid of *E. coli* that contains Tet^r (tetracycline resistant) marker and PstI cleavage site.

pKN80 A plasmid of *E. coli* that contains Amp^r (ampicillin resistant) marker and HindIII and HpaI cleavage sites.

pKO1 A plasmid of *E. coli* that contains Amp^r (ampicillin resistant) marker and SmaI and HindIII cleavage sites.

pKO11 A plasmid of *E. coli* that contains Amp^r (ampicillin resistant) marker and NruI, AvaI, SmaI, XmaI, BamHI, SalI, HindIII, and SphI cleavage sites.

pKO19 A plasmid of *E. coli* that contains Amp^r (ampicillin resistant) marker and HindIII, SmaI, BamHI, SalI, and EcoRI cleavage sites.

pKO100 A plasmid of *E. coli* that contains Amp^r (ampicillin resistant) marker and HindIII, SmaI, and HamHI cleavage sites.

pKOTW1 A plasmid of *E. coli* that contains Amp^r (ampicillin resistant) marker and NruI, EcoRI, AvaI, SmaI, XmaI, BamHI, SalI, HindIII, and SpeI cleavage sites.

pKT279 A plasmid of *E. coli* that contains Tet^r (tetracycline resistant) marker and PstI cleavage site.

pKTH601 A plasmid of *E. coli* that contains Amp^r (ampicillin resistant) marker and HindIII, BamHI, and PstI cleavage sites.

pKTH604 A plasmid of *E. coli* that contains Amp^r (ampicillin resistant) marker and HindIII, BamHI, and PstI cleavage sites.

PKU Abbreviation for phenylketonuria.

PL Abbreviation for 1. phospholipid and 2. placental lactogen.

PLA_1 Abbreviation for phospholipase A_1.

PLA_2 Abbreviation for phospholipase A_2.

pLA7 A plasmid of *E. coli* that contains Amp^r (ampicillin resistant) marker and BclII and AvaI cleavage sites.

Placebo An inactive substance with the identical appearance to the active one and used as a control in the testing protocol.

Placenta A structure within the pregnant uterus through which fetus takes in oxygen, nutrient, and removes CO_2 and waste products.

Placental Fementin A storage protein used for monitoring placental function from conception until delivery.

Placental Lactogen A protein hormone synthesized by the placenta that possesses lactogenic activity and is capable of stimulating production of progesterone.

Placidyl A trade name for ethchloroxynol, used as a sedative-hypnotic agent.

Plague An infectious disease caused by *Yersinia pestis*.

Plakalbumin A fragment of ovalbumin resulting from subtilisin cleavage that is more soluble in water than ovalbumin.

Plakoglobin A protein found in cell junction.

Planck's Constant A constant relating the energy and frequency of radiation equal to 6.265×10^{-27} erg-s or 1.58×10^{-34} cal-sec.

Plane of Symmetry A plane that divides a symmetrical structure into two mirror-image halves.

Planobispora A genus of bacteria (Actinomycetales).

Planococcus A genus of Gram-positive, asporogenous, catalase-positive, chemoorganotrophic, aerobic bacteria.

Planoconvex Describing a structure that is flat on one side and convex on the other.

Planogamete Referring to a motile gamete.

Planomonospora A genus of bacteria (Actinomycetales) that occurs in the soil.

Planospore A motile spore.

Plant Agglutinin A lectin from plants that is capable of agglutinating erythrocytes.

Plant Growth Substance Substances that influence plant growth and differentiation at low concentration (also known as plant growth hormone or plant growth regulator).

Plant/*nif* Gene Transfer A cloning procedure for transfer of *Klebsiella nif* genes (nitrogen fixation genes) into plants using Ti plasmid.

Plantibody Human antibody produced by genetically engineered tobacco plant.

PLAP Abbreviation for 1. peripheral laser angioplasty; 2. placental alkaline phosphatase.

Plaque 1. A clear spot on a bacterial lawn produced by the lysis of bacterial cells with bacteriophages; 2. A distinguishable region on a cell culture plate produced by the infection of culture with a virus.

Plaque Assay A quantitative assay for determining the relative concentration of a virus sample. It involves inoculation of virus onto a culture plate containing a layer of host cells, the number of spots (plaques) developed after a period of incubation represent the concentration of virus.

Plaque Forming Unit An unit of viral concentration determined by the plaque forming unit assay, the viral concentration is represented by the number of plaques per unit volume of the viral sample.

Plaque Mutant A mutant of bacteriophage that differs from wild type in the morphology of the plaque.

Plaquenil A trade name for hydroxychloroquine sulfate, used as an antimalarial agent.

Plaque Titer The concentration of a virus sample that is represented by the number of plaques per unit volume of the viral sample.

Plasbumin A trade name for normal serum albumin, used to expand plasma volume.

Plasma Blood without cells.

Plasma Amine Oxidase Synonym of amine oxidase (copper-containing).

Plasma Cells Antibody-producing cells differentiated from B lymphocytes.

Plasma Kallikrein A protease that catalyzes the selective cleavage of peptide bonds between arginine-lysine and arginine-serine.

Plasma Membrane The outer membrane of a cell.

Plasmablast The immature plasma cells or precursor of plasma cells.

Plasmacytoma A malignant tumor of plasma cells.

Plasmagene Genes present in the plasma, e.g., genes in the mitochondria and chloroplast.

Plasmalogen A glycerolphospholipid in which the glycerol moiety bearing an alkenyl ether group.

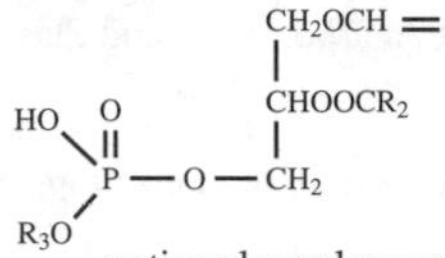

native plasmalogen phosphatide

Plasmanate A trade name for the 6% plasma protein fraction, used to expand plasma volume.

Plasmapheresis A method for separation of plasma protein from blood cells, cells resulting from the separation are suspended in saline and returned back to the individual.

Plasma-Plex A trade name for the 6% plasma protein fraction used to expand plasma volume.

Plasmaviridae A family of enveloped bacteriophage containing dsDNA, e.g., plasmavirus and bacteriophage MV-L2.

Plasmavirus A bacteriophage in the family of plasmaviridae.

Plasmid A linear or a covalently closed circular DNA present in the cytoplasm of a cell that replicates autonomously and can be transferred from one cell to another. Plasmids may carry antibiotic-resistant or toxin genes, changing phenotypes of the host cells and used as clonal vectors.

Plasmid Chimera A hybrid DNA resulted from *in vitro* joining of fragments of separate plasmids and functioning as a replicon when inserted into a cell.

Plasmid Cloning Vector A plasmid that functions as a vector in DNA cloning or DNA recombinant technology.

Plasmin The enzyme present in the serum that catalyzes the hydrolysis of fibrin leading to the dissolution of blood clots (also known as fibrinolysin).

Plasminogen Profibrinolysin or precursor of plasmin.

Plasminogen Activators Proteases that are capable of converting plasminogen to plasmin.

Plasmocid (mol wt 287) An antimalarial agent.

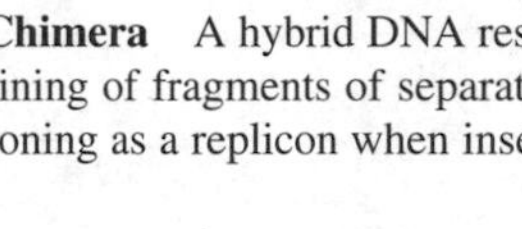

Plasmodesma A cytoplasmic connection between adjacent plant cells.

Plasmodium 1. A genus of protozoa. 2. A multinucleated motile mass of protoplasm that is variable in size and form and is the main vegetative form in members of Acarpomyxea and Plasmodiophoromycetes.

Plasmogamy The fusion of the cytoplasms of two or more cells.

Plasmolysis Shrinkage of a plant cell away from its wall when the cell is placed in a hypertonic medium.

Plastic Surgery Medical science that deals with reconstruction of deformed or damaged parts of the body.

Plastid Organelles in plant cells that function as sites of photosynthesis or nutrient storage, e.g., amyloplasts and chloroplasts.

Plastino AQ A trade name for cisplatin, an antineoplastic agent.

Plastocyanin A copper-containing protein in the chloroplast that serves as an electron carrier in noncyclic photophosphorylation.

Plastogen A self-replicating genetic entity associated with plastides.

Plastoglobuli The lipid-containing droplets or lipid globules found in the plastide.

Plastoquinol The reduced form of plastoquinone.

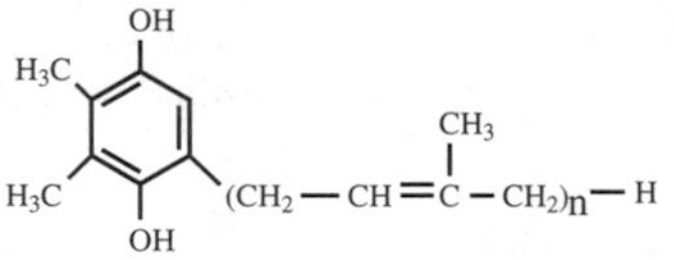

Plastoquinone A quinone present in the chloroplast that serves as an electron carrier in photosynthesis.

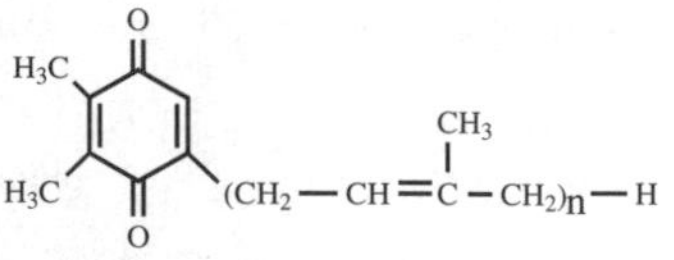

Platamine A trade name for cisplatin, used as an alkylating agent.

Plateaued Rat A slow-growing rat used for assaying growth hormone.

Platelet Flat disc-like, membrane-bound, colorless corpuscles present in mammalian blood that functions in blood coagulation and hemostasis.

Platelet Cofactor I Referring to antihemophilic factor.

Platelet Cofactor II Referring to the Christmas factor of the blood clotting system.

Platelet Derived Growth Factor A protein factor from platelets that promotes the proliferation of mouse fibroblasts and a variety of other cells.

Platelet Transfusion The intravenous administration of platelets to individuals who have a bleeding tendency due to insufficient platelets in the blood.

Plating Distributing an inoculum on the surface of a medium plate.

Platinol A trade name for cisplatin (cisplatinum), an alkylating agent that cross-links cellular DNA and interferes with transcription.

Platinum A chemical element with atomic weight 195, valences 2 and 4.

Platonin (mol wt 910) An immunomodulator.

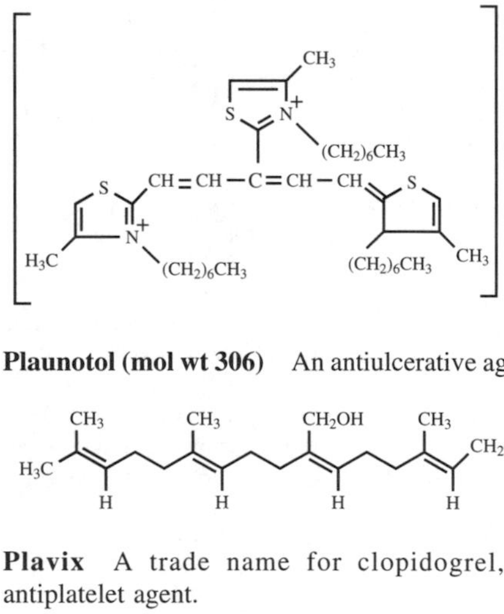

Plaunotol (mol wt 306) An antiulcerative agent.

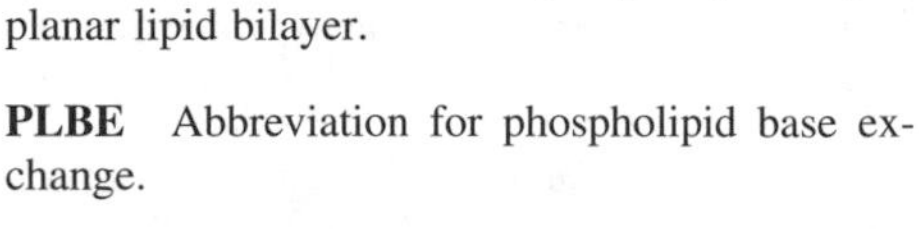

Plavix A trade name for clopidogrel, an antiplatelet agent.

PLB Abbreviation for 1. phospholipase B; 2. planar lipid bilayer.

PLBE Abbreviation for phospholipid base exchange.

PLC Abbreviation for 1. phospholipase C; 2. parenchymal liver cells.

PLD Abbreviation for phospholipase D.

PleI A restriction endonuclease from *Pseudomonas lemoignei* with the following specificity:

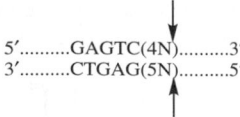

↓
5′..........GAGTC(4N)..........3′
3′..........CTGAG(5N)..........5′
↑

Ple19I A restriction endonuclease from *Pseudomonas lemoignei* 19 with the following specificity:

```
            ↓
5'........CGATCG........3'
3'........GCTAGC........5'
            ↑
```

Pleated Sheet A configuration of protein in which the polypeptide chains are fully extended and held together by interchain hydrogen bonds.

Pleckstrin A platelet protein and a substrate for serine/threonine phosphorylation by protein kinase C.

Plectonema A genus of cyanobacteria.

Plectonemic Coiling The antiparallel coiling of two polynucleotide chains in DNA so that they cannot be separated without unwinding the double helix.

Plectrovirus Bacteriophage of *Mycoplasma* and *Acholeplasma* (family of Inoviridae).

Plegine A trade name for phendimetrazine tartrate, used to promote transmission of nerve impulses by releasing stored norepinephrine from the nerve terminals in the brain.

Pleio- A prefix denoting 1. multiple; 2. excessive.

Pleiotropic Having multiple effects.

Pleiotropic Mutation A mutation that produces multiple phenotypic effects.

Pleiotropin A heparin-binding mitogenic protein.

Pleiotropism Producing multiple, phenotypic effects.

Plendil A trade name for felodipine, an antihypertensive agent that blocks the entry of calcium ions into vascular smooth muscle and cardiac cells.

Pleocytosis The presence of an abnormally large number of lymphocytes in the cerebrospinal fluid.

Pleomorphism The presence of different forms at different stages of the life cycle.

Pleromer The monomer in a polymer that can be replaced by another monomeric unit without a change in overall structural balance, e.g., cytosine and 5-methyl cytosine.

Plesiomonas A genus of Gram-negative bacteria of Vibrionaceae.

Pletal A trade name for cilostazol, a phosphodiesterase inhibitor that inhibits platelet aggregation.

Pleura The membrane lining the chest cavity.

Pleuripotent Stem Cell Cells capable of differentiating into different cell types.

Pleuritis The inflammation of pleura.

Pleurocentesis The insertion of a hollow needle into the pleural cavity through the chest wall to withdraw the needed fluid sample.

Pleurococcus A genus of unicellular green algae (Chlorophyta).

Pleuromutilin (mol wt 378) An antibiotic from basidiomycetes (*Pleurotus mutilus*).

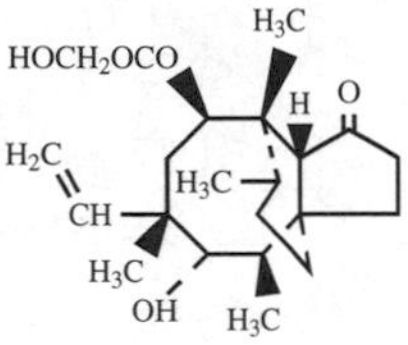

Pleuropneumonia Inflammation involving both the lung and pleura.

Pleuropneumonia-Like Organism Old term for *Mycoplasmas*.

Pleurotin(e) (mol wt 354) An antibiotic from fungus *Pleurotus griseus*.

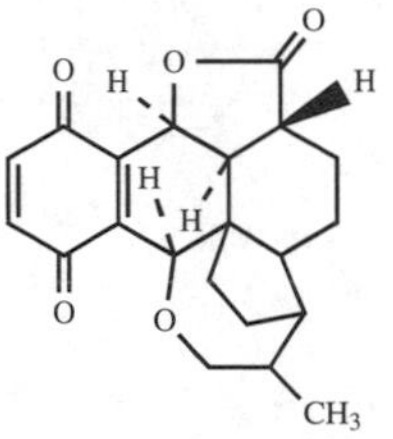

Plexiglass A transparent sheet made from polymethylmethacrylate.

PLF Abbreviation for placental fementin, a storage protein used for monitoring placental function from conception until delivery.

PLG Abbreviation for plasminogen.

PLI Abbreviation for phospholipase inhibitor.

Plicamycin (mol wt 1085) An oligosaccharide antibiotic with antineoplastic activity, produced by *Streptomyces argillaceus*.

PLK Abbreviation for polo-like kinase.

PLM Abbreviation for polarized light microscope.

-ploid A suffix meaning multiple chromosomes in the nucleus, e.g., diploid and polyploid.

Ploid Referring to the number of complete sets of chromosomes in a cell.

PLPC Abbreviation for 1-palmitoyl-2-linoleoyl-*sn*-phosphatidylcholine.

PLT Abbreviation for 1. primed lymphocyte test; 2. primed lymphocyte typing.

Plumbism Lead poisoning.

Plumbum Latin for lead.

Plumericin (mol wt 290) An antibacterial agent.

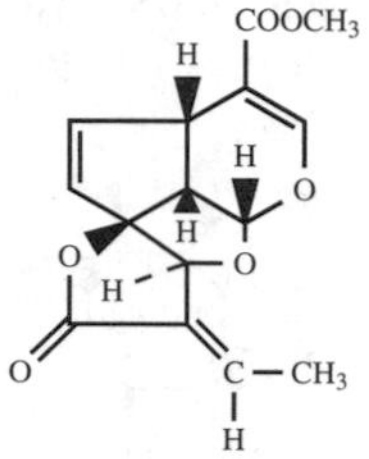

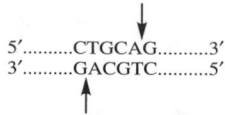

Pluripotent A cell capable of differentiating into more than one type of mature cell.

Pluronic Polyol F 127 A copolymer of ethylene oxide and polypropylene oxide used as a solidifying agent for culture medium.

Plus End The end of a microtubule or actin filament at which addition of monomers occurs readily.

Plus Strand A DNA strand thatserves as a template for synthesis of mRNA.

Plus Stranded-Virus RNA Viral RNA that is capable of acting as mRNA.

Plutonium (Pu) A chemical element with atomic weight 242, valence 3, 4, 5, 6, and 7.

PLV Abbreviation for panleukopenia virus.

PM Abbreviation for 1. plasma membrane; 2. picometer; 3. picomole.

Pm Symbol for promethium, a chemical element.

PM2 Phage A bacteriophage of the family Corticoviridae that infects *Pseudomonas*.

PMA Abbreviation for 1. *para*-methoxyamphetamine; 2. phosphomolybdic acid; 3. pyridylmercuric acetate.

PmaI (PstI) A restriction endonuclease from *Pseudonomas maltophila* with the same specificity as PstI.

Pma44I A restriction endonuclease from *Pseudonomas maltophila* with the following specificity:

5′..........CTGCA↓G..........3′
3′..........G↑ACGTC..........5′

PmaCI A restriction endonuclease from *Pseudonomas maltophila* CB50P with the following specificity:

5′..........CAC↓GTG..........3′
3′..........GTG↑CAC..........5′

PMACE Abbreviation for a combination drug containing prednisone, methotrexate, adriamycin, cytoxan, and etoposide.

pMAM17 A plasmid of *E. coli* that contains Amp[r] (ampicillin resistant) marker and EcoRI, BamHI, HindIII, and PvuII cleavage sites.

p42MAPK Abbreviation for 42-kD isoform of MAPK (mitogen-activated protein kinase).

p44MAPK Abbreviation for 44-kD isoform of MAPK (mitogen-activated protein kinase).

pMB9 A plasmid of *E. coli* that contains Tet[r] (tetracycline resistant) marker and BamHI, HindIII, SalI, and EcoRI cleavage sites.

PMBL Abbreviation for polymorphonuclear basophil leukocyte.

PMD Abbreviation for Pelizaeus-Merzbacher Disease.

PME Abbreviation for pectin methyl esterase.

pme Abbreviation for gene encoding pectin methyl esterase.

PmeI A restriction endonuclease from *Pseudomonas mendicina* with the following specificity:

5'........GTTT↓AAAC........3'
3.'........CAAA↑TTTG........5'

Pme55I A restriction endonuclease from *Pseudomonas mendicina* 55 with the following specificity:

5'........AGG↓CCT........3'
3'........TCC↑GGA........5'

PMEL Abbreviation for polymorphonuclear eosinophil leukocyte.

PMF Abbreviation for 1. plasma membrane fraction; 2. proton motive force.

pMF7 A plasmid of *E. coli* that contains Amp^r (ampicillin resistant) marker and EcoRI and SalI cleavage sites.

pMF517 A plasmid of *E. coli* that contains Amp^r (ampicillin resistant) marker and ClaI, HindIII, SalI, PstI, and EcoRI cleavage sites.

pMH621 A plasmid of *E. coli* that contains Amp^r (ampicillin resistant) marker and the BglII cleavage site.

PMI Abbreviation for phosphomannose isomerase.

PML Abbreviation for polymorphonuclear leukocytes.

PmlI A restriction endonuclease from *Pseudomonas maltophila* with the following specificity:

```
          ↓
5'........CACGTG........3'
3'........GTGCAC........5'
           ↑
```

pMLB1034 A plasmid of *E. coli* that contains Amp^r (ampicillin resistant) marker and BamHI, EcoRI, and SmaI cleavage sites.

PMM Abbreviation for 1. penta-methyl melamine; 2. phosphomannomutase.

PMN Abbreviation for 1. polymorphonuclear; 2. polymorphonuclear neutrophils.

PMNBL Abbreviation for polymorphonuclear basophil leukocyte.

PMNEL Abbreviation for polymorphonuclear eosinophil leukocyte.

PMNG Abbreviation for polymorphonuclear granulocytes.

PMNL Abbreviation for polymorphonuclear leukocyte.

PMNN Abbreviation for polymorphonuclear neutrophil.

PMNNL Abbreviation for polymorphonuclear neutrophil leukocyte.

pMOB45 A plasmid of *E. coli* that contains Cml^r (chloramphenicol resistant) marker and EcoRI, BamHI, HindIII, and ClaI cleavage sites.

pMOB48 A plasmid of *E. coli* that contains Cml^r (chloramphenicol resistant) marker and the BamHI cleavage site.

PMQ Abbreviation for pyrimethamine quinine.

PMR Abbreviation for proton magnetic resonance.

PMS Abbreviation for phenazine methosulfate, a dye used in histochemical detection of dehydrogenase activity.

PMS-Benztropine A trade name for benztropine mesylate, which blocks central cholinergic receptors.

PMS-Dimenhydrinate A trade name for the dimenhydrinate, used as an antiemetic agent.

PMS-Isoniazid A trade name for isoniazid, which inhibits bacterial cell wall biosynthesis.

PMS-Metronidazole A trade name for metronidazole, used as an antiprozoal agent.

PMS-Nylidrin A trade name for nylidrin hydrochloride, a vasodilator that stimulates beta receptors.

PMS-Prochlorperazine A trade name for prochlorperazine, maleate, a dopaminergic blocking agent used as an antipsychotic, antiemetic, and antianxiety agent.

PMS-Procyclidine A trade name for procyclidine hydrochloride, used as an antiparkinsonian agent.

PMS-Promethazine A trade name for promethazine hydrochloride, an antihistaminic agent that competes with histamine for H_1 receptors on effector cells.

PMS-Propranolol A trade name for propranolol hydrochloride, an antianginal agent that reduces cardiac oxygen demand.

PMS-Pyrazinamide A trade name for pyrazinamide, an antibacterial agent that inhibits the synthesis of folic acid in bacteria.

PMS-Sulfasalazine A trade name for sulfasalazine, an antibacterial agent that inhibits the synthesis of folic acid in bacteria.

PMS-Thioridazine E.C. A trade name for thioridazine hydrochloride, used as an antipsychotic agent.

PNA Abbreviation for 1. peptide nucleic acid; 2. peanut agglutinin; 3. *p*-nitroanilide.

***p*NA** Abbreviation for *p*-nitroanilide.

PNB (*p*NB) Abbreviation for *p*-nitrophenyl butyrate.

PNBT (*p*NBT) Abbreviation for para-nitroblue tetrazolium.

Pneumaturia The presence of air bubbles or other gas in the urine.

Pneumococcal Pneumonia A classical lobar pneumonia caused by certain strains of *Streptococcus pneumoniae.*

Pneumococci Referring to Gram-positive, pyogenic bacteria closely related to *Streptococci.*

Pneumococcus pneumoniae Referring to *Streptococcus pneumoniae.*

Pneumocystis A genus of parasitic protozoa, e.g. *P. carinii*, which causes pneumonia in immunodeficient individuals.

***Pneumocystis carinii* Pneumonia** A type of pneumonia that affects individuals whose immune defense has been compromised, e.g., in AIDS.

Pneumonia Inflammation of the lungs.

Pneumonitis Synonym for pneumonia.

Pneumotropic Having an affinity for the lungs.

Pneumovax 23 A trade name for a polyvalent pneumococcal vaccine.

Pneumovirus A virus of the family Paramyxoviridae that causes respiratory diseases in humans.

pNF4 A plasmid of *E. coli* that contains Kan^r (kanamycin resistant) and Amp^r (ampicillin resistant) markers and a BamHI cleavage site.

PNGase Abbreviation for peptide N-glycosidase.

PNGase F Abbreviation for peptide N-glycosidase F.

p[NH]ppA Abbreviation for adenosine 5'-[β-,γ-imido]triphosphate.

PNK Abbreviation for polynucleotide kinase.

PNNL Abbreviation for polymorphonuclear neutrophil leukocyte.

pNO1517 A plasmid of *E. coli* that contains Amp^r (ampicillin resistant) marker and SmaI, HpaI, BstEII, and SstII cleavage site.

pNO1523 A plasmid of *E. coli* that contains Amp^r (ampicillin resistant) marker and SmaI, HpaI, and SstI cleavage sites.

PNP Abbreviation for 1. para-nitrophenol and 2. purine nucleotide phosphorylase.

***p*NP (PNP)** Abbreviation for 1. para-nitrophenol; 2. para-nitrophenyl.

PNPase Abbreviation for polynucleotide phosphorylase.

PNPG Abbreviation for para-nitrophenyl beta galactoside.

PNPGB Abbreviation for *p*-nitrophenyl-*p*-guanidinobenzoate.

PNP-Glc (*p*NP-Glc) Abbreviation for para-nitrophenyl-beta-D-glucoside.

***p*NPP** Abbreviation for *p*-nitrophenyl phosphate.

***p*NPS** Abbreviation for *p*-nitrophenyl sulfate.

PNS Abbreviation for 1. peripheral nervous system; 2. post-nuclear supernatant.

Pnu-Imune 23 A trade name for polyvalent pneumococcal vaccine.

PNX Abbreviation for pneumothorax (air in the pleural cavity).

Po Symbol for polonium atomic wt 210 and valence 4, 2, and 6.

PO Abbreviation for 1. peroxidase; 2. pyruvate oxidase.

pO_2 Symbol for oxygen tension or partial pressure of oxygen.

POCC Abbreviation for a combination drug containing procarbazine, oncovin, cytoxan, and CCNU.

Pock 1. Referring to cutaneous pustules formed in the skin during small pox infection. 2. A lesion produced on the chorioallantoic membrane of chicken embryo by certain viruses, e.g., pox virus.

Podiatrist Individual who specializes in disorders of the feet.

Podiatry The medical science that deals with the feet.

Podocalyxin A sialated glycoprotein from kidney glomerulus.

Podofilox (mol wt 414) An anti-mitotic agent.

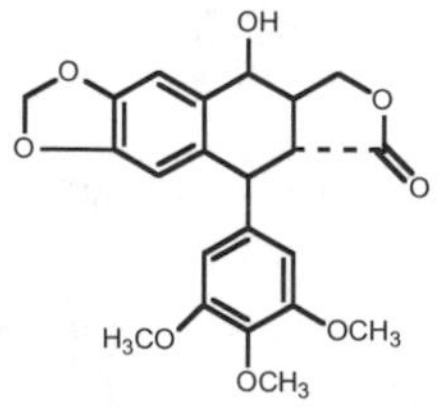

Podophyllinic Acid 2-Ethylhydrazide (mol wt 475) An antineoplastic agent.

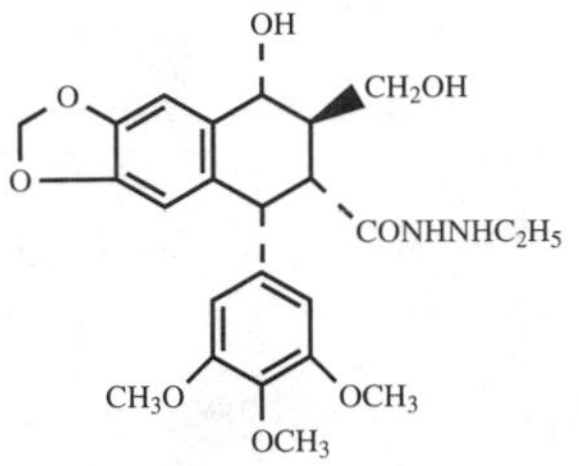

Podophyllotoxin (mol wt 414) An antineoplastic agent obtained from plant *Podophyllum peltatum.*

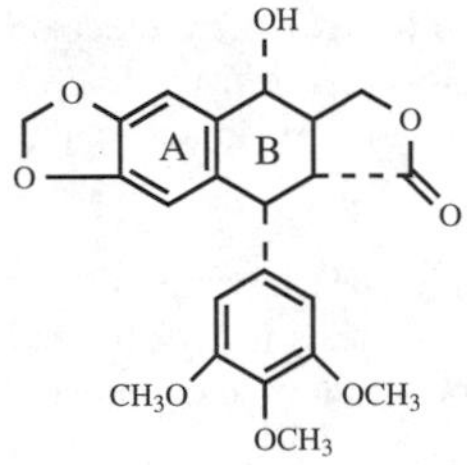

Podoviridae A family of DNA-containing bacteriophages with a short noncontractile tail, e.g., T_3,f 29, P22.

POFA Abbreviation for pancreatic onco-fetal antigen.

POG Abbreviation for 1-palmitoyl-2-oleoyl-*sn*-glycerol.

pOH The negative log of hydroxyl ion concentration in moles per liter.

-poiesis A suffix meaning synthesis or formation.

Poikilocyte An irregularly shaped red blood cell.

Poikilocytosis A disorder characterized by the presence of deformed red blood cells (poikilocytes).

Poikilotherm An animal whose body temperature tends to vary with the surrounding environment.

Point mutation A mutation caused by change of a single nucleotide

Poise A metric unit of viscosity.

Poission Distribution A mathematical equation used for calculation of % cells attacked by a different number of viral particles in a suspension of virus particles and host cells.

Pokeweed Mitogen (PWM) A lectin derived from pokeweed (*Phytolocca americana*) that stimulates both B and T lymphocytes.

pol Abbreviation for polymerase.

pol Abbreviation for the gene encoding of DNA polymerase.

Pol I Abbreviation for polymerase I.

Pol II Abbreviation for polymerase II.

Pol III Abbreviation for polymerase III.

***pol*A Gene** Gene that encodes DNA-polymerase I.

Poladex TD A trade name for dexchlorpheniramine maleate, used as an antihistaminic agent.

Polar 1. Electrically asymmetrical. 2. Water-soluble chemical group. 3. Hydrophilic.

Polar Amino Acid An amino acid that has a polar side chain, e.g., serine.

Polar Body A small cell produced during oogenesis that receives a disproportionately small amount of cytoplasm and subsequently degenerates.

Polar Bond A type of covalent bond in which two atoms have an unequal share in the bonding electrons.

Polar Compounds Compounds in which there is a polarized distribution of positive and negative charge due to uneven distribution of electrons. Polar compounds are water soluble.

Polar Covalent Bond A type of covalent bond in which the shared electrons are pulled closer to the more electronegative atom making it partially negative and the other atom partially positive.

Polar Mutation A mutation that reduces the expression of gene(s) down stream from the operator in the same operon.

Polar Nucleus One of two nuclei derived from each end of the angiosperm embryo sac, which fuses with a male nucleus to form the primary triploid nucleus that will produce the endosperm tissue of angiosperm seed.

Polar Solvent A solvent that contains charged groups and/or dipoles.

Polaramine A trade name for dexchlorpheniramine maleate, an antihistaminic agent that competes with histamine for H_1 receptors on effector cells.

Polargen A trade name for dexchlorpheniramine maleate, an antihistaminic agent that competes with histamine for H_1 receptors on effector cells.

Polarimeter An instrument for measuring the rotation of the plane of polarized light by an optically active substance.

Polarity 1. Nonuniform distribution of electrons in a molecule. 2. The distinction between the 5′ and 3′ end of a nucleic acid.

Polarized Light A light whose vibrations are in one plane only (vibrating in a defined pattern).

Polarizer A device used for the production of a plane of polarized light.

***pol*B Gene** The gene that encodes DNA polymerase II.

***pol*C Gene** The gene that encodes DNA polymerase III.

Poldine Methylsulfate (mol wt 452) An anticholinergic agent.

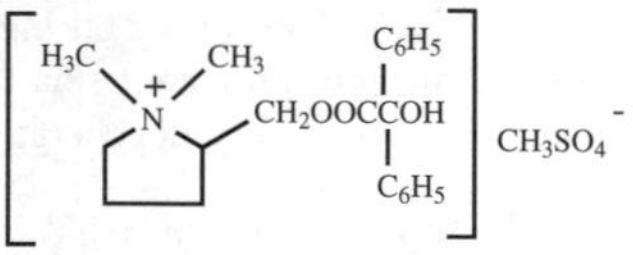

Polenske Number The number of milliliters of 0.1N alkaline required to neutralize the fatty acid in 5 grams of fat.

Polio Short for poliomyelitis.

Polioencephalitis Inflammation of the brain.

Poliomyelitis An acute inflammation of the gray matter of the spinal cord that is caused by a picornavirus.

Poliovirus A virus of the Picornaviridae that causes poliomyelitis.

Polivax A trade name for the inactivated poliovirus vaccine.

Pollen A collective term for pollen grains.

Pollen Grain A microspore in flowering plants that germinates to form male gametophytes that contain 3 nuclei, one fertilizes with ovum, the second fuses with two polar nuclei forming a 3N endosperm, and the third degenerates.

Pollen Mother Cell A diploid cell that produces four microspores by meiosis, which give rise to pollen.

Pollen Sac A cavity in the anther that contains the pollen grains.

Pollen Tube A tube formed following germination of the pollen grain that carries the male gametes into the ovule.

Polocaine A trade name for mepivacaine, an anesthetic agent.

Polonium A chemical element with atomic weight 210, valence 4.

Poly- A prefix meaning many.

Poly(A) Polymer of adenosine nucleotide.

Poly(1,4-N-Acetyl β-D-Glucosaminide Glycanohydrolase The systematic name for chitinase.

Polyacrylamide Gel An electrophoresis gel formed by polymerization of *N,N*-methylene bisacrylamide and acrylamide in the presence of polymerizing agents.

Polyacrylamide Gel Electrophoresis (PAGE) A type of electrophoresis in which a polyacrylamide gel serves as both a sieve and a supporting medium.

Polyadenylation Addition of a poly(A) tail to the 3′ end of a mRNA in eukaryotes by a mechanism that does not involve transcription.

Polyadenylate Referring to homopolymer of adenylate.

Polyadenylate Nucleotidyl Transferase The enzyme that catalyzes the following reaction:

$$\text{ATP} + (\text{adenylate})_n \rightleftharpoons \text{PPi} + (\text{adenylate})_{n+1}$$

Polyadenylation The process by which a poly(A) tail is enzymatically added onto the 3′-end of a mRNA.

Poly(ADP-Ribose) Synthetase Synonym of NAD^+ ADP-ribosyltransferase.

Polyamine A polycationic, long-chain aliphatic chain containing multiple amino groups and/or imino groups.

Polyamine-Methylene Resin An antacid.

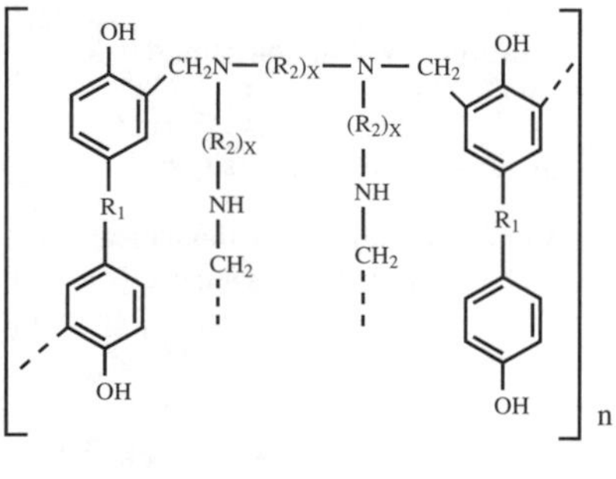

Polyamine Oxidase The enzyme that catalyzes the following reaction:

$$\text{Acetylspermine} + O_2 + H_2O \rightleftharpoons \text{Acetylspermidine} + \text{aminopropanal} + H_2O_2$$

Polyamino Acid A polymer of a single amino acid, e.g., polyglycine and polyglutamic acid.

Polyampholyte A polyelectrolyte that acts as either a proton donor or proton acceptor.

Polyanion A molecule that consists of a large number of negative charges.

Poly(A) Polymerase The enzyme that generates a poly(A) tail for mRNA from ATP.

Poly-A Tail Repetitive sequence of adenine nucleotide added posttranscriptionally to the 3′ end of a eukaryotic mRNA molecule.

Poly(C) A polymer composed of cytidine nucleotide.

Polycation A molecule that consists of a large number of positive charges.

Polycillin A trade name for ampicillin trihydrate, an antibiotic that inhibits bacterial cell wall synthesis.

Polycistronic mRNA A messenger RNA molecule that encodes more than one polypeptide product upon translation.

Polyclonal Activator A substance that activates large numbers of different clones of lymphocytes.

Polyclonal Antiserum A serum preparation that contains a heterogeneous population of antibodies, each antibody is specific for one antigenic determinant on an antigen.

Polyclonal Gammopathy The appearance in serum of a high level of immunoglobulins of many different specificities originating from many cell clones.

Polyclonal Hypergammaglobulinemia An increase in γ-globulin of various classes containing different H and L chains.

Polyclonal Mitogens Mitogens that activate large subpopulations of lymphocytes.

Polyclonal Proteins A group of molecules derived from multiple clones of cells.

Polycloning Site A DNA vector that is engineered to contain multiple cleavage sites for different restriction endonucleases.

Polycyclic Having more than one rings of atoms in the molecule.

Polycythemia A disorder characterized by the presence of an excessive number of red blood cells.

Poly(dA) Tail A polymer that consists of deoxyadenosine nucleotide used for joining DNA fragments containing single-stranded poly (dT) tail.

Poly(dC) Abbreviation for poly-deoxycytidylic acid or polydeoxycystidylate.

Polydeoxyribonucleotide An oligomer that consists of about 10 deoxyribonucleotides.

Poly-deoxy-Ribonucleotide Poly-deoxy-Ribonucleotide Ligase The systematic name for DNA ligase.

Poly(dG) Abbreviation for poly-deoxyguanylic acid or polydeoxyguanylate.

Polydipsia A condition of excessive thirst.

Poly(dT) Tail An oligomer that consists of polydeoxyribothymine nucleotides used for jointing DNA fragments containing single-stranded poly (dA) tail.

Polyelectrolyte A molecule that contains multiple charged groups.

Polyene A chemical compound that consists of many conjugated double bonds.

Polyene Antibiotics A group of antibiotics that contain conjugated double bonds in the lactone ring.

Polyenoic Fatty Acid Referring to polyunsaturated fatty acid.

Polyestradiol Phosphate An estrogen used for treatment of prostatic carcinoma.

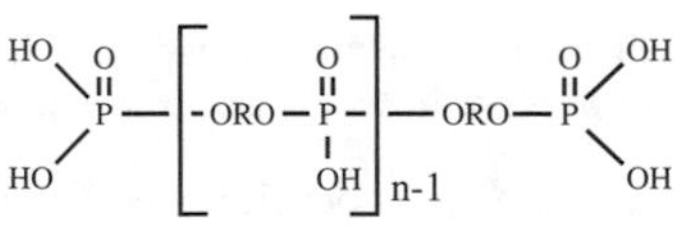

ORO represents the estradiol radical

Polyestrous Having several periods of estrus in a year.

Polyethylene A resin formed by polymerization of ethylene $(CH_2{=}CH_2)_n$.

Polyethylene Glycol A hydrophilic polymer of the general formula $H(OCH_2CH_2)_nOH$ that interacts with the cell membrane and promotes fusion of cells to produce hybrids.

Polyflex A trade name for a combination drug containing chlorzoxazone and acetaminophen, used as a muscle relaxant.

Polyfunctional Protein A protein that possesses multiple functions or enzyme activities, e.g., eukaryotic fatty acid synthetase.

Polygalacturonan Plant cell wall polysaccharide consisting of predominantly galacturonic acid.

Polygalacturonase The enzyme that degradates polygalacturonan.

Polygalacturonic Acid A polymer of galacturonic acid.

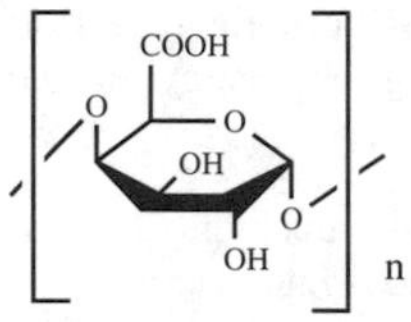

Polygene A group of genes that collectively control a quantitative character or modify the expression of a quantitative character.

Polygenic Character A variable phenotype that is governed by many pairs of alleles.

Polgenic Inheritance An additive effect of two or more gene loci on a single phenotypic characteristic.

Polygenic mRNA Synonym for polycistronic mRNA.

Poly-β-Glucosaminidase Synonym of chitinase.

Polygyny Mating system in which one male mates with more than one female.

Polyhydra Large inclusion bodies formed in the cells of insects infected by baculovirus or cytoplasmic polyhydrosis virus.

Polyhydric Any chemical compound that contains two or more hydroxyl groups per molecule.

Polyhydrin A virus-specific protein found in the polyhydra inclusion body.

Polyhydron A solid structure with many plane faces.

Polyhydrosis An insect disease caused by a nuclear polyhydrosis virus (Baculoviridae).

Poly(I:C) A synthetic double-stranded RNA that consists of one strand each of polyinosinic acid and polycytidylic acid that is an excellent interferon inducer.

Polykaryote A multinucleate cell.

Polylysine A polymer of lysine used to mediate adhesion of living cells to synthetic surfaces or particles.

Poly-β-D-1,4-Mannuronide Lyase The systematic name for poly-β-D-mannuronate lyase.

Polymer A high molecular weight substance that consists of many monomeric units, e.g., proteins and nucleic acid.

Polymerase The enzyme that catalyzes the formation of polymer, e.g., DNA polymerase and RNA polymerase.

Polymerase Chain Reaction A technique that employs a repetitive cycle of DNA amplification using heat-stable DNA polymerase from *Thermus aquaticus* to eliminate the need to add fresh enzyme after each heat denatured cycle.

Polymerization The process of forming a polymer.

Poly-Methoxy-L-Galacturonide Lyase The systematic name for pectin lyase.

Polymorphic Organism that exhibits polymorphism.

Polymorphism 1. The occurrence of multiple morphologically distinct forms of an organism. 2. The occurrence of multiple molecular forms of a protein in members of the same species. 3. The existence of two or more genetically different classes in the same interbreeding population.

Polymorphonuclear Neutrophil Polymorphonuclear granular leukocyte, that stains with neutral dyes.

Polymox A trade name for the antibiotic amoxicillin trihydrate, an antibiotic that inhibits bacterial cell wall synthesis.

Polymyxin A group of peptide antibiotics produced by *Bacillus polymyxa* that alter the permeability of the bacterial cell membrane.

Polymyxin B sulfate A peptide antibiotic of the polymyxin group that alters the permeability of the bacterial cell membrane.

Polynucleotide Linear polymer of nucleotide in which the nucleotides are linked by the phosphodiester bonds between 3′ position of the one nucleotide and the 5′ position of the adjacent nucleotide.

Polynucleotide 5'-Hydroxyl Kinase The enzyme that catalyzes the following reaction:

$$\text{ATP} + \text{5'-dephospho-DNA} \rightleftharpoons \text{ADP} + \text{5'-phospho-DNA}$$

Polynucleotide Kinase The enzyme that catalyzes the transfer of a phosphate from ATP to a 5′-hydroxyl terminal of a polynucleotide.

Polynucleotide Ligase The enzyme that joins two nucleotide fragments bearing a 5′-phosphate and a 3′-OH group.

Polynucleotide Phosphorylase The enzyme that catalyzes the following reaction:

$$RNA_{(n+1)} + Pi \rightleftharpoons RNA_n + \text{nucleoside diphosphate}$$

Polyol Referring to polyhydroxyl sugar alcohol.

Polyol Dehydrogenase Synonym of L-Iditol 2-dehydrogenase.

Polyomavirus A DNA-containing virus that causes tumors in new born mice (family Papovaviridae).

Polyoxins A group of antifungal antibiotics produced by *Streptomyces cacaoi*.

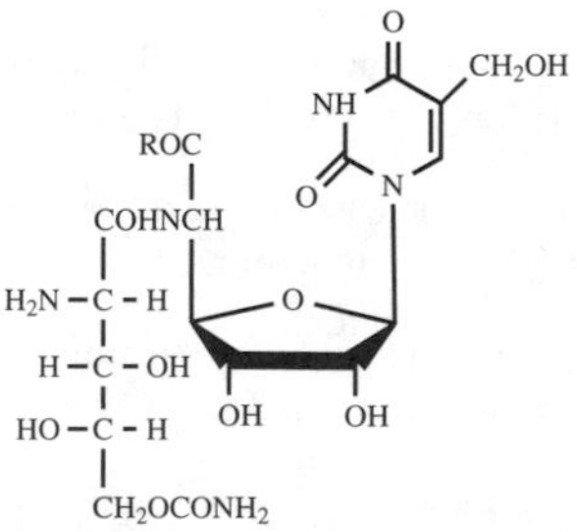

Different polyoxins have a different R group

Polyp A benign growth protruding from a mucous membrane.

Polypectomy Surgical removal of polyp.

Polypeptide A polymer composed of amino acid monomers joined by peptide bonds.

Polyphenol Oxidase The enzyme that catalyzes the oxidation of phenolic compounds in the presence of oxygen.

Polyphosphate A linear polymer of phosphate that occurs in bacteria, fungi, algae, protozoa, and certain higher eukaryotes.

Polyphosphorylase Synonym of phosphorylase.

Polyphyletic A group of species that consist of members that are derived from different evolutionary lines.

Polyploid Cell or organism that has more than two haploid sets of chromosomes.

Polypodial Pertaining to an ameba with several pseudopodia.

Polyposis A genetic disorder characterized by the development of a large number of adenomatous polyps in the large intestine with a tendency to become malignant.

Polyprotein A protein that is translated from a polycistronic mRNA and then cleaved into several functionally distinct polypeptides or proteins.

Polypus A growth (usually benign) that protrudes from a mucous membrane.

Polyribonucleotide A polymer of ribonucleotides (e.g., RNA).

Polyribonucleotide Nucleotidyltransferase The enzyme that catalyzes the following reaction:

$$RNA_{n+1} + \text{orthophosphate} \rightleftharpoons RNA_n + \text{a nucleoside diphosphate}$$

Polyribonucleotide Orthophosphate Nucleotidyl Transferase The systematic name for polyribonucleotide nucleotidyltransferase.

Polyribonucleotide Synthetase Synonym of RNA ligase.

Polyribosome A single molecule of mRNA with many ribosomes attached along its length; it is the functional complex of protein synthesis.

Polysaccharide A polymer composed of monosaccharides (e.g., cellulose and starch).

Polysaccharide Phosphorylase The enzyme that catalyzes the formation of monosaccharide phosphate from a polysaccharide.

Polyserositis Inflammation of the membranes that line the chest, abdomen, and joints.

Polysome See polyribosome.

Polysomy A condition in which some chromosomes are present in greater than diploid number.

Polyspermy Fertilization or penetration of an egg by more than one sperm.

Polysporin Ointment A trade name for a combination drug containing polymyxin B sulfate and zinc bacitracin, used as a local anti-infective agent.

Polystromatic Referring to a structure that is more than two cells thick.

Poly(T) Abbreviation for polythymidylic acid or polythymidylate.

Polytene Chromosomes Giant chromosomes produced by the successive replication of pairs of homologous chromosomes, joined together without chromosome separation.

Polytenization The process of the formation of the polytene chromosome by the repeated replication of a chromosome without separation of daughter chromosomes.

Polythiazide (mol wt 440) A diuretic agent that increases urine excretion of sodium and water by inhibiting sodium reabsorption in the cortical diluting site of the nephron.

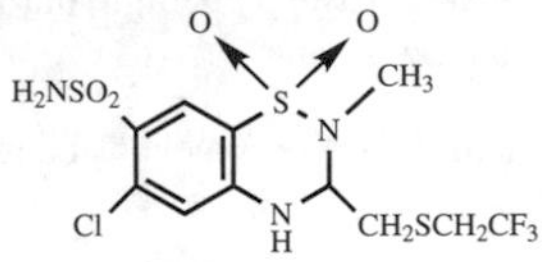

Polytrim Ophthalmic A trade name for a combination drug containing trimethoprim sulfate and polymyxin B sulfate, used as an ophthalmic anti-infective agent.

Poly(U) A polymer of uridine nucleotide.

Poly(U) Paper A sheet of paper that contains covalently linked polyurindylic acid that is used for isolation of mRNA with a poly(A) tail.

Polyunsaturated Fatty Acid Any fatty acid that contains many double bonds.

Polyuria The excretion of large quantities of urine.

Polyvalent Antiserum An antiserum that contains antibodies to a number of different antigens.

Polyvalent Vaccine A vaccine that contains the avirulent antigens and/or toxoids from each of the several different strains of one species of pathogen (also known as mixed vaccine).

pOM41 A plasmid of *E. coli* that contains Amp^r (ampicillin resistant) and $malT^+$ markers and a EcoRI cleavage site.

POMC Abbreviation for pro-opiomelanocortin.

Pompe's Disease A glycogen storage disease caused by the deficiency of α-(1,4)-glucosidase (also known as glycogen storage disease type II).

PON Abbreviation for paraoxonase, a serum esterase exclusively associated with high density lipoprotein.

Ponceau BS (mol wt 556) A dye.

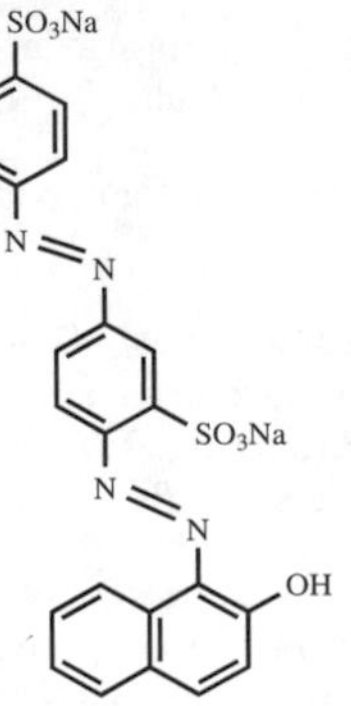

Ponceau S (mol wt 760) A dye.

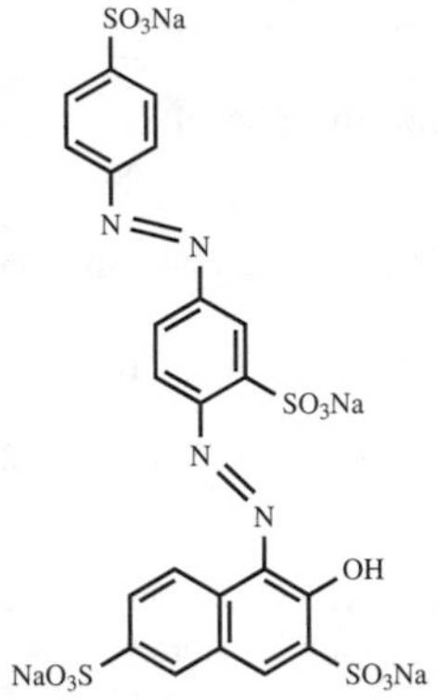

Ponderal A trade name for fenfluramine hydrochloride, used as a cerebral stimulant.

Ponderax A trade name for fenfluramine hydrochloride, used as a cerebral stimulant.

Pondimin A trade name for fenfluramine hydrochloride, used as cerebral stimulant.

Ponstan A trade name for mefenamic acid, used as an anti-inflammatory and antipyretic agent.

Ponstel A trade name for mefenamic acid used, as an anti-inflammatory agent.

Ponticulin An F-actin-binding transmembrane glycoprotein of plasma membranes of slime mould *Dictyostelium.*

Pontocaine Eye A trade name for tetracaine hydrochloride, used as a local ophthalmic anesthetic agent.

Pooled Plasma A mixture of plasma from many different donors.

pOP95-2 A plasmid of *E. coli* that contains a Tet^r (tetracycline resistant) marker and EcoRI, HindIII, and SalI cleavage sites.

pOP95-15 A plasmid of *E. coli* that contains Tet^r (tetracycline resistant) and Amp^r (ampicillin resistant) markers and EcoRI, HindIII, and SalI cleavage sites.

pOP203-2 A plasmid of *E. coli* that contains Tet^r (tetracycline resistant) and Amp^r (ampicillin resistant) markers and an EcoRI cleavage site.

pOP203-24 A plasmid of *E. coli* that contains Tet^r (tetracycline resistant) and Amp^r (ampicillin resistant) markers and an EcoRI cleavage site.

pOP203-27 A plasmid of *E. coli* that contains Amp^r (ampicillin resistant) and gpt^+ markers and an EcoRI cleavage site.

Poppy Any of the herbs of the genus *Papaver* (Papaveraceae). Opium is obtained from the fruit of the opium poppy (*P. somniferum*).

Porcine Pertaining to the pig family.

pORF2 A plasmid of *E. coli* that contains an Amp[r] (ampicillin resistant) marker and BamHI, BglII, and XmaI cleavage sites.

pORF5 A plasmid of *E. coli* that contains an Amp[r] (ampicillin resistant) marker and EcoRI, SmaI, BamHI, SalI, PstI, and HindIII cleavage sites.

Porfimer Sodium A light sensitive polyporphyrin oligomer with antineoplastic activity.

Porfiromycin (mol wt 348) An antibiotic produced by *Streptomyces ardus*.

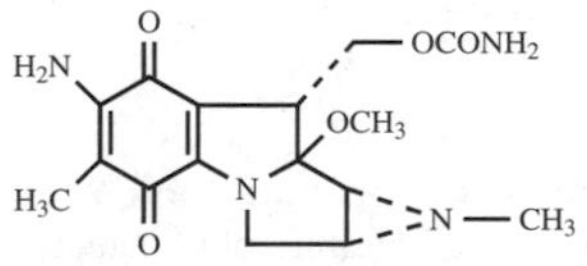

Porin A protein in the outer membrane of Gram-negative bacteria that forms a water-filled transmembrane channel or pore for the passage of ions or nonspecific molecules.

Porphin The parent cyclic tetrapyrrole ring structure.

Porphobilinogen (mol wt 226) A monopyrrole and precursor of porphyrins.

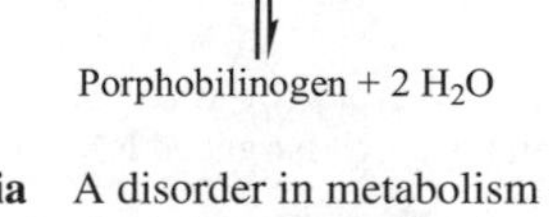

Porphobilinogen Synthetase The enzyme that catalyzes the following reaction:

2 5-Aminolevulinate

⇅

Porphobilinogen + 2 H_2O

Porphyria A disorder in metabolism of porphyrins and heme and characterized by the presence of a large quantity of porphyrins in urine.

Porphyrins A group of naturally occurring pigments containing chelated iron found in hemoglobin, cytochromes, and chlorophyll.

Porter Referring to transport agent.

Positive Catalysis A catalysis that leads to the increase of a chemical reaction or reaction product.

Positive Control A condition in which a regulatory protein is needed to initiate the transcription of structural genes (e.g., binding of regulatory protein to an operator to initiate transcription of structural genes).

Positive Cooperativity The binding of one ligand to the site on a macromolecule (e.g., enzyme) that increases the affinity for binding of a subsequent ligand to the other site on the same molecule.

Positive Effector A substance or metabolite that binds to the regulatory site on an allosteric enzyme initiating the binding of substrate to the enzyme.

Positive Feedback Initiation of a series of biochemical reactions by the presence of a metabolite or a specific substance.

Positive Regulation A control mechanism in which a pathway is activated by the presence of a specific substance or metabolite.

Positive Staining Staining of the specimen with electron dense material in electron microscopy so that the specimen appears darker than the background due to the binding of electron dense material to the specimen.

Positive Strand RNA Virus A single-stranded viral RNA that acts as a mRNA, e.g., polio viral RNA.

Positive Supercoil A coiled circular DNA molecule formed by a right-handed twist of a relaxed molecule.

Positron An elementary particle that is an antiparticle of the electron. It has the same mass of electron.

Positron Emission Tomography A technique of nuclear medicine used to evaluate the activity of tissues by measuring their uptake of the 2-deoxyglucose radioactively labeled with fluorine. The uptake and metabolism of 2-deoxyglucose produces positrons which collide with electrons and are annihilated with the production of gamma ray. Variation in tissue metabolic activities are measured by using a tomographic gamma camera.

Posology Science that deals with dosage of medicine.

Post Proline Cleaving Enzyme Synonym of prolyl oligopeptidase.

Post-Replication Modification Modification of DNA that occurs after the DNA replication has been completed, e.g., DNA methylation, DNA glycosylation, and hydroxymethylcytosine formation.

Post-Replication Repair DNA repair that occurs after the replication fork has passed that region to be repaired.

Post-Transcriptional Modification Enzymatic modification of transcriptional products to form a bioactive RNA, e.g., formation of the 5′-cap and 3′-poly(A) tail in mRNA.

Post-Translational Modification Enzymatic modification of proteins after translation is completed, e.g., glycosylation, removal of formylated methionine from the N-terminus.

Potaba A trade name for benzoate potassium, used as an antifibrosis agent.

Potassium (K) A chemical element with atomic weight 39, valence 1.

Potassium-40 The naturally occurring radioactive nuclide (^{40}K), emitting beta and gamma radiation with a half life of 1.28×10^9 years.

Potassium-42 The artificial radioactive nuclide (^{42}K), emitting beta and gamma radiation with a half life of 12 hours.

Potassium Acetate (mol wt 98) A reagent used to replace and maintain potassium levels in the body.

CH_3COOK

Potassium Bitartrate (mol wt 188) A laxative and diuretic agent.

$KO_2CCH(OH)CH(OH)COOH$

Potassium Carbonate (mol wt 132) A reagent used to replace and maintain potassium levels in the body.

K_2CO_3

Potassium Channel A type of structure in the membrane that controls the passage of potassium. There are three main types of potassium channels, namely 1. voltage-dependent or voltage gated K-channel; 2. calcium-activated K channel, and 3. receptor-coupled K channel.

Potassium Citrate (mol wt 306) An antiurolithic agent and antacid.

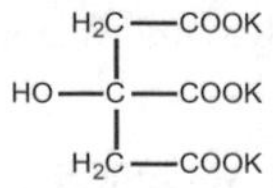

Potassium Chloride (mol wt 75) A reagent used to replace and maintain potassium levels in the body.

KCl

Potassium Gluconate (mol wt 234) A salt of gluconic acid used to replace and maintain potassium level in the body.

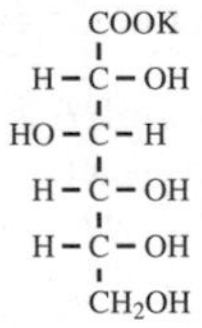

Potassium Iodide (mol wt 166) A reagent used to increase production of respiratory tract fluid to help liquefy and reduce the viscosity of thick secretions.

KI

Potassium Nitrate (mol wt 101) A reagent used as a diuretic agent.

KNO_3

Potassium Permanganate (mol wt 158) Substance used as an antimicrobial agent.

$KMnO_4$

Potassium Pump Generation of a concentration gradient of potassium across a membrane by K^+-dependent ATPase at the expense of ATP.

Potassium Rougier A trade name for potassium gluconate, used to replace and maintain potassium levels in the body.

Potato Spindle Tuber Virus A viroid that causes potato spindle tuber disease.

Potentiometer An instrument for measuring an electric potential difference of the constant polarity without drawing current from the circuit being examined.

Potexviruses A group of filamentous plant viruses containing single-stranded RNA, e.g., potato X virus.

Potyviruses A group of filamentous plant viruses containing single-stranded RNA, e.g., potato Y virus.

PovI (BclI) A restriction endonuclease from *Pseudomonas ovalis* with the following specificity:

```
        ↓
5′..........TGATCA..........3′
3′..........ACTAGT..........5′
                 ↑
```

Povidone Iodine An iodine-polyvinylpyrrolidone complex used as an anti-infective activity.

Powassan Encephalitis A tick-transmitted acute encephalitis caused by a flavivirus (flaviviridae).

POX Abbreviation for phenol oxidase.

Poxviridae A family of double-stranded DNA viruses, e.g., small pox and vaccinia.

Poyamin A trade name for vitamin B_{12}.

PP Abbreviation for 1. pancreatic polypeptide; 2. peroxisome proliferator; 3. phosphoprotein phosphatase; 4. protein phosphatase; 5. protoporphyrin IX.

PP Pathway Abbreviation for pentose phosphate pathway.

PPA Abbreviation for 1. phenylpropanolamine; 2. phenylpyruvic acid.

PpaI (Eco31I) A restriction endonuclease from *Pseudomonas paucimobilis* with the following specificity:

```
              ↓
5′..........GGTCTC(N)..........3′
3′..........CCAGAG(5N).........5′
                     ↑
```

PPACK A chloromethyl tripeptide and a selective inhibitor of thrombin.

D-phe-L-pro-L-Arg-CH_2Cl

PPAR Abbreviation for peroxisome-proliferator-activated receptor.

PPARE Abbreviation for peroxisome proliferator-activated receptor element.

PPase Abbreviation for 1. protein phosphatase; 2. pyrophosphatase.

ppb Abbreviation for parts per billion.

PP2C Abbreviation for protein phosphatase 2C.

PPCA Abbreviation for plasma prothrombin conversion accelerator.

PPCF Abbreviation for plasma prothrombin conversion factor.

pp[CH$_2$]pA Abbreviation for adenosine 5'-[α,β-methylene]triphosphate.

PPD Abbreviation for purified protein derivative, a preparation from *Mycobacterium tuberculosis* used for diagnosis of tuberculosis.

PPDME Abbreviation for protoporphyrin IX dimethyl ester.

PPE Abbreviation for porcine pancreatic elastase.

ppg Abbreviation for picopicogram.

PPGF Abbreviation for polypeptide growth factor.

ppGpp (mol wt 603) Abbreviation for guanosine tetraphosphate, a compound formed during amino acid starvation in the cell. The ppGpp blocks the synthesis of ribosomal RNA and tRNA.

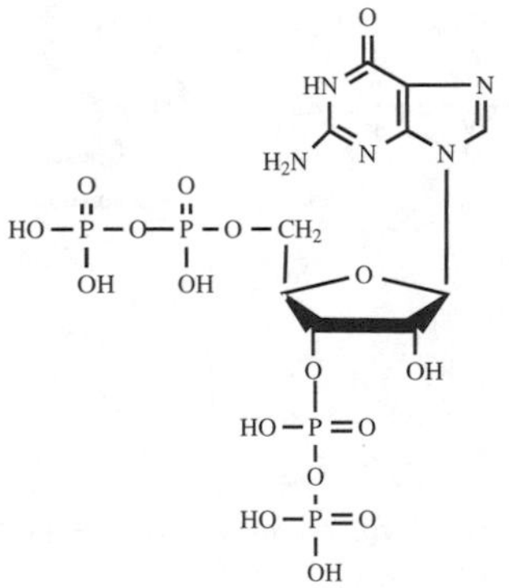

PPH Abbreviation for protocollagen proline hydroxylase.

PPHK Abbreviation for platelet phospho-hexose kinase.

PPi Abbreviation for pyrophosphate.

PPi Abbreviation for inorganic pyrophosphate.

PPL Abbreviation for penicilloylpolylysine.

PPLO Abbreviation for pleuropneumonia-like organism (former name for *Mycoplasma*).

ppm Abbreviation for parts per million or milligrams per liter.

PPM Abbreviation for phosphopentomutase.

PPNG Abbreviation for penicilinase-producing *Neisseria gonorrhoeae*.

PPO Abbreviation for pleuropneumonia organism.

PPP Abbreviation for pentose phosphate pathway, a pathway for production of NADPH from the metabolism of glucose.

pppA(2′-p-5′-A)$_n$ Abbreviation for an oligonucleotide of adenylate consisting of pppA(2′-p-5′-A)$_n$, synthesized by oligoadenylate synthetase in the presence of double-stranded RNA it activates the pre-existing RNAase L to degrade mRNA.

PP-Pathway Abbreviation for pentose phosphate pathway.

PPPG Abbreviation for post-prandial (after eating) plasma glucose.

pppGpp (mol wt 683) A guanosine pentaphosphate that inhibits transcription.

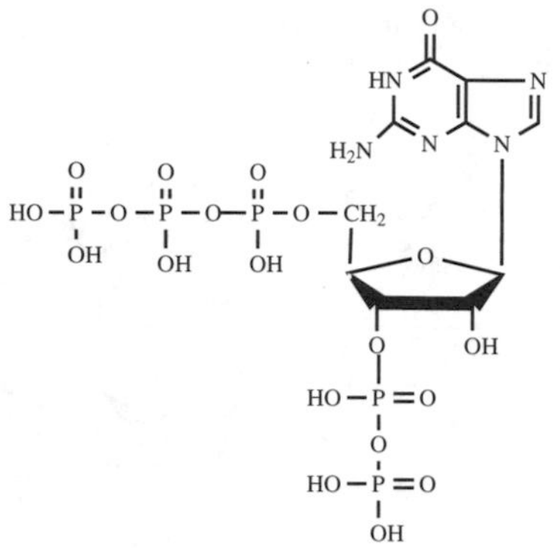

PPRE Abbreviation for peroxisome-proliferator regulatory element.

ppt Abbreviation for precipitate.

PpuI (HaeIII) A restriction endonuclease from *Pseudomonas putida* C-83 with the following specificity:

5′..........GGCC.........3′
3′..........CCGG.........5′

Ppu10I A restriction endonuclease from *Pseudomonas putida* RFL 10 with the following specificity:

5'........ATGCAT........3'
3'........TACGTA........5'

PpuMI A restriction endonuclease from *Pseudomonas putida* M with the following specificity:

5′..........PuGG(A/T)CCPy..........3′
3′..........PyCC(T/A)GGPu..........5′

PQ Abbreviation for plastoquinone, an electron carrier in photosynthesis.

PQH$_2$ Abbreviation for dihydroplastoquinone.

PQQ Abbreviation for pyrollo-quinoline quinone; it acts as a prosthetic group in the quinoprotein enzyme.

Pr Symbol for the chemical element praseodymium.

PR Abbreviation for 1. phenol red; 2. progesterone receptor.

PRA Abbreviation 1. plasma renin activity; 2. progesterone receptor assay.

Practolol (mol wt 266) An antiarrhythmic agent.

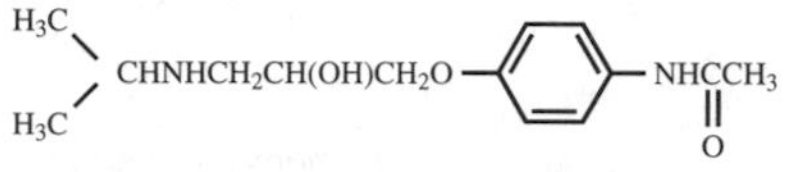

Prajmaline (mol wt 370) An antiarrhythmic agent.

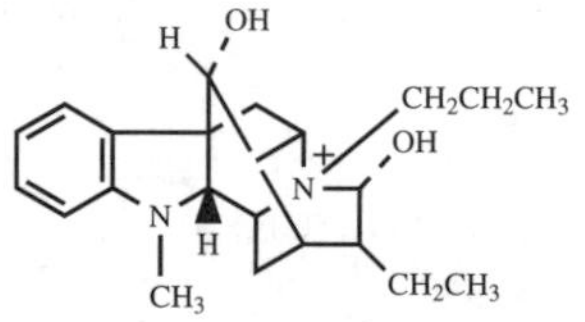

Pralidoxime Chloride (mol wt 173) A cholinesterase reactivator used to facilitate normal functioning of neuromuscular junctions.

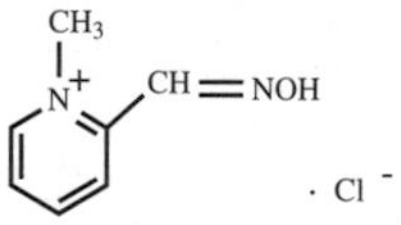

PrameGel A trade name for pramoxine, an anesthetic agent.

Pramiperxole (mol wt 269) A dopamine agonist used for treatment of Parkinson's disease.

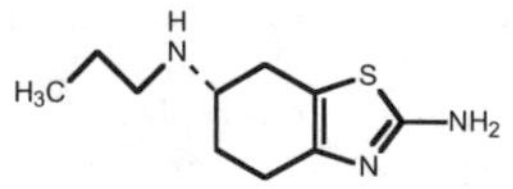

Pramiverin (mol wt 293) An antispasmodic agent.

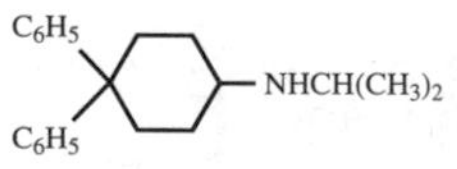

Pramoxine (mol wt 293) An anesthetic agent.

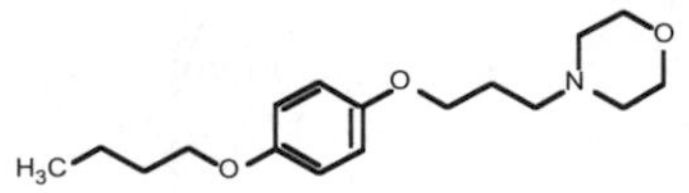

Prandase A trade name for acarbose, an anti-diabetic agent.

Prandin A trade name for repaglinide produced by DNA recombinant technology and used as an anti-diabetic agent for treatment of type 2 diabetes.

Pranlukast (mol wt 482) An antiasthmatic agent.

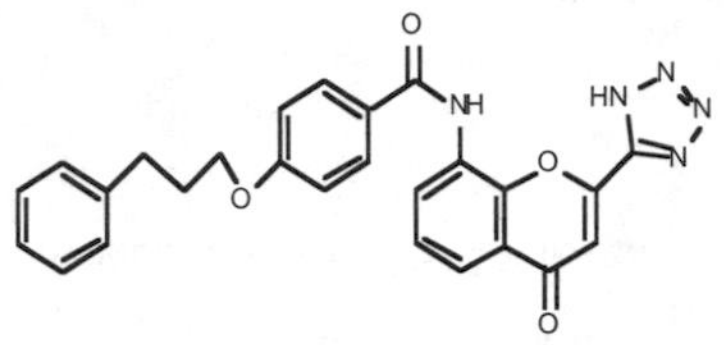

Pranoprofen (mol wt 255) An anti-inflammatory agent.

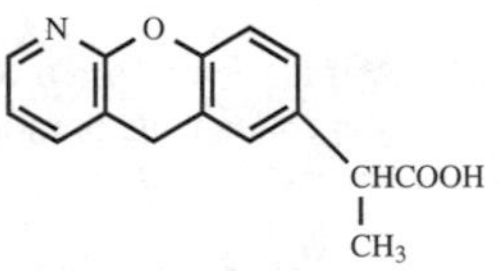

Praseodymium (Pr) A chemical element with atomic weight 141, valence 3 and 4.

Prasterone (mol wt 288) An androgen.

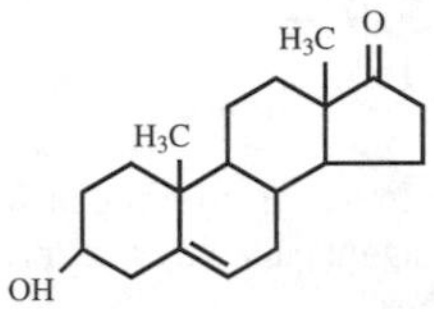

Prausnitz-Kustner Reaction A passive transfer by intradermal injection of serum-containing IgE antibodies from an allergic subject to a nonallergic recipient.

Pravachol A trade name for pravastatin sodium, an antilipemic agent that inhibits the synthesis of cholesterol.

Pravastatin Sodium (mol wt 447) An antilipemic agent that acts as inhibitor for 3-hydroxy-3-methylglutaryl-CoA reductase to inhibit the synthesis of cholesterol.

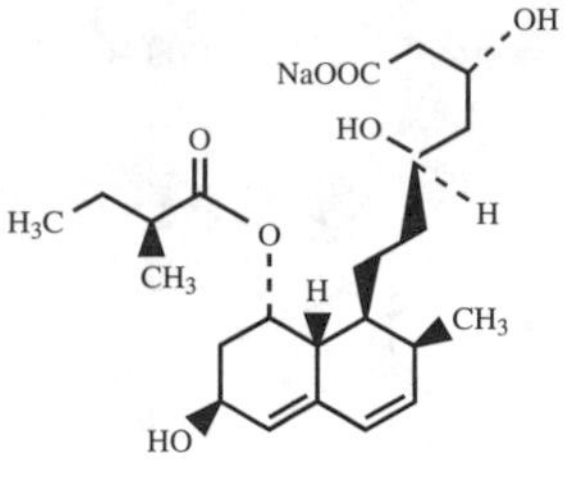

Prax A trade name for pramoxine, an anesthetic agent.

Prazepam (mol wt 325) An antianxiety agent that depresses CNS activity.

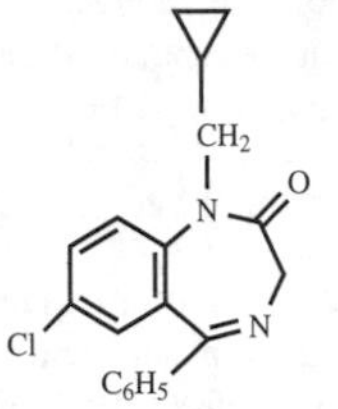

Praziquantel (mol wt 312) An anthelmintic agent that causes changes in the permeability of the cell membrane.

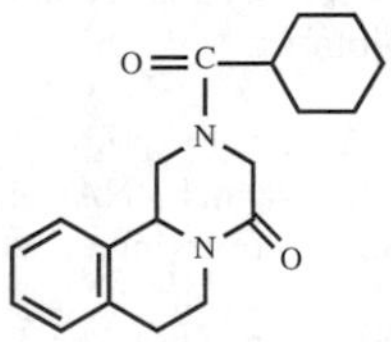

Prazosin (mol wt 383) An antihypertensive agent that relaxes both arteriolar and venous smooth muscle.

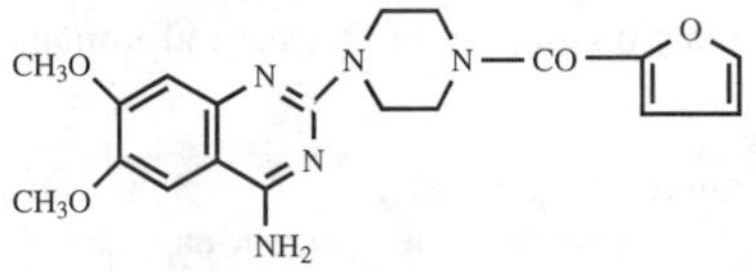

PRBC Abbreviation for packed red blood cells.

PRC Abbreviation for packed red cells.

PRD Abbreviation for proline-rich domain.

pRD111 A plasmid of *E. coli* that contains a Amp^r (ampicillin resistant) marker and BamHI, EcoRI, and HindIII cleavage sites.

PRDI Phage A phage of the family Tectiviridae.

PRE Abbreviation for positive regulatory element.

Prealbumin A serum protein whose electrophoretic mobility ranks ahead of serum albumin.

Pre-B Lymphocytes Immature B cells with diffuse cytoplasmic IgM but no membrane-bound surface immunoglobulin.

Precef A trade name for ceforanide, an antibiotic that inhibits bacterial cell wall synthesis.

Precipitation Reaction An antigen-antibody reaction that leads to the formation of visible precipitate.

Precipitin Precipitate resulting from antigen-antibody reactions (reaction between soluble antigen and soluble antibody).

Precipitin Curve A bell-shaped curve obtained by plotting the amount of antigen-antitody precipitate formed at constant concentration of antibody as a function of increasing amounts of antigen or vice versa.

Precipitin Reaction See precipitation reaction.

Precipitin Test A serological test in which reaction of soluble antigen with antibody forms a visible, insoluble antigen-antibody precipitin.

Precose A trade name for acarbose, an anti-diabetic agent.

Precursor mRNA An unfunctional, unprocessed and nonspliced mRNA that contains both exons and unspliced introns.

Precursor rRNA A large primary transcript from which different ribosomal RNAs are produced by posttranscriptional processing.

Precursor tRNA An unprocessed large precursor of tRNA.

Predaject A trade name for prednisolone acetate, an anti-inflammatory agent capable of suppressing inflammation and modifying normal immune responses.

Predalone A trade name for prednisolone acetate, an anti-inflammatory agent capable of suppressing inflammation and modifying normal immune responses.

Predate A trade name for prednisolone acetate, an anti-inflammatory agent capable of suppressing inflammation and modifying normal immune responses.

Predator An organism that feeds on another living organism.

Predcor A trade name for prednisolone sodium, an anti-inflammatory agent capable of suppressing inflammation and modifying normal immune response.

Pred-Forte A trade name for prednisolone acetate, an anti-inflammatory agent capable of suppressing inflammation and modifying normal immune responses.

Pred-G A trade name for a combination drug containing prednisolone acetate, gentamicin sulfate, chlorobutanol and petrolatum, mineral oil, and lanolin alcohol, used as an ophthalmic anti-infective agent.

Predicort A trade name for prednisolone sodium phosphate, an anti-inflammatory agent capable of suppressing inflammation and modifying normal immune responses.

Pred-Mild A trade name for prednisolone acetate, a hormone.

Prednicarbate (mol wt 489) A topical anti-inflammatory agent capable of suppressing normal immune responses.

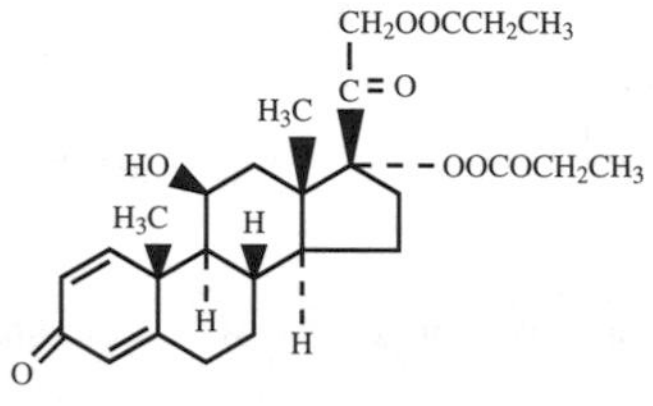

Prednicen-M A trade name for prednisone, an anti-inflammatory agent capable of modifying normal immune responses.

Prednimustine (mol wt 647) An antineoplastic agent.

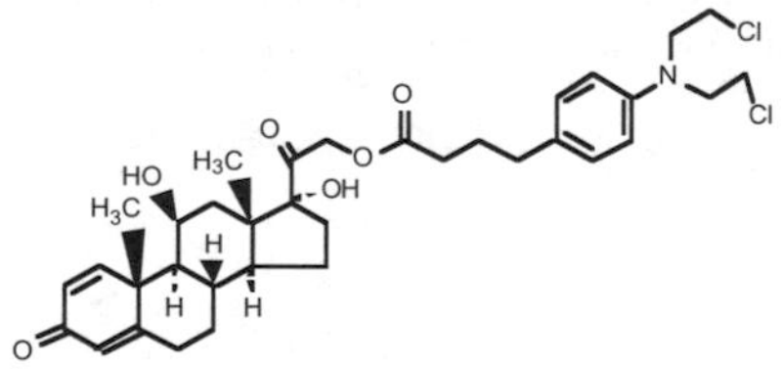

Prednisol A trade name for prednisolone, an anti-inflammatory agent capable of suppressing normal immune responses.

Prednisolone (mol wt 360) A hormone that suppresses inflammation and normal immune responses.

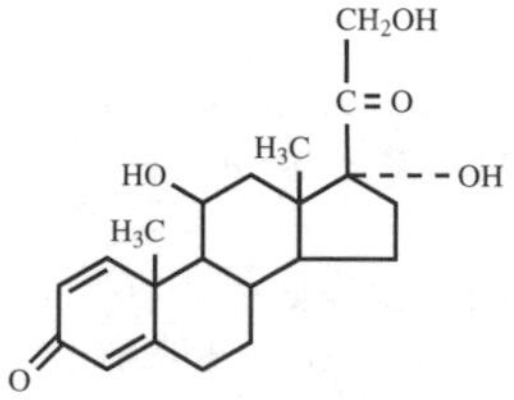

Prednisone (mol wt 358) A hormone that suppresses inflammation and normal immune responses.

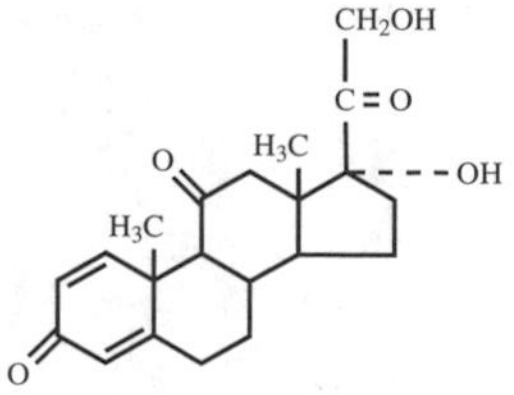

Prednisone Intensol A trade name for prednisone, a hormone used as an anti-inflammatory and immuno-suppressive agent.

Prednylidene (mol wt 372) An anti-inflammatory agent.

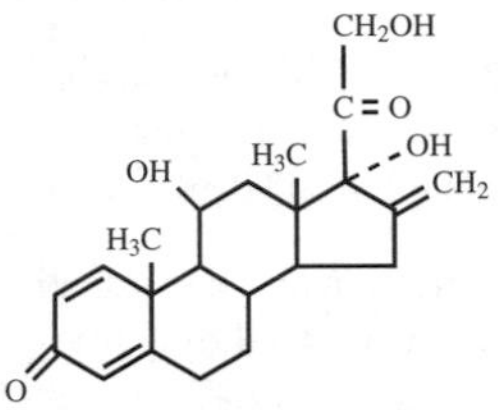

Predsol Eye Drops A trade name for prednisolone sodium phosphate solution, used for the treatment of inflammation of the eyes.

Prefrin-A A trade name for a combination drug containing phenylephrine hydrochloride, pyrilamine maleate, and antipyrine, used as an ophthalmic vasoconstrictor.

Pregnancy Associated α_2-Glycoprotein Synonym of pregnancy zone protein.

Pregnancy Test A serological test for pregnancy in which urine is examined for the presence of chorionic gonadotropin hormone by reaction with antibody to chorionic gonadotropin.

Pregnancy Zone Protein A homotetrameric disulfide-linked proteinase inhibitor of the α_2-macroglobulin family.

Pregnane (mol wt 289) The basic hydrocarbon skeleton of biologically and clinically important steroids.

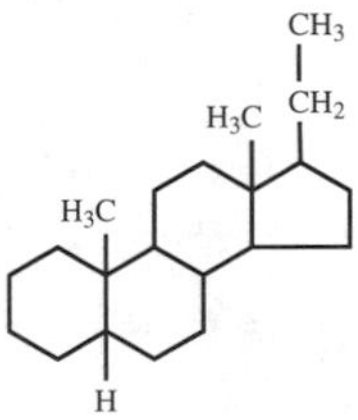

Pregnenolone A substance required for the synthesis of progesterone.

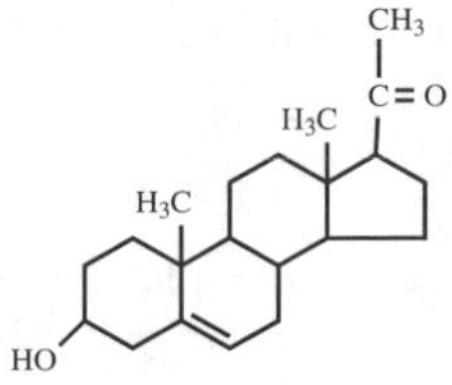

Pregnyl A trade name for the hormone gonadotropin, used to stimulate secretion of gonadal steroid hormones by stimulating production of androgen.

Prekallikrein A precursor of kallikrein.

Prelestone A trade name for betamethasone sodium, used as an anti-inflammatory agent.

Prelone A trade name for prednisolone, an anti-inflammatory agent capable of suppressing normal immune responses.

Prelu-2 A trade name for phendimetrazine tartrate, a cerebral stimulant that stimulates transmission of nerve impulses by releasing stored norepinephrine from the nerve terminals in the brain.

Preludin A trade name for phendimetrazine hydrochloride, a cerebral stimulant that stimulates transmission of nerve impulses by releasing stored norepinephrine from the nerve terminals in the brain.

Premarin A trade name for estrogen, which promotes the growth and development of female sex organs and maintains secondary sex characteristics in women.

Pre-Messenger RNA Gene transcript from which mRNA is formed by post-transcription processing.

Premphase A trade name for a combination drug containing estrogen and medroxyprogesterone used as a menopause drug.

Pre-mRNA Precursor of mRNA or non-processed mRNA.

Premsyn PMS A trade name for a combination drug containing acetaminophen, pamabrom, and pyrilamine used as an analgesic agent.

Prenalterol (mol wt 225) A cardiotonic agent.

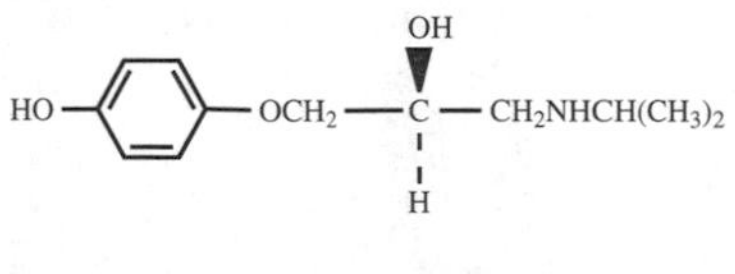

Prenoxdiazine Hydrochloride (mol wt 398) An antitussive agent.

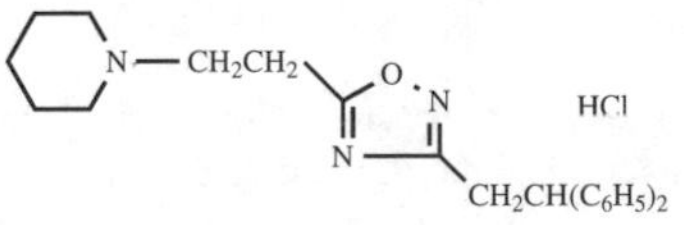

Prenyl Group Referring to the isoprene moiety.

Prenyl Pyrophosphatase The enzyme that catalyzes the following reaction:

$$\text{Prenyl diphosphate} + H_2O \rightleftharpoons \text{Prenol} + \text{PPi}$$

Prenylamine (mol wt 329) A coronary vasodilator.

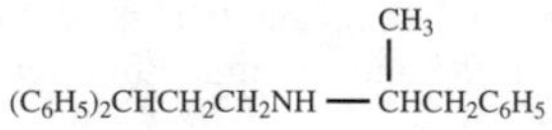

Prenylation Covalent attachment of an isoprenoid lipid group to a protein.

Preparative Method Any chemical method used from preparation of relatively large amounts of sample, e.g., isolation and purification of proteins and nucleic acids.

Preparative Ultracentrifugation An ultracentrifugation procedure for isolation and fractionation of macromolecules.

Prephenate Dehydratase The enzyme that catalyzes the following reaction:

$$\text{Prephenate} \rightleftharpoons \text{phenylpyruvate} + CO_2 + H_2O$$

Prephenate Dehydrogenase The enzyme that catalyzes the following reaction:

$$\text{Prephenate} + NAD^+$$
$$\updownarrow$$
$$\text{4-Hydroxyphenylpyruvate} + CO_2 + NADH$$

Prephenate Dehydrogenase (NADP$^+$-specific) The enzyme that catalyzes the following reaction:

$$\text{Prephenate} + NADP^+$$
$$\updownarrow$$
$$\text{4-Hydroxyphenylpyruvate} + CO_2 + NADPH$$

Prephenic Acid (mol wt 226) An intermediate in the synthesis of phenylalanine.

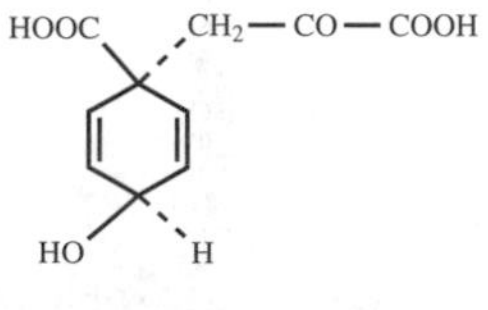

Prepidil Gel A trade name for dinoprostone, a prostaglandin used to stimulate the myometrium for the pregnant uterus to contract.

Preprimosome The precursor of primosome that contains no enzyme primase.

Preprocollagen A procollagen that contains signal peptide at the N-terminal.

Preproinsulin The newly synthesized insulin molecule that contains signal sequence at the N-terminal, which is a precursor of proinsulin.

Pre-Protein A protein that contains signal sequence to insert through the membrane for outward transport.

Prepulsid A trade name for cisapride, a GI drug used to improve GI mobility.

Pre-rRNA Precursor of ribosomal RNA.

Presequence Referring to the signal sequence in preproproteins or preproteins.

Presolol A trade name for labetalol hydrochloride, an antihypertensive agent that blocks alpha and beta stimulation and depresses renin secretion.

Pressyn A trade name for vasopressin, a hormone used to increase GI mobility.

Presumptive Test A test for evaluating water safety by inoculating a water sample into a lactate medium in a Durham tube to observe the formation of gas.

Presynaptic Pertaining to the part of the neuron that sends a signal across a synapse (the receiving cell is postsynaptic).

Presynaptic Cells The cell that releases neurotransmitter to stimulate the postsynaptic cell.

Pre-Transfer RNA Synonym of pre-tRNA.

Pre-tRNA Gene transcript from which tRNA is formed by post-transcription processing.

Pretz-D A trade name for ephedrine sulfate, a vasopressor, bronchodilator, nasal decongestant, and antiasthmatic agent.

Prevacid A trade name for lansoprazole, a protein pump inhibitor used as an antisecretory agent.

Prevalite A trade name for cholestyramine, an antihyperlipidemic agent.

Prevex B A trade name for betamethasone valerate, a hormone used as an anti-inflammatory and immuno-suppressive agent.

Previ-Dent A trade name for sodium fluoride, used to catalyze bone demineralization.

Previtamin Precursor of vitamin.

PRF Abbreviation for prolactin-releasing factor.

PRH Abbreviation for prolactin regulatory hormone.

PRI Abbreviation for phosphoribose isomerase.

Priadel A trade name for lithium carbonate that alters chemical transmitters in the CNS.

Pribnow Box The TATATT sequence on the promoter site of the bacterial cell centered at about –10 region, it is the site for binding sigma subunit of RNA-polymerase.

Pribnow Sequence See Pribnow box.

PRibPP Abbreviation for 5'-phosphoribosyl diphosphate.

Prickle Cells Cells with the cytoplasmic processes that form intercellular bridges.

Prifinium Bromide (mol wt 386) An antispasmodic agent.

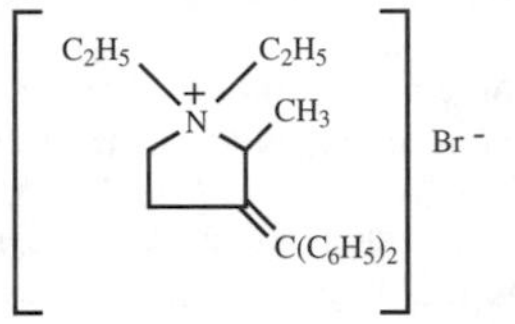

Priftin A trade name for rifapentine, an antituberculous antibiotic.

PRIH Abbreviation for prolactin-releasing inhibiting hormone.

Prilocaine (mol wt 220) A local anesthetic agent.

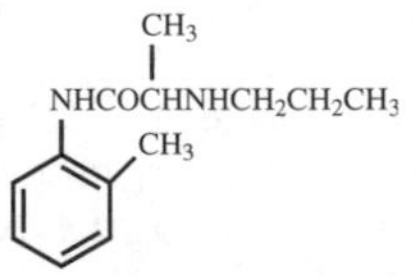

Prilosec A trade name for omeprazole, an antiulcer agent that inhibits activity of H^+/K^+- ATPase and blocks the formation of gastric acid.

Primaclone A trade name for primidone, an anticonvulsant.

Primacor A trade name for milrinone, a cardiotonic agent used for relaxation of smooth muscle.

Primaperone (mol wt 249) An antihypertensive agent.

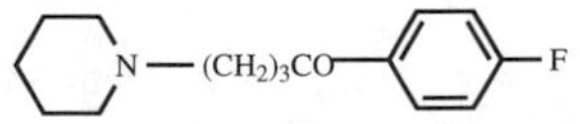

Primaquine (mol wt 259) An antimicrobial agent.

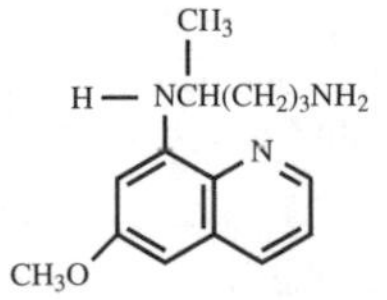

Primary Cell Culture A culture started from cells taken directly from living tissue.

Primary Cell Wall Cell wall of plant that is still capable of expanding, permitting cell growth.

Primary Cells Cells directly obtained from multicellular organisms and seeded onto culture plates.

Primary Fixation Initial step in the preparation of a specimen for microscopic examination that stabilizes the chemical components of the cells and hardens the specimen for further processing.

Primary Follicles Aggregates of lymphocytes in the cortex of lymph nodes and the white pulp of the spleen after antigenic stimulation. They are the sites of germinal center development.

Primary Hormone Any hormone that acts rapidly and uses a cyclic nucleotide as a second messenger.

Primary Immune Response Initial immune response to a particular antigen characterized by the production of large amounts of IgM antibodies.

Primary Lymphoid Organs Organs in which lymphocytes undergo a period of maturation and differentiation, e.g., bone marrow (or bursa of Fabricius in birds) and thymus.

Primary Lysosome Lysosome that has not yet fused with a vesicle or not yet engaged in digestive activity.

Primary Oocyte Cell derived from an oogonium by mitotic division that gives rise to an egg cell by meiosis.

Primary Oxaluria A genetic disorder characterized by the formation and deposition of oxalic acid in the kidneys or other tissue due to deficiency in a-ketoglutarate-glyoxylate carboligase.

Primary Pneumonic Plague A form of plague caused by invasion of the lungs by *Yersinia pestis*.

Primary Spermatocyte Cell that is derived from a spermatogonium by mitotic division and that gives rise to sperm cells by meiosis.

Primary Structure Referring to a sequence of amino acids in a polypeptide chain or sequence of nucleotides in a nucleic acid.

Primary Transcript Referring to the newly transcribed RNA from DNA without posttranscriptional modification or splicing.

Primary Tumor Initial malignant mass of proliferating cells in an organism.

Primary Wall See primary cell wall.

Primase A DNA-dependent RNA polymerase that synthesizes RNA primers required for initiation of replication of a DNA duplex.

Primatene A trade name for a combination drug containing theophyline and ephedrine, used as an anti-asthmatic agent.

Primatere Mist Solution A trade name for epinephrine solution, used as a bronchodilator.

Primaxin A trade name for imipenem/cilastatin sodium, used as an antibacterial agent that inhibits bacterial cell wall synthesis.

Primazine A trade name for promazine hydrochloride, a dopaminergic blocking agent used as an anti-psychotic agent.

Primed Referring to an individual who has had an initial immunological contact with a given antigen.

Primed Cells Referring to lymphocytes that have encountered an antigen.

Primer A short sequence of RNA or DNA that serves as starting point for synthesis of DNA.

Primer RNA A short sequence of RNA formed by primase that serves as starting point for DNA synthesis.

Primeverose (mol wt 312) A disaccharide.

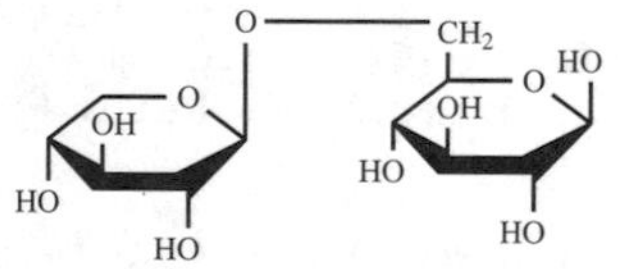

Primidone (mol wt 218) An anticonvulsant.

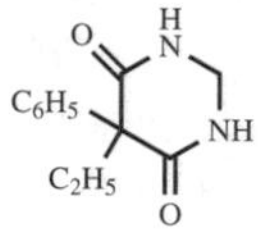

Priming 1. A process in which an individual becomes primed. 2. The initiation of DNA synthesis by synthesis of a primer RNA.

Primocarcin (mol wt 184) An antineoplastic antibiotic produced by the actinomycetes.

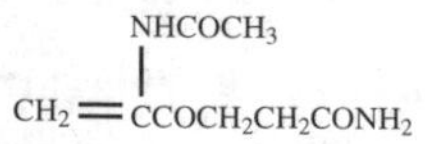

Primogyn Depot A trade name for estradiol valerate, used to stimulate the synthesis of DNA, RNA, and protein in the responsive tissue.

Primordial Cells Cells that form in the early stages of embryonic development.

Primosome Complex of proteins associated with primase that contains recognition factors required for synthesis of the primers in the initiation of DNA replication.

Principen A trade name for ampicillin, an antibiotic that inhibits bacterial cell wall synthesis.

Prinivil A trade name for lisinopril, an antihypertensive agent that inhibits angiotensin converting enzyme preventing conversion of angiotensin I to angiotensin II. It is also a potent vasoconstrictor.

Prinzide A trade name for a combination drug containing lisinopril and hydrochlorothiazide used as an antihypertensive agent.

Prions Infectious proteins that cause diseases such as scrapie in sheep and Kuru in humans.

Priscoline A trade name for tolazoline hydrochloride, used as a vasodilator.

PRIST Abbreviation for paper radio-immunosorbent test.

Private Antigen Histocompatibility antigen restricted to a specific given allele.

Private Specificities Antigenic specificity of MHC-encoded proteins that are unique to particular haplotypes.

Privine A trade name for naphazoline hydrochloride, used as a nasal agent to produce local vasoconstriction of dilated arterioles to reduce blood flow and nasal congestion.

pRL124 A plasmid of *E. coli* that contains a Amp^r (ampicillin resistant) marker and SphI, SalI, XmaI, SmaI, KpnI, and EcoRI cleavage sites.

PRM Abbreviation for 1. phosphoribose mutase; 2. photo-receptor membrane.

PRNT Abbreviation for plaque reduction neutralization test.

Pro- A prefix meaning 1. an inactive precursor, 2. a developmental stage that comes first, and 3. abbreviation for proline.

Pro-50 A trade name for promethazine hydrochloride, used as an antihistaminic agent.

Proaccelerin A blood clotting factor (factor V) that converts prothrombin to thrombin.

ProAmatine A trade name for midodrine hydrochloride, an antihypertensive agent.

Proaqua A trade name for benzthiazide, a diuretic agent that increases urine excretion of sodium and water by inhibiting reabsorption of sodium.

Probalan A trade name for probenecid, an antigout agent that blocks renal tubular reaborption of uric acid and increases uric acid excretion.

Pro-Banthine A trade name for propantheline bromide, an anticholinergic agent that blocks acetylcholine and decreases GI motility and inhibits gastric acid secretion.

Probe A sequence of labeled DNA or RNA that is used to locate and identify the sequences on a blot by hybridization under optimal conditions of salt concentration and temperature.

Proben-C A trade name for a combination drug containing probenecid and colchicine, used as an antigout agent.

Probenecid (mol wt 285) An antigout agent that blocks renal tubular reabsorption of uric acid, which increases excretion of uric acid.

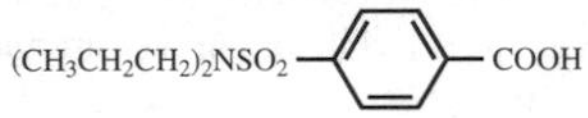

Probucol (mol wt 517) An antilipemic agent that inhibits cholesterol transport from the intestine, prevents oxidation of low density lipoproteins, and decreases synthesis of cholesterol.

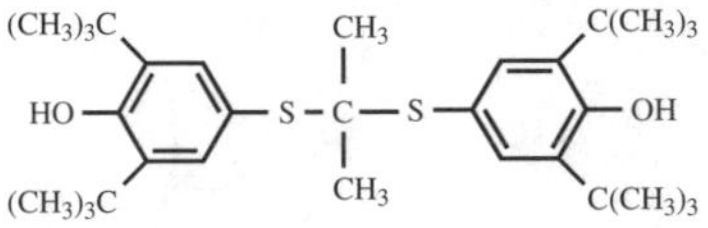

Procainamide Hydrochloride (mol wt 272) An antiarrhythmic agent that prolongs action potential.

Procaine (mol wt 236) A local anesthetic agent that blocks depolarization by interfering with sodium-potassium exchange across the nerve cell membrane.

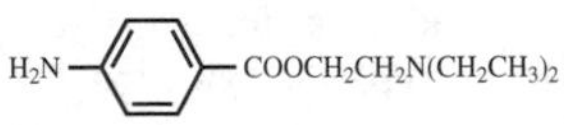

Procaine Esterase See carboxylesterase.

Procambium Primary meristem that produces the vascular tissue.

Procan SR A trade name for procainamide hydrochloride, an antiarrhythmic agent that prolongs action potential.

Procanbid A trade name for procainamide hydrochloride, an antiarrhythmic agent.

Procarb Abbreviation for procarbazine.

Procarbazine (mol wt 221) An antibiotic that inhibits the synthesis of DNA, RNA, and protein.

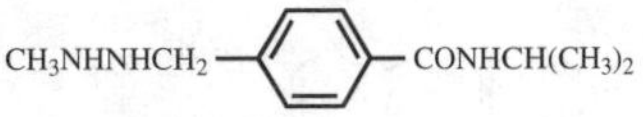

Procarboxypeptidase A precursor of carboxypeptidase that is converted to active carboxypeptidase by proteolytic enzymes.

Procarcinogen Chemical that acts as a carcinogen after activation by a biological or biochemical system.

Procardia A trade name for nifedipine, an antianginal agent that inhibits calcium ion influx across cardiac and smooth muscle cells and decreases cardiac oxygen demand.

Procaryon The nuclear region of a procaryotic cell.

Procaryote Synonym of prokaryote.

Procaterol (mol wt 290) A bronchodilator.

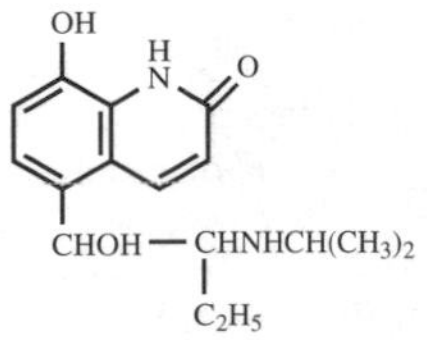

Procentriole An immature centriole

Processive Enzyme An enzyme that catalyzes a series of successive polymerization steps without releasing free enzyme.

Prochloraz (mol wt 377) An antifungal agent that inhibits sterol biosynthesis.

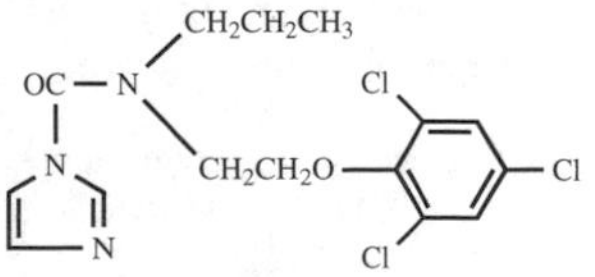

Prochlorperazine (mol wt 374) An antiemetic agent.

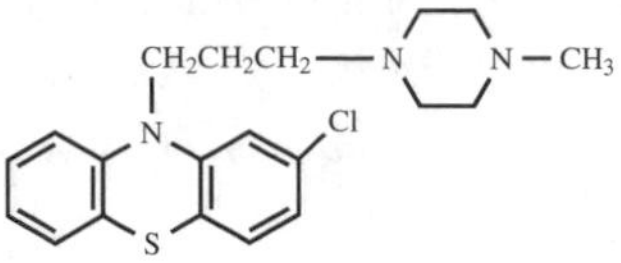

Procion Blue MX-R (mol wt 637) A dye for staining proteins in gel electrophoresis.

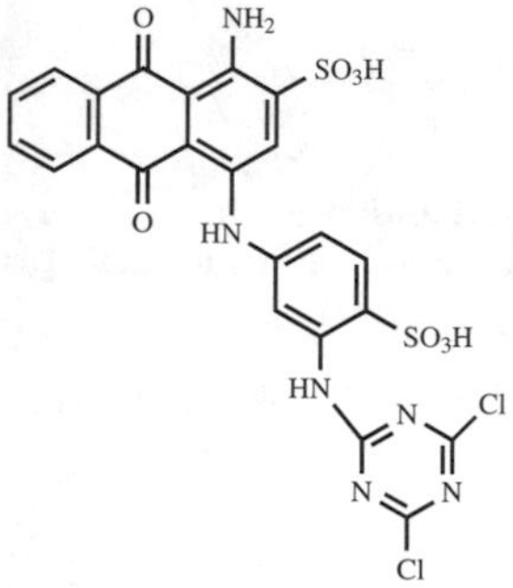

Procodazole (mol wt 190) An immunopotentiator.

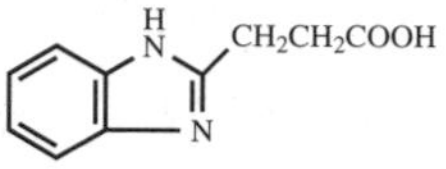

Procollagen The precursor of collagen that consists of triple-helical polypeptide chains with terminal extension peptides on both N-terminal and C-terminal regions.

Procollagen C-Endopeptidase The enzyme that catalyzes the cleavage of C-terminal propeptide from procollagen.

Procollagen N-Endopeptidase The enzyme that catalyzes the cleavage of N-terminal propeptide from procollagen.

Procollagen Galactosyltransferase The enzyme that catalyzes the following reaction:

UDP-galactose + procollagen-5-hydroxyl-lysine
⇅
UDP + procollagen- galactosyloxyl-lysine

Procollagen Glucosyltransferase The enzyme that catalyzes the following reaction:

UDP-glucose + 5-galactosyloxy-L-lysine-procollagen
⇅
UDP + 1,2,-D-glucosyl-5-D-galactosyloxy-lysine-procollagen

Procollagen Peptidase The protease that catalyzes the removal of the terminal extension peptides from procollagen.

Proconvertin The precursor of convertin.

Procrit A trade name for epoetin alfa (erythropoietin) that functions as a growth and cell differentiation factor.

Proctectomy Surgical removal of anus and rectum.

Proctitis Inflammation of the rectum.

Proctocolectomy Surgical removal of the rectum and colon.

Proctocolitis Inflammation of the anus, rectum.

Proctocort A trade name for hydrocortisone, a hormone that suppresses inflammation and normal immune responses.

Proctofoam-HC A trade name for hydrocortisone acetate, a hormone that suppresses inflammation and normal immune responses.

Proctogram X-ray photograph of the rectum taken after the introduction of a contrast medium.

Procyclid A trade name for procyclidine hydrochloride, an antiparkinsonian agent that blocks cholinergic receptors.

Procyclidine (mol wt 287) An antiparkinsonian agent that blocks the central cholinergic receptors, helping to balance cholinergic activity.

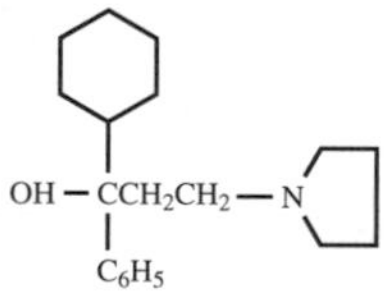

Procymate (mol wt 185) A tranquilizer.

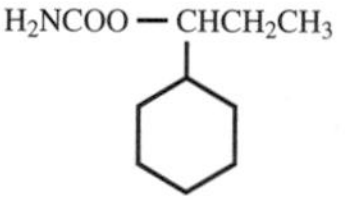

Procytox A trade name for cyclophosphamide, an alkylating agent that cross-links cellular DNA interferring with transcription.

Pro-Depo A trade name for hydroxyprogesterone caproate used to suppress ovulation.

Prodigiosin (mol wt 323) An antibiotic pigment produced by *Chromobacterium prodigiosum.*

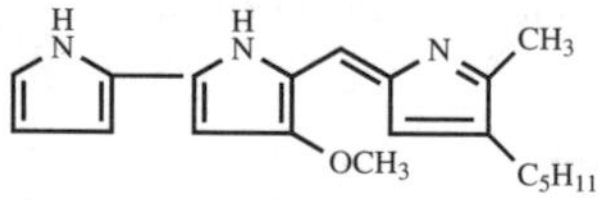

Prodilidine (mol wt 247) An analgesic agent.

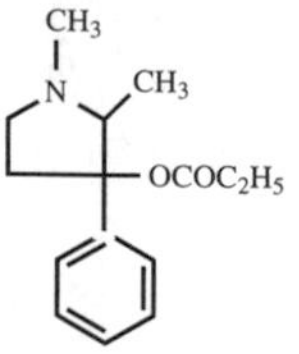

Prodipine (mol wt 279) An antiparkinsonian agent.

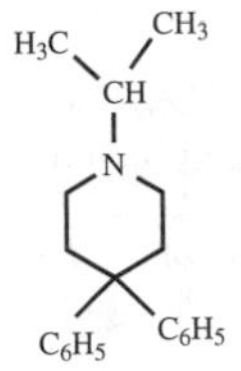

Prodium A trade name for phenazopyridine hydrochloride, an analgesic agent.

Prodrox A trade name for hydroxyprogesterone caproate, used to suppress ovulation.

Proelastase A precursor of elastase that can be activated by tryptic cleavage from the N-terminal.

Proenzyme The precursor of an enzyme that becomes fully functional after removal of the inhibitory sequence (also known as zymogen).

Proerythroblast The earliest recognizable precursor of erythrocytes (red blood cell).

Profasi HP A trade name for gonadotropin, used to stimulate ovulation.

Professional Phagocytes Referring to neutrophils and macrophages.

Profibrinolysin Precursor of fibrinolysin.

Profilactin Complex of profilin and actin.

Profilin An actin-binding protein that forms a complex with G-actin, preventing actin polymerization.

Profilnine SD A trade name for factor IX, an anti-hemophilic agent.

Proflavine (mol wt 209) A topical antiseptic agent, an inhibitor of DNA polymerase, and an acridine dye.

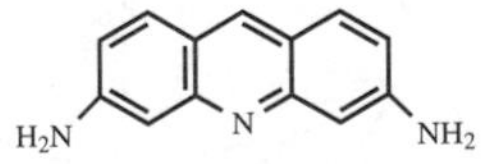

Progabide (mol wt 335) An anticonvulsant.

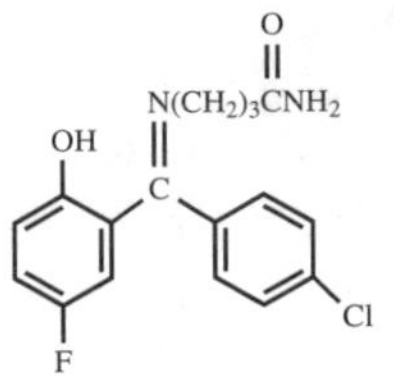

Pro-Gal-Sof A trade name for docusate calcium, used as a laxative.

Progenote Primitive organism presumed to be the phylogenetic progenitors for both prokaryotes and eukaryotes.

Progens A trade name for estrogen, used to reduce FSH and LH release from the pituitary.

Progeria A genetic disorder due to a deficiency in DNA repair.

Progestaject A trade name for progesterone, used to suppress ovulation.

Progestarsert A trade name for progesterone, a hormone that inhibits the secretion of pituitary gonadotropins.

Progesterone (mol wt 314) A 21-carbon female hormone that functions to maintain pregnancy and suppress ovulation.

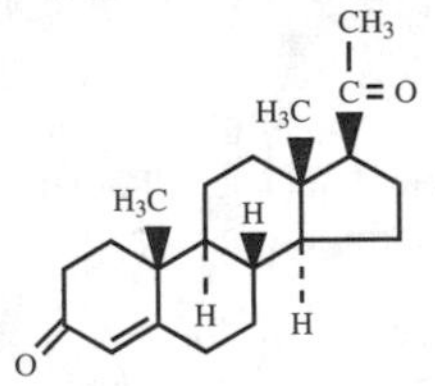

Progesterone Hydroxylase See progesterone monooxygenase.

Progesterone Monooxygenase The enzyme that catalyzes the following reaction:

$$\text{Progesterone} + AH_2 + O_2 \rightleftharpoons \text{Testosterone acetate} + A + H_2O$$

Progestilin A trade name for progesterone, used to suppress ovulation.

Progestin Substance with progesterone-like activity.

Progestogen Substance capable of inducing progestational changes in the uterus.

Proglumide (mol wt 334) An anticholinergic agent.

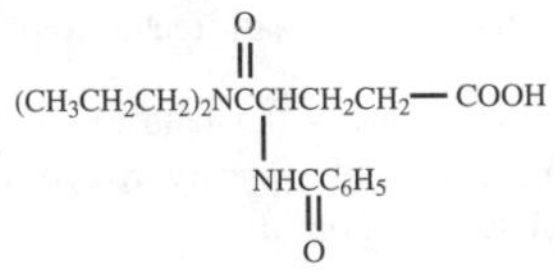

Proglycem A trade name for diazoxide, used to inhibit the release of insulin from the pancreas and decrease the peripheral utilization of glucose.

Prognosis A forecast of the course and final outcome of a disease.

Prograf A trade name for tacrolimus, an immunosuppressant.

Programmed Cell Death The notion that cells are destined to die at a specific stage through a programmed sequence of events. Also known as apoptosis.

Prohead The head of a bacteriophage that contains no DNA formed during the early stage of phage assembly.

ProHIBIT A trade name for hemophilus b vaccine.

Prohormone A precursor of peptide hormone that becomes active after proteolytic removal of the inhibitory sequence.

Proinsulin A precursor of insulin that becomes active after proteolytic removal of 33 amino acid residues.

Prokaryotes Unicellular organisms that lack membrane-bound nuclei, cell organelles, and contain a single chromosome.

Prokine A trade name for sargramostim, a granulocyte-macrophage colony stimulating factor.

Prokinetic Agent Any agent that induces intestinal activity.

Prolactin A peptide hormone secreted by the anterior lobe of the pituitary gland that initiates the growth of mammary glands and stimulates lactation.

Proladone A trade name for oxycodone pectinate, an analgesic agent that binds with opiate receptors altering both perception and emotional response to pain.

Prolamins Proteins that are soluble in 70-80% alcohol, insoluble in 100% alcohol, water, and other neutral solvents, e.g., zein (from corn) and gliadin (from wheat).

Prolastin An alpha-1 proteinase inhibitor.

Proleukin A trade name for interleukin-2 produced by DNA recombinant technology.

Prolidase The enzyme that catalyzes the removal of the amino acid proline or hydroxyproline from the N-terminal of a peptide.

Proliferin One of several proteins of the somatotropin/prolactin family that provide a growth stimulus to target cells in maternal and fetal tissues during the development of the embryo.

Proline (mol wt 115) A nonessential, heterocyclic and helix-breaking protein amino acid.

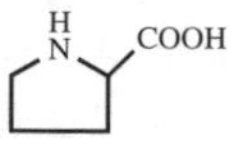

Proline Aminopeptidase The enzyme that catalyzes the removal of any amino acid linked to proline.

Proline Dehydrogenase The enzyme that catalyzes the following reaction:

Proline + acceptor + H_2O

⇅

1-Pyrroline-5-carboxylate + reduced acceptor

Proline Dipeptidase Synonym of X-pro dipeptidase.

Proline Iminopeptidase Synonym of prolyl aminopeptidase.

Proline Racemase The enzyme that catalyzes the following reaction:

L-Proline ⇌ D-Proline

D-Proline Reductase The enzyme that catalyzes the following reaction:

5-Aminopentanoate + lipoate

⇅

D-Proline + dihydrolipoate

Proline Specific Endopeptidase Synonym of prolyl oligopeptidase.

Proline tRNA Ligase The enzyme that catalyzes the following reaction:

ATP + L-proline + $tRNA^{pro}$

⇅

AMP + PPi + L-prolyl-$tRNA^{pro}$

Proline tRNA Synthetase See proline tRNA ligase.

Prolintane (mol wt 217) An antidepressant and a CNS stimulant.

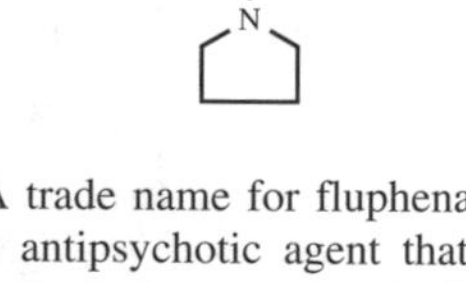

Prolixin A trade name for fluphenazine hydrochloride, an antipsychotic agent that blocks the postsynaptic dopamine receptors in the brain.

Proloid A trade name for thyroglobin, used to stimulate cellular oxidation.

Proloprim A trade name for trimethoprim, an antibacterial agent that inhibits bacterial folic acid synthesis.

Prolyl Aminopeptidase The enzyme that catalyzes the release of N-terminal proline from a peptide.

Prolyl Hydroxylase The enzyme that catalyzes the conversion of proline to hydroxyproline.

Prolyl Oligopeptidase The enzyme that catalyzes the hydrolysis of peptide bonds involving the carboxyl group of proline and alanine in oligopeptides.

Prolyl tRNA Synthetase See proline tRNA ligase.

Promazine (mol wt 284) An antipsychotic agent that blocks postsynaptic dopamine receptors in the brain.

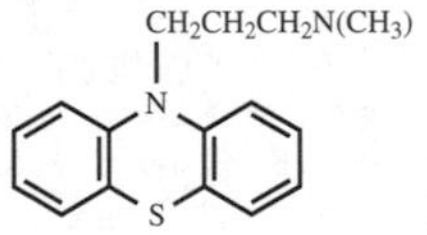

Promecarb (mol wt 207) An insecticide and an inhibitor of cholinesterase.

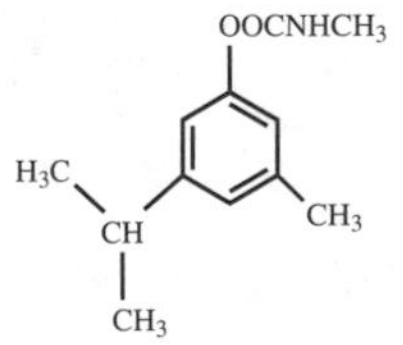

Promethazine (mol wt 284) An antihistaminic agent that competes with histamine for H_1 receptor sites on effector cells.

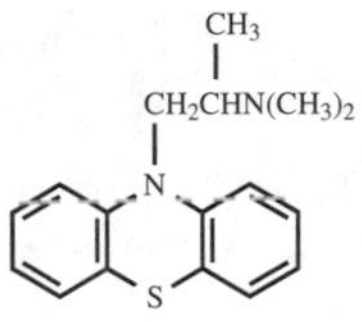

Promethegan A trade name for promethazine hydrochloride, an antihistaminic agent that competes with histamine for H_1 receptor sites on effector cells.

Promethium (Pm) A chemical element with atomic weight 147, valence 3.

Prometrium A trade name for progesterone, a hormone that inhibits the secretion of pituitary gonadotropins.

Promine A trade name for procainamide hydrochloride, an antiarrhythmic agent that prolongs action potential.

Promoter A DNA sequence to which RNA polymerase binds and initiates transcription.

Promoter Mutation A mutation that occurs in the promoter region of DNA that may lead to the decrease or increase of transcription or formation of new promoter.

Promoxolane (mol wt 188) A skeletal muscle relaxant.

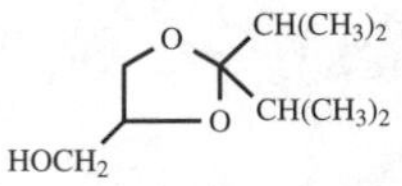

Pronase 1. A trade name for a mixture of various exo- and endo-peptidases obtained from *Streptomyces griseus* that hydrolyze protein to free amino acids. 2. A protease from *Streptomyces griseus*.

Pronestyl A trade name for procainamide hydrochloride, an antiarrhythmic agent that prolongs action potential.

Pronethalol (mol wt 229) An antianginal and antiarrhythmic agent.

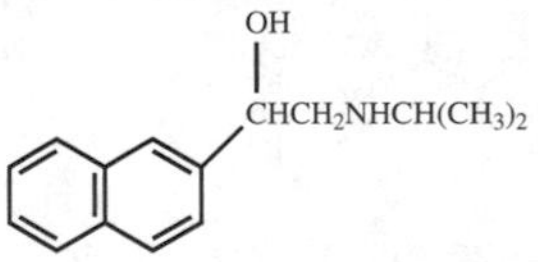

Pronto A trade name for pyrethrins, used as an antiparasitic agent.

Proof Reading A process of correcting errors during the process of DNA replication and transcription.

Pro(3-OH) Abbreviation for 3-hydroxyproline.

Proopiomelanocortin A polyprotein synthesized by the pituitary gland that yields multiple active hormones, e.g., ATCH and b-lipotropin.

Propacetamol (mol wt 264) An analgesic and antipyretic agent.

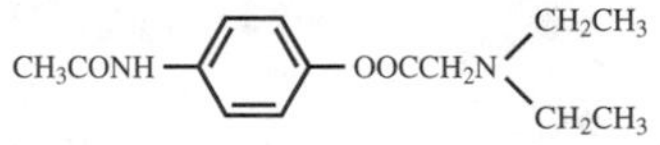

Propaderm A trade name for beclomethasone dipropionate, a hormone with anti-inflammatory activity.

Propagest A trade name for phenylpropanolamine hydrochloride, a nasal decongestant.

Propallylonal (mol wt 289) A sedative-hypnotic agent.

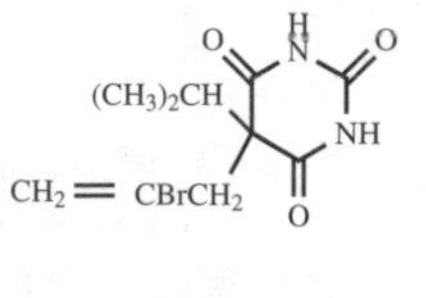

Propanidid (mol wt 337) An anesthetic agent.

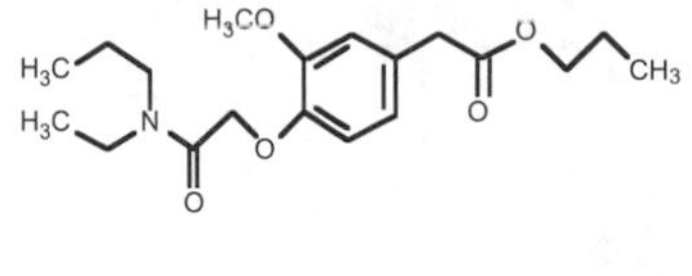

Propanocaine (mol wt 311) A local anesthetic agent.

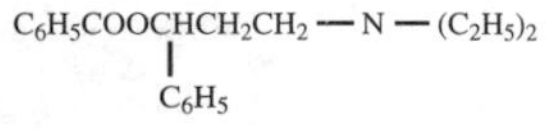

Propanthel A trade name for propantheline bromide, an anticholinergic agent that blocks acetylcholine, decreases GI motility, and inhibits gastric acid secretion.

Propantheline Bromide (mol wt 448) An anticholinergic agent that blocks acetylcholine, decreases GI motility, and inhibits gastric acid secretion.

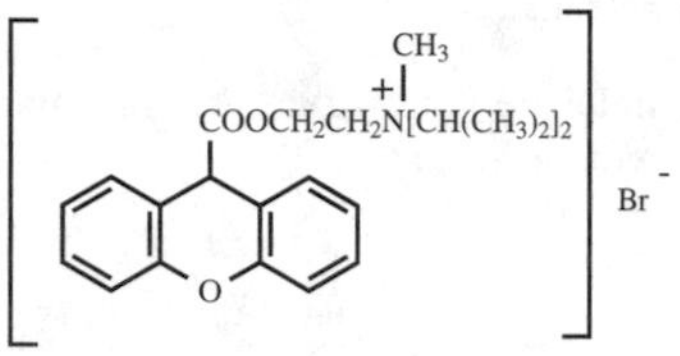

Proparacaine (mol wt 294) A topical ophthalmic anesthetic agent that produces anesthesia by preventing initiation and transmission of impulses at the nerve cell membrane.

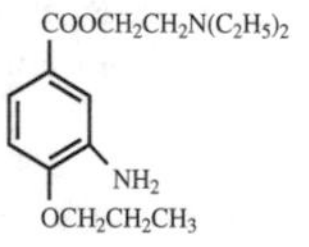

Propatyl Nitrate (mol wt 269) A coronary vasodilator.

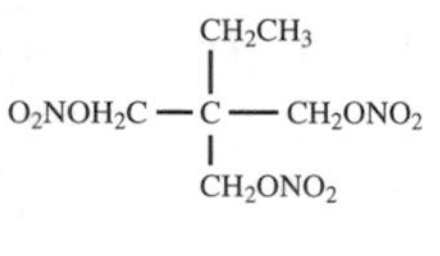

Propecia A trade name for finasteride, an inhibitor for androgen.

Propentofylline (mol wt 306) A vasodilator and inhibitor of cyclic AMP phosphodiesterase.

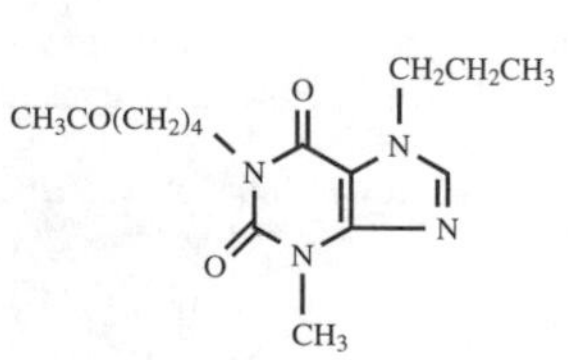

Propenzolate (mol wt 332) An anticholinergic agent.

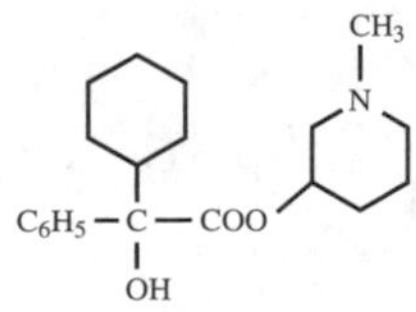

Properdin A basic serum protein involved in the alternative pathway of complement activation (also known as factor P).

Properdin Pathway Referring to the alternative pathway of complement activation.

Prophage The integrated phage genome on the host's chromosome that replicates as part of the bacterial chromosome during subsequent cell division.

Prophage Excision The excision of prophage DNA from the bacterial chromosome.

Prophage Immunity The resistance of lysogenic bacteria to further infection by a second phage of the same or similar type. The lysogenic infection established by P22 phage in *Salmonella typhimurium* causes glycosylation and modification of host receptors, preventing further infection by P22 and related phage.

Prophage Induction A process by which prophage DNA is excised from the bacterial chromosome leading to the production of phage particles and lysis of bacterial cells.

Prophage Integration The incorporation of DNA of the temperate phage into the bacterial chromosome.

Prophase The first stage of mitosis during which chromosomes are condensed but not yet attached to the mitotic spindle.

Prophylactic Preventive treatment for protection against disease.

Prophylaxis Preventive treatment for protection against disease, e.g., vaccination.

Propicillin (mol wt 378) A semisynthetic antibiotic related to penicillin that inhibits bacterial cell wall synthesis.

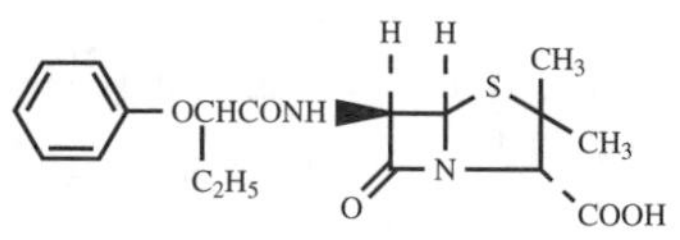

Propiconazole (mol wt 342) An agricultural fungicide.

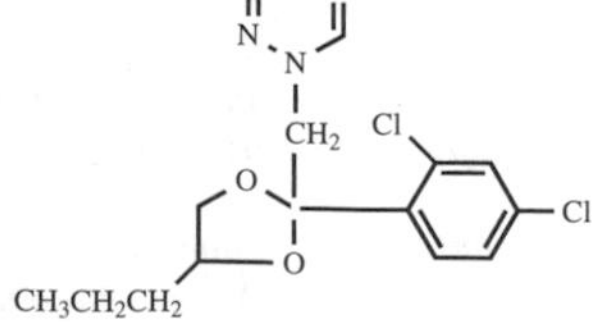

Propine A trade name for dipivefrin, used to treat eye disorders.

Propion A trade name for diethylpropion hydrochloride, used to promote transmission of nerve impulses.

Propionate-CoA Ligase The enzyme that catalyzes the following reaction:

ATP + propanoate + CoA
⇌
AMP + PPi + propanoyl-CoA

Propionate CoA-Transferase The enzyme that catalyzes the following reaction:

Acetyl-CoA + propanoate
⇌
Acetate + propanoyl-CoA

Propionibacterium A genus of Gram-positive rod-shaped bacteria (family of Propionibacteriaceae that produces propionic acid, acetic acid, or mixtures of organic acids by fermentation.

Propionic Acid (mol wt 74) A fermentation product of propionic bacteria, e.g., *Propionibacterium*

CH_3CH_2COOH

Propionic Acidemia A genetic disorder characterized by massive ketosis due to a deficiency in propionyl-CoA carboxylase leading to mental and physical retardation.

Propionic Acid Fermentation Pathway A metabolic pathway carried out by propionic bacteria for the production of propionic acid.

Propionigenium A genus of anaerobic, asporogenous bacteria.

Propionispira A genus of Gram-negative, obligately anaerobic, asporogenous, nitrogen-fixing bacteria.

Propionyl-CoA (mol wt 824) A substrate for synthesis of succinyl-CoA.

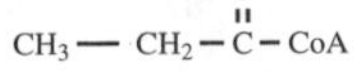

Propionyl-CoA Carboxylase The enzyme that catalyzes the following reaction:

ATP + propanoyl-CoA + CO_2
⇌
ADP + Pi + methylmalonyl-CoA

Propionyl-CoA Synthetase The enzyme that catalyzes the following reaction:

Propanoate + CoA + ATP
⇌
AMP + PPi + + propanoyl-CoA

Propionylpromazine (mol wt 341) A tranquilizer.

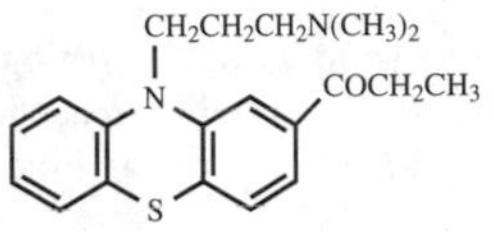

Propipocaine (mol wt 275) A local anesthetic agent.

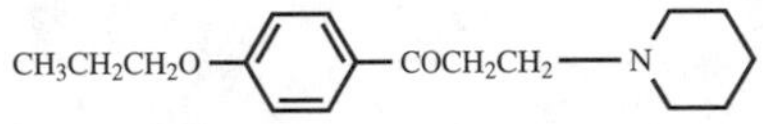

Propiram (mol wt 275) A narcotic analgesic agent.

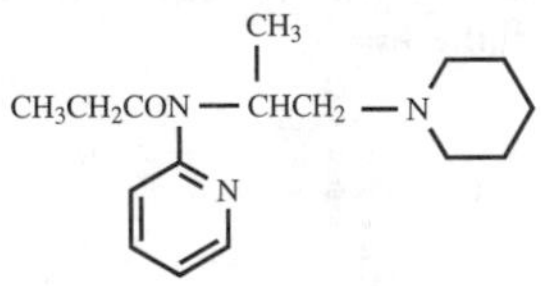

Propivane (mol wt 314) An antispasmodic agent.

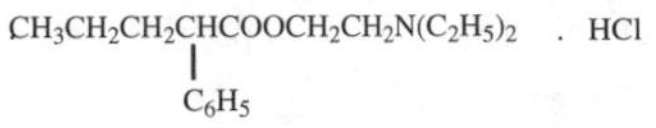

Propizepine (mol wt 296) An antidepressant.

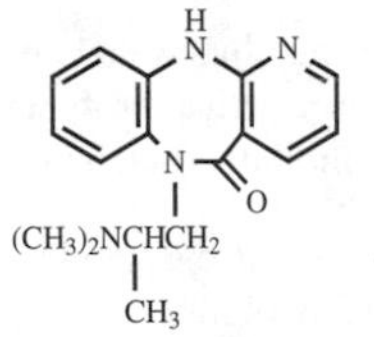

Proplastid A colorless, immature plastid that develops into a plastid under appropriate conditions.

Proplex T A trade name for factor XI complex, used to replace the deficient factor XI clotting factor.

Propofol (mol wt 178) An anesthetic agent.

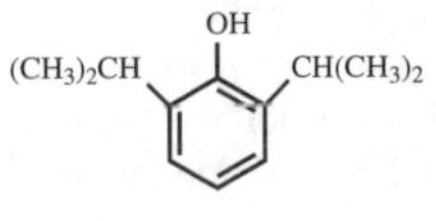

Pro-Pox A trade name for propoxyphene hydrochloride, an opiate analgesic agent that binds with opiate receptors in the CNS.

Propoxycaine Hydrochloride (mol wt 331) A local anesthetic agent.

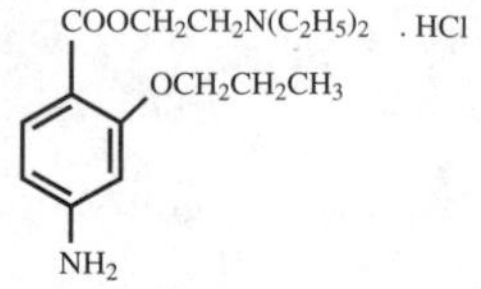

Propoxycon A trade name for propoxyphene hydrochloride, an opiate analgesic agent that binds with opiate receptors in the CNS.

Propoxyphene (mol wt 339) An opiate analgesic agent that binds to opiate receptors in CNS, altering perception and emotional responses to pain.

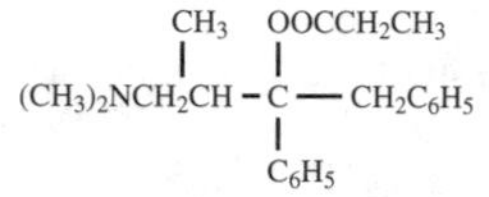

Propranolol (mol wt 259) An antianginal agent that reduces cardiac oxygen demand by blocking the catecholamine-induced increase in heart rate and blood pressure.

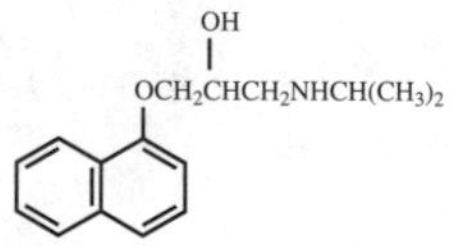

Propressophysin The polypeptide precursor of vasopressin and neurophysin.

Proprietary Name Synonym of trade name or brand name.

Pro-protein An unfunctional precursor of a protein that becomes active or functional after the removal of the inhibitory sequence by proteinase.

Propulsid A trade name for cisapride, a GI drug that improves GI mobility.

Propyl Gallate (mol wt 212) An antioxidant for food, fat, and oils.

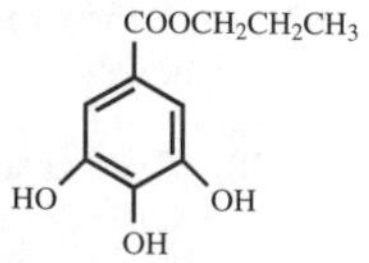

Propylhexedrine (mol wt 155) An adrenergic vasodilator and a nasal decongestant.

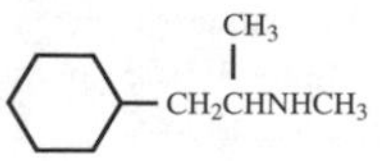

Propylmalate Synthetase The enzyme that catalyzes the following reaction:

3-Propylmalate + CoA

⇅

Pentanoyl-CoA + H_2O + glyoxylate

Propylthiouracil (mol wt 170) A substance that inhibits oxidation of iodine in the thyroid gland, blocking iodine's ability to combine with tyrosine to form thyroxine.

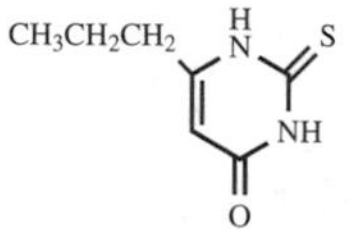

Propyl-Thyracil A trade name for propylthiouracil, used to inhibit oxidation of iodine and to prevent formation of thyroxine.

Propylxanthine (mol wt 194) An adenosine antagonist.

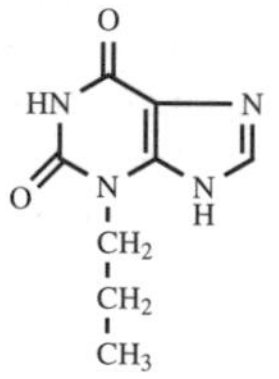

Propyphenazone (mol wt 230) An analgesic, antipyretic agent.

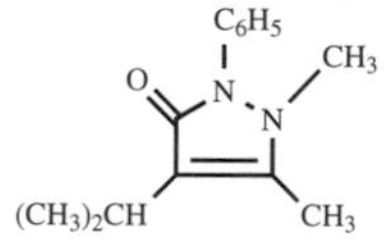

Propyromazine (mol wt 419) An anticholinergic and antispasmodic agent.

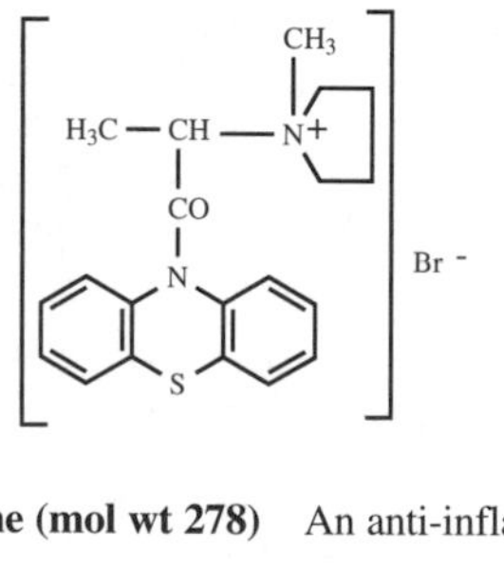

Proquazone (mol wt 278) An anti-inflammatory agent.

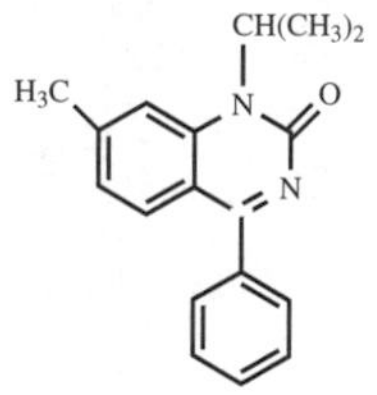

Prorex A trade name for promethazine hydrochloride, a dopaminergic blocking agent used as an antihistaminic and antiemetic agent.

Proscar (mol wt 373) An inhibitor for conversion of testosterone to androgen.

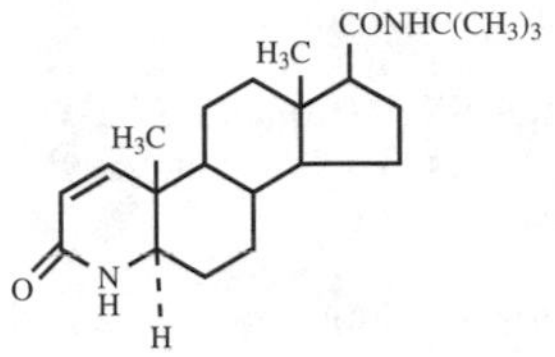

Proscillaridin (mol wt 531) A cardiotonic agent.

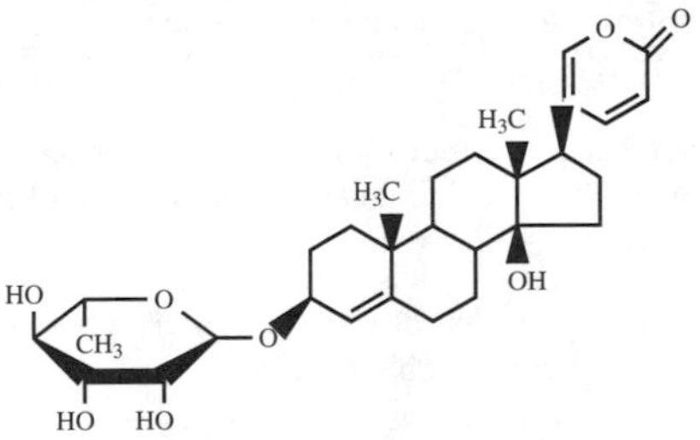

Pro-Sof A trade name for docusate sodium, a laxative and stool softener that promotes the incorporation of liquid into the stool.

ProSom A trade name for estazolam, a sedative-hypnotic agent.

Prosome A ribonucleoprotein particle present in the nucleus and cytoplasm of various types of eukaryotic cells that is involved in posttranscriptional regulation of gene expression.

Prostacyclin (mol wt 352) An inhibitor for platelet aggregation.

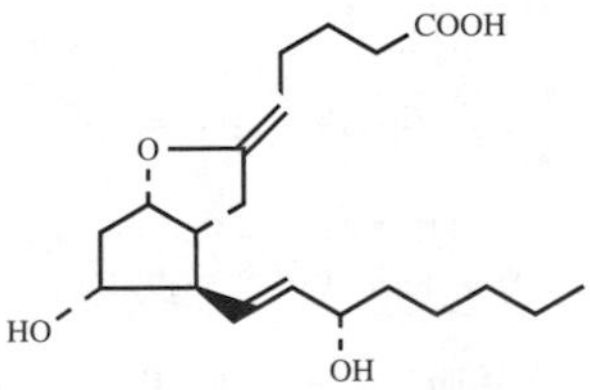

Prostacyclin Synthetase The enzyme that catalyzes the synthesis of prastacyclin.

Prostaglandin Bioactive lipids generated by the action of cyclooxygenase from arachidonic acid. Prostaglandins inhibit platelet aggregation, increase vascular permeability, and promote smooth muscle contraction.

Prostaglandin A1 (mol wt 336) A bioactive prostaglandin

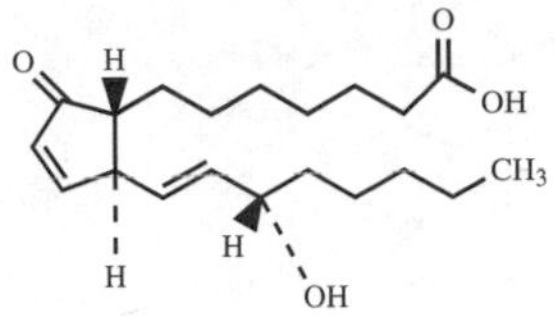

Prostaglandin A_2 (mol wt 334) A bioactive prostaglandin.

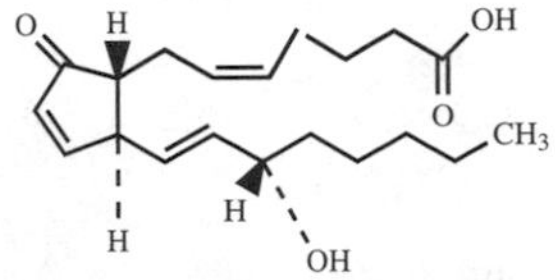

Prostaglandin B_1 (mol wt 336) A bioactive prostaglandin.

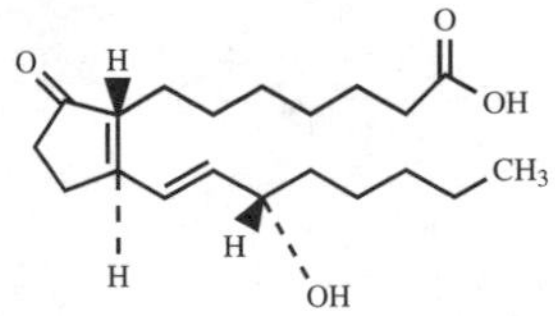

Prostaglandin B_2 (mol wt 334) A bioactive prostaglandin.

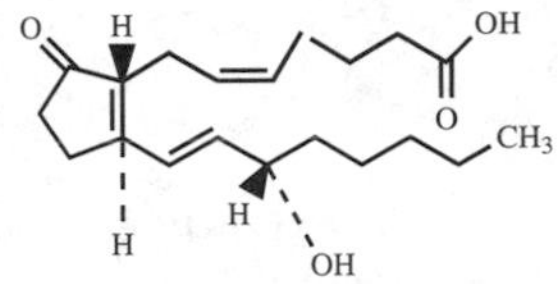

Prostaglandin D_2 (mol wt 352) A bioactive prostaglandin.

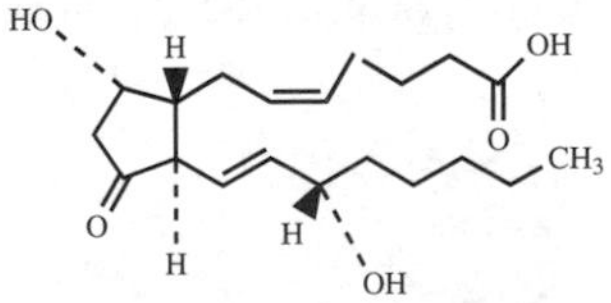

Prostaglandin E_1 (mol wt 354) A type of prostaglandin and a vasodilator.

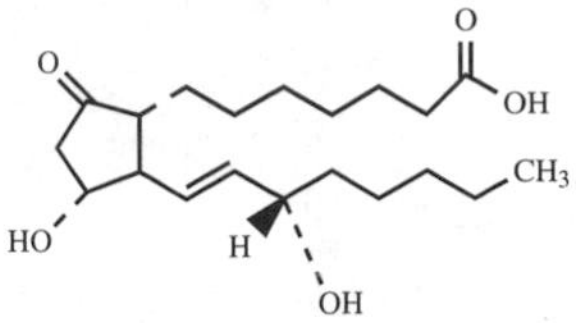

Prostaglandin E_2 (mol wt 352) A type of prostaglandin, an oxytocic, and abortifacient.

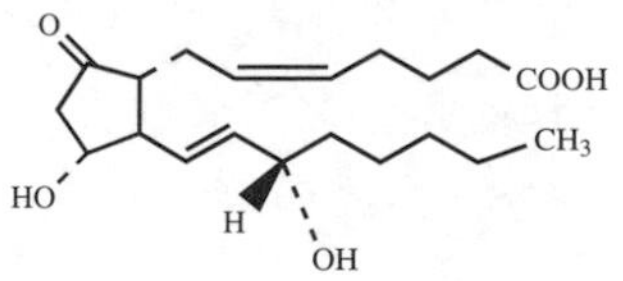

Prostaglandin E Synthetase The enzyme that catalyzes the following reaction:

(5Z,13E)-(15S)-9-α,11α-Epidioxy-15-hydroxyprosta-5,13-dienoate

⇅

(5Z,13E)-(15S)-11α,15-Dihydroxy-9-oxyprosta-5,13-dienoate

Prostaglandin F_{1a} (mol wt 357) An oxytocic and abortifacient.

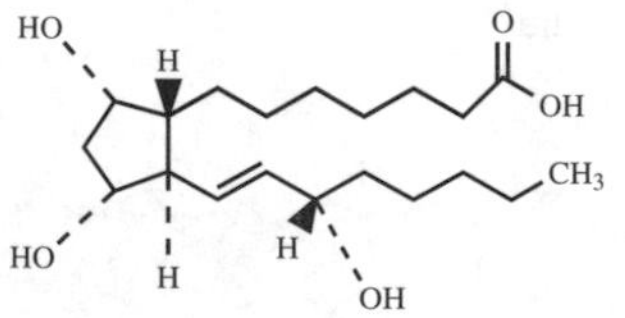

Prostaglandin F_{2a} (mol wt 354) An oxytocic and abortifacient.

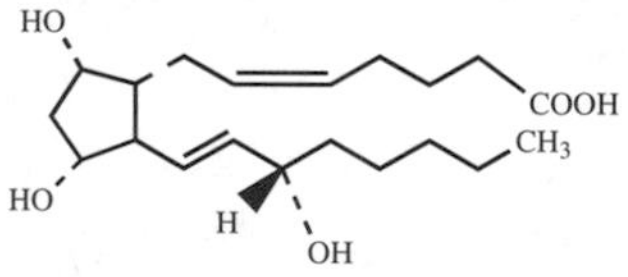

Prostaglandin G/H Synthetase Synonym of prostaglandin endoperoxide synthetase.

Prostaglandin-H_2 E-Isomerase Synonym of prostaglandin E synthetase.

Prostaglandin-I_2 (mol wt 352) A vasodilator.

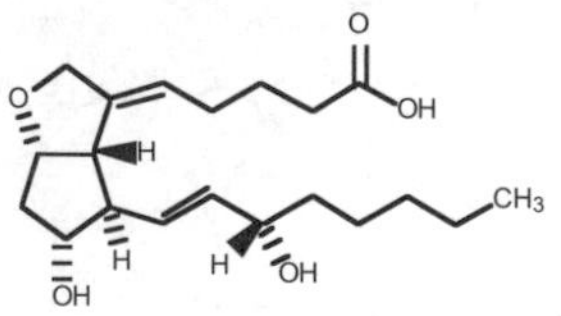

Prostaglandin-I Synthetase The enzyme that catalyzes the following reaction:

(5Z,13E)-(15S)-9α,11α–Epidioxy-15-hydroxyprosta-5,13-dienoate

⇅

(5Z,13E)-(15S)-6,9α-epoxy-11α,15-dihydroxyprosta-5,13-dienoate

Prostaphlin A trade name for oxacillin sodium, a penicillinase-resistant antibiotic.

Prostasin A protease with trypsin-like activity found in human seminal fluid.

Prostate A gland in the male that surrounds the urethra below the bladder.

Prostate-Specific Antigen A g-seminoprotein (human) related to kallikrein and used as a prostate cancer indicator.

Prostatectomy Surgical removal of the prostate gland.

Prostatitis Inflammation of the prostate gland.

ProStep A trade name for nicotine, a smoking deterrent.

Prosthecae A narrow extension or appendage bounded by cell wall and cytoplasm in a prokaryotic cell.

Prosthecomicrobium A genus of chemoorganotrophic, strictly aerobic bacteria.

Prosthetic Group Small organic molecule or metal on an enzyme that plays an indispensable role in the catalytic activity of the enzyme.

Prostigmin A trade name for neostigmine methylsulfate, a cholinergic agent that inhibits the destruction of acetylcholine.

Prostin E_2 A trade name for dinoprostone, an oxytocic agent that produces strong contractions of uterine smooth muscle.

Prosultiamine (mol wt 357) An enzyme cofactor and a vitamin.

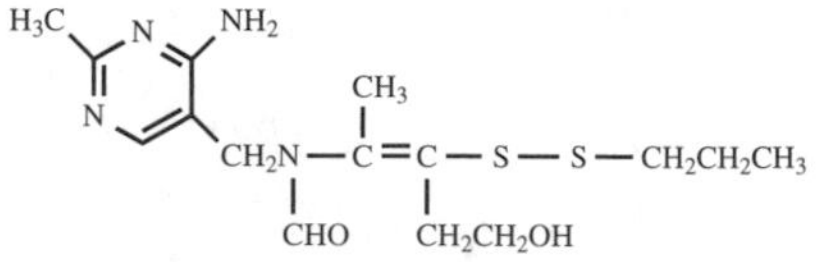

Protaminase The enzyme that catalyzes the release of C-terminal lysine or arginine.

Protamine Basic proteins or polypeptides that are soluble in water or NH_4OH but are not coagulated by heat.

Protamine Kinase The enzyme that catalyzes the following reaction:

$$\text{ATP + protamine} \rightleftharpoons \text{ADP + phosphoprotamine}$$

Protamine Zinc Insulin Suspension An insulin injection prepared by addition of zinc chloride and protamine sulfate. It is a long-lasting insulin and is absorbed slowly at a steady state.

Protanopia A form of partial color blindness in which the eye has difficulty distinguishing red from yellow or green.

Protease An enzyme that catalyzes the hydrolysis of peptide bonds in a protein.

Protease A A proteolytic enzyme from *Streptomyces griseus.*

Protease B A proteolytic enzyme from *Streptomyces griseus.*

Protease La The protease that catalyzes the hydrolysis of proteins in the presence of ATP.

Protease Pi The enzyme that catalyzes the cleavage of peptide bonds of oxidized insulin B chain.

Protecting Group Substance that binds to a functional group on a macromolecule, e.g., active site of an enzyme, preventing the functional group from destruction or participating in subsequent reactions.

Protective Antigen An antigen from a pathogenic organism that can elicit an immune response in individuals for protection against the pathogen.

Protegrin Any of a group of leukocyte antimicrobial peptides that are active against *E coli, Listeria monocytogenes*, and *Candida albicans in vitro.*

Protein A polymer of L-amino acids that folds into a conformation specified by the linear sequence of amino acids and functions as an enzyme, a hormone, an antibody, or a structural component of the cell.

Protein A A Protein derived from the Cowan strain of *Staphylococcus aureus* that is capable of binding with the Fc fragment of human IgG_1, IgG_2, and IgG_4; murine IgG_{2a}, and IgG_{2b}; and rabbit IgG.

Protein *N*-Acetylglucosaminyl Transferase The enzyme that catalyzes the following reaction:

UDP-*N*-acetylglucosamine + protein

⇅

UDP + *N*-acetylglucosaminyl protein

Protein Arginine Deiminase The enzyme that catalyzes the following reaction:

Protein-L-arginine + H_2O

⇅

Protein-L-citruline + NH_3

Protein Arginine Iminohydrolase The systematic name for protein-arginine deiminase.

Protein B A cell surfacc protein of group B *Sstreptococcus* species that binds to the Fc region human IgA.

Protein β-Aspartate O-Methyltransferase Synonym of protein-L-isoaspartate (D-aspartate) O-methyltransferase.

Protein Blotting A method for identification of electrophoretically separated proteins by radioactively labeled probe or enzyme-linked immunosorption assay. It involves transfer of electrophoretically separated proteins to a nitrocellulose paper or nylon-based membrane and followed by reaction with appropriate probe or enzyme-linked antibody.

Protein C A protease that catalyzes the degradation of blood coagulation factor Va and VIIIa.

Protein Conformation Referring to the three-dimensional structure of a protein.

Protein Disulfide Isomerase The enzyme that catalyzes the rearrangement of -S-S- bonds in proteins.

Protein Disulfide Reductase (Glutathione-dependent) The enzyme that catalyzes the following reaction:

2 Glutathione + protein-disulfide

⇅

Oxidized glutathione + protein-dithiol

Protein Disulfide Reductase (NADPH-dependent) The enzyme that catalyzes the following reaction:

NADPH + protein-disulfide

⇅

$NADP^+$ + protein-dithiol

Protein Fractionation Techniques for the separation of mixtures of proteins by means of salt precipitation, chromatography, centrifugation, and electrophoresis.

Protein G A cell wall protein from group G *Streptococci* that binds Fc region of a broader range of IgG molecules (similar to protein A).

Protein Glutamate O-Methyltransferase The enzyme that catalyzes the following reaction:

S-Adenosyl-L-methionine + protein-L-glutamate

⇅

S-Adenosyl-L-homocysteine + protein-L-glutamate methyl es

Protein Glutamine Amine Glutamyltransferase The systematic name for protein glutamine g-glutamyltransferase.

Protein Glutamine γ-Glutamyltransferase The enzyme that catalyzes the following reaction:

Protein glutamine + alkylamine

⇌

Protein N^5-alkylglutamine + NH_3

Protein L-Isoaspartate (D-Aspartate) O-Methyltransferase The enzyme that catalyzes the following reaction:

S-Adenosyl-L-methionine +
protein L-isoaspartate

⇌

S-Adenosyl-L-homocysteine +
protein L-isoaspartate methyl ester

Protein Kinase The enzyme that catalyzes the transfer of phosphate from ATP to the hydroxyl side chains of a protein causing changes in the functions of the protein.

Protein Kinase C Enzyme that phosphorylates specific serine and threonine residues on a variety of target proteins.

Protein Methylase The enzyme that catalyzes the methylation of a protein.

Protein Phosphatase Synonym of phytase.

Protein S A vitamin K-dependent protein in the blood that promotes binding of protein C to platelets and functions as a cofactor for the anticoagulant activity of the activated protein C.

Protein Tyrosine Kinase The enzyme that catalyzes the phosphorylation of tyrosine residues of a protein.

Protein Tyrosine Phosphatase The enzyme that catalyzes the removal of phosphate from tyrosine residues of a phosphorylated protein.

Protein Tyrosine Phosphate Phosphohydrolase The systematic name for protein tyrosine phosphatase.

Protein Z A major storage protein of barley endosperm.

Proteinase Synonym of protease or proteolytic enzymes that catalyze the hydrolysis of peptide bonds in a protein.

Proteinase K A nonspecific protease obtained from *Tritirachium album*.

Proteinoid Polymer formed by heat polymerization of amino acids.

Proteinoplast A protein storage organelle or structure in a cell.

Proteinuria The presence of an excessive amount of protein in the urine, which is a sign of kidney disorder.

Protenate A trade name for a plasma protein fraction used to expand plasma volume.

Proteoglycan High molecular weight complexes of proteins and polysaccharides that form ground substances in the extracellular matrix of connective tissue and serve as lubricants and support elements.

Proteohormone Synonym for protein hormone.

Proteoliposome Artificial membrane vesicle formed by the incorporation of specific proteins into a phospholipid bilayer.

Proteolysis Degradation of proteins.

Proteolytic Enzyme Synonym for protease or proteinase.

Proteoplast Synonym for proteinoplast.

Proteose A soluble product obtained from hydrolysis of protein that cannot be coagulated by heat but can be precipitated by ammonium sulfate.

Proteus A genus of Gram-negative bacteria.

Prothazine A trade name for promethazine hydrochloride, an antihistaminic agent that competes with histamine for H_1 receptors on effector cells.

Protheobromine (mol wt 238) A diuretic agent.

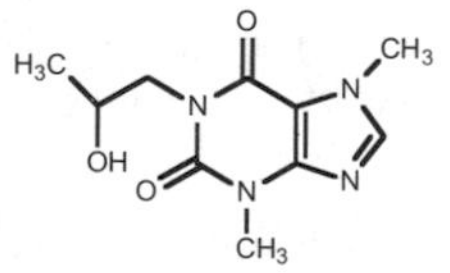

Prothrombase The enzyme that catalyzes the cleavage of peptide bonds in prothrombin at the position between arginine and threonine and arginine and isoleucine.

Prothrombin A precursor of thrombin.

Prothrombinase Synonym for prothrombase.

Prothymocytes Immature precursors of thymocytes within the thymus gland.

Proticins Bacteriocins from thymus *Proteus.*

Protilase A trade name for pancrelipase, an enzyme involved in digestion and absorption of fat.

Protiofate (mol wt 288) A topical fungicide.

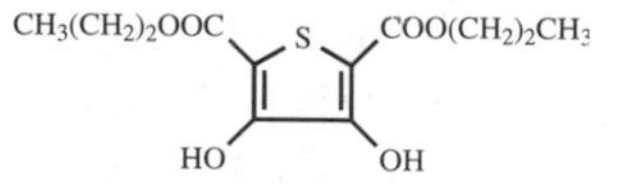

Protionamide (mol wt 180) An antibacterial agent.

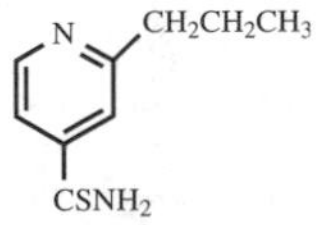

Protist The eukaryotic organisms of Protista including protozoa, algae, slime molds, and other groups that do not fit the definition of plant, animal, and fungi.

Protista One of the eukaryotic kingdoms including protozoa, algae, slime molds, and some groups that do not fit the definition of plant, animal, and fungi.

Protizinic Acid (mol wt 315) An anti-inflammatory agent.

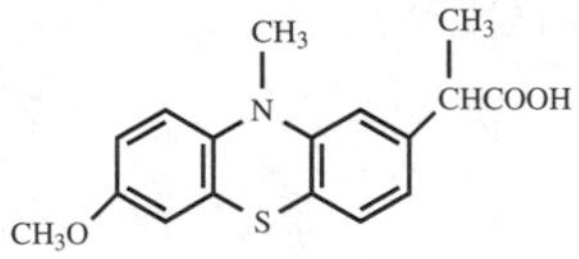

Proto- A prefix meaning first, e.g., protozoa, the first animal.

Protoanemonin (mol wt 96) An antibacterial agent from plant *Anemone pulsatilla* (Ranunculaceae).

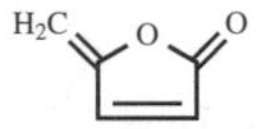

Protocatechuate Decarboxylase The enzyme that catalyzes the following reaction:

$$\text{3,4-Dihydroxybenzoate} \rightleftharpoons \text{Catechol} + CO_2$$

Protocatechuate Oxygenase Synonym of protocatechuate 3,4-dioxygenase.

Protocollagen An artificially synthesized collagen that contains no hydroxyproline and hydroxylysine.

Protocooperation The interaction between two microorganisms in which each organism benefits from the activities of the other but the interaction is not obligatory for either organism.

Protoderm The outermost primary meristem, which gives rise to the epidermis of roots and shoots.

Protofilament Polymer of subunits of tubulin that serves as the structural component of microtubules.

Protokylol (mol wt 331) A bronchodilator.

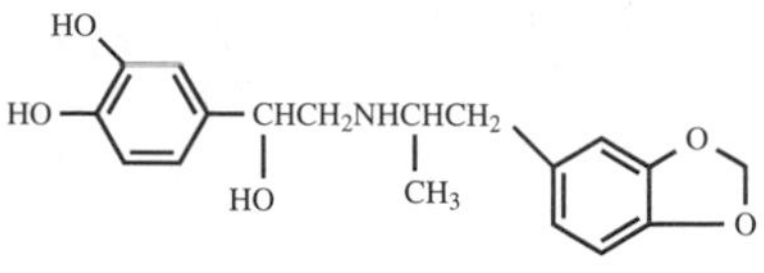

Protolignin A precursor of lignin that can be extracted from plants with ethanol or dioxane.

Protolysosome The lysosome that has not fused with phagocytic vesicles or has not been involved in digestive activity.

Protomonas A genus of methylotrophic bacteria.

Proton A particle that is identical to the nucleus of the hydrogen atom and has a mass of one atomic unit and an electric charge of +1.

Proton ATPase The ATPase that couples ATP hydrolysis to active transport of protons across an energy-transducing membrane.

Proton Gradient The difference between the hydrogen ion concentrations across an energy transducing membrane, e.g., the mitochondrial membrane it providing energy for phosphorylation.

Proton Motive Force A gradient of hydrogen ions across an energy transducing membrane, a form of potential energy stored in an electrochemical gradient for phosphorylation.

Proton PPase A proton pyrophosphatase that couples the hydrolysis of pyrophosphate to the energy-linked transport of protons across the membrane.

Proton Pump A mechanism for creation of a proton gradient by an active transport of hydrogen ions across an energy transducing membrane.

Proton Translocating ATPase Referring to F_0F_1-ATPase of mitochondria that generates a proton gradient.

Proton Translocator 1. Channel through which protons flow across a membrane 2. Ionophore that carries protons across a membrane.

Proto-Oncogene Oncogene sequences that appear within the genome of eukaryotic cells.

Protopam A trade name for pralidoxime chloride, an antidote.

Protophylline A trade name for dyphylline, used as a bronchodilator.

Protoplasm Cellular material within the plasma membrane of a cell.

Protoplast Cells of bacteria, fungi, or plants from which the cell wall has been completely removed by chemical or enzymatic treatment.

Protoplast Fusion Technique in which protoplasts are fused into a single cell.

Protoporphyrin IX (mol wt 563) A porphyrin ring structure and a precursor of blood and plant pigments.

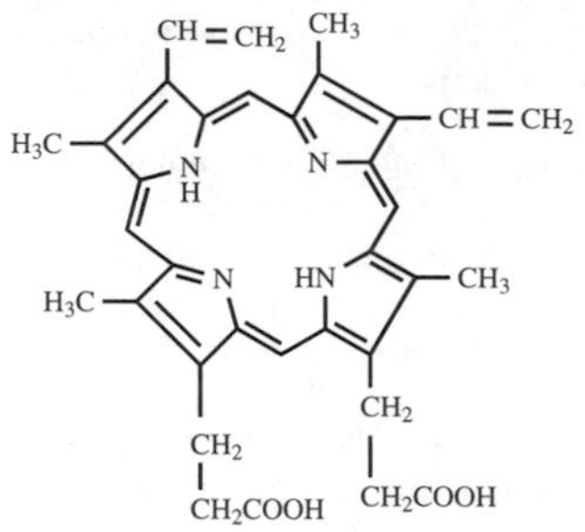

Protostat A trade name for metronidazole, used as an antiprotozoal agent.

Prototheria The egg-laying mammals.

Prototroph A nutritional wild-type; an organism that uses simple carbon sources (e.g., glucose) and requires no specific growth substance for metabolism and reproduction. An organism that requires growth nutrients not required by the prototroph is said to be a nutritional mutant or auxotroph.

Protozoa Unicellular, eukaryotic organisms that possess a distinct nucleus and cell organelles.

Protozoology The science that deals with protozoa.

Protran A trade name for chlorpromazine hydrochloride, an antipsychotic agent that blocks postsynaptic dopamine receptors in the brain.

Protrin A trade name for tco-trimoxazole, an antibacterial agent that inhibits bacterial dihydrofolate reductase.

Protriptyline (mol wt 263) An antidepressant that increases the amount of norepinephrine or serotonin in the CNS by blocking their uptake by presynaptic neurons.

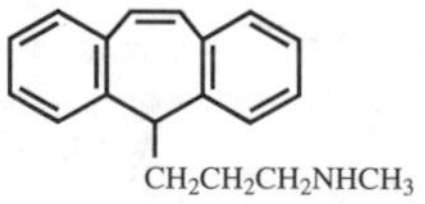

Protropin A trade name for somatrem, a pituitary hormone that stimulates growth of skeletal muscle and organs.

Prourokinase Precursor of urokinase.

Proventil A trade name for albuterol, used as a bronchodilator.

Provera A trade name for medroxyprogesterone acetate, used to suppress ovulation.

Providencia A genus of Gram-negative bacteria.

Provigil A trade name for modafinil, a CNS stimulant and narcolepsy agent.

Provirus Referring to viral DNA that becomes integrated into a host cell chromosome and is thus transmitted from one cell generation to another without production of viral progeny.

Provitamin The precursor of a vitamin.

Provitamin A Referring to b-carotene.

Proxazole (mol wt 287) A smooth muscle relaxant, analgesic, and anti-inflammatory agent.

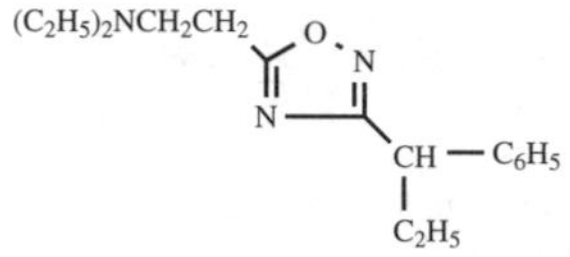

Proxibarbal (mol wt 226) A sedative-hypnotic agent.

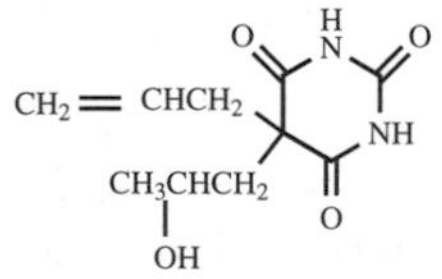

Proximal Located next or near to the point of reference, e.g., center of the body or point of attachment.

Proxyphylline (mol wt 238) A bronchodilator, vasodilator and smooth muscle relaxant.

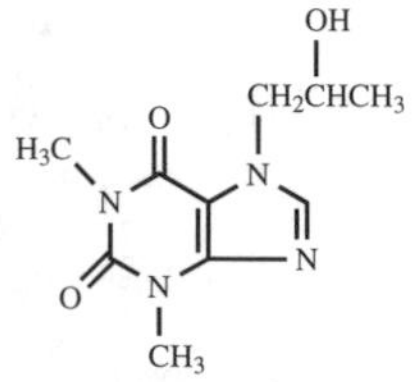

Prozac A trade name for fluoxetine hydrochloride used as an antidepressant.

Prozapine (mol wt 293) An antispasmodic and choleretic agent.

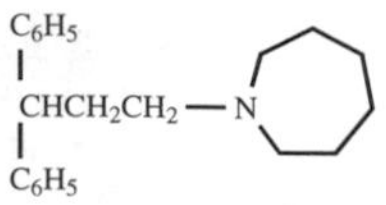

Prozone The inhibition of immunological agglutination or precipitation caused by the presence of high concentrations of antibody in the immunoprecipition reaction.

Prozone Effect The inhibition of agglutination or precipitation caused by the presence of a high concentration of antibody over antigen during the immunopreciptation reaction.

Prozone Phenomenon See prozone effect.

PRP Abbreviation for platelet-rich plasma.

PrP Abbreviation for prion protein.

PrP^c Abbreviation for normal cellular PrP (prion protein).

PRPP Abbreviation for phosphoribosyl-1-pyrophosphate, a substance used for biosynthesis of purine nucleotides.

PrP^{res} Abbreviation for normal protease-resistant PrP.

PrP^{Sc} Abbreviation for pathogenic isoform of prion protein.

p90rsk Abbreviation for 90-kDa ribosomal protein S6 kinase.

PRT Abbreviation for phosphoribosyl transferase.

Prulet A trade name for phenolphthalein, a laxative that promotes fluid accumulation in the colon and small intestine.

PRV Abbreviation for pseudo-rabies virus.

pRX-1 A plasmid of *E. coli* that contains an Amp^r (ampicillin resistant) marker and EcoRI, SacI, SmaI, BamHI, XbaI, SalI, HindIII, and ClaI cleavage sites.

PS Abbreviation for 1. phosphatidyl serine; 2. photosystem; 3. per second; 4. pico-second; 4. pantothenate synthetase.

p70S6 Abbreviation for 70-kDa ribosomal protein S6.

p90S6 Abbreviation for 90-kDa ribosomal protein S6.

PS I Abbreviation for photosystem I that occurs in both prokaryotic and eukaryotic photosynthetic cells.

PS II Abbreviation for photosystem II that occurs in eukaryotic photosynthetic cells.

PSA Abbreviation for 1. periodic acid Schiff; 2. polyethylene sulfonic acid; 3. prostate specific antigen.

pSBL-pKI A plasmid of *E. coli* that contains an Amp^r (ampicillin resistant) marker and BamHI and EcoRI cleavage sites.

pSC101 A plasmid of *E. coli* that contains a Tet^r (tetracycline resistant) marker and a unique EcoR1 cleavage site.

PSCT Abbreviation for peripheral stem cell transplantation.

PSD Medium Abbreviation for peptone-starch-dextrose medium.

PseI (AsuI) A restriction endonuclease from *Pseudoanabaena* species with the following specificity:

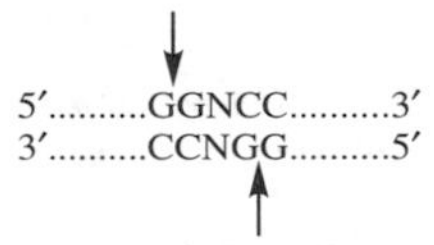

pSELECT1 A plasmid of *E. coli* that contains a Tet^r (tetracycline resistant) marker and ClaI, EcoRV, StyI, ScaI, and AatII cleavage sites.

Pseudanabaena A genus of filamentous cyanobacteria.

Pseudo- A prefix meaning false or temporary.

Pseudoalleles Genes that behave in the complement test as if they were alleles but that can undergo crossing over and recombination.

Pseudocaedibacter A genus of Gram-negative bacteria that occurs as endosymbionts in *Paramecium.*

Pseudocatalase A nonhemoprotein that is capable of catalyzing the breakdown of hydrogen peroxide into water and oxygen.

Pseudocholinesterase A nonspecific cholinesterase.

Pseudocilia Nonfunctional, immobile cilia.

Pseudocoagulase A protease produced by certain strains of *Staphylococcus* that mimics the effect of *Staphylocoaglase* thus causing a false-positive reaction in a coagulase test.

Pseudocow Pox A mild disease of cattle caused by parapox virus.

Pseudocumene (mol wt 120) A CNS depressant and respiratory irritant.

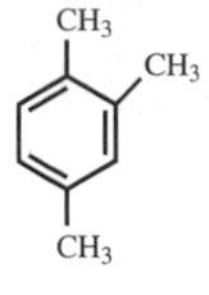

Pseudoephedrine Hydrochloride (mol wt 201) A nasal decongestant.

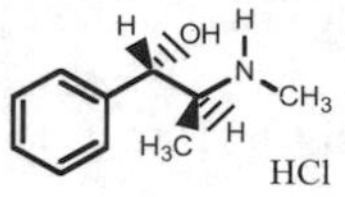

Pseudoflagellum An immobile, nonfunctional flagellum.

Pseudofructose (mol wt 180) Synonym for psicose.

Pseudogenes Noncoding DNA segments with sequences similar to functional genes but lacking signals necessary for gene expression.

Pseudoglobulin A globulin that is sparingly soluble in water.

Pseudo-Leucine Aminopeptidase Synonym of membrane alanyl aminopeptidase.

Pseudolysin A protease from *Pseudomonas.*

Pseudomonads Any bacteria of the genus *Pseudomonas.*

Pseudomonas A genus of Gram-negative, aerobic, chemoorganotrophic or facultative chemolithoautotrophic bacteria of the family Pseudomonadaceae.

Pseudomurein A component of the cell walls of *Archaebacteria* that is resistant to the actions of lysozyme and penicillin.

Pseudonigeran A polymer of 1,3-α-D-glucan from the cell wall of fungi, e.g., *Aspergillus nidulans.*

Pseudonocardia A genus of bacteria (Actinomycetales).

Pseudoplasmodium Multicellular, motile structure formed by the aggregation of amoeboid cells.

Pseudopod Extension of a cell formed by cytoplasmic streaming.

Pseudopodia Plural of pseudopodium.

Pseudopodium Cellular extensions of amoeboid cells for moving and feeding.

Pseudorabies A disease of pigs caused by herpes virus.

Pseudo-U Loop A loop in tRNA that contains pseudouridine nucleotide.

Pseudouridine (mol wt 244) An unusual nucleoside found in tRNA.

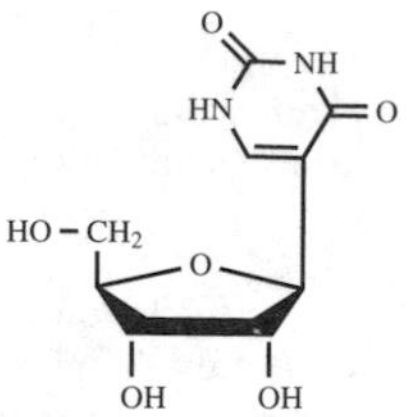

Pseudouridine Kinase The enzyme that catalyzes the following reaction:

$$\text{ATP + pseudouridine} \rightleftharpoons \text{ADP + pseudouridine 5}'\text{-phosphate}$$

Pseudouridylate Synthetase The enzyme that catalyzes the following reaction:

$$\text{Uracil + ribose phosphate} \rightleftharpoons \text{Pseudouridine 5}'\text{-phophate} + H_2O$$

Pseudouridylic Acid (mol wt 293) A nucleotide of pseudouridine found in tRNA.

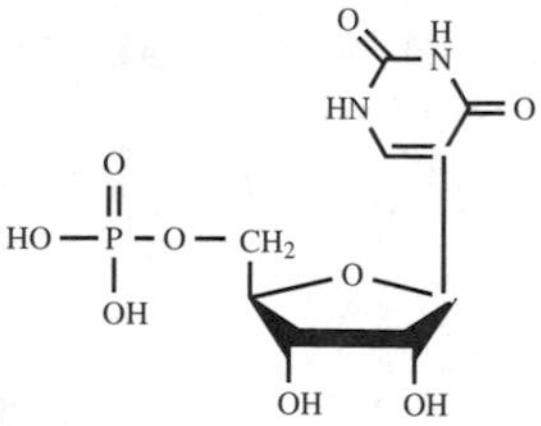

Pseudovirion An experimentally prepared virion in which protein capsid and viral genome are derived between two different viruses.

Pseudovitamin A substance that functions as a coenzyme but is not a dietary requirement.

psi (ψ) A letter in the Greek alphabet.

pSI4001 A plasmid of *E. coli* that contains an Amp[r] (ampicillin resistant) marker and HindIII, EcoRI, BamHI, SalI, and KpnI cleavage sites.

Psicofuranine (mol wt 297) A nucleoside antibiotic produced by *Streptomyces hygroscopicus* with antibacterial and antitumor activity.

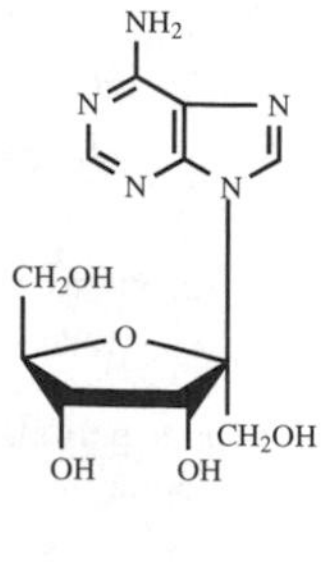

D-Psicose (mol wt 180) A nonfermentable monosaccharide.

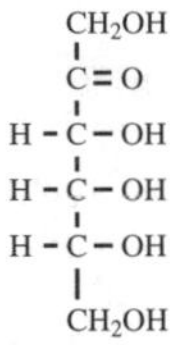

PSIFT Abbreviation for platelet suppression immuno-fluorescent test.

Psilocin (mol wt 204) A minor hallucinogenic component of Teonanacat, the sacred mushroom of Mexico (*Psilocybe mexicana*).

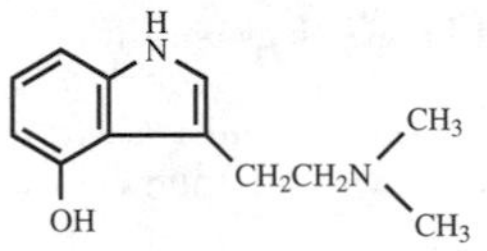

Psilocybin (mol wt 284) The major hallucinogenic component of Teonanacatle from the sacred mushroom of Mexico (*Psilocybe mexicana*).

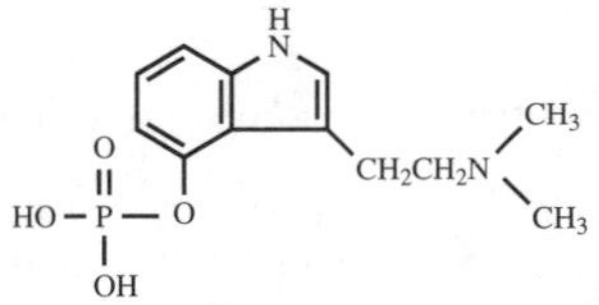

P-site A binding site on a ribosome that binds the tRNA carrying a growing polypeptide chain.

Psittacosis An infectious disease caused by *Chlamydia psittaci* transmitted by birds and characterized by fever, cough, and headache.

PSK A protein-bound polysaccharide from *Coriolus versicolor* (Basidiomycete) with immunostimulating and antineoplastic activity.

p70S6K (p70^{S6K}) Abbreviation for 70-kDa ribosomal protein S6 kinase.

p70S6-Kinase Abbreviation for 70-kDa ribosomal protein S6 kinase.

pSM6 A plasmid of *E. coli* that contains a Tet^r (tetracycline resistant) marker and an EcoRI cleavage site.

Psoralen (mol wt 186) A substance that binds with nucleic acid to form covalently linked photoproduct upon irradiation with UV light (365 mn).

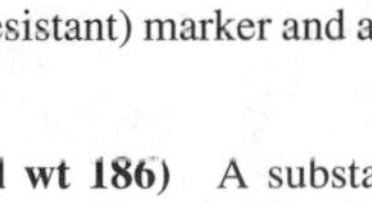

Psorion A trade name for betamethosone benzoate, an anti-inflammatory agent that suppresses inflammation and normal immune responses.

PSP Abbreviation for 1. pancreatic spasmolytic peptide; 2. parathyroid secretory protein.

PspI (AsuI) A restriction endonuclease from *Pseudoanabaena* species with the same specificity as AsuI.

Psp5II A restriction endonuclease from *Pseudomonas fluorescens* RFL 5 with the following specificity:

5'.......PuG↓G(A/T)CCPy.......3'
3'.......PyCC(T/A)G↑GPu.......5'

Psp61I (NaeI) A restriction endonuclease from *Pseudomonas* species MS61 with the following specificity:

5'..........GCC↓GGC..........3'
3'..........CGG↑CCG..........5'

Psp1406I A restriction endonuclease from *Pseudomonas* species RFL 1406 with the following specificity:

5'........AA↓CGTT........3'
3'........TTGC↑AA........5'

PspAI A restriction endonuclease from *Pseudomonas* species with the following specificity:

5'........C↓CCGGG........3'
3'........GGGCC↑C........5'

Psp124BI A restriction endonuclease from *Pseudomonas* species 124B with the following specificity:

5'........GAGCT↓C........3'
3'........C↑TCGAG.........5'

PspEI A restriction endonuclease from *Pseudomonas* species E with the following specificity:

5'........G↓GTNACC........3'
3'........CCANTG↑G........5'

PspLI A restriction endonuclease from *Pseudomonas* species L with the following specificity:

5'........CGTACG........3'
3'........GCATGC........5'

PspN4I A restriction endonuclease from *Pseudomonas* species N4 with the following specificity:

5'........GGNNCC........3'
3'........CCNNGG........5'

PspOMI A restriction endonuclease from *Pseudomonas* species OM2164 with the following specificity:

5'........GGGCCC........3'
3'........CCCGGG........5'

PspPI A restriction endonuclease from *Psychrobacter* species with the following specificity:

5'........GGNCC........3'
3'........CCNGG........5'

PspPPI A restriction endonuclease from *Pseudomonas* species PP with the following specificity:

5'........RGGWCCY........3'
3'........YCCWGGR........5'

R= A or G W= A or T Y= C or T

pSRBS A plasmid of *E. coli* that contains a Amp[r] (ampicillin resistant) marker and BalI, BamHI, BglI, BsaHII, ClaI, EcoRI, EcoRV, GdlII, HindIII, SacI, SauI, and Tth111I cleavage sites.

pSRW25 A plasmid of *E. coli* that contains Tet[r] (tetracycline resistant) and Kan[r] (kanamycin resistant) markers and EcoRI, XhoI, HindIII, and ClaI cleavage sites.

pSS9 A plasmid of *E. coli* that contains a Amp[r] (ampicillin resistant) marker and EcoRI, SmaI, BamHI, XbaI, and BglII cleavage sites.

pSS24 A phagemid of *E. coli* that contains an Amp[r] (ampicillin resistant) marker and KpnI, HindIII, PvuII, BamHI, and HpaI cleavage sites.

pSS25 A phagemid of *E. coli* that contains an Amp[r] (ampicillin resistant) marker and KpnI, PvuII, and BamHI cleavage sites.

PssI (DraII) A restriction endonuclease from *Pseudomonas* species with the following specificity:

5′..........PuGGNCCPy..........3′
3′..........PyCCNGGPu..........5′

PST Abbreviation for a combination drug containing penicillin, streptomycin, and tetracycline.

pST1800 A plasmid of *E. coli* that contains an Amp[r] (ampicillin resistant) marker and SmaI, BamHI, and EcoRI cleavage sites.

PstI A restriction endonuclease from *Pseudomonas stuartii* 164 with the following specificity:

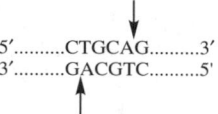

5′..........CTGCAG..........3′
3′..........GACGTC..........5'

PSTV Abbreviation for potato spindle tuber virus.

^{32}P-Suicide Loss of infectivity or activity by a biological entity (e.g., viral DNA) due to radioactive decay of the incorporated radioactive ^{32}P.

Psychedelic Drug Drugs that are capable of producing abnormal psychic effects (hallucination).

Psychiatry Medical science that deals with causes, treatment, and prevention of mental disorders.

Psychology The science that deals with behavior and function of the mind.

Psychoneuroimmunology Science that deals with interactions among the nerve, endocrine, and immune systems.

Psychro- A prefix meaning cold.

Psychroduric Microorganisms capable of surviving but not growing at low temperatures.

Psychrophile An organism that has an optimum growth temperature below 20° C that is associated with spoilage of refrigerated food.

Psyllium Laxative obtained from seeds of *Plantago psyllium* that absorb water and expand to increase bulk and moisture content of the stool.

P-System A blood group system in which blood group antigens are detected on erythrocytes, erythroblasts, platelets, megakaryocytes, and fibroblasts. They are designated P^k, P, and P1.

PT Abbreviation for 1. *para*-thyroid; 2. paratuberculosis; 3. para-typhoid; 4. phenytoin; 5. phototoxicity.

Pt Abbreviation for platinum, a chemical element.

PTA Abbreviation for 1. phosphotungstic acid; 2. plasma thromboplastin antecedent.

pTAC11 A plasmid of *E. coli* that contains an Amp^r (ampicillin resistant) marker and EcoRI, HindIII, and PstI cleavage sites.

pTAC12 A plasmid of *E. coli* that contains an Amp^r (ampicillin resistant) marker and a PvuII cleavage site.

pTAC12H A plasmid of *E. coli* that contains an Amp^r (ampicillin resistant) marker and HindIII, EcoRI, and PvuII cleavage sites.

pTACTERM A plasmid of *E. coli* that contains an Amp^r (ampicillin resistant) marker and SalI, BamHI, EcoRI, and HindIII cleavage sites.

PTB Abbreviation for phospho-tyrosine binding protein.

PTBD Abbreviation for phospho-tyrosine binding domain.

PTBP Abbreviation for phospho-tyrosine binding protein.

PTC Abbreviation for 1. phenylthiocarbamide; 2. phenylthiocarbamoyl; 3. phenylthiocarbarmyl.

PTC-Amino Acid Abbreviation for phenylthiocarbamyl amino acid.

Ptd Abbreviation for phosphatidyl group.

PtdCho Abbreviation for phosphatidylcholine.

PtdEt Abbreviation for phosphatidylethanol.

PtdIns Abbreviation for phosphatidylinositol.

PtdIns4P Abbreviation for phosphatidylinositol 4-phosphate.

PtdIns(4,5)P_2 Abbreviation for phosphatidylinositol 4,5-bisphosphate.

PtdOH Abbreviation for phosphatidic acid.

PtdSer Abbreviation for phosphatidylserine.

Pteridine (mol wt 132) A structural component of biopterin, folic acid, and riboflavin.

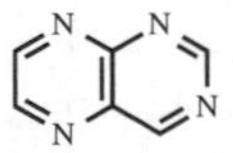

Pteroglutamic Acid Synonym for folic acid.

Pteroic Acid (mol wt 312) A structural component of folic acid.

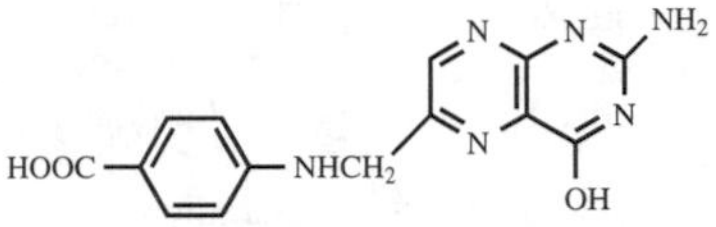

Pteroyl-Poly-γ-Glutamate Hydrolase Synonym of γ-glu-X carboxypeptidase.

PTF Abbreviation for plasma thromboplastin factor.

PTFE Abbreviation for poly-tetra-fluoro-ethylene.

PTH Abbreviation for 1. parathyroid hormone and 2. phenylthiodantoin.

PTI Abbreviation for pancreatic trypsin inhibitor.

PTK Abbreviation for protein tyrosine kinase.

pTK402 A plasmid of *E. coli* that contains an Amp^r (ampicillin resistant) marker and BamHI and PstI cleavage sites.

PTM Abbreviation for post-translational modification.

PTMA Abbreviation for phenyl-trimethyl-ammonium.

PTP Abbreviation for protein tyrosine phosphatase.

PTPase Abbreviation for phosphotyrosine phosphatase.

PTS Referring to phosphoenolpyruvate-dependent phosphotransferase system, a transport system present in both Gram-positive and Gram-negative bacteria in which sugar is phosphorylated and transported into the cell.

PTSA Abbreviation for *para*-toluene sulfonic acid.

PTTG Abbreviation for pituitary tumor-transforming gene.

pTTQ8 A plasmid of *E. coli* that contains an Amp^r (ampicillin resistant) marker and EcoRI, BamHI, SalI, PstI, HindIII, and SmaI cleavage sites.

pTTQ18 A plasmid of *E. coli* that contains an Amp^r (ampicillin resistant) marker and EcoRI, SstI, KpnI, SmaI, BamHI, XbaI, SalI, PstI, SphI, and HindIII cleavage sites.

pTTQ19 A plasmid of *E. coli* that contains an Amp^r (ampicillin resistant) marker and EcoRI, SstI, KpnI, SmaI, BamHI, XbaI, SalI, PstI, and SphI cleavage sites.

PTU Abbreviation for propylthiouracil.

PTX Abbreviation for 1. parathyroidectomy; 2. pertussis toxin.

Ptyalin A starch digesting enzyme secreted by salivary glands (also known as salivary a-amylase).

Ptyalism The excessive amount of saliva occurring during the early months of pregnancy or during mercury poisoning.

pTyr Abbreviation for phosphotyrosine.

pTZ Referring to plasmids and cloning vectors that contain T_7 promoter and *lac*Z genes (for blue/white colony screening), e.g., *p*TZ 18R, and *p*TZ18U.

Pu Abbreviation for purine.

Puberty Age at which the reproductive organs become functional.

Puberulic Acid (mol wt 198) An antibiotic produced by *Penicillium puberulum*.

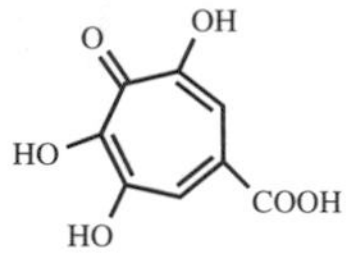

Puberulonic Acid (mol wt 224) An antibiotic produced by *Penicillium puberulum*.

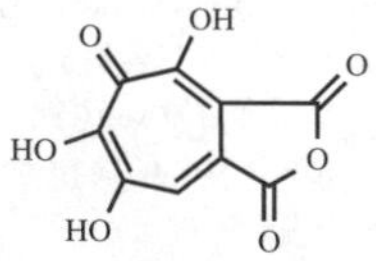

Public Antigen Antigenic determinant common to several distinct antigens.

pUC Vector Any of a series of plasmid cloning vectors bearing an ampicillin resistance gene and part of the *lac*Z gene.

pUC9 A plasmid of *E. coli* that contains an Amp^r (ampicillin resistant) marker and HindIII, PstI, SalI, AccI, HincII, BamHI, SmaI, XmaI, EcoRI, and HaeIII cleavage sites.

pUC18 A high copy-number plasmid cloning vector that contains β-lactamase gene and replication origin similar to that of pBR322.

pUCD9P A plasmid of *E. coli* that contains Amp^r (ampicillin resistant) and Kan^r (kanamycin resistant) markers and EcoRI, ClaI, HindIII, KpnI, SacI, PvuI, and PstI cleavage sites.

PUFA Abbreviation for polyunsaturated fatty acid.

Puff An expended or enlarged region of giant polytene chromosomes of the salivary gland of some Dipterans, which undergo active transcription (also known as chromosome buff).

pUH84 A plasmid of *E. coli* that contains an Amp^r (ampicillin resistant) marker and PstI, SalI, AccI, HincII, BamHI, and EcoRI cleavage sites.

pUK230 A plasmid of *E. coli* that contains an Amp^r (ampicillin resistant) marker and PstI and EcoRI cleavage sites.

Pullulan A linear polymer of D-glucan with α-1,6-glucosidic linkages that is synthesized by *Aureobasidium pullulans*.

Pullulanase The enzyme that catalyzes the hydrolysis of pullutan.

Pullulation Asexual reproduction by budding.

Pulmonary Referring to the lungs.

Pulmonary Circulation The circulation of blood between the heart and the lungs.

Pulmonary Edema The accumulation of fluid in the alveoli of the lung.

Pulmonary Embolism A blood clot or blockage of a lung artery.

Pulmozyme A trade name for enzyme DNase used for the treatment of cystic fibrosis.

Pulque A Mexican alcoholic beverage produced by fermenting juices of *Agave* species by *Saccharomyces cerevisiae* and *Zymomonas mobilis.*

Pulsating Ribosome Referring to the mechanism in which ribosomal subunits dissociate and reassociate repeatedly during the process of protein synthesis.

Pulse A regular beating caused by the contraction of the heart.

Pulse Chase Experiment A brief exposure of a bioactive system to a radioactively labeled substance (pulse) followed immediately by a high concentration of the unlabeled substance. Samples are taken at different time intervals to monitor the course of the labeled substance (chase).

Pulse Pressure The difference between arterial systolic and diastolic pressure of the heart cycle that is characteristic of the artery pulse.

Pulsed Field Gel Electrophoresis A type of gel electrophoresis in which large fragments of DNA can be separated by continuously altering the angle at which electric current is applied.

PUN Abbreviation for plasma urea nitrogen.

pUR222 A plasmid of *E. coli* that contains an Amp^r (ampicillin resistant) marker and PstI, SalI, AccI, HindIII, EcoRI, and BamHI cleavage sites.

Pure Culture A culture that contains the progeny of a single cell.

Purging To replace one type of gaseous environment by another.

Purification The processes of removal of nondesired substances and securing molecules of the same type through a series of preparative, biochemical techniques, e.g., ammonium sulfate fractionation, gel filtration, ion exchange chromatography, affinity chromatography, density gradient centrifugation and electrophoresis.

Purified Protein Derivative Referring to a protein preparation obtained from the culture of tubercle *Bacillus* grown on a synthetic medium that is used for diagnosis of tuberculosis.

Purine (mol wt 120) The nitrogen-containing parent structure of bases of adenine and guanine.

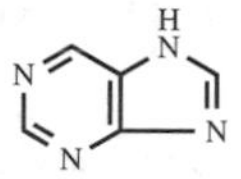

Purine Alkaloids Alkaloids that contain a heterocyclic purine ring, e.g., caffeine, theobromine, and theophylline.

Purine Antibiotics Antibiotics that contain a heterocyclic purine ring or a modified purine ring.

Purine Nucleosidase The enzyme that catalyzes the hydrolysis of purine nucleosides to D-ribose and purine base.

Purine Nucleoside Orthophosphate Ribosyltransferase The systematic name for purine nucleoside phosphorylase.

Purine Nucleoside Phosphorylase The enzyme that catalyzes the following reaction:

$$\text{Purine nucleoside} + \text{Pi} \rightleftharpoons \text{Purine} + \text{ribose 1-phosphate}$$

Purine Nucleotide Nucleotides that contain adenine or guanine.

Purine Nucleotide Cycle A metabolic cycle for interconversion of AMP to IMP.

Purinergic Nerves The neurons that use ATP as a neurotransmitter.

Purinethol A trade name for mercaptopurine, an agent that inhibits synthesis of RNA and DNA.

Purinol A trade name for allopurinol, an antigout drug.

Puromycin (mol wt 472) An antibiotic that binds to A-sites on the ribosome forming peptidyl–puromycin complexes, causing premature termination of protein synthesis.

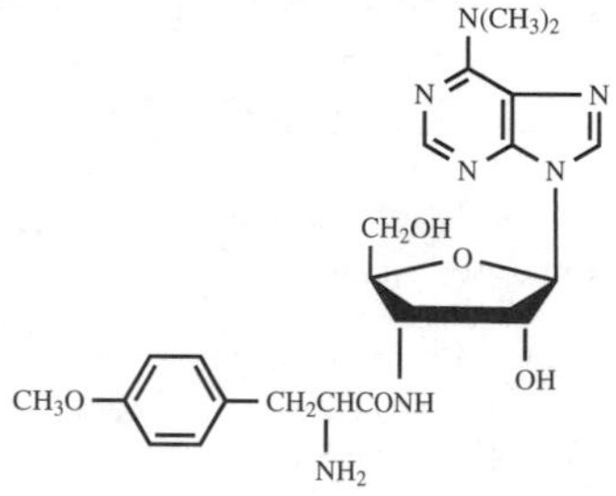

Purothionin A low molecular weight protein possessing antibacterial and antiyeast activity.

Purple Bacteria Bacteria of the Rhodospirillineae.

Purple Membrane The portion of the cytoplasmic membrane that contains bacteriorhodopsin, e.g., the membrane of *Halobacterium.*

Purple Sulfur Bacteria Photosynthetic bacteria of the family Chromatiaceae that use sulfur, sulfide, or hydrogen as electron donor.

Purpura A disorder characterized by patches of purplish discoloration of the skin due to hemorrhage of blood into the skin and mucous membrane.

Purpurin A heparin-binding protein found in cultures of chick neural retina cells.

Pus A thick, creamy-yellow or greenish-yellow product of inflammation formed at the site of infection that consists mainly of leukocytes, proteins, cell fragments, and dead pathogens.

Putrefaction The formation of food-smelling products, e.g., cadaverine and putrescine, by microbial degradation of proteinaceous materials.

Putrescine (mol wt 88) A product formed from decarboxyation of ornithine.

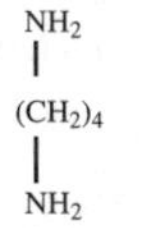

Putrescine Carbamoyltransferase The enzyme that catalyzes the following reaction:

Carbamoyl phosphate + putrescine ⇌ Pi + *N*-carbamoylputrescine

Putrescine Cycle The cycle for the synthesis of spermine, spermidine and putrescine.

Putrescine Oxidase The enzyme that catalyzes the following reaction:

Putrescine + O_2 + H_2O ⇌ 4-Aminobutanal + NH_3 + H_2O_2

Putrescine Oxygen Oxidoreductase Systematic name for putrescine oxidase.

PV Abbreviation for 1. penicillin V; 2. peripheral vessels; 3. plasma volume; 4. poliomyelitis vaccine; 5. polyoma virus.

PVA Abbreviation for 1. poly-vinyl acetate; 2. poly-vinyl-alcohol.

PVB Abbreviation for a combination drug containing platinol, velban, and bleomycin.

PVC Abbreviation for polyvinylcarbonate.

PVDF Abbreviation for polyvinylidene difluoride.

PVF-K A trade name for penicillin V potassium, an antibiotic that inhibits bacterial cell wall synthesis.

PVK A trade name for penicillin V potassium, an antibiotic that inhibits bacterial cell wall synthesis.

PVP Abbreviation for 1. Penicillin V potassium; 2. polyvinyl pyrrolidine.

PVP Iodine Abbreviation for polyvinyl pyrrolidine iodine (a disinfectant).

PVPI Abbreviation for polyvinyl pyrrolidine iodine.

PvuI A restriction endonuclease from *Proteus vulgaris* with the following specificity:

PvuII A restriction endonuclease from *Proteus vulgaris* with the following specificity:

PW Abbreviation for purified water.

PWM Abbreviation for pokeweed mitogen.

pWW84 A plasmid of *E coli* that contains Amp^r (ampicillin resistant), Kan^r (kannmycin resistant) markers and SacI, SphI, KpnI, XbaI cleavage sites.

pXf3 A plasmid of *E. coli* that contains Amp^r (ampicillin resistant) and Tet^r (tetracycline resistant) markers and EcoRI, ClaI, HindIII, BamHI, SalI, XorII, and PstI cleavage sites.

pXJ002 A plasmid of *E. coli* that contains Amp^r (ampicillin resistant) and Cml^r (chloramphenicol resistant) markers and SalI, BamHI, HpaII, and PstI cleavage sites.

PXM Abbreviation for projection X-ray microscope.

***p*XPRS** A shuttle vector that contains the replication origin of pBR322 for replication in prokaryotes, and SV40 replication origin for replication in eukaryotes.

Py Abbreviation for pyrimidine.

pY Abbreviation for phosphotyrosine.

pY2 A plasmid of *E. coli* that contains Amp^r (ampicillin resistant) and Tet^r (tetracycline resistant) markers and SalI, BamHI, and PstI cleavage sites.

Pyaemia Variant spelling of pyemia.

Pycnometer A device used for weighing liquid to determine its density or specific gravity.

Pycnosis The shrinkage and condensation of a nucleus caused by a viral infection or other conditions.

Pycocin See pyocin.

PYEG Medium Abbreviation for peptone-yeast-extract-glucose medium

pYEJ001 A plasmid of *E. coli* that contains Tet^r (tetracycline resistant) and Cml^r (chloramphenicol resistant) markers and PstI, SalI, BamHI, and EcoRI cleavage sites.

Pyelitis Inflammation of the renal pelvis.

Pyelolithotomy Surgical removal of stones from the kidney through an incision made in the pelvis of the kidney.

Pyelonephritis Inflammation of the kidney (nephritis) and of the renal pelvis (pyelitis).

Pyemia Infection of the blood stream by pyogenic bacteria.

PYG Medium Abbreviation for peptone-yeast-extract-glucose medium, used for culturing anaerobic bacteria; it contains peptone, yeast extract, cysteine-HCl, resazurin and glucose.

Pyknometer Variant spelling of pycnometer.

Pyknosis Variant spelling for pycnosis.

Pyo- A prefix meaning pus.

Pyocin Any bacteriocin produced by *Pseudomonas aeruginosa.*

Pyocyanine (mol wt 210) A pigment produced by *Pseudomonas aeruginosa.*

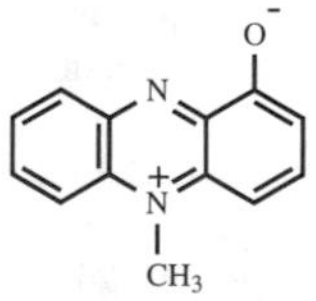

Pyoderma A pus-producing skin lesion.

Pyogen A pus-producing organism.

Pyogenic Capable of producing pus.

Pyogenic Microorganisms Microorganisms capable of producing pus.

Pyopen A trade name for carbenicillin disodium, an antibiotic that inhibits bacterial cell wall synthesis.

PYP Abbreviation for photoactive yellow protein.

PyP Abbreviation for pyrophosphate.

Pyr Abbreviation for pyridine.

Pyranose A six-member ring structure of monosaccharides, e.g., α-D-glucopyranose.

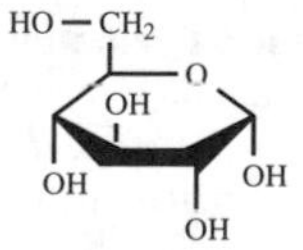

α-D-glucopyranose

Pyranose Oxidase The enzyme that catalyzes the following reaction:

$$\text{Glucose} + O_2 \rightleftharpoons \text{2-Dehydro-D-glucose} + H_2O_2$$

Pyranoside A glycoside involving a pyranose.

Pyrantel (mol wt 206) An anthelmintic agent that blocks the neuromuscular action of worms.

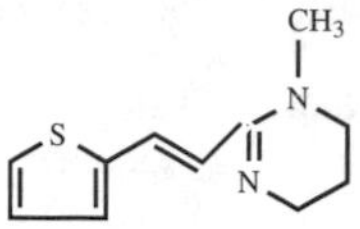

Pyrathiazine (mol wt 296) An antihistaminic agent.

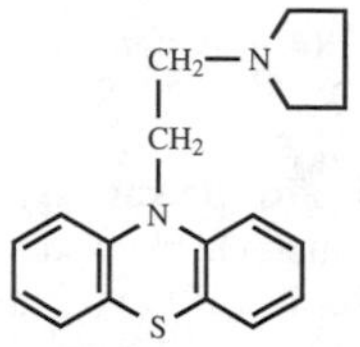

Pyrazinamide (mol wt 123) An antituberculostatic agent.

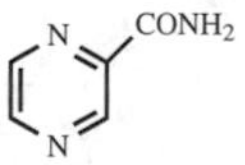

Pyrazodine A trade name for phenazopyridine hydrochloride, used as an analgesic and antipyretic agent.

Pyrazophos (mol wt 373) An antifungal agent.

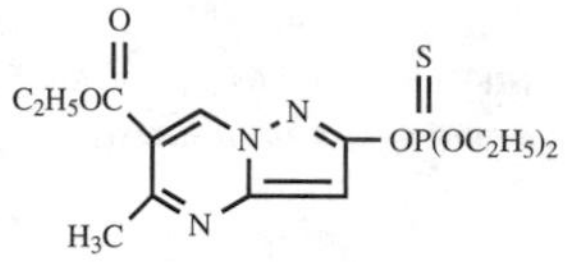

Pyregesic-C A trade name for a combination drug containing acetaminophen and codeine phosphate, used as an analgesic agent.

Pyrenoid A structure in the chloroplasts of certain algae that serves as a center for starch formation.

Pyrethrins A group of antiparasitic agents that disrupt the parasite's nervous system causing paralysis and death of the parasites.

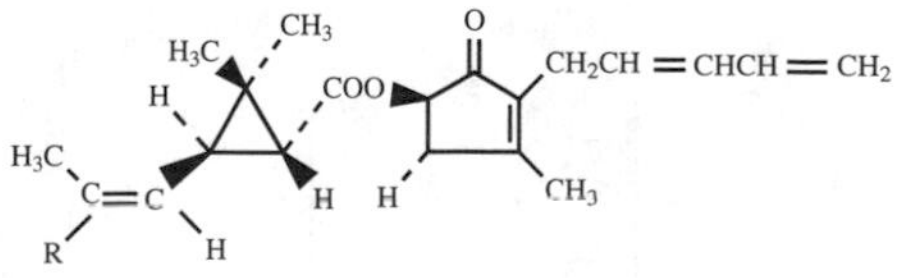

Pyretic Causing a fever.

Pyribenzamine A trade name for tripelennamine hydrochloride, an antihistaminic agent.

Pyridiate A trade name for phenazopyridine hydrochloride, used as an analgesic and antipyretic agent.

Pyridin A trade name for phenazopyridine hydrochloride, used as an analgesic and antipyretic agent.

Pyridine (mol wt 79) A solvent.

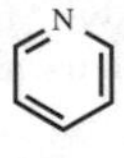

Pyridine Alkaloids Akaloids that contain a pyridine ring structure, e.g., nicotine.

Pyridine Nucleotide Referring to NAD^+, $NADP^+$, NADH, and NADPH.

Pyridine Nucleotide Coenzymes Referring to coenzyme NAD^+, $NADP^+$, NADH, and NADPH.

Pyridine Nucleotide Salvage Cycle A pathway in which nicotinamide derived from the breakdown of NAD or NADP is reused for synthesis of NAD or NADP.

Pyridinol Carbamate (mol wt 253) An antiarteriosclerotic agent.

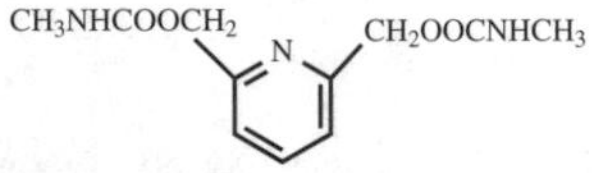

Pyridium A trade name for phenazopyridine hydrochloride, used as an analgesic and antipyretic agent.

Pyridomycin (mol wt 541) An antibiotic produced by *Streptomyces alblidofuscus*.

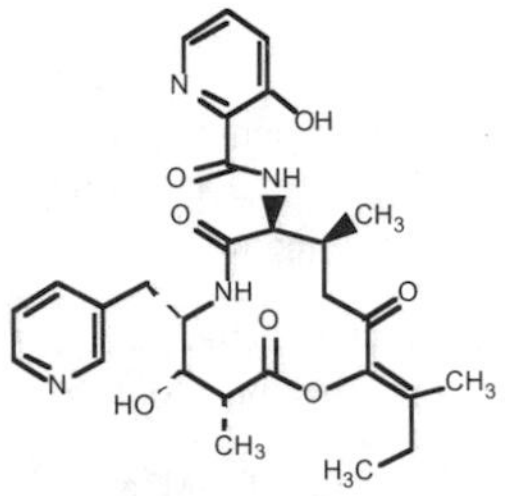

Pyridostigmine Bromide (mol wt 261) A cholinergic agent.

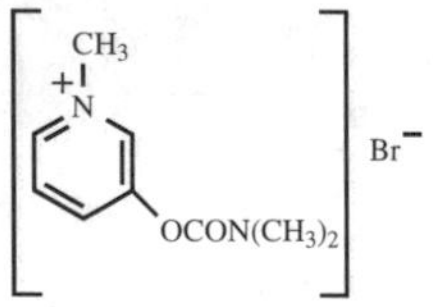

Pyridoxal (mol wt 167) A form of vitamin B_6.

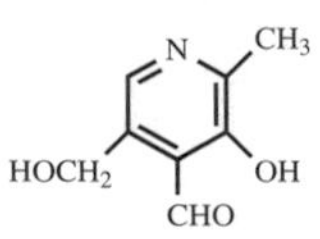

Pyridoxal Kinase The enzyme that catalyzes the following reaction:

ATP + pyridoxal ⇌ ADP + pyridoxal 5-phosphate

Pyridoxal 5-Phosphate (mol wt 247) A coenzyme (vitamin B_6) that participates in metabolism of amino acids.

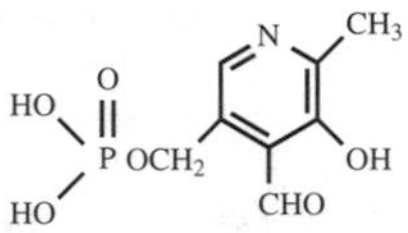

Pyridoxine (mol wt 170) A form of vitamin B_6.

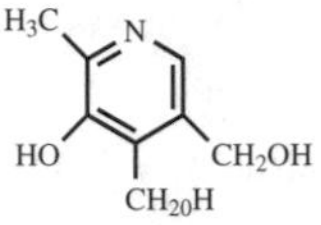

Pyridoxine 4-Dehydrogenase The enzyme that catalyzes the following reaction:

Pyridoxine + $NADP^+$ ⇌ Pyridoxal + NADPH

Pyrimethamine (mol wt 249) An antimalarial agent and an inhibitor of dihydrofolate reductase.

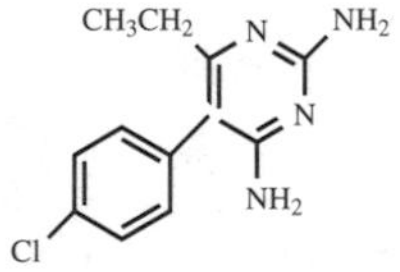

Pyrimidine (mol wt 80) The parent aromatic structure of the nitrogen-containing bases for cytosine, thymine, and uracil.

Pyrimidine Antibiotics Antibiotics that contain a pyrimidine or modified pyrimidine structure.

Pyrimidine Dimer DNA–Glycosylase Synonym of deoxyribopyrimidine endonucleosidase.

Pyrimidine Dimer Formation DNA damage caused by UV radiation in which a covalent bond is formed between two adjacent pyrimidine bases (e.g., thymine dimer), thereby blocking replication and transcription.

Pyrimidine Phosphorylase The enzyme that catalyzes the following reaction:

Thymidine + orthophosphate

⇅

Thymine + 2-deoxy-D-ruibose 1-phosphate

Pyrinoline (mol wt 416) A cardiac depressant.

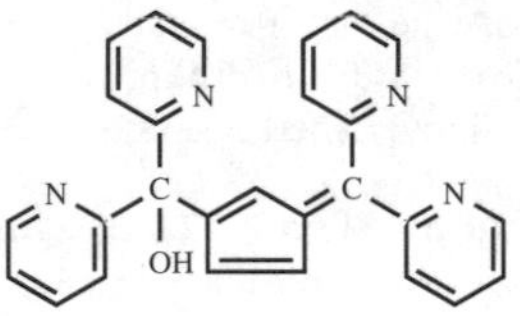

Pyrinyl A trade name for pyrethrins, used as an antiparasitic agent (scabicide and pediculicide).

Pyrithione (mol wt 127) An antibacterial and antifungal agent.

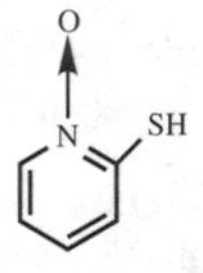

Pyrithyldione (mol wt 167) A sedative-hypnotic agent.

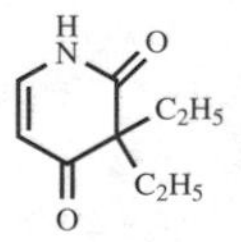

Pyrocatechol (mol wt 110) An antiseptic agent.

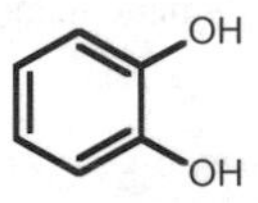

Pyrogen A fever-producing substance.

Pyrogenic Exotoxin C A toxin produced by *Staphylococcus aureus*.

Pyroglobulins Monoclonal immunoglobulins that precipitate irreversibly when heated to 56°C.

L-Pyroglutamic Acid (mol wt 129) An internal α-aminoglutaric acid lactam.

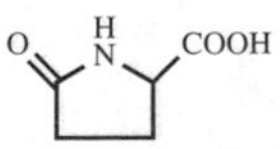

Pyroglutamyl Peptidase I The enzyme that catalyzes the release of an N-terminal pyroglutamyl group from a polypeptide (provided the next residue is not proline).

Pyrolysis Chemical transformation of a substance caused by heat, e.g., thermal isomerization or thermal decomposition.

Pyronine B (mol wt 1042) A dye used for staining bacteria, molds, and ribonucleic acid.

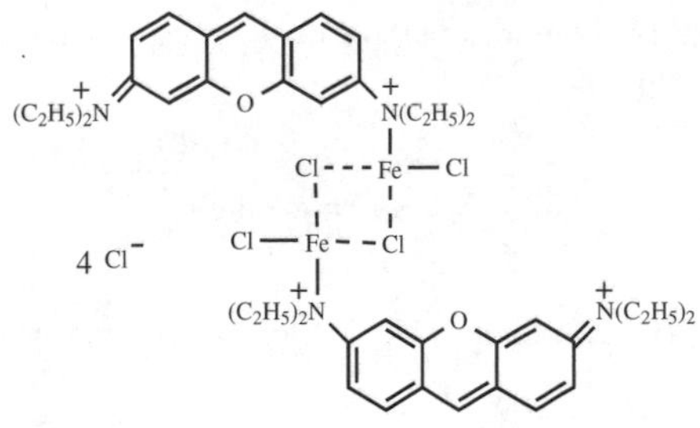

Pyronine Y (mol wt 303) A biological dye for staining RNA and bacterial cells in mammalian tissue.

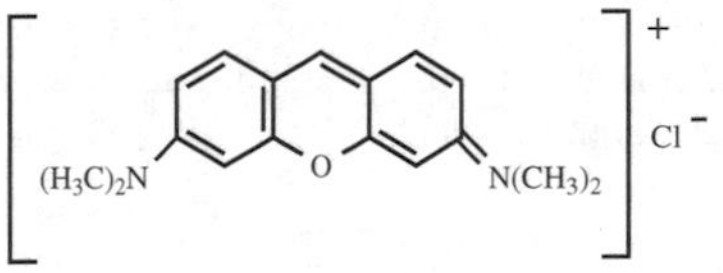

Pyronium A trade name for phenazopyridine hydrochloride, used as an antipyretic and analgesic agent.

Pyrophosphatase The enzyme that catalyzes the hydrolysis of pyrophosphate.

Pyrophosphate Cleavage The cleavage of a pyrophosphate group from a diphosphate or triphosphate nucleotide.

Pyrophosphate Fructose 6-Phosphate 1-Phosphotransferase The enzyme that catalyzes the following reaction:

Pyrophosphate + D-fructose 6-phosphate

⇅

Orthophosphate + D-fructose 1,6-bisphosphate

Pyrophosphate Phosphohydrolase Synonym of inorganic pyrophosphatase.

Pyrophosphoric Acid (mol wt 178) A diphosphoric acid.

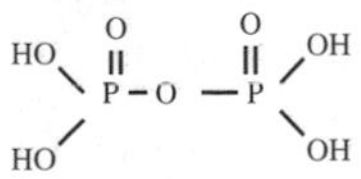

Pyrovalerone (mol wt 245) A CNS stimulant.

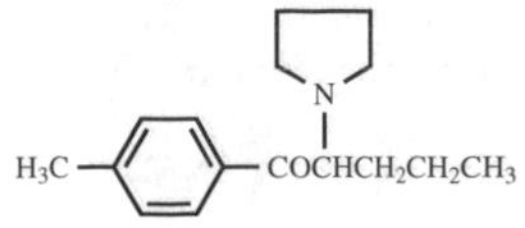

PyrP Abbreviation for pyridoxamine phosphate.

Pyrrobutamine (mol wt 312) An antihistaminic agent.

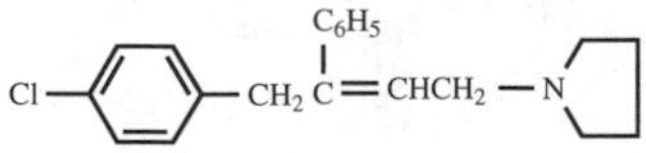

Pyrrocaine (mol wt 232) A local anesthetic agent.

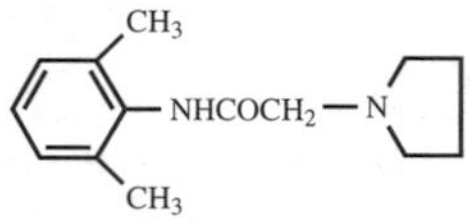

Pyrrole Ring A five-member heterocyclic ring that contains one nitrogen atom (part of the structure of porphyrins and heme).

Pyrrolnitrin (mol wt 257) An antifungal agent.

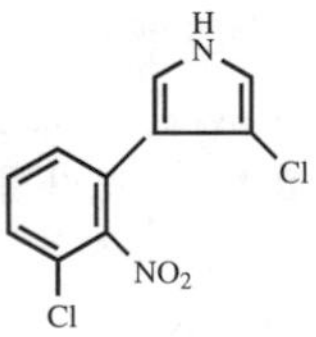

Pyruvate Carboxylase The enzyme that catalyzes the following reaction:

$$\text{ATP + pyruvate + } CO_2 \rightleftharpoons \text{ADP + Pi + oxaloacetate}$$

Pyruvate Decarboxylase The enzyme that catalyzes the following reaction:

$$\text{An } \alpha\text{-keto acid} \rightleftharpoons \text{An aldehyde + } CO_2$$

Pyruvate Dehydrogenase (NAD⁺) The enzyme that catalyzes the following reaction:

$$\text{Pyruvate + CoA + } NAD^+ \rightleftharpoons \text{Acetyl-CoA + } CO_2 \text{ + NADH}$$

Pyruvate Dehydrogenase (NADP⁺) The enzyme that catalyzes the following reaction:

$$\text{Pyruvate + CoA + } NADP^+ \rightleftharpoons \text{Acetyl-CoA + } CO_2 \text{ + NADPH}$$

Pyruvate Dehydrogenase Complex The enzyme complex that catalyzes the conversion of pyruvate to acetyl-CoA and carbon dioxide; it contains pyruvate dehydrogenase, dihydrolipoyl transacetylase, and dihydrolipoyl dehydrogenase.

Pyruvate Formate Lyase The enzyme that catalyzes the following reaction:

$$\text{Acetyl-CoA + formate} \rightleftharpoons \text{CoA + pyruvate}$$

Pyruvate Kinase The enzyme that catalyzes the following reaction:

$$\text{ATP + pyruvate} \rightleftharpoons \text{ADP + phosphenol pyruvate}$$

Pyruvate Lipoamide 2-Oxidoreductase The systematic name for pyruvate dehydrogenase.

Pyruvate $NADP^+$ 2-Oxidoreductase The systematic name for pyruvate dehydrogenase ($NADP^+$).

Pyruvate Oxidae The enzyme that catalyzes the following reaction:

$$\text{Pyruvate + Pi + } O_2 + H_2O \rightleftharpoons \text{Acetyl phosphate + } CO_2 + H_2O_2$$

Pyruvate Oxygen 2-Oxidoreductase (Phosphorylating) The systematic name for pyruvate oxidase.

Pyruvate Synthetase The enzyme that catalyzes the following reaction:

$$\text{Acetyl-CoA + } CO_2 \text{ + reduced ferredoxin} \rightleftharpoons \text{pyruvate + CoA + oxidized ferredoxin}$$

Pyruvic Acid (mol wt 88) A three-carbon organic acid and a key intermediate in the metabolism of glucose.

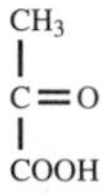

Pyuria The presence of pus in the urine.

PYY Abbreviation for peptide YY.

Pz Abbreviation for pancreozymin, a cytokin.

PZI Abbreviation for potassium zinc insulin, used as an antidiabetic agent.

PZP Abbreviation for pregnant zone protein.

Q

Q Abbreviation for 1. ubiquinone, 2. queuine, 3. queuosine, 4. coenzyme Q, and 5. glutamine.

Q_4 Referring to coenzyme Q_4 or ubiquinone-4 (mol wt 455).

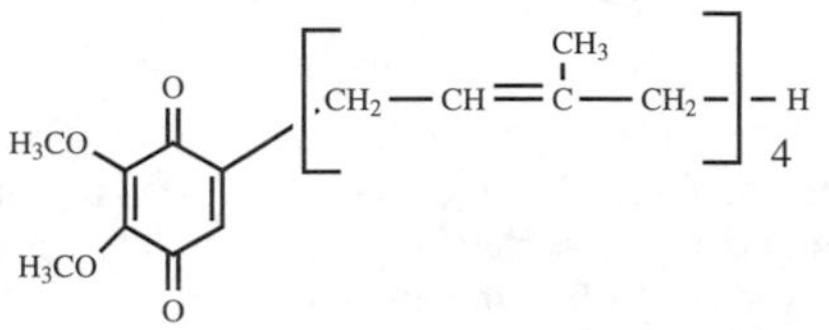

Q_9 Referring to coenzyme Q_9 or ubiquinone-9 (mol wt 795).

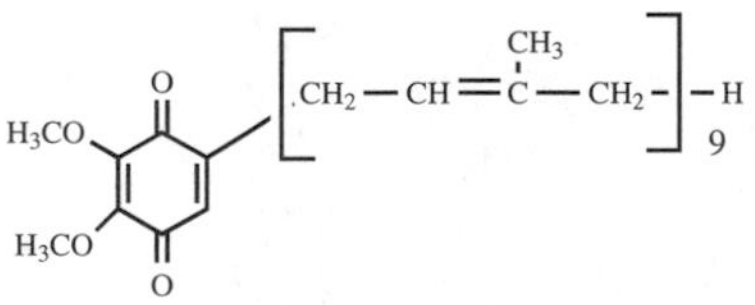

Q_{10} 1. Referring to coenzyme Q_{10} or ubiquinone-4 (mol wt 863).

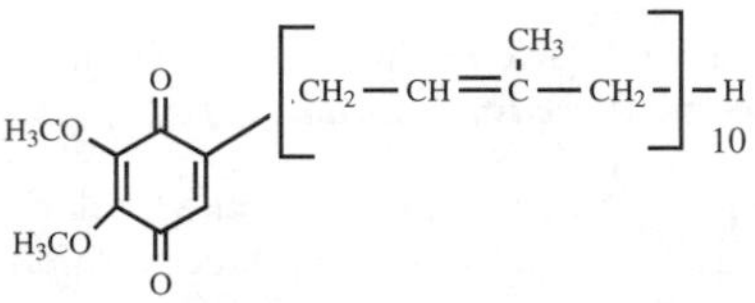

2. Change in the rate of a reaction or a process produced by raising the temperature 10° C.

Q Banding Technique A chromosome-banding technique in which a chromosome is stained with fluorochrome dye, e.g., quinacrine mustard or quinacrine dihydrochloride.

Q Bases Bases produced by modification of guanine, e.g., queuine.

Q Beta See QB.

Q Cycle A cycle of reactions for proton extrusion in the mitochondrial electron transport chain in which coenzyme Q undergoes cycles of reduction.

Q Enzyme Referring to the enzyme that catalyzes the transfer of a segment of 1,4-α-D-glucan chain to a 1,6-α-D-glucosidic linkage in a glucan.

Q Fever An acute disease in humans characterized by the sudden onset of headache, malaise, fever, and muscular pain caused by *Coxiella burnetii*; the reservoirs of infection are cattle, sheep, and ticks.

Q Value The total energy released from an atom when a nuclide is transformed to another nuclide.

Qa Antigen A class-I murine histocompatibility antigen found on certain T lymphocytes.

QAB Abbreviation for quaternary ammonium bases.

QAC Abbreviation for quaternary ammonium compound.

QAE-Sephadex Abbreviation for quaternary aminoethyl Sephadex, an anion exchanger for ion exchange chromatography.

QB Abbreviation for an icosahydron bacteriophage of the family Leviviridae that contains single-stranded RNA.

QB Bacteriophage See QB.

QB Replicase A viral-specific RNA-polymerase encoded by the RNA of QB bacteriophage.

QCT Abbreviation for quantitative computerized tomography.

QDHC Abbreviation for quinine dihydro-chloride.

Q-Enzyme See Q enzyme.

QF Abbreviation for quick freeze.

QH Abbreviation for semiquinone intermediate.

QH_2 Referring to the reduced form of coenzyme Q or the reduced form of ubiquinone.

Q-Pam A trade name for diazepam, used as an antianxiety agent.

QSP Abbreviation for quiescence-specific protein.

QT Abbreviation for quinine tannate.

Quadrant Streak A style of streaking for the isolation of a single colony of a microorganism on a solid medium plate that is divided into four areas or quadrants.

Quadri- A prefix meaning four.

Quadrinal A trade name for a combination drug containing theophylline calcium salicylate, ephedrine hydrochloride, potassium iodide, and phenobarbital, used as a bronchodilator.

Quadrivalent An association of four homologous chromosomes at diplotene, diakinesis, or metaphase I.

Quailpox Virus A virus of Poxviridae that infects bird.

Quaking Mutation A mutation in mice that produces a myelin-deficient animal.

Qualitative Analysis Analysis for the identification of constituents.

Quanta Plural of quantum.

Quantasome Variant spelling of quantosome.

Quantitative Analysis Analysis for the quantity of the constituents.

Quantitative Character A character whose phenotype can be numerically measured or evaluated.

Quantitative Inheritance An inheritable trait controlled by many alleles.

Quantosome A membrane-enclosed photosynthetic unit of the thylakoid membrane.

Quantum The energy of a photon, which is inversely proportional to the wavelength of the emitted radiation.

Quantum Efficiency The number of species that are reacted or decomposed per quantum of energy absorbed.

Quantum Yield The number of molecules of carbon dioxide fixed or oxygen evolved per quantum absorbed.

Quarantine The isolation of persons or animals suffering from an infectious disease in order to prevent transmission of the disease to others.

Quartz Referring to silicon dioxide (SiO_2) used for the manufacture of cuvettes for use in UV spectrophotometry.

Quarzan A trade name for clidinium bromide, an anticholinergic agent that blocks acetylcholine, decreases GI motility, and inhibits gastric acid secretion.

Quarternary Ammonium Compound A substance with a nitrogen atom bonded to four organic groups, thereby giving a positive charge to the nitrogen atom. Quarternary ammonium compounds are cationic surfactants and are used as disinfectants.

Quarternary Mixture A mixture of four different liquids in proportions such that all are mutually soluble in one another but will form phases when one of the components is added in excess.

Quarternary Structure The three-dimensional relationship of different polypeptide chains or subunits in a protein.

Quatrimycin (mol wt 444) An antibiotic prepared by epimerization of tetracycline.

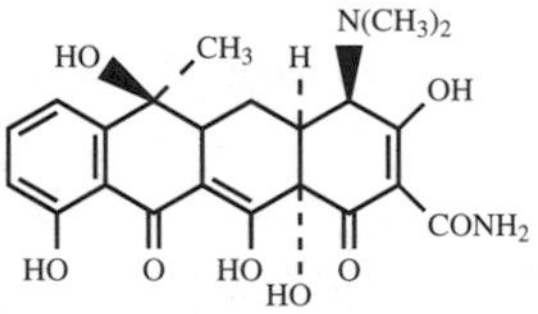

Quayle Cycle A metabolic pathway in some methylotrophic bacteria for assimilation of formaldehyde from ribulose monophosphate.

Quazepam (mol wt 387) A sedative-hypnotic agent that binds to specific benzodiazepine receptors in the CNS.

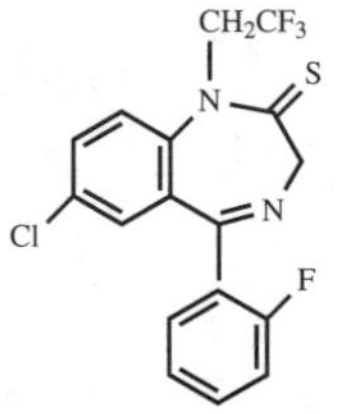

Que Abbreviation for queuine.

Queensland Tick Typhus A tick-borne disease caused by *Rickettsia australis*.

Quelicin A trade name for succinylcholine chloride, used as a neuromuscular blocker to prolong depolarization of the muscle end plate.

Quellung The swelling or enlargement of the capsules of *Pneumococci* by exposure to pneumococcal antibodies.

Quencher A substance capable of reducing or destroying luminescence through deactivation of an excited chemical species.

Quenching 1. Decrease in the counting efficiency in the liquid scintillation. 2. Stop in the discharge of radiation. 3. Decrease in fluorescence due to the absorption of emission energy. 4. Suppression of energy emission.

Quercetin (mol wt 302) An inhibitor of mitochondrial ATPase and glycolysis.

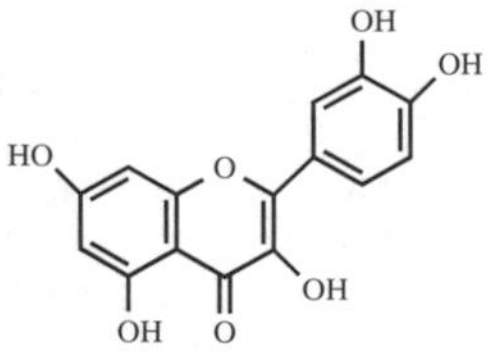

Quercetin 2,3-Dioxygenase The enzyme that catalyzes the following reaction:

Quercetin + O_2

⇅

2-Protocatechoylphloro-glucinolcarboxylate + CO

Quercitol (mol wt 164) An acorn sugar from various species of *Quercus.*

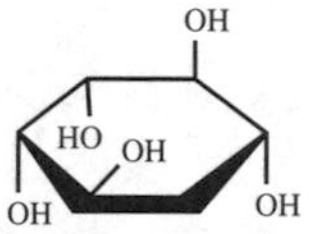

D-quercitol

Quercitrinase The enzyme that catalyzes the following reaction:

Quercitrin + H_2O

⇅

L-Rhamnose + quercetin

Questran A trade name for cholestyramine, an antilipemic agent that promotes utilization of cholesterol.

Queuine An unusual base found in tRNA that is a modified form of guanine.

Queuosine The ribonucleoside of queuine.

Quibron Capsules A trade name for a combination drug containing theophylline and guaifenesin, used as a bronchodilator.

Quibron Plus A trade name for a combination drug containing theophylline, ephedrine hydrochloride, guaifenesin, and butabarbital, used as a bronchodilator.

Quick Pep A trade name for caffeine, used as a cerebral stimulant.

Quiescent Not active.

Quiescent Stem Cells Referring to satellite cells that are capable of undergoing proliferation if muscle cells are damaged.

Quiess A trade name for hydroxyzine hydrochloride, used as an antianxiety agent.

Quin-2 AM (mol wt 830) A fluorescent calcium indicator capable of binding and releasing calcium.

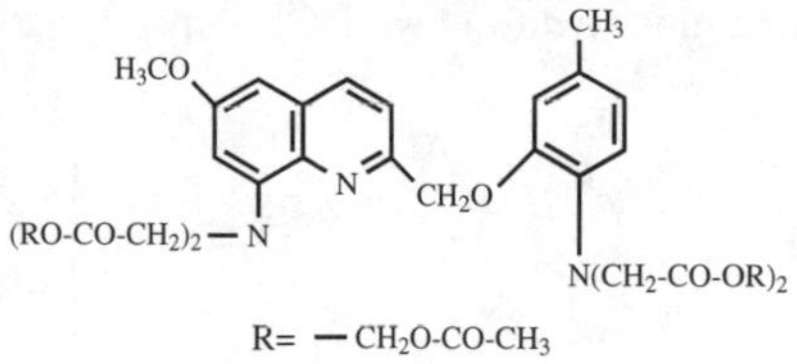

Quinacillin (mol wt 416) A semisynthetic antibiotic related to penicillin that inhibits bacterial cell wall synthesis.

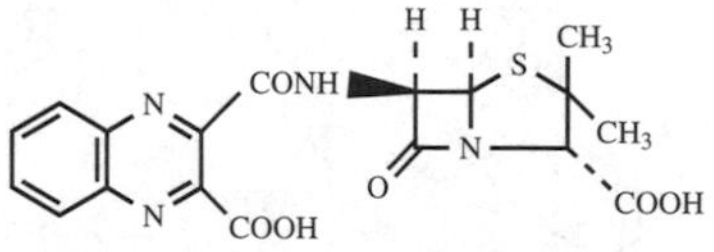

Quinacrine (mol wt 400) An anthelmintic agent that inhibits DNA synthesis in susceptible parasites.

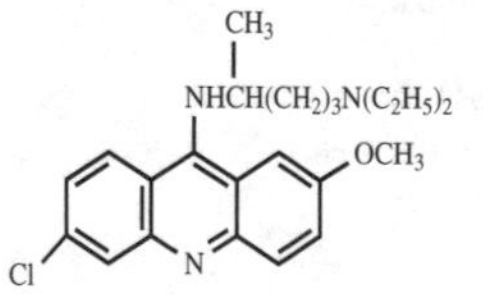

Quinacrine Mustard Dihydrochloride (mol wt 542) A fluorescent probe for labeling chromosomal DNA.

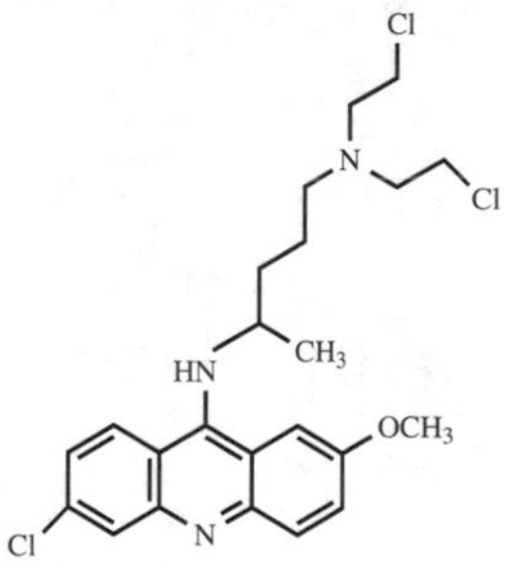

Quinaglute Dura-Tabs A trade name for quinidine gluconate, an antiarrhythmic agent that prolongs the action potential.

Quinalan A trade name for quinidine gluconate, an antiarrhythmic agent that prolongs the action potential.

Quinaldine Blue (mol wt 389) A histological stain.

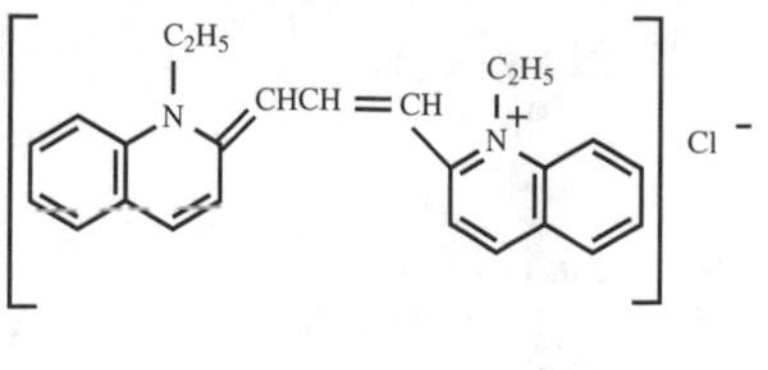

Quinaldine Red (mol wt 430) A pH indicator.

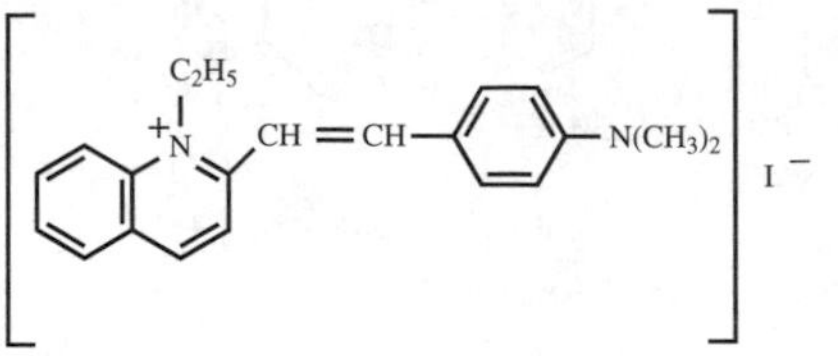

Quinapril (mol wt 439) An antihypertensive agent.

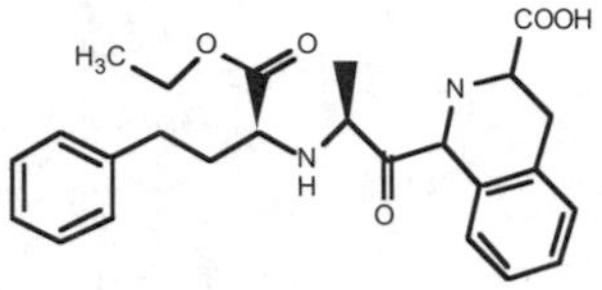

Quinate A trade name for quinine sulfate, used as an antimalarial agent.

Quinate 5-Dehydrogenase The enzyme that catalyzes the following reaction:

$$\text{Quinate} + \text{NAD}^+ \rightleftharpoons \text{5-dehydroquinate} + \text{NADH}$$

Quinbisul A trade name for quinine bisulfate, used as an antimalarial agent.

Quinbolone (mol wt 353) An anabolic hormone that promotes tissue-building processes.

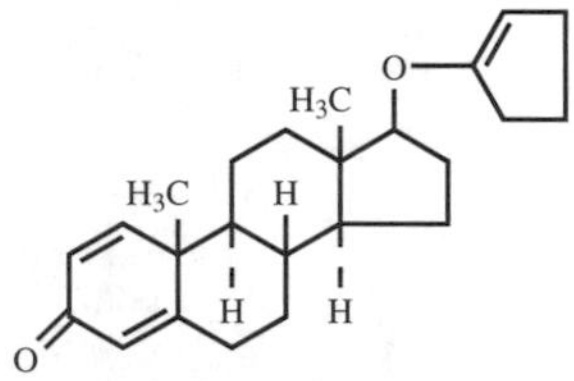

Quindan A trade name for quinine sulfate, used as an antimalarial agent.

Quine A trade name quinidine sulfate, an antiarrhythmic agent that prolongs the action potential.

Quinestradiol (mol wt 356) An estrogen.

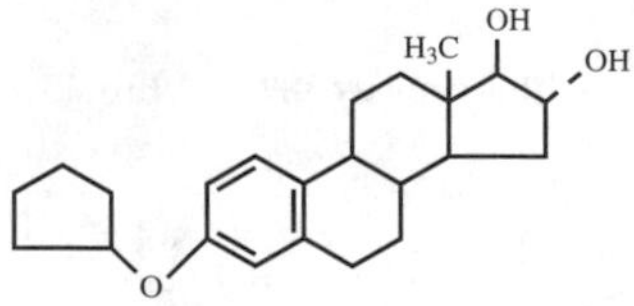

Quinestrol (mol wt 365) A hormone that increases the synthesis of DNA, RNA, and protein in the responsive tissue.

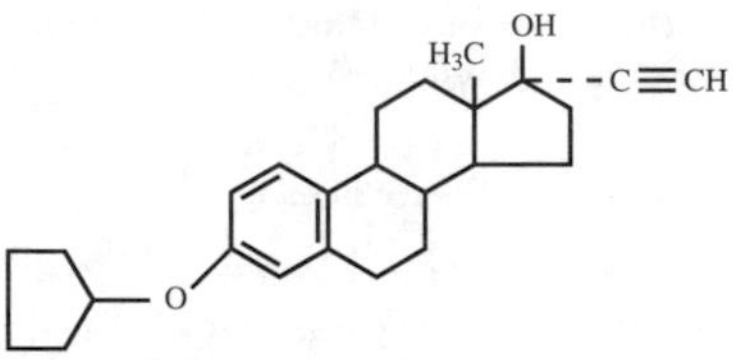

Quinethazone (mol wt 290) A diuretic agent that increases urine excretion of sodium and water by inhibiting sodium reabsorption.

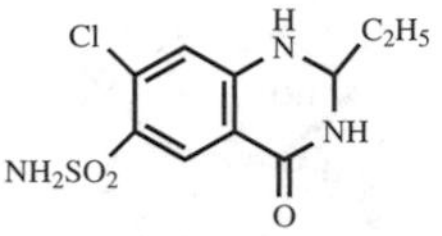

Quinfamide (mol wt 354) An antiamebic agent.

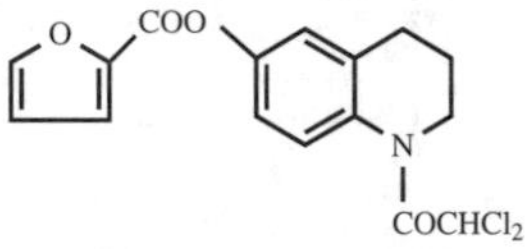

Quingestrone (mol wt 383) A progestogen.

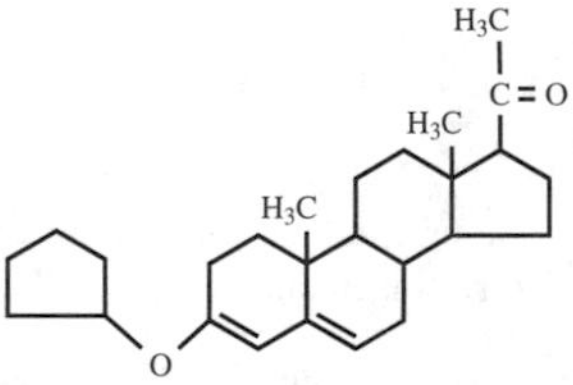

Quinidex Extentabs A trade name for quinidine sulfate, an antiarrhythmic agent that prolongs the action potential.

Quinidine (mol wt 324) An anti-arrhythmic agent that prolongs the action potential.

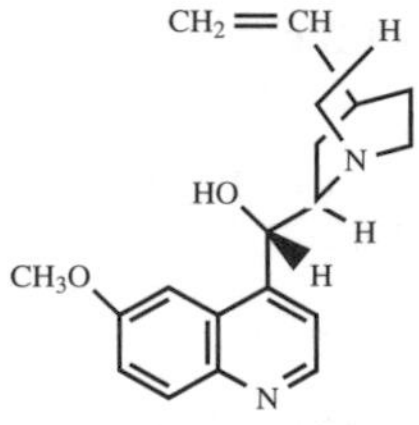

Quinine (mol wt 324) An antimalarial agent.

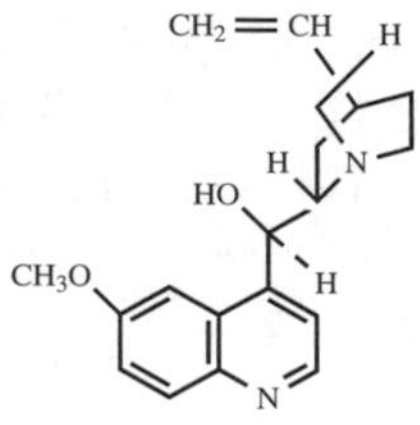

Quinoctal A trade name for quinine sulfate, used as an antimalarial agent.

Quinol Synonym for quinone.

Quinoline (mol wt 129) An antimalarial agent.

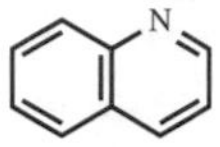

Quinoline Alkaloids Alkaloids that possess a quinoline structure, e.g., quinine.

Quinolinic Acid (mol wt 167) A metabolite of tryptophan.

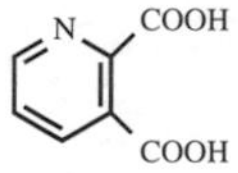

Quinolones A group of antibiotics that blocks normal DNA replication by interfering with DNA gyrase.

Quinone (mol wt 108) A substance involved in the oxidation-reduction system.

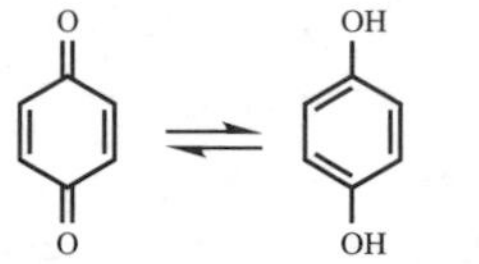

oxidized form reduced form

Quinone Reductase The enzyme that catalyzes the reduction of quinone to phenol using NADH or NADPH as electron donor.

Quinoproteins A group of enzymes that possess pyrroloquinoline quinone as a prosthetic group, e.g., methanol dehydrogenase.

Quinora A trade name for quinidine sulfate, an antiarrhythmic agent that prolongs the action potential.

Quinovose (mol wt 164) A monosaccharide (6-deoxy-D-glucose).

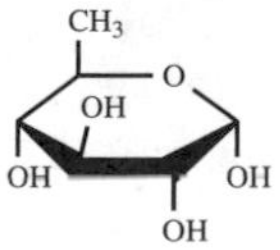

Quinsana Plus A trade name for undecylenic acid and zinc undecyclenate, used as a local antifungal agent.

Quintasa A trade name for mesalamine, an anti-inflammatory agent.

Quintozene (mol wt 295) An agricultural fungicide used for the treatment of seeds.

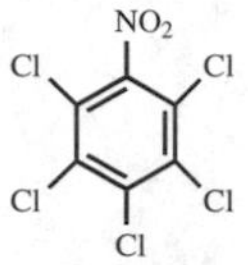

Quinupramine (mol wt 304) An antidepressant.

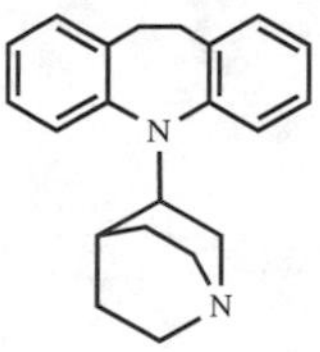

Quiphile A trade name for quinine sulfate, used as an antimalarial agent.

Quisqualic Acid (mol wt 189) An excitatory amino acid used to identify a specific subset of cell receptors.

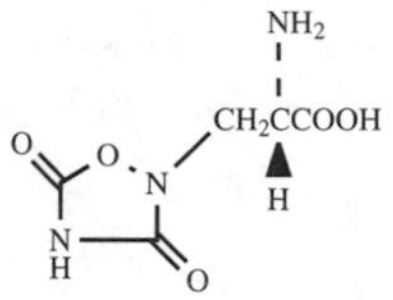

Quizalofop-Ethyl (mol wt 373) A herbicide.

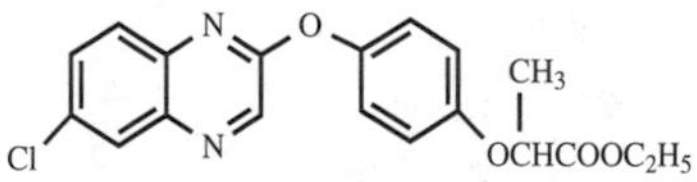

Q-Vel A trade name for a combination drug containing quinine sulfate, vitamin E, and lecithin, used as an antimalarial agent.

Qys A trade name for hydroxyzine, an anti-anxiety, antihistaminic and antiemetic agent.

R

R Abbreviation for 1. gas constant, 2. roengten, and 3. arginine.

R1 Abbreviation for a low copy-number of plasmids in *Enterobacteria* that contain Amp^r (ampicillin resistance), Cml^r (chloramphenicol resistant), and Str^r (streptomycin resistant) markers.

R6 A conjugative plasmid of *Enterobacteria* that contains chloramphenicol, streptomycin, sulphonamide, and mercury resistant markers.

R6K A multicopy conjugative plasmid of *E. coli* that contains Amp^r (ampicillin resistant) and Str^r (streptomycin resistant) markers.

R 17 An RNA bacteriophage of the family Leviviridae.

R_{18} A conjugative plasmid in *Pseudonomas* and *Enterobacteria* that encodes resistance to carbenicillin, kanamycin, and tetracycline.

R_{100} A conjugative plasmid of *Enterobacteria* that encodes resistance to chloramphenicol, streptomycin, sulfonamides, and mercury.

R199 A cloning vector that contains single-stranded DNA and a unique EcoRI cleavage site.

R208 A cloning vector that contains single-stranded DNA, Amp^r (ampicillin resistant), and Tet^r (tetracycline resistant) markers and HindIII, PstI, and SalI cleavage sites.

R229 A cloning vector that contains single-stranded DNA and a unique EcoRI cleavage site.

R Antigen A cell surface protein antigen from *Streptococci.*

R Factor 1. Referring to plasmids that carry drug resistant markers. 2. Abbreviation for releasing factor in protein synthesis.

R Gene 1. Regulatory gene. 2. Gene for drug resistant.

R Group The side chain of an amino acid.

R Glucan An alkaline-resistant glucan from cell wall of fungi.

R Plasmid Plasmids that carry drug resistant markers.

R Strain Strains of bacteria that produce rough or dull (nonsmooth, nonglossy) colonies on a culture plate.

R Strand DNA strand in the Adenovirus that transcribes from left to right.

Ra Abbreviation for the chemical element radium.

^{226}Ra Abbreviation for radioactive radium.

RA Abbreviation for 1. retinoic acid; 2. rheumatoid arthritis.

RA 27/3 A trade name for attenuated live rubella vaccine.

RAA Abbreviation for renin-angiotensin-aldosterone.

RAAg Abbreviation for rheumatoid arthritis agglutination.

RAANA Abbreviation for rheumatoid arthritis associated nuclear antigen.

RAANAg Abbreviation for rheumatoid arthritis associated nuclear antigen.

Rab Protein A family of GTP-binding proteins involved in membrane budding and fusion.

RabAvert A trade name for rabies vaccine.

Rabbit Reticulocyte System A cell-free system for protein biosynthesis from lysed rabbit reticulocytes in which the endogenous mRNA is destroyed by using calcium-dependent ribonuclease.

Rabeprazole (mol wt 359) An antiulcerative agent.

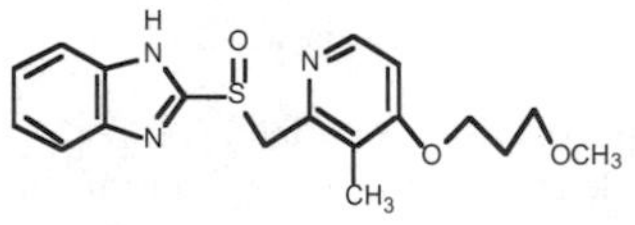

Rabid Affected by rabies.

Rabies An acute and usually fatal disease of humans, dogs, cats, bats, and other animals that is caused by the rabies virus (Rhabdoviridae) and commonly transmitted in saliva by the bite of a rabid animal.

Rabies Immune Globulin Immunoglobulin against rabies used for treatment of patients who have been exposed to rabies.

RABP Abbreviation for retinoic acid-binding protein.

RAC Abbreviation for rheumatoid arthritis cells.

RACC Abbreviation for receptor-activated Ca^{2+} channel.

RACE Abbreviation for rapid amplification of cDNA end.

5'-RACE-5' Abbreviation for rapid amplification of cDNA end.

Racefemine (mol wt 269) An antispasmodic agent.

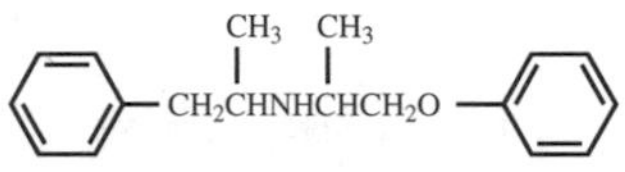

Racemase The enzyme that catalyzes the racemization reaction.

Racemate A mixture of equal parts of two optically active isomers (dextro- and levorotatory in equal concentrations) that neutralize the optical effect of each other.

Racemethorphan (mol wt 271) An antitussive agent.

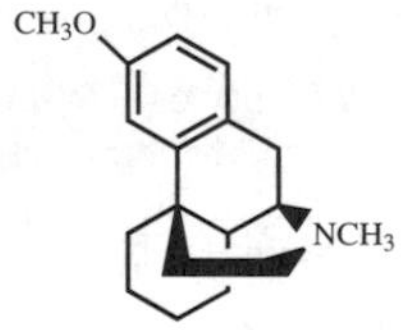

Racemization Conversion of an optically active substance into an optically inactive form in which the dextro- and levorotatory isomers are present in equal concentrations and incapable of rotating plane-polarized light.

Rad An unit of absorbed radiation dose (100 ergs per gram of irradiated tissue).

Radappertization Irradiation of food with radiation for inactivation or destruction of *Clostridium botulinum*.

Radial Chromatography A type of paper chromatography in which substances are allowed to migrate radially.

Radial Immunoassay A type of immunoprecipitation assay in which antigen is allowed to migrate from a well radially toward the antibody-containing agar to form an opaque circular antigen-antibody precipitate. The diameter of the antigen-antibody circle is related to the concentration of the antigen.

Radial Spoke The structure that links the nine microtubule doublets to the center pair of microtubules in the axoneme of a cilium or flagellum.

Radial Symmetry A type of symmetry in which two halves of a body are mirror images of each other regardless of the angle of the cut (cut along the center line).

Radian The angle subtended by an arc that equals to the length of the radius (one revolution equals 2π radians).

Radiation 1. The transfer of energy by electromagnetic waves. 2. The emission of particles or waves by radioactive materials. 3. The emission of light rays.

Radiation Absorbed Dose Referring to the absorbed radiation dose (100 ergs per gram of irradiated tissue).

Radiation Chemistry Science that deals with the chemical effects of radiation on living matters.

Radiation Sickness A disorder resulting from exposure to ionizing radiation, e.g., X-rays, and ultraviolet rays and characterized by nausea, vomiting, and weakness.

Radical Molecule or ion that has one or more unpaired electrons and hence extremely reactive, e.g., OH^-, NH_4^+, SO_4^-.

Radical Anion An anion that acts as a free radical.

Radical Cation A cation that acts as a free radical.

Radicidation Irradiation of food to inactivate or destroy certain nonsporing pathogens, e.g., *Salmonella*.

Radiculitis Inflammation of the root of a nerve.

Radio- A prefix meaning radiation.

Radioactive Having the property of undergoing nuclear disintegration with the emission of radiation.

Radioactive Dating A method that employs the half-lives of radioactive isotopes to determine the age of fossils and rocks.

Radioactive Decay The transformation of the nucleus of a radioactive atom leading to the emission of ionizing radiation, e.g., α particles, β particles, and γ-rays.

Radioactive Half-time The time required for an isotope to loose half of its radioactivity.

Radioactive Iodine Referring to ^{131}I or ^{125}I.

Radioactive Isotope An unstable isotope capable of emitting ionizing radiation, e.g., α particles, β particles, and γ-rays.

Radioactive Label A radioactive element used to label a compound to follow the course of the labeled compound in a biological system.

Radioactive Tracer A radioactive compound that can be traced throughout chemical or biological processes by its radioactivity.

Radioactivity The property of the spontaneous disintegration of an unstable nuclide that emits ionizing radiation, e.g., α particles, β particles, and γ-rays.

Radioallergosorbent Test (RAST) A type of radioimmunoassay for the detection of specific IgE antibody in the serum.

Radioassay Any assay that employs a radioactive isotope.

Radioautography A procedure in which a photographic film or plate is exposed to a radioactive sample to produce an image on the photographic film or plate.

Radiobiology Science that deals with the effects of radiation on biological systems.

Radiocarbon Dating The use of ^{14}C to establish the age of geological or biological remains by measuring its radioactive content.

Radiochemistry The science that deals with the properties and use of radioactive materials.

Radiochromium Referring to ^{51}Cr.

Radiodating See radioactive dating.

Radiodermatitis Inflammation of the skin after exposure to ionizing radiation.

Radiogold Referring to radioactive gold with a mass number of 198.

Radiograph The image produced by the activity of a radioactive substance on a photographic film or plate.

Radiography The production of a radiograph.

Radio-IEP Abbreviation for radio-immuno-electrophoresis.

Radioimmunoassay A highly sensitive serological technique used to assay specific antibodies or antigens employing a radioactive label to tag a reactant.

Radioimmunochemistry The science that deals with immunochemical techniques employing radioactively labeled components.

Radioimmunosorbent Test A solid-phase radioimmunoassay with great sensitivity.

Radioiodinated Protein A protein that is labeled with radioactive iodine (e.g., ^{131}I).

Radioisotope An atom that has an unstable nucleus that emits radiation as it decays, e.g., ^{14}C, ^{3}H, ^{32}P.

Radiolaria A group of free-living protozoa occurring in marine habitats.

Radiology The science that deals with radioactive substance and its application in diagnosis.

Radioluminescence Luminescence produced by radiation from a radioactive material.

Radiometer A device used for the measurement of the intensity of radiation energy.

Radiometric Analysis Analysis of unknown substances using radioactively labeled reagents.

Radiometric Clock The use of the known rate of decay of a radioisotope of an element to determine dates of events in the distant past.

Radiomimetic Substances Substances or drugs that produce physiological effects similar to that of ionizing radiation.

Radionuclide An unstable atomic species that decays and emits radiation.

Radiopaque A compound that is opaque to various forms of radiation, e.g., X-rays and γ rays.

Radiophosphorus Referring to ^{32}P.

Radiorespirometry A technique that employs radioactive isotopes for measurement of respiratory kinetics in the tissue of an organism.

Radiosensitizer A drug or substance that increases the efficiency of cell-killing by radiation.

Radiostol A trade name for ergocalciferol (vitamin D2) that promotes absorption and utilization of calcium and phosphate.

Radiostol Forte A trade name for vitamin D_2, which promotes absorption and utilization of calcium and phosphate.

Radiotherapy Treatment of a cancer by means of X-rays or γ rays.

Radium (Ra) A chemical element with atomic weight 226, valence 2.

Radixin A protein of the erythrocyte band 4.1 family that acts as an actin barbed-end capping protein.

Radon (Rn) Gaseous radioactive element with atomic weight 222, valences 2 and 4. A longest-lived natural isotope with half-life about 3.8 days that emits α-radiation.

RAF Abbreviation for rheumatoid arthritis factor.

raf An oncogene in chicken/mouse sarcoma that encodes protein kinase (serine/threonine).

Rafen A trade name for ibuprofen, used as an anti-inflammatory and antipyretic agent.

Raffinose (mol wt 504) A trisaccharide.

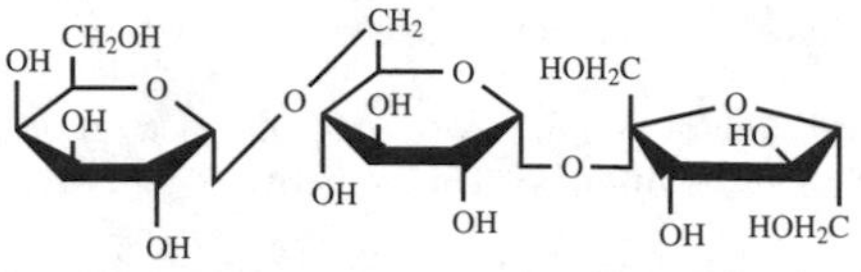

Rafoxanide (mol wt 626) An anthelmintic agent.

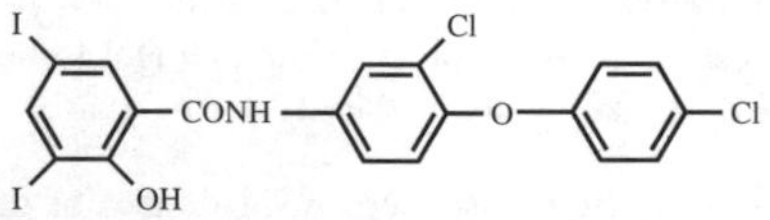

Rahnella A genus of bacteria of the family Enterobacteriaceae.

RAI Abbreviation for radioactive iodine.

RAIHS Abbreviation for radioactively iodinated human serum.

RAIHSA Abbreviation for radioactively iodinated human serum albumin.

RAIU Abbreviation for radioactive iodine uptake.

Raloxifene (mol wt 473) An anti-osteoporotic agent.

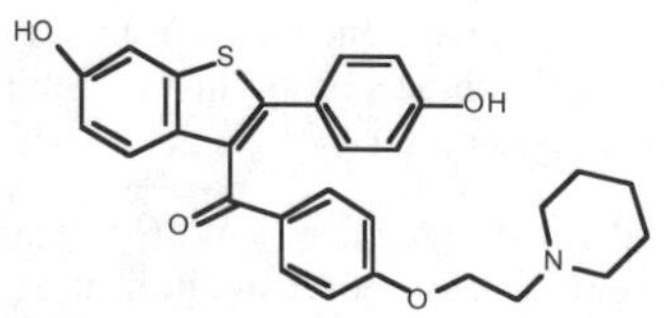

Ramifenazone (mol wt 245) An analgesic, anti-inflammatory, and antipyretic agent.

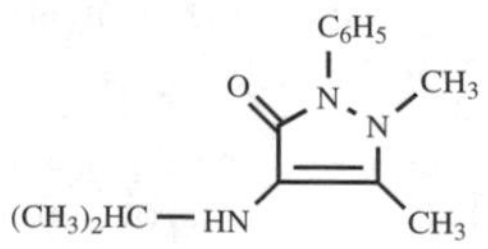

Ramipril (mol wt 417) An antihypertensive agent and a potent vasoconstrictor that inhibits angiotensin-converting enzyme, preventing conversion of angiotensin I to angiotensin II.

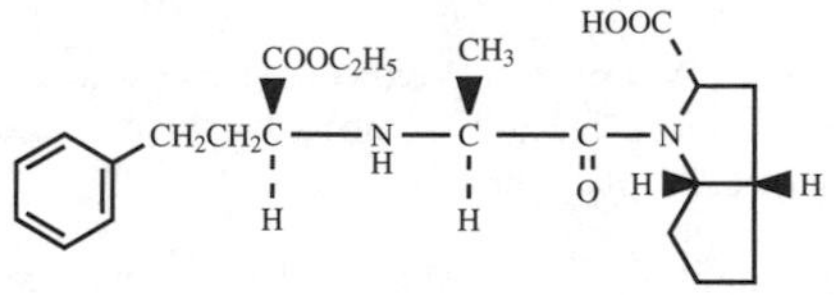

Ramon Titration A serological titration in which a constant volume of a given antigen is incubated with a series of dilutions of homologous antiserum to obtain a dilution of antiserum that gives detectable or nondetectable precipitin.

RAMP Abbreviation for 1. receptor activity modifying protein; 2. rifampicin-AMP.

Ramsay Hunt Syndrome A disease caused by the infection of herpes zooster virus and characterized by pain in the ear and face.

RAMT Abbreviation for rabbit anti-mouse thymocyte.

ran A gene encoding a small nuclear G-protein involved in regulating cell cycle progression and mRNA transport.

RANA Abbreviation for rheumatoid arthritis-associated nuclear antigen, an antigen from rheumatoid arthritis patients that can react with Epstein-Barr-immortalized lymphoid cell line.

Ranavirus A virus of the family Irridoviridae.

Rancidity The development of an unpleasant odor from decomposition of fat due to the liberation of butyric acid and other volatile fatty acids.

Ranimustine (mol wt 328) An antineoplastic agent.

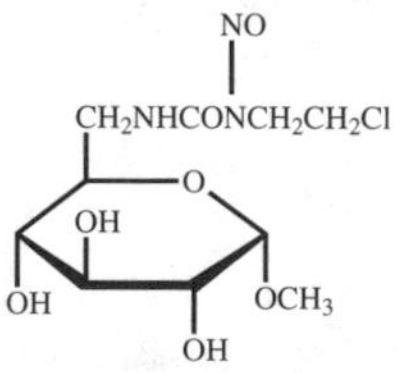

Ranitidine (mol wt 314) An agent that inhibits the action of histamine receptor sites and depresses gastric acid secretion.

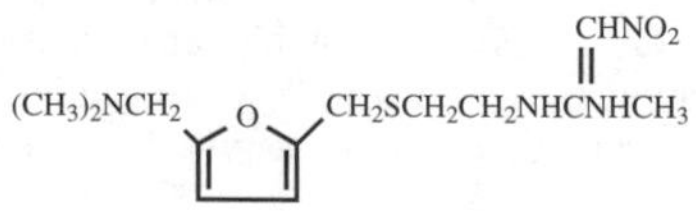

Ranolazine (mol wt 428) An antianginal agent.

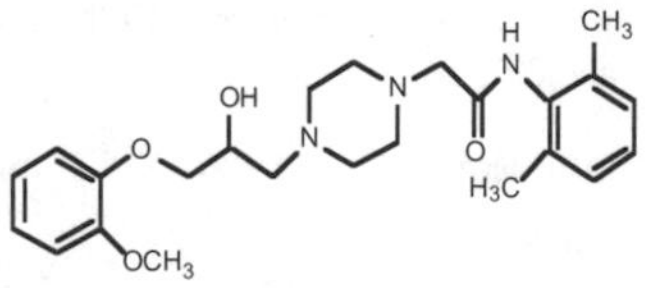

Raoult's Law The law states that the decrease of vapor pressure of the solvent by the addition of solute is proportional to the mole fraction of solute present in the solution.

RAP Abbreviation for receptor-associated protein.

Rapamycin (mol wt 914) A macrolide antibiotic isolated from *Streptomyces hygroscopicus* with immunosuppressant activity.

Rappaport-Vassiliadis Broth A broth medium containing tryptone, $MgCl_2$, and malachite green used for enrichment of salmonellae.

RAR Abbreviation for retinoic acid receptor.

RARE Abbreviation for retinoic acid response element.

Rare Amino Acids Referring to amino acids that occur in a few proteins, e.g., hydroxyproline and hydroxylysine.

Rare Gas Any of the 6 gases (helium, neon, argon, krypton, xenon and radon).

rARF Abbreviation for recombinant ADP-ribosylation factor.

RARLS Abbreviation for rabbit anti-rat lymphocyte serum.

RAS Abbreviation for 1. renin-angiotensin system; 2. rheumatoid arthritis serum.

***ras* Oncogens** A family of oncogens, originally discovered in rat sarcoma that occur in human, rodent, and *Saccharomyces* and encode proteins for conducting cellular signals, e.g., GTP-binding proteins.

***ras* Protein** A family of GTP-binding, GTP-hydrolyzing and autophosphorylating proteins encoded by *ras* genes that conduct cellular signals, e.g., relaying signals from cell-surface receptors to the nucleus.

RAST Abbreviation for radioallergosorbent test used for the detection of specific IgE.

Rast's Method A method for determination of molecular weight by measuring the depression of freezing point of a solvent by a known weight of solute.

Rate Constant A constant relating the concentration of a reactive species to the velocity of a reaction.

RATG Abbreviation for rabbit anti-thymocyte globulin.

RAU Abbreviation for radioactive uptake.

Rauhasine (mol wt 352) An antihypertensive agent.

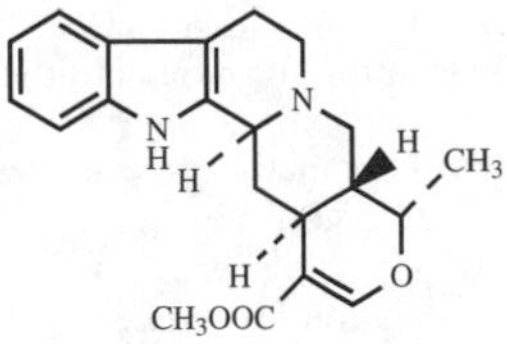

RAV Abbreviation for Rous-associated virus.

Raxar A trade name for grepafloxacin hydrochloride, an antibiotic.

Razoxane (mol wt 268) An antineoplastic agent.

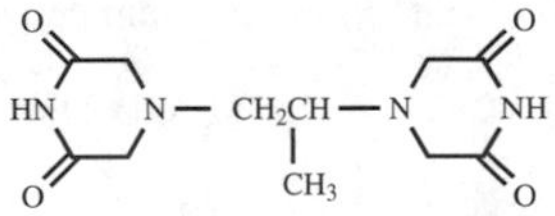

RB Abbreviation for retinoblastoma.

Rb Abbreviation for rubidium.

***rb* Gene** An antioncogen in retina cells whose product is responsible for suppression of retinablastoma.

RBA Abbreviation for rose bengal antigen.

RBC Abbreviation for red blood cell.

RBCC Abbreviation for red blood cell cast.

RBCM Abbreviation for red blood cell mass.

RBCV Abbreviation for red blood cell volume.

RBD Abbreviation for RNA-binding domain.

RBL Abbreviation for rabbit basal leukemia.

RBL Cell Abbreviation for rabbit basal leukemic cell.

RBP Abbreviation for 1. retinol-binding protein and 2. ribulose 1-5-bisphosphate.

rBV Abbreviation for recombinant baculovirus.

RcaI A restriction endonuclease from *Rhodococus capsulatum* with the following specificity:

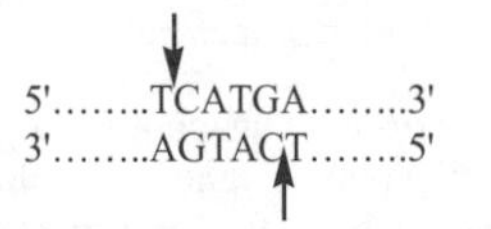

RCC Abbreviation for 1. red cell counts; 2. renal carcinoma cell.

RCF Abbreviation for relative centrifugal force.

RCFS Abbreviation for reticulocyte cell-free system.

RCG Abbreviation for radiocardiography.

rChromatin Chromatin that encodes ribosomal RNA.

RCIA Abbreviation for red cell immune adherence.

RCM Abbreviation for red cell mass.

R-Colony Abbreviation for rough colony.

RCP Abbreviation for riboflavin carrier protein.

RCS Abbreviation for reticulum cell sarcoma.

RCV Abbreviation for red cell volume.

RD Abbreviation for regulatory domain.

RDA Abbreviation for recommended daily allowance.

RDCSRV Abbreviation for Rhesus diploid cell strain rabies vaccine.

RDDA Abbreviation for recommended daily dietary allowance.

RDE Abbreviation for receptor destroying enzyme.

rDNA Abbreviation for 1. any DNA sequence that encodes ribosomal RNA; 2. recombinant DNA.

Rebamipide (mol wt 371) An antiulcerative agent.

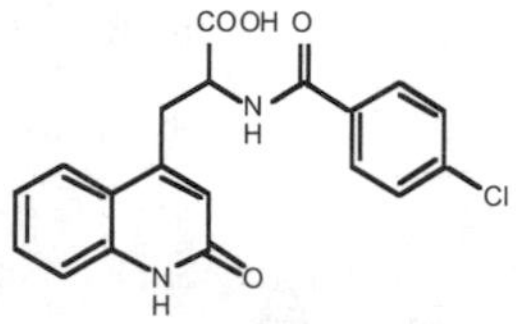

Re Abbreviation for rhenium.

RE Abbreviation for response element.

REA Abbreviation for radio-enzymatic assay.

Reactant Substance that initiates a reaction.

Reaction Center The photochemically active complex in the photosynthetic membrane of chloroplasts that receives the energy trapped by chlorophyll and accessory pigments and initiates electron transfer processes.

Reaction Mechanism The manner in which a chemical reaction proceeds.

Reaction of First Order A chemical reaction whose velocity depends on the concentration of the reactant.

Reaction of Second Order A bimolecular chemical reaction whose rate of reaction is proportional to the concentrations of two reactants.

Reaction Rate See reaction velocity.

Reaction Velocity The rate of a chemical reaction expressed either as the rate of disappearance of reactants or the rate of appearance of product.

Reactivation The restoration of an impaired bioactivity by processes such as genetic recombination, photoreactivation, chemical reaction, thermal reactivation, and the presence of helper elements.

Reactive Lysis The lysis of unsensitized erythrocytes by the binding of complement C5b and C6 onto the cell surface and followed by subsequent binding of C7, C8, and C9.

Reactive Oxygen A highly active oxygen species such as superoxide.

Reading Frame A triplet reading scheme of mRNA from initiation codon to termination codon, the triplets or interval between the initiation codon and termination codon is called open reading frame.

Reading Frameshift A change in nucleotide sequence of a mRNA due to insertion or deletion of a nucleotide by mutation, resulting in a change of reading frame.

Readthrough 1. Translation of mRNA into protein proceeds past the termination codon (an amino acid is inserted at the termination codon). 2. Transcription of DNA into RNA proceeds past the termination signal.

Readthrough Protein An abnormally large protein resulting from the readthrough of a termination codon in mRNA.

Reagent Substance used in a chemical analysis.

Reagin Referring to IgE antibody that mediates type-I hypersensitivity.

Reaginic Antibody Referring to IgE antibody.

Reannealing The formation of double-stranded DNA through complementary base pairing of two single-stranded DNA molecules or two DNA fragments.

Reassortant Virus A synthetically produced virion that contains genome segments and proteins from different viruses.

Reaumur Scale A temperature scale in which 0° is the freezing point of water and 80° is its boiling point.

rec⁻ Abbreviation for recombination-deficient mutant.

***rec*A Gene** A gene that encodes the *rec*A protein.

recA Protein A protein encoded by *rec*A gene that plays an essential role in genetic recombination and DNA repair.

Recalcitrant 1. A chemical that is resistant to microbial attack. 2. A disease that is not responsive to treatment.

Recanalization 1. The restoration of the passageway of a blood vessel that has been blocked by a blood clot. 2. The reunion of an interrupted channel of a body tube.

***rec*B Gene** The gene that encodes the *rec*B protein.

recB Protein A protein encoded by *rec*B gene involved in the recombination following conjugation in bacteria.

recBC Protein A multifunctional protein complex that consists of *rec*B protein and *rec*C protein that is involved in DNA recombination.

***rec*C Gene** The gene that encodes *rec*C protein.

***rec*C Protein** A protein encoded by *rec*C gene that is involved in DNA recombination in bacteria.

recDNA Abbreviation for recombinant DNA.

Receptor Proteins located either on the cell surface (membrane receptor) or within the cytoplasm (cytoplasmic receptor) that bind ligand initiating signal transmission and cellular activity.

Receptor Assay A method for the determination of receptor numbers and the dissociation constant of receptor-ligand binding.

Receptor Destroying Enzyme The enzyme that destroys specific cell receptors, e.g., neuraminidase of influenza virus.

Receptor Down Regulation The reduction of receptor activity on the cell membrane by treatment of cells with specific ligand, e.g., decrease of insulin-binding capability by treating cells with insulin.

Receptor-Mediated Endocytosis The uptake of substance by binding to a receptor and followed by the inward budding of membranous vesicles containing receptor-ligand complexes.

Receptor Protein Protein located either on the plasma membrane or within the cytoplasm that has a binding site for a specific ligand. The binding of a ligand to the receptor initiates signal transmission and cellular activity.

Receptosome A vesicular structure in the cytoplasm formed from receptor-mediated endocytosis that consists of a coated pit and receptor-ligand complex.

Recessive An allele whose phenotype is expressed only when in homozygous form.

Recessive Allele An allele that has no obvious phenotypic effect in a heterozygote; produces a phenotypic effect only when in homozygous condition.

Recessive Lethal An allele whose presence in homozygous condition causes death of cells or death of the organism.

RECG Abbreviation for radioelectrocardiogram.

Reciprocal Recombination Recombination that occurs as a result of crossing-over in which a symmetrical exchange of genetic material takes place, e.g., the genes lost by one chromosome are gained by the other and vice versa.

Recognition Sequence A sequence in the macromolecule that recognizes and binds to a particular sequence or site on the other macromolecule.

Recognition Site A site on a protein or nucleic acid to which a specific protein or ligand molecule binds.

Recombinant The progeny resulting from genetic recombination where the phenotype of the recombinant differs from that of the parent.

Recombinant DNA Any DNA molecule formed by joining DNA segments from different sources. Recombinant DNAs are widely used in gene cloning and genetic modification.

Recombinant DNA Technology Techniques for production of recombinant DNA *in vitro* and transfer of the recombinant DNA into cells where it may be expressed or propagated.

Recombinant Proteins Proteins resulting from cloned genes that are introduced into microorganisms or yeast cells for enhanced production of the gene product.

Recombinant Vaccine A vaccine containing antigen prepared by recombinant DNA technology.

Recombinase A collective term for enzymes involved in genetic recombination.

Recombination 1. The exchange of DNA by crossing over in meiosis. 2. Break and ligation of DNA fragments *in vitro* using purified DNA and enzymes.

Recombination Frequency Referring to the number of recombinant progeny divided by the total number of progeny, a parameter for measuring the distance between two loci or the locations of the genes on the chromosomes.

Recombination Repair A mechanism for repair of thymine dimer that involves the formation of gapped daughter strand opposite the dimer in the damaged parental strand. The gap in the daughter strand is repaired by recombination between gapped daughter strand and an undamaged homologous region of the strand complementary to the damaged parent strand.

Recombinational Hot Spot See recombinator.

Recombinational Regulation A type of genetic regulation in which intramolecular recombination acts as a mechanism for switching genes on and off.

Recombinator A DNA segment that is capable of enhancing general recombination within its vicinity.

Recombinatorial Germ Line Theory Theory states that variable-regions and constant-regions of the immunoglobulin genes are separated and rejoined at the DNA level.

Recombinogenic Element See recombinator.

Recombivax HB A trade name for hepatitis B vaccine.

Recon 1. The viable cell reconstructed by the fusion of a karyoplast with a cytoplast (also known as reconstituted cell). 2. Smallest segment of DNA that is capable of undergoing recombination.

Reconstituted Cell Synonym for recon.

rEC-SOD Abbreviation for recombinant extracellular superoxide dismutase.

Red Blood Cell Cells that specialize in transport of oxygen and have high concentration of hemoglobin in the cytoplasm (also known as erythrocytes).

Red Drop The decrease or reduction in photosynthetic activity that occurs when the wavelength of the incident light increases beyond 680 nm.

Red Muscle Dark skeletal muscle that contains a high concentration of myoglobin and cytochromes.

Red Shift Any shift of the peaks of a spectrum to a longer wavelength.

Redox Abbreviation of reduction–oxidation.

Redox Couple The reduced and oxidized form of a compound, e.g., NAD^+ and NADH, and lactate and pyruvate.

Redox Lipid A lipid that is capable of undergoing oxidation reduction reaction.

Redox Pair See redox couple.

Redox Potential The ability of a redox couple to accept or donate electrons. The standard redox potential of a redox couple at a concentration of one mole of each redox pair at pH 0 is referred to as E_0; at pH 7 is referred to as E_0'.

Redox Reaction A chemical reaction involving the transfer of one or more electrons from one compound to another (also known as oxidation-reduction reaction).

Redoxon A trade name for vitamin C (ascorbic acid).

Reduced Hemoglobin Hemoglobin that contains Fe^{++}.

Reducing Agent Substances that are capable of acting as electron donors (also known as reductants).

Reducing End The terminal of an oligo or a polysaccharide that carries a hemiacetal or a hemiketal group.

Reducing Power The capacity to function as a reducing agent.

Reducing Sugar A sugar that possesses an aldehyde or potential aldehyde group and is capable of reducing certain inorganic ions in solution.

Reductant A reducing agent that is capable of donating electrons.

Reductic Acid (mol wt 114) An antioxidant.

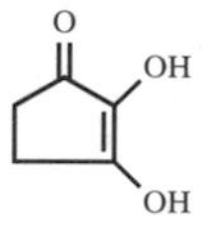

Reduction Potential See redox potential.

Reduction Reaction A chemical reaction involving the addition of electrons.

Reductive Amination The reaction of an α-carboxylic acid with ammonia to produce an amino acid.

Reductive Carboxylic Acid Cycle Synonym for reductive tricarboxylic acid cycle.

Reductive Pentose Phosphate Cycle Synonym for Calvin cycle.

Reductive Tricarboxylic Acid Cycle A pathway for CO_2 fixation in some bacteria in which one molecule of acetyl-CoA is synthesized for every two molecules of CO_2 consumed; it is essentially the reverse of Krebs Cycle.

Reed-Sternberg Cell A large binucleate cell that is characteristic of Hodgkin's disease.

REEG Abbreviation for radioelectroencephalograph.

Reese's Pinworm A trade name for pyrantel pamoate, an anthelminitic agent.

rEF Abbreviation for recombinant enhancing factor.

REF Abbreviation for renal erythropoietic factor.

Refludan A trade name for lepirudin, an anticoagulant.

Refractile Referring to cellular structures that refract light.

Refraction The deviation of a ray as it passes from one medium to another of different density.

Refractive Index A measure of the change in the velocity of light as it passes from one medium to another; it is a ratio of the velocity of light in a vacuum to the velocity of light in a given medium.

Refractometer A device used for measurement of the refractive index of a solution, e.g., Abbe's refractometer.

Refsum's Syndrome A metabolic disorder characterized by the accumulation of phytanic acid in the tissue and serum due to the deficiency of phytanate α-hydroxylase that produces serious neurological problems.

REG Abbreviation for radioencephalogram or radioencephalograph.

Regeneration 1. The process of repair or replacement of damaged tissues or structure. 2. Restoration of an ion exchanger to its original ionic form.

Regibon A trade name for diethylpropion hydrochloride, an anorexiant.

Regitine A trade name for phentolamine mesylate, an antihypertensive agent that blocks the effects of catecholamine on alpha-adrenergic receptors.

Reglan A trade name for metoclopramide hydrochloride, used as an antiemetic agent.

Regonol A trade name for pyridostigmine bromide, used as a cholinergic agent.

Regranex Gel A trade name for a platelet-derived factor produced by DNA recombinant technology and used for treatment of diabetic foot ulcers.

Regroton A trade name for a combination drug containing chlorthalidone and reserpine, used as an antihypertensive agent.

Regucalcin A hepatic calcium-binding protein and GTP-binding protein that modulates the hormonal regulation of plasma membrane Ca^+, Mg^+-ATPase.

Regulace A trade name for a combination drug containing docusate and casanthranol used as a laxative.

Regulator Gene 1. Gene that encodes the repressor for regulating transcription of structural gene(s) in a prokaryotic operon system. 2. A gene that controls the functions of other genes.

Regulatory Enzyme The enzyme that possesses an allosteric site or regulatory site for binding effector molecules in addition to the catalytic binding site. There are two types of regulatory enzymes, namely allosteric enzymes and covalently modified regulatory enzymes.

Regulatory Site Synonym for allosteric site.

Regulatory Subunit The protein subunit of an allosteric enzyme that binds effector molecules for regulating a metabolic pathway, e.g., CTP-binding subunit of aspartate transcarbamaylase.

Regulatory T Cell Referring to helper T cells.

Regulax SS A trade name for docusate sodium, used as a laxative.

Regulon A system in which the nonadjacent structural genes or structural genes of two or more operons are under control of a common regulatory gene product.

Regutol A trade name for docusate sodium, used as a laxative.

Reichert-Meissl Number The number of ml of 0.1N NaOH required for neutralization of the volatile fatty acids in 5 grams of fat.

Reidamine A trade name for dimenhydrinate, used as an antiemetic agent.

Rejection Response An immune response directed against transplanted tissue.

REKG Abbreviation for radio-electro-kardiogram or radio-electro-kardiograph.

rel An oncogene in turkey reticuloendotheliosis that encodes gene regulatory protein.

Rela A trade name for carisoprodol, used as a skeletal muscle relaxant that reduces transmission of nerve impulses from the spinal cord to skeletal muscle.

Relafen A trade name for nabumetone, an analgesic and anti-inflammatory agent.

Relapsing Fever A human disease caused by a *Borrelia* that is transmitted by ticks and characterized by recurrent fever.

Relative Centrifugal Force Centrifugal force expressed in multiples of gravitational force (g), e.g., 100,000 × g.

Relative Molecular Mass Mass of a molecule expressed as a multiple of the mass of a hydrogen atom.

Relaxant A substance that relieves muscular tension and produces relaxation.

Relaxation 1. Transition of supercoil circular DNA to a relaxed form with fewer superhelical turns. 2. Transition of constricted muscle to its resting stage.

Relaxation Enzyme Referring to the enzyme that catalyzes the conversion of supercoil DNA to relaxed DNA, e.g., topoisomerase.

Relaxed Circle DNA The double-stranded circular DNA that is not supercoiled.

Relaxed Control of Plasmid Replication Achievement of a high copy number of a plasmid under abnormal growth conditions (e.g., under the condition in which protein synthesis is inhibited).

Relaxed DNA DNA that is not supercoiled.

Relaxed Helix A nontwisted helix.

Relaxed Muscle A resting muscle.

Relaxed Plasmid A plasmid that can exceed its normal copy number only under abnormal growth conditions (e.g., under the condition in which protein synthesis is inhibited).

Relaxin A polypeptide ovarian hormone secreted by the corpus luteum during pregnancy that relaxes the pubic symphysis and dilates the cervix.

Relaxing Enzyme Synonym for DNA topoisomerase.

Releasing Factor Protein factor responsible for releasing mature polypeptide from the ribosome during protein synthesis.

Releasing Hormone A hormone that stimulates the release of another hormones.

Remeron A trade name for mirtazapine, an antidepressant.

REMI Abbreviation for restriction enzyme-mediated integration.

Remicade A trade name for chimeric $IgG1_k$ monoclonal antibody that consists of human constant and murine variable regions which it binds to tumor necrosis factor alpha.

Remineralization The restoration of the body's minerals that have been excreted during dehydration or lost through illness.

Remission A period during which the symptoms of a disease are abated.

Renacidin A trade name for a combination drug containing citric acid, glucono-delta-lactone, and magnesium carbonate used for the prevention and dissolution of calculi.

Renal Pertaining to the kidney.

Renal Glucosuria A disorder characterized by the excretion of glucose in the urine due to increased permeability of the kidney.

Renal Threshold A threshold concentration of a substance within the blood at which the substance starts to appear in the urine.

Renal Tubular Acidosis A kidney disorder characterized by the production of urine deficient in acidity due to a defective function in the renal tubules.

RenAmin An amino acid infusion for treatment of renal failure.

Renaturation 1. Restoration of a protein from a denatured state to the native conformation. 2. The reannealing of single-stranded DNA to form a duplex molecule.

Renese A trade name for polythiazide, a diuretic agent that increases urine excretion of sodium and water by inhibiting sodium reabsorption.

Renese-R A trade name for a combination drug containing polythiazide and reserpine, used as an antihypertensive agent.

Renin A protease that catalyzes the cleavage of the peptide bond involving the carboxyl group of leucine in angiotensinogen to generate angiotensin I.

γ-Renin The enzyme that catalyzes the cleavage of the leu-leu peptide bond in synthetic tetradecapeptide renin to produce angiotensin I.

Renin Angiotensin System A system involved in the control of blood pressure.

Renitec A trade name for enalapril maleate, an antihypertensive agent that inhibits angiotensin-converting enzyme, preventing conversion of angiotensin I to angiotensin II.

Rennet An extract from the stomach of young mammalians living on milk, it contains chymosin and rennin.

Rennin A highly specific aspartyl proteinase secreted by the kidney.

Renogram A photographic depiction of the course of renal secretion employing a radioactively labeled substance.

Renotropic Having a tendency to induce the enlargement of the kidney or to increase the activity of the kidney.

ReoPro A trade name for abciximab, an antiplatelet agent produced by recombinant DNA technology and used for reduction of acute blood clot-related complication and for high-risk angioplasty patients or for the patients undergoing coronary intervention.

Reoviridae A family of the double-stranded RNA viruses, e.g., reovirus.

Reovirus A virus of the family Reoviridae (respiratory enteric orphan virus) that contains double-stranded RNA and infects humans, birds, and cattle.

RepA A protein in *E. coli* encoded by *repA* gene involved in DNA replication.

***RepA* Gene** A gene in *E. coli* that produces repA protein.

Repaglinide (mol wt 453) An anti-diabetic agent that closes potassium channels in the beta cells of the pancreas which causes the opening of the calcium channels resulting in release of insulin.

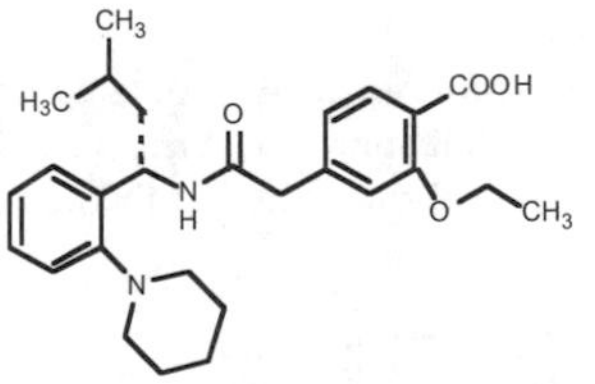

Repair Enzyme The enzyme that catalyzes the detection, removal, and repair of damaged DNA, e.g., DNA polymerase I.

Repair Synthesis Replacement of a damaged DNA base or segment by the synthesis of a new nucleotide segment using undamaged strand as template.

Repairosome A complex of enzymes and proteins involved in the repair of damaged DNA.

Repan A trade name for a combination drug containing acetaminophen, caffeine, and butalbital, used as an analgesic and antipyretic agent.

Repellent 1. A substance that repels insects or animals. 2. A substance that will not mix or blend with another substance.

Repetitive DNA A nucleotide sequence that is repeated many times in the genome.

Repirinast (mol wt 355) An antiallergic agent.

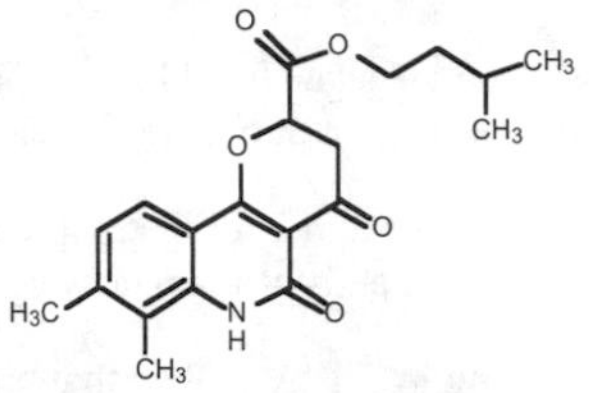

Replacement Vector A cloning vector that has a pair of cleavage sites flanking a dispensable sequence (known as stuffer); during the process of cloning, the stuffer sequence is removed and replaced with a foreign DNA.

Replica Method A method for preparating a replica of a specimen for transmission electron microscopy in which the freeze-fractured tissue is shadowed with metal, coated with carbon, and then the tissue is digested away to obtain a replica mold for examination under the electron microscope.

Replicase Referring to RNA-dependent RNA polymerase.

Replication 1. Multiplication of a microorganism. 2. Duplication of a nucleic acid from a template. 3. The formation of a replica mold for viewing by electron microscopy.

Replication Fork Y-shaped region of a replicating DNA molecule at which the two daughter strands are formed and separated.

Replication Origin A unique sequence in DNA where replication of DNA starts.

Replicative Form The double-stranded DNA or RNA formed during the replication of single-stranded DNA or single-stranded RNA.

Replicative Intermediate (RI) A replicative intermediate formed during replication of certain single-stranded RNA viral genomes that is partially double-stranded RNA with many single-stranded tails.

Replicon A segment of DNA that contains replication origin and serves as a replication unit. Eukaryotic chromosomes contain many replication units, each possessing a replication origin.

Replisome A complex of DNA polymerase, primase, helicase, and other proteins required for DNA replication at the replication fork.

Reposal (mol wt 262) A sedative, hypnotic agent.

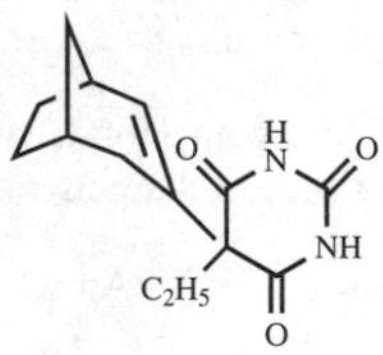

Reposans A trade name for chlordiazepoxide hydrochloride, used as an antianxiety agent.

Rep-Pred A trade name for methylprednisolone acetate, used as an anti-inflammatory agent.

Repressor Protein 1. A protein that binds to an operator gene blocking the transcription of structural genes under the control of that operator. 2. A protein that suppresses the replication of a prophage in lysogenic bacteria.

Repronex A trade name for menotropins, a purified preparation of gonadotropins extracted from the urine of postmenopausal women.

Reproterol (mol wt 389) A bronchodilator.

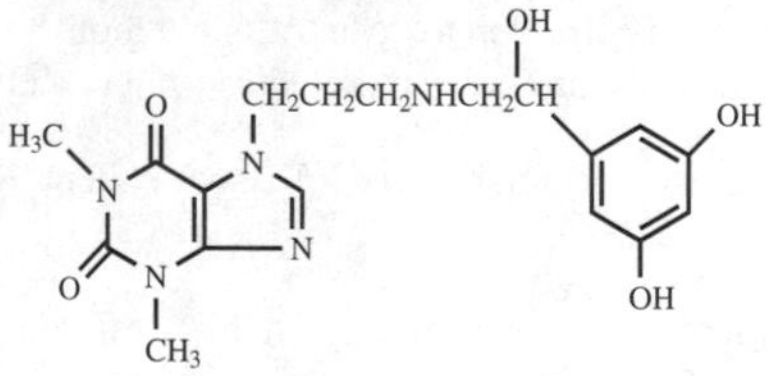

Requip A trade name for ropinirole hydrochloride, a dopamine receptor agonist used as an antiparkinsonism agent.

RER Abbreviation for rough endoplasmic reticulum.

RES Abbreviation for reticuloendothial system.

Resazurin (mol wt 229) A redox indicator dye.

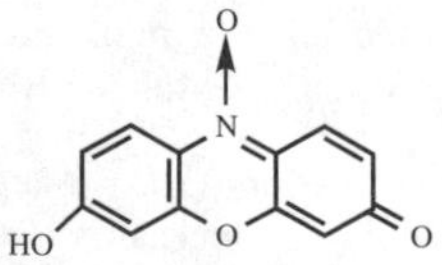

Resazurin Test A test for the presence of microorganisms in milk that employs resazurin as a color indicator.

Rescimetol (mol wt 591) An antihypertensive agent.

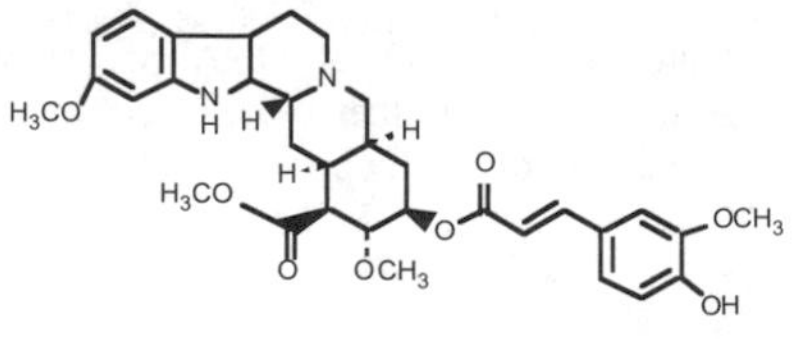

Rescinnamine (mol wt 635) An antihypertensive agent that inhibits norepinephrine release and depletes norepinephrine stored in adrenergic nerve endings.

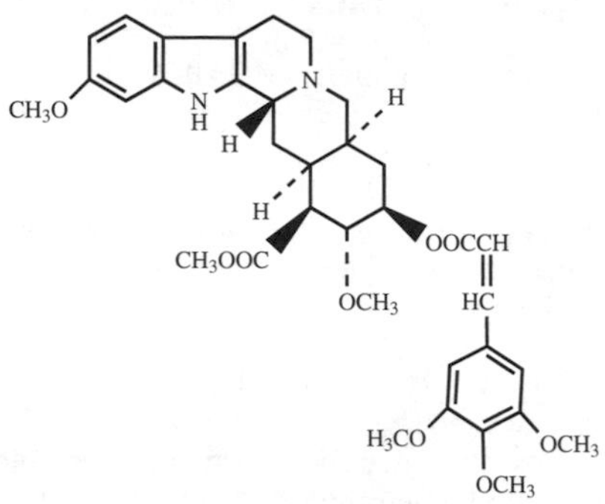

Rescriptor A trade name for delavirdine mesylate, used as an antiviral agent.

Resealed Ghost A membrane structure formed by resealing of the lysed membrane of the red blood cell in a defined medium.

Resectisol A trade name for mannitol, a sugar alcohol used as an osmotic diuretic agent.

Reserpiline (mol wt 412) An antihypertensive agent.

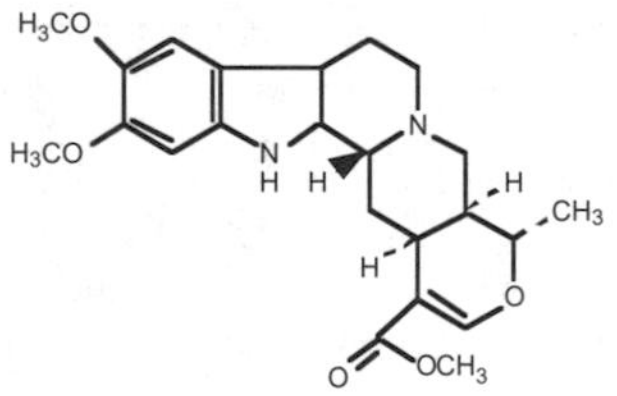

Reserpine (mol wt 609) An antihypertensive agent that inhibits norepinephrine release and depletes norepinephrine stored in the nerve ending.

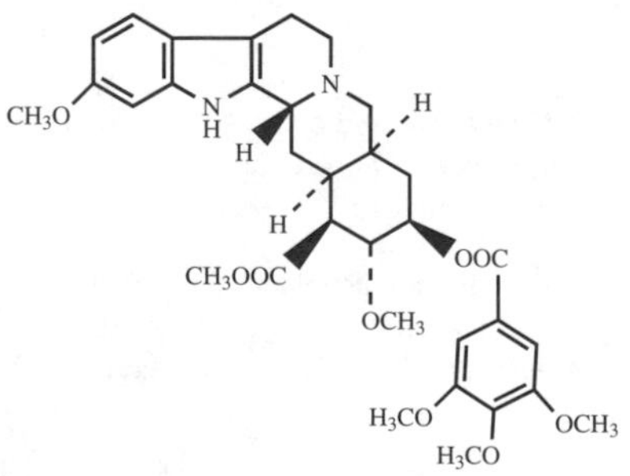

Residue The monomeric unit in a polymer without the atoms that are removed from it during polymerization, e.g., amino residue in a protein.

Resilin An elastic, rubberlike protein found in the tissue of insects.

Resin 1. Polymerized support used in ion-exchange chromatography, e.g., ion exchange resin. 2. A mixture of carboxylic acids, essential oils, and terpenes from plants.

Resistance Factor Referring to plasmids that carry drug-resistant genes (also known as R factor).

Resolving Power The ability of a given lens of a microscope or an optical system to distinguish minimum distance between two details.

Resonance Structures Structural formulas that differ from one another only by the position of electrons.

Resorcinol (mol wt 110) A keratolytic agent and a reagent used for the testing of ketohexoses.

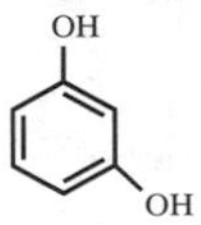

Respbid A trade name for theophylline, a bronchodilator that inhibits phosphodiesterase and increases cAMP concentration.

RespiGam A trade name for immune globulin antibodies against respiratory syncytial virus produced by DNA recombinant technology.

Respiration Oxidation of organic molecules and generation of ATP with O_2 as the terminal electron acceptor.

Respiratory Acidosis Acidosis caused by decrease in respiration and characterized by the increase of carbon dioxide and carbonic acid in the plasma.

Respiratory Alkalosis Alkalosis caused by an increase in respiration and characterized by the decrease of carbon dioxide and carbonic acid in the plasma.

Respiratory Burst A process by which neutrophils and monocytes kill microbial pathogens by conversion of oxygen to toxic oxygen products, e.g., to superoxide, singlet oxygen, and hydroxyl radicals.

Respiratory Chain Referring to mitochondrial electron transport chain (sequence of electron carrying proteins) in which oxygen is the terminal electron acceptor.

Respiratory Chain Phosphorylation Referring to oxidative phosphorylation (ATP-formation) in the mitochondrial electron transport chain.

Respiratory Complexes The multienzyme-protein systems in the inner mitochondrial membrane that carry out the reactions of electron transport.

Respiratory Control Referring to the regulation of oxidative phosphorylation and electron transport by the availability of ADP.

Respiratory Inhibitor Substances that inhibit the flow of electrons along the mitochondrial electron transport chain.

Respiratory Pigment Any of a number of pigmented proteins associated with respiratory processes such as cytochrome c or flavin protein.

Respiratory Quotient The molar ratio of carbon dioxide given off in respiration to that of oxygen consumed.

Respiratory Syncytial Virus A virus of the family Paramyxoviridae that forms syncytia in tissue and causes respiratory disorders.

Respirometer An instrument for measuring respiration.

Respirometry Measurement of the kinetics of respiration.

Respolin A trade name for albuterol, a bronchodilator that acts on alpha-2 adrenergic receptors.

Respolin Inhaler A trade name for albuterol sulfate, a bronchodilator that acts on alpha-2 adrenergic receptors.

Response Element A DNA sequence that is common to promoters/enhancers of genes whose expression is coordinately regulated.

Resprim A trade name for cotrimoxazole, an antibacterial agent that inhibits the formation of dihydrofolic acid from PABA in bacteria.

Resting Nucleus A nucleus that is not undergoing the process of cell division.

Resting Potential Potential difference across the plasma membrane of an unstimulated nerve cell.

Restoril A trade name for temazepam, a sedative-hypnotic agent.

Restriction Endonuclease An enzyme that catalyzes the cleavage of a double-strand DNA at a defined point with a specific sequence.

Restriction Enzyme See restriction endonuclease.

Restriction Fragment The DNA segments produced by the action of restriction endonuclease; different restriction endonucleases produce different numbers and sizes of fragments on the same DNA.

Restriction Fragment Length Polymorphism The fragment patterns produced by a specific restriction endonuclease on DNA from different individuals.

Restriction Mapping A method for determining the relationship between the genomes from different sources by comparison of the electrophoretic patterns of fragments of DNA resulting from treatment with a specific restriction endonuclease.

Restriction Modification System A system in bacteria in which cellular DNA is modified so that it cannot be degraded by its own endonuclease, e.g., modification by methylation.

Restriction Site A specific sequence in DNA at which a specific restriction endonuclease cleaves.

Restrictocin An antitumor polypeptide produced by *Aspergillus* spp.

Resyl A trade name for guaifenesin, an antitussive agent that increases production of respiratory tract fluids to liquify and to reduce the viscosity of tenacious secretions.

RET A human oncogene derived through fibroblast transfection with T-cell lymphoma DNA.

Retaplase A human tissue enzyme produced by recombinant DNA technology that converts plasminogen to the fibrinolysin for degradation of fibrin clots.

Retavase A trade name for retaplase, a human tissue enzyme produced by recombinant DNA technology that converts plasminogen to the fibrinolysin for degradation of fibrin clots.

Reticular Cell See reticulum cell.

Reticular Dysgenesis A congenital immune deficiency disorder characterized by the severe reduction in lymphocytes and phagocytes.

Reticulin A protein associated with connective tissue that occurs together with collagen and elastin.

Reticulocalbin A calcium-binding luminal protein of the endoplasmic reticulum.

Reticulocyte An immature red blood cell capable of hemoglobin synthesis.

Reticulocytosis The increase in the proportion of immature red blood cells (reticulocytes).

Reticuloendothelial System A former term for mononuclear phagocyte system, a diffuse system of phagocytes associated with the spleen, lymph nodes, and other lymphoid tissue.

Reticulum Cell Cells of the reticuloendothial system found in lymph node, bone marrow, and spleen.

Retina Layer of receptor cells in the eye that responds to light.

Retin-A A trade name for tretinoin (vitamin A acid or retinoic acid).

Retinal (mol wt 284) The aldehyde form of vitamin A.

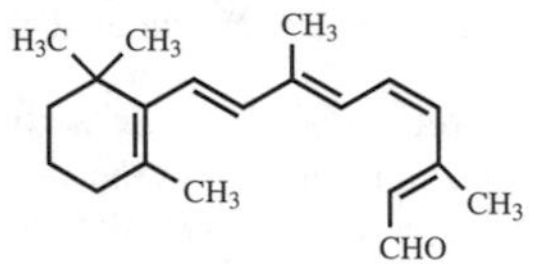

11-*cis* retinal

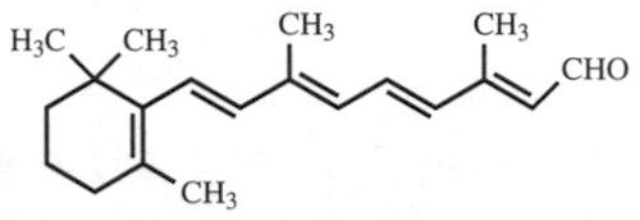

all-*trans* retinal

Retinal Dehydrogenase The enzyme that catalyzes the following reaction:

$$\text{Retinal} + \text{NAD}^+ + H_2O \rightleftharpoons \text{Retinoate} + \text{NADH}$$

Retinal Isomerase The enzyme that catalyzes the following reaction:

$$\text{all-}\textit{trans}\text{-Retinal} \rightleftharpoons 11\text{-}\textit{cis}\text{-Retinal}$$

Retinal Oxidase The enzyme that catalyzes the following reaction:

$$\text{Retinal} + O_2 \rightleftharpoons \text{Retinoate} + H_2O_2$$

Retinene Isomerase Synonym for retinal isomerase.

Retinitis Inflammation of the retina.

Retinoblastoma A cancer of the developing retina that affects only infants and young children.

Retinoic Acid (mol wt 300) A derivative of vitamin A.

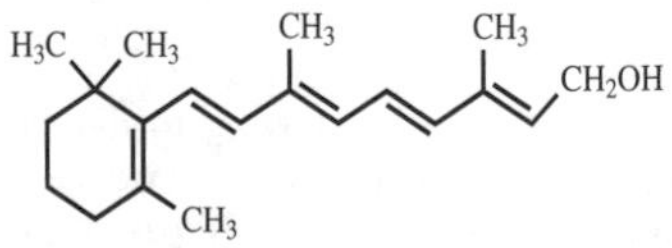

Retinol (mol wt 286) Synonym of vitamin A, a coenzyme necessary for retinal function, bone growth, and differentiation of epithelium tissue.

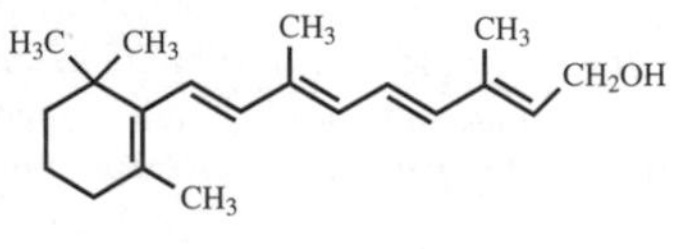

Retinol Binding Protein A plasma protein that binds and transports vitamin A.

Retinol Dehydrogenase The enzyme that catalyzes the following reaction:

$$\text{Retinol} + \text{NAD}^+ \rightleftharpoons \text{Retinal} + \text{NADH}$$

Retinol Isomerase The enzyme that catalyzes the following reaction:

$$\text{all-}\textit{trans}\text{-Retinol} \rightleftharpoons \text{11-}\textit{cis}\text{-Retinol}$$

Retinol O-Fatty-Acyltransferase The enzyme that catalyzes the following reaction:

$$\text{Acyl-CoA} + \text{retinol} \rightleftharpoons \text{CoA} + \text{retinyl ester}$$

Retinyl Palmitate Esterase The enzyme that catalyzes the following reaction:

$$\text{Retinyl-palmitate} + H_2O \rightleftharpoons \text{Retinol} + \text{palmitate}$$

Retrovir A trade name for AZT (azidothymidine), a drug used for treatment of AIDS.

Retrovirus An RNA virus that contains reverse transcriptase. Its RNA serves as a template for synthesis of cDNA, the synthesized cDNA is integrated into a chromosome of the mammalian host cell.

REV Abbreviation for reticulo-endotheliosis virus.

Reverse Electron Transport An energy-dependent movement of electrons along the electron transport chain found in some bacteria.

Reverse Gyrase A type-II topoisomerase that catalyzes the introduction of positive supercoiling in cccDNA (covalently closed circular DNA).

Reverse Transcriptase The enzyme that catalyzes the synthesis of DNA–RNA hybrid or single-stranded DNA using RNA as a template.

Reverse Transcription A process by which an RNA molecule is used as a template for synthesis of a single-stranded DNA copy or DNA–RNA hybrid.

Reversion Restoration of a mutant to its original phenotype.

Reversol A trade name for edrophonium chloride, an agent that inhibits the destruction of acetylcholine.

Revertase Synonym of RNA-directed DNA polymerase.

Revex A trade name for nalmefene hydrochloride, a narcotic antagonist that blocks the effects of opioids.

Rev-Eyes Trade name for dapiprazole hydrochloride, an ophthalmic agent for producing meiosis.

Revia A trade name for naltrexone hydrochloride, a narcotic antagonist.

Revimine A trade name for dopamine hydrochloride, used to stimulate dopaminergic, beta-adrenergic, and alpha-adrenergic receptors on the sympathetic nervous system.

Reviparin A low molecular weight heparin used as an antithrombotic agent.

Rexolate A trade name for sodium thiosalicylate, an analgesic and antipyretic agent that blocks the generation of pain impulses.

Rezide A trade name for a combination drug containing hydrochlorothiazide, hydralazine hydrochloride, and reserpine, used as an antihypertensive agent.

Rezulin A trade name for troglitazone, an oral antihyperglycemic agent.

RF Abbreviation for 1. replication factor; 2. replication form.

RF Abbreviation for 1. replicative form of single-stranded viral RNA or single-stranded viral DNA produced during the process of replication. 2. releasing factor in protein synthesis.

R_f Value The ratio of the migration distance traveled by a compound to the distance traveled by the solvent in paper chromatography.

RF_1 A releasing factor in protein synthesis that recognizes UAA and UAG termination codons and releases newly synthesized peptide from tRNA.

RF_2 Releasing factor in protein synthesis that recognizes UAA and UGA termination codons and releases newly synthesized peptide from tRNA.

RF_3 A releasing factor in protein synthesis that apparently stimulates the activities of RF_1 and RF_2.

R-Factor 1. A plasmid that contains one or more genes that encode resistance to antibiotics. 2. Releasing factor in protein synthesis that releases newly synthesized polypeptide from ribosome complexes.

RFC Abbreviation for rosette-forming cells.

RF-C Abbreviation for replication factor C.

RFLP Abbreviation for restriction fragment length polymorphism.

R-G Abbreviation for receptor-G-protein complex.

R-Ger Elixir A trade name for iodinated glycerol, an antitussive agent.

R-Group The side chain of an amino acid.

RGS Abbreviation for regulator of G-protein signaling.

Rh 1. Symbol for rhodium. 2. Abbreviation for rhesus factor or Rh factor.

Rh Blood Group System The use of Rh antigens for classification of blood groups, e.g., Rh positive and negative.

Rh D A type of Rh blood.

Rh Disease Referring to erythroblastosis.

Rh Factor See Rhesus factor.

Rh Incompatibility Type-II hypersensitivity reaction that occurs in a Rh negative mother bearing a Rh positive fetus.

Rha Abbreviation for rhamnose.

Rhabdomyoma A benign tumor derived from skeletal muscle.

Rhabdoviridae A family of helical, bullet-shaped, enveloped minus-stranded RNA viruses of animals and plants (e.g., rabies, vesicular stomatitis).

Rhabdovirus A virus of the family Rhabdoviridae that causes rabies.

L-Rhamnonate Dehydrase The enzyme that catalyzes the following reaction:

L-Rhamnonate

⇅

2-Dehydro-3-deoxy-L-rhamnonate + H_2O

Rhamnose (mol wt 164) A 6-deoxymannose.

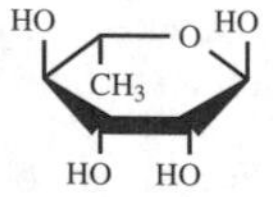

α-L-rhamnose

L-Rhamnose 1-Dehydrogenase The enzyme that catalyzes the following reaction:

L-Rhamnofuranose + NAD^+

⇅

L-Rhamno-1,4-lactone + NADH

L-Rhamnose Isomerase The enzyme that catalyzes the following reaction:

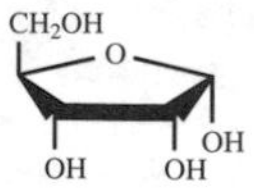

α-L-Rhamnosidase The enzyme that catalyzes the hydrolysis of terminal nonreducing α-L-rhamnose residues in α-L-rhamnoside.

β-L-Rhamnosidase The enzyme that catalyzes the hydrolysis of terminal nonreducing β-L-rhamnose residues in β-L-rhamnoside.

Rhamnoside Rhamnohydrolase The systematic name for α-L-rhamonsidase.

Rhamnulokinase The enzyme that catalyzes the following reaction:

ATP + L-rhamnulose

⇅

ADP + L-rhamnulose 1-phosphate

rHDC (RHDC) Abbreviation for rat histidine decarboxylase.

RheI(SalI) A restriction endonuclease from *Rhodococcus* species with the following specificity:

↓
5′..........GTCGAC..........3′
3′..........CAGCTG..........5′
↑

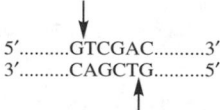

Rhenium (Re) A metallic element with atomic weight 186, valences 1 through 7.

Rheology The science that deals with the deformation and flow of matter.

Rheomacrodex LMD A trade name for low molecular weight dextran, used to expand plasma volume.

Rheometer An instrument for measuring flow of substances and electricity.

Rheotaxis A type of taxis in which the stimulus is a stream of fluid.

Rhesus Factor A group of red blood cell antigens that react with antibodies against rhesus monkey cells.

Rheum The discharge from mucous membranes of the eyes or nose.

Rheumacin A trade name for indomethacin, used as an anti-inflammatory agent.

Rheumatic Fever A disease characterized by fever, pain, and swelling around the joint due to complications from scarlet fever or streptococcal infection.

Rheumatism Disorders characterized by the inflammation or pain in muscle, joint, or fibrous tissue.

Rheumatoid Arthritis An autoimmune disorder marked by severe inflammation of joints due to the development of IgM autoantibody to IgG (rheumatoid factor). The deposition of IgM-IgG immune complexes in blood vessels and synovium cause severe inflammation.

Rheumatoid Factor An anti-immunoglobulin antibody directed against denatured IgG present in the serum of patients with rheumatoid arthritis and other rheumatoid diseases.

Rheumatrex A trade name for methotrexate sodium, used to prevent reduction of folic acid to tetrahydrofolic acid.

rHF Abbreviation for recombinant human ferritin.

Rhinalar Nasal Mist A trade name for flunisolide, used as a nasal agent to decrease inflammation.

Rhinall A trade name for phenylephrine hydrochloride, used to produce local vasoconstriction of dilated arterioles to reduce blood flow and nasal congestion.

Rhinitis Inflammation of nasal mucous membranes.

Rhinocort A trade name for budenoside, an anti-inflammatory agent.

Rhinosporidosis A chronic disorder characterized by the formation of polyplike growth on the mucosa of the nose and upper respiratory tract caused by fungus *Rhinosporidium seeberi.*

Rhinotracheitis Inflammation of the mucous membranes of the nose and trachea.

Rhinovirus A virus of the family Picornaviridae that causes the common cold.

Rhizobium A genus of Gram-negative, nitrogen-fixing bacteria and a symbiont in roots of leguminous plants.

Rhizoid 1. Rootlike structure. 2. Region of a cell or an organism that functions as a basal anchor to the substratum.

Rhizopuspepsin The enzyme from the *Rhizopus* species that catalyzes the hydrolysis of proteins with broad specificity.

rHLF Abbreviation for recombinant human lactoferrin.

rhLig Abbreviation for recombinant human DNA ligase.

rhLig I Abbreviation for recombinant human DNA ligase I.

Rho Abbreviation for rhodopsin.

Rho Factor The protein factor involved in the termination of synthesis of RNA molecule.

***rho* Gene** The gene that encodes the *rho* termination factor for transcription.

rho Independent Transcription Terminator A segment of DNA that signals the termination of transcription without the presence of the termination protein.

rho Protein Referring to the transcription termination protein factor.

Rhodacine A trade name for indomethacin, a nonsteroidal anti-inflammatory agent.

Rhodis A trade name for ketoprofen, a non-narcotic analgesic agent.

Rhodium (Rh) A chemical element with atomic weight 103, valence 1 through 6.

Rhodomicrobium A genus of photosynthetic bacteria (Rhodospirilaceae).

Rhodoplast The photosynthetic organelle of red algae.

Rhodopseudomonas A genus of photosynthetic bacteria of the family Rhodospirilaceae.

Rhodopsin A photoreceptor protein pigment consisting of protein and vitamin aldehyde, found in the rods of the retina of the eye, and used in the visual process of transducing photons of light.

Rhodopsin Kinase The enzyme that catalyzes the following reaction:

$$\text{ATP} + \text{rhodopsin} \rightleftharpoons \text{ADP} + \text{phosphorhodopsin}$$

Rhodospirillum A genus of photosynthetic bacteria (Rhodospirilaceae).

Rho-GAM A trade name for Rho(D) immune globulin.

Rhotral A trade name for betamethasone, a glucocorticoid.

Rhotrimine A trade name for acebutolol hydrochloride, a beta adrenergic blocking agent used as an antiarrhythmic and antihypertensive agent.

Rhovail A trade name for ketoprofen, a non-narcotic analgesic agent.

RhpI (SalI) A restriction endonuclease from *Rhodococcus* species with the same specificity as SalI.

rhPAH Abbreviation for recombinant human phenylalanine hydroxylase.

RhsI (BamHI) A restriction endonuclease from *Rhodococcus* species with the same specificity as BamHI.

rhsGC Abbreviation for recombinant human sGC (soluble guanylate cyclase).

rHTC Abbreviation for rat hepatoma tissue culture cell.

rhTNFα Abbreviation for recombinant human tumor necrosis factor alpha.

Rhythmin A trade name for procainamide hydrochloride, used as an anti-arrhythmic agent.

RI Abbreviation for replicative intermediate, an intermediate formed during the replication of single-stranded DNA or RNA that is a partially double-stranded structure with many single-stranded tails.

RIA Abbreviation for 1. radial immuno-assay; 2. radial immunodiffusion assay.

Rib Abbreviation for ribose.

Ribavirin (mol wt 244) An antiviral agent.

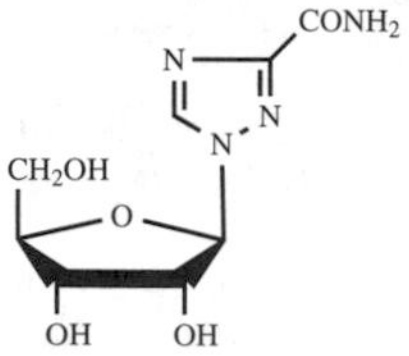

Ribitol (mol wt 152) A sugar alcohol derived from ribose (also known as adonitol).

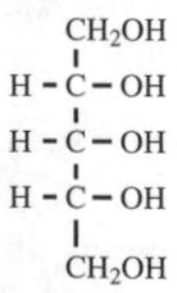

Ribitol Dehydrogenase The enzyme that catalyzes the following reaction:

$$\text{Ribitol} + NAD^+ \rightleftharpoons \text{Ribulose} + NADH$$

Riboflavin (mol wt 376) A nutritional factor commonly known as vitamin B_2.

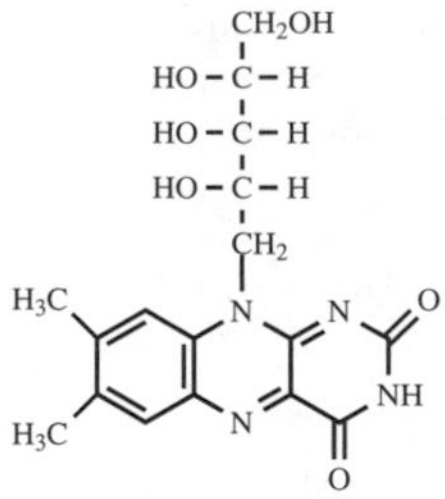

Riboflavin Hydrolase See riboflavinase.

Riboflavin Kinase The enzyme that catalyzes the following reaction:

$$\text{ATP} + \text{riboflavin} \rightleftharpoons \text{ADP} + \text{FMN}$$

Riboflavin Mononucleotide Synonym of flavin mononucleotide or riboflavin-5′-phosphate.

Riboflavin Phosphotransferase The enzyme that catalyzes the following reaction:

$$\text{Riboflavin} + \text{glucose 1-phosphate} \rightleftharpoons \text{Glucose} + \text{FMN}$$

Riboflavinase The enzyme that catalyzes the following reaction:

$$\text{Riboflavin} + H_2O \rightleftharpoons \text{ribitol} + \text{lumichrome}$$

Ribofuranose (mol wt 151) A ribose in the furanose form, e.g., α-D-ribose.

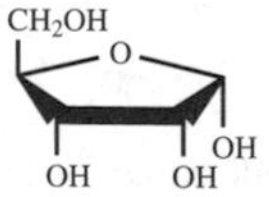

Ribokinase The enzyme that catalyzes the following reaction:

$$\text{ATP} + \text{D-ribose} \rightleftharpoons \text{ADP} + \text{ribose 5-phosphate}$$

Ribonic Acid (mol wt 167) An organic acid derived from ribose.

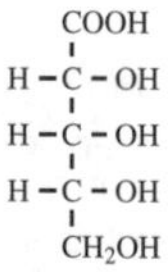

Ribonuclease A group of enzymes that catalyze the hydrolysis of phosphodiester bonds in RNA.

Ribonuclease I (Pancreatic Ribonuclease) An endoribonuclease that catalyzes the endonucleolytic cleavage of RNA to yield 3′-monophosphonucleotides and 3′-phosphooligonucleotides ending in Cp or Up with 2′,3′-cyclic phosphate intermediates.

Ribonuclease II An exonucleolytic enzyme that catalyzes the cleavage of RNA in the 3′- to 5′ direction to yield 3′-phosphomononucleotides (also known as exoribonuclease II).

Ribonuclease III The enzyme that catalyzes the endonucleolytic cleavage of double-stranded RNA, multimeric tRNA precursor, and rRNA.

Ribonuclease IV The enzyme that catalyzes the endonucleolytic cleavage of poly-A to fragments terminated by 3′-hydroxyl and 5′-phosphate groups.

Ribonuclease V The enzyme that catalyzes the hydrolysis of poly-A, forming oligonucleotides and 3′-AMP.

Ribonuclease IX The enzyme that catalyzes the endonucleolytic cleavage of poly-U and poly-C RNA to fragments terminated by 3′-hydroxyl and 5′-phosphate group.

Ribonuclease A The enzyme that catalyzes the endonucleolytic cleavage of RNA into mono- and oligonucleotides ending in 3′-pyrimidine nucleotides.

Ribonuclease Alpha The enzyme that catalyzes the endonucleolytic cleavage of *O*-methylated RNA to 5′-phosphomonoester.

Ribonuclease D The enzyme that catalyzes the endonucleolytic removal of nucleotides from tRNA precursor, producing 3′-terminal mature tRNA.

Ribonuclease E A ribonuclease involved in the formation of 5S rRNA from pre-rRNA.

Ribonuclease F The enzyme that catalyzes the endonucleolytic cleavage of RNA, leaving 5′-hydroxyl and 3′-phosphate groups.

Ribonuclease H (from Calf Thymus) The enzyme that catalyzes the endonucleolytic cleavage of RNA–DNA hybrids to yield 5′-phosphomonoester.

Ribonuclease M5 The enzyme that catalyzes the endonucleolytic removal of 21 and 42 nucleotides respectively from the 5′ and 3′-terminals of 5S rRNA precursor.

Ribonuclease P The enzyme that catalyzes the endonucleolytic removal of 5′-extranucleotides from a tRNA precursor.

Ribonuclease P4 The enzyme that catalyzes the endonucleolytic removal of 3′-extranucleotides from tRNA precursor.

Ribonuclease Protection A method for determination of the site on RNA that binds protein or enzyme by treating protein–RNA complexes with various ribonucleases to remove the RNA segment that is not protected by the protein or enzyme.

Ribonuclease S A ribonuclease derived from ribonuclease A by treatment with the protease subtilisin, which consists of two noncovalently linked fragments.

Ribonuclease T_1 The enzyme that catalyzes the endonucleolytic cleavage of RNA to yield 3′-phosphomononucleotides and 3′-phosphooligonucleotides ending in Gp with 2′,3′-cyclic phosphate intermediates.

Ribonuclease T2 The enzyme that catalyzes the endonucleolytic cleavage of RNA to yield 3′-phosphomononucleotides and 3′-phosphooligonucleotides with 2′-3′-cyclic phosphate intermediates.

Ribonuclease U2 The enzyme that catalyzes the endonucleolytic cleavage of RNA to yield 3′-phosphomononucleotides and 3′-phosphooligonucleotides ending in Ap or Gp with 2′,3′-cyclic phosphate intermediates.

Ribonuclease Y A ribonuclease that catalyzes the exonucleolytic cleavage of RNA in the 3′ to 5′ direction to yield 3′-monophosphonucleotides.

Ribonucleic Acid (RNA) Polymer of ribonucleotide that consists of ribonucleotides of adenine, guanine, cytosine uracil, and thymine (in tRNA); major types of RNA are tRNA, rRNA, mRNA, and viral RNA.

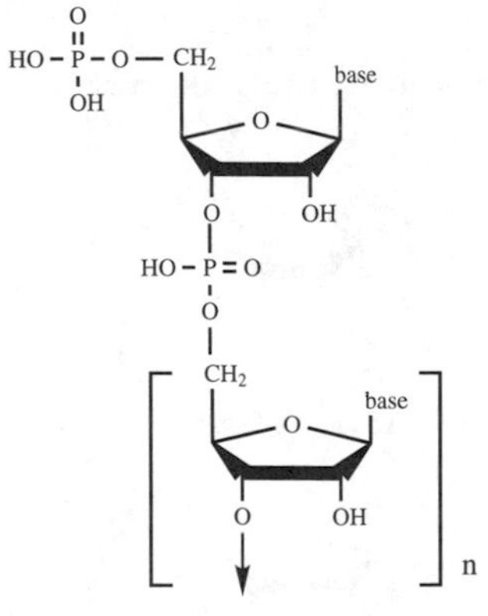

base = adenine, guanine, cytosine, uracil

Ribonucleoprotein Referring to a complex of RNA and protein.

Ribonucleoside Referring to ribonucleosides of adenine, guanine, cytosine, uracil, or thymine (in tRNA); a ribonucleoside consists of a ribose and a base.

Ribonucleoside Diphosphate Reductase The enzyme that catalyzes the following reaction:

2′-Deoxyribonucleoside diphosphate + oxidized thioredoxin + H_2O

⇌

Ribonucleoside diphosphate + reduced thioredoxin

Ribonucleoside Triphosphate Reductase The enzyme that catalyzes the following reaction:

2′-Deoxyribonucleoside triphosphate + oxidized thioredoxin + H_2O

⇌

Ribonucleoside triphosphate + reduced thioredoxin

Ribonucleotide Referring to ribonucleotides of adenine, guanine, cytosine, uracil, or thymine (in tRNA); a ribonucleotide consists of a ribose, base, and phosphate.

3'-Ribonucleotide Phosphohydrolase The systematic name for 3'-nucleotidase.

5'-Ribonucleotide Phosphohydrolase The systematic name for 5'-nucleotidase.

Ribonucleotide Reductase Synonym of ribonucleoside diphosphate reductase or triphosphate reductase.

Ribophorin A glycoprotein found in the endoplasmic reticulum that serves as ribosome binding sites.

Ribopyranose (mol wt 151) Ribose in the pyranose form.

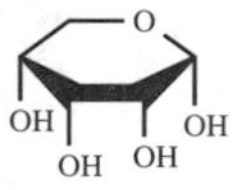

Ribose (mol wt 151) A 5-carbon aldosugar.

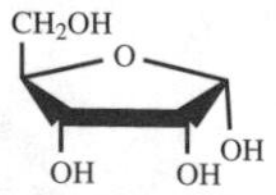

α-D-ribofuranose

Ribose Binding Protein A periplasmic binding protein from bacteria involved in the ribose transport system.

Ribose 1-Dehydrogenase ($NADP^+$) The enzyme that catalyzes the following reaction:

D-Ribose + $NADP^+$ + H_2O

⇌

D-ribonate + NADPH

Ribose Isomerase The enzyme that catalyzes the interconversion of ribose to ribulose.

Ribose 5-Phosphate (mol wt 320) The phosphate form of ribose.

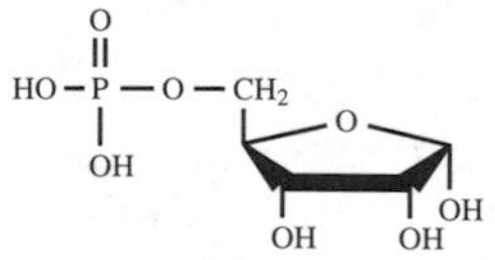

Ribose 5-Phosphate Adenyltransferase The enzyme that catalyzes the following reaction:

ADP + D-ribose 5-phosphate

⇌

ADP-ribose + Pi

Ribose 5-Phosphate Ammonia Lyase The enzyme that catalyzes the following reaction:

ATP + ribose 5-phosphate + NH_3

⇌

ADP + 5-phosphoribosylamine + Pi

Ribose 5-Phosphate Isomerase The enzyme that catalyzes the following reaction:

D-Ribose 5-phosphate

⇌

D-ribulose 5-phosphate

D-Ribose 5-Phosphate Ketol Isomerase The systematic name for ribose 5-phosphate isomerase.

Ribose Phosphate Pathway A cyclic pathway used by some methylotrophic bacteria for the assimilation of formaldehyde.

Ribose Phosphate Pyrophosphokinase The enzyme that catalyzes the following reaction:

ATP + D-ribose 5-phosphate

⇌

AMP + 5-phospho-α-D-ribose 1-diphosphate

Riboside A glycoside of ribose.

Ribosomal Maturation Nuclease Synonym of ribonuclease M5.

Ribosomal Proteins Proteins found in the ribosome or ribosomal subunits. The large 50S ribosomal subunit of prokaryotes contains 31 proteins designated L1 to L31 and that of the small 30S ribosomal subunit contains 21 proteins designated S1 to S21; the large 60S eukaryotic ribosomal subunit contains 49 proteins and that of 40S subunit contains 33 proteins.

Ribosomal RNA (rRNA) Several species of RNA that are present in the ribosome, e.g., 16S, 23S, and 5S RNA in prokaryotes and 5.8S, 5S, 28S, and 18S RNA in eukaryotes.

Ribosomal Subunits Ribonucleoprotein particles that form a functional ribosome for protein synthesis. There are two types of ribosomal subunits, namely, large subunit (50S of the prockaryote or 60S of eukaryote) and small subunit (30S of prokaryote or 40S of the eukaryote).

30S Ribosomal Subunit The small ribosomal subunit of prokaryotes that has a sedimentation coefficient of 30S (30×10^{-13} sec).

40S Ribosomal Subunit The small ribosomal subunit of eukaryotes that has a sedimentation coefficient of 40S (40×10^{-13} sec).

50S Ribosomal Subunit The large ribosomal subunit of prokaryotes that has a sedimentation coefficient of 50S (50×10^{-13} sec).

60S Ribosomal Subunit The large ribosomal subunit of eukaryotes that has a sedimentation coefficient of 60S (60×10^{-13} sec).

Ribosome Cellular structure composed of rRNA and protein and the sites where protein synthesis occurs within cells.

70S Ribosome The functional ribosome in prokaryotes that is the site of protein synthesis in prokaryotic cells, mitochondria, and chloroplasts. It has a sedimentation coefficient of 70S (70×10^{-13} sec).

80S Ribosome The functional ribosome in eukaryotes and the site of protein synthesis in eukaryotic cells that has a sedimentation coefficient of 80S (80×10^{-13} sec).

Ribostamycin (mol wt 454) An antibiotic related to neomycin that binds to 30S ribosomal subunit, inhibiting bacterial protein synthesis.

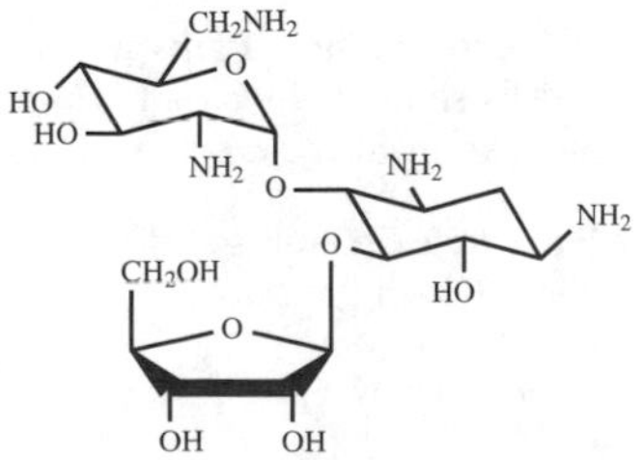

Ribosyladenine (mol wt 267) Referring to adenosine, a ribonucleoside.

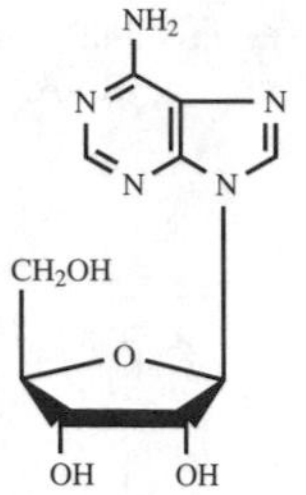

Ribosyladenylic Acid Synonym for adenosine phosphate, a ribonucleotide.

Ribosylcytidylic Acid Synonym for cytidine phosphate, a ribonucleotide of cytosine.

Ribosylcytosine (mol wt 243) A ribonucleoside of cytosine.

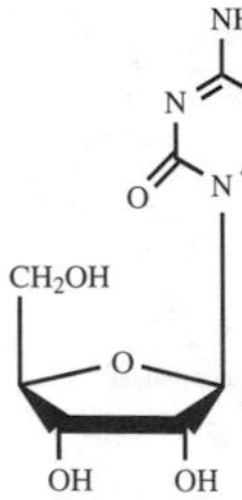

Ribosylguanine (mol wt 283) A ribonucleoside of guanine.

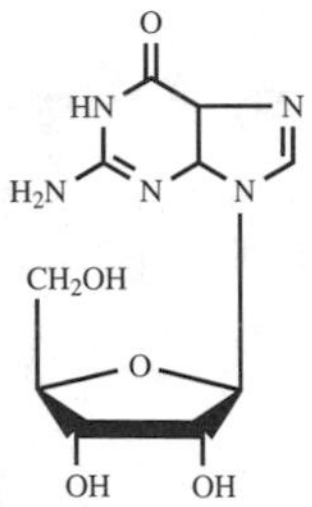

Ribosylguanylic Acid Synonym of guanosine phosphate, a riboucleotide of guanine.

Ribosylhomocysteinase The enzyme that catalyzes the following reaction:

$$S\text{-Ribosyl-L-homocysteine} + H_2O \rightleftharpoons \text{D-ribose} + \text{L-homocysteine}$$

Ribosylnicotinamide Kinase The enzyme that catalyzes the following reaction:

$$\text{ATP} + N\text{-ribosylnicotinamide} \rightleftharpoons \text{ADP} + \text{nicotinamide ribonucleotide}$$

Ribosylthymine (mol wt 242) A ribonucleoside of thymine that occurs in tRNA.

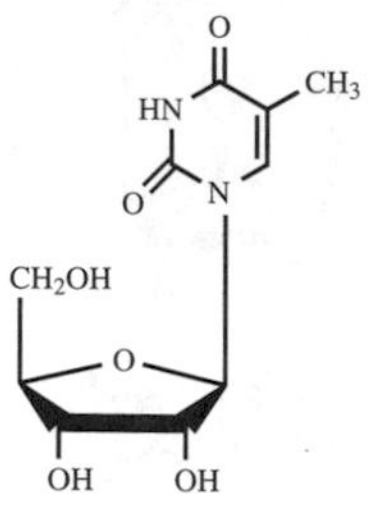

Ribosyluracil (mol wt 244) A ribonucleoside of uracil (see also uridine).

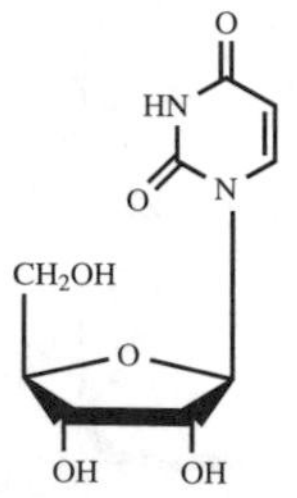

5-Ribosyluracil Synonym for pseudouridine.

Ribosyluridylic Acid Synonym for uridine phosphate, a ribonucleotide.

Ribothymidine See ribosylthymine.

Ribothymidylic Acid (mol wt 322) The ribonucleotide of thymine found in tRNA.

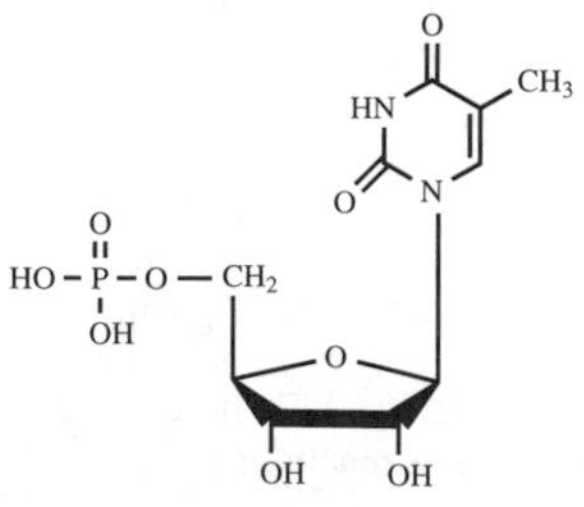

Ribotide Short for ribonucleotide.

Riboviruses Any virus with RNA as genetic material.

Ribozyme An RNA with catalytic activity.

Ribulokinase The enzyme that catalyzes the following reaction:

ATP + ribulose

⇅

ADP + ribulose 5-phosphate

D-Ribulose (mol wt 150) A 5-carbon keto-sugar.

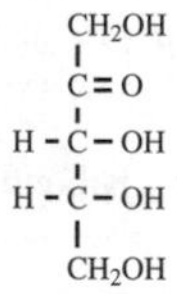

Ribulose 1,5-Bisphosphate (mol wt 310) A key intermediate in the Calvin cycle for production of sugar from photosynthesis.

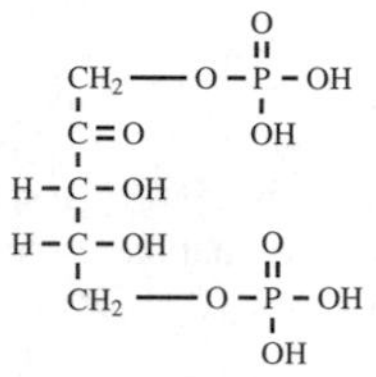

Ribulose 1,5-bisphosphate Carboxylase The enzyme that catalyzes the following reaction:

Ribulose 1,5-bisphosphate + CO_2

⇅

2 3-Phosphoglycerate

Ribulose 1,5-Bisphosphate Carboxylase-Oxygenase Synonym for ribulose 1,5-bisphosphate carboxylase.

Ribulose 1,5-bisphosphate Oxygenase The enzyme that catalyzes the following reaction:

Ribulose 1,5-bisphosphate + O_2

⇅

Phosophoglycolate + 3-Phosphoglycerate

Ribulose Diphosphate Synonym for ribulose bisphosphate.

Ribulose Diphosphate Carboxydismutase Synonym for ribulose-1,5-bisphosphate carboxylase.

Ribulose Monophosphate Pathway A cyclic pathway used by some methylotrophic bacteria for assimilation of formaldehyde.

Ribulose Phosphate Epimerase The enzyme that catalyzed the following reaction:

D-Ribulose 5-phosphate

⇅

D-xylulose 5-phosphate

Ribulose 5-Phosphate Kinase Synonym of phosphoribulokinase.

Ricin A toxic lectin from castor bean (*Ricinus communis*) that can be conjugated with monoclonal antibody to form an immunotoxin. It consists of a subunit for toxicity and a subunit for carbohydrate binding.

Ricinine (mol wt 164) A toxin from plant *Ricinus communis* that produces nausea, vomiting, hemorrhagic gastroenteritis, and hepatic and renal damage in human.

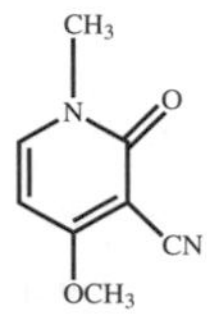

Ricinine Nitrilase The enzyme that catalyzes the following reaction:

$$\text{Ricinine} + H_2O \rightleftharpoons \text{3-Carboxy-4-methoxy-}N\text{-methyl-2-pyridone} + NH_3$$

Ricinoleic Acid (mol wt 298) An unsaturated, hydroxy fatty acid from oils derived from seeds of *Ricinus* spp.

$$CH_3(CH_2)_5CH(OH)CH_2CH{=}CH(CH_2)_7COOH$$

Ricinus Communis Agglutinin A toxic lectin from castor bean.

Rickets A disease caused by a deficiency of vitamin D and characterized by skeletal deformities.

Rickettsemia The presence of *Rickettsiae* in the blood.

Rickettsia A genus of Gram-negative bacteria of the family Rickettsiaceae causing a number of serious diseases in humans, including typhus and rocky mountain spotted fever.

Rickettsidsis Infection with *Rickettsia*.

RID A trade name for pyrethrins, used as a parasitic agent.

RIDA Abbreviation for radial immuno-diffusion assay.

Ridaura A trade name for auranofin, used as an anti-inflammatory agent for treatment of rheumatoid arthritis.

Ridogrel (mol wt 366) An antithrombotic agent.

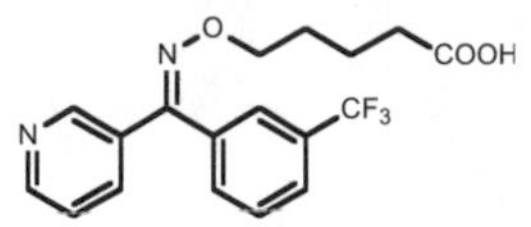

RIEP Abbreviation for rocket immunoelectrophoresis.

Rieske Protein Synonym of plastoquinol-plastocyanin reductase.

rif Abbreviation for rifamycin.

Rifabutin (mol wt 847) An antiviral agent that inhibits DNA-dependent RNA polymerase.

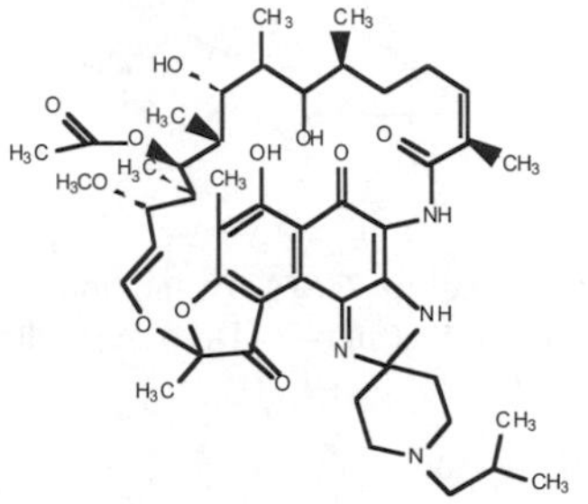

Rifadin A trade name for rifampin (rifampicin), used as an antitubercular and antileprotic agent.

Rifamate A trade name for a combination drug containing isoniazid and rifampin, used as an antitubercular and antileprotic agent.

Rifampin (mol wt 823) A semisynthetic antibiotic related to rifamycin.

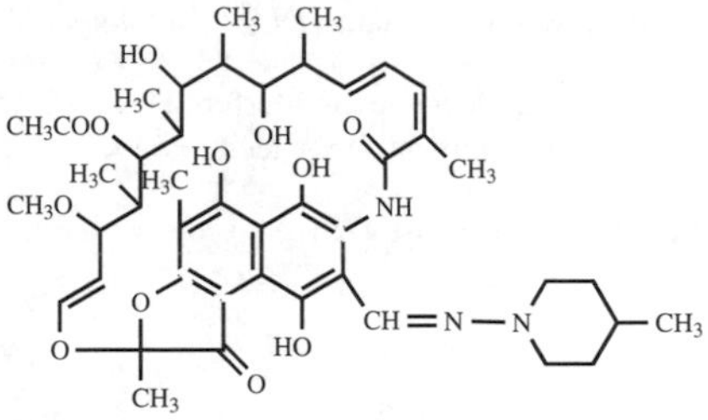

Rifamycin B Oxidase The enzyme that catalyzes the following reaction:

$$\text{Rifamycin B} + O_2 \rightleftharpoons \text{Rifamycin O} + H_2O_2$$

Rifamycin SV (mol wt 698) A semisynthetic antibiotic derived from rifamycin that inhibits DNA-dependent RNA polymerase.

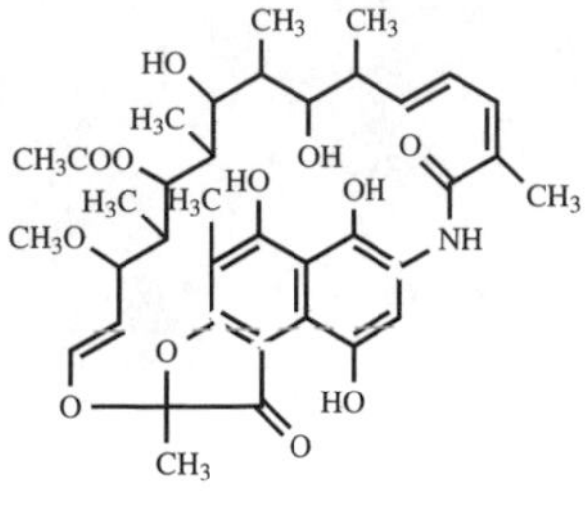

Rifamycins A group of antibiotics produced by *Streptomyces mediterranei* that inhibits DNA-dependent RNA polymerase.

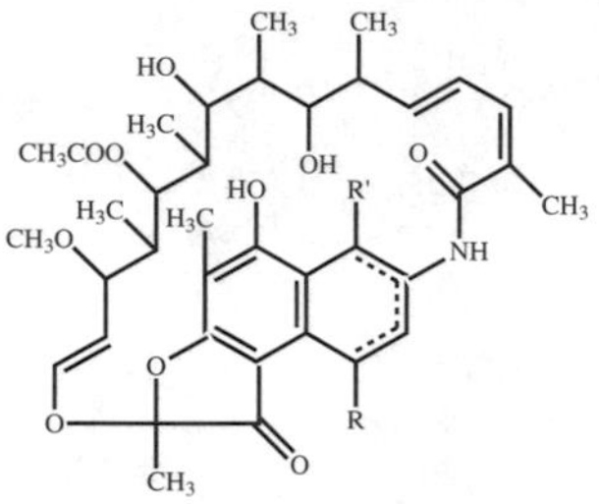

Rifaximin (mol wt 786) An antibiotic related to rifamycin SV that inhibits DNA-dependent RNA polymerase.

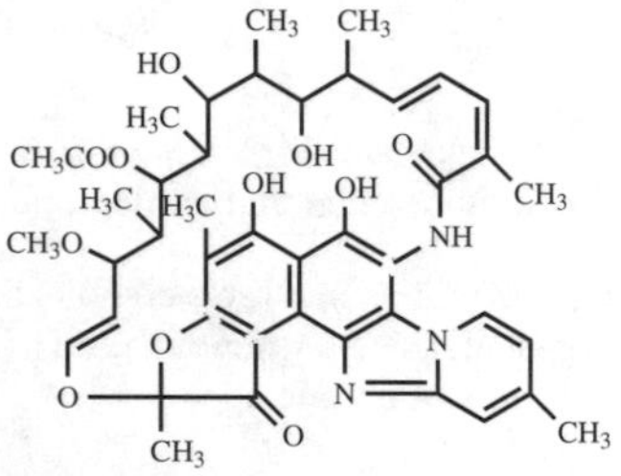

rIFN-A A trade name for interferon alpha-2A produced by recombinant DNA technology.

rIFN-B A trade name for interferon beta-2B produced by recombinant DNA technology.

Rilmazafone (mol wt 475) A sedative-hypnotic agent.

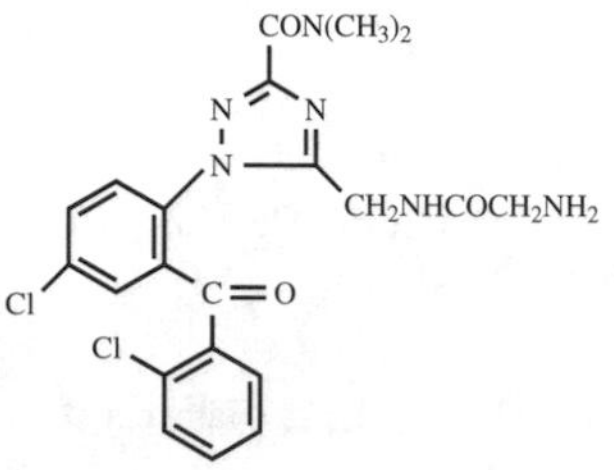

Rilmenidene (mol wt 180) An antihypertensive agent.

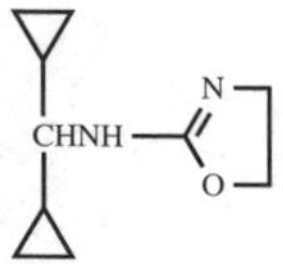

Rilutek A trade name for riluzole, a neuroprotective agent.

Riluzol (mol wt 234) A neuroprotective agent.

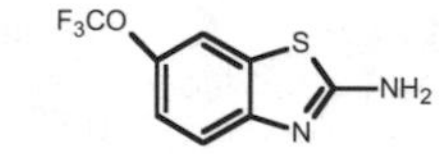

Rimactane A trade name for rifampin, used as an antibacterial agent.

Rimantadine (mol wt 179) An antiviral agent.

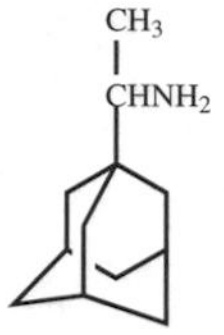

Rimazolium Metilsulfate (mol wt 362) An analgesic agent.

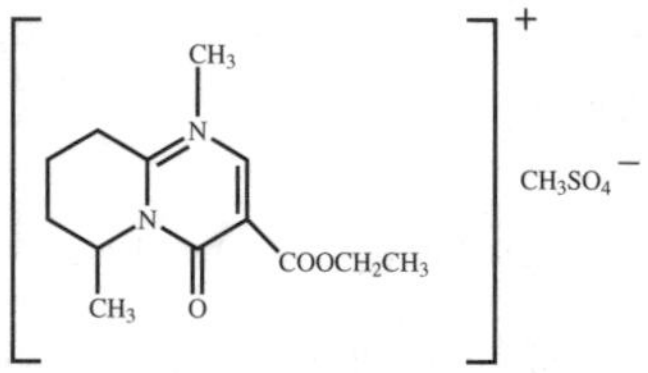

Rimexolone (mol wt 371) An anti-inflammatory agent.

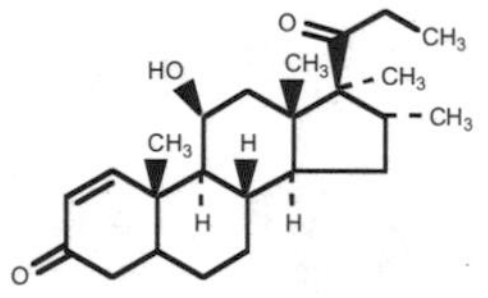

Rimiterol (mol wt 223) A bronchodilator.

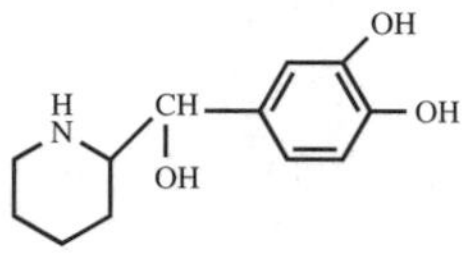

Rimocidin (mol wt 768) An antibiotic produced by *Streptomyces rimosus*.

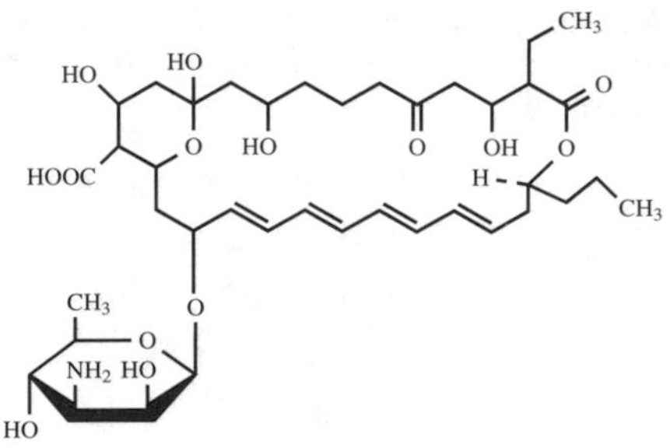

Rimycin A trade name for rifampicin, used as an antibacterial agent.

Ring Precipitation Test An antigen-antibody precipitation test in a small diameter Durham tube in which the antigen-antibody complex forms a ring at the interface between antigen and antibody.

Ringer's Solution A solution containing salts of sodium, potassium, and calcium (sometimes other salts) used in physiological experiments for maintaining cells or organs *in vitro.*

Riopan A trade name for magaldrate (aluminum-magnesium complex), an antacid.

Riophen A trade name for aspirin, an analgesic and antipyretic agent.

Rioprostil (mol wt 355) An antiulcerative agent.

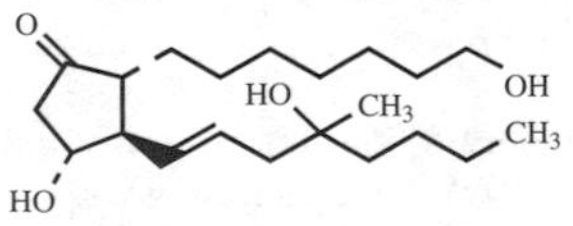

RIP Abbreviation for 1. radio-immuno-precipitation; 2. ribosome inactivating protein.

RIPA Abbreviation for radio-immuno-precipitation assay.

Riphen-10 A trade name for aspirin, an analgesic and antipyretic agent.

Riphenidate A trade name for methylprednisolone, a hormone that is responsible for anti-inflammatory and immunosuppressive effect.

RIPT Abbreviation for radio-immuno-precipitin test.

RISA Abbreviation for radioactively iodinated serum albumin.

Risedronic Acid (mol wt 283) A bone resorption inhibitor.

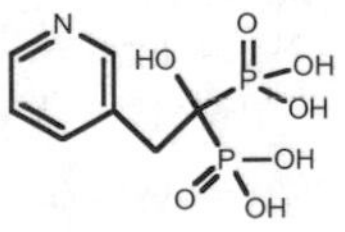

Risocaine (mol wt 179) A local anesthetic agent.

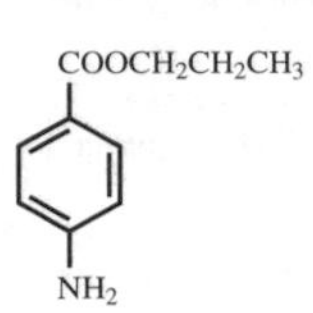

Risperdal A trade name for risperidone, an antipsychotic agent.

RIST Abbreviation for radioimmunosorption test, a serological test for detecting IgE antibody.

Ristocetin A glycopeptide antibiotic complex produced by actinomycetes (*Nocardia lurida*).

Ritalin A trade name for methylphenidate hydrochloride, a cerebral stimulant that promotes transmission of nerve impulses by release of stored norepinephrine from the nerve terminals in the brain.

Ritanserin (mol wt 478) An antidepressant and anxiolytic agent.

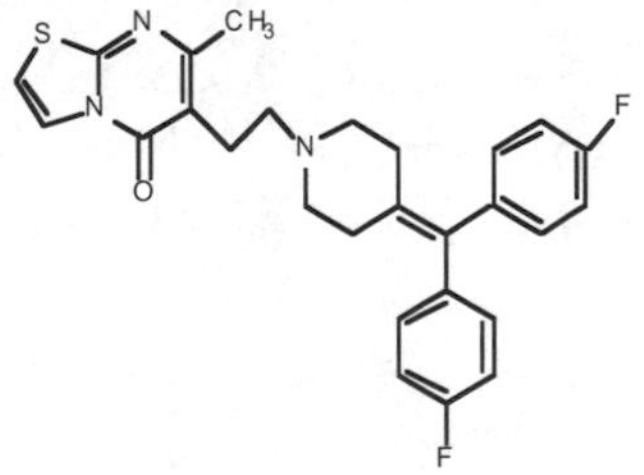

rITF Abbreviation for rat intestinal trefoil factor.

Ritipenem (mol wt 288) An antibacterial agent.

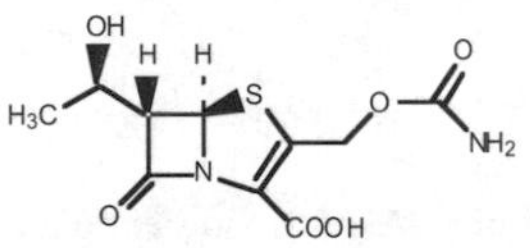

Ritodrine (mol wt 287) A beta-receptor agonist that stimulates beta-2-adrenergic receptors in the uterine smooth muscle inhibiting contractility.

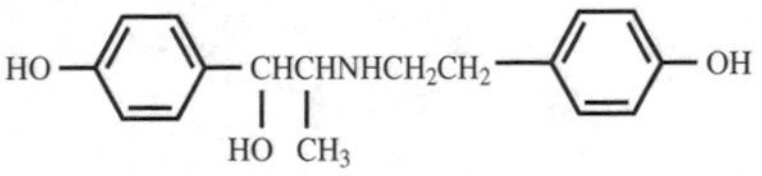

Ritonavir (mol wt 721) An antiviral agent that inhibits HIV protease activity leading to the decrease in production of HIV particles.

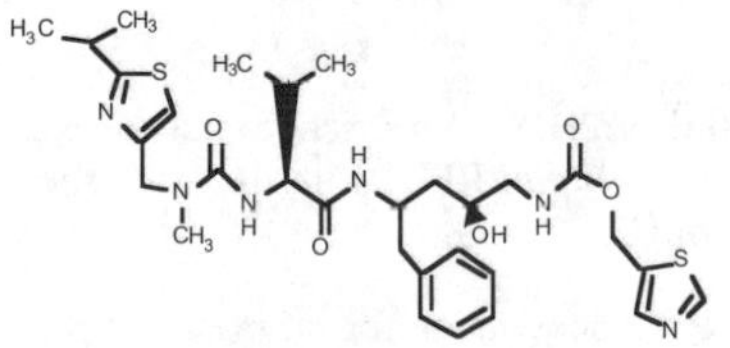

Rituxan TM A trade name for rituximab, a human/murine monoclonal antibody produced by DNA recombinant technology and used for the treatment of CD20 positive B cell non-Hodgkin's lymphoma.

Rituximab A murine/human monoclonal antibody to the antigen found on the surface of normal and malignant B cell leading to B cell lysis and used for treatment for CD20 positive B cell non-Hodgkin's lymphoma.

RIU Abbreviation for radioactive iodine uptake.

Rival A trade name for diazepam, an antianxiety agent.

Rivotril A trade name for clonazepam, an anticonvulsant.

RK Abbreviation for reductase kinase.

RKK Abbreviation for reductase kinase kinase.

RL Abbreviation for 1. *Renilla* luciferase; 2. *Renilla* luciferin; 3. Ringer's lactate.

RLLBP (rLLBP) Abbreviation for rat lens lipid-binding protein.

rLPP Abbreviation for rat lipid phosphate phosphohydrolase.

RluI (NaeI) A restriction endonuclease from *Rhizobium lupini* 1 with the following specificity:

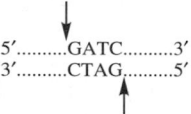

```
          ↓
5′..........GCCGGC..........3′
3′..........CGGCCG..........5′
          ↑
```

RluII (MboI) A restriction endonuclease from *Rhizobium lupini* RFL1 with the following specificity:

```
        ↓
5′..........GATC..........3′
3′..........CTAG..........5′
            ↑
```

Rlu3I (NlaIV) A restriction endonuclease from *Rhizobium lupini* RFL3 with the following specificity:

```
          ↓
5′..........GGNNCC..........3′
3′..........CCNNGG..........5′
          ↑
```

Rlu4I (Bam HI) A restriction endonuclease from *Rhizobium lupini* RFL4 with the same specificity as Bam HI.

rMCP Abbreviation for rat mast cell protease.

RME Abbreviation for receptor-mediated endocytosis.

RMP Pathway Abbreviation for ribulose monophosphate pathway, a cyclic pathway used by some methylotrophic bacteria for assimilation of formaldehyde.

RMRV Abbreviation for rhesus monkey rota virus.

RMS Abbreviation for root-mean square.

RMS Uniserts A trade name for morphine sulfate, an opiate analgesic agent that binds to opiate receptors in the brain altering the perception and emotional response to pain.

RMSD Abbreviation for root-mean square deviation.

RMUP (rMUP) Abbreviation for rat major urinary protein.

Rn Symbol for radon.

RNA Abbreviation for ribonucleic acid, a linear polynucleotide linked by phosphodiester bonds that is synthesized by RNA polymerase from DNA template and possesses 5′ and 3′ terminals.

RNA Blotting Transfer of electrophoretically separated RNA fragments from gel to a chemically reactive paper or film, e.g., nitrocellulose paper for subsequent detection of fragments by a radioactive probe or by enzyme-linked immunosorption assay.

RNA Coding Triplet Referring to codons on mRNA.

RNA Dependent DNA Polymerase Synonym for reverse transcriptase.

RNA Dependent RNA Polymerase A virus-encoded enzyme that catalyzes the synthesis of RNA from RNA template.

RNA Directed DNA Polymerase The enzyme that catalyzes the following reaction:

$$\text{Deoxyribonucleotide triphosphate} + \text{DNA}_n \rightleftharpoons \text{Pyrophosphate} + \text{DNA}_{n+1}$$

RNA Directed RNA Polymerase Synonym for RNA-dependent RNA polymerase.

RNA DNA Hybridization Association of a single-stranded DNA molecule with a complementary RNA molecule to form a hybrid molecule.

RNA Ligase The enzyme that catalyzes the following reaction:

$$\text{ATP} + (\text{ribonucleotide})_n + (\text{ribonucleotide})_m \rightleftharpoons \text{AMP} + \text{PPi} + (\text{ribonucleotide})_{n+m}$$

RNA Nucleotidyltransferase (DNA-Directed) Synonym of DNA-directed RNA polymerase.

RNA Phage Bacteriophages that contain RNA as genetic material, e.g., QB and ϕ6 RNA phage.

RNA 3′-Phosphate Cyclase The enzyme that catalyzes the following reaction:

$$\text{ATP + RNA-3'-terminal-phosphate} \rightleftharpoons \text{AMP + PPi + RNA-terminal-2'-3'-cyclic phosphate}$$

RNA Polymerase The enzyme that catalyzes the synthesis of RNA from triphosphate ribonucleotides using DNA as template.

RNA Primase The enzyme that catalyzes the synthesis of primer RNA for initiation of DNA replication.

RNA Primer A small fragment of RNA synthesized by RNA primase for the initiation of DNA replication.

RNA Replicase An RNA polymerase encoded by viral RNA that catalyzes the synthesis of RNA using RNA template (also known as RNA-dependent RNA polymerase).

RNA Splicing The process of removal of noncoding sequences from pre-mRNA molecules in the nucleus during formation of messenger RNA.

RNA Synthetase Referring to DNA-directed RNA polymerase.

RNA Tumor Virus Cancer-causing virus that contains RNA as the genetic material but uses double-stranded DNA as an intermediate for integration and propagation (see Retrovirus).

RNA Uridylyltransferase The enzyme that catalyzes the following reaction:

$$\text{UTP} + \text{RNA}_n \rightleftharpoons \text{PPi} + \text{RNA}_{n+1}$$

RNA Virus Any virus that contains RNA as genetic material.

RNAP Abbreviation for RNA polymerase.

RNA-pol Abbreviation for RNA polymerase.

RNAase Variant spelling for RNase.

RNase Any enzyme that catalyzes the cleavage of RNA, which may be either endoribonuclease or exoribonuclease (also known as ribonuclease).

RNase I See ribonuclease I.

RNase II See ribonuclease II.

RNase III See ribonuclease III.

RNase A See ribonuclease A

RNase BN An exoribonuclease that processes 3′-terminal extra-nucleotides of monomeric tRNA precursor.

RNase D See ribonuclease D

RNase E An endonucleolytic RNase in *E. coli* involved in the formation of 5S rRNA from pre-rRNA.

RNase F Synonym for ribonuclease F.

RNase H Synonym for ribonuclease H.

RNase I Synonym for pancreatic ribonuclease.

RNase M5 Synonym for ribonuclease M5.

RNase O Synonym for ribonuclease III.

RNase P See ribonuclease P.

RNase PH Synonym for tRNA nucleotidyltransferase.

RNase T An exoribonuclease that catalyzes tRNA *end turnover* and removal of terminal AMP residues from the 3′-CCA end of an uncharged tRNA.

RNase T1 Synonym for ribonuclease T_1.

RNase U2 Synonym for ribonuclease U2.

RNase Y The enzyme that catalyzes exonucleolytic cleavage of RNA in the 3′ to 5′ direction to yield 3′-phosphomononucleotides.

RNP Abbreviation for ribonucleprotein.

RNR Abbreviation for ribonucleotide reductase.

RNS Abbreviation for 1. post-nuclear supernatant; 2. reactive nitrogen species.

rNTP Abbreviation for ribonucleoside-5′-triphosphate.

Robafen A trade name for guaifenesin, an antitussive agent and expectorant that increases production of respiratory tract fluid, reducing the viscosity of tenacious secretion.

Robaxin A trade name for methocarbamol, used as a skeletal muscle relaxant.

Robaxisal A trade name for a combination drug containing methocarbamol and aspirin, a skeletal muscle relaxant that reduces the transmission of nerve impulses from the spinal cord to the skeletal muscle.

Robenecid A trade name for probenecid, an antigout agent that blocks the renal tubular reabsorption of uric acid.

Robicillin VK A trade name for penicillin V potassium, an antibiotic that inhibits bacterial cell synthesis.

Robidex A trade name for dextromethorphan hydrobromide, used as an expectorant and antitussive agent.

Robidrine A trade name for pseudoephedrine hydrochloride, an adrenergic agent that produces vasoconstriction.

Robigesic A trade name for acetaminophen, an analgesic and antipyretic agent that inhibits prostaglandin synthesis.

Robimycin A trade name for erythromycin ethylsuccinate, an antibiotic that inhibits bacterial protein synthesis.

Robinul A trade name for glycopyrrolate, an anticholinergic agent that inhibits acetylcholine action.

Robinul Forte A trade name for glycopyrrolate, an anticholinergic agent that inhibits acetylcholine action.

Robitet A trade name for tetracycline hydrochloride, an antibiotic that binds to 30S ribosomal subunits and inhibits bacterial protein synthesis.

Robitussin A trade name for guaifenesin, an expectorant and antitussive agent that promotes production of respiratory tract fluid helping to liquify and reduce viscosity of tenacious secretions.

Robomol A trade name for methocarbamol, a skeletal muscle relaxant.

Roborant An invigorating drug.

Rocaltrol A trade name for calcitrol, used to stimulate calcium absorption from the GI tract.

Rocephin A trade name for ceftriaxone sodium, an antibiotic that inhibits bacterial cell wall synthesis.

Rochalimaea A genus of Gram-negative bacteria (Rickettsieae).

Rochelle Salt Referring to potassium sodium tartrate.

$C_4H_4KNaO_6$

Rociverine (mol wt 340) A smooth muscle relaxant.

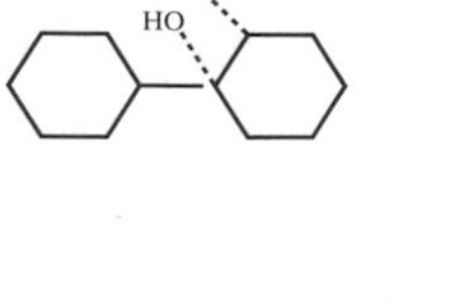

Rocket Immunoelectrophoresis A type of immunoelectrophoresis in which antigen or antigens are electrophoresed into an antibody-containing agarose gel. The antigen-antibody precipitin pattern resembles a rocket.

Rocky Mountain Spotted Fever A tick-borne human rickettsial disease caused by *Rickettsia rickettsii*.

Rocuronium (mol wt 529) A muscle relaxant.

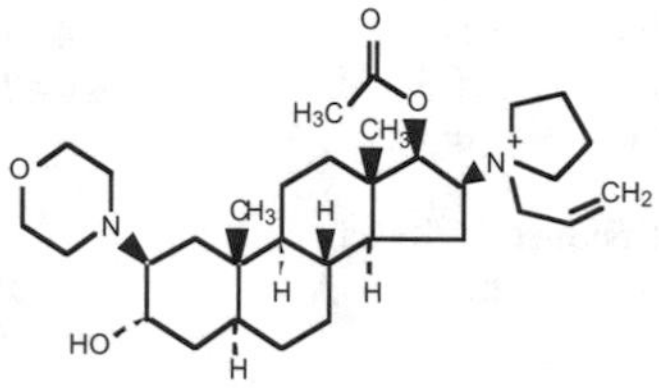

rODC (RODC) Abbreviation for rat ornithine decarboxylase.

Rodenticide Any substance that kills rats and other rodents.

Rodex A trade name for vitamin B_6.

Roentgen A quantity or dose of radiation for X-rays or γ-rays that produces one electrostatic unit of electricity of either sign in 1 cm^3 of dry air at 0°C.

Roentgen Ray Referring to X-ray.

Rofact A trade name for rifampicin, an antibiotic that inhibits DNA-dependent RNA polymerase in bacteria.

Roferon-A A trade name for interferon alpha-2a.

Rofocoxib (mol wt 314) A non-steroidal anti-inflammatory agent.

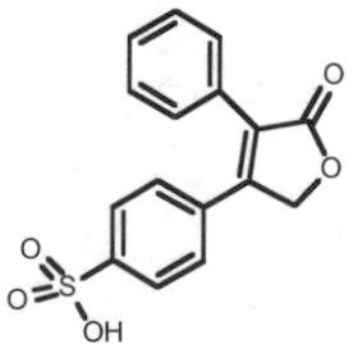

Rogaine A trade name for minoxidil, used to stimulate hair growth.

Rogesic A trade name for a combination drug containing acetaminophen, caffeine, and butalbital, an analgesic and antipyretic agent.

Rogitine A trade name for phentolamine mesylate, an antihypertensive agent and an alpha-adrenergic blocker that blocks the effects of catecholamine on alpha-adrenergic receptors.

ROI Abbreviation for reactive oxygen intermediate.

Rokitamycin (mol wt 828) An antibiotic produced by *Staphylococcus aureus.*

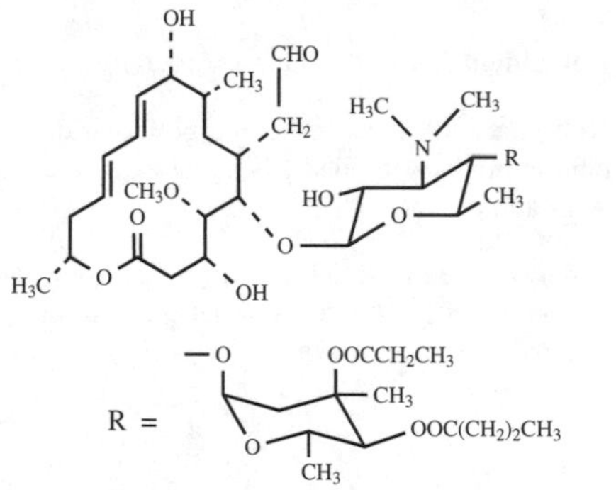

ROL Abbreviation for *Rhizopus oryzae* lipase.

Rolaids A trade name for dihydroxyaluminum sodium carbonate, used as an antacid to reduce total acid load in the GI tract.

Rolicyprine (mol wt 244) An antidepressant.

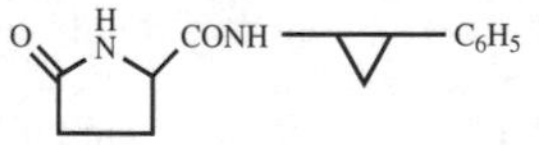

Rolipram (mol wt 275) An antidepressant.

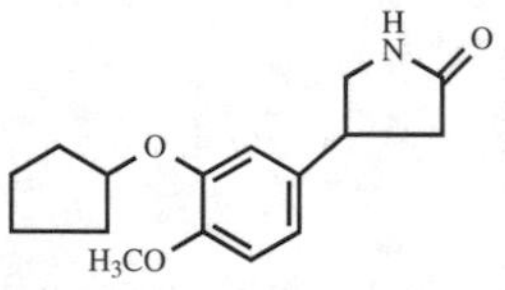

Rolitetracycline (mol wt 528) A semisynthetic antibiotic derived from tetracycline that inhibits bacterial protein synthesis.

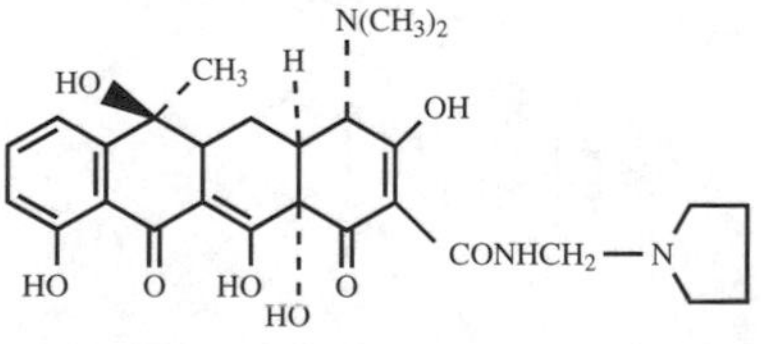

Roller Tube Method A technique used for propagation of anchorage-dependent cells in liquid culture in which the culture vessels are placed on a set of revolving rollers that rotate slowly in an incubator under a stream of carbon dioxide or nitrogen.

Rolling Circle Model A replication model of viral DNA in which a double-stranded circular DNA molecule is nicked at one strand and DNA synthesis proceeds using unbroken circular DNA as template, resulting in a double-stranded circular DNA with single-stranded displaced tail resembling the Greek letter σ. The displaced tail elongates to form a single-stranded linear DNA or may serve as template for Okazaki fragment synthesis.

Romanowsky Stain A range of composite stains, e.g., Giemsa's stain and Wright's stain, used for staining and identification of blood parasites.

Romazicon A trade name for flumazenil, a benzodiazepine receptor antagonist used as an antidote.

Ronase A trade name for tolazamide, an antidiabetic agent that stimulates the release of insulin from pancreatic beta cells.

Rondec A trade name for a combination drug containing pseudophedrine, carbinoxamine oxycodone, and acetaminophen, used as a decongestant and antihistaminic agent.

Ronidazole (mol wt 200) An antiprotozoal agent.

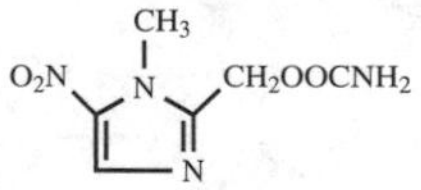

Ronifibrate (mol wt 378) An antihyperlipoproteinemic agent.

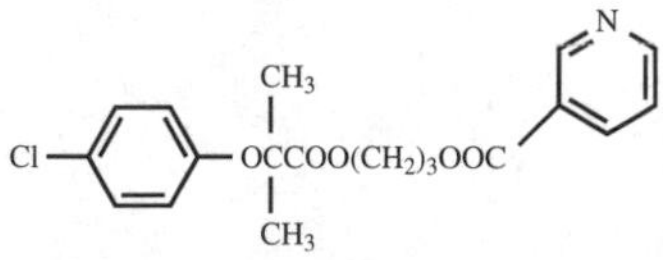

Rontgen Variant spelling of roentgen.

Root Nodule A gall-like structure in the roots of certain legume plants that contains endobiontic nitrogen-fixing bacteria for nitrogen fixation.

Ropinirole (mol wt 260) A dopamine receptor agonist used as an antiparkinsonism agent.

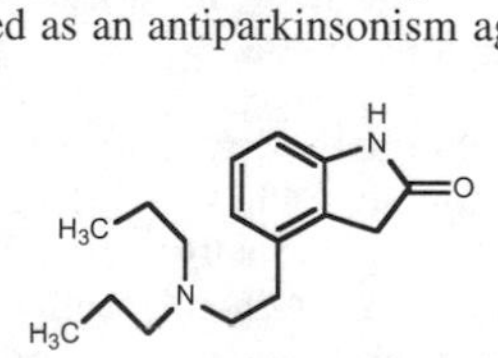

Ropivacaine (mol wt 274) An anesthetic agent.

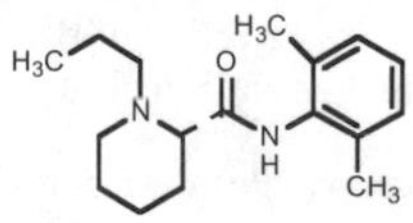

Roquinimex (mol wt 308) An antineoplastic agent.

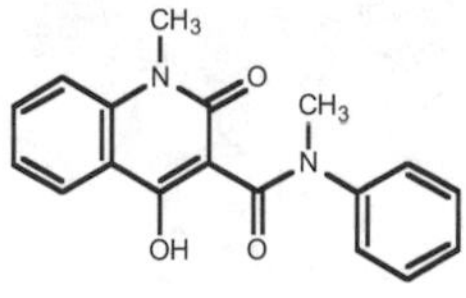

ROS Abbreviation for 1. reactive oxygen species; 2. radical oxygen species.

ROS A gene family encoding receptor-like tyrosine kinase, *v-ros* is the oncogene of the acutely transforming avian sarcoma virus UR2.

Rosaprostol (mol wt 298) A prostaglandin analog used as an antiulcerative agent.

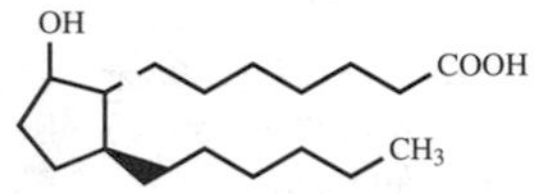

Rosaramicin (mol wt 582) A macrolide antibiotic from *Micromonospora rosaria.*

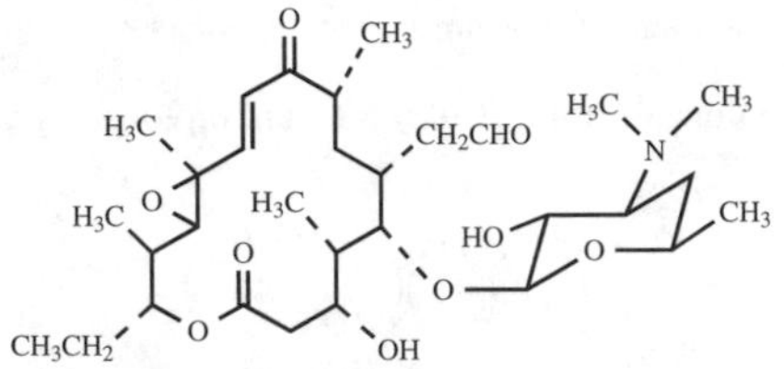

Rose Bengal (mol wt 1050) A biological dye.

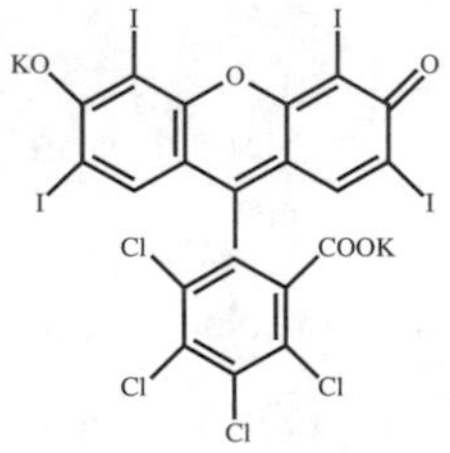

Rosette Multiple cellular structure formed by binding of cells to the surface of another cell, e.g., binding of erythrocytes to the surface of a lymphocyte to form an E rosette.

Rosette Technique A technique for detecting antigen or antibody on the cell surface employing antigen-coated or antibody-coated particulates (e.g., erythrocytes). Adhering antigen- or antibody-coated erythrocytes on the surface of the testing cells results in the formation of rosettes.

Rose-Waaler Test A type of passive hemagglutination test for the detection of rheumatoid factor that employs IgG-sensitized erythrocytes.

Rosoxacin (mol wt 294) An antibacterial agent.

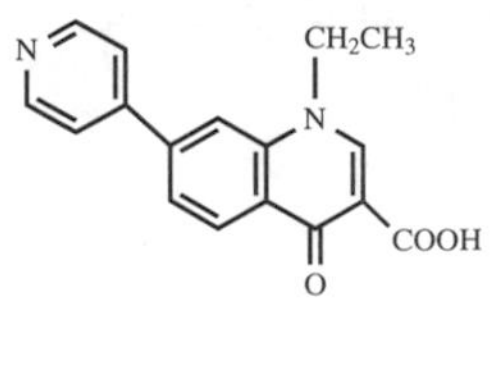

Rotary Evaporator A device to enhance the removal of solvent by evaporation under reduced pressure and rotating motion.

Rotashield A trade name for rotavirus vaccine.

Rotavirus A virus of the family Reoviridae containing double-stranded RNA that causes diarrhea in infants.

Rotenone (mol wt 394) An antiparasitic agent and an inhibitor of mitochondrial electron transport isolated from *Lonchocarpus nicou* (Leguminosae).

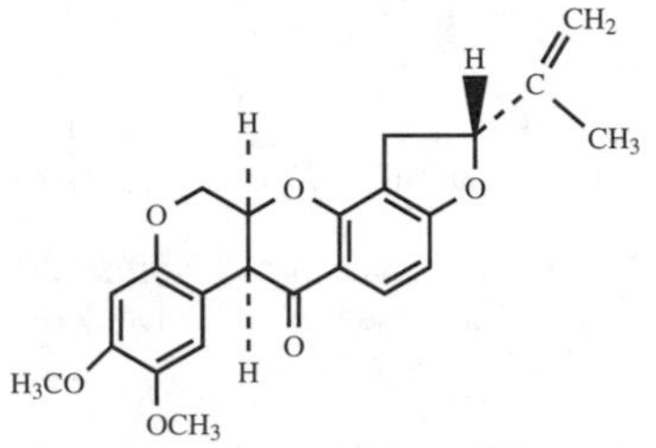

Rothera's Test A method for testing ketone bodies in urine based on the production of blue-purple color upon addition of sodium nitroprusside and ammonium hydroxide to the urine.

Rothia A genus of aerobic, facultative anaerobic, catalase-positive, asporogenous bacteria (Actinomycetales).

Rotometer A device for measuring the flow rate of a gas.

Rotor A device that holds centrifuge tubes during the process of centrifugation.

Rotraxate (mol wt 289) An antiulcerative agent.

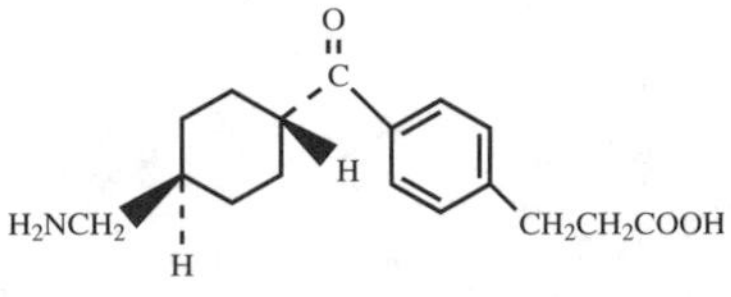

Rough Endoplasmic Reticulum (RER) Region of the endoplasmic reticulum with ribosomes attached.

Rough E R Short for rough endoplasmic reticulum.

Rough Smooth Variation A change of appearance of bacterial colonies (e.g., rough to smooth) due to a change in cell-surface composition.

Rough Strain A strain of bacteria that forms smooth or glossy colonies on the culture plate (e.g., rough strain of *Streptococcus pneumoniae*).

Rounox A trade name for acetaminophen, an antipyretic and analgesic agent that inhibits prostaglandin synthesis.

Rous Sarcoma A readily transplantable malignant fibrosarcoma of chickens caused by Rous sarcoma virus (Retroviridae).

Rous Sarcoma Virus An avian sarcoma virus of the family Retroviridae that causes Rous sarcoma in chicken.

Rowasa A trade name for mesalamine, an anti-inflammatory agent that inhibits prostaglandin synthesis.

Roxanol A trade name for morphine sulfate, an opiate analgesic agent that binds to opiate receptors in the CNS, altering the perception and emotional response to pain.

Roxatidine Acetate (mol wt 348) An antiulcerative agent.

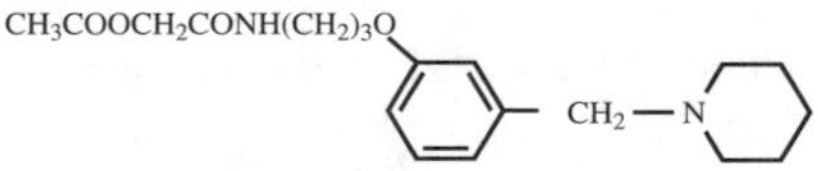

Roxicet A trade name for a combination drug containing oxycodone hydrochloride and acetaminophen, used as an opiate analgesic agent.

Roxicodine A trade name for oxycodone hydrochloride, an opiate analgesic agent that binds to opiate receptors in the CNS, altering the perception and emotion response to pain.

Roxidodone A trade name for oxycodone hydrochloride, an opiate analgesic agent that binds to opiate receptors in the CNS, altering the perception and emotional response to pain.

Roxithromycin (mol wt 837) A semisynthetic antibiotic derived from erythromycin.

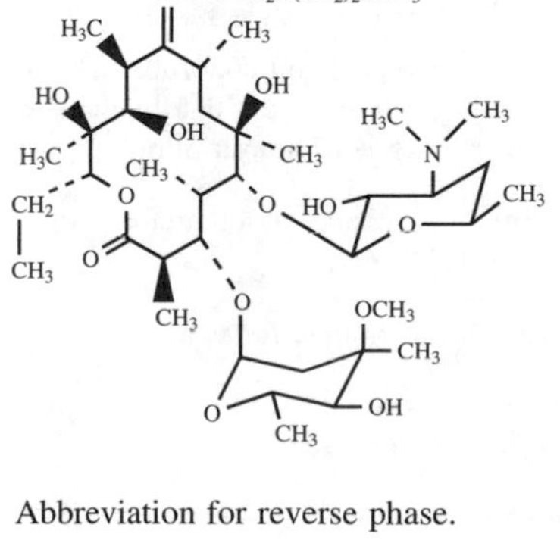

RP Abbreviation for reverse phase.

R5P Abbreviation for ribose 5-phosphate.

RPI Plasmid A conjugative plasmid of both *Enterobacteria* and *Pseudomonas* that contains resistant markers for carbenicillin, kanamycin, and tetracycline.

RP4 Plasmid A conjugative plasmid similar to that of RPI.

r-PA A trade name for retaplase, an enzyme produced by recombinant DNA technology that converts plasminogen to fibrinolysin for degradation of fibrin clots.

rPAH Abbreviation for recombinant phenylalanine hydroxylase.

RPC Abbreviation for reverse-phase chromatography.

RPE Abbreviation for retinol pigment epithelium.

RP-HPLC Abbreviation for reverse phase HPLC (high performance liquid chromatography).

rPLA$_2$ Abbreviation for recombinant PLA$_2$.

R-plasmid A plasmid that contains drug resistant marker(s).

rpm Abbreviation for revolution per minute.

rPP Abbreviation for recombinant protein phosphatase.

rPP2Cα Abbreviation for recombinant protein phosphatase 2Cα.

rpS6 Abbreviation for ribosomal protein 6 of a small ribosome subunit.

RQ Abbreviation for respiratory quotient.

RRA Abbreviation for radio-receptor assay.

RrhI (SalI) A restriction endonuclease from *Rhodococcus rhodochrous* with the following reaction:

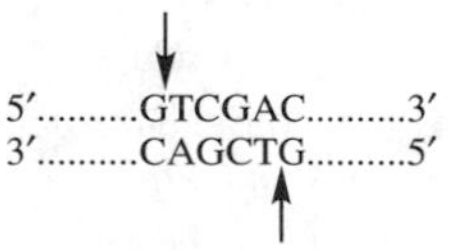

-rrhea A suffix denoting flow or discharge from an organ.

RRM Abbreviation for RNA recognition motif in protein that binds RNA.

rRNA Synonym for ribosomal RNA.

RroI (SalI) A restriction endonuclease from *Rhodococcus rhodochrous* with the same specificity as SalI.

rRyR Abbreviation for rabbit ryanodine receptor.

RS Abbreviation for Ringer's solution.

RSA Abbreviation for 1. rabbit serum albumin; 2. relative specific activity.

RsaI A restriction endonuclease from *Rhodopseudomonas sphaeroides* with the following reaction:

5′..........GT↓AC..........3′
3′..........CA↑TG..........5′

rSAA Abbreviation for recombinant SAA (serum amyloid A).

RshI (PvuI) A restriction endonuclease from *Rhodopseudomonas sphaeroides* with the following reaction:

5′..........CGAT↓CG..........3′
3′..........GC↑TAGC..........5′

RshII (CauII) A restriction endonuclease from *Rhodopseudomonas sphaeroides* with the same specificity as CauII.

RSMC Abbreviation for rat aortic smooth muscle cell.

rSP Abbreviation for rat spasmolytic polypeptide.

RspI (PvuI) A restriction endonuclease from *Rhodopseudomonas sphaeroides* with the same specificity as PvuI.

RspXI (BspHI) A restriction endonuclease from *Rhodococcus* species with the following reaction:

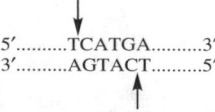

5′..........T↓CATGA..........3′
3′..........AGTAC↑T..........5′

RsrI (EcoRI) A restriction endonuclease from *Rhodopseudomonas sphaeroides* with the same specificity as EcoRI.

RsrII A restriction endonuclease from *Rhodopseudomonas sphaeroides* with the following reaction:

5′..........CG↓G(A/T)CCG..........3′
3′..........GCC(T/A)G↑GC..........5′

RST Abbreviation for reagin screen test for syphillis in which the antigen has been dyed blue, permitting microscopic examination.

R-Strain Synonym for rough strain.

r-Strand The strand in the DNA of Adenoviridae that is transcribed from left to right (right-ward transcription).

RSV Abbreviation for 1. respiratory syncytial virus; 2. rous sarcoma virus.

RSV-Ig A trade name for respiratory syncytial virus immune globulin.

RT Abbreviation for 1. reverse transcriptase; 2. room temperature.

R_t Abbreviation for retention time.

rt-PA A trade name for retaplase, an enzyme produced by recombinant DNA technology that converts plasminogen to fibrinolysin for degradation of fibrin clots.

rTPA (RTPA) Abbreviation for recombinant tissue plasminogen activator.

RT-PCR Abbreviation for reverse transcriptase PCR.

rTRP Abbreviation for rat transient receptor potential protein.

Ru Abbreviation for 1. ribulose; 2. ruthenium, a chemical element.

RU-486 A French birth control pill capable of causing abortion.

Rubella An infectious human disease caused by rubivirus of the Togaviridae (also known as German measles).

Ruberlysin A protease that catalyzes the cleavage of peptides bonds in the β-chain of insulin and angiotensin I.

Rubesol-1000 A trade name for cyanocobalamin (vitamin B_{12}).

Rubex A trade name for doxorubicin hydrochloride, an antineoplastic agent that intercalates with DNA, interfering with transcription.

Rubidium A chemical element with atomic weight 85, valence 1.

Rubion A trade name for vitamin B_{12}.

RuBisCO Short for ribulose 1,5-bisphosphate carboxylase-oxygenase.

Rubivirus A virus of the Togaviridae that causes rubella.

RuBP Abbreviation for ribulose-1,5-bisphosphate.

RuBP Carboxylase Ribulose-1,5-bisphosphate carboxylase.

RuBP Oxygenase Short for ribulose1,5-bisphosphate oxygenase, the enzyme that catalyzes the synthesis of phosphoglycolate and phosphoglycerate from oxygen and ribulose 1,5-bisphosphate.

Rubramin A trade name for vitamin B_{12}.

Rubredoxin Any of a group of bacterial iron-sulfur protein electron acceptors with molecular between 6,000 and 19,000.

Rubredoxin Reductase (NAD+) The enzyme that catalyzes the following reaction:

Reduced rubredoxin + NAD^+

⇅

Oxidized rubredoxin + NADH

Rubseol-A A trade name for vitamin B_{12a}.

Rufen A trade name for ibuprofen, an anti-inflammatory agent that inhibits prostaglandin synthesis.

Ruffini Endings Sensory receptors in the skin that are sensitive to touch.

Rufloxacin (mol wt 363) An antibacterial agent.

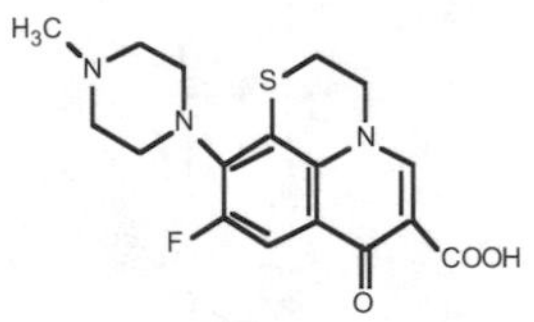

Rumen One of the four compartments of the stomach of a ruminant in which cellulose is digested by the action of symbionts.

Rumenitis Inflammation of rumen.

Rumination The process to rechew what has been chewed slightly and swallowed.

Ruminococcus A genus of Gram-positive, asporogenous, anaerobic, cellulolytic bacteria present in the rumen.

Rum-K A trade name for potassium chloride, used to replace and maintain potassium levels in the body.

RuMP Pathway Abbreviation for ribulose monophosphate pathway, a cyclic metabolic pathway used by some methylotrophic bacteria for assimilation of formaldehyde.

Runella A genus of pink-pigmented bacteria (Spirosomaceae).

Runting Disease A disorder, graft-vs.-host reaction, in the experimental animal resulting from transfer of allogeneic, immunocompetent cells.

RuP Abbreviation for ribulose 5-phosphate.

Ru5P Abbreviation for ribulose 5-phosphate.

Ru1,5P Abbreviation for ribulose1,5-bisphosphate.

Russell's Viper Venom The venom of Russell's viper (*Vipera russellii*), it is an enzyme source for phosphodiesterase, phospholipase, protease, and ribonuclease.

Russian Autumn Encephalitis Synonym for Japanese B encephalitis.

Russian Spring Summer Encephalitis A tick-transmitted acute human encephalitis caused by flavivirus (Flaviviridae).

Rust A plant disease caused by fungi of the order Uredianales, so-called because it forms rust-colored spores on the surfaces of the plants infected with many of the rust fungi.

Rusticyanin A blue copper-containing single chain polypeptide obtained from *Thiobacillus*.

Ruthenium (Ru) A chemical element with atomic weight 101, valence 1-8.

Ruthenium Red (mol wt 858) A dye used for staining mucopolysaccharide.

$(NH_3)_5Ru\text{-}O\text{-}Ru(NH_3)_4\text{-}O\text{-}Ru(NH_3)_5Cl_6$

Rutin (mol wt 611) A substance isolated from buck wheat used as a capillary protectant.

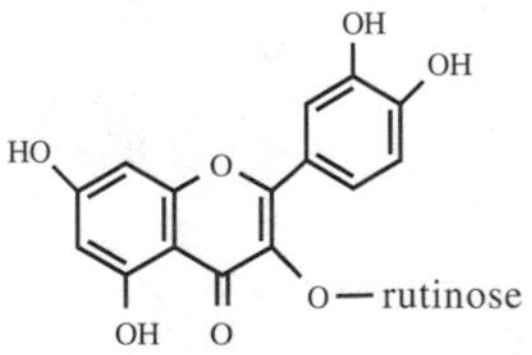

Rutinose (mol wt 326) A disaccharide.

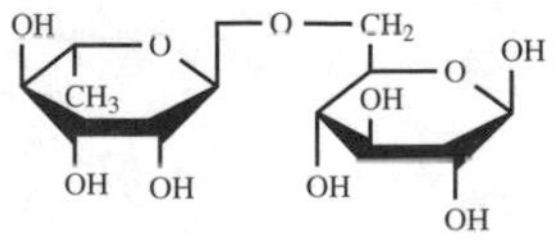

Ru-Tuss A trade name for a combination drug containing phenylepinephrine, propanolamine, chlorpheniramine, hysocyamine propanolamine and alcohol used as a decongestant and antihistaminic agent.

Ru-Vert-M A trade name for meclizine hydrochloride, an anti-emetic, anti-motion sickness, and antihistaminic agent.

RV Abbreviation for rabies virus.

RVV Abbreviation for 1. rubella virus vaccine; 2. Russell viper venom.

RWC Abbreviation for resting wandering cell.

RXR Abbreviation for *cis*-retinoid X receptor.

RXRα Abbreviation for retinoid X receptor alpha.

Ry Abbreviation for ryanodine.

Ryanodine (mol wt 494) An insecticide.

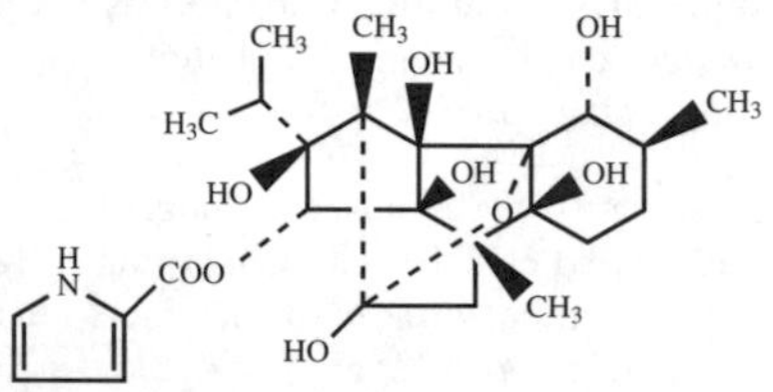

Rylosol A trade name for sotalol, an antianginal and antiarrhythmic agent.

Ryna-C A trade name for a combination drug containing codeine, pseudoephedrine, and chlorpheniramine, used as a decongestant and antihistaminic agent.

Rynatan A trade name for a combination drug containing azatadine maleate and pseudoephedrine sulfate, a long-lasting antihistaminic agent and decongestant.

Rynocrom A trade name for cromolyn sodium, a mast cell stabilizer used to inhibit degranulation of sensitized mast cells.

RyR Abbreviation for ryanodine receptor.

Rythmodan A trade name for disopyramide, used as an antiarrhythmic agent.

Rythmol A trade name for propafenone hydrochloride, an antiarrhythmic agent.

S

S Abbreviation for 1. amino acid serine; 2. sedimentation coefficient (1S = 1×10^{-13} sec); 3. a phase in cell division (period of DNA synthesis); 4. sulfur.

^{35}S Radioactive isotope of sulfur, a beta emitter with a half life of 87 days.

S12 A protein from small, 30S ribosomal subunit of and it is the determinant of streptomycin sensitivity in bacterial cells.

S100 A protein found in both central and peripheral nervous tissue and a marker protein for melanoma and other tumors.

S180 Abbreviation for sarcoma 180, a highly malignant mouse sarcoma cell line.

[S] Abbreviation for substrate concentration.

$[S]_{0.5}$ Referring to the substrate concentration that gives half maximum velocity in an enzyme-catalyzed reaction.

5S, 7S, 16S, 23S, 30S……..80S Referring to the sedimentation coefficient or Svedberg's unit ($1S=10^{-13}$ sec).

7S Antibody Referring to immunoglobulin G (IgG) with the sedimentation coefficient of 7S.

19S Antibody Referring to IgM.

S Colony Abbreviation for smooth colony.

S Layer The outmost layer of both Gram-positive and Gram-negative bacteria.

S Protein A protein in the human that prevents the generation of membrane attack complexes of complement.

S Region Referring to the region of the murine major histocompatibility complex that encodes class-III histocompatibility antigens.

5S RNA Referring to 1. RNA with 5 Svedberg units, e.g., 5S rRNA in 50S ribosome; 2. tRNA with a Svedberg unit of 5S.

16S RNA Referring to RNA with a Svedberg unit of 16S, e.g., 16S rRNA in 30S ribosome.

23S RNA Referring to RNA with 23 Svedberg units, e.g., 23 S rRNA in 50S ribosome.

30S Ribosome Referring to the small ribosomal subunit of the prokaryote.

40S Ribosome Referring to the small ribosomal subunit of the eukaryote.

50S Ribosome Referring to the large ribosomal subunit of the prokaryote.

60S Ribosome Referring to the large ribosomal subunit of the eukaryote.

70S Ribosome Referring to the complex of 30S and 50S ribosome of the prokaryote.

80S Ribosome Referring to the complex of 40S and 60S ribosome of the eukaryote.

S Strain A strain of bacteria (e.g., *Streptococcus pneumoniae*) that produces a smooth, glossy colony.

S Value Referring to Svedberg unit of the sedimentation coefficient.

SA Abbreviation for 1. salicylic acid; 2. sclerotic acid; 3. serum albumin; 4. specific activity; 5. surface antigen.

SAA Abbreviation for serum amyloid A.

SaaI (SacII) A restriction endonuclease from *Streptomyces alanosinicus* with the same specificity as SacII.

SAB Abbreviation for soluble amyloid β.

SAβ Abbreviation for soluble amyloid β.

Sabeluzole (mol wt 416) A nootropic agent.

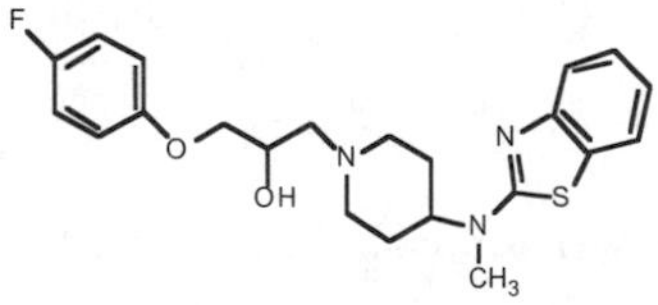

SabI (SacII) A restriction endonuclease from *Streptomyces albohelvatus* with the same specificity as SacII.

Sabin Feldman Dye Test A test for toxoplasmosis (infection of *Toxoplasma gondii).*

Sabin Vaccine An attenuated strain of poliovirus used as an oral vaccine against poliomyelitis.

SABP Abbreviation for secretory actin-binding protein.

SacI A restriction endonuclease from *Streptomyces achromogenes* with the following specificity:

5′..........GAGCT..........3′
3′..........CTCGA..........5′

SacII A restriction endonuclease from *Streptomyces achromogenes* with the following specificity:

5′..........CCGCGG..........3′
3′..........GGCGCC..........5′

SacAI (NaeI) A restriction endonuclease from *Streptomyces achromogenes* with the same specificity as NaeI.

Saccharase Synonym for β-fructofuranosidase or invertase.

Saccharated Iron Oxide Iron oxide particles employed for the determination of phagocytic activity of mononuclear phagocytes.

Saccharic Acid (mol wt 210) A six-carbon dicarboxylic acid derived from oxidation of glucose.

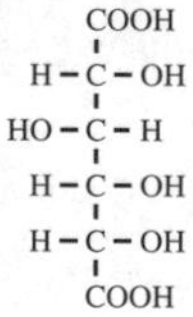

Saccharide Referring to carbohydrate.

Saccharimeter A device for measuring the concentration of sugar in a solution, e.g., polarimeter.

Saccharin (mol wt 183) A nonnutritive sweetener.

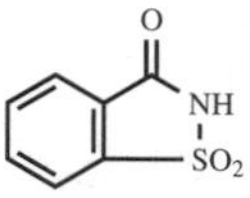

Saccharogen Amylase Synonym for β-amylase.

Saccharolysin Referring to the enzyme from *Saccharomyces* that catalyzes the cleavage of the peptide bonds (pro-phe, al-al) in a protein.

Saccharolytic Capable of splitting or degrading sugar compounds.

Saccharomonospora A genus of bacteria (order of Actinomycetales).

Saccharomyces A genus of yeasts that reproduce asexually by budding or sexually by conjugation. They are economically important in brewing and baking.

Saccharomycosis Infection by yeast, e.g., *Candida* or *Cryptococcus*.

Saccharopepsin A protease from *Saccharomyces* with broad specificity for hydrolysis of peptide bonds.

L-Saccharopine (mol wt 276) A precursor for lysine.

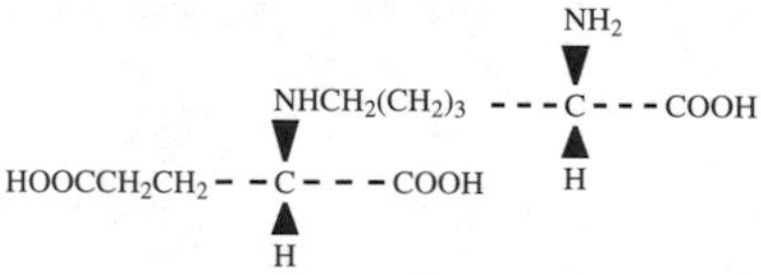

Saccharopine Dehydrogenase The enzyme that catalyzes the following reaction:

N^6-(L-1,3-Dicarboxypropyl)-L-lysine + $NADP^+$ + H_2O

⇅

L-lysine + α-ketoglutarate + NADPH

Saccharopinuria A genetic disorder characterized by mental retardation due to a deficiency of saccharopine dehydrogenase.

Saccharose Synonym for sucrose.

Sacculus Any small, saclike structure, e.g., murein succulus in the eubacteria.

SACE Abbreviation for serum angiotensin converting enzyme.

***sad* Mutants** Mutants that are defective in ribosomal subunit assembly.

***S*-adenosylmethionine (mol wt 400)** A methyl group donor.

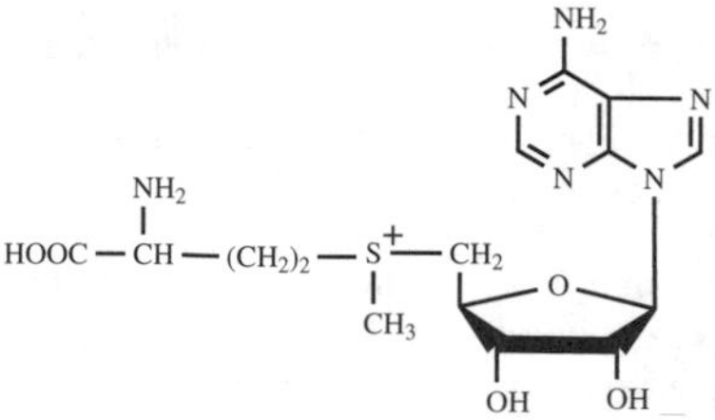

SAF Abbreviation for serum accelerator factor.

SAFA Abbreviation for soluble antigen fluorescent antibody.

Safranine O (mol wt 351) A biological dye.

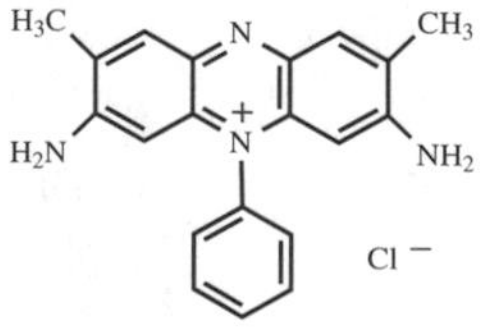

Safranine T Safranine O.

SAG Abbreviation for salicyl acyl glucuronide.

SAH Abbreviation for S-adenosyl-L-homocysteine.

Saizen A trade name for somatotropin produced by recombinant DNA technology and used for the treatment of growth hormone deficiency in children.

SakI (SacII) A restriction endonuclease from *Streptomyces akiyosinicus* with the same specificity as SacII.

Sakaguchi Reaction A reaction to test for arginine based on the production of red color following treatment of the arginine sample with α-naphthol and sodium hypochlorite.

SalI (PstI) A restriction endonuclease from *Streptomyces albus* with the following specificity:

↓
5′..........GTCGAC..........3′
3′..........CAGCTG..........5′
↑

SalAI (MboI) A restriction endonuclease from *Streptomyces albus* with the following specificity:

↓
5′..........GATC..........3′
3′..........CTAG..........5′
↑

Salacetamide (mol wt 179) An analgesic, antipyretic, and anti-inflammatory agent.

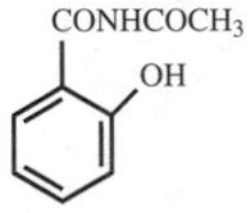

Sal-Adult A trade name for aspirin, an analgesic and antipyretic agent.

Salazopyrin A trade name for sulfasalazine, an antibacterial agent.

Salazosulfadimidine (mol wt 427) An antibacterial agent.

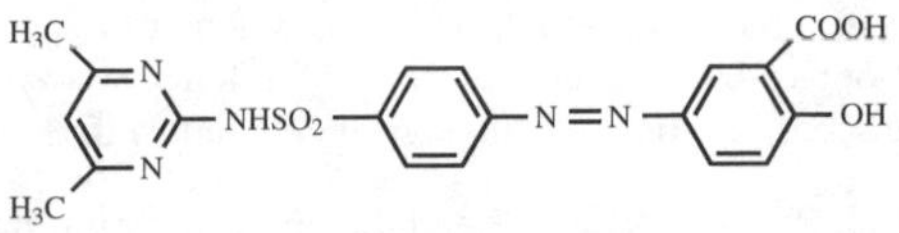

Salbutamol See albuterol.

SalCI (NaeI) A restriction endonuclease from *Streptomyces albus* with the same specificity as Nae I.

SalDI (NruI) A restriction endonuclease from *Streptomyces albus* with the following specificity:

↓
5′..........TCGCGA..........3′
3′..........AGCGCT..........5′
↑

Salflex A trade name for salsalate, used as an analgesic and antipyretic agent.

Salgesic A trade name for salsalate, used as an analgesic and antipyretic agent.

SalHI (MboI) A restriction endonuclease from *Streptomyces albulus* with the same specificity as MboI.

Salicin (mol wt 286) An analgesic agent and a substrate for β-glucosidase.

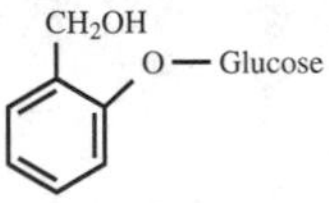

Salicyl Alcohol (mol wt 124) A local anesthetic agent.

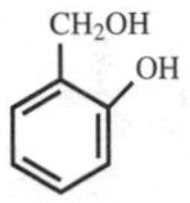

Salicylaldehyde (mol wt 122) A reagent used for reaction with the ε-amino group of lysine in protein.

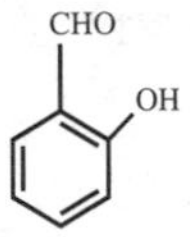

Salicylamide (mol wt 137) An analgesic agent.

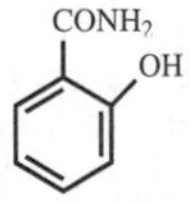

Salicylamide *O*-Acetic Acid (mol wt 195) An analgesic, antipyretic, and anti-inflammatory agent.

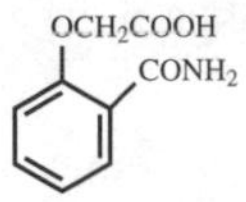

Salicylanilide (mol wt 213) An antifungal agent.

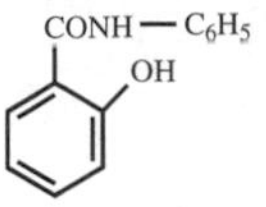

Salicylate 1-Monooxygenase The enzyme that catalyzes the following reaction:

$$\text{Salicylate} + \text{NADH} + O_2 \rightleftharpoons \text{Catechol} + \text{NAD}^+ + H_2O + CO_2$$

Salicylic Acid (mol wt 138) A topical keratolytic agent.

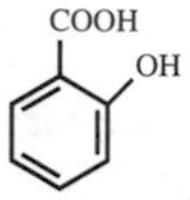

Salicylism Poisoning due to an overdose of aspirin or other salicylate-containing drugs.

4-Salicyloylmorpholine (mol wt 207) A choleretic agent.

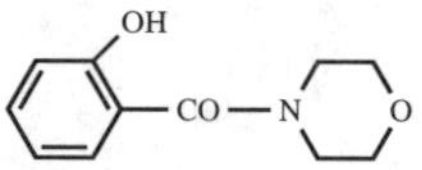

Salicylsulfuric Acid (mol wt 218) An analgesic and anti-inflammatory agent.

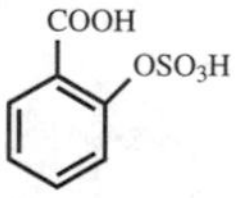

Salimeter A device used for measuring the specific gravity and concentration of sodium chloride in solution.

Salinazid (mol wt 241) An antibacterial agent (tuberculostatic).

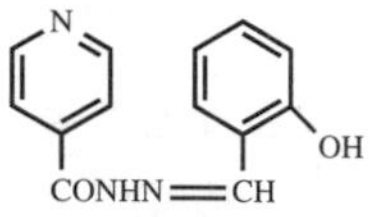

Saline An aqueous solution of 0.85% NaCl (physiological saline).

Saline Aggutination Synonym for auto-agglutination.

Sal-Infant A trade name for aspirin, an analgesic and antipyretic agent.

Salinity The amount of salt in water.

Saliva A mixture of secretions from mucous glands of the oral cavity.

Salk Vaccine The inactivated poliovirus used as a vaccine against poliomyelitis.

Salkowski Test A test for the presence of cholesterol.

Salmeterol (mol wt 416) A beta-2 adrenergic agonist used as an antiasthmatic agent and bronchodilator.

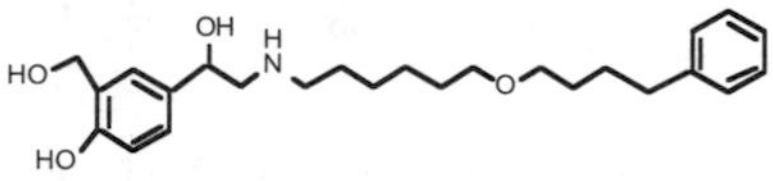

Salmine A protamine consisting of 32 amino acid residues isolated from salmon sperm.

Salmonella A genus of rod-shaped, motile, aerobic bacteria.

Salmonellosis Any disease of humans or animals caused by *Salmonella*, e.g., typhoid and paratyphoid fever (also known as gastroenteritis).

Salmonine A trade name for salmon calcitonin, a hormone and calcium regulator.

Salofalk A trade name for mesalamine, an anti-inflammatory agent.

Salometer Variant spelling of salimeter.

SAL-Plasmid A *Pseudomonas* plasmid that encodes the ability to metabolize salicylate.

Salsalate (mol wt 258) An analgesic and anti-inflammatory agent (a nonacetylated aspirin analog).

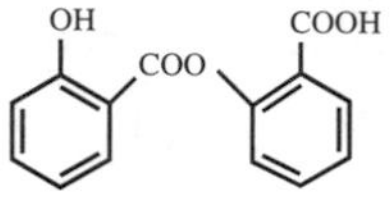

Salsitab A trade name for salsalate, used as an analgesic and antipyretic agent.

Salt The compound formed by replacing the hydrogen of an acid with a metal or its equivalent, e.g., NaCl.

Salt Bridge 1. Ionic bond formed by two oppositely charged ions. 2. Association of two oppositly charged groups or components.

Salt Fractionation Isolation of different proteins from a protein mixture by means of different concentrations of inorganic salt, e.g., ammonium sulfate.

Salt Precipitation Precipitation of proteins by neutral salt (see salt fractionation).

Salting In Increase solubility of a protein by addition of neutral salt.

Salting Out Rendering a protein insoluble by increasing the salt concentration of the solution. Different proteins can be precipitated by different concentrations of ammonium sulfate.

Saluron A trade name for hydroflumethiazide, a diuretic agent that increases urine excretion by inhibiting sodium reabsorption.

Salutensin A trade name for a combination drug containing hydroflumethiazide and reserpine, used as an antihypertensive agent.

Salvage Pathway A pathway that converts the catabolic products into bioactive molecules, e.g., conversion of adenine and guanine from the degradation of DNA or RNA into the corresponding nucleotides.

Salverine (mol wt 312) An analgesic agent.

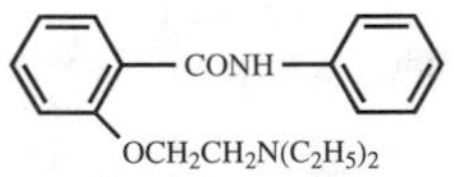

SAM Abbreviation for 1. S-adenosylmethionine; 2. scanning acoustic microscope; 3. surface active material.

Samaderins A group of antitumor agents isolated from the bark and seeds of *Samadera indica* (Simaroubaceae).

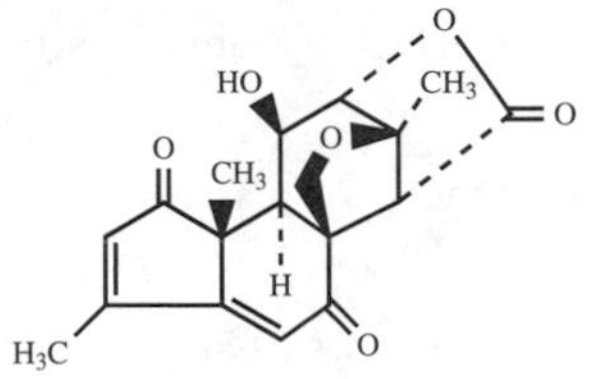

samaderin A

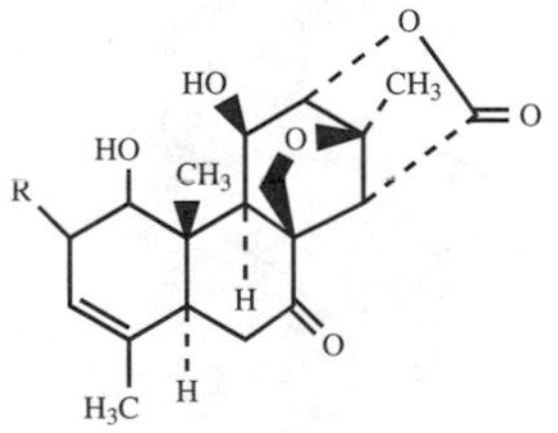

samaderins B and C

Samarium A chemical element with atomic weight 150, valence 2 and 3.

SAMDC Abbreviation for S-adenosylmethionine decarboxylase.

Samesense Mutation A point mutation that results in a change of nucleotide in a codon (usually third position in the codon) but the mutated codon still encodes the same amino acid.

Sam-Pam A trade name for flurazepam hydrochloride, a sedative-hypnotic agent.

Sample Gel The portion of electrophoretic gel that contains protein sample.

Sancycline (mol wt 414) A semisynthetic antibiotic related to tetracycline.

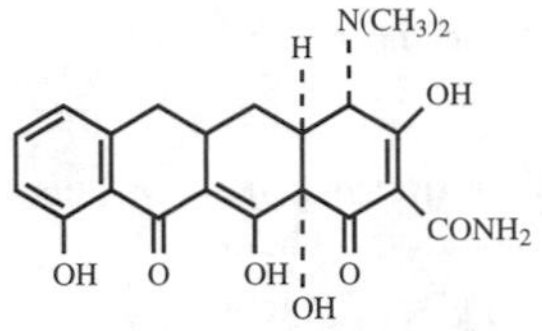

SanDI A restriction endonuclease from *Streptomyces* species with the following specificity:

5'........GG↓GWCCC........3'
3'........CCCWG↑GG........5'

Sandfly Fever A virus disease transmitted to humans by the bite of the sandfly (*Phlebotomus papatasii*).

Sandhoff's Disease A metabolic disorder due to a deficiency of hexosamidase leading to neurological deterioration.

Sandimmune A trade name for cyclosporin, an immunosuppressant that inhibits the proliferation of T lymphocytes.

Sandoglobulin A trade name for human immune globulin.

Sandostatin A trade name for octreotide acetate, an antidiarrheal agent that mimics the action of neural somatostatin.

Sanfillipo Syndrome A metabolic disorder caused by a deficiency of either heparan-*N*-sulfatase or *N*-acetyl α-D-glucosaminidase.

Sangcya A trade name for cyclosporine, a potent immuno-suppressant agent.

Sanger-Coulson Method A DNA sequencing technique in which the single-stranded DNA to be sequenced is used as template for DNA synthesis and incubated with DNA polymerase I, suitable primer, four deoxyribonucleotide triphosphates (one is labeled with ^{32}P) and a small amount of one of the four 2′,3′-dideoxyribonucleotide triphosphate as specific terminator of DNA synthesis for each experiment. Electrophoresis of the reaction mixtures for the four different dideoxyribonucleotide experiments reveal the sequence of the DNA.

Sanger's Reagent The reagent of 1-fluoro-2,4-dinitrobenzene used by Sanger for determination of N-terminal amino acids in a protein. The reaction of 1-fluoro-2,4-dinitrobenzene with N-terminal amino groups of a peptide forms a yellow stable dinitrophenyl derivative. The dinitrophenyl derivative of N-terminal amino acids can be identified after hydrolysis.

Sanguification Formation of blood (hematopoiesis).

Sanguinarine (mol wt 332) A poisonous alkaloid from bloodroot (a herb).

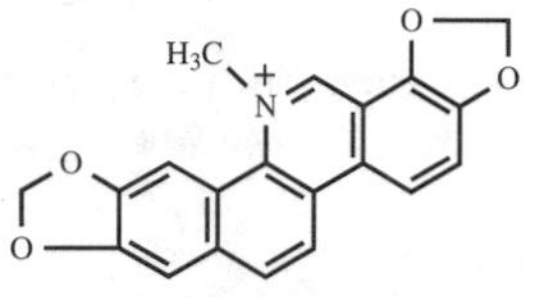

Sani-Supp A trade name for glycerin, a laxative.

Sanorex A trade name for mazindol, a cerebral stimulant.

Sans-Acne A trade name for erythromycin, an antibiotic that inhibits bacterial protein synthesis.

Sansert A trade name for methysergide maleate, an adrenergic blocker.

α-Santonin (mol wt 246) An anthelmintic agent isolated from dried flower heads of *Artemisisa maritima*.

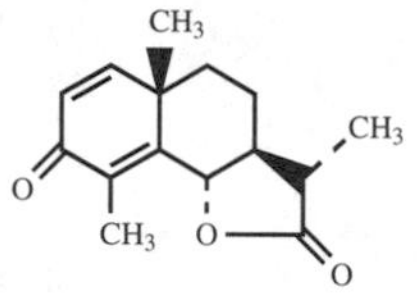

α-Santonin 1, 2-Reductase The enzyme that catalyzes the following reaction:

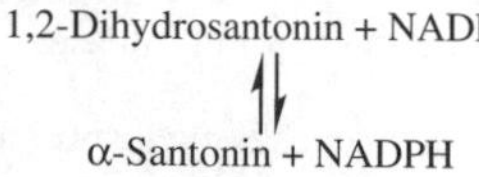

SaoI (NaeI) A restriction endonuclease from *Streptomyces albofaciens* with the same specificity as NaeI.

SAP Abbreviation for 1. secreted alkaline phosphatase; 2. serum alkaline phosphatase.

SapI A restriction endonuclease from *Saccharopolyspores* species with the following specificity:

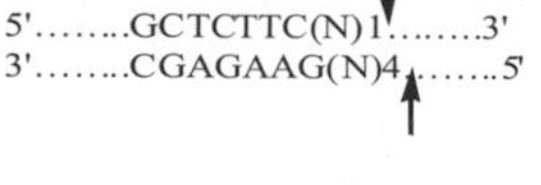

SAP Kinase Abbreviation for stress-activated protein kinase.

SAPK Abbreviation for stress-activated protein kinase.

Sapogenin The nonsugar portion of a saponin, which may be a steroid or a complex terpenoid.

Saponification The alkaline hydrolysis of fat that results in production of glycerol and salts of fatty acid (soap).

Saponification Number The number of milligrams of potassium hydroxide required to hydrolyze 1 g of a sample of fat.

Saponins A group of glycosides and potent surfactants and hemolytic agents consisting of sugar and sapogenin (aglucon moiety).

Saprodinium A genus of protozoa that feed on decaying organic matter.

Saprophytes Organisms that act as decomposers by absorbing nutrients from decaying organic matter.

Saprospira A genus of pigmented chemoorganotrophic gliding bacteria (Cytophagales).

Saprotroph An organism that obtains nutrient from nonliving organic matter.

SAPX Abbreviation for stromal ascorbate peroxidase.

SAR Abbreviation for 1. sarcosine; 2. structure-activity relationship.

Sarafloxacin (mol wt 385) An antibacterial agent.

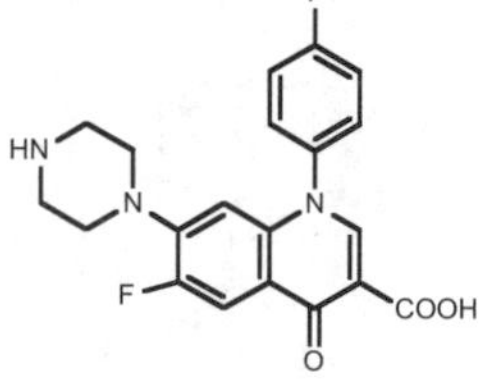

Saramycetin A polypeptide antifungal antibiotic produced by *Streptomyces saraceticus.*

Sarcolysis Lysis of muscular cells.

Sarcoma Tumor derived from connective tissue.

Sarcomatosis A sarcoma that has spread throughout the body.

Sarcomere Repeating unit of a myofibril in muscle cells, which is composed of an array of overlapping thick (myosin) and thin (actin) filaments between two adjacent Z discs.

Sarcoplasm Cytoplasm of a muscle cell.

Sarcoplasmic Reticulum Network of the internal membranes in the cytoplasm of a muscle cell that contains high concentrations of Ca^{2+} that is released into the cytosol during muscle excitation.

Sarcopticide Any agent that kills itch mites.

Sarcosine (mol wt 89) An nonprotein amino acid.

$H_3C—NH—CH_2—COOH$

Sarcosine Acceptor Oxidoreductase The systematic name for sarcosine dehydrogenase.

Sarcosine Dehydrogenase The enzyme that catalyzes the following reaction:

Sarcosine + acceptor + H_2O

⇅

Glycine + formaldehyde + reduced acceptor

Sarcosine Oxidase The enzyme that catalyzes the following reaction:

Sarcosine + H_2O + O_2

⇅

Glycine + formaldehyde + H_2O_2

Sarcosine Oxygen Oxidoreductase The systematic name for sarcosine oxidase.

Sarcosome Mitochondria of a striated muscle fiber.

Sarcotoxins Antibacterial proteins produced by the flesh fly (*Sarcophaga peregrinia*).

Sargramostim A genetically engineered glycoprotein of granulocyte-macrophage colony stimulating factor.

Sarin (mol wt 140) An extremely toxic chemical warfare agent.

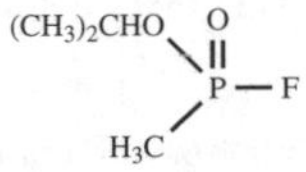

Sarisol No2 A trade name for butabarbital sodium, a sedative-hypnotic agent.

Sarkomycin A (mol wt 172) An antibiotic produced by *Streptomyces erythrochromogenes*.

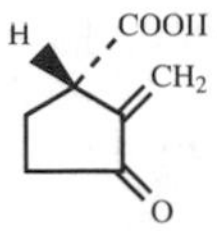

Sarkosyl An anionic detergent of sodium *N*-lauryl sarcosinate.

Sarmentose (mol wt 162) A sugar derivative and a component in some cardiac glycosides.

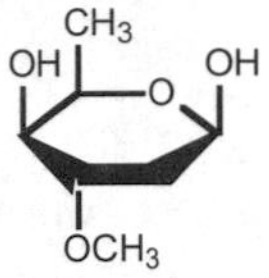

S.A.S. A trade name for sulfasalazine, an antibacterial agent that inhibits the formation of dihydrofolic acid from PABA.

Satellite DNA Highly repetitive fraction of nontranscribed DNA from eukaryotic chromosomes that differs from normal DNA in an unusual nucleotide composition.

Satellite RNA A small, self-splicing RNA molecule that may be encapulated within a specific plant virion (e.g., tobacco ring spot virus) or associated with plant ribosomes.

Satellite Virus A small virus that is replicated only in the presence of a specific helper virus.

Saturated Fatty Acid Fatty acid without double or triple bonds.

Saturated Molecule Molecule that contains only single covalent bonds between the carbon atoms.

Saturation Density The maximum number of cells attainable under specified culture conditions in a culture vessel.

Saturnism Referring to lead poisoning.

Sau96I (AsuI) A restriction endonuclease from *Staphylococcus aureus* PS96 with the following specificity:

↓
5′..........GGNCC..........3′
3′..........CCNGG..........5′
↑

Sau6782I (MboI) A restriction endonuclease from *Staphylococcus aureus* 6782 with the same specificity as MboI.

Sau3239I (XhoI) A restriction endonuclease from *Streptomyces aureofaciens* with the following specificity:

↓
5′..........CTCGAG..........3′
3′..........GAGCTC..........5′
↑

SauI A restriction endonuclease from *Streptomyces aureofaciens* with the following specificity:

5′..........CC↓TNAGG..........3′
3′..........GGANT↑CC..........5′

Sau3AI (MboI) A restriction endonuclease from *Staphylococcus aureus* 3A with the same specificity as MboI.

SauAI (NaeI) A restriction endonuclease from *Streptomyces aureofaciens* with the same specificity as NaeI.

SauBMKI (NaeI) A restriction endonuclease from *Streptomyces aureofaciens* with the same specificity as NaeI.

Saxicolous Growing on or in a rock or stone.

Saxitoxin (mol wt 299) A neurotoxin produced by dinoflagellates (*Gonyaulax catenella* and *G. tamarenesis*) that binds to the sodium channel, blocking the passage of the action potential.

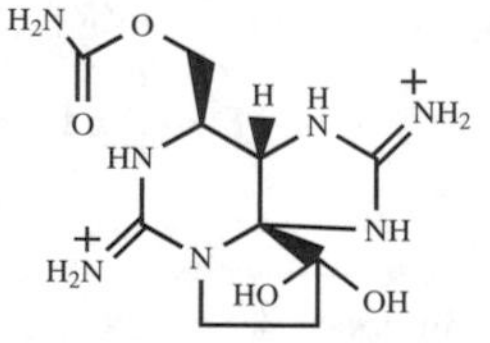

SB Abbreviation for 1. serum bilirubin; 2. southern blot.

SBA Abbreviation for soybean agglutinin.

SBF Abbreviation for serologic-blocking factor.

SbfI A restriction endonuclease from *Streptomyces* species Sb61 with the following specificity:

5'........CCTGCA↓GG........3'
3'........GG↑ACGTCC........5'

SBG Abbreviation for selenite brilliant green.

SblAI (StyI) A restriction endonuclease from *Salmonella blockley* YY156 with the following specificity:

5′..........C↓C(A/T)(A/T)GG..........3′
3′..........GG(T/A)(T/A)C↑C..........5′

SblBI (StyI) A restriction endonuclease from *Salmonella blockley* YY176 with the same specificity as StyI.

SblCI (StyI) A restriction endonuclease from *Salmonella blockley* YY242 with the same specificity as StyI.

SBMV Abbreviation for southern bean mosaic virus.

Sbo13I (NurI) A restriction endonuclease from *Shigella boydii* 13 with the following specificity:

5′..........TCG↓CGA..........3′
3′..........AGC↑GCT..........5′

SBP Abbreviation for 1. sedoheptulose 1,7-bisphosphate; 2. steroid-binding plasma; 3. steroid-binding protein.

SBPase Abbreviation for sedoheptulose 1,7-bisphosphatase.

SBPC Abbreviation for soybean phosphatidylcholine.

SBTI Abbreviation for soybean trypsin inhibitor.

SC Abbreviation for 1. sex chromosome; 2. sickle cells; 3. silicone coated; 4. squamous cancer; 5. stem cells; 6. sugar coated.

SCA Abbreviation for sickle cell anemia.

ScaI A restriction endonuclease from *Streptomyces caespitosus* with the following specificity:

5′..........AGT↓ACT..........3′
3′..........TCA↑TGA..........5′

Scabanca A trade name for benzyl benzoate lotion.

Scabene A trade name for lindane, a pesticide that inhibits neuronal membrane function of arthropods.

Scabicide An agent that kills itch mites causing scabies.

Scabies The contagious itch or mange caused by parasitic mites (e.g., *Sarcoptes scabiei*).

SCAD Abbreviation for 1. short chain acyl-CoA dehydrogenase; 2. short chain alcohol dehydrogenase.

Scaffolding Protein A protein or protein complex that provides a scaffolding (a temporary structure framework) for the assembly of bacteriophage heads but is absent from mature heads.

Scandium (Sc) A chemical element with atomic weight 45, valence 3.

Scanner 1. A device for measuring the distribution of color intensity or radioactivity. 2. A device for scanning the human body or an organism.

Scanning Measurement of color intensity, radioactivity or component distribution across an object or a body structure or whole human body.

Scanning Electron Microscopy (SEM) Technique of electron microscopy in which the specimen is coated with heavy metal and then scanned by an electron beam.

Scarlet Fever An infection or disease caused by *Streptococcus pyogenes* and characterized by sore throat, swelling of lymph nodes, nausea, vomiting, fever, rash, and strawberry-colored tongue.

Scarlet Red (mol wt 380) A stain for fat.

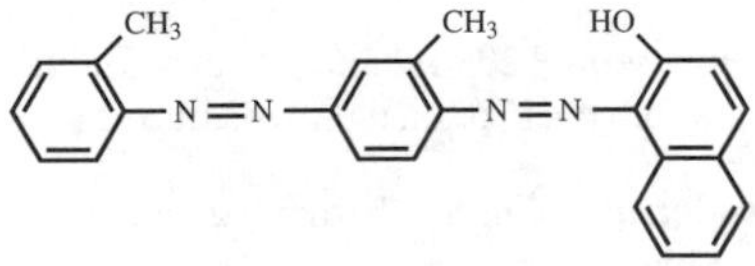

SCAT Abbreviation for sickle cell anemia test.

Scatchard Plot A method for analysis of reaction data for reversible ligand/receptor binding interactions and determination of association constant, binding affinity, and the number of binding sites on the receptor macromolecule.

Scatter Diagram A diagram in which data are plotted as points in a plane of rectangular coordinate to see if there is any correlation between the two plotted parameters.

Scattering The change in direction of a light beam due to collision of the particles or photons with the medium it passes through.

Scavenger A substance that reacts with or traps reactive intermediate in the chemical reaction.

SCC Abbreviation for 1. small cell carcinoma; 2. squamous cell cancer; 3. squamous cell carcinoma.

SCD Abbreviation for 1. sickle cell disease; 2. stearoyl-CoA desaturase.

SceI (FnuDII) A restriction endonuclease from *Synechococcus cedrorum* with the same specificity as FnuDII.

SCF Abbreviation for stem cell factor.

SCFA Abbreviation for short chain fatty acid.

SCHAD Abbreviation for short chain 3-hydroxyacyl-CoA dehydrogenase.

Schardinger Dextrins A group of cyclodexins formed from starch or glycogen by the action of cyclodexin glucosyltransferase from *Bacillus macerans*.

Schardinger Reaction A reaction for testing oxidase activity in milk by incubating milk with formaldehyde and methylene blue. The oxidation of formaldehyde by oxidase results in reduction of methylene blue and disappearance of blue color.

Schick Test A skin test used to determine the susceptibility of an individual to diphtheria by cutaneous injection of diluted diphtheria toxin, which causes reddening and induration in the injected area in the susceptible individual.

Schiff Bases Condensation products of an aromatic amine with aldehydes or ketones.

$$R-CH=N-C_6H_5$$

Schiff's Reagent A reagent used for colorimetric determination of aldehydes that consists of fuchsin and sulfurous acid.

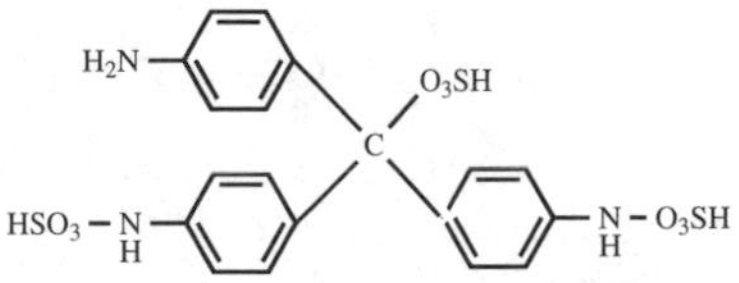

Schilling Test A test used to determine a patient's capacity to absorb vitamin B_{12} from the bowel.

Schistosoma A genus of trematode worms of the family Schistomatidae.

Schistosome Any trematode worm of the genus *Schistosoma*.

Schistosomiasis Infection or disease caused by parasitic *Schistosoma*.

Schistosomule Stage of the schistosome's life cycle that occurs shortly after penetration into a definitive host.

Schizogony A type of asexual reproduction in which the nucleus undergoes division many times resulting in a multinucleate schizont that gives rise to uninucleate cells.

Schizont A multinucleate cell in certain members of sporozoa that is reproduced by schizogony.

Schizonticide Agent that kills schizonts of sporozoan parasites (e.g., malaria).

Schlepper Referring to hapten carrier.

Schlesinger Test A quantitative test for urobilin in urine.

Schlieren Optical System An optical device for the measurement of boundary movement of macromolecules in an analytical centrifugation and in Tiselius electrophoresis.

Schultz-Charlton Test A test for detection of scarlet fever in which antibody to streptococcal erythrogenic toxin is injected into the skin. A localized blanching of the rash indicates a positive reaction.

Schultz-Dale Test An *in vitro* assay for immediate-type hypersensitivity in which smooth muscle is passively sensitized by cytotropic IgE antibody.

Schwann Cell Specialized cell in the peripheral nervous system that is responsible for the formation of the myelin sheath around a nerve axon.

SciI (XhoI) A restriction endonuclease from *Streptoverticillium cinnamonium* with the same specificity as XhoI.

SciAI (BstEII) A restriction endonuclease from *Synechocystis* species with the following specificity:

```
       ↓
5′..........GGTNACC..........3′
3′..........CCANTGG..........5′
              ↑
```

SciAII (PvuII) A restriction endonuclease from *Synechocystis* species with the same specificity as PvuII.

SCID Abbreviation for severe combined immunodeficiency disease.

Scillabiose (mol wt 326) A disaccharide consisting of glucose and rhamnose.

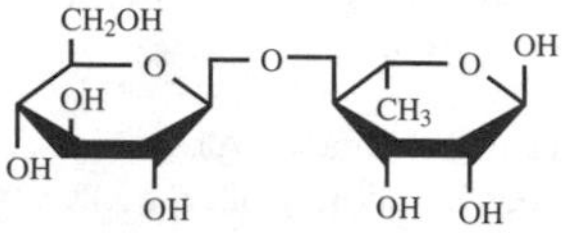

Scillarenin (mol wt 385) A cardiotonic agent.

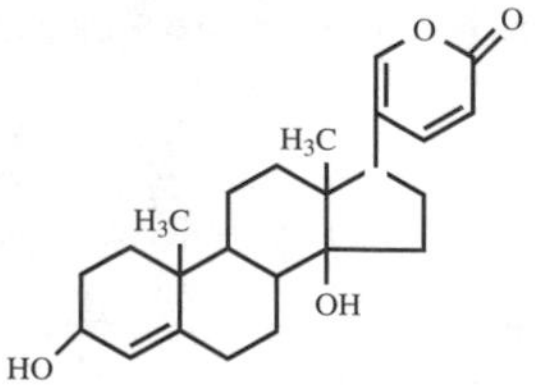

Scinderin A calcium-dependent cytosolic actin filament severing protein found in chromaffin cells, platelets, and a variety of secretory cells.

Scintillation A flash or pulse of light.

Scintillation Cocktail A solution of fluors used for liquid scintilation counting.

Scintillation Counter A device used to detect and register scintillations induced by incident ionizing particles.

Scintillator A substance that emits a scintillation upon interaction with radiation.

Scintillon Substance or structure that emits flashes of light upon acidification in the presence of oxygen.

SCK Abbreviation for serum creatine kinase.

SCLC Abbreviation for small-cell lung cancer.

Sclereid A type of sclerenchyma cell with a thick, lignified secondary wall with many pits.

Sclerenchyma Supporting plant tissue that consists of cells with lignified thick walls.

Sclerenchyma Cell Plant cells of variable form and size with thick, often lignified, secondary walls.

Scleroderma Chronic hardening and thickening of the skin.

Scleroglucan An uncharged microbial glucan found in *Sclerotium glucanicum*.

Scleroproteins A group of proteins that are insoluble in all neutral solvents and in dilute acids and alkalies and function as proteins of supportive tissues, e.g., collagen, elastin, and keratin.

Sclerosis Hardening of tissue due to overgrowth of fibrous tissue or increase in interstitial tissue.

Sclerotan Polymer of 1,3-β-D-glucan found in the sclerotia.

Sclerothrix Abnormal hardening of hair.

Sclerotia Plural of sclerotium.

Sclerotium A genus of fungi.

Sclerotium A compact mass of hardened mycelium.

ScoI (AacI) A restriction endonuclease from *Streptomyces coelicolor* with the same specificity as SacI.

Scoline A trade name for succinylcholine, a neuromuseular blocker that prolongs the depolarization of the muscle end plate.

Scoparone (mol wt 206) An anticholinergic agent.

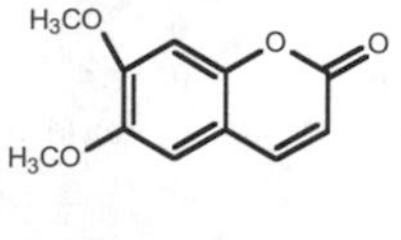

Scopolamine (mol wt 303) An anticholinergic agent that inhibits the muscarinic action of acetylcholine.

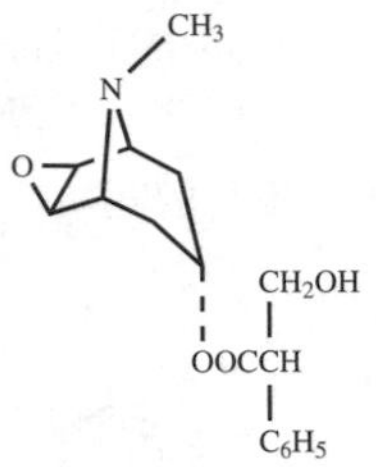

Scopolamine *N*-Oxide (mol wt 319) An anticholinergic agent.

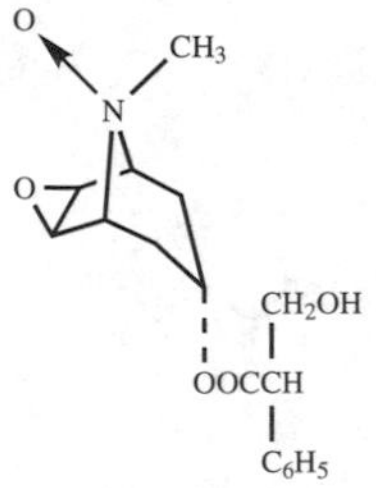

Scotophobin (mol wt 1581) A polypeptide isolated from brains of rats trained to avoid darkness that induces dark avoidance in untrained mice.

Scotopsin A protein found in retinal rods that combine with retinal to form rhodopsin.

Scot-Tussin Cough A trade name for dextromethorphan hydrobromide, a non-narcotic antitussive.

SCP Abbreviation for 1. serine carboxypeptidase; 2. single cell protein; 3. sterol carrier protein.

SCP1 Abbreviation for sterol carrier protein 1.

SCP2 Abbreviation for sterol carrier protein 2.

SCR Abbreviation for structurally conserved region.

Scrapie An infectious degenerative disease of the CNS of sheep and goat caused by prion (infectious protein).

ScrFI A restriction endonuclease from *Streptococcus cremoris* F with the following specificity:

5'........CC↓NGG........3'
3'........GGN↑CC........5'

scRNA Abbreviation for small cytoplasmic RNA associated with ribonucleoprotein particles in eukaryotic cells.

scRNP Abbreviation for small cytoplasmic ribonucleoprotein in eukaryotic cells.

Scrub Typhus An acute systemic disease of humans caused by *Rickettsia tsutsugamushi* and transmitted by the larval stage of mites.

SCT Abbreviation for 1. sickle cell trait; 2. stem cell transplantation.

SCTPA Abbreviation for single chain tissue plasminogen activator.

ScuI (XhoI) A restriction endonuclease from *Streptomyces cupidosporus* with the same specificity as XhoI.

Scurvy A disorder caused by the deficiency of vitamin C and characterized by spongy gum, loosening of the teeth, and bleeding into the skin and mucous membrane.

Scutelarin A protease from venom of the Tipan snake.

SCWP Abbreviation for soluble cell wall-associated protein.

Scytalidopepsin A A protease isolated from fungus *Soytalidium lignicolum* with activity similar to that of pepsin A.

Scytalidopepsin B A protease isolated from fungus *Soytalidium lignicolum* with broad specificity in cleavage of the B chain of insulin.

Scytonema A genus of filamentous cyanobacteria.

SD Antigen Abbreviation for serologically defined antigen, an MHC gene product.

SD Sequence Abbreviation for Shine-Dalgarno sequence (4 - 7 nucleotides) in the leader region of a mRNA that pairs with 16S rRNA and orients AUG initiation codon to a proper position on the ribosome.

S6D Mutation Abbreviation for a mutation in which serine in position 6 is replaced by aspartic acid.

SDA Abbreviation for serologically defined antigen.

SDAg Abbreviation for serologically defined antigen.

SDC Abbreviation for sodium deoxycholate.

SDF-1α Abbreviation for stromal cell-derived factor-1α.

SDH Abbreviation for 1. serine dehydrogenase; 2. shikimate 5-dehydrogenase; 3. sorbitol dehydrogenase; 4. succinate dehydrogenase.

SDS Abbreviation for sodium dodecylsulfate, a strong anionic detergent used in SDS-PAGE and for the solubilization of membrane proteins.

SDS-PAGE Abbreviation for sodium dodecyl sulfate polyacrylamide gel electrophoresis.

SDS-Polyacrylamide Gel Electrophoresis A separation technique in which proteins are treated with sodium dodecyl sulfate (SDS) and electrophoresed in SDS-polyacrylamide gel. The SDS-coated proteins migrate toward the anode at a rate inversely proportional to the molecular weights of the proteins.

SduI A restriction endonuclease from *Streptococcus durans* RFL3 with the following specificity:

5'........G(T/A/G)GC(T/A/C)↓C........3
3'........C↑(A/T/C)CG(A/T/G)G........5'

Se Symbol for selenium.

SE Abbreviation for secretor allele.

SEA Abbreviation for sheep erythrocyte agglutination.

***Sea* Genes** A gene family encoding membrane receptor tyrosine kinase, *v-sea* is an oncogene.

Sealase Synonym for DNA ligase.

SEAT Abbreviation for sheep erythrocyte agglutination test.

Sebaceous Glands The oil glands of the skin, which are associated with hair follicles that produce an oily, waxy secretion.

Sebacic Acid (mol wt 202) Decanedioic acid.

$HOOC(CH_2)_8COOH$

Seborrhea An abnormal increase of secretion and discharge of sebum that produces an oily appearance of the skin and forms greasy scales.

Sebum The secretion of sebaceous glands that contain unsaturated free fatty acids that act as antimicrobial agents.

SecI A restriction endonuclease from *Synechocystis* species with the following specificity:

5′..........C↓CNNGG..........3′
3′..........GGNNC↑C..........5′

SecII (HpaII) A restriction endonuclease from *Synechocystis* species with the same specificity as HpaII.

SecIII (SauI) A restriction endonuclease from *Synechocystis* species with the same specificity as SauI.

Secalonic Acid (mol wt 639) A toxic metabolite produced by mold.

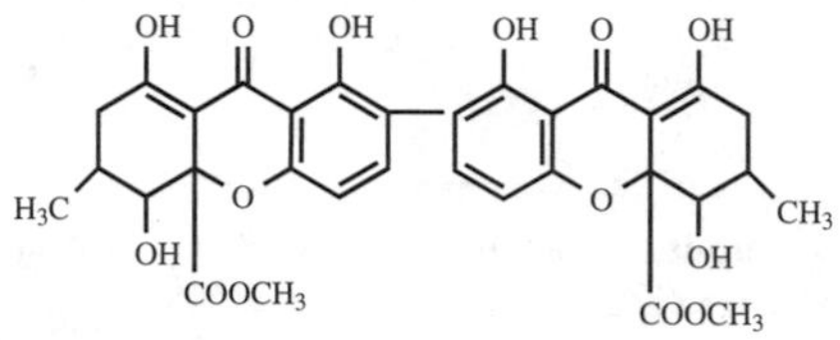

Secnidazole (mol wt 185) An antiprotozoal agent.

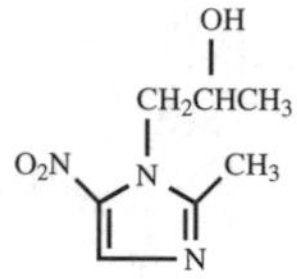

Seconal Sodium A trade name for secobarbital sodium, a sedative-hypnotic agent.

Second Fluor A second fluorescent agent that absorbs the fluorescent light emitted by the primary fluor and thus excited to emit light at a higher wavelength.

Second Law of Thermodynamics The law states that all physical and chemical changes proceed in such a direction that the entropy of the system increases until the equilibrium is reached.

Second Messenger Small molecule or ion generated in the cell in response to the binding of extracellular signal molecules to their receptors, e.g., AMP and IP_3.

Second Order Reaction See reaction of second order.

Second Set Graft Rejection The rejection of an allograft by a host who is sensitized to antigens contained in that graft due to a previous transplant of the same antigenic specificity.

Secondary Active Transport An active transport resulting from the accumulation of a substance across a membrane against a net electrochemical gradient without linkage to ATP hydrolysis.

Secondary Alcohol Oxidase The enzyme that catalyzes the following reaction:

$$\text{An secondary alcohol} + O_2 \rightleftharpoons \text{A ketone} + H_2O_2$$

Secondary Cells Cells arising from the proliferation of cultured primary cells.

Secondary Cell Wall Wall layer laid down by a plant cell on the inner surface of the primary wall when a plant cell has achieved its final size and shape. It is often impregnated with lignin.

Secondary Culture A culture derived from a primary culture.

Secondary Immune Response A rapid and pronounced immune response to a previously encountered antigen due to the presence of memory lymphocyles.

Secondary Lymphoid Organ Organs in which effector lymphocytes are located, e.g., lymph node and spleen.

Secondary Lysosome Cell organelle formed by the fusion of a primary lysosome with a phagocytic vesicle that contains enzymes for digestion of phagocytized materials.

Secondary Messenger See second messenger.

Secondary Metabolism Metabolism that is not essential for growth.

Secondary Products Metabolic products that are not vital to the organism.

Secondary Protein Structure Referring to the local helical, extended, or folded structure of a polypeptide chain (e.g., α helix, β sheet) that are formed by hydrogen bonds and other weak interactions (e.g., ionic bonds, hydrophobic bonds) with the neighboring amino acid residues.

Secondary Radiation Rays emitted by atoms or molecules as the result of the incident or primary radiation.

Secondary Spermatocyte Product of the first meiotic division of a primary spermatocyte from which sperm cells will eventually arise.

Secondary Tumor Malignant mass of cells in an organism originating from a primary tumor located elsewhere in the body.

Secondary Wall See secondary cell wall.

Secretin A basic peptidic gastrointestinal hormone produced by the duodenum of animals that stimulate pancreatic secretion.

Secretor An individual who secrets blood group antigen A, B, or AB into the saliva or other bodily fluids.

Secretor Gene A dominant gene in humans that controls the secretion of blood group antigen A, B, and AB into the saliva and other bodily fluids.

Secretory Cells Cells that specialize in secretion, e.g., epithelial cells.

Secretory Component A protein molecule associated with secretory immunoglobulin (e.g., IgA) that carries secretory immunoglobulin from the Golgi complex to the plasma membrane for exocytosis.

Secretory IgA A dimer of an IgA molecule with a sedimentation coefficient of 11S, containing J chain and secretory component.

Secretory Immune System A distinct immune system of external secretion that consists of predominantly IgA.

Secretory Piece See sceretory component.

Secretory Protein Protein destined for export from the cell in which it was synthesized.

Secretory Vesicle Membrane-bounded organelle in which molecules are made for secretion and stored prior to their release.

Sector Cell A sector-shaped cell used in analytical ultracentrifugation.

Sectrol A trade name for acebutolol, an antihypertensive agent that decreases myocardial contractility and reduces heart rate.

Secubarbiton A trade name for butabarbital, a sedative and hypnotic agent.

Secubutabarbital A trade name for butabarbital, a sedative and hypnotic agent.

Securin A protein that inhibits the transition from metaphase to anaphase in cell division (also called anaphase inhibitor).

Securopen A trade name for azlocillin sodium, an antibiotic that inhibits bacterial cell wall synthesis.

Se-Cys Abbreviation for selenocysteine.

Sedabamate A trade name for meprobamate, an antianxiety agent.

Sedapap A trade name for a combination drug containing acetaminophen and butabarbital, used as an analgesic agent.

Sedative A natural or synthetic therapeutic agent with the property of inducing relaxation and depressing the central nervous system.

Sedatuss A trade name for dextromethorphan hydrobromide, an antitussive agent.

Sedimentation Coefficient A quantitative measure of the rate of sedimentation of a given substance in a centrifugal field, expressed in Svedberg units (one Svedberg unit = 1×10^{-13} sec). Sedimentation coefficients can be calculated from the following equation:

$S = v/w^2x$ where
S = sedimentation coefficient
v = velocity of sedimentation
w^2x = centrifugal force.

Sedoheptulokinase The enzyme that catalyzes the following reaction:

ATP + sedoheptulose
⇅
ADP + sedoheptulose 7-phosphate

Sedoheptulose (mol wt 210) A seven-carbon ketosugar and an intermediate in the Calvin cycle and pentose phosphate pathway.

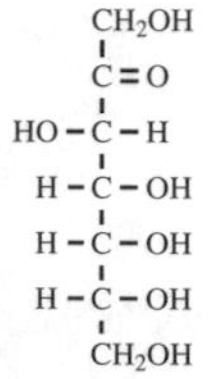

Sedoheptulose Bis-phosphatase The enzyme that catalyzes the following reaction:

Sedoheptulose-1,7-bisphosphate + H_2O
⇅
Sedoheptulose 7-phosphate + Pi

Segment Long Spacing Collagen An abnormal packing pattern of the collagen molecules (e.g., lateral aggregates or cross-striation) formed in acidic solution in the presence of ATP.

Segmented Genome A viral genome composed of several separate RNA molecules, e.g., influenza virus.

Segregation of Chromosome The separation of homologous pairs of chromosomes during meiosis so that only one from each pair is present in any single gamete.

SEH Abbreviation for soluble epoxide hydrolase.

Seldane A trade name for terfeniadine, an antihistaminic agent that competes with histamine for H_1 receptors on effector cells.

Selectable Marker A phenotype that can be used as a marker for identification of a mutant and selection of a recombinant or transformant, e.g., antibiotic resistant marker in a plasmid.

Selectins A group of glycoprotein adhesion receptors.

Selective Medium A medium used for selection of cells with specific growth character or nutritional requirement, e.g., HAT medium for selection of monoclonal antibody producing cells.

Selective Permeability The characteristic of a membrane that allows certain substances to pass through while other substances are excluded.

Selective Toxicity The toxicity of a substance that kills or inhibits one type of organism but not the other.

Selectins A family of cell-surface adhesion proteins.

Selegiline (mol wt 187) An antiparkinsonian agent.

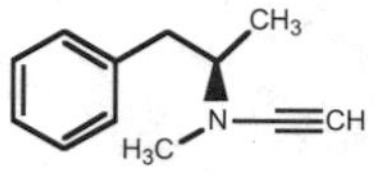

Selenitrace A trade name for selenium.

Selenium (Se) A chemical element with atomic weight 79, valence 2, 4, and 6.

Selenium-72 A radioactive nuclide of selenium (^{72}Se) emitting gamma radiation with a half life of 8.4 days.

Selenium-73 A radioactive nuclide of selenium (^{73}Se) emitting gamma and beta radiation with a half life of 7 hours.

Selenium-75 A radioactive nuclide of selenium (^{75}Se) emitting gamma radiation with a half life of 118.5 days.

Selenium-79 A radioactive nuclide of selenium (^{72}Se) emitting beta radiation with a half life of 6.5¥10^4 years hours.

Selenocysteine (mol wt 168) A naturally occurring nonprotein amino acid found in the active site of enzymes.

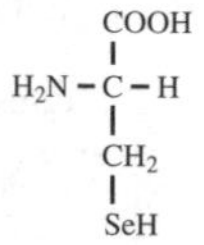

Selenocysteine Lyase The enzyme that catalyzes the following reaction:

L-Selenocysteine + reduced acceptor
⇅
selenide + L-alanine + acceptor

Selenocysteine Reductase Synonym for selenocysteine lyase.

Selenomethionine (mol wt 196) A selenoamino acid used as an anti-metabolite competing with methionine.

CH_3-Se-$[CH_2]_2$-CH$[NH_2]$-COOH

Selenomonas A genus of Gram-negative bacteria of the family Bacteriodaceae.

Selenoprotein Any protein or enzyme that contains selenocysteine or selenomethionine.

Selestoject A trade name for the hormone betamethasone sodium phosphate, an anti-inflammatory agent.

Self-Absorption Absorption of radiation by the sample from which it is emitted.

Self-Antigen The cells or cell products that are antigenic to one's own immune system. The clones of immune cells reactive with self-antigens are normally eliminated.

Self-Assembly The formation of a functional complex of macromolecules or supramolecular structures from its structural components in the absence of a template or parent structure.

Self-Assembly of Ribosomes The formation of functional 70S or 80S ribosomes from ribosomal subunits in the presence of Mg^{2+}.

Self-Cloning Experiment An experiment in which DNA to be cloned is derived from the source where the recombinant DNA is to be replicated.

Selfish DNA Various repetitive DNA sequences with no discernible cellular function (also known as junk DNA).

Self-Limiting A disease or infection that normally does not result in mortality and can be eliminated by the immune system of the host.

Self-Priming Replication of RNA without primer, e.g., replication of viral RNA by viral encoded RNA-dependent RNA polymerase (RNA replicase).

Self-Splicing Introns An intrinsic property of DNA to excise introns from genes without the presence of protein or enzyme.

Self-Tolerance The unresponsiveness of the im mune system to one's own immunogens. The self unresponsiveness was developed during fetal life.

Self-Transmissible Plasmid A plasmid that encodes all the functions needed for its intercellular transmission by conjugation (also known as conjugative plasmid).

Seliberia A genus of Gram-negative, iron-accumulating, budding bacteria.

Seliwanoff's Test A colorimetric test for ketohexoses that uses resorcinol.

SEM Abbreviation for scanning electron microscope.

Semecet A trade name for a combination drug containing acetaminphen, caffeine, and butalbital, used as an analgesic and antipyretic agent.

Semelparous Organism An organism that reproduces only once in its lifetime.

Semen The fluid containing sperm, produced by the male reproductive organ in mammals.

Semenogelin A major gel-forming protein of semen.

Semialdehyde An aldehyde produced by the conversion of one of the two carboxyl groups in a molecule to an aldehyde group, e.g., glutamate semialdehyde.

Semiautonomous Organelle Organelle, e.g., mitochondrion or chloroplast, that contains DNA and is able to encode some of its polypeptides but dependent on the nuclear genome to encode most of the essential products.

Semicarbazide Hydrochloride (mol wt 112) A reagent for assaying aldehydes and ketones.

$NH_2NHCONH_2 \cdot HCl$

Semiconservative Replication Mode of DNA replication in which the parent strands separate, and each strand serves as template for the synthesis of a new strands. Each daughter DNA molecule consists of one parent strand and one newly synthesized strands.

Semidiscontinuous Replication A mode of DNA replication in which one of the complementary new strands is synthesized continually as a leading strand while the other is synthesized discontinuously as Okazaki fragments.

Seminoma A malignant tumor of the testis.

Semiochemical Any chemical substance that delivers a message or signal from one organism to another.

Semipermeable Membrane A membrane that selectively permits passage of certain molecules or ions but not others.

Semipermissive Cells The fraction of a cell population that is permissive for lytic infection by a given virus.

Semisynthetic Referring to natural substances that are modified by chemical alteration.

Semliki Forest Virus A virus of the family Togaviridase that was first isolated from the mosquito in Uganda.

Semotiadil (mol wt 537) An antihypertensive and antianginal agent.

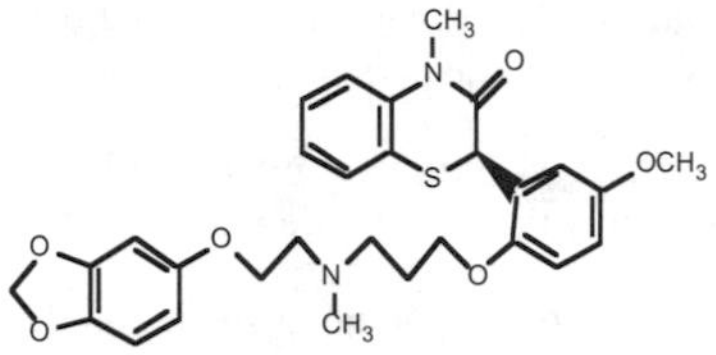

Semple Vaccine An antirabies vaccine, a phenol-inactivated, rabbit-fixed rabies virus (rabies virus that has passaged in rabbit brains).

Sendai Virus A virus in the family of Paramyxoviridae used for fusion of cells.

Senefen III A trade name for a combination drug containing acetaminophen and hydrocodone bitartrate, used as an analgesic agent.

Senescence Aging or deteriorative changes with aging.

Senescent Cell Antigen An antigen that appears on the surface of senescent erythrocytes.

Senna Dried leaflets of *Cassia senna.*

Senokot A trade name for senna, dried leaflets of *Cassia senna,* a laxative agent that promotes accumulation of fluid in the colon and small intestine.

Sense Codon A codon that specifies an amino acid.

Sense Strand Synonym for coding strand.

Sensitization A process in which specific IgE antibodies are synthesized in response to an allergen leading to the subsequent development of an anaphylactic response upon exposure to the same allergen.

Sensitized Erythrocyte An antibody-coated red blood cell.

Sensitized Lymphocytes Lymphocytes that have been exposed to antigen.

Sensor Any device that monitors the level of a substance under investigation using physical (e.g., heat or conductivity) or chemical (e.g., enzyme) parameters.

Sensorcaine A trade name for bupivacaine hydrochloride, a local anesthetic agent that interferes with sodium–potassium exchange and prevents generation and conduction of nerve impulses.

Sensory Cells 1. A nerve cell that transmits sensory impulses. 2. A peripheral nerve cell that receives sensory impulses.

Sensory Neuron A neuron leading from a sensory cell to the central nervous system.

Separation Gel The part of the polyacrylamide gel where separation of proteins or nucleic acids takes place during electrophoresis.

Sephacryl A trade name for covalently cross-linked dextrose beads used in gel filtration chromatography.

Sephadex A trade name for a group of covalenty cross-linked dextran beads used for gel filtration chromatography.

Sepharose A trade name for a group of agarose gels used for gel filtration chromatography.

Se-PHGP Abbreviation for selenium-dependent phospholipid hydroperoxide glutathione peroxidase.

Sepsis The spread of bacteria or bacterial products from a focus of infection to the blood or tissue.

Septate Divided by having a septum.

Septicemia Infection of the blood stream by a virulent microorganism from a focus of infection.

Septic Shock A disorder produced by endotoxin from the infection of Gram-negative bacteria and characterized by hyperpyrexia, rigors, and impaired cerebral function.

Septopal A trade name for gentamicin sulfate, an antibiotic.

Septra A trade name for cotrimoxazole, an antibacterial agent that inhibits the synthesis of dihydrofolic acid from PABA.

Septrin A trade name for the co-trimoxazole, an antibacterial agent that inhibits the synthesis of dihydrofolic acid from PABA.

Septum A membrane or wall between two structures, e.g., two cavities, two nuclei, or two chromosomes.

Sequenator An automatic device for determination of amino acid sequences in a protein through repeated cycles of Edman degradation.

Sequence Referring to the linear order of the different monomeric units in a polymer, e.g., amino acid sequence in a protein or nucleotide sequence in a nucleic acid.

Sequence Analysis Determination of the sequence of nucleotide bases in a DNA molecule or sequence of amino acids in a protein.

Sequence Homolog Referring to the identity of the sequences of the nucleotides in the nucleic acids or sequences of the amino acids in proteins from different sources.

Sequencing Gel Gel used for sequencing DNA or RNA (e.g., polyacrylamide gel).

Sequestered Antigen Antigens that are anatomically isolated from contact with the immune system, e.g., myelin basic proteins, sperm protein antigen, and lens protein antigen. Sequestered antigens activate the immune system and provoke immune responses after release from the sequestered locations.

Sequon An essential peptide sequence in a protein that is required for a specific function, e.g., the tripeptide sequence (asn-x-ser) for asparagine in a protein to act as attaching site for carbohydrate.

ser Abbreviation for serine.

SER Abbreviation for 1. sarcoplasmic- endoplasmic reticulum; 2. smooth muscle endoplasmic reticulum.

Sera Plural of serum.

Seralazide A trade name for a combination drug containing hydrochlorothiazide, hydralazine hydrochloride, and reserpine, used as an antihypertensive agent,

Seratrodast (mol wt 354) An antiasthmatic agent.

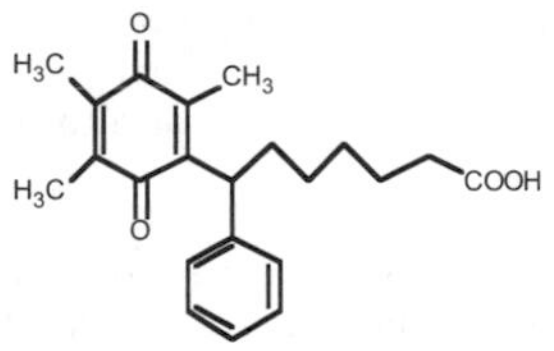

Serax A trade name for oxazepam, an antianxiety agent.

SERCA Abbreviation for sarcoplasmic/endoplasmic-reticulum Ca^{2+}-ATPase.

SERC-ATPase Abbreviation for sarcoplasmic/endoplasmic-reticulum Ca^{2+}-ATPase.

Sereen A trade name for chlordiazepoxide hydrochloride, an antianxiety agent.

Serentil A trade name for mesoridazine besylate, an antipsychotic agent that blocks postsynaptic dopamine receptors in the brain.

Serevent A trade name for salmeterol, a beta-2 adrenergaic agonist used as an anti-asthmatic agent.

Serial Dilution A set of progressive dilutions of a sample (e.g., antigen or antibody) used to determine the highest dilution (least concentration) that gives a positive reaction when incubated with a constant concentration of reactant.

Serial Passage A procedure for passage of a pathogen through different animal hosts or tissue cultures to attenuate the pathogenicity without altering its immunogenicity.

Sericin Proteins found in silk.

Serine (mol wt 105) A protein amino acid.

CH_2OH
|
$CHNH_2$
|
COOH

Serine Aldolase Synonym for glycine hydroxymethyltransferase.

Serine Carboxypeptidase The enzyme that catalyzes the release of a C-terminal amino acid with broad specificity.

D-Serine Dehydratase The enzyme that catalyzes the following reaction:

$$\text{D-serine} + H_2O \rightleftharpoons \text{Pyruvate} + NH_3 + H_2O$$

L-Serine Dehydratase The enzyme that catalyzes the following reaction:

$$\text{L-serine} + H_2O \rightleftharpoons \text{Pyruvate} + NH_3 + H_2O$$

Serine Dehydrogenase The enzyme that catalyzes the following reaction:

$$\text{L-serine} + H_2O + NAD^+ \rightleftharpoons \text{3-Hydroxypyruvate} + NH_3 + NADH$$

Serine Ethanolamine Phosphate Phosphodiesterase The enzyme that catalyzes the following reaction:

$$\text{Serine phosphoethanolamine} + H_2O \rightleftharpoons \text{Serine} + \text{ethanolamine phosphate}$$

Serine Glyoxylate Transaminase The enzyme that catalyzes the following reaction:

$$\text{L-serine} + \text{glyoxylate} \rightleftharpoons \text{3-Hydroxypyruvate} + \text{glycine}$$

Serine Hydroxymethylase The enzyme that catalyzes the following reaction:

5,10-Methylenetetrahydrofolate glycine + O_2
⇌
Tetrahydrofolate + L-serine

Serine Pathway A cyclic metabolic pathway used by some methylotrophic bacteria for assimilation of 1-C substances.

Serine Phosphoethanolamine Synthetase The enzyme that catalyzes the following reaction:

CDP-ethanolamine + L-serine
⇌
CMP + L-serine phosphoethanolamine

Serine Protease A group of proteolytic enzymes that contain an essential serine residue in the active site.

Serine Pyruvate Transaminase The enzyme that catalyzes the following reaction:

Serine + pyruvate
⇌
Hydroxypyruvate + L-alanine

Serine Sulfate Ammonia-lyase The enzyme that catalyzes the following reaction:

L-serine O-sulfate + H_2O
⇌
Pyruvate + NH_3 + sulfate

Serine Threonine Kinase Protein kinase that phosphorylates serine or threonine residues on the target protein.

Serine tRNA Ligase The enzyme that catalyzes the following reaction:

L-serine + ATP + $tRNA^{ser}$
⇌
AMP + PPi + L- seryl-$tRNA^{ser}$

Serine tRNA Synthetase See serine tRNA ligase.

Serine Type D-Ala-D-Ala Carboxypeptidase The enzyme that catalyzes the preferential cleavage:

$(Ac)_2$-L-lysine-D-alanine-↓D-alanine

The enzyme also catalyzes the transpeptidation of peptidyl-alanyl moieties that are *N*-acyl substituents of D-alanine.

Serine Type Carboxypeptidase The enzyme that catalyzes the release of C-terminal amino acids with broad specificity.

Sermorelin A human peptide growth hormone-releasing factor.

Serologic See serological.

Serological Pertaining to serology or method of serology.

Serologically Defined Antigens Antigens that can be sologically identined with antibodies and are present on membranes of all members of the same species and encoded by genes in the major histocompatibility complex (e.g., class-I major histocompatibility antigens).

Serological Reaction Referring to antigen-antibody reaction *in vitro*.

Serology Science that deals with serum and methods for study of antigen-antoHwdy interactions *in vitro*.

Seromucoid A glycoprotein in serum that is not coagulated by heat.

Seromycin A trade name for cycloserine, an antimicrobial agent that inhibits bacterial cell wall synthesis.

Serophene A trade name for clomiphene citrate, a hormone used as a fertility drug.

Seroquel A trade name for quetiapine fumarate, an anti-psychotic agent.

Serostim A trade name for somatropin, a growth hormone.

Serotherapy A type of passive immunity for treatment of infectious disease by injection of antiserum against a specific pathogen into an individual who has been exposed to the pathogen.

Serotonin (5-Hydroxytryptamine) (mol wt 176) A vasoactive amine that plays an important role in anaphylaxis. It is derived from tryptophan, induces contraction of smooth muscle, and enhances vascular permeability.

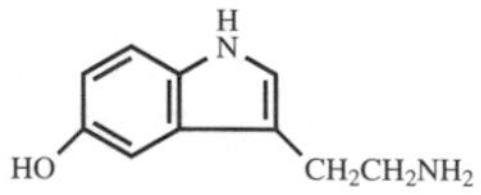

Serotype 1. A type of classification based on the variation of surface epitopes of microorganisms, e.g., serotypes of *Salmonella, Streptococci*, and *Shigella*. 2. The serologically distinguishable members of the same species.

Serotyping Serological identification of microorganisms based on serologically distinguishable epitopes on microorganisms.

Serous Pertaining to serum.

Serous Gland An exocrine gland that produced protein-rich secretions.

Serpalan A trade name for reserpine, an antihypertensive agent that inhibits release of norepinephrine.

Serpasil A trade name for reserpine, an antihypertensive agent that inhibits release of norepinephrine.

Serpazide A trade name for a combination drug containing hydrochlorothiazide and hydralazine hydrochloride, used as an antihypertinsive agent.

Serpens A genus of catalase-positive, oxidase positive, chemotrophic, Gram-negative bacteria.

Serpins A group of naturally occurring proteins that act as inhibitors for serine proteases.

Serralysin A protease from *Pseudomonas aeruginosa.*

Serratia A genus of Gram-negative bacteria (Enterobacteriaceae).

Sertan A trade name for primidone, an anticonvulsant.

Sertoli Cell Cells in the seminiferous tubules of the mammalian testis that surrounds and nourishes the developing sperm cells.

Sertraline (mol wt 306) An antidepressant.

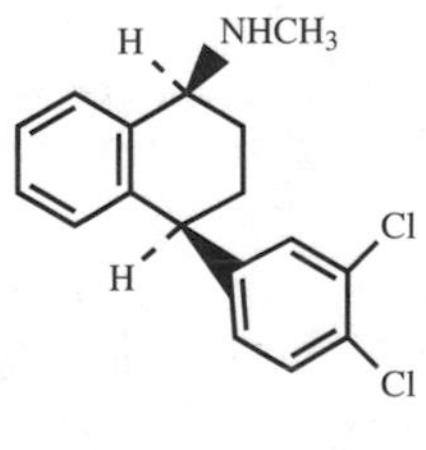

Serum Plasma without fibrin, the fluid remaining after blood coagulation.

Serum Albumin Water-soluble serum protein that serves to maintain the osmotic pressure of the blood.

Serum Carnosinase Synonym of β-Ala-His dipeptidase.

Serum Hepatitis A form of viral hepatitis (Heptadnavirus) transmitted through contaminated blood or blood products (also known as hepatitis B).

Serum Kallikrein Synonym of plasma kallikrein, a protease.

Serum Response Element A segment of DNA that regulates promoters for a group of genes that are activated by the addition of serum to the cell culture.

Serum Sickness A type of hypersensitivity reaction (type III) caused by the formation and deposition of soluble immune complexes in the tissues and characterized by rash, joint pain, and fever.

Seryl-tRNA Synthetase Synonym for serine-tRNA ligase.

Sesin (mol wt 311) A weed killer.

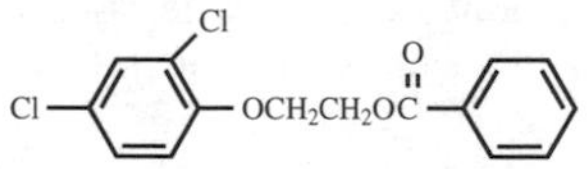

Serzone A trade name for nefazodone hydrochloride, an antidepressant.

Sessile Attachment of a structure directly to a base without a stalk, e.g., attachment of fruiting body or spore directly to the substratum without a stalk.

S-ester Abbreviation for thioester.

Setastine (mol wt 358) An antihistaminic agent.

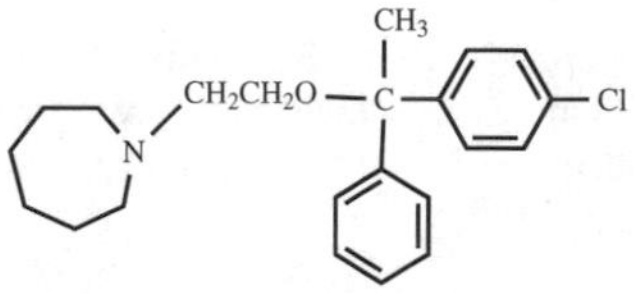

Sevag Method A procedure for deproteinization of nucleoprotein by treatment of the nucleoprotein sample with a mixture of chloroform and isoamyl alcohol.

Severe Combined Immunodeficiency A type of genetically determined immunodeficiency caused by the failure of stem cells to differentiate properly and characterized by the inability to mount an immune response.

Severin A calcium-dependent F-actin cleaving protein from *Dictyostelium discoideum* that binds irreversibly to the microfilament.

SexI (XhoI) A restriction endonuclease from *Streptomyces exfoliatus* with the same specificity as XhoI.

Sex Chromatin The condensed chromatin of the inactivated X-chromosome present in somatic cells of mammals (also known as Barr body).

Sex Chromosomes Chromosome that determines the sex of an animal, e.g., X and Y chromosome in humans (XX for female, XY for male). In birds the opposite is the case (XX for male, XY for female).

Sex Factor Referring to F-plasmid in *E. coli.*

Sex Hormone Hormones that are secreted by sex organs (e.g., gonads) that affect the growth and function of reproductive organs and development of secondary sex characteristics.

Sex Linkage Genes linked to a sex chromosome.

Sex Linked Character A phenotypic character carried on a sex chromosome of a eukaryote.

Sex Pheromone Substance released by an organism that encourages sex interactions between organisms.

Sex Pilus See *F. pilus.*

SexAI A restriction endonuclease from *Streptomyces exfoliatus* with the following specificity:

```
     ↓
5'........ACC(A/T)GGT........3'
3'........TGG(T/A)CCA........5'
                    ↑
```

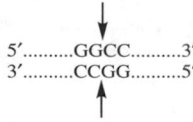

Sexduction Transfer of genetic material from one bacterium to another by conjugation.

Sexual Reproduction Reproduction by the fusion of gametes in which two parent organisms each contributes to the genetic information of the new organism.

Sezary Syndrome A T-cell lymphoma with prominent skin involvement that is caused by a retrovirus.

SF Abbreviation for 1. serum fibrinogen; 2. soluble factor; 3. sub-fragment; 4. Svedberg flotation.

Sf9 Abbreviation for *Spodoptera frugiperda* insect cell line.

Sf21 Abbreviation for *Spodoptera frugiperda* insect cell line.

SFA Abbreviation for saturated fatty acid.

SfaI (HaeIII) A restriction endonuclease from *Streptococcus faecalis* var *zymogenes* with the following specificity:

```
              ↓
5'..........GGCC..........3'
3'..........CCGG..........5'
              ↑
```

SfaGUI (HpaI) A restriction endonuclease from *Streptococcus faecalis* GU with the same specificity as HpaI.

SfaNI A restriction endonuclease from *Streptococcus faecalis* ND547 with the following specificity:

```
                      ↓
5'..........GCATC(N)5..........3'
3'..........CGTAG(N)9..........5'
                      ↑
```

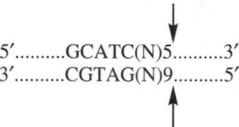

Sf9-BEV Abbreviation for Sf9-baculovirus expressing vector system.

SfcI A restriction endonuclease from *Streptococcus faecalis* with the following specificity:

```
          ↓
5'........CTPuPyAG........3'
3'........GAPyPuTC........5'
                ↑
```

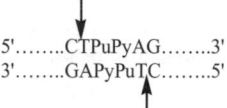

SF_1-Fragment A fragment produced by the treatment of myosin with papain.

SFHb Abbreviation for stroma-free hemoglobin.

SfiI A restriction endonuclease from *Streptomyces fimbriatus* with the following specificity:

```
                       ↓
5'..........GGCCNNNNNGGCC..........3'
3'..........CCGGNNNNNCCGG..........5'
                ↑
```

S-fimbriae See F fimbriae.

SflI (PstI) A restriction endonuclease from *Streptoverticillium flavopersicum* with the same specificity as PstI.

SfnI (AvaII) A restriction endonuclease from *Serratia fonticola* with the following specificity:

```
              ↓
5'..........GGATCC..........3'
3'..........CCTAGG..........5'
                  ↑
```

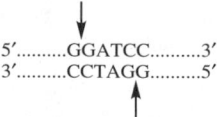

SfoI (NarI) A restriction endonuclease from *Serratia fonticola* with the following specificity:

```
              ↓
5'..........GGCGCC..........3'
3'..........CCGCGG..........5'
                ↑
```

S-Form Referring to bacterial colonies with a smooth appearance.

SfrI (SacII) A restriction endonuclease from *Streptovmyces fradiae* with the same specificity as SacII.

Sfr274I A restriction endonuclease from *Streptomyces fradiae* 274 with the following specificity:

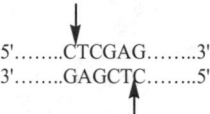

5'........CTCGAG........3'
3'........GAGCTC........5'

Sfr303I A restriction endonuclease from *Streptomyces fradiae* 303 with the following specificity:

5'........CCGCGG........3'
3'........GGCGCC........5'

SfuI A restriction endonuclease from *Streptomyces fulvissimus* with the following specificity:

5'........TTCGAA........3'
3'........AAGCTT........5'

SFV Abbreviation for Semliki forest virus.

SG Abbreviation for 1. serum globulin; 2. serum glucose; 3. soluble gelatin; 4. specific gravity.

SgaI (XhoI) A restriction endonuclease from *Streptomyces ganmycicus* with the same specificity as XhoI.

sGC Abbreviation for soluble guanylate cyclase.

SGF Abbreviation for skeletal growth factor.

SgfI A restriction endonuclease from *Streptomyces griseoruber* with the following specificity:

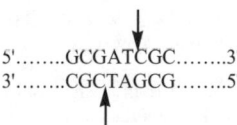

5'........GCGATCGC........3'
3'........CGCTAGCG........5'

SgoI (XhoI) A restriction endonuclease from *Streptomyces goshikiensis* with the same specificity as XhoI.

SGOT Abbreviation for serum glutamic oxaloacetic transaminase or serum glutamate oxaloacetate transaminase.

SGP Abbreviation for soluble glycoprotein.

SGPA Abbreviation for *Streptomyces griseus* protease A.

SGPT Abbreviation for serum glutamate-pyruvate transaminase.

SgrII (EcoRII) A restriction endonuclease from *Streptovmyces griseus* Kr20 with the same specificity as EcoRII.

SgrAI A restriction endonuclease from *Streptomyces griseus* with the following specificity:

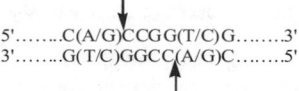

5'........C(A/G)CCGG(T/C)G........3'
3'........G(T/C)GGCC(A/G)C........5'

SH Abbreviation for 1. serum hepatitis; 2. sex hormone; 3. somatotropic hormone; 4. *Src* homology; 5. sulfhydryl.

SH Domain Abbreviation for *Src* homology domain.

SH Group Referring to the sulfhydryl group on a protein.

SH2 Domain Abbreviation for *Src* homology 2 domain, a protein module consisting of about 100 residues found in many proteins involved in signal transduction.

SH3 Abbreviation for *Src* homology 3 domain, it consists of approximately 50 amino acid residues that mediate protein–protein interaction.

SH4 Abbreviation for *Src* homology 4 domain.

SHAA Abbreviation for serum hepatitis–associated antigen.

SHAb Abbreviation for serum hepatitis-associated antibody.

SHAg Abbreviation for serum hepatitis-associated antigen.

Shadow Casting An electron microscopic technique used to study the morphology and dimensions of a structure or object (e.g., a virus particle) in which the object is covered by a thin layer of electron dense metal atoms deposited in a vacuum at a fixed angle to create a shadow of the object under the electron microscope, thus providing information on morphology and size of the object.

SHb Abbreviation for sickle hemoglobin.

SHBD Abbreviation for serum hydroxy-butyrate dehydrogenase.

SHBG Abbreviation for sex hormone binding globulin.

SHBP Abbreviation for sex hormone-binding protein.

SHE Abbreviation for standard hydrogen electrode.

Shear The force due to the variations in the velocities of flow of different layers relative to the parallel adjacent layers, e.g., forced flow of liquid through a capillary or Waring blender action for disintegration or homogenization of a suspended sample.

Shear Rate The rate of change of shear.

Shearing Degradation of matter as a result of shear, e.g., fragmentation of DNA by Waring blender treatment.

Sheathed Bacteria Bacteria whose cells occur within a filamentous sheath that permits attachment to solid surfaces.

Sheep Blood Cell Agglutination Test A method that employs sheep red blood cells or sheep red blood cells coated with antigen or antibody as carriers for testing antigen-antibody reactions by observing agglutination.

Shelf Life The time period during which a stored product remains effective, useful, or suitable for consumption.

Shemin Cycle A pathway for synthesis of tetrapyrrol from succinyl-CoA and glycine.

SHGSH Abbreviation for S-hexylglutathione.

Shift Experiment An experiment in which the conditions (e.g., medium, temperature) for growth of cells are precisely changed.

Shiga Toxin A protein neurotoxin produced by *Shigella dysenteriae* serotype I that inhibits protein synthesis by inactivating 60S ribosomal subunits in eukaryotes.

Shigella A genus of Gram-negative bacteria of the family Enterobacteriaceae.

Shigellosis Bacillary dysentery caused by bacteria of the genus *Shigella*.

Shikimate A salt of shikimic acid.

Shikimate Dehydrogenase The enzyme that catalyzes the following reaction:

$$\text{5-Dehydroshikimate} + \text{NADPH} \rightleftharpoons \text{Shikimate} + \text{NADP}^+$$

Shikimate Kinase The enzyme that catalyzes the following reaction:

$$\text{ATP} + \text{shikimate} \rightleftharpoons \text{Shikimate 5-phosphate} + \text{ADP}$$

Shikimate Pathway A pathway for synthesis of aromatic amino acids (see also shikimic acid pathway).

Shikimate 5-Phosphate (mol wt 254) The phosphate form of shikimic acid.

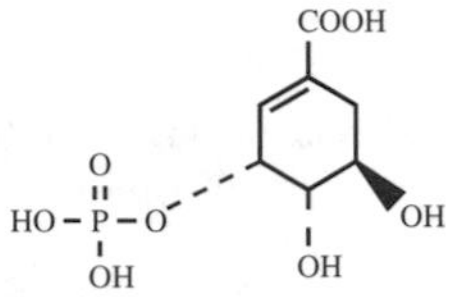

Shikimic Acid (mol wt 174) A precursor for synthesis of aromatic amino acids.

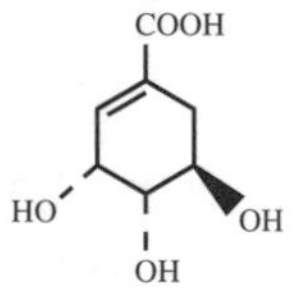

Shikimic Acid Pathway A pathway for synthesis of aromatic amino acids.

Shine Dalgarno Sequence A sequence of about four to seven nucleotides in mRNA upstream from the initiation codon that is complementary to the 3′-end of 16S ribosomal RNA. It serves as a binding site for ribosomes.

Shingles An acute inflammation of the peripheral nerves characterized by the formation of painful red, nodular skin lesions that are caused by reactivation of herpes varicella virus which remained latent after causing chicken pox in exposed individuals.

SHMT Abbreviation for serine hydroxymethyl transferase.

Shock Sensitive Permease A permease in the periplasmic space that is sensitive to osmotic shock treatment (inactivated by osmotic shock).

Shope Papilloma Virus An icosahydral virus of the family of Papovaviridae that contains DNA and produces papillomas in the rabbit.

Shotgun Method Random cloning of fragments of DNA from an entire genome to establish a clone library from which specific cloned fragments can be later selected.

Showdomycin (mol wt 229) A nucleoside antibiotic produced by *Streptomyces showdoensis*.

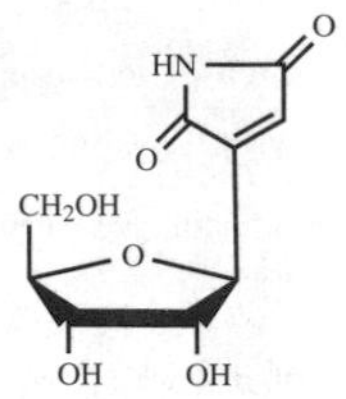

Shufflon Referring to a cluster of DNA segments that invert independently, resulting in complex rearrangement of the DNA genome.

Shunt A diversion from the normal, e.g., an alternative pathway in metabolism.

Shuttle Streaming Flow of cytoplasm that reverses direction with predictable periodicity.

Shuttle Vector A cloning vector that contains DNA sequences, permitting its replication in both bacterial and eukaryotic hosts.

ShyI (SacII) A restriction endonuclease from *Streptomyces hygroscopicus* with the same specificity as SacII.

Si Symbol for silicon.

SI Abbreviation for 1. serum iron; 2. soluble insulin; 3. sucrose isomaltase.

SI System Abbreviation for system international, a system based on seven basic units, namely meter (M) for length, kilogram (Kg) for mass or weight, ampere (A) for electric current, Kelvin (K) for thermodynamic temperature, candela (Cd) for luminous intensity, mole (M) for concentration of substance, and second (s) for time.

Sialadenitis Inflammation of the salivary gland.

Sialagogue Any agent that promotes the flow of saliva.

Sialate *O*-Acetylesterase The enzyme that catalyzes the following reaction:

N-acetyl-*O*-actylneuraminate + H_2O

$\rightleftharpoons$

N-acetylneuraminate + acetate

Sialic Acids Referring to derivatives of neuraminic acid, e.g., *N*- and *O*-substituted derivatives of neuraminic acid.

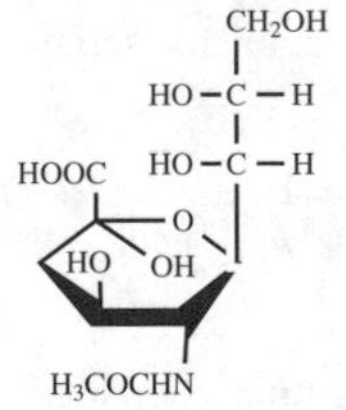

N-acetylneuraminic acid

Sialidase The enzyme that catalyzes the hydrolysis of α-2,3-, α-2,6-, and α-2,8-, glycosidic linkages (at a decreasing rate respectively) of the terminal sialic acid residues in oligosaccharides, glycoproteins, and glycolipid.

Sialoadenectomy Surgical removal of the salivary gland.

Sialoadhesin A macrophage-restricted sialic acid-dependent glycoprotein receptor that consists of 17 immunoglobulin-like domains. It recognizes specific oligosaccharide terminals in the glycan.

Sialoglycoprotein A glycoprotein that consists of sialic acid as a component.

Sialography X-ray examination of the salivary glands after introducing radiopaque material into the duct of salivary glands.

Sialolith Calculus that occurs in the salivary gland.

Sialolithiasis The formation or presence of calculus or calculi in the salivary gland.

Sialorrhea Excessive salivation.

Sialoyl Group Referring to the group resulting from removal of a hydroxyl group from the anomeric carbon of neuraminic acid or sialic acid.

Sialyltransferase Synonym of β-galactoside α-2,6-sialyltransferase.

Siamese Twins Identical twins who are physically joined together at birth.

Siblin A trade name for psyllium, a laxative that absorbs water and increases bulk and moisture content of the stool.

Sibling Species 1. Species that are similar and difficult to distinguish from one another. 2. Offspring from the same parents.

Sibutramine (mol wt 280) An anorexic agent.

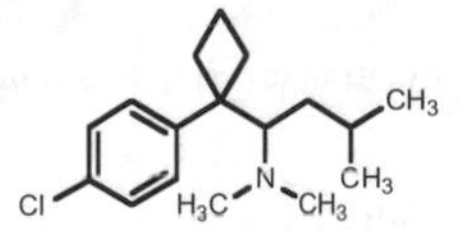

SICD Abbreviation for serum isocitrate dehydrogenase.

Sickle Cell An abnormal, crescent-shaped erythrocyte in patients with sickle cell anemia that contains sickle cell hemoglobin.

Sickle Cell Anemia An inherited disorder characterized by the presence of sickle cells in the blood and formation of abnormal hemoglobin, impairing the oxygen-carrying capability of the blood.

Sickle Cell Hemoglobin Abnormal hemoglobin from sickle cells in which the glutamic acid residue in the sixth position of the beta chain is replaced by valine.

Sickle Cell Trait A heterozygous condition of sickle cell anemia in which some erythrocytes tend to sickle but not enough to produce anemia.

Sickler An individual with sickle cell trait or sickle cell anemia.

Side Chain 1. A chain attached to the principal chain. 2. A chain attached to a ring structure.

Sideramines Synonym for siderophores or siderochromes.

Siderocapsa A genus of iron bacteria.

Siderochromes Synonym for siderophores.

Siderococcus A genus of bacteria found in soil.

Siderocyte An atypical erythrocyte containing iron that is not bound to hemoglobin.

Sideromycins Referring to iron-chelating antibiotics, e.g., albomycin and ferrimycin.

Sideropenia Iron deficiency.

Siderophillins A group of ferric iron-chelating glycoproteins, e.g., lactoferrin and transferrin.

Siderophores Referring to low molecular weight, ferric iron-chelating compounds produced by microorganisms for solubilization and uptake of iron (also known as siderochromes).

Siderosis 1. A lung disease caused by inhalation of iron particles, occurring often in iron workers. 2. Deposition of iron pigment in tissue.

SIDS Abbreviation for sudden infant death syndrome.

SIDV Abbreviation for simian immune deficiency virus.

Siemens A unit of electrical conductance, equal to the conductance between two points on a conductor when a potential difference of 1 volt between these points causes a current of 1 ampere to flow between them.

SIFR Abbreviation for sucrase-isomaltase factor repressor.

sIg Abbreviation for surface immunoglobulin on the B-cell.

sIgA Abbreviation for secretory immunoglobulin A.

sIgM Abbreviation for surface IgM.

Sigma (σ) A letter of the Greek alphabet.

Sigma Cycle A pathway for cycling sigma factor in transcription by attachment of sigma factor to core the enzyme of RNA-polymerase for initiation of transcription, followed by dissociation of sigma factor from the enzyme complex after completion of initiation and reassociation with core enzyme, starting another cycle of the initiation process.

Sigma Factor (σ Factor) The initiation factor for transcription and a subunit of RNA polymerase that functions by recognition of promoter and initiation of transcription.

Sigma Replication Synonym for rolling circle replication of DNA.

Sigma Structure A DNA structure formed during the process of rolling circle replication. It is a double-stranded circular DNA with displaced single-stranded tail resembling the Greek letter σ.

Sigma Subunit A subunit of bacterial RNA polymerase.

Sigma Virus A virus of the family Rhabdoviridae that contains RNA and infects *Drosophila melanogaster*.

Sigmoid Colon The S shaped terminal part of the descending colon, which leads to the rectum.

Sigmoid Kinetics An S-shaped curve obtained by plotting reaction velocity of an enzymatic reaction versus substrate concentration. It is characteristic of an allosteric enzyme or a cooperative binding interaction.

Sigmoidectomy Surgical removal of the sigmoid colon.

Sigmoidoscope An instrument used to inspect the interior of the rectum and sigmoid colon.

Signal Codons The codons that encode a signal peptide.

Signal Hypothesis The hypothesis that deals with the translocation of protein from its site of synthesis into or through a membrane. The protein destined to be transported is synthesized in a precursor form with an N-terminal signal sequence, which is essential for initiation of translocation across the membrane. The signal peptide sequence is highly hydrophobic and is cleaved off after the signal peptide extrudes through the membrane.

Signal Peptide Peptidase An endopeptidase that catalyzes the removal of signal peptide from a protein after polypeptide extrudes through the membrane.

Signal Recognition Particle A nucleoprotein particle that mediates the insertion of protein into or through the membrane.

Signal Response Coupling A cellular response initiated by a signal molecule at the outer surface of a cell membrane through signal-receptor binding action that frequently involves activation of GTP-binding protein and production of second messenger.

Signal Sequence A sequence of 15 to 30 amino acids at the N terminus of a secretory protein that directs the extruding of protein through a membrane (also called a signal peptide).

Signal Transduction A process by which cells convert an extracellular signal into cellular activity.

Silain A trade name for simethicone, an antacid that prevents the formation of mucus-surrounded gas packets in the GI tract.

Silanizing The conversion of an active silanol group to a less polar silyl ether by treatment with trichloromethylsilane to reduce the adsorption of sample to a glass surface in column chromatography.

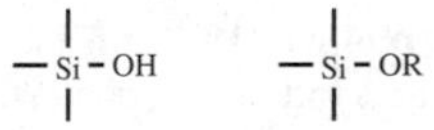

silanol silyl ether

Sildenafil Citrate (mol wt 667) An agent that prevents the breakdown of cGMP by phosphodiesterase, leading to increased cGMP levels and prolonged smooth muscle relaxation.

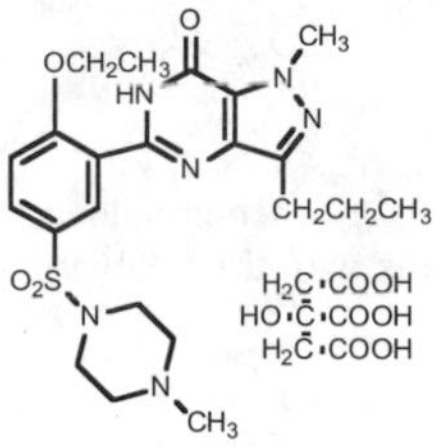

Silent Mutation A mutation that has no apparent effect on phenotype expression.

Silica Referring to silica dioxide.

Silica Gel Hydrated colloidal silicon dioxide used as adsorbent in column and thin-layer chromatography.

Silica Gel 40 A type of silica gel with particle size 0.015 to 0.035 mm.

Silica Gel 60 A type of silica gel with particle size 0.035 to 0.070 mm.

Silica Gel 100 A type of silica gel with particle size 0.063 to 0.2 mm.

Silica Gel Blue A type of silica gel containing a moisture indicator used as an active drying and adsorption agent.

Silicate Any of the widely occurring substances containing silicon, oxygen, and one or more metals without hydrogen. Silicon and oxygen may combine with organic groups to form silicate esters.

Silicoflagellates A group of unicellular planktonic marine algae that possess a netlike, siliceous endoskeleton.

Silicon A chemical element with atomic weight 28, valence 4, and 2.

Silicon Carbide (mol wt 40) A silicon–carbon complex used as an abrasive in dentistry.

$$CSi$$

Silicon Dioxide (mol wt 60) A substance used as an adsorbing and drying agent.

$$SiO_2$$

Silicones Any polymer of organosilicon oxide consisting of alternate silicon and oxygen atoms with various organic radicals attached to the silcon.

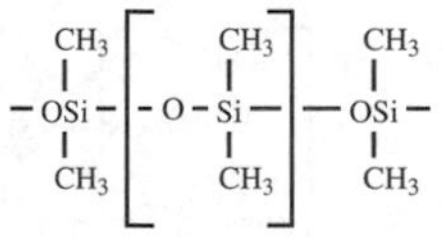

Siliconization Coating a surface with a thin film of silicone oil.

Silicosis Inflammation of the lung characterized by shortness of breath and caused by inhalation of silica that may lead to fibrosis.

Silk A natural protein fiber secreted as a continuous fibroin by silkworm (*Bombyx mori*).

Silk Fibroin Silk protein with antiparallel pleated sheet structure.

Silvadene A trade name for silver sulfadiazine, a local anti-infective.

Silver A chemical element with atomic weight 108, valence 1, and 2.

Silver Iodate (mol wt 283) A reagent used for determination of chloride in the blood.

$AgIO_3$

Silver Iodide (mol wt 235) A local anti-infective agent.

AgI

Silver Lactate (mol wt 215) A topical anti-infective and astringent.

$CH_3\ CH(OH)COOAg$

Silver Nitrate (mol wt 170) A topical anti-infective agent.

$AgNO_3$

Silver Protein Protein prepared from silver oxide and protein (e.g., albumin, gelatin or peptone) and used as an anti-infective and antiseptic agent.

Silver Stain A highly sensitive dye used for staining proteins in PAGE (polyacrylamide gel electrophoresis).

Silylation Introduction of a trimethylsilyl group [$-Si(CH_3)_3$] into an organic compound.

Simazine (mol wt 202) A herbicide.

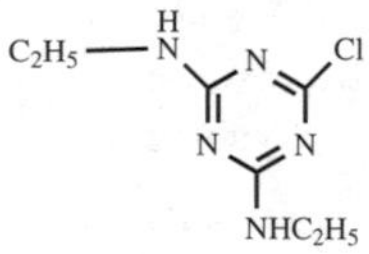

Simethicone An antacid with defoaming activity capable of preventing the formation of mucus-surrounded gas pockets in the GI tract.

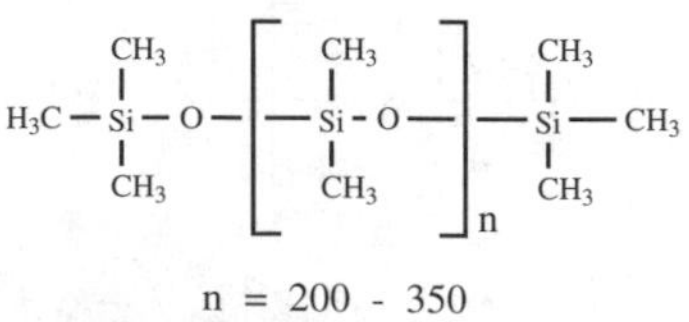

n = 200 - 350

Simetride (mol wt 499) An analgesic agent.

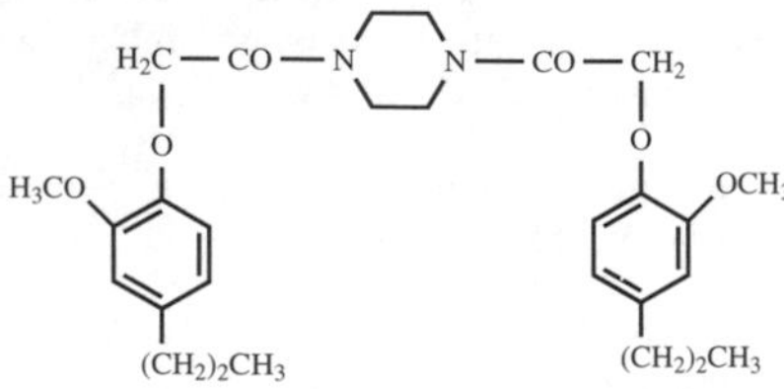

Simetryne (mol wt 213) An herbicide.

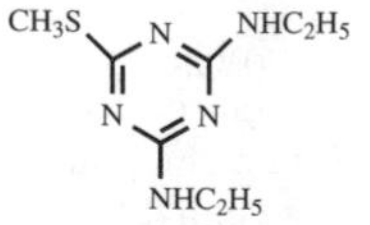

Simfibrate (mol wt 469) An anticholesteremic agent.

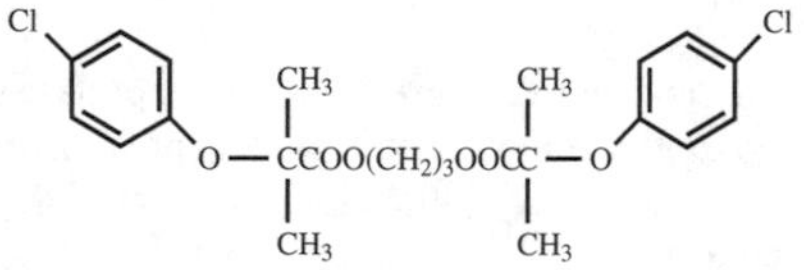

Simian AIDS A disease found in the macaque monkey closely related to human AIDS that is caused by a retrovirus.

Simian Virus 40 (SV40) A small DNA tumor virus of the family Papovaviridae that induces tumors in newborn hamsters (also known as SV40).

Simonsiella A genus of gliding bacteria found in the oral cavity of humans and other vertebrates.

Simple Diffusion A nonmediated transport of solute across a biological membrane.

Simple Protein Proteins that contain only amino acids.

Simple Sugar Synonym for monosaccharide.

Simple Triglyceride A triglyceride that contains only one type of fatty acid.

Simron A trade name for ferrous gluconate, an iron source.

Simulect A trade name for basiliximab, a monoclonal antibody used as an immunosuppressive agent.

Simvastatin (mol wt 419) An antilipemic agent that inhibits the synthesis of cholesterol.

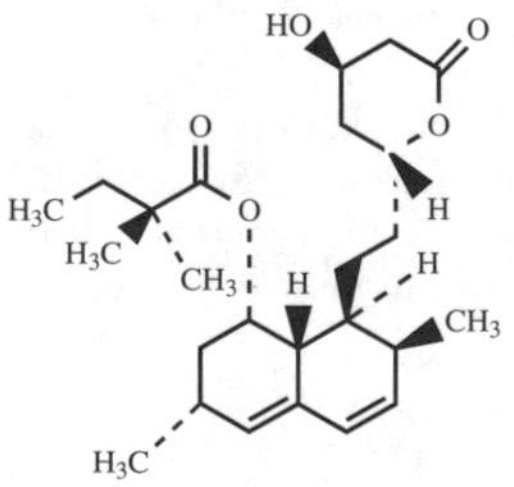

SinI (AvaII) A restriction endonuclease from *Salmonella infantis* with the following specificity:

5′..........GG(A/T)CC..........3′
3′..........CC(T/A)GG..........5′

SinAI (AvaII) A restriction endonuclease from *Salmonella infantis* YY163 with the same specificity as AvaII.

Sinarest Nasal A trade name for phenylephrine hydrochloride, an alpha adrenergaic agonist used to relieve pressure and promote drainage of the nasal passages.

SinBI (AvaII) A restriction endonuclease from *Salmonella infantis* YY190 with the same specificity as AvaII.

SinCI (AvaII) A restriction endonuclease from *Salmonella infantis* 85005 with the same specificity as AvaII.

SinDI (AvaII) A restriction endonuclease from *Salmonella infantis* 85020 with the same specificity as AvaII.

SinEI (AvaII) A restriction endonuclease from *Salmonella infantis* 85064 with the same specificity as AvaII.

Sine-Aid IB A trade name for a combination drug containing pseudoephedrine and ibuprofen used as a decongestant.

Sinefungin (mol wt 381) An adenine-containing antibiotic produced *by Streptomyces griseoleus*.

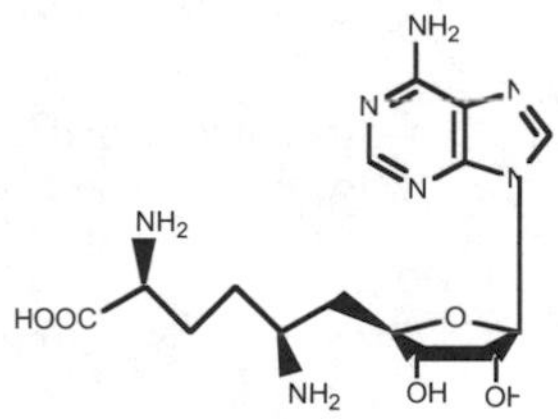

Sinemet CR A trade name for a combination drug containing carbidopa and levodopa, used as an antiparkinsonism agent.

Sinequan A trade name for doxepin, an antidepressant and anti-anxiety agent.

SinFI (AvaII) A restriction endonuclease from *Salmonella infantis* 85084 with the same specificity as AvaII.

SinGI (AvaII) A restriction endonuclease from *Salmonella infantis* 85144 with the same specificity as AvaII.

SinHI (AvaII) A restriction endonuclease from *Salmonella infantis* 85166 with the same specificity as AvaII.

Sinigrin (mol wt 397) A β-D-thioglucopyranoside and a substrate for thioglucosidase.

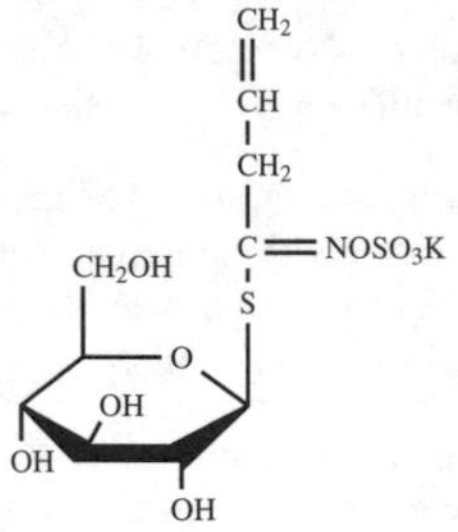

Sinigrinase Synonym of thioglucosidase.

SinJI (AvaII) A restriction endonuclease from *Salmonella infantis* 85325 with the same specificity as AvaII.

Sincomen A trade name for spironolactone, a diuretic agent that increases urine excretion of sodium and water.

Sindbis Virus An RNA-containing virus of the family Togaviridae that causes western equine encephalitis.

Sinemet A trade name for carbidopa-levodopa, used as an antiparkinsonian agent.

Sinequan A trade name for doxepin hydrochloride, an antidepressant that increases the level of norepinephrine or serotonin in the CNS.

Sinex A trade name for phenylephrine hydrochloride, used to produce local vasoconstriction of dilated arterioles to reduce blood flow and nasal congestion.

Single Bond Covalent bond formed between two atoms as a result of sharing a pair of electrons.

Single Burst Experiment A procedure for the quantitative determination of the burst size (number of bacteriophages produced per bacterial cell) resulting from a lytic infection.

Single Cell Protein Proteins derived from a single-cell organism grown on a large scale, e.g., bacteria, yeast, fungi, and algae for use as a source of protein in human or animal diet.

Single Diffusion An immunodiffusion method in which only one component of the antigen-antibody system diffuses through the gel, e.g., diffusion of antigen through an agarose gel containing antibody.

Single Radial Diffusion See radial immunoassay.

Single Strand Binding Protein A class of proteins that bind to single-stranded DNA near the replication fork during the process of replication, preventing unwound strands from rewinding.

Single Stranded DNA The DNA that consists of one polynucleotide chain, e.g., DNA of M13 phage.

Single Strand Exchange The pairing of one strand of a dsDNA with a complementary strand in another dsDNA molecule, displacing its homologous strand in the other duplex.

Singlet Oxygen A reactive but uncharged oxygen molecule produced from a respiratory burst in phagocytes (e.g., neutrophils) and is toxic to microbial cells.

Sintisone A trade name for prednisolone steaglate, a hormone used as an anti-inflammatory agent.

Sinufed A trade name for pseudoephedrine hydrochloride, an adrenergic agent that stimulates adrenergic receptors in the respiratory tract.

Sinulin A trade name for a combination drug containing acetaminophen and chlorpheniramine maleate and used as a decongestant and an analgesic, antihistaminic agent.

Sinusitis Inflammation of a sinus.

Sinusol-B A trade name for brompheniramine maleate, an antihistaminic agent that competes with histamine for H_1 receptors on effector cells.

Sinutab A trade name for a combination drug containing acetaminophen, chlorpheniramine, and pseudoephedrine, used as an analgesic and antipyretic agent.

S1P Abbreviation for sphingosine 1-phosphate.

Siphon A tubular device for drawing or removing fluids.

Siroheme An iron tetrahydroporphyrin serving as prosthetic group for nitrite reductase from *Neurospora crassa* and sulfite reductase from *E. coli.*

Sirohydrochlorin A siroheme from which iron has been removed.

SIRS Abbreviation for soluble immune response suppressor.

***sis* Gene** An oncogen originally identified as the transforming determinant of simian sarcoma virus. The viral *sis (v-sis)* product has nearly identical amino acid sequence to that of human platelet-derived growth factor.

Sisomicin (mol wt 448) An antibiotic produced by *Micromonospora inyoesis.*

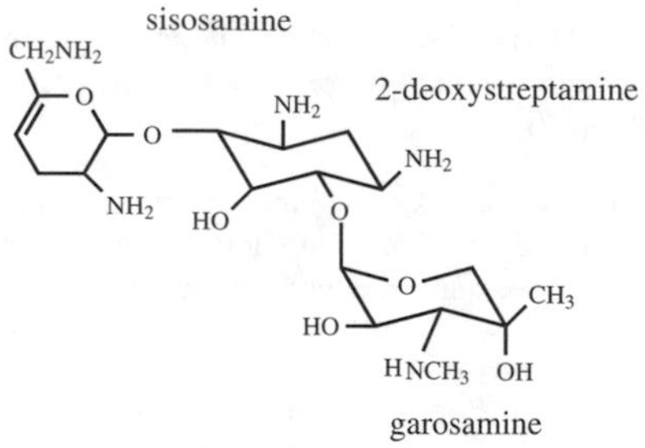

Sister Chromatid Exchange Crossing over between the sister chromatides of a meiotic tetrad or between a duplicated somatic chromosome.

Sister Chromatids Two nucleoprotein molecules formed by the replication of a chromosome and held together by a centromer.

Sistrand A translation unit in mRNA between initiation codon and termination codon.

SIT Abbreviation for serum inhibiting titer.

Site-Directed Mutagenesis Production of a highly specific, predetermined change of a DNA sequence in a given gene through the use of a chemically synthesized oligonucleotide containing the desired mutant base sequence. It involves *in vitro* synthesis and propagation of a mutant gene.

Site-Specific DNA-Methyltransferase (Adenine-specific) The enzyme that catalyzes the following reaction:

S-Adenosyl-L-methionine + DNA-adenine

⇅

S- Adenosyl-L-homocysteine +
DNA 6- methylaminopurine

Site-Specific DNA-Methyltransferase (Cytosine-Specific) The enzyme that catalyzes the following reaction:

S-Adenosyl-L-methionine + DNA-cytosine

⇅

S-Adenosyl-L-homocysteine +
DNA 5- methylcytosine

Site-Specific Endonuclease Synonym for restriction endonuclease.

Site-Specific Mutagenesis Synonym for site-directed mutagenesis.

Site-Specific Recombination A process by which two specific double-stranded DNA sequences in the same or different molecules are joined, e.g., integration and excision of λ prophage DNA.

SIV Abbreviation for simian immunodeficiency virus.

SK Abbreviation for 1. shikimate kinase; 2. squamous keratin; 3. streptokinase.

SkaI (NaeI) A restriction endonuclease from *Streptomyces karnatakensis* with the same specificity as NaeI.

SkaII (PstI) A restriction endonuclease from *Streptomyces karnatakensis* with the same specificity as PstI.

Skeletal Growth Factor A protein that stimulates bone growth.

Skeletal Muscle A muscle that is composed of striated muscle and is attached to the skeleton.

Skeleton A supporting framework of a vertebrate composed of bones, cartilage, and supporting soft tissue.

Skelex A trade name for a combination drug containing chlorzoxazone and acetaminophen, used as a skeletal muscle relaxant.

Skelid A trade name for tiludronate sodium, a biophosphonate.

SL Abbreviation for 1. sodium lactate; 2. streptolysin.

SlaI (XhoI) A restriction endonuclease from *Streptomyces lavendulae* with the same specificity as XhoI.

Slab Gel A sheet of polyacrylamide gel or starch gel used in electrophoresis.

Slant Culture A bacterial culture growing on a nutrient agar slant

SLAP Abbreviation for serum leucine aminopeptidase.

SLD Abbreviation for serum lactate dehydrogenase.

SLDH Abbreviation for serum lactate dehydrogenase.

SLE Abbreviation for systemic lupus erythematosus.

Sleep-Eze 3 A trade name for diphenhydramine hydrochloride, an antihistaminic agent that competes with histamine for H_1 receptors on effector cells.

Sleeping Sickness A human disease caused by *Trypanosoma* and characterized by fever, tremor, swelling of lymph nodes, and loss of weight; the disease is transmitted by tsetse flies.

SLEV Abbreviation for St. Louis encephalitis virus.

Sliding Filament Model A model to explain muscle contraction, accordingly, the contraction occurs in the sarcomere of striated muscle by the sliding of the thick filaments relative to the thin filaments.

Sliding Microtubule Model Model to explain microtubule-based motility; accordingly, model predicts that the length of microtubules remains unchanged but adjacent outer doublets slide past each other, causing a bending movement.

Slime Bacteria Bacteria of *Myxobacterales.*

Slime Layer Extracellular mucilage.

Slime Mold Referring to both acellular slime mold (e.g., *Physarum* of Myxomycetes) and cellular slime mold (e.g., *Dictyostelium* of Acrasidae).

Slimicide An agent that kills or inhibits slime-forming microorganisms.

Slo-Bid Gyrocaps A trade name for theophylline, a bronchodilator that inhibits phosphodiesterase and increases the concentration of cAMP.

Slo-Phyllin A trade name for theophylline, a bronchodilator that inhibits phosphodiesterase and increases levels of cAMP.

Slow Disease An infectious disease characterized by a long asymptomatic incubation period and a prolonged, progressive course of fatal consequence, e.g., disease caused by slow virus.

Slow Reacting Substance Referring to leukotrienes or peptidoleukotriens released from mast cells that cause a relatively slow contraction of smooth muscle compared to histamine.

Slow Virus A virus of the subfamily Lentivirinae (Retroviridae) that produces disease with a greatly delayed onset and protracted course.

SluI (XhoI) A restriction endonuclease from *Streptomyces luteoreticuli* with the same specificity as XhoI.

SM Abbreviation for sphingomyelin.

Sm Abbreviation for samarium, a chemical element.

SmaI A restriction endonuclease from *Serratia marcescens* with the following specificity:

```
          ↓
5′.........CCCGGG.........3′
3′.........GGGCCC.........5′
          ↑
```

Small Cytoplasmic Ribonucleoprotein Complex of small cytoplasmic RNA and protein found in eukaryotic cells.

Small Cytoplasmic RNA A class of small RNA molecules consisting of 100 to 300 nucleotides found in eukaryotic cells. These RNA molecules are associated with proteins to form small cytoplasmic ribonucleoproteins.

Small Nuclear Riboncleoprotein Particle A complex of enzyme and small nuclear RNA molecule that function in RNA splicing.

Small Nuclear RNA A class of small RNA that occur in the nucleus of eukaryotic cells that are about 100 to 300 nucleotides long and usually complexed with proteins to form small nuclear ribonucleoprotein particles.

Small Ribosomal Subunit Referring to the 40S ribosomal subunit in eukaryotes and the 30S ribosomal subunit in prokaryotes. The association of small ribosomal subunits with large subunits form a functional ribosome for protein synthesis.

Small T Antigen A tumor antigen encoded by polyomavirus genome that is found in the soluble cytoplasmic fraction of the cell and acts coordinately with large-T and middle-T antigen to induce transformation of cells.

Smallpox An acute contagious disease caused by a poxvirus, characterized by skin eruption with pustules, sloughing, and formation of permanent scars.

Smallpox Virus A brick-shaped virus of the family Poxviridae that causes smallpox in humans. The virus contains double-stranded DNA with covalently closed ends.

SMase Abbreviation for sphingomyelinase.

SMC Abbreviation for smooth muscle cell.

Smear Spreading material on a glass surface for microscopic examination, e.g., spreading bacteria on a glass slide for microscopic examination.

SMF Abbreviation for a combination drug containing streptozocin, mitomycin C, and 5-fluorouracil.

SmiI A restriction endonuclease from *Streptomyces milleri* S with the following specificity:

```
            ↓
5'........ATTTAAAT........3'
3'........TAAATTTA........5'
            ↑
```

SMMO Abbreviation for soluble membrane mono-oxygenase.

Smooth Endoplasmic Reticulum (Smooth ER) Endoplasmic reticulum that has no ribosomes attached and plays no direct role in protein synthesis but is involved in packaging secretory proteins and synthesizing lipids.

Smooth Muscle A type of muscle found in the walls of arteries and intestine and other viscera of the vertebrate body. It consists of long, spindle-shaped cells.

Smooth Muscle Antibody An autoantibody found in patients with chronic, active hepatitis.

Smooth to Rough Variation A change of morphology of bacterial colony from smooth (glossy) to rough due to changes in cell surface composition that occurs in both Gram-positive and Gram-negative bacteria.

Smooth Strain The virulent strains of bacteria that produce a smooth (glossy) colony on nutrient agar medium, e.g., *Streptococcus pneumoniae*.

SMP Abbreviation for submitochondrial particle.

sMtCK (SMTCK) Abbreviation for sarcomeric mitochondrial creatine kinase.

SN Abbreviation for streptonigrin.

Sn Symbol for tin, a chemical element.

sn Symbol for stereospecific numbering in lipid (e.g., *sn*-1 denoting C-1 of glycerol).

SnaBI A restriction endonuclease from *Sphaerotilus natans* with the following specificity:

```
           ↓
5'........TACGTA........3'
3'........ATGCAT........5'
           ↑
```

Snake Venom A mixture of toxic proteins produced by the venom gland of poisonous snakes, e.g., neurotoxins, cardiotoxins, protease inhibitors, and enzymes.

Snake Venom Phosphodiesterase The enzyme that catalyzes the removal of nucleotides from the 3′ end of polynucleotides.

SNAP Abbreviation for 1. S-nitroso-N-acetyl penicillamine; 2. synaptosomal-associated protein.

SNAP25 Abbreviation for synaptosome-associated protein of 25 kDa.

SNAR Abbreviation for synaptosome-associated protein receptor.

SnoI (ApaLI) A restriction endonuclease from *Streptomyces novocastria* with the following specificity:

↓
5′..........GTGCAC..........3′
3′..........CACGTG..........5′
↑

SNP Abbreviation for sodium nitroprusside

snRNA Abbreviation for small nuclear RNA in the nucleus of eukaryotic cells.

snRNP Abbreviation for small nuclear ribonucleoprotein particle.

S_1-Nuclease The enzyme from *Aspergillus oryzae* that catalyzes the hydrolysis of single-stranded DNA or single-stranded regions in double-stranded DNA.

SNV Abbreviation for spleen necrosis virus.

Soap The sodium or potassium salt of a fatty acid. The salts formed with heavy metal are called heavy-metal soap.

Sobuzoxane (mol wt 515) An antineoplastic agent.

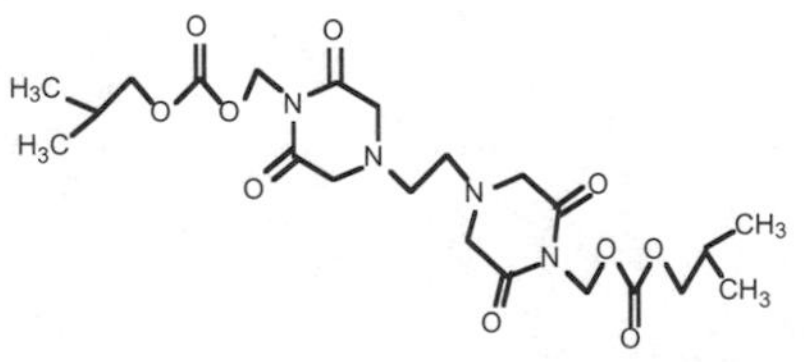

SOD Abbreviation for superoxide dismutase, an enzyme involved in the respiratory burst in neutrophils that converts superoxide to hydrogen peroxide.

Soda Lime A mixture of calcium oxide and sodium hydroxide (5-20%).

Sodium (Na) A chemical element with atomic weight 23, valence 1.

Sodium-22 A radioactive nuclide of sodium (^{22}Na) emitting beta and gamma radiation with a half life of 2.6 years.

Sodium-24 A radioactive nuclide of sodium (^{24}Na) emitting beta and gamma radiation with a half life of 15 hours.

Sodium Arsenate, Dibasic (mol wt 186) An antimalarial agent.

$$AsHNa_2O_4$$

Sodium Ascorbate (mol wt 198) The sodium salt of ascorbic acid and an antioxidant in food products.

Sodium Azide (mol wt 65) A potent vasodilator and antimicrobial agent.

$$NaN_3$$

Sodium Bicarbonate (mol wt 84) An antacid and an alkalizer.

$$NaHCO_3$$

Sodium Borate Solution A solution containing 1.5 grams sodium borate, 1.5 grams sodium bicarbonate, 0.3 ml liquefied phenol, and 3.5 ml glycerol per 100 ml of solution, used for washing mucous membranes.

Sodium Bromide (mol wt 102) A sedative-hypnotic agent and a convulsant.

$$NaBr$$

Sodium Cacodylate (mol wt 160) An arsenical substance used in the treatment of skin diseases.

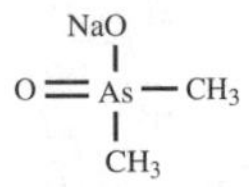

Sodium-Calcium Ion Exchanger A plasma membrane protein antiporter of muscle and nerve cells that promotes exchange diffusion of sodium and calcium.

Sodium Caseinate A food additive used as an emulsifier and a stabilizer that is prepared by dissolving casein in sodium hydroxide followed by evaporation.

Sodium Cellulose Phosphate A substance used for adsorption of calcium in the GI track and treatment of calcium urolithiasis.

Sodium Channel A transmembrane ion channel with negatively-charged interior to block the passage of anions. The channel is voltage gated, that is, it opens in response to a small depolarization of cells. It is the target of many of the potent neurotoxins.

Sodium Cholate (mol wt 431) A detergent and surface active agent.

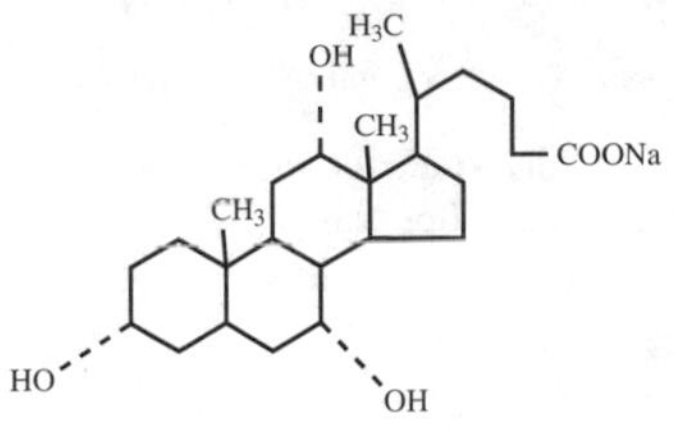

Sodium Cotransport Use of the highly exergonic inward transport of sodium ions to drive active symport of other organic solutes.

Sodium Deoxycholate (mol wt 415) A detergent and surface active agent.

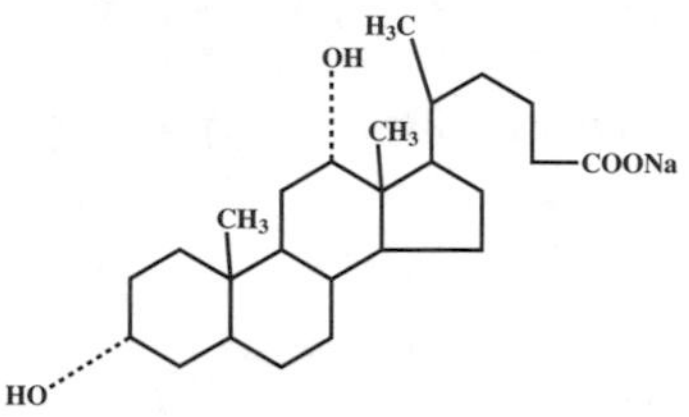

Sodium Dibunate (mol wt 342) An antitussive agent.

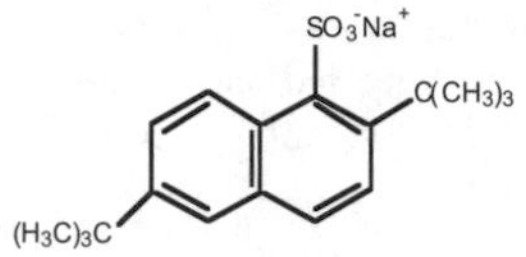

Sodium Dodecyl Sulfate (SDS) (mol wt 288) An anionic detergent used for solubilization of membrane proteins and SDS-gel electrophoresis.

$CH_3(CH_2)_{11}OSO_3Na$

Sodium Edecrin A trade name for ethacrynic acid, a diuretic agent.

Sodium Fluoride (mol wt 42) An agent that is capable of catalyzing bone remineralization.

NaF

Sodium Folate (mol wt 463) Sodium salt of folic acid, a water soluble hematopoietic vitamin (see folic acid for structure).

Sodium Gate See sodium channel.

Sodium Glutamate (mol wt 155) A flavor enhancer.

$HOOC(CH_2)_2CH(NH_2)COONa$

Sodium-Hydrogen Ion Exchanger A plasma membrane protein that functions as an Na^+/H^+ antiporter.

Sodium Hypochlorite (mol wt 74) A disinfectant and an agent used for bleaching.

ClNaO

Sodium Iodate (mol et 198) An antiseptic agent.

$NaIO_3$

Sodium Iodide, Radioactive Radioactive sodium iodide, e.g., $Na^{131}I$ or $Na^{125}I$.

Sodium Ionophore Sodium ion binder and transporter.

Sodium Ionophore I (mol wt 642) A neutral sodium ionophore.

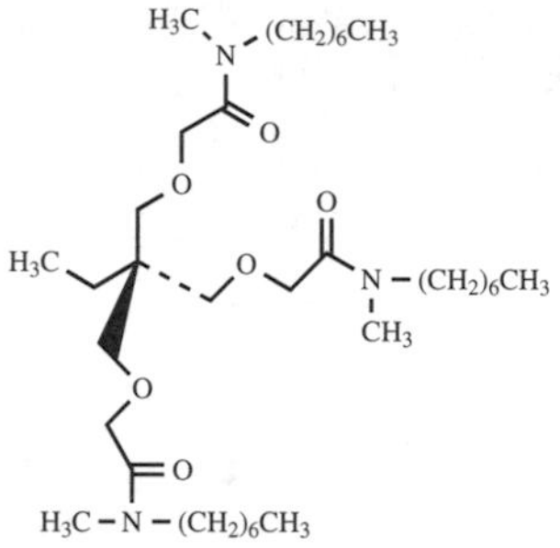

Sodium Ionophore II (mol wt 557) A neutral sodium ionophore.

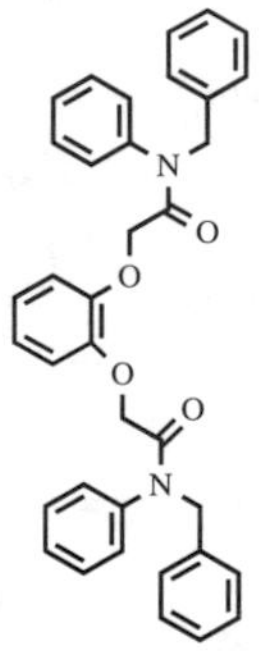

Sodium Ionophore III (mol wt 523) An ionophore for assaying sodium.

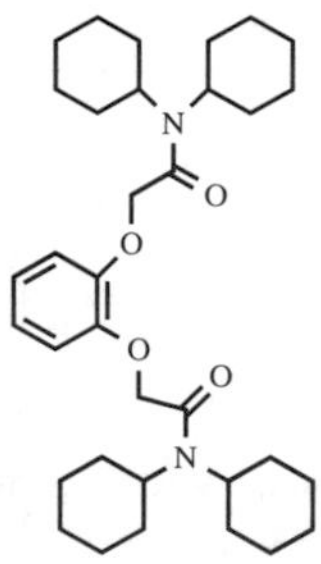

Sodium Ionophore V (mol wt 849) An ionophore for assaying sodium.

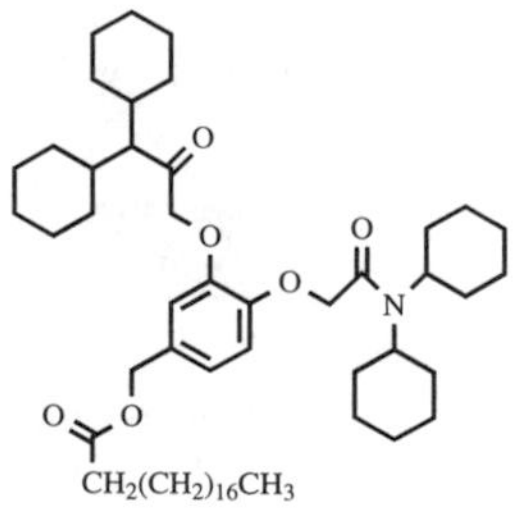

Sodium Lactate (mol wt 112) A substance capable of producing a buffering effect in the body.

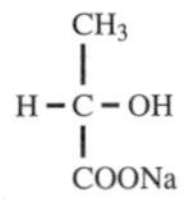

Sodium *N*-Lauroylsarcosinate (mol wt 293) A reagent used for solubilization of membranes and disruption of eukaryotic cells.

$$CH_3(CH_2)_{10}CON(CH_3)CH_2COONa$$

Sodium Lauryl Sulfate See sodium dodecyl sulfate.

Sodium Neural Amino Acid Cotransporter An integral membrane protein involved in sodium-dependent uptake of neutral amino acids.

Sodium Neurotransmitter Transporter Any of a family of integral membrane glycoproteins involved in sodium-dependent transport of neurotransmitters.

Sodium Polystyrene Sulfonate A potassium-removing resin that exchanges sodium ions for potassium ions in the intestine.

Sodium Potassium ATPase See Na+-K+ ATPase.

Sodium Potassium Pump See sodium pump.

Sodium Pump Referring to the active transport of sodium and potassium in animal cells mediated by Na^+-K^+-ATPase (a transmembrane carrier). It pumps Na^+ out of the cell and K^+ into the cell using the energy derived from ATP hydrolysis and creates concentration gradients of sodium and potassium across the membrane.

Sodium Salicylate (mol wt 160) An analgesic and antipyretic agent.

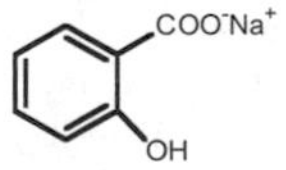

Sodium Sulamyd A trade name for a solution of sodium sulfacetamide, an ophthalmic anti-infective agent.

Sodium Tartrate (mol wt 194) A salt of tartaric acid.

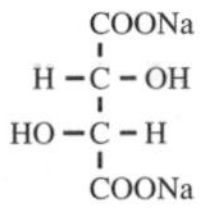

Sodium Tetradecyl Sulfate (mol wt 316) A sclerosing agent.

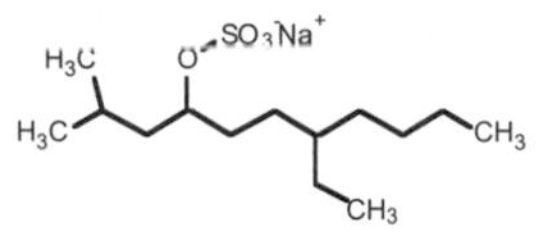

Sodol A trade name for carisoprodol, a skeletal muscle relaxant that reduces transmission of impulses from spinal cord to skeletal muscle.

SOD-PEG Abbreviation for superoxide dismutase polyethylene glycol.

Solan (mol wt 240) An herbicide.

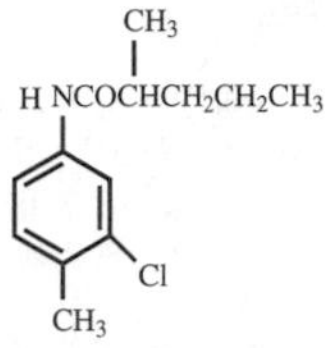

Solanine (mol wt 868) A substance obtained from *Solanum* species, e.g., fresh potato sprouts used as an agricultural insecticide.

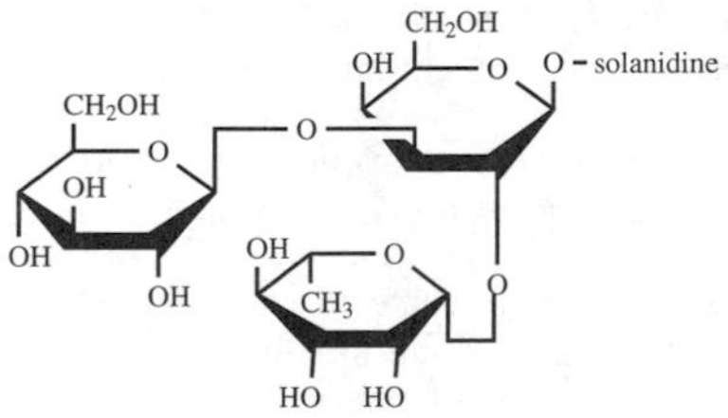

Solasulfone (mol wt 893) An antibacterial (Leprostatic) agent.

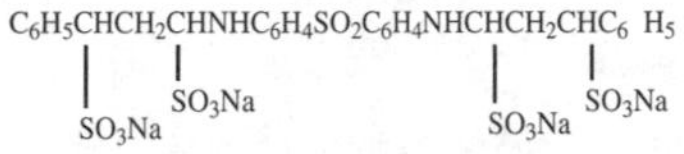

Solazine A trade name for trifluoperazine hydrochloride, an antipsychotic agent that blocks postsynaptic dopamine receptors in the brain.

Solenoid Helical coil of chromatin fiber that serves as an intermediate structure in chromosome condensation.

Solfoton A trade name for phenobarbital, used as an anticonvulsant.

Solganal A trade name for aurothioglucose, an anti-inflammatory agent.

Sol-Gel Transformation A change from the more fluid cytoplasm to a gel like cytoplasm, a proposed mechanism for ameba locomotion.

Solld Phase Radioimmunoassay A modification of radioimmunoassay in which antigen is linked to a solid particle or a surface. The immobilized antigen on the solid particles is allowed to capture antibody, followed by application of radiolabeled anti-Fc antibody specific for the Fc region of the captured antibody.

Solid Phase Technique A technique in which reagents are immobilized on a support and made insoluble.

Solid State Referring to a system in which current flow takes place entirely through solid materials such as a semiconductor.

Solium A trade name for chlordiazepoxide hydrochloride, an antianxiety agent.

Solone A trade name for prednisolone, an anti-inflammatory agent.

Solprin A trade name for aspirin, an antipyretic and analgesic agent that blocks the generation of pain impulses and inhibits the synthesis of prostaglandin.

Soluble Antigen Any antigen that is soluble in aqueous solution.

Soluble RNA Old term for tRNA.

Solu-Cortef A trade name for hydrocortisone sodium succinate, an anti-inflammatory agent.

Solu-medrol A trade name for methylprednisolone sodium succinate, an anti-inflammatory agent.

Solurex A trade name for dexamethasone sodium phosphate, an anti-inflammatory agent.

Solurex-LA A trade name for dexamethasone acetate, an anti-inflammatory agent.

Solute Any molecule that is dissolved in a liquid, the liquid is termed a solvent.

Solution A mixture of solute and solvent.

Solvent A liquid in which a solute is dissolved to produce a solution.

Solvent Demixing Separation of a solvent system into its constituent components.

Solvolysis The formation of a new substance from the interaction between solvent and solute.

Soma A trade name for carisoprodol, a skeletal muscle relaxant that reduces transmission of nerve impulses from spinal cord to skeletal muscle.

Soma Compound A trade name for a combination drug containing carisoprodol and aspirin, used as a skeletal muscle relaxant.

Somaclonal Variation Phenotypic variation, either genetic or epigenetic in origin, displayed among somaclones.

Somaclone Plants derived from cell culture of somatic cells.

Somatic Antigen 1. Antigen of a eukaryotic somatic cell. 2. Surface antigen of a prokaryotic cell (e.g., O-antigen).

Somatic Cell Any cell of a plant or animal other than a germ cell or sex cell.

Somatic Cell Genetics The genetics that deals with somatic cells.

Somatic Cell Hybrid The cell resulting from the fusion of somatic cells that differ genetically.

Somatic Cell Hybridization The *in vitro* fusion of somatic cells that differ genetically, e.g., fusion of somatic animal cells or plant protoplasts.

Somatic Crossing Over Crossing over during mitosis of a somatic cell leading to the segregation of heterozygous alleles.

Somatic Doubling Doubling of the diploid chromosome set.

Somatic Mutation A mutation occurring in the somatic cell that is not destined to become a germ cell.

Somatic Rearrangement Process whereby the DNA of somatic cells is rearranged, e.g., DNA rearrangement during the assembly of a complete V gene in the light or heavy chain of an immunoglobulin molecule.

Somatic Recombination See somatic rearrangement.

Somatocrinin A peptide (44 residues) with growth hormone releasing activity isolated from rat hypothalamus and some human pancreatic tumors (also known as somatotropin releasing hormone).

Somatoliberin A peptide growth hormone-releasing factor from hypothalmus that regulates somatotropin secretion.

Somatomedin A peptide hormone that is produced mainly in the liver and is released in response to somatotropin. It stimulates the growth of bone and muscle and also influences calcium, phosphate, carbohydrate, and lipid metabolism.

Somatoplasm Protoplasm of somatic cell.

Somatostatin Gastrointestinal and hypothalamic peptide hormone found in the gastric mucosa, pancreatic islets, nerves of the gastrointestinal tract, posterior pituitary, and in the central nervous system. It inhibits gastric secretion and release of somatotropin from hypothalamus.

Somatrem A growth hormone produced by DNA recombination technology that stimulates skeletal muscle and organ growth.

Somatotrophin Variant spelling of somatotropin.

Somatotropic Hormone Hormone secreted by the anterior lobe of the pituitary. It causes an increase in body growth and also affects carbohydrate and lipid metabolism.

Somatotropin Growth hormone that stimulates growth and synthesis of somatomedin.

Sominex A trade name for diphenhydramine hydrochloride, an antihistaminic agent that competes with histamine for H_1 receptors on effector cells.

Somogyi-Nelson Method A colorimetric method for determination of glucose in the blood based on the formation of blue color following treatment of deproteinized blood with copper sulfate and arsenomolybdate.

Somophyllin-CRT A trade name for theophyllin, a bronchodilator that inhibits phosphodiesterase and increases the concentration of cAMP.

Somophyllin-DF A trade name for aminophylline, a bronchodilator that inhibits phosphodiesterase and increases the concentration of cAMP.

Somophyllin-T A trade name for theophyllin, a bronchodilator that inhibits phosphodiesterase and increases the concentration of cAMP.

Sone 1. A trade name for prednisone, an anti-inflammatory agent. 2. A unit of loudness equal to a sound of 1000 cycles per second.

Sonication The disintegration or fragmentation of cells or molecules in a liquid medium by treatment with ultrasonic wave.

Sonicator A device that generates ultrasonic waves for disintegration and fragmentation of cells or molecules in a liquid medium.

Sonification See sonication.

Sonolysis Fragmentation of molecules or disintegration of cells by treatment with ultrasonic wave.

Soothe A trade name for tetrahydrozoline hydrochloride, an ophthalmic vasoconstrictor.

Sopamycetin A trade name for chloramphenicol, an antibiotic that binds to 50S ribosomal subunits, inhibiting bacterial protein synthesis.

Sophorolipid A glycolipid consisting of sophorose and hydroxy fatty acid.

Sophorose (mol wt 342) A disaccharide from *Sophora japonica* (Leguminosae).

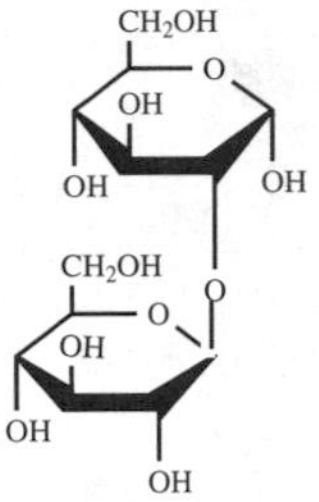

Soprodol A trade name for carisoprodol, a skeletal muscle relaxant that reduces the transmission of impulses from spinal cord to skeletal muscle.

Sorbent A substance capable of absorbing, adsorbing, or entrapping other substances.

Sorbic Acid (mol wt 112) A fungicide used as a food preservative.

$$CH_3CH{=}CHCH{=}CHCOOH$$

Sorbin A heat-stable peptide isolated from porcine intestine that increases water and sodium absorption in the intestine and gall bladder.

Sorbitol (mol wt 182) A sugar alcohol derived from glucose.

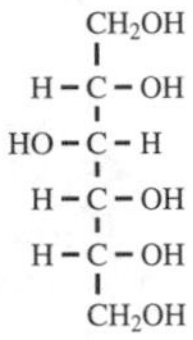

D-Sorbitol Dehydrogenase The enzyme that catalyzes the following reaction:

$$\text{D-Sorbitol} + \text{acceptor} \rightleftharpoons \text{L-sorbose} + \text{reduced acceptor}$$

Sorbitol 6-Phosphatase The enzyme that catalyzes the following reaction:

$$\text{Sorbitol 6-phosphate} + H_2O \rightleftharpoons \text{Sorbose} + \text{Pi}$$

Sorbitol 6-Phosphate Phosphohydrolase See sorbitol 6-phosphatase.

Sorbitrate A trade name for isosorbidedinitrate, an antianginal agent.

Sorbose (mol wt 180) A monosaccharide produced from sorbitol fermentation by *Acetobactor suboxydans.*

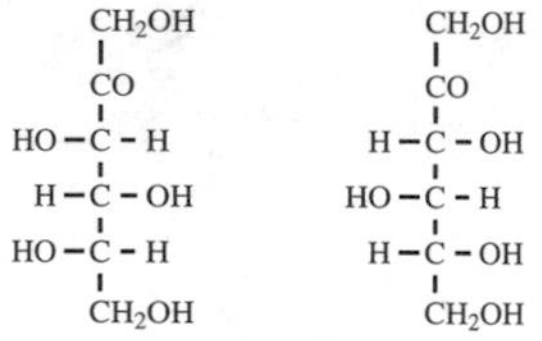

L-sorbose D-sorbose

Sorbose Dehydrogenase The enzyme that catalyzes the following reaction:

$$\text{L-Sorbose} + \text{acceptor} \rightleftharpoons \text{5-Dehydro-D-fructose} + \text{reduced acceptor}$$

Sorbose 5-Dehydrogenase ($NADP^+$) The enzyme that catalyzes the following reaction:

$$\text{L-Sorbose} + NADP^+ \rightleftharpoons \text{5-Dehydro-D-fructose} + NADPH$$

Sorbose Fermentation Fermentation of sorbitol for production of 2-ketohexose L-sorbose for manufacture of ascorbic acid by species of *Acetobacter* or *Gluconobacter.*

L-Sorbose Oxidase The enzyme that catalyzes the following reaction:

$$\text{L-Sorbitol} + O_2 \rightleftharpoons \text{5-Dehydro-D-fructose} + H_2O_2$$

Sorcin A calcium-binding protein encoded in multidrug-resistant cells.

Sore A localized spot on the body caused by tissue eruption or infection.

Sorivudine (mol wt 349) An antiviral agent.

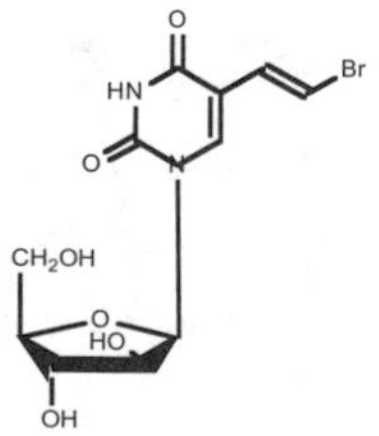

Sorption The process of sorbing or the state of being sorbed.

SOS Box A class of operator DNA sequences (about 20 nucleotide long) where the product of LexA gene binds to repress enzymes for the SOS repair system. Different SOS boxes have different affinities for LexA.

SOS Repair A repair mechanism induced as a result of the damage to DNA that involves the production of RecA protein, which cleaves LexA protein (repressor) resulting in activation of many genes involved in DNA repair.

SOS Response See SOS system.

SOS System A mechanism in *E. coli* for repairing damaged DNA in which the damaged DNA activates RecA protease that cleaves LexA protein (repressor for repair) leading to the expression of genes for enzymes involved in DNA repair.

Sotalol (mol wt 272) An antianginal, antiarrhythmic, and antihypertensive agent (a beta-adrenergic blocker).

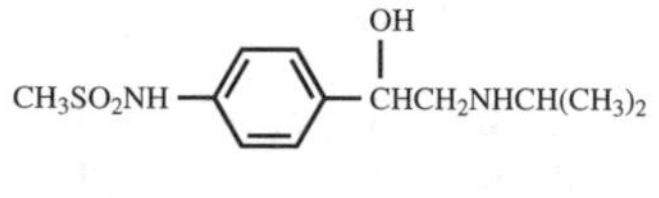

Soterenol (mol wt 288) A bronchodilator.

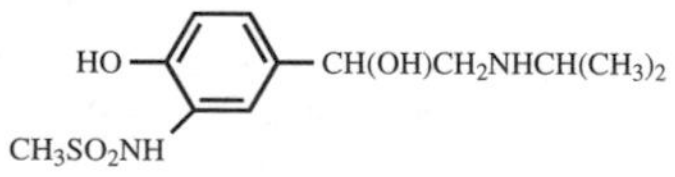

Sotradecol A trade name for sodium tetradecyl sulfate, a sclerosing agent.

Southern Blot Transfer of separated DNA fragments from electrophoretic gels to membrane filters such as nitrocellulose. The *blotted* DNA fragments are then detected by hybridization with radioactive DNA or RNA probes.

Southern Hybridization See Southern blotting.

SOX Abbreviation for sarcosine oxidase.

Soybean Trypsin Inhibitor A protein (180 amino acids) from soybean that forms an enzymatically inactive complex with trypsin.

sp Abbreviation for an unspecified species.

SP Abbreviation for 1. secretory piece; 2. split protein; 3. surface protein; 4. substrate P; 5. spasmolytic polypeptide; 6. surfactant protein.

S1P Abbreviation for sphingosine 1-phosphate.

S7P Abbreviation for sedoheptulose 7-phosphate.

SPA Abbreviation for scintillation proximity assay.

SP-A Abbreviation for surfactant protein-A.

[S]pA Abbreviation for adenosine 5'-thiophosphate.

Spacer Gel A small section of gel that lacks resolving power and is used to concentrate the samples being electrophoresed to the top of the resolving gel.

Spacer Sequence Transcribed sequence of nucleotides in an RNA molecule that is excised during RNA processing.

Span Nonionic surface active agents or detergents, esters of fatty acids and sorbitane.

Span 20 Referring to sorbitan monolaurate, a nonionic surface active agent.

Span 40 Referring to sorbitan monopalmitate, a nonionic surface active agent.

Span 60 Referring to sorbitan monostearate, a nonionic surface active agent.

Span 65 Referring to sorbitan tristearate, a nonionic surface active agent.

Span 80 Referring to sorbitan monoolearate, a nonionic surface active agent.

Span 85 Referring to sorbitan triolearate, a nonionic surface active agent.

SPAP Abbreviation for secreted placental alkaline phosphatase.

Sparassol (mol wt 196) An antibiotic substance produced by the fungus *Sparassis ramosa*.

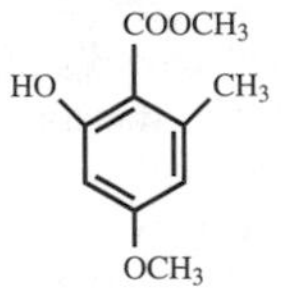

Sparfloxacin (mol wt 392) An antibacterial agent.

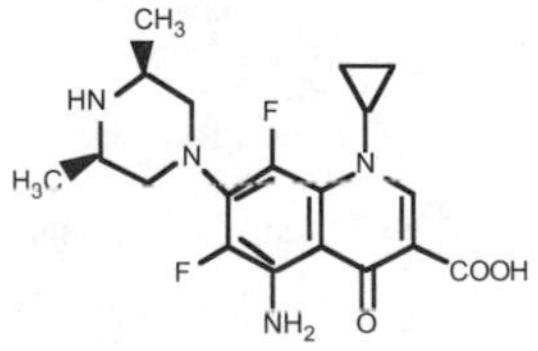

Sparine A trade name for promazine hydrochloride, an antipsychotic agent that blocks the postsynaptic dopamine receptors in the brain.

Sparsomycin (mol wt 361) A protein biosynthesis inhibitor with antibiotic and antitumor activity.

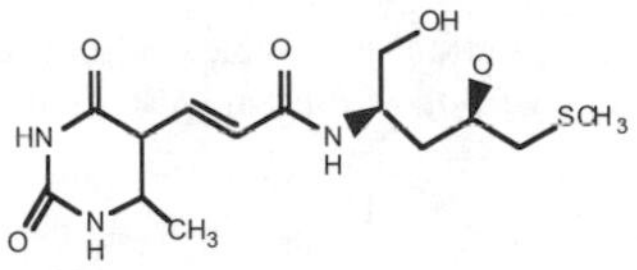

Sparteine (mol wt 234) An oxytocic agent from yellow and black lupin beans.

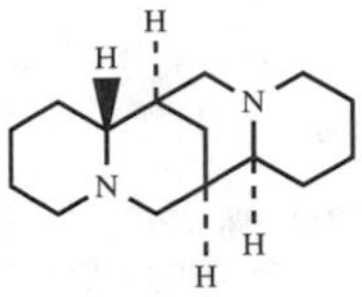

Spasm An involuntary and abnormal contraction of muscle or muscle fiber.

Spasmoban A trade name for dicyclomine hydrochloride, an anticholinergic agent.

Spasmolytic Capable of relieving spasm or convulsion.

Spasmolytol (mol wt 395) An antispasmodic agent.

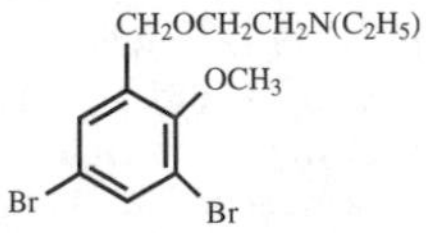

SpaXI (SphI) A restriction endonuclease from *Streptomyces phaeochromogenes* with the following specificity:

↓
5′..........GCATGC..........3′
3′..........CGTACG..........5′
↑

SP-B Abbreviation for surfactant protein-B.

SPBI Abbreviation for serum protein-bound iodine.

SP C Abbreviation for 1. surfactant protein C; 2. sphingosylphosphorylcholine.

***spc* Operon** An operon in *E. coli* that encodes genes for ribosomal proteins and genes involved in protein secretion.

SPCA Abbreviation for serum prothrombin conversion accelerator.

SP-D Abbreviation for surfactant protein-D.

SPE Abbreviation for serum protein electrophoresis.

SpeI A restriction endonuclease from *Sphaerotilus natans* with the following specificity:

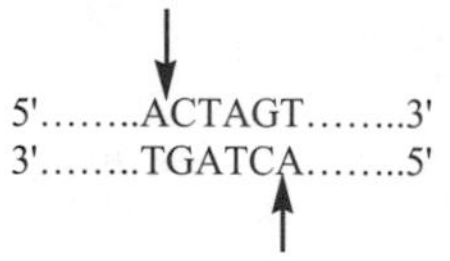

Specialized Transducing Phage A phage that transduces a unique sequence of bacterial DNA.

Specialized Transduction Transduction mediated by a temperate phage that transduces only genes adjacent to the sites where prophage integrates.

Species Category of biological classification ranking immediately below genus.

Specific Acid-base Catalysis The catalysis in which the catalysts are free protons (e.g., H^+, H_3O^+) and free hydroxyl ions. This type of catalysis is not affected by the other acidic or basic species present in the solution.

Specific Activity The number of activity units per unit of mass, e.g., number of enzyme units per milligram of protein or the number of microcuries per micromole of radioactive substance.

Specific Extinction Coefficient The extinction coefficient obtained when the concentration of the solution is expressed in mg/ml or g/L.

Specific Gravity The ratio of the density of a substance to the density of a reference substance, e.g., H_2O (density of water = 1).

Specific Growth Rate The number of grams of biomass formed per gram of biomas per hour.

Specific Heat The amount of energy that must be absorbed by one gram of a substance to raise its temperature by one degree centigrade. By convention, water is assigned a specific heat of one.

Specific Immunity An immunity specific for a given antigen.

Specific Rotation The observed rotation (in degrees) of the plane of polarized light by an optically active substance at 25° C with specified concentration and light path.

Specific Volume The volume of one gram of a substance (ml/g).

Specificity The ability of an enzyme or a receptor to discriminate among competing substrates or ligands.

SPECT Abbreviation for single photon emission computer tomography.

Spectazole A trade name for econazole nitrate, a local anti-infective agent that alters the permeability of the fungal cell wall.

Spectinomycin (mol wt 332) An antibiotic isolated from *Streptomyces spectabilis* that binds to ribosomes, preventing transfer of peptidyl-tRNA from the A site to the P site.

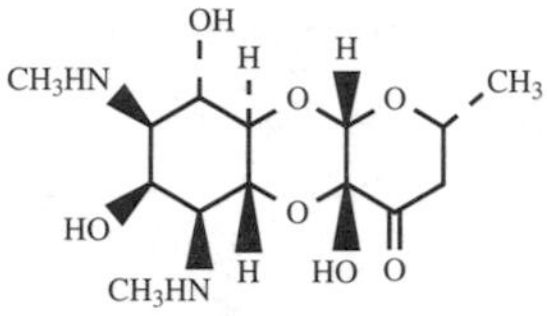

Spectrin A protein component of the erythrocyte membrane.

Spectrobid A trade name for becampicillin hydrochloride, an aminopenicillin that inhibits bacterial cell wall synthesis.

Spectrofluorometer An instrument used for measurement of fluorescence emitted by compounds.

Spectrophotometer An instrument used for characterization of the light-absorbing property of a substance (e.g., determination of absorption spectrum) and for determination of the concentration of a substance at a given wavelength.

Spectrophotometry Analysis of a substance by spectrophotometer.

Spectroscope An instrument for the separation of and examination of optical spectra.

Spectrum A graphic representation showing the extent to which light is transmitted or absorbed as a function of wavelength or a graph depicting a distribution of intensities as a function of the energy of radiation.

S-Peptide A small peptide from the N-terminal portion of ribonuclease consisting of 20 amino acid residues derived from subtilisin cleavage of ribonuclease.

Sperm Haploid gamete produced by the male.

Sperm Cell A male gamete or a male germ cell.

Spermact The sperm-activating peptide from the gelly coat of the egg of sea urchins.

Spermadhesin A protein found on the sperm surface that mediates sperm binding to the zona pellucida.

Spermatheca A sac structure in the female reproductive organs for the reception and storage of sperm.

Spermatids Haploid cells resulting from a second meiotic division in spermatogenesis that differentiate into mature spermatozoa.

Spermatocyte A cell that undergoes meiosis and gives rise to sperm.

Spermatogenesis The process of the formation of the male gamete, including meiosis and transformation of four spermatides into spermatozoa or sperm.

Spermatogonia Plural of spermatogonium.

Spermatogonium Undifferentiated germ cell that gives rise to primary spermatocytes.

Spermatozoa Plural of spermatozoon.

Spermatozoan Haploid male gamete produced by meiosis.

Spermatozoon The motile male gamete.

Spermaturia The presence of semen in the urine.

Spermicidal Any substance that kills sperm.

Spermidine (mol wt 145) A biogenic amine formed from putrescine.

$$NH_2(CH_2)_4NH(CH_2)_3NH$$

Spermidine Dehydrogenase The enzyme that catalyzes the following reaction:

Spermidine + acceptor + H_2O

⇅

1,3-Diamino-propane + 4-aminobutanal + reduced acceptor

Spermidine Synthetase The enzyme that catalyzes the following reaction:

S-Adenosylmethioninamine + putrescine

⇅

5′-Methylthioadenosine + spermidine

Spermine (mol wt 202) A biogenic polyamine formed from spermidine that occurs in all tissues of eukaryotes but not in prokaryotes.

$$NH_2(CH_2)_4NH — (CH_2)_4NH(CH_2)_3 — NH_2$$

Spermine Synthetase The enzyme that catalyzes the following reaction:

S-Adenosylmethioninamine + spermidine

⇅

5′-Methylthioadenosine + spermine

Sph Abbreviation for sphingosine.

SphI A restriction endonuclease from *Streptomyces phaeochromogenes* with the following specificity:

↓
5′..........GCATGC..........3′
3′..........CGTACG..........5′
↑

Sphaeroplast See spheroplast.

Sphaerosome Variant spelling of spherosome.

Sphaerotilus A genus of Gram-negative, rod-shaped, obligately aerobic, asporogenous bacteria.

S-Phase The synthetic phase in the eukaryotic cell cycle during which DNA is synthesized.

Spherophysine (mol wt 198) A ganglio blocking agent.

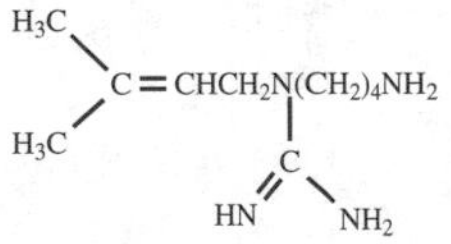

Spheroplast Gram-negative bacterial cell in which the cell wall has been partially removed by an enzymatic or chemical treatment.

Spherosome A lysosomelike, lipid-storage structure derived from endoplastic reticulum in plants.

Spherulin An antigen derived from the spherules of *Coccidioides immitis* used in skin testing for delayed hypersensitivity.

Sphinganine (mol wt 302) An intermediate in the biosynthesis of ceramide.

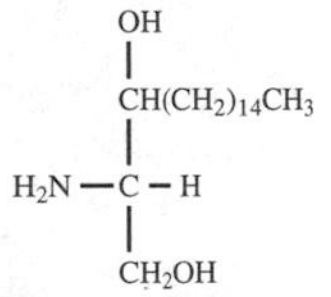

Sphinganine Kinase The enzyme that catalyzes the following reaction:

ATP + sphinganine

⇅

ADP + sphinganine 1-phosphate

Sphinganine 1-Phosphate Aldolase The enzyme that catalyzes the following reaction:

Sphinganine 1-phosphate

⇅

Phosphoethanolamine + palmitaldehyde

Sphingoglycolipids Sphingolipid that contains a carbohydrate moiety, e.g., ganglioside, cerebroside, or globoside.

Sphingolipidosis Genetic disorders characterized by the accumulation of various sphingolipids due to a deficiency of lysosomal enzymes for the degradation of sphingolipid.

Sphingolipids A class of lipids derived from sphingosine with long-chain fatty acid and polar alcohol attached. Sphingolipids are major components of membrane lipid.

Sphingolipid Storage Diseases Genetic disorders caused by the deficiency of enzymes for degradation of sphinoglycolipids, e.g., Tay-Sachs disease.

Sphingomyelin A group of sphingosine phosphatides found in the myelin sheath of nerves.

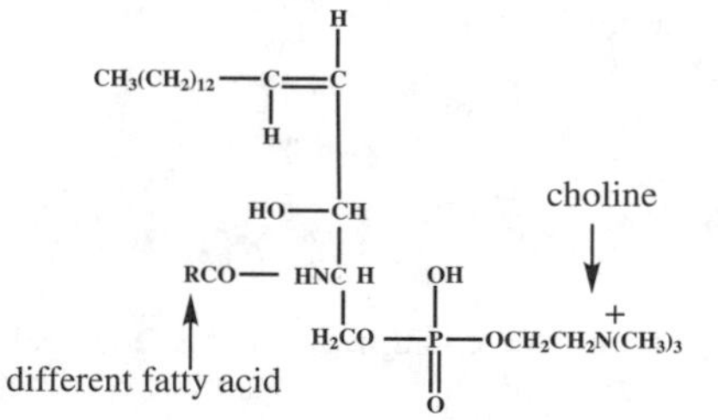

Sphingomyelin Choline Phosphohydrolase The systematic name for sphingomyelin phosphodiesterase.

Sphingomyelin Phosphodiesterase The enzyme that catalyzes the following reaction:

Sphingomyelin + H_2O
⇅
N-acylsphingosine + choline phosphate

Sphingomyelin Phosphodiesterase D The enzyme that catalyzes the following reaction:

Sphingomyelin + H_2O
⇅
Ceramide phosphate + choline

Sphingomyelinase The enzyme that catalyzes the following reaction:

Sphingomyelin ⇌ ceramide + choline

Sphingomyelinosis A disorder characterized by the accumulation of sphingomyelins caused by a deficiency in the enzyme sphingomyelinase.

Sphingophospholipids Phospholipids derived from sphingosine or sphingosine-related substances.

Sphingosine (mol wt 299) An amino alcohol that serves as backbone for sphingolipids.

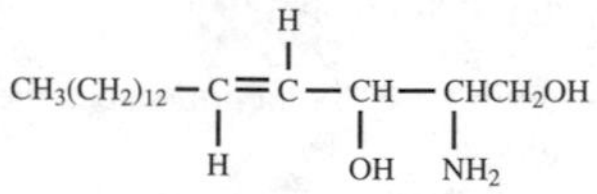

Sphingosine *N*-Acyltransferase The enzyme that catalyzes the following reaction:

Acyl-CoA + sphingosine
⇅
CoA + *N*-acylsphingosine

Sphingosine Choline Phosphotransferase The enzyme that catalyzes the following reaction:

CDP-choline + sphingosine
⇅
CMP + sphingosyl-phosphocholine

Sphingosine β-Galactosyltransferase The enzyme that catalyzes the following reaction:

UDP-galactose + sphingosine
⇅
UDP + psychosine

Sph-1P Abbreviation for sphingosine 1-phosphate.

SP-HPLC Abbreviation for straight phase HPLC (high performance liquid chromatography).

Sphygmomanometer An instrument used for measuring blood pressure.

Sphygmometer A device for measuring the strength of the pulse beat.

Sphygmus The pulse.

SPI Abbreviation for serum perceptible iodine.

Spikes Surface projections of varying lengths on the envelope of enveloped viruses, e.g., influenza virus.

Spin Label A free radical or a paramagnetic probe used in electron spin resonance spectroscopy for the study of molecular structure and biological function of bioactive molecules.

Spin Labeling The tagging of a part of a macromolecule with a group that has an unpaired electron that can be detected by magnetic measurements.

Spin Probe Referring to a protein that has a noncovalenty bound spin label.

Spinal Meningitis Inflammation of menings of the spinal cord.

Spindle Apparatus An array of microtubules that align and separate chromosomes in cell division.

Spindle Fiber An individual microtubular fiber associated with the mitotic spindle (spindle apparatus).

Spiperone (mol wt 395) An antipsychotic agent.

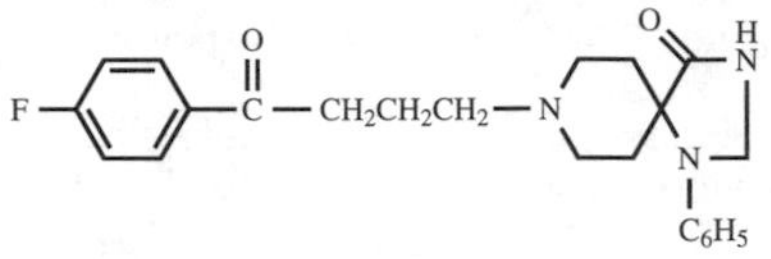

Spiramycin Antibiotics produced by *Streptomyces ambofacien.*

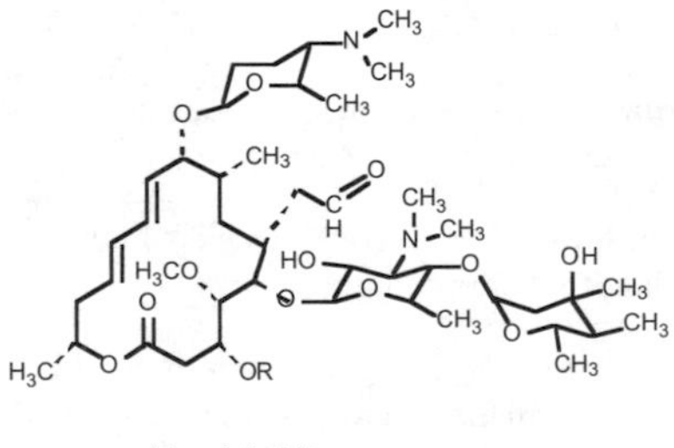

Spiramycin I: R=H
Spiramycin II: R=$COCH_3$
Spiramycin III: R=$COCH_2CH_3$

Spirilene (mol wt 394) An antipsychotic agent.

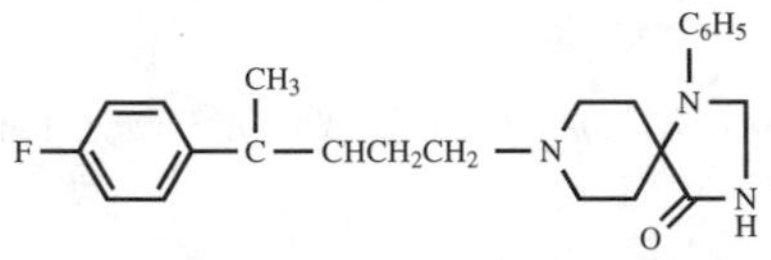

Spirilla Spiral-shaped bacteria (plural of spirilum).

Spirillospora A genus of bacteria (order of Actinomycetales).

Spirillum Spiral-shaped bacteria.

Spirillum A genus of Gram-negative, asporogenous bacteria.

Spirochaeta A genus of Gam-negative bacteria of the family Spirochaetaceae.

Spirochete Referring to clongated, spiral-shaped bacteria (members of Spirochaetaceae).

Spirochetemia The presence of spirochetes in the circulating blood.

Spirocheticide Any substance that kills spirochetes.

Spirochetosis Infection or disease caused by spirochetes.

Spirogermanium (mol wt 341) An antineoplastic agent.

Spirometer An instrument for measurement of air entering and leaving the lungs.

Spironazide A trade name for a combination drug containing spironolactone and hydrochlorothiazide, used as a diuretic agent.

Spironolactone (mol wt 417) A potassium-sparing diuretic agent that increases excretion of sodium and water.

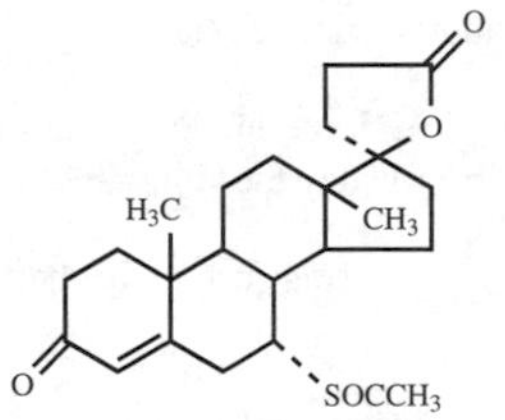

Spiroplasma A genus of facultative anaerobic bacteria (family Spiroplasmataceae).

Spiroplasmaviruses Bacteriophages that infect species of *Spiroplasma.*

Spirosoma A genus of yellow-pigmented bacteria (family Spirosomaceae).

Spirotone A trade name for spironolactone, a potassium-sparing diuretic agent that increases excretion of sodium and water.

Spirozide A trade name for a combination drug containing spironolactone and hydrochlorothiazide, used as a diuretic agent.

Spirulina A genus of filamentous cyanobacteria.

SPISA Abbreviation for solid phase immunosorbent assay.

Spizofurone (mol wt 202) An antiulcerative agent.

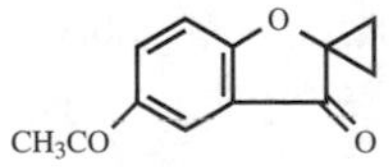

SplI A restriction endonuclease from *Spirulina platensis* with the following specificity:

5'........C↓GTACG........3'
3'........GCATG↑C........5'

SPLA$_2$ Abbreviation for secretory phospholipase-A$_2$.

Spleen Exonuclease The enzyme that catalyzes the exonucleolytic cleavage in the 5′- to 3′-direction to yield 3′-phosphomononucleotide.

Splenectomy Surgical removal of the spleen.

Splenocyte Mononuclear cells from the spleen.

Splenomegaly The abnormal enlargement of the spleen.

Splice Junction Segments containing a few nucleotides that reside at the ends of introns and function in excision and splicing reactions during the processing of transcripts.

Spliceosome A large nucleoprotein complex of eukaryotes involved in the splicing of pre-mRNAs.

Splice Site Referring to the site between intron and exon.

Splicing 1. Cutting and resealing of RNA transcript by precise breakage of phosphodiester bonds at the 5′ and 3′ splice sites (exon–intron junctions) to form active mRNA. 2. Cutting and resealing of DNA from different sources to form a recombinant DNA molecule.

Split Gene A gene whose nucleotide sequence is divided into exons and introns. After transcription, the introns are excised from the primary RNA and exons are joined together to form a functional mRNA.

SPO1 A bacteriophage.

Spoke Protein A protein associated with the filamentous structure of kinetochore microtubules in the mitotic spindle.

Spongioblast Embryonic cells of the supportive tissue found in the developing nervous system.

Spongiocyte The lipid-rich cells of the supportive tissue with a spongy appearance.

Spontaneous Mutation Mutation that occurs naturally in the absence of obvious external mutagen.

Spontaneous Reaction An exergonic reaction that is characterized by a negative free energy change and is therefore capable of proceeding spontaneously without an input of energy.

Spontaneous Transformation Transformation of a cell culture without deliberate addition of transforming agent.

Sporangium A specialized structure that houses spores.

Spore Any asexual reproductive cell capable of developing into an adult organism without gametic fusion.

Spore Mother Cell A diploid cell that undergoes meiosis and produces four haploid cells or four haploid nuclei.

Sporicide Any agent that kills spores.

Sporocytophaga A genus of gliding bacteria of the Cytophagales.

Sporogenesis The production of spores.

Sporolactobacillus A genus of Gram-positive, chemoorganotrophic, catalase-negative, endospore-forming bacteria.

Sporomusa A genus of Gram-negative, anaerobic, endospore-forming bacteria.

Sporophore Specialized mycelial branch upon which spores are produced.

Sporophyte Diploid generation in the life cycle of an organism that alternates between haploid and diploid forms. Sporophyte produces spores by meiosis.

Sporosarcina A genus of Gram-positive, chemoorganotrophic, obligately aerobic, endospore-forming bacteria.

Sporospirillum A genus of motile, rigid, helical bacteria.

Sporotrichosis Infection or disease caused by a fungus of the genus *Sporotrichum* and characterized by nodules, abscesses in the superficial lymph nodes, skin, and subcutaneous tissue.

Sporotrichum A genus of fungi of Hyphomycetes and causal agent of sporotrichosis.

Sporulation Production of spores.

Spot Desmosome The point of a tight adhesion between plasma membranes of adjacent cells.

[S]ppA Abbreviation for adenosine 5'-β-thiodiphosphate.

[S]pppA Abbreviation for adenosine 5'-β-thiotriphosphate.

Spreading Factors Products of certain microbial parasites that facilitate, or promote, their penetration of host tissues and spreading of infection, e.g., hyaluronidase.

SPRIA Abbreviation for solid phase radioimmuno-assay.

S-protein The large peptide resulting from subtilisin cleavage of ribonuclease, consisting of amino acid residues 21 through 124.

Sprx 105 A trade name for phendimetrazine tartrate, a cerebral stimulant that promotes transmission of nerve impulses by releasing stored norepinephrine from the nerve terminals in the brain.

SPS A trade name for sodium polystyrene sulfonate, used in the intestine to remove potassium.

SP-Sephadex Abbreviation for sulfopropyl Sephadex, a cation exchanger for ion exchange chromatography.

SPT Abbreviation for serine-palmitoyl transferase.

Sputum Expectorated matter consisting of saliva mixed with buccal and nasal mucus, cellular debris, and other substances.

SPV Abbreviation for Shope papilloma virus.

Squalene (mol wt 411) A precursor for synthesis of sterols that is found in large quantity in shark liver oil.

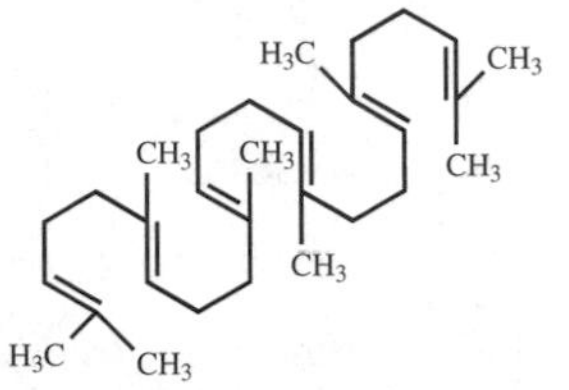

Squalene Epoxidase The enzyme that catalyzes the following reaction:

$$\text{Squalene} + O_2 + \text{NADPH} \rightleftharpoons \text{2,3-Oxidosqualene} + H_2O + \text{NADP}^+$$

Squalene Monooxygenase See squalene expoxidase.

Squama 1. A thin plate of bone. 2. A scalelike structure.

Squamous Cell Carcinoma A carcinoma arising from squamous cells.

Squamous Cells Flat, scaly cells derived from squamous epithelium.

Squamous Epithelium An epithelium in which the cells are flattened.

Square Bacteria Flat (often square), halophilic bacteria found in hypersaline environments.

Squiggle The symbol of ~ which is designated as a high energy bond, e.g., the bond involving phosphate groups.

Sr Symbol for the chemical element strontium.

SR Abbreviation for 1. sarcoplasmic reticulum; 2. sedimentation rate; 3. specific radioactivity.

^{90}Sr Abbreviation for strontium 90.

SRBC Abbreviation for sheep red blood cells, used for hemaglutination and complement fixation tests.

src An oncogen originally identified as transforminig determinant for Rous sarcoma virus (*v-src*). The cellular counter partner of *v-src* is designated as *c-src*. The gene product of v-src is a tyrosine-specific protein kinase.

SRE Abbreviation for 1. serum response element; 2. sterol-regulatory element.

SRF Abbreviation for 1. serum response factor; 2. somatotropin releasing factor.

SrfI A restriction endonuclease from *Streptomyces* species with the following specificity:

```
          ↓
5'........GCCC GGGC........3'
3'........CGGG CCCG........5'
              ↑
```

SRH Abbreviation for somatotropin releasing hormone.

SRIF Abbreviation for somatotropin release-inhibiting factor.

sRNA Abbreviation for soluble RNA.

SRP Abbreviation for signal recognition particle, a particle that is involved in the translocation of proteins across the endoplasmic reticulum.

SR-Protein Abbreviation for serine-rich protein.

SRS Abbreviation for slow-reacting substance (referring to leukotriene released from the degranu lation of mast cells and basophils).

SRS-A Abbreviation for slow-reacting substance of anaphylaxis, a leukotriene-mediated type-I hypersensitivity caused by degranulation of mast cells and basophils.

SRV Abbreviation for simian retrovirus.

sRyr Abbreviation for solublized RyR (ryanodine receptor).

ss Abbreviation for single stranded.

SS Abbreviation for 1. single-stranded; 2. starch synthetase; 3. supersaturated.

S-S Symbol for disulfide bond in protein resulting from linkage between two cysteine molecules.

SSII Abbreviation for starch synthetase II.

SS Agar Abbreviation for *Salmonella-Shigella* agar, an agar medium containing beef extract, lactose, thiosulfate, ferric citrate, bile salts, neutral red, and brilliant green, used for differentiation of *Salmonella* from *Shigella*.

SSA Abbreviation for 1. salicyl-salicylic acid; 2. skin sensitizing antibody; 3. succinic semialdehyde.

SSAT Abbreviation for 1. salicyl sulfonic acid test; 2. spermidine/spermine N-acetyltransferase.

SSB Abbreviation for single-strand breaks.

SSBP Abbreviation for single-stranded binding protein.

SSC Abbreviation for standard saline citrate solution (0.15M saline containing 0.015M sodium citrate, pH 7).

sscDNA Abbreviation for single-stranded circular DNA.

ssDNA Abbreviation for single-stranded DNA.

ssDNA Phage Any bacteriophage that contains single-stranded DNA as genetic material, e.g., M13 phage.

Sse9I A restriction endonuclease from *Sporosarcina* species 9 with the following specificity:

5'........↓AATT........3'
3'........TTAA↑........5'

Sse8387I A restriction endonuclease from *Streptomyces* species with the following specificity:

5'........CCTGCA↓GG........3'
3'........GG↑ACGTCC........5'

SseBI A restriction endonuclease from *Streptomyces* species with the following specificity:

5'........AGG↓CCT........3'
3'........TCC↑GGA........5'

S-Shaped Curve Synonym for sigmoid curve.

SsoI (EcoRI) A restriction endonuclease from *Shigella sonnei* 47 with the same specificity as EcoRI.

SsoII (ScrFI) A restriction endonuclease from *Shigella sonnei* 47 with the following specificity:

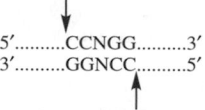

5′..........↓CCNGG..........3′
3′..........GGNCC↑..........5′

ssp Abbreviation for subspecies.

SspI A restriction endonuclease from *Sphaerotilus natans* with the following specificity:

5′..........AAT↓ATT..........3′
3′..........TTA↑TAA..........5′

Ssp1I (AsuII) A restriction endonuclease from *Streptomyces* species RFL1 with the following specificity:

5′..........TT↓CGAA..........3′
3′..........AAGC↑TT..........5′

SspII (AsuII) A restriction endonuclease from *Streptomyces* species with the following specificity:

5′..........TT↓CGAA..........3′
3′..........AAGC↑TT..........5′

Ssp2I (CauII) A restriction endonuclease from *Streptomyces* species RFL2 with the following specificity:

5′..........CC↓(C/G)GG..........3′
3′..........GG(G/C)↑CC..........5′

Ssp4I (XhoI) A restriction endonuclease from *Streptomyces* species with the same specificity as XhoI.

SspBI A restriction endonuclease from *Streptomyces* species with the following specificity:

5'........T↓GTACA........3'
3'........ACATG↑T........5'

SSR Abbreviation for succinic semialdehyde reductase.

SSRE Abbreviation for shear stress responsive element.

ssRNA Abbreviation for single-stranded RNA.

SstI (SacI) A restriction endonuclease from *Streptomyces stanford* with the following specificity:

5′..........GAGCTC..........3′
3′..........CTCGAG..........5′

SstII (SacII) A restriction endonuclease from *Streptomyces stanford* with the following specificity:

5′..........CCGCGG..........3′
3′..........GGCGCC..........5′

SstIV (BclI) A restriction endonuclease from *Streptomyces stanford* with the following specificity:

5′..........TGATCA..........3′
3′..........ACTAGT..........5′

S-strain Abbreviation for smooth strain, a virulent bacterial strain that produces colonies with a smooth appearance, e.g., S strain of *Streptococcus pneumoniae*.

SSV Abbreviation for small synaptic vesicle.

ST Abbreviation for sulphotransferase.

St. Louis Encephalitis An acute viral encephalitis caused by a virus of the family Flaviviridae (ssRNA-containing virus).

STA Abbreviation for serum thrombotic accelerator.

Stab Culture A bacterial culture made by a deep inoculation of a solid agar medium.

Stabilate A population of microorganisms which are maintained viable by freezing or freeze-drying.

Stabilizer Any substance that keeps a compound, mixture of compounds, or solution stable without changing the chemical nature.

Stable Factor Blood clotting factor VII (also known as proconvertin, autoprothrombin I, and cothromboplastin).

Stable Isotope A nonradioactive isotope of a chemical element.

Stachybotryotoxicosis A mycotoxicosis of animals and humans caused by the toxin of *Stachybotrys atra* (a mold).

Stachyose (mol wt 667) A nonreducing tetrasaccharide consisting of glucose, galactose, and fructose.

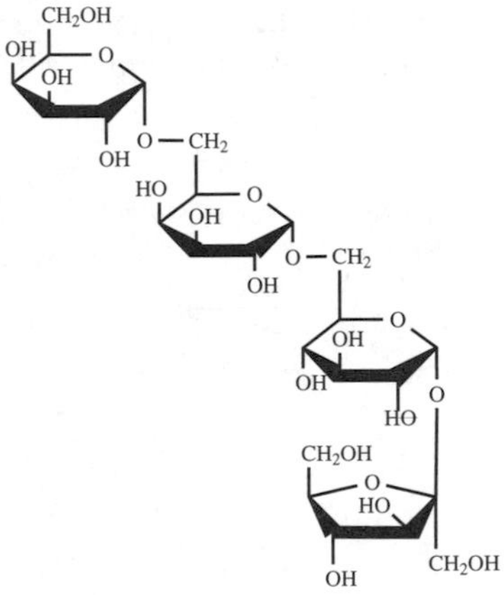

Stachyose Synthetase The enzyme that catalyzes the following reaction:

α-D-galactosyl-*myo*-inositol + raffinose
⇌
myo-Inositol + stachyose

Stacked Thylakoids Thylakoid membranes stacked on each other like a pile of coins (also known as grana).

Stacking Energy The free energy of stacking interaction between two base pairs in a double helix structure of DNA.

Stacking Gel The portion of gel used for concentrating of ionic components into a narrow region before entering the separation gel during gel electrophoresis.

Stacking Interaction The hydrophobic interaction between two base pairs that are in parallel planes in the interior of the double-stranded DNA.

Stadol A trade name for butorphanol tartrate, an analgesic agent that binds to opiate receptors in the CNS, altering both the perception and emotional response to pain.

Stage Micrometer A device on the microscope slide for measurement of a specimen.

Staggered Cut Breaking two strands of DNA at different positions.

Stain A dye used to color microorganisms, cells, or tissues as an aid to visual inspection.

Stain All (mol wt 560) A cationic carbocyanine dye that stains RNA (bluish-purple), DNA (blue), protein (red) and polysaccharide (blue to purple).

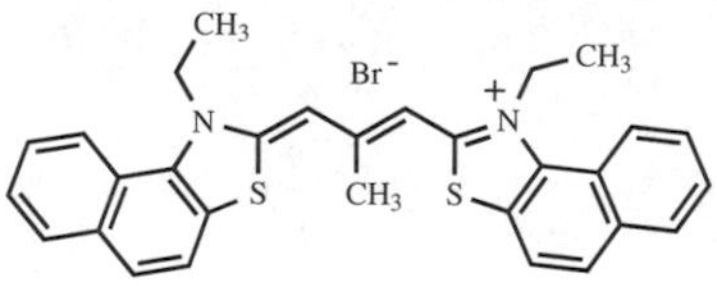

Stallimycin (mol wt 482) An antiviral and antitumor antibiotic produced by *Streptomyces distallicus*.

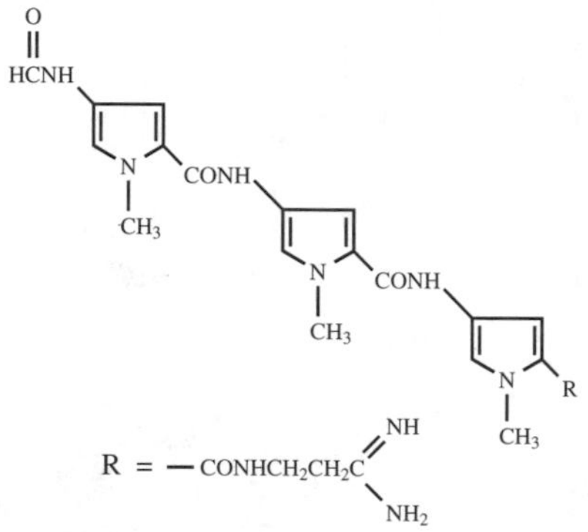

Standard Curve A graphic representation obtained by plotting absorbances at a given wavelength as a function of different concentrations of a light-absorbing compound, used for verification of Beer's law and determination of unknown concentration.

Standard Deviation A measure of variability in a population of items.

Standard Error A measure of variation of means.

Standard Electrode Potential (E^o) The electrode potential of a half reaction occurring at 25° C, 1 atm pressure, concentrations of reactant and product of 1 *M* and pH 0 (designated as E^o) or pH 7 (designated as $E^{o\prime}$).

Standard Free Energy Change (ΔG^0) The free energy change for a reaction occurring under a set of standard conditions, e.g., temperature of 25° C, 1 atm pressure, concentration of reactant and product at 1 *M* and pH 0 (designated as ΔG^0) or pH 7 (designated as $\Delta G^{0\prime}$).

Standard Hydrogen Half Reaction The half cell of the hydrogen ion (H^+) electrode in which H^+ at pH 0, 25° C and 1 atm pressure at equilibrium with H_2. The redox potential of the hydrogen half reaction is arbitrarily assigned as 0 volt.

Standard Oxidation Potential Referring to the standard electrode potential for an oxidation half reaction.

Standard Potential See standard oxidation reduction potential.

Standard Reaction Condition Referring to the reaction condition at 25° C, 1 atm pressure, concentrations of reactant and product at 1 *M* and pH 7.

Standard Redox Potential The ability of a redox couple to donate or to accept electrons, determined by measuring the voltage generated when a redox couple is connected with a hydrogen half cell electrode under the standard reaction conditions.

Standard Reduction Potential (E_0) Referring to the standard electrode potential for a reduction half reaction.

Standard Saline Citrate A saline solution (0.15M NaCl) containing 0.015*M* sodium citrate of pH 7.

Standard Solution A solution with known concentration used as a reagent in a chemical analysis.

Standard State See standard reaction condition.

Standard Temperature and Pressure An arbitrary reference conditions of °C temperature, a pressure of 1 atm (760 mm Hg), and a specified pH.

Stanolone (mol wt 290) An androgen used in the treatment of breast cancer.

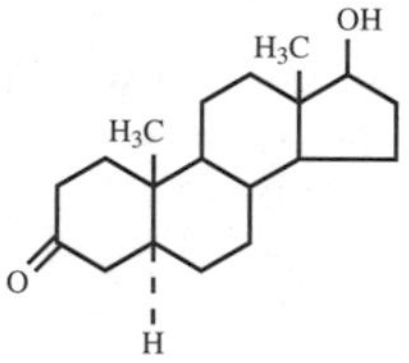

Stanozolol (mol wt 328) An anabolic steroid that promotes tissue-building processes.

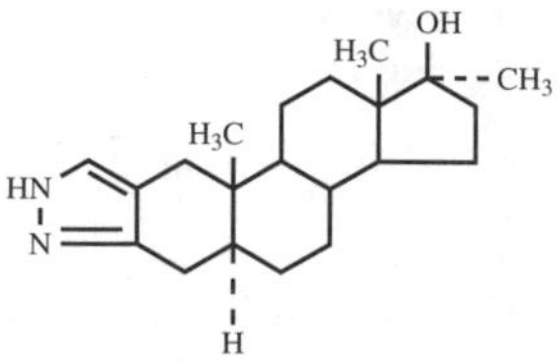

Staphcillin A trade name for methicillin sodium, a penicillinase-resistant antibiotic that inhibits bacterial cell wall synthesis.

Staphylocoagulase Coagulase produced by pathogenic *Staphylococci*.

Staphylococcal Protein A A protein derived from *Staphylococcus aureus* cell wall that binds IgG and stimulates B-cell activation.

Staphylococcemia The presence of *Staphylococci* in the blood.

Staphylococcins Referring to bacteriocins produced by *Staphylococci.*

Staphylococcosis Infection or disease caused by *Staphylococci.*

Staphylococcus A genus of Gram-positive, facultatively anaerobic, chemoorganotrophic, asporogenous, nonmotile, catalase-positive, commonly halotolerant bacteria. *Staphylococci* can be divided into coagulase-positive and coagulase-negative groups.

Staphylokinase A protein released by many strains of coagulative-positive *Staphylococcus* that activates profibrinolysin to fibrinolysin, causing fibrinolysis.

Staphylolysin A hemolysin produced by *Staphylococci.*

Staphyloslide A test for detection of the ability of strains of *Staphylococcus aureus* to clump fibrinogen-coated erythrocytes.

Staphylothermus A genus of heterotrophic, coccoid archaebacteria.

Starch Storage polysaccharide in plants, consisting of α-(1,4)-D-glucose repeating subunits and α-(1,6)-glucosidic linkages.

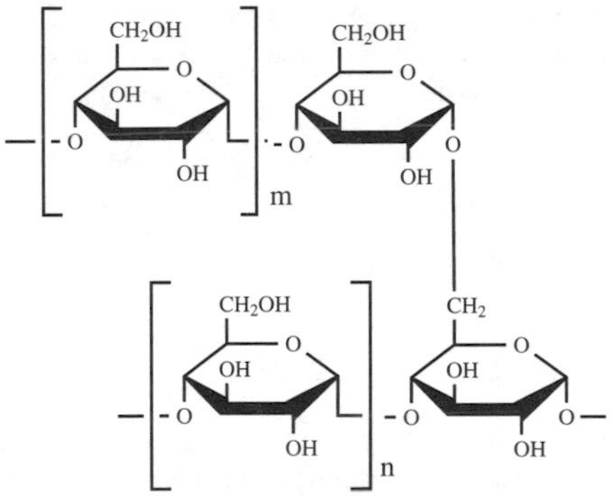

Starch Gel Electrophoresis A zonal electrophoresis techniques employing partially hydrolyzed starch as supporting medium.

Starch Granule The storage granule of starch occurring in the cytoplasm of plant cells, e.g., starch granule in the potato tuber cell.

Starch Phosphorylase The enzyme that catalyzes the following reaction:

$$(\alpha\text{-}(1,4)\text{-D-glucosyl})_n + \text{Pi} \rightleftharpoons (\alpha\text{-}(1,4)\text{-D-glucosyl})_{n-1} + \alpha\text{-D-glucose-phosphate}$$

Starch Synthetase The enzyme that catalyzes the following reaction:

$$(\alpha\text{-}1,4\text{-D-glucosyl})_n + \text{ADP-glucose} \rightleftharpoons \text{ADP} + (1,4\text{-}\alpha\text{-D-glucose})_{n+1}$$

Starria A genus of filamentous cyanobacteria.

Stasis 1. A disorder in which the flow of bodily fluid through a vessel is slowed or stopped. 2. A period of equilibrium during which change appears to be absent.

Stasis Dermatitis Inflammation of skin caused by the slowing or stoppage of the flow of blood to the affected area.

STAT Abbreviation for signal transducer and activator of transcription.

Statex A trade name for morphine sulfate, an analgesic agent that binds with opiate receptors in the CNS, altering both perception and emotional response to pain.

Staticin A trade name for erythromycin, an antibiotic that inhibits bacterial protein synthesis.

Stationary Phase 1. The stage of growth of a culture in which the number of viable cells remains constant. 2. The supporting material used in chromatography that selectively retards the flow of the sample.

Stationary State A reversible reaction in which the rate of the forward reaction equals the rate of the backward reaction (also known as state of equilibrium).

Statobex A trade name for phendimetrazine tartrate, a cerebral stimulant that promotes transmission of impulses by releasing stored norepinephrine from the nerve terminals in the brain.

Statospore A resting spore or cell.

Statrol A trade name for a combination drug containing neomycin sulfate and polymyxin B sulfate, used as an antibacterial agent.

Stavudine (mol wt 224) An antiviral agent that inhibits retrovirus replication.

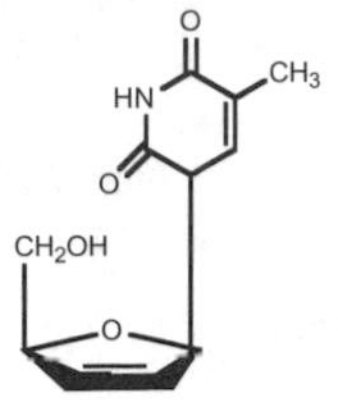

STCLV Abbreviation for simian T-cell leukemia virus.

STD Abbreviation for sexually transmitted disease.

Steady State The nonequilibrium condition of a system or reaction in which all components remain at constant concentration.

Steady State Kinetics The kinetics of an enzymatic reaction that proceeds under the steady state conditions.

STEAM Abbreviation for a combination drug containing streptonigrin, thioguanine, endoxan, actinomycin D, and mitomycin C.

Steapsin A lipase present in the pancreatic juice.

Stearic Acid (mol wt 284) An 18-carbon saturated fatty acid.

$$CH_3(CH_2)_{16}COOH$$

Stearin (mol wt 822) A glyceryltristearate (a triglyceride that consists of three stearic acids).

$$\begin{array}{l} H_2-C-O-CO(CH_2)_{16}CH_3 \\ \quad\;\; | \\ H-C-O-CO(CH_2)_{16}CH_3 \\ \quad\;\; | \\ H_2-C-O-CO(CH_2)_{16}CH_3 \end{array}$$

Stearyl Alcohol (mol wt 270) An alcohol derived from stearic acid used in various ointments.

$$CH_3(CH_2)_{16}CH_2OH$$

Stearoyl-ACP Desaturase The enzyme that catalyzes the following reaction:

$$\text{Stearoyl-ACP} + AH_2 + O_2 \rightleftharpoons \text{Oleoyl-ACP} + A + 2\,H_2O$$

Stearoyl-CoA Desaturase The enzyme that catalyzes the following reaction:

$$\text{Stearoyl-CoA} + AH_2 + O_2 \rightleftharpoons \text{Oleoyl-CoA} + A + 2\,H_2O$$

Steatitis Inflammation of fatty tissue.

Steatoblast Cells that give rise to fat cells (adipocytes).

Steatolysis The process of hydrolysis and emulsification of fats prior to their absorption from the intestine.

Steatorrhea The presence of an excessive amount of fat in the stools.

Steatosis The fatty degeneration in the liver.

Stelazine A trade name for trifluoperazine hydrochloride, an antipsychotic agent that blocks postsynaptic dopamine receptors in the brain.

Stella A genus of Gram-negative, asporogenous, nonmotile, flat, star-shaped bacteria found in water and sewage.

Stellacyanin A blue copper-containing mucoprotein of low molecular weight obtained from the Japanese lacquer tree, *Rhus vernicifera*.

Stellate Star-shaped.

Stellate Cell A cell with radiating cytoplasmic processes.

Stem 1. The main part of a plant bearing leaves, buds, and flowers. 2. The nonlooped portion of a tRNA. 3. The double-stranded region of a single-stranded DNA or RNA, e.g., a hair-pin structure.

STEM Abbreviation for scanning transmission electron microscope.

Stem and Loop Structure A type of secondary structure of DNA or RNA in which the complementary sequences within a strand form a stem structure while the noncomplementary sequences remain as a loop.

Stem Bromelain An endopeptidase from the stem of the pineapple plant.

Stem Cell Animal cells found in hematopoietic tissue (e.g., bone marrow) that are capable of undergoing differentiation and proliferation into particular cell types.

Stemetil A trade name for prochlorperazine, an antiemetic agent.

Stemonitis A genus of slime mold (class Myxomycetes).

Steno- A prefix meaning narrow.

Stenosalinic Pertaining to aquatic organisms that have restricted tolerance to changes in environmental salt water concentration.

Stenosis Narrowing a duct or canal.

Stenothermic Pertaining to organisms that have restricted tolerance to changes in environmental temperature.

Stenothermophiles Microorganisms that have restricted tolerance to changes in environmental temperature.

Stenoxenous Pertaining to organisms that have a narrow host range.

Stephanokont Cell with a ring of flagella at one pole.

Stepronin (mol wt 273) A mucolytic agent.

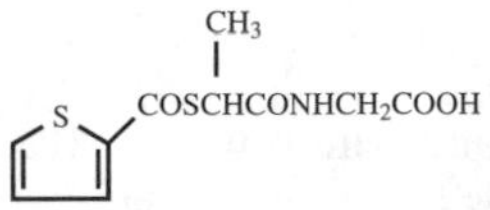

Stepwise Chromatographic Development A type of chromatographic technique in which the sample is repeatedly eluted from the column with different solvents.

Stepwise Elution A type of elution in which the composition of eluent is changed abruptly (not linearly) during the elution procedure.

Sterapred A trade name for prednisone, an anti-inflammatory agent.

Stereochemistry The science that deals with the three-dimensional spatial arrangements of atoms in molecules.

Stereoisomer Compounds that have identical chemical composition but differ from each other by the spatial arrangement of the atoms in the molecule.

Stereology The science that deals with the three-dimensional aspects of the morphology of structure or ultrastructure.

Stereoselective Reaction A reaction that generates or destroys selectively for one stereoisomer over another.

Stereospecific Reaction An enzymatic reaction that catalyzes the addition or removal of an atom or group of atoms to or from a particular stereoposition in a molecule.

Sterigmtocystin (mol wt 324) A carcinogenic metabolite from *Aspergillus*.

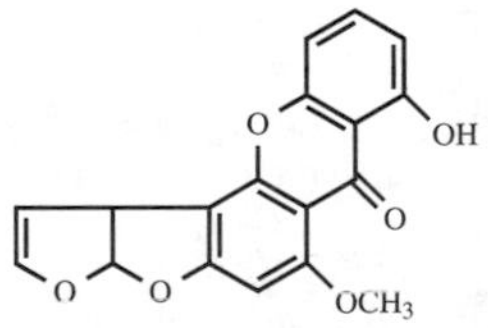

Sterilant Chemical agents that sterilize objects.

Sterile 1. Free of living microorganisms. 2. Unable to reproduce.

Sterilization The process of eradicating all viable microorganisms from culture media, glassware, and utensils.

Sterilizer Apparatus used for sterilization, e.g., autoclave and hot-air oven.

Sterilizing Immunity The immune response that provides the total elimination of a parasitic pathogen.

Sterine A trade name for methenamine mandelate, an anti-infective agent.

Sterne Strain A strain of *Bacillus anthracis* that lacks the poly-D-glutamate capsule that is used as a live vaccine against anthrax.

Steroid Numerous compounds that consist of a 17-carbon ring system (cyclopentanoperhydrophenanthrene), e.g., cholesterol, sterols, bile acid, estrogen, and testosterone.

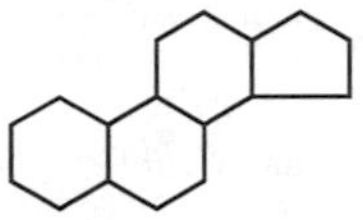

Steroid *N*-Acetylglucosaminyl Transferase The enzyme that catalyzes the following reaction:

UDP-*N*-Acetylglucosamine + estradiol-17-α 3-D-glucoside

⇅

UDP + 17α-(*N*-acetylglucosaminyl)-estradiol 3-D-glucuronoside

Steroid *O*-Acyltransferase The enzyme that catalyzes the following reaction:

Acyl-CoA + cholesterol ⇌ CoA + cholesterol ester

Steroid Alkaloids Referring to the nitrogen-containing steroids found in plants.

Steroid Cell Antibody An IgG autoantibody developed against antigens in the cytoplasm of steroid-producing cells in the ovary, testes, placenta, and adrenal cortex in patients with Addision's disease.

Steroid Conjugates Products formed by conjugation of products resulting from the breakdown of steroid with other organic molecules.

Steroid Diabetes A disorder due to the prolonged uptake of glucocorticoid that results in the production of glucose in the liver and inhibition of insulin activity.

Steroid Glycoside Steroids that contain nucleus structure of steroid and carbohydrate or carbohydrate derivative, e.g., saponin.

Steroid Hormone Any of the numerous hormones that have the characteristic ring structure of steroids, e.g., estrogens, androgens and corticoids.

Steroid Lactonase The enzyme that catalyzes the following reaction:

Testololactone + H_2O ⇌ Testolate

Steroid 11β-Monooxygenase The enzyme that catalyzes the following reaction:

A steroid + reduced adrenal ferredoxin + O_2
⇌
An 11β-hydroxysteroid + oxidized adrenal ferredoxin + H_2O

Steroid 17α-Monooxygenase The enzyme that catalyzes the following reaction:

A steroid + AH_2 + O_2
⇌
A 17α-hydroxysteroid + A + H_2O

Steroid 21-Monooxygenase The enzyme that catalyzes the following reaction:

A steroid + AH_2 + O_2
⇌
A 21-hydroxysteroid + A + H_2O

Steroid Receptor Cytoplasmic receptor protein that bind steroid hormone. The receptor-hormone complex then moves into the nucleus and binds to a specific site on the nuclear DNA and regulating gene activity.

Steroid Sulfotransferase The enzyme that catalyzes the following reaction:

3′-Phosphoadenylylsulfate + a phenolic steroid
⇌
Adenosine 3′,5′-bisphosphate + steroid 0-sulfate

Steroidogenesis The process of biosynthesis of steroids.

Sterol A steroid alcohol in which an alcoholic hydroxyl group and an aliphatic side chain with 8 or more atoms are attached to position 3 and 17 of the steroid nucleus, respectively.

Sterol O-Acyltransferase The enzyme that catalyzes the following reaction:

Acyl-CoA + cholesterol
⇌
CoA + cholesterol ester

Sterol Ester Synthetase See sterol *o*-acyltransferase.

Sterol Esterase The enzyme that catalyzes the following reaction:

A stearyl ester + H_2O ⇌ A sterol + a fatty acid

Steryl β-glucosidase The enzyme that catalyzes the following reaction:

Cholesteryl-β-D-glucoside + H_2O
⇌
Cholesterol + D-glucose

Sterol 3β-glucosyltransferase The enzyme that catalyzes the following reaction:

UDP-glucose + a sterol
⇌
A sterol 3-β-D-glucoside + UDP

Stethometer An instrument for measurement of chest expansion during breathing.

Stethoscope An instrument used for listening to sounds within the body.

STH Abbreviation for somatotropic hormone.

SthI (KpnI) A restriction endonuclease from *Salmonella thompson* with the same specificity as KpnI.

SthAI (KpnI) A restriction endonuclease from *Salmonella thompson* with the same specificity as KpnI.

SthBI (KpnI) A restriction endonuclease from *Salmonella thompson* YY106 with the following specificity:

```
              ↓
5′..........GGTACC..........3′
3′..........CCATGG..........5′
             ↑
```

SthCI (KpnI) A restriction endonuclease from *Salmonella thompson* YY150 with the same specificity as KpnI.

SthDI (KpnI) A restriction endonuclease from *Salmonella thompson* YY197 with the same specificity as KpnI.

SthEI (KpnI) A restriction endonuclease from *Salmonella thompson* YY209 with the same specificity as KpnI.

SthFI (KpnI) A restriction endonuclease from *Salmonella thompson* YY200 with the same specificity as KpnI.

SthGI (KpnI) A restriction endonuclease from *Salmonella thompson* YY217 with the same specificity as KpnI.

SthHI (KpnI) A restriction endonuclease from *Salmonella thompson* YY224 with the same specificity as KpnI.

SthNI (KpnI) A restriction endonuclease from *Salmonella thompson* YY148 with the same specificity as KpnI.

STI Abbreviation for soybean trypsin inhibitor.

Stibocaptate (mol wt 788) An anthelmintic agent.

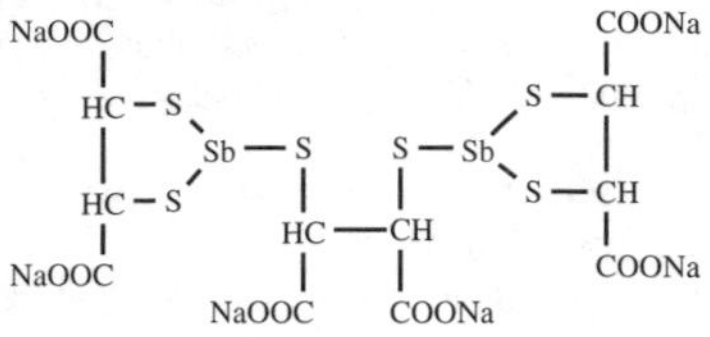

Stibophen (mol wt 895) An anthelmintic agent.

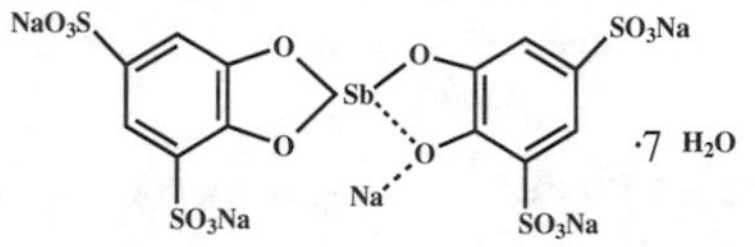

Stichonematic Pertaining to a eukaryotic flagellum bearing a single row of fine hairs along its length.

Stickland Reaction A metabolic reaction occurring in *Clostridium* in which oxidation of one amino acid is coupled to the reduction of another suitable amino acid.

Sticky Ends Complementary single-stranded termini of a double-stranded DNA molecule, e.g., the DNA of phage λ or DNA fragments produced by some restriction endonuclease.

Sticky Region A region in the nucleic acid molecule which is rich in G-C content (guanine and cytosine content).

Stigmasterol (mol wt 413) A plant sterol used for preparation of progesterone and other important steroids.

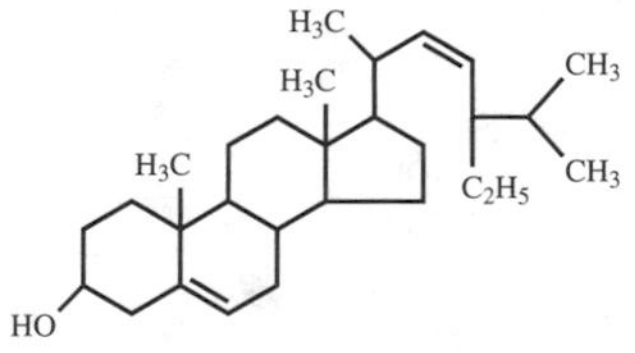

Stigmatellin (mol wt 515) An inhibitor of electron transport that binds to cytochrome b as well as other iron–sulfur proteins.

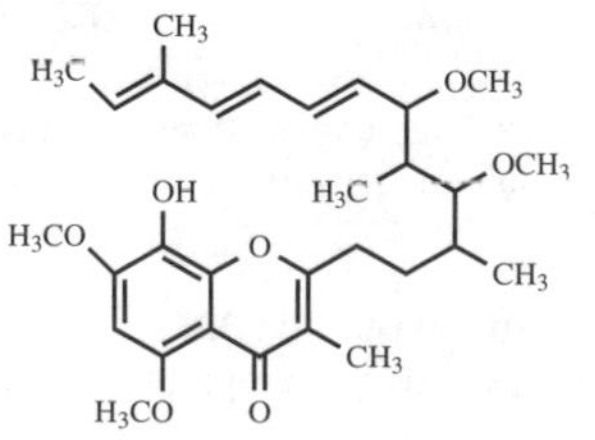

Stilbamidine (mol wt 264) An antiprotozoal agent.

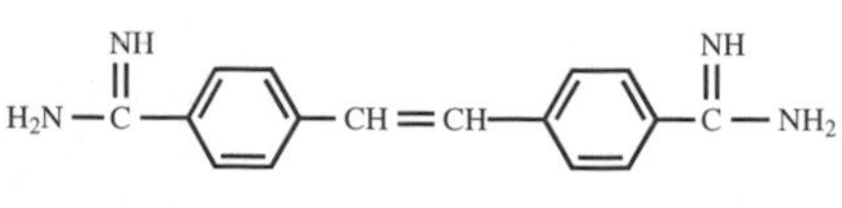

Stilbazium Iodide (mol wt 578) An anthelmintic agent (nematocide).

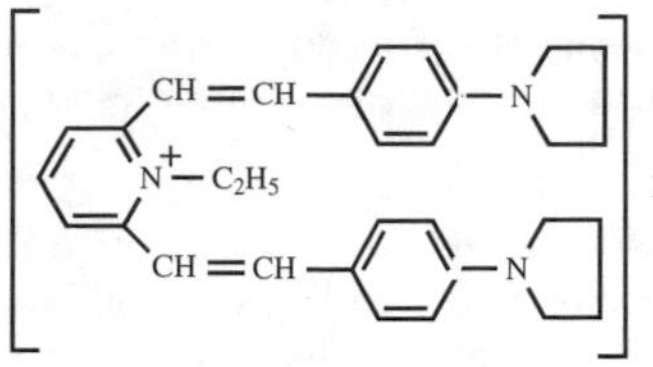

Still's Disease The rheumatoid arthritis of children.

Stilonium Iodide (mol wt 451) An antispasmodic agent and a ganglionic blocker.

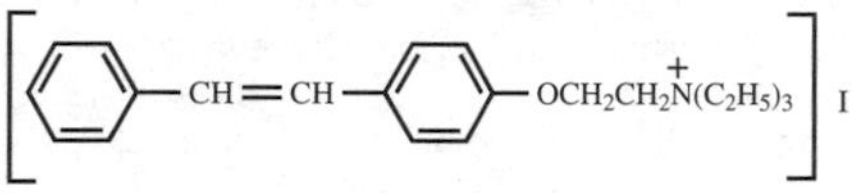

Stimate A trade name for desmopressin acetate, used to promote reabsorption of water and to produce a concentrated urine.

Stimulus Secretion Coupling The process of coupling the reception of a stimulus to the release of substance, e.g., coupling of membrane depolarization at the presynaptic terminal to the release of neurotransmitter into the synaptic cleft.

Stirofos (mol wt 366) An insecticide and a cholinesterase inhibitor.

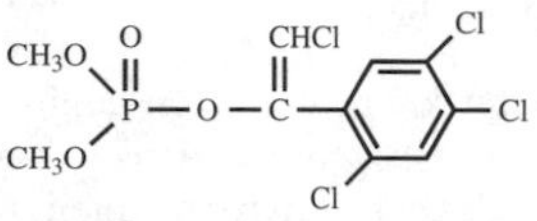

STLC Abbreviation for sulfated taurolithocholate.

STNV Abbreviation for satellite tobacco necrosis virus.

Stock Culture 1. A culture that is maintained as a source of authentic subcultures. 2. A pure culture from which working cultures are derived.

Stoichiometric Amount The quantity of a substance used in a chemical reaction.

Stoichiometry The science that deals with the quantitative relationship between elements in a compound or between reactants and the products in a chemical reaction.

Stoke A c.g.s. unit of viscosity for a fluid that has a viscosity of one poise and a density of one gram per cubic centimeter.

Stokes' Equation The equation describing the relationship of frictional coefficient (f) of a spherical particle to the radius (r) of the particle and viscosity of the solvent.

$$f = 6\pi\eta\rho$$

Stokes' Radius The radius of a perfect anhydrous sphere.

Stokes' Reagent A mixture of ferrous sulfate, tartaric acid, and ammonia used for testing of hemoglobin.

Stokes-Einstein Equation The equation describing the relationship of the diffusion coefficient (D) of the spherical particle to radius (ρ) of the spherical particle, temperature (T), and viscosity (η) of the solution.

$$D = KT/\ 6\pi\eta\rho$$

Stoma An opening, regulated by the guard cells, in the epidermis of a leaf or other plant part.

Stomata Plural of stoma.

Stomatitis Inflammation of tissues of the mouth.

Stomatology The medical science that deals with the mouth and its disorders.

Stone Cell A type of sclerenchyma cells in the plant.

Stop Buffer A buffer containing a component (e.g., EDTA) that stops an enzymatic reaction.

Stop Codon The triplet sequence of nucleotides on a mRNA molecule that signals the end of the translation, e.g., UAG, UAA, and UGA (also known as termination codon).

Stop Transfer Sequence The amino acid sequence that stops the translocation of a newly forming polypeptide chain through a protein-translocating channel in the membrane for anchoring the newly forming polypeptide or protein within the lipid bilayer.

Stopcock A valve that stops or regulates flow of a fluid through a tube.

Storage Granule A membrane-bound vesicle or structure containing food reserves or condensed secretory materials.

Storage Macromolecule Referring to a polymer that consists of one or a few types of subunits without specific order that serves as a storage polymer, e.g., starch and glycogen.

Storzolamide A trade name for acetazolamide, a diuretic agent that promotes the activity of carbonic anhydrase and renal excretion of sodium, potassium, bicarbonate, and water.

Stoxil A trade name for idoxuridine, an ophthalmic anti-infective agent that interferes with DNA synthesis.

STP Abbreviation for standard temperature and pressure.

Strain A cell or a population of cells of a given organism that possesses distinguishable characteristics with others within a given species or serotype.

Strand Displacement A replication mechanism in certain viruses in which one strand of the DNA is displaced as a new strand is being synthesized.

Strand-Specific Hybridization Probe The use of specifically designed single-stranded DNA or RNA as a probe for hybridization and detection of complementary sequences.

Stratified Epithelium An epithelium that consists of multiple layers of cell.

Streak Plate Technique A method of microbial inoculation whereby a loopful of culture is scratched across the surface of a solid culture medium so that single cells may be deposited at a given location.

Street Virus Referring to the virulent type of rabies virus isolated in nature from domestic or wild animals.

Strema A trade name for quinine sulfate, an antimalarial agent.

Strep Throat A sore throat caused by *Streptococcus pyogenes.*

Streptase A trade name for streptokinase.

Streptavidin A protein isolated from *Streptomyces avidinii* having high affinity for biotin.

Streptidine (mol wt 262) A component of certain aminoglycoside antibiotics.

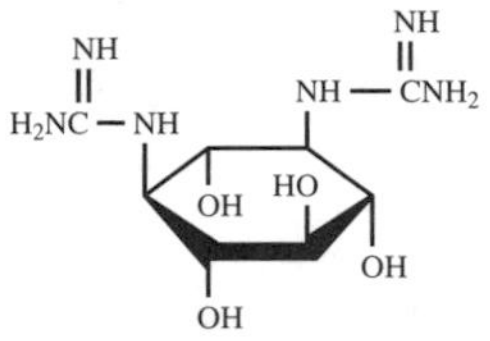

Streptidine Kinase See streptomycin 6-kinase.

Streptoalloteichus A genus of bacteria of the order Actinomycetales.

Streptobacillus A genus of anaerobic, facultative aerobic, chemoorganotrophic, Gram-negative bacteria.

Streptobiosamine (mol wt 337) A disaccharide and a component of the streptomycin molecule.

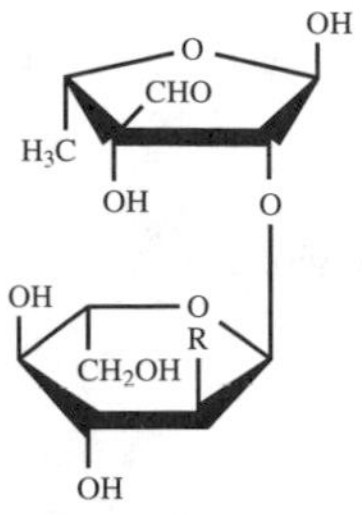

R = CH_3NH

Streptococcal M Protein A protein that protects cell walls found in the virulent strain of *Streptococcus pyogenes* that interferes with phagocytosis.

Streptococcal Toxin A group of hemolytic exotoxins produced by Streptococci, e.g., α-hemolysin, β-hemolysin, and γ-hemolysin.

Streptococcins Bacteriocin produced by *Streptococci.*

Streptococcus A genus of Gram-positive, asporogenous, chemoorganotrophic, facultatively anaerobic, and catalase-negative bacteria.

Streptodornase The extracellular nuclease produced by species of *Streptococcus.*

Streptogramins Referring to a group of closely related antibiotics, e.g., mikamycins, ostreogrycins, patricins, vernamycins, and virginiamycins.

Streptokinase The enzyme from *Streptococcus pyogenes* that catalyzes the following reaction:

$$\text{ATP + streptomycin} \rightleftharpoons \text{ADP + streptomycin 6-phosphate}$$

Streptolydigin Referring to antibiotics that bind to the β-subunit of eukaryotic RNA-polymerase, thus blocking the elongation of the nascent RNA chain.

Streptolysin Referring to toxins produced by Streptococci that cause hemolysis.

Streptolysin O A oxygen-labile, thio-activated hemolysin produced by *Streptococci* that causes hemolysis of cells with cholesterol in their membrane. Free cholesterol inhibits the hemolysis.

Streptolysin S An oxygen-stable hemolysin and leucocidin produced by *Streptococci* that causes β-hemolysis around streptococcal colonies on blood agar culture plates.

Streptomyces A genus of Gram-positive, aerobic bacteria (Actinomycetales) that occur in the soil and aquatic habitats.

Streptomycin (mol wt 582) An aminoglycoside antibiotic produced by *Streptomyces griseus* that inhibits protein synthesis by binding to the S12 protein of 30S ribosomal subunits, inhibiting proper translation.

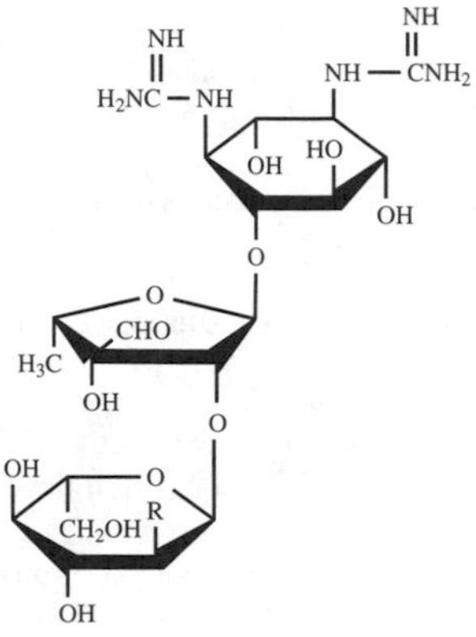

R = CH_3NH

Streptomycin Adenyltransferase The enzyme that catalyzes the following reaction:

$$\text{ATP + streptomycin} \rightleftharpoons \text{PPi + adenylylstreptomycin}$$

Streptomycin 6-Kinase The enzyme that catalyzes the following reaction:

$$\text{ATP + streptomycin} \rightleftharpoons \text{ADP + streptomycin 6-phosphate}$$

Streptomycin 6-Phosphatase The enzyme that catalyzes the following reaction:

$$\text{Streptomycin 6-phosphate} + H_2O \rightleftharpoons \text{Streptomycin + Pi}$$

Streptomycin-6-Phosphate Phosphohydrolase See streptomycin 6-Phosphate.

Streptonigrin (mol wt 506) An antineoplastic agent that causes strand breakage in DNA.

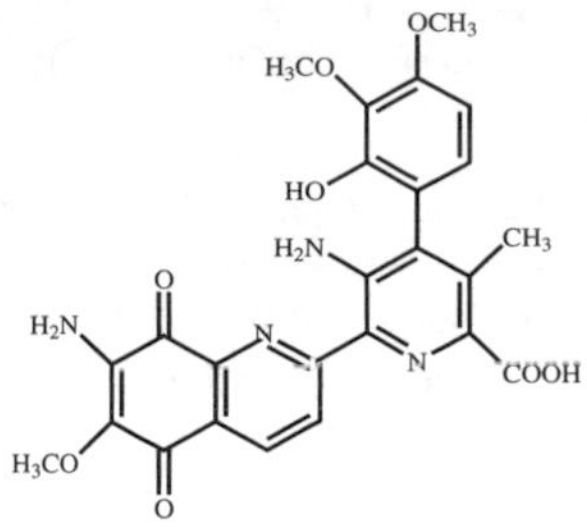

Streptopain A protease from *Streptococcus* that cleaves peptide bonds involving hydrophobic residues.

L-Streptose (mol wt 162) A sugar and component of the streptomycin molecule.

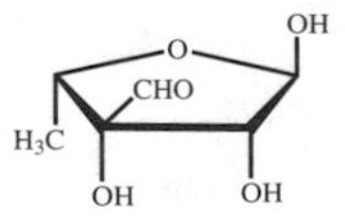

Streptosporangium A genus of bacteria (Actinomycetales).

Streptoverticillium A genus of aerobic bacteria (Actinomycetales).

Stress Fiber Bundles of contractile filaments resembling tiny myofibrils located in the cytoplasm of cultured fibroblasts. These bundles contain actin, myosin, and other cytoskeletal proteins.

Stress Protein A stress-induced protein that is synthesized following a heat-shock treatment (also known as heat shock protein).

Striated Muscle Muscle composed of transversely striped (striated) myofibrils, e.g., skeletal and cardiac muscles of vertebrates.

Strict Aerobes Synonym for obligate aerobes.

Strict Anaerobes Synonym for obligate anaerobes that cannot tolerate molecular oxygen and are inhibited or killed in the presence of oxygen.

Striction The reduction in volume resulting from mixing of two or more solutions as compared to the sum of the individual solutions due to the solute–solvent interaction.

Stringency Reaction conditions (e.g., temperature, salt, and pH) that dictate the annealing of two single-stranded DNAs, one single-stranded DNA and one RNA, and two single-stranded RNAs. At high stringency, duplexes form only between strands with perfect complementary; lower stringency allows annealing between strands with some degree of mismatch between bases.

Stringent Control A control mechanism in which the chromosome replication is dependent on the occurrence of protein synthesis.

Stringent Factor The gene product that is necessary for synthesis of ppGpp and pppGpp from GTP and ATP.

Stringent Plasmid A plasmid that replicates only with the bacterial chromosome and is present as a single copy or at most several copies per cell (also known as low copy number plasmid).

Stringent Response The reduction in synthesis of protein and RNA caused by the deprivation of an essential amino acids.

Stripped Hemoglobin A hemoglobin solution from which endogenous 2,3-bisphosphoglycerate has been removed.

Stripped Membrane Rough endoplasmic reticulum from which polysomes have been removed.

Stroke Sudden loss of consciousness and voluntary motion caused by rupture or obstruction of an artery of the brain due to the formation of an embolus or thrombus. A stroke may lead to paralysis, speech defect, or death.

Stroma Space enclosed by the envelope of a chloroplast containing enzymes for conversion of CO_2 to sugars.

Stroma Lamellae Membranes that connect thylakoid discs within the chloroplast.

Stromatin A structural protein present in the membrane of red blood cells.

Stromelysin A metalloprotease.

Strong Electrolyte An electrolyte that dissociates completely into ions in water.

Strongyloides A genus of parasitic nematode present in the intestine of various vertebrates.

Strongyloidiasis Infection or disease caused by the *Strongyloides* (parasitic nematode).

Strontium A chemical element with atomic weight 88, valence 2.

Strontium-90 Radioactive isotope of strontium resulting from a nuclear explosion, that is dangerous, especially for vertebrates, because it is taken up and incorporated in the construction of bone.

Strophanthus Dried ripe seeds of *Strophanthus kombe* used as arrow poison by African natives and also used as a cardiotonic agent.

Structural Gene A gene whose product is an enzyme, structural protein, tRNA, or rRNA, as opposed to a regulator gene whose product regulates the transcription of structural genes.

Structural Macromolecule Polymer that provides structure and mechanical strength to the cell, e.g., cellulose.

Structural Protein Proteins that serve as the structural component of cells or tissues.

Structural RNA Referring to ribosomal RNA.

Strychnine (mol wt 334) A highly toxic alkaloid from tropical plant (*Strychnos* species) used as a rodent poison.

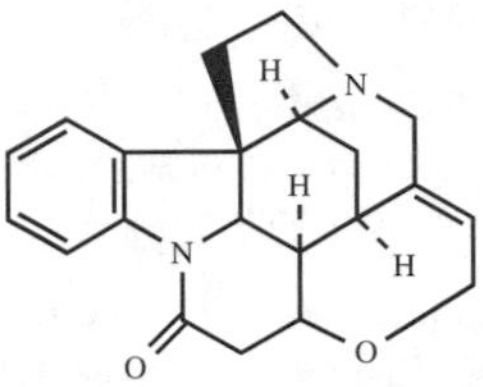

STS Abbreviation for 1. sequence-tagged site; 2. serological test for syphilis.

StuI A restriction endonuclease from *Streptomyces tubercidicus* with the following specificity:

↓
5′..........AGGCCT..........3′
3′..........TCCGGA..........5′
↑

Stuart Factor A blood coagulating factor that is activated by the Christmas factor in the intrinsic clotting pathway and activated by proconvertin in the extrinsic pathway.

Stukx A trade name for docusate sodium, a laxative.

Sturine A protamine found in sturgeon.

STV Abbreviation for simian T-virus.

StyI A restriction endonuclease from *Salmonella typhi* 27 with the following specificity:

↓
5′..........CC(A/T)(A/T)GG..........3′
3′..........GG(T/A)(T/A)CC..........5′
↑

Styloviridae A bacteriophage family with an elongated isomeric head and long, noncontractile tail.

Stylovirus Virus of the family Styloviridae.

Styragel A polystyrene support material used in gel filtration with lipophilic solvent.

Styramate (mol wt 181) A skeletal muscle relaxant.

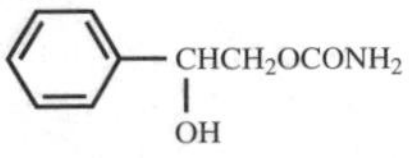

SU Abbreviation for thiouridine.

SUA Abbreviation for serum uric acid.

SuaI (HaeIII) A restriction endonuclease from *Sulfolobus acidocaldarius* with the same specificity as HaeIII.

Subacute Pertaining to disease that progresses more rapidly than a chronic disorder but less rapidly than an acute disorder.

Subacute Sclerosing Panencephalitis A chronic disorder of children associated with slow virus infection or with a persistent measles virus infection that involves demyelination of the cerebral cortex.

Subathizone (mol wt 271) An antibacterial substance used for treatment of tuberculosis.

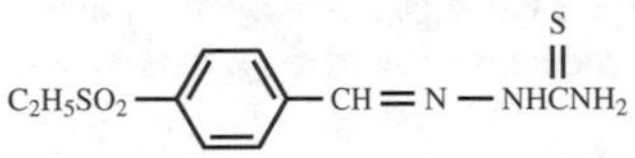

Subatomic Pertaining to the inside of the atom (smaller than atoms).

Subcapsular Pertaining to a structure occurring beneath or within a capsule.

Subcellular Referring to components or structures within the cell.

Subchondral Pertaining to a structure located beneath cartilage.

Subclinical Infection An infection without apparent symptoms.

Subcloning A method for obtaining a specific DNA fragment from a cloned DNA segment by DNA recombination technology.

Subculture 1. The process of reinoculation of a fresh culture medium with cells from an existing culture (e.g., from an overnight streak plate). 2. The process by which tissue or explant is first subdivided and then transferred to fresh medium.

Subcutaneous Beneath the skin.

Subcutaneous Tissue A layer of connective tissue that lies beneath the dermis (e.g., hypodermis).

Subcuticular Below the epidermis.

Subcutis The loose fibrous tissue below the skin.

Suberic Acid (mol wt 174) An eight-carbon dicarboxylic acid from caster oil.

$$HOOC(CH_2)_6COOH$$

Suberin A waxy material that covers the epidermis cell layer and serves as a waterproofing agent in higher plants. It contains long chain fatty acids and fatty esters and dicarboxylic acids.

Suberization Deposition of suberin by a plant at the site of the injury caused by fungal infection or insect damage.

Suberosis An allergy caused by the inhalation of cork dust contaminated with *Penicillium frequentans.*

Sublimation Transformation directly from a solid to a vapor state without the intermediate formation of a liquid.

Sublimaze A trade name for fentanyl citrate, an analgesic agent that binds with opiate receptors in the CNS, altering both perception and emotional response to pain.

Submerged Culture A type of culture in which cells reproduce below the surface of a liquid medium.

Submerged Culture Reactor A type of fermentor that uses forced aeration to maximize the rate of fermentation (e.g., vinegar production) in which bacteria grow in a fine suspension created by the air bubbles and the fermenting liquid.

Submetacentric A chromosome that appears J shaped at anaphase because the centromer is nearer one end than the other.

Submitochondrial Particle The small vesicles formed by sonification of mitochondria in which the inner mitochondrial membrane is inverted, producing an inside-out vesicles.

Submucosa A tissue layer directly under the epithelial lining of the lumen of the digestive tract.

Subnatant The liquid below another liquid or solid.

Subset A category within a particular type of cell, e.g., subset of a T cell or B cell.

Subspecies A taxonomic rank below species that describes a specific clone of cells.

Substance P A peptide neurotransmitter occurring in the brain and digestive tract that is capable of inducing inflammation of joints and stimulating synthesis of interleukines by monocytes.

Substituent An atom or a group of atoms that is introduced into a molecule by the replacement of another atom or another group of atoms.

Substitution Reaction A reaction that replaces an atom or a group of atoms attached to a carbon by another atom or another group of atoms.

Substitutional Vector A clonal vector in which a DNA segment can be excised and replaced by a foreign DNA fragment to be cloned.

Substrain A strain that has properties or markers not shared by all cells of the parent strain.

Substrate The substance on which an enzyme acts to form a product.

Substrate Adhesion Molecule Extracellular molecules that share a variety of sequence motifs with other adhesion molecules.

Substrate Analog A compound that resembles the true substrate for an enzyme and is capable of binding onto the active site of the enzyme but cannot be enzymatically converted.

Substrate Binding Site The site on an enzyme that binds substrate (also known as the active site).

Substrate Elution Chromatography A method for eluting enzyme from a chromatographic column by washing the column with solution containing substrate for the enzyme.

Substrate Induction The induction of synthesis of an enzyme by the presence of its substrate.

Substrate Level Phosphorylation Formation of ATP from ADP and inorganic phosphate resulting from the hydrolysis of a high energy compound without involvement of electron transport systems (e.g., formation of ATP from the hydrolysis of creatine phosphate).

Substratum The solid surface over which a cell migrates, grows, or adheres.

Subtillin A polypeptide antibiotic produced by *Bacillus subtilis.*

Subtilisin A protease that catalyzes the hydrolysis of proteins with preference for peptide bonds involving COOH groups of large uncharged amino acid residues.

Subtilopeptidase Synonym of subtilisin.

Subtilysin A hemolytic surfactant produced by *Bacillus subtilis*, a heptapeptide linked to a long chain fatty acid.

Subunits The smallest unit in a functional structure, e.g., LDH has 4 subunits.

Subviral Pathogen Referring to viroids and prions.

Succimer (mol wt 182) A chelating agent that forms a water-soluble complex with lead and increases its secretion in the urine.

$$HOOCCH(SH)CH(SH)COOH$$

Succinate A salt of succinic acid.

Succinate CoA Ligase (GDP-Forming) Synonym of succinyl-CoA ligase.

Succinate Coenzyme Q Reductase Enzyme complex of the electron transport chain involved in the transfer of electrons from succinate to coenzyme Q.

Succinate Dehydrogenase The enzyme that catalyzes the following reaction:

$$\text{Succinate} + \text{FAD} \rightleftharpoons \text{Fumarate} + \text{FADH}_2$$

Succinate Dehydrogenase (Ubiquinone) The enzyme that catalyzes the following reaction:

$$\text{Succinate} + \text{ubiquinone} \rightleftharpoons \text{Fumarate} + \text{ubiquinol}$$

Succinate Hydroxymethylglutarate CoA-transferase The enzyme that catalyzes the following reaction:

$$\text{Succinyl-CoA} + \text{hydroxymethylglutarate} \rightleftharpoons \text{Succinate} + \text{hydroxymethylglutaryl-CoA}$$

Succinate Pathway A pathway for synthesis of succinate from methionine, isoleucine, and valine.

Succinate Semialdehyde Dehydrogenase The enzyme that catalyzes the following reaction:

$$\text{Succinyl semialdehyde} + \text{NAD}^+ + \text{H}_2\text{O} \rightleftharpoons \text{Succinate} + \text{NADH}$$

Succinate Semialdehyde Dehydrogenase ($NADP^+$) The enzyme that catalyzes the following reaction:

$$\text{Succinyl semialdehyde} + \text{NADP}^+ + \text{H}_2\text{O} \rightleftharpoons \text{Succinate} + \text{NADPH}$$

Succinate Thiokinase The enzyme that catalyzes the following reaction:

$$\text{Succinate} + \text{CoA} + \text{ATP} \rightleftharpoons \text{Succinyl-CoA} + \text{ADP} + \text{Pi}$$

Succinic Acid (mol wt 118) An intermediate in the Krebs cycle.

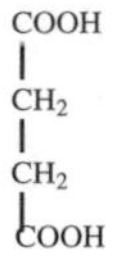

Succinic Anhydride (mol wt 100) A reagent used for the succinylation of proteins.

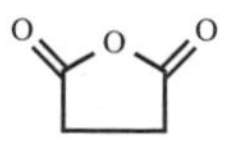

Succinic Dehydrogenase Synonym for succinate dehydrogenase.

Succinimide (mol wt 99) An antiurolithic agent.

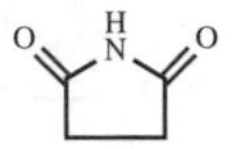

***N*-Succinimidyl 4-Maleimidobutyrate (mol wt 280)** A reagent used for enzyme immobilization and preparation of enzyme-hapten conjugates.

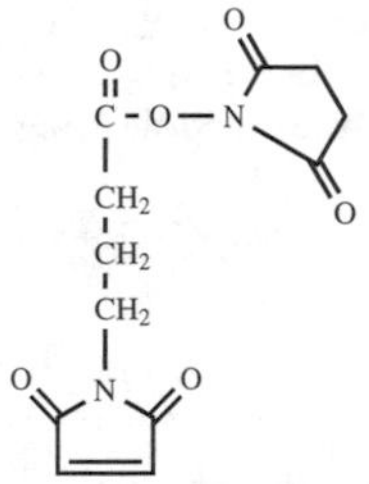

***N*-Succinimidyl 6-Maleimidocaproate (mol wt 308)** A reagent used for immobilizing and crosslinking proteins.

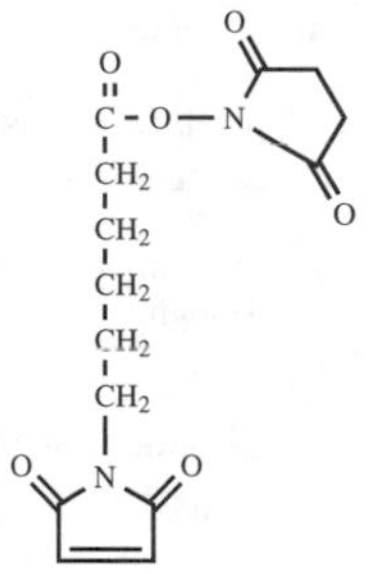

***N*-Succinimidyl 3-Maleimidopropionate (mol wt 266)** A reagent used for preparation of protein-hapten conjugates.

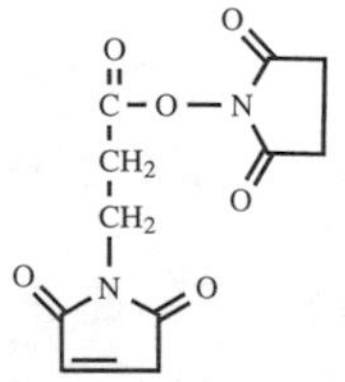

Succinimonas A genus of Gram-negative bacteria (family Bacteroidaceae).

Succinivibro A genus of Gram-negative bacteria (family Bacteroidaceae).

Succinyl-CoA (mol wt 868) An intermediate in the Krebs cycle.

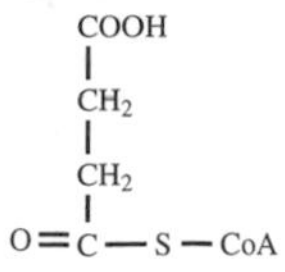

Succinyl-CoA Acylase See succinyl-CoA hydrolase.

Succinyl-CoA Hydrolase The enzyme that catalyzes the following reaction:

$$\text{Succinyl-CoA} + H_2O \rightleftharpoons \text{Succinate} + \text{CoA}$$

Succinyl-CoA Ligase (ADP-forming) The enzyme that catalyzes the following reaction:

$$\text{Succinate} + \text{ATP} \rightleftharpoons \text{Succinyl-CoA} + \text{Pi} + \text{ADP}$$

Succinyl-CoA Ligase (GDP-forming) The enzyme that catalyzes the following reaction:

$$\text{Succinate} + \text{GTP} + \text{CoA} \rightleftharpoons \text{Succinyl-CoA} + \text{Pi} + \text{GDP}$$

Succinyl-CoA Synthetase See succinyl-CoA ligase.

Succinyl-CoA Transferase Synonym of 3-oxoacid CoA-transferase.

Succinyl-Coenzyme A See succinyl-CoA.

Succinyl-Diaminopimelate Transaminase The enzyme that catalyzes the following reaction:

N-Succinyl-2,6-diamino-pimelate + α–ketoglutarate

⇅

N-Succinyl-2-amino-6-ketopimelate + L-glutamate

Succinyl-Glutathione Hydrolase The enzyme that catalyzes the following reaction:

$$S\text{-Succinylglutathione} + H_2O \rightleftharpoons \text{Succinate} + \text{glutathione}$$

Succinyl-Peroxide (mol wt 234) An antiseptic agent.

$$(HOOCCH_2CH_2CO)_2O_2$$

Succinylsulfathiazole (mol wt 355) An antibacterial agent.

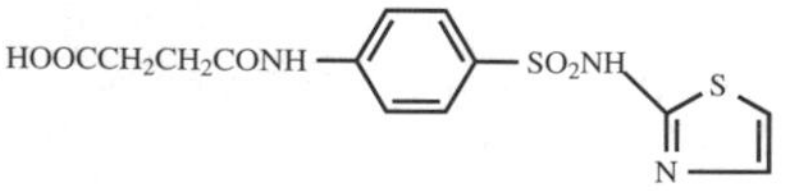

Succisulfone (mol wt 348) An antibacterial agent.

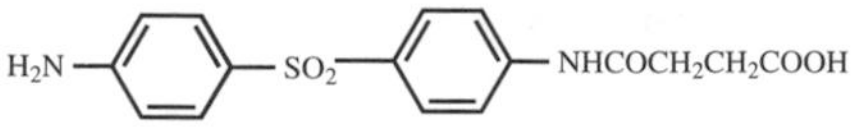

Suclofenide (mol wt 365) An anticonvulsant.

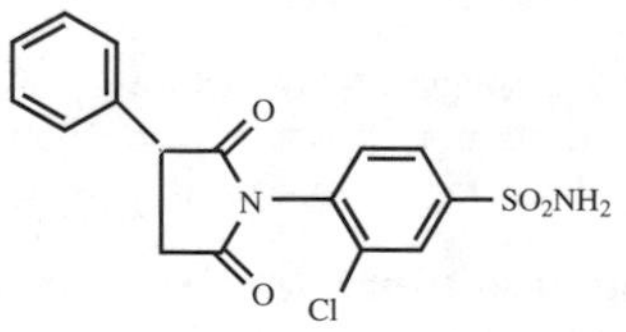

Sucostrin A trade name for succinylcholine chloride, a neuromuscular blocker that prolongs the depolarization of the muscle end plate.

Sucralfate (mol wt 2087) An antiulcer agent that protects the ulcer surface by forming a barrier.

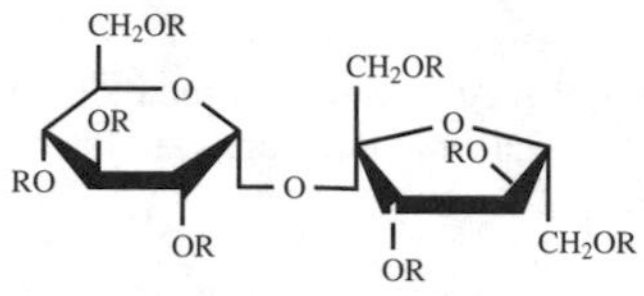

$$R = SO_3[Al_2(OH)_5]$$

Sucralose (mol wt 397) A chlorinated sucrose (tasting sweeter than sucrose itself).

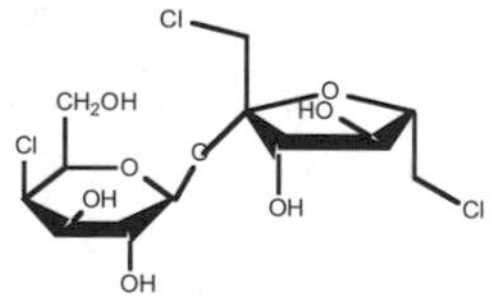

Sucrase The enzyme that catalyzes the following reaction:

$$\text{Sucrose} + H_2O \rightleftharpoons \text{Fructose} + \text{glucose}$$

Sucrets A trade name for dextromethorphan hydrobromide, an expectorant and an antitussive agent.

Sucrose (mol wt 342) A disaccharide consisting of glucose and fructose.

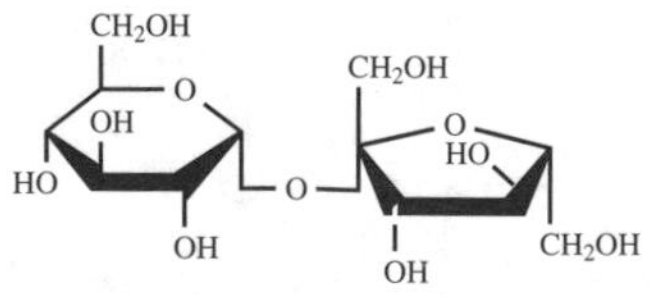

Sucrose Density Gradient A density gradient prepared with sucrose and used for density gradient centrifugation.

Sucrose Density Gradient Centrifugation A type of centrifugation technique using a density gradient prepared with sucrose for separation of macromolecules.

Sucrose α-Glucosidase The enzyme that catalyzes the hydrolysis of sucrose and maltose.

Sucrose Glucosyltransferase Synonym of sucrose phosphorylase.

Sucrose Isomaltase See sucrose α-glucosidase.

Sucrose Phosphatase The enzyme that catalyzes the following reaction:

$$\text{Sucrose 6-phosphate} + H_2O \rightleftharpoons \text{Sucrose} + \text{Pi}$$

Sucrose Phosphate Synthetase The enzyme that catalyzes the following reaction:

UDP-glucose + D-fructose 6-phosphate

⇅

UDP + sucrose 6-phosphate

Sucrose Phosphate UDP Glucosyltransferase Synonym of sucrose phosphate synthetase.

Sucrose Phosphorylase The enzyme that catalyzes the following reaction:

Sucrose + Pi

⇅

D-fructose + α-D-glucose 1-phosphate

Sucrose Polyester A nonabsorbable lipid and a substitute for fat in foods to reduce cholesterol levels.

Sucrose Synthetase The enzyme that catalyzes the following reaction:

UDP-glucose + D-fructose

⇅

UDP + sucrose

Sucrose UDP Glucosyltransferase Synonym of sucrose phosphate synthetase.

Sudafed A trade name for pseudoephedrine hydrochloride, an adrenergic agent that stimulates alpha-adrenergic receptors, producing vasoconstriction.

Sudan II (mol wt 276) A dye.

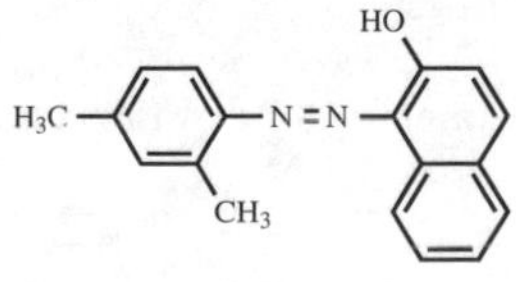

Sudan III (mol wt 352) A dye.

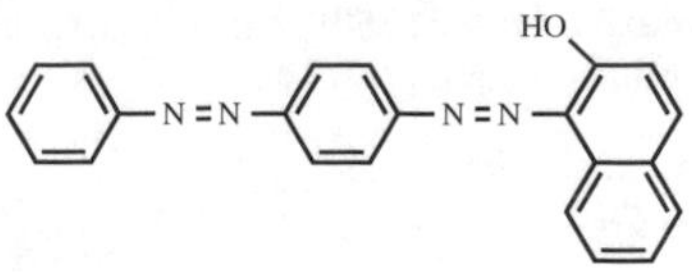

Sudan IV (mol wt 380) A dye.

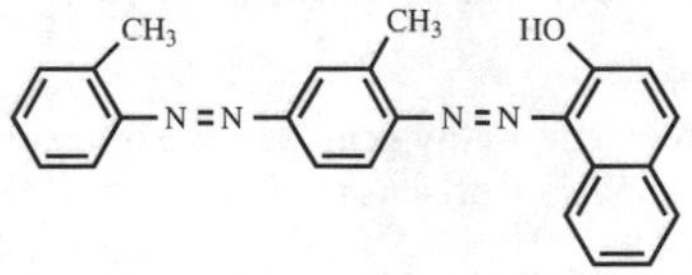

Sudan Black B Fat stain for animal tissue and microorganisms.

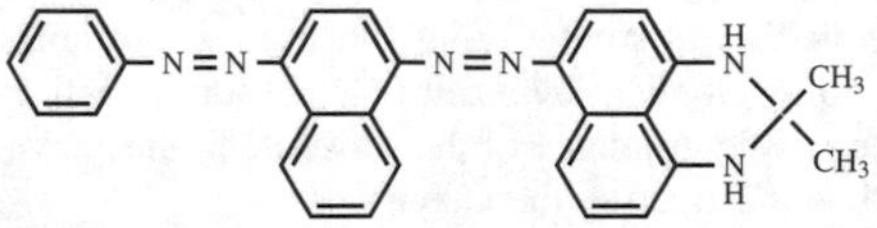

Sudden Infant Death Syndrome The unexpected death of an apparently healthy infant without any clinical evidence of disease. The medium-chain acyl-CoA dehydrogenase has been shown to be deficient in some of these infants.

Sufedrin A trade name for pseudoephedrine hydrochloride, an andrenergic agent that stimulates alpha-adrenergic receptors, producing vasoconstriction.

Sufenta A trade name for sufentanile citrate, an analgesic agent that binds with opiate receptors in the CNS, altering both perception and emotional response to pain.

Sufentanil (mol wt 387) An analgesic agent that binds with opiate receptor in the CNS, altering both perception and emotional response to pain.

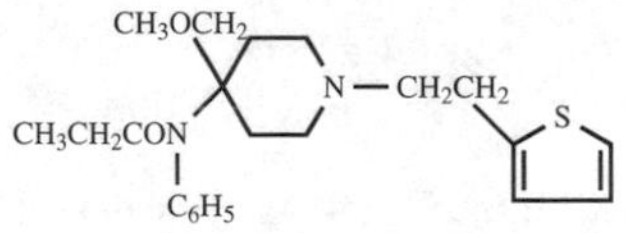

Sugar Referring to 1. sucrose and 2. monosaccharide or disaccharide.

Sugar Acid An acid derived from a sugar.

Sugar Alcohol An alcohol derived from a sugar.

Sugar Phosphatase The enzyme that catalyzes the following reaction:

$$\text{Sugar-phosphate} + H_2O \rightleftharpoons \text{Sugar} + \text{Pi}$$

Sugar Phosphate Backbone Referring to the chain of ester bonds formed between phosphoric acid and deoxyribose (in DNA) or between phosphoric acid and ribose (in RNA). This chain of ester bonds is the backbone of DNA and RNA.

Sugar Sulfate Sulfohydrolase The enzyme that catalyzes the following reaction:

$$\text{D-glucose 6-sulfate} + H_2O \rightleftharpoons \text{D-glucose} + \text{sulfate}$$

Sugar Transporter Any of various membrane proteins responsible for sugar transport into the cells.

Suicide Gene Any gene whose expression is lethal for cells or organisms.

Suicide Inhibitor A relatively inert molecule that is transformed by an enzyme into an active product that irreversibly inactivates the enzyme.

Suicide Substrate Any substrate whose uptake and metabolism by a cell is lethal for the cell due to covalent linkage of the substrate to the enzyme thus inactivating the enzyme.

Suipox Virus A virus of the Poxviridae that causes swinepox.

SulI (HaeIII) A restriction endonuclease from *Sulfolobus acidocaldarius* with the following specificity:

```
          ↓
5′..........GGCC..........3′
3′..........CCGG..........5′
          ↑
```

Sulbactam (mol wt 233) A semisynthetic β-lactamase inhibitor used in combination with β-lactam antibiotics.

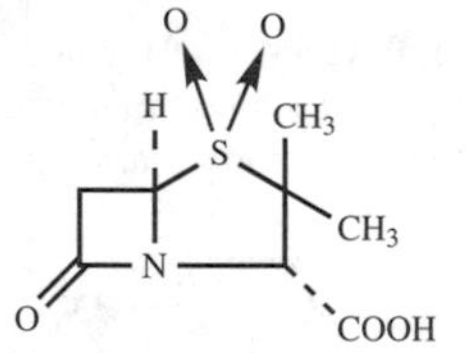

Sulbenicillin (mol wt 414) A semisynthetic antibiotic related to penicillin that inhibits bacterial cell wall synthesis.

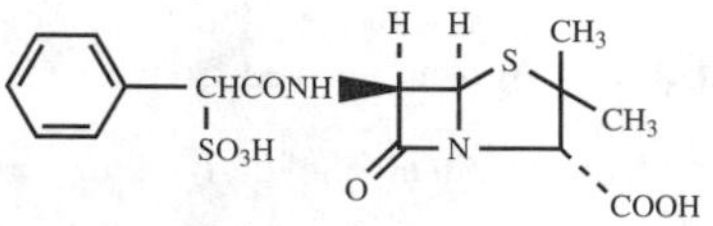

Sulbentine (mol wt 314) An antifungal agent.

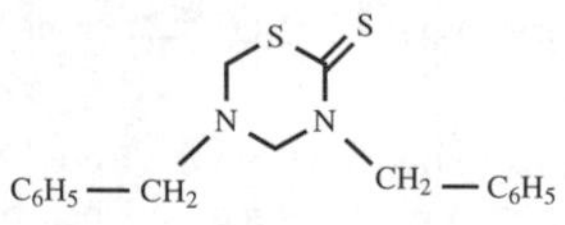

Sulconazole (mol wt 408) An antifungal agent.

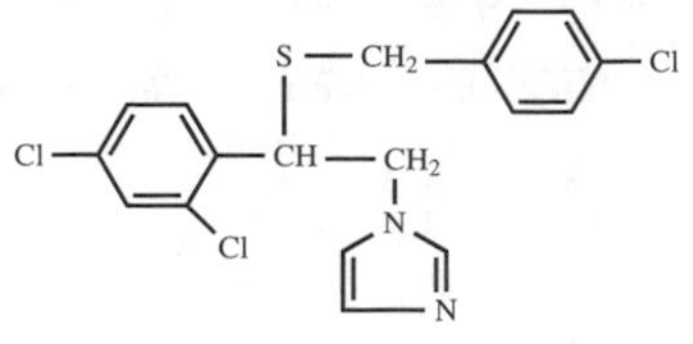

Sulcrate A trade name for sucralfate, an antiulcer agent.

Sulfa Drug Synthetic antibacterial agents characterized by having both sulfur and nitrogen (RSO_2NH_2). They are competitive inhibitors of *p*-aminobenzoic acid, which is required by bacteria for synthesis of folic acid.

Sulfabenzamide (mol wt 276) An antibacterial agent.

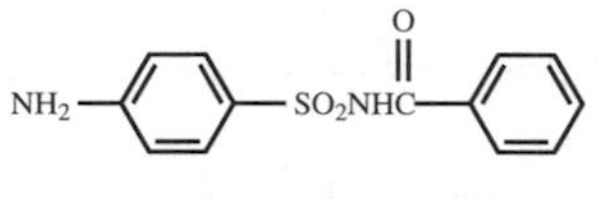

Sulfabromomethazine (mol wt 357) An antibacterial agent.

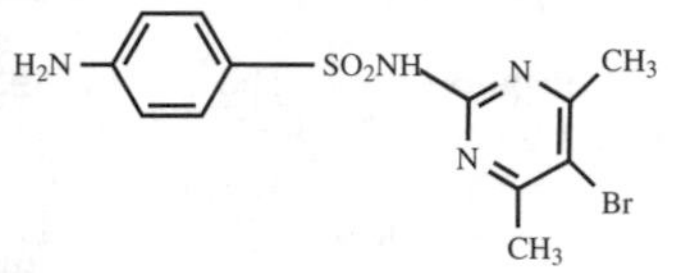

Sulfacetamide (mol wt 214) An ophthalmic, anti-infective agent.

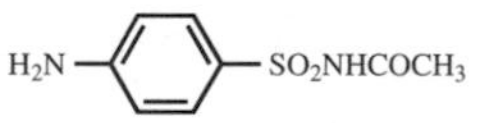

Sulfacet-R Lotion A trade name for a combination drug containing sulfacetamide sodium and sulfur, used as an anti-infective agent.

Sulfachlorpyridazine (mol wt 285) An antibacterial agent.

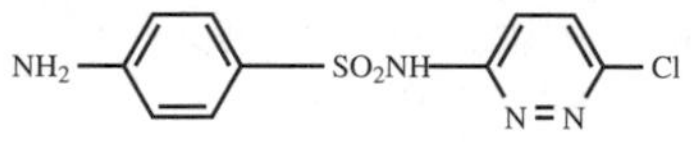

Sulfachrysoidine (mol wt 335) An antibacterial agent.

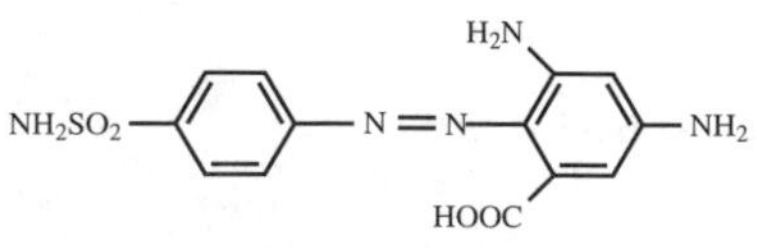

Sulfacytine (mol wt 294) An antibacterial agent.

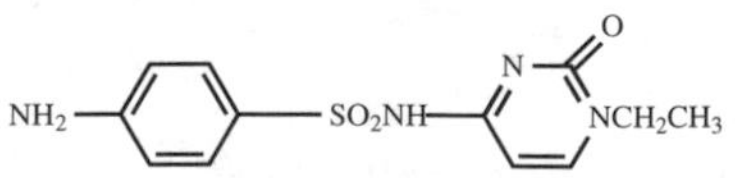

Sulfadiazine (mol wt 250) An antibacterial agent that inhibits bacterial folic acid synthesis.

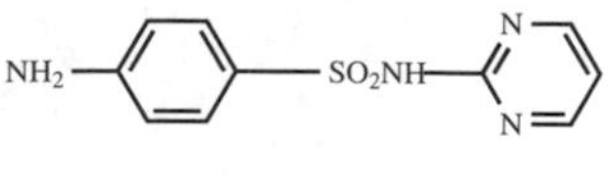

Sulfadicramide (mol wt 254) An antibacterial agent.

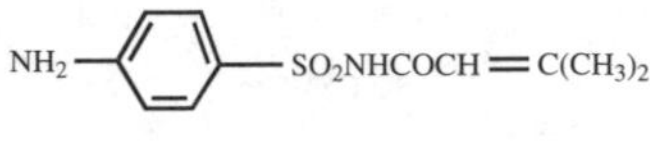

Sulfadimethoxine (mol wt 310) An antibacterial agent.

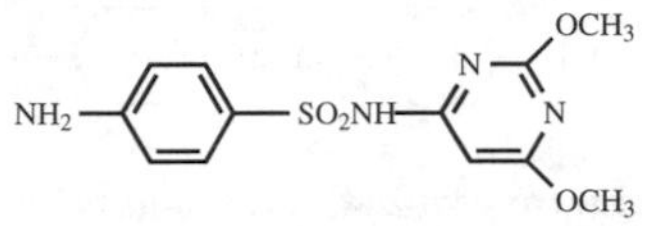

Sulfaethidole (mol wt 284) An antibacterial agent.

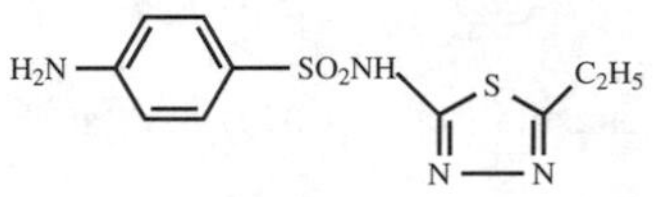

Sulfaguanidine (mol wt 214) An antibacterial agent.

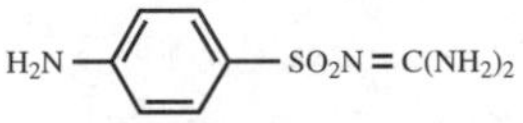

Sulfaguanole (mol wt 309) An antibacterial agent.

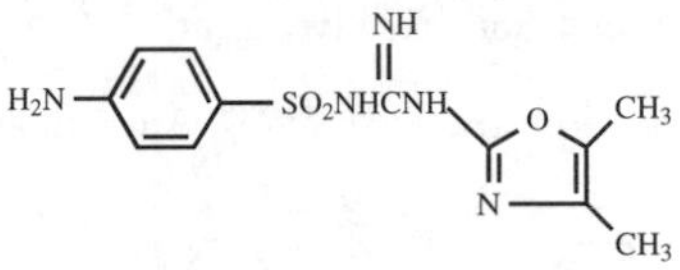

Sulfalene (mol wt 280) An antibacterial agent.

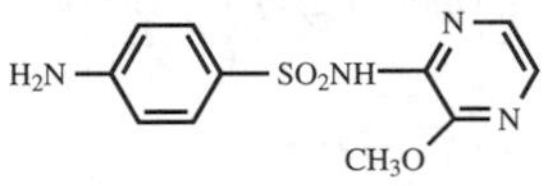

Sulfallate (mol wt 224) A herbicide.

Sulfaloxic Acid (mol wt 393) An antibacterial agent.

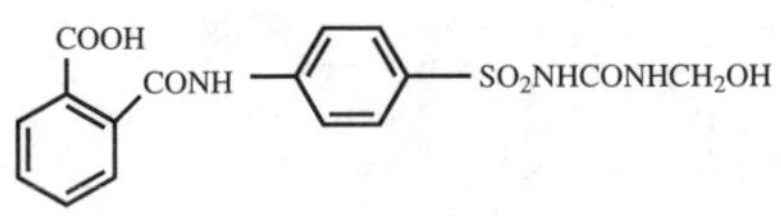

Sulfamerazine (mol wt 264) An antibacterial agent.

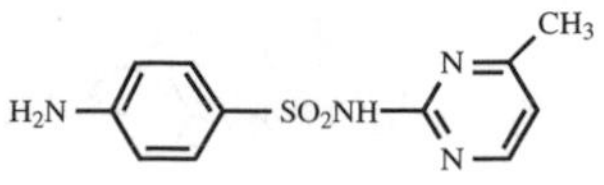

Sulfameter (mol wt 280) An antibacterial agent.

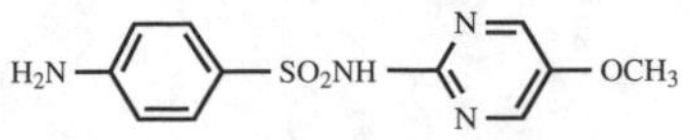

Sulfamethazine (mol wt 278) An antibacterial agent.

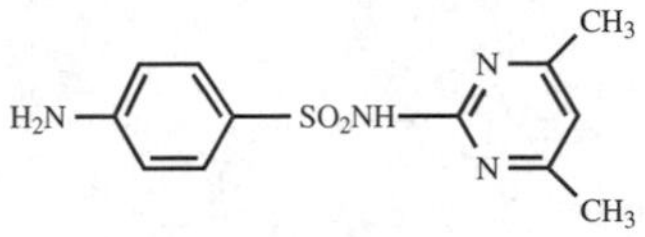

Sulfamethizole (mol wt 270) An antibacterial agent.

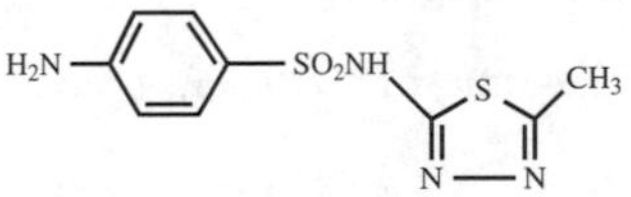

Sulfamethomidine (mol wt 294) An antibacterial agent.

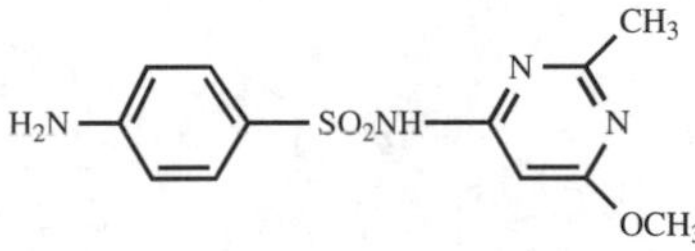

Sulfamethoxazole (mol wt 253) An antibacterial agent that inhibits bacterial folic acid synthesis.

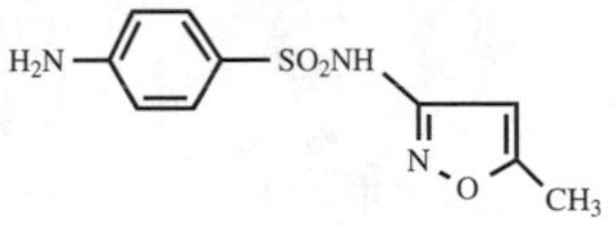

Sulfamethoxypyridazine (mol wt 280) An antibacterial agent.

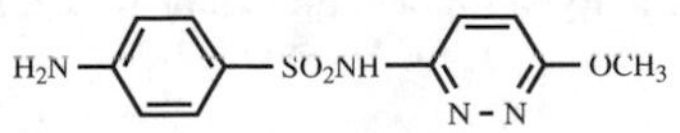

Sulfametrole (mol wt 286) An antibacterial agent.

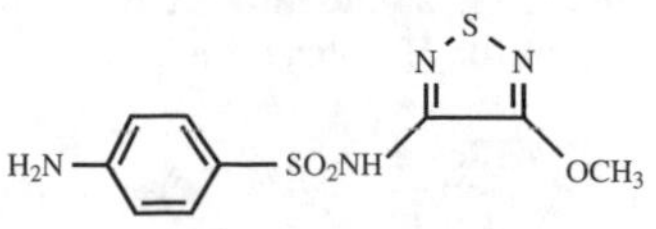

Sulfamidochrysoidine (mol wt 291) An antibacterial agent.

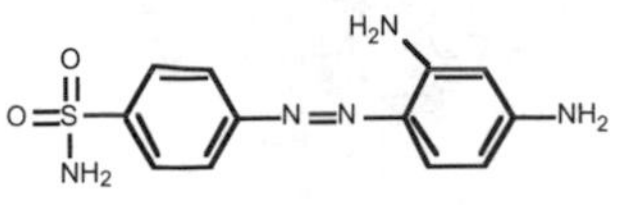

Sulfamipyrine (mol wt 319) An antiseptic and analgesic agent.

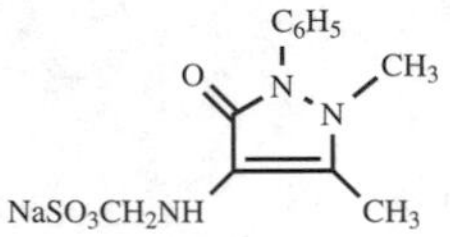

Sulfamoxole (mol wt 267) An antibacterial agent.

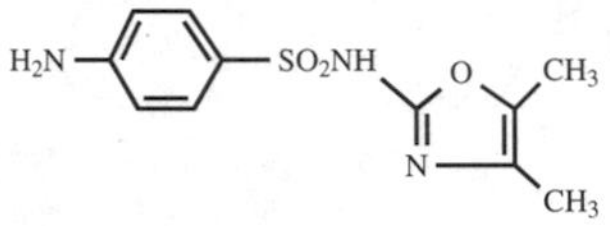

Sulfamylon A trade name for mafenide acetate, an anti-infective agent.

Sulfanilamide (mol wt 172) An antibacterial agent.

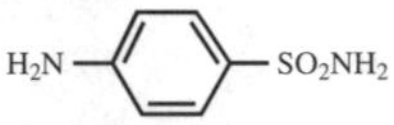

4-Sulfanilamidosalicylic Acid (mol wt 308) An antibacterial agent.

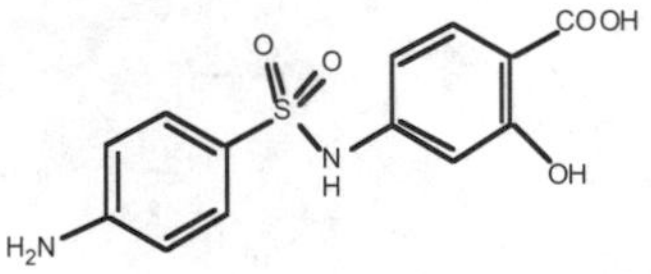

Sulfanilic Acid (mol wt 173) An antibacterial agent.

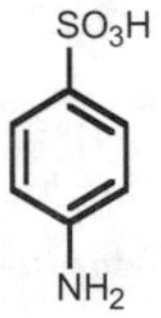

2-*p*-Sulfanilylanilinoethanol (mol wt 292) An antibacterial agent for *Mycobacterium.*

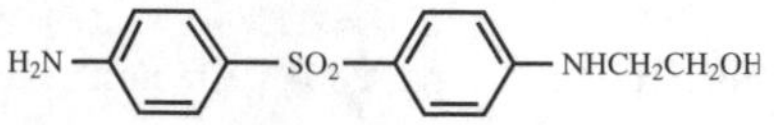

***p*-Sulfanilylbenzylamine (mol wt 262)** An antibacterial agent for *Mycobacterium.*

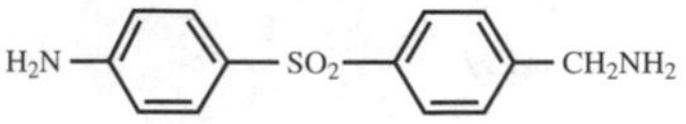

Sulfanilylurea (mol wt 215) An antibacterial agent.

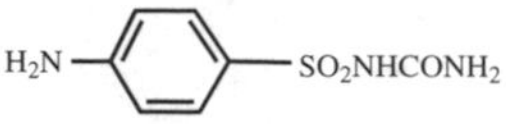

***N*-Sulfanilyl-3,4-xylamide (mol wt 304)** An antibacterial agent.

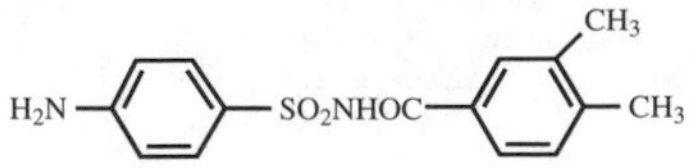

Sulfanitran (mol wt 335) An antibacterial agent.

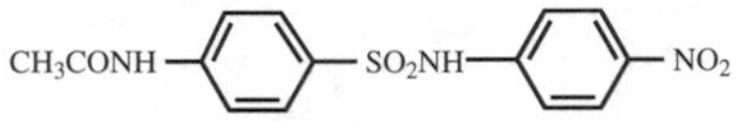

Sulfaperine (mol wt 264) An antibacterial agent.

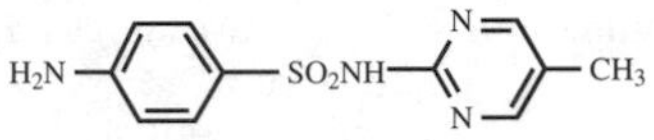

Sulfaphenazole (mol wt 314) An antibacterial agent.

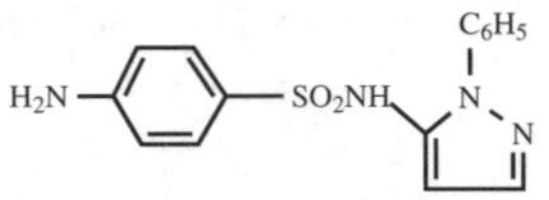

Sulfapred A trade name for a combination drug containing sulfacetamide sodium, prednisolone acetate, and phenylephrine hydrochloride, used as an ophthalmic anti-infective agent.

Sulfaproxyline (mol wt 334) An antibacterial agent.

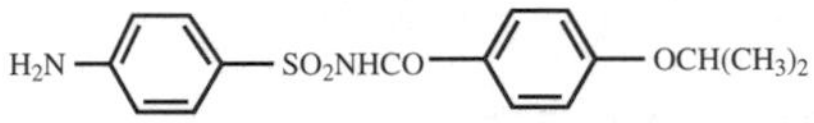

Sulfapyrazine (mol wt 250) An antibacterial agent.

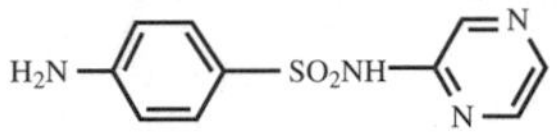

Sulfapyridine (mol wt 249) An antibacterial agent.

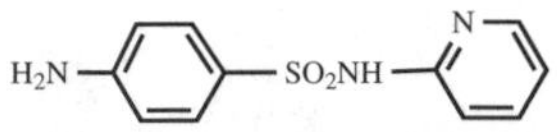

Sulfaquinoxaline (mol wt 300) An antibacterial agent.

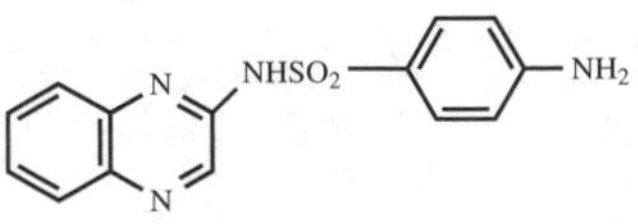

Sulfarsphenamine (mol wt 598) An antimicrobial agent (antisyphylitic).

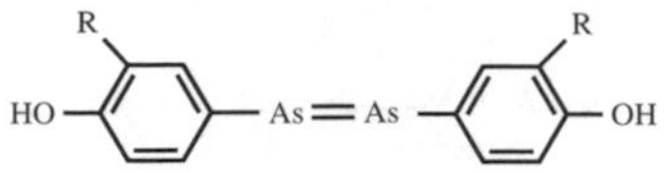

R = $NHCH_2SO_2ONa$

Sulfasalazine (mol wt 398) An antibacterial agent that inhibits formation of dihydrofolic acid from PABA.

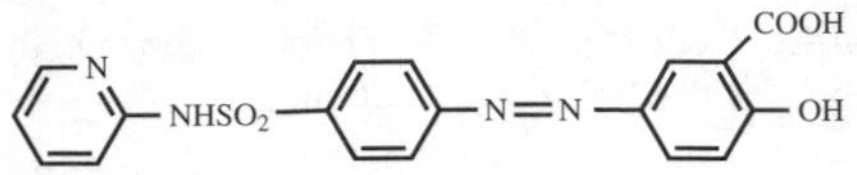

Sulfasomizole (mol wt 269) An antibacterial agent.

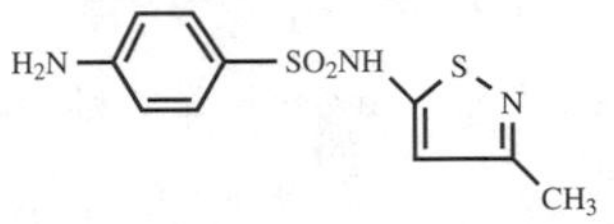

Sulfasymazine (mol wt 307) An antibacterial agent.

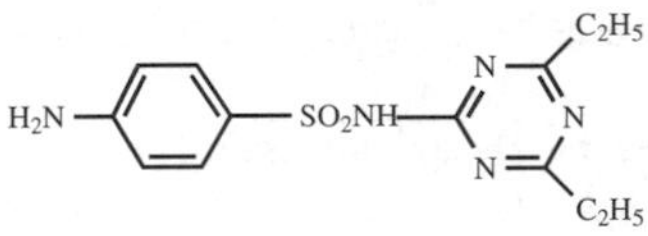

Sulfatase The enzyme that catalyzes the following reaction:

A phenol sulfate + H_2O $\rightleftharpoons$ A phenol + sulfate

Sulfate A salt of sulfuric acid, e.g., sodium sulfate.

Sulfate Adenylyltransferase The enzyme that catalyzes the following reaction:

ATP + sulfate $\rightleftharpoons$ PPi + adenylylsulfate

Sulfate Adenylyltransferase (ADP-specific) The enzyme that catalyzes the following reaction:

ADP + sulfate $\rightleftharpoons$ Pi + adenylylsulfate

Sulfate Assimilation The reduction of sulfate by bacteria and plants for biosynthesis of cysteine.

Sulfate Lipidosis A genetic disorder of childhood caused by the deficiency of cerebroside sulfatase and characterized by the increase of fats in the tissue of the central nerve system.

Sulfate Mineral A mineral in which the sulfur atom is linked to four oxygen atoms.

Sulfate Reducing Bacteria Bacteria capable of carrying out sulfate reduction, e.g., *Desulfovibrio.*

Sulfate Respiration Cellular respiration in which sulfate acts as the terminal electron acceptor to form secretory sulfide.

Sulfathiazole (mol wt 255) An antibacterial agent.

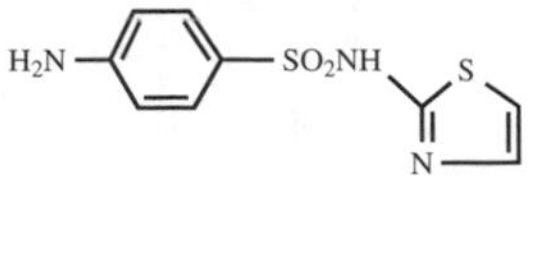

Sulfathiourea (mol wt 231) An antibacterial agent.

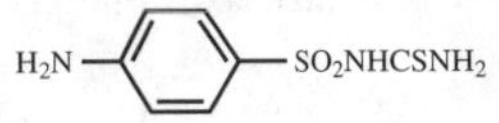

Sulfatide A glycosphingolipid containing a sulfate group.

Sulfatide Lipidosis A metabolic disorder caused by the deficiency of arylsulfatase A and characterized by the accumulation of sulfatide.

Sulfation The introduction of sulfate into a molecule or compound.

Sulfatolamide (mol wt 418) An antibacterial agent.

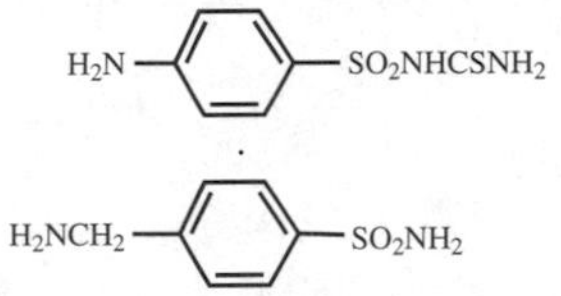

Sulfazamet (mol wt 328) An antibacterial agent.

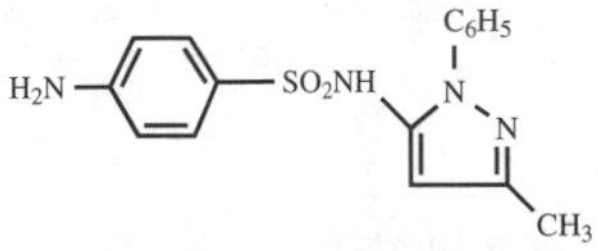

Sulfazecin (mol wt 396) An antibiotic produced by *Pseudomonas acidophila.*

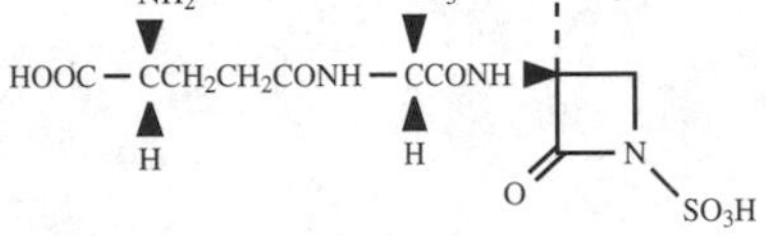

Sulfhemoglobin A green pigment formed by the reaction of hemoglobin with sulfide in the presence of oxygen or hydrogen peroxide.

Sulfhemoglobinemia The presence of abnormal sulfur-containing hemoglobin (sulfhemoglobin) in the circulating blood.

Sulfhydryl Group Referring to the -SH group found in cysteine and proteins.

Sulfhydryl Reagent Substances that are capable of reacting with the sulfhydryl group (-SH) of cysteine in proteins or peptides.

Sulfide A chemical compound that contains atoms of metal and sulfur.

Sulfinalol (mol wt 378) An antihypertensive agent.

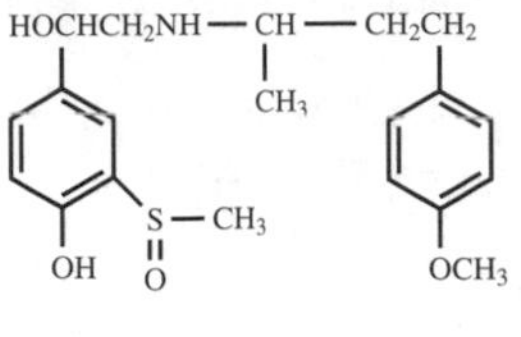

Sulfinpyrazone (mol wt 404) An antigout agent that blocks renal tubular reabsorption of uric acid.

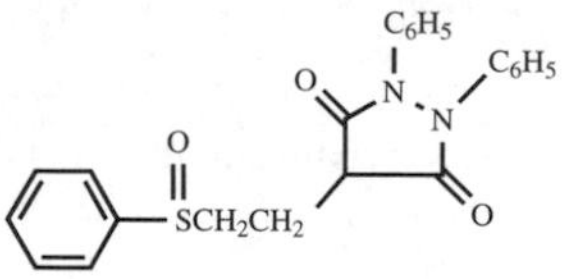

4′,4-Sulfinyldianiline (mol wt 232) An antibacterial agent.

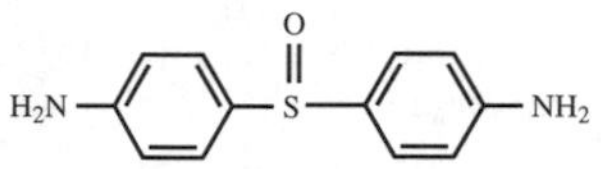

Sulfiram (mol wt 264) An antifungal agent.

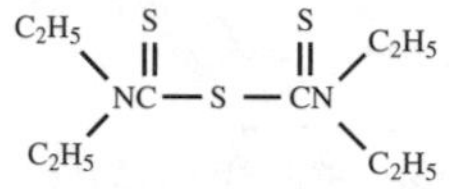

Sulfisomidine (mol wt 278) An antibacterial agent.

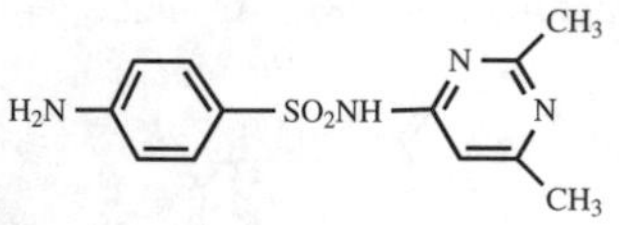

Sulfisoxazole (mol wt 267) An antibacterial agent that inhibits synthesis of dihydrofolic acid from PABA in bacteria.

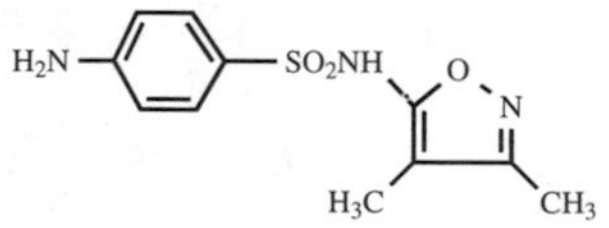

Sulfite A salt of sulfurous acid.

Sulfite Dehydrogenase The enzyme that catalyzes the following reaction:

Sulfite + 2 ferricytochrome c + H_2O ⇌ Sulfate + 2 ferrocytochrome c

Sulfite Oxidase The enzyme that catalyzes the following reaction:

Sulfite + O_2 + H_2O ⇌ Sulfate + H_2O_2

Sulfite Reductase The enzyme that catalyzes the following reaction:

Hydrogen sulfite + acceptor + $3H_2O$ ⇌ Sulfite + reduced acceptor

Sulfite Reductase (Ferredoxin) The enzyme that catalyzes the following reaction:

Hydrogen sulfite + 3 ferredoxin + 3 H_2O ⇌ Sulfite + 3 reduced ferredoxin

Sulfite Reductase ($NADP^+$) The enzyme that catalyzes the following reaction:

Hydrogen sulfite + 3 $NADP^+$ + 3 H_2O ⇌ Sulfite + 3 NADPH

Sulfitolysis The cleavage of a covalent bond by reaction with sulfite (SO_3^{-2}).

Sulfoacetaldehyde Lyase The enzyme that catalyzes the following reaction:

Sulfoacetaldehyde + H_2O ⇌ Sulfite + acetate

Sulfoalanine Decarboxylase The enzyme that catalyzes the following reaction:

3-Sulfino-L-alanine ⇌ Hypotaurine + CO_2

Sulfolipid Referring to sulfur-containing lipid.

Sulfonethylmethane (mol wt 242) A hypnotic agent.

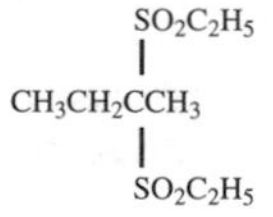

Sulfoniazide (mol wt 305) An antibacterial agent (tuberculostatic).

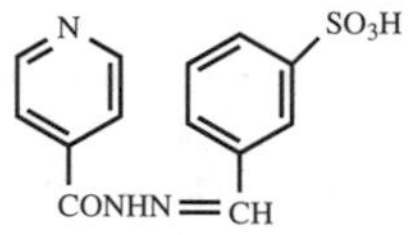

Sulfonic Acid An organic acid that contains one or more sulfo radicals ($-SO_3H$).

Sulfonmethane (mol wt 228) A hypnotic agent.

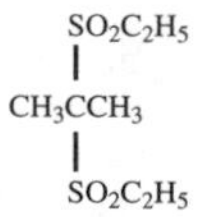

Sulforidazine (mol wt 403) An antipsychotic agent.

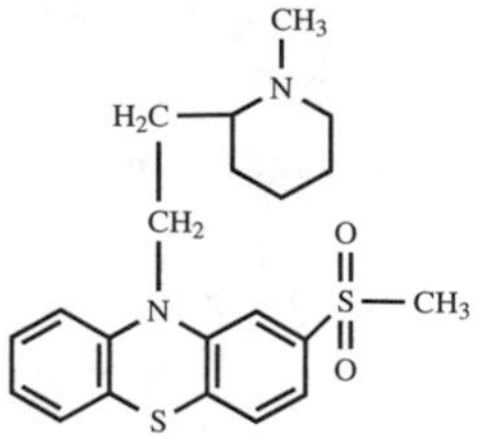

Sulfotep (mol wt 322) An insecticide.

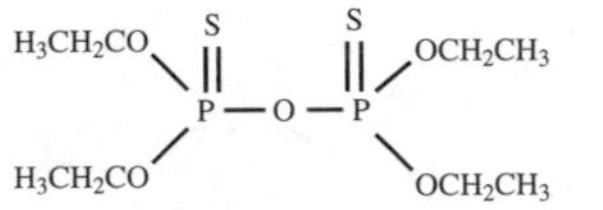

Sulfoxide (mol wt 324) An insecticide.

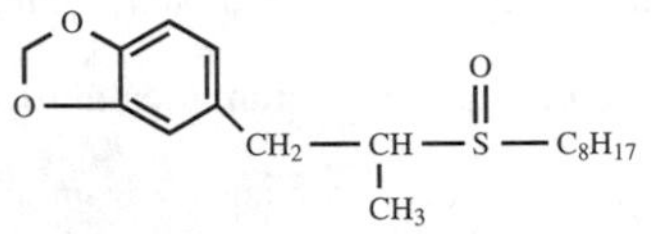

Sulfoxone Sodium (mol wt 448) An antibacterial agent.

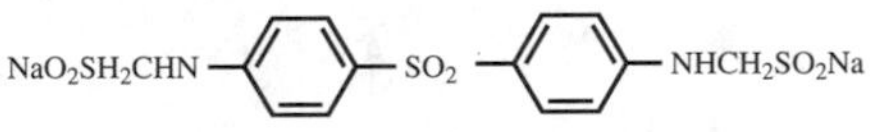

Sulfur A chemical element with atomic weight 32, valence 2, 4, and 6.

Sulfur-35 The radioactive nuclide of sulfur (^{35}S) emitting beta radiation and with a half life of 87 days.

Sulfur-38 The radioactive nuclide of sulfur (^{38}S) emitting beta and gamma radiation with a half life of 2.87 hours.

Sulfur Amino Acids Referring to amino acids that contain sulfur, e.g., cysteine and methionine.

Sulfur Cycle Biogeochemical cycle mediated by microorganisms that changes the oxidation state of sulfur within various compounds.

Sulfur Dioxide (mol wt 64) A pungent toxic gas used as a food preservative.

$$O_2S$$

Sulfur Dioxygenase The enzyme that catalyzes the following reaction:

$$\text{Sulfur} + O_2 + H_2O \rightleftharpoons \text{Sulfite}$$

Sulfur Oxidizing Bacteria The bacteria capable of oxidizing reduced sulfur compounds, e.g., *Thiobacillus thiooxidans*.

Sulfur Reducing Bacteria Bacteria capable of using elemental sulfur (S^0) as terminal electron acceptor in anaerobic respiratory metabolism.

Sulfur Reductase The enzyme that catalyzes the reduction of elemental sulfur or polysulfide to H_2S.

Sulfur Respiration An energy-yielding metabolism in which elemental sulfur (S^0) is used as terminal electron acceptor.

Sulindac (mol wt 356) An anti-inflammatory agent.

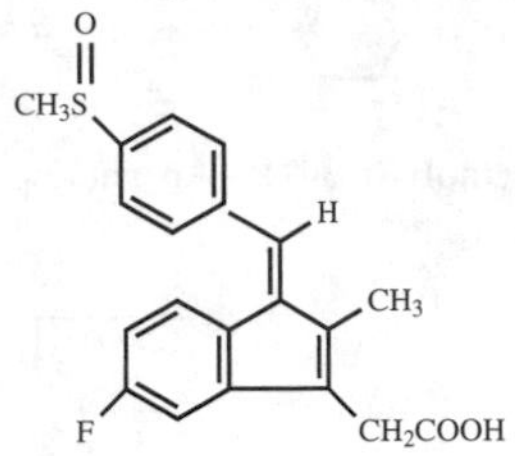

Sulisatin (mol wt 492) A laxative that stimulates motility of the large intestine.

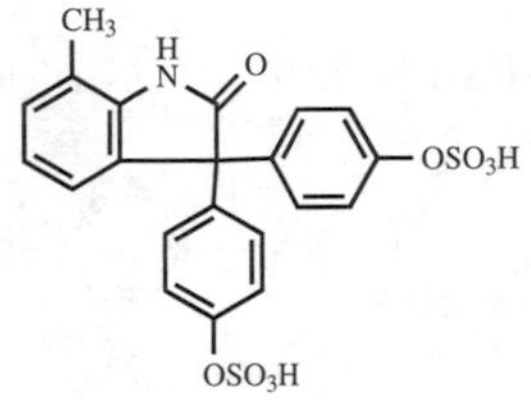

Sulkowitch Test A test for determination of calcium in the urine based on the measurement of turbidity after addition of oxalate to the urine.

Sullivan Test A colorimetric test for the determination of cysteine based on treatment of sample with 1,2-naphthoquinone-4-sulfonate and sodium sulfite.

Sulmarin (mol wt 352) A hemostatic agent.

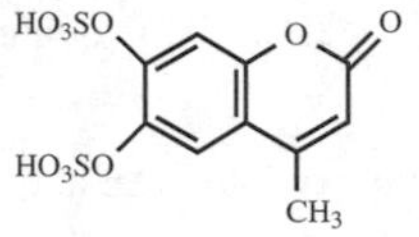

Sulmazole (mol wt 287) A cardiotonic agent.

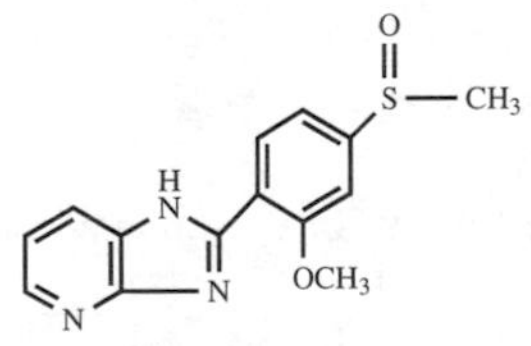

Sulmepride (mol wt 327) A neuroleptic agent that blocks presynaptic dopaminergic receptors.

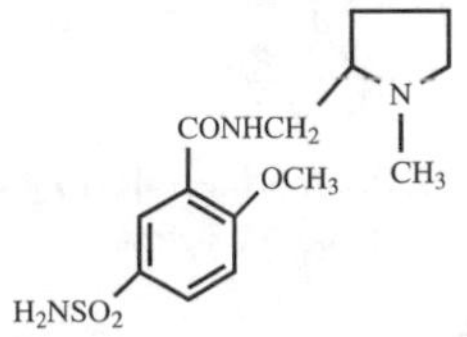

Sulmeprin A trade name for cotrimoxazole, an antibacterial agent.

Suloctidil (mol wt 338) A peripheral vasodilator.

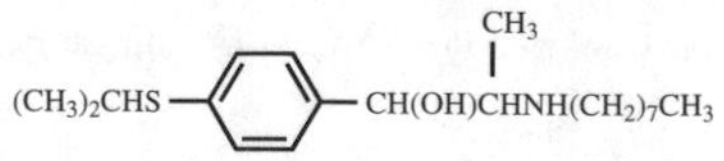

Sulpiride (mol wt 341) An antidepressant, antipsychotic agent.

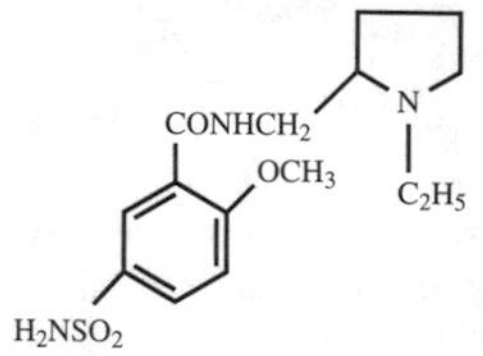

Sulprofos (mol wt 322) An insecticide.

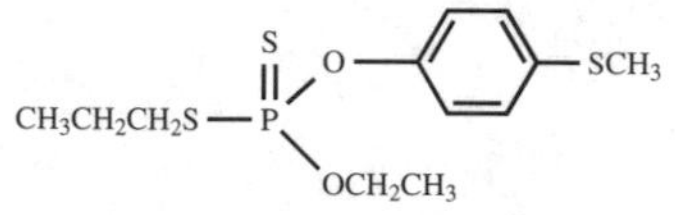

Sulprostone (mol wt 466) An abortifacient and an analog of prostaglandin E_2.

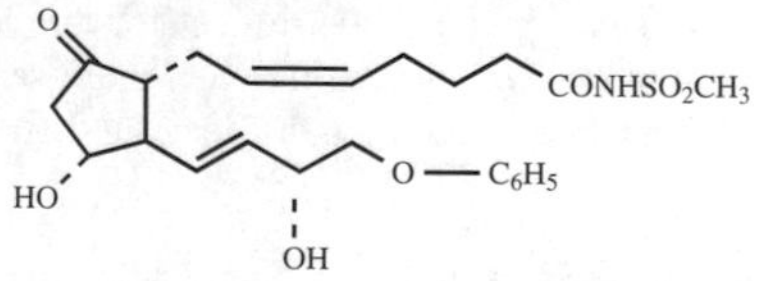

Sulquin A trade name for quinine sulfate, an anti-malarial agent.

Sulthiame (mol wt 290) An anticonvulsant.

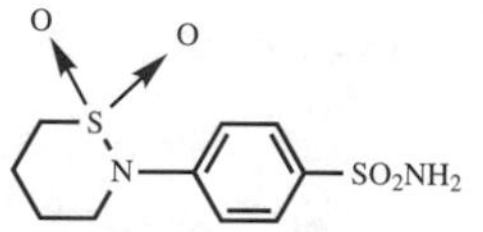

Sultopride (mol wt 354) An antidepressant.

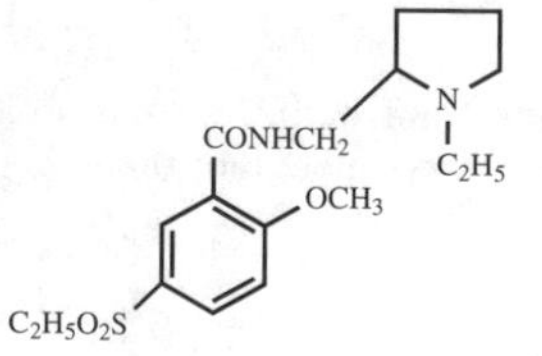

Sultosilic Acid, Piperazine Salt (mol wt 430) An antihyperlipoproteinemic agent.

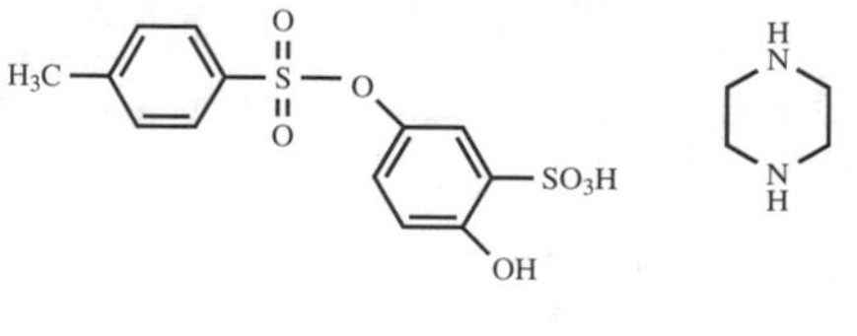

Sumatriptan (mol wt 295) An antimigraine agent.

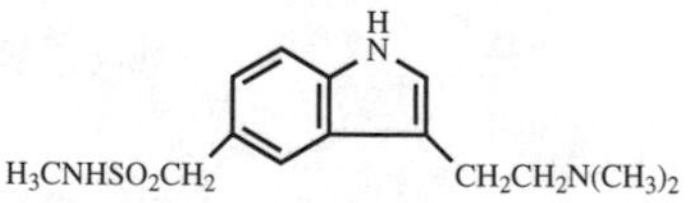

Sumycin A trade name for tetracyline hydrochloride, an antibiotic that binds to 30S ribosomal subunits, inhibiting bacterial protein synthesis.

SUN Abbreviation for serum urea nitrogen.

SunI A restriction endonuclease from *Synechococcus uniformis* with the following specificity:

```
        ↓
5'……..CGTACG……..3'
3'……..GCATGC……..5'
             ↑
```

Supasa A trade name for aspirin.

Superantigen An antigen capable of activating a large proportion of T lymphocytes in a given individual by interaction with various domains of antigen receptor on the T cell.

Superchar A trade name for activated charcoal.

Supercoil DNA A structure formed by the twisting of double-stranded DNA helix upon itself either in a circular DNA molecule or in a DNA loop anchored at both ends.

Superfamily Any group of genes and their cognate proteins that can be related by sequence homology.

Superhelix See supercoil DNA.

Superinfection The repeated infection of a bacterial culture that has been infected previously with a bacteriophage.

Superinfection Exclusion A phenomenon in which a virus-infected cell resists infection by a second virus of the same or similar type.

Superinfection Immunity See superinfection exclusion.

Supermolecule Referring to a protein complex that functions as an energy transducing unit in the inner membrane of mitochondria, e.g., complex of ATP synthetase, transprotonase, and transhydrogenase.

Supernatant See supernate.

Supernate The liquid fraction above sediment after centrifugation.

Superoxide Dismutase The enzyme that catalyzes the following reaction:

$$O_2^- + O_2^- + 2H^+ \rightleftharpoons O_2 + H_2O$$

Superoxide Radical A toxic free radical of oxygen (O_2^-).

Superrepressor A repressor that binds tightly and permanently to the operator, which therefore represses enzyme synthesis regardless of whether inducer is present.

Supersecondary Structure The specific clusters of secondary structure motifs in proteins, e.g., βαβ, ααα, and βαβαβ structure in the protein.

Supeudol A trade name for oxycodone hydrochloride, an analgesic agent that binds to opiate receptors in the CNS, altering both perception and emotional response to pain.

Suplatast Tosylate (mol wt 500) An antiallergic and antiasthmatic agent.

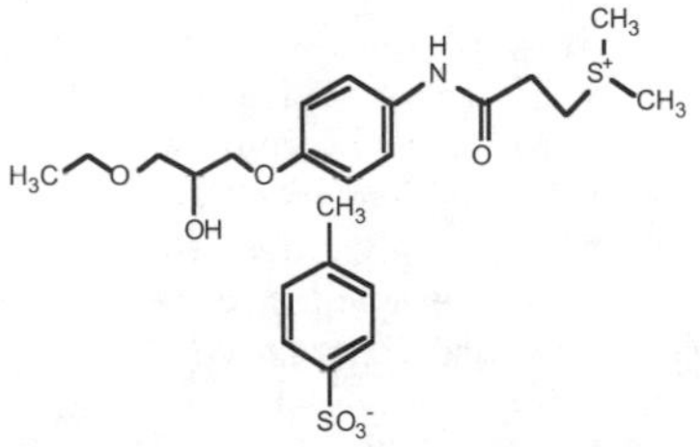

Suppap A trade name for acetaninophen, an antipyretic and analgesic agent.

Supprelin A trade name for histrelin acetate, which mimics the effects of gonatropin-releasing hormone.

Suppressor Gene A gene capable of reversing the effect of another mutation.

Suppressor Lymphocyte (T_s Cell) Subclass of T cells that suppresses the expression of cellular or humoral immune responses to antigen.

Suppressor Mutation Mutation that alters the anticodon of tRNA, causing it to insert an amino acid at a stop codon generated by another mutation.

Suppressor Sensitive Mutant A mutant that has the wild-type phenotype when a suppressor is present (designated *sus*).

Suppressor Strain Strain of bacteria with a mutant tRNA molecule that inserts an amino acid at the termination codon.

Suppressor T Cell The CD8+ T cells that suppress the immune response.

Suppressor tRNA A mutant tRNA molecule that can insert an amino acid at a termination codon.

Suppuration The formation of pus.

Supra- A prefix meaning above.

Supramolecular Structure Component of a cell consisting of macromolecules organized into a variety of multimolecular assemblies, e.g., ribosome, and multiple-enzyme complex.

Suprane A trade name for desflurane, an anesthetic agent.

Suprax A trade name for cefixime, an antibacterial agent that inhibits bacterial cell wall synthesis.

Suprazine A trade name for trifluoperazine hydrochloride, an antipsychotic agent that blocks postsynptic dopamine receptors in the brain.

Supres A trade name for hydralazine hydrochloride, an antihypertensive agent that relaxes arteriolar smooth muscle.

Suprofen (mol wt 260) An anti-inflammatory and analgesic agent.

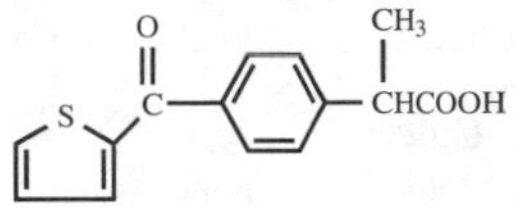

SUr Abbreviation for thiouracil or thiouridine.

Surface Active Compound A detergent-like molecule or substance that is capable of lowering the surface tension.

Surface Antigens Antigens associated with cell surfaces, e.g., O antigen of bacteria or hepatitis surface antigen.

Surface Chemistry The science that deals with the observation and measurement of forces acting at the surfaces of gases, liquids, and solids, or at the interfaces between them.

Surface Phagocytosis The enhancement of phagocytosis by entrapment of organisms on surfaces.

Surface Tension The cohesiveness of the surface of a liquid.

Surfactant Referring to a surface-active compound that reduces the surface tension.

Surfactin A surface-active lipopeptide produced by *Bacillus faecalis* consisting of a heptapeptide covalently linked to a hydroxy fatty acid.

Surfak A trade name for docusate calcium, a laxative that promotes incorporation of fluid into the stool.

Surgicel A trade name for oxidized cellulose.

Suriclone (mol wt 478) An anxiolytic agent.

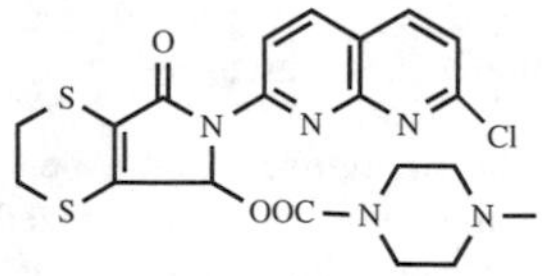

Surmontil A trade name for trimipramine maleate, an anti-depressant that increases the concentration of norepinephrine or serotonin in the CNS.

Survanta A trade name for beractant (natural lung surfactant).

Survivin A protein expressed in the G_2/M phase of the cell division cycle and associated with microtubules of the mitotic spindle.

Susceptibility A state of being open to disease or infection.

***sus* Mutant** A suppressor-sensitive mutant, e.g., amber mutant.

Suspension A colloid dispersion in which the particles are undissolved in the solution.

Suspension Culture A type of culture in which cells are suspended in a liquid medium.

Suspensor Cell A plant cell that links the growing embryo to the wall of the embryo sac in the developing seed.

Sus-Phrine A trade name for epinephrine hydrochloride, a bronchodilator that stimulates both alpha- and beta-adrenergic receptors within the sympathetic nerve system.

Sustaire A trade name for theophylline, a bronchodilater that inhibits phosphodiesterase and increases cAMP concentration.

Sustiva A trade name for efavirenz, a reverse transcriptase inhibitor used as an antiviral drug.

SUV Abbreviation for small unilamellar vesicle.

Suxethonium Bromide (mol wt 478) A neuromuscular blocker.

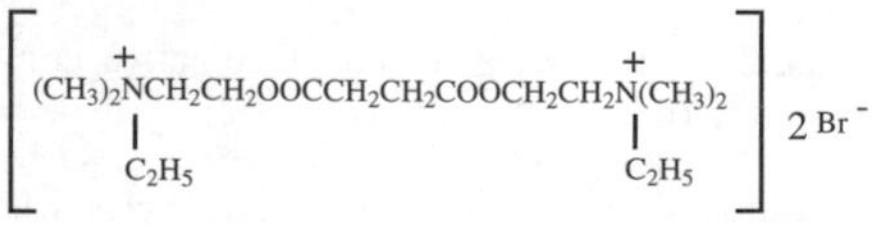

Suxibuzone (mol wt 438) An anti-inflammatory agent.

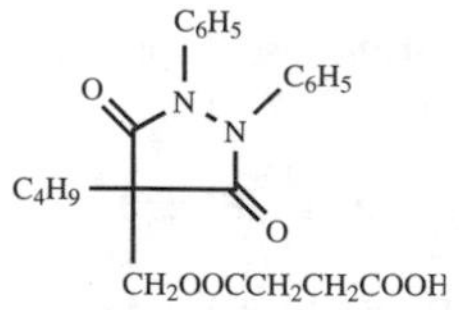

SV Abbreviation for 1. salk vaccine; 2. simian virus; 3. synaptic vesicle.

SV40 Abbreviation for simian virus 40 (SV40), a virus of the family Papovaviridae.

SV40 Tag Abbreviation for simian virus 40 encoded large tumor antigen.

S-Value Referring to Svedberg unit.

Svedberg Equation An equation that relates the relative molecular mass of a solute to its sedimentation velocity in an applied centrifugal field. The equation is as follows:

$$M = sRT/D(1\text{-}\nu p)$$

where

M = relative molecular weight
s = sedimentation coefficient
ν = particle specific volume
D = diffusion coefficient
p = density of the solution
T = absolute temperature
R = gas constant

Svedberg Unit A unit for expressing the sedimentation coefficient of biological macromolecules (1S = 10^{-13} second).

SV3TS Swiss 3T3 cells transformed by SV40.

SW13 Abbreviation for human adrenal adenosarcinoma cell line.

S_{w20} The sedimentation coefficient determined in water at 20°C.

SW480 Abbreviation for human colon carcinoma cell line.

SW620 Abbreviation for human colon carcinoma cell line.

Swelling A cardinal sign of the inflammatory response associated with the accumulation of fluid at the site of inflammation.

Swim Ear A trade name for boric acid, used to destroy bacteria in the ear canal.

Swinging Bucket Rotor A special type of centrifuge rotor used for density gradient centrifugation in which the tube holders swing out at right angles to the axis of rotation.

Swiss 3T3 Cells An immortal line of fibroblast-like cells derived from trypsinized embryo of the Swiss mouse.

Switch Referring to a change in synthesis of heavy chain isotypes during differentiation without affecting V-region expression.

Switch Cell A subset of T lymphocytes existing with Peyer's patches that governs isotype differentiation of B lymphocytes to ensure the formation of IgA-producing plasma cells in Peyer's patches.

Switch Gene A gene that causes an organism or cell to follow a different differentiation pathway.

Switch Peptide The flexible covalent connection between the variable region and constant region of an immunoglobulin.

Switch Region The regions within the immunoglobulin heavy chains that combine with each other to delete intervening DNA sequences, thus switching the class of immunoglobulin to be synthesized by the cell.

Switch Site The breakage points on a chromosome where gene segments unite during gene rearrangement.

Swivel Point The nick point at which DNA unwinds before a new strand can be formed during the replication of the DNA.

Swivelase Type-I topoisomerase.

Syllact A trade name for psyllium, a laxative that absorbs water and expands bulk and moisture content of the stool.

Sym- A combination meaning together.

Symadine A trade name for amantadine hydrochloride, an antiviral agent (anti-influenza).

Symbiogenesis The evolutionary development of chloroplasts and mitochondria from endosymbiotic microorganisms.

Symbiont An organism that lives in symbiotic association with another organism.

Symbiosis An intimate association of two dissimilar organisms in which both derive benefits from the association.

Symbiotic Nitrogen Fixation The fixation of atmospheric nitrogen by bacteria living in symbiotic association with plants, e.g., *Rhizobium* and leguminous plants.

Symclosene (mol wt 232) A topical anti-infective agent.

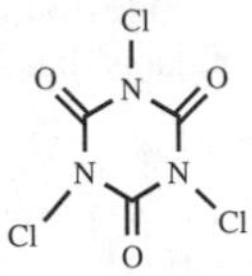

Symmetrel A trade name for amantadine hydrochloride, an antiviral agent (anti-influenza).

Symmetrical Replication Bidirectional replication of DNA.

Symmetric Transcription The synthesis of RNA from the complementary segments of the two strands of DNA by the core enzyme RNA polymerase *in vitro*.

Symmetrogenic Pertaining to longitudinal cell fission producing two daughter cells that are mirror images of each other.

Sympathetic Nerve A nerve of the sympathetic nervous system.

Sympathetic Nervous System The part of the vertebrate autonomic nervous system that uses epinephrine or norepinephrine as neurotransmitters. Its activities include increasing blood pressure and acceleration of the heart beat.

Sympatholytics Inhibiting the sympathetic nervous system.

Sympathomimetics Producing effects similar to those caused by the stimulation of the sympathetic nervous system.

Sympatric Living in the same environment or geographical region.

Symplast The continuous meshwork of the interior of living cells in the plant due to the presence of plasmodesmata.

Symport A membrane transport protein that carries two substances in the same direction across the membrane.

Symptom The subjective evidence of a disease or physical disturbance observed in plants, animals, and humans.

Symptomatology The study that deals with symptoms of diseases or disorders.

Syn- A prefix meaning together.

Synacort A trade name for hydrocortisone.

Synalar A trade name for fluocinolone acetonide, a dermatomucosal agent.

Synalbumin A peptide found in the blood of some diabetic individuals that acts as an inhibitor for insulin.

Synalgos DC A trade name for a combination drug containing aspirin, caffeine, and dihydrocodeine used as an analgesic agent.

Synapse The site of communication between two nerve cells (cell-cell junction) that allows signals to pass from one nerve cell to another. In a chemical synapse the signal is carried by a diffusible neurotransmitter.

Synapsin A protein that links the synaptic vesicle to the cytoskeleton and serves as substrate for Ca^{2+}-calmodulin and cAMP-dependent protein kinase.

Synapsis The pairing of homologous chromosomes during prophase I in meiosis.

Synaptic Cleft Gap between the presynaptic and postsynaptic membranes at the junction between two nerve cells.

Synaptic Transmission The process of the propagation of a signal from one cell to another via a synapse.

Synaptic Vesicle Membrane-bound structure filled with neurotransmitter molecules that are discharged exocytotically into the synaptic cleft upon arrival of a nerve impulse.

Synaptogenesis The formation of nerve synapses.

Synaptonemal Complex Structure that holds paired chromosomes together during prophase I of meiosis.

Synaptophysin An integral membrane glycoprotein found in presynaptic vesicles of neurons.

Synaptosome A preparation from nerve tissue that is rich in nerve endings.

Synaptotagmin An abundant integral membrane protein of synaptic vesicles.

Syncaryon Synonym for synkaryon.

Synchronous Culture A culture in which all cells pass through the same stage of the cell cycle at the same time.

Synchrony A state or condition of a culture in which all cells are dividing at the same time.

Syncytia Plural of syncytium.

Syncytium Mass of cytoplasm containing many nuclei enclosed by a single plasma membrane. It is formed as a result of either cell fusion or a series of incomplete division cycles in which the nuclei divide but the cell does not.

Syndecan An integral membrane proteoglycan that links the cytoskeleton to the interstitial matrix.

Syndrome A set of symptoms that occur together in a given disease.

Synechococcus A genus of unicellular Cyanobacteria.

Synechocystis A genus of unicellular Cyanobacteria.

Synemin A protein found in the Z-disc of skeletal and cardiac muscle cells.

Synemol A trade name for fluocinolone acetonide, a dermatomucosal agent.

Synephrine (mol wt 167) An adrenergic agent.

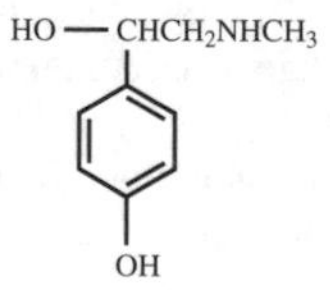

Syneresis The expulsion of a liquid from a gel due to shrinkage of the gel.

Synergism The greater activity or effect achieved by the interaction of two different agents, e.g., the effect of two or more antibiotics acting together on a given organism are greater than the additive effects of those antibiotics acting independently.

Synergistic Pertaining to synergism.

Synexin A protein that causes Ca^{2+}-dependent aggregation of isolated chromaffin granules.

Synflex A trade name for naproxen, an anti-inflammatory and analgesic agent.

Syngamy Fusion of gametes (also known as fertilization).

Syngeneic 1. Denoting the relationship that exists between genetically identical members of the same species, e.g., between identical twins. 2. Derived from genetically identical individuals.

Syngeneic Graft The graft between genetically identical individuals, e.g., between identical twins.

Synkaryon A hybrid cell resulting from the fusion of the nuclei it carries.

Synkavite A trade name for vitamin K_3 (menadione/menadiol sodium diphosphate).

Synonym Codons Codons that encode the same amino acid.

Synophylate A trade name for theophylline sodium glycinate, a bronchodilator that inhibits phosphodiesterase and increases cAMP concentration.

Synovia A transparent, sticky lubricating fluid secreted by a synovial membrane that acts as lubricant for joint, bursa, and tendons. It contains mucin, albumin, fat, and mineral salts.

Synovial Membrane Membrane of connective tissue that secrets synovial fluid.

Synovin Mucinous substance found in the synovia.

Synringic Acid (mol wt 198) An aromatic organic acid.

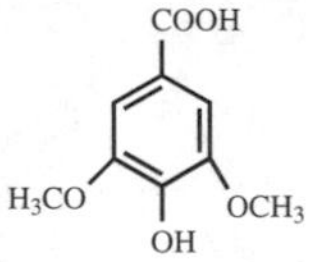

Synthermal Of the same temperature.

Synthetases The enzymes that catalyze the synthesis of molecules, their activities often coupled with hydrolysis of ATP.

Synthetic Medium A medium with known chemical composition and quantity.

Synthroid A trade name for levothyroxine sodium (a thyroid hormone), which stimulates cellular oxidation (also known as T_4).

Synthrox A trade name for levothyroxine sodium (a thyroid hormone), which stimulates cellular oxidation (also known as T_4).

Syntocinon A trade name for oxytocin, used for selective stimulation of uterine and mammary gland smooth muscle.

Syntrophism A phenomenon in which the extent of growth of an organism is dependent on the factors or nutrients provided by another organism growing in the vicinity.

Syntrophobacter A genus of Gram-negative, anaerobic, nonmotile bacteria.

Syntrophomonas A genus of Gram-negative, anaerobic bacteria.

Syntrophus A genus of Gram-negative, anaerobic bacteria.

Synuclein A brain presynaptic protein.

Syphilid The skin lesion caused by syphilis.

Syphilis A chronic, communicable, sexually transmitted disease of humans, caused by *Treponema pallidum* and characterized by a variety of lesions.

Syringaldazine (mol wt 360) A reagent for the detection of laccase and peroxidase.

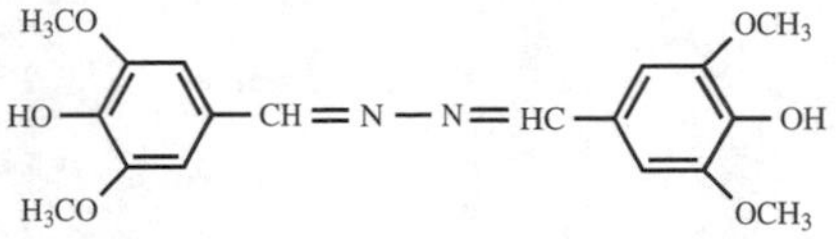

Syringaldehyde (mol wt 182) One of the components in plant lignin.

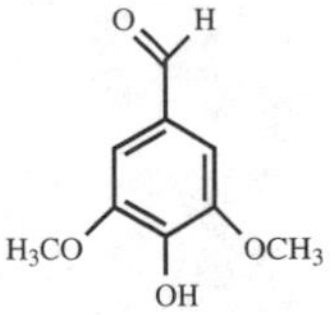

Systematics The system of scientific classification of organisms.

Systemic Infections The infections that are disseminated throughout the body via the circulatory system.

Systemic Lupus Erythematosus (SLE) An autoimmune disease resulting from the failure of the immune system to recognize self antigens. It is characterized by severe swelling of blood vessels and by kidney disorders. Antinuclear antibodies against DNA, histone proteins, and nonhistone protein bound to DNA are found in patients with this syndrome.

Systemin An 18-residue polypeptide isolated from tomato leaves, it induces two proteinase inhibitors in tomato and potato leaves.

Systole The tightening of the heart that drives blood forward in the circulation system.

Systolic Pressure The highest arterial blood pressure of a cardiac cycle.

T

t Symbol for Student's t test, a test used to test the difference between the means of two samples.

$t_{1/2}$ Symbol for half-time.

T Abbreviation for 1. thymine, thymidine, threonine, and tritium, 2. absolute temperature, 3. transmittance, 4. tesla, and 5. tensed configuration of an allosteric enzyme.

T1 A bacteriophage of the family Styloviridae with a long, noncontractile tail and containing double-stranded DNA.

T_2 Abbreviation for di-iodo-thyronine.

T2 A bacteriophage of the family Myoviridae with a long, contractile tail and containing double stranded DNA.

T3 1. A bacteriophage of the family Podoviridae with a short noncontractile tail and containing double-stranded DNA. 2. Symbol for triiodothyronine, a hormone that stimulates tissue metabolism by accelerating the rate of cellular oxidation.

T4 1. A bacteriophage of the family Myoviridae with a long contractile tail and containing double-stranded DNA. 2. Symbol for levothyroxine, a thyroid hormone that stimulates tissue metabolism by accelerating the rate of cellular oxidation.

T5 A bacteriophage of the family Styloviridae with a long, noncontractile tail and containing double-stranded DNA.

T6 A bacteriophage of the family Myoviridae with a long, contractile tail and containing double-stranded DNA.

T7 A bacteriophage of the family *Podoviridae* with a short, noncontractile tail and containing double-stranded DNA.

T24 Abbreviation for human bladder carcinoma cell line.

3T3 Symbol for a type of cell line.

293T An epithelium cell line.

$T_{1/2}$ Symbol for the time required for a substance to be diminished to one half of its original level of activity by a biological, chemical, or physical process.

T_{90} The time required for 90% mortality to occur for microorganisms exposed or treated with an antimicrobial agent.

T Antigen Tumor antigens encoded by *Papovaviruses* that are involved in the transformation of cells. SV40 has two T-antigens (small t and large T); polyoma virus has three T-antigens (large, small, and middle T).

T Cell Lymphocytes that are differentiated in the thymus and are important in cell-mediated immunity and modulation of antibody-mediated immunity.

T Cell Antigen Receptor A protein receptor on the surface of T cells that recognizes an antigenic determinant (also known as T-cell receptor).

T Cell Dependent Antigen The antigen that induces antibody production only in the presence of T cells.

T Cell Growth Factor Referring to interleukin 2.

T Cell Independent Antigen The antigen that is capable of inducing antibody production in the absence of T cells, e.g., polysaccharide antigens from bacteria. The antibody induced by T-cell independent antigens are of type IgM.

T Cell Receptor See T-cell antigen receptor.

T Cell Rosette A rosette formed between a T cell and erythrocytes (also known as E rosette).

T4 DNA Ligase DNA ligase from bacteriophage T4.

T Form Synonym of tense form.

T4 Ligase The enzyme isolated from T4 bacteriophage that catalyzes the ligation of the DNA fragments with blunt ends.

T Lymphocyte See T cells.

T4 RNA Ligase RNA ligase from T4 bacteriophage.

T Tubules Membrane channels that transmit action potentials from the surface of the skeletal muscle cell to the cell interior where T tubules make close contact with sarcoplasmic reticulum.

Ta Abbreviation for tantalum, a chemical element.

TA Abbreviation for 1. tannic acid; 2. titratable acid; 3. toxin-antitoxin; 4. transaldolase; 5. transplantation antigen; 6. tryptose agar; 7. tube agglutination; 8. tumor antigen.

TAA Abbreviation for 1. thyroid autoantibody; 2. transfusion-associated AIDS; 3. tumor-associated antibody; 4. tumor-associated antigen.

TAB An inactivated vaccine containing *Salmonella typhi and S. paratyphi* A and B.

TABT Abbreviation for a combined vaccine against typhoid A, typhoid B, and tetanus.

TABTD Abbreviation for a combined vaccine against typhoid A, typhoid B, tetanus, and diphtheria.

Tabtoxin A β-lactam-containing dipeptide produced by *Pseudomonas syringae pv tabaci* in infected tobacco plants.

Tabun (mol wt 162) A nerve gas and a potent cholinesterase inhibitor.

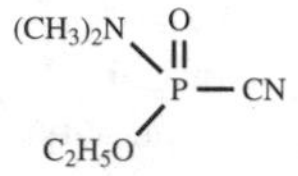

TAB-Vaccine Abbreviation for vaccine against typhoid A and typhoid B.

Tac-3 A trade name for triamcinolone acetonide, a corticosteroid used as an anti-inflammatory agent.

Tac Antigen A surface antigen on T cells associated with the receptor for interleukin 2.

***tac* Promoter** A hybrid promoter formed by fusion between elements of *lac* and *try* promoters of *E. coli.*

Tacaribe Virus A virus of the family Arenaviridae isolated from the South African bat.

Tacaryl A trade name for the methdilazine hydrochloride, an antihistaminic agent that competes with histamine for H_1 receptors on effector cells.

TACE A trade name for chlorotrianisene, a hormonal agent that increases the synthesis of DNA, RNA, and proteins.

Tachometer A device for measurement of angular velocity.

Tachykinins Referring to a group of neuropeptides.

Tachyphylaxis The diminished response of a tissue to a phamacologically active agent or substance following repeated exposure.

Tachypnea Increased rate of respiration.

Tackifiers Referring to substances used for making an adhesive more sticky.

Tackiness The property of being sticky or adhesive.

Tacrine (mol wt 198) A cholinesterase inhibitor and an antidote.

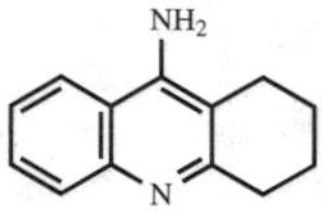

Tacrolimus (mol wt 804) An immuno-suppressant that inhibits T-cell activation.

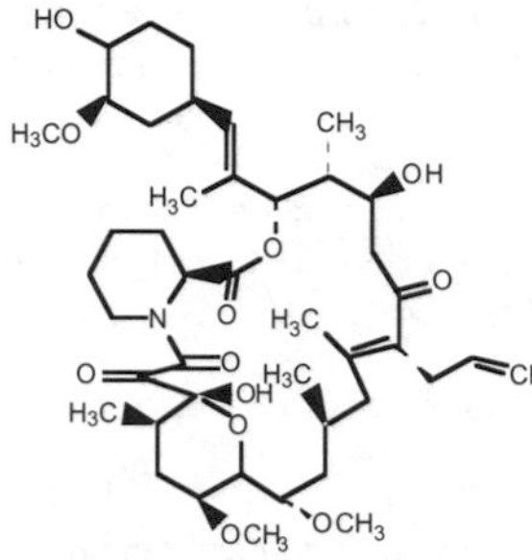

Tactic Pertaining to taxis.

Tactic Response Synonym for taxis.

Tactoid A paracrystalline aggregate that appears as a spindle-shaped structure under a polarizing microscope, e.g., aggregate of sickle cell hemoglobin in sickle cell anemia.

Tactophily A tendency to adhere or to grow on a solid surface.

TAD Abbreviation for a combination drug containing thioguanine, ara-C, and daunomycin.

Taenia A genus of large tapeworms, some of which are parasites of the human intestine.

Taeniasis An infection with tapeworms of the genus *Taenia.*

TAF Abbreviation for 1. TBP associated factor; 2. tissue angiogenesis factor; 3. toxoid antitoxin floccules; 4. trypsin aldehyde fuchsin; 5. tumor angiogenesis factor.

Tag To label a compound with a radioactive tracer or enzyme, e.g., labeling DNA with ^{32}P.

TAG Abbreviation for triacylglycerol.

Tagamet A trade name for cimetidine, an antiulcer agent that decreases gastric acid secretion.

Tagatose (mol wt 180) A six-carbon sugar.

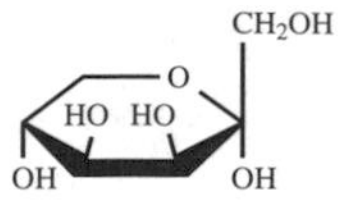

Tagatose 1,6-Bisphosphate The diphosphate form of tagatose, an intermediate in the metabolism of lactose in some bacteria.

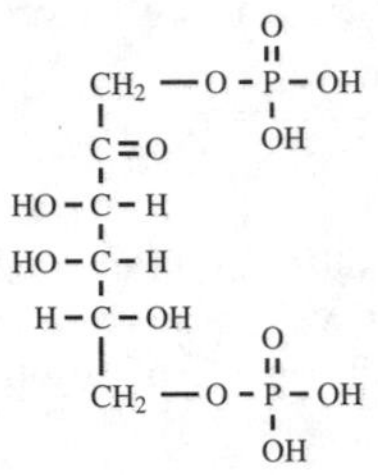

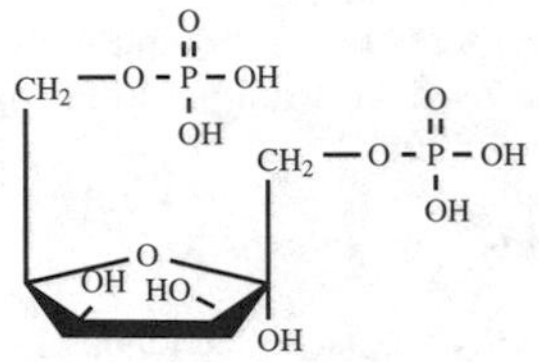

Tagatose Kinase The enzyme that catalyzes the following reaction:

$$\text{ATP} + \text{D-tagatose} \rightleftharpoons \text{ADP} + \text{tagatose 6-phosphate}$$

Tagatose 6-Phosphate Pathway A pathway for metabolism of lactose in some bacteria in which galactose resulting from hydrolysis of lactose is converted to tagatose 6-phosphate and tagatose 1, 6-bisphosphate.

Tagaturonate Reductase The enzyme that catalyzes the following reaction:

$$\text{D-altronate} + \text{NAD}^+ \rightleftharpoons \text{D-tagaturonate} + \text{NADH}$$

Taglutimide (mol wt 276) A sedative–hypnotic agent.

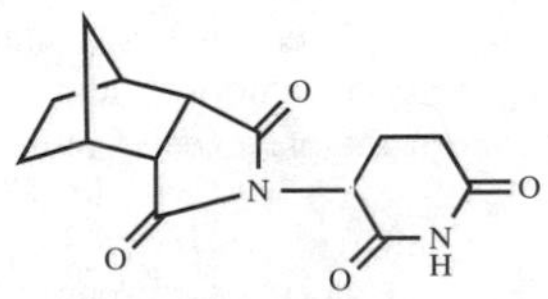

TAGVHD Abbreviation for transfusion-associated graft-vs-host disease.

Taka-Amylase Abbreviation for alpha amylase from *Aspergillus oryzae*.

Takadiastase An enzyme preparation from *Aspergillus* containing α-amylase.

Takatsi Technique A method for preparation of double dilutions using a specially designed device.

Talacen A trade name for a combination drug containing acetaminophen and pentazocine, used as an opiate analgesic agent.

Talampicillin (mol wt 482) A semisynthetic antibiotic related to penicillin that inhibits bacterial cell wall synthesis.

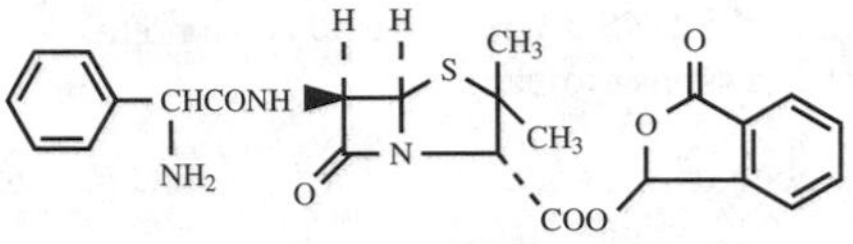

Talastine (mol wt 307) An antihistaminic agent.

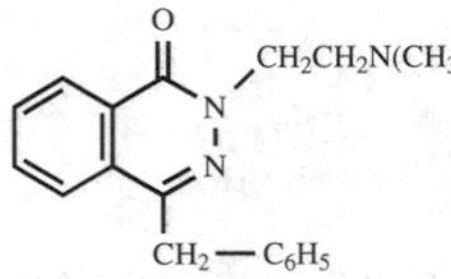

Talbutal (mol wt 224) A sedative–hypnotic agent.

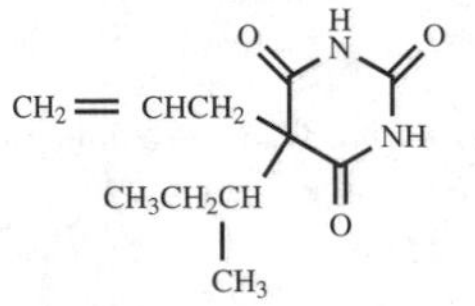

Talin A protein that links fibronectin receptors in the plasma membrane to the protein vinculin.

Talipexole (mol wt 209) An antiparkinsonian agent.

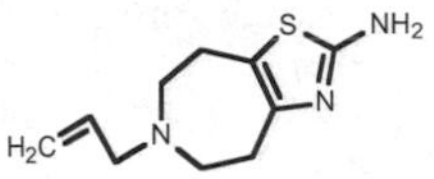

Talniflumate (mol wt 414) An anti-inflammatory and an analgesic agent.

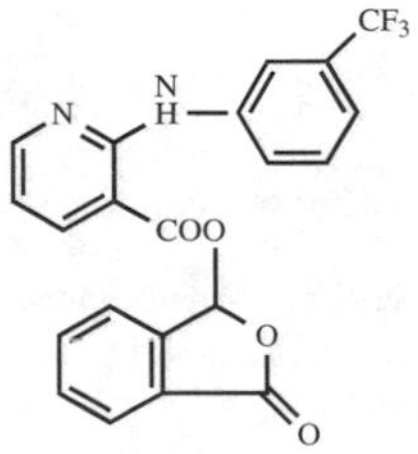

Talose (mol wt 180) An aldosugar.

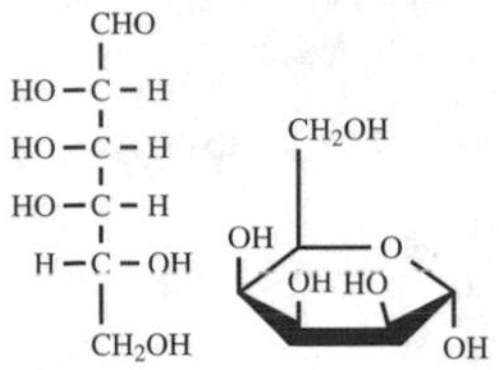

D-talose

Talwin A trade name for pentazocine hydrochloride, used as an opiate analgesic agent that binds with opiate receptors in the CNS, altering the perception and emotional response to pain.

Talwin Nx A trade name for a combination drug containing pentazocine hydrochloride and naloxone hydrochloride, which binds with opiate receptors in the CNS altering the perception and emotional response to pain.

Tambocor A trade name for flecainide acetate, an antiarrhythmic agent.

Tamfen A trade name for tamoxifen citrate, an antineoplastic agent that acts as an estrogen antagonist.

Tamm-Horsefall Glycoprotein The principal glycoprotein of human urine.

Tamone A trade name for tamoxifen citrate, an anti-estrogen.

Tamoplex A trade name for tamoxifen, an anti-estrogen.

Tamoxifen (mol wt 372) An antagonist of estrogen used as an antineoplastic agent.

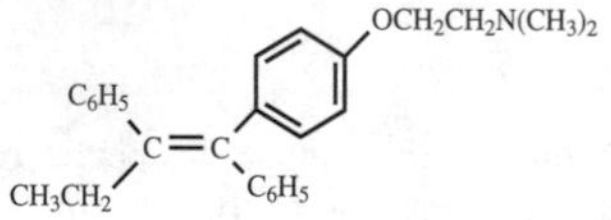

Tamsulosin (mol wt 409) An alpha adrenergic blocker that blocks the smooth muscle alpha-1 adrenergic receptors in the prostate, prostatic capsule, prostatic urethra, and bladder neck leading to the relaxation of the bladder and prostate.

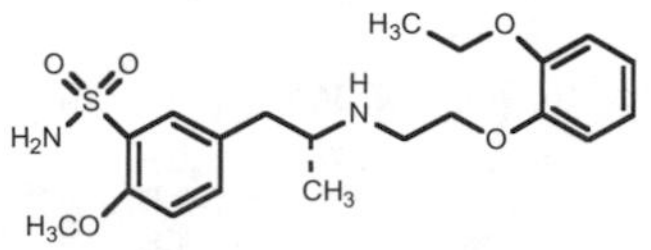

TAN Abbreviation for total ammonia nitrogen.

Tandem One after or behind another.

Tandem Duplication An aberration in which two identical chromosomal segments lie one behind the other in the same orientation.

Tandem Repeats When identical DNA segments lie one behind the other on the same chromosome and separated by a spacer.

Tandospirone (mol wt 383) An antidepressant.

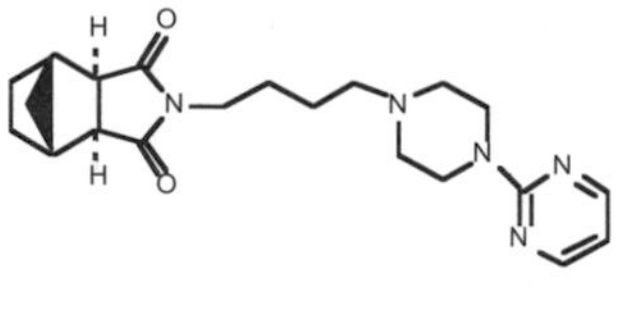

Tangier Disease A metabolic disorder characterized by the low concentration of plasma HDL and excessive deposition of cholesterol in tissues.

Tannase The enzyme that catalyzes the following reaction:

$$\text{Digallate} + H_2O \rightleftharpoons 2\ \text{Gallate}$$

Tanned Red Blood Cells Red blood cells that have been treated with tannic acid to render the surface of the cell capable of absorbing soluble antigen.

Tannic Acid-Agarose A complex of tannic acid and agarose used in affinity chromatography as supporting medium.

Tannic Acids See tannins.

Tannin Acylhydrolase See tannase.

Tannins A group of compounds with astringent taste obtained from the bark, fruits, and leaves of many plants (e.g., bark of oak). There are two groups of tannins: hydrolyzable tannins (esters of a sugar with one or more trihydroxybenzene carboxylic acids) and nonhydrolyzable tannins (derivatives of flavanols).

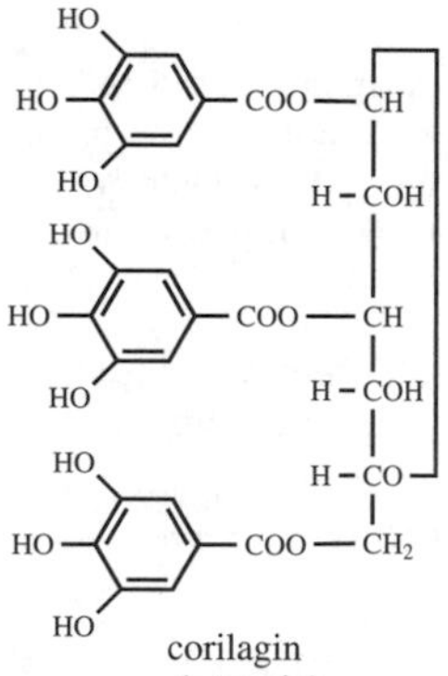

corilagin
(a tannin)

Tanret's Reagent A reagent that contains potassium iodide, mercuric chloride, and acetic acid used for testing of the presence of albumin in the urine.

Tanret's Test A test for the presence of albumin in the urine based on the formation of a white precipitate by addition of Tanret's reagent.

Tantalum (Ta) A chemical element with atomic weight 181, valences 2, 3, 4, and 5.

T-Antigen See T antigen.

TAP Abbreviation for tryptophan aminopeptidase.

Tapanol A trade name for acetaminophen, used as an analgesic and antipyretic agent that blocks the generation of pain impulses and inhibits prostaglandin synthesis.

Tapazole A trade name for methimazole, a thyroid hormone antagonist that blocks the ability of iodine to form thyroxine.

TAPDD TNF-associated protein with a death domain.

Taprostene (mol wt 399) An antithrombotic agent

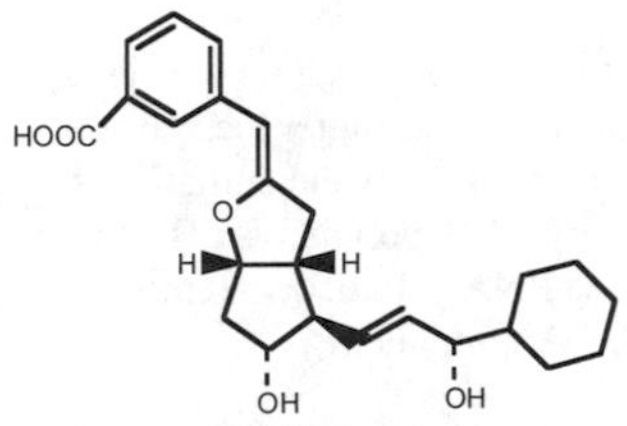

TAPX Abbreviation for thylakoid-bound ascorbate peroxidase.

Taq DNA Ligase DNA ligase from *Thermus aquaticus* (Taq).

Taq DNA Polymerase A heat-stable DNA polymerase from *Thermus aquaticus.*

Taq DP Abbreviation for *Thermus aquaticus* DNA polymerase.

TaqI A restriction endonuclease from *Thermus aquaticus* YTI with the following specificity:

```
          ↓
5′..........TCGA..........3′
3′..........AGCT..........5′
              ↑
```

TaqII A restriction endonuclease from *Thermus aquaticus* with the following specificity:

```
                        ↓
5′..........GACCGA(11N)..........3′
3′..........CTGGCT(9N)............5′
                      ↑
```

TaqXI (EcoRII) A restriction endonuclease from *Thermus aquaticus* with the following specificity:

```
             ↓
5′..........CC(A/T)GG..........3′
3′..........GG(T/A)CC..........5′
                   ↑
```

TAR Abbreviation for trans-activation response.

Tarabine PSF A trade name for cytarabine (cytosine arabinoside) that inhibits DNA synthesis.

Taractan A trade name for chlorprothixene, an antipsychotic agent that blocks postsynaptic dopamine receptors in the brain.

Tarassant A trade name for chlorprothixene, an antipsychotic agent that blocks postsynaptic dopamine receptors in the brain.

Target Cell A cell that has the appropriate receptors to bind and respond to a particular hormone or other chemical mediator.

Target Organ The organ that has hormone receptors and respond to the hormone action.

Target Sequence A short nucleotide sequence in recipient DNA where a transposon can be inserted. The target sequence is duplicated after transposon insertion so that the transposon is sandwiched between two copies of the target sequence.

Target Signal The amino acid sequence that directs the newly synthesized polypeptide or protein to its proper destination.

Target Theory The theory stating that the damage or death from radiation is caused by the inactivation of specific targets within the organism.

Target Tissue Tissue that has hormone receptors and respond to hormone action.

Targeted Gene Knockout The production of a null mutation employing an artificially designed DNA sequence that is introduced into the genome through homologous recombination and replacement of the normal allele.

Tarka A trade name for a combination drug containing trandolapril and verapamil used as an antihypertensive agent.

Taro-Sone A trade name for betamethasone dipropionate, a corticosteroid.

D-Tartaric Acid (mol wt 150) A dicarboxylic acid.

```
      COOH
       |
  HO — C — H
       |
   H — C — OH
       |
      COOH
```

DL-Tartaric Acid (mol wt 150) Racemic tartaric acid, a dicarboxylic acid.

```
COOH
 |
CHOH
 |
CHOH
 |
COOH
```

***meso*-Tartaric Acid (mol wt 150)** A dicarboxylic acid.

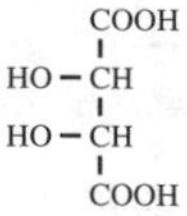

Tartrate Decarboxylase The enzyme that catalyzes the following reaction:

$$\text{Tartrate} \rightleftharpoons \text{D-glycerate} + CO_2$$

Tartrate Dehydratase The enzyme that catalyzes the following reaction:

$$\text{Tartrate} \rightleftharpoons \text{Oxaloacetate} + H_2O$$

Tartrate Dehydrogenase The enzyme that catalyzes the following reaction:

$$\text{Tartrate} + NAD^+ \rightleftharpoons \text{Oxaloglycolate} + NADH$$

***meso*-Tartrate Dehydrogenase** The enzyme that catalyzes the following reaction:

$$\text{meso-Tartrate} + NAD^+ \rightleftharpoons \text{Dihydroxyfumarate} + NADH$$

Tartrate Epimerase The enzyme that catalyzes the following reaction:

$$(R,R)\text{-Tartrate} \rightleftharpoons \text{meso-Tartrate}$$

Tartrazine (mol wt 534) A dye.

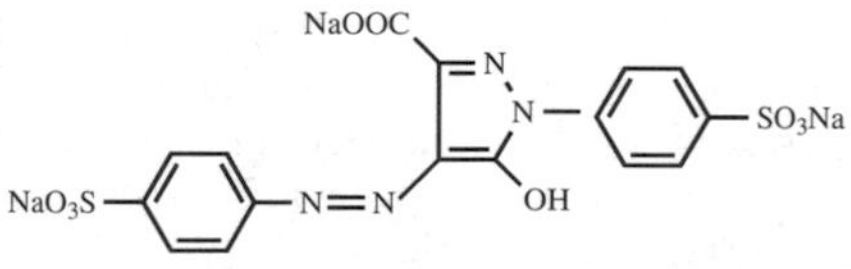

Tartronate Semialdehyde Carboxylase See tartronate semialdehyde synthetase.

Tartronate Semialdehyde Reductase See tartronate semialdehyde synthetase.

Tartronate Semialdehyde Synthetase The enzyme that catalyzes the following reaction:

$$2\ \text{Glyoxylate} \rightleftharpoons \text{Tartronate semialdehyde} + CO_2$$

Tartronic Acid (mol wt 120) A dicarboxylic acid.

$$HOCH(COOH)_2$$

Tarui's Disease A glycogen storage disease due to a deficiency in phosphorylase kinase.

TASA Abbreviation for tumor-associated surface antigen.

Tasmar A trade name for tolcapone, an antiparkinsonism drug.

TAT Abbreviation for 1. tetanus antitoxin; 2. total antitryptic activity; 3. tumor-associated trypsinogen; 4. tyrosine aminotransferase.

TAT-2 Abbreviation for tumor-associated trypsinogen-2.

TATA Referring to the nucleotide sequence of thymine, adenine, thymine, and adenine in DNA.

TATA Binding Protein A protein that binds to TATA box for stimulation of RNA polymerase II to initiate transcription.

TATA Box A consensus tetranucleotide sequence of the eukaryotic promoter located about 30 nucleotides upstream from the initiation site for transcription by RNA polymerase II. It is involved in positioning RNA polymerase-II molecules correctly on the DNA template.

TATA-Lu Abbreviation for TATA-luciferase.

Tatumella A genus of bacteria of the family Enterobacteriaceae.

tau (τ) A letter in the Greek alphabet.

***tau* Protein** Accessory microtubular protein that enhances the polymerization of tubulin subunits.

***tau* Protein Kinase** The enzyme that catalyzes the following reaction:

$$\text{ATP} + \text{tau-protein} \rightleftharpoons \text{ADP} + \text{phospho-tau-protein}$$

Taurine (mol wt 125) An aminosulfonic acid derived from cysteine that forms bile salt with bile acid and is a detoxifier.

$$NH_2CH_2CH_2SO_3H$$

Taurine Dehydrogenase The enzyme that catalyzes the following reaction:

$$\text{Taurine} + H_2O + \text{acceptor} \rightleftharpoons \text{Sulfoacetaldehyde} + NH_3 + \text{reduced acceptor}$$

Taurine Transaminase The enzyme that catalyzes the following reaction:

$$\text{Taurine} + \alpha\text{-ketoglutarate} \rightleftharpoons \text{Sulfoacetaldehyde} + \text{glutamate}$$

Taurochenodeoxycholic Acid (mol wt 500) A bile acid.

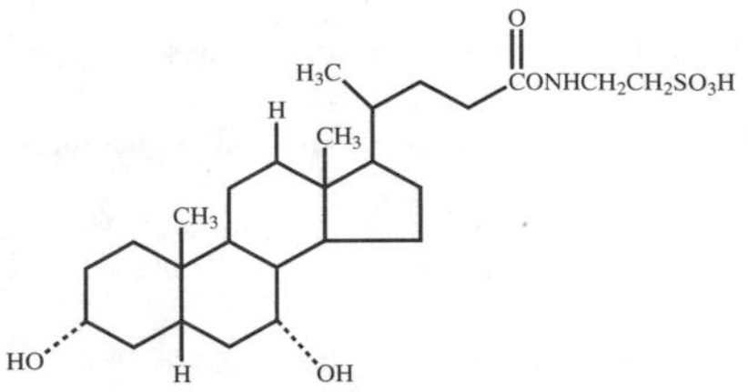

Taurocholic Acid (mol wt 516) Conjugation product of cholic acid with taurine and a choleretic agent.

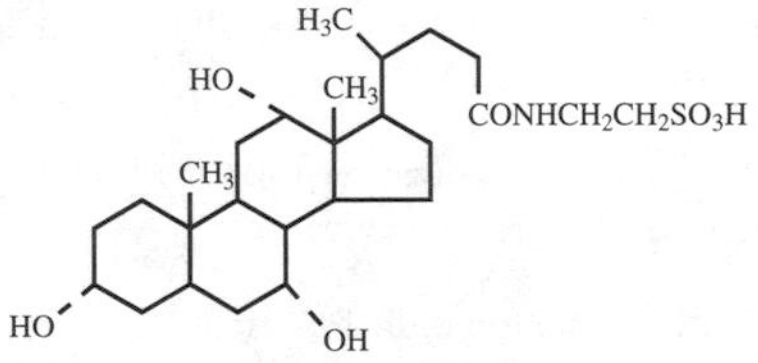

Taurocyamine Kinase The enzyme that catalyzes the following reaction:

ATP + taurocyamine

⇌

ADP + *N*-phosphotaurocyamine

Taurolidine (mol wt 284) An antibacterial agent.

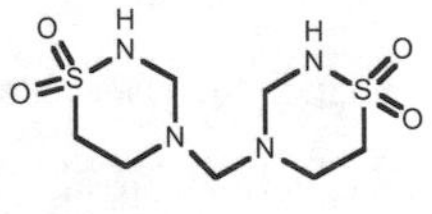

Tauropine Dehydrogenase The enzyme that catalyzes the following reaction:

Tauropine + NAD^+ + H_2O

⇌

Taurine + pyruvate + NADH

Tautomeric Shift A reversible change in the location of a hydrogen atom and double bond in a molecule that changes the molecule from one isomeric form to another and alters the chemical properties of the molecule.

Tautomerization Change in the location of a proton in a molecule leading to an alteration in the chemical property of the molecule.

Tautomers The alternate chemical forms of the same molecule that differ in their hydrogen-bonding properties.

Ta-Verm A trade name for piperazine citrate, an anthelmintic agent.

Tavist A trade name for clemastine fumarate, an antihistaminic agent that competes with histamine for H_1 receptors on effector cells.

Taxa Plural of taxon.

Taxis A directional locomotive response to a given stimulus exhibited by certain motile organisms or cells.

Taxol (mol wt 854) An antineoplastic agent isolated from the western yew evergreen tree (*Taxus brevifolia*) that inhibits mitosis.

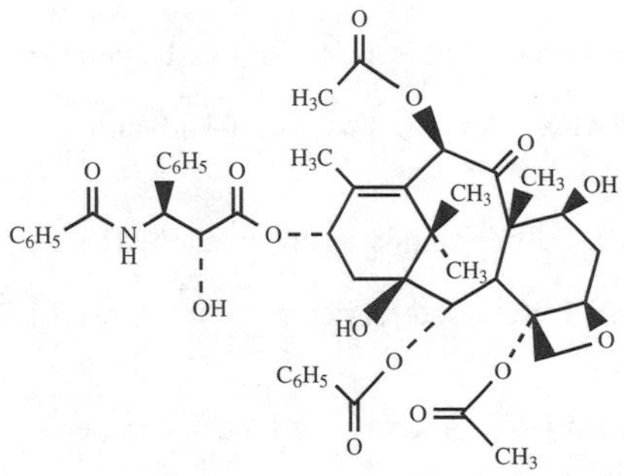

Taxonomy The science that deals with the classification or grouping of organisms according to their mutual affinities or similarities.

Taxotere A trade name for docetaxel, an antineoplastic agent.

Tay-Sachs Disease A disorder caused by the deficiency of the enzyme that breaks down gangliosides and characterized by retardation and early death.

Tazanolast (mol wt 289) An antiallergic agent.

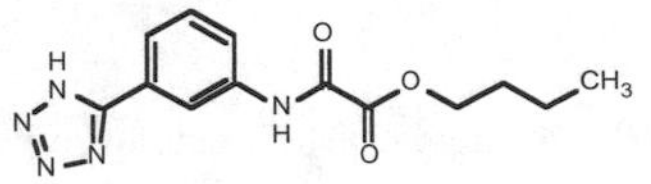

Tazarotene (mol wt 351) An anti-acne and antipsoriatic agent.

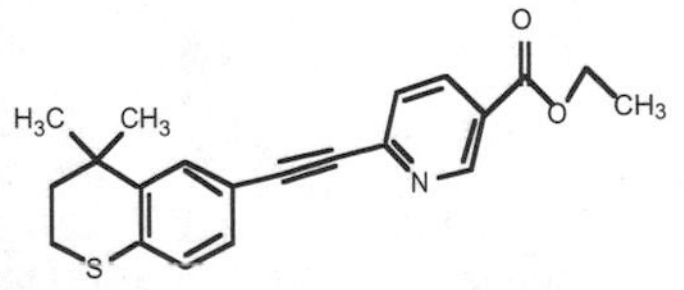

Tazicef A trade name for ceftazidime, an antimicrobial agent that inhibits bacterial cell wall synthesis.

Tazidime A trade name for ceftazidime, an antimicrobial agent that inhibits bacterial cell wall synthesis.

Taziprinone Hydrochloride (mol wt 458) An antitussive agent.

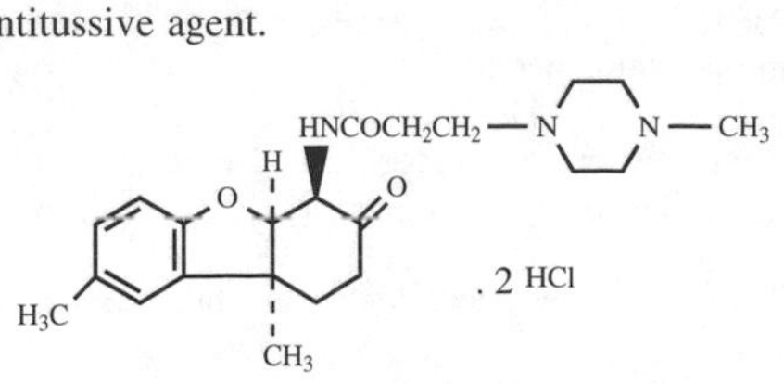

Tazorac Gel A trade name for a gel containing tazarotene used as an anti-acne and anti-psoriatic agent.

Tb Abbreviation for terbium, a chemical element.

TB Abbreviation for 1. tuberculin; 2. tuberculosis.

TB Fever Abbreviation for tick-borne fever.

TBARS Abbreviation for thiobarbituric acid reactive substances.

TBE Abbreviation for tris/borate/EDTA.

TBE Buffer Abbreviation for tris/borate/EDTA buffer.

TBE Virus Abbreviation for tick-borne encephalitis virus.

TBG Abbreviation for thyroxine-binding globulin.

Tbilisi Phage A bacteriophage of the family Podoviridase with a short, noncontractile tail and containing double-stranded DNA.

***t*Boc** Abbreviation for tertiary butyloxycarbonyl group, an amino group protecting agent used in solid phase protein synthesis.

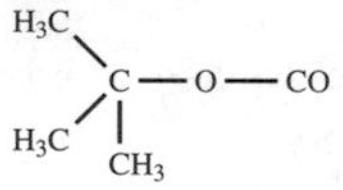

tB-OOH Abbreviation for t-butylhydroperoxide.

tBOQ Abbreviation for *t*-butylhydroquinone.

TBP Abbreviation for TATA-box binding protein.

TBPA Abbreviation for thyroxine-binding prealbumin.

TbpA Abbreviation for meningococcal transferrin-binding protein A.

TbpA+B Abbreviation for meningococcal transferrin-binding protein A+B.

TbpB Abbreviation for meningococcal transferrin-binding protein B.

TBS Abbreviation for tris-buffered saline.

TBSN Abbreviation for tris-buffered saline containing Nonidet P40.

TBST Abbreviation for tris-buffered saline containing Tween.

TBSV Abbreviation for tomato bushy stunt virus.

tBu Abbreviation for tertiary butyl.

TBUA Abbreviation for thio-barbituric acid.

Tc Abbreviation for technetium, a chemical element.

TC Abbreviation for 1. terminal cisternase; 2. tissue culture.

3TC A trade name for lamivudine, an antiviral agent that inhibits HIV reverse transcriptase.

TCA Abbreviation for 1. tricarboxylic acid; 2. trichloroacetic acid.

TCA Cycle Abbreviation for tricarboxylic acid cycle (see also Krebs cycle).

TCA-PF Abbreviation for trichloroacetic acid-precipitable fraction.

TCAR Abbreviation for thymus cell antigen receptor.

TψC Arm A small loop in tRNA that contains ribothymidine, pseudouridine, and cytidine nucleotides.

TCBS Agar Abbreviation for thiosulfate-citrate-bile salt agar used for isolation of *Vibrio* bacteria.

TCD_{50} Abbreviation for tissue culture dose 50%.

TCDC Abbreviation for taurochenodeoxycholate.

TCDD Abbreviation for tetra-chloro-dibenzo-dioxin.

TC-Detector Abbreviation for thermal conductivity detector used in gas-liquid chromatography for detection of chromatographically separated compounds.

TCE Abbreviation for trichloroethylene.

TceI (MboII) A restriction endonuclease from *Thermococcus celer* with the same specificity as MboII.

TCGF Abbreviation for T-cell growth factor.

$TCID_{50}$ Abbreviation for tissue culture infectious dose 50, an agent that causes infection in 50% of the inoculated tissue culture samples.

TCIS Abbreviation for T cell immunodeficiency syndromes.

TCL Abbreviation for T cell lymphoma.

TCM Abbreviation for tris-calcium-magnesium buffer.

TCMIF Abbreviation for tumor cell migration inhibition factor.

TCR Abbreviation for T-cell receptor.

T-Cypionate A trade name for testosterone cypionate, an anabolic steroid that stimulates tissue building processes.

tD Symbol for thermal death time, the time required for heat killing of a given population of microorganisms at a specified temperature.

TD Abbreviation for 1. tetanus/diphtheria; 2. threonine dehydrogenase; 3. thoracic duct; 4. thymus dependent; 5. tryptophan dehydrogenase; 6. tyrosine decarboxylase.

T47D Abbreviation for a human breast cell carcinoma cell line.

TD_{50} Abbreviation for toxic dose 50, a dose of toxic compound that kills 50% of the testing subject.

TD Antigen Abbreviation for thymus-dependent antigen.

TDC Abbreviation for 1. tryptophan decarboxylase; 2. tyrosine decarboxylase, taurodexoycholate.

TDE Abbreviation for tetrachlorodiphenyl ethane.

TDGA Abbreviation for tetradecylglycidic acid.

TDH Abbreviation for tryptophan dehydrogenase.

tDNA DNA that encodes tRNA.

T-DNA A segment of DNA in the Ti plasmid of *Agrobacterium tumefaciens* that can be integrated into a plant cell and cause tumor formation.

T7-DNAP Abbreviation for T7 DNA polymerase or DNA polymerase from bacteriophage T4.

TDP Abbreviation for thymidine diphosphate.

TDPA Abbreviation for tetradecanoyl-phorbol acetate.

Tdr Abbreviation for thymine deoxyriboside.

TdT Abbreviation for terminal deoxynucleotidyl transferase.

TE Abbreviation for 1. thioesterase; 2. transposible element.

Te Abbreviation for tellurium, a chemical element.

TE Buffer Abbreviation for tris-EDTA buffer.

TEA Abbreviation for 1. tetra-ethyl acid; 2. tetraethyl ammonium; 3. triethanolamine.

TEAC Abbreviation for tetra-ethyl ammonium chloride.

TEAE Cellulose Abbreviation for triethylaminoethyl cellulose, an ion exchanger used in ion exchange chromatography.

TEAEC Abbreviation for tetra-ethyl ammonium ethyl cellulose.

Tebamide A trade name for trimethobenzamide hydrochloride, an antiemetic agent.

TEBG Abbreviation for testosterone–estradiol binding globulin.

Tebrazid A trade name for pyrazinamide, an antimicrobial agent.

Tebuthiuron (mol wt 228) A herbicide.

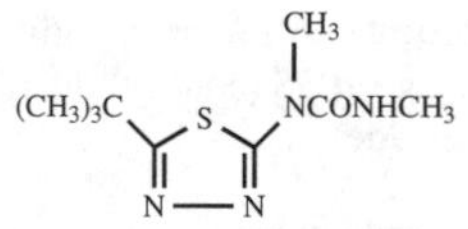

TEC A trade name for a combination drug containing zinc sulfate, copper sulfate, and magnesium sulfate, used for retinal function, bone growth, and differentiation of epithelial tissue.

Technetium A chemical element with atomic weight 98, valence 4 and 7 (usually).

Technetium-99 An isotope of technetium which emits gamma radiation and is used as a radioactive imaging agent.

Teclothiazide (mol wt 415) A diuretic agent.

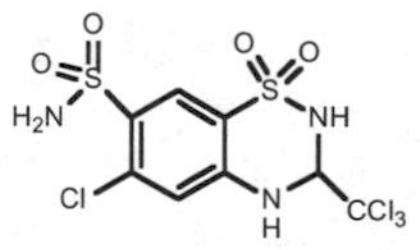

Tecnal A trade name for a combination drug containing aspirin, caffeine, and butalbital, used as an antipyretic and analgesic agent.

Tectibacter A genus of Gram-negative bacteria and a symbiont in the cytoplasm of *Paramecium aurelia.*

Tectiviridae A family of icosahydral, lipid-containing, dsDNA bacteriophages, e.g., phage PRD1.

Teczem A trade name for a combination drug containing enalapril and diltiazem used as an antihypertensive agent.

Tedelparin A fragment of haparin obtained by nitrous acid depolymerization that possesses antithrombotic activity.

Tedral A trade name for a combination drug containing theophylline, ephedrine hydrochloride, and phenobarbital, used as a bronchodilator.

TEE Abbreviation for tyrosine ethyl ester.

Teejel A trade name for choline salicylate, an analgesic and antipyretic agent.

Teflon A trade name for polytetrafluoroethylene, a plastic.

Tegamide A trade name for trimethobenzamide hydrochloride, an antiemetic agent.

Tega-Vert A trade name for dimenhydrinate, an antiemetic agent.

Tegison A trade name for etretinate, an anti-psoriatic agent.

Tego Compound A group of antimicrobial surfactants, e.g., tego 103S (dodecyl-diaminoethyl glycine hydrochloride).

Tegopen A trade name for cloxacillin sodium, a penicillinase-resistant penicillin that inhibits bacterial cell wall synthesis.

Tegretol A trade name for carbamazepine, an anticonvulsant.

TEIBQ Abbreviation for triethylene-imino-benzo-quinone.

Teichoic Acids A major component of bacterial cell wall consisting of long chains of either phosphoglycerol- or phosphoribitol-carrying sugars or amino acids.

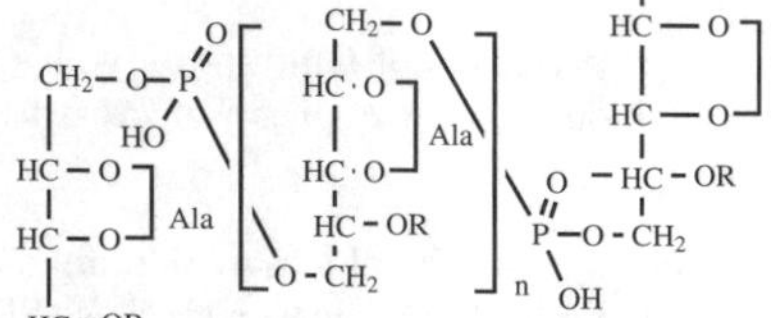

R = glycosyl
membrane teichoic acid

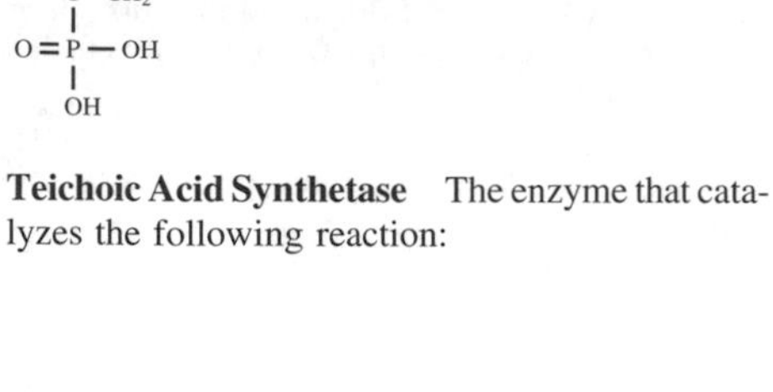

R = glycosyl
wall teichoic acid

Teichoic Acid Synthetase The enzyme that catalyzes the following reaction:

$$\text{CDP-ribitol} + (\text{ribitol phosphate})_n \rightleftharpoons \text{CMP} + (\text{ribitol phosphate})_{n+1}$$

Teicoplanin A glycopeptide antibiotic produced by *Actinoplanes teichomyceticus.*

Tektin Protein component of axonemal microtubules that is involved in organizing tubulin molecules into the A and B tubules of the doublet.

Telachlor A trade name for chlorpheniramine maleate, an antihistaminic agent that competes with histamine for H_1 receptors on effector cells.

Teladar A trade name for betamethasone, a glucocorticoid used as an anti-inflammatory agent.

Teldane A trade name for terfenadine, an antihistaminic agent that competes with histamine for H_1 receptors on effector cells.

Teldrin A trade name for chlorpheniramine maleate, an antihistaminic agent.

Teleology The doctrine stating that the structural development in living organisms tends to be determined by the purpose and function that they serve.

Teliospores Thick-walled, binucleate, resting spores of fungi (e.g., rusts and smuts).

TELISA Abbreviation for thermometric enzyme linked immunosorbent assay, an ELISA technique that measures the heat generated by the action of the enzyme.

Tellurium (Te) A chemical element with atomic weight 128, valence 2, 4, and 6.

Telmisartan (mol wt 515) An angiotensin II receptor antagonist used as an antihypertensive agent.

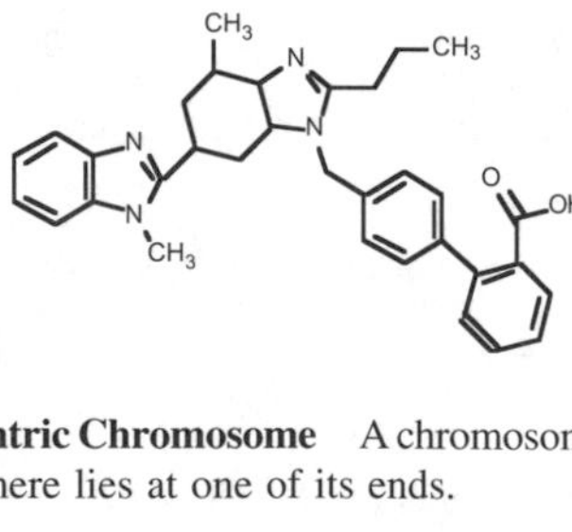

Telocentric Chromosome A chromosome whose centromere lies at one of its ends.

Telomerase An enzyme that catalyzes the addition of telomeric sequences to the ends of eukaryotic chromosomes without using template.

Telomere The terminal section of the eukaryotic chromosome containing a few hundred base pair that is involved in chromosomal replication and stability.

Telomycin A polypeptide antibiotic produced by *Streptomyces*.

Telophase Final phase of mitosis in which daughter chromosomes arrive at the poles of the spindle and begin to decondense, accompanied by the reappearance of the nuclear envelope and nucleoli.

TEM 1. Abbreviation for transmission electron microscopy. 2. Abbreviation for triethylenemelamine, an aziridine mutagen.

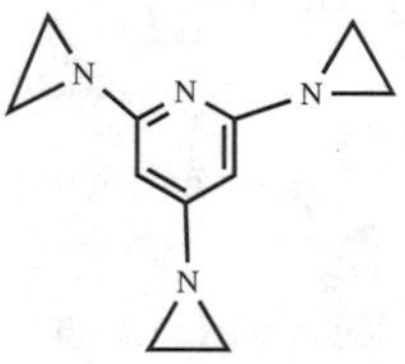

Temaril A trade name for trimeprazine tartrate, an antihistaminic agent that competes with histamine for H_1 receptors on effector cells.

Temaz A trade name for temazepam, a sedative–hypnotic agent.

Temazepam (mol wt 301) A sedative–hypnotic agent that acts on the CNS to produce a hypnotic effect.

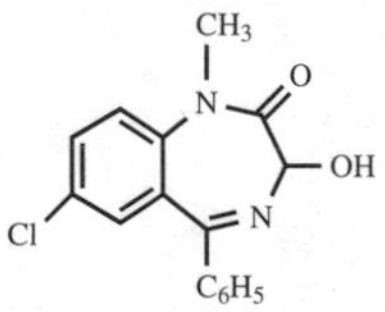

TEMED Abbreviation for *N,N,N′,N′*-tetramethylethylenediamine, a reagent used in the preparation of polyacrylamide gels.

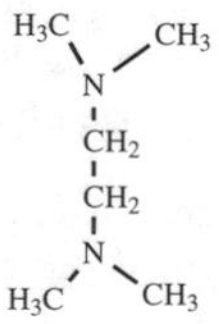

Temephos (mol wt 466) An insecticide.

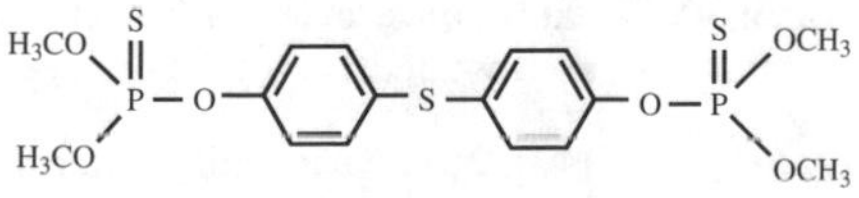

Temocapril (mol wt 477) An antihypertensive agent.

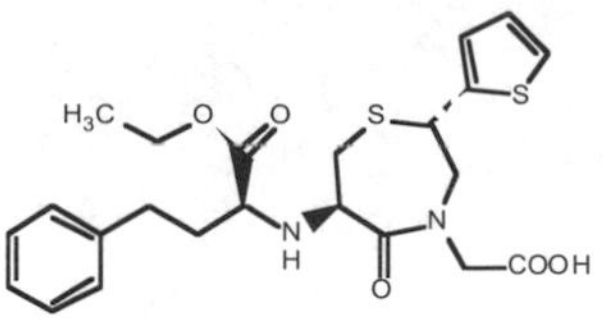

Temocillin (mol wt 414) A semisynthetic penicillin derivative with high activity against Gram-negative bacteria.

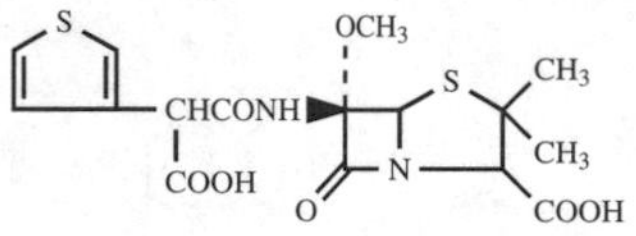

Temovate A trade name for clobetasol propionate, a topical corticosteroid.

Temperate Bacteriophage Bacteriophage capable of undergoing either lytic or lysogenic states. In the lysogenic state the phage DNA is integrated into the bacterial chromosome; In the lytic state the phage replicates and releases progeny by cell lysis.

Temperate Phage See temperate bacteriophage.

Temperate Viruses Viruses capable of undergoing either a lytic or lysogenic state.

Temperature-Sensitive Mutation Mutation that functions normally at one temperature (permissive temperature) but functions abnormally or not at all at another temperature (restrictive temperature).

Template The macromolecular mold for the synthesis of another macromolecule in a complementary fashion, e.g., DNA serves as a template (mold) for the synthesis of RNA.

Template Strand The strand of DNA that is transcribed into mRNA.

Template Switching A mechanism of DNA replication in which DNA polymerase I uses one strand as template and then shifts to use the second displaced strand as template.

Tempra A trade name for acetaminophen, an antipyretic and analgesic agent that inhibits prostaglandin synthesis and blocks the generation of pain impulses.

Tendinitis Inflammation of tendon.

Tendon A collagen-containing band of tissue that connects muscle with bone.

Tendonitis Variant spelling of tendinitis.

Tenex A trade name for guanfacine hydrochloride, an antihypertensive agent.

Tenidap (mol wt 321) An anti-inflammatory agent.

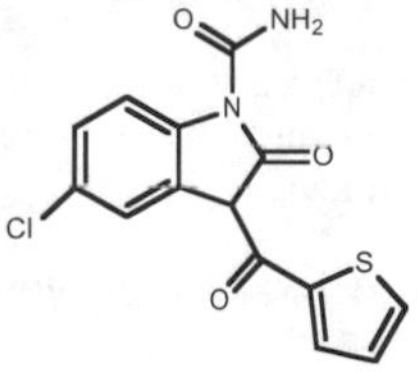

Teniloxazine (mol wt 289) An antidepressant.

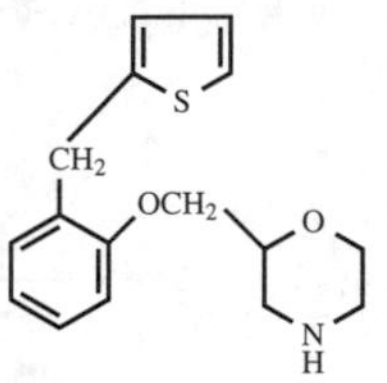

Teniposide (mol wt 657) A semi-synthetic derivative of podophyllotoxin that arrests cell mitosis.

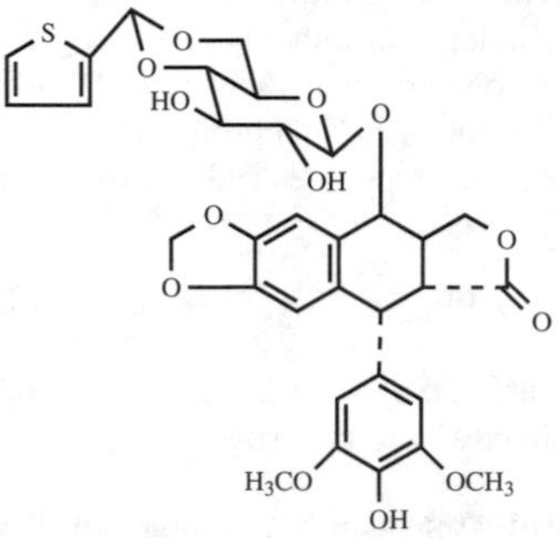

Tenol A trade name for acetaminophen, an analgesic and antipyretic agent that inhibits prostaglandin synthesis and blocks the generation of pain impulses.

Tenolin A trade name for atenolol, a beta adrenergic blocking agent used as an antihypertensive drug.

Tenonitrozole (mol wt 255) An antiprotozoal and antifungal agent.

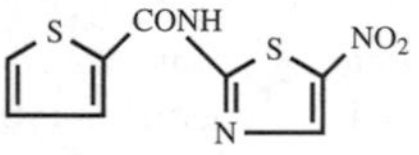

Tenoretic-50 A trade name for a combination drug containing atenolol and chlorthalidone, used as an antihypertensive agent.

Tenormin A trade name for atenolol, an antihypertensive agent that blocks the response to beta stimulation and decreases renin secretion.

Tenoxicam (mol wt 337) An anti-inflammatory and analgesic agent.

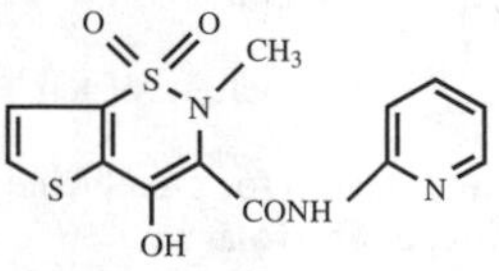

Tense Form The form of an allosteric protein that has lower affinity for a ligand.

Tensilon A trade name for edrophonium chloride, a cholinergic agent that inhibits the destruction of acetylcholine.

Tensiometer A device for measuring surface and interfacial tensions.

Tentoxin A cyclic tetrapeptide toxin produced by fungus *Alternaria* that induces chlorosis in plants and inhibits F_oF_1-ATPase.

Tenuate A trade name for diethylpropion hydrochloride, a cerebral stimulant.

Tenuazonic Acid (mol wt 197) An antineoplastic agent.

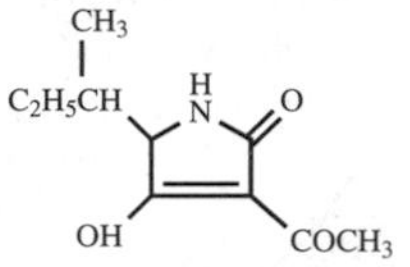

TEPA Abbreviation for triethylene phosphoamide.

Tepanil A trade name for diethylpropion hydrochloride, a cerebral stimulant that promotes the transmission of nerve impulses by releasing stored norepinephrine from nerve terminals in the brain.

TEPP Abbreviation for tri-ethyl-pyro-phosphate.

Teprenone (mol wt 331) An antiulcerative agent.

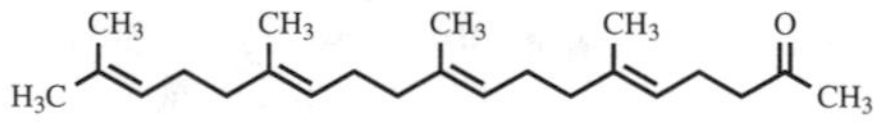

all *trans*-form

Ter- A prefix meaning three.

Tera- A prefix meaning 10^{12}.

Teramine A trade name for phentermine hydrochloride, a cerebral stimulant that promotes the transmission of nerve impulses by releasing stored norepinephrine from nerve terminals in the brain.

Teratocarcinoma Malignant tumor of primordial germ cells that is capable of differentiating into a variety of specialized type of cells.

Teratogen Any agent that induces abnormal development of embryos or causes birth defects.

Teratoma A neoplasma composed of more than one type of tissue.

Terazol 3 Vaginal Suppositories A trade name for terconazole, used as a local anti-infective agent.

Terazosin (mol wt 387) An antihypertensive agent that decreases blood pressure by enhancing vasodilation.

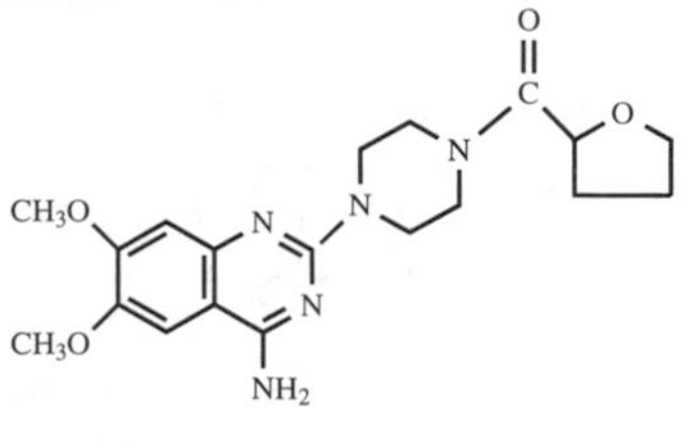

Terbacil (mol wt 217) A herbicide.

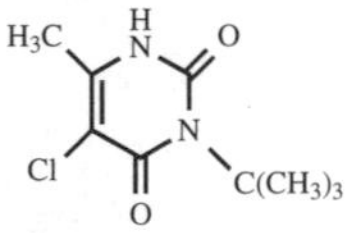

Terbinafine (mol wt 291) An antifungal agent.

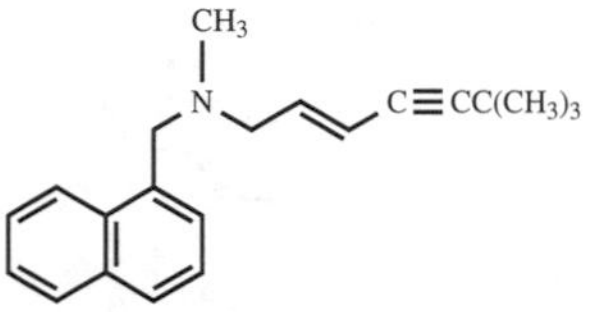

Terbium (Tb) A chemical element with atomic weight 159, valence 3, and 4.

Terbutaline (mol wt 225) A bronchodilator that relaxes bronchial smooth muscle by acting on beta-2 adrenergic receptors.

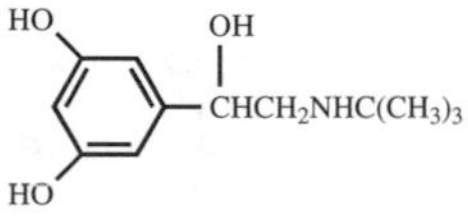

Terconazole (mol wt 532) An antifungal agent.

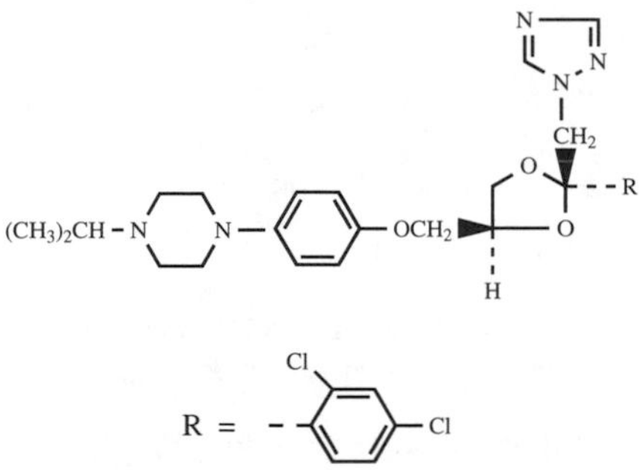

***ter*-Cutting** A cut made by the enzyme terminase on phage-DNA during packaging of DNA into the capsid of the λ bacteriophage.

Terfenadine (mol wt 472) An antihistaminic agent that competes with histamine for H_1 receptors on target cells.

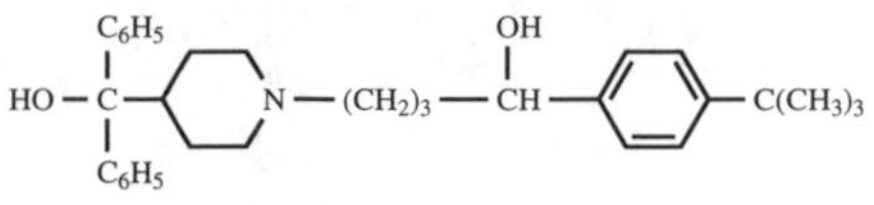

Terfluzine A trade name for trifluoperazine hydrochloride, an antipsychotic agent that blocks postsynaptic dopamine receptors in the brain.

Terguride (mol wt 340) An antiparkinsonian agent and a dopamine agonist derived from ergot.

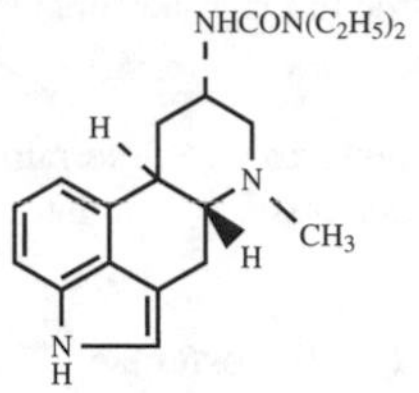

Teril A trade name for carbamazepine, an anticonvulsant.

Terlipressin A vasopressor peptide used for the treatment of uterine and esophageal bleeding.

$(Gly)_3$-Cys-Tyr-Phe-Gln-Asn-Cys-Pro-Lys-$GlyNH_2$

(S—S bridge between the two Cys residues)

3′-Terminal The end of a polynucleatide strand that carries either a free OH or a phosphorylated hydroxyl group at the 3′ position of the terminal ribose or deoxyribose.

5′-Terminal The end of a polynucleatide strand that carries either a free OH or a phosphorylated hydroxyl group at the 5′ position of the terminal ribose or deoxyribose.

Terminal Cisterna Regions of the sarcoplasmic reticulum that form transverse, connecting channels.

Terminal Deoxyribonucleotidyl Transferase The enzyme that catalyses the addition of the same type of deoxyribonucleotide to the 3′-terminal OH group of a DNA chain.

Terminal Desaturase The enzyme that catalyzes the addition of a double bond to a saturated fatty acyl-CoA.

$$CH_3(CH_2)_XCH_2CH_2(CH_2)_YCO\text{-}S\text{-}CoA \rightleftharpoons CH_3(CH_2)_XCH{=}CH(CH_2)_YCO\text{-}S\text{-}CoA$$

Terminal Electron Acceptor The molecule that is the final acceptor of electrons in a metabolic pathway (e.g., in aerobic respiration oxygen is the terminal electron acceptor).

Terminal Enzyme The enzyme that catalyzes the addition of the same type of ribonucleotide or deoxyribonucleotide to the terminal of an existing nucleic acid strand without the presence of a template.

Terminal Oxidase The enzyme that catalyzes the direct transfer of electrons to oxygen in the electron transport chain.

Terminal Redundancy The presence of identical sequences at both ends of a dsDNA molecule, e.g., DNA of T4 bacteriophage.

Terminal Ribonucleotide Transferase The enzyme that catalyses the addition of the same type of ribonucleotide to the 3′-terminal OH group of a DNA chain.

Terminal Transferase See terminal deoxyribonucleotidyl transferase or terminal ribonucleotide transferase.

Terminal Uridylyl Transferase The enzyme that catalyzes the addition of uridine ribonucleotide to an existing RNA.

Terminal Web The network of filaments located at the base of microvilli that contains actin, myosin, and several other cytoskeletal proteins.

Terminase The enzyme that catalyzes the cut of λ phage DNA and thereby initiates the packaging of phage DNA into the new phage head.

Termination Codon Referring to the three codons UAA, UAG, and UGA that terminate protein synthesis.

Termination Factor Protein factors that release the completed polypeptide from the ribosome (see releasing factor in protein synthesis).

Termination Signal A sequence on the DNA at the end of a transcription unit that determines the termination of transcription.

Termination Sites See termination signal.

Terminator A sequence of DNA that causes RNA polymerase to terminate transcription.

Terminator Sequence A sequence in DNA that signals the termination of transcription.

Terminator Stem A double-stranded structure formed in the RNA transcript that signals transcription termination.

Termolecular Reaction A chemical reaction that requires three reactants to form a product or products.

Ternary Having three elements, parts, or division, e.g., a molecule containing three different atoms.

Ternary Acid An acid that contains three different elements, e.g., acetic acid (CH_3COOH).

Ternary Fission An asexual process in which three cells are produced from one cell.

Ternary Mixture A mixture of three different liquids in proportions such that all are mutually soluble in one another but which will form two phases when one of the components is added in excess.

Terodiline (mol wt 281) An antianginal agent used for treatment of urinary incontinence.

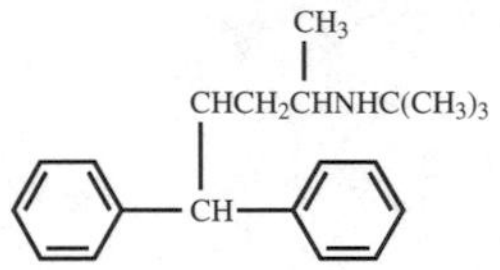

Terofenamate (mol wt 354) An anti-inflammatory and analgesic agent.

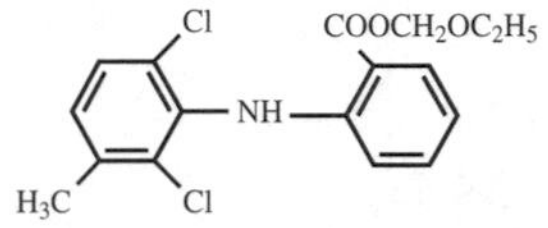

Terpene Organic hydrocarbon or hydrocarbon derivatives formed from recurring isoprene units (5-carbon compound).

Terpin (mol wt 172) An antitussive agent that increases production of respiratory tract fluid to help liquefy and reduce the viscosity of thick secretions.

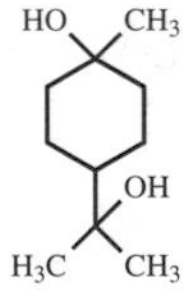

Terpineol (mol wt 154) An antiseptic agent.

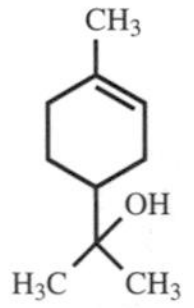

Terpolymers Polymers (usually linear) that consist of three different amino acids.

Terramycin A trade name for oxytetracycline, an antibiotic that binds to the 30S ribosomal subunit, inhibiting bacterial protein synthesis.

Terreaction See termolecular reaction.

Terreic Acid (mol wt 154) An antibiotic metabolite produced by *Aspergillus terreus*.

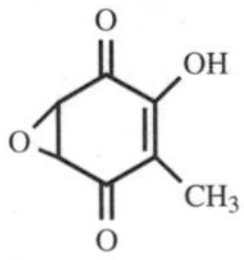

Tertiary Alcohol An alcohol in which the carbon atom to which the hydroxyl group is attached is also attached to the three other carbon atoms.

Tertiary Amine An amine in which three carbon atoms are attached to the amino nitrogen.

Tertiary Base Pairs The base pairs responsible for maintaining the three-dimensional structure of tRNA.

Tertiary Carbon The carbon atom that is joined to three other carbon atoms.

Tertiary Coiling Referring to DNA supercoiling.

Tertiary Hydrogen Bond Hydrogen Bonds formed between various hydrogen bond donor and acceptor groups in the three-dimensional structure of tRNA.

Tertiary Structure The three-dimensional folding of a polymer chain into a native folded conformation. The forces involved in the formation of tertiary structure are hydrophobic interactions, ionic bonds, and hydrogen bonds.

Tertroxin Synonym for T_3 (triiodothyronine).

Tesionate A trade name for testosterone cypionate, an anabolic steroid.

Tesla (T) A unit of magnetic field strength (1 tesla = 10,000 gauss).

Teslac A trade name for tesolactone, an antineoplastic agent.

Tessalon A trade name for benzonatate, an antitussive agent.

Test Cross Mating between one organism of unknown genotype with a tester organism carrying known homozygous recessive alleles to determine the unknown genotype.

Testa-C A trade name for testosterone cypionate, an androgen and an anabolic hormone.

Testamone A trade name for testosterone, an androgen hormone.

Testaqua A trade name for testosterone, an androgen.

Testectomy The surgical removal of the testis.

Tester An organism homozygous for one or more recessive alleles and used in a test cross.

Testex A trade name for testosterone propionate, an androgen.

Testicular Pertaining to the testes.

Testicular Feminization Syndrome A disorder caused by a mutation in a gene coding for androgen receptors in which XY males develop into women.

Testis The gamete-producing organ in male animals.

Testis-Determining Factor Genetic element on the mammalian Y chromosome that determines maleness.

Testoderm A trade name for testosterone, an androgen hormone.

Testoject-LA A trade name for testosterone cypionate, an androgen.

Testolactone (mol wt 300) An antineoplastic agent.

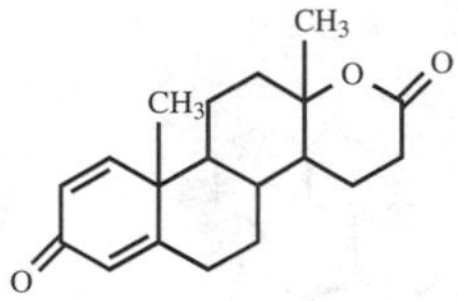

Testololactone Lactonohydrolase See steroid lactonase.

Testomet A trade name for methyltestosterone, an androgen.

Testone-LA A trade name for testosterone enanthate, an androgen.

Testosterone (mol wt 288) An androgen produced by the adrenol cortex and testes used in treatment of certain estrogen-dependent breast cancers.

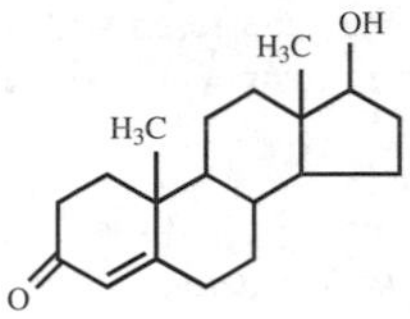

Testosterone 17β-Cypionate (mol wt 413) An androgen used for treatment of certain estrogen-dependent breast cancers.

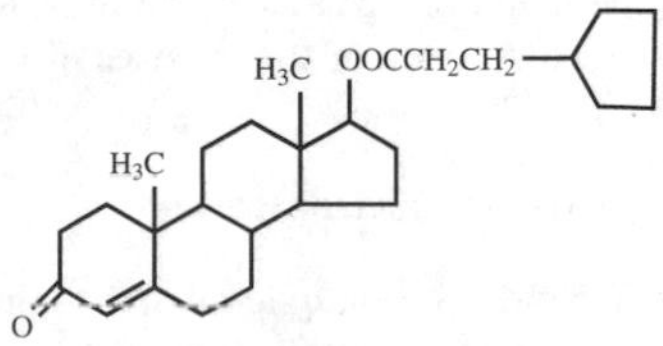

Testosterone 17β-Dehydrogenase The enzyme that catalyzes the following reaction:

$$\text{Testosterone} + \text{NAD}^+ \rightleftharpoons \text{Androst-4-ene-3,17- dione} + \text{NADH}$$

Testosterone Enanthate (mol wt 401) An androgen used for treatment of certain estrogen-dependent breast cancers.

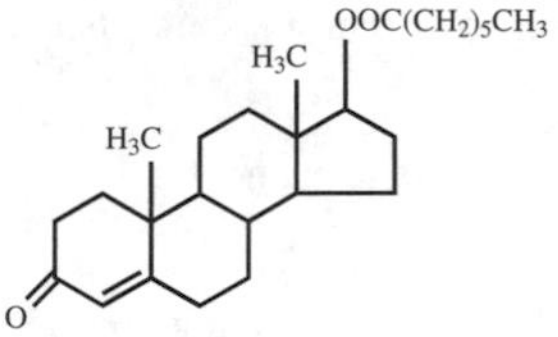

Testosterone Propionate (mol wt 344) An androgen used for treatment of certain estrogen-dependent breast cancers.

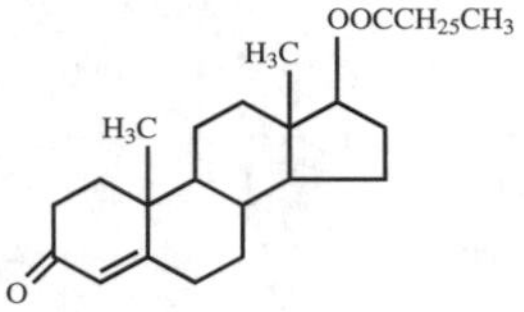

Testred A trade name for methyltestosterone, an androgen.

Testred Cypionate A trade name for testosterone cypionate, an androgen.

Testrin PA A trade name for testosterone enanthate, an androgen.

Tetanolysin A hemolysin produced by *Clostridium tetani.*

Tetanospasmin Neurotoxin produced by *Clostridium tetani* that interferes with the ability of peripheral nerves to transmit signals to muscle cells.

Tetanus An often-fatal disease caused by the bacterium *Clostridium tetani.*

Tetanus and Diphtheria Toxoids An active vaccine containing tetanus and diphtheria toxoid.

Tetanus Lockjaw A disease of humans and other animals caused by a neurotoxin produced by *Clostridium tetani* and characterized by sustained involuntary contraction of the muscles of the jaw and neck.

Tetanus Toxin See tetanospasmin.

Tetanus Toxoid A nontoxic tetanus toxin used for vaccination and promotion of immunity against tetanus toxin.

Tetany A disorder resulting from abnormal calcium metabolism and characterized by cramps, convulsions, and twitching of the muscle.

Tetra- A prefix meaning four.

Tetrabarbital (mol wt 240) A sedative–hypnotic agent.

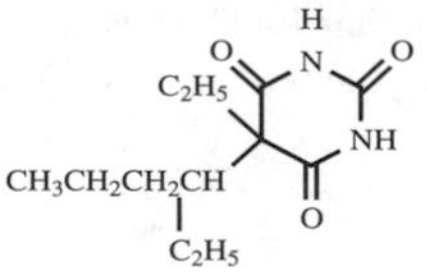

3,4,5,6-Tetrabromo-*o*-Cresol (mol wt 424) A fungicide.

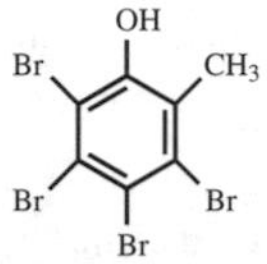

Tetracaine Hydrochloride (mol wt 301) A topical ophthalmic anesthetic agent that prevents initiation and transmission of impulses at the nerve-cell membrane.

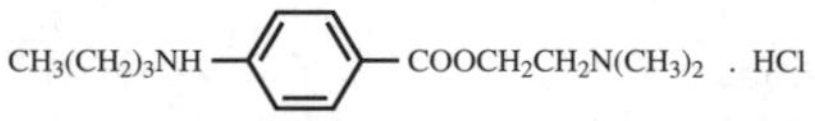

Tetracap A trade name for tetracycline hydrochloride, an antibiotic that binds to 30S ribosomal subunits, inhibiting bacterial protein synthesis.

Tetrachlormethiazide (mol wt 415) A diuretic agent.

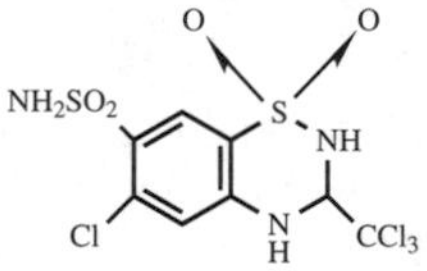

Tetracosa- A prefix denoting 24.

Tetracosapeptide A 24-residue peptide.

Tetracycline (mol wt 444) An antibiotic produced by *Streptomyces viridifaciens* that binds to 30S ribosomal subunts, preventing binding of aminoacyl tRNA to the ribosome A site.

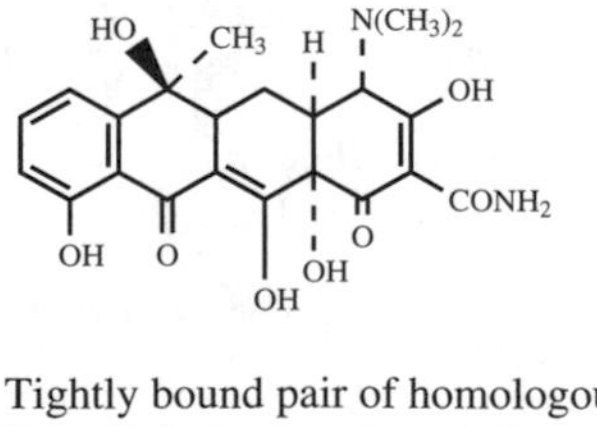

Tetrad Tightly bound pair of homologous chromosomes formed during prophase I of meiosis that consists of four chromatids.

Tetrad Analysis A method for the analysis of crossing over, linkage, and recombination using the four haploid products of single meiotic divisions.

Tetradeca A prefix denoting 14.

Tetradecadienoic Acid Any 14-carbon straight-chain fatty acid having two double bonds.

Tetradecapeptide A polypeptide containing 14 amino acid residues.

Tetraenoic Referring to an acid having four double bonds.

Tetraethylammonium Chloride (mol wt 166) A blocker of potassium channels.

$$(C_2H_5)_4NCl$$

Tetraethyl Pyrophosphate (mol wt 290) An insecticide.

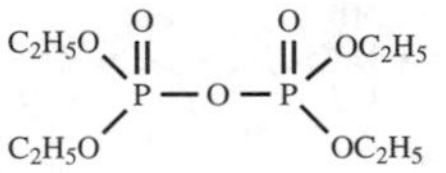

Tetrahydrate A compound that contains four molecules of water.

Tetrahydrobiopterin The reduced form of biopterin.

Tetrahydrocortisone (mol wt 364) A metabolite of cortisone.

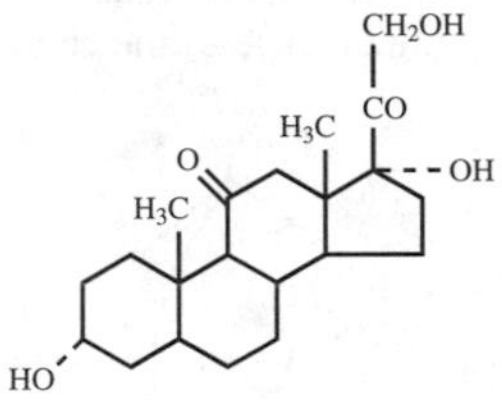

Tetrahydrofolate (mol wt 446) The metabolically active form of the vitamin folic acid, a coenzyme and carrier of one-carbon groups.

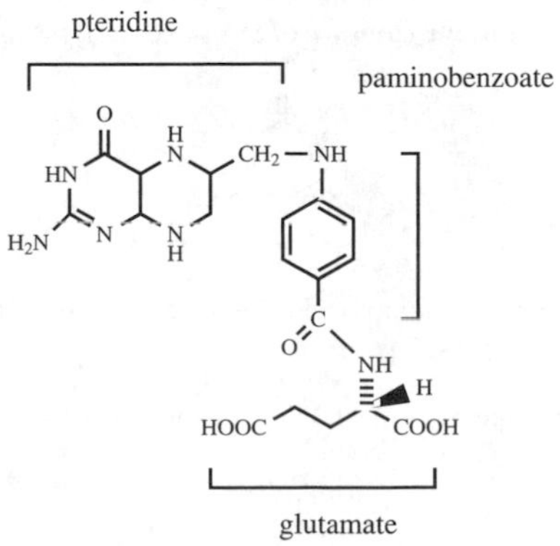

Tetrahydrofolate Dehydrogenase See dihydrofolate reductase.

Tetrahydrofolyl Polyglutamate Synthetase The enzyme that catalyzes the following reaction:

$$\text{ATP} + \text{tetrahydrofolyl-(glu)}_n + \text{L-glutamate} \rightleftharpoons \text{ADP} + \text{Pi} + \text{tetrahydrofolyl-(glu)}_{n+1}$$

Tetrahydrozoline (mol wt 200) An ophthalmic vasoconstrictor that produces vasoconstriction by local adrenergic action on the blood vessels of the conjunctiva.

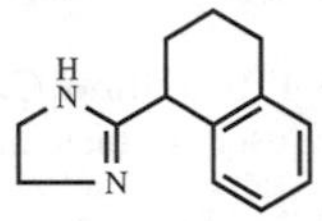

Tetrahymena A genus of ciliate protozoa.

Tetralan A trade name for the antibiotic tetracycline hydrochloride.

Tetralean A trade name for tetracycline hydrochloride, an antibiotic that inhibits bacterial protein synthesis.

Tetramer A compound that consists of four monomers.

Tetramethrin (mol wt 331) An insecticide.

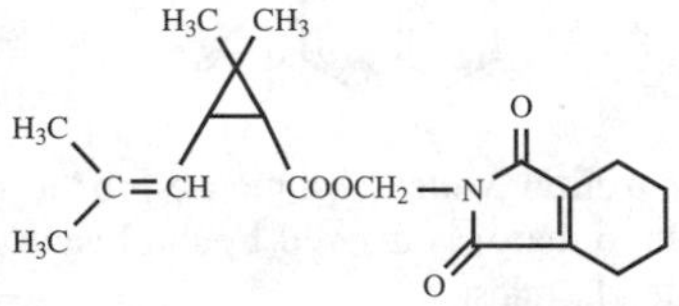

Tetramethylrhodamine B Isothiocyaninate (mol wt 444) A fluorochrome for labeling antibodies.

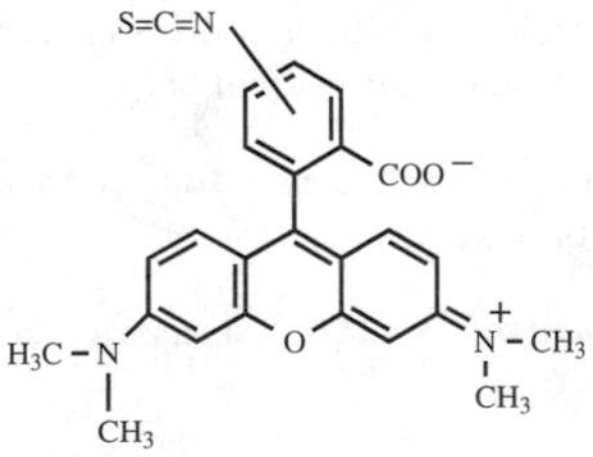

Tetramisole (mol wt 204) An anthelmintic agent and an immunostimulant.

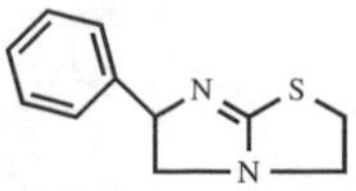

Tetramune A trade name for a vaccine containing diphtheria, tetanus toxoids, and whole cell pertussis with *Haemophilus influenzae b* conjugate.

Tetrandrine (mol wt 623) An analgesic and an antipyretic agent.

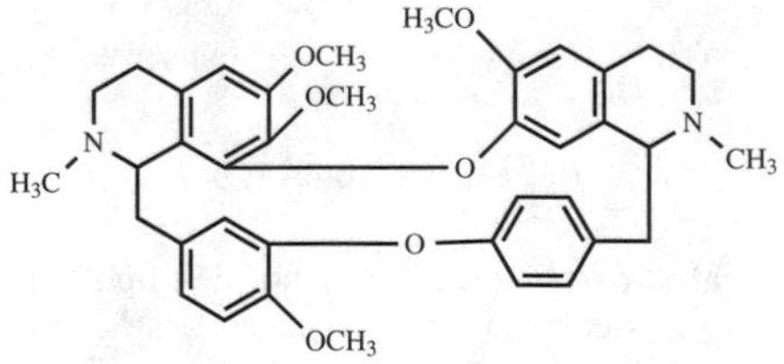

Tetranectin A tetrameric protein from human plasma that enhances plasminogen activation.

Tetranitroblue Tetrazolium Chloride (mol wt 908) A reagent used for histochemical detection of hydrolases.

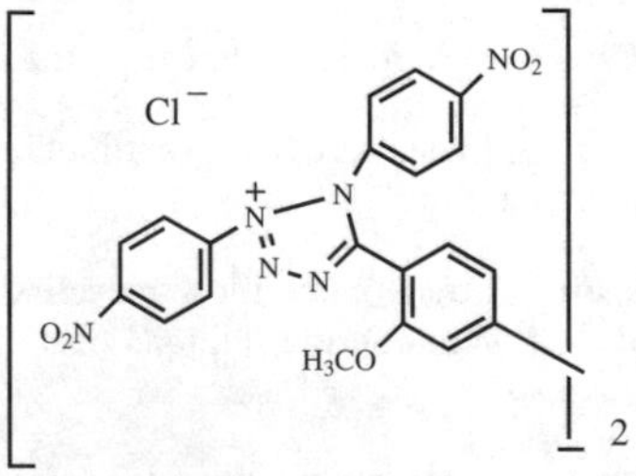

Tetrantoin (mol wt 216) An anticonvulsant.

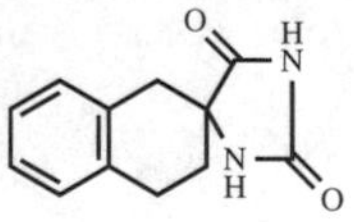

Tetraparental Mouse A mouse produced from an embryo that was derived by the fusion of two separate blastulas.

Tetraploid A cell that has four copies of haploid chromosome sets.

Tetrapyrrol A molecule consisting of four united pyrrol units, e.g., chlorophyll.

Tetrasine A trade name for tetrahydrozoline hydrochloride, a vasoconstrictor.

Tetrazepam (mol wt 289) Skeletal muscle relaxant.

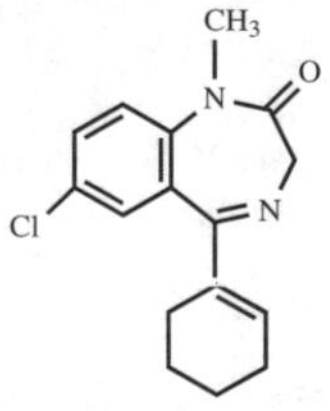

Tetrazole (mol wt 70) A coupling reagent used for automated synthesis of polynucleotides.

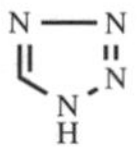

Tetrazolium Blue Chloride (mol wt 728) A reagent used for detection of oxidation–reduction of enzymes.

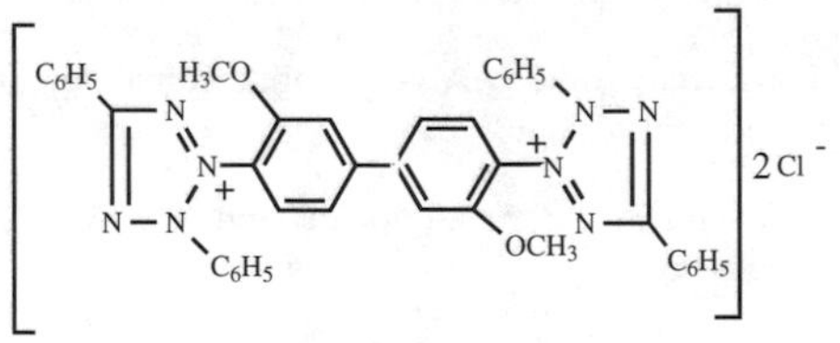

Tetrazolium Salt Referring to organic compounds that change color upon reduction with the formation of insoluble formazans.

Tetrin A group of antifungal antibiotics (e.g., tetrin A and tetrin B) produced by *Streptomyces* spp.

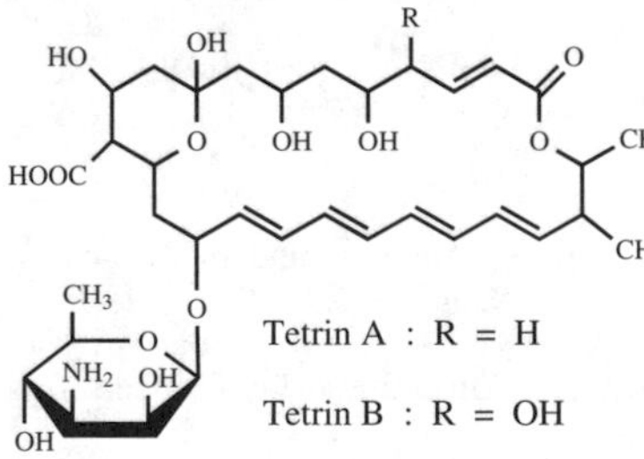

Tetrodotoxin (mol wt 319) A potent neurotoxin produced by ovaries and livers of many species of fishes of Tetraodontidae. It binds to sodium channels blocking the passage of the action potentials.

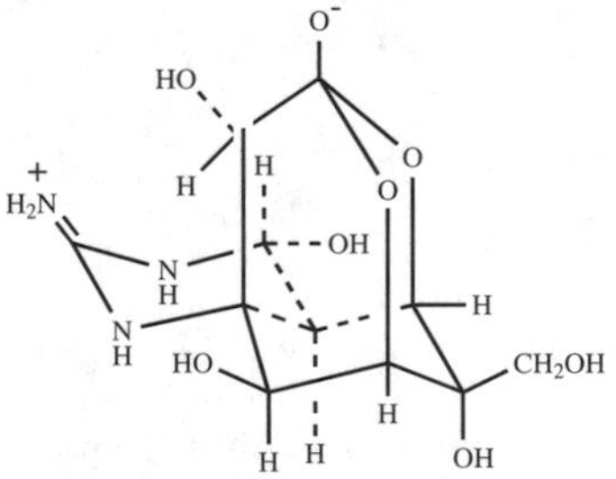

Tetroquinone (mol wt 172) A keratolytic agent.

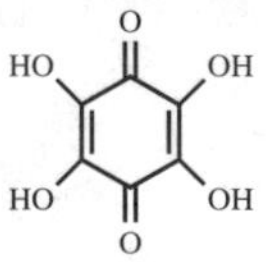

Tetrose A monosaccharide that consists of four carbon atoms, e.g., erythrose.

Tetroxoprim (mol wt 334) An antibacterial agent.

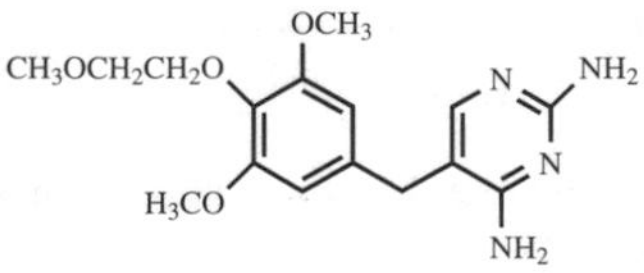

TeTx Abbreviation for tetanus toxin.

T-Even Phages Referring to bacteriophages of T2, T4, or T6.

Tevenel (mol wt 357) An anti-infective agent.

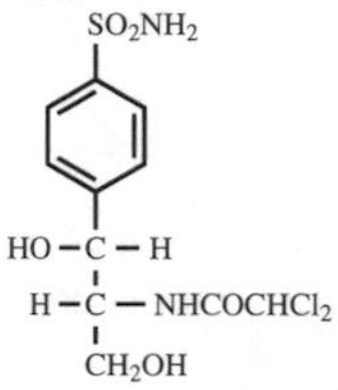

TF Abbreviation for 1. tissue factor; 2. transferrin; 3. transcription factor; 4. transfer factor.

TFA Abbreviation for trifluoroacetic acid.

TF-IIIA Abbreviation for transcription factor IIIA.

TFIIIB Abbreviation for transcription factor IIIB.

TFIIIC Abbreviation for transcription factor IIIC.

TFE Abbreviation for 1. trifluoroethanol; 2. trifluoroethylene.

TFF Abbreviation for trefoil factor.

TF-2H Abbreviation for transcription factor 2-H.

TflI(TaqI) A restriction endonuclease from *Thermus flavus* AT62 with the same specificity as TaqI.

TFMH Abbreviation for trifluoromethyl histidine.

TFMTR Abbreviation for trifluoromethylthioribose.

TFO Abbreviation for triplex-forming oligonucleotide.

TFPI Abbreviation for tissue factor pathway inhibitor.

TFT Abbreviation for trifluorothymidine.

TG Abbreviation for 1. thapsigargin; 2. thioguanine; 3. thyroglobulin; 4. transgenic; 5. transglutaminase; 6. triacylglycerol; 7. triglyceride.

6-TG Abbreviation for 6-thioguanine.

(TG)AL Referring to a tyrosine and glutamate polymer linked to the polylysine backbone through alanine residues used as a synthetic antigen for study of antigenic determinants.

TGase Abbreviation for transglutaminase.

TGF Abbreviation for transforming growth factor.

TGF-α Abbreviation for transforming growth factor α.

TGF-β Abbreviation for transforming growth factor-β, which is a potent regulator of cell proliferation and acts as a growth inhibitor of epithelial, hematopoietic, and endothelial cells.

TGF-β1 Abbreviation for transforming growth factor β1.

TglI(SacII) A restriction endonuclease from *Thermopolyspora glauca* with the same specificity as SacII.

TGN Abbreviation for *trans*-Golgi network.

TGN-p38 Abbreviation for trans-Golgi network protein of 38kD.

Th Symbol for the chemical element thorium.

T_H Abbreviation for helper T lymphocyte.

TH Abbreviation for 1. thyroid hormone; 2. tyrosine hydroxylase.

THA A trade name for tacrine hydrochloride, a cholinesterase inhibitor used as an Alzheimer's drug.

ThaI (FnuDII) A restriction endonuclease from *Thermoplasma acidophilum* with the following specificity:

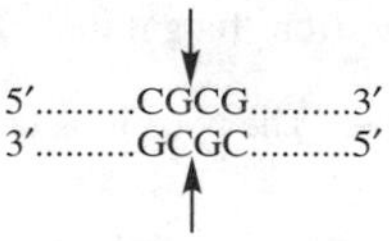

Thalamic Pertaining to the thalamus.

Thalamus An ovoid structure located on either side of the third ventricle of the cerebrum, serving as a relay center for sensory impulses in the cerebral cortex.

Thalassemia A genetic disorder of hemoglobin synthesis characterized by a reduced rate of synthesis of one or more of the globin chains in the hemoglobin leading to anemia.

α-Thalassemia A form of thalassemia caused by the decrease in rate of synthesis of the α peptide chain of hemoglobin. The homozygous form of α-thalassemia that causes death occurs shortly after birth.

β-Thalassemia A form of thalassemia caused by the decrease in the rate of the synthesis of β-peptide chain of hemoglobin (also known as Cooley's anemia).

Thalfed A trade name for a combination drug containing theophylline, ephedrine hydrochloride, and phenobarbital, used as a bronchodilator.

Thalicarpine (mol wt 697) A tumor inhibitory alkaloid.

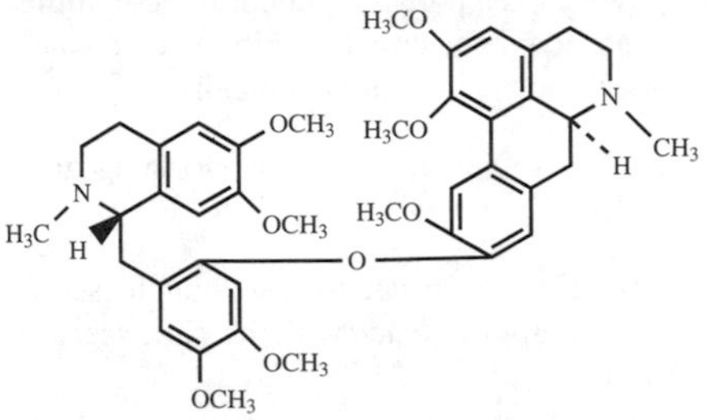

Thalidomide (mol wt 258) A sedative–hypnotic agent causing malformations of the fetus when taken between the 3rd and 5th week of pregnancy.

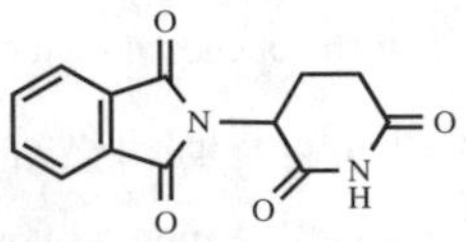

Thalitone A trade name for chlorthalidone, a diuretic agent that increases urine excretion of sodium and water by inhibiting the reabsorption of sodium.

Thallium A chemical element with atomic weight 204, valence 1 and 3.

Thallospore A spore developed by fragmentation or budding from fungal thallus.

Thallotoxicosis The poisoning caused by the intake of thallium salt.

Thallus The vegetative form of an organism that is not differentiated into root, stem, and leaf, e.g., certain fungi, algae and lichens.

THAM Abbreviation for tris (hydroxymethyl)-aminomethane, known as tris, a reagent used for the preparation of biological buffers.

Thanatochemistry The science that deals with the chemical reactions of tissues or organisms after death.

Thanatology The science that deals with phenomena of death.

Thapsigargin (mol wt 650) A tumor-promoting endoplasmic reticulum calcium-transporting ATPase.

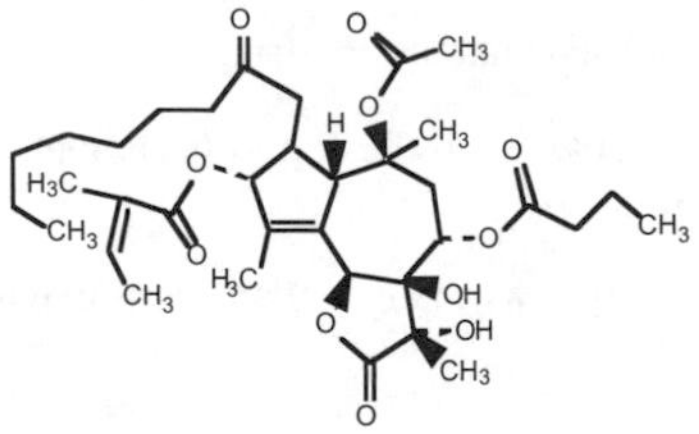

Thaumatin A sweet-tasting basic protein from fruit of the tropical plant *Thaumatococcus danielli.*

ThDP Abbreviation for thiamine diphosphate.

THE-1 Sequence A transposible element of DNA genome in human.

Thebaine (mol wt 311) An alkaloid found in opium that possesses a sharp astringent taste.

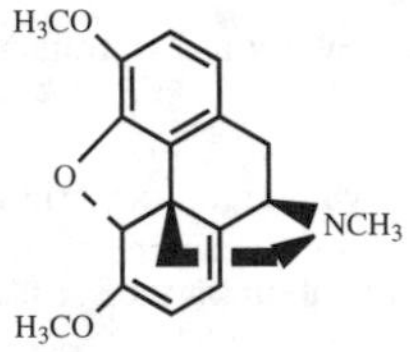

Theca 1. A sheath or an enveloping case of an anatomical structure, e.g., tendon sheath. 2. A type of cell wall in certain algal cells that lacks the microfibrillar structure.

Thecae Plural of theca.

Thecitis Inflammation of tendon sheath.

Theileria A genus of parasitic protozoa (Babesiidae).

Theileriosis Any disease or infection caused by *Theileria.*

Theliolymphocyte Intraepithelial lymphocytes associated with intestinal epithelial cells.

T-Helper Cells A class of T cells that enhance the activities of B cells and cytotoxic T cells.

Theo-24 A trade name for theophylline, a bronchodilator that inhibits phosphodiesterase and increases cAMP concentration.

Theo LA A trade name for theophylline, a bronchodilator used for relaxing bronchial smooth muscle.

Theobid Duracaps A trade name for theophylline, a bronchodilator that inhibits phosphodiesterase and increases cAMP concentration.

Theobromine (mol wt 180) An alkaloid from cacao bean that possesses diuretic and cardiotonic activity.

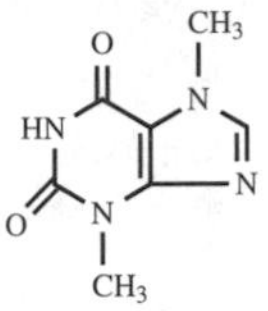

1-Theobromineacetic Acid (mol wt 238) A bronchodilator.

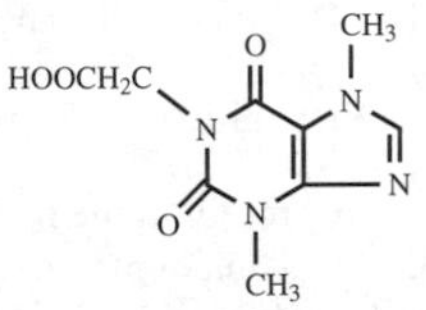

Theochron A trade name for theophylline, a bronchodilator that inhibits phosphodiesterase and increases cAMP concentration.

Theoclear-80 A trade name for theophylline, a bronchodilator used for relaxing bronchial smooth muscle.

Theodrine A trade name for a combination drug containing ephedrine, theophylline, and phenobarbital used as an anti-asthmatic agent.

Theo-Dur A trade name for theophylline, a bronchodilator that inhibits phosphodiesterase and increases cAMP concentration.

Theofibrate (mol wt 421) An antihyperlipoproteinemic, antithrombotic agent that also possesses platelet aggregation inhibitory activity.

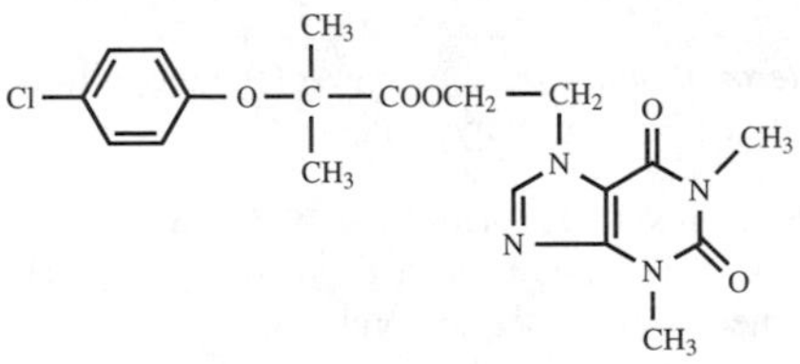

Theolair A trade name for theophylline, a bronchodilator that inhibits phosphodiesterase and increases cAMP concentration.

Theon A trade name for theophylline, a bronchodilator that inhibits phosphodiesterase and increases cAMP concentration.

Theophyl A trade name for theophylline, a bronchodilator that inhibits phosphodiesterase and increases cAMP concentration.

Theophylline (mol wt 180) A bronchodilator that inhibits phosphodiesterase, preventing degradation of cyclic AMP, resulting in relaxation of smooth muscle of the bronchial airways and pulmonary blood vessels.

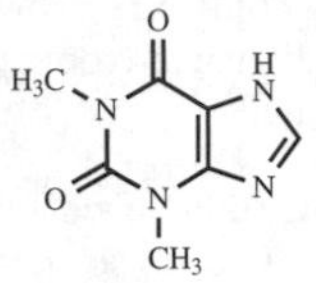

Theophylline Ethylenediamine A theophylline derivative and a bronchodilator that inhibits phosphodiesterase, preventing degradation of cAMP, thus resulting in relaxation of smooth muscle of the bronchial airways and pulmonary blood vessels.

Theophylline Sodium Glycinate A theophylline derivative and a bronchodilator that inhibits phosphodiesterase, preventing the degradation of cAMP, resulting in relaxation of the smooth muscle of the bronchial airways and pulmonary blood vessels.

Theospan SR A trade name for theophylline, used as a bronchodilator that inhibits phosphodiesterase and increases cAMP concentration.

Theo-Time A trade name for theophylline, used as a bronchodilator that inhibits phosphodiesterase and increases cAMP concentration.

Theovent A trade name for theophylline, a bronchodilator used for relaxing bronchial smooth muscle.

Thera Cys A trade name for *Bacillus* Calmette-Guerin (BCG), used as vaccine against tuberculosis.

TheraFlu A trade name for a combination drug containing pseudoephedrine, chlorpheniramine, and acetaminophen used as an antihistaminic agent and decongestant.

Thera-Flur A trade name for sodium fluoride, used to catalyze bone remineralization.

Theralax A trade name for bisacodyl, a laxative that promotes fluid accumulation in the colon and small intestine.

Therapeutic Index A measure of the safety of a drug expressed as the ratio of the maximum tolerated dose to the minium effective dose.

Thermal Pertaining to heat.

Thermal Chromatography A type of column chromatographic technique in which the eluting process is carried out at increasing temperature.

Thermal Conductivity Detector A type of detector used in gas chromatography for detecting inorganic gas and organic compounds by a thermal conductivity cell.

Thermal Death Point The lowest temperature that kills a population of microorganisms of a given species in 10 minutes.

Thermal Denaturation Denaturation of macromolecules by heat, e.g., separation of double-stranded DNA into single-stranded DNA.

Thermal Inactivation Point The temperature required to inactivate a suspension of virus particles in 10 minutes.

Thermal Melting Profile A plot of UV absorbance at 260 nm vs. temperature for a given sample of double-stranded DNA. The double-stranded DNA is progressively converted to single-stranded molecules as the temperature increases. The temperature at which one-half of the maximum change in absorbance is referred to as T_m (melting out temperature).

Thermal Polymer Heat-induced polymerization.

Thermazene A trade name for silver sulfadiazine, a local anti-infective agent.

Thermionic Emission The emission of ionic particles (e.g., electrons) from materials at high temperature.

Thermitase A serine-type endopeptidase used for hydrolysis of peptides and collagen.

Thermo- A prefix meaning heat.

Thermoacidophiles Microorganisms capable of growing in hot, acidic environment.

Thermoactinomyces A genus of thermophilic bacteria (Actinomycetales).

Thermoanesthesia The inability to feel the variation in temperature.

Thermobacteroides A genus of Gram-negative, obligately anaerobic, chemoorganotrophic, rod-shaped bacteria with an optimum growth temperature of 55-70° C.

Thermochemistry The science that deals with the interrelation of heat with a chemical reaction or state of physical change.

Thermochromism The reversible color change by a compound with change in temperature.

Thermococcus A genus of marine, chemolithoheterotrophic bacteria (Thermoproteales).

Thermodiscus A genus of bacteria with disc-shaped cells (Thermoproteales).

Thermoduric Capable of surviving high temperature.

Thermodynamics The science that deals with energy transformation, e.g., energy flow in closed and open systems.

Thermofilum A genus of chemolithoheterotrophic bacteria (order Thermoproteales).

Thermogenesis The production of heat in the body.

Thermogenic Pertaining to heat production.

Thermogenin A protein in the mitochondria of brown fat tissue that uncouples oxidative phosphorylation and enables this tissue to produce heat.

Thermohypoesthesia The diminished sensitivity to temperature.

Thermolabile Sensitive to heat or loss of characteristic properties upon heating.

Thermolysin A heat-stable proteinase produced by a strain of *Bacillus stearothermophilus* that catalyzes the hydrolysis of peptide bonds involving the hydrophobic amino acid residues at the N-terminal.

Thermolysis 1. Chemical decomposition by heat. 2. The dissipation of heat by the body.

Thermomicrobium A genus of aerobic, catalase-positive, chemoorganotrophic, thermophilic, Gram-negative bacteria.

Thermomonospora A genus of thermophilic bacteria (order Actinomycetales).

Thermoosmosis The flow of water or solvent across a membrane as a result of a temperature gradient across the membrane.

Thermophiles Organisms having an optimum growth temperature above 45° C.

Thermophilic Pertaining to thermophiles.

Thermoplasma A genus of aerobic, heterotrophic, archaebacteria (order Thermoplasmales).

Thermoreceptor A cell receptor or structure that responds to changes in temperature.

Thermorubin An antibiotic produced by *Thermoactinomyces antibioticus* that inhibits bacterial protein synthesis.

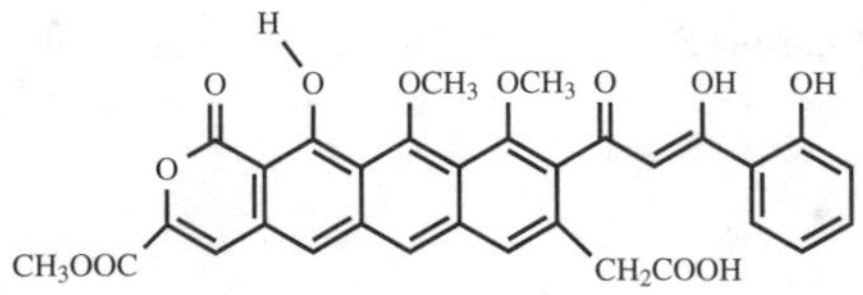

Thermoset A high polymer that solidifies irreversibly when heated.

Thermotaxis A type of taxis in which the stimulus is a temperature gradient.

Thermotitration The determination of the endpoint or equilibrium of a reaction by measurement of the heat of the reaction.

Thermotropism A type of tropism in which temperature gradient determines the orientation.

Thesaurosis A disorder characterized by the abnormal accumulation of cerebrosides in the body, e.g., in Gaucher's disease.

Theta (θ) A letter of the Greek alphabet.

Theta (θ) **Antigen** Alloantigen present on the surface of most thymocytes and peripheral T lymphocytes of nonhuman mammals (also known as Thy-1 antigen).

Theta (θ) **Replication** Bidirectional replication mode of double-stranded circular DNA (so-called because the intermediates look like the Greek letter theta).

Theta Structure An intermediate structure formed during the replication of a circular dsDNA molecule.

Thevetose (mol wt 178) A methylated sugar of quinovose.

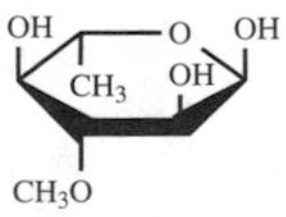

α-L-thevetose

THF Abbreviation for 1. tetrahydrofolate; 2. thymic humoral factor; 3. tetrahydrofuran.

THFA Abbreviation of tetrahydrofolic acid.

Thia A trade name for vitamin B_1 (thiamine hydrochloride).

Thiabendazole (mol wt 201) An anthelmintic agent and a fungicide.

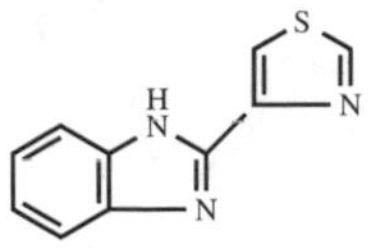

Thiacetazone (mol wt 236) An antibacterial agent.

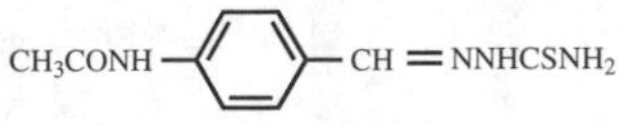

Thiacide A trade name for a combination drug containing methenamine mandelate and potassium phosphate.

Thialbarbital (mol wt 264) An anesthetic agent.

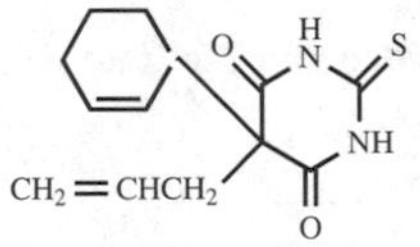

Thiambutene (mol wt 291) A narcotic analgesic agent.

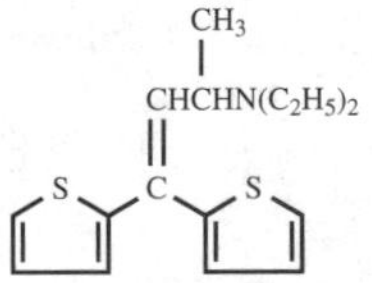

Thiaminase The enzyme that catalyzes the following reaction:

Thiamine + H_2O

⇅

4-Amino-5-hydroxymethyl-2-methylpyrimidine + 5-(2-hydroxyethyl)-4-methylthiazose)

Thiamine (mol wt 300) A vitamin and a coenzyme.

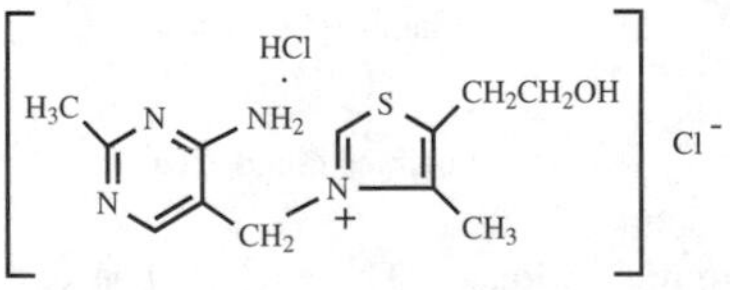

Thiamine Dehydrogenase See thiamin oxidase.

Thiamine Diphosphate Kinase The enzyme that catalyzes the following reaction:

ATP + thiamine diphosphate

⇅

ADP + thiamine triphosphate

Thiamine Disulfide (mol wt 563) A vitamin and enzyme co-factor.

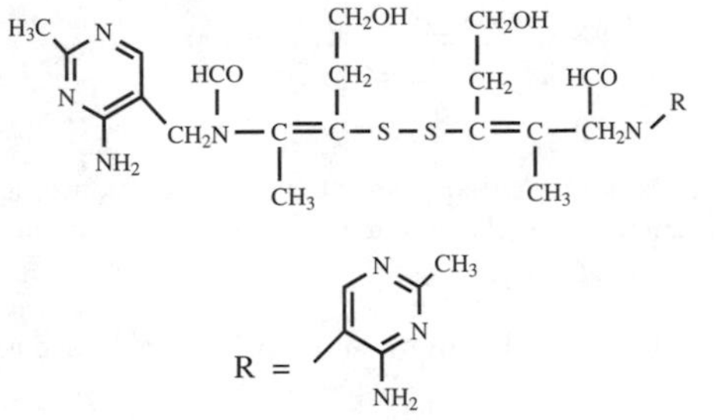

Thiamine Hydrochloride (mol wt 337) A vitamin and coenzyme.

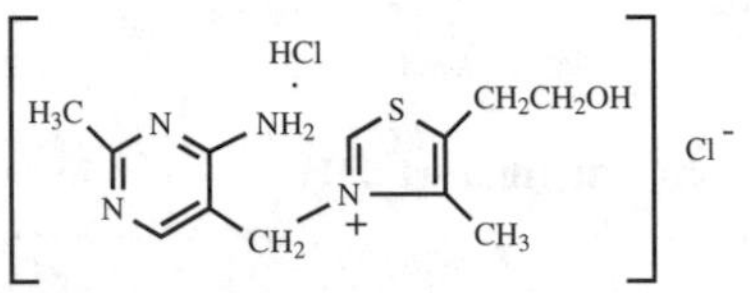

Thiamine Hydrolase See thiaminase.

Thiamine Kinase The enzyme that catalyzes the following reaction:

ATP + thiamine ⇌ ADP + thiamine phosphate

Thiamine Mononitrate (mol wt 327) A vitamin (B_1) and a coenzyme.

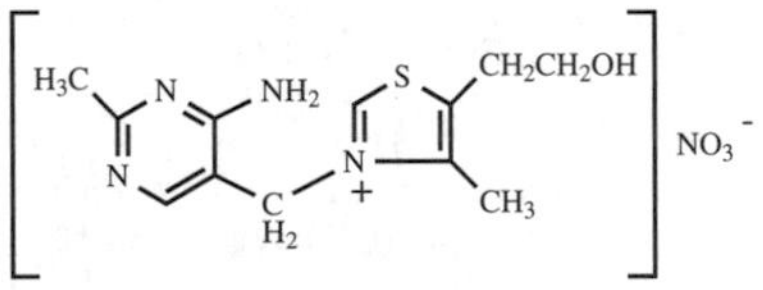

Thiamine Monophosphate Kinase The enzyme that catalyzes the following reaction:

ATP + thiaminephosphate
⇅
ADP + thiamine diphosphate

Thiamine Oxidase The enzyme that catalyzes the following reaction:

Thiamine + 2 O_2 ⇌ Thiamine acetic acid + 2H_2O

Thiamine Pyrophosphatase The enzyme that catalyzes the following reaction:

Thiamine triphosphate + H_2O
⇅
Thiamine diphosphate + Pi

Thiamine Pyrophosphate A coenzyme involved in the transfer of two-carbon units and other enzymatic reactions.

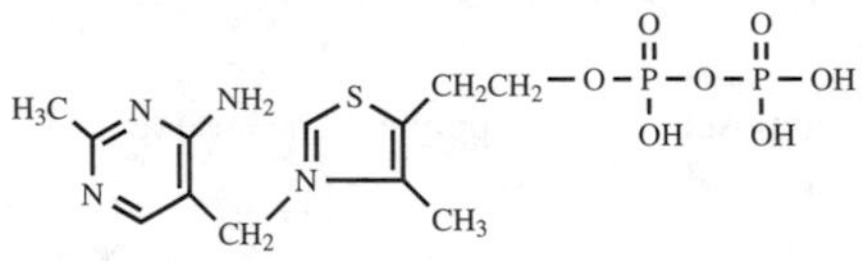

Thiamine Pyrophosphokinase The enzyme that catalyzes the following reaction:

ATP + thiamine ⇌ AMP + thiamine diphosphate

Thiamine Triphosphatase The enzyme that catalyzes the following reaction:

Thiamine triphosphate + H_2O
⇅
Thiamine diphosphate + Pi

Thiamine Triphosphate Phosphohydrolase See thiamine triphosphatase.

Thiamiprine (mol wt 292) An antineoplastic agent.

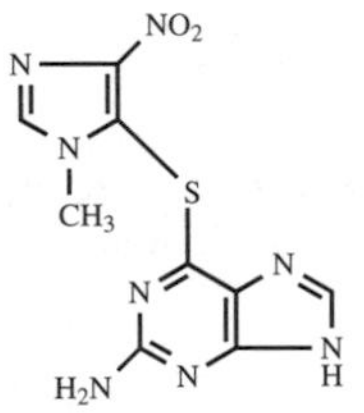

Thiamphenicol (mol wt 356) An antibacterial agent.

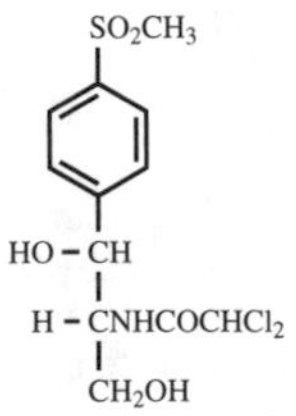

Thiazesim (mol wt 326) An antidepressant.

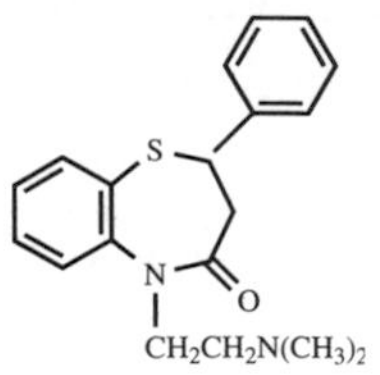

Thiazinamium Methylsulfate (mol wt 411) An antihistaminic agent.

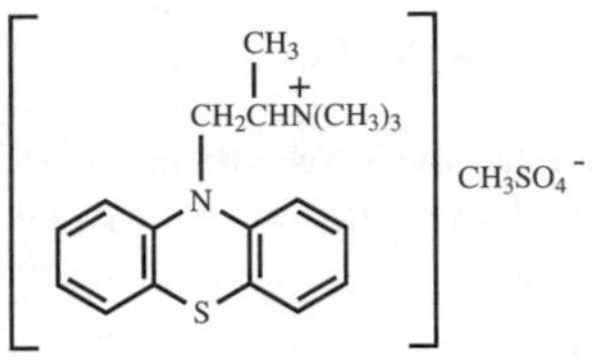

Thiazolinobutazone (mol wt 411) An anti-inflammatory agent.

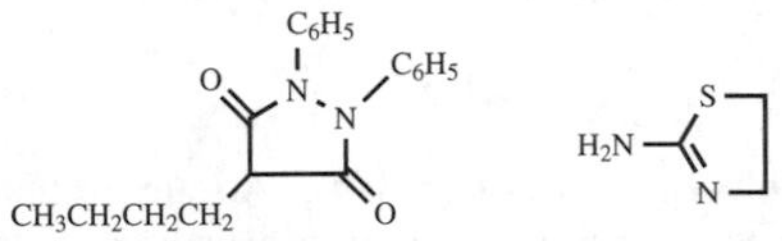

Thiazolsulfone (mol wt 255) An antibacterial agent.

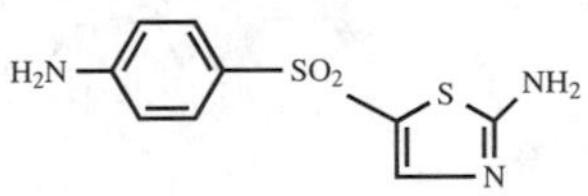

Thiazol Yellow G (mol wt 696) A biological stain and a reagent for determination of magnesium that is also a pH indicator (yellow at pH 11, red at pH 13).

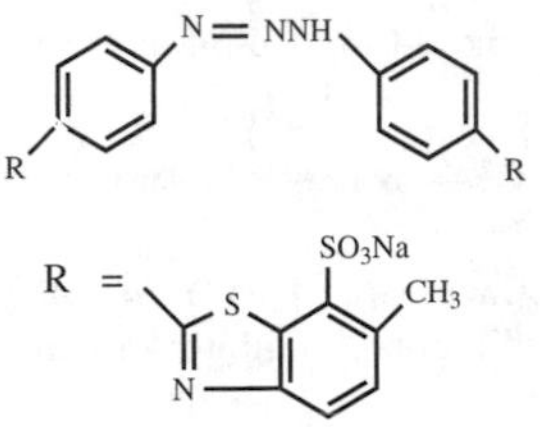

Thibenzazoline (mol wt 210) An antihyperthyroid agent.

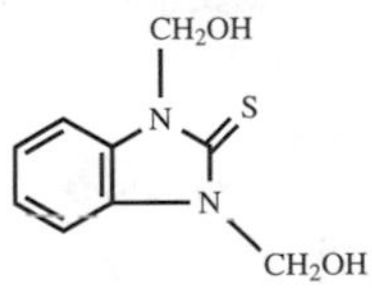

Thick Filament Myosin-containing filaments found in the myofibrils of muscle cells.

Thienamycin (mol wt 272) An antibiotic produced by *Streptomyces cattleya.*

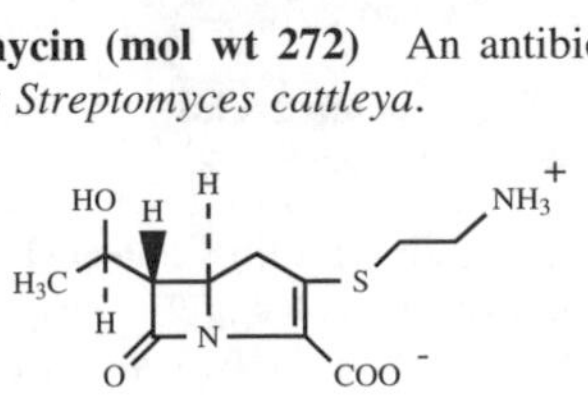

Thiery Staining A staining method for the detection of polysaccharide by electron microscopy.

Thiethylperazine (mol wt 400) An antiemetic agent that inhibits nausea and vomiting.

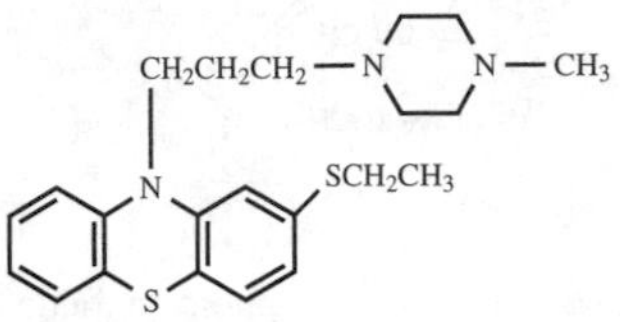

Thigmotaxis A type of taxis in which the stimulus is physical contact or touch.

Thihexinol (mol wt 322) An anticholinergic agent.

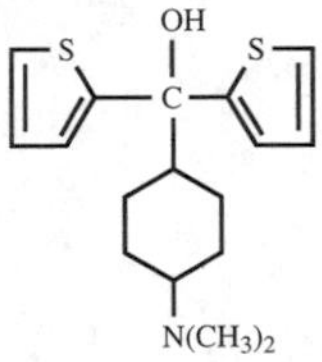

Thimerfonate Sodium (mol wt 441) A topical anti-infective agent.

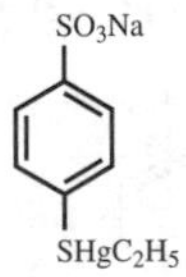

Thimerosal (mol wt 405) An anti-infective agent.

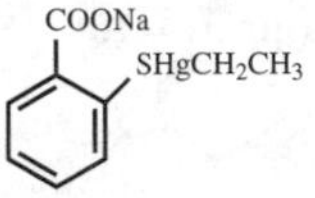

Thin Filament Actin-containing filaments found in the myofibrils of muscle cells.

Thin Layer Chromatography Chromatographic technique for separation of compounds on a thin layer of solid medium, e.g., a thin layer of silica on a glass plate.

Thin Layer Electrophoresis Electrophoretic separation of ionic components on a thin layer supporting medium, e.g., an agarose layer.

Thio- A prefix meaning sulfur.

Thiobacillus A genus of Gram-negative, obligately or facultatively chemolithoautotrophic bacteria.

Thiobarbital (mol wt 200) A thyroid inhibitor.

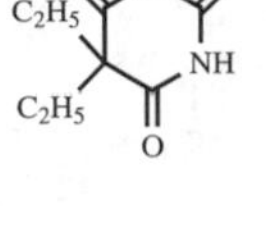

Thiobutabarbital (mol wt 228) An anesthetic agent.

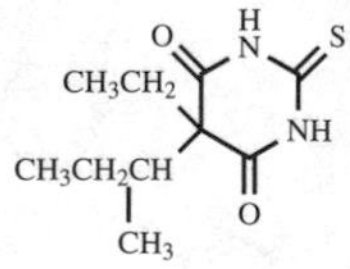

Thiocapsa A genus of photosynthetic bacteria (Chromatiaceae).

Thiocarbamizine (mol wt 516) An anti-amebic agent.

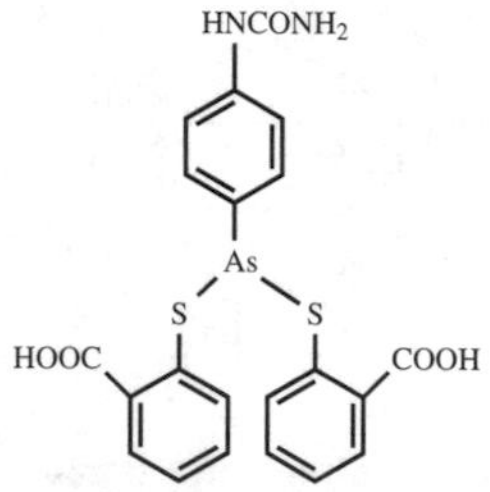

Thiocarbarsone (mol wt 392) An antiamebic agent.

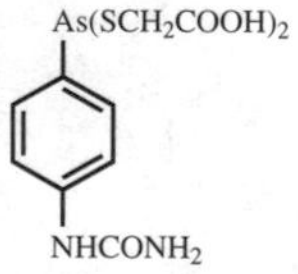

Thiocolchicine (mol wt 416) A muscle relaxant.

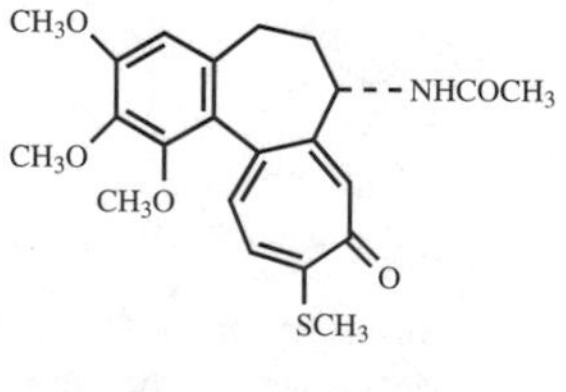

Thioctic Acid (mol wt 206) An antimicrobial agent (see also lipoic acid).

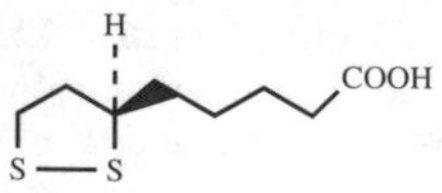

Thiocyanate Isomerase The enzyme that catalyzes the following reaction:

Benzyl isothiocyanate ⇌ Benzyl thiocyanate

Thiodicarb (mol wt 354) An insecticide.

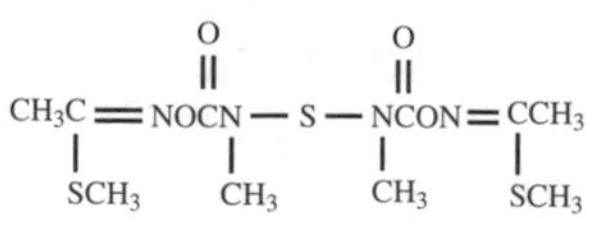

Thioester An ester formed between a carboxylic group and a sulfhydryl group.

R - CO- S - R′

Thioethanolamine *S*-Acetyltransferase The enzyme that catalyzes the following reaction:

Acetyl-CoA + thioethanolamine

⇅

CoA + S-acetylthioethanolamine

Thioether An ether with the general formula.

R - S - R′

Thioflavine T (mol wt 319) A stain for human reticulocytes in flow cytometric analysis.

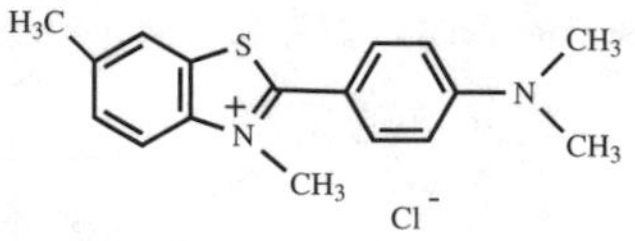

Thiogalactoside A complex of a thiol with a galactose.

Thiogalactoside Acetyltransferase The enzyme that catalyzes the following reaction:

Acetyl-CoA + β-D-galactoside

⇅

CoA + 6-acetyl β–galactoside

5-Thio-D-glucose (mol wt 196) A glucose derivative and a potent inhibitor of cellular glucose transport.

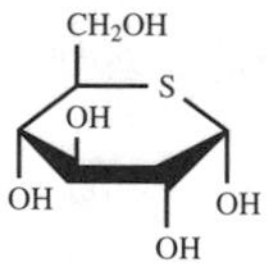

Thioglucosidase The enzyme that catalyzes the following reaction:

A thioglucoside + H_2O ⇌ A thiol + a glucose

Thioglucoside A complex of a sulfur-containing compound and glucose.

Thioglucoside Glucohydrolase The systematic name for thioglucosidase.

Thioglycerol (mol wt 108) Potential substitute for the unpleasantly smelling 2-mercapto-ethanol and a probe for the study of lymphocyte activation.

$HSCH_2CH(OH)CH_2OH$

Thioglycolic Acid (mol wt 92) A tryptophan protection agent.

$HS-CH_2-COOH$

Thioglycolic Acid Treatment A treatment with thioglycolic acid to break disulfide bonds in a protein.

Thioguanine (mol wt 167) A guanine derivative and an antineoplastic agent.

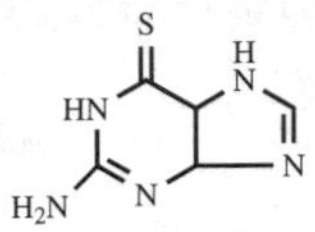

Thioguanosine (mol wt 299) A derivative of guanine nucleoside.

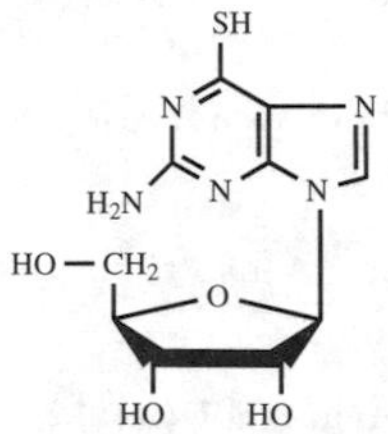

Thiokinase The enzyme that catalyzes the following reaction:

Fatty acid + ATP + CoA

⇌

Acyl-CoA + AMP + PPi

Thiol A compound that contains a sulfhydryl group with the general formula R-SH.

Thiol Activated Hemolysins A group of oxygen-labile proteins produced by certain bacteria that require reductive activation by thiols in order to function as hemolysins.

Thiol Endopeptidase A protease that has an active thiol group.

Thiol *S*-Methyltransferase The enzyme that catalyzes the following reaction:

S-Adenosyl-L-methionine + a thiol

⇌

S-Adenosyl-L-homocysteine + a thioether

Thiol Proteinase See thiol endopeptidase.

Thiolate An anion of a thiol.

Thiolation The introduction of a sulfhydryl group into a molecule.

Thiolutin (mol wt 228) An antibiotic produced by *Streptomyces albus.*

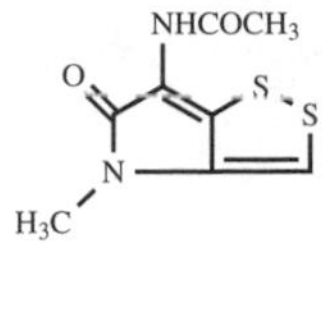

Thionicotinamide Adenine Dinucleotide Phosphate Monosodium (mol wt 782) A reagent for the study of the liver malic enzyme.

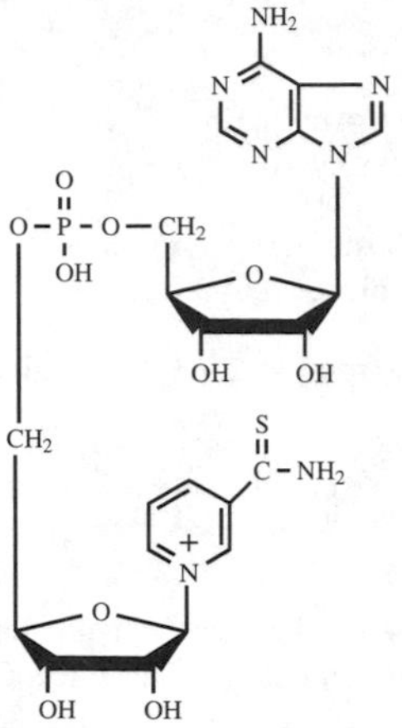

Thionins A group of toxic proteins from plants that are toxic to animals.

Thionucleoside Any nucleoside that contains a thiol group.

Thiopentone Sodium See thiopental sodium.

Thiophanate (mol wt 370) An anthelmintic agent.

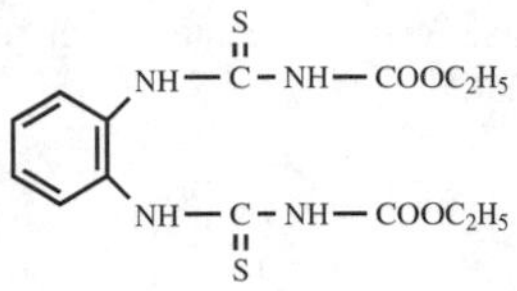

Thiophilic Growing best in the presence of sulfur compounds.

Thioplex A trade name for thiotepa, an antineoplastic agent.

Thioploca A genus of gliding bacteria (Cytophagales).

Thioporine A trade name for azathioprine, an immunosuppressant.

Thiopropazate (mol wt 446) An antipsychotic agent.

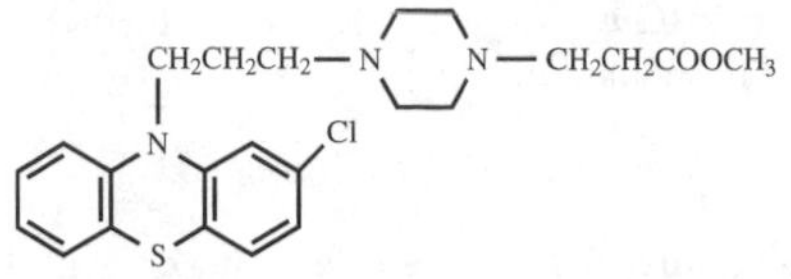

Thioproperazine (mol wt 447) A neuroleptic and an antiemetic agent.

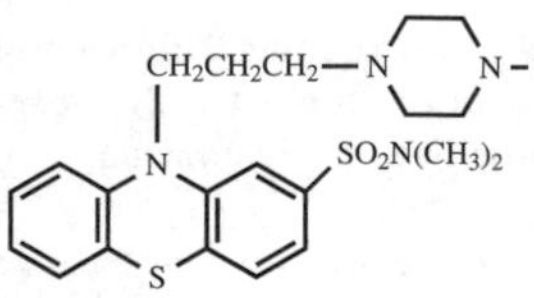

Thiopurine Any purine in which a thiol group has replaced a hydroxyl group.

Thiopyrimidine Any pyrimidine in which a thiol group has replaced a hydroxyl group.

Thioquinox (mol wt 236) A fungicide.

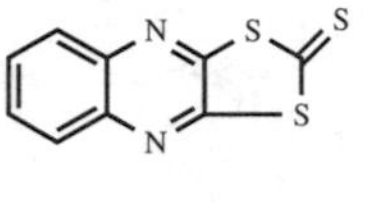

Thioredoxin A heat-stable hydrogen carrier involved in reduction of ribonucleotide to deoxyribonucleotide and a variety of other biochemical reactions.

Thioredoxin Reductase The enzyme that catalyzes the following reaction:

Oxidized thioredoxin + NADPH

⇅

$NADP^+$ + reduced thioredoxin

Thioridazine (mol wt 371) A dopaminergic blocking agent used as an anti-psychotic and antianxiety agent.

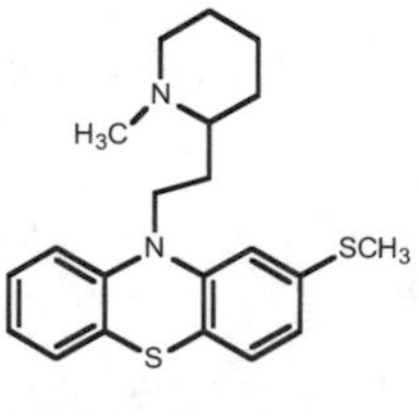

Thiosalicylic Acid (mol wt 154) A derivative of salicylic acid.

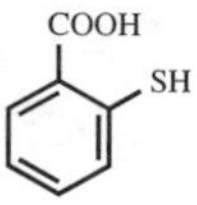

Thiospira A genus of Gram-negative, helical, polar flagellated bacteria.

Thiospirillum A genus of photosynthetic bacteria (Chromatiaceae).

Thiosulfate Salt of thiosulfuric acid.

Thiosulfate Dehydrogenase The enzyme that catalyzes the following reaction:

2 Thiosulfate + 2 ferricytochrome c

⇅

Tetrathionate + 2 ferrocytochrome c

Thiosulfate Sulfurtransferase The enzyme that catalyzes the following reaction:

Thiosulfate + cyanide

⇅

Sulfite + thiocyanate

Thiosulfate Thiol Sulfurtransferase The enzyme that catalyzes the following reaction:

Thiosulfate + 2 glutathione

⇅

Sulfite + oxidized glutathione + sulfide

Thiosulfate Thiotransferase See thiosulfate sulfurtransferase.

Thiosulfil Forte A trade name for sulfamethizole, an antibiotic.

Thiosulfuric Acid (mol wt 114) A highly unstable acid.

$H_2S_2O_3$

Thiotepa (mol wt 189) An antineoplastic and cytotoxic agent.

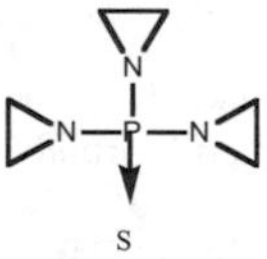

Thiothiamine (mol wt 296) An intermediate in the preparation of thiamine.

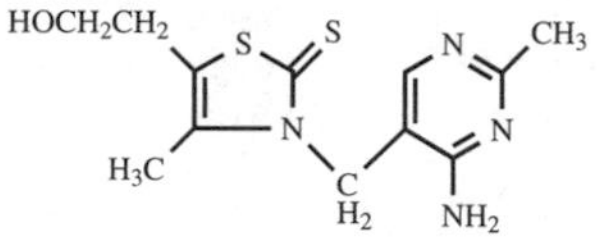

Thiothixene (mol wt 444) An antipsychotic agent that blocks postsynaptic dopamine receptors in the brain.

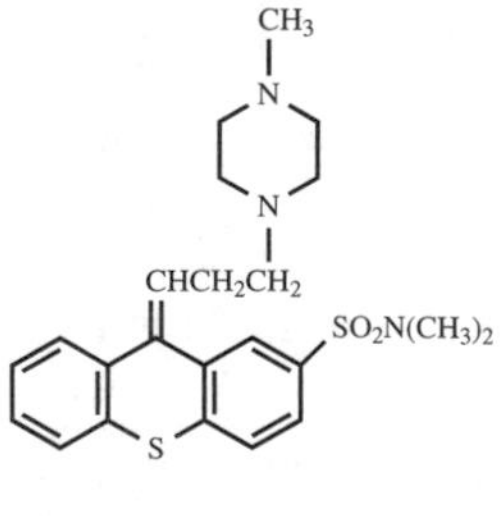

Thiothrix A genus of gliding bacteria (Cytophagales).

2-Thiouracil (mol wt 128) A thyroid depressant.

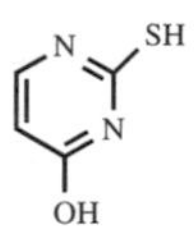

Thiourea (mol wt 76) An antithyroid substance.

H_2NCSNH_2

THIP (mol wt 140) Structural analog of muscimol with potent GABA agonist activity.

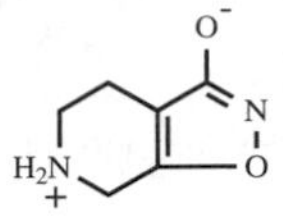

Thiphenamil (mol wt 328) An anticholinergic agent and a smooth muscle relaxant.

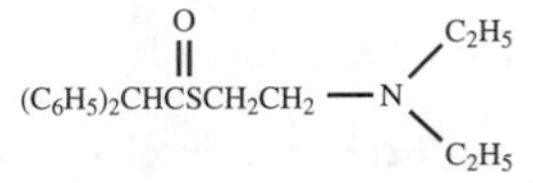

Thiram (mol wt 240) An antiseptic agent.

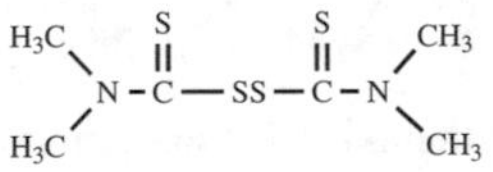

Third Order Reaction A reaction in which the velocity is proportional to the product of three concentration terms.

Thiuretic A trade name for hydrochlorothiazide, a diuertic agent that increases urine excretion of sodium and water by inhibiting sodium reabsorption.

Thoma's Counting Chamber A type of hemocytometer.

Thomson's Disease A glycogen storage disease caused by the deficiency in enzyme phosphoglucomutase.

Thonzylamine Hydrochloride (mol wt 323) An antihistaminic agent.

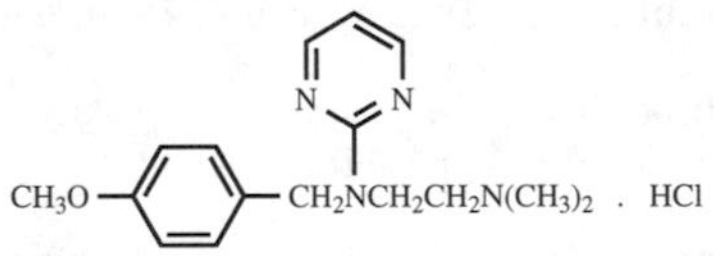

Thoracentesis The aspiration of sample fluid from the chest cavity by puncture.

Thoracic Pertaining to the chest cavity.

Thoracic Duct The major efferent lymph duct into which lymph from most of the peripheral lymph nodes drains. The lymphocytes in the lymph node return to the blood through the thoracic duct.

Thorax Synonym for chest.

Thorazine A trade name for chlorpromazine hydrochloride, an antipsychotic agent that blocks postsynaptic dopamine receptors in the brain.

Thorium A chemical element with atomic weight 232, valence 4.

Thor-Pram A trade name for chlorpromazine hydrochloride, an antipsychotic agent that blocks postsynaptic dopamine receptors in the brain.

Thozalinone (mol wt 204) An antidepressant.

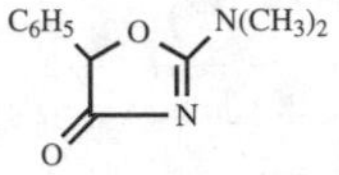

THP-1 A type of cell line.

ThPD Abbreviation for thiamine diphosphate.

ThPP Abbreviation for thiamine pyrophosphate.

Thr Abbreviation for 1. threonine and 2. threonyl.

Three Micron DNA Plasmid (3μ plasmid) A cccDNA (covalently closed circular DNA) plasmid in the strains of *Saccharomyces cerevisiae*.

Three Point Cross A series of genetic crosses for the determination of the order of the three nonallelic-linked genes on a single chromosome.

Threitol (mol wt 122) A sugar alcohol.

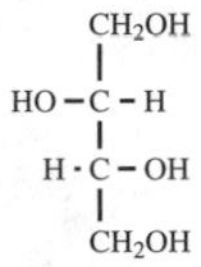

D-threitol

L-Threonate 3-Dehydrogenase The enzyme that catalyzes the following reaction:

$$\text{L-Threonate} + NAD^+ \rightleftharpoons \text{3-Dehydro-L-threonate} + NADH$$

Threonine (mol wt 119) An essential protein amino acid.

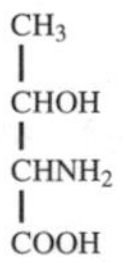

Threonine Acetaldehyde-Lyase See threonine aldolase.

Threonine Aldolase The enzyme that catalyzes the following reaction:

$$\text{L-Threonine} \rightleftharpoons \text{Glycine} + \text{acetaldehyde}$$

Threonine Deaminase The enzyme that catalyzes the following reaction:

$$\text{L-Threonine} + H_2O \rightleftharpoons \alpha\text{-Ketobutanoate} + NH_3 + H_2O$$

Threonine Dehydrase See threonine deaminase.

Threonine 3-Dehydrogenase The enzyme that catalyzes the following reaction:

$$\text{L-Threonine} + NAD^+ \rightleftharpoons \text{L-2-Amino-3-ketobutanoate} + NADH$$

Threonine Racemase The enzyme that catalyzes the following reaction:

$$\text{L-Threonine} \rightleftharpoons \text{D-Threonine}$$

Threonine Synthetase The enzyme that catalyzes the following reaction:

$$\textit{O}\text{-Phospho-L-homoserine} + H_2O \rightleftharpoons \text{L-Threonine} + \text{Pi}$$

Threonine-tRNA Ligase See threonine-tRNA synthetase.

Threonine-tRNA Synthetase The enzyme that catalyzes the following reaction:

$$\text{ATP} + \text{L-threonine} + \text{tRNA}^{thr} \rightleftharpoons \text{AMP} + \text{PPi} + \text{L-threonyl-tRNA}^{thr}$$

Threoninyl tRNA Ligase See threonine tRNA ligase.

Threoninyl tRNA Synthetase See threonine tRNA synthetase.

Threose (mol wt 120) A monosaccharide.

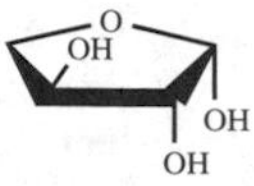

α-L-threose

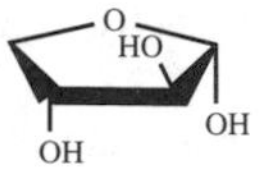

α-D-threose

Threshold Dose A dose of radiation below which it produces no detectable effect.

Threshold Potential The potential that an excitable cell membrane must reach for an action potential to be initiated.

Threshold Stimulus The stimulus that is able to depolarize the membrane and thereby initiate a nerve impulse.

THRF Abbreviation for thyrotropic hormone releasing factor.

THRH Abbreviation for thyrotropic hormone releasing hormone.

Thrombasthenia An inherited disorder of blood platelets in which the platelets lack a factor for blood clotting.

Thrombi Plural of thrombus.

Thrombin A serine proteinase that converts fibrinogen to fibrin, thus triggering the formation of blood clots.

Thrombinar A trade name for thrombin (a blood clotting factor).

Thrombinogen Synonym for prothrombin (factor II).

Thromboarteritis Inflammation of an artery with thrombus formation.

Thromboblast Precursor cell of platelets.

Thrombocytasthenia A platelet disorder characterized by abnormal adhesion and/or aggregation of platelets.

Thrombocytes Synonym for platelets.

Thrombocythemia A disorder in which an abnormal proliferation of platelet-producing cells occurs leading to the increase in the number of platelets in the blood and the tendency to form clots within the blood vessels.

Thrombocytopenia The decrease in number of platelets in the blood.

Thrombocytopenic Purpura An autoimmune disease in which antiplatelet autoantibody destroys platelets.

Thrombocytosis The increase in number of platelets in the blood.

Thromboembolism The blocking of a blood vessel by a dislodged blood clot.

Thrombogenic Tending to produce thrombi.

Thrombokinase A protease that catalyzes the conversion of prothrombin to thrombin.

Thrombolysis Lysis or dissolving of a blood clot.

Thrombolytic Enzymes The enzymes that catalyze fibrinolysis.

Thrombomodulin The specific endothelial cell receptor that forms a complex with thrombin, regulating blood clotting.

Thrombophlebitis Inflammation of the wall of a vein with the formation of a blood clot.

Thromboplastin A blood clotting factor from platelets released from injured tissue that initiates the extrinsic pathway of blood clotting.

Thromboplastinogen The antihemophilic factor.

Thrombopoiesis The formation of a blood clot.

Thrombosis The formation of a blood clot in the blood circulatory system.

Thrombospondin A major glycoprotein of the human platelet α-granules that is released in response to platelet activation by α-thrombin, which plays an important role in platelet aggregation.

Thrombostat A trade name for thrombin, a blood clotting factor.

Thrombosthenin A contractile protein occurring in blood platelets.

Thromboxanes A group of immunologically active substances derived from arachidonic acid involved in platelet aggregation, artery contraction, and other biological functions.

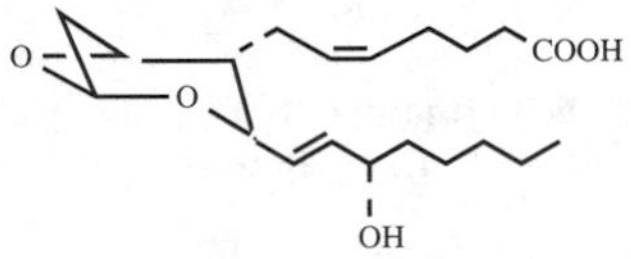

thromboxane A_2

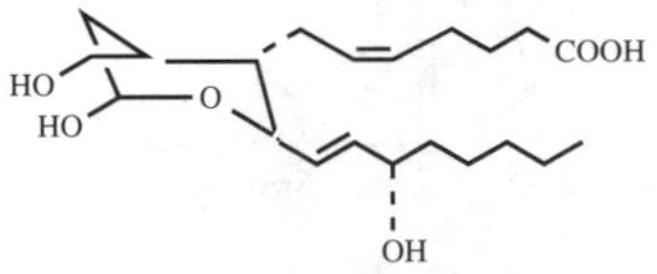

thromboxane B_2

Thrombus A blood clot formed within a blood vessel that remains attached to the wall of the vessel.

Thulium A chemical element with atomic weight 169, valence 3.

Thurfyl Nicotinate (mol wt 207) A topical vasodilator.

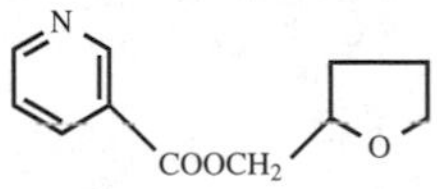

Thx Short for thyroxine.

Thy Abbreviation for thymine.

Thy-1 Antigen See thy-1 glycoprotein.

Thy-1 Glycoprotein Glycoproteins on the surface of murine T lymphocytes that serve as markers for T lymphocytes.

Thylakoid Disks Membranous sacs that form grana of the chloroplast that contain photosynthetic pigment, electron carriers, and ATP-forming enzymes, and are the sites of light reaction in photosynthesis.

Thymectomy Surgical removal of the thymus.

Thymic Pertaining to the thymus.

Thymic Alymphoplasia A severe combined immune deficiency transmitted as an X-linked recessive trait.

Thymic Aplasia A deficiency of T lymphocytes caused by the failure to develop a thymus.

Thymidine (mol wt 242) A deoxyribonucleoside.

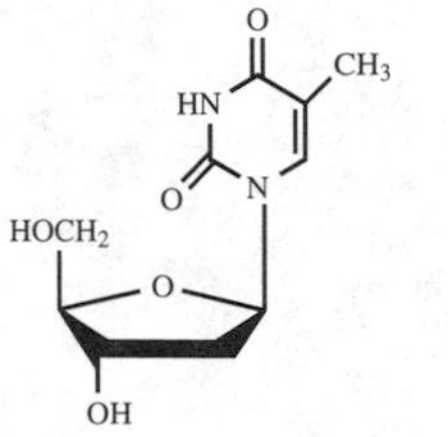

Thymidine 5′-Diphosphate (mol wt 402) The diphosphate form of thymine nucleotide.

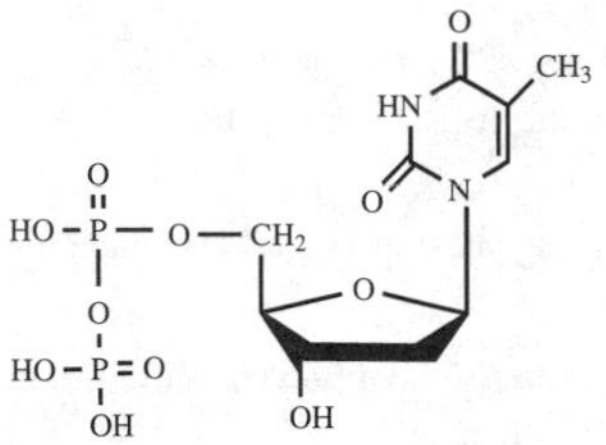

Thymidine Kinase The enzyme that catalyzes the following reaction:

ATP + thymidine

⇅

ADP + thymidine 5′- phosphate

Thymidine 5′-Monophosphate (mol wt 322) A nucleotide of thymine.

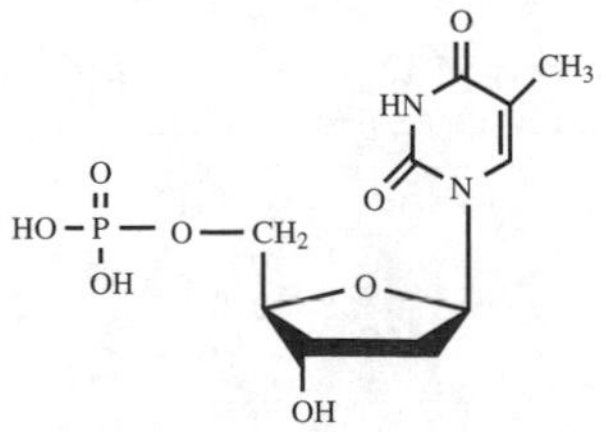

Thymidine Phosphorylase The enzyme that catalyzes the following reaction:

Thymidine + Pi

⇅

Thymine + 2-deoxy D-ribose 1-phosphate

Thymidine Triphosphatase The enzyme that catalyzes the following reaction:

$$\text{dTTP} + H_2O \rightleftharpoons \text{dTDP} + \text{Pi}$$

Thymidine 5′-Triphosphate (mol wt 482) The triphosphate form of thymine nucleotide.

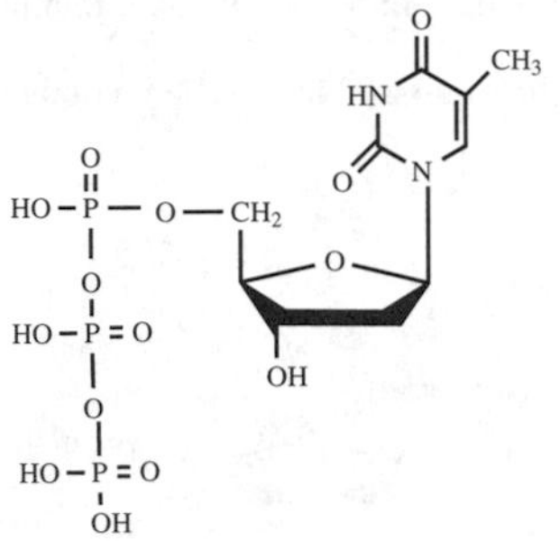

Thymidylate Salt of thymidylic acid.

Thymidylate 5′-Nucleotidase See thymidylate 5′-phosphatase.

Thymidylate 5′-Phosphatase The enzyme that catalyzes the following reaction:

$$\text{Thymidylate} + H_2O \rightleftharpoons \text{Thymidine} + \text{Pi}$$

Thymidylate 5′-Phosphohydrolase See thymidylate 5′-phosphatase.

Thymidylate Synthetase The enzyme that catalyzes the following reaction:

$$\text{5,10-Methylenetetrahydrofolate} + \text{dUMP} \rightleftharpoons \text{Dihydrofolate} + \text{dTMP}$$

Thymidylic Acid The deoxyribonucleic acid of thymine (e.g., thymidine 5′-monophosphate).

Thymine (mol wt 126) A pyrimidine nitrogenous base found in DNA and tRNA.

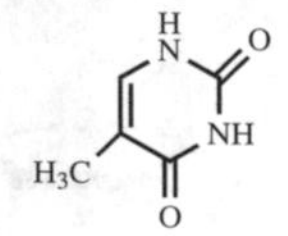

Thymine Dimer A type of dimer formed by covalent linkage between two adjacent thymines on the same strand of DNA, caused by exposure of DNA to ultraviolet light.

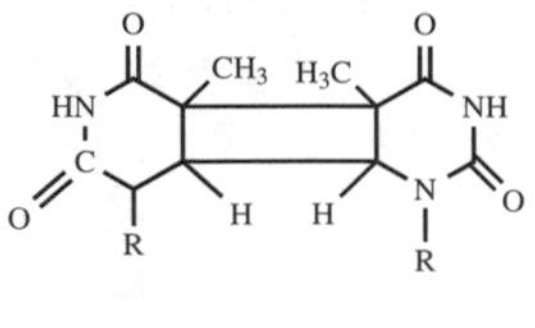

Thymine Dioxygenase The enzyme that catalyzes the following reaction:

$$\text{Thymine} + \alpha\text{-ketoglutarate} + O_2 \rightleftharpoons \text{5-Hydroxymethyluracil} + CO_2 + \text{succinate}$$

Thymitis Inflammation of the thymus gland.

Thymocytes Lymphoid cells present in the thymus.

Thymol (mol wt 150) An antiseptic and an anthelmintic agent.

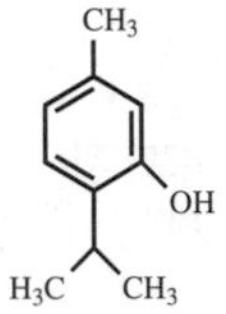

Thymol-Blue (mol wt 467) A dye and pH indicator (red at pH 1.2, yellow at pH 2.8).

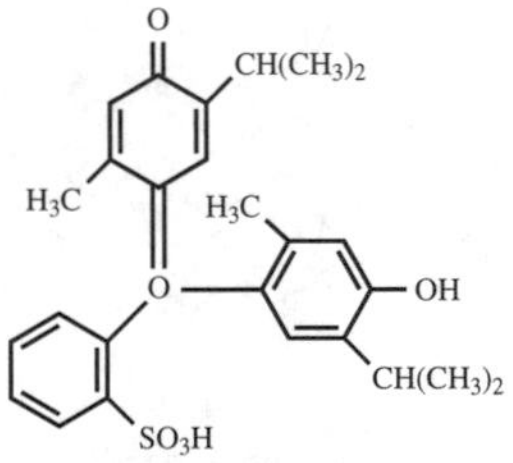

Thymol-Iodide (mol wt 550) An antifungal and an anti-infective agent.

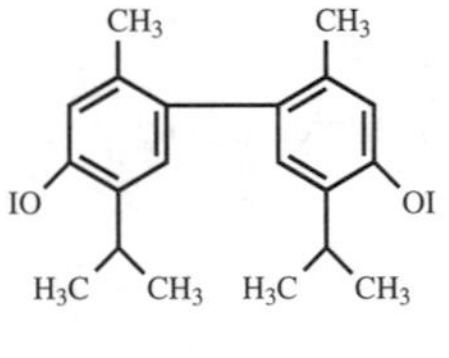

Thymolphthalein (mol wt 431) A pH indicator (colorless at pH 9.3, blue at pH 10.5).

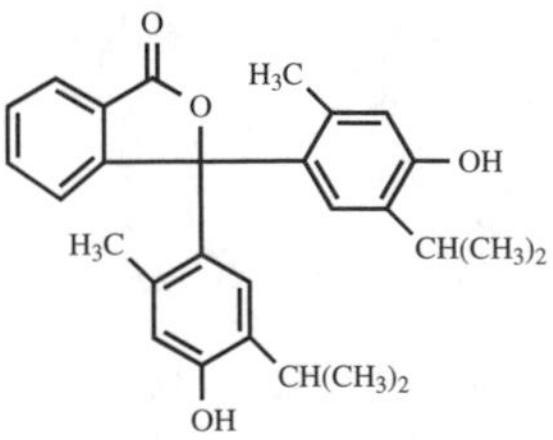

Thymol-Turbidity Test A liver function test based on the production of turbidity upon mixing thymol-barbiturate buffer with serum from individuals with hepatitis.

Thymoma A tumor arising from the tissue elements of the thymus.

Thymomodulin Protein hormone from calf thymus that acts as immunoregulator that possesses antileukopenic activity.

Thymopentin A pentapeptide thymic hormone analog consisting of arg-lys-asp-val-tyr that exhibits the full biological activity of natural thymic hormone and it also induces T lymphocyte differentiation.

Thymopoietin A polypeptide thymic hormone that induces T-cell maturation and inhibits B-cell differentiation. Thymopoietin occurs in the thymus of animals with autoimmune thymitis and myasthenia gravis and impairs neuromuscular transmission.

Thymosin A thymic hormone protein that can restore T-cell immunity in thymectomized animals.

Thymotaxin Referring to β_2 microglobulin.

Thymulin A thymic nonapeptide hormone known to induce thymic T cell differentiation.

Thymus The primary lymphoid organ, located in the thorax, which regulates the differentiation and maturation of T lymphocytes.

Thymus-Dependent Antigen Antigen that depends on T-cell interaction with B cells for production of antibody.

Thymus-Derived Lymphocytes Referring to T lymphocyte or T cell.

Thymus-Independent Antigen Antigen that induces antibody production without direct cooperation from T cells. The antibodies synthesized are of the IgM type.

Thymyl *N*-Isoamylcarbamate (mol wt 263) An anthelmintic agent.

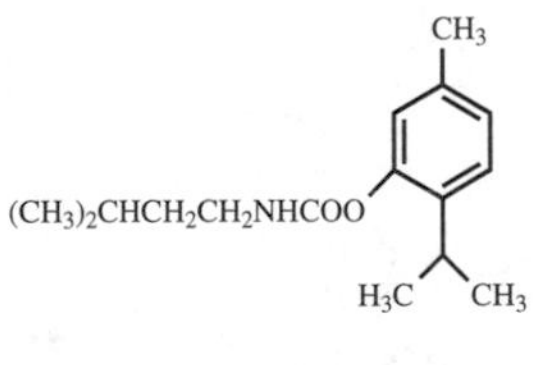

Thyrar A trade name for desiccated thyroid, which stimulates and accelerates the rate of cellular oxidation.

Thyrocalcitonin A bioactive peptide secreted by the thyroid gland.

Thyroglobulin The iodinated protein of the thyroid gland, from which thyroxine and triiodothyroxine are derived.

Thyroid 1. Pertaining to the thyroid gland. 2. A pharmaceutical preparation from thyroid gland of various animals used for the treatment of hypothyroid conditions.

Thyroid Antibody Autoantibody against thyroglobulin that is present in patients with Hashimoto's thyroiditis.

Thyroid Gland A two-lobed gland that produces the hormones thyroxine and triiodothyroxine.

Thyroid Hormone Referring to thyroxine and triiodothyroxine.

Thyroid Hormone Transaminase The enzyme that catalyzes the following reaction:

L-3, 5, 3′-Triiodothyroxine + α-ketoglutarate

⇅

3, 5, 3′-Triiodophenylpyruvate + L-glutamate

Thyroidectomy Surgical removal of the thyroid gland.

Thyroiditis Inflammation of the thyroid gland.

Thyroid-Stimulating Hormones Referring to thyrotropin, which activates cAMP production in the thyroid cell and promotes production and release of thyroid hormone by the pituitary.

Thyrolar A trade name for liotrix, a thyroid hormone.

Thyroliberin See thyrotropic-releasing hormone.

Thyromegaly Abnormal enlargement of the thyroid gland.

Thyropropic Acid (mol wt 636) An anticholesteremic agent.

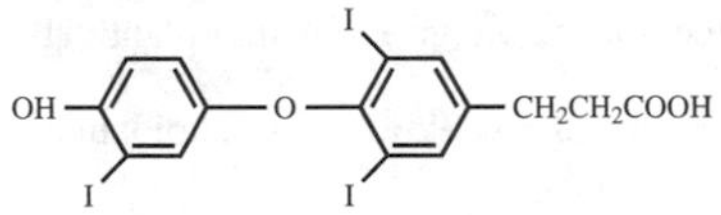

Thyro-Teric A trade name for desiccated thyroid.

Thyrotoxicosis Toxic condition caused by the overproduction of thyroid hormone.

Thyrotropic Hormone A hormone (e.g., thyrotropin) produced in the pituitary gland that stimulates the thyroid gland to produce the thyroid hormone (also known as thyroid-stimulating hormone).

Thyrotropin A thyroid-stimulating hormone that promotes thyroid hormone production by the anterior pituitary and also stimulates the uptake of radioactive iodine in patients with thyroid carcinoma.

Thyrotropin-Releasing Hormone A hypothalamic hormone that regulates the secretion of thyrotropin from the pituitary gland.

Thyroxine (mol wt 777) The hormone secreted by the thyroid gland that contains iodine and controls the rate of oxygen consumption and overall metabolism (also known as T_4).

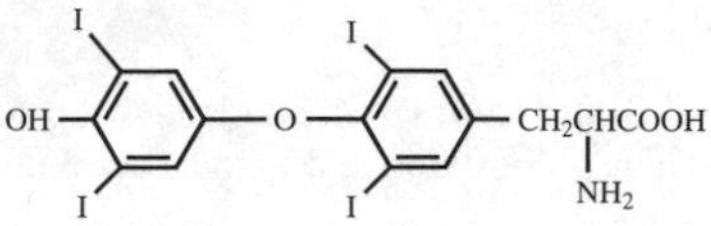

Thyroxine-Binding Globulin A glycoprotein in the plasma that serves as a major specific carrier for thyroxine.

Thyroxine-Binding Prealbumin A serum albumin that serves as a minor specific carrier for thyroxine.

Thyroxine-Deiodinase The enzyme that catalyzes the following reaction:

$$\text{L-Thyroxine} + AH_2 \rightleftharpoons \text{3, 5, 3}'\text{-Triiodo-L-thyronine} + \text{iodide} + A + H^+$$

Thytropar A trade name for thyrotropin, a thyroid hormone that stimulates thyroid hormone production in the pituitary gland.

Ti Symbol for the chemical element titanium.

TI Antigen Abbreviation for thymus-independent antigen.

Ti Plasmid A circular DNA plasmid of *Agrobacterium tumefaciens* that induces crown gall disease in dicotyledonous plants and carries T-DNA, which can be incorporated into host genone. Ti plasmid has been used as a cloning vector for introduction of foreign DNA into plant cells.

Tiadenol (mol wt 295) An antilipidemic agent.

$$HOCH_2CH_2S(CH_2)_{10}SCH_2CH_2 - OH$$

Tiagabine (mol wt 376) An anticonvulsant and anti-epileptic agent.

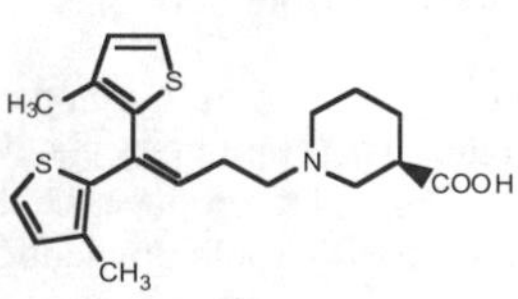

Tiamate A trade name for diltiazem hydrochloride, a calcium channel blocker used as an antianginal and antihypertensive agent.

Tiamenidine (mol wt 216) An antihypertensive agent.

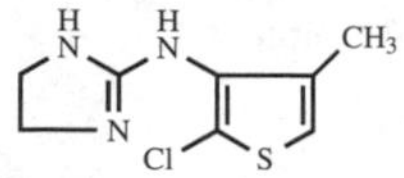

Tianeptine (mol wt 437) An antidepressant.

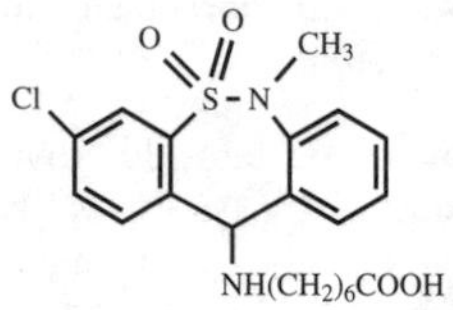

Tiapride (mol wt 328) An antidyskinetic agent.

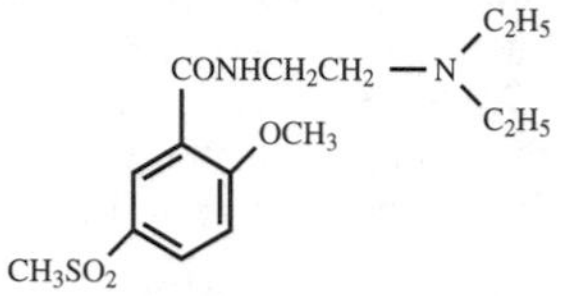

Tiaprofenic Acid (mol wt 260) An anti-inflammatory agent.

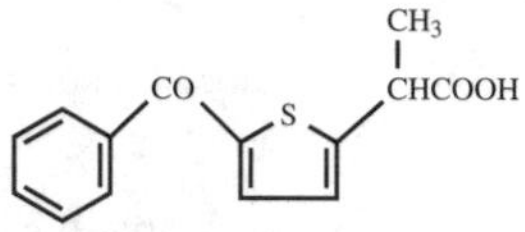

Tiaramide (mol wt 356) An antiasthmatic and anti-inflammatory agent.

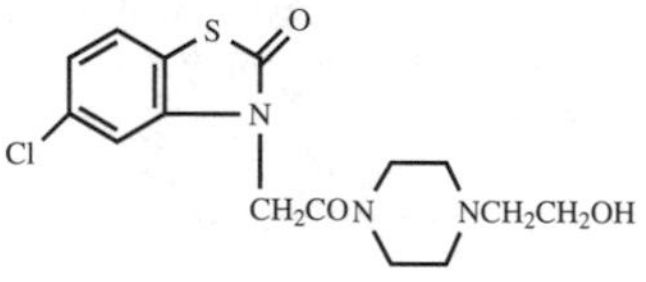

Tiazac A trade name for diltiazem hydrochloride, a calcium channel blocker used as an antianginal and antihypertensive agent.

TIB90 Abbreviation for a mouse fibrosarcoma cell line.

Tibezonium Iodide (mol wt 602) An antibacterial agent.

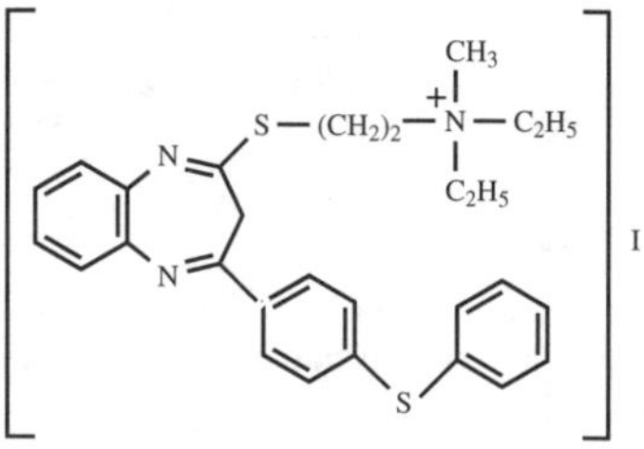

TIC Abbreviation for total ion chromatogram.

Ticar A trade name for ticarcillin disodium, an antibiotic that inhibits synthesis of bacterial cell wall.

Ticarbodine (mol wt 316) An anthelmintic agent.

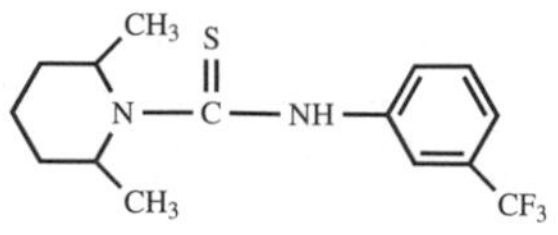

Ticarcillin (mol wt 384) A semisynthetic antibiotic related to penicillin that inhibits the synthesis of bacterial cell wall.

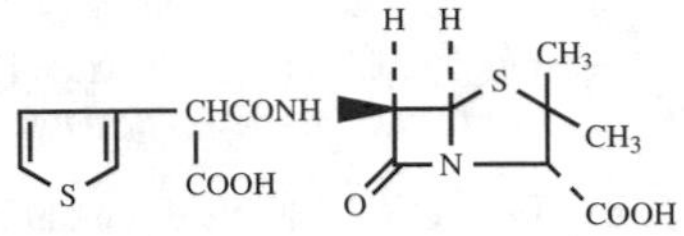

Ticillin A trade name for ticarcillin sodium, an antibiotic that inhibits bacterial cell wall synthesis.

Tick-Borne Disease Diseases that are transmitted by ticks.

Tickicide Any agent that kills ticks.

Ticlid A trade name for ticlopidine hydrochloride, an antiplatelet agent that blocks ADP-induced platelet fibrinogen and platelet–platelet binding.

Ticlopidine (mol wt 264) An antiplatelet agent that blocks ADP-induced platelet fibrinogen and platelet–platelet binding.

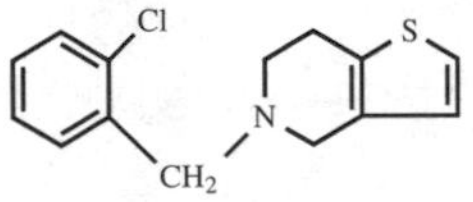

Ticon A trade name for trimethobenzamide hydrochloride, an antiemetic agent.

Ticrynafen (mol wt 331) A diuretic, uricosuric and antihypertensive agent.

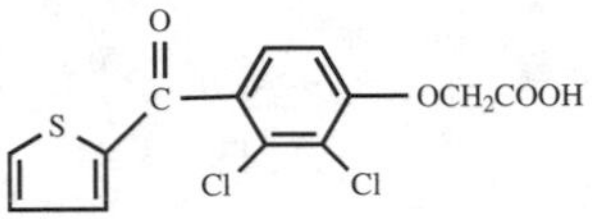

Tidal Air The air that passes in and out of the lung in a normal respiratory breath (about 500 cm^3 in a normal adult human male).

Tiemonium Iodide (mol wt 445) An anticholinergic and antispasmodic agent.

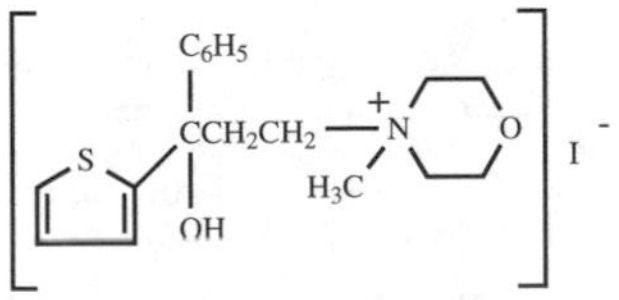

Tiered Metabolic Pathway A metabolic pathway in which the product of the first reaction activates the second reaction, and the product of second reaction initiates the third reaction, such as:

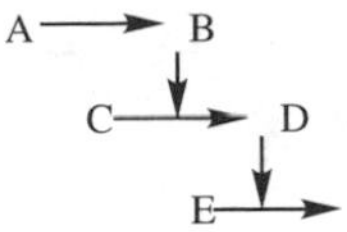

TIF Abbreviation for tumor inhibitory factor.

TIG Abbreviation for tetanus immune globulin.

Tigan A trade name for trimethobenzamide hydrochloride, an antiemetic agent.

Tigemonam (mol wt 437) An antibacterial agent.

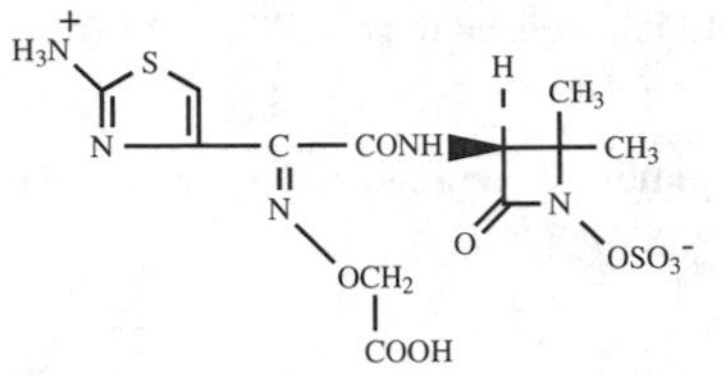

Tight Junction A type of cell junction in which the adjacent plasma membrane of the neighboring cells are tightly sealed, thereby preventing molecules from diffusing from one side of the epithelial cell layer to the other by passing through the space between adjoining cells.

Tigloidine (mol wt 223) A CNS depressant and an antiparkinsonian agent.

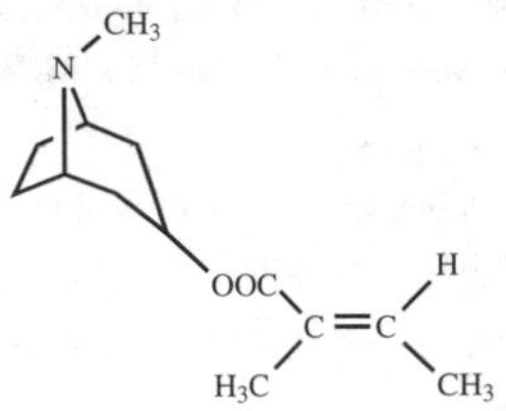

TIIV Abbreviation for trivalent inactivated influenza vaccine.

Tija A trade name for oxytetracycline, an antibiotic.

Tiject-20 A trade name for trimethobenzamide hydrochloride, an antiemetic agent.

TIL Abbreviation for tumor-infiltrating lymphocytes.

Tilade A trade name for nedocromil sodium, used to inhibit mast cell degranulation and release of vasoactive substance.

Tilidine (mol wt 273) A narcotic analgesic agent.

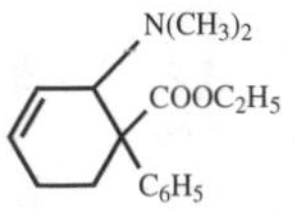

Tilorone (mol wt 411) An oral interferon inducer.

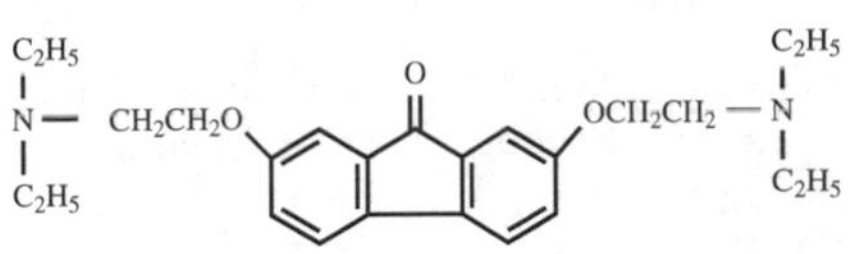

Tiludronic Acid (mol wt 319) A biophosphate used for the treatment of osteoporosis.

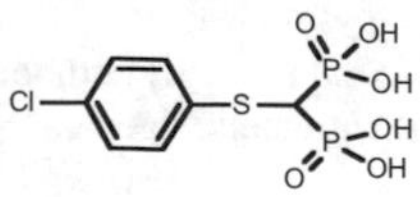

Timentin A trade name for ticarcillin disodium, an antibiotic related to penicillin that inhibits bacterial cell wall synthesis.

Timepidium Bromide (mol wt 400) An anticholinergic agent.

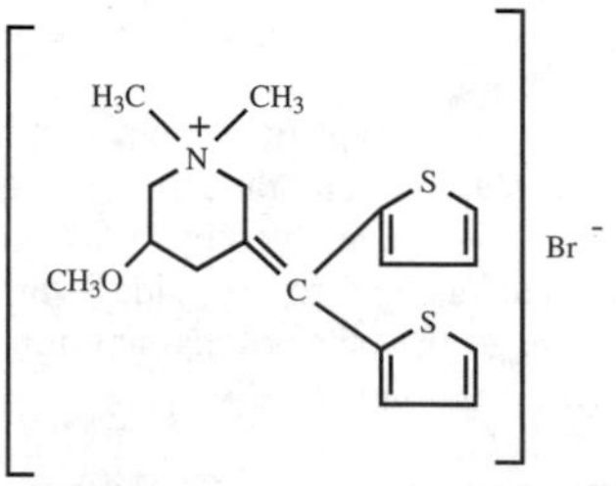

Timolide 10/25 A trade name for a combination drug containing timolol maleate and hydrochlorothiazide, used as an antihypertensive agent.

Timolol (mol wt 316) An antihypertensive agent that blocks the response to beta stimulation and depresses renin secretion.

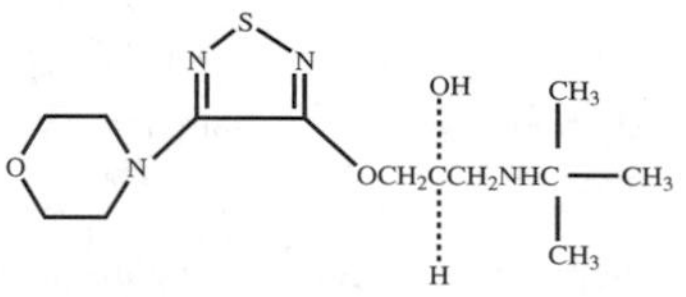

Timonacic (mol wt 133) A choleretic agent.

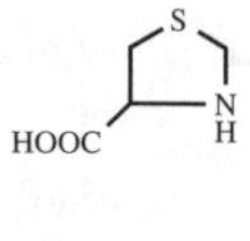

Timoptic Solution A trade name for timolol maleate solution, used as a beta blocker.

TIMP Abbreviation for tissue inhibitor of metallo-proteinase.

Tin A chemical element with atomic weight 119, valence 2 and 4.

Tinactin A trade name for tolnaftate, used as a local anti-infective against fungi.

Tindal A trade name for acetophenazine maleate, an antipsychotic agent that blocks the postsynaptic dopamine receptors in the brain.

Tine Test A test for tuberculosis in which tuberculin is introduced subcutaneously by a mechanical device, producing multiple punctures in the skin.

Tinea The skin lesions caused by fungi (e.g., *Trichlphyton*, *Microsporum*, and *Epidermophyton*).

Ting A trade name for tolnaftate, an antifungal agent.

Tinidazole (mol wt 247) An antiprotozoal agent.

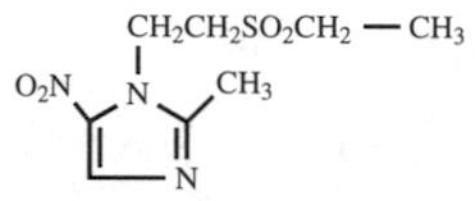

Tinnitus A sensation of noises in the ear.

Tinofedrine (mol wt 356) A cerebral vasodilator.

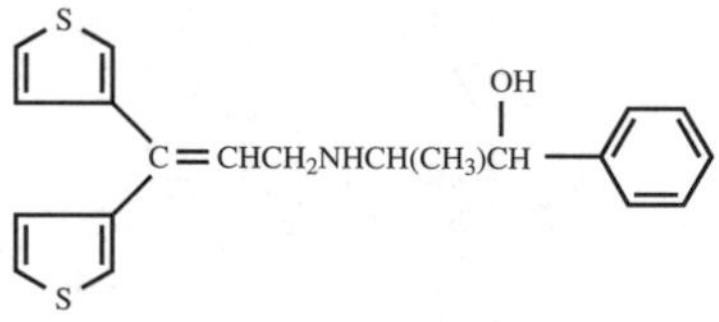

Tinoridine (mol wt 316) An analgesic, antipyretic, and anti-inflammatory agent.

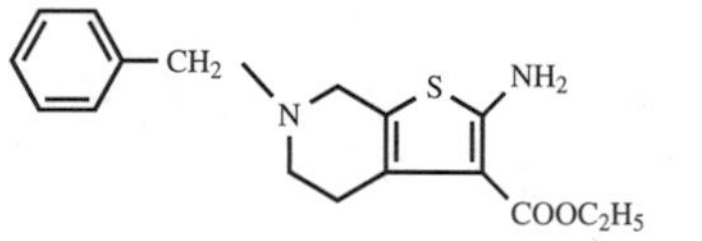

Tinsel Flagellum Flagellum of a eukaryotic organism that bears fine, filamentous appendages along its length.

Tiocarlide (mol wt 401) An antibacterial agent (tuberculostatic).

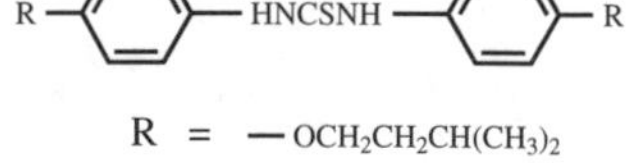

R = $—OCH_2CH_2CH(CH_3)_2$

Tioclomarol (mol wt 447) An anticoagulant.

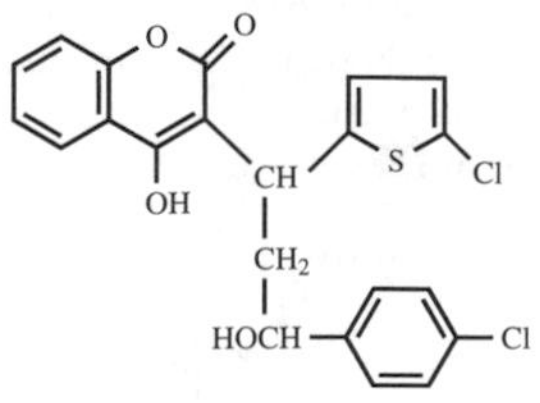

Tioconazole (mol wt 388) An antifungal agent that alters fungal cell wall permeability.

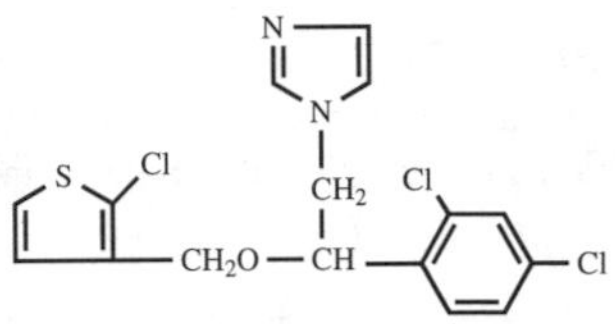

Tiomesterone (mol wt 451) An androgen.

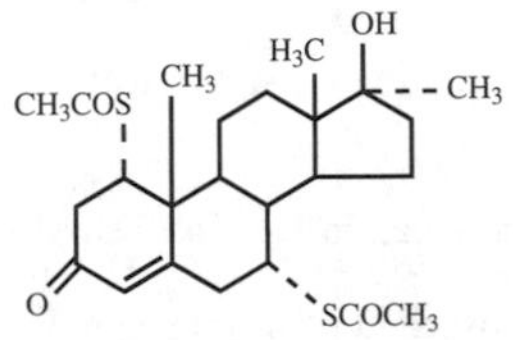

Tiopronin (mol wt 163) A substance used to prevent the formation of urinary cysteine stones.

$CH_3CHCONHCH_2COOH$
|
SH

TIP Abbreviation for 1. thermal inactivation point, 2. tumor-inducing principle, and 3. translation inhibitory protein.

Tipepidine (mol wt 275) An antitussive agent.

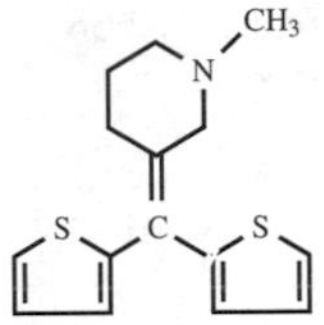

Tipula Iridescent Virus An insect virus.

Tiquizium Bromide (mol wt 410) An antispasmodic agent.

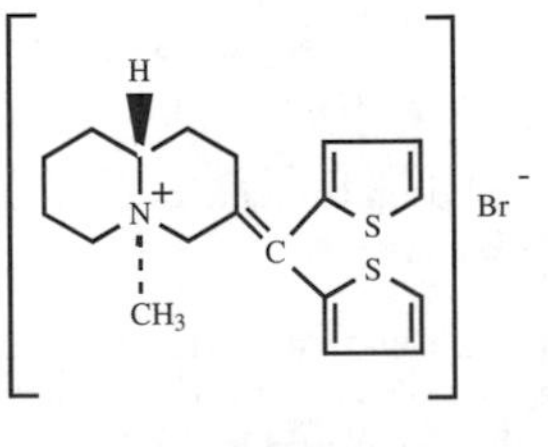

Tiratricol (mol wt 622) A substance used for thyroid replacement therapy.

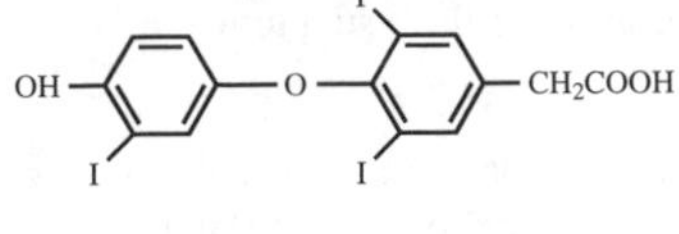

Tirend A trade name for caffeine.

Tirofiban (mol wt 441) An antiplatelet and antithrombotic agent.

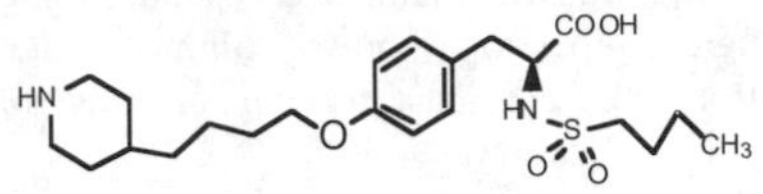

Tiropramide (mol wt 468) An antispasmodic agent.

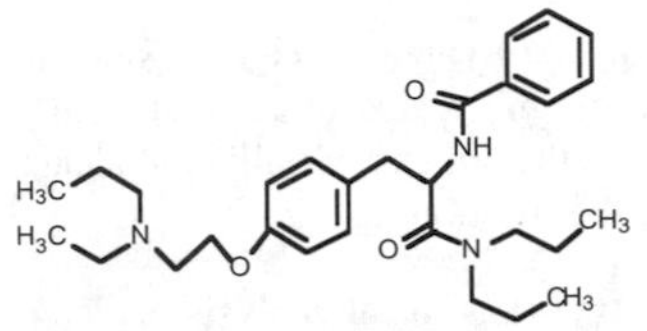

Tiselius Apparatus An apparatus used for performing moving boundary electrophoresis.

Tissue An integrated group of cells with a common structure and function.

Tissue Culture The maintenance and growth of isolated tissues *in vitro* from plants or animals.

Tissue Fixed Macrophage Referring to histiocytes.

Tissue Kallikrein 1. A protease that catalyzes the cleavage of peptide bonds involving the COOH group of arginine in a peptide. 2. A protease that catalyzes the selective cleavage of kininogen to release kallidin by hydrolysis of the peptide bond involving COOH groups of methionine and leucine.

Tissue Plasminogen Activator A serine protease that catalyzes the conversion of plasminogen to plasmin for initiation of local fibrinolysis by hydrolysis of the peptide bond between arginine and valine.

Tissue Typing The determination of MHC antigen compatibility between transplant donor and recipient.

Titanium A chemical element with atomic weight 48, valences 2, 3, and 4.

Titer 1. The relative concentration of a substance in solution. 2. The relative strength of an antiserum.

Titin Muscle protein that links thick filaments to the Z band of the sarcomere.

Titracid A trade name for calcium carbonate, used to reduce the total acid load in the GI tract.

Titralac A trade name for calcium carbonate, used to reduce the total acid load in the GI tract.

Titrant The solution that is added to a second solution in the process of titration.

Titration A volumetric analysis of strength or concentration of a test solution of a substance by addition of known volume of the standard solution through a pipet to a known volume of the test solution (e.g., acid-base titration or antigen-antibody neutralization).

Titration Curve A graphic representation obtained by plotting amount of acid or base added as a function of pH.

Titration Equivalent Weight Referring to the number of milligrams of fatty acid in a sample divided by the number of milliequivalents of alkali used in the titration.

Titrimetry Chemical analysis or determination by means of titration.

Tixocortol (mol wt 379) An anti-inflammatory agent.

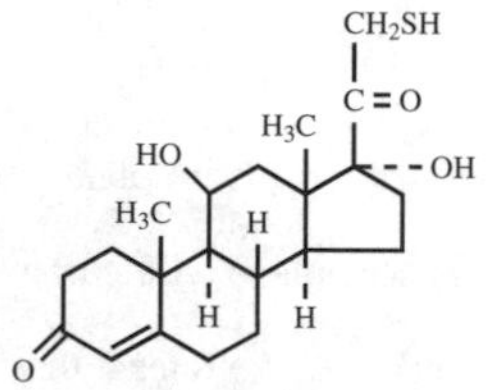

Tizanidine (mol wt 254) A skeletal muscle relaxant.

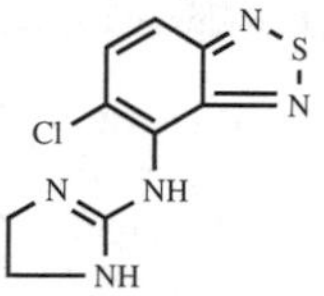

TK Abbreviation for thymidine kinase.

T-Kininogenase See tissue kallikrein.

TL Abbreviation for thymus leukemia.

TL Antigen Abbreviation for thymic leukemia antigen, a mouse antigen encoded by the TLa complex. TL antigen disappears as T cells mature but resurfaces as leukemia develops.

TLA Abbreviation for thymus leukemia antigen.

TLC Abbreviation for thin layer chromatography.

tld A gene encoding tolloid protein in *Drosophila* that is required for dorsal development.

TLE Abbreviation for thin layer electrophoresis.

TLI Abbreviation for total lymphoid irradiation.

TM Abbreviation for transmembrane.

Tm Abbreviation for thulium, a chemical element.

T_m The temperature at the midpoint of transition of a double-stranded DNA into single-stranded molecules.

TMA Abbreviation for tetramethylammonium.

TMAD Abbreviation for trimethylamine dehydrogenase.

TMADH Abbreviation for trimethylamine dehydrogenase.

TMB Abbreviation for tetramethyl benzidine.

TMD Abbreviation for transmembrane domain.

TMF Abbreviation for thymocyte mitogenic factor.

TMP A trade name for trimethoprim, an antibacterial agent.

TMϕ Abbreviation for thioglycolate-elicited macrophage.

7-TMR Abbreviation for seven-transmembrane-segment receptor.

TMR Spectroscopy Abbreviation for topical magnetic resonance spectroscopy.

TmuII (CauII) A restriction endonuclease from *Tuberoidobacter mutans* RFL1 with the same specificity as CauII

TMV Abbreviation for tobacco mosaic virus, a single-stranded RNA virus that infects and multiplies in tobacco or other plants.

Tn Abbreviation for transposon.

T_{n1} A transposon that carries genes encoding β-lactamase-like activity.

T_{n2} A transposon that carries genes encoding β-lactamase-like activity.

T_{n3} A transposon that carries genes encoding β-lactamase-like activity that tends to insert preferentially at the target sites in the AT-rich regions of DNA.

T_{n4} A transposon that carries genes encoding resistance to ampicillin, streptomycin, and sulfonamide.

T_{n5} A transposon that carries genes encoding aminoglycoside 3′-phosphotransferase, an enzyme that can phosphorylate karamycin, neomycin, and other aminoglycoside antibiotics.

T_{n9} A transposon that carries genes encoding resistance to chloramphenicol.

T_{n10} A transposon that carries genes encoding resistance to tetracycline.

T_{n21} A transposon that carries genes encoding resistance to sulphonamides, streptomycin, and mercuric ions.

T_{n501} A T_{n3}-like transposon that carries a gene encoding resistance to mercury.

T_{n554} A repressor-controlled transposon that carries genes for resistance to erythromycin and spectinomycin.

T_{n1681} A transposon that carries a gene for the heat-stable enterotoxin of enterotoxigenic *E. coli*.

TNA Abbreviation for total nucleic acid.

TNB Abbreviation for 5-thio-2-nitrobenzoate.

TNBS Abbreviation for trinitrobenzene sulfonate or trinitrobenzenesulfonic acid.

TNF Abbreviation for tumor necrosis factor.

TNFα Abbreviation for alpha tumor necrosis factor.

TNMR Abbreviation for tritium nuclear magnetic resonance.

TNP Abbreviation for 1. trinitrophenol, 2. trinitrophenyl.

TNT Abbreviation for trinitrotoluene.

TNV Abbreviation for tobacco necrosis virus.

TO Abbreviation for tri-oleoyl.

Tobacco Mosaic Virus See TMV.

Tobamoviruses A group of plant viruses in which the virion is a rigid filament consisting of linear, positive-stranded RNA as genetic material, e.g., tobacco mosaic virus.

TobraDex A trade name for a combination drug containing dexamethasone, tobramycin, chlorobutanol, mineral oil, and white petrolatum, used as an ophthalmic anti-infective agent.

Tobramycin (mol wt 468) An antibiotic produced by *Streptomyces tenebrarius* that inhibits bacterial protein synthesis.

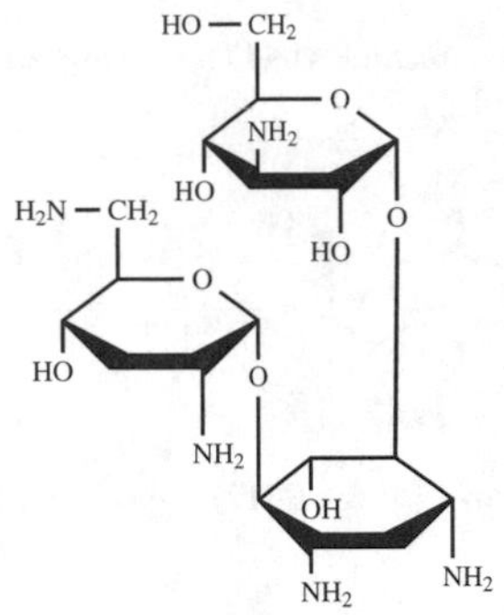

Tobraviruses A group of bipartite ssRNA-containing viruses with a wide host range.

Tobrex A trade name for tobramycin, which inhibits bacterial protein synthesis.

Tocainide (mol wt 192) A cardiac depressant and an antiarrhythmic agent that shortens the action potential.

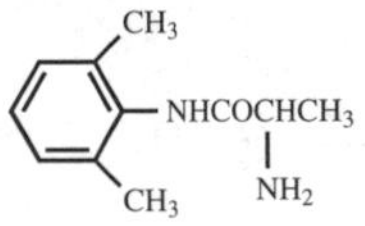

Tocamphyl (mol wt 424) A choleretic agent.

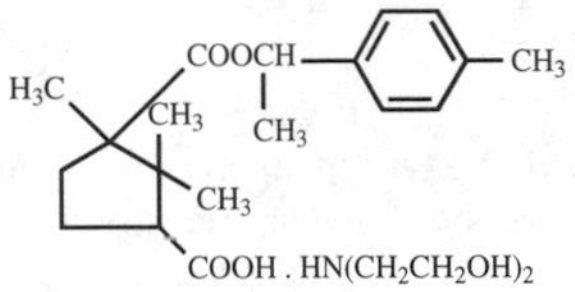

Tocol (mol wt 389) An antioxidant.

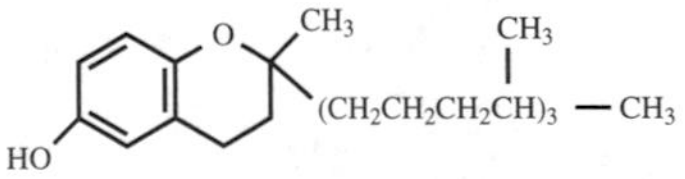

Tocolytic Drugs Drugs that suppress premature labor.

Tocopher A trade name for vitamin E.

α-Tocopherol (mol wt 431) The most active form of vitamin E, which protects unsaturated membrane lipid from oxidation.

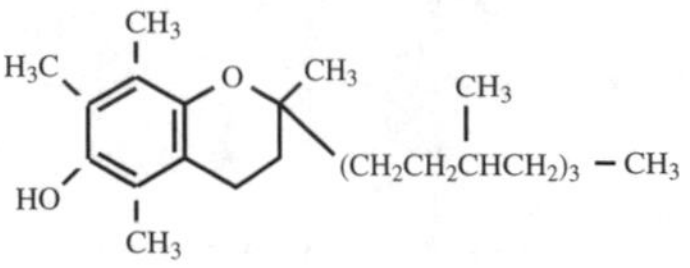

β-Tocopherol (mol wt 417) A substance related to vitamin E.

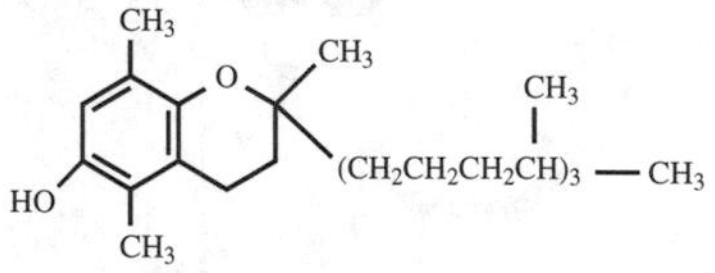

γ-Tocopherol (mol wt 417) A substance related to vitamin E.

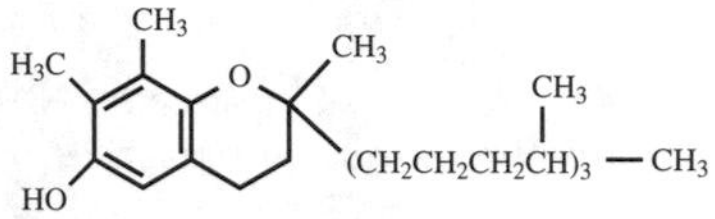

δ-Tocopherol (mol wt 403) A substance related to vitamin E.

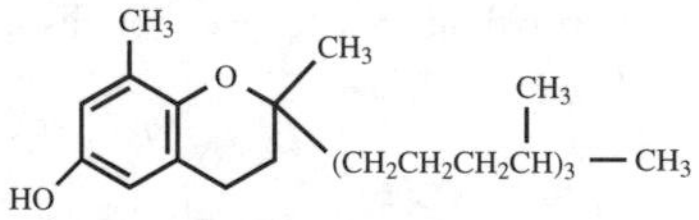

ε-Tocopherol (mol wt 411) A substance related to vitamin E.

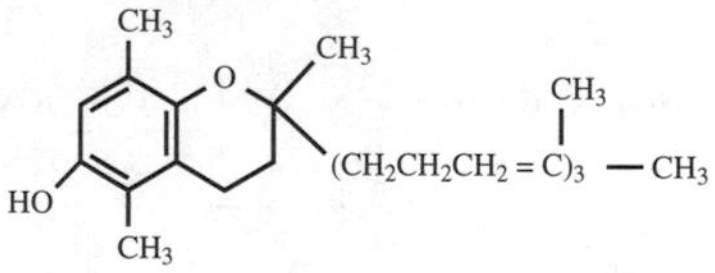

Tocopherol *O*-Methyltransferase The enzyme that catalyzes the following reaction:

S-Adenosyl-L-methionine + γ-tocopherol

⇅

S-Adenosyl-L-homocysteine + α-tocopherol

Tocopherols Forms of vitamin E.

T-Odd Phage Referring to bacteriophages of T1, T3, T5, and T7.

Todralazine (mol wt 232) An antihypertensive agent.

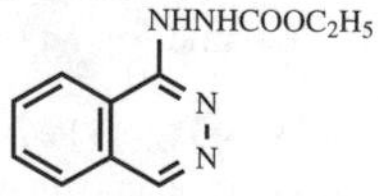

Tofenacin (mol wt 255) An antidepressant.

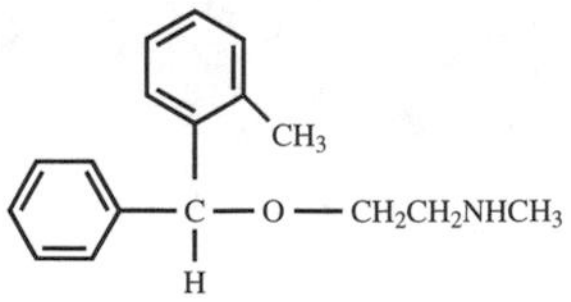

Tofisopam (mol wt 382) An anxiolytic agent.

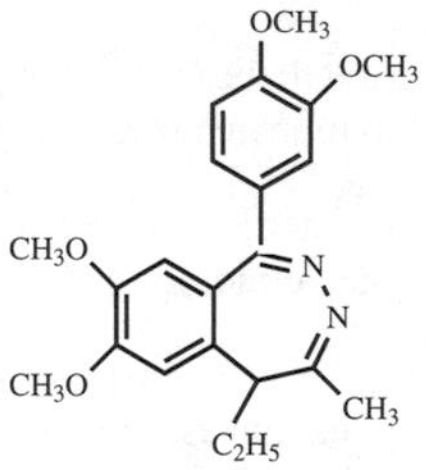

Tofranil A trade name for imipramine hydrochloride, an antihypertensive agent that increases concentrations of norepinephrine and serotonin in the CNS.

TOG Abbreviation for tri-oleoylglycerol.

Togaviridae A family of nonenveloped, icosahydral animal viruses containing plus-stranded RNA genome, e.g., polio virus and yellow fever.

Tolazamide (mol wt 311) A hormonal agent capable of stimulating the release of insulin from pancreatic beta cells.

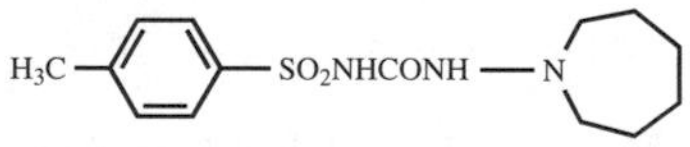

Tolazoline (mol wt 160) A vasodilator.

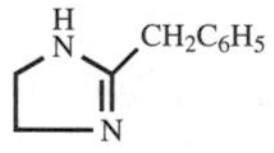

Tolboxane (mol wt 232) A tranquilizer.

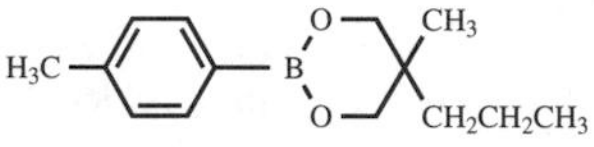

Tolbutamide (mol wt 270) A hormonal agent that stimulates the release of insulin from pancreatic beta cell and reduces glucose output by the liver.

Tolciclate (mol wt 323) An antifungal agent.

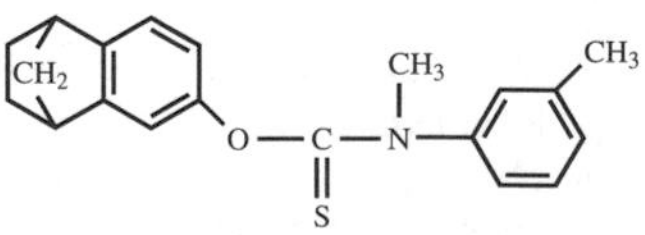

Tolcyclamide (mol wt 296) An antidiabetic agent.

Tolectin A trade name for tolmetin sodium, an anti-inflammatory, analgesic, and antipyretic agent.

Tolerance The development of an active state of specific immunological unresponsiveness to one or more particular antigenic determinants.

Tolerance Limit The point at which resistance to a poison or drug breaks down.

Tolerogen An antigen that is capable of inducing immunological tolerance.

Tolerogenic Antigen See tolerogen.

Tolfenamic Acid (mol wt 262) An anti-inflammatory and analgesic agent.

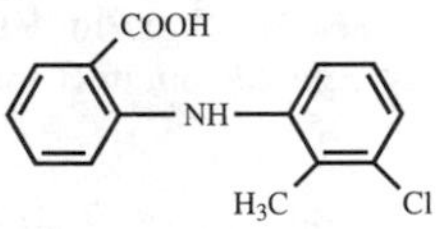

***tol*-G Protein** A protein of the outer membrane in Gram-negative bacteria encoded by *tol* G gene.

***tol*-Plasmid** A plasmid in *Pseudomonas* that encodes the capacity for metabolism of toluene and xylene.

***O*-Tolidine (mol wt 212)** A benzidine derivative.

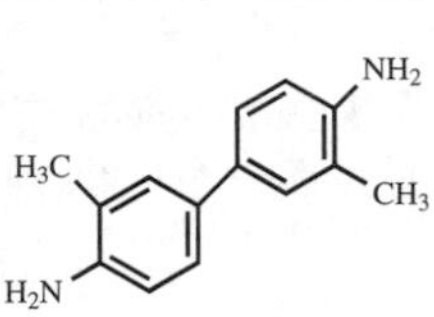

Tolinase A trade name for tolazamide, used as a hormonal agent for stimulating the release of insulin from pancreatic beta cells and reducing glucose output by the liver.

Toliprolol (mol wt 223) An antianginal and antihypertensive agent.

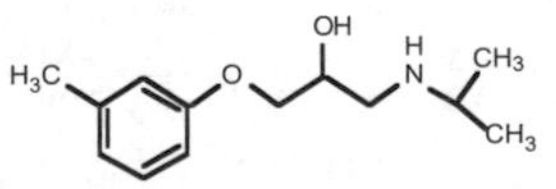

Tollen's Test A test for reducing sugar based on the reduction of silver ions to metalic silver in alkaline solution.

Tolloid A developmental protein in *Drosophila melanogaster* encoded by the *tld* gene.

Tolmetin (mol wt 257) An anti-inflammatory, analgesic, and antipyretic agent.

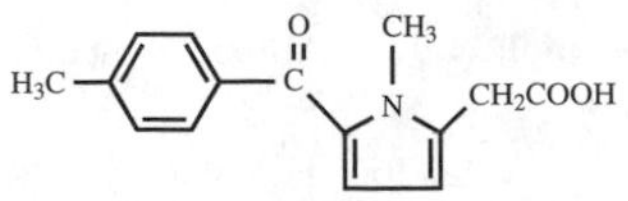

Tolnaftate (mol wt 307) An antifungal agent.

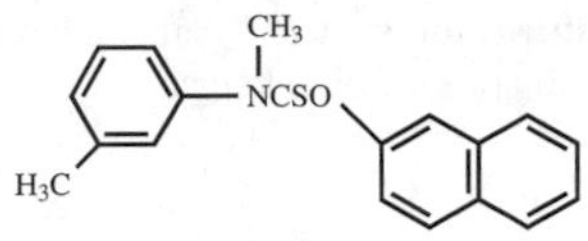

Tolonidine (mol wt 210) An antihypertensive agent.

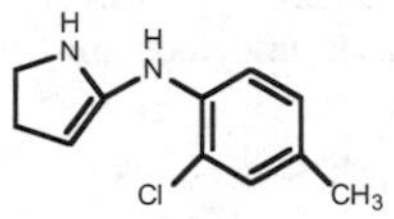

Toloxatone (mol wt 207) An antidepressant.

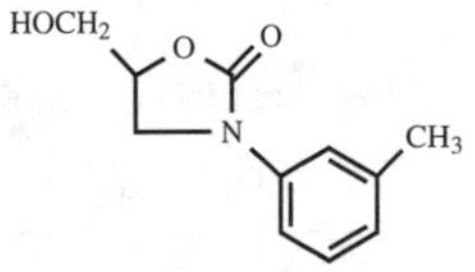

Tolperisone (mol wt 245) A skeletal muscle relaxant.

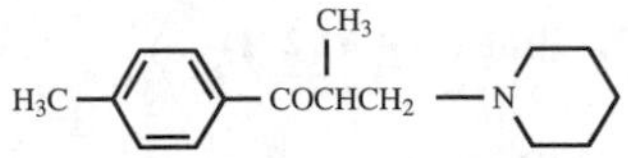

Tolpronine (mol wt 247) An analgesic agent.

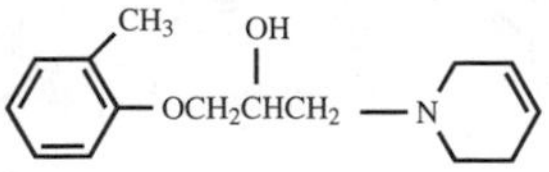

Tolpropamine (mol wt 253) A topical antihistaminic and antipuritic agent.

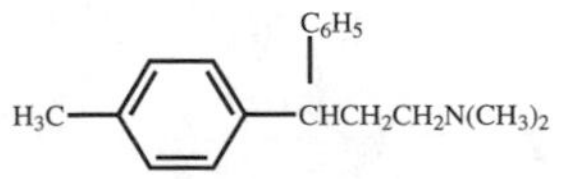

Tolrestat (mol wt 357) An inhibitor for aldose reductase that is capable of preventing the formation of sorbitol and galactitol from glucose and galactose, respectively.

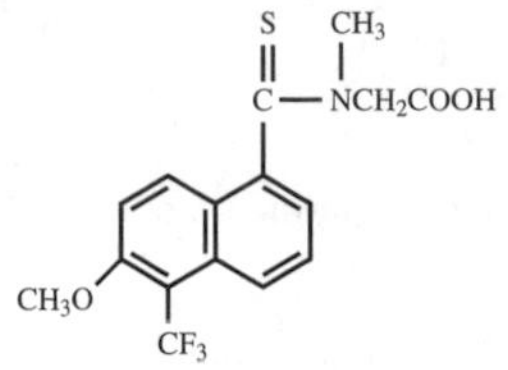

Toluene (mol wt 92) A toxic compound.

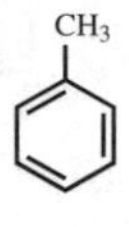

***O*-Toluidine (mol wt 107)** A methylaniline used for determination of blood sugar.

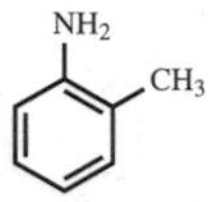

Toluidine Blue (mol wt 306) A metachromatic nuclear stain used in cytochemistry.

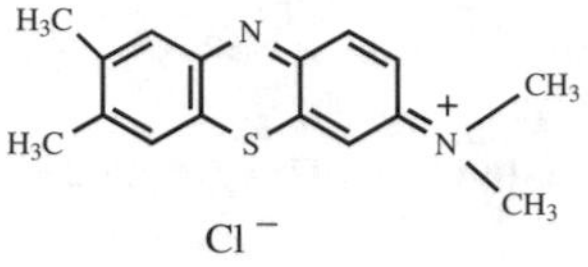

Toluidine Red (mol wt 307) A biological stain.

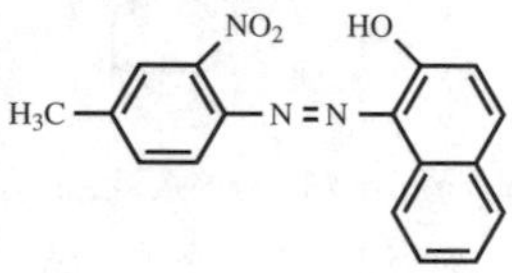

Toluylene Blue (mol wt 291) A biological dye.

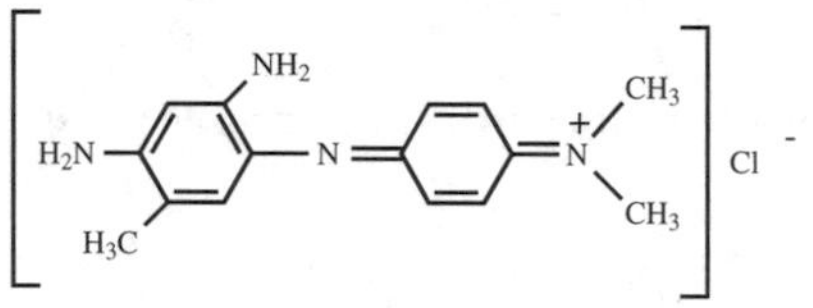

Toluylene 2,4-Diisocyanate (mol wt 174) A reagent used in cross-linking proteins.

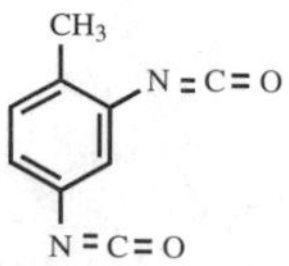

Tolycaine (mol wt 278) An anesthetic agent.

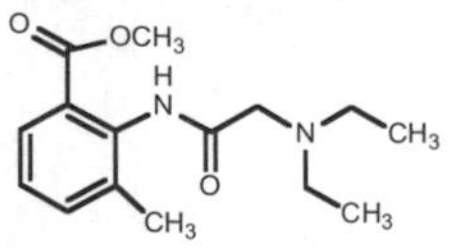

Tomatidine A steroid alkaloid from tomato leaves.

Tomatine A glycoalkaloid saponin found in the green tomato plant.

Tombusviruses A group of isosahydral, single-stranded, RNA-containing plant viruses, e.g., tomato bushy stunt virus.

Tomite A small motile, non-feeding stage in the life cycle of certain protozoa.

Tomography The technique of using a rotating source of X-rays to produce an image of structures at a particular depth within the body.

***ton*-A Protein** A protein in the outer membrane of *E. coli* encoded by ton-A gene that acts as receptor for colicin M, bacteriophage T1, and φ80.

***ton*-B Protein** A protein in the periplasmic region of *E. coli* involved in the energy-dependent transport of ferric iron-chelate complexes and vitamin B_{12} that is also involved in determining the susceptibility of cells to colicin B and bacteriophage φ80.

Tonic Agent A substance or drug that increases body tone.

Tonin A protease.

Tonocard A trade name for tocainide hydrochloride, an antiarrhythmic agent that shortens the action potential.

Tonofibril The fine fibril found in the cytoplasm of epithelium cells that gives support to the cell.

Tonofilament Keratin-containing intermediate filaments found in epithelial cells (the bundles of tonofilament form tonofibril).

Tonometer A device used for the measurement of tension or pressure.

Tonoplast A membrane-enclosed large central vacuole present in a plant cell.

Tonsilitis Inflammation of the tonsils or infection of tonsils by *Streptococcus pyogenes*.

Tonsillectomy Surgical removal of the tonsils.

Topamax A trade name for topiramate, an anticonvulsant and anti-epileptic agent.

Tophus An accumulation of urate crystals in the tissue, a characteristic of gout.

Topical Pertaining to the surface of a part of the body.

Topical Agent Substance or drug that is applied topically.

Topical Anesthesia Referring to the surface pain-killing produced by the application of anesthetic solution, gel, or ointment to the skin, mucous membrane, or cornea.

Topicort A trade name for desoximetasone, a corticosteroid that increases the concentration of liver glycogen.

Topicycline A trade name for tetracycline hydrochloride, an antibiotic that inhibits bacterial protein synthesis.

Topiramate (mol wt 339) An anti-epileptic and anticonvulsant agent.

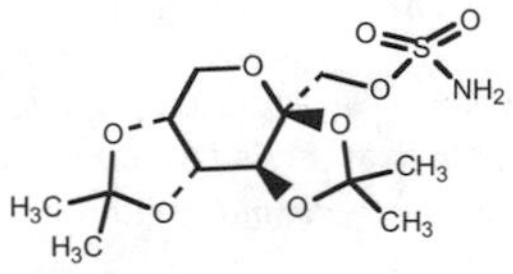

Topo- A prefix denoting place or region.

Topogenic Sequence A sequence in nascent polypeptide that is involved in getting the mature protein to its proper location in the cell.

Topoinhibition The inhibition of cell proliferation caused by the closely packed cell mass in a culture dish.

Topoisomerase A group of enzymes that catalyzes the conversion of DNA from one topological form to another by introducing transient breaks in one or both DNA strands.

Topoisomerase Type I The topoisomerase from *E. coli* that cuts one strand of the DNA and relaxes negatively supercoiled DNA but does not act on positively supercoiled DNA.

Topoisomerase Type II The enzyme that cuts both strands of DNA and increases the degree of negative supercoiling in DNA.

Topoisomers Forms of DNA with the same sequence but differing in their linking number (coiling).

Topological Isomers Molecules that differ only in their state of supercoiling.

Topology The science that deals with properties of the geometric configuration.

Toponacrosis The loss of sensation on a localized area of the skin.

Toposar A trade name for etoposide, a mitotic inhibitor used as an antineoplastic agent.

Topotaxis The response of a motile organism toward or away from a directional stimulus.

Topotecan (mol wt 421) An antineoplastic agent.

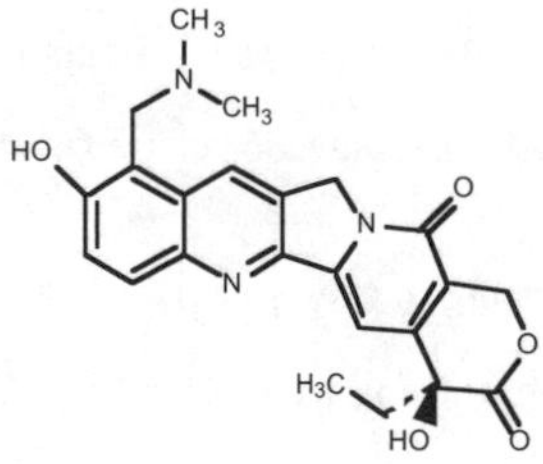

Toprol XL A trade name for metoprolol, an antihypertensive agent.

Topsyn A trade name for fluocinonide, a corticosteroid.

TOPV Abbreviation for trivalent oral polio vaccine.

Toradol A trade name for ketorolac tromethamine, an anti-inflammatory agent that inhibits prostaglandin synthesis.

Torasemide (mol wt 348) A diuretic agent.

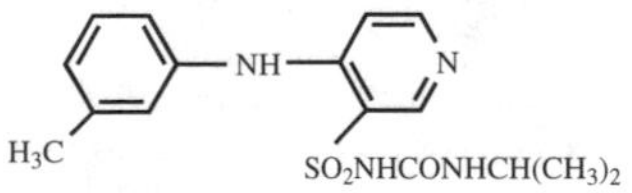

Torecan A trade name for thiethylperazine maleate, an antiemetic agent.

Toremifene (mol wt 406) An estrogen receptor modulator used as an antineoplastic agent.

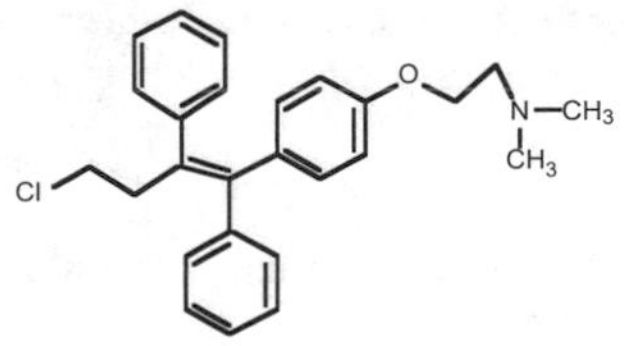

Tornalate A trade name for bitolterol mesylate, a bronchodilator.

Torofor A trade name for iodochlorhydroxyquin, an antifungal agent.

Torr A unit of pressure equal to 1/760 of 1 atm of pressure (a pressure of 1 mm of mercury).

Torsemide (mol wt 348) A diuretic agent.

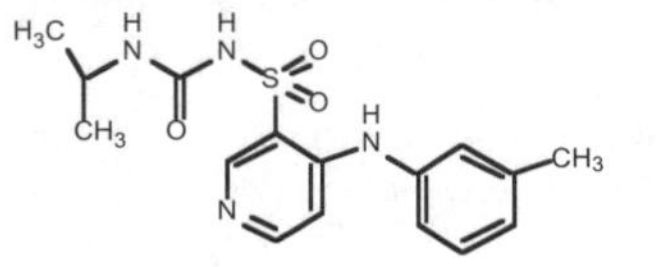

Torulaspora A genus of yeast (Saccharomycetaceae).

Torulopsis A genus of yeast (Hyphomycete).

Torulopsosis An infection caused by *Torulopsis glabrata*.

Tos Referring to tosyl group.

Tosufloxacin (mol wt 404) An antibacterial agent.

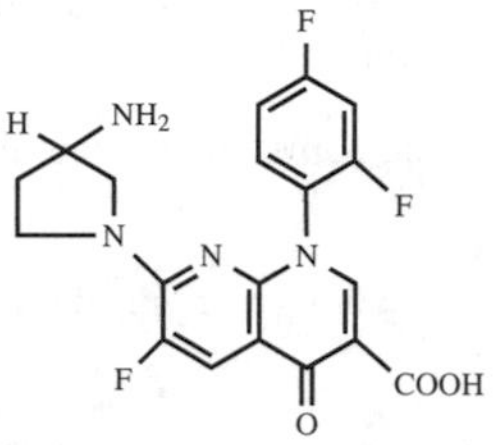

Tosyl Group The group

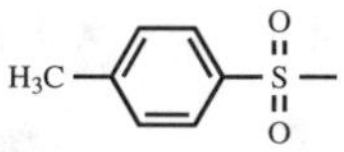

Tosylate Referring to a compound that contains a tosyl group that is used as a protecting agent for blocking amino groups in peptide synthesis.

Totacillin A trade name for ampicillin trihydrate, an antibiotic that inhibits bacterial cell wall synthesis.

Totacillin-N A trade name for ampicillin sodium, an antibiotic that inhibits bacterial protein synthesis.

Totipotent A cell that is not irreversibly committed to a single specific developmental fate.

Toxemia The presence of toxin in the blood.

Toxic Capable of producing a harmful effect to a living system.

Toxicant Any agent capable of producing toxic reactions.

Toxicosis Any disorder caused by the action of toxin or poison.

Toxin Poisonous substance produced by a cell, (sometimes encoded by a plasmid). There are two main classes of toxins, namely exotoxins and endotoxins. Exotoxins are toxins that are released from the cell, whereas the endotoxins are part of the cell surface.

Toxoflavin (mol wt 193) A toxic antibiotic produced by *Pseudomonas cocovenenans*.

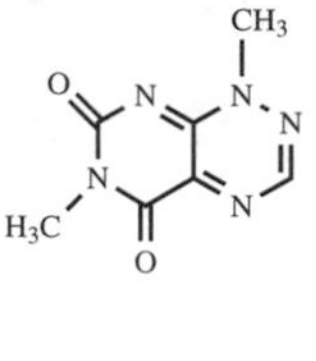

Toxoids Toxins that have been chemically modified to reduce or to eliminate their toxic effects while retaining their immunogenic and antigenic properties.

Toxophore The toxic moiety of a toxin molecule.

Toxoplasma A genus of parasitic, facultative, heteroxenous protozoa (suborder Eimeriorina) and the causal agent of toxoplasmosis.

Toxoplasma Dye Test A serological test for detection of *Toxoplasma gondii* that was alkaline methylene blue and quantitation of antibody to *Toxoplasma gondii* in a patient's serum.

Toxoplasmosis An acute or chronic disease of humans and other animals caused by the intracellular pathogen *Toxoplasma gondii*.

Toxothrix A genus of gliding bacteria that occur in cold, iron-containing springs.

Toyocamycin (mol wt 291) An antibiotic produced by *Streptomyces toyocaensis*.

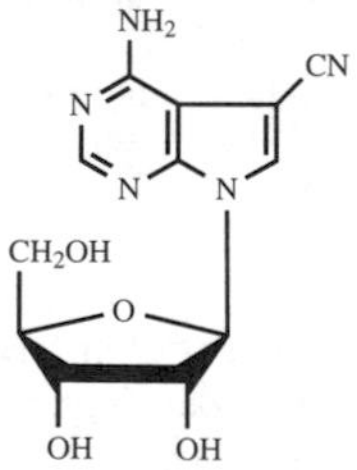

TP Abbreviation for 1. thymidine phosphorylase; 2. template primer; 3. total protein; 4. triphosphate; 5. tuberculin precipitation.

Tp-44 A T-lymphocyte receptor that regulates cytokine synthesis.

TPA Abbreviation for 1. tissue-type plasminogen activator; 2. tetra-decanoyl phorbol acetate; 3. a trade name for alteplase, a human tissue enzyme produced by DNA recombinant technology used for degradation of fibrin clots.

t-PA Abbreviation for tissue plasminogen activator.

TPBS Abbreviation for triton phosphate buffered saline.

TPCK Abbreviation for tosyl-L-phenylalanine chloromethyl ketone.

TPHA Test Abbreviation for *Treponema pallidum* hemaglutination test in which absorbed serum is tested for its ability to agglutinate tanned red blood cells sensitized with antigen from *P. pallidum*.

TPI Abbreviation for 1. *Treponema pallidum* immobilization test; 2. *Treponema pallidum* isomerase; 3. triose phosphate isomerase.

TPI Test Abbreviation for *Treponema pallidum* immobilization test in which the serum from a patient is tested for its ability to immobilize cells of *Treponema pallidum* in the presence of complement.

TPIA Abbreviation for 1. *Treponema pallidum* immobilization adherence; 2. *Treponema pallidum* immune adherence.

t-Plasminogen Activator See tissue plasminogen activator.

TPM Abbreviation for triphenyl methane.

TPMP$^+$ Abbreviation for triphenylmethyl phosphonium.

TPN$^+$ Abbreviation for the oxidized form of triphosphopyridine nucleotide (synonym for NADP$^+$).

TPNH The reduced form of TPN (triphosphopyridine nucleotide).

TPO Abbreviation for thyroid peroxidase.

TPP Abbreviation for thiamine pyrophosphate.

TPP$^+$ Abbreviation for tetraphenylphosphonium ion.

TPT Abbreviation for *Treponema pallidum* test.

TR Abbreviation for 1. tandem repeat; 2. tetrazolium reduction; 3. Texas red; 4. thyroid receptor; 4. thioredoxin reductase.

T_3R Abbreviation for thyroid hormone T_3 receptor.

***tra* Operon** See transfer operon.

Trace Elements The elements essential for life but required in extremely minute amounts.

Tracer Referring to isotopes (e.g., ^{14}C, ^{32}P) used to label a compound such as ^{14}C-labeled amino acid or ^{32}P labeled nucleic acid.

Trachea The tube that conducts air from the pharynx to the lungs.

Tracheid A water-conducting and supportive element of xylem, composed of long, thin cells with tapered ends.

Tracheitis Inflammation of the trachea.

Tracheobronchitis Inflammation of the trachea and bronchi.

Trachoma A communicable disease of the eye caused by *Chlamydia trachomatis*.

Tracking Dye A dye used as a migration indicator in gel electrophoresis.

Tracrium A trade name for atracurium besylate, a neuromuscular blocker that prevents acetylcholine from binding to receptors on the muscle end plate, blocking depolarization.

TRAF Abbreviation for TNF-receptor-associated factor.

TR-AIDS Abbreviation for transfusion related AIDS.

Trailer Sequence A nontranslated region at the 3′ end of a mRNA following the termination codon.

Trait Any detectable phenotypic variation of a particular inherited character.

Tral Filmtabs A trade name for hexocyclium methylsulfate, an anticholinergic agent that blocks acetylcholine, reducing GI motility and inhibiting gastric acid secretion.

Tramacort A trade name for triamcinolone acetonide, an anti-inflammatory agent.

Tramadol (mol wt 263) An analgesic agent.

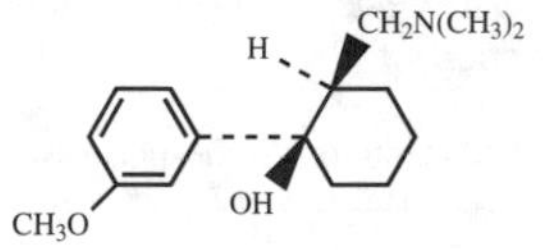

Tramazoline (mol wt 215) An adrenergic agent.

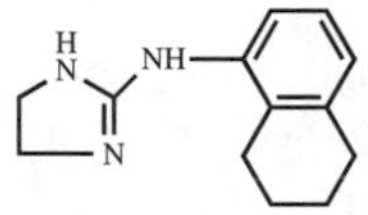

Trandate A trade name for labetalol hydrochloride, an antihypertensive agent that blocks the response to alpha and beta stimulation and depresses renin secretion.

Trandolapril (mol wt 431) An angiotensin converting enzyme inhibitor used as an antihypertensive agent.

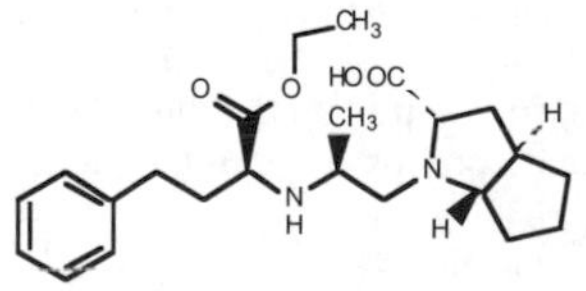

Tranilast (mol wt 327) An antiallergic agent.

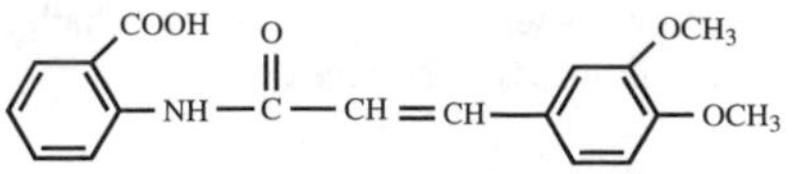

Tranmep A trade name for meprobamate, an antianxiety agent.

Tranquilizer A drug that relieves anxiety and mental tension.

Trans- A prefix meaning 1. configuration of a geometric isomer in which two chemical groups are attached to the opposite sides of a double bond formed by two carbon atoms, and 2. geometric configuration of mutant alleles across from each other on a homologous pair of chromosomes.

***trans* Face** The region of the Golgi complex that is opposite from the *cis* face (usually oriented toward the cell surface).

***trans* Golgi Network** Network of Golgi channels and vesicles located adjacent to the *trans* face of the Golgi complex.

Trans-Acting Referring to a genetic element, e.g., regulatory gene that acts on another genome through a diffusable gene product.

Transacylase The enzyme that catalyzes the transfer of an acyl group from one compound to another.

Transaldolase The enzyme that catalyzes the following reaction:

Sedoheptulose 7-phosphate + D-glyceraldehyde 3-phosphate

$\rightleftharpoons$

D-Erythrose 4-phosphate + D-fructose 6-phosphate

Transamidation The transfer of amide nitrogen of glutamine to another compound.

Transcapsidation 1. The virion resulting from a phenotypic mixing in which the genome of one virus is within the capsid of another virion. 2. A capsid that contains structural components derived from another virus.

Transcarboxylase See methylmalonyl-CoA carboxyltransferase.

Transconjugant A bacterial cell that has received DNA from a donor cell (male) during conjugation.

Transcortin A protein that binds and transports cortisol and corticosterone in the blood.

Transcribed Spacer The region of RNA that is transcribed but excised and discarded during maturation.

Transcript The single-stranded RNA chain transcribed from a DNA template.

Transcriptase Referring to DNA-dependent RNA-polymerase.

Transcription The process by which the information contained in the DNA is copied into a single-stranded RNA molecule by the enzyme RNA polymerase.

Transcription Factors Eukaryotic proteins that promote RNA polymerase to recognize promoters (analogous to prokaryotic sigma factors).

Transcription Unit The segment of DNA between the sites of initiation and termination, which may include more than one gene.

Transcriptional Activator Protein Protein factor that stimulates transcription by binding with particular sites on DNA.

Transcriptional Control The control of protein synthesis through the regulation of transcription.

Transcytosis Endocytosis of material into vesicles that move to the opposite side of the cell and fuse with plasma membrane, releasing the material into the extracellular space.

Transcytotic Vesicle The membrane-bound vesicle that shuttles fluid from one side of the endothelium to another.

Transderm-V A trade name for scopolamine, an anticholinergic agent.

Transderm-Nitro A trade name for nitroglycerin, used as an anti-anginal agent that reduces cardiac oxygen demand and increases blood flow.

Transderm-Scop A trade name for scopolamine, an anticholinergic and antimotion agent.

Transduced Element The DNA fragment that is transferred from one bacterium to another during transduction.

Transducer A device capable of transforming energy from one form to another, e.g., photocell which transforms light energy into electric energy.

Transducin A GTP-binding protein occurring in the membrane of retinal rods and cones that plays an important role in the visual excitation process by interacting with photoexcited rhodopsin.

Transducing Particle A bacteriophage that carries part of the host's genome.

Transducing Phage A bacteriophage that is capable of producing phage particles containing bacterial DNA. There are two types of transducing phages: 1. General transducing phages that produce tranducing phage particles containing nonspecific regions of bacterial DNA. 2. Specialized transducing phages that produce transducing particles containing specific regions of the bacterial DNA.

Transductant A cell that has been transduced.

Transduction The transfer of genetic material from one cell to another by a viral vector, e.g., transfer of a bacterial gene by bacteriophage or transfer of a eukaryotic genome by retrovirus.

Transfection The introduction of foreign DNA into eukaryotic cells.

Transferase An enzyme that catalyzes the transfer of a chemical group from one compound to another.

Transfer Factor Factor from sensitized lymphocytes capable of transferring delayed hypersensitivity to nonsensitized individuals.

Transfer Operon An operon that contains genes necessary for conjugation and production of F-pili phenotype in bacteria.

Transfer Origin The site on a plasmid DNA at which a nick is made for transfer of plasmid DNA into the recipient during bacterial conjugation.

Transfer RNA (tRNA) A small RNA molecule that recognizes a specific amino acid, transports it to a specific codon in the mRNA, and positions it properly on the ribosome during protein synthesis.

Transferin Receptor Plasma membrane protein involved in the transfer of iron into the cell.

Transferins A group of nonheme, iron-binding glycoproteins involved in iron transport in the developing red blood cell for hemoglobin synthesis.

Transform The conversion of a normal cell into a cancer cell.

Transformation 1. Introduction of exogenous DNA into a cell to give the cell a new phenotype. 2. Conversion of normal eukaryotic cells in tissue culture to a cancerlike state of uncontrolled division.

Transformed Cell Cell that has undergone tumor transformation.

Transforming Genes Genes that are capable of transforming cells, e.g., oncogen.

Transforming Growth Factor (TGF) Proteins secreted by transformed cells that stimulate the growth of normal cells. TGF-α has sequence similar to that of epidermis growth factor (EGF) and binds to EGF receptor and stimulates growth of microvascular epithelium cells. TGF-β is a homodimer of two peptide chains secreted by many different cell types and it stimulates wound healing.

Transforming Virus Virus capable of transforming animal cells, e.g., polyoma and retrovirus.

Transfusion Transfer of blood or plasma from a donor to the blood stream of a recipient.

Transfusion Reaction Hemolytic reaction that occurs following the administration of histoincompatible blood (wrong blood type) to an individual.

Transgene A foreign gene that is introduced into an organism by introducing the gene into a newly fertilized egg. The organism developed from the egg will carry the foreign gene and transmit it to progeny.

Transgenesis The process of incorporation of foreign DNA into an organism.

Transgenic Organism Referring to an organism that contains a foreign gene.

Transgenome A genome in which a foreign DNA has be incorporated.

Transglutaminase The enzyme that catalyzes the following reaction:

$$\text{Protein glutamine + alkylamine} \rightleftharpoons \text{Protein } N^5\text{-alkylglutamine} + NH_3$$

Transglycosylation The transfer of a sugar moiety to a hydroxyl group by glycosyl transferase.

Transgressive Variation The quantitative variation in a phenotypic characteristic of the offspring that exceeds that shown by either parent.

Transhydrogenase The enzyme that catalyzes the following reaction:

$$NADPH + NAD^+ \rightleftharpoons NADP^+ + NADH$$

Transin A protease secreted by carcinoma cells (also known as stromelysin I).

Transition Mutation A mutation in which a purine is replaced by a different purine or a pyrimidine by a different pyrimidine.

Transition State An intermediate stage in which an unstable and high-energy configuration assumed by the reactant prior to the formation of product.

Transition State Analog A stable molecule that resembles the transition state of the substrate, thereby acting as competitive inhibitor in the enzymatic reaction.

Transition State Inhibitor A molecule that is structurally similar to the transition state of the substrate, thereby capable of acting as a competitive inhibitor in the enzymatic reaction.

Transition Temperature The temperature at which transition of membrane phospholipid from one state to another occurs.

Transitional Endoplasmic Reticulum The region at the boundary of the rough endoplasmic reticulum and Golgi apparatus that is responsible for the transfer of secretory proteins from the rough endoplasmic reticulum to the Golgi system.

Transketolase An enzyme that catalyzes the following reaction:

Xylulose 5-phosphate + erythrose 4-phosphate

⇅

Fructose 6-phosphate + glyceraldehyde 3-phosphate

Translation The process whereby the genetic information present in a mRNA molecule is translated into a polypeptide.

Translational Amplification A mechanism for synthesis of a large quantity of a protein by prolonging the life time of the mRNA.

Translocase (EF-G) The elongation factor in prokaryotes for transfer of peptidyl-tRNA from A-site to P-site in the ribosome so that the next codon in mRNA is in position for translation.

Translocation 1. Chromosomal mutation in which a segment of chromosome is changed from one location to another either within the same chromosome or to another chromosome. 2. The motion of the ribosome along the mRNA to read the genetic codes during protein synthesis.

Transmembrane Protein Proteins that span the membrane with peptide chains exposed on both sides.

Transmembrane Protein Kinase Enzyme that catalyzes the phosphorylation of tyrosine, threonine, or serine residues of specific proteins on the cytoplasmic side of the plasma membrane when activated by the binding of a specific growth factor on the outer membrane surface.

Transmembrane Receptor A transmembrane protein with the extracellular portion of the protein having the ability to bind to a ligand and the intracellular portion having a specific activity upon ligand binding.

Transmethylation The process in which the methyl group is transferred from one compound to another.

Transmission Electron Microscope A type of electron microscopy in which the image is formed by varying the extent to which electrons are transmitted by different parts of the specimen.

Transmission Genetics The science that deals with the mechanisms of transfer of genes from one generation to the next.

Transmittance The ratio of the intensity of the transmitted light to that of the incident light, e.g., I/I_0, where I_0 is the intensity of incident light and I is the intensity of transmitted light.

Transmitter Substance A variety of molecules synthesized in the nerve axon terminals and released by the arrival of an action potential.

Transpeptidase See peptidyl transferase.

Transpeptidation The process of forming peptide bonds catalyzed by peptidyl transferase.

Transphosphoribosidase Synonym of hypoxanthine phosphoribosyltransferase.

Transphosphorylation The process of transfer of a phosphoryl group or pyrophosphoryl group from one organic compound to another.

Transpiration The passage of fluid through the skin.

Transplacental Pertaining to the movement of substances through the placenta.

Transplant An organ or part of an organ that is transplanted.

Transplantation The process of transfer of tissue or organ from one site to another within the same individual or from one individual to another.

Transplantation Antigens Class-I antigens, encoded by the major histocompatibility complex, that are expressed on the cell surface of all nucleated cells. These antigens induce graft rejection if the cells of the graft are not histocompatible with the recipient.

Transplantation Immunity The immunity against transplant.

Transplantation Reaction An immunological reaction observed in graft rejection that is mediated by cytotoxic T lymphocytes.

Transport The conveyance of substances across a cell membrane.

Transport Protein A protein that mediates the entry of specific substances into a cell.

Transport System A system that permits the influx or efflux of substances (e.g., nutrients, ions, waste products) across a biological membrane.

Transport Vesicle Vesicle formed by the pinching off of the rough endoplasmic reticulum, which is involved in the transfer of lipids and proteins synthesized in the rough ER of the Golgi complex.

Transposable Genetic Elements Specific segments of DNA that can undergo nonreciprocal recombination and thus move from one location to another.

Transposase The enzyme involved in the insertion of transposon at a new site.

Transposition 1. The movement of a transposon to a new site in the genome. 2. Insertion of a replica of a transposable element at a second site.

Transposition Immunity The ability of certain transposons to prevent others of the same type from transposing to the same DNA molecule.

Transposon Transposable genetic elements that move from one locus to another by nonhomologous recombination.

Transposon Tagging Insertion of a transposable element containing a genetic marker into a gene of interest.

Transprotonase The enzyme that catalyzes the transprotonation reaction.

Transprotonation Reaction The process of transport of protons across a biological membrane, e.g., creation of a proton gradient during electron transport in mitochondria.

Transsynaptic Pertaining to the transmission of nerve impulses across a nerve synapse.

Transthyretin A protein that carries vitamin A to the eye.

Transudate Plasma-derived fluid or electrolytes exuded during transudation.

Transudation The process of movement of fluid or electrolytes across a membrane or the interstices of tissue.

Transvection The ability of a gene to influence activity of an allele on the other homologue when two chromosomes are synapsed.

Transverse Diffusion Movement of a phospholipid or protein from one side of the membrane to the other.

Transversion Mutation A point mutation in which a purine is replaced by a pyrimidine or a pyrimidine is replaced by a purine.

Tranxene A trade name for clorazepate dipotassium, an antianxiety agent.

Tranxene-SD A trade name for clorazepate dipotassium, an antianxiety agent.

Tranxene-T-Tab A trade name for clorazepate dipotassium, an antianxiety agent.

Tranylcypromine (mol wt 133) An antidepressant that acts as a MAO inhibitor to promote accummulation of neurotransmitter by inhibiting MAO.

Trapidil (mol wt 205) A vasodilator.

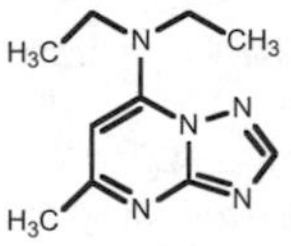

Trasylol A trade name for aprotinin, a hemostatic agent that forms complexes with plasmin, kallikreins to block the activation of the kinin and fibrinolytic systems.

Travamine A trade name for dimenhydrinate, an antiemetic agent.

Travs A trade name for dimenhydrinate, an antiemetic agent.

Traxanox (mol wt 300) An antiallergic agent.

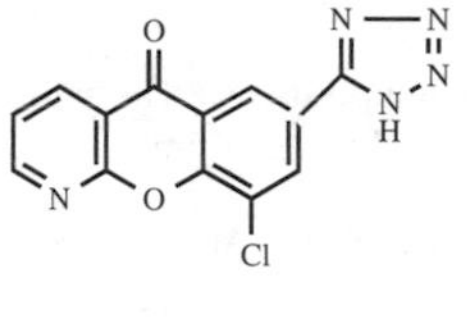

Trazodone (mol wt 372) An antidepressant that inhibits serotonin uptake in the brain.

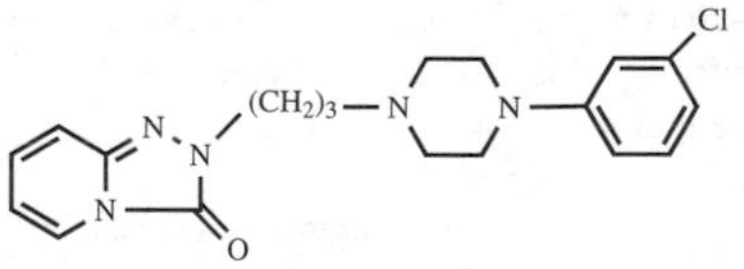

Trazon A trade name for trazodone hydrochloride, an antidepressant that inhibits serotonin uptake in the brain.

TRC Abbreviation for tanned red cells.

TRCH Abbreviation for tanned red cell hemagglutination.

TRE Abbreviation for thyroid response element.

Treadmilling Process whereby a microfilament is assembled on the plus end by polymerization of G-actin molecules and simultaneously disassembled at the minus end by dissociation of actin monomers; individual actin molecules are therefore continually transferred from the plus end of the microfilament to the minus end, even though the net length of the microfilament does not change.

Trebouxia A genus of unicellular, cocoid, nonmotile green algae.

Trecator-SC A trade name for ethionamide, an antimicrobial agent.

α,α-Trehalase The enzyme that catalyzes the following reaction:

α,α-Trehalose + H_2O ⇌ 2 D-Glucose

Trehalose (mol wt 342) A naturally occurring disaccharide consisting of glucose.

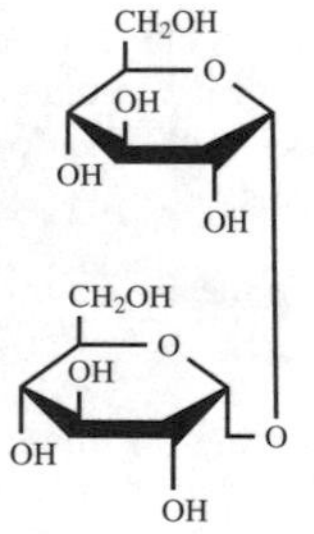

α,α-Trehalose Glucohydrolase See α,α-trehalase.

Trehalose Phosphatase The enzyme that catalyzes the following reaction:

Trehalose 6-phosphate + H_2O ⇌ Trehalose + Pi

Trehalose 6-Phosphate Phosphohydrolase See trehalose phosphatase.

α,α-Trehalose Phosphate Synthetase (GDP-forming) The enzyme that catalyzes the following reaction:

GDP-glucose + glucose 6-phosphate
⇅
GDP + α,α-trehalose 6-phosphate

α,α-Trehalose Phosphate Synthetase (UDP-forming) The enzyme that catalyzes the following reaction:

UDP-glucose + D-glucose 6-phosphate
⇅
UDP + α,α-trehalose 6-phosphate

α,α-Trehalose Phosphorylase The enzyme that catalyzes the following reaction:

α,α-Trehalose + Pi
⇅
D-Glucose + β-D-glucose 1-phosphate

Trematodes A group of parasitic flatworms, e.g., schistosomes and flukes.

Trenbolone (mol wt 270) An anabolic steroid that promotes tissue-building processes.

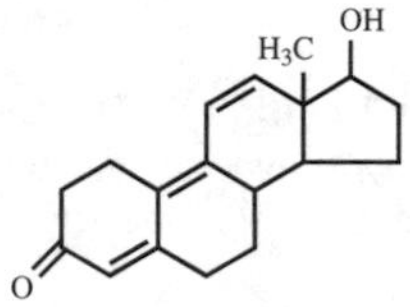

Trendal A trade name for pentoxifylline, used to improve capillary blood flow.

Trendar A trade name for ibuprofen, an anti-inflammatory and antipyretic agent.

Trengestone (mol wt 345) A progestogen.

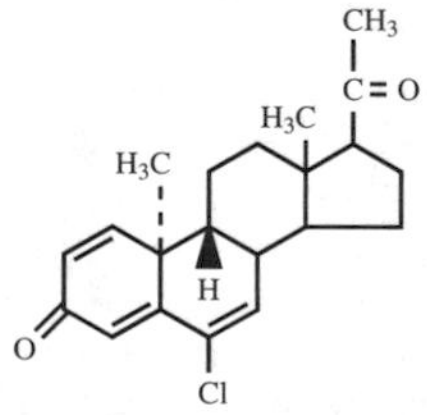

Trental A trade name for pentoxifylline, an agent used to decrease platelet aggregation.

Trepibutone (mol wt 310) A choleretic and an antispasmodic agent.

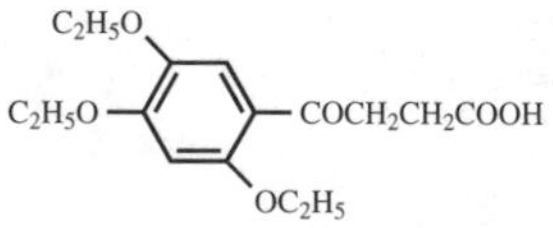

Treponema A genus of Gram-negative bacteria (Spirochaetaceae) that causes syphilis in humans.

Treponemal Test A test in which treponemal antigens or intact cells of *Treponema* are used to detect specific anti-*Treponema* antibodies in the patient suffering from syphilis.

Treponematosis Any disease caused by *Treponema*.

Treponeme A bacterium of the genus *Treponema*.

Treponemiasis See treponematosis.

Tretinoin A trade name for vitamin A acid.

Trexan A trade name for naltrexone hydrochloride, an opiate antagonist used to reversibly block the subjective effects of opioids.

TRF Abbreviation for 1. T-cell replacing factor; 2. thymus-replacing factor; 3. thyrotropin-releasing factor; 4. time-resolved fluorescence.

TRH Abbreviation for thyrotropin-releasing hormone or factor, it is a hypothalmic neurohormone that stimulates the release and synthesis of TSH (thyroid-stimulating hormone).

TRH Receptor Abbreviation for thyrotropin-releasing hormone receptor.

Tri A prefix meaning three.

TRI Abbreviation for tetrazolium reduction inhibition.

T_2RIA Abbreviation for di-iodo-thyronine radioimmuno-assay.

T_3RIA Abbreviation for tri-iodo-thyronine radioimmuno-assay.

T_4RIA Abbreviation for thyroxine radioimmuno-assay.

Triacet A trade name for triamcinolone acetonide, a corticosteroid.

Triacetate Lactonase The enzyme that catalyzes the following reaction:

$$\text{Triacetate lactone} + H_2O \rightleftharpoons \text{Triacetate}$$

Triacetyldiphenolisatin (mol wt 443) A cathartic agent.

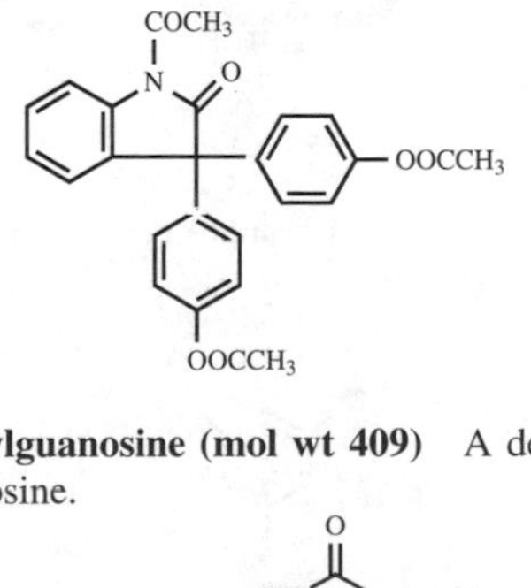

Triacetylguanosine (mol wt 409) A derivative of guanosine.

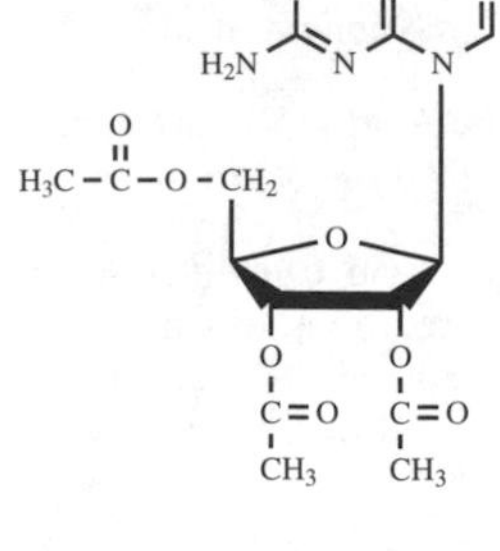

Triaconta- A prefix denoting 30.

Triacontanoic Acid A 30-carbon straight chain aliphatic acid.

Triacontanol (mol wt 439) A naturally occurring plant growth regulator.

$$CH_3(CH_2)_{28}CH_2OH$$

Triacontapeptide A 30-residue peptide.

Triacylglycerol (triglyceride) A lipid formed by esterification of three fatty acids to glycerol.

$CH_2 — O — CO — R_1$
|
$HC — O — CO — R_2$
|
$CH_2 — O — CO — R_3$

R_1, R_2, R_3 = different fatty acid

Triacylglycerol Acylhydrolase See triacylglycerol lipase.

Triacylglycerol Lipase The enzyme that catalyzes the following reaction:

$$\text{Triacylglycerol} + H_2O \rightleftharpoons \text{Diacylglycerol} + \text{a carboxylate}$$

Triacylglycerol Sterol *O*-Acyltransferase The enzyme that catalyzes the following reaction:

$$\text{Triacylglycerol} + 3\beta\text{-hydroxysterol} \rightleftharpoons \text{Diacylglycerol} + \text{a } 3\beta\text{-hydroxysterol ester}$$

Triad A trade name for a combination drug containing acetaminophen, caffeine, and butalbital, used as analgesic agent.

Triadapin A trade name for doxepin hydrochloride, an antidepressant that increases norepinephrine and serotonin in the CNS.

Triadimefon (mol wt 294) An agriculture fungicide.

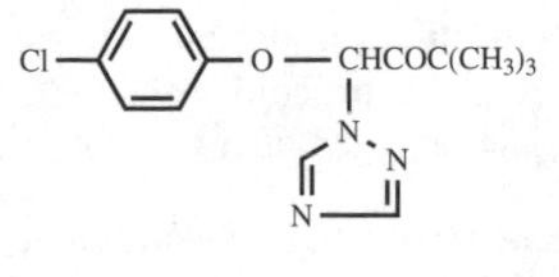

Trialodine A trade name for trazodone hydrochloride, an antidepressant.

Triam Forte A trade name for triamcinolone diacetate, a glucocorticoid hormone used as an anti-inflammatory and immuno-suppressive agent.

Triamcinolone Acetonide (mol wt 434) A glucocorticoid used as an anti-inflammatory agent.

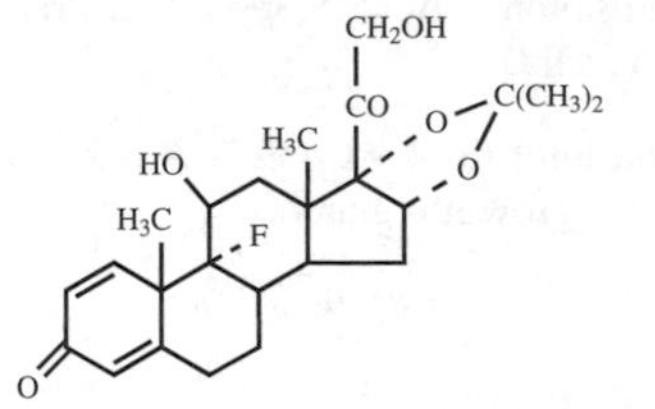

Triamcinolone Hexacetonide (mol wt 533) A glucocorticoid used as an anti-inflammatory agent.

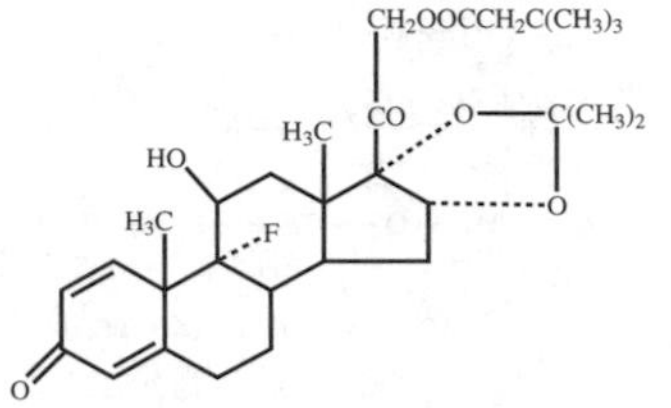

Triaminic A trade name for a combination drug containing phenylpropanolamine hydrochloride, pyrilamine maleate, and pheniramine maleate, used as a bronchodilator.

Triaminic Allergy A trade name for a combination drug containing chlorpheniramine, and phenylpropanolamine used as a decongestant and an antihistaminic agent.

Triaminic DM A trade name for a combination drug containing dextramethorphan and phenylpropanolamine used as a decongestant and an antihistaminic agent.

Triaminic TR A trade name for a combination drug containing pyrilamine, pheniramine, and phenylpropanolamine used as a decongestant and an antihistaminic agent.

Triaminicol A trade name for a combination drug containing phenylpropanolamine, chlorpheniramine, and dextromethorphan used as a decongestant and an antihistaminic agent.

Triamonide A trade name for triamcinolone, a glucocorticoid hormone used as an anti-inflammatory agent and an immuno-suppressant.

Triamonlone 40 A trade name for triamcinolone diacetate, a glucocorticoid hormone used as an anti-inflammatory agent and immuno-suppressant.

Triamterene (mol wt 253) A diuretic agent that inhibits sodium reabsorption and potassium excretion.

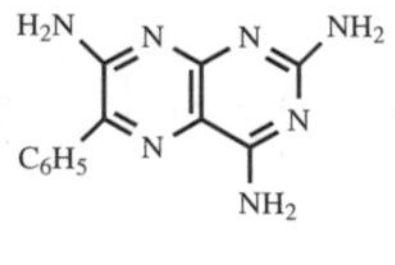

Triangulation Number The total number of the small equilateral triangles in an icosadeltahydron structure, e.g., a virion.

Triaphen-10 A trade name for aspirin, which blocks the generation of pain impulses and inhibits prostaglandin synthesis.

Triavil A trade name for a combination drug containing perphenazine and amitriptyline used as an antidepressant.

Triaziquone (mol wt 231) An antineoplastic agent.

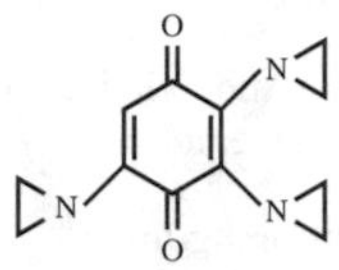

Triazolam (mol wt 343) A sedative–hypnotic agent.

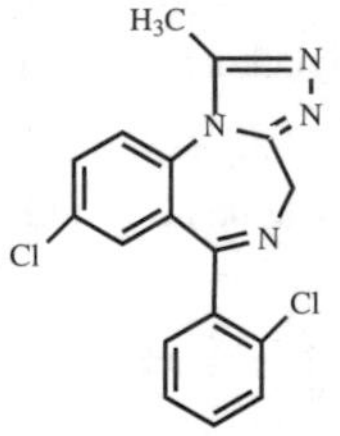

Tribasic A compound that has 3 hydrogen atoms replaceable by a metal (e.g., Na_3PO_4) or an acid that can provide 3 hydrogen ions (e.g., H_3PO_4).

Tribenoside (mol wt 479) A sclerosing agent.

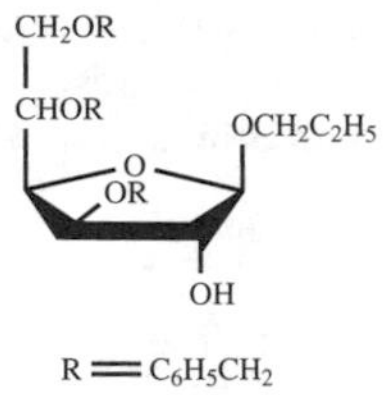

2,4,6-Tribromo-*m*-Cresol (mol wt 345) A topical antifungal agent.

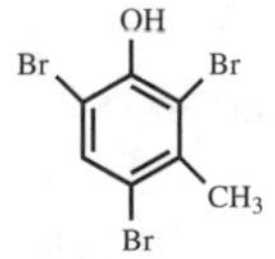

Tributyrase Synonym of triacylglycerol lipase.

Tricarboxylic Acid Any organic acid that contains three carboxylic acid groups.

Tricarboxylic Acid Cycle (TCA Cycle) A series of cyclic reactions in which acetyl-CoA derived from carbohydrate or fatty acid is oxidized to carbon dioxide, resulting in generation of ATP through the electron transport system and oxidative phosphorylation (also known as citric acid cycle, Krebs cycle).

Tricetamide (mol wt 324) A sedative agent.

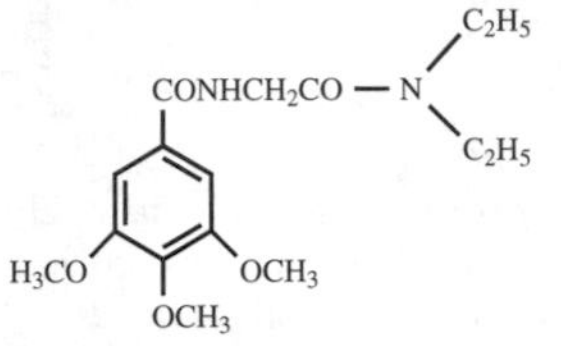

Trichinella A genus of nematode parasite that causes trichinosis in humans.

Trichinosis A disease caused by nematode *Trichinella spiralis* that is characterized by muscular stiffness and painful swelling.

Trichlorex A trade name for trichlormethiazide, a diuretic agent that inhibits the re-absorption of sodium and chloride.

Trichlorfon (mol wt 257) An insecticide and a cholinesterase inhibitor.

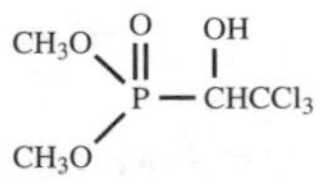

Trichlormethiazide (mol wt 381) A diuretic agent that inhibits re-absorption of sodium and chloride, thereby increasing excretion of sodium, chloride, and water by the kidneys.

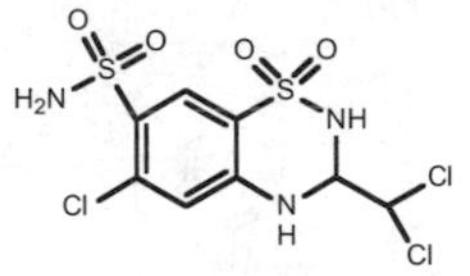

Trichloroacetic Acid (mol wt 163) A decalcifier and precipitant for proteins.

$Cl_3C\text{-}COOH$

2,2,2-Trichloroethanol (mol wt 149) A sedative agent.

CCl_3CH_2OH

Trichloroethylene (mol wt 131) An inhalation anesthetic agent.

$ClCH{=}CCl_2$

2,4,5-Trichlorophenol (mol wt 197) An antifungal and antibacterial agent.

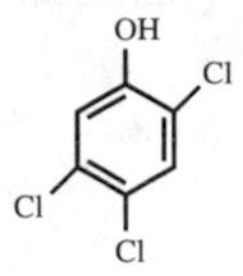

2,4,6-Trichlorophenol (mol wt 197) An antifungal and antibacterial agent.

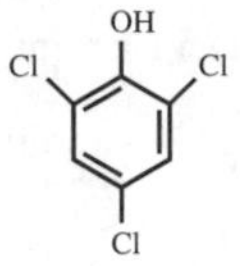

3′,4′,5-Trichlorosalicylanilide (mol wt 317) A topical antiseptic and antifungal agent.

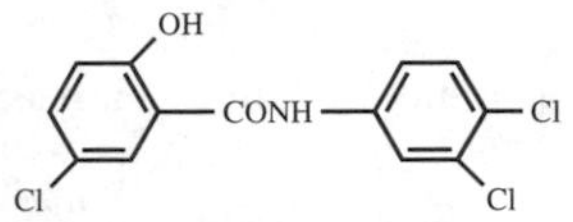

Trichlorourethan (mol wt 192) A sedative–hypnotic agent.

$NH_2COOCH_2CCl_3$

Trichocyst A threadlike organelle ejected from the surface of a ciliate, used both as a weapon and as an anchoring device.

Trichoderma A genus of fungi used for production of cellulase.

Trichodermin (mol wt 292) An antifungal metabolite from *Trichoderma viride*.

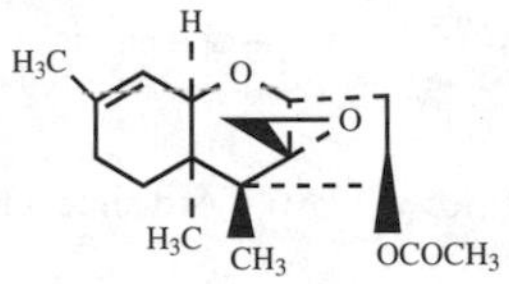

Trichogen Any agent that stimulates the growth of hairs.

Trichology The science that deals with hair and its diseases.

Trichomonas A genus of flagellate protozoa (order Trichomonadida).

Trichomoniasis Disease or infection caused by the species of *Trichomonas*.

Trichosporon A genus of fungi (Hyphomycete) that infect hair and skin.

Trichosporosis Any mycotic infection of hair caused by pathogenic *Trichosporon*.

Trichothecin (mol wt 332) A mycotoxin produced by *Trichothecium roseum*.

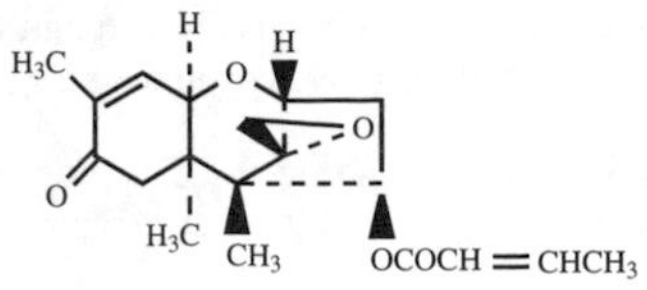

Trichuriasis Infection of intestine by roundworm (*Trichuris trichiura*).

Trichuris A genus of worms (Nematoda).

Triclabendazole (mol wt 360) An anthelmintic agent.

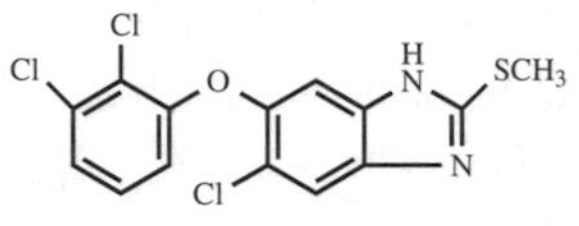

Triclocarban (mol wt 316) A disinfectant.

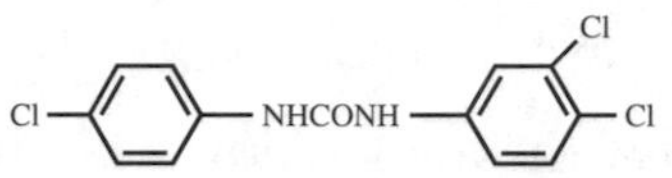

Triclodazol (mol wt 386) A tranquilizer.

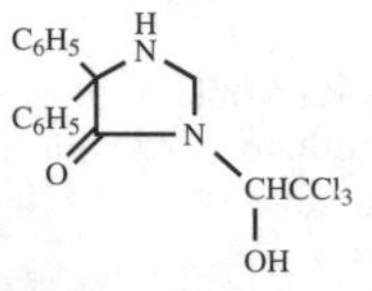

Triclofos (mol wt 229) A hypnotic–sedative agent.

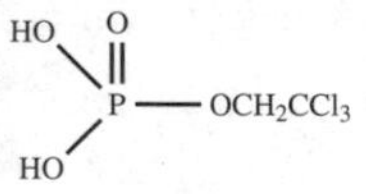

Triclosan (mol wt 290) A disinfectant.

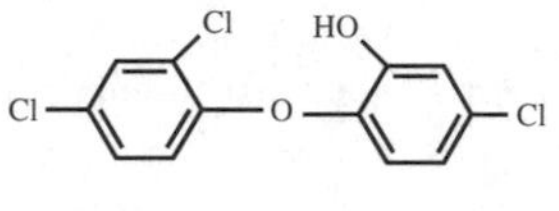

Tricor A trade name for fenofibrate, an antihyperlipidemic agent.

Tricosa- A prefix denoting 23.

Tricosapeptide A 23-residue peptide.

Tricromyl (mol wt 160) A coronary vasodilator and an antispasmodic agent.

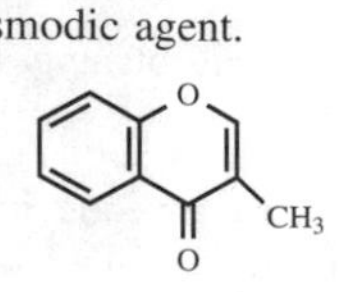

Trideca- A prefix denoting 13.

Triderm A trade name for triamcinolone acetonide, a glucocorticoid used as an anti-inflammatory agent and an immunosuppressive agent.

Tridesilon A trade name for desonide, a corticosteroid.

Tridihexethyl Iodide (mol wt 445) An anticholinergic agent.

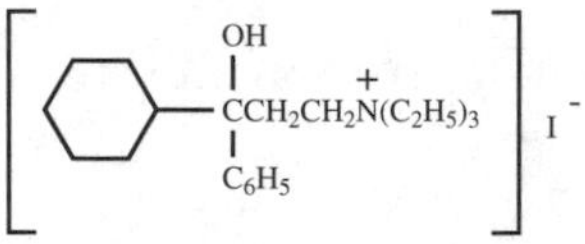

Tridil A trade name for nitroglycerin, an antianginal agent that reduces cardiac oxygen demand and increases blood flow.

Tridione A trade name for trimethadione, an anticonvulsant.

Trientine (mol wt 146) A chelating agent that chelates copper and increases its secretion.

$(NH_2CH_2CH_2NHCH_2\text{-})_2$

Trietazine (mol wt 230) A herbicide.

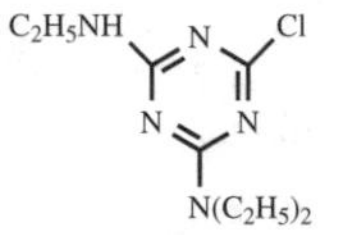

Triethanolamine (mol wt 149) An analgesic agent.

$N(CH_2CH_2OH)_3$

Triethylenemelamine (mol wt 204) An antineoplastic agent.

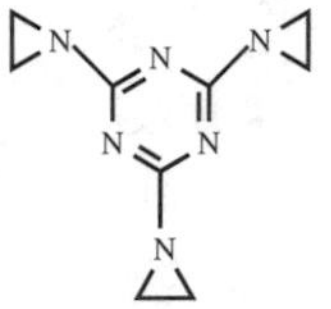

Triethylenephosphoramide (mol wt 173) An antineoplastic agent.

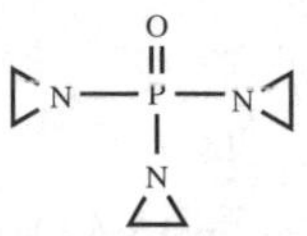

Triethylenethiophosphoramide (mol wt 189) An antineoplastic agent.

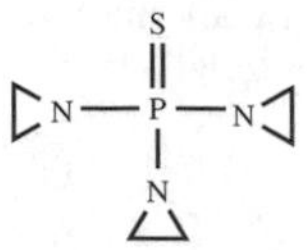

Trifluomeprazine (mol wt 366) A tranquilizer.

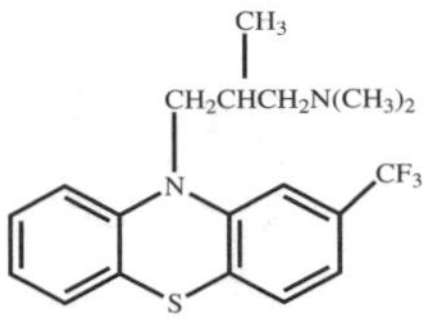

Trifluoperazine (mol wt 407) An antipsychotic agent.

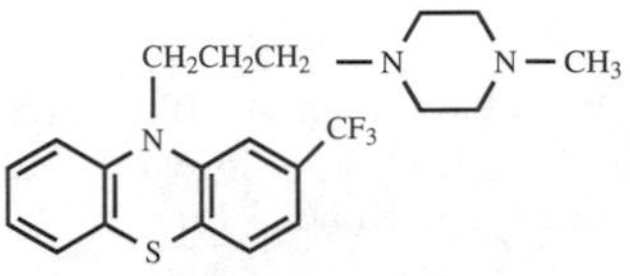

Trifluoromethylphenyl Isocyanate (mol wt 187) A reagent used for chemical modification of lysine.

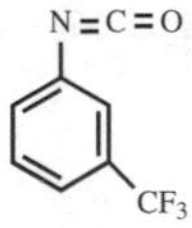

Trifluperidol (mol wt 409) An antipsychotic agent.

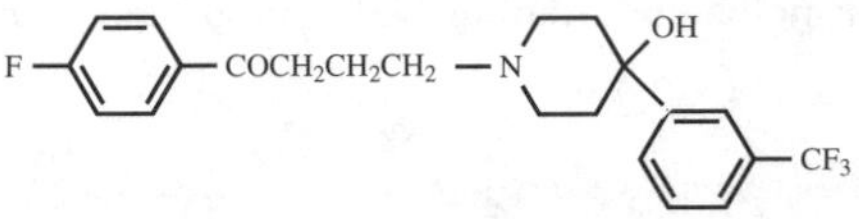

Triflupromazine (mol wt 352) An antipsychotic agent.

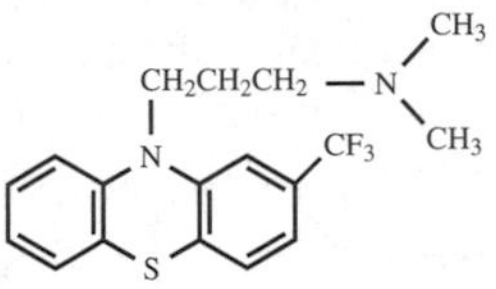

Trifluridine (mol wt 296) An anti-infective agent that interfers with DNA synthesis.

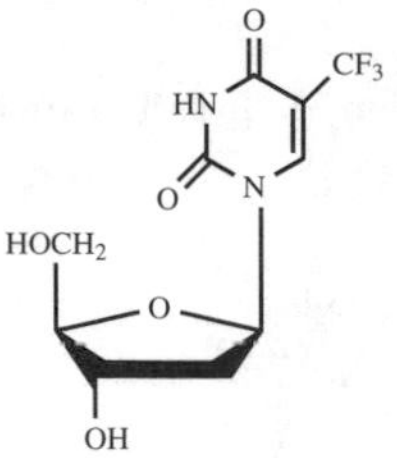

Triflusal (mol wt 248) An antithrombotic agent.

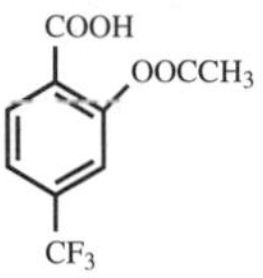

Trigesic A trade name for a combination drug containing acetaminophen and caffeine, used as an antipyretic and analgesic agent.

Trigger Protein A protein that accumulates rapidly during the G-1 phase of the cell cycle, causing the cell to pass into S phase.

Triglyceride See triacylglycerol.

Triglyceride Lipase Synonym of triacylglycerol lipase.

Trigonitis Inflammation of the urinary bladder.

Trihexane A trade name for trihexyphenidyl hydrochloride, an antiparkinsonian agent that blocks cholinergic receptors.

Trihexy-2 A trade name for trihexyphenidyl hydrochloride, an antiparkinsonian agent that blocks cholinergic receptors.

Trihexyphenidyl Hydrochloride (mol wt 338) An anticholinergic and antiparkinsonian agent.

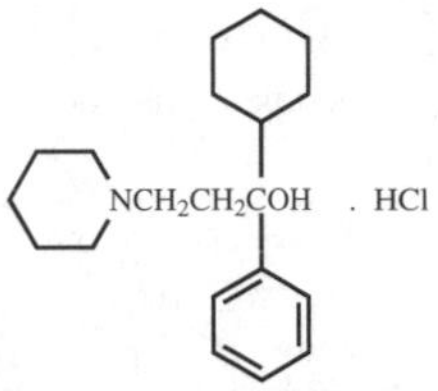

Trihybrid An organism heterozygous at three loci.

Trihydrate Substance that contains 3 molecules of water.

Trihydric Having three hydroxyl groups per molecule.

Tri-Hydroserpine A trade name for a combination drug containing hydrochlorothiazide, hydralazine hydrochloride, and reserpine, used as an antihypertensive agent.

Tri-Immunol A trade name for diphtheria, tetanus and pertussis vaccine.

Triiodobenzoic Acid (mol wt 500) An inhibitor of auxin transport in plants.

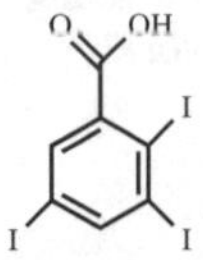

Triiodo-L-Thyronine (mol wt 651) A thyroid hormone with function similar to thyroxine (also known as T_3).

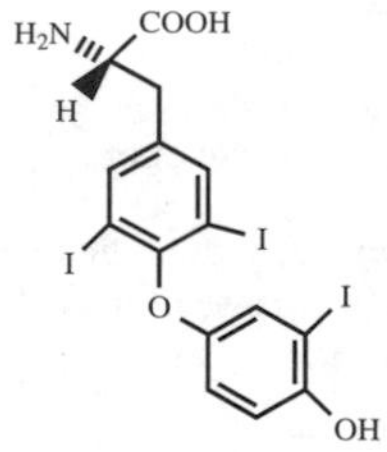

Trilafon A trade name for perphenazine, used as an antipsychotic agent that blocks postsynaptic dopamine receptors in the brain.

Tri-Levlen A trade name for ethinyl estradiol and levonorgestrel, used as an oral contraceptive to inhibit ovulation.

Trilisate A trade name for choline magnesium trisalicylate, used as antipyretic and analgesic agent.

Trilog A trade name for triamcinolone acetonide, a corticosteroide.

Trilone A trade name for triamcinolone diacetate, an anti-inflammatory agent.

Trilostane (mol wt 329) An adrenocortical suppressant and antineoplastic agent for treatment of breast cancer.

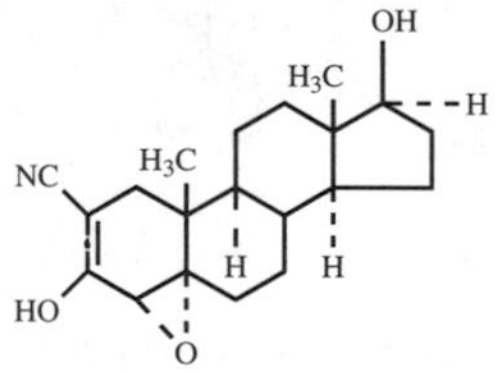

Trimazid A trade name for trimethobenzamide hydrochloride, an anti-emetic agent.

Trimazosin (mol wt 435) An antihypertensive agent.

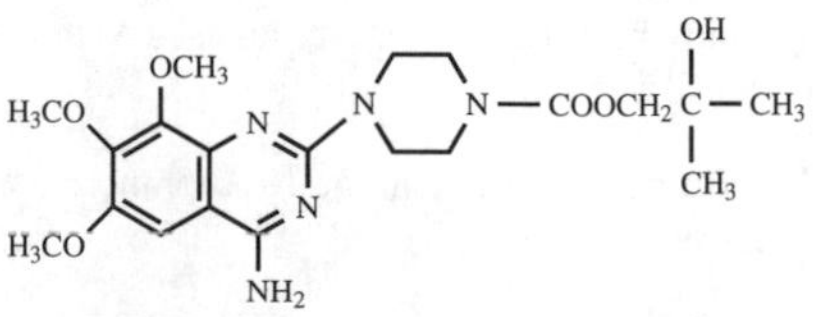

Trimebutine (mol wt 387) A drug that regulates intestinal motility by acting on intestinal opiate and serotonin receptors.

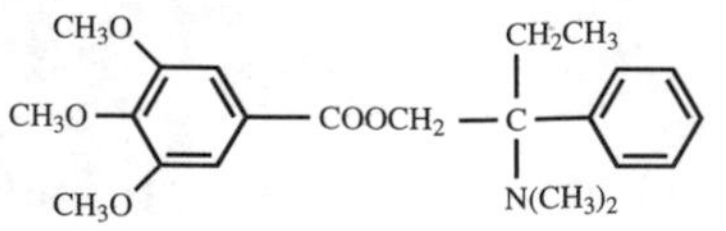

Trimeprazine (mol wt 298) An antihistaminic agent that competes with histamine for H_1 receptor sites on effector cells.

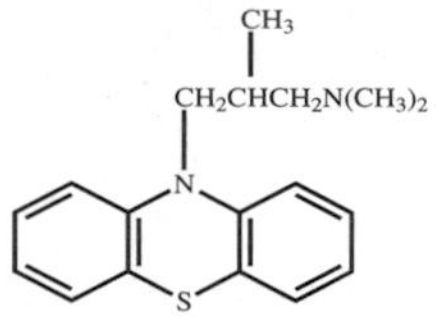

Trimer A compound that consists of three monomeric units.

Trimerelysin I The enzyme that catalyzes the cleavage of the B chain of insulin between positions 10 and 11, and 14 and 15.

Trimerelysin II The enzyme that catalyzes the cleavage of the B chain of insulin between the position 3 and 4, 10 and 11, 14 and 15.

Trimetazidine (mol wt 266) A coronary vasodilator.

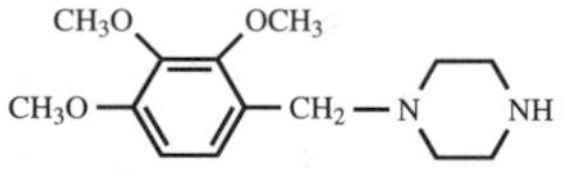

Trimethadione (mol wt 143) An anticonvulsant.

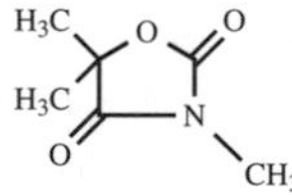

Trimethaphan Camsylate (mol wt 597) An antihypertensive agent that acts as a ganglionic blocker to stabilize postsynaptic membranes.

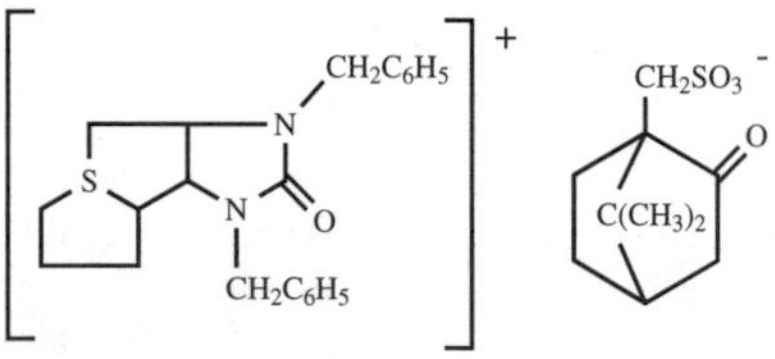

Trimethidinium Methosulfate (mol wt 491) An antihypertensive agent.

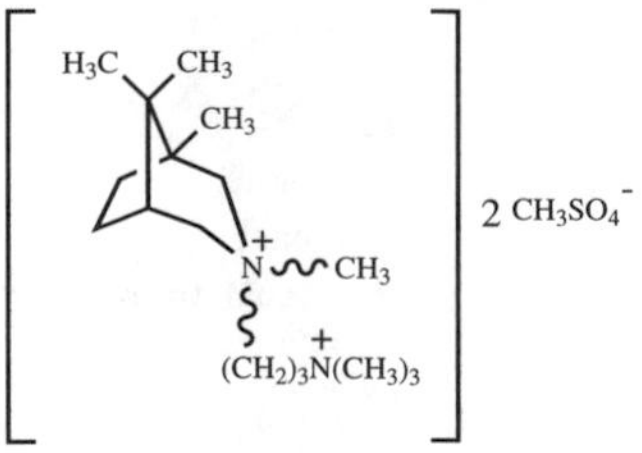

Trimethobenzamide (mol wt 388) An antiemetic agent.

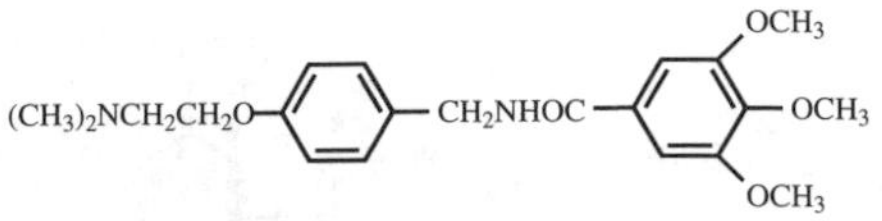

Trimethoprim (mol wt 290) An antibacterial agent that interfers with the action of dihydrofolate reductase, inhibiting bacterial synthesis of folic acid.

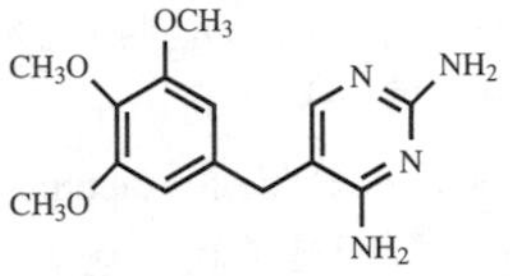

Trimethylamine (mol wt 59) A substrate for trimethylamine dehydrogenase.

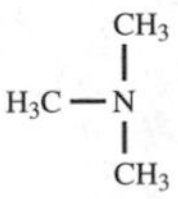

Trimethylamine Dehydrogenase The enzyme that catalyzes the following reaction:

Trimethylamine + H_2O + acceptor

⇅

Dimethylamine + formaldehyde + reduced acceptor

Trimethylamine *N*-Oxide Reductase The enzyme that catalyzes the following reaction:

NADH + trimethylamine-*N*-oxide

⇅

NAD^+ + trimethylamine + H_2O

Trimethyllysine (mol wt 187) A derivative of lysine.

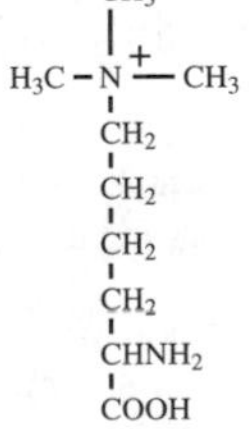

Trimethyllysine Dioxygenase The enzyme that catalyzes the following reaction:

N^6,N^6,N^6-Trimethyl-L-lysine + α-ketoglutarate + O_2

⇅

3-Hydroxy-^6N,^6N,^6N-trimethyl-L-lysine + succinate + CO_2

Trimethylolmelamine (mol wt 216) An antineoplastic agent.

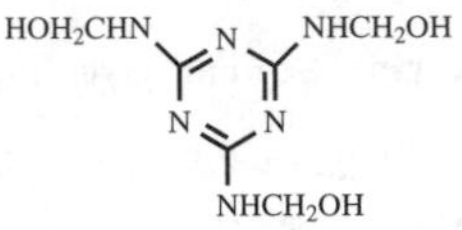

Trimetozine (mol wt 281) A sedative agent.

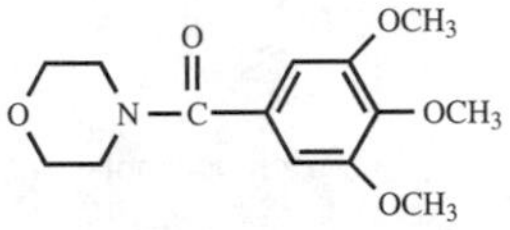

Trimetrexate (mol wt 369) An antineoplastic agent.

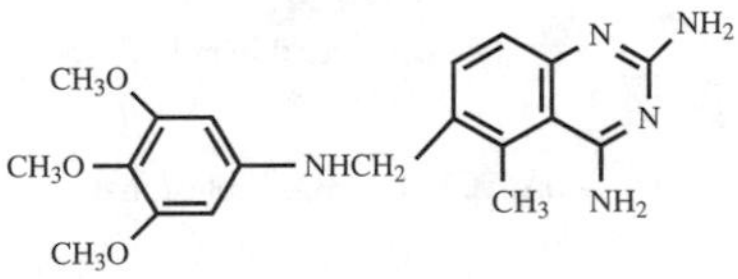

Trimipramine (mol wt 294) An antidepressant that increases the amount of norepinephrine or serotonin in the CNS by blocking their uptake by presynaptic neurons.

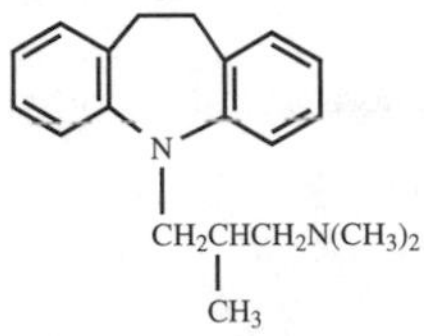

Trimoprostil (mol wt 379) An antiulcerative agent.

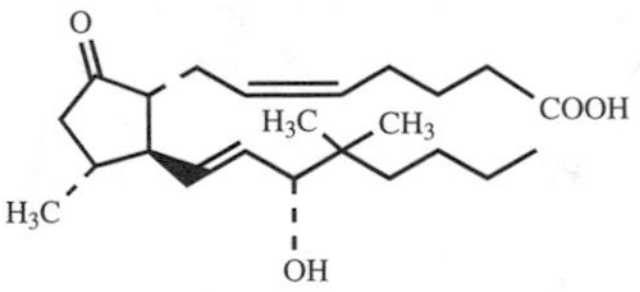

Trimox A trade name for amoxicillin trihydrate, an antibiotic that inhibits bacterial cell wall synthesis.

Trimpex A trade name for trimethoprim, an inhibitor of bacterial folic acid synthesis.

Trimstat A trade name for phendimetrazine tartrate, a cerebral stimulant that promotes the transmission of impulses by releasing the stored norepinephrine from the nerve terminals in the brain.

Trimtabs A trade name for phendimetrazine tartrate, a cerebral stimulant that promotes the transmission of impulses by releasing the stored norepinephrine from the nerve terminals in the brain.

Trinalin Repetabs A trade name for a combination drug containing azatadine maleate and pseudoephedrine sulfate used as an antihistaminic agent.

Trinitroglycerin See nitroglycerin.

Trinitrophenol (mol wt 229) A phenol derivative (see also picric acid).

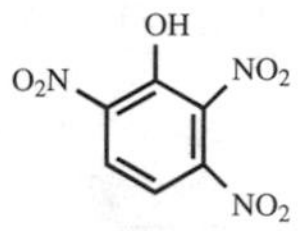

2,3,6-trinitrophenol

Triokinase The enzyme that catalyzes the following reaction:

ATP + D-glyceraldehyde
⇅
ADP + D-glyceraldehyde 3-phosphate

Triolein (mol wt 885) A triglyceride in olive oil.

$CH_2OCO(CH_2)_7CH = CH(CH_2)_7CH_3$
|
$CHOCO(CH_2)_7CH = CH(CH_2)_7CH_3$
|
$CH_2OCO(CH_2)_7CH = CH(CH_2)_7CH_3$

Triose A monosaccharide consisting of three carbon atoms.

Triose Kinase See triokinase.

Triose Phosphate Dehydrogenase Synonym of glyceraldehyde 3-phosphate dehydrogenase.

Triose Phosphate Isomerase The enzyme that catalyzes the following reaction:

D-Glyceraldehyde 3-phosphate
⇅
Glycerone phosphate

Triose Phosphate Mutase Synonym of triose phosphate isomerase.

Triostat A trade name for liothyronine sodium, a thyroid hormone.

Trioxsalen (mol wt 228) A photochemical probe for nucleic acid.

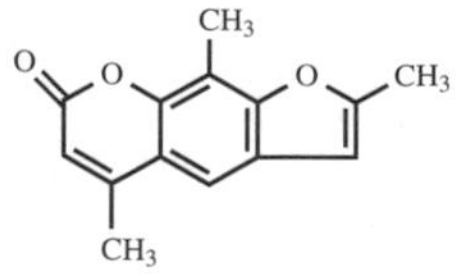

Tripamide (mol wt 370) An antihypertensive and a diuretic agent.

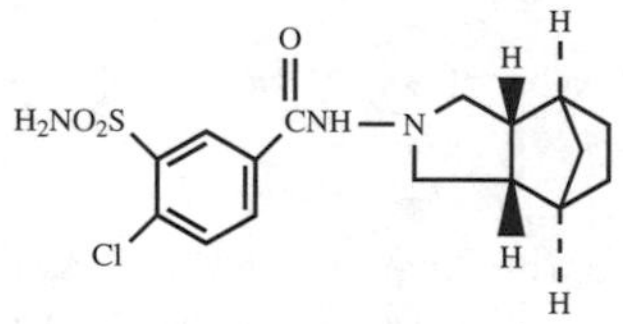

Triparental Recombination Referring to a bacteriophage that contains three marker genes derived from three different phages in a host infected by three different phages.

Tripedia A trade name for diphtheria and tetanus toxoids and pertussis vaccine.

Tripelennamine (mol wt 255) An antihistaminic agent that competes with histamine for H_1 receptor sites on effector cells.

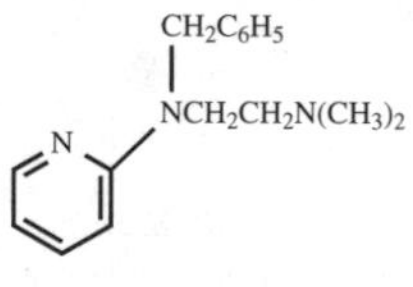

Tripeptide Aminopeptidase The enzyme that catalyzes the release of the N-terminal residue from a tripeptide.

Tripeptidyl-Peptidase The enzyme that catalyzes the release of an N-terminal tripeptide from a polypeptide.

Triphasil A trade name for ethinyl estradiol and levonorgestrel, used as oral contraceptives to inhibit ovulation.

Triphosphatase The enzyme that catalyzes the following reaction:

$$\text{Triphosphate} + H_2O \rightleftharpoons \text{Pyrophosphate} + \text{Pi}$$

Triphosphopyridine Nucleotide (TPN^+) Referring to $NADP^+$.

Triple Bond Chemical bond formed between two atoms as a result of sharing three pairs of electrons.

Triple Fusion In angiosperms, the fusion of the second male gamete, or sperm, with the polar nuclei, resulting in formation of a primary endosperm nucleus, which is most often triploid.

Triple Sulfa A trade name for a combination drug containing sulfadiazine, sulfamerazine, and sulfamethazine, used for inhibition of synthesis of bacterial folic acid.

Triplet Code A sequence of three nucleotides (a triplet) in mRNA that specifies an amino acid in a protein.

Triple X A trade name for pyrethrin, an antiparasitic agent.

Triplication Presence of three copies of a DNA sequence.

Triploblastic Having three primary germ layers in an animal embryo.

Triploid A cell or an individual organism with three complete sets of haploid chromosomes.

Triploidy The condition in which a cell or an organism possesses three haploid sets of chromosomes.

Triplopia A visual defect in which one object is perceived as three objects.

Tripramine A trade name for imipramine hydrochloride, an antihypertensive agent that increases the level of norepinephrine or serotonin in the CNS by blocking their uptake by presynaptic neurons.

Triprim A trade name for trimethoprim, an anti-infective agent that inhibits bacterial folic acid synthesis.

Triprolidine (mol wt 278) An antihistaminic agent that competes with histamine for H_1 receptors on effector cells.

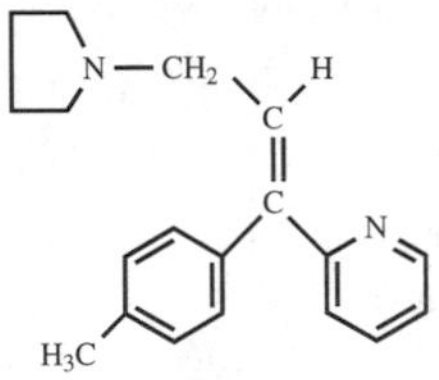

Triptil A trade name for protriptyline hydrochloride, an antidepressant that increases the concentration of norepinephrine and serotonin in the CNS by blocking their uptake by presynaptic neurons.

Triptone Caplets A trade name for dimenhydrinate, an antiemetic agent.

Tris (mol wt 121) Tris (hydroxymethyl) aminomethane, a substance used for the preparation of biological buffers (also known as THAM).

$$(HOCH_2)_3CNH_2$$

Trisaccharide A carbohydrate consisting of three monosaccharides.

Tris/DTT Buffer A buffer that contains 25 m*M* Tris-HCl, 1 m*M* DTT, and 20% (v/v) glycerol (pH 8.1 to 8.3).

Tris/EDTA Buffer A buffer that contains 10 m*M* Tris-HCl, 1.5 m*M* EDTA, 10% (v/v) glycerol and 10 m*M* monothioglycerol (or 1 m*M* DTT), pH 7.4.

Tris/Glucose Buffer A buffer that contains 25 m*M* Tris-HCl, 10 m*M* EDTA, and 50 m*M* glycerol (pH 8).

Triskelion A hexamer of clathrin consisting of three heavy chains and three light chains.

Tris/NaCl Buffer A buffer that contains 0.02 *M* Tris-HCl and 0.5 *M* NaCl (pH 7.5).

Trisomic A diploid cell with an extra chromosome.

Trisomic Cell A cell that has three copies of one chromosome instead of the usual two copies present in the diploid nucleus.

Trisphosphate A compound that has three phosphate groups.

Tris/Saline/Azide Solution A solution that contains 0.01 *M* Tris-HCl, 0.14 *M* NaCl, and 0.025% sodium azide (NaN_3), pH 8.

Tris/SDS Buffer A buffer that contains 91 g Tris base and 2 g SDS (pH 8.8) in 500 ml of solution.

Tristearin A triacylglycerol in which all fatty acids are steraoyl.

Tritanopia A disorder or inability to perceive the color blue.

Triterpenol Esterase Synonym of sterol esterase.

Tritiated Labeled with tritium.

Tritin A ribosome-inactivating protein from wheat.

Tritium (^{3}H) A radioactive isotope of hydrogen, with atomic weight 3, and half-life of 12.46 years.

Triton A trade name for a series of organic nonionic detergents.

Triton Lysis Buffer A buffer that contains 0.25 *M* Tris-HCl and 0.5% Triton X-100 (pH 7.8) used for lysis and solubilization of cells.

Triton X-100 A nonionic detergent used for solubilization of membranes.

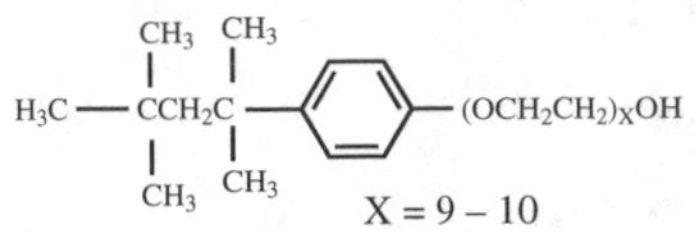

X = 9 – 10

Tritoqualine (mol wt 501) An antihistaminic agent.

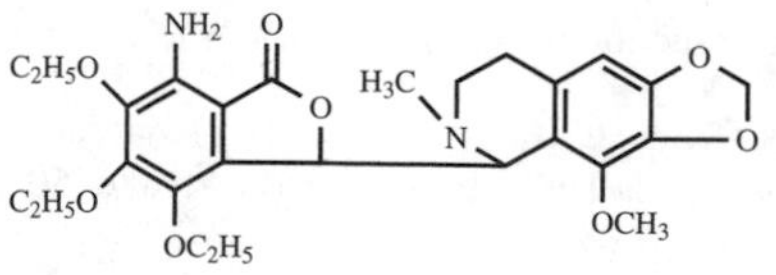

Trityl Group Referring to the triphenylmethyl group used for protection of amino groups in protein synthesis.

Trivalent Denoting an atom or radical with a valence of three.

Trivial Name A common name for an organism or a chemical.

Trivora A trade name for a combination drug containing levonorgestrel and ethinyl estradiol used as a contraceptive agent.

Trizma A trade name for tris (hydroxymethyl) aminomethane.

tRNA See transfer RNA.

tRNA Adenylyltransferase The enzyme that catalyzes the following reaction:

$$\text{ATP} + \text{tRNA}_n \rightleftharpoons \text{PPi} + \text{tRNA}_{n+1}$$

tRNA Cytidylyltransferase The enzyme that catalyzes the following reaction:

$$\text{CTP} + \text{tRNA}_n \rightleftharpoons \text{PPi} + \text{tRNA}_{n+1}$$

tRNA Intron Endonuclease The enzyme that catalyzes the endonucleolytic cleavage of pre-tRNA, producing 5′-hydroxyl and 2′,3′-cyclic phosphate termini.

tRNA Isoacceptor Different tRNAs that accept the same amino acid.

tRNA Nucleotidyltransferase The enzyme that catalyzes the following reaction:

$$\text{tRNA}_{n+1} + \text{Pi} \rightleftharpoons \text{tRNA}_n + \text{a nucleoside diphosphate}$$

tRNA Pseudouridine Synthetase The enzyme that catalyzes the following reaction:

$$\text{tRNA uridine} \rightleftharpoons \text{tRNA pseudouridine}$$

tRNA Sulfurtransferase The enzyme that catalyzes the following reaction:

$$\text{L-Cysteine} + \text{activated tRNA} \rightleftharpoons \text{L-Serine} + \text{tRNA containing a thionucleotide}$$

tRNA Synthetase See aminoacyl-tRNA synthetase.

tRNA Uridine Isomerase See tRNA-pseudouridine synthetase.

tRNA Uridine Uracilmutase See tRNA-pseudouridine synthetase.

tRNA^{al} Referring to tRNA for alanine or tRNA carrying alanine.

tRNA^{asp} Referring to tRNA for aspartic acid or tRNA carrying aspartic acid.

tRNA^{arg} Referring to tRNA for arginine or tRNA carrying arginine.

tRNA^{asn} Referring to tRNA for asparagine or tRNA carrying asparagine.

tRNA^{cys} Referring to tRNA for cysteine or tRNA carrying cysteine.

tRNA^{fmet} tRNA that carries formylmethionine, which is enzymatically formed after methionine is attached onto the CCA-terminal of tRNA.

tRNA^{gln} Referring to tRNA for glutamine or tRNA carrying glutamine.

tRNA^{glu} Referring to tRNA for glutamic acid or tRNA carrying glutamic acid.

tRNA^{his} Referring to tRNA for histidine or tRNA carrying histidine.

tRNA^{ile} Referring to tRNA for isoleucine or tRNA carrying isoleucine.

tRNA^{leu} Referring to tRNA for leucine or tRNA carrying leucine.

tRNA^{lys} Referring to tRNA for lysine or tRNA carrying lysine.

tRNA^{met} Referring to tRNA for methionine or tRNA carrying methionine.

tRNA^{phe} Referring to tRNA for phenylalanine or tRNA carrying phenylalanine.

tRNA^{pro} Referring to tRNA for proline or tRNA carrying proline.

tRNA^{ser} Referring to tRNA for serine or tRNA carrying serine.

tRNA^{thr} Referring to tRNA for threonine or tRNA carrying threonine.

tRNA^{try} Referring to tRNA for tryptophan or tRNA carrying tryptophan.

tRNA^{tyr} Referring to tRNA for tyrosine or tRNA carrying tyrosine.

Trobicin A trade name for spectinomycin dihydrochloride, an antibiotic that binds to 30S ribosomal subunits, inhibiting bacterial protein synthesis.

Troclosene Potassium (mol wt 236) A topical anti-infective agent.

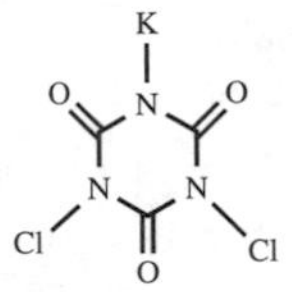

Trofosfamide (mol wt 324) An antineoplastic agent.

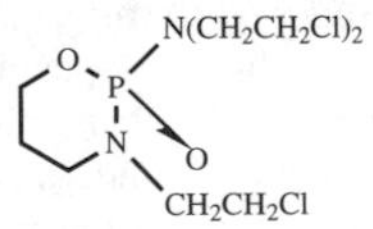

Troglitazone (mol wt 442) An anti-diabetic agent that stimulates insulin receptor sites to lower blood glucose and improve the action of insulin.

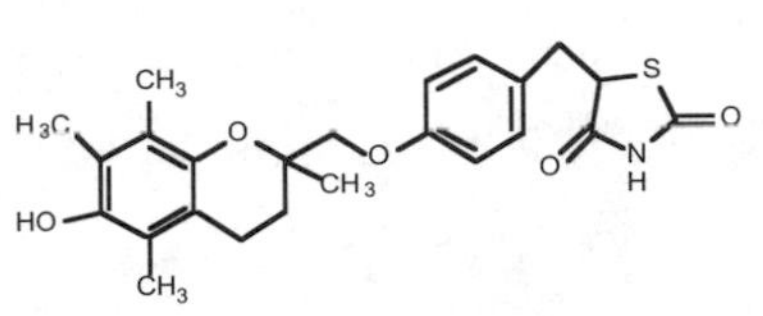

Trolnitrate Phosphate (mol wt 480) A vasodilator.

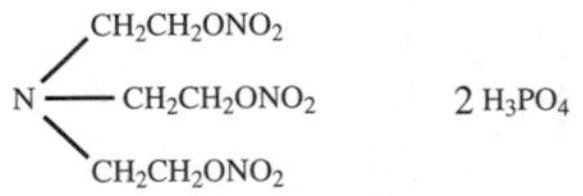

Tromantadine (mol wt 280) An antiviral agent.

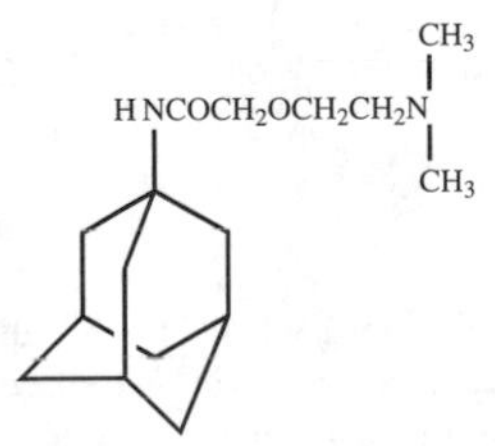

Trombicula A genus of mites capable of infecting humans.

Trombiculiasis Infestation by mites of the genus *Trombicula*.

Tromethamine (mol wt 121) An alkalinizer.

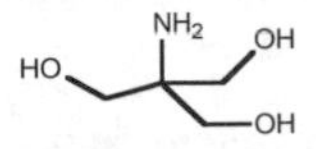

Tropacine (mol wt 335) An anticholinergic agent.

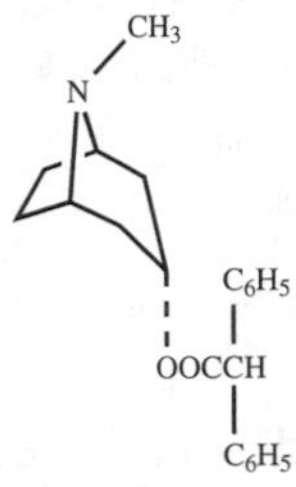

Tropaeolin O (mol wt 316) A biological stain and pH indicator; yellow at pH 11, orange-brown at pH 12.7.

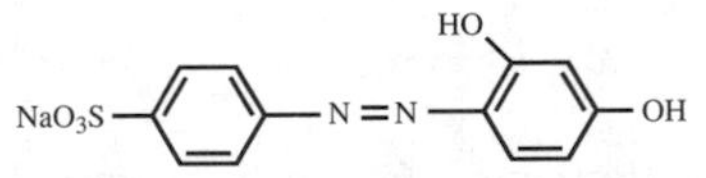

Tropaeolin OO (mot wt 375) A biological stain and a pH indicator; red at pH 1.4, yellow at pH 2.6.

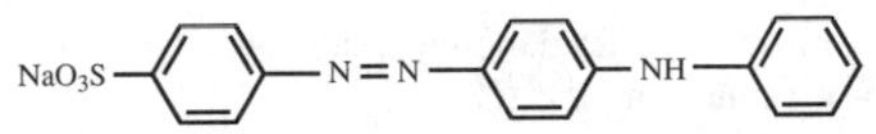

Tropenzile (mol wt 381) An antispasmodic agent.

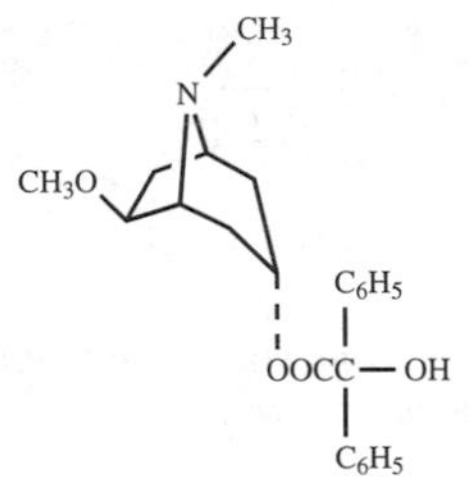

-troph A suffix meaning feeder, e.g., autotroph.

-trophic A suffix meaning related to nutrition.

Trophic Hormones Hormones that have other endocrine glands as their target.

-trophin A suffix denoting a trophic function.

Trophoblast The outer layer of the mammalian blastocyst that differentiates into extraembryonic membranes.

Trophoneurosis Tissue alteration caused by the interruption of nerve supply to that part of the tissue.

Trophophase A phase of a culture in which growth-directed metabolism is dominant over secondary metabolism.

Trophotropism Migration or movement of living cells toward or away from nutritive material.

Trophozoite A cell capable of feeding, e.g., the active feeding stage of a protozoan.

-trophy A suffix denoting 1. food or nourishment; 2. growth or size.

-tropic A suffix denoting a stimulatory function, e.g., gonadotropic.

Tropic Acid (mol wt 166) An aromatic organic acid obtained from hydrolysis of atropine.

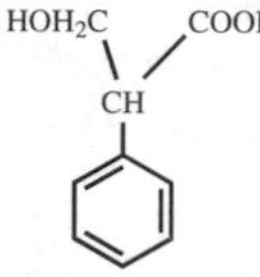

Tropic Hormones See trophic hormones.

Tropicacyl A trade name for tropicamide, used as an anticholinergic agent.

-tropin A suffix denoting a substance with stimulatory function.

Tropine (mol wt 141) A poisonous heterocyclic amine alcohol obtained from hydrolysis of atropine.

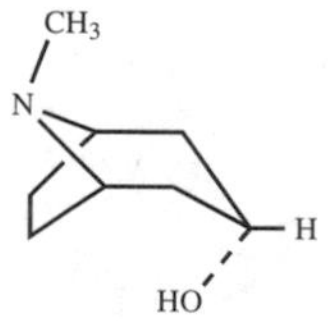

Tropine Dehydrogenase The enzyme that catalyzes the following reaction:

Tropine + $NADP^+$ ⇌ Tropinone + NADPH

Tropinesterase The enzyme that catalyzes the following reaction:

Atropine + H_2O ⇌ Tropine + tropate

Tropinone (mol wt 139) A substrate for tropine dehydrogenase.

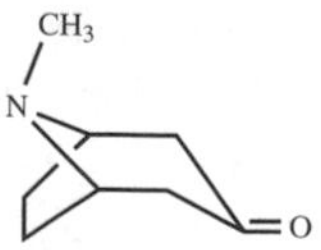

Tropism The involuntary orientation of an organism toward (positive) or away (negative) from a stimulus.

Tropocollagen The protein subunit of collagen fibrils that consists of a triple helix polypeptide chain.

Tropomodulin A cytoskeletal tropomyosin-regulatory protein that binds to the end of erythrocyte tropomyosin and blocks its head-to-tail association along actin filaments.

Tropomyosin Protein associated with actin filaments that blocks interaction between actin and myosin in the absence of calcium; it consists of two elongated a-helix polypeptide chains.

Tropomyosin Kinase The enzyme that catalyzes the following reaction:

ATP + tropomyosin

⇅

ADP + *O*-phosphotropomyosin

Troponin Protein associated with actin filaments that displaces tropomyosin in the presence of calcium. It consists of three polypeptide subunits (C, I, and T). Troponin T contains binding sites for tropomyosin; troponin I binds actin and inhibits interaction of myosin and actin; troponin C binds calcium ion and regulates the interaction of troponin I and T.

Trospectomycin (mol wt 374) An antibacterial agent.

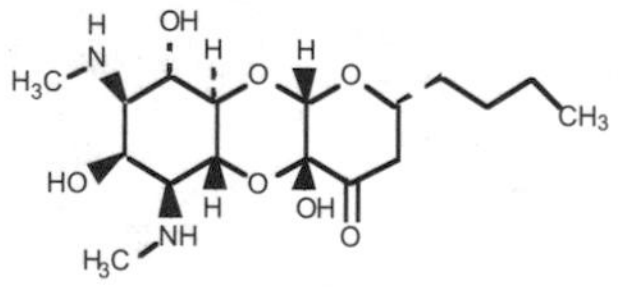

Trospium Chloride (mol wt 428) An antispasmodic agent.

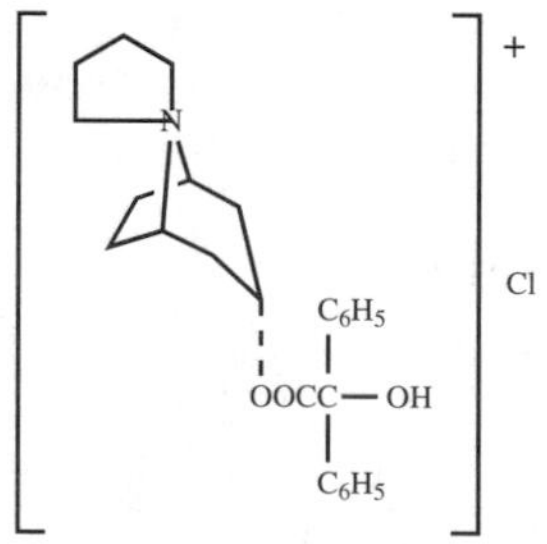

Trovafloxacin (mol wt 416) An antibiotic that interferes with DNA replication in susceptible gram-negative gram-positive aerobic and anaerobic bacteria.

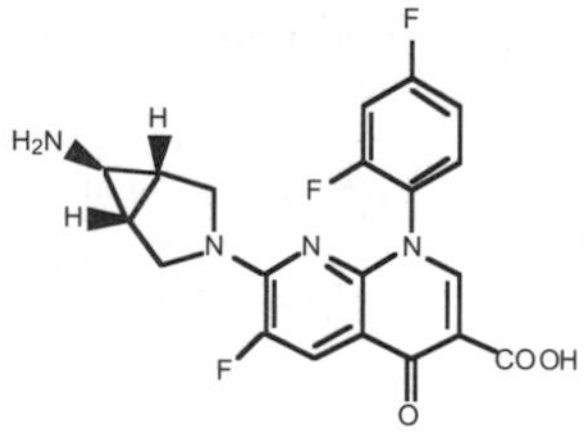

Trovan A trade name for trovafloxacin, an antibiotic that interferes with DNA replication in susceptible gram-negative and gram-positive aerobic and anaerobic bacteria.

Troxipide (mol wt 294) An antiulcerative agent.

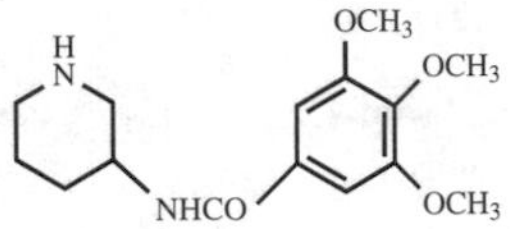

Trp Abbreviation for tryptophan.

TRP Abbreviation for transient receptor potential.

trp Genes A family of tryptophan biosynthetic genes found in *E. coli*.

***trp* Operon** See tryptophan operon.

TRPA Abbreviation for tryptophan-rich pre-albumin.

TRPL Abbreviation for transient receptor potential-like.

TRS Abbreviation for total reducing sugar.

TRSV Abbreviation for tobacco ring spot virus.

Trt Abbreviation for trityl group.

TruI (AvaII) A restriction endonuclease from *Thermus ruber* strain 21 with the same specificity as AvaII.

TruII (MboI) A restriction endonuclease from *Thermus ruber* strain 21 with the same specificity as MboI.

Tru9I A restriction endonuclease from *Thermus ruber* 9 with the following specificity:

↓
5'........TTAA........3'
3'........AATT........5'
↑

True Breeding Organisms that are homozygous for the traits under consideration.

Truphylline A trade name for aminophylline, a bronchodilator.

TRX Abbreviation for thioredoxin.

Trymegen A trade name for chlorpheniramine maleate, an antihistaminic agent that competes with histamine for H_1 receptors on effector cells.

Trypan Blue (mol wt 961) A biological stain to distinguish dead cells from viable cells.

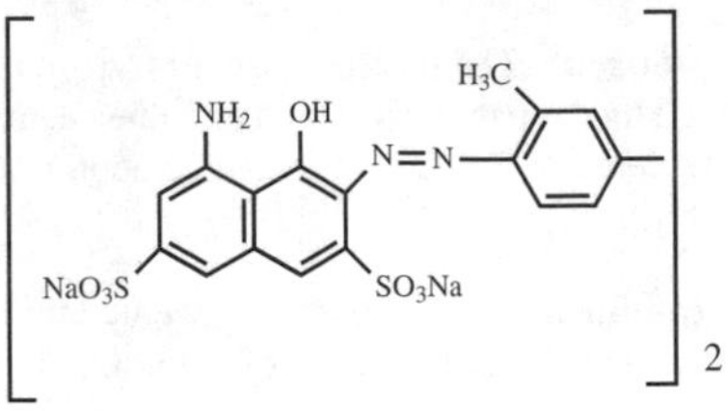

Trypan Red (mol wt 1003) A biological stain.

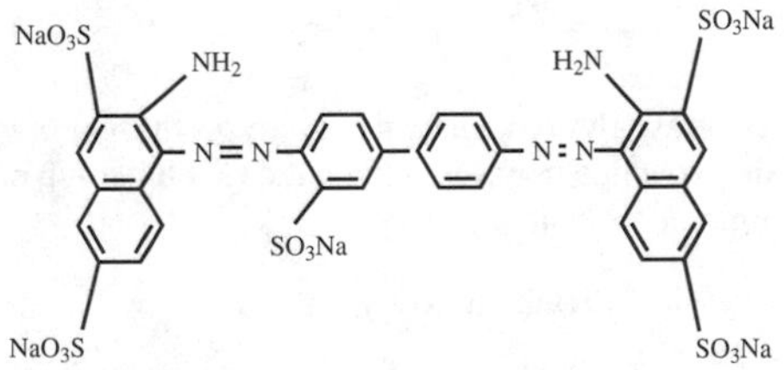

Trypanosoma A genus of parasitic flagellate protozoa (family Trypanosomatidae). Some species are pathogenic to human (e.g., causal agent of sleeping sickness).

Trypanosomatid Any member of the family Trypanosomatidae.

Trypanosome Any member of the genus *Trypanosoma*.

Trypanosome Adhesion Test A test for detecting the presence of antibodies homologous to a particular species of *Trypanosoma*.

Trypanosomiasis Human or animal diseases caused by a member of the genus *Trypanosoma*.

Trypanothione Reductase The enzyme that catalyzes the following reaction:

NADPH + trypanothione

⇅

$NADP^+$ + reduced trypanothione

Tryparsamide (mol wt 296) An antiprotozoal agent (Trypanosoma).

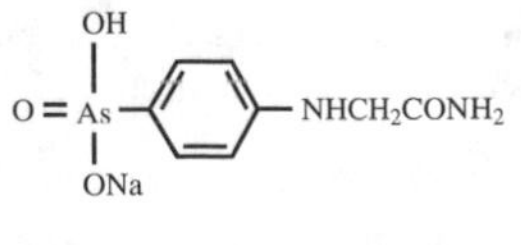

Trypomastigote Mature, infective form of trypanosomes.

Trypsin Proteolytic enzyme secreted by the pancreas that catalyzes the hydrolysis of peptide bonds involving the basic amino acids arginine and lysine.

Trypsinization Dissociation of tissue into cells by trypsin, a technique used for obtaining cells for cell culture.

Trypsinogen The inactive form of trypsin that is secreted by the pancreas, which becomes activated by the removal of N-terminal hexapeptide (val-$(asp)_4$-lys).

Tryptamine (mol wt 160) A biogenic amine resulting from decarboxylation of tryptophan.

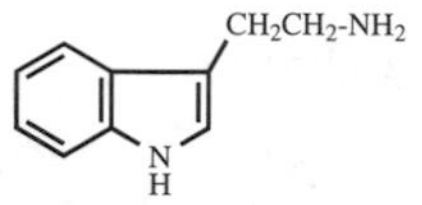

Tryptase A proteinase that catalyzes the hydrolysis of peptide bonds involving the COOH group of arginine or lysine.

Tryptic Pertaining to trypsin.

Tryptic Peptides Peptides produced as a result of tryptic digestion of a protein molecule.

Trypton A tryptophan-rich peptone produced by the action of typsin on casein.

Trypton Water A microbial medium containing 1-2% tryptone and 0.5% NaCl in water.

L-Tryptophan (mol wt 204) An essential protein amino acid.

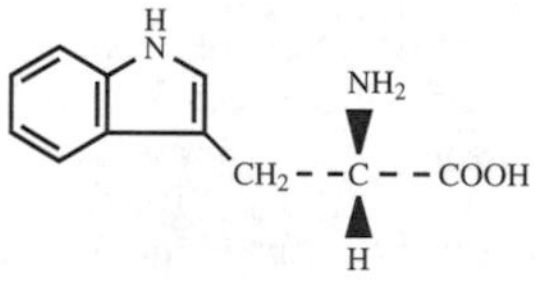

Tryptophan *N*-Acetyltransferase The enzyme that catalyzes the following reaction:

$$\text{Acetyl-CoA} + \text{D-tryptophan} \rightleftharpoons \text{CoA} + N\text{-acetyl-D-tryptophan}$$

Tryptophan Aminopeptidase The enzyme that catalyzes the preferential release of N-terminal tryptophan.

Tryptophan Decarboxylase The enzyme that catalyzes the following reaction:

$$\text{L-Tryptophan} \rightleftharpoons \text{Tryptamine} + CO_2$$

Tryptophan Dehydrogenase The enzyme that catalyzes the following reaction:

$$\text{L-Tryptophan} + \text{NADP}^+ \rightleftharpoons \text{Indole-pyruvate} + NH_3 + \text{NADPH}$$

Tryptophan 2,3-Dioxygenase The enzyme that catalyzes the following reaction:

$$\text{L-Tryptophan} + O_2 \rightleftharpoons \text{L-Formyl kynurenine}$$

Tryptophan 5-Hydroxylase See tryptophan monooxygenase.

Tryptophan Indol-Lyase The systematic name for tryptophanase.

Tryptophan *N*-Malonyltransferase The enzyme that catalyzes the following reaction:

$$\text{Malonyl-CoA} + \text{L-tryptophan} \rightleftharpoons \text{CoA} + N\text{-malonyl-tryptophan}$$

Tryptophan 2-Monooxygenase The enzyme that catalyzes the following reaction:

$$\text{L-Tryptophan} + O_2 \rightleftharpoons \text{Indole-3-acetamide} + CO_2 + H_2O$$

Tryptophan 5-Monooxygenase The enzyme that catalyzes the following reaction:

$$\text{L-Tryptophan} + \text{tetrahydrobiopterin} + O_2 \rightleftharpoons \text{5-Hydroxy-L-tryptophan} + \text{dihydrobiopterin} + H_2O$$

Tryptophan Operon Cluster of structural genes that code for enzymes involved in tryptophan synthesis and the control elements that regulate the synthesis of these enzymes in *E. coli*.

Tryptophan Oxygenase See tryptophan 2,3-dioxygenase.

Tryptophan Phenylpyruvate Transaminase The enzyme that catalyzes the following reaction:

$$\text{L-Tryptophan} + \text{phenylpyruvate} \rightleftharpoons \text{Indolepyruvate} + \text{L-phenylalanine}$$

Tryptophan Pyrrolase See tryptophan 2,3-dioxygenase.

Tryptophan Synthetase The enzyme that catalyzes the following reaction:

$$\text{Serine} + \text{indole 3-phosphate} \rightleftharpoons \text{L-Tryptophan} + \text{glyceraldehyde 3-phosphate} + H_2O$$

Tryptophan Transaminase The enzyme that catalyzes the following reaction:

L-Tryptophan + α-ketoglutarate
⇌
Indolepyruvate + L-glutamate

Tryptophan tRNA Ligase The enzyme that catalyzes the following reaction:

ATP + L-Tryptophan + $tRNA^{trp}$
⇌
AMP + PPi + L-tryptophan-$tRNA^{trp}$

Tryptophan tRNA Synthetase See tryptophan tRNA ligase.

Tryptophan $tRNA^{trp}$ Ligase See tryptophan tRNA ligase.

Tryptophan $tRNA^{trp}$ Synthetase See tryptophan tRNA ligase.

Tryptophanamidase The enzyme that catalyzes the following reaction:

L-tryptophanamide + H_2O
⇌
L-tryptophan + NH_3

Tryptophanyl Aminopeptidase See tryptophan amino peptidase.

Tryptophanyl-tRNA Synthetase See tryptophan tRNA synthetase.

Tryptophase The enzyme that catalyzes the following reaction:

L-tryptophan + H_2O
⇌
Indole + pyruvate + NH_3

Ts A component of the elongation factor in protein synthesis.

TS Abbreviation for 1. thromboxane synthetase; 2. toxic substance; 3. tryptophan synthetase; 4. threonine synthetase.

ts Abbreviation for temperature sensitive.

Ts Cell Abbreviation for suppresser T cell.

***ts* Mutant** Referring to temperature sensitive mutant, a mutant that is viable at one temperature and inviable at another temperature.

***ts* Mutation** A mutation that leads to the production of a protein functioning at one temperature but inactive at another temperature.

TSA Abbreviation for 1. tissue specific antigen; 2. toxic shock antigen; 3. trichostatin-A; 4. trypticase soy agar; 5. tumor-specific antigen.

T4SA Abbreviation for thyroxine specific activity.

TSAb Abbreviation for thyroid-stimulating antibody.

TSAP Abbreviation for toxic shock-associated protein.

TSB Abbreviation for tripticase soy broth.

TscI A restriction endonuclease from *Thermus* species 49 with the following specificity:

```
           ↓
5'........ACGT........3'
3'........TGCA........5'
          ↑
```

TSE Abbreviation for tris/sodium/EDTA buffer.

TseI A restriction endonuclease from *Thermus* species 93170 with the following specificity:

```
          ↓
5'........GCWGC........3'
3'........CGWCG........5'
   W= A or T ↑
```

TSF Abbreviation for 1. T-cell suppressor factor; 2. thrombopoietic stimulating factor; 3. thymocyte-stimulating factor.

TSG Abbreviation for tumor-specific glycoprotein.

TSH Abbreviation for thyroid-stimulating hormone, a glycoprotein hormone secreted by the pituitary gland to activate cAMP production.

TSH Releasing Hormone A tripeptide hormone produced by the hypothalamus to stimulate the release of TSH from the anterior pituitary.

TspI (Tth111I) A restriction endonuclease from *Thermus thermophilus* strain 110 with the same specificity as Tth 111I.

Tsp45I A restriction endonuclease from the *Thermus* species with the following specificity:

```
          ↓
5'........GT(C/G)AC........3'
3'........CA(G/C)TG........5'
                  ↑
```

Tsp509I A restriction endonuclease from the *Thermus* species with the following specificity:

```
          ↓
5'........AATT........3'
3'........TTAA........5'
              ↑
```

TspRI A restriction endonuclease from the *Thermus* species with the following specificity:

5'........NNCAGTGNN........3'
3'........NNGTCACNN........5'

TspZNI (HaeIII) A restriction endonuclease from *Thermus* species 2AZN with the same specificity as HaeIII.

TSTA Abbreviation for tumor-specific transplantation antigen.

T-Stat A trade name for erythromycin, an antibiotic that inhibits bacterial protein synthesis.

T-Suppresser Cells A class of T cells that suppress the immune response.

Tsutsugamushi Disease An infectious disease caused by *Rickettsia tsutsugamushi* and characterized by a painful swelling of lymphatic glands, fever, and headache (also known as scrub typhus).

***tsx* Protein** An outer membrane protein of *E. coli* that acts as receptor for T_6 bacteriophage.

TSY Abbreviation for trypticase soy yeast.

TT Abbreviation for tetanus toxoid or tetanus toxin.

TTC Abbreviation for triphenyl tetrazolium chloride.

TteAI (HaeIII) A restriction endonuclease from *Tolypothrix tenuis* with the same specificity as HaeIII.

TteI (Tth111I) A restriction endonuclease from *Thermus thermophilus* strain 110 with the same specificity as Tth 111I.

TTF-1 Abbreviation for thyroid-specific transcription factor-1.

TTF-1HD Abbreviation for thyroid-specific transcription factor-1 homoeodomain.

tTG Abbreviation for tissue transglutaminase.

tTGase Abbreviation for tissue transglutaminase.

Tth111I A restriction endonuclease from *Thermus thermophilus* strain 110 with the following specificity:

5′..........GACNNNGTC..........3′
3′..........CTGNNNCAG..........5′

TthHB8I (TaqI) A restriction endonuclease from *Thermus thermophilus* HB8 with the same specificity as TaqI.

TTM Abbreviation for tetrathiomolybdate.

TTP Abbreviation for thiamine triphosphate.

TTPA Abbreviation for triethylene thiophosphoamide.

TtrI (Tth111I) A restriction endonuclease from *Thermus thermophilus* strain 23 with the same specificity as Tth111I.

TU Abbreviation for tuberculin unit (one TU = 0.02 μg PPD (purified protein derivative) from the filtrate of a steamed culture of *Mycobacterium tuberculosis*.

Tu A component of the elongation factor in protein synthesis.

Tubarine A trade name for tubocuramine chloride, a neuromuscular blocker that prevents acetylcholine from binding to the receptors on muscle end plate.

Tube Cell 1. One of the two cells resulting from the germination of the microspore of flowering plants. 2. One of the cells of the mature male gametophyte of pine trees.

Tube Nucleus The haploid nucleus in a pollen tube that does not participate in double fertilization.

Tubercidin (mol wt 266) An antifungal and antibacterial agent.

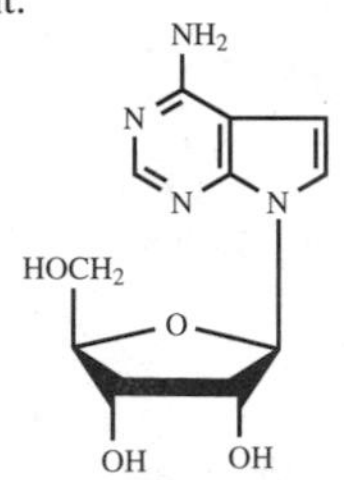

Tubercle 1. The lesion of tuberculosis. 2. A nodule on the skin.

Tubercle Bacillus Referring to *Mycobacterium tuberculosis* or any species of *Mycobacterium* that causes tuberculosis.

Tuberculin A preparation containing tuberculoprotein, obtained from filtrate of a steamed culture of *Mycobacterium tuberculosis*. The old tuberculin is an evaporated filtrate of a steamed culture. The purified form of tuberculin called PPD (purified protein derivative).

Tuberculin Purified Protein Derivative The purified form of tuberculin obtained from the filtrate of steamed culture of *Mycobacterium tuberculosis*.

Tuberculin Reaction The classic skin test for detecting tuberculosis in which a purified protein derivative of *Mycobacterium tuberculosis* is injected subcutaneously and the area near the injection site is observed for evidence of a delayed-hypersensitivity reaction.

Tuberculoprotein Protein derived from tubercle bacilli.

Tuberculosis An infectious disease in humans caused by *Mycobacterium tuberculosis* that infects any organ or tissue of the body, but usually the lungs.

Tuberculostatic Any agent that inhibits growth of tubercle baccilus.

Tuberin (mol wt 177) An antibiotic produced by *Streptomyces amakusaensis*.

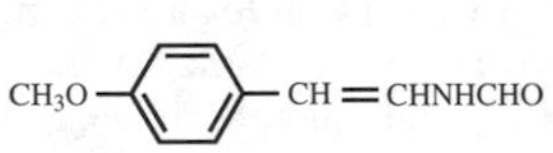

Tubersol A trade name for tuberculin purified protein used as a diagnostic antigen.

Tubulin Major protein component of microtubules that exists in microtubules as a dimer of α- and β-tubulin.

Tuftsin A tetrapeptide (threonine-lysine-proline-arginine) that enhances macrophage function.

Tularemia An acute or chronic systemic disease characterized by malaise, fever, and an ulcerative granuloma, caused by *Francisella tularensis*.

Tulobuterol (mol wt 228) A bronchodilator.

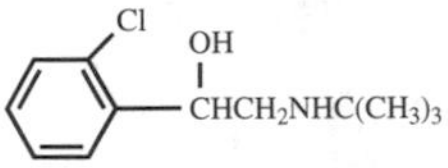

Tumefacient Causing swelling.

Tumefaction A swollen condition or a process of swelling.

Tumid Swollen.

Tumor An abnormal mass of tissue that forms within the normal tissue and caused by uncontrolled growth of transformed cells.

Tumor Angiogenesis Establishment of a vascular system within a tumor.

Tumor Angiogenesis Factor Substance released from a tumor that promotes vascularization of the mass of neoplastic cells.

Tumor Antigen See tumor-associated antigen.

Tumor Associated Antigens Cell-surface antigens found on transformed cells but not on normal cells.

Tumor Inducing Plasmid (Ti plasmid) Referring to the plasmid found in *Agrobacterium tumefaciens* that codes for proteins involved in the formation of plant tumors (galls) when this bacterium infects plants.

Tumor Infiltrating Lymphocyte A lymphoid cell that can infiltrate solid tumors.

Tumor Marker A substance produced by a tumor that can be used to monitor the size of the tumor and the effects of treatment.

Tumor Necrosis Factors Product of macrophages and lymphocytes that can exert a direct toxic effect on neoplastic cells.

Tumor Promoter A substance or agent that promotes progression of a transformed cells.

Tumor Specific Antigens Cell surface antigens that are expressed on malignant cells but not normal cell.

Tumor Suppresser Genes Genes that control unlimited cellular growth. The mutation of tumor suppresser gene causes the development of tumors.

Tumor Transformation A process in which a cell undergoes transformation from normal to the expression of the traits of a tumor cell.

Tumor Virus Virus that induces the formation of a tumor, e.g., retrovirus and polyoma virus.

Tumorigenesis The process of development of a tumor.

Tumorigenic Capable of causing tumors, e.g., carcinogen, radiation, or transformed cells.

Tumorigenicity Capacity to induce tumors in a susceptible animal.

Tums A trade name for the calcium carbonate, an antacid used to reduce total acid load in the GI tract.

TUNEL Abbreviation for terminal deoxyribonucleotidyl transferase-mediated dUTP nick end-labeling.

Tungsten A chemical element with atomic weight 184, valence 6, 5, 4, 3, and 2.

Tunicamycin A family of nucleoside antibiotics produced by *Streptomyces lysosuperificus*.

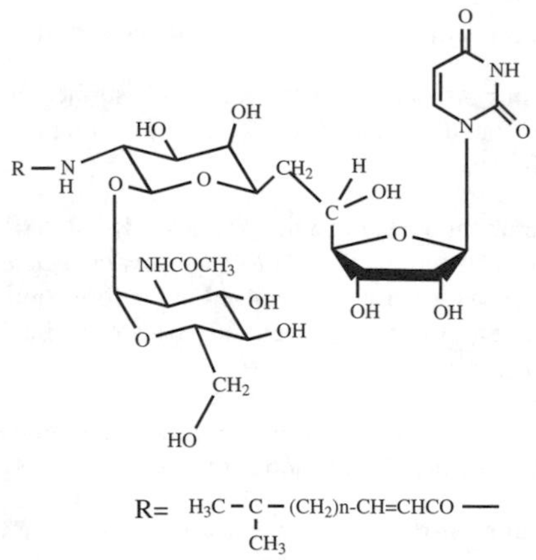

(t/u)-PA Abbreviation for (tissue/urinary)-plasminogen activator.

Turanose (mol wt 342) A disaccharide consisting of glucose and fructose.

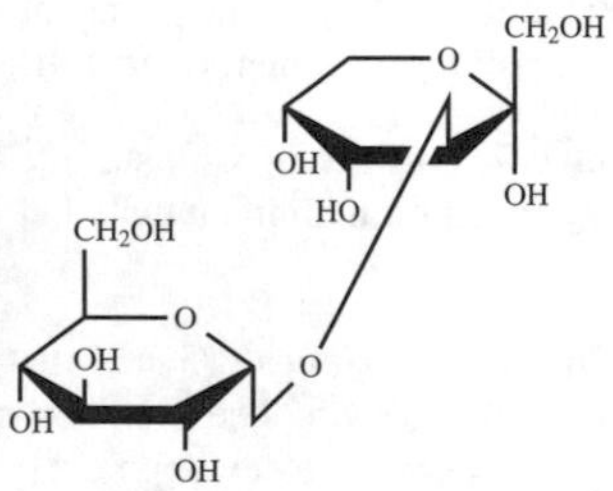

Turbidimetry A method of estimating the concentration of suspended particles in a solution, or determination of bacterial growth by measurement of the degree of opacity (or tubidity) caused by the presence of particles or bacteria.

Turbidity Referring to the opacity caused by suspended particles or bacteria in an aqueous system.

Turbidostat A system in which an optical sensing device measures the turbidity of the culture in a growth vessel and generates an electrical signal that regulates the flow of fresh medium into the vessel and the release of used medium and mature cells.

Turbinaire Decadron Phosphate A trade name for dexamethasone sodium phosphate, a glucocorticoid hormone used as an immunosuppressive and an anti-inflammatory agent.

Turgid Referring to the swelling of a cell due to the water uptake.

Turgor Pressure The pressure exerted by the contents of a cell against the cell membrane.

Turnaround Sequence The sequence between two inverted repeats in a DNA.

Turner Syndrome An abnormal human female phenotype produced by the presence of only one X chromosome (XO). Such individuals are female in phenotype but are sterile.

Turnover The continuous replacement of a substance in a metabolic pool.

Turnover Number The number of times an enzyme molecule transforms a substrate molecule per unit of time under optimal conditions (e.g., at the substrate concentration that gives maximum enzyme activity).

Tusal A trade name for sodium thiosalicylate, an analgesic and antipyretic agent that blocks the generation of pain impulses, inhibits synthesis of prostaglandin, and increases blood flow.

Tussend A trade name for a combination drug containing hydrocodone bitartrate, pseudoephedrine hydrochloride, and chlorpheniramine hydrochloride used as an antitussive agent.

Tussi-Organidin DM Liquid A trade name for dextromethorphen hydrobromide, an expectorant and antitussive agent.

Tussis Cough.

Tussive Relating to coughing or involved in coughing.

Tusstat A trade name for diphenhydramine hydrochloride, an antihistaminic agent that competes with histamine for H_1 receptors on effector cells.

Tutin (mol wt 294) Poisonous constituent of *Coriaria ruscifolia* or *C. japonica* (Coriariaceae).

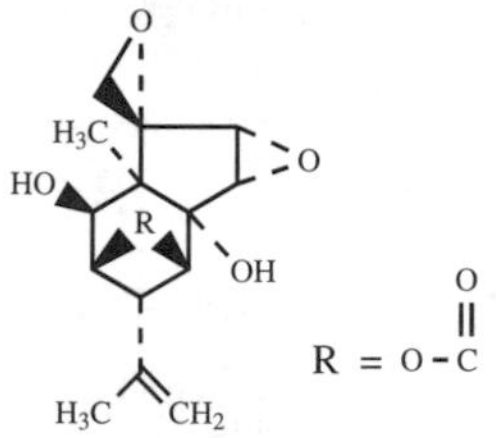

Tween A trade name for a group of nonionic detergents that are polyoxyethylene derivatives of fatty acid esters of sorbitan.

Tween 20 A trade name for a nonionic detergent, polyoxyethylene sorbitanmonolaurate.

Tween 40 A trade name for a nonionic detergent, polyoxyethylene sorbitanmono-palmitate.

Tween 60 A trade name for a nonionic detergent, polyoxyethylene sorbitanmono-stearate.

Tween 65 A trade name for a nonionic detergent, polyoxyethylene sorbitantristearate.

Tween 80 A trade name for a nonionic detergent, polyoxyethylene sorbitanmonooleate.

Tween 85 A trade name for a nonionic detergent, polyoxyethylene sorbitantrioleate.

Tween Hydrolysis Liberation of fatty acids from Tween by lipases, used for characterization and identification of certain bacteria, e.g., species of *Mycobacterium.*

Tween 20 Solution A 0.3% (v/v) Tween 20 solution in phosphate-buffered saline, pH 7.4.

Twilite A trade name for diphenhydramine hydrochloride, an antihistaminic agent that competes with histamine for H_1 receptors on effector cells.

Twisting Number of DNA The total number of turns in a dsDNA, which is the number of base pairs in the double helix divided by the number of base pairs per turn.

Two Five A Referring to a group of oligonucleotide with the formula of ppp-$(2'p5'A)_n$, which are formed from ATP and are involved in the interferon action against viral infection.

Two Gene One Polypeptide The participation of more than one gene in the production of one polypeptide or protein, e.g., formation of immunoglobulin chain by the genes for variable and constant regions.

2μ Plasmid A naturally occurring extragenomic circular DNA molecule found in some yeast cells with a circumference of 2 μ. Several types of cloning vectors are derived from this plasmid.

Two-Dimensional Chromatography A type of flat-bed chromatographic technique in which chromatographically separated substances on a flat bed (e.g., paper or thin layer) are rechromatographed following rotation of the chromatogram 90°.

Two-Dimensional Electrophoresis A type of flat-bed electrophoresis in which electrophoretically separated ionic components (e.g., proteins) on a paper or a slab gel are re-electrophoresed following rotation of the electrophoregram 90°.

Two-Carbon Fragment Referring to acetyl group.

Two-Point Cross A cross involving two loci.

Tx Abbreviation for thromboxane.

TX-100 Abbreviation for Triton X-100.

TxA_2 Abbreviation for thromboxane A_2, a potent inducer of platelet aggregation.

TxB_2 Abbreviation for thromboxane B_2.

Ty Caplets A trade name for acetaminophen, used as an analgesic and antipyretic agent that blocks the generation of pain impulses and inhibits synthesis of prostaglandin.

Ty Caps A trade name for acetaminophen, used as an analgesic and antipyretic agent that blocks the generation of pain impulses and inhibits synthesis of prostaglandin.

Ty Element A transposable element in the genome of *Saccharomyces cerevisiae.*

TyK Abbreviation for tyrosine kinase.

Tylax A trade name for a combination drug containing oxycodone and acetaminophen used as an analgesic agent.

Tylenol A trade name for acetaminophen, an analgesic and antipyretic agent that blocks the generation of pain impulses and inhibits synthesis of prostaglandin.

Tylenol PM A trade name for a combination drug containing acetaminophen and diphenhydramine used as a decongestant and an antihistaminic agent.

Tylose A methylated cellulose.

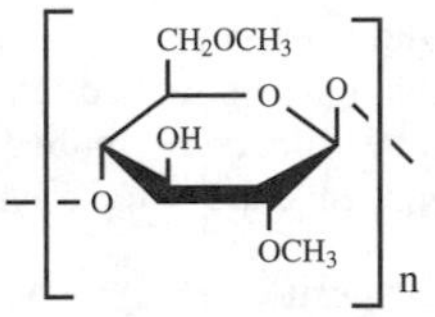

Tylosis Formation of callus.

Tylox A trade name for a combination drug containing acetaminophen and oxycodone hydrochloride, used as an antipyretic and analgesic agent.

Tymazoline (mol wt 232) A nasal decongestant.

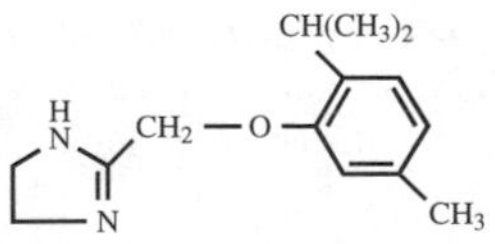

Tyndallization A process of fractional sterilization to eliminate endospores in which the material is heated to 80-100°C to kill all the vegetative organisms and followed by incubation at 37° C to allow endospores to germinate and develop into vegetative cells and then reheated to kill new vegetative cells.

Type Culture Referring to the culture of a type strain or any culture in type culture collection.

Type Culture Collections The centralized storage and preservation of all microbial species.

Type-I Analphylactic Hypersensitivity See type I hypersensitivity.

Type-I Antigen Referring to thymus-independent antigen.

Type-I DNA Topoisomerase A type of DNA topoisomerase that catalyzes the ATP-independent breakage of single-stranded DNA, followed by passage and rejoining.

Type-I Hypersensitivity An IgE-mediated hypersensitivity (also known as immediate-type hypersensitivity) caused by binding of allergen to mast cells, leading to mast cell degranulation and release of histamine.

Type-I Site-Specific Deoxyribonuclease The enzymes that catalyze the endonucleolytic cleavage of DNA to give random double-stranded fragments with 5′-terminal phosphates. ATP hydrolysis is required in the reaction.

Type-II DNA Topoisomerase A type of DNA topoisomerase that catalyzes the ATP-independent breakage of double-stranded DNA, followed by passage and rejoining.

Type-II Hypersensitivity An antibody-dependent hypersensitivity or antibody-dependent cytotoxicity, e.g., destruction of antibody-coated cells by antibody-dependent cytotoxic lymphocytes.

Type-II Site-Specific Deoxyribonuclease The enzymes that catalyze the endonucleolytic cleavage of DNA to give specific double-stranded fragments with 5′-terminal phosphates without hydrolysis of ATP.

Type-III Hypersensitivity A complex-mediated hypersensitivity or antigen-antibody complex-mediated hypersensitivity, e.g., serum sickness.

Type-IV Hypersensitivity A cell-mediated hypersensitivity (also known as delayed hypersensitivity), e.g., cell-mediated immunity to tuberculin vaccination.

Type-V Hypersensitivity A form of immediate-type hypersensitivity in which antibody reacts with cell surface components (e.g., a hormone receptor), leading to the stimulation of the cells (also known as stimulatory hypersensitivity).

Type-A Hepatitis An inflammation of liver caused by viral infection transmitted through mouth and intestine (also known as infectious hepatitis).

Type-B Hepatitis A type of hepatitis transmitted by means of punctures or transfusions of contaminated blood (also known as serum hetpatitis).

Type-B Oncoviruses Viruses of Retroviridae in which the viral particle has prominent surface spikes and electron-dense core located eccentrally within the envelope.

Type-C Oncoviruses Viruses of Retroviridae in which the viral particle has a spherical core with an electron-lucent center that buds through the plasma membrane to form an extracellular, enveloped viral particle.

Typhim Vi A trade name for typhoid Vi polysaccharide vaccine against typhoid fever.

Typhoid Pertaining or resembling typhus.

Typhoid Fever An acute infectious disease of humans caused by *Salmonella typhi* and characterized by fever and skin, intestinal, and lymphoid lesions.

Typhus Fever An acute infectious disease of humans characterized by a rash and high fever, transmitted by the body louse and fleas infected with *Rickettsia prowazekii*.

Typing Methods for characterization and distinguishing between closely related strains of a given microorganism, which exhibit minimal biological and biochemical differences.

Tyr Abbreviation for amino acid tyrosine.

Tyr Microtubules Abbreviation for microtubules rich in tyr tubulin.

Tyr Tubulin Abbreviation for tubulin dimer containing tyrosine residue at the C-terminal of its a tubulin.

Tyraminase See tyramine oxidase.

Tyramine (mol wt 137) An adrenergic agent.

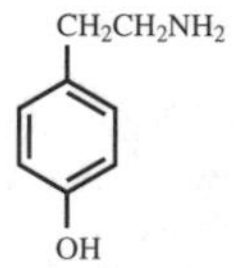

Tyramine *N*-Methyltransferase The enzyme that catalyzes the following reaction:

$$S\text{-Adenyl-L-methionine} + \text{tyramine} \rightleftharpoons S\text{-Adenyl-L-homocysteine} + N\text{-methyltyramine}$$

Tyramine Oxidase Synonym of amine oxidase.

Tyrimide A trade name for isopropamide iodide, an anticholinergic agent that blocks acetylcholine, decreases GI motility, and inhibits gastric acid secretion.

Tyrocidine A cyclic peptide antibiotic produced by *Bacillus brevis*.

[Val - Orn - Leu - D - Phe - Pro
Tyr - Glu - Asn - D - Phe - Phe]

Tyrocidine A

β-Tyrosinase See tyrosine phenol lyase.

Tyrosine (mol wt 181) An aromatic amino acid.

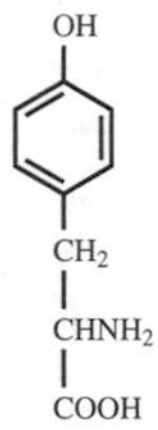

Tyrosine 2,3-Aminomutase The enzyme that catalyzes the following reaction:

L-Tyrosine
⇅
3-Amino-3-(4-hydroxyphenyl) propionate

Tyrosine-Arginine Ligase The enzyme that catalyzes the following reaction:

ATP + L-tyrosine + L-arginine
⇅
AMP + PPi + L-tyrosine- L-arginine

Tyrosine Carboxy-Lyase The systematic name for tyrosine decarboxylase.

Tyrosine Decarboxylase The enzyme that catalyzes the following reaction:

L-Tyrosine ⇌ Tyramine + CO_2

Tyrosine 3-Hydroxylase The enzyme that catalyzes the following reaction:

L-tyrosine + tetrahydrobiopterin + O_2
⇅
3,4-dihydroxy-L-phenylalanine + dihydrobiopterin + H_2O

Tyrosine Kinase Synonym of protein tyrosine kinase.

Tyrosine *N*-Monooxygenase The enzyme that catalyzes the following reaction:

L-Tyrosine + NADPH + O_2
⇅
N-Hydroxy-L-tyrosine + $NADP^+$ + H_2O

Tyrosine Phenol-Lyase The enzyme that catalyzes the following reaction:

L-Tyrosine + H_2O ⇌ Phenol + pyruvate + NH_3

Tyrosine Protein Kinase Enzyme that catalyzes transfer of the terminal phosphate group from ATP to a tyrosine residue in a target protein. Many growth factors are tyrosine-specific protein kinases. Several mitogen receptors also have tyrosine protein kinase activity.

Tyrosine Transaminase The enzyme that catalyzes the following reaction:

L-Tyrosine + α-ketoglutarate
⇅
4-Hydroxyphenylpyruvate + L-glutamate

Tyrosine-tRNA Ligase The enzyme that catalyzes the following reaction:

ATP + L-tyrosine + $tRNA^{tyr}$
⇅
AMP + PPi + L-tyrosyl-$tRNA^{tyr}$

Tyrosyl Protein Kinase Synonym of protein tyrosine kinase.

Tyrosine-tRNA Synthetase See tyrosine-tRNA ligase.

Tyrosinosis A genetic disorder in humans characterized by the excess excretion of *p*-hydroxyphenylpyruvic acid caused by the deficiency of enzyme *p*-hydroxyphenylpyruvate oxidase.

Tyrosyl-tRNA Synthetase See tyrosine-tRNA ligase.

TyrS Abbreviation for tyrosine-O-sulphate.

Ty-Sequence A transposable element in the genome of yeast.

Ty-Tab A trade name for acetaminophen, an analgesic and antipyretic agent that blocks generation of pain impulses and inhibits synthesis of prostaglandin.

Tyvelose (mol wt 148) A 3,6-dideoxy sugar found in O-antigen of Gram-negative bacteria.

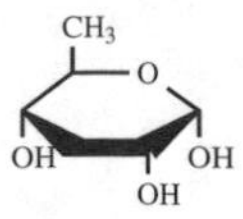

Tyzine Drops A trade name for tetrahydrozoline hydrochloride, used as a nasal agent to produce local vasoconstriction of dilated arterioles to reduce blood flow and nasal congestion.

U

U Abbreviation for 1. uracil or uridine and 2. uranium.

U1, U2, U3, U4, U5 Referring to small nuclear RNA present in the nucleus of eukaryotic cells that are so designated because of their high content of uridylic acid.

U87 Abbreviation for human astrocytoma cell line.

U251 Abbreviation for human glioma cell line.

U-937 Abbreviation for human promyelocytic leukemia cell line.

UA Abbreviation for 1. uric acid; 2. uronic acid.

UAA A stop codon on mRNA that terminates protein synthesis.

UAC A codon on mRNA that encodes the amino acid tyrosine.

UAG A stop codon on mRNA that terminates protein synthesis.

UAP Abbreviation for universal amplification primer.

UAS Abbreviation for upstream activation sequence.

UAU A genetic codon on mRNA that encodes the amino acid tyrosine.

Ub Abbreviation for ubiquitin.

Uba21I (HgiAI) A restriction endonuclease from an unidentified bacterium with the following specificity:

```
                    ↓
5′..........G(A/T)GC(A/T)C..........3′
3′..........C(T/A)CG(T/A)G..........5′
            ↑
```

Uba26I (BsmAI) A restriction endonuclease from an unidentified bacterium with the same specificity as BsmAI.

Uba44I (ApaII) A restriction endonuclease from an unidentified bacterium with the following specificity:

```
          ↓
5′..........GTGCAC..........3′
3′..........CACGTG..........5′
                ↑
```

Ubenimex (mol wt 308) A dipeptide and an antitumor antibiotic produced by *Streptomyces oliboreticuli.*

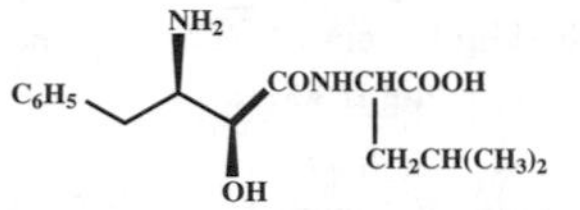

UBIP Abbreviation for ubiquitous immunopoietic polypeptide.

Ubiquinol The reduced form of ubiquinone (abbreviated as UQH_2).

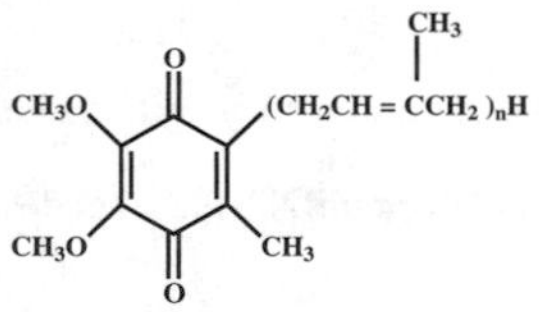

Ubiquinol-Cytochrome c Reductase The enzyme that catalyzes the following reaction:

$$QH_2 + 2 \text{ ferricytochrome c} \rightleftharpoons Q + 2 \text{ ferrocytochrome c}$$

Ubiquinone A lipid-soluble benzoquinone involved in electron transport in mitochondrial preparations (also known as coenzyme Q).

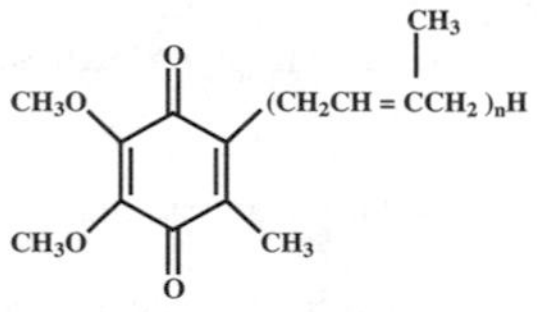

Ubiquinone Reductase The enzyme that catalyzes the following reaction:

$$\text{NADH} + \text{ubiquinone} \rightleftharpoons \text{NAD}^+ + \text{ubiquinol}$$

Ubiquitin A polypeptide (consisting of 76 amino acids) involved in ATP-dependent protein degradation (also known as ATP-dependent proteolytic factor).

Ubiquitin Activating Enzyme See ubiquitin-protein ligase.

Ubiquitin Calmodulin Ligase The enzyme that catalyzes the following reaction:

$$n\text{ATP} + n \text{ ubiquitin} + \text{calmodulin} \rightleftharpoons n\text{AMP} + n\text{PPi} + (\text{ubiquitin})_n - \text{calmodulin}$$

Ubiquitin Protein Ligase The enzyme that catalyzes the following reaction:

ATP + ubiquitin + protein-lysine
⇅
AMP + PPi + protein *N*-ubiquityllysine

Ubiquitin Thiolesterase The enzyme that catalyzes the following reaction:

Ubiquitin C-terminal thiolester + H_2O
⇅
ubiquitin + thiol

Ubiquitination Covalent linkage of ubiquitin to a protein.

UC Abbreviation for unesterified cholesterol.

UCA A codon on mRNA that encodes the amino acid serine.

UCC A codon on mRNA that encodes the amino acid serine.

UCDE Abbreviation for ubiquitin-conjugate degrading enzyme.

Ucephan A trade name for a mixture of sodium benzoate and sodium phenylacetate, used to reduce the formation of ammonia.

UCG A codon on mRNA that encodes the amino acid serine.

U-Cort A trade name for hydrocortisone, an anti-inflammatory hormone.

UCP Abbreviation for uncoupling protein.

UCU A codon on mRNA that encodes the amino acid serine.

UDN Abbreviation for ulcerative dermal necrosis.

Udo Referring to undecanoyl group CH_3-$[CH_2]_9$-CO-.

UDP Abbreviation for uridine diphosphate.

UDP-*N*-Acetylglucosamine 6-Dehydrogenase The enzyme that catalyzes the following reaction:

UDP-*N*-acetyl-D-glucosamine + $2NAD^+$ + H_2O
⇅
UDP-*N*-acetyl-2-amino-2-deoxy-D-glucuronate + 2NADH

UDP-*N*-Acetylglucosamine 4-Epimerase The enzyme that catalyzes the following reaction:

UDP-*N*-acetyl-D-glucosamine
⇅
UDP-*N*-acetyl-D-galactosamine

UDP-*N*-Acetylglucosamine 2-Isomerase The enzyme that catalyzes the following reaction:

UDP-*N*-acetyl-D-glucosamine
⇅
UDP-*N*-acetyl-D-mannosamine

UDP-*N*-Acetylglucosamine Pyro-phosphorylase The enzyme that catalyzes the following reaction:

UTP + *N*-acetyl-α-D-glucosamine 1-phosphate
⇅
PPi + UDP-*N*-acetyl-D-glucosamine

UDP-*N*-Acetylmuramate Alanine Ligase The enzyme that catalyzes the following reaction:

ATP + UDP-*N*-acetylmuramate + L-alanine
⇅
ADP + Pi + UDP-*N*-acetylmuramoyl-L-alanine

UDP-*N*-Acetylmuramate-L-Alanine Synthetase See UDP-*N*-Acetylmuramate-L-Alanine ligase.

UDP-*N*-Acetylmuramate Alanine-D-Glutamate Ligase The enzyme that catalyzes the following reaction:

ATP + UDP-*N*-acetylmuramoyl-L-alanine + glutamate
⇅
ADP + Pi + UDP-*N*-acetylmuramoyl-L-alanine-D-glutamate

UDP-*N*-Acetylmuramate Alanyl-D-Glutamyl-Lysine-D-Alanyl-D-Alanine Ligase The enzyme that catalyzes the following reaction:

ATP + UDP-*N*-acetylmuramoyl-alanyl-D-glutamyl-L-lysine + D-alanyl-D-alanine
⇅
ADP + Pi + UDP-*N*-acetylmuramoyl-L-alanyl-D-glutamyl-L-lysyl-D-alanyl-D-alanine

UDP-*N*-Acetylmuramoyl-L-Alanyl-D-Glutamate Synthetase See UDP-*N*-acetylmuramoylalanyl-D-glutamate ligase.

UDP-Arabinose 4-Epimerase The enzyme that catalyzes the following reaction:

UDP-L-arabinose ⇌ UDP-D-xylose

UDPG Abbreviation for 1. uridine diphosphate glucose and 2. uridine diphosphate galactose.

UDP-GA Abbreviation for uridine diphosphate-glucuronic acid.

UDP-Gal Abbreviation for uridine diphosphate galactose.

UDP-D-Galactopyranose Mutase The enzyme that catalyzes the following reaction:

UDP-D-galactopyranose
⇅
UDP-D-galacto-1,4-furanose

UDP-Galactose (mol wt 566) Abbreviation for uridine diphosphate galactose, an intermediate in the metabolism of galactose and synthesis of lactose.

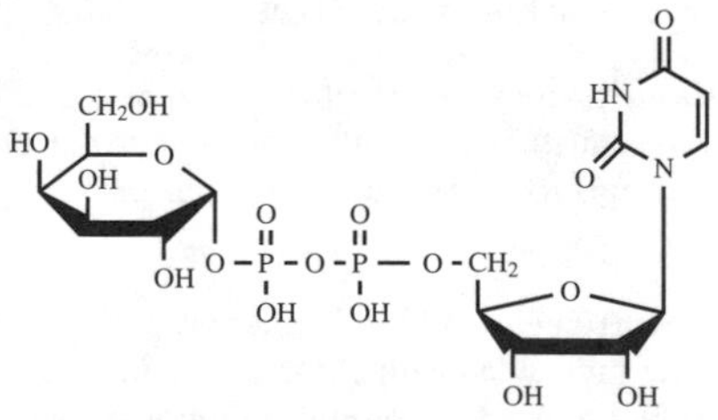

UDP-Galactose-Glucose Galactosyl Transferase Synonym of lactose synthetase.

UDP-Galactose 4-Epimerase See UDP-glucose 4-epimerase.

UDP-Galacturonate Decarboxylase The enzyme that catalyzes the following reaction:

UDP-D-galacturonate
⇅
UDP-L-arabinose + CO_2

UDP-Galacturonic Acid (mol wt 566) Complex of UDP and galacturonic acid.

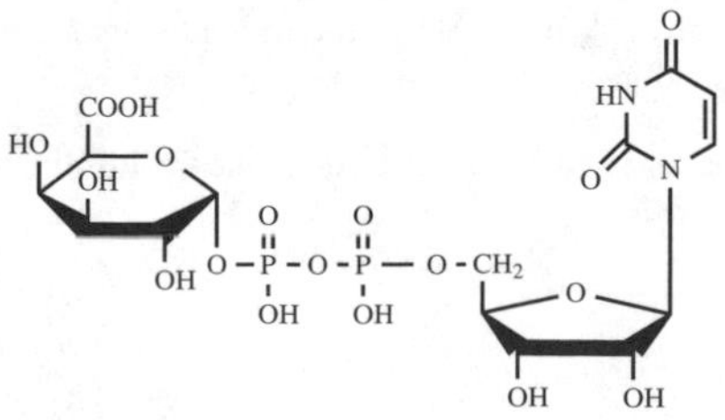

UDP-Glc Abbreviation for uridine diphosphate glucose.

UDP-GlcDH Abbreviation for UDP-glucose dehydrogenase.

UDP-Glucosamine 4-Epimerase The enzyme that catalyzes the following reaction:

UDP-glucosaminee ⇌ UDP-galactosamine

UDP-Glucose (mol wt 566) An intermediate in glucose metabolism and synthesis of lactose.

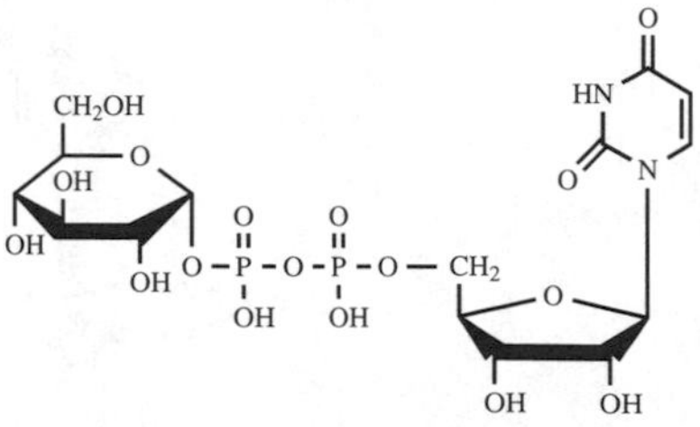

UDP-Glucose 4,6-Dehydratase The enzyme that catalyzes the following reaction:

UDP-glucose
⇅
UDP-4-dehydro-6-deoxy-D-glucose + H_2O

UDP-Glucose 6-Dehydrogenase The enzyme that catalyzes the following reaction:

UDP-glucose + 2 NAD^+ + H_2O
⇅
UDP-glucuronate + 2 NADH

UDP-Glucose 4-Epimerase The enzyme that catalyzes the following reaction:

UDP-glucose ⇌ UDP-galactose

UDP-Glucose Fructose Phosphate Glucosyltransferase Synonym of sucrose phosphate synthetase.

UDP-Glucose Glycogen Glucosyltransferase Synonym of glycogen synthetase.

UDP-Glucose-Hexose 1-Phosphate Uridylyltransferase The enzyme that catalyzes the following reaction:

UDP-glucose + α-D-galactose 1-phosphate
⇅
α-D-glucose 1-phosphate + UDP-galactose

UDP-Glucose Pyrophosphorylase See UTP-glucose-1-phosphate uridylyltransferase.

UDP-glucouronate ⇌ UDP-D-Xylose + CO_2

UDP-Glucuronate 4-Epimerase The enzyme that catalyzes the following reaction:

UDP-glucuronate ⇌ UDP-D-galacturonate

UDP-Glucuronate 5′-Epimerase The enzyme that catalyzes the following reaction:

UDP-glucuronate ⇌ UDP-L-iduronate

UDP-Glucuronic Acid (mol wt 598) An active glucuronic acid.

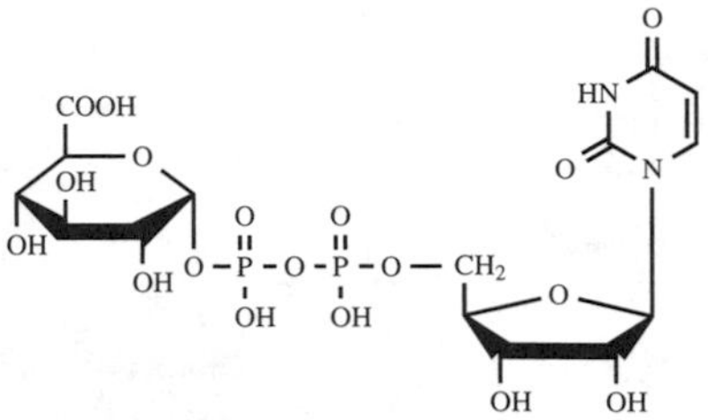

UDP-Glucuronosyltransferase Synonym of glucuronosyltransferase.

UGA A stop codon on mRNA that terminates protein synthesis.

UGC A codon on mRNA that encodes the amino acid cysteine.

UGG A codon on mRNA that encodes the amino acid tryptophan.

UGT Abbreviation for UDP-glucuronosyl transferase.

UGU A codon on mRNA that encodes the amino acid cysteine.

Ujothion (mol wt 282) An antifungal agent.

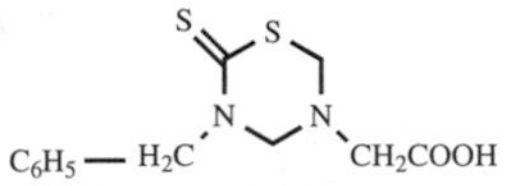

Ukidan A trade name for the enzyme urokinase.

Ulcer An inflamed lesion on the epithelium surface, e.g., skin or mucous membrane.

Ulcerate To form an ulcer.

Ulcerative Causing the formation of ulcers.

Ulcerogenic Capable of causing ulcers.

Uliginosins Antibiotics isolated from *Hypericum uliginosum* HBK, a woody herb found in Mexico and Central America.

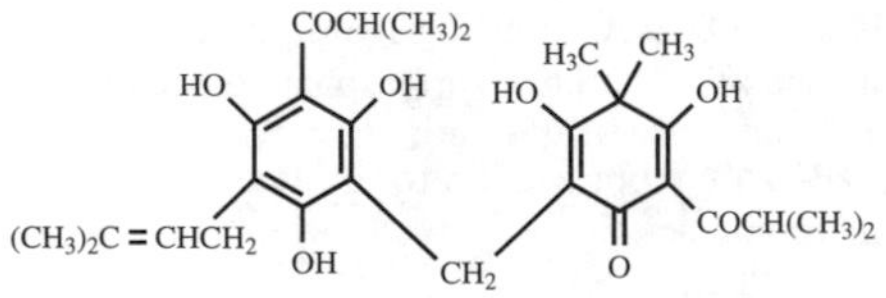

Uliginosin A

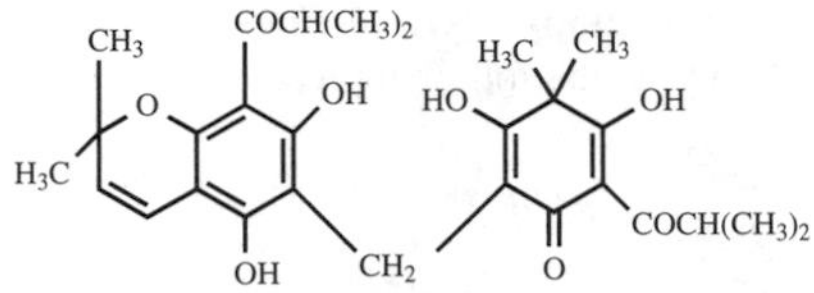

Uliginosin B

Ultiva A trade name for remifentanil hydrochloride, a narcotic analgesic agent.

Ultracef A trade name for cefadroxil monohydrate, an antimicrobial agent.

Ultracentrifugation Centrifugation performed in an ultracentrifuge for sedimentation and fractionation of macromolecules, subcellular particles, or viral particles.

Ultracentrifuge A high-speed centrifuge capable of generating a centrifugal force of approximately 500,000 × g, used for sedimentation and fractionation of macromolecules, subcellular particles, or viral particles.

Ultradol Synonym for etodolac, an anti-inflammatory agent.

Ultrafiltration A type of filtration used for concentrating macromolecules (e.g., DNA, protein) or removal of small molecules from a solution containing both micro- and macromolecules.

Ultragesic A trade name for a combination drug containing acetaminophen and hydrocodone bitartrate, used as an analgesic agent.

Ultralente Iletin I A trade name for insulin zinc suspension.

Ultram A trade name for tramadol hydrochloride, an analgesic agent.

Ultralente Insulin A trade name for insulin zinc suspension.

Ultramicroscope The microscope that uses ultraviolet light as source of illumination. It has twice the resolving power of the ordinary light microscope.

Ultramicrotome An instrument for cutting tissue or specimens into very thin sections (0.1 m or less) for electron microscopic examination.

Ultrase MT A trade name for pancrelipase, an enzyme capable of digestion of fat.

Ultrasonic A sound wave above the human ear's audibility limit of about 20,000 cycles per second.

Ultrasonic Cell Disintegration A technique used to disrupt cells by bombarding the cells with ultrasonic waves.

Ultrasonic Cleaning A method of cleaning by passing ultrasonic waves through a solution containing materials to be cleaned.

Ultrasonics The science that deals with the effects of ultrasonic waves.

Ultrasonification The use of ultrasound for disintegration of cells and tissues.

Ultrasonogram The record or profile made by ultrasonography.

Ultrasonography The examination of the body structure with ultrasound by measuring the reflection of ultrasonic waves directed into the tissue.

Ultrasound The sound vibration beyond the limit of audible frequencies (above 20,000 cycles per second).

Ultrastructure Referring to the cellular organization that is below the level of resolution of the light microscope.

Ultratard HM A trade name for insulin zinc suspension.

Ultratard MC A trade name for insulin zinc suspension.

Ultravate A trade name for halobetasol propionate, an anti-inflammatory agent.

Ultraviolet (UV) Radiation in the region of the electromagnetic spectrum from 200 nm to 390 nm.

Umbelliferone (mol wt 162) A pH-sensitive fluorescent indicator.

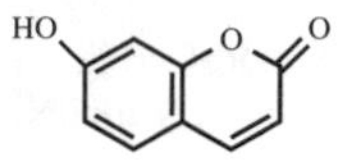

Umber Codon Referring to the UGA termination codon.

Umber Mutation A mutation that changes a normal codon to a termination codon (e.g., from a normal codon to UGA).

UMP (mol wt 324) Abbreviation for uridine monophosphate or uridine 5′-monophosphate.

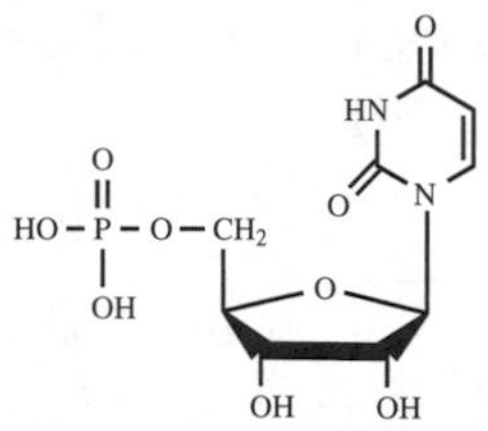

UMP Pyrophosphorylase See uracil phosphoribosyl-transferase.

uMtCK Abbreviation for ubiquitous mitochondrial creatine kinase.

umu Bacterial genes encoding proteins that are involved in error-prone SOS repair process.

Unasyn A trade name for ampicillin sodium/sulbactam sodium, an antimicrobial agent.

***unc* Mutant** A bacterial mutant that is defective in the generation of ATP from chemoisomatic coupling.

***unc* Operon** The operon involved in the synthesis of subunits of bacterial ATPase.

Uncharged tRNA A tRNA molecule that does not carry an amino acid.

Uncoating The removal of outer protein coat from a virion following infection, leading to viral replication and release of progeny virions.

Uncoating Enzyme The enzyme that catalyzes the removal of coat protein from a virus.

Uncoded Amino Acid Amino acid that occurs in protein but no codon existing on the mRNA (e.g., hydroxyproline or hydroxyline). Such amino acids are formed by post-translational modification of parent amino acids (e.g., proline to hydroxyproline, or lysine to hydroxylysine).

Uncompetitive Inhibition The inhibition of enzyme activity in which the inhibitor does not combine with free enzyme but only with one of the enzyme-substrate intermediate.

Uncoupler Any agent that uncouples electron transport from oxidative phosphorylation.

Uncoupling Agent A substance that uncouples phosphorylation of ADP from electron transport, e.g., 2,4-dinitrophenol.

Undecapeptide A peptide that consists of eleven amino acid residue.

Undecaprenol Kinase The enzyme that catalyzes the following reaction:

$$\text{ATP + undecaprenol} \rightleftharpoons \text{ADP + undecaprenyl phosphate}$$

Undecaprenyl Diphosphatase The enzyme that catalyzes the following reaction:

$$\text{Undecaprenyl diphosphate + } H_2O \rightleftharpoons \text{Undecaprenyl phosphate + Pi}$$

Undecylenic Acid (mol wt 184) A topical antifungal agent.

$$CH_2{=}CH(CH_2)_8COOH$$

Underwinding DNA The DNA produced by negative supercoiling.

Undulin A collagen associated with noncollagenous glycoprotein of the interstitial extracellular matrix.

Unequal Crossing Over Crossing over between two improperly aligned homologues, e.g., crossing over between the upstream copy in one chromosome and the downstream copy in the homologous chromosome.

Uniblue A (mol wt 506) A dye for staining protein.

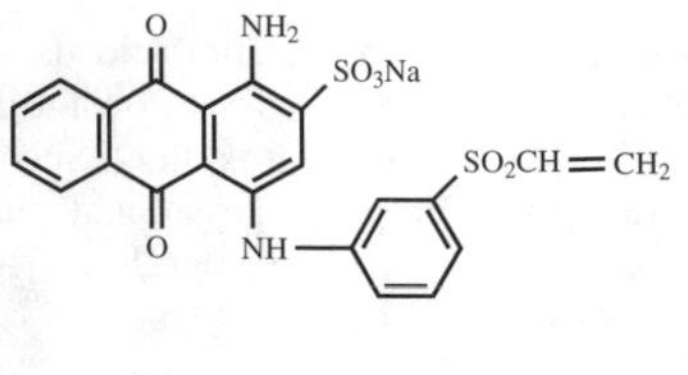

Unicellular Consisting of only one cell, e.g., bacteria and protozoa.

Unidirectional Replication The movement of a single replication fork from a given origin on a replicating DNA.

Uni-Dur A trade name for theophylline, a bronchodilator.

Unifast A trade name for phentermine hydrochloride, used as a cerebral stimulant that promotes transmission of impulses by releasing the stored norepinephrine from the nerve terminals in the brain.

Unilaminar Consisting of only one layer.

Unilateral Occurring only on one side.

Unilax A trade name for a combination drug containing docusate sodium and yellow phenolphthalein, used as a laxative.

Uniparental Inheritance The transmission of certain phenotypes from one parental type to all the progeny, e.g., extranuclear inheritance of a trait through cytoplasmic factors or organelles contributed by only one parent.

Uniparin A trade name for heparin, an anticoagulant.

Uniparin-Ca A trade name for heparin calcium, an anticoagulant.

Unipen A trade name for nafcillin sodium, a penicillinase-resistant antibiotic that inhibits bacterial cell wall synthesis.

Uniphyl A trade name for theophylline, used as bronchodilator that inhibits phosphodiesterase, increases cAMP concentration, and relaxes bronchial airway and pulmonary blood vessels.

Uniport A transport system that carries only one solute from one side of the membrane to another (distinct from cotransport).

Unipres A trade name for a combination drug containing hydrochlorothiazide, reserpine, and hydralazine hydrochloride, used as an antihypertensive agent.

Unique DNA A fragment of DNA without repeat or duplicate.

Unique Sequence A DNA sequence that is present in only one copy per haploid genome.

Uniretic A trade name for a combination drug containing moexipril and hydrochlorothiazide used as an antihypertensive agent.

Unit Complement The quantity of complement that causes lysis of 50% of sensitized red blood cells.

Unit Membrane Referring to any membrane based on the lipid bilayer model.

Uni-Tussin A trade name for guaifenesin, an expectorant.

Univalent 1. Chromosome that remains unpaired at the first division of meiosis. 2. Substances or ions with a combining power of one.

Univalent Antibody Referring to antibody or antibody fragment that contains only one binding site, e.g., Fab fragment resulting from papain digestion.

Univasc A trade name for moexipril, an angiotensin converting enzyme inhibitor used as an antihypertensive agent.

Universal Donor Referring to the individual with type O blood.

Universal Recipient Referring to the individual with type AB blood.

Univol A trade name for a combination drug containing aluminum hydroxide and magnesium carbonate as co-dried gel and magnesium hydroxide used as an antacid.

Unprimed Animal Referring to an animal that has not been exposed previously to a particular antigen.

Unsaturated 1. A solution capable of dissolving more solute at a given temperature. 2. Organic compound possessing double or triple bonds.

Unsaturated Acyl-CoA Hydratase Synonym of enoyl-CoA hydratase.

Unsaturated Fatty Acid Fatty acid that contains one or more double bonds.

Unscheduled DNA Synthesis Any DNA synthesis occurring outside the S phase of the eukaryotic cell.

Unstable Mutation A mutation that has a high frequency of reversion, e.g., a mutation caused by the insertion of a controlling element whose subsequent exit produces a reversion.

Unstacked Thylakoids Membrane-bound channels that pass from one stack of thylakoid membrane to another.

Untwisting Enzyme Referring to DNA topoisomerase.

Unusual Bases Referring to bases in addition to adenine, cytosine, guanine, and uracil, found in tRNA or other RNA.

Unwinding Proteins Proteins that unwind the DNA helix ahead of the replicating fork during DNA replication.

UOD Abbreviation for urate oxidase.

Up Mutation See up promoter mutation.

Up Promoter A promoter sequence that increases the rate of transcription initiation (also known as strong promoter).

Up Promoter Mutations A mutation at the promoter sequence that increase the frequency of initiation of transcription.

uPA Abbreviation for urokinase-type plasminogen activator.

UPAR Abbreviation for 1. urinary-type plasminogen activator receptor; 2. urokinase type plasminogen activator receptor.

U5'ppGal Abbreviation for uridine diphosphogalactose.

U5'ppGlc Abbreviation for uridine diphosphoglucose.

UPRE Abbreviation for unfolded protein response element.

Upsilon (υ) A letter in the Greek alphabet.

Upstream Located toward the 5′-end of the DNA strand lying opposite the strand that serves as the template for transcription.

Upstream Activation Site A sequence of the eukaryotic DNA (about 50-300 bp) upstream from promoter that regulates transcription.

Upward Flow The upward flow of eluent in a chromatographic column to minimize the compression of the column and allowing better flow rate.

UQ Abbreviation for ubiquinone.

UQH_2 The reduced form of coenzyme Q (dihydroubiquinone).

Ur Abbreviation for uridine.

Ura Abbreviation for uracil.

Urabeth A trade name for bethanechol chloride, a cholinergic agent that binds to acetylcholine receptors, mimicking the action of acetylcholine.

Uracel-5 A trade name for sodium salicylate, an antipyretic and analgesic agent that blocks the generation of pain impulses and inhibits synthesis of prostaglandin.

Uracil (mol wt 112) A pyrimidine base that occurs in RNA.

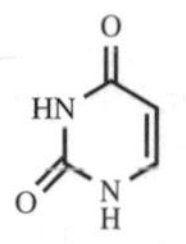

Uracil 5-Carboxylate Decarboxylase The enzyme that catalyzes the following reaction:

$$\text{Uracil 5-carboxylate} \rightleftharpoons \text{Uracil} + CO_2$$

Uracil Dehydrogenase The enzyme that catalyzes the following reaction:

$$\text{Uracil} + \text{acceptor} \rightleftharpoons \text{Barbiturate} + \text{reduced acceptor}$$

Uracil DNA Glycosidase The enzyme that catalyzes the removal of uracil resulting from deamination of cytosine in DNA.

Uracil Mustard (mol wt 252) 1. An alkylating agent that cross-links strands of DNA and interferes with transcription. 2. An antineoplastic agent.

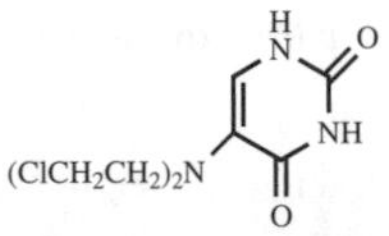

Uracil Oxidase See uracil dehydrogenase.

Uracil Phosphoribosyltransferase The enzyme that catalyzes the following reaction:

$$\text{UMP} + \text{PPi} \rightleftharpoons \text{Uracil} + \text{5-phospho-}\alpha\text{-D-ribose 1-diphosphate}$$

Uracil Pyrophosphorylase See Uracil Phosphoribosyl-transferase.

Uranium (U) A chemical element with atomic weight 238, valences 6, 5, 4, and 3. There are three naturally occurring isotopes: ^{238}U, ^{235}U, and ^{234}U.

Uranyl Acetate (mol wt 388) Reagent used for staining specimens for electronmicroscopic examination.

$$UO_2(OCOCH_3)_2$$

Urapidil (mol wt 387) An antihypertensive agent.

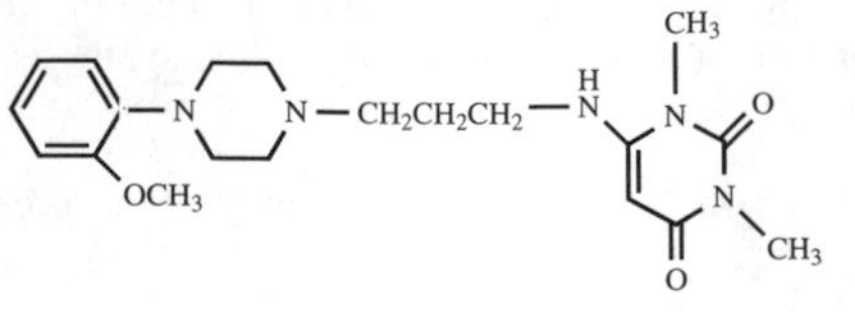

Urasal A trade name for methenamine, an anti-infective agent.

Urate A salt of uric acid.

Urate Oxidase The enzyme that catalyzes the following reaction:

$$\text{Urate} + O_2 + H_2O \rightleftharpoons \text{Allantoin} + H_2O_2 + CO_2$$

Urate Ribonucleotide Phosphorylase The enzyme that catalyzes the following reaction:

$$\text{Urate-D-ribonucleotide} + \text{Pi} \rightleftharpoons \text{Urate} + \text{D-ribose 1-phosphate}$$

Uraturia The presence of urate in the urine.

Urd Abbreviation for uridine.

UrdP Abbreviation for uridine phosphate.

Urd2'P Abbreviation for uridine 2'-phosphate.

Urd2'3'P Abbreviation for uridine 2',3'-phosphate.

Urd3'5'P Abbreviation for uridine 3',5'-phosphate.

Urd5'P Abbreviation for uridine 5'-phosphate.

Urd5'PP Abbreviation for uridine 5'-diphosphate.

UrdPPGal Abbreviation for uridine 5'-diphosphogalactose.

UrdPPGlc Abbreviation for uridine 5'-diphosphoglucose.

Urd5'PPP Abbreviation for uridine 5'-triphosphate.

Urea (mol wt 60) A diuretic agent that increases the osmotic pressure of glomerular filtrate, inhibiting tubular reabsorption of water and electrolytes.

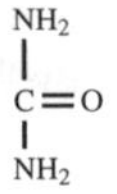

Urea Amidohydrolase See urease.

Urea Carboxylase The enzyme that catalyzes the following reaction:

$$\text{ATP} + \text{urea} + CO_2 \rightleftharpoons \text{ADP} + \text{Pi} + \text{urea 1-carboxylate}$$

Urea Cycle A metabolic pathway in the ureotelic animal for the synthesis of urea from amino groups and carbon dioxide. It occurs in the liver.

Urea Hydrogen Peroxide (mol wt 94) A disinfectant.

$$CO(NH_2)_2 \cdot H_2O_2$$

Ureacin A trade name for urea used as a diuretic agent.

Ureaphil A trade name for urea, used as a diuretic agent.

Ureaplasma A genus of urease-positive, aerophilic bacteria (family Mycoplasmataceae).

Urease The enzyme that catalyzes the following reaction:

$$\text{Urea} + H_2O \rightleftharpoons CO_2 + 2NH_3$$

Urecholine A trade name for bethanechol chloride, a cholinergic agent that binds to cholinergic receptors, mimicking the action of acetylcholine.

Urechysis Escape of urine into the tissue due to the rupture of the bladder.

Uredema An edematous condition resulting from the infiltration of urine into the tissue.

Uredepa (mol wt 219) An antineoplastic agent.

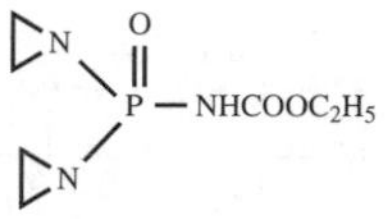

Ureidoglycolate Dehydrogenase The enzyme that catalyzes the following reaction:

$$\text{Ureidoglycolate NADP}^+ \rightleftharpoons \text{Oxalureate} + \text{NADPH}$$

Ureidoglycolate Hydrolase The enzyme that catalyzes the following reaction:

$$\text{Ureidoglycolate} + H_2O \rightleftharpoons \text{Glyoxylate} + 2\,NH_3 + CO_2$$

Ureidoglycolate Lyase The enzyme that catalyzes the following reaction:

$$\text{Ureidoglycolate} \rightleftharpoons \text{Glyoxylate} + \text{urea}$$

β-Ureidopropropionase The enzyme that catalyzes the following reaction:

$$N\text{-Carbamoyl-}\beta\text{-alanine} + H_2O \rightleftharpoons \beta\text{-Alanine} + CO_2 + NH_3$$

Ureidopropionic Acid (mol wt 132) An intermediate in pyrimidine catabolism in animals (also known a *N*-carbamyl-β-alanine).

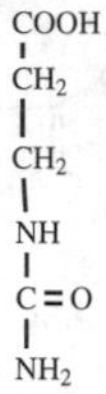

Ureidosuccinase The enzyme that catalyzes the following reaction:

$$N\text{-Carbamoyl-L-aspartate} + H_2O \rightleftharpoons \text{L-Aspartate} + CO_2 + NH_3$$

Uremia Disorder caused by the accumulation of waste products in the blood that are normally excreted in the urine, e.g., products from protein metabolism.

Ureolysis Breakdown of urea into carbon dioxide and ammonia.

Ureotelic Referring to the organisms that secrete excess nitrogen in the form of urea.

Ureter A long tube that conveys urine from the kidney to the bladder.

Ureterectomy Surgical removal of the ureter.

Ureteritis Inflammation of the ureter.

Ureterolithiasis The presence of calculus in the ureter.

Ureteroscope An instrument used to examine the lumen of the ureter.

Urethan (mol wt 89) An antineoplastic agent.

$$NH_2COOC_2H_5$$

Urethra The canal that conveys urine from the bladder to the exterior of the body.

Urex A trade name for furosemide, a diuretic agent that inhibits reabsorption of sodium and chloride.

Urex-M A trade name for furosemide, a diuretic agent that inhibits reabsorption of sodium and chloride.

URF Abbreviation for unidentified reading frame.

-uria A suffix meaning the presence of a substance in the urine.

Uric Acid (mol wt 168) A product in purine metabolism. Deposition of uric acid in the joints causes arthritis.

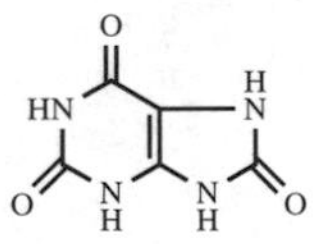

Uricase See urate oxidase.

Unicort A trade name for hydrocortsone.

Uricosuria The presence and passage of uric acid in the urine.

Uricosuric Drug Drug that relieves the pain of gout or to increase the elimination of uric acid.

Uricotelic Referring to organisms or animals that secret nitrogen in the form of uric acid.

Uridine (mol wt 244) The ribonucleoside of uracil, a component of RNA.

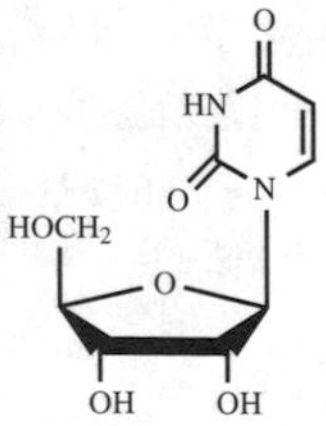

Uridine 5′-Diphosphate (mol wt 404) A diphosphate form of uracil nucleotide.

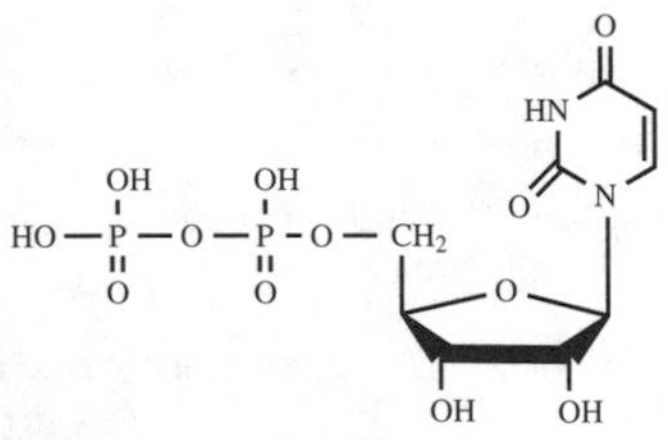

Uridine Diphosphate Glucose (mol wt 566) A coenzyme involved in conversion of galactose-1-phosphate to glucose-1-phosphate.

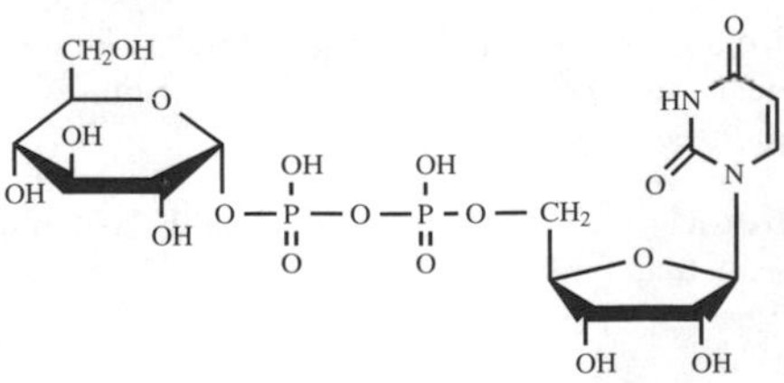

Uridine Kinase The enzyme that catalyzes the following reaction:

$$\text{ATP} + \text{uridine} \rightleftharpoons \text{ADP} + \text{UMP}$$

Uridine 5′-Monophosphate (mol wt 324) The monophosphate form of the ribonucleotide of uracil (also known as uridylic acid).

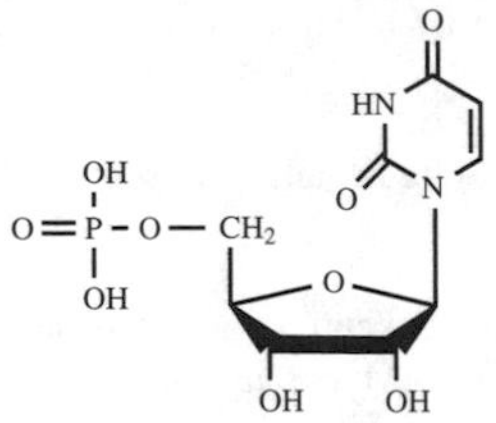

Uridine Nucleosidase The enzyme that catalyzes the following reaction:

$$\text{Uridine} + H_2O \rightleftharpoons \text{Uracil} + \text{D-ribose}$$

Uridine Phosphorylase The enzyme that catalyzes the following reaction:

$$\text{Uridine} + \text{Pi} \rightleftharpoons \text{Uracil} + \text{D-ribose 1-phosphate}$$

Uridine Ribohydrolase See uridine nucleosidase.

Uridine 5′-Triphosphate (mol wt 484) The triphosphate form of ribonucleotide of uracil.

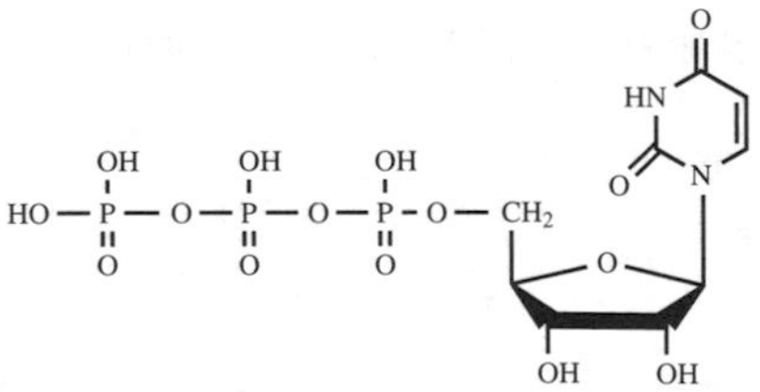

Uridon A trade name for chlorthalidone, a diuretic agent that inhibits urine excretion of sodium and water by inhibiting sodium reabsorption.

Uridrosis The presence of excessive amounts of urea in the sweat.

Uridyl Transferase See UDP-glucose-hexose-1-phosphate uridylyltransferase.

5′-Uridylic Acid (mol wt 324) Referring to the ribonucleotide of uracil, e.g., uridine 5′-monophosphate.

Uridylylation The reaction of transfer of UMP from UTP that is catalyzed by uridylyltransferase.

Urinalysis Chemical and physical analysis of urine.

Urine The fluid excreted by the kidney, stored in the bladder, and discharged through the urethra that consists of chiefly water and urea (about 96% water and 4% urea).

Urinogenous Production of urine.

Urinometer A device for measuring the specific gravity of urine.

Urisedamine A trade name for a combination drug containing methenamine mandelate and hyoscyamine, used as an antibiotic.

Urishiol Oxidase The enzyme the catalyzed the following reaction:

$$4\text{ Benzenediol} + O_2 \rightleftharpoons 4\text{-Benzosemiquinone} + 2H_2O$$

Urispas A trade name for flavoxate hydrochloride, a spasmolytic agent.

Uri-Tet A trade name for oxytetracycline hydrochloride, an antibacterial agent that binds to 30S ribosomal subunits, inhibiting bacterial protein synthesis.

Uritol A trade name for furosemide, a diuretic agent that inhibits sodium reabsorption.

Uroacidimeter A device for determining the acidity of urine.

Urobilinemia The presence of urobilin in the blood.

Urobilinogen Precursor of urobilin.

Urobilinogenuria The presence of an excess amount of urobilinogen in the urine.

Urobilins Bile pigment found in feces and urine.

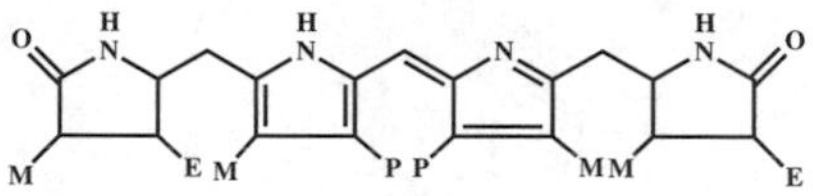

M = CH_3
E = C_2H_5
P = CH_2CH_2COOH

Urobiotic-250 A trade name for a combination drug containing oxytetracycline hydrochloride, sulfamethizole, and phenazopyridine hydrochloride, used as an antimicrobial agent.

Urocarb A trade name for bethanechol chloride, a cholinergic agent that binds to the cholinergic receptors, mimicking the action of acetylcholine.

Urochloralic Acid (mol wt 326) A sugar derivative from urine after digestion with chloral hydrate.

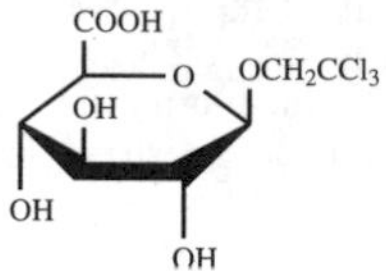

Urodine A trade name for phenazopyridine hydrochloride, an analgesic and an antipyretic agent.

Urodynia Pain and discomfort from urination.

Urofollitropin A follicle stimulating hormone used as a fertility drug.

UroGantanol A trade name for a combination drug containing sulfamethoxazole and phenazopyridine hydrochloride, used as an antibacterial agent.

Urogesic A trade name for phenazopyridine hydrochloride, an analgesic and antipyretic agent.

Urokinase The enzyme that catalyzes the specific cleavage of the peptide bond between arg-val in plasminogen to form plasmin.

Urolene Blue A trade name for methylene blue, an antiseptic dye.

Urolite Synonym for urolith.

Urolith The presence of calculus in the urinary tract.

Urolithiasis The formation of calculus in the urinary tract.

Urology The science that deals with disease, diagnosis, and treatment of the urinary tract.

Uro-Mag A trade name for magnesium oxide, used as an antacid.

Uromodulin A glycoprotein with immunosuppressive property found in normal human urine.

Uronate Dehydrogenase The enzyme that catalyzes the following reaction:

$$\text{D-Galacturonate} + NAD^+ + H_2O \rightleftharpoons \text{D-Galactarate} + NADH$$

Uronic Acid Compound derived from a monosaccharide with general formula:

$$CHO(CHOH)_nCOOH$$

Uronic Isomerase (Glucuronate Isomerase) The enzyme that catalyzes the following reaction:

$$\text{D-Glucuronate} \rightleftharpoons \text{D-Fructuronate}$$

Uronolactonase The enzyme that catalyzes the following reaction:

$$\text{D-Glucurono-6,2-lactone} + H_2O \rightleftharpoons \text{D-Glucuronate}$$

Uro-Phosphate A trade name for a combination drug containing methenamine and sodium acid phosphate, used as an antibacterial agent.

Uroplania The escape of urine into the tissue.

Uroplus DS A trade name for co-trimoxazole, an antibacterial agent that inhibits the synthesis of dihydrofolic acid from PABA.

Uroplus SS A trade name for co-trimoxazole, used as an antibacterial agent.

Uropoiesis The process of the formation of urine.

Uroquid Acid A trade name for a combination drug containing methenamine mendelate and sodium acid phosphate, used as an anti-infective agent.

Urothion (mol wt 325) A constituent of human urine.

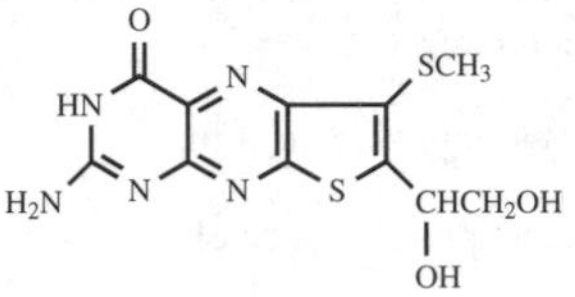

Urozide A trade name for hydrochlorothiazide, a diuretic agent.

URSO A trade name for ursodiol, a gallstone solubilizing agent.

Ursodiol (mol wt 393) A naturally occurring bile acid used to suppress hepatic synthesis and secretion of cholesterol as well as intestinal cholesterol absorption.

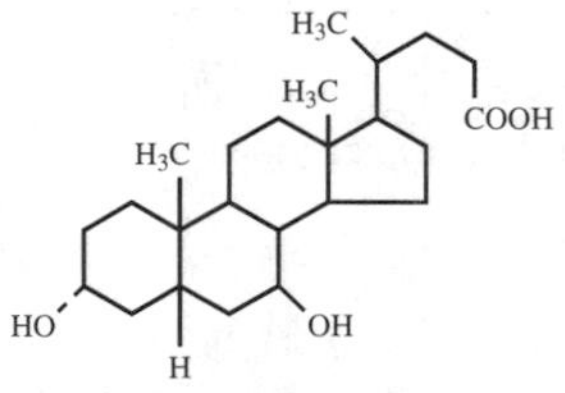

Ursofalk A trade name for ursodiol, a gallstone solubilizing agent.

Ursolic Acid (mol wt 457) An emulsifying agent.

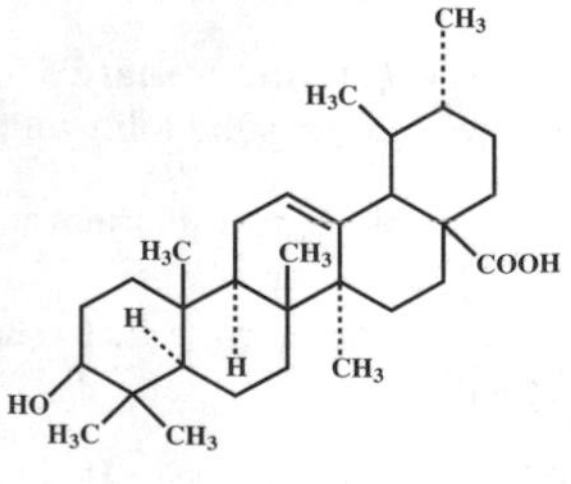

Urticant Any agent that causes itching or stinging.

Urushiol The main constituent of the irritant oil of poison ivy (*Toxicodendron radicans*). It is a mixture of derivatives of catechol with unsaturated 15-carbon or 17-carbon side chains.

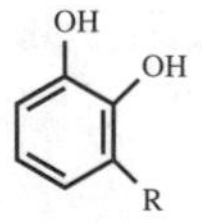

I R = $(CH_2)_{14}CH_3$

II R = $(CH_2)_7CH{=}CH(CH_2)_5CH_3$

III R = $(CH_2)_7CH{=}CHCH_2CH{=}CH(CH_2)_2CH_3$

IV R = $(CH_2)_7CH{=}CHCH_2CH{=}CHCH{=}CHCH_3$

V R = $(CH_2)_7CH{=}CHCH_2CH{=}CHCH_2CH{=}CH_2$

Uscharidin (mol wt 531) An African arrow poison and cardiac glucoside produced by *Calotropis procera* (Asclepiadaceae).

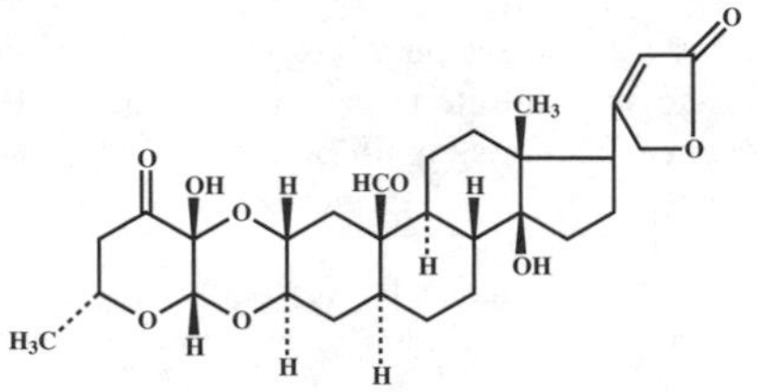

USE Abbreviation for upstream sequence element.

USF Abbreviation for upstream stimulating factor.

USF-ARE Abbreviation for upstream-stimulating factor antioxidant response element.

Usnic Acid (mol wt 344) An antibacterial agent produced by lichens.

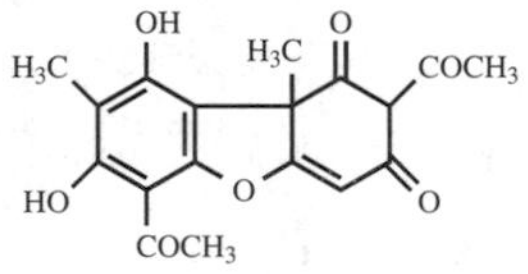

Usn-RNA Abbreviation for uridylate-rich small nuclear RNA.

USP Abbreviation for United State Pharmacopeia denoting a chemical that meets the specifications of the US Pharmacopeia.

Uticort A trade name for betamethasone benzoate, a topical corticosteroid.

Utimox A trade name for amoxicillin trihydrate, an antibiotic that inhibits bacterial cell wall synthesis.

UTP Abbreviation for uridine triphosphate, a triphosphate form of the ribonucleotide of uracil.

UTP-Glucose-1-Phosphate Uridylyltransferase The enzyme that catalyzes the following reaction:

$$\text{UTP} + \alpha\text{-D-glucose 1-phosphate} \rightleftharpoons \text{PPi} + \text{UDP-glucose}$$

3'-UTR Abbreviation for 3'-untranslated region.

UTS Abbreviation for untranslated sequence.

UUA A codon on mRNA that encodes the amino acid leucine.

UUC A codon on mRNA that encodes the amino acid phenylalanine.

UUG A codon on mRNA that encodes the amino acid leucine.

Uur960I A restriction endonuclease from *Ureaplasma bacterium* RFL 44 with the following specificity:

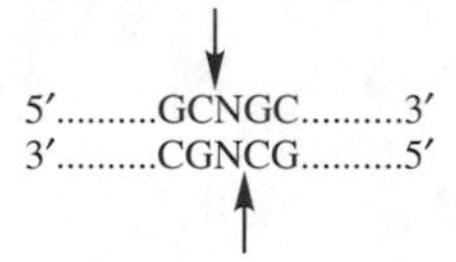

UUU A codon on mRNA that encodes the amino acid phenylalanine.

UV Abbreviation for ultraviolet.

Uvadex A trade name for methoxsalen, a pigmentation agent capable of increasing the synthesis of melanin.

UV-B Abbreviation for UV light of medium wavelength (≈280 to 320 nm).

UV-C Abbreviation for UV light of short wavelength (≈260 nm).

UVDRP Abbreviation for UV-damaged DNA-recognition protein.

UVR Abbreviation for UV radiation or ultraviolet radiation.

***uvr*A Gene** The gene that encodes subunit-A of *uvr*-ABC endonuclease in *E. coli*.

***uvr*A Protein** Product of *uvr*A gene in *E. coli*.

***uvr-ABC* Excinuclease** Abbreviation for *uvrABC* gene-encoded enzyme complex involved in the detection and removal of pyrimidine dimer (TT-dimer) in the DNA produced by UV radiation.

***uvrABC* genes** Abbreviation for genes of *uvr*-A, *uvr*-B, and *uvr*-C, that encoded enzyme complex involved in the detection and removal of pyrimidine dimer (TT-dimer) in the DNA produced by UV radiation.

UVRB Abbreviation for UV resistant gene B.

***uvr*B Gene** The gene that encodes subunit-B of *uvr*-ABC endonuclease in *E. coli*.

***uvr*B Protein** Product of *uvr*B gene in *E. coli*.

UVRC Abbreviation for UV resistant gene C.

***uvr*C Gene** The gene that encodes subunit-C of *uvr*-ABC endonuclease in *E. coli*.

***uvr*C Protein** Product of *uvr*C gene in *E. coli*.

Uzarin (mol wt 699) An antidiarrheal agent.

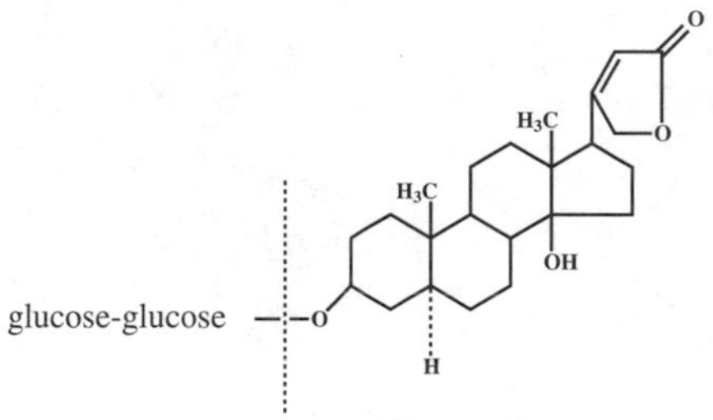

Uzarigenin

V

V Symbol for 1. amino acid valine and 2. vanadium.

^{48}V A radioactive isotope of vanadium with half-life of about 16 days.

$\bar{v}$ Symbol for particle specific volume (e.g., ml/g or volume/mass).

V Abbreviation for velocity or reaction rate.

V_0 Abbreviation for initial velocity.

V Antigens Virally induced antigens that are expressed on viruses and virus-infected cells.

V Gene The gene that encodes a variable region of an immunoglobulin chain.

Vaccenic Acid (mol wt 282) A fatty acid found in butterfat and animal fats, it acts as a growth promoting factor in rats.

$$CH_3(CH_2)_5CH{=}CH(CH_2)_9COOH$$

Vaccination The administration of an inactive or attenuated form of a pathogen to the body to prevent the infection by the virulent strain of the same pathogen, e.g., administration of vaccinia for the prevention of small pox.

Vaccine A nonvirulent antigenic preparation used for vaccination to stimulate the recipient's immune defense mechanisms against a given pathogen or toxic agent.

Vaccinia A virus of Poxviridae derived from cowpox that is a useful cloning vector for insertion of foreign DNA into eukaryotic cells.

Vacciniin (mol wt 284) A glucose derivative.

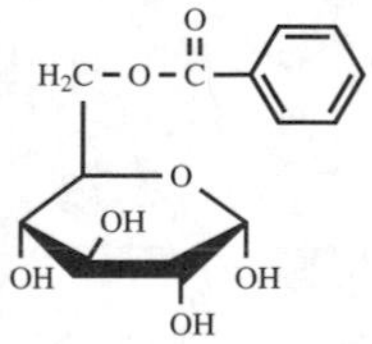

Vacuant An agent that promotes the emptying of bowels.

Vacuole A membrane-bound structure within a cell involved in digestion, secretion, storage, or excretion.

VAD Abbreviation for a combination drug containing vincristine, adriamycin, and dexamethasone.

VAFAC Abbreviation for a combination drug containing vincristine, amethopterin, FU, adriamycin, and cytoxan.

Vaginitis Inflammation of the vagina.

Vaginomycosis Fungal infection of the vagina.

Vagistat A trade name for tioconazole, an antifungal agent that alters fungal cell permeability.

Vahlkampfia A genus of amebae (order Schizopyrenida).

VAIPP Abbreviation for vaso-active intestinal polypeptide.

Val Abbreviation for the amino acid valine.

Valacyclovir (mol wt 324) An antiviral agent that inhibits viral DNA replication and deactivates viral DNA polymerase.

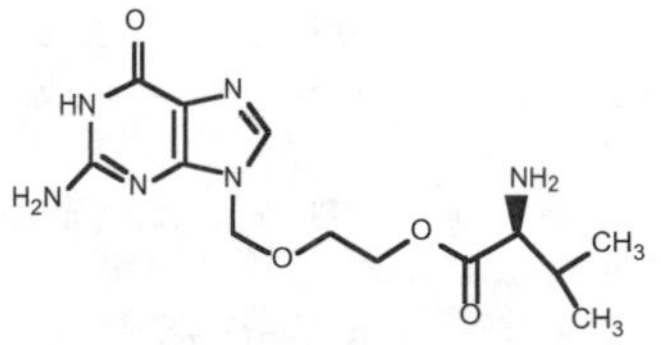

Valadol A trade name for acetaminophen, an analgesic and antipyretic agent that prevents generation of pain impulses and inhibits prostaglandin synthesis.

Valdrene A trade name for diphenhydramine hydrochloride, an antihistaminic agent that competes with histamine for H_1 receptors on effector cells.

Valence 1. The combining power or number of charges on an ion, e.g., Mg^{++} has a valence of +2; Cl^- has a valence of –1. 2. The number of antigenic determinants on an antigen or the number of binding sites on an antibody molecule.

Valence Shell The outermost shell of an atom, containing the valence electrons involved in chemical reactions of that atom.

Valergen A trade name for estradiol cypionate, which increases the synthesis of DNA, RNA, and protein.

Valerian The dried rhizome and root of *Valeriana officinalis* with sedative/hypnotic activity.

Valeric Acid (mol wt 102) A 5-carbon acid.

$$CH_3(CH_2)_3$$

Valine (mol wt 117) A protein amino acid.

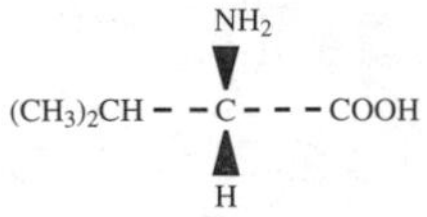

L-valine

Valine Carboxy-lyase See valine decarboxylase.

Valine Decarboxylase The enzyme that catalyzes the following reaction:

$$\text{L-Valine} \rightleftharpoons \text{2-Methylpropanamine} + CO_2$$

Valine Dehydrogenase (NADP$^+$) The enzyme that catalyzes the following reaction:

$$\text{L-Valine} + H_2O + NADP^+ \rightleftharpoons \text{3-Methyl-2-ketobutanoate} + NH_3 + NADPH$$

Valine Pyruvate Transaminase The enzyme that catalyzes the following reaction:

$$\text{Valine} + \text{pyruvate} \rightleftharpoons \text{3-Methyl-2-ketobutanoate} + \text{L-alanine}$$

Valine tRNA Ligase The enzyme that catalyzes the following reaction:

$$ATP + \text{L-valine} + tRNA^{val} \rightleftharpoons AMP + PPi + \text{L-valyl-}tRNA^{val}$$

Valine tRNA Synthetase See valine tRNA ligase.

Valinemia A genetic disorder due to a deficiency of valine aminotransferase.

Valinomycin (mol wt 1111) A depsipeptide antibiotic produced by *Streptomyces fulvissimus* that acts as an ionophore for uniport of Rb^+, K^+, Cs^+, and NH_3^+.

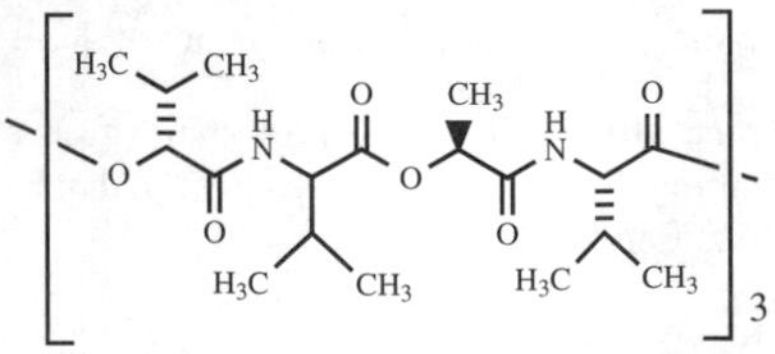

Valinyl tRNA Ligase See valine tRNA ligase.

Valisone A trade name for betamethasone valerate, a topical steroid.

Valium A trade name for diazepam, an antianxiety agent.

Valnoctamide (mol wt 143) A tranquilizer.

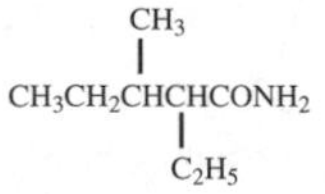

Valorin A trade name for acetaminophen, an analgesic and antipyretic agent that blocks the generation of pain impulses and inhibits prostaglandin synthesis.

Valosin A bioactive peptide that stimulates pancreatic secretion.

Valpin 50 A trade name for anisotropine methylbromide, an anticholinergic agent that blocks acetylcholine, decreases GI motility, and inhibits gastric acid secretion.

Valproic Acid (mol wt 144) An anticonvulsant.

$$(CH_3CH_2CH_2)CHCOOH$$

Valpromide (mol wt 143) An anticonvulsant.

$$(CH_3CH_2CH_2)CHCONH_2$$

Valrelease A trade name for diazepam, an antianxiety agent.

Valrubicin (mol wt 724) An inhibitor of topoisomerase.

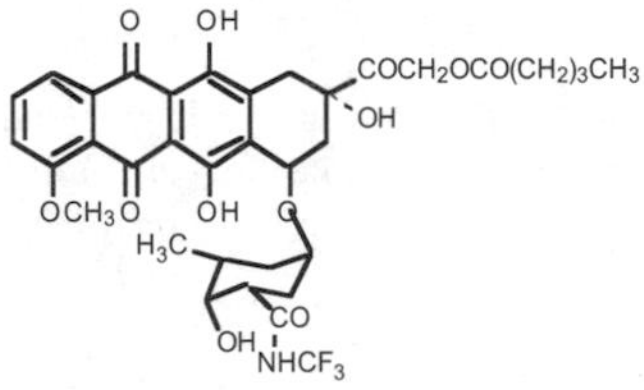

Valsartan (mol wt 436) An angiotensin II receptor blocker used as an antihypertensive agent.

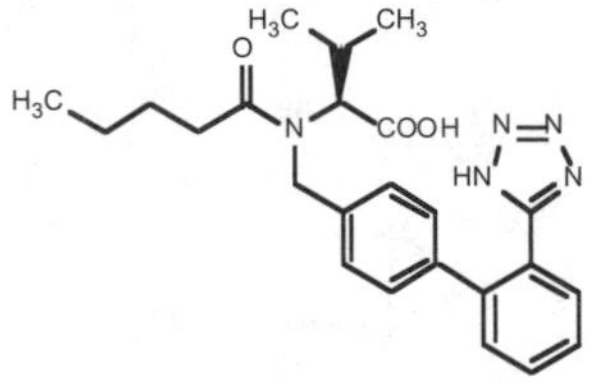

Valtrex A trade name for valacyclovir, an antiviral agent.

Valyl-tRNA Synthetase See valine tRNA ligase.

Vamate A trade name for hydroxyzine pamoate, an antianxiety agent.

VAMP Abbreviation for 1. vesicle-associated membrane protein; 2. a combination drug containing vincristine, actinomycin, methotrexate, and prednisone.

Vampirovibrio A genus of Gram-negative bacteria.

Van Abbreviation for vanadate, a chemical element.

VanI (BglI) A restriction endonuclease from *Vibrio anguillarum* with the same specificity as BglI.

Van91I A restriction endonuclease from *Vibrio anguilarum* with the following specificity:

5'........CCA(N)$_4$↓NTGG........3'
3'........GGTN↑(N)$_4$ACC........5'

van den Bergh's Test A test used to distinguish the hemolytic jaundice from congenital unconjugated hyperbilirubinemia by determining the excess bilirubin in the blood is conjugate or unconjugated.

Van der Waals Attraction The weak force of attraction between atoms, ions, and molecules. It is active only at short distances caused by the interaction of varying dipoles.

Vanadium (V) A chemical element with atomic weight 50, valence 2, 3, 4, and 5.

Vanadium Protein Referring to enzymes that contain vanadium as a prosthetic group essential for enzyme activity.

Vancenase A trade name for beclomethasone dipropionate, used as a nasal spray or nasal inhaler.

Vanceril A trade name for beclomethasone dipropionate, used as an anti-inflammatory agent.

Vancocin A trade name for beclomethasone dipropionate, used as a nasal spray or nasal inhaler.

Vancoled A trade name for vancomycin hydrochloride, an antibiotic that inhibits bacterial cell wall synthesis.

Vancomycin A glycopeptide antibiotic produced by *Streptomyces orientalis* that inhibits bacterial mucopeptide biosynthesis.

Vanillin (mol wt 152) Substance occurring in vanila.

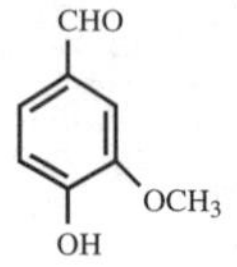

Vanilmandelic Acid (mol wt 198) A catecholamine metabolite.

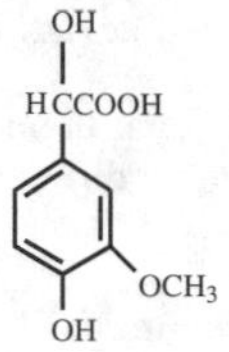

Vanitiolide (mol wt 253) A choleretic agent.

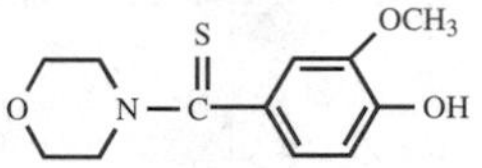

Vanoxide A trade name for a benzoyl peroxide lotion that possesses antimicrobial and comedolytic activity.

Vanponefrin A trade name for epinephrine, an alpha and beta adrenergic agonist used as a cardiac stimulant, vasopressor, bronchodilator, nasal decongestant, and antiasthmatic agent.

Vanquish A trade name for a combination drug containing aspirin, acetaminophen, caffeine, aluminum hydroxide, and magnesium hydroxide, used as an analgesic and antipyretic agent.

Vansil A trade name for oxamniquine, an anthelmintic agent.

Vantin A trade name for cefpodoxime, an antibiotic that inhibits bacterial cell wall synthesis.

VAP Abbreviation for a combination drug containing velban, actinomycin D, and platinol.

Vapocet A trade name for a combination drug containing acetaminophen and hydrocodone bitartrate, an analgesic agent.

Vapo-Iso A trade name for isoproterenol, a bronchodilator.

Vaqta A trade name for hepatitis A vaccine.

Variable Arm The extra arm in the tRNA.

Variable Domain Discrete region of an antibody molecule that is formed from the variable regions of peptide chains of the immunoglobulin and functions as binding sites for antigens.

Variable Number of Tandem Repeats Short repeat DNA sequences whose variability between individuals forms the basis for DNA fingerprinting.

Variable Region N-terminal region of the light or heavy chain of an immunoglobulin molecule that differs greatly in amino acid sequence among different immunoglobulin chains.

Variance A statistical measure of the variation of values from a central value, calculated as the square of the standard deviation.

Variant An individual organism or protein that possesses recognizable differences from the arbitrary standard type.

Varicella Referring to chicken pox.

Varicella Zoster Virus A virus of the Herpesviridae and the causal agent of chicken pox and shingles.

Variety A subgroup of species that displays an identifiable set of variations.

Variola Referring to smallpox.

Variolation Inoculation with live smallpox virus in order to protect individuals against subsequent attacks by the smallpox virus.

Varivax A trade name for varicella live virus vaccine.

Vascor A trade name for bepridil hydrochloride, a calcium channel blocking agent that inhibits calcium ion influx across the cardiac and smooth muscle cells and used as an antianginal agent.

Vascular Pertaining to vessels that conduct fluid, e.g., blood vessels.

Vascular Bundle Vascular tissue in a plant that is composed of xylem and phloem.

Vascular Bundle Sheath A sheath of cells surrounding the vascular bundle in C_4 plants.

Vascular Bundle Sheath Cells The cells in the vascular bundle sheath of the C_4 plants that contain the C_3 pathway.

Vascular Tissue Tissue for internal transport, such as xylem and phloem in plants and blood and lymph tissues in animals.

Vascularization The formation of blood vessels.

Vasculitis Inflammation of blood vessels.

Vasepam A trade name for diazepam, an antianxiety agent.

Vaseretic A trade name for a combination drug containing enalapril maleate and hydrochlorothiazide, used as an antihypertensive agent.

Vaso- A prefix meaning blood vessel.

Vasoactive Having an effect on blood vessels, usually inducing constriction or dilation.

Vasocidin Ophthalmic Ointment A trade name for an ophthalmic drug containing sulfacetamide sodium (10%), prednisolone acetate (5%), and phenylephrine hydrochloride (0.125%) used as an ophthalmic anti-infective agent.

Vasocidin-A Ophthalmic Solution A trade name for a combination drug containing naphazoline hydrochloride (0.05%) and antazoline phosphate (0.5%), used as an ophthalmic vasoconstrictor.

Vasoclear A trade name for naphazoline hydrochloride, an ophthalmic vasoconstrictor.

Vasoconstriction A reduction of the lumen of a blood vessel.

Vascoconstrictor Any agent capable of causing constriction of a blood vessel.

Vasoconstrictor Nerve A nerve that causes constriction of a blood vessel.

Vasodilan A trade name for isoxsuprine hydrochloride, a vasodilator.

Vasodilatation An increase of lumen of a blood vessel.

Vasodilator An agent capable of causing dilation of blood vessels.

Vasodilator Nerve A nerve that causes dilation of a blood vessel.

Vasoinhibitor Substance or drug that inhibits the action of vasomotor nerve.

Vasomotor Nerve Nerves that cause constriction or dilatation of blood vessels.

Vasopressin An antidiuretic peptide hormone (ADH) secreted by the posterior lobe of the pituitary that causes constriction of blood vessels. Arginine vasopressin and ornithine vasopressin have amino acid arginine and ornithine at position 8, respectively.

Vasopressin Tennate A derivative of vasopressin that promotes reabsorption of water and produces concentrated urine.

Vasopressor Any substance which causes constriction of blood vessels and increases blood pressure.

Vasoprine A trade name for isoxsuprine hydrochloride, a vasodilator.

Vasostatin A 76-residue peptide released by proteolytic cleavage of chromogranin-A. It lowers vascular tension.

Vasostimulant Agent or substance capable of causing dilatation or constriction of blood vessels.

Vasosulf A trade name for a combination drug containing sulfacetamide sodium and phenylephrine hydrochloride, used as an ophthalmic anti-infective agent.

Vasotec A trade name for enalaprilat maleate, an antihypertensive agent that inhibits the conversion of angiotensin I to angiotensin II.

Vasotocin A cyclic peptide hormone secreted by the posterior lobe of the pituitary gland of birds and reptiles with a function similar to vasopressin.

VASP Abbreviation for vasodilator-stimulated phosphoprotein.

VATD Abbreviation for a combination drug containing vincristine, ara-C, thioguanine, and daunomycin.

V-ATPase Abbreviation for vacuolar H^+-ATPase.

VB Abbreviation for a combination drug containing velban and bleomycin.

VBA Abbreviation for a combination drug containing velban, BCNU, and adriamycin.

VBAP Abbreviation for a combination drug containing vincristine, BCNU, adriamycin, and prednisone.

VBC Abbreviation for vitamin B complex.

VBM Abbreviation for a combination drug containing vinblastine, bleomycin, and methotrexate.

VBMCP Abbreviation for a combination drug containing vincristine, BCNU, melphalan, cytoxan, and prednisone.

VBP Abbreviation for a combination drug containing vinblastine, bleomycin, and platinol.

VBS Abbreviation for veronal (barbital) buffered saline.

VCA Abbreviation for viral capsid antigen.

VCAM Abbreviation for vascular cell adhesion molecule.

VCAP Abbreviation for a combination drug containing vincristine, cytoxan, adriamycin, and prednisone.

VCF Abbreviation for a combination drug containing vincristine, cytoxan, and fluorouracil.

V-cillin K A trade name for penicillin V potassium, an antibiotic that inhibits bacterial cell wall synthesis.

VCMP Abbreviation for a combination drug containing vincristine, cytoxan, melphalan, and prednisone.

VD Abbreviation of venereal disease.

VDAC Abbreviation for voltage-dependent anion channel, a porin channel in the mitochondrial outer membrane.

VDC Abbreviation for valine decarboxylase.

VDG Abbreviation for venereal disease gonorrhea.

VDH Abbreviation for valine dehydrogenase.

VDP Abbreviation for a combination drug containing vincristine, daunomycin, and prednisone.

VDR Abbreviation for vitamin D receptor.

VDRE Abbreviation for vitamin D_3 response element.

VDRL Test Abbreviation for Venereal Disease Research Laboratory Test, a test for detecting syphilis that is performed on a slide employing fixed volumes of test antigen and inactivated patient serum.

VDS Abbreviation for venereal disease syphilis.

V_e Abbreviation for elution volume, a volume of solvent required to elute the solute from a column in gel filtration chromatography.

Vector 1. A plasmid or viral DNA molecule into which another DNA molecule can be inserted without disruption of the ability of the molecule to replicate itself. 2. An organism that acts as carriers of pathogens and is involved in the spread of disease from one individual to another.

Vectorial Group Translocation The directional translocation of a given molecular or ionic species across a membrane due to the presence of a fixed pathway across the membrane.

Vectorial Synthesis Synthesis of protein destined for export from the cell.

Vectrin A trade name for minocycline hydrochloride, a semisynthetic derivative of tetracycline.

Vecuronium Bromide (mol wt 638) A neuromuscular blocker that prevents acetylcholine from binding to receptors on the muscle end plate, thus blocking depolorization.

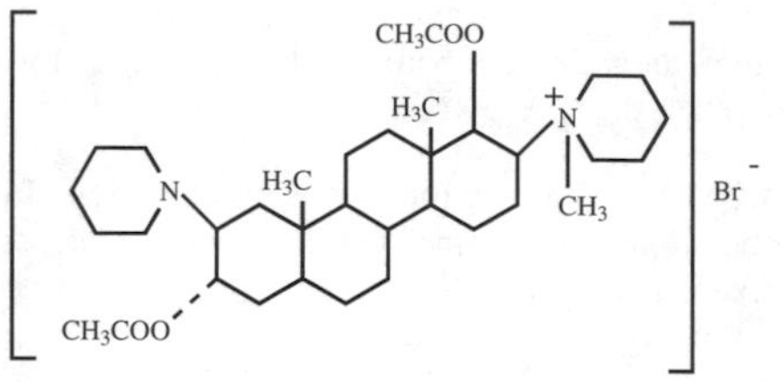

VEE Abbreviation for Venezuelan equine encephalomyelitis, an acute disease caused by an alphavirus.

Veetids A trade name for penicillin V potassium, an antibiotic that inhibits bacterial cell wall synthesis.

VEEV Abbreviation for Venezuelan equine encephalomyelitis virus.

Vegetable Black A more or less pure carbon produced by the incomplete combustion of vegetable matter or wood.

Vegetable Dye Referring to dyes that are derived from vegetable sources, e.g., logwood and indigo.

Vegetative Cells Cells that are not specialized in reproductive or dormant function.

Vegetative DNA The phage DNA that has not yet been packed into the capsid or assembled into the viral particle.

Vegetative Reproduction An asexual reproduction, e.g., cloning of plants by asexual means.

VEGF Abbreviation for vascular endothelial growth factor.

Vehicle Plasmid A plasmid containing a piece of passenger DNA, used in recombinant DNA experimentation.

Veillonella A genus of Gram-negative bacteria (family Vellonellaceae).

Veins Vessels that return blood to the heart.

Velban A trade name for vinblastine sulfate, a mitotic inhibitor.

Velbe A trade name for vinblastine, a mitotic inhibitor.

Velocity of Reaction An amount of reactant transformed or of product produced per unit of time.

Velosef A trade name for cephradine, an antibacterial agent.

Velosulin A trade name for zinc insulin.

Velsar A trade name for vinblastine, a mitotic inhibitor.

Veltane A trade name for brompheniramine maleate, an antihistaminic agent that competes with histamine for H_1 receptors on effector cells.

Vena Cava A large vein that delivers blood to the right atrium of the heart.

Venereal Disease Sexually transmitted diseases such as syphilis or AIDS.

Venereology The science that deals with venereal disease.

Venezuelan Equine Encephalomyelitis An acute disease of humans caused by an alphavirus.

Venlafaxine (mol wt 277) An antidepressant that inhibits serotonin, norepinephrine, and dopamine re-uptake leading to prolonged stimulation at neuroreceptors.

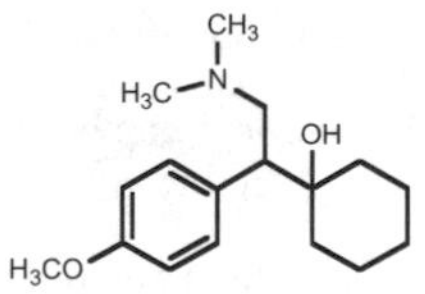

Venom A toxin secreted by snakes or other animals.

Venombin A A protease that catalyzes the selective cleavage of peptide bonds involving the COOH of arginine in fibrinogen to form fibrin and release fibrinopeptide A.

Venombin AB A protease that catalyzes the selective cleavage of peptide bonds involving the COOH group of arginine in fibrinogen to form fibrin and release fibrinopeptide A and B.

Venom Exonuclease The enzyme that catalyzes the exonucleolytic cleavage in the 3′- to 5′-direction to yield 5′-phosphomononucleotide.

Venom Phosphodiesterase See venom exonuclease.

Venom RNase The enzyme that catalyzes the two-stage endonucleolytic cleavage to yield 3′-phosphomononucleotides and 3′-phosphooligonucleotides ending in Ap or Gp with 2′, 3′-cyclic phosphate intermediate.

Venopressor Any agent that increases venous blood pressure by venoconstriction.

Vent Polymerase A thermostable DNA polymerase of *Thermococcus litoralis* isolated from a submarine thermal vent.

Ventodisk A trade name for albuterol, a beta-2 adrenergic agonist used as a bronchodilator and antiasthmatic agent.

Ventolin A trade name for albuterol, a beta-2 adrenergic agonist used as a bronchodilator and antiasthmatic agent.

Ventricle 1. A heart chamber that pumps blood into arteries. 2. Fluid-filled cavity in the brain.

Ventriculitis Inflammation of the lining of the ventricles of the brain.

Venturicidins An anti-fungal antibiotic.

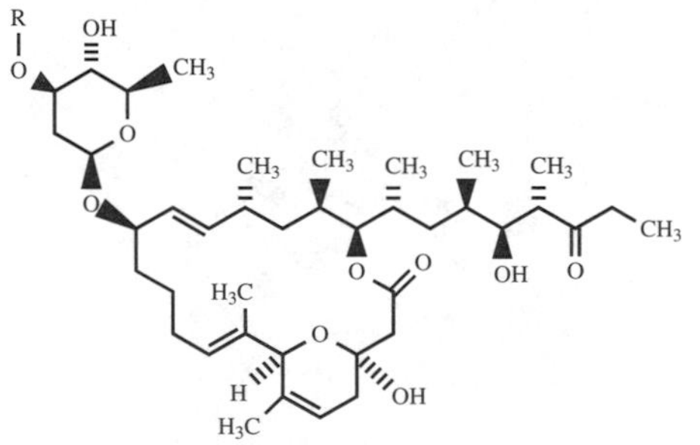

Venturicidin A : R = NH_2CO
Venturicidin B : R = H

Venule Small vein in the circulatory system.

VePesid A trade name for etoposide, a mitotic inhibitor.

Veradil A trade name for verapamil hydrochloride, an antianginal agent that inhibits calcium influx across cardiac and smooth muscle cells.

Veralipride (mol wt 383) A substance used in the treatment of menopausal disorders.

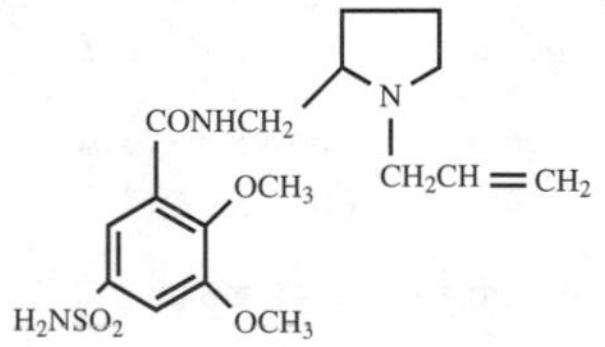

Verapamil (mol wt 455) An antianginal agent that inhibits calcium ion influx across cardiac and smooth muscle cells, thus decreasing myocardial contractility and oxygen demand, and dilating coronary arteries and arterioles.

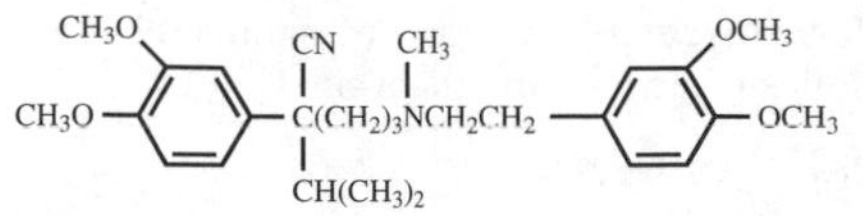

Verazide (mol wt 285) An antibacterial agent.

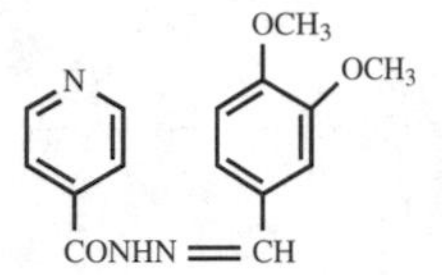

Verbascose (mol wt 829) An oligopolysaccharide isolated from roots of mullein (*Verbascum thapsus*).

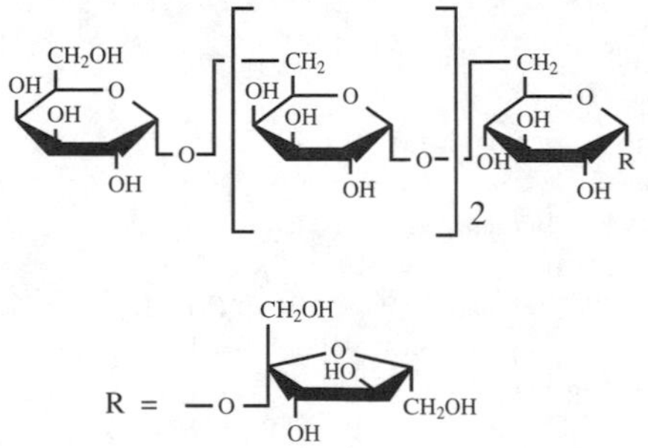

***v-erbB* Gene** A viral oncogen that encodes a truncated version of the EGF-receptor, which lacks the EGF-binding domain but retains its transmembrane segment and its protein kinase domain.

Verelan A trade name for verapamil hydrochloride, an antianginal agent that inhibits calcium influx across cardiac and smooth muscle cells and decreases cardiac oxygen demand.

Veriga A trade name for piperazine citrate, an anthelmintic agent.

Vermicide An agent capable of killing intestinal worms.

Vermicular Wormlike.

Vermifuge An agent capable of expelling intestinal worms.

Vermirex A trade name for peperazine citrate, an anthelmintic agent.

Vermizine A trade name for peperazine citrate, an anthelmintic agent.

Vermox A trade name for mebendazole, an anthelmintic agent.

Vernamycin B A group of antimicrobial antibiotics related to virginiamycin.

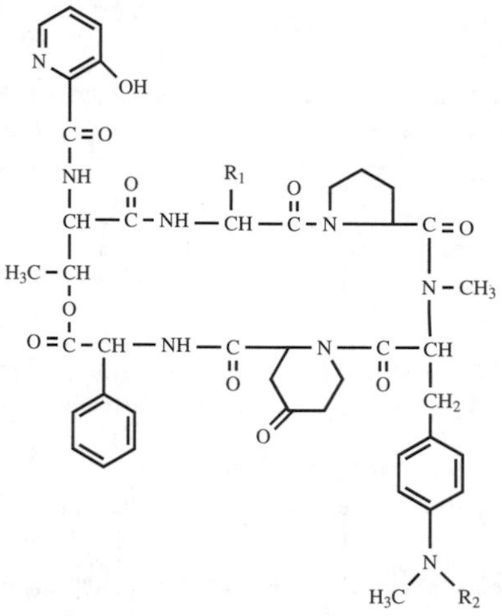

Vernamycin B_α : $R_1 = CH_2CH_3$, $R_2 = CH_3$

Vernamycin B_β : $R_1 = CH_2CH_3$, $R_2 = H$

Vernamycin B_γ : $R_1 = CH_3$, $R_2 = H$

Vernolate (mol wt 203) A herbicide.

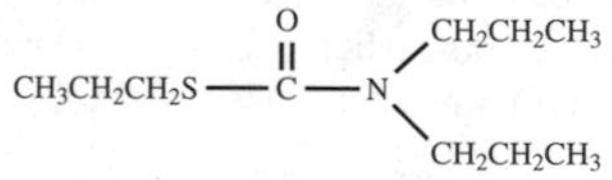

Vernolepin (mol wt 276) An antineoplastic agent.

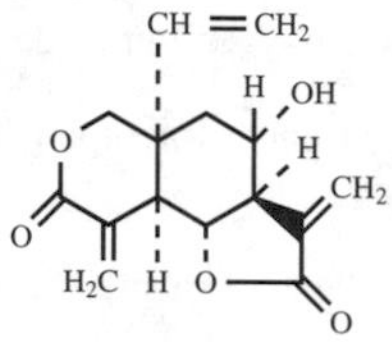

Vero An established cell line derived from kidney of an African green monkey.

Veronal A trade name for barbital.

Verrucarins A group of antifungal substances with cytostatic activity.

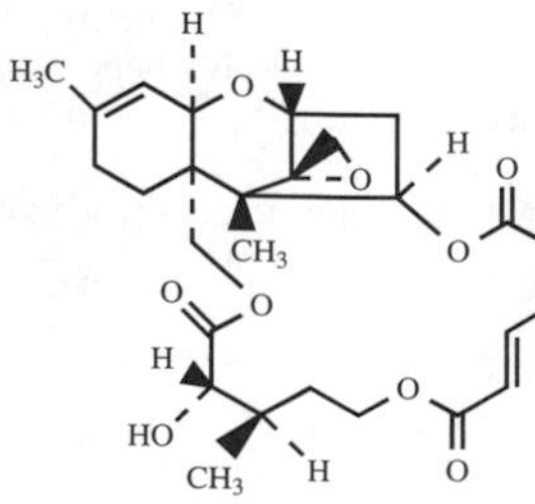

verrucarin A

Verrucosis A condition characterized by the presence of multiple warts or wartlike structure.

Versed A trade name for midazolam hydrochloride, a sedative–hypnotic agent.

Versene A trade name for sodium salt of EDTA.

Versican A protein that plays an important role in the intracellular signaling.

Vertical Transmission The transmission of a pathogen or disease from parent to offspring via egg, placenta, or genetic inheritance.

Verticillins An antibiotic produced by fungus *Verticillium*.

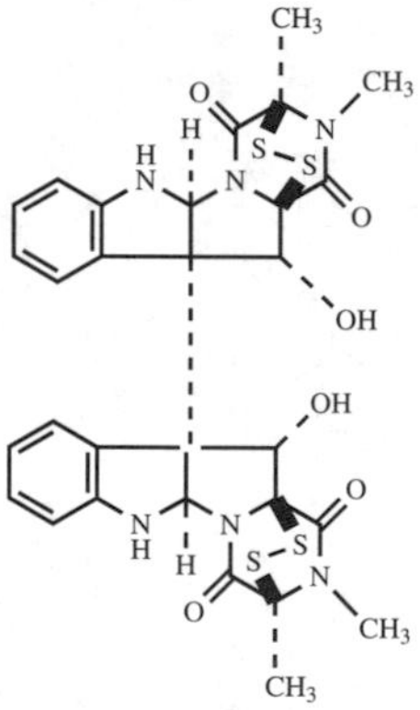

verticillin A

Very High Density Lipoprotein (VHDL) The plasma lipoprotein with a density above 1.21 g/ml that consists of approximately 57% protein, 21% phospholipid, 17% cholesterol and cholesterol esters, and 5% triglycerides.

Very Low Density Lipoprotein (VLDL) The plasma lipoprotein with a density of 0.95–1.006 g/ml that consists of approximately 10% protein, 18% phospholipid, 22% cholesterol and cholesterol esters, and 50% triglycerides.

Vesanoid A trade name for tretinoin, an antineoplastic agent.

Vesication Formation of blisters.

Vesicles Small, membrane-bound structure formed by budding from the membrane.

Vesiculin A protein (about 10,000 daltons) found in the synaptosome vesicle.

Vesiculovirus A virus of the Rhabdoviridae, e.g., vesicular stomatitis virus that causes an acute, infectious disease in animals.

Vesprin A trade name for triflupromazine hydrochloride, a dopaminergic blocking agent used as an anti-psychotic, anti-emetic, and anti-anxiety agent.

Vestigial Small and nonfunctional biological structure.

Vetrabutine (mol wt 313) A uterine relaxant.

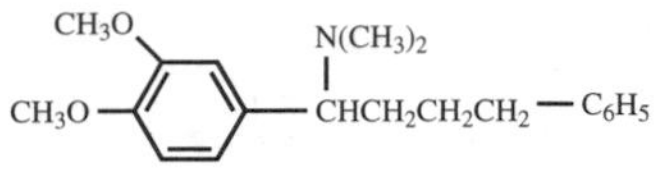

VFA Abbreviation for volatile fatty acid.

***v-fos* Gene** A retrovirus oncogen.

VH Abbreviation for viral hepatitis.

V_H Symbol for variable region of the heavy chain of an immunoglobulin molecule.

VhaI (HaeIII) A restriction endonuclease from *Vibrio harveyi* with the same specificity as HaeIII.

Vha464I A restriction endonuclease from *Vibrio harveyl* 464 with the following specificity:

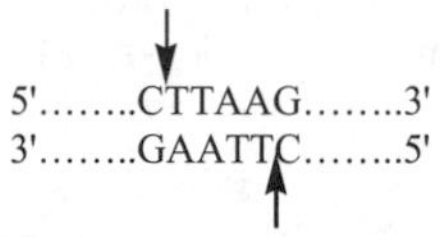

V-H⁺ATPase Abbreviation for vacuolar proton pumping ATPase.

VHDL Abbreviation for very high density lipoprotein.

V-H⁺PP Abbreviation for vacuolar proton pumping pyrophosphatase.

Vi Symbol for velocity of an enzyme-mediated reaction in the presence of an inhibitor.

Vi Antigen Polysaccharide cell-surface antigen in some enterobacteria, e.g., *E. coli* and *Citrobacter freundii.*

Viability The ability to grow and to reproduce.

Viable Capable of living.

Viable Plate Count The method for the enumeration of bacteria whereby serial dilutions of a bacterial suspension are plated onto a suitable solid growth medium, the plates incubated, and the number of colony-forming units counted.

Viagra A trade name for sildenafil citrate, an impotence agent that prevents the breakdown of cGMP by phosphodiesterase, leading to the increased cGMP levels and prolonged smooth muscle relaxation, promoting the flow of blood into the corpus cavernosum.

Vibramycin A trade name for doxycycline, an antibiotic that binds to 30S ribosomal subunits, inhibiting bacterial protein synthesis.

Vibramycin IV A trade name for the antibiotic doxycycline hydrochloride.

Vibra-Tabs A trade name for the antibiotic doxycycline hydrochloride.

Vibrio A genus of Gram-negative bacteria of the family Vibrionaceae.

Vibrio cholerae The causal agent of cholera that produces cholera toxin (Vibrionaceae).

Vibriosis Infection or disease of humans caused by *Vibrio* spp.

Vibriostatic Any agent that is capable of inhibiting or killing *Vibrio.*

Vicianose (mol wt 312) A disaccharide consisting of arabinose and glucose.

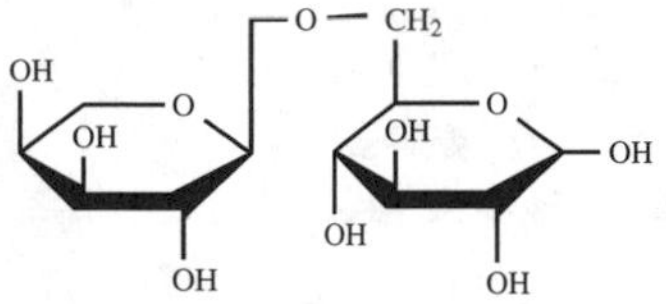

Vicilin A major storage protein in the seeds of *Pisum sativum* (pea).

Vicks DayQuil A trade name for a combination drug containing dextromethorphan, pseudoephedrine, acetaminophen, and guaifenesin used as a decongestant and an antihistaminic agent.

Vicks Formula 44 A trade name for a combination drug containing dextromethorphan, chlorpheniramine, and alcohol used as a decongestant and an antihistaminic agent.

Vicks Nyquil A trade name for a combination drug containing dextromethorphan, doxylamine, pseudoephedrine, and acetaminophen used as a decongestant, an antitussive, and an antihistaminic agent.

Vicodin A trade name for a combination drug containing acetaminophen and hydrocodone bitartrate, an analgesic agent.

Vicoprofen A trade name for a combination drug containing hydrocodone and ibuprofen used as an analgesic agent.

Victoria Blue B (mol wt 506) A dye.

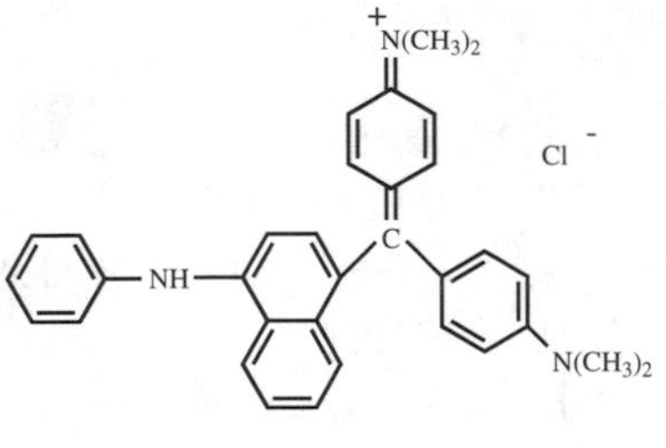

Victoria Blue R (mol wt 458) A dye.

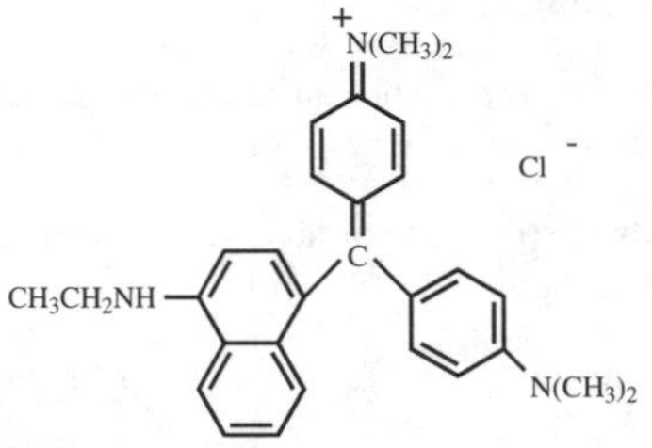

Vidarabine (mol wt 285) A purine nucleoside and an anti-infective agent (antiviral) that interferes with DNA synthesis.

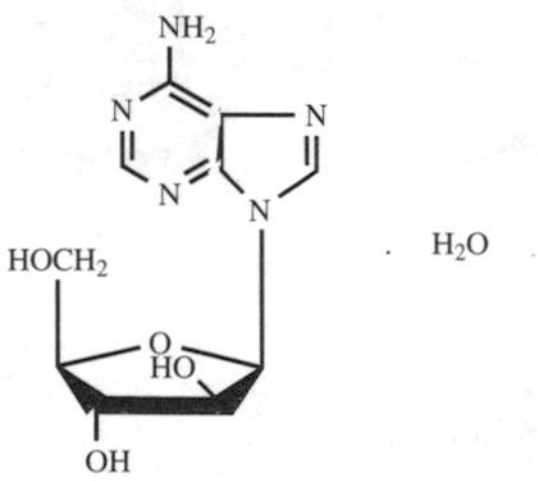

Videx A trade name for didanosine, an antiviral agent that inhibits replication of HIV.

VIG Abbreviation for vaccinia immune globulin.

Vigabatrin (mol wt 129) An anticonvulsant.

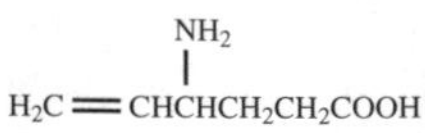

Villi Plural of villus.

Villikinin A gastrointestinal hormone that regulates the movement of the villi.

Villin A protein from intestinal villi.

Villus A small, vascular, hairlike projection from the surface of a membrane, e.g., inner mucous membrane of the intestine.

Vimentin Major protein component of the intermediate filaments found in the connective tissue and other cells of mesenchymal origin.

Viminol (mol wt 363) An analgesic agent.

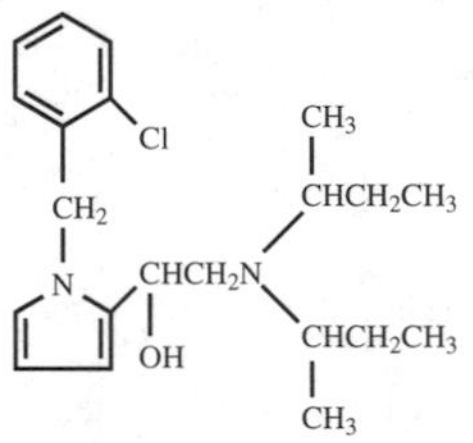

Vinbarbital Sodium (mol wt 246) A sedative–hypnotic agent.

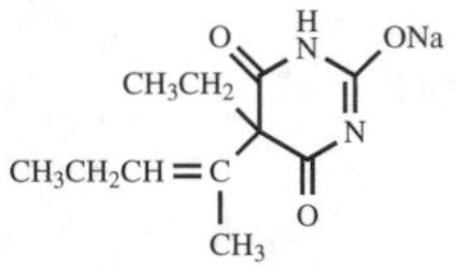

Vinblastine (mol wt 811) An antineoplastic agent that arrests mitosis and blocks cell division.

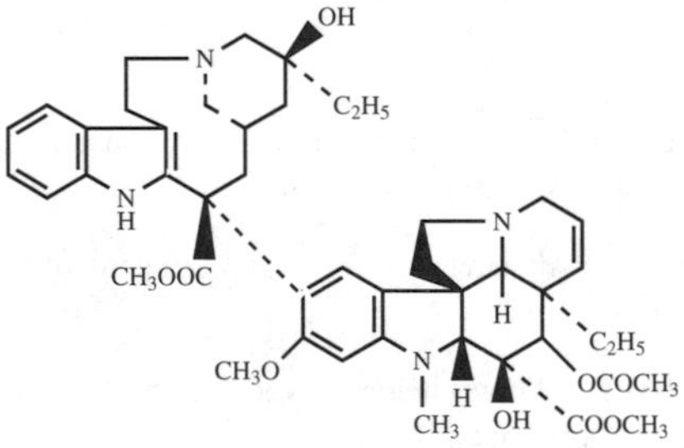

Vinca Alkaloids Alkaloids products of Madagascan periwinkle (*Vinca rosea*), e.g., vinblastin.

Vincamine (mol wt 354) A vasodilator.

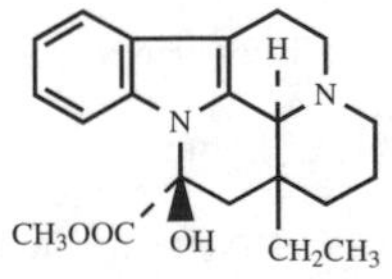

Vincasar PFS A trade name for vincristine sulfate, an antineoplastic agent that inhibits mitosis.

Vincent's Angina An ulcerative disorder of the tonsils and gums caused by bacteria (*Treponema vincentii*).

Vincent's Powders A trade name for aspirin, an analgesic and antipyretic agent that blocks the generation of pain impulses and inhibits prostaglandin synthesis.

Vinclozolin (mol wt 286) An antifungal agent.

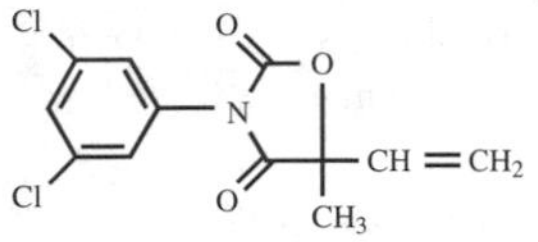

Vincristine (mol wt 825) An antitumor alkaloid isolated from *Vinca rosea* (Apocynaceae).

Vinculin A protein involved in linking the actin cytoskeleton to the cell membrane; this protein is highly phosphorylated by tyrosine kinase from Rous sarcoma transformed cells.

Vindesine (mol wt 754) A mitotic inhibitor that arrests mitosis at metaphase, blocking cell division.

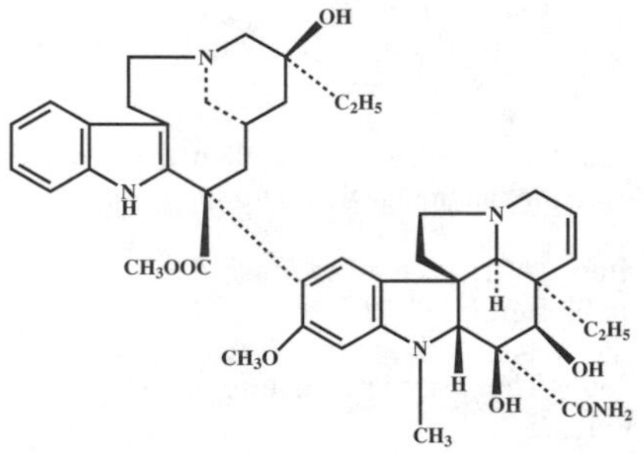

Vindoline (mol wt 457) An alkaloid from the leaves of *Vinca rosea* (Apocynaceae).

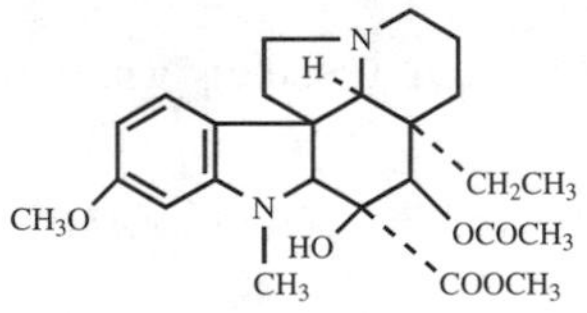

Vinorelbine (mol wt 779) A mitotic inhibitor used as an antineoplastic agent.

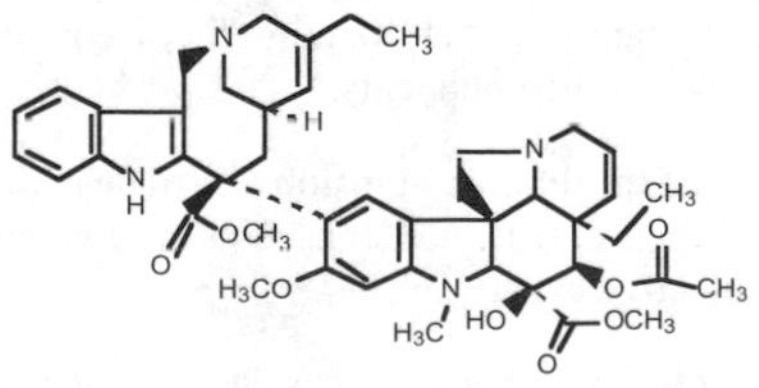

Vinpocetine (mol wt 350) A cerebral vasodilator.

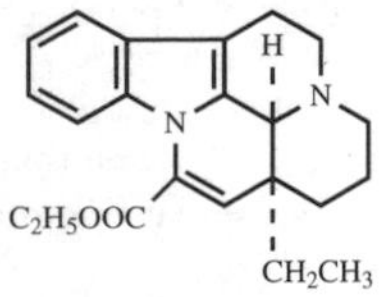

Vintiamol (mol wt 413) A vitamin and enzyme cofactor.

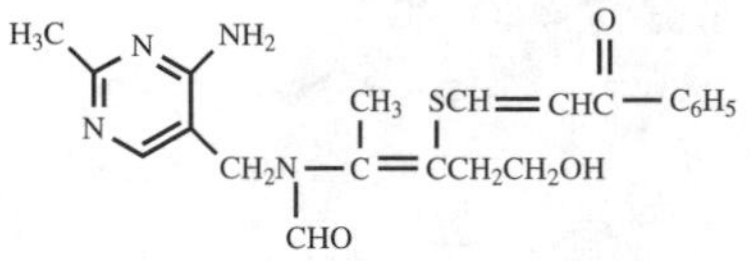

Vinyl Chloroformate (mol wt 107) A reagent for amino groups protection in the peptide synthesis.

$CH_2CHOCOCl$

Vioform A trade name for iodochlorhydroxyquin used as a local antifungal agent.

Viokase Powder A trade name for pancrelipase, used as a digestant.

Viokase Tablets A trade name for pancrelipase, used as a digestant.

Violacein (mol wt 343) A pigment formed from tryptophan by *Chromobacterium.*

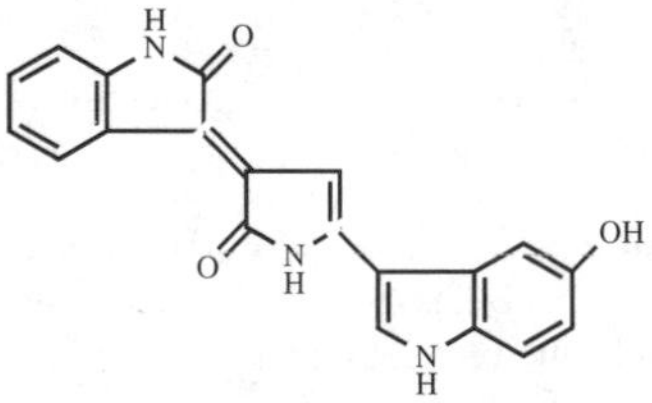

Viomycin Kinase The enzyme that catalyzes the following reaction:

$$\text{ATP} + \text{viomycin} \rightleftharpoons \text{ADP} + O\text{-phosphoviomycin}$$

Viomycin Phosphotransferase See viomycin kinase.

Viomycin Sulfate (mol wt 784) A peptide antibiotic produced by *Streptomyces peniceus*.

Vioxx A trade name for rofecoxib, a nonsteroidal anti-inflammatory agent.

VIP Abbreviation for vasoactive intestinal polypeptide, a gastrointestinal hormone capable of relaxation of the vascular smooth muscle and stimulation of secretion of insulin and cAMP formation in the intestine.

Vipoma A tumor of islet cells of the pancreas that produces vasoactive intestinal polypeptide.

VIPST Abbreviation for vasoactive intestinal polypeptide secreting test.

Viquidil (mol wt 324) A vasodilator and antiarrhythmic agent.

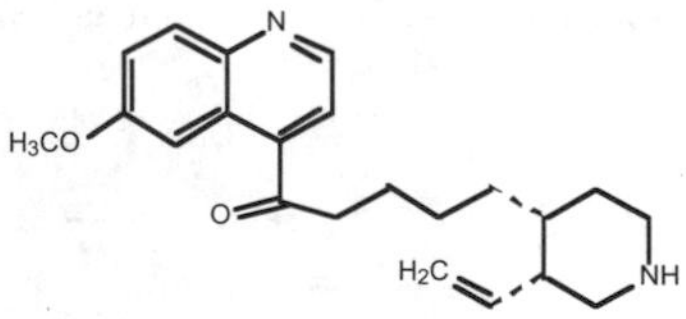

Vira-A A trade name for vidarabine monohydrate, an antiviral agent that interferes with DNA synthesis.

Viracept A trade name for nelfinavir mesylate, an antiviral agent that inhibits HIV-1 protease resulting in production of immature, non-infectious virus.

Viral Antigen Referring to viral encoded protein antigens.

Viral Hepatitis Referring to hepatitis A, hepatitis B, and non-A, non-B hepatitis.

Viral Oncogene (*v-onc*) A cancer-causing gene present in the viral genome, e.g., *src* gene of Rous sarcoma virus.

Viral Transformation Malignant transformation induced by a virus.

Viramune A trade name for nevirapine, an antiviral agent that binds to HIV-1 reverse transcriptase and blocks the replication HIV.

Virazole A trade name for ribavirin, an antiviral agent.

Viremia Presence of virus in the blood.

Virginiamycin An antibiotic complex produced by *Streptomyces virginiae* that consists of two principal components, virginiamycin M_1 and S_1.

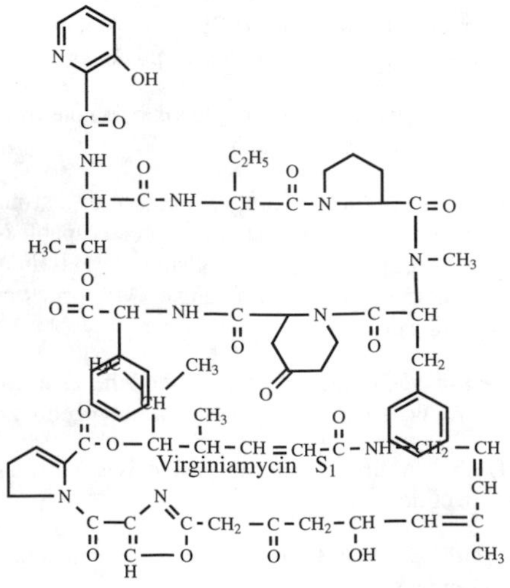

Virginiamycin M_1

Virgin Lymphocyte A lymphocyte that has not yet encountered the antigen for which it possesses receptors.

Viricides Chemicals capable of inactivating viruses.

Viridicatin (mol wt 237) An antibiotic produced by *Penicillium viridicatum.*

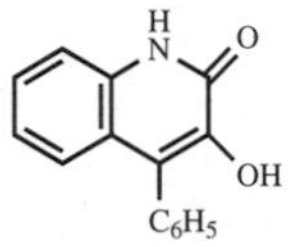

Viridin (mol wt 352) An antifungal antibiotic produced by *Gliodadium virens*.

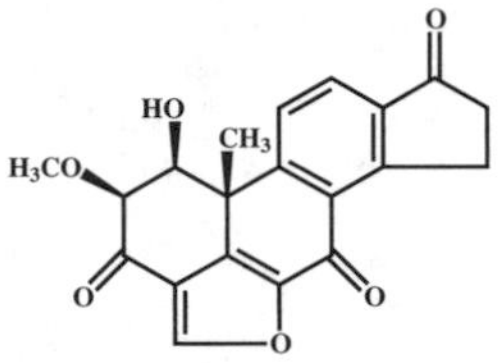

Viridium A trade name for phenazopyridine hydrochloride, an analgesic and antipyretic agent.

Virilon A trade name for methyltestosterone, an anabolic steroid.

Virion The complete mature infectious virus particle.

Viroid Infectious subviral particles consisting of nucleic acid without a protein capsid, e.g., potato spindle tuber virus.

Virology Science that deals with viruses and viral diseases.

Viropathic Tissue damage caused by a viral infection.

Viropexis The process by which viruses enter cells.

Viroplasm A type of viral inclusion body, an electron dense structure in the cytoplasm of the virus-infected cell that contains the accumulated virus and viral components.

Viroptic Ophthalmic Solution A trade name for 1% solution of trifluridine, used as an ophthalmic anti-infective agent.

Virosome A liposome that incorporates viral proteins.

Virulence The degree of pathogenicity or ability to cause disease.

Virulence Factors Factors that enhance the pathogenicity of disease-causing microorganisms, allowing them to invade tissue and to disrupt normal body functions.

Virulent Bacteriophage Bacteriophage that always undergoes lytic growth in a host cell, leading to the production of progeny bacteriophages and resulting in death and lysis of the host cell.

Virulent Pathogen An organism with ability to cause disease.

Viruria The presence of viruses in the urine.

Virus An infectious intracellular parasite capable of reproduction only inside living cells, consisting either of DNA or RNA (never both), which is usually surrounded by the protein capsid. The capsid in some viruses is enclosed in a membranelike envelope.

Visceral Muscle Smooth muscle found in the walls of the digestive tract, bladder, arteries, and other internal organs.

Viscerol A trade name for dicyclomine hydrochloride, an anticholinergic agent.

Viscometer A device used for measuring the viscosity of a fluid.

Viscose A colored aqueous solution of sodium cellulose xanthogenate.

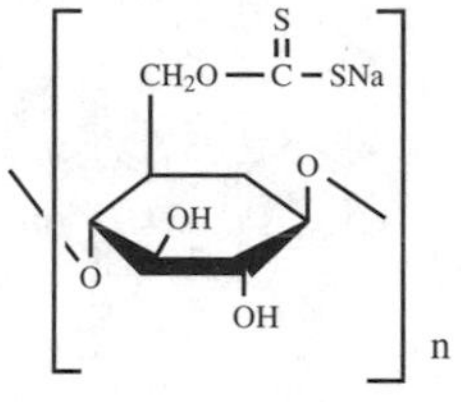

n = 200 - 400

Viscosimeter Variant spelling for viscometer.

Viscosin A peptide antibiotic produced by *Pseudomonas viscosa*.

Viscosity The physical property of a fluid that is determined by the size, shape, and property of the molecules in the fluid.

Viscotaxis A taxis in which the stimulus is a change in viscosity.

Visible Light Referring to the wavelength in the range of 400–750 nm.

Visible Spectrum Referring to the electromagnetic spectrum in the range of 400–750 nm.

Visine A trade name for tetrahydrozoline hydrochloride, an ophthalmic vasoconstrictor.

Visinin A calcium-binding protein specific to cone photoreceptors.

Visken A trade name for pindolol, an antihypertensive agent that blocks the response of beta stimulation.

Visnadine (mol wt 388) A coronary vasodilator.

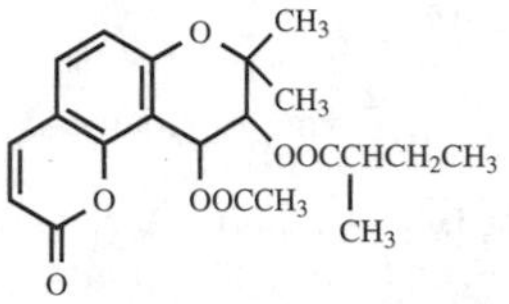

Vistacon A trade name for hydroxyzine hydrochloride, an antianxiety agent.

Vistaject A trade name for hydroxyzine hydrochloride, an antianxiety agent.

Vistaquel A trade name for hydroxyzine hydrochloride, an antianxiety agent.

Vistaril A trade name for hydroxyzine hydrochloride, an antianxiety agent.

Vistazine A trade name for hydroxyzine hydrochloride, an antianxiety agent.

Vistrax-10 Tablets A trade name for a combination drug containing oxyphencyclimine hydrochloride and hydroxyzine hydrochloride, used as an anticholinergic agent.

Visual Purple Referring to rhodopsin.

Visual Yellow Referring to *all-trans*-retinal.

Visuoauditory Pertaining to both vision and hearing.

Vitacarn A trade name for L-carnitine, used to facilitate the transport of fatty acids.

Vital Stain A stain that stains living cells without destructive effects on the cells.

Vital Staining The staining of living cells or their components by vital stain.

Vitamer 1. One or two or more forms of a vitamin, e.g., vitamin A_1 and A_2. 2. Substance with a vitamin function.

Vitamin A group of structurally unrelated organic compounds that an organism cannot synthesize itself but nevertheless requires in small quantity for normal growth and metabolism.

Vitamin A (mol wt 287) A fat-soluble vitamin structurally related to carotenes that is required for normal development of bone and production of visual pigment. A vitamin A deficiency causes night blindness. It exists in a number of derivatives, e.g., vitamin-A aldehyde and vitamin-A acid (also known as retinol).

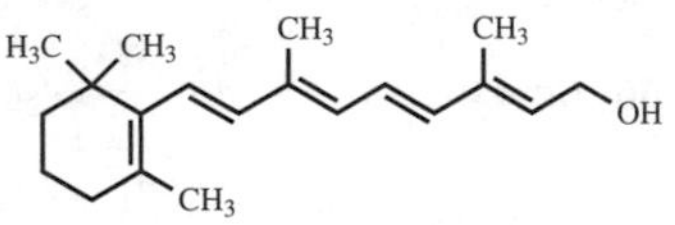

Vitamin A_1 Referring to vitamin-A alcohol.

Vitamin A_2 (mol wt 284) A derivative of vitamin A that combines with opsin to form visual pigment.

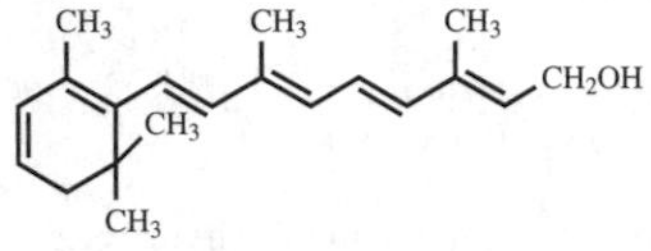

Vitamin A Acetate (mol wt 329) A vitamin-A derivative.

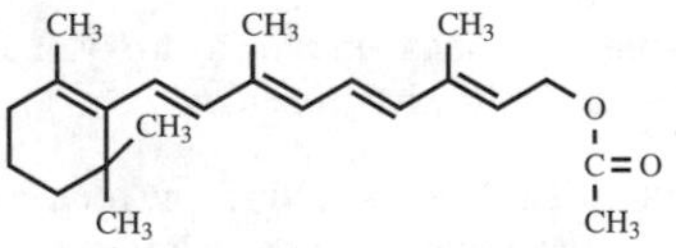

Vitamin A Acid (mol wt 300) A vitamin A derivative.

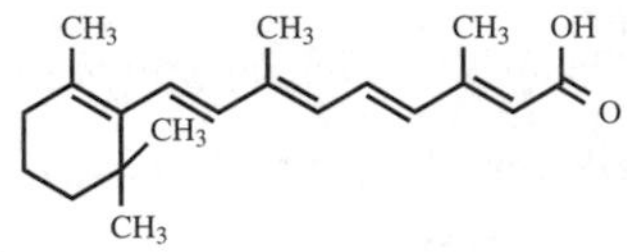

Vitamin A Alcohol (mol wt 286) A vitamin-A derivative and a fluorescent probe for membrane lipid (also known as vitamin-A_1).

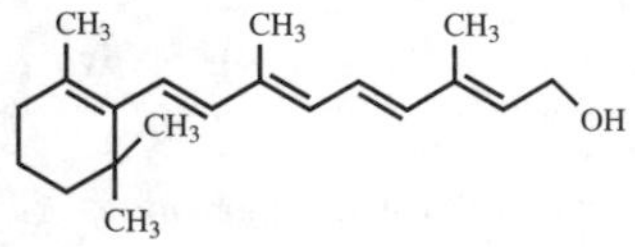

Vitamin A Aldehyde (mol wt 284) A vitamin A derivative known as retinal.

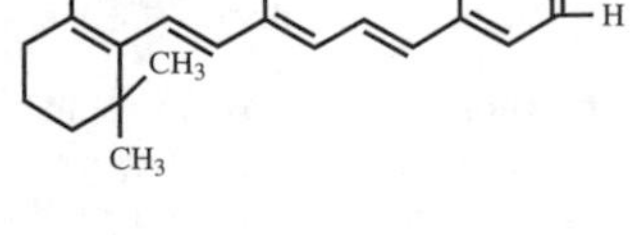

Vitamin A Palmitate (mol wt 528) A vitamin-A derivative.

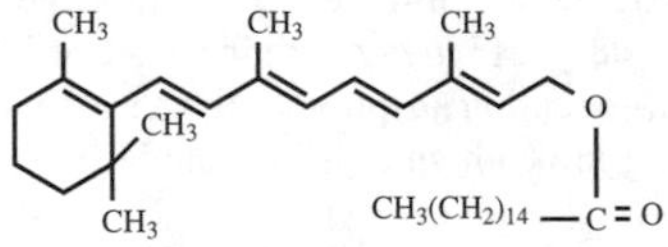

Vitamin B_1 (mol wt 301) Referring to thiamine that acts as coenzyme for carbohydrate metabolism.

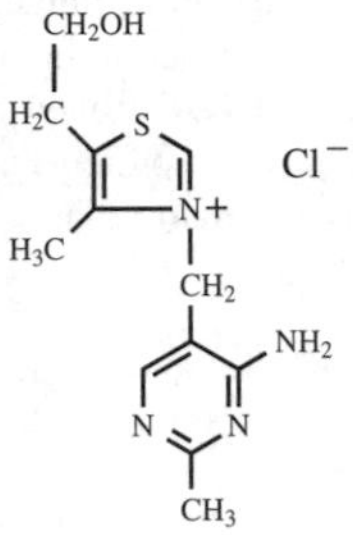

Vitamin B_2 (mol wt 376) Referring to riboflavin that acts as coenzyme and is needed for normal tissue respiration.

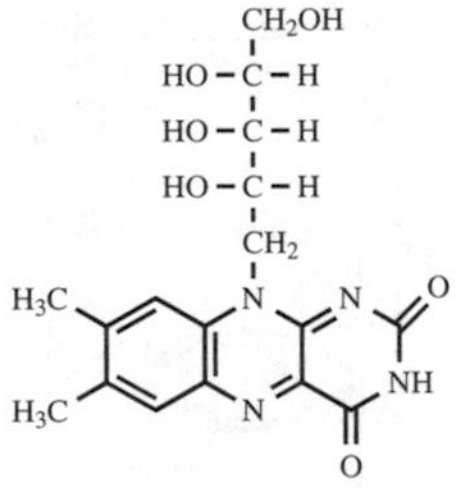

Vitamin B_3 Referring to niacin (nicotinic acid) that is needed for lipid metabolism, tissue respiration, and glycogenolysis. It also decreases synthesis of low-density lipoprotein and inhibits lipolysis in the adipose tissue.

Vitamin B_6 (mol wt 170) Referring to pyridoxine hydrochloride that acts as coenzyme for a number of metabolic reactions.

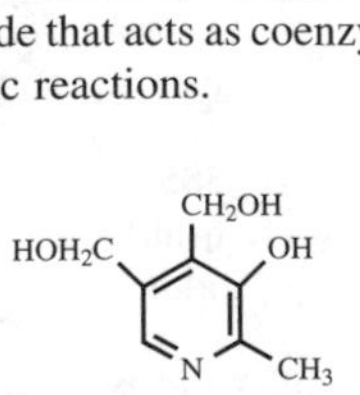

Vitamin B_9 Referring to folic acid that is needed for normal erythropoiesis and nucleoprotein synthesis (see folic acid for structure).

Vitamin B_{12} Referring to cyanocobalamin that acts as coenzyme for a number of metabolic reactions and is also needed for cell replication and hematopoiesis.

Vitamin B_{12a} Referring to hydroxocobalamin that acts as coenzyme for a number of metabolic reactions and is also needed for cell replication and hematopoiesis.

Vitamin B_{12r} Reductase The enzyme that catalyzes the following reaction:

$$NADH + 2\ Cob(II)alamin \rightleftharpoons NAD^+ + 2\ Cob(I)alamin$$

Vitamin C (mol wt 176) Referring to ascorbic acid that is needed for collagen formation, tissue repair, and oxidation-reduction reactions throughout the body.

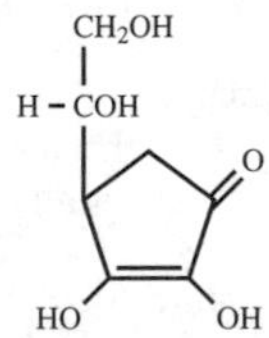

Vitamin D Vitamins derived from cholesterol that play an important role in control of calcium and phosphorus metabolism.

Vitamin D_1 (mol wt 793) A 1:1 molecular compound of lumisterol and vitamin D_2.

Vitamin D_2 (mol wt 396) The synthetic form of vitamin D, prepared by irradiation of ergosterol, that promotes absorption and utilization of calcium and phosphate (also known as ergocalciferol).

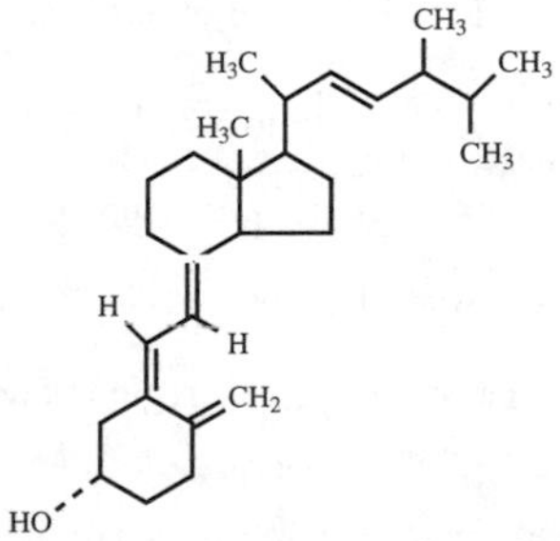

Vitamin D_3 (mol wt 385) A vitamin that mediates intestinal calcium absorption and bone calcium metabolism (also known as cholecalciferol).

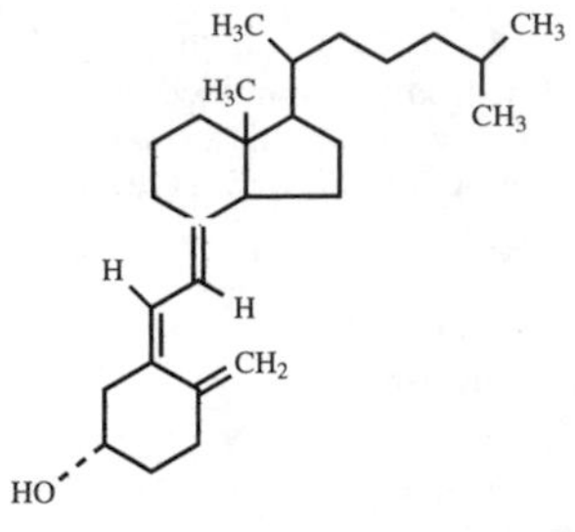

Vitamin D_4 (mol wt 399) A vitamin obtained by irradiation of 22:23-dehydroergosterol with light from a magnesium arc.

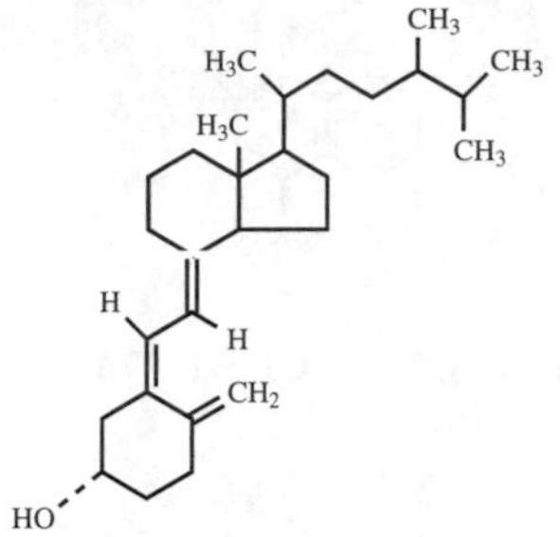

Vitamin E (mol wt 431) Natural α-tocopherol that acts as an enzyme cofactor and an antioxidant.

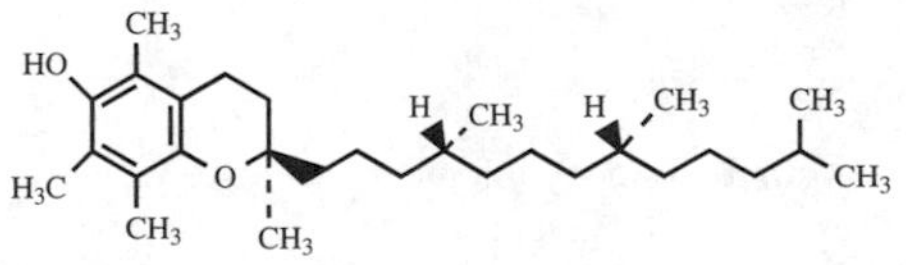

Vitamin F Obsolete designation for thiamine.

Vitamin G Obsolete designation for riboflavin.

Vitamin H Synonym for biotin.

Vitamin I Obsolete designation for B_7 (carnitine).

Vitamin K Fat-soluble vitamins involved in blood clotting.

Vitamin K_1 (mol wt 451) An antihemorrhagic vitamin that promotes the formation of active prothrombin.

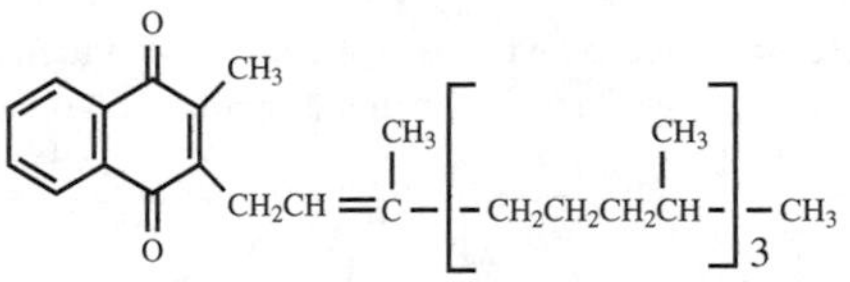

Vitamin K_2 An antihemorrhagic vitamin possessing side chains that vary in length.

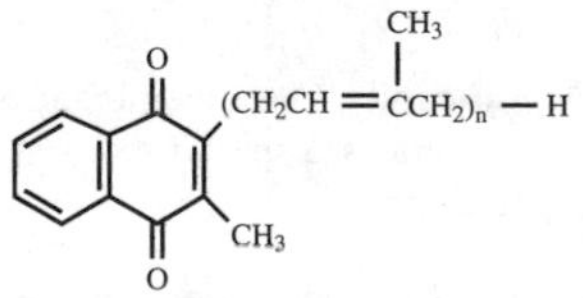

Vitamin K_5 (mol wt 173) An antihemorrhagic vitamin.

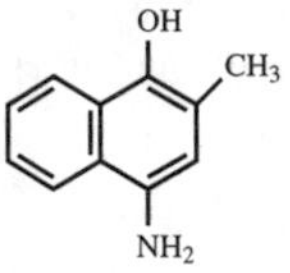

Vitamin K_6 (mol wt 172) An antihemorrhagic vitamin.

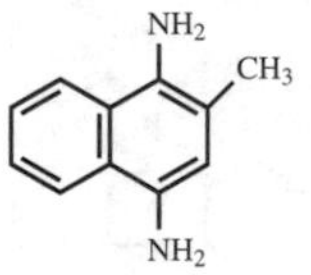

Vitamin K-S(II) (mol wt 276) An antihemorrhagic vitamin.

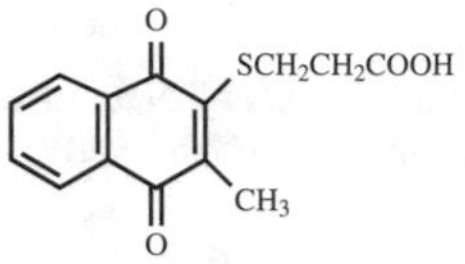

Vitamin L Adenylthiomethylpentose, which is necessary for lactation.

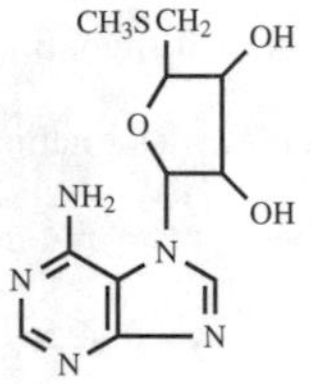

Vitamin M Referring to folic acid, which is necessary for a number of metabolic reactions.

Vitamin N Obsolete designation for nicotinic acid.

Vitamin P Referring to bioflavonoids.

Vitamin U (mol wt 200) An antiulcer vitamin from cabbage leaves, a methionine derivative.

$$[(CH_3)_2 + CH_2CH_2CH(NH_2)COOH]Cl$$

Vitamin V Referring to NAD^+ and NADH.

Vitellin The major protein of egg yolk.

Vitelline Resembling egg yolk or producing egg yolk.

Vitelline Membrane Outer membrane of the egg of marine invertebrates, a specialized coat that only sperm can penetrate.

Vitellogenesis Oogenesis (egg development) in which yolk forms.

Vitellogenic Hormone Referring to an insect hormone that is needed for vitellogenesis.

Vitellogenin A protein in the egg yolk of various vertebrates.

Vitellus The yolk of an egg.

Vitravene A trade name for fomivirsen sodium used for treatment of CMV in AIDS patients who cannot tolerate other treatments for CMV retinitis.

Vitreoscilla A genus of gliding bacteria (Cytophagales).

Vitreous Humor The transparent gelatinous secretion filling the posterior cavity of the eye.

Vitrification The process of converting a siliceous material into an amorphous, glassy form by melting and cooling.

Vitronectin An adhesion protein that promotes adhesion of cells in tissue culture.

Vitrosin An outdated term for collagen.

Vivactil A trade name for protriptyline hydrochloride, an antidepressant that increases the amount of norepinephrine and serotonin in the brain.

Vivarin A trade name for caffeine, used as a cerebral stimulant.

Vivelle A trade name for estradiol, an estrogen hormone.

Viviparous Reproduction in which fertilization of the egg and development of the embryo occur inside the mother's body.

Vivol A trade name for diazepam, an antianxiety agent.

Vivotif Berna Vaccine An oral typhoid vaccine containing killed Ty-2 strain of *Salmonella typhi*.

V-J Joining The joining of a variable gene (V gene) and a joining gene (J gene) in the process of formation of a functioning immunoglobulin gene.

***v-jun* Gene** A viral oncogen.

V_κ Abbreviation for variable region of an immunoglobulin κ chain.

V_L Symbol for the variable region of light chains of an immunoglobulin molecule.

V_λ Abbreviation for variable region of an immunoglobulin λ chain.

VLB A trade name for vinblastine sulfate, a mitotic inhibitor used as an antineoplastic agent.

VLCAD Abbreviation for very long chain acyl-CoA dehydrogenase.

VLCADH Abbreviation for very long chain acyl-CoA dehydrogenase.

VLCFA Abbreviation for very long chain fatty acid.

VLDL Abbreviation for very low density lipoprotein.

VLDLR Abbreviation for receptor of very low density lipoprotein.

VLIA Abbreviation for virus-like infectious agent.

VLP Abbreviation for a combination drug containing vincristine, leucogen, and prednisone.

VM-26 Referring to teniposide that arrests cell mitosis.

V_{max} Symbol for maximum velocity, the velocity of an enzymatic reaction occurring when the enzyme is saturated with substrate.

VMCP Abbreviation for a combination drug containing vincristine, methotrexate, cyclophosphamide, and prednisone.

VneI (ApaLI) A restriction endonuclease from *Vibrio nereis* with the following specificity:

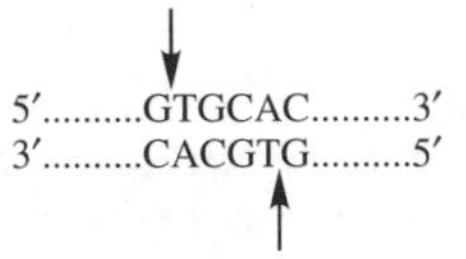

VneAI (DraII) A restriction endonuclease from *Vibrio nereis* with the following specificity:

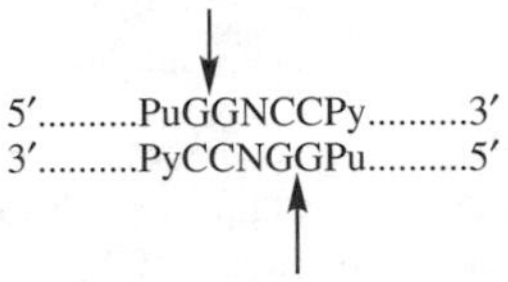

VniI (HaeIII) A restriction endonuclease from *Vibrio nigripulchritudo* with the same specificity as HaeII.

VNTR Abbreviation for variable number tandem repeat.

V_o Abbreviation void volume.

Voges-Proskauer Test A test used to determine the ability of an organism to form acetoin based on the production of red color by the reaction of acetoin with α-naphthol (Barritt's method).

Void Volume The volume of solvent or eluant in the interstitial space of a chromatographic column.

Vol Abbreviation for volume.

Volatility The tendency of a solid or liquid material to pass into the vapor state at a given temperature.

Volemitol (mol wt 203) A sugar alcohol found in plants, fungi, and lichens. It is used as a sweetening agent.

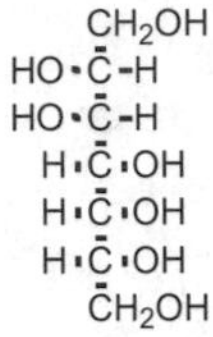

Vollmer Patch Test A type of tuberculin test in which a patch of tuberculin-impregnated gauze is fastened to the skin with adhesive tape for detection of tuberculosis.

Volmax A trade name for albuterol, a beta-2 adrenergic agonist used as a bronchodilator and antiasthmatic agent.

Volt The potential difference required to send a current of 1 ampere through a resistance of 1 ohm.

Voltage Clamp Technique in electrophysiology in which a microelectrode is inserted into a cell, and current injected through the electrode so as to hold the membrane potential at some predefined level.

Voltage-Gated Channel A channel in the cell membrane whose opening is governed by the membrane potential.

Voltage Gradient The potential difference across a plasma membrane.

Voltage Sensitive Gate An ion channel in the membrane of excitable cells that allows only certain ions to pass under specific changes of electric potential.

Voltammetry Analysis of chemical reactions by the measurement of current generated by electrolysis as a function of the applied voltage.

Voltaren A trade name for diclofenac sodium, an anti-inflammatory agent.

Volulin Synonym for polyphosphate.

Volume Receptor The receptors in the CNS that respond to changes in the volume of blood.

Volumetric Analysis A quantitative analysis that employs precise volumes of the standard solution that react with the sample under investigation, e.g., titration.

Volutin Granule Polyphosphate storage granules found in cyanobacteria, yeast, and other microorganisms.

Vomitoxin (mol wt 296) A mycotoxin from *Fusarium roseum*.

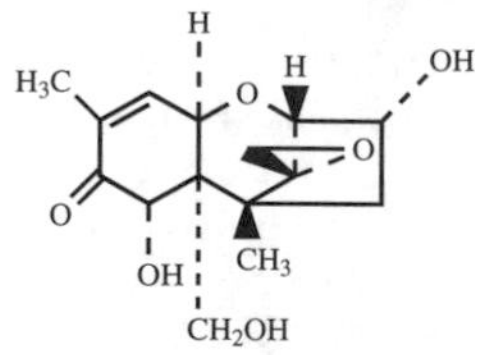

von Willebrand Disease An autosomal dominant platelet disorder caused by a deficiency in blood coagulation factor VIII.

von Willebrand Factor A plasma factor involved in platelet adhesion through an interaction with Factor VIII.

v-onc Symbol for viral oncogene acquired by the virus from a eukaryotic cell.

Vortex A device for mixing solutions in a tube by a rapid whirling action.

Voxsuprine A trade name for isoxsuprine hydrochloride used as a vasodilator.

VP Abbreviation for 1. vasopressin and 2. viral protein.

VP-16 Referring to etoposide used as an antineoplastic agent for inhibition of cell mitosis.

VPC Abbreviation for volume of packed cells.

VPCG Abbreviation for vapor-phase chromatography.

V_{pg} A protein that is covalently linked to 5′-viral RNA, priming the viral RNA replication.

VPL Abbreviation for a combination drug containing vincristine, prednisone, and leucogen.

V-PPase Abbreviation for vacuolar pyrophosphatase.

VPRC Abbreviation for volume of packed red cells.

Vorticella Genus of ciliate Protozoa (subclass Peritrichia).

VRE Abbreviation for vancomycin-resistant *Enterococcus*.

V-Region The region of the immunoblobulin chain near the N-terminal where the amino acid sequence shows great variation between molecules (see also variable region).

vRNA Abbreviation for viral RNA.

VSIEF Abbreviation for vertical slab isoelectric focusing.

***v-sis* Gene** An oncogen of simian sarcoma virus.

VSM Abbreviation for vascular smooth-muscle.

VSMC Abbreviation for vascular smooth-muscle cell.

VspI A restriction endonuclease from *Vibrio* sp with the following specificity:

↓
5′..........ATTAAT..........3′
3′..........TAATTA..........5′
↑

v-src Rous sarcoma oncogen.

VSV Abbreviation for vesicular stomatitis virus.

VSVG Abbreviation for vesicular-stomatitis-virus glycoprotein.

V_t Abbreviation for total bed volume (in gel filtration chromatography).

Vulvovaginitis Inflammation of the vulva and vagina, usually caused by *Candida albicans*, herpes viruses, *Trichomonas vaginalis*, or *Neisseria gonorrhoeae.*

Vumon A trade name for teniposide, which inhibits cell mitosis.

VV Abbreviation for vaccinia virus.

v/v Referring to the concentration of a solution expressed in terms of volume per unit volume, e.g., number of ml of a solute per 100 ml of solution.

VX (mol wt 267) Symbol for methylphosphonothioic acid, a cholinesterase inhibitor.

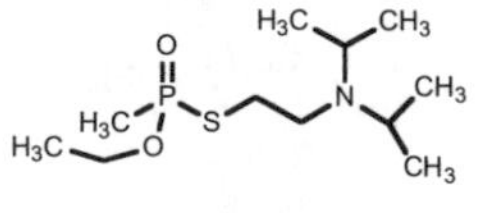

VZ Abbreviation for varicella-zoster.

VZIg Abbreviation for varicella-zoster immune globulin.

VZV Abbreviation for varicella-zoster virus.

W

W Abbreviation for 1. tryptophan and 2. watt.

W-138 A type of human embryonic fibroblast cell line.

Waldenström's Disease A disorder characterized by the presence of a high concentration of macroglobulin (IgM).

Waldenström's Macroglobulinmia See Waldenström's disease.

Wandering Cells Cells capable of amebic movement, e.g., free macrophages.

WAP Abbreviation for whey acid protein.

Warburg Apparatus An instrument used for the study of cellular respiration of tissue slices and suspended cells by measuring the O_2 uptake and CO_2 production.

Warburg Effect The inhibition of CO_2 fixation in photosynthesis by the presence of a high concentration of oxygen.

Warburg Method A manometric method for study of cellular respiration of tissue slices and cells with the Warbury apparatus.

Warfarin (mol wt 308) An anticoagulant that inhibits vitamin K-dependent activation of clotting factors, II, VII, IX, and X.

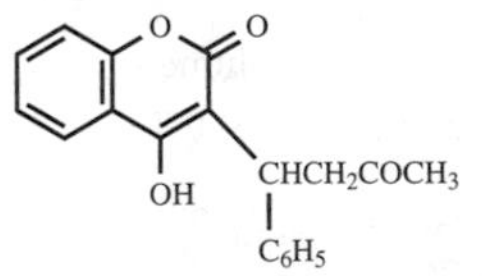

Warfilone See warfarin.

Waring Blender A trade name for a homogenizer or blender used for preparation of tissue homogenates.

Warm Antibody Antibody that has optimum activity at elevated temperature (e.g., 37°C).

Wart Small, benign tumors of the skin, caused in humans by the human papillomavirus.

Wassermann Antibody Antibody in the Wassermann reaction.

Wassermann Reaction 1. A complement fixation test for diagnosis of syphilis in which antibody (Wassermann antibody) undergoes a complement fixation reaction with cardiolipin-lecithin-cholesterol antigen. 2. Any serological reaction used for diagnosis of syphilis.

WAT Abbreviation for white adipose tissue.

Water Activity Referring to the amount of "free" or "available" water in a given substrate, it may be defined as 1/100th the relative humidity of the air in equilibrium with the substrate.

Water Blue (mol wt 738) A water soluble dye.

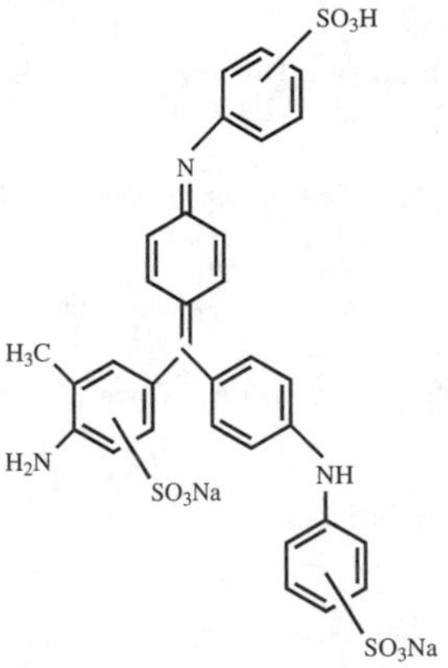

Water Borne Disease transmitted by contaminated water.

Water of Hydration Water molecules that bind to a molecule to form a hydrated state.

Watson-Crick Base Pair Referring to the base pair between guanine and cytosine, and adenine and thymine in the double-stranded DNA.

Watson-Crick Model Referring to the right-handed, antiparallel, double helix structure of DNA in which two antiparallel polynucleotide chains coil around the same axis forming a double helix. The purine-pyrimidine bases are held together by hydrogen bonds between the complementary bases (e.g., guanine–cytosine base pair and adenine–thymine base pair).

Watt (W) The amount of electric power produced by one volt with one ampere of current.

Wattage The amount of power expressed in watts.

Wavelength The distance between successive peaks of a wave train, such as electromagnetic radiation.

Wave Number The number of waves per unit of length; the reciprocal of the wavelength.

Wavetrain A series of waves along the same axis.

Wax The ester of fatty acid with a long chain alcohol.

Wax D A high molecular glycolipid or peptidoglycolipid extracted from *Mycobacterium tuberculosis.*

Wax Ester Acylhydrolase See wax ester hydrolase.

Wax Ester Hydrolase The enzyme that catalyzes the following reaction:

$$\text{A wax ester} + H_2O \rightleftharpoons \text{A long chain alcohol} + \text{a long chain carboxylate}$$

Wax Synthetase The enzyme that catalyzes the following reaction:

$$\text{Acyl-CoA} + \text{long chain alcohol} \rightleftharpoons \text{CoA} + \text{long chain ester}$$

WB Abbreviation for 1. Western blot; 2. whole blood.

WBAP Abbreviation for winged-bean acidic protease.

WBC Abbreviation for 1. white blood cells; 2. white blood count.

WBCD Abbreviation for white blood count and differentiation.

WBC/hpf Abbreviation for white blood cells per high-power field.

WBF Abbreviation for whole blood folate.

WBH Abbreviation for whole blood hematocrit.

WBHC Abbreviation for whole blood hematocrit.

WBS Abbreviation for whole blood scan.

WBV Abbreviation for willow-brook virus.

WC Abbreviation for 1. white cells; 2. white count; 3. whooping cough.

WCC Abbreviation for white cell count.

WCHP Abbreviation for wood-chuck hepatitis virus.

WE Abbreviation for 1. Western encephalitis; 2. Western encephalomyelitis.

WEE Abbreviation for 1. Western equine encephalitis; 2. Western equine encephalomyelitis.

WEEV Abbreviation for 1. Western equine encephalitis virus; 2. Western equine encephalomyelitis virus.

Wehamine A trade name for dimenhydrinate, an antiemetic agent.

Wehdryl A trade name for diphenhydramine hydrochloride, an antihistaminic agent that competes with histamine for H_1 receptors on effector cells.

Wehless A trade name for phendimetrazine tartrate, a cerebral stimulant that promotes transmission of nerve impulses by releasing stored norepinephrine from the nerve terminals in the brain.

Weightrol A trade name for phendimetrazine tartrate, used as a cerebral stimulant that promotes transmission of nerve impulses by releasing stored norepinephrine from the nerve terminals in the brain.

Weil-Felix Test A serological test for diagnosis of certain rickettsial diseases based on the reaction (agglutination) of rickettsial antibody with cells of *Proteus vulgaris.*

Wellbutrin A trade name for bupropion hydrochloride, an antidepressant.

Wellcovorin A trade name for leucovorin calcium, a vitamin (reduced form of folic acid).

Westcort Cream A trade name for hydrocortisone valerate, a topical corticosteroid.

Western Blotting A technique to identify particular proteins in a mixture by separation on polyacrylamide gels, followed by blotting onto nitrocellulose, and detected by radiolabeled or enzyme-labeled probes.

Western Transfer Technique See Western blotting.

Wetting Agent Any surface-active agent that promotes spreading of a liquid on a solid surface.

WGA Abbreviation for wheat germ agglutinin.

Wheal and Flare A local, cutaneous anaphylactic reaction to the intracutaneous injection of allergen. It is a form of type I-hypersensitivity and characterized by the appearance of local swelling (wheal) and redening (flare).

Wheat Germ The embryonic plant at the tip of the seed of wheat.

Wheat Germ Agglutinin A nonmitogenic plant lectin from wheat germ that binds to oligopolysaccharide containing *N*-acetylglucosamine or neuraminic acid residues.

Wheat Germ System The *in vitro* protein synthetic system employing components extracted from wheat germ.

Whey The fluid fraction of milk after precipitation of casein by acidification that contains about 4–5% lactose, 0.8% protein (whey protein), and 0.2–0.8% lactic acid.

Whey Proteins The noncasein milk protein obtained by acidification (pH 4.7) of skim milk and removal of precipitated casein.

Whiplash Flagella The smooth flagella of algae and fungi.

Whipple's Disease A disorder of intestinal malabsorption, characterized by excess facial fat, anemia, swelling of the joints, and infiltration of intestinal mucosa by macrophages.

Whipworm A parasitic worm (*Trichuris trichuria*) that causes trichuriasis in the intestine.

White Adipose Tissue A highly specialized tissue in mammals and birds that stores fat.

White Blood Cells Referring to basophils, eosinophils, neurtophils, monocytes, and lymphocytes; collectively termed as leukocytes.

White Corpuscles Referring to leukocytes of the blood.

White Muscle Skeletal muscle with relatively low content of hemoglobin and cytochromes. It is almost devoid of mitochondria, obtaining ATP from glycolysis and capable of only short bursts of activity.

White Thrombus The thrombus that consists of mainly platelets and fibrin, without erythrocytes or hemoglobin.

Whitlow Inflammation of finger tip.

Whole Plasma The nonfractionated plasma.

Whole Serum The nonfractionated serum.

Whooping Cough (pertussis) An acute respiratory tract disease occurring mainly in children, caused by *Bordetella pertussis* and characterized by paroxysms of coughing.

WI38 (Wistar Institute 38) A diploid cell line derived from a female embryonic lung tissue.

Widal Test An agglutination test for the diagnosis of typhoid fever in which serum is tested for the presence of agglutinin to the H and O antigens of *Salmonellae*.

WIGA Bacterium Referring to *Legionella bozemanii.*

Wigraine A trade name for a combination drug containing ergotamine and caffeine used as an antimigraine.

Wildfire Toxin (mol wt 289) Toxin produced by *Pseudomonas tobaci* in plants responsible for wildfire disease of tobacco. It is also highly toxic to bacteria algae, plants, and animals.

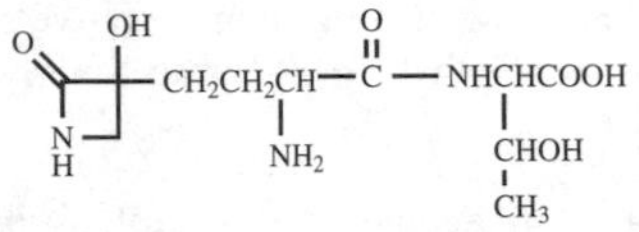

Wild Type The genotype or phenotype that is found in nature or in the standard laboratory stock for a given organism.

Wilm's Tumor A childhood kidney cancer caused by the inactivation of an antioncogene.

Wilson and Blair's Agar An agar medium for isolation of *Salmonella* strains that contains peptone, beef extract, disodium phosphate, ferrous sulfate, brilliant green, bismuth sulfate, and agar.

Wilson's Disease An genetic disorder in humans caused by a deficiency in ceruloplasmin, leading to an increase of copper in the brain and liver.

WinGel A trade name for a combination drug containing aluminum hydroxide and magnesium hydroxide, used as an antacid.

Win-Kinase A trade name for urokinase, which activates plasminogen.

Winpred A trade name for prednisone, an anti-inflammatory agent.

Winsprin Capsules A trade name for aspirin, an analgesic and an antipyretic agent that blocks the generation of pain impulses and inhibits prostaglandin synthesis.

Winstrol A trade name for stanozolol, an anabolic steroid that promotes tissue-building processes.

Wiskott-Aldrich Syndrome An X-linked recessive combined immune deficiency disorder characterized by defects in leukocyte functions, leading to thrombocytopenia, eczema, increased levels of IgE and IgA, and decreased levels of cell-mediated immunity.

WLF Abbreviation for whole lymphocyte fraction.

WNFV Abbreviation for West Nile fever virus.

WNV Abbreviation for West Nile virus.

Wobble Base Referring to the base at the 5′ end of an anticodon that can align itself with more than one base in the codon of mRNA.

Wobble Hypothesis The flexibility in base pairing between the third base of a codon and the complementary base of its anticodon.

Wohl-Zemplen Degradation The removal or degradation of the carbon atom of the reducing group in an aldosugar by treatment with hydroxylamine, producing an aldose with one less carbon.

Wolbachia A genus of Gram-negative bacteria (Rickettsiaceae) that grows intracellularly in the arthropod host.

Wolfina A trade name for rauwolfia serpentina, an antihypertensive agent.

Wolinella A genus of oxidase-positive, Gram-negative bacteria (family Bacteroidaceae).

Wolman Disease An inherited disorder due to the deficiency of lysosomal acid lipase.

Wood Alcohol Referring to methanol or methyl alcohol.

Woodward's Reagent K (mol wt 253) A reagent used for modification of carboxylic groups in enzymes.

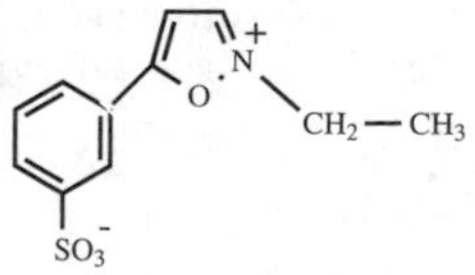

Woodward's Reagent L (mol wt 240) A reagent used in peptide synthesis.

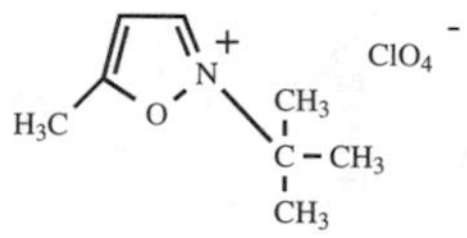

Wood-Werkman Reaction Referring to the carboxylation of pyruvate to oxaloacetate.

Wool Fat Lanolin obtained from sheep's wool used for preparation of ointment.

Woolf-Augustinson Plot A plot of velocity (v) of an enzyme-mediated reaction as a function of v/[S] where v is velocity and [S] is substrate concentration.

WP Abbreviation for washed platelets.

WR Abbreviation for Wassermann reaction.

WRC Abbreviation for washed red cells.

Writhing Number The number of times that the axis of a DNA duplex crosses itself by supercoiling, which is a measure to estimate the degree of supercoiling.

WT Abbreviation for wild type.

WT33 A protein associated with Wilm's tumor.

WT-GDP Abbreviation for wild type GDP (glucose dehydrogenase).

WTR Abbreviation for wild type receptor.

Wuchereria A genus of parasitic nematodes (Filarioidea).

Wuchereriasis Infestation with nematodes of the genus *Wuchereria*.

w/v Abbreviation for a weight/volume solution in which the concentration is expressed in terms of weight per unit of volume, e.g., number of grams of a solute per 100 ml of solution.

w/w Abbreviation for a weight/weight solution in which the concentration is expressed in terms of weight per unit of weight, e.g., number of grams of a solute per 100 grams of liquid.

Wyamine A trade name for mephentermine sulfate, an adrenergic agent that stimulates both alpha- and beta-adrenergic receptors.

Wyamycin S A trade name for erythromycin stearate, an antibiotic that inhibits bacterial protein synthesis.

Wycillin A trade name for penicillin G procaine, an antibiotic that inhibits bacterial cell wall synthesis.

Wydase A trade name for hyalurnidase, which promotes diffusion of fluid in the tissue.

Wygesic A trade name for a combination drug containing acetaminophen and propoxyphene hydrochloride, an analgesic agent.

Wymox A trade name for amoxicillin trihydrate, an antibiotic that inhibits bacterial cell wall synthesis.

Wyo Abbreviation for wyosine, a minor nucleoside in tRNA.

Wyosine A highly modified nucleoside found in tRNA.

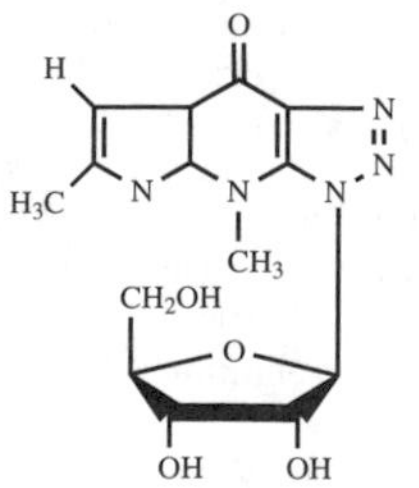

Wytensin A trade name for guanabenz acetate, an antihypertensive agent.

Wyvac A trade name for rabies virus vaccine.

W,X,Y Boxes Three conserved sequences found in the promoter region of HLA-Dr α-chain gene.

W, Z Chromosomes Sex chromosomes in species where the female is the heterogametic sex (WZ).

X

X Abbreviation for 1. xanthine, 2. xanthosine, and 3. cross with.

X and Y Linkage The inheritance pattern of genes found on both the X and Y chromosomes.

X Bacteria Gram-negative, endosymbiotic bacteria occurring in the cytoplasm of certain strains of *Amoeba proteus*.

X Body A type of inclusion body that is presumed to be the site of virus replication (also known as viroplast, viroplasm, and viroplasmic matrix).

X Chromosome A chromosome associated with sex determination that is present in two copies in the homogametic sex and in one copy in the heterogametic sex.

X Factor A growth factor required for growth of *Haemophilus* species.

X Hyperactivation A process in *Drosophila* by which the structural genes of the male X chromosome is transcribed at the same rate as the two X chromosomes of the female combined.

X Inactivation The random cessation of transcriptional activity of one X chromosome in mammalian females.

X Linkage The inheritance pattern of genes found on the X chromosome but not on the Y chromosome.

Xaa Symbol for an unspecified amino acid residue in a protein.

XAFS Spectroscopy Abbreviation for X-ray absorption fine structure spectroscopy.

Xalatan A trade name for latanoprost, an ophthalmic agent.

XamI (SalI) A restriction endonuclease from *Xanthomonas amaranthicola* with the same specificity as SalI.

Xamoterol (mol wt 339) A cardiotonic agent.

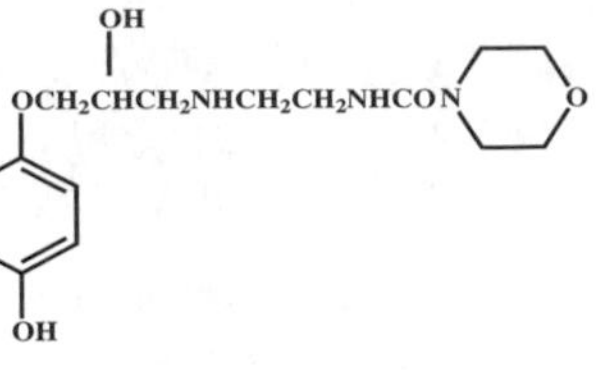

Xan Abbreviation for xanthine.

Xanax A trade name for alprazolam, an anti-anxiety agent.

Xanomeline (mol wt 281) A cholinergic and nootropic agent.

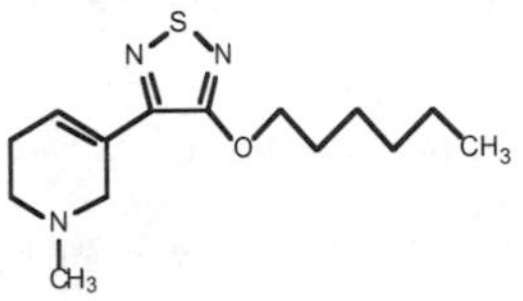

Xanthatin (mol wt 246) An antimicrobial agent from *Xanthium strumarium.*

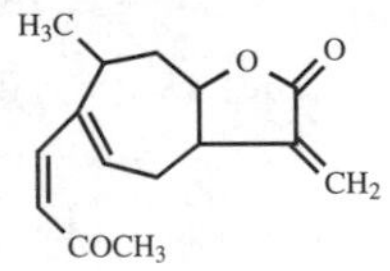

Xanthene (mol wt 182) Basic structure for many dyes.

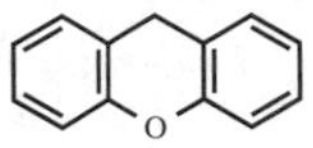

Xanthine (mol wt 152) A nitrogenous base found in tRNA that is the product from the deamination of guanine.

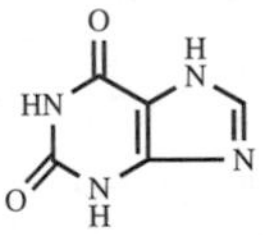

Xanthine Dehydrogenase The enzyme that catalyzes the following reaction:

$$\text{Xanthine} + \text{NAD}^+ + H_2O \rightleftharpoons \text{Urate} + \text{NADH}$$

Xanthine Oxidase The enzyme that catalyzes the following reaction:

$$\text{Xanthine} + O_2 + H_2O \rightleftharpoons \text{Urate} + H_2O_2$$

Xanthine Phosphoribosyltransferase The enzyme that catalyzes the following reaction:

$$\text{5-Phospho-D-ribose 1-phosphate} + \text{xanthine} \rightleftharpoons \text{D-Ribosylxanthine 5}'\text{-phosphate} + \text{PPi}$$

Xanthinol Niacinate (mol wt 434) A vasodilator.

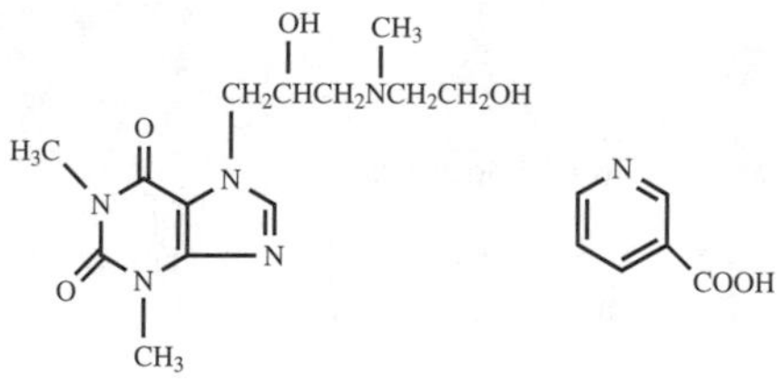

Xanthinuria A genetic disorder characterized by the presence of excessive amounts of xanthine in the urine.

Xantho- A prefix denoting yellow color.

Xanthobacter A genus of aerobic, catalase-positive, Gram-negative bacteria.

Xanthomas Yellow skin plaque or nodule characterized by the deposition of cholesterol in the skin because of a defect in fat metabolism.

Xanthomatosis The presence or development of multiple xanthomas.

Xanthomonas A genus of Gram-negative, obligately aerobic, chemoorganotrophic, plant pathogenic bacteria (family Pseudomonadaceae).

Xanthophyll (mol wt 569) A yellow carotenoid found as an accessory pigment in photosynthesis.

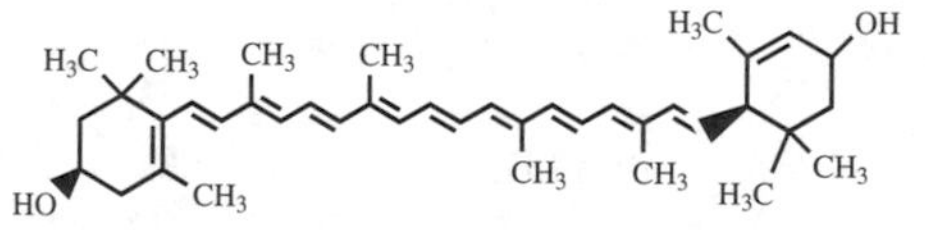

Xanthopterin (mol wt 179) Pigment found in insects, e.g., wings of butterflies.

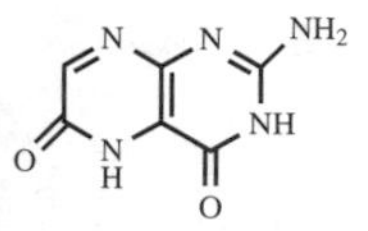

Xanthosine (mol wt 284) A nucleoside of xanthine found in tRNA.

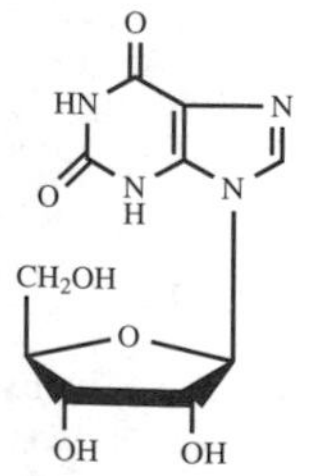

Xanthosine 5′-Diphosphate (mol wt 444) Diphosphate nucleotide of xanthine.

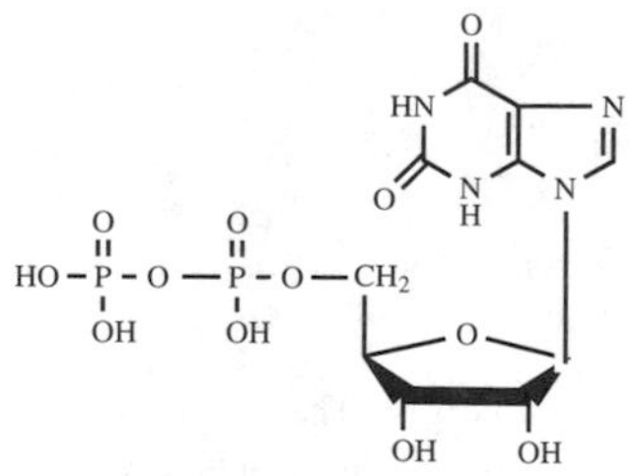

Xanthosine 5′-Monophosphate (mol wt 364) A nucleotide of xanthine found in tRNA as a minor nucleotide.

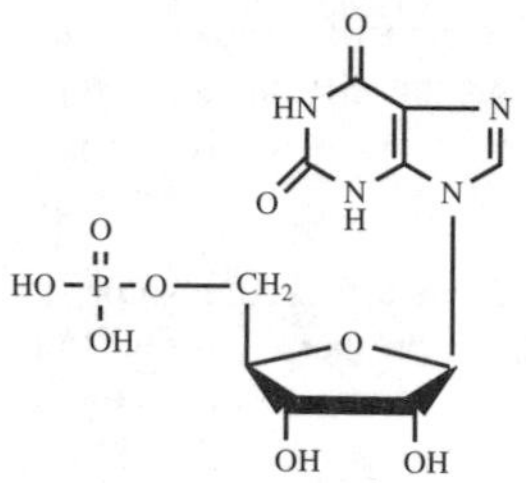

Xanthosine 5′-Triphosphate (mol wt 524) Triphosphate nucleotide of xanthine.

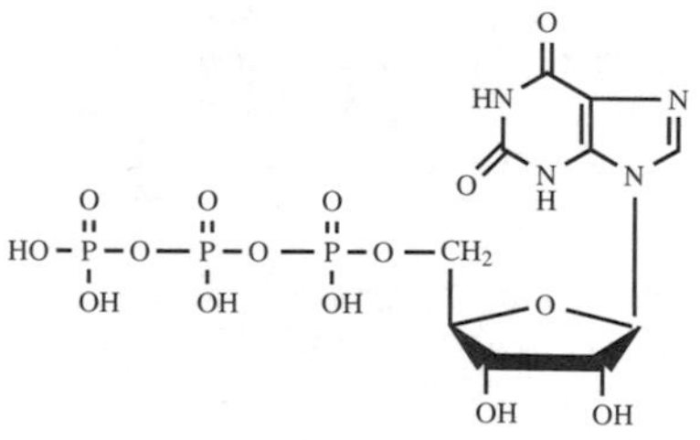

Xanthosis The yellow discoloration of skin.

Xanthotoxin (mol wt 216) A reagent for treatment of vitiligo.

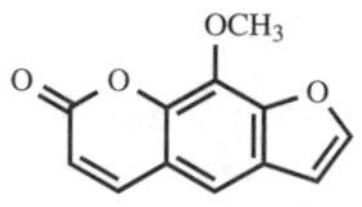

Xanthurenic Acid (mol wt 205) A metabolic intermediate that accumulates during pregnancy and in vitamin B_6-deficiency individuals.

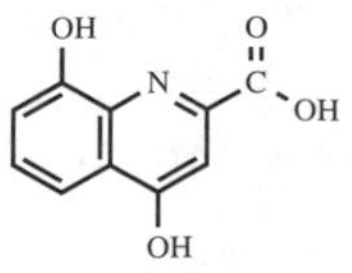

Xanthurenic Aciduria A genetic disorder in humans due to a deficiency of the enzyme kynureninase in tryptophan metabolism.

Xanthydrol (mol wt 198) Reagent used for the determination of urea concentration.

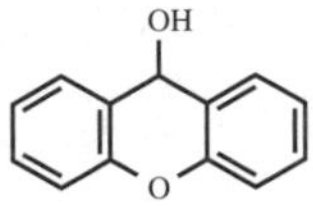

Xanthylic Acid Referring to nucleotide of xanthine.

Xao Abbreviation for xanthosine.

Xao5'P Abbreviation for xanthosine 5'-phosphate.

Xao5'PP Abbreviation for xanthosine 5'-diphosphate.

Xao5'PPP Abbreviation for xanthosine 5'-triphosphate.

XAS Abbreviation for X-ray absorption spectroscopy.

XbaI A restriction endonuclease from *Xanthomonas badrii* with the following specificity:

↓
5′..........TCTAGA..........3′
3′..........AGATCT..........5′
↑

XBD Abbreviation for xylan-binding domain.

XBP Abbreviation for xylan-binding protein.

XcaI (SnaI) A restriction endonuclease from *Xanthomonas campestris* with the following specificity:

↓
5′..........GTATAC..........3′
3′..........CATATG..........5′
↑

XciI (SalI) A restriction endonuclease from *Xanthomonas citrii* with the following specificity:

↓
5′..........GTCGAC..........3′
3′..........CAGCTG..........5′
↑

XcyI (SmaI) A restriction endonuclease from *Xanthomonas campestris* with the following specificity:

↓
5′..........CCCGGG..........3′
3′..........GGGCCC..........5′
↑

XD Abbreviation for xanthine dehydrogenase.

XDH Abbreviation for 1. xylitol dehydrogenase; 2. xanthine dehydrogenase.

XDP Abbreviation for xanthosine diphosphate.

Xe Abbreviation for xenon, a chemical element.

XECTG Abbreviation for xenon-enhanced computed tomography.

Xeloda A trade name for capecitabine, an antineoplastic agent.

Xenazoic Acid (mol wt 375) An antiviral agent.

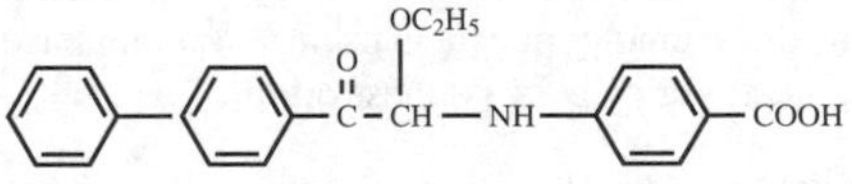

Xenbucin (mol wt 240) An antihyperlipoproteinemic agent.

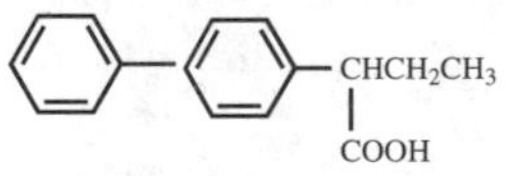

Xenical A trade name for orlistat, a lipase inhibitor used as a weight loss agent.

Xenin A 25-residue peptide from human gastric mucosa, it stimulates exocrine pancreatic secretion.

Xeno- A prefix meaning foreign.

Xenoantibody Any antibody, raised in one species, whose homologous antigen is derived from a different species.

Xenobiotic Referring to a compound that is found in the environment but is not formed by natural biosynthetic processes, e.g., pesticides in the environment.

Xenococcus A genus of unicellular cyanobacteria.

Xenodiagnosis A method for diagnosis of a vector-transmitted disease in which a pathogen-free vector is allowed to suck blood from a patient and examined for the presence of pathogen in the vector after a proper incubation period.

Xenogeneic Relationship between members of genetically distinct species.

Xenograft A tissue or organ graft between members of different species.

Xenoma A tumor-like lesion on the tissue infected with certain pathogenic parasites that consists of hypertrophied tissue within which parasite lives.

Xenon (Xe) A chemical element with atomic weight 131.

Xenon-133 (^{133}Xe) A gamma-emitting radioactive gas with half-life of about 5 days, used to measure blood flow and regional pulmonary ventilation.

Xenopus A genus of African clawed toads.

Xenorhabdus A genus of motile bacteria (family Enterobacteriaceae).

Xenosome A bacteria-like endosymbiont in certain marine protozoan, e.g., *Parauronema acutum*.

Xenotropic Referring to an endogenous retrovirus that cannot replicate in its host of origin but can replicate in cells of nonhost origin.

Xenytropium Bromide (mol wt 537) An antispasmodic agent.

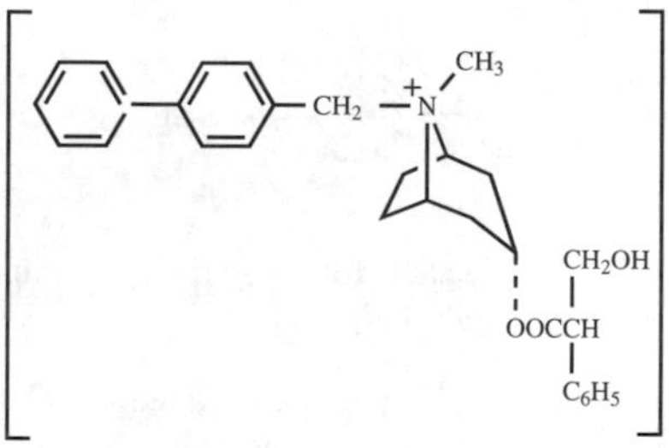

Xero- A prefix meaning dry.

Xeroderma Pigmentosum An inherited defect characterized by the extreme sensitivity to sunlight and development of multiple skin cancers. The disorder is due to a defect in the repair of UV-damaged DNA.

Xerogel Dried gel.

Xerophic Referring to organisms capable of growing at low water activity.

Xerophthalmia A pathological disorder of the cornea of the eye, characterized by a dry and lusterless cornea caused by deficiency of Vitamin A.

Xerosis Abnormal dryness of skin or mucous membrane.

Xerotolerant An organism capable of growth at low water activity.

XET Abbreviation for xyloglucan endotransglycosylase.

xg Symbol for the multiplicity of earth gravity force; $1 \times g = 980$ cm/sec^2.

X_{gal} Referring to 5-bromo-4-chloro-3-indolyl-β-D-galactoside, a substance used for assaying β-galactosidase in bacteria that forms blue colonies when growing on a medium containing X_{gal}.

XhoI A restriction endonuclease from *Xanthomonas holcicola* with the following specificity:

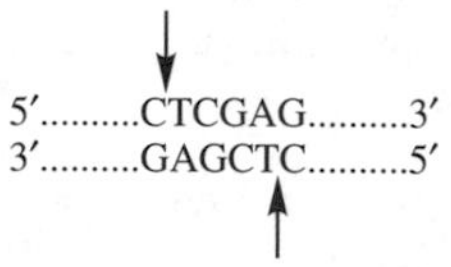

XhoII A restriction endonuclease from *Xanthomonas holcicola* with the following specificity:

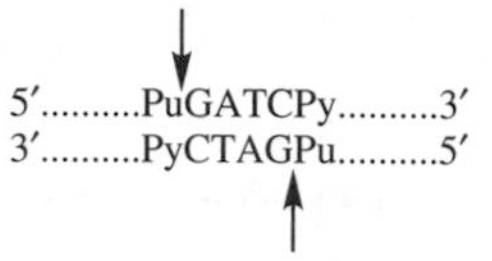

Xi (ξ) A letter in the Greek alphabet.

Xibenolol (mol wt 251) An antiarrhythmic agent.

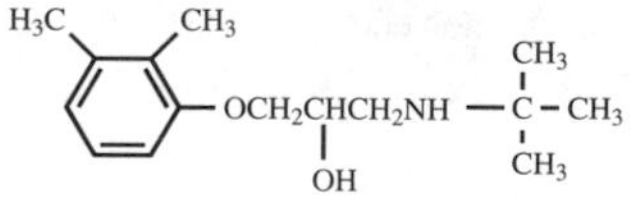

Xibornol (mol wt 258) An antibacterial agent.

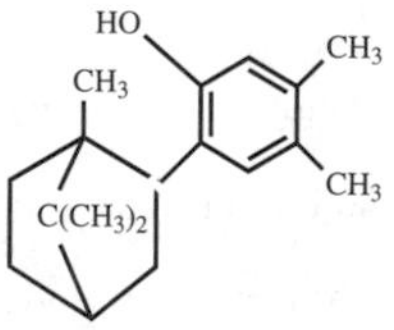

Ximoprofen (mol wt 261) An anti-inflammatory agent.

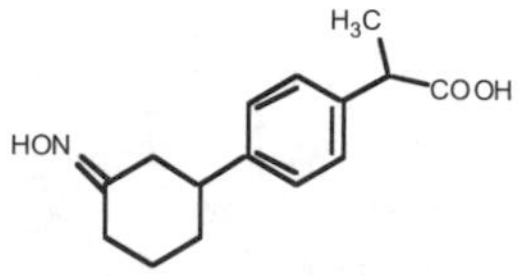

Xipamide (mol wt 355) An antihypertensive agent.

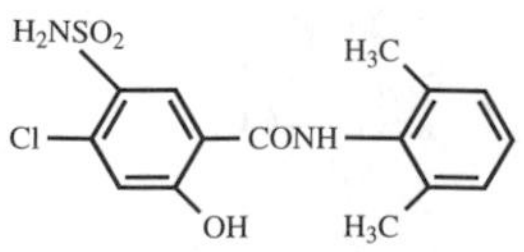

***xis* Gene** A gene in bacteriophage λ that is responsible for excision of λ prophage from bacterial chromosomes.

***xis* Protein** Referring to the enzyme excisionase for exicision of λ prophage from bacterial chromosomes.

XLD Agar Abbreviation for xylose/lysine/deoxycholate agar.

X-linked Disease Any inherited disease whose controlling gene is located on an X chromosome, e.g., hemophilia.

X-linked Gene Genes located on the X chromosome.

X-linked Inheritance The pattern of hereditary transmission of genes located in the X chromosome.

xln Abbreviation for a gene encoding xylanase.

XlnA Abbreviation for xylanase-A.

XlnB Abbreviation for xylanase-B.

XlnC Abbreviation for xylanase-C.

XLP Syndrome An X-linked lymphoproliferative syndrome in which infection of Epstein-Barr virus leads to a fetal infectious mononucleosis.

XmaI (SmaI) A restriction endonuclease from *Xanthomonas malvacearum* with the same specificity as SmaI.

XmaII (PstI) A restriction endonuclease from *Xanthomonas malvacearum* with the same specificity as PstI.

XmaIII A restriction endonuclease from *Xanthomonas malvacearum* with the following specificity:

```
         ↓
5′..........CGGCCG..........3′
3′..........GCCGGC..........5′
                 ↑
```

XmaCI A restriction endonuclease from *Xanthomons malavacearum* strain C with the following specificity:

```
         ↓
5'........CCCGGG........3'
3'........GGGCCC........5'
              ↑
```

XmnI A restriction endonuclease from *Xanthomonas manihotis* 7AS1 with the following specificity:

```
                ↓
5′..........GAANNNNTTC..........3′
3′..........CTTNNNNAAG..........5′
                ↑
```

XMP Abbreviation for xanthosine monophosphate.

XMP Pathway Abbreviation for xylulose monophosphate pathway, a cyclic metabolic pathway for assimilation of formaldehyde by yeast growing on methanol medium (also known as dihydroxyacetone pathway).

XniI (PvuI) A restriction endonuclease from *Xanthomonas nigromaculans* with the same specificity as PvuI.

XO Abbreviation for xanthine oxidase.

XOR Abbreviation for xanthine oxidoreductase.

XorI (PstI) A restriction endonuclease from *Xanthomonas oryzae* with the same specificity as PstI.

XorII (PvuI) A restriction endonuclease from *Xanthomonas oryzae* with the same specificity as PvuI.

xotch A transmembrane early embryo protein in *Xenopus laevis*.

XP Abbreviation for 1. *Xeroderma pigmentosum*; 2. xylene poisoning.

XPA Abbreviation for *Xeroderma pigmentosum* group A.

XpaI (XhoI) A restriction endonuclease from *Xanthomonas papavericola* with the same specificity as XhoI.

XPB Abbreviation for *Xeroderma pigmentosum* group B.

XphI (PstI) A restriction endonuclease from *Xanthomonas phaseoli* with the same specificity as PstI.

X-prep Liquid A trade name for senna, used as a laxative.

XPS Abbreviation for X-ray photoelectron spectroscopy.

X-Rays An energetic form of electromagnetic radiation having wavelengths shorter than ultraviolet radiation.

X-Ray Crystallography A physical method for the analysis of X-ray diffraction patterns of a crystalline compound used to determine the molecule's three-dimensional structure.

X-Ray Diffraction Pattern The pattern of spots, arcs, and rings obtained when X-rays are reflected from the atoms of a crystal.

X-Ray Film A photographic film coated with a sensitive emulsion on both sides, used in X-ray crystallography.

X-Ray Microanalysis Analysis of concentrations of elements within a cell or tissue by bombarding sections of freeze-dried specimen with electrons and then analyzing the X-rays produced by various metallic and nonmetallic elements in the specimen.

XRE Abbreviation for xenobiotic responsive element.

XRF Abbreviation for X-ray fluorescence.

XRFS Abbreviation for X-ray fluorescence spectrometry.

XTP Abbreviation for xanthosine triphosphate.

X-Trozine A trade name for phendimetrazine tartrate, a cerebral stimulant that stimulates transmission of nerve impulses by releasing stored norepinephrine from the nerve terminals in the brain.

Xul Abbreviation for xylulose.

Xul5P Abbreviation for xylulose 5-phosphate.

Xyl Abbreviation for 1. xylene; 2. xylose.

Xylan Polysaccharide of xylose occurring in the plant cell wall.

Xylan Endo-1,3-β-Xylosidase The enzyme that catalyzes the internal hydrolysis of 1,3-β-D-xylosidic linkage in 1, 3-β-D-xylan.

1,4-β-D-Xylan Synthetase The enzyme that catalyzes the following reaction:

$$\text{UDP-D-xylose} + (\text{1,4-}\beta\text{-D-xylan})_n \rightleftharpoons \text{UDP} + (\text{1,4-}\beta\text{-D-xylan})_{n+1}$$

1,3-β-D-Xylan Xylanohydrolase The systematic name for xylan endo-1,3-b-xylosidase.

Xylan 1,3-β-Xylosidase The enzyme that catalyzes the successive hydrolysis of xylose residues from the nonreducing termini of 1, 3-β-D-xylan.

Xylan 1,4-β-Xylosidase The enzyme that catalyzes the hydrolysis of 1,4-β-D-xylan so as to remove successive D-xylose residues from the nonreducing termini.

Xylanase The enzyme that catalyzes the internal hydrolysis of xylan.

Xylazine (mol wt 220) A sedative, analgesic, and muscle relaxant.

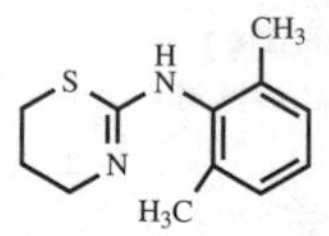

xylE A gene in *E. coli* encoding a xylose transport protein.

Xylem A vascular tissue that transports water and dissolved minerals upward through the plant.

Xylene (mol wt 106) Any of three flammable toxic isomeric aromatic hydrocarbons, used as organic solvent.

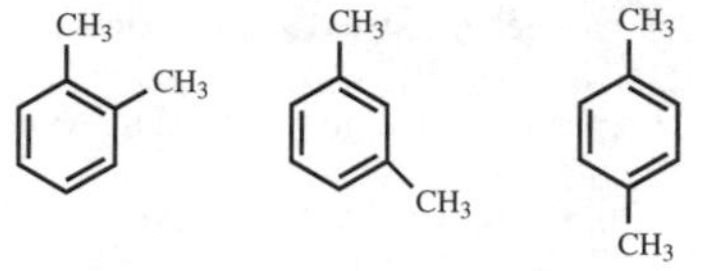

o-xylene *m*-xylene *p*-xylene

Xylenecyanol FF (mol wt 539) Tracking dye in electrophoresis for DNA sequencing.

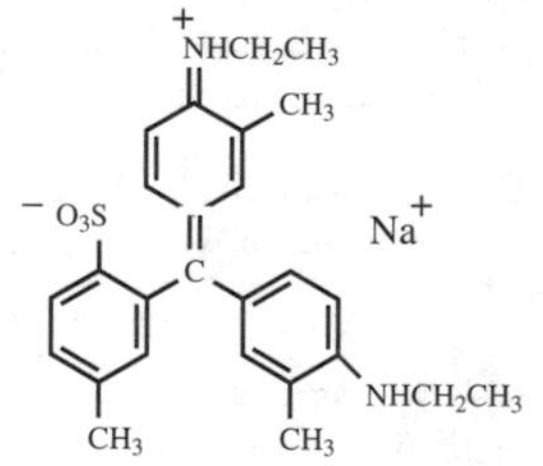

Xylenol Blue (mol wt 411) A pH indicator dye.

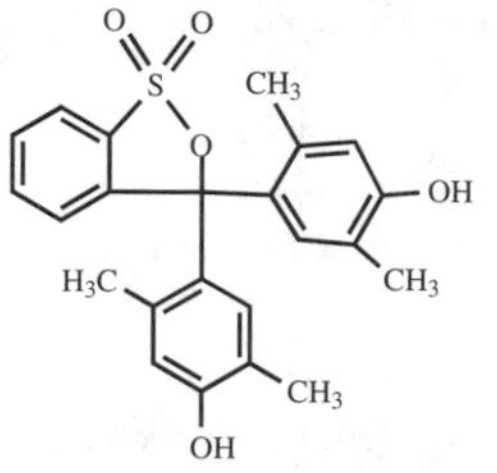

Xylenol Orange A dye.

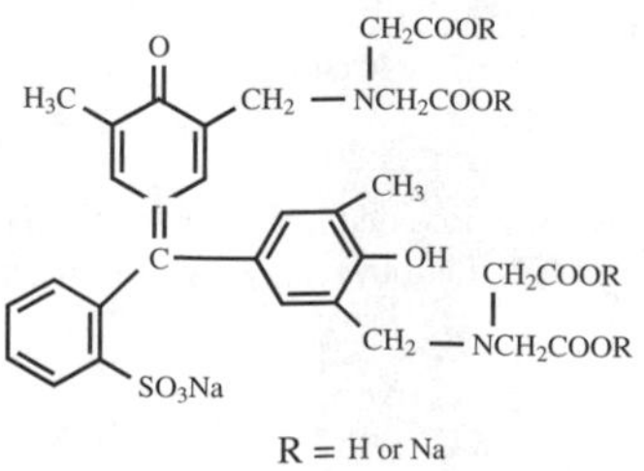

R = H or Na

Xylidyl Blue (mol wt 514) A dye used for the determination of magnesium in plasma and urine.

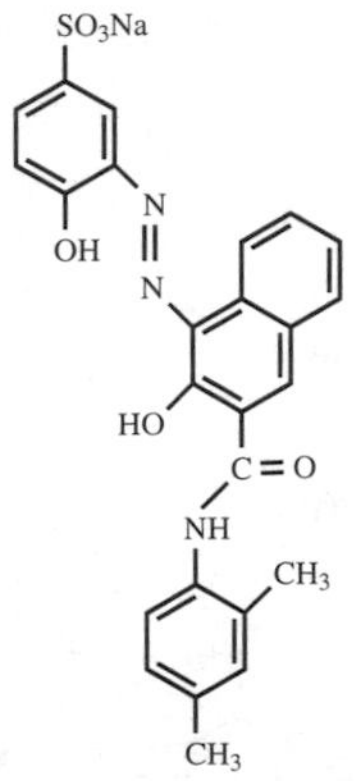

Xylitol (mol wt 152) A sugar alcohol.

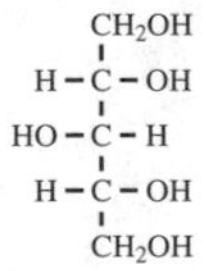

XylNAc Abbreviation for N-acetyl-xylosamine.

Xylocaine A trade name for lidocaine hydrochloride, an antiarrhythmic agent that shortens the action potential.

Xylocard A trade name for lidocaine hydrochloride, an antiarrhythmic agent that shortens the action potential.

Xylometazoline (mol wt 244) A nasal agent that produces local vasoconstriction of the dilated arterioles to reduce blood flow and nasal congestion.

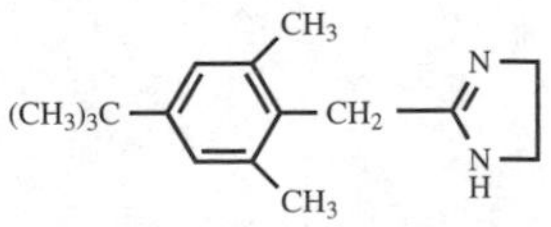

Xylonate Dehydratase The enzyme that catalyzes the following reaction:

$$\text{D-xylonate} \rightleftharpoons \text{2-Dehydro-3-deoxy-D-xylonate} + H_2O$$

Xylopropamine (mol wt 163) An adrenergic agent.

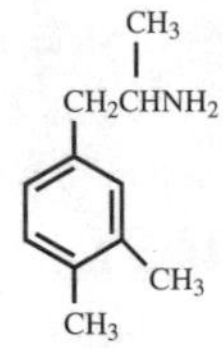

Xylose (mol wt 150) A monosaccharide.

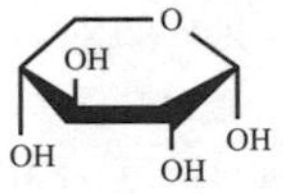

α-D-xylose

D-Xylose 1-Dehydrogenase The enzyme that catalyzes the following reaction:

$$\text{D-Xylose} + NAD^+ \rightleftharpoons \text{D-xylonolactone} + NADH$$

D-Xylose 1-Dehydrogenase (NADP⁺ Specific) The enzyme that catalyzes the following reaction:

$$\text{D-Xylose} + NADP^+ \rightleftharpoons \text{D-Xylono-1,5-lactone} + NADPH$$

L-Xylose 1-Dehydrogenase The enzyme that catalyzes the following reaction:

$$\text{L-Xylose} + NADP^+ \rightleftharpoons \text{L-Xylono-1,4-lactone} + NADPH$$

Xylose Isomerase The enzyme that catalyzes the following reaction:

$$\text{D-Xylose} \rightleftharpoons \text{D-Xylulose}$$

Xylobiase Synonym of xylan 1,4-β-xylosidase.

Xylose Ketol Isomerase The systematic name for xylose isomerase.

Xyl5P Abbreviation for xylose 5-phosphate.

Xylulokinase The enzyme that catalyzes the following reaction:

$$\text{ATP} + \text{D-xylulose} \rightleftharpoons \text{ADP} + \text{D-xylulose 5-phosphate}$$

L-Xylulokinase The enzyme that catalyzes the following reaction:

ATP + L-xylulose

⇅

ADP + L-xylulose 5-phosphate

Xylulose (mol wt 150) A monosaccharide keto-sugar.

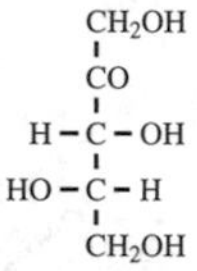

L - isomer

Xylulose Monophosphate Pathway A cyclic metabolic pathway for assimilation of formaldehyde by yeast growing on methanol medium.

D-Xylulose Reductase The enzyme that catalyzes the following reaction:

Xylitol + NAD^+ ⇌ D-xylulose + NADH

L-Xylulose Reductase The enzyme that catalyzes the following reaction:

Xylitol + $NADP^+$ ⇌ L-xylulose + NADPH

Xylyl Chloride (mol wt 141) A powerful lacrimator.

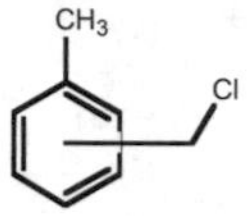

Y

Y Abbreviation for 1. amino acid tyrosine; 2. the chemical element yttrium.

Y79 Abbreviation for human retinoplastoma cell line.

Y Chromosome A sex chromosome present only in the heterogametic sex (XY).

Y Fork The DNA replication fork, a point at which a DNA molecule is being replicated and the two newly synthesized double strands form the arms of the Y-shaped structure.

Y Linkage The inheritance patterns of genes found on the Y chromosome but not on the X chromosome.

Yabapox A smallpox of nonhuman primates.

YAC Abbreviation for yeast artificial chromosome.

YAD Abbreviation for yeast alcohol dehydrogenase.

YADH Abbreviation for yeast alcohol dehydrogenase.

YAG Laser Abbreviation for yttrium-aluminum-garnet laser, a type of laser used for cutting tissue.

Yaws An infectious tropical disorder characterized by ulcerating sores on the body surface, caused by a spirochete of the genus *Treponema* (e.g., *T. pertenue*).

Yb Abbreviation for ytterbium, a chemical element with atomic weight 173 and valences 2, 3.

Y-Chr Abbreviation for Y chromosome.

Yeast Unicellular, saprophytic, eukaryotic fungi, capable of fermenting a range of carbohydrates.

Yeast Artificial Chromosome A cloning vector in yeast that can accept very large fragments of DNA.

Yeast Extract Water soluble preparation from brewers' yeast, rich in amino acids, vitamin B, and peptides, and used as a component of the culture medium.

Yellow AB (mol wt 247) A dye.

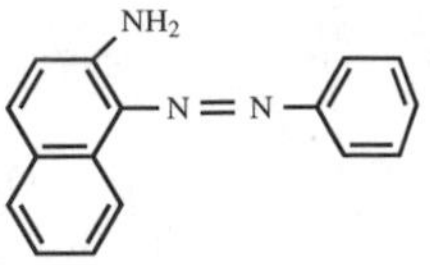

Yellow Enzyme Referring to the enzyme that contains a yellow flavin prosthetic group.

Yellow Fever An acute, systemic disease caused by a togavirus (Togaviridae) and transmitted to humans by mosquitoes.

Yellow Jacket Venom A toxin from bee, wasp, or ant that induces anaphylactic shock that may lead to death.

Yellow OB (mol wt 261) A dye.

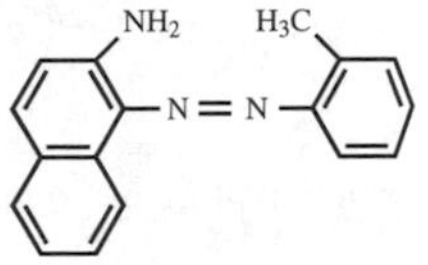

YenAI (PstI) A restriction endonuclease from *Yersinia enterocolitica* with the same specificity as PstI.

YenBI (PstI) A restriction endonuclease from *Yersinia enterocolitica* 08 Bi1212 with the same specificity as PstI.

YenCI (PstI) A restriction endonuclease from *Yersinia enterocolitica* 08 Bi3995 with the same specificity as PstI.

YenDI (PstI) A restriction endonuclease from *Yersinia enterocolitica* 08 Bi9534 with the same specificity as PstI.

YenEI (PstI) A restriction endonuclease from *Yersinia enterocolitica* 08 85-775 with the same specificity as PstI.

YenI (PstI) A restriction endonuclease from *Yersinia enterocolitica* 08 A 2635 with the same specificity as PstI.

YEPD Abbreviation for yeast extract-peptone-dextrose. A complex medium for culturing yeast.

Yersinia A genus of Gram-negative bacteria of the Enterobacteriaceae.

Yersiniosis Infection caused by *Yersinia.*

yes A gene encoding non-receptor tyrosine kinase, *v-yes* is an oncogene, and its cellular counterpart is *c-yes*.

YF-Vax A trade name for yellow fever vaccine, an attenuated 17D yellow fever virus.

Y-Junction The point of active DNA replication where the double helix opens up so that each strand can serve as a template for synthesis of daughter DNA molecules.

Y-linked Genes Genes located in the Y chromosome.

YMA Medium Yeast extract-mannitol-agar medium.

Yocto- A prefix denoting 10^{-24}.

Yodoquinal A trade name for iodoquinol, an amebicide.

Yodoxin A trade name for iodoquinol, an antiprotozoal agent.

Yoghurt Variant spelling of yogurt.

Yogurt Food made by fermenting milk with a mixed culture of *Loctobacillus bulgaricus* and *Streptococcus thermophilus*.

Yohimbine (mol wt 354) An indole alkaloid with α-adrenergic blocking activity.

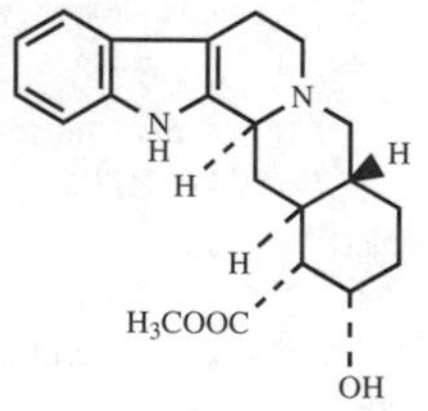

Yolk The nutrient portion of an egg that supplies food to the developing embryo.

Yolk Sac The highly vascularized extraembryonic membrane that serves as the site for the earliest hemopoiesis during ontogeny.

Yolk Stalk The narrow passage between the intraembryonic gut and yolk sac.

Yomesan A trade name for niclosamide, an anthemintic agent.

Yotta- A prefix denoting 10^{24}.

YPC Abbreviation for yeast pyruvate carboxylase.

YPD Abbreviation for yeast peptone dextrose.

YS Abbreviation for yolk sac.

YSC Abbreviation for yolk sac carcinoma.

Ytterbium A chemical element with atomic weight 173, valences 2 and 3.

Yttrium (Y) A chemical element with atomic weight 89, valence 3.

Yttrium-90 A radioactive isotope of yttrium used for the treatment of breast and prostatic cancer.

Yutopar A trade name for ritotrine hydrochloride, used to stimulate β_2 adrenergic receptors in the uterine smooth muscle, inhibiting contractility.

Z

Z Abbreviation for 1. atomic number; 2 amino acid glutamine or glutamate; 3. average net charge of an ion; 4. proton number.

(Z) A prefix in chemical nomenclature denoting a geometric isomer in which the highest priority substituent groups are determined according to the Cahn-Ingold-Prelog rules.

Z Disc Thin, dense structure that runs down the middle of each I band in a skeletal muscle myofibril, the structure into which actin filaments are inserted.

Z DNA A form of DNA in which the two antiparallel polynucleotide chains form a left-handed double helix. It consists of about 12 residues per turn and has been shown to be present along with B-DNA in chromosomes and may have a role in regulation of gene expression.

Z Line See Z disc.

ZA Abbreviation for zinc acetate.

Zadine A trade name for azatadine maleate, an antihistamine that competes with histamine for H_1 receptors on effector cells.

Zaditen A trade name for ketotifen fumarate, used to stabilize mast cells.

Zafirlukast (mol wt 576) A leukotriene receptor antagonist and an antiasthmatic agent.

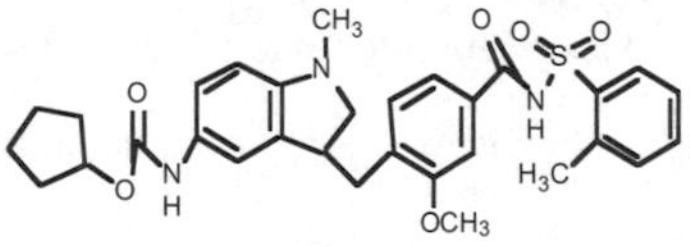

ZAG A soluble glycoprotein present in the serum and other body fluids. It is so named because it precipitates zinc salt and has an electrophoretic mobility in the region of α_{2}- globulin. ZAG stimulates lipid degradation in adipocytes and causes extensive losses of fat associated with advanced cancers.

Zagam A trade name for sparfloxacin, an antibiotic.

Zaleplon (mol wt 305) A hypnotic agent.

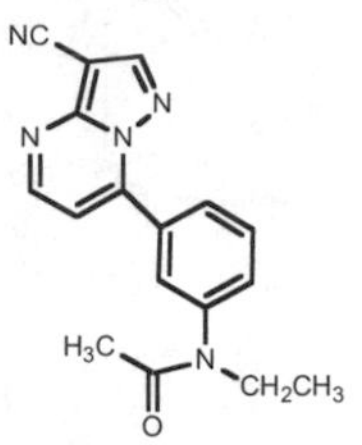

ZanI (EcoRII) A restriction endonuclease from *Zymomonas anaerobia* with the following specificity:

↓
5′..........CC(A/T)GG..........3′
3′..........GG(T/A)CC..........5′
↑

Zanaflex A trade name for tizanidine, an antispasmodic and sympatholytic agent.

Zanflo An anionic polysaccharide produced by certain soil bacteria.

Zanosar A trade name for streptozocin, an alkylating agent that cross-links cellular DNA, interfering with transcription.

Zantac A trade name for ranitidine hydrochloride, an antiulcer agent that decreases gastric acid secretion.

Zapex A trade name for oxazepam, an antianxiety agent.

Zarontin A trade name for ethosuximide, an anticonvulsant.

Zaroxolyn A trade name for metolazone, a diuretic agent that increases urine excretion of sodium and water by inhibiting sodium reabsorption.

Zatebradine (mol wt 457) An antianginal agent.

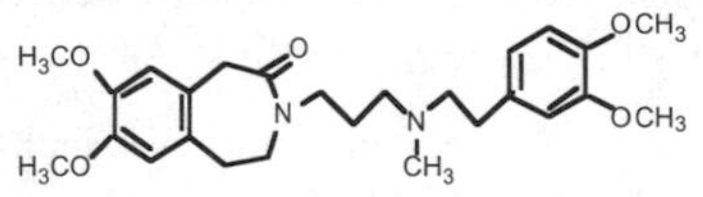

Zearalenone (mol wt 318) An anabolic agent isolated from mycelia of *Gibberella zeae*.

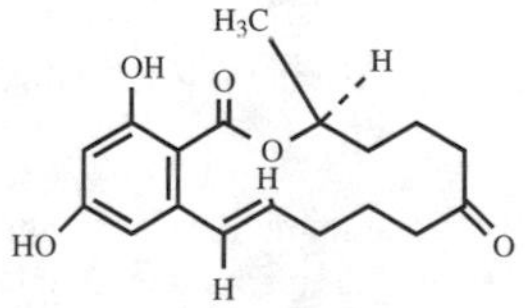

Zeasorb-AF A trade name for tolnaftate, a local anti-infective agent.

Zeatin (mol wt 219) A plant growth hormone or cytokinin isolated from sweet corn.

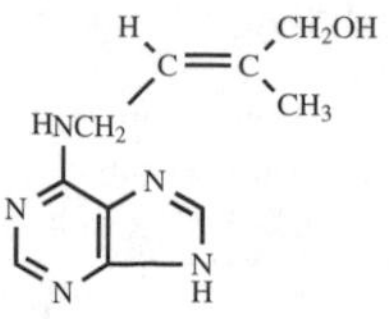

Zeatin *O*-β-D-Glucosyltransferase The enzyme that catalyzes the following reaction:

UDP-glucose + zeatin

$\rightleftharpoons$

UDP + *O*-β-D-glucosylzeatin

Zeatin *O*-β-D-Xylosyltransferase The enzyme that catalyzes the following reaction:

UDP-D-xylose + zeatin

$\rightleftharpoons$

UDP + *O*-β-D-xylosylzeatin

Zeatin Reductase The enzyme that catalyzes the following reaction:

Dihydrozeatin + $NADP^+$ $\rightleftharpoons$ Zeatin + NADPH

Zeaxanthin (mol wt 569) A carotinoid alcohol isolated from yellow corn and certain seaweed.

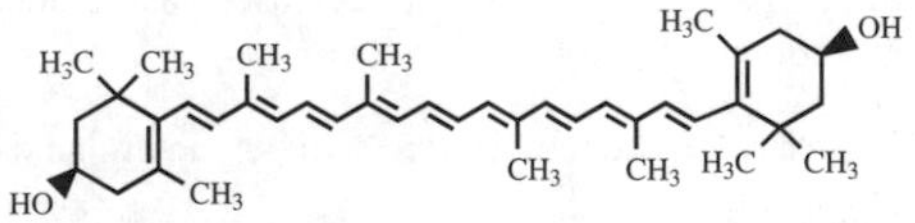

Zebeta A trade name for bisoprolol, a beta adrenergic blocking agent used as an antihypertensive drug.

Zebutal A trade name for a combination drug containing butalbital, acetaminophen, and caffeine.

Zefazone A trade name for cefmetazole sodium, an antibacterial agent that inhibits bacterial cell wall synthesis.

Zein A seed protein of corn (prolamine) that is soluble in 70 to 80% alcohol.

Zemuron A trade name for rocuronium bromide used as an anesthetic agent.

Zendole A trade name for indomethacin, an antiinflammatory agent.

Zenepax A trade name for daclizumab, a monoclonal antibody that reacts specifically with exposed receptor sites on the activated T lymphocytes.

Zenker's Fluid A fixative used in histological techniques consisting of mercuric chloride (5 g), potassium dichromate (2.5 g), and 5 ml glacial acetic acid per 100 ml of solution.

Zeolites Hydrated alkaline earth aluminum silicates used as an ion-exchanger in ion-exchange chromatography and as a molecular sieve material in gel filtration.

Zepto- A prefix denoting 10^{-21}.

Zerit A trade name for stavudine, an antiviral agent that inhibits replication of some retroviruses.

Zero Order Kinetics The kinetics of zero order reactions.

Zero Order Reaction The reaction in which the velocity of a reaction is independent of the concentration of substrate or reactant.

Zero Time-Binding DNA The DNA duplex formed at the beginning of a reassociation reaction because of the intramolecular reassociation of inverted repeats.

Zestoretic A trade name for a combination drug containing lisinopril and hydrochlorothiazide, used as an antihypertensive agent.

Zestril A trade name for lisinopril, an antihypertensive agent that prevents the conversion of angiotensin I to angiotension II.

Zeta (ζ) A letter in the Greek alphabet.

Zeta Potential The potential across the interface of all solids and liquids, e.g., the potential across the diffuse layer of ions surrounding a charged colloidal particle.

Zetran A trade name for diazepam, an antianxiety agent.

Zetta A prefix denoting 10^{21}.

ZF Abbreviation for zinc finger.

ZFM Abbreviation for zinc-finger mutant.

ZGM Abbreviation for zymogen granule membrane.

Ziac A trade name for a combination drug containing bisoprolol and hydrochlorothiazide, used as an antihypertensive agent.

Ziagen A trade name for abacavir, an antiviral agent that inhibits reverse transcriptase.

Zidovudine (mol wt 267) An antiviral agent; a synonym for AZT.

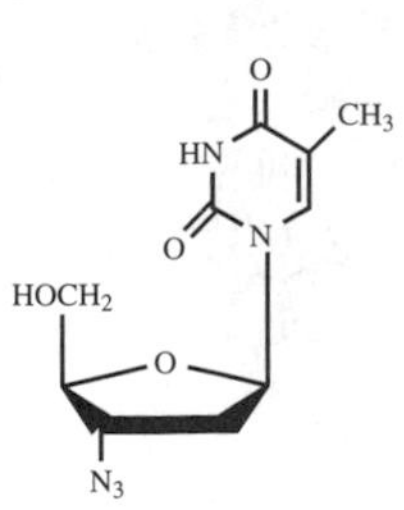

Ziehl-Neelsen's Stain An acid-fast stain (carbofuchsin) for staining acid-fast bacteria.

ZIG Abbreviation for zoster immune globulin.

ZIGV Abbreviation for zoster immune globulin vaccine.

Zileuton (mol wt 236) A leukotriene receptor antagonist used an antiasthmatic agent.

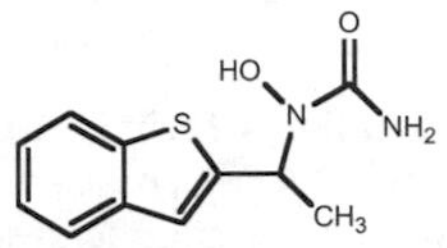

Zimeldine (mol wt 317) An antidepressant that inhibits 5-hydroxytryptamine uptake.

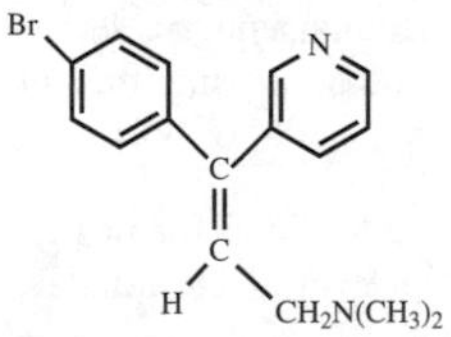

Zinacef A trade name for cefuroxime sodium, an antibacterial agent that inhibits bacterial cell wall synthesis.

Zinamide A trade name for pyrazinamide, an antibacterial agent.

Zinc A chemical element with atomic weight 65, valence 2.

Zinc-62 The artificial radioactive nuclide of Zinc (^{62}Zn), emitting beta and gamma radiation with a half life of 9 hours.

Zinc-65 The artificial radioactive nuclide of Zinc (^{65}Zn), emitting beta and gamma radiation with a half life of 244 days.

Zinc-72 The artificial radioactive nuclide of Zinc (^{72}Zn), emitting beta and gamma radiation with a half life of 2 days.

Zinc Bacitracin A zinc-bacitracin complex prepared from zinc salts and bacitracin, used as an antimicrobial ointment.

Zinc Carbonate (mol wt 125) A zinc salt used as an antiseptic agent.

$ZnCO_3$

Zinc D-Ala-D-Ala Carboxypeptidase The enzyme that catalyzes the cleavage of the peptide bond of the following sequence:

↓

(Ac)$_2$-L-Lysyl-D-Alanyl-D-Alanine

Zinc Finger A structural motif of DNA-binding proteins in which finger-like loops in the protein are stabilized by interactions with zinc atoms. In cysteine-histidine zinc fingers, two cysteines and two histidines bind the zinc atom, while in cysteine-cycteine zinc fingers four cysteines bind the zinc atom. Zinc fingers are involved in the regulation of gene transcription.

Zinc Finger Protein See zinc finger.

Zincfrin A trade name for a combination drug containing phenylephrine hydrochloride and zinc sulfate, an ophthalmic vasoconstrictor.

Zinc Iodate (mol wt 415) A zinc salt used as a topical antiseptic agent.

$ZI(IO_3)_2$

Zinc Oxide (mol wt 81) A mild astringent and antiseptic agent.

ZnO

Zinc Permanganate (mol wt 303) An antiseptic and astringent agent.

Mn_2O_8Zn

Zinc Peroxide Mixture A mixture of zinc peroxide, zinc carbonate, and zinc hydroxide, used as a local disinfectant, astringent, and deodorant.

Zineb An agricultural fungicide.

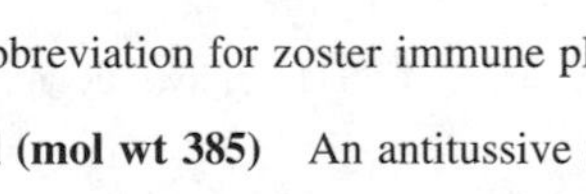

Zinostatin An antitumor acidic antibiotic consisting of protein components and a nonprotein chromophore isolated from *Streptomyces carcinostaticus* var F-41.

ZIP Abbreviation for zoster immune plasma.

Zipeprol (mol wt 385) An antitussive agent.

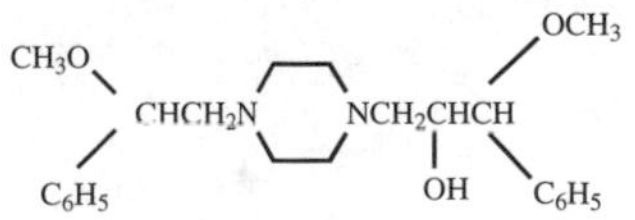

Zippering The formation of double helix DNA or RNA from the complementary strands.

Zirconium A chemical element with atomic weight 91, valence 4 and 3.

Zithromax A trade name for azithromycin, an antibacterial agent that binds to 50S ribosomal subunits, inhibiting bacterial protein synthesis.

Zn Symbol for the element zinc.

Zn^{2+} G Peptidase See zinc D-Ala-D-Ala carboxypeptidase.

ZnPP Abbreviation for zinc protoporphyrin.

ZO Abbreviation for zinc ointment.

Zocor A trade name for simvastatin, an antilipemic agent that inhibits 3-hydroxy-3-methyl glutaryl-CoA reductase, which is involved in cholesterol synthesis.

Zofran A trade name for ondansetron hydrochloride, an antiemetic agent.

Zoladex A trade name for goserelin acetate, an antineoplastic agent.

Zolamine (mol wt 291) An antihistaminic and anesthetic agent.

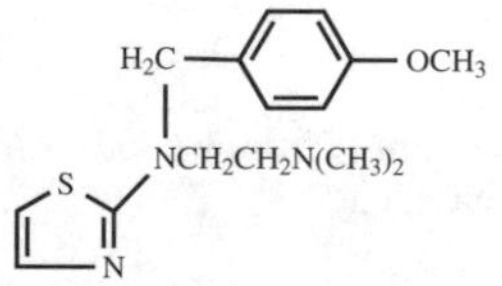

Zolicef A trade name for cefazolin sodium, an antibiotic.

Zolimidine (mol wt 272) An antiulcerative and a nonanticholinergic gastroprotective agent.

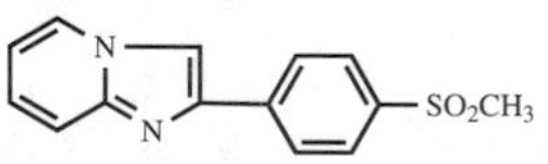

Zollinger-Ellison Syndrome A disorder characterized by gastric hypersecretion, hyperacidity, and severe peptic ulcers caused by the occurrence of gastrinomas (gastrin-producing tumor).

Zoloft A trade name for sertraline hydrochloride, an antidepressant.

Zolpidem (mol wt 307) A hynotic agent.

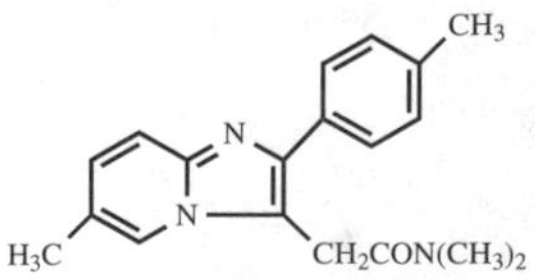

Zomepirac (mol wt 292) An analgesic and an anti-inflammatory agent.

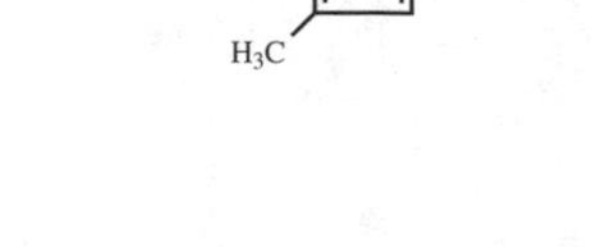

Zometapine (mol wt 274) An antidepressant.

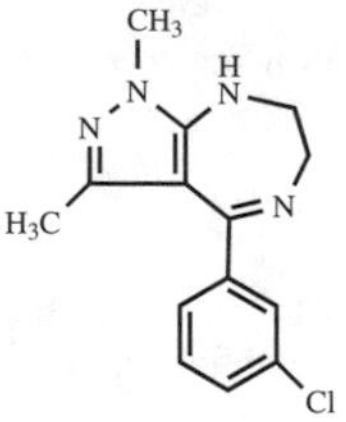

Zonal Centrifugation Technique of density gradient centrifugation in which macromolecules or subcellular components are fractionated into discrete zones or bands in a centrifuge tube.

Zonal Electrophoresis An electrophoretic technique in which ionic components are separated into zones or bands in a solid supporting medium (e.g., paper, cellulose acetate, starch, or polyacrylamide).

Zonalon A trade name for doxepin hydrochloride, an antidepressant and antianxiety agent.

Zone Electrophoresis See zonal electrophoresis.

Zone Equivalence See equivalence zone.

Zonegran A trade name for zonisamide, an antiepileptic agent.

Zonisamide (mol wt 212) An antiepileptic agent.

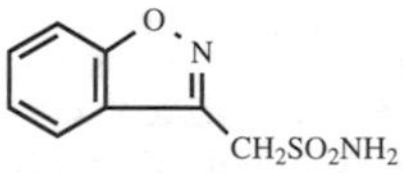

Zonula Adherens The intercellular junction in which the adjacent cell membranes have a 15–20 nm space into which microfilaments are inserted.

Zonula Occludens The intercellular junction in which the adjacent membranes are fused or separated by only a 1–2 nm space.

Zoo Blot The use of Southern blotting to test the ability of a DNA probe from one species to hybridize with the DNA from the genome of a variety of other species.

Zoobiont A animal symbiont.

Zoochlorella A symbiotic, green-pigmented algal cell living in the tissue of an animal host.

Zoogloea A genus of Gram-negative, aerobic, chemoorganotrophic bacteria (family Pseudomonaceae).

Zooid Motile spore.

Zoonosis Infection or disease transmitted from animal to man.

Zooparasite An animal parasite.

Zoophagous Feeding on animals.

Zoosporangia Plural of zoosporangium.

Zoosporangium A sporangium that produces zoospores.

Zoospore A flagellated, asexual reproductive spore or a motile, flagellated cell.

Zoosteroid Steroid of animal origin.

Zootoxin Toxin of animal origin.

Zopiclone (mol wt 389) A sedative–hypnotic agent.

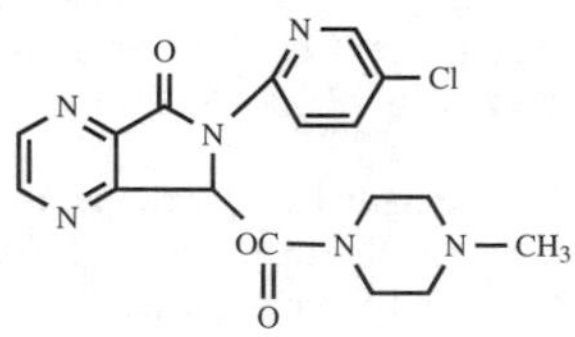

Zorubicin (mol wt 646) A semisynthetic antibiotic related to daunorubicin.

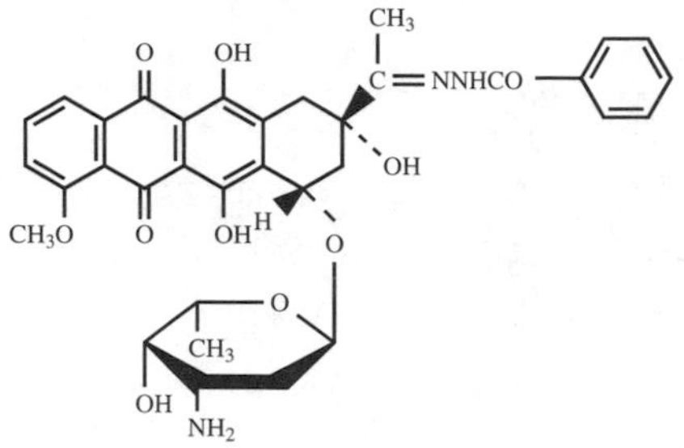

Zotepine (mol wt 332) An antipsychotic agent.

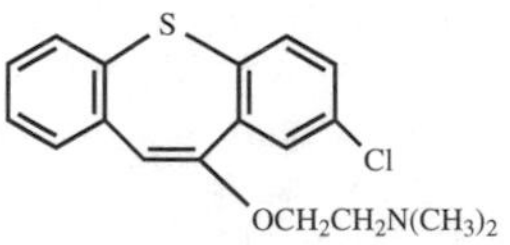

Zovia A trade name for a combination drug containing ethynodiol diacetate and ethinyl estradiol used as a contraceptive.

Zovirax A trade name for acyclovir sodium, an antiviral agent that inhibits viral DNA synthesis by incorporation into the DNA.

Zoxaphen A trade name for a combination drug containing chlorzoxazone and acetaminophen, a skeletal muscle relaxant.

Zoxazolamine (mol wt 169) A skeletal muscle relaxant and uricosuric agent.

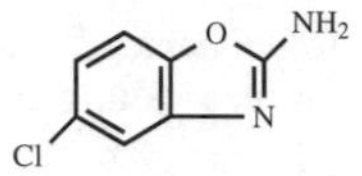

Z-Pathway See Z scheme.

Zr Symbol for zirconium.

Z-Scheme The oxygenic, noncyclic mode of photosynthesis in plants in which two photosystems, PSI and PSII, function together to drive electrons from water to generate NADPH and ATP.

Zwitterion A dipolar ion that has a positive charge on one portion of the molecule and a negative charge on the other, (also known as dipolar ion).

Zy DNA The DNA synthesized during the zygotene stage of meiosis.

Zyban A trade name for bupropion, an antidepressant and a smoking deterrent.

Zydone A trade name for a combination drug containing hydrocodone and acetaminophen.

Zyflo A trade name for zileuton, a leukotriene receptor antagonist used as an antiasthmatic agent

Zygomycetes A class of fungi of the Zygomycotina.

Zygomycosis Diseases or infections caused by fungi of Zygomycete.

Zygonema (Zygotene Stage) The stage of first meiotic prophase during which homologous chromosomes pair.

Zygosis Fusion of two unicellular cells.

Zygospores Thick-walled resting spores formed after gametangial fusion by members of the zygomycetes.

Zygote A diploid cell formed by fusion of two gametes.

Zygotene A stage of meiotic prophase I in which the homologous chromosomes synapse and pair along their entire length, forming bivalents.

Zygotic Induction The activation of a prophage to a virulent state when a chromosome containing a prophage in the lysogenic bacterial cell is transferred into a nonlysogenic bacterium through conjugation. It occurs as a result of the absence of repressor protein in the nonlysogenic recipient cell.

Zygotic Meiosis Meiosis that proceeds to the formation of haploid vegetative cells in a life cycle in which haploid phase predominates.

Zyloprim A trade name for allopurinol, an antigout agent that reduces uric acid production.

Zymase A trade name for pancrelipase, a digestive enzyme used to help with the digestion of lipids.

Zymenol A trade name for mineral oil, used as a laxative.

Zymogen The inactive precursor of an enzyme, e.g., pepsinogen for pepsin.

Zymogenesis The process of formation of an active enzyme from a proenzyme or zymogen.

Zymogen Granules The intracellular vesicles for storage of zymogen, e.g., intracellular vesicles in the pancreas for storage of pancreatic zymogen.

Zymogram The patterns of the electrophoretically separated and histochemically stained isozymes of a given enzyme, e.g., five histochemically stained isozymes of LDH in an agarose gel or polyacrylamide gel. A zymogram can provide information for taxonomy and clinical diagnosis.

Zymoid Resembling an enzyme.

Zymolysis Chemical process caused by an enzyme.

Zymomonas A genus of oxidase-negative, catalase-positive, chemoorganotrophic, Gram-negative bacteria.

Zymophore The active portion of an enzyme molecule, e.g., active site.

Zymosan A protein polysaccharide complex derived from yeast cell wall.

Zymotype A biotype characterized and distinguished on the basis of a zymogram.

Zyprexa A trade name for olanzapine, a dopaminergic blocking agent used as an anti-psychotic agent.

Zyrtec A trade name for cetirizine, a histamine H_1 receptor antagonist that inhibits histamine release and eosinophil chemotaxis leading to reduced swelling and decreased inflammatory response.